2009
Rapid Excavation and Tunneling Conference
PROCEEDINGS

Edited by
Gary Almeraris
and Bill Mariucci

Published by

Society for Mining,
Metallurgy, and Exploration, Inc.

Society for Mining, Metallurgy, and Exploration, Inc. (SME)
8307 Shaffer Parkway
Littleton, Colorado, USA 80127
(303) 948-4200 / (800) 763-3132
www.smenet.org

SME advances the worldwide mining and minerals community through information exchange and professional development. With members in more than 50 countries, SME is the world's largest association of mining and minerals professionals.

Copyright © 2009 Society for Mining, Metallurgy, and Exploration, Inc.

All Rights Reserved. Printed in the United States of America.

Information contained in this work has been obtained by SME, Inc. from sources believed to be reliable. However, neither SME nor its authors guarantee the accuracy or completeness of any information published herein, and neither SME nor its authors shall be responsible for any errors, omissions, or damages arising out of use of this information. This work is published with the understanding that SME and its authors are supplying information but are not attempting to render engineering or other professional services.

No part of this publication may be reproduced, stored in a retrieval system, or transmitted in any form or by any means, electronic, mechanical, photocopying, recording, or otherwise, without the prior written permission of the publisher. Any statement or views presented here are those of individual authors and are not necessarily those of the SME. The mention of trade names for commercial products does not imply the approval or endorsement of SME.

ISBN: 978-0-87335-304-5

CONTENTS

Preface xi

Executive Committee xii

Session Chairs xiii

International Committee xiv

Part 1: Conventional and Rock Tunneling

High Speed Excavation by Drill & Blast with Mechanized Mucking System—Mitholz Railway Tunnels, Switzerland ▪ *Stig Eriksson* 2

Lake Dorothy Hydroelectric Project Lake Tap and Tunnel, Juneau, AK ▪ *Jason Morrison, Ben Roberds, Chris Hickey* 12

Underground Construction for a Combined Sewer Overflow System in Providence, Rhode Island ▪ *John L. Kaplin, Jeffrey P. Peterson, Philip H. Albert* 22

Modern Caverns in Gotham—Geotechnical and Design Challenges for Large Rock Caverns in Manhattan ▪ *Timothy P. Smirnoff* 35

Tunneling Under the Harlem River ▪ *Jacek B. Stypulkowski, James P. Mooney Jr., Hugh S. Lacy* 48

Part 2: Design and Planning

Bay Tunnel—Design Challenges ▪ *R. John Caulfield, Isabelle Pawlik, Johanna I. Wong* 60

The Selection of Excavation Methods for the Detroit Upper Rouge Tunnel CSO Control Project ▪ *Jean Habimana, Wern-Ping Chen, Robert Barbour, Pamela Turner, Gary Stoll Jr., Mirza M. Rabbaig* 72

Planning and Design Features of the Waller Creek Tunnel, Austin, Texas ▪ *Gary Jackson, Stan Evans, Tom Saczynski, Peter Jewell, Prakash Donde* 86

NATM Strategies in the U.S.—Lessons Learned from the Initial Support Design for the Caldecott 4th Bore ▪ *Bhaskar B. Thapa, Thomas Marcher, Michael T. McRae, Max John, Zuzana Skovajsova, Mahmood Momenzadeh* 96

The Price Is Right—Planning Large Water Tunnel Contracts In New York ▪ *Christian Maguire, Eric Cole* 108

Daylighting Thorn Creek Tunnel into Chicago's TARP Thornton Composite Reservoir ▪ *Cary Hirner, Kevin Fitzpatrick, Marcella Landis, Faruk Oksuz* 123

Part 3: Difficult Ground

Times Square Connection: "Supporting the Cross Roads of the World" ▪ *Victor F. Paterno Jr., Robert W. Fonkalsrud* 134

Gibe II Tunnel Project—Ethiopia: 40 Bars of Mud Acting on the TBM "Special Designs and Measures Implemented to Face One of the Most Difficult Events in the History of Tunneling" ▪ *Antonio De Biase, Remo Grandori, Pierfrancesco Bertola, Martino Scialpi* **151**

Design and Construction of the Lenihan Dam Outlet Tunnel and Shaft ▪ *Hemang Desai, Brett Mainer, Mike Murray, Shawn Spreng* **171**

Daniel Island Surprise—Sand Lens Lurking In Cooper Marl, Charleston, SC ▪ *Ray Brainard, Paul Smith, Larry Drolet* **185**

San Vicente Pipeline Tunnel—Reach 4W, 3, and 2: Case History ▪ *Michael Jatczak, Brett Zernich, Steven Short, William Monahan* **195**

A Review of Tunneling Difficulties in Carbonate Sedimentary Rocks ▪ *D.P. Richards, T.W. Pennington, J.A. Fischer, G. Garcia* **215**

Part 4: Geotechnical

Geotechnical Baseline Reports—A Review ▪ *T. Freeman, S. Klein, G. Korbin, W. Quick* **232**

Ground Characterization for CSO Tunnels in Washington, D.C. ▪ *Maurice A. Ponti Jr., Steven B. Fradkin, Xiaohai Wang, Moussa Wone, Ronald E. Bizzarri, Edward J. Cording, Roger C. Ilsley, Qamar A.O. Kazmi* **242**

Actual vs. Baseline Tracking During TBM Tunneling in Highly Variable Glacial Geology ▪ *Ulf G. Gwildis, Leon E. Maday, John E. Newby* **250**

Assessing Ground Ahead of TBM Tunnel Using Low-Interruption Wireless Seismic Reflector Tracing System ▪ *Takuji Yamamoto, Yasuhiro Yokota, Jozef Descour, Matthew Kohlhaas* **263**

Ground Characterization and Feasibility Evaluation of Tunneling Methods for Mather Interceptor ▪ *Mohammad R. Jafari, Michael D. Middleton, Bruce J. Corwin, Andrew Page* **272**

Part 5: Ground Modification

North 27th Street ISS Extension: Unique Owner/Contractor Agreement Settles Major Disputes ▪ *Donald J. Olson, Roger J. Maurer, Martin Vliegenthart* **286**

Brightwater Conveyance System: Ground Freezing for Access Shaft Excavation Through Soft Ground ▪ *Joseph M. McCann, David K. Mueller, Paul C. Schmall, James D. Nickerson* **297**

New Approach of ASFINAG for Tunnel Construction Monitoring of the Tauern Tunnel Project in Austria ▪ *Robert Schnabl, Josef P. Mayr* **308**

Research in Soil Conditioning for EPB Tunneling Through Difficult Soils ▪ *Rory P.A. Ball, David J. Young, Jon Isaacson, Jeffrey Champa, Christopher Gause* **320**

An Analysis Method for Modeling Compensation of Settlements Due to Tunnel Driving by Grouting Cement Suspensions ▪ *Christian Wawrzyniak, Wolfgang Krajewski* **334**

Part 6: Innovation

Uetliberg Tunnel: Soft Ground Excavation and Premiere of New Tunnelling Machine: World's First Tunnel Bore Extender Excavated by Undercutting ▪ *S. Maurhofer, H.P. Müller* **346**

Extensible Conveyor Systems for Long Tunnels Without Intermediate Access ■
Dean Workman 363

New Cutter Soil Mixing (CSM) Technology Used to Construct Microtunneling Shafts
for Mokelumne River Crossing ■ *Matthew Wallin, Mary Asperger*........................ 370

An Introduction to Virtual Design and Construction (VDC) ■ *Matthias Rheinlander,
Rachel Arulraj, Alan Hobson, Chris Graham* 383

Placement of Concrete Lining for Water Tunnel No. 3, Manhattan Portion ■
Robert Labbe Jr., Michael Gorski.. 396

Part 7: International

The Hallandsås Dual Mode TBM ■ *Werner Burger, Francois Dudouit* 416

Effective Planning of Underground Space—Planning and Implementation of the First
Underground Water Reservoirs in Hong Kong ■ *T.H. Chan, Derek Arnold, Edwin K.F.
Chung, Chris C.W. Chan* 438

Hobson and Rosedale Tunnels—New Technology In Auckland ■ *Harry Asche,
Tom Ireland, Mike Sheffield, Mike Bonnette* 450

Feasibility and Implementation of Shield Machine Tunnel Passing Through
an Operating Airport Runway ■ *Hongjie Yang, Xiuzhi Wang, Hongbo Liu*...................... 463

Experience Gained in Mechanical and Conventional Excavations in Long Alpine
Tunnels in Switzerland ■ *Y. Boissonnas* 471

Part 8: Las Vegas

Design and Construction of Lake Mead Intake No. 3 Shafts and Tunnel ■ *Jon Hurt,
Jim McDonald, Gregg Sherry, A.J. McGinn, Luis Piek* 488

Project Delivery Selection for Southern Nevada's Lake Mead Intake No. 3 ■
Michael Feroz, Erika P. Moonin, James McDonald 503

Design and Subsurface Construction at Yucca Mountain, Nevada ■
John F. Beesley, Jaime A. Gonzalez 516

Feasible Tunnel Construction Options for the Systems Conveyance and
Operations Program Reach 3 Tunnel ■ *S.H. Jason Choi, Scott Ball, Sean Tokarz,
Jim Devlin*... 525

What Happens in Vegas: The Apex Tunnel Geologic Investigation ■
Ann L. Backstrom, Joel G. Metcalf, Stephen McKelvie 534

The Cost and Benefit of the Phase 2 Investigation for the Reach 4 Tunnel, How a
Roll of the Dice Came Up Big in Las Vegas ■ *Ray Brainard, Racheal Johnsen,
Jim Devlin, Eugene Smith, Tom Knox, Jim Werle, Alston Noronha* 548

Part 9: Microtunneling

Microtunneling 1.2-Mile, 72-in RCP with Crossings of NJ Turnpike and CSX
Railroads ■ *Zhenqi Cai, Alberto G. Solana, Neal O'Connor, Philip Lloyd*.................. 558

The Longest Drive—Portland's CSO Microtunnels ■ *Christa Overby, Matt Roberts,
Craig Kolell* 570

Microtunnels vs. EPB Risk-Based Selection ■ *Michelle L. Ramos, Kimberlie Staheli* 581

Microtunneling Challenges in Soft Ground of Downtown Hartford, CT ■
William Bergeson, Verya Nasri, James Sullivan, Alan Pelletier 589

Microtunneling for Utilities Under Harold Railroad Interlocking ■ *William Reininger, Wai-shing Lee, Richard Pociopa* 602

Part 10: Mining

Technical Challenges in Mine Rehabilitation ■ *Don Dodds, David A. Clayton, Terry L. Johnson* 616

Open Pit TBM Driven Drainage Tunnel—Ok Tedi Mine ■ *Tony Peach, Nigel Sudgen* 629

The Deep Underground Science and Engineering Laboratory and the Construction of Physics Megacaverns ■ *C. Laughton* 638

Subsurface Repository Ventilation Design ■ *Edward Thomas, Jaime Gonzalez* 646

Part 11: New Projects 1

Planning New Metro Subways—Los Angeles, California ■ *Amanda Elioff, David Perry, Girish Roy, Pierre Romo* 660

Slurry TBM Tunnel in Rock, the Modified Detroit River Outfall No. 2 ■ *William H. Hansmire, Pamela Turner, Frederic Mir, Parvez Jafri* 675

Port of Miami Tunnel Update—A View from Design Builder's Engineer ■ *Wern-Ping Chen* 687

Design Considerations and Evaluation Process for a New Tunnel and Ocean Outfall Project ■ *Steve Dubnewych, Michael Torsiello, Jon Kaneshiro, David Haug* 700

MBTA Silver Line Phase III—Completes Boston's Newest Transit Line ■ *Gregory Yates, Mary Ainsley, William Gallagher* 713

Part 12: New Projects 2

Design of NATM Tunnels and Stations of Silver Line Phase III Project in Boston ■ *Verya Nasri, Kosmas Vrouvlianis, Irwan Halim* 728

Geotechnical and Structural Design Challenges of the Fremont Central Park Subway for the BART Warm Springs Extension ■ *Mitchell L. Fong, M. Shing Owyang, Thomas S. Lee* 742

Atlanta North-South Tunnel ■ *Samer Sadek, Hugh Caspe, Tim Heilmeier, Darryl D. VanMeter, John D. Hancock* 757

Proposed Contracting Practices for the Caltrain Downtown Extension ■ *Derek J. Penrice, Bradford F. Townsend* 766

Part 13: New York City

Alternative Final Cavern Linings for the East Side Access Transit Project ■ *Colin Barratt, William Cao* 780

Continuing the Legacy: An Update on the Construction of the New Second Avenue Subway ■ *Jaidev Sankar, Christopher K. Bennett, David Caiden, Anil Parikh, Thomas F. Peyton* 790

No. 7 Subway Extension Crossing Under an Existing Subway Station: Challenges and Integration of Underpinning into the Design of New Tunnels ■ *Aram Grigoryan* **802**

Railroad Interface Management for MTA East Side Access Project Tunnels and Structures ■ *Michael A. Piepenburg, Daniel A. Louis, Robert Magnifico, Augustine Juliano* **814**

Construction of the MCUA Tunnel and Force Mains Under the Raritan River, New Jersey: A Case History ■ *Bob Rautenberg, Julian Prada, Frank Perrone, Donato Tanzi* **824**

Part 14: Risk Management

Hindsight Is 20/20—Reverse Engineering Tunnel Risk Analyses ■ *Lee W. Abramson*... **836**

Getting the Engineer's Estimate Right ■ *Tom Martin, Joe O'Carroll, Keith Caro, Tom Peyton* **845**

Transfer of a Project Risk Register from Design into Construction: Lessons Learned from the WSSC Bi-County Water Tunnel Project ■ *R.J.F. Goodfellow, P.J. Headland* **854**

The Delivery of Underground Construction Projects in the UK: A Review of Good Practice ■ *Andy Alder, Mike King*... **861**

Using Risk Analysis to Support Decision Making on the Central Subway Project ■ *Joe O'Carroll, Noel Berry, Arthur Wong* **871**

Short Tunnels = High Risk?: Pipeline Construction Involving Open-Cut and Tunnel Segments ■ *Michael Gilbert, Michael Schultz* **881**

Part 15: SEM

Case History of the Wachovia–Knight Theater Pedestrian Tunnels ■ *Eric Eisold, Prakash Donde, Ivona Tarchala* **894**

Boggo Road Busway Project, Brisbane, Australia ■ *Ted Nye, Maxwell Kitson, Ravin Chinniah* **906**

Loosening and Face Stability with Shallow Overburden in the "SITINA Tunnel," Bratislava, Slovakia ■ *Chikaosa Tanimoto, Kimikazu Tsusaka, Toshihiko Aoki, Masahiro Iwano* **916**

Innovative NATM—Design for a Large Shallow Cavern at Stanford ■ *Thomas Marcher, Max John, Steffen Matthei, Zuzana Skovajsova* **927**

ADECO as an Alternative to NATM: How It Works, Why It Works ■ *Fulvio Tonon* **942**

Part 16: Shaft

New Technology Changes Blind Shaft Drilling ■ *Alan Zeni* **970**

Tamerlane Hoist and Vertical Belt Project ■ *Shawn Collins* **975**

A Small Diameter Shaft Design Alternative ■ *Brian E. Gombos, David D. DiPonio* **986**

Kansas River Tunnel Shaft Drilling ■ *Clay Haynes, Cary Hirner, Clay Griffith*................. **998**

Design Considerations for the Use of Slurry Walls as Permanent Walls for Deep Rectangular Shaft Structures in Seismic Areas—Silicon Valley Rapid Transit Project ■ *Michael J. Lehnen, Ching Wu, Mike Wongkaew, James Chai* **1005**

Part 17: Slurry and EPB 1

Selection, Design, and Procurement of North America's Largest Mixshield TBM for Portland, Oregon's East Side CSO Tunnel ▪ *Christof Metzger, Greg Colzani, Gary Irwin* **1022**

Construction of Drilled Shafts for the Upper Northwest Interceptor, Sections 1&2 Project—Sacramento, CA ▪ *Jeremy Theys, Chris Schäfer* **1037**

Port Authority of Alleghany County North Shore Connector Project Tunnels and Station Shell Case History Contracts 003 and 006 ▪ *Paul Zick* **1051**

Construction of the North Dorchester Bay CSO Storage Tunnel in Boston ▪ *J. Davies, K. Chin, J. Ohnigian, J. Stokes* **1062**

High Risk Tunneling Adjacent to Large Water Tank on the UNWI Sections 3&4 Project ▪ *Andrew Finney, John Wong, Craig Vandaelle, Cody Painter, Steve Cano* **1074**

Construction Works of Large-Section Vertically Parallel Twin Tunnels in Close Proximity ▪ *Kentaro Kuraji, Masami Morita, Naoyuki Araki, Yoshihir Taniguchi* **1083**

A Practical Approach for Precast Concrete Segmental Ring Selection ▪ *Steve Skelhorn, Laura McNally* **1102**

Part 18: Slurry and EPB 2

Big Walnut Outfall Augmentation Sewer—Part II: TBM Case History ▪ *Tom Szaraz, Gary Bulla, Chris Smith* **1114**

EPB Tunnelling Through Cohesionless Saturated Ground Under Very Shallow Cover—Perth New MetroRail City Project ▪ *Hiroshi Yamazaki, Oskar Sigl, Fumihiro Aikawa, Raghvendra Bhargava* **1124**

Sao Paulo Metro Project—Control of Settlements in Variable Soil Conditions Through EPB Pressure and Bicomponent Backfill Grout ▪ *Lorenzo Pellegrini, Pietro Perruzza* **1137**

Planning and Preparation for Tunneling at Brightwater West ▪ *Mina M. Shinouda, Glen Frank, Greg Hauser* **1154**

Brightwater East—A Case History ▪ *Luminita Calin, Tony Hupfauf* **1171**

Gotthard-Base Tunnel, Section Faido Previous Experience with the Use of the TBM ▪ *Martin Herrenknecht, Olivier Böckli, Karin Bäppler* **1182**

Part 19: TBM Case Studies 1

TBM Tunneling at the Ashlu Hydropower Project, Squamish, BC ▪ *Serge Moalli, Steve Redmond, Dean Brox, Peter Procter, Daniel Jezek, Michelle van der Pouw Kraan, Richard Blanchet, Robert Kulka* **1208**

TBM and NATM Combined Solution for a Very Deep Tunnel—The "Pajares" Case ▪ *Enrique Fernández, Paz Navarro, Alejandro Sanz* **1218**

8 m Diameter 7 km Long Beles Tailrace Tunnel (Ethiopia) Bored and Lined in Basaltic Formations in Less than 12 Months ▪ *Antonio Raschillà, Francesco Bartimoccia* **1235**

Construction of Louisville Water Company's Riverbank Filtration Tunnel and Pump Station Project ▪ *S. Holtermann, W. Klecan, K. Ball* **1248**

Technical Considerations for TBM Tunneling in the Andes ▪ *Dean Brox, Guido Venturini* **1261**

Robbins 10m Double Shield Tunnel Boring Machines on Srisailam Left Bank
Canal Tunnel Scheme, Alimineti Madhava Reddy Project, Andhra Pradesh, India ▪
William Brundan **1275**

Part 20: TBM Case Studies 2

Madiq Tunnel, Lebanon: TBM Tunneling vs. Karst Geology ▪ *William D. Leech,
Issam Bou Jaoude, Nicholas Ghanem* **1294**

Onsite Assembly and Hard Rock Tunneling at the Jinping-II Hydropower Station
Power Tunnel Project ▪ *Stephen M. Smading, Joe Roby, Desiree Willis* **1308**

Double Shield TBM in Challenging, Difficult Ground Conditions—A Case Study
from Zagros Long Water Transfer Tunnel, Iran ▪ *Jafar Khademi Hamidi,
Kourosh Shahriar, Jamal Rostami* **1321**

Impacts of Ground Convergence on TBM Performance in Ghomroud Tunnel ▪
Ebrahim Farrokh, Jamal Rostami **1334**

TBM Data Management and Quality Assurance for the Brightwater Conveyance
Project ▪ *Jeffrey Mitsopoulos, Frank Stahl, James Wonneberg, Kenneth Rossi*....... **1354**

Construction of the East Side Access Manhattan Tunnels ▪ *Don Hickey*............... **1365**

Index   1373

PREFACE

The 2009 Rapid Excavation and Tunneling Conference returns to Las Vegas, Nevada, and begins with the always interesting topic of Risk, followed by 20 technical sessions totaling 107 papers over four days. These sessions in this 19th RETC cover every aspect of underground construction, design, and planning as presenters share their trials, tribulations, and triumphs. The authors of these papers come from a wide spectrum of the industry and bring both innovative ideas and unique perspectives to the presentations.

RETC has become a truly international conference as our world and industry becomes smaller due in part to increased sharing of knowledge. We have 40 international papers being presented, and our conference committee maintains an international subcommittee representing approximately 16 countries.

RETC is a firm believer of student participation and the promotion of pursuing careers in underground construction. To that extent, we sponsor student attendance at RETC and are looking to expand the scholarship program for students in related studies. These proceedings are a valuable source of knowledge for all who attend, but we should endeavor to pass these proceedings and "lessons learned" on to the next generation of tunnelers. It is up to all of us to mentor our engineers, superintendents, designers, supervisors, and craftspeople. They are our future and we owe it to them because we all have someone who inspired us to become part of this exciting industry.

The program chairs wish to thank the session chairs, co-chairs, authors, and other members of the RETC Executive Committee for their dedication to this conference. Additionally, the chairs wish to thank the dedicated staff of SME for their hard work, patience, and enthusiastic support. Finally, we thank all the participants for joining us in this global conference, which will not only provide technical knowledge and networking but "good, old-fashioned camaraderie."

EXECUTIVE COMMITTEE

James Marquardt ('09)
J F Shea Company Inc.
New York, NY

Bill Hansmire ('09)
Parsons Brinckerhoff Quade & Douglas, Inc.
Detroit, MI

Red Robinson ('11)
Shannon & Wilson Inc.
Seattle, WA

Dave Rogstad ('11)
Frontier Kemper
Seattle, WA

John Hutton ('13)
S. McNally Intl.
Hamilton, Ontario, Canada

John Townsend ('13)
Hatch Mott MacDonald
Pleasanton, CA

Michael T. Traylor ('15)
Traylor Brothers Inc.
El Segundo, CA

Gary Almeraris ('15)
Skanska USA
Civil NorthEast, Inc.
Whitestone, NY

Bill Mariucci ('17)
Kiewit Constuction
Portland, OR

Victor Romero ('17)
Jacobs Associates
San Francisco, CA

Steve Redmond ('19)
Frontier Kemper
Seattle, WA

SESSION CHAIRS

Bill Austell
Coluccio Construction

Luminita Calin
Kenny Construction

Carl Christensen
J.F. Shea

Eric Eisold
Bradshaw Construction Corp.

Matthew Fowler
Parsons Brinckerhoff

Robert Goodfellow
Black & Veatch

Jean Habimana
Jacobs Engineering Group

Greg Hauser
Jay Dee

Paul Helsop
Arup

Don Hickey
Judlau

Scott Hoffman
Skanska

Jon Hurt
Arup

Gary Irwin
City of Portland

Heather Ivory
URS Corp

Lonnie Jacobs
Frontier Kemper

Lars Jennemyr
Skanska USA

Marc Jensen
Southern Nevada Water

Jon Kaneshiro
Parsons

John Kennedy
Atkinson Construction

Mike Kucker
Shannon & Wilson

Isabel Lamb
Jacobs Associates

Darrel Liebno
Obayashi

Andrew Liu
Hatch Mott MacDonald

Calvin Locke
King County DNRP

Jim McDonald
Impregilo

Laura McNally
McNally Construction

Tony O'Donnell
Kiewit

Shemek Oginski
J.F. Shea

Ryan Ostoyich
Atkinson Construction

Berry Roberts
Parsons Brinckerhoff

Brett Robinson
Traylor

Paul Roy
AECOM USA

Mike Ryan
Schiavone

Steve Skelhorn
McNally Construction

James Smith
Brierley Associates

Sarah Wilson
Jacobs Associates

Scott Wimmer
Kiewit

Terry Yokota
Frontier Kemper

David Young
Hatch Mott MacDonald

Brett Zernich
Traylor Brothers Inc.

INTERNATIONAL COMMITTEE

Australia	**Tony Peach** Terratec Asia - Pacific Pty., Ltd Tasmania, Australia
Austria	**Manfred Jaeger** Batloggstrasse 95 Schruns, Austria
Canada	**Rick Lovat** Lovat Tunnel Equipment, Inc. Etobicoke, Ontario, Canada
England	**Alan P. Finch** Mott MacDonald Croydon, UK
France	**Bernard Gautrais** Cedex, France
Germany	**Otto Braach** Wayss & Freytag AG Kelkheim, Germany
Hong Kong	**Ian McFeat-Smith** IMS Tunnel Consultancy Ltd. Tai Hang, Hong Kong
Italy	**Remo Grandori** SELI Societa Esecuzione Lavori Rome, Italy
Japan	**Katsuji Fukumoto** Obayashi Corp. Tokyo, Japan
Korea	**Nam-Seo Park** Daeduk Consulting & Construction Co. Seoul, Korea
Mexico	**Roberto Gonzalez Izquierdo** Moldequipo Internacional, SA Tepotzotlan, Mexico
Norway	**Thor Skjeggedal** Skjeggedal Construction Services Stabekk, Norway
Spain	**Enrique Fernández Gonzales** Dragados O.P. Madrid, Spain
Sweden	**Stig Eriksson** Skanska ID Solna, Sweden
Switzerland	**Fredric Chavan** Prader AG Tunnelbau Switzerland

PART 1

Conventional and Rock Tunneling

Chairs

Ryan Ostoyich
Atkinson Construction

Lars Jennemyr
Skanska USA

HIGH SPEED EXCAVATION BY DRILL & BLAST WITH MECHANIZED MUCKING SYSTEM—MITHOLZ RAILWAY TUNNELS, SWITZERLAND

Stig Eriksson ▪ Skanska Infrastructure Development AB

BACKGROUND

In Switzerland two major railway systems as shown below in Figure 1 has recently been completed or is under construction:
- Lötschberg, south of Bern which was commissioned in 2007
- Gotthard, south of Zurich which is scheduled to be commissioned in 2016

The aim is to increase the transport capacity of the railway system and to transfer more transport of goods to the railway by offering transport of trailers and lorries on the new railway systems. It will have a positive environmental impact but also improve the congested traffic situation of the highways in north-south direction.

The Gotthard tunnel is totally 57 km (35 mile) long and the Lötschberg tunnel is 35 km (22 mile) long. Both tunnel systems consist of two parallel single track tunnels connected with cross tunnels at every 333 m (1 093 ft). Additionally there are cross-over tunnels and caverns for installations and operation. In the Lötschberg tunnel, the majority of the excavation has been by Drill & Blast and in Gotthard tunnel the major part is being excavated by TBM.

MITHOLZ PROJECT

The Mitholz Project is the northern contract of Lötchsberg tunnel which consists of three major contracts (Mitholz, Ferden and Steg-Raron) as shown below. The Mitholz project was constructed by the large joint venture, Satco. The JV consisted of the following companies; Strabag, Austria, Skanska, Sweden, Vinci Construction, France, Rothpletz and Walo, Switzerland.

Satco signed the Construction Contract in February, 2000. The Contract covers the construction of:
- North-East railway tunnel, 7 787 m (25 548 ft) long and concrete lined
- South-East railway tunnel, 9 019 m (29 590 ft) long and concrete lined
- South-West railway tunnel, 9 278 m (30 440 ft) long and only primary support consisting of bolts, shotcrete and mesh
- Other works such as cross tunnels, caverns etc

The 9.6 km (31 500 ft) long investigation tunnel, the 1.5 km (4 900 ft) long access adit tunnel and the junction area were constructed earlier under separate Contracts. Due to the high rock cover up to 1500 m (4 900 ft) and risk of squeezing rock, the tunnels were excavated by Drill & Blast. Totally 26 091 m (85 600 ft) of main tunnels were excavated from July 2000 to April 2004.

The Junction area is shown in Figure 3 and consists of two caverns, several tunnels and some shafts.

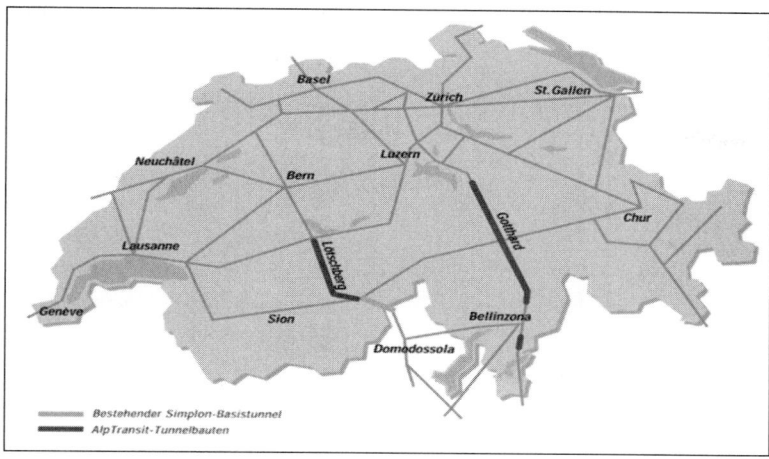

Figure 1. Switzerland

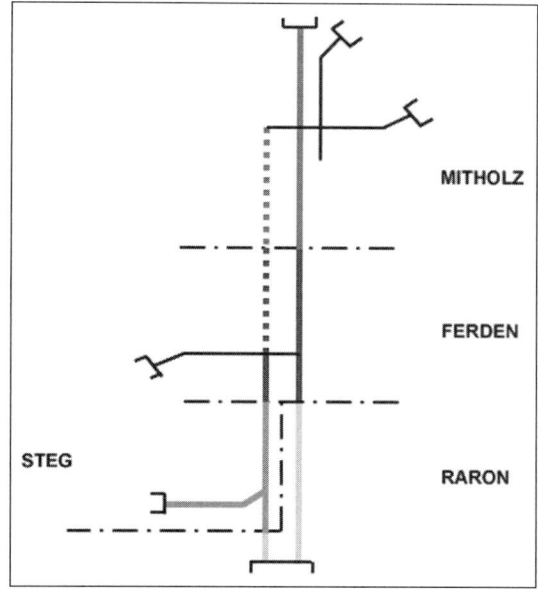

Figure 2. Lötschberg tunnel

The cross section of the main tunnels varied from 63 m² (677 ft²) to 90 m² (967 ft²) depending on the rock quality. Below shows the section used in good rock to the left and the one used for poor rock to the right with an arch invert.

The excavation was done from the junction and from the Junction area two parallel main tunnels were excavated towards South and one main tunnel towards North (except for the last 800 m (2600 ft)), where also two parallel main tunnels were excavated). In

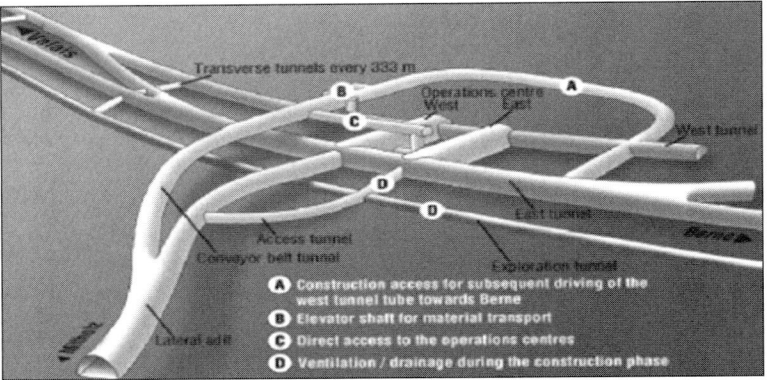

Figure 3. Junction area (main railway tunnels in red)

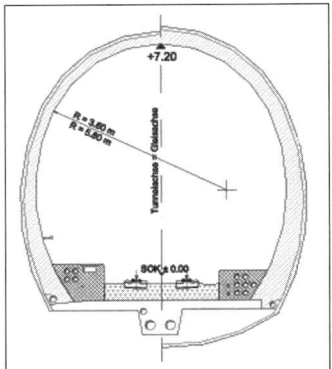

Figure 4. Cross section

the caverns located at the junction area a batching plant and workshop was erected to serve all three tunnels. Additionally there were workshop and stores above ground as well as office, canteen and sleeping quarters.

CONTRACT

- Client: BLS Alptransit AG.
- Contract: Unitprice with > 10,000 unit rates
- Contract Period: February 2000 till June 2006.
- Main tunnels: 26 091 m (85 600 ft)
- Concrete lining: Around 17 000 m (55 800 ft)

The Contract was based on the Swiss norm, SIA. The Bill of Quantity and the coding of the Bill were in accordance with Swiss standards. The Bill of Quantity was very detailed and contained more than 10 000 unit rates. In regards to rock excavation, the following conditions were valid:

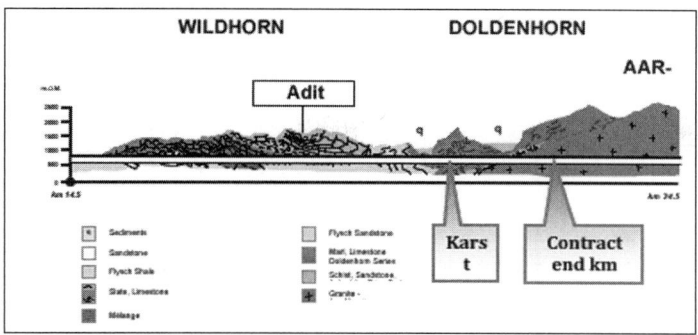

Figure 5. Geology

- Unit rates for standstills caused by ingress of water, gas etc.
- Unit rates for probe drilling
- Unit rates for grouting (ground treatment)
- 8 Rock classes with defined temporary support with separate unit rates

GEOLOGY AND ROCK CLASSES

In Mitholz project, the tunnels go through three different geologic parts as shown below. A simplified description of these is:

- Wildhorn consists of mainly slates and flysh with various disposition and metamorphic level. Partly it consists also of mica schist, sandstone and limestone.
- Doldenhorn consists of different types of limestone with risk for Karst.
- Aar is mainly granite and gneiss.

Rock support was installed in accordance with the contractual requirements of the rock classes as shown below. Rock class 1 corresponds to very good rock, rock class 2 to good rock, rock class 3 to fair rock etc. The capacity for each class is undisturbed capacity excl. disturbances for probe drilling, cross tunnels and other disturbances. The time schedule was regulated every month based on the actual encountered conditions compared to the predicted ones.

Rock class	Pull/ Advance	Bolt	Shotcrete	Other Info
1	> 4.0 m	2.5–3.5 m long 4–8 no./tunnel meter	8 cm	Capacity = 12,5 m/d (41 ft/d)
2	4.0 m	2.5–4.0 m long 6–10 no./tunnel meter	10 cm (ev. mesh)	Capacity = 11,5 m/d (38 ft/d)
3	2,5–3.0 m	3.5–4.5 m long 10–15 no./tunnel meter	15 cm (ev. mesh)	Capacity = 8,5 m/d (28 ft/d)
4	2.0 m	3.5–5.0 m long 0–4 no./tunnel meter	21 cm (with mesh)	Steel rib HEB 120 1 no./meter , 4,5 m/d
5	1.5 m	4.0–6.0 m long 0–4 no./tunnel meter	27 cm (with mesh)	Steel rib HEB 160 1–2 no./meter, 3,5 m/d

Rock class	Pull/ Advance	Bolt	Shotcrete	Other Info
6 (gallery + bench)	1.5 m	3.5–8.0 m long 15–20 no./tunnel meter	25 cm (with mesh)	Steel rib of Omega type 1–2 no./meter, 1.5–2 m/d Slots for deformations
7 (gallery + bench)	1.0 m	3.5–6.0 meter long 0–6 no./tunnel meter	27 cm (with mesh)	Steel rib HEB 160 1–2 no./meter, 1.5 m/d Spiling l=3–4 m (5 ft/d) Face support
8 (rock burst)	4.0 m	2.5–3.5 m long 10–15 no./tunnel meter	10 cm (ev. mesh)	Capacity = 10 m/d (33 ft/d)

The client had geologists at site covering 24 hours per day and the geologists classified the rock and decided the actual rock class.

Convergence points were installed at suitable locations and monitored continuously to verify that the rock support was adequate.

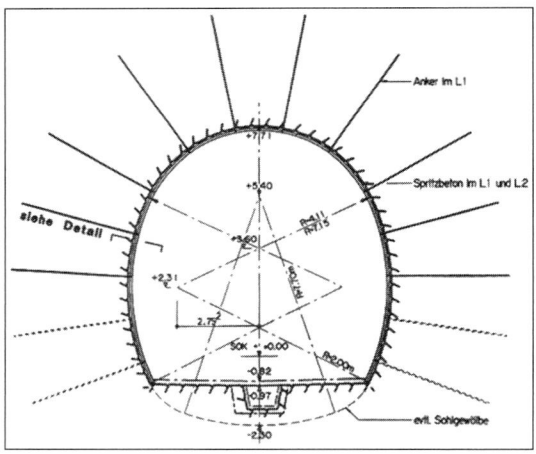

Figure 6. Rock class 3

Figure 7. Tunnel face

During the works, the rock support was revised for some classes to suit the actual conditions and new classes were defined such as 3A, 3B, 6A, 6B, 6K etc. According to the Swiss norm the rock support should generally be installed latest 25 m (80 ft) from the face. Below shows rock class 3, which was the most common and 50 % of the excavation was in rock class 3. Invert arch excavation was only occasionally used in rock class 3.

PRODUCTION

Each tunnel was driven as a single-front with its own equipment and crew. The production was going on 24 hours per day with 3 shifts of 8 hours each, 7 days a week and stopped only during some main holidays such as Christmas, Easter etc. Workshop and batching plant were common resources serving all three tunnels, located in the junction at the end of the adit. At the junction there was some spare equipment such as drilljumbo, shotcreterobot, loader, excavator, etc.

When choosing the production method for Mitholz the main aim was to find a method that could manage a high capacity for long tunnel drives. Many different options were discussed such as rail bound mucking, conveyor systems, different types of crushers, varies mucking systems with electric excavators, tunnel loaders to pay loaders.

The final choice was the following for each tunnel:

- Drilling: Atlas Copco XL 3C
- Charging: Dyno Nobels emulsion and Noneelectric detonators
- Mucking: Side dumping GHH L.F and CAT 966G.
- Crushing: Mobile DBT impact crushers
- Platform: Rowa Tunnelling Logistics.
- Tunnel Conveyor: Continental Conveyor.
- Scaling: Liebherr R-932.
- Shotcrete: Meyco Robojet Suprema.

Additionally there were 2 no Atlas Copco L 2 drilljumbos as spare units and to excavate cross tunnels and other minor excavations.

The unique components in the mechanized mucking system that was used consisted of the following main items:

- The mobile crusher that was located around 100 m (330 ft) behind the face
- The hanging platform with the conveyor
- The main tunnel conveyor, maximum length 9 km (29 500 ft)
- Bunker station at the junction area

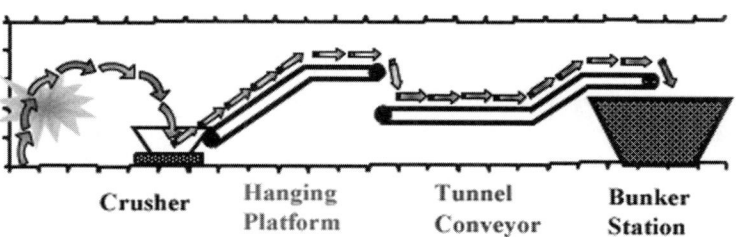

Figure 8. Conceptual mucking system

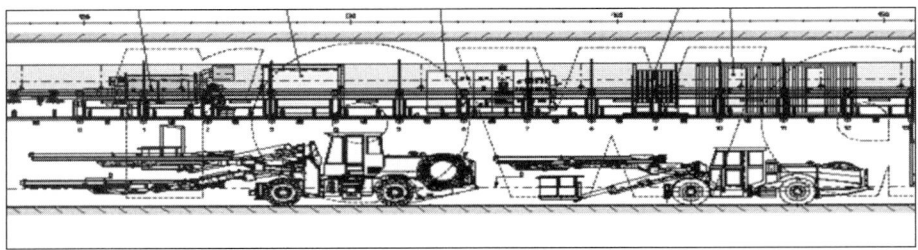

Figure 9. Hanging platform

Figure 10. Crusher

The mobile crusher was a track mounted impact crusher that was located around 100 m (330 ft) from the tunnel face. After blasting the tunnel muck was carried from the face by the scoop tram and/or loader to the crusher where it was processed to a size < 200 mm (8") before going on to the conveyor of the hanging platform. The hanging platform was one very important part of the production system and resembles of a backup to a TBM.

This type of platform has earlier only been used for some few long Drill & Blast tunnels. It is a mobile platform that travels on rails suspended from the roof by the assistance of bolts and chains. The free height under the platform was around 4.5 m (15 ft). The more than 100 m (330 ft) long platform housed the following.

- Conveyor, around 300 m (1000 ft) long (longer than the platform as it was overlapping the tunnel conveyor)
- Dust collector (similar type as for a TBM)
- Electric cable drum.
- Transformers.
- Compressor
- Ventilation duct cassette for 100 m (330 ft) for blowing ventilation
- Fan for suction. Ventilation duct cassette for suction ventilation
- Air Cooling equipment.
- Personnel cabin, office and Toilet

Figure 11. Bunker station

Figure 12. Hanging platform—bolt and chain supporting the rails to the right

- Safety equipment and rescue-container
- Workshop and stores

The mobile crusher and the hanging platform followed the production and were advanced 15 to 20 m (50 to 67 ft) during blast drilling (daily in rock class 1 and 2 but less frequent in the other rock classes).

Figure 10 shows the mobile crusher with the hanging platform behind. From the crusher, the conveyor was climbing for the first 20 m (65 ft) up to level of the platform level. The chain curtain shown above was protecting the platform when blasting. The short conveyor on the platform was overlapping the long tunnel conveyor by 165 m (540 ft) for the northern tunnel and 333 m (1100 ft) for the southern tunnels (equals to the spacing between the cross tunnels).

The muck was transported by the shorter conveyor on the hanging platform to the long main tunnel conveyor that transported it to the Bunker station at the junction. The Bunker Station had a capacity corresponding to two full rounds of muck and from this point the transfer of the muck to the surface was the responsibility of the Client.

At the Bunker station the muck was transferred to another conveyor which took it via the Adit Tunnel to the surface where it was stockpiled or processed for concrete aggregate. The aggregate for concrete was transported by another conveyor from the crushing plant outside to the batching plant at the junction area.

CONVENTIONAL AND ROCK TUNNELING

Figure 13. Platform from below—tunnel conveyor on the floor to the left

The mobile platform traveled on rails suspended from the roof by the assistance of bolts and chains. Below the bolts and the chains can be seen that were supporting the rails on which the platform is travelling.

For the two Southern tunnels (East and West), only one large main tunnel-conveyor located in the West tunnel was handling all muck from the two South tunnels up to the Bunker station located in the junction area. From the crusher in the South-West tunnel it was an around 300 m (1000 ft) long straight conveyor on the hanging platform that overlapped and fed the main tunnel conveyor. The muck from the conveyor on the hanging platform in South-East tunnel was fed via a cross-conveyor located in the first or second cross tunnel to the main tunnel conveyor located in South-West tunnel.

By this arrangement the South-East tunnel was kept free from transports and concrete lining was started and followed about 2 km (6500 ft) behind the excavation.

The first 300–400 m in each tunnel was excavated conventionally with mucking by loaders and dumpers to achieve sufficient space for the installation of the mobile crusher, hanging platform and the main tunnel conveyor. The installation and commissioning of the mobile crusher, hanging platform and conveyors was completed in around one month period for each tunnel and after another 4 months of training and fine tuning of the system, the crew and the system was working as expected.

CAPACITIES (ROCK EXCAVATION)

The capacity depends of course on the rock conditions and the conditions varied in these long tunnels. The best capacities are shown below (the best capacities were mainly achieved in rock classes 1 and 2 but the monthly capacity may include also some rock class 3):

Best Capacities		Day	Week	Month
South	West tunnel	19 m (62 ft)	104 m (341 ft)	338 m (1109 ft)
	East tunnel	18 m (59 ft)	94 m (308 ft)	303 m (994 ft)
North	East tunnel	18 m (59 ft)	103 m (338 ft)	342 m (1122 ft)

The average monthly capacity for the northeastern tunnel was around 240 m (790 ft) when using the mechanized mucking system.

The average capacities for the southern tunnels were lower as more poor and extremely poor rock was encountered in those and all rock class 6 (with flexible steel ribs to allow for deformations) was encountered in the southern tunnels. Some parts of the Carbon zone had to be re-excavated due to very large convergence. Hence the average capacity was lower but was still around 170 m (560 ft) per month.

CONCLUSION

After more than 3 years of rock excavation it can be concluded that the chosen production method with the mechanized mucking system was functioning as expected or even better than expected.

The final distribution of rock classes were as follows:

Rock Class	RC 1	RC 2	RC 3	RC 4–5	RC 6	RC 8
Total (m)	7079	3851	13 238	139	1159	625
(ft)	23 225	12 635	43 430	455	3800	2050
%	27	15	50	1	5	2

The high contractual capacity of up to 12.5 m/day (41 ft/day) in rock class 1 has been met and the actual capacity in rock classes 2 and 3 were higher than planned and the rock excavation was hence completed earlier than planned.

The production method with the mechanized mucking system gave the following advantages:

- High capacities, as planned or better
- High capacities on mucking due to short haulage = short cycle time.
- Short stops for extension of services (ventilation, electricity etc.)
- Simplified logistics for supply and maintenance
- All recourses and service were always available on the hanging platform close to the face
- Less heavy traffic in the tunnel. Positive for both safety and working environment
- Good working environment in regards to air quality and temperature.

LAKE DOROTHY HYDROELECTRIC PROJECT
LAKE TAP AND TUNNEL, JUNEAU, AK

Jason Morrison ▪ J.S. Redpath Corporation

Ben Roberds ▪ J.S. Redpath Corporation

Chris Hickey ▪ J.S. Redpath Corporation

ABSTRACT

The Norwegian Lake Tap Method was successfully used to tap into the side of Lake Dorothy 36.5 m (120') below the lake surface. The lake tap allows water to flow into an access tunnel from the lake lowering the water elevation below its natural outfall. When the lake reaches a predetermined elevation below the outfall, a 1.5 m (60") valve in the tunnel will be shut and discharge from Lake Dorothy will stop temporarily allowing construction of a dam located 2.9 km (1.8 miles) downstream at Bart Lake.

The lake tap and tunnel were one phase of the $64 million Lake Dorothy Hydroelectric Project, owned and operated by Alaska Electric Light & Power Company (AEL&P), a private utility located in Juneau, AK. The Lake Dorothy portal site and tunnel are located approximately 25.7 km (16 miles) southeast of Juneau, AK, in the remote Tongass National Forest at an elevation of 701 m (2,300') above sea level in the mountains above the Taku Inlet. Thirty degree snow covered mountainsides and limited access by helicopter were just two of the many factors which challenged this unique project.

The lake tap and tunnel development consisted of 250 m (820') of horizontal tunnel, averaging 3.6 m x 3.6 m (12' × 12'), excavated using conventional drill and blast methods in highly competent Granodiorite and Tonalite rock, placement of one 3.9 m (13') long concrete plug, or bulkhead, located at the tunnel midpoint, and one 3 m (10') diameter lake tap into the side of Lake Dorothy.

This paper will provide an analysis of the logistical and engineering challenges involved in the development of one of only a handful of lake taps in North America. AEL&P and contractor J. S. Redpath Corporation worked together to complete this critical project phase safely and on schedule, in a location accessible only by helicopter and plagued by constantly changing, harsh weather conditions. A strict schedule had to be maintained to ensure the timely operation of the downstream dam and power plant.

INTRODUCTION

In order to provide clean, renewable energy to the City of Juneau, Alaska, Alaska Electric Light & Power Company (AEL&P) has several small hydroelectric projects tied together on an independent power grid. This power grid is not connected with other communities or power grids outside of the Juneau zip code. Due to electrical load growth within the community and extension of the power grid to uninterruptible customers, such as local mines and cruise ships, any substantial shortage in the water supply at existing hydroelectric projects puts a strain on the energy available to the community. The Lake Dorothy Hydroelectric Project (LDH) is designed to meet the rising demand for a dependable power source, minimizing reliance on diesel generated electricity

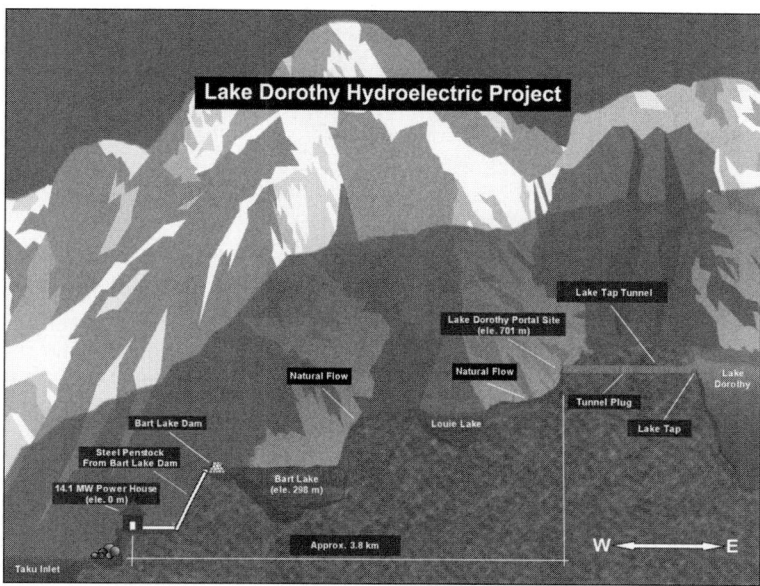

Figure 1. Lake Dorothy Hydroelectric Project overview

when either water reserves are low or energy demand is high. The LDH will provide a large reserve of clean, sustainable, energy to customers in the Juneau area.

LDH consists of 5.6 km (3.5 miles) of transmission line, a 14.1 megawatt powerhouse (located at sea level on the Taku Inlet), a 2.5 km (8,200') steel penstock from the powerhouse to a new dam at Bart Lake (elevation 298 meters (978') above sea level), and a lake tap into Lake Dorothy (elevation 701 meters (2,300') above sea level). (See Figure 1.) In order to build the dam at Bart Lake, the water elevation at Lake Dorothy must be lowered to temporarily stop the flow of water into Bart Lake. This required an access tunnel with a lake tap connected to a system of piping using a regulator and a valve to draw water from the lake during the long winter months when the lake is frozen. The inline valve and regulator will also allow AEL&P to control the flow of water from the tunnel as necessary. Flow control is critical for construction of the Bart Lake Dam and later when the powerhouse is in full operation.

In the spring of 2007 AEL&P hired a project chain consisting of underground mine contractor J.S. Redpath Corporation (Redpath), engineering, consulting, and project management firms Lachel Felice & Associates, Inc. (LF&A) and Norconsult of Sandvika, Norway. This team's job was to establish a portal site, excavate the lake tap tunnel, perform the lake tap, and begin installation of the piping system which would carry water out of the tunnel and begin lowering the elevation of Lake Dorothy. North Pacific Erectors (NPE) of Douglas, Alaska, was also contracted to complete all concrete and construction work in the tunnel and at the portal site.

The Lake Dorothy tunnel and lake tap were completed over two separate construction seasons: May to November in 2007 and April to October in 2008. Due to various weather, logistical, and permit related delays, Redpath was unable to begin the 2007 construction season until July, 2007. Construction during the winter months of December to March was not feasible given the amount of snowfall at the portal site.

LIMITED ACCESS AND PROJECT LOGISTICS

Unlike most cities and towns, there are no roads connecting Juneau to other parts of the state or North America. As a result, normal freight delivery by truck was not an option for this project. All equipment and materials were shipped in and out of Juneau from Seattle, Washington or from elsewhere in Alaska by barge or by air freight. The only "next day" or "rush" deliveries were provided by Alaska Air Lines Gold Streak service. Barge shipments to and from Juneau require a minimum of one week transit time. Given indeterminate weather windows, constant planning and advanced scheduling were crucial to ensure that materials and equipment were on site and available to meet the critical construction schedule.

Once materials and equipment arrived in Juneau, getting to the jobsite was the first major challenge. Portal access was limited from mid November to early April due to heavy snow fall. Dense fog and high winds limited daily access for the remaining construction season. To reduce the number of long helicopter trips from Juneau to the portal site, a staging area was setup at the Lake Dorothy powerhouse near the shoreline. Bulk loads of equipment and materials were brought to the staging area by barge from Juneau and stored waiting for good weather to be flown to the portal. Transporting materials from the staging area to the portal site 3.8 km (2.4 miles) away was always difficult. The portal site was constantly plagued by unpredictable weather. Poor unpredictable weather could limit safe helicopter access for as long as two weeks. As much material as was possible was flown to the portal site when weather permitted; however, storage capacity was very limited due to the small site footprint of the storage area.

ENVIRONMENTAL CONSIDERATIONS

The U.S. Forest Service, Alaska Department of Environmental Conservation, Environmental Protection Agency and the Federal Energy Regulatory Commission provided environmental oversight for the project.

The proximity to Dorothy Creek required special measures to avoid hydrocarbon spills and excavated material from entering the stream. Oil and fuel totes were stored in lined containments. Absorbent booms were placed in any area where hydrocarbons could potentially seep into the creek. Drainage was constructed so that all discharge water from the tunnel and surface areas was pumped to an oil water separator and skimmed as required. Fiber Web Ty-par was placed on four foot lift intervals on the development rock waste pad and road to filter any potential silt that might make its way to the creek. Storm drainage around the site was directed to a settling sump also lined with Fiber Web Ty-par. No incidents or spills requiring agency notification occurred during the project.

Wildlife was another consideration at this remote site and bears were naturally a primary concern. The solution was to remove any leftover food and excess garbage from the site on a daily basis. Waste and other miscellaneous materials were stored in bear proof dumpsters and also removed from the site on a regular basis.

During the start of the 2008 construction season avalanches were a significant danger at the portal site. Redpath used controlled blasting to intentionally trigger several small avalanches directly above the portal site preventing larger avalanches from covering the project area and destroying the facilities.

REMOTE SITE MOBILIZATION

When Redpath arrived at the project site, 10.6 m (35′) of snow had to be removed prior to starting work. Additionally, as there was no location to land and assemble large equipment, using larger equipment for the initial portal work was impractical. Snow removal began on June 2, 2007 using four 10 cm (4″) inch gasoline and diesel powered

Figure 2. Clearing a landing site using handheld equipment

Figure 3. Small excavator and air track drill used to establish portal site

pumps as hydraulic monitors to melt and remove the snow. Larger pumps were considered but could not be transported to the site due to helicopter lifting restrictions. Crews worked for 21 days to remove enough snow to begin development of a small landing site. Using hand held jackleg drills and blasting larger boulders, a small landing area was cleared to begin delivering mechanized equipment. (See Figure 2.) During project mobilization and start up, crews were limited to smaller equipment which would fit at the portal site. One Hitachi ZX35 excavator and one Ingersoll LM100 track drill were disassembled and flown in using a Bell 214 helicopter with a lifting capacity of 5,500 pounds. (See Figure 3.) Once at the portal site, the equipment was reassembled for use.

A small staging area and portal face were then cut into the thirty degree sloped rock face using the LM100 drill and the ZX35 excavator. Other mining equipment was then flown in to this small, remote staging area. Equipment size and weight were major logistical challenges during the entire project. Columbia Helicopters' Boeing 234 Chinook with a lift capacity 11,793 kg (26,000 lbs.) and a Boeing/Kawasaki Vertol 107-II with a lift capacity 3,629 kg. (8,000 lbs.) heavy lift helicopters were used to transport large equipment. Crews dismantled the larger pieces of mining equipment to meet weight restrictions of these helicopters. As work progressed more room for equipment and materials became available. Larger equipment could be brought to the site

Figure 4. Larger equipment arriving to site

enabling the construction of a short road and switch back along with more lay down area, a temporary shop, and an emergency overnight shelter. (See Figure 4.) Utilities at the portal site consisted of one Ingersoll Rand 260 kw and one Ingersoll Rand 290 kw diesel powered generators and two Ingersoll Rand 375 diesel powered air compressors. Service water was taken directly from nearby Lake Dorothy creek.

Larger heavy lift helicopters were only available approximately 10% of the time throughout the project. All remaining lifts had to be done using a smaller A-Star B-2 helicopter, lifting capacity of 816 kg (1,800 lbs.). The A-Star B-2 was the primary support helicopter and was used for everything from transporting crews to and from the portal site to hauling trash dumpsters and other required supplies.

Limited helicopter access created a challenging working condition for the mining crews. In the event of a medical emergency, the helicopter had to immediately access the project site providing transport of an injured worker. Therefore, if a helicopter was unable to fly to or from the project site, all work had to stop. In the case of extended weather prohibiting access, crews ceased work and stayed in the emergency shelter until helicopter access was restored. Although helicopter traffic was discouraged during darkness, the helicopter was equipped with night flying instruments, and all pilots were night rated.

Another logistical problem was the number of contractors depending on helicopter support. During the peak of construction, three separate project contractors, all in different locations required helicopter support to complete their individual tasks. AEL&P project management was responsible for coordinating helicopter activities to ensure that all contractors had a coordinated access to the helicopters as needed.

GEOLOGY AND GROUND CONDITIONS

Geotechnical analysis, tunnel design, and ground support requirements were provided by LF&A. The rock mass was a highly competent Tonalite and Granodiorite with granitic intrusions (the granite showed up as dikes in the tunnel). Overall the rock had a Rock Quality Designation (RQD) of 98 and Class I Rock Mass Rating (RMR). (See Figure 5.) Every 24 m (80′) of tunnel advance, a 30 m (100′) long probe hole was drilled to search for any unexpected water course that would require grouting and to help identify rock conditions ahead. No significant water inflows were encountered during tunnel development.

After the portal was advanced approximately 15 m (50′), engineers were able to visually inspect the rock conditions in greater detail. They determined that ground

Figure 5. Photo of rock in the portal entrance

support requirements originally proposed would not be necessary for the remainder of the tunnel except in larger openings. Daily inspections were done confirming this decision. Only one area in the drift required 2.4 m (8') long bolts to be installed on a 1.5 m × 1.5 m (5' × 5') pattern. This area was slightly blocky and fractured and only extended 33.5 m (110'). Controlled drilling and blasting helped define the outer perimeter of the tunnel and minimize over break.

The rock and ground conditions proved to be an asset when drilling the final lake tap rock plug. Due to the strength of the rock, crews were able to drill the final lake tap round to within 20.3 cm (8") of the edge of the lake without the holes collapsing or leaking. No grout curtains or other type of rock stabilization were needed prior to drilling the final lake tap round. Initial development plans called for an extensive probe hole operation to help define the nature of the rock near the edge of the lake. However, as the tunnel advanced, rock conditions determined that several of these holes were not necessary. Thirteen probe holes into the lake bottom from the tunnel were used to determine the actual distance to the lake and helped engineers determine the physical profile of the edge of Lake Dorothy. Based on the probe holes, engineers discovered the lake bottom was actually 3.6 m (12') closer to the end of the tunnel than original plans indicated.

MINING CYCLE

During the 2007 construction season, mining crews were assigned a day and night shift rotation lasting 12 hours each. A typical mining crew was made up of one lead miner, one miner, and one mechanic. The tunnel project had three separate mining crews: one master mechanic, a surface operator and a project superintendent, for a total of 12 workers. An electrician was assigned to the project on an as needed basis only. Crew changes were made at the portal or at the face to help minimize downtime. Due to the complexity of the mechanical installations mining crews worked day shift only during the 2008 construction season.

The strength and density of the rock allowed mining crews to expedite development of the tunnel. When weather was reasonable and crews could easily access the portal, advance rates of up to 12 m (40') per day were achieved. Typical advance rates averaged 7.6 m (25') per day, using one Tamrock Quasar single boom drill jumbo and two Tamrock JS-220 (2-cy) LHD's. However, poor weather conditions and access to the site prevented mining crews from achieving steady advance rates. Fog and wind constantly hampered the mining crews and forced an early exit from the portal, or work

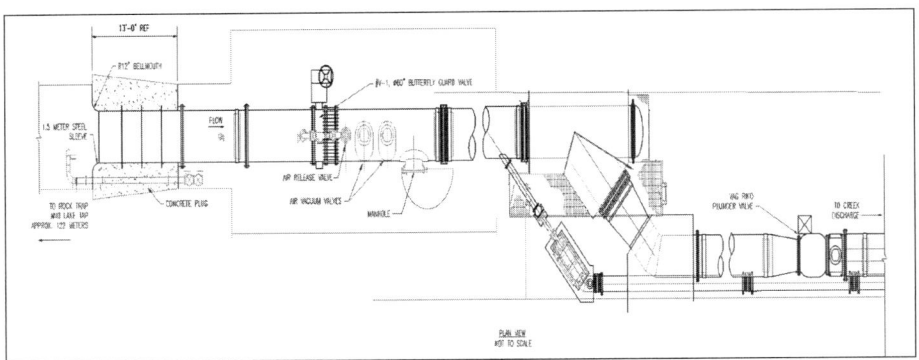

Figure 6. Tunnel plug and piping

stoppage, when a helicopter could not access the portal. A total of 35 weather related delay days were encountered during the 2007 construction season and 33 weather related delays during the 2008 season. Crews were able to establish the portal site, complete excavation of the 250 m tunnel, including the 415 m^3 (14,664 ft^3) rock trap excavation, install lake bottom ground support, and complete drilling of the lake tap, in approximately six (6) months.

TUNNEL PLUG AND PIPING

Following completion of tunnel development a 3.9 m (13′) long concrete tunnel plug was poured at the tunnel midpoint to support installation of a 1.5 m (5′) diameter butterfly type valve (guard valve) and steel pipe which would carry water out of the tunnel. A 1.5 m (5′) diameter steel sleeve with a bolted flange on the front side and smooth bell mouth on the back side was poured into the concrete plug. The guard valve was attached to the sleeve and approximately 128 m (420′) of steel pipe was connected to the guard valve. A 91 cm (36″) VAG RIKO Plunger Valve (flow control valve) was installed near the pipe discharge. The plunger valve precisely regulates pressures in the piping system and guarantees a constant supply of water at any time. Flow control from Lake Dorothy is a critical element for construction of the Bart Lake dam. (See Figure 6.)

Extensive contact grouting with cementitious and chemically based polyurethane and epoxy grouts were used at the concrete plug/rock interface to ensure the plug would be water tight. The chemical grouting process was a non-stop procedure, and once the grouting process began, crews worked a continuous 30 hour shift until the process was complete. Crews then began to install the 1.5 m (60″) diameter pipe using a monorail system that had been set up during the concrete plug construction. The overhead monorail system was used for transport and installation of pipe in the tunnel, and allowed miners to safely and easily handle the 6 m (20′) segments of 1.5 m pipe.

NORWEGIAN LAKE TAP METHOD

A lake tap is a method of physically accessing a body of water from below the surface without first lowering the water's elevation or using coffer dam construction. The principal in lake tapping involves driving a tunnel under a lake to within a few short meters of the edge of the lake leaving a short rock plug between the tunnel and lake bottom. (See Figure 7.) The final plug is then blasted piercing the lake floor from below. Rock from the plug settles into a rock trap leaving an unobstructed path for water to

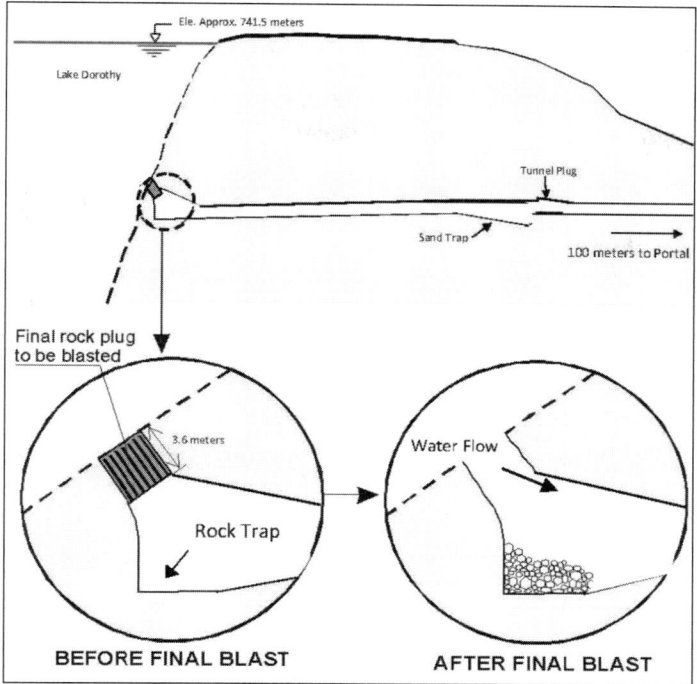

Figure 7. Principles of the lake tap method

flow. Lake taps are unique because there is only a single chance of a successful lake piercing. Effects of an unsuccessful lake tap include significant schedule delays and expenses to recover the lake piercing. Successful lake taps require trained specialists having practical experience in these challenging works. Norconsult was responsible for design of the tunnel, from the tunnel plug to the lake tap, the lake tap piercing, and explosives used for the blast. Norconsult arrived at the portal site on August 12, 2007 to assist and oversee the lake tap operations. The final lake tap blast took place on August 19, 2008.

The end of the tunnel was excavated with a 415 m³ rock trap 9 m high with the rock plug located above. To access the rock plug, mining crews drove in on the bottom level then moved back and breasted down top cuts, working on top of the muck pile as they ramped up. Once the tunnel was within 3.6 m (12') of the lake bottom, crews set up to drill out the final rock plug and install ground support around the lake piercing. (See Figure 8.) This support was intended to secure the perimeter of the lake tap after the blast. Ground support consisted of 14 each 3 m (10') galvanized cable bolts fully encapsulated in grout and 14 each 3 m (10') galvanized #8 Dywidag bar point anchored and also fully encapsulated in grout, installed around the perimeter of the lake tap penetration.

Prior to blasting the final rock plug the back half of the tunnel was filled with water using two 43 kw (58 hp) pumps leaving a small air pocket directly in front of the rock plug. This air pocket was pressurized using a 5 cm (2") diameter high density polyethylene (HDPE) pipe that was installed from the tunnel plug to the end of the rock trap. At the tunnel plug the HDPE pipe was connected to a 5 cm (2") diameter stainless steel pipe that was poured into the concrete. The purpose of the air pocket is to counteract,

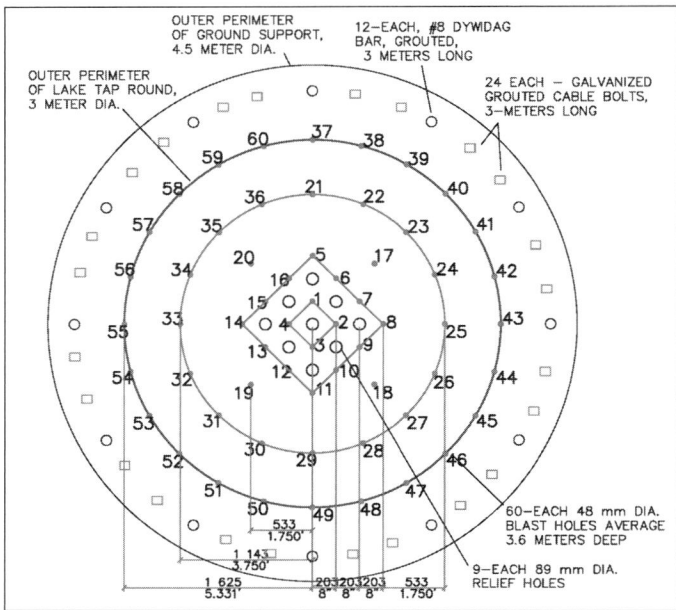

Figure 8. Lake tap round with ground support

or soak up, energy given off from the blast and to help isolate the explosives from water prior to detonation. The volume and pressure of the air pocket had to be precisely calculated and closely monitored as the tunnel was being filled with water. A system of float sensors and a single pressure transducer installed directly in front of the rock plug was use to monitor the air pressure and water elevation inside the rock trap. A communication cable connected to the float sensors and pressure transducer was attached to the HDPE air pipe and penetrated the tunnel plug through a separate 5 cm (2") stainless pipe.

Several probe holes drilled directly through the face of the final rock plug helped engineers determine specific depths for each of the 69 blast holes that were drilled. A detailed blast hole drill pattern, listing hole depth and angle, was provided by Norconsult and was carefully followed for maximum accuracy.

With water temperatures averaging 3.3°C (38°F) and head pressures of 4 bar (58 psi.), drilling probe holes into Lake Dorothy required concentration and attention to detail. First, a smaller 47 mm (1.8") diameter hole was drilled approximately 1.2 m (4') deep and then reamed out to 89 mm (3.5") in diameter. A 1.2 m (4') long packer having an inner diameter of 50 mm (2") was then installed and secured. Crews were then able to drill through the packer, recording the hole depth, until breaking into the lake. Due to the head pressure against the drill steel, crews depended on the jumbo's hydraulic centralizer to safely remove the drill steel from the probe hole. After the drill steel was removed, a valve on the end of the packer was shut, stopping the inrush of icy lake water. All probe holes penetrating the lake were grouted with Portland Type I-II cement using a one-half to one (½:1) water to cement (W/C) ratio.

Loading and blasting the final rock plug had to be completed within a 24-hour time frame to eliminate the risk of moisture exposure to the nitroglycerin (NG) explosive. To load the final lake tap round, crews worked from scaffolding which had been set up after the rock trap was mucked out. To save time while loading the round, the NG based explosive was preloaded into PVC pipes. Each pipe was cut to a specific length and marked for its intended use in a specific blast hole. The PVC pipes were preloaded outside in good ventilation, reducing exposure of the NG explosive to the miners. Each blast hole had two non-electric (nonel) detonators, or was "double primed," to ensure the round went off as planned. All 138 nonel leads were tied into groups of 5 until only 2 nonel leads remained. The two remaining leads tied into two (2) electric blasting caps and the firing cable. The firing cable was strung through the tunnel plug, outside of the tunnel to the edge of Lake Dorothy. The scaffolding was then disassembled and carried out of the tunnel before being filled with water. The lake tap was initiated from the Lake Dorothy shoreline so all individuals involved were able to witness the air pocket and energy from the blast emerge from below the water line.

CONCLUSION

Completing a remote construction project in Alaska is a unique challenge. Supporting a remote project entirely with helicopters, and working under constantly difficult environmental conditions adds to this challenge. Being limited to less than a one acre footprint for storage for equipment and supplies requires equipment to be versatile and planning to be precise. Ordinary planning and scheduling does not suffice. Construction personnel must be willing to work under ever changing weather conditions, be flexible and innovative, and must properly sequence all portions of the work to maintain project schedules. Co-operation between all parties involved is an absolute necessity to allow work to be accomplished in a fashion which yields quality products.

ACKNOWLEDGMENTS

J.S. Redpath Corporation would like to thank Lachel Felice and Associates, Alaska Electric Light and Power Company, and Norconsult for the opportunity to work together on this unique project and for their individual contributions and support while preparing this manuscript.

UNDERGROUND CONSTRUCTION FOR A COMBINED SEWER OVERFLOW SYSTEM IN PROVIDENCE, RHODE ISLAND

John L. Kaplin ▪ Gilbane

Jeffrey P. Peterson ▪ Jacobs Associates

Philip H. Albert ▪ Narragansett Bay Commission

ABSTRACT

The underground construction for the Phase I Combined Sewer Overflow (CSO) Project is on an impressive scale by any measure, appearing more so in a midsize city like Providence, RI. The Narragansett Bay Commission recently completed $350 million in construction to improve the water quality in Narragansett Bay. This project included a number of serious challenges, delays, and disputes. Looking back on this complex project completed within budget, it is hard to consider the work as anything other than a success. Along the way, this outcome did not always appear certain.

INTRODUCTION

With a sewer system dating back to the 1870s and a treatment plant dating back to the early 1900s, Providence and the surrounding communities have a long history of keeping up with population growth and technological advances in handling the city's sewage and stormwater. The Narragansett Bay Commission (NBC), which has existed since 1982, provides sewerage treatment for 10 communities, with 360,000 residents and 8,000 businesses in the Providence area. Many upgrades were performed on NBC's system over the years. Since its inception, NBC has improved the function of the Fields Point Wastewater Treatment Facility from one of the nation's worst-rated treatment system in the 1980s to being designated by the EPA as the nation's best secondary treatment system in 1995. However, despite an effective secondary treatment system, Providence is burdened with combined sewer and stormwater overflows to the upper Narragansett Bay on an average of 71 times annually. The Bay is a water body of national estuarine significance, and the overflows have major impacts on shellfish harvesting and other commercial and recreational uses.

The Combined Sewer Overflow (CSO) system was mandated by the federal Clean Water Act to end overflows. A preliminary design was first developed in 1995 to address the overflows. Since then, the design has developed into a three-phase program that has spread construction out over 20 years.

When the $350 million plus Phase I program began, it was the largest civil works project in the history of Rhode Island. Although its scope is impressive in its own right, it is amplified when considering the project was constructed in the smallest state in the union by a midsize wastewater agency. Prior to Providence, deep tunnel storage projects for CSO abatements in the U.S. had been built predominantly in major metropolitan areas by larger agencies such as Milwaukee, Chicago, and Atlanta. The NBC managed to overcome difficult funding and public affairs hurdles to start Phase I. In terms of scale, this is a relatively small owner tackling a large project.

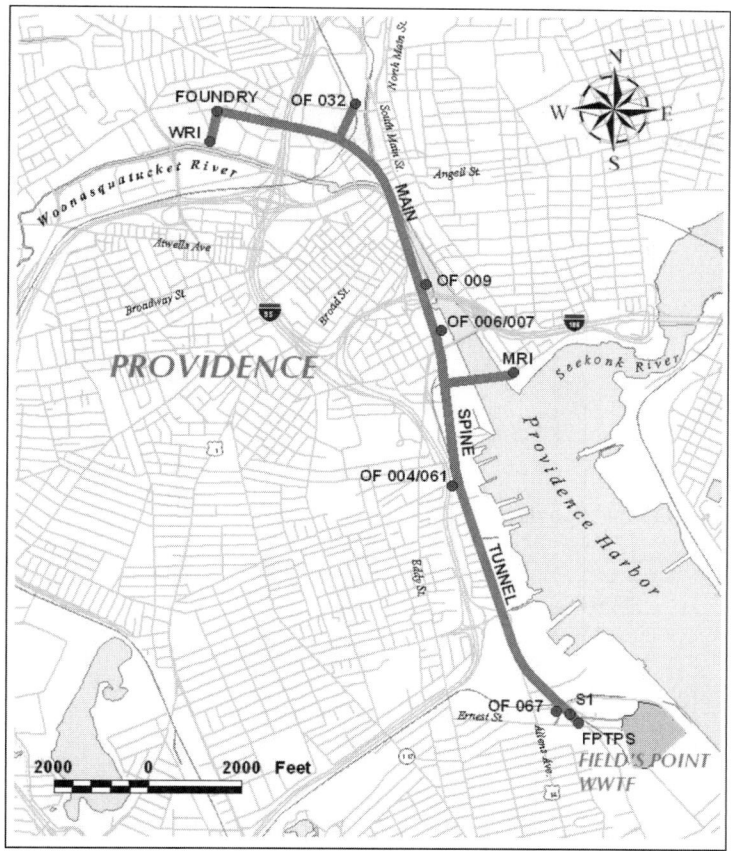

Figure 1. Location plan

Figure 1 shows the Phase I layout. Starting in 2001, nine construction contracts were issued to build the Phase I CSO system. The subject of this paper is one of these contracts, Contract 6, which was the largest and covered the underground works. It was completed between 2002 and 2007 by Shank/Balfour Beatty (S/BB) with a final contract value of $173 million, 6% above the original bid. The scope of the contract is summarized in Table 1.

Other Phase I contracts outside the bounds of this paper include six construction packages that tie the existing CSO system to the tunnel. Each of these packages consists of diversion structures, connecting conduits, gate and screening structures, and drop and vent shafts drilled down from the surface. The construction of these shafts is discussed in Castro et al. (2007). There also is a $54 million contract in the final stages of completion to fit-out an underground pump cavern with a 189 ML/day pump station, construct associated support buildings, and complete the mechanical, electrical, and instrumentation work. As of this writing, all of this work is sufficiently completed such that the bulk of Providence's storm overflows within the Phase I area is now collected in a tunnel rather than polluting the bay. Effluent is pumped from the tunnel after storms to the existing Fields Point Treatment Plant at the south end of Providence.

Table 1. Contract 6 scope of work

Work Element	Description
Shafts (six total)	S-1: main work shaft servicing MST; screening shaft for permanent works. Utility and Access: servicing the pump cavern. Foundry: north terminus of MST; vent shaft for permanent works. 067: drop and vent shafts for a CSO connection (six other locations constructed under other contracts; not in scope of this paper).
Main spine Tunnel (MST)	4,963 m (16,284 ft) long, 7.9 m (26 ft) diameter, 70 m (230 ft) deep. This is the main CSO storage tunnel, which can hold 235 million liters (62 million gallons).
Adits (seven total)	2.7 to 4.8 m (8.75 to 15.75 ft) diameter deaeration chambers; 2.4 m (8 ft) diameter adits. Adits inclusive of chambers total 1,237 m (4,057 ft) in length. Convey CSO flows from drop shafts to MST.
Pump cavern	A 189 ML/day (50 MGD) Field's Point Pump Station housed in a 92 m (300 ft) deep rock cavern with dimensions of 35.7 m × 18.6 m × 20.7 m (117 ft × 61 ft × 68 ft). S/BB constructed the cavern, floor, and roof; all other fit-out work was performed in subsequent contract. The Contract 6 pump cavern work is not covered in this paper (see Hughes et al. 2008).

Table 2. Shaft excavation data

	S-1 Shaft m (ft)	Utility Shaft m (ft)	Access Shaft m (ft)	Foundry Shaft m (ft)	OF-067 Drop Shaft m (ft)	OF-067 Vent Shaft m (ft)
Excavation diameter	Soil: 15.2 (50) circular Rock: 10.4 (34) square	11 (36)	4.3 (14)	11 (36)	3.7 (12)	1.8 (6)
Finish diameter	7.9 (26)	9.8 (32)	3.4 (11)	7.9 (26)	2.7 (9)	1.2 (3.9)
Depth to water table	4.6 (15)	6.4 (21)	6.4 (21)	7 (23)	4 (13)	4 (13)
Depth to top of rock	50 (165)	48 (157)	53 (174)	46 (152)	50 (164)	50 (164)
Depth of shaft	89 (292)	74 (242)	74 (242)	78 (257)	71 (233)	71 (233)
Soil support method	Ground freezing					
Initial and final shaft liners	Cast-in-place concrete, slip-lining method					

Three bids were received for Contract 6 on September 25, 2001, ranging from $163.5 million to $227.4 million. As low bidder, S/BB was awarded the contract in December 2001. S/BB mobilized to the site in early 2002.

SHAFTS

First, three shafts were sunk at the S-1 site at the southern terminus of the project, near the Field's Point Treatment Plant at the Port of Providence. The S-1 was sunk first, starting in 2002, followed by the Utility and Access shafts. Construction on the Foundry shaft at the northern end of the Main Spine Tunnel (MST) started later, in 2003, as it would not be needed until the MST hole-through. The 067 drop and vent shafts were frozen and raise-bored near the tail end of Contract 6. Table 2 summarizes pertinent facts about these six shafts.

The S-1 is the main work shaft from which the MST was excavated. As noted in Table 2, the depth to bedrock is quite deep at 50 m (165 ft). The freezewall constructed to provide soil support for this shaft encountered a substantial delay, which impacted the critical path schedule of the project and the overall Phase I program. It was originally estimated that the freezewall would close in 38 days from the start of the freeze, and by 50 days the freezewall thickness would be sufficient for excavation to start. The actual duration for closure was 122 days, and excavation started at 138 days. Further time was lost during the shaft excavation due to the increased ice growth inward during the initial freeze delay. This additional frozen ground had to be chipped by impact hammer, prolonging the excavation far beyond the time it would have taken had the bulk of the core remained unfrozen as planned. These events led to considerable efforts to determine cause and mitigate delays. The contractors worked aggressively to resolve the freezing problems with a variety of techniques, including additional freeze pipes, instrumentation, grouting, and pump tests.

After the freezewall closed and soil excavation was completed, the contractor and freezewall subcontractor submitted a claim that contended that three differing site conditions (DSCs) had interacted to cumulatively delay the freeze. They claimed that "free-phase" gasoline contamination delayed the freeze near the top of the shaft and was remedied by the installation of shallow freeze pipes. This problem masked two deeper "windows" at 37 m and 47 m (120 ft and 155 ft) depths caused by higher-than-anticipated permeability and groundwater flow up from bedrock through preconstruction grout holes that had been incompletely sealed by a previous NBC contractor.

The owner's team was not in complete agreement with the contractors' explanations as to the causes for the freezewall problems. For example, no "free-phase" gasoline was ever discovered. When freezing was initiated, the shaft core had already been excavated nearly to the groundwater table, which was then exposed to hot summer temperatures that apparently overwhelmed the desired effect of the freeze pipes on groundwater temperature. Thus, warm water, not gasoline, was the most likely culprit of the shallow freeze problems, and this conclusion was supported by direct temperature measurements. As for the deep windows, the owner's team concurred that there likely was leakage up from rock into the shaft core but found the flow could have come through bedrock fractures, a contractor's own observation well, or grout holes as asserted. Soil permeability was found locally to be higher than anticipated; however, the gradient of the groundwater table was no steeper than described in the contract documents.

There also was some debate about contract provisions, specifically whether the geotechnical information included was sufficient to determine groundwater flow for the purposes of designing a freezewall, and whether the contractor needed to perform investigations. This was never done, with the freezewall contractor arguing that the information already included was adequate.

Sorting out responsibility for delays on a freezewall claim with multiple problems is certainly not clear cut. In the end, agreement was reached as part of a global settlement. The lessons learned are that closer attention must be paid when determining whether groundwater flow at a site is an issue for ground freezing, who will conduct those investigations, and how it is accounted for in the design of the freezewall. Secondly, a shaft collar installed and soil excavated to near the groundwater table within the core also must be properly accounted for in the design of the freezewall, especially during summer months when elevated air temperatures may persist very close to the ground that is to be frozen. The causes and mitigation measures to resolve this problem are described in greater detail in Schmall et al. (2007), which presents the perspective of the ground-freezing contractor.

The ground freezing at the other five shafts proceeded largely as planned. Cast-in-place (CIP) liners were used for longer-term soil support by means of the slip-lining method at all shafts, following which the ground-freeze systems were shut off. Excavation of rock in all the shafts was by drill and blast. The exception was the raiseboring of the 067 shafts. All shaft rock excavation proceeded without any noteworthy problems. However, there were some issues at the Foundry shaft that deserve mention, as briefly described below.

During the drilling for Foundry shaft pregrouting from the surface and the subsequent installation of freeze pipes, heavy water losses and grout takes were observed in the upper 6–9 m (20–30 ft) of bedrock. No clear cause of permeable ground was detected in the core borings, so there naturally was cause for concern with shaft sinking through this ground. A decision was made to increase the length of freezewall pipes to a depth from surface of 56 m (185 ft), well into rock. When the shaft was excavated through this zone, the occurrence of 12–50 mm (0.5–2.0 in.) wide fractures filled with a clean, fine-to-medium sand of probable glacial origin were discovered at a depth of 6 m (20 ft) below the top of rock. Grout had penetrated some of the fractures, but incompletely. These fractures would have been heavy water makers had the ground not been frozen and subsequently sealed by the cast-in-place liner. Even then, chemical grouting through the CIP liner was required to cut off silt laden inflows that occurred through shrinkage cracks.

These fractures also were observed to a lesser extent in the Utility shaft, where some remedial work was required. In fact, the authors are aware of other shafts in New England where this problem has been encountered. The common characteristics of these fractures are their occurrence in the upper 6 m (20 ft) of bedrock, an aperture of up to several inches, a filling of clean sand, and the potential to produce high inflows at depths below the water table. They often are undetected by the single boring typically used for shaft exploration and require grouting, time, and money so that shaft sinking can proceed. Perhaps these features are unique to shaft sinking at locations with a record of glacial activity.

MAIN SPINE TUNNEL EXCAVATION

It has been a practice of the M.L. Shank side of the S/BB team to design and build its own tunnel boring machines (TBMs), and the Contract 6 machine was a scaled-up version of their previous smaller TBMs. Other than the cutterhead, main shield, and tail shield being fabricated by Hitachi of Japan, the fabrication and assembly of the TBM and trailing gear were performed by S/BB forces.

Under a value-engineering proposal, the tunnel liner was a composite consisting of 25.4 cm (10 in.) thick precast segments and a cast-in-place liner 30.5 cm (12 in.) thick, placed after mining was completed. Installation of the precast segments was within the TBM shield. As the TBM advanced forward and the four-piece, 1.2 m (4 ft) wide precast ring emerged from the tail shield, hydraulic jacks were engaged in a 0.6 m (2 ft) wide open key gap at the crown to expand the segments against the rock. Struts filling the key gap were installed thereafter to hold the segments in place.

The cutterhead was 9.1 m (30 ft) in diameter and equipped with 65 43-cm (17-in.) diameter, back-loading cutters. The machine employed 20 jacks to propel and steer off precast segments, with a total available thrust of 28,024 kN (6,300,000 lb). The cutterhead was rotated at 3 to 5 rpm and was driven by 20 hydraulic motors rated at a total of 1,837 kW (2,465 hp). Power for the thrust, head rotation, conveyors, and segment erection was electric over hydraulic.

The TBM was launched from a starter tunnel at the S-1 shaft on March 8, 2004 (Figure 2). Within a month, production was in the 12.2 to 13.7 m/day (40 to 45 ft/day) range. Mining was performed on one long shift, from 6:30 a.m. to 6 p.m., with maintenance performed on the back shifts. The basic mining cycle consisted of mining through 1.2 m (4 ft) of rock in 25 to 40 min, separated by time to install segments and wait for

Figure 2. TBM in starter tunnel

trains. Haulage by rail was performed by three locomotives hauling five muck cars and one segment car. Two rail switches were used—a fixed one at the S-1 shaft and a second, forward one that was periodically advanced toward the heading to improve rail transit. A muck car roll-over dump, a vertical conveyor in the shaft, and a radial stacking conveyor at the surface were used to move tunnel muck to a surface pile, where it was trucked off site for commercial uses. As with the TBM, most of this equipment was designed and fabricated by S/BB forces.

The rock encountered in all the underground works was a weakly metamorphosed sequence of folded sedimentary rocks of the Rhode Island Formation that includes sandstone, shale, siltstone, graphitic shale, and conglomerate. With the exception of the graphitic shale, the rocks were stable with good stand-up time.

A ground classification system for tunneling was included in the Geotechnical Baseline Report (GBR) and made part of the contract documents, as a two-pass lining system was permitted in the bid documents. The classification system was based on criteria that characterized the rock behavior and was intended for use in the selection of ground support. Type I ground required rock bolts and shotcrete; Type II ground called for steel sets, lagging, and shotcrete. S/BB's precast segments installed immediately behind the TBM made the need to classify ground unnecessary. However, the classification system did become a source for a differing site condition (DSC) claim related to TBM penetration rate. S/BB anticipated that the TBM would penetrate weaker, broken ground faster than less-fractured, more-competent ground, relying upon the ground classification quantities provided in the GBR for its estimate. This was not an unreasonable approach, although certainly not the one intended by the authors of the report. Considerably more Type I ground was encountered in the MST than the 65% predicted by the GBR, and the actual TBM penetration rate was lower than bid. The claim was eventually settled as part of a global settlement.

None of this should overshadow the overall performance of the TBM, which operated with a high degree of reliability and made steady progress. The TBM holed through at the Foundry shaft in early December 2005 (Figure 3). Production on the single long shift, five days a week, yielded 274–427 m (900–1,400 ft) per month, with a daily production between 12.2 m and 18.3 m (40 ft and 60 ft) per mining day. Table 3 provides a summary of performance data for the TBM. One factor that affected monthly production was the number of days lost to mining to perform pre-excavation grouting, as discussed next.

Table 3. TBM production data

Description	Quantity	Unit
Single-shift mining	10.5	h
Average TBM penetration	2.1 (6.8)	m/h (ft/h)
Average cycle time (mine 1.2 m [4 ft], erect segment ring; wait for next train at heading)	54	min
Mining	36	min
Segment erect	11	min
Wait for train at heading	7	min

Figure 3. TBM cutterhead following hole-through at the Foundry shaft (photo by Sue Bednarz)

PRE-EXCAVATION GROUTING

It was anticipated that pre-excavation grouting would be required to maintain a total steady-state groundwater inflow from the MST, adits, and shafts below 12,490 Lpm (3,300 gpm). All inflow except that from the pump cavern was pumped from a dewatering station at the bottom of the S-1 shaft to a settling pond and pretreatment system at the surface, then to the Field's Point Treatment Plant for discharge. Contract specifications called for grouting if probe holes through the tunnel face, which had to provide continuous, overlapping coverage, produced greater than 190 Lpm (50 gpm). Once triggered, grouting would be performed using Type III or microfine cement grout.

During the early stages of mining, inflow into the tunnel increased at a greater rate than anticipated, despite the use of the pre-excavation grouting program. By the time 25% of the MST had been mined, the inflow already had reached 4,540 Lpm (1,200 gpm), or 36% of the total anticipated inflow. For this reason, the grouting program was modified a number of times, and it is worth noting the more significant changes to the program based on conditions encountered. A summary is shown in Table 4. Figure 4 is a graphic of the inflow and the grout-event history.

Table 4. Pre-excavation grouting program summary

Criteria	Original Program	Modified Program	Discussion
Grout trigger	190 Lpm (50 gpm)	57 Lpm (15 gpm)	Trigger was changed several times: first down to 26 Lpm (7 gpm), then up to 57 Lpm (15 gpm), which was the predominant trigger during mining.
Grout pressure	3.5 MPa (500 psi)	2.4 MPa (350 psi)	At the higher pressures, grout takes were in the range of 142–170 m³ (5,000–6,000 ft³) per event. It was judged that a considerable portion of the grout was pushed through fractures outside the tunnel envelope, perhaps due to hydrojacking. After the pressure was reduced, grout takes were lowered by a factor of two to four, with no apparent loss in effectiveness, as is visible in Figure 4 from Station 80+00 and beyond.
Grout holes	two	four	Based on inspection of the face, it was judged that two holes were insufficient for a 9.1-m (30-ft) diameter face.
Equipment	Complete mobilization of all grouting equipment from surface to heading for each grout event, probing and grouting through one hole at a time.	Rolling gantry behind the TBM trailing gear was fabricated; included equipment to allow rapid mob/demob for grout events and ability to probe/grout two holes simultaneously. A second grouting/mixing plant was also fabricated to boost grouting capacity.	No mining could be accomplished when pre-excavation grouting was performed. Initially, mob/demob time resulted in two lost mining days per grout event. These equipment changes, which the owner paid for through a change order, reduced the lost mining days from two to one while increasing the number of holes that could be grouted at a time.
Grout	Microfine and Type III	Same. However, with probe flow of greater than 190 Lpm (50 gpm), Type III would be started with to reduce cost.	This was done to reduce cost by initially filling heavy water-making fracture systems that took large quantities of grout at low pressure with the less-expensive Type III cement.
Cost	$5.8 million paid using unit price items	$10 million using unit price items with quantities replenished through change order.	All parties universally agreed that inflow had to be reduced by pre-excavation grouting to a point where it was manageable for mining and final CIP lining, and that cut-off grouting (attempting to grout off inflows after exposed in the mined opening) was not a reasonable approach. A total of 3,364 m³ (4,400 yd³) of grout was pumped over 44 grouting events, or 59 working days.

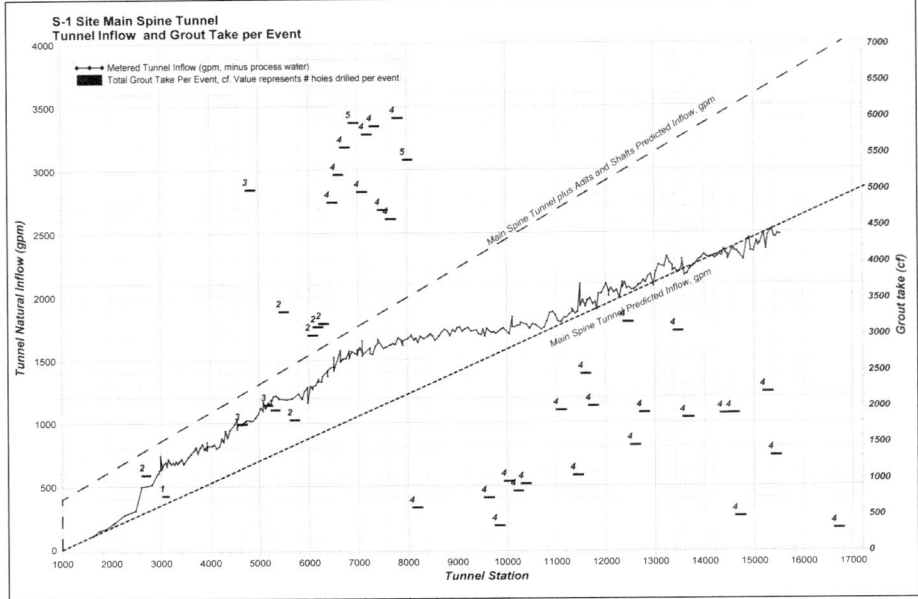

Figure 4. Water inflow and grout take per event in the Main Spine Tunnel

There was relatively little decay observed in the inflow rates over time in the MST. This assessment was based on direct measurement of sustained inflows. The overall effectiveness of pre-excavation grouting can therefore be analyzed in rough terms by comparing probe inflow to tunnel inflow in grouted versus ungrouted portions of the tunnel. The results are shown in Table 5.

It is critical to understand the purpose of a grouting and water handling program in a rock tunnel. In this case, the intention was to keep inflow at a level manageable for tunneling equipment operation and successful placement of the CIP liner without washout. Following placement of the liner, subsequent contact grouting and a final round of chemical grouting at a few select locations, the long-term, steady-state inflow from the MST, all adits, and connecting shafts was approximately 1,140 Lpm (300 gpm).

ADIT CONSTRUCTION

Seven adits were excavated off the MST to connect the tunnel to the drop and vent shafts previously installed by other NBC contractors. To mitigate the schedule loss of a year's time that occurred prior to the start of MST mining, all adits were excavated and lined concurrently with the excavation of the MST. The exceptions were the 067 and WRI adits, which were driven from the base of the S-1 and Foundry shafts, respectively. Elevated decks were constructed in the MST at each of the adit portals to service the adit excavation and concrete lining. This allowed access to the adits, which intersected the MST at the spring line while allowing MST train traffic to pass unimpeded beneath the deck (Figure 5).

Excavation was by drill and blast and proceeded on a 24-hour basis. Most of the blasting was performed on the second and third shifts so as to avoid interference with the daytime MST mining operation. Considerable outreach helped project neighbors understand the program and allayed common fears that come with blasting. Convincing

Table 5. Pre-excavation grouting performance summary. Grouted versus ungrouted intervals of the Main Spine Tunnel.

	Interval Length, m (ft)	Increased Inflow in Tunnel, Lpm (gpm)	Cumulative Probe Inflow for Interval, Lpm (gpm)	Ratio of Increased Tunnel Inflow/ Probe Inflow
Grouted intervals	2,425 (7,957)	4,546 (1,201)	12,343 (3,261)	0.37
Ungrouted intervals	2,275 (7,465)	5,435 (1,436)	1,154 (305)	4.71

Figure 5. Elevated deck for construction of the adit from the Main Spine Tunnel

authorities to permit 24-hour blasting was also simplified because few businesses above the adits operated at night and there were few residences.

One of the adit alignments is beneath the main power plant in the city, causing concern with the local hospital, which had previously experienced power outages unrelated to construction. Through meetings and test blasts, these concerns also were addressed, and adit excavation proceeded as planned.

Ground support consisted of swellex bolts and shotcrete in competent rock (Type I ground). Steel sets, lagging, and shotcrete were used where weak graphitic shale was encountered (Type II ground). Probing ahead of the excavation face was performed, yet pre-excavation grouting was triggered rarely and water inflow was generally not a problem in the adits. Table 6 provides a summary of pertinent adit/chamber data.

The deaeration chamber connections to the drop and vent shafts installed under other NBC contracts proceeded as planned with one exception. A failure of the shaft liner was experienced just above the intersection with the chamber at one adit where Embedded Cylinder Pipe (ECP) had been used for a shaft liner. In just a few hours, the mortar lining spalled off the bottom 6 m (20 ft) of the shaft in dramatic fashion, and the thin steel lining ballooned inwards under the groundwater pressure, pinching the shaft diameter to a few feet. It was apparent that water had leaked in between the inner layers of the pipe, causing it to burst. The failed pipe section was removed and repaired with mesh and shotcrete. In hindsight, a steel transition section for the bottom 3 m (10 ft) of each shaft where cylinder pipe was selected would have been far more robust and able to withstand the rigors of the chamber excavation and shaft installation. For three other locations that also had ECP pipes used for shaft liners, procedures were revised and all were completed without problems.

CONVENTIONAL AND ROCK TUNNELING

Table 6. Adit/deaeration chamber summary

Adit	Adit Length m (ft)	Chamber Length m (ft)	Total Length m (ft)	Adit Excavated Diameter m (ft)	Adit Finished Diameter m (ft)	Chamber Excavated Diameter m (ft)	Chamber Finished Diameter m (ft)	Ratio Type I/II Ground %	Shifts to Excavate
067	3.0 (9.7)	36 (117.5)	39 (127.2)	3.4 (11.3)	2.4 (8)	5.8 (19)	4.8 (15.75)	100/0	69
004/061	1.5 (5)	20 (65.5)	22 (70.5)	3.4 (11.3)	1.4 (4.5)	3.7 (12)	2.7 (8.75)	100/0	28
MRI	521 (1,710)	20 (65.5)	541 (1,776)	3.4 (11.3)	2.4 (8)	3.7 (12)	2.7 (8.75)	86/14	386
006/007	96 (314)	20 (65.5)	116 (380)	3.4 (11.3)	2.4 (8)	3.7 (12)	2.7 (8.75)	100/0	81
009/010	40 (132)	36 (117.5)	76 (250)	3.4 (11.3)	2.4 (8)	5.8 (19)	4.8 (15.75)	100/0	83
032	226 (740)	20 (65.5)	246 (806)	3.4 (11.3)	2.4 (8)	3.7 (12)	2.7 (8.75)	100/0	152
WRI	189 (619)	20 (65.5)	209 (685)	3.4 (11.3)	2.4 (8)	4.3 (14)	3.2 (10.5)	33/67	143

Figure 6. Gantry for bulkhead construction at downstream end of Main Spine Tunnel forms

Concrete lining and contact grouting of the adits and deaeration chambers was accomplished by pumping from a surface pump fed by ready-mix trucks through slicklines run down the drop shafts.

MAIN SPINE TUNNEL—FINAL LINING

The final cast-in-place MST liner was constructed from the Foundry shaft to the S-1 shaft by pumping concrete from the surface at four locations. A fleet of ready-mix trucks delivered 27.6 MPa (4,000 psi) concrete with 10 mm (⅜ in.) stone from a batch plant constructed and operated by S/BB at the S-1 site. The maximum pumping distance was 1,006 m (3,300 ft) from the pump to the forms. Additives included new generation full-range and high-range water reducing admixtures, retarder, fly ash, and an air-entraining admixture. A total of 44,340 m^3 (58,000 yd^3) of cast-in-place concrete was placed in the MST for the final lining.

Once beyond the initial learning curve, 50 m/day (165 ft/day) placements were achieved. This required a sustained delivery of 443 m^3 (580 yd^3) over eight hours. A bulkhead was constructed at the downstream end (Figure 6) for each day's pour, and concrete was typically filled to a height of approximately 80% full or higher, leaving a sloping construction joint. Within the forms, a slickline at the crown was dragged along hanging rollers as the pour progressed.

When the liner was nearly complete, displacement was observed to have occurred along this construction joint at about half of the 100 bulkhead locations. In the most severe case, a concrete block approximately 0.6 m × 1.2 m (2 ft × 4 ft) fell to the invert. An investigation revealed that concrete in the crown area was displacing away from the precast segments along the sloping joints, possibly as a result of increasing pressure from the recovering groundwater table. A repair plan was developed and implemented to remove the displaced concrete, install dowels, and backfill with shotcrete. Pressure relief holes were added to the problem areas. Repairs were completed without impact to the project schedule, and costs to perform this work were split between the NBC and S/BB. An inspection performed a year after the work was completed revealed that the repairs were successful. In the future, the authors suggest that the use of sloping cold joints in large diameter cast-in-place liners should be avoided and full bulkhead pours stipulated as a requirement.

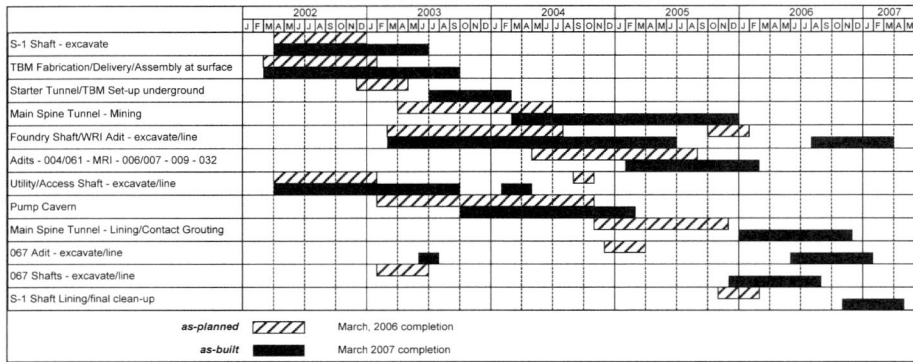

Figure 7. Comparison of planned and actual schedules

CLOSING

A year was lost on the schedule prior to the start of MST mining. Of this, six months is accounted for on the S-1 shaft freeze and excavation. A concurrent and continuing delay associated with TBM delivery, fabrication, and setup in the starter tunnel resulted in an additional four-month slippage. S/BB developed a recovery plan to mine and line six of the seven adits concurrently with MST mining. This plan was successful in preventing further schedule loss and was well managed by S/BB. Figure 7 depicts a comparison of the planned and actual project schedules.

Contract 6 was completed within budget and at a final contract price within 6% of the original bid price. Although some delays also occurred on the shaft/surface contracts, the overall Phase I program also was completed within budget and within 7% of the total original bid price.

Prior to the Phase I CSO, the last major tunneling project completed in Providence was over 50 years ago, and it was a relatively short bus tunnel. Phase I remained largely out of the local press, allowing the project to achieve its objectives without negative reviews. Multiple tunneling technologies were used to improve the environment, in this case Narragansett Bay. This project is an example of how a public agency in a midsize city like Providence, without significant tunneling experience, can tackle a large and complex underground project for the benefit of its citizens and do it within budget. Tackling problems early and head on is one of the keys to success, particularly with large projects.

REFERENCES

Castro, R., Vincent, F., Hughes, G., and Albert, P. 2007. Drop shafts for Narragansett Bay Commission CSO Abatement Program. In *Proceedings of the Rapid Excavation and Tunneling Conference*, ed. M.T. Taylor and J.W. Townsend. Society for Mining, Metallurgy, and Exploration.

Hughes, G., Kaplin, J., Halim, I, Albert, P. 2008. Design and construction of the Fields Point Tunnel Pump Station for the Narragansett Bay Commission CSO Abatement Program, Providence, Rhode Island. In *Proceedings of the North American Tunneling Conference*, ed. M.T. Taylor and J.W. Townsend. Society for Mining, Metallurgy, and Exploration.

Schmall, P., Corwin, B., Spiteri, P. 2007. Ground freezing under the most adverse conditions: Moving groundwater. In *Proceedings of the Rapid Excavation and Tunneling Conference*, ed. M.F. Roach, M.R. Kritzer, D. Ofiara, and B.F. Townsend. Society for Mining, Metallurgy, and Exploration.

MODERN CAVERNS IN GOTHAM—GEOTECHNICAL AND DESIGN CHALLENGES FOR LARGE ROCK CAVERNS IN MANHATTAN

Timothy P. Smirnoff ▪ Parsons Brinckerhoff

ABSTRACT

Sustaining the region's mobility needs, achieving projected economic growth, and maintain Midtown Manhattan as a center of regional, national and global importance has led to continued investment in commuter rail and transit infrastructure in New York City and resulted in design and construction of large underground caverns. The current plans for stations in Manhattan include New York Penn Station Expansion (NYPSE) caverns for Trans Hudson Express (THE) Tunnel Project for New Jersey Transit (NJ Transit) and the Port Authority of New York and New Jersey (PANYNJ), Metropolitan Transportation Authority Capital Construction Long Island Railroad (MTACC/LIRR) twin caverns for East Side Access (ESA) Project, and station caverns of the No. 7 Line Subway Extension Project and Second Avenue Subway Project for Metropolitan Transportation Authority Capital Construction/New York City Transit (MTACC/NYCT).

This paper provides a general overview of these large caverns now in design or construction in Manhattan, the area's geologic setting, the geotechnical factors considered and approaches commonly applied for cavern design.

INTRODUCTION

The continued growth of commuter volume in Manhattan has posed a challenge and provided unique opportunities for the region's transit agencies to improve mobility within the area. Limited rail capacity beneath the Hudson River and at the existing New York Pennsylvania Station (PSNY) was recognized as a significant barrier to sustaining future economic growth for the New York and New Jersey metropolitan regions. This resulted in THE Project of NJ Transit/PANYNJ, which would increase trans-Hudson rail capacity and provide redundancy in trans-Hudson rail operations through development of a new, two-track rail tunnel beneath the Hudson River and construction of a new and expanded terminal station (NYPSE) under West 34th Street in Manhattan (Figure 1).

The Long Island Rail Road (LIRR) presently provides passenger service from ten branch lines on Long Island through Amtrak's tunnels under the East River to the west side of Manhattan into New York Penn Station. The East Side Access (ESA) Project will allow LIRR to provide direct service to the east side of Manhattan enabling commuters from Long Island to directly access mid-town Manhattan (Figure 2). This service will connect the LIRR main lines through the Queens Segment, the lower level of the existing 63rd Street East River tunnel and new bored tunnels and caverns into a new terminal station to be constructed beneath Grand Central Terminal (GCT) in Manhattan and under mid-town Manhattan's densely populated business district (Figure 3).

The existing No. 7 Subway Line, ("A" Division, IRT Service), constructed in the early 1900s as the Queensboro Subway, currently provides service between Main Street in downtown Flushing in Queens and the Times Square Station in Manhattan. The No. 7 Subway Line currently terminates in an east-west orientation beneath 41st Street

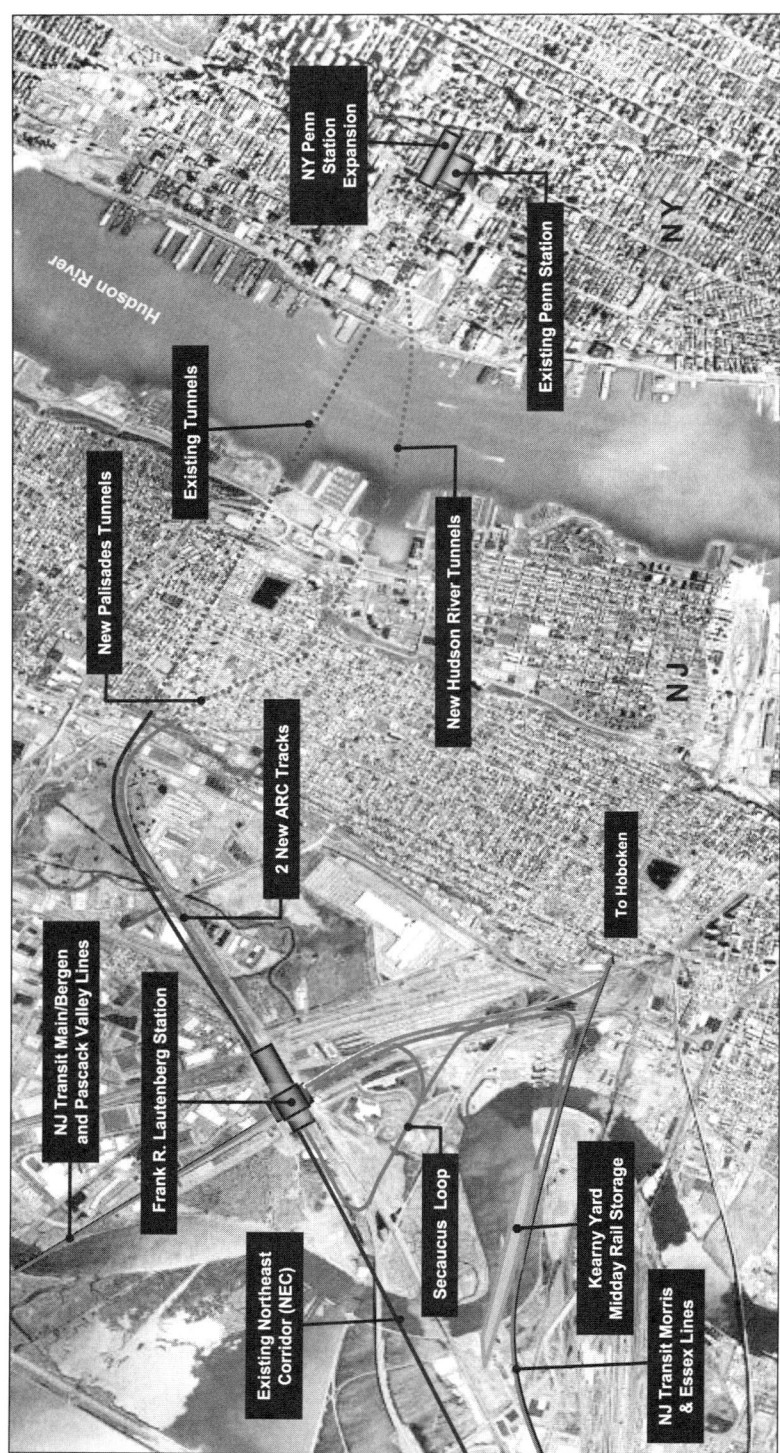

Figure 1. THE tunnel project

CHALLENGES FOR LARGE ROCK CAVERNS IN MANHATTAN 37

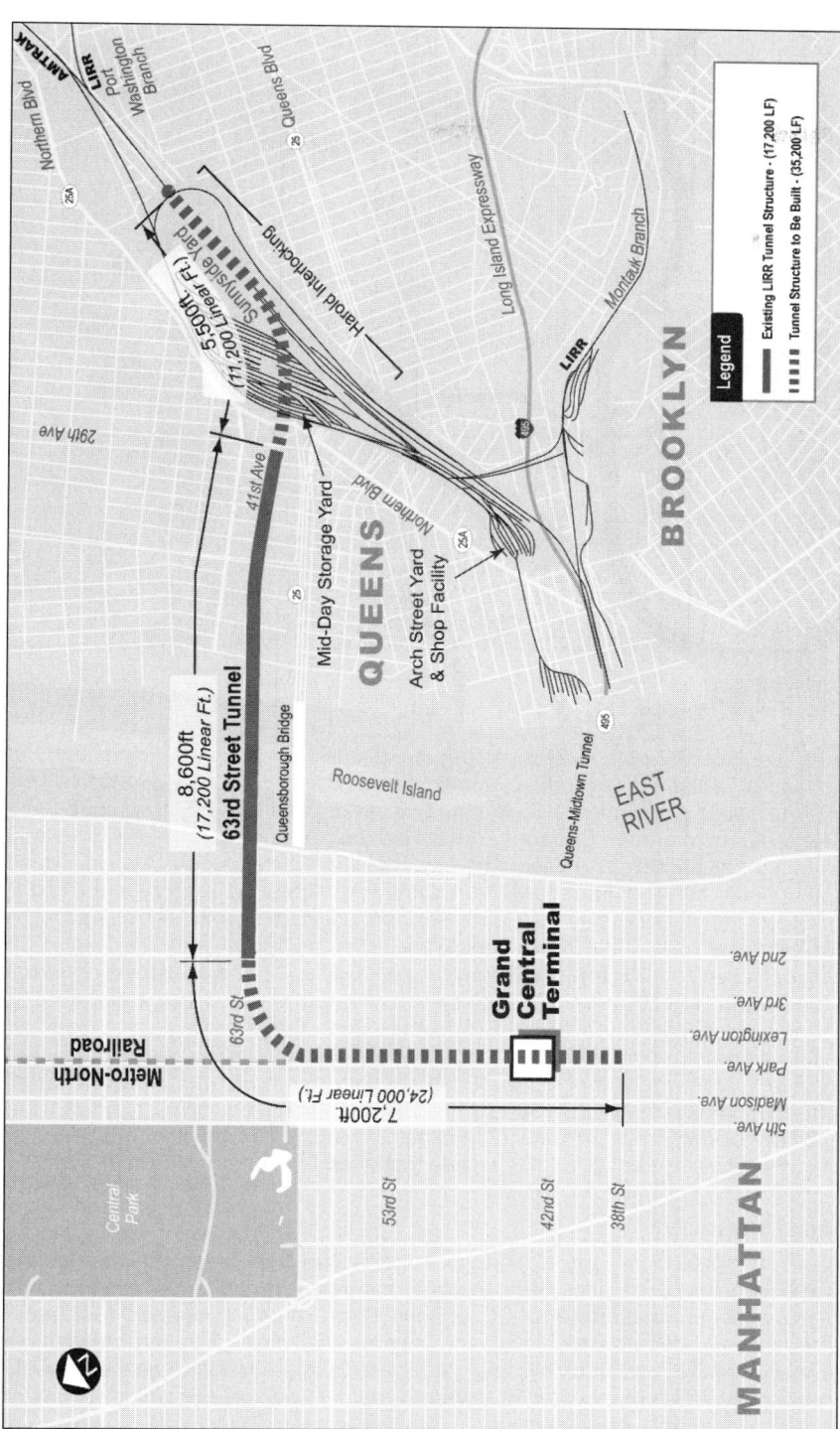

Figure 2. East side access project

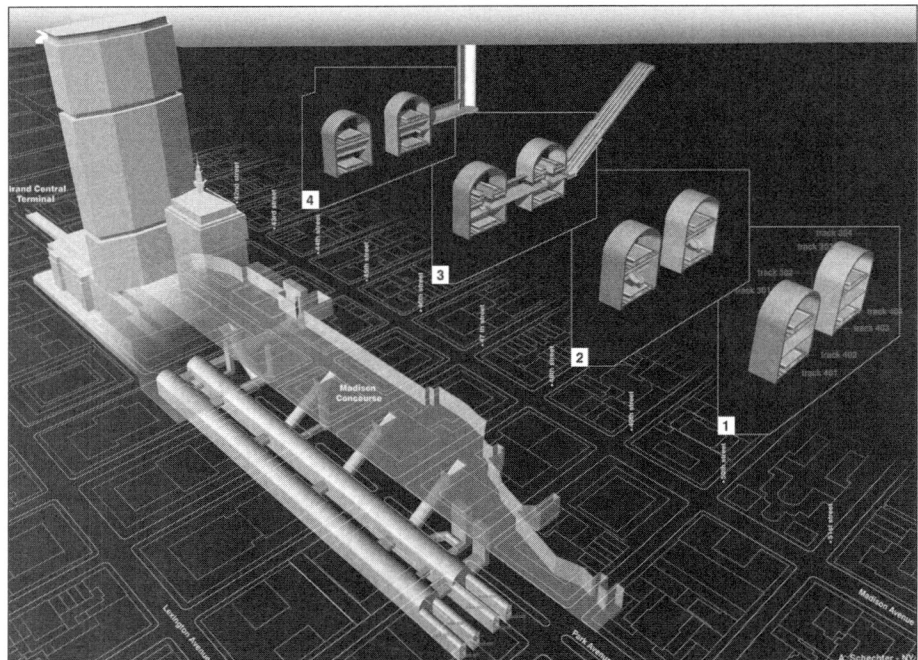

Figure 3. ESA caverns

between Seventh and Eighth Avenues. Concurrent with New York City Department of City Planning proposal for the redevelopment and revitalization of the Hudson Yards area, MTA prepared a preliminary design study to extend the No. 7 Subway Line from its current terminus at Times Square into the Hudson Yards area (Figure 4). Several alignment alternatives were analyzed. The selected alternative extends the No. 7 Line westward under West 41st Street and then southward along Eleventh Avenue, a distance of approximately one mile. The extension provides for a 34th Street Station on Eleventh Avenue at West 34th Street (Figure 5).

The Second Avenue Subway will be New York City's first major expansion of the subway system in over 50 years. When fully completed, the line will extend 8.5 miles along the length of Manhattan's East Side, from 125th Street in Harlem to Hanover Square in Lower Manhattan. In addition, track connections to the existing 63rd Street and Broadway Lines would enable the Second Avenue Subway Line to provide direct service from East Harlem and the Upper East Side of Manhattan to West Midtown via the Broadway express tracks. Phase One construction will include station caverns at 72nd and 86th Streets, each one housing two tracks and a center platform (Figure 6).

Manhattan Caverns

THE Project 2245-foot long NYPSE caverns extend from west of Ninth Avenue to west of Broadway constituting the eastern terminus of the project alignment. They are situated within Manhattan's 34th Street right-of way, approximately 90 feet (measured at the crown) beneath street level. The caverns will consist of several sections to be excavated entirely in bedrock by enlarging the four single track tunnels previously excavated by TBMs. The main station cavern (Figure 7) will include two platforms at

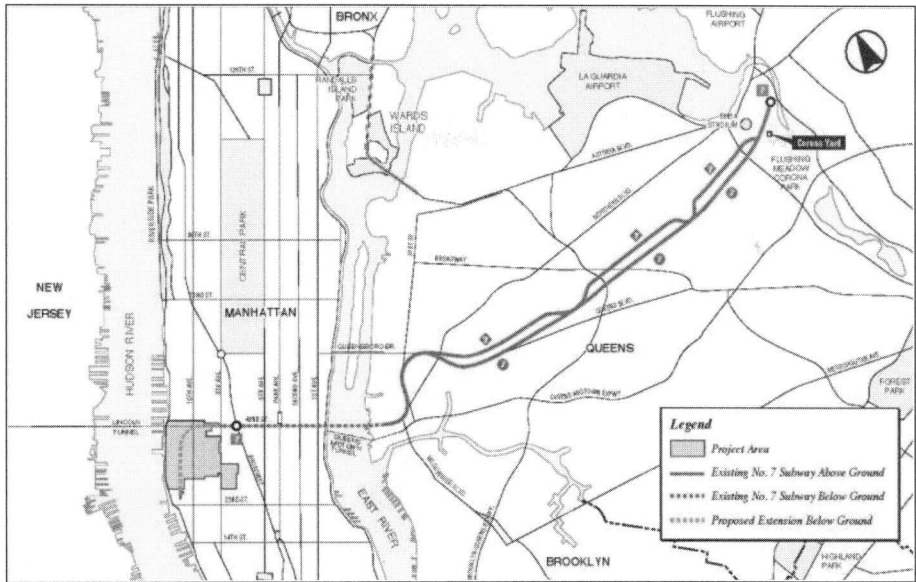

Figure 4. No. 7 line subway extension project

Figure 5. No. 7 line subway extension alignment

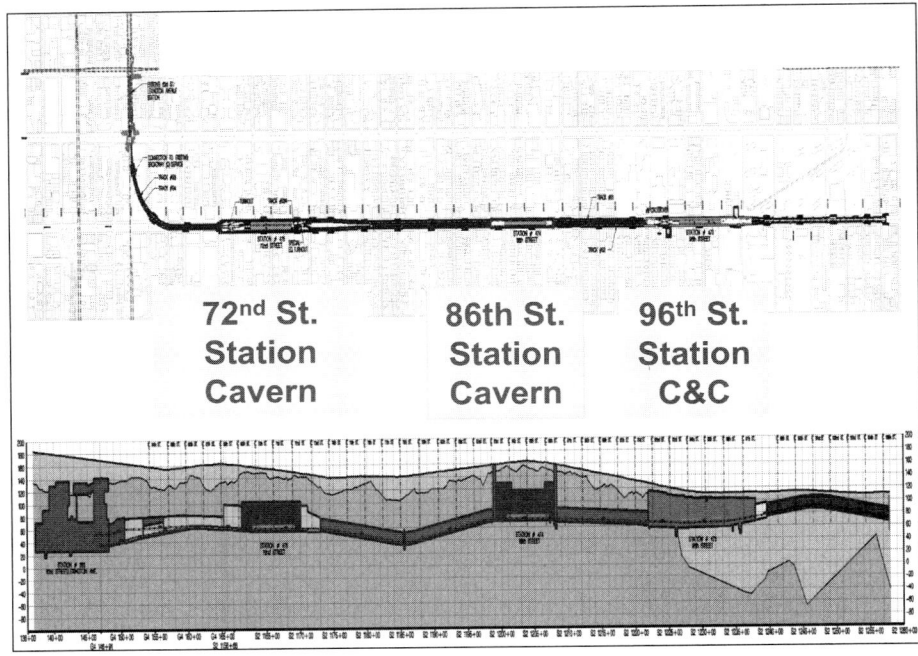

Figure 6. Second Avenue subway (Phase I)

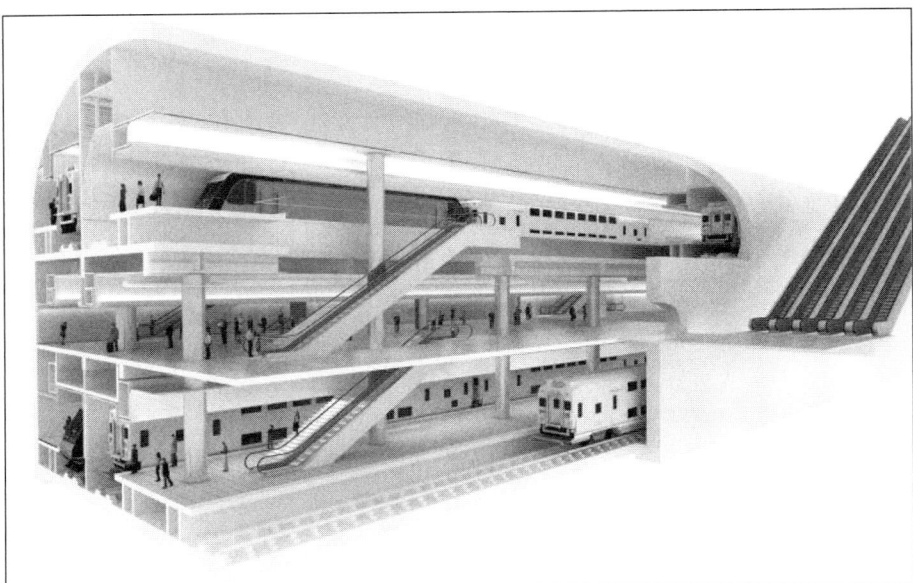

Figure 7. NYPSE cavern

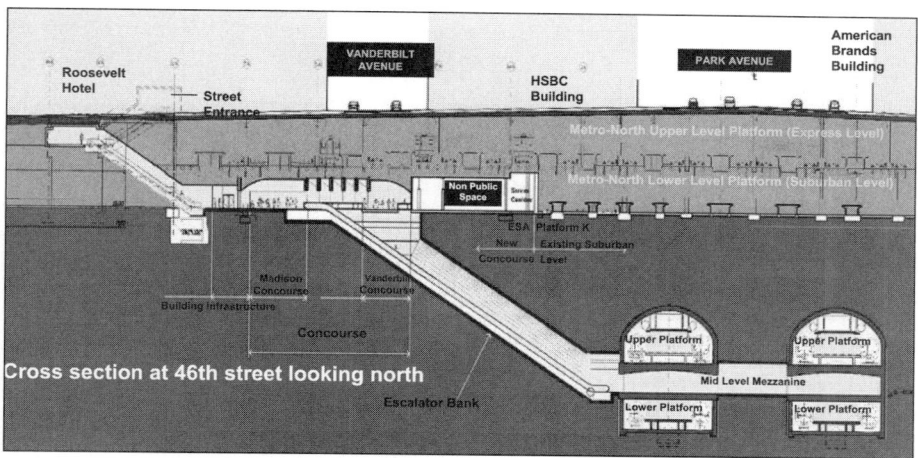

Figure 8. ESA caverns

each track level, one along the north side, and one along the center of the cavern. The maximum overall dimensions of this cavern are 96 feet wide by 90 feet high. The cavern profile will follow an upward grade of 0.2% from west to east. Entrance and utility tunnels will intersect the cavern below its springline to connect with the mezzanine area.

The ESA station caverns consist of two parallel caverns approximately 58 feet wide by 66 feet high and 1155 feet long. The twin caverns have eight platform tracks (two over two in each cavern) served by four 1020-foot long center-island platforms serving 12-car train consists. Each cavern houses a lower two-train chamber, an upper two-train chamber and four non-continuous mezzanine areas in between the upper and lower chambers. The mezzanines are positioned for direct connection with the Madison Avenue Concourse by three banks of four escalators and one five-escalator bank (Figure 8).

The 34th Street Station Cavern of the No. 7 Line Subway Extension Project is approximately 960 feet long, 54 feet high and approximately 66 feet wide. Including the north and south interlocking caverns, the total length of the enlarged cavern complex will be 1210 feet. The cavern will be constructed in bedrock approximately 60 feet (measured to the crown) beneath the Eleventh Avenue viaduct and within Eleventh Avenue right-of-way. Several utility and entrance tunnels connect with the station cavern at the mezzanine level (Figure 9).

MANHATTAN GEOLOGY

The topography of Manhattan is largely controlled by bedrock geology, and Manhattan's elongate ridges generally trend northeast, parallel to the established street grid. A map of Manhattan's topography and drainage, prepared in by Egbert Viele in 1865 before heavy urbanization, shows former stream channels generally following zones of weakness in the underlying bedrock. Most of these channels were filled during Manhattan's urban development and are no longer reflected in ground surface topography, but they are valuable indicators of underlying geologic structures.

Bedrock underlying Manhattan is a deeply eroded assemblage of folded and faulted Proterozoic- to Ordovician-age metamorphic and igneous rocks, as shown in the geologic map of midtown Manhattan in Figure 10. Surficial materials overlying

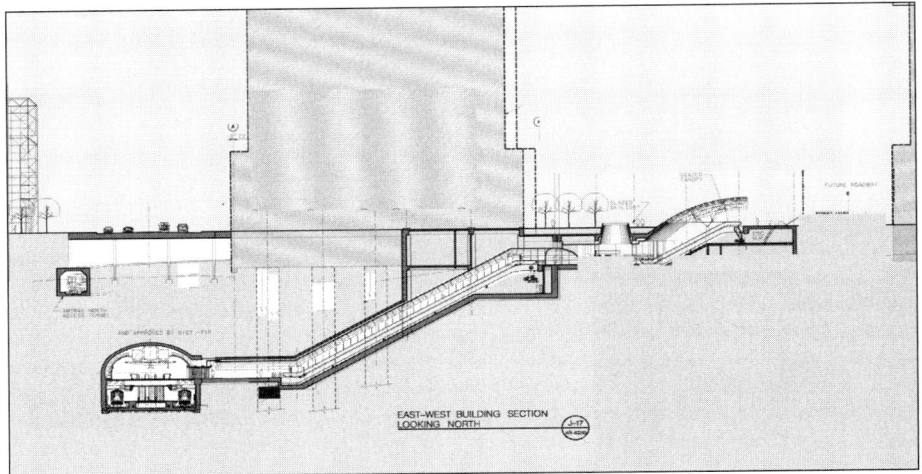

Figure 9. 34th Street Station cavern

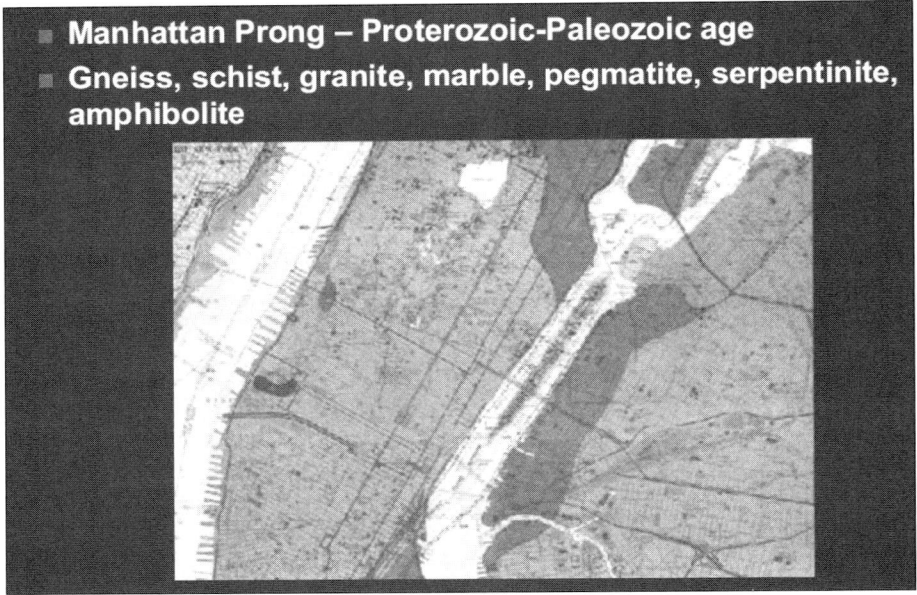

Figure 10. Bedrock geologic map of midtown Manhattan (from Baskerville, 1994)

Figure 11. Rock mass discontinuities in a surface excavation

bedrock in Manhattan include glacial, fluvial, lacustrine, and estuarine deposits along with artificial fill.

Geologic structure, lithology, and stratigraphy of rock and soils in Manhattan are complex and reflect a complex sequence of tectonic, erosional, and depositional events. Five major phases of deformation affected the region, and the oldest rocks show effects of faulting, folding, intrusion, and hydrothermal alteration. The complex system of joints and faults and the well-developed metamorphic foliation are the geologic features most affecting stability of underground openings in Manhattan.

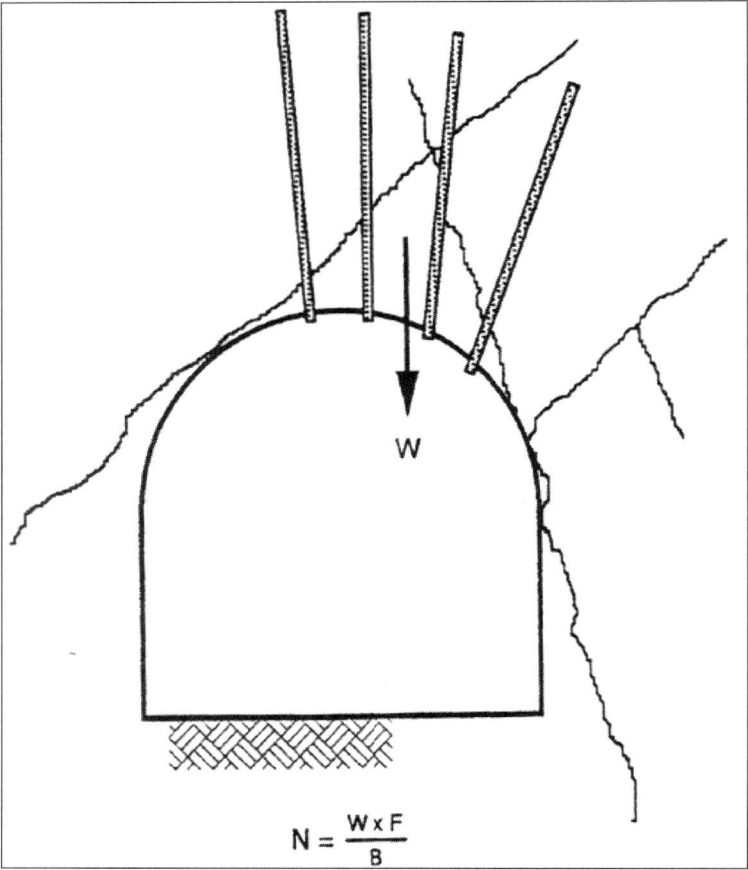

Figure 12. Gravity wedge analysis to determine anchor orientaions (from USACE, 1977)

PRINCIPLES OF CAVERN DESIGN

The stability of rock caverns at shallow depths is controlled by the discontinuities in the rock mass, including joints, fractures, foliation planes, contacts, and faults. Discontinuities visible in a surface excavation are highlighted in Figure 11. Intersections of these discontinuities allow formation of potential wedges or blocks which may loosen as the state of stress within the rock mass is altered by excavation. Gravity-driven movements along combinations of discontinuities daylighting into the excavation are typically seen as wedge fallout around the perimeter of the opening (Figure 12).

Foliation forms a penetrative planar fabric in the metamorphic rock units underlying Manhattan. It is the single most influential type of discontinuity affecting stability of underground openings. Rock wedges formed by adversely oriented intersections of foliation and other types of discontinuities generally control the stability of cavern crown and sidewalls. Wedge size is a function of the span opening and the spacing and persistence of discontinuities.

CHALLENGES FOR LARGE ROCK CAVERNS IN MANHATTAN

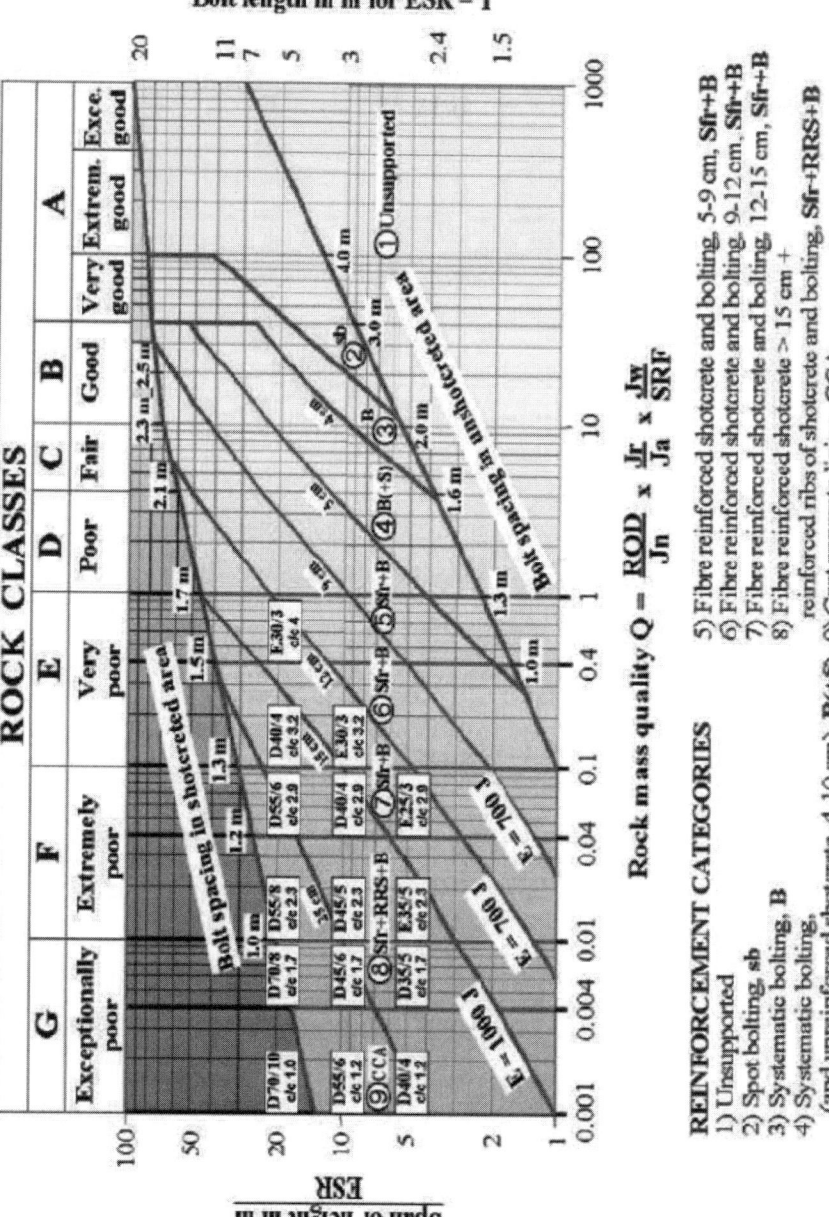

Figure 13. Estimated support catergories based on the tunneling quality index, Q (after Grimstad and Barton, 1993)

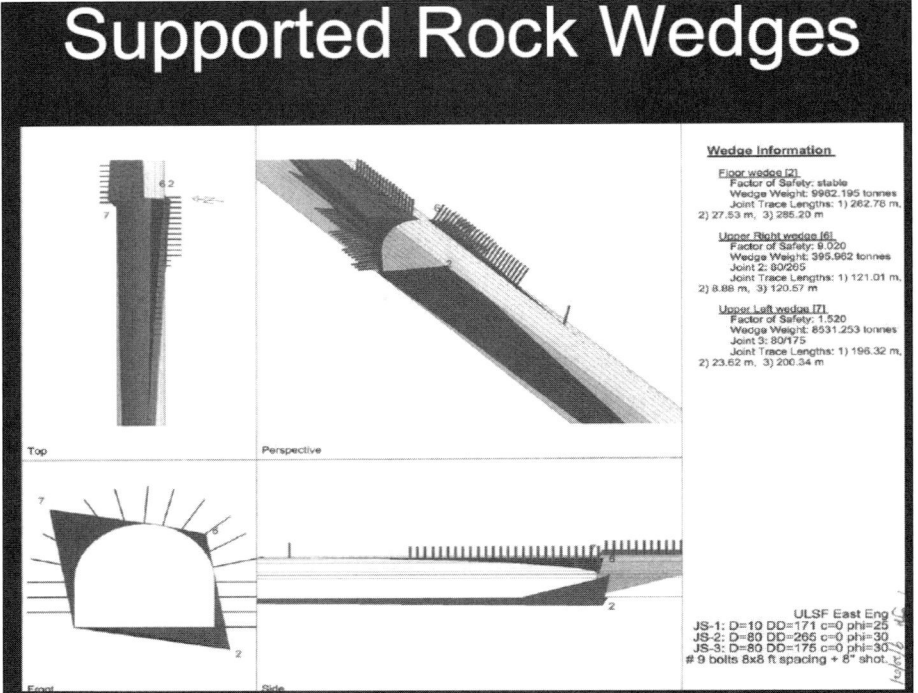

Figure 14. Sample 3D stability analysis and visualization for underground excavation in rock with intersecting discontinuities (Rocscience Inc. 2003)

The most effective and prudent means to control such wedge movements is to apply initial support as close and as soon as possible to the opening of the rock face. Rock bolts, shotcrete, and lattice girders used as a coordinated support provide an effective means to control wedge movements. Control of heading size and sequence of excavation are important for limiting fallout, especially in blocky or closely jointed ground.

Often the first attempt at design of rock support for large caverns is based on empirical design methods derived from the Terzaghi classification or from the more recently developed RMR and Q systems, shown in Figure 13. These classifications have been used successfully to design initial and permanent rock support, including the development of rules for bolt lengths and spacing.

The mechanisms of rock mass response to excavation are generally controlled by the planes of the primary geologic structures. These planes effectively control the size of individual blocks that may undergo independent movement and interact with the excavation profile to develop the principal stability mechanisms. The planes of the secondary structures control smaller scale interactions, notably overbreak effects and minor fall outs, and will reduce the stiffness of the individual blocks bounded by the primary structure.

Computer-based numerical and 3D vector stability analyses of wedge and block geometries and loadings have become staples of modern cavern design. Several common computer programs calculate wedge and block loading and allow input designs for rock bolts and shotcrete singularly or in combination. A sample output from one such program, UNWEDGE, is shown in Figure 14. Generally, the designer includes simulations of the worst case rock mass conditions within each modeled zone.

CONCLUSION

The sizes of the caverns to be constructed in these four projects are large and without precedent in New York City. Advancements in rock mass characterization and numerical and analytical design methods have greatly enhanced our ability to develop economical and stable support of such large openings.

REFERENCES

Baskerville, C. A., 1994. Bedrock and Engineering Geologic Maps of New York County and Parts of Kings and Queens Counties, New York, and Parts of Bergen and Hudson Counties, New Jersey, U.S. Geological Survey, Miscellaneous Investigations Series Map I-2306, scale 1:24,000.

Grimstad, E. and Barton, N., 1993. Updating the Q-System for NMT. Proc. Int. Symp. on Sprayed Concrete – Modern Use of Wet Mix Sprayed Concrete for Underground Support, Fagernes. 46-66. Oslo: Norwegian Concrete Assn.

Rocscience Inc. 2003. *Unwedge Version 3.0 - Underground Wedge Stability Analysis.* www.rocscience.com, Toronto, Ontario, Canada.

U.S. Army Corps of Engineers (USACE), 1977. Engineering and Design, Tunnels and Shafts in Rock, EM 1110-2-2901, May.

Viele, Egbert L., 1865. Sanitary & Topographical Map of the City and Island of New York, prepared for the Council of Hygiene and Public Health of the Citizens Association, under the direction of Egbert L. Viele, Topographical Engineer, Ferdinand Mayer & Co. Lithographers, New York, scale 1:12,000.

TUNNELING UNDER THE HARLEM RIVER

Jacek B. Stypulkowski ▪ MRCE

James P. Mooney Jr. ▪ Consolidated Edison Company of New York, Inc.

Hugh S. Lacy ▪ MRCE

ABSTRACT

This paper discusses the challenges faced by the design team of the under-river crossings of electric transmission feeders in New York City, a densely populated urban area. The under-river crossing of a 345 kV feeder was initially designed as Horizontal Directional Drilling (HDD), but was redesigned as a utility tunnel because of site constraints on both sides of the river. Multiple studies addressed location, available technologies and costs. Multiple soil investigation phases addressed design aspects of HDD, as well as tunneling. Contractual issues varied from a design build alternative with Geotechnical Data Report (GDR) to design tender approach with GDR & Geotechnical Baseline Report (GBR).

INTRODUCTION

Project History HDD

The Harlem River Crossing is a part of the 9.5 mile long 345 kV primarily underground transmission circuit, running north-south from Sprain Brook Substation in Yonkers NY to Inwood in upper Manhattan. The first crossing location of the river was selected along the existing right-of-way where retired 138 kV feeders remained buried in the river bottom. The design team suggested two technically feasible alternate options (1 & 2) for HDD as shown in Figure 1.

The owners of the properties affected appeared very concerned with the proposed combination of the technology and the location of construction activities. This led to another study in the vicinity of the crossing. Alignments on the west side of the bridge were considered, however, a feasibility study determined that they would require a lengthy permitting process associated with the NYC Parks requirements and a full environmental assessment because of the potential impact on wetlands, so they were quickly ruled out.

A workable HDD alignment was found east from the bridge, similar to option 1 above, and the contract documents were prepared while waiting for the real estate department to finalize access and easement agreements. All technical, as well as bureaucratic, challenges which led to this point were described in detail in the recently published paper by Mooney & Stypulkowski, 2008. Those studies led to the conclusion that tunneling technologies needed to be considered in this case. The combination of technical, political, and environmental issues facing HDD led the design team to introduce a new crossing concept: the utilidor.

Project History Tunnel

To determine which tunneling technology could be considered, internal diameters of the tunnel had to be established at first. While the original tunnel concept was only

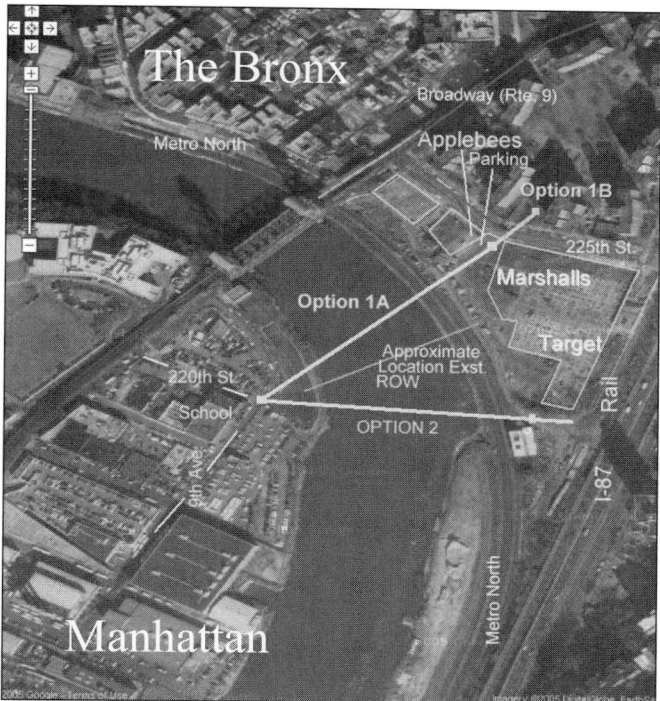

Figure 1. Harlem River crossing first study site plan

to house the 345 kV feeder, the design was quickly expanded to accommodate other Con Edison utilities. The final design contains high pressure gas, 13.8 kV distribution feeders, as well as space for future 345 kV transmission feeders. Each of these separate utilities has individual and unique design considerations for installation, emergency repair, maintenance, etc. The design team worked with subject matter experts for each utility type to determine and design for these special considerations. To accommodate specific needs of utility, a traditional shaft-tunnel layout had to be customized not only to facilitate sweeps in the conduit pipes but also to provide splicing space in the bottom of the shaft and in the tunnel. A small tunnel shown in Figure 2 was initially considered, since it could be potentially excavated using micro-tunneling technology; however, this option would not allow for splicing and repairs inside and was ruled out early in the process. An internal diameter of 7'2" could be achieved inside a large diameter concrete pipe jacked from the shaft using the micro-tunneling boring machine in the front. In addition, access for inspection was not possible in this confined environment.

During the review of the tunnel's operational requirements for various users, it appears that most demanding was installation and splicing of the 345 kV cable inside the tunnel. It was determined that all required utilities could not coexist in the small tunnel. Based on numerous discussions and several initial designs, a medium-size tunnel was developed as shown in Figure 3. Tunnel internal dimensions were set to 13' high and 12' wide which could be excavated using NATM technology in soil or D&B in rock.

50   CONVENTIONAL AND ROCK TUNNELING

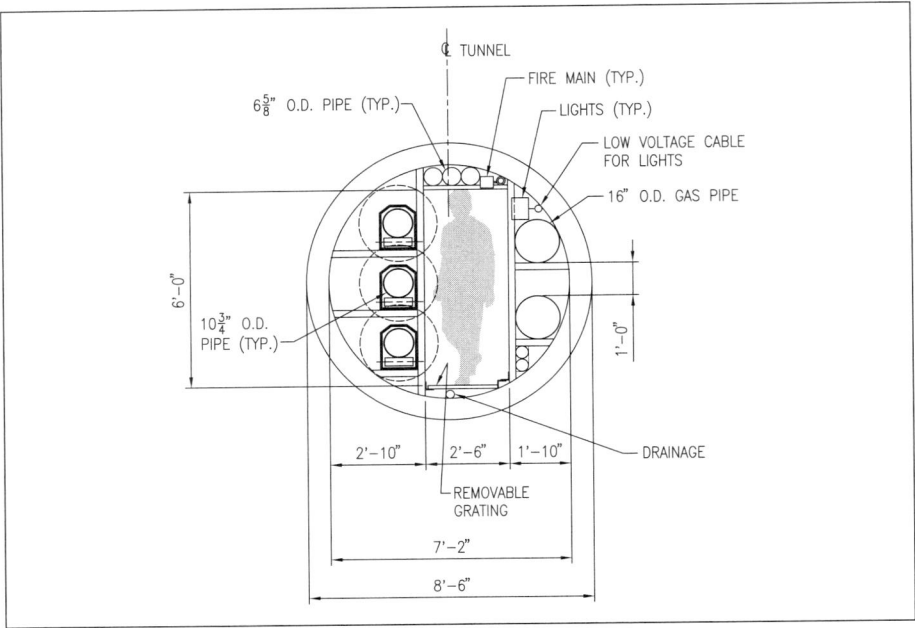

Figure 2.   Small tunnel

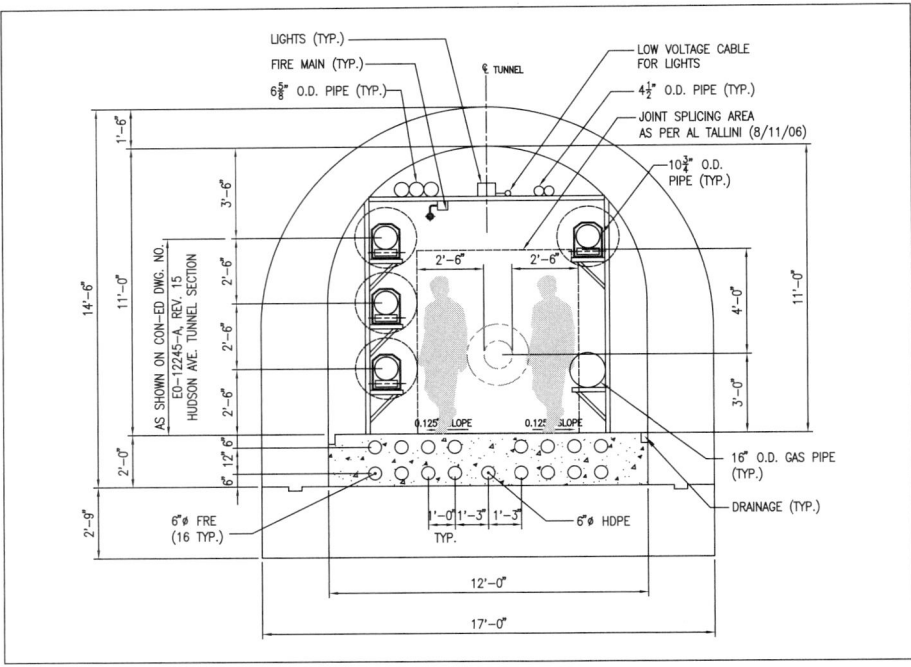

Figure 3.   Mid-size tunnel

SITE DESCRIPTION

General

New York City is a deep water port connected to various estuarine courses to the east, west, and north. Manhattan, an island, is the most densely populated part of the city. The Harlem River is a tidal strait separating the borough of Manhattan from the Bronx. At the time of European discovery, the Harlem River was approximately 900 to 1,000 feet wide in the area of the proposed crossing, as opposed to its current width of approximately 425 feet. Utilities supplying Manhattan must cross one of these water bodies. The electricity supply provided by Consolidated Edison Company of New York, Inc. (Con Edison) is only one of many utilities facing this challenge.

Field Investigations

A key element in the design of the river crossing is adequate knowledge of sub-surface conditions, as well as sub-aqueous characteristics. Multiple phases of design resulted in collection of data which was published in the GDR. In general, the main goal of those investigations, before the decision on tunneling option was made, was to define top of the rock and determine rock quality along proposed HDD alignments. Since some of the proposed crossings were quite deep, the depth of borings was adequate for the tunneling design which followed. This approach resulted in numerous borings, defining top of the rock, but not necessarily providing all the required information for tunnel design until late into the process.

River bottom was mapped using hydrographic and geophysical survey techniques involving a Trimble DGPS system interfaced with Maretrack II vessel trackline control and data logging navigation system. A side scan sonar system to map the river bottom in the corridor was used to identify debris on the river bottom. An EdgeTech GeoStar Chirp sub-bottom profiling system with 2–16 kHz was run simultaneously with the acquisition of side scan sonar information. The Chirp system was used to collect subsurface sediment profiles.

A relatively shallow depth of the bedrock and numerous man-made structures buried in soil (described below) convinced the design team early in the process that a rock tunnel crossing would carry the lowest risk.

Since size of the tunnel, as well as the means of excavation, wasn't predetermined, the laboratory testing program had to take into account various possibilities including roadheader specific tests performed at the rock testing laboratory of VOEST-ALPINE Bergtechnik GmbH, Zeltweg, Austria.

Due to the high cost of in situ testing, after the decision was made to tunnel under the river, three additional boreholes included in-situ measurements of the density, frequency, and orientation of bedrock fractures. Acoustic televiewer (ATV) logging was performed using Sopris MGX II of Matrix digital logging system in exploratory boreholes. The permeability of the rock was evaluated in the same boreholes by performing packer permeability tests in accordance with the procedures specified in the Bureau of Reclamation Earth Manual Designation E-18.

Geotechnical core logging methods and procedures developed for the International Society of Rock Mechanics (ISRM) were used to describe all cores and prepare borehole logs. A customized discontinuity log was prepared to provide input into the Q and RMR system for all rock cores.

After televiewer data was correlated with discontinuity logs and water pressure testing results were analyzed, it became apparent that results were consistent and the coverage was adequate for a 700 ft long tunnel.

Since the proposed structures included two shafts to depths of 150 feet below grade, two in-situ Downhole Seismic (DS) tests were performed. This geophysical test determined shear and compressional wave velocity profiles of soils and bedrock. The above properties were used to classify the site as per code requirements, perform seismic site-specific response analyses, and determine seismic pressures induced on the proposed structure by the surrounding soil and rock.

Site Geology

Figure 4 illustrates the location of the borings, the selected alignment and a profile showing the soils and rock encountered.

The site is located in the New England Upland, a division of the Appalachian Highlands that includes all of New England and parts of New York, New Jersey and Pennsylvania. The area consists primarily of metamorphic and igneous rock. The Upland in the New York area consists of two prongs that extend from northeast to southwest. The smaller of these prongs lies to the east and is known as the Manhattan Prong. The prong, that underlies New York City and extends to the southern tip of Manhattan Island, consists predominantly of highly metamorphosed rock together with some igneous rock. The site lies on a subdivision of the Manhattan Prong called the Inwood Lowland, which consists of Inwood Marble. The units have been subjected to folding and faulting. The Inwood Marble consists of calcite and dolomite marbles and calcareous schist.

Across the site, the rock is overlain by a relatively thin, but fairly continuous, layer of glacial basal till. In Manhattan and under the Harlem River, the till is overlain by a thick, continuous deposit of glacial lake varved silts and clay. In the Bronx, the varved soil and till is generally not present. Either it was never deposited, or more likely it was eroded away when a layer of clean outwash sand was later deposited in its place. In Manhattan the varved soil is overlain by a stratum of glacial sand, probably outwash. Under the river, the varved soil is overlain by a layer of river bottom sand. Further to the north in the Bronx, the corresponding outwash is overlain by silts and fine sands typical of a low energy environment, such as a lake or abandoned stream meander. It's not clear if the Bronx outwash sand, Harlem River sand and Manhattan sand were deposited as a continuous body, or as separate units. The composition of the present river bottom sand reflects reworking by the river currents. Above this lies a layer of soft organic silty clay and peat that was deposited as a result of post-glacial sea level rise.

Site Challenges

At the time of European discovery, this part of the Harlem River wasn't passable by vessels with significant draft. In the late 19th century, a decision was made to construct a Ship Canal to provide a navigable connection to the Hudson River via Spuyten Duyvil Creek. At the site area, the channel was excavated through soft alluvial and sand deposits and stone-filled timber crib walls at the channel's edge were constructed. Metro North Railroad (MNR) commuter train tracks were built on top, on the north side of the channel. From the beginning, the rail line received heavy use bringing thousands of daily commuters from north of NYC to Grand Central Terminal. The depth of the crib wall wasn't ever established and the clearance between top of the rock and bottom of the crib was estimated to be about 30 ft.

Other obstructions identified at this site included the original Marble Hill Station. In 1975, the present elevated platform structures were constructed on the other side of the Broadway Bridge and the original station abandoned. The original platforms were removed. However, foundations of the platforms and elevator pit remained. The platform foundation consisted of anchorage piles and steel cables or steel rods. The steel cables or steel rods span from the platform to the U.S. pier and bulkhead line.

In addition, a 54-inch Intercepting Sewer has been identified in the project area. The sewer runs parallel to MNR tracks from Broadway to a pump station located south of the Applebee Restaurant. The invert is located approximately 25 feet below ground surface. Due to the presence of organic soils at this site, this sewer was founded on timber piles.

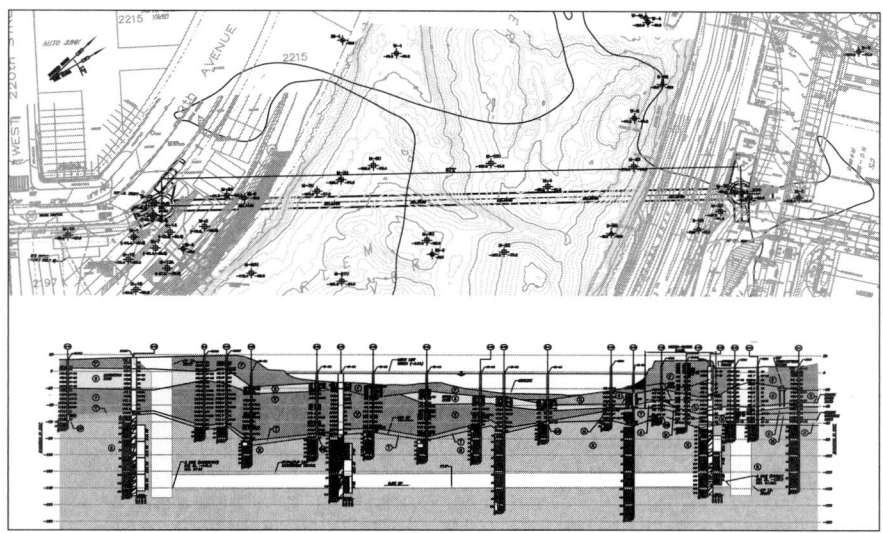

Figure 4. Tunnel plan and profile

Major man-made structures in combinations with poor soil conditions made the Bronx shaft a challenging location for a deep shaft.

UTILITY TUNNEL

General

Utilidors are tunnels housing different utilities and are usually built in densely populated areas or in areas with catastrophic environmental hazards, like tsunamis. They provide access for maintenance and inspection and they do not require digging to look for problems. Their costs, while high initially, become competitive if numerous utilities are placed in them and long-term maintenance costs are considered.

The 345 kV feeder has been placed just under the ground surface for most of the 9.5 mile distance. Occasionally, it has to be attached to the existing bridges or go under shallow utilities. The typical river crossing using HDD technology normally involves just a pull between terminal manholes located on each side of the river. The non-HDD solution required re-evaluation of the layout, 345 kV pipes have a maximum bending radius of 25 feet. In the tunneling option, placing pipes in vertical shafts and horizontally on brackets within the tunnel is a different installation. In addition, splicing needs to be done on the bottom of the shaft and pulling across 700 ft has to be done between the shafts.

Tunnel Layout

The proposed Harlem River Tunnel (HRT) crossing consists of two shafts, with an inside diameter of 24 feet and approximately 150 feet deep and tunnel about 700 feet long. To take advantage of the economy of scale a tunnel provides, additional facilities were designed to be included in the HRT. The facilities include several additional 345 kV and 13.8 kV feeders, and a high pressure gas header. Smaller conduit was designed to accommodate communications and control wiring. The shafts and tunnel were sized to accommodate all the proposed additional Con Edison facilities. The

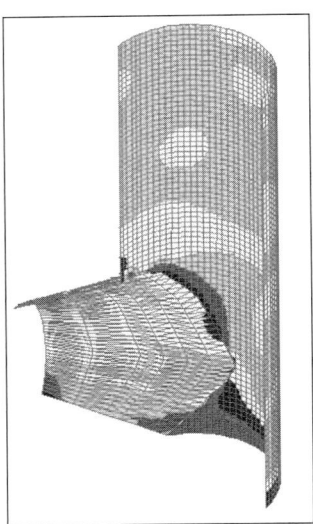

Figure 5. Transition structure between the shaft and the tunnel

Bronx Shaft is located in the parking lot adjacent to Metro-North railroad tracks. The Manhattan Shaft is located in the parking lot adjacent to Ninth Avenue. A headhouse will be located in the vicinity of the Bronx shaft to house ventilation, electrical, and communications equipment.

The tunnel to carry the new Con Edison facilities will be horseshoe-shaped, and concrete-lined. It will also be waterproofed to minimize water in-leakage. It will cross under the Harlem River, which has a water depth of between 17 and 25 feet and is approximately 450 feet wide in the vicinity of the tunnel.

To facilitate all the required cables smooth transition from the shaft to the tunnel, a 3D portal enlargement solution has been developed as shown on Figure 5 below. The portal height tapers from 32 ft at the shaft to 13 ft where the tunnel height becomes constant.

Design Philosophy

A rock mass classification system was used to define the support elements of the initial lining in different rock classes. Typical initial support systems were designed as ground reinforcement or direct support to establish a stable opening for continuing construction operations and consist of pattern bolting in the shafts and running tunnel. Shotcrete in the crown of the tunnel was added to provide additional long term support against deterioration of rock mass. Smoothing criteria for shotcrete was added to satisfy permanent waterproofing placement criteria.

Initial support was designed for application as close to the working face as practicable both to protect construction personnel and to control ground movements. The initial lining was designed for the rock above the opening by considering rock structure interaction, treating the surrounding rock as part of the support system. In poor ground conditions, steel sets and lagging assisted with grouting were designed for temporary tunnel support. Long term support will be provided by the cast in place reinforced concrete.

Cut-off grouting, to reduce ground water inflow from the rock formations into the tunnel and shaft excavations, will be required. Grouting in the tunnel will be performed in advance of the tunnel excavation (from within the tunnel) before potential water

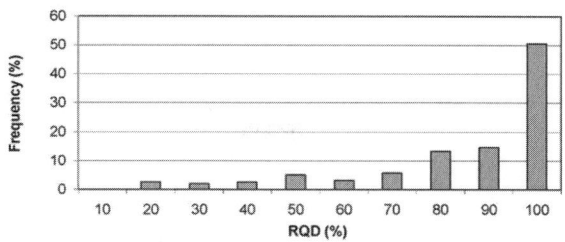

Figure 6. Frequency histogram of RQD from all borings

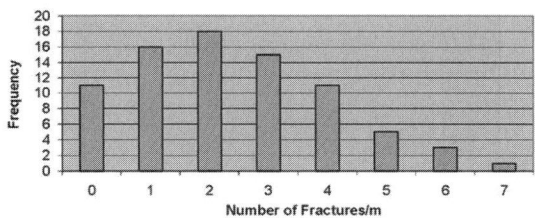

Figure 7. Fracture frequency histogram for 3 selected borings (# of fractures/m)

bearing zones are exposed. Modification of the possible fault material by means of chemical and/or cement grout was assumed to be feasible to allow tunneling to proceed below with heavier support like ribs and lagging, spiles or shotcrete. The grouting would need to both decrease the permeability of the ground and increase the strength.

Excavation Considerations

Roadheader use was encouraged in the Contract Documents due to their high flexibility to cut profiles and to reduce opening of joints in the rock compared to the impact of drilling and blasting in this tunnel with limited rock cover and a river above. The rock testing program included 13 UCS tests ranging from 4,000 to 36,000 psi. Tensile strength testing was performed on 9 samples with results ranging from 330 to 1,000 psi. Two rock samples were sent to Austria for specialized testing. One was a medium-crystalline marble with mica laminations which tested UCS = 13,000 psi and the other one was a coarse crystalline marble UCS = 22,000 psi. However, pegmatite veins found in the core tested as much as UCS = 36,000 psi. A summary of RQD's are shown on Figure 5, and fracture frequency for three selected borings are shown on Figure 6.

We concluded, looking at the results as summarized above, that coarse crystalline marble (22,000 psi) when it is massive, is only marginally cuttable with the largest sized roadheader that could be used in a tunnel of this size. The contractor needed to have more flexibility in choosing adequate excavation technology.

The D&B method is often not considered suitable for excavation at shallow depths below flowing watercourses. However, due to the relatively short length of the tunnel, the highly variable rock strength, the general good quality of the rock and the requirement to perform thorough grouting ahead of the rock face, it was allowed and ultimately was chosen by the Contractor as the preferred method.

Design Build

A Request for Proposal (RFP) for Design Build Construction of M-29 Utility Tunnel below the Harlem River was issued in April 2007. The following documents were provided to the invited bidders:

- Conceptual Design Drawings
- Performance/Technical Specifications
- Geotechnical Data Report

The Design/Build Contractor was required to provide the prime contractor; designers and all proposed subcontractors. The major elements of the project included civil, mechanical, electrical as well as security elements including:

- Two 24 ft ID shafts.
- A water tight tunnel and shafts.
- Embedded steel plates/rings capable of supporting utilities.
- An elevator in the Bronx shaft and stairs in each shaft.
- Concrete lined shallow vaults (pulling chambers) connected to the shafts.
- Above ground head house adjoining the Bronx shaft.
- A sump below the floor in the Manhattan shaft including permanent pumps with controls extending through the tunnel to the head house above the Bronx shaft.
- Below grade ventilation and cable chamber.
- Equipment hatches and equipment hoist.
- The installation of 6-inch diameter fiberglass conduits in the base of the tunnel.
- Terminal Manhole for Distribution System.
- The security access to the Manhattan shaft and the Bronx head house.

This contract, as described above, attracted limited attention in the tunneling industry. It was perceived that coordination of all the design aspects of highly specialized engineering fields was too risky for a contractor and no bids were ever submitted and the RFP was withdrawn.

Design–Tender–Build

A decision was made in late summer 2007 to complete the design and re-tender the package. In September 2007 final design of the tunnel commenced with a very aggressive schedule. In addition to the tunnel and shaft construction, the engineering design package consisted of several elements: Electrical, Mechanical, Plumbing, Ventilation, Elevator, Civil, Communication, Fire Detection, and Security. The design team quickly recognized that the project should be divided into three phases. To get the construction started, civil works were separated from all the other disciplines. Taking highly specialized items like elevators, cables, ventilation, security, etc. out of the tunneling contract helped to get more tunneling contractors interested and, as a result, more bidders submitted bids. The utilities were also designed separately.

The complete construction package included drawings and specifications to permit Con Edison to obtain the necessary building permits, competitive pricing, and to complete the construction of the tunnel, head structures and associated systems. Final Design Package for civil works was delivered in late November 2007. The contract was bid and it was awarded in July 2008 to Kiewit.

Contract Documents

The Geotechnical Data Report (GDR) was prepared to describe the anticipated subsurface conditions and geotechnical properties of subsurface soil and rock strata for the construction of the tunnel beneath the Harlem River. The GDR was included with RFP for Design Build as well as with Build packages.

The decision was made to prepare a Geotechnical Baseline Report (GBR) after the design-build bid failed to attract enough bidders. It became evident that bidding contractors expected risk to be shared with the Owner. Therefore, the report expressly identified baseline statements with respect to ground and groundwater conditions. The GBR provided an interpretation of the geotechnical data and addressed anticipated ground conditions that might be encountered during the performance of construction, as well as their design basis. It also included discussions of geologic and manmade features of engineering design and construction significance and monitoring requirements of the existing structures and utilities.

The following geotechnical parameters and conditions are baselined in this GBR and provided in this document:

- Top of Rock
- Cobbles and Boulders
- Rock Domain Quantities
- Engineering Properties of Intact Rock
- Rock Mass Joints and Fractures Orientations
- Rock Mass Classifications
- Soil Parameters
- Groundwater Conditions

The GBR was included with the other Contract Documents: Contract Drawings and Specifications.

ACKNOWLEDGMENTS

The writers acknowledge contribution of the following companies to the successful completion of the design: JD Hair and Associates, BGA, LLC, Hager-Richter Geoscience, Inc., Ocean Surveys, Inc. TECTONIC Engineering & Surveying Consultants P.C. and Washington Group International.

CONCLUSIONS

At the time of article submittal, shafts in the Bronx and Manhattan are under construction, with the estimated deadline for civil works to be completed by February 2010.

REFERENCE

Mooney, J, Stypulkowski, J "Two HDD crossings of the Harlem River in New York City" in Underground Infrastructure of Urban Areas Wrocław , Poland, 22–24/10/2008, Edited by C. Madryas, B. Przybyla, A. Szot, Wrocław University of Technology, CRC Press, Taylor & Francis Group.

PART 2

Design and Planning

Chairs

Jon Hurt
Arup

Sarah Wilson
Jacobs Associates

BAY TUNNEL—DESIGN CHALLENGES

R. John Caulfield ■ Jacobs Associates

Isabelle Pawlik ■ Jacobs Associates

Johanna I. Wong ■ San Francisco Public Utilities Commission

ABSTRACT

The primary water supply system for the City of San Francisco and adjacent peninsula areas consists of aging and deteriorated pipelines crossing under the San Francisco Bay (the Bay). This system is highly vulnerable to seismic damage from the nearby San Andreas and Hayward fault zones.

The Bay Tunnel Project will replace the existing system with a 8-km-long (5-mi-long) tunnel underneath the Bay. The tunnel will be constructed using earth pressure balance methods and will have a two-pass lining system of precast concrete bolted, gasketed segments for initial support, and a welded steel pipe final lining. The project challenges include tunneling through both soft ground and rock conditions without any intermediate shaft access over the entire alignment. Seismic performance criteria include maintaining service flows after a large earthquake.

INTRODUCTION

In November 2002, the San Francisco Public Utilities Commission (SFPUC) launched a $4.3 billion Water System Improvement Program (WSIP) to repair, replace, and seismically upgrade its water system (Figure 1), consisting of 320 km (200 mi) of aging pipelines, tunnels, reservoirs, and treatment systems serving 2.4 million customers in the greater Bay Area. More than 75 WSIP projects in San Francisco and the surrounding region will be completed by the end of 2015.

Preliminary studies commissioned by the SFPUC determined that two major pipeline arteries in the system—Bay Division Pipelines 1 and 2—would be particularly vulnerable during a seismic event. Their replacement or upgrade was defined as a key element of WSIP and resulted in the creation of the Bay Division Pipelines Reliability Upgrade Project. The project consists of constructing the 33.8-km (21-mi) Bay Division Pipeline No. 5 (BDPL No. 5) from Irvington Tunnel Portal in Fremont to Pulgas Tunnel Portal near Redwood City, including a tunnel (Bay Tunnel) under the Bay. The project will provide seismic and delivery reliability to SFPUC's customers and contribute to its goal of ensuring a secure, high-quality water supply.

FACILITY LAYOUT

The Bay Tunnel segment of BDPL No. 5 will consist of a 2,743-mm (108-in.) internal-diameter (ID) pipeline extending 8 km (5 mi) from the Newark Shaft site in the City of Newark to the Ravenswood Shaft site in the City of Menlo Park. The alignment crosses under the Bay, adjacent marshlands, and salt ponds (Figure 2). It will be constructed while the existing pipelines remain in service. BDPL Nos. 1 and 2 and the new BDPL No. 5 then will be tied in to the tunnel at both ends during short outage windows.

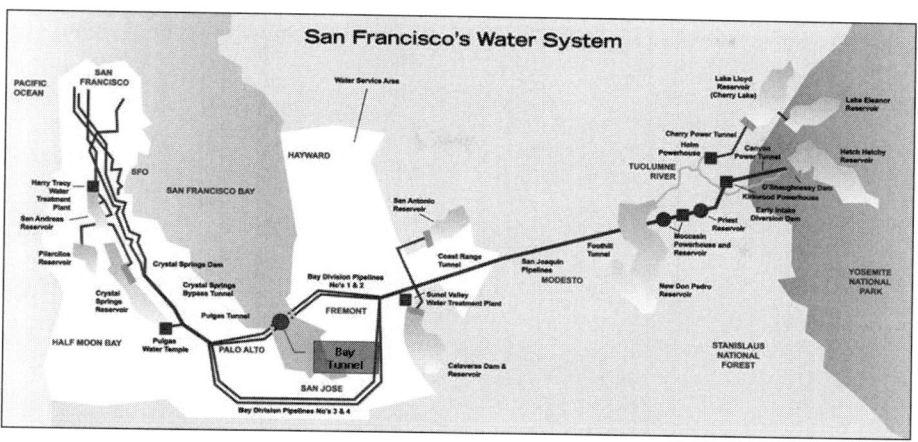

Figure 1. Diagram of SFPUC's water distribution system

The presence of environmentally sensitive habitats on the Bay margins requires that the entire 8-km-long (5-mile-long) tunnel be constructed only with launching and receiving shafts; there will be no intermediate construction shafts. The shafts will be constructed using diaphragm slurry wall or sunken caisson construction methods, with tremied structural concrete slabs tied into the shaft bottoms. The tunnel boring machine (TBM) launching and tunnel construction shaft, located at the Ravenswood site on the west side of the Bay, will be approximately 17.7 m (58 ft) in diameter and 33.5 m (110 ft) deep. The receiving shaft, located at the Newark site on the east side of the Bay, will be approximately 8.5 m (28 ft) in diameter and 22.5 m (74 ft) deep. The tunnel depth ranges from about 21.3 to 30.5 m (70 to 100 ft) below the mean surface (sea level) of the Bay and was selected to maintain minimum clearances under the bottom of the Bay and situate the tunnel in materials favorable for tunnel excavation and seismic performance while locating it as shallow as possible to reduce construction cost.

The Bay Tunnel will be constructed as a two-pass system. In the first pass, the tunnel will be excavated with a TBM, and a precast concrete segmental lining will be erected as initial ground support immediately behind the TBM. In the second pass, the 2,743-mm (108-in.) ID welded steel pipeline will be installed inside the tunnel and cellular concrete used to backfill the annular space between the outside of the pipe and the initial support.

GEOLOGIC CONDITIONS

Field Investigations—Geotechnical Program

The field investigations included a series of marine-based borings across the Bay and land-based borings and cone penetrometers on the Newark and Ravenswood site locations. The initial phase of the field geotechnical investigation, completed in September 2006, included 28 borings; 13 land-based and 15 marine-based. The borings were drilled along the alignment and at the proposed shaft locations. In February and October 2007, additional field investigations better defined the areal extent and depth of a buried bedrock ridge encountered west of the Newark shaft and reduced a geological information gap under the salt ponds due to access restrictions from the landowners. This second phase of investigations included four additional land-based

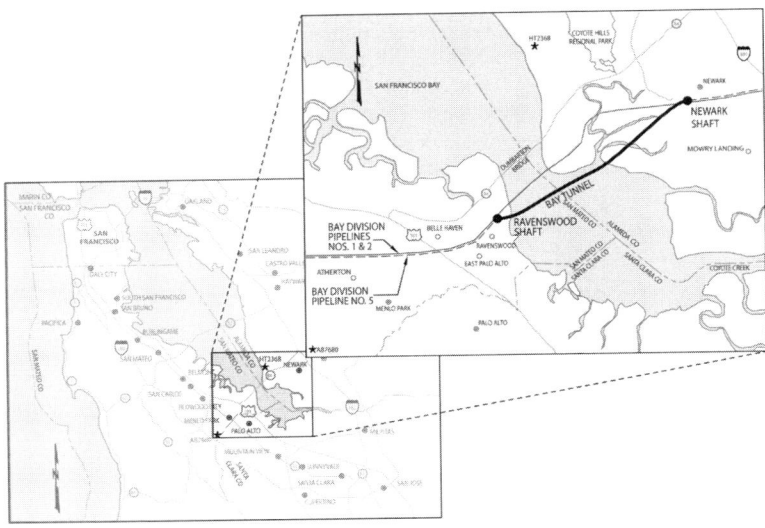

Figure 2. Proposed Bay Tunnel horizontal alignment

borings west of the proposed Newark shaft location and two marine borings drilled within a tributary to the Bay.

Geotechnical exploration included drilling and sampling at each of the referenced boring locations; downhole seismic suspension logging at nine boring locations; downhole vane shear and pressuremeter testing at two boring locations; installation of five vibrating wire piezometers; and installation of a standpipe piezometer.

Field Investigations—Geophysics

In addition to conventional exploration, an extensive marine- and land-based geophysics program along the majority of the proposed alignment corridor helped the design team make interpretations between boreholes that assisted with the mapping of subsurface stratigraphic horizons within the proposed tunnel corridor.

For the geophysics along the eastern end of the tunnel alignment, seismic data were collected using a Digipulse AWD-100 accelerated weight drop as the energy source and a Seistronix 24-bit, 120-channel recording system. Both seismic reflection and refraction survey data were collected at the Newark Shaft site and the adjacent salt pond levees.

The marine seismic reflection survey mapped the soil and rock horizons beneath the offshore part of the tunnel corridor across the Bay to a depth beneath the lowest elevation of the tunnel envelope. Two systems were used to collect the reflection data: (1) an 8-channel boomer provided higher resolution of the near-surface (15 to 23 m [50 to 75 ft] depth range) horizons, and (2) a 24-channel minisparker provided reflections down to subsurface depths of 46 m (150 ft) or more. Approximately 89 km (48 nautical miles) of marine seismic reflection data were collected along the survey lines that were arranged in a grid over the proposed tunnel alignment.

Utilization of land and marine seismic data for site characterization of the proposed Bay Tunnel required the integration of the geophysical data with geotechnical boring and downhole velocity data sets. Onshore, the seismic refraction and reflection data were integrated to provide a surface-to-depth image of the subsurface soil and rock horizons in the Newark area. Seismic velocity data from suspension logging and reflection record processing procedures were used to convert seismic reflection record travel

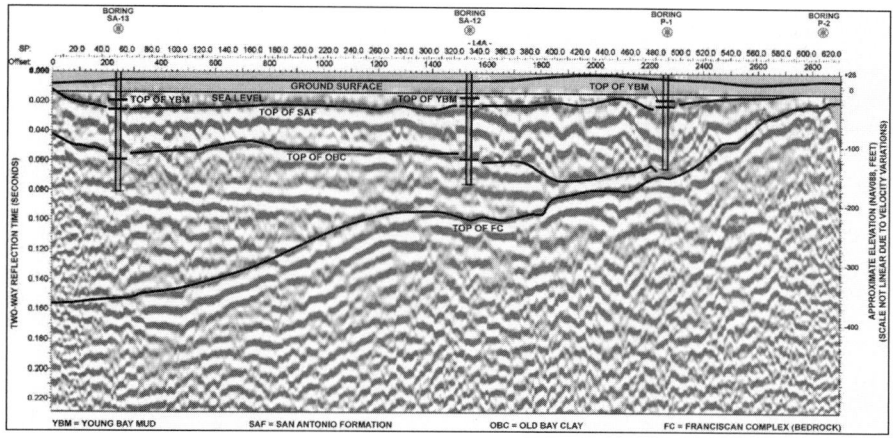

Figure 3. Typical Integration of seismic reflection and geotechnical data

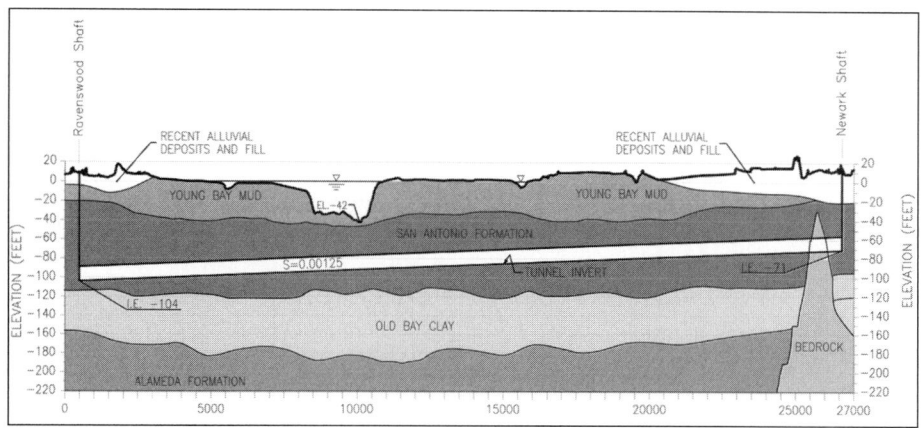

Figure 4. Geologic profile along the proposed Bay Tunnel alignment

times to depths. Geotechnical explorations were used to characterize the soil and rock units at specific sites and provide correlations with the geophysics information. This process was employed to identify and map the lateral extent and thickness of soil and rock units between the geotechnical exploration sites.

A typical interpreted seismic reflection record from Line 4A in the Newark Shaft area (Figure 3) shows four geotechnical borings with depths to the various soil and rock units converted to reflection times. The corresponding reflections are associated with the interfaces of the specific soil and rock units.

Site Geology

After the geophysical and geotechnical investigation and associated laboratory testing were complete, a detailed geologic cross section was generated showing the stratigraphy along the proposed alignment. The anticipated geology based on the results of these studies is summarized on Figure 4.

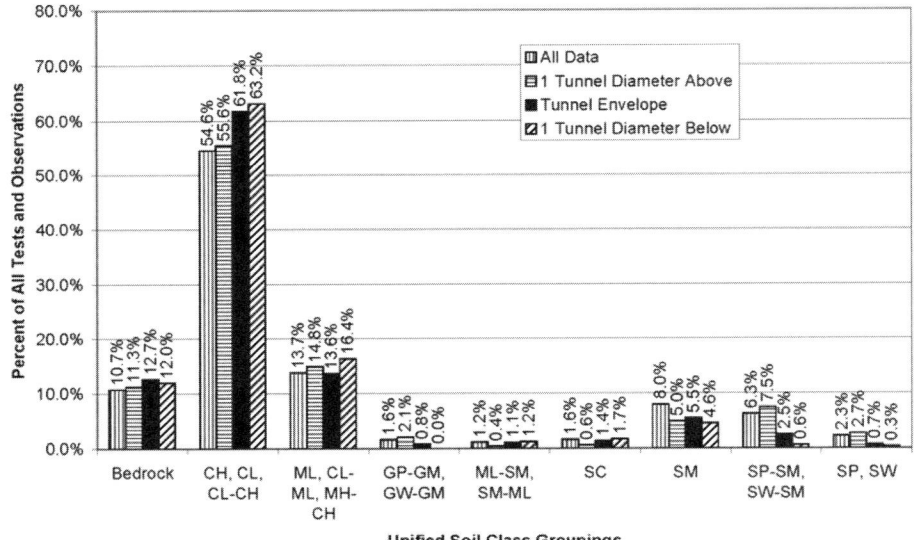

Figure 5. Distribution of USCS soil types along the proposed tunnel alignment (highest fines content on the left; lowest on the right)

Soil deposits along the tunnel alignment can be grouped into six primary units: artificial fill, Young Bay Mud (YBM), San Antonio (SA) Formation, Old Bay Clay (OBC), Alameda Formation, and Franciscan Complex bedrock.

ANTICIPATED TUNNELING CONDITIONS

Design studies determined that the optimal location for the tunnel alignment was primarily within the San Antonio Formation (see Figure 4), which consists of interbedded clays, silts, and sands. A 213-m-long (700-ft-long) reach of the tunnel will also traverse a buried ridge of highly weathered Franciscan Complex bedrock containing basalt, sandstone, shale, serpentinite, and chert that can be very hard and abrasive. Due primarily to environmental and property-access restrictions, the Newark Shaft site could not be relocated to the west of the buried bedrock ridge.

The entire tunnel alignment is under the water table, subject to approximately 3.2 bar (48 psi) of hydrostatic pressure. The proposed tunneling methods needed to account for both the soil and hard rock conditions, as well as the hydrostatic pressures expected.

To select and specify tunneling methods, it was important to quantify the amount of soil with high fines content (clays and silts) and coarse materials (sands and gravels). In addition, it was important to characterize the stratigraphic distribution of the soil types. Laboratory testing, including extensive sieve and hydrometer testing, determined the gradation curves for the various soil types sampled in the borings. These data, along with Atterberg Limits testing and visual-manual soil classification of logged samples, were used to classify soil by the Unified Soil Classification System (USCS) (Figure 5).

Figure 5 provides a statistical summary of the Unified Soil Classification groupings from the laboratory testing program. Data from one tunnel diameter above and below also were included to provide a more complete picture of soil types and variability that may exist within the tunnel horizon.

BAY TUNNEL—DESIGN CHALLENGES

MECHANIZED EXCAVATION CONSIDERATIONS

Tunneling Methods

A pressurized-face TBM was determined to be well suited for the Bay Tunnel due to the soil conditions, presence of a high groundwater table, and varying permeabilities and strengths of the materials along the alignment. The silts and sands existing along the alignment could present running or flowing ground conditions if the face is not pressurized during excavation. There also is the potential for squeezing behavior of clay deposits. This was determined by computing the overload factor (stability factor) along the proposed alignment depth. It was determined that approximately 70 percent of the tunnel alignment could experience some moderate to extreme squeezing conditions. This requires control and regulation of the face pressure.

The two most common types of pressure-balancing TBMs are earth pressure balance (EPB) and slurry shield. Due to the large percentage of silts and clays and the cohesive nature of the soils, an EPB TBM was judged to be the most appropriate mechanical excavation method. Almost 90 percent of the materials encountered within the borings at tunnel level will be fine grained and cohesive (Figure 5). The other 10 percent will be mixed soil conditions or, rarely, a full face of sandy soil. For the Bay Tunnel, the cutterhead also will need to be fitted with cutting tools that can excavate through the bedrock reach in the Franciscan Complex. The cutterhead must include scrapers to bring loosened material into the cutterhead openings, rippers (drag bits), and disc cutter mounts that can be back-loaded.

Along the body of the TBM, squeezing forces can occur as the result of relaxation of material into the steering (i.e., overcut) gap created by the cutterhead. To compensate for large frictional forces that could be imposed on the TBM shield, specific equipment and techniques must be implemented. Included among these are sufficient overcut on the cutterhead, the ability to inject and control the pressure of the bentonite in the steering gap along the shield, the ability to measure pressure along the shield, adequate TBM thrust capability, and limits on stoppages through areas identified as being particularly susceptible to squeezing conditions.

Ground Conditioners

Developments in ground conditioning have steadily improved over the past decade. Therefore, EPB TBMs now are viable for applications for which only slurry TBMs had been considered before. With an EPB TBM, the tunnel face is supported by pressure from the mass or plug of remolded soil (muck) within the cutterhead and screw conveyor. For soils with high permeability or high clay content, ground conditioners are required to achieve a stable soil mass or plug. In addition to supporting the face, ground conditioners lower friction and permeability, thus preventing or reducing water inflows through the screw conveyor. Ground conditioners also enhance excavation efficiency by creating muck that is more plastic.

Ground conditioners will be required to treat both fine-grained cohesive soils and coarser permeable soils. Fine-grained soils will require the addition of conditioners such as foam and polymer to lubricate the muck and prevent it from sticking. Coarse-grained soils will require the addition of polymers to prevent the flow of water through the TBM screw and maintain stability at the face.

TBM Cutterhead Wear

Ground conditioners also are used to reduce abrasion wear and extend cutting tool longevity. This is important for the Bay Tunnel, which has a long TBM drive with no

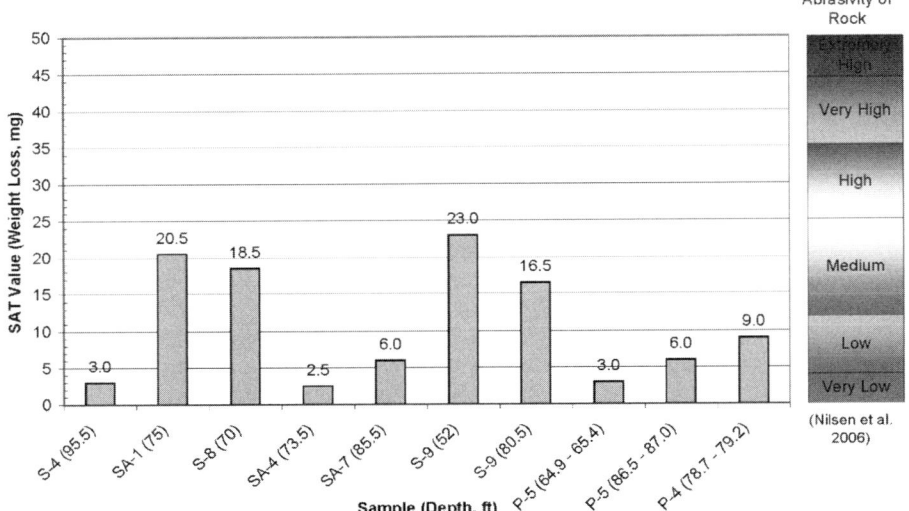

Figure 6. Soil abrasion test results of soils at the tunnel horizon from geotechnical investigation

access to the machine through intermediate shafts. During tunnel excavation, control of wear will need to be carefully monitored by regular inspection of the cutterhead.

Soil Abrasion Testing (SAT) was used to evaluate the relative abrasiveness of soil samples. Figure 6 shows the results of SAT for the Bay Tunnel and the relative abrasiveness of the soil samples from west to east along the alignment. The SAT evolved from the hard rock TBM boreability test methods developed by the Norwegian University of Science and Technology (NTNU) and NTNU's Foundation for Scientific and Industrial Research (SINTEF) (Nilsen et al. 2006). The SAT method is relatively new, and test values currently are correlated to the abrasion value (AV) for rock (Nilsen et al. 2007). Therefore, they cannot be used directly to accurately estimate cutter wear. The SAT values do, however, provide valuable information about areas of higher abrasion and areas where more frequent cutterhead tool inspections are recommended.

The Bay Tunnel alignment passes through different abrasive zones, including low-to-medium abrasive soils. For most of the alignment, it is anticipated that the TBM will encounter sticky soils, where clogging can impact advance rates and exacerbate abrasion. Therefore, these soils will require consistent use of surfactant foaming agents to combat adhesion and stickiness caused by high fines content. In the two areas identified as medium abrasive, lubricating polymer-enriched foaming agents can help reduce abrasion wear and prevent clogging of the excavation chamber. Consistent and appropriate ground treatment combined with regular cutterhead inspections will maximize TBM performance and minimize unexpected delays due to blockages, worn out tools, or damage to the cutterhead structure.

SEISMIC ANALYSIS AND DESIGN

Seismic Performance Criteria and Design Earthquake

The proposed Bay Tunnel will be located between two of the most seismically active faults in the Bay Area—the San Andreas and Hayward faults. The SFPUC

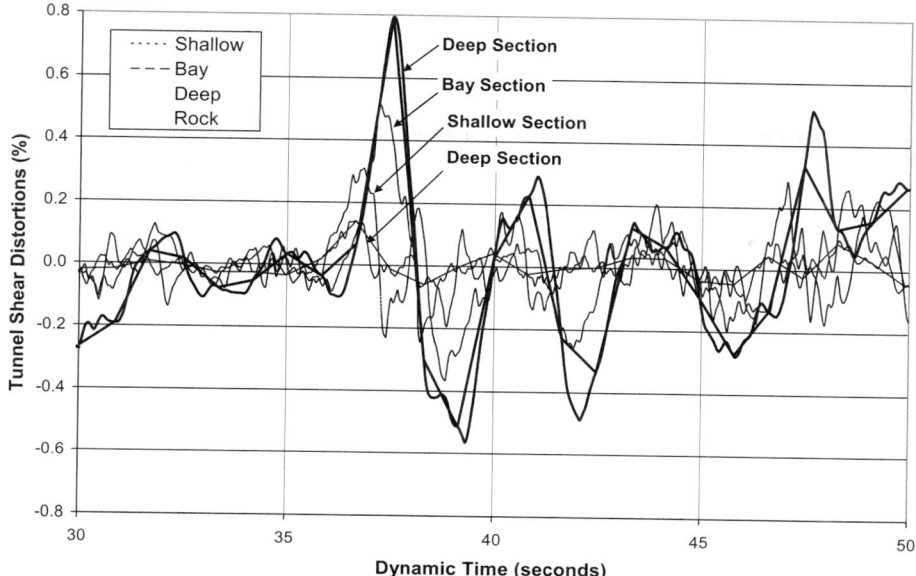

Figure 7. Time histories of final lining shear distortions at various sections

program criteria indicate that the water system must deliver a winter-day demand of 215 MGD within 24 hours of a major earthquake, so evaluation of the response of the tunnel facilities due to extreme seismic activity was an essential design activity.

The design earthquake for the Bay Tunnel was established using a probabilistic seismic hazard analysis (PSHA) with two major controlling earthquakes: a M7.9 earthquake on the San Andreas fault and a M7.1 earthquake on the Hayward fault. A deterministic ground motion analysis also was performed to define the limit of the ground motions for the design earthquake. The selected design earthquake has a 5 percent probability of exceedance in 50 years (975 year approximate return period) and a horizontal peak ground acceleration (PGA) of approximately 0.6 g.

Tunnel Analyses and Results

Two different types of two-dimensional analyses for the tunnel and the shafts were performed for the Bay Tunnel. These analyses considered the transverse effects of the design earthquake, and their primary focus was assessing the magnitude of the shearing distortions and hoop stresses induced in the final lining by the vertically propagating shear waves, and their potential impact on final lining performance.

The tunnel analyses were performed using three methods: closed-form solutions, numerical racking analysis, and a full nonlinear dynamic analysis. The racking and dynamic analyses were performed using the FLAC computer program. Four tunnel sections were selected to represent the range of overburden depths and ground conditions along the tunnel alignment, referred to as the Shallow, Bay, Deep, and Rock Sections.

The critical ground deformations were the shear distortions. The shear distortions defined here are the relative horizontal displacements between the tunnel crown and invert. They are transient, following similar cycles of ground displacements. The magnitude of the shear distortions was found to be highly dependent on the amplitude of shear waves and ground conditions (Figure 7).

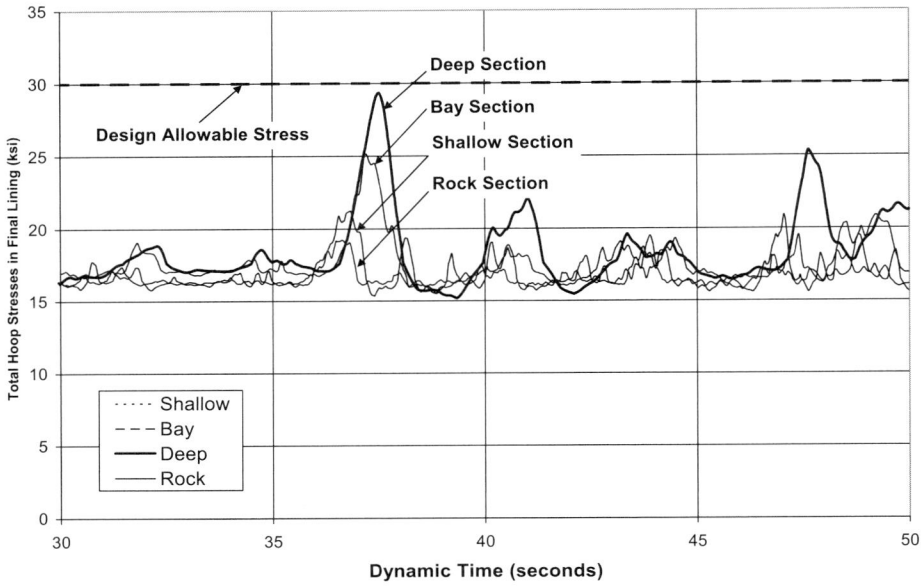

Figure 8. Time histories of total hoop stresses in final lining at various sections

Input ground motion time histories vary along the tunnel alignment. Among the four tunnel sections analyzed, the highest peak ground velocity (PGV) is associated with the Deep Section, and the lowest with the Shallow and Rock sections. This explains why the calculated shear distortion at the Deep Section is larger than that at the Shallow Section, even though there is greater ground cover at the Deep Section.

Time histories of the calculated hoop stresses that developed in the final lining are shown in Figure 8 for the four tunnel sections. These stresses include the effects of ground loads, groundwater pressures, and internal pressure. The final lining along the entire tunnel alignment is in tension due to internal pressure. Similar to shear distortions, the hoop stresses are dependent on the amplitude of shear waves and ground conditions. The maximum tensile hoop stress was predicted to be about 200 MPa (29 ksi), occurring at the Deep Section. This stress is below the allowable design stress of 207 MPa (30 ksi) for the grade 40 steel specified for the final lining. Stresses are transient, and the maximum stress occurs only once during the design earthquake (Figure 8).

The segmental lining was included in the FLAC racking and dynamic analyses. Its stiffness was accounted for, but its function for the long-term performance of the Bay Tunnel is not required. In addition, a separate analysis to review seismic performance during the construction period was performed using a 100-year-return-period earthquake on the segmental lining. The analysis was performed based on a racking analysis using FLAC. Results of the analysis indicate that a 229-mm-thick (9-in.-thick) precast segmental lining using 41.4-MPa (6,000-psi) concrete reinforced by either steel rebars or steel fibers as the initial support will be adequate for the Bay Tunnel during construction under combined static- and seismic-loading conditions.

Shaft Analyses and Results

The shaft analyses included the seismic performance of both shafts and the connecting pipelines. One of the key areas of the analyses was the examination of

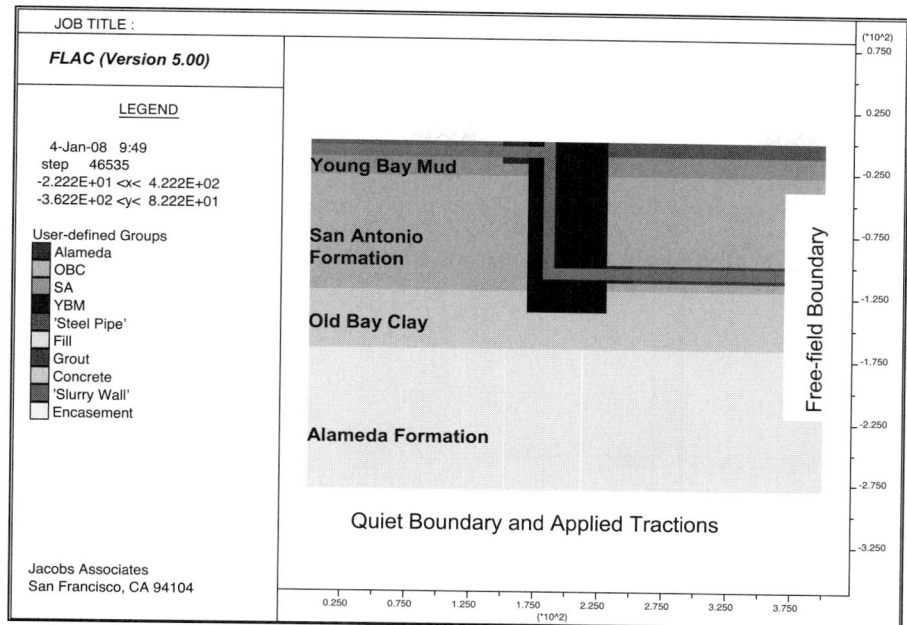

Figure 9. FLAC model with boundary conditions for Ravenswood Shaft

the bending stress concentrations in the final lining near the intersections with the Ravenswood and Newark shafts and the effectiveness of mitigation measures that could be used to reduce the magnitude of the stress concentrations. The stress concentrations are primarily the result of the rigid-body rocking of the shaft during ground shaking.

Shaft analyses were initially performed with a FLAC3D model. This model identified specific focus areas at the shaft/pipeline intersections that required further analysis. These additional analyses were performed with the FLAC program and were based on a two-dimensional dynamic approach. Separate analyses were generated for the Ravenswood (Figure 9) and Newark shafts because of their different sizes and ground conditions. The models each represent a unit-thick slice of the corresponding shaft intersected by the tunnel initial segmental lining and final lining.

The bending stresses are shown in Figure 10. The highest stress concentrations in the final lining are about 241,3 MPa (35 ksi); these occur about 3 to 6 m (10 to 20 ft) outside the slurry wall of the Ravenswood Shaft. Similar observations were made at the Newark Shaft. The stress levels in these localized areas exceed the allowable design stress of 206.8 MPa (30 ksi) for Grade 40 steel, although in all other areas stress levels are within acceptable limits.

Mitigation measures to reduce the high seismic stress levels induced in the final lining near the intersections with the two shafts included the use of higher grade steel, such as grade 60 steel instead of grade 40, and a structural concrete encasement for the final lining at this location. To examine the effectiveness of concrete encasement for the final lining, additional FLAC analyses were carried out. In these analyses, the cellular concrete backfill in the zone extending 9.1 m (30 ft) from the shaft slurry walls was replaced with structural concrete. The results from the FLAC model for the Ravenswood Shaft with concrete encasement are also shown in Figure 10. Use of a 9.1-m-long (30-ft-long) structural concrete encasement reduces the stress levels to

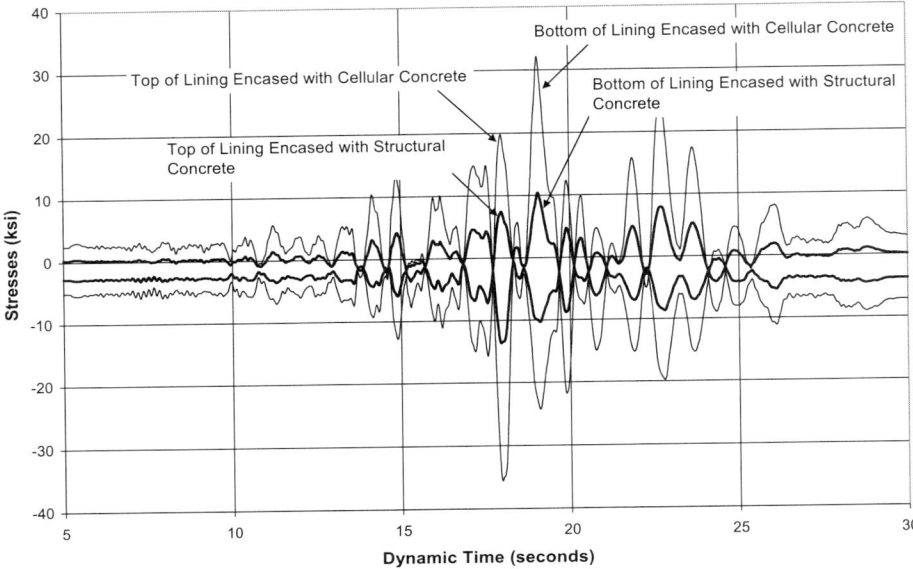

Figure 10. Time histories of axial stresses in final lining near intersection outside Ravenswood Shaft

about 69 MPa (10 ksi) and maintains them within the allowable stress of 206.8 MPa (30 ksi) for the grade 40 steel. In addition, elevated axial stress levels in the final lining outside the concrete encasement regions remain below the design allowable stress.

PROJECT SCHEDULE

The final design for the Bay Tunnel is currently underway and is expected to be complete in July 2009. The Final Environmental Impact Report (FEIR) was released to the public for comments in December 2008. The environmental documentation is being performed in parallel with the design and is expected to be completed in July 2009. The project is expected to begin construction in February 2010, with an anticipated completion date in 2014.

SUMMARY AND CONCLUSIONS

The proposed 8-km-long (5-mi–long) Bay Tunnel is a critical lifeline water supply facility for the City of San Francisco and the San Francisco peninsula communities. It will be the first tunnel excavated by a TBM under the San Francisco Bay.

It is expected to encounter interlayered medium stiff to hard silt and clay, with dense sand lenses within the San Antonio Formation along the majority of the tunnel alignment. Due to the large percentage of silts and clays, and the cohesive nature of the soils, in conjunction with a high groundwater table, an EPB TBM is the most appropriate excavation method. The cutterhead also will be fitted with tools that can excavate a short reach of hard Franciscan Complex bedrock. Ground conditioners will be required to treat both fine-grained cohesive soils and coarser permeable soils.

The proposed tunnel is located in an area of high seismicity and will need to remain serviceable after a large earthquake on nearby active faults. Earthquake-induced shear strains and hoop stresses were found to be below allowable values and are expected to have no significant impact on the final lining performance. High seismic stress concentrations occurring near the pipeline/shaft intersections will make it necessary to use higher grade steel and structural concrete encasement in these areas.

ACKNOWLEDGMENTS

The authors acknowledge the support provided by the San Francisco Public Utilities Commission and Jacobs Associates for the preparation of this publication.

REFERENCES

Caulfield, J., Romero, V., Wong. J. 2007. Planning and Design of the Bay Tunnel. In *2007 RETC Proc.*

Erickson, L., Raleigh, P., and Romero, V. 2008. Geotechnical Conditions and TBM Selection for the Bay Tunnel. In 2008 *NAT Proc.*

Jacobs Associates and URS Corporation. May, 2008. *Bay Division Pipeline Reliability Upgrade (CUW36801), Bay Tunnel Project, Final Technical Memorandum No. 7: Seismicity.*

Jacobs Associates and URS Corporation. May, 2008. *Bay Division Pipeline Reliability Upgrade (CUW36801), Bay Tunnel Project, Geotechnical Data Report.*

Nilsen, B., Dahl, F., Holzhaeuser, J., Raleigh, P. 2007. New Test Methodology for Estimating the Abrasiveness of Soils for TBM Tunneling. In *2007 RETC Proc.*

Nilsen, B., Dahl, F., Holzhaeuser, J., Raleigh P. 2006. Abrasivity of Soils in TBM Tunnelling. In *Tunnels and Tunnelling International Proc.*

Sun, Y., Klein, S., VanGreunen, J., and Louie, P. 2008. Seismic Design Evaluation of the Bay Tunnel. In *2008 NAT Proc.*

THE SELECTION OF EXCAVATION METHODS FOR THE DETROIT UPPER ROUGE TUNNEL CSO CONTROL PROJECT

Jean Habimana ▪ Jacobs Engineering Group

Wern-Ping Chen ▪ Jacobs Engineering Group

Robert Barbour ▪ Jacobs Engineering Group

Pamela Turner ▪ Detroit Water and Sewerage Department

Gary Stoll Jr. ▪ Detroit Water and Sewerage Department

Mirza M. Rabbaig ▪ Detroit Water and Sewerage Department

ABSTRACT

Complex and distinct geological units spread along the 7-mile long 30-ft inside diameter Upper Rouge Tunnel CSO Control Project in Detroit, Michigan. Its construction is divided in two tunnel construction packages: The South Tunnel, to be driven in finely laminated weak TG Limy Shale, will be excavated by a shielded TBM with precast concrete segmental lining, while the North tunnel, predominantly in a relatively more competent geologic formation Antrim Shale, will be excavated by a main beam TBM with a two-pass lining system. The adits and deaeration chambers will be excavated by sequential excavation method. This paper addresses the process of selecting excavation methods and lining systems for shafts, tunnels, connecting adits, and deaeration chambers.

PROJECT OVERVIEW

The Rouge River in Southeast Michigan does not currently meet state water quality standards due to frequent point and non-point wet weather sources. The combined sanitary and storm sewers in this area serve most of the Rouge Valley watershed including parts of Detroit and most of the western suburbs in the watershed. The Northwest Interceptor (NWI), which services the western Detroit and the near-west suburbs, is overwhelmed during heavy, intense storm events and the excess combined sewage is discharged untreated into the Rouge River. In order to improve the quality of the discharge into the Rouge River, the Upper Rouge Tunnel (URT) CSO Control Project is expected to totally capture all but one storm event on an annual basis.

The URT project will capture and store up to 825 million liters (218 million gallons) of combined sewer overflow (CSO) from 28 outfalls in Detroit, Dearborn Heights and Redford Township during wet-weather events. After storm events, CSO captured by the storage tunnel will be dewatered by a pumping station and routed to the NWI when it has the capacity to receive flow for conveyance for treatment at the downstream Detroit Wastewater Treatment Plant.

The alignment of the deep tunnel is predominantly within land owned by the City of Detroit and managed by the Detroit Recreation Department. Surface works sites are,

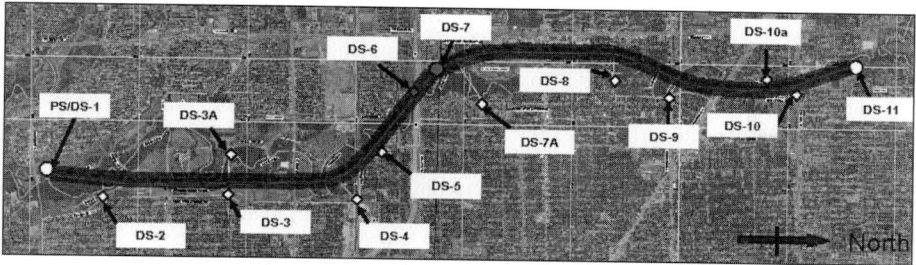

Figure 1. Project location plan

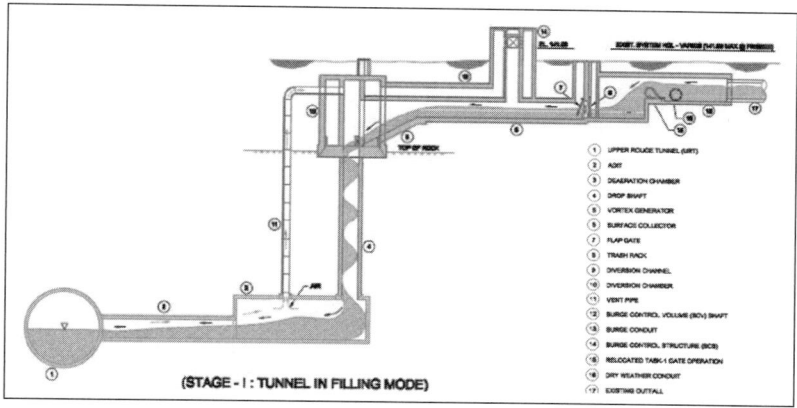

Figure 2. Typical components of URT at each outfall structure

by necessity, located along the alignment of existing outfalls to be intercepted. The specific surface works sites have been selected to minimize the impact on wetlands and, to the extent practical, remain outside the 100-Year flood plain. The main tunnel consists of a 9.14 m (30 foot) finished diameter rock tunnel, which extends along an approximately 11.28 km (7 mile) alignment, generally parallel to the Upper Rouge River. In addition to the main tunnel, the project includes a 100 MGD capacity pump station; three work shafts, including the pump station shaft; 13 near surface interceptor (diversion) structures and appurtenant facilities; 11 drop shafts, connecting adits and deaeration chambers; three collector sewers; and outfall modifications. The location of the URT's primary components and their relationship to their surroundings is illustrated in Figure 1, and Figure 2 shows the typical components of the URT at each outfall structure.

GEOLOGIC SETTING

The project site is located along the southeasterly edge of an extensive, circular shaped depression called the Michigan Basin that is a series of concentric circles with layers of sedimentary rocks slopping or dipping gently inward from the rims toward the center of the basin. The regional strike of the bedrock is northeast-southwest, with beds dipping gently to the northwest at typical less than 10 m per kilometer (50 ft per mile).

Overburden Geology

Most of the soils in Wayne County are of glacial origin (drift), deposited either directly by glacial ice (glacial till), by glacial melt water streams (glaciofluvial), or in glacial lakes (glaciolacustrine deposits). The glacial drift consists of clayey fills, outwash sands and gravels, glaciolacustrine silts and clays, and glacial till that overlie the buried bedrock surface. The depth of the overlying soil thickness varies from 11 m to approximately 30 m (35 to 95 feet), generally thickening from south to north. These soils typically overlie an over-consolidated clay (glacial till) layer locally referred to as hardpan that ranges from 1.5 to 5.5 m (5 to 17.5 feet) in thickness. Occasionally, there is a layer of sand or silt above the hardpan which typically ranges from 1 to 6.5 m (3 to 21 feet) in thickness.

Bedrock Geology

The underlying bedrock consists of very shallow dipping sedimentary rocks comprising shales, limestones, and dolomites. Figure 3 shows the geological profile along the proposed tunnel alignment with three major geologic formations: the Antrim Shale (AS), the Traverse Group (TG), and the Dundee Limestone in order of the youngest to the oldest.

The Antrim Shale is comprised of two sub-formation layers:

- The AS Carbonaceous Shale, which is a dark brown to black, moderately hard to hard, brittle laminated shale comprising mostly illite clay and quartz.
- The AS Gray Shale, which is weak, soft, slightly to moderately calcareous and consists of predominantly illite clay, a trace of chlorite, and has little to no quartz.

Traverse Group (TG) is divided into five geologic formations that are described below, from top to bottom layers:

- The TG Dolomite, which is characterized by gray, strong, hard to very hard, fine to medium grained Dolomite and Dolomitic Limestone having an average thickness of about 3.5 to 4.5 m (12 to 14 feet).
- The TG Breccia, which is a rehealed Chert/Dolomite Solution Breccia with an average thickness ranging from 2 to 3.5 m (6 to 12 feet). This unit comprises chert nodules and angular fragments of chert generally well bonded in a matrix of aphanitic to fine grain siliceous dolomite and dolomitic limestone. The Solution Breccia is separated into two to three layers comprising fresh breccia; a slightly to moderately weathered, finely porous, gray/light brown mottled breccia; and a layer comprising chert and dolomite fragments in a matrix of dark gray mudstone. The upper portion of this unit is often pitted and vuggy, and frequently exhibits trace of hydrocarbon staining.
- The TG Upper Limestone, which comprises a dark gray, strong, hard dolomitic to calcareous mudstone mottled with occasional light gray chert nodules and irregular, fine grain dolomite and/or dolomitic limestone layers.
- The TG Limy Shale, which is a thick unit of gray to dark gray, low strength, medium-hard to soft, argillaceous (clay-rich), calcareous Limy Shale. Analyses show the shale to consist of a mix of an exceptionally fine carbonate (calcite) mass and clays that are mostly illite, with significantly lesser amounts of chlorite and kaolinite. The TG Limy Shale is finely laminated, fissile, and readily separates disks apart along laminar surfaces after drying out or upon handling.

DETROIT UPPER ROUGE TUNNEL CSO CONTROL PROJECT

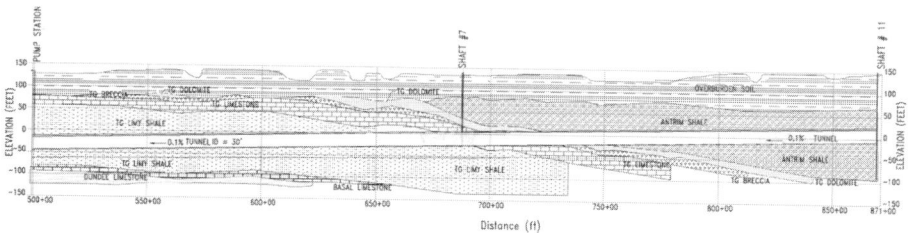

Figure 3. Simplified geologic profile along URT alignment

- The a TG Basal Limestone, which is 11 to 14 feet thick and comprises a sound, moderately hard to hard, highly calcareous, fossiliferous mudstone that grades to a hard, light gray, fine crystalline limestone.

The deepest formation investigated is the Dundee Limestone, which comprises a sound, hard, light gray to buff-color, fine to medium grain, crystalline limestone with occasional fossil shell horizons, numerous stylolites, and thin finely porous to pitted horizons. Historically, the Dundee is an oil and gas producer, and within its occasionally highly fractured upper reaches, often contains groundwater under artesian pressure. The URT tunnel alignment was purposely selected such that this formation is not to be encountered.

GEOTECHNICAL INVESTIGATIONS AND GROUND CHARACTERIZATION

History of Rock Tunneling in the Detroit Area

Rock tunnel construction in the Detroit area has a history of encountering explosive gas (methane), hazardous gas (hydrogen sulfide) and high groundwater inflows. Two previous local tunnel projects, the Detroit River Outfall Tunnel No.2 and the Dearborn CSO Retention Treatment Tunnel, have been flooded and abandoned. Both tunnels were excavated in the Dundee Limestone formation, which is prone to solution activity (gypsum and anhydrite) and washout behavior (numerous joints filled with clay, silt, or even sand). The encountered construction problems were linked to three characteristics of Dundee Limestone:

- High groundwater inflows during construction. In addition, with the presence of vertical joints that were either partially open or clay filled. The clay infilling can be washed away once it is unconfined from tunnel excavation, which leads to the dramatic increase of the permeability of these joints.
- Artesian pressure in Dundee Limestone contributes to high groundwater inflows.
- Dissolved gases (methane and hydrogen sulfide) in groundwater resulting in hazardous condition.

Geotechnical Investigation Program

Geotechnical investigations for the URT Project have been carried out under three different design contracts as described below:

The Feasibility Study Investigation Program. Led by NTH Consultants in 1997 and 1998, this investigation performed a total 56 vertical borings and 3 incline borings to define general subsurface conditions along the alignment and at proposed near surface structures and connections. In situ tests that include packer tests and 3 pumping tests were performed in bedrock, at soil/rock interface, and in overburden. The Feasibility

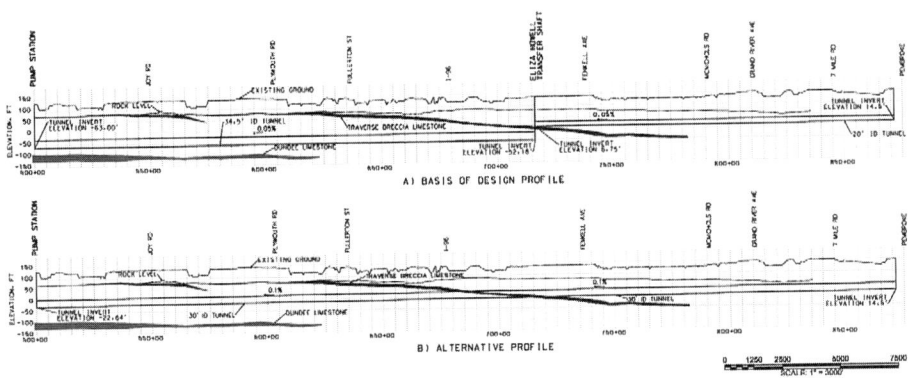

Figure 4. Vertical split alignment and vertical single alignment as considered by preliminary design

Study depicted the TG Breccia as highly porous with direct hydraulic connection to the top of the rock zone that was considered highly fractured and weathered with artesian pressure condition due to confinement by the above silty clay. The TG Breccia layer was then considered to be susceptible of producing significant groundwater inflow during construction, and thus should be avoided. The Feasibility Study then recommended a vertical split alignment.

The Preliminary Design Investigation Program. Led by Parsons Brinckerhoff and Somat Engineering in 2005, the Preliminary Design performed a total 55 additional borings along the tunnel alignment and or at adjacent to selected drop shaft/ diversion control structures, along connection adits and sewers. In situ tests performed during this investigation program include packer tests and borehole geophysical logging. Although these tests performed under this investigation program did not confirm the highly permeable nature of the TG Breccia as depicted by the Feasibility Study, the Preliminary Design maintain the recommendation of a vertical split alignment as a mean to minimize the tunneling risk through this rock unit with the North Tunnel being excavated primarily in Antrim Shale and the South Tunnel in the TG Limy Shale. A transfer shaft would then be excavated through the TG Breccia unit to connect the two tunnels, basically allocating all the large groundwater inflow risk into a vertical shaft element, which can be pre-grouted prior to the excavation. The Preliminary Design, however, left the option of a single vertical alignment open if warranted by the results of a supplemental investigation program to be performed during Final Design. Figure 4 illustrates the preferred option of vertical split alignment and an optional single vertical alignment considered during the Preliminary Design.

The Supplemental Investigation Program. Undertaken by Jacobs Engineering and Somat Engineering in 2006 and 2007 during the Final Design phase, this program includes 10 inclined borings and two horizontal borings for a total length of about 610 and 460 m (2,000 and 1,500 feet) that targeted the TG Breccia to characterize its geomechanical and hydromechanical properties. In situ tests performed during this program consisted of Packer tests and Acoustic Televiewer (ATV) surveys and hydraulic fracturing. In addition, 85 shallow borings to support the design of near surface structures or facilities associated with the pumping station and the 17 outfall structures and 10 additional deep vertical borings were drilled near the shafts and adits. Results of the Supplemental Investigations (Habimana et al. 2008) allowed a better characterization of the TG Breccia, which is not highly porous as depicted by previous investigations and no vertical joints were observed within the investigated area. The final design team

Table 1. Design properties for different rock units along URT

Rock Type	UCS (psi)	Modulus (psi)
AS Carbonaceous Shale	6,500	1.0×10^6
AS Gray Shale	8,000	3.0×10^6
TG Dolomite	16,000	6.4×10^6
TG Breccia (Fresh)	18,000	9.6×10^6
TG Breccia (Weathered)	3,000	1.6×10^6
TG Upper Limestone	15,000	5.9×10^6
TG Limy Shale (clay rich layers)	2,200	0.5×10^6
TG Limy Shale (limestone & wackestone beds)	13,600	4.2×10^6
TG Basal Limestone	11,000	6.5×10^6
Dundee Limestone	16,000	8.5×10^6

then selected a single vertical profile with the intent of achieving a shallower, ease for Pump Station operation and maintenance, and better-performing tunnel with a steeper grade and less grit accumulation in the tunnel system (Chen et al. 2007). The single alignment raises the invert of the Pump Station, which results in a thicker buffer zone, of more than 15 m (50 feet), between the Pump Station invert to the top of the Dundee Limestone formation, which is known to be a difficult tunneling ground and should be avoided.

Ground Characterization

The results of ground investigations and ground characterization were summarized in Habimana et al. 2008. Table 1 provides a summary of the unconfined compressive strength and deformation modulus adopted for the design for various rock types along the URT alignment.

Based on the nature of the different rock formations encountered, the characterization of the rock mass permeability along the tunnel alignment was grouped into three different reaches:

- The South Reach that is almost exclusively within the TG Limy Shale (from Sta 500+00 to Sta 680+00);
- The Middle/Transition Reach, which is 1,000 m (3,500 ft) long and where the tunnel will be excavated within the TG Limestone, Breccia, and Dolomite (from Sta 680+00 to Sta 735+00);
- The North Reach that is in Antrim Shale (Sta 735+00 to Sta 870+00). Rock mass permeability distributions for the three reaches are presented in Table 1 and illustrated in Figure 5.

Expected Ground Behavior Along South Reach. The South Tunnel reach consists of tunneling through predominantly TG Limy Shale formation with frequent occurrences of thick clay seams, or clay filled joints within the shale, or where at least the upper third of tunnel perimeter lies within this formation. Some of the better ground conditions in this reach generally occur when there are occasional presences of stronger, harder wackestone layers and limestone beds within the generally weaker and softer shale. The stability of the excavation in the Limy Shale will be governed by the following properties:

- Low to very low shale durability characteristics, ranging from rock like to a complete breakdown after two cycles of the slake durability test with the residual material degrading into a soil like material.

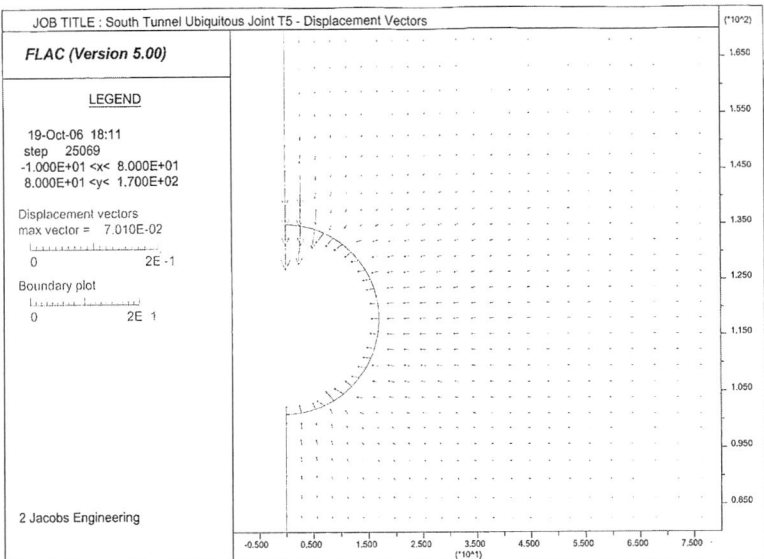

Figure 5. Slabbing behavior as simulated by the FLAC ubiquitous joint model

- Fissile behavior along partings when subjected to air drying or handling, or a general propensity toward disking or parting along bedding planes.
- Low strength of the frequent thick clay seams within the rock mass.
- Presence of stronger, harder wackestone and limestone beds that can occur throughout the section.
- Anisotropic behavior for the generally horizontal beds of the shale layers with a strength anisotropy ratio of approximately 1.5 was estimated for the horizontal versus vertical directions, based on point load test data on the TG Limy Shale.

In order to simulate the behavior of the TG Limy Shale, a laminar rock, i.e. ubiquitous joint, model was used for the initial tunnel support design in TG Limy Shale using the two dimensional FLAC (Fast Lagrangian Analysis of Continua) computer program from Itasca (Version 5.0). This model was able to simulate the horizontal bedding planes, where the shear strength properties (cohesion and friction angle) along the bedding can be combined with the intact strength properties of the rock matrix. The failure mechanism in this type of formation, if not adequately supported, is anticipated to initiate along the tunnel crown and upper sidewall areas in the form of shearing movements and tensile failures along the bedding (Figure 5). If left unsupported, this failure could propagate upward in a "chimney" fashion until the progressive collapse of the tunnel crown occurs.

Rock mass permeability in this reach is expected to be relatively low and estimates for groundwater inflows for the 5,750 m (18,865 feet) long tunnel South Reach are approximately 1,900 liters per minute (l/min) (500 gallons per minutes (gpm)) for the main tunnel based on the 2005 version of the Heuer method.

Ground investigations found traces of hydrogen sulfide at measurable levels and methane gas was observed at concentration levels that were less than the 100% of Lower Explosive Limit (LEL). The excavation in this reach was classified as potentially gassy operation in accordance with OSHA 1926.800 (1989) regulations.

Expected Ground Behavior along Middle/Transition Reach. The Middle Reach consists of tunneling through the limestone and Breccia/Dolomite formations. The ground conditions in these units are generally considered good and stable except in the highly fractured and weathered TG Dolomite and Breccia, where localized failure of rock blocks and wedges at the tunnel crown and/or falling rock conditions can be expected.

The rock mass in this reach is considered to be more permeable than the rest of the tunnel and pre-excavation grouting is required. The groundwater inflow estimates in this reach were performed for both ungrouted and grouted rock assuming that grouting operations will reduce the groundwater inflow by about 50%. Estimates of groundwater inflow within this 1,000 m (3,500 feet) tunnel reach are 7,600 and 3,800 l/min (2,000 and 1,000 gpm) for ungrouted and grouted conditions, respectively.

Expected Ground Behavior Along the North Reach. The remainder of the North Tunnel consists of tunneling through predominantly Antrim Shale formation, which has relatively better quality than that of the TG Limy Shale, with medium high to very high slake durability indices, less propensity for disking apart along the bedding, and higher Unconfined Compressive Strength as shown in Table 1. Similar to the TG Limy Shale, but to a much lesser degree due to stronger rock and shear strength along bedding, the main failure mechanism in this tunnel is anticipated to be shear failures and movements along the bedding planes, which could lead to progressive slabbing failures along the tunnel crown, if not immediately and adequately supported. In addition, some areas of the northern portion of the tunnel are anticipated to have shallow rock cover and will require a more robust support system.

This reach is considered to be relatively low permeability and groundwater inflow estimates of 3,800 l/min (1,000 gpm) are expected for ungrouted conditions. The excavation in this reach is also classified as potentially gassy.

THE SELECTION OF TUNNEL EXCAVATION METHOD AND LINING SYSTEM

South Tunnel

Given that no significant groundwater inflows are expected within the Limy Shale, an open face Tunnel Boring Machine (TBM) was recommended for the South Tunnel. In addition, due to the Limy Shale high degree of fissility and low slake durability, immediate protection after excavation is required to maintain the rock integrity.

The following options were thus considered for the South Tunnel:

1. An open face main beam TBM with rock bolts and wire mesh as initial support for good ground and steel sets within poor ground. The wire mesh was to be installed tightly against the rock surface to prevent initiation of raveling of the rock. The final lining would have consisted of cast-in-place concrete lining.

2. A special open face single shield TBM that is able to erect expanded precast segments as initial supports with a cast-in-place concrete final lining overlay. The expanded segmental lining role is to provide immediate protection and support for the rock behind the TBM. The TBM advances via jacking against the completed and expanded precast segmental ring behind it.

3. A special main beam TBM that facilitates a shotcrete operation near the tunnel face. In this scenario, the tunnel perimeter would have been protected immediately by shotcrete and rock bolts or steel ribs behind the cutter head. The final lining will consist of cast-in-place concrete lining.

4. An open face single shield TBM that employs a one pass precast concrete segmental lining as both initial support and final lining of the tunnel. The segments would be bolted and gasketed for watertightness. The TBM would

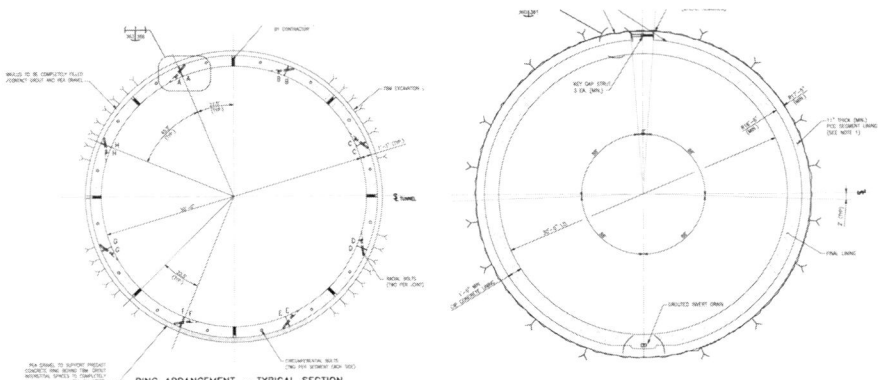

a) One pass precast bolted and gasketed concrete segmental lining

b) a two-pass lining system with expandable concrete segments and a cast in place final lining

Figure 6. Two lining options for the South Tunnel

be advanced via jacking against the completed segmental lining. The gap between the segmental lining and the ground should be contact grouted as quickly as practically possible through the tail seal of the TBM shield.

These options were evaluated using several criteria that include cost, schedule, constructability and risk assessment. Two important factors were considered during evaluation of excavation methods: First, systematically applying shotcrete or sealants for the entire length of the tunnel was considered but judged to be inefficient because it would significantly impact the TBM performance and cause other maintenance problems. Second, the design team recognized a risk of progressive chimney failure above the crown that could have impacted construction progress in case of the use of rock bolts as initial support.

After a comparative review of these options, the design team selected an open face TBM with two options of lining system (Figure 6):

- A one-pass precast bolted and gasketed precast concrete segmental lining
- A two-pass lining system with initial lining consisting of expanded precast concrete segments with cast-in-place concrete final lining.

This package was advertised in December 2007 and the two bids received were both for the one pass precast bolted and gasketed precast concrete segmental lining. The Kenny Construction / Obayashi Joint Venture was awarded the contract on October 27, 2008 for about $316 million. At the time of the writing of this paper, preparation for the excavation lunch shaft is underway.

North Tunnel

In the early stages of the Final Design, recognizing the benefits of a single vertical alignment as outlined above, three options were considered based on the geologic conditions as described in the Preliminary Design phase. The tunnel excavation method and lining system were then to be selected based on the results of the supplemental investigation program. The three options considered were:

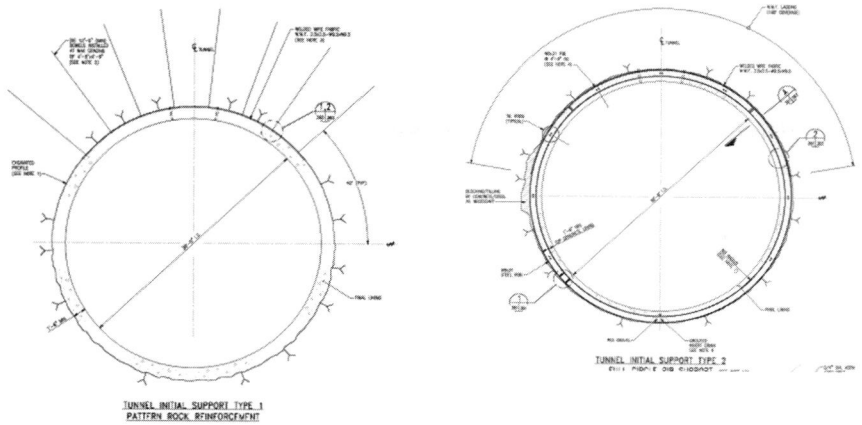

a) Rock dowel and welded wire fabric   b) Steel ribs and welded wire fabric lagging

Figure 7. Two initial support systems for the North Tunnel

1. A main beam TBM with rock bolts and wire mesh as initial support for good ground and steel sets within poor ground. The wire mesh should be installed tightly against the rock surface to prevent initiation of raveling of the rock.
2. A hybrid TBM that could have operated in closed mode for the first 3500 ft and then in open face mode for the rest of the tunnel.
3. A combined approach of Sequential Excavation Method (SEM) for the Middle/Transition Reach and then an open main beam TBM for the rest of the tunnel.

Based on the results of the supplemental investigation that allowed a better geo-mechanical and hydromechanical characterization of the TG Breccia, the design team recommended a main beam TBM that is equipped with finger shields behind the cutterhead, as a protection for personnel working near the tunnel face and to allow immediate installation of initial rock dowels support. As mentioned above, because of the potential of encountering high groundwater inflows in the first 1,000 m of tunneling through the TG Breccia and Dolomite units, continuous probing and pre-excavation grouting ahead of the TBM is required along this Middle/Transition Reach to control the groundwater infiltration during tunneling.

The tunnel initial support system will need to be installed concurrently with the excavation and immediately behind the TBM to prevent rock blocks/wedges from loosening and progressive slabbing to take place above the tunnel roof. Two types of initial support were considered (Figure 7):

- Type 1 initial support consists of patterned rock dowels in the tunnel crown, supplemented by Welded Wire Fabric (WWF) when tunneling through the weathered TG Breccia,, fractured TG Dolomite, and the Antrim Shale Formation.
- Type 2 initial support for ground with low, i.e. less than 1.5 tunnel diameter, rock cover consists of regularly spaced full-circle steel ribs with WWF lagging over the tunnel crown. Precast concrete invert segments can be used in lieu of full-circle steel for invert protection.

Table 2. Dimension and location of different connecting adits and deaeration chambers along URT

Tunnel Reach	Ground Condition Reach	Location	Connecting Adit		Deaeration Chamber	
			Finished Diameter (ft)	Length (ft)	Finished Diameter (ft)	Length (ft)
South Tunnel	South Reach	DS-2	15.0	274.7	32.0	129.0
		DS-3	15.5	257.9	33.0	132.0
		DS-3A	7.0	1,441.2	13.5	99.0
		DS-4	9.0	1,037.6	18.0	131.0
		DS-5	9.0	43.9	17.0	124.0
		DS-6	7.0	226.2	11.0	80.0
North Tunnel	Middle/Transition Reach	DS-7A	9.5	1,812.0	19.0	140.0
	North Reach	DS-8	7.0	785.6	13.0	98.0
		DS-9	18.5	564.80	40.0	159.0
		DS-10A	7.0	438.0	8.0	32.0
		DS-10	8.0	80.20	15.5	115.0

The baseline specifies Type 1 and 2 initial supports for 85% and 15% of the tunnel length, respectively. As a contingency measure, additional rock support elements including additional rock dowels, shotcrete, steel straps and channels were specified to be used as warranted by the ground conditions encountered during excavation.

THE SELECTION OF EXCAVATION METHODS AND LINING SYSTEMS FOR CONNECTING ADITS AND DEAERATION CHAMBERS

Table 2 summarizes the finished inside diameter and length of each connecting adit and deaeration chambers for both the South and North Tunnels and expected ground conditions. Because of surface constraints and environmental issues, all the connecting adits and deaeration chambers excavations and muck disposal will be staged from the tunnels.

The excavation geometries for the connecting adits and deaeration chambers associated with both the South and North Tunnel Contracts are of horseshoe shapes. They will be constructed by drill and blast method with the initial lining installed as close as practicable to the tunnel heading. Smaller connecting adits and deaeration chambers are expected to be excavated in a single drift. The largest size deaeration chambers, i.e. 9.75 to 10-m (32 to 33-ft) finished diameter chambers in the South Tunnel and 12-m (40-ft) diameter in the North Tunnel) are to be excavated in multiple headings using the sequential excavation method with multiple headings and benches (see Figure 8). The support system consists of an initial support with a cast-in-place concrete final lining. The initial support system will consist of a combination rock dowels and shotcrete for all connecting adits and deaeration chambers, except the deaeration chamber at DS-9 site, which has a shallow rock cover and lattice girders as initial support will be employed. While the maximum spacing of rock dowels was maintained at a same spacing, 1.2-m × 1.2-m (4 × 4 ft) for all connecting adits and deaeration chambers, their lengths and tunnel unsupported spans was assigned based on the excavation dimensions.

Provision for additional support measures (i.e. rock dowels and shotcrete) were also adopted to deal with the potential of encountering weaker than expected bedding layers or joints, i.e. thick clay seams or clay filled joints.

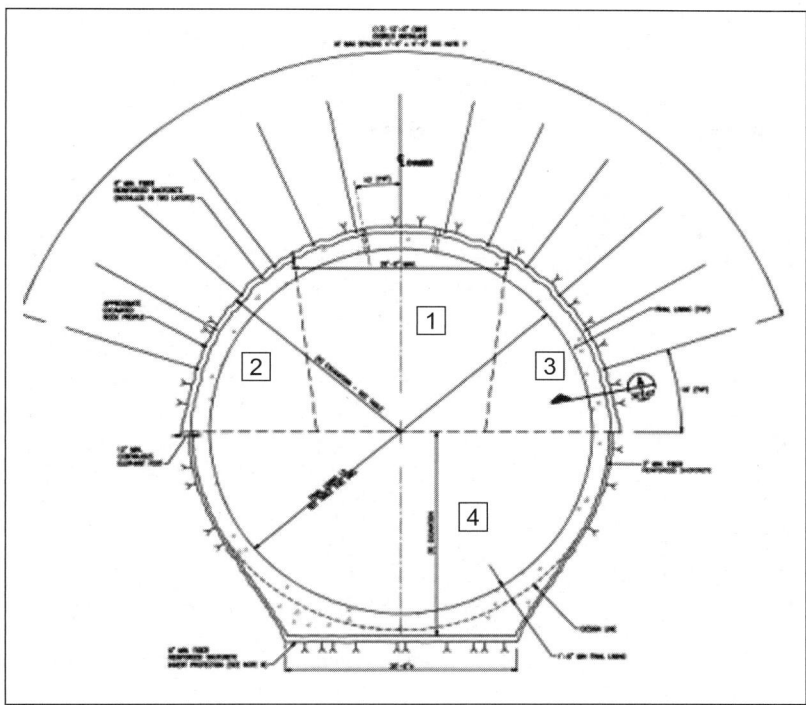

Figure 8. Excavation sequences and initial support for deaeration chambers at DS-2 and DS-3

THE SELECTION OF EXCAVATION METHODS AND LINING SYSTEMS FOR THE SHAFTS

The URT has a total of 3 work shafts and 11 drop shafts and their associated ventilation and surge control volume shafts. Finished diameters range from 14 to 24 m (47 to 80 feet) for work shafts, 1.5 to 5.5 m (5 to 18.5 feet) for drop shafts, and 0.6 to 2.5 m (2 to 8 feet) for vent shafts. Their depth ranges from 45 to 55 m (150 to 180 feet). This section will focus on the larger diameter work shafts only, since the construction and final lining methodologies for other smaller diameter shafts are similar to the work shafts.

Shaft Excavation in Overburden

Some of the important overburden soil characteristics that will potentially impact the shaft excavations are:

- The potential of encountering zones of soil with relatively high hydraulic conductivity.
- The occurrences of cobbles and boulders in Glacial Till.
- The potential of encountering artesian pressure that can be as high as 1.8 m (6 ft) above ground surface with methane/hydrogen sulfide gas in solution at the soil-rock interface and trapped within the bedrock.
- The presence of man-made obstructions in the upper portion of soil, the fill layer

The excavation support system is required to penetrate below the weathered rock zone to prevent instability at the bottom of excavation and soil migration into the excavation. In addition, surface grouting program around the shaft perimeter is required before the shaft excavations to reduce the rock mass permeability. The initial support installation in the shaft soil section shall be able to deal with the following issues:

- Provide excavation wall and face stability at all times
- Resist earth and water pressure
- Provide watertightness

For Work Shafts DS-7 and DS-11, the selection and the design of the support of excavations in overburden were left to the contractor based on its means and methods in accordance with the specified design criteria. For the Work Shaft at the Pump Station, a 39-inch thick reinforced concrete slurry wall was required, because of its 24-meter (80-ft) larger than normal shaft diameter.

Shaft Excavation in Rock

The excavation of shafts in the rock by drill and blast or mechanical methods are feasible for the expected ground conditions. The 24-m (80-ft) ID Pumping Station Shaft will be excavated almost entirely in TG Limy Shale, except for the top 6 m (20-ft) of rock that will be excavated in TG Upper Limestone. Rock excavation at the 14-m (47-ft) ID Work Shaft at DS-7 will be in Antrim Shale and TG Dolomite, Breccia, Upper Limestone and Limy Shale at the bottom of the shaft. Rock excavation at the 20-m (64-ft) ID Work Shaft at DS-11 will be entirely in Antrim Shale.

The mechanical excavation of the Pump Station Shaft is expected to be performed in low strength rock sections, where drill and blast is not necessary to loosen the rock. Initial ground support in the shafts consist of pattern rock dowels with steel straps, 20-cm (8-in) of fiber reinforced shotcrete, and lattice girders. The shotcrete is to be applied in two layers with the first 7.5-cm (3-in) applied to the rock surface immediately after excavation before the installation of rock bolts or lattice girders to protect the shale. The allowable excavation depth before installing the support is limited to 1.8-m (6-ft). This unsupported depth is reduced to 1-m (3-ft) if highly fractured rock, a high percentage of clay-filled joints, clay beds several inches thick and decomposed shale layers are encountered and the monitoring indicates excessive movement. The reduced initial support spacing will mainly be applied in poor to very poor quality rock, which is expected for 20% of the shaft excavation in TG Limy Shale.

Due to the close proximity of Pump Station Shaft invert to the underlying Dundee Limestone, which is a known aquifer, and the potential for high artesian pressure condition within this zone, shaft excavation bottom stability have been carefully evaluated. Contingency glass fiber dowels are specified as tie-downs in the shaft excavation bottom to stabilize the uplift groundwater pressure based on piezometer pressure monitored during construction. Furthermore, the groundwater in the Dundee Limestone is known to contain dissolved methane and hydrogen sulfide gas. Therefore penetrating within 2.5 m (8 feet) distance from the Dundee Limestone is strictly forbidden, because perforating the top of this confined groundwater horizon would allow the gas-laden groundwater to flow into the shaft excavation.

Ground conditions in rock for Work Shafts DS-7 and DS-11 are in the shale and limestone formations. They are expected to be generally stable and self-supporting, except occasional areas where thick clay-filled joints maybe present along the sub-horizontal bedding planes, which could lead to some movements and localized rock mass failure at the excavation face. In the weathered and highly fractured TG Dolomite and Breccia zones, localized block failures and falling rocks should also be anticipated. In these cases, shotcrete application shall be immediately following the excavation face. The initial support consists of pattern rock dowels with 7.5-cm (3-in) and 10-cm (4-in) minimum fiber reinforced shotcrete for shafts at DS-7 and DS-11, respectively.

CONCLUSION

The selection of excavation methods for the two tunnel construction contracts of the URT was primarily based on the two distinct geologic formations that are expected along the URT alignment. The Supplemental Investigation program undertaken during the Final Design has targeted specific geological formations that have been depicted by early studies as potentially risky to tunnel through. The selection of a single vertical alignment, at the conclusion of this investigation program, was judged to be optimum for both construction and operation standpoints. It was finally selected, as opposed to the vertical split alignment recommended by earlier studies. For the South Tunnel contract, an open shielded TBM with precast concrete segmental lining was selected to address the relatively low strength and fissile nature of the TG Limy Shale. A main beam TBM equipped with finger shield was selected for the North Tunnel that will be driven in relatively more competent rock formations. Given the possibility of significant groundwater inflow in the first portion of the North Tunnel, probing and pre-excavation grouting are required for that portion. Initial support for the North Tunnel will consist of rock dowels and welded wire fabric for 85% of the tunnel while a more robust initial support with steel sets will be used for portion of low rock cover that correspond to 15% of the tunnel length. Excavation for adits and deaeration chambers and shafts in rock are to be performed by drill and blast with initial support to be installed immediately behind the excavation face. Both tunnel excavations are classified as potentially gassy ground operations.

REFERENCES

Chen, W., Rabbaig, M., Barbour, R., Habimana, J., and Liu, J. 2007. "Upper Rouge Tunnel CSO Contro," the Rapid Excavation and Tunneling Conference, 627–635.

Habimana, J., Halim, I., Chen, W., Rabbaig, M. and Barbour R. 2008. "Geotechnical Considerations for the Final Design of the Upper Rouge Tunnel CSO Control Project" the North American Conference, San Francisco, 302–311.

Heuer R., 2005. Estimating Rock Tunnel Water Inflow–II, Rapid Excavation and Tunneling Conference, 394–407.

PLANNING AND DESIGN FEATURES OF THE WALLER CREEK TUNNEL, AUSTIN, TEXAS

Gary Jackson ▪ City of Austin

Stan Evans ▪ City of Austin

Tom Saczynski ▪ Kellogg Brown & Root

Peter Jewell ▪ Kellogg Brown & Root

Prakash Donde ▪ Jenny Engineering Corporation

ABSTRACT

Preliminary engineering for the Waller Creek Tunnel was completed in 2001, but the project was held in abeyance until the summer of 2007. This paper describes several features of the revised design, which required significant changes from the original concept. Most notable is the expected replacement of a tunnel boring machine (TBM) by conventional tunneling. The considerable influence of geological conditions on the tunnel vertical alignment is discussed, in particular the impact of the zones of Eagle Ford Shale with swelling potential. Other features which are described include the wide span tunnels and large diameter shaft requirements.

INTRODUCTION

The Waller Creek watershed is the most developed tributary watershed of the Colorado River within the city limits of Austin. The lower reach of Waller Creek traverses the City of Austin's (COA) downtown corridor, including portions of the State Capitol complex and the University of Texas campus. The flood plain is as wide as 800 ft. Moreover, flood events have periodically inundated developed areas above the banks of Waller Creek, resulting in the loss of life, property and damage to infrastructure.

The COA has long desired to improve flood control of Waller Creek. Reducing the flood plain to within the banks of the creek will allow approximately 92,903 m^2 (1,000,000 ft^2) of previously unusable land to be fully utilized. In addition, 42 commercial, public and residential structures as well as portions of 12 roadways would no longer be subjected to periodic flooding.

To that end, the COA has conducted several studies for the purpose of identifying strategies for flood control and enhancing the water quality of Waller Creek. The purpose of the project is, therefore, to capture and divert floodwaters into a tunnel that will convey the flood flows to the Colorado River via Lady Bird Lake, thereby reducing the floodplain to within the banks of Waller Creek.

INITIAL PLANNING

The goals of the project are to:
- Capture a 100-year flood event and safely divert it to the Lake;
- Improve the water quality in the Creek; and
- Improve recreational opportunities along and near the Creek.

Such an accomplishment would foster redevelopment of public and private properties that are currently within (or near) the floodplain of Waller Creek.

COA selected a team of consultants to perform engineering and design of the Waller Creek Tunnel. The design team, a Joint Venture of Kellogg Brown & Root and Espey Consultants, with a number of subconsultants, including Jenny Engineering Corporation, was authorized to proceed with initial design in October 2007.

The design team recommended that an inlet capture facility be located at Waterloo Park. The Inlet will be capable of capturing the 100-year flood event and, with a tunnel alignment under Sabine Street, deliver captured storm water to the proposed outlet facility adjacent to the mouth of Waller Creek.

The recommended alignment of the tunnel is primarily under existing right-of-way and mostly within the Austin Limestone. The tunnel depth not only meets the design requirements, but it is deep enough to provide a tunnel diameter of rock over the tunnel excavation.

Preliminary engineering in 2001 revealed that tunnel excavation by several means was feasible, including the use of a tunnel boring machine (TBM), but a particular method was not recommended at that time. Lining of the tunnel would likely be by cast-in-place concrete techniques, but precast concrete segmental liner was feasible.

Since the tunnel would be normally filled with water from the lake, it would be necessary to recirculate the water in the tunnel to prevent stagnation. Additionally, recirculation of water from the tunnel could yield a source of additional base flow in the creek during the dry seasons, and could improve aesthetics. Although recirculation entails mechanical, electrical and piping considerations, they are beyond the scope of this paper.

To limit the flood flows in the creek well below the top of bank to a depth of about 4 feet, it would be necessary to capture flows into the creek between the proposed inlet and outlet sites. As a result, two additional capture points were proposed along the creek. These structures would consist of side-overflow weirs with shafts and lateral connections to convey the flows to the main tunnel.

A public workshop held in November 2007 generated additional ideas and influences were contributed. After evaluation and incorporation of accepted suggestions, the cost of the construction is estimated at about $100 million with a construction duration of about 48 months.

WALLER CREEK HYDROLOGY AND RESULTING TUNNEL HYDRAULICS

The key requirement of the project is to protect the lower part of the basin from flooding, thereby facilitating redevelopment of this part of downtown Austin. Accepting the upper basin flows into the tunnel and conveying the flows to the Lake will accomplish this.

The Waller Creek watershed includes approximately 12.23 km (7.60 mi) of urbanized stream reach that drains approximately 5.60 km^2 (14.50 mi^2) into the Colorado River basin. The basin lies parallel to and west of IH-35, and is rather steep in the downtown, urbanized lower part of the basin where the tunnel will be located.

The creek drops about 12.2 m (40 ft) in elevation below the main Inlet at Waterloo Park over approximately one mile to the Outlet at Lady Bird Lake. The location of the main inlet at Waterloo Park allows the proposed project to accept flows from the upper 85% of the creek basin. Figure 1 shows the preliminary tunnel alignment with respect to the creek and Lady Bird Lake.

The meteorological characteristics of central Texas, along with the influence of the Balcones Escarpment, produce conditions conducive to large rainstorms in the area. Many of the highest rainfall intensities in the world have occurred in central Texas:

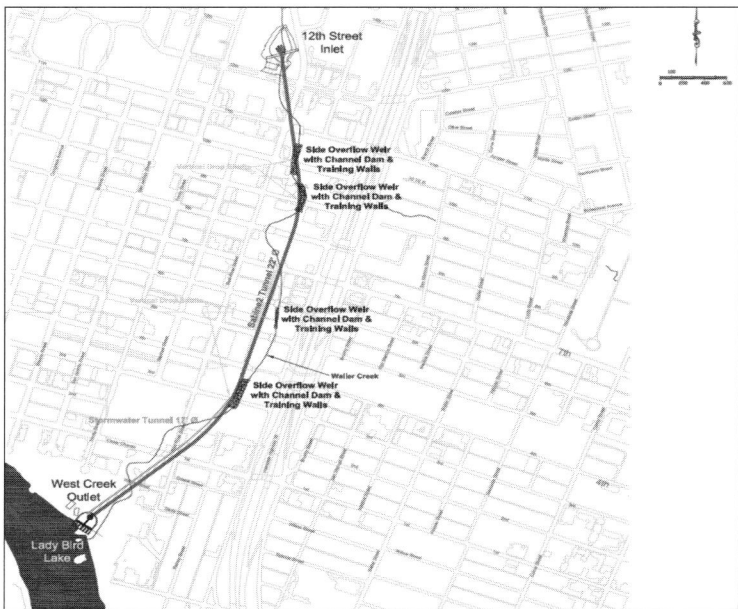

Figure 1. Tunnel alignment

- 1921 storm in Thrall (NE of Austin) produced 813 mm (32 in) 32 of rain in 12 hours;
- 1935 storm at D'Hanis (SW of Austin) produced 559 mm (22 in) in 2 hours 45 minutes;
- 2007 storm at Marble Falls (W of Austin) produced 203 mm (8 in) in one hour.

The most notable events in the modern era to strike Waller Creek were the November 2004 and October 1973 events where 76 mm (3 in) of rain fell in approximately 90 minutes in each storm event. The 2004 storm produced the largest recorded peak rate on Waller Creek at the 23rd Street gage (upstream of the tunnel inlet) of 148 m^3 (5,233 ft^3/s), which is between a 20 year and 50 year event. A severe storm in May 1981 within the Shoal Creek basin resulted in 254 mm (10 in) of rain in 3 hours. This basin is adjacent to the Waller Creek basin. The 1981 event produced the fifth largest recorded peak flow at the 23rd Street gage on Waller Creek of 3,333 ft^3/s/s.

The design storms for the COA are based on depth of rain in a 24 hour period as shown in Table 1.

About 76% of the surface of the Waller Creek basin is comprised of impervious material. Thus much of the precipitation cannot penetrate the ground but runs off rapidly into the creek.

The main inlet within Waterloo Park will be designed to accept the calculated 1% annual chance (100 year) storm event flow of 217 m^3/s (7,666 ft^3/s). However, the tunnel will be designed to accept and convey an even greater flow, since additional flows will be routed into the tunnel between the inlet and the outlet locations.

In addition to accepting the flows from the upper watershed, there is a requirement for the project to provide a water surface in the creek that will be maintained at typically no more than 1.22 m (4 ft) above the normal base flow water surface during a

Table 1. Recurrence interval, probability, and rainfall for design rain storm

Event Recurrence (Years)	Probability (over 1 year)	Precipitation (mm/24 hours)	Precipitation (in/24 hours)
2	0.50	87	3.44
10	0.10	155	6.10
25	0.04	194	7.64
100	0.01	270	10.62

storm event (up to the 100-year event). The base flow will be controlled by release of water from the main inlet facility and will normally be 2.13 to 3.05 m³/s (7 to 10 ft³/s). Additionally, two intermediate creek side inlets will be constructed to limit the rise of the water in the creek during storm events caused by overland flows and the contributions from the existing stormwater systems. These intermediate inlets will be located at 8th Street and 4th Street.

Each intermediate inlet will consist of a low-flow, in-channel dam within the creek; a side overflow weir; and a passive screening facility to capture debris (bag or net). The captured flows will be conveyed to the main tunnel via drop shafts and connector tunnels.

These dams will span the width of the creek and, during dry conditions, will create a permanent pool approximately 1 m (3 ft) deep. Each dam will allow up to 2 m³/s (60 ft³/s) to pass over the spillway without engaging the side overflow weir. However, during flood conditions, that is, flows in excess of 2 m³/s (60 ft³/s), each dam will induce a tailwater condition to maximize flood flow diversion to the side overflow weir while keeping flow passed by the dam spillway to a minimum. The pool water surface will increase by only one foot at the maximum design condition. The dams will also serve as a pedestrian walkway during dry weather conditions, and the pools will allow for formation of biological habitat.

The creek side inlets will each have a 61 m (200 ft) long lateral weir that will allow flows over 2 m³/s (60 ft³/s) to be diverted from the creek. When the water surface elevation in the pond increases above the weir elevation, the flow will enter the creek side inlet facility and will go through passive trash collectors, and then fall into the drop shaft and be conveyed via a connecting tunnel to the main tunnel. These two intermediate inlets will each accept design peak flows of 20 m³/s (700 ft³/s) while a peak of 6 m³/s (200 ft³/s) will pass over the dam and continue downstream in the creek.

To accommodate additional inflows from the weirs at 8th street and 4th street, the tunnel diameter varies to define 3 separate reaches. From the main Inlet shaft just above 12th Street to the first intermediate inlet shaft at about 8th Street, the tunnel diameter is 20 feet. The diameter increases to 6.7 m (22 ft) where the connector tunnel from the 8th Street connector joins the main tunnel. Similarly, where the 4th Street intermediate inlet connector tunnel joins the main tunnel, the main tunnel diameter increases to 7.9 m (26 ft) and remains so until it reaches the Outlet shaft.

For the connector tunnels, a minimum diameter for hydraulics requirements was established as 4.3 m (14 ft). There are 2 tunnel components of the 4th Street connector, a north arc and a south arc. Whereas the south arc will convey water, the north arc is only to accommodate tunnel excavation operations and will be backfilled. The contractor will have the flexibility to increase the diameters of the 4th Street connector to suit support requirements. The 8th Street tunnel connector is straight in plan, and impinges upon the main tunnel at a 45 degree angle. A summary of tunnel dimensions for the hydraulics requirements is shown in Tables 2 and 3.

Table 2. Variations in tunnel diameter and reach length—main tunnel

Structural Feature	Outlet to 4th St.	4th St. to 8th St.	8th St. to Inlet
Reach	1	2	3
Diameter	7.9 m (26 ft)	6.7 m (22 ft)	6.1 m (20 ft)
Increment Length	725 m (2,380 ft)	445 m (1,460 ft)	508 m (1,668 ft)

Table 3. Tunnel diameter and segment lengths—connector tunnels

Segment	4th St. Connector		8th St. Connector
	South Arc	North Arc	
Minimum Diameter	4.3 m (14 ft)	4.3 m (14 ft)	4.3 m (14 ft)
Length	65.6 m (215 ft)	33.2 m (109 ft)	30.5 m (100 ft)

DESIGN CONSIDERATIONS FOR TUNNELS AND SHAFTS

The hydraulic and hydrologic considerations essentially dictate the horizontal and vertical alignments of the project, as well as the diameter and shape of the tunnels. Rights-of-way issues required initial adjustments to shaft locations, as well as to horizontal curves in the tunnel alignment. With the project geometry established, the design of the tunnels and shafts becomes a matter of excavation and support, which in turn must take account of the geology.

Through value engineering examinations by the design team and the Owner's staff in the initial design phase undertaken during 2008, several key decisions were reached that resulted in a consensus that the tunnel would be excavated by conventional methods, that is, by roadheader or drill-and-blast techniques.

Eliminating TBM excavation from consideration affords two important advantages. The first is a schedule benefit as time waiting for a TBM to be manufactured or refurbished, delivered to the site and inserted into the ground is avoided. Secondly, it allows for concurrent construction for the inlet site and the outlet site since these shafts would not be required for insertion and removal of the TBM.

To allow tunnel, excavation, and support, it is envisaged that one of the creek side inlet shafts would become the primary operational shaft. Since the tunnel excavation would not be limited to the diameter of a TBM, the tunnel diameter could be varied to accommodate increases in flow from the intermediate inlet shafts.

Ground Characterization

Geotechnical investigations indicate the tunnel and shafts are primarily located within Austin Limestone. Although most of the inlet and outlet shaft vertical alignments are within the Austin Limestone, the lower portions are located within the Eagle Ford Shale, as is the main tunnel where it connects to these shafts.

The data collected in the geotechnical investigations indicate that the rock quality is typically "good" to "excellent" because Rock Quality Designations (RQD; Deere & Deere, 1988) range primarily from 90 to 100. This generalization applies to the Austin limestone as well as to the Eagle Ford Shale.

Based upon the investigation and analysis to date, the rock would appear to be "tight," as with experience on other tunnel projects in Austin. Packer tests indicate very low permeability of the rock mass. Additionally, although there are some faults along the alignment, experience has demonstrated that faulting in the area does not transmit water.

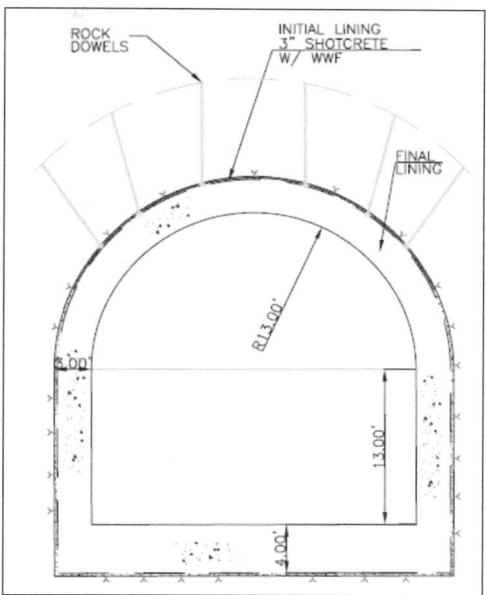

Figure 2. Horseshoe

Sufficient exploration and laboratory analyses have been accomplished to date to determine the general behavior of tunnel and shaft excavations. Whereas geotechnical work is essentially complete, additional work is still being considered for specific, isolated areas of interest. However, it would be overly optimistic at this stage to conclude that the project is without geotechnical challenges.

Certain members of the Eagle Ford Shale exhibit significant swelling potential (ISRM, 1989, 1994). The consequence of swelling on lining thickness is that the lining must withstand loads from swell in addition to hydrostatic forces and the nominal rock load. Initial testing indicates that loads from swell can be as high as 0.69 MPa (100 psi) over the hydrostatic and nominal rock loads.

The Eagle Ford Shale, which includes the swelling rock, is also susceptible to slake. As the rock mass dries out and disintegrates, joints and bedding planes become wider and stability problems develop. Therefore rapid application of shotcrete will be required to preclude slaking, as well as to provide immediate support of the opening.

Another challenge stems from the relatively low strength of the rock. With uniaxial compressive strength for both limestone and shale at approximately 10.3 MPa (1,500 psi), the unsupported excavated width might entail limitations, such as requiring multiple drift excavation.

Low strength rock might also present other stability challenges at intersections of the tunnel with the main shafts and with the oblique intersections with the connecting tunnels. Further analysis in the final design phase will investigate the potential.

Tunnel Design

There are two loading conditions under evaluation for the tunnel design. These are the construction stage (short term), which defines criteria for initial support, and the service stage (long term), which defines criteria for final tunnel lining.

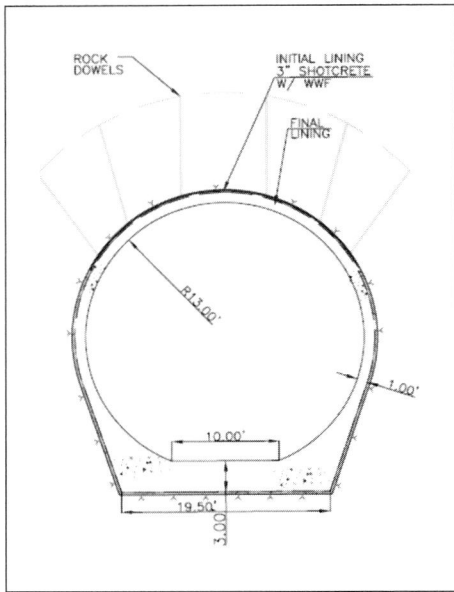

Figure 3. Modified horseshoe with circular lining

Initially, the excavated cross-section for conventional tunneling was envisioned as a horseshoe (i.e., a semicircle over a rectangle, Figure 2). The tunnel will, however, undergo dewatering frequently over the design life. In a dewatered state, hydrostatic pressure would induce large bending stresses in the vertical sidewalls. The result would be an increase the thickness of the lining, as well as in the amount of reinforcing steel, over the requirements for a circular opening.

Establishing stability of the excavated opening is the primary short term consideration. The contractor will be allowed considerable latitude in determining the shape of the excavated opening. Whereas a flat invert would be desirable for traficability and a circular opening for the final lining, to best capture the benefits of both these systems something approaching a modified horseshoe is envisioned for the excavated opening (Figure 3).

Flexibility in the selection of initial support will also be a feature of the design. Rock Mass Rating and Q ratings will be assessed for intervals along the entire alignment. Initial support will be designed by analyzing tunnel cross-sectional configurations at various locations using FLAC 3D software which is based on finite difference modeling.

Excavation. Conventional and TBM excavation were investigated. Conventional methods comprise excavating a single "round," for example 10 feet, with the application of initial support immediately following the excavation. Either drill and blast or roadheader methods would be appropriate for excavation. Some advantages and disadvantages of each are summarized in Table 4.

Practical considerations limit TBM application. Primarily, an excavated diameter of 29 feet would almost certainly require a custom manufacture, which has additional cost and schedule implications. Additionally, only the maximum of the three diameters could be excavated, thereby creating about 20% more excavated material and requiring 10%more concrete and steel for the liner. Further, the inlet and outlet shafts would be required for launching and retrieval, and one of the two shafts would be required for tunneling operations, thereby extending the schedule even further. Finally, right

Table 4. Comparison of excavation methods: Advantages and disadvantages

Drill and Blast	
Advantages	**Disadvantages**
Considerably lower equipment costs	Blasting vibrations
Shorter mobilization time	Public reaction to blasting
Easier surveying	Safety and security issues
Easier inspection of the working face	Higher potential for overbreak
Relatively easy drilling	Blasting plan requirements
Roadheader	
Advantages	**Disadvantages**
Relatively easy cutting	Will require mechanical dust extraction
Less overbreak and scaling	Difficult to control opening dimensions

of way considerations require tight radius horizontal curves, which further limit TBM applicability.

Initial Support. The design presently envisions initial tunnel support as consisting primarily of rock dowels and fiberglass reinforced shotcrete. Further investigation and analysis might, however, indicate that some tunnel intervals require additional support. Additional support would take the form of lattice girders or structural steel shapes.

The large size of the tunnel cross section is not likely to allow a full face excavation of the tunnel and a multiple drift, or heading and bench, approach would therefore be utilized. It is expected that the top heading would be excavated using a roadheader. It would be followed by the excavation of the bench by again a roadheader or ripping the bench as is frequently done in limestone excavations. This excavation sequence will be modeled numerically in the analysis of the tunnel excavation for design purposes.

Additionally, in the event that the investigation and analysis of the final design determines the presence of intervals of rock with relatively low rock mass ratings (RMR; Bienawski, 1989) and Q values (Barton et al., 1980). SEM methodology would be required. Such measures could include frequent precision convergence measurement, multiple drift excavation, lattice girders, steel sets, and additional dowels and shotcrete.

Final Lining. Because of the large finished diameters and the smoothness required from hydraulics, the final lining will be cast in place concrete. At this writing, lining thickness is anticipated as 300 mm to 450 mm (12 to 18 in) with a single layer of reinforcing steel, Further analysis could demonstrate that thicker lining and more reinforcement are necessary, particularly for tunnel intervals in swelling shale and at intersections.

Shafts

There are four shafts planned for the following locations:
1. Inlet at the upstream terminus to capture and control all water that would enter the desired area from upstream sources;
2. Outlet at Lady Bird Lake (formerly Town Lake); and
3. Two intermediate shafts for weirs that capture the surface water between the Inlet and Outlet.

Table 5 lists the expected diameters, thicknesses of Alluvium and rock and total shaft depths.

Primarily for scheduling considerations, the inlet and outlet shafts will not be used for excavation of the main tunnel. Rather, the 4th Street shaft will support tunnel

Table 5. Summary of principal shaft dimensions

Shaft	Diameter	Alluvium Thickness	Depth of Shaft in Rock	Total Shaft Depth
Outlet	12.2 m (40 ft)	6.7 m (22 ft)	12.2 m (40 ft)	18.9 m (62 ft)
Inlet	9.1 m (30 ft)	3.0 m (10 ft)	21.6 m (71 ft)	24.7 m (81 ft)
4th Street	9.1 m (30 ft)	6.7 m (22 ft)	17.1 m (56 ft)	23.8 m (78 ft)
8th Street	6.1 m (20 ft)	6.1 m (20 ft)	17.4 m (57 ft)	23.5 m (77 ft)

operations. At roughly mid way, excavation and support of the main tunnel could proceed from this shaft in two directions, both up-tunnel and down-tunnel.

Excavation. In all cases, the shafts will be excavated through overburden soils before rock. Conventional excavation is anticipated for removing overburden soils.

Below the overburden, conventional rock excavation will be possible in the Austin Limestone and Eagle Ford Shale. Previously, excavation of these rock types has been successful with pilot/auger/ream methods for shafts up to 32 feet excavated diameter. Alternative methods, especially for larger shafts, would include ripping, roadheader excavation or drill and blast would be effective.

Excavation Support. Excavation support for the soil would normally result from the contractor's choice of means and methods and would most likely consist of one or several of the following systems:

- Ring beams and liner plate;
- Ring beams and lagging;
- Secant piles;
- Tangent piles;
- Sheet piling; and
- Slurry wall.

Obviously, secant piles require more time, but they offer the advantage of a watertight shaft. Additionally, the rates for both soil and rock excavation appear optimistic. The matter will be thoroughly examined during the final phase.

Owing to the proximity of Lady Bird Lake, which establishes the ground water table around the outlet shaft, the excavation support will be required to cut off the ground water. Similarly, the water table elevation and flood events will influence the support designs and the resulting ability of the shafts to effectively cut off water. Because flooding of any of the shafts would seriously jeopardize the project, prescriptive methods to ensure that the shafts remain watertight will be specified and have sufficient freeboard.

At this point in time, a ring of secant piles socketed into Austin Limestone rock is considered as the most effective means to support the ground, as well as of reasonable cost. This will be more fully explored during the final design phase.

Excavation support in rock will consist of shotcrete and supplemented with rock dowels, as required. Whereas shotcrete application to the Eagle Ford Shale will be mandatory, consideration might evolve for change order work to eliminate shotcrete cover of the Austin Limestone where stability is not an issue.

Final Lining. Final lining for the shafts will consist of reinforced concrete of about 24 inches in thickness. The lining thickness might be increased for the Eagle Ford Shale because of swelling stresses. This will be determined in the final design.

CONCLUSIONS

After several years of conceptual planning and budgeting, COA is now poised to finalize the design of the Waller Creek Tunnel project and proceed to construction. The concept for the project has been updated and the paper has made note of the principal features which will address the 100-year flood event, by means of an inlet capture facility, a tunnel to divert flood flows, an outlet facility adjoining the mouth of the existing Waller Creek and two capture points for intermediate flows to be diverted into the tunnel. The paper has also introduced the revised design concepts for tunnels and shafts.

Completion of the Waller Creek Tunnel within the next few years will remove the threat of periodic flooding inundation to many properties in the existing flood plain in the heart of downtown Austin, and will help stimulate renewed economic growth of an area of some 92,903 m^2 (1,000,000 ft^2) of prime development.

REFERENCES

Barton, N., Løset, F., Lien, R. and Lunde, J. 1980. Application of the Q-system in design decisions. In *Subsurface space,* (ed. M. Bergman) 2, 553–561. New York: Pergamon.

Bieniawski, Z.T. 1989. *Engineering rock mass classifications.* New York: Wiley.

Deere, D.U. & Deere, D.W. 1988. The rock quality designation (RQD) index in practice. In *Rock classification systems for engineering purposes,* (ed. L. Kirkaldie), ASTM Special Publication 984, 91–101. Philadelphia: Am. Soc. Test. Mat.

ISRM Commission on Swelling Rock 1994. Comments and Recommendations on Design and Analysis Procedures for Structures in Argillaceous Swelling Rock. *Int. J. Rock Mech. Min. Sci. & Geomech. Ahstr. 31*(5), pp. 535–546. London: Pergamon.

ISRM Commission on Swelling Rock 1989. Suggested Methods for Laboratory Testing of Argillaceous Swelling Rock. *Int. J. Rock Mech. Min. Sci. & Geomech. Abstr, 26*(5), pp. 415–426. London: Pergamon.

NATM STRATEGIES IN THE U.S.—LESSONS LEARNED FROM THE INITIAL SUPPORT DESIGN FOR THE CALDECOTT 4TH BORE

Bhaskar B. Thapa ▪ Jacobs Associates

Thomas Marcher ▪ ILF Consultants

Michael T. McRae ▪ Jacobs Associates

Max John ▪ Tunnel Consultant

Zuzana Skovajsova ▪ ILF Consultants

Mahmood Momenzadeh ▪ California Department of Transportation

ABSTRACT

The design of the Caldecott 4th Bore, located along State Route 24 in Oakland, California, is based on the principles of the New Austrian Tunneling Method (NATM). Typical NATM initial support design practices used in Europe were adapted for this project to account for U.S. conditions and requirements, such as degree of experience with NATM construction, the prevailing contractual environment, and preferences for contractual simplicity. Key design features are: (1) support selection criteria based on ground behaviors and ground conditions; (2) a prescriptive design with allowance for support adjustments based on observations during construction; and (3) organization of support requirements into only four major support categories, while permitting some adjustment for variations in ground behaviors and conditions using a few subtypes and additional support measures.

INTRODUCTION

Project Background

The existing Caldecott tunnels consist of three bores along State Route 24 (SR 24) through the Berkeley Hills in Oakland, California. The California Department of Transportation (Caltrans) and the Contra Costa Transportation Authority (CCTA) propose to address congestion on SR 24 near the existing three Caldecott tunnels by constructing a fourth bore that will provide two additional traffic lanes. The length of the proposed fourth bore is 1,036 m (3,399 ft). The project will include short sections of cut-and-cover tunnel at each portal, seven cross-passageway tunnels between the fourth bore and the existing third bore, and a new Operations and Control Building.

The fourth bore includes two 3.6-m (12-ft) traffic lanes and two shoulder areas that are 3-m and 0.6-m (10-ft and 2-ft) wide. The horseshoe-shaped mined tunnel is 15-m (50-ft) wide and 9.7-m (32-ft) high. A typical section of the tunnel is shown in Figure 1. The tunnel includes a jet fan ventilation system, a wet standpipe fire protection system, and various operation and control systems, including closed-circuit television (CCTV) monitoring, heat and pollutant sensors, and traffic monitoring systems.

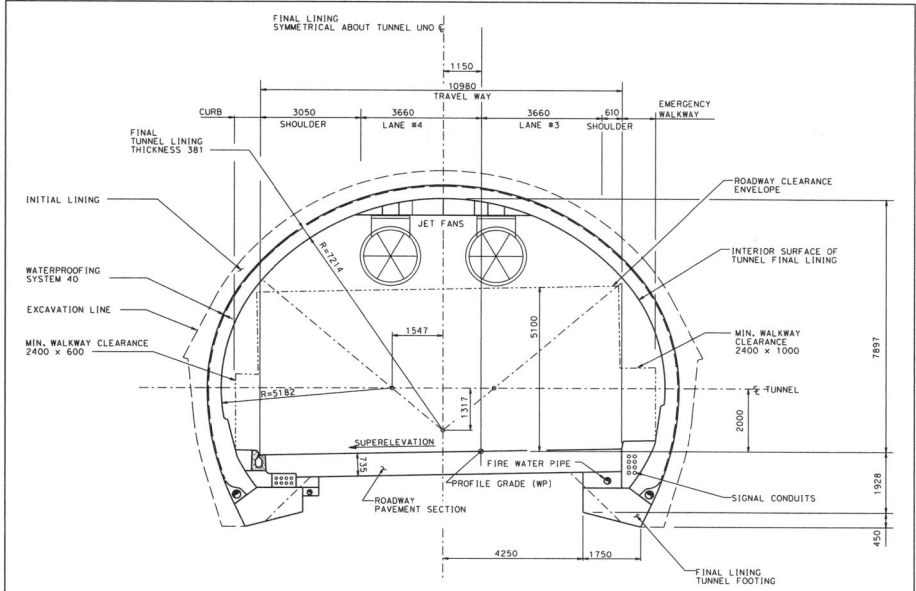

Figure 1. Typical section

Details on ground conditions and initial support design have been described in previous papers (Thapa et al. 2007, 2008) and are not repeated here. This paper reviews general aspects of NATM design used on recent projects, identifies some key design issues regarding NATM practice in the U.S., and describes how these issues have been treated in the design for the 4th Bore.

RECENT NATM PROJECTS

This section is a review of some recent NATM projects in the U.S. and Europe that define aspects of NATM tunneling considered in the design for the Caldecott 4th Bore.

United States

Devil's Slide Tunnel. The Devil's Slide Tunnel consists of twin one-lane roadway tunnels on California Route 1, just south of the City of Pacifica near San Francisco. The length of each tunnel is approximately 1,250 m (4,101 ft). The tunnels are 9-m (30-ft) wide and 6.8-m (22.3-ft) high.

Ground conditions along the alignment consist of a granitic block and two sedimentary rock (claystone, siltstone, sandstone, and conglomerate) blocks separated by faults. Other faults produce disturbed zones within these major blocks. Amini et al. (2005) describe the geotechnical design of the Devil's Slide Tunnel based on identification of rock mass types, behavior types, and support categories. Four behavior types and five support categories were developed to address the range of anticipated ground conditions. Criteria for application of the support categories were provided as part of the contract documents in terms of ground conditions and behaviors. Criteria for additional support measures were provided in relation to warning and alarm level tunnel convergence, support performance, and ground condition criteria. Support-selection decisions are made on a daily basis in meetings between the contractor and engineer, with the contractor proposing the appropriate support category selection and the engineer approving the proposed support. Support elements used in the NATM design include fiber-reinforced shotcrete, lattice girders, rock dowels, spiling, a pipe canopy, and an invert arch. Monitoring is being used during construction to verify the design.

Beacon Hill. The Beacon Hill Station and tunnels include one mile of twin tunnels, an underground station, portals, and ancillary works. Ground conditions consist of variable glacial soil deposits ranging from soft, water-bearing sands to stiff, slickensided clays (Phelps et al. 2005). A large number of geologic units have been grouped into five major ground types to reduce the complexity of the geologic profile.

The NATM (or Sequential Excavation Method, SEM) design of the station tunnels was based on a three-stage approach (Field. et al. 2005) consisting of (1) a prescriptive design for excavation and support sequences, dimensions, and shotcrete support for each excavation element of the station complex; (2) prescriptive ground conditioning and presupport at critical locations; and (3) a "tool-box" of additional support measures to be used as required. Daily meetings between the contractor and engineer were used to confirm support requirements (Akai et al. 2007).

Europe

Strenger Tunnel, Austria. The Strenger Tunnel consists of a twin two-lane road tunnel in Austria that was driven through highly squeezing ground, as described by John et al. (2005). Ground conditions consist of quartz phyllonites and quarzitic schists of low permeability, with a strike at an acute angle to the tunnel axis and steeply dipping. According to the Austrian Standard (ÖNORM B 2203), support categories are differentiated for the top heading, bench, and invert. Eight categories are defined for the top heading, six categories for the bench, and four for the invert. At each excavation stage, the ground behavior is evaluated and the applicable support category adopted. This allows for maximum flexibility with regard to changing ground conditions between the crown and invert. The project also utilized a sophisticated system for the adjustment of payment for support elements that accounts for the impact of support installation time on advance rates. Employing this remuneration system enabled payment for support elements on a unit-price basis.

NATM DESIGN ISSUES

This section discusses some key considerations regarding NATM design that were identified as requiring adaptation to suit U.S. tunneling practices.

Experience with NATM Construction

United States. Considering the size of the U.S tunneling industry, there have been relatively few tunnels constructed in the U.S using NATM. In addition to the Devil's Slide and Beacon Hill projects, other recent NATM projects in the U.S. (reported in the 2005 and 2007 RETC Proceedings) include the Stanford Linear Accelerator (LINAC) Tunnel (Halim et al. 2007), Dulles Corridor Metrorail Project (Rudolf et al. 2007), a reach of the San Vicente Tunnel (Krulc et al. 2007), the Michigan Street Pedestrian Tunnel (Madsen et al. 2007), the Dulles International Airport People Mover (Frandina 2005), and the San Diego Mission Valley East Extension (Field et al. 2005).

The limited experience with NATM in the U.S is important because NATM is an observational method that involves verification/finalization of excavation and support designs during construction, and effective application of the method requires experienced construction personnel within both the owner's and contractor's organizations. These personnel are required to observe ground conditions and behaviors and adjust support accordingly. Although the widely used (in the U.S.) ground classification scheme by Terzaghi (Proctor and White 1968) does consider some behavior modes, it is mainly focused on ground conditions. On the other hand, NATM ground classification considers a wide range of ground behaviors. The limited U.S. experience in working with ground behavior observations/evaluations, which are a key part of NATM support selection/design verification, was considered a limitation to be accounted for in the design layout and preparation of the Caldecott 4th Bore contract documents.

Europe. In contrast to the U.S., numerous tunnel projects have been constructed throughout Europe using NATM and, therefore, there is a wealth of experience with both owners and contractors. This wealth of experience translates to well-trained miners and site supervisors with experience identifying key ground conditions and behaviors that allows immediate decisions at the face on support requirements. Workers are well acquainted with the procedures for handling and installing all support elements, which enables prompt switching of support categories with minimal impacts on productivity. Finally, complex regulations for measurements of bid items are standard as this procedure has been built upon for years. For example, in Austria each support element is remunerated by a separate pay item and all elements are clearly defined for each support category. Adjustments of support elements during construction are handled by payment for the actual number and type of elements installed. Impacts of support adjustment on advance rates are addressed by adjustment of contract support category advance rates established during bidding using evaluations of time consumed for installation of adjusted support elements. For instance, if additional dowels are to be installed, the advance rate will be reduced. This would be addressed by an adjustment for construction time and remuneration of costs.

Comparison of Contractual Practice

Contracting for NATM tunnels in Europe allows optimization of construction to achieve cost and schedule efficiencies by placing the designer in the construction manager role to provide continued validation, back-analysis, and design adjustments (Field et al. 2005). In contrast, the U.S. practice tends to be oriented towards developing a set of clearly scoped contract documents for competitive bidding and does not allow the designer to fill the construction manager role. The risk to owners using the European NATM approach is accepted because differing site conditions are the responsibility of the owner in any case, and the owners are experienced in managing the risk by participating in the decision-making process at the face. By comparison, the risks to a U.S. owner entailed in using the European approach to NATM contradict the general risk management approach to construction projects that is prevalent in the U.S. Thus, it is necessary to adapt the European approach to NATM construction by simplifying and translating European NATM practice such that the bids are a meaningful basis for selecting the contractor and that the bid price can be fairly adjusted if conditions are different than anticipated.

In Europe it is practice for the NATM initial support design to be optimized for variations in ground behaviors so as to achieve the most efficient tunnel production system possible. This is accomplished by using a large number of support categories with each support category suited to a narrow range of ground behavior and providing bid items for the use of additional support measures as required. This approach is inconsistent with the general contractual practice in the U.S. of using only a few support categories that group a range of ground behaviors and provide a clear basis for bids. The Devil's Slide Tunnel is an example of the U.S practice on a recent U.S. NATM tunnel project. A similar approach was considered for the Caldecott Project so as to attract the largest number of bidders and promote competitive bidding.

CALDECOTT 4TH BORE NATM DESIGN FEATURES

This section describes strategies used in the design of the 4th Bore NATM initial support to address NATM adaptation issues identified above.

Site Investigation

An extensive site investigation program involving over 1,245 m (4,100 ft) of borings, geological field mapping, in situ testing, and laboratory testing was undertaken to characterize materials along the alignment, The borings cover over 90% of the alignment. Additionally, construction records from the existing three bores were reviewed in assessing ground conditions along the alignment. These data provided a detailed understanding of the range of ground conditions expected along the alignment. The site investigation information was evaluated to identify Rock Mass Types (RMTs), which are rock units with similar mechanical properties, and ground classes having similar excavation and support requirements (Thapa et al. 2007, 2008). Results of the site investigation indicate feasible excavation methods include use of a roadheader, excavator with cutter-head attachment, and drill-and-blast methods.

Geotechnical Design

In order to address the NATM support selection criteria based on ground behaviors and also accommodate the typical U.S. practice of using ground conditions, the design defined ground classes, which are groups of RMTs having similar predominant behaviors in an unsupported opening (Thapa et al. 2007). Seven behaviors were defined in terms of failure modes and manifestations, as summarized in Table 1 and depicted on Figure 2. Support categories consisting of a set of excavation and support requirements were developed for each ground class on a one-to-one basis.

Support Requirements

A prescriptive approach to the specification of the excavation and initial support requirements was adopted to implement NATM construction of the 4th Bore. Excavation and support requirements for each support category include overall excavation and construction sequence, restrictions on advance lengths, drift dimensions, arrangement and dimensions of support elements, as well as alternative schemes where applicable (Figures 3a, 3b and 3c). The construction sequence consists of a top heading, bench and invert excavation sequence. The design of the support system includes the following measures: fiber-reinforced shotcrete; drill and grout, as well as self-drilling rock dowels; lattice girders; invert arch; drill and grout, as well as self-drilling spiles, pipe canopy, face dowels, and a sloping core for face support.

In keeping with general tunneling practice in the U.S., it was decided to minimize the number of support categories on the 4th Bore to simplify the tunnel production operational requirements. The initial support design was organized into: (1) standard support consisting of four major support categories and three subtypes, each having a separate pay item; and (2) additional support elements on a unit-price basis (including time-dependent costs such as impacts on advance rates) to be used for local ground conditions/behaviors, as required. Table 2 summarizes the key support elements for the four major support categories, and Figure 3 shows the arrangement of support elements and support installation requirements for one of the support categories.

Additional support measures are supplementary to the standard support measures. These additional measures are required to address observed or measured local ground conditions or behaviors. They will be installed when measured convergence exceeds warning levels or when specific ground conditions or support system behaviors are observed, as defined in the contract documents. Estimated quantities

Table 1. Ground behaviors*

Behavior	Description of Failure Modes and Manifestations in an Unsupported Tunnel
Block failures	Discontinuity-controlled, gravity-induced failure of rock blocks that manifests as falling and sliding of blocks.
Raveling	Progressive, discontinuity-controlled failure of small rock blocks within the general rock mass at or near the excavation surface. Raveling is manifested as successive fallout of small rock blocks and can ultimately result in a significant overbreak.
Shallow shear failure	Shallow shear failures result from overstressing of the ground within 0.25D to 0.5D of the tunnel perimeter (D=tunnel diameter) and may be enhanced by the potential for discontinuity and gravity-controlled failure modes. Shallow shear failure is manifested by moderate inward movement of the tunnel perimeter, including invert heave, and possibly by movement of rock into the tunnel opening along discontinuities.
Deep shear failures	Deep-seated shear failures result from overstressing of the ground beyond 0.25D to 0.5D from the tunnel perimeter. Deep-seated shear failure manifests as large radial convergence of the tunnel perimeter, including invert heave.
Slaking/softening	Slaking is the deterioration and breakdown of intact rock upon exposure by excavation and manifests as slabbing of material from the crown and sidewalls. The severity of this behavior is assessed on the basis of slake durability tests performed according to ASTM Test Method 4644. Softening, which is dependent on wetting and exposure by excavation, is the reduction of intact rock strength at the invert or elsewhere and manifests as the development of a muddy or unstable invert or sloughing along segments of the tunnel perimeter elsewhere.
Swelling	Swelling occurs due to absorption of water by clay minerals in rock upon excavation-induced unloading. Swelling manifests as movement of the ground into the tunnel opening or additional tunnel support loading.
Crown instability due to low cover	Excessive crown geological overbreak and chimney-type failure will occur due to lack of confinement under low-cover reaches at portals. It manifests as block fallout and raveling above the crown.

*Modified from Austrian Society for Geomechanics, 2004.

of additional support measures included in the contract are based on an assessment of variations in ground conditions expected using the results of the site investigation program. Additional support elements include spiling, rock dowels, shotcrete, lattice girders, and an invert arch.

Numerical models were used to analyze the behavior of the support system during the excavation process, including evaluation of forces, moments and rotations in the shotcrete lining, and forces in the rock reinforcement support elements (Thapa et al. 2007, 2008). The design for the support categories was based on numerical calculations of typical sections along the alignment. The possible variation of support category application location, extent, and sequence from the design prognosis (Figure 4) is clearly stated on the contract drawings and Geotechnical Baseline Report (GBR) so as to clarify that the construction impact of such variations must be accounted for in the bids. The total quantities of support categories and the number of changes between support categories to be used for bidding purposes were stated in the contract documents. The "Variation In Quantities" Clause of the specifications allows for renegotiation of unit rates of support category payment should the bid quantities vary by more than 25%, or by a lineal meter threshold for items with a low estimated quantity.

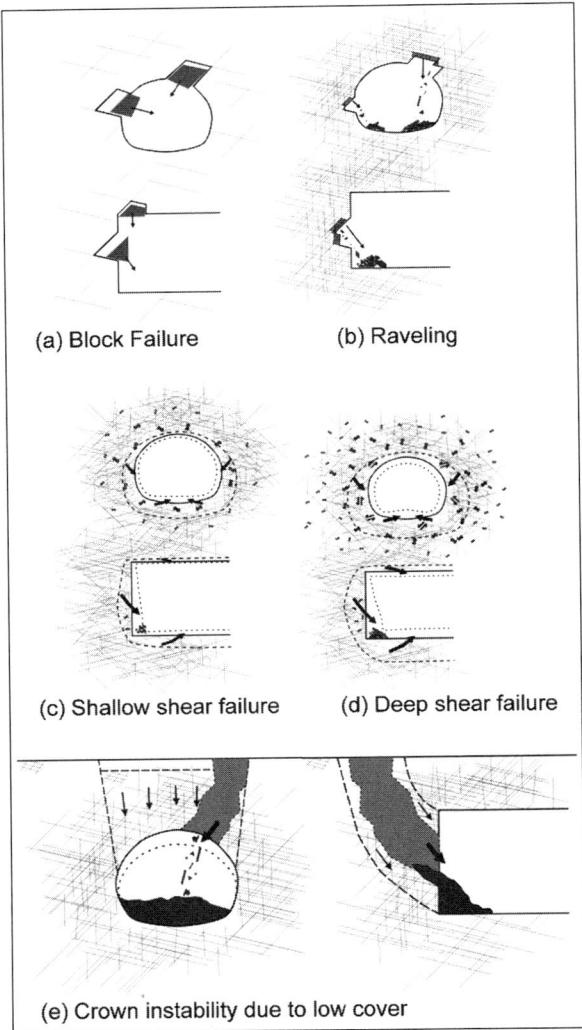

Figure 2. Ground behaviors

Support Selection Criteria

The definitions for each ground class, which are a key element of the support category selection criteria, were developed using consistent terminology based on the U.S. Bureau of Reclamation (USBR,1998) and Geological Strength Index (Marinos et al. 2005). Rock mass descriptions are summarized in the plans (see Figure 5) using the Geological Strength Index system and expanded upon in the GBR using the USBR terminology. Ground class behaviors also were defined consistently using the definitions shown in Table 1 and Figure 2. Expanded discussions of behaviors in the GBR clarify which behaviors will be directly observable and which ones will not, the locations and special conditions associated with specific behaviors, and the relations between ground behaviors and support elements required to control the behaviors. Criteria for

NATM STRATEGIES IN THE U.S.

Table 2. Summary of support categories

Support Category	Max. Advance Length, m (ft)	Presupport	Face Support	Shotcrete Thickness, cm (in.)	Avg. Dowel Spacing, m (ft)	Invert Arch
I	1.8 (6)	None	SC IA: face dowels/ sealing fiber reinforced shotcrete (FRS) as required SC IB: systematic face dowels/ sealing FRS	20.3 (8)	1.8 m (6)	None
II	1.4 (4.5)	SC IIA: none SC IIB: spiles	Face dowels/ sealing FRS or sloping core/ sealing FRS	25.4 (10)	1.5 m (5)	None
III	1 (3.3)	Spiles	Sloping core/ sealing FRS	30.5 (12)	1.2 m (4)	SC IIIA: none SC IIIB: top heading and bench
IV	1 (3.3)	Pipe canopy	Sloping core/ sealing FRS	30.5 (12)	None	Top heading and bench

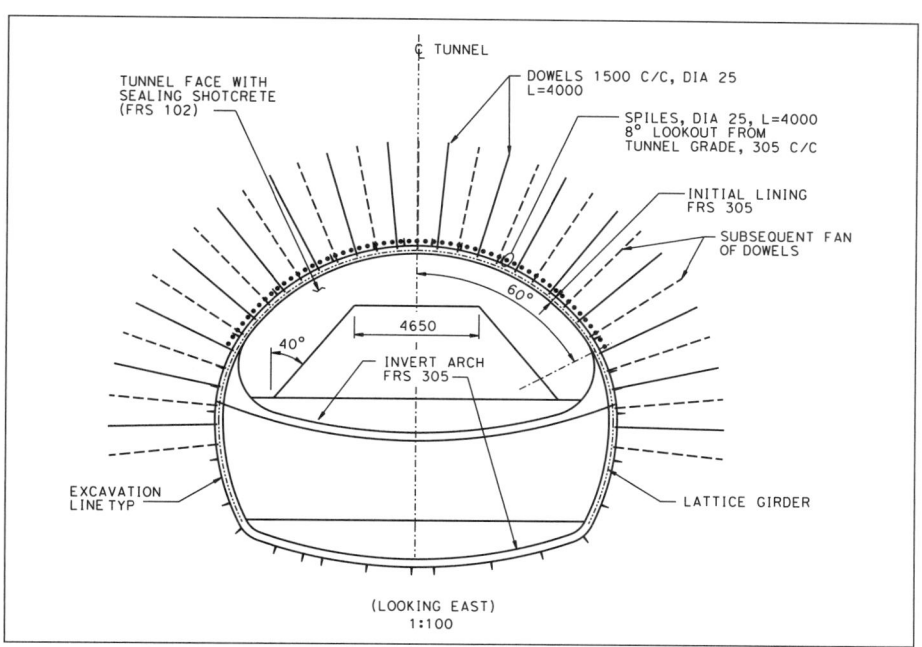

Figure 3a. Example of support category requirements, typical excavation cross section

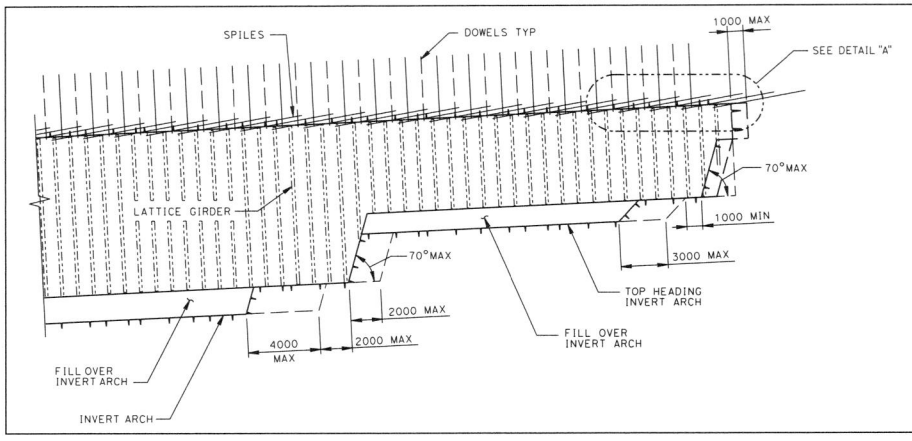

Figure 3b. Example of support category requirements, section

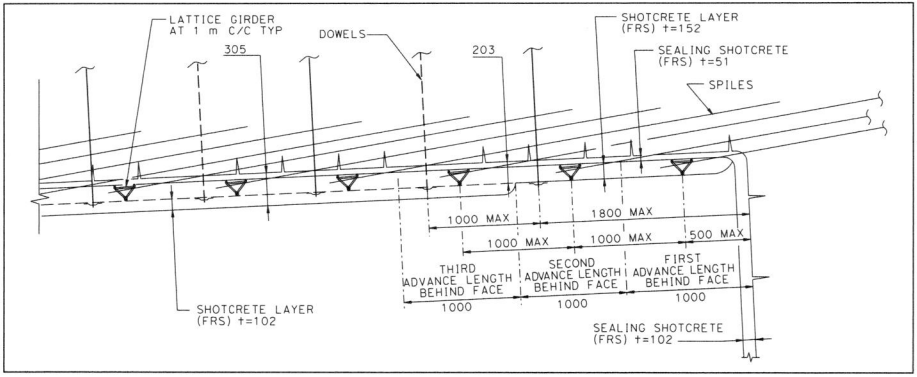

Figure 3c. Example of support category requirements, detail "A" of section

application of additional support measures were defined in the plans in terms of local ground conditions and behaviors, including warning and alarm levels of tunnel convergence (Figure 6).

Roles and Responsibilities

The roles and responsibilities for monitoring of the tunnel performance and support selection are detailed in the technical specifications. The contractor is required to collect monitoring bolt data within 6 hours after installation in the last excavation round, and provide data and interpretations to the engineer within 12 and 24 hours, respectively, after taking readings. The engineer's role is to evaluate the information provided by the contractor in making independent assessments of support requirements. The contractor and engineer are to make independent assessments of support requirements and performance based on mapping of exposed excavation surfaces, probe drilling results, observations, and evaluations of ground behavior and the monitoring data. Daily meetings (or more as required) between the contractor and engineer are to

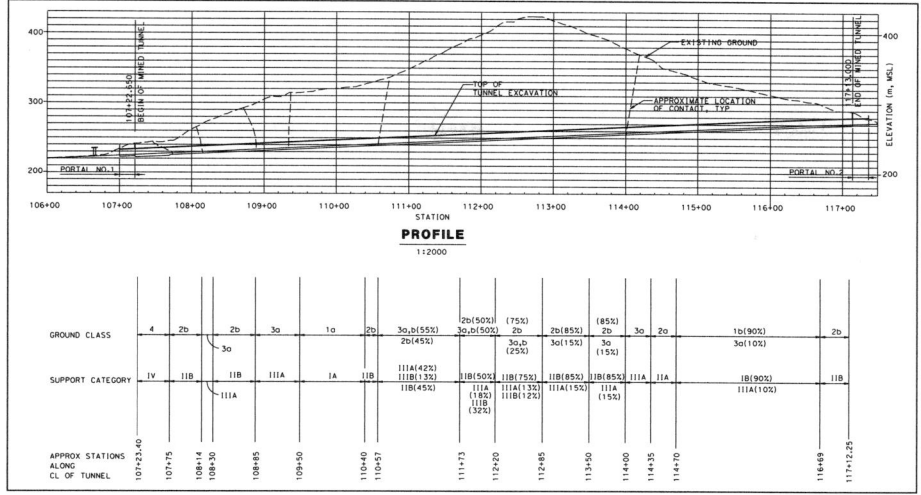

Figure 4. Design prognosis of ground classes and support categories along the alignment

be used to make decisions on excavation and support requirements. Decisions are as proposed by the contractor and as approved by the engineer.

CONCLUSIONS

Several aspects of NATM practice in Europe were modified to suit U.S. tunneling practices in developing the design of the excavation and initial support for the Caldecott 4th Bore. The role of ground behavior in NATM support selection was clarified and integrated with the traditional ground condition criteria used for support selection in the U.S. A prescriptive design, including detailed criteria and procedures for construction decisions, was used to develop a biddable contract package while maintaining flexibility for the typical NATM practice of design verification and adjustment based on observations made during construction. Finally, the design was laid out to strike a balance between contractual simplicity and design optimization by minimizing the number of major support categories, subtype support categories, and use of additional support measures.

ACKNOWLEDGMENTS

The authors would like to acknowledge AMEC (Geomatrix Consultants) for its work on the site geology and Marlène Villeneuve of Jacobs Associates for her work on the ground behavior figures. The contents of this paper were reviewed by the State of California's Business, Transportation and Housing Agency; the California Department of Transportation; and the Contra Costa Transportation Authority. The contents of this paper reflect the views of the authors, who are responsible for the facts and accuracy of the data presented herein, and do not necessarily reflect the official views or policies of the State of California or the Contra Costa Transportation Authority. This paper does not constitute a standard, specification, or regulation.

DESIGN AND PLANNING

GROUND CLASS	SUPPORT CATEGORY	PREDOMINANT ROCK MASS TYPES AND GSI DESCRIPTIONS	OBSERVED OR ANTICIPATED ROCK MASS BEHAVIOR	
			PREDOMINANT	SECONDARY
2b	IIB	Tsp: BLOCKY/DISTURBED/SEAMY ROCK MASS STRUCTURE AND POOR TO FAIR DISCONTINUITY SURFACE CONDITIONS Tss-1: BLOCKY/DISTURBED/SEAMY TO VERY BLOCKY ROCK MASS STRUCTURE AND POOR TO FAIR DISCONTINUITY SURFACE CONDITIONS Tcs-2: BLOCKY/DISTURBED/SEAMY TO VERY BLOCKY ROCK MASS STRUCTURE AND FAIR DISCONTINUITY SURFACE CONDITIONS Tc-1: VERY BLOCKY ROCK MASS STRUCTURE AND FAIR DISCONTINUITY SURFACE CONDITIONS Tc-2: BLOCKY/DISTURBED/SEAMY ROCK MASS STRUCTURE AND FAIR DISCONTINUITY SURFACE CONDITIONS Tc-3: VERY BLOCKY ROCK MASS STRUCTURE AND FAIR DISCONTINUITY SURFACE CONDITIONS Tc-4a: DISINTEGRATED TO BLOCKY/DISTURBED/SEAMY ROCK MASS STRUCTURE AND FAIR DISCONTINUITY SURFACE CONDITIONS Tc-4b: DISINTEGRATED TO BLOCKY/DISTURBED/SEAMY ROCK MASS STRUCTURE AND FAIR DISCONTINUITY SURFACE CONDITIONS Tor-3: BLOCKY/DISTURBED/SEAMY ROCK MASS STRUCTURE AND POOR TO FAIR DISCONTINUITY SURFACE CONDITIONS	PREDOMINANT GROUND BEHAVIOR CONSISTS OF: • DISCONTINUITY-CONTROLLED BLOCK FAILURES	SECONDARY BEHAVIORS INCLUDE: • RAVELING • SHALLOW SHEAR FAILURE • SLAKING • SOFTENING • SWELLING

STANDARD SUPPORT APPLICATION CRITERIA

Figure 5. Example of support category selection criteria

ADDITIONAL SUPPORT - APPLICATION CRITERIA AND MEASURES				CONTINGENCY MEASURES REQUIRED	
DISPLACEMENT WARNING LEVEL AND OBSERVATIONS OF SUPPORT PERFORMANCE	ADDITIONAL SUPPORT FOR WARNING LEVEL AND OBSERVATIONS OF SUPPORT PERFORMANCE	OBSERVED LOCAL ROCK MASS CONDITIONS AND ANTICIPATED BEHAVIORS REQUIRING ADDITIONAL SUPPORT	ADDITIONAL SUPPORT FOR OBSERVED CONDITIONS	ALARM LEVEL	OBSERVED CONDITIONS
RADIAL DISPLACEMENT: 50 CRACKS IN SHOTCRETE	ADDITIONAL FAST SETTING CEMENT GROUTED ROCK DOWELS. APPLY ADDITIONAL FRS (t=51) TO INITIAL LINING.	DISINTEGRATED ROCK MASS WITH POOR INTERLOCKING. BEHAVIORS: RAVELING (FAST)	4000 LG GROUTED SPILES AND LATTICE GIRDERS AS REQUIRED.	RADIAL DISPLACEMENT: 70	CRACKS WITH OFFSET GREATER THAN 3 mm, OR WITH AN OPENING GREATER THAN 1.5 mm APPEAR IN SHOTCRETE
		DISCONTINUITIES FORM UNSTABLE BLOCK AT TUNNEL FACE (APPLIES TO SUPPORT CATEGORY IIB ALTERNATIVE 1) BEHAVIORS: DISCONTINUITY CONTROLLED BLOCK FAILURES	APPLY LOCAL FIBERGLASS FACE DOWELS (25 DIA, 6000 LG) AND FRS (t=51) TO FACE AS REQUIRED.		
		PORTIONS OF SLOPING CORE EXHIBITS RAVELING/SHALLOW SHEAR FAILURE BEHAVIORS. (APPLIES TO SUPPORT CATEGORY IIB ALTERNATIVE 1)	APPLY FRS (t=51) AND FIBERGLASS FACE DOWELS (25 DIA, 6000 LG) TO SLOPING CORE AS REQUIRED.		

Figure 6. Example of additional support application criteria

REFERENCES

Akai, S., Murray, M., Redmond, S., Sage, R., Shetty, R., Skalla, G., and Varley, Z. 2007. Construction of the C710 Beacon Hill Station using SEM in Seattle—"Every chapter in the book." In *2007 RETC Proc.*, Toronto, Ontario, Canada, June 10–13. Littleton, CO: SME.

Amini, M., John, M., Sander, H., and Wang, Y.N. 2005. Geotechnical design of Devil's Slide Tunnel. In *2005 RETC Proc.*, Seattle, WA, June 26. Littleton, CO: SME.

Austrian Society for Geomechanics. 2004. "Guideline for the geomechanical design of underground structures with conventional I excavation." Draft English translation.

Field, D. P., Hawley, J., and Phelps, D. 2005. The North American Tunneling Method—Lessons learned. In *2005 RETC Proc.*, Seattle, WA, June 26. Littleton, CO: SME.

Frandina, F.P., Hirsch, D.R., Weeks, C.R., and Field, D.P. 2005. Design of shallow tunnels for Washington Dulles International Airport people mover. In *2005 RETC Proc.*, Seattle, WA, June 26. Littleton, CO: SME.

Halim, I., Vincent, F., and Taylor, J. 2007. NATM Design for Stanford LINAC Coherent Light Source Tunnels. In *2007 RETC Proc.*, Toronto, Ontario, Canada, June 10–13. Littleton, CO: SME.

John, M., Spöndlin, D., Ayadin, N., Huber, G., Westermayer, H., and Mattle, B. 2005. Means and methods for tunneling through highly squeezing ground: A case history of the Strenger Tunnel, Austria. In *2005 RETC Proc.*, Seattle, WA, June 26. Littleton, CO: SME.

Krulc, M. A., Murray, J.J., McRae, M.T., and Schuler, K.L. 2007. Construction of a mixed face reach through granitic rocks and conglomerate. In *2007 RETC Proc.*, Toronto, Ontario, Canada, June 10–13. Littleton, CO: SME.

Madsen, P.H., Younis, M.A., Gall, V. and Headland, P.J. 2007. NATM through clean sands—The Michigan Street experience. In *2007 RETC Proc.*, Toronto, Ontario, Canada, June 10–13. Littleton, CO: SME.

Marinos, V., Marinos, P., and Hoek, E. 2005. The Geological Strength Index: Applications and limitations. *Bull. Eng. Geol. Environ.* 64, 55–65.

ÖNORM B 2203-1: *Underground Works. Part 1: Cyclic Driving.* Works contract, Issue 2001-12-01.

Phelps D.J., Gildner, J., and Tattersall, C. 2005. Design and risk management strategy for the Sound Transit Beacon Hill Station and tunnels. In *2005 RETC Proc.*, Seattle, WA, June 26. Littleton, CO: SME.

Proctor, R.V., and White, T.L. 1968. *Rock Tunneling with Steel Supports, with an "Introduction to Tunnel Geology" by Karl Terzaghi.* Youngstown, OH: Commercial Shearing and Stamping Company.

Rudolf, J., and Gall, V. 2007. The Dulles Corridor Metrorail Project—Extension to Dulles International Airport and its tunneling aspects. In *2007 RETC Proc.*, Toronto, Ontario, Canada, June 10–13. Littleton, CO: SME.

Thapa, B.B., McRae, M.T., and Van Greunen, J. 2007. Preliminary design of the Caldecott Fourth Bore. In *2007 RETC Proc.*, Toronto, Ontario, Canada, June 10–13. Littleton, CO: SME.

Thapa, B.B., Van Greunen, J., Sun, Y., McRae, M.T., and Law, H. 2008. Design analyses for a large-span tunnel in weak rock subject to strong seismic shaking. In *2008 NAT Proc.*, San Francisco, CA, June 8–11. Littleton: CO: SME.

USBR (U.S. Bureau of Reclamation). 1998. *Engineering Geology Field Manual*, 2nd Ed., U.S. Washington, D.C.: Government Printing Office.

THE PRICE IS RIGHT—PLANNING LARGE WATER TUNNEL CONTRACTS IN NEW YORK

Christian Maguire ▪ URS

Eric Cole ▪ AECOM

ABSTRACT

Tunnel contracts in New York have in recent years had few bidders with the bids received being much higher than anticipated. Since 9/11, security has become a key issue in planning key infrastructure projects in New York. This is especially true in the case of large water supply tunnels. Striking the right balance between cost and security was the major factor in determining the best configuration for the Kensico-City Tunnel. This paper outlines the efforts made to determine realistic probable bid costs for the tunnel contracts, and the innovative approaches taken to deal with security issues.

INTRODUCTION

The Kensico-City Tunnel (K-CT) represents a major investment by NYCDEP in the dependability of the key stretch of the conveyance system downstream of Kensico Reservoir. About 90% of the potable water supplies for New York City are conveyed from Kensico Reservoir via the Catskill Aqueduct (CAT) and the Delaware Aqueduct (DEL). At present and projected levels of demand, it would be too risky to take either the CAT or the DEL out of service for any significant period of time. Thus although these two aqueducts are aging, it is not possible for them to be inspected and repaired without the additional capacity that the K-CT would provide.

The DEL and CAT are in close proximity at Kensico and Hillview reservoirs (see Figure 1) and will be inter-connected at the Eastview site when the CAT/DEL UV facility comes on line within the next few years. Thus 90% of the water supplies to New York City pass through several common points, which is a cause for concern from the security viewpoint. Severe economic, social, and political consequences would arise out of any significant shortfalls in the supplies of water to New York City. It is for this reason and faced with the vulnerabilities discussed above that NYCDEP gives a high priority to adding redundancy to the conveyance system downstream of Kensico Reservoir through the implementation of the K-CT.

The justification for the high investment in the K-CT, estimated at about $3 billion at 2006 price levels, is thus increased system reliability and reduced vulnerability. It is not surprising therefore that risks of failure from natural causes and acts of deliberate damage were major factors in the determination of the size and configuration of the K-CT.

Following a systematic screening process, two main contending options for the K-CT emerged. These are shown on Figure 2, and may be summarized as follows:

▪ **East Corridor Reference Project**

  The tunnel in this option follows an easterly alignment that runs directly from an intake (Site 4) on the Rye Lake branch of Kensico Reservoir to an existing underground valve chamber, the Van Cortlandt Valve Chamber (VCVC). On this alignment, the tunnel would be located in a broad band of Manhattan

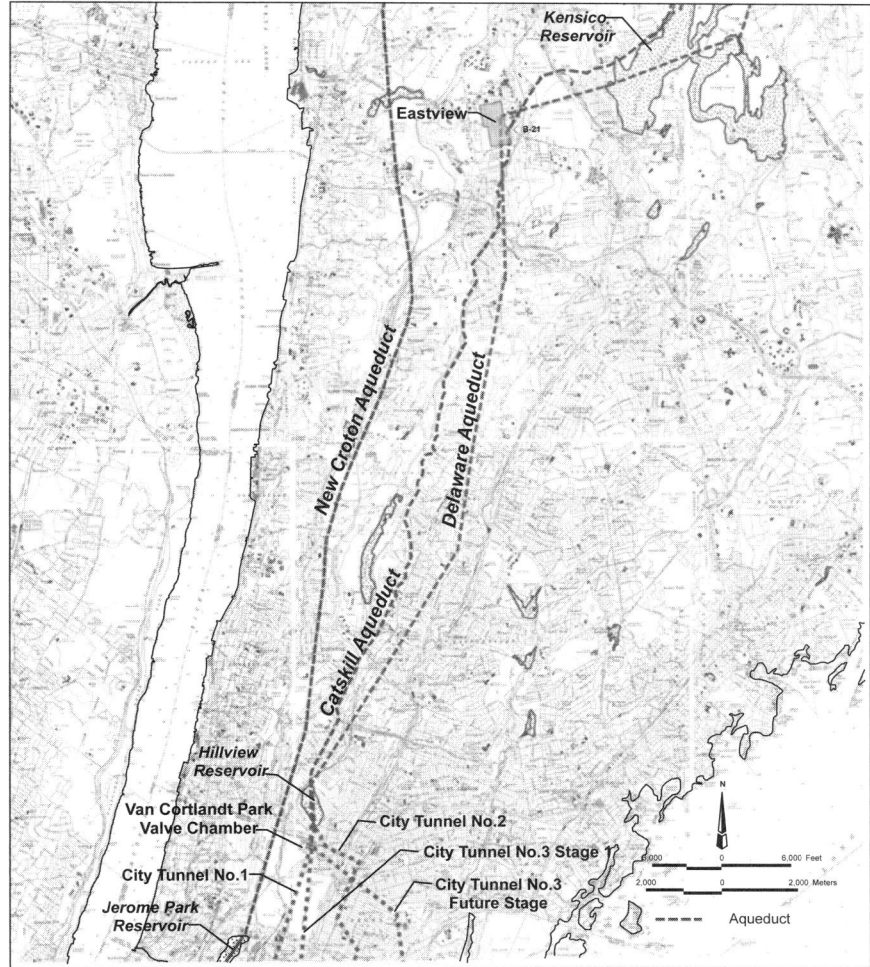

Figure 1. Existing conveyance system in project area

Schist for most of its length. An additional UV Facility would be required at Intake Site 4 for this option.

- **West Corridor Reference Project**

    For this option the tunnel passes from an intake (Site 2) on the main branch of Kensico Reservoir through the Eastview Site to take advantage of the CAT/DEL UV facility under construction at the site. Downstream of Eastview, the tunnel follows a fairly direct route to the VCVC, passing through a complex geological structure made up of alternating bands of metamorphic rocks.

Both alignments corridors are located within the Manhattan Prong Geological Province, which is characterized as a narrow, northeast-trending belt of tightly folded, high grade metamorphic rocks that extends northward from Staten Island to Danbury,

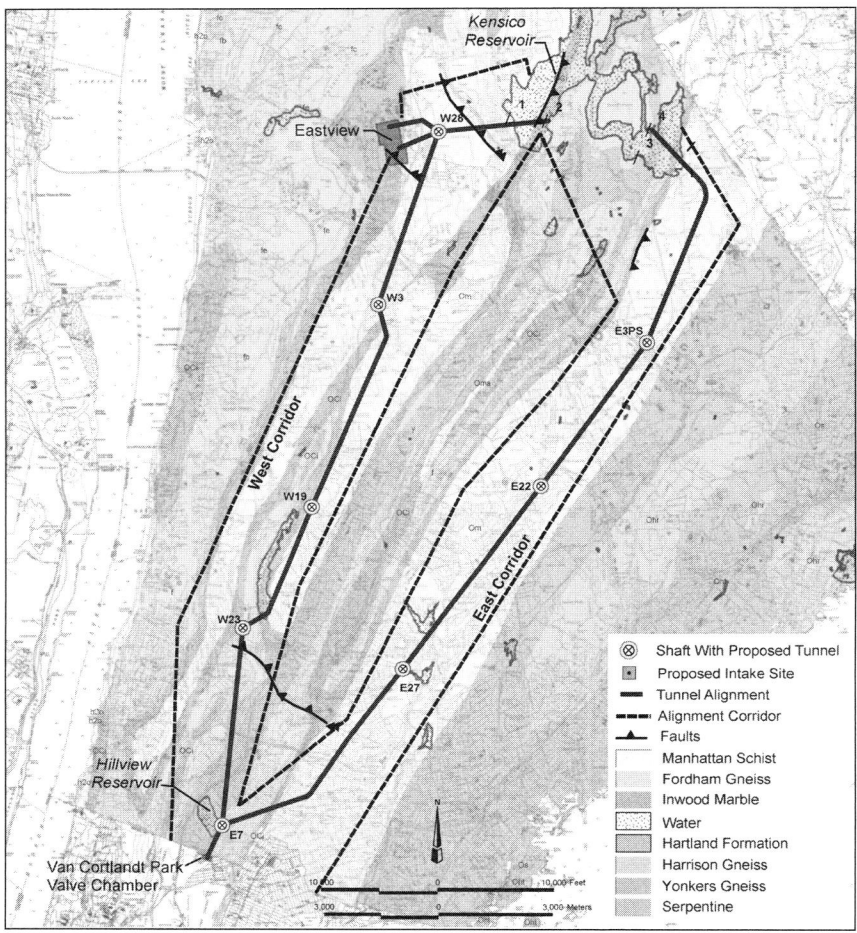

Figure 2. K-CT reference project alignments

Connecticut. The bedrock of the Manhattan Prong is comprised of metamorphic rocks that are Precambrian to Ordovician in age forming a landscape of rolling hills and valleys. The specific rock formations in the K-CT alignments consist of the Manhattan Schist, Inwood Marble, Yonkers Gneiss, Fordham Gneiss and the Lowerre Quartzite. Through the entire length of the tunnel, the alignments are expected to cross numerous minor faults, sheared and crushed zones, and three major faults, all of which will have significant impacts on TBM advance rates and tunnel costs.

The theme of this paper is the interplay between project costs and the protection of the system from the consequences of natural failures and acts of sabotage, and how these factors have interacted to determine the main characteristics of the K-CT and the comparison of the competing options.

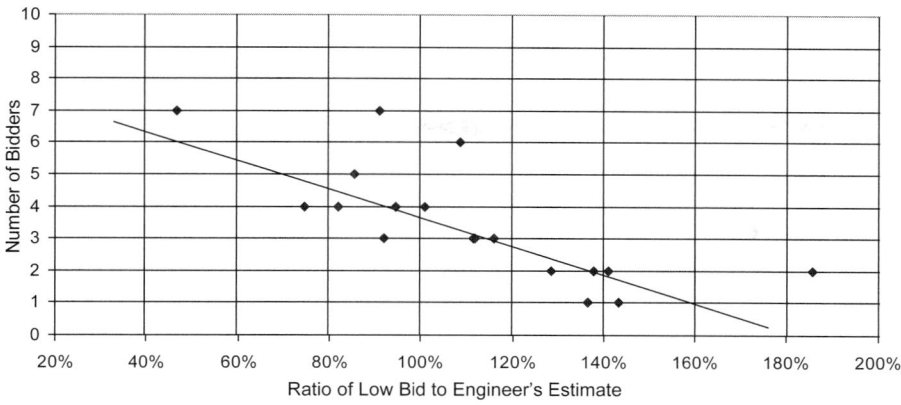

Figure 3. Impact of number of bids

COST DETERMINATIONS

Recent Experience

Engineers have had difficulty in recent years in estimating costs for large infrastructure projects in the New York area. There has been a marked tendency to underestimate costs significantly. This may be attributable to many factors including, depending too much on standard price databases, the impact of steep rises in commodity prices internationally, and to supply-side market constraints that have impacted on bid prices in recent years. A good example of the latter point is NYCDEP experience in the bids received for the various contracts for a major waste water treatment plant in the period 2000 through 2007. This is illustrated on Figure 3 which gives the ratio of low bid to Engineer's Estimate, for contracts in excess of $20 million (ranging up to $500 million), as a function of the number of bids received. The trend is unmistakable. For any contract receiving less than three bids, there is a marked risk of the bids being significantly above the Engineer's estimate, by a margin of 30%–80%.

TBM tunneling costs in New York have not been typical of the country as a whole. This is readily apparent from the data taken from TBM tunnel contracts across the country over the past 20 years as shown on Figure 4. The data on this figure have been adjusted to 2006 price levels, but are otherwise not adjusted for varying tunnel lengths, geological conditions and unusual design aspects. Despite this the data indicates that New York costs have been significantly higher than those for the US as a whole.

Resource Based Estimating for Tunnel Construction

Resource-based estimates, in which construction costs are built up from an analysis of labor, equipment, materials and consumables, were used for deriving costs of tunnels and shafts for the K-CT. The tunneling costs derived in this way were checked against the bid prices for the water tunnels that have been constructed or are about to be constructed by TBMs in the New York City area. The structure of the cost model is shown on Figure 5.

Resource-based cost estimate models were developed for typical TBM tunnel contracts for three different internal diameter tunnels, namely, 4.3 m (14 feet), 7.0 m

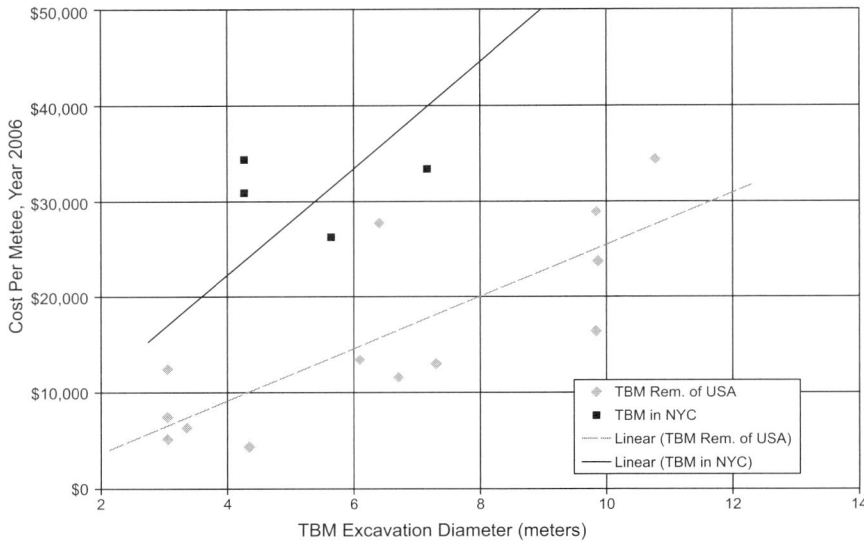

Figure 4. TBM tunnel costs in NYC and US

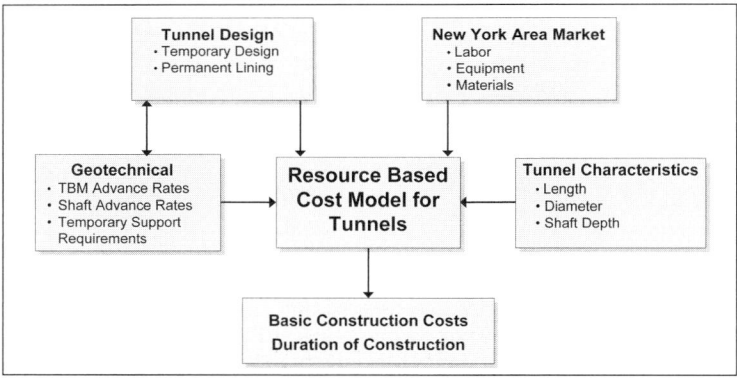

Figure 5. Structure of resource-based cost model

(23 feet) and 8.2 m (27 feet). These models were based on the construction practices and the labor wage levels for the New York City area. The estimates covered the following major activities:

- Mobilization and de-mobilization
- Shaft construction in overburden and rock
- Starter tunnel and chamber excavation
- Materials for tunnel excavation, including temporary support
- TBM assembly and disassembly
- Permanent lining, including materials and forms
- Time-related overhead costs.

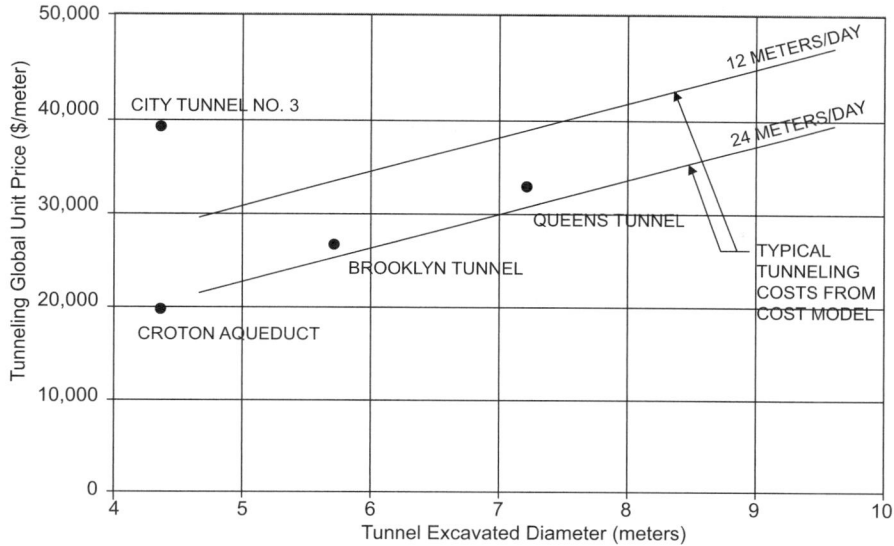

Figure 6. TBM water tunnel costs in NYC area

The cost models were structured so that several major parameters could be varied to reflect the particular characteristics of each construction contract making up the K-CT Reference Projects. These basic parameters are:

- Construction shaft depth
- Tunnel length
- Breakdown of tunnel length into various classes of temporary support
- Shaft advance rates in each rock type
- Tunnel advance rates in each rock type
- Average permanent lining advance rate.

There has been limited but significant TBM experience for water supply tunnels in the New York area. Most of this was for the construction of various reaches of City Tunnel No. 3 (CT3) which conveys water from Hillview Reservoir to the various city boroughs. The CT3 construction differs significantly from transportation tunnel construction in that the CT3 is founded much deeper in the bedrock than the normal transportation tunnels, and is provided with an un-reinforced in-situ concrete lining, in contrast to the pre-cast segment linings commonly used in transportation tunnels. The unit cost data for the various water tunnels that have been constructed by TBM in the New York area are shown on Figure 6. This data has been updated to 2006 price levels, and adjusted to remove the impact of varying construction durations and tunnel lengths. The resulting cost data shows fairly consistent trends, with the exception of the Manhattan leg of CT3, for which a single bid was received. For comparison typical outputs from the resource-based cost modeling are shown on the figure.

TBM Advance Rates and Performance Review

A series of studies were performed to establish the fundamental geological/geotechnical and physical properties of the rock, and deduce TBM performance statistics, and characteristics of operation. This was then used to determine the operational

Table 1. Review of gneissic layering

Rock Type	Unfavorable Layering
Fordham gneiss	33%
Yonkers gneiss	80%

values to be inserted into the tunnel cost models. The study was performed using data from the K-CT site investigations and laboratory test programs, as well as information from previous studies of the area such as geological surface mapping, earlier borehole information, and the excellent geological logging done in the 1940s for the Delaware aqueduct tunnel.

Three main lithological constraints were evident:

- Rock layering
- Garnet concentrations
- In situ rock stresses.

The effect of gneissic layering on TBM performance was documented during the construction of City Tunnel No. 3. This unexpected and unfavorable lithological orientation resulted in slow TBM progress rates in parts of the tunnel. In some cases the unfavorable layering reduced penetration rates by a factor of six. Faster penetration rates are achieved when the layering approaches 90 degrees to the direction of tunneling, however when it approaches 0° (degrees) penetration rates drop with an associated increase in stress for the TBM. The effect of this layering is particularly pronounced in the gneissic rock, leading to an increased compressive strength mode of failure due to cutting. An analysis of the rock anisotropy produced the results shown in Table 1.

Garnet is very hard and in concentrations of up to 50%, as reported in other New York tunnels running through similar rock, results in dense, hard, abrasive rock. As confirmed by the Cherchar abrasivity measurements and Cutter Life Index Testing, these high garnet concentrations will lead to high cutter wear and slow TBM progress in these zones. Clusters of high concentrations of garnet are expected in the Manhattan schist and gneissic formations.

High in-situ rock stress can have an adverse impact on tunnel construction. Stress concentrations in the tunnel perimeter can overstress the rock and result in spalling, popping, and general blocky conditions. The magnitude of the impact is a function of the level of in-situ rock stress, the quality of the rock mass, the orientation of the excavation relative to the in-situ rock stress, and the method of tunnel construction. A literature search of data regarding rocks stress in the K-CT study area indicated that the Yonkers granitic gneiss formation is highly stressed and characterized by "popping" from these stress conditions. Specific measurements during construction of the Van Cortland Valve Chamber recorded in-situ rock stress up to 10 MPa (1,450 psi).

The tunnel contracts making up the K-CT options were broken down into specific geologic zones, with associated sets of data specific to each zone that were used to calculate TBM performance. Some of the functions used were Q–system, RQD, angle of rock layering, density, porosity, I_{50} point load testing, cutter life index, TBM cutter force, in-situ stress, quartz content, tunnel diameter, TBM utilization gradient, and other basic rock properties. The advance rate (AR) of the TBM was then computed for each rock zone using the penetration rate (PR), and the TBM utilization (U) obtained from the above analyses.

Utilization is sometimes harder to determine and in this case maximum utilization was used as the base for calculation. Time (T) and a gradient of deceleration were used to determine U, from the relationship:

$$U \approx T^m$$

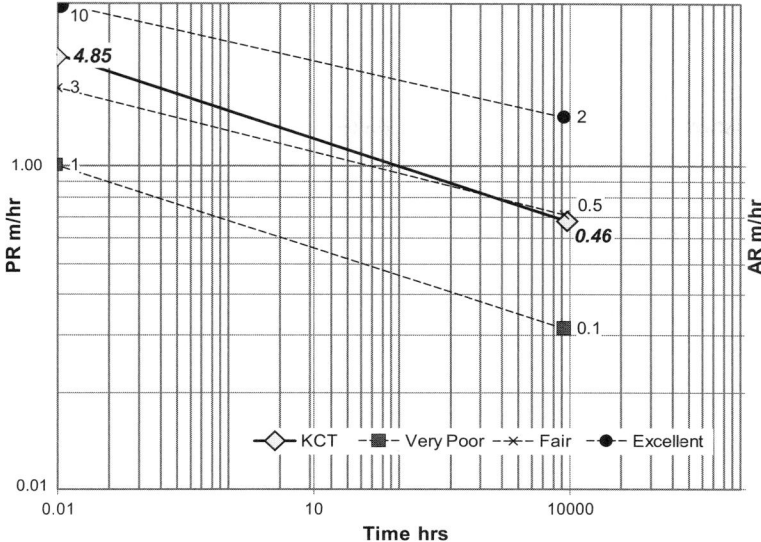

Figure 7. Typical gradient of TBM penetration rate (PR), and advance rate (AR)

This gradient 'm' depends on several factors such as rock mass, TBM design, tunnel length etc. Typical gradients for the K-CT are in the range of –0.16 to –0.25 for normal tunneling, rising to a range of –0.5 to –0.8 for areas with rock stability problems implying low utilization for example in fault zones. The performance of each TBM assessed in this way can be represented graphically as shown on Figure 7, giving a comparison with a wide range of empirical data.

In good rock conditions large diameter tunnels can be driven more efficiently than smaller tunnels, typically however more delays will occur in larger tunnels as the rock quality reduces. This experience is reflected in the 'm' values selected for the gradient of deceleration for different size tunnels. Other diameter sensitive factors which modify the gradient of deceleration include cutter life index, quartz content, and porosity.

The TBM advance rates computed for each tunnel contract were inserted into the tunnel cost model to determine contract costs. Typical cost data for the contracts making up the K-CT in the West Corridor, for tunnel diameters of 7.0 m (23 feet) and 8.2 m (27 feet) are shown on Figure 8.

These costs are at December 2006 prices levels and do not include contingencies and incidental costs such as steel liners, distribution shaft costs etc.

Cost and Schedule Risk Assessments

A comparative risk analysis of the K-CT Reference Projects was conducted to assess and quantify the cost and schedule risks associated with the projects. Comprehensive risk registers were developed covering design, construction, environmental, geotechnical, market and labor, permitting and stakeholder, and political risks. Typical project cost distributions resulting from the analyses are shown on Figure 9. Contingency levels over base-line costs were estimated be in the range of 25% to 50%, depending on the project alignment and type of construction.

Tornado Charts were developed for each project to illustrate the events that are likely to have the largest impact on project cost and schedule. The factors affecting cost most were assessed to be:

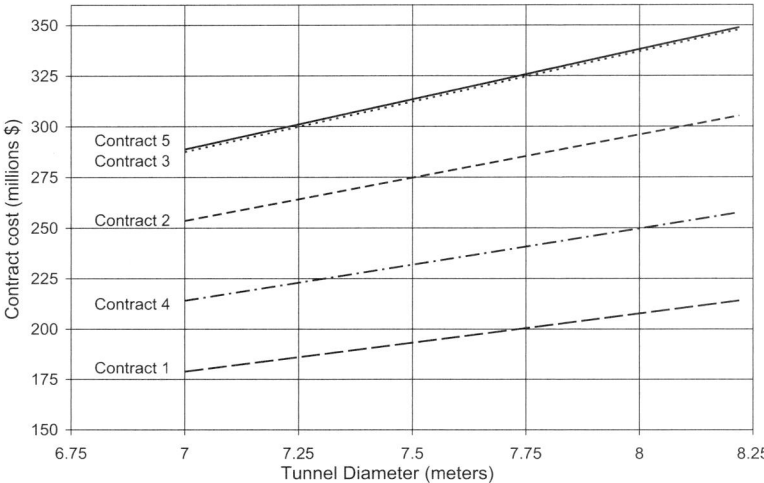

Figure 8. Typical tunnel cost variation with diameter

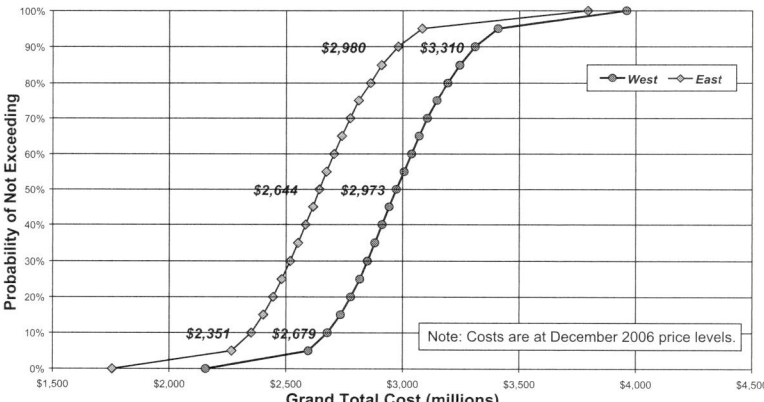

Figure 9. Project cost distribution

- Community and stakeholder objections to shaft sites and mucking routes
- Need for additional temporary support measures
- Lack of competitive bidding.

RELIABILITY AND SECURITY CONSIDERATIONS

Conceptual Approach

One of the main objectives of the K-CT is to provide additional discharge capacity between Kensico and Hillview Reservoirs to assure supplies in the event of a failure (unplanned outage) of any component of the conveyance system, and to enable the other components of the conveyance system in this reach to be taken out of service

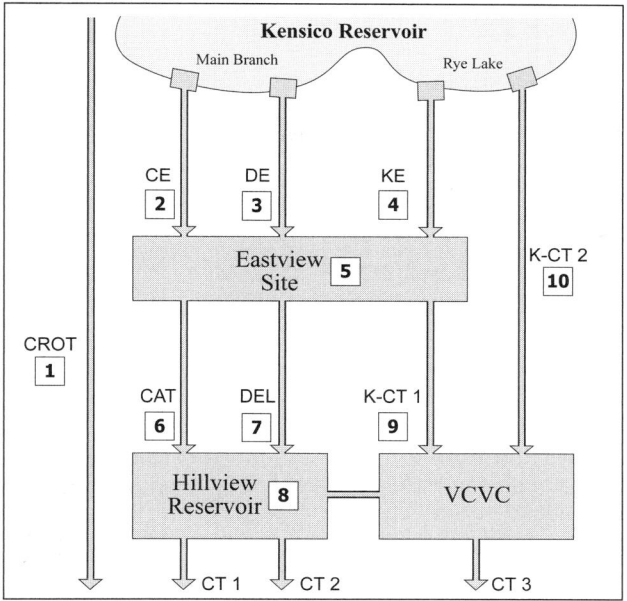

Figure 10. Reliability model structure

(planned outage) for inspection and, if necessary, repair. In general, unplanned outages are the result of unusual circumstances, such as structural collapse, unusual operating conditions or accidents, or even deliberate sabotage. It is not certain how frequently unplanned outages will occur, if at all. Typically for such critical infrastructure as the NYC water supply system, the occurrence of an unplanned outage is an infrequent event.

Any significant unplanned outage will result in water shortages, or deficits. The larger the capacity of the K-CT the smaller will be the deficits resulting from outages. The approach taken in the planning of the K-CT was to compare the incremental cost of providing additional K-CT capacity to the probable reductions in water deficits and thereby determine if the additional investment is justified. To assist in this process a Reliability Model was developed for the conveyance system downstream of Kensico Reservoir, as shown schematically on Figure 10. The model consists of ten components as defined in Table 2.

Components 1, 2, 3, and 5 through 8 represent the main features of the existing water supply system between Kensico and Hillview reservoirs, as shown schematically on Figure 6. Components 4, 9 and 10 represent two categories of K-CT options, those that pass through the Eastview Site (4 and 9) and those that do not (10).

System reliability is a product of the reliability of each of the conveyance system components. Each one of the ten components can either be available for operation or be out of operation either on unplanned outage or planned outage. The probability of an unplanned outage is assigned to each component. There are 1,024 possible combinations of the availability of the ten components. Each one of these combinations may be considered possible states of the system, each with a probability of occurrence derived from the reliability levels of the system components, and each associated with a resulting system capacity.

For each state of the system, the probability of occurrence associated with that state can be determined as well as the resulting capacity of the system, which will

Table 2. Reliability model components

No.	Conveyance System Component	Code	Nominal Capacity (mgd)	Nominal Capacity (m³/s)
1	Croton Aqueduct	CROT	290	12.7
2	Catskill Aqueduct between Kensico and Eastview	C-E	700 /1,000	30.7/43.9
3	Delaware Aqueduct between Kensico and Eastview	D-E	1,400	61.4
4	Kensico Aqueduct between Kensico and Eastview	K-E	To be determined	
5	Eastview Site	EV	Not applicable	
6	Catskill Aqueduct between Eastview and Hillview	CAT	700	30.7
7	Delaware Aqueduct between Eastview and Hillview	DEL	1,400	61.4
8	Hillview Reservoir	HV	Not applicable	
9	Kensico Aqueduct between Kensico and the VCVC	K-CT1	To be determined	
10	Kensico Aqueduct between Eastview and the VCVC	K-CT2	To be determined	

vary with the capacity selected for the K-CT. These system capacities can then be compared with system demand and the water deficit computed. This permits probable deficits to be computed as a function of the capacity of the K-CT and thus the reliability model provides a means of placing an economic value on the K-CT reliability benefits.

Important inputs to the model are the estimated reliability levels of the existing conveyance components, and the economic impact of water deficits.

Impact of Deficits and Value of Water Supply Reliability

Over the past two decades NYCDEP has undertaken a program of conservation measures aimed at reducing water demands. This program has been very successful as demonstrated by Figure 11 which presents historic and projected annual water demand in NYC. It will be noted that over the last 15 years, water demands have fallen by 25%, despite a growing population. For the next 40 years demands are expected to increase modestly in line with trends population and per capita consumption.

The safe yield of the NYC's water sources will just about meet the projected long term annual demands. Despite this, failures in the conveyance system could lead to significant water deficits in NYC. Downstream of Kensico Reservoir the conveyance system is required to convey peak day capacities. A reduction in system conveyance capacity of say 15% would result in deficits during a few days each year, which could be handled by normal drought management policies. The loss of either the CAT or the DEL would lead to more serious deficits, and the simultaneous loss of both aqueducts would be catastrophic.

A few demographic and economic facts about New York are pertinent in this context. At about 8.3 million, New York City's population is over twice that of the next largest US city, Los Angeles. The population of the New York metropolitan area is about 19.8 million. The city is one of the leading financial centers of the world, and is the world headquarters for numerous leading financial services companies. The New York area, with a gross metropolitan product of over $1.1 trillion, is the largest regional economy of the US. It is against this background that the economic impacts of water deficits should be assessed.

Several US studies have been undertaken into the value of water reliability and how much investment should be made to increase reliability. Studies into value of water reliability in Orange County, CA, which is a highly populated area with a very productive economy, indicate that the value of reducing water deficits would be in the range of $0.03 to $0.08/litre ($0.10 to $0.30/gallon), depending on the mix of business and residential impacts. The base value used in the K-CT studies was $0.05/litre ($0.20/gallon), which is about 60 times the average water rate. Loss of the DEL would result in an average annual water deficit of about 33%. During peak day demands,

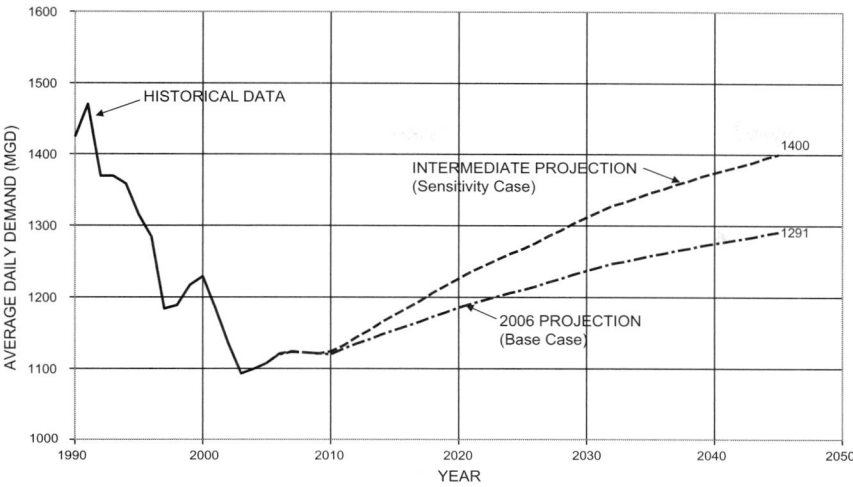

Figure 11. NYC average day demand projections

the deficit level would rise to about 45% to 55%. At a value of $0.05/litre, these deficit levels would be valued at about $50 billion per year. Similarly the economic impact of losing Hillview Reservoir would be about $85 billion per year.

Natural Failure Events

Based on detailed risk analyses of other portions of the Delaware Aqueduct undertaken by NYCDEP, the annual risk of failure from natural causes of the portion of the aqueduct downstream of Kensico Reservoir was taken as $1/1000$. Comparative assessments, taking into account type of construction and age, place the annual risks of failure of the Catskill and Croton aqueducts at $1/200$ and $1/100$ respectively. Note that all these failure rates are high in comparison to target reliability levels for this type of infrastructure, and is a reflection of the aging of the aqueducts. It is to be expected that without inspection and maintenance the reliability levels would continue to fall. The impact of adopting pessimistic values for the reliability of the existing conveyance system components is to increase the required size of the K-CT.

Vulnerability to Deliberate Damage

The impact of the terrorist attacks on September 11, 2001, and subsequent attacks elsewhere has heightened the need to protect key infrastructure systems, including water supply systems. Tunnel infrastructure is regarded as particularly vulnerable, and some engineers have proposed that for potential target areas such as New York and Washington DC, deliberately placed explosives should be considered a loading case in tunnel design. In general this would lead to increased lining thicknesses and steel reinforcement percentages, and other measures depending on the type and function of the tunnel.

For the New York City water supply system, NYCDEP has deployed security measures for the vital components of the system, which include policed security perimeters at surface installations. Hardening of existing tunnels and shafts downstream of Kensico Reservoir to resist explosions would be costly and time consuming, and would require an additional tunnel to be put into operation, namely the K-CT to ensure the continuity of water supplies. The addition of a new tunnel raises the possibility of adopting an additional security strategy, namely geographic dispersion, which would make it much more difficult to inflict significant water deficits on New York City.

Specialist security consultants were retained to examine the vulnerability associated with alternative tunnel alignments and configurations, and to review the risks of acts of deliberate damage, including terrorist attacks, against the NYC water system. The annual probability of a terrorist attack on the NYC water supply system was assessed to be in the range of $1/100$ to $1/1000$.

IMPACT ON PROJECT CHARACTERISTICS

Discharge Capacity of the K-CT

Comparisons were made between the cost of increasing the size of the K-CT, and the economic impact of deficit reductions, using the reliability model described earlier, taking into account the risks of natural failure and deliberate damage attempts. It was concluded that the K-CT should have a diameter of 7.0 m (23 feet), which would give a discharge capacity of about 88 m^3/s (2,000 mgd) when operating directly off Kensico Reservoir, and about 70 m^3/s (1,600 mgd) when interconnected hydraulically with Hillview Reservoir. At this size, the K-CT would assure the water supplies to New York City in the event that Hillview Reservoir and connecting aqueducts are unavailable, and would give sufficient additional conveyance capacity to enable NYCDEP with confidence to take each of the existing aqueducts out of service for inspection and repair.

Project Location and Configuration

The plan and hydraulic profile of the K-CT in the East Corridor is shown on Figure 12.

From the security viewpoint the advantage of this option is its separation from the existing aqueducts. The main objective of the planning studies was to develop a completely independent system. Thus the option in this corridor will have its own UV facility close to the intake structure, and if in due course it becomes necessary to filter the CAT/DEL supplies, a portion of the filtration plant could be located at the intake area. Because of this, intake Site 4 which is located on a large tract of land owned by NYCDEP was selected. The other advantage of the East Corridor option is the more uniform geological conditions along this corridor.

The plan and profile of the K-CT in the West Corridor is shown on Figure 13. The advantage of this configuration is that use can be made of the UV facility already under construction at Eastview. Also, should it be necessary in the future to filter the CAT/DEL supplies, a full filtration plant could be accommodated at the Eastview site.

This option has potential security issues, in that all three aqueducts, the DEL, the CAT, and the K-CT, would follow similar alignments and would be in close surface proximity at the Eastview Site. Security modeling was undertaken to mitigate the security risks as much as possible. This showed that the intake structure should be located at Site 2 rather than Site 1. The modeling also showed that the significant additional investment for an off-site bypass at Eastview would be justified. A typical option for such a bypass is shown on Figure 13. In this case the bypass structure would be located remotely from Eastview, on the opposite side of a major highway. The bypass structure would be a subsurface secure installation.

Despite these efforts to improve the security of the West Corridor option it is still regarded as intrinsically less secure than the completely independent K-CT option in the East Corridor. In addition the geological conditions in the West Corridor are more problematical. Finally the cost of the East Corridor option was estimated to be about 30% lower than the West Corridor option. Thus security, geological risk, and cost considerations have lead to the recommendation to build the K-CT in the East Corridor.

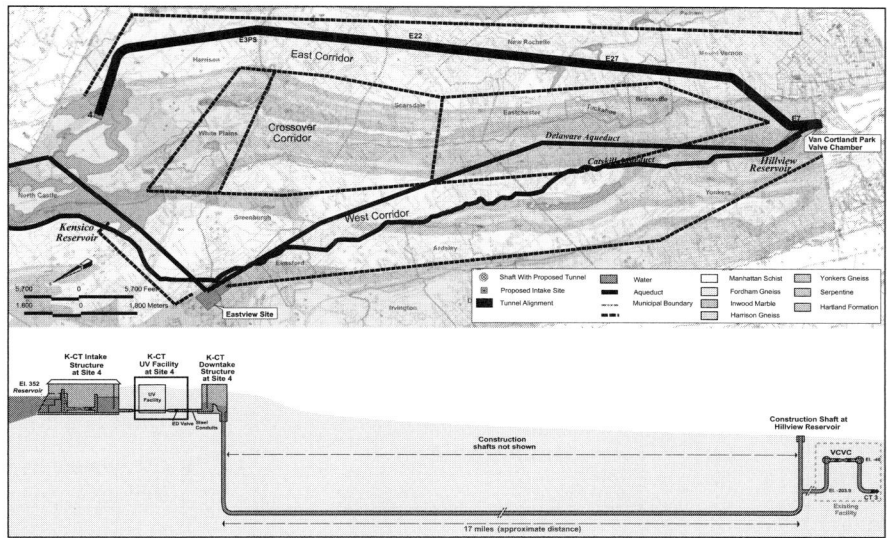

Figure 12. K-CT in East Corridor

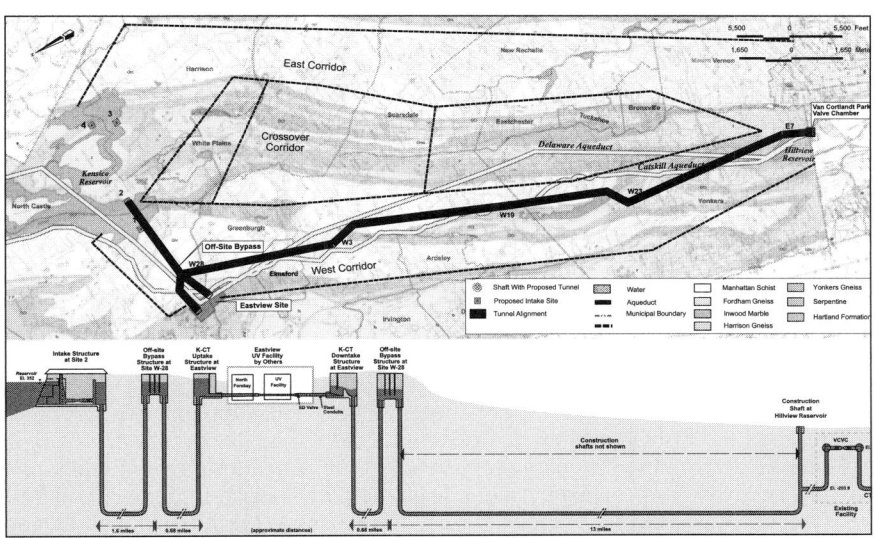

Figure 13. K-CT in West Corridor

Project Costs and Benefits

The tunnel for the recommended project in the East Corridor would be about 25 km long (15.5 miles), with a finished diameter of 7.0 m (23 feet). The project would be broken down into a number of construction contracts, including four tunnel contracts, with the objective of not exceeding $500 million at December 2006 price levels for any one

Table 3. Project cost estimate

Contract	Estimated Cost (Million $)
Intake Structure	404
UV Facility (Stages 1 and 2)	529
UV Connections and K-CT Downtake	129
Tunnel Contract No. 1 (includes shaft)	420
Tunnel Contract No. 2 (includes shaft)	315
Tunnel Contract No. 3 (includes shaft)	355
Tunnel Contract No. 4 (includes shaft)	471
Conversion to Distribution Shafts	160
Total Construction	2,783

Table 4. K-CT security and reliability benefits

Loss Event	K-CT Benefit (Million $)
Natural Events	
Loss of Hillview Reservoir	798
Loss of DEL downstream of Eastview	694
Loss of DEL upstream of Eastview	71
Loss of CAT downstream of Eastview	281
Loss of CAT upstream of Eastview	0
Loss of Croton Aqueduct	0
Loss of Key Structures at Eastview	798
Deliberate Damage Event	
Loss of Hillview, Eastview, or Kensico Intakes	1,553
Total Security and Reliability Benefits	4,195

Note: Benefits are at December 2006 price levels, discounted to the first year of operation (2024).

contract. In this way it is hoped that there will be at least three bidders for each contract. The cost estimates for the contracts are given in Table 3.

The main benefits of the K-CT are derived from increased levels of reliability of the conveyance system downstream of Kensico Reservoir, and reductions in system vulnerability to deliberate damage. Other benefits include improvements in water quality, and additional flexibility in system operations. As discussed earlier, the main benefits have been evaluated by assigning probabilities of unplanned outages to the various components of the conveyance system downstream of Kensico Reservoir, and determining the economic impact of the water deficits associated with unplanned outages. The main potential benefits of the K-CT are the avoidance of the economic consequences of probable deficits as indicated in Table 4.

For comparison, construction costs at December 2006 price levels discounted to the year 2024 amount to about $3,610 million. Thus for the assumptions made the K-CT would have a benefit/cost ratio of 1.16.

ACKNOWLEDGMENT

The authors would like to thank William Meakin of DEP New York for his contribution to this paper.

DAYLIGHTING THORN CREEK TUNNEL INTO CHICAGO'S TARP THORNTON COMPOSITE RESERVOIR

Cary Hirner ■ Black & Veatch

Kevin Fitzpatrick ■ Metropolitan Water Reclamation District of Greater Chicago

Marcella Landis ■ Metropolitan Water Reclamation District of Greater Chicago

Faruk Oksuz ■ Black & Veatch

ABSTRACT

The Metropolitan Water Reclamation District of Greater Chicago (District) is in the final phase of completing the Calumet System component of the Tunnel and Reservoir Plan (TARP) by bringing the Thornton Composite Reservoir online by 2014. The Thornton Composite Reservoir will provide nearly 30 billion liters (7.9 billion gallons) of additional storage for minimization of combined sewer overflows (CSO) and flooding in the Chicago metropolitan area.

The project requires that the existing 6.7 m (22 ft) diameter Thorn Creek tunnel conveying overbank flood flows is re-routed and connected to the Thornton Composite Reservoir. This paper presents the components of the Thornton Composite Reservoir and specifically the detailed evaluation of a drill-blast tunnel. Other project features include an access shaft, control gate, energy dissipating structure, bulkheads, and live tunnel connection. The construction contract will be procured in late 2009.

INTRODUCTION

The Thornton Composite Reservoir is being constructed by the District to minimize waterway pollution by combined sewer overflows (CSOs) in the Calumet service area and reduce flood damages by providing an outlet for Thorn Creek floodwater. The reservoir site is being constructed in the expanded North Lobe of the Thornton Quarry which is located adjacent to I-80/I-294 in southern Cook County, Illinois (Figure 1). The north lobe of the quarry, which is currently being mined for the future Thornton Composite Reservoir, will ultimately provide 30 billion liters (7.9 billion gallons or 24,200 acre-ft) of storage capacity as part of the District's TARP.

Three construction contracts are planned to complete the reservoir:
1. Contract 04-201-4F: Tollway Dam, Grout Curtain, and Quarry Plugs:
   - Groundwater Protection System (Grout Curtain)
   - Tollway Dam Grout Curtain
   - RCC Gap Dam
   - Quarry Haul Tunnel Plugs
2. Contract 04-202-4F: Connecting Tunnels and Gates:
   - Indiana Avenue Connecting Tunnels, Gates and Shaft
   - Reservoir Inlet/Outlet Structure and Energy Dissipation

Figure 1. Thornton Quarry site map

3. Contract 04-203-4F: Final Reservoir Preparation:
   - Thorn Creek Connecting Tunnel and Gate
   - Abandon Transitional Reservoir
   - Convert Thorn Creek Diversion Tunnel to a Drainage Adit

Contract 04-203-4F is the focus of this paper and primarily entails making a live connection to the existing Thorn Creek Diversion Tunnel at about 76 m (250 ft) below ground and diverting flow to the Thornton Composite Reservoir. A portion of the existing tunnel going below the I-80/I-294 highway will be abandoned and converted into a drainage adit to intercept seepage through the tollway rock wall dividing the north (Composite Reservoir) and main lobes of the quarry. Black & Veatch is responsible for the design of the new connection tunnel and drainage adit conversion. At the time this paper was prepared the project was in the preliminary design phase, which included tunnel alignment alternatives evaluation.

EXISTING THORN CREEK DIVERSION TUNNEL

An interim storage facility, the Thornton Transitional Reservoir (TTR), was created in the West Lobe of the Quarry with the objective of providing floodwater diversion and detention for the residents affected by overbank flooding from Thorn Creek (Figure 2). The TTR was placed into operation in 2003 and has captured billions of gallons of floodwaters over 20 plus fill events. At the upstream end of the tunnel, there is a diversion

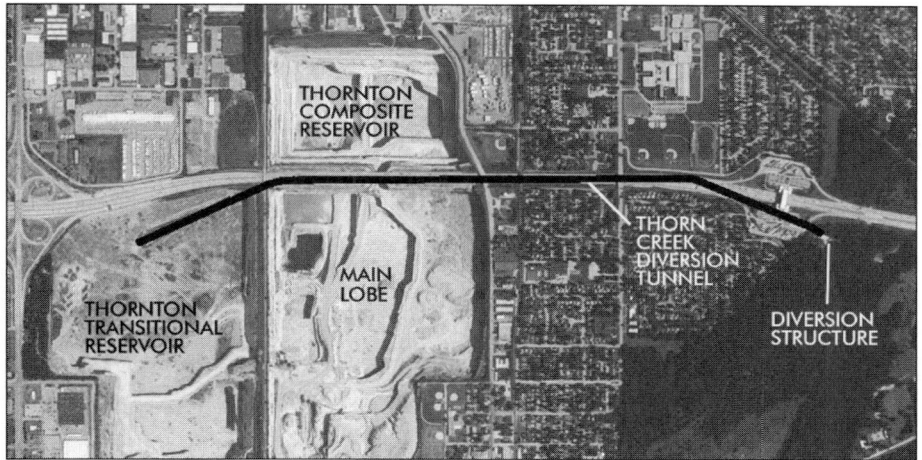

Figure 2. Thorn Creek diversion tunnel alignment

Figure 3. Thorn Creek diversion structure

structure on Thorn Creek. It is comprised of a weir and three 3.7 × 3.7 m (12 × 12 ft) electrically operated sluice gates that control flow from the creek to the TTR (Figure 3). A trash rack and screen are located in front of the weir to limit debris entering the tunnel. The diversion structure is designed for a maximum flow of 175 m^3/sec (6,200 cfs). Downstream of the gates flow drops down 70 m (230 ft) through a 7.3 m (24 ft) diameter drop shaft leading to the 2,440 m (8,000 ft) long Thorn Creek Diversion Tunnel (Figure 4). The tunnel daylights into a tapered trapezoidal outlet channel in the floor of the TTR. The water level in the TTR is not allowed to exceed El. −119 CCD, since above that elevation floodwater would flow into the active Main Lobe of the quarry. After the storm passes, the reservoir is drained through part of the diversion tunnel and an 8-foot diameter tunnel that connects to the Indiana Avenue Leg of Calumet TARP. Two 106.7 cm (42-in) diameter cones valves control the flow and help to dissipate hydraulic energy while the TTR is dewatered into TARP tunnel.

Figure 4. Thorn Creek diversion tunnel

THORN CREEK CONNECTION TUNNEL

By the end of 2014, when the Composite Reservoir is scheduled to be on line, the Transitional Reservoir will be decommissioned and returned to the quarry operator. To complete the decommissioning of the TTR, a new tunnel will be constructed to connect the existing Thorn Creek Diversion Tunnel to the Composite Reservoir, and allow the existing diversion tunnel outlet into the TTR to be abandoned.

The new Thorn Creek Connection Tunnel will allow the system to operate in essentially the same manner as the existing system with the following exceptions:

- The floor of the Composite Reservoir will be approximately 15 m (50 ft) lower than the TTR so the new tunnel outlet in the Composite Reservoir must incorporate an energy dissipating structure;
- Since the Composite Reservoir will also accept CSO flow from the Calumet TARP System, the new connection tunnel will need means to prevent CSO backflow from the reservoir back into the un-lined Thorn Creek tunnel and adversely impacting groundwater;
- Because the Composite Reservoir will be allowed to fill to a much higher level, i.e., El. –5 CCD compared to El. –119 CCD in the TTR, the diversion tunnel can potentially be pressurized during operation;
- The mode of dewatering the Composite Reservoir will be via a new TARP tunnel connection near the northeast corner of the reservoir that has gates for flow control.

PRELIMINARY DESIGN

The Final Reservoir Preparation contract includes a number of project facilities and components to be constructed to make the Composite Reservoir operational. They include:

- Tunnel connection to the Composite Reservoir consisting of approximately 305 m (1,000 ft) of 6.7 m (22 ft) diameter rock tunnel;
- Measures to control reservoir water from impacting groundwater resources by either installing a large wheel gate in the new tunnel connection or concrete lining the entire portion of the existing Thorn Creek Diversion Tunnel that will remain operational;

- Energy dissipation structure to prevent reservoir floor damage caused by the 15 m (50 ft) drop of water;
- Plug part of the Thorn Creek Diversion Tunnel and convert it into a drainage adit, including drilling of drain holes to intercept seepage from the tollway rock wall;
- Provide for a method to dewater the converted drainage adit through use of a shaft and pump station, drain holes or use of existing facilities;
- Decommission the TTR including removal of rock bolts at the Thorn Creek Diversion Tunnel inlet/outlet structure in the floor of the TTR and the cleaning out sediments accumulated over time;
- Highwall stabilization measures and floor drainage improvements for the Composite Reservoir;
- Improvements to quarry (reservoir) access road and ramps;
- Fencing and landscaping along the grounds of the Thornton Composite Reservoir.

GEOLOGIC SETTING

Thornton Quarry lies on the eastern flank of the Kankakee Arch where the strata dip eastward at approximately 5 m (15 ft) per mile. The Kankakee Arch is an anticlinal structure trending northwestward across Indiana and Illinois, connecting the Cincinnati Arch with the Wisconsin Arch, and separating the Illinois Basin from the Michigan Basin. The bedrock consists of about 1,220 m (4,000 ft) of sedimentary rocks, ranging in age from Cambrian to Silurian, which are overlain by Quaternary glacial and lacustrine sediments. The sedimentary rock is underlain by Precambrian granitic rock.

A thin layer of Quaternary glacial drift overlies bedrock at the Thornton Quarry. The Racine Formation, the youngest and most lithologically varied stratigraphic unit of the Silurian System, ranges from about 73m (240 ft) to 91 m (300 ft) thick in the quarry. The Thornton Reef structure occupied the project area during Racine deposition and largely controls the local bedrock structure. Three generalized rock facies are recognized in the Racine: reef, reef-flank, and inter-reef facies. Underlying the Racine Formation is the Sugar Run and Joliet formations, which primarily consist of dolomite beds with some shale partings. These formations comprise the rock currently exposed in the quarry. Shales of the Maquoketa Group create the regional aquitard underlying dolomite limestone that limits vertical groundwater seepage, allowing conversion of the quarry into a storage reservoir.

TUNNEL CONFIGURATION

The major component of this project is the new connection tunnel to the reservoir. Multiple horizontal and vertical tunnel alignments were evaluated during preliminary design along with control options to prevent reservoir water from impacting local and regional groundwater. Key criteria used in this evaluation included:

- Operation and functionality of the future works
- Cost
- Constructability
- Interfaces with other project components and contracts
- Interface with the quarry operator
- Schedule including limiting disruptions to the operation of the existing works
- Ground conditions and easement requirements
- Maintenance of the future works

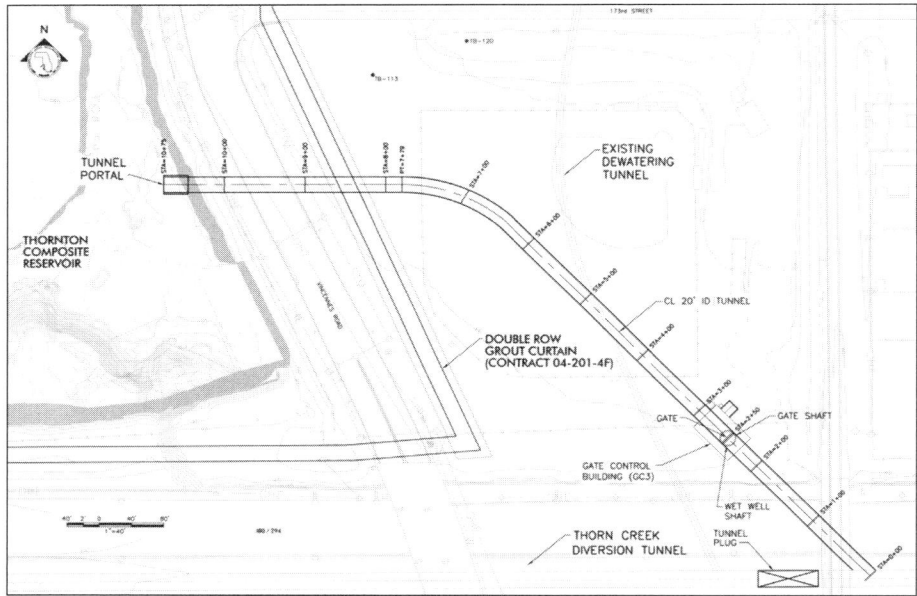

Figure 5. Thorn Creek connection tunnel preliminary alignment

The tunnel alignment corridor on the southeast corner of the reservoir quarry deems to be the most desirable because it allows the existing diversion tunnel in the tollway dam to be converted into a drainage adit. Furthermore, the reservoir highwalls are excavated using pre-split blasting in this area and minimal impacts are anticipated on the grout curtain to be installed in Contract 04-201-4F. There is also convenient access to the shaft, if one is necessary. This location allows for the possibility of using the existing dewatering tunnel located east of the Composite Reservoir to empty out the drainage adit system; thereby eliminating the need for an additional shaft and pump station (Figure 5). To use the existing dewatering tunnel to drain the drainage adit the connection to the existing tunnel will need to occur east of the dewatering tunnel. In this scenario, the new connection tunnel must incorporate a drop of approximately 12 m (40 ft) to 15 m (50 ft) using a vertical shaft prior to crossing the dewatering tunnel alignment.

MEASURES TO PROTECT GROUNDWATER RESOURCES

The District is addressing the potential for infiltration into and exfiltration out of the reservoir by constructing a full perimeter grout curtain in the Silurian dolomites as part of Contract 04-201-4F. This grout curtain is intended to protect existing groundwater to be impacted by the CSO waters that are to be contained in the reservoir. Because the existing Thorn Creek Diversion Tunnel is unlined, engineered controls will need to be constructed with the new connection tunnel to prevent water from the reservoir backing up into the unlined tunnel and potentially seeping out into the bedrock. Two control measures currently being considered during preliminary design include installing a large wheel gate in a shaft that would be constructed as part of the new tunnel connection and concrete lining the portion of the Thorn Creek Diversion Tunnel that will continue to provide floodwater conveyance.

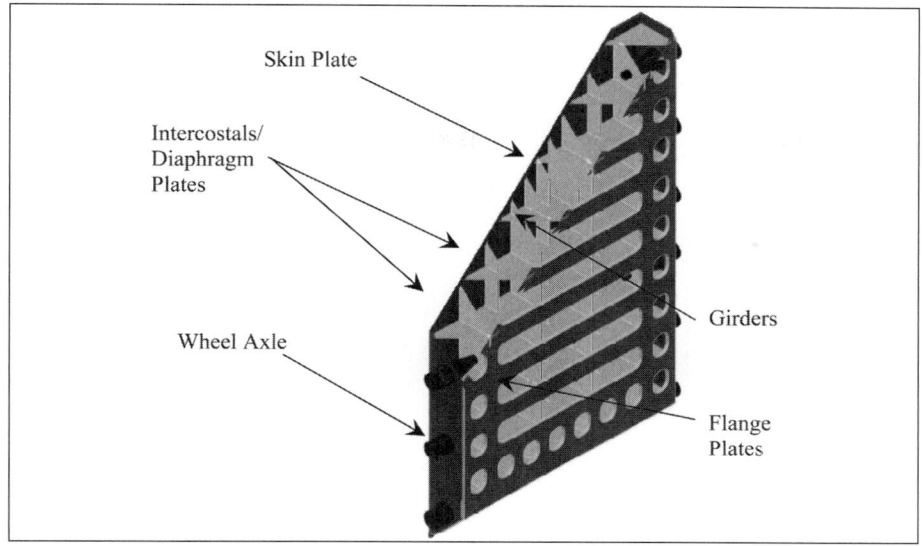

Figure 6. Connection tunnel gate schematic

Roller Gate

Gate alternatives were evaluated during the early stages of preliminary design and a large wheel roller gate was recommended for further evaluation. The maximum size of the steel roller gate would be 6 m by 6 m (20 ft by 20 ft, gate net opening) and the gate would be operated by a hydraulic cylinder (Figure 6). The gate and cylinder would be located in an excavated shaft from the tunnel to the surface. The connection tunnel will need to transition from a circular to square cross-section to accommodate the gate structure. To provide removal of the gate for maintenance or repair and maintain the ability of the Thornton Composite Reservoir to receive combined sewer inflows, a stoplog would be inserted and removed under balanced head conditions to allow for removal of the roller gate. Located above the gate shaft would be a gate control building to house the gate hydraulic power unit, motor control center and local control panel. The gate would be capable of manual or automatic control based on water level in TCR, feedback signal from the Thorn Creek diversion gates, and pressure in the diversion tunnel when the gate is closed.

Concrete Tunnel Lining

A less maintenance intensive measure also being considered during preliminary design to protect groundwater would be the installation of a cast-in-place concrete liner in the tunnel from the Thorn Creek Diversion drop shaft to the new reservoir outlet. This would entail lining the new connection tunnel and approximately 1,220 m (4,000 ft) of existing tunnel. Upon liner installation, secondary grouting of the bedrock would be conducted behind the liner to decrease the rock mass permeability. Depending upon the final alignment selection, the potential exists to eliminate the need for an access shaft during construction by mining and concrete lining the tunnel from the quarry reservoir floor using an access ramp.

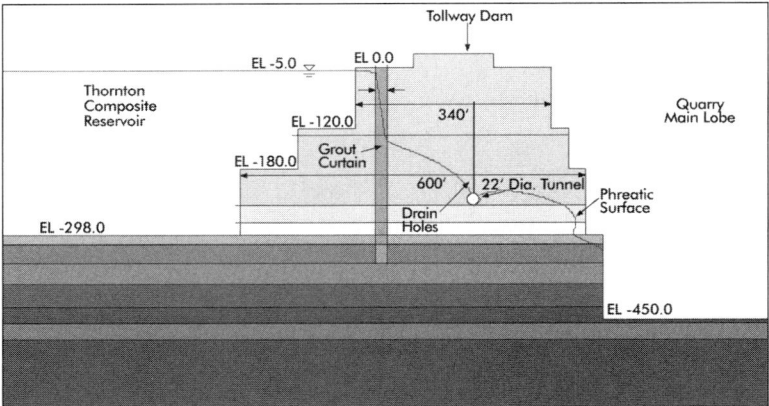

Figure 7. Tollway Dam drainage adit

DRAINAGE ADIT CONVERSION

A portion of the existing Thorn Creek Diversion Tunnel will be transitioned into a drainage adit to reduce seepage pressure in the rock dam to increase stability. The drainage adit will collect seepage water within the tollway dam downstream of the Composite Reservoir and the grout curtain, effectively lowering the phreatic surface and therefore the pore pressure exerted within the downstream face of the dam (Figure 7). The existing tunnel/drainage adit is parallel to the long axis of the rock dam at the base of the Racine Formation. To most efficiently intercept seepage drain holes will be installed in a vertical plane parallel to the tunnel axis from the tunnel crown and likely the tunnel invert. Drain holes will be drilled upward from the tunnel at an angle up to 30 degrees from vertical at a spacing of 12 m (40 ft) to efficiently cross the joints with the drain holes. The upward directed drain holes would be drilled to the top of the phreatic surface modeled in the seepage analysis. This is about 20 m (65 ft) above the crown of the diversion tunnel or at El. −160 CCD. Actual drilling length of the upward drain holes would be 23 m (75 ft), taking into account the inclination of 30 degrees from vertical for the drain holes.

Drainage Adit Dewatering

Water collected by the drainage adit will need to be removed and returned to the Thornton Composite Reservoir or conveyed directly to the TARP tunnel. To do this, three alternatives are being analyzed in preliminary design.

Pump Drainage Adit Water

A shaft would be constructed in the tollway dam that is connected to the drainage adit. A pump station would be installed to pump the water back to the reservoir.

Gravity Drain Water to the Main Quarry

Outflow holes would be drilled through the south face of the tollway dam from the west end of the drainage adit above the existing sump in the northwest corner of the Main quarry lobe. Water collected in the quarry sump would be pumped back over the tollway dam and into the Composite Reservoir.

Gravity Drain Water Using Dewatering Tunnel

Using existing infrastructure, locate the Thorn Creek tunnel plug east of the dewatering tunnel to keep the dewatering tunnel functional. This would allow the drainage tunnel to be dewatered by gravity using the dewatering tunnel to convey flows to the TARP tunnel. This alternative is currently considered to be the most favorable to dewater the drainage adit and likely will be recommended for final design.

CONCLUSIONS

TARP was selected by the Metropolitan Water Reclamation District of Greater Chicago in 1972 as the Chicago area's plan to cost effectively comply with federal and state water quality standards. The Thornton Composite Reservoir is part of the second phase of TARP that will reduce flooding and further minimize waterway pollution by CSO. The Thorn Creek Tunnel connection and drainage adit conversion of the Final Reservoir Preparation contract are an integral part of commissioning this visionary component of TARP.

ACKNOWLEDGMENTS

The authors acknowledge contributions of many District staff, engineers, vendors and contractors on this truly marvelous project that continues to improve water quality and the environment.

PART 3

Difficult Ground

Chairs

Andrew Liu
Hatch Mott MacDonald

Brett Robinson
Traylor

TIMES SQUARE CONNECTION: "SUPPORTING THE CROSS ROADS OF THE WORLD"

Victor F. Paterno Jr. ▪ Skanska USA Civil

Robert W. Fonkalsrud ▪ Skanska USA Civil

ABSTRACT

In December of 2007, a joint venture of J.F Shea, Skanska and Schiavone referred to as S-3II, was awarded a $1.15 billion (€ 886,880,000) contract to construct the running tunnels and station structures for the Number 7 Line Extension Project in New York City by the Metropolitan Transportation Authority Capital Construction (MTACC) and New York City Transit (NYCT). As part of the contract, a technically complex and critical connection of the new tunnel to the existing, operational 7 Line subway was required.

This paper describes the innovative design approach that was conceived to carry out the underpinning and excavation works both within the east-west 7 Line subway tunnel and beneath the north-south, 4 track, 2 level Times Square Station for the A, C & E trains. This underpinning provided temporary support to the subway substructure as the grade of the tunnel invert was lowered up to 10 feet (3 meters) into the rock sub-grade to allow the future subway to pass under the basement of the world's busiest bus terminal.

In addition, the creation of a new portal connecting the existing subway tunnel and the subterranean level of the Port Authority Bus Terminal by means of underpinning structural footings and horizontal spiling/poling of existing ground will be discussed.

INTRODUCTION

New York City Transit (NYCT) and MTA Capital Construction (MTACC), which are both public benefit corporations, are the sponsors of the Construction of the Running Tunnels & Station Structure for the Number 7 Line Extension in the Borough of Manhattan (see Figure 1). The project begins on the west side of Manhattan at 11th Ave. & 25th Street. It progresses north below 11th Ave. and turns west under 41st Street to connect to the existing 7 Line terminus below the A, C and E Line Station at 8th Ave. & 41st Street. The subway station platform is about 30 feet (9 meters) below the street grade.

This project consists of two 22 ½ foot (6.86 meters) diameter tunnels, each approximately 4,025 feet (1,227 meters) in length, to be excavated utilizing tunnel boring machines and lined with pre-cast concrete segmental liners. The new station structure including the interlocks is approximately 1,200 feet (366 meters) in length. It starts just north of 32nd Street and ends at 36th Street. The station is constructed in a rock cavern with access shafts and adits that are being constructed using conventional drill and blast methods. The project also includes approximately 300 linear feet (91.4 meters) of retrofit. The retrofit consists of lowering the invert of the existing tail tracks of the 7 Line 10 ft (3 meters) into the rock sub-grade, underpinning centerline columns to support the A, C, and E Line that cross over the top of the tail tracks, and the installation of temporary steel

TIMES SQUARE CONNECTION

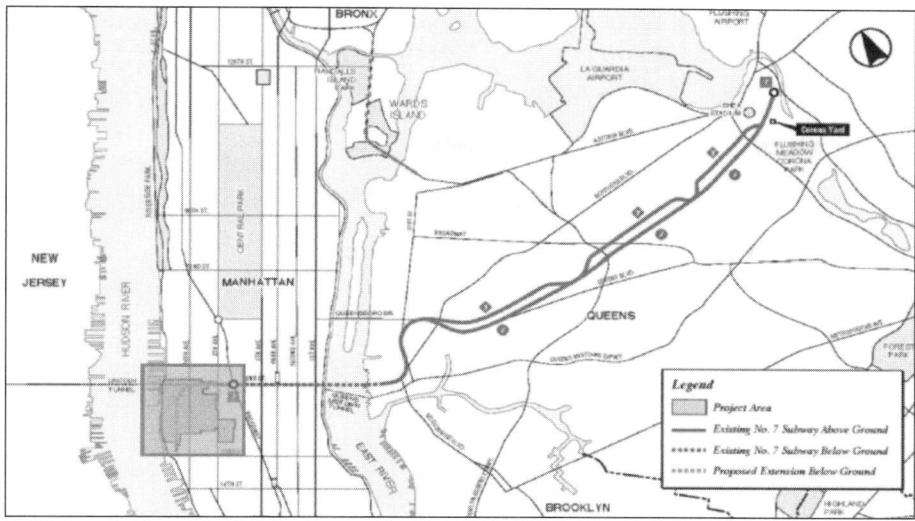

Figure 1. Project location map

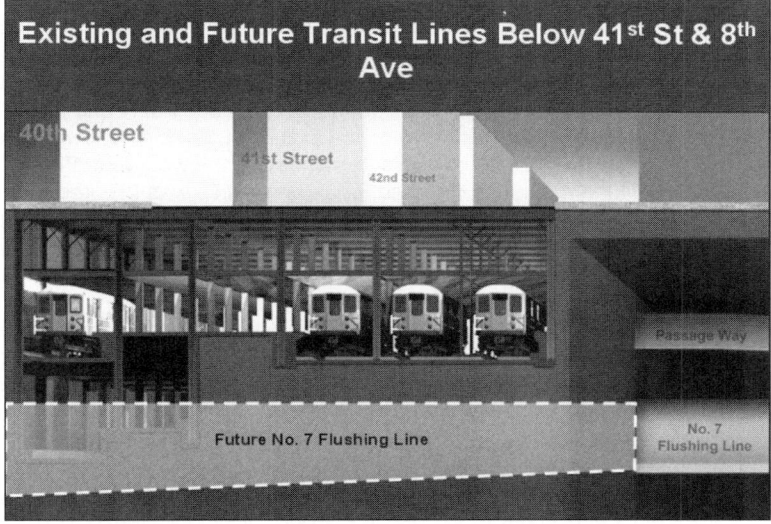

Figure 2. Existing and future subway lines in Times Square work area

bracing to facilitate structural supports for the new configuration through the abandoned 8th Ave. Line (see Figure 2).

The location of the 300 linear feet (91.4 meters) of retrofit is below one of the busiest intersections in Manhattan, one of NYC Transit's busiest subway stations and the world's busiest bus terminal operated by Port Authority of New York & New Jersey. There are three quarters of a million people that travel through this area on a daily basis. It is one of the major intermodal transportation facilitates in the city. The 7 Line portion of the retrofit section of the project was put into operation March of 1927. The

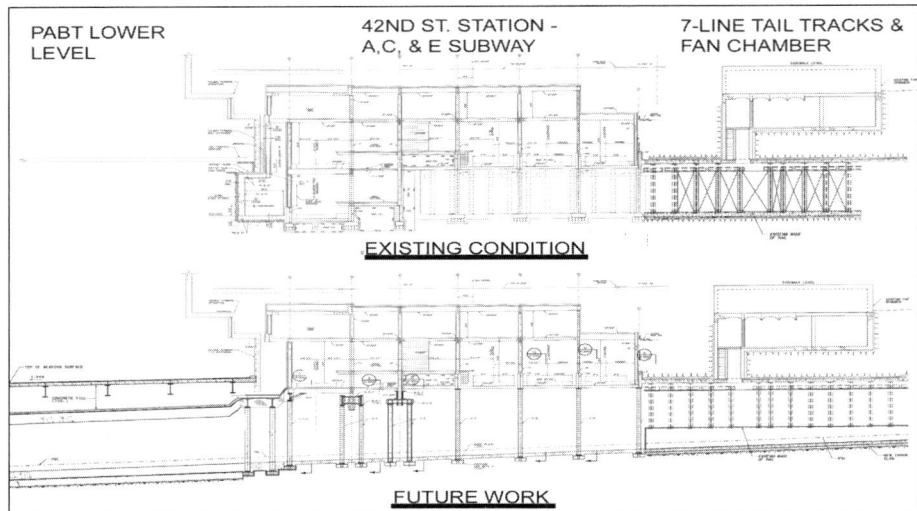

Figure 3. Existing condition of 42nd Street station and New 7 Line extension in lower level

A, C, and E Line, along with an abandoned 8th Ave. Station directly below, was constructed in 1932, but the lower level single platform was not completed with finishes and open to the public until 1959. This single track lower level platform, downtown service only, was for odd services like the Aqueduct Racetrack trains and for rush hour E Line service. This lower level platform service closed in 1981. The lower level platform and track blocked the 7 Line from being extended to the west (see Figure 3). It is often thought that 7 Line owned by IRT and BMT companies was blocked to the west by Independent Subway System (IND) Company because they wanted to crush the competition in any cross town line. This lower level platform, after it's shut down, has been use by film producers for several movies the most famous was "Ghost" starring Patrick Swayze and Demi Moore.

In order to extend the 7 Line west and not to disturb one of the 7 Line's most traveled stations at 42nd Street and 7th Avenue, the abandoned 8th Avenue platform and track needed to be demolished to meet the elevation of the TBM tunnels coming in from the west.

TIMES SQUARE CONSTRAINTS

There were only two ways to access the work areas but neither one provided access into both zones. One access point was a door on the active platform level of the downtown 8th Ave. A, C, and E lines which is part of two level active station roughly 40 feet (12 meters) below street elevation of 8th Ave. & 41st Street. The other access point to the work area in the 7 Line tail tracks was west of the last stop in the Times Square Station, which is below the intersection of 7th Avenue and 41st Street, about 70 feet (21.34 meters) below street grade. The drawback of using this access point is that it requires special permission from NYC Transit. This involves a request for a track outage and flagman assistance that at a minimum took six weeks to be approved.

The work zones are divided by structural walls, which were installed at the time of their construction. Equipment access is being made available only by work trains that either originated out of the 38th Street Yard Facilities in Brooklyn or the Corona Yard

Figure 4. Top of material access shaft and muck removal compound at West 41st Street and 8th Avenue

Facilities in Queens. Both locations are about 12 miles (19 kilometers) from Manhattan but the time to get to the work zone is between 45 to 55 minutes without train traffic. The dimensions of the work train flat car pose yet another restriction. The useable surface on the flat cars that transport the equipment are 8' wide × 40' long (2.5 m wide × 12 m long) with a 7'-6" (2.3 meters) height restriction. Work train availability is either week nights or weekends. The weekend work train requests are very limited as week nights are preferred by the client. Material movement in and out of the work area also falls under these requirements with a minimum six week work train notification required by NYCT.

The physical size of work areas also limits the choice of equipment. The abandoned platform is only 12'-7" (3.8 m) wide with a 12 ft. (3.6 m) height. The abandoned track area for future NYC Transit worker facilities is only 16'-7" (5 m) wide by 14 ft. (4.3 m) in height. The 7 Line tail tracks is the largest of the work areas at 11'-9" (3.58 meters) and a height up to 15 ft. (4.6 m) but this too is limited by crash walls between C1 and C2 track and duct banks on the outside walls. This required the team to come up with innovative ideas to overcome the conditions associated with working in such a restrictive work environment.

Even the staging area for the project on the street is restricted due to construction of a new high rise office tower on the east side of 8TH Ave. between 41ST and 42ND Street (see Figure 4).

Not only do we need to coordinate our work with an active railroad to gain access, we now have to coordinate with a high rise building construction operation that has no direct relationship to the outcome of our work.

The craft labor rules with jurisdiction requirements are another factor that determines how we do our work. Table 1 is a matrix of the craft labor and the work they perform.

Table 1. Existing and future subway lines in Times Square work area

Craft Labor	Concrete Demolition	Rock Excavation	Temporary Steel	S.O.E.	Permanent Steel	Concrete Forming	Hvac	Mechanical	Electrical
Laborers Local 731	X	X		X		X			
Drill Runners Local 29	X	X							
Operating Engineers Local 14	X	X		X	X	X			
Operating Engineers Local 15	X	X	X	X	X	X			
Operating Engineers Srurvey15D	X	X	X	X	X	X			
Carpenters Local 608						X			
Dock Builders Local 1456			X	X					
Timbermen Local 1536				X					
Iron Workers Local 40			X		X				
Metallic Iron Rebar Local 46						X			
Masons Local 780					X				
Electricians Local 3									X
Plumbers Local 1								X	
Mill Rights Local								X	
Tin Knockers Local 28							X		
Teamsters Local 282	X	X	X	X	X	X	X	X	X

Since the New York City hourly rate is one of the highest in the country, we must plan and coordinate work to maximize the field operations with multiple work areas. These areas all have common work themes; concrete demolition, drill and split rock, lift and load debris, underpinning, waterproofing installation as well as form and pour structural concrete. Additional work will include electrical, mechanical and HVAC. The work areas provide the flexibility to perform multiple work disciplines simultaneously as opposed to moving trades in and out.

The first order of business is to link the abandoned 8th Avenue. platform with the tail tracks of the 7 Line. Then analyze what are the largest types of equipment we could bring in and, above all, overcome the limitations of working in and around an active railroad operation. The active railroad limited our access time to the work areas to off hours, which is nights and weekends.

In order to create access between the 7 Line tail tracks and the abandoned 8th Avenue platform an access way was cut between the five foot steel bents. This was done early in the project with electric drills and chippers during our mobilization period.

The project team started planning for the removal of the third rail power (600 volt DC) and relocation of the train signal stops and insulating joints at the track bumper blocks which are located at the end of the existing 7 Line. The relocation of the signal equipment and signal stops have to be approved by NYC Transit Signal Department and must be implemented by a signal engineer who also has to be approved by NYCT. This involves adjusting signal timing of the trains in and out of the last station stop which can only be done on the weekends. Performing this work involves a number requests for weekend track outages.

The project team came up with the idea to install an access shaft adjacent to an existing fan plant air shaft on the corner 41st Street & 8th Avenue. Though this provided access to the tail tracks for movement of material in and out, it was limited with the size of material that could be lowered down the shaft (see Figure 5).

The creation of the shaft triggered the need for more DOT permits. When performing any work on the streets of New York City, DOT permits are required. We now had to prepare engineered critical pick drawings and calculations for our application to place a cherry-picker on the street along with permits to occupy two lanes of a three lane street (see Figure 6). Along with those permits, we needed to obtain a work agreement with the building contractor constructing a new high rise on 8th Avenue between 41st & 42nd Streets.

Table 2 is a list of equipment used to assist with the construction operations in the subway retrofit portion of the project.

7-LINE TAIL TRACK UNDERPINNING

Existing Site Conditions

As briefly described in previous sections, the first area of temporary underpinning in the Times Square Connection work area covered approximately 190 linear feet (57.9 m) of existing double arched tunnel located adjacent to the contract limit where the new work ties in to the existing rail system. All as-built plans of the 7-line tunnel in this area show the concrete arches to be unreinforced concrete with steel center columns spaced at 5'-0" (1.5 m) on center between the two track beds (see Figure 7). The existing ground layers above the tunnel consist of 10 ft. (3 m) of fill material and approximately 21 ft (6.4 m) of rock. West 41st Street runs directly over the path of the tunnel and was open to traffic for the duration of the project.

Planned Final Condition

The contract design in this area specified the existing invert slab to be removed and the rock subgrade to be lowered from 4–8 ft (1.2–2.4 m). A new concrete reinforced track invert slab is to be installed with a 3.00–3.75% vertical curve in order for the new extended tunnel to clear the lower level of the Port Authority Bus Terminal located to the west. The steel center columns were required to remain in place and are designed to be supported by a longitudinal concrete reinforced pedestal between tracks in the permanent condition (see Figure 7).

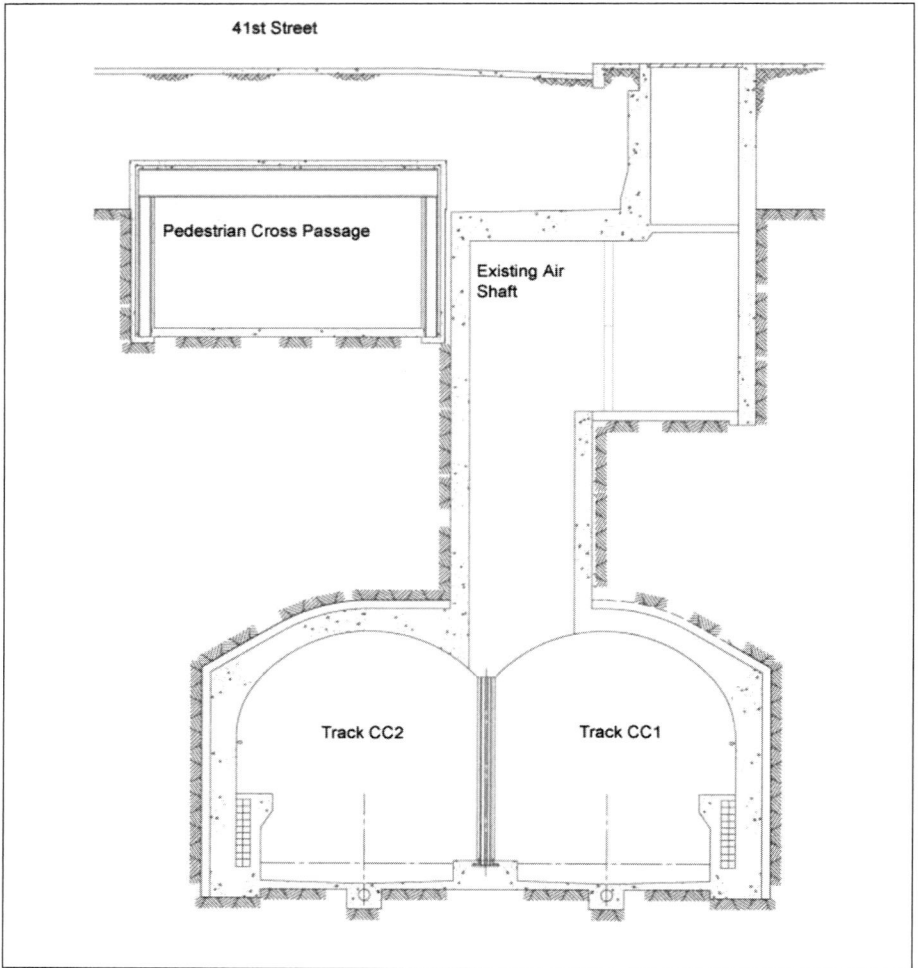

Figure 5a. Sections of 7 Line tail tracks showing before installation of the material access shaft on West 41st Street

Temporary Underpinning

The Joint Venture recognized from the onset the challenge of supporting the existing tunnel structure in place during the excavation phase in light of the limited work area in the tunnel environment and the restricted access of heavy equipment for underpinning and rock removal. In addition, the underpinning system needed to be installed without disrupting the existing 8th Ave. subway lines and 42nd St. station which are located directly above the 7-line extension work area. Finally an underpinning system that could be installed at one time for the duration of the project was crucial in order to meet schedule and continuity of work constraints. For this reason a support system of back to back channel shaped steel members was selected to "sandwich" the existing structural steel center columns of the tunnel and transfer the load through foundations into the rock sub-grade below the final invert elevation (see Figure 8).

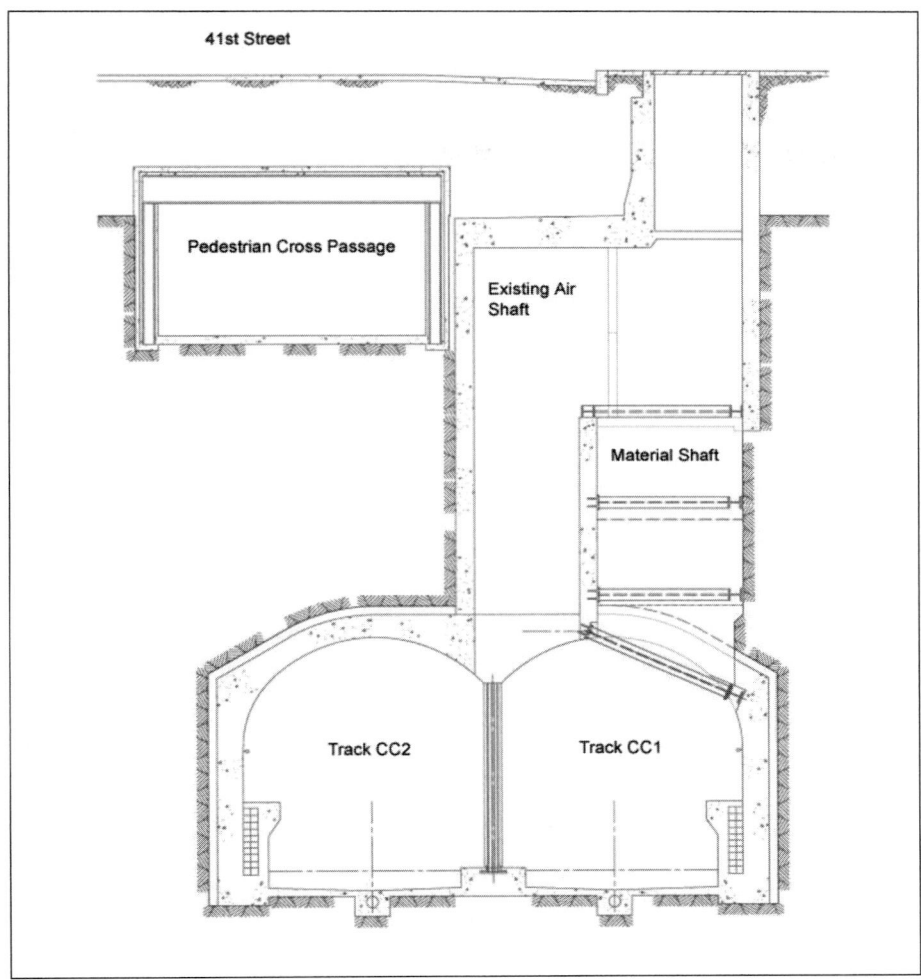

Figure 5b. Sections of 7 Line tail tracks showing after installation of the material access shaft on West 41st Street

Geological boring data from adjacent sites indicate the rock below the existing 7-line tunnel to be predominantly competent schist with zones of pegmatite-schist as well. Due to this favorable rock condition and the fact that the partners of the Joint Venture possess extensive experience in rock drilling, it was decided that small diameter caissons would be the most suitable means to distribute the underpinning loads into the rock sub-grade.

The small diameter caissons consist of 9 5/8" (24.5 cm) diameter, 80 ksi, steel pipe installed into 12" (30.5 cm) diameter drilled holes. Three and one half inch diameter, 75 ksi threaded bar was inserted full depth into the hole and the entire caisson was filled with 6,000 psi grout. The caissons range from 14 ft.–17 ft. (4.3–5.2 m) in depth and are spaced at 10 ft. (3 m) on center between the existing tunnel center columns.

The back to back channels gripping the existing center columns were fabricated from W24 members and distributed the underpinning load into the caisson through

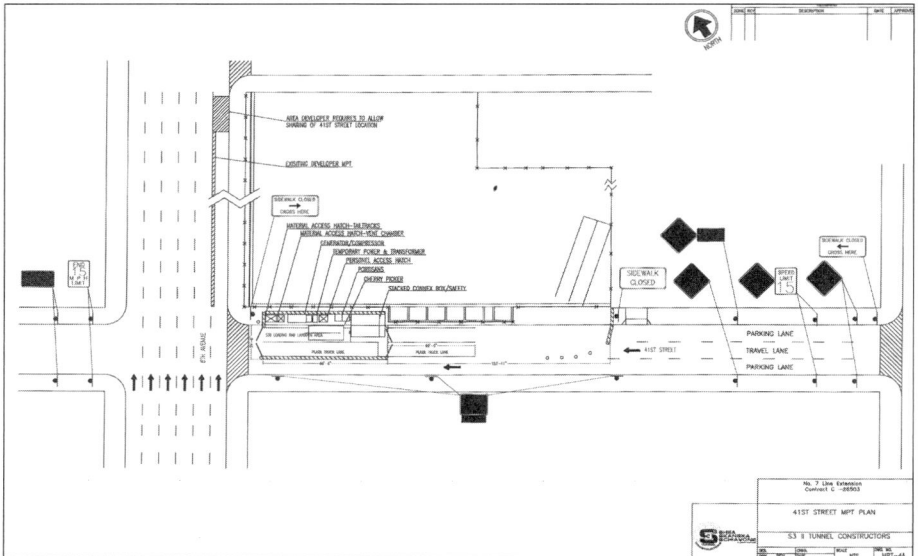

Figure 6. West 41st Street and 8th Avenue traffic plan

Table 2. Equipment list for underpinning and rock removal

	Pneumatic Supply	Concrete & Rock Removal	Loading & Moving Material	Concrete & Rock Removal	Rock Removal	Rock Removal	Rock Bolt Installation	Caisson Installation
Equipment Task	Air Compressor	Hydraulic Excavator	Skid Steer	Demolition Robot	Air Track	Air Track	Drill Rig	Down-The-Hole Hammer
Equipment	Ingersoll Rand 915	Bobcat 435	Bobcat A300	Brokk 250	IR LM 100	IR ECM 350	Perfora Rock Buggy	Davy DK515
Power Supply	Diesel	Diesel	Diesel	Electric	Air	Air	Diesel	Electric

localized W12 cap beams. In addition horizontal braces connecting the cap beam to the tunnel side wall were required at each caisson in order to prevent a moment build up at the underpinning steel to caisson connection.

As mentioned, the contract design called out for the removal of the existing tunnel invert slab which in addition to providing a track bed served as a diaphragm preventing lateral movement in the tunnel side walls. In order to temporarily support these walls during excavation, 8 ft. (2.4 m) long Swellex PM24 rock bolts were placed in the tunnel side walls at 4 foot (1.2 m) on center and continued in levels as the rock excavation proceeded on a typical 4 ft. × 4 ft. (1.2 m × 1.2 m) pattern. The installation of rock bolts in lieu of conventional lateral struts was a critical factor since construction access in the tunnel could be maintained throughout the excavation process.

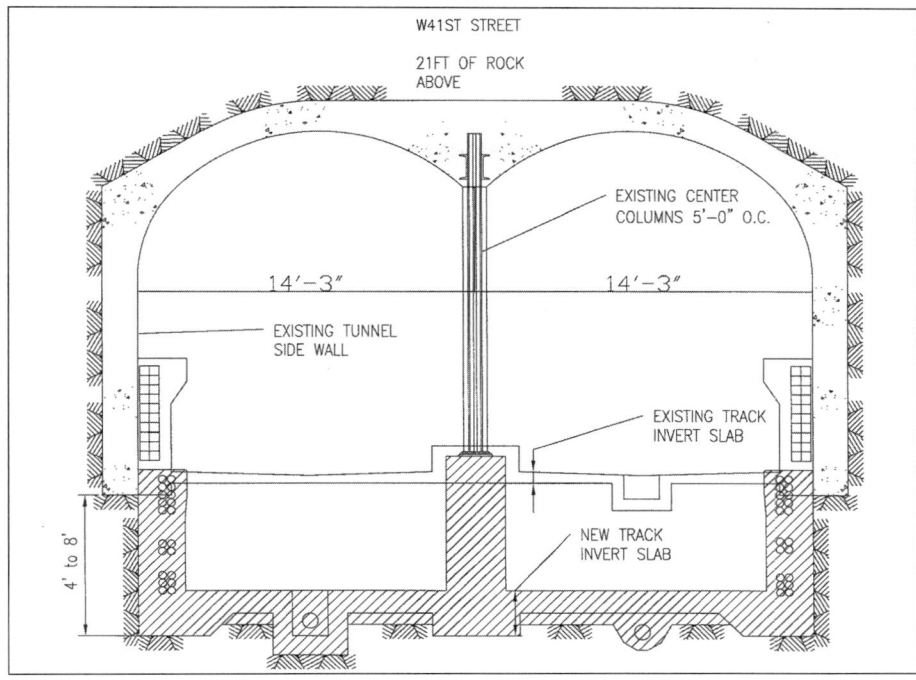

Figure 7. Section of 7 Line tail tracks showing existing and new work

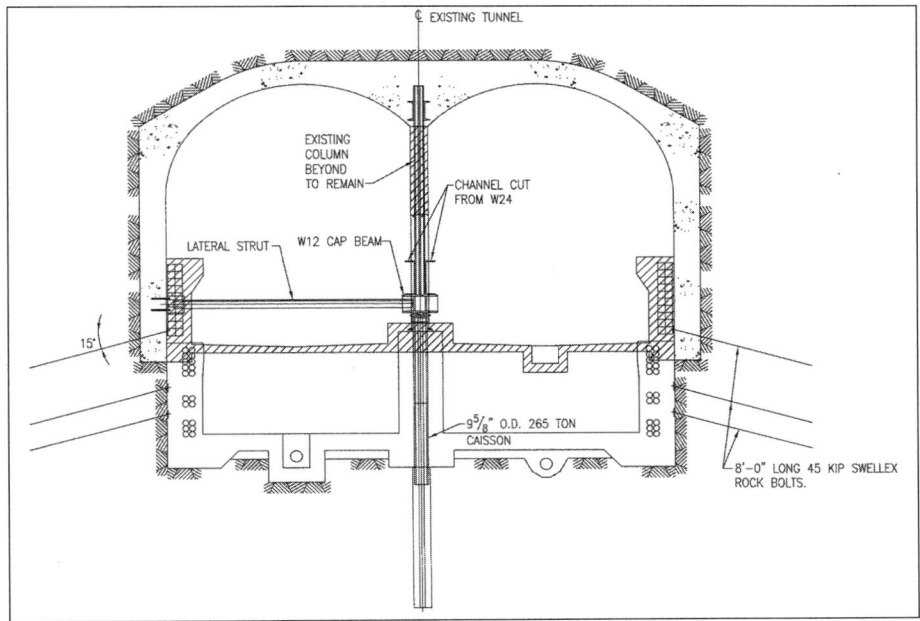

Figure 8. Section of 7 Line tail tracks showing temporary underpinning

Figure 9. Installation of small diameter caissons with Davey DK 515 down-the-hole hammer

Temporary Underpinning Installation

Due to the aforementioned site access restrictions, the rock drill rig for installing the small diameter caissons had to meet three prerequisites:

- Size—7 ft. (2.1meters) or smaller to meet clearance for mobilization train
- Capacity—power to drill 12" (30.5 cm) diameter holes in rock
- Emissions—pneumatic/electric rig (not diesel) due to air quality restrictions in the tunnel work environment

The compact yet powerful Davey Drill DK515 with a 12" (30.5 cm) down the hole hammer fulfilled all of these requirements and had proven to be effective on similar NYCT subway tunnel projects. The compressed air for the rig was supplied from a mobile compressor located on 41st Street and the connecting hose was run through the temporary access shaft. An additional benefit of the DK515 was the ability of the mast to fit in between the tightly spaced center columns (see Figure 9). After the DTH had reached the designed depth, the rock quality was verified and recorded by video camera.

The limited head room above the caisson locations and the material access constraints prevented a continuous pipe from being installed into the drilled holes for the caissons. For this reason as well as difficulties in welding in the tunnel environment, the pipe segments were fabricated in 5 ft. (1.5 m) lengths with threads at each end to enable flush joint splices. Although the threaded pipes resulted in higher initial material costs, the Joint Venture was able to achieve an excellent production rate for the installation of the pipe segments with the assistance of the DK rig. This is significant in New York City where labor rates are considerably higher than the national average.

In order to place the 6,000 psi grout in the caissons, a mobile hydraulic grout plant was located above the tunnel on 41st Street and a slick line was run down the shaft to the tunnel. Although the street location removed the equipment further from the work area, it was better suited than in the tunnel due to potential dust control issues when batching the bagged mix and mobilization difficulties of the equipment and material in the tunnel.

The W24 channel members were fabricated in manageable 18 ft. (5.5 m) lengths and all of the bolted connection points were pre-drilled by the fabricator to expedite the field installation. The underpinning steel was lowered into the tunnel via the access shaft with a 35 ton (31.8 metric ton) Tadano mobile crane and once in the tunnel, the

Figure 10. Installation of underpinning steel on existing center columns

Dockbuilder crew rigged the underpinning steel into position adjacent to the center columns with a Bobcat skid steer and 5 ton (4.5 metric ton) chain falls (see Figure 10). Careful preparation to grind and level of the top of the caissons was critical in order to assure a uniformly distributed load transfer between the underpinning steel and the caissons. After all channel members were bolted in place and the cap beam installed, 4" (10.2 cm) steel wedges were driven into position with hammers and this completed the load transfer from the center columns to the temporary underpinning system.

In conjunction with the underpinning, the Joint Venture designed and installed an elaborate system of instrumentation to monitor the major existing structural elements in the tunnel. Survey information was provided to a fully automated theodolite continuously and the results were posted "real time" on a secure internet web site.

The rig selected to drill the 1 ⅞" (4.76 cm) diameter, 8 ft. (2.43 m) deep holes for the rock bolt support in the tunnel side walls was the Perfora Rock Buggy for its compact size (fit on mobilization train), ability to drill at near horizontal angles and high production rate. The use of Swellex PM 24 rock bolts offered numerous advantages as well including:

- Light weight—able to be handled by craft
- Straightforward installation—hydraulically inflated to full bearing capacity
- Quality assurance—built in to design of bolts

ROCK REMOVAL

As described in earlier sections, the existing 7 Line tunnel in the Times Square work area is situated under approximately 21 ft. (6.40 m) of rock. Following the contract design documents approximately 3,000 CY (2,294 m^3) of rock subgrade were required to be removed in order to lower the invert for the future track elevation. The Joint Venture considered numerous means to remove this rock efficiently however traditional methods such as blasting were ruled out due to the following reasons:

- the close proximity of structural supports for active subway traffic above the work area
- the existing NYCT ventilation system for the 42nd St. Station is open to the demolition work area and potential blast air over pressure and fumes

- existing active electrical and mechanical utilities are exposed in the work area
- mandated off shift blasting times (2 AM–5 AM) within the NYCT structure
- blasting protection would require small mats installed by way of SNATCH blocks and chainfalls since the low tunnel headroom limited the use of heavy equipment

After initial mobilization on the site and probing the rock in different locations in the tunnel, a combination method of hammering the rock in fractured areas in conjunction with drilling and splitting the rock in the competent areas was developed.

The hammer chosen to break up the rock was the remote controlled Brokk 250 demolition rig. The Brokk is an extremely powerful hammer in relation to its compact size and proved to be durable in the tunnel demolition environment. An additional benefit of the Brokk is that it is electrically powered which eliminated emission issues associated with traditional diesel powered excavators.

Due to the limited size of the tunnel and the tight clearance for mobilization trains, hydraulic drill rigs were not an option for the drill and split operation. However the pneumatic driven Ingersoll Rand ECM 350 rock drill was a dependable rig and was able to meet production requirements. The drill hole pattern varied from 8" on center to 12" on center and drill depths ranged from 4 ft.–6 ft. (1.20 m–1.80 m) depending on the condition of the rock.

Following the drilling operation and the creation of an open relief for the rock, a hydraulic splitter which generates a break out force of more than 26,000 psi fractured the rock adjacent to holes. The splitter operates through a single electric/ hydraulic power pack which can power up to 4 breaking cylinders. Each cylinder is equipped with 11 pistons that extend 1" (2.54 cm) when hydraulic pressure is applied. Due to the size of the cylinders 3 ½" (8.90 cm) diameter holes were required to be drilled in advance of the splitting operation. Although the drill and split operation proceeded in a methodical manner, it advanced the rock removal safely and efficiently.

PABT OPEN CUT—SUBWAY INTERFACE

Existing Site Conditions

One of the most challenging areas of engineering and construction on the 7 Line extension project was connecting the new open cut excavation area in the basement of the Port Authority Bus Terminal with the existing lower level of the 42nd St. subway station (see Figure 11). Although only 13 ft. (3.90 m) of rock separates the two work zones, the existing loading condition above the interface presented a complex scenario for the design of the temporary support system. The following loads were factored into the design:

- Wheel loads from the PABT bus traffic
- Dead load from the bus terminal structural east wall
- Dead load from the rock pillar above the interface
- Station load from the overhanging mezzanine level of the 42nd St. station
- Utility loads from major 8th Ave. sewer lines
- Street load from 8th Ave.

Furthermore due to the slenderness of the interface area, conventional arching effects of the rock overburden to assist in distributing loads outside the new tunnel area could not be relied upon.

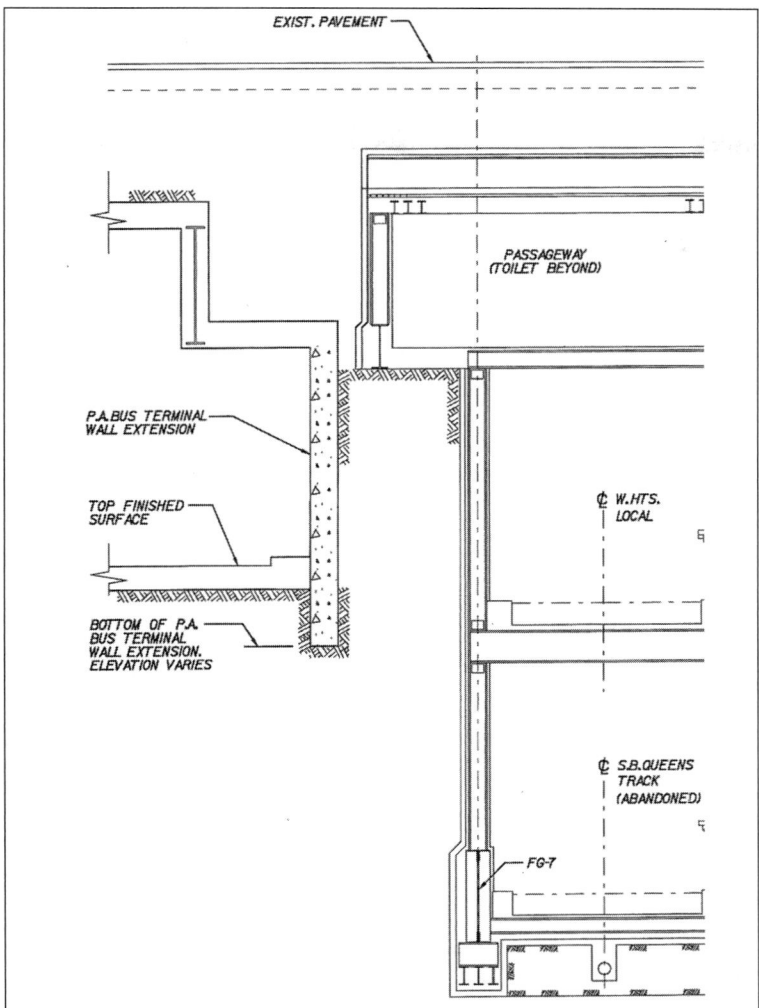

Figure 11. Existing condition in PABT-Subway interface area

Initial research on the project unveiled that both the bus terminal and subway station structures had undergone several renovations and expansions by drill and blast methods in this area since the original construction. This suggested potential over break in the rock and an existing condition scenario more complex than showed on the "as-built" drawings.

Planned Final Condition

The permanent design for the new tunnel section in this area connecting the existing subway structure with the open excavation specified a reinforced concrete structural "box" supplemented with a structural steel frame to assist in the distribution of the previously described existing loads.

One note of interest in the interface area involves the permanent waterproofing system. The new work to the east followed the traditional NYCT method of relieving hydrostatic pressure with weep holes through the track invert slab. However all new 7 Line work to the west of this point was specified to be fully enclosed in a membrane waterproofing system. The permanent waterproofing design therefore played a strong influence on the temporary support system that the Joint Venture planned in this area.

Temporary Underpinning

A temporary support system to facilitate the construction of the new 7 Line in the interface area was designed by the Joint Venture, along with our design consultants, Mueser Rutledge Consulting Engineers. In addition to supporting the existing vertical loads, the design of the system was influenced by the equipment and material which could be mobilized in the tight site conditions.

One of the first activities performed by the Joint Venture after receiving permission to proceed with work in the bus terminal was to acquire more information on the geological conditions in the underground directly above the interface. Horizontal and angled cores were taken through the concrete east wall of the PABT into the rock pillar between the subway and PABT. The cores confirmed that rock was present in this zone however as presumed, the concrete walls were much wider than shown on the as-built drawings due to the previous excavation over break.

Several support systems including methods of bridging as well as dense patterns of rock bolts were considered but the slender rock pillar above the interface presented a particular challenge. Ultimately a system of horizontal spiling was developed which transferred loads onto different structural members depending on the stage of the construction sequence (see Figure 12).

The spiling consisted of twelve 7 ⅝" (19.37 cm) diameter, 80 ksi, structural pipe spaced horizontally across the width of the future tunnel. In order to prevent settlement in the structural components above the spiling, a second 5 ½" (13.97 cm) steel pipe was inserted into the larger pipe to provide additional stiffness. The annular space in the drill hole between the exposed rock and pipe as well as the inside of the pipes was filled with 5,000 psi grout to further mitigate settlement potential.

The limited clearance and the aforementioned mobilization difficulties on the subway side of the interface, dictated the drilling operation for the pipe spiling to take place from the bus terminal side of the work zone. The 40 ft. (12.20 m) wide open excavation under the PABT provided an excellent opportunity for a large hydraulic drill rig to position itself directly adjacent to western rock face of the spiling. One caveat however for the drilling operation was that a temporary decking system was required to be installed over the open excavation in this area so that critical long distance bus traffic could be maintained during 7 Line construction. The decking system presented a problem for the spiling since the eastern most decking girder was located directly in front of the planned spiling holes. However the Joint Venture working in cooperation with the MTACC engineers was permitted to reengineer the permanent design of the interface by lowering the elevation of the tunnel roof without sacrificing the required MTACC clearances. The spiling was lowered as well and thereby able to clear the W40 decking girder.

One limitation of the pipe spiling was the inability of the system to support lateral loads resulting from hydrostatic and surcharge pressure on the PABT east wall. The footing for the east wall was keyed into the rock below the road surface and the distributed lateral loads into the subgrade. However this stabilizing rock subgrade was specified to be removed in conjunction with the open cut excavation. In order to temporarily support the lateral influences from the wall, a long reinforced concrete member spanning over the width of the excavation was installed which acting as a beam in the horizontal direction distributed all lateral loads into the rock subgrade on both sides of

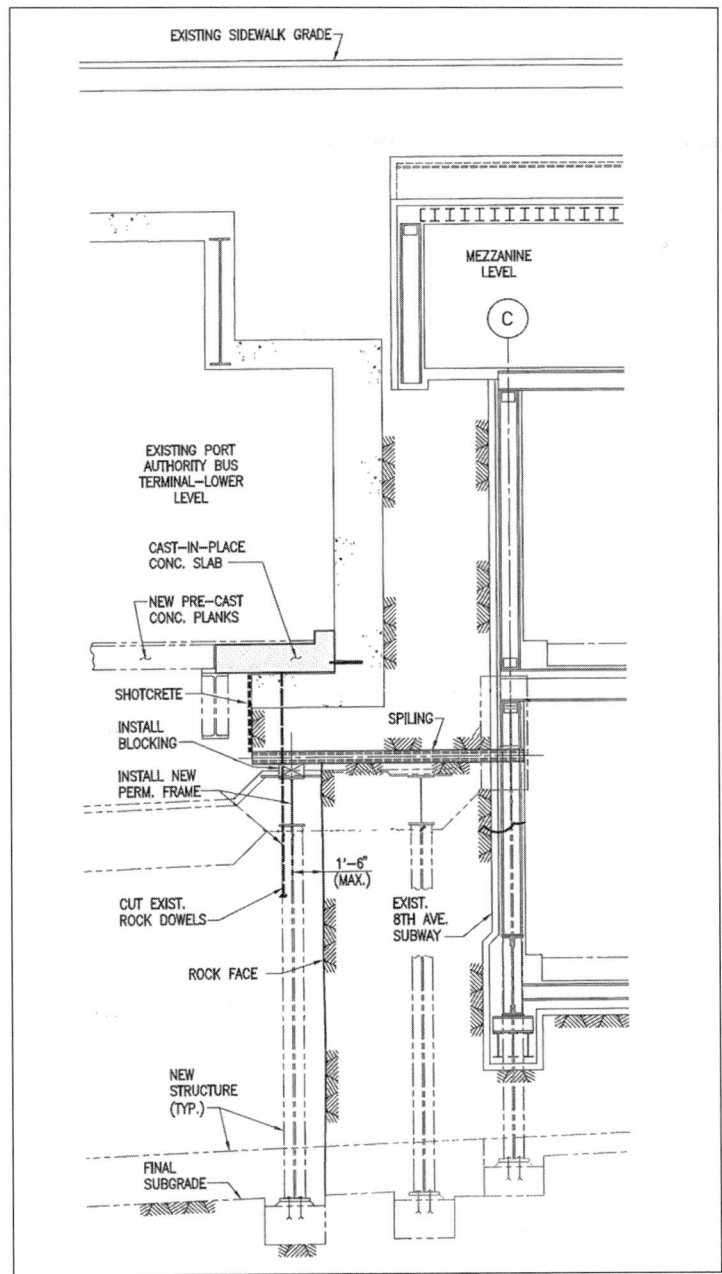

Figure 12. Temporary pipe spiling in the PABT- subway interface area

the cut. This concrete beam also served as a road surface for the heavy bus traffic during the construction phase.

CONCLUSION

The Number 7 Line Extension Project is a technically complex and critical connection of new tunnels to the existing operational 7 Line Subway in the heart of Manhattan. This paper discussed the challenges and unique methods developed to overcome these challenges in the Times Square and PABT work area where the new tunnel connects to the existing subway system.

An intricate system of temporary shoring was developed that could support active subway traffic without interruptions and more importantly from the constructor's perspective was constructible in a productive fashion in the restricted tunnel environment. A system of traditional structural underpinning on caissons was devised in addition to more innovative methods such as large diameter pipe spiling in the area between the bus terminal and the lower level of the subway. In terms of rock removal, a conventional drill and split operation was enhanced with new technology offered by a powerful 26,000 psi hydraulic splitter.

Logistical and mobilization obstacles for construction of the 7 Line in the urban Times Square work environment were overcome through meticulous planning and cooperation of the MTACC. The reliance on less than predictable work trains to supply the site with material and equipment was resolved by the creation of a new shaft directly over the work area offering 24/7 hour access.

Finally this paper also highlights the complexities in planning work with respect to maintaining continuity of work for the multiple trade unions that are involved in this project. Over 16 different unions contribute to the construction of the new tunnel and each trade is a critical piece of the puzzle.

ACKNOWLEDGMENTS

We would like to take this opportunity to acknowledge the organizations that helped S3-II to achieve our accomplishments. First of all we thank the NYCT and MTACC authorities for choosing the joint venture team of S3-II Tunnel Constructors (J.F. Shea, Skanska, Schiavone) to take on this challenge. In addition, we express our gratitude to the Port Authority of NY & NJ who supported the efforts needed to accomplish the 7 Line extension work in the lower level of their bus terminal. Finally we acknowledge the efforts of the owner's construction consultant management team, HLH7, a joint venture team of Hill International Inc., LiRo Engineers, Inc. and HDR/Daniel Frankfurt) and the owner's design consultant, Parson Brinckerhoff.

GIBE II TUNNEL PROJECT—ETHIOPIA: 40 BARS OF MUD ACTING ON THE TBM "SPECIAL DESIGNS AND MEASURES IMPLEMENTED TO FACE ONE OF THE MOST DIFFICULT EVENTS IN THE HISTORY OF TUNNELING"

Antonio De Biase ▪ SELI Spa

Remo Grandori ▪ SELI Spa

Pierfrancesco Bertola ▪ Lombardi SA

Martino Scialpi ▪ SELI Spa

ABSTRACT

In October 2006, a 7m Double Shield TBM, boring through very poor volcanic formations, was pushed back by fluid mud, which presence had not been detected during previous field investigations, due to the high covers characterizing the majority of the tunnel axis.

The present work describes the investigations made to gain better knowledge of the conditions in front and all around the TBM, the special measures implemented, the exploratory and bypass tunnels excavated, the extraordinary occurrences came true during the execution of these tunnels, the drainage campaigns, the design of the chamber excavated to free the TBM, the dismounting and restarting of the TBM.

INTRODUCTION

The Gilgel Gibe II Hydroelectric Power Plant is the second one after Gilgel Gibe I HPP. It is located approximately 240km southwest of Addis Ababa, in Ethiopia, between the Gibe and the Omo Rivers.

The Project layout mainly consists of a concrete lined headrace tunnel (Figure 1) spanning from the Gilgel Gibe River valley to the Omo River valley, a penstock and a power-house located on the right bank of the Omo River.

SALINI Spa and SELI Spa respectively are the Contractor and the Specialized Subcontractor for the excavation of the 25,8km of tunnel.

The tunnel is currently excavated by two Double Shield TBMs: the former started at the Intake Tunnel Portal on August 2005 and stopped on October 2006 at the ch.: 4+196 due to an exceptional geological event. Then the excavation started again on August 2008 along a new tunnel alignment; the latter started from the Outlet Portal on November 2005.

The excavation with both, Intake and Outlet TBMs, is almost completed.

DESIGN CONSIDERATION

The Main data of the water conveyance project are listed in the following Tables 1, 2 and 3.

Table 1. Water conveyance project main data

Purpose	HPP (432MW)
Tunnel length	25,8 km
Internal tunnel diameter	6,3 m
Design discharge	100 m³/s
Design internal pressure	min.2 bar max.7 bar

Table 2. Lining geometrical characteristics

Segment type	n.4 Hexagonal
Segment thickness	25 cm
Segmental lining outside diameter	6.800 mm
Segmental lining final internal	6.300 mm
Segment width	1.600 mm

Table 3. TBM main characteristics

Cutterhead excavation diameter	6.980 mm (new cutters)
Cutterhead structure	No. 5 parts—heavy structure-bolted
Cutters	17 inch DISC (431,8 mm) backloading
Number of cutting discs	44
Maximum recommended average cutterload	222 kN
Maximum recommended cutterhead thrust	222 kN × 39 cutters = 8.659 kN
Cutterhead drive type	Electric Variable Frequency
Number of drives	6 with individual frequency controls
Installed power	6 × 315 = 1890 kW
Rotation speed	0–6,9 RPM
Torque 0–6,9 rpm	2590 kNm
Front shield outer diameter	6.930 mm (Bolted structure)
Gripper and tail shield outer diameter	6.896 mm (Bolted structure)
Main thrust cylinders	No.10 × 1.400 mm stroke
Auxiliary thrust cylinders	No.8 × 2.200 mm stroke

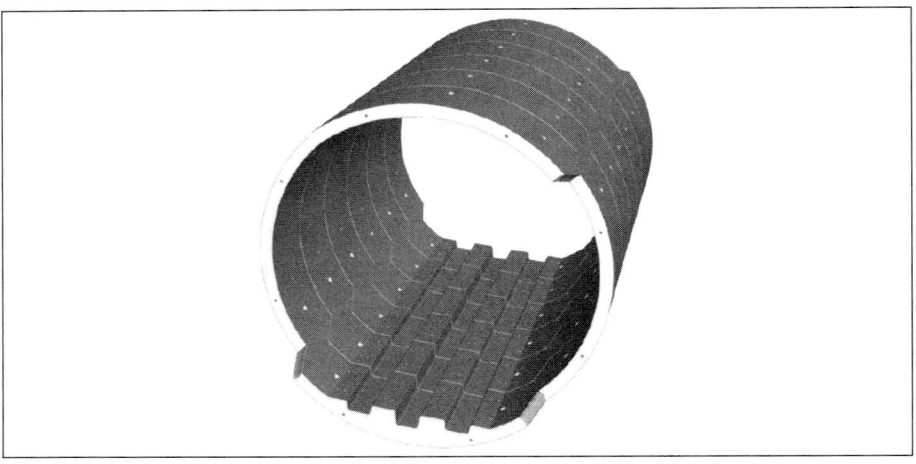

Figure 1. Typical lining section

The initial geological investigations mainly revealed 5 rock formations along the tunnel alignment. In particular, the rock mass mainly consists of tertiary volcanic rocks, rhyolite, trachyte, basalt and some dykes close to the surge shaft. Some fault zones were identified thanks to site analysis performed through aerial photo interpretation; they are mapped along the geological longitudinal profile (Figure 2).

According to the foreseen geological profile, the TBM tunnel, from section ch.: 0+000 to approx ch.: 11+000 (Intake drive), is in the Omo basalt formation, comprising massive, amygdaloidal and scoriaceous layers, inter-bedded with some weak layers of sedimentary tuff, and brick-red paleosoil. The Intake TBM drive encountered the Omo formation since the beginning of the excavation, which was mainly composed of the weak layers of sedimentary tuff and paleosoil formation.

EVENT AT THE CHAINAGE 4+196 FROM INTAKE HEADING

Description of the Event

The Omo rock formation was excavated by the Intake TBM during September and October 2006. The geological and geomechanical characteristics (at the chainage 3800 ÷ 4100) can be resumed as follows:

- Tunnel overburden: about 670m
- Volcanic unit = Omo Vulcanites (Basalt flows with intercalation of Miocene Trachyte layers)
- Rock description:
  - *Weathered brecciated basalt*: greenish grey, dark to light grey and brown incolor, highly joined and fractured, moderately to highly weathered, weak to moderately strong and slickensided basalt with reddish brown, highly weathered to decomposed. Joint planes are filled with calcite and clay infillings
  - *Decomposed Basalt:* reddish brown, highly weathered to decomposed, weak and slickensided basalt, grey to greish grey, smooth to slickensided highly weathered tuff-clay
- Rock Mass Rating between 17–19
- Rock Mass Class V—Very poor
- No ground water was encountered. The rock was only damp
- The temperature was varying from 42 to 53°C

At the end of October 2006, the TBM was stopped (chainage 4100) as consequence of the sudden extrusion and collapse of the tunnel front face against the cutterhead and the front shield.

Two times it was attempted to excavate a bypass tunnel in the tunnel crown in order to free the front shield and cutterhead. In both occurrences the bypass collapsed due to the very high pressure of the rock mass.

The extraordinary ground pressure was mainly coming from the tunnel face (semi-horizontal) and, in particular, from its left side (inclination of 30°). The rock mass moved towards the TBM 40–60mm/hour. No support was able to stop this movement.

The TBM has been pushed back and displaced laterally more than 40cm. All the free spaces were filled by the plastic rock mass till a new static equilibrium was reached. As consequence, severe damages occurred to the shields, the cylinders and the last 7 segment rings installed behind the TBM.

Figures 3, 4 and 5 show the results of the TBM backward displacement (bent shield and cylinders, damaged lining segments).

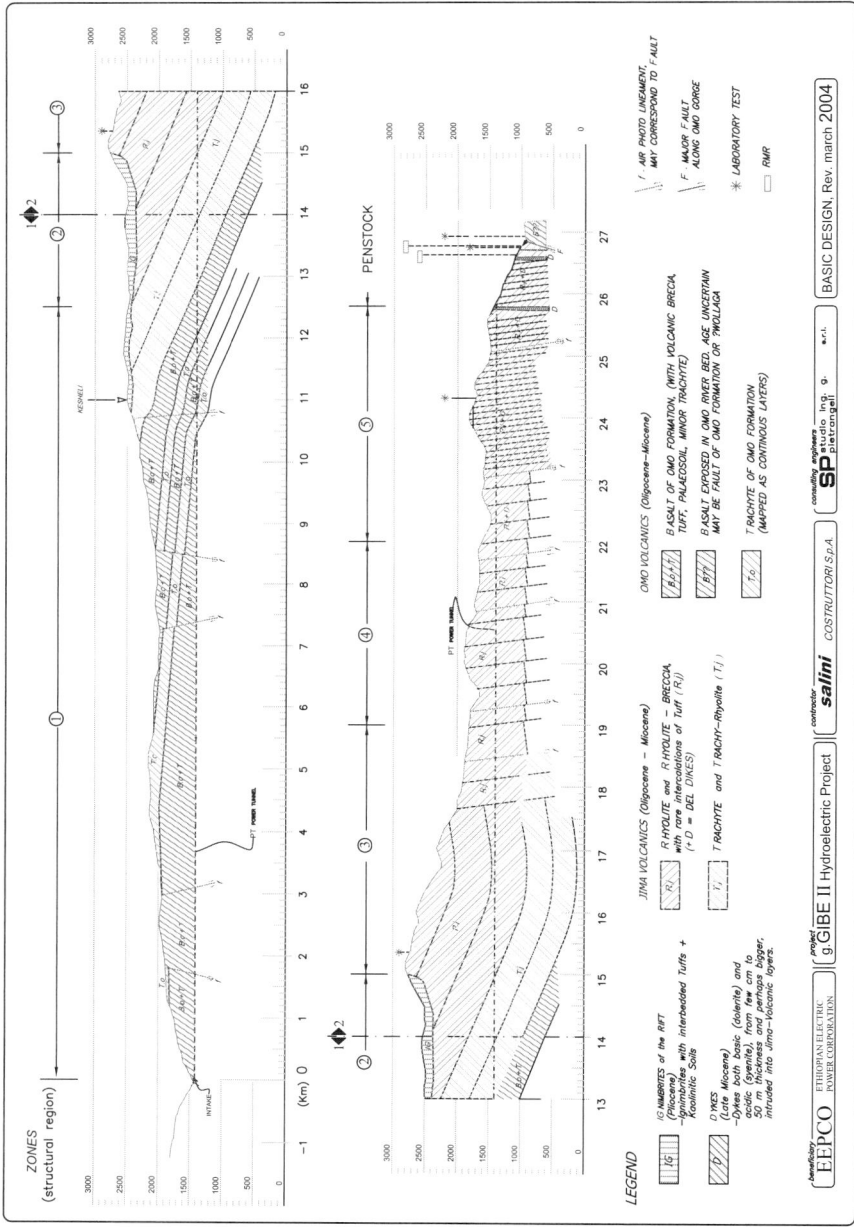

Figure 2. Geological profile

Figure 3. Tail TBM shield damaged and bent

Figure 4. Lining segments moved and severely damaged

Figure 5. Overview of the TBM shield left part and installed lining segments

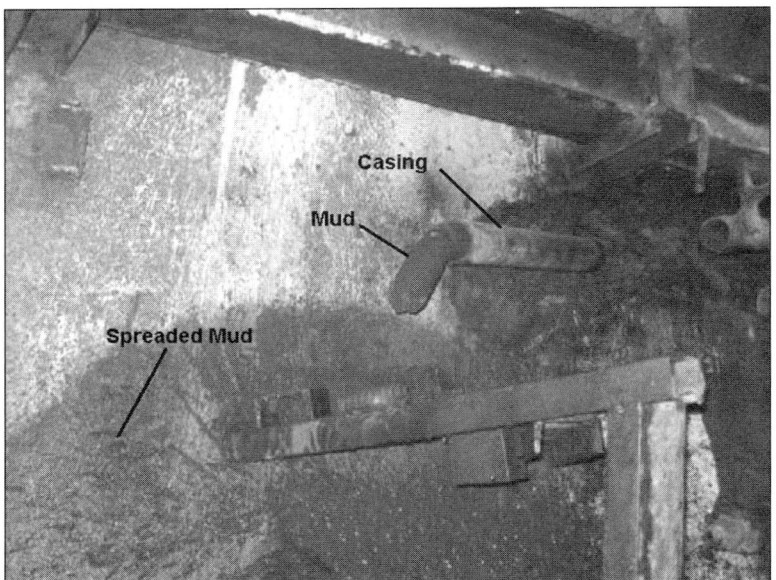

Figure 6. Mud poured out from one of the boreholes

For this particular event, Lombardi Consulting Engineers was involved. Really both SELI and Lombardi Consulting Engineers never encountered before a similar behaviour of the Rock Mass in a TBM tunnel drive.

Some boreholes, drilled through the front shields after the event, were performed with extreme difficulties, sometimes blocked at 17–18m and, in some cases, the rod bars were pushed out by the high pressure of the rock (clay and mud poured out with a temperature of about 40° and the pressure raised to 40 bar—Figure 6).

Nature of the Exceptional Event

The front face instability was not due to gravity failure, generally with brittle behaviour and sudden collapses. In the observed zone the rock mass had an elasto-plastic behaviour with long deformation (squeezing).

The rock behaviour can be assumed as a consequence of the poor properties of the rock mass, characterised by low strength and plastic deformation (viscosity), the effect of water drainage either on the rock conditions (passage from undrain situation at short time to a drained situation at mid-long term) and on its properties, with the decomposition of the sound rock in clay or mud.

Figure 7 shows the rock mass at the front face at the beginning of the decomposition and few hours before the TBM displacing.

EVENT STORYBOARD

On January 2007 was designed and planned the strategy to overcome the event in the shortest time:

- Additional campaign of investigations (boreholes and geophysical survey) in order to characterise geometrically and geo-mechanically the fault zones
- Monitoring plan

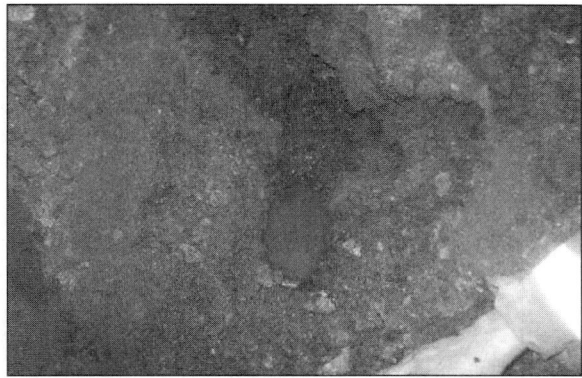

Figure 7. Rock mass at the front face before the TBM displacing

- Rehabilitation works to repair the damaged lining rings
- Construction of a rescue chamber (Back Chamber) surrounding the TBM Shields to repair the TBM and the Back-Up
- Excavation of an Exploratory Adit on the left side of the power tunnel

The new investigations had to be carried out with no effect on the Back Chamber works; both had to be performed by using site materials.

Figure 9 shows the Back Chamber aimed to free, repair and re-align the TBM.

Its design, including the shape, the support structures and the excavation phases, was performed by Lombardi Consulting Engineers and carried out through a back analysis of the event for estimating the geomechanical properties of the rock mass and the loads have been acting on the TBM.

The Back Chamber construction, with the excavation works and the installation of the support structures, developed in four next phases depending on the Chamber part (Figure 8 and 9):

- Top heading
- I bench
- II bench
- Invert

These works were executed with typical mining methods.

Continue monitoring was executed in order to prevent unexpected events and improve the Back Chamber design step by step with the new data and the relative interpretation. Daily contact between SELI's staff on site and Lombardi in Switzerland was maintained. SELI's daily reports were analysed and discussed in Switzerland and, when necessary, new proposal were transmitted directly to the site by e-mail. Continuous contacts with SELI's staff in Rome were also maintained.

At the end of May 2007:

- Almost 60% of the Back Chamber was built
- The left Exploratory Adit reached the fault zone
- About 550m of investigation boreholes were completed

Figure 10 shows the situation at that time.

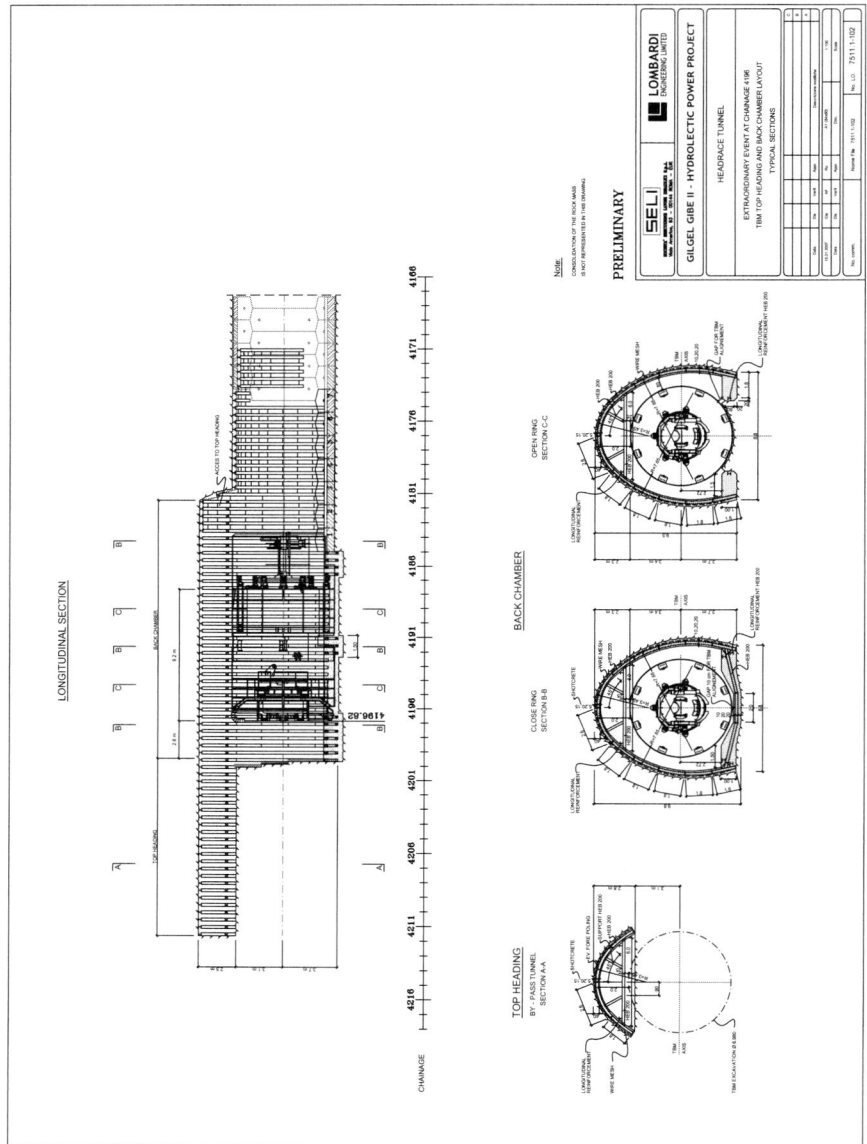

Figure 8. Back chamber layout

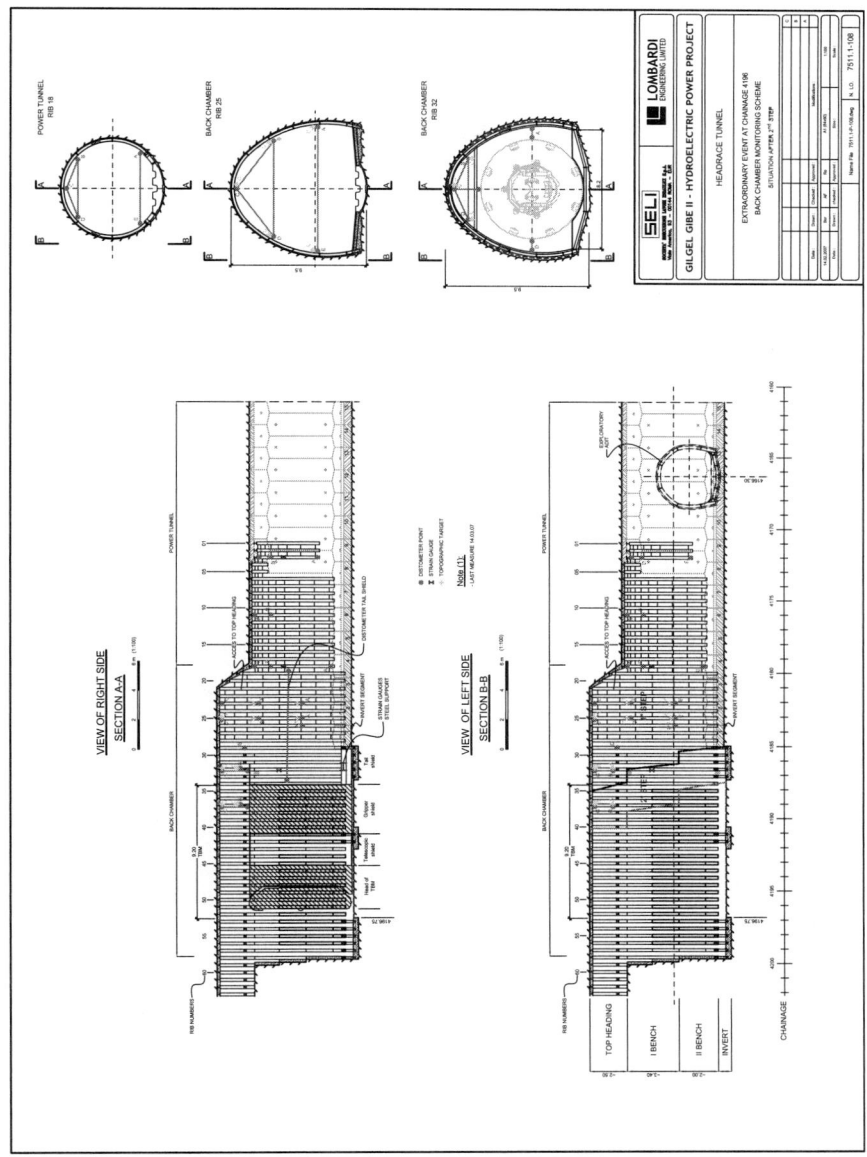

Figure 9. Monitoring stations in the back chamber

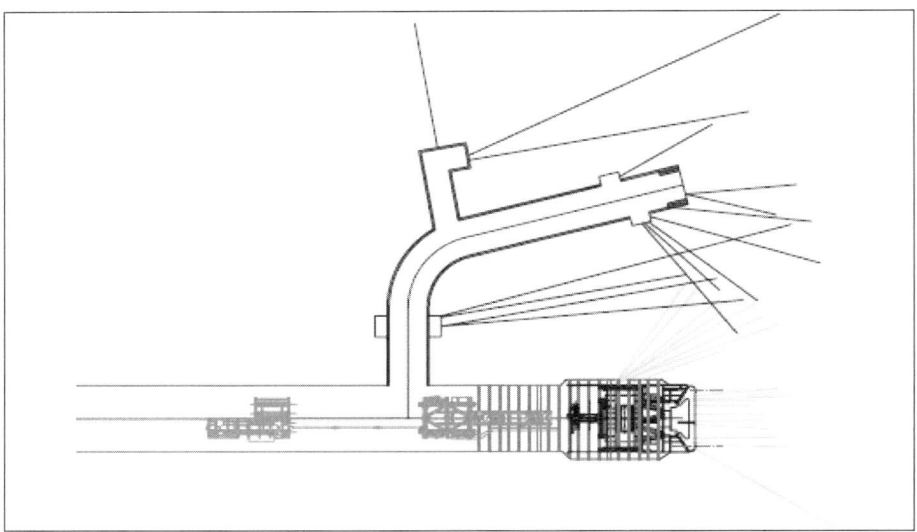

Figure 10. State of the intervention on May 2007 (ch. 4+196)

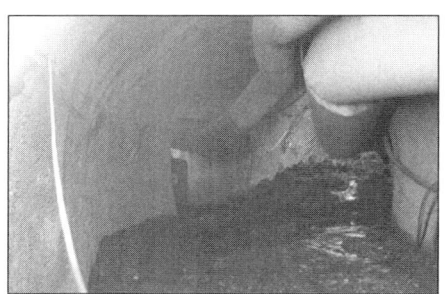

Figure 11. Extruding mud in the Exploratory Adit

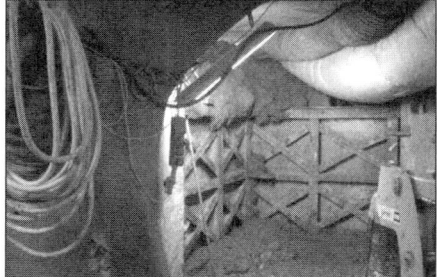

Figure 12. Extruding mud at the Adit Portal

In June 2007 a collapse of the front face started at the left Exploratory Adit, producing three controlled leaching of about 3,500m^3 of mud, which filled the all Adit and about 80m of the power tunnel (Figures from 11 to 16).

At the same time, the monitoring devices (strain gauges applied on the steel structure and distometers) showed an increasing of the loads and displacements in the Back Chamber structure and on the TBM itself (see Figures 17 and 18). In particular, Figures 17 and 18 describe the TBM and the Gripper Shields position monitoring respectively.

A new strategy was then defined:

- The Mud was removed from the power tunnel and partly only from the left Exploratory Adit (a plug wall was erected in the middle of the left Exploratory Adit)
- The Back-Up of the TBM was extracted from the tunnel
- The construction of the Back Chamber was temporary suspended

Figure 13. Extruding mud in the Power Tunnel

Figure 14. Extruding mud filling the Back-Up

Figure 15. Extruding mud in the Power Tunnel crown

Figure 16. Consolidated mud filling the Power Tunnel

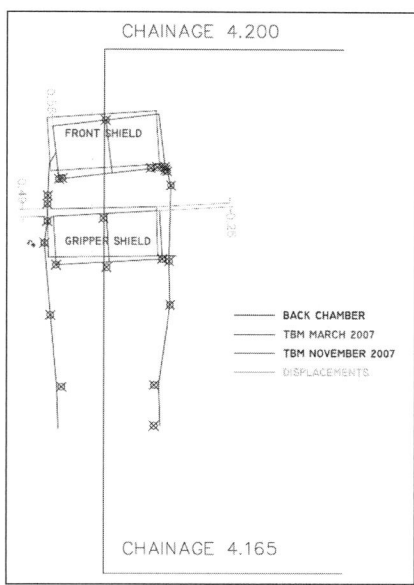

Figure 17. Topographic monitoring of TBM displacements

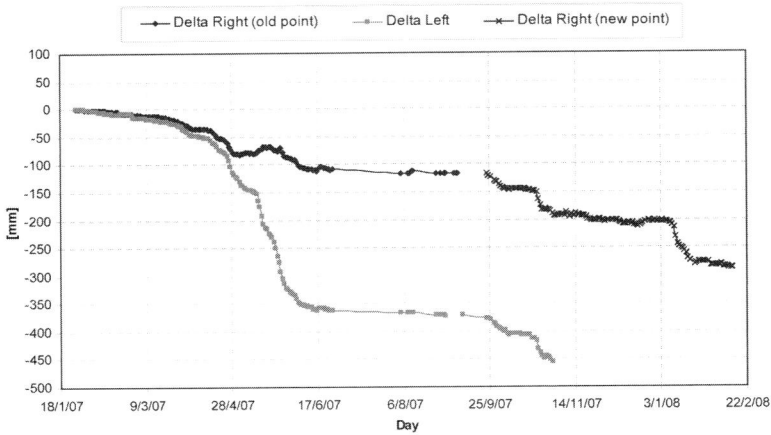

Figure 18. Distometer measurements of Gripper Shield backing

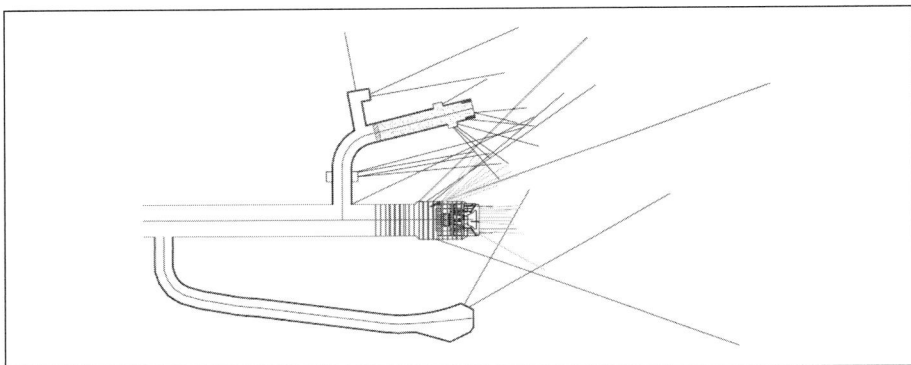

Figure 19. State of the intervention at the end of September 2007

- A more powerful drilling equipment was positioned behind the TBM shields (having removed the Back-Up of the TBM)
- The construction of a new Exploratory Adit on the Right side (40m behind) then started

The state of the works at the end of September 2007 is shown in Figure 19.

New boreholes were made behind the shields by the more powerful drilling equipment, which allowed reaching the basalt formation behind the fault zone. Then the water drainage started, lowering the pressure surrounding the TBM head and the constructing Back Chamber. Consequently, the excavation of the Back Chamber re-started (October 2007).

A new attempt to reach the TBM cutterhead was carried out from the right side Exploratory Adit but the monitoring of loads and deformations suggested renouncing (Figure 20).

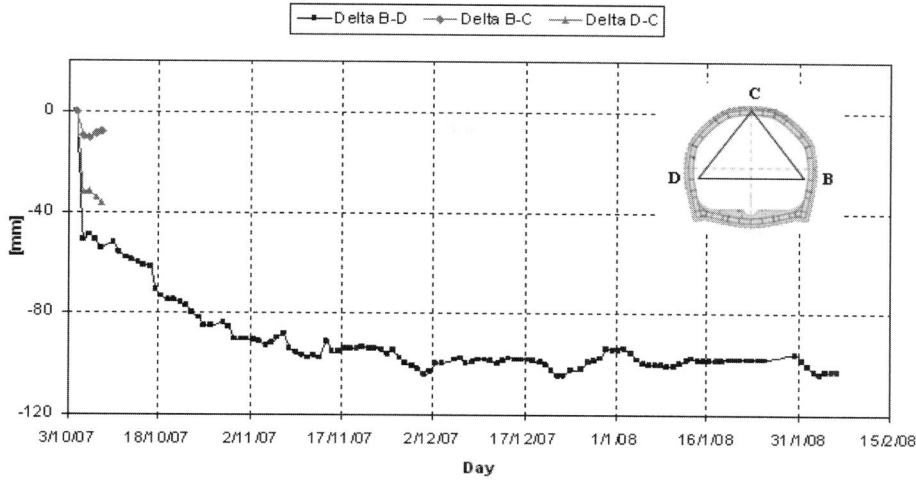

Figure 20. Convergence measurements in the Right Adit (steel rib n°39 close to the TBM Cutterhead)

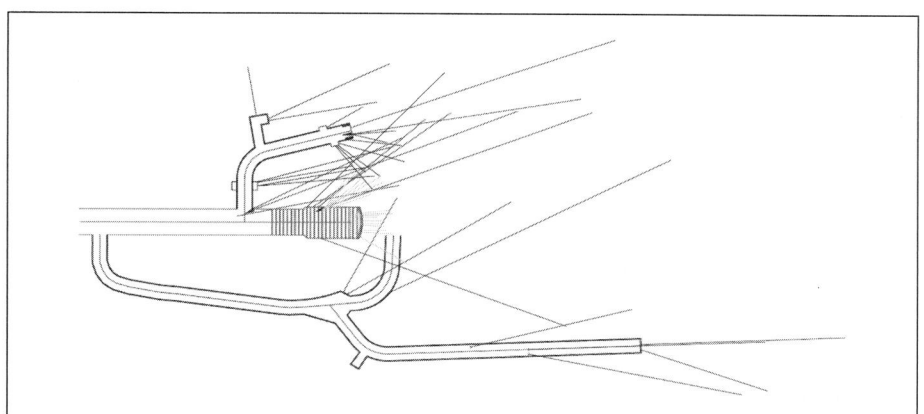

Figure 21. State of the intervention at the end of February 2008

The right Exploratory Adit was then continued along a bifurcation, crossing completely the fault zone and entering into the basalt formation (sound rock). The excavation of the right Exploratory Adit was always anticipated by exploratory and drain boreholes. Finally, the chainage 4+270 was reached (about 70m in front of the TBM).

On December 2007, monitoring measurements have shown a further lowering and next stabilization of the rock stresses, while the right Exploratory Adit was crossing the fault zone. The intervention at the beginning of 2008 is shown in Figure 21.

A further new strategy was considered:

- Completion of the Back Chamber through a concave shot-concrete wall, reinforced with horizontal steel ribs HEB200 (see Figure 22)
- TBM Dismantling inside the Back Chamber and carrying outside

Figure 22.   3D Model of the Rescue Chamber realised to recover the TBM

- Refurbishing and reconstruction of the TBM (in the external yard)
- Construction of a new assembly chamber and TBM launching chamber
- Segmental lining dismantling and casting of a concrete plug in the Power Tunnel
- Resuming of the Intake drive excavation along a new alignment at ch. 3+805

The recovery of the TBM was successfully completed and at the end of February 2008.

Bypass Power Tunnel

The new Power Tunnel alignment was chosen maintaining a minimum distance from the explorative Right Adit, satisfying the hydraulic requirements of the main tunnel and considering the minimum radius of curvature achievable with the TBM (600 m) (Figure 23).

For the Assembly and Launching Chamber was reached a zone of class II (ch. 3760–3805). In this area the main tunnel section has been enlarged by drill and blast method. In the transition area (totally 160m) the segmental lining was removed.

In order to improve the TBM steering, to contain the backfilling volumes and to reinforce the pillar between the two tunnels, in the transition zone the old tunnel has been filled with concrete (Figure 24).

The TBM has been refurbished on site and pre-assembled on the external yard between March and May 2008.

In order to reduce squeezing effects acting around the shields in the rock mass with plastic behaviour, the excavation diameter has been enlarged from 6,980mm to 7,074mm while the Shields diameter has been maintained according to the original TBM SELI design. For this purpose, peripheral cutter housings (from n°37 to n°44) have been moved and re-positioned as shown in Figure 25.

The assembly chamber was completed during the month of May 2008 installing three monorails for trolley and lifting hoists anchored to the crown (Figure 26).

The TBM was re-assembled (Figure 27) and the excavation along the new alignment restarted on August 1st, 2008 as planned.

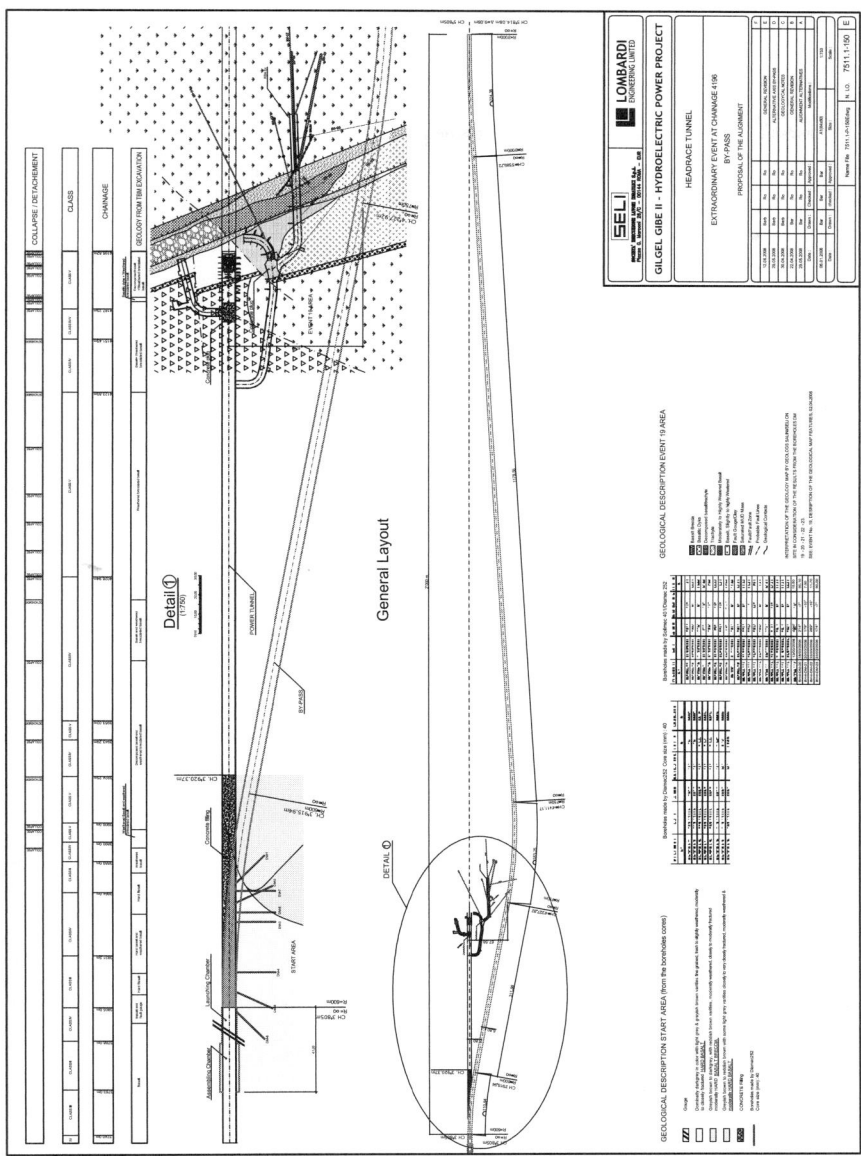

Figure 23. Position of TBM assembly-launching chamber and by-pass general layout

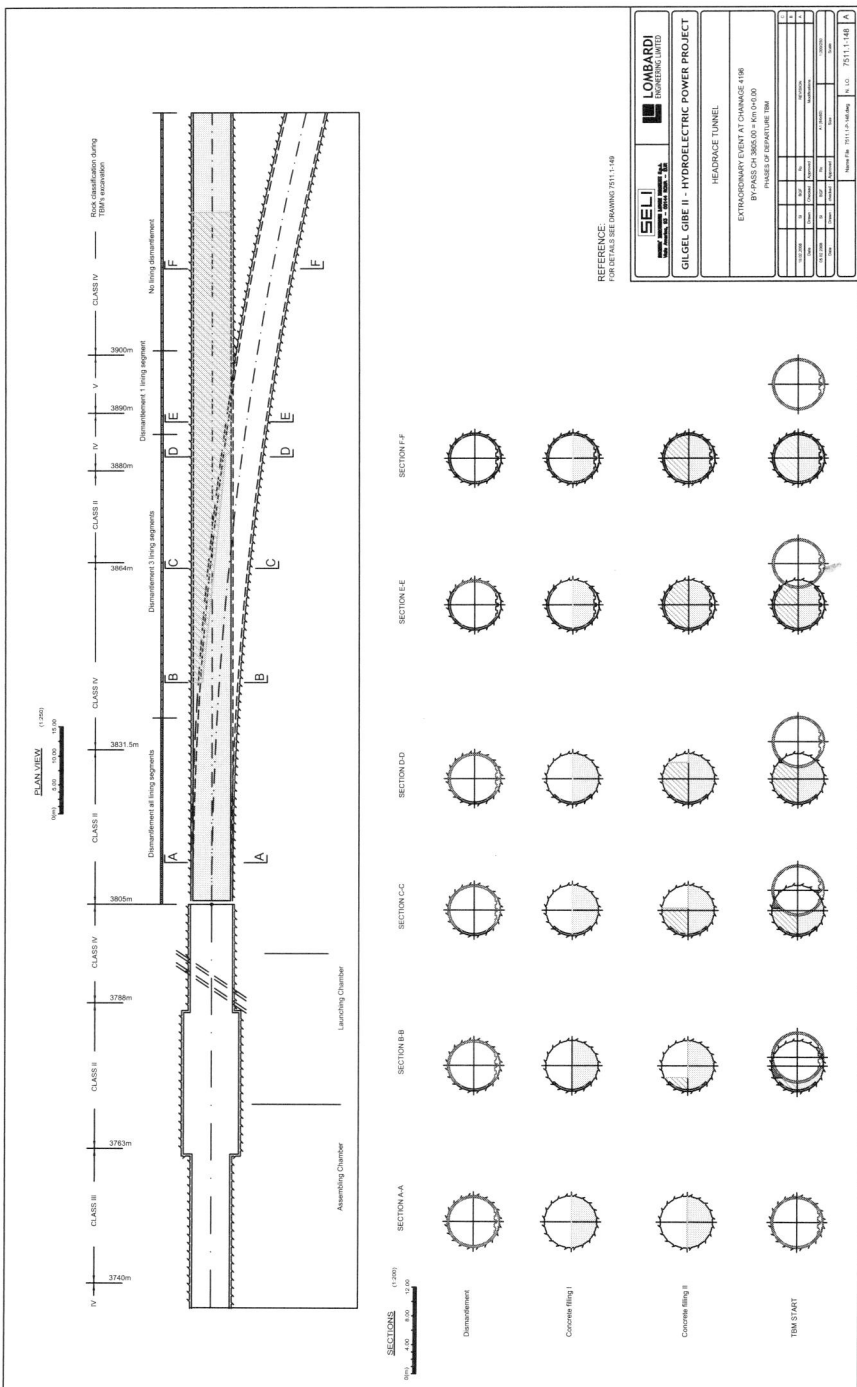

Figure 24. Restarting of excavation and plug in the transition zone TBM repair, modification and assembly

GIBE II TUNNEL PROJECT

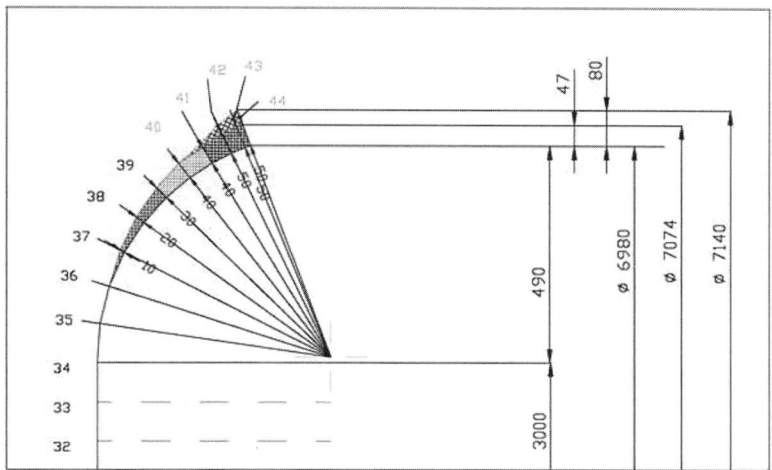

Figure 25. TBM cutting profile after modification

Figure 26. Monorails in the assembly chamber

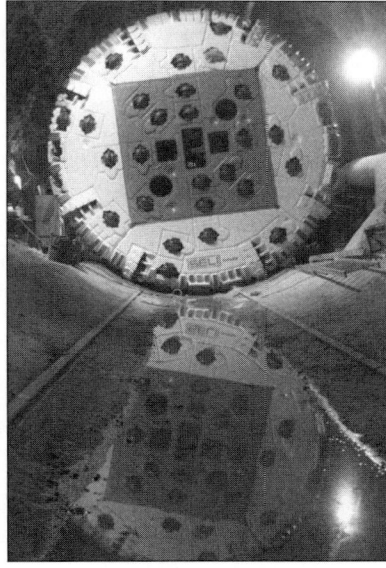

Figure 27. TBM cutterhead

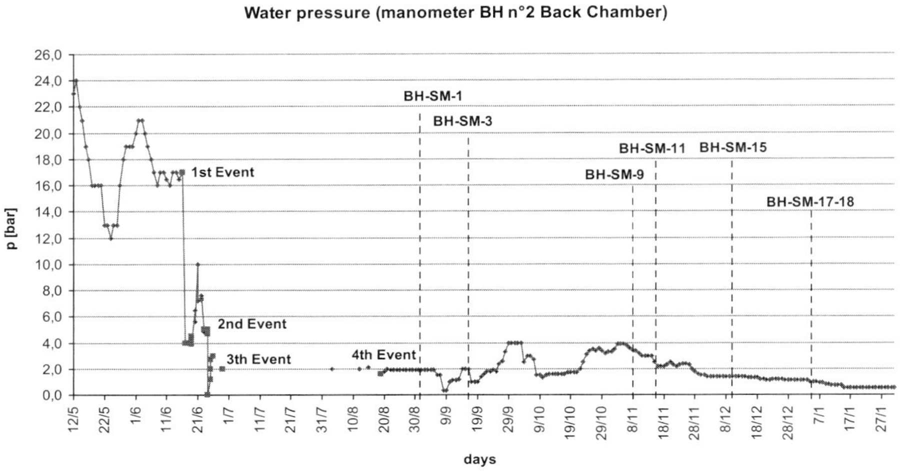

Water pressure (manometer BH n°2 Back Chamber)

Figure 28. Pressure variation measured in the Back Chamber

FINAL CONSIDERATIONS

The success of the intervention was possible only after the releasing of the pressure and the stresses acting in the area and surrounding the TBM.

Some numbers are significant for the Event n°19:

- 3,500m^3 of mud flowed and removed
- 39,600m^3 of water drained during the intervention
- 230m of Exploratory Adit excavated with traditional method
- 1,600m of investigation boreholes

Figure 28 shows the variation of the pressure in the rock mass, measured some meters in front of the TBM. The three events represent the most important leaching of mud in the left Exploratory Adit. Each event was immediately followed by a sudden reduction of the acting pressure and convergence (Figures 29 and 30). Important pressure reductions were also obtained after August 2007, when the new boreholes crossed the fault zone, draining water (Boreholes BH- SM- 1 to BH-SM-18).

The intensive campaign of investigation allowed shedding light the geology in the fault zone (Figure 31).

The TBM crossed the fault zone along the new alignment during the month of October 2008, performing later excellent production. Figure 32 resumes the monthly TBM excavation from August 2008.

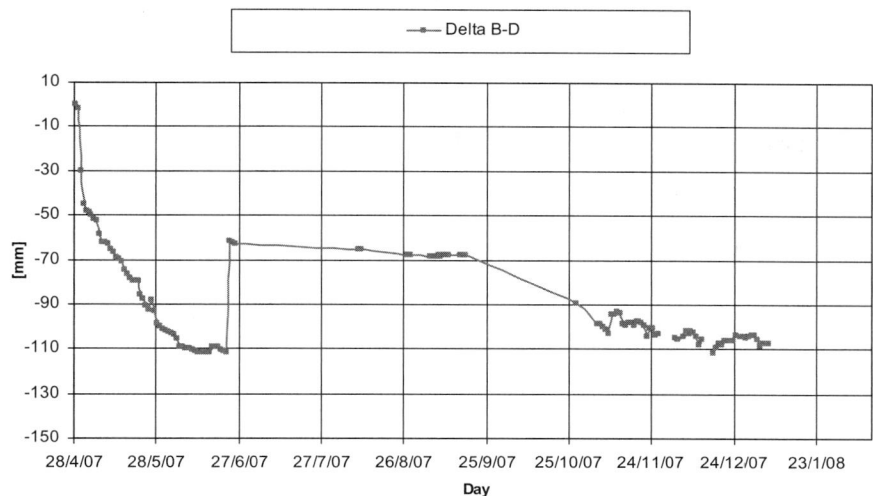

Figure 29. Convergence measurements in the Back Chamber (steel rib n°41)

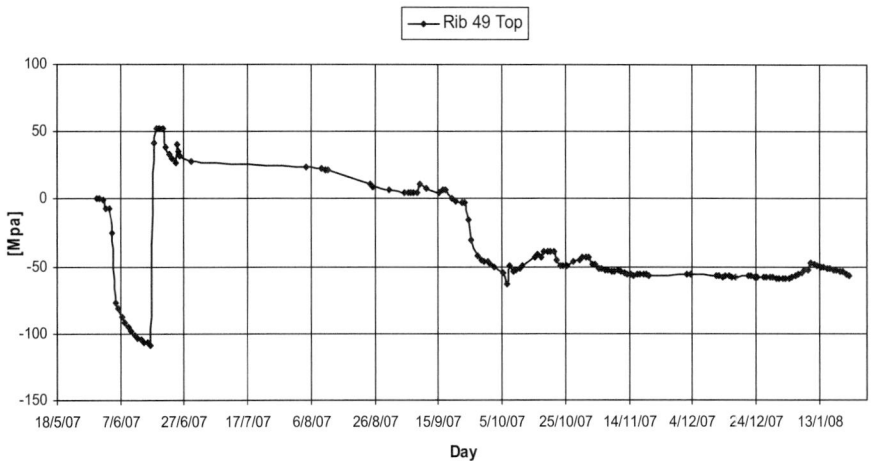

Figure 30. Strain gauges measurements in the Back Chamber (steel rib n°49)

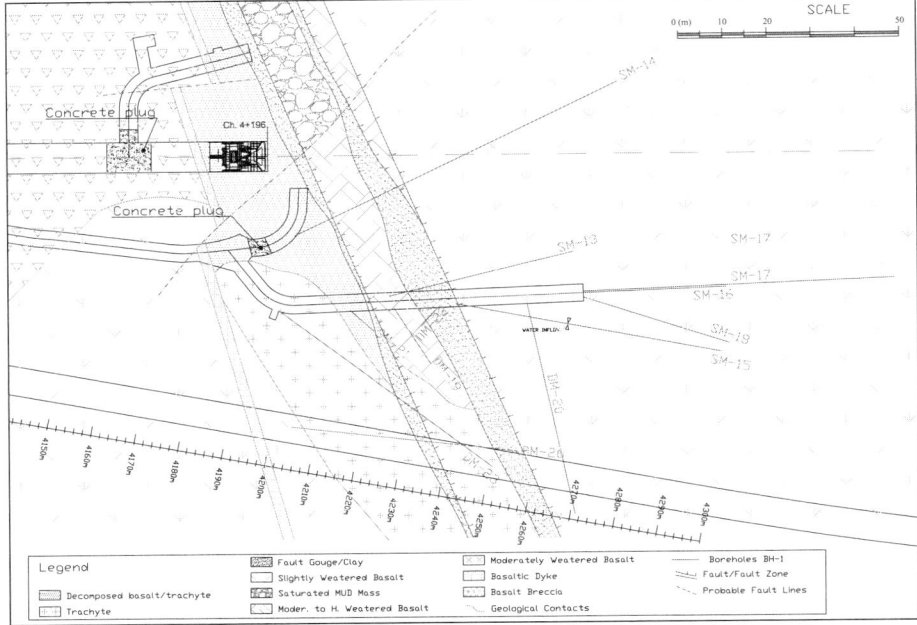

Figure 31. Geological model of the area

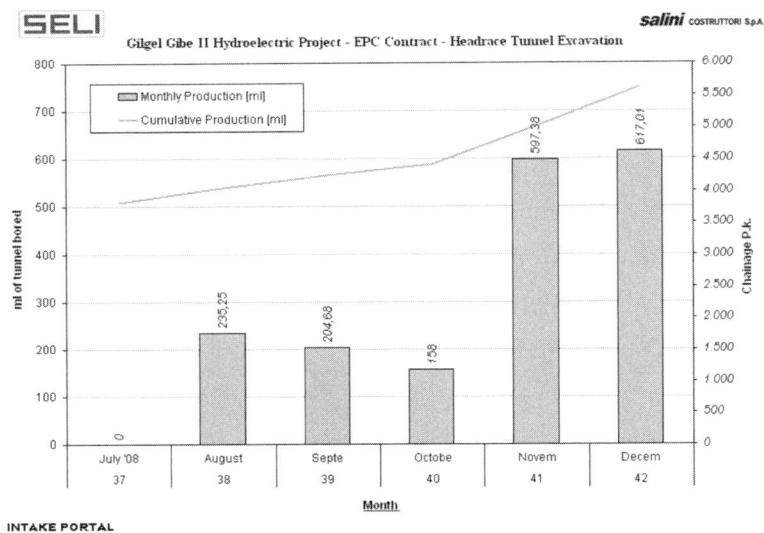

Figure 32. TBM production chart after the excavation resuming

DESIGN AND CONSTRUCTION OF THE LENIHAN DAM OUTLET TUNNEL AND SHAFT

Hemang Desai ▪ Santa Clara Valley Water District

Brett Mainer ▪ Drill Tech Drilling and Shoring

Mike Murray ▪ Hatch Mott MacDonald

Shawn Spreng ▪ Jacobs Associates

ABSTRACT

A damaged outlet pipe beneath Lenihan Dam has restricted reservoir discharges, severely constraining the owner's groundwater recharge program. The solution, a new outlet pipeline through the dam's right abutment, includes an approximately 12.2-m-deep (40-ft-deep) shaft and 618.7-m-long, 4.3-m-diameter (2,030-ft long, 14-ft-diameter) tunnel mined in Franciscan melange composed of hard blocks of rock and soft clayey matrix. Alternating roadheader with drill and blast operations and using steel ribs and shotcrete have proven a flexible combination for handling the variable ground conditions. Excavating the shaft and connecting it to the tunnel require lowering the reservoir for five months, during which time careful construction sequencing and use of a tunnel plug eliminate the risk of an uncontrolled discharge.

INTRODUCTION

Lexington Reservoir is one of ten reservoirs owned and operated by the Santa Clara Valley Water District (District). Lenihan Dam (formerly Lexington Dam) is a rolled earthfill dam that impounds the Lexington Reservoir along Los Gatos Creek. It is located on the east side of State Highway 17, approximately 2.4 km (1.5 mi) south of the town of Los Gatos, California. Construction of Lenihan Dam was completed in 1952. It is approximately 59 m (195 ft) high, and has a crest width of 12.2 m (40 ft) and crest length of 255.7 m (830 ft). The dam, constructed across Los Gatos Creek, impounds water into Lexington Reservoir for the purpose of groundwater recharge.

The Lexington Reservoir drains an area of approximately 95.6 km^2 (36.9 mi^2), and was constructed primarily to capture winter storm runoff and replenish the groundwater basin underlying the Santa Clara Valley. The reservoir's only water source of water is runoff within the watershed. It has a capacity of approximately 23,490,000 m^3 (19,044 acre-ft) and a surface area of approximately 1.9 km^2 (475 acres). Under normal operation, the Lexington Reservoir typically fills during the winter and early spring, and is lowered by late fall. The District tries to maintain its reservoir level as high as possible during the summer months to allow for recreational uses, provided that groundwater recharge derived from reservoir releases is not adversely affected.

The Lenihan Dam low level outlet pipe was originally installed in 1952 and is approximately 479 m (1,570 ft) long. The pipeline extends from a gated intake structure upstream of the dam to a free discharge structure near its downstream toe. The outlet pipe is composed of 410.2 m (1,346 ft) of 127 cm (50 in.) diameter steel pipe encased in 30.4-cm (12-in.) reinforced concrete, and 68.3 m (224 ft) of 122-cm-diameter

(48-in.-diameter) reinforced concrete pipe, also encased in concrete. The existing outlet works at Lenihan Dam has a history of problems. On several occasions, sections of the steel outlet pipe liner have buckled. Emergency repair work on the outlet pipe was recently completed in March 1999. The old outlet works will be replaced by new outlet works under the current Lenihan Dam Outlet Modification project. The new facilities include an approximately 128-m-long (420-ft-long) inclined 137-cm-diameter (54-in.-diameter) pipeline with multiple intake ports on the bank of the reservoir. The pipeline drops down a 12.2-m-deep (40-ft-deep intake) shaft and is fixed to concrete pedestals within a new 609.6-m-long (2,000-ft-long) access tunnel located through the right abutment of the dam. The pipeline terminates at an outlet structure with energy dissipation chambers along Los Gatos Creek at the downstream tunnel portal. The intake profile is shown in Figure 1.

DESIGN

Geology

The project site is located in the Santa Cruz Mountains within the California Coast Ranges geomorphic province. The San Andreas fault bisects the Santa Cruz Mountains and is located approximately 3 km (1.9 mi) southwest of the site. Northeast of the San Andreas fault, and underlying the project area, basement rocks of the Coast Ranges consist of the Franciscan Complex, which includes heterogeneous melange bodies of marine sedimentary and volcanic rocks of Jurassic to Cretaceous age (Geomatrix Consultants, 2006). Franciscan lithologies consist primarily of greywacke sandstone and interbedded shale, with lesser amounts of limestone, radiolarian chert, pillow basalt, meta-basalt or greenstone, and associated tabular intrusive bodies of serpentinite (Geomatrix Consultants, 2006).

The tunnel and shaft are located entirely within Franciscan melange rocks. The Franciscan melange is characterized by hard and strong blocks of sandstone, serpentinite, shale and greenstone, embedded in a matrix of crushed and sheared shale and serpentinite. The blocks at the project site range from sand sized particles to masses over one hundred feet in greatest dimension, while the matrix tends to be severely weathered and often clayey (Spreng et al. 2008).

Franciscan melange rocks are typically complexly deformed and intermixed due to faulting, fracturing, shearing, and crushing associated with ancient tectonic processes. The bedding, foliation and structures observed in blocks of sandstone and shale in the melange are not generally representative of the overall structure of the melange rock mass because the individual rock blocks and the surrounding matrix have been subject to random and chaotic movements during the evolution of the Franciscan. Blocks and also the rock matrix are very closely fractured, and joints/fractures are generally randomly oriented, although systematic joints do occur locally. Overall, the matrices often exhibit foliations that appear to wrap around the melange blocks with sheared contacts.

Ground Conditions

Based on the assessment of the ground conditions, the tunnel was anticipated to encounter zones comprised primarily of friable to very strong rock (blocks), zones comprised primarily of crushed rock and soil (matrix), and mixed face conditions where various properties of blocks and matrix are encountered in any given location. The unconfined compressive strength of the intact rock can range up to 241.3 MPa (35,000 psi) for greywacke and 275.8 MPa (40,000 psi) for greenstone although strengths of the fractured rock mass are generally much lower. Block shapes range from generally ellipsoidal to irregular and appear to follow a 2-to-1 ratio of major to minor axes.

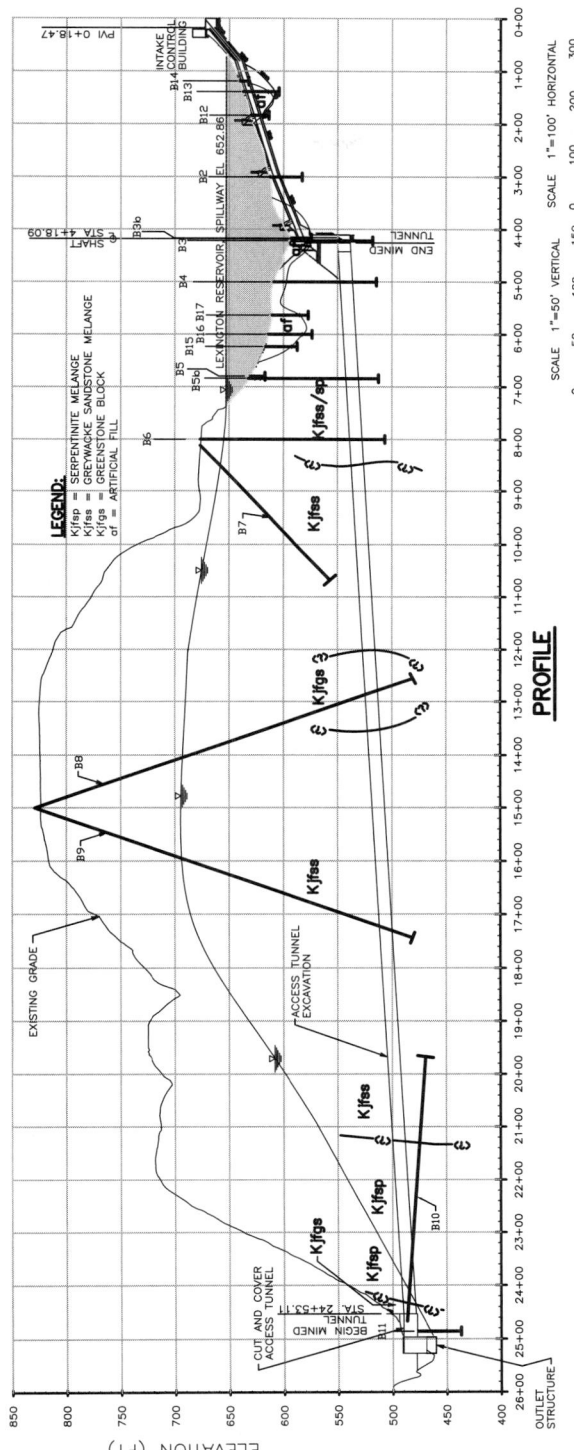

Figure 1. Intake profile

The irregularity and wide variation in rock type and block size has resulted in a material ranging from dense and hard soil, to rock that is friable to very strong. The main consequence of this geologic mixture, especially in regard to the prediction of ground conditions for tunneling, is that the melange typically lacks spatial continuity and therefore exhibits frequent and abrupt variations in geomechanical characteristics, which makes characterizing the ground difficult. Both Bieniawski's Rock Mass Rating (RMR) system (Bieniawski 1989) and Barton's Tunneling Quality Index (Q) (Barton et al. 1974) rely on the rock mass's Rock Quality Designation (RQD) (Deere 1989) as a primary index for classification. Because discontinuities are ubiquitous in the melange, RQD is typically zero and therefore does not allow sufficient differentiation of the rock mass properties. However, the limited RQD combined with a qualitative assessment of the rock mass did allow estimates to be made of the percentage of ground that could be classified by the Terzaghi classifications: "very blocky and seamy," "crushed," and "squeezing" ground. Table 1 provides a modified version of the Terzaghi classification in Proctor and White (1968), along with a summary of the percentages of ground conditions encountered in the geotechnical borings. Although the distribution of melange materials is chaotic, it was assumed that the amounts of materials (blocks and matrices of the various rock types) encountered during construction would be approximately proportional to the percentage of materials logged from the boreholes.

TUNNEL INITIAL SUPPORT

The melange materials were classified on the basis of the Terzaghi ground conditions as shown in Table 1. Three ground types were developed: Type 1, Type 2, and Type 3, roughly corresponding to blocky and seamy, crushed, and squeezing ground, respectively. However, the ground types typically overlapped so that Type 1 ground, for example, might include blocky and seamy ground and a certain percentage of crushed ground. Factors such as overburden pressure, anticipated rock mass permeability, lithology, and average strength along the tunnel alignment affected the distribution of ground types. As an example, squeezing ground was classified as Type 3 only when the overburden pressures exceeded about 1.5 times the average strength of the rock mass.

Because the ground conditions at the Lenihan Tunnel were expected to vary widely, the initial support system was designed to be flexible and adaptable. This was achieved by developing a basic initial support design that could be augmented without making substantial changes to the geometrical cross section of the excavation or creating a need to remove or change any structural components if loads/deformations were found to be higher than anticipated at the beginning of excavation.

Ground support types correspond directly with the ground types, thus three types of initial rock support were designed for the variable rock conditions. Type 1 consists of blocked steel ribs spaced at 1.2 m (4 ft) on-center, with up to 7.6 cm (3 in.) of steel-fiber-reinforced shotcrete lagging applied as needed to stabilize the excavation surface and control localized instabilities. Type 2 support consists of steel ribs spaced at 1.2 m (4 ft) on-center with 15.2 cm (6 in.) of steel-fiber-reinforced shotcrete lagging. Type 3 support consists of steel ribs spaced at 1.2 m (4 ft) on-center with 22.9 cm (9 in.) of steel-fiber-reinforced shotcrete lagging. Type 3 could be split-spaced with jump sets at 0.6 m (2 ft) if the ground loading was expected to be particularly heavy. Typical initial support configuration is shown in Figure 2.

Both Type 2 and Type 3 supports require the use of a reinforced concrete invert slab to act as a strut to carry side loads and prevent heave of the invert. In order to maintain a consistent excavation cross section throughout the support types, a 30.5-cm (12-in.) slab thickened to 70 cm (24 in.) at the tunnel centerline was used for all support types. Although the invert slab is thicker than necessary for Type 1 and Type 2 initial

Table 1. Ground conditions and rock loads

Ground Condition	Definition	Percentage of Core	Terzaghi Rock Load Classification[1]	
			Vertical Rock Load[2]	Side Pressure
Blocky and Seamy	Rock mass mainly consisting of imperfectly interlocked, angular to subangular blocks or fragments of rock. Rock mass contains occasional seams of silty, sandy clay and clayey sand. Individual blocks or fragments of rock are chemically intact or nearly intact are typically smaller than one foot. Where individual blocks are larger than one foot, the ground is termed moderately blocky and seamy.	22%	(0.35 to 1.1)C	Little or no side pressure
Crushed	Rock mass consisting of poorly interlocked, heavily broken rock fragments that are chemically intact or nearly intact and gravel sized or smaller. Rock mass is frequently interspersed with zones and seams of silty, sandy clay and clayey sand.	39%	1.1 C	Considerable side pressure
Squeezing	Soil-like ground that is predominantly fine-grained, consisting mainly of clayey, silty, and sandy gouge. Non-blocky due to soil-like consistency and closely-spaced shears and foliation. Contains gravel-sized or smaller rock fragments.	36%	(1.1 to 2.1)C	Heavy side pressure

1. From Terzaghi (Proctor and White, Table 3, 1968).
2. C = B+H, where B = tunnel width, and H = tunnel height. Formulas are valid for any consistent set of units. Multiply by unit weight to get load per unit area (pressure).

support, the added thickness is not wasted. High hydrostatic groundwater loading on the final lining necessitates the use of a thick final concrete invert. The reinforced concrete invert used for initial support was designed to be incorporated into the final lining to provide permanent ground support.

SHAFT AND TUNNEL CONNECTION DESIGN

An approximately 12.2-m-deep (40-foot-deep), 4.9-m-diameter (16-ft-diameter) vertical shaft connects the sloping intake structure to the mined access tunnel. Shaft-site location was determined by several factors: planned reservoir operating levels during construction period, local topography, and required elevation for the lowest intake port. The circular shaft configuration (Figure 3) includes a rectangular excavation at the outboard edge in order to construct the lowest intake ports, potentially below the reservoir water level. An approximately 3.7-m. (12-ft) section of tunnel connects the shaft to the access tunnel. This section is backfilled with concrete and serves as a plug between the active reservoir and the access tunnel.

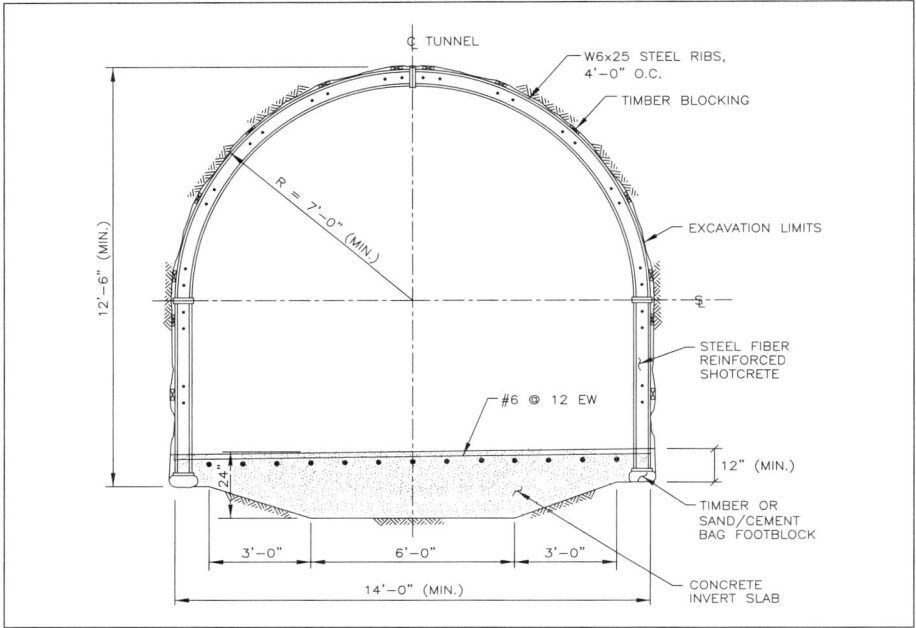

Figure 2. Typical initial support

The shaft and tunnel connection was designed to be constructed independently of the remainder of the access tunnel. The shaft and connector tunnel were anticipated to be constructed and backfilled prior to completion of the access tunnel excavation in order to avoid the possibility of inundation from the reservoir into the tunnel. The access tunnel could have been excavated just short of the shaft, allowing the tunnel plug to be constructed from the tunnel side.

TEAM ORGANIZATION

Field trailers for the general contractor, subcontractor, and construction management staff were located in a public car park at the dam. The District's Project Manager was located in the field trailer along with the construction management staff from Hatch Mott MacDonald (HMM) and design support staff from Jacobs Associates (JA). The convenient location helped encourage daily communications among the field staff. Project partnering was written into the contract and adopted by the parties.

Tunnel excavation and shaft construction were undertaken by Drill Tech Drilling and Shoring (DTDS) with Flatiron West (FCI), formerly FCI, providing logistical support as well as muck disposal. DTDS provided a full-time project engineer along with experienced superintendents and tunnel walkers. FCI constructed the intake structure and the final tunnel concrete lining. At the time of writing, preparations are in progress for the installation of the 20.3-cm (8-in.) low-flow pipe and 137-cm (54-in.) outfall pipe in the tunnel.

HMM provided experienced tunnel staff, including inspectors, to cover the continuous shift working. FCI was responsible for quality control inspections and testing, and HMM was responsible for quality assurance inspections and testing. JA provided a full-time field engineer who carried out the daily field mapping, confirmed the tunnel

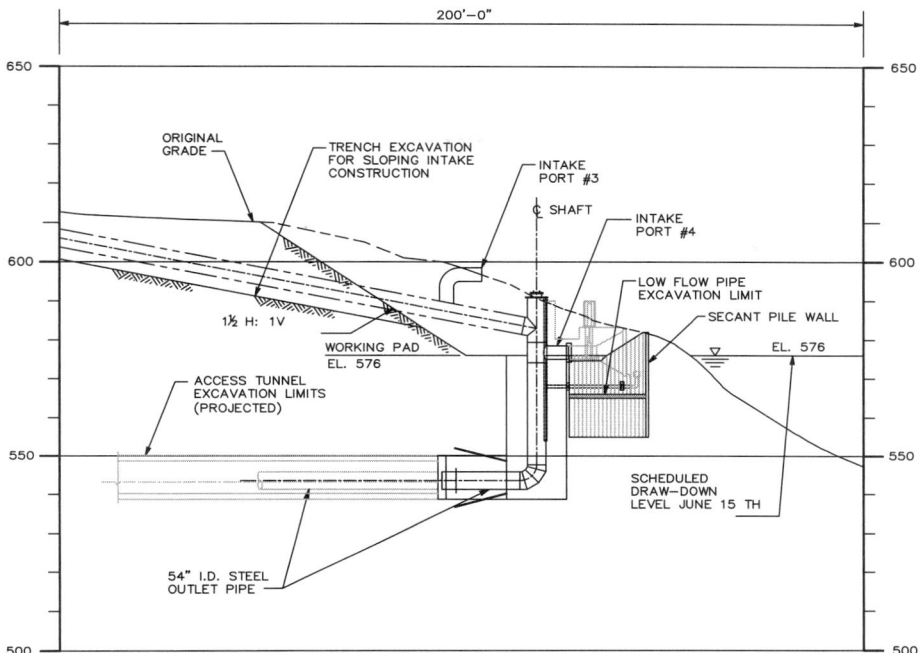

Figure 3. Lenihan shaft configuration

support class, and reviewed submittals and Requests for Information. The tunnel support class determination was confirmed using Required Excavation Support Sheets countersigned by the parties following face mapping and occasional joint inspections at the tunnel face.

The contract was set up so that unresolved disputes between the parties would be referred to the Disputes Review Board (DRB). The DRB meets quarterly for status updates, and only on one occasion has required an informal hearing.

The Division of Safety of Dams (DSOD) has been involved with the project since the design stage. During construction, regular inspections took place by the DSOD field engineer, who reviewed the construction activities against the approved design. Close coordination with DSOD was necessary to ensure that DSOD received three days' notice to attend field inspections.

CONSTRUCTION

Portal Development

A primary challenge to constructing the Lenihan Dam Outlet Tunnel was the access to the tunnel portal, which was located in a steep ravine adjacent to the creek channel at the downstream base of the dam. Prior to construction, the available level area adjacent to the tunnel portal was only suitable for parking pick-up trucks in a single file line. There was no available room for equipment and material laydown, let alone muck transfer areas, therefore, the first order of work on the job was to create as much usable space as possible.

Figure 4. Development of tunnel portal

Figure 5. Portal area during mining activities

The contractor designed a temporary soil nail and shotcrete shoring wall to support the box-like excavation that was made into the hillside to create the tunnel portal. An anchored steel post and chain link rockfall catchment fence was installed on top of the soil nail wall to protect the portal area from loose material on the steep talus area above the portal. Figure 4 shows a photograph of the tunnel portal during construction. As the portal was being excavated, the existing outlet pipe at the base of the dam was extended past the tunnel portal and fill was placed over the creek bed until the ground surface was level from the tunnel portal to the opposite side of the ravine.

Portal preparation took approximately six weeks. At the end of this time, approximately 1,012 m^2 (0.25 acre) of level space had been created adjacent to the tunnel portal where air compressors, generators, equipment, con-ex boxes, a water treatment facility, materials, and tunnel muck had to be placed. Access to this area was via a 20% grade road down the face of the dam that included a tight switchback turn. Above the dam, more than an acre of laydown space was available. These access constraints meant that only a small quantity of the materials required for the project could be stored at the portal. Typically, only several days' worth of fan line, steel sets, and premixed bags of shotcrete could be staged at the tunnel, and only a day's worth of tunnel muck could be stockpiled near the portal (Figure 5). Therefore, for the duration of the tunneling, there was a relatively constant stream of material being ferried down the steep hill,

Figure 6. Looking downstream around the first turn of the tunnel

with muck being transported up the hill to the staging area above the dam for offsite disposal. To maximize the space available, the water treatment tanks were placed on top of the con-ex boxes used for tool storage and the shotcrete bags were stacked four-high inside a shed to protect them from rain.

Tunnel Excavation

Tunnel excavation began on November 7, 2007, and was performed using both roadheader and drill and blast methods. An AM50 roadheader was selected to excavate the softer portions of the tunnel. Due to the potential for naturally occurring asbestos and silica dust, the roadheader excavation crew wore half-face respirators and the heading was kept damp by a constant spray of water from nozzles mounted near the cutting head on the roadheader. Ventilation was maintained in the tunnel with double 75 HP Jet Air fans extracting air through a 66-cm. (26-in.) fanline fabricated by Tumsco, Co. Airflow in the tunnel was maintained between 45.7 and 76.2 linear m/min (150 and 250 linear ft/min). No naturally occurring toxic or explosive gases were encountered.

A custom-built, two boom drill jumbo mounted on a low-profile forklift was used to drill blast holes, spiling holes, probe holes, and pre-excavation grout holes. This could be done without removing the roadheader from the tunnel because the drill jumbo was able to reach past the roadheader.

A 2.7-m (3.5-yd) Wagner ST-3 scooptram was used to transport muck out of the tunnel. Initial cycle times for the scooptram were negligible, but the wait for the scooptram's return became noticeable as the tunnel reached completion. However, the delay costs in scooptram cycle time more than offset the costs of an additional operator and the creation of turnouts to allow for multiple muck transport vehicles.

For full-face drill and blast operations, the drill pattern typically consisted of about 60 holes. Perimeter holes were spaced on 0.6-m (2-ft) centers, and tunnel delays were used with a combination of nitroglycerin-based and non-nitroglycerin-based explosives provided by WA Murphy, Co. The blast muck pile was removed with the scooptram, and tunnel sets were brought to the heading and erected with a low-profile forklift.

The steel sets (W6x25) were provided by Heintzmann Corporation through Antraquip in four pieces consisting of two posts and two arch segments. Steel sets were brought to the heading on a forklift, and the arch segments were lifted into place with the roadheader and bolted together by the miners. Tunnel alignment was maintained with the use of pipe lasers set for line and grade as daily survey checks. In curved portions of the tunnel, each set was surveyed in place to maintain accuracy.

Figure 6 shows a view looking around the first curve of the tunnel. The sets were then blocked, with spacing not exceeding 0.9 m (3 ft). Additional timber lagging was provided to afford rockfall protection so that excavation could proceed with the application of shotcrete, typically once a day over several sets.

The shotcrete initial lining was applied by hand-held nozzles pumping from a custom-built mobile concrete pump that towed a modified trailer mounted Cemen-Tech batch plant filled with premixed shotcrete. In areas of limited stand-up time or particularly wet ground, dry-mix shotcrete with powder accelerator was applied from a Reid Pot. Extensive spiling and forepoling also was required to protect tunnel advancement in areas of crushed and very blocky and seamy ground. Spiling typically consisted of 3.7-m (12-ft), number 9 rebar driven into holes drilled by the boom drill jumbo.

Convergence monitoring was performed at monitoring points placed in areas of squeeze-susceptible ground. The maximum convergence recorded was 4.6 cm (1.8 in.), or about 1.1% of the tunnel diameter.

Tunnel excavation work was performed in two twelve-hour shifts, five days per week, with tunnel invert preparation on Friday nights and invert pours on Saturdays. Invert preparation could not be performed simultaneously with tunnel excavation because of the rebar placement required for the invert slab. A 10.3 MPa (1,500 psi) minimum strength was required prior to traveling equipment over the concrete invert. The concrete invert was placed using a high-pressure concrete pump that pumped ready-mix concrete from the portal through a 12.7-cm (5-in.) slickline to the heading. In areas of stable ground and when the week's advancement had been slow, the invert was placed every other week.

Tunnel excavation was completed in just over eight months on July 15, 2008. Tunnel advancement proceeded at an average of approximately 18.2 m (60 ft) per week with a best day of advancement of 8.5 m (28 ft) in 24 hours. There were no lost time injuries, which is a tremendous accomplishment given the tight quarters and dangerous ground through which the tunnel was driven.

Pre-Excavation Grouting

Throughout the tunnel excavation, it was required that a minimum of two probe holes be kept a minimum of 6.1 m (20 ft) out in front of the heading. This was typically accomplished by drilling two 18.2-m-deep (60-ft-deep) probe holes every 12.2 m (40 ft) of tunnel advancement. The purpose of the probe holes was to detect groundwater and hazardous gasses and to provide information on the ground ahead. Should any of the probe holes produce more than 0.38 L/min (0.1 gal/min) per foot of hole, it was required that the contractor drill an additional two holes and pressure grout the four holes to refusal. The contractor then drilled two additional holes; if these additional holes exceeded the water threshold, the process was repeated. Grout consisted of Type III cement and water mixed with a colloidal grout plant and pumped through grout-through packers at a pressure not to exceed 1.1 psi per foot of rock cover at that point in the tunnel.

In only one location did the grouting program have limited success. Water inflows in excess of 378.5 L/min (100 gal/min) were experienced while the heading was located in a rapidly softening clay zone, preventing the grout packers from holding the required pressure. Shotcrete was applied to stabilize the face, additional holes were drilled, and pressure grout pipes were installed with post grout valves grouted in place spaced well ahead of the face. Pressure grout was pumped through these pipes, and the water inflow was brought to under 189.3 L/min (50 gal/min). It was decided to drill four bypass holes into the water feature and pipe the remaining flowing water around the heading and out of the tunnel. Tunneling was cautiously resumed with full spiling and Type III ground support. Just beyond the clay zone the tunnel was excavated through a block of greenstone well over 30.5 m (100 ft) in extent—the water had likely been impounded within the greenstone block by the impermeable clay zone. Within a week, the tunnel had reached the hardest rock encountered in the tunnel and conditions were dry.

As the tunnel approached the reservoir, significant groundwater inflows were expected. However, due to clay-filled seams in the rock and a lowered reservoir, major inflows were not encountered. The tunnel was constantly wet and dripping and shotcrete adhesion became difficult; however, the water inflows were only slightly above the grouting threshold and were easily controlled by the grouting program.

Shaft and Tunnel Plug Construction

The tunnel/shaft connection was designed with a concrete plug in place so that the tunnel would not be susceptible to inundation by the reservoir through an open shaft. Since the tunnel arrived at the shaft site before reservoir operations allowed access to complete the shaft, the concrete plug was constructed from the tunnel side. During construction of the plug, 4.27-m (14-ft.) lengths of the 137-cm (54-in.) and 20.3-cm (8 in.) pipes were set in place in the upstream end of the tunnel. The face of the tunnel plug was formed, and concrete was pumped from the portal to the face to create the solid concrete plug. The shaft was then completed, and pipes were joined to the previously placed pieces. The pipes made a 90° turn and extended vertically out of the shaft. The shaft was then backfilled with concrete.

The intake shaft and associated low flow pad consisted of a 1.4-m^2 (15-ft^2) rectangular excavation about 4.9-m (16-ft) deep connected to a 4.9-m (16-ft) inside diameter circular shaft extending about 15.2 m (50 ft) below original grade. The shaft and low-flow excavation were to be constructed immediately adjacent to the edge of the lowered reservoir. Due to the District's water supply constraints, the reservoir could only be lowered for shaft construction between June 15 and November 15, 2008.

The contract drawings specified the use of an extensive pre-excavation curtain grouting program, overlapped 20.3 cm-diameter (8-in.-diameter) micropiles with interior bracing around the low flow pad, and a gasketed liner plate supported, drill and blast excavated shaft. The contractor submitted a value engineering design and proposal to line both the low flow pad and the shaft with 7.6-cm-diameter (36-in.-diameter) drilled secant piles. The larger, overlapping concrete piles eliminated the need for interior bracing around the low flow pad and, most importantly, eliminated the need for curtain pressure grouting to control water inflow. All parties determined that the value engineering proposal eliminated the environmental risk of grout frac-out into the reservoir.

The secant pile drilling was performed using a Bauer RTG25 drill rig (Figure 7) capable of drilling 7.6-cm-diameter (36 in.-diameter) holes into rock with compressive strengths in excess of 137.9 MPa (20,000 psi). After the secant piles were in place, the drill was used to "Swiss cheese" the interior of the shaft to eliminate the need for blasting. The rock inside the low flow pad and the top portion of the shaft was excavated with a large excavator, and then a midsize excavator was placed on the low flow pad to dig the shaft as far as it could reach. The last 3 m (10 ft) of the shaft excavation, beyond the reach of the excavator, was accomplished by hand, with the miners hand-loading the excavated material into a muck bucket transported out of the shaft by an tending crane (Figure 8).

Intake Structure

The intake structure consists of approximately 128 linear m (420 linear ft) of 137-cm (54-in.) welded steel pipe encased in reinforced concrete, with four 106.6-cm (42-in.) hydraulic intake gates at various elevations. Construction started by excavating the existing reservoir slope to rock, placing lean concrete base to grade then forming a footing

Figure 7. Bauer RTG25 used to drill at shaft

with small saddles on which to set the pipe. After the footing was poured the pipe was placed, welded, and strapped down with U-shaped rebar and bar-lock couplers to prevent pipe floating during concrete placement. The rebar was installed, and the main pipe encasement portion was formed up and poured around the pipe. A 76.2-cm-tall (30-in.-tall) wall was poured along the top of the structure to place 2.54-cm (1-in.) hydraulic lines to control the hydraulic gates and the 20.3-cm (8-in.) low-flow inlet. Over 1,128 linear m (3,700 linear ft) of 2.54-cm (1-in.) welded stainless steel pipe was installed to control the gates. The pipe terminates at a new intake control housing houses the hydraulic power unit. The work on the shaft and intake structure was required to be completed within a very tight window of construction. Due to operational constraints on the reservoir, the District could only guarantee the contractor the water level would remain below elevation 576 between June 15 and November 15, 2008. The shaft and intake work, however, were successfully completed without reservoir operations impinging on production.

Tunnel Final Lining

The final concrete lining consists of a 45.7-cm (18-in.) reinforced concrete invert and a minimum 30.5 cm-thick (12-in.-thick) reinforced concrete arch. The invert was completed first, working from the shaft towards the portal. The invert rebar was installed at about 30.5 linear m (100 linear ft) per day, and the invert work was completed in seven separate concrete pours. After the invert was completed, the reinforcing steel for the arch was brought into the tunnel. The rebar was installed from the portal towards the plug/shaft. After a week of rebar installation, the final arch concrete lining started at the portal. The final tunnel lining was placed using a 18.3-m (60-ft) traveling form system with a 9.1-m (30-ft) form carrier manufactured by Wausau-Everest Form Systems. A night shift set and prepared the form for concrete and advanced the slick line and utilities. A day shift poured anywhere from 52 to 96 m^3 (68 to 125 yd^3), depending on tunnel overbreak, in each pour. The 35 arch pours were completed in 40 days.

Figure 8. Shaft excavation by hand mining

Outlet Pipe

The outlet pipe is approximately 640 linear m (2,100 linear ft) of 137-cm (54-in.) mortar-lined welded steel pipe. The outlet pipe connects to the intake pipe at the tunnel plug with a dismantling joint. The 54-in. pipe sits on concrete saddles 0.3 m (1 ft) off the invert, spaced at 6.1-m (20-ft) centers along the tangent portions of the tunnel and 3-m (10-ft) centers along the curved portions. The pipe came in 12.20-m (40-ft) lengths for the tangent sections and 20-ft sections for the curves with shop-welded mitered joints approximately half way along each pipe segment to account for the curvature of the tunnel. The 40-ft sections weigh 5,443 kg (12,000 lb; 6 tons), making it difficult to move the pipe in such a confined tunnel. The size of the tunnel did not allow for the installation of the pipe saddles prior to bringing in the pipe segments. One change in the design proposed by the contractor was to replace the cast-in-place pipe saddles with precast saddles, allowing the contractor to set them just prior to installing in the pipe. The work sequence was comprised of a swing shift for drilling anchor holes for the precast saddles and setting the epoxied anchors. The day crew would bring two saddles in with a small forklift and align them over the anchors. The pipe would then be delivered via a specially designed pipe handler. The pipe handler has four-wheel steering capability with rollers to allow the contractor to line the pipe up to saddles and launch the pipe onto the saddles. Another crew would fit up and tack weld the pipe together followed by a crew welding the pipe. The pipe terminates at an outlet building containing an arrangement of valves to allow controlled discharge into Los Gatos Creek.

ACKNOWLEDGMENTS AND CONCLUSION

The open partnering approach adopted by the parties as well as the regular informal communications were key factors in helping the project achieve the enviable status of being on schedule, within budget, displaying good quality, and having a good safety record. The authors acknowledge the hard work and commitment of other key members of the project team, including Jeff Wells, Project Manager of FCI; Mike Cox, Project Engineer of DTDS; Kevin Bolle and Harlan Opp, mining foremen of DTDS; Donna Collins, Senior Project Manager of the District; as well as the miners and workers who helped make this project such a success story.

REFERENCES

Geomatrix Consultants, Inc. 2006. *Final Geologic and Geotechnical Data Report, Lenihan Dam Outlet Modification Project.* Prepared for Jacobs Associates.

Spreng, S.P. et al. 2008. Tunnel Design and Construction in a Franciscan Melange. In *Proc. 42nd U.S. Rock Mechanics Symposium* and *2nd U.S.-Canada Rock Mechanics Symposium,* San Francisco, 2008.

Bieniawski, Z.T. 1989. *Engineering Rock Mass Classifications.* New York, NY: John Wiley & Sons, Inc.

Barton, N.R. et al. 1974. Engineering classification of rock masses for the design of tunnel support. *Rock Mechanics.* 6(4) 189–236.

Deere, D.U. 1989. *Rock Quality Designation (RQD) After 20 Years.* U.S. Army Corps of Engineers Contract Report GL-89-1. Vicksburg, MS: Waterways Experimental Station.

Proctor, R.V., and T.L. White. 1968. *Rock Tunneling with Steel Supports.* 2nd ed. Youngstown, OH: Commercial Shearing and Stamping Company.

DANIEL ISLAND SURPRISE—SAND LENS LURKING IN COOPER MARL, CHARLESTON, SC

Ray Brainard ▪ Black & Veatch Corporation

Paul Smith ▪ Black & Veatch Corporation

Larry Drolet ▪ Charleston Water System

ABSTRACT

Charleston Water System's Daniel Island Extension Tunnel, initially thought to be the simplest of the four replacement tunnels to design and build, became the most challenging after the discovery of an unexpected 30-ft thick lens of sand in the normally consistent Cooper Marl while sinking a shaft on Daniel Island. This paper will explore the challenges involved with determining the extent of the unforeseen lens of sand, the accelerated boring program, and the redesign of the vertical alignment—keeping the alignment between the sand layer and the bottom of the Cooper River. It will describe how the challenges were overcome, and how the Owner, Engineer, and Contractor were able to team together to achieve the desired results.

INTRODUCTION

When the Contractor for Charleston Water System's Daniel Island Extension Tunnel encountered an unexpected thick layer of running sand while attempting to sink the shaft at the Daniel Island Wastewater Treatment Plant, it was crucial for clear and decisive direction to be provided by the Project Construction Management (PCM) team in a timely manner. The Owner, Charleston Water System, and the Engineer/ Construction Manager, Black & Veatch Corporation in association with Hussey, Gay, Bell and DeYoung, Inc. (B&V/HGBD), moved quickly to determine a viable solution for this problem, one that was previously not encountered in numerous tunnels and deep foundations in the region. The Daniel Island Extension Tunnel was a negotiated contract with the Contractor, and in an effort to expedite and strengthen the decision making process for this problem, the PCM added the Contractor, Affholder, Inc., to the team. This paper will describe the challenges involved with determining the extent of the unknown lens of sand, the accelerated boring program, the redesign of the vertical alignment, how the challenges were overcome, and how the Owner, Engineer, and Contractor were able to team together to achieve the desired results.

HISTORICAL CHARLESTON

Charleston, South Carolina is an important city in American history, rich in Southern culture and heritage. Originally settled in 1670 by the English, Charleston has over 100 buildings listed on the National Register of Historic Places and was intimately involved with the American Revolution and the American Civil War (See Figure 1). Based on its historic significance and charm, Charleston has experienced an extended boom of tourists and growth with over 1.5 million people visiting Charleston each year (Drolet, McKelvey, and Swartz 2005). Tourism is a major economic engine, driving the local

Figure 1. Photo of historic Charleston

economy and benefiting the area businesses. A disruption in services or the inability to provide those services, notably wastewater conveyance, could cause an economic disaster for any community associated with tourism and growth. Charleston faced this problem when the existing wastewater tunnel system began to fail.

Recognizing the severity of the problem, CWS engaged the services of B&V/ HGBD to develop a new Wastewater Tunnel Master Plan, and to design and manage the construction of a replacement wastewater tunnel system. Heavy civil construction in urban settings, especially in historic communities such as Charleston, requires careful and detailed planning. Many hours were spent during the design phase to minimize the impact to historical properties, roadways and neighborhoods. This work included archeological studies, environmental studies, building/structure assessment studies, traffic control studies, and numerous mitigation meetings with neighborhood and business associations (Drolet, McKelvey, and Swartz 2005).

ORIGINAL SYSTEM

Prior to the Water Quality Act of 1965 Charleston, like other municipalities in that era, discharged wastewater directly into adjacent bodies of water. For the Charleston peninsula, this included the Ashley River, Cooper River, and Charleston Harbor. Charleston's original wastewater tunnel system and Plum Island Wastewater Treatment Plant (PIWWTP) were constructed in the late 1960s and early 1970s after the Water Quality Act of 1965 and Clean Water Act of 1972 were signed into law. The original wastewater tunnel system included the Harbor Tunnel, Ashley River Tunnel, Cooper River Tunnel, and West Ashley Tunnel.

When completed, the original system was 12.3 km (40,300 ft) long and consisted of 2.1 m (7 ft) diameter tunnels constructed approximately 36.6 m (120 ft) below ground. A typical cross section of the original tunnel consisted of ribs and lagging boards as the initial support and reinforced concrete pipe 30 to 76 cm (12 to 30 in.) in diameter as the tunnel liner or carrier pipe. Previous experience with water tunnels north of Charleston led to the presumption that new tunnels in the same impervious geological formation would be stable for many years (Drolet, McKelvey, Smith, and Webb 2006). This presumption led to the decision and design of filling the annulus space between the initial support and carrier pipe with water in lieu of grout. Figure 2 is a tunnel section detail from the original drawings that indicates the carrier pipe, steel rib, wood post, wood chocks, and the annulus space to be filled with water.

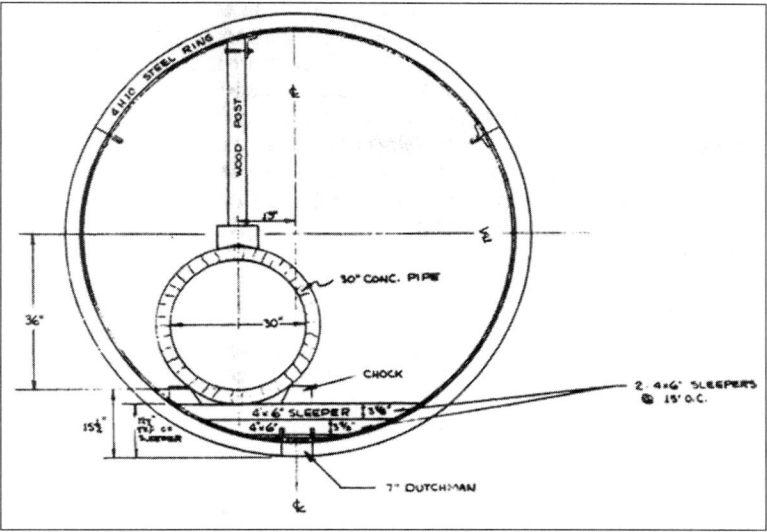

Figure 2. Original tunnel section detail

The original wastewater tunnel system included numerous inlets and drop shafts that collected wastewater from near surface gravity sewer lines and dropped to the tunnel system for conveyance to the PIWWTP. Each inlet structure had a vortex designed to funnel wastewater in a circular motion down the drop shaft pipe, allowing entrained air to be released from the wastewater. The original wastewater tunnel system also included small diameter (15 to 20 cm or 6 to 8 in) ventilation pipes installed in the tunnel system that were designed to expel any remaining trapped air. These ventilation pipes proved to be problematic and eventually became clogged with solids and floatables in the wastewater. Air that could not be released from the system combined with water and hydrogen sulfide gas (H_2S) to produce sulfuric acid, which corroded the reinforced concrete carrier pipes. Ultimately the crown of the carrier pipes, particularly near drop shafts, was breached, and wastewater entered the surrounding water. This in turn led to corrosion of the steel arch tunnel supports and eventually portions of the tunnel collapsed as shown in Figure 3 (Drolet, McKelvey, Smith, and Webb 2006). The costly blockages that occurred after the collapses began led to the timely development of a replacement system.

REPLACEMENT SYSTEM

The new replacement wastewater tunnel system was specifically designed for a 100 yr life cycle, and in a calculated effort to combat the harsh atmosphere of wastewater and hydrogen sulfide gas, the design incorporated the latest corrosive resistant materials available such as centrifugally cast fiberglass reinforced polymer mortar (CCFRPM) pipe, stainless steel fittings, fiberglass grating, epoxy lined concrete, polymer concrete, and HDPE pipe. Larger ventilation pipes and drop pipes were incorporated into the design along with improved vortexes, approach channels, odor control structures, and a SCADA monitoring system. Also, incorporated into the design was the same initial support design from the original tunnels (ribs and lagging boards) with one important change—the annulus space between the initial support and carrier pipe was filled with a low density cellular concrete grout (See Figure 4).

Figure 3. Photo of collapsed tunnel

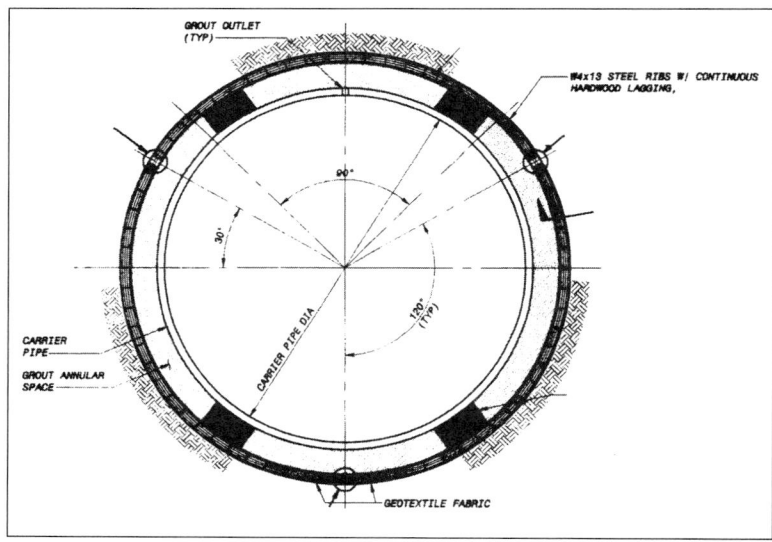

Figure 4. Replacement tunnel section detail

The new replacement wastewater tunnel system, as shown in Figure 5, includes Phase I—Harbor Replacement Tunnel, Phase II—Ashley River Replacement Tunnel, Phase III—Cooper River Replacement Tunnel, Phase IV—Daniel Island Extension Tunnel, and Phase V—West Ashley Replacement Tunnel. When the final scheduled phase, Phase V, is completed, the replacement wastewater tunnel system will be 18.8 km (61,600 ft) long. To date 16.3 km (53,400 ft) of wastewater tunnel has been completed, including the Phase IV Daniel Island Extension Tunnel. The excavated tunnels range from 1.5 to 2.4 m (5 to 8 ft) in diameter, and the CCFRPM carrier pipes installed range from 51 to 137 cm (20 to 54 in) in diameter. HDPE was utilized for the shaft drop pipes and ventilation pipes, and they range from 51 to 76 cm (20 to 30 in) in diameter installed.

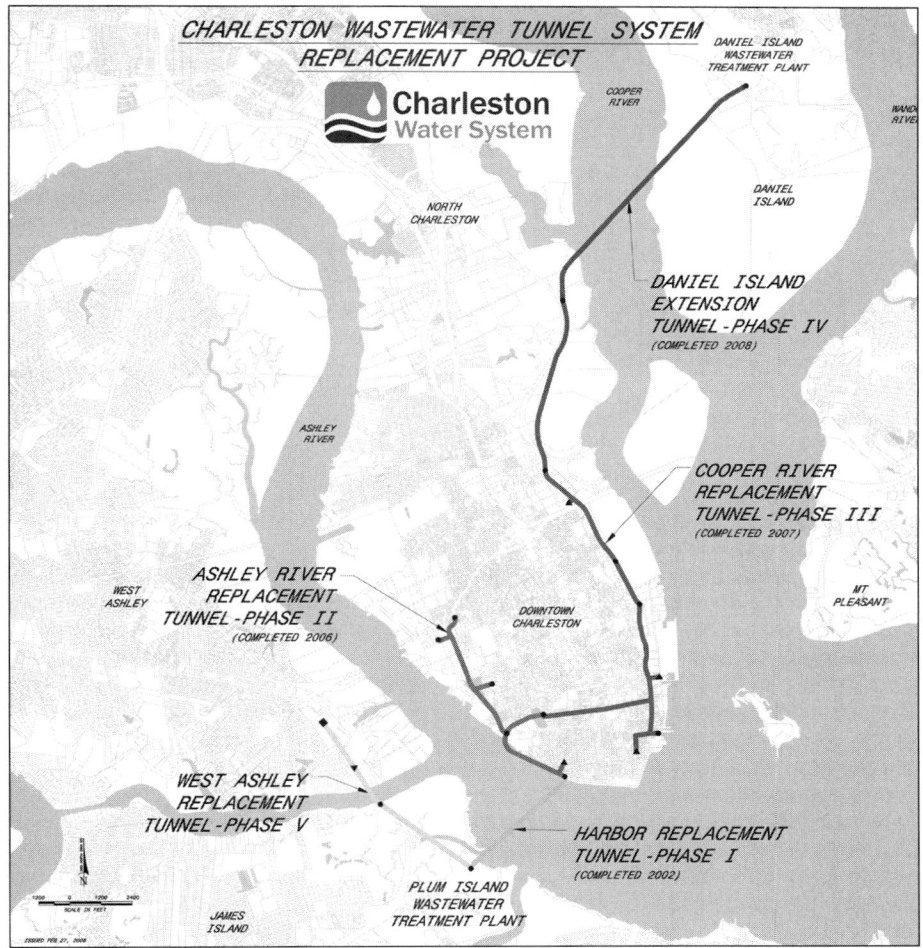

Figure 5. New replacement wastewater tunnel system

Geology

The Charleston Peninsula and Daniel Island were deposited as part of an estuary influenced by the combination of marine and continental processes. The estuarine surficial soils range in thickness from 10.7 to 18.3 m (35 to 60 ft) and consist of artificial fill, cohesive soils, and noncohesive soils. The artificial fill consists mostly of sandy soils and manmade rubble. The cohesive surficial soils, classified as silty clay with sand and sandy clay, includes the zero shear strength formation locally known as "Pluff" mud. The noncohesive surficial soils, located in layers and lens throughout the overburden, are characterized as silty sand, poorly-graded sand, and sand with silt.

Lying below the surficial soils is the Eocene and Oligocene age Cooper Marl. Cooper Marl is a highly calcareous, clay and silt-sized stratum characterized as clayey silt, silty clay and sandy silt (MH, CH, ML, CL, MH-CH, ML-CL) per ASTM D 2487. Based on previous tunnel construction in the area and the geology, the replacement tunnels were located approximately 30.5 to 36.6 m (100 to 120 ft) below the surface within the Marl. The Marl provides a nearly optimum tunneling medium due to its very

Table 1

	Ashley River Tunnel		Cooper River Tunnel		Daniel Island Tunnel	
	Min	Max	Min	Max	Min	Max
Moisture Content (%)	29	56	27	59	29	75
Dry Unit Wt (kg/m3)	1121	1297	1121	1458	913	1490
Wet Unit Wt (kg/m3)	N/A	N/A	1682	2018	1538	1922
Liquid Limit (%)	39	88	41	125	32	155
Plastic Limit (%)	N/A	N/A	23	44	23	64
Plasticity Index (%)	15	57	8	91	4	91
Wash #200 (% passing)	52	92	31	96	18	86
Carbonate Content (%)	54	68	21	75	8	77
Direct Shear c(kPa)/Φ	0/22	79/32	65/28 (1 test)		0/22	76/35
UCS (kPa)	22	294	51	175	69	155
UU Triaxial (kPa)	86	170	98	202	68	278

low permeability, relative softness, sufficient stand-up time for initial support installation, and low abrasivity.

To verify tunnel alignments, a subsurface investigative boring program was conducted along the proposed alignment. Regionally the Marl becomes weaker in a southeastward direction from inland areas to the Charleston Peninsula, but across the Peninsula the range in geotechnical properties is very uniform. Also across the region, the Marl has an erosional surface with down cut valleys representing paleo-channels. The subsurface investigation was conducted primarily to confirm the top of the Marl and to confirm soil characteristics. Table 1 represents a summary of the results of the Cooper Marl laboratory testing for the Ashley River, Cooper River, and Daniel Island Tunnel alignments.

During the subsurface investigation for the Daniel Island Extension Tunnel, a deep channel into the Marl was discovered just to the east of the Cooper River that would reduce Marl cover over the crown of the tunnel to 3.0 to 4.6 m (10 to 15 ft). Several borings were completed near both banks of the Cooper River, while only a single boring tagging the top of the Marl was completed near the center of the channel, due to heavy ship traffic on the River. Following the investigation, it was decided to lower the vertical alignment of the Tunnel (Figure 6). This was to avoid any chance of excavating out of the Marl if the bottom of the deep incised channel was deeper than seen in the borings and to increase the Marl cover over the crown of the tunnel under the Cooper River. Generally, a layer of very loose saturated sand is found immediately overlying the Marl, and excavating through that material would require a different Tunnel Boring Machine (TBM) than what the Contractor had available on site.

This change in alignment put the proposed tunnel below most of the borings, including all the borings on Daniel Island. Discussions were held to decide whether additional boring information was needed because of the lack of data at tunnel depth. Because this was a fast tracked project with a hard deadline, and the uniformity of the Marl had been established with fairly high certainty, it was decided to move forward without any further investigations.

Shaft Sinking

Analyzing shaft sinking methods through the surficial soils determined that caisson construction for large diameter shafts and steel cased drilled construction for small diameter shafts would minimize the probability of settlement and groundwater

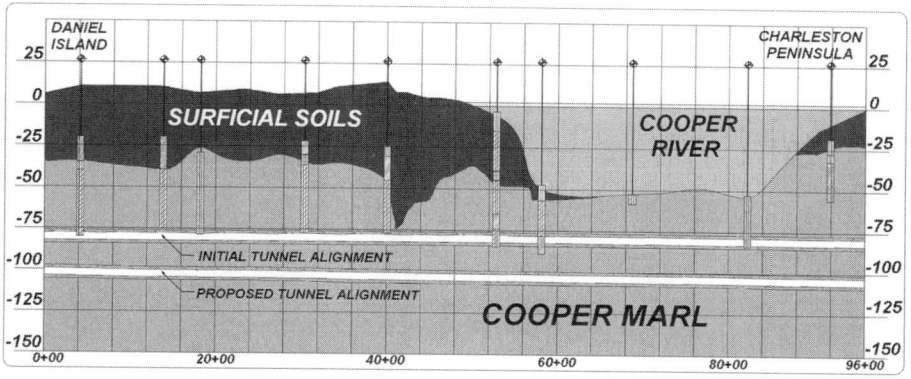

Figure 6. Initial and proposed vertical tunnel alignment

drawdown of surrounding soils at the shaft sites. End terminal shafts, 6 m (20 ft) in diameter, were designed as full depth seismically resistant caissons and sized to provide storage for hydraulic flushing of the tunnel system. Drop shafts, 3 m (10 ft) in diameter, were designed to be flexible and incorporated a final lining of AWWA C-300 reinforced concrete or polymer concrete cylinder pipe, with steel or stainless steel gasketed joints, stacked vertically. The temporary caisson support for the drop shafts was designed and sized by the Contractor. The annulus between the temporary caisson support and the final shaft lining was backfilled with pea gravel as a means of achieving the desired differential movement during a seismic event (Drolet, McKelvey, Smith, and Webb 2006).

Cracks in the Caisson and Running Sand

Caisson and drilled shaft sinking during the Ashley River and Cooper River Replacement Tunnel Contracts was uneventful and went rather smoothly. The Daniel Island caisson shaft sinking proved to be otherwise. The Daniel Island shaft was originally designed as a full depth end terminal shaft with 61 cm (24 in) thick walls. During the sinking operation, with the caisson into the marl, substantial cracks developed in the walls at elevation −18.0 m (−59.0 ft) or 19.7 m (64.5 ft) deep resulting in an unsafe condition if the caisson sinking continued. An investigation into the concrete and the caisson sinking method led to a non-conclusive explanation for the cracks. Subsequently, the shaft was redesigned and resized to allow for ribs and lagging board support down to final grade of −36.8 m (−120.8 ft) and to construct the final shaft lining at a smaller diameter of 4.9 m (16.0 ft). The shaft sinking continued smoothly until the Contractor encountered saturated running sand at elevation −26.8 m (−88.0 ft), approximately eight feet lower than the initial geotechnical investigation. The Contractor notified the PCM team of the situation and was directed to continue the shaft sinking and grouting of the annulus outside of the ribs and lagging board support in an effort to control the running sand. The Contractor struggled with little or no advancement for several days. During that period, the PCM team quickly determined that supplemental borings were required to determine the limits of the saturated sand. Also during that same period, the PCM team requested the Contractor join forces in an effort to solve the problem in an efficient and expeditious manner. The Contractor brought another level of expertise to the table with intuitive ideas and methods for solving the problem and completing the work.

Figure 7. Photo of river boring from barge

While the details of a solution and plan were being developed, the Contractor attempted to advance the caisson down by cutting the lagging boards between ribs to shorter lengths in an effort to minimize the amount of ground exposed. In addition, the Contractor procured liner plate as a substitute for the lagging broads. The Contractor also drove 20 cm (8 in) steel channel spilling along side the ribs and attempted to grout the annulus behind the spilling.

Accelerated Boring Program

The PCM team moved quickly and engaged the services of the same drilling subcontractor utilized for the initial subsurface investigation to complete the supplemental subsurface investigation. The drilling subcontractor was able to mobilize very quickly after locations of the borings were determined. The supplemental subsurface investigation included eight additional borings that were located along the alignment. Work at the shaft stopped after the first two borings near the Daniel Island Shaft revealed that the extent of the sand lens was much deeper than originally anticipated. Two of the eight additional borings were located in the Cooper River, and the accelerated mobilization for the river borings proved to be more difficult to execute than the borings themselves. The logistics of working near a commercial shipping channel involved detailed scheduling, planning, and approvals from multiple agencies including the U.S. Army Corps of Engineers, U.S. Coast Guard, and the South Carolina Department of Health and Environmental Control. Specified dates and work hours were provided by the governing agency. These criteria and the local tide charts dictated how the marine subcontractor and drilling subcontractor were able to plan and complete the two river borings from a barge (See Figure 7).

Redesign of Vertical Alignment

After each boring was completed, the boring log information was added to the geotechnical profile. As the information was added, it became apparent that the sand lens was approximately 9 to 12 m (30 to 40 ft) thick and was dipping southwest from the Daniel Island Shaft to the Cooper River. Preliminary observations indicated the sand lens was not directly linked to the Cooper River. After the borings were complete, a meeting was held to discuss the two options available and decide on the best one moving forward. Everyone was reminded that the Daniel Island Extension Tunnel schedule was driven by the Daniel Island Wastewater Treatment Plant mandated milestone to

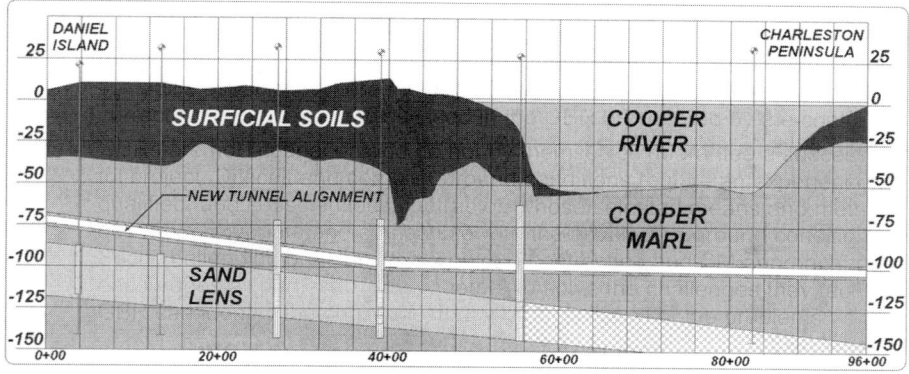

Figure 8. Redesigned vertical tunnel alignment

discharge into the tunnel by a specific date. The connection between the plant and tunnel was completed by others.

The first option would involve continuation of the caisson advancement down to design grade, modification of the TBM to operate in Earth Pressure Balance (EPB) mode, and changing the initial support to a closed system such as liner plates or gasketed segments. This option would significantly reduce caisson and TBM advance rates, and would have increased costs for the changed initial support and schedule impacts. The second option would involve no further caisson advancement, redesign of the vertical alignment, and relocation of the working shaft and tunneling operation from the Daniel Island Shaft to the Greenleaf Shaft. The redesign of the vertical alignment would allow the tunnel to be constructed over the sand lens, but still remain in sufficient marl cover under the deep channel in the marl and under the Cooper River. This option would have increased costs associated with relocating the tunneling operation to the new shaft location and schedule impacts. The second option was chosen because analysis indicated this option was the most cost effective and offered the most time saving to the Owner. Figure 8 indicates the redesigned vertical alignment and the previously unknown sand lens.

CONCLUSION

The second option proved to be the correct decision. The logistics of relocating the tunneling operation were completed very quickly and efficiently. The schedule impacts for tunnel excavation and pipe installation were corrected by adding an additional hour to both shifts and working double shifts on Saturdays as needed. This acceleration brought the Contract back on schedule with three months remaining. The relationship between the Owner, Engineer and Contractor proved to be the most important part of the whole process and manifested itself from the previous tunnel contracts. Without this relationship and teamwork, the process of acknowledging the difficult ground as a differing site condition and developing the right direction and compensation would have taken much longer. That could have potentially impacted the schedule beyond the final completion date and could have cost the Owner a substantial amount in penalties.

The discovery of this previously unseen lithology deep in the Cooper Marl astonished the local contingent of Engineers and Contractors experienced with the ground conditions in the region. There is an old important lesson to be learned again from this experience: Never assume that uniform ground conditions are consistent enough to not warrant additional subsurface investigations when changes are made to the alignment.

The cost of a well planned subsurface investigation is great insurance against surprises such as the sand lens lurking in the Charleston Cooper Marl.

REFERENCES

Drolet, Larry; McKelvey, James and Swartz, Jason. (2005). *Planning and Fast Track Design in Historical Downtown Charleston.* Rapid Excavation and Tunneling Conference Proceedings.

Drolet, Larry; McKelvey, James; Smith, Paul; and Webb, Ross. (2006). *Tunnel Construction in Historical Downtown Charleston.* North American Tunneling Conference Proceedings.

BIBLIOGRAPHY

Black & Veatch Corporation in association with Hussey, Gay, Bell & DeYoung, Inc. (2003). *Wastewater Tunnel Master Plan.* Commissioners of Public Works of the City of Charleston, SC.

Black & Veatch Corporation. (2006). *Geotechnical Baseline Report Daniel Island Extension Tunnel.* Commissioners of Public Works of the City of Charleston, SC.

Black & Veatch Corporation. (2006). *Geotechnical Data Report Daniel Island Extension Tunnel.* Commissioners of Public Works of the City of Charleston, SC.

SAN VICENTE PIPELINE TUNNEL—
REACH 4W, 3, AND 2: CASE HISTORY

Michael Jatczak ▪ Traylor Brothers, Inc.

Brett Zernich ▪ Traylor Brothers, Inc.

Steven Short ▪ Construction & Tunneling Services, Inc.

William Monahan ▪ Traylor Brothers, Inc.

ABSTRACT

The San Vicente Tunnel Project is an 11-mile water tunnel constructed for the San Diego County Water Authority. One of the many challenges of the project included tunneling a 22,000-foot single-heading tunnel on the western half of the job in varying geology using a 3.66m (12ft) diameter open shield excavator or "digger shield." This paper will discuss the design considerations for the temporary liner, the tunneling, and ancillary equipment. It will also highlight the challenges faced while mining this geology, explain the methods implemented to drive this section of tunnel, and discuss the experiences while performing mechanized tunneling in an extensive and varied San Diego geology.

INTRODUCTION

The San Vicente Pipeline Tunnel (SVPT) is a 17.45 km (57,230 ft) tunnel that is part of a major undertaking by the San Diego County Water Authority (SDCWA) to create emergency storage for San Diego's regional raw water supply. Once complete, this Emergency Storage Project (ESP) will create a system of reservoirs, pipelines, and pump stations to comprise a storage and distribution system for an approximate six-month water supply for the San Diego region should a natural disaster disrupt water deliveries from Northern California or the Colorado River. The ESP includes the construction of the Olivenhain Reservoir and Pipeline, Lake Hodges Pipeline and Pump Station, San Vicente Dam Raise, San Vicente Pump Station, and San Vicente Pipeline. A regional map showing the project's locations is shown in Figure 1. (Emergency Storage Project, 2007)

The SVPT project was awarded to a Joint Venture between Traylor Bros., Inc., of Evansville, IN and J.F. Shea Construction, of Walnut, CA. The Traylor Shea Joint Venture (TSJV) was given notice to proceed on July 14, 2005. The initial contract value was $198,266,900 and the original contract duration was 1,250 calendar days. The designer of record was Jacobs Associates, of San Francisco, CA and Construction Management was contracted to Parsons, of Pasadena, CA.

The pipeline tunnel is located in a highly variable geological formation and intersects all three rock types: sedimentary, igneous, and metamorphic. The predominant of these along the drive is sedimentary, which includes sandstone, claystone, siltstone, and conglomerate. The tunnel alignment is broken into six reaches distinguished by the geology, Reach 1 through 6. Reach 4 is further subdivided into East and West, split by the location of a Central Shaft. A 7 km (22,000 ft) section of tunnel was driven from

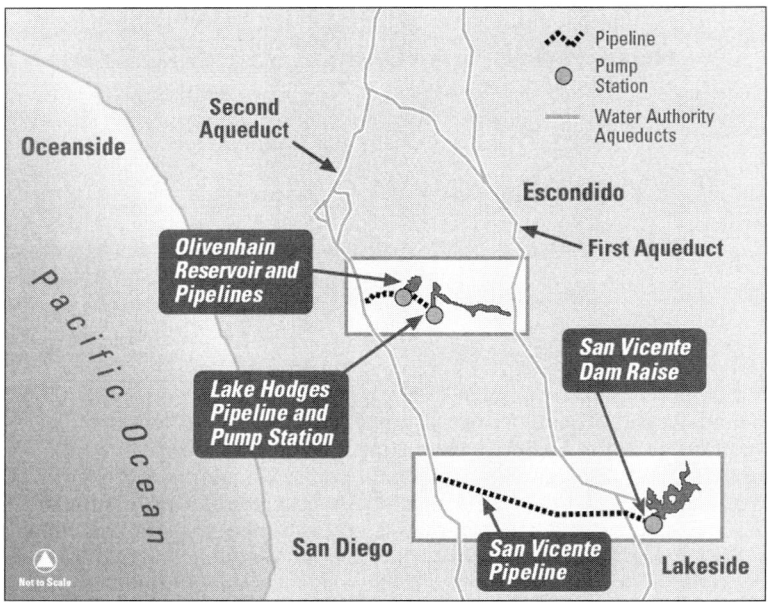

Figure 1. Regional map and project locations

the Central Shaft towards the West Shaft and encountered 3 different rock types in Reaches 4W, 3, 2 and the final portion of Reach 1. A cross section of the overall tunnel is shown in Figure 2.

This paper will highlight the use of a digger shield to mine this over four-mile section of tunnel and describe the geology as presented in the Contract versus actual conditions. This portion of the tunnel completed mining in August 2008 and the overall project is expected to be completed in late 2010. Further background information on the project can be seen in a 2008 NAT paper. (Campbell, et al., 2008)

GEOLOGICAL BACKGROUND

The project is located in the Peninsular Range geomorphic province, which extends from the Los Angles basin to the tip of Baja California. In the San Diego region the basement rocks, Cretaceous igneous and Jurassic metamorphic, are overlain with sedimentary rock formations. This tunnel drive was broken up into three distinct sections: Reach 4W (3.64 km/11,950 feet), Reach 3 (168 m/550 feet) and Reach 2, (2.88 km/9,450 feet) in the heading direction respectively. The Central Shaft and Reach 4W lay entirely within the Stadium Conglomerate formation and ranged in depth to invert from 40 to 300 feet. Reach 3 was a relatively short section at the top of a much larger granitic rock formation near the midpoint of the drive that ranged in depth from 67 m (220 ft) to 94 m (310 ft). Reach 2 was entirely within the Friars Formation, more specifically the Fine-Grained Friars Formation, and ranged in depth from 36 m (120 ft) to 88 m (290 ft). Groundwater levels were expected to range from 4.6m (15 ft) to 26 m (85 ft) above the tunnel invert. The geological profile is shown in Figure 3 for this western section of tunnel.

SAN VICENTE PIPELINE TUNNEL

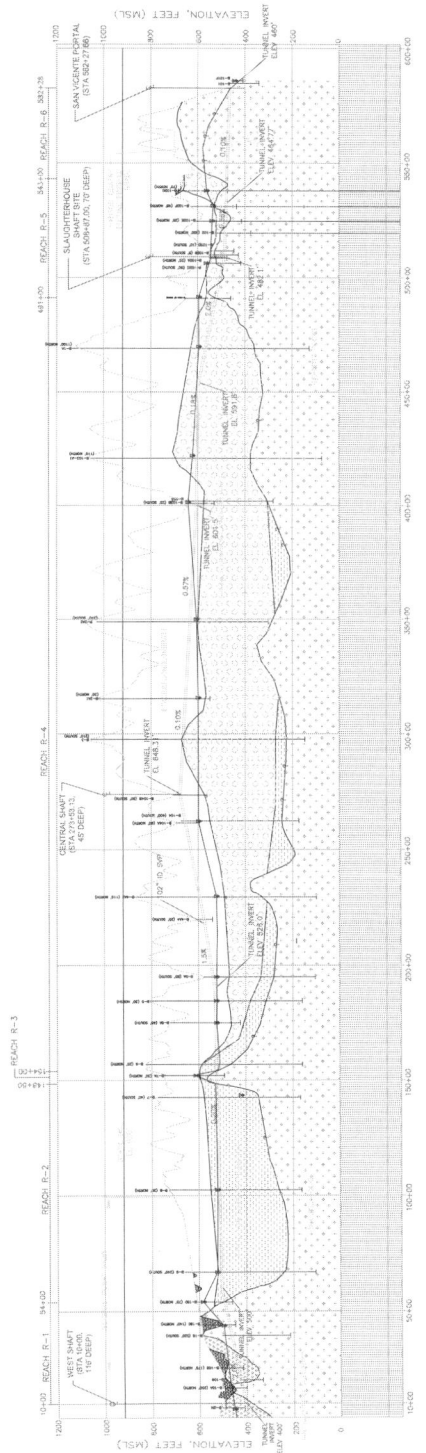

Figure 2. Tunnel geological cross section

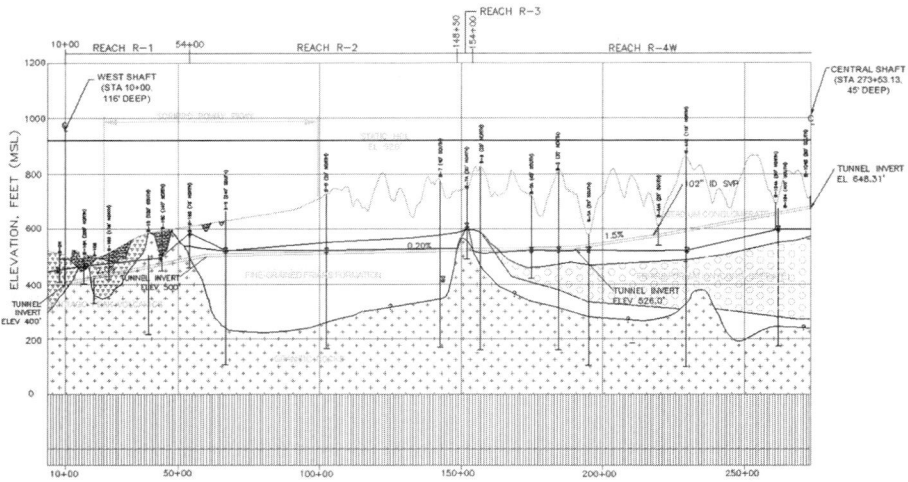

Figure 3. Western tunnel geological profile

Stadium Conglomerate—Reach 4 West

The Geotechnical Baseline Report (GBR) did not distinguish Reach 4 West from Reach 4 East, but defined the entirety of Reach 4 as Stadium Conglomerate and Friars Formation Conglomerate. The GBR summarized that both conglomerate formations were expected to be nearly the same and that differences between the two conglomerates would be difficult to distinguish.

The Stadium Conglomerate is a fairly vast formation in the San Diego region and can be seen in quarries and cuts throughout the area. The GBR defined the formation by referencing a 1975 Publication from the California Division of Mines and Geology as a "massive cobble conglomerate with a weak sandstone matrix." The formation also contains layers of sandstone, lenses of fine-grained sediments, and boulders. The GBR went on to define the matrix as "generally uncemented, although in some places it is weakly to strongly cemented with calcium carbonate." The average clast size was described as cobble (3 to 12 inches), however clast sizes up to 1.8 m (6 ft) and larger could be expected. The GBR described Reach 4 specifically for baseline purposes as follows:

> "Ground conditions in this reach will be controlled by the strength of the matrix and clasts, the grain size distribution, and groundwater conditions. The ground in this reach is expected to exhibit physical characteristics ranging from very dense soil to a weak to very weak rock. Firm to slow raveling ground is anticipated above groundwater table. Slow to fast raveling ground is anticipated below groundwater table. Where the conglomerate matrix is uncemented, cobble and boulder clasts will not be firmly held in place by the matrix, and it is expected that the clasts will fall from the tunnel roof, sidewalls, and face if not supported. The high strength and abrasive characteristics of the clasts in this reach is an important consideration for tunnel excavation equipment."

Granitic Rocks—Reach 3

The short stretch of "hard rock" covered by the sedimentary deposited Fine Grained Friars Formation was referred to as Reach 3, Granitic Rock. This rock mass is part of the Southern California Batholith and is the peak of this basement rock mass that lies beneath the entire tunnel alignment. This formation would have been familiar to TSJV after having seen the formation in Reach 1 and 6.

This stretch of granite was referred to by the GBR as mostly massive, moderately jointed rock; however the contact zones would be "very blocky and seamy." The GBR indicated the rock could range with Unconfined Compressive Strength (UCS) between 12,000 to 41,000 psi, with an average strength of 26,500 psi. The GBR indicated that the maximum UCS had reached 69,000 psi at the nearby Cowles Mountain job in the mid-1990s.

Fine-Grained Friars—Reach 2

Reach 2 was classified as the Fine Grained Friars Formation. This reach extended 2.88 km (9,450 ft) from the contact of Reach 1 Granitic Rock to the contact between Reach 3, as described above. The Fine Grained Friars is a sedimentary deposit that consisted of much smaller grain size distribution than that of the Stadium Conglomerate. The GBR described the formation as consisting of "primarily weak siltstone interbedded with layers of claystone, sandstone, and conglomerate." The GBR indicated that the thickness of these sedimentary seams could vary, but that over half of the drive would be mostly siltstone, splitting a little less than half the drive equally between claystone, sandstone, and conglomerate. The formation was classified as extremely weak to weak rock. The GBR also stated that the formation would have an average strength of 250 psi for 80% of the drive and 2,000 psi for the remaining 20%. The Fine Grained Friars formation could be described as mostly massive with some discontinuities between the various layers.

SHAFTS

The western portion of the Project contained two shafts. The Central shaft, located approximately halfway between each end of the entire tunnel, defined the boundary between Reaches 4 West and 4 East. From this shaft, the first digger shield mined west. The West Shaft, at the western terminus of the Project, was the starting point for a Robbins TBM to drive east into Reach 1, and the removal point for the digger shield.

Central Shaft

Location. Located in the middle of a new housing development, the Central Shaft was the most suburban of the San Vicente sites. The only third-party issues during construction consisted of a single homeowner complaining of noise during the night, certain individuals using a construction access road as a short-cut, and a residential complaint about the shade of brown of the site fencing.

The location had been a difficult one for the Water Authority to obtain, and certain agreements were necessary with the adjacent developer. One important one was the requirement that all tunnel muck remain on site. The muck removed from most of the western portion of the Project would become compacted fill to turn the Central shaft site 'bowl' into usable, marketable acreage. As a result, maximizing stockpile area was a consideration in setting up the site plant and facilities.

Situated as it was along the tunnel alignment, the Central shaft was a key access point for much of the job's activities. It was from here that Reaches 4 West, 3, and 2 were driven, and it was the point from which all the liner pipe for the west portion of the job was installed. Additionally, digger #2 would eventually hole-thru into this shaft, and much of the cellular grout placed around the liner pipe, in both directions, originated from here.

Figure 4. Top of the central shaft

Design. The bid documents required that the shaft dimensions be at least 10.7 m (35 ft) in diameter if round, and approximately 13.7 m (45 ft) deep, to later accommodate a permanent riser. An option was presented for a ramp down to tunnel elevation in lieu of a shaft, essentially creating a portal, and eliminating the need for cranes. As this shaft would later serve as the entry point for liner pipe, up to 15.24 m (50 ft) long if steel were chosen, a round shaft of minimum diameter was out of the question. A rectangular shaft, 24.4 m × 6.1 m (80 ft × 20 ft), and the ramp option were both evaluated at bid time. The ramp option was eventually rejected as it eliminated too much usable space in the yard.

After the eventual Award of the Project, TSJV proposed to deepen the shaft approximately 7.6 m (25 ft), thereby reducing the grade in the adjacent 4 West tunnel from 1.9% to 1.5%. This would aid in the selection of rail haulage equipment. The Owner accepted this proposal. The final shape of the shaft was changed from a rectangle to an ellipse, approximately 13.3 m × 9.1 m (60 ft long × 30 ft) wide. This size and shape allowed for the installation of the 15.24 m (50 ft) long sections of pipe, and used more economical ground support methods.

The ground conditions were expected to be 1.5 m to 3 m (5 ft–10 ft) of fill, followed by 1.5 m (5 ft) of alluvium, and the remainder Stadium conglomerate. It was noted that portions of the Stadium may be so hard as to require blasting. The design method chosen consisted of one course of ribs and lagging, below which, ribs and shotcrete were used. Ribs were 10×60 Gr. 50 on 1.5 m (5 ft) centers, and shotcrete was 4" thick, 5,000 psi with Sika hooked-end steel fibers. Support near shaft bottom was modified slightly to incorporate an eye at each end to allow the exit of one digger shield, and the entrance of the other. Actual ground conditions essentially matched those predicted. Figure 4 shows the top of shaft.

Construction. The excavation was initially accomplished with a John Deere 225C excavator, which loaded a muck bucket hoisted by crane. When the Stadium conglomerate was reached, and limited portions of it proved too hard to dig, a small Link Belt excavator with breaker was added. Eventually, most of the excavation was too hard for even the Link Belt breaker, and a larger Cat 315 with breaker was used. This machine was capable of removing the strongly cemented conglomerate, with no blasting required. Figure 5 shows the John Deere digging Stadium conglomerate.

Shotcrete was applied by hand using a trailer mounted Reed 2050 diesel pump, 2,400 psi, 5" outlet. Accelerator, SA-160, was dosed with a Watson Marlow Bredel SPX squeeze pump. Steel fibers were added on site to Ready-mix trucks. Due to the

Figure 5. Central shaft excavation

Figure 6. Tunnel eye

requirement to cover the 10″ ribs, thick portions of shotcrete under the ribs tended to sag and fall out. Here 4x4 welded wire fabric was placed to help provide support.

Upon reaching the bottom, a short length of tunnel was driven through the west eye, to allow for the 'socketing' of the digger shield during assembly and startup. This ~4.3 m (14 ft) section was excavated using the Cat 315, and supported by several self-drilling spiles above the brow, and shotcrete on the interior walls. See Figure 6.

The shaft bottom was prepared for mining and installation of the digger shield, which included landings for the Alimak, utilities, sumps, and large concrete pedestals to ease the later transitions between bottom configurations for setup, mining and pipe installation. Muck guides were added later after the shield was installed.

West Shaft

Location. The West shaft is located at the western terminus of the Project. This shaft was used to provide access to the Robbins TBM which mined Reach 1, and allow removal of the digger shield components after completion of Reach 2. It is located in the busiest, most visible portion of the job, cornered between the I-15 freeway, and Scripps Poway Road.

Design. The design of this shaft was similar to the Central Shaft and the Slaughterhouse Shaft, in that ribs, plus either lagging or shotcrete, provided ground support in the conglomerate above, while bolts plus shotcrete were used to support the rock below. The depth of the shaft was approximately 30.5 m (100 ft) with a diameter of 10.4 m (34 ft).

Construction. The surface fill and conglomerate sections, more extensive than at the other shafts, were excavated using the John Deere 225C excavator and muck bucket. Shotcrete was applied with the same system used at Central shaft. Once rock was encountered, drill & shoot methods were used. After blasting, bolts were installed and a thin layer of shotcrete applied to the walls. Once reaching depth, a 122 m (400 ft) long starter tunnel was excavated in anticipation of the Robbins TBM delivery from Reach 6.

TUNNEL LINING

The tunnel was designed as a two-pass system consisting of initial ground support followed by a 262 cm (103 in) Welded Steel Pipe (WSP) that forms the final liner. The WSP had to be backfilled with cellular grout and mortar lined on the inside. Because of the varying geology across the alignment, several ground support methods were implemented depending on tunneling methods. However, for Reaches 4W, 3, and 2 a precast concrete segmental liner was chosen. The table below gives the critical dimensions for the project and the various mined tunnels. The only reach not included in this table is Reach 5, which was drilled and shot using NATM techniques for ground support. (See Krulc, et al. RETC, 2007)

The precast segmental liner was made of steel fiber reinforced concrete. This liner was designed using a 6-piece trapezoidal ring with 2 bolts across the longitudinal joint and 12 dowels between the circumferential joint. The annular space around the ring was backfilled with grout and after reaching strength the bolts were removed and reused. The segments were cast by Traylor Shea Precast in Littlerock, CA and designed by Halcrow Group Ltd, of England. Of note is the fact that these rings were cast vertically using self-compacting concrete without mechanical compaction techniques. A previous paper (King and Hebert, 2007) better describes the segment design considerations for this project and discusses the advantages and disadvantages of this design. The geometry of the ring is shown in Figure 7.

TSM GENERAL ARRANGEMENT

The CTS Tunnel Shield Machines (TSMs) were open-faced rippers with enhanced features intended to engage the anticipated unique geology of the San Vicente Pipeline Project. The design was robust due to the length of the drives and anticipated variations in this geology. The machines can be described in three main shield sections along with the backup and ancillary equipment. The shields were tapered, or reduced in diameter, starting from the cutting face rearward to minimize the potential of getting stuck in heavy ground.

The TSM was designed by Construction & Tunneling Services, Inc. of Auburn, WA and manufactured at Jesse Engineering facilities in Tacoma, WA. The two machines were assembled one at a time for staggered delivery to the site.

Forward Shield. The first section, the forward shield, was configured with a hardened leading edge for cookie cutting. The edge was backed up by articulation rams capable of transferring full shield jacking loads. This available edge cutting force exceeds standard open-shield practice. The crown of the forward shield contained six (6) extensible breasting doors intended to restrain potential running material. This feature is enhanced by a shield hood rake angle of 12 degrees, further intended to slow

Table 1. Critical project dimensions and lengths

Shaft	Station	Depth (ft.)
West	10+00	116
Central	273+53	70
SlaughterHouse	508+87	70
Portal	582+28	0

Length (ft.)	Reach	Station
		10+00
4,400	Reach 1	
	Rock TBM	54+00
9450	Reach 2	
	Segment Liner	148+50
550	Reach 3	
	Segment Liner	154+00
11,953	Reach 4 West	
	Segment Liner	273+53
21,747	Reach 4 East	
	Segment Liner	491+00
5,200	Reach 5	
	Drill and Shoot	543+00
3,920	Reach 6	
	Rock TBM	582+28

Total Pipe Length	
Feet	Miles
57,228	10.84

Machine and Tunnel General Dimensions		
In.	Ft.	Description
148.5	12.38	Digger Shields CTS cut
97.4	8.12	CTS Front Shield-Top Length
47.4	3.95	CTS Front Shield-Bottom Length
146.25	12.19	OD CTS Mid Shield
143.75	11.98	ID CTS Mid Shield
252	21.00	CTS Middle Shield
146	12.17	OD Tail Shield
143	11.92	ID Tailskin
113.6	9.47	CTS Tail Shield
463	38.58	CTS Total Length
126	10.50	ID of Segments
140	11.67	OD of Segments
7	0.58	Segment Thickness
4.25	0.35	Nominal Backfill Grout
102	8.50	ID Pipe
48	4.00	Length of Segment
138	11.50	TBM Robbins
196.75	16.40	Front of Cutter Head to End of Roof Support
266.75	22.23	Front of Cutter Head to End of Fingers
62.75	5.23	Propel Stroke
568.25	47.35	Cutterhead Face to Center of Gripper Carrier, Retracted
631	52.58	Cutterhead Face to Center of Gripper Carrier, Extended
1169	97.42	Cutterhead face to deck #1
4752	396.00	Total Length of TBM & Trailing Gear
129	10.75	ID of Ring Steel

loose material ingress. The lower part of the interior of the forward shield is comprised of a heavy steel muck apron that contains mined material to be pulled out of the heading and onto the leading end of the conveyor. Figure 8 is a view of the forward shield.

Mid-Shield. The mid-shield, the second main section of the machine, contains the major working elements for mining. These are the cutting tool mounted to the boom, the conveyor, the operator's control stations and the propel rams. This shield is split into two sections, upper and lower, for manufacturing and shipping purposes. Figure 9 shows the 2 shield sections being fitted in the shop.

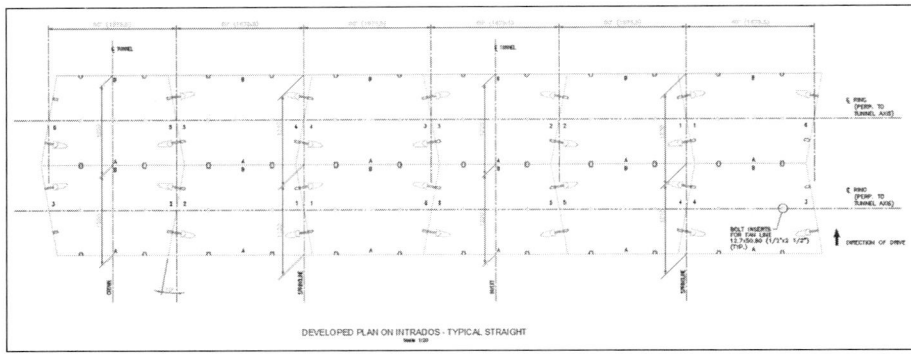

Figure 7. Precast segment geometry—plan view

Figure 8. Digger shield

Figure 9. Shield assembly

Figure 10. Shield with Krummenauer head

Boom. Predicted variations in geology led to a machine design that could utilize two uniquely different cutting tools in different situations. The primary tool was a single-toothed ripper with loading paddle that was mounted to the end of the boom (see Figure 8). The system was designed to apply high-penetration loads to the single tip in order to rip the matrix and loosen the exposed cobbles. The single tip can be rotated infinitely about a 360-degree axis and locked into place for ripping. Ripped material that falls onto the muck apron is then raked up the muck apron and into the conveyor by means of an integral paddle.

The secondary tool is a road header bit designed and manufactured by Krummenauer of Germany. The road header can be exchanged for the ripper while utilizing the same drive motor and mounting surface. Figure 10 shows the Krummenauer head mounted to the boom during manufacturing.

The heavy boom was pinned to a turntable using hardened, shrink fit pins backed up by high capacity bushing material. The turntable rotates on a set of large bronze alloy bushings that are mounted within a slide plate. The slide plate in turn translates fore and aft through the length of the shield on hard steel ways with replaceable surfaces. Boom swing and lift functions allow the cutting tools to engage the face. The slide plate travel function moves the entire assembly fore and aft to reach out to the face and to enable loading of loose material onto the belts. A large Hagglund hydraulic drive motor imbedded in the boom provides rotation for the selected cutting tool.

Traylor's past experience dictated that each boom function—boom swing, boom lift, slide plate travel and tool rotation—should be independently and significantly powered. This application of high, independent horsepower to each function helps ensure that the machine will not stall under duress. As a result, each of the four mining motions is backed up by its own independent 100 Hp (74 kW) electric motor driving a Gearoil variable volume pump. Proportional control of these functions is via joysticks mounted to the arm rests of the operator's chair, similar in concept to control of a hydraulic excavator.

Conveyors. The conveyor system, from heading to loading area (2 conveyors total), is designed to handle heavy discharge volumes and to pass 60 cm (24 in) boulders intact. The 90 cm (36 in) wide belt is oversized for machines of this diameter. The conveyors include high horsepower drive pulleys for extra torque to carry large volumes of loose material from the heading. Troughing idlers are all heavy-duty impact idlers and are spaced more closely than standard design dictates, again to handle large volumes and oversized stones.

Figure 11. Operator station

Control. The TSM operator's control station is mounted in the right hand side of the mid-shield (See Figure 11). In addition to the joysticks mentioned above, the operator has access to a variety of control and monitoring screens selected from an Allen-Bradley touch screen controlled by the A-B ControLogix PLC. The operator can select between the basic mining modes (ripper vs. road header), start and stop all main systems and review critical operating parameters. A secondary operator station is located in the rear of the mid-shield for independent control of the propel rams during ring build.

Information on machine location is displayed on the separate screen of the ZED guidance system. As with most recent projects, operating and guidance parameters are bundled and transmitted for viewing on surface PCs in real time. Select users can also access this information on the internet.

Propulsion. A total of twelve (12) propel rams; two for each segment, provided forward push forces and segment holding during ring build. The TSMs were designed for very high jacking loads to advance the shields in falling ground and to provide high edge loads for cookie cutting with the forward shield. In practice this was rarely used in R4W due to the hard material encountered.

Tail shield. The last shield section of the TSM is the articulated tail. This shield encloses the segment build and grout injection areas. The shield is manufactured from high-strength low-alloy steel plate. The rear of the shield encompasses two rows of wire brush tail seals at the rear that are greased via piping that is integral with the shield skin.

Erector. The segment erector is mounted from the back of the mid-shield and extends into the tail. Segments are moved forward from the unloading area at backup deck 1 and fed to the erector. The rotary erector picks and places an individual segment in the order advised by the ZED system. Each segment in turn is held in place by extension of two propel cylinders under low pressure until the entire ring is built and bolted together. The erector is controlled by a battery-operated hand-held remote radio unit.

Backup. The backup train supports the TSM utilities, segment unloading, ventilation and grout injection system. The backup decks are open-bottom frames that straddle the tunnel rail with polyurethane tread wheels rolling on the tunnel segments. A walkway runs the full length along one side of the decks with all utilities mounted to the other side or over the top. Muck haulage is via rail with the conveyor discharge approximately halfway along the length of the backup. As described earlier, the conveyor system was designed to convey 60 cm (24 in) boulders intact for discharge into the muck train.

Muck Handling. Muck was removed from the TSM using five 8-cubic-yard muck boxes pulled with 16-ton Brookville-24in gauge locomotives. Electric 10-ton Brookville locomotives were introduced later in the drive to limit horsepower within the airstream after the second switch was installed at the bottom of the 1.5% grade, about 2.7 km (9,000 ft) from the shaft. The 16-ton locomotives pulled this grade while the electric locomotives handled the 0.2% grade quite nicely, thus reducing diesel emission concentrations inside the trailing gear during mucking. The empty trains would swap with the loaded trains at the switch. A third and final switch was eventually added about 2.1 km (7,000 ft) after the second.

Ventilation. The main tunnel was ventilated using a 100 cm (38 in) spiral metal duct and 16 Spendrup 125 HP fans. The reason that so many fans were required was that a ventilation shaft was not allowed along the drive. The fans were spaced quite close together, approximately 450 m (1,500 ft). These fans were designed to deliver to around 43,000 cfm of airflow. The fans were all controlled using variable frequency drives (VFD) and could be controlled via a programmable logic controller using software on the surface to monitor their status. This system was designed to control start up so that the fans could be started in a synchronized manner to prevent a single fan from stalling. The system was also used to turn fans down to a minimal speed during off hours to save electricity, while maintaining a smaller airflow than that needed during mining.

Besides the main tunnel ventilation, the TSM was equipped with its own negative pressure booster fan system to ventilate the trailing gear and prevent dust from entering the shield. An auxiliary system was used and maintained inside the shield to minimize exposure to dust particulates around the shield operator and other employees.

Tunnel Backfill Grout. The segments were backfilled using a two-component system that has been used on several other Traylor jobs with huge success. The grout is pumped from a surface plant in liquid form and activated with an accelerator at the point of entry behind the segments using a two-component injection packer. The grout plant was built by Team Mixing of Canada to Traylor Shea's specifications. The grout was pumped to the heading using a 75 mm (3") peristaltic pump. This pump was able to deliver grout to the heading for the entire 7 km (23,000 ft) drive. Some problems were encountered during production from time to time, which forced crews to have to clean the 10 cm (4 in) pipe out several times; however pumping grout this length of distance was viewed as an accomplishment.

Grout was injected behind the segments using other peristaltic pumps in the heading which pumped grout from a 2000-liter agitator tank located on the back-up trailing gear of the TSM. The grout pumps were controlled from their own ControLogix screen and could be adjusted as necessary depending on mining speed. This information could then be logged from the office using Allen Bradley trending software to track grout takes and pressures. Also of note, was that the segments were not gasketed. This created some concern that grout would leak out between the joints, however this grout system performed quite well and could be pumped in a controlled fashion to avoid significant leakage.

TUNNEL PRODUCTION

Startup

Delivery of the digger shield to the Central shaft site occurred mostly in May 2006. The various bits and pieces were arranged so that startup of tunneling could occur with only the shield and first conveyor in the hole. The remainder of the critical decks would be connected via umbilicals, remaining on the surface, until room allowed.

As has been mentioned previously, the West end of the Central shaft was equipped with an eye capable of accepting the shield for startup. It was approximately 14' long,

Figure 12. Shaft shield assembly

and lined with shotcrete. The eye served two purposes: it allowed moving the shield into the tunnel and somewhat out of the shaft so as to make room for other items; and also to anchor the shield during the initial excavation with the boom. Without some means of holding the shield stable, the boom had the ability to lift the shield off the ground, similar to an excavator lifting its own tracks when digging hard.

The general sequence of installation was as follows:

- Lower and assemble the shield prior to pushing it into the eye. (See Figure 12)
- Start with just the shield in the eye as there was only enough room in the shaft for the machine with #1 conveyor, and one small muck box. Figure 13 shows the forward and mid-shields assembled in the eye.
- Advance the shield until room was available to add the segment feeder and deck 1, which contained the segment unloading trolley, greatly improving ring handling and installation.
- Mine with one box until the shield and deck 1 were buried.
- Set the grouting decks in the back of the shaft, elevated on beams, to allow proper encasement of the early rings.
- Mine for 180 ft to allow a secondary short mode, with the hydraulic motors, main power decks, PLC and control decks, and the #2 conveyor.
- Mine with this arrangement until the shield was buried 380 ft.
- Add all the remaining accessory decks with non-crucial gear, fan deck, trombone, tool box decks, grout decks, transformer etc.

As is usually the case when umbilicals are part of a startup, difficulties were had with fluid pressures and flows; electrical signals and controls; troubleshooting; and connecting, disconnecting, reconnecting. Eventually, all of the trailing gear decks were installed and functioning, while the focus could move on to digging the ground.

STADIUM CONGLOMERATE R4W

Ground Conditions

Even before the entire machine was installed and functioning, attentions naturally turned to the ground. The machine had been having difficulties excavating the Stadium conglomerate, but initially this was attributed to startup issues. It soon became evident, though, that the biggest problem was the degree and amount of cementation of the conglomerate matrix.

Figure 13. Forward and middle shield assembled in the shaft

The GBR outlined the expected ground conditions as relates to degree and quantity of cementation of the conglomerate. An eventual Differing Site Condition was declared based on exceeding these values over the full length of Reach 4 West. The actual ground conditions encountered at any given point varied between extremely hard and well cemented, to very loose sand.

The anticipated excavation methods included hogging out the center of the face, and then using the shield edge to push into the remaining ground, cutting the desired perimeter. In reality, the actual method normally used entailed physically digging 100% of the face, and pushing the shield into the open hole. The eventual overall mining rate for Reach 4W was an average of approximately 45 ft/day, though this number was much lower early in the reach.

Initial Problems

In the early portions of mining, much downtime had been incurred fixing boom hydraulic hoses. Sticking the pick in the hard ground and stalling the boom caused the hydraulic oil on the pressure side of the system to compress. Once the bite of ground that the pick was engaged with broke free, a shock wave was produced from the release of energy stored in the compressed hydraulic oil, breaking hoses and causing leaks on the return side of the hydraulic system. The return side plumbing had not been designed to handle such high pressures and flows, which were not expected. Additionally, the violent action of the boom in attempting to break up hard ground caused increased wear on the outer surfaces of the hoses. These problems were overcome by upgrading the return side of the hydraulic system to bigger hoses and hard lines.

In August 2006, while mining in hard ground, it was discovered that the point of connection for the pick was bent and fractured. The fractures were repaired by welding. In September 2006 the pick again broke off the boom. Shortly thereafter, the boom travel hydraulic cylinder broke. When the pick broke again for the third time, operations were shut down to perform the first of several major upgrades to the boom.

Reinforce Boom

The arduous conditions presented by the harder than anticipated matrix required the TSM operator to mine the face in a different fashion from what was originally anticipated. TSM design was based on being able to penetrate the matrix using a high tip force and then rake the face in a sweeping motion to dislodge the cobbles and boulders. This is a single motion requiring the use of the boom swing and/or lift cylinders.

Dislodged material would then be raked to the belt using the boom paddle actuated by the slide plate pull back cylinder.

In softer formations resembling the anticipated geology this in fact was the technique utilized. In the harder formations it proved very difficult to even penetrate the matrix and impossible to simply sweep the face to mine. The operators developed a technique whereby a slight depression in the face could be leveraged by applying full boom or swing force at maximum system pressure and then "popping" out a piece of rock by rapidly actuating the slide plate pull back cylinder.

This concentrated application of energy to the face resulted in an explosive release of rock, albeit in small pieces. At times heated pieces of the ripper tip in a molten state could be seen falling from the ripper tip. The unanticipated combined loading to the boom and the shock loads created when the face released took its toll on the machine. In particular the boom was subjected to higher than anticipated combined bending loads followed immediately by stress reversals during each rebound from the face. This was magnified by the significant increase in the number of ripping cycles required to mine out the length of one segment.

As a result, Traylor Shea and CTS undertook a program to reinforce the in-situ booms and re-design a replacement boom based on the current realities. Figure 14 shows an early attempt to strengthen the boom 'elbow' by adding 4-inch plate.

The boom already underground was subjected to constant inspections to detect weld cracking. When cracks were found, mining was stopped to effect repairs. These ongoing repairs highlighted those areas of the boom most subject to distress. These areas were reinforced as much as practical by welding on overlay plates and additional gussets.

At the same time, a finite element analysis was performed by inputting the enhanced loading that the boom was seeing. Using the results from the analysis, a new boom design evolved that was strengthened in critical areas. The boom was made stronger by trading off space requirements of the roadheader for added steel structure. Eliminating the roadheader excavation option produced more clearances between machine components and the boom, thereby creating room for the boom structure to grow.

Changes to Pick/Paddle

The same conditions that conspired to damage the boom also took a heavy toll on the ripper with the following detrimental effects:

- The high excavation forces created friction against the hard matrix and wore the pick tips rapidly. At times, as described above, the tips could actually be seen to be red hot. The result was rapid wear exacerbated by loss of hardness and heat treatment properties of the ripper tooth.

- The combined action of the boom lift/swing and slide plate travel cylinders created the same bending overloads and rebound issues as the boom saw. The result was a repeated breaking of the shank that retains the ripper tooth.

Experiments were conducted using a wide variety of tooth shapes from various manufacturers. It was found that a sharp pick point would penetrate the ground more readily, a fact appreciated by the operators. The drawback is that a sharp point has less material, so this style of tip wore more rapidly. A chisel shaped tip was found to be the opposite of the sharp tip, less penetration per cycle but longer life. Between these two diggers, this Project became one of the biggest (if not the biggest) consumers of ripper and dredging teeth in the country.

In general the trend has been towards ever larger tips, from the original D8 size up to D10. Nothing larger will fit into the small tunnel. Variation between sharp and chisel point continued depending on prevailing ground conditions at any given time and to some extent depending on availability of various products.

Figure 14. Boom reinforcement

To combat the chronic breakage of the tooth shank, various bolt-on designs were tested. This led to a design where breakage, when it occurred, was on the shank and not at the bolt flange. This allowed spare shanks with fresh tips to be stored on the machine for rapid tip replacement when required.

A series of other improvements were implemented when it was clear that the number of operating cycles required for the excavation would be a large multiple of the as-planned cycle time. These included armor plating of contact surfaces and reconfiguring welded mucking paddles into bolted replaceable units.

The improvements were an ongoing process, accomplished in many steps. The following shows some of the incremental improvements to the digger:

- As-Designed: Heavy steel design based on GBR and normal ripper pick arrangements
- Added independent base plate, with groove, to strengthen the connection of the shank and avoid snapping (breaking). This change recognized that shank failure required a quicker method to change out.
- Fabricated a dual dowel pin connection of the shank to base plate. This allowed quicker changeover of damaged parts. Also increased proportions of several components; pick, paddle, and stiffeners.
- Extended shank entirely through base plate and welded periphery of connection, using a D9 base plate.
- Switched shank base material to 4140 heat treated steel.
- Reduced the length of dowel holes into shank base, from 3" to 1.5". Also countersunk the D9 shanks into base and added welding. Eliminated the dowel hole stress riser.
- Used D10 generation shank employing a thicker and wider base plate. Continued to countersink the shank into the base.
- Modifications to dollar plate to make use of D10 shank.

Eventually the boom transformed into something similar in function, but much stronger and more readily repairable than the original. Compare Figure 15 to the earlier photo of the boom tip. Figure 15 actually shows the boom of digger #2 before mining Reach 4 East, however the first digger's configuration was essentially identical.

Granitic R3

The geology and planning for construction has been addressed in detail in a NAT paper (Campbell, 2008), and only a relatively brief description will be included here. As

Figure 15. Pick-paddle evolution

described above, at approximately the midpoint of the western portion of the Project, a 'hump' of granitic rock intrudes up into the tunnel alignment. This hump is approximately 168 m (550 ft) long, and is referred to as Reach 3. Reach 3 lies between the mostly Stadium Conglomerate of Reach 4 West, and the Fine Grained Friars of Reach 2. The practicalities of construction demanded that this granite be excavated in front of the existing CTS #1 digger shield, which had just completed Reach 4 West, in an East to West direction.

Ground Conditions. Encountered geology in this reach was very similar to that defined by the GBR. The initial 24 m (80 ft), and final 28 m (91 ft) were able to be excavated with the digger shield, simply as if it were advancing through hard conglomerate. The 116 m (379 ft) between these areas consisted of very hard rock and, thus, were blasted.

Methods. The details regarding expected blast patterns, ground support, and muck removal are covered in Campbell, 2008 and won't be fully rehashed here. The general methodology, however, consisted of the following. The face would be drilled for the next round. These rounds were variably full-face or top/bottom, and the depths ranged from 1.2 m to 2.4 m (4 ft to 8 ft), depending on the rock conditions. The holes were loaded and wired, protective blast mats were hung in the front of the shield, and the round shot. Ground support, which chiefly consisted of Swellex bolts, was then installed off the muck pile. Mucking occurred while the machine was pushed forward, and a new segmental ring built behind the shield.

An early plan for excavation of the reach involved installing a small mucker/loader in front of the shield, and leaving the shield stationary while the drill & shoot face advanced the full 168 m (550 ft). Theoretically, the mucker would deliver shot rock to the shield's boom and conveyor system for removal. Unfortunately, when the time came to actually find a piece of equipment that could be put through the digger shield, and be of adequate size to carry muck, the plan was forced to change. In addition, when the safety aspects associated with potential fallout and rib/lagging ground were considered, it seemed logical to bring the machine with the excavation, use it as a component of initial support, and be able to promptly install the final support, precast concrete rings.

Production. Excavation of Reach 3 ground began September 6, 2007. The initial 24 m (80 ft) of TSM-diggable ground took approximately 3 workdays to mine. The final 28 m (91 ft) took approximately 4 workdays. The remaining 115 m (379 ft) between took approximately 79 days, for an average of 4.8 ft/day. The reach overall had a 6.4 ft/day average. When compared to the estimated average of 6.6ft/day, particularly with the added safety of using the digger shield as protection from fallout, this construction should be considered a success.

REACH 2

Besides starting 3.8 km (12,500 ft) from the production shaft, Reach 2 proved to be less significant in terms of mining challenges than Reach 4W. The ground was much softer than that which was experienced in the strongly cemented Stadium Conglomerate and the mining shield was able to better penetrate and hold a tighter profile than in Reach 4W. A particular challenge was that the claystone, sandstone and siltstone were layered, dipping fairly shallow, which often presented a mixed face condition. This meant that there could be hard sandstone in the bottom and soft siltstone at the top or any manner of a face mixed up along those lines. The tunneling machine proved quite adaptable to most of the geology presented using its single pick and paddle configuration.

Production. Reach 2 was mined using 3 shifts, 24 hours per day, 5 days per week, with maintenance, when required, on Saturday. Usually, the crews were able to mine and build a ring in less than an hour depending on the strength of the face. The overall average daily production was 66.3 feet with a best daily production of 117 feet. During the drive, the crews had 14 days over 100 feet and 18 days in the 90 to 100 foot range. Reach 2 was mined out in 29 weeks.

Roadheader. The discussion of using the roadheader attachment was heavily analyzed during the initial mining stages of Reach 2. In the end, it was determined by job management that the benefits of mining with the single pick would outweigh the downtime needed to swap out to the roadheader. Although the roadheader would have worked quite well within the Fine Grained Friars, it did not make enough sense to change the attachment knowing that it would have to be changed back for the final couple hundred feet of mixed face granitic rock at the contact zone between Reach 1 and 2. Simply put, the potential production improvement did not outweigh the downtime necessary to make the two changes.

SURVEY

The project possessed some very difficult challenges from a surveying standpoint over this long alignment. On the western half of the project, the tunnel traveled beneath suburban neighborhoods, under roads and power line easements. On the eastern half of the project, the alignment traveled under an open space preserve spanning canyons and mountains. Initially, this meant that the shaft sites would have to be tied in together using GPS survey to avoid the long land traverses and, because boreholes were not allowed along the alignment, this further meant that no control loop could be tied into the surface from any locations but the shaft sites.

The western half of the tunnel had to be completely driven from the shaft transfer at West and Central Shaft. Control monuments were specified as fixed locations in the contract at each shaft site. TSJV included these points as fixed in their control network and set temporary points that were floating as well at each site. TSJV later performed conventional surface traverses from each site to verify and further develop the overall network. The horizontal points were transferred vertically at the shafts using Taylor-Hobson spheres and Baechler prisms above the shaft in three locations.

Tunnel. The machine tunnels were driven using Leica total stations with control points set behind the trailing gear, transferred to a temporary point as close to the heading as possible. The total station was set up and tied into the machine using a ZED interface unit and a ZED infrared target unit. ZED also provided the Windows-based software for the guidance screen and ring build software. The system was fairly simple and could be integrated with existing ZED control units and targets that Traylor already owned. The control points were set as reasonably close to the back-up trailing gear as possible and brought down the tunnel by locating points on both sides of the tunnel in a triangular and boxed zigzag fashion.

Furthermore, TSJV had Towill Surveying, Mapping, and GPS Services, of Englewood, CO perform a total independent survey at the Central Shaft as an additional layer within the project network. Towill performed their own shaft transfer and also used a DMT Gyromat 2000 gyroscopic-theodolite to verify azimuths independent of the shaft transfers. During the course of the tunnel drive, TSJV also used their own Sokia GP1 gyro-theodolite to verify azimuths and add extra measurements to the already robust network.

The TSM holed into the dead-end of Reach 1 on August 22nd, 2008. The end of the Reach 1 drive was accurately surveyed prior to the Robbins TBM removal and adjusted in the TSM DTA for the digger shield to find. After hole through, the convergence point ended up differing by 1″ in elevation and 1.5″ horizontally. This was viewed as quite a success for such a long drive.

AFTERWORD

At the time of this writing, pipe is currently being installed in the western half of the project. Mining continues with digger 2 on the eastern half and is expected to finish at the end of 2009. Another paper is bound to be written about the saga at the San Vicente Pipeline Tunnel, if not a couple of papers. Shall we say, to be continued…

REFERENCES

Jacobs Associates, November 29, 2004, "Geotechnical Baseline Report San Vicente Pipeline Tunnel," San Vicente to Second Aqueduct Pipeline Project, Prepared for the San Diego County Water Authority.

M. King and C. Hebert, 2007, "Steel-Fiber Reinforced Self-Compacting Concrete on the San Vicente Aqueduct Tunnel," pp 1243, *RETC Proceedings*.

M. Krulc, J. Murray, M. McRae, K. Schuler, 2007, "Construction of a Mixed Face Reach Through Granitic Rocks and Conglomerate," pp 928, *RETC Proceedings*.

B. Campbell, G. Revey, B. Garrod, 2008, "San Vicente to 2nd Aqueduct Pipeline-Tunnel: Drill and Blast in Advance of an Open-Face Digger Shield," pp 740, *NAT Proceedings*.

http://www.sdcwa.org/infra/cip-esp.phtml, December 2008, Emergency Storage Project Factsheet, San Diego County Water Authority.

Suggested Reading

S. Boone, A. Poschmann, A. Pace, C. Pound, 2001, "Characterization of San Diego's Stadium Conglomerate for Tunnel Design," pp 33, *RETC Proceedings*.

G. Raines and R. Wright, 2001, "San Diego's Tunnels," pp 15, *RETC Proceedings*.

S. Klein, D. Hopkins, M. McRae, Z. Ahinga, 2005, "Design Evaluations for the San Vicente Pipeline Tunnel," pp 804, *RETC Proceedings*.

B. Zernich, M. Krulc, and B. Robinson, 2005, "Northeast Interceptor Sewer Case History," pp 357, *RETC Proceedings*.

A REVIEW OF TUNNELING DIFFICULTIES IN CARBONATE SEDIMENTARY ROCKS

D.P. Richards ▪ Parsons Brinckerhoff

T.W. Pennington ▪ Parsons Brinckerhoff

J.A. Fischer ▪ Geoscience Services

G. Garcia ▪ GilCo Group Inc.

ABSTRACT

Tunneling in carbonate sedimentary rocks such as limestones and dolomites can be difficult when solution channels or voids are anticipated. Addressing the risk aspects associated with these formations can be done effectively through the Geotechnical Baseline Report (GBR) by properly characterizing the ground and ensuring that the Owner's and the Contractor's interests are represented and protected. This paper presents a review of tunneling difficulties in these rocks and includes a review of site investigations, geologic characterization, and presentation in the GBR. Key construction issues are also discussed, including appropriate excavation technology, mechanisms of face support and groundwater control, and quantifying grout takes.

CHARACTERISTICS OF CARBONATE SEDIMENTARY ROCKS

Carbonate sedimentary rocks can exhibit wide variations in engineering parameters and mechanical behavior that are often indicative of the complex geological history of the formation. These variations can be attributed to of alternating cycles of deposition, where slight changes in climate produced facies ranging from unconsolidated, soil like material to intact and sound rock. Other variations can be attributed to secondary effects, such as metamorphism, folding, faulting, weathering, dissolution and erosion, where weakness zones and voids often develop. Therefore, where carbonate sedimentary rocks are discussed in this paper, they are assumed to include those rocks that have been undergone these secondary effects, as many of the engineering difficulties encountered in these rocks are intimately related to the conditions resulting from secondary effects.

For the tunnel engineer, carbonate sedimentary rocks can also present a wide range of difficulties relating to both design and construction, including:

- Mixed-faced conditions consisting of hard, or consolidated, lenses in an otherwise soft ground face
- Open or soil-filled solution cavities
- Differential weathering
- Variable development of any of the above

KARST AND KARSTIC GROUND

Carbonate sedimentary rocks commonly exhibit karst, or karstic features. To facilitate the discussion of these features and their impact on engineering difficulties, the following definitions are presented for karst based on Hatheway (2005):

Karst. A geological process or condition in which carbonate or other water soluble terrane is or has been subjected to the passage of groundwater of relatively low total dissolved solids content and in which a variety of subsurface voids or enlarged joints and other discontinuities are found as openings, sometimes filled with soil material. Karst implies a proven condition of potential subsurface instability and represents the potential for karstic features to compromise the integrity and function of engineered works.

Karstic Terrane. A broad area exhibiting surface or subsurface evidence, from place to place, of the presence of karst processes. It is not appropriate to declare such ground "karst" and expect to encounter surficial or subsurface evidence of dissolution everywhere in or across an area broadly exhibiting such features. Karstic implies that topographic or subsurface evidence exists that suggests the presence of carbonate dissolution processes that may affect site suitability.

Based on these definitions, the term "karst" denotes a specific geologic condition, or process, that has been verified through direct observation, either through surface or subsurface detection techniques. "Karstic Terrane" is a more general term used to describe a broad area which exhibits characteristics of karst, but has not been directly observed.

KARST DEVELOPMENT IN SOUTH FLORIDA

Lane (1986) has noted that karst formation in Florida involves primarily weathering and erosion of carbonate sedimentary rocks, with the extent of karst development being a function of formation porosity and permeability. For example, he has categorized the south Florida area around Miami-Dade and Broward Counties, as "bare or thinly covered limestone—sinkholes are few, generally shallow and broad, and develop gradually. Solution sinkholes dominate."

Sowers (1996) has also noted typical observations on the potential for karstic conditions in south Florida, which are worthy of consideration at any site with a significant karst potential. For example, primary porosity of well-indurated, unweathered oolitic limestone underlying much of Miami, Florida, above the water table averages approximately 15% (void ratio = 0.18). Below the water table, the porosity can be as high as 75% (void ratio = 3.0), an increase of 500%. Also, loss of solids through dissolution in the bedrock is proportional to increase in porosity and can result in the decrease of rock strength; the unconfined compressive strength often decreases by as much as 75% to 90%.

Sowers' assessment should be compared with the observations made during the subsurface exploration for tunnel projects located in a known karstic area, in which vuggy to extremely vuggy materials are observed in core samples. It is also consistent with frequently observed difficulties encountered during drilling with lost circulation of the drilling fluids. Obviously, lost circulation is an excellent indicator of high porosity and/or extensive cavities.

Types of karst development are a function of geologic and environmental conditions. For consistency, terminology relating to karst (i.e., void) development should follow the characterization by Atkinson (1986). These terms are summarized in Table 1.

The types of karst development have also been classified by Kutzner (1996) and can be used as basic guidelines in evaluating the extent of karst at a proposed tunnel construction site. Kutzner's classification is provided below:

TUNNELING DIFFICULTIES IN CARBONATE SEDIMENTARY ROCKS

Table 1. Classification of voids in soluble rocks

Void Type	Modal Size	Remarks	Hydrogeological Function
Intergranular	10^{-3} to 10^{-1} mm	(a) Intercrystal boundaries (b) Intergranular pores that have not been filled with diagenetic cement. (c) Skeletal voids ranging up to several cm in late Tertiary and younger coral reef limestones.	3-dimensional network of interconnecting voids. Porosity varies from very low values in (a) up to 40%. Laminar (Darcian) flow.
Fractures	10^{-1} to 10 mm	Planar or curving fractures such as bedding planes, joints and minor faults, with or without minor enlargement by solution.	3-dimensional network of intersecting planar voids. May have strong preferred orientation in directions of dominant joint sets. Laminar (Darcian) flow.
Fissures	10 to 100 mm	(a) Fractures that have been mechanically widened. (b) Washouts of fine-grained sediments from shale or mudstone seams between massive carbonate beds. (c) Solutionally widened fractures, with tubular or anastomosing openings.	(a) and (b) affect local drainage. (c) may form local or regional networks of interconnecting fissures which channel ground water flow.
Conduits	100 mm to 10 m	(a) Network caves consisting of enlarged fissures within a single group of strata. (b) Sponge-like dimensional networks of interconnected caverns. (c) Dendritic or trellised networks of water filled tubes or freely draining caves and shafts.	(a) and (b) are uncommon, although (a) are found in thin limestones interbedded with sandstone. (c)-type conduits form underground river systems analogous to surface rivers, ranging in length from 100 m to 100 km.

Source: Adapted from Atkinson, 1986.

- Type 1—Large caverns extending up to several hundreds of meters (unlikely to not be detected in a tunnel site investigation)
- Type 2—Small caverns extending up to several meters, open or filled
- Type 3—Pipe-like open channels within a largely impermeable rock mass
- Type 4—Open or filled joints of various width, created by dissolution of minerals—grouted by common means, with borehole arrangement, pressure and grout material being a function of the degree of karst development
- Type 5—Porous rock—similar grout treatment as Type 4 noted above

It is important to note that early phases of karst development (Types 4 and 5) often encounter field hydraulic conductivities ranging from about 10^{-1} to 10^{-4} cm/s, or higher. These values should also be considered as a general condition, with actual in-situ hydraulic conductivity potentially up to several orders of magnitude higher locally (up to 10^1 cm/s as an upper bound).

Table 2. Typical karst indicators

Observation	Potential Karstic Condition
Drilling rod drops, sudden drop in SPT N-values	Open cavities or soft soil in-filled cavities
Repeated loss of drilling fluid and high measured hydraulic conductivities	Open and interconnected dissolution features; possibly only dissolution-enlarged bedrock discontinuities
Zones of poor recovery and RQD	Dissolution features
Through-flow of groundwater low in total dissolved solids	Indications of ground water flow through the site, as measured by ground water tracers
Ridges and clefts Top-of-Rock Surface	Indicator of paleo-geologic dissolution activity.
Dissolution-Affected Bedrock	Zones of relatively poor recovery and RQD directly below intervals of rod drops in stiff overburden soils

Source: Hatheway, 2005.

IDENTIFYING KARSTIC FEATURES

In accordance with the previous definitions by Hatheway (2005) and comments by Lane (1986), and including observations from detailed subsurface explorations at a proposed tunnel construction site, the site can be considered to be karstic terrane if it exhibits suggestive evidence of karst development. Even though no significant large-scale karst features may be detected during site exploration (i.e. dissolution cavities or large voids), observations made during exploration may suggest karst development at the site. Possible karst indicators are summarized in Table 2.

Site Investigations

Site investigations for tunnel projects commonly utilize standard exploration techniques in overburden soils and bedrock. In overburden, Standard Penetration Tests (SPT) provide disturbed samples for index testing as well as providing a relative indication of soil consistency/stiffness. In bedrock, conventional coring using a split barrel is a must. In very weak, friable, poorly cemented materials, it is often necessary to use triple tube core barrels and/or larger core barrels to improve recovery. If these improvements are not done, the low core recovery and poor RQD values may provide misleadingly low indications of rock quality.

The size and extent of karstic features can be difficult to estimate using conventional exploration techniques, i.e. auger or sonic drilling. In rotary wash drilling techniques, using water or a thin drilling mud and recording the depths and quantities of drilling fluid losses should be used as an aid in defining the presence, depth and approximate size of encountered karstic features. Extent of karst development can also be estimated by noting the occurrence of rod drops. Backfilling the boreholes with a thin, low shrinkage grout can aid in interpreting the subsurface conditions as well as protecting against sinkhole formation at the borehole location.

Furthermore, use of a split barrel in coring solutioned carbonate rocks (despite a driller's reluctance) should be considered. One has only to imagine a driller striking the side of the core barrel to release the rock "stick" and seeing zero RQD gravel pour out of the bottom of the barrel.

Standard Penetration Testing. Driving samplers in poorly cemented materials, while often appropriate for obtaining samples of in-situ material for visual classification, occasionally breaks down the more competent interbeds of limestone, and the number of blows recorded during SPTs may not be representative of the in-situ strength. When this is a concern, a Denison-type sampler may be necessary. Regardless, SPT

TUNNELING DIFFICULTIES IN CARBONATE SEDIMENTARY ROCKS 219

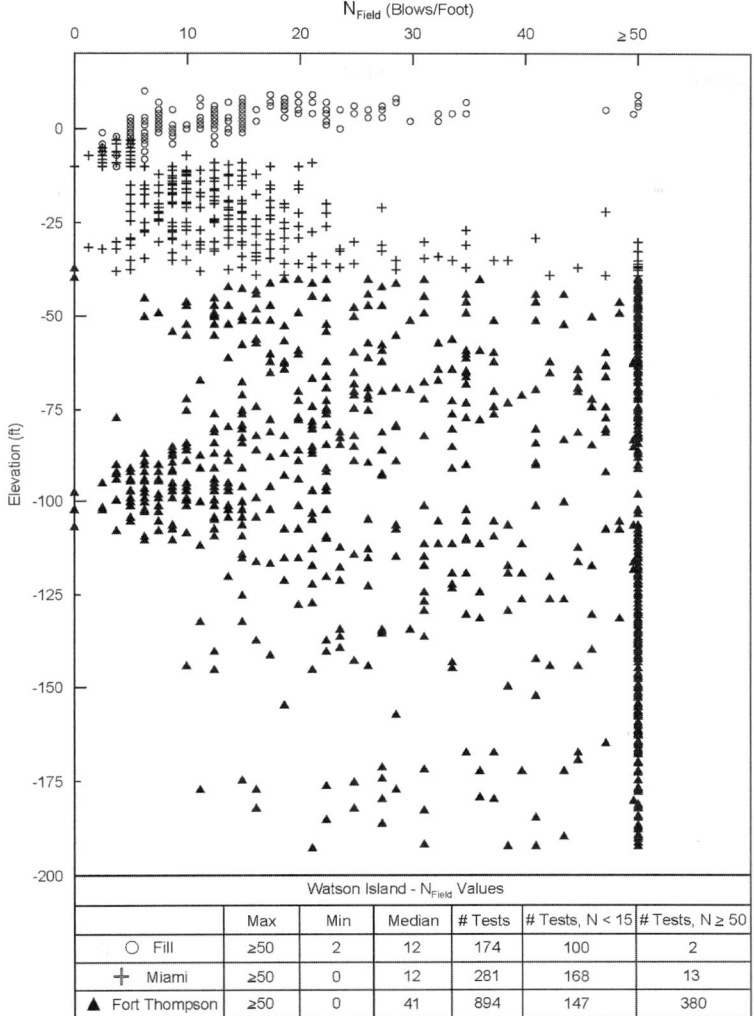

Figure 1. Representative SPT N-values from south Florida site

sampling in non-cemented or non-indurated materials generally provides good recovery and strength correlations using N-values are considered appropriate.

Unconsolidated (i.e., soil-like) zones are sometimes encountered in karstic prone limestone formations. Data from a site in south Florida where karstic prone, oolitic limestones (Miami and Fort Thompson Formations) were encountered is presented in Figure 1. The widely variable SPT values indicate the potential for these unconsolidated zones and provide a good illustration of the variability of formational consistency and strength with depth. It is evident weaker zones are present but a discernable pattern to SPT N-value distribution is difficult to identify. Also, stronger zones are most dominant in the Fort Thompson Formation as indicated by a median SPT N-value of 41 blows/foot; however unconsolidated zones are present at nearly all elevations throughout the formation.

Rock Coring. Primary geotechnical parameters obtained from rock coring include core recovery and RQD. From the measured RQD values, other engineering parameters are commonly inferred or interpreted. However, these commonly employed correlations may not always be valid. For instance, RQD is defined as the total length of all "hard and sound" core pieces greater than 4 inches in length, expressed as a percentage of the total length of core run. Deere and Deere (1988) recommended that core not meeting the "hard and sound" requirement should not be counted towards the determination of RQD. Deere and Deere recommended rock meeting ISRM (1978) weathering Grade I—Fresh and Grade II—Slightly Weathered, can be assigned the traditional RQD value, while Grade III—Moderately Weathered should be reported by qualifying the RQD by means of an asterisk (RQD*). Rock with Grade IV—Highly Weathered and Grade V—Completely Weathered rating should not be counted towards determination of RQD.

The "hard and sound" criteria can be subjective but it is important for carbonate sedimentary rocks where weathering or solutioning are encountered. In these rocks, the RQD values determined often deviate from standard application and hence negate common engineering correlations between rock mass classification and tunnel excavation and support requirements.

Variable data from subsurface explorations should be fully recognized when making engineering interpretations and conclusions with respect to RQD values. Use of RQD as an engineering parameter should be applied with caution and judgment and should be considered only as one measure of the variability of subsurface conditions. As illustrated in Figure 2, RQD usually provides a good identification of the range of rock quality and presence of relatively strong or hard strata in relation to relatively weak or soft strata.

REMOTE DETECTION OF KARSTIC FEATURES

Despite the amount of subsurface investigation often performed for a tunnel project, it can be difficult to detect, in advance of construction, subsurface cavities large enough to influence construction, but not so large that they are easy to identify. Experience indicates that it is often impractical, if not impossible (both budget and schedule considered) to detect all potential subsurface cavities within the tunnel envelope along the full length of a proposed tunnel alignment. An example is a cavern in Saudi Arabia (Grosch et al. 1987) that took over 11,000 m^3 of grout to partially backfill in the zone of construction. One of the initial 23 borings detected it, but at the peak of the rubble pile where the roof had collapsed leaving only a 3-inch open void. During the site investigation this void was attributed to being a large vug, which had been identified in multiple places in all of the borings and were common in the open excavation sidewalls.

Aerial photos and site reconnaissance often aid in locating potential problem zones prior to drilling. This first step is not only cost-effective, but can aid in interpreting the results of future explorations.

Geophysics

In many cases, geophysics can be an excellent adjunct to any exploration process and in identifying potential problematic areas that may require further investigation by conventional exploration techniques. To gain the most from geophysical surveys, the data must be correlated with direct investigation techniques, such test pits, probes or test borings, to provide a realistic interpretation of the data considering the inherent vagaries of karst.

Geophysical techniques considered appropriate for karstic terrane have been outlined by Fischer et al. (2004) who note that geophysical investigations often include,

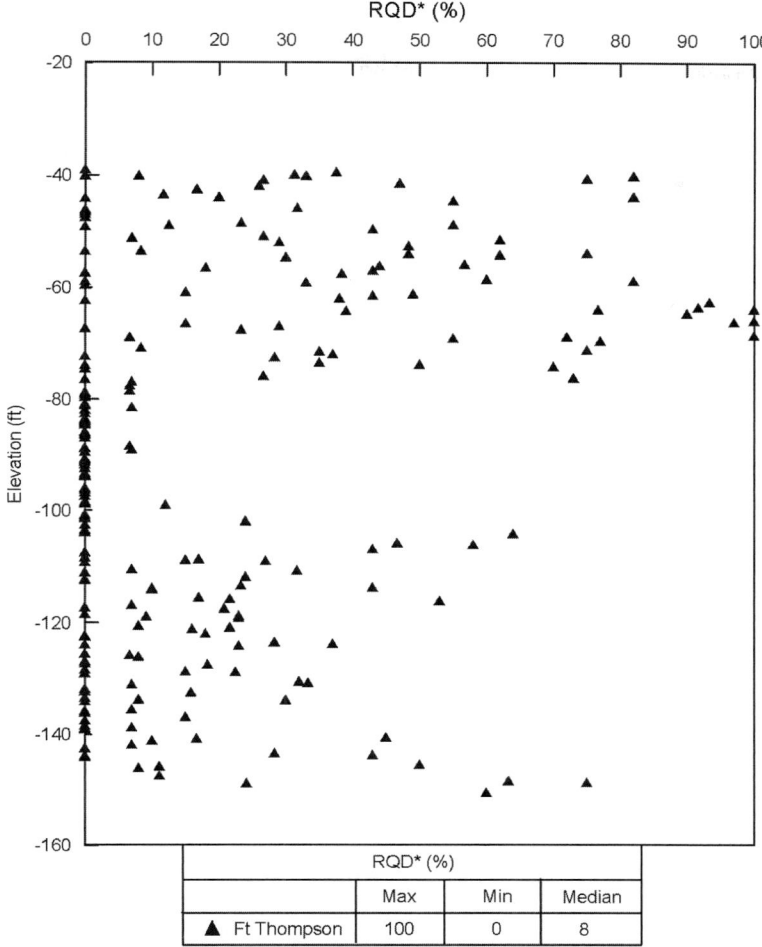

Figure 2. Measured RQD vs. depth for carbonate rocks from south Florida site

depending upon site specific ground conditions, ground penetrating radar, resistivity, spontaneous potential, electromagnetics, seismic reflection, and gravity methods. The selection of the most appropriate equipment is highly dependent upon the type of karst, i.e., recent coralline, flat-lying, Cambro-Ordovician (e.g., the mid-continent U.S.), or folded and faulted Cambro-Ordovician (e.g., Appalachian carbonates).

Using a combination of complementary geophysical methods is generally recommended to increase the likelihood of a successful investigation. It is possible to identify medium- to large-scale karst features when detailed geophysical investigations using multiple methods are integrated with drilling or other direct sampling methods. However, shallow, small, karst features may still be beyond resolution unless directly encountered by a drill hole or test excavation. Thus, interpretation of the geophysical data always requires direct observation or indication from conventional exploration techniques to identify or eliminate various diagnostic features.

The fact that the drill holes and geophysics did not identify any solution features does not mean that there are none present, especially small-scale features. Most geophysical investigations are not intended to identify small, but potentially significant, solution features. The costs of increased resolution must be balanced with the need for identifying specific features versus having a reasonable understanding of the geologic model along the tunnel alignment. An elaboration of some of the problems encountered at an Appalachian karst site is described in Connor, et al (2008).

Fischer, et al. (2004) have further noted factors that can adversely affect the identification of karst features include overburden thickness and composition, the presence of a thick weathered/saprolite layer, size and orientation of the karst feature, and whether the feature is clay-filled, water-filled, or air-filled. The geophysical results should be integrated with other geological and geotechnical data to delineate subsurface anomalies. It is in the authors' opinion that geophysical investigations for engineering applications are not "stand alone" studies. Hard data from borings is necessary to aid in interpreting the geophysical study results.

LABORATORY TESTING

In most site investigations for underground construction projects, the recovered samples are tested for both index properties and for strength and deformation characteristics. When in complex carbonate geology, it can be difficult to determine representative values of material properties. Any laboratory testing program must recognize that often only the good materials have been recovered, and that samples may be disturbed and not representative of in-situ conditions.

Unconfined Compressive Strength

As with both the SPT and RQD discussion previously, determining reliable unconfined compressive strength (UCS) data in carbonate rocks has similar difficulties. Method of drilling can affect sample integrity and resulting strength. Often, test samples can break during sample preparation. It is likely these samples exhibited limited strength due to lack of cementation, increased weathering and/or high porosity.

UCS of carbonate rocks is generally controlled by the degree of cementation and the extent of porosity, whereby an increase in cementation will tend to result in higher compressive strengths while an increase in porosity will tend to result in decreased compressive strength. Similarly, as unit weight decreases (often a function of porosity), compressive strength tends to decrease. The UCS data presented in Figure 3 provides an example of this strength relationship for the stronger, more competent samples of carbonate rocks, rather than the weaker materials, such as very weakly cemented limestone and interbeds of soil, or soil-like material. For these weak zones, the SPT N-value will provide a better representation of the strength and consistency (or density) of the material.

Elastic Properties

As with UCS testing, determination of elastic properties is difficult due to wide-ranging rock quality of limestone. Again, only the more competent samples can be sampled in an undisturbed condition and subsequently tested, resulting in a bias in the data set, as test results don't capture the weaker materials not being of sufficient quality to allow such physical property testing.

The measured elastic modulus (ε) and Poisson's ratio (μ) can vary widely, even amongst relatively intact and undisturbed samples. For more competent samples, ε can often be expected to be in the medium to high range (Deere and Miller, 1966). But with wide variations in rock quality, such values for the more competent samples are

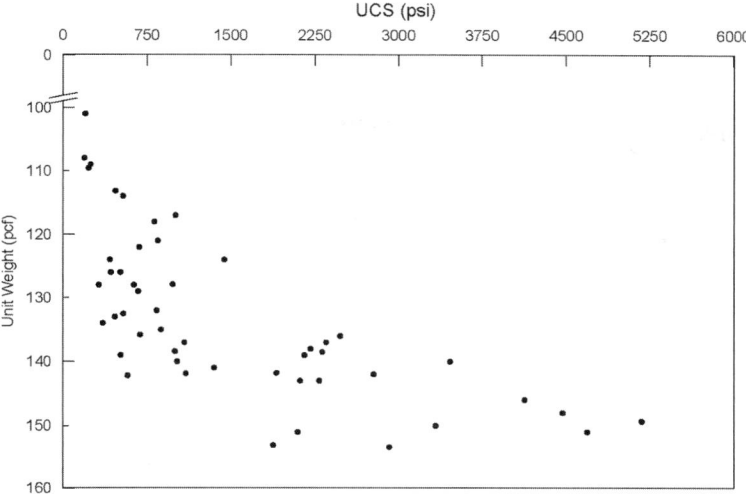

Figure 3. Compressive strength vs. unit weight from south Florida site (Fort Thompson Formation)

not indicative of the behavior of the less competent components of the same formation. In such cases, it is often considered appropriate to have two separate representative values, one for more competent materials and one for less competent materials. For the less competent materials, even the lower value should be considered as an upper bound since most of the weaker materials are sometimes not recovered, or testable if they were recovered.

Similar observations can be made for the measured values of Poisson's ratio, with the value dependent upon sample quality and integrity, which governed the sample failure mode during testing. As with the ε values, it may be appropriate to present two representative μ values, one for the more competent materials and one for the less competent materials. Again, the representative value for less competent material should be considered an upper bound, since most of the weaker materials will be either not recovered, or not testable if they were recovered.

The purpose of considering more than a single value of elastic properties is to represent two different material types in the tunnel. This is particularly critical for larger excavations, where an excavation may encounter both strong and weak (variable elastic properties) materials in the face at once, effectively presenting a "mixed-face" condition even though the rock type and formation is the same. Also, selected elastic properties should consider relevant scale effects to account for differences between intact and rock mass behavior.

Particle Size Distribution

Most samples of highly weathered limestone or calcareous sand interbeds are tested for gradation and usually classify as silts or sands. However, disturbances imposed by the SPT sampling procedures may cause natural cementation within sand and limestone inter-beds to break down, resulting in gradations representative of a "broken-down" aggregate, rather than a particulate sand or silt.

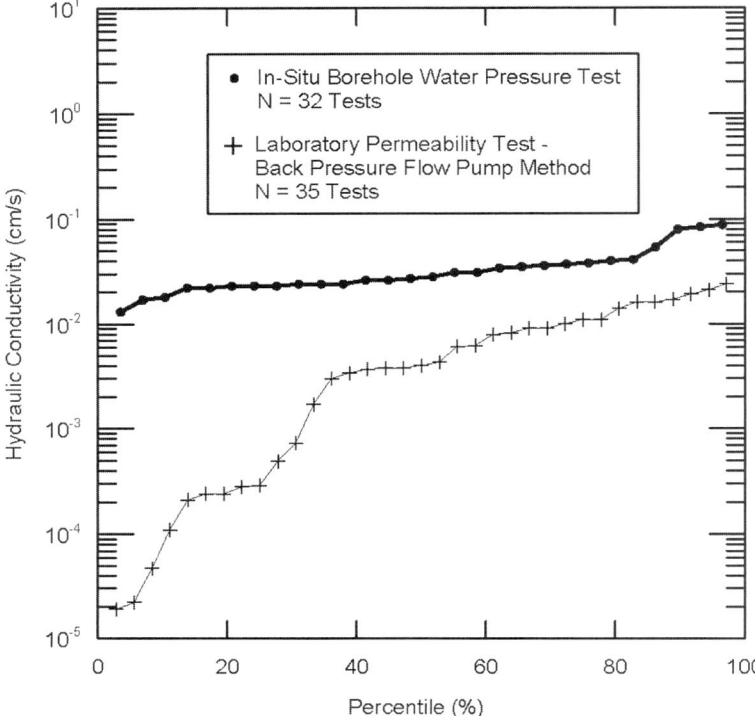

Figure 4. Cumulative distribution of hydraulic conductivity (k) from south Florida site

For such disaggregated materials, the gradations of material excavated by a tunnel boring machine (TBM) may be different than those gradations resulting from disaggregation by the SPT sampling process, due to the cutting action of the TBM and different amount of energy imparted on the material during excavation. In this situation, the gradation curves may not reflect the natural gradation present in-situ prior to the sampling process, but rather represent a partial disaggregation of the cemented material. Therefore, the sampled material can be assumed to be a crude representation of the disaggregation process that would take place during the excavation by a TBM.

Hydraulic Conductivity

In most subsurface exploration programs, borehole water pressure tests (packer or slug tests) and laboratory permeability tests are used to develop an estimate of the in-situ hydraulic conductivity. An example of such tests carried out at a site in south Florida is shown below in Figures 4 and 5. Measured field (in-situ) hydraulic conductivity was commonly in the range of 10^{-1} to 10^{-2} cm/s, while hydraulic conductivity from laboratory permeability tests were underestimated by up to three orders of magnitude. Note karst development at this site was consistent with Type 4 and 5.

Results from this site suggest laboratory tests may need to be subject to greater scrutiny when applied for engineering purposes. In-situ testing will likely provide a more reliable estimate of hydraulic conductivity of large-scale features, as interconnectivity of voids and varying porosity are likely to be captured.

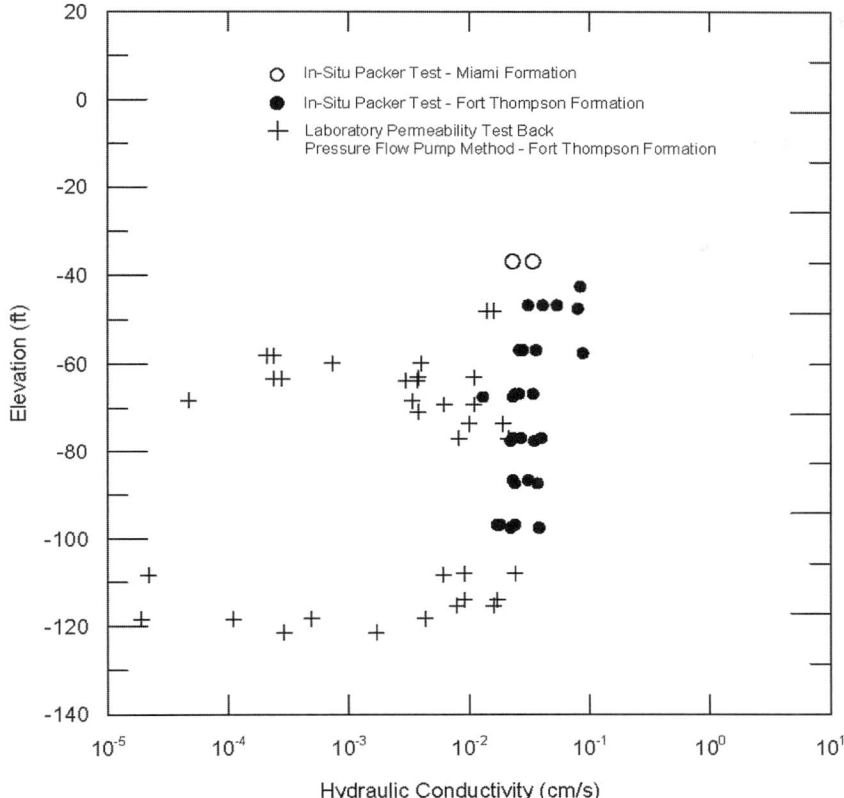

Figure 5. Hydraulic conductivity (k) vs. depth from south Florida site

Another consideration when estimating hydraulic conductivity of carbonate sedimentary rocks is the size and degree of interconnectivity of the pore spaces. Field tests, such as in-situ borehole packer tests or pumping tests, are useful for capturing the hydraulic properties of aquifer. However, in the absence of in-situ data, Atkinson's (1986) conceptual model relating pore size to porosity and hydraulic conductivity can provide general guidance (Figure 6). Note the relationships proposed by Atkinson assume laminar flow within the aquifer and consider both intergranular (secondary) and fracture (primary) flow. For further discussion on primary and secondary hydraulic conductivity of carbonate sedimentary rock, see Braithwaite (2005).

CONSIDERATIONS FOR TUNNELING

For tunnels constructed in carbonate sedimentary rocks, several important factors should be considered when evaluation the potential means and methods of tunnel construction including: (1) elevation of static ground water level, (2) presence of karst features or other highly permeable zones, (3) presence of weathered or weak, soil-like zones. Note tunnel construction considerations in well-developed karst are not discussed herein. For a more detailed discussion of these conditions and their impact on tunneling, refer to Milanovic (2004).

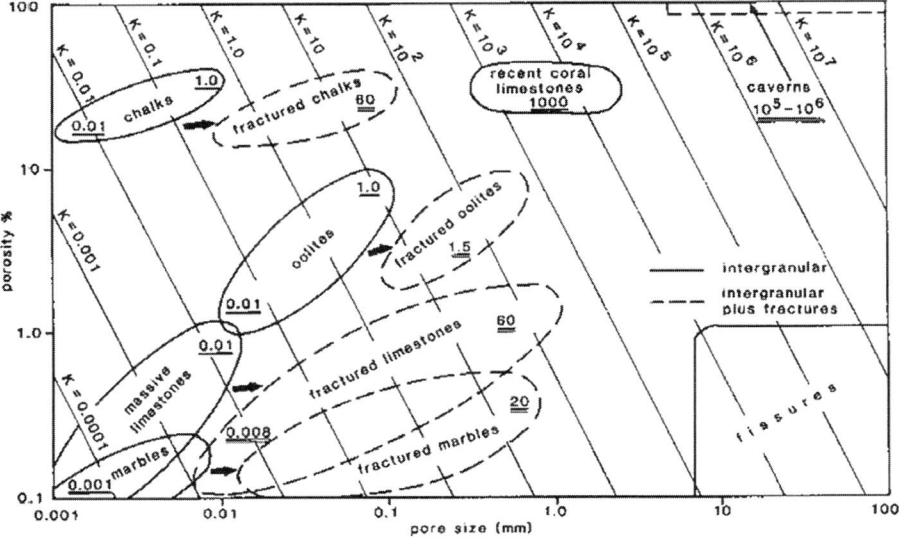

Source: Atkinson, 1986.

Figure 6. Relationships between porosity, pore size, and theoretical hydraulic conductivity (k) of carbonate rocks. Note k is presented in m/day.

Face Stability

One of the primary contributing factors to successful excavation of a bored tunnel is the ability to maintain face stability during the excavation process so that groundwater inflow and excessive ground loss is minimized. This is anticipated to be an important factor in all geological horizons to be encountered in carbonate sedimentary rocks, but will be more critical in zones of weakly cemented or loose, highly permeable granular materials. If such ground loss is not prevented, loss of face stability and a chimney to the ground surface is a major risk. Where heterogeneous conditions are expected, earth pressure balance (EPB) and slurry shield TBMs can provide the required face stability and are generally considered suitable for carbonate sedimentary rocks.

For earth pressure balance (EPB) TBMs, the proper selection and proportioning of conditioning agents is critical to the successful operation. In addition to facilitating muck removal and providing face stability, conditioning agents will help to achieve the plug in the screw conveyor. In carbonate sedimentary rocks, the gradation of the excavated material can be variable and difficult to predict based on site investigations alone and maintaining a sufficiently "impermeable" plug may be difficult. A range of conditioning agents may need to be considered. Also managing face pressure during the tunnel drive will be challenging and will require adequate definition of weak (unconsolidated) or highly permeable (karst) zones to allow sufficient time for face pressure to be increased or decreased prior to encountering the weak or highly permeable zone.

For slurry shield TBMs, the key risk is the loss of slurry in porous, and highly permeable zones, resulting in a reduction in face pressure and increase in face instability or uncontrolled inflow of water and soil into the cutting chamber. If site exploration boring logs indicate that there are zones within the tunnel horizon where drilling fluid circulation was lost, it is likely a similar condition will be encountered during tunneling with slurry shield TBM, with the magnitude of slurry loss a function of tunnel diameter and extend of the highly permeable zone.

For a bored tunnel in carbonate sedimentary rocks, it is generally advantageous for the TBM to be equipped to probe ahead of the face in order to detect weak or highly permeable zones. This is particularly critical when interventions for cutterhead maintenance are planned, as the presence of a large cavity or highly permeable zone will impact the ability to maintain a stable face under compressed air.

Formation Grouting

Planning a grouting program will depend on the nature and extent of karstic features and should consider:

- Some karstic terrane includes the presence of relatively intact, massive rock with widely open fissures and voids. Widely varying grout takes are typical.
- Cavities are often difficult to detect, particularly when the karst is in an initial stage of development, and the openings are widely distributed.
- Karstic cavities are occasionally filled with a soil like material that can be cohesionless and erodible. This is a very unfavorable situation, since these fillings are difficult to grout or replace.

Formation grouting executed ahead of an advancing TBM should be considered for the typical ground conditions encountered in carbonate sedimentary rocks. The necessity for and extent of formation grouting will be a function of the prevailing geologic conditions and development of karstic features, as well as the means and methods of construction selected by the contractor. These grouting requirements will be developed based upon sufficient interpretation of the geological, geohydrological, and geotechnical conditions.

As noted by Fischer et al. (1997, 2003, 2006), the type of grouting appropriate for karstic features will be a function of the type and extent of karst development. Different or multiple stages of grouting are often required, including a first stage of permeation grouting with a high mobility grout; if the grout take is excessive, this is followed by a second stage using low mobility grouts and/or the use accelerators to limit grout penetration. These considerations also apply to tail void grouting as discussed below. Generally, low mobility (compaction) grouting is effective in high permeability granular soils, but not economical in the residual clays often found atop the older, harder carbonates of much of the U.S.

Tail Void Grouting

Geologic variability of carbonate sedimentary rocks can often result in inconsistent or non-uniform tail void grouting into the surrounding ground and produce higher than anticipated grout takes. Limiting the grout penetration distance before grout set-up is often required. It is also probable that the cutting action of the TBM will cause overbreak along the perimeter of the excavation in weakly cemented materials causing higher tail void grout volume.

The amount of excess grout will also be a function of formation porosity, the grout mix design, injection pressure and the set-up time. Given the wide variation in porosity and hydraulic conductivity, it would not be considered unusual for total grout volumes to exceed 150% of the theoretical annular tail void grout volume. Total grout volume should be estimated based upon a knowledge of the site specific conditions, previous construction experience if available, and the contractor's proposed means and methods of construction.

LESSONS LEARNED FROM CONSTRUCTION

Alafia River Tunnel, Tampa, FL

The Alafia River Tunnel included approximately 900 feet of 96-inch outside diameter pipeline excavated by EPB TBM with ground water pressures as high as 2.5 bars. Subsurface conditions included limestone bedrock known as the Florida Aquifer. Details regarding construction in addition to those provided below can be found in Garcia (2006).

The TBM selected for the project was designated a hybrid EPB since it combined design components of a hard rock tunnel boring machine, EPB, and slurry shield machine. The TBM utilized 17-inch disc cutters with a maximum cutterhead rotational speed of 12 rpm. The cutterhead incorporated face ports to allow drilling and grouting ahead of the face. The TBM was designed to operate with a face pressure of 3 bars and was not equipped with a soil conditioning system, although the mucking system was closed and pressurized. Despite not having a soil conditioning system, the addition of polymers to the excavated muck (after secondary crushing) was required to improve its fluidity and create a plug flow within the auger screw conveyor.

Even with the addition of polymers, and the secondary crushing, it was not possible to convert the excavated limestone into a matrix that was suitable for extrusion and conveying with the screw auger. Therefore a field modification was made to the TBM that provided a closed mucking system, with the screw auger discharge directly connected to the "slurry" line that pumped the muck to the access shaft, where a second stage pump transferred it to the ground surface to a settlement tank prior to disposal. After this field modification, advance rates increased to about 40 feet per 9½-hour shift, almost double that experienced prior to the modification.

Other lessons learned on this project included the following:

- In many of the carbonate sedimentary rocks of Florida, rock quality does not improve significantly with depth, and tunneling difficultly generally increases due to higher hydrostatic pressures and buoyancy considerations.

- Voids and cavities in Florida carbonate sedimentary rocks are rarely larger than 48 inches, and can possibly be filled during tunneling due to the rotational and mixing action of the cutterhead.

- Soil conditioning was not applicable since the excavated and crushed weak rock was often poorly graded and had insufficient fines content to form a paste. A closed mucking system was preferable.

- Penetration of the cutterhead must be carefully controlled since it is possible to embed cutting discs too deep, forcing rotation of the shield, causing damage to other mechanical components.

CONCLUSIONS

For tunnels in carbonate sedimentary rocks, unique measures must be taken during design and construction, such that every reasonable effort is taken to develop reliable predictions of ground conditions and ground behavior during construction. These reasonable efforts include:

- Desk study of available subsurface geological and geohydrological information in the project area, as well as related tunnel case histories

- Site reconnaissance using aerial photography

- Comprehensive subsurface investigation program; preferably more extensive than what would have been done for less complex geological conditions

- Geophysical explorations to "fill in the gaps" (if appropriate) from the boring program
- Detailed laboratory testing program with careful consideration of sample integrity and evaluation of test results
- Risk assessment of geotechnical issues and need for additional explorations to better define or mitigate those risks
- Search for case histories of similar underground projects completed in similar ground conditions, even if none were completed in the project area
- Use of judgment and experience to establish realistic representation of subsurface conditions based upon all available project and related data
- A reasonable and clearly defined risk sharing strategy in the contract documents (e.g., through a geotechnical baseline report) so that neither the contractor or the owner are expected to carry more that a reasonable share of the construction risk

REFERENCES

Atkinson, T.C., 1986, "Soluble Rock Terrains," *A Handbook of Engineering Geomorphology*, edited by P.G. Fookes and P.R. Vaughan, Surry University Press, Glasgow.

Braithwaite, C.J.R., 2005, *Carbonate Sediments and Rocks*, Whittles Publishing, Glasgow.

Connor, J.G., M.J. McMillen, R.W. Greene, J.A. Fischer, J.G. McWhorter, D.L. Jagel, 2008, "Electrical Resistivity in Northeastern U.S. Karst—A Case History," *Sinkholes and the Engineering and Environmental Impacts of Karst, Proc. of the 11th Multidisciplinary Conference*, ASCE, pp. 71–80.

Deere, D.U. and D.W. Deere, 1988, "The Rock Quality Designation (RQD) Index in Practice," *Rock Classification Systems for Engineering Purposes*, ASTM STP 984, pp. 91–101, L. Kirkaldie, ed., Philadelphia.

Deere, D.U. and R.P. Miller, 1966, Engineering Classification and Index Properties of Intact Rock, Air Force Weapons Laboratory, Technical Report No. AFNL-TR-65-116, New Mexico.

Fischer, J.A., J.J. Fischer and R.J. Canace, 1997, "Geotechnical Constraints and Remediation in Karst Terrane," *Proc. of the 32nd Symposium Engineering Geology and Geotechnical Engineering*, Boise, Idaho.

Fischer, J.A., J.J. Fischer and R.S. Ottoson, 2003, "Grouting in Karst Terrane—Concepts and Case Histories," *Grouting and Ground Treatment: Proc. of the 3rd International Conference*, ASCE Special Geotechnical Publication #120, ASCE, New Orleans, LA.

Fischer, J.A., D.L. Jagel, J.J. Fischer and R.S. Ottoson, 2004, "The Expectations and Realities of Geophysical Investigations in Karst," *Proceedings of 55th Highway Geology Symposium*, Kansas City, MO, pp. 94–103.

Fischer, J.A., J.J. Fischer, J.J., J.G. McWhorter and R.S. Ottoson, 2006, "Geotechnical and Geohydrological Investigations in Eastern United States Karst," *Underground Construction and Ground Movement: Proceedings of the GeoShanghai Conference in Shanghai, China*, ASCE Special Geotechnical Publication #155, ASCE, Reston, VA.

Garcia, G., 2006, "The Challenge of Florida Limestone," *Tunnel Business Magazine*, June, pp. 33–39, Peninsula, Ohio.

Grosch, J.J., F.T. Touma, and D.P. Richards, 1987, "Solution Cavities in the Eastern Province of Saudi Arabia," *Karst Hydrogeology: Engineering and Environmental Impact : Proc. of the Second Multidisciplinary on Sinkholes and the Environmental Impacts of Karst*, A.A. Balkema, Rotterdam.

Hatheway, A.W., 2005, "Karstic May Not be Karst—When is it Safe for a Landfill," Engineering Geology Perspectives #21, Association of Engineering Geologists Special Publication No. 13.

ISRM, 1978, "Suggested Methods for the Qualitative Description of Discontinuities in Rock Masses," Commission on Standardization of Laboratory and Field Tests, *International Journal of Rock Mechanics and Mining Sciences*, Vol. 15, pp. 319–368.

Kutzner, C. 1996, *Grouting of Rock and Soil*, A.A. Balkema, Rotterdam.

Lane, E., 1986, Karst in Florida, Special Publication No. 29, State of Florida, Department of Natural Resources, Division of Resource Management, Bureau of Geology, Tallahassee, Florida.

Milanovic, P.T., 2004, *Water Resources Engineering in Karst*, CRC Press, Boca Raton, Florida.

Sowers, G.F., 1996, *Building on Sinkholes*, American Society of Civil Engineers, New York.

PART 4

Geotechnical

Chairs

Calvin Locke
King County DNRP

Mike Kucker
Shannon & Wilson

GEOTECHNICAL BASELINE REPORTS—A REVIEW

T. Freeman ▪ GeoPentech

S. Klein ▪ Jacobs Associates

G. Korbin ▪ Independent Consultant

W. Quick ▪ Independent Consultant

ABSTRACT

In the past ten years, the Geotechnical Baseline Report (GBR) has arguably become the key document for tunnel construction. This report not only allocates much of the risk involved with the work, it serves as the basis for bid preparation and is used extensively in resolving disputes during construction. This paper discusses some important issues related to the GBR and presents suggestions for improving these vital reports.

INTRODUCTION

The use of Geotechnical Baseline Reports (GBRs) for contractually defining anticipated ground conditions has become a widely accepted practice in the tunneling industry. The importance and the critical nature of these reports have increased the scrutiny they receive to unprecedented levels. The basic premise of a contractual geotechnical baseline has been well developed and communicated to the industry in the Underground Technology Research Council's guideline document titled *Geotechnical Baseline Reports for Underground Construction* (UTRC, 1997) and in the updated version of this document (UTRC, 2007).

Despite the acceptance of GBRs, there is room for improvement. Engineers and geologists struggle to develop specific numerical baselines from a myriad of geotechnical properties, especially where geologic conditions are highly variable. Contractors are frustrated because they are not always provided with the baselines they need. Owners feel taken advantage of when baselines are used to justify claims in a manner not intended or the baselines are not respected in the dispute resolution process.

The intent of this paper is to provide suggestions for improving the effectiveness of GBRs, based on our collective experience with many tunnel projects and their GBRs. It is intended that these suggestions will complement the UTRC guidelines and result in better, more useful GBRs. Our suggestions fall into four categories:

1. Establishing baselines
2. Ground behavior/performance assessments
3. Construction considerations
4. Use of the GBR during construction

ESTABLISHING BASELINES

Two of the greatest challenges in preparing a GBR are determining the ground conditions that need to be baselined and how to quantify these baselines. As stated in the UTRC guidelines, the goal of the GBR is to "translate the results of geotechnical investigations and previous experience into clear descriptions of anticipated subsurface conditions upon which bidders may rely."

Baselines were intended to provide contractors a mechanism whereby they were not held responsible for unlimited risks involved with unforeseen ground conditions. Contractors have the right to expect that the baselined ground conditions presented in the GBR are reasonable. Ultraconservative baselines (i.e., attempts to shift unreasonable risk to the contractor) may not be successful, and this practice is discouraged because it may distort the bidding process and is contrary to the intent of the GBR.

Owners have the right to expect that contractors will consider the full spectrum of baselined ground conditions when determining appropriate construction means and methods and preparing their bids. Furthermore, owners have the right to expect that Dispute Review Boards (DRBs) will evaluate differing site condition claims against the actual baselines presented in the GBR.

Developing accurate or representative numerical baselines can be difficult, especially considering the natural variability of most geologic formations. In addition, the geotechnical investigations conducted for most tunnel projects only sample and test a small fraction of the soil/rock mass that will be encountered in the tunnel, typically much less than one percent. Considering the variability of Mother Nature and our limited geologic database, it is understandable that developing baselines presents some significant challenges. The following paragraphs discuss issues that have been problematic and suggestions for overcoming these issues.

Appropriate Baseline Topics

It is important to recognize that each baseline presented in the GBR establishes a potential target for a claim. Many GBRs baseline an extensive array of geotechnical parameters (such as unit weight, Atterberg Limits, blow counts, etc.), including parameters not relevant to the tunnel construction. Although this increased level of detail creates the illusion of a comprehensive GBR, and seemingly serves to protect the owner from differing site condition (DSC) claims, in some cases it has come back to haunt owners during construction. In order to justify a DSC claim, some contractors search for a deviation in one of the baselined variables, regardless of its importance in tunnel construction, and devise a creative explanation for how this variable impacted their means and methods and/or progress rate. Therefore, as a general rule it advisable to only baseline the soil/rock properties that are necessary for a contractor to evaluate tunnel excavation means and methods, estimate production rates, and design temporary or initial supports, when required.

After contract award, the GBR may not receive much attention until problems arise and DSC claims appear. Evaluations of DSC claims should be based on a comparison of the actual encountered ground conditions against the ground conditions baselined in the GBR. Accordingly, it is desirable for the baselines to be selected and quantified in terms that are readily measurable or verifiable in the tunnel, considering the means and methods that are likely to be used to construct the tunnel. For example, it can be very difficult to measure the frequency, orientation, and aperture of rock mass discontinuities along the tunnel (certainly not continuously) if the contractor elects to use a shielded tunnel boring machine (TBM) with a segmental concrete lining. Limited access to observe in situ ground conditions hampers both the contractor and owner. The contractor has difficulty demonstrating that ground conditions deviate from the baseline, and the owner cannot easily verify that ground conditions are consistent with the baseline. In some cases, it is better to baseline the end result, such as groundwater inflow into the tunnel, rather than a soil/rock property, such as permeability, which may vary widely and is difficult to measure in the field during construction. In Europe, partly to overcome these limitations, TBM boreability parameters (i.e., penetration rate and cutter wear) are sometimes used to classify the rock mass; however, these parameters can be difficult to interpret if machine operation becomes an issue. As discussed below, in many cases the end result can be significantly influenced by the contractor's selected means and methods and execution of the work; therefore, care is required to ensure that baselines are as independent of the contractor's means and methods as possible.

Quantifying Baselines

After the variables to be baselined have been selected, the next step is to quantify them. As discussed above, quantifying baselines is often a difficult process due to inevitable limitations in the geotechnical database, and these difficulties are even more acute when geologic conditions are complex.

Engineers and geologists tend to baseline the full range of possibilities, arguing that as long as actual conditions fall within that range, a DSC will not be encountered. However, in cases where a physical property can have a wide range, such as rock strength, it is difficult for contractors to select a single or average value on which to base their bid. Contractors must recognize that most tunnels have to be constructed through a range of geologic conditions, often a wide range of conditions. The commonly applied "average value" of a certain ground characteristic may not even be meaningful. It may just be the midpoint between two extremes. Owners must realize the importance of carrying out sufficient geotechnical investigations to develop a statistically representative sampling of the ground prior to bid. If a dispute arises, a similar statistically representative data set of the encountered ground conditions should be collected to allow for a valid comparison between anticipated and actual conditions.

The distribution of the various ground conditions along any given tunnel reach also are generally described by a range of conditions, and if possible, with specific identifiable features, such as fault zones. Similar to the disadvantages of baselining too many individual ground characteristics, as discussed above, dividing the tunnel into too many reaches also can be a disadvantage.

Furthermore, in complex geology the orientation and characteristics of many faults or shears can change significantly horizontally and vertically such that predicting their specific subsurface locations and conditions is quite uncertain. This raises an important question. If a fault zone, anticipated between certain limits in a hard rock tunnel, is described as "poor ground," is it possible to also baseline some additional randomly located poor ground elsewhere in that reach to accommodate the likely presence of faults and shears that cannot be specifically located? Of course, there is no rational reason why "poor ground" cannot occur randomly in other portions of a reach that is outside specifically delineated fault zones. This condition certainly can be covered by baselining a percentage of the reach as poor ground to accommodate anticipated ground that cannot be specifically delineated. Although, if characterized in this manner, this randomly located poor ground would have to be similar in character to the poor ground associated with the specifically delineated fault zones and be limited to some maximum width.

Average Baselines

As an example of typical geologic variability and the significance of average or mean baselined values, consider the unconfined compressive strength (UCS) data shown in Figure 1 for two sandstones from two projects in which the authors were

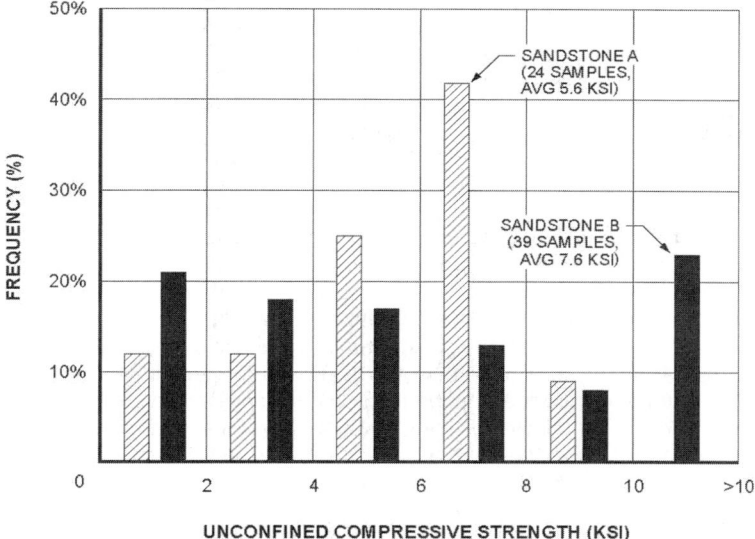

Figure 1. Comparison of UCS data from two sandstones

involved. Sandstone A has strength distribution that roughly resembles the bell-shaped Gaussian distribution that you might expect for a material subject to natural statistical variability. Sandstone B, on the other hand, has an almost uniform distribution across this UCS range, with about 20 percent of the samples having a strength of less than 2 kips per square inch (ksi) and about 25 percent with an UCS higher than 10 ksi. There are valid reasons for this type of distribution, such as variations in the mineralogy or texture of the rock, differences in cementation, healed joints, or other small scale defects in the samples. The key point is that for Sandstone A, over 60 percent of the data is within 2 ksi (±) of the mean strength, whereas for Sandstone B, only about 20 percent of the data is within 2 ksi (±) of the mean value. Baselining the average UCS value for Sandstone A appears to be a reasonable way to represent the mean strength of this rock. However, for Sandstone B it would be better to baseline a strength distribution by histogram or a distribution similar to its histogram (i.e., baselining the percentage of the rock greater than 10 ksi and less than 2 ksi as well as some intermediate values).

Another approach might be to determine the average rock strength from each boring in the formation and either use a weighted average approach or have the baseline average rock strength vary along the tunnel alignment similar to the borings, provided this is a valid trend based on the data and the data are not adversely influenced by features such as fault zones. Setting the baseline using an approach other than the strict arithmetic mean of the test data, which may or may not be representative, should be acceptable as long as the basis is logical and clearly explained.

Baselines Not Consistent with Test Data

Care must be used to assign baselines that are reasonably consistent with the statistical database. For example, Gould (1995) describes a project where the engineer increased the baseline rock strength to account for the difference between laboratory rock strength tests and in situ rock strength resulting from sample disturbance. However, the reason for this adjustment was not fully explained in the GBR, and the

DRB ruled for the contractor, indicating that this adjustment appeared to be a strategy to protect the owner and not a proper baseline for bidding. When the baseline deviates from the test data, it may be important to explain the basis or justification for the adjustment in the GBR so it is clear that the baseline is not being established arbitrarily or unfairly. However, in the authors' opinion, a baseline is a baseline, even without such a justification.

Groundwater Control

More recently, the issue of tighter control of groundwater inflows has been coming up more frequently for a variety of reasons, including potential third-party impacts. This issue becomes most challenging in complex geological environments, particularly in hard rock with high groundwater levels, where the seemingly unpredictable nature of rock mass discontinuities controls the inflow into the tunnel. To define the range of conditions expected, two baselines are often employed: one for flush flows and steady-state inflows from untreated ground to define in situ conditions, and a second for flush and steady-state inflows from grouted ground. Sophisticated numerical groundwater models are useful tools that can be used to help quantify the inflows. However, extreme care is necessary in setting the baseline for treated ground as the amount and type of grouting needed to achieve this condition will be difficult to determine in advance of construction, and will also be highly dependent on the contractor's grouting means and methods.

In cases where the ground or groundwater conditions are poorly defined and have the potential to adversely impact the work, the owner often elects to direct pre-excavation grouting measures to achieve desired reduction in groundwater inflows and to control costs. However, this approach can impact the contractor in several ways that complicate the work. Contractors typically want to be free to do the grouting they feel is necessary and to be paid for this grouting. With these competing interests, whether it is better for the owner or contractor to control such grouting remains an unresolved issue in the tunneling industry. The best way to handle this issue will vary by project depending on the technical requirements of the work; the risk or uncertainties involved; the contracting/delivery method being utilized; and the relationships between the owner, contractor, and any third parties. From a GBR perspective, it is important to recognize that the baselines that are ultimately established need to be consistent with the contracting approach and specified requirements.

Conflicting Baselines

Conflicting baselines result when more than one soil/rock property with the same general purpose are baselined. For example, with regard to TBM boreability and disc cutter wear, the abrasiveness of hard rock can be evaluated by using the Cutter Life Index (CLI) of the SINTEF method, or by using the Cerchar Index (Johannessen, 1988). If both parameters are baselined and they do not result in the same predicted cutter life, the GBR has provided two baselines for the same purpose that do not agree. While in this instance the contractor would be free to select either baseline, it is clearly not an ideal situation and one that can easily be avoided by providing, in this case, only one baseline for cutter wear.

Similar problems arise in determining the completeness of the baseline. The CLI requires information on mineralogy (rock type) and quartz content to evaluate cutter wear. If only the CLI is baselined, but not quartz content, the baseline is incomplete.

Conservative Baselines

Some GBRs seem to be written to protect the owner from any risk, attempting to put all of the risk on the contractor. There is little advantage, either to the owner or to the contactor, in artificially making baselines too conservative. Although technically the owner has the right to do so, it serves no purpose to make the baseline so conservative that the owner pays for conditions that are unlikely to occur. Also, when baselines are made overly conservative it can backfire when the "law of unintended consequences" comes into play. In general, baselines should be realistic and, as close as possible, reflect the actual expected ground conditions. It should be recognized that there is a natural tendency to be conservative due to the uncertainties involved with a limited database and our imprecise knowledge of geologic variability. This underscores the importance of completing adequate geotechnical investigations in the first place.

Ground Classification

In classifying the ground, the GBR commonly employs terms and definitions that are standard to the industry. Occasionally, however, it is advantageous to develop new definitions or modify existing ones. For example, many tunnel projects in Atlanta, Georgia, have employed a weathering classification system solely developed for the region and with terminology familiar to local contractors. In other situations, established classification systems, such as the Terzaghi's rock classification system (Proctor and White, 1968), have been found to be too narrow. For example, Terzaghi's definition of "crushed" rock indicates it is comprised of "chemically intact rock" with the "character of a crusher run." If, however, the crushed material is weathered or altered, resulting in a clay fraction, the clay can act as a weak cohesive binder. Therefore, to include this material in the crushed rock category, a modification to the definition is required. Contrary to some recent DRB decisions, the authors believe such modifications or project-specific definitions are acceptable as long as the definitions are clear and unambiguous and the reasons for making these changes are also explained in the GBR.

Terzaghi also uses the phrase "chemically intact" in his description of blocky and seamy rock (Proctor and White, 1968). Some GBRs have used Terzaghi's rock classification system and baselined the weathering condition of the rock mass separately. This has been criticized by some as being inconsistent because the term "chemically intact" does not appear to be consistent with a weathered rock mass. However, Terzaghi did not intend to limit the term "blocky and seamy rock" to only "chemically intact" rock, as he also indicates:

> This condition is encountered in both closely jointed and badly broken rock. The joints may be narrow or wide, empty or filled with the products of rock weathering.

In dealing with this apparent contradiction, the engineer has two options: (1) modify the standard Terzaghi definition, as discussed above; or (2) develop a project-specific classification system. Using a project-specific ground classification system avoids the contradiction by defining each ground class in terms selected to specifically communicate the unique aspects of the ground conditions to the contractor.

GROUND BEHAVIOR/PERFORMANCE

Predicting ground behavior in the tunnel excavation is one area of the GBR where engineers and geologists often have difficulty in making an accurate assessment. This is because ground behavior ultimately depends on both the in situ properties of the soil or rock mass and the contractor's applied means and methods. Ideally, a discussion of potential ground behavior in the GBR should only reflect the unmodified, in situ ground conditions and not relate to effects of the applied construction means and methods. While selected means and methods will modify the behavior of the ground, this is the contractor's domain and, as has been demonstrated on many projects, the engineer is rarely correct when making assumptions about the contractor's means and methods or workmanship, unless the methods are specified.

For example, subhorizontal foliation combined with high angle intersecting joints can result in overbreak within the crown area. In TBM excavations, fallout could occur some distance behind the cutterhead only in response to displacements due to stress redistribution around the excavation opening. However, if drill and blast methods are used, all of the overbreak can occur simultaneously with excavation of each round. Should the GBR explain this? While in our opinion the answer is yes (as discussed below), the magnitude of the effects are primarily related to the contractor's particular means and methods, and thus the specific amount of overbreak should not be baselined.

In ground with short standup time, if the contractor delays the installation of adequate initial support, more overbreak can occur than if installed in a timely manner; the key words being "adequate" and "timely." In this regard, great care is necessary in preparing the GBR by limiting the discussions to the relevant facts, mainly unmodified ground conditions. Baselines regarding ground behavior that are heavily influenced by the applied means and methods may not be appropriate.

CONSTRUCTION CONSIDERATIONS

Some have argued it is not appropriate for GBRs to address construction considerations and the report should focus entirely on baselining the anticipated ground conditions. This may have to do with concerns regarding the level of detail provided when the contractor's means and methods are unknown and there is a potential of introducing extraneous or possibly misleading material. On the other hand, the GBR is the only interpretive geotechnical report provided to the contractor and the consequences of the anticipated ground conditions on the construction work should be explained, *but not be baselined*, to enhance the contractor's understanding of the project requirements.

Many claims deal with the relationship between the ground conditions and the construction process. Examples include slaking or softening of the ground, overbreak, timeliness of ground support, TBM gripper problems, face stability, abrasive rock conditions, and the need for pre-support, to name a few. The value of discussing construction considerations (or potential construction problems) is that this provides the contractor with key information in planning the work and may help avoid claims and delays during construction (Waggoner et al., 1969; USNCTT, 1984). Unless the specifications require the contractor to specifically address the issue, the contractor may not have a contractual requirement to provide a solution but at least they are alerted to the situation and have the opportunity to take appropriate action.

Another advantage of discussing construction considerations in the GBR from a legal perspective involves the differing site conditions clause. This clause recognizes two types of DSCs: Type 1 where conditions differ materially from those indicated in the Contract; and Type 2 which involves unusual physical conditions that differ from those ordinarily encountered or generally recognized as being inherent in the work (UTRC, 1997, 2007). Violating a GBR baseline usually would be considered a Type 1 DSC. If the GBR were silent on a certain construction problem/difficulty (i.e., no discussion of construction considerations involved with the anticipated ground conditions) then the contractor could possibly allege a Type 2 DSC.

After the engineer identifies a potential construction problem or issue, there are three general approaches that can be adopted in the GBR (and design documents):

1. Be silent on solving the problem/issue and let the contractor decide how to address it.

2. Provide a recommended solution(s) and allow the contractor to select the solution.
3. Specify a solution that will work and require the contractor to implement that approach.

It is difficult to identify which approach is best. This depends on the issue, the potential consequences in terms of risk and cost, and in some cases, the owner's preference. When the problem/issue is left entirely up to the contractor to solve (No. 1 above), then it must be assumed that the low bidder will probably try to implement the least expensive solution. No claims should result as long as this solution works. This approach may offer the owner potential cost savings because the contractor is free to select the approach that is most economical, but it could involve some significant risk for the owner if this approach does not work.

The second approach is to not specify means and methods, but instead to discuss concerns and appropriate solutions through recommendations provided in the GBR. Presenting relevant discussions, recommendations, and/or suggestions in the GBR as to appropriate means and methods (such as the need for certain features on a TBM) provides the contractor with an assessment of how ground conditions could impact the work and would indicate intent, but would not be binding. In that case, the contractor is free to accept the recommendations or ignore them. However, if they are ignored and this issue becomes a factor in a dispute, the DRB has been provided with specific indications for consideration.

The last approach often is used when there are difficult or risky ground conditions and also when construction could impact third parties. Rather than try to define or baseline conditions that are too complex to convey accurately in words, an approach that is believed to be the appropriate (and usually somewhat conservative) is specified. In response, contractors often indicate that the resulting specifications are overly prescriptive. Design engineers argue that it is important to define requirements they perceive as important for success; however, this results in a potential for conflicts as the contractor wants to be able to select the means and methods of construction. Again, discussions of construction considerations in the GBR can be useful in explaining the designer's logic for prescriptive requirements.

USE OF THE GBR DURING CONSTRUCTION

Contractors need to carefully study the GBR and understand the implications of the baselines before preparing their bid. Some contractors retain an engineering geologist or geotechnical engineer to assist them in assessing the GBR during their bid preparation. Others only review the GBR well into the construction and then retain engineering geology or geotechnical engineering support when they want to pursue a DSC claim. Possibly this has to do with the limited time frame for preparing the bid. Whatever the reason, this later approach is counterproductive and is not in keeping with the intent of the GBR.

Likewise, the owner's construction management (CM) team also should study the GBR at the outset and not wait until the contractor submits a DSC claim to get familiar with the document. Extra care should be exercised during the review of the contractor's submittals to ensure that the construction considerations discussed in the GBR have been addressed and incorporated into the contractor's selected means and methods. Recommendations and/or warnings in the GBR, not addressed by the contractor, should be pointed out in the CM team review comments.

During construction, when reviewing DSCs it is most important that DRBs render their decisions based on entire contents of the GBR, not just on a narrow interpretation of the report. For example, if the condition of the ground (strength, joint frequency or characteristics, weathering, etc.) varies from the GBR descriptions but the behavior is the same as indicated in the report, does this constitute a valid DSC? The only way to find a DSC in this situation is for the impact on the contractor to be due to the ground characteristics that varied from the GBR, not the behavior.

The authors recently experienced a case where the GBR warned the contractor of raveling ground conditions in "highly altered" rock; however, a DSC was awarded by the DRB for more highly altered ground even though the actual behavior of the ground was consistent with the baseline. This questionable decision was not accepted by the owner.

It is important for DRBs to recognize that a DSC claim is not automatically valid just because a baseline has been violated. There must also be some impact to the contractor, and the impact must be "solely" due to the property or characteristic that has differed from the baseline (Cibinic et al., 1995). Failure to evaluate and determine if this connection exists is inappropriate. Where the contractor was warned of potential problems in discussions of construction considerations and these warnings were ignored, even if certain baselines were violated, the contractor may have some responsibility to address the condition.

RECOMMENDATIONS

In a relatively short time, only about ten years, GBRs have become an accepted practice in the tunneling industry. Therefore, it is important to make these reports as effective as possible. The following recommendations should be considered:

- Reasonable baselines need to be incorporated into the GBR. Avoid unnecessary or superfluous baselines. Overly conservative baselines or attempts to place unreasonable risk on the contractor may not be successful and are not recommended. Other approaches for incorporating contingency measures in the contract to deal with adverse conditions should be utilized.

- It must be recognized that geotechnical investigation needs for construction often exceed the needs strictly for design (Gould, 1995). Preparing realistic baselines starts with having a statistically adequate geologic database.

- Quantification of the baselines should consider how the properties will be measured in the tunnel, and in general, the baselined values should be consistent with the geologic database or explain any deviations.

- Ground performance or baselines related to ground behavior should be avoided to the extent they are affected by the contractor's selected means and methods. However, discussions of construction considerations/difficulties should be included in the GBR to communicate to the contractor the potential consequences of the baselined ground conditions. Whether recommendations are provided for handling these issues or specific contract requirements are included in the specifications depends on the potential impact, the risk in leaving it up to the contractor (and the marketplace), and the potential for third-party impacts.

- Contractors should give attention to discussions of construction difficulties/problems described in the GBR and consider these issues in the selection of their means and methods.

- Owners should actively review the GBR while it is in the developmental stage and make sure they understand the cost/risk ramifications. For major projects, an independent opinion from a consultant board/panel is recommended. The implications of the baselined conditions should be fully addressed by the CM team in reviewing the contractor's submittals during construction.

- DRBs should respect the baselines indicated in the GBR and consider the GBR in its entirety in deciding the merit of DSC claims. It is not reasonable for the contractor to base a claim solely on one word, phrase, or sentence in the GBR, especially where the potential construction difficulties/problems are correctly recognized and discussed in the GBR but ignored by the contractor.
- GBRs are not a panacea and do not by themselves guarantee a successful project. Difficult and complex ground conditions require an experienced, qualified, and conscientious contractor, whose selected means and methods are appropriate and compatible with the anticipated ground conditions. Such challenging projects also require a contracting and CM approach that encourages all parties to work together to solve the challenges they face. A GBR can identify the challenges but will not solve all of the problems.

REFERENCES

Cibinic, J. and Nash, R.C., 1995, *Administration of Government Contracts*, 3rd edition, George Washington University, National Law Center, Washington, D.C., 1560 p.

Gould, J.P., 1995, "Geotechnology in Dispute Resolution," *Journal of Geotechnical Engineering*, ASCE, July, pp. 523–534.

Proctor, R.V. and White, T.L., 1968, *Rock Tunneling with Steel Supports*, with an "Introduction to Tunnel Geology" by K. Terzaghi, Commercial Shearing, Youngstown, OH, 296 p.

Johannessen, O., 1988, *Hard Rock Tunnel Boring*, Project Report 1-88, University of Trondheim, Norwegian Institute of Technology, Trondheim, Norway, 183 p.

Waggoner, E.B., Sherard, J.L., and Clevenger, W.A., 1969, "Geologic Conditions and Construction Claims on Earth- and Rock-Fill Dams and Related Structures," in *Engineering Geology Case Histories*, Number 7, Geological Society of America, pp. 33–43.

U.S. National Committee on Tunneling Technology (USNCTT), 1984, *Geotechnical Site Investigations for Underground Projects*, National Academy Press, Washington, D.C., Vol. 1, 182 p.

Underground Technology Research Council (UTRC), 1997, *Geotechnical Baseline Reports for Underground Construction*, Technical Committee on Contracting Practices of the UTRC, ASCE, Reston, VA, 40 p.

Underground Technology Research Council (UTRC), 2007, *Geotechnical Baseline Reports for Construction*, Technical Committee on Geotechnical Reports of the UTRC, ASCE, Reston, VA, 62 p.

GROUND CHARACTERIZATION FOR CSO TUNNELS IN WASHINGTON, D.C.

Maurice A. Ponti Jr. ▪ Camp Dresser & McKee

Steven B. Fradkin ▪ Hatch Mott MacDonald

Xiaohai Wang ▪ Camp Dresser & McKee

Moussa Wone ▪ Hatch Mott MacDonald

Ronald E. Bizzarri ▪ District of Columbia Water and Sewer Authority

Edward J. Cording ▪ University of Illinois at Urbana-Champaign

Roger C. Ilsley ▪ RI Geotechnical

Qamar A.O. Kazmi ▪ Schnabel Engineering

ABSTRACT

The D.C. Water and Sewer Authority is implementing the Anacostia River Projects component of its Long-Term CSO Control Plan, including 21 kilometers of 3.7- to 7-meter-diameter CSO near-surface diversion storage/conveyance structures and flood relief, pressurized-face, soft ground tunnels at a maximum depth of 65 meters. This paper describes a project-specific soil grouping system that is primarily based on soil characteristics for tunneling derived from tests including Atterberg limits, grain size, swelling, mineralogy, triaxial, pressuremeter, and consolidation, which differentiates the project from past area tunnel projects using descriptive systems based primarily on area geology.

INTRODUCTION

The proposed Anacostia River Projects portion of the Long-Term CSO Control Plan (LTCP) for the D.C. Water and Sewer Authority (authority) includes 21 kilometers of 3.7- to 7-meter-diameter CSO storage, conveyance, and flood relief tunnels. The LTCP is being implemented in accordance with the requirements of a federal consent decree. A facility plan was developed for the Anacostia River Projects to establish a viable, cost effective approach to integrate consent decree and functional hydraulic requirements while limiting design, construction and long-term operational risks to the authority.

A preliminary subsurface investigation program consisting of approximately 70 widely spaced borings was performed in several phases to support the facility plan level conceptual design. The results of the program establish a feasible tunnel alignment and develop a preliminary description of the anticipated soil types to be encountered, and to serve as a basis for subsequent design evaluations. The recommended tunnel system alignment begins in the southwest part of Washington, D.C. (D.C.) at the authority's Blue Plains Advanced Wastewater Treatment Plant, trends northward along the east banks of the Potomac and Anacostia Rivers, crossing the Anacostia River

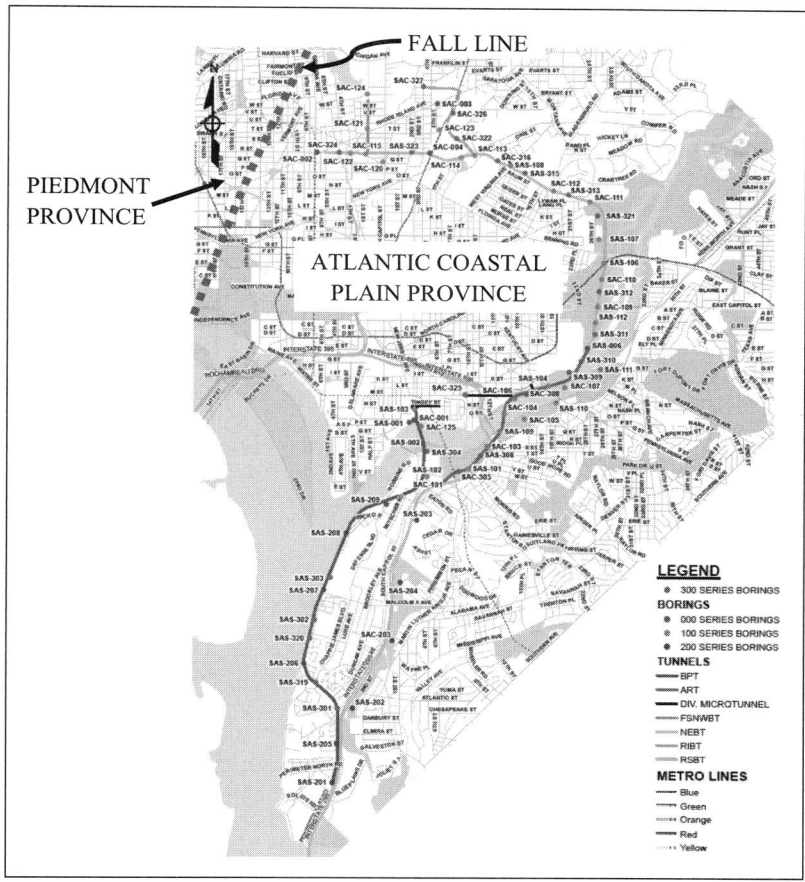

Figure 1. Proposed tunnel alignment and geotechnical boring plan

at the 11th Street Bridges, trending northward near the west bank of the Anacostia River, and finally curves to the west through northeast D.C. to a termination point at 6th Street, NW at Rhode Island Avenue (Figure 1). The proposed soft ground tunnels are anticipated to be excavated using pressurized-face tunnel methods at depths to 65 meters with associated shafts, near-surface diversion structures, and conveyances. The focus of this paper is to describe a project-specific soil grouping system developed for the soils within the proposed tunnel zone that is based on standard soil test results such as Atterberg Limits, gradation, pressuremeter and consolidation/triaxial tests, with consideration of the characteristics of mineralogy, stickiness, swell potential and abrasion. The proposed alignment occurs within the Potomac Group soils at a greater depth than most of the existing Washington Metropolitan Area Transit Authority (WMATA) tunnels. The soil descriptions used by WMATA (1967) appear to be generally applicable to the overlying Terrace and Alluvium deposits along the proposed alignment.

The proposed soil grouping system utilizes soil characteristics and is reflective of the complex, dissected, coastal plain depositional environment. It is aimed to assist in the evaluation of the soil-related performance of pressure-faced Tunnel Boring Machines (TBMs).

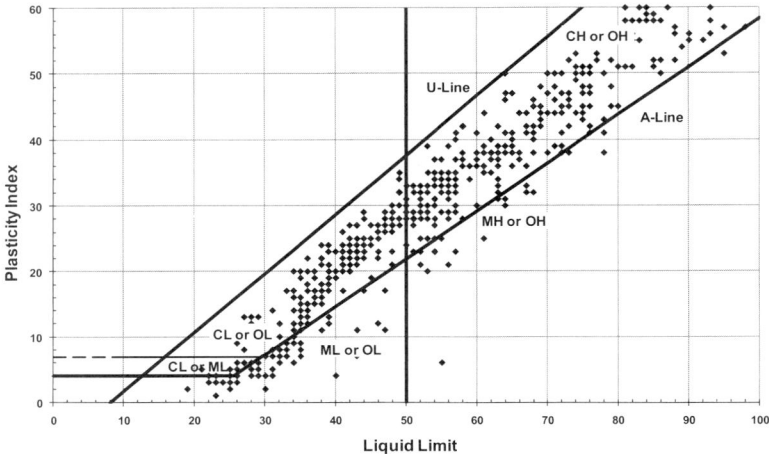

Figure 2. Atterberg limits Potomac Group test data

GEOLOGICAL SETTING

The project area is located near the western edge of the Atlantic Coastal Plain Physiographic Province and differentiated from the adjacent Piedmont Physiographic Province by a southwest-northeast trending "fall zone" in the D.C. area (Figure 1). The bedrock in the project area consists of gneiss and schist that is overlain by a few hundred feet of unconsolidated Cretaceous and Quaternary Period deposits. The uppermost Cretaceous deposits are assigned to the Potomac Group; these are overlain, successively, by Tertiary and Pleistocene Terrace deposits of reworked Potomac sand, gravel and alluvium, by Recent Alluvial deposits of clay/silt/sand/ gravel, and by fill deposits of various compositions mixed with manmade materials (WMATA 1967). The Potomac Group is subdivided into the upper Patapsco/Arundel and lower Patuxent Formations. The Patapsco/Arundel Formation consists primarily of silt and clay layers with minor sand layers, and the underlying Patuxent Formation consists primarily of silty/clayey sands, with minor silt and clay layers, and occasional lignite bearing seams. Most of these geologic units represent fluvial deposits or dissected /braided deposits of clay, silt, sand, and gravel layers of various thickness and lateral extent. The Potomac Group fine grained deposits generally exhibit over-consolidated behavior with several hundred feet of overburden believed to have been eroded away (WMATA 1967).

SUBSURFACE CHARACTERIZATION

The subsurface investigation program was conducted using conventional drive and wash techniques with selected Standard Penetration Test (SPT) and Shelby tube samples. Sonic drilling methods were also used at a number of drilling locations to obtain a more continuous vertical soil profile to evaluate depositional homogeneity. Based on 841 Sieve Analysis tests and 636 Atterberg Limit tests, the soils were classified in accordance with the Unified Soil Classifications System (USCS). Evaluation of the data resulted in a significant overlap between several of the referenced geologic formations and individual soil strata behavior at various depths within the tunnel area; highly plastic soils with liquid limits exceeding fifty percent were observed at several locations within the Potomac Group (Figure 2). Generally, the swelling potential is a function of the moisture adsorption capacity of the plastic clay particles, and the stickiness potential is a function of the adherence of plastic clay material to metal or TBM machinery. The clay size particles that occur in clayey sands exhibit a low to high plasticity range, which indicates the swelling potential and stickiness are not uniformly related to grain size, but may be attributed to mineralogy.

The results of the 12 mineralogy tests indicate that the sand portions of the samples tested contain high to very high percentages (65% to 95%) of Quartz and low percentages (2% to 15%) of Potassium (K) feldspar. Manufacturers of soil additives indicate high quartz and/or K-feldspar content within a TBM slurry mix generally exhibit moderate to high abrasivity on cutter tools, bearings, and TBM hardware (Nilsen et al. 2006). The clay portions tested contain low percentages of Chlorite, Kaolinite, and Illite micaceous minerals, and significant percentages (up to 69%) of Smectites, which are a group of clay minerals with expansive properties similar to bentonite and montmorillonite. Highly expansive clay materials such as smectite have a high stickiness potential and may exhibit an increased frictional resistance against the TBM shield resulting from an increased swelling from an increase in moisture content of the in-situ clays. This ground condition may become important at the contact between predominantly clayey and predominantly sandy strata within the tunnel zone, such as the Patapsco/Arundel clays and the saturated Patuxent sand interface zones.

Results of 14 swell tests conducted indicate the Potomac Group clay samples from the southern portion of the alignment exhibit significantly higher swell pressures than the Potomac Group clay samples from the Anacostia River area. All of these results correlate well with the limited number of clay mineralogy tests performed to date.

A total number of 31 pressuremeter, 28 consolidation, and 15 triaxial tests reveal several hard clay zones at various elevations that correspond empirically to several hundred feet of eroded overburden, and indicate over-consolidated soils with a lateral earth pressure coefficient (K_0) greater than 1.0 at several locations within the upper Patapsco/Arundel Formation.

The subsurface investigation and laboratory test results indicate a varied and overlapping depositional environment. A soil grouping system based on soil characteristics appears to be an appropriate method to describe anticipated ground conditions and will assist in the evaluation of the performance of pressure-faced TBMs, as it relates to ground conditions.

SOIL GROUPING SYSTEM

The encountered soil materials have been divided into six groups according to physical characteristics and engineering properties. A brief description of the soil groups follows:

- Group 1 (G1), silty clay and silt/clay mixtures (CH, MH);
- Group 2 (G2), clayey silt and silt/clay mixtures (CL, ML, CL-ML);
- Group 3A (G3A), silty/clayey sand (SM, SC, SM-SC);
- Group 3B (G3B), silty clayey gravel (GM, GC, GC-GM);
- Group 4 (G4), fine to coarse sands, non-plastic (SP, SP-SM, SP-SC, SW, SW-SM, SW-SC);
- Group (G5), well graded and poorly graded gravel, non-plastic (GP, GP-GM, GP-GC, GW, GW-GM, GW-GC).

The results of the soil grouping system are presented in detail on Table 1.

The Atterberg Limits of soil groups G1, G2, and G3 are presented in Figure 3. This figure generally shows that the G1 group is highly plastic and has a Liquid Limit (LL) over 50. The G2 and G3 groups exhibit an LL less than 50, and overlap with an LL

Table 1. Soil grouping system description

Group Symbol	Description	Criteria	USCS Symbol
G1	Generally consists of clay, silt and silt/clay mixtures of high plasticity that may contain minor amounts of fine sand. Usually hard and over-consolidated. Some stress relief may have occurred where exposed to weathering. A major component of the Patapsco/Arundel Formations. Also present in layers within the underlying Patuxent Formation at some locations. These soils have a high stickiness potential and are considered to be low to non- abrasive.	More than 50% passing the #200 sieve. The liquid limit is higher than 50.	CH, MH
G2	Generally consists of clay, silt and silt/clay mixtures of medium to low plasticity that may contain subordinate amounts of sand. Usually hard and over-consolidated; some stress relief has occurred where exposed to weathering, a major component of the Patapsco/Arundel Formations. Lignite may be present within the Patapsco/Arundel and Patuxent Formations. These soils have a moderate to high stickiness potential and are considered to have a low abrasive potential.	More than 50% passing the #200 sieve. The liquid limit is lower than 50.	CL, ML, CL-ML
G3 (G3A & G3B)	Generally consists of non-plastic silty and/or clayey sand, or sand/silt clay mixtures of low plasticity clay (G3A) or non-plastic silty and/or clayey gravel, or gravel/silt clay mixtures of low plasticity clay (G3B). Usually very dense/hard with cohesive materials potentially over-consolidated except where exposed to weathering. Locally contains fragments, seams or layers exhibiting cementation to various degrees often in close association with the presence of lignite. Contains some lignite at various locations. The crisp texture of the lignite often yields pieces characterized as sand in the particle size analyses that can be easily crushed to finer material. This condition, along with any associated cementation could affect the particle size tests resulting in dispersion of similar materials over multiple soil groups. A major component of the Patuxent Formation. It is also common within the overlying Patapsco/Arundel Formations of the Potomac, but less widely distributed. These soils have low to high abrasive potential with low to high stickiness potential.	**G3A** Less than 50% passing the #200 sieve and more than 12% passing No. 200 sieve. The percent sand is more than the percent gravel. **G3B** Less than 50% passing the #200 sieve and more than 12% passing No. 200 sieve. The percent gravel is more than the percent sand.	SM, SC, SC-SM GM, GC, GC-GM
G4	Generally consists of fine to coarse sands. May also contain subordinate amounts of fine gravel. Lignite fragments and seams may also be present. Usually very dense. Locally contains fragments, seams or layers exhibiting cementation to various degrees. A major component of the Patuxent Formation. Also present within the overlying Patapsco/Arundel Formation of the Potomac at some locations, but less widely distributed. These soils are non-plastic, have a high abrasive potential, and are considered to be non-sticky.	Less than 12% passing the #200 sieve. The percent sand is more than the percent gravel.	SP, SP-SM, SP-SC, SW, SW-SM, SW-SC
G5	Generally consists of very dense gravel and gravel/sand mixtures. Cobbles and potentially boulders may also be present. Near the base of the Patuxent Formation and above the underlying bedrock at the far west and north edges of the investigation area. These soils are non-plastic, have a high abrasive potential, and are considered non-sticky.	Less than 12% passing the #200 sieve. The percent gravel is more than the percent sand.	GP, GP-GM, GP-GC, GW, GW-GM, GW-GC

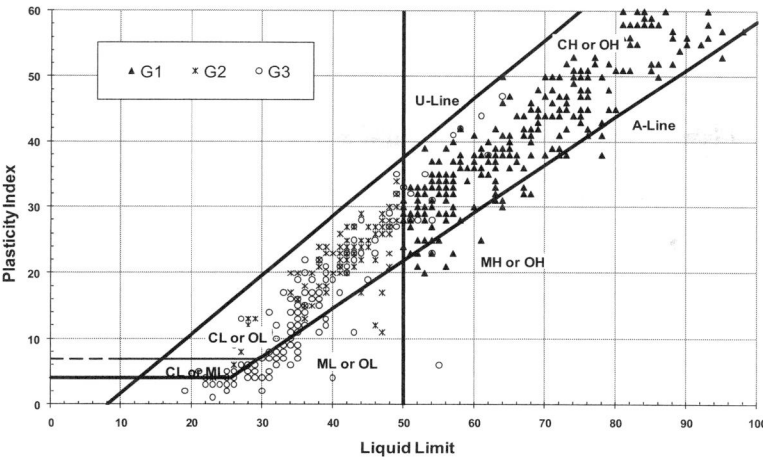

Figure 3. Atterberg Limits test data for groups G1, G2, and G3 soils

between 35 and 50. This overlap may be attributed to reworked mineralogical source areas.

SOIL ADDITIVES

Currently it is anticipated that two types of pressurized-face TBMs may be considered to excavate the tunnel: earth pressure balance (EPB) and slurry machines. Generally, the stickiness of fine grained soils and the abrasiveness of the coarse grained soils are significant issues affecting advance rate and machine wear. The most common types of soil additives attempt to create a flowable, toothpaste consistency with respective ground types. The additives that would potentially be used with project soil groups are as follows: highly plastic, swelling, smectite type clays (G1) may require foaming agents; layered, clayey silts, with minor sand percentage (G2) may require foam and an abrasivity reducer; clayey/silty sands/gravels (G3A/B) may require foam and polymers to reduce heat buildup and abrasivity. Clean sands and gravel mixtures (G4, G5) may require foam, polymer and lubricating gels to improve uniform consistency and reduce abrasivity and friction.

CONCLUSIONS

The selected USCS classifications appear to correlate well with the six project soil groups that occur in the varied coastal plain deposits located in the project area. A comparative summary of the project soil groups, their USCS classifications, a summary soil description and soil additive ground types, is shown on Table 2.

To evaluate the impact of the ground conditions on the performance of pressure-faced TBMs on tunnel projects located within complex, inter-layered, and discontinuous soil deposits, a quantitative approach is necessary. The method used should differentiate the potential effects of the ground conditions on the performance of EPB and slurry machines. The project specified soil grouping system described in this paper is a quantitative approach that generally meets these objectives. It is based on the *in-situ* tests and laboratory testing results from over eight hundred soil samples. By developing the project soil group system, both the physical characteristics of the existing soil and the anticipated soil behavior during tunneling operations are taken into consideration.

Table 2. Project soil group and ground type summary

Project Group Symbol	USCS Classifications	Soil Description*	Soil Additive Ground Type
G1	CH, MH	Plastic clay	Swelling clays, plastic
G2	CL, ML, CL-ML	Silty clay	Layered sandy silts/clays
G3A	SM, SC, SC-SM	Clayey/silty sand	Clayey/silty sands
G3B	GM, GC, GC-GM	Clayey/silty gravel	Clayey gravel
G4	SP, SP-SM, SP-SC, SW, SW-SM, SW-SC	Clean sand, minor gravel	Clean sands
G5	GP, GP-GM, GP-GC, GW, GW-GM, GW-GC	Gravel, minor sand	Clean gravels

* Soil descriptions are generalized and not in accordance with the USCS classification system.

Based on the test data, the clay mineralogy appears to be a significant indicator of swell potential, plasticity, and stickiness, particularly with respect to G3 soils (clayey/silty sands), as well as G1 and G2 soils. Mineralogy tests identify the potential swelling characteristics of fine grained materials and should be included as part of further subsurface investigation programs assessing ground conditions for tunneling.

BIBLIOGRAPHY

American Society for Testing and Materials. 2004. Annual Book of ASTM Standards, Section Four, vol. 04.08, Soil and Rock (I): D 420-D 5779.

ASTM D 2487, Standard Classification of Soils for Engineering Purposes (Unified Soil Classification System).

Darton, N.H. 1950. Configuration of the Bedrock Surface of the District of Columbia and Vicinity. United States Geological Survey (USGS) Professional Paper 217.

Drake, A. A. Jr. 1989. Metamorphic Rocks of the Potomac Terrane in the Potomac Valley of Virginia and Maryland: Washington, D, C., American Geophysical Union, 28th International Congress Field Trip Guidebook T202, 22p.

Fleming, A.H., A.D. Drake Jr., and L. McCartan. 1994. Geologic Map of the Washington West Quadrangle, District of Columbia, Montgomery and Prince Georges Counties, Maryland, and Arlington and Fairfax Counties, Virginia.

Froehlich, A.J. and J.T Hack. 1975. Preliminary Geologic Map, District of Columbia. U.S. Geological Survey Open File Report 75-537.

McCartan, L. 1990. Geologic Map of the Coastal Plain and Upland Deposits in the Washington West 7.5 Minute Quadrangle, Washington D.C., Maryland and Virginia. U.S. Geological Survey Open File Report 90-654.

Mixon, R. B., et. al. 1989. Geologic Map and Generalized Cross Sections of the Coastal Plain and Adjacent Parts of the Piedmont, Virginia. U. S. Geological Survey Miscellaneous Investigations Series I-2033.

Nilsen, B., F. Dahl, J. Holzhäuser, and P. Raleigh. 2006. Abrasivity of soils in TBM tunneling. Tunnels and Tunneling International, 38(3): 36–38

Southworth, S., and D. Denenny. 2005. Geologic Map of the National Parks in the National Capital Region, Washington, D.C., Virginia, Maryland, and West Virginia. United States Geological Survey (USGS) Open File Report 1331 (Southworth, 2005).

Tenbus, F.T. 2003. Lithologic Coring in the Lower Anacostia Tidal Watershed, Washington, D.C., July 2002. United States Geological Survey (USGS) Open File Report 03-318.

Thewes, M. 1999. "Adhäsion von Tonböden beim Tunnelvortrieb mit Flüssigkeitsschilden"(Adhesion of Clay Soils in Tunnel Driving Using Slurry Shields). Bergische Universität Gesamthochschule Wuppertal, Bodenmechanik und Grundbau, No. 21.
Washington Metropolitan Area Transit Authority (WMATA). 1967. Washington Metropolitan Area Rapid Transit Authorized Basic System—B&O Route. Final Report—Subsurface Investigation—Volume 3. By Mueser, Rutledge, Wentworth & Johnston Consulting Engineers, New York, N. Y., December 1967.
Washington Metropolitan Area Transit Authority (WMATA). 1969. Washington Metropolitan Area Rapid Transit Authorized Basic System—Benning Route and A Portion of the Pentagon Route. Final Report—Subsurface Investigation—Volume 4. By Mueser, Rutledge, Wentworth & Johnston Consulting Engineers, New York, N. Y., July 1969.
Washington Metropolitan Area Transit Authority (WMATA). 2006. Adjacent Construction Design Manual, Office of Joint Development & Adjacent Construction, Washington Metropolitan Area Transit Authority, Washington, DC, April 2006 (Revision 2).
Williamson, L., L. Benson, D. Girard, and R. Bizzarri. 2007. The DC WASA Anacostia River CSO Control Tunnel Project. In *Rapid Excavation and Tunneling Conference Proceedings 2007*, Toronto, Canada, June 10–13. Society for Mining, Metallurgy, and Exploration

ACKNOWLEDGMENTS

The authors want to extend their gratitude to Leonard Benson of the District of Columbia Water and Sewer Authority, Randall Essex, Larry Williamson, Mike Schultz, François Bernardeau, Michael Gilbert, Gary Shaughnessy, David Field, Andrew Stone and all others who contributed to the development of this paper.

ACTUAL VS. BASELINE TRACKING DURING TBM TUNNELING IN HIGHLY VARIABLE GLACIAL GEOLOGY

Ulf G. Gwildis ▪ CDM

Leon E. Maday ▪ King County DNRP

John E. Newby ▪ CDM

ABSTRACT

The Brightwater Conveyance System includes 21 km of tunnels mined by two earth pressure balance Tunnel Boring Machines (TBMs) and two slurry TBMs. The probabilistic baseline approach defined in the Geotechnical Baseline Report (GBR) can representatively be verified by use of a comprehensive tracking system to identify face conditions that allows comparison with the baselines. The glacial geology includes a high lateral variation in soil types that, when combined with the pressurized face tunneling methods, creates challenges for documenting face conditions. Baseline tracking using spoil samples and TBM operational parameters provides information to document actual conditions and support resolution of differing site condition claims.

INTRODUCTION

The Brightwater Conveyance System will connect a third wastewater treatment plant in the growing Seattle metropolitan area—the plant being currently under construction—to a marine outfall into the Puget Sound through a 21 km tunnel. The tunnel with design I.D.s between around 4 m and 6 m includes a gravity driven effluent pipe for treated wastewater and also includes sections where untreated wastewater is conveyed towards the treatment plant by means of a pump station. Additional conveyance system connections to the existing sewage system with design I.D.s between 1.2 m and 2.1 m are part of the construction contracts and are constructed by open cut as well as microtunneling methods. The tunnel construction project consists of three contracts with the tunnel alignment being divided into four sections. These four tunnel sections are partially completed by means of two Herrenknecht Mix-Shields (Slurry TBMs) and two Lovat Earth Pressure Balance TBMs (EPB TBMs).

The pronounced topography of the project area required up to 60 m deep launch and receiving shafts resulting in an overburden thickness up to 150 m and in a hydrostatic head above the tunnel invert up to 7.3 bar (Figure 1). Vertical borings with an average spacing of 150 m were carried out to investigate the subsoil conditions along the alignment. The variability of the glacial and non-glacial sediments of several glaciation cycles—evidence has been found for at least six continental glaciations in the Puget Sound area—allows correlation between the borings with regard to groups of deposits of similar geological history but not for geological facies units or units of similar geotechnical properties. In response to this complexity, a probabilistic geotechnical baseline approach was chosen. This approach categorizes the geological conditions into Tunnel Soil Groups (TSGs) with specified composition and engineering properties. Single TSGs or combinations of TSGs constitute typical face conditions defined in the GBR. For each Brightwater tunnel the geotechnical baseline allocates percentage

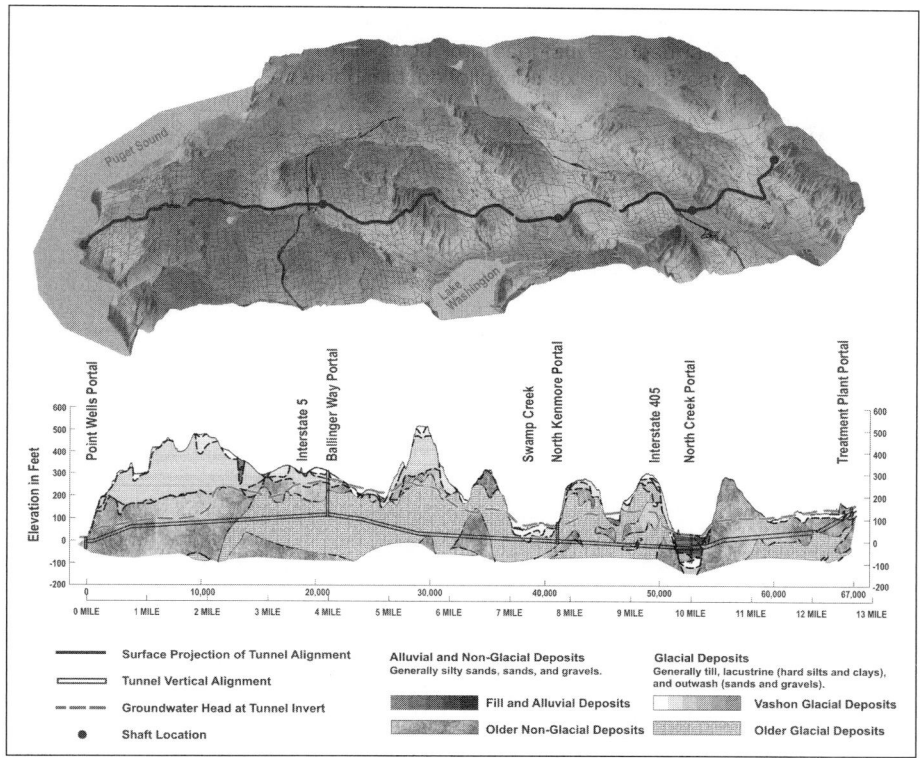

Figure 1. Block diagram showing tunnel alignment and shaft locations

ranges of the tunnel length to each typical face condition. This concept lends itself to actual-versus-baseline tracking during mining to provide comparisons to the baselines.

GEOLOGY AND GEOTECHNICAL BASELINE

The Brightwater Conveyance System is located within the Puget Sound Trough, a structural basin situated between the Olympic Mountains to the West and the Cascade Range to the East, which was formed in conjunction with the Juan de Fuca oceanic plate being thrust beneath the North American continental plate. The more recent geologic history of the Puget Sound Trough is dominated by a succession of at least six continental glaciation cycles. In the foreland of the glacier advances from the North, large lakes were formed in which fine-grained lacustrine sediments were deposited. As the ice sheets advanced farther to the South, the sediment supply became coarser and the lowlands were filled with glaciofluvial sand and gravel. When the ice front reached the geographic latitude of the project area, sub-glacial melt water and ice reworked the deposits re-depositing them farther south. Subsequent warming of the climate and receding of the glaciers resulted in deposition of the entrained sediment over the uncovered landscape. Each successive glaciation partially eroded the pre-existing ground surface and deposited a new depositional sequence. A trench with multi-level benching that was excavated at the site of the planned wastewater treatment plant for the purpose of seismic investigation reveals the lateral and vertical variability of the glacial deposits due to facies transitions as well as erosional contacts (Figure 2 and 3).

Figure 2. Glacial deposits exhibited in a trench excavated for seismic investigation of the treatment plant site

Figure 3. Erosional contacts as a typical characteristic of glacial sequences

King County, the owner, commissioned a team of geologists and geotechnical engineers led by CDM to investigate the geological and hydrological conditions along the tunnel alignment, provide the geotechnical design, and coordinate the geotechnical baselines for the tunnel contracts with the county and the conveyance design team. During the subsoil investigation phase, the deposits of at least three glacial and three inter-glacial sedimentation cycles were identified. Correlation of glacial and non-glacial sedimentary deposits over horizontal and vertical distances between the 150 m spaced borings was an important tool to evaluate the potential for continuity in soil conditions, or lack thereof. Soil types of similar composition, with similar engineering properties, and partly products of similar geologic forming processes, were combined in groups (TSGs), which were described using a color code. Those groups were the basis for the contractual description of typical face conditions, which consist of either single or combinations of more than one TSG (Figure 4). The TSGs are furthermore the basis for identifying the anticipated soil adhesion and soil abrasivity. Specific soil abrasivity testing was carried out and baselines were established for the TSGs to allow the contractors to plan for TBM system selection and operation and maintenance requirements.

In order to provide a baseline of the number and sizes of boulders to be encountered along the tunnel alignment, the geotechnical design team carried out geological mapping of regional outcrops of those glacial deposits that had been identified at the

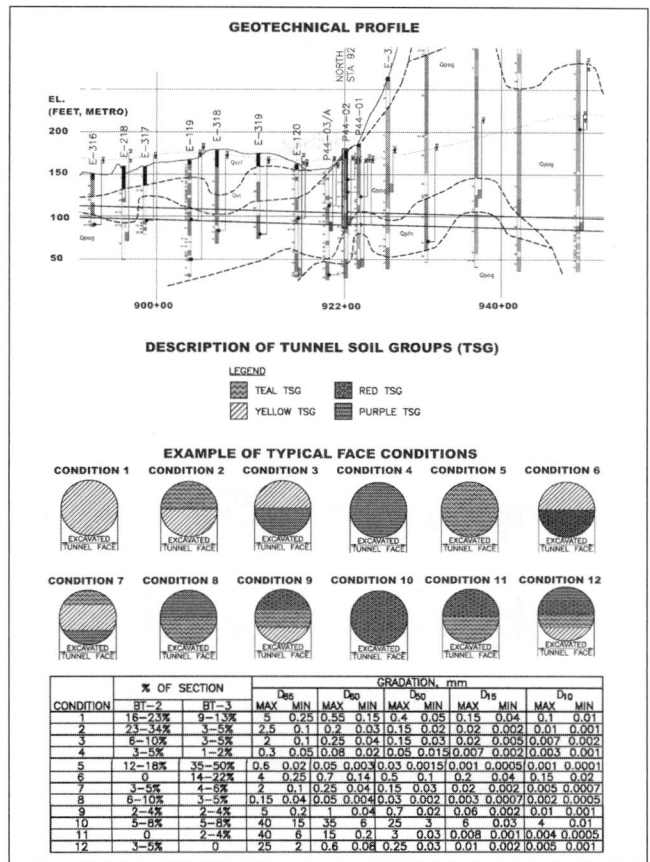

Figure 4. Geotechnical section and categorization of tunnel face conditions

tunnel elevation at the boring locations. This information was integrated with empirical methods to provide the baseline values. The contract documents provide baseline numbers for boulders for each tunnel, divided into categories based on size. Similar to the TSGs, no prediction was made as to where along the alignment the boulders would be encountered. However, the geologic classification of the soils to be mined through allows areas where encountering boulders is more likely or less likely to be identified.

For investigating the hydrogeological conditions, borings were equipped with groundwater monitoring instrumentation at elevations where aquifers had been encountered. Grouted-in vibrating wire piezometers (VWPs) as well as standpipes equipped with VWPs were installed. The VWPs were connected to data loggers. During the design period of a few years, hydrostatic data were collected often in half-hourly to two-hourly intervals. Based on those measurements, the hydrostatic head at the tunnel invert was calculated and baselined for the project design and for the contractors' planning of their tunneling operations.

For each of the three tunnel contracts, baseline values for the geotechnical conditions at the shaft locations and along the tunnel alignments were provided in the GBR. In addition, all the data collected during the subsoil investigation phase—boring logs; geotechnical, geological, and mineralogical laboratory test results; pump test data;

Figure 5. EPB TBM installation in the IS launch shaft

Figure 6. EPB TBM before removal from the treatment plant portal receiving pit after tunneling 4,282 m

groundwater monitoring data; and geophysical test results—were compiled in the form of a Geotechnical Data Report (GDR), which is part of each contract.

CONSTRUCTION AND DATA COLLECTION

In January 2006, King County awarded the East Contract for the easternmost tunnel section to the joint venture of Kenny Shea Traylor (KST) with an offer of $130.9 million. This first of the four Brightwater tunnels (BT-1) is a segmentally lined tunnel with a design I.D. of 5.87 m and a length of 4,282 m. Also included in the construction contract are the 22.5 m deep launch shaft (IS) at the North Creek Portal with an I.D. of 24.3 m and a second shaft for a pump station (IPS), consisting of two overlapping cells each with an I.D. of 25.6 m and a depth of 25.3 m. Furthermore, the contract includes a 741 m long influent microtunnel with an I.D. of 1.8 m as well as two interconnecting tunnels between the shaft structures of 3.66 m I.D. each. For the construction of BT-1 KST chose an EPB TBM manufactured by Lovat (Figure 5). After installing the shaft structures by two-phase slurry wall construction method using a hydromill and by pouring tremie slabs, tunneling started in November 2007. The TBM reached the receiving pit at the treatment plant site in 13 months on schedule in November 2008 (Figure 6).

Figure 7. Mix-Shield installation in the North Kenmore Portal launch shaft

Figure 8. Mix-Shield cutter head tools during inspection stop

The two tunnels of the Central Contract (BT-2 and BT-3) were awarded in July 2006 to the joint venture of Vinci Parsons Frontier-Kemper (VPFK) based on an offer of $209.8 million. Both tunnels use the same launch shaft (North Kenmore Portal), BT-2 heading eastwards to the IS shaft (North Creek Portal) has a length of 3,536 m and BT-3 heading westwards to the receiving shaft at Ballinger Way Portal has a length of 6,126 m. BT-2 and BT-3 are segmentally lined tunnels with a design I.D. of 5.12 m. The construction contract includes the 27 m deep launch shaft with an I.D. of 15.8 m and the receiving shaft for BT-3 which is 61 m deep with an I.D. of 7.31 m. The contract includes an influent line for connection of existing sewage flows to the North Kenmore Portal that consists of a 549 m open cut section and two microtunnel sections with a combined length of 488 m and design I.D.s of 1.2 m and 1.8 m. VPFK used a clamshell excavator and two-phase slurry wall construction method including depressurization of the aquifer for the North Kenmore Portal shaft and ground freezing for excavation support for the Ballinger Way Portal shaft. VPFK chose two Mix-Shields manufactured by Herrenknecht (Figure 7). Tunneling started for BT-2 in October 2007 and for BT-3 in March 2008 (Figure 8).

The West Contract includes the fourth Brightwater Tunnel (BT-4), a 6,431 m long segmentally lined tunnel with a design minimum I.D. of 4 m that connects the launch pit (Point Wells Portal) at Puget Sound with the Ballinger Way Portal shaft, the launch pit itself, and a 132 m long microtunnel connection (I.D. 2.1 m) between the launch pit and the marine outfall. This construction contract was awarded in February 2007 for $102.5 million to the joint venture of Jay-Dee Coluccio Taisei (JCT). JCT chose a Lovat EPB TBM for its tunneling operations, which started in September 2008.

During mining operations, the contractors continuously monitor and collect specified TBM operational parameters and provide those parameters to the county in real-time. Using project-specific visualization software, the county can monitor the TBM operations and evaluate the operational parameters with regard to the geotechnical conditions encountered. Additional documentation and reporting by the contractors include shift and ring reports as well as information regarding tools and products used. Inspection and maintenance stops are reported in detail regarding tool changes and tool wear measurements.

For each TBM, the contractors provide representative tunnel spoil samples on a daily basis to the county. For the EPB TBMs this task is accomplished in a relatively simple manner by taking samples out of the muck cars after they have been filled via the TBM conveyor screw and moved to the portal. The task of obtaining a representative sample is more complicated for the slurry TBM operations. With slurry systems the configuration and operation mode of the separation plant needs to be taken into consideration that can result in several samples from different separation steps and the need to quantify those steps with regard to the material flow. Slurry mix composition and the possibility of accumulation of solids in the slurry tanks needs to be taken into account.

Before the start of construction activities, the network of groundwater monitoring points that had been installed by CDM during the subsoil investigation phase was expanded to provide further capabilities of construction impact monitoring. Depending on the positions of the TBMs and other construction activities, selected vibrating wire piezometers connect to data loggers with typically half-hourly recording intervals, whose data are made accessible to the county, the designers, and the contractors via an online Project Instrumentation Database (PID). In addition to the groundwater monitoring data, the geodetic and geotechnical measurements taken by the contractors—settlement survey data, extensometer data, inclinometer data—are also uploaded to the PID.

DATA INTERPRETATION AND ACTUAL VS. BASELINE COMPARISON

Identifying the tunnel face conditions encountered during mining can be based on (1) the daily TBM spoil samples taken by the contractors, (2) the continuously recorded TBM operational parameters, and (3) construction reports including information regarding tools, products, special observations (shift reports, ring reports, inspection and maintenance stop reports, etc.). The TBM spoil samples provide a representative indication of the tunnel face conditions at specific locations while the other data sources provide additional details regarding the system behavior. The data can then be used for further evaluation of the geotechnical conditions, for example with regard to material changes between the tunnel stations where the spoil samples were taken. Statistical analysis used for evaluating correlation between certain TBM operational parameters and interpreted tunnel face conditions can serve as a validation tool.

The initial determination is whether the TBM spoil material corresponds to one or more of the TSGs. Along the tunnel alignments of BT-1 through BT-4 the soils encountered in the investigation phase were categorized as fine-grained, plastic soils (Teal TSG), fine-grained, non-plastic soils (Purple TSG), predominately fine to medium sand (Yellow TSG), predominately coarse sand and gravel (Red TSG), and in the area of a post-glacial valley filling as plastic, very soft, organic silt (Tan TSG). Each TSG is contractually described with regard to composition and engineering properties, for example

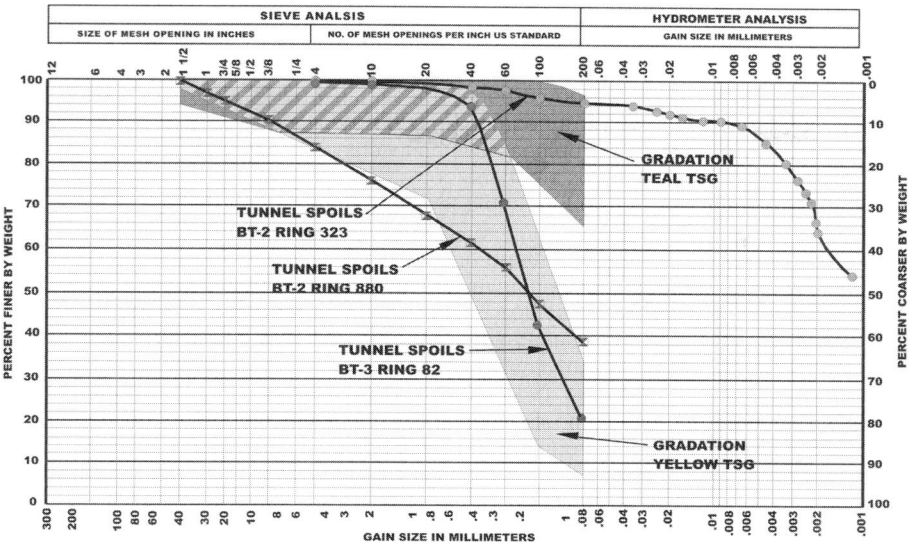

Figure 9. Grain size distribution curves of TBM spoils overlain on gradation bands for Yellow and Red TSG

by grain size distribution bands and plasticity. Each of the TSGs may include layers of different TSGs up to 0.305 m (1 ft) in thickness. For the tunnel excavation diameters of BT-1 through BT-4, a layer thickness of 1 ft corresponds with an average soil volume of about 5 %, plus or minus a few percentage points depending on the position of the soil layer within the tunnel face. For example, if chunks of plastic clay are found within a sample of otherwise fine to medium grained sand, this would mean the existence of either one or two TSGs depending on the volume percentages. Grain size analyses carried out with the TBM spoil sample material allow more thorough identification of the sample. These analyses allow checking if the grain size distribution curve falls within the boundaries of the grain size distribution band of a TSG. Gap-graded grain size distribution curves are an indication of a composite of two or more TSGs (Figure 9 and 10).

In some instances discontinuities of continuously recorded TBM operational parameters such as the abrupt change of cutterhead torque in conjunction with cutterhead rotation speed and changes in the amplitude of chamber pressure fluctuations may provide indications of changes in tunnel face conditions near the tunnel station where the discontinuity was recorded (Figure 11). Other instances may indicate a more gradual transition from one tunnel face condition to another (Figure 12). In cases where the TBM operational parameters or other data sources do not provide any indication, the transition between two typical face conditions is determined by either linear or non-linear interpolation or extrapolation between and from sampling locations.

Plotting certain recorded TBM operational parameters, for example the advance rate per ring in mining mode, over certain observed properties of the daily spoil samples, for example the percentage ratio of fine grained to coarse grained components, can help identify trends and correlations (Figure 13). The results of these statistical analyses can be used to better interpret TBM system behavior between sampling points. Statistical analysis is also a valuable tool for checking or validating data interpretation.

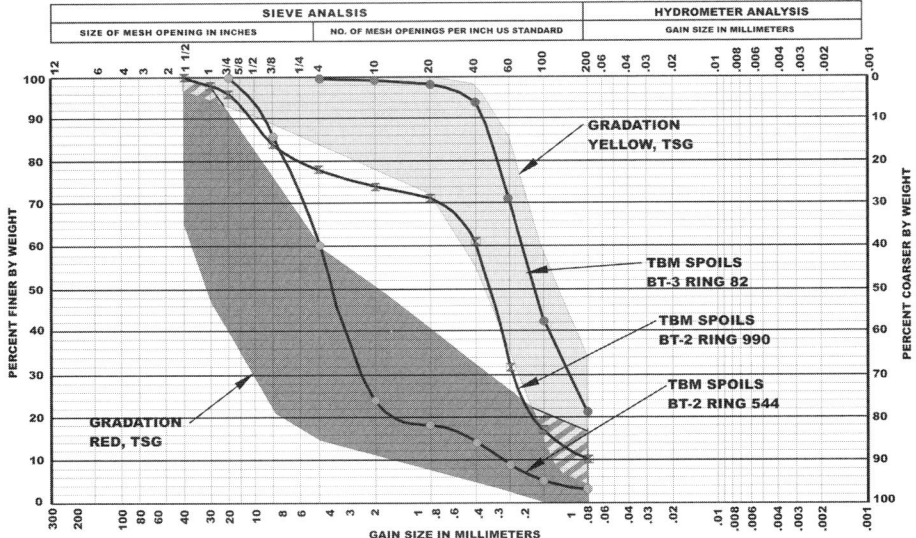

Figure 10. Grain size distribution curves of TBM spoils overlain on gradation bands for Yellow and Teal TSG

An example is the comparison of TBM advance rates of the two Mix-Shields used for BT-2 and BT-3 in different tunnel face conditions at the first reaches of each tunnel drive (Figure 14). However, the analysis also needs to consider differences in the geological setting of tunnel sections and the differences in the installed electrical power of the two TBMs as well as modifications to system components like the separation plant after the start of mining operations.

Actual-versus-baseline tracking can involve the contractor's own personnel if so desired by the contractor. For the tunnels where the contractor chooses to track, evaluate, and document the tunnel face conditions encountered, the county and the contractor can agree on a procedure that includes the exchange of each party's tracking results and discussion of the interpretations on a periodic basis. In case of discrepancies, the data sources can be compared against each other and if there is the prospect of useful additional information, further investigations, for example laboratory testing, can be carried out. Each party can update its summations of interpreted tunnel face conditions as percentage of the tunnel length and track the results with the percentage ranges provided in the GBR.

The groundwater monitoring data and the geodetic and geotechnical measurements, which are uploaded to the online PID, primarily serve for identifying and evaluating impacts of the construction activities on the aquifers or existing structures. In addition, the contractors utilize the groundwater monitoring data for planning of their TBM operations. During hyperbaric interventions for TBM maintenance, repair, and tool changes, the groundwater monitoring points in the vicinity of the tunnel face allow observation of the change in the pressure field around the TBM (Figure 15). These data provide a better understanding of the interaction of TBM operations and ground conditions and can also serve to evaluate average soil properties, for example by back-calculating hydraulic conductivities.

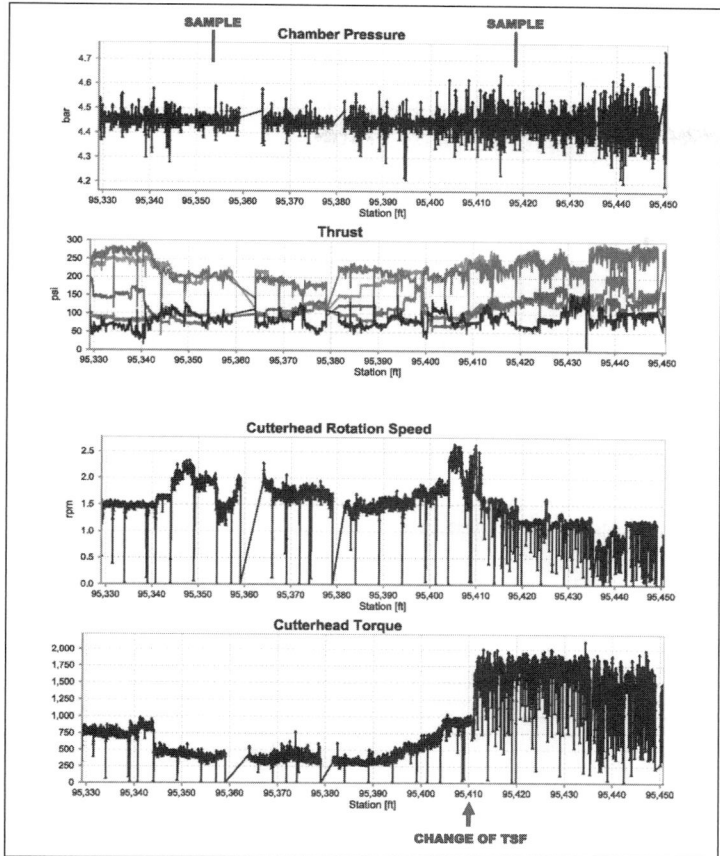

Figure 11. Recording of TBM operational parameters indicating a rather abrupt change in tunnel face conditions

CONCEPT EVALUATION

The Geotechnical Baseline Reports provide geotechnical baselines for each of the four Brightwater tunnels. The geotechnical baseline approach includes the description of tunnel face conditions and allocates percentage ranges of the length of the tunnel alignment. No specifics are identified as to where along the alignment a specific tunnel face condition will be encountered. The Geotechnical Data Reports include the data that were collected during the subsoil investigation phase. This approach provides the contractors information that can assist them in the selection of equipment and planning for operation and maintenance. Tracking of the actual conditions encountered and comparison with the baseline tunnel face conditions provides a basis for allocation of responsibility.

Tracking of the actual tunnel face conditions to date has shown the existence of frequent and abrupt changes in soil composition between the boring locations as had been anticipated during the investigation and design phase. Those anticipated changes were the reason for choosing the probabilistic baseline approach described. On-going tracking of actual conditions by the contractors and the county may result in interpretations that differ to a certain degree. In the case of discrepancies between the

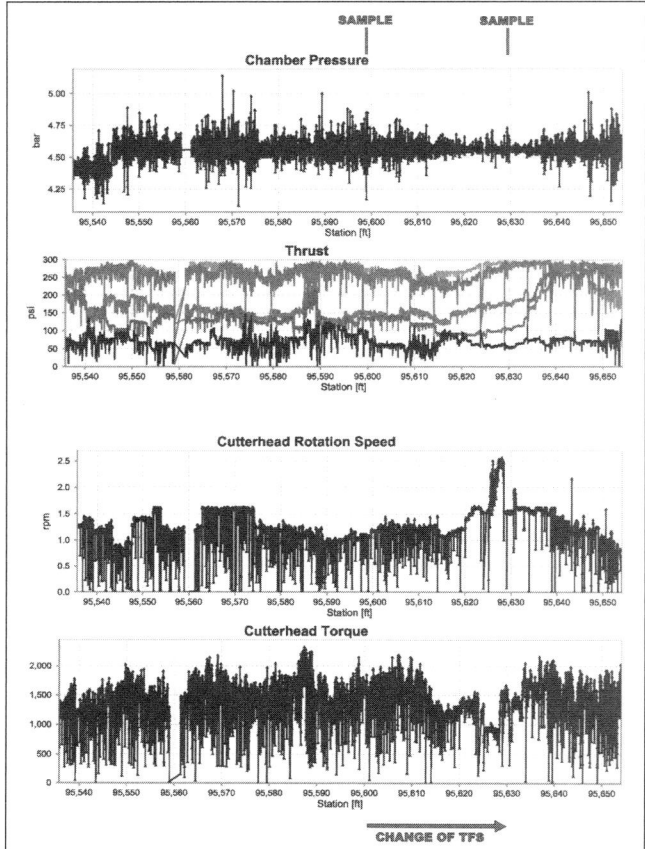

Figure 12. Recording of TBM operational parameters indicating a rather gradual change in tunnel face conditions

county and contractor interpretations, verification of data sources and reproducibility of the interpretation process will be the key elements for a successful technical resolution.

The systematic data collection inherent to the described procedures will also contribute to on-going efforts of gaining a better understanding of the complex system behavior like the effect of abrasive soils on TBM wear for the given geological, hydrogeological, and hydrological conditions and the specific equipment, operation, and maintenance regime choices made by each of the contractors. Further insight on that topic bears the promise of contributing to increased planning reliability and reduced risk during construction of future tunnel projects.

BIBLIOGRAPHY

ASCE, 1997. "Geotechnical Baseline Reports for Underground Construction, Guidelines and Practices," prepared by The Technical Committee on Geotechnical Reports of the Underground Technology Research Council, Randall J. Essex, Editor, ASCE, 1997.

ACTUAL VS. BASELINE TRACKING DURING TBM TUNNELING

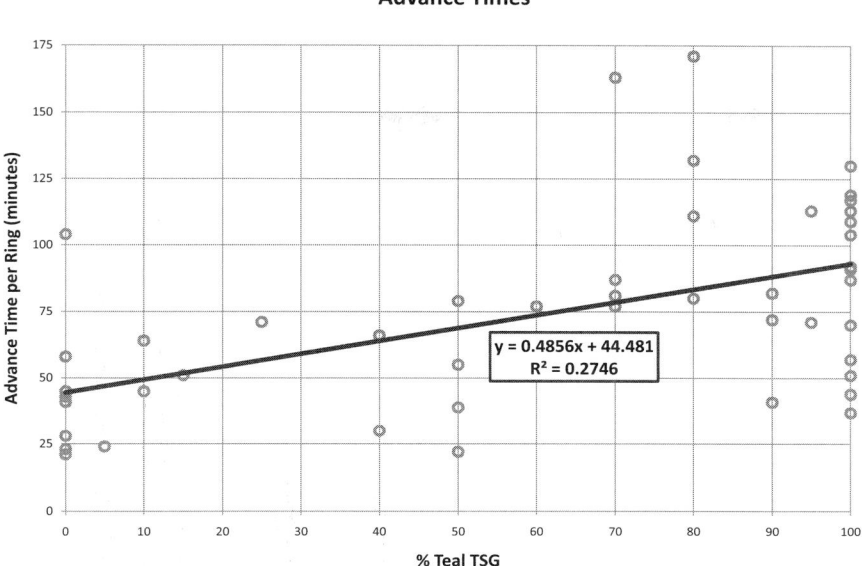

Figure 13. Correlation of mining time per ring and fines content of the spoils for a tunnel section

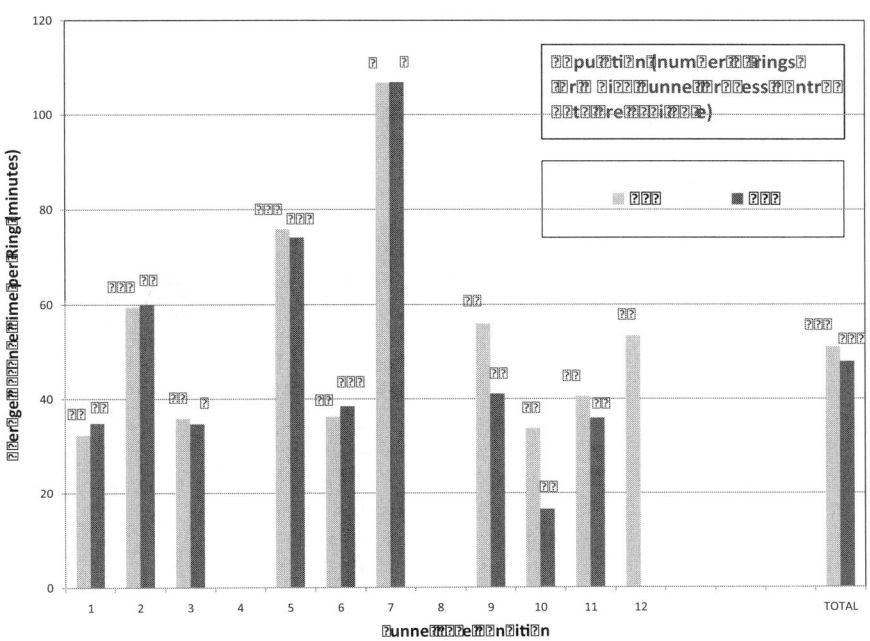

Figure 14. Comparison of BT-2 and BT-3 advance rates for the tunnel face conditions encountered at the first reaches of each tunnel drive

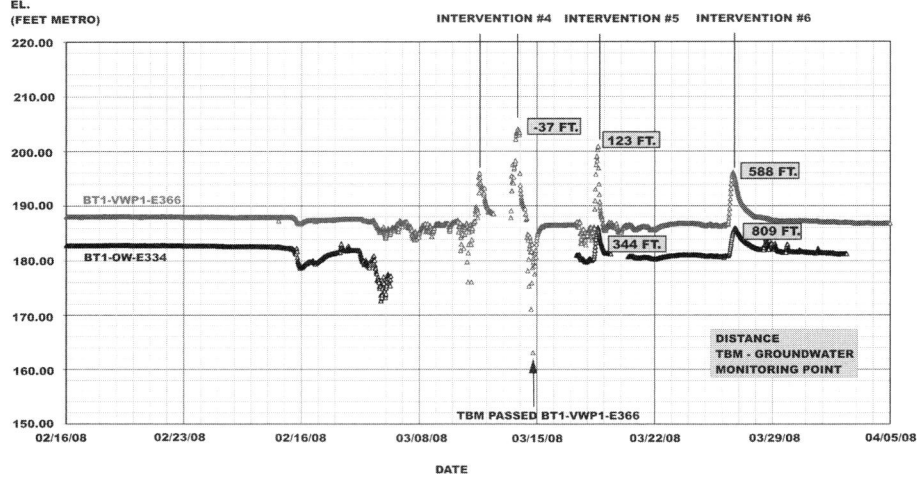

Figure 15. Hydrograph curves showing the impact of hyperbaric interventions in the vicinity of the groundwater monitoring instruments

ASCE, 2007. "Geotechnical Baseline Reports for Construction: Suggested Guidelines." Prepared by the Technical Committee on Geotechnical Reports of the Underground Technology Research Council, Randall J. Essex, Editor, ASCE, 2007

Booth, D. B., Troost, K. G., 2003. "Puget Lowland Geologic Framework," Training Course for King County Wastewater Treatment Division, University of Washington, Department of Earth and Space Sciences, Seattle, Washington, January 2003.

Gilbert, M.B., Perrone, V.J., Maday, L. E., 2005. "Exploration Decision Analyses for Brightwater Tunnels," Proceedings Rapid Excavation and Tunneling Conference, Seattle, Washington, June 2005.

Gwildis, U.G., Sass, I., 2008. "Maschineller Tunnelbau in wechselhafter glazialer Geologie - Ein Ansatz probabilistischer Baugrundbeschreibung und baubegleitender Ueberpruefung", Geotechnik v.4/2008, December 2008.

Newby, J.E., Gilbert, M.B., Maday, L.E., 2007. "Brightwater Conveyance System will expand Seattle's Wastewater Treatment," Tunneling and Underground Construction, June 2007.

Newby, J.E., Gilbert, M.B., Maday, 2008. "Establishing Geotechnical Baseline Values for Deep Soft Ground Tunnels," Proceedings North American Tunneling Conference, San Francisco, California, June 2008.

ASSESSING GROUND AHEAD OF TBM TUNNEL USING LOW-INTERRUPTION WIRELESS SEISMIC REFLECTOR TRACING SYSTEM

Takuji Yamamoto ▪ Kajima Technical Research Institute

Yasuhiro Yokota ▪ Kajima Technical Research Institute

Jozef Descour ▪ C-Thru Ground, Inc.

Matthew Kohlhaas ▪ Sawtooth Wireless Sensors, LLC

ABSTRACT

High-speed TBM excavation requires rapid ground assessment ahead of excavation, particularly in complex grounds. The Tunnel Reflector Tracing technology (TRT) developed by authors, uses seismic reflections to produce volumetric images of ground anomalies. To further minimize interruption in the TBM excavation the authors modified surveying technique switching from cable (wired) to radio (wireless) data acquisition system. This paper presents the outline of the Wireless TRT system, and compares the results for the wired and the wireless surveys. The comparison shows that the wireless system offers high precision data, is more efficient and economical and significantly less intrusive in tunneling operations.

INTRODUCTION

Local ground conditions greatly affect the efficiency of tunnel construction. An unexpected change in the ground can result in a rupture of the tunnel face, collapse of the tunnel top and walls, or a massive outbreak of underground water. Any of these failures can significantly slow down tunnel excavation, increase its cost, and jeopardize the safety of tunneling personnel. Therefore it is extremely important to be able to accurately predict the ground conditions sufficiently far in front of an advancing tunnel face and around the excavated tunnel.

The authors have developed three-dimensional ground imaging technology, "TRT" (Tunnel Reflector Tracing) (Ashida and Sassa, 1993; Neil et al., 1999; Aoki et al., 2003; Descour et al., 2005; Yamamoto et al., 2007). TRT uses reflected seismic waves recorded in an advancing tunnel, for detection and delineation of features in the ground which can potentially disrupt tunneling operations. The processing of acquired seismic data generates three-dimensional images of anomalies in the ground (Figure 1), and can be usually done in less than 3 hours.

So far this technology has been used in more than 70 tunnel excavations around the world, including sites in Japan, United States, Great Britain, Laos, and China.

Unlike some other seismic ground imaging techniques, the TRT system does not require blasting to generate seismic waves. Instead, a sledge-hammer or swept-frequency source is applied directly to the tunnel walls at a number of selected points. When seismic waves encounter boundaries of changing ground conditions (type of rock, fractured ground, cracks carrying water, etc.), part of their energy is reflected (Aki and Richards, 1989; Watters 1978; Yamamoto et al., 2007). The reflected seismic

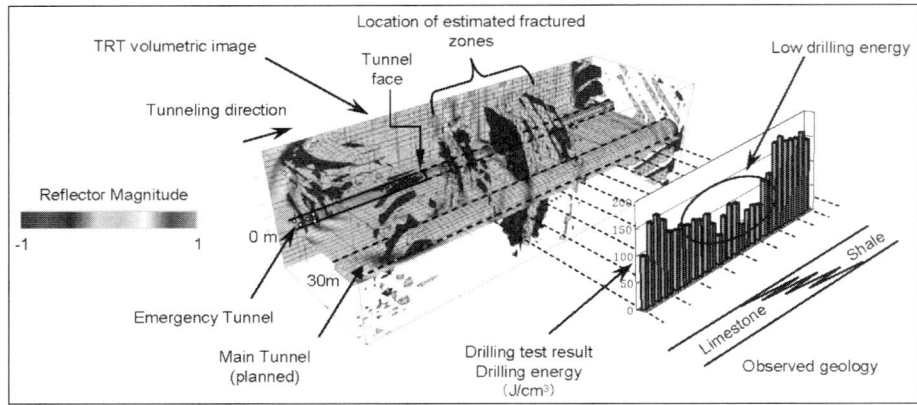

Figure 1. Volumetric image of features in the ground corresponds well with rock assessment using drilling energy and matches geology observed by advancing tunnel excavation

waves are detected using small and very sensitive accelerometers that are also attached directly to the tunnel walls.

The type of sources and a quick coupling of both sources and accelerometers to the tunnel walls allow significant saving of data acquisition time for TRT surveys.

However, in the original, wired equipment design the accelerometers are linked to the seismic data acquisition equipment (digital seismograph) via a network of coax cables.

In an effort to further reduce the time required for TRT surveys, improve reliability and fidelity of acquired seismic data, and reduce cost and weight of the equipment, the authors developed a new Wireless TRT surveying system (Kohlhaas and Descour, 2008). The new system not only replaces cables with radio communication, but also shifts a number of functions performed by the seismograph to the Remote Modules close to individual accelerometers.

This paper presents an overview of the new Wireless TRT system, and the results of its testing side-by-side with the Wired TRT system.

WIRELESS TRT SURVEYING SYSTEM

Improving Surveying Operations

The TRT data acquisition system is typically composed of ten accelerometers coupled mechanically to the tunnel walls, and arranged in a specific pattern (Neil et al. 1999; Yamamoto et al., 2007). The accelerometers produce electric signals in response to vibrations caused by passing seismic waves.

Figure 2 presents the block-schematic of the Wired TRT surveying system, and the picture of data acquisition equipment. In that system each accelerometer is connected to a seismograph using a separate 30-meter long coax cable. Each cable carries power to the analog amplifier on board of the connected accelerometer. The same cable carries back to the seismograph the analog signals that represent vibrations caused by seismic waves and detected by that accelerometer. The seismic source activation triggers the process of simultaneous recording by the seismograph of the signals from all accelerometers to form a data file.

LOW-INTERRUPTION WIRELESS SEISMIC REFLECTOR TRACING SYSTEM

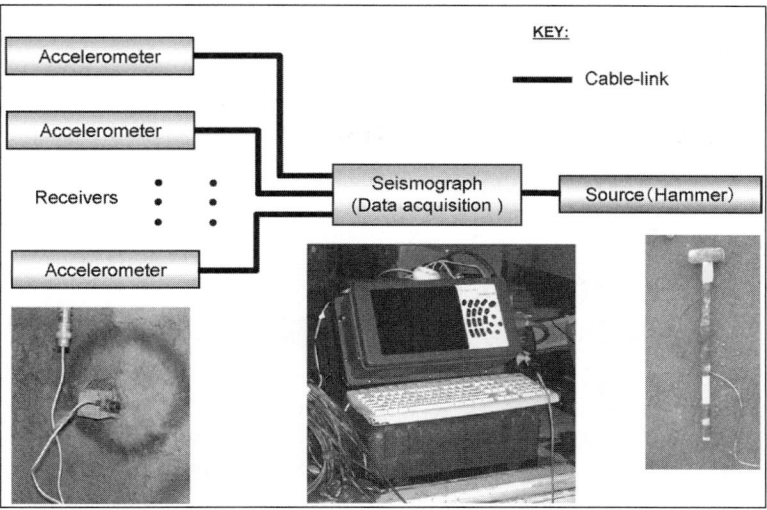

Figure 2. Block-schematic of the Wired TRT surveying system and pictures of major equipment components

Figure 3 presents the block-schematic of the wireless TRT system, and pictures of the equipment installed in a tunnel. For the Wireless TRT surveying system each accelerometer is connected via a short signal cable to its own Remote Module. The Module not only powers the accelerometer, but also continuously receives, digitizes, and buffers seismic signals from that accelerometer. On the trigger signal sent via radio when the source is activated, the pre-selected length of buffered signal from each Module is sent back to the notebook computer for recording as part of the data file for that source. A ruggedized notebook computer replaces the cumbersome seismograph. It is used to setup the system, and controls the acquisition of seismic data.

By replacing cables with the proper radio-link technology, the Wireless TRT system either eliminated or significantly reduced many of the problems associated with its wired and otherwise reliable predecessor, such as:

- Time consuming installation and retrieval of cable lines;
- Delicate, prone to damage cables;
- Interference from electromagnetic noise associated with tunneling operations;
- Electric noise caused by ground-loops via cable lines due to multiple grounding points.

Technical Development Tasks

Synchronization. All accelerometer signals for the same source are sent sequentially from Remote Modules to the notebook computer. These signals have to be properly synchronized to form a coherent data file.

This task is being accomplished by using the trigger signal produced when the source is activated. At that moment the trigger signal is sent via cable to the computer, and is broadcast via radio to all Remote Modules. This marks the beginning of the data transmission cycle via radio to the computer.

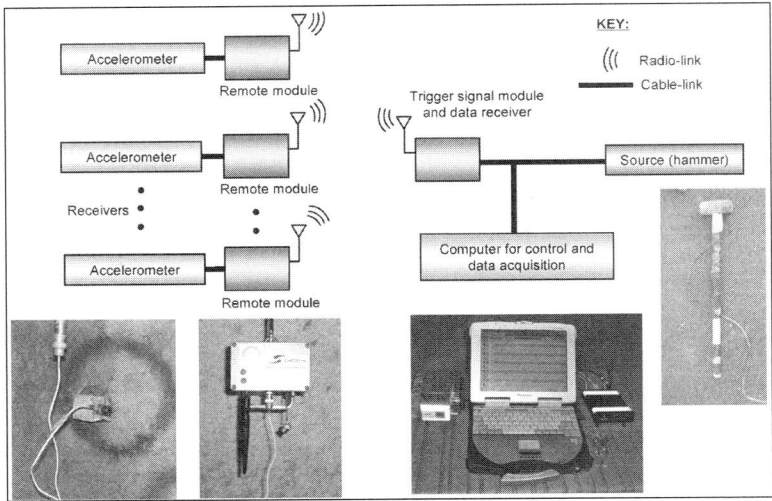

Figure 3. Block-schematic of the wireless TRT surveying system and pictures of the major equipment components

Communication Interference. Even short sections of underground tunnels can be difficult for a reliable radio communication. Electromagnetic noise from external sources, loss of the "line-of-site" due physical obstructions in the tunnel profile, and multipath fading caused by multiple reflections of radio waves from tunnel walls and from the tunneling equipment, all contribute to degraded communication.

The issues associated with interference and multiple reflections were resolved by using FHSS (frequency-hopping spread-spectrum) technology in a narrow frequency band near 2.4GHz. The selection of FHSS for wireless communications has many benefits (Dixon 1994) such as:

- High resistance to narrowband interference;
- Ability to share a frequency band with many types of conventional transmissions with minimal interference.

In addition, most countries set aside the 2.4GHz frequency band for communication systems to operate below certain power thresholds without the requirement of a site license (FCC, 2007).

WIRELESS TRT FIELD EVALUATION

The field test was designed to verify fitness and reliability of the Wireless TRT surveying system versus the Wired TRT system in possibly the toughest and most disruptive conditions for radio communication.

Evaluation Outline

An initial evaluation of the Wireless versus Wired TRT survey system was conducted in a larger cross-section (80 m^2) NATM tunnel A (Figure 4). This evaluation has proven that in relatively large underground openings the radio communications for the Wireless system are equally reliable as the cable transmission for Wired TRT system.

The recent evaluation was conducted in a small cross-section (6.15 m^2), highly obstructed TBM tunnel B (Figure 5). This test was the most critical for assessing

Figure 4. No significant obstructions for radio communications in large cross-section NATM tunnel A (cross-section 80 m^2)

Figure 5. Obstructions (trailing gear) block 30 to 40% of the cross-section in small size TBM tunnel B (diameter 2.8 m, cross-section 6.15 m^2)

performance of the radio communications in the Wireless TRT system due to ferromagnetic construction equipment (trailing gear) blocking the tunnel profile, electromagnetic interference from the power lines, and other sources.

Site Characteristics

The length of the extracted tunnel B at the time of survey was 4.2 kilometers. Tunnel B is excavated by a TBM with a cutting diameter of 2.8 meters (Figure 6). The tunnel path cuts mostly through claystone interrupted sporadically by sandstone deposits.

The preliminary seismic survey done from surface along the projected tunnel alignment yielded P-wave velocities between 3.8 and 4.6 km/s. This led to prediction of relatively competent rock and good geological conditions for tunneling. However, the investigation also detected some lower-velocity zones, possibly fractured ground. An intrusion of very competent gabbro into the tunnel path was considered possible as well.

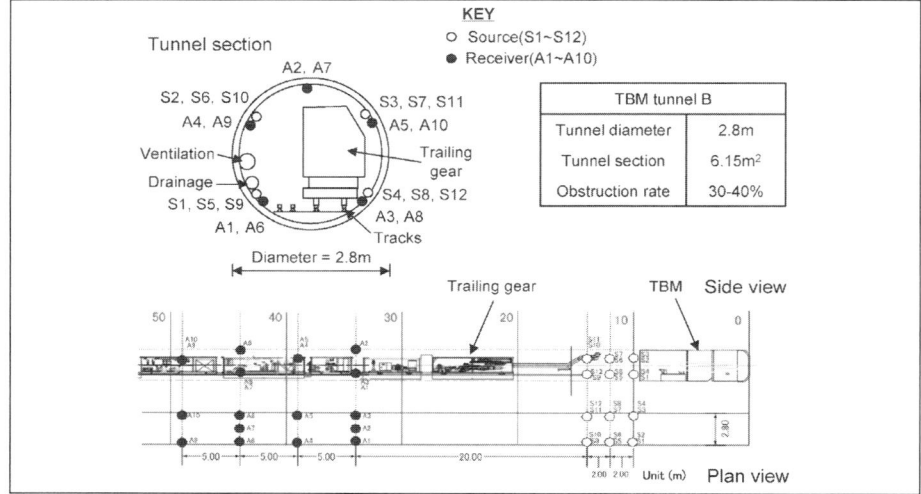

Figure 6. Configuration of seismic sources and receivers (accelerometers) for TRT survey in TBM tunnel B to compare wired and wireless communication systems

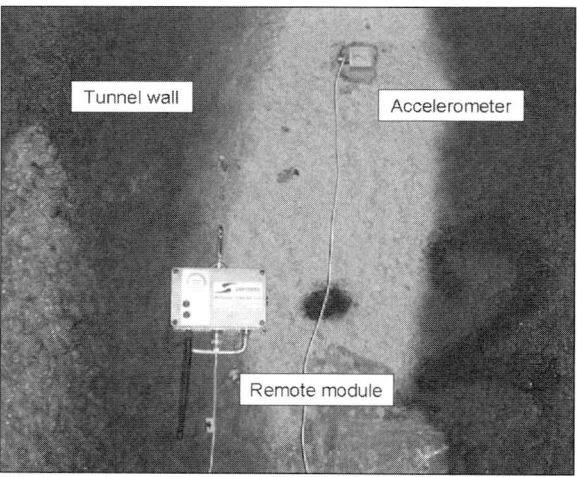

Figure 7. A light 3-m long cable delivers power to accelerometer, and carries back to the wireless remote module the analog signals generated by accelerometer in response to passing seismic waves

Survey Layout

The layout of seismic sources and receivers comprises 12 source points and 10 receiver points (Figure 6). The Wired TRT system test was followed by the test of the Wireless TRT system. Each system used the same layout of sources and receivers.

Figure 7 shows the wireless Remote Module connected to an accelerometer coupled to the tunnel wall. A 3-m long cable connecting accelerometer to the Remote Module allowed some flexibility in positioning the Module further from the ferromagnetic

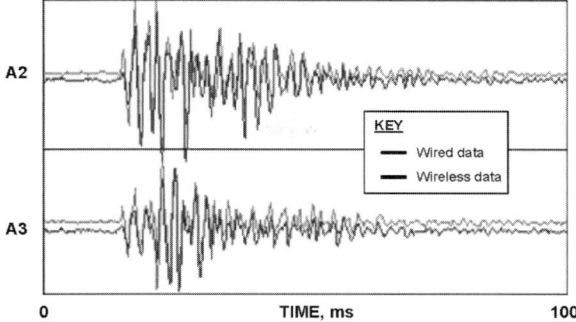

Figure 8. Nearly perfect match between waveforms of seismic signals recorded or two accelerometer locations (A2, and A3) as part of performance evaluation test for wired and wireless TRT surveying systems

components in the tunnel profile, and closer to the optimum line-of-sight to mitigate possible disruption of radio communications with the data acquisition equipment. This was particularly important as the steel structure of the trailing gear occupied approximately 30 to 40% of the tunnel profile (Figures 5 and 6).

Survey Results

In spite of the small diameter of the tunnel, which was additionally reduced by the components of the tunneling equipment, there were no indications of any difficulties in radio communication between the components of the Wireless TRT survey system. Multipath fading, electromagnetic radiation and other sources of radio interference did not adversely affect the seismic records produced during the wireless TRT survey.

Figure 8 shows a comparison between waveforms acquired using the Wireless TRT system (gray) and waveforms acquired using the Wired TRT system (red). The records appear very similar with no difference in the travel times measured for the direct waves coming from the same source points. The matching travel times also confirmed that all the Remote Modules were precisely time synchronized.

Figure 9 shows the volumetric image of the ground produced by processing seismic records acquired by the wireless TRT survey. The image is compared with the geology observed, and with the "Rock mass rating" and "Estimated rock mass strength" parameters defined as the tunnel advanced. The weaker mostly negative anomalies correspond with the fractured rock zone. The strong mostly positive anomalies appear to match a competent gabbro formation.

CONCLUSIONS

The Wireless TRT surveying system was developed to improve efficiency, precision and reliability compared to the Wired TRT system. These objectives were tested and confirmed in highly adverse conditions of a small diameter TBM tunnel.

The radio communications were also effective even with the total cross-section of the small tunnel reduced by 30–40%, and the line-of-sight disturbed due to obstacles in the tunnel profile.

The use of the trigger signal to synchronize data acquisition of seismic signals for all recording receivers was proven very effective and precise.

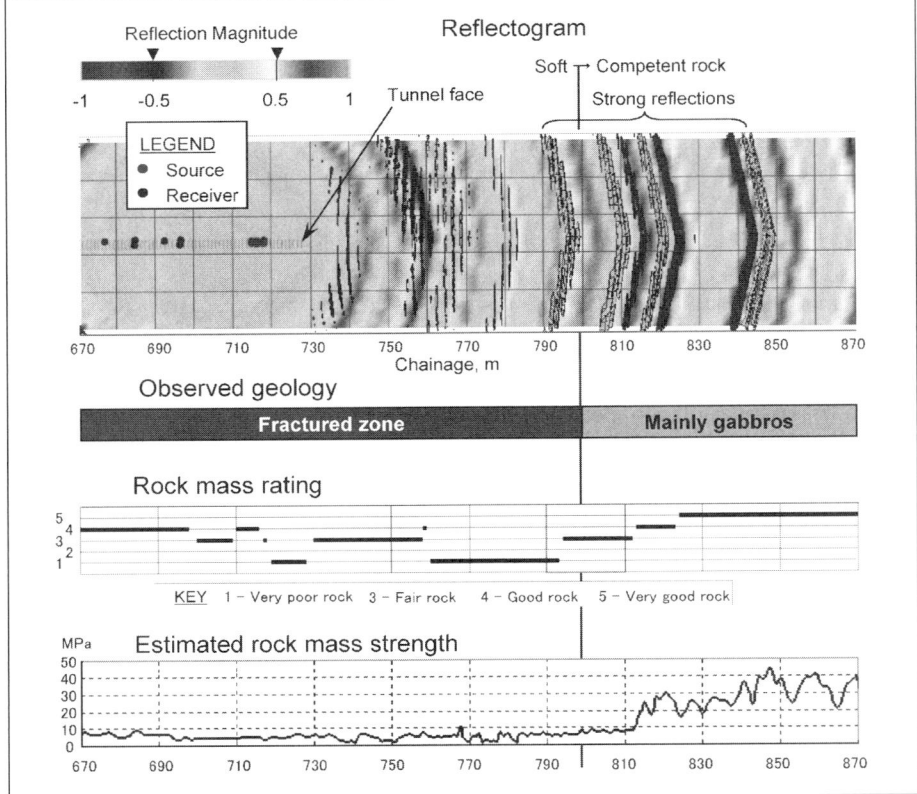

Figure 9. Volumetric image of reflective anomalies in the ground (reflectogram) ahead of tunnel B appear to correctly identify the boundary between fractured soft rock zone and competent gabbro formation

The short (3 meter) cable connecting each accelerometer to its Remote Module provided additional flexibility, when needed, to mitigate negative effects on radio communication caused by ferromagnetic obstacles in the tunnel profile.

The main objective for developing the Wireless TRT surveying system was to reduce time and manpower required for:

- Deploying and retrieving the system;
- Conducting the survey; and,
- Maintaining the system integrity.

The field test confirmed that this objective was accomplished.

REFERENCES

Aki, K, and P. G. Richards. 1980. Quantitative seismology. Theory and methods. Vol. 1. W. H. Freeman and Co., N. York.

Aoki, K., Mito, Y., Yamamoto, T., and Shirasagi, S. 2003. Advanced support design system for TBM tunnels using the seismic reflective survey and TBM driving data. *Proceedings of the International Symposium on the Fusion Technology of Geosystem Engineering, Rock Engineering and Geophysical Exploration, Seoul, Korea, November 18–19.*

Ashida, Y., and K. Sassa, 1993. Depth transform of seismic data by use of equi-travel time planes. *Exploration Geophysics 1993 (23):341–6.*

Descour J. T. Yamamoto and K. Murakami, 2005. Improving 3D imaging of unknown underground structures. In *Proceedings (CD-ROM) of FHWA Unknown Foundation Summit, Lakewood, CO. November 15–16, 2005.*

Dixon, R. C., 1994, Spread spectrum Systems with Commercial Applications, 3ed, John Wiley & Sons, New York

FCC, October 1, 2007, Code of Federal Regulations CFR 47, Part 15, section 247.

Kohlhaas, M, and J. Descour, 2008. Monitoring the Structural Integrity of Our Nation's Bridges. ASNT NDE/NDT for Highways and Bridges: Structural Materials Technology (SMT) 2008 topical conference, Oakland, CA. September

Neil, D. M., K. Y. Haramy, J. Descour, and D. Hanson, 1999. Imaging ground conditions ahead of the face. *World Tunneling, 12(9):425–429.*

Watters, K. 1978. *Reflection seismology. A tool for energy resource exploration.* John Willey & Sons, New York.

Yamamoto T., S. Shirasagi, K. Murakami, and J. Descour, 2007. Evaluation of Geological Conditions Ahead of TBM Using Seismic Reflector Tracing and TBM Driving Data. Proceedings of Rapid Excavation and Tunneling Conference. Toronto, Canada, June 10–11, 2007.

GROUND CHARACTERIZATION AND FEASIBILITY EVALUATION OF TUNNELING METHODS FOR MATHER INTERCEPTOR

Mohammad R. Jafari ▪ Camp Dresser & McKee

Michael D. Middleton ▪ Camp Dresser & McKee

Bruce J. Corwin ▪ Camp Dresser & McKee

Andrew Page ▪ Sacramento Regional County Sanitation District

ABSTRACT

The Camp Dresser & McKee (CDM) has been retained by the Sacramento Regional County Sanitation District (SRCSD) to design the Mather Interceptor, a portion of the South Interceptor and Mather Interceptor program. The Mather Interceptor consists of 14,667 feet of 54 to 72 inch gravity interceptor. Approximately 90 percent of the alignment will be constructed via tunneling. Cobbles and boulders in a sand/gravel matrix dominate the ground conditions combined with seasonal perched groundwater table at some of the deeper reaches of the alignment, impacting design consideration. This paper will present the results of a comprehensive geotechnical investigation conducted for this project and in-depth technical evaluation of various trenchless techniques feasible for tunneling in the cobble and boulder laden formations of this project. This paper will also present the risks and issues of the design alternatives and selection criteria.

INTRODUCTION

Sacramento Regional County Sanitation District (SRCSD) provides wastewater conveyance and treatment services to the cities of Sacramento, West Sacramento, and Folsom as well the Sacramento Area Sewer District (SASD). SASD provides sewer collection services to unincorporated areas of Sacramento County and the cities of Citrus Heights, Rancho Cordova, and Elk Grove. All told, SRCSD conveys and treats the sewage of approximately 1.3 million residents in the greater Sacramento area through approximately 200 miles of sewer interceptors. All sewage is conveyed to, and treated by, the Sacramento Regional Wastewater Treatment Plant (SRWTP) located in Elk Grove which has a permitted capacity of 181 million gallons per day average dry weather flow.

CDM has been retained by SRCSD to design the Mather Interceptor, a portion of the South Interceptor and Mather Interceptor (SIAMI) program. Serving the Mather sewer sheds, Aerojet (AJ) sewer sheds, and Laguna Creek Area 5, the Mather Interceptor will convey wastewater flow from the Chrysanthy Pump Station near Chrysanthy Boulevard and Sunrise Boulevard, to the Bradshaw Interceptor Section 7C where it crosses Zinfandel Drive.

The Mather Interceptor consists of 14,667 feet of 54 to 72 inch gravity interceptor. The site is located in the city of Rancho Cordova and Sacramento County. Figure 1 provides an overview of the alignment and the major features of the project.

EVALUATION OF TUNNELING METHODS FOR MATHER INTERCEPTOR 273

Figure 1. Proposed Mather Interceptor location plan

The proposed alignment extends west from the upstream end of the proposed pipeline which is located on Chrysanthy Boulevard about 50 feet east of the intersection with Sunrise Boulevard. The alignment then trends north along Sunrise Boulevard, west along Douglas Road, and northwest along the future alignment of Zinfandel Drive, as shown in Figure 1. A proposed pipeline stub to the future Aerojet 2 pipeline will cross Sunrise Boulevard from the proposed junction structure. The proposed Mather interceptor pipeline will be 72 inches in inside diameter between Stations 10+20 and 114+50 and 54 inches in inside diameter between Stations 114+50 and 156+68. The Aerojet 2 stub pipe line will also be 72 inches in inside diameter.

Various manholes will be located at regular interval along the alignment. The current proposed Mather Interceptor pipeline has invert depths that are about 30 to 68 feet below ground surface (bgs) along the present and future alignment of Zinfandel Drive, about 46 to 70 feet bgs along Douglas Road except at the Folsom South Canal where the pipeline invert is about 21 feet below the canal invert, and about 30 to 62 feet bgs along Sunrise Boulevard and Chrysanthy Boulevard. The proposed Aerojet 2 pipeline stub invert depth is approximately 45 feet bgs at the Mather Junction Structure (MJS) and approximately 42 feet bgs at the north end. No structure is planned at the north end of the Aerojet 2 pipeline stub.

Open trench construction is currently planned for the northern portion of the present Zinfandel Drive alignment between Stations 10+20 and 14+00. The remainder of the pipeline will be constructed using either 2-pass tunneling or pipe jacking methods employing open shield.

Figure 2. Geologic map and boring location plan

REGIONAL GEOLOGICAL SETTING

The project site is situated in the eastern portion of Sacramento County, California and within the southern portion of Sacramento Valley. The generalized distribution of geological earth units in the site area is mapped by Helley and Hardwood (1985) as shown in Figure 2. This published geologic mapping was modified based on aerial photographic evidence and subsurface exploration data. A total of six earth units are mapped within the project area including (from oldest to youngest): Laguna Formation (Tl), Riverbank Formation (Qrl), alluvium (Qal), dredge tailings (t), landfill (lf), and artificial fill (af). Figure 2 presents the approximate aerial extent of these units in conjunction with the locations of borings drilled for this project and previous studies along the proposed alignment.

In addition to the geological units mapped by Helley and Harwood (1985), the California Department of Water Resources (DWR, 1974) maps several buried stream channels intersecting the proposed pipeline alignment. These buried stream channel deposits are identified within the Laguna Formation materials beneath the site and include relatively permeable sands and gravels surrounded by less permeable silt and clays forming "a network of meandering tubular aquifers hydraulically isolated from the deeper aquifers," (DWR, 1974). The stream channel deposits are believed to have been formed by streams draining the Sierra Nevada during glacial and interstadial periods and were later buried by more recent alluvium and basin deposits. As shown on Figure 2, the project area is mostly underlain by Laguna Formation. Towards the central portion of the site area, alluvium is inferred along the seasonal stream that is tributary to Mather Lake. A relatively small area of Riverbank Formation occurs along Eagles Nest Road, but is likely a thin veneer overlying the Laguna Formation based on its relatively small aerial extent. In the northern portion of the project area, dredge tailings have been well documented extending several thousand feet south of White Rock Road. Much of the dredge fields were leveled and/or reclaimed for development. One such development is the Village of Zinfandel located on the north end of the alignment where as much as 30 feet of tailings were removed and replaced with engineered fill to facilitate residential development. A brief description of each geological earth unit known to exist within the site vicinity is provided below:

Laguna Formation (Tl). Alluvial deposits consist primarily of interbedded arkosic gravel, sand, and silt forming a matrix with pebbles and cobbles composed of metamorphic clasts. According to United States Bureau of Reclamation (USBR) the Laguna Formation exhibits properties to claystone at the canal bottom elevation near Douglas Road. Perched groundwater conditions may also develop within or above these layers.

Riverbank Formation (Qru and Qrl). These alluvial terraces consist of weathered reddish gravel, sand, and silt with minor clay. These deposits often include cobbles and small boulders. Riverbank Formation can be differentiated as the upper and lower members, with the upper members typically being more granular (cobbles, gravels, and sands with lesser silts and clays) than the lower member (silts and clays with lesser coarse deposits). The upper member appears to be more prevalent at the site.

Alluvium (Qal). This is typically unconsolidated sand, silt and clay that has been deposited by stream flows during the last thousand years.

Dredge Tailings (t). Dredge tailings consist predominantly of rounded cobbles derived from Laguna and Riverbank Formations. Sluicing on board the dredges resulted in separation of coarse and fine-grained materials and subsequent deposition of rows of sand, gravel, and cobble tailings. The fine-grained materials separated from the granular gravel, and cobbles were discharged into the dredge ponds.

Landfill (Lf). Landfill sites occur east of the Mather Field Airport that are associated with the former Mather Air Force Base. Landfill gas monitoring wells are situated surrounding these landfills. The content and thickness of both the debris and cap material is unknown.

Artificial Fill (af). Area where cuts and fills were made during recent grading operations. The fill is composed mainly of reclaimed dredge tailings, quarry wastes, and Riverbank Formation materials. Cuts within these areas expose mainly Riverbank Formation materials. Within the Village of Zinfandel Development on the north end of the alignment, engineered fills are found.

SUBSURFACE EXPLORATIONS

In support of the design and construction of Mather Interceptor, a two-phase subsurface exploration program was conducted to delineate the soil, rock and groundwater conditions at the site. The Phase 1 program consisted of drilling 44 test borings at 36 locations along the alignment between July 2 and November 5, 2007, while the Phase 2 program included the drilling of eight additional test borings along the proposed pipeline (tunnel) and was performed between September 15 and 22, 2008. Figure 2 shows the location of test borings along the pipeline (tunnel) alignment. The drilling of borings throughout the alignment was performed using three techniques as described below:

- **Hollow Stem Auger Borings:** A truck-mounted CME 85 or CME 75 drill rig equipped with 8 inch diameter hollow stem augers and a rock bit.
- **Roto-Sonic Borings:** A truck-mounted roto-sonic drill rig equipped with a 25 foot long, 4.5 or 6 inch inside diameter core barrels with 6 or 7 inch diameter outer casings with carbide button bits. For the roto-sonic borings, an inner core barrel and outer casing were advanced simultaneously using rotation and counterbalanced vibrator. The core barrel was retrieved at nominal 7 to 10 foot intervals providing continuous soil sampling. The continuous core permitted observation and detailed logging of the entire soil stratigraphy.

Figure 3. Large diameter flight auger in cobble laden ground

- **Large Diameter Borings:** Truck-mounted Soilmec T-108 drill rig equipped with 30 to 36 inch diameter flight auger with carbide rock teeth. A bucket auger was used for drilling below the groundwater level.

Drill refusal was typically encountered in the hollow stem auger borings prior to reaching the target depths (2 tunnel diameter below the tunnel invert). The drill refusal appeared to be caused by gravel and cobble materials. In some cases the auger was able to penetrate these materials with some difficulty. A high degree of bit wear was noted following drilling due to the abrasive nature of the materials. Where premature refusal of hollow stem auger borings was encountered, a roto-sonic or large diameter boring was performed in order to reach the target depth.

Roto-sonic borings were able to penetrate the subsurface materials and recover continuous samples of subsurface soils. Due to the vibration associated with roto-sonic drilling, the recovered samples are disturbed, preventing evaluation of the stiffness or relative density of the soils encountered. However, a 140 pound automatic trip hammer falling 30 inches was used in several of the test borings to obtain driven sampler blowcounts and intact samples in addition to the core samples.

The 30 to 36 inch diameter flight auger rig was able to penetrate the gravel and cobble bearing layers with moderate drilling effort. A few zones of hard drilling were encountered where larger cobbles were present, although the auger was able to lift the cobbles from the hole as shown in Figure 3.

Groundwater Monitoring Wells

Groundwater monitoring wells (drilled-in-place stand-pipe piezometer) were constructed in 11 of the test borings. Groundwater level measurements were made periodically following monitoring well installation.

Geotechnical Laboratory Test Program

Geotechnical laboratory testing was performed on selected samples obtained from the borings. The laboratory testing program included index tests and strength and deformation tests to classify soil/rock into similar geologic groups and to measure the engineering properties of each geologic unit to support engineering analyses. The main laboratory test program can be summarized as follows:

EVALUATION OF TUNNELING METHODS FOR MATHER INTERCEPTOR

- Soil Index Tests
  - Visual Classification
  - Water Content
  - Unit Weight
  - Grain Size Distribution
  - Atterberg Limits
  - Specific Gravity
- Soil Strength and Consolidation Tests
  - Unconsolidated Undrained (UU) Triaxial Compression
  - Consolidated Undrained (CU) Triaxial Compression
  - Direct Shear
- Rock Testing
  - Uniaxial Compressive Strength
  - Cerchar Abrasivity Index
  - Thin-Section Petrographic Analysis

GENERAL GROUND BEHAVIOR AND SUBSURFACE CHARACTERIZATION

Table 1 summarizes the strata that are anticipated to be encountered within the bored tunnel zone (within approximately one tunnel diameter below and two tunnel diameters above the tunnel alignment) and their behavior with respect to tunneling activities based on Tunnelman's Ground Classifications. Figure 4 also shows the subsurface profile along Mather Interceptor tunnel alignment.

The ground conditions evaluated for tunneling include the zones located between two tunnel diameters above and one tunnel diameter below the tunnel crown and invert, respectively. Soil conditions encountered along the tunnel alignment are highly variable in composition but are generally overconsolidated and relatively firm/dense. Table 2 presents the type of soils to be encountered along the tunnel zone and the blowcounts recorded during driving of Standard Penetration Test and California Samplers at various stations.

Table 3 presents the general range of total unit weight and cohesion for various soil groups to be encountered along the tunnel alignment.

The cobbles encountered in the borings drilled as shown in Figure 5 for this study are generally very hard and abrasive. The cobble sizes generally vary from 3 to 8 inches; however, some cobbles up to 12 inches in maximum dimensions were encountered in several of the borings. The subsurface explorations to date have not encountered boulders, but previous underground projects performed in the area indicate that boulders up to 24-inches may be encountered (Castro et al. 2001).The predominant rock types are meta-andesite and quartzite (meta-sandstone) with minor meta-sedimentary and other metamorphic rocks. Although generally not as abrasive as the quartzite clasts, the meta-andesite clasts in many cases exhibited the highest compressive strength. Cobble samples obtained from borings throughout the site indicated the meta-andesite clasts have compressive strengths ranging from 18,000 to 38,000 psi in most samples with one sample as high as about 81,000 psi. The quartzite (meta-sandstone) clasts tested exhibited compressive strengths ranging between 31,000 to 37,000 psi. Compressive strength test results on miscellaneous clasts ranged between 11,000 to 20,000 psi. Cerchar abrasion indices (CAI) on meta-andesite clasts ranged from 3.51 to 4.19 for the samples tested. Cerchar abrasion indices on quartzite (meta-sandstone) clasts ranged from 4.4 to 4.68.

Table 1. Anticipated ground behavior

Soil Type	USCS Symbol	Anticipated Ground Behavior	
		Above Groundwater	Below Groundwater
Silt and Sandy Silt	ML	Firm	Cohesive running to slow raveling. Slow groundwater inflow
Elastic Silt	MH	Firm	Slow to fast raveling. Very slow groundwater inflows
Fat Clay and sandy fat Clay	CH	Firm. Swelling if wetted by construction process.	Same as MH
Lean Clay and sandy Lean Clay	CL	Firm. Slight swelling if wetted by construction process	Same as MH
Poorly graded sand and silt	SP-SM	Cohesive running to fast reveling	Flowing to cohesive running. Slow to moderate groundwater inflows
Poorly graded	SP	Running to fast raveling	Flowing. Moderate groundwater inflows
Silty Sand	SM	Slow to fast raveling	Cohesive running to fast raveling. Slow groundwater inflows
Clayey Sand	SC	Firm to slow raveling	Cohesive running to fast raveling, moderate to fast groundwater inflows
Well and Poorly graded gravels with cobbles	GP, GW	Cohesive running	Flowing to cohesive running. Fast groundwater inflows
Poorly graded gravel with silt and cobbles	GP-GM	Cohesive running	Flowing to cohesive running. Fast groundwater inflows
Poorly graded gravel with clay and cobbles	GP-GC	Cohesive running	Flowing to cohesive running. Fast groundwater inflows
Clayey Gravel with cobbles	GC	Firm to slow raveling	Cohesive running to fast raveling, moderate to fast groundwater inflows
Silty Gravel with cobbles	GM	Firm to fast raveling	Cohesive running to fast raveling, moderate to fast groundwater inflows

EVALUATION OF TUNNELING METHODS FOR MATHER INTERCEPTOR

As currently planned, approximately 36% of the tunnel alignment (5,100 LF) for Mather Interceptor Project will be constructed using two-pass tunneling method and the remaining (9,700 LF) via one-pass pipejacking.

The two-pass tunnel construction consists of a 10 ft OD tunnel. The type of tunneling machine for two pass tunneling method can range from an open/partial face excavation machine (digger shield) to a convertible EPB machine or mixed face tunnel boring shield machine. The digger shield is composed of a fabricated, open-ended steel cylinder fitted with an excavator, breasting system, muck apron, muck conveyor, and thrusting system. An important feature on this type of shield is the forward projection of the crown of the shield to form a protective hood that can help reduce ground loss. The tunnel support for two pass method consists of steel ribs with wood lagging or steel liner plates. The 72 inch RCP carrier pipe will be installed inside the tunnel support with the space between tunnel support and the carrier pipe filled with grout.

EVALUATION OF TUNNELING METHODS FOR MATHER INTERCEPTOR

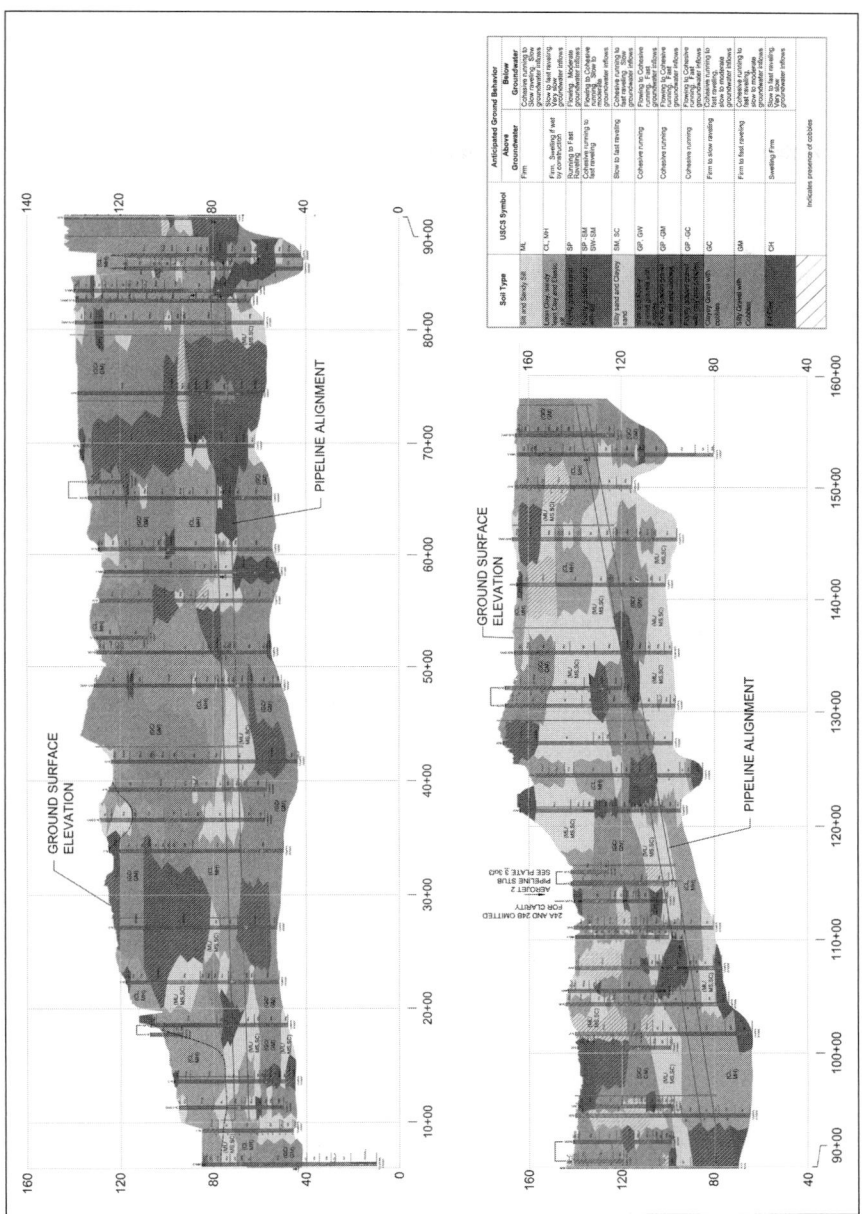

Figure 4. Subsurface soil profile along the tunnel alignment

Table 2. Blow counts for SPT and California sampler

Station (tunnel reach)	USCS Soil Type	Subsurface Soil Condition	SPT (blow counts per foot)*	California Sampler (blow counts per foot)*
14+00 to 65+00	CL, CH	Firm to very hard silts and clays with varying amounts of sand, gravel, cobbles, and relatively dense sands and gravels with varying amounts of silt, clay and cobbles.	8 to 40	9 to 61
	ML, MH		—	40 to 73 for 9¼"
	SP, SP-SM, SP-SC, SM, SC		30	37 to 78
	GP, GP-GM, GP-GC, GM, GC		14, 50 for 6"	54 to 50 for 5"
65+00 to 88+00	CL, CH	Soft to very hard silts and clays with varying amounts of sand, gravel, cobbles, and relatively dense sands and gravels with varying amounts of silt, clay and cobbles.	—	20
	ML, MH		—	77 to 81
	SP, SP-SM, SP-SC, SM, SC		—	24 to 50 for ¾"
	GP, GP-GM, GP-GC, GM, GC		14 to 50 for 6"	49 for 6" to 50 for ¾"
88+00 to 119+00	CL, CH	Firm to very hard silts and clays with varying amounts of sand, gravel, cobbles, and relatively dense sands and gravels with varying amounts of silt, clay and cobbles.	16 to 46	20 to 50 for 2"
	ML, MH		11	77 to 81
	SP, SP-SM, SP-SC, SM, SC		53	31 to 50 for ¾"
	GP, GP-GM, GP-GC, GM, GC			53 to 50 for 4"
119+00 to 157+40	CL, CH	Soft to very hard silts and clays with varying amounts of sand, gravel, cobbles, and relatively dense sands and gravels with varying amounts of silt, clay and cobbles.	—	17 to 50 for 2"
	ML, MH		—	—
	SP, SP-SM, SP-SC, SM, SC		33 to 50 for 6"	28 to 50 for 5"
	GP, GP-GM, GP-GC, GM, GC		42	88 to 50 for 3"

*Blowcounts are uncorrected

The type of machine to be used for one-pass method will likely be an open face pipejacking machine. The diameter of the carrier pipe for the project varies from 72 inch to 54 inch. The variation in diameter will require the use of two machines, upsizing of the TBM skin and cutterhead, or standardization of the diameters.

A primary tunneling methodology issue that can lead to disputes between the owner and the contractor is a mismatch of tunneling methods and the ground conditions encountered. To manage this risk the contract documents need to provide for a system that is workable and provides the most flexibility with regards to the ability to overcome the obstacles presented by the selected tunnel methodology.

The construction techniques to be utilized for the Mather Interceptor were selected and ranked based on the following criteria.

Applicability to Subsurface Conditions

The recommended construction method and excavation equipment should be compatible with subsurface conditions to be encountered during the construction program. Inappropriate methods of construction or machine selection would result in project delay and construction budget overrun. Good examples include the Portland TriMet project where the tunnel boring machine used to excavate the first of 15,000-ft-long

EVALUATION OF TUNNELING METHODS FOR MATHER INTERCEPTOR 281

Table 3. Summary of soil total unit weight and cohesion

Material Type	Total Unit Weight (pcf)	Ultimate Cohesion (psf)	Recommended Ultimate Cohesion for Design (psf)
Clay (CL, CH)	115 to 135	1,000 to 4,500	1,000
Silt (ML, MH)	85 to 110	150 to 4,500	250
Silty Sands (SM)	120 to 130	150 to 3,000	250
Well and Poorly Graded " Clean" Sands (SW, SP)	120 to 130	0	0
Sandy Gravels and Cobbles (GW, GP)	125 to 140	0	0
Silty Gravels and Cobbles, (GM)	125 to 140	150 to 3,000	250
Clayey Gravels and Cobbles, (GC)	125 to 140	500 to 4,500	1,000

Note: The higher range ultimate cohesion values presented above are for cemented soils.

Figure 5. Cobbles and boulders in a sand/gravel matrix

twin transit bore took over 60 weeks and the second bore took only about 20 weeks for the same length of tunnel after major modifications to the TBM were undertaken between drives (Abramson, 1998). Another example is the Chamber Creek Tunnel near Tacoma, WA where the construction started with a closed-face machine that was stopped by a higher concentration of boulders/cobbles and then was completed successfully with a digger shield machine (Abramson, 1998).

Cobbles. For the Mather Interceptor ground condition, cobbles will not stay in a soil matrix when encountered. As the cobbles fall to the bottom of the face where they will collect and these nested cobbles will clog and eventually stop the advancement of the tunnel if not cleared out on a continuous routine basis.

The two-pass system with a 10-ft OD has the room at the heading to allow for a digger shield machine. The pipejacking alternatives with 72-in pipe provide less room, but can accommodate a digger shield. The shield for 54-in pipe may be too small to use a digger shield and may instead utilize a rotating cutter head to excavate the face and process material for conveyance out of the tunnel.

The use of pipejacking may carry a slightly higher risk from excessive cobbles. Two-pass tunneling reduces this potential risk.

Boulders. Boulders can be pushed away from the tunnel face or can be broken depending on cutter type, rock composition and strength, boulder shape, orientation, and soil matrix strength. Cutter types might include; drag bits, picks, roller cutters, and single/multi-disk cutters.

The existence of frequent cobbles and possible boulders will generate an uneven and irregular surface around the carrier pipe which may cause damage to the RCP pipe during pipejacking operations. This adverse environment is mitigated to some degree by the lubricating material used in the pipejacking process but the lubrication is not designed to eliminate that effect; it is only a beneficial byproduct of lubrication. However, a 10-ft OD two pass tunnel provides a designed and known environment for pipe installation with no contact between carrier pipe and the ground.

Open-face digger-shield type TBMs allow access to the face for boulder removal. The digger shield type TBM can ingest boulders up to approximately half of the excavated diameter, but mucking large boulders will require splitting. This will reduce the risk of the TBM machine becoming stalled or the need for excavating a rescue shaft or drilled hole to rescue the TBM or to retrieve the boulder respectively. Using a digger shield machine would eliminate the risk of cutter head wear or damage due to existing cobbles and boulders, saving significant time to repair damaged or worn cutterheads. In general application of open face pipejacking machine is limited to grain size less than 40% to 50% of the size of the cutterhead by direct ingestion. As previously mentioned, small boulders up to 24-inches may possibly be expected for the Mather project. This is larger than the 20-inch boulder size limit for a 54-inch (approximately 66-in OD) pipejacking machine. However, the open face shield design allows for workers to remove larger rocks by splitting or other methods that do not require rescue shafts.

The use of pipejacking carries a slightly higher risk from boulders. Two pass tunneling risks are lower for boulders.

Groundwater. A closed-face pipejacking alternative would not require ground modification techniques such as soil grouting for a small portion of the alignment that is identified to be below groundwater. The digger shield excavation with an open face will require grouting to provide adequate face stability in this same reach of the tunnel.

The use of a two-pass digger shield carries a higher risk from groundwater. A closed-face pipejacking machine reduces this potential risk. An open-face pipejacking machine carries the same risk as a two-pass digger shield machine with regards to groundwater. However, inflows at the face of the 2-pass digger shield could be proportionally larger than inflows at the face of the smaller pipejacking shield.

Alignment

Application of two-pass tunneling method for a portion of the alignment will enable the pipeline to be installed in a curving alignment following the future Zinfandel Drive extension, saving the expense of additional right-of-way procurement. While the one-pass pipejacking method will require shafts every 600 to 1,000 ft, current cost estimates predict that this method is far less expensive than the two-pass method given the relatively short linear footage in question.

Ground Deformation and Instrumentation

Tunnel over-excavation using an open-face pipejacking machine with rotating cutters can be a source of ground deformation and excessive settlements, however a two-pass digger shield may also experience over excavation at the face. A two-pass digger shield reduces the risk of settlement through installation of the initial support behind the TBM, whereas a pipejacking machine has a larger radial overcut that is not supported to allow jacking of the trailing pipe string. This annular space is grouted after completion of each pipejacking drive.

Maintenance Access to the Heading

A 10 ft OD two pass tunnel provides greater space for faster muck handling, lighting, and installation of temporary ventilation system, than the pipejacking alternative.

Scheduling Impact

In general, a two pass tunneling method is slower than one pass pipejacking. However, for difficult ground conditions, especially for cobble and boulder laden formation, application of two pass tunnel methods can be faster than one pass trenchless technique such as pipejacking or microtunneling. This would be negated by the concurrent use of two different sized pipejacking machines for the 54- and 72-inch pipe reaches.

Schedule risk is similar for the two methods.

History Lessons of Tunneling Methods in the Local Area

In 1998 the Construction Method Proving Project (CMPP) (Black and Veatch Construction Inc. 1998) was conducted in similar ground formation in close vicinity of Mather Interceptor project (approximately 4 miles east of Sunrise Boulevard) but at shallower depth (20 ft below ground surface) to assess the feasibility of two trenchless methods for the construction of Folsom East 2 (FE-2) project in Sacramento California. The open face shield pipejacking machine with rotating cutter bar and Microtunneling boring machine (MTBM) had been selected for installation of a 66 inch OD pipe. The results of this trial indicated that both techniques were unsuccessful and both failed as viable construction options for the encountered subsurface condition due primarily to settlements related to over excavation. However, the US Highway 50/Folsom Blvd tunnel was successfully constructed approximately ½ mile east of the CMPP, using a closed face TBM with access doors. Some settlements were observed, but these settlements were much smaller than those observed on the CMPP.

Our interpretations of applicable lessons learned from this project are that:

- The allowable tunneling method for ground conditions similar to the Mather Interceptor should provide enough room during tunneling and flexibility to execute possible mitigation plans such as ground treatment or boulder extraction.
- The allowable tunneling method and excavation technique should also provide adequate flexibility for swift machine modification and repair with no major impact on project schedule and cost.
- The selected excavation method should minimize the associated tunnel face loss or excessive ground loss.
- The selected tunneling method and excavation technique should minimize the down time of machine operation due to damage and repair.

CONCLUSIONS

The tunnel construction for Mather Interceptor project will be performed through challenging mixed ground consisting of mainly dense silty sand and gravel with some reaches passing through silt/sandy silt and lean clay/sandy lean clay and frequent cobbles and possible boulders. As currently planned, approximately 36% of the tunnel alignment will be constructed using two-pass tunneling method and the remaining via one-pass pipejacking.

ACKNOWLEDGMENT

The authors would like to acknowledge the collaboration of several individuals for preparing this paper. Particular thanks are due to Kenneth Sorensen (Kleinfelder) and Jeff Cabrera (CDM Inc.).

REFERENCES

Abramson I.W. (1998). Root Causes of Disputes on Tunnel Construction Projects. *Proceedings 1998 North American Tunneling Conference*, pp. 55–62.

Castro R., Webb R., and Nonnweiler J. (2001). Tunneling Through Cobbles in Sacramento, California. *Proceedings 2001 Rapid Excavation and Tunneling Conference*, pp. 907–918.

Black and Veatch Construction, Inc., and Bennett/ Staheli Engineers, Inc. (1988). Construction Method Proving Project.

Helley, E.J., and Harwood D.S. (1985). *Geologic Map of the Late Cenozoic Deposits of the Sactamento Valley and Northern Sierran Foothills, California*. U.S. Geological Survey Miscellaneous Field Studies Map MF-1790.

PART 5

Ground Modification

Chairs

Jon Kaneshiro
Parsons

Paul Roy
AECOM USA

NORTH 27TH STREET ISS EXTENSION: UNIQUE OWNER/ CONTRACTOR AGREEMENT SETTLES MAJOR DISPUTES

Donald J. Olson ▪ CH2M Hill

Roger J. Maurer ▪ MMSD

Martin Vliegenthart ▪ J. F. Shea

ABSTRACT

The North 27th St. ISS Extension tunnel will, under present plans, be the last constructed section of Milwaukee Metropolitan Sewerage District's (MMSD's) existing deep tunnel system. The 6.5 m (21 ft) finished diameter tunnel is approximately 3,260 m (10,590 ft) long and is located in NE Milwaukee County. When disputes over the processes to be used in owner-directed pre-excavation grouting in rock threatened the completion date stipulated between MMSD and the State of Wisconsin, the MMSD and J.F. Shea, the contractor, entered into a unique agreement. Shea accepted responsibility for all shaft and tunnel grouting and water control during the project in exchange for settlement of claims related to grout quantities and perceived delays. This significantly reduced risk for the MMSD and helped ensure the stipulated completion of the project.

HISTORY

The Milwaukee Metropolitan Sewerage District's (MMSD's) Contract C05013C01, the N. 27th Street ISS Extension (N.27th ISS Ext) was designed as the last in a long series of projects for its Inline Storage System (ISS) of tunnels in rock 300 feet beneath the region's three major rivers. The ISS was conceived in the late 70s after the MMSD was sued by various state and local governments for polluting Lake Michigan. The MMSD was forced to implement a massive public works program to reduce combined sewer overflows (CSOs) to Milwaukee's rivers and streams, all of which were tributaries to Lake Michigan. That program, known at the time as the Water Pollution Abatement Program (WPWP), was completed in the mid-1990s at a cost of approximately $2.5 Billion, more if one considers the interest on money borrowed to complete the work. Work performed under the WPAP resulted in: (a) the design and construction of nearly 30 km (20 miles) of ISS tunnels, known locally as the Deep Tunnel System, to prevent nearly all CSOs to Milwaukee's rivers and the lake; (b) massive improvements to the existing system of conveyance pipelines and tunnels that carry sewage flows to the region's two main treatment plants; and (c) significant enhancements to these plants which increased their sizes and flow capacities, among other tasks.

Several of the tunnel projects included in the WPAP have been discussed at length in RETC papers dating back to the early 1980s. The WPAP has been highly praised by numerous organizations and agencies over the last two decades for having met the needs of the MMSD and the people of the region by reducing CSOs to the rivers and lake from 50 to 60 per year to just over two per year.

Since the completion of the WPAP in the mid-1990s, MMSD has added one tunnel to the ISS, the MMSD's Northwest Side Interceptor, a 10 km (7 mi.) tunnel to store and convey sanitary sewage from the suburban communities in the northwest quadrant of the MMSD's jurisdiction. Other tunnels, most much shallower than the ISS, have also been constructed to continuing the work of controlling contaminated waters in Milwaukee.

DESIGN

In 2003, MMSD was working toward a goal of designing and building two interceptor sewer tunnel projects, one to the west and one to the north of downtown. These tunnels would convey flows to the treatment plants. The Wisconsin Department of Natural Resources (WDNR), in its review of these projects, recommended that they be consolidated into one final addition to the ISS. WDNR felt it was critical to enhance the storage capacity of the ISS. The storage capacity of the ISS was approximately 1 billion liters (494 million gallons). The desire was to increase this capacity by about 5%. MMSD opted to consolidate the design projects for the proposed north and west interceptors into the design of what would become the N. 27th Street ISS Extension.

The result of this change was the design of a 3,260 m (10,590 ft) long, 6.5 m (21 ft) diameter tunnel which would be an extension of MMSD's North Shore Interceptor. The North Shore Interceptor comprises about half of the original ISS. It is comprised of: a) the NS Phase 1A tunnel, 8,125 m (26,000 ft) of nominal 9.8 meter (32 ft) diameter tunnel in rock beneath the Milwaukee River; b) the NS Phase 2A tunnel, 4,875 m (15,000 ft) of nominal 5.5 m (19.5 ft) diameter tunnel under W. Hampton Avenue and N. 32nd Street, and a connecting tunnel, 1,630 m (5,200 ft) of nominal two meter diameter beneath W. Hampton Ave. from N. 32nd to N. 51st Streets. Figures 1 and 2 show these tunnels and the N.27th ISS Ext. tunnel alignment.

The N.27th ISS Ext. was designed to store about 102 million liters (27 million gallons) of sanitary flow in tunnel from the intersection of W. Hampton Avenue and N. 30th Street in Milwaukee north to a point at W. Mill Road in Glendale, WI. It would run beneath N. 27th Street in Milwaukee for a portion of its length. From this came the name of the project. The northern terminus on W. Mill Road was chosen in order to receive flows from a series of MMSD relief sewers that had been installed in the 1970s and 80s. Several of these sewers meet at this location. However, during design a planned drop shaft on Mill Road was removed from the project for cost reasons. Thus, the tunnel was finally designed to accept flows only from its southern connection to the Ph 2A tunnel at Hampton Ave. After storing water accumulated during wet weather CSO flows from the ISS, this water would drain from the tunnel via this same connection at its south end. All ISS waters flow by gravity to the MMSD's Inline Pump Station beneath the Jones Island Treatment Plant south of downtown Milwaukee.

The design was amended several times prior to being released for bidding. The Hampton Ave. shaft was relocated about 50 m (150 ft) to the east of its originally planned location when geotechnical borings showed contaminated soils would be encountered during excavation at the original location. Later, study of geologic maps developed after excavation of the NS Ph 2 tunnel provided evidence a faults or shear zones which might intersect the base of the shaft. Extra support elements were included in the Contract for use as needed if rock conditions were found to be less than favorable.

There were several items of concern raised during design. The existence of an abandoned landfill some 300 m (1,000 ft) west of the alignment's northern end could be a problem. The possibility of migration of contaminated water from the landfill to the tunnel was considered. This could be a problem for workers during excavation and might be cause for concern by environmental agencies. Several groundwater monitoring wells were installed to allow engineers to observe water levels and quality. If needed, actions could be taken to mitigate migration.

The rock types to be excavated were anticipated to be similar to those excavated in other ISS tunnels, particularly the NS Ph 2A tunnel. The tunnel would start

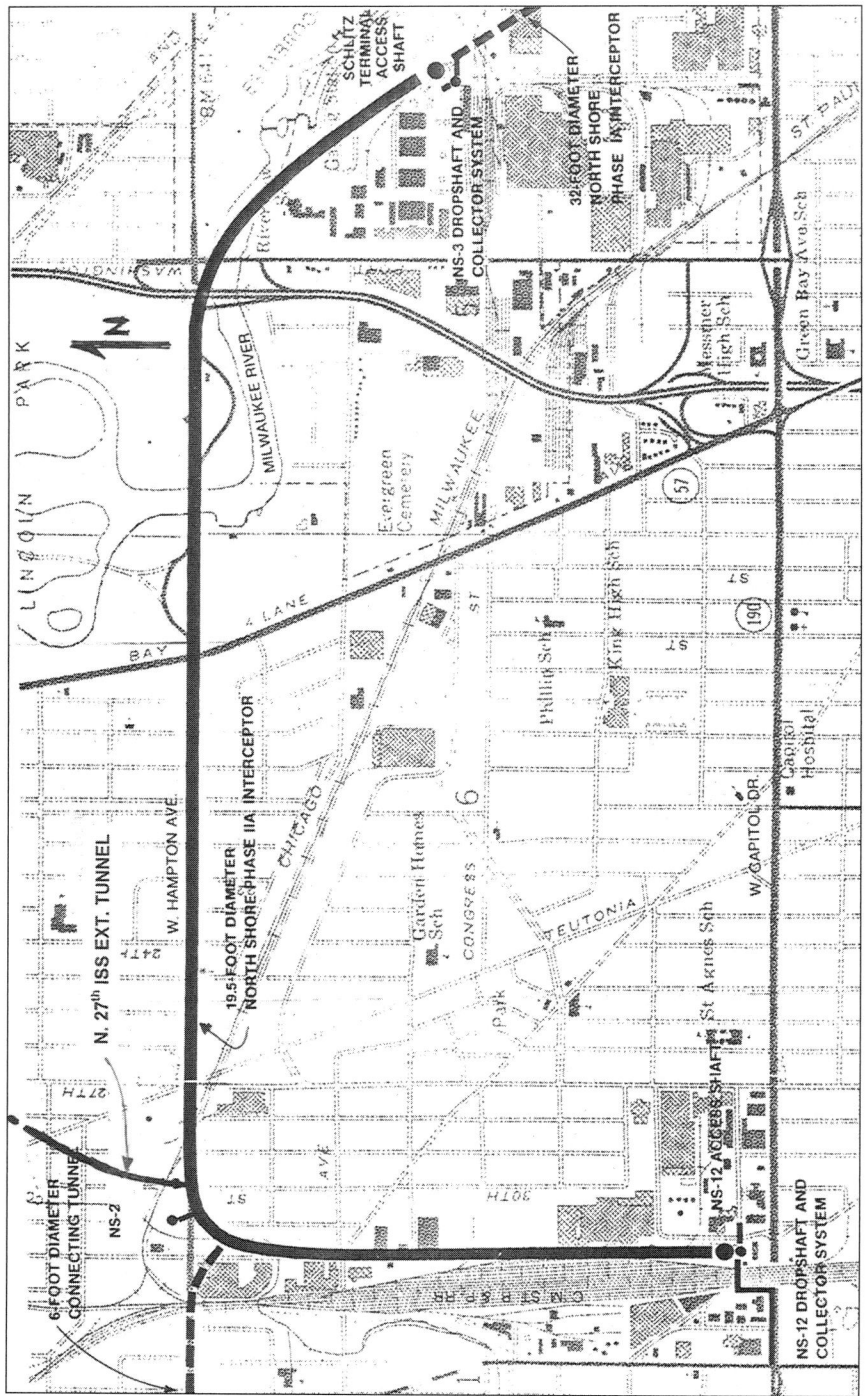

Figure 1. Vicinity map

NORTH 27TH STREET ISS EXTENSION: AGREEMENT SETTLES DISPUTES 289

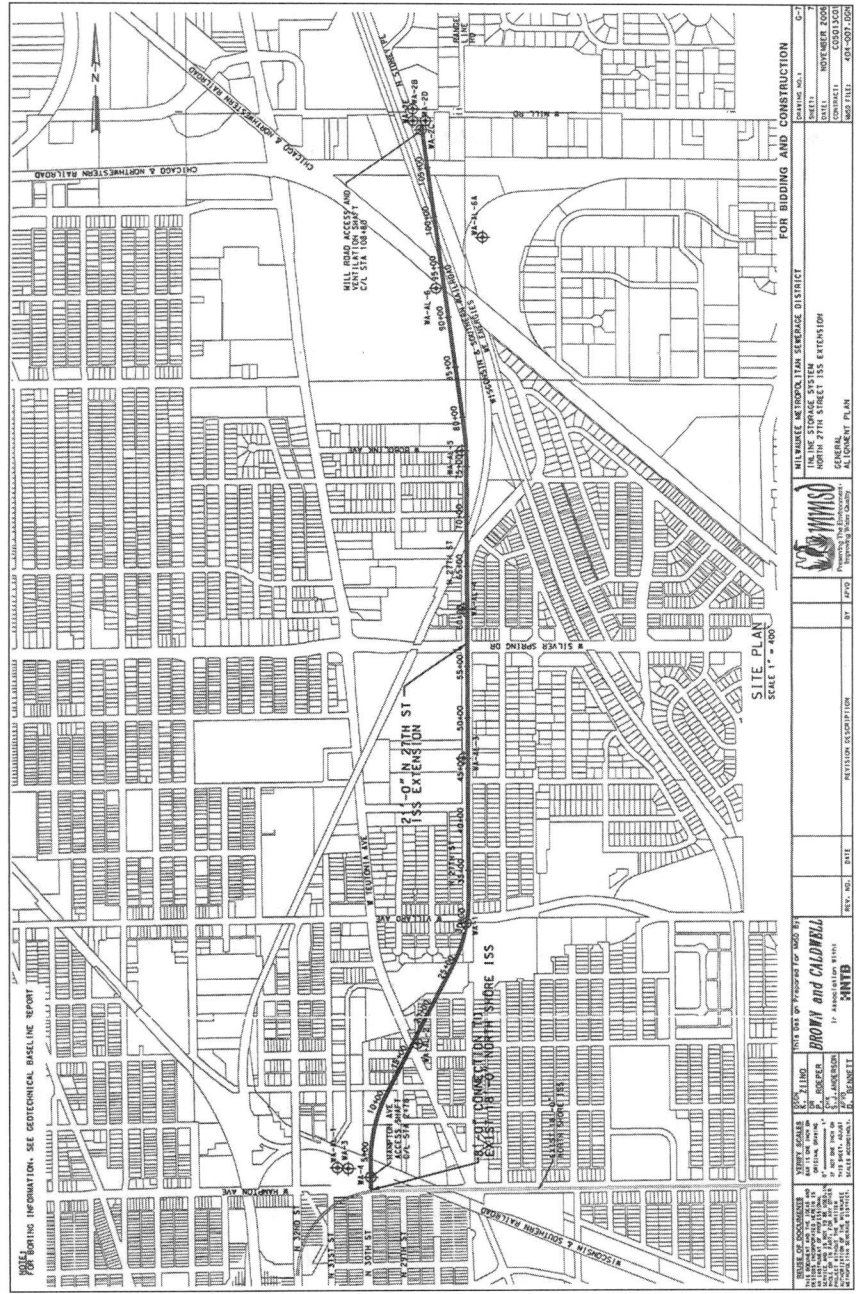

Figure 2. Alignment map

at its southern end in the center of three sub-divisions of the Waukesha Formation. As excavation progressed to the north, it would move into the upper sub-division of the Waukesha. Most of the northern half of the tunnel would be mined with the tunnel invert in the Waukesha and the tunnel crown within or close to the overlying Racine Formation. These are Silurian Age dolomites, generally close to medium bedded sediments with varying numbers of vugs, some filled, some not. Some areas of these formations contain faults, shear zones and solution cavities. Primary tunnel support methods and materials were specified that would allow the successful bidder to handle virtually all rock conditions.

Groundwater inflow to the new tunnel was another area of concern during design. The NS Ph 1A tunnel excavation in the late 1980s was seriously impacted by poor rock conditions and by greater-than-expected groundwater inflows. The NS Ph 2A tunnel was also quite wet during excavation especially along its north-south alignment. However, it was anticipated that improved ability of contractors to probe ahead of the tunnel face and perform pre-excavation grouting would help the situation. Probing ahead of the face was used extensively in Milwaukee during later stages of the WPAP. The design of the N.27th ISS Ext. included use of new grout techniques and materials that would reduce inflows to manageable amounts. Water inflow to the tunnel was anticipated to be as high as 21,000 liters per minute (5,500 gallons per minute) in the tunnel prior to final lining and grouting.

Of major importance during design was the time of construction. The MMSD was under significant pressure to assure that the N. 27th ISS Extension was completed on time. The completion date of the work had been defined long before the start of design. In fact, a completion date for the north and west interceptor tunnels mentioned above, which morphed into the N. 27th ISS tunnel, had been ordered by the courts. In the 1990s, the WDNR had sued MMSD for alleged infractions of discharge rules. As part of the settlement of that case in Milwaukee County District Court, MMSD agreed to stipulated dates of completion of these two interceptors. When the N.27th ISS Ext. tunnel replaced these interceptors, MMSD agreed to accept the stipulated completion date of the interceptors for the N.27th ISS Ext. project.

BID

Bids for MMSD Contract C05013C01 were accepted on January 30, 2007. The low bid ($65,360,000) was submitted by Affholder, Inc. They were awarded the Contract on March 12, 2007, but immediately informed the MMSD that they would prefer to not proceed with the work. Affholder made an internal decision, apparently after bidding, to remove itself from work in the tunnel construction industry, and they wished to be released from the contract.

This notice was a complete shock to the MMSD. MMSD staff had no idea that Affholder was considering such actions. MMSD had to quickly develop and consider several options. One option was to force Affholder or its bonding company to perform the work. Realizing the complications that could be wrought by signing a contract with a business entity which was effectively shutting down, MMSD considered other options, including taking Affholder's $3 million bid bond and going out again for new bids for the work. MMSD decided to approach SKJV, who had been second low bidder on the project. SKJV's bid had been about $3 Million higher than Affholder's bid. MMSD, SKJV and Affholder then met. An agreement was reached to develop a method of assigning the contact to SKJV. A "novation agreement" removed Affholder from the contract altogether. SKJV would complete the work for Affholder's bid price but would bill MMSD using its own bid breakdown. SKJV would maintain Affholder's Small, Women's, and Minority Business Enterprise (S/W/MBE) goals and adopt Affholder's S/W/MBE subcontractors. Affholder would give to SKJV several elements of equipment and other consideration; specifics of this part of the agreement were not made known to the MMSD. SKJV would become the second party in the contract, accepting all requirements and duties of the contractor as binding upon them.

This solution was viewed favorably by the MMSD. Without such an agreement, they might be forced to re-bid the job. The time involved in such a change would prevent the MMSD from completing the project in compliance with the stipulated court date. In addition, MMSD gained some comfort with signing SKJV, a "known commodity" so to speak. MMSD had worked with SKJV and its staff on numerous projects in the past. MMSD had some assurance that the work would be completed appropriately.

DISPUTES

A Notice to Proceed was issued to SKJV on May 15, 2007. They provided a preliminary schedule and an initial set of submittals including the plans for the tunnel boring machine (TBM) and shaft excavation. They would utilize a former Robbins TBM that had been used on one of the MMSD's early ISS projects, the Kinnickinnic/Lake Michigan tunnel in the late 1980s, and again on the MMSD's Northwest Side Interceptor project, which was completed in 2004. Shaft excavation would proceed in fundamentally the same way that most of SKJV's shafts were constructed. The nominal 25 m (80 ft) of soil at each shaft would be frozen prior to excavation. Prior to soil freezing, the rock beneath would be grouted with Portland cement.

On most MMSD contracts during the previous 25 years, grouting of rock to reduce permeability of the rock mass has been specified as work to be directed by the Engineer, acting on behalf of the MMSD. Pursuing the grouting under Engineer direction is not unique in the industry. Owners often feel more confident that their interests will be served if the grouting is controlled by them or their agents. There have been projects in the past which specified that the contractor would effectively self-supervise his operations and the owner would pay using bid prices for the various grout components. This type of arrangement had some history of being advantageous for the contractor without significant benefit for the owner. MMSD believed this type of program simply wasteful and wanted no part of such a procedure. Engineer-directed grouting would be best.

Consistent with previous MMSD contracts, the specifications for the N.27th ISS Ext. defined in great detail how the grouting work would be accomplished. Open joints, bedding planes, faults, and any other discontinuities in the rock would be intercepted by drilled holes. The holes would be tested for permeability and filled with a slurry of cement, water and additives as needed. The slurry would then harden into a solid mass, creating a barrier impeding the flow of groundwater into the excavation. This general concept of filling water-producing features in rock with grout had been used for many years in the tunnel industry.

MMSD had not called for the use pre-excavation grouting during tunnel excavation during the first years of construction of the ISS. Generally contractors were left to their own devices in terms of control of water inflow to the tunnels during construction. If contractors chose to allow groundwater to flow into the excavation, then pump and treat it prior to discharge, and address the consequences later, this was allowed in most cases. Eventual reduction of groundwater inflow to stated limits was to be accomplished using post excavation grouting, either before or after concrete lining of the tunnels was installed.

However, on these early ISS projects, the MMSD paid a heavy price for this approach. In many locations, groundwater inflow to the tunnels was quickly followed by lowering of the groundwater levels in rock and soil aquifers. In some cases this led to settlement of and distress to buildings and other structures above the tunnels. Pre-excavation tunnel grouting processes were developed as modifications to these early contracts in order to mitigate damage to near-surface structures. Eventually, pre-excavation grouting was instituted as a requirement of all contractors who pursued work beneath the streets of Milwaukee. Concurrently, SKJV became a strong proponent of this approach.

However, SKJV had routinely opposed some of the elements of pre-excavation grouting as specified by MMSD in the late 1990s. This became a major stumbling block on the N. 27th ISS Ext. project. This opposition can be summarized by the defining the technique for pre-excavation grouting specified in recent MMSD documents.

SPECIFIED GROUTING TECHNIQUES

The MMSD has tended in recent years to incorporate techniques in its tunnel contracts which generally follow the concepts of the Grout Intensity Number (GIN) method forwarded by Prof. Giovanni Lombardi and others in the early 1990s.

Grouting methods commonly used prior to development of the GIN method routinely called for the injection of thin grout mixes of Portland cement and water, typically starting with 4:1 water/cement ratios, and proceeding to thicker mixes incrementally if grout takes and pressures remained constant. Depending on grout take and pressures used, grout mixes would be thickened up to 1:1 water/cement ratios. In particularly large fissures and cavities, mixes with even smaller water/cement ratios might be used, or additives for increasing the bulk or the viscosity of the mix might be employed. The intent was to fill with cement all nearby fissures that might transmit water. Once filled, the cement would set up and form a practically impermeable barrier to water flow.

The GIN method utilizes the same principle of filling voids with solids. However, the method calls for creating thick mixes (low water/cement ratios) and supplementing the mixes to develop very low viscosities so that they might travel more easily into the voids, filling them more quickly and efficiently. Super-plasticizers were often specified to create such thick mixes with very low viscosities. The reduction of viscosities would allow these mixes to work their way deeper into openings in the rock. A significant advantage of this method would be the high solids content of the grouts. This would minimize most of a common trait in neat cement grouts: separation of solids from the water in the injected slurry after the slurry is delivered to the fractures in the rock. Solids separation, also known as bleed, often would reduce the effectiveness of thin, neat-cement grouts. After injection, the cement particles would remain in the fractures, but if the water bled out, the fracture would once again be capable of transmitting water. Permeability would increase if the water could bleed away. Bleed of water from the mix would decrease if "stable and balanced" mixes, as defined in the GIN method were used.

In addition to the concepts in the GIN method, use of ultra-fine cement was also encouraged in the MMSD specifications as a means of filling smaller cracks and voids in the rock mass.

The grouting specification in the N.27th ISS Ext. contract was intended to maximize grouting efficiency. Creation of a grout "curtain" near the eventual "A-line" of the opening was desired. There would be no need to fill all voids within the A-line because this rock (and the grout within this rock) would eventually be removed and thus wasted. Ostensibly a "cylinder" of impermeable rock and grout around the tunnel perimeter would restrict inflow after the rock inside the cylinder was removed. If this method was successfully executed, a significant reduction in groundwater flow to the tunnel could be achieved without filling with grout every water-producing feature within the "A Line." This could save the contractor and owner in both materials and time if executed properly.

Also, the contract defined limits on the extent of pre-excavation grouting ahead of the tunnel face to enhance efficiency. The Engineer was to direct pre-excavation grouting ahead of the tunnel face only if groundwater flow from any 30 m (100-foot) long section of probe holes exceeded 250 liters per minute (60 gallons per minute). Probe holes producing less than this flow rate would be slugged off with a thick grout mix not expected to travel into water-producing features. The Contractor would then continue to mine immediately. Whatever flow volumes were encountered at the walls of the tunnel would have to be dealt with by the Contractor during mining and lining. These flows would be addressed during cut-off grouting of the tunnel after lining.

In early discussions with the Engineer, it was obvious that SKJV did not agree with the concepts specified in the documents. As mentioned above, SKJV had a history of preferring to perform extensive grouting before shaft and tunnel excavations were undertaken. But they also wanted to do this work using older methods proven on previous projects. The GIN method was not a method that they agreed with. They disagreed with the use of super-plasticizers to reduce the viscosity of thick grout mixes. They argued that attempting to create a grout "curtain" would not be successful; trying to save a few sacks of cement from working deep into the joints inside the "A Line" would be an exercise in futility. Filling every conceivable joint with solids would be much more likely to cut off inflows. Starting with thin mixes of neat cement grout (4:1 water/cement ratio) and progressively working to thicker mixes would work again as it had in the past. Use of expensive modern super-plasticizers would simply stand as an extra cost to them and to the Owner; such grouts would travel far from tunnel alignment and be useless in the pursuit of reduction of water inflow.

SKJV added other notes of disagreement and distain for the directions concerning grouting. The specifications required that only one vertical grout hole could be drilled at a time at the shafts. Any grout hole was to be drilled and grouted individually and completely before any other grout holes were drilled. SKJV said this was completely inappropriate: it would prevent the communication of grout between grout holes. They perceived that drilling numerous holes (at least four), into water-bearing features, placing packers in each hole and then grouting through one packer, carefully observing the other packers for the production of clear water or a slurry of cement and water (from the grouted hole) was the better way to go. Communication between holes was, in their opinion, critical.

They also argued that the specified pattern of shaft grouting holes, starting with an outer ring of holes, 8 feet outside the A-line and then installing to a second ring 5 feet outside the A-line, would lead to poor results. They preferred to start drilling and grouting near the center of the shaft, with communication between holes as the desired result. They would then drill and grout another ring of holes closer to (but still inside) the A-Line, hoping for communication between holes. Possibly then they would drill and grout several holes in a ring outside of the A-line. Eventually a test hole would be drilled at the center of the shaft and pumping test would be run; extracting water from this hole would give an indication of the total groundwater that might be seen after excavation was complete.

The Hampton Avenue shaft was grouted under the Engineer's direction generally in accord with the specifications. A ring of grout holes was drilled and grouted outside of the A-line. A second ring was drilled and grouted. Based on grout takes and other considerations, three added holes were drilled within the A-line. Then in concert with freeze hole drilling, a burn hole was drilled to invert and a pumping test was accomplished. It indicated that likely the rock portion of the shaft would produce less than 10 GPM when the shaft was fully excavated.

Note that several elements of the specified grouting method were *not* called for by the Engineer. For instance, the Engineer did not impose the requirement for using super-plasticizers in the grout mixes. Laboratory testing of various grout mixes by the Engineer resulted in significant bleed of water from the mixes with and without the super-plasticizer. The viability of the method was in question. The Engineer's tests showed that grout mixes as thin as 2:1, the specified starting point for grout injection, would always bleed. Half of the mix water would separate from the solids whether additives were used or not. The specification inferred that such mixes with super-plasticizers would not bleed. Later the Engineer learned from outside experts that ALL such thin mixes would ALWAYS bleed. Only thick mixes (water/cement ratios of 1:1 or smaller) would resist bleeding, and this was independent of the addition of super-plasticizers. The Engineer also tested the addition of bentonite powder to the cement in an effort to reduce bleed. Results were varied; sometimes the bentonite seemed to help. However, of the many experts in grouting consulted, the Engineer found none who would advocate the use of bentonite. Thus it was removed from consideration. In the end, the Engineer allowed use of neat cement grout as preferred by SKJV because the specification could not reasonably work.

The specified "one at a time" grout hole drill/grout procedure was found to be very time consuming. Eventually the Engineer allowed multiple holes to be drilled prior to the injection of grout into first one and then the other holes. This seemed to be an efficient approach.

Grouting of the Hampton Shaft ran from mid-June to mid-September 2007. This three month period was close to the time shown in SKJV's preliminary schedule. However, they were not happy with the time used. With the completion of the Hampton shaft grouting, SKJV inquired pointedly about how the Mill Road Shaft would be grouted. The Engineer and the MMSD said there was no intention to change the methods more than had occurred at Hampton Shaft. SKJV also asked if the tunnel would be similarly grouted using hole patterns, materials, criteria and schedules defined in the specifications. The Engineer was bound to the Contract Documents and the MMSD was not inclined to significantly modify the specified method.

SKJV was confident that groundwater conditions at the Mill Road shaft and in the tunnel would be reasonably close to those defined in the Contract. They felt they could handle all the water that the tunnel might throw at them, assuming they could address inflows on their own terms. If they were provided control over the groundwater and grouting operations, they were sure that they would be able to complete the project with no further claims for water. Otherwise, they and MMSD would lose significant amounts of money and the schedule constraints imposed by the courts would not be met.

Therefore, just after the start of drilling at the Mill Road Shaft, SKJV was compelled to make a bold proposal. If the MMSD would agree to allow SKJV to self direct the pre-excavation grouting, SKJV would accept total responsibility for all grouting, all water control, all water handling, all treatment of water to be disposed off site, and all other aspects of groundwater. They would file no further claims related to water. SKJV would agree to install grouting bid items as needed, and would request payment for bid items only up to the quantities defined in the Bid. If more grout units (sacks of cement, drilled holes, volume of mixed grout, crew hours for drilling and grouting, etc.) than were shown in the bid were in fact needed, SKJV would perform that work without cost to the MMSD. And they would make no more claims related to water.

Note that SKJV was not as sure about the rock quality and support needs in the tunnel as they were about the grouting. They were not prepared to accept responsibility for any unknown conditions related to rock quality, soil conditions, or other such differing site conditions.

As part of the proposal SKJV insisted that one element of the grout bid elements could not be included in the "no pay beyond bid quantities" agreement. They had already made a claim related to the 130 grout connections for shaft pre-excavation grouting in the bid. To SKJV the quantity shown in the Bid seemed to be inappropriately low. The Engineer reviewed this and agreed that the bid quantity seemed low. Discussion with the Designer showed that his interpretation of the term "connection" was apparently different than the interpretation that had historically been employed by SKJV. The Engineer agreed with SKJV. A significant increase in the number of connections seemed appropriate. Eventually, the over-run in quantities amounted to just over

200% of the bid quantity. The Contractor was paid for this increase using the unit price in the bid. It amounted to an increase in the Contract Amount of just under $500,000.

MMSD was favorably impressed with the proposal. The Engineer provided draft language for a modification order to adopt the proposal. SKJV assumed all responsibility for grouting the second vertical shaft on the project (Mill Road Shaft) and the entire length of tunnel using their own methods. Both parties agreed to and signed Modification Order No. 1 to the Contract in January 2008. It included a negotiated non-compensable time extension of 30 days to account for about half of the delay time claimed by SKJV.

RESULTS TO DATE

Results for the work as of January 12, 2009, are as follows:

1. Both shafts have been excavated and lined. At its completion, the Hampton Shaft produced a bit over 20 liters per minute (five gallons per minute) of inflow after a three month long Engineer-directed pre-excavation grouting program and limited cut off grouting after lining. About 2300 sacks of cement were installed prior to excavation.
2. The Mill Road shaft produced about 85 liters per minute (20 gallons per minute) after a nominal 2.5 month long Contractor-directed pre-excavation grouting program and some cut off grouting after lining. About 3,500 sacks of cement were installed prior to excavation.
3. A bit less than half of the tunnel has been excavated on a schedule close to the schedule presented at NTP by SKJV. The groundwater inflows to the tunnel are less than 500 GPM. Slightly more than half of the bid quantity for tunnel pre-excavation grouting paid as crew hours to account for the costs of labor, equipment, supervision, power, supplies, etc., have been paid expended. About 25% of the bid quantity of cement for tunnel pre-excavation grouting has been expended.
4. In general, the contentious disagreements of the earliest days of the contract have been replaced with general accord and agreement on most items elements of the project. The MMSD remains very concerned about the overall schedule of the work. The potential reaction of the Milwaukee County Circuit Court and/or the WDNR if the stipulated completion date is not met looms large over every aspect of the project. The most recent update to the schedule indicates that underground work *can* finish on time. Full compliance with all aspects of Substantial Completion can reasonably be expected before the end of 2009.

CONCLUSION

At the Hampton Ave. Shaft, the Engineer-directed pre-excavation rock grouting program resulted in 75% less groundwater inflow than did the Contractor-directed program at the Mill Road Shaft. On the other hand, about 50% more cement was injected into Mill Road Shaft's slightly deeper rock section, and it took the Contractor about 15% less time to inject the greater amount of cement at the Hampton Shaft. With all variables considered, the authors conclude that the results at the two locations are comparable. There is no clear "winner" between Engineer-directed and Contractor-directed operations.

As it is for any RETC paper written prior to the completion of a project, no final conclusions can be drawn as yet about the tunnel. If significantly more groundwater inflow is encountered during mining, the schedule may suffer. No float exists in the schedule. If things go well, both parties and the overseeing governmental agencies will be happy. If poor progress is made in the second half of the tunnel, these agencies may not exhibit feelings of joy. And I know the MMSD will be unhappy with SKJV, the Engineer, the Designer and anyone else connected to the project. We hope to report the job complete and all parties happy in San Francisco in 2011.

ACKNOWLEDGMENTS

The authors acknowledge the input and help of Bob Plecash, Joni Johnson, Mark Pospyhalla and Mel Russell of MMSD; Lenny Postregna and Milan Jovanovich of J.F. Shea; Scott Anderson and Dave Bennett of Brown and Caldwell; Roger C. Ilsley of RI Geotechnical, Inc.; Samer Sadek of HNTB; Chuck Pape of WDNR; and employees and staff of these and other related firms and agencies. In addition the input and advice of the Disputes Review Board (Ed Plotkin, Tony Stewart and Jim Mahar) and the MMSD's Tunnel Review Board (Ron Heuer, Skip Hendron and Andy Merritt) is appreciated. Finally the invaluable input of Dr. Farrukh Mazhar of Brown and Caldwell and grouting expert Masrour H. Kizilbash of Montgomery Watson Harza is highly appreciated. The insight provided concerning grouting and the GIN method was huge.

BIBLIOGRAPHY

Lombardi, G. and Deere, D., 1993, Grouting Design and Control Using the GIN Principle, International Water Power & Dam Construction.

Brown and Caldwell in association with HNTB and RI Geotechnical, 2006, North 27th Street ISS Extension Geotechnical Baseline and Data Reports.

Brown and Caldwell in association with HNTB and RI Geotechnical, 2007, North 27th Street ISS Extension Design Report.

Milwaukee Water Pollution Abatement Program, 1992, North Shore Interceptor—Phase IIA Lining Report No. 2.

IN MEMORIUM

Author Roger J. Maurer was Section Leader for MMSD on the N. 27th St. ISS Ext. project. He was central to the claims and final agreement about which this paper is written. He alone proposed the paper to his co-authors and remains the inspiration for our work on the project. We are very sad to report that Roger died on December 16, 2008 after battling cancer for over a year. He was a good engineer, but more importantly he was a good man, dedicated to public works and public service. We will all miss him.

BRIGHTWATER CONVEYANCE SYSTEM: GROUND FREEZING FOR ACCESS SHAFT EXCAVATION THROUGH SOFT GROUND

Joseph M. McCann ▪ Moretrench

David K. Mueller ▪ Moretrench

Paul C. Schmall ▪ Moretrench

James D. Nickerson ▪ Frontier-Kemper Constructors, Inc.

ABSTRACT

The Brightwater regional wastewater treatment facility, under construction in King County, WA, includes an extensive conveyance system that will feature deep, soft-ground tunnels and shafts. Excavation of the Ballinger Way Receiving Portal to 65.8 m (216 ft) below grade through challenging geology was accomplished by means of artificial ground freezing. This paper discusses freeze design development, system installation and monitoring, and the contractual and construction challenges that were overcome, including work sequencing and excavation means and methods, together with lessons learned.

INTRODUCTION

The Brightwater regional wastewater treatment facility, due to be completed in 2010, includes a new treatment plant and extensive conveyance system consisting of more than 21 km (13 miles) of influent and effluent conveyance lines and five portals. The majority of the conveyance system lines are mined by tunnel boring machines (TBMs) and microtunnel boring machines (MTBMs). Tunnel and portal construction is divided into three main contracts. The Ballinger Way Receiving Portal was constructed under the Central Tunnel Contract. The portal has a 7.3 m (24 ft) finish ID and extends to a depth of 65.8 m (216 ft) below grade through a complex and challenging soil profile characterized by multiple groundwater tables (Newby et al. 2007).

The joint venture team for the Central Tunnel Contract Ballinger Way Portal considered either slurry wall techniques or ground freezing for earth support during excavation of the portal and liner installation. Taking into account the vertical depth, cost, past practice, geology, water tables and the risk of contaminated soils, ground freezing was selected. A demonstrated successful work history and the experience of key personnel was a significant factor in the selection of the ground freezing contractor. The frozen ground would also act as a hydraulic barrier to groundwater. Extensive pre-production laboratory testing, including frozen soil testing, was conducted by the owner to establish parameters for the development of the ground freezing system design.

The freeze bottomed out in a substantially thick stratum consisting of predominantly clay, but with some inclusions of permeable soil which indicated significant groundwater pressures. In order to ensure stable conditions at the bottom of the shaft, the ground freezing contractor, which also specializes in dewatering, designed and installed a perimeter array of deep wells to temporarily depressurize the permeable soils. This unusual and challenging combination of techniques was necessary because there was no impermeable and structurally competent stratum within practical reach into which the frozen ground could be keyed to effectively complete groundwater cut-off.

THE GROUND FREEZING TECHNIQUE

Deep excavation through soft ground for shaft sinking typically includes the dual requirement of support of excavation and groundwater control. Perimeter ground freezing to create a cofferdam within which shaft excavation and completion can take place meets both of these objectives in a single operation and is the most widely used and best known ground freezing application. Ground freezing can be accomplished in the full range of soils, from clays to cobbles and boulders, and in pervious or fissured rock as the frozen walls can be formed around underground structures or obstructions. The technique is ideally suited for mixed ground conditions varying from highly permeable sands and gravels to clay and rock, and difficult ground conditions commonly encountered at the soil/rock interface where the geology is generally the most challenging and where displacement or improvement of the ground by other techniques is impractical. Compared to other groundwater cut-off or excavation support methods, a frozen wall is easily connected to the underlying bedrock and will also perfectly conform to adjoining subsurface installations, if necessary, to provide a composite cut-off structure. By providing a complete groundwater cut-off, freezing does not impact the surrounding groundwater regime (Schmall et al. 2004).

To create a frozen earth cofferdam, a coolant, usually calcium chloride brine, is introduced into closed-end freeze pipes that are inserted into holes drilled in a pattern consistent with the shape of the area to be stabilized and the design thickness of the cofferdam wall. As the brine moves through the system, extracting heat from the soil, frozen earth forms around the pipes in the shape of vertical, elliptical cylinders. The brine is returned to the refrigeration plant through an insulated header and, after re-cooling, is re-circulated within the closed system. With heat extraction continuing at a rate greater than heat replenishment, the frozen ground cylinders gradually enlarge with time until they intersect to form a continuous wall. Once the frozen wall has achieved its design thickness, the freeze plant may be operated at a reduced rate to maintain the condition during shaft excavation and liner placement (Powers et al. 2007).

Strategically located monitoring pipes instrumented with temperature sensors provide data to allow the ground freezing contractor to ensure that the desired conditions are maintained during freeze formation and the system maintenance period. Following excavation of the soils within the frozen cofferdam and completion of the intended construction, refrigeration is discontinued, allowing the ground to thaw and return to its pre-freeze temperature.

GEOTECHNICAL CONSIDERATIONS

The geotechnical engineering firm selected by the King County Department of Natural Resources and Parks, Wastewater Treatment Division (WTD) prepared a Geotechnical Baseline Report (GBR) for the Central Contract, in conjunction with WTD and the final design team. The geotechnical contract documents also included a Geotechnical Data Report (GDR) that provided details on laboratory and geologic testing procedures pertinent to ground freezing.

Subsurface Conditions

Based on geotechnical exploration, representative subsurface conditions at the Ballinger Way Portal were divided into the following six soil units, categorized as follows in the GBR (GBR 2006):

- Soil Unit 1: Vashon Recessional Outwash, consisting of medium dense to dense, moist, slightly silty, sandy gravel and gravelly sand to slightly silty sand.
- Soil Unit 2: Vashom Meltout Till and Vashon Lodgement Till consisting of very dense, moist, gravelly, slightly silty to silty sand with cobbles. Interfingered zones of relatively clean, very dense sand in upper portion.
- Soil Unit 3: Vashon Advance Outwash, consisting of dense to very dense, moist to wet sand and slightly silty to silty sand with occasional layers of sandy silt.
- Soil Unit 4: Glaciolacustrine and non-glacial Lacustrine deposits consisting of hard, moist to wet silt to fine sandy silt, hard, moist clayey silt, and hard, moist, low to high plasticity clay.
- Soil Unit 5: Pre-Fraser Glacial Till, Meltout Till and Diamicton, consisting of very dense, moist to wet, gravelly, silty sand, very dense, moist, sandy, silty gravel with cobbles, and hard, moist to wet, gravelly sandy silt. Also layers and lenses of gravel, sand, slightly silty sand, and silty, gravelly sand.
- Soil Unit 6: Pre-Fraser Glaciolacustrine and Glaciomarine deposits, consisting of hard, moist, low to high plasticity clay with trace to slight amounts of gravel and sand.

The presence of boulders was anticipated throughout the soil profile. As is typical of the geology in the Seattle area, multiple groundwater tables were encountered, as shown in figure 1 (GBR 2006).

GROUND FREEZING DESIGN

While many North American shaft freeze projects have been very successful, and indeed sometimes have been the only viable solution, there have been incidences of freeze failure on the West Coast, underscoring the need for close attention to be paid during the ground freezing operations with respect to ground temperatures, piezometric levels and overall system performance.

In the complex groundwater and soil conditions at the Ballinger Way Portal, the ground freezing contractor ensured that experienced oversight of the freezing operation was present on site throughout freeze formation and shaft sinking, and that close attention was paid to physical conditions as they were actually uncovered so that immediate adjustments could be made to the system if necessary.

Structural and Thermal Design

For the Ballinger Way Portal, structural analyses were based on a maximum excavated shaft diameter of 9.75 m (32 ft), an excavation depth of 65.8 m (216 ft), and a freeze pipe installation depth of 73.15 m (240 ft). Thermal analyses additionally considered a freeze pipe circle diameter of 14 m (46 ft), a theoretical freeze pipe spacing of 1.05 m (3.44 ft) center to center, and 42 freeze pipes.

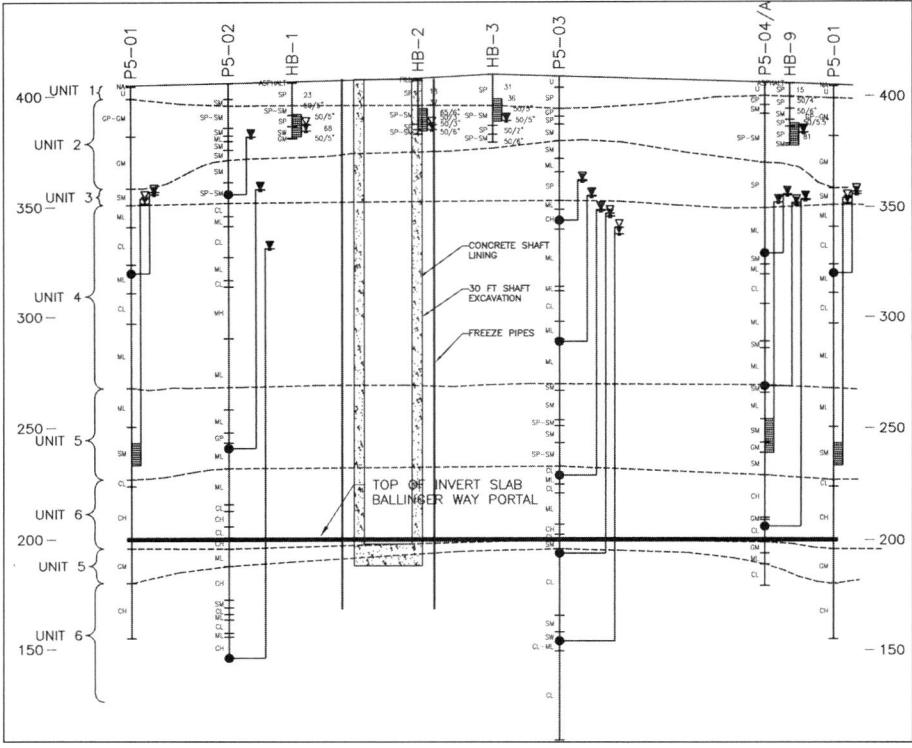

Figure 1. Ballinger Way subsurface profile from the GBR

Structural design. The structural design provides the minimum dimensions of the frozen wall. The highly complex geologic and groundwater conditions at the Ballinger Way Portal dictated that the ground freezing contractor develop designs in various soil strata and pay particular attention to the evaluation of groundwater gradients.

As well as the subsurface geology, elements of the structural design for the Ballinger Way Portal freeze included calculation of stresses and deformations in the circular frozen wall using analytical closed-form solutions.

The final structural design called for an average freeze wall temperature of −10°C (+14°F) for Soil Units 1 through 5 and −15°C (+5°F) for Soil Units 6 where lateral pressures were high. The design freeze wall thickness, including a factor of safety, ranged from 1.5 m (5.0 ft) in Soil Unit 2 to 4.11 m (13.5 ft) in soil Unit 6. Excavation was anticipated to begin after approximately 10 weeks of freezing, at which time a minimum frozen wall thickness of 1.5 m (5.0 ft) would have developed in Soil Units 1 through 5. Total stand-up time for the frozen face was anticipated to be six months for Soil Units 1 through 5 and three months for Soil Unit 6 since less time was required for bottom-up placement of the final concrete lining.

Thermal design. Thermal analyses were performed using finite element modeling to determine the requisite time to form the frozen wall, freeze plant capacity, refrigeration plant operation during maintenance freezing, and temperature development and distribution in the soils. Finite element modeling was also used once ground freezing was underway to calibrate the model and confirm design assumptions.

Figure 2. Well flushing

The thermal design analysis resulted in a refrigeration requirement of three plants for the first 10 weeks, two plants for the next 10 weeks, and a single operating plant for the remainder of the ground freezing operations.

Closure was anticipated to occur at approximately 10 weeks in Soil Unit 6, assuming a maximum freeze pipe spacing of 2.4 m (8 ft) that accounts for drilling deviations. Full structural formation was shown to occur at approximately 20 weeks in Soil Unit 6.

PRESSURE RELIEF DEWATERING

Pressure relief is typically not required for frozen shaft excavation since frozen walls, by design, typically key into an underlying cut-off. However in this case, with no competent cut-off stratum within practical reach, deep well pressure relief was essential. With shaft subgrade encountering either soil units 5 or 6, and groundwater levels significantly above subgrade elevation, shaft bottom stability during excavation activities was a concern. To depressurize units 5 and 6, three 152-mm (6-in.) diameter deep pressure relief wells were installed to 67 m (220 ft) below grade around the exterior of the shaft to cater to a calculated total system flow of 200 L/min. (50 gpm). Figure 2 shows well development.

It is important to note that movement of the groundwater generated by pressure relief can be detrimental to the freeze formation if not performed properly. The experience of the ground freezing contractor with understanding and controlling groundwater is paramount in the success of this unusual and difficult combination of operations.

The deep wells were installed during the freeze pipe system installation but not activated until freeze closure had been achieved. They remained operational until the portal had been completed.

FREEZE SYSTEM INSTALLATION AND OPERATION

The 42 freeze pipe locations were spaced at approximately 1 m (3.4 ft) on center in a circular array 14 m (46 ft) in diameter. At each location, a temporary surface casing was installed to a depth of approximately 9.1 m (30 ft) to prevent vertical migration of surficial contamination during drilling operations. Mud rotary drilling techniques were then used to advance the drill holes to full design depth. Directional surveys of all freeze pipes were made using an inclinometer to measure verticality. From the freeze pipe head assemblies, coolant supply and return hoses connected to custom-built, insulated manifolds that distribute the chilled calcium chloride brine solution pumped

Figure 3. Installed and operating ground freezing system

from the refrigeration plants. Three, self-contained, trailer-mounted refrigeration plants were mobilized for the project, each operating at approximately 500 HP. The operating ground freezing system is shown in Figure 3.

The freezing operation is a continuous process where the temperature of the circulating brine gradually decreases with time. This, in turn, reduces the average temperature of the soil surrounding the freeze pipes and increases the size of a frozen, cylindrical structure. Groundwater exclusion is the first milestone in this process: closure occurs when the cylinders become continuous between all adjacent freeze pipes. Structural formation comes later when the thickness of the frozen wall is adequate to support an open shaft sinking operation at its current depth.

Of the six separately identified soil units at the Ballinger Way Portal, the deepest in the series required the thickest and coldest structural support system. But even if this was in place early, it would not be fully stressed for many weeks after starting shaft excavation from the surface. The freeze design was therefore integrated with the shaft sinking schedule. A start to the excavation can be made some time after closure is confirmed at all depths. Then, full-powered freezing (all available plant capacity) continues to expand the dimensions of the frozen wall to resist increasing soil and groundwater pressures as the shaft excavation is deepened. Also, during this period, increasingly cold brine temperatures improve the inherent strength of the frozen soil.

The finite element analyses guided the contractor along this path. Ultimately, the cold brine delivery temperatures attained with two of the three available freeze plants proved sufficient to achieve the target strengths and dimensions required in the Unit 6 soils well before excavation would take place. Soil Unit 6 was the last zone to achieve closure, which was verified after approximately 30 days of continuous operation. The finite element model, incorporating actual data from the field, concluded that full structural thickness was achieved in Soil Unit 6 after approximately 110 days of continuous operation.

INSTRUMENTATION AND MONITORING

A comprehensive quality control program was instituted to ensure that freeze formation progressed as designed and that, ultimately, full closure and thickness of the peripheral cofferdam was achieved. A central pressure relief well was installed and screened through the water-bearing soils to relieve water pressure build-up within the unfrozen core of the shaft as the freeze continued to grow inward. Data logging probes installed in the relief well monitored groundwater elevations. Four monitoring pipes were located within and outside of the freeze pipe array and equipped with thermocouple wire temperature sensors installed at select intervals. Temperature data was recorded by an automated data acquisition system, a custom-built unit consisting of thermocouple scanning modules and a data recorder. The data obtained allowed real-time analysis of freeze progression, temperature distribution within the freeze wall, and freeze wall thickness.

PORTAL CONSTRUCTION CONSIDERATIONS

The Ballinger Way Portal is the terminus for the BT3 Leg of the Central Contract and the BT4 terminus of the West Contract. The shaft diameter was a Contractor's option and the JV team selected 9.85 m (32.33 ft) to facilitate removal of the Central Project TBM and the West Contract TBM.

Although off the critical path of the project schedule, the portal had to be completed before the TBMs arrived. Thus the owner imposed a hefty liquidated damages clause to reflect this requirement. Other owner-imposed restrictions included a 114 L/min. (30 gpm) allowable discharge to the existing city sewer system and work hours limited to between 7:00 am and 10:00 pm due to permitting requirements with the City of Shoreline that also included noise regulations not to exceed 65 decibels.

Consideration also had to be given to a 9.1 m (30 ft) thick zone at the top of the shaft where the soils were contaminated with pesticides, hydrocarbons, and dry cleaning fluids. All excavation crews underwent training in preparation for working in this environment.

EXCAVATION AND LINER INSTALLATION

Prior to freeze pipe installation, an existing facility had to be removed, the remaining site cleared and grubbed, environmental controls instituted, and site utilities set up.

Drilled holes for the freeze pipes were cased for the top 9.1 m (30 ft) and grouted, in order to isolate the contaminated zone. Used drill slurry was maintained in designated muck bins to facilitate testing and disposal of drill cuttings. The air and soil was continually sampled and tested to ensure no risk to employees and environment. The excavated material was tested by a local environmental contractor and hauled to a contaminated disposal site.

Following the 10 weeks required for the frozen wall to develop, the shaft collar, handrail, and survey controls were installed, together with a customized head frame, and work deck assembly operated from the surface by means of four Bayard winches. The deck assembly complied with OSHA requirements for a guided personnel hoist to be used below 21.3 m (70 ft).

Excavation Through Frozen Ground

A 200-ton crane was utilized for pulling a 10-BCY sinking bucket, loaded by a 12,700-kg (28,000-lb) excavator fitted with multiple quick attachments that included a bucket, a rotary head miller, and a hoe ram. The vast majority of excavation was accomplished by hoe-ramming (Figure 4), with the milling head only used in limited zones of particularly hard or frozen ground. The shaft alignment was monitored by using plumb-bob measurements and tape extensometers to measure convergence. In addition, four inclinometers were installed around the circumference of the shaft to accompany this monitoring.

The JV team had elected to install 8 × 31 ring steel sets with 63.5 mm (2.5-inch) thick lagging for the shaft (Figure 5). This would provide additional safety protection against potential sloughing of material; a mechanism to install a full-length PVC liner

Figure 4. Excavation of frozen ground underway with hoe ram

Figure 5. Ring steel set

incorporated in the design to satisfy the Owner's water leakage criteria and provide a barrier in the contaminated zone in the upper 9.1 m (30-ft) reach; an additional insulation barrier for the ground freeze wall; and additional ground support in the upper 9.1 m (30 ft) for crane loading.

Excavation was conducted in two consecutive phases, with the upper 9.1 m (30 ft) contaminated section excavated and ring steel sets installed in order to pour a monolithic annulus grout in this 9.1-m (30-ft) zone to create a barrier between the shaft and contaminants. The remaining 56.7 m (186 ft) to the shaft bottom was cycled in 1.5-m (5-ft) increments.

For the first phase, initial excavation advanced to a depth of 4.6 m (15 ft) from ground surface while three 1.5-m (5-ft) vertical sections of steel set units were preassembled on the surface and then installed. Excavation in 1.5-m (5-ft) increments and steel ring set assembly from inside the shaft continued similarly to 9.1 m (30 ft). At an excavated depth of 10.7 m (35 ft) the ground was tested to determine if it was contaminated. Annulus grouting with a bentonite grout mix was poured monolithically for the full 9.1 m (30 ft) depth. Grouting was accomplished by drilling holes into the lagging and installing packers through which grout was pumped from the surface. A cross-section of the excavation and steel set installation is presented in Figure 6.

For the second phase, beneath the contaminated zone, shaft excavation continued in 1.5-m (5-ft) increments, with intermittent stops at 4.6 m (15 ft) intervals in order to place the annulus grout in conjunction with the shaft excavation. Excavation and grouting continued in this manner until the design depth of 216 feet below ground surface was reached. Figure 7 shows removal of excavated soils at depth with the sink bucket.

Production challenges. Ground conditions posed a problem with rate of production. Over consolidated glacial till encountered in Soil Units 1 and 2 was difficult to split with a hoe ram and would break in very small pieces, making this part of the excavation more time-consuming than originally anticipated.

In the upper reaches of the shaft, within the water tables, progress was slow due to the contaminated soils in conjunction with start-up learning curves. In the silt and clays progress improved drastically only to slow once more when reaching the bottom reach of the shaft where the longer freeze time had resulted in greater encroachment of the frozen ground into the excavation and the ground was more dense.

Shaft Lining

Once the excavation was complete, a 150-mm (6-in.) non-structural work slab was poured. The steel sets support was then removed from the bottom of the shaft to facilitate reinforcement of the two tunnel eyes. The PVC liner was then installed from the top down using a work deck assembly (Figure 8).

With the liner in place, the reinforced invert was poured and work progressed with the shaft liner at the tunnel eyes. This section was necessarily more robust than usual, requiring additional rebar reinforcement due to the loading of the tunnel interfaces. The height of this area was 8.2 m (27 ft) and was poured in three individual 2.7-m (9-ft) lifts.

Pouring concrete at depth in a frozen shaft is a challenge due to the hydration of the concrete reacting to the cold temperatures and the resulting condensed vapors. Additional ventilation was installed in order to overcome reduced visibility and facilitate completion of the shaft lining.

Once the concrete lifts for the tunnel eyes portion were completed, a 3-m (10-ft) compression shaft form was lowered and 18 sequential lifts were poured until completion. The maximum excavation and concrete deviation from design alignment throughout the shaft length was 25 mm (1 in.).

The freeze was maintained the entire duration of the excavation, and turned off almost at the conclusion of concrete liner installation.

CONCLUSION

Ground freezing at depth through complex soil conditions characterized by multiple groundwater levels presented a number of challenges, both in the freeze design and implementation, the need for deep well pressure relief, and in the excavation and completion of the shaft. These challenges were overcome with careful planning by all parties and by comprehensive instrumentation and monitoring during the freeze formation and shaft excavation and completion. The experience of the ground freezing contractor and the general contractor, and close cooperation throughout the work, were instrumental in the success of this project.

306 GROUND MODIFICATION

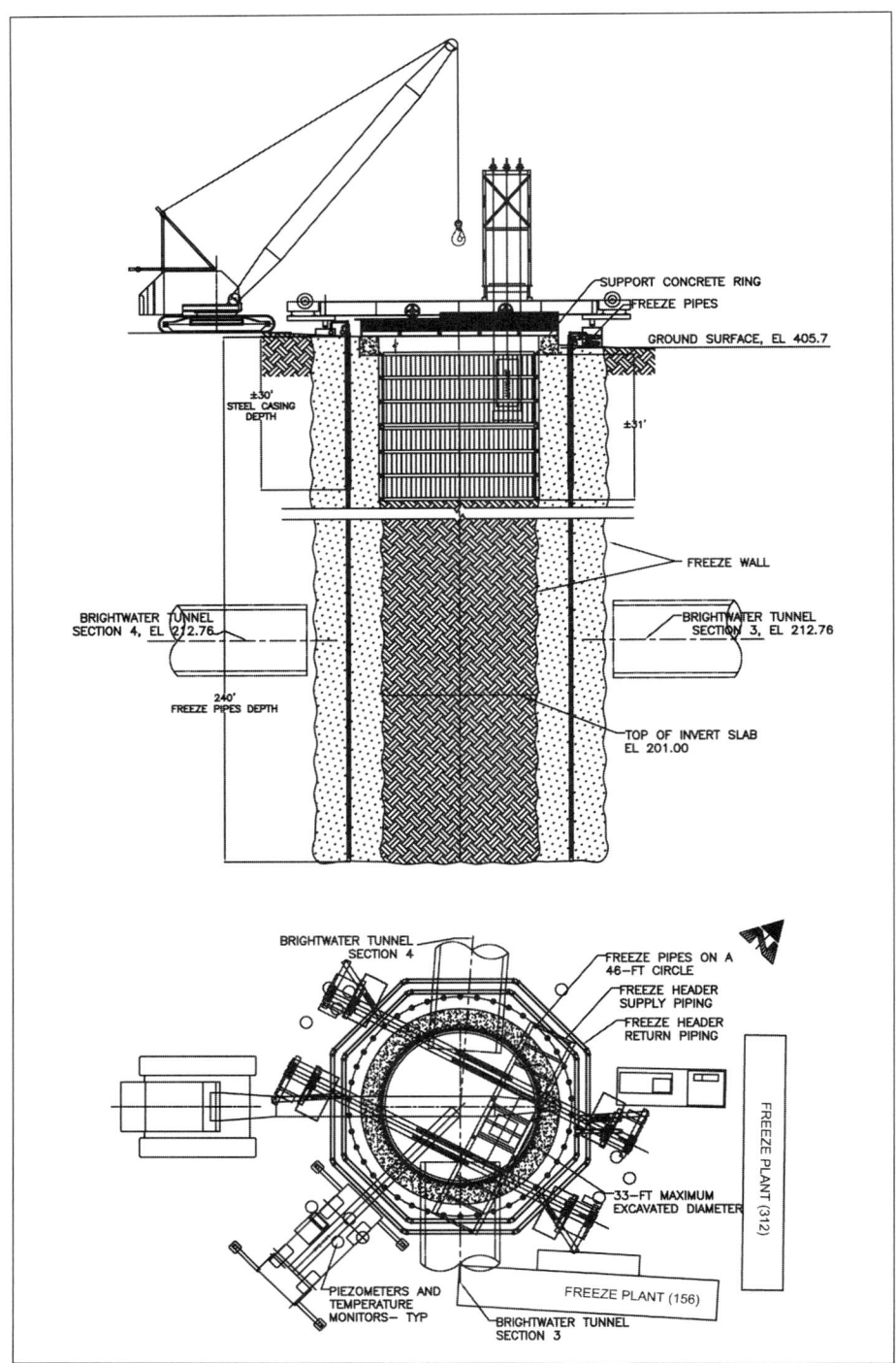

Figure 6. Cross-section and plan view of the Ballinger Way Portal

Figure 7. Deep excavation underway within the steel set liner

Figure 8. Installed PVC liner

ACKNOWLEDGMENTS

The authors extend their thanks and appreciation to the staff and management of Moretrench and Frontier-Kemper Constructors for their cooperation and support in the production of this paper, and to the staff of VPFK, Bernd Braun and Derek Maishman, consultants to Moretrench, and Steve Spenser of GZA for their contribution to the successful completion of the Ballinger Way Portal ground freezing project.

REFERENCES

Geotechnical Baseline Report, Brightwater Conveyance Syste Central Contract (2006). Prepared for King County Department of Natural Resources and Parks Wastewater Treatment Division by CDM, Bellevue, WA.

Newby, J.E., Gilbert, M.B. and Maday, L.E. (2007). "Brightwater's Conveyance system will expand Seattle's wastewater treatment." Tunneling and Underground Construction, Volume 1 Number 2, June 2007, 31–38.

Powers, J.P., Corwin, A.B., Schmall, P.C. and Kaeck, W.E. (2007). *Construction Dewatering and Groundwater Control, 3rd Ed.* John Wiley & Sons, New York, NY.

Schmall, P. C, Maishman, D., McCann, J.M. and Mueller, D.K. (2004). "Ground Freezing for Urban Applications." Proceedings of North American Tunneling Conference 2004, Atlanta Georgia.

NEW APPROACH OF ASFINAG FOR TUNNEL CONSTRUCTION MONITORING OF THE TAUERN TUNNEL PROJECT IN AUSTRIA

Robert Schnabl ▪ ASFINAG

Josef P. Mayr ▪ DIBIT Messtechnik GmbH

ABSTRACT

ASFINAG is responsible for the construction and operation of the Austrian highway system. It is approx. 2,400 km of highways and 65 tunnels with a total length of 200 km including the Tauern Tunnel.

In order to improve the documentation of the tunnel construction, ASFINAG chose to use a state of the art tunnel scanning system to record the various stages of construction: excavation, initial shotcrete lining, smoothing layer and final cast in place concrete lining. The geometric and photogrammetric documentation is provided on a daily basis to all parties involved. This innovative concept provides accurate & complete information in real time.

The tunnel scanner documentation of the 2nd tube at the 6.4 km long Tauern Tunnel will be presented in this paper as a part of the survey contract.

INTRODUCTION—THE PROJECT

The project is located in the center of the Federal Republic of Austria, about 100 km (62 miles) south of the city of Salzburg. The existing highway A10 represents one of the most important north south highways of Austria, connecting Germany and Italy. The A10 highway is built and operated by ASFINAG.

The A10 highway is part of the E55 highway which is starting in Sweden at Helsingborg and ending in Greece at Kalamata. As the A10 highway crosses the eastern Alps more than 60% of the Highway consists of bridges and tunnels. The main tunnel of the A10 is the 6.700 m long Tauern tunnel (first tube).

The construction of the A10 was started in the early 1970s. The road consists of two divided lanes with 2 lanes for each direction. As the predicted traffic at that time did not succeed the traffic capacity per day for one tube, it was decided to build only a single tube for the Tauern tunnel. Nevertheless the portals including approx. 70 m of tunnel for the second tube where already constructed at that time.

Traffic increased significant in the 1990s. In May 1999 a tragic accident happened in the first tube of the Tauern Tunnel, when a truck crashed into a second truck, which was loaded with chemical sprays. The chemicals exploded and a deadly fire burned inside the tunnel and killed 12 people.

After this accident ASFINAG decided to increase tunnel safety in all ASFINAG's tunnels according to the latest health and safety regulations. ASFINAG initiated a tunnel building program for more than 20 tunnels to be built within the next 15 years. For this program ASFINAG optimized and focused their personal and technological capabilities for tunnel construction within the ASFINAG Management GMBH.

As ASFINAG is also responsible for the operation of the highway grid it is ASFINAG's intention to assure high quality of all infrastructure to be built and therefore to employ best practices and highest standards for the future tunnels to be built.

Figure 1. Liesertal Bridge on A10

Figure 2. North Portal Tauern tunnel

Construction works for the second Tauern Tunnel tube commenced on September 15, 2006 with NATM top headings in the already existing short tunnel tubes at both portals. For the complete civil construction 42 month construction time is scheduled. All final lining including concrete road surface shall be finished by the end of 2009. In fall 2010 the second tube shall be opened to the traffic. After opening of the second tube ASFINAG commences with the rehabilitation of the first tube which shall be finished by the end of 2011.

Total cost is estimated at EUR 216 Mill. This cost also includes the rehabilitation of the first tube, which was constructed in the 1970s.

STRUCTURE OF THE SITE ORGANIZATION

ASFINAG's concept to manage construction projects foresees a classic design-bid-build process. The projects are managed by project teams of ASFINAG Construction Management GMBH, which is a subsidiary of ASFINAG. Each ASFINAG project team consists of an experienced project manager and his assistant. Within the last 10 years ASFINAG spent an average of 700 Mill EUR/year into new infrastructure.

310 GROUND MODIFICATION

The project consists of the construction of 6.546 m of main tunnel, 17 walk able cross passages, 3 drive able cross passages and 6 drive able cross passages for access with rescue vehicles. The tunnel is excavated according to the principles of NATM in three excavation stages (top heading, bench and invert). The size of cross sections for the main tunnel vary between 80 m^2 and 90 m^2 depending on the rock classes.

Figure 3. NATM excavation stages, top heading and bench

It is ASFINAG policy to contract out all works necessary for the project to experienced designers, contractors and experts. The main function of the ASFINAG project team is to manage all contracts and to make final decisions on design and construction.

It is also ASFINAG policy to have a site organization in place, which is capable to guarantee for efficient usage of the provided funds and for high quality of the built infrastructure. The later is very important to ASFINAG as the four service companies who operate the highways, are geared to minimize the operating cost for repair an rehabilitation of the infrastructure.

To guarantee cost efficiency and high quality ASFINAG established a QM system, whereas all parties involved into the construction process are controlled by other parties and all parties work is structured into contracts so that no conflict of interest shall occur. All contracts are with ASFINAG Construction Management GMBH. According to ASFINAG's QM system the contracts for the civil construction phase are structured as follows:

- Design Contract
- Construction Management Contract
- Civil Construction Contract
- Survey Contract
- Geology Contract
- Project Controlling Contract

For tunnel projects the actual ground conditions are very crucial to construction cost and in many cases the reason for claims of the contractor. ASFINAG therefore separates the data collection for ground conditions from the civil construction contract in order to obtain impartial results about the ground conditions. The data collection of ground conditions therefore is assigned to experts for surveying and geology. The geometric quality and quantity control is assigned to the survey contract. The results of the survey and geology experts are provided to all parties involved in the construction process.

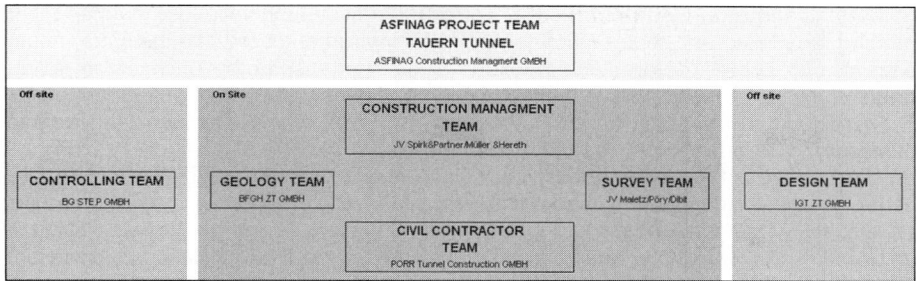

Figure 4. The site organization

During the civil construction phase the following parties are permanently on site with their teams:
- Construction Management (CM)Team
- Civil Contractor Team
- Geology Team
- Survey Team

All other parties visit the site on request by the CM Team or for the scheduled meetings. Meetings are organized on a daily, weekly and monthly basis.

The **daily** meetings are organized by the CM team during the excavation phase of the tunnel and are attended by representatives from the civil contractor, the geology team and the survey team. During this meeting all observations and data collected on this day by the attendees are provided to all partners and discussed within the site team. Based on the observations and the collected data the CM team on behalf of the owner and the civil construction team decide on all open questions with regard to NATM excavation and support.

The **weekly** project meetings are organized by the CM team throughout the complete project duration and are attended by the representatives of the site teams as well by the ASFINAG project team to discuss technical questions, to review project progress and quality of work performed. Within the weekly project meetings commercial questions are prepared for the monthly project meeting. If necessary one member of the design team attends the weekly meeting to answer all questions with regard to design.

The **monthly** project meetings are organized by the CM team throughout the complete project duration and are attended by the attendees of the weekly meetings plus the project controlling team. The scope of the meeting is to discuss all open commercial and technical questions and to solve disputes.

SCOPE OF THE SURVEY CONTRACT

The scope of the survey contract is to provide all necessary data to judge the behavior of the ground and the quality and quantity of the performed work. All data is collected and provided according to a given time schedule. All collected data is compiled into standardized formats. The results are stored onto the site server of the construction management team, where also the geological documentation is stored. All site teams have access rights to view the results on the site server.

The survey contract is structured into three major sections.
   a. Geotechnical monitoring (GTM)
   b. Construction survey control (CSC)
   c. Documentation with tunnel scanner (DTS)

The scope of the **Geotechnical Monitoring** section is installation, maintenance, reading, analyzing and reporting of geotechnical instrumentation. The instrumentation consists of optical 3D deformation monitoring points, extensometers, inclinometers and pressure cells. The instruments are installed in monitoring cross sections.

Readings and results are performed and provided to the site teams on a daily basis for the daily meeting in standardized formats.

The scope of the **Construction Survey Control** is to perform independent survey control for the underground and aboveground survey grid established by the contractor on request by the construction managment team.

Above ground survey control is scheduled at the beginning of civil construction. The main confirming underground surveys are scheduled into periods with no tunnel excavation like Christmas holidays or Easter holidays.

Results are provided to the site team 14 days after executed survey in standardized reports.

The scope of the **Documentation with Tunnel Scanner** is to provide visual and geometrical documentation of defined construction phases like excavation, shotcrete lining, smoothing layer and final lining, to evaluate the geometry compared with the design, to prove quality like shotcrete thickness, shotcrete smoothness and thickness of final lining and to provide accurate volume calculation for excavation, shotcrete and final lining.

Preliminary results documenting excavation profile and supporting shell (shotcrete thickness) are first provided to the field engineers inside the tunnel at the tunnel face after scanning to organize immediate reaction if the results are not within the limits.

Final results of the excavation profile and the supporting shell are provided on a daily basis for the daily meeting.

Results for final shotcrete lining, smoothing layer or final lining are provided within 48 hours after scanning.

The types of deliveries, the digital data format and the amount of printouts is clearly defined in the contract.

IMPLEMENTATION OF DOCUMENTATION WITH TUNNEL SCANNER ACCORDING TO CONTRACT

The scope of tunnel scanning is to document all construction phases for verification of ground condition, excavation geometry, shotcrete and final lining geometry as well as layer thickness of shotcrete and final lining and smoothness of smoothing layer. For this reason the following construction phases are documented with the tunnel scanner system. Figures 5 through 8 show various forms of this documentation.

TYPES AND FORMAT OF DELIVERIES

Figures 9 through 11 show types and format of deliveries.

For all plots the same color code is used to signalize construction management if limits are exceeded. Figure 12 shows these calculations. The color code is according to the stop light colors. Green [light gray in Figure 12] is within the limits. Red [dark gray in Figure 12] is without the limits. Yellow indicates that the trigger values have been exceeded.

PLACE OF DELIVERIES

On Site (preliminary results)

Results of excavation profile and shotcrete thickness are provided after scanning to the field engineers. The results are displayed on the field computer in cross section view or tunnel surface view.

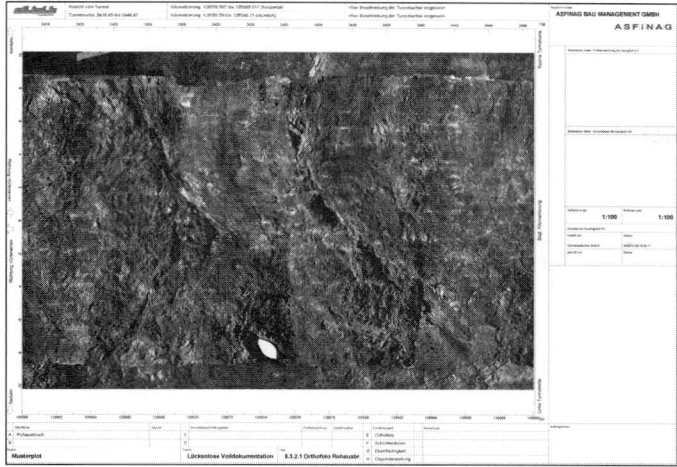

The time for scanning of one round is limited to 5 minutes in order to minimize interruption of the excavation process. Scanning is performed after completion of mucking. At the same time the geologist is at the tunnel face to verify the ground conditions. After scanning the 3D model of the excavated rock is calculated. On the display of the field computer the actual profile in comparison with the design is displayed.

Figure 5.   Excavated rock

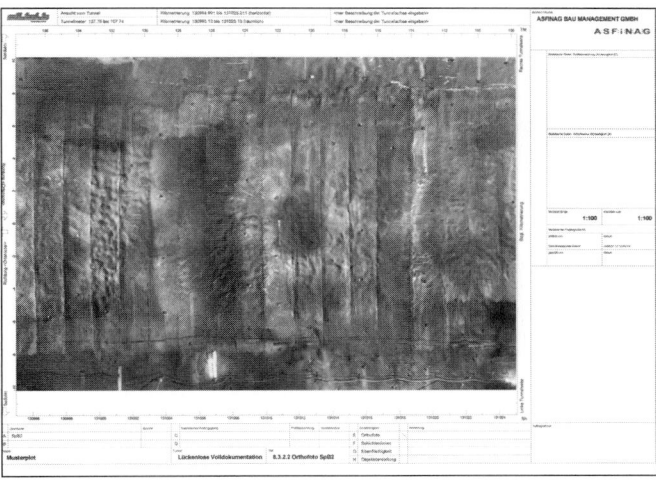

The daily production of shotcrete is scanned during face drilling for the next round behind the drill jumbo. This procedure does not interrupt the tunnel construction process. After all deformation is finished the shotcrete lining is scanned again in order to find all tights before the smoothing layer is installed. Scanning of the deformed shotcrete lining is performed in sections of more than 300 m.

Figure 6.   Shotcrete lining after application and after finish of deformation

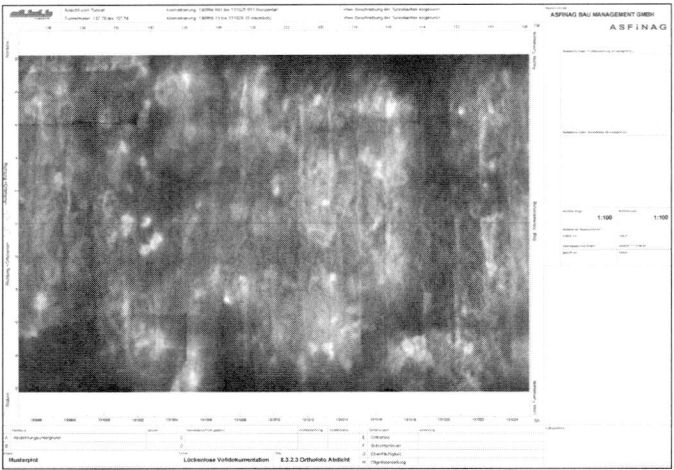

Before installation of the waterproofing membrane, the smoothing layer is scanned in order to detect any tights and to control the smoothness of the smoothing layer. Installation of smoothing layer cannot be started before prove of quality of smoothness. Scanning of the smoothing layer is performed in sections of more than 300 m.

Figure 7. Smoothing layer before installation of water proofing membrane

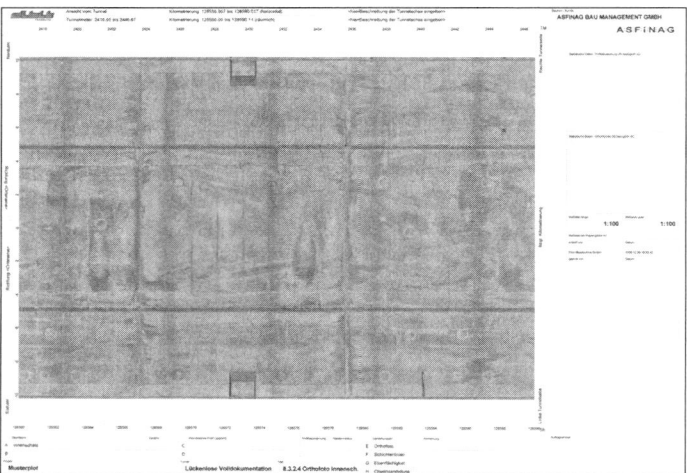

After installation of the final lining, the final lining is scanned in order to control the profile and the thickness of the final lining. Scanning of the final lining is done on a weekly basis in order to receive early results about profile and thickness.

Figure 8. Final concrete lining

TUNNEL CONSTRUCTION MONITORING OF TAUERN TUNNEL 315

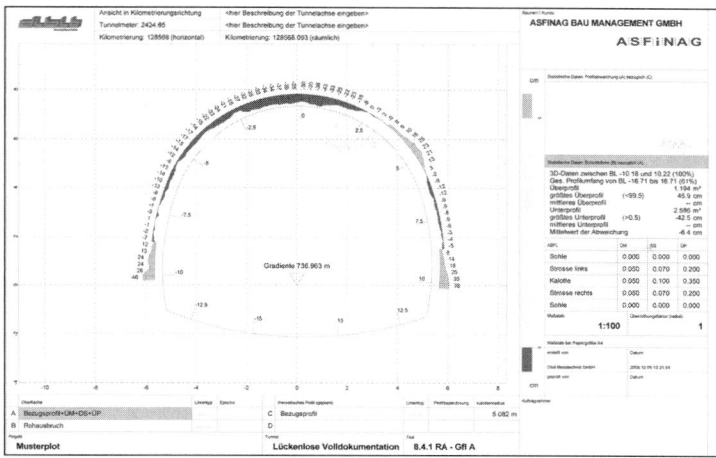

Cross section plots show the actual profile of the scanned layer. The profile of the actual layer can be compared with the designed profile for this layer or with another actual layer. The plots show statistic values like area of over and under break, average thickness of shotcrete or final lining etc. Cross sections are produced at a typical spacing of 0,5 m.

Figure 9.   Cross section plots

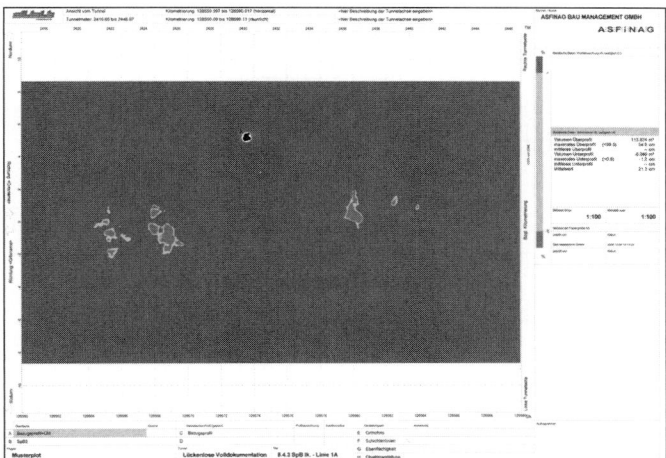

Tunnel surface plots are used to visualize the tunnel surface of a defined layer according to a chosen criteria and style. Criteria's are:
- Image of the surface.
- Deviation from a defined geometry.
- Deviation from another real tunnel surface.
- Indication of quality of the tunnel surface.
- Display of defined and captured objects of the tunnel surface.

Figure 10.   Tunnel surface plots

316 GROUND MODIFICATION

3D models are textured with black and white or colored images of the tunnel surface depending on the type of scanning system utilized. Textured 3D models are used to visualize critical tunnel sections for better understanding of the local situation by the project team.

Figure 11. 3D Models

ARGE GTM TT2R
Volumen Innenschale
(Volumen der Nischen wurden NICHT berücksichtigt)
Abschnitt: Kalotte + Strosse

Bereichs Name	Bereichs Beginn	Bereichs Ende	Bogen Länge Beginn	Bogen Länge Ende	Volumen Ist AUG zu Soll AUG hochgerech.	Auswerte- prozentsatz	Fläche Soll Innenschale	Innenschalen Stärke IS [cm]	Volumen Soll Innenschale	Vollumen Innenschale gesamt
Block 423	-0,01	11,35	-10,75	10,75	49,41	96,30	8,579	35	97,4574	146,8721
Block 422	11,34	15,75	-10,75	10,75	23,77	98,86	6,620	25	29,1942	52,9617
Block 421	15,74	27,75	-10,75	10,75	65,32	98,49	6,620	25	79,5062	144,8296
Block 420	27,74	39,75	-10,75	10,75	63,24	95,91	6,620	25	79,4996	142,7393
Block 419	39,74	51,75	-10,75	10,75	61,43	98,35	6,620	25	79,5062	140,9336
Block 418	51,74	63,75	-10,75	10,75	57,36	98,38	6,620	25	79,5062	136,8641
Block 417	63,74	75,76	-10,75	10,75	85,57	97,11	6,620	25	79,5724	165,1473
Block 416	75,75	87,76	-10,75	10,75	37,33	97,87	6,620	25	79,5062	116,8411
Block 415	87,75	99,76	-10,75	10,75	44,04	92,80	6,620	25	79,4996	123,5393
Block 414	99,75	111,76	-10,75	10,75	46,39	97,62	6,620	25	79,5062	125,8935
Block 413	111,75	123,76	-10,75	10,75	46,82	99,02	6,620	25	79,5062	126,3229
Block 412	123,75	135,76	-10,75	10,75	43,49	91,98	6,620	25	79,5062	123,0005
Block 411	135,75	147,76	-10,75	10,75	54,18	97,49	6,620	25	79,5062	133,6884
Block 410	147,75	159,76	-10,75	10,75	46,86	98,95	6,620	25	79,4996	126,3630
Block 409	159,75	165,76	-10,75	10,75	26,36	98,52	6,620	25	39,7862	66,1428

Calculations provide exact statistic values about the documented construction phases. Results are:
- Volume of eg. overbreak, shotcrete, final lining.
- Average deviation of profile.
- Areas of same thickness or with need for profiling.

All calculations are provided in Excel format for easy further computation.

Figure 12. Calculations

In the Field Office

All results are provided in the field office as a single print out to the construction management team. In addition all printouts are plotted into a digital plot file (pdf) and stored on the site server of the construction management.

EXAMPLES OF DELIVERIES

Figures 13 through 16 show examples of deliveries.

SUMMARY

The experience of ASFINAG with the concept to collect all project relevant data by independent experts to receive an impartial documentation of the ground conditions and the work performed is very positive. The major advantages are:

- Complete documentation and fast delivery of results enable for immediate reaction on changed ground conditions or improper workmanship.
- Independent and impartial documentation of ground condition and tunnel structure is a valuable tool to solve disputes fast and fair.
- Providing actual results to all partners involved into the construction process enables them to optimize their work and to reduce construction cost.
- The QM system leads to higher quality and reduced future operation cost.

For future projects it is intended to utilize the tunnel scanning system to support the geological documentation of the tunnel face.

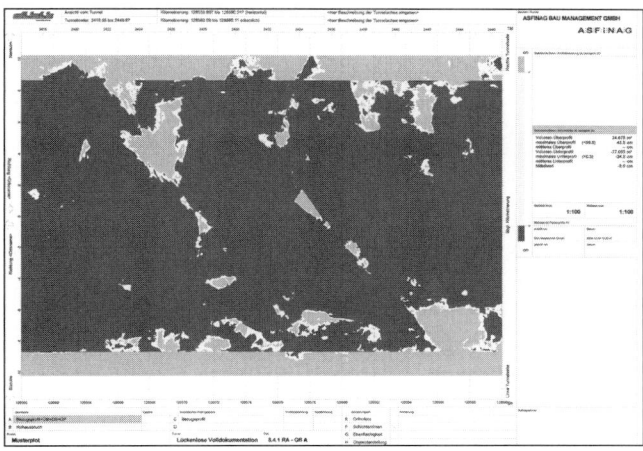

The actual excavation profile is indicated in false colors by comparison of the excavated rock surface with the designed excavation profile, plus the over excavation profile. The over excavation profile is defined by construction management and geology team according to the total actual deformation measured and on experience.

Figure 13.   Excavation profile

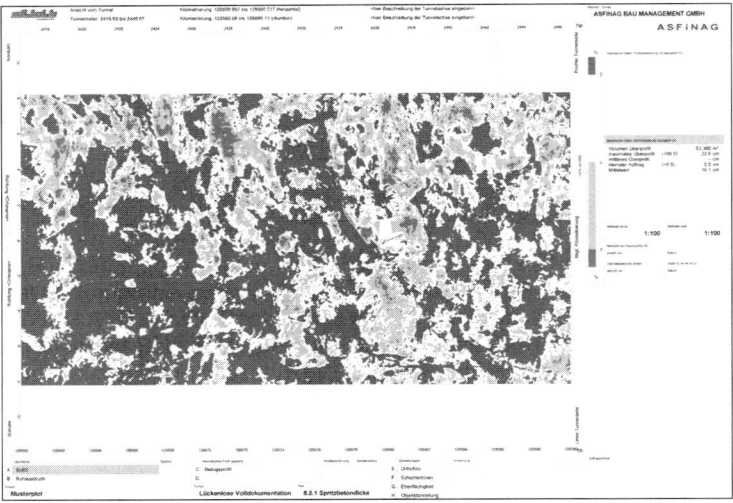

The actual thickness of the shotcrete lining is proven by comparison of the surface of the excavated rock and the surface of the shotcrete layer. A separate calculation is performed to judge, if the quality criterias are met, as minor undercuts of the designed shotcrete layer are accepted by the shotcrete specification.

Figure 14. Shotcrete thickness

Shotcrete smoothness is tested according to the wave method or bowl method. The tunnel surface is displayed in ortho picture style. Areas which do not met the criterias are highlighted in red color. Before installation of the waterproofing membrane these areas are scanned again to prove quality.

Figure 15. Shotcrete smoothness

ASFINAG BAU MANAGEMENT GMBH
ASFINAG
1:100
1:100
Musterplot
Lückenlose Volldokumentation

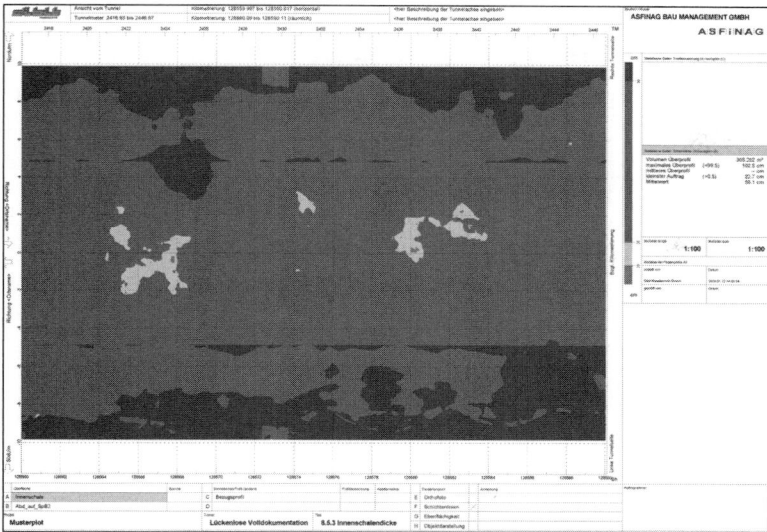

The actual thickness of the final lining is proved by comparison of the surface of the final smoothing layer with the surface of the final lining. A separate calculation is performed to estimate the actual concrete volume for each block and surplus concrete for geological overbreak.

Figure 16. Thickness of final lining

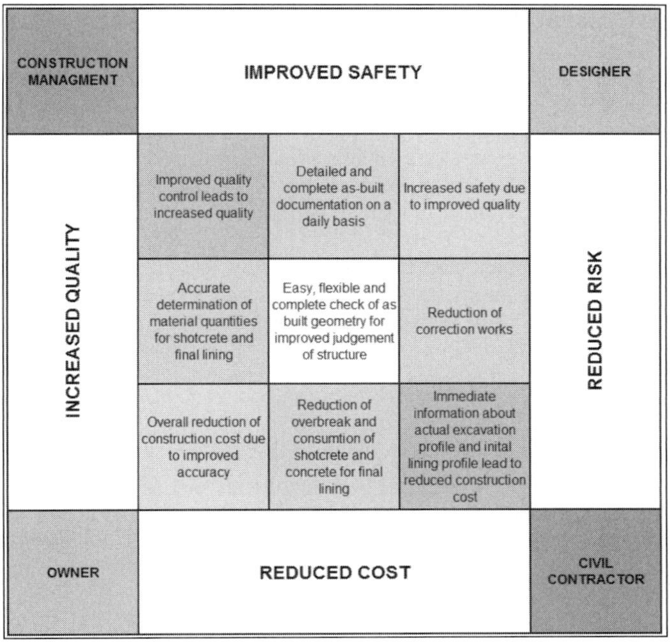

CONSTRUCTION MANAGMENT
IMPROVED SAFETY
DESIGNER
INCREASED QUALITY
Improved quality control leads to increased quality
Detailed and complete as-built documentation on a daily basis
Increased safety due to improved quality
REDUCED RISK
Accurate determination of material quantities for shotcrete and final lining
Easy, flexible and complete check of as built geometry for improved judgement of structure
Reduction of correction works
Overall reduction of construction cost due to improved accuracy
Reduction of overbreak and consumtion of shotcrete and concrete for final lining
Immediate information about actual excavation profile and inital lining profile lead to reduced construction cost
OWNER
REDUCED COST
CIVIL CONTRACTOR

Figure 17. Advantage matrix

RESEARCH IN SOIL CONDITIONING FOR EPB TUNNELING THROUGH DIFFICULT SOILS

Rory P.A. Ball ▪ Hatch Mott MacDonald

David J. Young ▪ Hatch Mott MacDonald

Jon Isaacson ▪ Hatch Mott MacDonald

Jeffrey Champa ▪ BASF Construction Chemicals, LLC

Christopher Gause ▪ BASF Construction Chemicals, LLC

ABSTRACT

The excavation of difficult soils by means of an Earth Pressure Balance Machine (EPBM) creates the potential for a reduced advance rate and increased downtime. EPBMs typically use soil conditioning to modify soil behavior to reduce abrasion, reduce cutterhead torque, control water, and ensure control of the spoil passing through the screw conveyor. Recent research is presented concerning two difficult soils for EPBM tunneling, which are sticky soils and coarse grained soils with low fines content. Soil conditioning tests were performed on several samples of difficult soils, which provide insight into how different polymers and additives can modify soil behavior to improve the performance of EPBM mining. During testing, the use of high density slurry to augment soils with low fines content was investigated. Results of investigation and a review of available literature is presented to help predict the potential for machine clogging and to help select useful soil conditioning agents in soils with low fines.

INTRODUCTION

The use of soil conditioning agents may be necessary in EPBM tunneling to modify soil behavior to control face pressure while excavating, to reduce abrasion, reduce cutterhead torque, control water, and ensure control of the spoil passing through the screw conveyor. The most difficult soils are those with an insufficient content of fine-grained material or those that clog the EPBM.

Samples of difficult soils from the Silicon Valley Rapid Transit (SVRT) project in San José, California, and from the North Dorchester Bay Tunnel (NDBT) project in Boston, Massachusetts, were tested in BASF laboratories in Cleveland, Ohio. The methods of identifying the difficult soils, the methods of testing and the results are included in this paper.

CLOGGING SOILS

The excavation of cohesive soils by means of an EPBM creates the potential for adhesion at the cutterhead, excavation chamber surfaces, in the screw conveyor and in the mucking system (belt conveyor or muck cars). If enough soil adhesion occurs, transport passages may become clogged, leading to delays and slower advance rates.

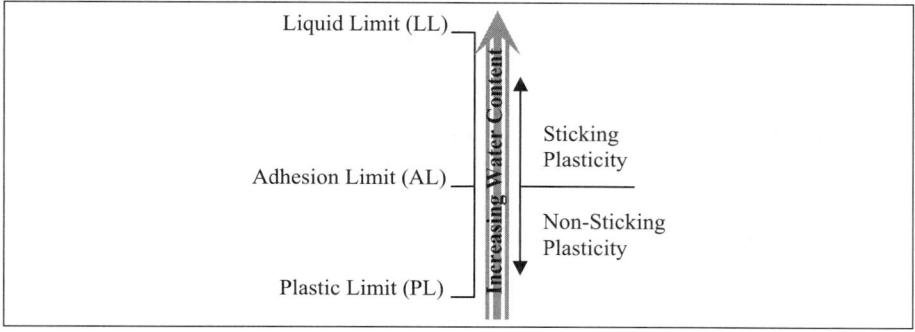

Figure 1. Relationship of increasing water content to the adhesion limit

Identifying Clogging Soils

There are at least two methods available for evaluating the clogging potential of cohesive soil during EPBM tunneling. Each method is examined below:

- Analyzing in-situ soil moisture content in relation to Atterberg Limits; including the adhesion limit (Atterberg, 1974)
- Analyzing the relationship between soil consistency and soil plasticity (Thewes, 2004)

Adhesion Limit

The adhesion limit test was first described by Atterberg in his 1911 paper (published in English in 1974). Several of the limits described in this paper were subsequently made into ASTM 4318 and are referred to as "Atterberg Limits." The adhesion limit test was left out of the ASTM standard, but has been used in tunnel engineering to assess clogging potential. The more common application of adhesion limit is in the agricultural engineering field for which it was originally developed.

The terms stickiness and adherence are typically given to soil that adheres to metal. The sticky limit, or adhesion limit, is the lowest water content at which soil adheres to a nickel spatula when drawn lightly across the soil paste's surface. Atterberg divides the plasticity range into a sticking plasticity and a non-sticking plasticity. This is illustrated in Figure 1.

The Soil Consistency Versus Plasticity Relationship

Detailed research in Thewes (1999) identifies the interaction of four mechanisms for the clogging potential:

- The adhesion of clay particles on a component surface; the most important single effect mechanism
- The bridging of clay particles over openings in the path of the spoil transport
- The cohesion of clay particles, sticking to each other
- The low tendency of clay towards dissolving in water

Thewes (2004) uses empirical data from slurry machine tunnel projects to highlight three categories of clay clogging potential based on the consistency (I_c) and plasticity (I_p) indices, and shows where each category exists on a plot of soil consistency versus plasticity illustrated in Figure 2. Soil samples from the field can be tested in the lab for Atterberg Limits and water content and then be plotted on the figure to find the associated category of clogging potential. The following equations show how each value is calculated:

$$I_c = (LL - \omega_c) / (LL - PL)$$

$$I_p = LL - PL$$

Where, LL = Liquid Limit
PL = Plastic Limit
ω_c = Water Content

Categories of clogging potential published by Thewes (2004) with proposed changes to apply to EPBM tunneling are as follows:

- Soils with **high clogging potential** lead to substantial problems during excavation and require daily cleaning works. Machine modifications only lead to a reduction, not a solution to the problem.

- Soils with **medium clogging potential** can be mastered after a number of mechanical modifications to the shield machine and soil transport system, along with changes in the operation of the machine.

- Soils with **low clogging potential** require a reduction in the advance rate, but making major alterations to the EPBM is unnecessary.

Differences Between Prediction Methods

As an example to illustrate the difference between predictions based on adhesion limit and predictions based on the consistency/plasticity relationship, consider the high plasticity clay soil sample taken from approximately 50 foot depth on SVRT. Data for the same soil sample is plotted in Figure 2 for both prediction methods.

In this sample, the natural moisture content lies below the Adhesion Limit (AL) and above the PL; therefore indicating sticking behavior. Whereas, the empirical relationship graph by Thewes indicates a "high clogging risk." The natural moisture content is expected to be altered during actual EPBM excavation because water is introduced as a component of foam soil conditioning through the injection ports. As water is mixed with the soil in the excavation chamber, the moisture content increases, thereby lowering the consistency. It is intuitive that this should lower the clogging risk as indicated by the Thewes graph, but this would move the soil into the sticking plasticity range using the adhesion limit prediction method (see Figure 2 for illustration). Through continued addition of water and foam, the conditioned spoil would ultimately reach the liquid limit where cohesion is rapidly reduced and clogging will become less of a concern as the spoil behaves more like a fluid.

The simplicity of the adhesion limit test leads to a drawback for predicting EPBM clogging potential. Research published by Kooistra (1998) and Ziminik (1999) has shown that real world effects such as high normal stress, contact time between clay and steel, steel roughness, differential pore pressure/capillary pressure, and variations in clay mineral composition are relevant to clogging of an EPBM, and these factors are not considered in the adhesion limit test.

Assessing the Suitability of Anti-Clay Soil Conditioners for Clogging Soils

Once potentially clogging soils have been identified on the tunnel alignment, the next step is to find a soil conditioner that will reduce or eliminate the clogging risk. The various soil conditioner suppliers have each developed procedures for evaluating anti-clay conditioners and the correct dosage. Standard practice has been for the soil

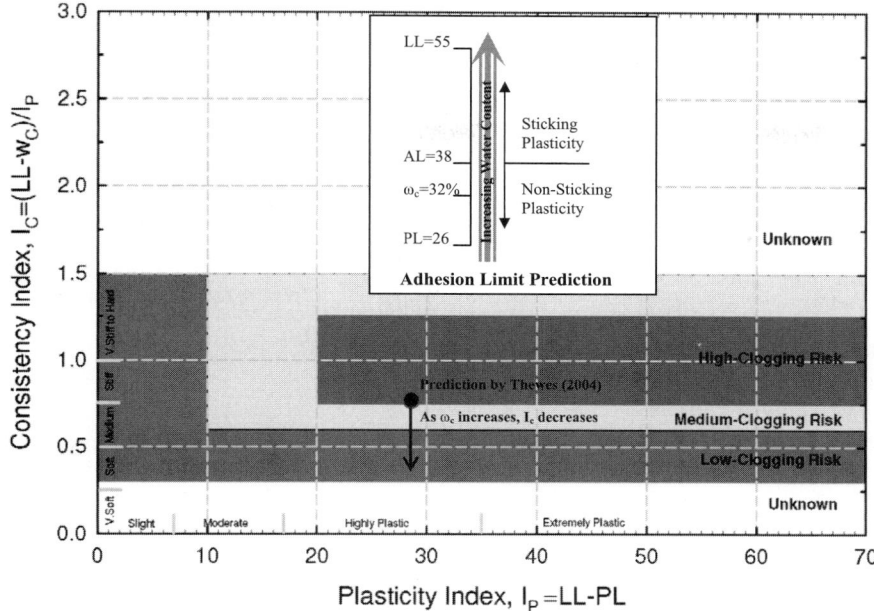

Figure 2. Comparison of prediction methods for clogging soils

conditioner suppliers to grab soil from the site (usually from early shaft excavations) that is assumed to represent the conditions along the tunnel alignment and to perform their own testing, and then recommend a product and a dosage rate. On the SVRT project which is currently in the design phase, extra soil samples from the design phase geotechnical investigations at tunnel depth were made available for this testing. Laboratory testing for the SVRT project and also for the NDBT project is described below.

Sample Selection for SVRT

The SVRT extension of the existing Bay Area Rapid Transit (BART) system includes five miles of twin bore tunneling by EPBM through San José. During the subsurface investigation soil samples were supplied to BASF laboratory, at their request, and testing was performed to establish suitable quantities of soil conditioners for estimating purposes. The offer is available to other soil conditioner suppliers who may request soils samples. Soil samples were tested for the effectiveness of soil conditioners on different soil types to be encountered during tunneling. The soil samples sent to be tested were taken from a geotechnical sonic boring. In order to mimic the mixing of materials within the excavation chamber of the EPBM machines, some soils samples were blends of soils from different intervals along the boring.

A potentially clogging soil sample was identified within the sonic boring for testing. The vast majority (~80%) of this clay was taken from a layer of CH at a shallow depth. The remaining portion (added to increase volume) was taken from a marine clay layer (CL type material) at approximately 90 feet depth. The samples were aggregated and mixed in the sample container. Trace amounts of fine sand were present in the CH material, and one piece of fine gravel was discovered in the mixer during testing at the lab. The total sample volume was approximately 2.5 gallons.

Soil as received + 7.5% Water   7.5% Water+1.5% R211   7.5% Water+1.5% R211+ FIR 30%

Figure 3. Soil conditioning for clogging soil

Clogging Soil Testing

No dilatancy or water segregation was noted in the sampled soils prior to testing. The initial soil material was very cohesive and therefore no slump testing was conducted during the testing. A small, measured quantity of soil was placed in a Hobart Model N-50A mixer for testing and soil condition agents were then added. Methods to reduce clogging include the addition of water and the use of foam and anti-clay agents.

Two tests were conducted on this material. For the first test, no water was initially added to the clay soils during mixing. The clay stuck to the mixing paddle stiffly, and the adhesion had fully obscured the windows through the paddle completely. A 30% Foam Injection Ratio (FIR) of Meyco SLF 30 foam [at a Foam Expansion Ratio (FER) of 7] was added to the sample and the result was a reduction in the clogging of the mixing paddle, and a slightly visible reduction in the adhesion of the clay to the paddle. The basics of soil conditioner mix design are described by the European federation dedicated to specialist construction chemicals and concrete systems (EFNARC), (2005) and this document can be referenced for additional information. Next, 1.5% by weight of an anti-clay agent (Rheosoil 211) was added to the mix and the resulting clay soil mix was still sticky and the paddle remained thoroughly loaded with adhered clay soil, but mixer motor effort was audibly reduced and the mix became more plastic in appearance. After the addition of water, the resulting soil mix had lost its ability to bridge the gaps in the mixing paddle, becoming fluid like a thin pudding. The original foam matrix had been destroyed prior to this point due to the amount of previous mixing that had taken place, so a second test was run, with the foam added last.

For the second test of this material, illustrated in Figure 3, 7.5% by weight of water was initially added to the second sample and mixed thoroughly. The resulting clay mix was still very sticky, but was slightly more plastic than the sample in the first round at this point. Next, 1.5% of Rheosoil 211 (R211) was added (by weight) while mixing. The effect on adhesion was immediate, and was at a similar state to where the first test ended exhibiting low bridging ability of the clay across the paddle openings. Motor effort was also audibly reduced. Then, 15% FIR of the SLF 30 foam was added to the mix, which resulted in visible fluffing (bulking). The mix was now more fluid and required even less torque from the mixer to agitate. An additional 15% FIR was then added, bringing the total to 30% by volume. The soil mix now had the appearance and viscosity of a mousse, and could be described as very fluffy (notable air entrainment) almost immediately. It required very little mixing effort. The soil mix also now had a dilatant quality when shaken, and vibrated like mousse or jello in a mold.

The addition of water with soil using the small lab mixer seemed to improve the ability to mix the other ingredients. The Rheosoil 211 appeared effective in small quantities at reducing the adhesion of the clay. During actual tunneling, Rheosoil 211 and the foam would be mixed and introduced at the same time. The mixer did not allow clods of material to be present at the time of addition, and it is unknown how capable this chemical is of reducing clods that may occur in the cutting chamber. The occurrence of clods of clay material may require higher dosing of Rheosoil 211 during mining than that presented in lab testing if the agent is capable of reducing clod size. The addition of the foam served to fluidify the mix, reduce torque for mixing, while bulking and aiding in making the mix very homogeneous.

Laboratory Testing for NDBT

Part of the NDBT excavation was through the Boston Blue Clay (BBC), a marine clay deposit. Laboratory tests were performed to determine the effects of foam and anti-clay agents on the clay. Included in these tests were rheology measurements using an Anton Parr Physica MCR 301 type rheometer with a 12 mm measuring bob. Initially, shear rate and torque measurements were made on four mixtures to illustrate the effects of foam addition. The foaming agent concentration (CF) of the foam solution was 10%, the FER was 10, and the FIR was variable. Testing was run in the following order:

1. BBC as received
2. BBC, 10% addition of water
3. BBC, 10% water, Meyco SLF 30 ($0.5kg/m^3$ of soil), R211 ($0.5kg/m^3$ of soil)
4. BBC, 10% water, Meyco SLF 30 ($1.5kg/m^3$ of soil), R211 ($0.5kg/m^3$ of soil)

The addition of water decreased the torque required to move the bob through the clay but did not significantly reduce stickiness. The addition of foam lowered the torque further and lessened the stickiness of the clay. The shear rate used in these tests was limited due to the amount of torque needed to test the clay as received; it rapidly reached the limit of the Rheometer. A graphic representation of the tests is shown in Figure 4.

To illustrate the effects of anti-clay agent (Rheosoil 211), tests were made using three mixtures. The FIR was kept low to better show the effects of the dispersant.

1. BBC, 10% water, Meyco SLF 30 ($0.5kg/m^3$ of soil), R211 ($0.5kg/m^3$ of soil)
2. BBC, 10% water, Meyco SLF 30 ($0.5kg/m^3$ of soil), R211 ($0.75kg/m^3$ of soil)
3. BBC, 10% water, Meyco SLF 30 ($0.5kg/m^3$ of soil), R211 ($1.0kg/m^3$ of soil)

The shear rate on these mixtures was increased by three decades (each decade is a ten fold increase in the shear rate) as the torque values were lowered due to the soil conditioning. The incremental increase of Rheosoil did decrease the torque of the mixtures. A graphic representation of the results is shown in Figure 5.

The results show the value of foam and Rheosoil on the rheology of the clay. Both lessened the adhesion of the clay and lowered the torque required to move through the clay as measured by the Rheometer.

Both the foam and clay dispersant technologies were utilized in the NDBT.

Case Histories

Analyzing empirical data from completed EPBM tunnel drives is one of the most valuable methods to evaluate the usefulness of the adhesion limit and the Thewes method to predict clogging.

Although the Thewes paper does not specifically describe its application to EPBMs, there is limited other published empirical data found in this regard. Langmaack (2007) applies the Thewes paper to the soil data from the Toulouse Metro Line B EPBM drive completed in France in 2007. During the EPBM drive on this project the machine experience significant clogging even working in dry ground conditions under compressed air. The difficult soil, "Toulouse molasses," plots in the medium to high clogging risk

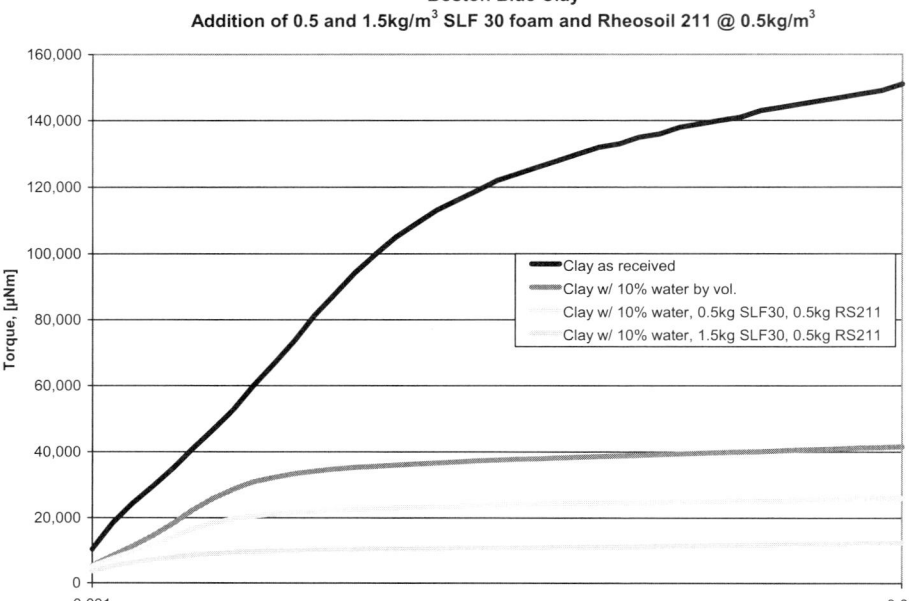

Figure 4. Rheometer testing in Boston Blue Clay

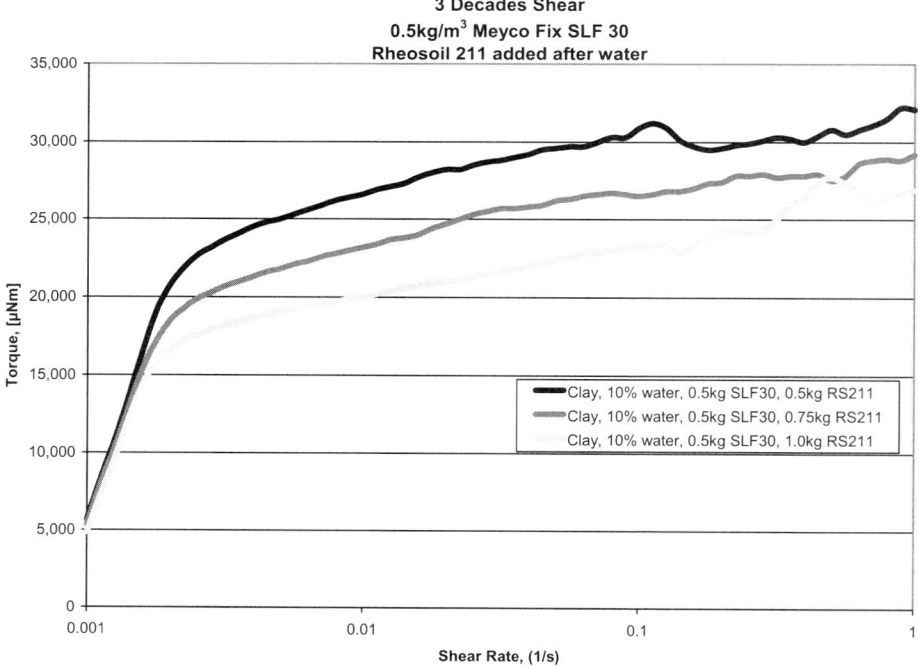

Figure 5. Rheometer testing in Boston Blue Clay with three decades of shear

zones. By the addition of water, foam and an anti-clay agent to the soil mix, the project was able to stay on track by reducing clay clogging problems.

The Beacon Hill Project, recently under construction, was a tunnel excavated by EPBM primarily in glacial till. During this project, adhesion limit tests were performed along with Atterberg Limits and the Natural Moisture Content (NMC). Each test plotted the NMC between the plastic limit and the adhesion limit, and below the liquid limit. Based on this data alone, the clay at its NMC is not prone to sticking in the manner the test was done. The project Geotechnical Baseline Report (GBR) baselined 25% of all very stiff to hard clay soils to be sticky clays based on the designer's related tunneling experience and the owner's allocation of risk rather than the adhesion test results alone. Although it was not considered at the time of writing the GBR, the Thewes method does indicate a high potential of clogging for some soils in the tunnel drive. When construction took place, EPBM clogging was apparent during the Beacon Hill drives, lending credibility to the Thewes method. Clay adhesion was reported in the cutterhead, screw and muck conveyance equipment along with other surfaces the clay came into contact with.

Until more soil properties, including adhesion limit, from a large database of past tunnel projects can be interpreted into meaningful relationships to predict EPBM clogging, the tunnel engineer's tools are limited. The Thewes method is one tool that does provide a relationship between soil properties and case histories of TBM clogging. While this method was originally developed based on slurry TBM experience, the basic principles making this method valid for assessing slurry TBM drives are also valid for EPBM drives. EPBM case histories from Beacon Hill and Toulouse Metro appear to confirm this.

SOILS WITH LOW FINES

Another difficult soil is coarse-grained soil with insufficient fines combined with free water for EPBM tunneling. In order for an EPBM to properly control face pressure while excavating, it must dissipate the face pressure along the length of the screw conveyor. "Toothpaste" is a term often used to describe the ideal consistency of conditioned soil mixture for an EPBM. The material in the screw must be a stiff viscous fluid like toothpaste in order to properly dissipate the face pressure. Some sands and gravels have insufficient fines to achieve the consistency of toothpaste. Instead, they tend to drain free water and segregate, which are undesirable spoil characteristics for EPBM spoil. Sands and gravels that segregate and drain free water do not behave like a viscous fluid, and could not be expected to dissipate pressure along a screw conveyor. How much fines are needed is a point of discussion. In the British Tunneling Society (BTS) guideline for closed face tunneling, a minimum value of 10% is recommended (BTS, 2005), but this would rely on the addition of polymer. Without the addition of polymer, 20% fines is considered a minimum.

Identifying Soils with Low Fines

Identification of low fines soils can be done with sieve analyses during the subsurface investigation that measure particles smaller than 75 µm or particles able to pass through a No. 200 sieve. Fines contents of soil layers in the face are averaged because the EPBM cutterhead, cutterhead arms and screw conveyor will at least partially mix the soil from each soil layer. It is this average fines content of soil in the excavation and screw conveyor that influences the ability to control face pressure while excavating.

Mitigating Soils with Low Fines

In soils with low fines, sole use of foam may not be enough to provide the desired plasticity to the soil. Segregation risks persist, even with stable foam. Various polymers can be combined with foam technology to control segregation and improve plasticity and work by binding the water in the soil matrix and decreasing the slump. However, if the fines content is less than 10%, additional fines may need to be added in the excavation chamber because extensive polymer addition to achieve the desired plasticity may prove uneconomical. The addition of fines along with foam and a polymer can provide a more economical solution. The addition of fines, made at the cutterhead, changes the characteristics of the soil in the working chamber, providing sufficient cohesion to the soil. This is traditionally achieved using bentonite slurry. An alternative to bentonite slurry is High Density Limestone Slurry (HDLS) which has a number of advantages over bentonite. As well as being capable of achieving higher density than bentonite slurry, HDLS is more economical and does not require the special handling and disposal measures for the spoil that is required for bentonite contaminated spoil.

HDLS, comprised of approximately 70% pulverized limestone by weight, when combined with a fluidifier to control bleed, produces a viscous fluid which can be pumped and delivered to the excavation chamber. Higher percentages of pulverized limestone may result in nozzle clogging.

Using this method, the excavation chamber spoil is treated with enough HDLS and polymer to enhance the fines content and consistency of the spoil such that:

- The permeability is significantly reduced to prevent free water flow through spoil in the screw conveyor
- The spoil has sufficient cohesion and friction within the screw conveyor for plug formation to resist excavation chamber pressure without induced slippage within the screw conveyor
- Spoil texture is consistent and exhibits enough cohesion to prevent free flowing or very low slump material on the conveyor or spoil removal system

The amount of HDLS being added, and when it is needed, will generally be controlled through visual observation at the screw conveyor discharge point. A slump test can be used, but careful attention must be paid to how well the spoil holds water in the soil matrix. Sole use of slump testing can be misleading due to aggregate angularity. Addition of HDLS to achieve a fines content between 15% and 20% in silty gravels may effectively improve consistency to a point where the screw conveyor and EPBM could efficiently function. Due to the water portion of the HDLS, a point of diminishing returns will be reached where consistency will no longer exhibit significant change. Testing of alignment soils and conditioning using HDLS is recommended to find the approximate point of optimal effectiveness.

Laboratory Testing on Soils with Low Fines

Testing of soil with low fines at BASF laboratories included two laboratory manufactured soil samples with low fines and a sample from SVRT. The SVRT low fines sample was assembled from two separate layers that appeared to contain the least fines of the layers present in the boring. The sample would be classified as sandy gravel with 4 to 6% fines.

Laboratory Testing on BASF Samples

To illustrate the conditioning of soils with low fines content, laboratory tests were conducted on a sand sample, and a mixture of gravel and sand. The tests used the following materials and parameters:

Figure 6. Photos of foam only soils conditioning on BASF sand sample with low fines

- Conditioning foam: Meyco SLF 30 @ C_f of 2.5%, FER 10%, FIR was varied
- HDLS; addition rate of 20% by weight of sample
- Polymer: Meyco SLF P2

The HDLS was made using Marble White 200 pulverized limestone [98% passing the No. 200 (75μm) sieve] and water, at a solids content of 73.5%. No stabilizer was added to this HDLS due to the short time between creating the slurry and adding it to the soil.

Sand 0 to 4 mm

This sand contains >17% fines and EPBM tunneling could be done with only foam for conditioning. The addition of foam improved the appearance and consistency of the sand. The sample was more homogeneous. The slump increased as the FIR increased. To decrease the slump, Meyco SLF P2 was added and provided improved soil structuring for better handling. Figure 6 illustrates the change in the soil with the associated soil conditioning.

Sand & Gravels 0 to 25 mm

The mixture of sand and gravel has <5% fines and required the addition of HDLS. Figure 7 illustrates the change in the soil with the associated soil conditioning.

Figure 7. Photos of soils conditioning on BASF mixture of sand and gravel with low fines

The addition of foam to the sand and gravel mixture improves the consistency of the materials but it remained porous. The addition of the HDLS improved the consistency and appearance of the mixture and produced excellent rheological properties. The addition of the polymer modified the rheology of the mixture by decreasing the slump, thus improving the soil structuring for EPBM excavation.

This laboratory testing was performed and reported as part of a study for the NDBT project where the EPBM was designed to incorporate the addition of fines to the face using HDLS. Although fully equipped, the need to incorporate the HDLS never occurred.

To further illustrate the ability of added fines to improve a coarse soil, Langmaack (2000) cites laboratory tests on a porous soil from Boulevard Peripherique Nord de Lyon (BPNL). The addition of a clayey silt and polymer to the sample decreased the permeability of the soil.

Laboratory Testing on SVRT Low Fines Soil

The SVRT low fines sample, with an initial sample moisture content of 13% and a degree of saturation of 100% prior to soil conditioning treatment, was mixed using a Kol Mixall mixer. The sample possessed a 7.5 inch slump. No significant separation of water from soil solids was observed during this slump test. Figure 8 illustrates the sequence of testing.

Untreated, 7.5" slump

10% HDLS added by wt, 8.5" slump

FIR 30%+50ml P2+HDLS, 6.5" slump

FIR 30%+65ml P2+LS slurry, 6.5" slump

Figure 8. Photos of soils conditioning on SVRT sample with low fines

The addition of the HDLS at 10% of the initial sample weight created a color change that was immediately apparent, as well as a visual change in the consistency of the soil mix, which began to behave more uniformly as soil slurry. The addition of the water in the limestone slurry had affected the slump of the base soil material, but the uniformity of the mix may also have contributed to the increase in slump.

The soil was then treated by adding a 30% FIR and 50ml of P2 polymer to bind the water and reduce the slump. However, after approximately four minutes of mixing, the soil mixture was found to have a 6.5 inch slump. Additional polymer was added (considered to be excessive), but the slump did not decrease. It was then noticed the polymer being used had an expired shelf life, drastically reducing the effectiveness. The soil consistency at this point was very uniform, homogeneous, and slightly to moderately cohesive.

A side test was performed that consisted of a clear polycarbonate tube (5.25 inch ID) placed vertically and seated over a No. 8 sieve. A portion of the soil mix used in the soil conditioning tests was poured into the tube to a height of 8.5 inches above the base of the tube. Then, 15 inches of water head was surcharged above the plug of soil to test for permeability and soil seal against the container. Upon completion of all testing (2.5 hours later), no measurable leakage through the low fines sample plug and No. 8 Sieve had occurred.

Laboratory Testing Conclusions

It is clear from the testing on the SVRT sample that the same effect could not be produced as the NDBT samples. Since the polymer used was beyond the shelf-life, being biodegradable, there was a considerable effect on the potency/efficacy of the polymer. Tracking of and adhering to age limits is important when using these polymers.

Addition of the first 10% by weight of HDLS (to supplement the soil's fine content) appears to be important in gaining consistency. HDLS and foam added together are not sufficient because they result in too much slump. Polymer needs to be added with the foam to control the slump of the spoil plus fines mixture. Before testing, determine the ideal and maximum material slump for the proposed conveyor system. In addition, determine the percentage of free water in the muck that the conveyor can handle before problems occur.

All tests at the BASF lab were performed at atmospheric pressure with mixers that may be more effective in mixing the soil more thoroughly than would take place in the EPBM excavation chamber. Dosage rates determined during these tests are considered to be a general guideline and may need to be adjusted in the field under actual tunneling conditions.

CONCLUSIONS

The soil conditioning laboratory testing allowed insight into how different additives could modify soil behavior to improve the performance of EPBM mining. Using soil conditioners expands the applicability of EPBMs into the realm of soils formerly considered to require a slurry machine and steps towards the goal of a universal machine that can handle a broad range of conditions. The tests resulted in several important conclusions.

For soils with low fines:

- Addition of the polymer appears to have significant benefits if the fines content is below 20%, as it binds free water in the soil
- If the fines content is below 10%, the use of a polymer and a HDLS appears to improve soil behavior
- Addition of the high density HDLS appears to have significant benefits at a 10% by weight application for improved consistency, cohesion and lower permeability
- The slump test may be used to test the effectiveness of soil conditioning, but careful attention should be paid to the amount of free water that drains from the spoil
- The addition of polymer appears to bind the additional water introduced into the mix in the slurry (preventing adverse slump effects)
- The stated shelf life of chemical additives is important to treatment efficacy. Additives to be used in additional soil conditioning tests must have a manufacture date that puts the proposed test usage within the applicable shelf life of the additive.

For soils that are potentially clogging:

- The Adhesion Limit test does not have enough empirical data to support its use in estimating soil clogging in an EPBM
- Addition of water prior to addition of an anti-clay agent or foam substantially increased efficacy of additives

- Addition of an anti-clay agent and a foam additive resulted in significant reductions in adhesion and required torque, while creating a uniform consistency
- The higher the FIR of the foam additive, the more likely the pressure change at the screw conveyor will induce effervescing of the material and could promote soil bypassing through a stationary conveyor that may be undesirable during mining operations
- Unlikely that the EPBM excavation chamber can provide similar mixing effort as the mixing unit used in the laboratory
- A FIR somewhere between 15 and 30 percent seems to be the most effective to establish a good mix in the EPBM chamber
- Atterberg limits and NMC are proposed on the pre-treated sample to assess the anticipated clogging potentials
- Design excavation chamber layout to maximize mixing effort applied to spoil

REFERENCES

Atterberg A. (1974), *Plasticity of Clays*. International Reports on Soil Science. Vol. 1, 1911, pp.40–43. Reproduced in English by the U.S. Joint Publications Research Service for U.S. Army Cold Regions Research and Engineering Laboratory. September.

BTS (2005), *Closed-Face Tunnelling Machines and Ground Stability*. British Tunnelling Society (Closed-Face Working Group) in association with the Institution of Civil Engineers: Thomas Telford Publishing, London, 77 p.

EFNARC (2005), *Specification and Guidelines for the use of Specialist Products in Mechanized Tunnelling (TBM) in Soft Ground and Hard Rock*. April.

Kooistra A. (1998), *Appraisal of stickiness of natural clays from laboratory tests*. Engineering Geology and Infrastructure, pp. 101–113. Ingeokring.

Langmaack L. (2000), *Advanced Technology of Soil Conditioning*. North American Tunnelling Congress, Boston 2000 A.A Balkema, Rotterdam, Brookfield, 2000, p. 525.

Langmaack L. and Martinotto A. (2007), *Toulouse Metro Lot 2: soil conditioning in difficult ground conditions*. ITA Tunnel Congress, Prague. May.

Thewes M. (1999), *Adhäsion von Tonböden beim Tunnelvortrieb mit Flüssigkeitsschilden*. Berichte aus Bodenmechanik und Groundbou Vol. 21, Bergische Universität Wuppertal Fachbereich Bauingenieurwesen. September.

Thewes M. and Burger W. (2004) *Clogging risks for TBM drives in clay* Tunnels & Tunnelling International, pp.28–31. June.

Zimnik A. (1999) *The adherence of clay to steel surfaces*. Memoirs of the Centre of Engineering Geology in the Netherlands, No. 187.

AN ANALYSIS METHOD FOR MODELING COMPENSATION OF SETTLEMENTS DUE TO TUNNEL DRIVING BY GROUTING CEMENT SUSPENSIONS

Christian Wawrzyniak ▪ CDM Consult GmbH

Wolfgang Krajewski ▪ University of Applied Sciences

ABSTRACT

An analysis method for modeling compensation grouting based on the finite element method is presented. By this method the design parameters of compensation grouting, grouting pressure and grouting volume, can be evaluated. Also the subsidence of a building due to tunnel driving as well as the lifting of the building by compensation grouting can be simulated. The stress changes in the tunnel lining and in the foundation of the building are computed as well. The application of the analysis method is demonstrated on two examples for compensation grouting in combination with construction of tunnels.

INTRODUCTION

In connection with the construction of inner city tunnels settlement sensitive buildings are often crossed under. In many cases the distance between the tunnel and the foundation of the building amounts to a few meters only. Moreover the ground conditions are often difficult with the tunnel lying in layers of low stiffness and strength.

In cases when additional constructive measures are not sufficient, the subsidence of settlement-sensitive buildings due to tunnel driving is often balanced by compensation grouting. Further this technique can be used to set back buildings that are inclined due to settlement differences or to compensate mining induced subsidence of structures.

Usually compensation grouting is designed and carried out using empirical analysis methods and experience. However, the displacements and stress changes in the subsoil, tunnel lining, or the foundation of the building resulting from the compensation grouting cannot be evaluated using such methods. In this paper, an analysis method for modelling compensation grouting based on the finite element method is presented.

TECHNIQUE OF COMPENSATION GROUTING

In Germany, the first use of compensation grouting was carried out in the early 1980s during the undercrossing of a production hall with highly sensitive machines by the new subway of the city of Essen in the Ruhr-area.

Compensation grouting belongs to the group of the so-called crack or fracture injection techniques. Contrary to fill injections where the voids of the subsoil are filled, during fracture injection artificially created joints are filled with a hardening cement based suspension. Fracture injection leads to a volume increase in the subsoil and causes a lifting of the ground surface.

ANALYSIS METHOD FOR MODELING COMPENSATION OF SETTLEMENTS

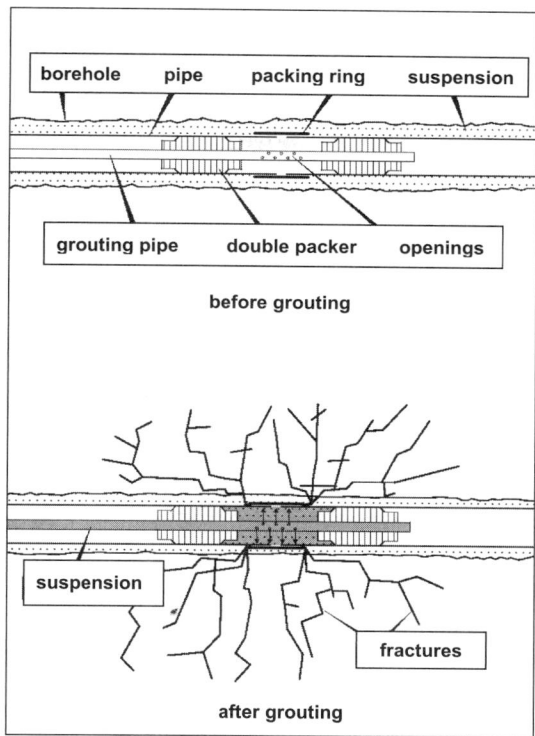

Figure 1. Technique of grouting

The method of compensation grouting is based on repeated injections of small volumes of grout in regularly distributed boreholes. The total lifting of a building is the result of many partial heaves in the millimeter range [Raabe & Stockhammer, 1995].

The injection boreholes usually are located to produce a fan-like distribution in the subsoil below the building that is going to be lifted. After the drilling of the boreholes is finished, steel or plastic pipes with packing rings made of rubber are installed into the boreholes. The annular space between the wall of the borehole and the pipes is filled by a hardening suspension (Figure 1).

For the injection a double packer is placed at a chosen packing ring. The double packer is fastened by compressed air and suspension is injected into the internal space between the double packers. After opening the packing ring, and breaking the hardened suspension in the annular space, fractures in the subsoil are created and filled with suspension. The injection volume is limited and the injection pressure results from the injection rate.

Compensation grouting under a building in connection with a tunnel heading consists of different stages. The first stage is named contact grouting and includes the injections until the pores in the ground are filled and the first uplifting is measured. The next stage is compensation grouting which includes the uplifting of the building. Depending on the total amount of the tunnel induced subsidence as well as on the permissible subsidence of the building the stage of compensation grouting can be subdivided into a stage of pre-heaving before the tunnel heading takes place and a compensation grouting after the tunnel heading.

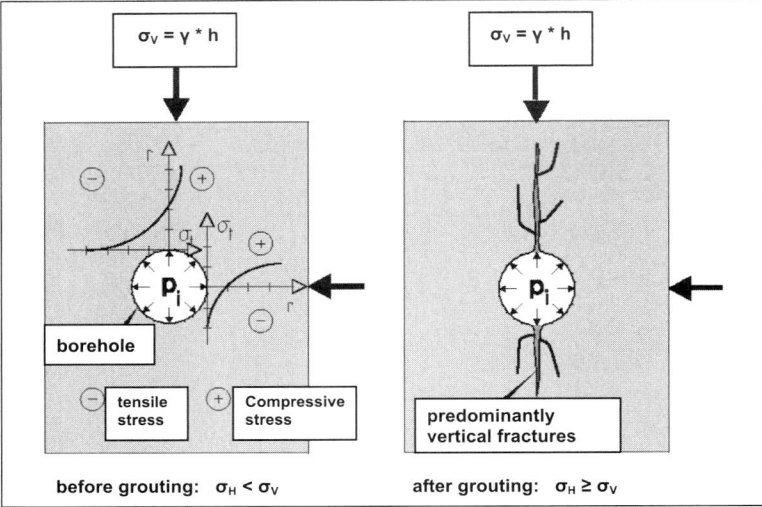

Figure 2. Fractures in the subsoil during contact grouting

Electronic hydrostatic tube balances usually measure the heaves and the subsidence of settlement-sensitive buildings. The hydrostatic tube balances, which are known for a high standard of measuring accuracy, are connected on-line with the control desk. By that method the heaves and the subsidence of the building are continuously interpreted. The results are the basis for further injections.

NUMERICAL SIMULATION OF COMPENSATION GROUTING

The numerical model is developed to fulfil a variety of demands. The design parameters, injection volumes, and grouting pressure used in the model must agree with the construction. Also the subsidence of the building due to tunnel driving as well as the lifting of a building by compensation grouting should be simulated realistically. The stress changes in the tunnel lining and in the foundation of the building should be computed as well.

One of the fundamentals of the model is the question of the direction in which the fractures occur in the subsoil. In order to answer this question, a single borehole is considered (Figure 2). The initial stress state in a consolidated ground is that the horizontal principal stress σ_H is smaller than the vertical principle stress σ_V, which results from the load of the overlying ground. As a result of the grouting pressure inside the borehole, tensile stress occurs in tangential direction in the subsoil around the borehole. In the top and the bottom of the borehole, the tensile stress is higher than beside the borehole. Therefore, predominantly vertical fractures develop. Due to the vertical fractures the subsoil becomes compressed in a horizontal direction and the principle horizontal stress increases until a secondary stress state is reached, in which $\sigma_H > \sigma_V$. The injections that lead to the secondary stress state are the so-called contact grouting.

After reaching the secondary stress state, the borehole is strained again by injection pressure, and the highest tensile stress occurs in the subsoil beside the borehole. Predominantly horizontal fractures then develop. These injections are the essential heave grouting (Figure 3).

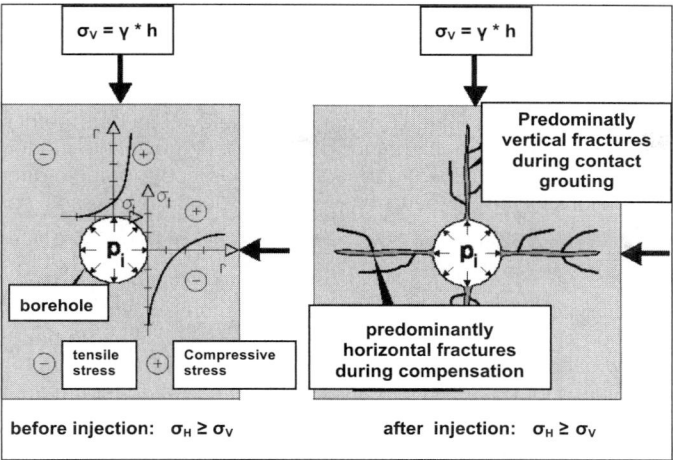

Figure 3. Fractures in the subsoil during compensation grouting

Another fundamental question is the length of the fractures caused by the injections. Assuming elastic conditions, the length of the fractures can be determined by the following equation which was developed by Griffith for the evaluation of the width w (r) of a fracture being under an inner pressure p_i.

$$w(r) = \frac{2(1-v^2)}{E} p_i \sqrt{a^2 - r^2} \quad \text{with } 0 \leq r \leq a \tag{1}$$

The volume V_E of the fracture results from integrating the width over the length of the fracture.

$$V_E = 2 \cdot \int_0^a \left[\frac{2(1-v^2)}{E} \cdot p_i \cdot \sqrt{a^2 - r^2} \right] dr = \frac{\pi(1-v^2)}{E} \cdot p_i \cdot a^2 \tag{2}$$

The length of the fracture a can be determined by solving the equation for a.

$$a = \sqrt{\frac{V_E \cdot E}{\pi(1-v^2) \cdot p_i}} \tag{3}$$

From the above mentioned fundamentals the analysis model with horizontal and vertical fractures of determined length can be derived [Wawrzyniak, 2002].

In the numerical model based on the finite element method the grouting umbrella consists of horizontal and vertical elements. To simulate the process of grouting the grouting pressure is activated as an uniformly distributed load at the inside of the fracture elements. During this step in the modeling process a very low stiffness is assigned to the fracture elements in the grout injection zone. The grouting volume results from the widening of the fracture elements and the hardening of the cement based suspension is simulated by increase of the stiffness in the following calculation step.

EXAMPLES OF APPLICATION

Crossing Under a High Stack Warehouse by the Limburg Tunnel

The analysis method mentioned above is applied to an example for compensation grouting carried out during the construction of the Limburg tunnel, a part of the Cologne-Frankfort high-speed railway line.

The high speed railway line Cologne-Frankfort has a central position in the middle European railway net of high speed trains and is one of the largest transportation projects of the last 10 years. By this link the German railway net gets a new connection to the cities of London, Brussels, Paris and Amsterdam. With the start-up of the line the travel time between Cologne and Frankfort is shortened from 2 h 13 min to 58 min. Moreover the airports of Cologne and Frankfort are connected to each other by the new high speed link which means an extraordinary increase in the quality of interconnection between the different traffic systems.

The new 105 mi long railway line between Cologne and Frankfort runs parallel to the interstate highway A3, minimizing the impact to the nature and the loss of forest and agricultural areas. The track was designed as a pure passenger traffic line, which allows a high speed up to 200 mph and at the same time a small radius of curvatures and a comparatively high slope of 4% at maximum. By this conception the number of bridges and tunnels could be reduced as far as possible.

Nevertheless several bridges and tunnels were required to be built. The middle lot has 20 tunnels with a total length of more than 20 mi. One of these tunnels is the Limburg tunnel.

The Limburg tunnel has a total length of 2,395 m and crosses underneath an industrial park in the city of Limburg. The industrial property of the TetraPak-company with several halls and a high stack warehouse was driven under by the tunnel of a length of 450 m (Figure 4). The distance of the tunnel roof to the foundation of the structures amounts to 10 to 12 m.

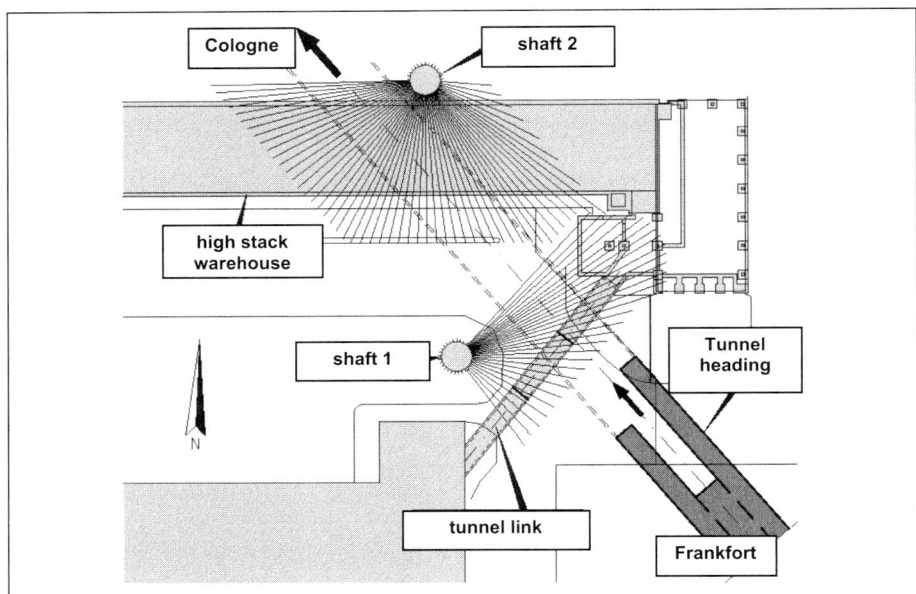

Figure 4. Crossing under of a high stack warehouse by the Limburg tunnel

ANALYSIS METHOD FOR MODELING COMPENSATION OF SETTLEMENTS 339

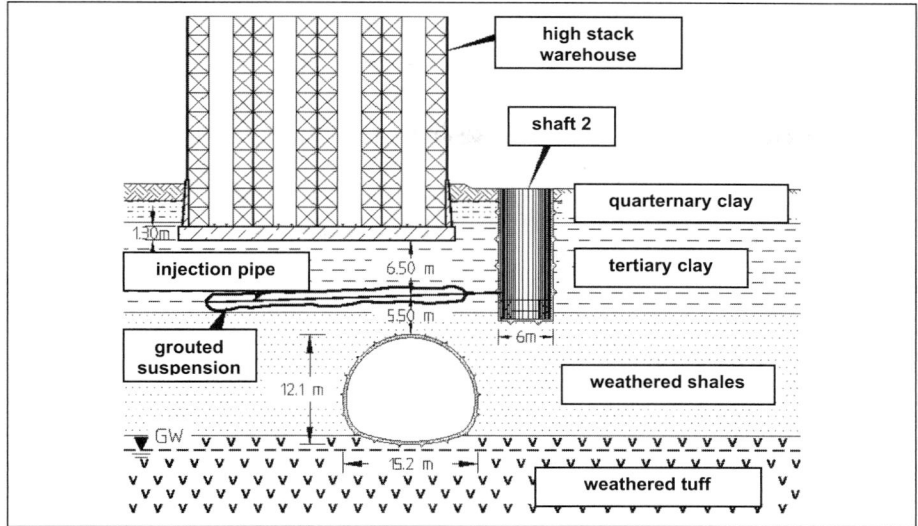

Figure 5. Cross section of the Limburg tunnel and the high stack warehouse

The surface the underground consists of quarternary and tertiary clays on which the foundations of the buildings are based. Below the clay the ground is comprised of weathered shale and tuff, in which the tunnel is lying.

The entire tunnel except the portal areas was driven using an excavator and shotcrete method. To increase the stability of the tunnel face and reduce the subsidence of the surface, the cross section was divided into two sidewall-headings and a middle core heading. The outer shotcrete lining was 35 cm thick and supported by systematically suited anchors. To increase the stability of the tunnel face a certain number of spiles were brought in after every excavation step. Following completion the tunnel is supported by an inner lining of 60 cm reinforced concrete.

The high stack warehouse is fully automated and is run by computer-controlled robots. For this reason the building itself is highly settlement-sensitive. Every settlement of the structure more than 15 mm would cause a disruption of the working process and lead to a financial loss of the TetraPak company which had to be funded by the railway company.

During design it became clear that additional measures to compensate the tunnel induced settlements had to be carried out to avoid a disruption of the working process in the high stack warehouse. The designer recommended grouting of cement based suspension.

In the first step a shaft of 6 m diameter was excavated and supported by a reinforced shotcrete lining (Figure 5). From the bottom of the shaft a number of horizontal injection drillings with a length up to 45 m was bored. Each borehole was lined by a pipe with packing rings. The grouting umbrella has a size of about 2,100 m² and is placed right between the foundation slap and the tunnel roof.

For the verification of the analysis model the excavation of the tunnel and the compensation of the settlements by grouting cement suspension are simulated through a numerical calculation according to the finite element method.

The two-dimensional finite element mesh consist of 5,790 elements with 17,509 knots and simulate the different layers of the subsoil, the tunnel heading, the high stack warehouse, and the grouting area (Figure 6).

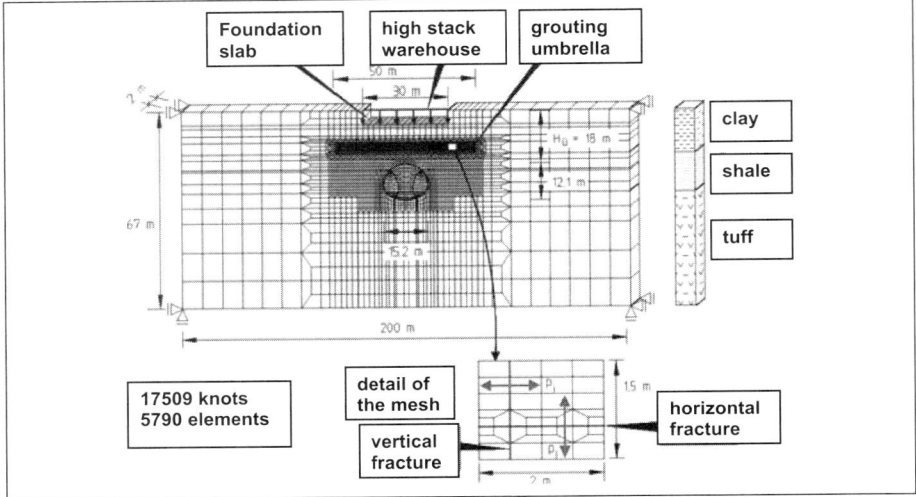

Figure 6. Finite element mesh

The calculation steps of the numerical simulation copy the different stages of the construction which alternate between grouting and tunnel heading [Wawrzyniak, 2002]. In the first step the initial stress state is calculated (step1). Before the excavation of the tunnel the contact grouting and the pre-heave grouting are simulated (step 2 to step 8). With the pre-heave grouting the high stack warehouse is lifted up before the tunnel heading arrives. In the next steps excavation and support of the side wall headings are simulated (step 9 and step 10). Afterwards the compensation grouting I compensates the settlements and the building is lifted up again before the tunnel heading of the core takes place (step 11 to step 15). The following calculation steps include excavation and support of the middle core of the tunnel (step 16 to step 19). In the last steps the second stage of the compensation grouting is simulated. The whole simulation takes 25 calculation steps.

To consider the three-dimensional effect of the displacements at the tunnel face each step of excavation and support is divided into two steps (pre-relaxation and excavation/support). The relaxation of the ground is simulated by reducing the stiffness in the excavation cross section. In the shown example the reduction of the stiffness amounts to approximately 50%.

As it is shown in Figure 7 and Figure 8 the calculated grouting pressure and the calculated grouting volume approximately agree with the effective grouting pressure and the realized grouting volumes on the construction site.

Furthermore the calculated displacements of the high stack warehouse as well as the tunnel itself match quite well with the measured displacements on the construction site.

The calculated displacements of a hydrostatic tube balance installed on the foundation slab outrange the measured displacements by a small amount. The reason for this difference is the different stages of grouting and tunnel heading overlay each other on the construction site a certain amount of time while the computer model calculates each stage for itself (Figure 9).

The comparison of the deformation of a measuring point of an extensometer 2 m above the tunnel roof shows almost exact conformance of the calculated and the measured displacements (Figure 10).

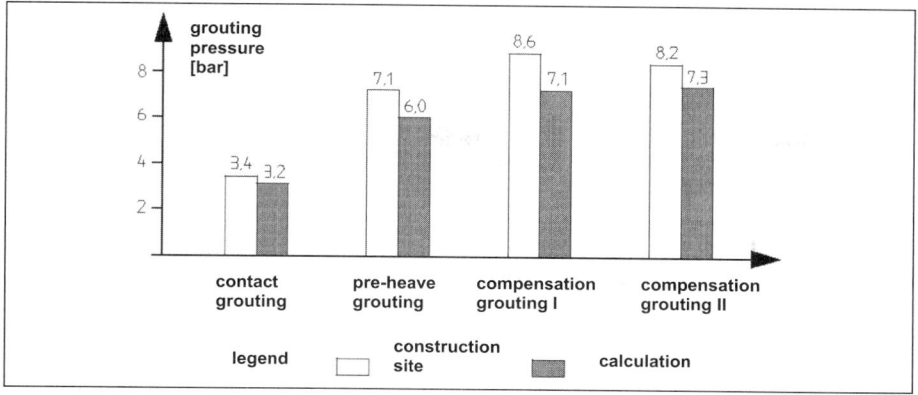

Figure 7. Grouting pressure—Comparison between building site and calculation

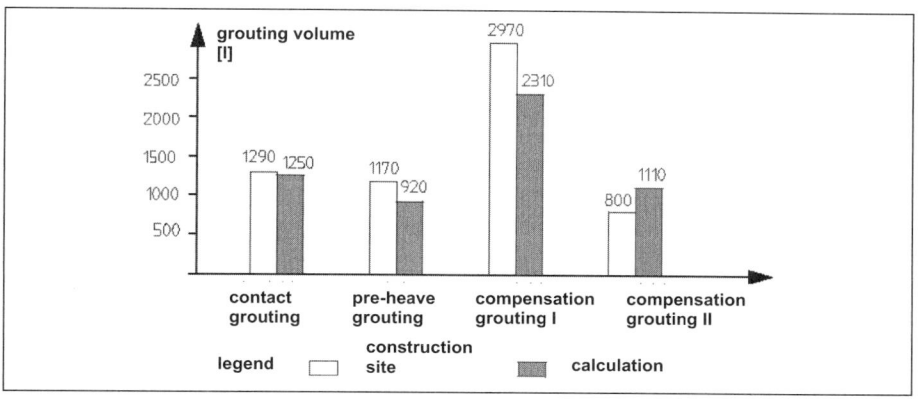

Figure 8. Grouting volume—Comparison between building site and calculation

Another result of the calculation is that the stress in the shotcrete lining increases due to grouting. In the tunnel roof the bending moments are almost twice as high as without an effect of grouting. In the side walls of the tunnel the bending moments increase about 25%.

Crossing Under of the Juridicum-Mall Building by the "City-Tunnel" Leipzig

In the city of Leipzig, a hydro-shield full-face cutter will drive the new railway "City-Tunnel." The tunnel will cross under numerous buildings and will consist of two parallel tubes. As the distance between tunnel lining and foundations is rather small at several estates, technical means have to be chosen to support the buildings and to avoid damage. At the Juridicum-Mall building near to the market place of Leipzig, an extraordinary difficult situation has to be solved. The building consists of several upper floors and four underground floors. The bearing construction is very complex due to several halls, which interrupt the concrete slabs. At the foundation level, the tunnel passes within a distance of only 1.4 m. The subsoil consists of river gravels and the so-called Bitterfeld sands.

It is agreed by contract that differential settlements of more than 2 cm resp. >1:800, due to the tunnel driving, have to be avoided or compensated by technical means. The

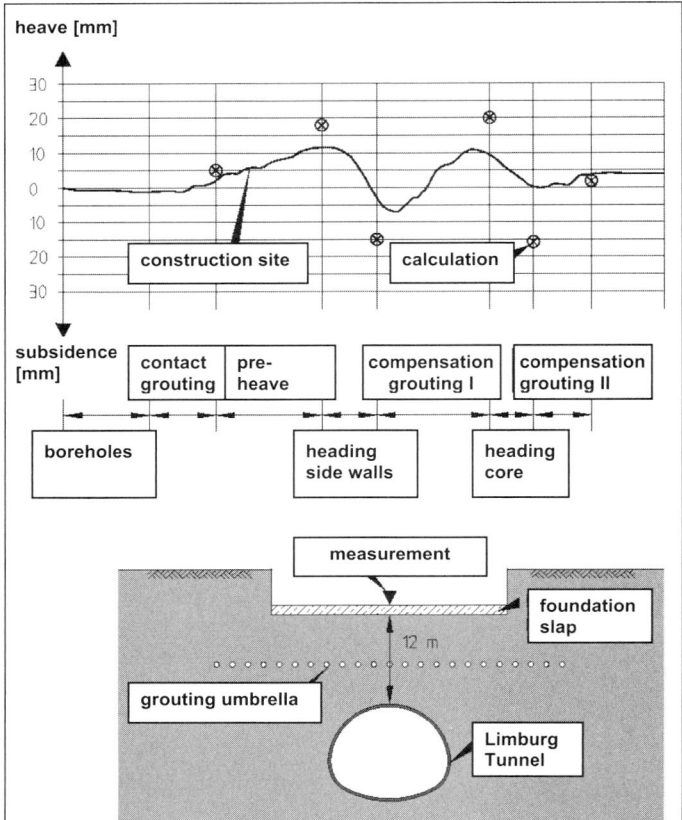

Figure 9. Comparison of the deformation measured on the foundation slab

probability of inadmissible deformations and the effect of technical countermeasures, especially the advantage of heave grouting, have been investigated by extensive three-dimensional Finite-Element calculations.

As a characteristic result of the investigations Figure 11 shows that the vertical stresses below the overhanging part of the building are reduced significantly and may be critical for the stability of the building, reducing the necessary bedding of this part of the foundation almost completely. The investigation of the concrete structure shows inadmissible loadings.

In a further calculation step, the effect of heave grouting has been evaluated. The injections will be performed using a horizontal grouting curtain, which is already installed between the roof of the inner tunnel tube and the overhanging part of the building. The grouting causes a pre-stressed zone between tunnel and building. Hence the necessary bedding of the foundation can be generated (Figure 12).

Moreover, no significant settlements of the building occur as the deformations due to tunnel driving are compensated by the injections. However, this positive result is accompanied by a considerable additional loading of the tunnel lining. The area between tunnel and grouting curtain is characterized by large settlements due to the injection pressures, which cause corresponding large bending moments in the tunnel lining.

ANALYSIS METHOD FOR MODELING COMPENSATION OF SETTLEMENTS 343

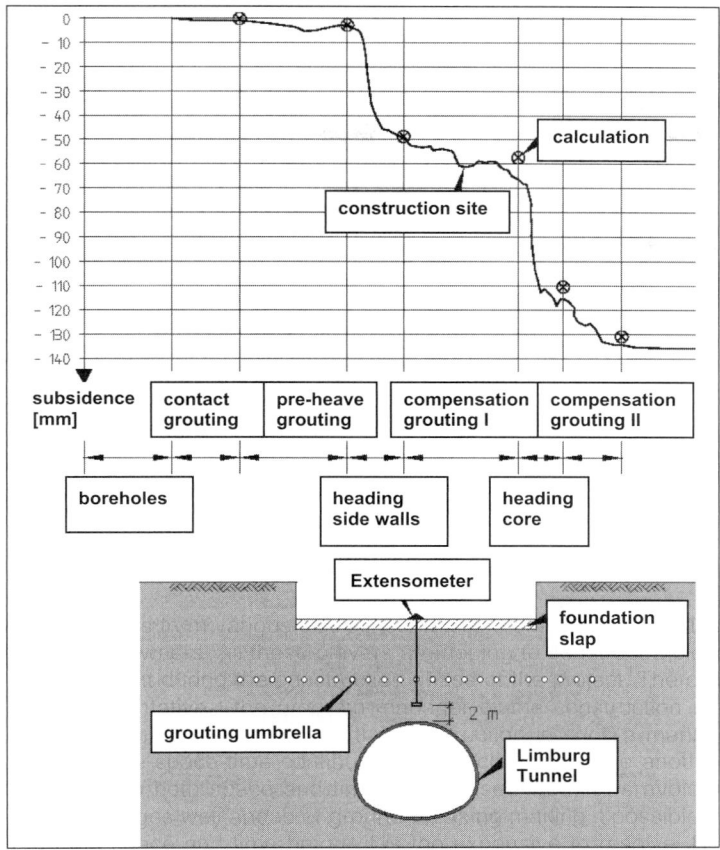

Figure 10. Comparison of the deformation measured 2 m above the tunnel

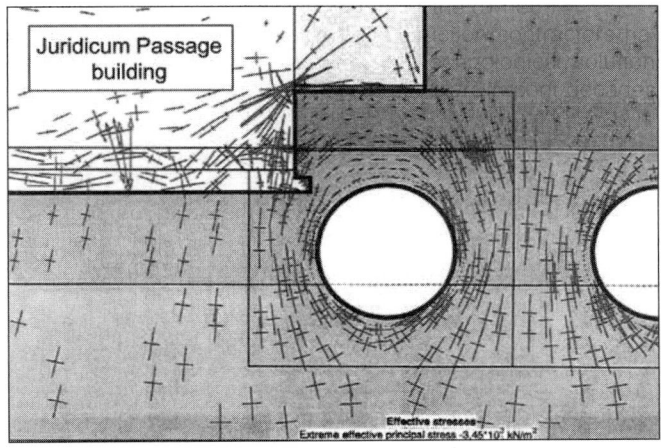

Figure 11. Normal stress without compensation grouting

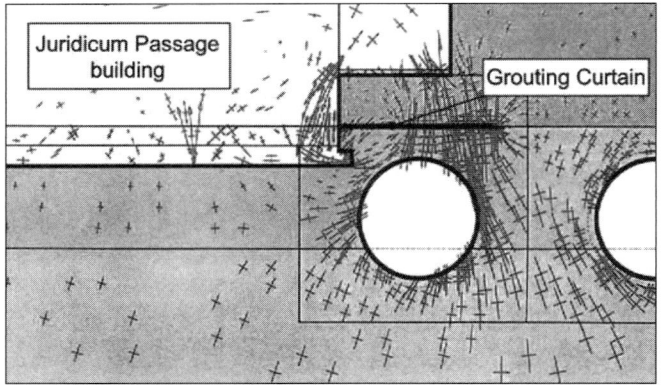

Figure 12. Normal stress with compensation grouting

REFERENCES

Krajewski, W. 2004. Numerical calculations in tunnel engineering, Festschrift zum 60. Geburtstag von Herrn Prof. Dr.-Ing. P. Vermeer, Balkema- Verlag, Rotterdam, Netherlands

Raabe, E.W. & Stockhammer, P. 1995. Einsatz von Soilcrete und Soilfrac im Tiefbau—Möglichkeiten und Grenzen beider Injektionstechniken. 10. Christian Veder Kolloquium, Graz.

Wawrzyniak, C. 2002. Simulation von Hebungsinjektionen durch numerische Berechnungen. WBI Print 12, Hrsg. Prof. Dr.-Ing. W. Wittke Beratende Ingenieure für Grundbau und Felsbau GmbH, Aachen, Verlag Glückauf, Essen.

PART 6

Innovation

Chairs

James Smith
Brierley Associates

Paul Helsop
Arup

UETLIBERG TUNNEL—SOFT GROUND EXCAVATION AND PREMIERE OF NEW TUNNELLING MACHINE: WORLD'S FIRST TUNNEL BORE EXTENDER EXCAVATED BY UNDERCUTTING

S. Maurhofer ▪ Amberg Engineering Ltd.

H.P. Müller ▪ Amberg Technologies Ltd.

PROJECT DESCRIPTION AND OVERVIEW

The Uetliberg Tunnel in Switzerland is the longest tunnel in Zurich's new Western Bypass Expressway, connecting the Birmensdorf bypass in the west with the existing Zurich-Chur national motorway in the east. Thus the tunnel links the Zurich-West motorway interchange with the Zurich-South interchange (shown in Figure 1).

The Uetliberg Tunnel project comprises two parallel tubes, each about 4.4 km long. The two tunnel tubes are connected every 300 m by a transverse walkway and every 900 m by a transverse roadway (Figure 2). The SOS niches are spaced 150 m apart. Portal stations with machinery rooms are located at the west and east portals. The tunnel falls with a gradient of 1.6% from the west portal at Wannenboden to the east portal at Gänziloo.

The Project costs are about CHF 1.12 billion.

The carcass construction started in March 2001. The Opening of the Uetliberg Tunnel is planed in spring 2009.

GEOLOGY

From west to east, the Uetliberg Tunnel passes under two parallel ranges of hills (Ettenberg and Uetliberg) with the Reppisch Valley in between. Maximum overburden of the tunnel under the Uetliberg is about 320 m.

Before the molasse sections of Eichholz (500 m) and Uetliberg (2,800 m) were reached, three sections of soft ground (Gjuch, Diebis and Juchegg) had to be driven.

Detailed Geology of the Molasse Sections

The upper fresh water molasse consists of interchanging strata comprising hard sandstone outcrops and soft marl layers plus a variation in the two strata. The marl layers split the sandstone outcrops. In addition, various marker beds can be encountered. These marker beds primarily consist of pale-reddish freshwater calcium, accompanied by black, carbonaceous, sandy marls and marly sandstone ranging in thickness between a few centimetres and a few decimetres. No actual bentonite horizon has been encountered. In addition, thin black horizons with carbonaceous substance were encountered.

Generally, the strata are flatly bedded. Separating planes which occur include stratification planes, bedding planes and cracks in the form of fissures. The marl layers feature strata thicknesses in the range of millimetres to centimetres and the sandstones feature strata thicknesses in the range of decimetres to metres. No cracks are observable in the marls. This contrasts with the sandstones where the fissures are

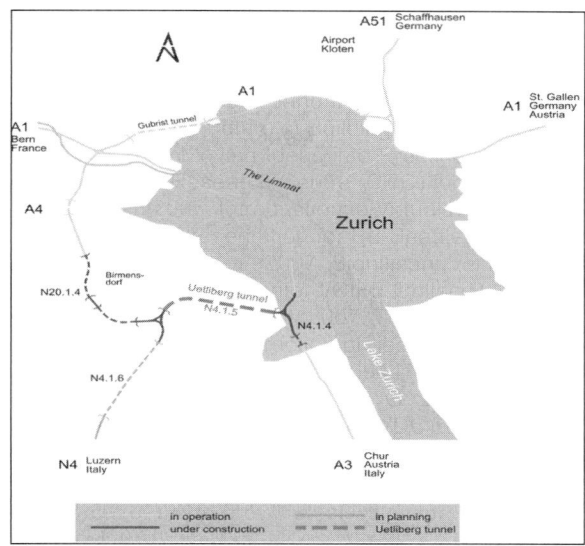

Figure 1. National highways in the Zurich area

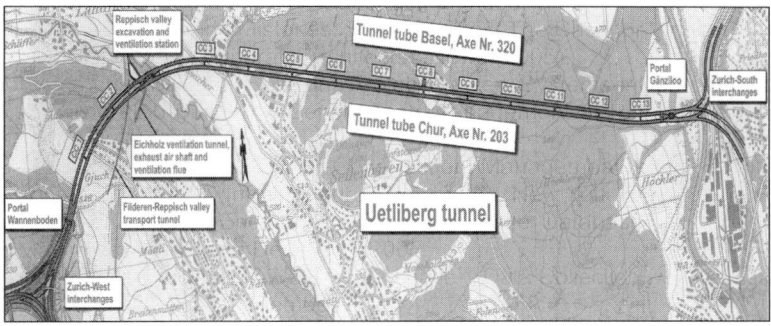

Figure 2. Overview of the Uetliberg Tunnel N4.1.5 project

closed or open only a few millimetres. There are also far larger fracture openings in the slope areas. Overall, plate-form fissure structures can be anticipated. It is not possible to preclude the possibility of subsequent fractures in the crown area in the marls.

The permeability of the marls is low. They can be considered as more or less impervious. The sandstones have a low porosity but moderate fissure water may occur owing to the fissures. Overall however, it is possible to assume a moderate permeability.

TUNNEL CROSS-SECTION

Tunnel Cross-Section for Soft Ground and Drill-and-Blast

The horseshoe cross-section used in all of the soft ground sections (SG) and for the drill-and-blast excavation in the Eichholz molasse section (MO-EIC) is about 14.70 m wide and about 12.70 m high. The face area is roughly 143 to 148 m². All of the soft ground sections are driven with the core method of tunnel construction. Steel

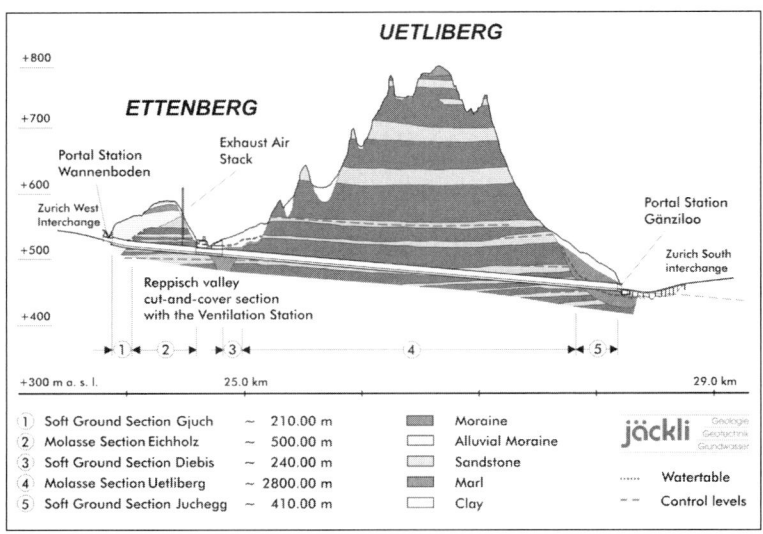

Figure 3. Geologic profile along the tunnel

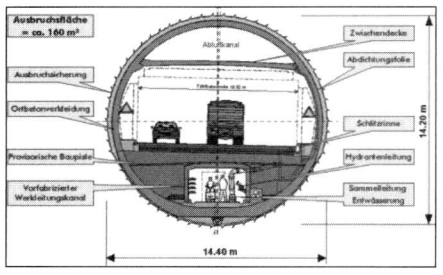

Figure 4. Standard cross-section for the soft ground sections/Eichholz molasse

Figure 5. Standard cross-section for Uetli molasse

arches (HEM-180 girders spaced 1 m apart) are used for excavation support in these sections together with a 25 cm thick layer of steel-fiber-reinforced shotcrete. The section under the Ettenberg (Eichholz molasse section), which is about 500 m long, is excavated by blasting. Three drill-and-blast divisions are used: crown, bench and base.

Tunnel Cross-Section for TBE-Excavation

The standard cross-section (see Figures 4 and 5) for the Uetliberg molasse section is 14.40 m wide and 14.20 m high. The face area is about 160 m².

Tunnel excavation is carried out with a pilot tunnel boring machine (TBM, diameter 5.00 m) followed later by a tunnel bore extender (TBE) employing the undercutting technology. The TBE extends the pilot tunnel to the final cross-section diameter of 14.00 to 14.40 m. Excavation support, which comprises Swellex rock bolts, cable bolts, mesh and shotcrete (use of steel supports where necessary), is installed right behind the boring head. Waterproofing, concrete invert, service duct (prefabricated elements), a cable conduit block, and side backfilling were installed beneath the back-up

WORLD'S FIRST TUNNEL BORE EXTENDER EXCAVATED BY UNDERCUTTING

The face is divided here into these seven cross-sectional components:

Upper sidewall galleries	2 × 17.35 m²
Lower sidewall galleries	2 × 22.55 m²
Crown	24.30 m²
Core	26.66 m²
Base	16.84 m²
Total	147.60 m²

Figure 6. Core method for all soft ground heading

Figure 7. Division of the face

equipment behind the TBE suspended from rock bolts in the ceiling. The concrete ring of the inner shell and the intermediate ceiling were installed successively later on in the rear areas.

SOFT GROUND HEADING WITH THE CORE CONSTRUCTION METHOD

All soft ground sections were driven with the core construction method (see Figure 6). Normally, temporary support measures consisted of steel arches (HEM-180 girders spaced 1 m apart) and a 25 cm thick layer of steel-fibre-reinforced shotcrete. The experience gathered in the three soft ground sections already driven is summarized in Figure 7.

After 180 m in the Basel tube and 230 m in the Chur tube, so in the Diebis soft ground section, the headings struck the moraine material and transition zone to the molasse. A standard support consisting of steel arches and shotcrete was installed continuously until the base of the upper side wall galleries had penetrated into the unweathered molasse. From this point onwards, steel support was restricted to the upper side wall galleries with crown legs underneath. Excavation was accomplished with a small backhoe, sometimes assisted by pneumatic picks and blast charges. For the lower half of the face, which entered the molasse, drill-and-blast was employed. This work was divided into two sections, benches and base.

In the Juchegg soft ground section are two significant differences. Near the portal, more than half of the tunnel cross-section lies in "Uetliberg loam," which rises above the roof. This short section was traversed using four pipe screen sections, each 12 m long.

As expected, the Gjuch section proved to be the toughest of the three soft ground sections. Fortunately, this very heterogeneous "end moraine" complex was made to order for the exceptionally flexible core construction method. Before excavation began, about 15 spring wells were drilled from the surface to lower the groundwater table to the level of the tunnel invert or slightly above it.

Surveying

Heading was controlled by motor lasers (Figure 8). For this purpose, Amberg Technologies AG's TMS SetOut plus software was installed on Leica's TPS 1100 series total stations. During the start-up phase pipe arch construction is controlled using this system; it is then used for heading the 8 parallel galleries. Setting-out work for the pipe arch, excavation and arch installation was carried out independently by the foremen, as well as inspecting the application of the shotcrete (Figure 9). They were supported by 1–2 surveyors, who moved and checked the motor laser as required. The

Figure 8. Motor laser in side-wall gallery with blasting protection

Figure 9. Foreman inspecting a steel arch

final inspection of profile precision and the volume calculations were performed by the surveyors using TMS ProScan software.

In order to record the deflection curve during construction, 3D convergence was measured simultaneously with driving. The interval of the sections with 5–6 points was between 6 m and 20 m. Convergence was visualised using graphic diagrams and published directly on the Amberg Technologies AG GEOvis internet platform. Authorised personnel thus had immediate access to all necessary information.

The motor laser method and operation by the heading team proved a fruitful choice. Although the surveyors needed to be on site 24 hours a day during the start-up phase without the motor laser, it was possible to later limit this presence to one surveyor on the day shift and freed up time for other work.

Summary of the Soft Ground Excavation

Positive results can be reported after 1,900 meters of tunnelling in soft ground and the installation of 10,000 tonnes of steel. During this two-year period, about 1,800 meters of tunnel has been driven successfully through soft ground. The excavation rate specified in the contractor's bid, 1.2 m advance per day over the entire cross-section of 148m^2 was achieved and even exceeded. The systematic installation of steel supports has made it possible to organize efficient and predictable tunnelling progress practically free of unpleasant surprises and complications (first of all without a cave-in) despite these exceptionally heterogeneous geological formations. Overall support at the working face and support of the partial cross-sections was insured at all times with face anchors and cement injections.

DRILL-AND-BLAST EXCAVATION OF THE EICHHOLZ MOLASSE SECTION

Coming from the west, the Gjuch soft ground section is followed by the Eichholz molasse section. The 500 m long Eichholz molasse section was excavated by drill and blast with round lengths of 3 m, divided into crown, bench and base. Rock support generally consisted of mesh reinforced shotcrete and rock bolts. Initially, the tunnel crown was driven over the entire length of the molasse stretch. As soon as breakthrough of the crown was achieved, work started on excavation of the core and the base (Figure 10).

The drill jumbo was controlled by TMS SetOut in aiming laser mode. The aiming laser data were integrated into the drill jumbo's control systems. Profile surveying

WORLD'S FIRST TUNNEL BORE EXTENDER EXCAVATED BY UNDERCUTTING

Figure 10. Excavation of the Eichholz molasse section

following each drilling stage was carried out using the Bever system installed on the drill jumbo. For profile evaluation the measured data were transferred via an interface to TMS ProFit and visualised. All additional tasks such as arch installation and shotcrete inspections were done using TMS SetOut andTMS ProScan as for a soft ground heading.

The Eichholz molasse section underpassed a 140-year-old railway tunnel still being in use by the Zurich S-Bahn system. The two tunnels intersected at an oblique angle only 7 meters apart. In the areas of the crossing, the support was reinforced with steel. The zone of the crossing was investigated geomechanically in great detail. It was necessary to strengthen the original structure to make sure that the existing tunnel suffers no damage and to safeguard railway operation at all times. A comprehensive monitoring program was initiated for the tunnelling period. 3D convergence measurements and precision levelling were primarily carried out in the old and new tunnels.

BORING OPERATIONS WITH TBM AND TBE

Uetliberg Molasse Section of the Pilot Tunnel

The TBM for the pilot tunnel bored two 5.00 m diameter tunnels in both tubes.

Rock support used consisted of steel fiber reinforced shotcrete and glassfiber reinforced rock bolts 2.5 m long.

Surveying was done by means of fixed laser and passive target sight, and profile measurements and evaluations with TMS ProScan and TMS ProFit from Amberg Technologies AG.

Tunnel Bore Extender (TBE) Employing Undercutting

Starting in November 2002, the TBE was assembled in the Diebis soft ground section and in the starting pit. The rock was excavated with the undercutting technique, known to be an effective cutting method ever since the early days of tunnel boring machines. With this technique, the cutting rollers work against the rock's tensile strength, which is much lower than its compressive strength (Figure 13 and 14).

The TBE enlarged the pilot tunnel of 5.00 m to a diameter of 14.00 to 14.40 m. The head of the TBE consisted of a two-piece cutterhead body with six cutter arms (Figure 15). It rotated on the inner kelly, which was braced and positioned in the pilot tunnel and in the extended tunnel cross-section. The cutters, which were offset both

Figure 11. TBM-Excavation of the pilot tunnel

Figure 12. Breakthrough of first pilot tunnel

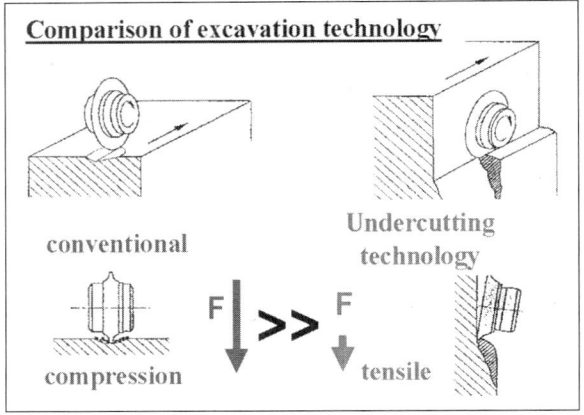

Figure 13. Principle of the undercutting technique

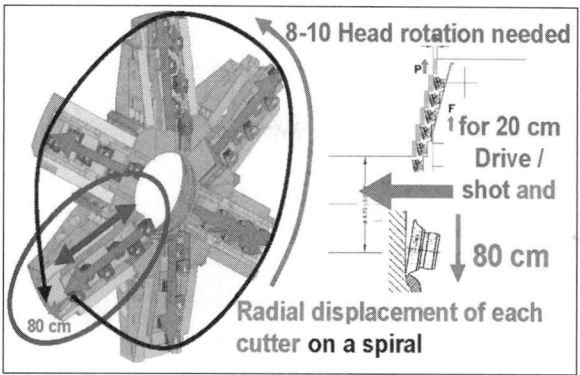

Figure 14. Operating principle of the TBE

Figure 15. The TBE boring head

axially and radially from the tunnel axis, were mounted on radially movable slides on the cutter arms.

When the six-arm boring head rotated, the rollers were shifted from an inner to an outer boring diameter. When the nominal boring diameter was reached, the slides were retracted. Then the continuously rotating head was shifted by one 20 cm advance towards the face, and boring of the next round could begin. The erection of the TBE with its back-up equipment (total length about 180 m, Figure 16) was completed by mid-April 2003. Following initial test runs, the extension of the pilot tunnel to the final cross-section of 160 m^2 began after Easter 2003.

Experiences with the TBE-Driving

Back-cutting technology. The successful enlargement of the two tubes showed that the back-cutting technology functions well. It was possible to reduce the energy

Figure 16. TBE back-up equipment

Figure 17. Cutterhead seen from the pilot tunnel

output by some 50% through working against the rock's tensile strength rather than its compressive strength. The TBE started extending the pilot tunnel bore on April 11, 2003. Its progress over the first 400 m showed that the undercutting principle works well. The rock was cut away smoothly. After advancing 40 m, the TBE reached the first cross passage. Here it was possible to observe the cutting action closely out of the bypass (Figure 18). The stepped traces of the cutting rollers were clearly visible on the tunnel face. The cross-sectional accuracy was very good. Excavation support, comprising cable bolts, Swellex rock bolts, nets and shotcrete, was installed right behind the cutting head.

The offset traces of the cutter rollers could clearly be identified on the face as well as the exact removal of material in accordance with the profile. Dust, vibrations and water meant that the complex electronic control facilities in the cutterhead, which were installed for the back-cutting technology, had to be maintained intensively. Even when individual cutterhead components failed, which meant that only three of the six cutter arms could be extended, it was still possible to continue excavating. Through ensuring that the diverse technical components for the cutterhead were set up in a redundant manner for the second tunnel tube, the TBE's availability could be constantly geared to a high level albeit with a substantial maintenance requirement. As the forces in each case act in a diametrically opposed manner when the slides with the roller bits are extended radially and consequently nullify each other, only slight forces are generated

Figure 18. TBE-cutterhead passing an accessible cross-passage

from the main bearing. The abolition of the major contact forces (thrust) and the related high torque loads on the tunnel boring machine was also positive.

Block-form of the excavated material. The block-form of the backcut excavated molasse material (flat lying sandstone and marl layers) differed considerably from the chip form of traditional TBM drives. The grading curve of the material is more similar in fact to that of a drill and blast operation. On account of its larger block-form the material had to be reduced at the end of the back-up by means of a crusher in order to adhere to the maximum permissible edge length of 30 cm for conveyor belt transportation. A powerful dedusting plant together with its casing was installed at the crusher to protect the crew's health. The resultant block-form made it possible to reutilise the excavated material for backfilling at both sides of the service duct after sieving and adding a hydraulic binder.

Possibility of overcutting. A circular profile of 7.0 m radius was excavated by the roller bits on the six radially extendable cutterhead slides. Thanks to the three additional extendable roller bits (Figure 19) installed in the cutterhead an overbreak with a maximum radius of 20 cm could be attained in order to provide more room for supporting the excavation. The overbreak was only required in the vault area because the completed base invert was placed some 80 m behind the face so that the temporary excavation support there could be reduced. Thanks to targeted overbreak, it was on the one hand possible to save on material transport and on the other on concrete and backfill requirements.

Applied supporting media. The supporting friction bolts that were used for overhead protection could be installed at the earliest some 4 m behind the face slightly inclined forwards and 4.5 m behind the face vertical to the tunnel axis.

In zones with geological overbreak for example on account of the molasse's horizontal bedding, it was possible to react flexibly at local level to spalling thanks to the chosen support system (friction and rope anchors, netting and shotcrete, Figure 20). In this connection, the anchors and netting were placed in the cavities that ensued and subsequently filled with shotcrete. Where potential areas of spalling were identifiable (e.g., coaly base levels) additional anchors with l = 3.60 m were placed in addition to the anchorage system (l = 5.0 or 6.0 m in a 1.20 × 1.20 to 1.50 × 1.50 m grid). They were set up in such a way that they penetrated possible areas prone to spalling that were identified in order to additionally nail the fissured zones. A row of the anchor system was installed every 1.20 tm in such geologically tricky zones and a row of additional anchors offset by 0.60 m placed every 1.20 m.

Figure 19. Overbreake roller bit on the TBE cutterhead

Figure 20. Excavation support in L1 of the TBE (with spalling above the cutterhead)

This flexible system of rock support did stand the test. However it demanded of all the involved a high attention and a professional competence to react correctly and in time to the identified hazard scenarios.

Deformations of the rock support. The deformations measured (3D convergence measurements following installation of the support means were of about 1–2 cm. The deformation occurred around the entire vault periphery, mostly radial to the excavated surface towards the cavity. Extensometers in the base showed normally about 1 mm of enhancement, at most 3 mm because of swelling after contact with water.

Ventilation. As far as the TBE application was concerned the prior production of the centric pilot tunnel represented a prerequisite in order to provide pre-tensioning for the enlargement. Alongside some invaluable geological findings from the pilot tunnel excavation, the pilot tunnel served for surveying purposes during the TBE drive and especially for ventilation. The dust that occurred during the drive or rather during the drilling of the anchors (dry) was suctioned off by fans stationed at the end of the pilot tunnel, cleansed by means of dedusters and released into the open. As a result, it was possible to largely minimize dust nuisance in all TBE working sectors.

Figure 21. Service duct and backfilling behind the TBE

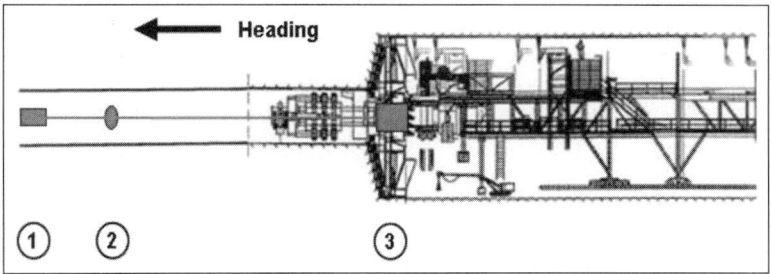

Figure 22. (1) Laser, (2) aperture, (3) passive target

On the basis of the three-shift operation, the average rate of advance for the entire cross-section amounts to approx. 29 m/week (including the pilot tunnel and the completion of the base area in the back-up zone, Figure 21).

Surveying. The TBE was controlled by means of rearward fixed laser and aperture from within the pilot tunnel. The fixed laser and the aperture were placed on a special mounting in the bore centre. The passive target sight was mounted directly in the TBE's Kelly shown in Figure 22. The laser was moved approximately every 200 m.

COMPARISON OF MECHANICAL AND DRILL AND BLAST HEADING

Because of scheduling considerations, a counter drive was undertaken in the tube Chur, starting from the east. It has already been completed. The counterdrive is about 700 m long. Its crown was excavated by drill-and-blast, the core with pneumatic picks, and the side wall benches and the invert by cutting to the exact profile. Because the same stretch was driven in the other tube with the TBE, this provided a good opportunity to compare the two heading methods.

In the following explanations the term "drill-and-blast" will be used for the whole counterdrive-section.

In both tubes, the tunnel was lined with a double shell and a full seal. The standard TBE cross-section measures 160 m^2, which is about 12 m^2 larger than the standard drill-and-blast profile. Moreover, the TBE's cutterhead produces a virtually circular cavity while the profile produced by drill-and-blast is rather mouth-shaped (Figure 23). Of

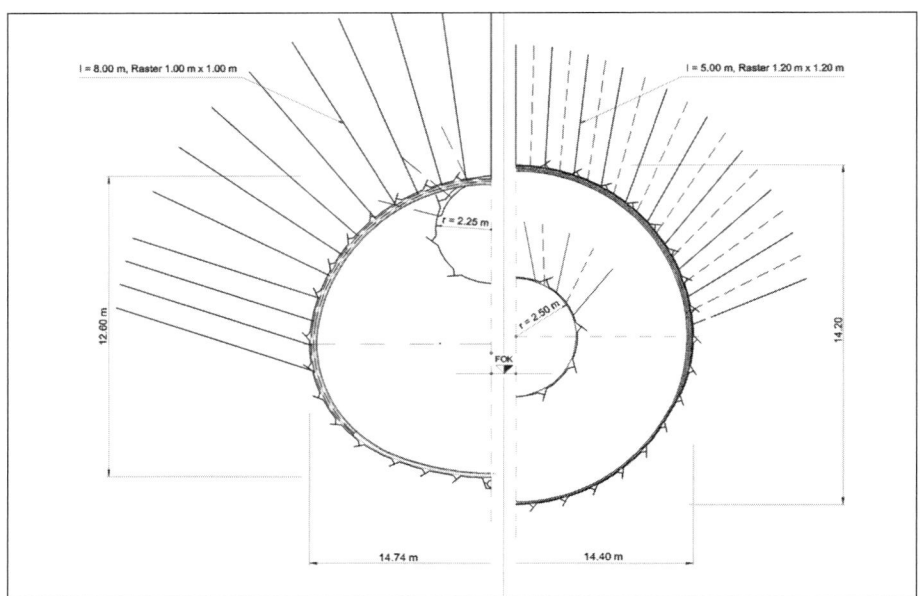

Figure 23. Comparison of drill & blast (left side) and TBE (right side)

the two, the TBE profile form is more favorable in terms of structural analysis. On the other hand, the greater excavation volume of the TBE profile means that transportation of the debris out of the tunnel is more costly.

During the TBE drive in the molasse, rock loosening as great as 2 m^3 per meter of advance occurred above or close to the cutterhead in zones with carbonaceous horizontal layers. Even though the molasse layers ran horizontally, most of this loosening occurred on one side only, namely the left side of the cross-section in the drive direction. During back-cutting, as the cutterhead rotates clockwise, the slides with the cutting rollers are shifted radially outward. This results in a propagation of pressure as the rock is cut away and thereby produces tension forces in the rock. The tension forces act horizontally on the left-hand half of the cross-section, and this, in combination with the existing vertical clefts, might well have caused the rock loosening phenomena. So the possibility that the clockwise cutterhead rotation could have contributed to the one-sided rock loosening cannot be ruled out.

Hardly any rock loosening occurred during drill-and-blast, even though this excavation method would appear at first glance to be more concussive than TBE heading. Drill-and-blast's fundamentally different excavation process with only very brief effects during blasting—compared with the TBE's continuous rock cutting and removal—may help explain this.

During excavation of the core with pneumatic picks and precise cutting of the side wall benches and the invert, the frequently hard sandstone layers greatly reduced the rates of advance. The TBE drive was unaffected by these hard zones.

When the different excavation support types were established during the final design phase, the circular profile of the TBE drive, with its more favorable structural analysis, was compared with the mouth-shaped drill-and-blast cross-section. As a result, a somewhat lighter excavation support type was chosen for the TBE than for the drill-and-blast drive, shown here for the excavation support type 1. The rock-loosening risk pattern with the TBE in the L1 zone was one factor considered in choosing the

Type of rock support	Rock bolts / steel vaults			Shotcrete		Meshing	
	L = ...[m]	a_L = ...[m]	a_Q = ...[m]	Vault d = ...[cm]	Invert d = ...[cm]	Pitch/Ø [mm]	Meshing type [Steel S500]
1	5.0	1.5	1.5	20		150/150/6/6	K 188
TBE 1	5.0	1.5	1.5	20	10	150/150/6/6	K 188
2	5.0	1.2	1.2	25		150/150/6/6 150/150/6/6	K 188 K 188
TBE 2	5.0	1.5	1.5	20	10	150/150/6/6	K 188
3	8.0	1.0	1.0	30		100/100/6/6 100/100/6/6	K 283 K 283
TBE 3	5.0	1.2	1.2	25	10	100/100/6/6 100/100/6/6	K 283 K 283
4	8.0	1.0	1.0	35		100/100/6/6 150/150/8/8	K 283 K 335
TBE 4a	6.0	1.2	1.2	30	10	100/100/6/6 100/100/6/6	K 283 K 283
TBE 4b	8.0	1.2	1.2	30	10	100/100/6/6 100/100/6/6	K 283 K 283
TBE 5	Steel vaults TH 29 / 58			30	20	150/150/8/8	K 335

Figure 24. Comparison of the types of rock support in the official proposal and in the TBE version (1..4 = types of rock support, official proposal; TBE 1..TBE 5 = types of rock support, TBE)

excavation support means. The result was a systematic arrangement of friction tube bolts with reinforcement nets for immediate overhead protection, interspersed with cable mortar bolts. Compared with the mortared bolts, the immediately load-bearing friction tube bolts reduce further loosening and therefore help prevent the formation of larger cleft blocks. Besides the circular profile, this was an additional reason for the relatively light TBE excavation support means compared with those planned for drill-and-blast, which are steel mortar bolts only.

The finally used excavation support types break down as follows for the drill-and-blast and TBE drives (Figure 24).

As you can see, the experience gained during TBE heading, with some fairly large rock loosening right above the cutterhead, finally made it necessary to employ mainly excavation support type 3.

Hardly any friction tube bolts were set during drill-and-blast heading. One reason was that rock loosening hardly ever occurred in this case. Furthermore, it was possible to remove loose material at the periphery of the cavity with heavy equipment immediately following an advance round. This procedure essentially took into account the possibility of rock loosening in the secured part of L1.

The large difference in the thickness of the invert excavation support between TBE and drill-and-blast is explained by the fact that the finished invert arch including the service duct and backfilling was installed right in the back-up section of the TBE after about 100 m, while the drill-and-blast invert arch was installed only a year or more after installation of the excavation support.

The deformations measured following installation of the support means were of about the same order of magnitude in TBE and drill-and-blast, usually about 1–2 cm. The deformations occurred around the entire profile periphery, mostly radial to the excavation surface towards the cavity.

During the year 2003, TBE heading took place on five days per week in two-shift operation: from 07.00 o'clock in the morning to 22.00 o'clock in the evening during 14 hours. From 23.00 o'clock in the evening to 03.00 o'clock in the morning rock support and maintenance work were done. Daily advances were between 5 m and 7 m. The peak daily advance was 12 m. Starting at the beginning of 2004, three-shift operation was adopted with a reduced third shift responsible for support and maintenance work. So TBE-heading took place also on five days per week, but from 06.00 o'clock in the morning to 22.00 o'clock in the evening without a break, remaining support and maintenance work from 22.00 o'clock in the evening to 06.00 o'clock in the morning. This new working schedule yielded average daily advance ranging from 9 m to 11 m. The peak daily advance was 16.5 m. In the back-up zone, all of the work on the invert—which means invert arch, service duct and backfilling—was carried out simultaneously with heading.

The previously completed pilot tunnel of 5.0 m in diameter was driven in three-shift operation with an average advance of 20 m/day. The peak advance rate was 42.6 m/day.

This yields an average advance of mechanical driving for the entire cross-section of about 6.6 m/day, including finishing of the invert in the back-up zone.

Drill-and-blast heading was done on a two-shift basis, five days per week from 06.00 o'clock in the morning to 22.00 o'clock in the evening, followed by mucking and completion of maintenance work. Crown heading regularly produced advances of 6 m/day, with a peak daily advance of 8 m. Core, side wall benches and invert excavation followed completion of the crown heading with an advance rate of 8 m/day. This yields an average advance of drill-and-blast heading over the entire cross-section of 4 m/day, without completion of the invert installations.

The cost comparison for excavation and support, not including seal supports or installations, shows that TBE heading of the 700-m-odd stretch was less costly than drill-and-blast. If the cost of removing the 7,000 m^3 of additional excavation debris is taken into account, produced by TBE heading as a result of the larger cross-section, the drill-and-blast's cost disadvantage is cut in half.

INTERIOR WORKS

When the tunnel driving work was completed in the first soft ground sections, the interior works began in the Diebis soft ground section in summer 2002.

Sealing

Sealing system used in the soft ground sections and in the Eichholz molasse section. On the basis of the hydrogeological investigations and the sealing specifications, all three soft ground sections in the vicinity of existing bodies of flowing and/or stagnant groundwater and the short Eichholz molasse section between them were sealed with an all-round, pressure-resistant full seal (sealing sheet 3 mm TPO). In order to be able to repair any leakage of the seal and also to satisfy the required service period of 100 years for the structure as a whole, these pressure-resistant areas were provided with an additional post-injected system. It consists of an additional TPO pastille sheet (pastille height 0.3 mm) laid on top of the sealing sheet. The resulting cavities between the sealing sheet and the pastilles can be filled later on (if necessary) with an injected substance based on PU. The tightness of the system can be checked by means of special field control nozzles arranged on each block. To divide the seals into individual injection fields, each block seam (every 12.5 m) is fitted with a six-rib post-injectable seam tape welded all around to the sealing sheet.

Setting-out the joints and joint tapes was done using a combination of TMS SetOut and an inclined rotary laser in the joint plane. This allowed the formwork car team to work generally autonomously and guaranteed a high degree of safety and precision.

Figure 25. Sealing of the base in the Diebis soft ground section

Figure 26. Backfilling of the service duct

The inner concrete lining profile inspections were carried out using TMS ProScan. In special areas laser scanning was carried out using the TMS Tunnelscan system and Amberg Profiler 5003. Measurements on the seal support were compared with the measurements on the final lining and an analysis of the exact concrete thickness made.

Type of sealing system used in the Uetliberg molasse section. The Uetliberg molasse section is sealed all around with a single-layer full drainage seal. Effective drainage is assured by a pressure-relieving, flushable seepage line laid outside the sealing sheet in the base area. When the first pilot tunnel was driven, the geological and hydrogeological predictions were confirmed, i.e., it was found that the entire core zone of the Ueltiberg Tunnel was dry, so that—in contrast to all of the soft ground sections—there was no need for pressure-resistant sealing.

Base Concrete/Service Duct

In their tender, the contractor "ARGE Uetlibergtunnel" proposed that the 2 × 2,800 m long service duct be backfilled with processed molasse debris from the TBE operations (Figure 26). In the summer of 2002 the client agreed with this proposal.

Positive results from earlier preliminary tests and ecological considerations prompted the client to set up a central processing plant for molasse debris over the past two years and to backfill all service ducts in the tunnel sections of the west bypass with the processed excavation material. For the main driving operation and backfilling of the 2 × 2,800 m long molasse section, the contractor will process his own molasse material and backfill it within the back-up part of the boring machine.

CONCLUSION

The mechanical heading with the TBE in the Uetliberg tunnel allowed reducing the excavation support, compared to drill and blast heading.

It was feasible to mechanically drive large cross-sections in the molasse without aid of concrete segments by constantly assessing the rock conditions. To be successful with this method, we needed a selection of pre-defined types of rock support. The rock support had to be placed well forward, behind the cutter-head.

In fact there was more excavated material to remove because of the larger cross-section, but a part of this material could be used to backfill all service ducts in the tunnel sections.

All things considered the mechanical heading with the TBE proved to be an economic way to drive through the Uetliberg molasse section.

For further information:

- www.westumfahrung.ch (client)
- www.uetlibergtunnel.ch (project coordinator)
- www.arge-uetli.ch (subcontractors)

EXTENSIBLE CONVEYOR SYSTEMS FOR LONG TUNNELS WITHOUT INTERMEDIATE ACCESS

Dean Workman ▪ The Robbins Company

ABSTRACT

Long extensible conveyor systems are an important option for efficient tunnel excavation where intermediate access is impractical. The conveyor system designed for the Pula Subbaiah Veligonda Tunnel #2 in Andhra Pradesh, India will eventually extend to 19.35km. The Ø10.0m water diversion tunnel will be excavated under India's largest tiger sanctuary using a Robbins Double Shield TBM. Once extended, the conveyor will be one of the longest single-flight systems ever installed in India, and is expected to offer increased system availability over comparable systems using muck cars. This paper will examine the engineering challenges of long-flight extensible conveyor systems in the areas of belt selection, component design, and equipment location in confined assembly and launch areas.

OVERVIEW

As extensible conveyors continue to gain acceptance as the standard muck removal system for tunnel boring machines, the length of the muck conveyor required has also continued to increase based on the project specifications. In some projects, intermediate access to the tunnel, either by a vertical shaft or a sloping adit, has allowed the muck to be removed from the tunnel at a point other than the starting portal. Utilizing these designed-in features results in a shorter and therefore less expensive conveyor system, with the trade-off being an increase in labor costs due to the time necessary to disassemble and then re-install the system after the TBM has advanced beyond the access. However, there are instances where intermediate access to the tunnel by a shaft or an adit would either be impractical or cost-prohibitive. Contributing factors preventing intermediate access include (a) environmentally sensitive areas, such as a national park or nature preserve; (b) geographic locations, such as a dense metropolitan area or a large body of water; or (c) geological limitations, such as the depth of the tunnel or the stratigraphy of the overlying ground. In these situations an extensible conveyor can still be the most economic muck removal method, despite the increase in its ultimate length and the resulting conveyor system requirements. This paper will highlight the engineering challenges and solutions faced in one such example.

TUNNEL DESCRIPTION

The Pula Subbaiah Veligonda Tunnel #2 in Andhra Pradesh, India will connect the existing Srisailam Reservoir on the Krishna River with eighteen irrigation projects in the districts of Prakasam, Nellore, and Kadapa, and is in the same general location as tunnel projects AMR #1 and AMR #2, which will also connect with the reservoir. However, this Ø10.0M water diversion tunnel will be bored under the Nallamala Hills and will eventually reach a length of 19.2 kilometers. The Nallamala is home to the Nagarjunasagar Reserve, India's largest tiger sanctuary, making intermediate access to the tunnel from within the Reserve environmentally complicated and potentially

Figure 1. Veligonda portal area

dangerous. This constraint required that the Veligonda tunnel be bored in one continuous operation, and with a minimum of perceivable noise and vibration on the surface. Access to the tunnel would only be from the portal, faced-up at the end of a long, narrow starting trench cut into the hillside. There are no curves along the tunnel centerline, which has a +.082% grade (see Figure 1).

TBM SELECTION

A Robbins Company double shield TBM with a complete back-up system was chosen by the Hindustan Construction Company (HCC) / Coastal Projects Pvt. Ltd. Joint Venture in October 2007 to bore the tunnel through a mixture of quartzite, shale, and phyllite. A double shield machine was specified to allow for continuous installation of the lining segments, as well as being a duplicate of the two TBMs currently in service at the AMR projects. With a mining stroke of 1,700mm and a minimum cycle time of 21 minutes, the TBM could theoretically generate 917 metric tons of tunnel muck per hour. The back-up system features an articulated conveyor tail pulley and a conveyor running structure assembly station on separate cars, rather than a conventional advancing tailpiece on one car. This arrangement reduces the overhung load on the cars and reduces the additional reinforcement necessary to withstand the increase in belt tension.

CONVEYOR SYSTEM

The Robbins Company was also awarded the contract for an extensible continuous conveyor system with a design capacity of 800 metric tons per hour at a belt speed of 3.05 meters per second. With a peak volume capacity of over 1,500 tph, the 914mm belt-width conveyor can easily handle the 'under ideal conditions' TBM output for limited periods of time. Despite the tunnel length, a single flight conveyor was preferred over a series of smaller cascading systems since it would not require the construction of any intermediate transfer stations and the loss of time in disassembling, relocating, and re-installing the major conveyor components (see Figure 2).

The key component of any belt conveyor is obviously the belting itself. For this system, an ST-1600 steel cable belt was selected. The belting has an ultimate strength of 1,600kN/m and a recommended operating strength of 240kN/m. Analysis of the conveyor at its full 19.35km length by a proprietary computer program revealed that under natural load sharing, the conveyor would require 758kw at starting and 482kw during normal running. Belt tension at starting would peak at 36,044kg, while belt tension at running would decrease to 24,255kg, and the maximum belt tension at the articulated tail pulley in the back-up car would be 17,507kg and occur during starting. During normal running conditions at its full length, the *AutoBelt 2* computer program also indicated that the conveyor would require a belt with an operational rating of 260kN/m, which still fell within the ST-1600 operating parameters. The steel cable core is sandwiched between 5mm thick, flame resistant top and bottom covers. Permanent vulcanized splices were also specified as new belting is added into the system to maintain the integrity of the conveyor. An in-line splicing station between the main drive and the belt cassette is dedicated to this requirement.

The belt drives for this system were designed with modular construction to allow for simplicity during manufacturing and interchangeability of component parts. The power units are self-contained and shaft-mounted utilizing a 300kw variable speed motor and a right angle hollow-shaft reducer. This configuration allows for easier shipping and faster assembly at the job site. The main drive (See Figure 3) consists of two power units (600kw total) with the drive pulleys mounted in the traditional tandem configuration and features an elevated, boom-mounted discharge pulley. The drive pulleys are also equipped with ceramic lagging to reduce belt slippage. The power units are mounted on the same side of the main frame to reduce the overall width and thereby minimize the footprint required by the main drive in the narrow confines of the starting trench.

The tripper-booster drives (See Figure 4) also use the same 300kw power unit as the main drive. Two of the booster drives are rated at 600kw; however, booster #3 has only a single 300kw power unit. This is due to the spacing of the tripper-boosters along the tunnel length. The booster drives will be equally spaced at 5.66km intervals from the main drive. This positions booster #3 only 2.75km from the end of the tunnel and reduces the load it is required to carry. The tripper-boosters share nearly the same design advantages as the main drive, but feature a loading hopper and an impact bed instead of a discharge boom and pulley. Again, the 600kw boosters have the power units mounted on the same side to minimize the amount of space in the tunnel cross-section required to mount the booster.

No booster drive was required for the return side of the belt due to the absence of any curves along the tunnel centerline which would have induced a considerable amount of additional drag for a conveyor of this length. Had curves been present, Robbins' patented curve carrying idlers and patented return hold-downs would have been specified, and the increase in return tension would have been reflected in the *AutoBelt 2* printout. As it is, the running tension in the return side of the conveyor is only 12,827kg and is easily within the operating parameters of the ST-1600 belt. This also assumes that regular scheduled conveyor maintenance will keep the belt properly tracked as misalignment of the belt can and will increase belt tension as well as consuming horsepower.

A 620m capacity belt storage cassette (See Figure 5) was specified to accept the 500m rolls of new belting while still having enough extra take-up capacity to accommodate the belt stretch at startup. This is especially critical in a conveyor of this length, and was also a deciding factor in selecting steel cable belting, as its modulus of elasticity is dramatically lower than that of fabric carcass belting. The cassette has a 10:1 ratio of main carriage movement vs. belt taken up and features ten intermediate carriages to maintain proper belt separation, thereby further reducing the drag on the return side of the conveyor. This feature also optimizes the ability of the cassette to react to the various loading conditions the conveyor is subjected to during a typical mining shift. An equally crucial component in this operation is the 3900kg electric drum winch which either winds in the Ø13mm wire rope to pull the main carriage back and maintain proper belt tension, or releases it as the TBM advances and pulls belting out of the cassette.

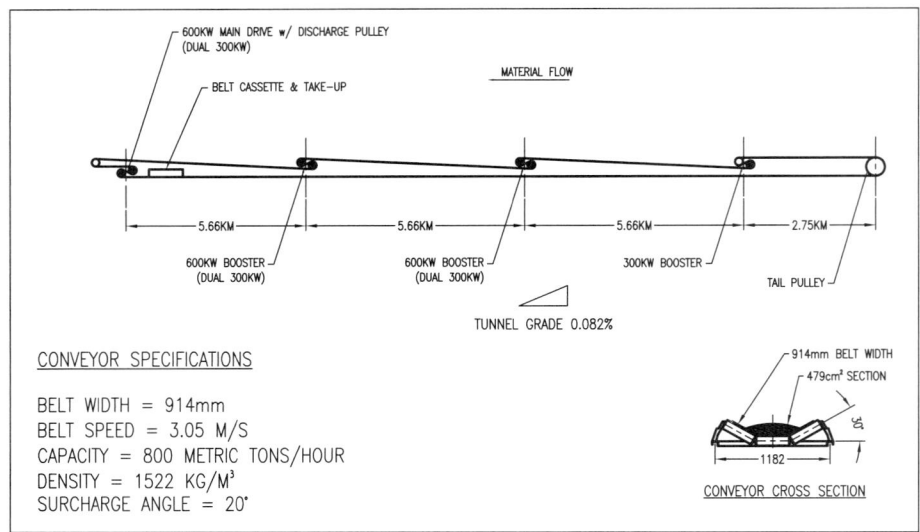

Figure 2. Belt conveyor profile

Additionally there are times when the winch drum must remain stationary—simply holding tension on a running belt while the TBM is idle. To accomplish this, the winch features a variable speed, constant torque motor that has its own internal cooling fan, and a hydraulically applied clutch-type brake integral with the right-angle reducer input shaft. The cassette also has a 10:1 rope sheave ratio to facilitate the take-up operation.

While the belting may be the key component of any conveyor, its heart is the control system. Each electric motor in the system has its own air-cooled VFD which has a 200% × 60sec overload rating, and each belt drive has its own PLC. A master PLC is located at the tunnel portal. Touch-screen stations at each PLC, plus one in the TBM operator's cab, allow the conveyor system, or any individual belt drive, to be operated from a variety of points to assist in maintenance and trouble-shooting. The only exception to this is the cassette winch, which can only be operated in the AUTOMATIC mode or from its own PLC. Programming for the PLCs is proprietary and is based on extensive experience in this area. Safety for the miners is assured by E-stop switches spaced at 100m intervals along the tunnel conveyor. Each switch is identified by its own internal transmitter, enabling instant identification when tripped, and resulting in a much faster response time. Combined with the features integrated into the PLC touch screens, this system has a Safety Integrity Level of 3 / Category 4. Additionally the conveyor is constantly monitored with belt slip and belt misalignment switches at the main drive and at each booster.

CONCLUSION

When fully extended, the Pula Subbaiah Veligonda Tunnel #2 will be one of the longest single-flight conveyors ever installed in India. As TBM-excavated tunnels continue to grow in length and complexity, extensible continuous belt conveyors can still become the most viable method of muck removal. Through proper engineering consideration and detailed communication between all the parties involved, from the customer to the component manufacturer, there is conceivably no limit to what can be accomplished.

EXTENSIBLE CONVEYOR SYSTEMS

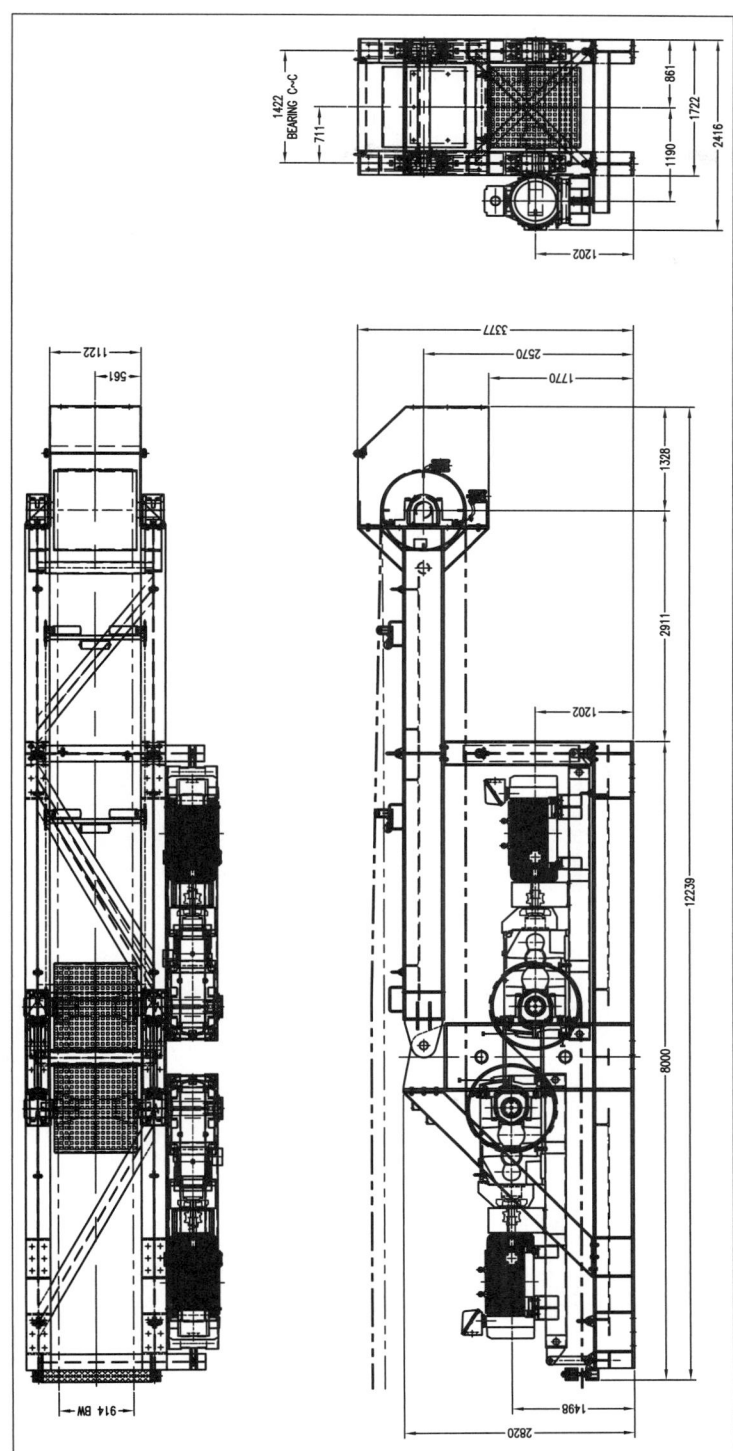

Figure 3. Main belt drive

368 INNOVATION

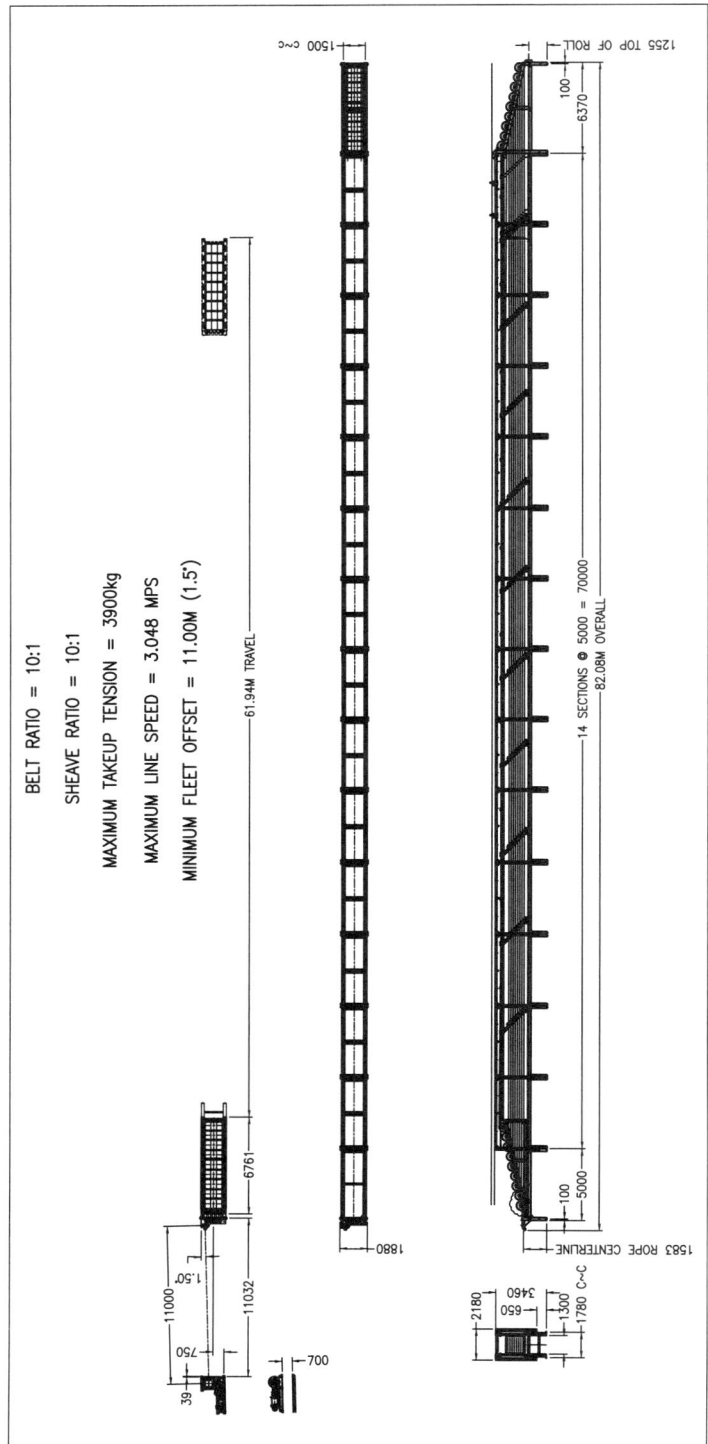

Figure 4. Belt cassette and take-up winch

EXTENSIBLE CONVEYOR SYSTEMS

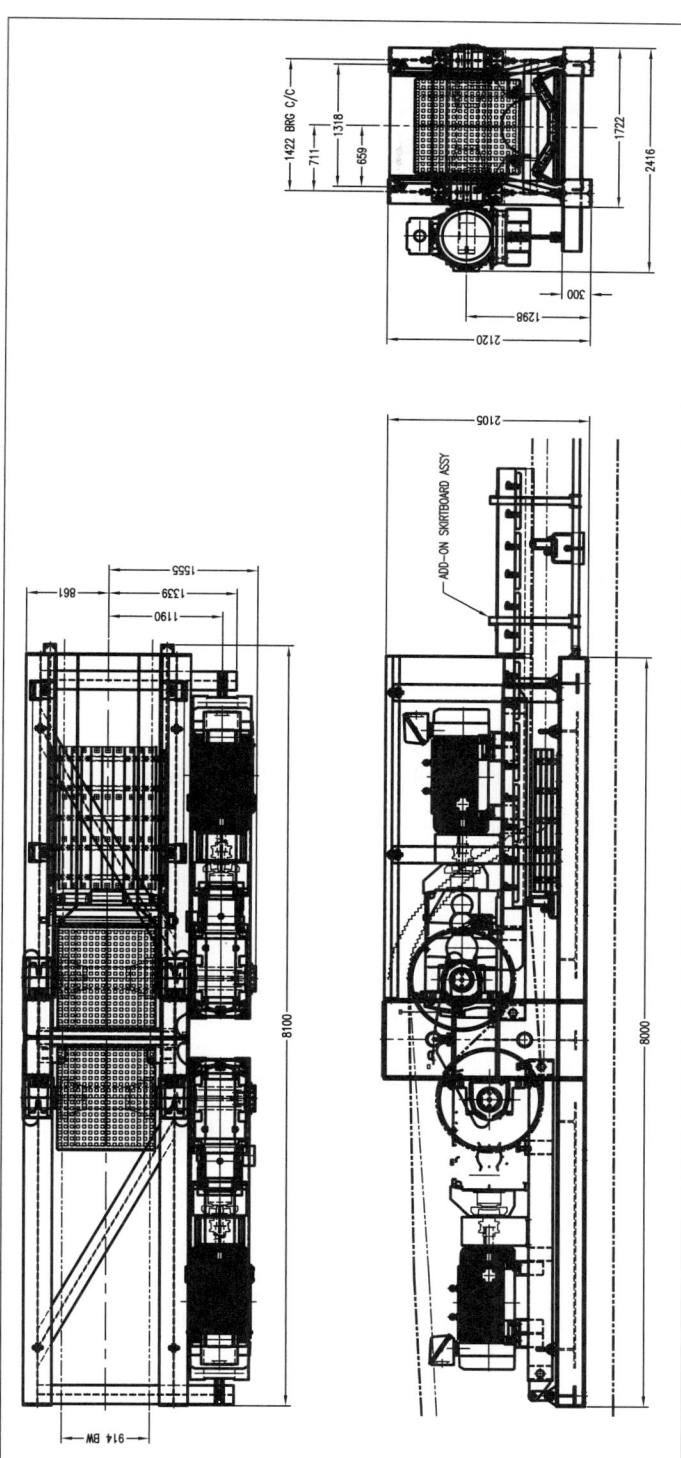

Figure 5. Tripper-booster belt drive

NEW CUTTER SOIL MIXING (CSM) TECHNOLOGY USED TO CONSTRUCT MICROTUNNELING SHAFTS FOR MOKELUMNE RIVER CROSSING

Matthew Wallin ▪ Bennett Trenchless Engineers

Mary Asperger ▪ Bennett Trenchless Engineers

ABSTRACT

A part of the East Bay Municipal Utility District's Folsom South Canal Connection Project, 430 feet of raw water pipe was installed using microtunneling beneath the Mokelumne River. The access shafts for this crossing were constructed 50 and 35 feet deep through very dense sands and soft sandstone and siltstone. High groundwater and the immediate source of recharge from the river required watertight shaft construction methods. The shaft contractor, Drill Tech of Antioch, CA, used a brand new Bauer cutter-soil mixing (CSM) rig to construct the large shafts. The structural concept is similar to secant piles, but instead of creating cylinders of soilcrete, the CSM rig creates interlocking rectangular panels. This technology has been used previously in Europe, Asia, and Canada, but this was the first project to use CSM technology in the United States. Cutter-soil mixing technology has other applications beyond shaft construction, including retaining wall construction and levee reinforcement.

INTRODUCTION

In 2007 the East Bay Municipal Utility District (EBMUD) began construction of its Folsom South Canal Connection (FSCC) Project. This project adds a major component to EBMUD's raw water delivery system. The FSCC Project consists of 19 miles of 72-inch raw water transmission main carrying flow from the southern terminus of the Folsom South Canal (southwest of Clay, California) to the existing Mokelumne Aqueducts (southwest of Wallace, California) as shown in Figure 1. The alignment crosses mainly open agricultural land following existing roads and was installed primarily using traditional open-cut construction. However, five trenchless crossings were required to pass beneath two state highways, a county road and flood control levee, a wetland area, and the Mokelumne River. It was at the river crossing, shown in Figure 2, where the cutter soil mixing (CSM) construction method was used for the first time in the United States to construct access shafts for a microtunneled drive.

EBMUD began final design of the FSCC project in early 2005. Fugro West was hired by EBMUD to perform a comprehensive geotechnical investigation and to provide design recommendations for the pipeline and associated structures. Fugro West subsequently contracted with Bennett Trenchless Engineers for assistance with the design of the trenchless crossings on the project.

EXISTING CONDITIONS

The FSCC pipeline crosses beneath the Mokelumne River immediately downstream of EBMUD's Camanche Dam and Reservoir. The upstream end of the pipeline crossing is located in an open pasture north of the river and south of Buena Vista Road.

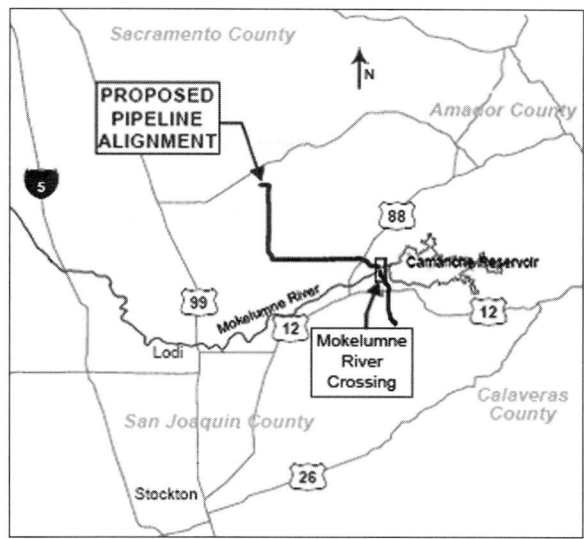

Figure 1. Folsom South Canal Connection Project vicinity map (Bennett-Staheli, 2007)

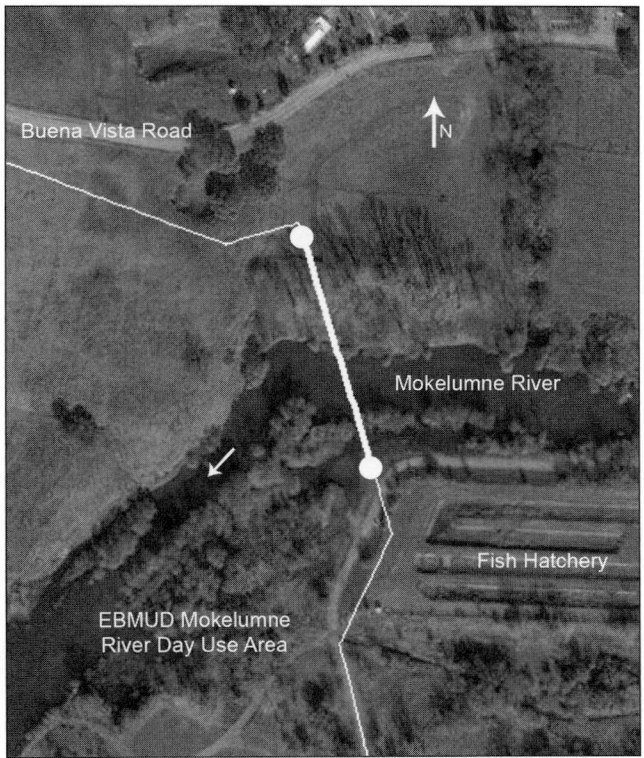

Figure 2. Aerial photo of the Mokelumne River crossing

Figure 3. The Mokelumne River at the trenchless crossing

The alignment crosses south, exiting in the Mokelumne River Day Use Area, just west of the Mokelumne River Fish Hatchery. This section of the river is an environmentally sensitive area that provides habitat for salmon and steelhead. Protecting the area from damage during construction was a critical issue during design and construction.

At the proposed pipeline crossing location, the active river channel measures approximately 150 feet wide and is only 10 to 12 feet deep during normal flow conditions. However several factors would contribute to making the seemingly simple crossing design quite challenging. The surface conditions on either side of the river included sensitive habitat that could not be disturbed. Both river banks support riparian habitat (see Figure 3) including elderberry bushes, which are protected habitat for the federally threatened Valley Elderberry Longhorn Beetle. Additionally, the northern bank of the river contained a grove of mature dogwood, willow, and box elder trees that were to be protected due to their size. These restrictions caused the length of the crossing to be extended to approximately 430 feet.

From a profile standpoint, the surface elevation on the northern side of the crossing is approximately 110 feet. The southern side was located within the flood plain of the river at elevation 96 feet. The bottom elevation at the deepest portion of the river was approximately 82 feet. To keep the risk of hydrofracture of drilling fluid into the river to an acceptably low level, a minimum clearance of 10 feet was required below the river bottom. These factors led to a pipeline invert elevation of 65 feet, requiring the north and south shafts to be approximately 50 and 35 feet deep, respectively.

The subsurface conditions at the crossing location would provide yet another challenge. Six borings were advanced by Fugro at the crossing, including two in the river channel, to investigate the subsurface conditions. The north side of the river had approximately 15 feet of stiff sandy clay and silt overlying dense to very dense sand and silty sand with some fine gravel. The sandy soils graded into soft rock materials of the Mehrten Formation at a depth of approximately 35 feet. The rock primarily consisted of soft, friable, sandstone with unconfined compressive strengths of 300 psi up to approximately 1,000 psi. However, within the sandstone were lenses of siltstone, claystone, mudstone, and agglomerate that had strengths up to approximately 2,000 psi. Of particular concern during design was the possible presence of cobble or boulder-size clasts within the agglomerate that might present possible obstructions to the tunneling equipment, or complicate shaft construction. While the borings did not encounter such clasts, the tunnel profile was selected to avoid the known agglomerate layers to the extent possible to minimize the obstruction risk.

At the south shaft location, the surficial soils consisted of channel deposits including sands and gravels with cobbles for the first few feet. Below the gravels, the soils were very similar to the northern side with clean and silty sands with fine gravel overlying the soft Mehrten Formation bedrock. Given the sandy soil conditions and the proximity to the river, groundwater levels at both shafts were approximately equal to the river level. Additionally, recharge from the river through the permeable sands and gravels meant that attempts to dewater the ground outside the shafts would be very difficult, if possible at all.

DESIGN CONSIDERATIONS

The large diameter of the pipeline combined with the high groundwater levels and immediate source of recharge made the choice of microtunneling as the specified tunneling technique simple. The fully sealed excavation face of a microtunnling machine (MTBM) is designed for use in unstable ground and below the water table. Additionally, MTBM's can be fitted with cutterheads specifically designed for cutting rock of various types. This project would call for a combination style cutterhead intended primarily for cutting soft rock, but also fitted with features to improve the efficiency of excavation of soft ground, if lenses of non-cemented sand were encountered.

Given the high internal pressure of the raw water transmission line, combined with the anticipated need for high jacking force capacity needed on this bore, reinforced concrete cylinder pipe (RCCP, AWWA C300) was chosen as the carrier pipe to be installed by direct jacking. The mortar-lined and dielectric tape-coated carrier pipe used for the open-cut portions of the project could have been installed inside a steel casing in a second pass, but it was decided that keeping the excavated diameter smaller would increase the number of available MTBM's and contractors that could bid on the project.

The shaft design requirements would be even more challenging than the tunneling requirements. The shaft methods to be employed by the contractor needed to provide a safe, dry, excavation large enough for both the microtunneling equipment during installation and the sweeping 90 degree bends in the final pipeline. The shafts would have to be installed through potentially flowing sands, some cobbles and small boulders, and competent soft rock with strengths as high as 2,000 psi. Additionally, the shafts had to provide not only watertight walls, but a sealed base that could resist the buoyant uplift loads when the shaft was complete and the interior dry. The design assumed that the contractor would excavate the shaft interior in-the-wet, place an underwater concrete seal slab by tremie, and then dewater the sealed interior.

The watertight shaft methods considered during design included interlocking steel sheetpiles, auger drilled shafts, sunken concrete caissons, and secant piles. However, steel sheetpiles were eliminated from consideration early on due to concerns related to the ability to successfully drive the sheets through the cobbles and boulders anticipated on the south side, and into the rock present at both locations, and to maintain the integrity of the interlocks to ensure a watertight seal. There was also concern about the production rates that would likely be achieved advancing a sunken caisson in the difficult soils. Finally, the shaft diameters necessary to accommodate the pipeline appurtenances were at the edge of what was feasible for auger drilled shafts. These concerns left secant pile shafts as the most likely shaft construction method. However, traditional secant pile construction would also have been challenging, with slow drilling likely in the difficult ground.

CUTTER SOIL MIXED (CSM) METHOD

In May of 2007 EBMUD awarded a contract for construction of the southern portion of the FSCC pipeline to Sundt, Inc. Sundt's construction team included Drill Tech

Figure 4. Counter-rotating cutterwheels of the Bauer CSM rig

Drilling and Shoring, Inc. (Antioch, CA) for shaft construction at the Mokelumne River, and Nada Pacific Corporation (Caruthers, CA) as the microtunneling subcontractor. As an alternative to the methods called out in the specifications, Drill Tech submitted a proposal for employing cutter soil mixing (CSM) to build the access shafts. This process was little known to the Owner, and the Engineers had no experience working with the method. However, CSM is similar in principle to the accepted secant pile method where columns of strengthened material overlap to form watertight walls that support the ground in ring compression. Unlike traditional secant pile construction where the native soil is replaced with concrete columns, CSM forms walls by mixing the native soil in place with bentonite and cement to form soilcrete panels. The overlapping panels are then formed into a shaft, in whatever pattern is needed for the particular project.

Soil mixing techniques have been used in the United States since the mid-1930s for applications such as improving soil for road and dam construction. These applications consisted primarily of mixing the native soil with various mixtures of bentonite, cement, lime, and other admixtures using common construction equipment. As the technology evolved it became general practice to inject cementitious grout into the soil via the tip of a drill rig's auger. ("Soil Mixing Technology Overview," 2005.) This process, combined with experience with diaphragm wall cutters for the construction of retaining walls, was further adapted by BAUER Machinen GmbH in 2003 into the CSM technique seen at the Mokelumne River crossing (Mainer, 2008).

The difference between CSM and more traditional soil mixing is in the rig used to create the soilcrete panels. Instead of a single- or triple-axis auger rig where the augers turn around a vertical axis, the mixing tools of a CSM machine rotate around a horizontal axis. Two counter-rotating cutterwheels, as shown in Figure 4, are mounted on a Kelly bar that can control their position in all three dimensions (see Figure 5). A primary advantage to this configuration is that the cutter gear drives are located on the end of the Kelly bar and advance with the drilling head. This configuration allows for inclinometers to be used to provide real-time tracking of the cutter's position during drilling, allowing for more accurate panel construction. The data collected regarding cutter head location, as well as cutterwheel RPM, down pressure, penetration rate and fluid injection rates and pressures are monitored and tracked on a control screen in the operator's cabin. There are a variety of mixing tools available for use with soils ranging from non-cohesive sands to cohesive clays and harder materials. The end product from all of these types of tools is a rectangular panel of soilcrete, as opposed to the cylindrical column created by a traditional auger.

Panels can be created using either a single-step or two-step process. In the two-step process, as was used on this project, the ground is fluidized by injecting bentonite slurry as the cutter wheels progress downward. When the final depth is attained, the cutter wheels are retracted and cement slurry is pumped and mixed in with the fluidized bentonite and cuttings from the downstroke. The single-step process simply combines the two steps and is suitable for applications in softer ground where progress will be quick so that the risk of the soilcrete setting while the rig is still drilling is low. When multiple panels are to be drilled, CSM uses an alternating method of primary and secondary drilling as is common for secant piles. The primary panels are drilled first consisting of every other panel in a continuous wall. After the primary panels have had sufficient time to cure, secondary panels are drilled, overlapping between the pairs of primary panels.

The CSM method has been used extensively in Europe and Asia to form retaining walls, cut-off walls, foundations, levee reinforcement, and excavation supports. However, this project was not only the first use of CSM in the United States, but was also the first time that a Bauer CSM rig was used to create access shafts for a microtunneling drive. (Mainer, 2008)

Figure 5. The Bauer CSM rig

SHAFT CONSTRCUTION

Drill Tech's design, shown in Figure 6, for the Mokelumne River shafts called for twelve panels, 94 inches in length and 30 inches wide, overlapping to approximate a circular shaft with a finished inside diameter of 24 feet. The design provided for a minimum overlap at the outside edges of 1 ¾-inches, with an overlap of approximately 18 inches on the inside faces. The final excavated depths of the north and south shafts were 51 feet and 38 feet below grade, respectively. To provide groundwater cutoff, however, the panels were designed to extend to 70 feet and 51 feet below the excavated depth, respectively.

Structurally, the shaft was designed to work in ring compression like a traditional secant pile shaft or sunken caisson. The 30-inch thick panels overlapped to form a 24-inch wide continuous ring of soilcrete. The contractor's design calculations demonstrated that the compression ring alone would have been sufficient to withstand the soil and groundwater loads for this project, given a soilcrete panel strength of 300 psi. However, an additional level of conservatism was included by the contractor. While the soilcrete panels were still wet, a wide flange beam was embedded in each panel. Additionally, several levels of ring steel were then installed during excavation. At the north shaft, three permanent sets and one temporary set of ring steel (left in place until the bottom slab was cured) was installed (see Figure 7). The south shaft had two sets of ring steel.

A unique feature of Drill Tech's design was the breakout areas. At both the jacking and receiving shafts where the MTBM would come through the shaft wall, the contractor constructed a sealed rectangular chamber by drilling four additional overlapping panels.

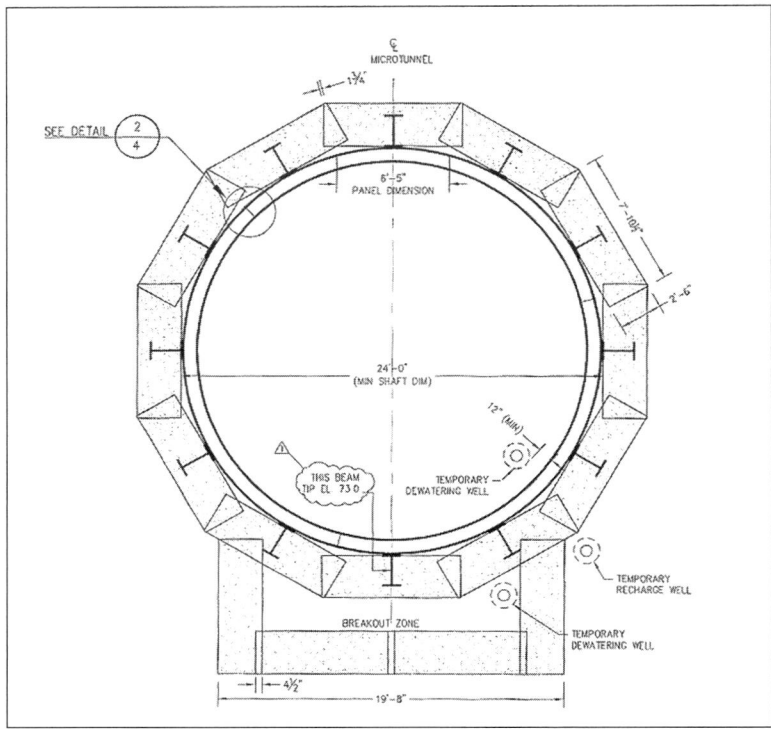

Figure 6. Plan view of the north shaft

These panels created a small, isolated zone which could be dewatered, eliminating the need to do any ground improvement outside the shafts.

QC testing to verify that the panels would attain the proper strength was conducted by wet sampling the soilcrete from every other panel. Samples were obtained by using a custom-designed box sampler that was open on the top and had a one-way flap on the bottom. This sampler, shown in Figure 8, allowed wet soil-cement to enter through the bottom when lowered into the panel, but would seal when the sampler was withdrawn. Thus, samples could be obtained from specific depths. The samples were cast into 3-inch diameter cylinders and 5-inch square beams for testing in compression and bending.

Once construction began, drilling proceeded relatively smoothly at both shafts. The layout of the drill site at the north shaft is shown in Figure 9. Drilling of the 16 panels for the north shaft took 13 total days, averaging approximately 85 linear feet of drilling per day. The shallower south shaft took only 11 days, however the drilling average was only 75 linear feet per day due to harder drilling in the cobbles near the surface and the higher percentage of Mehrten Formation material overall. Drilling rates averaged around 3 minutes per foot during the downstroke. Grouting on the upstroke was much faster at approximately 1 minute per foot. Instantaneous drilling rates varied widely depending on type of material encountered. In the softer materials at the top of the panels, penetration rates varied from 1 to 5 minutes per foot of progress. However, near the bottom, in the fine-grained rocks and agglomerate, penetration rates were as slow as 8 to 16 minutes per foot of drilling. Overall, Drill Tech averaged approximately 1.25 panels per day at the north shaft and 1.5 panels per day at the shallower south

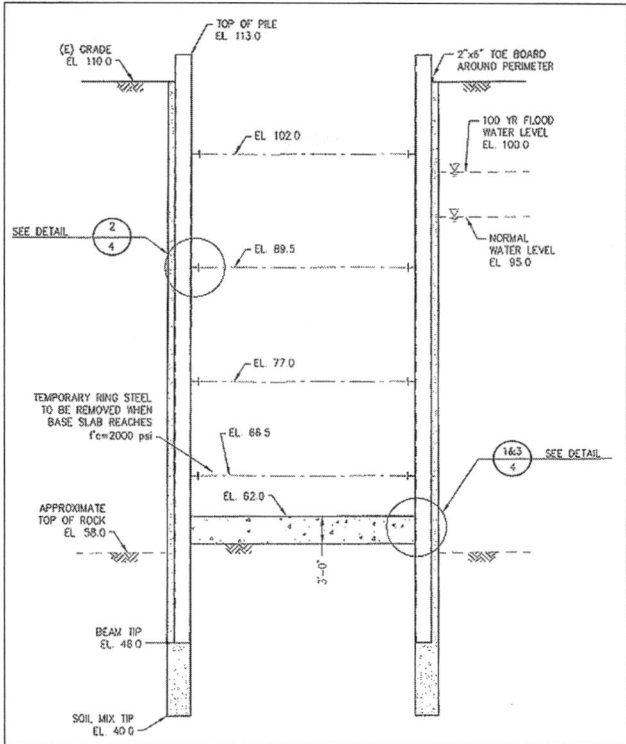

Figure 7. Section view of the north shaft

shaft. Volumetrically, approximately 60 yd^3 of material could be treated in one day, including panel setup, beam placement, and QC testing.

While the panels were curing, two dewatering wells and a temporary recharge well were drilled at each shaft location. One dewatering well was placed inside the breakout zone and another inside the main shaft. See Figure 6, above, for the dewatering well layout. The shaft well allowed for water trapped within the panels to be removed ahead of the excavation and re-injected through the recharge well installed outside the shaft. Prior to launch of the MTBM the same process was conducted in the breakout zone.

Initial tests on the north shaft soilcrete samples showed significant scatter between individual breaks. The wet grab samples contained primarily well blended soil cement, but also contained some small lumps of unmixed native soil, typically no larger than one inch in diameter. It was presumed that the test scatter was due primarily to these anomalies in the samples. While the effects on the strength of the 30-inch panels from these discontinuities would likely be negligible, they had significant effect on the small cylinders. The contractor felt that on future projects typical 6-inch or larger diameter concrete cylinder molds would provide for more representative testing of the panel strength.

Once the soilcrete panels had achieved sufficient strength, excavation began in each shaft. Excavation was accomplished using a hydraulic excavator for the upper 20 to 25 feet. Once the work extended beyond the reach of the hydraulic excavator, an electric mini-excavator was mobilized to the site and the remainder of the excavation proceeded using a crane and muck buckets, as shown by Figure 10. The ring steel was

Figure 8. Drill Tech taking a sample from a CSM panel

Figure 9. The north shaft site

Figure 10. Excavation at the north shaft

Figure 11. The completed south shaft

installed as excavation progressed and seal slabs were placed at the bottom. Figure 11 shows one of the completed shafts.

During excavation one section was discovered in the north shaft where the panels did not overlap as intended. This was remedied by drilling out the untreated area and grouting this location, providing a continuous wall. Additionally, there were a few small leaks discovered at the bottom of each shaft which were repaired by spot grouting.

MICROTUNNELING

Nada Pacific began setup of the microtunneling equipment in mid-January of 2008. The MTBM selected for the project, shown in Figure 12, was an Akkerman SL72 with an 88-inch diameter upsize kit to match the RCCP jacking pipe. The original electric drive motor had been upgraded to a 250 horsepower model to drive the 89.5-inch diameter cutterwheel. The custom-designed cutterwheel was a combination-type rock head with a row of double disc cutters across the face, triple-disc cutters and inset

Figure 12. The MTBM used for the Mokelumne River crossing

picks around the perimeter, and two rows of inset spades to help scoop material into the crushing chamber.

The MTBM was set into the shaft on January 30th, setup continued on the 31st, and tunneling began on February 1st. Over 14 days the MTBM made very good progress in the Mehrten Formation rock, especially in the large pockets of sandstone. Progress rates as high as 62 feet per day were achieved. However, an overall average rate of 31 feet per day was achieved over the entire 430-foot drive, including a few stoppages for typical mechanical issues and delays.

Two intermediate jacking stations (IJS's) were required in the specifications due to a prohibition on recovering the MTBM from the surface if it became stuck beneath the river. However, the stable ground conditions led to very low jacking forces overall. Over the course of the drive, jacking forces typically ranged between 100 to 125 tons. The maximum jacking force recorded was 210 tons at just past 300 feet of advancement. However, through aggressive lubrication the contractor was able to lower jacking forces back to 100 tons by 400 feet. The drive was finished with under 150 tons of jacking force for a very low overall average unit friction of 0.015 tons/ft^2, and the IJS's were never needed.

Surprisingly, wear on the cutterwheel was light to moderate despite concerns about the abrasive sandstone. The circumferential picks had significant wear, and the triple-disc cutters were also fairly worn between the carbide inserts. However, most of the tooling on the face was very lightly worn, and almost no wear was visible in the crushing chamber. Overall, the tunneling was very successful with few issues.

CONSTRUCTION COST

The total bid cost for the Mokelumne River crossing was approximately $3.5M. The bid for the shafts was $1.4M or approximately $15,000 per foot of depth. Calculated by shaft volume the bid price was approximately $900 per cubic yard. This cost places these CSM shafts within the order of magnitude typically seen for secant piles or more complex auger drilled shafts, but below the typical cost of sunken concrete caissons. While this cost is significantly higher than typical sheetpile shafts, it should be noted that the ground conditions at this location would have made sheetpile construction much more costly, if even possible.

The total microtunneling bid was approximately $2.75M; however this included both the 430-foot river crossing and a 240-foot crossing of State Highway 12 in soft ground. Based on the added risk associated with the river crossing in full-face rock, it is assumed that approximately $2M of the total microtunneling bid was attributed to the Mokelumne River tunnel. If so, the unit cost for tunnel would be approximately $4,650 per foot, very high for typical microtunneling. However, the high level of perceived risk involved with constructing a river crossing in full-face rock, combined with the low overall tunneling footage on the project, is what likely skewed the unit costs higher than normal.

LESSONS LEARNED

On any project using a new method for the first time, there are inevitably a few issues encountered and lessons learned that get carried forward to the next project. Each of the issues discussed below were addressed proactively by the contractor to ensure that the project would be a success.

While CSM creates far less spoil than traditional secant pile construction due to the in-situ mixing of the native soil rather than replacement, there is still a significant volume of wet spoil to contain. For this project the contactor excavated a trench approximately three feet deep around the perimeter of the shaft where the panels would be drilled. The trench was used to collect the excess slurry and spoil that was created during the mixing process. The excess slurry was then alternately moved by excavator and front-end loader, or pumped with a concrete pump, to a sedimentation basin excavated on site. A small separation plant was brought in to remove the sand from the slurry and create drier spoil that was easier to haul offsite. The fluids were then transferred to a second, plastic-lined, basin for limited reuse and later disposal. The slurry handling operation was fairly messy and containment storage and disposal of this material should be addressed during the design phase of future projects, specifically during layout of temporary construction easement.

Accurate layout of the shaft panels was a crucial aspect of the construction to ensure that proper overlap was achieved. Control points were established by survey in the field for the center point of the shaft, the centerline of the pipeline crossing, and 50-foot offsets of the shaft center. Thereafter, the contractor used these control points to place pins marking the inside center point of each panel manually. The mixing head was lined up to each pin and then adjusted to ensure it was perpendicular. Subsequent panels on each side were laid out by transcribing a circle on the ground from the shaft center pin at the intended inside diameter of 24 feet. The chord length from the inside center point of one panel to the center of the adjacent panel was taken from the contractor's design drawings and measured in the field to set the center pin for each subsequent panel. This method proved to be challenging to complete in the field with high accuracy, especially with the added complication of the spoils trench excavated around the shaft perimeter. The center pins had to be offset one to two feet inside the planned shaft circumference. The layout difficulties resulted in one significant and a few very minor locations where complete overlap was not achieved. For the larger gap, the contractor drilled the area out and placed grout to provide a continuous wall. The minor areas were spot grouted to seal minor leaks. On future projects, it would be well worth the expense of having all necessary control points surveyed to ensure accurate wall placement. A set of four pins for each panel, located outside the spoils trench footprint, would allow for stringlines to be used to locate the panel centerline and inside panel face, greatly simplifying rig setup.

Productivity was generally very impressive during the panel drilling. The CSM rig was very efficient at drilling through the soft ground near the surface and the very dense sand and soft sandstone that accounted for the majority of the panel volume. However, the penetration rate dropped significantly in the siltstone, claystone, and agglomerate layers near the bottom of the panel depth. The upper materials could often be drilled at 20 to 40 feet per hour before slowing to as low as 5 feet per hour in the layers where both unconfined compressive strength and induration of the rock increased.

The stiffer material in the bottom third of the panel profile caused another minor issue during construction. Overall, the Kelly-mounted mixing system provided for a very well-controlled and plumb panel, however in the harder rock material where significant crowd was necessary to achieve penetration, the Kelly bar had a tendency to "walk" away from the rig slightly. This tendency was evident during the placement of the H-piles into the panels. For panels drilled from inside the shaft footprint, the H-pile would begin to hang up on unmixed native ground on the inside panel face near the bottom. Had all the panels been drilled from outside the shaft footprint, this issue would not have been cause for concern as any walking of the panels would have been inward toward the shaft center, and would only have resulted in slightly greater overlap, albeit a slightly smaller shaft diameter. However, on this project one side of the north shaft was located very close to the environmentally sensitive riparian area that could not be encroached on. The lack of working area on one side forced the contractor to drill half of the panels from inside the shaft, meaning that the tendency to "walk" would reduce the theoretical panel overlap. Despite some concern, the walking effect did not seem to cause any issue with the integrity of the finished shaft. On future projects, some additional design overlap, as well as ensuring working space to drill from outside the shaft footprint, could alleviate this concern altogether.

CONCLUSION AND ACKNOWLEDGMENTS

The FSCC Mokelumne River crossing was ultimately a very successful trenchless project thanks to the efforts of all involved parties. Drill Tech's use of the innovative CSM-method of shaft construction for the first time in the United States set the stage by efficiently providing a safe, dry, excavation in very challenging ground. Further, the CSM method has many applications beyond shaft construction including larger excavation support, foundation construction, curtain walls, levee and dam improvement, and others. A second CSM project has already been completed immediately on the heels of the Mokelumne River crossing by a second west coast contractor who has purchased a Bauer CSM rig. We expect to see CSM used on many projects in the future, trenchless and otherwise.

The authors would like to thank our clients EBMUD and Fugro West for the opportunity to work on this innovative project. Additionally, we would like to thank the construction team of Sundt, Drill Tech, and Nada Pacific for their hard work making this project a success.

SELECTED READING

Bennett-Staheli Engineers. (2007). Geotechnical Baseline Report, Folsom South Canal Connection Project, Specification 1946 (Southern Reach). February 22, 2007.

Fugro West, Inc. (2007). Geotechnical and Geologic Study, Folsom South Canal Connection Pipeline. January 2007.

Mainer, B., and Gerressen, F. (2008). First Use in the USA of the CSM Method for use as Microtunnel Sending and Receiving Shafts. Proceedings of 2008 Deep Foundations Institute Annual Conference on Deep Foundations, New York, New York, October 15–17, 2008.

Shallow Soil Mixing Technology Overview. (2005). Retrieved January 14, 2009, from http://www.soil-mixing.com/Soil-Mixing-technology/Soil-Mixing-technology.asp

AN INTRODUCTION TO VIRTUAL DESIGN AND CONSTRUCTION (VDC)

Matthias Rheinlander ▪ Parsons Brinckerhoff

Rachel Arulraj ▪ Parsons Brinckerhoff

Alan Hobson ▪ Parsons Brinckerhoff

Chris Graham ▪ Parsons Brinckerhoff

ABSTRACT

The paper will describe Virtual Design and Construction (VDC) as a tool which will enable us to build a project "virtually" on a computer before constructing and operating it in the real world. VDC computer models are built, using information from various sources including survey, 3D laser scanning, CAD and GIS to represent the project and its components in a coordinated and consolidated infrastructure model. This model is reviewed and evaluated by, and shared among project teams including owners, contractors, designers, engineers, construction managers, stakeholders and operators. It allows teams to fix costly design and construction issues before construction on site. Benefits and lessons learned from global projects will be discussed.

INTRODUCTION

In today's Architectural, Engineering and Construction (AEC) industry there is an ever increasing need to provide owners, contractors, designers, engineers, architects, construction managers, stakeholders and operators with highly detailed information regarding different aspects of a project. Managing public concern over the impact on the surrounding environment and community forms an integral part of any project.

Implementation of VDC modelling can provide an effective means of adequately addressing these requirements (refer Figure 1—VDC Beneficiary Diagram). VDC is a tool enabling the construction of a project in a digital environment or "virtually" using specific computer modelling software. The virtual model has the ability to integrate multidisciplinary information required for the project including design elements, survey, 3D laser scanning, CAD, GIS information, aerial photography, construction sequencing information, critical path method (CPM) schedule, and cost/budget information. Through the integration of all these various areas of information, a highly accurate virtual model of the project is created.

For example, the application of VDC in the tender phase for a design-build-operate (DBO) project provides project team members with transparent and reliable information to improve interdisciplinary coordination so more timely and informed decisions can be made. Also, the integrated design can be evaluated before construction and operation in the real world to fix costly design and construction errors before they happen on the construction site. The final model can also be handed over to the operator after construction has been completed as a tool used during the concession period in the operation, maintenance and management of the constructed project.

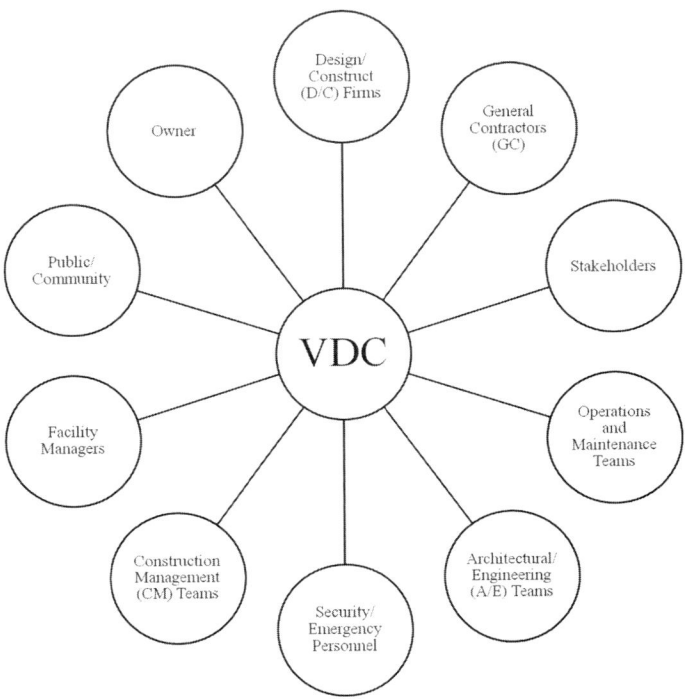

Figure 1. VDC beneficiary diagram

The best medium in conveying information in an easy to understand, detailed and accurate manner to the prospective audience is through a Visualisation Information System (VIS). Correct use of VDC technology provides a wealth of information which can be readily utilised for the creation of visualisations intended for stakeholders, including government agencies and authorities, financing groups and community groups. Because of the organisation and accuracy of the geometry information provided within the VDC technology, this information is also readily available for photo-realistic visualisation solutions, making the process cost effective. Also by establishing the visualisation environment as a model, information can be updated as required by importing new data which reflects the status in the VDC technologies.

VDC embedded as a tool in any standard project delivery system application and in any collaborative planning, design and construction process, can significantly improve performance, both from the owner's perspective (reduced cost and time, change orders and improved quality) and from the design-contractor's perspective (increased profitability, reduced risk and liability exposure, and improved safety records).

WHAT IS VDC?

VDC is an approach that lets us build a project in a virtual world (model) on a computer before constructing and operating it in the real world. To do this, 3D and 4D computer models are built to represent the project and its individual components in a coordinated and consolidated virtual project model (Arulraj, 2008).This model can then be reviewed and tested by and shared among several project participants to evaluate design and construction options. Conflict resolution is enhanced through the ability to see the project from any point of view or point in time.

The approach of VDC supports data centric decision-making rather than a document centric perspective. This represents a digital project delivery where purpose built software are integrated in a systematic manner. The success of VDC is that integration occurs throughout all design, planning and construction stages.

The Centre of Integrated Facility Engineering (CIFE) at Stanford University defines Virtual Design and Construction (VDC) as "the use of multidisciplinary performance models of design-construction projects, including the *product* (i.e., facilities), *organisation* of the design-construction-operation team, and work *processes*, to support explicit and public business objectives" (Fischer and Kunz 2004).

VDC allows teams to find and fix costly design and construction errors before they happen and provides useful tools and information to improve communication and coordination during construction. With input from the field, VDC also will yield an accurate 3D as-built model of the entire facility that can be used for training and for better management of longer-term operations and maintenance. 3D Models are assembled from the various contractors or individual 3D models created by the project team from incoming 2D CAD data. The VDC modeling process (inputs and outputs) will be incorporated into a data-centric repository and tied to the current design review, documentation, mark-up, comment capture, transmittal and reporting workflows. A real-time visualization system comprising a high-performance computer graphics workstation and high resolution computer projector is normally used to support formal and informal review and coordination meetings, working sessions, ad-hoc collaborations and presentations.

The typical tools used in a VDC approach of *product, organisation* and *process* are cutting-edge technologies that meet the increasing demands of the AEC market. In the typical VDC framework these technologies would include:

- *Product* modelling tools—3D design technology such as AutoCAD, ADT, Revit;
- *Process* modelling tools—such as location based schedules and CPM;
- *Organisation* and *process* modelling tools—such as SIMVision and VDT; and
- *Product* and *process* modelling tools—4D technologies such as CommonPoint Project 4D, NavisWorks Timeliner
  (Fischer 2006; Khanzode, Fischer, Reed and Ballard 2006)

The VDC process can assess the 3D design performance against targets as well as improving the process of cost estimation, construction planning and supply chain management. The approach facilitates better coordination of design teams, subcontractors and trades on the construction site. (Jongeling, Olofsson and Norberg 2007)

HISTORY OF VDC APPLICATIONS BY CIFE AND PARSONS BRINCKERHOFF (PB)

CIFE introduced the term 'Virtual Design and Construction' in 2001 (Kunz and Fischer 2008) and the methods of VDC have been taught since that time. The CIFE has also been involved in research and project opportunities that support a VDC framework. These include:

- 1998: Paradise Pier in Disney's California Adventure—First use of 4D modelling to review all construction schedule alternatives for a major project.
- 2000: Walt Disney Concert Hall in Los Angeles—First use of 4D modelling for planning the construction of a highly complex building.

- 2001: 600-seat Lecture Hall at the Helsinki University of Technology—First application of Building Information Modelling in practice.
- 2005: Fulton Street Transit Centre Project in New York City—First large-scale application of 4D modelling for an urban infrastructure renewal project.
- 2003—ongoing: Piloting and support for GSA National BIM Program (Fischer and Hartmann 2007)

PB has been involved with implementing VDC in projects from a commercial and technical perspective, both in the USA and internationally over a period of five years. These projects are in large infrastructure applications and include:

- rail/transit tunnel new build
- rail/transit tunnel rehabilitation
- multimodal transportation hubs new build
- highway new build/rehabilitation
- bridge and viaduct new build/rehabilitation
- power generation plant new build
- rail station new build/rehabilitation
- airport new build
- road tollway rehabilitation
- large infrastructure/commercial buildings new build

GENERAL BENEFITS

Currently in the AEC industry a sizeable proportion of projects are being delivered late, over budget (up to 40%) and fail to meet quality expectations. The greatest losses in profit and time throughout the execution of projects are due to:

- Rework and delays: Up to approx. 30% of construction work is rework (i.e. due to clashes resulting from poor planning, design, collaboration and quality)
- Poor configuration of design and construction procurement
- Inefficient project scheduling, poor pre-construction planning and construction look-ahead planning
- Poor project coordination between project-wide design disciplines, project design teams and external design firms
- Fragmentation of design disciplines (performing of design work in isolation from other designers)
- Project team members not communicating effectively and "not working together" and thus, collaborative design and planning process including a transparent and cooperative exchange of information does not exist
- Safety incidents and claims: Construction workers are exposed to hazards on the construction site causing accidents and injuries
- Inadequate construction sequencing, and insufficient value engineering and constructability reviews
- Ineffective communication across the project teams (off-site including stakeholders and 3rd parties)

- Poor project workforce utilization and inconsistent workflow: Approx. 30 to 40% of the manpower used on construction sites can be wasted due to poor utilization

By adopting VDC, project coordination, integration of information and decision making will greatly improve the planning, review and communication processes during the design and construction life cycle, therefore minimising the aforementioned losses of profit, and time. Contractors, designers, engineers, architects, construction managers, stakeholders and operators all benefit from VDC implementation through collaboration, and visualization opportunities which lead to shorter project duration and thus, earlier hand-over to the end-user or owner and greater return on investment.

DESIGN COLLABORATION IN A COLLABORATIVE DESIGN-BUILD (ALLIANCE) ENVIRONMENT

VDC framework enables the multidisciplinary team to predict the performance of project tasks before they are constructed, adapt to any design changes quickly, perform what-if analyses, simulate scenarios and create visualisations to provide effective means in optimising the overall design and construction strategies, align design, planning and construction, distribute design and construction documentation of a superior quality, and provide an interactive communication environment. Early collaboration of the multidisciplinary design and construction team enables the creation of a variety of designs. These variations in designs can be quickly and easily assessed from all aspects. Additionally, important information contained in the virtual model can be extracted to assist in earlier decision making and more cost-effective project delivery.

To aid in the communication and data transfer between the multidisciplinary parties involved in the project, a digital environment is established. The digital environment provides team coordination across the various multidisciplinary personnel required throughout the design and construction period of the project. This environment enables team members the ability to access:

- The overall model of the project and/or specific project areas
- Schedules which display the allocation of quantity take-off boundaries; coordination and integration responsibilities; and scope of technical project team disciplines, manufacturers/suppliers and sub-contractors
- Property data and ownership restrictions
- 3rd Party information in regard to city, authority, council and town planning aspects
- Geotechnical and hydrogeological information
- Existing building and infrastructure information
- Project-wide design development for ventilation, emergency egress, fire engineering, durability, traffic modelling etc.
- "As-Built" information from previous projects and/or adjacent projects
- Links to the project-wide management and document control system

With this environment proposed designs can be quickly analysed in-depth by benchmarking the operation of differing designs by performing simulations from differing aspects, enabling improved and innovative solutions which creates an overall better design. The digital environment also enables prompt resolution of construction and scheduling clashes across all disciplines resulting in more concise allocation of onsite resources and reliable workflow.

In summary, the integrated multidisciplinary virtual model provides the following specific benefits during the bid/tender phase, pre-construction phase, design and construction phase, post-construction phase and O+M/concession period:

- Rapid pre-determination of spatial conflicts, interferences and collision detections (i.e. predict unplanned or unforseen field conditions and scenarios)
- Configuration of schemes relating to design, planning, procurement and construction
- Integration of project control for scope, schedule, cost/budget and estimation
- Extraction of quantities
- Integrated assembly and itemised take-off
- Support of early procurement of critical components or "long-lead" items (i.e. Procurement of Tunnel Boring Machine (TBM): sizing of TBM diameter including tunnel space-proofing check or procurement of water pumps in a water desalination plant: sizing of pumps and check of equipment location and tie-ins)
- Management of design changes made by different technical disciplines and design teams
- Alternative design and construction sequencing and constructability analysis with respect to:
  - Size of the workforce according to availability (within the "home offices" and on the project)
  - Speed of construction
  - Location of equipment and site material set down
  - Level of off-site pre-fabrication ("outsourcing") that can be achieved or will be required as per the design and construction schedule
- Better production quality—documentation output is flexible and exploits automation
- Faster and more effective processes—information is more easily shared, can be value-added and reused
- Integration of planning and implementation processes—government, industry and manufacturers/suppliers have a common data protocol
- Tool for general marketing, business development and tender/bid presentations
- Management of resources, equipment, and subcontractors
- Reduction of re-design and re-work
- Assistant for day-to-day design and construction execution, and coordination including safety-in-design (SID) reviews, schedule look-ahead meetings, design coordination and integration meetings etc.
- Tool for general conflict resolution, alternative dispute resolution and litigation process
- Reduction of correspondence, request for information (RFI), design errors and omissions (E&O), risk and liability exposure, change order requests, and claims
- Excellent return on investment (ROI) (refer Figure 2—ROI Diagram)

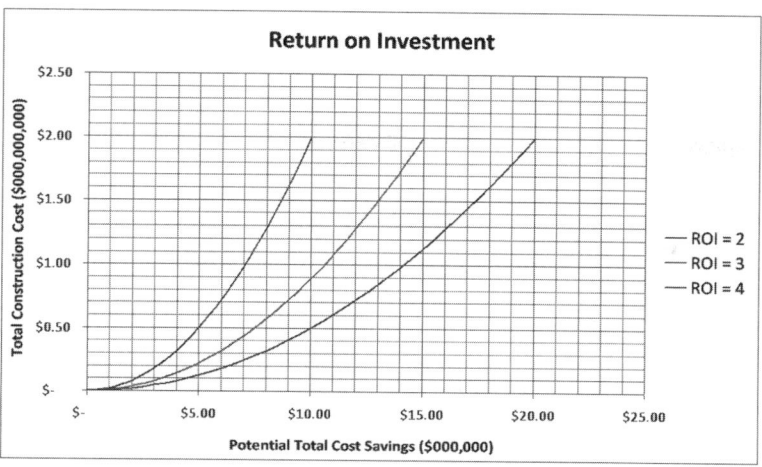

Figure 2. ROI diagram

ROI Assumptions:
- Example: Average size Public-Private Partnership (PPP)/Design-Build-Operate (DBO) Infrastructure Project in Australia is approximately AUS$2 billion (Total Construction Cost)
- Initial VDC installation cost is approximately 0.25% of Total Construction Cost (approximately $5 million)
- Potential total cost savings could be in the range of approximately $15 million
- Return on Investment (ROI) could be 2 to 4 times of the initial investment

VDC is extremely helpful on larger DBO projects where the awarded design-contractor is responsible to furnish a concept design (approx. 30% design level or re-confirmation of the tender/bid design) and where the M+E design-contractor firm (subcontractor) is responsible to develop the detailed design to a 100% level. Checking of the original design intent, fit-for-purpose and compatibility checks, technical review including space proofing and fit-out planning will be much easier with such a model established.

LIFE CYCLE INFORMATION AND POST-CONSTRUCTION USE

By using VDC, the integration of all aspects of the design enables the designed performance of the project to be tested in relation to its environmental impact and energy requirements before construction. Therefore the life-cycle running costs can be determined and optimised which can be compared and assessed according to the environmental and performance design criteria.

After construction owners, O+M teams and operators can benefit from the virtual model. The model can be used as a tool by owners, O+M teams and operators to aid in the allocation of resources such as maintenance, security, operations and emergency personnel during the concession period or after project completion. The model can be shared easily and also reused, overall creating faster and more effective processes. At project completion, the as-built model shall be part of the hand-over documentation (Contractor to O+M team, operator or facility manager). The model can support

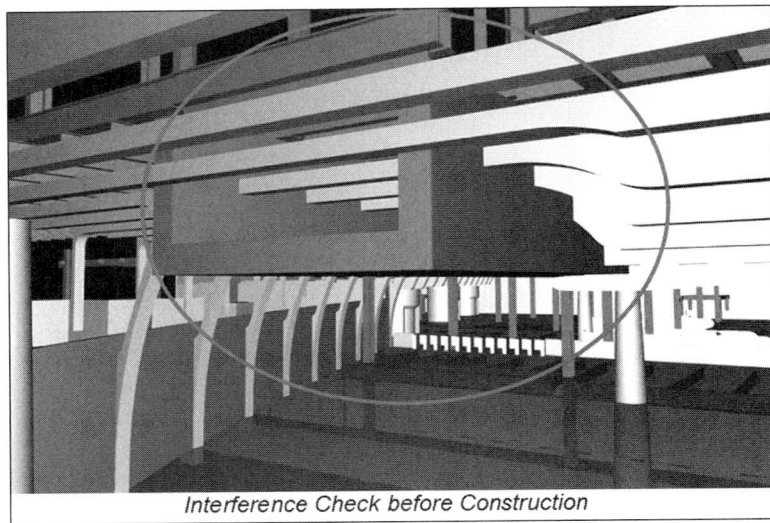

Interference Check before Construction

Figure 3. 4th Dimension decision support system

day-to-day routine maintenance and operation activities such as assessing safe access and egress into/from underground stations and can provide valuable information to the local fire and rescue departments in case of emergency operations.

Some public agencies are pushing the acceptance of 3D models as a digital deliverable. Many industries have successfully used 3D models as digital deliverables for decades. Only recently, 3D models are becoming acceptable as digital deliverables in the AEC market. As engineering designers move toward a total 3D/4D design scenario, it will become more cost efficient to the facility owner to reuse the 3D/4D models for construction management and eventually for long term operations and maintenance.

Infrastructures around the world have a huge stock of as-built records. These "as-builts" are mostly in 2D format. Several infrastructure owners have transformed their 2D hardcopy maps into 2D digital files. However, they may not reflect the situation of the facility "as-maintained" unless the agency has rigorously updated the 2D digital files to reflect maintenance updates. With the advent of 3D laser scanning, many agencies are capturing their existing infrastructure in accurate 3D point cloud data to augment or supplement their current as-built records (Arulraj, 2007). The December 1, 2008 issue of Engineering News-Record published by McGraw-Hill covers "Brave New Worlds Building teams on the brink of virtual design and construction." The jump from 2D to 3D is imminent.

Project requirements, design, construction and operational information can be utilised in facility management, providing an accurate understanding of the life-cycle and environmental information.

PRODUCT AND PROCESS MODELLING—4D TECHNOLOGY

In normal parlance the fourth dimension connotes temporality. That includes the three dimensions of space (x, y, and z) and one dimension of time. Such a space is called the Minkowski space and is the space used in Einstein's theories of special relativity and general relativity. Moving beyond scientific concepts into the ground realities of a construction project, the fourth dimension (time) plays a vital role in the development of an object in 3 dimensions. Most civil infrastructure development projects

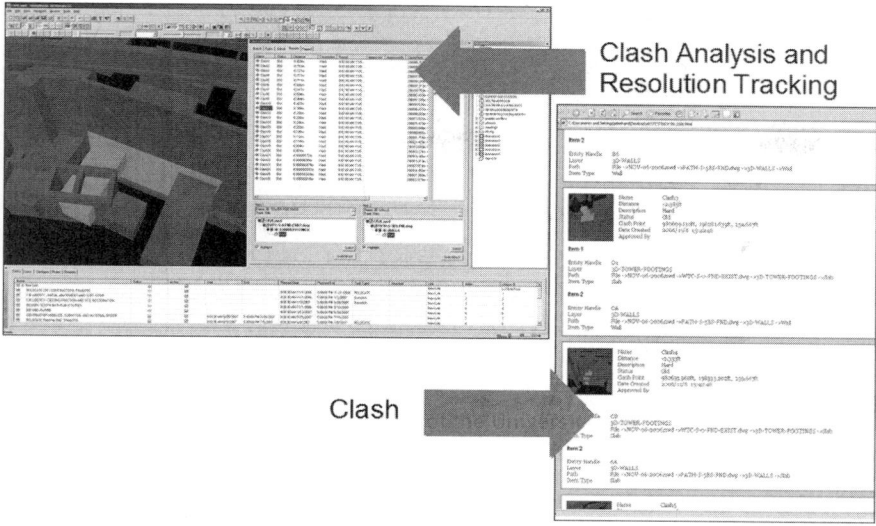

Figure 4. Automatic clash report generation from the 4D model

include several players from various disciplines, multiple stakeholders and sometimes a conglomeration of owners. All these players are hoping to construct a piece of infrastructure at some point during the construction phase of the project. The owner of the facility or his representative—the program manager or the construction manager, is tasked with the responsibility to oversee the construction in such a manner that the multiple players in the space-time continuum do not clash with one another either in space or in time or in both space and time. VDC based workflows avoid the occurrence of such challenges well ahead of ground breaking making it possible to find and fix costly errors in space and time before it is too expensive to resolve. Below is an example of a potential error in the fourth dimension—when the permanent structure (light grey) will be erected before a temporary structure (dark grey) is removed. Since the issue was seen in the model before the two contractors discovered the challenge on site, an expensive change order was avoided for the owner.

For centuries designers, engineers and construction contractors have dealt with these kinds of situations without dependence on technological solutions. They had read and understood construction sequence laid in more recent decades by scheduling software like Primavera or Microsoft Project. What is now possible with 4D software such as Autodesk's NavisWorks, Intergraphs's SmartPlant Review, etc., is that others can see the situation and make timely decisions without having to rely on the interpretation of piles of 2D CAD drawings and scheduling documents. This has made it possible to be more collaborative in decision making and realistic in construction sequencing. The technological framework provided by software of recent years has made it possible to share the complexity of construction challenges with people who might not have the background to envision the situation.

The term fourth-dimensional (4D) modelling describes the three-dimensional CAD models that can show changes over time. The 4D (3D + time) computer models are built to represent the project and its individual components in a coordinated and consolidated virtual project model. The 'time' component represents sophisticated software used for complex construction scheduling. The '3D' component represents the design geometric information. Figure 5 illustrates that when 3D and 4D modeling methods are linked together, the construction schedule is animated with objects appearing as they will during construction.

Since the tools are integrated, a change in one will trigger a change in the other. This is a useful feature of the 4D environment because a change in construction sequence can be readily visualised.

The 4D model gives explicit explanations of construction operations. The 4D technology can be used for clash detection by comparing 3D files from contractors, architects, engineers, designers or other services. Geometric clashes will be exposed and reported. The process is shown in Figure 5. An example of a clash between two files being compared and resultant reports is shown in Figure 4.

VISUALISATION FACILITATES COLLABORATION AND CONSULTATION

VDC provides images and information that improve communication and coordination during design or construction. All participants see or experience the project in a highly visual, consistent and interactive manner, and individual teams can drill deeper into the modelling database to evaluate specific project elements.

The AEC industry can benefit from the provision of visualisation techniques by enabling the evaluation and presentation of design options, acceleration of decision-making and approvals/permitting, and increase public outreach and build stakeholder consensus.

When a project has been designed in a 3D environment the geometry information is readily available to be used within sophisticated visualisation technologies. This is illustrated in Figure 5. The 3D visualisation models would emulate the same level of accuracy as the 3D design environment. A typical 3D visualisation model could be built on a platform that allows the export of visualisation products such as walk-, drive- and fly-throughs as well as stills.

Recent trends in 3D visualisation technologies indicate that there is a shift towards real-time game engines. The advantage of these environments is that the visualisation models are fully interactive in real-time where a design can be "experienced" as though one were immersed within the proposed infrastructure. These models can export the same animations mentioned above but with the difference that they are more photo-realistic. In addition, these technologies can export sections of the model as self-executing files which embed the same real-time read-only functionality that is within the source model. These self-executable files are useful for consultation where community stake holders can 'explore' the proposed new infrastructure from their own particular vantage points. This is achieved without requiring access to the full real-time visualisation model which is dependent on a high-end bulky computer.

The products available from a visualisation model would include:

- 3D Real-time executable files;
- Photo-realistic 3D and 4D visualisation animation; and
- Photo-realistic 3D and 4D visualisation stills.

These are briefly defined below.

3D Real-Time Executable File

The 3D real-time visualisation outputs include a range of deliverables based on the technology being used. Two particularly useful ones are firstly, 3D real-time solutions and secondly, 3D real-time high-definition solutions.

3D Real-Time Solutions. Interactive technologies create detailed 3D environments that run in real-time. 3D models form the basis of an immersive environment, with complex interactions, analysis and other information added to create a comprehensive

AN INTRODUCTION TO VIRTUAL DESIGN AND CONSTRUCTION

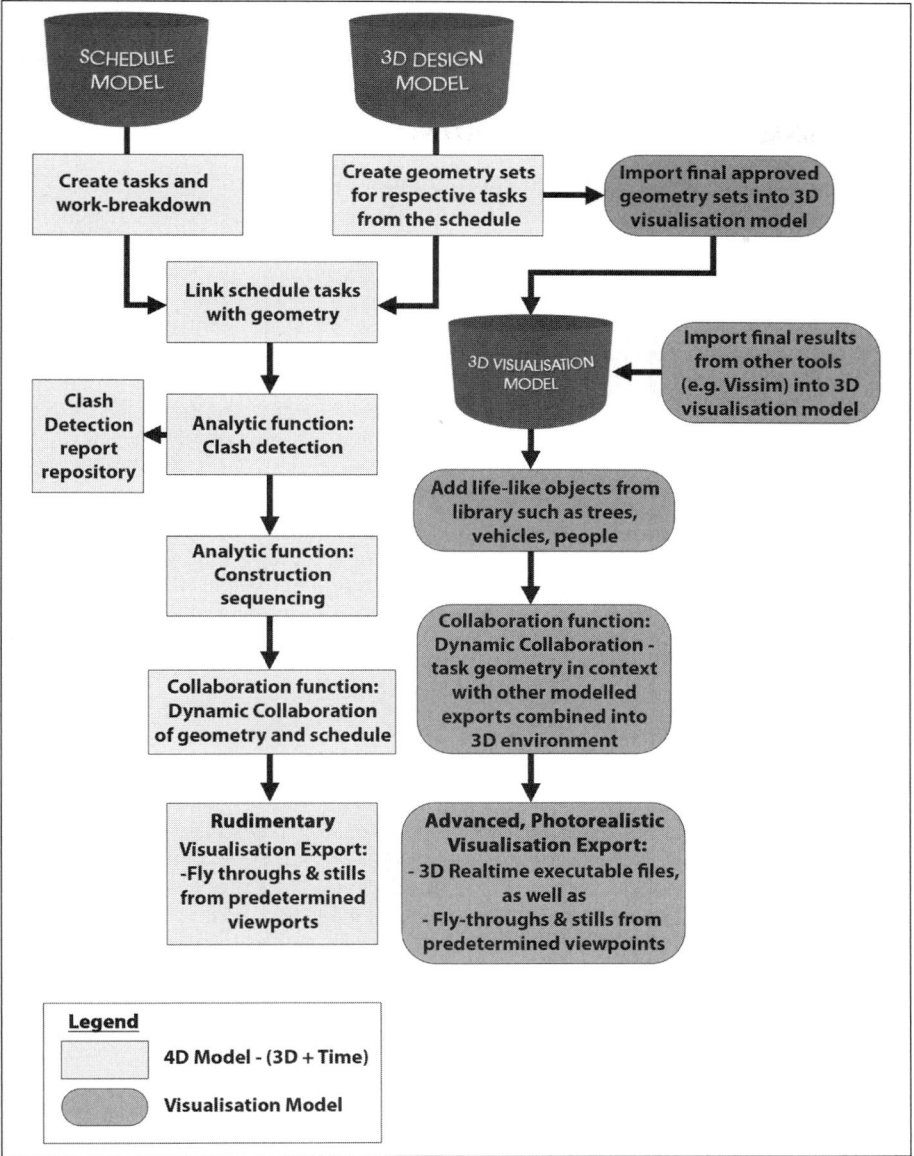

Figure 5. Workflow in a typical 4D model showing links to a visualisation model

experience. The user is able to navigate around the scene with freedom. Using custom user interfaces allows for a great deal of interactivity and usability without any training on the user's part. This simplicity and accessibility allows anybody to experience the simulation by moving the cursor within the 3D environment. They can even be distributed via the web using free and readily available software.

3D Real-Time High-Definition Solutions. 3D high-definition real-time technologies allow the user to experience an environment in the most realistic way technology can provide in real time. The level of interactivity can range from having the user interact with some part of the environment to offering complete freedom of movement and actions. Typically this technology is use in training simulation settings. 3D real-time models include simulated environmental factors such as shadows, daytime, night time as well as the inclusion of scale correct objects to convey greater realise.

Photo-Realistic 3D and 4D Visualisation Animation

3D animation is a tool used for visualising a future (or existing) condition in motion. Animations can provide controlled walking, driving or flying tours, incorporating pre-determined information. These tours are created in a computer-modelled environment, giving the viewer a set-path virtual tour.

4D animation links 3D models in a set sequence which can convey complex staging in an understandable manner. Depending on the target audience, 4D animations can be coarse-level steps or down to detailed steps. The animations are well suited to revealing project risks or opportunities to promote the project design or construction methods.

Photo-Realistic 3D and 4D Visualisation Stills

3D visualisation is a rendered computer model that uses the CAD design geometry to show what the project would look like once constructed. Within the model the camera is moved to selected points and a still snap-shot is made of the line of sight. The camera can be moved to multiple vantage points resulting in many rendered static views and the result is often integrated with actual photography to see the design in context. 4D visualisation stills are aimed at showing the planned construction sequencing. The series of stills display the constructability staging.

SUMMARY AND CONCLUSION

Virtual reality is not reserved for science fiction alone. VDC is virtual reality in action or in other words it is "Real Virtuality"!

Major infrastructure projects around the world are now used to the benefits offered by computing technologies to make VDC based digital project delivery a possibility. We can view, analyse, walk through and experience the proposed infrastructure before actual construction. If the public has to choose from a set of alternatives, it is possible to provide 3D/4D models to them over the web or at an information centre—so they can make a visually informed decision. These models being used in collaboration would be photo realistic 3D visualisation models which are more cost effective to produce because they use the same geometric information as being used in the product and process 4D models. The owners are in a better position to manage risk and revenue. The designers, engineers, architects and construction contractors are in a better position to explain their occupational complexities to their clients. Thus the VDC approach to building and infrastructure construction provides a win-win situation for all involved.

REFERENCES

Arulraj, R. 2008. Virtual Design and Construction: An Introduction, PB Network #67, Volume XXIII, Number 1 (Page 93) http://www.pbworld.com/news_events/publications/network/issue_67/pdf/67_37_Arulraj_VirtualDesignAndConstruction.pdf. Accessed January 2009.

Arulraj, R. 2007. Emerging Geospatial Technologies for 3D Visualization, Proceedings of the Geospatial Information Technology Association's Annual Conference 2007, San Antonio, Texas, March, 2007.

Arulraj, R. 2007. Emerging Geospatial Technologies: Focus on 3D Laser Scanning, PB Network #66, Volume XXII, Number 2 (Page 43) http://www.pbworld.com/news%5Fevents/publications/network/issue_66/pdf/66_22_Arulraj_EmergingGeospatial.pdf. Accessed January 2009.

Fischer, M. 2006. Leaving Today's Future of Building Behind (Page 4). Presented at the Clients Driving Innovation: Moving Ideas into Practice Conference (12–14 March 2006) for the Cooperative Research Centre (CRC) for Construction Innovation.

Fischer, M. and Hartmann, T. 2007. Introduction to Virtual Design and Construction and 3D/4D Modeling. (Page 6) Presentation for PB Seattle.

Gilligan. B., Kunz, J. 2008. VDC Use in 2007: Significant Value, Dramatic Growth, and Apparent Business Opportunity. Stanford/CIFE Online Publications (2008).

Jongeling, R., Olofsson, T., and Norberg, H. 2007. The Virtual Design and Construction Process. (Page 23) http://construction.project.ltu.se/main.php/ VC%202%20VirtualConstructionProcess.pdf?fileitem=7569447. Accessed November 2008.

Khanzode, A., Fischer,M., Reed, D. and Ballard, G. 2006. A Guide to Applying the Principles of Virtual Design & Construction (VDC) to the Lean Project Delivery Process. (Page 8) Center for Integrated Facility Engineering (CIFE) Working Paper #093. Stanford, CA: Stanford University.

Kunz, J. and Fischer, M. 2008. Virtual Design and Construction: Themes, Case Studies and Implementation Suggestions. (Page 2) Center for Integrated Facility Engineering (CIFE) Working Paper #097. Stanford, CA: Stanford University.

Lichtig W.A. 2005. Ten Key Decisions to A Successful Construction Project—Choosing Something New: The Integrated Agreement for Lean Project Delivery. American Bar Association—Forum on the Construction Industry.

Lingerfelt, M. 2005. Avoiding Mistakes, Saving Money: The Case for Virtual Design and Construction. AIA—Corporate Architects eNews (2005).

Sanvido, V.Konchar, M. 1999. Selecting Project Delivery Systems—Comparing Design-Build, Design-Bid-Build and Construction Management at Risk. Project Delivery Institute (pdi) 1999.

Reseigh, C and Arulraj, R.; 2008. Construction Management in the Digital Age, Presentation at the 3D Infrastructure Symposium, New York. December 2008.

American Society of Civil Engineers (ASCE)—Website Database—Data Collection. www.asce.org. Accessed December 2008.

Construction Industry Institute—Website Database—Data Collection. www.construction-institute.org. Accessed December 2008.

Design-Build Institute of America (DBIA)—Website Database—Data Collection. www.dbia.org. Accessed December 2008.

Engineers Australia—Website Database—Data Collection. www.engineersaustralia.org.au. Accessed December 2008.

U.S. Department of Labour, Bureau of Labour Statistics—Website Reports—Data Collection. www.bls.gov. Accessed December 2008.

PLACEMENT OF CONCRETE LINING FOR WATER TUNNEL NO. 3, MANHATTAN PORTION

Robert Labbe Jr. ■ Schiavone Construction Company

Michael Gorski ■ Schiavone Construction Company

ABSTRACT

The final tunnel lining for this project involved pumping vast quantities of concrete over large distances under the extremely busy streets of lower Manhattan. Approximately 68,000 cubic meters (89,000 cubic yards) were pumped from various locations over 152 meters (500 feet) below ground (Figure 1) to create nearly 14.5 kilometers (9 miles) of 3.05 meter (10 foot) diameter tunnel. Total distances pumped exceeded 1,645 meters (5,400 feet) horizontally after a vertical difference of over 152 meters (500 feet). The urban setting of the operation, coupled with the long distances over which concrete was pumped, created many problems which had to be dealt with along the way. A three-shift labor operation was implemented to allow sequential pouring of tunnel sections nearly every day. The fine tuning of the mix design was critical to make this operation successful.

INTRODUCTION

Introduction to Water Tunnel No. 3, Stage 2

When placed into service, Stage 2 of City Water Tunnel No. 3 will be an integral part of the growth and prosperity of New York City for generations to come. The ambitious nature of creating over 14.5 kilometers (9 miles) of tunnel over 152 meters (500 feet) below ground was made more impressive by the intensely urban setting in which the work was to be constructed. The Borough of Manhattan presented Schiavone/ Shea/ Frontier-Kemper with truly unique and challenging engineering problems, which could only be overcome with creative solutions.

Stage 2 is a continuation of the ongoing effort by New York City to meet an ever-increasing demand for fresh, clean drinking water. In the city that never sleeps, it is imperative that the utilities are capable of keeping up with the fast pace of city life. The New York City Department of Environmental Protection (NYCDEP) has been working for more than a century to ensure that this is the case—not only now, but in the future as well.

New York City Department Environmental Protection Water Supply Background

The History of the New York City Water Supply can be traced back as far as 1677, when the first public well was dug in front of the old fort at Bowling Green. By the late 1700s, a reservoir system was constructed, using a series of hollow logs to distribute the water. More wells were sunk in order to keep up with the demand and to keep the reservoirs full. Eventually, the wooden mains were replaced with cast iron pipes, but as time passed, this system proved to be inadequate and the wells became polluted. The completion of City Tunnel No. 3 will provide additional capacity to all areas of the

Figure 1. Starter tunnel (dill and blast) area at bottom of shaft 26B looking south

city and allow the dewatering and inspection of tunnels that have been in continuous operation for many decades.

New York City saw a need to increase its water supply, so they looked to the Croton River. In 1842, the Old Croton Aqueduct was placed into service. This, along with the addition of more reservoirs, allowed for the supply of 90 million US gallons per day. By the late 1800s, the New Croton aqueduct was placed into service to provide additional capacity from new reservoirs in the Croton Watershed. With a population still growing, the New York City Board of Water Supply was created in 1905, to find new sources of water and finance the construction of reservoirs and aqueducts to convey water from the new upstate sources. The Board of Water Supply identified the Catskill Mountains as another source. This led to the construction of the Catskill Aqueduct system (circa 1915), which used City Tunnel No. 1 to deliver water to Manhattan. In the 1930s, the Delaware Aqueduct system was built along with City Tunnel No. 2 to increase the water supply to New York City.

Background of Project

By the late 1960s, anticipated increases in demand for water required construction of a new water tunnel to support the demands of an expanding New York. City Tunnel No. 3 (Figure 2) is the largest capitol construction project in New York City's history and one of the world's engineering marvels. Construction began in 1970 and is projected to be completed by 2020. The tunnel is being built in four stages and, when completed, will total more than 96.5 kilometers (60 miles) in length, with costs exceeding $5 billion.

The first Stage of City Tunnel No. 3 originates in the Hillview Reservoir in Yonkers and continues on to Queens. It was constructed by the drill and shoot method, which is the same method used to construct the Croton, Catskill and Delaware Aqueducts. Advances in technology led to Stage 2 being constructed utilizing a Tunnel Boring Machine (TBM), which greatly increased the speed of the excavation and greatly improved safety. The first part of stage 2 is the Brooklyn/Queens Tunnel, which was completed in 2001 and is projected to be placed in service by 2009. The second part of Stage 2 is the Manhattan Legs, which total 14.5 kilometers (9 miles) running from Central Park down to lower Manhattan and across to the East Side. It is projected that this section will be placed in service by the 2012. Stage 3, currently in the final stages of planning, will be a 25.75 kilometer (16 mile) run from the Kensico Reservoir to the Valve Chamber in the Bronx. Finally, Stage 4 will service the eastern parts of the Bronx and

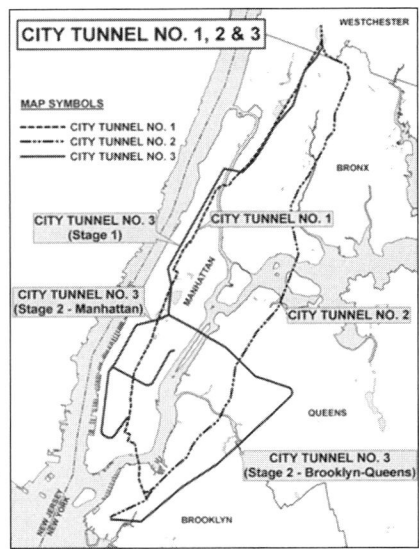

Figure 2. Map of City Tunnels No. 1, 2 & 3 within the limits of New York City

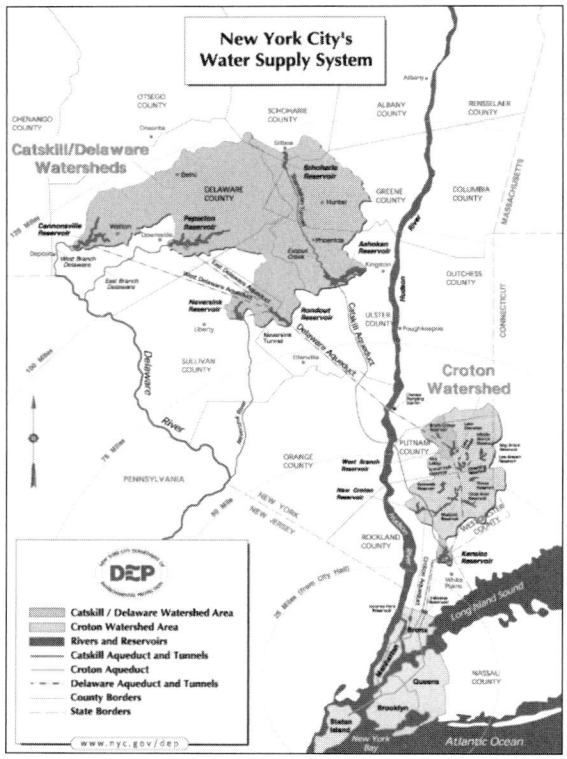

Figure 3. Map of the New York City water supply system

Queens with a 22.5 kilometer (14 mile) long tunnel traversing from the Valve Chamber in the Bronx and into Queens by going under the East River.

The lowest point of City Tunnel No. 3 is roughly 244 meters (800 feet) below ground and the highest point is about 122 meters (400 feet) below ground. The gradual slopes of the tunnel utilize the differential elevations of the reservoirs and shafts to create a gravity delivery system to deliver the water to NYC. The system is so efficient that water can be delivered to the sixth floor of most buildings from the reservoirs 100 miles upstate without pumping. The finished diameters of Tunnel No. 3 will range from 3.05 meters (10 feet) to 7.32 meters (24 feet). The eventual operation of Tunnel No. 3 will allow inspection and repair of Tunnels No. 1 and No. 2 for the first time since they were put in service in 1917 and 1936, respectively.

Thanks to this colossal municipal undertaking, New York City has the largest unfiltered surface water supply in the world. Every day, some 1.3 billion US gallons of water from this vast system are delivered to more than eight million people. The New York City Water Supply System originates from a watershed of roughly 3,160,000 hectares (2,000 square miles) across eight counties both north and west of the City and ends with the tap water that is so easily accessible throughout the Five Boroughs and nearby counties (Figure 3). With the future completion of the remaining planned stages, this supply will only grow.

1st Contract for Water Tunnel No. 3, Manhattan Portion

The joint venture of Schiavone/ Frontier-Kemper /Shea started on the first of the projects on the Manhattan portion in January 2002. This project consisted of: uncovering and dewatering shaft 26B, which had previously been sunk by Frontier-Kemper under a previous contract, to a depth of 176 meters (580 ft), the North/ South starter chamber and the mining of the South tunnel, which was approximately 5,500 lineal meters (18,000 lineal feet) of 3.84 meter (12.6 foot) diameter. The mining operation was performed utilizing a Robbins model 1215-257 Tunnel Boring Machine. In addition to the mining, the drilling and blasting of the East starter tunnel was performed in preparation for the setting up of the tunnel boring machine for the second contract.

2nd Contract for Water Tunnel No. 3, Manhattan Portion

The joint venture of Schiavone/ Shea /Frontier-Kemper started on the second project, which consisted of mining the North and East tunnels, concrete lining the South, North and East tunnels, sinking nine distribution shafts up to 177 meters (580 feet) deep (Figure 4), and the connecting of the shafts to the tunnels with adits up to 53.34 meters (175 feet) long, support of excavation for construction of the distribution chambers included Raito walls, soldier beam & lagging, ground freezing to depths up to 33.53 meters (110 feet) deep and the installation of two 1.22 meter (four foot) diameter stainless steel riser pipes to bring water into the distribution chambers. The restoration, returning the sites to their preconstruction condition, of these shaft sites was also a part of the contract. After a separate future contract is performed to install the valves and other permanent equipment in the shafts, these shaft sites are planned to be parks for their communities.

The project mainly consisted of construction of the three tunnel sections, the South 5.5 km (18,000 lf), North 4.3 km (14,000 lf) and East 4.6 km (15,000 lf) legs, mined at a 3.84 meter (12.6 foot) diameter and lined to a finish of 3.05 meter (10 foot) diameter. Water Tunnel No. 3 is a gravity feed system, with Shaft 26B designed to be the low point (drainage shaft) if ever required in the future. So the South, North and East tunnels are pitched back to shaft 26B at 0.5%, 1.4% and 1.0% respectively.

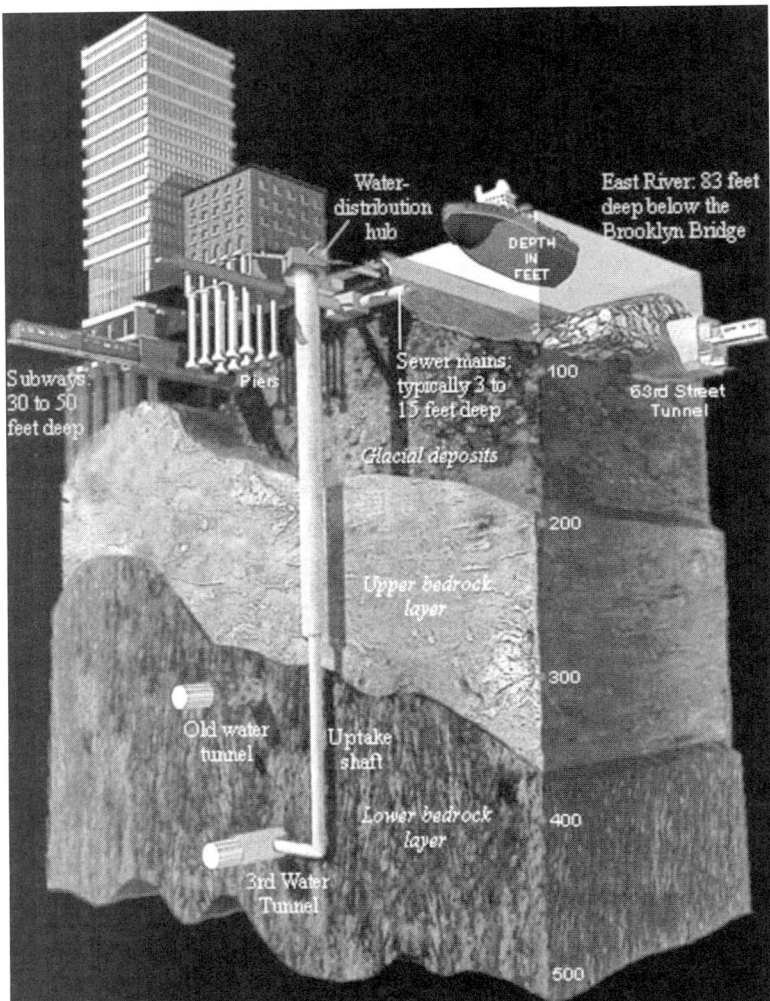

Figure 4. Illustration of the Water Tunnel No. 3 under New York City

This project also included the construction of ten shafts around Manhattan (Figure 5), nine of which were required to be sunk under this contract. While the South tunnel consists of 5 shafts, the North and East tunnels consists of only 2 shafts each. The central shaft, which was previously sunk and lined by another contractor, was the access point to all 3 legs and was utilized for personnel, equipment and material access for the majority of the tunneling work for both contracts.

The shaft construction itself required a significant scheduling, coordination and public relations effort. The shafts were located in many affluent areas of Manhattan, where some of the most expensive real estate in the world can be found. Special precautions were taken to prevent damage and complaints from the neighbors in the adjacent residential, office, religious and educational structures.

CONCRETE LINING FOR WATER TUNNEL NO. 3, MANHATTAN PORTION 401

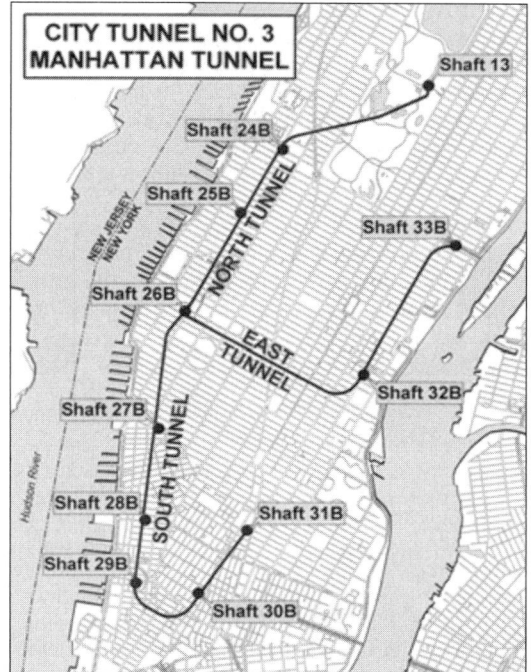

Figure 5. Map of the Manhattan portion of City Tunnel No. 3

SCOPE OF THE CONCRETE LINING OPERATION

The concrete lining operation required the placement of over 70,000 cubic meters (92,000 cubic yards) of 4,500 psi concrete for the 3.05 (10 foot) and 6.10 meter (20 foot) finished tunnel diameters, as well as, two transition sections for 3.05 meter (10 foot) to 6.10 meter (20 foot) and tying into ten shafts. Over 14.5 km (47,000 lf) of 3.05 meter (10 foot) diameter tunnel had to be lined, which was over 68,000 cubic meters (89,000 cubic yards). There were 122 lineal meters (400 lineal feet) of 6.1 meter (20 foot) diameter tunnel that connected the North and South tunnels, which consisted of approximately 2,700 cubic meters (3,500 cubic yard) of concrete. The two transition sections telescoped the tunnel from a 3.05 meter (10 foot) to a 6.1 meter (20 foot) diameter tunnel over a fifty foot span.

This would be the first time that concrete was placed to these depths and distances by pumping from street level in New York City. In the beginning, the concrete operation had many issues with the segregation of the mix, but after many adjustments (innovative ideas, constant attention to all of the details, adjusting the amounts of fine aggregate, course aggregate and cement, constantly monitoring the amounts of super plasticizer and a hydration controlling admixture) the concrete pumping proved to be more efficient than the traditional method of dropping concrete and using agitator cars to transport the concrete. Basically, the concrete lining operation was performed by pumping the concrete through the slickline from street level down a 0.36 meter (14 inch) diameter utility hole and then horizontally to the placement location.

Once the concrete was placed into the pump's hopper, it was then pumped through a short section of slickline followed by a 90 degree elbow connected to a vertical ("drop") pipe. At the bottom of the drop pipe, there was another elbow and a guillotine which minimized the risk of concrete sliding down the vertical section when the concrete pump was stopped periodically. After the guillotine, the pipe would typically have a few sweeps to position the pipe to the center of the tunnel or along the tunnel wall. The slickline was on the invert and/or the walls until it reached the concrete forms. Once the slickline reached the formwork, it would have a series of sweeps until the pipe attached to the inside (walking area) of the concrete forms and then to the placer car. The placer car allowed the concrete to enter the formwork through a port in the arch of the formwork.

This operation was performed utilizing three labor shifts. The goal was to pump concrete during day shift and have the other two shifts set forms and prepare for the next days concrete. This operation was safer and much more efficient than the previous method using agitator cars to transport the concrete, since the issue of agitator car derailments was eliminated.

PROGRESSION OF THE WORK

During a typical day, concrete was pumped almost 180 meters (600 feet) vertically down to tunnel and approximately 730 meters (2,400 feet) horizontally to the placement location. The typical total volume of concrete, while pumping, in the slickline at any time was approximately 11.5 cubic meters (15 cubic yards) of concrete. A typical concrete placement consisted of approximately 275 cubic meters (360 cubic yards).

South Tunnel

The concrete lining operation began in the South tunnel, which is over 5.5 kilometers (18,000 feet) long. This portion of the tunnel, when completed, will deliver water to the neighborhoods of The East Village, Chinatown, Little Italy, SoHo (South of Houston Street), TriBeCa (Triangle beneath Canal Street) and Greenwich Village.

During the concrete lining placement for the South tunnel, sequencing of the concrete operations at the south shafts was as follows:

- 31B—East Village
- 30B—Little Italy
- 29B—TriBeCa
- 28B—SoHo
- 27B—Greenwich Village
- 26B—West side, the main access shaft for Construction Activities

North Tunnel

The second tunnel on the project to be concrete lined was the North tunnel. The North tunnel is over 4.3 kilometers (14,000 feet) long going North from Shaft 26B to Shaft 13B in Central Park. This portion, when completed, will mainly deliver water to the Westside and Hell's Kitchen neighborhoods. The concrete lining operation started at the northern most part of the North tunnel and worked south to Shaft 26B.

During the placement of the concrete lining for the North tunnel, sequencing of the concrete operation at the north shafts was as follows:

- Utility Hole—midway between shaft 13B and 24B
- 24B—Midtown West adjacent to Fordham University
- 25B—Main Access Shaft in Hell's Kitchen for Post Construction Activities— Future Tunnel Inspection and Access
- 26B—West side, the main access shaft for Construction Activities

Figure 6. East starter tunnel with transformer pocket

East Tunnel

The final tunnel to be concrete lined was the East tunnel. The East tunnel is over 4.6 kilometers (15,000 feet) long and runs east from Shaft 26B to Second Avenue and then north the Queensboro (59th Street) Bridge. This portion, when completed, will mainly deliver water to many locations on the Eastside.

During the placement of the concrete lining for the East tunnel, sequencing of the concrete operation at the east shafts was as follows:

- 33B—Queensboro Bridge
  - Corner of East 59th Street and 1st Avenue
- 32B—Entrance to Midtown Tunnel
  - East 35th Street Between Tunnel Approach and 2nd Avenue
- Utility Hole
- 26B—Main Access Shaft for Construction Activities
  - West 30th Entrance to the Lincoln Tunnel
  - Corner of West 30th Street and 10th Avenue

Bottom of Shaft 26B

The concrete at the bottom of Shaft 26B was the last to be placed, to maintain all of the available areas to stage materials and equipment for as long as possible. The areas that were left until the end of the project were 55 lineal meters (180 lineal feet) of 3.05 meter (10 foot) diameter tunnel in the South tunnel, fifty lineal feet 3.05 meter (10 foot) to a 6.1 meter (20 foot) diameter transition in the North and South tunnels and 61 lineal meters (200 lineal feet) of 6.1 meter (20 foot) diameter tunnel in the North and South tunnels. These last placements took the longest in terms of time resulting from the number of different size forms required to build the project as designed. The tunnel at this location was excavated by means of drill and shoot method to allow for the installation of the TBM, therefore the volumes were larger than the TBM areas-and setting the forms took longer due to the irregularities in the rock surface (Figure 6). The quantities for these concrete placements were calculated using surveyed cross sections of the rock surface.

DIFFICULTIES ENCOUNTERED

In the business of dangerous and often unpredictable underground construction, the work requires extensive amounts of planning to maximize efficiency and minimize risk. Just think of some of the potential roadblocks that were faced in Manhattan. Some of the difficulties encountered were the mix design, sensitivity of the required low heat of hydration (type II modified cement), Canadian cement deliveries, batch plants outside Manhattan, wet and dry batch plants, intense traffic on the extremely busy city streets of Manhattan, the inability to stage trucks outside of the jobsites, shaft site locations, small jobsites ranging from 604 square meters (6,500 square feet) to 2,500 square meters (27,000 square feet), sharing the jobsites with other crews and/or with subcontractor's crew, community oversight of the activities at these shaft sites, the presence of fault zones, reinforcement required at various critical locations, periodic plugging of the slickline, pumping concrete great (long horizontal) distances, continuous three shift operation, breakdowns in communication, the room in the tunnel to fit all of the equipment required, weather (hot, cold, stormy, etc.) and potential equipment breakdowns.

Mix Design

The specifications called for a 4,500 psi design mix to include 38 millimeters (1 ½ inch) stone as the course aggregate. This size aggregate, if in the concrete, would not be–able to be pumped through a 125 millimeters (5 inches) diameter slickline. After many trial batches were performed the desired strengths and shrinkages were achieved and the NYCDEP allowed the use of a design mix utilizing 19 millimeters (¾ inches) aggregate to be used for this project. Throughout the concrete operation, mix designs were modified to get the best quality and performance.

Sensitivity of the Cement

One of the biggest obstacles faced was the required use, by the NYCDEP; of type II modified cement in the approved mix. This cement was specified because of its low heat of hydration, which theoretically would allow for a slower cure and less cracks than regular type II cement. The reduction of cracks would result in less water leakage from the tunnel in the future once it is in service. This cement was found to be extremely sensitive and would react differently from day to day when the same amounts of chemicals and water were added to the mix. The level of admixtures would have to be properly balanced to ensure that the concrete would be able to be pumped (slump and workability) completely through the slickline to the placement location without causing any quality issues.

With no history within the city or the joint venture of successfully pumping this type of cement at these depths and distances, the concrete pumping operation was on a learning curve of its own. The chemical composition of the cement was determined to be extremely sensitive requiring constant monitoring of the cement's reaction to other variables in the "mix" (amount of water, amount of super plasticizer, amount of a hydration controlling admixture, temperature of the water, air temperature, etc.). Monitoring of the switching of "burns of cement" and the random testing of loads at the concrete plant had to be performed, so it could be determined whether the use of more or less super plasticizer or a hydration controlling admixture was required to be able to successfully pump the concrete.

Canadian Cement Deliveries

The cement was imported from Canada with an average consumption rate of about six truck loads per day. The concrete supplier's limited storage space required immediate notification to the concrete plant of any issues that would affect their delivery schedule. Cement deliveries to the concrete plant needed to be coordinated with a two workday notice in order to assure that there was a sufficient supply of this low heat of hydration cement.

Concrete Batch Plants

A major obstacle in placing concrete in Manhattan is getting the concrete to the jobsite; constant communication was required between site supervision and the dispatcher. There are no concrete plants in Manhattan and site batching is not a viable option due to environmental (pollution, vibration, noise and dust) restrictions in the congested neighborhoods of Manhattan. The concrete for this project came from one of the major suppliers in New York City, which had four concrete batch plants within the city limits. There were six 91 metric ton (100 ton) capacity silos, eight 68 metric ton (75 ton) silos and one 113 metric ton (125 ton) capacity storage blimp totaling an available storage capacity of 1,200 metric tons (1,325 tons) of cement at the two wet batch concrete plants which were located in Queens. There were nine 68 metric ton (75 ton) capacity silos and two 113 metric ton (125 ton) capacity storage blimps totaling an available storage capacity of 840 metric tons (925 tons) of cement at the two dry plants which were located in Brooklyn. One silo at each of their four plants and all three 113 metric ton (125 ton) capacity storage blimps were reserved for the type II modified cement for this project.

The decision on which concrete batch plant was utilized for any particular concrete placement was determined based on several factors; location of the shaft site in relation to the batch plants, size of the concrete placement, the time of day of the placement, quantity of the required materials at the batch plant, the condition of the batch plants, breakdowns and maintenance issues of all equipment, which are more common during extreme temperatures and made more important when the schedule did not allow for weather related shutdowns, the quantity and location of other projects that our concrete supplier was servicing, the quantity and location of the other concrete placements for our project including subcontracted work, etc. The effect of other projects around the city also turned out to be a major issue. This work was performed at a time when the city of New York City like the rest of the country was experiencing a building boom.

Wet vs. Dry Batch Plants

Both wet (commonly referred to as central mix) and dry batch plants were utilized for this operation, so it was important for the field supervision to closely monitor the consistency and quality concrete in order to spot potential problems. Batching and material concerns were just other variables that needed to be identified as early as possible in order to avoid quality and productivity problems. During some concrete placements, both wet and dry plants were utilized in order to keep up with production.

The wet plant loads the raw materials into a central mix barrel which mixes the materials and loads them the same way into every truck making each load more consistent than loads from a dry batch plant. The loading process was done in less than three minutes per truck and provided a very consistent quality product. There was a designated individual responsible for batching all the concrete for the job each day. The central mix plant is better for production and quality, as it is better controlled.

A dry batch plant measures and loads the raw materials into the trucks in a similar fashion each time, but the mixing of the concrete is then left to the driver to handle. Each driver is then responsible for mixing the load to the proper slump required on the job site, which required all drivers to be specially trained due to the sensitivity of this mix design. The dry plant loads a truck in less than four minutes but the truck then needs to mix for an additional six minutes once it arrives to the jobsite.

Intense Traffic Issues

Even by New York City standards, many of these shaft sites were in high traffic areas, which made it very difficult to stage trucks when dealing with the occasional problems on the jobsite. Since rush hours last longer in Manhattan than many other cities and some of the shaft sites were located near the Lincoln Tunnel, Holland Tunnel, Midtown Tunnel and the Queensboro Bridge, during gridlock trucks could be only blocks away, but not arrive for another 15 minutes. An accident in, on or near the major bridges or tunnels would greatly effect the delivery time of the concrete, the trucks and/ or other vehicles would then be rerouted or having to just deal with the increased volume of vehicles on their regular routes. The travel time from the plant to the jobsite was inconsistent due to traffic caused from the extended morning and evening rush hours and accidents. Television and movie shoots around the city also affected travel time, as New York City prides itself on being a sought after location for the television/ movie industry. When the concrete placement took longer to perform, it would effect the next day's production since the other shifts would have less time to get their work done. At times, it took over 2 hours from batching for a truck to reach the jobsite, which also meant that any concrete still in the slickline was just sitting there getting older. The large quantities of super plasticizer and the hydration controlling admixture in the concrete prevented the concrete from setting up in the slickline.

Staging of Trucks

New York City regulations stated that concrete trucks were not allowed to line up outside of the jobsite idling or double parked, which was inconvenient at times. Even when there was physically room outside of our worksites to park concrete trucks, the neighboring business would have deliveries making it inconvenient for the operation to say the least.

Shaft Locations and Conditions

The locations of the shafts in Manhattan were in very busy and diverse residential and commercial neighborhoods. In all circumstances, there were many environmental (pollution, dust, noise and vibration) restrictions placed by the NYCDEP in addition to the difficulties that are inherent to work of this nature.

These small shaft sites range in size from 604 square meters (6,500 square feet) to 2,500 square meters (27,000 square feet) with the average being 1,700 square meters (18,000 square feet). These shaft sites were usually shared by a mechanical subcontractor in an effort to shave time off of the project schedule. This operation also had to work around many items (Figure 7 & 8); such as NYCDEP field offices, crew (operators, sandhogs and laborers) shanties required by our union labor agreements, shaft equipment (cranes, compressors, concrete formwork, etc.), Contractor/ Subcontractor field office, other crews performing operations which were occasionally placing concrete as well, materials (reinforcing steel, stainless steel pipes up to 16.75 meters (55 feet) long and ranging from 1.22 meters (4 feet) to 2.44 meters (8 feet) diameter, construction debris dumpsters, existing structures with contaminated materials to be demolished, etc.

Community Oversight

These small shaft sites needed to have enough room for the following tools and equipment required for the concrete operation: Putzmeister BSA 14000 diesel powered concrete pump, water pump, portable compressor, connex box with small tools and equipment, concrete testing station (including concrete cylinder curing boxes).

Figure 7. Shaft 33B (adjacent to Queensboro Bridge at the corner of East 59th St. and 1st Ave.) looking east at concrete pump while shaft is also being constructed

Figure 8. Shaft 33B (adjacent to Queensboro Bridge at the corner of East 59th St. and 1st Ave.) looking north at concrete pump while shaft is also being constructed

The shaft sites were also under the constant watch of many local community boards. The NYCDEP made many promises to the community, such as; monitoring and enforcing environmental restrictions (noise and dust), temporary jobsite shutdowns during moving in or out of Fordham University students, accelerating the return of certain shaft sites back to the community, etc. Handling the neighbors' complaints became a vital key to keeping peace in the neighborhoods.

Influences Immediately Outside of the Jobsite

Since the jobsites were nestled into neighborhoods that were already very congested and busy, the concrete operation was affected at times by the obstacles and events going on in the neighborhoods. Many of these events included the double parking of vehicles making it impossible for concrete trucks to pass on some of the narrow streets, local businesses receiving deliveries blocking traffic and/ or taking up room that was needed for waiting concrete trucks. Construction activity taking place at adjacent structures was commonplace and local college dormitories that had student move in and out days restricted what activity was allowed in certain areas per the contract specifications. Other unique problems arose when there were tourist amazed at the Manhattan sights and locals disregarding and/ or unaware of the moving concrete trucks, truck routes changing when there was a United Nations General assembly,

Figure 9. Steel reinforcement at the intersection of Shaft 25B and the South Tunnel

the President or other high ranking officials were in town or the immediate area and emergency utility repairs immediately in front of our site entrances. These issues were in addition to the normal conflicts and personality clash on a typical construction site.

Reinforcement Required at Various Locations

The tunnels did not have steel reinforcing except for the intersections (Figure 9) at the shaft adits and the two faults zones. When steel reinforcing was required, it slowed down the progression of the tunnel lining work. The reinforcing work slowed down production for the following reasons: (1) the three shifts were each used to doing everything but placing steel reinforcement, (2) steel reinforcement was needed to be transported into the tunnel pre-bent due to lack of room in the tunnel, (3) layout was required to get the proper spacing for the rebar, (4) supports and spacers were required due to the rock not having a mined finish, and (5) bars with various lengths and radius were required to be installed at a location with virtually no lay down area. The steel reinforcing required varied from a single layer of #6 bars all the way up to 2 mats of #11 reinforcing bar.

Fault Zones

Fault zones were encountered at two main locations. At these fault zones, the placement of steel reinforcement was required. These fault zone locations not only slowed our production down initially during the rock excavation, but then also slowed the concrete operation down.

Plugging Slickline

Each truckload of concrete was visually inspected prior to the pumping of that particular load of concrete. This was done to better assure that it would make it to its desired placement location without problems. The slickline became plugged at times due to air being introduced to the slick line and materials segregating in slick line. When this occurred it was very time and labor intensive as well as consuming process. When the slick line plugged, there was anywhere from 4 cubic meters (5 cubic yards) to 13 cubic meters (17 cubic yards) that needed to cleared from the slick line and eventually removed from the tunnel. Cleaning out the slickline was performed by placing a rubber cleanout "bullet" and pumping water behind the bullet. After the slickline was cleaned out, the concrete needed to be either scooped up with a skid steer when possible or placed in sand bags. The concrete was loaded up in muck trains and brought back to Shaft 26B to be removed from the tunnel.

Pumping at Great Distances and Depths

This was the first time in NYC history that concrete was pumped the distances and depths that took place during this project. Pumping concrete great depths, as we found out, requires an extremely focused team which paid attention to the details and rarely deviated from the developed system. When waiting for trucks and the concrete pump is shut down, there is a risk that the concrete can slide down the slickline enough to cause an air pocket which can result in the concrete being unable to be pumped through the slickline to the formwork.

The equipment used to successfully place the concrete tunnel lining included a Putzmeister BSA 14000 diesel concrete pump, hydraulic guillotine to reduce the likelihood of the creation of air pockets in the slickline, miles of slickline and a placer car to extend and retract the slickline to the necessary section of the concrete formwork, which changes throughout the pour.

The typical concrete placement had over a 150 meter (500 foot) vertical difference from street (concrete pump) level and a horizontal distance of over 450 lineal meters (1,500 lineal feet). This pumping equipment was used to successfully pump concrete up to 950 meters (3,100 lineal feet) horizontally after an approximately 150 meter (500 foot) foot difference in elevation.

Additional equipment was introduced to the operation for pumping distances from 900 (3,000) to 1,800 meters (6,000 feet) which included additional transformers, a rail mounted electric concrete pump and an electric jumbo trough for pump to a pump situations. The concrete pump at street level pumped the concrete to a remixer, which then utilized a Putzmeister 2107 electrical concrete pump to transfer the concrete to the concrete forms. The concrete had to be remixed and redosed with chemicals when required by pumping conditions. This method was utilized to pump concrete over one mile after it was pumped down a difference in elevation of almost 150 meters (500 feet).

Communication Challenges

With limited access to a tunnel that is almost 180 meters (580 feet) below street level, communication can also be a pretty big obstacle. Since there are so many variables and things can go wrong, proper communication between the topside crew and the tunnel crew is essential. Communication was especially important during the pump to a pump situations. These situations required constant communication in order to synchronize the pumps to successfully pump concrete at great distances.

Small Tunnel Challenges

This was only a 3.05 meter (10 foot) diameter finished tunnel with a concrete liner of approximately 0.38 meters (15 inches). The formwork would breakdown and travel inside of itself to get to the other end of the forms to be set up again. There was very little room to use as a lay down area. Smaller diameter tunnels have clearance issues magnified in comparison to the clearance issues present in bigger tunnels. For example, when larger tunnel forms collapse and travel inside each other there is a small amount of clearance, which could be up to 76 millimeters (3 inches). For smaller diameter forms the amount of theoretical clearance is then less which could be as little as to 25.4 millimeters (1 inch).

Weather

When is it a good time to order concrete? In the nicer weather, everybody is ordering concrete for their projects. During the hotter weather experienced in the summer months, ice and a hydration controlling admixture was a critical part of the success of this concrete pumping operation.

When the weather is cold (freezing), not too many people are ordering concrete, but the concrete trucks and plants are experiencing more breakdowns. During the colder weather, more equipment breakdowns and weather related issues occur than usual; such as formwork carrier, water lines, and fire hydrants.

Since the placement location of the concrete is underground, the stormy weather above ground does not directly effect how many concrete forms are set for the next day unless the concrete pour runs late or into the next shift due to weather related issues top side. In any event, the concrete operation was not stopped due to the weather conditions, as the weather related obstacles were just another part of the operation that needed to be handled. There were even times when the temperature was in the single digits for so long that many other contractors were canceling concrete orders and we became the only project for our concrete supplier.

Equipment Breakdowns

Another major obstacle was equipment breakdowns, which is certainly common in Tunnel Construction. The condition that the equipment is subjected to in a tunnel environment certainly puts additional or premature wear on the equipment. When equipment breaks down on a project above ground it is a major inconvenience, but when the broken equipment is 180 meters (580 feet) below ground and over 4.8 km (3 miles) in from the heading, it creates tremendous challenges. The mechanic at the heading has to communicate the details over the phone and hope to receive the proper parts to get the machine back up and running in order to minimize the lost time and keep production going. The equipment required to keep these operations running is very specialized and when parts break, they must be onsite to allow timely repairs to be made and keep the operation going. Equipment in this environment usually requires more maintenance than usual and it is vital to stay current with all maintenance of the key equipment; such as the formwork carrier, concrete pump, changing out worn slickline, guillotines, compressors, personnel elevators, rail mounted electric concrete pump for pump to a pump situations, electric jumbo trough for pump to a pump situations, trains taking crews to work underground, etc.

Forming Equipment

There were many different materials, sizes and shapes of forming required to build this project. All of the formwork designed for this project, with the exception of the wood forms, was fabricated by Everest Equipment Co. The sizes of the major formwork equipment were 3.05 meter (10 foot) diameter (Figure 10), 3.05 meter (10 foot) to a 6.1 meter (20 foot) transition formwork (Figure 11), 6.1 meter (20 foot) diameter formwork (Figure 12) and custom preassembled wood forms for sump pits. The major steel formwork equipment utilized for the 3.05 meter (10 foot) diameter locations were round telescoping forms used with a formwork carrier (Figure 13) and handset forms used in adits and handset forms used to tie in adits to tunnel. The major steel formwork

CONCRETE LINING FOR WATER TUNNEL NO. 3, MANHATTAN PORTION

Figure 10. 3.05 meter (10 foot) diameter formwork in the East Starter Tunnel

Figure 11. 3.05 meter (10 foot) to a 6.1 meter (20 foot) diameter transition formwork in the East Starter tunnel

Figure 12. 6.1 meter (20 foot) dia. formwork in the North Starter tunnel

Figure 13. Shop testing of 3.05 meter (10 foot) diameter formwork with carrier

equipment utilized for the 3.05 meter (10 foot) to 6.1 meter (20 foot) diameter transition locations had walkways installed so there was no form carrier, since the forms were a unique shape and were not intended to be stripped and moved to another location for immediate use. The major steel formwork equipment utilized for the 6.1 meter (20 foot) diameter locations were round forms used with a formwork carrier. There was also a "saddle" form that was used for the forming of the intersection of the 6.1 meter (20 foot) diameter shaft and the 6.1 meter (20 foot) diameter tunnel. The formwork used for the sump and pump pits at the bottom of Shaft 26B was custom wood forms by Contractor's Engineer preassembled prior to lowering down the shaft.

The steel formwork was lowered down the shaft in the biggest pieces practical for each particular size form. After it was determined how to put the pieces together safely and efficiently, the exact sequence of the formwork installation was decided upon. The two separate sections of the wood formwork was lowered down Shaft 26B and then placed into position at the sump and pump pits at the bottom of Shaft 26B. As shown in the various photos, additional consideration had to be made to account for the forms due to the fact that were not always centered on the tunnels.

Bulkhead Materials

Vertical bulkheads between consecutive placements were required to be stainless steel since the tunnel will carry New York City drinking water. These bulkheads were made from an expanded metal material (stay–form mesh type material) and steel angles rolled to the mined tunnel radius and bolted into rock to hold the stainless steel mesh.

CONCLUSION

In summary, the long distance pumping method was chosen over the conventional tunnel lining concrete placement technique used in previous New York City tunnel work to address the unique problems presented by the project was initially very challenging, but became successful due to the modifications and adjustments that were made. Constant monitoring of the project is necessary to identify potential problem areas before they become impacted production. Selection of the right mix will help to reduce many possible problems and the right equipment can be the difference between making and not making production. The work schedule was three shifts a day five days a week and all potential obstacles needed to be resolved to reduce the potential massive costs and schedule impacts from occurring. If it takes as long as two minutes extra to pump each truck, the amount of forms that were set for the next day's placement and the day's schedule that could easily affected, considering there have been as many as 50 concrete trucks during a placement. Given all of the difficulties encountered on this project, this method proved to be a success.

PART 7

International

Chairs

Heather Ivory
URS Corp

Isabel Lamb
Jacobs Associates

THE HALLANDSÅS DUAL MODE TBM

Werner Burger ▪ Herrenknecht AG

Francois Dudouit ▪ Skanska–Vinci JV

ABSTRACT

The Hallandsås project with a long history, first started in 1991, is well known in the tunneling community. In the past it has been stopped twice for technical and environmental reasons. The third attempt started in 2004 using a dual mode rock TBM. The heterogeneous highly fractured and abrasive rock mass as well as large areas of water bearing zones present most difficult subsurface conditions. The tunneling concept is based on the use of watertight segmental lining and a TBM able to operate in open mode, possibilities for extensive pre excavation grouting and high pressure closed mode operation up to 13 bar to cover exceptional conditions.

THE HALLANDSÅS PROJECT

The Hallandsås Railway Tunnel Project is a major infrastructural project presently under construction in southern Sweden. It is part of a large investment aiming to expand and re-build the west coast railway line between Gothenburg and Malmö. Once completed and taken into operation the new twin-track rail link, designed for fast trains, will shorten the travel time between the two cities by two hours. Furthermore, the overall capacity of the rail link will increase from 4 trains/hr to 24 trains/hr.

Two parallel 8.6 km long tunnels are presently being excavated using a 10.6 m dual mode Mix-Shield rock TBM.

The tunnel construction began first in 1991/92 but was suspended in 1997. By this time, 5.5 km of the tunnel, or a third of it's total length of 18.6 km, had been completed. The tunnels were originally scheduled to open in 1997, but the project has been dogged by problems. The first contractor unsuccessfully used an open gripper TBM which excavated only 18 m of tunnel. After changing to drill and blast method the contractor finally aborted the project leaving only 20% of the total length completed.

When the second attempt started in 1996 using traditional methods water leakage in the tunnel became a major problem. The water leakage limits for the project where exceeded. Alternative sealing methods where evaluated, leading to the testing of Rhoca Gil, a chemical sealing agent. This caused acylamide, a toxic additive, to leak into the tunnel and spread into nearby wells and watercourses. Tunnel workers experienced health problems, and cows in the vicinity became lame. These events caused great alarm and strong reactions, and the National Railway Administration (Banverket) left it up to the Government to decide the future of the tunnel. Between 1998 and 2000, on commission by the Government, the National Railway Administration conducted intensive investigations regarding possible working methods, environmental impact and the costs of continuing the project. A shielded TBM has been recommended for the tunnel construction, but traditional blasting techniques may also be used. In either case the rest of the tunnel is to be lined with concrete. In 2001 the Swedish Parliament and Government gave the go-ahead for the tunnel construction to proceed.

 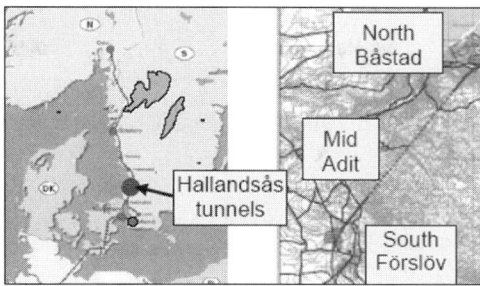

Figure 1. First attempt, gripper TBM "Hallborr" ready to go

Figure 2. Location of the Hallandsås project

The project started for the third time in 2004 (to excavate the remaining 2 times 5,500 m of tunnel) using a much more advanced technique including a dual mode hard rock Mix-Shield TBM and a watertight segmental lining to control water ingress. The client is the Swedish Railway Administration (Banverket) and the contractor a joint venture between Swedish company Skanska Sverige AB and French company Vinci Construction Grand Projets (Skanska-Vinci HB). The contract is a design and built contract amounting (based on end of 2001 numbers) to almost 430 M€.

The project is considered to have a high risk profile and three significant circumstances particular to the project are:

- the history, including two unsuccessful previous attempts and a public debate on the legitimacy of the project.
- the very complex geological and hydro geological situation.
- the high environmental demands, including comprehensive chemical evaluation of all chemicals used within the project and tight restrictions on water ingress.

End of December 2008, about one third of the remaining 11,000 m has been excavated by the TBM.

GEOLOGICAL DESCRIPTION

The rock mass in the area consists of precambrian gneiss and amphibolite. The Hallandsås ridge is a horst which is topographically very conspicuous. The horst is a result of major uplifting that occurred approximately 70 Million years ago. The rock mass within the horst is strongly affected by the pronounced jointing and faulting, implying very complex geological conditions. Three major tectonic zones, several hundreds of meters wide, are located along the tunnel alignment.

The uplifting movement especially caused severe fracturing and crushing along each side of the horst, today seen as the Northern and Southern Marginal Zones. These zones have also been subjected to a relatively strong deep-weathering and parts of the rock mass are completely disintegrated into clay. Within the Southern Marginal Zone triassic sediments, mainly siltstones, claystones and unconsolidated sandstones, overlay the weathered basement rock. The tunneling conditions within the Southern Marginal Zone are very difficult as the rock mass partly exhibits raveling or running ground with short stand-up time (less than 1 hour).

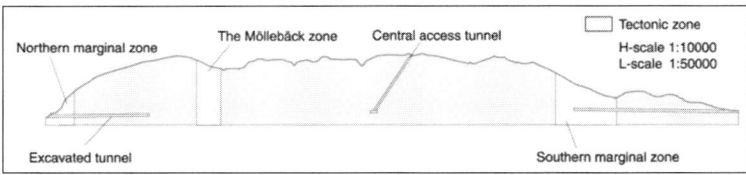

Figure 3. Schematic geological profile of the Hallandsås ridge with in 1997 already excavated tunnel sections

Inside the horst another major zone, the Mölleback Zone, is encountered. This zone differs from the marginal zones by being less weathered and therefore extremely permeable.

Pre-investigations comprising different geophysical methods, percussion and core drilling were initially carried out during 1989–1990.

The rock mass consists mainly of gneiss intruded by amphibolite dykes (generally with unknown geometry) and by dolerite dykes, which have been well defined during the site investigations. Unixial compressive strength of the unfractured rock can reach 250 MPa, abrasivity is generally above 4.5 Cerchar, but values up to 5.9 have been measured.

Due to the intense fracturing the horst is completely water saturated and the ground water pressure at the tunnel level reaches 15 bars. In this case the major problems for the tunnel excavation were expected to be related to ingress of water.

ENVIRONMENTAL DEMANDS

Following the two unsuccessful attempts, very strict environmental legal demands, set by the Swedish Environmental Court, were finally applied to the Hallandsås project.

The flow of ground water coming out from all tunnel excavations in the ridge shall, during the construction period, not exceed the following limits:

- over a rolling period of 30 days: 100 l/s
- over a rolling period of 7 days: 300 l/s
- at any time: 400 l/s

The water discharged from the project shall be sufficiently treated in order to comply to the defined quality criteria:

- suspended material: 65 mg/l for discharge in the sea and 30 mg/l for discharge in the surrounding rivers
- concentration of hydrocarbons: 5 mg/l

Any breach of the demands set by the Swedish Environmental Court will result in legal proceedings. In addition, Ph adjustment is required.

Before being brought to the project and used, any chemical product must have its full composition known, which in most cases requires secrecy agreements with the suppliers, and its impact on the environment must be duly assessed. Approval to use a product is granted by the client and by the Swedish authorities. Each product is only approved for a particular application and for a defined quantity. For example a product approved on the South site, cannot be used on the North site, since the recipient is not the same, as long as a new environmental assessment has not been carried out. More than 430 chemical products have been already assessed.

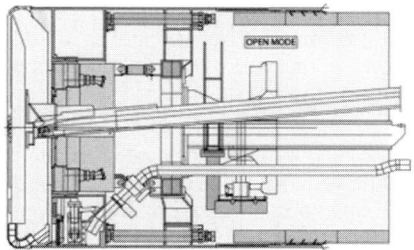
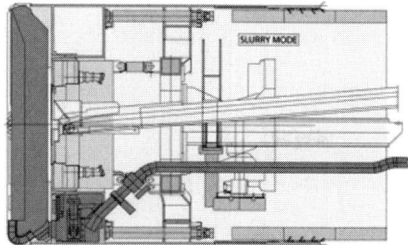

Open mode with dry primary muck discharge system (TBM conveyor)

Closed mode with hydraulic muck discharge system (slurry)

Figures 4 and 5.   Dual mode Herrenknecht Mix-Shield

TBM—BASIC CONCEPT DEVELOPMENT

In light of the project history and the high profile project requirements from both the subsurface ground conditions as well as environmental aspects the first conceptual developments started at Herrenknecht in the year 1999/2000 even in a pre tender stage based on the key requirements:

- excavation of a hard and abrasive rock mass
- zones of soft soil and mixed face conditions
- potential of high water inflow along the total length of the tunnel
- static water pressure above 10 bar along the majority of the alignment
- strict environmental (legal) restrictions on water inflow volume
- strict environmental restrictions and approval procedures on materials and methods used

At the time of this first conceptual design stage this set of requirements was asking for a technical solution close to or even beyond the state of the art of TBM design at the time. It became obvious quite soon that if any mechanized solution could be feasible only a shielded machine with watertight segmental lining would be able to address the requirements to their full extend. The TBM itself should be able to operate in open as well as in closed mode to overcome worst case conditions, with anticipated maximum water pressure and ground conditions leaving the slurry method as only feasible closed mode option to be available as a "last resort."

Being aware that high pressure closed mode excavation under hard rock and mixed face conditions is a most difficult way of operation extensive possibilities for pre excavation grouting from within the machine should be incorporated to support open mode operation. For such a concept of a dual mode hard rock Mix-Shield the available modes of operation could be:

- Open mode with dry primary muck discharge system (TBM conveyor)
- Open mode with (cyclic) pre excavation grouting
- Open mode with (cyclic) pre excavation grouting in closed static conditions
- Closed mode with hydraulic (slurry) muck discharge system under reduced face pressure
- Closed mode under full face pressure and potential for positive face support

After the contract for the Hallandsås tunnel was awarded to the Skanska Vinci JV an intensive twelve month pre design period for the TBM started whilst the project still was waiting for the final rulings of the environmental court. The decision to invest and use the "waiting time" for an intensive design work program involving the owner's representatives, the contractor and the TBM manufacturer was taken in light of the difficult nature of the project and finally proved to be very positive for the overall concept development process. The state of the art at the time of the pre design for the different aspects of the technical concept were:

- Large diameter rock excavation—the at the time ongoing excavation of the new Alpine tunnels at Lötschberg and Gotthard already provided results and "lessons learned" of cutter and cutterhead developments that had been done previously.
- Pre excavation grouting, inflow water management systems—the developments done for the difficult Arrowhead project in California and their success on site also including 10 bar static seal systems.
- Dual mode Mix-Shields—with the first large diameter dual mode Mix-Shield being employed at the Grauholz project in Switzerland in the early nineties the second generation of that machine type already had finished the Thalwil project in Zürich.
- High pressure operations—having the Westershelde project in the Netherlands finished long term experience for dynamic seal systems as well as process experience for saturation diving in tunneling for pressure ranges 6–8 bar were available.

With the most important mandatory requirement to control the water inflows within the limits set by the Swedish Environmental Court, the following technical means were planned to be used:

- grouting ahead of the TBM, with cementituous grout, rather at the periphery than at the face, in order to reduce the permeability of the rock and consequently the future ground water inflows, in water bearing zones.
- excavation in closed mode with a face pressure up to 8 bars, in order to reduce the water inflows entering the cutterhead chamber. However, maintenance on the cutterhead was planned to still take place under atmospheric conditions. It was estimated originally that 20 % of the tunnel will have to be excavated with the TBM operated in closed (slurry) mode at maximum 8 bar.
- installation of a segmental lining, equipped with gasket and capable to withstand 15 bars of water pressure, within the tail of the TBM.

These technical means should be combined with all the conventional requirements of hard rock tunneling with a TBM.

In addition, in order to handle possible zones of unstable ground (soil like conditions) under a high water pressure, the TBM should be capable to be operated with 13 bars of face pressure including the preparation for the use of saturation diving methods for hyperbaric face access.

TBM—DESIGN CONCEPT

The dual mode Mix-Shield TBM supplied for the Hallandsås project has on board the full size equipment for open mode single shield hard rock TBM excavation as well as for closed mode slurry operation with positive face support up to a maximum dynamic face pressure of 13 bar. Additional equipment and installation for extensive pre excavation grouting from within the machine as well as the handling of high water inflows and water laden muck in open mode operation.

Figures 6 and 7. The S-246 Hallandsås TBM pre assembled in the workshop

Figure 8. Trailing gear in front of the tunnel portal

CUTTERHEAD

At the design stage of the Hallandsås cutterhead five major design arguments were identified each of them causing its own design requirements. Some of these requirements contradicted each other and could only be combined accepting compromises.

Hard and Abrasive Rock Conditions

To address the strength and abrasivity of the rock mass the strongest structural design and cutter housing implementation had to be applied. The design principles and load assumptions had to follow to full extend the baseline and experience developed for the Alpine tunnels. The at the time proven 17" backloading disc cutter technology as used on the alpine projects was applied instead of the at that time still "experimental" 19" systems. Face cutter spacing of 85 mm followed a more conservative approach as the Alpine machines with 90 mm.

Occasionally Blocky Rock

Expected sections of blocky rock conditions in the vicinity of fault zones were addressed by integrating the full range of improvements and findings from the blocky rock sections at the Lötschberg tunnel, thus anticipating the cutterhead to operate as a rock crusher for limited length of tunnel as already experienced before at Lötschberg.

High Water Inflows in Open Mode

A new developed type of grids arranged along the muck discharge channels was realized, after real size testing and optimizing during workshop assembly of the TBM.

Wear protection on the cutter-head;

⇨ Wedges to protect the cutters.

➡ Wear plates for the face

⇨ Wear protection at cutter-head periphery.

➡ Grill bars to protect the cutterhead periphery.

⇨ Grill bars to protect the scrapers from big blocks and control maximum block size to enter the cutterhead.

Figure 9. Wear protection features and details on the original Hallandsås cutterhead

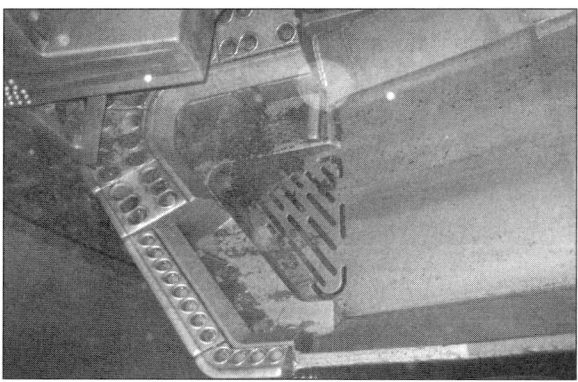

Figure 10. Dewatering grids along muck channel

The water management / discharge system layout integrated a flushing circuit with gantry arranged treatment plant for fines handling.

Dual Mode Operation (Center Belt—Slurry)

The basic design principle of the dual mode cutterhead is the same as on earlier convertible cutterheads using the structural arms of the cutterhead also as open muck channels to transport the muck to the center arranged belt conveyor in open mode. In closed mode (slurry mode) the open rear structure between the arms allows the muck to pass through the submerged wall opening to the crusher-suction pipe area. The permanent presence of two entirely different mucking versions caused some compromises for cutter arrangement and access compared to a single mode conventional hard rock cutterhead.

Potential Soil Like Ground Behavior in Closed Mode (Clogging)

A still acceptable amount and size of center and face openings in the cutterhead front had to be realized to address potential clogging risks that would potentially

Figure 11. Original size shop test of the main drive seal system

cause major problems and need for manual intervention in closed mode. The need for increased muck openings leaves fewer degrees of freedom for disc cutter arrangements especially in the center area causing a more "star type" disc cutter arrangement, less favorable for hard and blocky rock conditions.

CUTTERHEAD DRIVE

Due to the anticipated wide range of ground conditions and the potential of high water inflows the decision was taken in favor of a hydraulic cutterhead drive system. The fact of being able to combine the potential of very high torque at low speed (soft and mixed ground conditions) with a typical high revolution hard rock operation took advantage over the 20% better power efficiency of a VF drive. For the same reason of variable ground conditions and to maintain the maximum amount of operational options the installation of a fully articulated cutterhead drive system was decided with the possibility of longitudinal adjustment and angular tilting for eccentric overcut. A heavy three axis roller bearing, drive pinions with independent supports and a fully equipped lube oil circuit is installed as state of the art solution for today's heavy duty cutterhead drive systems.

Special care and further development was necessary for the bearing seal systems since they have to withstand a maximum dynamic pressure of 13 bar. The cascade seal system design that was first used successfully at the Elbetunnel in Hamburg and the Westersheldetunnel was further developed towards an automatic controlled pressure cascade system. Due to the importance of this system an original size dynamic shop test up to 15 bar pressure was performed to optimize and verify the settings of the automatic control system. In addition an activated system to isolate the main seals from the chamber pressure in standstill mode is installed to release the dynamic seal system from pressure during extended standstill periods for example during pre-excavation grouting periods.

SHIELD STRUCTURE

The multi piece bolted shield structure is designed to withstand a theoretical water pressure of 15 bar and associated ground loads for conditions as described in the geotechnical documents. In addition to the water and ground loads operational loads as well as special load cases following potential worst case scenarios from pre-excavation grouting are considered.

Figure 12. Underground assembly of the shield structure

The main shield body is built from six sections with bolted and sealed flanges including submerged wall, pressure bulkhead and erector support frame. A submerged wall gate and all required installation and connection ports for a full size slurry circuit as well as a jaw crusher are foreseen. All required connections and precautions are foreseen for a potential use of saturation diving including a pre chamber for a future transport shuttle connection.

The fixed double shell tailskin is supplied in three sections and welded on site to the shield. A wire brush tail seal is installed in a three chamber configuration (three rows of wire brush and rear row as five layer spring steel plate) and a 360 degree spring steel excluder to the outside. Grout lines and all necessary tail seal mastic and sampling lines are integrated in the dual layer tail plate.

The need for underground assembly of the shield structure also provided size and weight limits to be addressed during an early design stage.

MUCKING SYSTEM

A full size mucking system for open and closed mode operation is installed. The capacity of the open mode belt conveyor system being 1,000 to/h and the closed mode slurry circuit and treatment plant for a nominal flow rate of 1,800 m^3/h.

The open mode belt conveyor system consists of a TBM and gantry conveyor as well as the advancing tailpiece, transfer point and belt extension area for the tunnel conveyor. When changing the mode of operation from open to closed the muck hopper including the TBM conveyor in the cutterhead center can be hydraulically retracted. The front plate of the muck hopper then closes and seals the center opening. Special features are foreseen for the handling of water laden muck. Dewatering areas and drainage basins along the machine and gantry conveyor are installed including a flushing circuit with on board treatment plant for 600 m^3/h.

The slurry circuit for closed mode operation has besides the typical slurry shield requirements to address the wide variation of potential face pressures and pressure

Figure 13. Water laden muck on gantry conveyor

levels. Also the typical muck grain size distribution from a hard rock excavation process has to be taken into account for its layout.

SEGMENT ERECTION AND BACKFILL

The supply of segments and backfill material is via train into the closed deck trailing gear. Segment handling is with vacuum gripping systems for the segment crane and the erector. A segment feeder with a storage capacity for one complete ring consisting of 8 pieces is installed.

The machine is prepared to use either mortar or pea gravel backfill material or a combination of both. Handling systems and storage capacities for both are installed on the trailing gear. Also equipment as well as material handling and storage facilities for second stage grouting is provided.

PRE-EXCAVATION GROUTING

With the pre-excavation grouting being identified as one of the major tools to handle the difficult ground conditions the machine is prepared and equipped with extensive permanent installation for drilling and grouting.

Multiple inclined channels in the shield skin as well as a large number of ports in the bulkhead, submerged wall and corresponding cutterhead openings allow a dense drill pattern around and in front of the tunnel face:

- 30 channels through the shield skin in two different lockout angles of 10° and 13°
- 26 outer face positions
- 7 inner face positions

A total of three permanent installed drill rigs installed are located on a 360° carrier ring behind the erector able to reach the periphery and outer face positions. Two more

Figure 14. Permanent drill installation at the front end of the trailing gear behind the erector

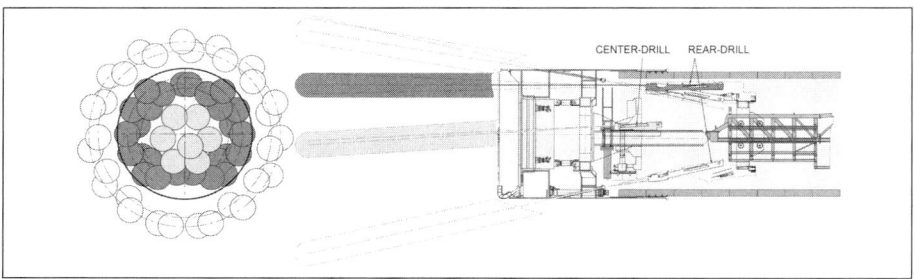

Figure 15. Possible drill pattern ahead of the tunnel face

permanent drills can be installed in the front of the center shield area if required. A temporary drill installation is possible in the shield center and on the erector.

A permanent pre-excavation grout plant and materials storage and handling area is provided on the gantry.

HYPERBARIC FACE ACCESS

In order to be prepared to inspect the cutterhead in worst case conditions with high water pressure and unstable face the machine is equipped with all necessary basic installation for chamber access in saturation mode. In front of the fact that along the entire tunnel alignment the water table is in the range of 100–130 m above the tunnel the machine is prepared for face access with saturation method only. No permanent man lock for decompression within the machine is foreseen as used for applications with a maximum of 3–5 bar chamber pressure. Besides piping and connections required for saturation access a permanent pre chamber is installed in the shield to which the transport shuttle can be connected. A transport shuttle is available on site and all means of transport and passage of the shuttle through the gantry to the pre chamber are foreseen and have been tested during workshop commissioning.

Figure 16. Test for shuttle handling during shop assembly

EXPERIENCED ROCK CONDITIONS AND CONSEQUENCES

After site assembly of the shield and gantries excavation started in November 2005. The shield was assembled in an undergound cavern at the end of the previous drill and blast excavation at the south portal, the gantry was assembled in front of the portal and then transferred through the drill and blast section and connected to the shield.

From the beginning of the excavation with the TBM, in addition to the difficulties associated with handling of the ground water inflows, difficulties associated with the rock mass have been experienced:

- Repetitive block instabilities. Significant overbreaks at the face and amounts of blocks were observed in such conditions
- High variability of the rock mass

The consequences of the blocks instability have been dramatic on the cutterhead and mucking systems, the belt conveyor system in open mode and on the slurry circuit in closed mode. From the start of the excavation until the arrival of the TBM in the underground cavern of the mid-adit, which is an intermediate access excavated in 1995–1996 to allow four additional drill and blast faces, approximately 14% of the then excavated distance of 2,540 m has been mined in closed mode.

CONSEQUENCES ON THE CUTTERHEAD

In blocky rock conditions (block instabilities at and in front of the face), the blocks falling from the face are of a size and strength that major rock crushing takes place in front of the cutterhead. The rock excavation does not proceed from the controlled cutting action of the disk cutters, but from the rather uncontrolled crushing actions of the entire cutterhead front.

Routine face inspections to record at the face overbreaks and percentage of cutters tracks are performed basically every one or two rings in order to know the percentage of cutters working in a regular cutting mode.

Observation of the damages on the cutterhead and on its tools (cutters and scrapers) has confirmed the disastrous effects of the blocks. The crushing actions are associated with peak loads on the cutters, which can exceed by at least a factor of 5–10 times the nominal load of 250 kN and which are variable in direction. In particular, they result in damages on the cutters, on their fixations and housings. The percentage of cutter changes due to normal wear reaches only about 40%, the rest being replaced due to

Table 1. Technical data Herrenknecht TBM S-246

Machine type	Mixshield, dual mode
Manufacturer	Herrenknecht AG
Excavation diameter	10.60 m
Total length	250 m
Total weight	3,100 to
Total power	8,600 kW
Cutterhead	Hard rock, dual mode, articulated
Cutters	17", backloading (19" second cutterhead)
Power	4,000 kW
Speed	0–5 rpm, hydraulic drive
Torque	20.3 MNm / 26.0 MNm
Thrust	22,000 kN (42,000 kN in high pressure mode)
Shield	⌀ 10,53 m
Max. pressure	15 bar
Thrust	139,500 kN (188,500 kN in high pressure mode)
Mucking system open mode	1,000 to/h (cont. Conveyor in tunnel)
Mucking system closed mode	1,800 m³/h, rock crusher (STP at portal)
Segment backfilling system	Two systems, mortar and pea gravel
Flushing system	600 m³/h on-board STP
Probing / grouting	3 permanent drills (5 possible)
Drill pattern	30 periphery positions, 33 face positions
Trailing gear	11 trailers, closed deck, train supply

blockage / damage (generally caused by premature failure of the bearing, cracks in the cutter ring, damaged hub). Damages in the fixing elements and housings result in additional difficulties in replacing worn or damaged cutters.

Duration for daily maintenance on the cutterhead increases from 0.75 hour / tunnel meter to 0.9–1.65 hour/m, as the boring conditions move to highly blocky conditions.

CONSEQUENCES ON THE BELT CONVEYOR SYSTEMS

The size of the blocks, transported by the mucking system is a consequence of the size and geometry of the openings provided in the cutterhead and of deflectors and grain size limiters installed in these openings. The size of the blocks can be reduced by closing the openings, but additional crushing actions take place and consequently more damages occur on the cutterhead front and its cutting tools and buckets.

Transport of blocks on the belt conveyors system is associated with blockages at transfer points between two belt conveyor units and with belt damage, being repetitively punched or suddenly cut by blocks, trapped in a hopper.

Figure 17. Typical tunnel muck showing crushed blocks from open mode operation

Figure 18. Typical tunnel muck from closed mode operation (rounded blocks)

CONSEQUENCES ON THE SLURRY CIRCUIT

In closed mode operation, the blocks, entering through the cutterhead openings, have to pass through the suction grill / jaw crusher combination in front of the discharge pipe inlet. The suction grill limits the grain size allowed to enter the discharge pipe to 150 mm. Larger blocks stay in front of the grill and will be crushed by the jaw crusher to a suitable size.

After less than 50 m of tunnel excavated in closed mode, severe damages to the slurry circuit have been experienced: several elbows and steel pipes being pierced by the repetitive impacts of the blocks. Such observations were even made on the special reinforced components of the slurry circuit installed in the TBM and trailing gear. The destructive actions of the blocks, worsened by the abrasivity of the transported rock materials, were observed also on other components such as slurry pumps, valves, and telescope pipes.

Despite time consuming repairs, maintenance programme on the slurry circuit and replacement of some components by ones with a higher wear resistance (made of steel with 500 HB on the complete thickness), severe difficulties were experienced to maintain the slurry circuit operational. The typical crushed blocks initially angular in front of the suction grill were found at the slurry treatment plant rounded after their transport through the slurry discharge circuit.

In blocky ground conditions the jaw crusher capacity often became the determining factor for the achievable mining speed. Depending on the severity of the situation significant reduction of the TBM advance speed or even to stop it in order to give the time to the crusher to absorb the blocks were found necessary. In the worst situations the blocks piling up in the chamber were obstructing the submerged wall opening to an extent that a proper hydraulic connection between the cutterhead and air bubble chamber was lost. The effect followed by this was over-pressurisation of the cutterhead chamber.

Due to the unfavourable shape and size of the rock particles the flow in the return line had to be increased from 1,800 to 2,000 m^3/hr (16 inches diameter pipes) to achieve a higher flow speed and reduce the risk of settlement and blockages in the discharge line.

ACTIONS TAKEN TO ADAPT TO THE ROCK CONDITIONS

The extremely difficult rock conditions and tunnel face behaviour resulted in a number of modifications of operational and mechanical equipment aspects resulting from a necessary re-adjustment of some of the basic projects requirements finally dictated by the in situ ground conditions.

TBM DRIVING CLASSES

In order to find the best compromise between spot progress and damages to the cutterhead, a back analysis exercise was launched very early, involving study of:

- TBM excavation parameters such as cutterhead rotation speed, penetration step of the cutters, average load on the cutters
- Face inspections (geological description, details of the overbreak, percentage of the cutters marks visible at the face)
- Records pertaining to the maintenance on the cutterhead

A strategy regarding selection of the TBM excavation parameters related to the degree of blockiness has then been adopted for the original 17" cutterhead. Three different TBM driving classes have then been defined respectively in the gneiss, amphibolite and dolerite (see Table 2).

Based on the parameters recorded during the probe drilling exercise (three holes are systematically drilled at the periphery ahead of the face), the geotechnicians are making a prognosis of the TBM driving classes, which is given to the TBM driver. The final selection of the TBM parameters is based on the observation of the blocks on the belt conveyor and on the face inspection results. Due to the high variability of the geology, it is very frequent that several changes of TBM driving classes are necessary during the same day.

CUTTERHEAD MAINTENANCE

The daily maintenance routine on the cutterhead was modified and a higher frequency of inspections was adopted following a defined procedure (segment ring length is 2.2 m):

- Ring n: Check of all tools (cutters and bucket lips) and wear measurement. Re-tightening of cutter fixing bolts and change of the worn or damaged tools
- Ring n+1: Re-tightening of the fixing bolts of the cutters changed at ring n
- Ring n+3: Visual check of all tools and change of the worn / damaged ones
- Ring n+4: Re-tightening of the fixing bolts of the cutters changed at ring n+3
- Ring n+5: As ring n.

Table 2. Parameter table of the three TBM driving classes

TBM Driving Class Ga/Aa/Da:
- Chips on belt conveyor
- More than 80% of the cutter marks visible at the face

TBM operation parameters:
- 4–5 rpm
- 200–250 KN/cutter
- 10 mm/rev

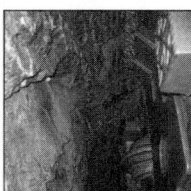

TBM Driving Class Gb:
- Chips and small blocks (<200 mm) on belt conveyor
- More than 20% of the cutter marks not visible at the face

TM operation parameters:
- 2,5–3,5 rpm
- 200 KN/cutter
- 12 mm/rev

TBM Driving Class Gc/Ac/Dc:
- blocks on belt conveyor
- Cuttermarks not visible at the face

TBM operation parameters:
- 1–1,5 rpm
- 60 KN/cutter
- 20 mm/rev

A follow up of the cutterhead and tool damages has shown that most damage is concentrated to a certain area of the outer face likely to correspond to intense crushing taking place at this location as a consequence of block instabilities in the face.

In addition to the daily cutterhead maintenance carried out from inside and behind the cutterhead a approximately 500m interval has been established for heavy cutterhead maintenance including access possibility to the cutterhead front to maintain and replace front wear protection and block deflectors. Whereas comparable heavy maintenance procedures are required every 3,000–5,000m of tunnel for the Alpine hard rock TBMs the blocky rock conditions in the Hallandsås project require far shorter intervals in order to keep the cutterhead fully operational.

SECOND CUTTERHEAD

With the view to reduce the impact of the cutterhead maintenance on the production, and aware of the fact that repetitive heavy underground welding operations may finally have a negative effect on the basic structure, decision was taken in autumn 2006 to launch the design for a second cutterhead. Also the access possibility of the mid-adit ahead about halfway of the first tube as well as the chance to do design modifications in order to adapt to the in situ experienced rock conditions and face behaviour were major arguments for the decision.

Design of the second cutterhead should take due consideration of the experience gained through the various conditions already encountered along the route of the east tunnel, in particular the blockiness, and all the modifications regarding wear protection and block deflectors / grill bars already done to some extend on the first cutterhead.

Since meanwhile reliable experience in the use of 19 inches cutters for backloading cutterheads existed, a decision was taken to use these heavier cutters for the second cutterhead. In the specific case of the Hallandsås project, the benefit is rather

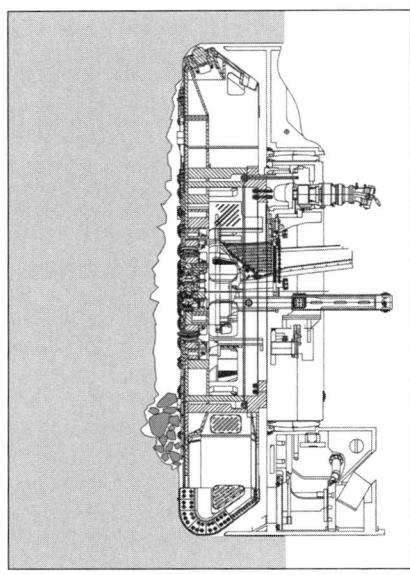

Figure 19. Schematic sketch of rock crushing area in front of the cutterhead

Figure 20. Front wear protection in worn conditions before replacement and repair

Table 3. Readjustment of cutterhead requirements based on in situ experience

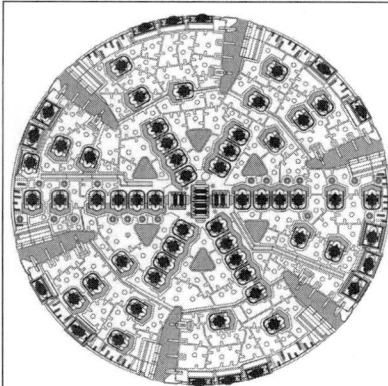 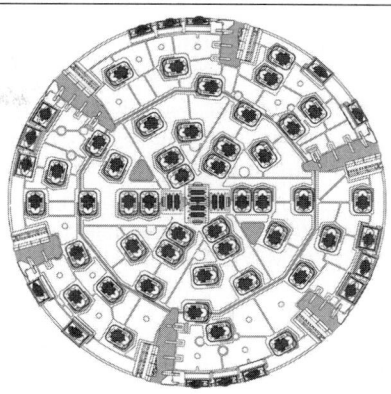

Original cutterhead requirements:	Adapted requirements for second cutterhead:
Conventional hard rock type with closed mode possibility	Closed mode possibility less important but still required
Hard and abrasive rock conditions	Increased importance
Occasional blocky rock	Blocky rock conditions as part of the regular situation, rock crushing in front of the cutterhead
High water inflows	No changes
Dual mode operation (center belt—slurry)	Slurry option less important but still required
Potential soil like ground behavior in closed mode (clogging)	Less important

thought to be in the heavier and more rigid structure of the cutters and of their fixation systems for withstanding the crushing impact loads than in the higher performance associated to the increase of the nominal load from 250 to 320 kN. In addition, the larger diameter of the 19 inches cutters permits to move the fixation system 25 mm backwards away from the face and more inside the basic cutterhead structure and therefore better protected.

The risk of clogging of the cutterhead in soil like conditions identified originally on the project has been re-evaluated and some openings on the cutterhead have been removed. This move permitted a better arrangement of the cutters in the face area.

Protections have been designed for easier refurbishment and some new developed backloading crushing tools have been added along the most affected outer face area.

Experience, since the re-start from the mid-adit, shows that the new cutterhead has permitted to reduce significantly the daily cutterhead maintenance. This has in return led to an increase of the daily production, which is in the range of 11.6 m/day outside the periods of ground treatment, barriers construction, heavy maintenance on the cutterhead. These periods amount to 55% of the time.

Overall the progress of the TBM is equivalent to the one achieved in the past by five simultaneous drill and blast headings, without considering the fact that the tunnels were unlined at that time.

BELT CONVEYOR SYSTEMS

The following measures have been taken in relation to the belt conveyors system:

- Modification of the geometry of transfer points between two belt conveyors to improve passage of blocks
- Strengthening of impact bars at muck chutes
- Additional personnel to watch transport of blocks at strategic positions to limit the risk of damages and blockages
- Installation of a block separating unit at the entrance of the tunnel in order to protect the outside belt conveyors. Such installation could not be done on the TBM back-up due to geometrical restrictions.

SLURRY CIRCUIT

Due to the severe damages and to the difficulties to maintain the slurry circuit operational in the encountered ground conditions, decision was taken to abandon progressively the use of closed mode to control the ground water inflows and put priority on the use of the alternative excavation in open mode supported by pre-excavation grouting.

In addition, the time consuming cutterhead maintenance, experienced on the project, reduces dramatically the benefit of the closed mode as a measure to control the ground water inflows, since cutterhead maintenance is carried out under atmospheric conditions.

Heavy maintenance on the slurry circuit, replacement of components with ones having an even higher wear resistance at strategic positions have been implemented in order to be still in a position to operate the TBM in closed mode for exceptional cases if required.

GROUND WATER INFLOW

With the strict limitation of groundwater inflow for the overall project being one of the essential mandatory project requirements, the limitation of inflow quantity in the active heading is of major influence for the total inflow water quantity. Ground water inflows are controlled by mainly the use of two techniques:

- Ground treatment ahead of the face (pre-excavation grouting) to limit the future inflows
- Optimised method of backfilling with frequent construction of water barriers behind the segmental lining

Works relative to control of the water inflows are time consuming. On a monthly basis, between 6 and 66% of the time is spent in such works. Consequently the monthly figure of tunnel excavation is highly depending on the ground water inflow conditions.

In order to permit advance of the TBM through water bearing zones, ground treatment is carried out ahead of the shield. The target is to seal off the rock mass sufficiently to permit progress of the TBM in open mode without violating the maximum limits set by the Swedish Environmental Court ruling.

Ground treatment is triggered by the water inflows measured during the forward probe drilling campaigns, which are systematically performed with the requirement to maintain a minimum of a 10m overlap between subsequent probe holes.

Drilling of grout holes is performed under atmospheric conditions, however depending on the water inflows, particularly the 30-day average value and the risk of grout leakage at the face, grouting is carried out either under atmospheric pressure or under pressurised conditions. To achieve pressurised conditions the muck hopper in

Figure 21. TBM arriving at mid-adit in April 2008 with the second cutterhead waiting in front

the centre is retracted thus isolating and sealing off the excavation chamber from the tunnel. Under such conditions the chamber pressure is increased until the balance with the surrounding ground water pressure has been reached and no more water inflows measured. The typical chamber pressure to achieve a static (balanced) situation is found in the range of 9–11 bar. Microfine cement is used as grouting material.

Experience, which has been gained along the first 2,500 m of tunnel, shows that better results are obtained by drilling through the face than at the periphery, confirming also experiences from other jobsites as for example the Arrowhead tunnels.

Difficulties have been encountered in backfilling behind the lining due the ground water inflows. Despite the use of PP fibres, anti-washout agent and accelerator, washout of the mortar, as backfilling material, could not be totally prevented. In order to complement the backfilling of the loss resulting from the washout phenomenon, second stage backfilling about 15 m behind the shield has been introduced.

The backfilling method, originally based on mortar, has been modified to introduce the use of the pea-gravel. The lining is backfilled with a controlled quantity of pea-gravel from the invert to the spring line. Mortar is injected in the upper part. Since the water is circulating through the lower pea-gravel matrix, washout of the mortar in the top is avoided. Construction of barriers is required to stop this circulation. Results in terms of movement of the heavy segmental lining behind the shield are very satisfactory using the combined technique pea-gravel and mortar.

CONSTRUCTION OF BARRIERS BEHIND THE SEGMENT LINING

Various methods have been tested to build efficient barriers behind the segmental lining, preventing ground water inflows from behind the watertight segmental lining along the shield and into the cutterhead chamber. The usual phases for construction of a barrier are the following ones:

- Installation of special segments equipped with drainage valves in the invert, backfilling being done with pea-gravel behind these segments

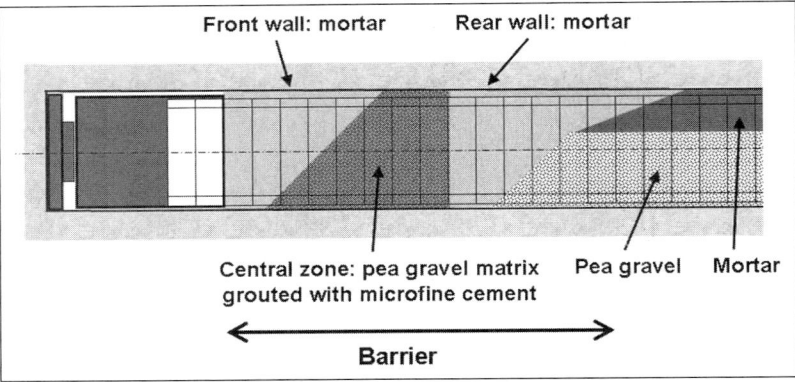

Figure 22. Barrier construction

- Backfilling with mortar of the following 5 rings in order to create a "rear wall"
- Backfilling with pea-gravel of the following 3 rings
- Excavation without backfilling of the following 4 rings. The design of the segmental lining includes strong dowels between rings which permit to omit backfilling without significant movements between the rings
- Pressurisation of the shield to balance the surrounding ground water pressure and achieve static conditions
- Backfilling of the last four rings with mortar
- Grouting of the pea-gravel matrix contained between the two mortar walls with a microfine cement grout
- De-pressurisation of the shield.

Results in terms of reduction of the water inflows vary in function of:

- Local geological conditions prevailing around the barrier two mortar walls (in particular characteristics of the rock mass fractures: orientation, opening, infill). It has been observed at various occasions that the ground water was circulating in the rock and parallel to the lining.
- Amount of water inflows and ground water pressure prevailing in the surrounding rock mass
- Overbreaks
- Length of the barrier
- Ability to balance strictly the ground water pressure

The two tools "pre-excavation grouting" and "barrier construction" are combined in order to control the water inflow within the given limits. Pre-excavation grouting is used to limit the increase of water inflow when mining through a water bearing zone. Barriers are used to reduce significantly the inflow from behind after passage of that zone.

Ground treatment and barrier construction is time consuming. The objective remains to limit them in time and consequently to maximize the available mining time of the TBM.

Figure 23. Lined tunnel

INFLOW WATER FLUSHING SYSTEM

The already installed inflow water management system with a flushing circuit including an on-board treatment plant has been upgraded and modified for several details to be able to handle short term high inflow volumes in order to support the tendency for as much as possible open mode operation.

The flushing circuit has been extended to pass as well through the invert of the air bubble chamber to evacuate water and fines direct from behind the cutterhead, involving as well the slurry circuit and above ground treatment plant also in open mode if required. An additional dewatering section along the gantry conveyor has been installed with a flat belt section, 7 m long and a large flushing basin below.

This upgrades and system adjustments enable the machine now to technically maintain open mode operation with water inflows from the face as high as 240 l/sec.

CONCLUSION

The Hallandsås Tunnel Project with its history of more than 15 years in total now presents one of the most demanding tunneling projects currently underway. At the end of 2008 more than 3,600 m of difficult waterbearing rock have been excavated by a dual mode hard rock Mix-Shield followed by a clean and dry final tunnel product. No further environmental issues have occurred and public acceptance could be reestablished to a large degree.

Appropriate technical solutions able to deal with the conditions may not have been available on the market at the time of its first start. The lessons learned for the tunneling industry from the third attempt of the project are definitely that difficult projects need a close and focused partnership and collaboration of all involved parties, the owner, the contractor and the TBM supplier to succeed.

The TBM has successfully excavated more than 1,000 m since the restart at the mid-adit with all the system adjustments and modifications in place. It is still a long and hard road to go to the end, but there are true reasons to be optimistic.

EFFECTIVE PLANNING OF UNDERGROUND SPACE— PLANNING AND IMPLEMENTATION OF THE FIRST UNDERGROUND WATER RESERVOIRS IN HONG KONG

T.H. Chan ▪ Water Supplies Department

Derek Arnold ▪ Black & Veatch

Edwin K.F. Chung ▪ Black & Veatch

Chris C.W. Chan ▪ Black & Veatch

SYNOPSIS

As the University of Hong Kong and their advisors planned their new Centennial Campus it became clear that they would have to re-provide existing water reservoirs on an adjacent site with new salt water reservoirs located underground in two new caverns. The existing reservoirs are owned by the Water Supplies Department who provide salt water for sanitary flushing and a separate potable water supply for regular use. Although underground reservoirs had not previously been adopted locally, the planning has been completed to a tight schedule and the construction is well advanced. The need to preserve historic buildings imposed constraints, as did the requirement to minimise potential construction nuisance for the local community and environment.

This paper presents the background to the project and the work carried out to develop an innovative, sustainable and environmental acceptable solution. Construction progress to date is also presented.

INTRODUCTION

The University of Hong Kong (HKU) is developing a Centennial Campus in order to facilitate their vision of achieving the highest levels of academic excellence while enhancing their reputation as one of the world's leading institutions for higher education. The development is timed to coincide with HKU's centenary in 2011, and to prepare for the implementation of a new four-year (increased from three) undergraduate degree curriculum in 2012.

A Millennium Master Plan (MMP) was completed in 2000 and further feasibility studies followed until December 2005 when Black & Veatch (B&V) were commissioned to provide engineering services for the infrastructure necessary to support the proposed Centennial Campus.

Hong Kong is a densely populated city and land for development is scarce. Furthermore, HKU wanted their new campus to be located as close as possible to the existing campus to minimise logistical issues and to foster a sense of unity for students and staff alike. The only available site that can meet these criteria is located to the west of the existing Main Campus; however, it is currently occupied by existing potable and salt water reservoirs and the associated waterworks facilities owned by the Water Supplies Department (WSD) of the Hong Kong Special Administrative Region (HKSAR). These facilities form an essential part of the water supply systems for the north-western part of Hong Kong Island. In order to conserve potable water in Hong

**Figure 1. Proposed Centennial Campus of the University of Hong Kong
(Source: HKU Centennial Campus (http://www.hku.hk/cecampus/))**

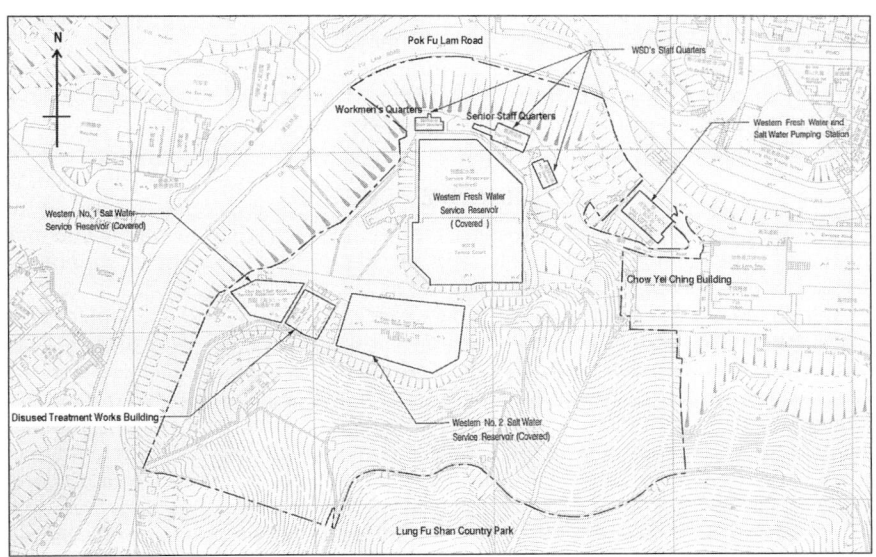

Figure 2. Layout of existing waterworks facilities on the proposed site

Kong, salt water is used as a separate supply for sanitary flushing purposes. Some of the older facilities on the site known as the Elliot Treatment Works are no longer used but are considered to be of historical significance. In particular, the Senior Staff Quarters, Workmen's Quarters and Elliot Treatment Works are listed as historic buildings under local regulations.

In order to make way for the new Centennial Campus, the existing potable and salt water reservoirs have to be re-provided without interrupting existing supplies. An aerial photograph showing the location of the proposed Centennial Campus is shown in Figure 1. The layout of existing waterworks facilities on this site are shown in Figure 2. The site is bounded by the existing Chow Yei Ching University Building to

Figure 3. General view of the existing site (looking southeast) with the Chow Yei Ching University Building in the background

the east and Pokfulam Road to the north and west. The southern portion of the site is a steep natural hillside bounded by Lung Fu Shan Country Park.

This paper describes the project background and the planning carried out to achieve an innovative, sustainable and environmental friendly solution for the re-provisioning of the waterworks facilities. The construction completed to date is also described.

ALTERNATIVE SCHEMES

Prior to 2005, various alternatives for the re-provisioning of the waterworks facilities and the subsequent development of the new campus were studied. A general view of the existing waterworks site is shown in Figure 3.

Among the options studied, decking over existing facilities allowing them to remain in service with re-provided new reservoirs on the existing slopes alongside Pokfulam Road, or on new open-cut site formations formed in the Lung Fu Shan hillside were favoured by early studies. The preferred arrangement at that time was an open-cut site formation as shown in Figure 4. This scheme comprised the construction of replacement potable water reservoirs on the southern portion of the site adjacent to the existing salt water reservoirs, and it required massive earthworks and felling of numerous trees to create level ground. Moreover, the three historic buildings mentioned above would have to be demolished as part of the proposal. A photomontage showing the scheme upon completion is given in Figure 5.

In 2006 B&V undertook a full review of the previous infrastructure studies and worked with HKU and WSD to update guiding principles for the Centennial Campus development. This review considered the comments raised by various interested parties during previous stages of study, and the increasing awareness and expectation of the community and stakeholders regarding the adoption of sustainability concepts for new developments. To meet the objectives it was soon agreed that an innovative solution with the re-provision of salt water reservoirs inside caverns within the Lung Fu Shan hillside would be adopted. The area currently occupied by the existing salt water reservoirs would be made available for potable water reservoirs. Calling this an innovative solution is considered to be justified because underground water reservoirs have not previously been built in Hong Kong.

The proposed alternative scheme for the re-provision of waterworks facilities (referred as "cavern scheme."hereafter) includes the following major components:

EFFECTIVE PLANNING OF UNDERGROUND SPACE 441

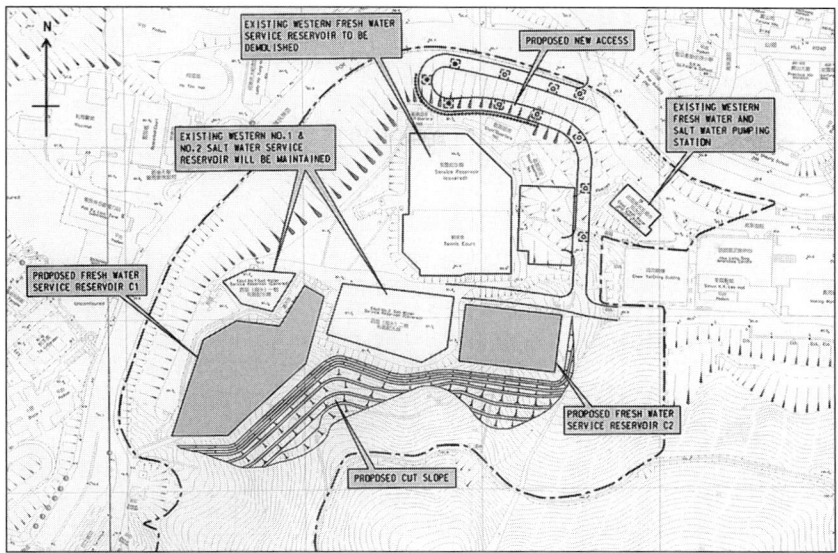

Figure 4. Preferred layout in 2005 with open-cut site formation for re-provided water reservoirs

Figure 5. Photomontage showing general view of the open-cut alternative with large cut slopes and new water reservoirs (new water reservoirs highlighted for identification purposes only).

- Construction of two caverns within Lung Fu Shan hillside to the south of the site. The caverns will house two salt water reservoirs with a total storage capacity of 12,000 m^3;
- Construction of two potable water reservoirs with a total storage capacity of 26,500 m^3 in the area currently occupied by the existing salt water reservoirs;
- Diversion, removal and construction of water supply pipework and associated facilities;

- Construction of a pipe gallery housing water supply pipework and associated facilities;
- Construction of a vehicular access road connecting the existing HKU Main Campus and the new Centennial Campus; serving as the maintenance access for the re-provided waterworks facilities and as the Emergency Vehicular Access (EVA); and
- Stabilisation and improvement works for adjacent slopes and retaining structures.

A layout plan showing the preliminary cavern scheme is given in Figure 6. A horseshoe shape cavern with a span width of 16.7 m and a height of 14.4 m was proposed. The length of each salt water reservoir is about 50 m. A typical section of the proposed salt water reservoir in cavern is shown in Figure 7.

The decision to locate salt water reservoirs instead of the potable water reservoirs in caverns is the result of a careful review of spatial constraints, water supply considerations, operation and maintenance, and tunnel engineering requirements. The total required storage capacity of salt water for this location is 12,000 m^3 which allows the sectioned cavern size to be within the bounds of past experience and with reasonable engineering economy. The length of the caverns can also be limited to suit the space available within the Lung Fu Shan hillside without encroaching into Lung Fu Shan Country Park above, which is protected from development by law. The excavation of caverns for salt water with a total storage capacity of 12,000 m^3 instead of potable water with a total storage capacity of 26,500 m^3 is also preferable from construction, cost and programme perspectives. In addition, placing salt water rather than potable water in caverns also eliminates the risk of contamination to drinking supplies by contaminated groundwater infiltration into the cavern.

To mitigate the risk of water leakage from the underground water reservoirs to the surrounding ground, structures are designed as water retaining structures in accordance with relevant design standards and in accordance with conventional WSD practice. This requirement limited the risk of groundwater contamination. To facilitate the future operation and maintenance of the water reservoirs in caverns and to secure the safety and health of maintenance personnel, careful consideration of ventilation requirements has also been made. As the corrosive nature of salt water is much higher than potable water, corrosion resistance of the works has also been considered. First of all, a forced ventilation system with adequate number of air changes per hour will be provided. A rate of air change of 6 times per hour has been adopted. Moreover, appurtenances inside the salt water reservoir caverns will be non-metallic or made from high grade stainless steel to minimise maintenance requirements.

The cavern scheme, which avoids massive excavations and extensive tree felling, is widely supported by interested parties; in particular, agreement with the Water Supplies Department and the Geotechnical Engineering Office (GEO) of the Civil Engineering and Development Department (CEDD) was essential and was obtained within a remarkably short period of just nine months from conception.

In addition to the benefits already mentioned, the proposal facilitates the preservation of the historic buildings and structures within the site. It was decided at an early stage to conserve the Senior Staff Quarters (Grade II Historic Building) and the Workmen's Quarters (Grade III Historic Building) located at the northern portion of the site. The two staff quarters will become part of the main entrance plaza for the future Centennial Campus. To support the strong commitment of the Government and HKU to heritage conservation, it was also decided to retain the Elliot Treatment Works (Grade III Historic Building), located at the heart of site. Substantial effort have been made by the whole project team to overcome numerous physical and technical constraints imposed by the preservation of Elliot Treatment Works and to work out a feasible options for the

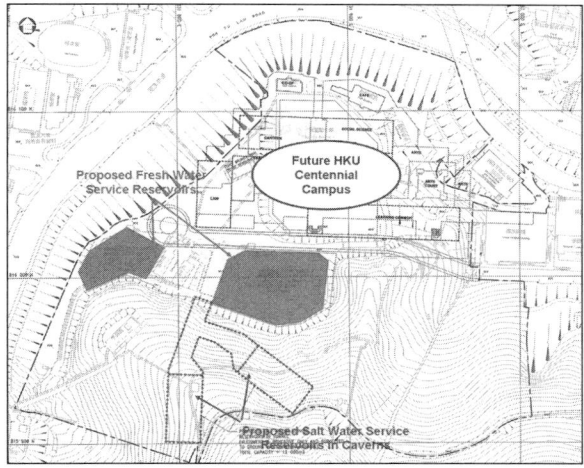

Figure 6. Proposed layout of the re-provision of waterworks facilities with salt water reservoirs in two underground caverns

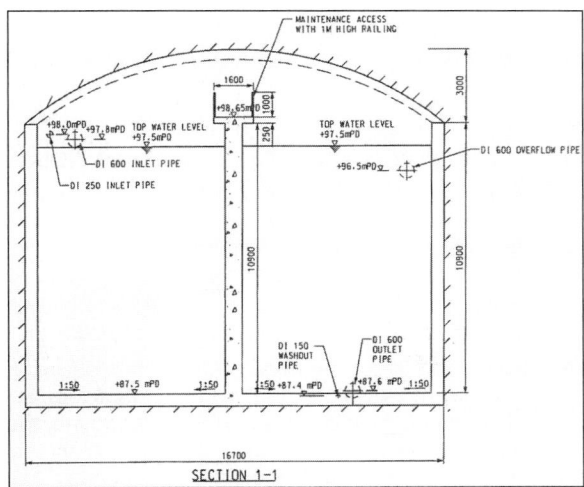

Figure 7. Typical cross section of the salt water reservoir in cavern

waterworks re-provisioning scheme without the demolition of the buildings. This led to potable water reservoirs with higher water retaining height, a more constrained site with limited space at the cavern portal for future operation and maintenance, and constrained access to the west of the potable water reservoirs. The future use of the three preserved historic buildings will be further studied by HKU's architect and heritage conservation consultants together with WSD and the community aiming for them to be reused in some way upon the completion of Centennial Campus.

Further considerations which have been taken into account in the formulation of the proposed cavern scheme are discussed below.

GEOLOGICAL AND GEOTECHNICAL ENGINEERING SETTING OF THE SITE

A comprehensive review on the geological and geotechnical engineering setting of the cavern scheme site was carried out based on available maps, publications, ground investigation information and site reconnaissance. A brief summary is presented as follows.

Based on the geological survey carried out by GEO as presented in The Pre-Quaternary Geology of Hong Kong (GEO, 2000), the main site consists predominantly of coarse ash crystal tuff with intercalations of eutaxite and sandstone beds. These are named as the Mount Davis Formation and are grouped in the Repulse Bay Volcanic Group. The northeastern side of the site is bounded by medium grained Kowloon Granite. The Kowloon Granite forms a sub-circular biotite monzogranite pluton centred on Kowloon and Hong Kong Island. Moreover, granite intrusions can be observed from the rock cores showing meta-sandstone / granite and coarse ash crystal tuff / granite contacts. Irregular and finger-like granite intrusions are also found within the site. Furthermore, the formations are considered to have possibly been subjected to thermal metamorphism due to the granite intrusion.

Rock head in the Lung Fu Shan hillside immediately to the south of the existing salt water reservoirs is found at relatively shallow depths making the scheme viable from cost and program perspectives. The layout of the proposed caverns was designed to maximise excavation in competent rock and at areas with adequate rock cover despite the complex geological setting.

The proposed cavern scheme takes advantage of design opportunities provided by the existing landform. It is formulated to suit the unique setting of the site and to maximise land use without compromising the environment. The proposed layout is arranged like two fingers and the caverns are sized to provide adequate width for maintenance access at the cavern portal to facilitate future operation and maintenance by WSD.

SUSTAINABILITY AND ENVIRONMENTAL CONSERVATION INITIATIVES

By housing the replacement salt water reservoirs in caverns, substantial slope cutting has been avoided together with extensive site formation works for the salt water reservoirs, which in turn minimised modifications to the existing landform. The cavern scheme also eliminates the loss of about 6,000 m^2 of secondary woodland on the southern portion of the site.

Design objectives for the cavern scheme included the following:

1. Minimise Modification to Existing Landform.
    Landforms including natural woodlands with adequate soil and rock mantle are essential to support living resources. Once destroyed, they are irreplaceable so one of the considerations leading to the formulation of the cavern scheme was to avoid modification to the existing landform. The initial open-cut site formation scheme would have involved significant modifications and changes to the existing landform. The proposed works would have resulted in changes to the slope, elevation, grade and composition of the existing landform, none of which is desirable.

    The cavern scheme takes advantage of design opportunities provided by the existing landform. In particular, an existing cut rock slope behind the southwestern corner of the existing water reservoir allows the cavern portal to be formed with minimal effort. The cavern scheme protects the land base including its form, soils, rocks, and associated natural processes. This in turn minimises disruption to the natural form of the site and its ecological integrity. It also protects biophysical features and maintains the visual character and identity of the natural landscape. An aerial photograph showing the site and

Figure 8. Aerial photograph (looking southwest) showing the conserved landform

indicating the extent of existing landform conserved by the cavern scheme is shown in Figure 8.

2. Minimise Ecological Impact by Avoidance of Substantial Tree Felling and Preservation of Woodland Habitat of Natural Hillside.

Another consideration leading to adoption of the cavern scheme was to minimise ecological impact. A major benefit of the scheme is that woodland habitat can be retained, preserving the ecology of the site, in particular the natural hillside to the south.

Concerning habitats and vegetation, woodland habitat were located mainly on the southern fringe of the site immediately north of Lung Fu Shan Country Park and the slopes to the south of the existing water reservoirs. The secondary woodland is fairly natural and of moderate to mature age. The canopy reaches a height of 8 to 15 m and is dominated by a number of native species. The understorey is densely vegetated by a variety of tree seedlings, shrubs and herbs all worthy of preservation.

Concerning birds, a variety of bird species were recorded including the Crested Goshawk Accipiter trivirgatus and Black Kite Milvus lineatus, which are of conservation concern. These bird species are recognised as Class II protected species in the People's Republic of China.

In addition to plants and birds, the secondary woodland is also recorded as the habitat for invertebrates and herpetofauna. For example, some uncommon butterflies, like Constable Dichorragia nesimachus and White Commodore Parasarpa dudu, were discovered at the site.

In summary, the ecological value of the woodland is high due to its naturalness, moderate to mature age, and diversity and presence of species of conservation interest. Moreover, the habitat characteristics and species composition would be difficult to recreate. By adopting the cavern scheme, the ecological impact due to the development is minimised and no adverse ecological impact associated with habitat loss is anticipated.

3. Minimise Visual Impact.

The open-cut scheme originally proposed would have involved the formation of steep man-made cut slopes (typically 80 degrees in rock and 55 degrees in soil) over 50 m high. The formation of large and steep cut slopes in the natural hillside would create severe visual impact, particularly affecting the residents of adjacent high-rise residential developments including

Figure 9. Elliot Treatment Works

the Belcher's apartment buildings as well as Country Park visitors and future users of the Centennial Campus. In view of the steepness of the proposed man-made slopes it was anticipated that, even if the cut slopes were adequately landscaped, the visual impact mitigation measures would not have been able to restore the original natural landscape character of the hillside.

For the cavern scheme, the natural landscape in the woodland at the southern portion of the site as shown in Figure 8 above will be retained.

4. Minimise Heritage Impact by Preserving Historic Buildings.

The planning of the cavern scheme also considered the conservation of heritage resources. The graded historic buildings within the site are to be preserved in-situ and properly protected during the waterworks reprovisioning project. A photograph of one of the historic buildings is shown in Figure 9.

Elliot Treatment Works was constructed between 1930 and 1931. It was a rapid gravity filtration plant. The building was constructed in red brick and cement plaster with blue painted iron framed glass windows. Concrete filter beds are constructed on both sides of the valve house. The Elliot Treatment Works was decommissioned in 1993. The primary importance of the Treatment Works Building lies in the machine and equipment inside the buildings as it demonstrates not only the history of the site but also the history of water supplies in Hong Kong.

The cavern scheme facilitates conservation of the three graded historic buildings and helps demonstrate the effort made by relevant parties to support heritage conservation. The future use of the historic buildings will be studied by HKU, WSD and the community for appropriate reuse in the future.

5. Minimise Other Environmental Impacts.

In addition to the geological, ecological, visual and heritage aspects, the cavern scheme is also environmental advantageous because dust and noise impacts during construction have been minimised because many of the construction activities are underground.

The cavern scheme also has advantages for waste management. The open-cut site formation scheme would have required over 280,000 m^3 of soil and rock excavation while the cavern scheme involves excavation of less than 40,000 m^3. The cavern scheme significantly reduces the amount of construction and demolition material that would have to be disposed at locations a long distance from the site. As a result the transportation of excavated materials off-site has been greatly reduced, minimising potential traffic and secondary environmental impacts.

Furthermore, the cavern was excavated using non-explosive drill-and-break methods. This avoided potential vibration disturbance and other potential safety hazards.

CONSTRUCTION

A detailed design and construction contract was awarded to Gammon Construction Limited in March 2007. The waterworks re-provisioning contract started in April 2007 with the cavern excavation commencing in August 2007.

The Design and Build Contractor proposed some modifications to the arrangement of the proposed salt water reservoirs for the final design; the span width of the horseshoe shape caverns is 17.6 m with an overall height of 16.8 m. The length of each of the main cavern is about 50 m and the water retaining height of the proposed salt water reservoir is 9 m.

The method of excavation for the caverns was as follows:

1. For weak rock, a mechanical breaker mounted on a backhoe was used for excavation without the need for pre-excavation drilling.
2. For stronger rock, a drilling jumbo was used to drill 50 or 100 mm diameter holes and then mechanical breaking followed using a hydraulic breaker and where necessary a hydraulic splitter.
3. In very strong rock, a non-explosive demolition agent (Bristar) was used.

Thirteen months after commencement, the excavation work with a total volume of 33,000 m^3 was completed by the drill-and-break method. Favourable rock conditions made this fast-track excavation programme feasible.

The sequence adopted for different sections included advance probing followed by:

- **Initial access tunnel**—Full face excavation of the 8.8 m span excavation with typical height of around 8.4 m (horseshoe shape at access tunnel and tunnel junction). The tunnel under construction is shown in Figure 10.
- **Full span cavern**—Header and bench excavation was adopted for the full span 17.6 m excavation with typical height of 16.8 m (rectangular base, arch roof at main cavern for water reservoir). The header excavation was approx 8 m high (see Figure 11) followed by a bench approximately 4 m high.

Typical advance for each round of excavation was 1 to 2 m for both the initial access tunnel and the main 17.6 m span caverns. The total volume of soil and rock excavated was about 33,000 m^3.

Temporary support was designed using the Q-system. This typically resulted in steel arch ribs and fibre reinforced shotcrete for the initial excavation in mixed ground or areas with shallow rock cover, i.e. less than half span rock cover. For excavation at locations with adequate rock cover, the excavation was supported by rock bolts and fibre reinforced shotcrete. Expansion type Swellex rock bolts were used to support the crown of the excavation. Typically 3 m long pattern bolt for 8.8 m span excavation and typically 5 m long pattern bolts for the 17.6 m span excavation, with centre to centre spacing between 1.5 m and 2.0 m were adopted. Traditional drill and grout type rock bolts (25 or 32 mm dia high yield steel bars, 3 m or 5 m long) were mainly used for supporting vertical face in the lower part of the main cavern.

Construction plant used included two Atlas Copco Drilling Jumbo equipped with rock drill and / or hydraulic splitter (see Figure 12); backhoes equipped with hydraulic breakers in various sizes; a loader for spoil removal; two Meyco Potenza shotcrete robots for the application of wet mix shotcrete; a Cherry Picker; and a Meyco Piccola

Figure 10. Excavation using mechanical breaker for the initial tunnel

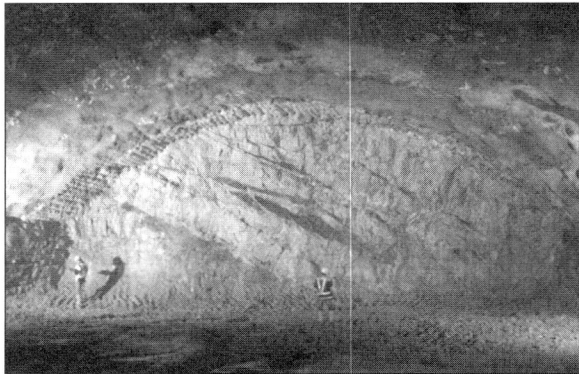

Figure 11. The main cavern first bench—17.6 m span by approx. 8 m high

shotcreting machine for the application of a dry mix final finishing shotcrete for the permanent shotcrete in the arch roof.

Final lining for the arch roof portion comprises 50 mm diameter 1 m long drain holes and drainage strips. A drainage composite has been installed behind the drainage strip to enhance drainage performance. A spray-on waterproofing membrane was then applied and permanent lining will comprise further steel fibre reinforced shotcrete. A final smoothing layer will then be applied to finish the lining.

For other vertical areas, 50 mm diameter 1 m long drain holes and drainage geotextile have been installed. A drainage composite has been installed behind the drainage geotextile to enhance drainage performance. A waterproofing membrane was then applied. Permanent lining will then be constructed using traditional reinforced concrete methods.

The total volume of concrete used was approximately 7,000 m^3 with 1,400 tonnes of reinforcement bar. The quantity is substantially higher than many tunnelling projects due to the tall sectional shape of the cavern required to house the water reservoir, and shallow rock cover at a few locations. In addition lining that forms the water reservoir is designed to be water retaining which imposes stringent crack width requirements.

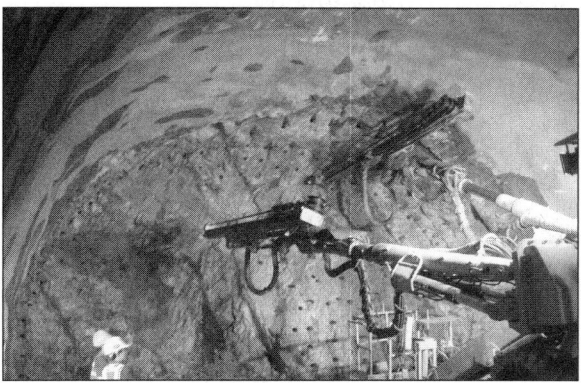

Figure 12. Drilling with the jumbo

CONCLUSIONS

In order to provide space for the development of HKU Centennial Campus to the west of the existing HKU Main Campus, the project team came up with an innovative, sustainable and environmental friendly solution proposing the re-provision of salt water reservoirs in caverns within the Lung Fu Shan hillside, with the area currently occupied by the existing salt water reservoirs made available for the potable water reservoirs.

The placing of a water reservoir in a cavern is unprecedented in Hong Kong and the construction of the project is currently in progress. By taking this approach, substantial slope cutting has been avoided, which in turn minimised the need for modification of the existing landform. The cavern scheme reduces significantly the numbers of mature trees that would need to be felled, avoids disturbance to the existing habitat and minimises visual impact to the neighbourhood. The cavern scheme also facilitates in-situ preservation of the three historic buildings. The approach is also environmental beneficial because dust and noise impacts and the generation of waste materials during the course of the construction works have been reduced.

It is hoped that the implementation of the cavern scheme to house water reservoir will serve as a good example for the increased use of underground space for public facilities in Hong Kong. In this way valuable surface land resources can be preserved and the impact on the environment can be minimised.

REFERENCE

Geotechnical Engineering Office (2000). The Pre-Quaternary Geology of Hong Kong, Hong Kong Geological Survey, Hong Kong Government.

ACKNOWLEDGMENTS

The authors wish to express their sincere thanks to the University of Hong Kong and the Water Supplies Department of the HKSAR Government, PRC for their kind permission to publish this paper.

HOBSON AND ROSEDALE TUNNELS— NEW TECHNOLOGY IN AUCKLAND

Harry Asche ▪ Connell Wagner

Tom Ireland ▪ Connell Wagner

Mike Sheffield ▪ Watercare Services Limited

Mike Bonnette ▪ McConnell Dowell Constructors

ABSTRACT

The last 12 months has seen the construction of the first Earth Pressure Balance (EPB) TBM excavated tunnel in New Zealand on the Project Hobson sewer storage tunnel for Watercare Services Limited, with the concurrent excavation of the second EPB tunnel on the Rosedale Outfall project for North Shore City Council.

Together these tunnels involve a total length of 6km and follow an 8 year pause in TBM activity in Auckland. TBM performance on these two projects is thus of significant interest as there are four major tunneling projects totaling over NZ$8B being planned for Auckland, in similar ground conditions.

Auckland's bedrock is a weak sandstone/siltstone (East Coast Bays Formation), subject to infilled paleovalleys. The geology in the region has been significantly influenced by volcanic activity; 49 volcanoes are present within 20km of the city centre.

The paper describes the design and construction challenges for the projects, including TBM performance, 'sticky spoil' management, and risks associated with basalt lava flows and two paleovalleys. Watercare's Project Hobson also saw the successful implementation of a Geotechnical Baseline Report—one of the first uses of this type of document in Australasia. Rosedale Outfall is designed for high internal water pressures with varying hydraulic flows, with the risk of exfiltration explicitly addressed.

INTRODUCTION

This paper describes the first implementation of Earth Pressure Balance (EPB) TBM technology in Auckland, following an 8 year pause in TBM activity in the region. Problems on previous projects have concerned public sector underground project sponsors, and therefore successful implementation would enhance confidence in the major tunneling projects presently planned for Auckland.

This paper describes the geological challenges presented by Project Hobson and the Rosedale Outfall project, and addresses in detail the performance of the TBM on Project Hobson.

BACKGROUND

Auckland Geological Setting

The basement rocks in the Auckland region comprise greywacke and argillite. These are overlain by the Waitemata Group which is comprised of a number of distinct subgroups and formations. The most common subgroup is the East Coast Bays Formation (ECBF) which occurs throughout the Auckland region and comprises shallow dipping alternating beds of extremely weak to weak sandstone and siltstone. In addition, there are occasional interbedded lenses of Parnell Grit, a weak to moderately strong sandstone. The Waitemata Group sediments were deposited during the Miocene period in shallow seas and have been subsequently uplifted and eroded to form the foundation of the modern landscape.

During the Pleistocene period, material derived from the ECBF rocks were deposited in a shallow marine environment to form the Tauranga Group, a firm to stiff clay which is often found infilling paleovalleys above the ECBF.

Intermittent volcanic activity has occurred throughout the region resulting in basalt cones and flows and several explosion craters that give rise to local depressions and tuff deposits. With respect to Project Hobson two main volcanic centers have affected the Hobson Bay area: the Orakei Basin, which formed from an explosive eruption and coinciding ash and tuff deposits; and the basalt lava flow from the Little Rangitoto volcano.

The ash and lapilli deposits consist of unconsolidated beds of angular to rounded, well sorted, dense to vesicular basalt fragments. When weathered, the ash deposits behave like soft to firm, silt and clay. The tuff consists of thin graded beds of angular to rounded, well sorted, clay and sand sized ejecta comprising pre-volcanic material and basalt fragments. The tuff may be very soft, compacted or welded. When weathered, it can exhibit behavioral characteristics similar to loose to dense sandy or silty gravels.

The basalt lava flow from Little Rangitoto, comprises very strong rock, with a fine grained crystalline texture. The top and bottom surfaces of the lava are highly fractured, behaving as a vesicular gravel and cobbles with soft clay and silt infilling.

Recent alluvium comprising loose silty sand to very soft silty clay to silty sand, and older alluvium comprising firm to stiff silts and clays of medium to high plasticity and/or dense sands and gravels, has been deposited in the low lying areas.

Previous TBM Tunneling in Auckland

Excluding a number of relatively small scale pipe-jack projects, the only other use of TBM technology within Auckland was on the Vector CBD Reinforcement Project, where a 3.56 diameter double-shield TBM was used to construct a 6.185km long tunnel. This project was completed in October 2000, after being delayed by poor performance of the TBM. One of the main issues that delayed production was the sticky nature of the ECBF material when excavated. This became the subject of a significant Unforeseen Physical Conditions Claim (UPCC) submitted by the contractor.

The outcome on this project caused project sponsors to doubt the cost certainty of tunneling in Auckland, and in some cases to question the viability of TBM tunneling in the prevailing ECBF material. As outlined below there is a significant volume of tunneling work proposed for Auckland in the medium term, and therefore Watercare's Project Hobson represents an important benchmark for proving the viability and cost certainty of TBM technology in Auckland.

Future TBM Tunneling Projects Proposed in Auckland

Like many modern cities, Aucklanders have realized that further development requires improved infrastructure, and that surface constraints will necessitate major tunneling projects. At present, four major projects are in varying phases of planning, and these future projects rely on the successes of Project Hobson and Rosedale to demonstrate that ECBF rock and paleochannels of Pleistocene and recent alluvium can be successfully handled by modern equipment.

Central Interceptor. Watercare is planning a 25km tunnel system to intercept flows from the CBD and nearby suburbs. The tunnels will be largely in ECBF with diameters ranging from 3m to 7m, depending on the hydraulic requirements of conveyance and storage. Watercare have recently requested expressions of interest from consultants to take this project through to the preliminary design and environmental approval stage.

Waterview Connection. A missing link in Auckland's freeway system, the Waterview Connection connects the unfinished SH20 (the freeway from the Airport) to the existing SH16 (the North Western Motorway). Originally proposed as a surface motorway, increasing lengths of this motorway have been proposed as cut and cover sections of tunnel. The difficulty with cut and cover solutions has been that a basalt flow overlays the ECBF along much of the motorway route. Furthermore, open cut excavation of the basalt with blasting or hammering is seen as a major environmental concern. As such, the current proposal is a driven tunnel beneath the basalt, staying in the ECBF, which uses the best of the geology, with soft rock cutting in the ECBF and the basalt above providing relief from surface settlement.

Currently, the scheme comprises twin 3km TBM tunnels with either two or three lanes in each tube. Environmental approvals are due to be sought in early 2009, with the preparation of contract documentation expected later in the year.

CBD Rail Link. The upgrading of Auckland's passenger rail system is currently underway with duplication of suburban lines and electrification of the network. However it is clear that further improvements will require the capacity of the CBD network to increase. The agencies responsible for planning and constructing passenger rail in Auckland (ARTA and Ontrack) have identified a route from the current terminus at Britomart to Eden Park station, with approximately 3km of tunnel, all in ECBF, and two new underground stations. They have called for tenders from consultants to take this project through to the preliminary design and Notice of Requirement, which will reserve the route.

Waitamata Harbour Crossing. Recent studies have considered how best to augment the existing crossings of the Waitemata Harbour. The studies concluded that the best solution comprised tunneled crossings of the harbor including both road and rail tunnels. This is expected to involve 4 tunnels using EPB machines. The geology is mainly ECBF with palaeochannels of Pleistocene and recently deposited alluvium. The NZ Transport Agency has recently called for consultants to tender for submitting a Notice of Requirement, which will reserve the route for both road and rail.

PROJECT HOBSON DESCRIPTION

Project Overview

Watercare is New Zealand's largest company in the water and wastewater industry, responsible for supplying water to the Auckland region, and operating the regional wastewater network. A key component of the wastewater network, conveying a quarter of the Greater Auckland wastewater flows, is the existing 90 year old Hobson Bay pipeline which is however nearing the end of its economic life. The $118M Project Hobson replaces the existing pipeline with a tunnel beneath Hobson Bay that will offer several benefits to Watercare and the local and regional communities:

- Provide capacity to meet projected growth in the region
- Virtually eliminate wastewater overflows into Hobson Bay and the Waitemata Harbour
- Open up the bay for recreational purposes

The new tunnel will receive flows from the Orakei Main Sewer and branch sewers and convey the flows to a new pump station (PS64), which will pump to the Eastern

Interceptor. The bored tunnel is approximately 3km long and has an internal diameter of 3.7m and external diameter of 4.2m. The tunnel is being constructed using an earth pressure balance tunnel boring machine capable of operating in open and closed modes. The ground support is provided by a single pass, bolted and gasketted precast concrete segmental lining.

The existing sewer is located just above sea level in Hobson Bay, and falls east, at a 1 in 2000 grade, towards the existing pump station (PS16). The replacement sewer is to be in tunnel, which requires the construction of inlet vortex drop shafts at each of the junctions (8m and 5m diameter) and a 22m deep pump station. The horizontal alignment of the tunnel commences on the west shoreline of Hobson Bay adjacent to Logan Terrace. Figure 1 illustrates the upstream connection to the existing network. The tunnel crosses the western side of Hobson Bay to the Victoria Avenue Peninsula, passes close to the existing foreshore to connect with Victoria Avenue Drop Shaft, and then continues across the eastern side of Hobson Bay, below the Orakei ridgeline to a new pump station (PS64) within the Orakei Domain adjacent to PS16.

Geological Challenges

A key decision for the project was the depth of tunnel. The tunnel traverses two palaeochannels within Hobson Bay, which are filled with sediments above the palaeosurface of the ECBF. During the environmental assessment the initial concept was a deep tunnel constructed in the ECBF; thereby avoiding the alluvial channels. At that stage, only preliminary geotechnical investigations had been completed; however, conventional tunneling methods such as roadheader excavation were envisaged. To ensure the tunnel was wholly located within the ECBF required a pump station with excessive depth and committed Watercare to higher operating costs throughout the life of the facility. Thus the decision was made to lift the alignment as high as possible, whilst remaining below the most adverse ground conditions—a basalt flow from Little Rangitoto on the eastern side of Hobson Bay.

The strength and consistency of the basalt flow is variable, creating potential for the rock to be water charged. Therefore the risks, and TBM performance and configuration requirements associated with tunneling through the basalt, were not considered to be acceptable. A tunnel elevation was therefore established to pass below the underside of the rock.

With the decision to use a higher alignment, the geological risk profile for the project increased. The roadheader tunnel would pass through ECBF only—a relatively consistent and tunnel friendly material. Whereas the TBM alignment would need to negotiate the following materials:

- ECBF
- Ash/tuff
- Recent (Holocene) alluvium
- Older (Pleistocene and Pliocene) alluvium

There was also the overriding risk of intersecting the Little Rangitoto basalt lava flow with the higher alignment. Whilst significant geotechnical investigation were undertaken to define the underside of the basalt flow, this is a complex task, as significant variations in basalt depth are possible as the lava flows fill the erosional surface at the time of the eruption.

The other risks associated with the higher alignment were:

- Tunneling through mixed face conditions at the unit boundaries
- Tunnel face stability in alluvial units
- Significant water inflows within the ash/tuff units

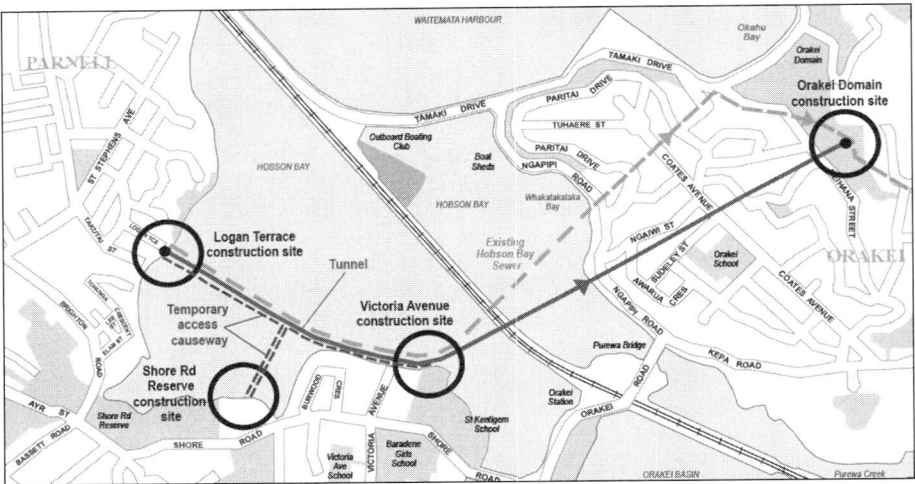

Figure 1. Project Hobson tunnel alignment

- Risk of volcanic ejecta (called "bombs"), which can be made from hard material

Risk Management

The approach to the management of the ground risk on this project, in order to achieve the objectives of cost and program certainty, included several key measures described below.

Geotechnical Baseline Report. Watercare elected to manage and share tunneling risk through, among other key measures, the implementation of a Geotechnical Baseline Report (GBR). This approach addressed one of the key issues that arose on the previous TBM tunnel in Auckland, whereby in a highly competitive bidding environment the low bidder made overly optimistic interpretations of the data, and used a second hand machine that was unsuitable for the ground conditions. When the actual conditions encountered were more adverse than the bidders interpretation significant claims were submitted. A key benefit from the implementation of the GBR on this project is that the risk profile adopted by each bidder was the same, and competitive advantage was not gained by a bidder through elevating the owner's risk.

In addition the geotechnical risks taken by the bidders was limited where it was judged that better project value was achieved through Watercare taking certain risks. The prime example related to the depth of basalt flow, and this key baseline was set in the GBR (i.e., the basalt will not be encountered within the tunnel horizon). Forward priced variations covered the risk of encountering the basalt. One of the geological units expected to be encountered is known to contain volcanic ejecta (called "bombs"), which can be made from hard material. A baseline was set to limit the number and size of bombs to be priced. Forward priced variations were included for encountering more than the baseline number.

Risk management utilizing the GBR was also undertaken through the specification of tunnel construction methods and risk mitigations, such as mandatory closed mode operation in certain locations and forward probe drilling to locate the basalt lava flow. For example, where the tunnel passes through the East Paleovalley which formed part of Hobson Bay, it was expected that mixed face conditions comprising weathered

ECBF, volcanic deposits such ash and scoria and the older (Pleistocene deposits) muds would be present. It was specified that the TBM should be operated in closed mode within this section. For the more consistent sections, where the tunneling face was expected to consist of entirely ECBF, the contractor was given the option of open or closed mode.

The implementation of a GBR on project Hobson has been very successful, with the TBM working as planned, the contractor finishing ahead of program, and no expected tunneling related ground condition claims or disputes.

TBM Specification. As part of the risk management undertaken by Watercare for this project, a detailed TBM specification was prepared so that the machines offered by each of the bidders adequately addressed the geological risks presented by the higher alignment.

An Earth Pressure Balance Machine (EPBM) is being used on the project, although the specification also allowed the use of Slurry Machines. The main potential benefit offered by a Slurry Machine was the improved performance tunneling though basalt. Whilst the TBM needed to be designed to tunnel though the basalt, this risk was considered unlikely, and even if required it would be for a short distance.

The requirement for tunnelling through the alluvium in the paleo-valleys beneath Hobson Bay meant a closed face machine was required, although the opportunity for both open-mode and closed-mode operation existed on the project.

The fundamental requirements of the TBM to address the geological risks outlined above were:
- Face pressure controlled to match the in-situ ground and groundwater pressures
- Means to measure volumes of excavated material to match advance
- Continuous grouting of the annular void
- Minimizing of overcutting and loss of ground resulting from steering and directional control
- Automatic controls to achieve face pressure control, control of excavated volumes, and tail void grouting
- Tools and cutting head configurations to cover both the expected ground conditions, and the basalt lava flow

Other specified requirements included:
- Cutting tools replaceable from the rear of the cutter head
- Soil conditioning system designed to inject water, foam or another lubricant such as a polymer to the face
- Computerized laser guided guidance system
- Guillotine doors to seal within 20 seconds the excavation bulkhead when encountering running ground or flowing water
- Compressed air system to allow tool changes within soft ground areas
- Variable drive system for the cutterhead
- Screw conveyor design to accommodate volcanic block/bombs of 180mm diameter
- Probe drill to allow 20m to be drilled ahead of the face
- TBM data collection system

The key to managing the risks presented by the higher alignment related to the possibility of intersecting the basalt lava flow. Our approach was to specify a TBM that could tunnel through the basalt using disc cutters if it was encountered. The other

Figure 2. Earth pressure balance TBM during factory testing

key risk was to confirm the depth of the basalt prior to reaching the basalt so that the tools could be changed from rippers to discs in free air. The depth of the basalt was confirmed using probe drilling. Although there are forward price variations within the contract for head interventions under compressed air, from a time, cost and safety perspective this was to be avoided if at all possible, and in the event this was not required. Figure 2 shows the TBM during the factory commissioning and testing prior to shipping.

ROSEDALE OUTFALL PROJECT

General

North Shore City Council operates its own wastewater system and treatment plant for the north shore suburbs of the Auckland region. Wastewater flows by gravity to a series of pumping stations which convey the water by rising mains to the Rosedale Waste Water Treatment Plant (WWTP), located in the suburb of Rosedale in the centre of North Shore City.

At the Rosedale WWTP, the wastewater is treated to tertiary level, concluding with UV treatment either occurring naturally by sunshine in the summer months, or through a UV plant at other times if required. The treated wastewater is then conveyed by gravity to Castor Bay with a pipeline including a 600m marine section of pipe. This outfall is at the end of its economic life, it is the major constraint for flows out of the WWTP, and its short marine length does not meet current needs for distance from bathing beaches.

The project consists of a new outfall with a 100 year design life, a specified capacity of $6m^3/s$ and improved diffusion characteristics. Figure 3 shows the project location.

The project commences with a new connection to the UV plant. The flow then falls to a 3km EPB tunnel. This tunnel was sized by the contractor to be the minimum size for economical EPB tunnel driving and has an internal diameter of 2.8m and external diameter of 3.2m. This tunnel takes the flow to 600m offshore of Mairangi Bay. A riser takes the flow up to a 1477mm internal diameter HDPE marine pipeline, 1800m long. A series of diffusers, spread over 300m completes the project.

The EPB tunnel is located entirely in ECBF rock. Near the UV plant and down at Mairangi Bay, there are surficial deposits of Tauranga Group (a Pleistocene formation of clay). The potential for settlement caused by groundwater drawdown in this formation is a risk for the project.

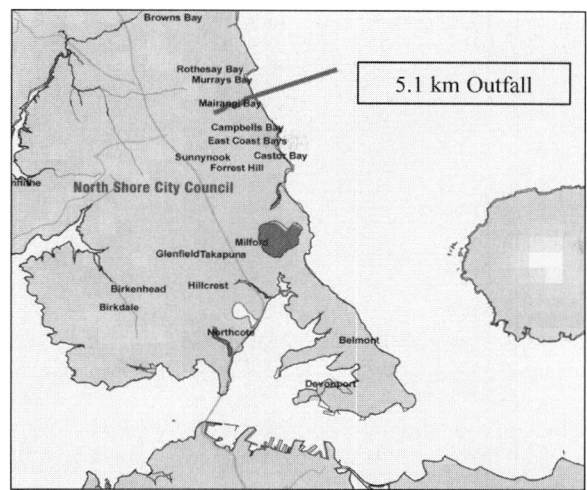

Figure 3. Rosedale outfall project location

Hydraulics

The average flow from the WWTP is currently less than 1m³/s. Upgrades to the WWTP have been planned up to 4m³/s. The outfall is specified to carry a maximum of 6m³/s.

Looking at the layout of the project, it can be seen that almost all of the resistance to flow is located in the HDPE section due its small ID. For a given flow, the hydraulic grade line (HGL) is almost flat along the driven tunnel section, then falling in the HDPE pipeline to the diffuser section, where the HGL is equal to sea level plus the outlet loss. The HGL in the driven tunnel varies from +3m (i.e., beach level) at 1m³/s to +22m at 6m³/s.

Hydrofracture

The driven tunnel is constructed with a gasketted segmental lining. The seal provided by the gaskets relies on the confinement of the surrounding ground. Calculations show that the ECBF has the capacity to keep the gaskets compressed under full internal pressure.

However, the nature of segmental precast lining construction means some leakage is possible. If water leaks from the inside to a gap on the outside of the segments, and the pressure in the gap increases, there is a risk that the confinement provided by the surrounding ground is reduced. In effect, the problem becomes similar to an unlined pressure tunnel, typical of hydro power schemes (Hoek and Moy 1995).

Therefore the design alignment has been chosen such that the internal pressure is lower than the minimum in-situ principal stress in the ECBF rock. Testing for in-situ stresses was carried out for the project by the hydrofracturing technique (Enever 1993). Whilst the majority of the readings showed horizontal stresses somewhat higher than the vertical stress, some values showed horizontal stresses slightly lower. The vertical alignment has been set so that the factored internal pressure is less than the minor horizontal principal stress which has a ratio of around 0.9 when compared to the vertical in-situ stress. Should the tunnel leak due to high internal pressures the insitu stress field will prevent hydro-fracturing of the ground and any risk of leakage to the surface.

Table 1. Project Hobson TBM performance statistics

	Number of Rings (1.2m long)	Distance (m)
Total drive length	2417	2900
Weekly average since commissioning	90	108
Top shift performance	16	19.2
Top daily performance	31	37.2
Top weekly performance	127	152.4
Top monthly performance	500	600

TBM PERFORMANCE

McConnell Dowell Constructors is undertaking the tunneling on both projects described above.

For Project Hobson the contractor has procured a new Lovat RME170SE 23700 TBM with a nominal cut diameter of 4.353m. The TBM length is 8m and the total length is 70m. The TBM weight is 175t and the total weight is 250t.

The Rosedale Outfall project is using a similar new Lovat RME131SE TBM with a nominal cut diameter of 3.352m. The TBM length is 9m and the total length is 73m. The TBM weight is 88t and the total weight is 160t.

Performance Figures

At the time of writing this paper the Project Hobson machine has 250m to go and will be complete in the next three weeks which is two weeks ahead of program. The Rosedale Outfall machine has just been launched, and therefore, this section describes the performance of the TBM on Project Hobson.

The machine has performed very well and has lived up to the reputation of Lovat machines for robustness and reliability. Refer to Table 1 for project performance statistics, which relates to a TBM availability of over 90%.

Sticky Spoil Management

As discussed in Section 2 the sticky nature of the ECBF material when excavated was the subject of an unforeseen conditions claim on a previous project in Auckland. This section describes how this issue was overcome on Project Hobson.

ECBF Properties. The ECBF material is an interbedded sandstone/siltstone. The rocks are typically extremely weak to weak, exhibiting both rock and soil strength characteristics. Petrographic analysis indicates that the sandstone and siltstone samples contain both lithic clasts (i.e., rock fragments) and mineral clasts (individual mineral grains or crystals) set in a matrix of predominately clay. The difference is that the clay matrix predominates in the siltstones ($>\frac{2}{3}$ of the rock mass) and is subsidiary in the sandstone ($<\frac{1}{3}$ of the rock mass). The sandstone is friable because diagenesis of the cement is incomplete (diagenesis is the process of sediments becoming rocks).

In its natural state the water content of the ECBF is close to the plastic limit of the pulverized fraction. Thus, the mechanized excavation process and the application of water becomes a trigger for clay stickiness. Destructuring occurs because silt-sized particles—some of which are aggregations of clay particles, are relatively weakly bonded. Under the influence of severe mechanical working the clay aggregations may be disaggregated and release clay mineral particles (Atkinson et al 2003). The resulting plasticity and swelling behavior depends on the mineralogy of the released clay, which for ECBF is Smectite. This has implications for the operation of a closed mode machine using a screw conveyor, as the cutterhead openings, working chamber and

Table 2. East Coast Bays Formation properties

Test Type	Mean Data
Water content	20%
Atterberg Limits—Liquid Limit	50%
Atterberg Limits—Plasticity Index	24%
Uniaxial Compressive Strength (UCS)	4 MPa
Sandstone beds clay content	15–30%
Siltstone beds clay content	60–75%

screw flights can rapidly become clogged with swelling clay. Table 2 summarizes the salient properties of insitu ECBF.

Solution. The solution to dealing with sticky spoil was to condition the ground appropriately. The contract specification required the TBM to have a ground conditioning system with the ability to inject four conditioning agents to the cutterhead, working chamber and screw conveyor.

Various spoil conditioning agents were trialed. It was found, however, that basic foam effectively prevented clogging. The TBM removes spoil using the screw conveyor in all modes of operation, and it was necessary to add a large volume of water to produce a wet foam, and spoil of a 'porridge' like consistency. If the mix became too dry, the screw conveyor torque would increase, and the rate of removal would decrease. The capacity of the spoil conditioning pumps was increased from 100l/min to 160l/min to provide the optimal consistency. It was found that only relatively small amounts of air were required (i.e., a low Foam Expansion Ratio).

The additional of foam was essential. When excavation was attempted without foam, the screw conveyor blocked immediately. The ease of excavation was also sensitive to the percentage of soap added with screw conveyor blockages resulting if the soap concentration or total foam flow rate was reduced. It was necessary to adjust the flow rate depending on the amount of water found in the rock. In dry areas it was also necessary to add water directly to the cutterhead chamber.

These observations indicate that in this case the foam is an effective surfactant in the clay rich soils. A surfactant molecule is made up of a water soluble (hydrophilic) and a water insoluble (hydrophobic) component. The surfactant absorbs at the surface of the solid particles with the hydrophilic ends of the molecules oriented towards the water phase. The surfactants form what amounts to a protective coating around the suspended material, while the hydrophilic ends associate with neighboring water molecules. This mechanism allows the spoil to be turned into slurry thereby reducing sticking and clogging both between particles and to metallic surfaces such as the cutterhead and cutterhead chamber.

One disadvantage of this ground conditioning approach is that the resultant spoil is a wet mixture with a moisture content of approximately 37%. The ECBF material is suitable for use as engineered fill, but has an optimum moisture content of around 18%. This means it needs to be windrowed to dry prior to use as engineered fill, which requires multiple handling and reduces the materials economic viability as a fill source. Table 3 summarizes the tunnel spoil characteristics found on Project Hobson.

Groundwater Control

For Project Hobson, the first kilometer of tunneling was under residential properties, and the last 700m was adjacent to the 90 year old operational main sewer that the tunnel is to replace when completed. For this reason, it was necessary to limit groundwater ingress, and an extensive monitoring program (as required by the project

Table 3. East Coast Bays formation tunnel spoil properties

Test Type	Mean Data
In situ Bulk Density	2.09 t/m^3
In situ Dry Density	1.74 t/m^3
Water Content	20%
Spoil Bulk Density	1.58 t/m^3
Bulking factor	1.32
Spoil Water Content	37%
Spoil Loose Density	1.26 t/m^3

consents) was implemented. A total of 17 piezometers were monitored during construction. Several of these piezometers have been monitored since 2003, which provided a good indication of the seasonal variation. Even so, the dry summer of 2007/8 in the months proceeding tunneling caused several piezometers to dip to their lowest recorded levels. Fortunately this lowering of the water table occurred prior to tunneling, and therefore, could not be confused as a tunneling related effect.

Although the tunnel has been constructed using a closed face machine with a watertight gasketted segmental lining installed behind, there is still 12m of exposed ground between the excavation face, and where the annular grout is installed behind the shield. If the TBM passes through a crush zone, water inflows from any high permeability joints in the rock are drained into the tunnel spoil at the face until the grouted segmental lining passes the joint. At the average advance rate this would take 13 hours, however crush zones could be intersected during a machine breakdown or over a weekend stoppage. One known crush zone was intersected during excavation, and the inflow was measured to be approximately 40l/min. Due to a machine breakdown and weekend stoppage it took five days to clear the crush zone. During that time an adjacent piezometer recorded a drop in the groundwater level within the ECBF material of 2.25m. This recovered by 1m within a week. This lowering of groundwater level within the ECBF is insignificant in terms of ground surface settlement. Figure 4 displays the drop in groundwater levels as the TBM intersected the crush zone, and the subsequent recovery.

At Rosedale, the tunneling related groundwater disturbance is expected to be similarly benign. The greatest cause of groundwater drawdown is the sinking of shafts. The construction and drop shaft at the UV plant caused significant groundwater drawdown.

Ground Settlement

Due to the location of the Project Hobson tunnel beneath residential properties for the first one kilometer the project consents require extensive settlement monitoring. This ranged from measurements every 6 months of all monitoring points to weekly measurements in the vicinity of the tunnel face, and daily measurements of 13 arrays to determine the settlement profile in relation to the excavation face.

The ground conditions beneath the residential area were self supporting ECBF rock, and tunneling was undertaken in open mode. The monitoring points consisted of survey pins installed in kerbing and the like. Both the monitoring points and the benchmarks were subject to seasonal movements of up to 5mm that were independent of the tunneling operation. Fortunately the monitoring points were established four years prior to the commencement of tunneling, and were monitored every six months in the two years before tunneling.

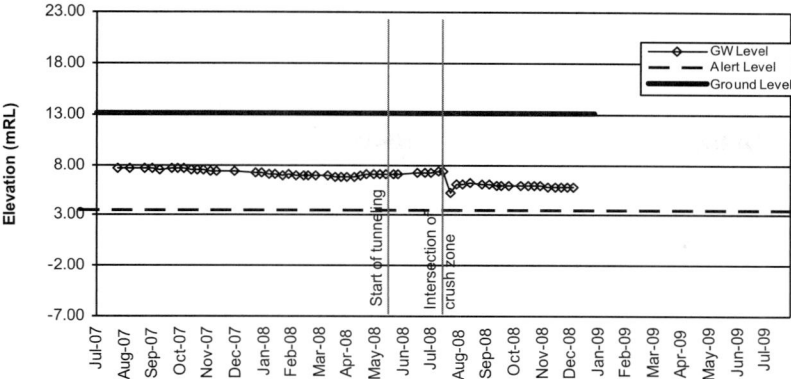

Figure 4. Piezometer showing influence of crush zone on groundwater table

The tunneling induced settlement was less than the tolerance of the survey, and within the range of seasonal movements. It was estimated to be from 1–2mm for the daily measurements taken, which is equivalent to a volume loss of around 0.5%.

The tunnel also passed under an operational surface railway, and daily monitoring of the tracks was undertaken when the TBM was within 50m of the railway. Fortunately the tunnel intersected the railway at the location of the basalt lava flow, and as the tracks were supported by a large volume of solid basalt no movement was recorded.

At Rosedale, settlement associated with the construction shaft drawdown has occurred in the Tauranga Group up to 12mm. However, there has been no damage observed in the closely monitored buildings around the construction shaft. The settlement caused by groundwater drawdown extends over 100m, but is very gradual and the buildings experience no more than a few mm of differential settlement each.

ACHIEVEMENTS

Watercare's Project Hobson has proved that EPB tunneling can be undertaken in Auckland with a large degree of cost and program certainty. This should give confidence to project sponsors who are currently planning several major underground projects for Auckland.

The implementation of a GBR by Watercare on Project Hobson has been a very successful risk management tool, and has contributed to making the project a success for all parties involved.

East Coast Bays Formation material has proved to be an excellent tunneling medium. The beneficial characteristics are:

- Material is consistent
- Low strength allows rapid excavation with minimal TBM wear
- Ground is self supporting leading to negligible settlement
- Ground is typically of low permeability leading to minimal impact on the groundwater table
- Potential stickiness can be overcome with appropriate spoil conditioning

The main remaining opportunity is to investigate an alternative spoil conditioning regime that minimizes the water content of the spoil in order to improve the economic viability of using the spoil as engineered fill.

ACKNOWLEDGMENTS

Thanks to Watercare Services Limited and North Shore City Council for permission to publish details of these projects.

REFERENCES

Atkinson, J.H., Fookes, B.F., Miglio, B.F., and Pettifer, G.S. 2003. Destructuring and disaggregation of Mercia Mudstone during full-face tunnelling. *Q. J. Eng. Geology Hydrogeology*. 36:296–303.

Enever, J.R., 1993. Experience with Hydraulic Fracture Stress Measurement in Australia. Comprehensive Rock Eng. Vol 3, Chapter 23, Pergamon Press.

Hoek E and Moy D 1995. Design of Large Powerhouse Caverns in Weak Rock. Comprehensive Rock Engineering Vol 5, ed John A Hudson, Pergamon Press.

FEASIBILITY AND IMPLEMENTATION OF SHIELD MACHINE TUNNEL PASSING THROUGH AN OPERATING AIRPORT RUNWAY

Hongjie Yang ▪ Shanghai Shentong Railway Transit Research & Consultancy Co. Ltd.

Xiuzhi Wang ▪ Shanghai Shentong Railway Transit Research & Consultancy Co. Ltd.

Hongbo Liu ▪ Shanghai Shentong Railway Transit Research & Consultancy Co. Ltd.

ABSTRACT

The No. 10 metro line of Shanghai city is 36 kilometers in length and will pass through the operating airport runway by using tunnel boring machine with earth pressure balance mode. Although there are examples through world that tunnels pass through airport runway and tunnelling underground is not a difficult undertaking technically, there is rare experience in China for the similar projects especially in Shanghai saturated soft soil region. Shanghai soft earth is famous for its high compressibility, high rate of water content, weakness in strength, large deformation and different characteristics in earth strata. Besides, the whole construction process will be under ultra-large airplane dynamic load on the surface ground over the tunnel, which increases the risks of construction process. This paper evaluates the feasibility of building the metro tunnel under such an extremely serious condition, especially the subsidence of ground surface caused by advancing of tunnel boring machine, which also is the most important control target for the safe operating of the runway. Furthermore, this paper discusses the effective measures to keep the settlements in the permissible level during the whole working period for the tunnel from aspects of plan, design and construction.

INTRODUCTION

Shanghai Hongqiao Airport (SHA) is located at the west suburb of Shanghai city, accommodating about 22 million passengers in 2008, SHA is the second busiest airport in Shanghai city. There is a runway at present which is 3400 meters in length and 57.6 meters in breadth and another new one will be built soon to improve the operating capacity of the whole airport. For the convenient communication between the airport and downtown of city, the No 10 metro line of Shanghai city which is 36 kilometers in length, was started to be built in 2004 and SHA is the west terminal of the whole metro line. The No 10 metro line consists of twin precast concrete tunnels, cut and over excavation at either portal to the tunnel. The airport portion of No 10 metro line includes two precast concrete tunnels that are machine bored through soils under flight operations at SHA. This portion lies between Konggang Yi Road station and Hongqiao East station, with Konggang Yi Road station on the east and Hongqiao East station on the west, about 1.8 kilometers in length for the single tunnel, as shown in Figure 1.

Since the underground tunnel will pass through the parking apron and runway, and the tunneling process is not permitted to interrupt their normal services, the strict safety requirements of the operating airport must make a tough construction process technically and a vast quantity of money must be spared for that. Hence, this paper discusses the feasibility and effective implementation measurements for the project.

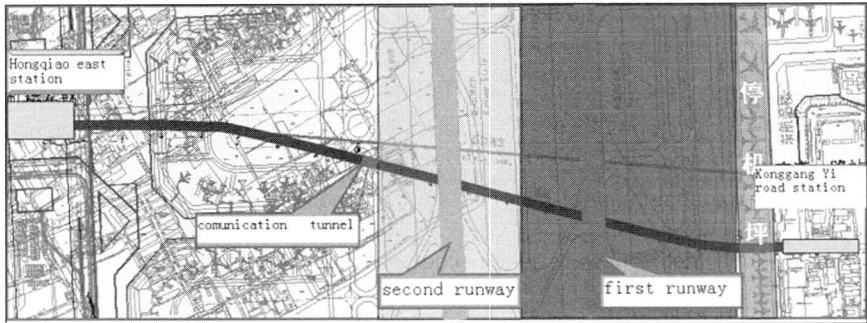

Figure 1. Site plan of metro tunnel and the airport runway

FEASIBILITY OF THE PROJECT

To evaluate the feasibility of the project, the geological condition is the first key factor in consideration. The geological report indicates within 50 meters in depth beneath airport runway, the soil mainly consist of clay, silt and sand, generally distributed in horizontal stratification. The planned twin tunnels beneath the airport runway mainly buried in ④2 and ⑤2 earth stratum. From the ground surface of runway downwards, ①,②1,②3,③, ④1, ④2 and ⑤2 earth stratum are encountered respectively, as shown in Figure 2. The serial number of earth stratum ①, ②1, ②3, ③, ④2 and ⑤2 stand for backfill earth, clay, sandy silt with silty clay, sullage clay, silty sand with silty clay, sandy silt stratum respectively. These different earth strata have different mechanic characteristics; experiments data about their mechanical characteristics in Table 1 indicate that these soil strata are soft, almost completely saturated, and weak in strength with high compressibility. Under such a complicated geological condition, TBM machine seems to be the only choice to the project. Generally speaking, alternative tunneling methods can not provide continuous face support in combination with a watertight lining, which are thought to be an unacceptable risk to the normal operation of the airport. And also other excavation methods can not provide the efficient control over ground movements. The most efficient, safe, and watertight alternative proved to be a TBM-installed precast tunnel with bolted, gasket connections.

Risks of disruption to airport operations due to excavation and construction in such difficult geological conditions beneath SHA are the most important consideration to the feasibility of the project. SHA airfield clearance restrictions (about 60 to 100m) for ensuring an obstacle free volume of space surrounding runway eliminates construction process with large ground surface construction equipment. Further more, the continuous operation of the airport requires that the only runway must be kept intact. That is to say, any earth improvements measure including grouting and lowering underground water can not be conducted for the project. In fact, the subsidence caused by tunneling through the soft untreated soil is the most dangerous risk that might threaten the normal operation of the runway. Hence, subsidence control is the key problem for the feasibility of the project. Fortunately for us, there are enough experiences for tunnel construction in Shanghai soft soil, a large quantity of field monitoring data have been accumulated during past twenty years of engineering practices.

We can conduct a rough evaluation about the controllable degree for ground surface subsidence of tunnelling beneath airport runway according to these experiences.

In Shanghai soft region, the ground surface subsidence usually is evaluated through a simple formula related to land loss rate as follows.

$$Lo = S*H/(pi*R^2) \tag{1}$$

Table 1. Mechanical parameters of earth strata beneath airport runway

Serial Number of Earth Stratum	Name of Earth Stratum	Water Content (%)	Gravity (KN/m³)	Saturation Degree (%)	Void Ratio (e)	Cohesion (KPa)	Internal Friction Angle (°)
②1	clay	32.3	18.5	96	0.925	17	15
②3	sandy silt with silty clay	28.8	18.8	96	0.829	1	31
③	sullage clay	39.8	17.6	96	1.129	11	15
④2	silty sand with silty clay	31.0	18.4	95	0.884	5	21
⑤2	sandy silt	30.5	18.4	94	0.874	2	29.5

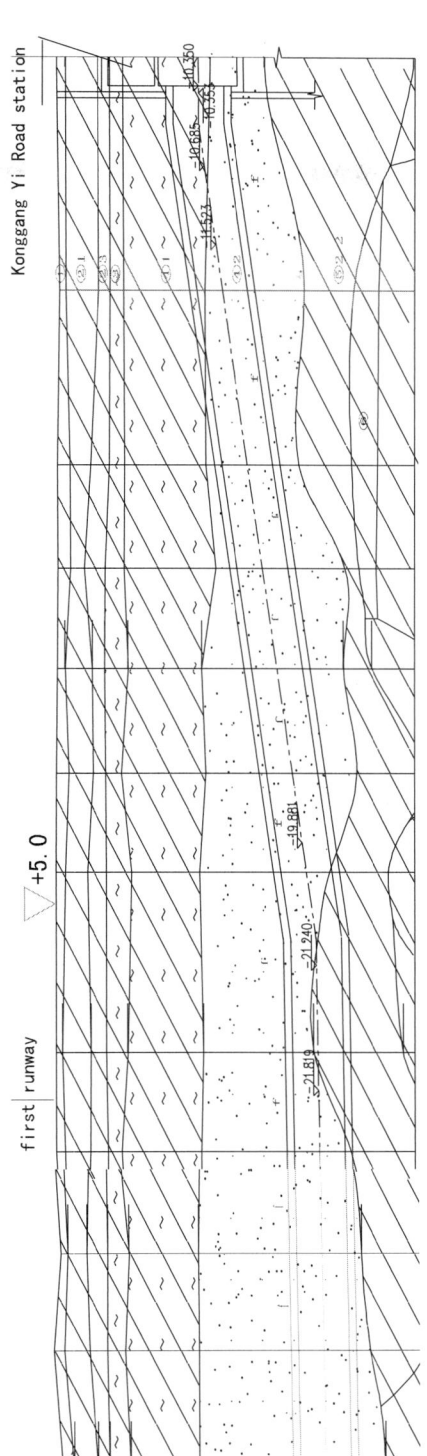

Figure 2. Soil profile at Shanghai Hongqiao Airport

In which Lo stands for land loss rate caused by shield machine advancing in soil, S stands for the maximum subsidence happened at the ground surface over center line of tunnel, H stands for the covered soil depth from ground surface to the center of tunnel lining. R stands for outer radius of shield machine. In most cases, Lo can be controlled within the scope of 5‰ to 2%. For example, for a tunnel excavated by shield of 3.17m in radius, with 10m depth of soil covered, 1% of land loss rate corresponds to maximum subsidence of 30mm. Of course the formula only reflects subsidence happened immediately after the construction of tunnel by using shield machine with EPB mode, it will continue to increase after finishing of the construction because of the creep characteristic of Shanghai soft soil and usually subsequent subsidence is more than that happened immediately after construction. However, if we take active and effective measures the subsequent subsidence can also be controlled in the permissible level. Hence subsidence happened during construction is the first factor must be considered to assure the safety of runway operating. According to the requirements of aviation rules, slope of subsidence of airport runway must be not more than 1.5/1,000 and subsidence curvature must be not more than 1/30,000, which is a very strict standard in considering the poor geological condition and airplane dynamic load on the surface ground over the tunnel. According to formula (1), the tunnel should be buried in soil as deep as possible in order to decrease the subsidence during construction. Suppose the depth is 20m, for a conventional land loss rate of 1% formula (1) gives subsidence of 16mm and corresponding slope of subsidence of airport runway is 0.8/1,000, which can meet the requirement of the aviation rule. Hence, the depth of buried tunnel should be more than 20 meters. However, equation (1) can only estimate the subsidence for the common case without effect of dynamic load of airplane. It can not judge whether the maximum subsidence curvature of runway meets the requirement of aviation rule, not more than 1/30,000. Hence a further numerical analysis is conducted to evaluate the feasibility of excavation beneath airport runway.

To simplify the analysis and spare the time for calculation, a two dimensional finite element model was constructed, as shown in Figure 3. Suppose the tunnel is buried 20 meters beneath the airport runway, the mesh region is 66 meters in breadth and 78 meters in height, with the twin tunnels 6.2 meters in outer diameter. Calculation results (as shown in Figure 4) indicate that the maximum subsidence caused by the excavation of tunnel is about 1.84mm and the corresponding curvature of runway face is about 1/36,407, less than 1/30,000. That is to say, the excavation can satisfy the requirement of aviation rule. Besides, the dynamic loads of airplane are also considered in the calculation. It can cause as much as 5mm subsidence for runway surface as shown in Figure 5, however, most of which is elastic and can resume to the initial shape as before dynamic load is exerted. About 0.05 MPa maximum additional vertical earth pressure is produced on the crown of tunnel because of airplane wheel load, which must be considered when set the earth pressure in the earth cabin of TBM. This additional earth pressure will cause disadvantageous effect for the shield machine excavation with EPB mode especially in the saturated sandy soil bearing water pressure. Under such circumstance, possibility of underground water coming into earth cabin of shield machine will increase, which can destroy the EPB working mode of excavation and result in the collapse of working face before cutter head of TBM. However, such a disadvantageous case is not insurmountable, much more related experience has been accumulated in the Shanghai soft earth region. Active and efficient measures will be discussed in the next part of the paper.

From the analysis above we can draw a conclusion that excavation of tunnel beneath the airport runway can be conducted theoretically. Subsidence caused by tunnelling process can be controlled in the permissible scope and will not disrupt the normal operation of runway, which indicates that the project is feasible technically.

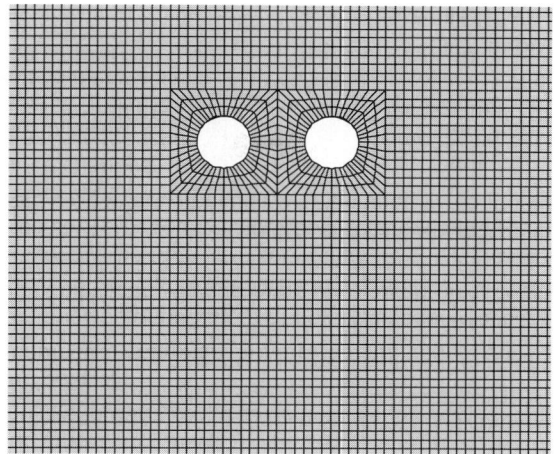

Figure 3. Finite element mesh used in the analysis

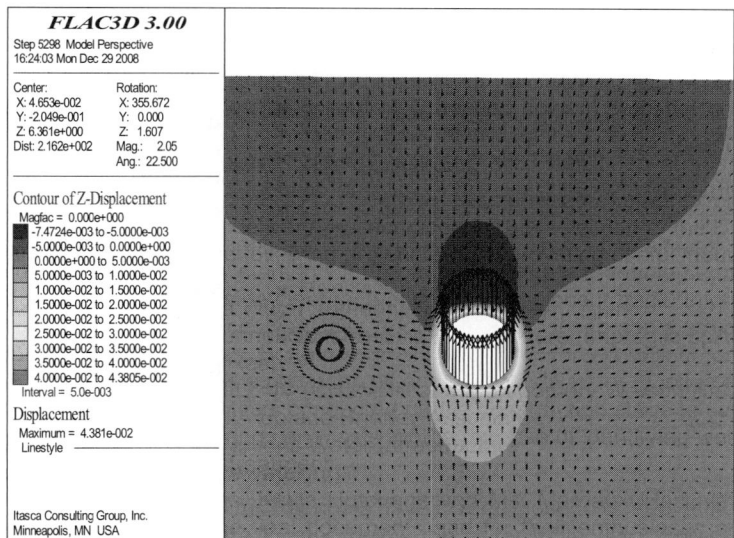

Figure 4. Vertical subsidence caused by shield excavation

ACTIVE MEASURES IN IMPLEMENTATION OF TUNNELLING

Although the project is feasible theoretically, there are great risks in the practical implementation thus active measures including aspects of plan, design and construction are discussed as follows.

Determining the Route of Twin Tunnels

Since there are strict requirements of deformations for operating runway from owner of airport, main attention is centered on decreasing the disadvantage effect to

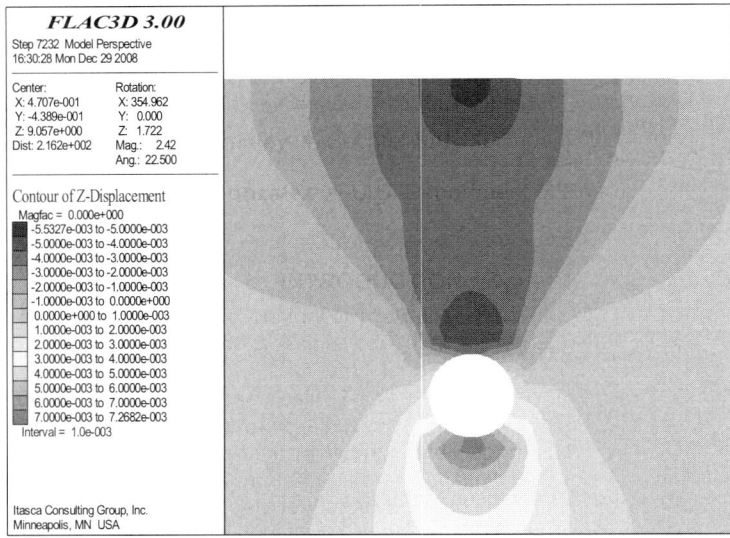

Figure 5. Vertical subsidence caused by airplane load

the airport runway when the twin tunnels route are planned. When the twin tunnels pass through the runway the route adopted straight style in horizontal plan to make the excavation of shield machine more easily, which can avoid more land loss when excavation in the curve tunnel route resulted from rectifying deviation of TBM body. At the same time vertical alignment of tunnels beneath the runway is planned in the slope angle as small as possible for the same purpose as horizontal plan. If only consider tunnel route style, it should not be buried in too deep soil. However, to assure the safe construction during the excavation process, an appropriate depth is needed. The different depth of tunnel will produce different effects to subsidence during excavation, load bearing capacity of tunnel structure, costs of project and so on. To decrease the mutual effect from airplane and underground metro, the twin tunnel should be buried in the soil as deep as possible, which needs to increase the depth of east and west portals at same time. This measure can assure the depth of tunnels beneath runway is not less than 20 meters.

According to such a alignment plan principle, the tunnel route is designed in detail as follows, in horizontal plan, after the tunnel extends out of the east portal (KonggangYi Road), it turns to the north in a curve with a radius of 1,200 meters, passes beneath the existed airport parking apron, taxiway, then passes through the existed runway (the first runway) and the new built runway (the second runway, will be built in April, 2009), then turns to the south in the curve with a radius of 1,200 meters and changes into straight route when passes beneath new built taxiway till at last passes through new built airport building reaching the west portal Hongqiao East Station. In the vertical plan, the tunnels start from east portal Konggang Yi Road with center elevation of −10.35 meters, goes downwards with a declivity slope angle of −2‰ to connect a vertical curve with a radius of 3,000 meters, then goes downwards with a slope angle of −28‰ to connect another vertical curve with a radius of 5000 meters reaching the position beneath airport runway with center elevation of −21.823 meters (the ground runway has an elevation of +4.407 meters). Then the tunnel goes downwards to reach the lowest point with an elevation of −23.614 meters where there is a communication tunnel to connect the twin tunnels, then goes upwards with a slope angles of 14.803‰

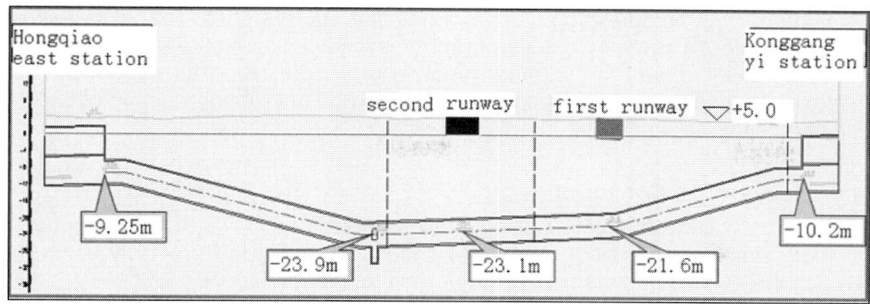

Figure 6. Vertical tunnel route view

and 25‰, till at last reaches the west portal Hongqiao East Station with tunnel center elevation of −10.41 meters. The designed route results in the tunnel buried depth of 11.28 meters at east portal Konggang Yi Road and 23.13 meters at the existed airport runway, as shown in Figure 6.

Structural Design of Twin Tunnels

The precast concrete tunnel lining system with strength over 55 MPa will be adopted for the whole project. The tunnel lining is composed of about 3,000 reinforced precast concrete rings that are connected together to achieve the whole designed length of the two TBM tunnels. A precast concrete ring is consists of five rectangular segments and one key. After assembled, the precast concrete segments will form a 1.2 meters wide ring, or circular tunnel with an inside diameter of 5.5 meters and outside diameter of 6.2 meters, each tunnel lining segment is 0.35 meters thick. For considerations of safe excavation and service life as long as 100 years, the precast concrete lining must meet some strict specifications such as good waterproof performance and high strength. All concrete segments are connected by metal bolts in longitudinal and circumferential directions.

In the structural designs of other underground tunnels there is only one back grouting hole in each segment except the key on the crown of tunnel. For the tunnel excavated beneath the airport runway, three back grouting holes are designed for each segments except the key, there are fifteen holes in all for each precast concrete ring, which can make the grouting process more uniformly.

Because of excavation beneath the airport runway, there are ultra-large airplane dynamic loads over the concrete tunnels, which make the joints of concrete segments open and close frequently. Such a continuous progress will be a very disadvantageous effect to the leakproof performance of tunnel structure, especially for tunnels in Shanghai saturated sandy soil. Hence, design of compression gaskets installed between joints of concrete segments is optimized to improve its impervious function and endurance. Ethylene-propylene-diene mischpolymer and materials which can expand when meet water are compounded to form the compression elastic gasket used in tunnel segment joints. New technology for gasket with porous cross section by compounding Ethylene-propylene-diene mischpolymer and materials which can expand when meet water is an advanced craft to produce compounds by one-off extrusion out of mold and microwave sulfuration, which overcomes the shortcomings once existed in such methods as mechanically embedding one material into another, for example, difficulty of embedding the watertight rubber into segment joints, easy exposure to the air and falling because of being affected with damp. Special material characteristics and specially designed porous cross section of the compression gaskets can completely take advantage of both compression leakproof function and material expanding leakproof function. Besides, a stripe of watertight rubber is installed along the outer brim of segment joints to increase the leakproof function of concrete ring further. All materials applied to produce the compression gasket and watertight rubber must be of excellent quality and endurance.

Active Measures in Construction

The excavation process by using TBM beneath the runway is important in the implementation of the project. The TBM should be controlled precisely in order to decrease the degree of disturbing earth around tunnel, thus decrease the subsidence as much as possible to meet the specification of airport owner. EPM mode is important during excavation to keep the stability of working face before blade, which can decrease compression and looseness of soil. Axis of shield machine should be in accord with the designed axis of tunnel, which can decrease the correcting deflection quantity of shield machine body and corresponding shearing disturb to earth, also can decease the gap between shield machine and earth around it. However, this gap is inevitable for the tunnels excavated by TBM since the diameter of machine is lager than outer diameter of precast concrete ring of tunnel, which is one of main factors for subsidence. Hence, this architectural gap should be filled with synchronized back grouting in time. According to experience of tunnel construction in Shanghai, quantity of grouting should be in the scope of 100% to 150% times of that theoretic gap. After extrusion of shield machine tail, the second grouting process should be started according to quantity of subsidence monitored by instruments through fifteen back grouting holes in the segments. Geotechnical instrumentation is located on the ground surface directly over the excavation to monitor any construction induced subsidence and vibrations. Grouting applied in this project contains cement, flyash, lime, addictive and water, which has a small bleeding, compression and a high initial strength. This special grouting can control the subsidence of ground surface and rising of tunnel efficiently.

According to geology report, silty sand and sandy silt will be encountered during excavation. When compressed by blade, friction between earth and blade will increase evidently, sometimes blade torsion moment can increase so high that exceeds the rated value. Hence, special slurry or water should be injected to the working face before blade to improve fluidity of earth.

Thrust velocity of shield machine is also important in controlling the subsidence during excavation. The slower thrust velocity will decrease disturb to earth and make the grouting more uniformly along the back of segments, thus decrease the subsidence of ground surface. The thrush velocity is decided to be 2~3 cm per minute when TBM advancing beneath the runway.

CONCLUSION

The project is located in a busy, operating airport environment. The excavation process will be subject to daunting ground surface conditions, stringent aviation rules and requirements. This paper discussed the feasibility and implementation in detail. Years of TBM tunnel construction experience in Shanghai soft earth region combined with a numerical analysis for the project indicate that shield machine tunnel passing beneath the airport runway is feasible technically without disturbing the normal service of runway. The practical implementation of the project was discussed from aspects of tunnel route plan, precast concrete segments design and detailed construction methods or skills. At present the TBM has started to excavation from the portal of Kongang Yi Station and tunnel construction beneath runway will be conducted in June, 2009.

EXPERIENCE GAINED IN MECHANICAL AND CONVENTIONAL EXCAVATIONS IN LONG ALPINE TUNNELS IN SWITZERLAND

Y. Boissonnas ▪ Amberg Engineering Ltd.

INTRODUCTION

The transalpine rail routes in Switzerland are well over one hundred years old. As the established routes no longer meet the demands of the continually increasing volumes of rail traffic between north and south, two new routes through the Alps were planed. The old Gotthard rail line is in fact a mountain railway. The northern and the southern access ramps—with a maximum speed of 80 km/h and a maximum slope of 2.2%—reach the old Gotthard rail tunnel at an elevation of approximately 1,100 m above sea level. This is approximately 900 m higher than the city of Milan.

In 1992, the Swiss electorate people voted with an overwhelming majority in favor of the project to build the new base routes through the Alps. The Swiss voted again in 1998 in favor of the financing proposal for the new infrastructure.

Swiss Federal Railways together with the BLS Lötschbergbahn were commissioned for the realization and management of the Gotthard and Lötschberg routes. AlpTransit Gotthard Ltd., founded by Swiss Federal Railways, was given the task of managing the design and realization of the Gotthard route until the start of regular service.

In the meantime the shorter Lötschberg Base Tunnel was open to traffic in 2007.

OVERVIEW OF THE NEW GOTTHARD ROUTE

The transalpine rail route "Gotthard" will connect the city of Zurich with the city of Milan (see Figure 1), benefiting approximately 20 million people in Germany, Switzerland and Italy. Shorter travelling times—one hour less between Zurich and Milan—will enable rail service across the Alps to compete with aircraft (improved modal split for rail) and permit optimized connections.

The realization of three important tunnels is required on the Gotthard route: the Ceneri Base Tunnel to the south (15 km in length), the Zimmerberg Base Tunnel to the north (20 km) and the Gotthard Base Tunnel at the heart of the project (57 km).

The most impressive structure of the Gotthard route is the Gotthard Base Tunnel (GBT), which is designed to handle high speed passenger trains travelling at speeds of up to 250 km/h and freight trains travelling at up to 160 km/h. This will be achieved through with a minimum radius of 5,000 m and a maximum slope of 0.70%. It will have an elevation of approx. 500–550 meters above sea level and a maximum overburden of approx. 2,400 m. When completed, the GBT will be the longest infrastructure tunnel in the world.

OVERVIEW OF THE TUNNEL DESIGN

The GBT stretches from Erstfeld in the north to Bodio in the south (Figure 2). It consists of two parallel single-track tubes with a diameter varying from 9.0–9.5 m,

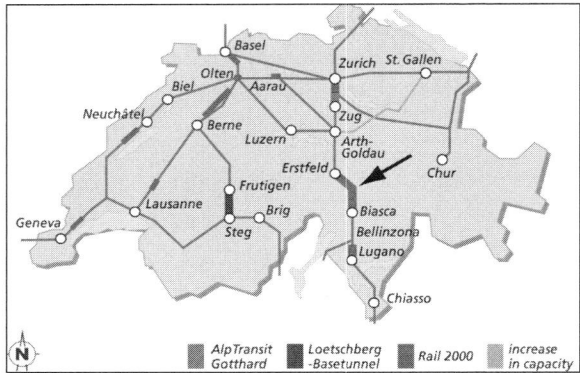

Figure 1. New transalpine rail routes through Switzerland (Graphic: Alp Transit Gotthard AG)

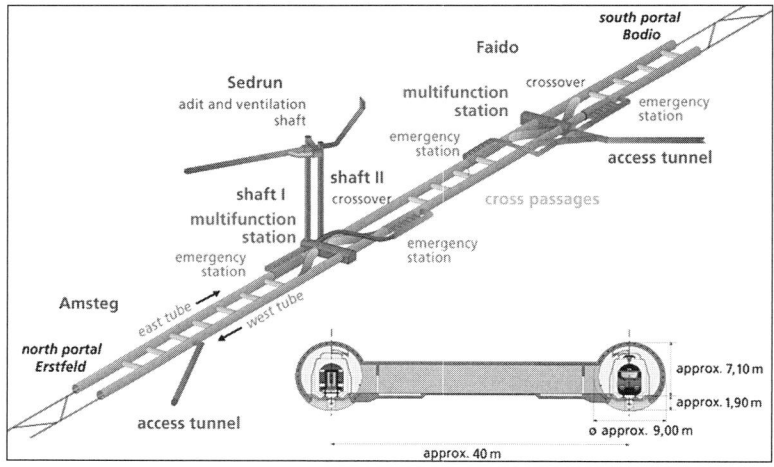

Figure 2. Layout of the tunnel system (Graphic: Alp Transit Gotthard AG)

which are linked by cross-passages every 312 m. Multifunction stations (MFS) are located at two locations one-third and two-thirds along the length of the tunnel. These will be utilized for the diversion of trains to the other tube via crossover tunnels, to house technical infrastructure and equipment, and as an emergency station for the evacuation of passengers.

Detailed and sophisticated studies demonstrated that this tunnel system concept was the most suitable for long alpine tunnels. To shorten construction time and for ventilation purposes, the tunnel will be excavated from several sites simultaneously. To this end, the tunnel has been divided into five sections. Excavation will take place from the portals as well as from three intermediate attacks located in Amsteg, Sedrun and Faido.

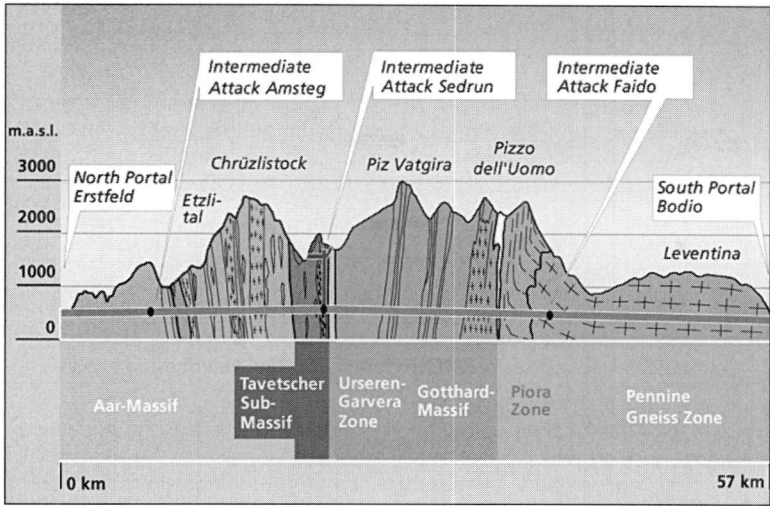

Figure 3. Geological profile of the Gotthard Base Tunnel (Graphic: Alp Transit Gotthard AG)

GEOLOGICAL CONDITIONS

From north to south, the 57 km long Gotthard Base Tunnel passes through mostly crystalline rock, the massifs which are interrupted by narrow sedimentary tectonic zones. The three crystalline rock sections include the Aare massif to the north, the Gotthard massif and the Pennine gneiss zone to the south (see Figure 3). These massifs consist mainly of high strength igneous and metamorphic rock. More than 90% of the total tunnel length consists of these types of rock. The main hazard is the risk of rock burst caused by high overburden, the instability of rock wedges and water inflow.

Much of the tunnel will have a very high overburden: more than 1,000 m overburden over approximately 30 km of the tunnel, more than 1,500 m over 20 km and more than 2,000 m over approx. 5 km. Overburden is an important parameter in determining the tunnel excavation method and rock support design.

The most difficult section of the new tunnel from the viewpoint of geology is expected to be the (old crystalline) Tavetsch intermediate sub-massif in the Sedrun section. Located between the Aar-massif and the Gotthard-massif, it is one of the about 90 different isolated short fault zones along the 57 km. It consists of a steeply-inclined, sandwich-like sequence of soft and hard rock. Exploratory drillings in the early nineties indicated extremely difficult rock conditions for about 1,100 m of the tunnel (see Figure 4). As well as compact gneiss, there are also intensively overlapping strata of schistose rock and phyllite.

In the Faido section, the Piora syncline was intensively investigated in the second half of the 1990s.

The initial planning phases concentrated on aligning the tunnel through the various fractured zones at their narrowest points wherever possible.

The high mountain overburden of up to 2,400 m means that operating temperatures in the tunnel can reach 35–50°C (see Figure 5). In order to maintain the required air temperature in the various working areas, an air conditioning system is required.

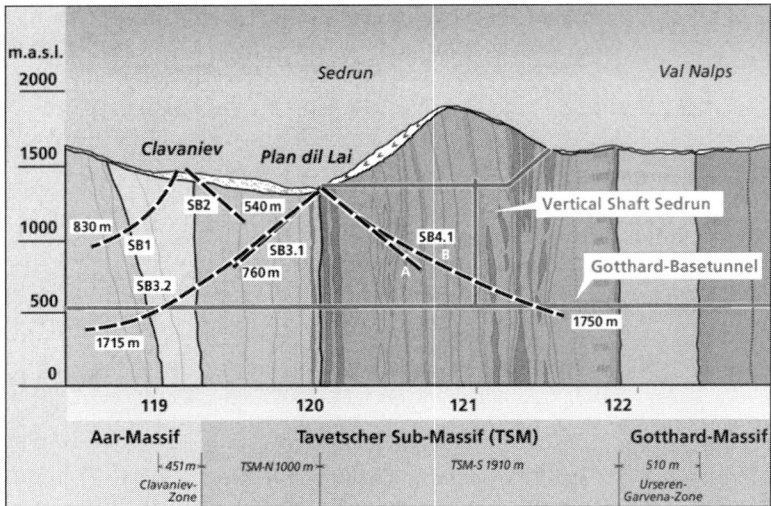

Figure 4. Geological investigations in the Sedrun section of GBT (Graphic: Alp Transit Gotthard AG)

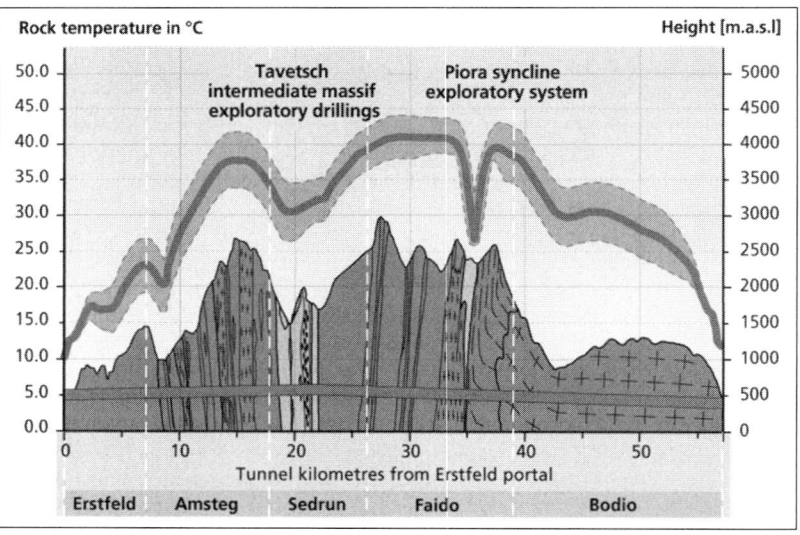

Figure 5. Forecast of rock temperatures at tunnel level (Graphic: Alp Transit Gotthard AG)

CONSTRUCTION PROGRESS

Construction of the Gotthard Base tunnel has been proceeding for many years, beginning in Sedrun in 1996. All five sections of the tunnel are currently under construction including the portals and intermediate attacks. Up to 1st of Oktober 2008, about 123.0 km of tunnels and galleries have been excavated, or 80.21% of total 153.5 km (see Figure 6).

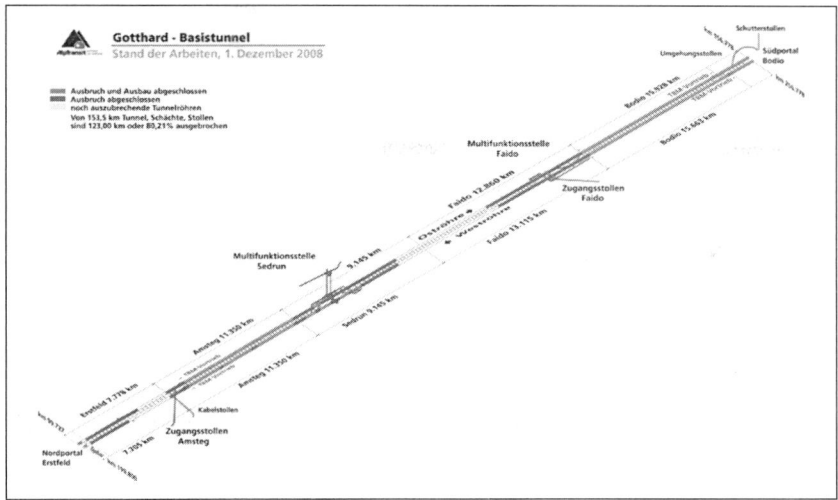

Figure 6. Gotthard Base Tunnel, construction status as of December 1, 2008 (Graphic: Alp Transit Gotthard AG)

Based on the progress achieved to date in all sections, the current overall time schedule indicates that tunnel excavation works will be completed in 2012, and the first train will pass through the GBT end of 2017.

Sedrun Section

Sedrun is the most complex section of the base tunnel due to logistical and geological reasons. The adit (1 km), the inclined ventilation shaft (450 m), the cavern at the top of the first shaft and both 800 m-deep vertical access and ventilation shaft were completed in several preparatory construction lots. The entire logistic supply for the heading of a total tunnel length of 2 × 9 km (4 simultaneous tunnel drives, together with others minor drives) has to be conducted through the access tunnel and the double shaft system. Up to 6,000 tonnes of excavated material have to be handled every day. Shaft drives were installed for a double floor hoisting cage with a 5 MW rating, providing a 60 km/h hoisting speed for an 80 t load.

Because of the complex geological conditions, different auxiliary support measures are to be applied during excavation, e.g. investigations ahead of the tunnel face with geophysical systems and/or percussion and core probe drillings. The drillings must be preventer-protected against water inflow and debris under high pressure. In the GBT Sedrun, for example, the core drilling rigs must be equipped to water pressure strikes as high as 200 bar. The triple preventer systems used offer three independent systems. If water under high pressure is struck unexpectedly, the drilling rod can be clamped (pipe ram) or even cut off (shear preventer). The third system, Roto Pac, makes it possible to keep on drilling even in zones with high water pressure. This preventer technology comes from the oil and gas drilling industry.

Shaft 1, with a diameter of 7.90 m was excavated in full section from the cavern at the shaft top by drilling and blasting. To sink Shaft 2, two extra caverns had first to be excavated at the top and bottom levels. In a first phase, a pilot bore with a bit diameter of 43 cm was drilled (daily heading rate 11.07 m). In a second phase, the pilot bore was enlarged with a raise-boring machine to a diameter of 1.80 m (daily heading rate 25.83 m). In the third and last phase, Shaft 2 was enlarged to the final diameter of

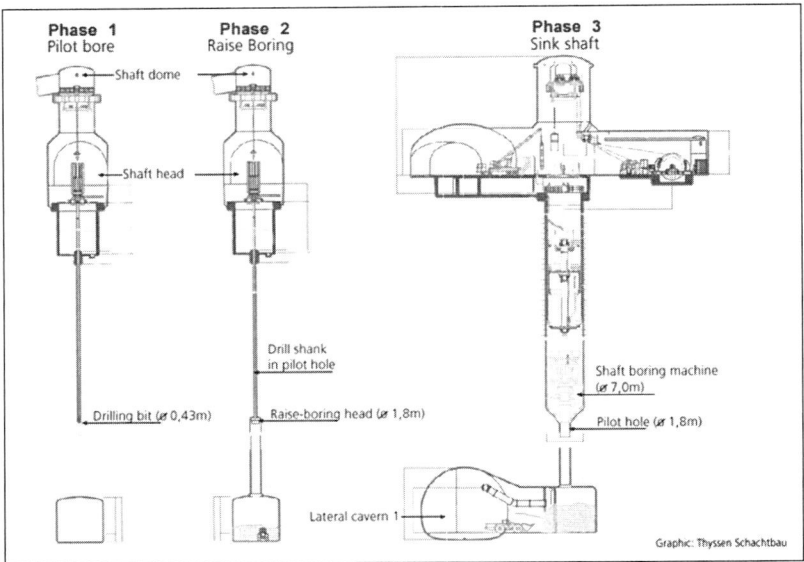

Figure 7. Shaft Sedrun, sink of shaft II

7.0 m with a shaft top down boring machine (daily heading rate 5.05 m). Phases: see Figure 7.

The northern section was driven in squeezing rock in the TZM North. The encountered geology complied with the forecast. With regard to deformations, the geotechnical behaviour of the kakiritic gneisses and slates encountered turned out to match the forecast. A rapid cessation of the deformations (occurring only some 20 to 30 m behind the face) from the perpendicular bedding made the transformation from the deformation to resistance principle easier. The friability was extremely high in some areas. In this regard, the influence of the strike of the layers vis-à-vis the tunnel axis is of major significance. In the case of upright layers there is a higher need for supporting the face (shotcrete with net reinforcement, excavation in stages, and installation of spiles), whereas in the case of flat lying layers, the danger at the edge of the excavation and deformation is increased. All situations occurring so far have been mastered by adapting the means of support and the length of advance.

The deformations in squeezing rock occur radially at the anticipated average magnitudes of 20 to 80 cm. The deformations in the cross section can occur in a very different manner (asymmetrical deformations). These asymmetrical deformations are reacted by corresponding additional supports (grouted anchors).

The cross section of the single tubes increases from about 80 m² up to more than 130 m² (see Figure 8). To summarize, it can be concluded that the support system used in the TZM North proved itself effective. The installed steel mining arches have the capacity to deform to the permissible 80 cm radial deformations as planned.

Following the conclusion of the learning phase up to the end of March 2005, average desired rates of approx. 90 cm/work day have been attained.

The southern drives in both tunnel tubes in the Gotthard Massif gneisses have been undertaken by means of highly mechanized excavation and transport installations with drill and blast since spring 2005. The driving operations for both single-track tunnels are roughly 10 months ahead of schedule as compared with the contract construction program. This is due to the technically easier penetration of the Urseren Garvera

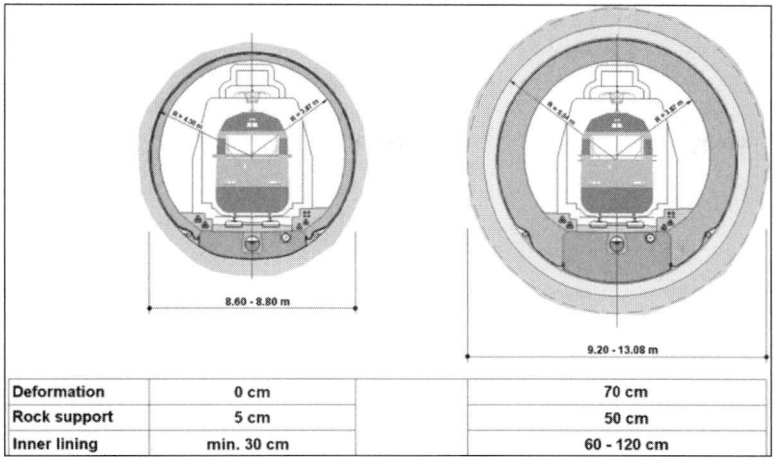

Figure 8. GBT Sedrun, typical cross sections

Zone, which was tackled prior to entering the Gotthard Massif in summer 2005. Further driving operations over the original lot boarder towards Faido will be run until the final breakthrough expected to be end of 2011. Starting in the second half of 2006, vault supporting work in the core area of the MFS commenced in the tunnel change-over enlargements followed by the lining operations for the adjoining single-track tubes.

Faido Section

In Faido, the GBT is accessed up via a 2.7 km long sloped tunnel (declination of 12.7%). Construction of this tunnel began in December 1999 and was constructed by conventional drill and blast methods. Close to the access tunnel portal, extensive material processing plants for recycling of a part of the muck and for the production of concrete aggregates as well as conveyor belt systems for the transportation of debris were installed. A 5 km long conveyor belt transports some of the muck for permanent disposal in an old quarry.

The Faido multifunctional station (MFS Faido) lies at the end of the Faido access tunnel. It consists of the two tunnel tubes that are connected with each other by two tunnel crossovers in the MFS, the cross cavern for housing the railway infrastructure equipment, the cross-cuts required for construction operations, the cavern for handling logistics in the construction phase, the two emergency train stops that are connected by side galleries, and a ventilation tunnel system. Altogether there are about 10 km of tunnels and galleries with cross-sectional areas ranging from 40 m^2 to more than 300 m^2. The tunnel crossroads were arranged north and south of the cross cavern (Figure 9).

In summer 2002 an intensive fault zone was tunnelled into unexpectedly. It passes through the large caverns of the multifunctional station at a very acute angle. An intensive investigation program was carried out to define the best location of the large cross over caverns with cross sections up to 330 m^2. As a result of these investigations, it was possible to adapt the layout of the MFS Faido with the aim of placing the large caverns in good rock conditions in the southern part. Therefore the emergency stations and the cross over caverns where moved 600 meter to the south in better rock conditions.

Besides the layout of the MFS and the station's critical cross-sections, the geology encountered also made it necessary to carry out a critical review of excavation

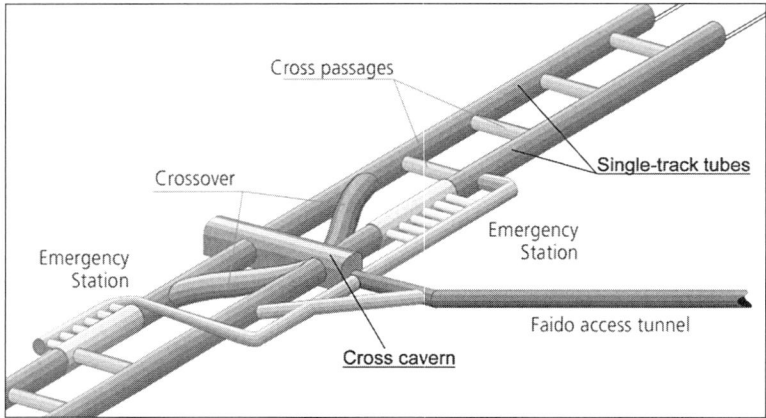

Figure 9. Primary disposition of the MFS Faido (Graphic: Alp Transit Gotthard AG)

Figure 10. Rebuilding of the HEM stretch

support means to be applied in the relevant cross-sections. Heavy steel sections Type HEM 180 were used, backed by about 40 cm of concrete. Initial support with rock bolts and shotcrete was planned to ensure safe working during the first deformation phase prior to installation of the steel supports.

In May 2004 it became necessary to cease heading operations in the north eastern tunnel because of excessive deformations in the rear areas. The steel arches in these areas were severed completely, and safe working could no longer be ensured (Figure 11).

The HEM stretch had to be rebuilt, because underbreak had been suffered over the entire length of 125 m and it also affected the invert. The HEM arches were dismantled and, following additional excavation, replaced by new supports (heavy flexible arches TH 44). In the meantime, this support system with movable TH arches and slitted shotcrete has been implemented successfully.

Rock bursts have been registered in various drives since March 2004. Additionally, a number of micro-quakes were registered by the Swiss Seismic Service (SED) in the Faido region. The times correlate to the rock bursts observed during the MFS drive. In order to be able to understand these incidents better, a working group was set up

Figure 11.   TBM switching by low-loading truck

Figure 12.   Reassembling of the Faido TBM

involving SED experts. It was proved that the tunnel excavation provoked the microquakes with intensity up to 2.4 on the Richter scale.

The main conventional drives in the MFS Faido were concluded in summer 2007. The problems relating to major de-formations in and around the fault zone have been resolved and the rock burst incidents were countered through appropriate means. Up to seven driving crews were engaged at 10 working points in order to prepare the MFS Faido so that the TBM from Bodio could arrive unimpeded. The TBMs reached the MFS Faido in the second half of 2006.

For the heading of the single track tubes in northern direction open hard rock TBMs are used. These TBMs have already excavated the twice 16 kilometres singletrack tubes of the Bodio section. After completion of the Bodio headings the TBM were dismantled, switched by low-load trucks through the approximately 2,500 m long MFS Faido (Figure 12) and reassembled in assembling caverns at the end of the MFS Faido (Figure 13). The excavation diameter was enlarged from 8.83 m in the Bodio section to 9.43 m for driving the Faido section.

The TBM in the east tube started the excavation in July 2007, the west tube TBM in October 2007.

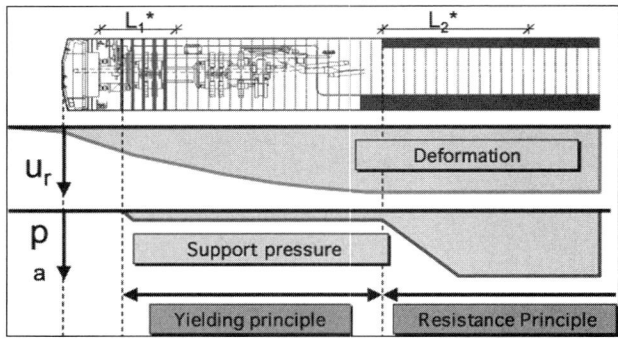

Figure 13. Rock support concept

The TBMs have a total length of 450 m each (424 m back-up installations) and a diameter of 9.43 m. The cutter head is equipped with 66 discs (load capacity 16,000kN) and has a weight of approx. 240 tons. For the excavation of the difficult horizontally layered Lucomagno gneiss with large overburden of up to 2,000 meters the excavation diameter was enlarged up to 9.53 m by shifting with installation of additional distance washers.

The excavation support in L1* (Figure 4) must be strong enough to stabilize the rock until the TBM has moved in order that this rock section has reached L2* (Figure 14), where the remaining excavation support is installed.

The supporting means used in L1* (only rock support installed in L1* has an effect on the advance rate) must be such that a maximal rate of advance can be achieved. In zones of squeezing ground, the excavation support must be realized in two steps: Immediately behind the cutter head, flexible steel arches with yielding friction clutches are put in place. These friction clutches allow deformations of the steel arches, until the lining resistance is equal to the rock pressure. In L2*, dependent upon the amount of deformations / relaxation which have already occurred, the remaining necessary lining resistance must be applied, because later, there will be no space for further application of rock support. Within the TBM and its back-up equipment, the zones L1*and L2* must be therefore spacious enough to enable supporting measures to be carried out as fast as possible without any obstructions of the heading.

On their way to Sedrun the TBMs have to pass through the Piora syncline with an overburden of about 1,800 m. It was a key point of the geology of the GBT since the structure and extent of the Piora syncline were initially unclear. The Piora syncline is a Triassic deposit of Dolomite in the Gotthard Massif, which is inserted between Lucomagno Gneiss and Medelser Granite. Above the base tunnel level this Dolomite is disintegrated into a matrix of fine sand and solid Dolomite blocks in water. An extensive exploration was necessary in order to ascertain the real presenting conditions on base tunnel level for the tunnel construction (Figure 15).

This exploration was realized by means of a 6 km long exploratory gallery reaching close to the Piora syncline and 300 m above the level of the GBT. Four inclined test bores from the Piora exploratory gallery down to GBT level indicated that conditions at that depth are solid Dolomite-Anhydrite formations with no water pressure or circulation. So for the main tunnel excavation favorable tunneling conditions where expected. However, this fault zone had to be handled with attention and respect and extensive investigation and monitoring measures where planed. The Piora Syncline was successfully crossed by the east tunnel in fall 2008 confirming the elaborated design.

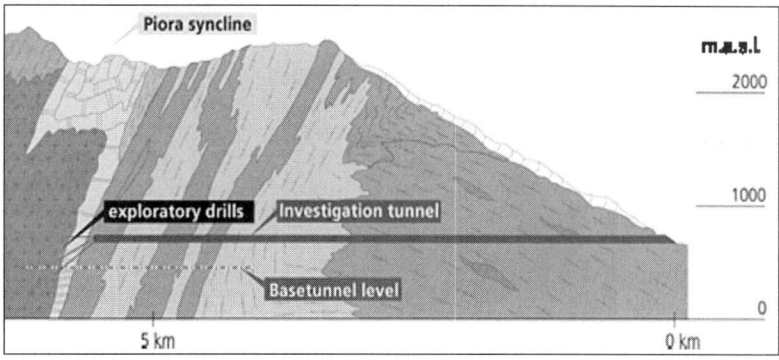

Figure 14. Investigations of the Piora Syncline (Graphic: Alp Transit Gotthard AG)

Bodio Section

Extensive construction work was realized at the southern Portal in Bodio before autumn 2002 in preparation for the heading of the main tunnel section. These were:

1. A bypass-tunnel—headed with conventional drilling and blasting, with a total length of 1,200 m—to bypass the very delicate and time-consuming loose rock zone at the portal of the base tunnel. The bypass was excavated in good rock conditions. The two 15 km tunnel tubes in Bodio, leading north in the direction of Faido started in the cavern at the end of the bypass.
2. The 3.2 km long mucking tunnel that houses a conveyer belt. The 3.7 km long conveyor belt system will transport 6 million tons of muck to the neighbouring Blenio valley. The mucking tunnel was excavated by means of a 5m-diameter TBM. The work was finished in April 2001.
3. The opencast constructed tunnel (380 m long), including the portal (see Figure 16). The construction works has been completed successfully in July 2003.
4. The tunnel section in the loose rock material that follows the opencast construction, which is also close to 410 m long. This tunnel section was excavated through very delicate rock fall and loose rock material (see Figure 15). To drive the heading through this zone, auxiliary measures were taken such as putting a pipe screen umbrella in place and undertaking a large amount of grouting. The heading of this section was completed successfully in June 2003, on time and on budget.
5. The processing plants that recycle muck into concrete aggregates as well as the conveyor belt systems that transport the surplus muck through the mucking tunnel to the Buzza of Biasca stone quarry or to the construction site of the open sections of the new railway route.

The main excavation work by TBM started in January 2003 and the 14 kilometer long section was completed by October 2006.

In 2003 the roughly 600 m of equipment for the rear "inner vault" construction site—the so-called "worm"—was assembled at the Bodio yard in front of the tunnel. The concreting of the lining support and of the lining itself was started in the west tube in April 2004. A specially designed vault formwork, which possesses some degree of adaptability, is being utilized. It is used to optimize the support taking into account any TBM deviations and the actually installed supporting elements. In September 2008 the last lining of the Bodio section was put in place.

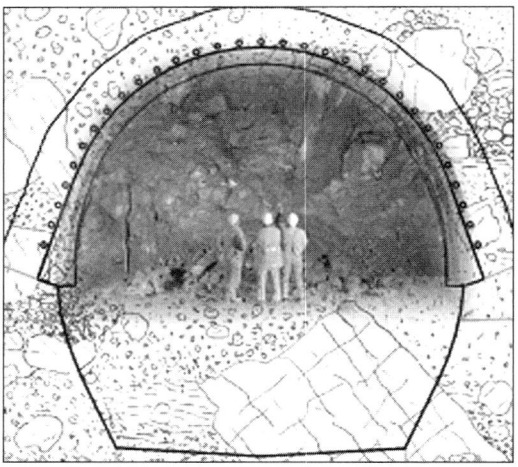

Figure 15. GBT Bodio, tunnel section in loose rock material

Figure 16. Portal Bodio, inner lining equipment before moving through the opencast constructed tunnel

Actually the sidewalks are installed in the west tunnel while the east tube is used for the transport logistic of the Faido excavations.

ENVIRONMENTAL CONSIDERATIONS

The Swiss population is in general very sensible towards environmental issues and the Swiss government sets a high priority on the preservation of land and water resources. Several federal laws prescribe measures to protect people and nature from contamination, pollution, noise and different kinds of waste. The realization of a significant recycling of materials is therefore a priority. Various concepts and technical solutions have been adopted within the project to reduce as much as possible temporary and permanent effects caused by the construction works and by the final deposit of the muck.

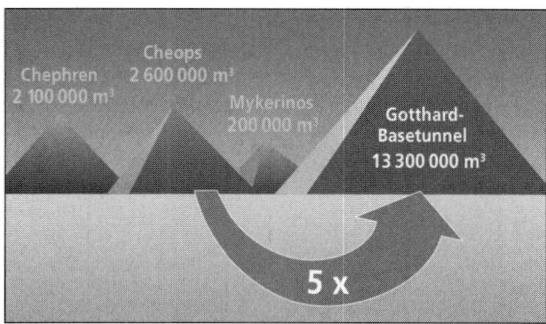

Figure 17. Gotthard Base Tunnel, volume of excavated material (Graphic: Alp Transit Gotthard AG)

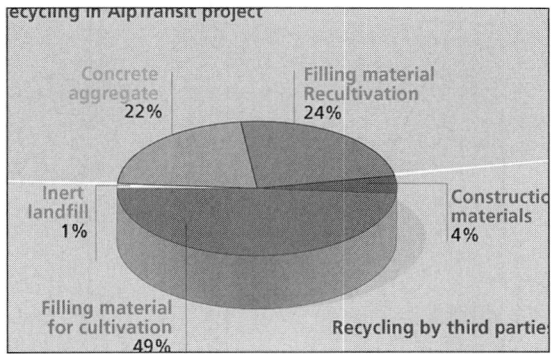

Figure 18. Gotthard Base Tunnel, recycling of excavated material (Graphic: Alp Transit Gotthard AG)

The excavated volume of the entire Gotthard Base Tunnel (including galleries and multifunctional stations) is estimated at 13.3 million cubic meters; with the entire volume of muck produced, it would be possible to build a pyramid five times the size of Cheops (Figure 17). On all construction sites, material processing plants have been built to recycle the muck and produce concrete aggregates. Depending on the petrography and the properties of the excavated material and according to the predicted geology, about 28% of entire muck will be processed. The expected recycling rate is around 85% (concrete aggregate production about 22% of entire muck). A part of the muck is used as fill material and the rest is transported to permanent disposal sites or to an inert landfill (Figure 18).

Muck contaminated through the loss of oil from the machines (especially the TMBs), waste products from the use of explosives, chemicals and heavy metals from rebounded shotcrete cannot be deposited in a "conventional" inert landfill. The requirements for "non-contaminated" muck are very high. Currently, approximately 1% of the excavated material of section Bodio has been transported to a special landfill for contaminated waste. Contaminated waste incurs intensive costs due to the extremely high level of deposit and treatment prices. To reduce noise and pollution, muck is transported with conveyor belt systems above ground and is sprayed with water to avoid dust production.

Figure 19. Construction site at Bodio, treatment of wastewater from the tunnel

Figure 20. Construction site at Amsteg, encapsulated plants and conveyors belt systems reduce noise

Wastewater from the tunnel and from several technical facilities outside the tunnel is treated to reduce the acidity (pH), to neutralise chemicals and to separate oils and particles in suspension. After treatment, the temperature of the cleaned water has to be reduced as necessary prior to the introduction in a river, before being recycled for industrial use or being used as liquid for the cooling system in the tunnel (Figure 19, treatment of maximum 200 l/s, recycling of about 5÷10 l/s). The sludge resulting from the treatment of wastewater is pressed to reduce water content and is handled with the same criteria as the muck. In general, the quality of the cake is "contaminated," especially due to the concentration of oils.

To reduce the impact of noise from the construction sites upon residents living nearby, the required standard of some equipment such as tunnel train wagons have been set very high. The concrete production plant, material processing plant and conveyor belt systems have been encapsulated (Figure 20). Temporary dams and absorbent walls have been specially erected to avoid the propagation of extremely noisy activities.

CONCLUSIONS

Construction of the Gotthard Base Tunnel is preceding full speed ahead. Up to now, some sections with poor geological conditions have delayed the excavation schedule of the Faido and Bodio sections. But most of the seven predicted difficult geological zones have been completed successfully.

As things stand today the GBT should be completed on schedule in the year 2017 and it will be a milestone for the realization of the New Alpine Transverse in Switzerland. A total investment of about 11 billions Swiss francs is going to be invested to ensure a future with sufficient transport capacity for the ever-increasing volume of rail traffic between northern and southern Europe. The GBT is a tunnel with an elevated standard of safety and technology, with several "new land" and "world-première" solutions. It is an important contribution to the preservation of the Alps.

PART 8

Las Vegas

Chairs

Marc Jensen
Southern Nevada Water

Jim McDonald
Impregilo

DESIGN AND CONSTRUCTION OF LAKE MEAD INTAKE NO. 3 SHAFTS AND TUNNEL

Jon Hurt ▪ Arup

Jim McDonald ▪ Vegas Tunnel Constructors

Gregg Sherry ▪ Brierley Associates

A.J. McGinn ▪ Brierley Associates

Luis Piek ▪ Arup

INTRODUCTION

The entire scope of new facilities for Lake Mead Intake No. 3 includes a submerged intake structure, a deep tunnel beneath Lake Mead, a tunnel access shaft, Intake Pumping Station No. 3, and connections to an existing intake and water treatment facility, as shown in Figure 1.

In March 2008 Vegas Tunnel Constructors (VTC), a joint venture of Impregilo SpA and SA Healy were awarded a $447m Design-Build contract by SNWA for a section of this work (Contract 070F 01 C1), including the 185 m deep tunnel access shaft, the 4.7 km long, 6.1 m diameter tunnel and the submerged intake. Arup, supported by Brierley Associates, is the Design Engineer for VTC.

The scope of the work is shown in Figure 2. This paper will describe the design and construction on this contract, with a focus on the access shaft which is the most complete section of the work. The shaft is being sunk using drill-and-blast methods with the final lining placed as excavation proceeds. The shaft passes through a major fault zone, and extensive pre-excavation grouting for water control is required for much of the depth of the shaft.

PROJECT BACKGROUND

The need for the new intake is driven by the declining levels of Lake Mead caused by drought and increased regional demand. Created by the construction of Hoover Dam in the 1930s, the 180 km Lake Mead lies on the Nevada-Arizona border about 50 km southeast of Las Vegas. It is supplied by the Colorado River and is the largest man-made reservoir in the U.S. Along with Lake Powell, it serves 25 million people in seven states, including the residents of Las Vegas and Phoenix. Prolonged drought conditions in the U.S. Southwest have strained the lake and it is currently only filled to around 50 percent of the full reservoir capacity. At its fullest in 1983, the level of the lake stood at 373 m (1,224 ft) above mean sea level (amsl). In July 2008 the level stood at 337 m (1,105 ft). Intake No. 3 will permit drawing lake water at elevations as low as 305 m (1,000 ft) amsl—a level at which Intakes No.1 and 2 would be unusable.

A more complete description of the project is provided in other RETC papers (Feroz et al. 2007; Feroz et al. 2009).

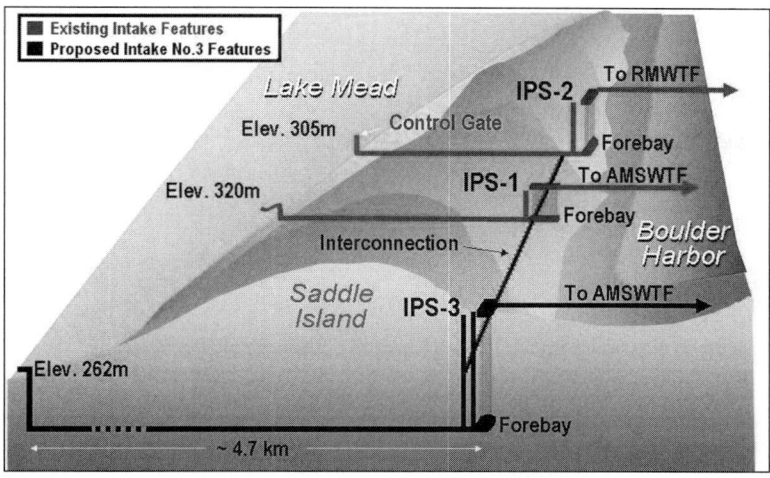

Figure 1. Lake Mead Intake No. 3

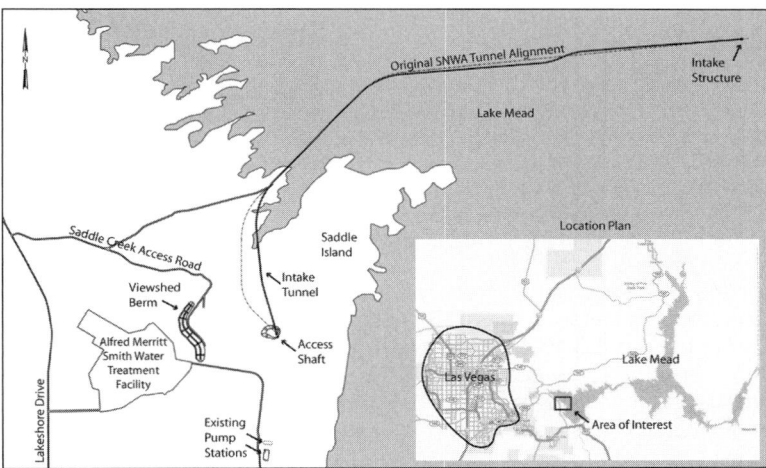

Figure 2. Project layout for Contract 070F 01 C1

SHAFT DESIGN

Shaft Layout

The 9.1 m (30 ft) internal diameter Tunnel Access Shaft extends 185 m (607 ft) below the final ground elevation of 382 m (1,254) amsl. At the bottom of the shaft a 60 m long, 14 m wide by 10.5 m high cavern will be excavated to allow launching and operation of the TBM. A short stub tunnel, oriented 180 degrees from the main tunnel, is also required at the shaft bottom to allow the option for future expansion of the system.

At elevation 247 m (811 ft) amsl, a short 26 m length of 6m wide by 6m high horseshoe configuration stub tunnel is being constructed, known as the IPS-3 Stub Tunnel, which will connect with the adjacent underground pump cavern which is being

Figure 3. Shaft layout

built under a separate contract. During construction, this will be used to house pumps for the slurry circuit and pump water discharge system. Another, slightly smaller stub tunnel is being constructed at elevation 311 m (1020 ft) amsl to also house slurry and water pumps, such that the head on each lift of the pumps is limited to around 70 m. A rendering of the shaft is shown in Figure 3.

Anticipated Geologic Conditions

Subsurface conditions for the Access Shaft were derived from boring LB-15, which was located approximately 8 m (26 ft) southeast of the center of the shaft. The rock mass encountered at LB-15 was subdivided into six distinct zones. These zones are summarized in Table 1. For design purposes, geologic structure encountered at LB-15 was translated to the center of the shaft. This allowed the designers to estimate the depth to geologic contacts and zones of varied rock quality. Actual conditions are currently being mapped as the excavation is advanced and images of the Detachment Fault (Zone B) and Lower Plate (Zone C) are shown in Figures 4 and 5.

Rock Structure

Numerous joints, joint sets, shears and faults were observed in the rock core and in surficial outcrop mapping across the shaft site area. The rock mass is highly fractured within approximately 15 to 30 m of the Detachment Fault. At the shaft, the Detachment Fault itself strikes approximately N 40 to 50° E, and dips 30 to 40° NW. In the Upper Plate rocks the joints and shear sets strike from N 75 W to N 60 E, and dips range from low angle to high angle. In the Lower Plate, the strikes range from N 80 W to N 45 E, and dips are generally steeper, from moderate to high angle. The Lower Plate rocks (Zones C through F) also manifest varying degrees of foliation having a prominent NE strike with dips of 10 to 80° NW and SE. A number of more significant shear/fault zones within the Lower Plate rocks were encountered.

Table 1. Summary of anticipated geologic condition

Zone	El. (m amsl)	Description	REC	RQD	Anticipated Rock Mass Behavior
A	375 to 363	Saddle Island Upper Plate: Pre-Cambrian rocks above the Detachment Fault, which consists primarily of blocky, seamy, variably slightly weathered to highly weathered-decomposed, very weak to strong Tertiary volcanic intrusive rocks, including dacite and basalt along with some pegmatite	57[1]	13[1]	Blocky to very blocky and seamy, with up to 45% of this zone being intensely fractured and highly weathered to decomposed soil-like conditions, which presents potentially raveling conditions on exposure without restraint. Unconfined compressive strengths of blocky fragments, core stones, and intact rock from this unit were measured to range between 13.4 to 68.9 MPa (1,950 to 10,000 psi)
B	363 to 335	Saddle Island Detachment Fault: Up to 75% of the rock profile is highly weathered to decomposed, or intensely fractured to crushed. Non-faulted rock in this zone comprises approximately 90% highly fractured Upper Plate rocks (basalt / dacite), and about 10% of strongly foliated chlorite phyllonite rock, most likely derived from the underlying Lower Plate amphibolite rock.	—	—	Varies from blocky/disturbed/seamy to laminated and intensely sheared. Raveling and unstable shaft wall conditions are expected in this zone where encountered below the water table [2]. Non-faulted rock is anticipated be blocky and seamy, variably slightly weathered to highly weathered-decomposed, generally weak to moderately strong, soft, and friable
C	335 to 271	Saddle Island Lower Plate: Mostly gneiss and chlorite-biotite schist with a significant shear zone, approximately 0.6 to 3 m wide (N49E, 39 NW), is anticipated to be encountered between El. 308 and 296 m.	40 to 100 98[1]	12 to 100 59[1]	Blocky to very blocky with localized shear zones. Blocks and wedges formed by intersecting joints and shears will be present causing wedge slipping and toppling conditions.
D	271 to 259	Saddle Island Lower Plate: Highly foliated chlorite-biotite schist that contains several intersecting shear and fault zones ranging from 1 cm to 3-m thick.	45 to 100 79[1]	0 to 35 13[1]	Very blocky to seamy and occasionally laminated and sheared. Some localized raveling rock conditions might occur
E	259 to 236	Saddle Island Lower Plate: Chlorite-biotite schist/ amphibolite gneiss with a more intensely fractured area projected to be encountered between El. 245 and 236 m	90 to 100 99[1]	20 to 95 55[1]	Blocky to very blocky ground conditions.
F	236 to 189	Saddle Island Lower Plate: Predominantly amphibolite gneiss with occasional interlayers of amphibolite schist/biotite-chlorite schist/pegmatite. Between El. 229 and 223 m, a more fractured area is projected to be encountered	98 to 100 99[1]	78 to 100 93[1]	Below El. 236 m, mostly massive to blocky with only occasional very blocky to seamy areas adjacent to localized shears and fractures.

NOTES:
(1) Average value. (2) At the beginning of construction the water level in Lake Mead was at El. 336 m. For construction-phase design, the maximum lake level is assumed to be at El. 345 m

Figure 4. Exposed shaft wall (left) and core (right) in Detachment Fault (Zone B)

Figure 5. Exposed shaft wall (left) and core (right) in Lower Plate (Zone C)

Table 2. Access shaft design joint sets

Joint Number	Dip	Dip Direction
1	26	310
2	44	081
3	12	174
4	71	096
5	58	244

Table 2 lists five design joints sets that were developed from the joint orientation information obtained at LB-15. Table 3 lists the orientation of major discontinuity planes.

Rock Mass Permeability

Hydraulic packer testing was performed to assess rock mass permeability values with depth. Interpreted values ranged from 2.6×10^{-6} to 2.7×10^{-4} cm/sec. A pumping test was conducted at Saddle Island. For design, this test was assumed to be representative of potential high end ungrouted rock mass ground water inflow conditions through rock mass features such as joints, fractures, shear zones and faults. The high hydraulic conductivity measured during the pumping test (2.4×10^{-3} cm/sec), as

Table 3. Orientation of discontinuity major planes

Plane Number	Pump Adit		IPS-3 Stub Tunnel	
	Dip	Dip Direction	Dip	Dip Direction
1	73	305	39	111
2	39	319	39	295
3	53	064	64	288
4	59	094	3	257
5	14	185	35	89
6	—	—	77	112

compared to the much lower packer test values, likely resulted from discrete fracture flow features rather than intact rock permeability.

Initial Support Design

Design of initial support employed two rock mass classification systems; Rock Mass Rating (RMR) and Geologic Strength Index (GSI). These systems were used to evaluate rock mass strength and deformation parameters and to evaluate rock reinforcement spacing, lengths, and size with empirical, structural, and numerical evaluations in general accordance with applicable codes.

The strength and deformation parameters, discussed previously, were input into elasticity and plasticity equations to evaluate the zone of plastic deformation with and without an applied internal pressure associated with initial support. If required, the internal pressure was increased according to available empirical correlations (Hoek, 2000) for conventional support type, e.g., bolts and shotcrete. Based on these analyses, it was determined that the rock mass would be self-supporting and would require minimal initial support prior to placement of the final lining.

Individual wedge instabilities were evaluated using the rock structure information listed in Tables 2 and 3 and the commercially available software, Unwedge. Rock reinforcement lengths and installation patterns and shotcrete thicknesses were adjusted until a safety factor of 1.3 was achieved for each potential wedge combination. The performance of the initial support systems developed through the empirical design methods and wedge stability analyses were then evaluated using commercially available finite element software Phase 2 and Examine 3D. The influence of variations in strength and deformation parameters was assessed by varying effective friction, effective cohesion, and rock mass modulus and calculating the predicted bending moment and thrust in the initial support system.

Individual wedge instability controlled the initial support type for Zones C, D, E, and F. In these zones, the initial support separated into three categories based on ranges of GSI values. The Shaft Design Engineer's representative (SDER) and Contractor evaluate the rock mass independently and then meet to assign one of the three support types after each excavation lift

- Type 2A initial support—spot rock reinforcement (dowels or bolts) and a flash coat of shotcrete.
- Type 2B initial support—spot rock reinforcement (dowels or bolts) and shotcrete.
- Type 2C initial support—1.8 m long pattern SS-46 Split Sets rock dowels, spaced 1.5 m on center vertically and horizontally, and shotcrete with a single layer of welded wire fabric or chain-link fence.

Shaft Final Lining Design

The final lining was evaluated for the specified loading conditions for the 100 year design criteria. The Contract placed a number of requirements on the shaft final lining in addition to the need for the design to meet the anticipated ground and water loads:

- The lining was required to be a minimum of 450 mm thick and to have a minimum concrete cylinder strength of 27.5 N/mm^2 (4,500 psi).
- Maximum inflow limits for the shaft and intake tunnel at the completion of tunneling were set at 125 l/s (2,000 gpm) with specific limits on inflows from discrete locations.

Given that pre-excavation grouting was also required to allow shaft construction to proceed, it was decided to utilize the grouting as the primary means of limiting water inflows and to design the shaft lining as a drained structure with reduced external groundwater pressures, rather than as a tanked structure. Below the elevation at which a 450 mm thick lining is sufficient to resist the hydrostatic water pressure, 296 m amsl, nine 50 mm diameter drain holes installed on an effective 3 m by 1.5 m grid are provided through the lining. These drainage holes extend a minimum of 0.6 m into the rock mass. The design philosophy is to provide a lining permeability higher than the grouted rock permeability such that the water pressure on the lining is significantly reduced. It should be noted that the most onerous design case is for a future dewatering of the tunnel coinciding with a maximum lake elevation. During construction, with the lower lake elevation, and operation, with internal and external water loads balanced, loads on the lining are significantly less.

The final lining was modeled as a thick concrete cylinder that resists lateral rock, ground water, surcharge, and seismic loads. For the final lining, lateral rock loads were evaluated based on the estimated zone of plastic deformation around the shaft, which is a function of the rock mass strength, and was determined to be zero for the shaft. The major design load for the shaft was the external water load that develops when the shaft is dewatered for inspection and maintenance.

For design purposes, a reduced seepage pressure equal to 50% of the hydrostatic head at 296 m amsl was adopted for the design of the cast-in-place final lining for the shaft. At the junctions between the pump adit and IPS-3 tunnel, the design pressure was reduced to 25% of this hydrostatic head. The reduced seepage pressure was based on modeling of the groundwater pressure and flow, accounting for the increased number of drainage holes installed through the shaft lining and the permanent shotcrete linings that will support the pump adit and IPS-3 tunnel.

Structural design was performed in accordance with ACI 318-08 to consist of plain (unreinforced) concrete throughout the shaft, reducing durability concerns that may arise with the use of reinforcement. At the two junctions, a steel rib and post frame was provided within the concrete lining for additional robustness. Joints between concrete pours are contact grouted with a cementitious grout with no requirement for a waterstop.

Load Factor on Water Loads

The main design code used for the design of the underground concrete structures is ACI 318 "Building Code Requirements for Structural Concrete," which as the name implies is generally written for the requirements of building structures. The load factors used are based on ASCE/SEI 7-05: Minimum Design Loads for Buildings and Other Structures, which specifies a load factor of 1.4 to be applied to fluid loads, which would require significant capacity to be provided for structures under 180 m of water load. To enable an efficient, yet robust design, a reduced load factor of 1.2 was agreed for use on the following basis.

LAKE MEAD INTAKE NO. 3 SHAFTS AND TUNNEL

The commentary to ASCE/SEI 7-05 regarding fluid load explains that the load factor was selected because "emptying and filling causes fluctuating forces in the structure, the maximum load may be exceeded by overfilling; and densities of stored products in a specific tank may vary." Compared with this commentary, the groundwater loading on deep underground structures, such as at Lake Mead is much less variable because:

- The rate of change of fluid load is very slow, so no fluctuating forces will be exerted on the structures
- The maximum load cannot be exceeded by overfilling; in this case the design load is based on a maximum lake elevation defined by SNWA of 376 m (1234 ft) amsl, which is above the spillway elevation of the Hoover Dam 372.3m (1221.4 ft)
- The density of the lake water will not vary
- The number of load cycles is low compared to a storage tank.

A more appropriate design approach is contained in ACI 357 "Guide for the Design and Construction of Fixed Offshore Structures," which states in Section 3.1 that dead loads include "external hydrostatic pressure" and in Section 4.4.1.1 provides a load factor of 1.2 for these dead loads. This is in line with the approach taken by international standards, which more appropriately address the situation of chiefly non-fluctuating and clearly defined fluids. A few examples are provided below:

- The British Standard 8110:1-1997 "Structural Use of Concrete" makes allowances for loads such as this by providing a reduced load factor of 1.2 (Section 2, Table 2.1) for water pressure where *"the maximum credible level of water can be clearly defined. If this is not feasible, a factor of 1.4 should be used."*
- The British Standard 6349-6:1989 "Maritime Structure: Code of practice for general criteria" recommends a partial factor of 1.1 for hydrostatic pressures under normal loading and a 1.0 factor elsewhere.
- The International Organization for Standardization document ISO 19903:2006 states: Section 6.4 PARTIAL FACTORS FOR ACTIONS: *"For external hydrostatic pressure, and for internal pressures resulting from a free surface, an action factor of 1.2 may normally be used, provided that the action effect can be determined with normal accuracy."*

Pump Adit/IPS-3 Final Design

A three-dimensional model of the shaft at the junctions with the pump adit and IPS-3 stub tunnel was employed to assess the stress distribution around the two openings in the dewatered condition. The shaft walls were modeled as concrete shell elements with a nominal thickness of 450 mm.

The pump adit and IPS-3 stub tunnel will be supported with a 125 mm. thick layer of fiber-reinforced shotcrete and fiberglass rock bolts. This support will also serve as the permanent lining for the tunnels. The linings were designed for as safety factor of 1.3 for 20 combinations of potential wedges each bounded by 3 discontinuities subjected to the design seepage pressure. Since the rock mass will be pre-grouted for ground water control, the improvement of the grout on the joint characteristics was incorporated into the wedge stability analysis.

SHAFT CONSTRUCTION

The Intake Access Shaft was located on a gently to steeply sloping mountainside. The Owner specified a final top of shaft elevation of 382 m amsl, along with a

Figure 6. Initial excavation for shaft collar

working access pad at that elevation for future operations. As the original ground was at 378 m ±, a fill would be required. Since material was not initially available and due to the area needed to support the future tunnel boring operations, a pad was cut into the hillside to elevation 375 m, and the collar of the shaft was constructed at that elevation. At the end of construction, the shaft will be extended up to elevation 382 m and the area filled to its final configuration.

VTC took full advantage of the Design-Build delivery system to start the shaft excavation early. The large cut to elevation 375 m was phased so that the center area was excavated first. This was done concurrent with the collar design so that the collar excavation, as shown in Figure 6, could be done before the pad was completely excavated and graded.

Because of the shaft collar doors that would be placed over the shaft, the shaft collar was a subcollar type. That is, a 12.2 m (40-ft) inside diameter reinforced structure was built to a depth of 4.9 m below the working pad elevation. At the bottom of this collar, a 1.5m concrete platform was constructed with an inside diameter of 9.1 m (30 feet) to match the final shaft configuration. The completed collar prior to placement of the shaft collar doors and headframe is shown in Figure 7. The subcollar was intended to provide working space for construction utilities, survey control, and secondary winches and tuggers. At that point, excavation, initial support, concrete lining, and pre-excavation grouting of the 9.1 m diameter shaft could proceed.

Excavation

Excavation of the shaft has been by means of drill-and-blast. A three-boom Tamrock shaft jumbo (Figure 8) is used to drill blast holes. A Cat 939 track loader loads 6 m^3 sinking buckets. Initially, hoisting was accomplished with a 180 tonne (200-ton) Link Belt crane. Later, a 26 m tall headframe was erected over the shaft to service the work. The work was supported with an Ingersoll-Rand double drum main hoist with 1250 hp, 36 tonne (80,000 lb) single line pull, and maximum rope speed of 245 m per minute. This hoist will also support the tunnel excavation. In addition to the double drum main hoist, four New Era stage winches, with 22.5 tonne (50,000 lb) line pull, are installed to support and hoist the work decks and concrete forms and to provide guides for the skips and buckets. A two-deck Galloway is suspended just above the working face to support all shaft activities.

A typical round pulls 3 m, with average 150 drill holes and about 270 kg (600 lb) of powder. Because of the placement of the shaft jumbo, a V-cut is drilled for the center

Figure 7. Completed shaft collar

Figure 8. Tamrock Shaft Jumbo

zone. Nonel detonators are used in all cases. ANFO was used above the water table, and stick powder was used below the water table.

The Pump Adit and the IPS-3 Stub Tunnel will be excavated in one pass, drilling with two Tamrock 200 Commando single boom jumbos.

Concrete and Final Linings

VTC selected the top down approach to the concrete lining to provide earlier permanent support and additional ability to control groundwater inflows. A blast-proof shaft form with nominal height of three meters was obtained from Everest Equipment. The forms were configured with a 750 mm curb ring and the main section of 2.25 m.

Figure 9. Shotcrete initial support, shaft form and completed lining

The curb ring is suspended from the previous placement by means of all-thread bars. Scribing pins and steel mesh are used to form the bottom of the curb. Concrete is then placed into the curb ring. While that sets, the rest of the form is lowered, set, and poured.

Concrete placements follow immediately behind the excavation. Generally, as shown in Figure 9, there is less than 6 m of excavated shaft below the most recent concrete placement. This minimizes exposure and allows enough room for the jumbo to drill the perimeter holes beneath the lining above. Contact grouting is performed after the concrete reaches its design strength.

In the drill-and-blast adits and the TBM chamber, the initial support is intended to be the final lining. In these areas, the support has to be able to achieve the 100-year design life. For this, a system of fiberglass rockbolts and steel fiber reinforced shotcrete has been designed as discussed above. A structural concrete invert is also placed in each area.

Pre-excavation Grouting

The groundwater table at the Access Shaft is directly related to the lake level in Lake Mead. Consequently, at the time of construction, the lake level will be approximately 150 m above the bottom of the shaft. Combined with numerous open zones that are predicted in the geotechnical investigations, high water inflows are anticipated. To reduce these inflows, a program of probe drilling and pre-excavation grouting is required.

Four probe holes are to be maintained at least 9 m (30 ft) ahead of the shaft excavation. If more than 3 l/s (50 gpm) of groundwater inflow is observed through the probes, then a pattern of grout holes is drilled and grouted in order to reduce the inflows to below this limit.

The grouting program designed by Arup and VTC is based on a 36 m (120-foot) deep grout curtain, with 9 m of overlap, in order to obtain an approximately 3 m thick

Table 4. Segmental liner information

No. of Segments	5 Segments + 1 Key
Internal Diameter:	6.1 m (20 feet)
Segment Thickness:	356 mm (14 inches)
Type of Construction	Bar reinforcement—WWR 51.7 kN/m² (75ksi) yield
Reinforcement Quantity:	78 kg/m³ (225 lb/cy)
Ring Type:	Universal Double Taper
Ring Width:	1.82 m (72 inches)
Average Segment Weight:	5,900 kg
Concrete Compressive Cylinder Strength:	40 N/mm² (6000 psi)
Radial Joint Connections:	Spear bolts w/ Guide Rods
Circumferential Joint Connections:	Spear bolts w/ Ball Joints
Gasket Groove and Gasket:	EDPM gasket rated to 38 bar
Number of Rams on TBM:	16 rams, 1 per shoe
TBM Ram Loads:	70,368 kN (7,910 tons)—Force on TBM Face 100,528 kN (11,300 tons)—Maximum Jacking Force 33 N/mm² (4,750 psi)—Max Pressure Under Each Shoe
Size of TBM Ram Shoe:	900 mm × 340 mm
Maximum Hoop Thrust:	7,400 kN/m

curtain around the shaft. Primary holes were spaced at approximately 1.25 m centers around the perimeter of the shaft, looking out at 5 degrees. Secondary holes were spaced between the primary holes, looking out at 2 degrees.

Three meter long standpipes were drilled and grouted in place at the drilling elevation. These were then fitted with valves and gages. Primary grout holes were drilled to a certain depth, and then pressure grouted. They were then redrilled and, if the results were satisfactory, deepened to the next elevation. When the primary holes did not return the desired result, secondary holes were drilled and grouted as well. Grouting continued in this manner until the 36 m depth was reached. Four interior probe holes were then drilled to verify the results.

The grout mix has depended on the ground conditions as well as the results, and has evolved as the shaft is extended. Cementitious materials included Type II/V cement, Type III cement, and ultrafine cements. Similarly, refusal was redefined as the program progressed, and was a function of achieved pressure or injection volume.

TUNNEL DESIGN

The intake tunnel alignment is approximately 4.7 km (15,400 feet) long and passes through variable geology including metamorphic, sedimentary, and volcanic rocks. While designing a tunnel lining through this changing geology can be challenging, the consideration of the high water head made the design even more so. Further complicating the design, the high water head requires a very high thrust ram capacity to advance the TBM, resulting in very high jacking loads on the liner. In the final state, the full hydrostatic pressure of the lake will be imposed on the tunnel liner, resulting in high hoop loads and therefore high bursting loads.

Some of the key features of the liner and TBM are included in Table 4.

The segmental lining has been designed as a universal tapered ring. Each segment ring consists of four rhomboidal segments, a trapezoidal counterkey, and a key. The relatively high segment Length/ID ratio of 0.3 was chosen to reduce the number of joints along the tunnel. In comparison with five foot long segments, this arrangement reduces the total joint length by 12%.

The segment ring is provided with a taper of 51 mm and the segment ring can be rotated in sixteen different positions. The taper is arranged such that the key is on the widest portion of the ring, and allows an absolute minimum turning radius of 220 m. To limit the occurrence of cruciform joints, which are more difficult to seal against external pressure, certain orientations on adjacent rings will be avoided wherever possible, and consequently the minimum radius will be around 275 m.

The tunnel design utilized several analysis methods and tools such as empirical relationships, closed-form continuum solutions, Finite Element (FE) models in SAP2000, and LS-DYNA, and numerical modeling of the ground in Plaxis, Unwedge, and beam-spring models. The multiple approaches were used to cross check results and ensure a robust design.

It was recognized early in the design process that the segment to segment joints would be a critical design feature. Due to the external water pressure, the lining is carrying a high compressive load during construction and in the final state when the tunnel is dewatered. The high compressive load has the most impact at the joints, where bursting forces are introduced as the section thickness reduces to provide the gasket groove and stress relief recesses. The joint bursting stresses were analyzed using a 2-D FEM plane-strain SAP analysis and the closed form solution by Leonhardt.

The response of the segmental tunnel lining to ground and water loads is often only modeled using a plane-strain loading model, which does not take into account 3-dimensional effects. However, due to the very high hoop loads it was decided to model the segment to segment interaction using an advanced numerical analysis. Detailed analyses of stresses and potential damage in the tunnel lining due to segment misalignments were carried out using the 3D nonlinear finite element program LS-DYNA which is particularly well suited to treat soil/structure interaction and contact problems. A model of a 4-ring tunnel section with its surrounding rock was first developed. The full tunnel construction sequence was simulated in order to investigate the effect of segment misalignment under a representative overall loading. The segment loads and displacements calculated in the 4-ring coarse model were then applied to a 2-segment fine mesh model more suitable for stress analysis. Pressure from exterior rock and pore water as well as jacking axial loads were taken into account. The analysis also included the effect of rotation of the segments or tunnel squat on the joint faces, as the longitudinal joints are inclined and therefore do not hinge around a line parallel to the tunnel axis, as would occur if the segments were rectangular. This effect can only be captured with an advanced 3D model and was reviewed in the design.

Of the three methods used to model bursting, the results obtained from LS-DYNA, which modeled the most onerous combination of misalignments, predict the highest bursting stress, though in general all the results agree. To design the segment to segment interface, steel reinforcement ties were designed to confine the busting stresses, with no benefit to the concrete tensile strength given. The concrete tensile strength of $6.0\sqrt{(f'c)}$ has only been included in calculations to check the handling loads of the segments during manufacture.

INTAKE DESIGN

To meet the SNWA criteria, the intake structure required an inlet area of 46 square meters and a riser pipe of approximately 5 m diameter. The design had to allow for a bulkhead to be used during construction which could also be refitted in the future to allow draining of the tunnel for maintenance. If a steel structure was used for the intake, stainless steel was required for section extending above the lake bed.

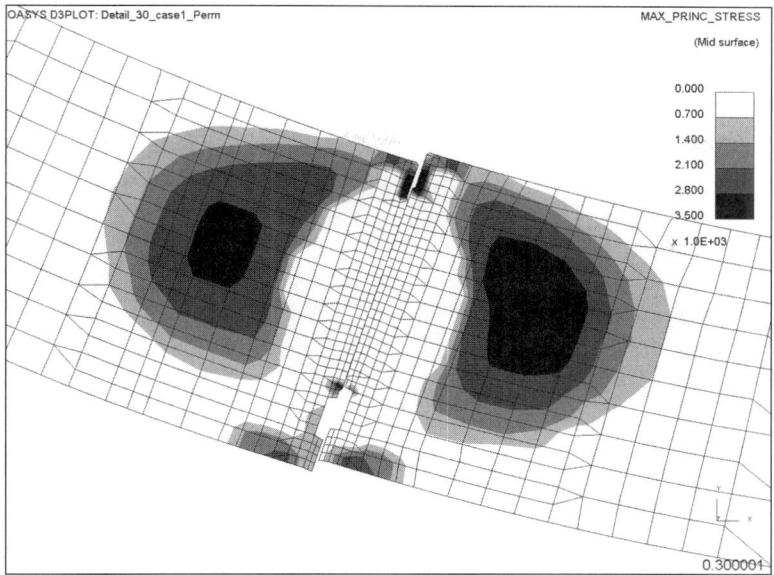

Figure 10. LS-DYNA model

There are two types of basalt in the vicinity of the intake, as well as flow breccia, which are anticipated to vary widely in composition and other physical factors. A total of 10 boreholes are available in the area, which show that there is continuous and persistent jointing going through all three types of rock. The jointing can be grouped into three regular sets of joints, together with numerous random joints. Basalt, both vesicular and non-vesicular, is usually a homogenous rock type with no cleavage or natural foliation. The observed jointing is therefore most likely related to local or more regional tectonic processes. The two steeply dipping possible fault zones give indications of activity that can have contributed to the uniformity of the observed jointing.

The excavation of the lake bed for the positioning of the pre-assembled caisson of the Intake Riser will be a challenging operation, because of the extreme depth of the water.

A number of significant health and safety risks were identified relating to the construction of the intake structure, including:

- Risk of flooding the tunnel while making the connection between the tunnel and the intake.
- Saturation diving techniques would be required for any manual underwater work (water depths between 90 and 110 m)
- Difficulties of working on the lake (it was anticipated being unable to work approximately one month per year due to rough lake conditions)

In addition, making the connection between the tunnel and the intake is on the critical path, at the end of the construction schedule.

During the bid design process a conventional drilled shaft, or dry tap, arrangement was developed. This consisted of placement of a 5 m diameter steel riser shaft into the lake bed. The tunnel would be bored under the riser, and a connection made by excavating between the tunnel and the riser. Before the tunnel arrived at the intake location,

Figure 11. Intake structure

this would require a sequence of ground improvement, drilling a large diameter shaft, and placing a riser shaft and grouting.

Recognizing the risks, the VTC team looked at alternative configurations and construction methods. The wet tap method (where the tunnel is bored first and then the connection to the lake made by blasting out the remaining section of material above the tunnel) was ruled out because there would have been high inflows before the connections was made as the rock is relatively permeable. The method would also result in a longer schedule. The chosen solution, shown in Figure 11, is to utilize an intake structure that could be fabricated close to the shore and then be prepositioned into the lake bed using immersed tube techniques, which would serve as a location in which to "dock" the TBM at the end of the drive. Once the TBM has entered the intake structure, and a seal made between the TBM skin and the structure through grouting, or freezing if necessary, the TBM can be partially dismantled and a final concrete lining placed. For the bid, a steel structure was designed, but a concrete alternative is currently being developed.

ACKNOWLEDGMENTS

The authors acknowledge the permission of the Southern Nevada Water Authority to publish this paper.

Project Team:
- Owner—Southern Nevada Water Authority
- Owner's Engineer—MWH and CH2MHill in joint venture
- Construction Manager—Parsons Water Infrastructure
- Contractor—Vegas Tunnel Constructors, a joint venture of Impregilo SpA and SA Healy Co.
- Contractor's Design Engineer—Arup, with Brierley Associates and Sneegeoconsult.

REFERENCES

Feroz, M., Jensen, M. and Lindell, J.E, 2007, The Lake Mead Intake 3 Water Tunnel and Pumping Station, Las Vegas, Nevada, USA, RETC 2007, Toronto, Canada.

Feroz, M., Moonin, E. and McDonald, J, 2009, Project Delivery Selection for Southern Nevada's Lake Mead Intake No. 3, RETC 2009, Las Vegas, NV.

PROJECT DELIVERY SELECTION FOR SOUTHERN NEVADA'S LAKE MEAD INTAKE NO. 3

Michael Feroz ▪ Parsons Water and Infrastructure

Erika P. Moonin ▪ Southern Nevada Water Authority

James McDonald ▪ Vegas Tunnel Constructors JV

INTRODUCTION

Severe drought in the Colorado River Basin over the past 10 years has caused water levels to drop in Lake Mead by more than 113 feet. The current lake level is at elevation 1,112 feet mean sea level (msl). Lake Mead is the primary source of water supply for greater Las Vegas. Water for Las Vegas from Lake Mead is accessed through two existing submerged intake shaft and tunnel systems. If the water levels continue to drop in the lake, the Intake No. 1 will become inoperable below water level elevation of 1,050 feet msl. To sustain the current levels of water delivery to all member agencies in the southern Nevada region, the Southern Nevada Water Authority (SNWA) has decided to construct a third intake at a deeper depth in the lake to ensure better quality water and enhanced reliability.

PROJECT DESCRIPTION

The Project will be the deepest sub-aqueous tunnel constructed with a pressurized-face tunnel boring machine (TBM) in the world to date. The TBM will have the potential of operating under 17 bars pressure. The contract will be completed towards the end of the year 2012.

The general project area is located in southern Nevada (Clark County), approximately 20 miles east of Las Vegas along the western shoreline of Lake Mead's Boulder Basin. The project location is shown in Figure 1. The area is entirely within the Lake Mead National Recreation Area (LMNRA), managed by the National Park Service (NPS), U.S. Department of Interior.

Areas within the LMNRA have been developed by SNWA for the existing intakes and pumping stations (IPS-1 and IPS-2) on the south half of Saddle Island and the Alfred Merritt Smith Water Treatment Facility (AMSWTF).

The project will be located on the northern third of Saddle Island and will extend to the northeast into an area presently inundated by Lake Mead. The pumping station will be situated on the western slope of Saddle Island, approximately 500 feet north of the existing causeway in an undeveloped area sparsely covered by native grasses, cacti, and scrub brush. The intake tunnel will be approximately 15,000-feet long and will extend from a 600-foot deep vertical shaft near the pumping station area, around the northwest tip of Saddle Island, to an intake area located in Lake Mead, northeast of the access shaft. The tunnel alignment and intake area will be within a corridor specified by the SNWA Requirements.

SNWA presently operates two water intakes and pumping stations at Saddle Island on the western shore of Lake Mead, approximately five miles northwest of Hoover Dam and approximately 20 miles east of Las Vegas. The existing facilities are

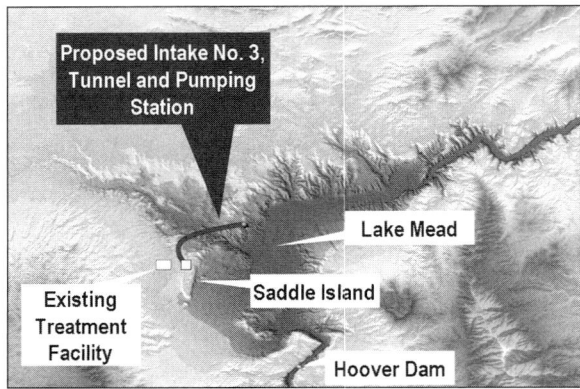

Figure 1. Project location

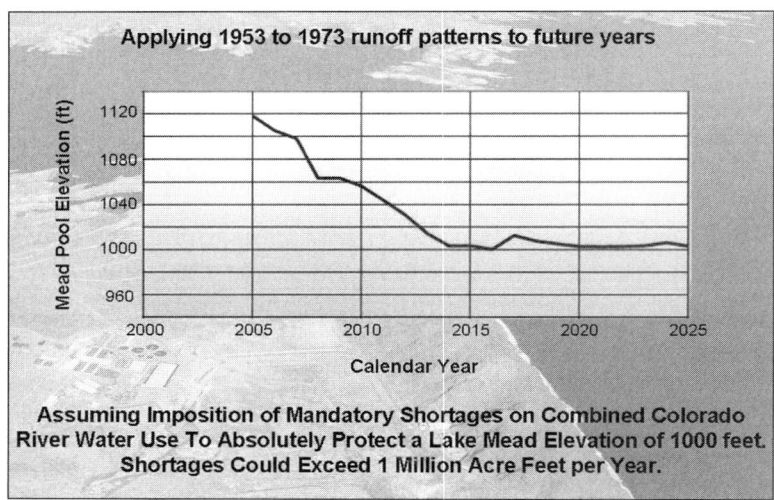

Figure 2. Possible Lake Mead water levels

designated as Intake Pumping Station No. 1 (IPS-1) and Intake Pumping Station No. 2 (IPS-2). IPS-1 serves the AMSWTF adjacent to Saddle Island and IPS-2 serves the River Mountains Water Treatment Facility (RMWTF) located approximately five miles west of Lake Mead.

Severe drought has caused declining water levels in Lake Mead during recent years. The current lake elevation is approximately 1,112 feet msl and the lake is expected to decline even more over the next several years as projected in Figure 2. The SNWA will construct a third deep-water intake in Lake Mead to protect the existing water system capacity against the potential inoperability of IPS-1 should lake levels fall below elevation 1,050 feet msl. The secondary purpose will be to improve water quality and system reliability and operational flexibility. The Figure 3 graphic displays water level impacts.

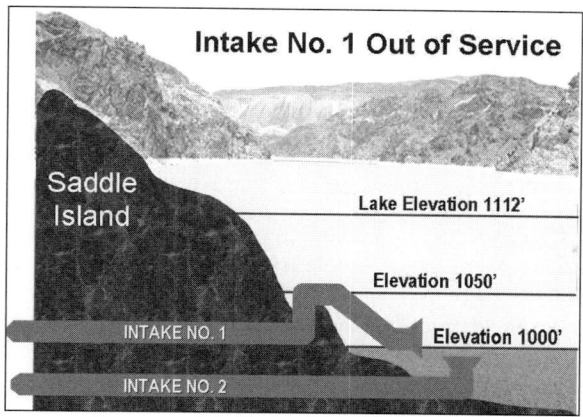

Figure 3. Lake Mead water level impacts

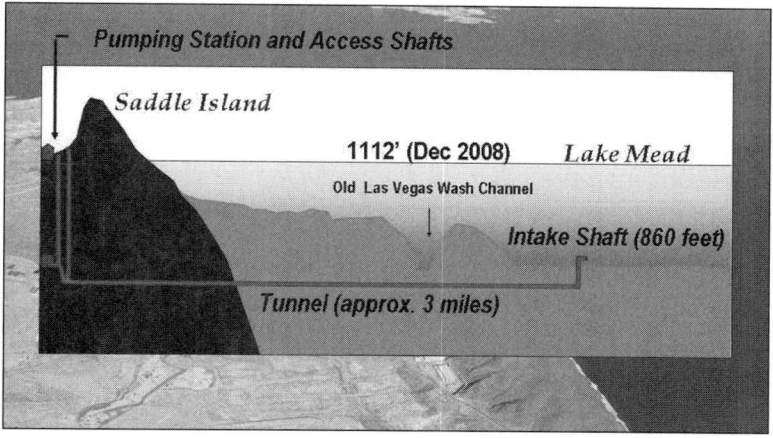

Figure 4. Profile of proposed Intake No. 3 system

The Lake Mead Intake No. 3 project will include the following major components:
- A deep-water intake riser and inlet structure;
- An 20-ft diameter, 1,5000 ft long intake tunnel driven beneath the lake and a portion of Saddle Island;
- A pumping station on Saddle Island (IPS-3);
- A discharge pipeline from IPS-3 connecting to the AMSWTF; and
- A tunnel connecting the IPS-3 facilities with the existing IPS-2.

The basic concept of the Lake Mead Intake No. 3 project is to draw water from below the Lake's thermocline through an intake structure located northeast of Saddle Island as demonstrated in Figure 4. Water will flow from the intake structure through an intake tunnel to a pump forebay located under Saddle Island. The location of the new intake was selected based on water quality modeling completed in conjunction with the

Clean Water Coalition. The pumping station will be an at-grade structure located on the western slope of Saddle Island, north of the existing pumping station facilities. Well shafts will extend from the ground surface of the pumping station to the underground forebay and will house vertical turbine pumps. The vertical turbine pumps will pump the water into a discharge pipeline to convey the water to AMSWTF located on the west side of Saddle Cove.

The major components of the Lake Mead Intake No. 3 project are divided into six major work efforts:

- Design-Build of a 20-ft diameter, 15,000 ft long intake tunnel driven beneath the Lake and a portion of Saddle Island and a deep-water intake riser and inlet structure (Contract No. 070F 01 C1)
- Design-Bid-Build of a Pumping Station Underground IPS-3 (Contract No. 070F 02 C1)
- Design-Bid-Build of a Pumping Station Superstructure IPS-3 (Contract No. 070F 02 C2)
- Design-Bid-Build Power Supply Facilities (Administered by CRC)
- Design-Bid-Build of a discharge pipeline from IPS-3 connecting to the AMSWTF (Contract No. 070F 04 C1)
- Design-Bid-Build of a shaft and tunnel (Intake 2 Connection and Modifications) connecting the IPS-3 interconnecting tunnel to the existing IPS-2 (Contract No. 070F 05 C1)

Summary of Key Project Features of the Design-Build Contract

Intake Tunnel

The intake tunnel will be approximately 15,000 lineal feet, sloping upward from the base of the tunnel access shaft to the intake riser(s). The actual vertical and horizontal alignment, grade and length will be determined by the Design-Builder to suit their operations but will be within a prescribed alignment corridor as shown in the SNWA Requirements. The excavated and finished dimensions will be also be determined by the Design-Builder to meet the prescribed design and operational requirements. The prescribed tunnel alignment corridor is defined by SNWA for the following purposes: to limit the amount of tunnel excavation in Saddle Island Upper Plate (Pcu); to preclude tunneling within the altered Saddle Island Volcanics (Tvsi); and to maximize the proportional length of tunnel in Tertiary sedimentary rock formations. The vertical limits of the alignment corridor extend from the existing ground surface and lake bed to elevation 600 feet below the surface. General elevations of the new intake system are shown in Figure 5.

Intake Riser and Inlet Structure

The intake riser (or risers) will be located within a prescribed area in the Boulder Basin of Lake Mead, northeast of Saddle Island and the Las Vegas Wash. The riser(s) will be a lake tap structure, similar in concept to the existing SNWA intakes on Saddle Island, and will connect the intake inlet structure to the intake tunnel. The number of risers and their excavated and finished dimensions and lining details will be determined by the Design-Builder to meet the prescribed design and operational requirements and suit the Design-Builder's operations. The intake inlet structure will be located on top of the intake riser and will be configured to draw water horizontally at a centerline elevation

PROJECT DELIVERY SELECTION FOR LAKE MEAD INTAKE NO. 3   507

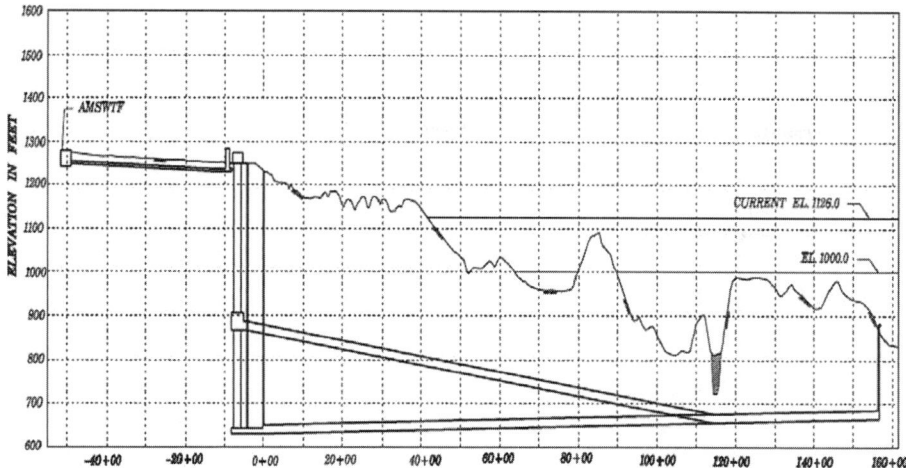

Figure 5. Elevations of Intake No. 3

of 860 feet msl. The inlet will include provisions for a future connection to extend the intake inlet to deeper waters northeast of the planned location by underwater pipeline.

Tunnel Access Shaft and Connections

The tunnel access shaft will be located approximately 500 feet north of the planned IPS-3 pumping station site. The shaft will provide access for constructing the intake tunnel and also function as one of the elements to contain surge during operation of the project. The access shaft will include a short stub tunnel for connection to the planned IPS-3 Pumping Station project (Contract No. 070F 02 C1). The IPS-3 stub tunnel will intersect the tunnel access shaft at approximately elevation 811 feet msl. The tunnel access shaft will also include provisions for future expansion to an additional pumping station in an area east of the shaft and north of the proposed IPS-3 location. The provisions for future expansion will be a short stub tunnel located at the base of the shaft. The stub tunnel will include a steel bulkhead to isolate the stub tunnel from the IPS-3 system.

EVALUATION OF ALTERNATIVE CONTRACTS

SNWA evaluated the following three delivery methods for the Intake No 3 facilities:
1. Competitive Bid (Design/Bid/Build)
2. Construction Manager At Risk (CMAR)
3. Design-Build (D-B)

Advantages and disadvantages for each delivery method were evaluated by SNWA as described below.

Competitive Bid (Design/Bid/Build)

This delivery method has been the most common for Public Work Projects. Final Plans and Specifications are completed by the Owner's Design Engineers. The Contractors bid the project exactly as designed and the responsive bidder with the lowest price is selected for the work.

The major advantages of this delivery method were as follows:
1. This delivery method is most commonly used and is simpler to manage
2. This method provides the opportunity for the lowest initial price
3. The scope of the project is completely defined for design and construction
4. Since the award is primarily based on lowest price, this method provides more opportunities for bidders
5. Owner has more control of the design and construction as compared to other delivery methods.

The main disadvantages of this delivery method were as follows:
1. Contractor is not selected based on experience and may increase potential for a lower quality project
2. This method is based on linear process and normally results in longer schedule duration as compared to other delivery processes
3. Generally results in adversarial relationships between the designer, the contractor and the owner
4. This method increases the potential for disputes/claims, and change orders resulting in higher final project cost.

Construction Manager At Risk (CMAR)

This delivery method allows the owner to select a designer and builder (construction manager/general contractor-CM/GC), based on qualifications and experience that is best suited for developing the design and bidding documents. CM's fee is agreed by the owner up front. The design consultant team is also selected by the owner. The CM and the design team develop the design and construction cost estimate. The CM provides a guaranteed maximum price to the owner with identified contingencies. The CM advertises project elements to receive proposals from subcontractors and awards the work. The final construction price is determined by adding the subcontractor price, the agreed upon CM's Fee, CM over Head, and contingencies. All contingencies that are not used at the end of the project revert back to the Owner.

The major advantages of this delivery method were as follows:
1. CM/GC team is selected based upon qualifications and experience
2. CM/GC provides design phase assistance in planning
3. Cost is developed as design progresses and, as such, continuous budget control possible
4. The screening of subcontractors allows Owner and CM/GC to select quality subcontractors for quality work
5. This method provides better control for handling changes in scope and project design
6. This method allows to fast track construction without additional cost because overlapping project phases are possible
7. The owner and CM/GC can develop guaranteed maximum price as compared to other project delivery methods
8. Prescriptive specifications allow better quality standards as compared to other delivery methods
9. This method reduces the potential for change orders and claims.

The major disadvantages of this delivery method were as follows:
1. It is difficult for the owner to evaluate whether the project is completed at the best price
2. Project costs may increase for details not identified in CM/GC's guaranteed maximum price
3. This method offers lesser opportunity for competitive price for CM/GC's Fee, Over Head and Subcontractor costs.
4. This method is not permitted by the Nevada Revised Statutes

Design-Build (D-B)

The Design-Builder (D-B) would be hired by the owner as one entity to deliver the complete project. The D-B would be set up to provide a firm fixed price at the time of bid based on the owners requirements identified in the Request for Proposal (RFP) documents. The D-B would be required to identify all its major subcontractors. The D-B would meet all of the owner's requirements and develop detailed design using the 30 percent conceptual drawings and performance specifications provided to the D-B team in the RFP.

The major advantages of this method were as follows:
1. The D-B could be selected based on qualifications of the team, and experience with similar projects
2. Design and construction would be accomplished through a single point of responsibility
3. The Owner has the benefit of obtaining the firm fixed price at the time of bid
4. The D-B can provide innovative design phase assistance in project planning and construction
5. The D-B delivery process offers faster project delivery compared to other available alternatives
6. Change orders due to errors and omissions are minimized in D-B delivery process
7. Allows continuous execution of design and construction
8. The D-B delivery process offers great advantage for projects with critical schedule.

The major disadvantages of this delivery method were as follows:
1. The Owner has little to no input on the selected design team
2. There is a potential for reduced quality due to lesser control by the Owner
3. There is higher potential for conflicts between the D-B and the Owner
4. At the completion of the project the Owner cannot determine whether the project has been delivered at the best price.
5. Due to a firm fixed price, changes are difficult and expensive to make once construction begins.
6. There is potential for higher cost as the D-B may build higher risk contingencies in his bid.
7. The D-B may direct the designer to implement cheaper alternatives which may compromise quality of the end products
8. Review time for design submittals for the Owner may be compromised as the D-B may require quick decisions. This can increase the potential for unsafe design or project.

SNWA'S SELECTED ALTERNATIVE

SNWA compared the advantages and disadvantages of each of delivery methods discussed in the preceding paragraphs for the Intake No. 3 Project. Schedule is of utmost importance to SNWA to build the Intake No. 3 in a timely manner. In order to achieve this goal SNWA embraced the Design-Build (D-B) delivery method for this project. Design Build (D-B) under Nevada Law (as revised October 2005) and Nevada Revised Statutes (NRS) 338 Public Workwas used to select the most qualified Design-Builder to meet the project goals. Based on the law, the winner was selected based on both qualifications and total price.

The engineering and construction requirements for this project required a D-B with extensive experience in mining a tunnel under a lake or similar conditions. Letters for expression of Interest (LOI) were issued to national and international firms and several firms expressed interest in the project. The LOI's were followed by issuing a Request for Preliminary Proposal (RFPP) to interested firms in accordance with NRS requirements. A short list of qualified firms was developed and a Request for Final Proposal (RFP) was issued in accordance with NRS requirements. Prior to receiving the proposal from the short listed firms, proprietary meetings were held individually with each of the firms. The finalists were requested to submit a Technical Proposal and a Cost Proposal in separate sealed submittals. The finalists were first evaluated on their technical proposal by the SNWA committee. The evaluating committee were assisted by a group of tunneling experts. Subsequent to scoring of the Technical proposals, the Cost Proposals were opened and evaluated in detail. The combined scores were used for selecting the Design Builder based on the points identified in the RFP and in accordance with NRS.

The D-B contract was awarded in March 2008 for approximately $447 million to Vegas Tunnel Constructors (VTC), a joint venture of Impregilo SpA and S.A.Healy. VTC has selected Arup as the Design Engineer. The selected Design Builder has mobilized to the site and excavation for the tunnel access shaft is currently under construction. The construction is currently under way with scheduled completion of the project in December 2012.

This section describes how SNWA balanced compliance with Nevada State Law and achievement of practical goals in requesting and evaluating design-build proposals.

SNWA Considerations for Design-Build Contract

- Design of the project is strongly related to Design-Builder's means and methods
- This approach puts responsibility for design and construction on one entity, the Design-Builder
- Synergy between Design-Builder and Owner's Team saves time
- This approach may increase construction costs or discourage participants if risks are high or are not predictable
- The Design-Builder can apply creative design solutions to tough construction challenges
- This contract is of longest duration of all the other project elements for Intake No. 3 and based on D-B means and method anticipated and contractual completion dates can be set by the D-B
- Based on NRS requirements a two step process was used for procuring this Contract:
    1. Preliminary Proposals
    2. Final Proposals

The design-build schedule including milestones was conducted as outlined in the following table:

Design–Build Schedule and Milestones

- Board approved Design-Build Method — 10-19-06
- Issue Request Statement of Interest — 10-19-06
- Issue Request for Preliminary Proposal — 12-18-06
- Conducted Pre-proposal Meeting — 01-19-07
- Received Preliminary Proposals — 05-11-07
- Board Approved Short List — 05-22-07
- Issued Request for Final Proposals — 06-22-07
- Conducted Meetings with Proposers — Jul/Sep 2007
- Opened Final Proposals — 11-02-07
- Board Awarded DB Contract — 03-20-08
- Construction Complete — December 2012

Request for Preliminary Proposals (RFPP)

- Request for preliminary proposals was advertised in December 2006
- SNWA held a Pre-Proposal Meeting
  - Provided overview of project scope
  - Reviewed selection criteria & process
  - Provided opportunity for attendees to ask questions
  - Conducted a tour of the project site
- Issued RFPP Addenda in response to questions received at the prebid
- Received four Preliminary Proposals

Preliminary Proposal Review

- The four proposals were reviewed by a Evaluation Committee in accordance with the following criteria:
- Preliminary Proposal Review Criteria:
  - Criteria Required by NRS (Minimum Requirements for Eligibility)
  - Qualifications and Experience relative to the Project requirements

  Three of the proposers were selected as Finalists to move to the next phase—Request for Final Proposals

Request For Final Proposals (RFP)

- The RFP was advertised on June 22, 2007
- SNWA initiated one on one proprietary meetings with the three Finalists
  - Series of optional meetings were also held at the request of each Finalist
  - Information discussed was confidential since the Finalists technical approach was considered proprietary
  - SNWA evaluated information obtained from each Finalist and in some cases issued Addenda to modify the RFP

Purpose of Proprietary Meetings

The purpose of the proprietary meetings was to receive feedback from the Finalists on the RFP requirements, including the contract agreement and allocation of risk issues considered during RFP Process:

- Appropriate and fair allocation of risk between D-B and Owner
- Technical requirements
- Use of allowances to compensate Contractor for unforeseen conditions for specific Unit Items
- Differing site conditions clause applied after allowance item is fully used for specific Unit Items
- Providing more flexibility for Finalists to determine appropriate means and methods
- Allowing the Finalists to establish the contractual completion date by specifying the number of days in proposal
- Providing additional time for Finalists to complete and submit proposal

Final Proposal Review

SNWA received two final proposals. The Proposals were reviewed and rated by the Evaluation Committee using the following criteria:

- Technical Proposal: Maximum Score = 60 Points
- Price Proposal: Maximum Score = 35 Points
- Nevada Contractor Preference: Score = 5 Points

Criteria Used in Evaluation of Final Design-Build Proposals:

1. **Technical Proposal: 60 Points**
   a. The Technical Proposal was evaluated based upon the following sub-factors:
   b. Project Management Plan: 15 points
   c. Risk Management Plan: 15 points
   d. Design and Construction Plan: 30 points

2. **Price Proposal: 35 Points**
   a. Price Proposals were scored and ranked on the basis of the Total Proposal Price and in relation to the lowest Total Proposal Price, with scoring as follows:
      i. The Finalist submitting the lowest Total Proposal Price was awarded the maximum number of points (35).
      ii. The next-lowest Total Proposal Price was awarded points based on the product of: (a) the ratio of the lowest Total Proposal Price divided by the next-lowest Total Proposal Price; and (b) 35 points (i.e., the points awarded for the lowest Total Proposal Price), with such product rounded to the nearest one hundredth of a point.

3. **Nevada Contractor Preference: 5 Points**

PROJECT DELIVERY SELECTION FOR LAKE MEAD INTAKE NO. 3

Price-Technical Trade-off Evaluation

After the Evaluation Committee evaluated each Proposal and assigned the scoring, the highest-scored Proposal had its Price Proposal compared to each of the other Proposals.

If the differential between the highest-scored Proposal and any other Proposal was ten percent (10%) or greater, and the highest-scored Proposal had the higher-priced Proposal, then SNWA conducted a price-technical tradeoff between the highest-ranked Finalist and such lower-priced Finalist(s).

If the differential was less than ten percent (10%), then there would be no price-technical tradeoff process.

Finalists with Price Proposals having less than a ten percent (10%) differential with the highest-ranked Finalist would not be eligible to participate in the price-technical tradeoff process.

The purpose of the price-technical tradeoff was to determine whether the combination of technical and price factors justify a recommendation of award to a Finalist other than the initially highest-ranked Finalist. The following process was applied in conducting the price-technical tradeoff evaluation:

- Each Proposal involved within the price-technical tradeoff process was to be compared to the other, in separate pairings.
- SNWA analyzed the significance of the differences in the Technical Proposal ratings as indicated by each Technical Proposal's strengths, weaknesses, and risks for each evaluation factor. The strengths, weaknesses, and risks were considered in light of the relative importance of each factor stated in the RFP.
- SNWA assessed the best mix of cost and non-cost benefits and determined whether the strengths of higher rated proposals were worth the price premium.

Based on Evaluation of Technical Proposal and Cost Proposal the SNWA Board awarded the Design-Build Contract to:

- Vegas Tunnel Constructors, a Joint Venture of S.A.Healy and Impregilo SpA
- Contract Price: $447,085,629
- Scheduled Completion: December 2012

Advantages of Awarding the D-B Contract:

- Partnering affords superior performance and quality
- Design-Builder responsible for design and construction as single source
- Design-Build team selected based upon qualification of experience of the designer and the contractor
- Firm Fixed Price was provided in the final proposal
- Project Delivery Schedule is faster than the traditional D/B/B or CMAR
- Risk Management by risk sharing between the Owner and the Design-Builder minimizes building contingencies in the Design-Builder's Firm Fixed Price
- Allows the Design-Builder to apply creative design solutions to tough construction challenges
- Change Orders in traditional D/B/B due to Errors and Omissions by Consultant eliminated

DESIGN BUILD UNDER NEVADA LAW
(AS REVISED OCTOBER 2005)

The State Law Allows Allows for the Following Design-Build Projects:
- Construction of a park and appurtenances thereto,
- Rehabilitation or remodeling of a public building, or
- Construction of an addition to a public building

Requirements of a Design Build Team:
- Assume overall responsibility for ensuring that the design and construction of the public work is completed in a satisfactory manner
- Consist of at least one General Engineering or Building Contractor AND Architect (if for building) or Professional Engineer
- Able to obtain performance and payment bonds
- Carry appropriate licenses
- Able to obtain general liability and errors and omissions insurance
- During previous 5 years, not have been found liable for breach of contract on a previous project
- Not have been disqualified under NRS 338 prequalification process

Selection Procedure:

1st Public body has to approve the use of Design-build for a specific Public Work

2nd Advertise at least 30 days a "Request for Preliminary Proposals" which includes:
  a. A statement whether or not team that is selected as finalist but not awarded contract will be partially reimbursed for the cost of preparing a final proposal, and if so, an estimate of the amount of the partial reimbursement. (May reimburse each unsuccessful finalist up to 3% of the total amount to be paid to the team.)

3rd After receiving Preliminary Proposals, select at least two but not more than four finalists from preliminary proposals received by conducting an evaluation of the qualifications of each team including:
  a. Performance history;
  b. Safety programs established and safety records accumulated;
  c. Management plan and details of ability of team to design and construct the public work; and
  d. Professional qualifications and experience

4th Issue "Request for Final Proposals" to finalists selected which includes list of factors to be used in evaluation including relative weight assigned to each factor.
  a. 5% weight given to in-state preference certificate;
  b. At least 30% weight given to proposed cost.

5th Select, at a public meeting, the most cost-effective and responsive final proposal, or reject all bids.

6th Make available to the public results of evaluations of both preliminary proposals and final proposals.

CONCLUSIONS

- Early integration of Specialty Contractor's during the RFPP phase proved to be very useful.
- Partnering between the Owner and the Proposers during the RFP process provided excellent exchanges and RFP documents were modified to incorporate those changes.
- SNWA held proprietary meetings with the short listed Finalists to develop risk sharing concerns. These were documented in the Risk Management Register and incorporated via Addenda to the RFP document.
- Balancing compliance with the State laws and project goals was a challenge and it was resolved during the proprietary meeting discussions.
- Limiting number of contracts for various components eased interfaces, and scheduling conflicts.
- Design-Build approach provides the vehicle through which Schedule, Cost, Safety and Quality are managed best.
- Balancing compliance with the State laws and project goals was a challenge and it was resolved during the proprietary meeting discussions.
- Limiting number of contracts for various components eased interfaces, and scheduling conflicts.
- Design-Build approach provides the vehicle through which Schedule, Cost, Safety and Quality are managed best.

ACKNOWLEDGMENTS

The authors are very gracious to the Southern Nevada Water Authority for granting the permission to discuss the delivery process used for the selection of the D-B team. We also wish to thank the following project team in providing useful input:

Owner: Southern Nevada Water Authority
Design-Builder: Vegas Tunnel Constructors Joint Venture (Impregilo spA and S.A.Healy)
Owner's Engineer: MWHILL JV (CH2MHill and Montgomery Watson Harza)
Design-Builder Engineer: Arup and Brierley Associates
Construction Manager: Parsons Water Infrastructure

Our special thanks to Ms. Kayla Hutchens and Ms. Margaux Macchiaverna for formatting and developing the final documents and for their enthusiastic and timely assistance.

REFERENCES

Feroz, M., Jensen, M. and Lindell, J.E., The Lake Mead Intake 3 Water Tunnel and Pumping Station, Las Vegas, NV, USA, RETC 2007, Toronto, Canada
Nevada Revised Statutes (NRS), NRS 338.1721, NRS 338.1723, NRS 338.1725, NRS 338.1727, Procedure for Qualifying Bidders, October 2005

DESIGN AND SUBSURFACE CONSTRUCTION AT YUCCA MOUNTAIN, NEVADA

John F. Beesley ▪ Bechtel SAIC Company, LLC

Jaime A. Gonzalez ▪ U.S. Department of Energy

INTRODUCTION

This paper discusses the design of the subsurface nuclear waste repository at Yucca Mountain, Nevada; and possible concepts for construction of the subsurface facility. Subsurface construction at Yucca Mountain is planned to last for 20 or more years, and may be accomplished using tunnel boring machines (TBM), drill and blast methods, raise bore machines, and roadheaders.

The design of the subsurface facility includes approximately 107 km (66 miles) of tunnels and drifts. The vast majority of the repository will be excavated with TBMs. Design and construction concepts for the subsurface repository discussed in this paper are based on the License Application that was submitted to the U.S. Nuclear Regulatory Commission (NRC) by the U.S. Department of Energy (DOE) on June 3, 2008. The License Application was subsequently docketed by the NRC in September 2008. When a Construction Authorization is received from the NRC, notices to proceed with subsurface construction at Yucca Mountain will be issued by DOE.

The Yucca Mountain site is located in the Mojave Desert approximately 150 km (90 miles) northwest of Las Vegas, Nevada. The site is on federal land in southern Nevada and is shown in Figure 1. Yucca Mountain is a ridge of volcanic rocks composed primarily of solidified volcanic ash.

Previous underground construction work at Yucca Mountain included excavation of two tunnels using two different TBMs. The first tunnel, the Exploratory Studies Facility (ESF), is a 7.62 m (25 ft) diameter tunnel that is 7,877 m (25,843 ft) long and was described by Morris and Hansmire at the Rapid Excavation and Tunneling Conference (RETC) in 1995. The second tunnel, known as the Enhanced Characterization of the Repository Block (ECRB), was a TBM drive of 5.0 m (16 ft, 5 in.) in diameter and 2,681 m (8,796 ft) long. The second tunnel was excavated in 1998 and was described by Fulcher, Eastlund and Copeland at RETC 1999 and is named the East–West Cross Drift in that paper. (These previously excavated tunnels are shown in Figure 2.) In the volcanic rock formation where the vast majority of the planned repository will be constructed, previous tunneling conditions were generally very good.

DESIGN AND CONSTRUCTION—MINED EXCAVATIONS

The proposed subsurface facility at Yucca Mountain is designed to hold spent nuclear fuel (SNF) from commercial power plants, SNF from U.S. naval operations, SNF from DOE, and high-level radioactive waste (HLW) from national defense activities. Most of the subsurface repository would be composed of emplacement drifts (tunnels) where containers of SNF and HLW would be placed. The subsurface facility includes about 68 km (42 miles) of emplacement drifts. The emplacement drifts are intended to be the final resting place for the SNF and HLW that are placed in the subsurface.

DESIGN AND SUBSURFACE CONSTRUCTION AT YUCCA MOUNTAIN 517

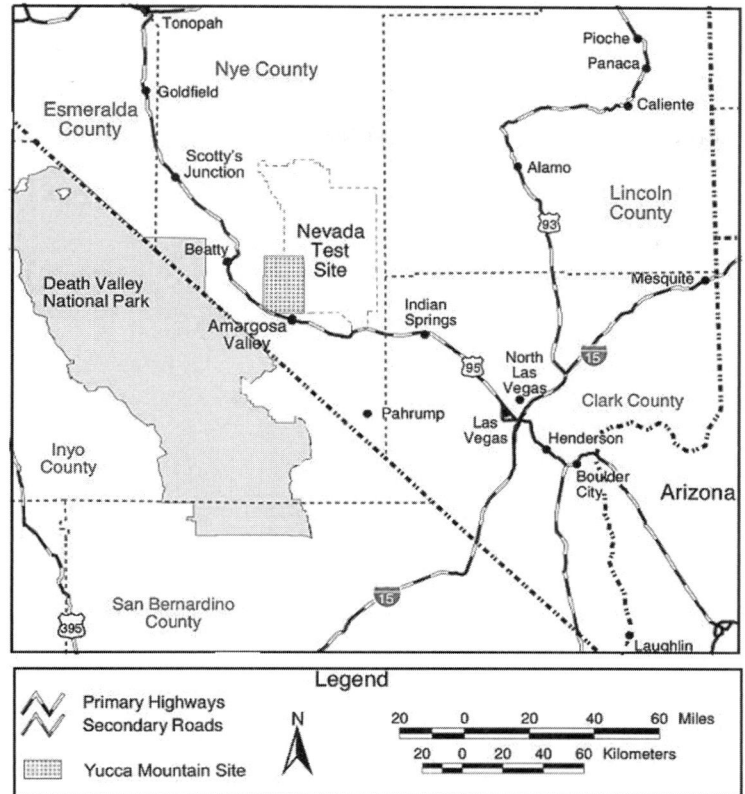

Figure 1. Location of the Yucca Mountain site

The waste packages that are placed in the drifts will be handled by remotely operated equipment. The overall subsurface facility layout is shown in Figure 2.

The repository design includes 108 emplacement drifts. The emplacement drifts are designed to be excavated with 5.5 m (18 ft) diameter hard rock TBMs. Each emplacement drift begins with a 60 m (200 ft) radius curve that ensures that the access mains in the underground facility will not be in the direct radiation shine path from waste packages in the emplacement drifts. Another key feature of the emplacement drifts is that they are relatively short when compared to most TBM drives. Most of the emplacement drifts are only about 600 m (2,000 ft) long. An overview of the arrangement of the emplacement drifts that will be constructed first, located in Panel 1 in the underground design, is shown in Figure 3.

Each emplacement drift has a TBM launch chamber and starter tunnel at the beginning of each drive. Since the emplacement drift drives are relatively short and begin with an initial curve section, the emplacement drift TBM will have a unique design. The machine will also need to be more mobile than most TBMs since it may be relocated three to five times a year. The TBM will also need design features to ensure that loss of any organic fluid is minimized.

Because of concerns with protecting the site to ensure that long term waste isolation characteristics are not compromised, other unique requirements have also been

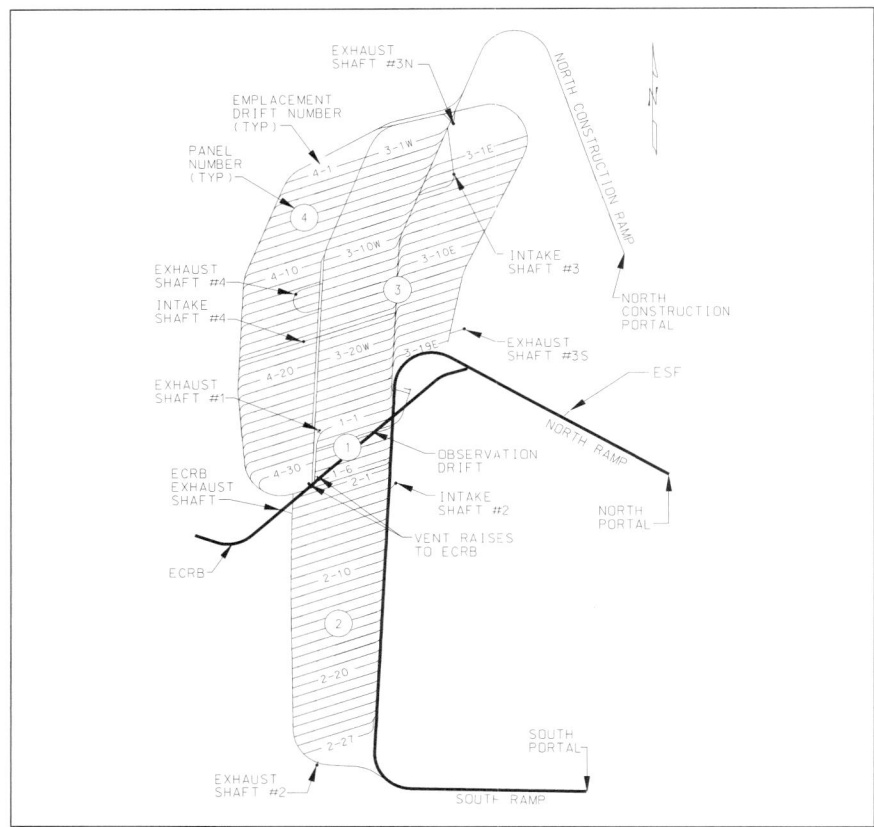

Figure 2. Subsurface repository layout

established. For example, use of water, hydraulic fluids, and concrete will be closely monitored. In some cases, such as with concrete, its use is prohibited in emplacement areas. The designers of the repository were not allowed to consider using concrete in the emplacement drifts because of concerns that it could affect radionuclide transport. Cement products may have negative impacts over long time periods because of chemical interactions between the cement and the materials in the emplacement drifts. Concrete use is allowed in the access mains and perimeter drifts where waste packages will not be emplaced.

The primary intake and exhaust ventilation access mains (tunnels) in the design are constructed by 7.62 m (25 ft) diameter TBMs. These mains are generally on the perimeter of the subsurface facility, though some tunnels do cut through the center of the repository. The design includes about 23 km (14 miles) of 7.62 m (25 ft) diameter mains. Table 1 lists the major excavation types included in the design for Yucca Mountain, the expected excavation method, and approximate excavation dimensions.

Excavation methods in Table 1 were selected due to requirements that were established during development of the design. Due to concerns with blasting residues, overbreak, and fracturing; excavation of the emplacement drifts by TBM has been part of the project planning basis for over a decade. Likewise, roadheader excavation may be required in some areas if mechanical excavation is mandated where, otherwise,

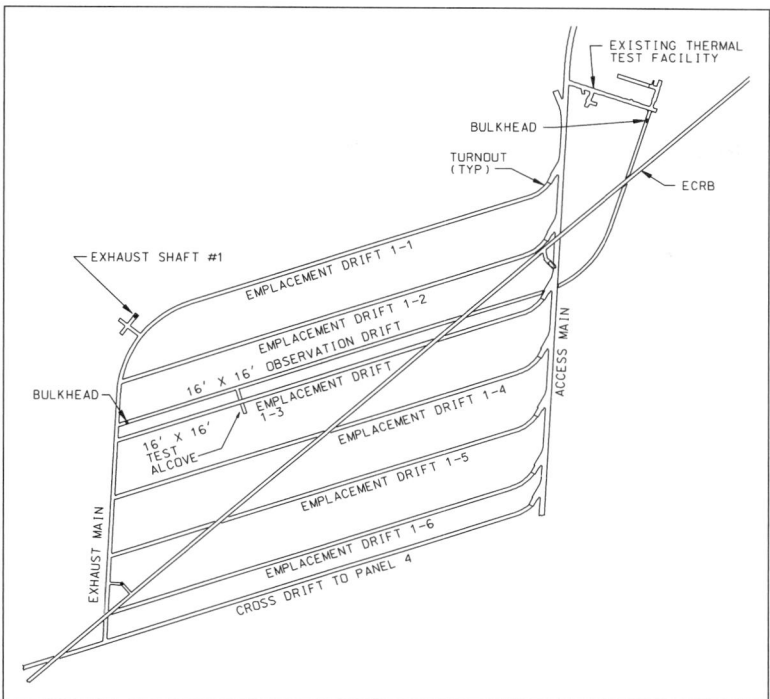

Figure 3. Panel 1, initial underground construction area

Table 1. Mined excavations at Yucca Mountain

Excavation Type	Excavation Method	Approximate Size Estimated Length
Ramps, access mains, and exhaust mains	TBM	7.62 m (25 ft) dia. 23 km (14 mi.)
Shafts	Drill and blast, raise bore, or raise and slash	5–8 m (16–26 ft) dia. (Excavated dimension) 3,134 m (10,281 ft)
Turnouts	Drill and blast, roadheader, and TBM	6 × 7 m (20 × 23 ft) 15 km (9 mi.)
Emplacement drifts	TBM	5.5 m (18 ft) 68 km (42 mi.)

drill and blast would have been the excavation method of choice. Excavation requirements have not been fully developed yet. Contractor input will be needed to ensure that requirements placed on excavation and construction processes can be reasonably implemented.

The subsurface repository has been designed to allow for phased construction of emplacement areas. The design includes 5 separate panels and those panels will likely be further subdivided and completed in phases concurrent with the emplacement of waste.

The exhaust main for a particular panel will usually be constructed prior to excavation of the emplacement drifts. This will allow the emplacement drift TBM to finish the emplacement drift drive by breaking into a larger tunnel. In a few cases, the emplacement drift TBM will need to complete its drive in a dead end heading, or it may break into a heading that is also only 5.5 m (18 ft) diameter. By designing the repository so that the 5.5 m (18 ft) diameter emplacement drift TBM usually breaks into a 7.62 m (25 ft) diameter ventilation main, the emplacement drift TBM can be broken down and relocated more easily for a subsequent drive.

In addition to the TBM excavations explained previously, the design also includes about 15 km (9 miles) of miscellaneous excavations for ventilation and access to the emplacement areas. Most of the additional excavations are expected to be excavated using drill and blast methods or roadheaders. Most of these additional excavations are the starter tunnels and emplacement drift turnouts that are needed at the beginning of each emplacement drift. Since the emplacement drift turnout and starter tunnel include a curve, rail haulage of muck is the likely option for muck handling during each emplacement drift drive.

DESIGN AND CONSTRUCTION—FINISHING ACTIVITIES

Excavation of the repository at Yucca Mountain is only one part of the work needed to complete the subsurface facility. The subsurface construction scope will include installation of a robust ground support system, a rail transportation system, and an appropriate ventilation system. The final design and construction of these systems will incorporate stringent requirements that are needed for handling and storing nuclear wastes. Finishing activities will occur in two primary areas. The areas are the access main tunnels, and the emplacement drifts.

Access Main Completion

The plan for access main completion during the first phase of construction includes placing a concrete cap on top of the existing concrete inverts. The concrete inverts were installed when the original 7.62 m (25 ft) diameter tunnel was excavated in 1994–1997. The new invert slab will be 0.46 m (18 in.) thick. The slab design includes top and bottom mats of reinforcing bar. During the first phase of construction it is estimated that about 3,700 m (12,000 ft) of access main will need to be completed. This will require placement of approximately 9,200 cu m (12,000 cy) of concrete. The crane rail for the transportation and emplacement vehicle will be placed on the new slab and would have a gauge of 3.35 m (11.0 ft). The access main to the first emplacement area and the first panel of six emplacement drifts will require that crews install about 4,200 mt (4,600 tons) of crane rail. A typical access main invert for the initial construction phase is shown in Figure 4.

The preliminary design for the access mains in the subsurface repository also specifies that the tunnel be fitted with a permanent and robust ground support system. The rock bolt layout design for the access main is shown in Figure 5. The fully grouted rock bolts will be installed on a square 1.25 m (4.1 ft) pattern. The ground support design for much of the access main also includes a 100 mm (4 in.) thick layer of shotcrete. Some areas, depending on localized rock conditions, may require installation of lattice girders or supplemental ground support components such as additional rock bolts, wire mesh, or steel sets.

Emplacement Drift Completion

The emplacement drifts are where waste packages will be placed. The waste packages are containers made of special corrosion resistant materials. The waste packages hold SNF and HLW. Waste packages will be handled by specially trained crews using remotely operated equipment. Construction personnel will be separated from waste

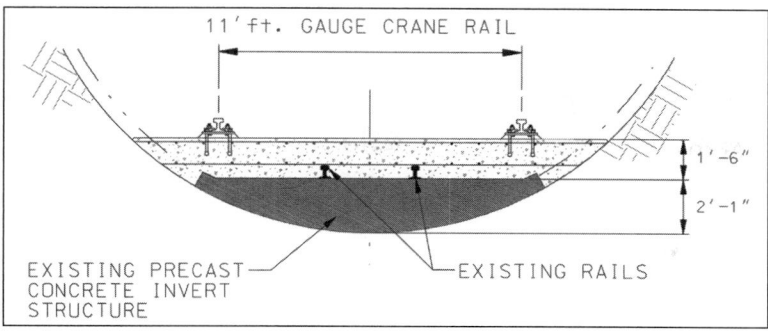

Figure 4. Access main invert and rail design

handling crews by a system of physical barriers and controls. Once construction operations have completed a section of the subsurface facility, a formal process will be used to start-up and commission the emplacement areas for that section.

After emplacement areas are completed, construction crews will not have access to, nor will they be in the vicinity of, those sections of the repository containing highly radioactive materials. The construction side of the facility will also have a separate ventilation system and separate security systems from those used on the emplacement side of the subsurface. Though some interfaces will have to be established, the subsurface construction and emplacement operations are intended to be independent from each other. The design of the subsurface repository has been developed to ensure that waste emplacement operations can remain totally isolated and separated from construction operations.

The design of the steel invert and rail emplacement system for the emplacement drifts is shown in Figure 6. The steel invert system in the emplacement drifts includes W8 and W12 structural beams that will support the transportation and emplacement vehicle rail system. The steel invert has an average weight of about 840 kg/m (560 lbs/ft). For completion of Panel 1, the first area to be completed in the subsurface repository, this will require installation of approximately 2,900 mt (3,200 tons) of structural steel. The steel invert system was selected because, as discussed previously, use of a concrete invert system was not appropriate due to concerns with the potential chemistry between cement and other materials in the emplacement drifts.

During the construction process, crushed TBM muck or a similar material will be returned to the emplacement drifts and installed in the invert as ballast. It may be installed after all of the steel has been placed, or it may be placed in conjunction with the steel installation. The primary function of the ballast is to provide support to the waste packages in the far future when the steel beams in the drift have degraded. This effort will require that approximately 15,000 cu m (20,000 cy) of material be moved to the subsurface and placed in the emplacement drift inverts using a controlled process for compaction, gradation, composition, and moisture conditioning.

At the beginning of each emplacement drift, an emplacement access door will control access into the drift. Due to the presence of SNF and HLW in waste packages, human entry into an emplacement drift will not be allowed. Systems at the beginning of each drift will control airflow, monitor vehicle movement, collect data, and will transmit information to the central control center facility on the surface.

The ground support system in the emplacement drifts is also shown in Figure 6. The system includes 3 m (10 ft) stainless steel friction-type rock bolts installed on a square 1.25 m (4.1 ft) pattern and perforated stainless steel sheets. These items were selected since they are expected to remain intact and continue their function

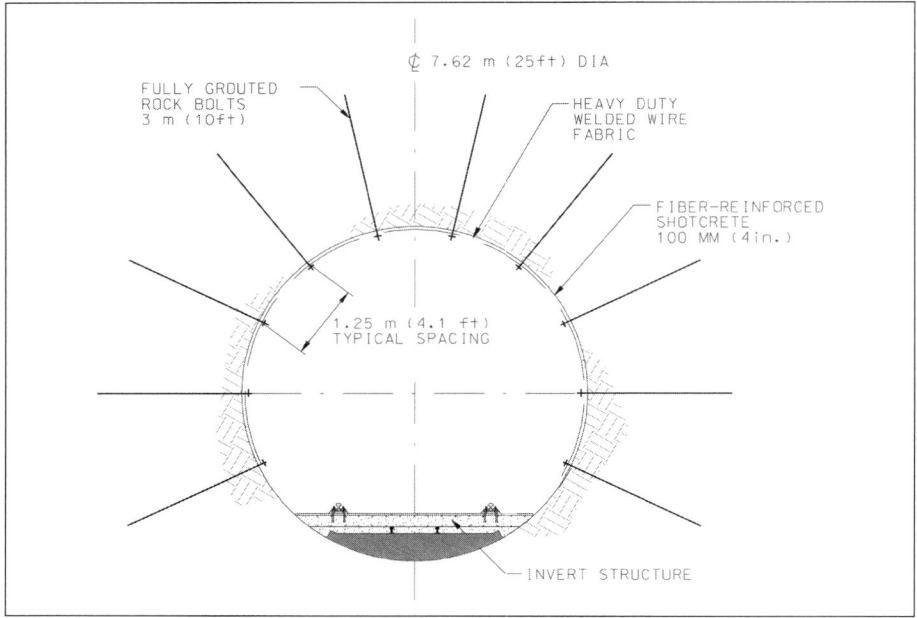

Figure 5. Typical access main cross section

of supporting the ground for the entire time period that the repository is open and being monitored by future caretakers. The repository may be monitored and open for 300 years. Expandable friction bolts were selected as the preferred bolt type over mechanical anchor-type rock bolts because of natural voids and cavities that occur in the volcanic rock. Grouted rock bolts were not selected because cement-based products will not be used in emplacement areas.

Stainless steel sheets that are rolled to fit the emplacement drift diameter will cover the upper two-thirds of the drift. The sheets are perforated so that any moisture that may accumulate around the sheets will be quickly evaporated. Evaporation is driven by the combined effects of heat generated by the nuclear waste and by forced ventilation circulating through the drifts. Because concrete usage is not allowed in the emplacement drifts, a stainless steel lining appears to be the best choice for full ground support coverage. Stainless steel was selected because of its durable properties and the material complies with scientific requirements.

Initial ground support will be installed as necessary to provide worker safety until the final ground support system shown in Figure 6 is installed. The initial ground support system will consist of split sets and wire mesh. Field engineering will determine the extent of the initial ground support. Installation of the final ground support system will occur after completion of the mapping activities.

DESIGN AND CONSTRUCTION—VERTICAL SHAFTS

The preliminary design of the subsurface facility includes nine vertical shafts that are excavated either 4.9 m (16 ft) or 7.9 m (26 ft) in diameter. Shafts will be lined with about 0.3 m (1 ft) of concrete. These shafts are for ventilation and no personnel hoisting systems for routine transportation of men and materials during repository emplacement operations are anticipated. The supply shafts will be available for emergency exit

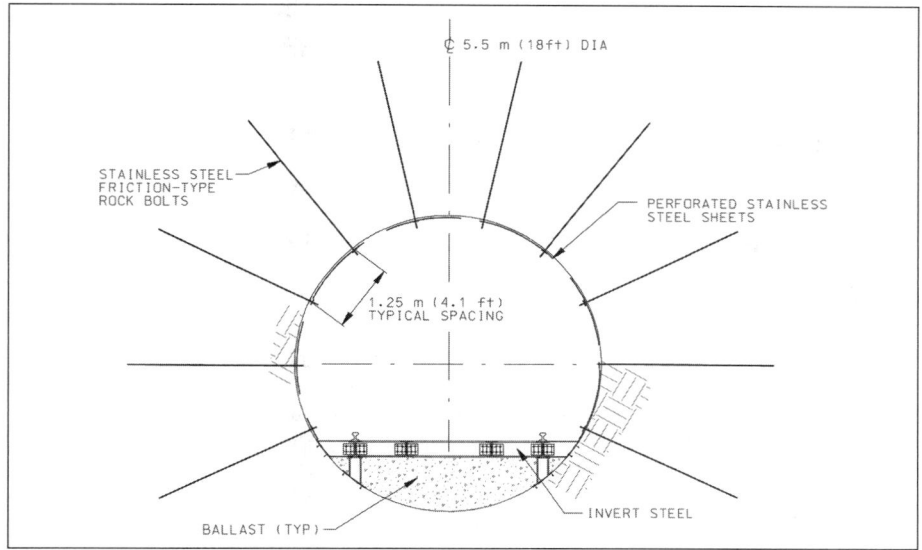

Figure 6. Typical emplacement drift cross section

during repository construction. Most of the shafts will be approximately 300 m (1,000 ft) deep and originate on the ridges above the repository.

The location for one of the ventilation shafts is shown in Figure 7. Figure 7 is typical of the conditions at most of the shaft sites. Access roads to some of the shaft sites are currently fairly steep; i.e., greater than 17 percent. New roads to the shaft sites are expected to have maximum grades of about 8–12 percent. However, construction may need to start prior to upgrading the roads and shaft contractors may need to mobilize equipment that can safely negotiate grades greater than 17 percent.

At least two of the nine shafts need to be completed before emplacement operations can begin. Until further planning is finalized, the final construction method for the shafts can not be determined. The shafts may be raise bored and then slashed using drill and blast techniques. Alternatively, they may be sunk blind from the top also using conventional drill and blast methods. Water infiltration is not expected during shaft construction since the repository is well above the water table. Future mining activity will not be affected by significant inflow of water. Natural inflow of water is not expected to affect any subsurface construction work at Yucca Mountain.

DESIGN AND CONSTRUCTION—SUMMARY

Future subsurface construction at Yucca Mountain, Nevada will be a large and significant effort that will last for decades. Crews performing drill and blast or roadheader operations will need to prepare numerous openings each year. Most of that work will be followed by TBM operations. After TBM excavations have been completed, concrete and steel travel ways and transportation systems will be constructed. Table 1 summarizes the major excavations at Yucca Mountain and their approximate dimensions.

During construction, the repository will be fitted with numerous utilities, including electrical power, water lines, compressed air lines, personnel communication, data acquisition, video monitoring, and command and control systems. Many ventilation structures will be installed, and barriers will be erected to separate the waste

Figure 7. Typical surface conditions at shaft sites

emplacement areas from subsurface construction locations. Shafts will be fitted with large ventilation fans and ancillary equipment to support ventilating the subsurface areas.

Timing for the exact start of construction has not yet been determined. As work packages are developed and discrete work activities are planned, contract solicitations will be issued to solicit contractors to perform the work. Construction will not begin until the NRC has completed their review of the design and authorizes DOE to start repository construction. The review and authorization process is expected to last 3–4 years after docketing of the license application, which occurred in September 2008.

ACKNOWLEDGMENTS/DISCLAIMER

The authors thank Bechtel SAIC Company, LLC and DOE for their support in preparing this paper. Though the authors have made every attempt to present information that is consistent with the official DOE design that was submitted to the NRC in June 2008, this paper must not be considered official. Officially approved design and construction documents are readily available to the public on both DOE and NRC websites. The Government (DOE) reserves for itself and others acting on its behalf, a nonexclusive, paid-up, irrevocable, world-wide license for Governmental purposes to publish, distribute, translate, duplicate, exhibit, and perform any of this work. The views expressed herein are those of the authors and do not necessarily represent the views of the DOE or Bechtel SAIC Company, LLC.

REFERENCES

Morris, James P. and Hansmire, William H. *TBM Tunneling on the Yucca Mountain Project.* 1995 Proceedings Rapid Excavation and Tunneling Conference. Chapter 51, pp. 807–822.

Fulcher, Brian; Eastlund, John; and Copeland, J.B. *Construction of the East-West Cross Drift Tunnel at the Yucca Mountain Site.* 1999 Proceedings Rapid Excavation and Tunneling Conference. Chapter 57. pp. 1011–1033.

FEASIBLE TUNNEL CONSTRUCTION OPTIONS FOR THE SYSTEMS CONVEYANCE AND OPERATIONS PROGRAM REACH 3 TUNNEL

S.H. Jason Choi ▪ Jacobs Associates

Scott Ball ▪ MWH Americas, Inc.

Sean Tokarz ▪ MWH Americas, Inc.

Jim Devlin ▪ Clean Water Coalition

ABSTRACT

The Systems Conveyance and Operations Program (SCOP) is being implemented to convey tertiary-treated effluent to Lake Mead near Las Vegas, Nevada. The Reach 3 Tunnel Project is part of SCOP. The tunnel's planned location along and across the Las Vegas Wash creates challenging tunneling conditions. A number of feasible excavation and initial support methods are being considered during the final design stage. This paper describes how the tunnel construction options were refined based on anticipated tunneling conditions and compatibility between tunnel excavation and initial support methods. A brief description of the most significant characteristics of the anticipated ground conditions is included for the purpose of discussion.

BACKGROUND

The SCOP project was initiated by the Clean Water Coalition (CWC) to address the need for a new discharge point in Lake Mead near Las Vegas, Nevada. SCOP will collect tertiary-treated effluent from various member agencies (Clark County Water Reclamation District, City of Las Vegas, City of North Las Vegas, and City of Henderson) and convey the effluent to the proposed diffuser facility in Lake Mead. SCOP includes approximately 22.5 km (14 mi) of pipeline and tunnels, a 291 million L/day (77 million gal/day) pump station, a power-generating station, and diffuser pipelines. The Reach 3 Tunnel Project, a part of SCOP, consists of two shafts, about 1,701 m (5,580 ft) of 120-in. inner diameter (ID) lined pressure tunnel, and about 122 m (400 ft) of pipeline constructed by cut and cover. Figure 1 shows the location of the Reach 3 Tunnel Project.

The Reach 3 Tunnel will be constructed between two shafts: the Drop Shaft at the upstream (west) end and the Access Shaft at the downstream (east) end. The horizontal alignment includes 304.8-m-radius (1,000-ft-radius) curves and straight sections. The invert elevation of the downstream end of the tunnel is El. +1375, and the invert elevation of the upstream end of the tunnel is El. +1392. The tunnel invert elevates at a constant slope of 0.0028 from the downstream end to the upstream end. Figure 2 shows the horizontal alignment of the tunnel, and Figure 3 shows the vertical profile of the tunnel.

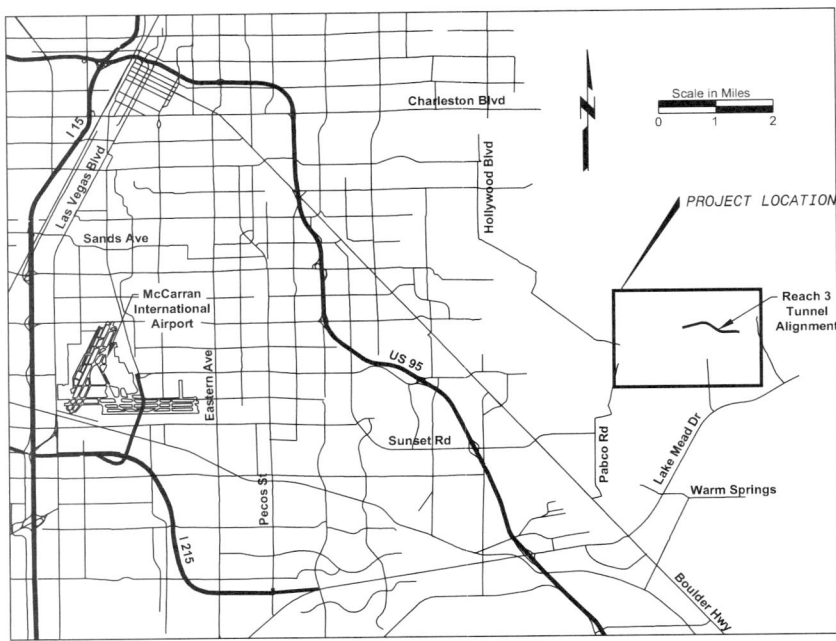

Figure 1. Location map

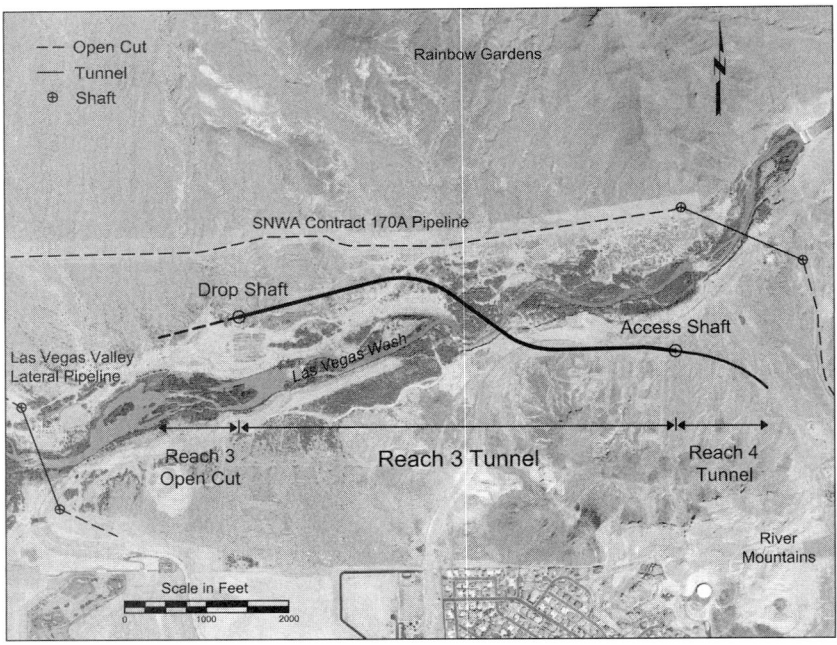

Figure 2. Horizontal alignment

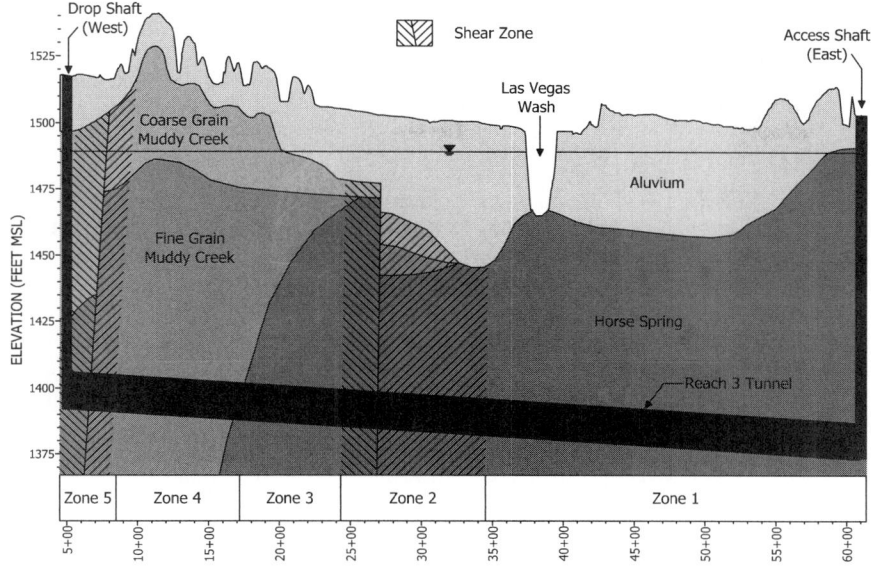

Figure 3. Tunnel and geologic profile

GEOLOGIC CONDITIONS

The Reach 3 Tunnel is located in the northwest-trending Las Vegas Valley, bounded on the northwest by the Sheep and Las Vegas ranges, on the west by the Spring Mountains, on the south by the McCullough Range, and on the east by Sunrise-Frenchman Mountain and the River Mountains. Two geologic formations—the Muddy Creek Formation and the Horse Spring Formation—are anticipated to be encountered during tunnel excavation. These geologic formations are Tertiary sedimentary rocks. Figure 3 shows the geologic profile along the tunnel.

Muddy Creek Formation

The Muddy Creek Formation consists of a coarse-grained unit overlaying a fine-grained unit. The fine-grained Muddy Creek Formation is expected to be encountered in the western quarter of the Reach 3 Tunnel alignment. The formation consists of poorly cemented mudstone, claystone, and sandstone that are thickly bedded to massive, and weak to strong with low hardness. It includes gypsum veins or gypsum healed joints. Bedding in the Muddy Creek Formation generally dips to the southwest at shallow angles. Bedding spacing varies from medium to thick (0.15 m to 1.8 m [6 in. to 6 ft]). Jointing in the Muddy Creek Formation is characterized as very widely spaced (>1.8 m [>6 ft]). In some cases, the formation is described as a soil because of its weak and poorly cemented characteristic.

Horse Spring Formation

The Horse Spring Formation is expected to be encountered in the eastern three quarters of the Reach 3 Tunnel alignment. This formation is described as very thinly-to-medium bedded (0.01 to 0.61 m [0.03 to 24 in.]), weak-to-strong calcareous sedimentary rock consisting of siltstone, claystone, and sandstone. Bedding in the

Horse Spring Formation generally dips southwest to west at low dipping angles. The Horse Spring Formation also includes two joint sets with high dipping angles. The primary joint set trends east-west, with joint spacing ranging from 0.08 to 0.30 m (3 to 12 in.). The secondary joint set trends north-south, with joint spacing ranging from 0.61 to 1.5 m (2 to 5 ft).

Shear Zones

Two shear zones, believed to be associated with splays of the Sunrise-Frenchman Mountain fault, are expected along the Reach 3 Tunnel alignment. The first shear zone is expected to be encountered approximately 106.7 m (350 ft) from the west end of the tunnel, within the Muddy Creek Formation. The second shear zone is expected to be encountered in approximately 310.8 m (1,020 ft) in the central portion of the tunnel alignment at the immediate north side of the Las Vegas Wash, crossing within the Horse Spring Formation (Figure 3).

Groundwater

The entire tunnel alignment will be below the groundwater table. The groundwater level is expected to be between 21.3 and 30.5 m (70 and 100 ft) above the tunnel invert, which translates to a maximum hydraulic pressure of 310.3 kPa (45 psi). During the geotechnical investigation, the groundwater solution concentrations of the common chemical constituents (iron, sulfate, and chloride, as well as Total Dissolved Solids) turned out to be generally above the regulatory limits for discharge into waterways, as set by the Nevada Division of Environmental Protection (NDEP). Discharge of groundwater removed from the Reach 3 Tunnel excavations into the ground or the Las Vegas Wash is prohibited. Also, groundwater at the Reach 3 Tunnel site is considered corrosive. According to the tunnel inflow estimate based on the methodology described by Heuer (1995 and 2005), the groundwater inflow could vary significantly between shear and non-shear zones. Groundwater controls and provisions such as pre-excavation grouting programs and/or the use of a pressurized-face method and water-sealed initial support system will be required in the shear zones.

ANTICIPATED GROUND BEHAVIOR

Anticipated ground behavior along the Reach 3 Tunnel alignment is described using Terzaghi's Rock Load Classification system (Terzaghi 1946; Deere et al. 1970; Rose 1982), the Tunnelman's Ground Classification method for soft ground (Heuer and Virgens 1987; Brandt 1970), Rock Mass Quality (Q), and Rock Mass Rating (RMR) assessment based on the borehole logs. Although the Tunnelman's Ground Classification was developed for excavation in soil, it can be used to appropriately describe the fractured and weak nature of the rock mass present in the shear zones (i.e., as soil). The fact that it was created as a classification system assumes an open-face excavation method. Nevertheless, the Tunnelman's Ground Clasification also is useful for evaluation of pressurized-face methods, as unstable ground conditions are an indication that ground modification methods may be needed and whether pressurized-face methods are applicable.

In weak rock, the influence of ground stresses must be considered in describing ground behavior in a tunnel. A fundamental aspect of the behavior of weak rock in a tunnel is the tangential stresses developed around the tunnel excavation as compared to the strength of the rock. The modified overload factor (OFM) was developed to evaluate the potential for overstressed conditions in terms of the tangential stress (σ_θ) and the uniaxial compressive strength (UCS) by the following equation (Deere et al. 1969):

OFM = σ_g/UCS

Overstressed rock conditions develop around a tunnel when the OFM is greater than one, at which point the behavior of the rock depends on the stress-strain characteristics of the rock. The identification of weak rock following the overstress criteria for the Reach 3 Tunnel was considered for the shear zones and in the Muddy Creek Formation along the alignment.

The Reach 3 Tunnel alignment is divided into five zones where similar ground behavior during construction is anticipated. Each zone is described below, from east to west in direction of the tunnel drive. Also refer to Figure 3.

Zone 1 extends approximately 817 m (2,680 ft) in length and consists of the Horse Spring Formation. Q values range from 0.2 to 17.6. RMR values range from 39 to 74. The ground in Zone 1 is expected to be moderately blocky and seamy in terms of Terzaghi's Rock Load Classification. Slabbing, spalling, and fallout/sliding block behavior are anticipated as the rock is bedded and jointed. Exposure of this rock to flowing water and air will result in slight degradation or slaking. The modified overload factor (OFM) in this zone is expected to be less than one; therefore, overstressed rock conditions are not anticipated.

Zone 2 extends approximately 311 m (1,020 ft) in length and consists of the Horse Spring Formation in the shear zone. Approximately half of Zone 2 is comprised of fractured, very weak sedimentary rock and crushed rock in a clayey matrix, with thicknesses up to 4.6 m (15 ft). Q and RMR assessments were not applicable for this ground because of low-quality, weak rock. Based on the Tunnelman's Ground Classification, raveling, squeezing, and swelling behavior are expected in this ground in open-face conditions. Exposure of this ground to flowing water and air will result in slight degradation or slaking. OFM for this ground is expected to be greater than one, and an overstress condition is expected. The other half of Zone 2 is characterized as blocky and seamy in terms of Terzaghi's Rock Load Classification. Rock behavior in this ground will be similar to that of Zone 1. Implementation of a groundwater control program will be required within Zone 2.

Zone 3 extends approximately 207 m (680 ft) in length and consists of the Horse Spring Formation. Ground behavior is anticipated to be similar to that of Zone 1.

Zone 4 extends approximately 290 m (950 ft) in length and consists of the fine-grained Muddy Creek Formation. The rock is similar to that in Zone 5, described below, but is expected to be of higher quality and behave more like rock than soil. Q values range from 1.3 to 1.8. RMR values range from 61 to 71. The ground in Zone 4 is anticipated to be blocky and seamy in terms of Terzaghi's Rock Load Classification. Slabbing, spalling, and fallout/sliding block behavior are expected as the rock is bedded and jointed. Exposure of this rock to flowing water and air will result in slight degradation or slaking. OFM in Zone 4 is expected to be less than one; therefore, overstressed rock conditions are not anticipated.

Zone 5 extends approximately 107 m (350 ft) in length and consists of the fine-grained Muddy Creek Formation in the shear zone. Q and RMR assessments were not applicable for the rock in Zone 5 because its ground behavior is more like soil. Based on the Tunnelman's Ground Classification, raveling, squeezing, and swelling behavior are expected throughout Zone 5. Exposure of this ground to flowing water and air will result in high degradation or slaking. OFM for the fine-grained Muddy Creek Formation in the shear zone is expected to be greater than one, and overstressed conditions are expected in Zone 5. Implementation of a groundwater control program will be required within this zone.

CONSTRUCTION ISSUES

In addition to the ground conditions, there were three construction issues considered when establishing and refining the feasible construction options for the Reach 3 Tunnel Project. The first issue was water disposal. As mentioned earlier, the water removed from the Reach 3 Tunnel excavations can not be discharged into the ground or the Las Vegas Wash. Wastewater from excavations (comprised of groundwater inflows and construction water) can be discharged into an evaporation pond. Then, the wastewater discharge to the evaporation pond must be controlled so that it does not exceed the maximum capacity of the evaporation pond during construction. The second issue was available construction staging area at each shaft site. Very limited staging area will be available at the Drop Shaft (west) site. An appropriate staging area will only be available at the Access Shaft (east) site to accommodate the tunnel construction equipment/materials and the evaporation pond. Consequently, only the Access Shaft is considered suitable for the tunnel construction shaft. The last issue was construction-induced vibrations. The potential impact of vibration on sensitive cultural resources is an important consideration during construction. The construction vibration levels at a cultural resource must be monitored and maintained below the specified limits during construction.

EXCAVATION METHODS

In order to establish feasible tunnel construction options for the Reach 3 Tunnel, the first step was to evaluate available excavation methods and narrow the selection for the project. For the Reach 3 Tunnel, the available excavation methods evaluated were drill-and-blast, roadheader (open-face and shielded), digger shield, main-beam tunnel boring machine (TBM), wheel-type TBM, single- and double-shielded TBMs, and earth-pressure-balance (EPB) TBM. The following should be considered in the selection of feasible excavation methods for the SCOP Reach 3 Tunnel should:

- Excavation will occur through weak-to-strong sedimentary rocks and weak sedimentary rocks, including clay matrix in shear zones;
- The potential exists for pressurized groundwater inflow;
- Low bearing capacity for TBM grippers;
- The potential exists for slaking materials;
- Water-disposal capacity;
- Vibration requirement for cultural resources;
- Available staging areas;
- The need for face stabilization in very closely fractured rock, soil-like Muddy Creek Formation, and shear zones; and
- 1,700.8 m (5,580 ft) long, 3.96 to 4.27-m (13 to 14 ft) diameter tunnel.

Drill-and-blast in such weak rock will disturb and loosen surrounding rock mass more than other methods and lead to significant overbreak beyond the intended excavation lines. In addition, it will be difficult to control an unstable face using drill-and-blast when raveling and flowing grounds are encountered within the shear zone. Combined with the potential for pressurized groundwater inflow, the weak ground conditions of the Reach 3 Tunnel would delay the tunnel advance rate significantly with the drill-and-blast method. Generally, this method is not likely to be cost effective for a rock tunnel longer than 1,219.2 m (4,000 ft). Other disadvantages of the drill-and-blast method are greater vibration and noise impacts.

Roadheader excavation, either with or without a tunnel shield, is considered feasible for the excavation of the Reach 3 Tunnel, provided that face support, presupport, or pre-excavation grouting is implemented in the shear zones. Since the tunnel length is near the trade-off point for economic viability of TBM versus conventional excavation, roadheader methods should be feasible for the tunnel excavation. However, conventional roadheader excavation would be competitive with TBM (single heading) excavation only if it can be performed with two headings from both shafts. Since only the Access Shaft can serve as tunnel excavation shaft and an appropriate water disposal system is not available at the Drop Shaft, conventional roadheader excavation for the Reach 3 Tunnel has to be single headed and may not be as economical as TBM excavation.

Digger shields are used mainly to excavate soft ground tunnels. A digger shield would be suitable for the weak Muddy Creek Formation and the shear zone materials, but not for the strong Horse Spring Formation in the Reach 3 Tunnel. For the same reason, the wheel-type TBM is precluded. Wheel-type TBMs are commonly designed for tunnels with a small diameter (<2.4 m [<8 ft]) and a short drive (<609.6 m [<2,000 ft]) in weak ground that requires small torque. Wheel-type TBMs would not be suitable for the high-quality Horse Spring Formation and for some of the cohesive materials that require high torque in the Reach 3 Tunnel.

Overall, the main-beam TBM is not considered suitable for the Reach 3 Tunnel due to the tunnel's anticipated ground conditions. It may be suitable for the limited high-quality Horse Spring Formation, but not for the majority of the Reach 3 Tunnel ground with its relatively short standing time. Also, the bearing capacity of the anticipated ground is expected to be very low in some locations, and the use of a main-beam TBM with grippers may not be feasible.

Shielded TBM methods, such as single-shield TBMs and EPB TBMs, are considered appropriate for the anticipated ground conditions. Shielded TBMs can be adapted to both weak-to-strong sedimentary rocks and cohesive soil conditions as they are equipped with interchangeable, rear-loading, low-profile disc cutters and drag teeth; adjustable muck bucket openings; and a mixed-ground cutterhead, providing adequate face support. The biggest advantage of the double-shield TBM is the flexibility it provides in varying the initial support system, saving costs on initial support. However, its high machine cost may offset this initial support cost saving, and an interchanging initial support system during shielded TBM operation is not practical. Also, the double-shield TBM has a higher risk of being wedged in the ground. Its relatively longer shield requires a longer tunnel section to be supported by the shield for a longer time. If the ground has a shorter standing time or squeezing behavior, the double-shielded TBM can be wedged in. Thus, double-shield TBM methods would not be beneficial for the Reach 3 Tunnel because of the varying ground conditions, including squeezing and short standing time.

In this study, available tunnel excavation methods were evaluated, and the feasible tunnel excavation methods for the Reach 3 Tunnel were refined to the shielded (circular) roadheader method, the single-shielded TBM method, and the EPB TBM method.

INITIAL SUPPORT SYSTEMS

The next step was to select initial support systems that are compatible with the selected excavation methods and suitable for the considerations listed in the previous section. For the anticipated ground conditions, feasible initial support systems for the Reach 3 Tunnel include rock reinforcement, shotcrete with lattice girders or steel ribs, shotcrete with rock reinforcement, and precast concrete segmental linings. Rock reinforcement is applicable to Zones 1 and 3. If combined with shotcrete, rock reinforcement can be used in portions of Zone 4, where blocky rock conditions are encountered. Shotcrete with lattice girders or steel ribs is suitable in Zones 2, 5, and portions of Zone 4. However, shielded roadheader and shielded TBM methods require robust initial support systems, regardless of the ground conditions, because of the need for a

Table 1. Feasible Reach 3 Tunnel construction options

Option	Excavation Method	Zone 1 512.1 m (1,680 ft) HSF	Zone 2 310.9 m (1,020 ft) SZ/HSF	Zone 3 207.3 m (680 ft) HSF	Zone 4 289.6 m (950 ft) MCF	Zone 5 106.7 m (350 ft) SZ/MCF
1	Single-shield TBM or Roadheader	EP	EP	EP	EP	EP
2		BGP	BGP	BGP	BGP	BGP
3	EPB TBM	BGP	BGP	BGP	BGP	BGP

HSF: Horse Spring Formation
MCF: Muddy Creek Formation
SZ: Shear Zone
EP: Expandable precast concrete segmental lining
BGP: Bolted and gasketed precast concrete segmental lining

proper reaction for the thrust jacks. Shotcrete is not recommended with any TBM or shielded roadheader operation because it is difficult to apply shotcrete close enough to the tunnel heading to be effective. In addition, the cathodic protection for the final pipe lining was considered for initial support selection. Continuous timber lagging was precluded by planned use of an impressed current corrosion protection system because timber lagging resists the electric current. A shielded TBM or shielded roadheader may use steel ribs with continuous timber lagging as the initial support system, but the use of continuous timber lagging with an impressed current system is not recommended. A possible alternative lagging system would be steel liner plates, which may not be cost effective, or shotcrete lagging, which may not be preferred in association with shielded machine operation. Therefore, steel ribs are precluded for shielded machine methods. Expandable and non-expandable (bolted and gasketed) precast concrete segmental linings are feasible with single-shield TBMs or shielded roadheaders for the entire Reach 3 Tunnel alignment. Table 1 summarizes feasible Reach 3 Tunnel construction options.

Finally, the Reach 3 Tunnel feasible tunnel construction options, presented in Table 1, were qualitatively evaluated, as shown in Table 2.

CONCLUSIONS AND FUTURE REFINEMENT

Ground conditions, project geometry, construction considerations, and environmental restrictions of the Reach 3 Tunnel provide multiple construction options that are feasible. The challenge for tunnel engineers is to refine the feasible tunnel construction options and to select the most suitable and economical one. During the final design stage of the Reach 3 Tunnel project, available tunnel excavation methods and initial support methods were evaluated, and the most probable tunnel construction options that would be selected by the contractor for the Reach 3 Tunnel were established and qualitatively evaluated. Up to this stage, technical aspects such as ground conditions, construction considerations, and environmental restrictions have been considered. In the next stage, further refinement of the tunnel construction options can be performed for economic aspects and construction-performance aspects. Then, a single construction method can be presented in the final bid documents. Otherwise, multiple tunnel construction options will be presented in the final bid documents leading the contractor to select the most suitable and economical tunnel construction method based on detailed construction cost estimates, schedule, and the safety during construction.

Table 2. Advantages and disadvantages of tunnel construction options

Options	Advantages	Neutral	Disadvantages
1	■ High support installation rate (only minor contact grouting behind initial support) ■ Low initial support cost	■ Moderate excavation rate ■ Moderate equipment cost ■ Precast concrete segment plant and storage requirements	■ Unsuitable for ground load sharing with final lining ■ High groundwater control cost
2	■ High support installation rate ■ Suitable for ground-load sharing with final lining	■ Moderate excavation rate ■ Moderate equipment cost ■ Moderate groundwater control cost ■ Precast concrete segment plant and storage requirements	■ High initial support cost ■ Requires grouting behind initial support
3	■ High excavation rate ■ High support installation rate ■ Low groundwater control cost ■ Suitable for ground-load sharing with final lining	■ Precast concrete segment plant and storage requirements	■ High equipment cost ■ High initial support cost ■ Requires grouting behind initial support

ACKNOWLEDGMENTS

The authors wish to acknowledge the support provided by the Clean Water Coalition, MWH America, Inc., and Jacobs Associates for the preparation of this publication.

REFERENCES

Brandt, C. T. et al. 1970. *A Systems Study of Soft Ground Tunneling.* U.S. Dept. of Transportation Report No. DOT-FRA-OHSGT-231.

Deere, D.U., R.B. Peck, J.F. Monsees, and B. Schmidt. 1969. *Design of Tunnel Liners and Support Systems.* Report prepared for U.S. Department of Transportation. OHSGT Contract 3-0152. NTIS.

Deere, D.U., R.B. Peck, H.W. Parker, J.F. Monsees, and B. Schmidt. 1970. *Design of Tunnel Support Systems.* Highway Research Record, No. 339, pp. 26–33.

Heuer, R E. 1995. Estimating rock tunnel water inflow. In *Proceedings, Rapid Excavation and Tunneling Conference.* Ed. G. E. Williamson and I. M. Gowring; p. 41–60. Littleton, CO: Society for Mining, Metallurgy, and Exploration, Inc.

Heuer, R.E. 2005. Estimating rock tunnel water inflow-II. In *Proceedings, Rapid Excavation and Tunneling Conference.* Ed. J. D. Hutton and W. D. Rogstad; pp. 394–407. Littleton, CO: Society for Mining, Metallurgy, and Exploration, Inc.

Heuer, R.E. and D.L. Virgens. 1987. Anticipated behavior of silty sands in tunneling. In *Proceedings of the Rapid Excavation and Tunneling Conference.* Littleton, CO: Society of Mining Engineers, Inc.

Rose, D. 1982. Revising Terzaghi's tunnel rock load coefficients. In *Proceedings, 23rd U.S. Symposium Rock Mechanics*, pp. 953–960. New York, NY: AIME.

Terzaghi, K. 1946. Rock defects and loads on tunnel support. In *Rock Tunneling with Steel Support*, ed. R. V. Proctor and T. White, pp. 15–99. Youngstown, Ohio: Commercial Shearing Co.

WHAT HAPPENS IN VEGAS: THE APEX TUNNEL GEOLOGIC INVESTIGATION

Ann L. Backstrom ▪ Kleinfelder

Joel G. Metcalf ▪ Kleinfelder

Stephen McKelvie ▪ HDR Engineering, Inc.

ABSTRACT

The Apex Tunnel is part of the Southern Nevada Water Authority's proposed Clark, Lincoln, and White Pine Counties Groundwater Development Project. The project is intended to develop unused Nevada groundwater for meeting the water needs of Southern Nevada and would involve over 322 km (200 mi) of water transmission facilities. The Apex Tunnel, as currently conceived by ongoing planning and environmental analyses, would extend 2804 m (9,200 ft) through a structurally complex terrain of Paleozoic carbonate rocks and younger basin fill deposits and will traverse both public and private lands. Several faults cross the tunnel study area, juxtaposing rocks of varying lithology and degree of folding, and creating intervals of differing rock conditions along the tunnel alignment. This paper discusses how the geotechnical program was developed to be flexible and allow for rapid implementation of revisions based on new knowledge that became available during program execution, resulting in more effective decision making and enhanced value from the geologic investigation during the planning phase of the project.

INTRODUCTION

The Southern Nevada Water Authority's (SNWA) proposed Clark, Lincoln, and White Pine Counties Groundwater Development Project would involve the design and construction of over 322 km (200 mi) of buried pipeline extending through east-central Nevada and terminating in the Las Vegas Valley. The proposed Apex Tunnel was identified during the pipeline planning process as a means to meet hydraulic requirements. It would be located near the southern end of the proposed pipeline alignment in the Las Vegas Range as the pipeline alignment enters the northern Las Vegas Valley. A location map is presented on Figure 1. The tunnel is approximately 2,804 m (9,200 ft) in length and is preliminarily planned as a nominal bore 3 m (10 ft) in diameter for 183 cm (72 in.) to 229 cm (90 in.) welded steel pipe.

The Project Owner and Client is SNWA, supported by Parsons Infrastructure as Program Manager. The civil designer, HDR Engineering Incorporated, and the geotechnical consultant, Kleinfelder Incorporated, were contracted directly to SNWA for the duration of the work described in this paper.

The work described in this paper was performed as part of the Phase I and II explorations for the Apex Tunnel, consisting of geologic reconnaissance and mapping, seismic refraction, eleven core borings ranging from about 38 m (125 ft) to 152 m (500 ft) deep drilled using track-mounted and helicopter-portable rigs, and laboratory testing of core samples. Later phases of exploration are expected to include work

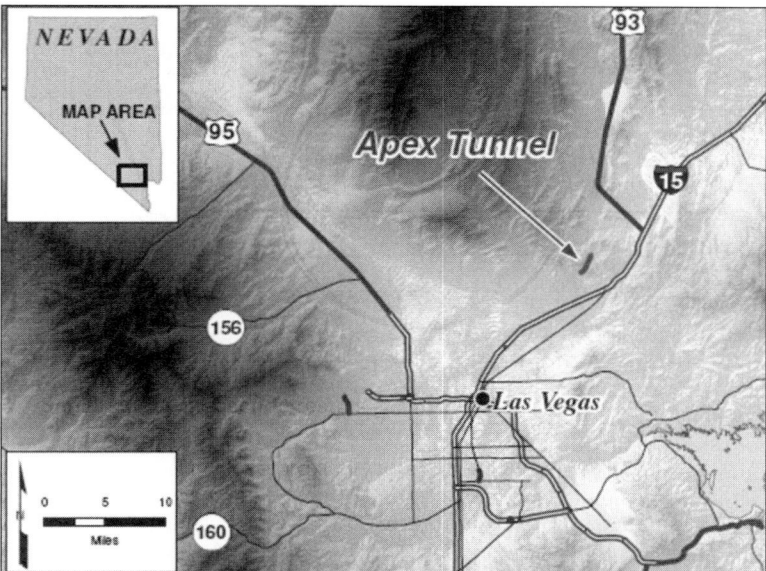

Figure 1. Location of the proposed Apex Tunnel alignment

within the public right-of-way pending permit approvals, as well as possible additional exploration on private lands.

An iterative approach was used during the planning and execution of the fieldwork to take advantage of our evolving knowledge of subsurface conditions along the tunnel alignment. By committing in advance to flexibility during the site exploration, the project team was able to make several "on-the-fly" adjustments to the drilling plan:

- An eleventh core boring was added to the initial set of ten borings to constrain the location of a low-velocity anomaly detected by seismic refraction.
- Two borings were extended by 18 m (60 ft) and 30 m (100 ft) to penetrate unexpectedly deep basin-fill deposits.
- The locations and angles of two borings were shifted to better characterize faulting along the alignment.
- Down-hole natural gamma logging was added to the final seven holes to aid in defining lithology and stratigraphic relationships in the bedded carbonate rocks.
- One hole was completed using smaller diameter core to facilitate hole completion in highly broken rock.
- A grouting program was developed to measure and compare corehole grout takes in the variable ground conditions encountered.

These modifications significantly enhanced the information gained during the exploration program and were incorporated as a result of a team approach founded on communication and interaction between the geotechnical consultant, the design team, and the Owner.

SITE DESCRIPTION

The proposed Apex Tunnel alignment is located in the southeastern Las Vegas Range in Clark County, Nevada, approximately 24 km (15 mi) north of Las Vegas (Figure 1). The northern half of the Apex Tunnel alignment (not yet explored) occupies public land and is characterized by open, low-relief terrain with few areas of rock outcrop. The southern half of the Apex Tunnel alignment traverses privately-owned land beginning in a localized, alluvium-covered area of low relief (Phase I area) then continuing south into rugged, rocky terrain with abundant outcrop (Phase II area) before finally ending at the north edge of the Las Vegas Valley.

Regional Geology

The southern Las Vegas Range defines the northeastern margin of the Las Vegas Valley and is located in the Basin and Range Province, a region of north-trending mountains and elongate intermountain basins. The Las Vegas Range is separated from the Sheep Range on the west by Yucca Forest Valley, and from the Arrow Canyon Range on the northeast by the southern extension of Hidden Valley. The elongate, north-south trending basin-range structure of the region is interrupted at the margin of the Las Vegas Valley by the northwest-striking, right-lateral Las Vegas Valley shear zone (Beard and others 2007).

Rocks in the southern Las Vegas Range are mostly Paleozoic marine strata laid down in a carbonate shelf depositional environment along a broad continental margin, or in a deeper water carbonate slope to basin depositional setting (Page and others 2005). The first major structural episode that affected the rocks in the Las Vegas Range was the development of thrust faults and folds related to the Sevier orogeny in Cretaceous time. Late Tertiary extension in southern Nevada resulted in regional strike-slip and normal faulting, and localized reverse faulting (Duebendorfer and Simpson 1994). The major extensional structure near the Las Vegas Range is the west-northwest-striking, right-lateral Las Vegas Valley shear zone (LVVSZ) to the south, the effects of which include oroflexural bending of the Las Vegas Range (Page and others 2005). As a result, structures in the southern Las Vegas Range trend northeast, rather than north-south.

Preliminary Studies

Kleinfelder performed a pipeline reconnaissance study for the Clark, Lincoln, and White Pine Counties Groundwater Development Project that covered the area of the Apex Tunnel alignment. The preliminary study included a review of published geologic maps and literature, new and archival aerial photographs, and reconnaissance geologic mapping.

Published geologic mapping by the U.S. Geological Survey (Beard and others 2007) indicates rocks exposed in the vicinity of the project site are part of the Bird Spring Formation, which consists of a variety of carbonate rock types including limestone, micrite, chert, mudstone, and shale. A long, northeast-striking fault lies west of and subparallel to the southern part of the tunnel alignment; the tunnel alignment crosses this fault near the location where the tunnel route bends northward. Northeast of its intersection with the tunnel alignment, the fault is shown abruptly branching into three strands. Northeast along the strike of these faults and filling the area between the three strands is a broad, northeast-trending swath of Tertiary-age alluvium and basin fill deposits.

Numerous lineations are visible in the aerial photographs of the project site, including northeast-trending linear features that correspond to the faults mapped by Beard and others (2007). Two additional sets of faults that strike northwest and north-northeast

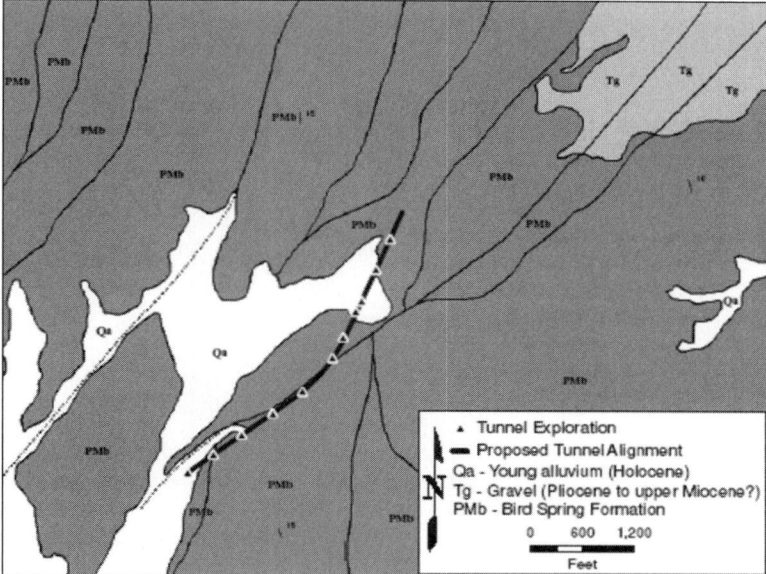

Figure modified from Beard and others 2007.

Figure 2. Geologic map of the Apex Tunnel vicinity

are exposed on the southeast side of the tunnel alignment. Although faults are clearly visible both in overhead imagery and on the ground, there is no evidence of recent faulting along the Phase I and II parts of the tunnel alignment.

FIELD EXPLORATION

Site Reconnaissance and Geologic Mapping

The tunnel exploration was conducted in phases to allow for early exploration of the private land part of the route while pursuing the application and review for the public-land access permit. An alluvium covered area accessible to track-mounted drill rigs at the north end of the private land portion of the alignment was designated the Phase I exploration area. An initial geotechnical concern was the depth of alluvium and the nature of the underlying rock. The rugged, rocky terrain south of Phase I is accessible only to helicopter-transported drill rigs and was designated the Phase II exploration area. The initial geotechnical concerns in the Phase II area related to the presence of the northeast-striking fault subparallel to, and intersected by, the tunnel alignment.

Although useful in planning and critical to gaining an understanding of the regional geology, the published geologic maps (Page and others 2005; Beard and others 2007) are at such a large scale that little insight is afforded at the level of resolution needed to address most geotechnical issues. The photo-geologic analysis provided insight into the geomorphology and underlying geologic structure of the site, but field observations were required to develop the stratigraphic and structural picture and allow effective placement of the exploratory coreholes.

Site reconnaissance confirmed the presence of faults shown on the published geologic maps, as well as several additional faults corresponding to lineations seen in the aerial photographs. Geologic mapping was performed along a corridor approximately

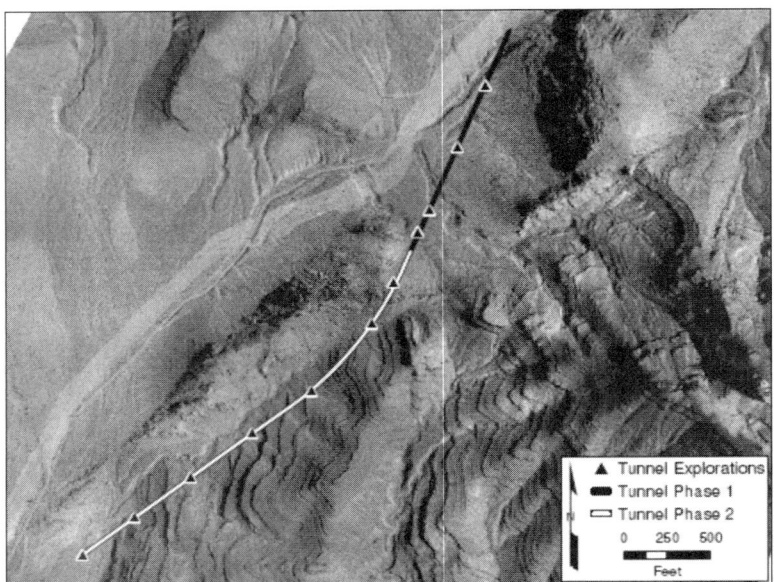

Figure 3. Aerial photograph of the southern half of the Apex Tunnel alignment showing Phase I and II corehole locations

305 m (1,000 ft) wide centered on the tunnel alignment to characterize surface exposures of the principle geologic units, to identify major geologic features such as faults, and to aid in the interpretation of the corehole results. Geologic mapping showed that the Phase I and II exploration areas comprised three distinct intervals: the alluvium-covered Phase I area; the north end of the Phase II area with steeply-dipping, thick-bedded limestone; and the south end of the Phase II area with gently-dipping, thin-to-thickly bedded carbonate rocks displaying a distinctive ledge-and-slope morphology.

Core Drilling

The preliminary geotechnical study provided a level of understanding of the site geology sufficient to develop a corehole exploration plan. Coreholes were initially placed roughly 120 m (400 ft) to 150 m (500 ft) apart, a compromise between increasing the potential volume of subsurface data and controlling exploration costs. The hole locations were close enough together to provide stratigraphic overlap over the tunnel route, provided large displacement faults were not present. The planned depth for each hole extended to 9 m (30 ft) below the lowest invert elevation being considered for tunnel design.

A shallow seismic refraction survey across the Phase I area was performed prior to drilling by Zonge Geophysics. The objective of the survey was to measure the depth of alluvium and the p-wave velocity of the underlying rock. The results of the seismic survey showed a zone of relative low velocity near the south edge of the alluvium-covered area. To investigate whether the low-velocity zone was a fault an additional corehole was added to bracket the zone. Three angle holes were planned, including the additional corehole: one to find the basin limits and explore the nature of the alluvium/bedrock contact at the basin margin; a second to cross a greater stratigraphic thickness in the zone of steeply-dipping strata; and a third to cross the northeast-striking fault that intersects the tunnel alignment. Figure 3 is an aerial photograph of the south half of the

Figure 4. Oblique aerial photograph of the southern half of the Apex Tunnel alignment showing corehole locations

alignment showing the Phase I and II exploration areas and corehole locations. Figure 4 is a low altitude, oblique aerial photograph of the same area.

Several other changes were made in the field to the original drilling plan. All of these changes were executed without significant delay, even when cost adjustments were required because of increased scope of work. A key change to the originally proposed drilling was to extend two borings by 18 m (60 ft) and 30 m (100 ft) to reach bedrock after unexpectedly deep basin-fill deposits were encountered. The discovery of a small, filled basin straddling the tunnel route has important implications for tunnel construction and extending the coreholes was crucial in helping to define the extent and geometry of the basin. Another useful change made after completing the first four holes was the addition of down-hole natural gamma logging in the remaining seven holes. Lithologic logging of the first four holes suggested that textural differences in the carbonate rocks encountered might be expressed as differences in the amount of natural gamma ray emissions. The natural gamma logs were key in determining the stratigraphic relationships between the cores collected from the southern part of the tunnel alignment.

Other changes to the drilling plan that improved the usefulness of the resulting data involved shifting the locations and increasing the angles of two holes to intersect two different faults. The surface locations of both of these faults were determined only a short time before the start of drilling of the holes in question, so rapid approval of changes was required to capture the desired data. In one of these holes the drilling plan was altered again when the drill encountered a large void and fluid loss from the HQ-size drill stem could not be controlled. Smaller diameter NQ-size drilling tools were brought in overnight from another location and "telescoped" inside the HQ rods, which acted to case the void and associated fault zone, allowing the hole to be completed to the target depth. Finally, in this same hole, the grouting plan was modified so that the grout take of the void could be measured. The altered grouting plan was extended to all of the remaining holes to improve our understanding of the potential for grout loss into the formation.

Core RQD, Fracture Frequency, and Cobble-Boulder Distribution

Rock Quality Designation (RQD) is a modified core recovery percentage in which all pieces of sound core over four inches in length are summed and divided by the length of the core run (Deere and Deere 1989). RQD values shown on the core logs or as histograms were obtained for the Paleozoic bedrock units only. With a few rare exceptions, we interpret the separations seen in the core of the basin-fill deposits as mechanical breaks (i.e., induced by drilling and core recovery) rather than pervasive discontinuities intersecting the core.

Fracture frequency measurements were also obtained in Paleozoic bedrock units as part of the core evaluation. The number of fractures per foot of core length was counted and recorded up to a maximum of six per foot, or an average spacing of two tenths of a foot. In general, core intervals of low RQD have high fracture frequency and intervals of high RQD have low fracture frequency.

The evaluation performed on the core from the basin-fill deposits included measuring the size and depth of large cobbles and boulders that may impact mining of the proposed tunnel. Individual clasts with a core length of 12 cm (0.4 ft) to 24 cm (0.8 ft) were recorded as cobbles, and clasts with a core length of greater than 24 cm were recorded as boulders, following the scale commonly used in sedimentary rock classification. The cored length of any clast is the minimum possible cobble/boulder size; if the core passes through the edge of a large clast, the maximum clast dimension could be several times larger than the cored length.

Preliminary Results

The geologic investigation for the proposed Apex Tunnel is nearly complete for the southern half of the tunnel alignment where it crosses private land. Eleven coreholes totaling 978 m (3210 ft) were drilled with an average core recovery of 97 percent. A database of rock properties and corehole rock characteristics was developed from a combination of core RQD, fracture frequency, discontinuity surveys, and laboratory testing.

Two general rock types are present in the vicinity of the Apex Tunnel alignment: marine carbonate rocks of the Paleozoic Bird Spring Formation, and much younger basin-fill deposits. Most of the rock exposed along the tunnel alignment is part of the Bird Spring Formation, which consists of a variety of carbonate rock types, including limestone, cherty limestone, and micrite, along with small amounts of mudstone, siltstone, and shale. The four coreholes drilled during the Phase I exploration revealed the presence of a deep, previously unknown, fault-bounded basin filled mostly with conglomerate of probable Tertiary age. The basin is topped with a veneer of older alluvium which hides most of the underlying basin-fill deposits, although a small area of basin-fill deposits is exposed northeast of the tunnel alignment. The basin is floored by carbonate rocks of the Bird Spring Formation.

Much of the basic geologic structure along the southern tunnel alignment was resolved and a preliminary geologic model was produced which will be used in making predictions of conditions at tunnel depth beyond the immediate vicinity of the coreholes. Geologic mapping and surface geophysics were used to locate faults.

Core lithology was used to define a simplified stratigraphy relevant to tunneling, and the results of corehole geophysical logging using natural gamma and optical televiewer tools were used to help correlate the stratigraphy between holes. The natural gamma log records the level of radioactivity of the rock. The optical televiewer results

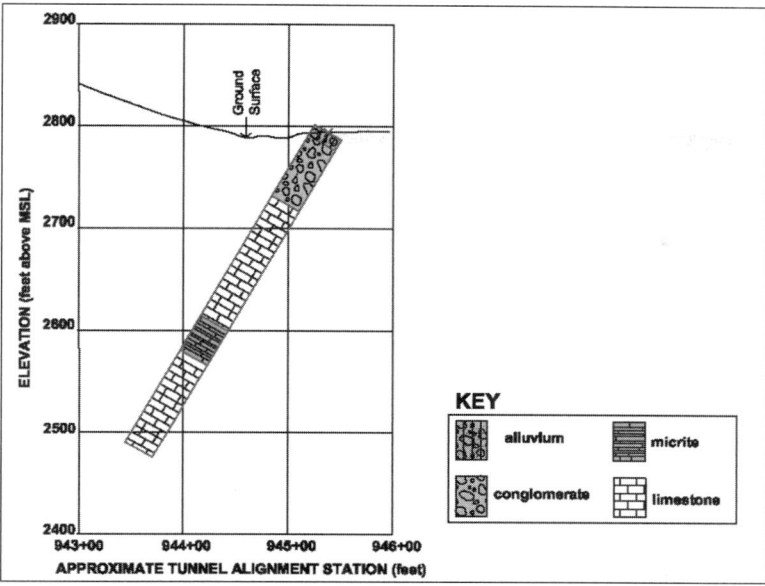

Figure 5. Simplified stratigraphy of bedrock and overlying basin-fill deposits at the basin margin

provide the orientation of bedding and discontinuities at depth in the logged coreholes. Figure 5 shows an example of the simplified stratigraphy from a hole near the margin of the basin. Figure 6 shows the corresponding bedrock RQD and fracture frequency for the same hole. Figure 6 also shows the distribution of cobbles and boulders in the basin-fill deposits overlying the bedrock.

The significance of gamma logging in the explorations for the Apex Tunnel is that the micrite, mudstone, siltstone, and shale rocks yield higher natural gamma radiation than the limestone rocks, presumably due to their higher concentration of terrigenous sediments. The strength of the correlation of beds between coreholes was greatest where the gamma logs intersected more than one marker bed. The most useful marker beds proved to be rare, thin mudstone beds with a very high relative gamma count in the logs and a distinctive reddish brown or olive brown appearance in the core. The gamma logs show overlapping intervals that are sufficiently similar in pattern and magnitude of signal that they can be correlated with reasonable confidence. An example of stratigraphic correlation between adjacent coreholes based on core lithology and natural gamma logs is presented on Figure 7.

CORE LABORATORY TESTS AND DISCONTINUITY MEASUREMENTS

The laboratory testing program was developed to achieve a representative sampling of test data on the varying lithologies expected to be encountered along the tunnel alignment. Preference was given to test sample selection within the range of possible tunnel depths within each core boring. However, because of bedding dip and a strike more or less perpendicular to the tunnel route, cored Paleozoic beds above the tunnel zone were assumed to be representative of conditions along the alignment between the cored locations, and testing was therefore not restricted to tunnel depth at the cored

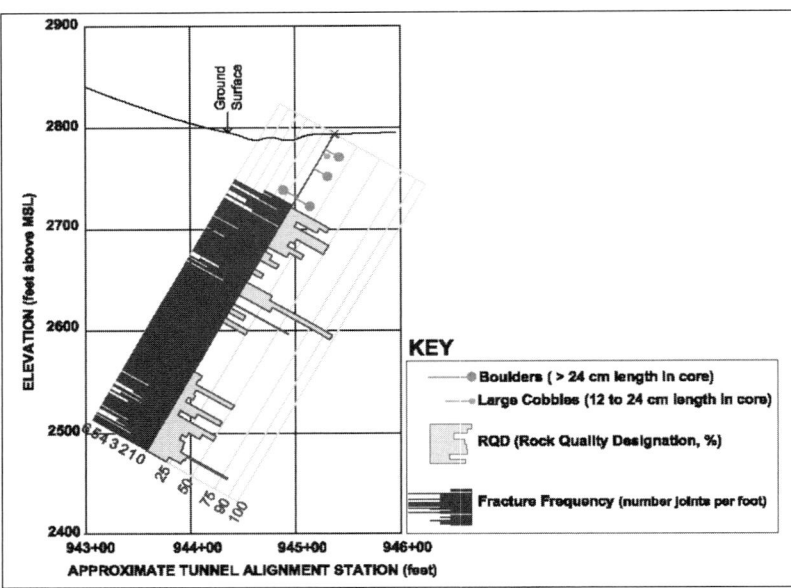

Figure 6. RQD and fracture frequency of bedrock, and cobble-boulder distribution of overlying basin-fill deposits at the basin margin

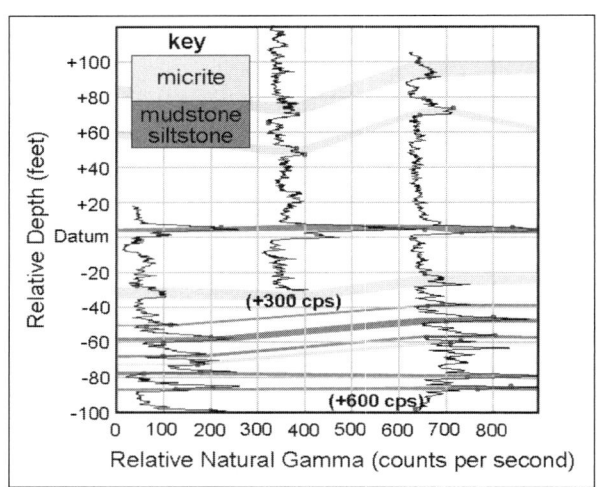

Figure 7. Example stratigraphic correlation between adjacent coreholes based on core lithology and natural gamma logs

locations. The laboratory tests generally fell into three categories: strength, drillability, and classification.

Unconfined compression, confined compression, and direct shear tests were performed at the Kleinfelder Rock Mechanics Laboratory. Additional direct shear tests were performed by Advanced Terra Testing. Brazilian tensile, unconfined compression,

Table 1. Summary of laboratory testing

Test Type	Number	Test Type	Number
Unconfined Compression	25	Sievers "J"	8
Confined Compression	10	Cerchar Abrasion	11
Direct Shear	5	Punch Penetration	6
Brazilian Tensile	12	Petrographic Analysis	14
Point Load	194	Grain Size Analysis	10
Brittleness	4	Atterberg Limits	10
Abrasion	4	Unit Weight	6

direct shear, abrasion, Cerchar abrasion, brittleness, and Siever's "J" tests, and petrographic analyses, were performed by the University of Texas at Austin Department of Civil Engineering. Punch penetration and Siever's "J" tests were performed by the Earth Mechanics Institute at the Colorado School of Mines. Point load testing, grain-size distributions, Atterberg Limits, and unit weight determinations were performed by Kleinfelder at the core storage facility and the soil testing laboratory in Las Vegas, Nevada. The type and number of each test performed is summarized in Table 1.

Rock Strength Testing

The strength parameters assessed in the Apex Tunnel Phase I and II exploration testing program included the unconfined compressive strength, σ_c or UCS, obtained from the unconfined compression strength test; the tensile strength, σ_t, which is obtained from the Brazilian tensile or splitting tensile test; and the corrected point load index, $I_{s(50)}$, which is obtained from point load test data.

Sample length constraints imposed by several of the testing methods introduce a bias toward testing stronger samples in weak rock. Samples tested for strength were tested at moisture conditions as received in the testing laboratory. The strength data was grouped according to lithology into the general categories of limestone, micrite, limestone/micrite with chert, limestone breccia, calcareous mudstone and conglomerate.

A total of 25 Phase I and II samples were subjected to unconfined compression testing. Categorizing the UCS by the ranges given in Table 2, nine samples of conglomerate and limestone breccia are categorized as weak; ten samples of limestone, limestone with chert, micrite, and limestone breccia are categorized as moderately strong; five samples of micrite and limestone/micrite with chert are categorized as strong; and one sample of micrite is categorized as very strong.

Twelve samples were subjected to Brazilian tensile testing. Intact samples relatively free from structural defects (most samples contained at least a few thin calcite veins or bedding laminations) were sought for testing; of the twelve tests performed, nine failures were judged to be nonstructural. The strength of the rock can be categorized as shown in the following Table 3 by assuming the tensile strength, σ_t, is approximately an order of magnitude lower than σ_c (Derringh 1998). The nine, nonstructural tensile failures were approximately evenly divided between the weak to moderate (3 results), strong (3 results), and very strong (3 results) tensile strength categories. When both structural and nonstructural failures are considered, half the samples tested fall within the weak and moderate tensile strength categories and half fall within the strong and very strong categories.

Point load testing was performed as part of the warehouse core evaluation. An average of approximately one test per 6 m (20 ft) of core was performed, uniformly distributed throughout the full length of each core. Both axial and diametrical point load

Table 2. Field estimation of strength based on rock grade

Grade	Description	Approximate Range of σ_c (psi)	Approximate Range of σ_c (MPa)
S1-S6	Soil	<50	<0.50
R0	Extremely Weak	50–150	0.25–1.0
R1	Very Weak	150–750	1.0–5.0
R2	Weak	750–3,500	5.0–25
R3	Moderate	3,500–7,500	25–50
R4	Strong	7,500–15,000	50–100
R5	Very Strong	15,000–35,000	100–250
R6	Extremely Strong	>35,000	>250

Source: Modified from ISRM 1978 and 1981

Table 3. Tensile strength classification derived from rock grade

Grade	Description	Approximate range of σ_t (psi)	Approximate range of σ_t (MPa)
S1-S6	Soil	<5	<0.05
R0	Extremely Weak	5–15	0.025–0.1
R1	Very Weak	15–75	0.1–0.5
R2	Weak	75–350	0.5–2.5
R3	Moderate	350–750	2.5–5
R4	Strong	750–1500	5–10
R5	Very Strong	1,500–3,500	10–25
R6	Extremely Strong	>3,500	>25

Source: Modified from ISRM 1978 and 1981

tests were performed to assess anisotropic behavior where bedding was expressed. Point load tests were also performed on the trimmed ends of unconfined compression test samples to develop a project-specific database for correlation of point load test results and unconfined compressive strength.

A total of 191 individual breaks were performed on samples from depths ranging from 0 to 134 m (440 ft). Seven samples (three limestone with chert, two limestone, and two micrite) exceeded the limits of the point load testing machine with $Is_{(50)}$ values near or in excess of 7.0 MPa (1.0 ksi). The weakest samples tested were conglomerates with $Is_{(50)}$ = 0.12 MPa (0.017 ksi). In general, the conglomerate, limestone breccia, and calcareous mudstones have mean $Is_{(50)}$ values that fall within the low to medium strength designations as presented in ASTM D5731. Mean $Is_{(50)}$ values for limestone and micrite fall within the high strength designation and the mean $Is_{(50)}$ values for limestone with chert fall into the very high strength designation (ASTM D5731).

The strength anisotropy index, $I_{a(50)}$, is defined as the ratio of the mean $I_{s(50)}$ values measured perpendicular and parallel to planes of weakness. The point load test data includes 25 pairs of data representing these two failure modes; each data pair is from approximately the same core interval in the Bird Spring Formation rocks. Two of the data pairs are in calcareous mudstone; the remaining 23 are in limestone and micrite. The strength anisotropy indices, considering test data from the limestone and micrite samples both separately and combined, range between approximately 0.9 and 1.1, suggesting quasi-isotropic behavior. Limited data from the calcareous mudstones suggests anisotropic behavior with an $I_{a(50)}$ value of approximately two.

A correlation, C, between σ_c and $I_{s(50)}$ was estimated from comparisons within the same lithology. Typical average values reported in the literature are in the range of 20 to 25. The Phase I and II data include 15 pairs of σ_c and $I_{s(50)}$ data representing tests on samples from approximately the same depth interval. Where possible, point load tests were performed on cut ends from σ_c test samples; where cut ends were unsuitable, point load testing was performed on an adjacent sample of core judged to be of similar lithology. A simple mean calculation developed from the quotient of the pairs of σ_c and $I_{s(50)}$ data yields a C value of 31 for the conglomerate; a C value of 24 for the Bird Spring Formation rocks, and an average value of 27 for all data sets combined.

Drillability Testing

Brittleness, Siever's J, and Abrasion testing were performed on samples from the Phase I and II explorations to aid in the assessment of drillability. Project Report 13A-98 Drillability Test Methods (NTNU-Anleggsdrift 1998) describes these laboratory methods and their use in the development of indices for indirect measures of the drillability of rock. Punch Penetration and Cerchar Abrasivity testing was performed to aid in the assessment of TBM cutter wear. A brief description of these tests and a summary of the results obtained are presented in the following paragraphs.

Brittleness. The Brittleness Value test provides a measure of the ability of the rock to resist crushing by repeated impact. Meaningful interpretation of the results requires that the test be performed on a homogeneous material; tests were therefore performed on micrite breccia and limestone but not on the conglomerate. The test results in a Brittleness Value S_{20}, equal to the percentage of material passing a 11.2 mm mesh after 20 impacts of a specified weight dropped from a specified height, and the Compaction Index, a qualitative assessment of the increasing degree of compaction after impact on a scale of 0 to 3. The reported Brittleness Value S_{20} for the four samples of Bird Spring Formation rocks tested ranged between 48 and 55, with a Compaction Index of 1 in each test.

Sievers' J-value. The Sievers' is a miniature drill test which gives a measure for the surface hardness of the rock. The Sievers' J-value is the mean value drill hole depth after 200 revolutions of the miniature drill, averaged from 4 to 8 drill holes, measured in $\frac{1}{10}$ mm. Typical reported values for the Bird Spring Formation rocks were in the range of 30 to 60, with the exceptions being one limestone value in excess of 100 and a chert sample with a reported value less than 5.

Abrasion. The abrasion value is a time dependent measure of abrasion on steel from crushed rock powder. Crushed rock powder less than 1 mm is passed under a test bit of steel from a new cutter ring. The reported abrasion value, AVS, is the weight loss in milligrams after 20 revolutions on a rotating steel disc in a one minute time period. Abrasion values for tests of four samples of the Bird Spring Formation rocks ranged from 1 to 3.

Punch penetration testing. The punch penetration device applies a load to the sample and measures the corresponding penetration. The punch penetration test evaluates the toughness or brittleness of the rock with respect to indentation (Ozdemir and Nilsen 2007). The force-penetration curve is used in TBM design to estimate the required energy input exerted by the TBM to form significant "chips" or rock fragments liberated from the cutting face. The peak slopes for the punch penetration testing range 641 MPa (93 kips/in.) to 1303 MPa (189 kips/in.) with an average value of 972 MPa (141 kips/in.).

Cerchar abrasivity testing. Four samples were selected for Cerchar Abrasivity testing. Two tests were performed on limestone clasts from the conglomerate and two tests were performed on samples of micrite and micrite breccia, respectively. The Cerchar Abrasiveness Index, CAI, is used to estimate the life expectancy of cutting heads in the TBM. The sample is scratched with a small metal pin over a distance of 10 mm and the CAI value is calculated from the average diameter of the pin to the nearest tenth of a millimeter divided by the scratching distance. For each test, a total of five scratching cycles are performed with five different steel pins for each cycle. Reported CAI values ranged from 0.6 to 4.5, with an average of 2.5.

Core Discontinuity Survey

A discontinuity survey was performed on each of the cores as part of the core evaluation to provide quantitative data for a by-core-run assessment of Q-value (Barton 1974) and Rock Mass Rating (RMR; Bieniawski 1989). Discontinuities are assumed to part easily and are identified as those features that influence tunnel excavation, stand-up time, and choice of permanent support. Discontinuities encountered in this project include shears (discontinuities showing evidence of offset), joints (discontinuities not showing any evidence of offset), and partings along bed contacts (bedding plane joints). Veins or healed joints not causing separation of the core were not described in the discontinuity survey. Discontinuity characteristics were recorded as outlined by Piteau and Martin (1977) to determine the following discontinuity characteristics:

- Type (shear, joint, or bedding),
- Inclination relative to the core axis (where possible),
- Width (distance between rock surfaces, measured normal to the plane of the discontinuity),
- Aperture (i.e., openness),
- Infilling material and thickness (if present),
- Roughness (i.e., small-scale surface asperities), and
- Planarity (i.e., larger-scale surface asperities).

Discontinuities were also not measured in the basin-fill deposits. Excluding slightly over half of the 978 m (3,210 ft) of core drilled (~28% of core in basin-fill deposits; ~23% of core with greater than six discontinuities per foot, or core loss), a total of 3391 discontinuities were examined.

SUMMARY

The project team adopted a flexible approach to aid project planning and execution. As a result, the team was able to make rapid adjustments to the exploration plan based on information gathered during the field investigation and increase the value of the geologic investigation to other concurrent planning activities.

The geologic investigation completed thus far for the Apex Tunnel has revealed much of the complex geology along the tunnel alignment. A key finding of the geologic investigation completed thus far is the discovery of a deep basin along the tunnel route. The presence of the basin has a potentially significant impact on the tunnel construction because the basin fill represents a change of material from the carbonate bedrock encountered elsewhere, and because of anticipated structural complexity at the basin margins. The success in resolving the site stratigraphy along much of the alignment investigated thus far means that it should be possible to make reasonably accurate forecasts of expected tunneling conditions, however, the variability in rock types and rock quality will present challenges to both tunnel design and construction.

REFERENCES

ASTM D 5731-07. 2007. *Standard Test Method For Determining Point Load Index Of Rock And Application To Rock Strength Classifications*. American Society for Testing and Materials International, West Conshohocken, PA. Available from www.astm.org.

Barton, N.R., Lien, R., and Lunde, J. 1974. Engineering classification of rock masses for the design of tunnel support, *International Journal of Rock Mechanics and Mining Sciences*, 6(4), p. 189–239.

Beard, L.S., Anderson, R. E., Block, D.L., Bohannon, R. G., Brady, R. J., Castor, S. B., Duebendorfer, E. M., Faulds, J.E., Felger, T. J., Howard, K. A., Kuntz, M. A., and Williams, V.S. 2007. Preliminary Geologic Map of the Lake Mead 30' X 60' Quadrangle, Clark County, Nevada and Mohave County, Arizona, U.S. Geological Survey Open-File Report 2007-1010, 1:100,000.

Bieniawski, Z.T. 1989. *Engineering Rock Mass Classifications*, John Wiley & Sons, New York.

Deere, D.U., and Deere, D.W. 1989. *Rock Quality Designation (RQD) after Twenty Years*, US Army Corps of Engineer Waterways Experiment Station report GL-89-1, Vicksburg, MS.

Derringh, E. 1998. *Computational Engineering Geology*, Prentice Hall, Upper Saddle River, New Jersey, p. 323.

Duebendorfer, E.M. and Simpson, D.A. 1994. Kinematics and timing of Tertiary extension in the western Lake Mead region, Nevada, *Geological Society of America Bull.*, v. 106, p. 1057–1073.

International Society of Rock Mechanics, 1978, Suggested Methods for the Quantitative Description of Discontinuities in Rock Masses, Int. J. Rock Mech. Min. Sci. & Geomech. Abstr., Vol. 15, pp. 319–368.

International Society for Rock Mechanics, 1981, ISRM suggested methods, Rock characterisation, testing and monitoring—ISRM suggested methods. Oxford: Pergamon.

NTNU-Anleggsdrift. 1998. *Project Report 13A-98 Drillability Test Methods*, Norwegian University of Science and Technology (NTNU, Norges Teknisk-Naturvitenskapelige Universitet). Available from www.drillability.com.

Ozdemir, L., Nilsen, B., Recommended Laboratory Rock Testing for TBM Projects, reviewed on the world wide web on 09/15/08 at: http://www.mines.edu/academic/mining/ research/emi/10_publications.html

Page, William R., Lundstrom, Scott C., Harris, A.G., Langenheim, V.E., Workman, J.B., Mahon, S.A., Paces, J.B., Dixon, G.L., Rowley P.D., Burchfiel, B.C., Bell, J.C., and Smith, E.I. 2005. Geologic and Geophysical Maps of the Las Vegas 30' × 60' Quadrangle, Clark and Nye Counties, Nevada and Inyo County, California, U.S. Geologic Survey Scientific Investigations Map 2814, 1:100,000.

Piteau, D.R., and Martin, D.C. 1977. Detail line engineering geology mapping method, in: *Rock Slope Engineering*, Part D, Federal Highway Administration Reference Manual, FHWA-TS-97-208, Federal Highway Administration, Washington, D.C.

THE COST AND BENEFIT OF THE PHASE 2 INVESTIGATION FOR THE REACH 4 TUNNEL, HOW A ROLL OF THE DICE CAME UP BIG IN LAS VEGAS

Ray Brainard ▪ Black & Veatch

Racheal Johnsen ▪ University of Nevada Las Vegas

Jim Devlin ▪ Clean Water Coalition

Eugene Smith ▪ University of Nevada Las Vegas

Tom Knox ▪ Black & Veatch

Jim Werle ▪ Converse Consultants

Alston Noronha ▪ Black & Veatch

ABSTRACT

The subsurface investigation for the Reach 4 Tunnel was expected to be completed in a single phase. This assumed that data from two nearby previous tunnels, combined with limited new subsurface information and surface mapping would be sufficient to characterize the rock behavior. Unforeseen rock conditions and complex faulting led to the recommendation for a second phase. This decision was not reached easily, as much of the follow-on work required helicopter support, and associated costs. The second phase of drilling found significantly more variability than the first phase and led to a very different and better understanding of the geologic and geotechnical conditions, and rock characterization.

INTRODUCTION

The wastewater agencies within the Las Vegas Valley have operated treatment facilities since the early 1950s. Each facility operated with the sole purpose of receiving wastewater, treating the water to the highest quality possible, and reusing the water in an efficient manner. As the population of the Las Vegas Valley increased, excess treated wastewater was returned to the Las Vegas Bay area of Lake Mead via the Las Vegas Wash. This return flow augments Las Vegas' allocation of raw Colorado River water, as the Colorado River Commission permits an additional gallon of water to be withdrawn for every gallon of highly treated effluent returned to the Lake.

Effluent flows will continue to increase in the future from the current average annual flow of about 6,570 l/s (150 mgd) currently to a projected 17,520 l/s (400 mgd) in 2050. The mixing capacity of the Las Vegas Bay will reach its limit as the effluent flows increase, and the nutrient loadings of the reclaimed water will rapidly approach the maximum limits allowed under current regulations. Therefore, the flows to the Lake need to be introduced in a manner that promotes sustainability (of the resource and environment) and maintains the appropriate mixing capacity.

Currently, there are three participating entities discharging effluent into the Las Vegas Wash: the City of Las Vegas, the Clark County Water Reclamation District, and the City of Henderson, collectively referred to as the Dischargers.

To deal with the pending water quality issues in Lake Mead, the Dischargers implemented the Alternate Discharge Study which led to the development of the System Conveyance and Operations Program (SCOP). SCOP is a multifaceted program for managing the highly treated effluent generated at the Dischargers' facilities with the goal of keeping Lake Mead in compliance with the permits issued by the Nevada Division of Environmental Protection. Evaluation of current water quality in Las Vegas Wash, Lake Mead, and Lake Mohave using water quality models, examination of potential process enhancements utilizing existing and emerging technologies, and evaluation of alternative receiving areas within Lake Mead led to the conclusion to construct a pipeline to gather and transport the effluent through the River Mountains and out into Lake Mead, including diffusers that discharge near the Boulder Islands. The project has been divided into multiple design packages, of which the Reach 4 Tunnel is one. The design and construction packages that make up SCOP are listed below, and can be seen in Figure 1.

- Reach 1 cut-and-cover pipeline—2,440 m long
- Reach 2 cut-and-cover pipeline—4,120 m long
- Reach 3 cut-and-cover pipeline—1,940 m long
- Reach 3 tunnel—1,570 m long
- Reach 4 tunnel—13,394 m long
- Pressure reducing/power generation station
- Reach 5 cut-and-cover pipelines—five pipes, each 1,830 m long
- Boulder Islands outfall—five pipes ranging from 2,560 m to 4,435 m long
- City of Henderson pump station
- City of Henderson cut-and-cover force main and tunnel—2,000 m long
- City of Henderson tunnel—380 m long
- Supervisory control and data acquisition system

PROJECT MANAGEMENT

Formed as a joint powers authority in November 2002, the Clean Water Coalition (CWC) was established to carry out the SCOP project. Members of the CWC include the three Dischargers and the city of North Las Vegas. The CWC has kept staffing at a minimal level, limited to the Program Administrator, Finance/Contracts Manager, Engineering Manager, Water Quality Manager, and an Administrative Secretary. Almost all technical work is outsourced. To manage the design and construction of the system, the CWC contracted with Black & Veatch to perform the Lead Design Engineering (LDE) services, and with CH2M Hill to perform Construction Management (CM) services.

The primary role of the LDE is to perform initial design (30%) of all the SCOP features; complete the geotechnical evaluation for the entire project including investigations, geotechnical data reports and geotechnical design memorandum for all project features; perform a system wide hydraulic and transient analysis; and manage final design engineers. The CM's primary responsibilities include developing and updating project schedules and cost estimates, constructability reviews of all design packages, and managing construction.

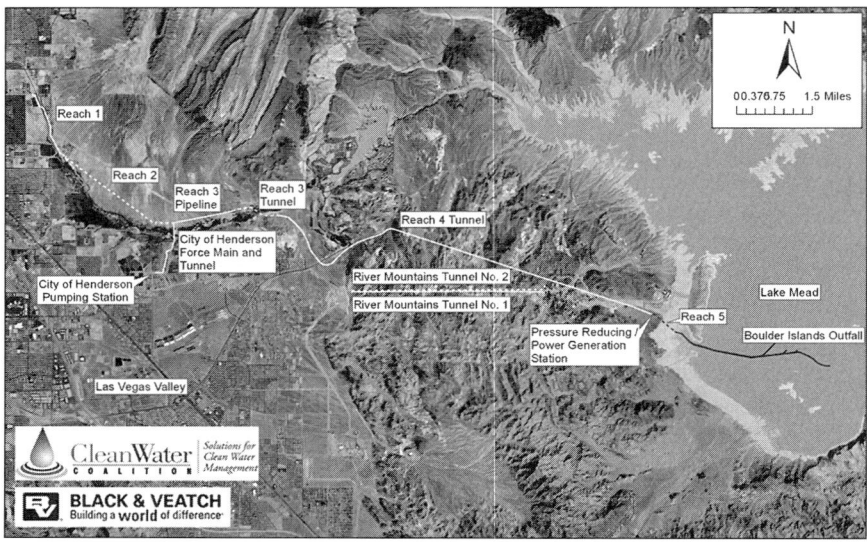

Figure 1. Plan of SCOP project

GEOLOGY

The project is located in the Basin and Range physiographic province, which is characterized by linear block faulted and tilted mountain ranges separated by broad valleys filled with sediments derived from the ranges. This topography formed in response to Miocene age east-west extension resulting in north-south oriented normal faults bounding the ranges and valleys (Faulds et al. 2001).

From west to east, the Reach 4—River Mountains Tunnel No. 3 (Reach 4 Tunnel) crosses the eastern edge of the Las Vegas Valley (basin), the River Mountains (range), and the western edge of the Boulder Basin that includes Lake Mead (Figure 1). The geology that will be encountered by the tunnel varies across these geomorphic boundaries. The Las Vegas Valley contains Tertiary age sedimentary rocks of the Muddy Creek and Horse Spring Formations. The River Mountains are predominantly comprised of Tertiary age volcanic rocks. The tunnel will pass through unconsolidated and weakly cemented Quaternary age alluvial fan deposits in the Boulder Basin. The geologic units are described below from youngest to oldest; their spatial distribution can be seen in Figure 2.

Alluvium and Alluvial Fans. These unconsolidated deposits of Quaternary age generally consist of gravel, sand, and clay.

Fanglomerate. Fan deposits of Quaternary age are present below surficial soil at the eastern end of the alignment. Cementation is variable, ranging from absent to well-cemented. Due to this variability, the strength ranges from friable to moderately strong. Clasts are strong and hard igneous rock (>75 MPa) observed in core samples up to 3 feet in diameter.

Muddy Creek Formation. The primary lithologies of the Muddy Creek Formation are mudstone, siltstone, and sandstone. Most of the units are gypsiferous. The gypsum is present as cement, veinlets, and in gypsum beds up to 45 feet thick.

Horse Spring Formation. The primary rock types of the Horse Spring Formation include calcareous siltstone, sandstone and shale; and limestone.

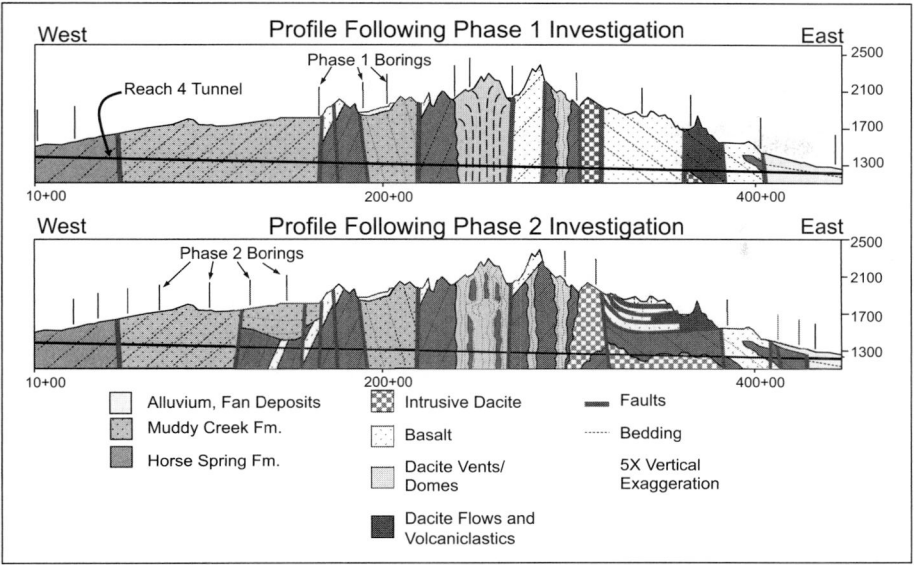

Figure 2. Profile along Reach 4 Tunnel after Phase 1 investigation (upper profile), and Phase 2 investigation (lower profile)

Powerline Road Volcanic Rock Suite. The majority of the River Mountains along the alignment are comprised of rocks of the Powerline Road Volcanic Suite. The primary lithologies are dikes, intrusive dacite, basalt, and dacite; as described below.

- Dikes of basalt, dacite, rhyolite, and andesite composition up to 30 feet in thickness were observed in surface exposures and the core samples obtained in the project area. These dikes cut across all the rock types except for Fanglomerate. The dikes have variable strike and dip orientations ranging from near horizontal to vertical, and can be seen in outcrop to change orientation over short distances.
- The Intrusive Dacite was intruded into the basalt and dacite volcanic units at very shallow depths, as evidenced by the very fine grain size and high glass content. This is the strongest and hardest unit on the alignment. It is characterized by a dacite matrix with basalt inclusions.
- Basalt is found in three textural forms: massive, agglomerate, and breccia. The massive and agglomerate basalts are commonly interbedded, and represent the interior and exterior portions of a single flow, respectively. The agglomerate contains angular clasts of scoriaceous or vesicular basalt. Commonly the matrix is altered to a much higher degree than the clasts causing a large difference in strength, the clasts being stronger. The basalt breccia is more uniformly altered, and the clasts are not scoriaceous or vesicular.
- The most common Powerline Road rock type is dacite, which was deposited in two ways; as near-vertical vents or domes, and as sub-horizontal flows. Several textural forms of dacite have been identified, including:
  - Massive dacite is generally strong, hard, and dense with no apparent bedding or clasts within it. This unit is generally found as flows, or occasionally associated with domes and vents as the core unit.

- Flow banded dacite is generally strong, hard, and dense and is found as the core unit of domes and vents. The banding is generally vertical and the unit forms a cylinder in the center of the dome and vent deposits. This unit represents flow of the lava up from the magma chamber.
- Perlitic dacite is predominantly glass and is generally black, but sometimes grades to gray. Where it is found associated with domes and vents, perlite can be seen encircling the flow banded dacite at the core of the vent. It also occurs as flows.
- Brecciated dacite is the most variable textural unit of dacite. It ranges from moderately strong, hard, and dense, to weak and light. This variation is a function of degree of alteration and pumice content. Brecciated dacite occurs as the outermost ring of the vent deposits, and as flows.
- Volcaniclastic dacite are deposits of reworked and water laid dacite. Bedding and grading are common. Generally, these deposits are moderately strong and moderately hard with a low density. Deposition ranges from mud and debris flows to alluvium.

PREVIOUS TUNNELING EXPERIENCE

Two tunnels were previously constructed through the River Mountains, River Mountain Tunnel (RMT) No. 1 and RMT No. 2 (Figure 1). Both tunnels convey water, one treated and one raw, from Lake Mead to the Las Vegas Valley and are owned and operated by the Southern Nevada Water Authority.

RMT No. 1 was completed in 1970, with RMT No. 2 constructed in 1997. The tunnels follow the same alignment with RMT No. 2 approximately 150 ft north of, and paralleling, RMT No. 1. They are located less than one mile to the south of the Reach 4 Tunnel, but the Reach 4 Tunnel is about 180 m (600 ft) lower in elevation. Surface geology, based on the Nevada Bureau of Mines published Quadrangles, is essentially the same across the area of the three tunnels through the volcanic rocks of the River Mountains.

The RMT No. 1 and 2 tunnels traversed similar geology consisting of about 650 ft of conglomerate assigned to the Muddy Creek Formation along at the western end. East of the conglomerate both tunnels passed through a broad, faulted antiform consisting of four lava flows belonging to the Powerline Road Volcanic Suite. Based on the generally accepted terminology established by the published quadrangles, the flows were dacite (McCormick, et al, 1997). Geologic mapping by Black & Veatch of RMT No. 2 during construction differentiated the dacite into several textural forms, including breccia, perlite, agglomerate, and volcaniclastic deposits. A good correlation was found between the surface mapping and tunnel mapping.

Over 300 faults were mapped along the tunnel length of 20,000 feet. Neither tunnel penetrated a groundwater table. A considerable amount of geologic and engineering data was obtained for both tunnels during investigations and construction that was available for our use.

PHASE 1 INVESTIGATION

The philosophy in developing the original scope of investigation for the Reach 4 Tunnel was that it could be done in a single phase. This was based on the assumption that data from the previous tunnels in the River Mountains could be used to help characterize the ground conditions. The idea was to have a limited drilling and testing program, combined with a focused geologic mapping effort aimed at confirming published

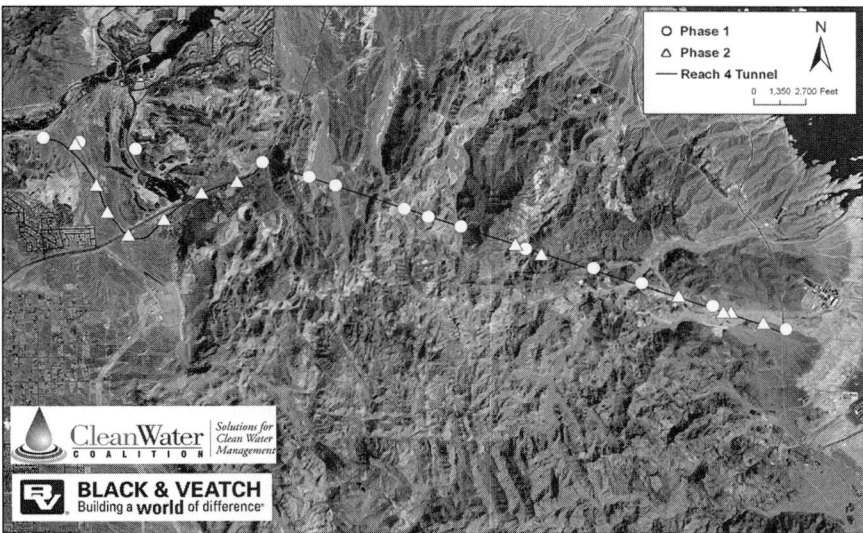

Figure 3. Plan of borings

maps of the area by the Nevada Bureau of Mines (Bell and Smith 1980, Smith 1984). This data could then be combined with the data from the previous tunnels to provide a sufficient data set.

The Phase 1 investigation included 13 core borings (Figure 3) for an average spacing of over 1,100 m (3,650 ft). The total footage of the 13 borings was about 2,160 m (7,100 ft). Packer testing, downhole televiewer, and a typical suite of laboratory strength and abrasion testing were also performed.

The total subconsultant cost for the Phase 1 investigation was $1,200,000, with Crux Subsurface, Inc. contracted to do the drilling, and Colorado School of Mines Earth Mechanics Institute to do the laboratory testing. The field work was performed from October 2006 to July 2007.

Geologic mapping carried out simultaneously with the drilling identified a group of volcanic vents or domes near the center of the alignment. Prior to this, the Nevada Bureau of Mines quadrangles had only a single vent mapped close to station 300+00. This led to the addition of a boring to help characterize these rocks.

Based on our understanding of the geology following the Phase 1 investigation, the upper profile in Figure 2 was developed. Additional investigation results included:

- Only limited groundwater was encountered at the very western end of the tunnel alignment.
- Intrusive dacite was unexpectedly found at the bottom of one of the borings, and a single unconfined compressive strength (UCS) test yielded a strength of 250 MPa (36,400 psi), more than double the next highest value.
- Surficial soils on the east end of the alignment covered very weakly cemented Fanglomerate at tunnel grade, rather than the stronger Muddy Creek Formation anticipated.

PHASE 2 INVESTIGATION

Several factors contributed to the decision to conduct a second phase of investigation, including the following.

- The horizontal alignment for the western third of the tunnel was changed due to easement problems.
- The new alignment parallels two major faults including the boundary fault between the Las Vegas Basin and the River Mountains.
- The two borings drilled to test basalt encountered heavily altered and weakened rock at one location, and highly sheared rock at the other due to encountering the River Mountains/Boulder Basin boundary fault. These borings were not considered to be representative of the basalt based on surface outcrop. No basalt was encountered in the previous tunnels, so no other data was available.
- The very strong intrusive dacite was not encountered in the earlier tunnels and required additional testing to fully characterize it.
- The boring added later in the Phase 1 program to test the volcanic vents was interpreted to have penetrated the outer rim of a vent due to the extreme variability seen through the core, and was not considered to be representative.
- A privately owned water well near the new alignment was found, which had reported groundwater levels above the proposed tunnel.
- Large differences in geology were observed between the two previous tunnels and the Reach 4 Tunnel, including the vertically emplaced vents, basalt, intrusive dacite, and an overall higher degree of alteration that made reliance on the data from the RMT No. 1 and No. 2 tunnels more difficult and riskier.

From this, it was determined that a more detailed mapping effort would be made utilizing Dr. Eugene Smith (the author of the Bureau of Mines geologic quadrangles, and now a professor at UNLV) and two students, Racheal Johnsen and Shirley Robinson. Additionally, seven core borings were proposed, two of which would require helicopter access. The borings were targeted at the new alignment on the west end (3 borings), the interior of a volcanic vent (1 boring), intrusive dacite (1 boring), basalt (1 boring), and the very weak Fanglomerate (1 boring).

Costs for the drilling increased primarily due to lower productivity than expected in Phase 1. This was mostly the result of drilling above groundwater and the presence of open fractures that resulted in the loss of all drilling fluids, usually very quickly, and sticking of the drill string. The initial boring plan of 3,150 feet was contracted for $700,000, including two helicopter locations.

Surprises arose almost immediately, as volcanic rock was found at tunnel grade on the west side under the Muddy Creek Formation, and on the east side under the Fanglomerate. This caused the addition of four borings on the west side and two on the east side, raising the total number of borings in Phase 2 to thirteen (Figure 3).

Geologic mapping determined that the fault bounded block of basalt near the center of the alignment (station 275+00) was a half graben (bounded by a fault on only one side), and the basalt would not project down to tunnel grade. Additionally, a series of low angle normal faults were discovered on the east end of the alignment that suggested that the surface geology had no bearing on what to expect at tunnel depth (lower profile in Figure 2). A further review of the heavily altered rock in the boring near station 350+00 resulted in a new interpretation that dacite rather than basalt was present at depth.

Observations made nearby, but off the alignment, by the geologic mappers indicated that the intrusive dacite is generally found in subhorizontal tabular bodies, hence the interpretation seen in Figure 2 (lower profile).

Based on all the information gathered in the Phase 2 investigation, a new profile was developed along the tunnel alignment (Figure 2 lower profile).

CONCLUSIONS

Although at considerable expense, the addition of the Phase 2 investigation increased our knowledge considerably of the ground conditions anticipated for the Reach 4 Tunnel, and thereby reduced the project risk. Some of the most important findings from the Phase 2 investigation were:

- An additional 2,000 m (6,500 ft) of tunnel was found to be in harder and stronger volcanic rock, rather than sedimentary rock.
- The tunnel excavation will be within the groundwater table for about 35 percent of the tunnel length.
- The very high UCS value (36,400 psi), in the lone sample of intrusive dacite tested in Phase 1 investigation, was considerably higher than all testing performed in Phase 2 on the same rock type (average 8,500 psi). This led to a better characterization of this unit and baseline parameters.
- Detailed geologic mapping and the Phase 2 borings established the overall complexity of the volcanic rocks, including rapid changes in textural variation and degree of alteration. This led to the lumping of most of these rocks into a single engineering unit for baselining purposes.
- The major faults that are paralleled by the revised alignment do not have significant gouge or fracture zones, and should not require special excavation and support requirements.
- There were significant differences between the previous tunnels and Reach 4, limiting the use of data from RMT No. 1 and 2. This difference is primarily a result of the volcanic vents that have much more variable geology than flows, and caused considerably more alteration of the rocks.

REFERENCES

Bell, John W., and Smith, Eugene I., 1980, Map 67, "Geologic Map of the Henderson Quadrangle, Nevada," Nevada Bureau of Mines and Geology, 1:24,000.

Faulds, J.E., D.L. Feuerbach, C.F. Miller, and E.I. Smith 2001. Cenozoic evolution of the northern Colorado River extensional corridor, southern Nevada and northwest Arizona, Utah Geological Association Publication 30—Pacific Section American Association of Petroleum geologists Publication GB78, p. 239–263.

McCormick, Bill, E. Gilmore, R. Tudor, B. Bush, and G. Korbin 1997. Analysis of TBM performance at the record setting River Mountains Tunnel No.2, *Proceedings of the 1997 Rapid Excavation and Tunneling Conference 1997*, Carlson and Budd (eds.), p. 135–149.

Smith, Eugene I., 1984, Map 81, "Geological Map of the Boulder Beach Quadrangle, Nevada," Nevada Bureau of Mines and Geology, 1:24 000.

PART 9

Microtunneling

Chairs

Gary Irwin
City of Portland

Bill Austell
Coluccio Construction

MICROTUNNELING 1.2-MILE, 72-IN RCP WITH CROSSINGS OF NJ TURNPIKE AND CSX RAILROADS

Zhenqi Cai ▪ Hatch Mott MacDonald

Alberto G. Solana ▪ Northeast Remsco Construction

Neal O'Connor ▪ Hatch Mott MacDonald

Philip Lloyd ▪ Hatch Mott MacDonald

ABSTRACT

The ongoing Overpeck Valley Parallel Sewer Project of Bergen County Utilities Authority (BCUA), New Jersey, is a US $65 million CSO improvement program to meet a State DEP consent order by 2010. One major program component includes a total of 1.8 km (1.2 mile) long, 1,800 mm (72 in) diameter reinforced concrete pipe (RCP) installed by a Herrenknecht AVND1800AB MTBM. The alignment includes drives through 850 m (2,788 ft) of extremely soft glaciolacustrine varved clay, 980 m (3,214 ft) of loose to medium dense mixed soils containing silty sands, gravels, and boulders, and crosses under the New Jersey Turnpike, US Highway Route 46, and a CSX Railroad Intermodal Yard and a branch mainline. Two types of MTBM cutterheads are being used, one configured for the weak varved clay, and one for the mixed soil conditions. The MTBM is also equipped with a double-steering system consisting of a standard front steering joint and a second trailing steering joint to assist grade control should adequate steering not be achieved by the front primary steering in extremely weak soil conditions. This paper discusses the significant engineering aspects and challenges related to site settings, ground characterization, key MTBM features and performance, and required ground treatment.

INTRODUCTION

Bergen County Utilities Authority (BCUA) was mandated by a New Jersey Department of Environmental Protection Consent Order to eliminate by 2010 wet weather overflows through a 50-year-old trunk sewer system (1,520 mm, 60 in RCP) into surrounding waterways. As a result, BCUA commissioned the design and construction of a new relief sewer installed in parallel to the existing system. This relief sewer line is part of an overall scheme being undertaken by BCUA to provide additional capacity and redundancy for sewage conveyance from its service areas to its wastewater treatment plant (WWTP). It consists of the installation of 8.45 km (5.3 miles) of new interceptor pipelines that vary in diameter from 1,050 mm (42 in) to 2,400 mm (96 in) through one of the most densely industrialized areas in the region.

The project site and ground conditions required several different construction techniques, including conventional open-cut, microtunneling, and subaqueous installation. The entire project was contracted in three separate packages—two open-cut only contracts and one contract with a combination of microtunneling, subaqueous installation and open-cut. BCUA retained Hatch Mott MacDonald as the Engineer for the planning, design and construction management of the entire project. The construction of the

MICROTUNNELING CROSSINGS OF NJ TURNPIKE AND CSX RAILROADS 559

Figure 1. Schematic microtunnel alignment and site aerial view

microtunnel (combined with subaqueous construction and open-cut) was awarded to Northeast Remsco Construction. This paper focuses on the major engineering aspects of the microtunneling portion of the project including site settings, ground characterization, key MTBM features and performance, and required ground treatment.

MICROTUNNEL ALIGNMENT AND SITE SETTINGS

The schematic microtunnel alignment and an aerial view of the surrounding settings is shown in Figure 1. The alignment consists of ten (10) individual drives with varying drive distances totaling approximately 1.8 km (6,000 ft). These drives are summarized in Table 1. Each drive is formed of strings of 3 m (10 ft) long reinforced concrete pipes with a 1,830 mm (72 in) ID and a 2,270 mm (89 in) OD. The selection of the microtunneling method to install the sewer was governed by the heavy density of utilities and the crossings of major transportation facilities. From downstream, the microtunnel begins from the east shore of the Hackensack River (a subaqueous crossing leading to the WWTP) and extends upstream with crossings of the following major sites.

- Drive #1—CSX Railroad Intermodal Yard

    Drive #1 runs at a depth of 6 m (20 ft) under a CSX Railroad Intermodal Yard that consists of 12 railroad tracks and a shipment storage area. As illustrated in Figure 2, this drive is advanced from the MTBM jacking shaft (S2) to the exit shaft (S1) by the east shore of the Hackensack River. Although packaged in the same contract, the river crossing is installed by subaqueous means in consideration of the low cover above the WWTP intake elevation and extremely weak soil conditions in the riverbed, both of which are considered to be unsuitable for microtunneling. Under the railroad yard the microtunnel runs in parallel to the existing trunk sewer and crosses under an existing water main with a separation of slightly less than one tunnel diameter at the intersection.

- Drives #2 & 3—PSE&G (Public Service Electric and Gas) Property

    These two drives are advanced from S2 to S3 (Drive #2) and from S4 to S3 (Drive #3) within the property limits of the utility provider. In addition to the

Table 1. Summary of microtunnel drives and site crossings

Drives	Direction of Drive	Drive Distance m (feet)	Major Crossings and Site Settings
#1	S2 to S1	256 (839)	Under-crossing of CSX Intermodal Rail Yard; In close proximity to existing 760 mm (30 in) force main; In parallel to existing 1,520 mm (60 in) trunk sewer
#2	S2 to S3	198 (650)	Electric and Gas Utility; In close proximity to existing 1,520 mm (60 in) trunk sewer; Numerous buried utilities, above ground electrical equipment and overhead HV power lines
#3	S4 to S3	242 (794)	
#4	S4 to S5	187 (613)	Under-crossing of New Jersey Turnpike (NJTP); At-grade and elevated sections of highway; drainage channels; In vicinity of viaduct piles
#5	S6 to S5	205 (674)	Heavy industrialized zone; In close proximity to existing 1,520 mm (60 in) trunk sewer and a number of underground chambers and pipelines; In vicinity of industrial complexes
#6	S6 to S7	69 (227)	
#7	S8 to S7	308 (1,010)	
#8	S8 to S9	157 (515)	
#9	S10 to S11	148 (487)	Under-crossing of US Highway Route 46
#10	S12 to S13	52 (170)	Under-crossing of CSX Railroad branch mainline

1,520 mm (60 in) RCP sewer, a variety of other facilities including electrical transmission towers, overhead power lines, numerous gas and water mains, and buried communication cables, lie within the influence zone of the microtunnel.

- Drive #4—New Jersey Turnpike (NJTP)

Drive #4 is the most critical drive of the project, as shown in Figure 3. It crosses under a major expressway in the region—the New Jersey Turnpike (NJTP), located west of New York City. This drive is situated at a depth of 6–7 m (20–23 ft) below grade and is advanced from S4 to S5. Along the line of crossing, the NJTP is composed of at-grade lanes, an elevated section supported on battered piles, and ramps and drainage channels. The line of crossing is selected to provide adequate separation from the piles supporting the abutments of the elevated section. A comprehensive review on the construction history of the NJTP at this location was performed to ensure that all potential obstructions were identified and avoided in the vicinity of the crossing.

- Drives #5, 6, 7 & 8—Heavily Industrialized Zone

Drives #5 through #8 are located to the east of NJTP at a shallower grade starting at shaft S5. These segments of the alignment run through a heavily industrialized zone occupied by electrical facilities, trucking transports, storage buildings, and other industrial complexes. The existing BCUA trunk sewer and its service chambers are also located below the streets in this area. Of particular consideration is the narrow clearance between Drive #7

MICROTUNNELING CROSSINGS OF NJ TURNPIKE AND CSX RAILROADS

Figure 2.   Drive #1—Microtunnel drive crossing CSX Railroad Yard

Figure 3.   Drive #4—Microtunnel drive crossing New Jersey Turnpike (NJTP)

(S8 to S7) and the existing trunk sewer that is less than one tunnel diameter for a stretch more than 10 m (33 ft) and reduces to 1 meter (3 ft) near the end of the drive (at S7). Additionally, the existing trunk sewer was placed using the trenching method with timber lagging and a gravel bedding, both of which would pose unfavorable conditions for microtunneling operations at the face.

Many of the utility facilities can only be taken out of service for a very limited time window as required by owners' operations. The potential elevated cost, schedule delay, and disruption of traffic associated with relocation of utilities prohibited an economical open-cut method of installation in this industrial community.

- Drive #9—US Highway Rt. 46

Drive #9 crosses under US Highway Rt. 46, a major regional throughway providing access to and from New York City. At the line of crossing, the highway consists of a 9 m (30 ft) high embankment constructed in the 1930s, and the tunnel crown is approximately 4.5 m (15 ft) below the toe elevation of the embankment.

- Drive #10—CSX Railroad Mainline

  Drive #10 is a relatively short drive under a branch mainline railroad owned by CSX. This segment of the pipeline was initially designed as a conventional jack and bore installation. However, the MTBM was later used due to the presence of saturated silty sand and clay exposed in test pits, which would be difficult to manage at the excavation face using jack and bore.

  At the time of writing this paper, Drives #3 through #10 have been completed successfully.

GEOTECHNICAL EXPLORATION AND GROUND CHARACTERIZATION

General Geological Descriptions

The ground conditions along the microtunnel alignment feature two distinct geological settings, divided, coincidently, by the NJTP. To the west of the NJTP, there exist deep beds of glaciolacustrine deposits (Qhkl) underlain by a highly compacted dense till (Qt). The natural surface soil in this area is an estuarine and salt marsh deposit (Qm) formed in past tidal marsh environments, which is now covered by variable amounts of artificial fill (af). Immediately below the marsh deposit lies a thin layer of alluvium containing variable amounts of organic matter. These soils are underlain by a Qhkl deposit of alternating thin layers of silt and clay, reaching a thickness of 18 m (60 ft). The microtunnel to the west of NJTP runs within this soft and varved soil deposit. These clays and silts were deposited in a glacial lake environment exhibiting a very soft consistency with sporadic "weight of rod" or "weight of hammer" and low SPT counts during sampling. The glacial clays and silts give way to a very compact lodgment till composed of clays, sands, silts, and gravels with cobbles and boulders.

As the alignment crosses the NJTP eastward, the Qhkl deposit is intersected by a gradual rise of deltaic deposit (Qhk). This deposit becomes the dominant soil type east of the NJTP. Typical of Qhk deposits, the gradation of this stratum becomes finer with depth. The upper portion of this deposit predominates as a loose to medium dense fine grained silty sand which gives way to a sequence of interbedded stiff silt and clay. Published surficial geology maps estimate this deposit to be as much as 15 m (50 ft) thick. The microtunnel to the east of NJTP runs through this mixed soil deposit.

Geotechnical Exploration and Testing

In consideration of the variable and difficult ground conditions and achieving safe crossings of the major highways and railroads, a comprehensive geotechnical exploration program was undertaken and documented in a contractual Geotechnical Data Report. The exploration program consisted of the following elements:

- 31 standard soil borings that were evenly distributed along the alignment to obtain soil samples, estimate field soil properties, and establish subsurface stratigraphy;
- 7 directional geoprobes to detect potential obstructions at locations where timber piles and gravel beddings were used for construction of the existing trunk sewer;
- environmental sampling at 12 locations along the alignment;
- a standard suite of soil laboratory tests conducted for soil classification, estimate of engineering properties and design parameters, and assessment of the feasibility of microtunneling through these soils; and

Table 2. Summary of ground characterization for microtunnel drives

Drives	Direction of Drive	Drive Distance m (ft)	Ground Characterization
#1	S2 to S1	256 (839)	Glaciolacustrine Deposits (Qhkl) ■ Soft varved clay ■ Low plasticity CL ■ Baseline undrained Cu = 20 kPa (400 psf) ■ Sticky and squeezing behavior ■ Wetting and remolding lowers strength ■ Scattered "weight of rod" & "weight of hammer" SPTs.
#2	S2 to S3	198 (650)	
#3	S4 to S3	242 (794)	
#4 (half)	S4 to S5	187 (613)	
#4 (half)			Deltaic Deposits (Qhk) ■ Mixed soil conditions ■ Low to medium dense fine silty sand ■ Baseline strength φ=32° ■ Flowing behavior when not supported ■ Gravel bedding and timber piles ■ Defined boulders 750 mm (30 in) in diameter with UCS=175 MPa (25 ksi)
#5	S6 to S5	205 (674)	
#6	S6 to S7	69 (227)	
#7	S8 to S7	308 (1,010)	
#8	S8 to S9	157 (515)	
#9	S10 to S11	148 (487)	Estuarine Deposit (Qm) and Deltaic Deposit (Qhk) ■ Soft marsh ■ Mixed soil conditions ■ Soft to medium stiff silty clay ■ Highway embankment containing rock fills ■ Defined rock fill sizes 750 mm (30 in) in diameter with UCS=175 MPa (25 ksi)
#10	S12 to S13	52 (170)	Deltaic (Qhk) ■ Mixed soil conditions ■ Loose to medium dense silty sand ■ Soft to medium stiff silty clay

- 15 cone penetration tests conducted during construction stage to supplement original soil borings and tests for the purpose of identifying weak soil locations for ground treatment.

Ground Characterization

The field borings, laboratory tests and analyses confirmed the general geological settings characterized by the two distinct soil distributions—the glaciolacustrine varved clay and mixed ground conditions divided geographically by NJTP, across which the microtunnel transitions from one soil medium to the other. A contractual Geotechnical Baseline Report was developed to characterize the ground conditions, including subsurface stratigraphy, baseline soil properties, expected soil behaviors, and obstructions. A summary of the generalized ground characterizations for all the drives is given in Table 2.

Varved Clay

Drives #1, 2 and 3 as well as one half of Drive #4 to the west of NJTP (see Figures 1 & 3) run exclusively in the glaciolacustrine varved clay deposit. Figure 4 shows the

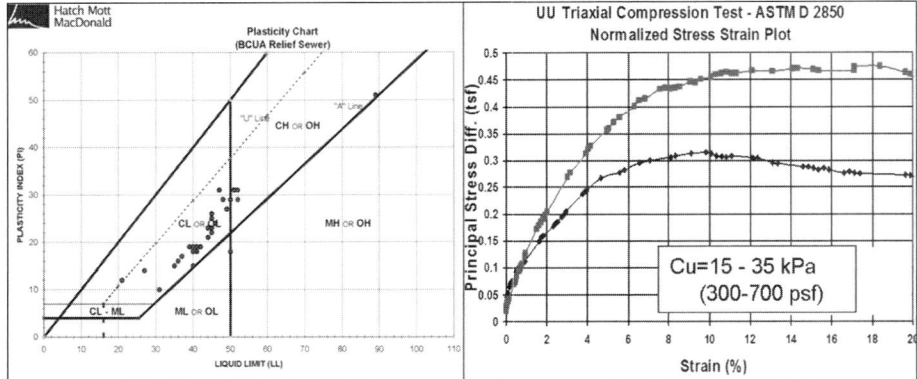

Figure 4. Varved clay (Qhkl deposit) plasticity chart and example triaxial test results

laboratory results of soil plasticity tests and selected triaxial tests. Atterberg limits indicated that the varved clay has a low plastic to semi-liquid consistency at the majority of locations where samples were taken. Laboratory soil tests subjected to triaxial loading exhibited highly nonlinear and strain softening behavior and resulted in relatively low undrained peak shear strengths ranging from 15 kPa (300 psf) to 35 kPa (700 psf).

All field observations and laboratory data appeared to demonstrate that this varved clay deposit would undergo large settlement upon loading, e.g., the self weight and steering thrusts of the MTBM. If excessive settlement occurs under the advancing MTBM, there exist the risks of MTBM "sinking," loss of line and grade, and severe distortion of pipe joints. The GBR spelled out these associated risks and as a result, two specific actions were taken as risk mitigation measures:

1. a corresponding set of specific requirements were specified for the MTBM design and manufacture that would allow the microtunneling operation to accommodate the anticipated soil conditions—this is further elaborated in Section 4; and
2. a ground treatment Allowance was incorporated in the contract to provide the necessary soil stabilization where required during the construction stage— this is further elaborated in Section 6.

Mixed Soils

All drives to the east of NJTP consist of mixed ground conditions. From the eastern portion of Drive #4 under the NJTP through Drive #10 (see Figure 1) the alignment runs primarily in mixed cohesionless soils composed of sands, silty sands, gravels, and scattered artificial fills containing debris, blocks and stones from previous constructions. Drives #9 and #10 also contain soft to medium stiff silty clay within the tunnel horizon. Figure 5 shows a collection of gradation curves of samples from both the deltaic and the glaciolacustrine deposits.

Boulders

To overcome risks of encountering boulder obstructions within the tunnel horizon, the assumed amount of boulders with a defined size and UCS strength were baselined (see Table 2) in the mixed soil drives. This baseline was deemed a necessary requirement considering the nature of geological setting and construction history of this industrial area. Although not measured in size and quantity, boulders were indeed

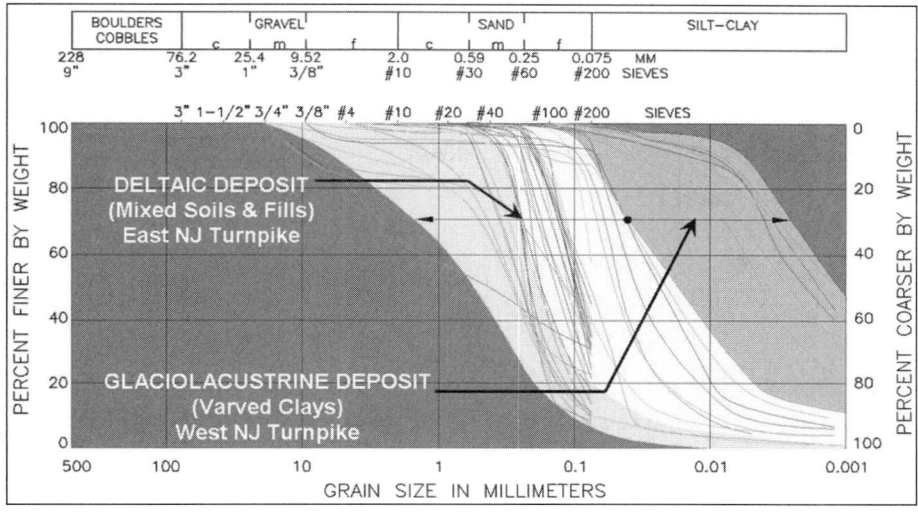

Figure 5. Gradation curves of soil samples

encountered and excavated successfully, with one boulder within Drive #7 taking nearly four hours to mine through. Drive #9, which crosses under US Highway Rt.46 with a construction history of rock fill below the embankment base, was also baselined to encounter boulders (see Table 2). This was proven to be also a prudent requirement as the MTBM cutterhead mined through at least one boulder during the course of this crossing.

Groundwater

As the project site lies within the immediate vicinity of the Hackensack River and the Overpeck Valley Creek (see Figure 1), and based on previous local construction experience, the groundwater table was baselined at 300 mm (1 ft) below the ground surface.

KEY MTBM FEATURES

The principal considerations given to the specification of the MTBM include:

- Pressurized closed-face shield for the highway and railroad crossings and drives in the immediate proximity of existing utilities and structures;
- Cutterhead(s) capable of excavating the soft varved clay and mixed soils containing the defined obstructions including boulders; and,
- Steering capability to maintain line and grade in the soft varved clay deposit.

Achieving the steering required to control line and grade in the soft varved clay would be difficult if the steering actions induce excessive ground settlement, or "sinking," below the MTBM. It was initially contemplated to specify lateral mechanical "plows" or "grippers" to counteract the front weight of the MTBM, but consultation with the Contractor and the MTBM manufacturer resulted in the adoption of a second active steering joint incorporated between the power pack section and the airlock section of the MTBM. This is shown in Figure 6. When line and grade cannot be maintained with the standard steering joint, the second steering joint can then be activated. In

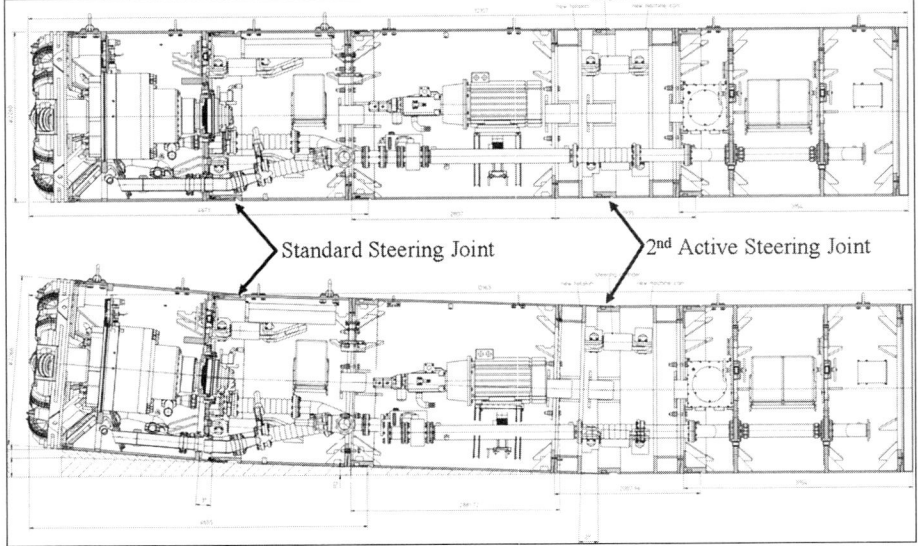

Figure 6. Herrenknecht AVND 1800AB MTBM with second steering joint

its activated status, a greater area of the MTBM will be positively inclined to achieve a more stable upward reaction with the MTBM engaging seven times as much soil volume for steering reaction as with the standard joint. This provides an important recourse needed to correct the MTBM back to grade in case of excessive downward movement, or "sinking."

The cutterhead that was supplied with the MTBM was configured with disk cutters to mine through specified boulders, as shown in Figure 7(a). It has proven to be more suitable for the mixed ground conditions, having successfully completed Drives #4 through #10. As described earlier, boulders have been encountered and successfully excavated within Drives #7 and #9.

The activation of the second active steering joint was required to assist grade control during the crossing of Drive #9 under Highway Rt. 46 when the soft silty clay portion of the drive was encountered. For the soft varved clay portion of Drive #4 under NJTP, jet grouting (see Section 6) was deemed necessary to ensure the safe crossing of this critical regional expressway.

While these approaches had been proven successful, a further risk mitigating measure was adopted, as an amendment to the contract, by procuring a more open and less heavy cutterhead, as shown in Figure 7(b), that was considered more suitable to mine through the soft varved clay drives (#1, 2 & 3). The use of the "soft" ground cutterhead reduced the front weight of the MTBM from 7,700 kg (17 kips) to 4,550 kg (10 kips), thus improving steerability, excavation efficiency, and factor of safety against bearing failure below the front section of the MTBM. This measure resulted from a cautious approach and cooperative effort between the Engineer and the Contractor to proactively managing project risks. Currently, Drive #3 in the soft varved clay deposit has been successfully completed with the use of the "soft" ground cutterhead and assisted by the second steering joint to navigate through scattered weak zones.

(a) "Mixed" ground cutterhead
Weight=7,700 kg (17 kips)

(b) "Soft" ground cutterhead
Weight=4,550 kg (10 kips)

Figure 7. MTBM cutterheads

Figure 8. Jacking pipe

JACKING PIPE

The adopted jacking pipe is a reinforced concrete pipe in compliance with ASTM C76 with an internal diameter of 1,830 mm (72 in), a wall thickness of 215 mm (8.5 in), and an outside diameter of 2,270 mm (89 in). Each RCP pipe segment is 3 m (10 ft) in length. The pipe joint is of the flat-faced bell-spigot type, with a rubber gasket confined between the cast-in bell and spigot steel rings. Typical pipe segments in storage and during installation are shown in Figure 8.

The allowable jacking load of the pipe section is 10 MN (1,200 tons) with a factor of safety of 3. Recorded jacking load during production driving has ranged from 1 MN (110 tons) to 3 MN (330 tons) with injected bentonite lubricant. To inhibit potential hydrogen sulfide corrosion, non-surcharged pipe segments are lined with an internal PVC liner cast-in around the top 270° circumference of the pipe section.

GROUND TREATMENT

To implement a "blanket" ground treatment to solidify the soft glaciolacustrine varved clay deposit would not be an economically viable approach to eliminate the risks of losing line and grade or MTBM "sinking." Therefore, a supplemental set of 15

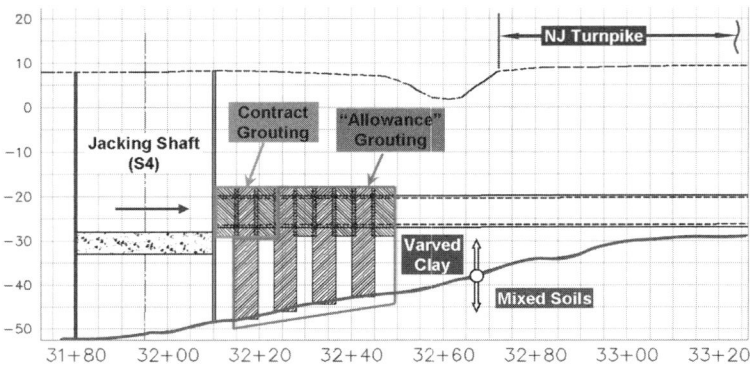

Figure 9. Jet grouting MTBM approach to NJTP crossing

Figure 10. MTBM launching in jacking shaft

field Cone Penetration Tests (CPT) were conducted to further identify "weak spots" where ground treatment would be merited. The field CPTs resulted in soil strength values consistently higher than the laboratory results. The adoption of the "soft" ground cutterhead and the assistance of the second active steering joint have also improved further the steering performance of the MTBM. With these new data and risk mitigating measures, the original belief that a substantial portion of the soft varved clay drives would have to be treated was re-assessed. As a result, theoretical analyses indicated additional ground treatment beyond the tunnel "eye" zones would be required at only one key location: the approach leading to the crossing of the NJTP, as illustrated in Figure 9. The jet grouted area extended as much to the right-of-way of NJTP as site access allowed, covering the entire body length (12 m, 40 ft) of the MTBM beyond the jacking shaft (S4). This additional grouted area is compensated for by the ground treatment Allowance and is beyond the MTBM entry grout specified by contract for each jacking shaft. The jet grouting measure ensured a stable MTBM launch and approach into the soft varved clay portion of the NJTP drive and a subsequent successful crossing of the expressway.

MICROTUNNELING PRODUCTION

To date, Drives #3 through #10 have been completed within schedule. The microtunneling operations are staged from six jacking shafts. A typical example is shown in Figure 10. Generally, it takes two 10-hour days to launch all the MTBM sections and a third day to install the first few pipes that contain the cooling pipes, the tunnel slurry pump, and an Intermediate Jacking Station. It is not until day four when efficient productivity starts. For Drives #5 through #10, the best productivity is 8 pipes (24 m, 80 ft) in a 10-hour shift. For Drive #4 crossing the NJTP, a continuous 24-hour work schedule was adopted to ensure the uninterrupted MTBM advance, and a record productivity of 11 pipes (33 m, 110 ft) during a 12-hour period has been achieved. The crossing of the CSX Railroad Yard has also been planned for a 24-hour work schedule.

CONCLUSIONS

A large majority of the BUCA relief sewer line has been successfully installed by microtunneling in challenging ground conditions, in close proximity to dense utilities, and crossing under major highways and railroads. A site exploration program involving targeted tests followed by a thorough engineering analysis resulted in soil-specific MTBM requirements and risk mitigating measures that proved to be the key success factors to date. These included supplemental field CPT tests to identify weak soil zones that would require ground treatment, the incorporation of a specially designed second steering joint as well as a lightweight "soft" ground cutterhead for the MTBM to steer and excavate efficiently in soft silty clays, and a "mixed" ground cutterhead equipped to mine through mixed soils containing sands, gravels, and boulders. As a result of these proactive approaches, the completed drives have been accomplished on schedule and without the occurrence of any unexpected events, and the remaining projects are projected to be completed by May, 2009.

ACKNOWLEDGMENTS

The authors would like to acknowledge Bergen County Utilities Authority for permission to publish this paper. They also wish to express their gratitude to Ms. Rebecca Carmine and Mr. Eric Prantil, both recently completed their internships at Hatch Mott MacDonald, for their help in preparing the manuscript.

THE LONGEST DRIVE—PORTLAND'S CSO MICROTUNNELS

Christa Overby ▪ Bureau of Environmental Services

Matt Roberts ▪ Kiewit-Bilfinger Berger

Craig Kolell ▪ Jacobs Associates

ABSTRACT

As part of the City of Portland, Oregon's East Side Combined Sewer Overflow Tunnel Project, nine separate 2,130 mm (84-inch) diameter reinforced concrete pipe microtunnel drives totaling approximately 2,380 m (7,808 ft) are being constructed to divert flows from several of the existing outfalls to the main tunnel. The drives were constructed in challenging ground conditions that include soft soils under the groundwater table, large amounts of wood and metal debris, and close proximity to sensitive structures. While all of the drives will be addressed, the focus of the paper will be on the 931 m (3,055 ft) continuous drive connecting Outfall 46, currently the longest microtunnel driven in North America.

INTRODUCTION

In 1994, the City's Bureau of Environmental Services (BES) entered into an Amended Stipulation and Final Order (ASFO) for combined sewer overflow (CSO) abatement with the Oregon Department of Environmental Quality (DEQ). The agreement required the City to control 55 combined sewer outfalls by December 1, 2011, with intervening major deadlines to complete specific parts of the work. The East Side CSO Tunnel Project, the last major component of the City's $1.4 billion program, is required to meet the regulatory milestone of December 1, 2011 for control of outfalls along the east side of the Willamette River. The project consists of a 9,144 m (30,000 ft) long, 6,700 mm (22 ft) finished inside diameter tunnel, ranging in depth from 30 to 50 m (100 to 160 ft), which will function as a storage and conveyance conduit for the captured flows. The tunnel alignment will intercept a series of gravity conduits and drop structures that connect to 13 existing combined sewer outfalls. The project also contains nine microtunnel drives, totaling approximately 2,380 m (7,808 ft) that will divert flows from several of the existing outfalls to the main tunnel to an existing pump station that transports 832 million liters per day (220 MGD) of CSO flow to the City's existing wastewater treatment plant. Each microtunnel drive consisted of 2,130 mm (84-inch) reinforced concrete jacking pipe, the longest of which was a 931 m (3,055 ft) drive that runs along the east side of the Willamette River.

GEOTECHNICAL CONDITIONS

A Geotechnical Baseline Report (GBR) developed for the project included data from a variety of sources: geologic maps and publications; historical maps of the Portland area; previous investigations in the project area including foundation investigations for bridge structures, buildings, and other facilities; and the specific subsurface

Table 1. Drive lengths and geological formations encountered

Drive by Outfall (OF) Number	Length of Drive m (ft)	Geological Formation(s) Encountered	Depth Below Groundwater (GW) Table mm (ft)
OF 28	275 (900)	100% Troutdale Formation (Tt)	300 to 3700 (1 to 12)
OF 36	35 (116)	100% Artificial Fill (Qaf)	600 to 900 (2 to 3)
OF 37-1	42 (138)	100% Sand/Silt Alluvium (Qal and Qff)	4600 to 5800 (15 to 19)
OF 37-2/38	220 (720)	80% Sand/Silt Alluvium (Qal and Qff), 15% Artificial Fill (Qaf), 5% Gravel Alluvium (Qfc)	900 to 4600 (3 to 15)
OF 40	335 (1,100)	90% Troutdale Formation (Tt), 5% Sand/Silt Alluvium (Qal and Qff), 5% Artificial Fill (Qaf)	0 to 1800 (0 to 6)
OF 41	88 (290)	100% Troutdale Formation (Tt)	Above the GW
OF 44A	427 (1,400)	50% Sand/Silt Alluvium (Qal and Qff), 50% Gravel Alluvium (Qfc)	0 to 6100 (0 to 20)
OF 46	931 (3,055)	75% Sand/Silt Alluvium (Qal and Qff), 25% Artificial Fill (Qaf)	300 to 6400 (1 to 21)
OF 46-2	27 (89)	100% Sand/Silt Alluvium (Qal and Qff)	6700 (22)

exploration program for the project. Geotechnical reports and construction records were also available from the recently completed West Side CSO Tunnel Project. All of the microtunnel drives, except for one, were below the groundwater table. The four geological units encountered on this project are described below. Table 1 and Figure 1.

Artificial Fill (Qaf) mostly consists of gravel, sand, sandy silt, and silt with organic debris. However, in this case, it also contained building debris, abandoned steel rails and timber railroad ties, concrete, logs, and wood waste including sawdust, branches, wood chips and fragments. Fill composition varied within the project area depending on previous site use.

Sand/Silt Alluvium (Qal and Qff) predominately comprises interbedded sandy silt and silty fine sand deposited as recent alluvium (Qal) and late Pleistocene fine-grained catastrophic flood deposits (Qff). Qal and Qff are typically non-plastic to low plasticity, with occasional zones of moderate to high plasticity elastic silt found. Gravel lenses are also found in this unit, but are not typical. The alluvium is typically stratified with alternating layers of fine sand, sandy silt, silt, and clayey silt. The consistency of the alluvium is often described as soft to medium stiff for fine-grained layers or loose to medium dense for coarse-grained layers. Organic material commonly consisting of organic silts, wood fragments, logs, and wood debris were found.

Gravel Alluvium (Qfc) is typically described as dense to very dense poorly graded gravel, cobbles, and scattered boulders in a sand matrix. Scattered lenses of silt and sand are interbedded with the coarser material, although silt and fine sand from the overlying Sand/Silt Alluvium often form the matrix of the upper portion of the Gravel Alluvium deposits. The sandy interbeds are described as medium dense to dense. The gravel, cobbles, and boulders consist predominantly of basalt with lesser amounts of andesite and quartzite. Gravel and cobble-sized particles are generally rounded.

The Troutdale Formation (Tt) generally consists of very dense poorly graded gravel in a matrix ranging from sand to clayey silt with cobbles and scattered boulders. The clayey or sandy silt matrix material sometimes provides weak cementation within the formation; however, the unit is best described as gravel rather than weak rock.

Figure 1. Geological formations encountered

DESIGN CONSIDERATIONS

A slurry pressure microtunnel boring machine (MTBM) was required to minimize the possibility of soil running or flowing uncontrollably into the machine face. Slurry microtunnelling minimized the groundwater drawdown and provided positive face support, thereby minimizing the effects of ground movement on nearby structures and utilities. The MTBM was specified to be capable of supporting the soil exposed in portions of the face while cutting harder layers and concretions in other parts. The selected MTBM was equipped with a stone crusher and disc cutters to break up boulders with compressive strengths up to 380 MPa (55,000 psi). A compressed air chamber was included which allowed for hyperbaric interventions for removal of obstructions, and to change cutterhead tools if required during tunneling.

The microtunnels used a one pass pipe support system comprised of Class IV 2,130 mm (84-inch) reinforced concrete jacking pipe with rubber gaskets that were designed for this project. All of the microtunneled outfall connections were designed using 2,130 mm (84-inch) pipe so that the same MTBM machine could be utilized. The use of a uniform pipe size was determined to be the most cost effective option to control the City's ASFO design storm rather than varying the size of each microtunneled connection.

The OF46 drive was from 9 to 15 m (30 to 50 ft) below the ground surface with up to approximately 6,400 mm (21 feet) of hydraulic head above the planned pipe crown. Several options were considered in designing the connection of Outfall 46 to the main tunnel. These included:

- Constructing a deep tunnel connection shaft near the outfall
- Microtunneling the connection pipeline over 914 m (3,000 ft) from OF46 to Port Center Shaft.
- Open cutting the connection pipeline over 914 m (3,000 ft) from OF46 to the Port Center Shaft.

The Port Center shaft is a 15 m (49 ft) diameter shaft being constructed and is located at the north end of the East Side CSO tunnel. At the bottom of this shaft a connection will be made to the completed West Side CSO tunnel's Confluent shaft discharging to the Swan Island pump station. Due to property restrictions, cost and constructability, microtunnelling was chosen as the best option. The outfall connection was determined to be best-located at the end of the outfall near the Willamette River. Locating this further east away from the river was not feasible due to nearby structures (cement silos and their concrete access roads exist directly east), property restrictions (Union Pacific Railroads main rail yard is to the east), and business activities. The alignment was then determined based on the location of the connection from this structure to the jacking shaft on the Port Center site, avoiding the railroad property to the east.

The original design was to drive two microtunnels of 590 m (1,934 ft) and 302 m (991 ft) respectively, with an intermediate shaft. During construction the contractor proposed eliminating the intermediate shaft. It should be noted that the construction contract is a cost-plus-fixed-fee contract, where labor, equipment, and materials are paid by the owner as reimbursable costs. The City accepted the higher calculated risk of the long drive based on the potential cost savings of eliminating the intermediate shaft, and on the City's previous experience with the microtunnelling on the West Side CSO Project in similar ground conditions with a similar machine. The advantage to the contractor was a savings to the construction schedule and the elimination of the shaft construction in a difficult location with limited space. Design modifications were also made to the MTBM system to support the 931 m (3,055 ft) long alignment. These modifications included changes to the slurry system, intermediate jacking stations (IJS) controls, guidance system, and additional electrical equipment.

EQUIPMENT

An AVND 2000AB machine built by Herrenknecht Tunnelling Systems was purchased for the project. The 2,655 mm (104.5-inch) diameter cutting wheel rotates in both directions. This offers the possibility to correct or compensate for roll of the machine that can occur as reaction to the torque of the cutting wheel. Situated in the shield, directly behind the cutting wheel, the excavation chamber processes all excavated material, which gets crushed to a grain-size suitable for transportation by the closed-loop slurry circuit. The excavated and crushed material is transported in a slurry suspension. A feed pump pushes this suspension through a feed line into the excavation chamber, where it mixes with the excavated material. Afterwards this mixture is pumped via slurry pumps to the surface and into a separation plant. The separation plant separates the solids from the fluid, which is then returned to the MTBM for reuse. Finally, adding bentonite can vary the density and viscosity of the slurry suspension to adapt the slurry to the changing geological conditions.

The MTBM's AVND-mode enables face stabilization and effective control of settlement in sensitive ground with high water tables. A pressure wall separates the shield pressure chamber from the excavation chamber, which is completely filled with pressurized slurry. The pressure corresponds to the earth and groundwater pressure and thereby prevents uncontrolled penetration and loss of stability at the tunnel face. For control reasons, a submerged wall divides the extraction chamber behind the cutting wheel. While bentonite suspension and some extracted material completely fills the excavation chamber, it only partially fills the pressure chamber behind the submerged wall, which also contains an air cushion situated in the top area of the chamber. A compressed air control unit regulates the cushion in order to adjust the exact supporting pressure.

To separate the muck from the return slurry, A Type MAB 400 slurry separation plant was procured from Schauenburg-MAB, Inc. The plant was equipped with a horizontal classifying-dewatering screen, and a double stage hydro-cyclone system. Figure 3.

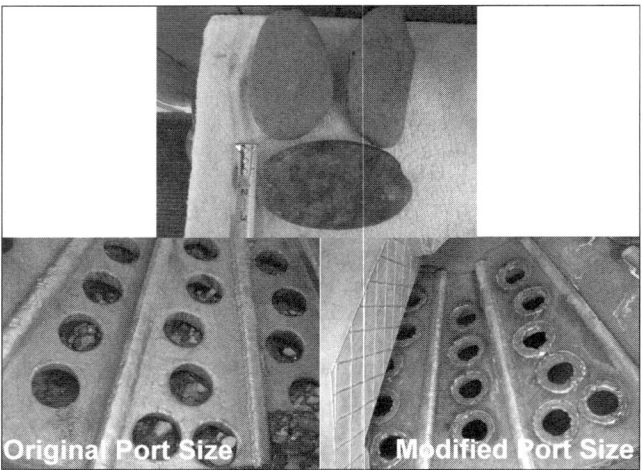

Figure 2

Figure 3. MTBM slurry separation plant

The first drive was excavated entirely through the Troutdale Formation. Fish shaped cobbles with an approximate length to width ratio of two to one caused numerous slurry line plugs. Two changes were made after the completion of the first drive to help eliminate this problem. First, longer elbow sweeps were installed in the slurry circuit. Second, the ports located in the rock crushing area of the machine were slightly reduced to allow the cobbles to be broken down into smaller pieces. These modifications resolved the plugging issue and the Schauenburg plant was well suited for separation of large grained materials and cobbles found in the gravel alluviums and troutdale formation. Figure 2.

All of the remaining drives except one were primarily excavated through sand/silt and artificial fills. The silts and clays were clogging the shaker screen on the plant so modifications to the separation plant were required. Brandt, Inc. was contracted to design a "bolt on" solution to the existing Schauenburg separation plant. A decanting

Figure 4. Gumbo box

Figure 5. Centrifuge with discharge conveyor

centrifuge and a gumbo box were added to the plant to assist in separating the fine-grained material while still utilizing the screens and cyclones on the existing separation plant. The return line from the tunnel would first dump into the gumbo box which was mounted to the top of the plant. Figure 4. The gumbo box's chain conveyor would separate out the "balled up" silt and clay materials and dump this material directly into the muck bin. The remaining wetter material would gravity feed back to the separation plant shaker screen and cyclones to be processed. The fine-grained material that was not removed would pass into the settlement tank, and then would be pumped to the centrifuge. The centrifuge would then separate out the remaining material and return clean water to the mining water tanks. The material coming out of the centrifuge would discharge directly into the muck bin via a short conveyor. Figure 5.

Figure 6. Secant pile and large diameter slurry jacking shafts

CONSTRUCTION

Jacking and receiving shafts for the project were constructed using secant piles, solider piles with wood lagging, and circular ribs with shotcrete and wood lagging. Some of the drives were also driven from or jacked into the larger diameter main tunnel shafts where the tunnel dropshafts are to be located. Figure 6.

Break-in/break-out ground support was used to mitigate ground loss, and allow an even bearing surface across the face for the cutterhead. Where slurry wall shaft construction was utilized, fiberglass rebar was placed in the shaft wall and additional controlled density fill (CDF) panels were constructed outside the shaft to form a CDF block. The block had a minimum length of 3m (10 ft) while the other side of the block varied due to the curvature of the shaft. The block extended 1.5 m (5 ft) five feet below the invert of the pipeline and was filled with CDF to surface. For shafts constructed using secant piles or soldier piles, either jet grouted columns or an additional row of secant piles filled with CDF were used to provide additional ground support outside the shaft wall.

On some of the shafts a "square up" concrete block was poured inside the shaft to provide uniform bearing surface for the cutterhead and attachment of an entrance or exit seal. A steel bulkhead containing three gaskets for containing the groundwater and slurry in the annular space was integrated into the "square up" block pour. Most of the break-ins to shafts above the water table were completed with no outside ground support and no inside support. Break-ins to shafts below the water table were completed with the above stated ground support outside the shaft and additional break-in support was provided inside the shaft by a steel top hat fabricated with grout ports (Figure 7). The top hat was filled with CDF and the cutterhead was driven into the top hat. The annular space around the machine was then be grouted with polyurethane grout to limit the inflow of water. Once the flow of water was reduced to manageable amounts, the top hat was removed and the machine was pushed into the shaft and removed.

One of the receiving shafts was located underneath an elevated section of the Interstate 5 Freeway. Given the low overhead clearance, a hydraulic boom gantry, and large capacity air chain hoists were used to remove the MTBM. Figure 8.

For the 931 m (3,055 ft) OF 46 drive, the jacking shaft constructed was a 7.3 m (24 ft) diameter, 16.5 m (54 ft) deep secant pile shaft comprised of 34 interlocking 1 m (39 inch) diameter secant piles. The secant piles were reinforced with steel rebar except where the break-in or break-out of the machine was to occur. The bottom of the shaft was sealed using overlapping jet grout columns. Jet grout columns were also

Figure 7. Break-in top hat

Figure 8. Hydraulic boom gantries hoisting the MTBM

used to construct a 3 m (10 ft) thick by 4.9 m (16 ft) high break-in and break-out block at this shaft. In addition, jet grout columns were also installed to construct a 1.5 m (5 ft) thick by 8.5 m (28 ft) tall jacking reaction block.

Due to space limitations at the bottom of the shaft, a platform was erected 3 m (10 ft) above the shaft bottom for the slurry pump. An 1,100-ton jacking frame was selected to handle the 3 m (10 ft) pipe sections that were used for the microtunnel operations.

The OF46 receiving shaft is located where the outfall is diverted into the microtunnel. The walls of the structure were constructed first and the MTBM was removed from within a 4,300 mm (14 ft) wide by 5,500 mm (18 ft) long chamber in the structure. A 3,400 mm by 3,400 mm (11 by 11 ft) opening was left in the diversion structure wall to allow for the break-in.

There were many logistical challenges that had to be addressed prior to the start of the drive. One of these challenges included electrical service for the MTBM and slurry pumps in the tunnel. Due to the length of the drive, two options were available. The first option was to enlarge the electrical cables feeding the machine and add extra cables. The second option was to run high voltage into the tunnel to a transformer in the pipe and transform the high voltage into the service needed for the machine. The project team evaluated both options and decided to install the second option.

Figure 9. Alignment of the 931 m (3,055 ft) OF 46 drive

Special electrical modifications had to be made to the control container to accommodate this electrical system. A special transformer and skid were manufactured for the transformer and two special variable frequency drives (VFDs) and skids were manufactured by Herrenknecht specifically for this drive. Once the machine was advanced 46 m (150 ft) with the standard electrical system, it was switched over to the high voltage system in the pipe. A 13.2 kVA high voltage line was run down the jacking shaft to a special disconnect box at the bottom of the shaft. At the disconnect box, 91 m (300 ft) long high voltage cables were connected and disconnected for every pipe push. This system proved to be very effective and through training of every employee and development of special high voltage handling procedures, no electrical incidents were encountered during the connection and disconnection of pipe approximately 305 times (305 pipes) during the drive.

Another critical planning aspect of the drive included cutter head tool and head hardfacing. Through experiences on previous drives, the decision was made to hardface the entire head and commence the drive with a combination of carbide and hardfaced tools on the cutter head. This proved to be the right decision as an inspection of the cutter head under atmospheric conditions approximately half way through the drive found minimal tool wear. Once the machine holed through after driving 931 m (3,055 ft) through difficult ground conditions, the cutterhead wear was still minimal.

Tunnelling began on February 14, 2008 and was completed on April 19th, 2008. At the start of the drive the high silt content in the ground required some modifications to the separation plant equipment and handling procedures. The material required lime to be added to the wet material coming out of the separation plant to stiffen it up so it could be loaded into trucks. One concern with the very soft material was the machine grade (sinking). Keeping the machine on grade turned out not to be an issue. Controlling the line was more difficult while pushing the pipe string from the jacking shaft. This problem was overcome through more frequent survey checks. Due to the length of the OF 46 drive, the standard laser guidance system would not work for machine guidance. A north seeking gyro to control line and water level system to control grade was employed to control machine line and grade for the long drive. The standard laser guidance system was used for the first 500' then the rented gyro / water level system was installed for the remainder of the drive. With the gyro system and due to the length of the drive, survey checks were required more often towards the end of the long drive. These survey checks proved well worth the time. The machine holed through well within the specified tolerances for line and grade at the reception shaft.

A total of seven IJS's with a 1100 ton jacking capacity each were used. The first IJS was installed 43 m (140 ft) behind the machine, with all subsequent IJS's installed approximately every 128 m (420 ft). The jacking forces slowly increased to approximately 900 to 1,000 tons towards the end of the drive. On most pushes only one or two IJS's were needed. At times, advancing as much as 550 m (1,800 ft) of pipe string with as little as 500 tons of jacking forces using only the shaft jacks was achieved.

The GBR identified the potential for the tunnel to intersect an abandoned dry dock. Contingency plans were made in the event the MTBM encountered steel sheet piles or any other buried construction materials from the dry dock. Actual conditions were monitored more closely as the MTBM approached where the old dry dock was suspected to be located however, it was never encountered. An old railroad trestle was also identified in the geotechnical review. The MTBM did in fact encounter the old railroad trestle towards the end of the drive. Old debris consisting of wood piles, steel spikes, and other debris was encountered. The machine advance rate was reduced to as low as 25 mm per minute in order to process this debris but never caused a full stoppage of the machine advance. Through earlier drives in wood pile debris, it was determined that if the advance rate of the machine was reduced, the machine was better able to process the wood material which helped to avoid plugging the head and the slurry lines with the wood debris. Because of this earlier lesson learned, we reduced the rate of machine advance through the wood piles resulting in minimal slurry line plugs and no downtime due to plugging of the head. Production rates were also reduced when wood piles, miscellaneous pieces of metal including large spikes, nails, and bolts caused the slurry lines to plug. One small sinkhole directly over the centerline line of the tunnel occurred in a location where a large volume of this wood and metal was encountered. The sinkhole was backfilled with CDF and did not cause any damage to the surrounding structures. The daily production averaged 16.5 m (54 ft) per day with the best day being 46 m (152 ft). No hyperbaric interventions were required and no cutterhead tools were replaced. This 931 m (3,055 ft) drive, like most rewarding achievements, was completed with a lot of hard work and long hours by a team of laborers, operators, engineers, and managers. Most importantly, it was completed safely, installing 306 pieces of undamaged pipe with no injuries.

The US $426 million East Side CSO Tunnel Project, designed by Parsons Brinckerhoff and is being constructed by the joint venture of Kiewit-Bilfinger Berger, is on schedule to reach completion in 2011. Parsons Brinckerhoff assists the owner, Portland's Bureau of Environmental Services, with design services for the project. Jacobs Associates assists the owner with construction management services, including contract administration and resident engineering. A big part of the ongoing success for the microtunneling work on this project is attributed to the participation of the crew in a formal lessons-learned session after each drive. Some of the major lessons learned were as follows:

Figure 10. MTBM after holing through on the OF 46 drive

- Reduced the size of the MTBM intake ports to better the cobbles.
- Installed longer elbow sweeps in the slurry circuit to prevent plugging of the slurry lines.
- Modifiied the entrance seal construction process by pouring the entrance seal in the "square up" block.
- Modified the separation plant by adding a gumbo box and decanting centifuge to better handle the fine-grained materials.
- Reduced the advance rate while excavating through wood and other fill debris
- Fabricated a small circular jacking frame ("push pup") to advance the IJS lead pipe further ahead to give provide more room to accommodate the IJS can and the special trailing pipe.
- The crews also implemented a lot of safety improvements to reduce the risk for slips, falls, pinch points, and electrical hazards at the sites.

REFERENCE

Geotechnical Baseline Report, East Side CSO Tunnel Project, Parsons Brinckerhoff in association with CH2MHill and Tetra Tech/KCM, February 10, 2006.

MICROTUNNELS VS. EPB RISK-BASED SELECTION

Michelle L. Ramos ▪ Staheli Trenchless Consultants

Kimberlie Staheli ▪ Staheli Trenchless Consultants

INTRODUCTION

The Ballard Siphon was built in 1935 and consists of two 91.4 cm (36-inch) diameter wood-stave pipes. These pipes convey sewer flows under the Ship Canal from northwest Seattle and the Ballard Regulator to the North Interceptor and the West Point Treatment plant. The siphons were constructed using open cut construction, are now buried in the sediment on the floor of the canal, and have severely limited ability to be dewatered, inspected, or cleaned. The existing siphon is 394.7 m (1,295 feet) long from the junction structure on the southern shoreline to the regulator structure on the north. The location of the Ballard Siphon is shown in Figures 1 and 2.

The area around the Ballard Siphon has become more populated and developed since the siphon was built. Additionally, the Ballard Siphon is subject to high flows that are greater than the capacity of the pipelines. This has resulted in predictable Combined Sewer Overflows (CSO's) whenever a large, long duration rain event occurs. The aged pipeline of uncertain viability has become a critical piece of the sewer system infrastructure.

Because of the age of the pipelines and their importance, a sonar survey was completed on both barrels of the Ballard Siphon in December of 2005. The sonar images appeared to indicate crown intrusion of the wood-stave pipes, which suggested that the pipes were in imminent danger of collapse. As a result, an Emergency Declaration was executed by King County and the Ballard Siphon Replacement Project was initiated.

Staheli Trenchless Consultants was retained to perform feasibility studies to determine the viability of using microtunneling and horizontal directional drilling (HDD) for replacement of the Ballard Siphon. For the evaluations, the preferred site location was determined to be at or near the site of the existing Ballard Siphon in order to minimize the ancillary sewer work which would accompany moving to a new site. To meet flow demands and hydraulic requirements, the replacement siphon needed a minimum equivalent diameter of 182.9 cm (72 inches). This could be accomplished with a number of different siphon diameters and configurations including, but not limited to, three 106.7 cm (42-inch) diameter pipelines, two 152.4 cm (60-inch) pipelines, or two 91.44 cm (36-inch) pipelines and a 121.9 cm (48-inch) pipeline.

HDD FEASIBILITY

Although the preferred product for the new pipelines was HDPE pipe, at the time of the evaluation, 106.7 cm (42-inch) diameter HDPE had not been installed using HDD at the lengths required to cross the Ship Canal, about 609.6 m (2,000 feet).

If steel casing pipe was installed with HDD methods with a product pipe inserted into the casing, the set-back requirements for sufficient bore geometry would have resulted in a 609.6 m (2,000 foot) pipeline that would preclude use of the existing regulator facilities. Additionally, concerns were raised regarding drilling triple 106.7 cm

Figure 1. Location map of the Ballard Siphon project

Figure 2. Ballard Siphon detail

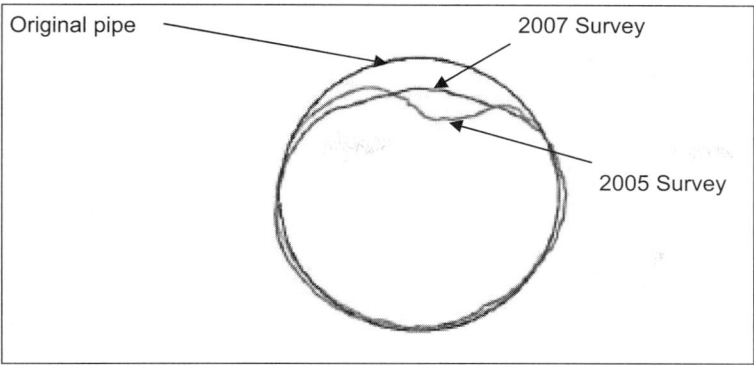

Figure 3. Sonar survey inspection results

(42-inch) diameter pipes in close proximity to existing critical utilities and structures that line both sides of the Ship Canal.

HDD posed a large risk of near-surface hydrofracture, settlement of historic structures, and binding during pull back since sufficient layout area was not available to preassemble the pipeline. HDD was therefore not recommended for replacement of the existing siphons (Staheli and Ramos August 2006).

CHANGE IN PROJECT STATUS

During HDD consideration, both barrels of the siphon were cleaned and another sonar inspection was completed. Data from the subsequent inspection were in conflict with the 2005 inspection data, and did not indicate crown intrusion (Figure 3). Analysis of the sonar images could not resolve the discrepancies between the two inspections. One of the siphon barrels was partially dewatered, allowing a CCTV inspection. The visual inspection revealed that the crown intrusion shown on the 2005 sonar survey was actually grease deposits on the crown of the pipe.

This new information allowed the emergency status of the project to be lifted. Since King County was still interested in increasing the capacity of the siphons to accommodate a 20-year storm event and considering the age of the siphon, the inability to accurately asses the structural stability of the wood-stave pipes, and the benefits gained by installing additional capacity, it was determined that the project would continue to move forward. The focus of the project then changed to include both a replacement of current flow capacity and a reduction in CSO's.

REVISED PIPELINE CONFIGURATIONS

King County wanted to develop a pipeline configuration that was adequate to carry daily flows but also have the capacity to handle large storm events. After extensive hydraulic analysis, the desired equivalent diameter was set at 213.4 cm (84 inches). To accommodate both flow regimes, it was decided that new dual pipelines would be installed beneath the Ship Canal. For the new siphons, one of the siphon barrels had to be maintained at a diameter relatively close to the existing 91.44 cm (36-inch) size to maintain adequate flow velocities across the Ship Canal during normal daily flow events. However, the second barrel would have to be significantly larger to accommodate a 20-year storm event.

Table 1. Equivalent inner diameter [cm (inches)] nominal sizes

Desired Equivalent	Low Flow Barrel	Barrel 2	Barrel 3	Barrel 4	Equivalent Diameter
213.4 (84)	76.2 (30)	167.6 (66)	167.6 (66)	0	222.2 (87.5)
	76.2 (30)	152.4 (60)	167.6 (66)	0	213.1 (83.9)
	76.2 (30)	137.2 (54)	182.9 (72)	0	216.2 (85.1)
	76.2 (30)	137.2 (54)	137.2 (54)	137.2 (54)	212.3 (83.6)
	91.44 (36)	167.6 (66)	167.6 (66)	0	225 (88.6)
	91.44 (36)	152.4 (60)	167.6 (66)	0	216.2 (85.1)
	91.44 (36)	137.2 (54)	182.9 (72)	0	219.2 (86.3)
	91.44 (36)	137.2 (54)	137.2 (54)	137.2 (54)	215.4 (84.8)

King County developed a myriad of pipeline diameter combinations, including two, three, and four barrel alternatives that would meet their requirements for a desired equivalent of 213.4 cm (84 inches) (Crawford 2006). Table 1 presents some of the pipeline dimensions considered for the new siphons.

All of the configurations above assumed that a single siphon would be used for low-flow and that additional siphons would be used as flows increased. With any of the pipeline configurations listed above, it was necessary to have either multiple crossings of the Ship Canal or to construct a large casing that would accommodate more than one pipe.

Since the 2007 sonar survey indicated that the existing siphons were in adequate condition, Staheli Trenchless Consultants suggested slip-lining the existing siphons with HDPE pipe. This would result in using both slip-lined siphons to accommodate low-flow, and constructing only one additional crossing that would serve as the overflow siphon. Staheli Trenchless was tasked to perform a feasibility study for slip-lining the existing siphons to determine if this potentially cost-saving alternative was viable.

SLIP-LINING FEASIBILITY

The general factors that affect slip-lining feasibility are the number of bends and deflection angles of the host pipeline, the outer diameter of the slip-lined pipe compared to the inner diameter of the host pipe, the potential for blocking debris in the pipeline, and the available area to construct access pits. All of these factors were evaluated in detail along with cost and schedule impacts.

The preliminary analysis indicated that the only unknown factor was the potential for blocking debris in the wood-stave pipes. To raise confidence in the success of the method, it was decided to pig the existing siphons while carefully measuring the force required to move the pig through the pipelines.

Polyethylene pigs were used to clean the siphon. These pigs were sequentially upsized after each run from 60.96 cm (24-inch) to 66.04 cm (26-inch) to 74.93 cm (29.5-inch) diameters. Force readings were collected to evaluate the force necessary to move the pig through the pipeline. There was no significant difference in pull forces when the pig was increased from a 66.04 cm (26-inch) diameter to a 74.93 cm (29.5-inch) diameter, with maximum forces near 800 pounds. The results of the pigging clearly indicated that the host pipe did not contain any significant obstructions and that slip-lining the pipe with 76.2 cm (30-inch) OD HDPE pipe was feasible.

The slip-lining alternative was further evaluated and was found to provide a potential $3.1 million cost savings over constructing multiple crossings of the Ship Canal (Staheli and Ramos 2007). The project was subsequently moved forward based on

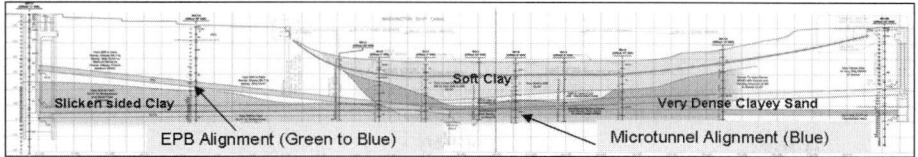

Figure 4. Generalized cross section

slip-lining the existing siphons with 67.31 cm (26.5-inch) ID (76.2 cm (30-inch) OD) pipelines and constructing a single pipeline beneath the channel using tunneling methods.

TUNNELING METHODS AND FEASIBILITY

Site conditions were analyzed to determine the feasibility of using various tunneling methods to cross the Ship Canal. The primary factor affecting feasibility was the geotechnical conditions at the preferred crossing location.

A geotechnical investigation was completed along the preferred alignment. The soils adjacent to and beneath the Ship Canal were identified, tested, and categorized by geologic unit. In agreement with the previously conducted historical geotechnical review, soil conditions along the alignment consisted of fill/recent alluvium over glacially consolidated silts and clays. The fill/recent alluvium consisted of loose to medium dense silty sand. The glacially consolidated silts and clays were very stiff to very dense clayey silt and silty clay. Very loose/soft soils (with blow counts of less than 3 blows per foot) were identified within the canal in the uppermost 3.048 to 9.144 m (10 to 30 feet). On the south side of the canal, slicken-sided clays were also identified. Groundwater was identified at approximately the elevation of the water in the Ship Canal. Figure 4 shows a geotechnical cross section of the alignment.

It was concluded that tunneling was a feasible crossing method due to the dense soils present along the alignment. However, the soft/loose soils on the canal floor present significant challenges for tunneling as they do not provide adequate bearing capacity to support the machine and will not allow steering. It was therefore necessary to choose a tunnel horizon that did not traverse these soils. In addition, it was desirable to select an alignment depth that avoided the slicken-sided clays to prevent excessive soil loading on the tunnel. Based on the soil and groundwater conditions and input from a Value Engineering panel, it was decided to further investigate microtunneling and Earth Pressure Balance (EPB) tunneling for the canal crossing.

TUNNEL METHOD SELECTION

A number of factors were considered when determining the preferred tunneling method. These factors included geotechnical conditions, alignment and shaft locations, pipeline elevation, site constraints, shaft construction methods, available pipe materials, jacking force considerations, liner design consideration, construction risks, estimated costs, and schedule. While many factors contributed to the final selection, the impact of perceived and actual risk was an overwhelming factor.

For the microtunneling alternative, a 472.4 m (1,550-foot) tunnel with 213.4 cm (84-inch) flow diameter pipe (one-pass tunneling) was considered. This represented the minimum straight-line distance across the canal with the set-back necessary to avoid structures near the edge of the canal. The shafts and tunnel were placed at depths sufficient to minimize geotechnical risks, resulting in shaft depths of approximately 30.5 m (100 feet) on each side of the canal.

For the EPB alternative, the preferred alignment was a 609.6 m (2,000-foot) tunnel with a minimum diameter of approximately 264.2 cm (104 inches). It was assumed that a bolted and gasketed concrete segmental liner would be used as the tunnel support in which a 213.4 cm (84-inch) flow diameter pipe would be placed upon tunnel completion. The alignment included a vertical curve that reduced the number of required easements as private property could be avoided. The EPB alternative contained a vertical curve that allowed shallower shafts than the microtunneling alternative while maintaining the vertical clearance beneath the very soft soils in the canal.

The minimum straight-line distance alignment considered for microtunneling required obtaining difficult easements on the south side of the canal due to the highly developed nature of the property and the multiple property owners. Lengthening the alignment to approximately 609.6 m (2,000 feet) allowed the south shaft to move away from the shore and the highly developed property. However, this resulted in an extremely long microtunnel. Although lengths in excess of 457.2 m (1,500 feet) are possible for microtunneling and have frequently been completed outside of the United States, the number of United States contractors with microtunneling experience on these drive lengths is very limited. Lengthening the tunnel to 609.6m (2,000 feet) to avoid difficult easements was not concerning for EPB tunneling methods as tunnels well in excess of this length are commonly completed.

There was significant concern over the risk of hitting an obstruction that would require retrieval of the microtunneling machine from beneath the canal. Although microtunneling machines of this diameter can have face access with an air-lock, few existing machines have this capability and the access is limited due to space restrictions. In addition, the vast majority of microtunneling contractors in the United States do not have experience with accessing the face through an air-lock system. This was of significant concern to King County due to the ever-present risk of encountering large boulders in glacial soils and their previous experiences with microtunneling which have resulted in a number of unplanned machine retrievals. Large boulders are much less of a concern for EPB machines due to the soil removal mechanism and the ability to access the face of the machine during tunneling.

RISK EVALUATION

The selection of the preferred tunneling alternative was driven by the analysis of risks. The first step in the risk evaluation was to determine risk events which would be considered for both tunneling methods. Over 50 risk events were identified which could impact the design, bid or construction process. These potential risk events varied from impacts to schedule, permit acquisition, right-of way easements, construction, material procurment, and slip-lining failure. In order to gain input from as many perspectives as possible, potential risks were suggested by all disciplines of the project team. This included input from the right-of-way acquisition team to the schedule team, from the hydraulics engineer to the maintenance lead, and from the designers to the construction manager.

It is important to note that with many of the risk events, there is potential for a "cascading effect" of impacts that cross traditional engineering disciplines and fields. To avoid this potential, the risk events were vetted within a multi-disciplinary brainstorming session.

Construction risks, such as encountering obstructions beneath the Ship Canal, encountering soft soils that cause grade loss, and settlement of the critical structures above the tunnel alignment were considered. Other risks included acquiring necessary easements; property acquisition; and permitting risks.

Once brain storming for potential risk events was complete, the multitude of possible risk events were reduced by focusing on those events which could cause changes

Table 2. Risk allocated cost matrix

	Microtunnel			EPBM		
	Probability	Impact	Risk Allocated Cost	Probability	Impact	Risk Allocated Cost
Soft material causes grade loss	15%	$8,700,000	$1,305,000	15%	$250,000	$37,500

to cost or schedule. Each remaining risk event was discussed and a probability ranking was assigned for that event. This probability ranking was qualitatively determined as Low, Medium, or High for the individual risk. Low probability items were given a likelihood of 15 percent, medium probability events were assigned a likelihood of 50 percent, and high probability events were assigned a likelihood of 85 percent.

Once the probability of an event was established, the cost impact of that event was evaluated. The cost impacts were quantitatively assessed within a multidisciplinary group setting. Each risk event was evaluated using similar criteria for both microtunneling and EPBM tunneling. This ensured that the same methodology was applied to compare the two construction types.

One of the identified risk events was encountering unexpected soft soil which may cause the tunnel to lose grade. For both construction techniques, the probability of encountering soft soil is the same. As shown in Table 2, the probability of encountering unexpected soft material was determined to be 15%.

The anticipated cost impact to deal with loss of grade due to soft soil is greater for microtunneling than it is for EPBM tunneling due to the required actions to correct the problem. With EPBM tunneling with a bolted segmental liner, creating curves is common practice. If the machine were to lose grade due to soft soils, a curve would be introduced once the machine entered competent ground. With microtunneling, curved alignments require specialized guidance equipment and there are limitations on the amount of curve that can be introduced by the jacking joint. Therefore, recovery of line and grade was considered unlikely. For microtunneling, the impacts included deepening the reception shaft or abandoning the tunnel and constructing a new microtunnel.

Once the probability of a risk event and the cost impact of that risk event are determined, the Risk Allocated Cost can be determined by multiplying the Probability and the Impact. In this manner all the risk events can be evaluated and the Risk Allocated Cost for each determined (Table 2). Based on this cost-based risk analysis, microtunneling had approximately two times the Risk Allocated Cost as EPBM tunneling.

The Risk Allocated Cost for microtunneling and EPBM tunneling were then added to the corresponding estimated construction cost for microtunneling and EPBM tunneling. The resulting risk-weighted construction cost was evaluated by the County and EPB tunneling was selected as the preferred alternative. The EPBM tunneling alternative has now been moved into final design and is slated for bid in the first quarter of 2009.

CLOSING

In summary, the Ballard Siphon Replacement Project has gone through a number of different permutations over the initial three years of the design process. After HDD was eliminated as a feasible replacement method, the County chose to slip-line the existing dual wood-stave siphons to provide adequate low flow capacity, and to tunnel one supplemental crossing to provide overflow capacity for a 20-year storm event. Of the potential trenchless methods investigated for the additional crossing, EPB tunneling was chosen as the preferred alternative as the County was unwilling to accept the risks associated with microtunneling. These risks included the unusually long drive length, the extremely deep shafts required, and the high potential for unplanned machine retrieval. This project is currently in the 100% design stage, and is expected to go out to bid in the first quarter of 2009.

REFERENCES

Crawford, B. and Swarner, B. (2006) Ballard Siphon Capacity and Sizing, Technical Memorandum, July 11, 2006.

Staheli, K. and Ramos, M. (2006) HDD Feasibility Study for Ballard Siphon, Technical Memorandum, August 21, 2006.

Staheli K, and Ramos, M. (2007) Sliplining Assessment for the Ballard Siphon, Technical Memorandum, February 7, 2007.

Staheli, K, and Ramos, M. (2007) Microtunneling Assessment for the Ballard Siphon, Technical Memorandum, April 27, 2007.

Staheli, K. and Ramos, M. (2007) Earth Pressure Balance Tunneling—30% Design Recommendations, Ballard Siphon Replacement Project, Technical Memorandum, January 7, 2008.

MICROTUNNELING CHALLENGES IN SOFT GROUND OF DOWNTOWN HARTFORD, CT

William Bergeson ▪ AECOM

Verya Nasri ▪ AECOM

James Sullivan ▪ AECOM

Alan Pelletier ▪ Hartford Metropolitan District Commission

ABSTRACT

This paper presents the challenges of the Homestead Avenue Interceptor Extension (HAIE) project in downtown Hartford, Connecticut. The new 72-inch (1,830 mm) sewer pipeline project consists of installing approximately 3,010 linear feet (920 m) of PVC lined reinforced concrete pipe using pressurized face microtunneling. The entire alignment is located within soft to very soft, varved silt and clay. To the extent possible, the alignment was selected to avoid bedrock, glacial till, miscellaneous fill, buried steel piles, and utilities. Challenges for this project include low ground cover, crossing major transportation routes, and limiting settlement to prevent damage to historical buildings and critical utilities. The machine was selected for the ground conditions. Design calculations were performed for selecting the reinforced concrete pipe class and for evaluating settlement. The potential for settlement damage was evaluated using both conventional analysis and finite element analysis.

INTRODUCTION

This paper presents the microtunneling challenges for the Homestead Avenue Interceptor Extension (HAIE) project in downtown Hartford, Connecticut. The Metropolitan District Commission is the project owner. AECOM provided final design and construction management services. Northeast Remsco Construction won the contract. The new pipeline will divert storm water from an existing combined sewer system and eliminate overflow discharges into the Connecticut River.

The microtunnel alignment runs about 3,010 lf (920 m) through a highly urbanized section of the city as shown in Figure 1. This new 72-inch (1,830 mm) reinforced concrete storm sewer pipeline crosses major transportation routes, encroaches upon critical utilities, and passes in close proximity to historically significant buildings. Four building footprints are located within 20 feet (6 m) of the alignment. Important transportation routes crossed by the alignment include interstate I-84 with multiple traffic lanes and an Amtrak rail corridor with 4 sets of tracks. Three locations along the alignment have been identified where multiple utilities are within close proximity to the alignment.

The sewer will be installed by microtunneling in saturated, soft to very-soft, varved silt and clay with about two tunnel diameters of ground cover for much of the alignment; however, there are limited reaches where the ground cover is limited to about one tunnel diameter. To the extent possible, the alignment was selected to avoid mixed ground conditions with much stiffer materials to include bedrock, glacial till, and miscellaneous fill. The groundwater table is typically about 7 to 10 feet (2.1 to 3 m) below ground surface.

The six drives are performed under existing city owned streets from three jacking pits and four receiving pits. The longest drive is about 1,100 feet (335 m). Since settlement is critical along the entire alignment, continuous pressurized face microtunneling is necessary for balancing the face pressure and minimizing ground loss.

Damage criteria were established for the historic buildings, the rail crossings, and the critical utilities encountered along the alignment. For the historic buildings, the potential for damage was evaluated based on limiting criteria for both angular distortion and horizontal strain as suggested by Boscardin and Cording (1989). For the rail crossings, allowable settlements were provided in Amtrak's MW-1000 documents that are based on chord length and the track travel-speed classification. For the critical utilities, the tensile strains computed in the ground around the utility were compared with the limiting tensile strains as published in the literature for the primary materials of the utility. Since interstate I-84 rested on deep foundations that were avoided as part of alignment considerations by the design team, settlement criteria was not developed for this crossing.

Design calculations were performed to select the class of reinforced concrete pipe and to evaluate settlement due to tunneling. The jacking pipe was designed for operating loads; while the temporary design loads from jacking were left for design by the contractor. An estimate of the time needed to dissipate excess pore pressures was also performed to determine how long it will take for consolidation settlements to take place to determine if construction induced settlements will occur within the liability period of the contractor. Good microtunnel installation practices are required as part of the contract documents to include closely balancing the face pressure, filling of overcut annulus with lubricant, and contact grouting of the pipeline immediately after installation.

GEOLOGY

The project site lies within the Connecticut Valley. The sedimentary rocks at the basement of this valley consist of conglomerates, feldspar-rich sandstone (arkose), and red and black shales. Glacial sediments consisting of till, floodplain and lacustrine (lake) deposits overlay the bedrock. Miscellaneous fill is the topmost layer.

The alignment is located entirely within the lacustrine deposit which is composed of varved silt and clay material that formed approximately 13,000 years ago as melting glaciers formed an ancient lake. During the warmer months, fine-grained material was flushed into the lake through the process of erosion. The larger silt sized particles settled quicker than the clay sized particles and were deposited in greater proportion at the lake bed. Later in the year when the lake surface was frozen and the erosion process had been cut off, the fines continued to settle but that remaining suspended over the winter now had a greater portion of clay sized particles. As this process occurred over many annual cycles of thawing and freezing at the lake surface, well-laminated varves were formed. This deposit is some 10 ft to 30 ft thick at the project site.

The varved silt and clay at the project site is bounded on the bottom by glacial till, which is characterized as very stiff to hard reddish brown sandy and gravelly silts and clays; the varved silt and clay is bounded on top by miscellaneous fill, which is characterized as highly variable with mixtures of sand, clay, gravel, and boulders, as well as some construction debris.

SUBSURFACE INVESTIGATION

To establish the subsurface conditions, available information was collected and reviewed from a nearby project; new borings were drilled during three distinct stages of design; in-situ and laboratory testing programs were implemented; and geophysical exploration methods were performed.

Figure 1. HAIE sewer alignment in downtown Hartford, Connecticut

Data was available from a nearby interstate highway (I-84) project that crossed paths with the new microtunnel alignment. The existing data contained drawings that showed the locations of piles for bridge abutments and piers.

In the preliminary design stage, 12 new exploratory boreholes were drilled. Glacial till was encountered in 6 of the 12 borings; and bedrock was not encountered in any of these borings. At the 30% design stage, 10 additional test borings were drilled. Falling head tests were conducted in 7 of the 10 borings. Monitoring wells were installed in 8 of the 10 borings. Glacial till was encountered in 7 of the 10 borings. Rock was encountered unexpectedly in 2 of the 10 borings. At the 60% design stage, the vertical alignment was modified to avoid the bedrock that was encountered; and 4 additional borings were drilled to determine the depth of glacial till and bedrock in locations where the geophysical testing methods indicated the possible presence of very stiff materials.

The borings of the final design stage were drilled with solid augers to the water table; afterwards, steel casings were driven to support the hole. Coring was performed when bedrock was encountered. Continuous split-spoon samples were collected to the water table as well as throughout the projected tunnel horizon. Undisturbed (Shelby tube) samples were also collected; all together, 170 split spoon samples and 16 undisturbed samples were performed.

Laboratory testing was performed on 37 samples to determine the moisture content, grain size distribution, Atterberg limits, oedometer, unconfined compression, unconsolidated undrained triaxial, consolidated undrained triaxial, electrical resistivity, chlorides, and sulfates.

In addition to the borings and laboratory testing program, geophysical methods were used to map the top of glacial till and bedrock along the alignment. The geophysical methods included seismic refraction, multi-channel analysis of surface waves (MASW) seismic, and low frequency ground penetrating radar (GPR). Typical profiles are shown in Figure 2 for MASW and Figure 3 for GPR. Seismic refraction was not useful because of interference between the pavement layer and the equipment.

MASW data were collected along two lines using a 48 channel exploration seismograph and 4.5-Hz OYO geophones on a land streamer array. The geophone array was attached to the seismograph unit via a seismic cable that relayed the motion-induced electrical signals from individual sensors. The electrical signals were recorded in the seismograph unit as 32-bit integer data.

GPR data were collected using a digital acquisition system, which displayed the data on a color monitor and recorded the data to a hard drive. GPR antennas with both 100-MHz and 40-MHz center frequencies were used for the survey. The very high

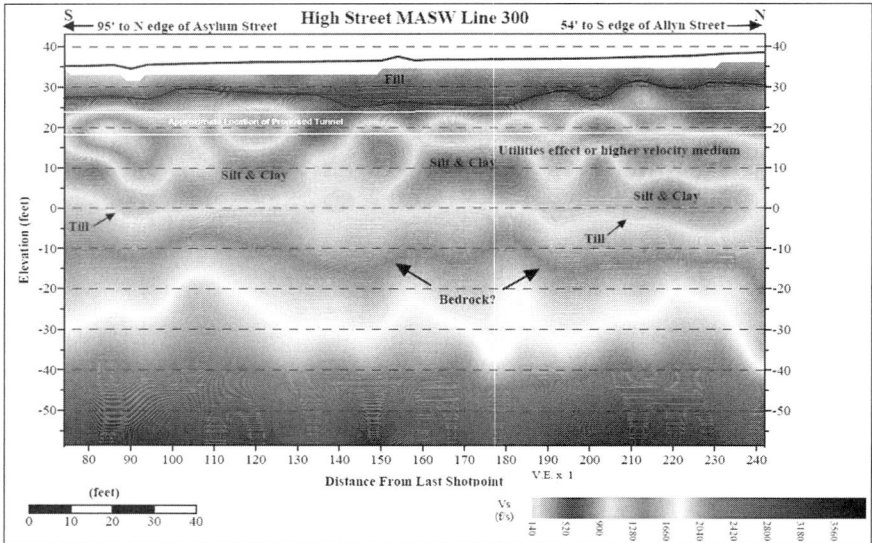

Figure 2. Profile with multi-channel analysis of surface waves for HAIE

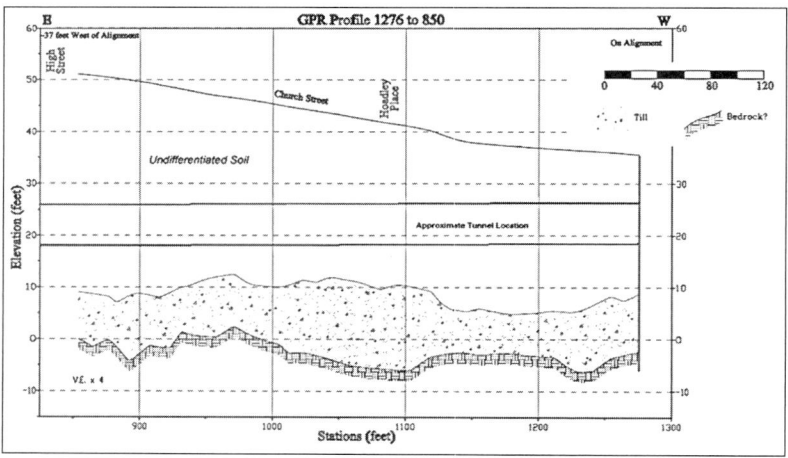

Figure 3. Profile with ground penetrating radar for HAIE

power bi-static 100-MHz antenna reached estimated depths of 60 to 70 feet (18.3 to 21.3 m), and the 40-MHz obtained penetration as high as 100 feet (30.5 m). The data collection time ranges were between 100 and 600 nanoseconds, Hager GeoScience, Inc. (2008).

ALIGNMENT SELECTION

The alignment was selected to minimize the length of the pipeline, provide adequate ground cover, avoid installation impediments/obstructions, minimize impact at crossings of major transportation routes, and position the pipeline beneath city owned streets/properties, Bergeson (2008).

The elevation difference between the two tie-in end connection points that helped to establish the vertical alignment of the sewer was minimal for a gravity flow system; and the length of the sewer between these points had to be minimized for flow considerations. The finalized invert slope was only about 0.0014 ft/ft (m/m), as such, the concrete pipe had to be installed with a PVC liner, which also served to protect the concrete and minimize leakage at the joints.

The ground cover varies along the alignment from 10 ft to about 20 ft. Where the ground cover is less than about two tunnel-diameters, the settlement effects will be more pronounced; and the machine operator will have more difficulty in balancing the face pressure. Temporary road closures may be necessary due to roadway instability, sink holes, ground heave, as well as inadvertent returns when using a slurry type machine.

Installation impediments include mixed face with stiff ground; while obstructions include buried building foundation slabs in the fill, steel piles that support the bridge piers and abutments of interstate I-84, and the existing utilities located beneath almost all of the streets. Entering mixed ground conditions with the MTBM is not advisable since the varved silt and clay deposit is much softer than the glacial till, bedrock, and large portions of the miscellaneous fill. Misalignment of the tunnel and over excavation will most certainly be the result if these conditions are encountered. Obviously, obstructions are avoided to the extent possible for successful installation of the tunnel.

The major transportation routes that cross the project alignment serve regions of the country; and damage to these routes would result in major consequences. When crossing under railways, design standards from AREMA are generally followed, while AASHTO design standards are generally followed for the interstate highway crossing. Both of these standards recommend that the utility cross their respective alignments at roughly perpendicular angles. This was accommodated by adjusting the shafts locations along the microtunnel alignment. For the interstate I-84 crossing, the roadway was elevated by bridge piers and abutments resting on steel H-piles driven to glacial till and/or bedrock; therefore, the microtunnel alignment was selected to cross at locations where piles have not been driven. For the Amtrak crossing, which is not supported by deep foundations, the ground cover had to be sufficient in order to limit settlement at the location of the tracks; therefore, the alignment was adjusted vertically to provide as much ground cover as possible whenever the at-grade crossing locations had less than two tunnel-diameters of ground cover.

Local ordinances restrict activities on important city streets; therefore, jacking pits and receiving pits were generally located on streets where the local ordinances allowed for some surface disruption. Since microtunneling drives are routinely performed upwards to 1,000 feet (300 m), the more restrictive streets were simply avoided.

PIPE DESIGN

The pipe was designed for embedment loads reflecting the long-term operating conditions of the pipeline. The design of the pipe for the jacking loads was left to the contractor since specifying a strength class of pipe beyond that required for embedment loading may add unnecessary costs to the project. Intermediate jacking stations, lubricants, and a host of other mechanisms are commonly used to increase the distance jacked; however, the jacking loads were analyzed for evaluating contractor submittals pertaining to use of pipe design for jacking loads as well as the thrust wall design of the jacking pit.

Embedment loads for the reinforced concrete pipe were calculated according to the procedures outlined by the American Concrete Pipe Association (ACPA) (2005).

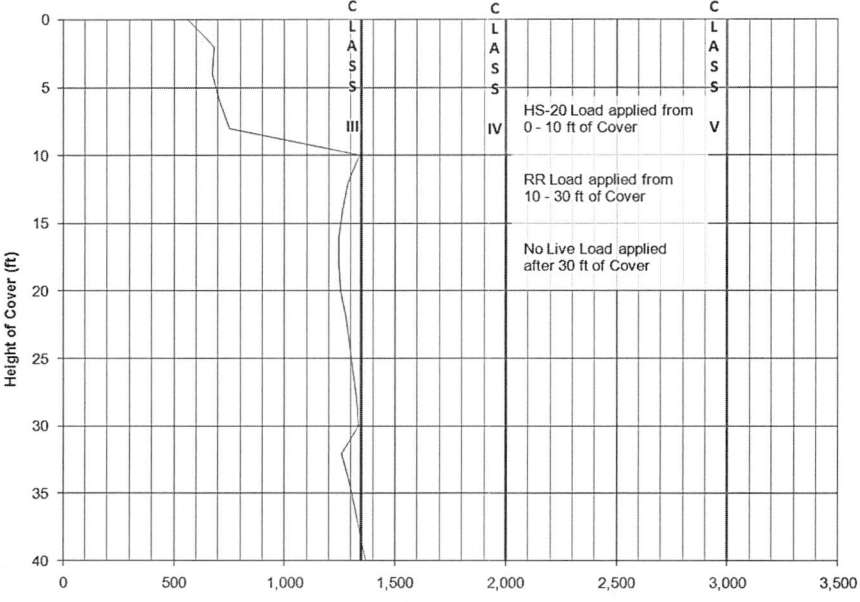

Figure 4. Embedment loads

Embedment loads (D_load) for a given diameter (D_0) of reinforced concrete pipe are based on earth loads (W_E), live loads (W_L), self-weight of pipe (W_P), and weight of the sludge (W_F) within the pipe under operating conditions. The manual also provides appropriate values for the factor of safety (FS), bedding factor (B_f), and live load bedding factor (B_{fLL}).

$$D\_load = \left(\frac{W_E + W_F + W_P}{B_f} + \frac{W_L}{B_{fLL}}\right) \cdot \frac{FS}{D_0} \text{ lbs per linear ft of pipe}$$

Both AREMA and AASHTO allow the designer to utilize ACPA design standards, which take into account axel loadings, to include impact, from Cooper E 80 for track crossings and from HS-20 truck axel loadings, to include impact, for highway crossings. Figure 4 shows that Class III reinforced concrete pipe was adequate for all the embedment design loading requirements, however, this is the lowest class of pipe that is allowed by the mentioned transportation authorities even if analysis indicates that a lower strength class would otherwise be satisfactory.

Axial loads are determined for estimating the pipe wall thickness, the need for intermediate jacking stations and lubrication requirements, types of jacking systems, and evaluating the thrust block design of the jacking shafts. Axial loads result from pressure at the tunnel face, steering loads, and skin resistance along the perimeter of the installed pipe. These loads can also be influenced by misalignment, dewatering, and work stoppages. Both the pressure at the face and the skin resistance were calculated according to the method outlined by the Pipe Jacking Association (1995). An estimate of the steering loads was based on the fact that only half of the steering jacks, at most, can be used to perform a steering function into the heading after overcoming the face pressure. As a result, the maximum steering load probably can not exceed

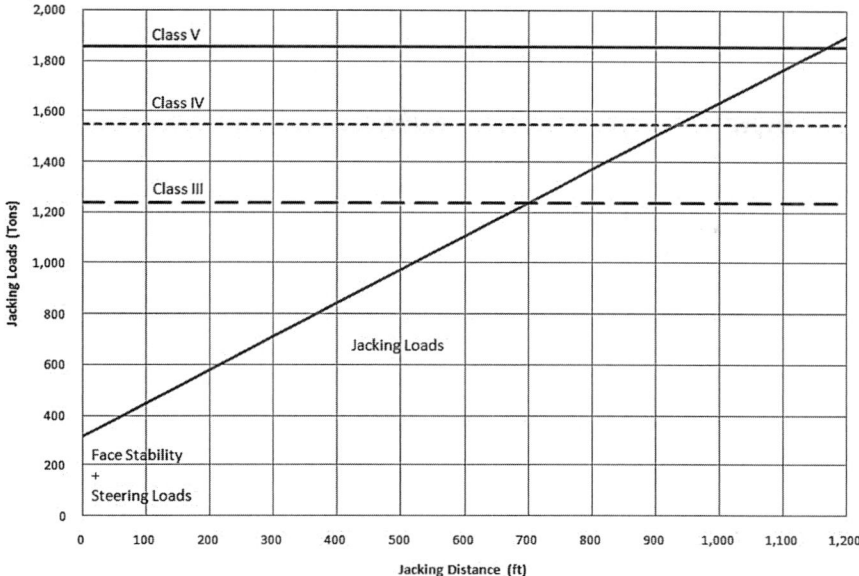

Figure 5. Jacking loads

about one-half the net capacity of the steering jacks after taking into account the face pressure. Misalignment, dewatering, and work stoppages were not addressed in the design calculations; however, the contract documents have been written to ensure that the contractor takes these items into consideration as part of the pipe design submittal requirements for axial loading.

$$P = \frac{1}{2} \cdot \frac{0.85 \cdot \phi \cdot f_c'}{LF_j} \cdot A_p$$

The axial capacity of the pipe was calculated using the method outlined by ACPA. Figure 5 shows that standard reinforced concrete jacking pipe is adequate for jacking distances up to 700 ft (215 m) with Class III, 925 ft (280 m) with Class IV, and 1,150 ft (350 m) with Class V for the ground conditions at the site; however, additional pipe reinforcement can be added to increase the jacking distances.

The contact area between joint packing and concrete surface with no joint separation (A_p), design compressive strength of concrete (f_c'), strength reduction factor for compressive strength (ϕ), and load factor due to jacking thrust eccentric load (LF_j) are used to compute the allowable jacking loads (P) for the reinforced concrete pipe. Concrete pipes can be fitted with steel sleeves to confine the concrete at the joints and increase the axial capacity.

MACHINE SELECTION

The Japanese Society of Civil Engineers (1996) indicates that both EPBM (with and without soil conditioning) and slurry machines are adequate for silt and clay type soils with SPT N = (0 – 2). These machines are also fairly well suited for dealing with low ground cover and the potential for mixed ground conditions while maintaining line and grade, controlling settlement, avoiding obstructions and steering in the vicinity of

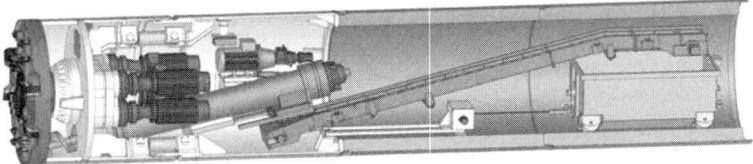

Figure 6. EPB MTBM—AE Herrenknecht

critical utilities. These machine types, which differ principally in their method of stabilizing the excavated face, each have advantages and disadvantages; however, it is important not to limit one particular type of machine without sufficient justification.

The machine used on this project has to be nearly neutrally buoyant to limit any sinking within the very soft ground since an extensive ground treatment program was not being implemented; however, the balance point of any tunnel machine will be forward of center because of the heavy weight of the cutting head and the main drive. As a result, a longer stable base has to be formed to increase the bearing area, which will allow for better control of the machine. This can be accomplished by connecting the machine, machine cans, and following pipes. Secondary steering articulation at the machine aft is also available to more effectively steer the machine in the soft ground conditions.

EPB machines are simpler to learn, maintain, and operate. Production rates are higher; capital costs and consumption of additives tend to be lower. Launch sites and shafts can be reduced in size. Face collapse generally results in less ground volume loss; and the face is stabilized using a controlled removal process that takes advantage of the self-supporting nature of the ground. Controlling the face pressure requires careful synchronization between the excavation and advance rates; and results in less uniform face support. After conditioning agents are mixed into the face, which produces a low permeability, viscous muck, ground that is conditioned to a state resembling near normally consolidated clay provides ideal tunneling conditions for EPB machines. Figure 6 shows an EPB type machine.

Slurry machines are considered more versatile and can be used within a wider variety of grounds. These machines perform well in ground with hydraulic conductivity values ranging from 1E–6 to 1 cm/s, as well as under variable groundwater pressures. The excavated face is sealed by a mud cake, which allows the slurry to be pressurized. Using mixshield principles that incorporate a cushion of air behind the face bulkhead, the face can be balanced within carefully controlled limits. In areas of low ground cover, very good operating techniques will be necessary in order to prevent inadvertent returns of the slurry to the surface, especially in areas where the more permeable miscellaneous fill resides. A fairly extensive separation plant with centrifuge and vertical clarifier will be needed for removing the suspended fine material from the slurry. The generated muck volume may also be greater than with the EPB because of the loose wet nature of the muck. For economics, the mucking and separating systems has to keep pace with the excavation process. The slurry machine may not advance as rapidly within the fine grained material unless high pressure backward pointing water jets are installed to clear the crusher of blocked clay. These water jets can double the production rate through sticky clay deposits. Figure 7 shows a slurry type machine.

Northeast Remsco owns an AVND 1800AB Herrenknecht machine. Similar machines have performed well in projects with comparable ground conditions worldwide.

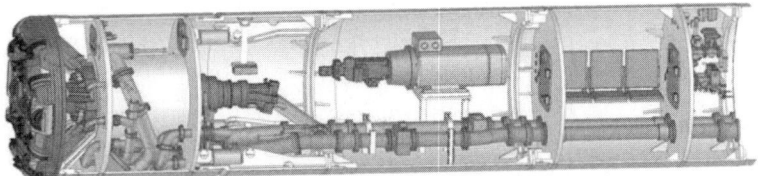

Figure 7.   Slurry MTBM—AVND Herrenknecht

MICROTUNNELING IN SOFT GROUND

Microtunneling within the soft ground will generate surface settlements which must be realistically determined in order to evaluate the impacts to historic buildings, major transportation routes, and critical utilities. Construction of a shallow tunnel produces a settlement trough resembling an inverse Gaussian curve.

The settlements within the settlement trough, perpendicular to the alignment, can be expressed in relation to the maximum surface settlement and the distance to the inflection point of the settlement trough. This trough tends to be of broader extent in clays than in sands. The volume of soil contained in the settlement trough is sometimes estimated as a percentage of the volume of the excavated tunnel; when this is done, it is referred to as volume loss, which can be highly dependent on construction methods and workmanship. Leca (2000) provides some values for the volume loss parameter as observed on various projects under different ground conditions. Closed face shields may have ground losses that are generally less than about 0.5% for sands; whereas in soft clays, the ground loss can be about 1% to 2%; and even larger volume losses can be expected when encountering mixed face conditions.

Balasubramaniam and Musa (1993) report experience from Singapore in very soft marine clay (SPT N=0) with about 20 ft (6 m) of ground cover using 5 ft (1,500 mm) slurry shields where contractors achieved advance rates of about 32 ft/day (10 m/day) and had average long-term ground settlements of about ¾ inch (20 mm) using high face pressures that initially resulted in about $7/_{16}$ inch (11 mm) of heave in front of the machine. Long term settlements were recorded for 90 days, with the data showing little further settlement after 25 days. The machine operators had difficulty balancing the face pressure in the very soft ground as any excess pressure would result in heave, while any pressure deficiency would result in free-flow of the marine clay into the machine. The deviation from alignment was about 2 inches (50 mm) in both the horizontal and vertical directions over about 4,450 lf (1,354 lm) of sewer pipe installation.

SETTLEMENT ANALYSES

Conventional settlement analyses were performed to obtain initial estimates of the ground movements in accordance with Peck (1969). Conventional methods assume that the ground is homogenous; and it is generally not possible to account for the benefits provided by soil-structure interaction. In reality, the surrounding ground and the existing structures at the site influence both the magnitude and the distribution of the ground displacements. Furthermore, the effects of ground loss at tunnel level do not always propagate fully to the ground surface. In order to overcome some of these limitations with conventional methods, finite element analysis has been performed using Plaxis.

Plaxis geotechnical engineering finite element software was used to explicitly take into account the heterogeneity of the ground, the cross sectional geometry at selected locations along the alignment, as well as to account for the material properties of the

Figure 8. Historic buildings along alignment in downtown Hartford, Connecticut

soil. Plaxis supports various material models; the linear elastic and Mohr-Coulomb constitutive models were used, respectively, to model the structural elements and the soils. The linear elastic model represents Hooke's Law of isotropic linear elasticity, which is completely defined using two stiffness parameters, Young's Modulus and Poisson's Ratio. The Mohr-Coulomb is a basic linear-elastic-perfectly-plastic model that makes use of five input parameters to include the two mentioned stiffness parameters as well as strength parameters to include cohesion, friction, and dilatancy. The Mohr-Coulomb model can account for a linear increase of stiffness with depth but can not account for any stress path dependency such as unload/reload. The principal strain is composed of an elastic part and a plastic part.

All material models implemented in Plaxis are based on the relationship between effective stress rates and strain rates; therefore, effective stiffness parameters are entered and not total stiffness parameters. For simulating undrained behavior, Plaxis automatically assumes an implicit undrained bulk modulus when the material type behavior is set to undrained. This is referred to as undrained effective stress analysis, which may be performed using either effective strength parameters or undrained strength parameters, Brinkgreve (2002). The liner contraction feature is used in the program to simulate volume loss resulting in tunnel induced settlements.

The settlement distributions generated by Plaxis generally agreed with the distributions predicted by conventional analyses; however, the magnitudes of settlement in Plaxis were only about ½ the values from conventional analysis where it is noted that many settlements observations in the field are only about ½ to ⅓ of those predicted during the design stage using conventional analyses.

Plaxis was also used for estimating the time period for primary consolidation to ensure that the settlements would occur within the liability period of the contractor. Primary consolidation will be substantially complete within 2 weeks of installing the pipeline by microtunneling, which is consistent with the experience as reported from Singapore under generally similar conditions.

Historic Buildings

Four subject buildings have been identified within 20 feet of the tunnel alignment. The responses of the buildings due to tunnel excavation include rigid body translation as well as rotation of the structure; however, the responses of the individual structural members are generally more critical with respect to the potential for building damage. Buildings that are situated adjacent to excavations are generally less tolerant to

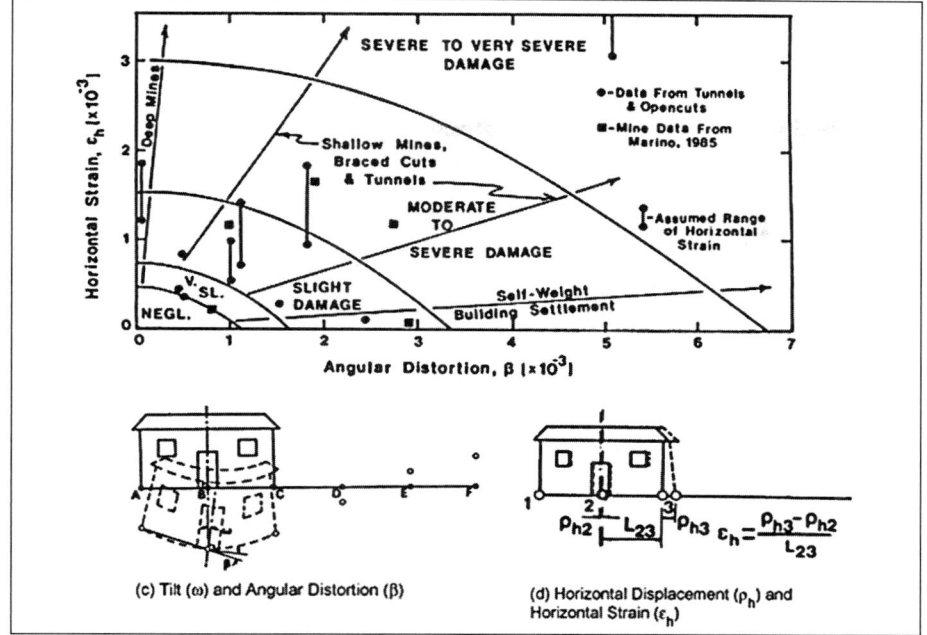

Figure 9. Summary of building damage criteria, Boscardin and Cording (1989)

excavation induced differential settlements than similar structures that settle as a result of their self weight. This is because lateral strains develop in response to most excavations. The potential for building damage has been evaluated based on angular distortion and horizontal strain using the method suggested by Boscardin and Cording (1989).

Analyses of the buildings indicate that differential settlements will be about 0.2 inches, while horizontal deformations will be about 0.15 inches. For buildings that are placed on spread footings spaced at 15 foot centers, the angular distortion will be limited to about 1/1000, while the tensile strains will be limited to about 0.083%, which are below the limits specified in the project design criteria.

Rail Crossing

There are four tracks that cross the microtunnel alignment. The crossing takes place over about 215 linear feet along the alignment of the new pipeline. At this location, the elevation difference between the highest and lowest track is almost 5 feet. Because of this elevation difference, the lower tracks will experience more settlement due to the shallower ground cover; however, the settlement of these tracks are not allowed to exceed one-half the maintenance limits for the assigned track class, which is based on the travel-speed of the passing trains as provided in Table 1.

The maximum settlement from tunneling under the tracks will be about 0.25 inches; the width of the settlement trough will be slightly more than 75 ft. Table 1 shows that ¼ inch settlement is acceptable for any track class using a 62 ft chord deviation from profile for one-half the maintenance limits.

Table 1. Amtrak's maintenance limits as contained in MW-1000

TRACK CLASS	MAX. PASSENGER SPEED (MPH)	CROSS LEVEL (INCHES) The Difference in Cross Level Between Any Two Points Less Than		DEVIATION FROM PROFILE (INCHES)		DEVIATION FROM HORIZONTAL ALIGNMENT INCHES	DEVIATION FROM HORIZONTAL ALIGNMENT INCHES
		10'	62'	31' CHORD	62' CHORD	31' CHORD	62' CHORD
MAINTENANCE LIMITS							
1	15	1	2 1/4	2 5/8	2 1/4	N/A	3 3/4
2	30	1	1 5/8	2 1/4	2	N/A	2 1/4
3	60	1	1	1 1/2	1 5/8	N/A	1 1/4
4	80	1	1	1 1/8	1 1/2	N/A	1
5	90	1	1	3/4	1	N/A	1/2
6	110	3/4	1	3/4	3/4	3/8	1/2
7	125	3/4	1	3/4	3/4	3/8	3/8
8	160	3/4	1	1/2	3/4	3/8	3/8
9	200	3/4	1	3/8	1/2	3/8	3/8
1/2 MAINTENANCE LIMITS							
1	15	1/2	1 1/8	1 5/16	1 1/8	N/A	1 7/8
2	30	1/2	13/16	1 1/8	1	N/A	1 1/8
3	60	1/2	1/2	3/4	13/16	N/A	5/8
4	80	1/2	1/2	9/16	3/4	N/A	1/2
5	90	1/2	1/2	3/8	1/2	N/A	1/4
6	110	3/8	1/2	3/8	3/8	3/16	1/4
7	125	3/8	1/2	3/8	3/8	3/16	3/16
8	160	3/8	1/2	1/4	3/8	3/16	3/16
9	200	3/8	1/2	3/16	1/4	3/16	3/16

Utilities

For the utilities that were deemed critical, the tensile strains were computed in the ground around the utility instead of attempting to model the engineering behavior of each utility, especially since obtaining sufficient information is impractical. In reality, the utilities will reinforce the ground to some degree and the magnitude of tensile strains in the utilities will be less than those calculated in the ground, which would make the results from this approach slightly more conservative. Leblais (1995) provides some published tensile strain values for the service limit condition as well as the ultimate limit condition for the utilities to include 0.3% and 1% for concrete and cast iron, 0.5% and 1% for steel, 1% and 2% for ductile iron, and 6.7% and 20% for plastics; however, gravity draining utilities also need to be checked to ensure that settlement will not impair the flow characteristics of the utility.

Strain contours were developed from the tunnel outward based on both conventional analysis and finite element analysis. The strain levels at the location of the utilities were not found to exceed the mentioned limits for the most part; however, two sewer manholes having multiple pipe connections will likely need to be supported or repaired as a result of the tunneling operations taking place in close proximity. A few other utilities subject to lower amounts of tensile strain will have to be monitored during tunnel installation.

SUMMARY AND CONCLUSIONS

The new sewer pipeline projects consists of about 3,010 linear feet (900 m) of 72-inch (1,830 mm) reinforced concrete pipe installed using pressurized face microtunneling in downtown Hartford, CT with the potential to impact several historical buildings, critical utilities, passenger rail tracks, city streets, and interstate I-84; however, some of these impacts were mitigated to the extent possible by alignment considerations. The alignment is located within varved silt and clay with areas of low ground cover. The response of the machine is improved in the soft ground by using a long stable base, formed by connecting the machine, machine can, and trailing pipe as well as by using secondary steering articulation at the machine aft. Evaluating the impact of settlement to historic buildings, rail crossing, and utilities has been described.

REFERENCES

American Concrete Pipe Association. "Concrete Pipe Design Manual."March 2005.

Amtrak National Railroad Passenger Corporation "Amtrak Specifications for Construction and Maintenance of Track, M.W.—1000."

Balasubramaniam, K. and Musa, M. A. "Mechanised Shields for Trenchless Sewer Construction in Singapore. VIII Australian Tunneling Conference. August 1993.

Bergeson, W., Nasri, V., and Sullivan, J. "Design of Hampstead Avenue Interceptor Extension in Downtown Hartford Connecticut." ISTT Moscow. June 2008.

Boscardin, M.D., and Cording, E.J. "Building Response to Excavation-Induced Settlement." Journal of Geotechnical Engineering. ASCE, 115(1), 1–21. 1989.

Hager GeoScience, Inc. "Geotechnical Report Homestead Avenue Interceptor Extension—Hartford, Connecticut." January 2008.

Leblais, Y. "Settlements Induced by Tunneling." Association Française des Tunnels et de l'Espace Souterrain. October 1995.

Leca, E, Leblais, Y., and Kuhnhenn, K. "Underground Works in Soils and Soft Rock Tunneling." GeoEng 2000: Invited Papers. June 2000.

Japanese Society of Civil Engineers. "Standard Specifications for Tunnels." 1996.

Pipe Jacking Association. "Guide to Best Practices for the Installation of Pipe Jacks and Microtunnels." June 1995.

Peck, R. "Deep Excavations in Soft Ground." Proceedings of 7th International Conference Soil Mechanics and Foundation Engineering. Mexico City. 1969.

Brinkgreve, R. B. J. "PLAXIS 2D Version 8 Manual." Balkema Publishers, the Netherlands. 2002.

MICROTUNNELING FOR UTILITIES UNDER HAROLD RAILROAD INTERLOCKING

William Reininger ▪ STV, Incorporated

Wai-shing Lee ▪ Parsons-Brinkerhoff

Richard Pociopa ▪ STV, Incorporated

INTRODUCTION

Microtunneling provides a necessary solution for multiple utility crossings associated with the New York Metropolitan Transportation Authority Capital Construction Company's (MTACC) Eastside Access Project for the Long Island Rail Road. Construction of four TBM tunnels and their approach structures at the eastern end of the project in Sunnyside, Queens required significant modification and expansion of the railroad infrastructure at this location. Maintenance of railroad traffic during project construction is a principal requirement. As a consequence fourteen (14) utility crossing segments are being installed under active traffic by microtunneling methods. Microtunnels were installed in a variety of conditions including depths, types of soils and construction restraints.

PROJECT OVERVIEW

The East Side Access project will connect the Long Island Rail Road's (LIRR) Main and Port Washington lines in Queens to a new LIRR terminal beneath Grand Central Terminal in Manhattan as an addition to existing service to Pennsylvania Station. In Queens three new TBM tunnels and a yard access TBM tunnel surface in existing Harold Interlocking, which provides multiple interlocked switch movements to the approaches of the original East River tunnels to Penn Station. With the addition of the new tunnel approaches the complexity of Harold increases with the addition of tracks and more than doubling the amount of switches. The reconfigured Harold Interlocking requires significant additions to railroad system and control power and modifications to the sewer systems. At fourteen locations utility casing pipes are required to be installed under multiple active track locations.

Currently, over 670 Long Island Rail Road commuter trains and Amtrak's Northeast Corridor trains to New England pass through Harold Interlocking on a typical weekday and over 830 trains daily on the western approaches to Harold, which include Amtrak and NJ Transit trains looping through Sunnyside Yard. Train traffic is required to be maintained throughout the project. Due to this train service requirement installation of utility casings by tunneling methods was mandated for most of the crossings. Microtunneling provides the precision and productivity necessary for the project.

Construction contract CH053 is the first of four civil/structures third party contracts planned for the improvements to Harold. All microtunnels planned for Harold are to be built in CH053 to facilitate scheduling and to reduce mobilization costs. The $139 million contract was awarded to Perini Corporation Peekskill, NY in December 2007. Cruz Contractors LLC was engaged as the subcontractor to do microtunneling for CH053. A new microtunnel boring machine (MTBM) was fabricated for the project. The first bore is to be Run 9 and is scheduled to begin in January 2009.

DESCRIPTION OF UTILITY CROSSINGS

Fourteen utility crossings were indicated for installation by microtunneling methods for the project. The contract documents mandated that eight of the fourteen crossings be constructed using a microtunneling boring machine (MTBM). Required alignment precision and/or construction control were considered to require MTBM construction at these eight runs. Other construction methods (primarily jacking) were allowed with review at the remaining locations. A description of the fourteen utility crossings follows including a discussion of construction constraints.

Runs 1 to 4—60-inch Casing Under Long Island Mainline

These four utility crossings carry electrical conduits under the Long Island Rail Road Mainline west of the Honeywell Street Bridge. Conduits primarily are used for DC Traction Feeder and Return cables although conduits for control and AC power are also installed. (See Figure 1)

Runs 1 to 3 utilize a 60-inch steel casing. Run 4 utilizes a 48-inch reinforced concrete pipe (RCP) casing. Run 4 required fewer conduits; the use of RCP was made to reduce material cost while maintaining a nominal 60-inch outside diameter for the MTBM for all four runs. Runs 1 and 2 cross entirely under the Harold Interlocking (LIRR Mainline) from south to north embankment slopes with approximate lengths of two hundred forty (240) feet. (See Figure 2A). Runs 3 and 4 cross under the eastbound tracks from the south embankment slope approximately one hundred forty feet to a middle area of the interlocking.

A review of Figure 2A and 2B indicates a number of items of interest during construction. A jacking pit (JP4) is installed at the south side of the Harold Embankment. Support of excavation of JP4 is sized to accommodate microtunneling operations for Runs 1 through 4 as well as Runs 1G to 4G discussed below. Bulkheading compartments in JP4 are anticipated to decrease dewatering requirements and accommodate permit restrictions on pumping rates. Given the groundwater elevation dewatering is a relatively minor issue for Runs 1 through 4; more so for Runs 1G to 4G. Figure 2A also illustrates future elements of construction of the Eastside Access Project. The Westbound Bypass (WBY) and Eastbound Reroute (ERT) approach structures are cut and cover structures to be built in 2011–2012. TBM tunnels B/C and D are to be constructed in 2010–2013. Clearance distances to these future structures dictated relatively tight tolerances for the installation of the conduit casings. These tolerances and the soil factors imposed by the glacial till at the site dictated the use of a MTBM for these runs.

Runs 1G to 4G—60-inch Casing Under Amtrak Sunnyside Yard Loop Tracks

These four utility crossings carry electrical conduits from LIRR DC Traction Substation G02 under Amtrak Sunnyside Yard Loop Tracks west of the Honeywell Street Bridge. Conduits primarily are used for DC Traction Feeder and Return cables although conduits for control are also installed. (See Figure 1)

Runs 2 & 3 utilize a 60-inch steel casing. Runs 1 & 4 utilize a 48-inch reinforced concrete pipe (RCP) casing. All four runs are eighty (80) feet long. Runs are designed with a profile grade of 10% constrained by the design to keep lower vault lever of G02 above ground water and to maintain at least a 5'-0" cover to Amtrak Loop 2 track. Use of MTBM method installation was not required for these runs.

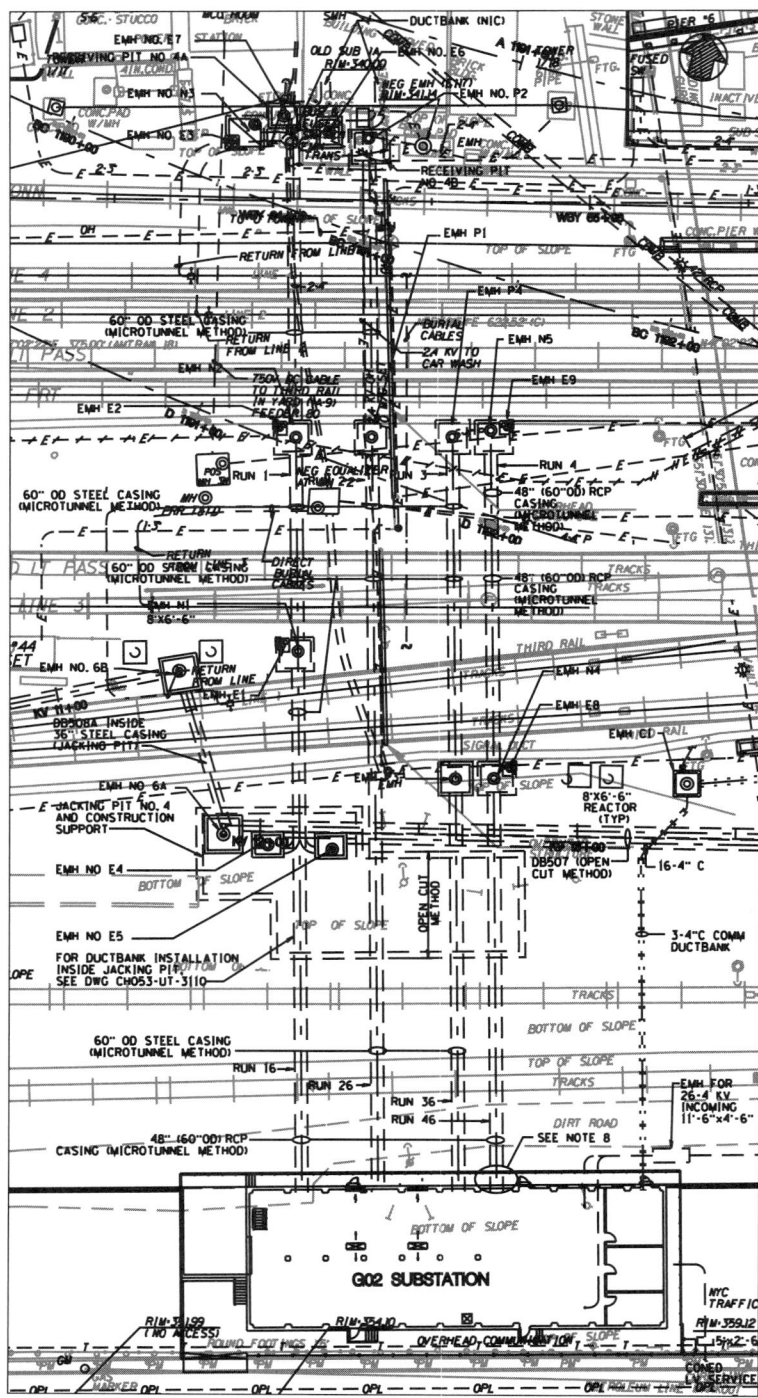

Figure 1. Plan Runs 1 through 5 and Runs 1G through 4 G

Run 5—36-inch Casing Under Long Island Mainline

Utility crossing provides electrical conduits under two eastbound Long Island Rail Road Mainline tracks west of the Honeywell Street Bridge adjacent to Runs 1 to 4. Conduits primarily are used for AC Traction Feeder cables to a substation. (See Figure 1).

Run 5 was designed with a 36-inch steel casing. Cruz Contractors LLC was allowed a substitution of a 48-inch reinforced concrete pipe (RCP) casing; which permitted use of the nominal 60-inch (1,575 mm) MTBM for the run. Run 5 has a length of fifty-one feet.

Run 6–7—60-inch Casing Under Long Island Mainline

This utility crossing casing carries electrical conduits in a sub-casing under the Long Island Rail Road Mainline west of the 39th Street Bridge. Conduits are primarily are used for AC Traction Feeder although conduits for control and AC power are also installed. (See Figure 3). In addition, this casing also carries a gravity storm sewer drain for the three Harold Tunnel Approach structures.

Run 6–7 utilizes a 60-inch steel casing. Two 24-inch sub-casings are used AC conduits and a 16-inch ductile iron pipe is used for storm sewer. Run 6–7 crosses entirely under the Harold Interlocking (LIRR Mainline) from south to north embankment slopes with a length of three hundred six (306) feet. (See Figure 4). Note design Run 7 was combined with Run 6 at the request of MTBM sub-contractor Cruz Contractors, LLC. This permitted a joint jacking pit with Run 8. Profile tolerances and the soil factors imposed by the glacial till at the site dictated the use of a MTBM for Run 6–7.

Run 8—36-inch Casing Under 39th Street

Utility crossing provides electrical conduits under a four lane local street. Conduits are used for AC Traction Feeder cables to a traction power substation. (See Figures 3 & 4). New York City Department of Transportation restrictions on maintenance of traffic on 39th Street discouraged use of open-trench ductbank construction at this location.

Run 8 was designed with a 36-inch steel casing. Cruz Constructors LLC was allowed a substitution of a 48-inch reinforced concrete pipe (RCP) casing; which permitted use of the nominal 60-inch (1,575 mm) MTBM for the run. Run 8 has a length of one hundred seventy-two (172) feet. Casing profile has a rise of twenty five feet over this length. Profile was influenced by requirement for clearance to drilled and grouted tie-backs to adjacent bridge abutment (See Figure 4).

Run 9—36-inch Casing Under Amtrak Acela Shop Leads

This utility crossing carries a sewer under multiple yard switch tracks leading to Amtrak's Acela High-Speed trainset Maintenance Shops.

Run 9 was designed with a 36-inch steel casing. Cruz Contractors LLC was allowed a substitution of a 48-inch reinforced concrete pipe (RCP) casing; which permitted use of the nominal 60-inch (1,575 mm) MTBM for the run. Run 9 has a length of sixty-four (64) feet (See Figure 5). A 20-inch fiber-reinforced, polymer mortar pipe is installed in the casing as a carrier pipe. Double pipes were considered required to provide precision to meet an existing sewer and a shallow gradient on entire sewer run. These tolerances dictated the use of a MTBM for this run.

Runs 10 and 11—60-inch OD Casing Under Long Island/Amtrak Mainline

These two utility crossings carry electrical conduits under approach tracks to Amtrak's East River Tunnels and the Long Island Rail Road tracks to Long Island City

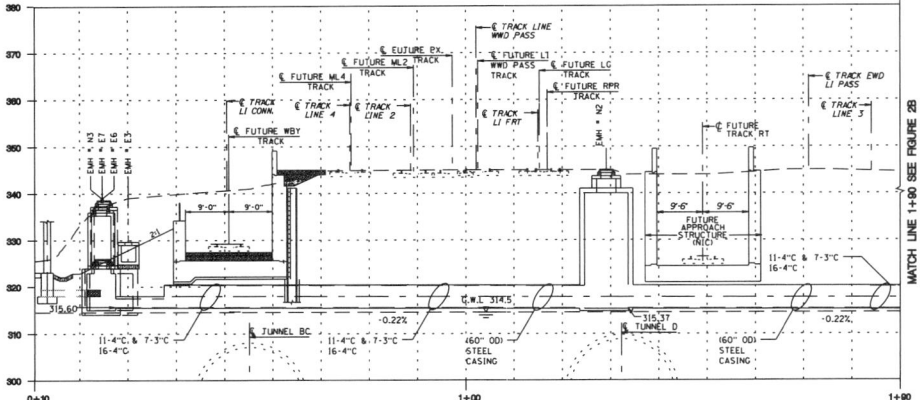

Figure 2A. Profile Run 1, Part A

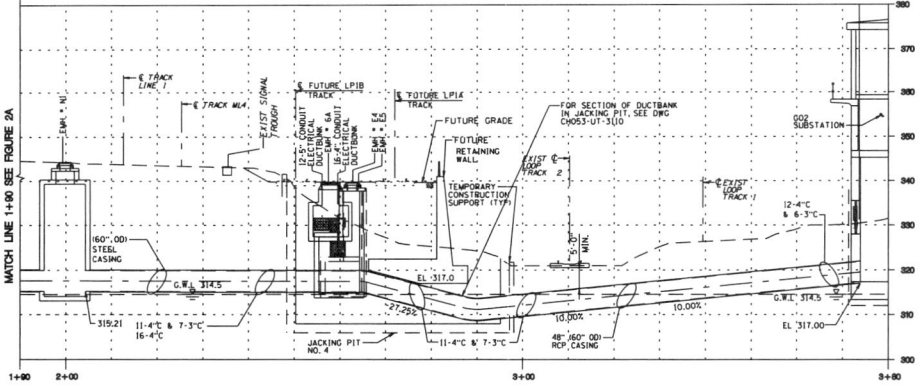

Figure 2B. Profile Run 1, Part B

Yard west of the Thompson Avenue Bridge. Conduits primarily are used for DC Traction Feeder and Return cables although conduits for control cables are also installed. (See Figure 6)

Runs 10 and 11 utilize a 48-inch reinforced concrete pipe (RCP) casing. The use of 48-inch RCP with a nominal 60-inch outside diameter maintains the typical diameter MTBM used for the contract. Runs 10 and 11 have an approximate length of one hundred ninety (190) feet. These runs are constructed in the poor peat and organic silts and clay soils at the west of the site (See Figure 7). Depth of groundwater in the excavation and settlement control sensitivity at the site dictated the use of a MTBM for these runs.

GENERALIZED SUBSURFACE CONDITIONS ON SITE

The subsurface information for the project site was interpreted from over three hundred borings drilled on the East Side Access Project site in Harold/Queens.

The subsurface strata expected in the microtunneling area in general sequence by increasing depth are Miscellaneous Fill, Organic Deposits, Mixed Glacial Deposits

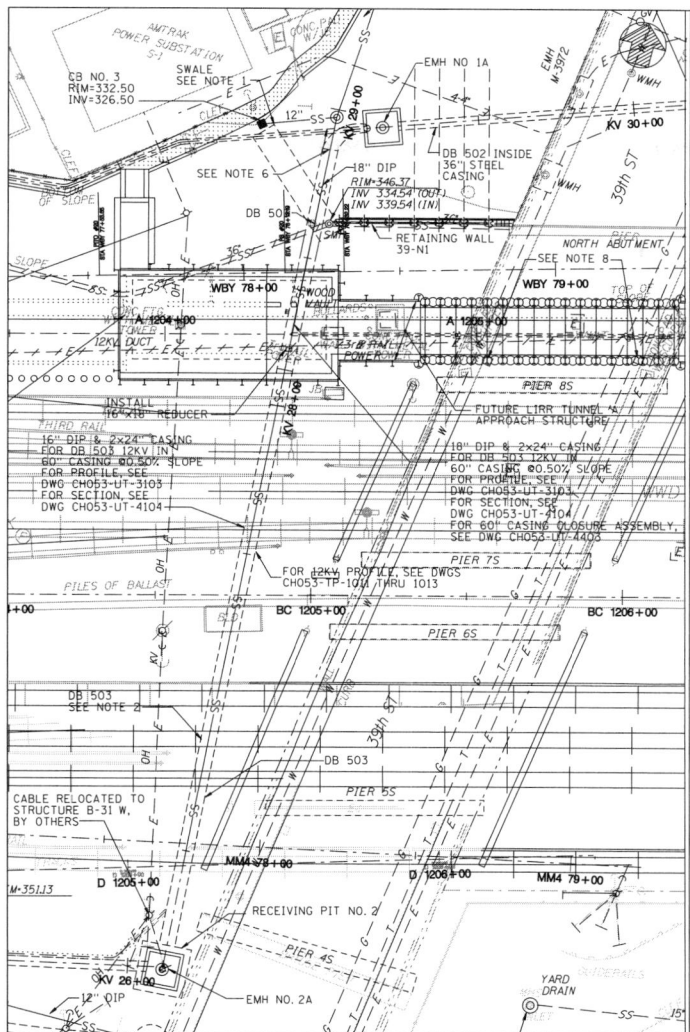

Figure 3. Plan Run 6–7

Glacial Till/Reworked Till/Outwash, Gardiners Clay, Jameco Gravel, Raritan Clay Decomposed Rock and Bedrock. However, most of the microtunneling work is expected to be conducted within Miscellaneous Fill, Glacial Till/Rework Till/Outwash and Organic Deposits only. Generally, the fill occupies the top 5 to 15 feet of the soil profile. Below the fill is glacial till with thicknesses ranging from 40 feet to greater than 95 feet.

Miscellaneous Fill

The Fill materials consist of a heterogeneous mixture of sand, silts, gravel, cobbles and boulders with varying amounts of debris such as cinders, brick fragments, wood, metal pieces, concrete and other rubble. Most of this fill is believed to be material excavated elsewhere and reused. Therefore, it is often difficult to distinguish this material

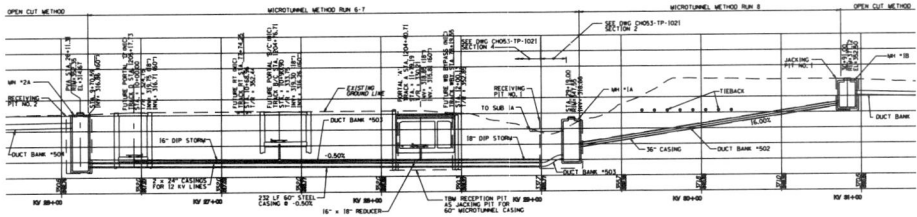

Figure 4. Profile Runs 6–7 & 8

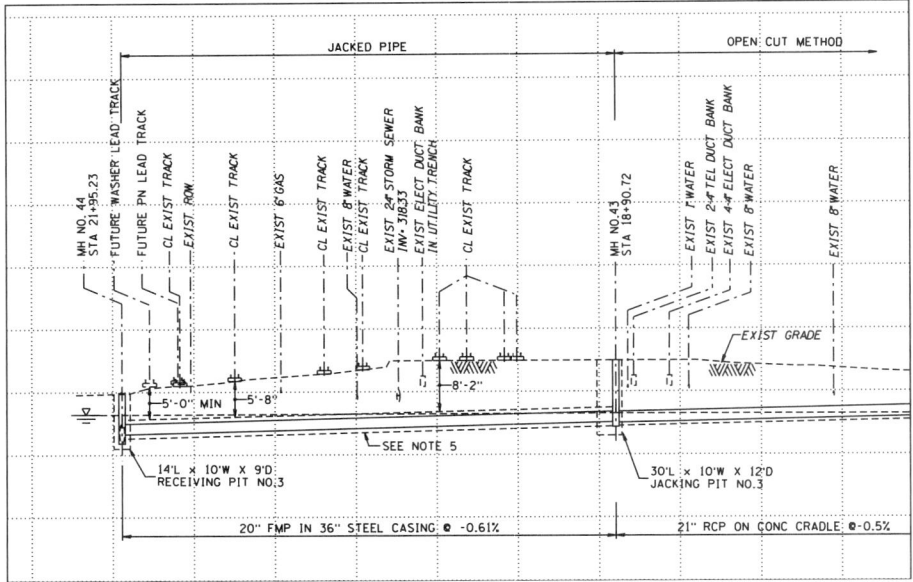

Figure 5. Profile Run 9

from the natural glacial deposits in the absence of foreign objects such as cinders or other manmade objects. Standard Penetration Test N-values generally ranged from 3 to over 100 blows per foot.

Glacial Till/Reworked Till/Outwash Deposits

The glacial till materials consist of a poorly sorted heterogeneous mixture of sand, gravel, cobbles and boulders, mostly with less than 30 percent cohesive binder material. Sand is the major component in most samples collected in this layer. Boulders were encountered within this layer, with the largest found being 4 feet in the vertical dimension. Based on borings drilled in the area, more frequent boulder occurrences are expected within this layer than the fill layer, and boulders with size up to 15 feet are expected. The boulders are generally randomly distributed, but can be concentrated or "nested" at some locations.

Intermixed with the till are materials that were deposited in melt-water channels, as outwash within and under the glacier, and as the result of reworking of glacial till. Re-advances of the glacier in some locations scraped off previously deposited soils

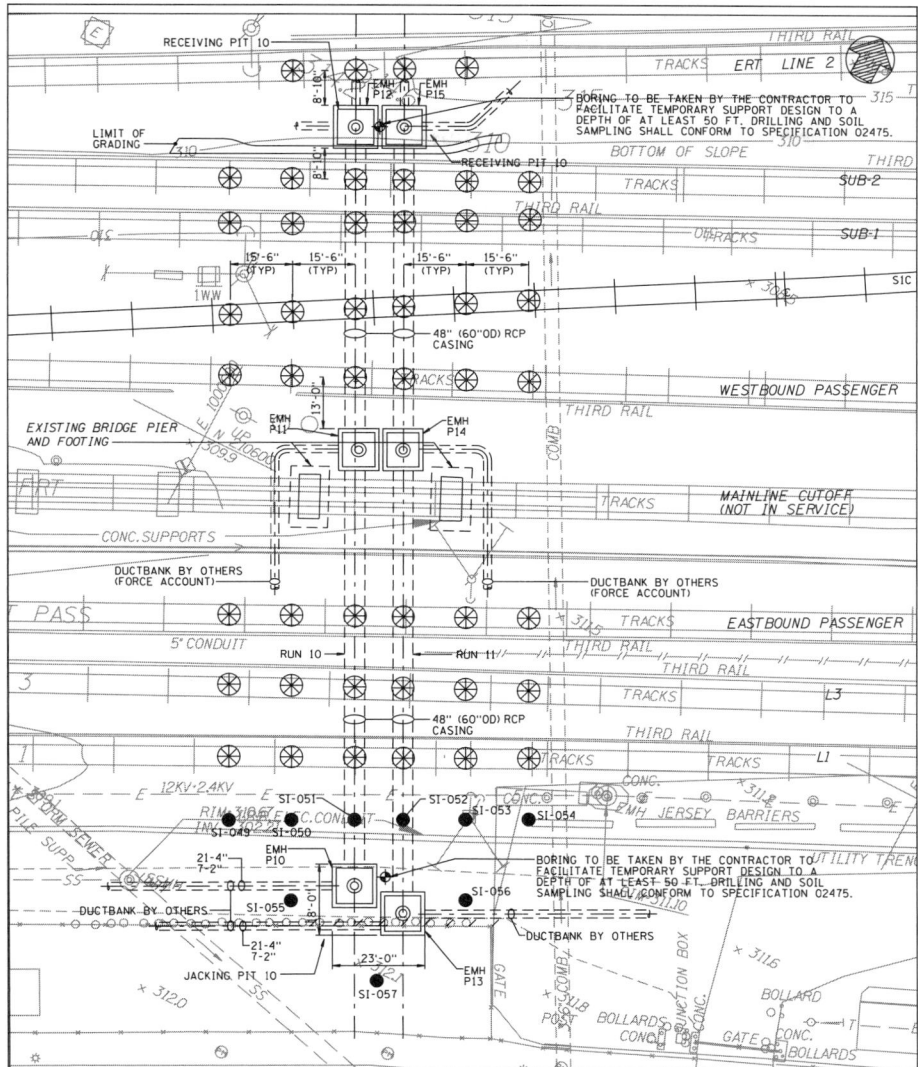

Figure 6. Plan Runs 10 & 11

and reworked and re-deposited the materials, sometimes removing fine particles in the process. Because of this depositional process, these materials exhibit a high degree of interlayering within the till material and are difficult to delineate. These materials were not delineated from the till. However, the outwash material tends to have lower SPT N-values and fewer fines. The SPT N-values generally ranged from 7 to over 100.

Organic Deposits

This layer, when found, lies beneath fill. It consists of highly fibrous to woody peat and organic silts and clays. This deposit is not continuous and was encountered mostly in the western part of the project site, which lies in the vicinity of the former location

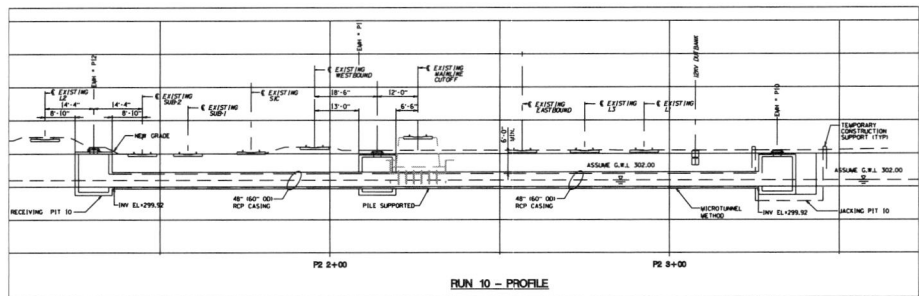

Figure 7. Profile Run 10

Dutch Kills Creek. This area was created by filling of the old creek in the early 1900s. When encountered, the thickness is approximately 6 to 22 feet. SPT N-values generally range from 0 to 6.

Groundwater

Much of the microtunneling work is expected to be conducted above or with the invert scraping the groundwater. Groundwater control is a greater issue at the jacking pits of Runs 1 through 4, Run 9, Run 10 and Run 11. Generally, groundwater elevations have a seasonal fluctuation of two feet.

EXPECTED SOIL PROFILES TO BE ENCOUNTERED BY THE MICROTUNNELS

Microtunnel Runs 1 to 5

These five runs will be bored mostly within glacial till below the Mainline tracks. However, for Runs 1G to 4G, they will be bored within fill from the Loop Track area to the GO2 Substation, and possibly with the inverts scraping the top of glacial till when they are approaching the substation.

Microtunnel Runs 6–7 and 8

The portion of Run 6–7 below the Mainline embankment is expected to be conducted entirely within glacial till. For the portion that extends north of the Mainline embankment and for Run 8, they are expected to be constructed in fill.

Microtunnel Run 9

Run 9 is expected to be bored mostly within fill. At some locations however, the invert may scrap a few feet into glacial till.

Microtunnel Runs 10 and 11

Runs 10 and 11 are located at the western edge of the project area and there are no existing borings covering this area. The Contractor is required to drill additional borings to confirm the assumed ground conditions. Based on the borings drilled east of these two runs, they will likely be constructed mostly within the organic deposits, perhaps with the crown a few feet into the fill at some locations.

DESIGN AND CONSTRUCTION CONSIDERATIONS

Given the large area of the project site, the number of utility crossings to be constructed, varied geological conditions and the sensitivity of working adjacent and under railroad facilities each of the microtunnel runs required specific considerations to facilitate construction as follows.

Microtunneling Under Mainline Railroad

Runs 1, 2, 3, 4, 5, 6–7, 10 and 11 are installed under critical mainline tracks. Interruption of "rush-hour" (6–10AM, 3–8PM) rail traffic was to be considered unacceptable which generally dictated the use of microtunneling construction methods. Further, the provisions for the maintenance and protection of rail traffic resulted in stringent requirements for monitoring and correction of settlement discussed in a later section. In addition, ince the duration of mainline track outages are very limited, the construction of vertical relief shafts for the recovery of a disabled MTBM was precluded. This was a concern given the varied nature of the glacial till and also the potential for encountering manmade obstructions that might impact the MTBM. Provisions were made in the specification requiring the use of MTBM with accessible face both for disk replacement, but also to increase access for obstruction removal. Risk sharing provisions discussed below were also made part of the special conditions.

Caisson Reception Pit/Manholes

Runs 3 and 4 terminate partially across the width of the Harold embankment. A manhole is installed at this location to route DC traction power cables to the surface. Location is tightly constrained by existing tracks and a future approach structure. As can be seen by Figure 2A microtunnel invert is some 25 feet deep at this location. Insufficient clearance existed for temporary support of excavation shoring and the required manhole in the space available. Specialized caisson construction method was designed to function as the manhole and also as the MTBM reception pit for these runs. Manhole/pit is constructed by starting with a precast concrete manhole segment with steel cutting shoe placed outside clearance envelope of adjacent track. Interior soil is removed and segment advances vertically. Additional precast segments are placed as the caisson/manhole advances. Bottom segment has unreinforced cut-out for break-in by MTBM. Manhole is sized as a reception pit and then modified for function as electrical manhole.

Groundwater Control and Dewatering

Approximately half of the microtunnels have inverts close to or above the groundwater level. The remainder are located below the water table with consequent construction issues concerning groundwater control and dewatering. A common jacking pit is shared for Runs 1 through 5 and Runs 1G through 4G. This north end of Runs 1G through 4G is completely below the water table. Permit restrictions limit the dewatering of the pit to less than 45 gpm. Compartmentalizing the jacking pit and use of Tremie concrete jacking pit slabs in conjunction with ring seals are utililized to restrict inflow to meet these dewatering restrictions. Boulders in the glacial till soils present some issues with the installation of the pit temporary support of excavation. Sheet piling for jacking pit cofferdam was utilized and driven until it met resistance from an obstruction. Adjacent sheets were generally could be advanced to a greater depth. Partial excavation was performed; the obstruction removed and the sheeting advanced in this manner to required depth. This method was effective given the relatively shallow depth of the water.

Shallow Casing Cover

Depth of cover for MTBM operations are generally preferred to 2 to 2.5 the bore diameter. Contract documents indicated a minimum depth of cover or a specific profile to meet design parameters, but permitted latitude for the Contractor flexibility in construction where practical to optimize construction. Runs 1G through 4G, 9, 10 and 11 were design with shallow cover. Different design parameters led to differing construction adjustments.

Runs 1G through 4G generally had a profile fixed by a requirement that the lower vault of Substation G02 be above ground water. Shallow cover was assumed to require ground improvement to permit use of MTBM operation. As this paper is written the feasibility of installing these runs by open-cut excavation during successive weekend outages on a loop track is being considered by the parties. Rail traffic would be maintained during weekend on single adjacent track with all service restored for Monday AM.

Run 8 profile could not be adjusted as it is a gravity storm sewer intercepting an existing sewer. Cover was considered more than adequate for much of the length. Soils permitted MTBM installation with minimal soil improvement.

Runs 10 and 11 profile was permitted to be lowered at the request of contractor. These runs are ductbanks that did not have profile constraints. Increase cover was particularly beneficial given the soft soils at this location. Soft soils and peat layer at the site of these runs facilitated sheeting installation. Cofferdam sheeting for pits made dewatering for the increased depths easier.

Risk Sharing

The owner, MTACC, recognized there were elements of risk potential in the project's MTBM construction that should not be borne exclusively by the contractor/subcontractor. On the other hand MTACC did not want to assume risk that would normally be considered contractor risk for production rates and means and methods. Special conditions and bid items were prepared that limited contractor cost exposures to those special work restrictions that were particular to the Eastside Access project. Given the nature of the glacial till and potential of man-made obstructions use of a MTBM with an accessible face was mandated. Contract documents provided that crew hours expended in excess of a base allowance used in internally accessing the head to clear obstructions, replace cutters or make obstruction caused repairs would be compensated. In addition, contract documents provided that man-made obstructions that required an external intervention to the MTBM would be cause for additional compensation.

MICROTUNNELING BORING MACHINE SYSTEM UTILIZED

The most important criterion that the MTBM has to satisfy for this project is to minimize the ground disturbance and therefore, allow the train traffic to operate without interruptions. Based on the ground conditions and the Contract requirements, the Contractor has selected the Herrenknecht AVN1200TB Microtunneling System. The major components of this system include:

- Articulated shield tunneling machine with front face access for cutting tools replacement.
- Slurry System
- Control Container, Steering & Controlling Unit.
- Laser Guidance System.
- Bentonite Lubrication System.
- Gas Detection System.

Access to Excavation Chamber

The Herrenknecht machine allows the possibility to enter into the excavation chamber to inspect the cutting head and change worn cutting tools without recovery of the tunneling machine from a reception shaft. Also, the access to the excavation chamber enables the removal of possible obstacles/obstructions, consequently reduces the likelihood of any intervention from the ground surface, which is strongly undesirable, given the congested train traffic above ground and could be precluded at some locations.

Steering and Controlling Unit

With the aid of a steering and controlling unit, the drive will be accurately steered, important data can be measured, controlled, monitored and recorded and faults will be displayed on a screen monitor.

The steering and controlling unit consists of a control tableau that contains all necessary switches, buttons and displays, a computer unit and a monitor. Via this computer unit, all data, e.g. pressures of hydraulic oil or strokes of several hydraulic cylinders, are measured and monitored. Furthermore, a graphical display of the active position of the tunneling machine occurs. This gives the possibility to operate via corresponding buttons the steering cylinders and thus actively steer the tunneling machine. Measured data is checked, compared with the scheduled data, e.g. temperatures and pressures of hydraulic oil, and, in case of an exceeding of tolerance values, error messages will be displayed or the tunneling machine will be automatically stopped. Thus, the formation of defects and failures of the tunneling machine can be limited or avoided and the pipe run can be placed in the ground as accurate as possible. Additionally, with controlling units data from each drive can be stored for later evaluation, interpretation and analysis.

Alignment Control

The design of the MTBM system allows a laser to be set in the jacking pit. A target mounted on the MTBM machine enables the Contractor to determine the exact alignment at all times (horizontal & vertical). The information from the laser and target are available in the operator's booth inside the control room. Since the machine is steerable in all directions, adjustments can be made as necessary.

PROCEDURE FOR EXCAVATION AND CONVEYANCE OF THE VARIOUS SOIL/BOULDERS/COBBLES

A water bentonite suspension mix will be used as flushing water (slurry) in order to stabilize the ground soil around the cutter head and avoid material being transported from settlings. This system mixes the drill cuttings with the flushing water (water or water bentonite suspension) and then transported by the conveyor pump(s) through the tunnel slurry lines to the separation/screening plant.

The screening plant removes the solids from the slurry and recycles the water back into the system. Water that is not pumped back into the system will be pumped into settling tanks or settling basins for later disposal.

The AVN1200TB is equipped with a crusher so that the boulders/cobbles in the pipe route can be broken down to a transportable size. The crusher works like a cone crusher according to the principle of a "coffee grinder." Inside the machine there is an internal cone with crusher bars. The crusher arms on the back of the cutting wheel form the counterpart to it. The material is broken up between the stationary crusher bars of the internal cone and the rotating crusher arm of the cutting wheel.

DERRICK TBSS-225 SEPARATION PLANT

Derrick TBSS-225 Separating Plant (Removes solids from slurry to be pumped back into the system). The plant is able to process 225 cubic meter of flow per hour and able to separate 60 Tons of Solids per hour. The soil is mixed with the flushing water (water or water bentonite suspension) and then transported by the conveyor pump(s) through the tunnel slurry lines to the separation system.

INSTRUMENTATION AND MONITORING PROGRAM FOR RAILROAD

Maintenance of rail traffic is a critical element of the project. LIRR and Amtrak required that a comprehensive track instrumentation and monitoring program be in place before the start of tunneling operations. Similar measures are implemented for both the MTBM construction and the TBM bores for train tunnels to follow.

Project specifications identify displacement alert and action limits for tracks and other facilities. In the case of tracks these displacement tolerances are all set below speed class reduction limits for the railroads, which are at or below tolerances set by the Federal Railroad Administration (FRA). The intent of the program is early detection of any tunneling induced settlement and the implementation of changes to boring operations and maintenance surfacing of railroad tracks. With these early actions the intent is that track tolerances that would mandate reductions in train speed are never reached. Contract documents allow for the stopping of MTBM for settlement induced displacements that are below the FRA limits.

It is thought that the microtunneling work under the should not typically induce track displacements so as to trigger action responses. However, some microtunneling will be conducted in bouldery ground and with potential existence of other buried man-made obstructions, particularly at shallower depths. A monitoring program is therefore placed for detecting and forecasting the trend of any excessive settlements caused by the work and the need for any immediate remediation.

At some locations, the microtunnels will be bored underneath up to ten tracks and switches. The settlements will be monitored by ground settlement points and track dynamic profile points. Due to the rigidity of the rail/tie, the rail may not reflect the actual ground settlement below it. A track dynamic profile point is a short wooden rod installed directly underneath a rail and will register any gap between the ground and the rail/tie system once a train has passed over the rod.

A set of ground settlement point and track dynamic profile point is installed under each rail/tie directly above the tunnel centerline. Additional sets are placed at 15.5 ft (a quarter of a 62-ft chord) intervals on both sides of the centerline up to a distance of approximately 1.5 times the depth of the microtunnels. For a microtunnel with a depth of about 30 ft, sets of monitoring points are placed at centerline, and 15.5, 31 and 46.5 ft on both sides of the centerline. The monitoring will be performed by the Construction Manager's team, at every 3 to 4 hours. Prior to any microtunneling activities, the railroad track profiles will be baselined. Any significant deviation from the reference profile will be corrected.

PART 10

Mining

Chairs

Terry Yokota
Frontier Kemper

Lonnie Jacobs
Frontier Kemper

TECHNICAL CHALLENGES IN MINE REHABILITATION

Don Dodds ▪ North Pacific Research

David A. Clayton ▪ Clayton & Associates

Terry L. Johnson ▪ Construction Management Consultant

IRON MOUNTAIN MINE REHABILITATION PROJECT

Brief History

The Iron Mountain Mine is located in Northern California approximately 12 miles northwest of Redding. The ore body is a large, very dense pyritic intrusion, discovered in 1860, and initially minded for silver. The silver played out quickly and 34 years later the ore body was mined for copper. Copper mining was scheduled to cease in 1939, but WWII prolonged the closure. After WWII, the pyrite ore was mined to produce sulfuric acid. In 1949 when large amounts of cheaper sulfuric acid were being generated by the oil industry the underground mine shut down.

Little was done with the site conditions until 1970 when the U.S. Bureau of Mines began monitoring the pH of the effluent. The initial pH readings were around 2.0. This was lower than anyone expected. Laboratory tests showed that when the ore from the mine was submerged in water, the pH of the water would stabilize above 4.0. However, the pH got steadily lower. In 1989 with the ph around 0.7, the U.S. Environmental Protection Agency (EPA) required the owner to build a capture and treatment facility. During this initial period, the majority of the effort was spent on the treatment plant. The relationship between the mine owner and the EPA was basically an adversarial relationship from the very beginning. In 2001, North Pacific Research was hired to upgrade the capture and control facility. The general layout of the Iron Mountain Mine is shown in Figure 1.

ENVIRONMENTAL CHALLENGES

Extreme Heat and Humidity

The mine, especially behind the Five Way, presented hot, 58° C and humid, 100%, conditions. These conditions made it extremely difficult for small work parties to work away from the cool ventilation air.

Acidic Water

In the Richmond Adit, the ph ranged from 2.7 near the Five Way to 7.0 around station 3+00. A pH of 2.7 is about the same as diluted lemon juice. Behind the Five Way, the ph was initially about 0.5, which is about ten times more acidic than battery acid. These extremely acidic conditions are poorly understood. Acid's effect on materials was a key issue. Behind the Five Way, a 1-inch rebar placed in the invert rusted in two pieces within 36 hours. Steel sets in the low pH area of the Richmond Adit required them to be replaced about every 4 years. Stainless steel sets were experimented with

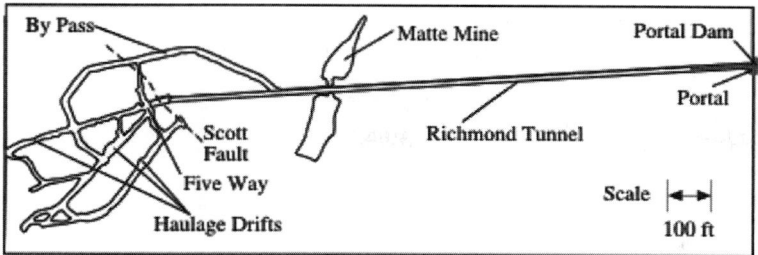

Figure 1. Iron Mountain mine general layout

before we arrived on the job. The acid's effect on concrete was little better. The acid attacked the lime-based product and quickly rendered it useless.

A second major impact of the acidic water was on the construction equipment. We had two different areas on the job where the equipment could be hosed down; one at the portal and the other near the shop. Frequent use of these facilities slowed the deterioration but did not stop it. The cost of maintaining the heavy equipment was considerably higher than on normal projects.

The pH levels measured inside the mine are somewhat controversial. In one study, the research team reported pH as low as –3.0. However, this pH was measured using hydrogen reactivity, which is more repeatable, but requires an equation dependent on pH to compare it to pH measured by concentration. At very low levels of measurement the modification equations changes rapidly. The accuracy of that equation is questionable, especially at very low pH. We monitored the pH continuously over the nearly three-year construction period. The lowest pH reading measured by the construction team was 0.0 and was measured in a pool directly below the level, which produced the –3.0 measurement. We were unable to identify why the pH would decrease so drastically in a few vertical feet. See Figure 2, the small dark pool in the middle ground is the pool where we took our measurements. The water was stratified, the upper water with a pH of about 1 was clear, while the lower water had a pH of 0.0 and had a dark color shown in the photograph.

TECHNICAL CHALLENGES

Poor Access

The mine had been completely neglected for over 50 years. The access to the Richmond Adit was blocked by the portal dam. Personnel could scramble over the dam but any equipment and materials had to be carried over the dam by hand. Once in the adit, the access was only about 1.3 meters wide. This allowed the luxury of using a wheelbarrow until we reached the Scott Fault Bunker. At this point in the mine, the clearance was reduced to about 1.5 meters high and less than 1 meter wide. Figure 3 is a photograph of the pump station early in the construction. The Scott Fault Bunker is in the background. Note the condition of the steel sets.

Six Months to Design and Award Job

The nature of the contract required a very short design period. The design, specifications, and award of the job were required to be completed within six months after the award of the design contract. This allowed no time for exploration and testing.

Figure 2. B drift 20 feet directly under –3.0 pH reading

Figure 3. Scott Fault Bunker from the Richmond Adit

Very Limited Geotechnical Data

The only information available for the design and construction of the project was 55-year old drawings, which may or may not have been updated near the end of operation. We did have the luxury of walking the adit itself and seeing pretty much what we had to deal with except in areas that were supported. The blocking and lagging hindered observation in those areas where observation was most needed.

Zero Release of Mine Effluent to the Environment

During the entire construction project the EPA required zero release and imposed heavy fines if any effluent reached the environment.

APPROACH

Design/Build/Maintain

Since the client was also responsible for maintenance, the project was conceived as a design, build and maintain project. The addition of maintenance to the mix added efficient maintenance as a goal. Thus, both ease of construction and maintenance became important in making the design decisions. Rather than approach the maintenance problem in a reactive manner, fixing failures as they occur, a proactive approach was adopted to significantly improve the state of the Richmond Adit. The design was based on installing active rock support systems, increasing the minimum excavated opening dimensions to 10.5 by 10.5 horseshoe shaped tunnel and included capturing the floor drainage, a portal system that would pass vehicular traffic, provide adequate ventilation air and other utilities.

Risk Sharing

Very often in lump sum or unit priced contracts the contractor and the engineer are in an adversarial relationship. The contractor makes money if the product meets the specifications as he has priced them, but does not if he is expected to exceed the specifications or the scope. This type of contract requires tight specifications, good geotechnical data and constant surveillance. The more the contractor can cut corners the more money he makes. The more corners that are cut the greater the maintenance costs.

Underground work is notorious for large overruns in cost and large expensive claims. A primary factor in the run away cost of underground work revolves around uncertainty. Obviously much uncertainty surrounded the Iron Mountain Mine project. In most underground construction, the risks are passed to the contractor. Successful contractors do not lose money. The result is the contractor must increase the price to cover the risk for unknowns by adding a large contingency.

Following this approach, the outcome is that if the project is less difficult, the contractor makes extra profit. If the job is more difficult, the contractor resorts to claims and avoiding costly specification requirements. The end result of this process is that the owner always pays more for the job and sometimes gets an inferior product. Successful contractors have a large legal staff.

As the project moved forward, the Owner decided that assuming all the risk would allow them to get a better product, reap any reward and reduce the project price. The final contract was set up such that the contractor supplied men and equipment and the owner's oversight team controlled all work. All the risk was on the owner with a bonus for the contractor if the job was finished under budget and ahead of schedule. In any case, the decision was right and the owner won.

Team Building

The success of this effort was directly dependent on the ability of the overall project manager to build a successful team between design and construction. Many of the owner's key people were from contracting backgrounds, a factor that may have been pivotal in this decision. Team building is often talked about between design and construction but rarely achieved. Entire books have been written on team building. But it boils down to good communication and taking care of the people below you and them taking care of you. It requires trust and no surprises. Immediate disclosure must be encouraged not punished. The emphasis must be on fixing the problem not the blame. How does a highly educated person become part of a team of miners? Simple, enjoy getting your hands dirty and swear a lot. Might sound funny but it works.

Crew Training

As part of our team building, we held classes for the miners on active support, smooth wall blasting and proper shotcrete application. This may appear to be a waste of time, but it worked for us. I have worked with miners for over 40 years and while I will agree that they are not part of the intellectual elite, they are definitely not stupid. They possess considerable knowledge about rock that us so-called "intellectuals" never get, knowledge that is of great value to the success of the project. The problem with teaching miners new tricks resides with the instructor. These classes were not theoretical but practical, explained in terms they were familiar with and in a way they could evaluate and understand. Talking neither up nor down but straight at them. By the end of this job, they all believed in smooth wall blasting and active support because they understood how it worked and saw that it worked for them.

Contractor Selection

Because of the short lead-time before construction, we decided to let the contract using a modified RFQ and RFP procedure similar to consultant selection. Six firms were selected from those who responded to the RFQ and were asked to present a Proposal. The proposal was based on the 70% design drawings and specifications. It consisted of their labor and overhead costs, equipment rental rates, a fixed profit, and supporting data. Three contractors were qualified to bid selected based on their written proposals and site walks. One of those three was selected as the apparent successful contractor based on allocating,

- 30 percent to price
- 20 percent to similar experience and personnel and equipment resources
- 20 percent to understanding of the site conditions
- 20 percent to approach to the work
- 10 percent to innovation

Site Visit

The site visit was held with one contractor at a time. It was specifically designed to protect the contractor's confidential ideas and provide the engineer with the information necessary to determine the experience level of the contractor's personnel. All key personnel were required to attend. The questions during this visit were designed to allow the contractor to demonstrate their knowledge of site conditions, their ability to innovate, and general tunneling knowledge. The contractors were asked to defend prices that were considerably out of line with other candidates.

Specific areas judged during the interviews were

1. General underground knowledge
2. Experience with similar conditions, such as
   a. Smooth wall blasting
   b. Shotcrete
   c. Low pH water
3. Innovation
4. Willingness to listen and participate

The scope of the work was defined as all the work necessary to get from the portal to the limit of the work, regardless of what we faced. If negotiations failed with the primary selection, we would then approach the second choice and so on until a contract was signed. The contract essentially made the contractor the hands and legs of the designer. The contract guaranteed that the only risk taken by the contractor was a diminished profit if the job ran considerably over the schedule.

Maximum Design Done at the Face

Again, because of the short lead-time and the poor geotechnical data, most of the design was to be done at the face. This is often done in mines but rarely done in tunnels. There is a subtle difference between mining and tunneling, mining cares most about what's in the hole; tunneling cares more about what surrounds the hole. In this case, a combined team consisting of the project manager, a rock mechanics consultant, mining engineer, and tunnel superintendent along with the shift boss made all decisions after each blast on changes, if necessary, to the round, temporary support and other pertinent matters.

SAFETY

This project was an extremely dangerous job in a very dangerous profession. Tasks such as removing sets and liner plate were extremely dangerous particularly in the Portal and Mattie mine areas. Previous contractors had refused to remove sets because of the danger. Also, since the Richmond level had been mainly used for haulage during mining operations, 38 chutes and two winzes had to be cleaned, stabilized, and plugged. Due to the skill, and vigilance of the crew and supervision, no serious accidents or injuries occurred. Other than normal cuts, bruises and the occasional splash of low pH water in the eyes the job was injury free on a three year, 240-hour workweek project.

WORK BREAKDOWN

Water Control

The first order of business was to remove the old portal dam. This removal required two temporary dams and changing the pipe from a combination of 18-inch and 24-inch pipes to a single 8-inch pipe and a high flow pump. We then alternately stored and pumped the water around construction until we could build a 1-meter high portal dam with a 1.25% upstream slope, a 6% downstream slope and a 10 m long 35 cm wide intake. The new portal dam offered no restriction to loaded trucks entering and exiting the portal, was designed to contain twice the maximum history flow and acted as an equipment wash station. See Figure 4.

The portal dam and floor drain were made out Dynastone, a concrete-like material made from fly ash, which relies on silica based cement rather than a calcium based cement. The cement was used to make highly acid resistant pipe. A call to the manufacturer early in the project and a discussion with the developer of the product, revealed an interest in using large aggregate but they could not find an application. Even though Dynastone had not been used in this type of an application nor been poured in place before, six months later two technicians from the manufacturer were on the site to supervise the mixing and placement of the Dynastone. The material was mixed on site in a standard mix truck, containing the aggregate and water.

Dynastone does not set up like concrete. It takes an initial set in around 2.5 hours and rapidly becomes sticky and difficult to finish. This material also cures differently than concrete. The manufacturer recommends that Dynastone be cured at a temperature of 48°C. This presented a problem at the portal, less in the Richmond Adit but was no problem behind the Five Way where the temperatures were ideal for it setting.

Figure 4. Portal dam looking upstream

Figure 5. Steel sets before rehabilitation

Widen 400 Meters of the Richmond Adit

Once the portal dam was finished we widened the full 400 meter tunnel from an 8'×8' to a 10.5'× 10.5' wide horseshoe. Along the way, we pulled all the sets and replaced them with shotcrete. We then drove the By Pass Drift around the Scott Fault Bunker to gain vehicle access to the haulage drifts.

The tunnel environment presented several challenges that required untested technology. The low (2.5±) pH water in the Richmond Adit oxidized steel sets and also attacked the calcite bond in the concrete. Wood sets seemed little effected by water with a pH below 2.5. Unfortunately, a previous contractor had removed all the wood sets and replaced them with steel sets. Since these sets, not surprisingly deteriorated quickly, the same contractor split spaced these sets with stainless steel sets. These expensive sets lasted a few years more, but at the time of our work were looking fairly ragged. Figure 5 is a close up of the set conditions existing in the Mattie mine area before the work began.

Support of the openings and control of the low pH water required the use of engineering materials resistant to acid attack. The unsatisfactory performance of steel and concrete seriously limited what could be done underground. We knew what didn't work,

Figure 6. Drainage blanket

but not what would work, especially behind the Five Way where the pH was 15 times lower.

We decided to use polypropylene fiber reinforced, silica flume enhanced shotcrete for support in the Richmond Adit. These fibers were acid resistant and would eliminate the steel mesh. Another advantage of the shotcrete was that it could be applied in variable thickness on the surface of the tunnel allowing the engineer in the field to tailor the support to the rock conditions. Under low pH conditions, the disadvantage of shotcrete is that it is composed of a calcite-based cement. By limiting the exposure to the water, decreasing the calcite content of the aggregate, replacing some of the cement with silica flume and making the mix as dense as possible to eliminate water infiltration, the team was able to minimize the effects of the acid ground water.

The shotcrete performed well in all dry areas. However, shotcrete is notorious for refusing to stick to the rock surface where water is flowing. In these areas, we placed a waterproof barrier between the rock and the shotcrete. Unfortunately, this barrier was designed to be nailed to the rock. This works if the water is not acidic. We knew that prolonged exposure to low pH water would dissolve the steel nails. We reduced the failure in the areas where the pH was above 2.5 by covering the nails with thick epoxy mastic before applying the shotcrete. Figure 6 shows the drainage blanket with the mastic-covered nails in the Richmond Adit before placement of shotcrete. The nails allowed the shotcrete to set and the mastic plugged the nail hole when the nail dissolved. Figure 7 is a photo of the finished Richmond Adit. Compare this figure with Figure 5.

Need for the By Pass Drift

The Scott Fault Bunker was a result of the adversarial relationship between the owner and the EPA. The EPA's contractor felt that the Scott Fault might move and sever the main pipeline for effluent out of the deep mine, a reasonable concern. The Owner apparently didn't think it was that much of a threat. The resolution of the dispute ended unsuccessfully for both parties. The Owner, forced to build a bunker that wouldn't fail, constructed a virtual barrier to the Five-Way. The contractor constructed a support of the Scott Fault by lining the tunnel with steel liner plate reinforced with .75 meters of concrete. The result in an 8 by 8 horseshoe tunnel was that access was limited to about 1.5 by 1.0 meters. This allowed personnel only to pass and they were required to stoop if they were taller than 5 feet. Nobody won.

Figure 7. The finished Richmond Adit

The only notable thing during the construction of the By Pass Drift was the lack of the Scott Fault. The path of the By Pass Drift was such that it again had to pass through the dreaded Scott Fault. Miners, by their nature, soon had wagers on where we would hit it. After we reached the Haulage drifts, the fault was so innocuous we had to get a committee of geologists, rock mechanics, and mining engineers to comb the Bypass Drift to identify and agree on its location. In conclusion, one cannot fault either side for their positions, but the problem was the adversarial relationship set up by the Environmental Protection Act.

Haulage Drifts

Once access was established to the haulage drifts using the By Pass Drift, it was necessary to remove the muck that filled these existing drifts from the invert to the back, remove the muck from the abandon chutes and backfill all chutes with concrete, and construct a maintainable mine water catchment facility.

Once in the haulage drift, we had to turn and muck down grade, back to the Five Way to establish air flow. This meant that the acid water flooded the face and sometimes rose as high as 0.5 meters. Besides the normal problems with working a drowned face, the pool of acidic water increased equipment maintenance.

To our surprise, the tunnel muck was not normal tunnel muck. The gradation showed 90% passing a 200 sieve and only 1% larger than 30 cm. This was certainly not your mothers tunnel muck and definitely not produced by mining. Figure 8 shows the non-typical tunnel muck, notice the circle around the foot caused by tapping the toe.

Whatever the process, it also produce grain size small enough to be quick, which increase the loading and hauling problems. Too much vibration in the loader bucker or in the back of the truck could turn the material instantly into liquid and it would run out of the bucket or bed.

The need for proper ventilation in the 100 percent humidity and 58 °C temperature was for the most part a routine problem, requiring only the use of brattices and shifting the cool ventilation air on the work area. In the areas away from normal ventilation, care had to be taken that core body temperature remained normal. In these cases, we used a cooling vest that did not use ice as the cooling agent. Instead it used a liquid with a melting point around 10 °C. This allowed two advantages. One, the long-term exposure to a 10 °C substance is less uncomfortable than to a 0 °C substance; and two, the coolant can be recharged by immersing the cooling agent in ice water.

TECHNICAL CHALLENGES IN MINE REHABILITATION   625

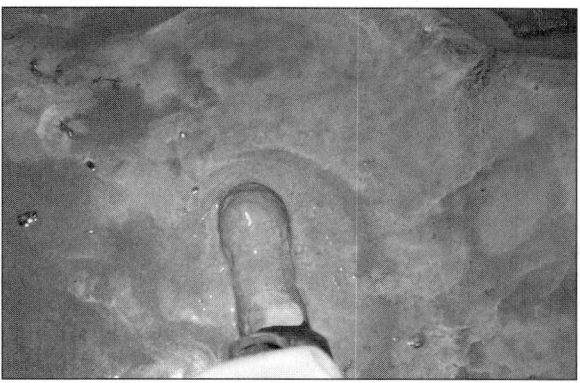

Figure 8. Haulage drift quick material

Plugging the 38 chutes, a notoriously dangerous job in active mines, proved to be fairly benign in this abandoned mine. The chutes came in four varieties: clean, running, hung, and plugged. The clean chutes were plugged with a concrete plug 1.5 times as deep as the maximum diameter of the chute. Heavy steel mesh and shotcrete was used as a form to retain the concrete. This allowed the construction of the form with minimum exposure of personnel below the open chute. Running chutes were mucked until clean and then treated as clean chutes. In the hung chutes, several attempts were made to remove the obstruction. If it was removed, the chute was empted and treated as a clean chute. If the obstruction could not be removed, a concrete plug was installed. The chutes that were plugged were filled with a material that had been naturally cemented by ground water chemicals. An attempt to tap into the plugged chutes was made to remove the ground water. This was not always successful.

Shotcrete was used for support in the dry areas behind the haulage drifts. In the wet areas fiberglass bolts were used. Most of the area behind the Five Way was stable or dry. Fiberglass bolts were confined to a short section of the Haulage Drift that was driven through country rock outside of the ore body.

THE RESULTS

We finished a 3-year job with No Deaths, no lost time injuries, 1% over the budget, and on schedule with no claims. The credit for this accomplishment belongs to the entire team. We were most thankful and pleased that no one was killed or injured during this extremely dangerous job. Second, we were also proud that a job with so many unknowns and problems could be completed on budget and without any claims. Claims are almost a way of life in the underground world. We believe the tunneling profession can learn from the contractual model for this job.

The difference between the mine before and after is great and can be seen by the difference between before and after photos. Figure 9 is a shot of the Five-Way near the beginning of the job and Figure 10 is the same location near the end of the project.

UNRESOLVED QUESTIONS

Acid Producing Bacteria

Coincidentally while we dealt with the excavation problems, the University of California at Berkley was studying a bacteria found in the mine water. The results of this

Figure 9. The Five-Way before

Figure 10. The Five-Way after the job was complete

study showed that the nature of this material was due to the processing of the pyrite by microbes.[1] These microbes lived in total darkness, in 0.7 pH water and 58°C temperature, and lived on pyrite. Figure 11 is a micrograph provided by the Berkley research team showing both the size of the grains and the organically produced markings.

The research showed that these bacteria break the iron and sulfur molecules apart and extract energy by striping off on electron from the Fe_2. The waste products from this process are Fe_3 and a sulfate radical SO_4. The SO_4 combines with Hydrogen in the water to produced $H2SO_4$. The result is the ultra low pH. Ironically, it seems that it is this rare and endangered species of bacteria, not the mining, which is producing the major environmental catastrophe. This leaves us with a dilemma, do we kill the endangered species to save the river environment or spend millions annually to repair their damage, an environmental and logical nightmare that is still being avoided. If not, should we be spending millions of dollars trying to eradicate the bacteria responsible for small pox, aids or other disease? Do humans have sufficient knowledge to be able to correctly

 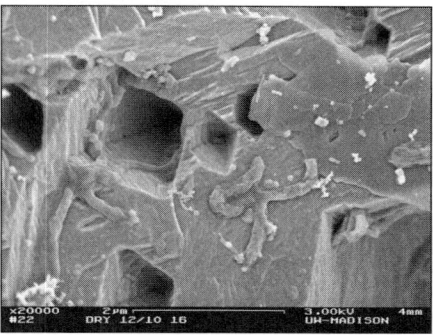

Figure 11. Grain size—bar is 200 mm; close up—bar 2 mm

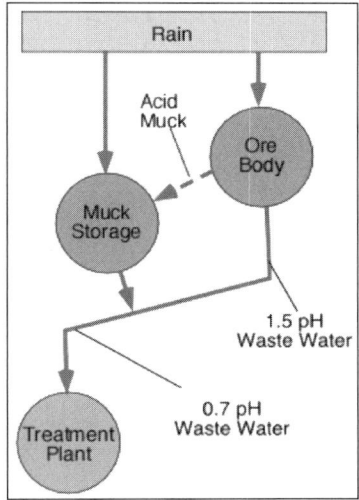

Figure 12. Still in the loop

make decisions that direct global evolution? The work at Iron Mt may rattle dishes as far away as Washington D.C.

The Acid Loop

After cleaning all the Haulage drifts our routine measurement of pH at the Five Way showed a definite upward trend. Something we were doing was increasing the pH. The management team decided that it made sense to clean out more drift and see what occurred. We cleaned out another 400 meters of drift, which increased the pH at the Five Way from 0.5 to 1.2—a 700% improvement in two months. This was good news except that the pH at the plant remained constant. Although considerable time was spent investigating, the contradiction could not be explained until six months after job completion.

After the project ended a routine schematic of the process illuminated a possible answer. See Figure 12. Could the answer to the inconsistency be that we had removed the acid producing material out of the mine, but we had not removed it from the loop.

We simply placed the same material back in the loop further down the line. The muck we removed was still draining high pH water into the pipeline only now the entry point was below the Five Way. Even though the cell was sealed at the finish of the job, it is reopened periodically to accept more material during routine maintenance. We have no recent data on the pH of the plant to see if sealing the cell lowers the plant pH.

LESSONS LEARNED

- Problem caused by bacteria
- Contract used was instrumental in reducing claims
- Adversarial relationships set up by Environmental Protection Act aggravated the problems.

REFERENCE

1. Katrina J. Edwards, Philip L. Bond, Thomas M. Gilhring, Jillian F. Banfield, An Archaeal Iron-Oxidizing Extreme Acidophile important in Acid Mine Drainage. March 10, 2000,Science, vol 287 No 5459 pages 1701–1876.

OPEN PIT TBM DRIVEN DRAINAGE TUNNEL— OK TEDI MINE

Tony Peach ▪ Terratec Asia Pacific Pty. Ltd.

Nigel Sudgen ▪ Sugden Consulting

PROJECT DESCRIPTION

The Ok Tedi Mine is located approximately 18 km east of the Papua and New Guinea-Indonesia border, on the southernmost extremity of the Star Mountains Range, it is situated almost equidistant from the north and south coastlines. The headwaters of the Fly River system are in the Star Mountains.

The open pit mining activities are on Mt Fubilan with a (pre-mining) elevation of 2,095 metres (6,871 Feet) in the Western Province. The mine is in an area characterised by very steep topography and dense tropical rainforest. The pit has been free draining but will cease to be so in the first quarter of 2009. Ok Tedi Mining Limited (OTML) required a drainage tunnel to be driven under the mine, if possible before free drainage was no longer possible.

The tunnel was required to be at least 5 m in diameter and 4,200 m (13,776 ft) long, with a possible extension of 1,100 m (3,608 Feet) if the mine decided it was required for additional drainage and underground exploration. The tunnel is being driven up grade at 0.67%, and will be connected to the pit by the raising boring of two 1.1 m (3 Ft. 6 In.) diameter shafts from the tunnel level 250 m (820 Feet) below to the pit.

In Tabubil, the adjacent township at an elevation of 640 m (2,100 feet), air temperatures range from a mean at night of 20 (68°F) degrees centigrade to a daytime average of 27 (81°F) degrees. Extremes of 12 and 39 (54°F–102°F) degrees centigrade have been recorded. Rain falls, on average, 339 days each year and totals 8,000 mm (315 inches), and have been known to approach 12,000 mm (472 inches) making the area one of the wettest places on earth. The portal site is at an elevation of 1,200 m or 4,000 ft and the air temperatures are a little cooler and the rainfall a little greater.

Access to the mine is by air for personnel and light freight and by ship for heavier goods and materials. The port is at Kiunga on the Fly River 153 km from the portal site.

The program for the project was tight, the decision to purchase the required TBM was made in July 2007 and the final decision to go ahead with the construction of the tunnel was made in November 2007.

MAJOR RISKS

At the planning stage of the project OTML conducted workshops at which the major risks to the project were identified (or most of them). The principal risks identified are listed below:

- TBM related risks including late delivery, assembly delays, unsuitability, and unreliability;
- Transport logistics
- Inexperienced Contractor and inexperienced personnel
- High rainfall delays to site establishment and operations

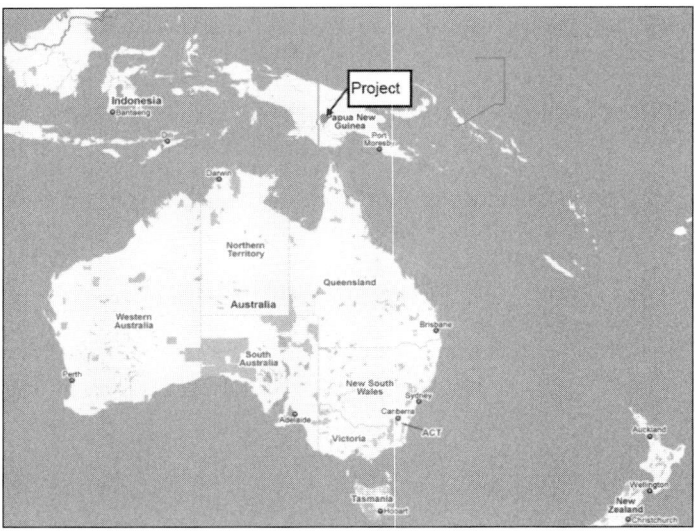

- High ground water flows in tunnel
- Adverse ground conditions, particularly faults across the tunnel line with water pressures up to 37 bar (cover over the tunnel was up to 500+ meters).
- The Potential of disputes between OTML and the Contractor.

GEOLOGY

The geology along the tunnel alignment from the geotechnical baseline was:
- CH 0-2000 siltstones/mudstones
- CH 2000-2330 an igneous intrusion of monzodiorite
- CH 2330-3430 high grade metamorphosed siltstones
- CH 3430-3800 monzodiorite
- CH 3800-4200 high grade metamorphosed siltstones

Either side of the intrusion the siltstones are highly to moderately metamorphose. The rock strengths vary from around 20 MPa at the portal to around 300 MPa in the highly metamorphosed siltstone. The tunnel alignment follows the axis of an anticline, the ground has also been faulted, with faults of up to 2.5 m widths expected in the alignment.

The faults were expected to hold water at considerable pressure of up to 37 Bar. The number of faults of significant width is expected to be relatively small. The geotechnical baseline report for the project stated that two wide faults, requiring ground treatment prior to taking the TBM through, were expected. The tunnel was expected to be wet with high quantities of ground water draining into the tunnel as it advanced.

TBM RELATED RISKS

To mitigate the risks associated with the TBM procurement OTML engaged a suitably qualified mechanical engineer to help manage the process.

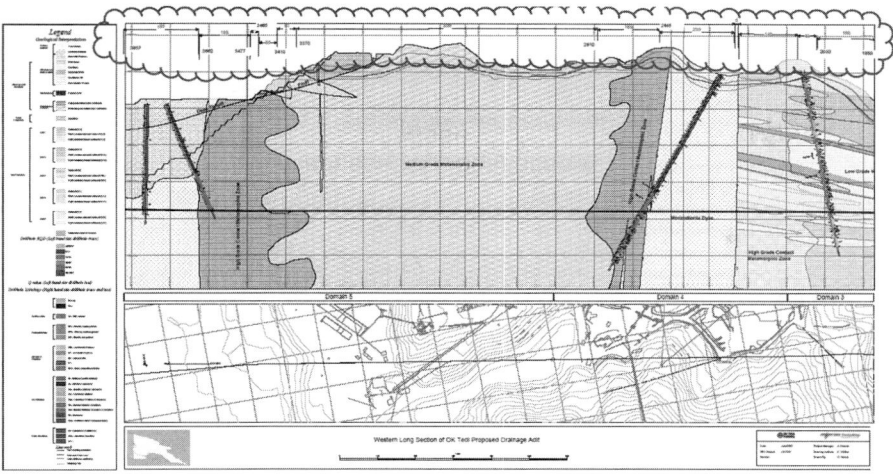

Figure 1

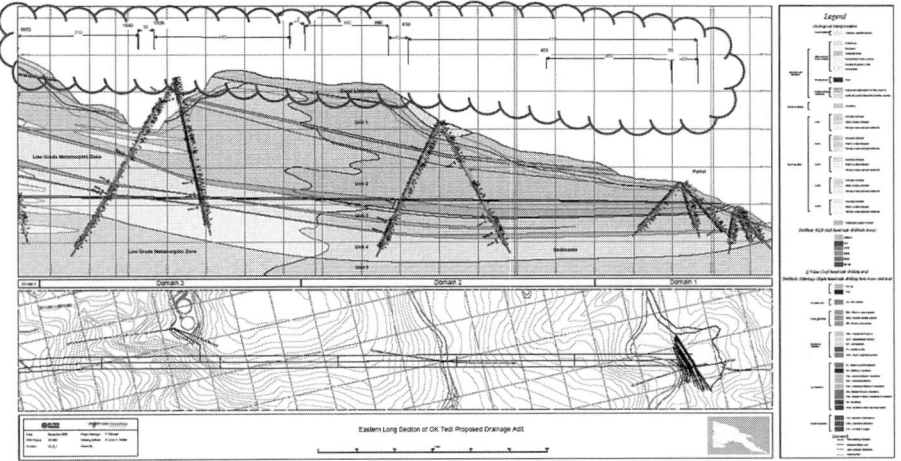

Figure 2

TBM DELIVERED TOO LATE

If the TBM was not delivered and commissioned by the middle of 2008 it would not be possible to achieve the required completion date for the project. This risk could be broken up into, the TBM:

- was ordered too late;
- took too long to manufacture, refurbish;
- took too long to assemble and commission

It was clear from the first discussions held with OTML that there would not be enough time to have a new TBM manufactured for the project. It was also clear that to ensure that a TBM would be on site in time OTML would have to procure the TBM and

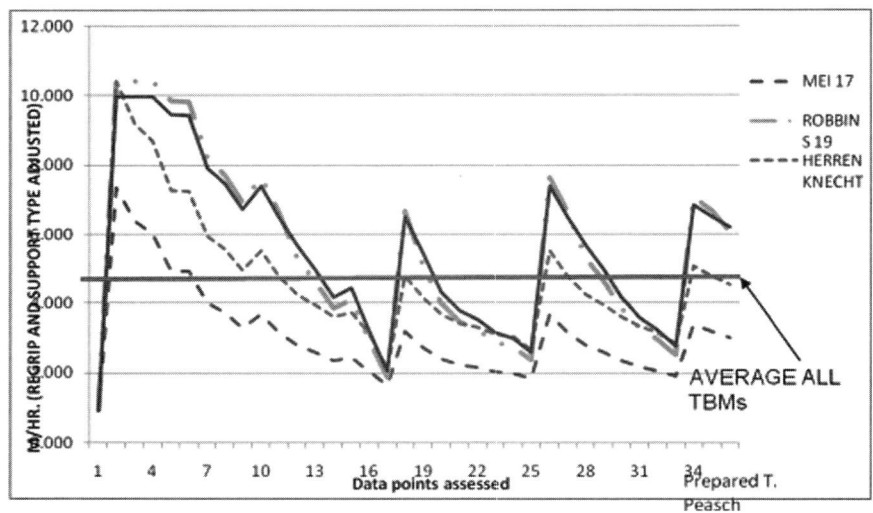

Figure 3. TBM preformance at chainage

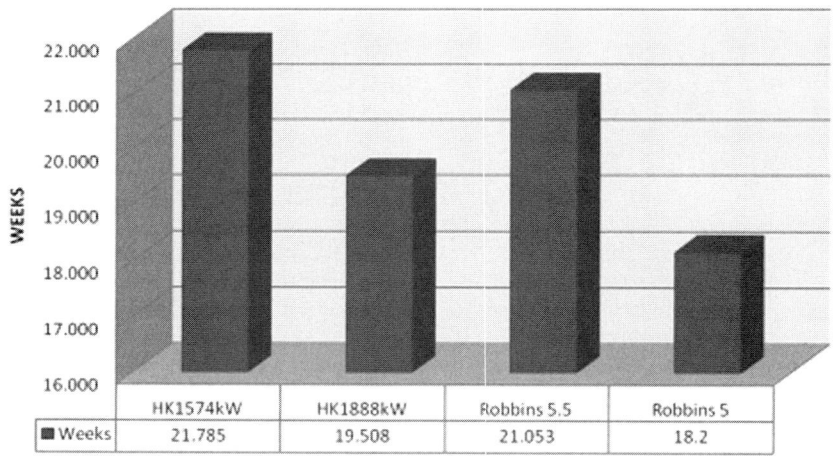

Figure 4. Comparison of potential TBMs, estimated weeks to complete main tunnel

provide it to the contractor for the tunnel construction. The contract for the supply of the refurbished TBM included liquidated damages for late delivery.

The geology was evaluated and the resulting available information was modeled to arrive at suitable TBM specifications. This revealed after investigation that four possible candidate TBMs may be suitable.

The obvious choice due to the critical nature of completion was to select a machine with an excess of power with appropriate ground support systems to be able to advance as rapidly as possible. Basically there was no limit to the provision of electrical power for the TBM as the mine operates its own hydro and diesel generating plants.

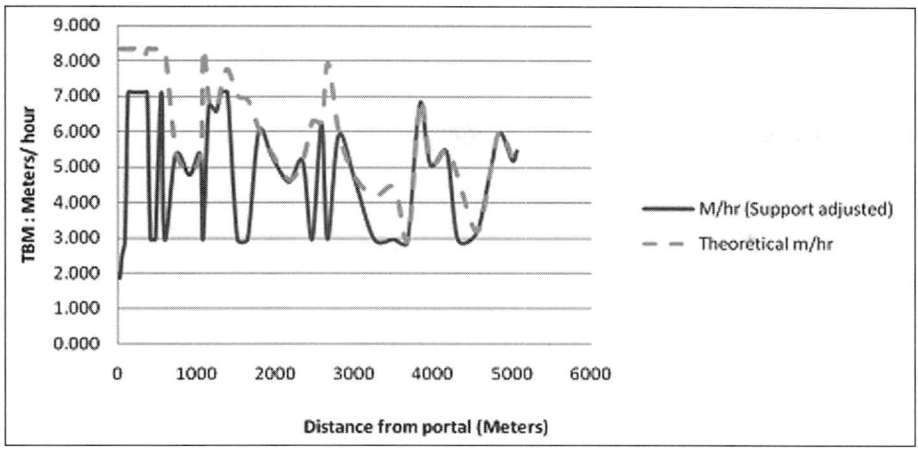

Figure 5. OTML drainage drive main tunnel ground support adjusted TBM production expected

TBM UNSUITABLE

The risk that the TBM would not be suitable for the tunnel was firstly mitigated by producing a specification for the machine. The re-manufacturers were asked to quote for the supply of the machines refurbished to OTML's specification. The Contract of the supply of the TBM was eventually awarded to The Robbins Company.

Once the TBM had been selected a more detailed evaluation of the potential performance of the TBM including the anticipated ground support was performed.

The results were formatted to provide the industry accepted weekly production histogram.

The risk that the TBM refurbishment would not be of the required standard was increased by the decision to shorten the delivery time by only carrying out a partially assemble the TBM at the supplier's workshops. The risk was mitigated by having regular inspections of the refurbishment while it was in progress. The supplier was given an incentive to provide a quality TBM by the provision of a bonus payment if the TBM availability was 85% or greater and the TBM excavated the tunnel by the required key date (End of February 2009). A detailed sub assembly criteria was developed and generally if components were NEW to THIS TBM and had a mass in excess of 300 Kg (660 lb.) it was required that they be assembled. Unfortunately the required assembly was not carried out as components comprising the new ground support systems had not arrived at the supplier's workshop by the required shipping date and a trial assembly was never completed in the workshop. This caused significant delays to the site assembly as parts were installed and modifications were either necessary or alterations mandated by the Contractor to mitigate operational safety concerns.

It was disappointing that, because there was a relaxation of the workshop trial assembly, there also appeared to be a relaxation of the diligence in insuring that all major fasteners (Structural bolts etc.) was not present and accounted during the packaging process. This lack of foresight created delays at assembly as the suppliers personnel hunted for parts they expected to be present but were not. There were resultant delays as components were air lifted from USA to the remote project site in Papua New Guinea. The supplier had been cautioned on this exact possibility, but the message appeared not to have been relayed to the personnel responsible for packaging and shipping.

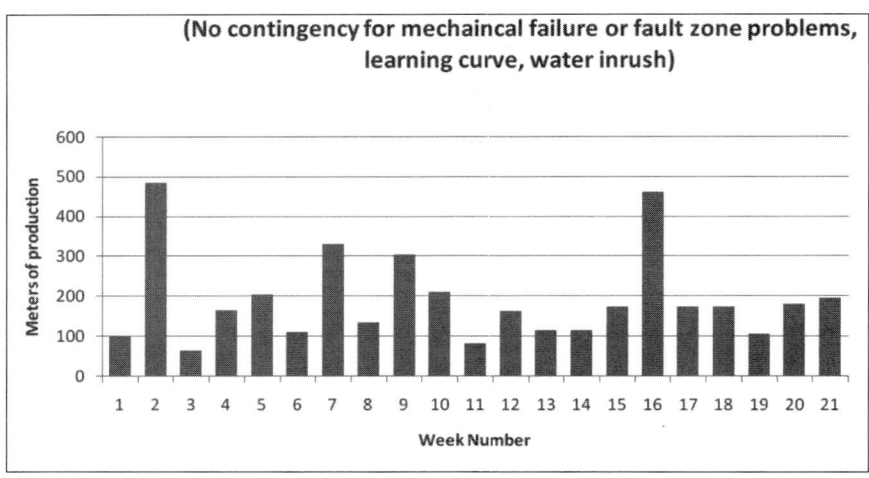

Figure 6. Predicted weekly production by TBM including support

Figure 7. TBM during site assembly (Note rain covers and water on the ground)

HIGH RAINFALL RESULTING IN DELAYS TO TBM ASSEMBLY

The portal site experiences between 8 and 12 meters of rainfall per year. Through the TBM assembly period it rained almost 100% of the time. The original intention was to construct a shelter over the TBM assembly after the major elements of the TBM had been positioned. What eventually transpired was the erection of a combination of small shelters and tarpaulins.

The personnel were almost continually working in heavy rain and were regularly soaked to the skin. However the temperatures were such that the crew was not too intimidated and the resultant delay caused by the wet conditions was minimal.

GENERAL RISKS

Logistics

There is no road access to the Ok Tedi Mine and thus the project. Twenty five kilometers from the mine there is an airport at Tabubil, the village built to house the mine workers. The port, Kiunga, is on the Fly River approximately 160 km from the mine. The port and mine are connected with an unsealed road. OTML has a logistics organization that is set up to supply the mine. It is based in Brisbane and was heavily involved with organizing and overseeing the transport of the TBM, and all the other equipment and materials to site. The time that was required in the program to get something to site by sea, from the time it was ordered, was 3 months. Thus if a machine part, or a particular material, was required and was not on site, it could not be brought to site inside 3 months without using air freight and incurring the resultant cost increases.

Inexperienced Contractor and Inexperienced Personnel

A major risk in hard rock tunneling is that the contractor and, (or), the personnel on site do not have the required expertise.

OTML sent requests for expressions of interest to a large number of tunneling contractors both within Australia and international. Unfortunately only two contractors expressed an interest and finally bid. The Contractor who was finally awarded the work was not as experienced as was initially considered necessary. The experience of the critical roles for the Contractor's team on site was specified in the contract documents which helped the on-site team but did not improve the backup from the Contractor's head office.

A comment from a number of the contractor's who did not bid was that they were not interested as OTML had procured the TBM, both reducing the size of the contract and the contractor's control over the project outcomes. This was not a risk identified during discussions leading up to the decision to procure the TBM.

High Ground Water Flows

High ground water flows were predicted in the Geotechnical Baseline for the tunneling. The flows would have made a back up train system in the tunnel invert unworkable. The decision was taken at the time the TBM was specified to use an invert segment 500 mm high to lift the rails out of the water.

Faults Across Tunnel Alignment

It was predicted in the Geotechnical Baseline that the tunnel would encounter faults of between 2 and 3 meters width in the alignment. It was also predicted that the water pressure in these faults could be up to 37 bar.

It was decided that to mitigate the risk of running the TBM into a fault containing crushed material and water at 37 bar, probing would be carried out in front of the TBM.

To mitigate the risk that the faults would result in the grippers not being able to engage the tunnel walls adequately invert thrusters were specified to push off the invert segments.

Contractual Disputes

The Contractor that was engaged to do the construction was a relatively small contractor who was principally involved with the mining industry in Australia. That industry was booming and almost all of the contract work in it had gone to cost plus when the Ok Tedi project was being set up. The Contractor was very enthusiastic to maintain a cost plus arrangement for this project. OTML were not enamored with the concept and finally an 'Alliance' agreement was negotiated. This had a number of beneficial effects including reducing the risk of contractual problems due to the procurement of the TBM by OTML. The perennial problem of latent ground conditions for a tunneling project was mitigated by OTML providing a Geotechnical Baseline for the project and making it clear that OTML took the risk of the ground being worse than that report indicated. Some risks associated with using an Alliance arrangement were identified during the workshops but were not adequately managed due to the limited time available.

CONCLUSIONS

TBM

The project was successfully set up in a reasonably short period. From the time the decision was made to go to the time that the TBM started boring was just over a year. This was a delay of around two months from the program.

The time set out for delivery of the TBM in the TBM supply Contract was 6 months after the signing of the LOI and this was not sufficient. The authors would recommend that delivery times of at least 8 months are included in project programs regardless of what the people selling you the machine say. It is better to build the correct period into your program from the start.

Delivery of the TBM was not completed until 4 months after the required delivery date. The quality of the refurbishment of the TBM was not adequate, and this and the partial delivery resulted in a TBM assembly that took over 3 months. The authors believe a full assembly and testing of the TBM must be specified for the refurbishment of any machine and that the purchaser should consider programming inspections by experienced mechanical/electrical personnel at biweekly intervals during the refurbishment. An allowance should be made in the budget for 100% presence by the purchaser's inspectors during the time of trial assembly and testing in the manufacturer's workshop. There is a major risk (almost certain) that any time saved by skipping steps in the manufacturer's workshop will result in major delays during the assembly/commissioning of the machine.

The packing of the TBM for shipment should also be specified by the purchaser to avoid problems with the TBM packing delaying the assembly process. The purchasers' representatives should be present during the entire packing process.

To date the TBM has handled the ground conditions well and made good progress once it was commissioned and the tunneling crews learned how to use it and made the necessary adjustments to get the required performance.

General

The very tight program for this project meant that some of the problems that should be avoided in any project were not. The project took longer and cost more than it would have done if more time had been available to set it up. Owners should be aware that compressing any project program can only be done with a costs penalty which can be quite large.

The logistics backup to the project from OTML was very good; there were minimal delays due to shipping problems, though some delays were experienced due to inadequate stock control on site. The project was fortunate that the logistics department had spent the previous year supplying a large construction project at the mine and thus the systems had been put in place and the personnel concerned had the required experience.

The lack of experience in the Contractor's project management team (the Contractor did not manage to hire a project manager) both on site and in the head office resulted in unnecessary delays and costs. However the specification of required experience levels for the supervisors and TBM operators meant that there were adequate experienced people around to train the national workforce at same time as the TBM was being driven forward. OTML retained 4 Robbins personnel to assist with the driving of the TBM during the tunnel excavation. The authors recommend that required experience levels be specified in the contracts for all tunneling works.

There was not enough time available to set up the Alliance Agreement and educate the parties to the Agreement as to what was required. This resulted in less than optimum management of the project as the Alliance Agreement allows for too many degrees of freedom to the participants. The Alliance method did not work particularly well for OTML. The Contractor tended to treat the project as a cost plus contract and the Alliance agreement does not have the controls necessary to manage a Contractor who takes this approach. This resulted in delays and cost overruns that were only partly recovered by OTML through the pain/gain share inherent in Alliance contracting.

At the time of writing the TBM had advanced over 2,000 m (6,600 ft) and the ground conditions had been more or less, what was expected from the geotechnical baseline, two faults had been encountered that required sets as ground support. The probe drilling was of limited value for mitigating the risk associated with these encounters. The ground was so variable that the conditions encountered by the probe only a few meters outside the tunnel perimeter could be quite different from those encountered by the TBM. No high pressure water had been found in the faults encountered.

THE DEEP UNDERGROUND SCIENCE AND ENGINEERING LABORATORY AND THE CONSTRUCTION OF PHYSICS MEGACAVERNS

C. Laughton ■ Fermi Research Alliance

ABSTRACT

The Deep Underground Science and Engineering Laboratory (DUSEL) is to be constructed at the defunct Homestake Mine, in South Dakota. As a cornerstone of DUSEL's research program, it is intended to conduct experiments into the fundamental nature of neutrino and proton particles. The experimental apparatus required to conduct these experiments will be housed in large rock cavern(s) sited at depths of over a kilometer. As conceived, these cavern(s) are planned to have mined volumes of up to half a million cubic meters, and spans of approximately 50 meters.

The paper will describe key end-user requirements, and outline the in situ rock conditions and mine site constraints that are likely to play determinant roles in identifying candidate sites and developing cost-effective design options(s).

INTRODUCTION

The Deep Underground Science and Engineering Laboratory (DUSEL) is a multi-disciplinary Laboratory designed for construction within the footprint of the Homestake Gold Mine, located in South Dakota, United States of America. DUSEL will provide international science and engineering communities with a dedicated underground facility capable of supporting a broad spectrum of fundamental and applied research at depth in the earth's crust. Research partners of the facility include physicists, biologists, geo-chemists, rock mechanics and rock engineers.

To perform new experiments, researchers are calling for the development of laboratory facilities and the study of large, undisturbed rock volumes. One major physics experiment proposed for DUSEL is the Long Baseline Experiment. This experiment requires the construction of single or multiple, large-span rock caverns, sited at depths well in excess of a kilometer. This experiment will allow for the performance of "frontier research" into the fundamental behavior of neutrino and proton particles. To support this new research initiative, the rock engineer must design access tunnels, infrastructure and cavern opening(s) that meet demanding requirements for long-term environmental and structural stability, delivered at an affordable price and tolerable level of risk.

THE HOMESTAKE MINE AND DUSEL SITE

The Homestake mine is located in the town of Lead. Lead is located in the Black Hills of South Dakota, some 60 km northwest of Rapid City. Gold was extracted from the mine for over a century. The inventory of existing openings is large, including over five hundred kilometers of development and haulage tunnel, and interconnecting ramps, shafts and winzes that provide access to a large rock mass volume to depths of approximately 2.4 kilometers.

Figure 1. Homestake Shaft Head House

The State of South Dakota is currently performing initial shaft re-entry work to re-establish access to shallow and intermediate mine depths, install a dewatering system and perform an initial round of basic tunnel rehabilitation tasks. The University of California at Berkeley is responsible for the preliminary design of the full facility under a 15 million dollar grant from the US National Science Foundation.

THE PLANNED UNDERGROUND FACILITIES

As currently envisaged, researchers intend to identify, develop and occupy a number of research campuses, strategically placed at depth within the reopened areas of the mine. The shallowest sites will be accessed by drive-in portal. The rehabilitated shafts, and tunnels will provide access to laboratory sites, planned for intermediate and deep mine levels. Levels are designated as feet below ground level.

On-going work at the mine is focused on re-opening mine shafts to regain access to the 4850 level. At this level, a combination of rehabilitated mine workings and new excavations will be used to house an initial suite of experiments. A head house for one of the shafts that will provide access to depth is shown in Figure 1. The installation and operation of these experiments will be funded and managed by the State of South Dakota. At the time of writing, the 4850 level is under water, but it is anticipated that it will be dewatered in the near future as part of the on-going rehabilitation program.

REQUIREMENTS FOR UNDERGROUND PHYSICS

As currently envisaged, designers of the Long Baseline cavern(s) can usefully draw on experience gained building similar facilities at the European Particle Physics Laboratory (CERN), Switzerland; Fermilab, USA, the Gran Sasso Laboratory, Italy; and laboratories sited in the Creighton, Kamioka, Pyhasalmi, Soudan mines, located in Canada, Japan, Finland, and the USA respectively. Underground requirements for physics research can also be usefully compared to those adopted at other underground sites where an underground environment suitable for permanent human occupancy and the operation of electrical-mechanical systems must be accommodated, e.g., road, rail, hydro-electric and public use facilities. Indeed, proven engineering solutions have been developed to address many commonly encountered underground requirements for access, equipment and infrastructure, storage, maintenance, egress, refuge and decommissioning.

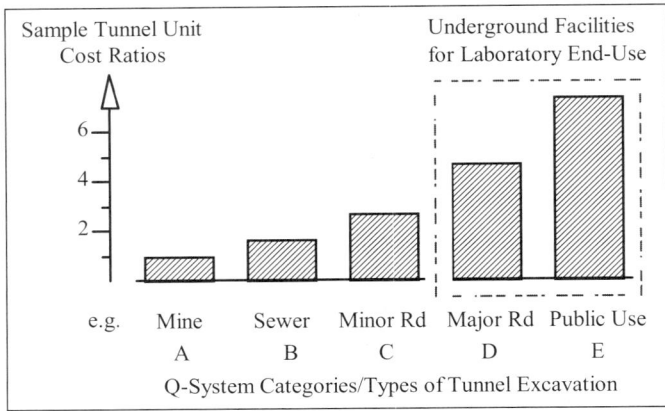

Figure 2. Sample unit cost ratios for Q-System categorized tunnels

However, in some instances physic research calls for the adoption of requirements that are more stringent than those demanded by other, more prosaic users. The physics end user may impose strict tolerances or specifications for excavation alignment, surface cleanliness, air quality and climate control (dust, temperature, humidity, etcetera), gas exclusion (radon), groundwater control, fluid containment, radiation shielding, and floor movement. In particular, a high premium may be placed on long-term, maintenance-free stability. In the context of a physics experiment, stability is not simply an issue of limiting ore dilution or retaining larger falls of ground for short periods.

Within laboratory space, even comparatively minor vibration, convergence, stress- or gravity-driven fall-out occurring over an extended operating period (decadal) could be highly disruptive to research activities. Installed apparatus is typically unique, high value, and highly sensitive to the impacts of ambient vibration or deformation. Damage scenarios can result in major repair and recalibration, potentially leading to additional project cost and delay that in turn can undermine the timeliness and relevance of the research products and negatively impact the development of follow-on experiments. To ensure that facilities can successfully host research, additional experiment-specific design mitigations may be required. Such mitigations typically result in increments in construction cost relative to facilities built for other end-use, as indicated in Figure 2.

The figure was developed to provide early guidance on cost as a function of end-use. As shown here, the figure is not intended to be used directly for estimating purposes but rather serves to underline the major unit cost differences that may exist between underground spaces built for different purposes. A recent study conducted in Finland identifies a lab space to raw space unit cost ratio (Euros/cubic meters) of four-to-one for hard rock openings excavated in an operating mine (Peltoniemi, 2006). At larger spans excavation costs will also be strongly influenced by factors that limit excavation productivity, including the size and layout of access tunnels, infrastructure and rock removal capacities, the opening shape, and above all the site geology.

HOMESTAKE GEOLOGY

The Homestake Mine is sited in heavily-folded, meta-sedimentary hard rock units, consisting primarily of schists, phyllites and amphibolites. These units contain faults, fracture zones and cross-cutting dike intrusions. The stress and temperature gradients at the mine are significant, with seismic activity, high-humidity and hot water inflows all

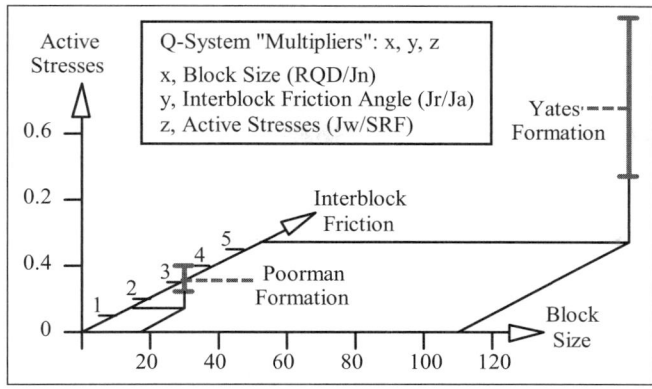

Figure 3. Q-Plots for the Yates and Poorman formations

reported at depth over the life of the mine. Mine descriptions of the geostructure, stress regime and hydrologic all indicate that a relatively wide range rock mass properties and behaviors may be expected at the mine site.

Rock engineers will need to prepare a site investigation plan that can provide for adequate characterization of site-specific rock mass parameters (intact strength, anisotropy, stress, fracture, water conditions, etc.) before key siting and design decisions can be made in confidence.

PRELIMINARY ROCK MASS CHARACTERIZATION ON THE 4850 LEVEL

Pending re-entry, evaluation of ground conditions and support adequacy has focused on the assessment of the Yates and Poorman rock units. These units will host the new and rehabilitated excavations planned on the 4850 level. Preliminary characterization of these units has been based on the study of existing data sets that were collected over the life of the mine (Steed and Carvalho, 2006). The data was used to calculate the three multiplying factors that represent, block size, inter-block friction and active stresses within the context of the Q-System rock mass classification (Barton and Grimstad, 1994). The three parameters are depicted in 3-D space in Figure 3. The figure provides for a visual assessment of the relative rock mass strengths and weaknesses for the Yates and Poorman formations, as weighted in the Q-system.

The block size (RQD/Jn) and inter-block friction (Jr/Ja) parameters shown in the figure were estimated by reference to rock core logs and local outcrop. Stress Reduction Factors SRF's were calculated using mine-specific strength and stress measurements at the crown and sidewalls of the excavations.

A joint water factor (J_w) of one was adopted. Although the proposed 4850 campus level is currently under water the "dry" J_w value was judged to best represent mine conditions during operation. Dewatering operations on the upper levels have begun and it is anticipated that the rock mass will be fully-drained below the 4850 level well prior to the commencement of the Long Baseline construction activities.

To estimate ground support needs an Excavation Support Ratio of 1 was used. This ratio reflects the demands for permanent occupancy and long-term stability, described above.

DETECTOR CAVERN—CANDIDATE SITE IDENTIFICATION

In rock engineering practice, the siting of underground excavations is often constrained by non-geotechnical factors e.g., function, land-ownership, rights-of-way, water and mineral rights, environmental acceptability, politics, etcetera. At DUSEL, there are no such constraints and the rock engineer is afforded the luxury of being able to seek-out the best ground conditions within a relatively large, rock mass volume. A start-up list of siting criteria is offered below:

1. Adequate shielding depth to ensure the detector serves a multi-functionary role.
2. Within the Yates formation where desktop Q-studies indicate that the Yates rock mass material is superior to that of the Poorman.
3. Outside the zones of influence of major mapped geo-structures where excavation wall and crown rock materials could be weakened or local stress and/or displacement anomalies present.
4. Outside the zones of influence of rock contacts where significant stiffness/stress contrasts could be present.
5. Outside the zones of influence of mine stopes and other old mine workings where zones of deterioration, overstress or destress may be more frequent and severe than in virgin ground.
6. Outside the zones of influence of existing or planned DUSEL excavations where higher levels of blast vibration and in situ stress superposition can be expected.
7. In close proximity to rehabilitated openings and infrastructure where costs to support the investigation, construction and operation can be shared.

Based on the above criteria, the 4850 level has been determined to be the best level upon which to initiate a search for candidate Long Baseline detector site(s). Siting the detector on the 4850 level will allow the Long Baseline project to benefit from the shared use of rehabilitated openings and ready access to in-place infrastructure. If adverse conditions were found on 4850 back-up sites could also be investigated on adjacent levels.

EARLY SITE CHARACTERIZATION OBJECTIVES

The criteria outlined above emphasize the need for the early acquisition of site investigation data to support the siting process. For the Long Baseline detector, an initial site investigation and characterization effort is needed to develop a shortlist of candidate sites and establish a credible baseline concept that can meet the design goals, in a timely manner, at an affordable price and tolerable level of risk.

The initial phase of investigation will likely focus on the review, validation and extension of an existing mine model, including a refined representation of the Yates formation. Although the Yates has been identified as the preferred host rock mass material, it was not a primary focus of the geologists during mining. The Yates is gold-barren and was, in all probability, under-mapped relative to neighboring gold-bearing units. Comprehensive remapping of the existing excavations will ensure that key rock mass features are more reliably characterized and located in 3-D across the campus site.

Once the 3-D mine model has been updated and cavern siting criteria confirmed and weighted, the model can be reviewed by geologists and rock engineers to identify a shortlist of best available site(s). It is these sites that will be investigated first.

KEY ELEMENTS OF THE INITIAL SITE INVESTIGATION WORK

After the mine has been dewatered below 4850, mine drifts re-stabilized and critical infrastructure installed, initial site investigation work will begin. During the first phase of field activity it will be important to prioritize the acquisition of data to support an improved characterization of the Yates rock mass and in situ stress regime on the 4850 and adjacent levels. This phase will likely include field and laboratory tasks commonly used in the investigation of initial rock excavation sites.

For the early study of large rock mass volumes, it is proposed to include the use of geophysical methods to cost-effectively screen large rock volumes. Geo-tools and sensors can reoccupy existing exploratory holes and tunnels to better locate major discontinuities, and develop 3-D models of material properties and stress distributions. Local conditions within candidate rock mass volumes can be further explored and models validated using core drilling, logging, down-the-hole testing and cross-hole tomography. Given the SRF ratios projected for openings on the 4850 level, the measurement of stress within the Yates formation will be a key component of the early site investigation campaign.

Once geotechnical attributes of candidate site(s) have been identified, appropriate analytical model(s) can be selected to provide for first estimates of ground response and support requirements. These early sensitivity studies should focus on identifying and studying the ground and construction factors that will most strongly influence stability and cost and identify the key design aspects that merit further investigation. Follow-on investigation work and detailed modeling should ultimately allow for optimization of the design to include the detailed consideration of factors such as excavation size and shape, orientation, drift and bench sizing and sequencing, ground treatment, reinforcement, and liner options. Modeling accuracy should be evaluated and refined during the earlier excavation phases of DUSEL excavations, and the large cavern facilities themselves comprehensively monitored during construction and operation.

SEEKING THE BEST VALUE DETECTOR-CAVERN OPTION

At this point in time, there appears to be no intrinsic reason why large-deep caverns cannot be built at DUSEL. However, this statement of feasibility is based on limited site data and a minimal engineering effort. Key issues of affordability and risk, which may ultimately determine whether the facility is built or not, remain to be addressed in more detail. Pending the acquisition of site-specific data and a substantial design effort, the full scope of the project is yet to be accurately determined. However, for the scale of facility envisaged here, costs are conservatively estimated to be between a quarter and half a billion dollars and the duration for design and construction between five and ten years. Such numbers are based on inadequate geo-data sets and must necessarily be accompanied by high contingencies. Rock masses are inherently variable and, in the author's experience, surprises encountered during investigation and construction rarely lead to a cost under-run or early delivery.

Given the above commentary, it is premature to speculate on the merits of various detector-cavern options. However, an early effort will be needed to collect the data necessary to achieve "best value" in design, construction and operation at an appropriate time during the design process. Key technical criteria that need to be considered in developing an underground design are shown schematically in Figure 4.

The figure is intended to draw the end-users' attention to the added complexity that an underground site brings to a project. Working at depth in an old mine and within a natural material brings a broad spectrum of constraints and risks to the Long Baseline project. Developing a safe and cost-effective execution plan for an underground project like the Long Baseline requires the adoption of a multidisciplinary approach and, in particular, an upfront allocation of significant resources for investigation and

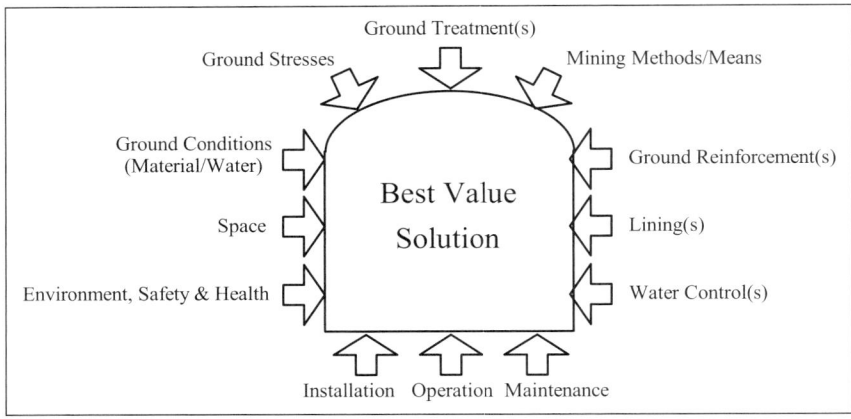

Figure 4. Underground engineering design criteria

geo-engineering work. Facets of an underground design process that may be undervalued or completely overlooked by project management teams less familiar with the vagaries and diverse challenges of underground work (Laughton, 2004).

OPPORTUNITIES FOR ROCK ENGINEERING RESEARCH

Given the scope of the Long Baseline facilities there are clear incentives to consider including the use of targeted engineering research tasks within the design and construction framework of the Project. Physicists and practicing engineers should consider research in a number of areas including: safety, site investigation, rock mass characterization, risk management, geological and geo-mechanical modeling, excavation, support, and ground instrumentation and detector technologies.

CONCLUSIONS

Large-deep caverns are needed to pursue "frontier research" into the behavior of protons and neutrinos. To support this research, the rock engineer must design and deliver large-deep cavern(s) that meet demanding requirements for long-term stability.

Pending investigation, it has been assumed that rock masses can be found at DUSEL in which large caverns can be excavated and supported cost-effectively. However, an early site characterization effort is critical in order to confirm the validity of this basic assumption, and support the initial investigation and short-listing of best-qualified sites. Even if rock mass conditions are highly conducive to the construction of large caverns the cost will undoubtedly represent a major investment of research funds.

To use these research funds to best effect, the design must be subject to a rigorous optimization process that takes into account the full range of engineering factors that drive costs underground, including not only a consideration of rock mass conditions and laboratory requirements, but also the influence of mine site constraints and construction practicality. The quest for best value should also actively consider the benefits of research, performed in partnership with industry, that can not only enhance the viability of the Long Baseline project but also have broader impact across the mining and civil engineering communities.

REFERENCES

Peltoniemi, J., 2006. "Finnish Underground Lab." EL SUD—LAGUNA Meeting, Garching, Germany, April 24, pages 16. http://www.e15.physik.tu-muenchen.de/research_and_projects/lena/laguna_meeting_april_2006/

Barton, N; Grimstad, E., 1994. "The Q-System following twenty years of application in NMT support selection." Felsbau, Volume 12, Number 6, page 428–436.

Steed, C.M. and Carvalho, J.L., 2006 "Geotechnical analysis of proposed laboratory excavations at the former Homestake Mine, Lead, South Dakota." Golder Associates Report 06-1117-014, pages 59. http://www.lbl.gov/nsd/homestake/conceptualdesign.html

Laughton, C., 2004. "Drawing from past experience to improve the management of future underground projects." North American Tunneling, Conference. Atlanta, Georgia, USA. Ed. L. Ozdemir, page 15–20.

SUBSURFACE REPOSITORY VENTILATION DESIGN

Edward Thomas ▪ Bechtel SAIC LLC

Jaime Gonzalez ▪ U.S. Department of Energy

YUCCA MOUNTAIN OVERVIEW

The U.S. Department of Energy's Yucca Mountain Project is a geologic repository project designed to receive, handle, package, transport, and emplace spent nuclear fuel and high level radioactive waste. The planned repository would be located in Yucca Mountain on the western edge of the Nevada Test Site, about 90 miles northwest of Las Vegas, Nevada, and would include both surface and subsurface areas. Yucca Mountain rock formations consist of successive layers of volcanic rocks (called tuffs) approximately 14 to 11.5 million years old, formed by eruptions of volcanic calderas to the north. With the volcanic rock the potential for methane emission is not an issue and the subsurface repository is a non-gassy (methane) classification.

This paper will provide a summary of the development ventilation system design that provides fresh air for subsurface construction activities, and the emplacement ventilation system design that removes heat generated by the waste packages. The subsurface layout, illustrated in Figure 1, contains 108 emplacement drifts, one Performance Confirmation (PC) observation drift, nine intake/exhaust shafts, and three access ramps. The existing Yucca Mountain facilities, shown in bold lines in Figure 1, include a 7.8 km long, 7.62 m (25 ft) diameter U shaped tunnel that is formed by the North and South Ramps at both ends of the Exploratory Studies Facility (ESF) drift and a 1.5 km long, 5.0 m (16.4 ft) diameter Enhanced Characterization of the Repository Block drift (ECRB) that were constructed from 1994 through 1998. These drifts were used for site characterization activities and have been incorporated into the subsurface repository configuration. The unexcavated perimeter access main and exhaust main drifts are typically 7.62 m (25 ft) diameter TBM excavations and the emplacement drifts are 5.5 m (18 ft) diameter TBM excavations. The turnouts leading into the emplacement drifts are to be excavated by either mechanical, or drill and blast methods and serve as the launch chambers for the emplacement drift TBM excavations.

SUBSURFACE VENTILATION SYSTEM OVERVIEW

The subsurface repository ventilation system consists of two operationally independent and separate systems: the development ventilation system and the emplacement ventilation system. The development ventilation system supports the excavation of subsurface openings and construction activities such as installing emplacement drift inverts, turnout bulkheads, isolation barriers, and the like. The emplacement ventilation system supports thermal management goals by ventilating and cooling emplacement drifts during the forced ventilation and closure periods. Isolation barriers separate the two ventilation systems.

Fans located on the surface circulate the ambient surface air throughout the subsurface development and waste package disposal areas. The development side fans are configured to force air into the subsurface while the emplacement side fans are configured to draw heated air from the subsurface. This configuration ensures that if

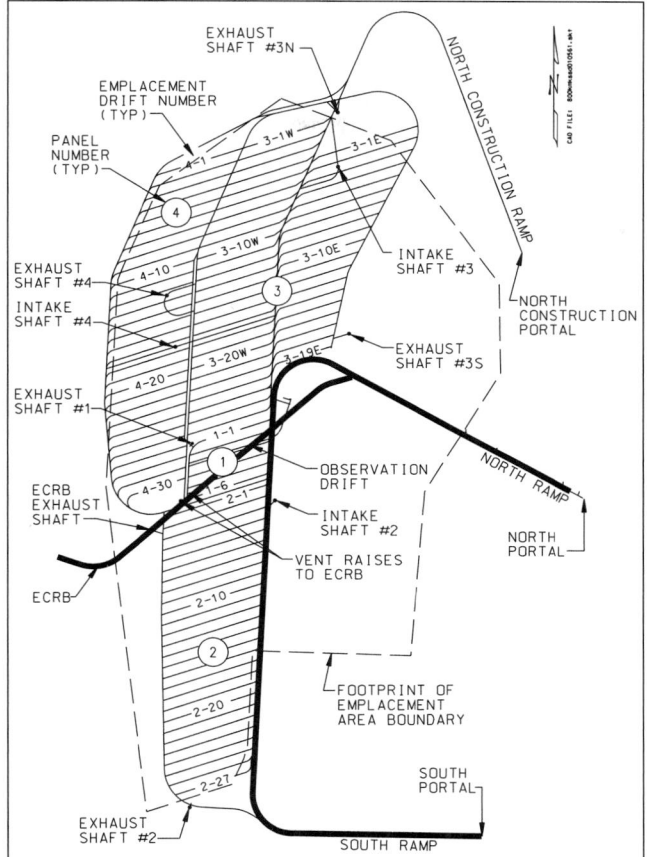

Figure 1. Subsurface repository layout configuration

one system shuts down, the flow direction of potential leakage across the isolation barriers is maintained toward the emplacement areas.

The subsurface ventilation system operates throughout the development, waste emplacement, the post emplacement monitoring, and repository closure operations; a period of up to 100 years. The subsurface emplacement ventilation system operates after final waste emplacement to achieve the repository thermal conditions for the postclosure phase.

Isolation Barriers

As noted above isolation barriers separate the two ventilation systems and are considered a common structure to both systems. The two separate ventilation systems allow for concurrent development of emplacement drifts on one side of the isolation barriers and waste emplacement in operational emplacement drifts on the other side of the isolation barriers. The separate ventilation systems create distinct fire areas, and the isolation barriers have a fire rating of three hours. Isolation barrier monitoring functions include door status, door control, and pressure differential.

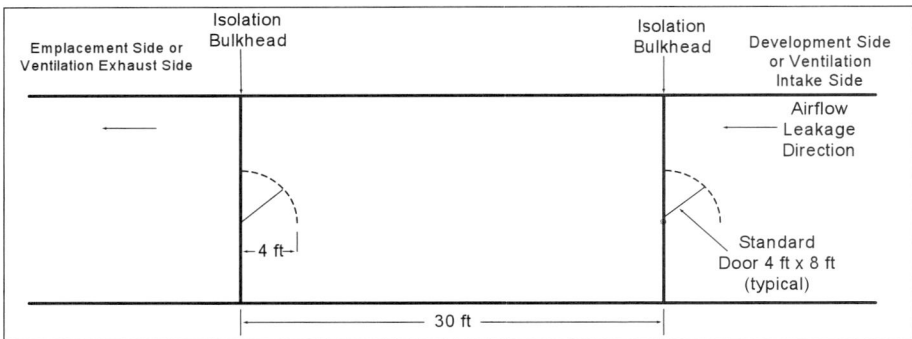

Figure 2. Type B isolation barrier plan view

The emplacement side of the repository is regulated by the Nuclear Regulatory Commission (NRC), and as such, development personnel egress through select barriers is restricted to emergency conditions and these isolation barriers will contain perimeter intrusion detection and assessment system controls. There are three types of isolation barriers used in the repository:

- Type A—Isolation barrier between the development and emplacement ventilation systems that would not permit emergency access.
- Type B—Isolation barrier between the development and emplacement ventilation systems that would permit emergency access.
- Type C—Permanent isolation barrier between emplacement intake airflow and exhaust airflow.

During development the Type A and Type B isolation barriers are installed in the exhaust mains and access mains and separate the development ventilation system from the emplacement ventilation system. As the subsurface is developed into a fully-loaded repository the Type A and Type B isolation barriers are relocated to maintain separation between the two ventilation systems. The Type C barriers are permanent because they remain in place for the forced ventilation period after final emplacement.

The Type B isolation barrier structure consists of two bulkheads, airlocks, and related operating and monitoring components spaced a sufficient distance apart to form an airlock chamber between the two bulkheads. Figure 2 illustrates a plan view of a typical Type B isolation barrier concept with an access door/air lock installation. A cross section of a typical isolation barrier bulkhead is shown in Figure 3. The dual bulkhead provides a configuration that assures at least one bulkhead remains available and effective if the other bulkhead is damaged or being maintained.

The Type A and Type C isolation barriers are similar to the Type B isolation barrier but located on the exhaust access drifts and prevents access to the very high radiation areas.

DEVELOPMENT VENTILATION SYSTEM

The development ventilation system supports construction of the subsurface repository by providing fresh air for a safe work environment, and removing or diluting potential contaminants such as radon, dust, and fumes. The development ventilation system will be modified as emplacement drifts are completed and turned over for waste emplacement. Ventilation systems described herein are common to the mining industry

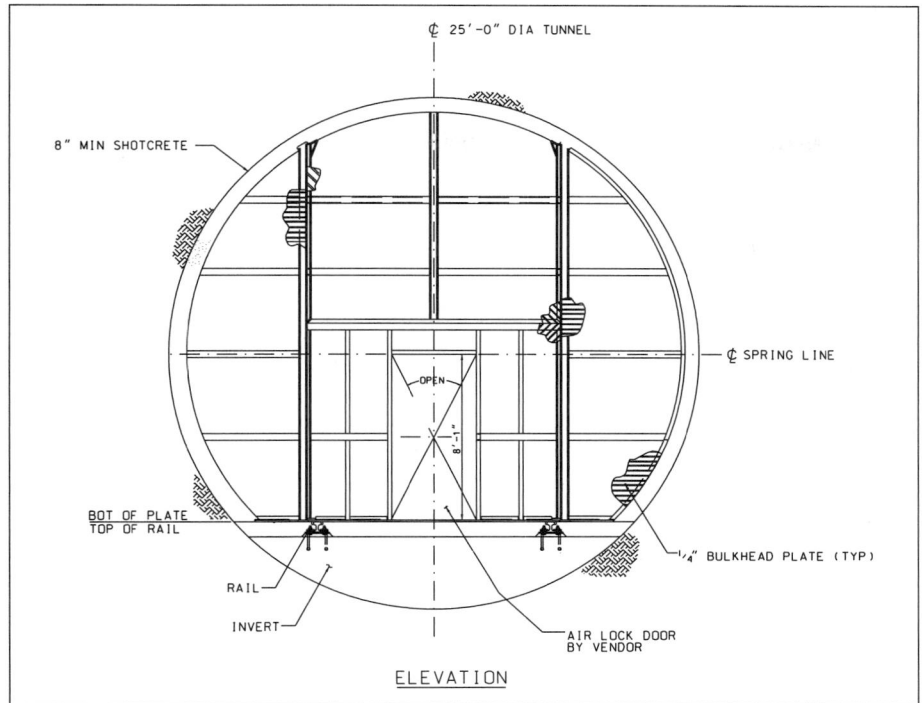

Figure 3. Isolation barrier bulkhead section view

and final construction specifications and design details would be coordinated with the contractor.

Except during the initial construction phase, the development ventilation system uses fans installed on the intake shafts to force air into the subsurface development areas. Auxiliary fans and ducting direct fresh air into the advancing excavation fronts and dust collectors will remove and clean the used air. Conventional shaft excavations are ventilated by a local ventilation system that intakes fresh air from the surface.

Dust Control

With the water table located approximately 210 m (690 ft) below the repository horizon and Yucca Mountain's desert climate with limited rainfall, the repository horizon conditions are extremely dry. The process of subsurface excavation generates dust and therefore with the dry conditions, efficient dust control systems will be a critical and integral part of tunnel excavation equipment. The use of multiple methods may be required to effectively control dust. Supplemental dust scrubbers, similar to those used during the original ESF construction, may be utilized in the return air stream.

Development Ventilation Design Interests

Water Usage. Water is considered the primary method used to mitigate dust so various design issues must be addressed for systems that use water. Water sprays are typically used to wet the broken material and capture airborne dust particles. Wetting the broken material will also help reduce dust generated during secondary handling processes, such as conveyor belt transport, or muck haul car dumping.

Spray nozzles are used to capture a portion of airborne dust particles and their effectiveness is determined by a well-engineered design. Use of high-pressure water sprays to control dust will reduce water consumption at the same time.

Scientific constraints severely restrict the use of organic materials in the underground since residue may impact radionuclide transport and may influence postclosure performance. This eliminates the potential use of commercially available surfactants, foams, or encrusting agents that would assist in meeting the dust control regulation.

Tunnel Boring Machine Dust Control. The majority of the repository tunnels are to be excavated by tunnel boring machines (TBM) and stringent dust control measures for TBM operations are planned. An efficient TBM dust collection and suppression system design is essential, and multiple stages of dust filtration may be required. The TBM cutting head design creates a contained area suited for dust capture, mitigation, and extraction. TBM generated dust is commonly suppressed using water sprays at the cutting head, a scrubber mounted on the trailing gear that will extract and filter dust-laden air from the cutter head area, and by water sprays at conveyor discharge points.

The TBM specification will indicate the performance requirements related to dust control and will include requirements for water sprays, dust scrubbers, cabs, and booster fans to handle local airflow requirements. A refuge chamber is planned as part of the TBM trailing gear. The successful contractor will submit plans prior to start of excavation to demonstrate ability to meet performance specifications.

Based on the ESF construction experience with conveyor belts, a covered muck rail car system is being considered for muck haulage to surface in place of conveyor belts.

Radon. Radon and its decay products are naturally occurring in the subsurface repository area and emanate continuously into the airflow. Mechanical ventilation (dilution) and other protective techniques (respirator or tunnel lining) can control the exposure to radon. The air volumes moving through the subsurface are designed to control, within acceptable limits, the concentration of radon and its decay products.

Heat Stress During Development. Based on historical ESF meteorological records and experience during the ESF construction, mechanical refrigeration will not be necessary for normal repository construction. Historical average air temperatures and humidity at the repository horizon are $75°F_{td}$, $52°F_{tw}$, and 19% relative humidity. However, the design of development ventilation systems will include considerations for equipment heat loads added to the ambient air temperature and consider supplemental cooling. The TBMs driving the long access mains, using multiple booster fans, may incorporate an integral cooling unit in the trailing gear to cool the work area near the TBM face.

TBM Ventilation Design Concept

This section describes the TBM ventilation concept for a 5.5 m (18 ft) diameter emplacement drift. The larger 7.62 m (25 ft) diameter access mains would use a similar system, scaled accordingly. The average emplacement drift excavation length is approximately 630 m (2,066 ft) and the longest emplacement drift excavation length is 808 m (2,651 ft). The emplacement drift TBM ventilation system is designed for the longest drift plus a corresponding turnout length, for a total length of approximately 910 m (3,000 feet). The 7.62 m (25 ft) access main drift excavation lengths vary from 2,885 m (9,460 ft) long to approximately 6,400 m (21,000 ft) long.

Conceptually the emplacement drift excavations will use two TBMs in a leapfrog process for excavation where one will be excavating and a second one is being moved or readied in the adjacent emplacement drift. When the excavation in the 1st drift is

SUBSURFACE REPOSITORY VENTILATION DESIGN

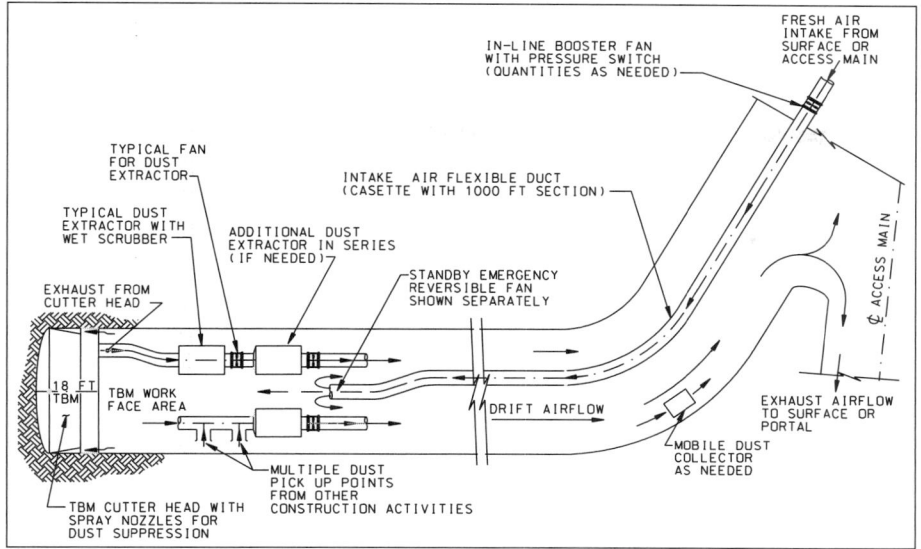

Figure 4. Emplacement drift TBM ventilation concept

complete, the tunneling crew will then move to the next drift where the TBM has been readied for use. The TBM that has completed the finished excavation is backed out of the tunnel, service and maintenance is performed, and the TBM is then moved to the next drift to be excavated.

Ventilation Duct. The ESF was constructed with a suction ventilation system using 20' sections of steel duct with fans placed in series along the length of the tunnel. A positive pressure system (blowing) using ventilation bags of approximately 305 m (1,000 ft) lengths provides limited joints, is considered for the repository TBM excavations.

With a bag line system the absence of joints minimizes leakage, pickup and recirculation of construction dust. Air leakage from a positively pressurized supply duct results in fresh air to the work area. The ventilation bag system does not place a burden on the supply handling system.

Figure 4 illustrates a typical ventilation system arrangement for an 18-ft diameter emplacement drift TBM excavation. As illustrated in this figure, water spray nozzles are provided at the cutter head and dust collectors are used to reduce airborne dust in the excavation areas and behind the TBM where personnel are deployed performing multiple duties.

Initial Construction Design Concepts

The initial construction effort includes all work necessary to establish the first three emplacement drifts in Panel 1 for emplacing waste. Panel 1 consists of six 5.5 m (18 ft) diameter emplacement drifts, an exhaust main, and Exhaust Shaft #1. An Observation Drift, used for instrumentation and performance confirmation purposes, is located under and offset from emplacement drift 1-3 and excavated with 5 m × 5 m (16 ft × 16 ft) dimensions. The Panel 1 excavation sequence is from north to south and is broken into two phases. Once waste emplacement is initiated, separate development and emplacement ventilation systems are maintained until the entire repository has been developed, a planned period of approximately 20 years.

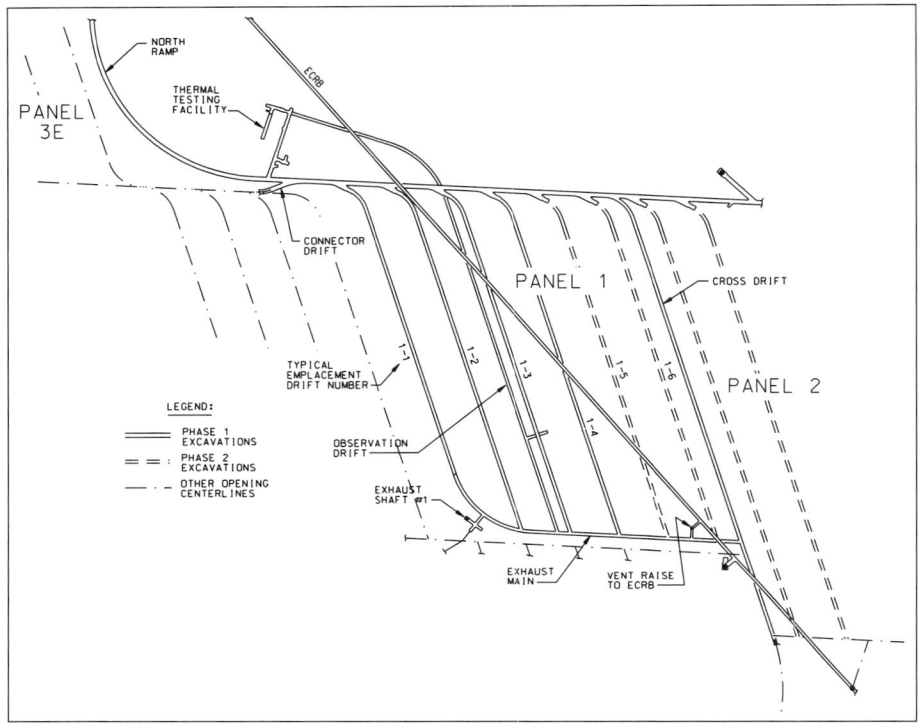

Figure 5. Initial excavation phases and corresponding openings

The following information summarizes the methodology used to construct Panel 1 in two separate phases, with Figure 5 identifying the excavations related to the two development phases. Phase 1 includes excavation of openings necessary to emplace waste in Panel 1 drifts 1-1, 1-2, and 1-3. The Phase 1 excavations are identified by solid lines in Figure 5 and include the Panel 1 exhaust main, connector drift, the Cross Drift to Panel 4, the Observation Drift, Panel 1 turnouts, Exhaust Shaft #1, and emplacement drifts 1-1 through 1-4. Emplacement drift 1-4 excavated in Phase 1, provides construction access and ventilation support for the Phase 2 effort.

After the emplacement drift excavation is complete the drift invert, rail, and ventilation structures are installed in drifts 1-1 through 1-3. As the outfitting of emplacement drifts nears completion, isolation barriers are installed to separate the first three emplacement drifts from the remaining Phase 2 development activities. The completed facilities are then turned over for commissioning and emplacement operations. At that time the emplacement side operations are NRC controlled and subject to applicable regulatory standards.

Phase 2 excavations, identified by the dotted lines in Figure 5, complete the Panel 1 development and initiate excavation of Panel 2. The ECRB Shaft and Intake Shaft #2 are excavated to support Phase 2 work, and a 7.62 m (25 ft) diameter TBM is launched from the bottom of the South Ramp to excavate the Panel 2 exhaust main. Once the Phase 1 isolation barriers are in place, flow-through ventilation from the North Portal to the South Portal is not possible and the ECRB Shaft is used to supply the ventilation air for Phase 2 development.

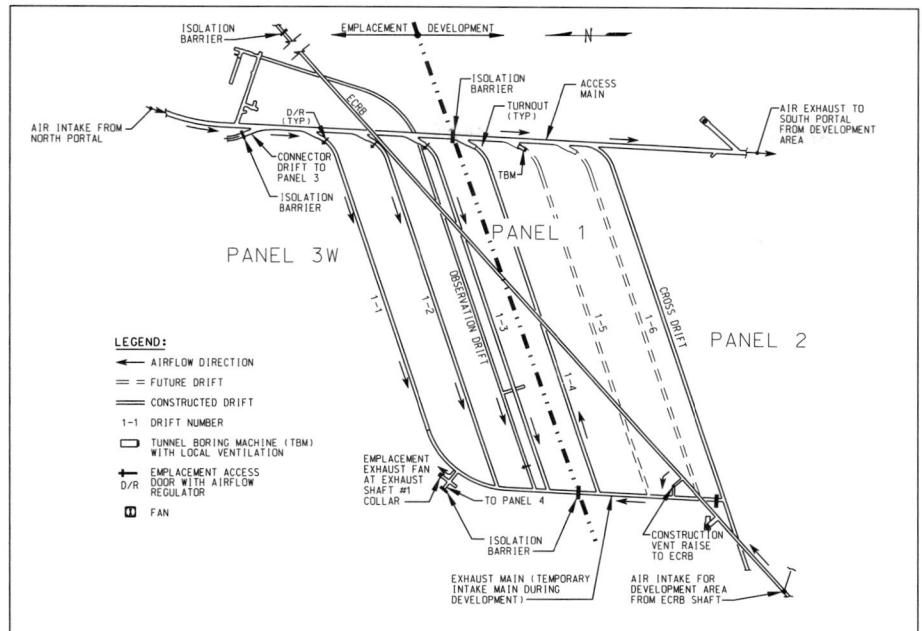

Figure 6. Ventilation configuration at Phase 1 transition to Phase 2

Figure 6 illustrates the ventilation configuration at the Phase 1 to Phase 2 transition. At each transition stage where emplacement drifts are turned over to accept waste, isolation barriers are relocated, and an extra drift provides the construction access and ventilation support for the next phase. This construction method of developing the repository in blocks of drifts is followed for the rest of the repository.

EMPLACEMENT VENTILATION SYSTEM

Thermal management is the limiting consideration for the emplacement ventilation system in terms of airflow rates and duration of ventilation. The emplacement ventilation system operates continuously after the last waste emplacement to remove enough heat generated by the waste packages to ensure the initial thermal conditions for the postclosure safety analyses are established. The nominal airflow rate to a fully loaded emplacement drift is 15 m^3/s (32,000 cfm) for thermal line loads up to 2.0 kW/m. This airflow volume and duration have been demonstrated to provide efficiencies of 85% to 90% in terms of heat removal during the preclosure period, for the thermal loading plans proposed for the repository.

The emplacement ventilation system components include ventilation fans, turnout bulkheads containing emplacement access doors and airflow regulators, isolation barriers, and instrumentation for control and monitoring of the system. The emplacement ventilation system uses fans installed on the exhaust shafts to draw ambient air down intake shafts or ramps, across the emplacement drifts, then exhausted to the surface. The repository's airflow distribution is controlled using a bulkhead and regulator located in each emplacement drift turnout. The system operates with the ambient intake air characteristics, and to handle both normal and off-normal situations in the operational phases of the repository.

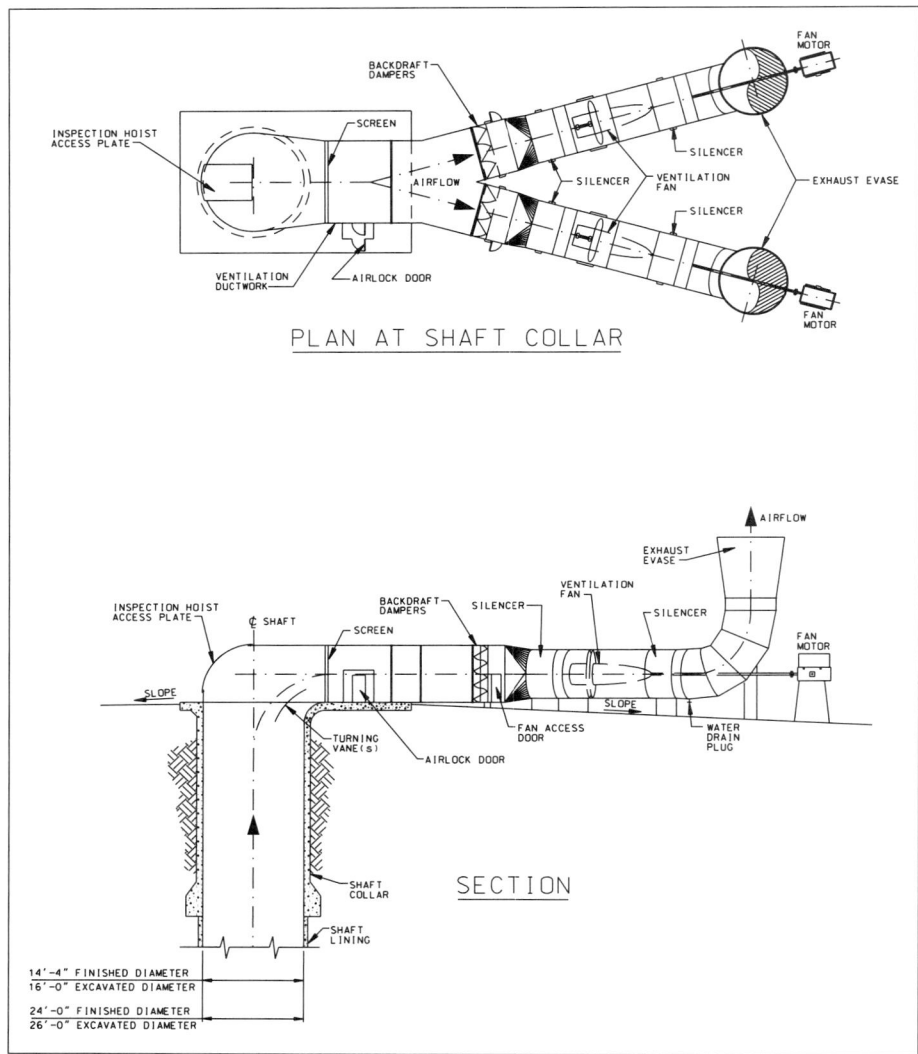

Figure 7. Emplacement exhaust fan configuration

Main Fans

The main fans provide the means to produce and control the repository airflow for the emplacement ventilation system. As illustrated in Figure 7 the emplacement fans are located at the exhaust shaft collars and contain two fans, typically operating in parallel to provide the desired airflow. The fully emplaced repository uses six exhaust shafts to provide approximately 2,100 m³/s (4,450,000 cfm) of airflow through the repository during normal operations. The total fan power required at each exhaust shaft is approximately 1,800 hp, or 900 hp per fan.

The layout configuration, shaft quantities, and subsurface opening dimensions provide operating pressures and airflows that are within the design capacity of axial flow fans. The fans are variable pitch, axial flow, and are driven by variable-speed motors. These characteristics allow flexibility to vary the air volumes as emplacement operations advance over time as the number of loaded drifts increases. With multiple ventilation shafts, each having two fans, a single fan down for maintenance does not have a major impact on the repository airflow volume. When one of the fans is down for maintenance, the second fan can be isolated to maintain a reduced airflow in the effected area.

Emplacement Ventilation Configuration

The integrity of the emplacement ventilation system and the goal to deliver the desired airflow through each emplacement drift supports the repository thermal and human safety objectives. The ambient air drawn through the repository enters the emplacement drift, via airflow regulators (regulator) that control the airflow quantities, then travels to the exhaust mains, and exhausts to the surface through the exhaust shafts. As air passes over the waste packages, it removes heat from waste packages and expands in volume. The exhaust mains are inaccessible by humans due to potentially high radiation and elevated air temperatures.

The airflow distribution throughout the repository is continuously monitored in real time and can be controlled by the digital control and management information system. Both the position and status of the regulator and the emplacement access doors are monitored remotely. The airflow volume through each regulator is also remotely monitored and can be adjusted remotely or manually. As the airflow requirements vary over time or as changes in performance are detected, changes in system's operating points can be evaluated and implemented remotely through the digital control and management information system.

The turnout bulkhead occupies the full cross-sectional area of the turnout and is located near the junction of the turnout and the access main, as close to the access main as possible in order to minimize the potential radiation dose rate at the location. The turnout bulkhead contains the emplacement access doors and airflow regulator. The emplacement access doors provide the means to restrict access to the emplacement drift, and the regulator provides the means to control the emplacement drift airflow volume. Figure 8 illustrates the turnout bulkhead location and emplacement access doors with respect to the access main and waste packages.

The emplacement access doors function as the access point to the very high radiation area of the emplacement drift, and as such, remain locked, except during periods of equipment access, to prevent unauthorized or inadvertent access to the very high radiation area within the emplacement drift and turnout. A Central Control Center on the surface provides supervisory control of the access door interlocks in order to maintain security and positive control of each emplacement drift.

Figure 9 illustrates an isometric view of the turnout bulkhead, emplacement access door, and airflow regulator. A nominal design volume of 15 m^3/s (32,000 cfm) per emplacement drift satisfies thermal requirements. A design range of 15 ±2 m^3/s (32,000 ± 4,000 cfm) per emplacement drift is based on thermal calculations. The range would help minimize the automated regulators searching for and continually adjusting to the operating points. The regulators and fan system allow for airflow to a single emplacement drift to be varied from zero to a maximum of 47 m^3/s (100,000 cfm).

The emplacement access doors are a counter-opening configuration and accommodate the transport and emplacement vehicle entrance and exit. The emplacement access door operations are instrumented to provide real time monitoring of door indicator functions such as door position, status, alarms, interlocks, and local overrides with safety controls.

The airflow regulator is interlocked with the emplacement access door such that when an emplacement access door is opened, causing a temporary airflow anomaly,

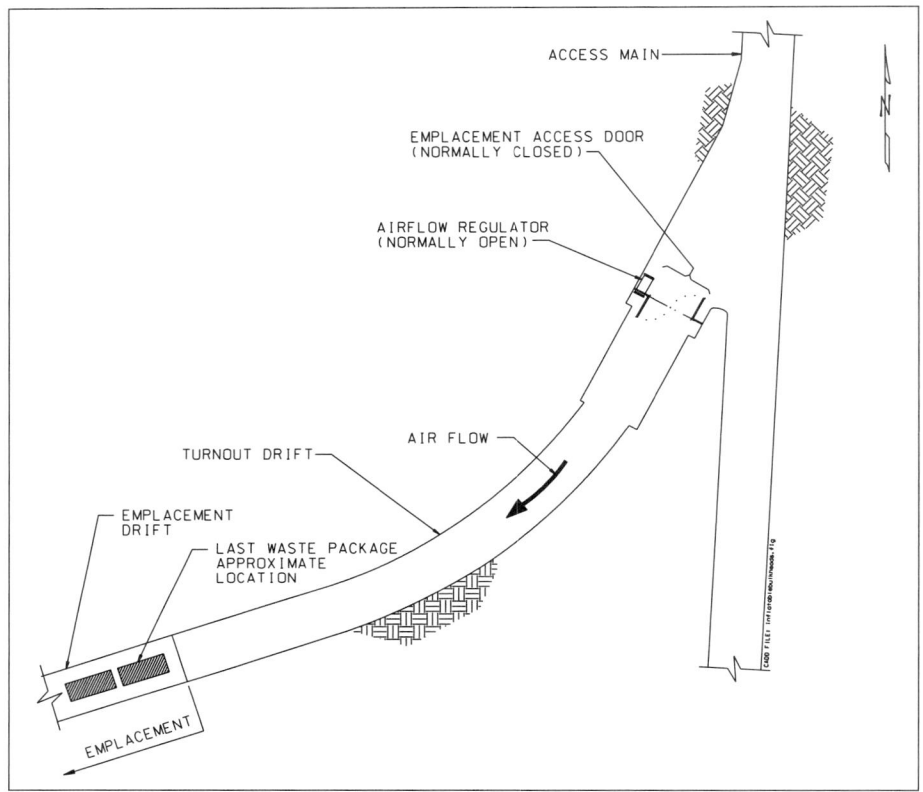

Figure 8. Typical turnout and ventilation control arrangement

the regulator louvers do not adjust for the perceived decrease in airflow. The airflow regulator monitoring provides operational information, supports performance confirmation, and enables the system components to be adjusted if repository requirements change.

The overall ventilation system is monitored to provide component operational information, system airflow conditions, and establish a baseline to confirm thermal management goals. The main intake and underground monitoring parameters include airflow velocity, barometric pressure, air temperature, humidity, airborne particulates, carbon monoxide, and radon. Airflow, temperature, relative humidity, and particulate monitors at the exhaust shaft collars provide performance information.

Normal operations in the subsurface facility involve the transport and placement of waste packages that are closed and sealed. Since the breach of a waste package and leakage of radioactive material from a waste package has been identified as beyond Category 2 (non-credible event) during preclosure, the emplacement ventilation system is not required to mitigate releases from the subsurface and no filtration is included in the subsurface ventilation system design. Potential airborne releases of radioactive materials during normal operations of the subsurface facility are possible due to resuspension of radioactive contamination from external surfaces of the emplaced waste packages, or neutron activation of ventilation air or dust as air passes through the emplacement drifts. The methodology and results for these potential releases have

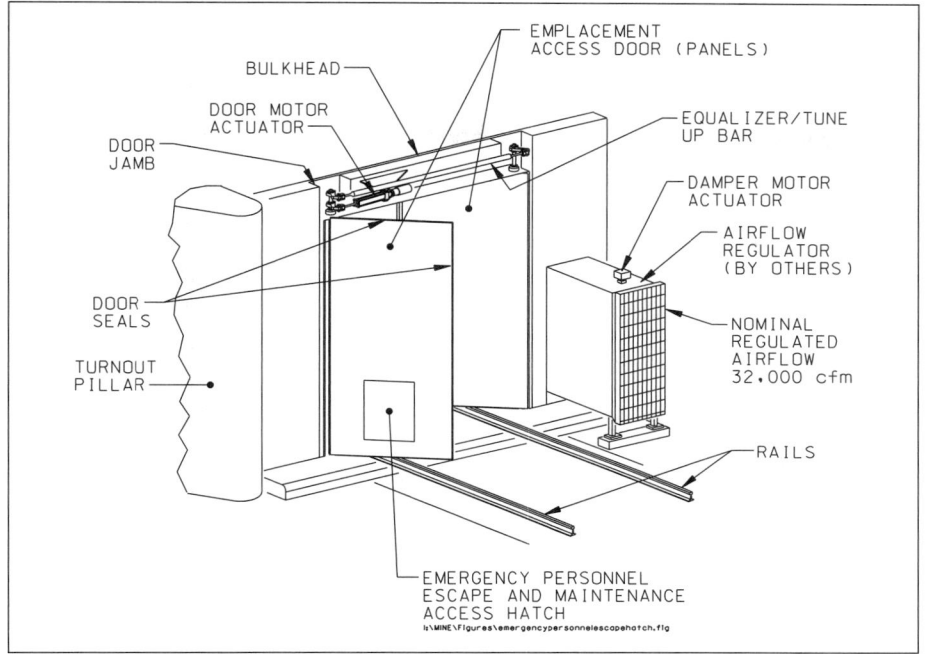

Figure 9. Emplacement access door isometric illustration

been analyzed, and it has been determined that any resulting radiological doses would not exceed the regulatory limits.

Off-Normal Conditions. The subsurface ventilation system is operated to maintain repository temperatures at acceptable operational levels during the preclosure period and to maintain an adequate margin. Analyses demonstrate that thermal limits are not exceeded during system shutdowns of limited duration, up to 30 days, resulting from power failure, fan failure and replacement, or from partial airflow blockage due to the unlikely event of rockfall. The analyses show that there is sufficient time to restore the system or portions of the system to normal operating conditions before exceeding the thermal temperature limits.

Maintenance. With a nominal 100-year design life for structures and components, routine maintenance, refurbishing, and replacement will be required. The radiological dose rate does not influence the basic design of the turnout bulkhead, but would require consideration for maintenance.

Equipment or components that require maintenance, such as electrical or instrumentation equipment are placed near the intersection of the access mains and turnouts to reduce potential radiation exposure.

Natural Ventilation Pressure

The waste packages add a significant heat load to the ventilation system that causes the air in the exhaust system to be less dense than the air in the intake system, creating a natural imbalance in the ventilation system. This effect is referred to as natural ventilation pressure (NVP).

A sustained power outage to ventilation fans could result in loss of forced ventilation to the subsurface facility. The NVP will maintain airflow even if the exhaust fans are off due to a power outage or out of service for repair. Various analyses of the natural convection potential of a fully loaded drift during the preclosure period indicate that in the absence of forced ventilation, airflow rates of about 6 to 10 m^3/s can be maintained across the emplacement drifts.

Since the WP thermal energy reduces over time, the corresponding NVP will reduce according to the waste package decay heat. The NVP also varies due to the ambient natural diurnal cycles and seasonal temperature changes.

ACKNOWLEDGMENTS/DISCLAIMER

The authors thank Bechtel SAIC Company and DOE for their support in preparing this paper. Though the authors have made every attempt to present information that is consistent with the official DOE design that was submitted to the NRC in June 2008, this paper is not an official DOE/NRC document and cannot be relied on for purposes other than that for which it was developed. Officially approved design and construction documents are readily available to the public on both DOE and NRC websites. If this document conflicts with the official DOE/NRC documents on file, the official DOE/NRC document supersedes. The Government (Department of Energy) reserves for itself and others acting on its behalf, a nonexclusive, paid-up, irrevocable, world-wide license for Governmental purposes to publish, distribute, translate, duplicate, exhibit, and perform this article. The views expressed herein are those of the authors and do not necessarily represent the views of the Department of Energy or Bechtel SAIC Company, LLC.

REFERENCE

BSC (Bechtel SAIC Company) 2008. Subsurface Construction and Emplacement Ventilation. 800-KVC-VU00-00900-000-00C. Las Vegas, Nevada: Bechtel SAIC Company.

PART 11

New Projects 1

Chairs

Matthew Fowler
Parsons Brinckerhoff

Tony O'Donnell
Kiewit

PLANNING NEW METRO SUBWAYS—
LOS ANGELES, CALIFORNIA

Amanda Elioff ▪ PB Americas, Inc.

David Perry ▪ MACTEC

Girish Roy ▪ LACMTA

Pierre Romo ▪ MACTEC

ABSTRACT

Success of the Gold Line Eastside Extension rail tunnels in Los Angeles has prompted Metro's planning for two new rail lines to include underground alternatives. The Westside Extension Transit Corridor alternatives include up to 17 miles of subway to be constructed in soft ground. The routes will require construction in Los Angeles' "Potential Methane Zone," not considered for subway construction until recently. Slurry face TBMs will likely be specified. The Regional Connector, through the downtown area, includes one underground alternative with 1.7 miles of tunnel as well as three underground stations in soft ground and rock. Challenges will include tunnel and station construction under narrow streets. This paper will describe the alternatives, planning, geologic conditions and tunneling concepts for both planning studies.

BACKGROUND

The Los Angeles County Metropolitan Transportation Authority's (Metro) existing rail transit system includes 18 miles of Heavy Rail (HRT) subway, 55.7 miles of Light Rail (LRT) and currently has 14.6 miles of LRT under construction. In addition, a fixed guideway (Orange Line Bus Rapid Transit—BRT) extends the system 14 miles in the San Fernando Valley (Figure 1).

The Los Angeles County Metropolitan Transportation Authority, (Metro) has recently completed Alternatives Analysis (AA) studies for new rail lines (January 2009): the Westside Extension and the Regional Connector. Both studies included subway as well as aerial sections for the analyses. A heavy rail subway for the Westside of the Red and Purple Lines will provide a viable alternative to driving in the heavily congested Westside Extension Transit Corridor. The Regional Connector will add 1.7 miles of light rail through downtown Los Angeles to connect existing light rail lines and their extensions now under construction. Stations along the new downtown segment will serve areas of the city's core not currently easily accessible to Metro's Red/Purple Line. The new connection will allow 37 miles of continuous rail from Pasadena to Long Beach, and 15.5 miles from East Los Angeles to Culver City. Construction of the Regional Connector will also reduce transfers and the potential for overcrowding at the existing LRT terminal stations—7th/Metro and Union Station.

Underground alternatives have, in part, been considered due to the recent success of Metro's Gold Line Eastside Extension (MGLEE), which includes almost two miles of LRT in tunnels with two underground stations. This new line will go into revenue service as planned in the summer of 2009. Tunnels were constructed in soft ground using EPB

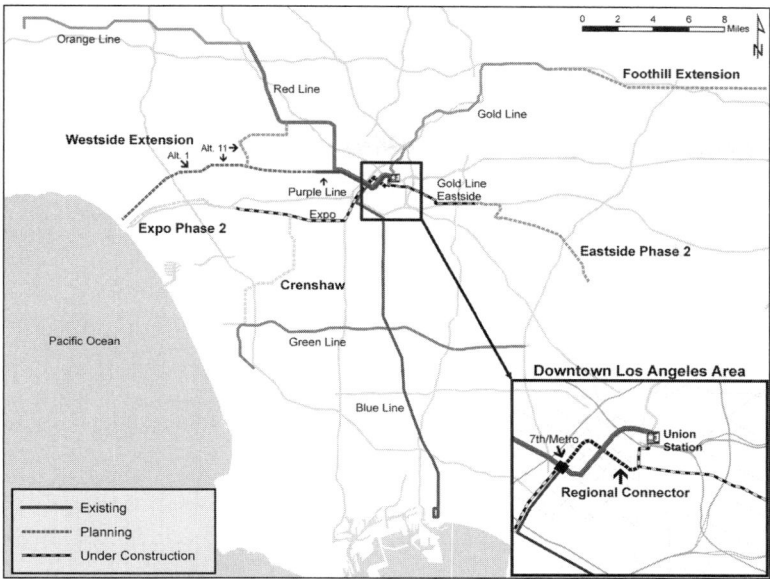

Figure 1. Metro's existing rail system and rail lines under planning and construction

technology, a single pass (double gasketed) precast liner, and were completed with virtually no surface settlement (Choueiry et al., Robinson and Bragard, RETC 2005).

Heavy rail had not been considered along Wilshire Boulevard, the major east-west street connecting downtown to the westside city of Santa Monica, since 1997, when all rail projects in planning and design were suspended due to funding issues and a perception that tunneling was not suitable in Los Angeles' subsurface conditions, including gassy ground. This perception evolved from tunneling issues that occurred during the Red Line construction (1986 to 2000) and the presence of subsurface gases in the area around the La Brea tar pits.

Now that the MGLEE tunnels have been completed—with minimal disruption to the community—the Mayor of Los Angeles, and other elected officials have announced their support for major extensions to the rail transit system, including and extension to the existing Red Line (now Purple Line) Subway to the west, and the Regional Connector, the 1.7 mile connection of two separate light rail lines through Downtown Los Angeles. Support of these and other transportation projects has been affirmed by the residents of Los Angeles County, who in November 2008, passed Measure R, an initiative to fund transportation projects with by raising the local (Los Angeles County) sales tax by 0.5 percent. Over a 30 year period, beginning July 1, 2009, the tax revenues are projected to generate $40 billion for congestion relief. This paper describes tunneling aspects of the two new studies and the challenges for future phases of design.

WESTSIDE EXTENSION

In 2007, Metro initiated the AA study to begin the project definition process. The AA study is the first step in defining a project in the Federal Transit Administration's New Starts Program. The analysis included an evaluation of a wide range of transit alternatives (alignments and transit modes), screening the alternatives against criteria, and selecting those most promising to proceed into an environmental review. For the

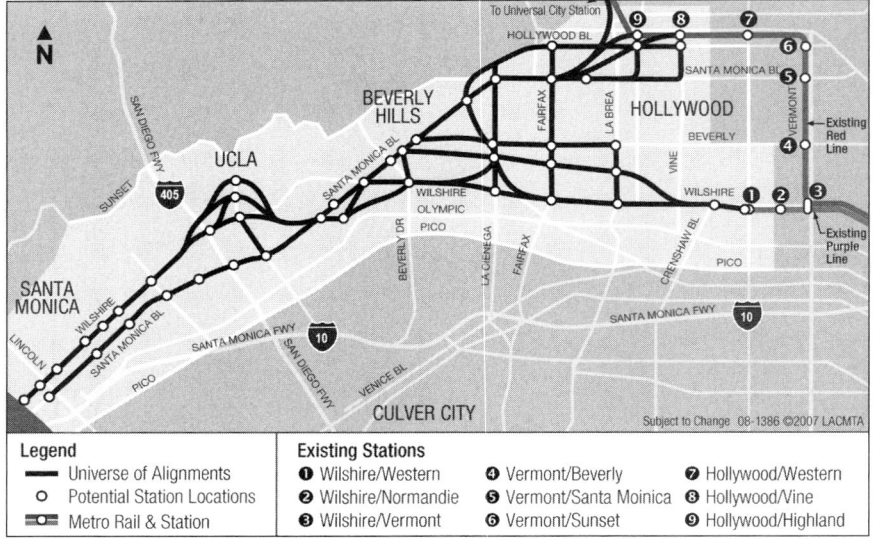

Figure 2. Seventeen initial alternatives for screening

Westside Study, HRT in subway emerged from the alternatives analysis as the best transit mode. The initial set of alternatives included 17 alternative alignments known as the "Universe of Alternatives" to extend the existing Red and Purple Lines from either the existing Wilshire/Western station or Hollywood/Highland station or both. All alternatives ended in the City of Santa Monica (Figure 2).

A series of 17 community meetings were held to present alternatives and receive public comments. Results were that the communities overwhelmingly supported transit improvements, favored a Wilshire Boulevard alignment, and also supported a line to West Hollywood. There was limited support for aerial, monorail, or BRT alternatives. Ridership demands and travel times over the 16 mile distance from downtown to Santa Monica showed that the area would best be served by the HRT transit mode. Preliminary models estimated travel times from downtown to Santa Monica at 27 minutes, a trip that can currently take over an hour by car in rush hour traffic. Two alternates, Alternatives 1 and Alternative 11 of the original 17 (Figure 3 and Figure 4) were selected to proceed to the next phases of study—Environmental Analysis and Advanced Conceptual Engineering (ACE).

As noted above, the study included subways previously prohibited from funding in areas along Wilshire Boulevard due to legislation passed in 1986. At that time, the United States Congress enacted legislation that funded the initial Red Line segment, but prohibited use of federal funds for subway construction in Los Angeles' High Potential Methane Zone (Methane Zone), deemed an unsafe area for tunneling after a serious non-tunneling related methane explosion occurred in 1985. In 2007, Congress repealed the 1986 legislation given advances in technology and demonstrated successes in underground construction projects including those in Los Angeles. Key to the repeal of this legislation were conclusions by a panel of experts assembled by the American Public Transportation Association (APTA). The panel reviewed new data and evaluated advances in worldwide tunneling technology and the safety of buildings related to operating transit tunnels in the methane hazard zone along Wilshire Boulevard (Figure 5).

The panel concluded that advances in technology and tunneling practice in the last 20 years would permit tunneling—such that it "could be undertaken at no greater

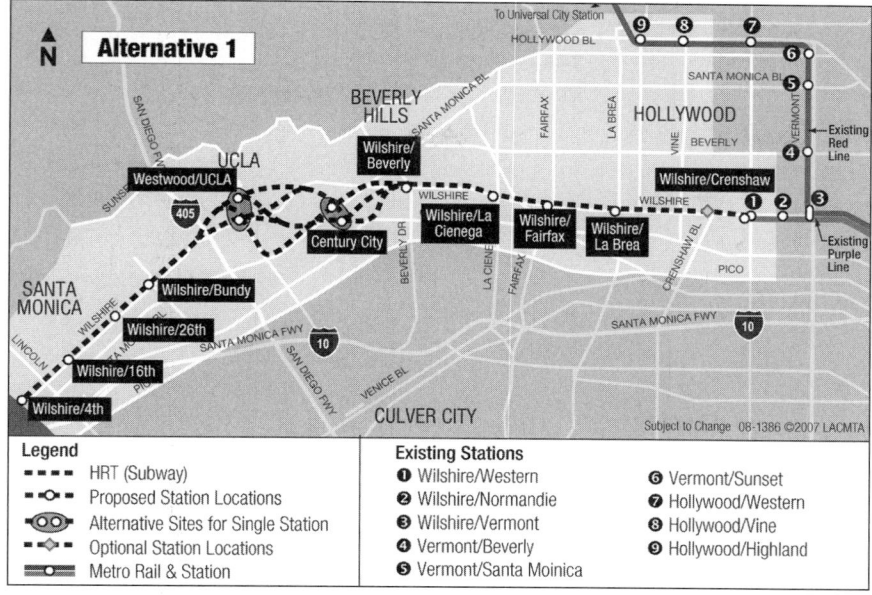

Figure 3. Westside Extension Alternative 1

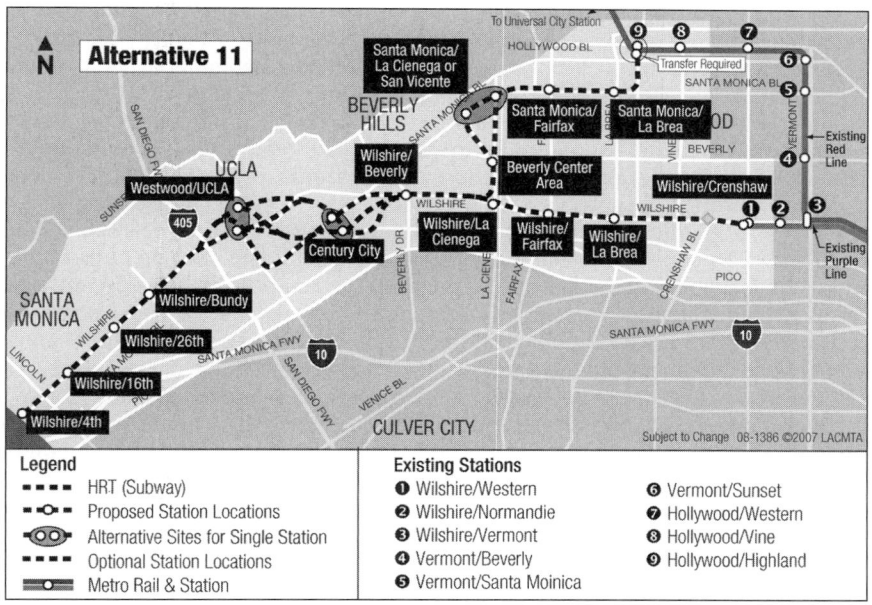

Figure 4. Westside Extension Alternative 11

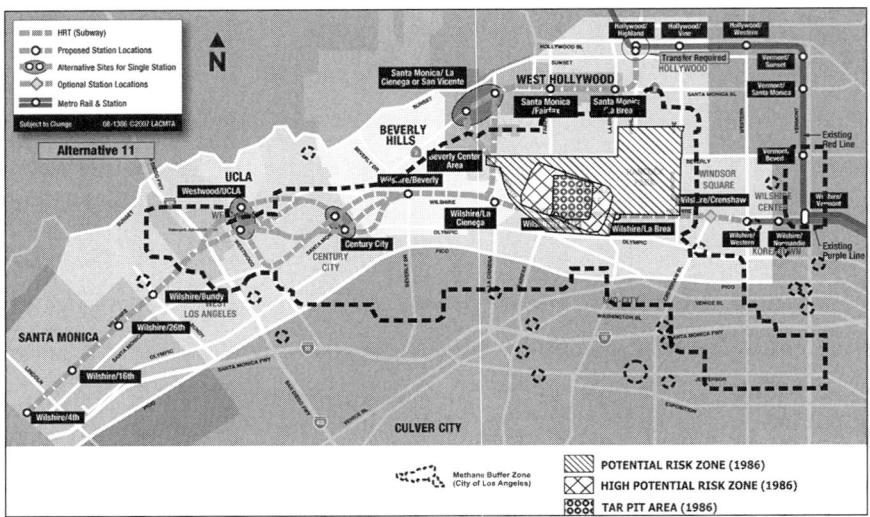

Figure 5. Methane Zones, Los Angeles Westside

risk than other subway systems in the United States." Supporting the APTA panel's conclusions were that since 1986 there have been: advances in Tunnel Boring Machine (TBM) technologies such as use of pressure face TBMs, improvements in gas measurement and instrumentation technology, more local tunneling experience, and successful operation of the existing Metro system.

With these general conclusions, the panel also made recommendations for future design and construction that involved: use of pressure face (slurry) TBMs in high gas risk areas, minimizing construction in the gas and tar bearing formations as much as possible, particularly the San Pedro Formation's unsaturated zones, incorporating lessons learned from other projects as information becomes available, and developing procedures for repair of sealing systems—should seismic events or fires occur.

For the AA study, special tasks were undertaken to follow-up on recommendations of the APTA panel, to obtain additional geologic data, and develop further recommendations for future study and design phases. These would be based on obtaining more detailed geologic information as well as additional experience in Metro construction since the 2006 study, in particular, successful completion of the MGLEE tunnels and cross-passages constructed through the former Boyle Heights oil field. Among the study tasks was a literature review to compile previous Metro work and a survey of projects world-wide that could have similar gas conditions. Much of this previous work supported APTA's recommendations to use slurry face TBMs (SFM), examine ground treatment methods to reduce hazards from gas encounters, and consider shallow stations to avoid geologic formations having the high gas concentrations (Elioff, et al., 1995, Jacobs et al. 1999).

Geologic Conditions

Geologic conditions and profiles for the initial study alternative alignments were assessed from research of over 230 geotechnical reports in Metro and the Consultant team's files as well as other published data. Data was collected on geologic formations and their properties, groundwater, gas conditions and seismicity, summarized below.

Geologic Setting of Study Area

The study alignments are located in the northern portion of the Los Angeles Basin, approximately ½ to 3 miles south of the eastern portion of the Santa Monica Mountains. The geomorphic surface along the alternative alignments was largely formed by the aggradation of sediments shed out from the mountain front, subsequently uplifted, and later modified by the erosion of broad channels into the alluvial surface. Regionally, the alignment is located near the boundary between the Peninsula Ranges and the Transverse Ranges, two major geomorphic and structural provinces of southern California. The Santa Monica and Hollywood faults are considered the boundary between the two geomorphic provinces within the area of the alignments under study.

Geologic Conditions

The Wilshire Boulevard tunnel profile (Alternatives 1 and 11) will encounter several geologic units that range in age from Miocene to Holocene. From oldest to youngest in geologic age, are the Miocene-age sedimentary bedrock of the Puente Formation, Pliocene-age sedimentary strata of the Fernando Formation, Pleistocene-age San Pedro and Lakewood Formations, Pleistocene-age (older) alluvium, and Holocene-age (younger) alluvium. Pleistocene- and Holocene-age alluvial deposits comprise the surficial geologic units along the alignment. The San Pedro, Fernando, and Puente formations would be encountered beneath the Holocene and late Pleistocene sediments in the subsurface along portions of the Wilshire Boulevard alignment.

The Holocene-age sediments generally consist of interlayered silty clay, silty sand, and sandy silt. These surficial deposits are underlain by variably thick late-Pleistocene-age older alluvial deposits and semi-consolidated continental and marine sediments of the late Pleistocene-age Lakewood Formation consisting generally of gravel, sand, and silty sand with some interlayers of silt and clay. The lower portion of the undifferentiated older alluvial sediments/Lakewood formation consists primarily of interlayered silts and sandy clays with some silty sand and is anticipated to be the primary geologic unit that would be encountered at tunnel depth along the Wilshire tunnel alignments under study.

The Lakewood Formation materials are underlain by sediments of the early Pleistocene-age San Pedro Formation. These materials generally consist of stratified sand with some fine gravel, silt and clay layers The Pleistocene age alluvial deposits are underlain by Tertiary age sedimentary rocks of the Fernando and Puente Formations. The Fernando Formation, where encountered in prior borings along the alignment, mainly consists of massive yellow brown to olive-gray siltstone and claystone with few sandstone interbeds. The claystone and siltstone was described as friable and weak (CWDD/ESA/GRC, 1981). The Puente Formation is comprised of both massive, light brown to medium-brown siltstone and zones of claystone that are interbedded with thin laminae of sandstone and siltstone. The Tertiary age sedimentary rocks deepen in the subsurface toward the west and are anticipated to be encountered primarily in the eastern portion of the alignment at tunnel depth. These formations were tunnelled for the existing Metro system (to the east of the study area) using open-shields.

Groundwater

Exploratory borings drilled along Wilshire Boulevard between Western and Fairfax Avenues in 1980 to 1981 for the Metro Rail project encountered shallow groundwater, probably perched, between approximately 10 to 35 feet below ground surface. Locally, ground water as shallow as 5 to 10 feet below ground surface has been reported in borings drilled along Wilshire Boulevard in the Wilshire/Fairfax area.

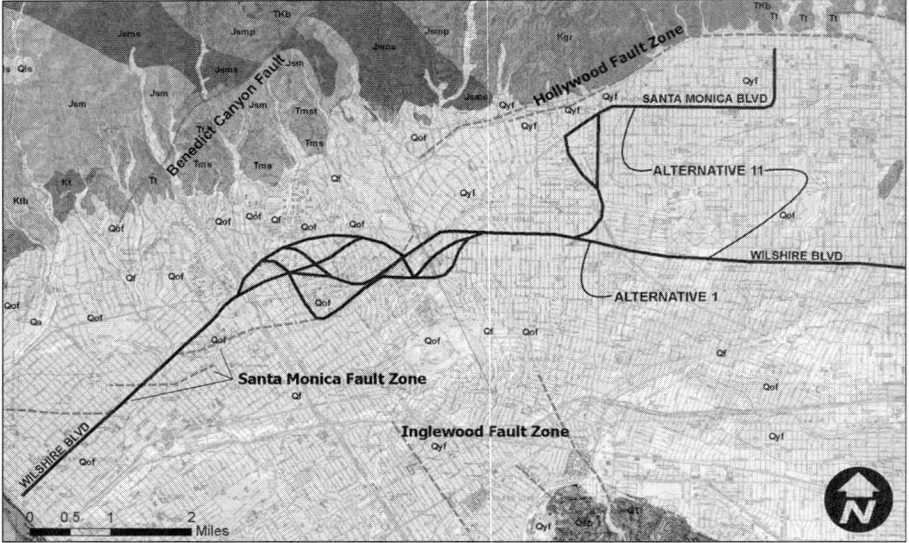

Figure 6. Geologic map showing Santa Monica fault traces

Seismic Considerations and Faulting

The numerous faults in Southern California include active, potentially active, and inactive faults. The Santa Monica and Hollywood fault zone form a portion of the active Transverse Ranges Southern Boundary (TRSB) fault system. The Santa Monica fault zone (SMFZ) is the western segment of the Santa Monica-Hollywood fault zone. The fault zone trends east–west from the Santa Monica coastline on the west to the Hollywood area on the east (Bryant, 2005, and U. S. Geological Survey, 2005). Urbanization and development within the greater Los Angeles area has resulted in a poorly defined lateral extent, location, and rupture history of the SMFZ. Thus, the Wilshire tunnel alignment may traverse the Santa Monica Fault Zone (Figure 6).

Gas Conditions

Of primary concern are subsurface gas conditions in the 1986 Methane Risk Zone, the area surrounding the La Brea tar pits. The tar pit area is northeast of the intersection of Wilshire Boulevard and Fairfax Avenue, a planned subway station area (also shown in Figure 5). It has long been documented that in the Mid-Wilshire area tar seeps and associated gases (methane and hydrogen sulfide) are present. These gases migrate upward to the surface from deeper Miocene Formations and are usually associated with the many oil fields of the Los Angeles Basin. In the Mid-Wilshire area, the gases, primarily methane with some amounts of hydrogen sulfide, are mainly found in the San Pedro and Lakewood Formations at approximately 10 to 50 feet below the ground surface—or at least to typical depths of foundation excavation. In some areas near the La Brea tar pits, methane can reach up to 90 to 100 percent by volume. Additionally, hydrogen sulfide, has been found in the range of 10 to 600 parts per million (ppm) in the Wilshire/Fairfax area. Methane is explosive in the range of 7 to 14 percent in air, and hydrogen sulfide is considered unsafe at concentrations of over 10 ppm.

Historically, there have been occasions when the gas has accumulated. In 1985, there was an explosion at the Ross Dress for Less store at Fairfax Avenue and 3rd Street (about ½ mile north of Wilshire Boulevard and Fairfax Avenue) where methane

PLANNING NEW METRO SUBWAYS—LOS ANGELES, CALIFORNIA

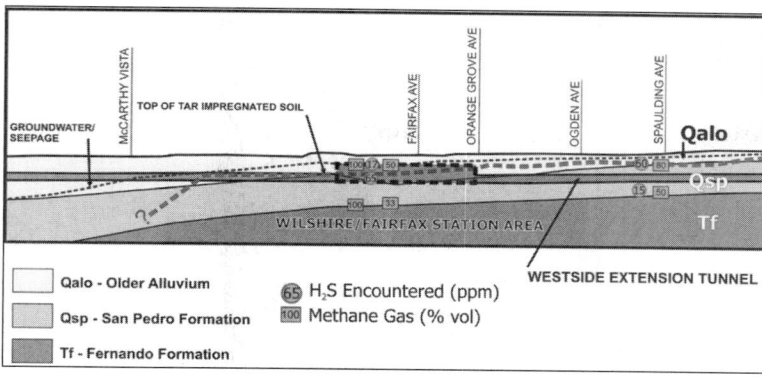

Figure 7. Conceptual profile Wilshire/Fairfax area

had accumulated in the basement of the store. Because of this and the knowledge of the presence of methane in the city, the City of Los Angeles has implemented special building code provisions for "Methane Risk Zones" and "methane buffer zones" within the city to address this natural occurrence and provide mitigation. These measures include proper investigation, construction of barriers/liners, passive venting systems beneath building slabs, special HVAC requirements, and detection and alarm systems. Since 1986, the City has revised the Methane Risk Zone boundaries (Figure 5) to correspond to oil field maps and reported gas seepage.

Figure 7 shows the conceptual profile of the Wilshire/Fairfax station area developed during the Westside study. Gas data has been collected over the last 30+ years, beginning with the original Metro explorations in 1985 to the present time, as monitoring wells have been maintained. Methane and hydrogen sulfide measurements were plotted on the profiles, indicating 0 to 100 percent methane by volume (maximum) and hydrogen sulfide gas concentrations at over 50 parts per million (ppm), in the area west of Crenshaw Boulevard to about San Vicente Boulevard.

Construction Methods in Gas Risk Areas

Tunnel Construction

The existing Metro Red Line tunnels were built with open face tunneling shields in both in soft ground and siltstones, including ground with high methane levels and some hydrogen sulfide present. The new extension will traverse areas with similar methane concentrations, but higher gas pressures (up to about 200 inches of water). Figure 8, presents gas measurements taken in the 1980s along the existing alignment and those measured in the current study area.

From previous Metro studies and APTA recommendations, use of SFMs is now recommended for relatively high gassy ground areas. For other tunnel reaches, the choice between Earth Pressure Balance (EPB) and SFM may be left to the contractor. In the higher gassy ground reaches (as a minimum), a SFM should be required for the following reasons:

- Gas in hazardous concentrations is known to be present in this reach. Using SFMs, the gas is always confined in a pressurized system in the tunnel making the tunnel safer for workers.
- Spoil treatment is provided at the slurry treatment plant. Gas can be more easily monitored, safely controlled and treated at this plant than within the tunnel.

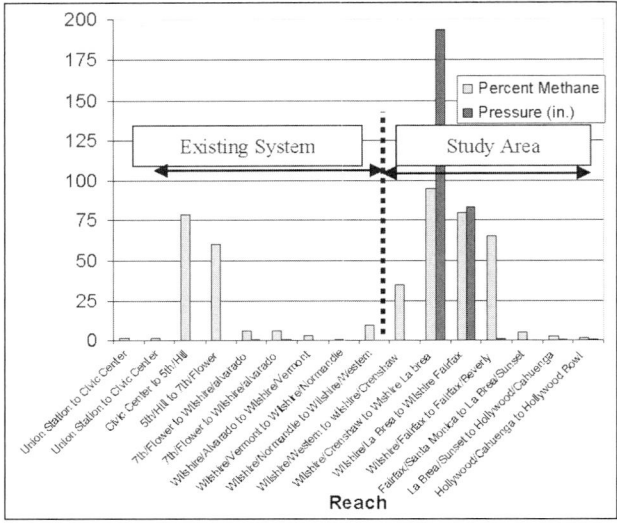

Figure 8. Gas measurements (1980s) for Metro Red Line

- APTA panel, and previous Metro study findings (Elioff et al. 1995), lead to recommendations SFMs in areas with high hydrogen sulfide concentrations.

Other reaches requiring SFMs could be identified during later design efforts after additional geotechnical explorations. Other tunneling studies during subsequent phases of design will follow up on experience of slurry face TBM projects constructed in gassy ground including; slurry treatment and processing related to presence of gas, measurement of gas inflows through installed liners, and monitoring of the slurry treatment plant, its discharges, and work areas. To date SFMs have not been used for hydrogen sulfide mitigation, however, tunnels in Detroit, Michigan—known to have similar hydrogen sulfide gas issues (Hansmire 2008)—are now under construction, and lessons learned from this project will be used for advances in treatment and handling and possible application on the Westside Extension.

Station Construction

To date, Metro stations have been constructed safely using cut and cover methods. These methods were also used for construction of existing deep foundations and parking garages built in the Methane Risk Zone, specifically in the Wilshire/Fairfax area. Some were constructed according to city codes developed after defining of the Methane Risk Zone, others prior to that time. Metro reviewed this information and researched additional deep parking garages, having up to five levels of underground parking on one site. Findings were that of the garages surveyed, some reported no incidences of measured gas intrusions (gas alarms); others occasionally dealt with seepage of tar through fine cracks in the structures. Seeping tar material is collected and disposed of by specialty environmental firms. During recent excavation for an underground parking garage project in the vicinity, hydrogen sulfide gas pockets were occasionally encountered.

Metro considers its existing subway system and operations a success, but there will be additional challenges for the Westside Extension. Because of the presence of hydrogen sulfide and methane gases, tar sands, and high groundwater, construction

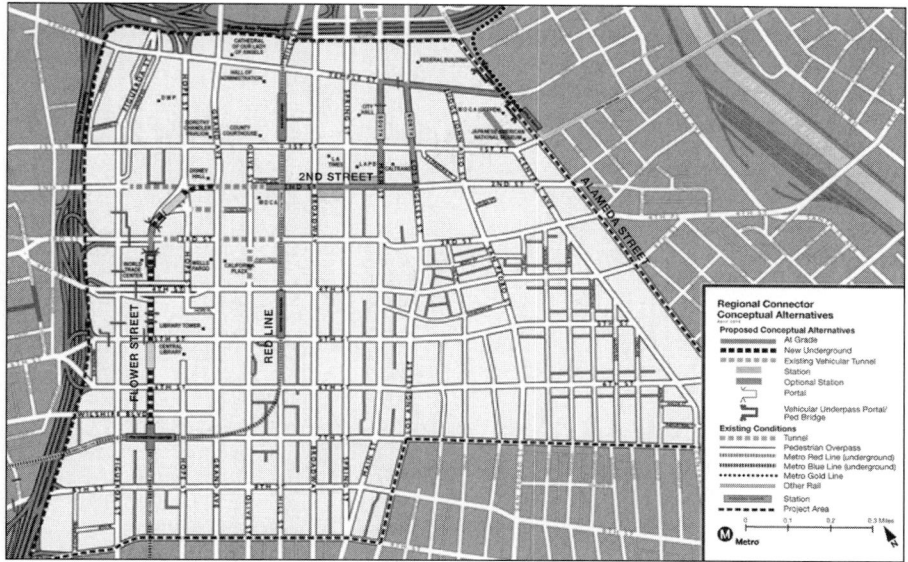

Figure 9. At-grade emphasis alternative (Couplet A)

and operation of the Westside Extension design and construction will be approached with considerable caution and an added degree of conservatism. The approach for future design and successful accomplishment of construction and operations in the higher gas risk areas will incorporate; redundancy in design, focus on construction quality control, state-of-the-art instrumentation for leak detection; activation of emergency operations and repairs, and design of ventilation systems along with back-up systems.

REGIONAL CONNECTOR

Metro's Regional Connector will link the 1.7 mile gap between the existing 7th Street/Metro Center Blue Line station and the Little Tokyo/Arts District Gold Line station. This connection will allow connecting service between several light rail service lines in operation or in construction; i.e., the Metro Gold Line to Pasadena, the Metro Gold Line Eastside Extension, the Metro Blue Line, and the Metro Expo Line. The connection will greatly broaden and improve the region's public transit, mobility, and accessibility. For example, upon completion of the Regional Connector, riders will be able to travel from Pasadena to Long Beach, 37 miles, without a transfer. Currently this trip's travel time, with two transfers, is about 115 minutes. When the Regional Connector is complete, the trip will be shorter and is estimated to take 95, minutes, therefore, a time savings of about 15 minutes. In addition, the elimination of transfers will make the ride more attractive to transit patrons. Given the relatively short alignment length, travel time savings to be realized, and new riders attracted, the Regional Connector has a favorable cost effectiveness rating not common for underground projects.

The study area is primarily within the Central Business District of the Los Angeles (Figure 1). During the AA Study, three alternatives currently referred to as "Promising Alternatives" were identified to be studied further in the Environmental and ACE phases. These alternatives are presented in plan view in Figure 9 and Figure 10, and are named Couplet Alternative A, Couplet Alternative B, the at-grade emphasis alternatives, and the Underground Emphasis Alternative. Couplet B is similar to Couplet A

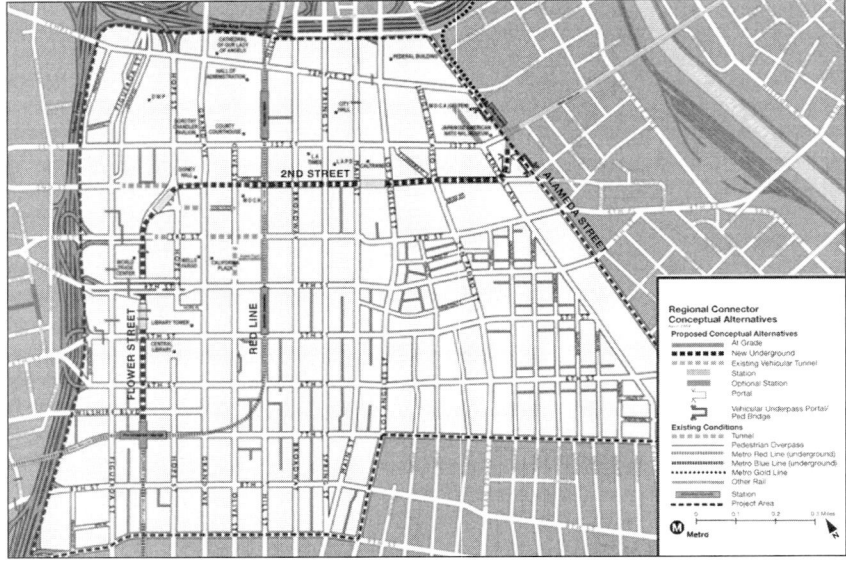

Figure 10. Underground emphasis alternative

except that it has an at-grade station on Flower Street. The underground alternative would be in a tunnel for almost the entire 1.7 miles, have three underground stations and a portal just west of Alameda Street. Connection with the at-grade MGLEE would be south of the existing Little Tokyo station. A grade separation (underpass) of Alameda Street is also planned. The at-grade alternative options (Figure 10) must also begin at existing 7th/Metro station, but would portal to the surface along Flower Street. Couplet A has an underground station on Flower Street between 6th and 5th streets, and portals south of 3rd Street. Couplet B portals south of 4th Street and has an at-grade station between 4th and 3rd Streets. From 3rd Street north to 2nd Street, the at-grade alternatives are the same. Both alternatives would go underground again for a short distance under Bunker Hill to connect into the existing 2nd Street tunnel. The at-grade emphasis connects to the MGLEE north of the Little Tokyo station, and would be grade separated on Temple Street.

Geology

The Regional Connector tunnel alignment (and at-grade alternatives) will traverse the south-eastern end of the Elysian Park Hills, west of downtown, and the ancient floodplain of the Los Angeles River. The Elysian Hills comprise the low-lying hills west of the Los Angeles River and southeast of the eastern end of the Santa Monica Mountains. The Elysian Hills are comprised largely of Miocene age sedimentary rocks with Pliocene age rocks flanking the south-eastern edge of the hills. The project alignment is located on the south-western flank of the northwest trending Elysian Park anticline. This fold is a major geologic structure of the Elysian Hills and trends south-easterly as mapped by Lamar (1970). In the vicinity of the project alignment, bedding within the Fernando and Puente formations strike approximately east-west to slightly north of east and dips moderately to steeply to the south (Soper and Grant, 1932, Lamar, 1970, and Dibblee, 1991).

PLANNING NEW METRO SUBWAYS—LOS ANGELES, CALIFORNIA

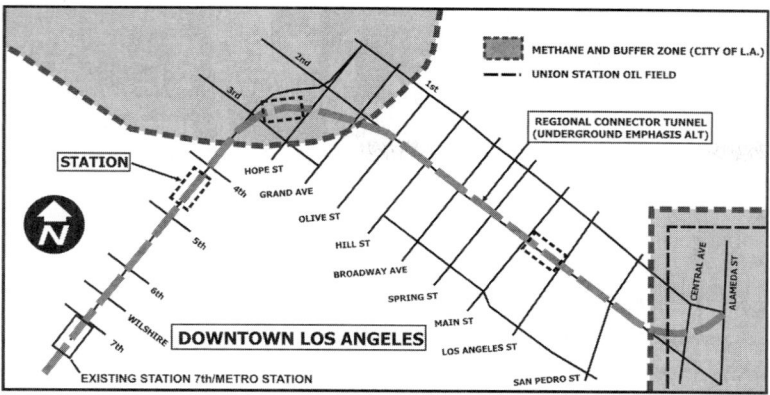

Figure 11. Methane Risk Zone Area, downtown Los Angeles

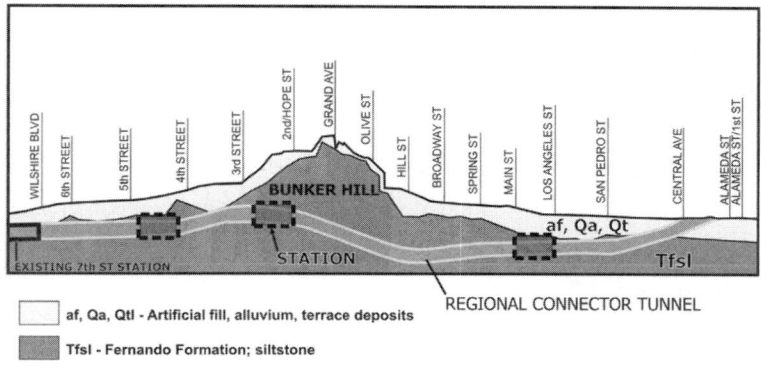

Figure 12. Geologic profile (underground emphasis alternative)

The proposed tunnel alignment along the Flower to 2nd Street alternative will encounter several geologic units that range in age from Miocene to Holocene. The geologic units that would be encountered within the proposed tunnel alignment, from oldest to youngest in geologic age are; the Miocene-age sedimentary bedrock of the Puente Formation, Pliocene-age sedimentary strata of the Fernando Formation, Pleistocene-age alluvium (older alluvium), and Holocene-age alluvium. The Pliocene and Miocene formations are exposed in the Bunker Hill area in the northern part of downtown Los Angeles. These formations are overlain depositionally by Pleistocene and Holocene age alluvial sediments. The Pleistocene sediments are present as remnant depositional terraces in the Bunker Hill area, whereas, the more recent alluvial sediments are present at lower elevations. Artificial fill has been placed at various locations along the alignment including areas overlying both existing and abandoned tunnels. Portions of the study area are also included in the City's Methane Risk Zone, as shown in Figure 11.

The geologic profile prepared for the conceptual engineering study, Figure 12, illustrates the interpretation of the subsurface contacts between the geologic units. Geologic data was obtained from 75 existing borings, drilled over a period of many years along or adjacent to the alignment.

Tunneling and Underground Stations

Tunneling is anticipated to be through use of pressure face TBMs and by cut and cover methods where reaches are relatively short and shallow. Major challenges for underground construction will include all of those typically associated with tunneling in dense urban areas: work site constraints, major utility relocations, subsurface obstructions, and protection of existing structures, some of them among the oldest in Los Angeles. In addition, the conceptual design includes a deep mined station under Bunker Hill to allow the tunnels to pass under existing Metro Red Line tunnels along Hill Street. Specific design elements to be addressed in the next study phases include:

Protection of Existing Utilities

The downtown area has numerous utilities both known and unknown due to the age of the area. Major storm drains exist along Flower, 2nd and Alameda Streets that will require relocation or support during tunnel and station construction. The existing Metro Red Line tunnels cross the alignment on Hill Street and will likely require design of protection.

Protection of Existing Buildings

2nd Street is relatively narrow—40 feet from curb line to curb line in some locations. The tunneled sections must pass close to the building lines and in some cases directly under building sections or vaults extending beyond the property lines.

Subsurface Obstructions

With new high-rise buildings along Flower Street, other obstruction such as tie-backs used for temporary support during construction of underground parking garages, are expected along Flower and 2nd Streets. In addition there are abandoned tunnels, originally used for the Pacific-Electric Line (former trolley) tunnel, crossing Flower Street—directly in-line with all alternatives.

Worksite Areas Downtown

Construction staging areas for each of the station sites as well as the tunnel portals will need to be identified. Sufficient space to carry out efficient operations will need to be identified.

Cut and Cover Construction along Flower Street

Flower Street from 7th/Metro to 3rd Street is wide; however, the existing storm drains will constrain design and construction methods. The existing, 7th/Metro station is a side platform station, such that a transition structure will be required if a center platform station is designed for the underground station between 4th and 5th Streets. Conceptual designs include side platforms for the Couplet B shallow underground station. In addition, for the underground alternative, the short radius line section south of the Bunker Hill area may not be suitable for excavation using a TBM. Cut and cover, or possibly mined methods may need to be used there as well.

Break-in to 2nd Street Tunnel

Couplet Alternatives A and B have a 2nd underground reach at Bunker Hill and must break into the existing, 2nd Street tunnel (Figure 13 and Figure 14). This work will require specialized design and construction to support the existing tunnel during the break-ins and for the final structure.

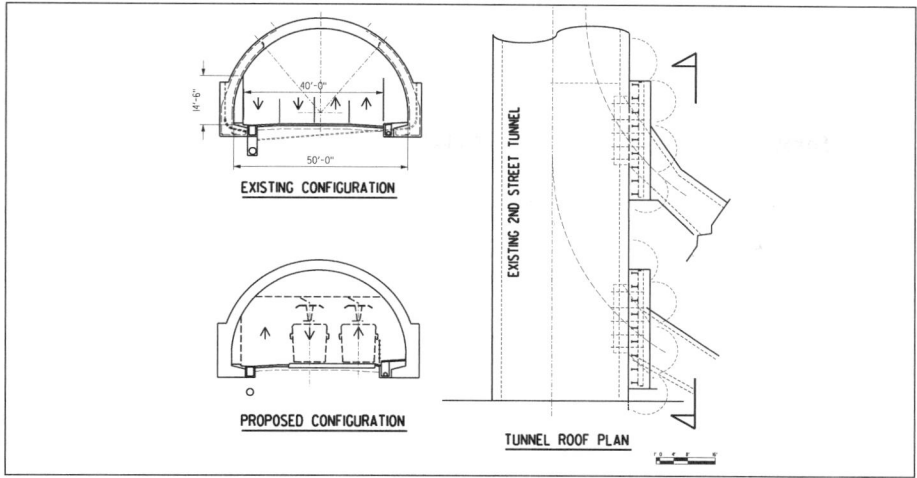

Figure 13. Break-into 2nd Street Tunnel

Figure 14. 2nd Street Tunnel, East Portal

Mined Station

The underground alternative has a deep profile to reduce potential impacts to future development at Grand Avenue and 2nd Streets, and to pass under the existing Metro Red Line Tunnels. A mined station is currently planned for the station in the Bunker Hill area. This method will need detailed analysis for station design and to determine the construction sequence and monitoring program.

CONCLUSIONS

Underground transit has again become the transportation mode of choice in Los Angeles for a number of new rail lines. Continued traffic congestion, fuel prices, environmental disturbance and right-of-way costs have led to consideration of subways as the preferred system. The two new subway sections planned for Los Angeles will face design and construction challenges typical for urban areas: minimal work areas, subsurface obstructions, protection of existing buildings, utility relocations as well as construction methods to limit community inconvenience. In addition, Metro will need to address new design challenges—some unique to Los Angeles—for construction in gassy ground, tar impregnated sands, and potential for fault crossings. The underground construction industry has seen proven advances in the past 20 years in pressure face TBM technology, tunnel lining systems, excavation support systems, and barrier methods to build systems safely with fewer community impacts. These advances have opened opportunities to expand rail transit in Los Angeles.

ACKNOWLEDGMENTS

The authors acknowledge Metro's planning teams for the Westside Extension Corridor and Regional Connector Studies led by David Mieger and Dolores Roybal-Saltarelli. We also thank Tom Jenkins, Jim Monsees, and Perry Maljian for their assistance in preparing this paper.

The consultant team for the Westside Extension Transit Corridor Study was led by Parsons Brinckerhoff with MACTEC as the geotechnical consultant. The Regional Connector consultant team was led by CDM with Parsons Brinckerhoff and MACTEC as engineering sub consultants.

REFERENCES

American Public Transportation Association (APTA), Final Report of Peer Review Panel for the Los Angeles County Metropolitan Transportation Authority, November 2006

Bryant, W. A., (compiler), 2005, Digital Database of Quaternary and Younger Faults from the Fault Activity Map of California, Version 2.0, California Geological Survey Web Page, http://www.conrv.ca.gov/CGS/information/publications/QuaternaryFaults_ver2.htm, 8-13-07

Choueiry et al., Planning and Construction of the Metro Gold Line Eastside Extension Tunnels, Los Angeles, CA, Proceedings, RETC 2007

Converse Ward Davis Dixon, Earth Science Associates, Geo-Resource Consultants, (CWDD/ESA/GRC), 1981, Geotechnical Investigation Report, Volume I; Volume II—Appendices 1 and 2, for Southern California Rapid Transit Metro Rail Project

Dibblee, T. W., 1991, Geology of the Los Angeles Quadrangles, Los Angeles County, California, Dibblee Foundation Map #DF-22, map scale 1:24,000.

Elioff et al., Geotechnical Investigations and Design Alternatives for Tunneling in the Presence of Hydrogen Sulfide Gas, Proceedings, RETC 1995

Hansmire and Jafri, "Slurry-Face Rock TBM Outfall Tunnel in Detroit, MI, USA," Proceedings, World Tunnelling Congress, 2008, Agra, India

Jacobs et al., Hydrogen Sulfide Controls for Slurry Shield Tunneling in Gassy Ground Conditions—A Case History, Proceedings RETC 1999.

Lamar, D. L., 1970, Geology of the Elysian Park-Repetto Hills Area, Los Angeles County, California, California Division of Mines & Geology Special Report 101

Soper, E. K., and Grant, U. S. IV, 1932, Geology and Paleontology of a Portion of the Los Angeles, California: Geological Society of America Bulletin, Vol. 43, pp 1,041–1067.

Robinson and Bragard, "Los Angeles Metro Gold Line Eastside Extension—Tunnel Construction Case History," Proceedings, RETC 2007

U.S. Geological Survey, 2005, "Preliminary Geologic Map of the Los Angeles 30′×60′ Quadrangle, Southern California," Open File Report 2005-1019, Version 1.0, http://pubs.usgs.gov/of/2005/1019, map compilation by R. F., and Campbell, R. H., 2005, prepared by U. S. Geological Survey in cooperation with California Geological Survey.

SLURRY TBM TUNNEL IN ROCK, THE MODIFIED DETROIT RIVER OUTFALL NO. 2

William H. Hansmire ▪ Parsons Brinckerhoff

Pamela Turner ▪ Detroit Water and Sewerage Department

Frederic Mir ▪ Vinci Construction Grands Projets

Parvez Jafri ▪ Detroit Water and Sewerage Department

ABSTRACT

The Modified Detroit River Outfall No. 2 (MOD DRO-2) will discharge treated wastewater to the Detroit River. Construction of a 6.4-m-diameter, 1.9-km-long tunnel was halted in 2003 by flooding during tunneling with an open face hard rock TBM. A new design for a higher rock tunnel, a "Modified" design, was completed in 2007 and Notice to Proceed to the tunnel contractor was given in November 2008. Project duration is four years for completion of tunneling and start of outfall tunnel operation. This tunneling project has high groundwater pressure (5 bar), pervious limestone rock of up to 140 MPa (20,000 psi) strength, and hydrogen sulphide in the groundwater. The new tunnel will be connecting to access shafts and outfall diffuser riser shafts previously built for the now-abandoned tunnel. The new tunnel design construction contract, and initial planning to use a pressurized-face (slurry) hard rock TBM are presented.

WASTEWATER SYSTEM IN DETROIT—A NEED FOR TUNNELS

The requirements for preventing or minimizing pollution have become more stringent with implementation of the Clean Water Act in 1972. A major action by the Detroit Water and Sewerage Department (DWSD) to achieve regulatory compliance has been to improve the capability and increase the capacity of the wastewater treatment plant (WWTP). That work has been underway for several years and is nearing completion. The plant is one of the largest in the United States with primary treatment capacity of 6.4 billion L/day (1,700 million gallons per day [mgd]) and a secondary treatment capacity of 3.5 billion L/day (930 mgd). Detroit's central wastewater treatment plant discharges treated effluent to the Detroit River. As in many older cities, the sanitary wastewater system in Detroit is combined with surface drainage. In times of no rainfall (dry weather flow), wastewater to the treatment plant is mostly sanitary flow. However, in times of high rainfall, the combined sanitary and storm-water flow greatly increases the total flow that reaches the plant.

The first outfall tunnel, now identified as Detroit River Outfall No. 1 (DRO-1), was constructed in the 1930s primarily by shield tunneling in the clay stratum that overlies rock. With installation of increased capacity of the treatment plant, the existing outfall tunnel has insufficient capacity for critical conditions and a second outfall tunnel was required. The second outfall tunnel can be considered the final measure of upgrading the capabilities of the treatment plant.

PROJECT SCHEDULE

Tunnel Design and Bid Period

Tunnel design for the Modified Detroit River Outfall No. 2 (MOD DRO-2) started October 2006 and was completed at the end of 2007. Parsons Brinckerhoff (PB), working under DWSD Contract No. CS-1448, is the prime consultant for design and construction services. Tunnel contractor prequalifications were started November 2007 and completed February 2008 with prequalification of two contractor teams. The bid period started in March 2008. With time extensions, the bid period was approximately 3 months with bids taken 5 June 2008. One bid in the amount of $299 million was received from Vinci/Frontier-Kemper JV (VFK), a joint venture of Vinci Construction Grands Projets of Montreal, Canada, and Frontier-Kemper Constructors, Inc. of Evansville, Indiana. The contract was awarded to VFK and Notice to Proceed (NTP) was 17 November 2008.

Construction Contract Duration

In establishing the duration of the tunnel construction contract, the intent was to meet a date for wastewater outfall tunnel operation that was acceptable to the regulators and be practical to construct. Considering the past problems and attempt to construct this tunnel (see section following), there was great emphasis on the part of the owner (DWSD) and the regulatory agency (MDEQ) on having high assurance of completing the tunnel and placing it into operation. At the time of bid, the contract duration had been established at 39 months. In recognition of the risks the tunnel contractor has for timely completing and tunnel ground conditions, in particular the uncertainty for need for grouting to implement interventions for cutter inspection and replacement, a 48 month contract duration was established for Substantial Completion. This requires all tunneling and related underground work to be completed, flooding of the tunnel, and start-up testing of elements of the operating system including gates, instrumentation, and controls.

Construction Schedule

Key dates are summarized as follows:

Notice to Proceed (PC-771)	November 17, 2008
TBM Delivery to Site	January 2010
Complete TBM Mining	June 2011
Substantial Completion	December 16, 2012
NPDES Permit Operation Date	December 31, 2012
Final Completion	December 23, 2013

FIRST TUNNEL DESIGN AND CONSTRUCTION EXPERIENCE

Design of the Detroit River Outfall No. 2 (DRO-2) took place in the 1990s and the tunnel construction contract started November 1, 1999. Geologic exploration included borings on land along the alignment, and borings in the Detroit River. The new outfall tunnel connects to the treatment plant through a drop shaft (entrance shaft), which was also the construction access for tunneling.

Shaft sinking and preparations for tunneling are presented in detail by Traylor et al (2003). During the initial two years of work the entrance shaft was constructed by first sinking a reinforced concrete caisson 24 m through the clay and then by conventional shaft sinking in rock by drill and blast to the total depth of 91 m at tunnel level. Shaft size was 10.4 m (34 ft) ID in the caisson through soil and 9.2 m (30 ft) ID in rock. The

SLURRY TBM TUNNEL IN ROCK, THE MODIFIED DETROIT RIVER OUTFALL 677

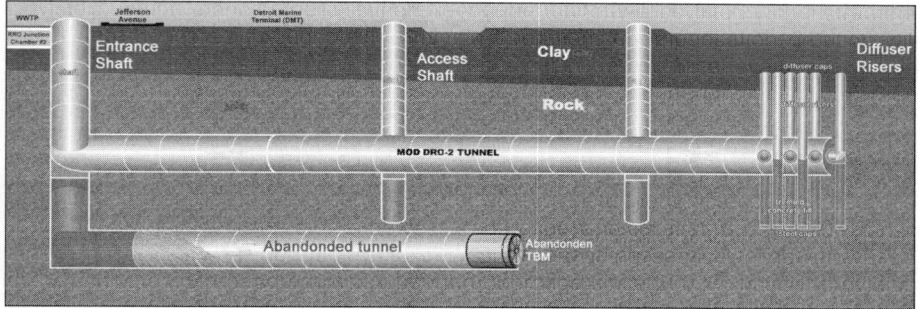

Figure 0. Profile of Detroit River Outfall No. 2 showing abandoned tunnel and new tunnel (not to scale)

rock was grouted from the ground surface before sinking the shaft. Two 2.4-m-ID (8 ft) access shafts with a 22-mm-thick (⅞ in.) steel lining, and six 3-m-ID (10 ft) offshore outfall diffuser riser shafts with 32-mm-thick (1-¼ in.) steel lining were constructed by large-diameter rotary drilling.

The tunnel contract required tunnel boring machine (TBM) excavation and a one-pass segmental precast concrete tunnel lining. Based on the site exploration, some reaches of tunnel were expected to have substantial ground water inflow. Grouting ahead of the TBM (pre-excavation grouting) was required and the tunnel construction contract had specific payment provisions for grouting.

Ground conditions were good at the start of tunneling and permitted construction of the starter tunnel in substantially dry conditions. At the time of preparation of the paper by Traylor (January 2003), on the first attempt, tunneling was projected to be completed in 2004. In the next few months, however, tunneling encountered large volumes of ground water and substantial delays to construction resulted from the grouting efforts needed to deal with water inflow. Evaluation of the conditions and a variety of measures to improve tunneling progress were in process when a catastrophic flooding occurred on April 23, 2003 and, despite significant concentrations of hydrogen sulphide, work crews were able to safely evacuate the tunnel. Alternatives for recovery were evaluated, but ultimately the tunnel was officially abandoned and a new tunnel design was required.

NEW TUNNEL DESIGN—MAJOR DESIGN AND CONSTRUCTABILITY ISSUES

The new design, termed the Modified DRO-2 (MOD DRO-2), required not just a re-issue of the prior plans, but a complete review and validation of the past design plus new designs to deal with several constructability issues.

Hydraulics

Two major hydraulic requirements must be met: outfall capacity of 3.8 million L/day (1,000 mgd) for the critical 100-year design conditions, and discharge velocity to achieve rapid mixing of effluent in the Detroit River. For the critical condition of high river elevation, the driving elevation head is very low, on the order of 0.5 m (20 in.). With the small available driving head, all unnecessary sources of head loss needed to be avoided. Hydraulic head losses had to be minimized and special attention was given to minimizing head loss for connections. Trade-offs were required among hydraulic efficiency, risk to construct, cost, and overall constructability. Even a 25 mm (1 in.) head loss represented approximately 5% of the available head.

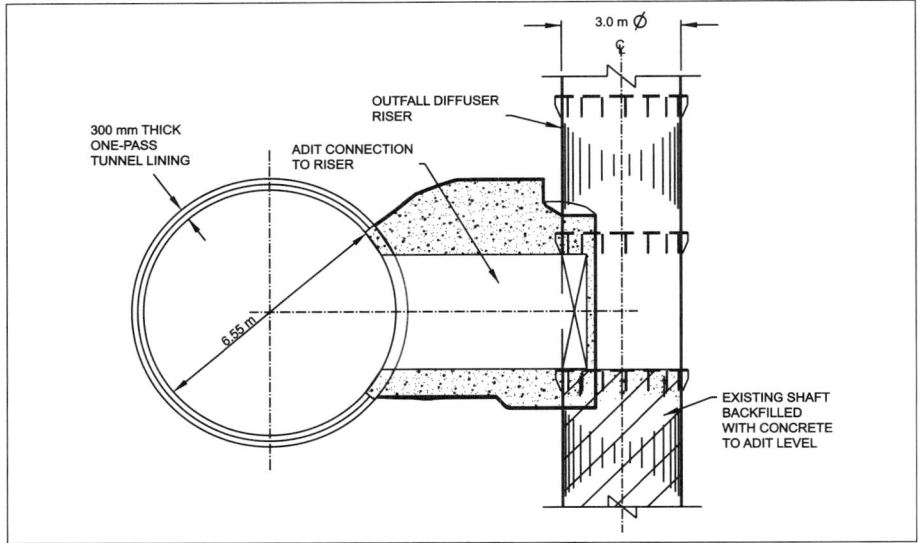

Figure 1. New design for connection to outfall tunnel diffuser riser shafts

Using Existing Constructed Works

As much as possible of the relevant parts of structures that were previously constructed are to be used. The enatrance shaft is to be dewatered after plugging the existing tunnel. The two access shafts and six diffuser risers were originally designed to be connected by tunneling underneath them. A new design was required to connect between the tunnel and shafts from the side of the tunnel. New constructability concerns had to be dealt with, see below. Substantial precast tunnel lining segments (6.4 m [20 ft] internal diameter) left on site are to be demolished since the new tunnel is slightly larger and a more robust, thicker lining for use with a slurry TBM is required.

Connections to Access Shafts

For a new tunnel at a higher elevation, tunneling on exactly the same alignment as originally intended presents a serious constructability problem. If left in place, the 22-mm-thick (⅞ in.) steel lining of the shaft becomes a man-made obstruction for the TBM, particularly for a closed-face slurry TBM. To avoid the hazards and extraordinary constructability problems of trying to make the connection as previously designed, the tunnel was offset from the shaft. The connection was designed to be made with a horizontal adit, similar to what is shown in Figure 1 for the diffuser risers.

Connections to Existing Diffuser Risers

Connections to the diffuser risers (shafts in which the effluent "rises" and is discharged through a "diffuser" at the top of the riser at the bottom of the river) were deemed to be substantially more risky than for the access shafts on account of being under the river and the larger size of the connecting adit. Any work from the surface would have to be done from a barge on the Detroit River. These diffuser risers were installed to a depth just above the crown of the DRO-2 tunnel, but now will be connected at a much higher elevation. See Figure 3. The new connection will cut through the 3-m-dia, 32 mm-thick wall of the steel diffuser riser shaft. Even though the distance from tunnel to diffuser riser is relatively short, ranging from 5 to about 15 m, stable rock conditions and controllable ground water inflow are essential. Done by hand mining, these short excavations are required to have pre-excavation grouting from the tunnel. After grouting but before hand tunneling, ground freezing is required. These measures are intended to do more than mitigate the risk of flooding of the tunnel, but avoid the risk entirely.

Tunnel

Large volumes of groundwater inflow with generation of unacceptable volumes of hydrogen sulfide gas from the groundwater in previous tunnel construction was a determining factor in the flooding and ultimate tunnel abandonment. Evaluation of alternatives to complete the tunnel concluded that a pressurized-face tunnel boring machine was required. Accordingly, the new design has been based on the mandatory use of a pressurized-face rock tunnel boring machine (slurry TBM), which must be designed to operate under the full head of groundwater pressures.

Major Non-Tunnel Scope of Project

This DWSD construction contract (PC-771) is largely a tunnel contract, but the contract also includes civil site work and structures, and all mechanical, electrical, instrumentation, and control systems to connect this tunnel to the WWTP. New motorized gates for hydraulic control are to be constructed along with tie-ins for electric power and controls for the mechanical equipment. The waste water is disinfected by injection of chlorine solution thru diffusers in the conduits and then dechlorinated with sulfur dioxide thru diffusers in the tunnel before discharge into the Detroit River. Sampling systems and other instrumentation are required to acquire data to monitor regulatory compliance for wastewater treatment. A new junction chamber (Junction Chamber No. 2) is to be constructed that ties into an existing box structure known as the Rouge River Outfall (RRO). The RRO serves as the emergency discharge from the WWTP and by permit requirement, can only be decommissioned during a five-month period of time from October thru February of successive years, which imposes a significant construction schedule constraint.

GEOLOGIC CONDITIONS

Subsurface Exploration

Geologic conditions along the alignment comprise overburden soils (glacial deposits of till and outwash) ranging in thickness from 25 to 30 m overlying bedrock. The tunnel horizon is in limestone and dolomite bedrock from the middle Devonian age Detroit River Group. The project presented the challenge of assimilating the geologic information from past and new subsurface exploration programs. See Figure 2.

Subsurface investigation (18 borings) had been undertaken for the initial DRO-2 tunnel design in 1995 and 1996. More subsurface investigation (8 borings plus pump tests) was done when tunneling problems were experienced in 2003. For the new design in 2007, an additional 6 borings (3 vertical and 2 inclined on land, and one barge-based in the Detroit River) were drilled. Purposes of the new borings were to collect geotechnical data between previous design borings where widely spaced (some were spaced ≥300 m apart), to further characterize the ground conditions in the new tunnel horizon, and to identify high angle fracture systems in the vicinity of the previous tunnel failure.

A basic premise of the new site investigation was to concentrate on doing more in a few bore holes rather than less in a larger number of borings. Improving the understanding of the geologic conditions in each new boring was judged to be preferable to having many more borings with less detailed information. Borehole lengths were up to 70 m (230 ft) in order to extend to about two tunnel diameters below the preliminarily

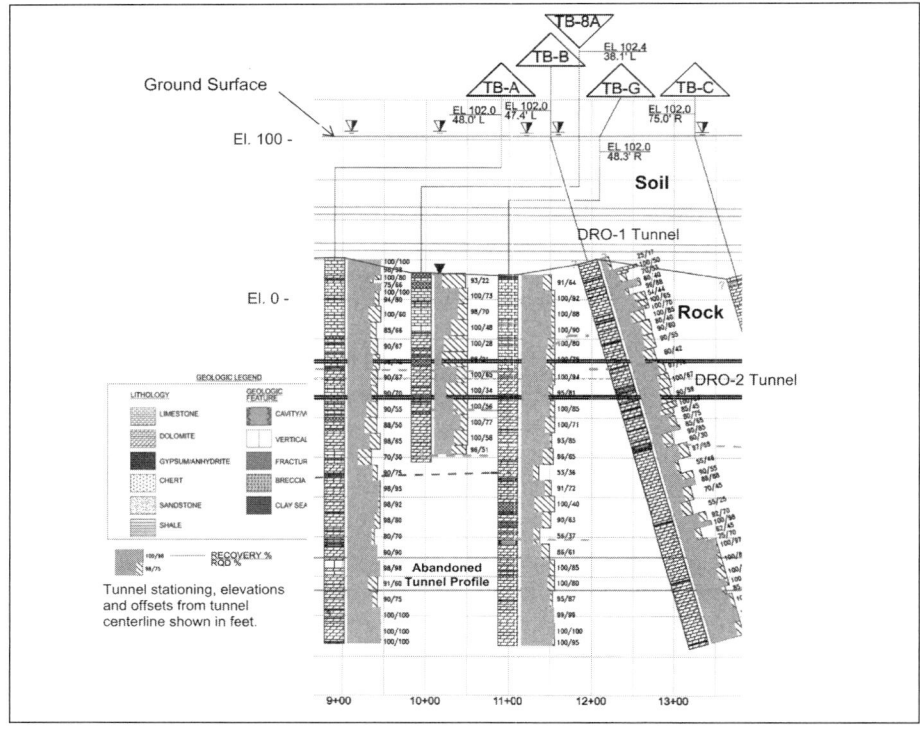

Figure 2. Typical geologic profile showing subsurface investigation

established new tunnel invert. Lugeon water pressure testing and downhole geophysical logging was carried out within the entire rock-cored length of each borehole. The water pressure testing was done on 5 m (15 ft) spacing. Downhole geophysical tests included acoustic and optical televiewer; acoustic caliper; natural gamma ray; fluid temperature; fluid resistivity; electrical logs including spontaneous potential, single point resistance, and resistivity; and heat-pulse flowmeter (3 vertical borings only). Three borings had multi-level vibrating wire piezometers installed.

Geologic profiles prepared from a combination of the information from both the old and new borings. Clarity of data presentation was dealt with by preparing two profiles. One profile presented the core recovery and RQD data for each boring and stratigraphy (see example in Figure 3). The companion profile presented permeabilities calculated from water pressure testing and groundwater observations. The stratigraphic column portrayed the significant geologic features greater than 150 mm (6 in.) in dimension including, lithology (rock type); cavities; anhydrite or gypsum layers; chert layers or nodules; high angle joints and fractures; highly fractured rock; and clay seams, sandstone or sandy dolomite, shale, and brecciated zones or layers.

Rock Properties

Rock properties important to hard-rock TBM tunneling were determined on the basis of laboratory tests. All data from the new borings were used, but the data from the earlier borings included only rock samples collected above approximately the same elevation as the bottom of new borings. Baseline values of rock parameters

Table 1. Intact rock properties baselines (expressed in percent of excavated rock mass—Lucas dolomite)

Property	Range	Average (Mean)	Baseline
Unconfined Compressive Strength (MPa)	27–158	75	95% of rock—138 MPa or lower 5% of rock—Greater than 138 MPa
Shore Rebound Hardness	20.2–42.0	29.7	90% of rock—35 or lower 10% of rock—Greater than 35
Abrasion Hardness	0.15–2.14	0.66	90% of rock—1.6 or lower 10% of rock—Greater than 1.6
Cerchar Abrasivity	0.5–1.9	0.8	95% of rock—1.5 or lower 5% of rock—Greater than 1.5

are summarized in Table 1. These parameters were the baseline values to be used in planning and procuring tunneling equipment for the project. No single rock property value is representative for the entire tunnel. To illustrate the range of data, histograms of the data were prepared as shown for unconfined compressive strength of the rock in Figure 3. Rock mass quality was baselined in a similar manner. For the tunnel, Rock Mass Rating (RMR) and Tunnel Quality Index (Q) were baselined as 70% of the tunnel would have RMR 50 or greater (Q = 1.95 or greater).

Groundwater

Groundwater is under artesian conditions. The tunnel site, located at the Detroit River, is topographically low compared to the surrounding area and the soil overlying the rock is an aquaclude. Ground surface is approximately El +30 m (+100 ft), but maximum groundwater head measured by piezometers in the rock was El +36 m (+118 ft) (artesian) and became a geotechnical baseline. Boreholes cased through the soil would flow steadily if not capped. It was known from the DRO-2 tunneling that the rock was in some locations very pervious. Lugeon-type water pressure testing indicated permeabilities from low to high, which reflects the nature of the limestone rock that, in some places is intact rock, in other places highly jointed. Rock permeability for tunneling has to be interpreted in the context of the specific geologic conditions along the tunnel. As an example, rock mass permeability was baselined such that 50% of the rock would be in the range 10^{-4} to 10^{-3} cm/s.

Toxic and/or explosive gases are present as hydrogen sulfide, with some methane in limited volumes, and occur throughout the rock mass along the entire tunnel alignment. Hydrogen sulfide (H_2S) is toxic at concentrations greater than 10 ppm and is combustible in air at concentrations above 4.3% by volume. Methane gas (CH_4) is explosive in air at concentrations in the range of 5 to 15%. As measured at the borehole collar, hydrogen sulfide gas was observed in most of the site investigation borings in concentrations exceeding 100 ppm (0.01%); methane was observed in three of the borings drilled in 2007 in concentrations of 1 to 7%.

SLURRY TBM TUNNELING IN ROCK

Project Conditions and State-of-the-Art TBM Tunneling

State-of-the-practice tunneling technology will be required to successfully construct this tunnel. The flooded tunnel was constructed with an open-faced TBM with a one-pass precast segmental lining at a depth of 91 m (300 ft) and water pressure of approximately 9.5 bar, when artesian ground water levels are considered. As of 2007,

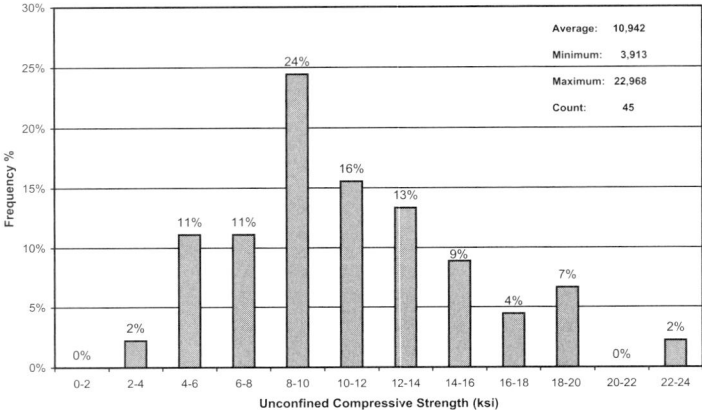

Figure 3. Unconfined compressive strength of rock (138 MPa = about 20 ksi)

only one tunnel in the world (Hallandsas Tunnel in Sweden) was being constructed with a pressurized-face rock TBM for high ground water pressure (designed for 13.5 bar). In the US, the Lake Mead Intake 3 Water Tunnel is being planned to tunnel with a closed-face TBM with pressures ranging from 13 to 17 bar (Feroz, et al 2007).

In configuring the new tunnel, having the tunnel as high as possible was desirable so that the lowest possible groundwater pressures would have to be dealt with. In balance with other factors, the tunnel profile was also selected to provide as much rock cover as possible for tunneling and for adit connection construction. There is no ideal elevation for the tunnel due to the pervasive occurrence of open fractures and cavities throughout the rock mass, and the unlimited volume of artesian groundwater that is charged with dissolved hydrogen sulfide gas. The tunnel is situated to be as high as possible to minimize hydrostatic pressures during construction, not because better rock was expected at shallower depth, but as a choice among minimizing groundwater pressures, having rock cover above the tunnel profile, and maintaining clearance to a zone of rock observed to contain several partially filled to open cavities which lies approximately 6 to 12 m below the tunnel invert. The tunnel slopes back toward the Entrance Shaft and avoids connecting to the existing diffuser risers at the bolted joints between fabrication sections.

A piezometric head of Elev +36 m (+ 118 ft) (tunnel invert is at about El. −19 m, −62 ft) was baselined for which the tunneling equipment must operate and must be anticipated to be present at any location along the tunnel alignment. This corresponds to about 5.5 bar pressure at the tunnel invert (5 bar at tunnel centerline). "Interventions" (interruption of TBM mining and accessing the cutter head for inspection, maintenance, or repair) will be required as tunneling takes place. An intervention is made especially difficult by the pervious rock and gassy conditions. If the rock is tight with minimal ground water inflow, an intervention could be relatively straight forward as long as ventilation deals with gassy conditions. If permeability is high, pregrouting is indicated to be required to create a safe working condition for the intervention. The relatively high ground water pressures are too high to be dealt with by conventional compressed air, and a mixed-gas system would be needed. Risk associated with interventions is the contractor's. In order to provide more time for tunnel construction, including time for interventions, the contract duration was increased during bidding.

SLURRY TBM TUNNEL IN ROCK, THE MODIFIED DETROIT RIVER OUTFALL

A 7.5 m (24 ft-8 in.) Herrenknecht "Mixshield" using 17 in. cutters is planned for use. Selection of the TBM and associated Slurry Treatment Plant and Water Treatment Plant has been made considering several aspects as summarized in the following.

TBM Main Aspects

The slurry TBM is designed to:

- Deal with rock: The cutterhead is similar to a standard 7.5 m (25 ft) diameter soft rock TBM cutterhead with the exception that the rock material has to be carried through the slurry circuit. This is achieved by maintaining a high flow and velocity in the slurry circuit. Due to the rock properties, but also based on the experience of the first attempt in the same rock formation, the cutter consumption and the abrasion in the slurry circuit is expected to be very low.

- Deal with the hydrostatic pressure: the project specifications require the use of a Slurry TBM. Slurry TBMs are commonly used in soft ground to stabilize the ground face in front of the TBM. In this case, the hard rock face will be stable and the purpose of the slurry circuit is to prevent water and H2S inflow into the tunnel. In an ideal situation, the slurry pressure balances the hydrostatic pressure so the groundwater inflow is expected to be minimal, leading to minimal amount of H2S in the slurry circuit.

- Deal with cutter changes and cutterhead maintenance: Periodic maintenance interventions in the cutterhead are expected, about once every 100 m. With a pressurized TBM, these interventions are normally performed in closed mode in compressed air where the air pressure balances the hydrostatic pressure. Hyperbaric interventions can normally be performed up to 5 bars. But unlike soft ground, hard rock is a rigid material and sudden fractures can occur, leading to sudden pressure loss in the cutterhead chamber. Furthermore, unlike in soft ground, the slurry can not be expected to "seal" the face of the excavated tunnel. In this situation, hyperbaric interventions cannot be safely performed for the project and regular cutterhead maintenance is therefore planned to be performed at atmospheric pressure (in open mode).

- Deal with water in the cutterhead during interventions: This represents one of the biggest challenges of the project. As the maintenance work is performed at atmospheric pressure, the permeability of the rock leads to water inflow in open mode from 0 to 200 L/s (3,400 gpm). It is expected that for 90% of the tunnel drive, the groundwater flow will be below 30 L/s (500 gpm). The TBM is designed to handle such a flow of water (and H_2S) in the cutterhead with pumps. For the balance of the tunnel drive, the groundwater inflow needs to be reduced using pre-grouting to permit safe work in the cutterhead chamber. The TBM is equipped with 11 grout ports around the tail skin plus 15 grout ports directly through the cutterhead. It is equipped with 2 dedicated drill rigs installed in the crown just behind the segment erector. Pre-excavation grout operations are to be performed at atmospheric pressure, each grout hole being equipped with packers.

Slurry Treatment Plant and Water Treatment Plant

The Slurry Treatment Plant and Water Treatment Plant are designed to:

- Deal with H_2S in the atmosphere inside the tunnel: A high flow of fresh air is forced into the tunnel at all times though ventilation tubes with an emergency capacity of up to 45 m³/s (100,000 cfm). In addition, during interventions, secondary fans provide fresh air directly in the cutterhead chamber.

- Deal with H_2S in water: The highest potential to be in contact with H_2S occurs during the maintenance interventions (also during shaft connections) when significant groundwater flow enters the TBM cutterhead at atmospheric pressure. The water is pumped to the surface and is treated at the surface in a water treatment plant that precipitates the H_2S using hydrogen peroxide. This method has been successfully used on several projects in the Detroit area where H_2S is naturally present in several rock formations.

- Deal with H_2S in the slurry: Although the flow of groundwater in the slurry line is expected to be low when the system is balanced, a minimum amount of H_2S will have to be dealt with. As the slurry circuit is a closed circuit, the concentration of H_2S will increase with time and will have to be neutralized. One characteristic of H_2S is that it stays dissolved in water when the pH is above 9, preventing offgassing that can become a health issue at concentrations above 10 ppm. A high pH will also prevent corrosion of the steel. Constant monitoring of the pH is performed at the Slurry Treatment Plant. An evaluation of several options to reduce the H_2S in the slurry leads to the conclusion that the most effective way is to use a scavenger, zinc or iron based, or hydrogen peroxide that can still be efficient at high pH. The Slurry Treatment Plant is designed to include H_2S and pH monitoring as well as the introduction of lime, caustic soda, hydrogen peroxide or scavengers. Another precautionary measure is taken at the Slurry Treatment Plant in case the pH drops and offgassing occurs: the separation plant tanks are enclosed; the enclosure is kept at negative pressure with vacuum pumps; and, the extracted air can be filtered for H_2S with photo catalysis (oxidation).

Tunnel Lining

A robust tunnel lining is required for the construction condition, which will readily accommodate the relatively smaller long-term loading conditions for the perpetually submerged tunnel during outfall operation. The tunnel lining is designed as 300 mm thick (12 in.), 6.55 m (21 ft-6 in.) ID, one-pass, 6-piece segmental precast concrete. The contractor is responsible for checking the adequacy of the lining for the slurry TBM, shield jacks, transport, and erection procedures to be used. The contractor's selected segment configuration includes particular features to cope with lining erection and shield jack forces on the lining when operating in closed mode at 5 bar pressure: no packers between segments and rings to increase concrete-to-concrete friction, use of bolts instead of dowels between segment rings, and "rectangular" segments. All these features have been successfully used in another slurry TBM project in Seattle in similar pressure conditions in order to minimize the potential for movement of segments during the erection and advancing the machine.

Access Shaft and Diffuser Riser Shaft Connections

The access and diffuser riser shafts were installed as part of the first tunnel works. There are no connections to the abandoned tunnel. As shown in Figure 2, these tunnel-to-shaft connections were designed as lateral adits. Due to concerns with rock quality and groundwater flows, mandatory ground improvement is required for construction. For the Access Shafts, which are on land, ground improvement by pre-grouting is required. For the Diffuser Riser Shafts in the Detroit River, which are larger and considered to have higher risk of tunnel flooding, ground improvement by pre-grouting followed by freezing is required.

CONDITIONS OF CONTRACT

Specific efforts were made to implement contract terms and conditions to deal with tunneling risk. Some of the contract features deemed necessary were already planned by the Owner prior to start of design, while others were implemented as the contract documents were developed. Because many of these measures are interrelated, or in varying degrees dependant on each other, a suite of risk-mitigating measures was implemented during design, as follows:

Prequalification of Tunnel Contractors

Prequalification was considered an essential requirement in order to get contractor teams that had the experience to deal with the many project risks and the required use of a state-of-the-art slurry TBM. Prequalification started in November 2007 as the design was being completed. Notice of prequalification was made February 2008 to two joint venture teams.

Geotechnical Baseline

The MOD DRO-2 tunnel design was required by the owner to have a GBR. Preparation of the GBR followed the *Geotechnical Baseline Reports for Construction* (Essex, 1997, 2007).

Inclusion of Geotechnical Data as Contact Document

The geotechnical data obtained for the original design (1990s) and the new data obtained under the CS-1448 design (2007) were kept as separate documents but collectively defined as a Geotechnical Data Report (GDR). The GDR was made a contract document and linked to the GBR as a source for the development of the baseline conditions. Contract documents state that the GBR presents the final interpretation of the data provided in the GDR, and that the GBR takes precedence over the GDR or any other document for baseline purposes. This practice follows ASCE's guideline referred to above.

Differing Site Conditions Clause

A Differing Site Conditions (DSC) clause in the contract was considered an essential complement to the use of other contractual risk-mitigation measures, specifically the GBR.

Escrow of Bid Documents

Escrow Bid Documents (EBD) were required for this contract and follows industry guidelines (ASCE 1989, 1991). Only the lowest and next to low bidders were required to submit.

Payment to Contractor for Major Costs for Tunnel Boring Machine

The tunnel construction contract has provision to pay separately for the large costs of the tunnel boring machine (TBM) and related equipment for the slurry system. Contractors' financing costs and the large upfront payment for the manufacture of the TBM can not be adequately covered in a "mobilization" bid item or in the early construction work activities. A separate bid item for payment of the TBM and related support equipment was established. A payment schedule was developed on the basis of approved submittals, manufacturing invoices, manufacturing plant verification visits, and final shipping and delivery charge. This payment procedure was intended to reduce the cost of financing up-front costs that the contractor would otherwise have to spread through other bid items as a cost to the Owner.

ACKNOWLEDGMENTS

This is a Detroit Water and Sewerage Department (DWSD) project (PC-771). DWSD Director (Interim) is Pamela Turner. Assistant Director of Engineering Services, responsible for design and construction, is Ramesh Shukla. Parvez Jafri, Head Engineer for Water Systems, is DWSD's Project Manager for the consultant contract (CS-1448) for design and construction services. K. V. Ramachandran, Head Engineer (Field), is in charge of the construction contract (PC-771). Regulatory agencies include the Michigan Department of Environmental Quality (MDEQ), the U. S. Army Corps of Engineers, and the U. S. Coast Guard. Parsons Brinckerhoff Michigan, Inc. (a PB Americas, Inc. company) is the prime consultant (DWSD Contract CS-1448) for design and construction services. Bill Hansmire, Malcolm Hudson, and Zephaniah Varley are Project Manger, Design Manager, and Resident Engineer for construction, respectively. Consultant teammates are Applied Science, Inc.; Brierley Associates, LLC; CDM Michigan, Inc; Hazen & Sawyer, Inc.; Hinshon Environmental Consulting, Inc.; Integrated Management Services, Inc.; METCO Services, Inc.; Sigma Associates, Inc.; and SOMAT Engineering, Inc. Thierry Portafaix is Project Manager (Director) for Vinci/Frontier-Kemper JV. Frederic Mir is Technical Manager for Vinci Construction Grands Projets, and Dave Rogstad is the principal representing Frontier-Kemper Constructors, Inc. on the JV board. At the 2008 World Tunnel Congress in Agra, India, the technical paper by Hansmire and Jafri (2008) presented this project as of the time of bidding and award of the tunnel contract. That paper has been adapted for RETC 2009 by updating with construction planning, in particular TBM design, as of January 2009, which was two months after Notice to Proceed for the tunnel construction contract. The authors thank Ramesh Shukla for his contribution to this paper.

REFERENCES

ASCE. 1989. *Avoiding and Resolving Disputes in Underground Construction*, Technical Committee on Contracting Practices, Underground Technology Research Council, 24 p + appendices.

ASCE. 1991. *Avoiding and Resolving Disputes During in Underground Construction*, Technical Committee on Contracting Practices, Underground Technology Research Council, ASCE, 82 p.

Essex, R. J, ed. 1997. *Geotechnical Baseline Reports for Underground Construction*, Technical Committee on Geotechnical Reports, Underground Technology Research Council, ASCE, 40 p.

Essex, R. J. ed. 2007. *Geotechnical Baseline Reports for Underground Construction*, Technical Committee on Geotechnical Reports, Underground Technology Research Council, ASCE, 62 p.

Feroz, M, Jensen, M, Lindell, J. E. 2007 "The Lake Mead Intake 3 Water Tunnel and Pumping Station, Las Vegas, Nevada, USA," Proceedings, 2007 Rapid Excavation and Tunneling Conference," Traylor, M. T. and Townsend, J. W. eds, SME, pp. 647–662.

Hansmire, W. H. and Jafri, P. 2008. *Slurry-face Rock TBM Outfall Tunnel In Detroit, Michigan, USA*, Proceedings of the World Tunnel Congress—2008, Agra, India, Kanjlia, V. K. et al eds., Vol. 3, pp. 1575–1584.

Traylor, M., Jatczak, M, Robinson, B. 2003, "Construction of the Detroit River Outfall No. 2," Proceedings, 2003 Rapid Excavation and Tunneling Conference," Robinson, R. A, & Marquardt, J. M. eds., SME, pp. 450–482.

PORT OF MIAMI TUNNEL UPDATE—A VIEW FROM DESIGN BUILDER'S ENGINEER

Wern-Ping Chen ▪ Jacobs Engineering

ABSTRACT

Technical and non-technical challenges exist in each mega infrastructure project. It's no exception to the first US underground PPP Port of Miami Tunnel (POMT) project in Miami, FL. The POMT will connect two man-made islands, the Dodge and the Watson Islands, under the Biscayne Bay within highly porous limestone formation. Permeability as high as 0.1 cm/sec has been identified. Its 41-ft outside diameter, if constructed, is the largest precast segment lining in the US. Besides its geological and hydrogeological challenges and risks, local community influence and impacts have controlled the fate of this project. This paper describes its tender design enhancements and the evolving of its slow advancement in resolving local community issues.

INTRODUCTION

The Florida Department of Transportation District 6 (FDOT D6) began a study in 1987 for a master plan to improve the traffic circulation and congestions between the Port of Miami and downtown Miami. This study includes a tunnel connection between two man-made islands, the Watson and the Dodge Islands. Though a Finding of No Significant Impact (FONSI) was completed in 1992, the project was on hold for about 10 years. In 2002, Florida Turnpike took over this project, re-evaluated its benefits to the local communities, determined it feasibility (specifically the tunnel options), and established a probable cost.

In 2005, FDOT D6 regained its charge over the project and concluded that a bored tunnel, under Biscayne Bay, is feasible. Its primary components are:

1. Widening of the MacArthur Causeway Bridge, to accommodate truck traffic to the Port of Miami
2. A tunnel connection between Watson Island and Dodge Island, where the Port of Miami (POM) is located
3. Connections to the POM roadway system, either for the traffic from the City of Miami or for the truck traffic from I-395 to the POM

The Project will improve access to and from the POM and provides a dedicated roadway connector linking the POM with the MacArthur Causeway (State Route 41/1) and I-395, specifically for the truck in and out of the POM without going through downtown Miami. As identified in the Project Information Memorandum, PIM, (FDOT), its primary objectives are to:

- Improve access to the POM, helping to keep it competitive and ensuring its ability to handle projected growth in both its cruise and cargo operations
- Improve traffic safety in downtown Miami by removing POM traffic, trucks and buses, from the congested downtown street network; and in so doing
- Facilitate ongoing and future development plans in and around downtown Miami.

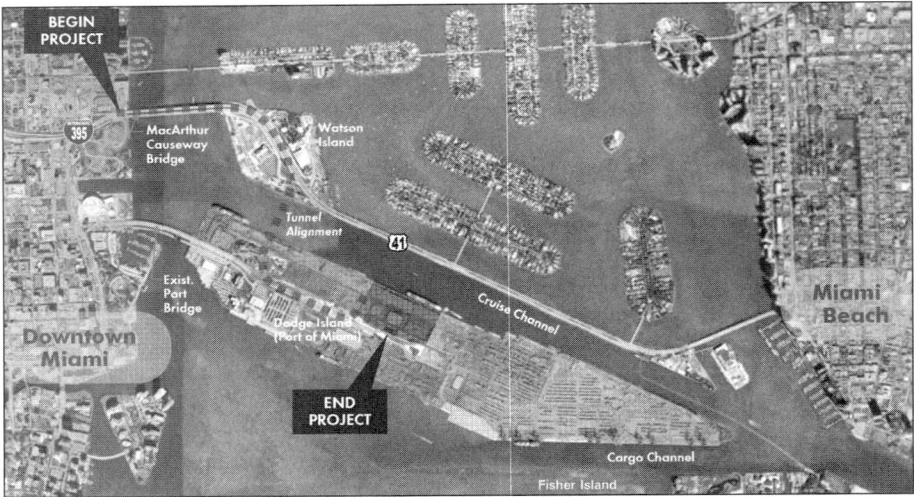

Figure 1. Project location plan and alignment

Project Location Plan and Alignment

The Project is located adjacent and east to downtown Miami. It alignment extends from MacArthur Causeway Bridge, Watson Island, under the Cruise Channel in Biscayne Bay, and to Dodge Island, where the POM is located. Its schematic project location plan and alignment is shown on Figure 1.

Purpose of This Paper

The purpose of this paper is to document the process and development of the first US underground project utilizing Public-Private Partnership (PPP) project delivery model. It will address the following topics:

- Project funding and financing
- Regional geology
- FDOT indicative designs
- Tender design performed by a consortium, the Miami Access Tunnel (MAT)
- Project developments in a chronological order from 2006 to 2008
- Conclusions and lesson learned

The MAT consists of the contractor Bouygues Travaux Publics of France, the investment bank Babcock & Brown of Australia, the tunnel operation and maintenance controller Transfield Services of Australia, and the design engineer Jacobs Engineering with its subconsultant Langan Engineering and Environmental Services, Miami. Contractually, the design team is a subconsultant to the contractor Bouygues.

Opinion and interpretation presented in this paper are sole perspectives from the author, who was the Tunnel Design Manager of MAT during tender design phase. They are not official views from MAT.

PROJECT FUNDING AND FINANCING

The Project is being undertaken in cooperation with FDOT, Miami-Dade County, MDC, the Port of Miami (a Department of MDC), the City of Miami, and other local stakeholders. The funding source is jointly by FDOT, $457 M, MDC, $402.5 M, and the City, $50 million cash, $5 million in right-of-way on Watson Island. As described earlier, the Project is to be developed through a PPP delivery system. A concession team, the Concessionaire, is selected by FDOT as the best value Proposer and then enters into the Concession with FDOT. The Concessionaire is responsible for the Design/Build/Finance/Operation/Maintenance of the facilities, O/M only for the tunnel portion, for a Concession period of 35-years, which include design and construction.

In accordance with FDOT, the primary objectives for pursuing the Project as a PPP are to:

- Achieve the most efficient possible design, construction and maintenance of the Project
- Receive a high-level of quality, availability, upkeep, safety, and user service
- Share risks with a private partner(s) that is experienced in mitigating such risks
- Agree to a long-term, guaranteed cost structure for the Project
- Facilitate a predictable and efficient implementation process

Milestone Payments

During construction, the Concessionaire is responsible for privately financing the Project and will receive Milestone Payments, $100 Million, upon completion of the associated works:

- $20 Million—Completion of the tunnel designs, excluding mechanical, electrical and plumbing components
- $40 Million—Tunnel boring machine is at work in the first bore
- $25 Million—Tunnel boring machine is at work in the second bore, but in no event prior to completion of the first bore
- $15 Million—Substantial completion of construction work on MacArthur Causeway

In addition, FDOT and its funding partners will provide approximately $300 million in "Construction Milestone Payments" at the completion of the construction.

Changed Geotechnical Conditions Payment

A $180 Million geotechnical contingency fund is setup to mitigate Extra Work Costs and Delay Costs arising out of Changed Geotechnical Conditions during construction. Its mechanisms are:

- The first $10 Million—Borne solely by Concessionaire
- The next $150 Million—Borne solely by FDOT
- The last $20 Million—Borne solely by Concessionaire

Extra Work Costs and Delay Costs for Changed Geotechnical Conditions that exceed $180 Million is deemed Extraordinary Geotechnical Losses.

Maximum Availability Payments

After the Concessionaire's completion of construction and the commencement of operations, FDOT will begin making periodic payments to the Concessionaire. These

Table 1. Typical geotechnical parameters—Miami Limestone and Fort Thompson formation

Parameters	Miami Limestone	Fort Thompson Formation
Low to Median RQD	0 to 5%	8 to 14 %
Low UCS	No Test	0.1 to 35 Mpa (40 to 5,000 psi)
High Permeability	1.0 E-1 cm/sec baseline; locally can be 1 cm/sec	

Figure 2. Good Fort Thompson Formation rock sample

"Availability Payments" will be based on the availability of the below-grade portions of the Project, the Tunnel, to provide vehicular access to the POM, as well as the Concessionaire's conformance with other criteria established in the RFP.

REGIONAL GEOLOGY

The following geological, hydrogeological, and geotechnical information are excerpted from the Geotechnical Baseline Report (GBR) developed by Parsons Brinckerhoff in 2006 for this Project. Detail regional geology adjacent to the project site and subsurface exploration programs are addressed in the GBR and are not to be repeated here.

The topography in the area of Biscayne Bay is flat lying, varying in elevation from approximately 1.5 to 4.3 meters (5 to 14 feet) above Mean Sea Level (MSL), based on National Geodetic Vertical Datum (NGVD), except for the ship channel area.

Along the proposed tunnel alignment, only the Miami Limestone and the Fort Thompson Formations interfingered with the Anastasia and Key Largo Formations are expected to be encountered by the planned tunnel construction, with the exception of the recent sediments and man-made fill which overly the bedrock formations as a thin veneer. Typically this thin veneer is with mostly dredge-placed fill composed of very loose sand and lime-rock overlying naturally occurring loose silty fine sand to sandy silt. The Miami Limestone is very porous and permeable due to the dissolution of carbonate by ground water migration. The Fort Thompson Formation, underlies the Miami Limestone Formation, generally contained alternating units of sands, marls, shells, and sandy fossiliferous limestones. The GBR has been assumed to be at elevation –40 feet. Furthermore, the Key Largo and Anastasia Formations are interfingered within the Fort Thompson Formation. Some representative geotechnical parameters of the Miami Limestone and Fort Thompson Formation are listed in Table 1 to get a flavor of the ground condition to be encountered during tunneling.

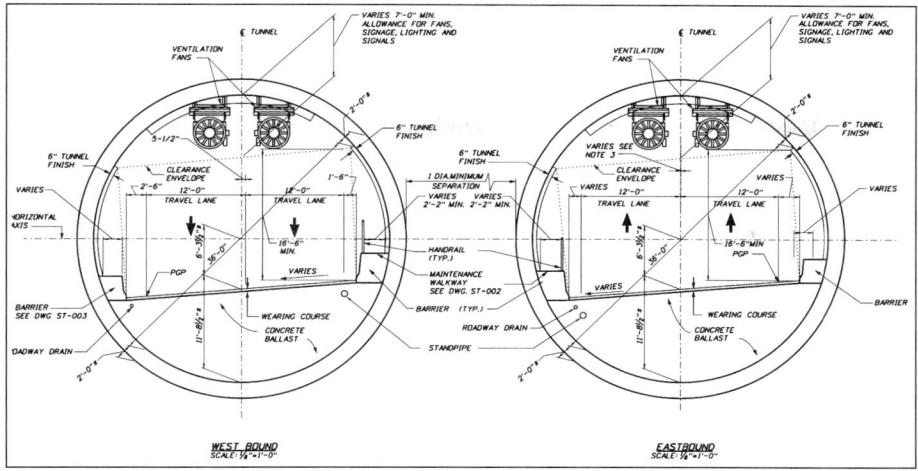

Figure 3. FDOT indicative tunnel cross section

The high porosity of the ground to be encountered can be illustrated by a typical high quality Fort Thompson rock sample shown in Figure 2.

In general, ground water elevations were measured between +0.6 and +1.5 meters (+2 and +5 feet) NGVD with extreme measurements of –0.3 and +1.68 meters (–1 and +5.5 feet).

FDOT INDICATIVE DESIGNS

The proposed tunnel is an 11-meter (36-ft) inside diameter twin bored two-lanes, each, highway tunnel to be excavated by a closed face earth Pressure Balance Machine or a Slurry Shield machine. Its lining is a 61-cm (2-ft) thick bolted and gasketed precast concrete segmental lining. Figure 3 shows FDOT indicative tunnel cross section. The determination factor for the size of the tunnel cross section is a functional requirement to satisfy the vehicular clearance, the space for tunnel ventilation system, the width of the walkway and life/safety issues, the tunnel stability, the buoyancy, and the security reason.

The vertical tunnel alignment of the proposed tunnel is limited by the accesses on both sides of the two manmade islands, the Watson Island and the Dodge Island, the maximum grade to meet the current highway tunnel industry standard, the tunnel clearance to the bottom of the future ship channel deepening, and the manmade obstructions, such as pile foundations. Figure 5 shows FDOT indicative vertical tunnel alignment.

To facilitate tunnel design, bases for bidding purpose, and basis for differencing Site Condition (DSC), the GBR provides a geological profile baseline along the indicative tunnel alignment. It divides the geological profile in to four (4) primary zones: 1) Fill (F), 2) Miami Limestone (M), 3) Transition Zone (T), and 4) Fort Thompson Formation (FT) in an order from top to bottom as illustrated in Figure 5.

MAT DESIGNS

Design philosophies of the MAT team are to minimize the risks, geological or political, anticipated to be encountered, maximize the efficiency of tunnel constructions,

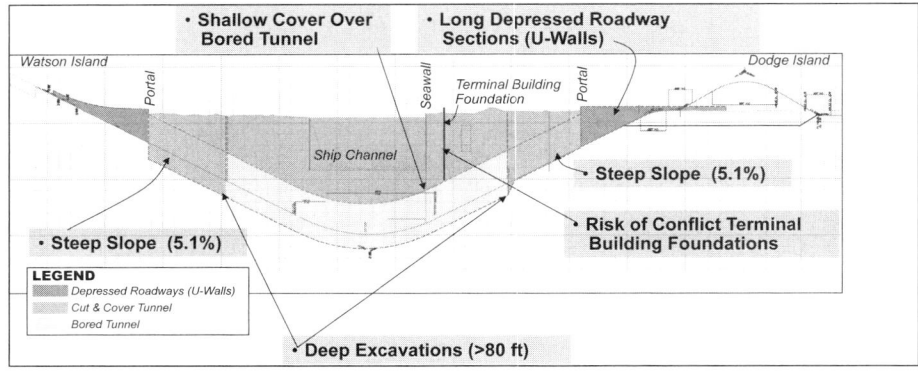

Figure 4. FDOT indicative vertical tunnel alignment

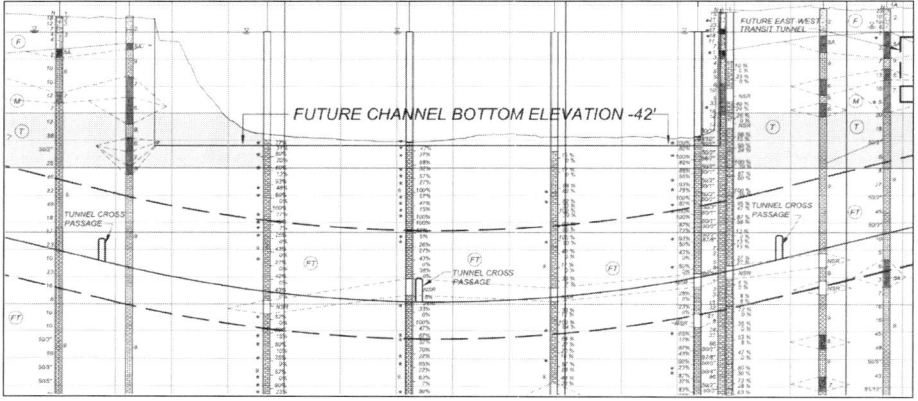

Legend—F: Fill; M: Miami Limestone; T: Transition zone; FT: Fort Thompson Formation

Figure 5. FDOT indicative geological profile

maximize the safety for surface and underground traffic operations, and minimize the cost associated with the future operation and maintenance of the tunnel facilities. Many design enhancements to FDOT indicate designs were investigated and evaluated. This paper only provides limited typical design enhancement examples, including:

- Geological Profile—A more detailed geological profile along the proposed tunnel alignment to reduce potential geotechnical risks for TBM Tunnelling and machine selection
- Tunnel Design—Extending the bored tunnel length and minimize the depth of cut-and-cover tunnel—cost effective TBM tunneling and minimize the risk of deep support of excavation
- Traffic Movement—Watson Island

Geological Profile

A detailed subsurface geological profile along the proposed tunnel alignment was developed. It consists of eight (8) different layers, instead of four layers by FDOT,

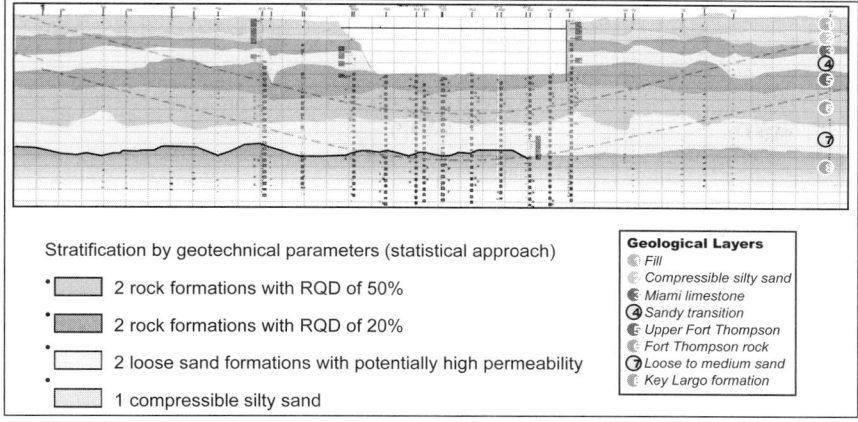

Figure 6. MAT geotechnical profile

including (1) Fill, (2) Compressible silty sand, (3) Miami Limestone, (4) Sandy Transition, (5) Upper Fort Thompson, (6) Fort Thompson Rock, (7) Loose to medium sand, and (8) Key Largo Formation. Figure 6 shows the developed detail geological profile.

Table 2 makes a comparison of geotechnical profile interpretations made by FDOT and MAT. Interpretations of Ground properties of each MAT layer were also made in Table 2.

The detailed geologic profile interpretation facilitates MAT team to select its tunnel construction means and methods, such as:

- The EPBM has to be operated in a closed face mode, be able to perform grouting from inside the TBM when tunneling under the ship channel
- Ground improvements at the invert of the TBM are required on either end of the ship channel to prevent the "stepping-down" of the TBM during tunnel excavation, since loose to medium sand are expected at the bottom of the TBM. These ground improvements can be performed from ground surface
- Condition agent selection, either polymer or form, for each different ground zones to be encountered

Tunnel Design

The strategies are to maximize bored tunnel length and decrease the depth and length of both cut-and-cover tunnels and U-walls. Tunnel grade is flattened in Dodge Island and ground improvements were provided near tunnel portal areas such that the bored tunnel can be performed in much shallow depth in comparison to FDOT indicative design. See Figure 7 for a comparison between MAT and FDOT indicative designs. MAT design eliminated FDOT designed bifurcation point, a traffic hazard, in the East Bound cut-and-cover tunnel section. Also, because of the deeper alignment, the existing cruise line recreation center that is on the FDOT proposed alignment does not required to be relocated.

After detail tunnel electrical and mechanical designs, the MAT team concluded that the tunnel cross section designed by FDOT is difficult to house all required tunnel system components; therefore, MAT design increased the inside tunnel diameter from 11 to 11.3 meters (36 to 37 feet). Both its cross section and lining system are similar

Table 2. MAT vs. FDOT geological layers

MAT Layer #	MAT Designation	FDOT Designation	Ground Properties—MAT Interpretation
1	Fill	Fill	Sand and gravel
2	Compressible silty sand		Soft plastic cohesive soil
3	Miami Limestone	Miami Limestone	Limestone—soft rock, very weakly cemented (soil type behavior); low porosity; fairly consistent
4	Sandy transition	Sandy Transition	Sand with limestone—highly permeable soil like behavior with inclusions and interbedded zoned of limestone
5	Upper Fort Thompson		Limestone with some sand—porous soft rock with sand zones
		Fort Thompson Formation	
6	Fort Thompson Rock		6a: Cemented sand/shell, with some sand—very porous rock with sand, well cemented 6b: Cemented sand/shell—very porous, well cemented, consistent rock
7	Loose to medium sand		Sand with inclusion and interbedded zones of sandstone—can be present 1) very loose soil, 2) potentially voildy condition, 3) vuggy soil filled zones, and 4) isolated zones of hard rock
8	Key Largo Formation		Sandstone, interbedded with sand lenses, zones, seam and occasional sand/silty sand pocket—very porous, well cemented, interbedded soft rock with sand

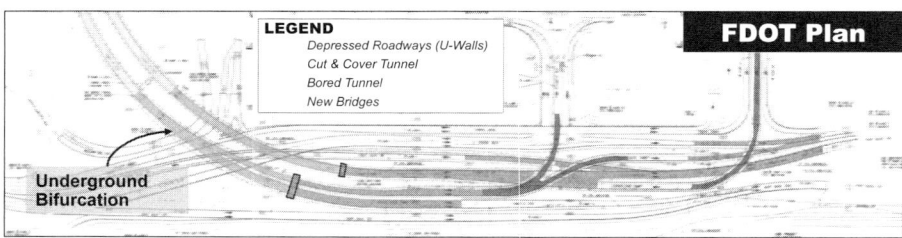
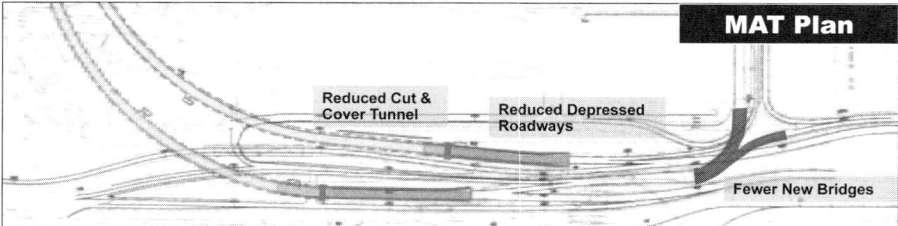

Figure 7. MAT vs. FDOT plans—Dodge Island

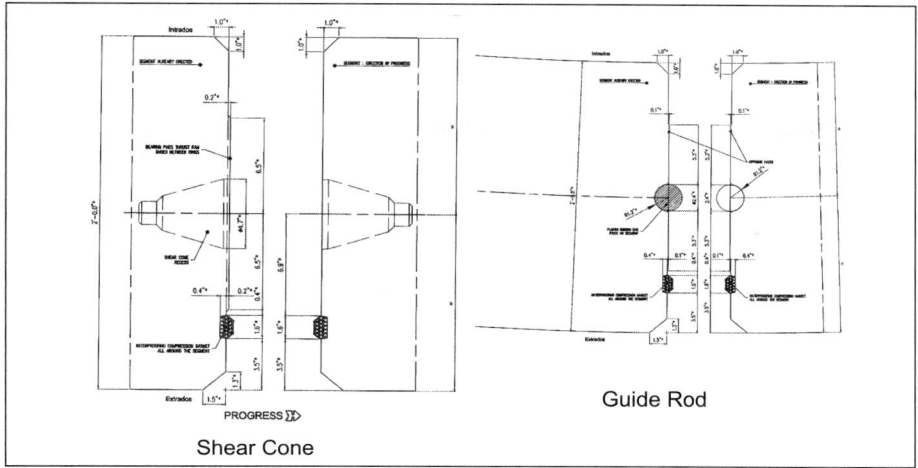

Figure 8. Shear cone and guide rod

to that designed by FDOT, as shown in Figure 3. It consists of a 61-cm (2-ft) thick precast concrete segmental lining as required for security reason; however, its connection details for radial and circumferential joints are revised from that designed by FDOT to facilitate segment erections. Shear cones were utilized in the circumferential joints and guide rods were used for radial joints. See Figure 8 for the proposed shear cone and guide rod.

Several Support of Excavation (SOE) systems were evaluated, including Cutter Soil Mixing (CSM) wall, slurry wall, and secant pile wall. Slurry wall was first excluded, since it has a potential of slurry leaking and loosing soil confinement when porous ground condition is encountered. CSM was seriously considered in the early design phase; however, it was also excluded because of strength capacity reason for this site. Secant pile was selected as the SOE for cut-and-cover tunnels and for deep U-wall sections. For shallow U-wall sections, steel sheet piles were selected as the SOE systems.

Figures 9 and 10 show typical cut-and-cover tunnel and U-wall sections, respectively.

Tremie seals are required both for the cut-and-cover tunnel and U-wall sections to against the uplift hydrostatic pressure. Their construction sequences are (1) Installed secant pile wall, (2) Excavate to the top of groundwater level and install cross lot bracing as needed, (3) Excavating the remaining portion in side the secant pile wall in wet condition until to the bottom of the tremie seal location, (4) Install tension piles, (5) Cast tremie seal and dewater the pit, (6) Finalize the permanent structures within the pit.

Cross passage constructions will be under ground freezing conditions. Tunnel break-in scheme will utilizes concrete block inside the launching pit to provide TBM confinement, instead of ground improvement outside the pit. Tunnel break-out scheme will utilize water pressure balance method, i.e., during tunnel break-out, the TBM receiving pit will be flooded prior to the TBM break through.

Traffic Movement—Watson Island

As stated in the beginning of this paper, the primary purpose of this Project is to divert truck traffics to the POM from I-395 to Watson Island, through the tunnel, and then to POM, instead of through downtown Miami. This movement, by FDOT design, has the potential of interference between the truck traffics to POM, through I-395, with

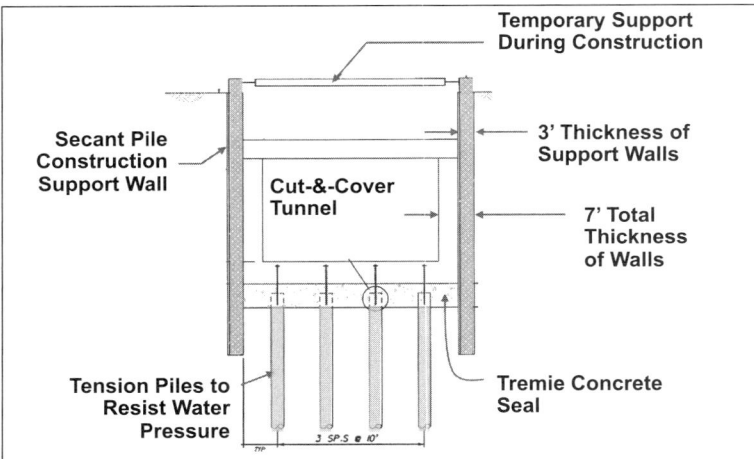

Figure 9. Typical cut-and-cover section

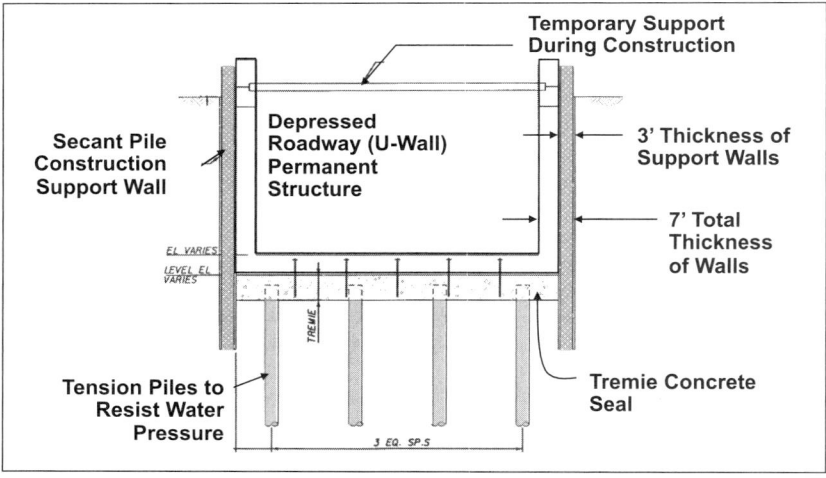

Figure 10. Typical U-wall section

the passenger traffics, through Route 1, from downtown Miami to Miami Beach. MAT design extended the bored tunnel length in Watson Island and at the same time eliminated the traffic interference weaving movement. Figures 11 and 12 illustrate these traffic movements.

CHRONOLOGICAL PROJECT DEVELOPMENTS

Any underground project at this magnitude would take a long time from planning to design and to construction. There is no exception to the Port of Miami Tunnel Project. This section provides the final and most critical development stage of this Project in a chronological order, specifically from the years 2006 to 2008. It is summarized in Table 3.

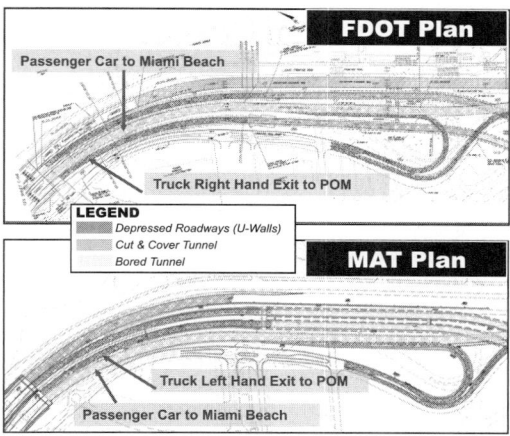

Figure 11. Traffic movement improvements—Watson Island

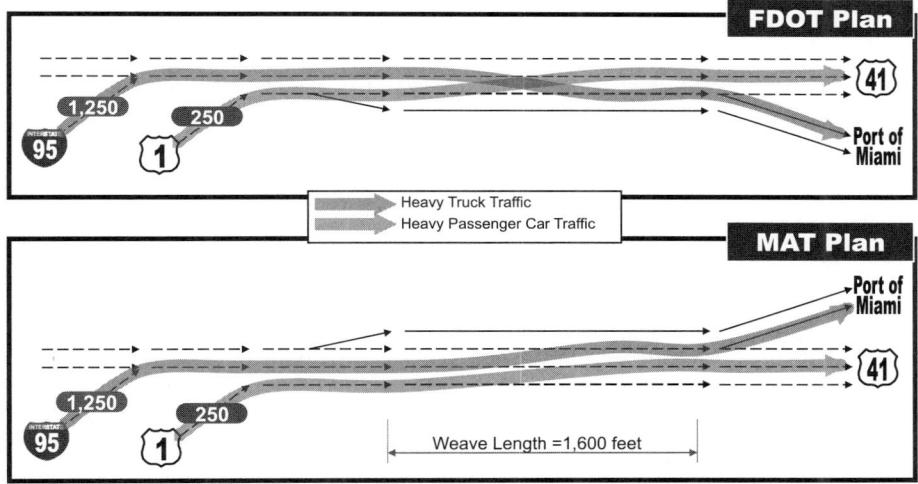

Figure 12. Traffic movement comparisons

Major reasons for FDOT to terminate the Project are:
- Agreement of the construction cost of the Project can't be made on time. Though MAT has started its pricing negotiation with FDOT since February 2008, mutual agreement was not finalized. The cost proposal presented by MAT in 2006 requires an adjustment to cover material and labor escalations, since the award is almost one 91) year late. Interpretation of the GBR between FDOT and MAT is also a hurdle during the negotiation.
- The financial sponsor Babcock & Brown can no longer support this Project after the financial downturn in the last quarter of 2008.

Table 3. Chronological project developments

Date	Significant Events/Activities
February 17, 2006	FDOT publishes a Request for Qualifications (RFQ) for a DFBOM through a Concession Agreement.
April 12, 2006	FDOT receives Statement of Qualifications (SOQ) from potential proposers
April 28, 2006	FDOT announces Short-list of qualified proposers, Miami Mobility Group (Dragados / Odebrecht / Parsons), Miami Access Tunnel (Bouygues/ABN-AMRO), and FCC/Morgan Stanley.
June 2006	FDOT published its schedule as follows: 1) Prepare Proposal: 7/06–11/06, 2) Submit Proposal: 11/1/06, 3) Award Concession: 1/4/07, 4) Notice To Proceed: 3/1/07, 4) Permitting Phase: 3/07–12/07, 5) Construction Phase: 9/07–4/12, 6) Open to Traffic: 2012, and 6) Concession Term: 2010–2045
August 2006	Procurement schedule was identified: 1) Early October, 2006—FDOT publishes Final Request for Proposals (RFP), 2) Mid January, 2007—FDOT receives Proposals, 3) Late February, 2007—FDOT makes Final Selection and Awards Contract, 4) five (5) Year Construction Period beginning in 2007
2 PM, March 4 2007	Actual date Tender Design Submitted, which is three (3) moths behind original schedule
May 2, 2007	FDOT posted a Notice of Intent to Award to MAT
July 24, 2007	Miami-Dade County commissioners approved its funding contribution to the Project
August 1, 2007	City of Miami commissioners voted to oppose using the City Redevelopment Fund on the tunnel unless the city wants to be repaid that money over the years from tolls that might be charged at the tunnel.
February 15, 2008	FDOT awarded MAT consortium as the winning bidder for the Port of Miami Tunnel and the City of Miami agreed its funding contribution to the project.
December 12, 2008	FDOT pulling plug on the Project

CONCLUSIONS AND LESSON LEARNED

The Port of Miami Tunnel project is well known for its engineering challenges, including design and construction in a very difficult ground condition and under ship channel. Risks from geological condition, hydrogeological condition, construction means and methods, permits, utilities, and funding commitments (from the State, the County, and the City) were well addressed. Though most of the challenges and risks were eventually, or will be, overcome, the risk of the funding from the private financial sponsor was under-estimated, which is a primary reason this 20-year long planning project. Near the end of its journal, FDOT almost secured an opportunity to fund, to design, to construct, and to operate and maintain this facility in a very creative contract delivery mechanism, which is a common model nowadays in the other part of the world. Several lessons are learned from the tender design of this PPP project, including:

- Financial Risk of the Concessionaire can be unpredictable and risky
- High Quality End Products—It is different from a Design-Build project that the concessionaire is not just deign and built the agreed facilities. It also has to operate and maintain its constructed facilities for an agreed concession terms, which can be as long as over 30 years. In this type of contractual mechanism, the contractor (and its design team) of the concessionaire has do its best to design and construct the facilities in a high and efficient standard, since the concessionaire is paid only when the facilities is open and functional

- Innovation—Like the Design-Build project, the PPP encourages the cooperation between the contractor and the engineer through both design and construction phases. This can result in a creative engineering product that cost less for construction and achieve higher functional standards for end users
- The Purpose of the GBR—We all know this document has to be clear and concise and shall not be a document as a mechanism to shield the liabilities of the engineer of an Owner. Like a Design-Build project, obtaining a consensus of this document can be difficult between the Owner's representative/engineer and the Concessionaire, since the baselines can be ambiguous and the Concessionaire may select a scheme that alters the original alignment proposed by the Owner. The purpose and the utilization of this document, especially for Design-Bid-built contract, is well known for the US underground design and construction communities; however, this may not be the case for foreign tunneling communities. It is the Author's opinion that the rule and its use for this document shall be clearly defined in bidding stage, which can avoid the disputes and arguments for detail parameters.
- Consensus Baseline Approach—We still have a need to resolve the issue how we define the geotechnical baselines for Design-Build and PPP projects. Consensus approach does not exist.

REFERENCES

FDOT, 2006. Project Information Memorandum—Port of Miami Tunnel and Access Improvement Project.

Parsons Brinckerhoff, 2006. Geotechnical Baseline Report, Port of Miami Tunnel & Access Improvement. Prepared Florida Department of Transportation, August 2006.

DESIGN CONSIDERATIONS AND EVALUATION PROCESS FOR A NEW TUNNEL AND OCEAN OUTFALL PROJECT

Steve Dubnewych ■ Jacobs Associates

Michael Torsiello ■ Jacobs Associates

Jon Kaneshiro ■ Parsons Corporation

David Haug ■ Sanitation Districts of Los Angeles County

ABSTRACT

The Joint Water Pollution Control Plant, operated by the Sanitation Districts of Los Angeles County, treats wastewater generated by over three million people. A new tunnel and ocean outfall is being considered to meet future hydraulic demands and provide long-term redundancy for the Districts' existing tunnels, portions of which were constructed as early as the 1930s.

The detailed process used to evaluate the feasibility of various tunnel and outfall locations is described. Challenges include a geologic profile with mixed face and squeezing ground conditions; high water pressures; active fault crossings; gassy and contaminated ground conditions; and liquefaction, slope stability, and lateral spreading concerns in the area of the riser and diffusers.

INTRODUCTION

The Sanitation Districts of Los Angeles County (Districts) are 24 independent special districts serving approximately 5.3 million residents in Los Angeles (LA) County. Seventeen of the districts that furnish sewerage services to metropolitan LA are signatory to a Joint Outfall Agreement that provides for a regional, interconnected system of facilities known as the Joint Outfall System (JOS). The JOS serves an area that encompasses 73 cities as well as unincorporated territory and parts of the City of LA. The JOS provides wastewater collection, treatment, reuse, and disposal for residential, commercial, and industrial users, and it includes seven treatment plants, the largest of which is the Joint Water Pollution Control Plant (JWPCP), located in the City of Carson. Currently, secondary effluent from the JWPCP is conveyed through two parallel tunnels 2.4 and 3.7 m (8 and 12 ft) in diameter. The tunnels interconnect at a manifold structure at Royal Palms State Beach on the Palos Verdes (PV) Peninsula, from which two operational seafloor outfalls extend offshore. Both tunnels are required to be in service at all times and have not been inspected since 1958. The new tunnel and ocean outfall system, if constructed, will provide additional capacity and long-term redundancy, and will allow inspection, maintenance, and repair of the existing tunnel and outfall system.

In June of 2006, feasibility studies and preliminary engineering for this project was awarded to the Parsons Corporation in association with Jacobs Associates, which is leading the underground design efforts. Currently, the team is undertaking an alternative selection process to choose the alignment alternatives to consider during the preliminary engineering phase of this project.

NEW TUNNEL AND OCEAN OUTFALL PROJECT

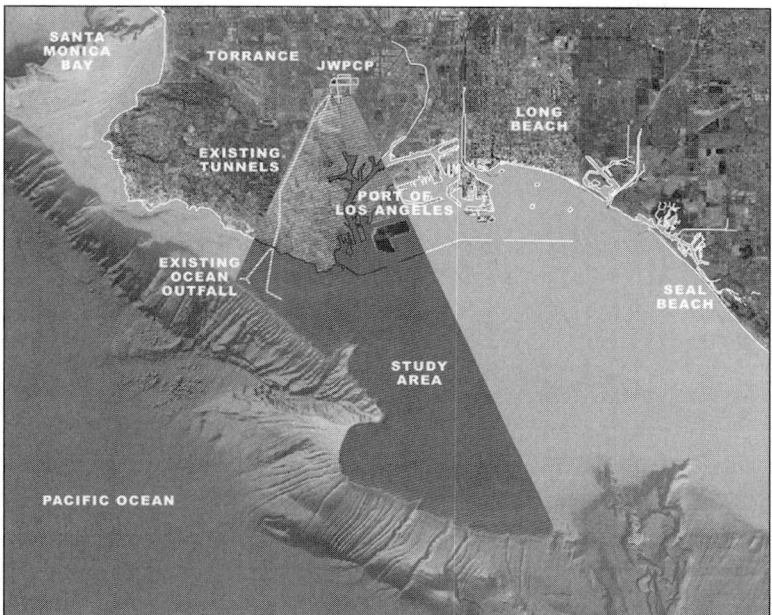

Figure 1. Project study area

STUDY AREA

The new tunnel will commence with a shaft at the JWPCP. From the JWPCP, an onshore tunnel will be constructed in a southward direction to a shaft at the shoreline and then will continue offshore. The offshore alignment will begin with a tunnel that will either continue to a diffuser as a tunnel or as a seafloor pipeline.

The onshore study area (Figure 1) includes multiple alignments from the JWPCP to the shoreline—from Royal Palms State Beach on the PV Peninsula to the far eastern boundary of the Port of LA. The offshore study area extends from the shoreline, southwards to the San Pedro and PV Shelves.

AREA SITE CONDITIONS

Geologic and Seismic Setting

The study area lies on the southwest boundary of the LA Basin and straddles the paleo-LA River and two prominent geomorphic features—the PV Hills and the Newport-Inglewood Uplift (Figure 2). The dominant structural feature is the PV fault, which forms the northwest-trending Gaffey syncline and anticline. The Cabrillo fault, a splay of the PV fault, although not as pronounced, is nevertheless a significant geomorphic feature (Figure 2). Another pervasive feature of the area is that it is underlain by the Wilmington Oil Field, a well-developed, northwest-trending anticline structure of oil-bearing rock.

The stratigraphy of the LA Basin and the PV Hills is generally characterized as follows:

- Quaternary age: Surficial deposits include Holocene sediments consisting of fill, alluvium, sand dunes, and terrace deposits underlain by Pleistocene sediments, including the Lakewood and the San Pedro formations. The

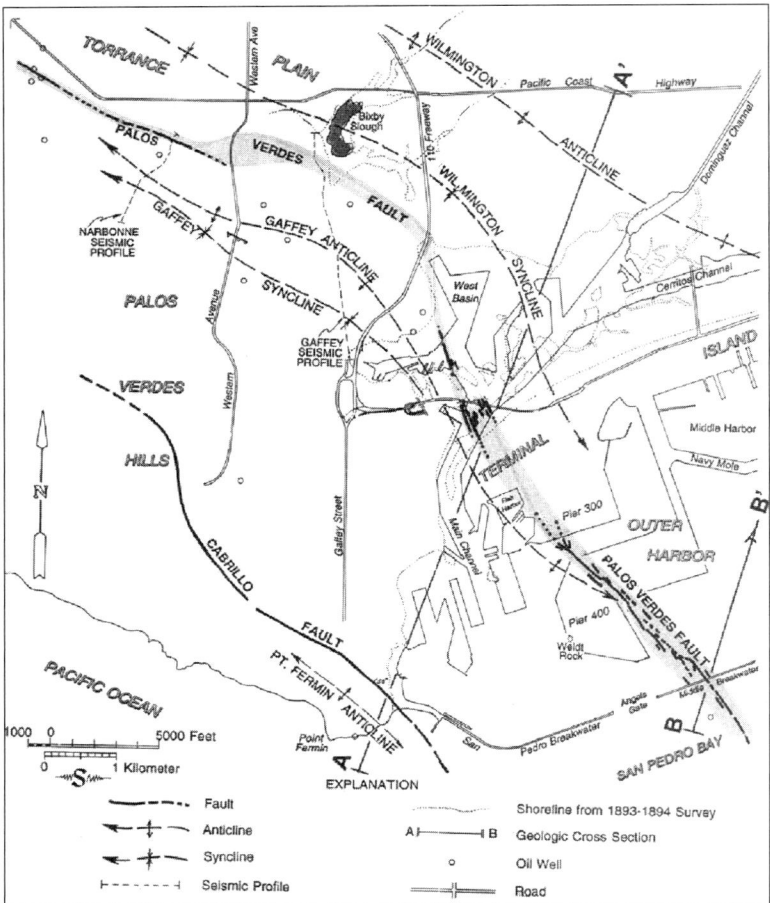

Figure 2. Geologic structures in the project area

Lakewood Formation contains the PV sand member. The San Pedro Formation contains the San Pedro sand, the Timms Point silt, and Lomita marl members. These formations range from primarily unconsolidated sediments to strong soil and very weak rock.

- Tertiary age: Pliocene sediments include the Fernando Formation, consisting of the Pico and oil-bearing Repetto members, which are underlain by the Miocene-age Malaga Mudstone and the Monterey Formation. The Monterey Formation consists of Valmonte Diatomite, and Altamira Shale. These formations range from strong soil to weak and moderately strong rock.
- Jurassic age: Basement rock is Catalina Schist, metamorphic hard rock that varies from moderately strong to very strong.

Generally, the onshore tunnel alignments west of the PV fault and toward the PV peninsula are primarily Tertiary-age ground, with the possibility of encountering Jurassic-age rock at depth and Quaternary-age sediments as the tunnel alignments

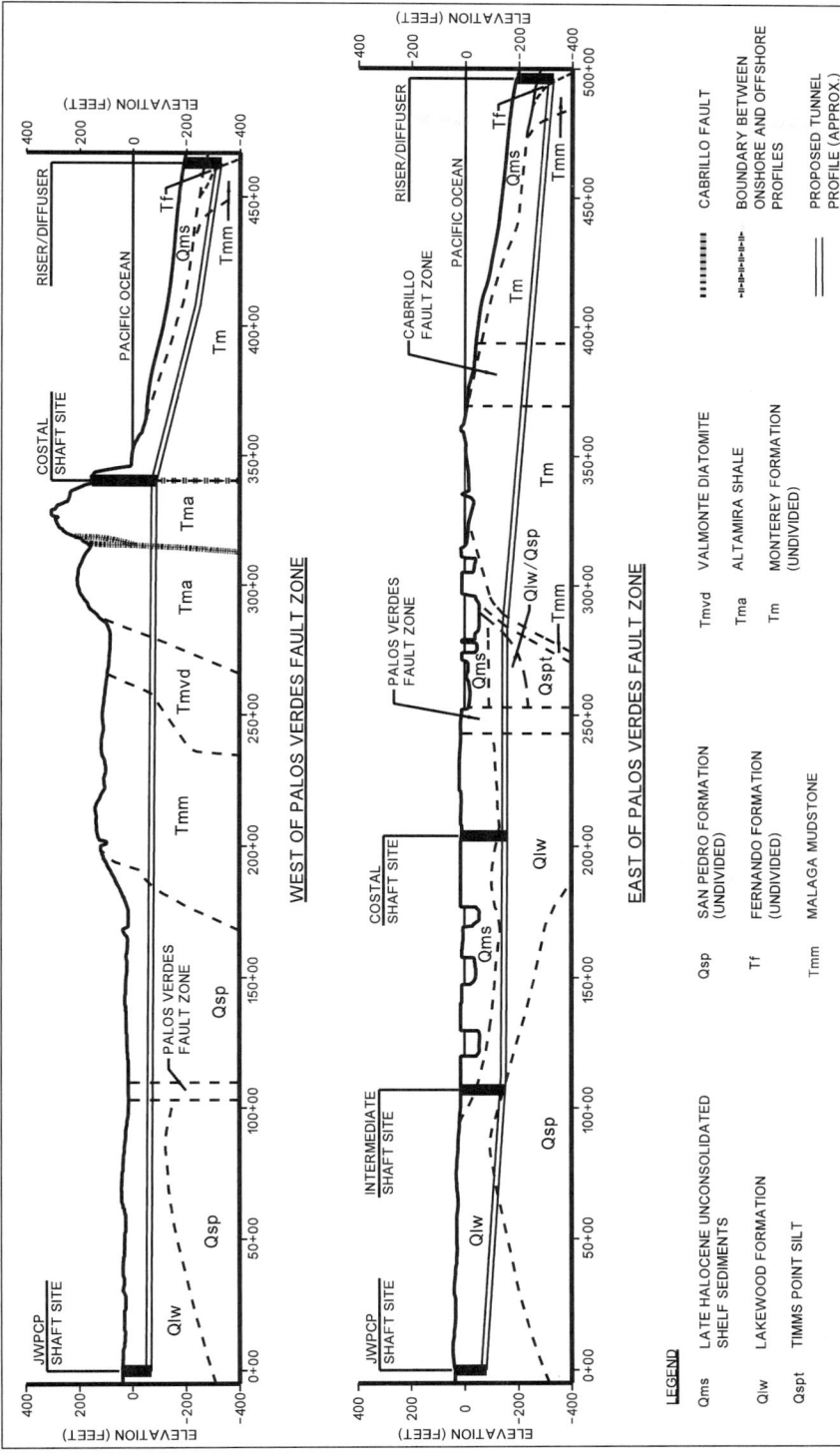

Figure 3. Geologic profiles illustrating differences in subsurface conditions on either side of the PV Fault

drift/trend away toward the east flanks of the peninsula. The onshore alignments east of the PV fault are entirely in Quaternary-age sediments.

Offshore, the tunnel would be mostly west of the PV fault to reach the proposed diffuser areas. These portions of the alignment are located in Tertiary-age ground. Where the offshore tunnel alignments are east of the PV fault (primarily within the Port of LA Harbor area), the tunnel alignments will encounter Quaternary-age sediments before transitioning to Tertiary-age ground.

Example geologic profiles for alignments both east and west of the PV Fault are shown in Figure 3. The differences in the onshore geology, as noted above, can be seen in these two alignments.

A significant earthquake is expected within the next 30 years within Southern California, and there is significant risk of a M7. Palos Verdes fault offset is also a major concern, for which mitigation measures are discussed in "Design Considerations," below.

GROUNDWATER CONDITIONS

Groundwater onshore varies from sea level near the shoreline and rises toward JWPCP to 17 m (55 ft) below the ground surface. The Gaspur Aquifer represents runoff down the Los Angeles River from the San Bernardino Mountains and into deep aquifers and is a source of groundwater. A series of deep injection wells prevent seawater intrusion into this aquifer. Between the ground surface and the permanent groundwater level, perched water or trapped virgin water is expected to be encountered in lenses or pockets. At tunnel depth, however, groundwater pressure conditions are expected to reflect those of the permanent groundwater table. Groundwater offshore and pressure head are expected to be consistent with sea level.

PROJECT COMPONENT AREAS

To aid in establishing and evaluating the preliminary alternatives for a potential tunnel and ocean outfall system, five component areas were developed and independently analyzed. These include JWPCP shaft sites, coastal shaft sites, onshore alignments, marine alignments, and diffuser areas. For each area, preliminary options were developed based on a set of initial criteria, as described below.

Shafts

Shafts will be necessary to construct the tunnel; however, the total number of shafts is dependent on the alignment alternative selected and whether it is feasible to construct additional intermediate shafts. Regardless of the alternative selected, at least one mining shaft will be necessary, and additional shafts may be used for access and/or retrieval. The proposed shafts range in depth, depending on where they are located along the alignment alternatives, with potential depths of 75 m (245 ft) below the ground surface. Additionally, the proposed diameters of the shafts range from approximately 9 to 18 m (30 to 60 ft), depending on the expected use for each shaft. Working shafts are expected to be at least 12 m (40 ft) in diameter and up to 18 m (60 ft) if a shaft supports two headings. For example, a shaft at the shoreline may be used to mine both onshore and offshore headings. The diameter of access shafts is expected to be approximately 9 m (30 ft).

Given that these shafts could be excavated in a range of geologic materials, several shaft excavation and temporary support systems are discussed. For excavations in soil below the groundwater table, watertight excavation methods will be used. The watertight methods of excavation support that are considered most feasible include

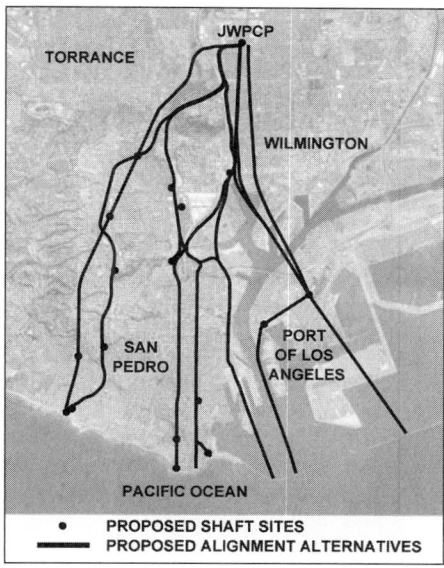

Figure 4. Proposed alignments and shaft sites

both slurry diaphragm walls and ground freezing. Two nonwatertight methods include sequential excavation and soldier piles with lagging. The nonwatertight methods are expected to be used for shafts in sedimentary rock.

JWPCP Shaft Sites

Two potential shaft sites at the JWPCP have been identified and are being evaluated—they are known as the JWPCP West and East Shaft sites. A location at the JWPCP is necessary to house the permanent structure, which will connect the existing treatment facilities to the new tunnel and could also be used for mining. Location criteria for the JWPCP shaft site are that the site must be within the confines of the JWPCP property boundaries; avoid conflicts with current facilities or planned future facilities; have a minimum area of 1.6 ha (4 acre); be roughly rectangular to square in shape and relatively flat; and have access for equipment, ventilation systems, and personnel, as well as long-term access for excavated material removal on a continuous basis.

Coastal Shaft Sites

The remaining potential shaft sites are considered coastal shaft sites and fall into one of three categories: working shafts, retrieval shafts, or access shafts. Locations near the coast along the different alternatives could be used to mine the offshore portion of the tunnel. A shaft close to the shoreline will reduce the length of the tunnel drives if it could be constructed and used as a mining shaft. An intermediate shaft site not used for mining could still be beneficial for maintenance and ventilation and other access needs. A site of 1.6–2 ha (4–5 acre) was determined to be the minimum size necessary for a working shaft, but a 2.4-to-3.2-ha (6–8 acre) site is preferred.

Retrieval shafts require approximately 0.8 ha (2 acre). This is adequate for routine access for personnel ingress and egress, as well as for the removal of salvageable portions of the tunnel boring machine (TBM) at the project's conclusion. Retrieval shafts will not be used to remove excavated material during the tunneling operations.

Access shafts will require the least area and are limited to personnel ingress and egress, as well as connection with the overall tunnel's ventilation system. Public land is preferred over private land for the selection of coastal shaft sites. The site has to be relatively flat and near rectangular or square in shape. Based on these criteria, locations were identified as potential coastal shaft sites (Figure 4).

Onshore Alignments

Initial evaluations of construction methods indicated traditional open-cut construction was not a viable option due to the disruption that would occur to the surrounding communities and the depth of excavation required—in excess of 18 m (60 ft). Therefore, the onshore alignments must be built by tunneling methods. Using long, continuous public streets between the JWPCP and Pacific Ocean or existing easements, and a minimum radius of 305 m (1,000 ft), criteria were established for developing the onshore alignments. Twenty-two potential tunnel alignments were identified for further analysis (Figure 4).

Tunnel drives up to 10.7 km (6.6 mi) in length are being considered for the onshore drives. The excavated diameter of the tunnel could be up to 7 m (23 ft). Tunnel excavation may require a combination of soft ground and rock excavation methods due to the variable soil conditions that may be encountered. Challenges include long tunnel drives under very high groundwater heads. Groundwater pressures up to 6 bar could be encountered, depending on the alignment alternative selected. A pressurized-face TBM is ideally suited for this project due to the expected presence of high groundwater pressures combined with the varying permeability and strength of the soil units, including mixed-face conditions (i.e., both rock and soil in the excavation face) along the proposed alignment corridors.

A significant factor that helps differentiate which type of TBM to select is the geology expected along the alignment. With the data currently available, a definitive conclusion cannot be made on the most suitable tunnel excavation method. Nevertheless, the advantages and disadvantages of using Slurry TBMs, earth pressure balance (EPB) TBMs, or hybrid TBMs are discussed under "Design Considerations," below.

Independent of machine type, there are other geotechnical considerations that also must be evaluated, including the potential for ground squeezing, naturally occurring hydrocarbons, and faults in the project area that must be crossed. Given the expected ground and groundwater conditions, a watertight bolted gasketed liner system will be required for ground support and for advancing the TBM. This initial support also could be used as the final lining of the tunnel, or a two-pass lining could be used.

Marine Alignments

The marine alignments, which consist of an offshore tunnel, a seafloor pipeline, or a combination of the two, connect the onshore alignments and the diffuser areas to create complete viable alternatives for further evaluation. The vertical connection from the tunnel to the seafloor pipeline or diffuser site will be made with a riser. The establishment of preliminary options for the marine alignments was deferred until after the establishment of preliminary options for the other four component areas.

Tunnel and Pipeline

Similar tunneling conditions, tunnel diameter, TBM types, and geotechnical and liner considerations are required for offshore tunnel alignments, except the maximum drive lengths are generally longer than onshore alignment drive lengths—the offshore drives could be as long as 17 km (10.5 mi). Additionally, external hydrostatic pressures could approach 10 bar.

Table 1. Marine pipeline constraints

	Reinforced Concrete Pipe ID	Pipe Laying Depth
State of practice	1.5 to 3.7 m (5 to 12 ft)	<60 m (<200 ft)
State of art	>4.3 m (>14 ft)	>60 m (>200 ft)

Table 2. Marine riser constraints

	Maximum Depth	Maximum Diameter
State of practice	30 m (100 ft)	5 m (16 ft)
State of art	60 m (200 ft)	5 to 6 m (16 to 20 ft)

Fundamental to the selection of the preferred offshore tunnel alignment is the consideration of feasible seafloor pipelines based on geotechnical considerations, ship-traffic considerations, and constructability. Based on a review of case histories, the design team developed constraining criteria for layout of deep marine pipelines (Table 1).

Riser

The design of offshore structures is largely driven by construction considerations; this applies particularly to a riser that connects a tunnel to the seabed portion of the outfall system. Among the major considerations for offshore shafts and risers are:

- Seabed bathymetry and water depth
- Riser height and diameter
- Geologic conditions
- Offshore waves and currents
- Shipping and navigation patterns
- Construction equipment limitations

Tunnel construction constraints directly affect design of the riser, as riser height is controlled by the elevation of the tunnel relative to the seabed, and the depth of water at the riser is a function of the tunnel length and alignment. Water depth, riser diameter, and riser height are perhaps the most important factors affecting the riser design, construction costs, and risks, particularly for the potentially large-diameter structures being considered on this project. As the tunnel length increases to extend under deeper water, the associated costs of the shaft and riser increase, along with construction risks.

Construction of a large riser in the ocean is delicate and risky. Generally, the construction of the riser proceeds independently of and prior to tunnel construction. When both the riser and tunnel are completed, the connection between the two is made from the tunnel. This is the riskiest part of the operation, and its success is determined largely by the accuracy of the planning during the design. The design has to address all that could go wrong with the different methods as applied to the specific site geology, and a procedure that has built-in safety redundancy must be selected. Based on a review of case histories, the design team developed the constraining criteria for layout of deep marine risers (Table 2).

Diffuser Areas

An underlying criterion for the proposed diffuser is that the system must perform equal to or greater than the existing outfalls. To achieve this, initial parameters

- Tunnels requiring multiple drives to reduce the total project delivery time

However, some very long tunnels have been excavated in soil and rock as a single heading without the use of intermediate shaft sites, including:

- Submarine tunnels
- Tunnels in urban environments with limited shaft-site availability
- Tunnels in remote areas with difficult terrain or limited surface access

Based on a review of case histories, the design team developed the constraining criteria for layout of long tunnels (Table 3). Some long tunnel drives in soil include the Westerschelde Tunnel in the Netherlands and the South Bay Ocean Outfall in California, which had drives of approximately 6,600 m and 5,900 m (21,650 ft and 19,500 ft), respectively. In rock, some of the Channel Tunnel drives between the United Kingdom and France were over 22,000 m (72,000 ft) in length. An example of a long

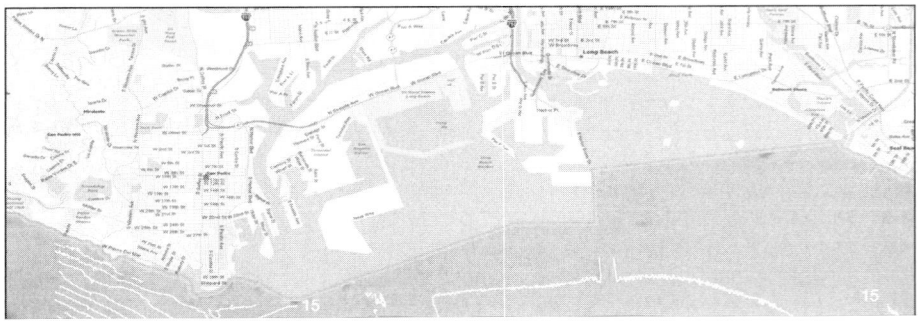

Table 3. Tunnel length constraints

	Maximum Length in Soil	Maximum Length in Rock
State of practice	4,600 m (15,000 ft)	7,600 m (25,000 ft)
State of art	6,700 m (22,000 ft)	15,900 m (52,000 ft)

outfall tunnel in rock is the Deer Island Outfall in Boston, which is just over 15,000 m (49,500 ft).

The maximum external pressures for the case histories reviewed ranged from 7 to 11 bar. Also, the lengths provided in Table 3 are approximate within 10% of what is feasible/practical.

Depending on the alignment selected, both the onshore and offshore tunnels are considered at the State of the Art (SOA) for tunnel length. Preliminary onshore tunnels could be over 10,700 m (35,000 ft), with the preliminary offshore portion being over 16,800 m (55,000 ft). Both the onshore and offshore tunnel alignments will likely be excavated in a single heading because of the lack of intermediate shaft site availability onshore and the subaqueous nature of the offshore portion. In all alignment alternatives being considered, there is at least one intermediate shaft site near the coast, which will separate the onshore and offshore tunnel drives.

Long tunnel drives increase the amount of design difficulty because the entire drive must be excavated with the same TBM. For example, the geology is extremely variable within the project area of the JWPCP tunnel and ocean outfall, and due to the lack of intermediate shafts, each entire drive (onshore and offshore) will need to be mined with one TBM. Additionally, as tunnel length increases, ventilation requirements become greater. Lastly, TBMs may need to be outfitted with special provisions to provide for stopping in potentially bad ground conditions for cutterhead maintenance.

Geological Considerations

Geotechnical considerations that also must be evaluated include the potential for ground squeezing, naturally occurring hydrocarbons, and faulting. Squeezing ground occurs when ground slowly advances into the tunnel excavation, which could occur in both cohesive soils and weak rock. In areas where the alignments cross either the PV or Cabrillo fault, squeezing conditions could be encountered if they are not designed against or mitigated properly.

Additionally, alignments passing through the Tertiary-age formations could encounter naturally occurring hydrocarbons during excavation. Both methane gas and hydrogen sulfide gas may be encountered in these formations. In these conditions, a Slurry TBM would be advantageous since it operates in a "closed circuit," minimizing workers' exposure to gas underground. Also, a Slurry TBM provides more safety for the expected high pressures, especially in cohesionless soils. It is costlier, however, to separate the bentonite slurry, especially in cohesive soils.

An EPB TBM would discharge the excavated material into muck cars underground and transport it by rail to the surface. During transport, gases could be released and could be dangerous. A number of provisions would need to be in place to mitigate dangers, such as increased ventilation and restrictions on electrical equipment. An EPB TBM could be outfitted with a "closed circuit" muck-removal system similar to a slurry system to prevent discharge of gas into the tunnel environment. An EPB TBM can offer economic tunneling in cohesive soils and weak rock/strong soils, although safety is compromised in cohesionless soils, especially in pressures over 3 bar. In recent years in Japan, the use of EPB has greatly outpaced that of slurry machines except where safety issues are overwhelming (controlled face tunneling in cohesionless soils under high head over 3 bar).

A hybrid TBM may offer the best compromise between slurry and EPB methods. They are equipped with slurry pumping injection and removal systems and a screw conveyor for removal of muck. Hybrid TBMs have been most common in Japan and Europe for over 15 years. To the authors' knowledge, two small projects have been completed using hybrid TBMs in the U.S.: one in Miami Dade County, Florida, and one in San Mateo County, California. The Southern Nevada Water Authority's Intake Pump Station No. 3 Tunnel will be excavated by a Hybrid TBM. Hybrid TBMs may excavate in slurry mode when required for safety reasons to prevent the uncontrolled inflow of ground and groundwater. When in good ground, a Hybrid TBM may operate using EPB open-mode tunneling for efficient and economical tunneling.

Seismic Considerations

Another design challenge is the crossing of active faults. Fault crossing strategies are being evaluated, including:

- Specially designed one-pass tunnel lining in fault zone (with joints able to withstand some displacement while retaining their water-tightness).
- Two-pass tunnel lining in fault zone (may reduce cross-sectional area if excavated diameter is kept constant). This could accommodate more displacement by leaving an annular gap between the two linings.
- Shallow tunnel profile in fault zone would make repair in the event of rupture more feasible and economical.

Shaft Construction Considerations

Soil and Groundwater Conditions

Deep shafts excavated in soils with high groundwater heads will be the most challenging ground conditions for shaft construction. Potential shaft sites at the coast in and around the Port of Los Angeles could be excavated more than 50 m (165 ft) below grade entirely in soil. In these locations, the groundwater table is at or just below the existing grade, making for large groundwater pressures at the bottom of the excavation. Additionally, lateral pressures exerted on the shafts excavated in soil will be higher than those excavated in rock because the rock shares some of the load. Shaft excavation in saturated soils is fairly routine at depths of 30 m (100 ft) or less; however it becomes challenging at depths of greater than 45 m (150 ft).

Achieving tight construction tolerances needed to obtain a watertight structure with a deep shaft will be an issue when installing slurry wall panels or ground freeze pipes. Maintaining verticality during installation of both slurry walls and ground freeze pipes becomes more difficult with depth. Potential horizontal deviations of 0.5–1.0% of the vertical distance (depth) could occur. Practical construction tolerances also may be an issue when improving ground for break ins or break outs (e.g., with jet grouting).

While slurry walls and ground freeze walls are effective in controlling groundwater seepage through the shaft walls, groundwater also can enter the excavation from the bottom. With groundwater heads as high as those expected at shafts near the coast, shaft bottom seals become a critical consideration in shaft design. Some methods being considered to mitigate this include establishing groundwater cut off by extending the shaft wall below the base of the excavation, shaft excavation in the wet followed by the construction of a tremie concrete bottom slab designed to resist the hydrostatic pressures at the bottom of the shaft, and temporary aquifer depressurization (if acceptable).

Sensitive Receptors and Land Use

The current land use of the potential shaft sites is important because some sites are being used for recreational purposes and some are already in industrial areas. Construction traffic, noise, light, and dust can create bad public relations with the surrounding community if not properly mitigated. Even when mitigated, these issues may affect a shaft being selected in close proximity to sensitive receptors. Additionally, a site located adjacent to a major freeway or truck route would decrease impacts on the community and traffic flow in general.

Postconstruction access is desirable to allow maintenance and inspection and to house any permanent structures. Because it will be desirable to have permanent access to the tunnel via the shaft sites after construction, it is also important to consider the future land use of the sites.

Outfall Construction

Just as the threat of an earthquake provides design challenges to the tunnel fault crossings, it also adds challenges to the riser and diffuser design and site selection. The seismic hazards are those due to wave propagation damage (shaking) and permanent ground deformation (PGD). Shaking produces higher tunnel liner or pipe liner stresses and strains (hoop, axial, curvature bending, buckling, and racking). Shaking also may produce hydrodynamic forces, which create a water-hammer effect (overpressure) on the system components. In loose or soft-to-medium dense soils like those at some of the potential diffuser locations, PGD consists of liquefaction-induced settlements and liquefaction-induced lateral spreading and slope failure. Differential settlements at the seafloor surface or an entire slope failure could have catastrophic effects for either a seafloor pipeline or a riser and diffuser structure.

The potential for seismic-induced slope instability and liquefaction-induced lateral spreading will have a significant impact on the location of the marine facilities, which will in turn impact the location of the tunnel alignment. The challenge is to be able to predict the potential for PGD in the large offshore study area with limited geotechnical and geophysical data.

CONCLUSIONS

During the preparation of the feasibility studies and screening of alternative alignments (which is underway), evaluations and review of construction case histories of tunnels and marine pipelines, risers, and diffusers were conducted. Criteria were developed by the design team for maximum length, depth, and diameter of constructible alternatives. These criteria may be useful for the planning of other mega projects.

ACKNOWLEDGMENTS

The authors thank their respective employers for permission to publish this paper. Appreciation is also due to the design team: G. McBain, D. Yankovich, L. Meiorin, C. White, S. Min, M. Sinha, S. Bache, and M. Strugis of Parsons; M. McKenna, J. Yao, and S. Klein of Jacobs Associates; and C. Jin, T. Sung, C. Boehmke, S. Highter, and J. Houghton of the Districts. Black & Veatch is assisting on the interconnection to the JWPCP to the tunnel and ocean outfall. The geotechnical exploration program was led by Fugro West, Inc.

MBTA SILVER LINE PHASE III—COMPLETES BOSTON'S NEWEST TRANSIT LINE

Gregory Yates ▪ AECOM USA

Mary Ainsley ▪ Massachusetts Bay Transit Authority

William Gallagher ▪ URS Corporation

INTRODUCTION

Project Description

The Silver Line is a Bus Rapid Transit (BRT) Line and is Boston's newest transit service. Silver Line Phase I extends along Washington Street entirely on the surface in priority bus lanes from Dudley Station in Roxbury through the South End and Chinatown neighborhoods to downtown Boston. The Phase I opened in July 2002. Silver Line Phase II consists of an underground transit tunnel from South Station to the South Boston Waterfront with underground stations at Courthouse and World Trade Center to Silver Line Way east of D Street. Phase II continues through the Ted Williams Tunnel to all the terminals at Logan Airport, and on the surface to the Boston Marine Industrial Park and City Point. Phase II opened in December 2004.

Silver Line Phase III is a one mile tunnel connection to the existing Phase II tunnel in the vicinity of South Station and will eventually portal on Tremont Street just north of Marginal Road. After leaving the portal, the buses will use dedicated bus only lanes on the surface roadway network to connect to the Phase I service. Figure 1 shows the alignment of the Silver Line upon completion of the Phase III project.

Physical Alignment

The Phase III tunnel begins where a cut and cover connection will be made to the Phase II tunnel in the Atlantic Avenue/Essex Street intersection, The cut and cover tunnels will then continue westerly on Essex Street to South Street where the new tunnels will be mined under the existing 72 inch diameter New East Side Interceptor Sewer (NESI). West of the NESI sewer the twin mined tunnels of the Silver Line Phase III will continue westward along the alignment of Essex Street passing underneath the Central Artery tunnel. The mined tunnels continue to a cut and cover pit near Harrison Avenue which will facilitate the construction of the vertical circulation elements between the Silver Line and the existing Orange Line at Chinatown Station. Continuing westward along the alignment of Essex/Boylston Street, Phase III will then create a transfer connection between the Silver Line and the existing Green Line at Boylston Station. Westward from Boylston Station, a loop will be constructed for operational purposes to allow buses to loop back and make the return trip to South Station. Approximately two-thirds of the buses will use the turnaround loop, and one-third of the buses will leave Boylston Station and make a left turn into new Phase III tunnels under the alignment of Charles Street South. They will continue to the new portal on Tremont Street where the BRT vehicles will join and connect to the Phase I service on Washington Street.

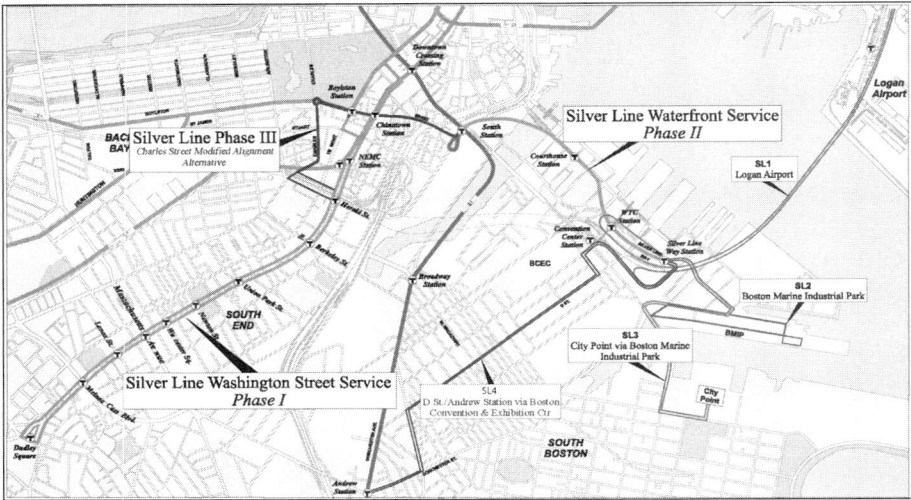

Figure 1. Silver Line Phases I, II and III

Owner and Stakeholders

The owner and operator of the Silver Line, along with all of the other transit lines within the city of Boston, is the Massachusetts Bay Transportation Authority. The Silver Line in particular joins the communities of Roxbury, the South End and Chinatown to downtown via the newly redeveloped Washington Street Corridor. With the Reconstruction of Washington Street and the Silver Line, the corridor has seen the opening of many new businesses and residences. The new Silver Line also serves Tufts Medical Center and the Downtown Crossing area of downtown Boston. Phase II also services Logan Airport and the new Boston Convention and Exhibition Center as well as the new development ongoing along the Boston Waterfront. The Phase III project will provide a connection from the communities south of Boston and downtown to the emerging employment centers along the South Boston Waterfront and the airport.

Timeline

The Silver Line Phase III project was approved to enter Preliminary Engineering under the Federal Transit Administration's New Starts Program in late 2002. A contract for professional services was awarded by the MBTA in early 2003 for development of a Supplemental Environmental Impact Statement and Preliminary Engineering. The originally planned portal location on Washington Street across from New England Medical Center (now Tuft's Medical Center) was determined to be undesirable. Several alternatives were evaluated prior to selecting the present alignment with the portal on Tremont Street. Interim Preliminary Engineering (30% design) was completed in October 2008. Preliminary Engineering (60% design) is currently scheduled to be completed in 2009 and Final Design in late 2010. Construction is set to start in late 2010/early 2011. Revenue Service is scheduled to begin by the end of 2016.

ALIGNMENT CONSIDERATIONS/IMPACTS

The Phase III project connection was planned for within the design of the Phase II tunnel. A specific location was designed to allow for the connection within the Atlantic

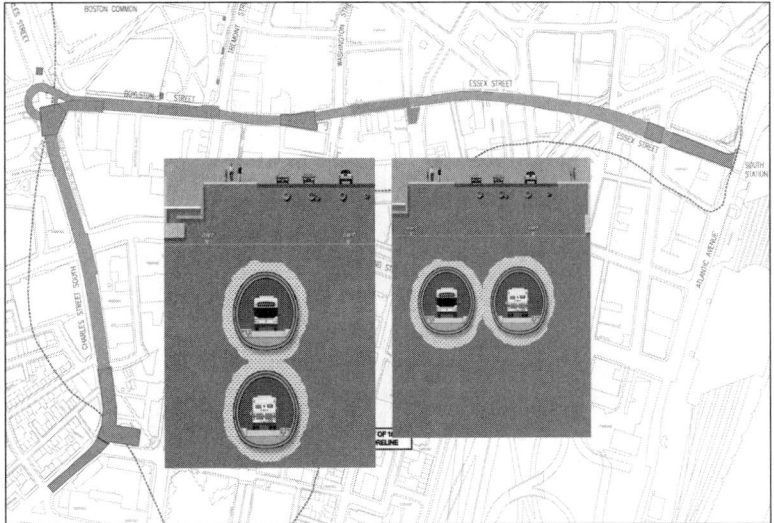

Figure 2. Side-by-side and stacked tunnel configurations

Avenue/Essex Street intersection. From the connection the Phase III alignment follows the public right of way under Essex Street. The tunnels are in a side by side configuration to facilitate the actual connection with the Phase II tunnel, however, to stay within the public right of way the tunnels are "rolled" into a stacked configuration. See Figure 2. The mined tunnels immediately begin to descend at 5.6% grade. The lower (or Inbound Tunnel) continues to descend while the upper (or Outbound Tunnel) levels off to allow for the stacked configuration. The depth of the stacked tunnels is predicated on the fact that they must pass under NESI sewer, the southbound Central Artery Tunnel, the Orange Line subway tunnels and the Green Line LRT tunnels. The depth of the Outbound Tunnel invert is approximately 60 feet below the surface and the Inbound Tunnel invert is approximately 100 feet below. The tunnels are approximately 26 feet in diameter allowing for the dynamic envelope of the BRT vehicle and the required appurtenances within the tunnel. The BRT vehicles operate off of an overhead contact system at 650 vdc within the tunnels. Emergency tunnel ventilation, emergency lighting, communications, and fire protection facilities run system-wide within the tunnels.

To make the physical connection of the Silver Line Chinatown Station to the existing Orange Line Station, two large and full depth cut and cover shafts are constructed on either side of the existing Orange Line tunnels. The deep cut and cover sections for the vertical circulation cores down from both the Inbound and Outbound sides of the existing Orange Line Station. See Figure 3.

The close proximity of the existing Green Line tunnels within the alignment of Boylston Street to the north and the historic buildings located on the south side, require that the configuration of the Silver Line Boylston Station and its connection to the Green Line be different. The two running tunnels are each widened to form a mined cavern approximately 35 feet across. The station cavern holds the platform, vertical circulation elements and the busway. A shallow cut and cover section at the east end of the station allows for the transfer connection between the Green Line and the new Silver Line. See Figure 4.

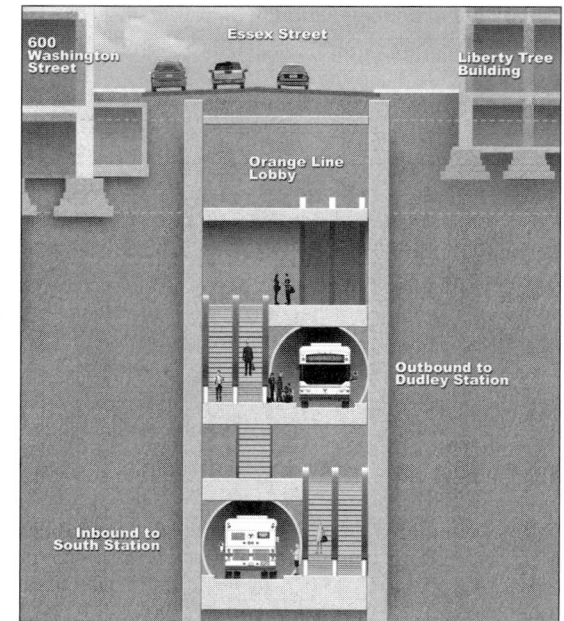

Figure 3. Vertical circulation core at Chinatown

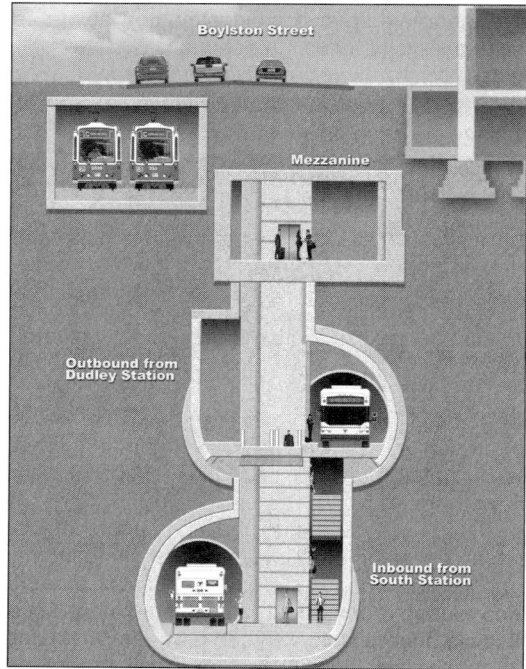

Figure 4. Boylston Station configuration

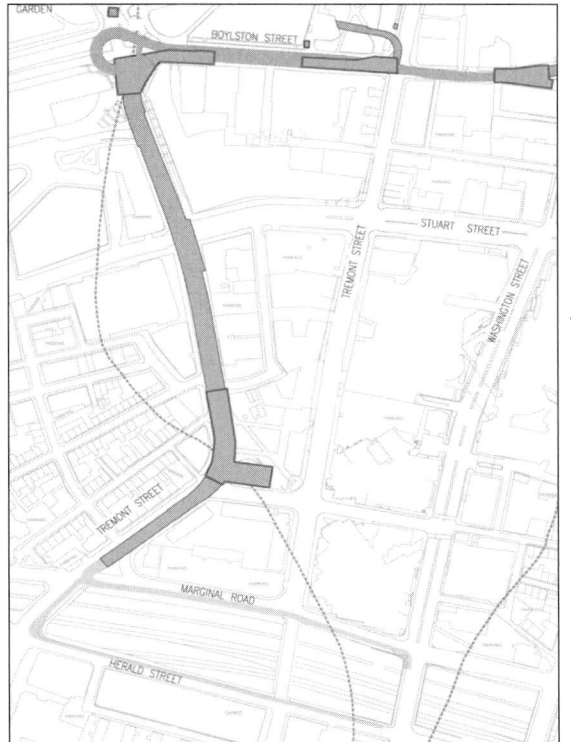

Figure 5. Alignment along Charles Street South to the portal at Tremont Street/Marginal Road

The Silver Line Boylston Station has one new station entrance that is located at the far west end of the station constructed in the plaza at the intersection Boylston and Charles Street South.

The turn-around loop is a mined helical tunnel connecting the upper outbound tunnel to the lower inbound tunnel and descends at a 7.5% grade. This grade is acceptable due to the fact that it is a non-revenue move (buses will not be carrying passengers) around the loop, and will be descending from the outbound tunnel back to the inbound tunnel toward South Station.

As the alignment turns south on Charles Street, the tunnels begin to roll out of their stacked arrangement into a side-by-side configuration as they approach the portal and "boat section" transition to grade at the Tremont Street/Marginal Road intersection. See Figure 5.

SUBSURFACE CONDITIONS

The subsurface profile from investigation along the Silver Line Phase III tunnel is shown on Figure 6, also depicting the current inbound (lower) and outbound (upper) tunnel alignments. Starting at the ground surface, the profile consists of the following layers: fill, organic soil (in limited areas outside the original Shawmut Peninsula), marine clay and sand, glacial till, and bedrock. Marine clay and sand layers were typically encountered beneath the fill and/or organic soil (when present). This layer ranged from about 11 feet to 117 feet thick, generally became thinner as the tunnel alignment progressed from its starting point at Marginal Road to its end point at Atlantic Avenue. In general, the individual sub-layers of clay and sand were discrete and relatively homogeneous. However, finely stratified clay and sand, consisting mostly of clay but with sand layers a few millimeters thick up to several inches thick, were encountered in several areas.

Glacial till was encountered beneath the marine sand and clay, and overlying bedrock, at most locations. The maximum till thickness was about 48 feet, and generally the top of the till layer rose in elevation, and the till thickness increased, as the tunnel alignment progressed from its starting point at Marginal Road to its end at Atlantic Avenue. The depth to bedrock ranged from 72 to 134 feet, generally decreased as the tunnel alignment progressed from its starting point at Marginal Road to its end point at Atlantic Avenue.

The groundwater levels measured in shallow observation wells installed in the fill and upper marine clay and sand layers along the project alignment typically ranged between 10 to 30 feet below existing ground surface, the variation of which generally attributed to the local drainage conditions. The groundwater levels measured in the deeper observation wells installed in the lower portion of clay, glacial till, and bedrock were typically several feet lower than those measured in the shallow wells, indicating a slight downward flow gradient. Therefore, except for the boat section and short cut-and-cover tunnel section on the Charles Street South end of the alignment, the remaining Silver Line Phase III tunnels are completely below the groundwater table.

CONSTRUCTION METHODS

Project Environment

The Silver Line Phase III project will be constructed in some of the most congested and busiest areas in downtown Boston. The cut and cover connection of the Phase III tunnels to the existing Phase II tunnel occurs at the intersection of Atlantic Avenue and Essex Street. This intersection/work zone is bounded by South Station, One Financial Center, Two Financial Center (which is currently under construction and will be completed prior to the start of Silver Line Phase III construction) and the Plymouth Rock building.

Essex Street between Kingston and Washington Streets and Boylston Street between Washington and Charles Streets are narrow streets with an approximate ROW of less than 50 feet in width.

On Boylston Street between Tremont and Charles Streets the campus of Emerson College is located on the south side of Boylston (Piano Row—Historical Buildings) and the Boston Common (National/Local historic landmark) is located on the north side. Although the Silver Line tunnels are quite deep along Boylston Street, they will run partially under and parallel to the existing historic Green Line tunnels.

The turn-around loop within the Boylston/Charles Streets intersection is bounded by the Boston Common, the Public Garden, the Four Seasons hotel and the west end of the Piano Row historic district.

The running tunnels along Charles Street South ascend at a 5.6% grade that requires a transition to cut and cover construction in the vicinity of Warrenton Place because of minimum depth of ground cover over the tunnels. The portal and boat section are being constructed in Tremont Street between Bay Village and the Mass Pike Towers residences.

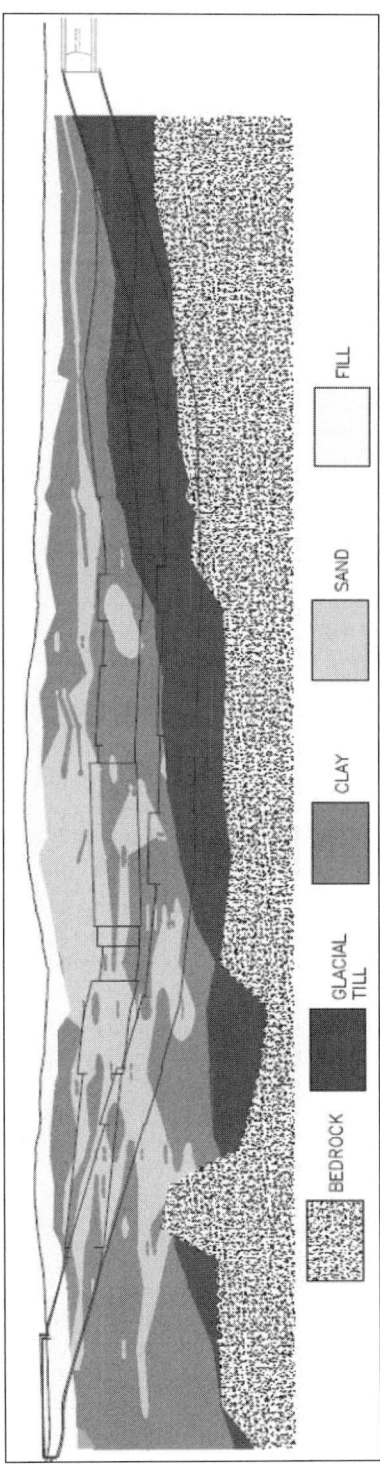

Figure 6. Geological profile

Construction Techniques

An initial evaluation was made as to the construction of the tunnels using either a Tunnel Boring Machine or Sequential Excavation (NATM) techniques.

The actual "running tunnel" configuration exists primarily at three locations. They are:

- From the Access Shaft n the western side of the NESI pipe at South Street to the eastern most edge of the east vertical circulation core of Chinatown Station—approximately 1,100 feet of running tunnel.
- From the western most edge of the west vertical circulation core of Chinatown Station to the east end of Boylston widening point of the tunnels where the cavernous shape to incorporate the platforms of Boylston Station begins—approximately 350 feet of running tunnel.
- From the southern most edge of the cut and cover junction section the Boylston and Charles Street intersection to approximately Warrenton Place where the tunnels must enter the cut and cover section to the portal— approximately 1,000 feet of running tunnel.

The widening of the tunnel into a more cavernous section is necessary at Chinatown and Boylston Stations such that the platforms may be incorporated into the tunnels. The widened shape of the Loop tunnel will accommodate the tight radius curve and the dynamic envelope of the BRT vehicle. The requirement for varying tunnel sections makes the use of a tunnel boring machine much less efficient and practicable.

In addition to the project configuration and relatively short length of "running tunnel," below is the summary of the advantages and disadvantages of each methodology.

Advantages and Disadvantages of a TBM Construction Methodology. They are:
Advantages:

- There is less of a need to provide significant soil improvement and ground water cutoff.
- TBM creates the finished tunnel in one pass and at a significantly faster rate.

Disadvantages for Silver Line Phase III project:

- Neither the Atlantic Avenue nor Tremont portal ends of the project have a staging site of sufficient size to fully support a TBM operation.
- There is not an adequate source of electrical power at either location that would be required to operate the TBM.
- Notwithstanding the lack of adequate staging areas to support a TBM, the compound horizontal and vertical geometry of the tunnel alignment will require an articulated machine causing somewhat slower progress. Additionally, to enter a receiving pit in the Loop area, dismantle the TBM and turn it to re-launch down the Charles Street South alignment would require approximately 6 months.
- The "just in time" delivery approach required due to insufficient TBM staging areas will expose the Project to schedule and cost overruns.
- The thrusting pressures created as the TBM advances through the soil would place stresses on the adjacent tunnel, and possibly the utilities and existing structures above, therefore requiring a greater separation between tunnels and greater clearances below existing structures. A criterion of a full tunnel diameter of separation between the longitudinal tunnels would be required, forcing the depths of the stations deeper.

- The relatively short length of the overall Silver Line Phase III project would require that only one TBM be used for the entire project thus requiring the entire tunneling be contained in a single construction contract.

Advantages and Disadvantages of a Sequential Excavation (NATM) Construction Methodology. They are:

Advantages:
- The mining operation allows for the telescoping and/or transitioning of one tunnel diameter or configuration to another.
- Allows for multiple construction contracts thereby allowing for multiple locations for the mining of the tunnels to occur simultaneously.
- The mining technique allows for better control of the stresses from the construction of one tunnel on the adjacent tunnel thereby permitting the tunnels to be constructed closer together. In critical areas, the mined tunnel design will allow for the tunnels to get within a third of a tunnel diameter of each other.
- Allows construction of the tunnels to much smaller radius curves and more complex compound horizontal and vertical geometry.

Disadvantages on the Silver Line Phase III project:
- Significant attention to the design of the soil improvements and ground water control will be necessary.
- Although the mining may occur in multiple locations simultaneously, significant outreach to the construction industry to make sure that available mining expertise is available for the project. Please reword
- The mining operation is much slower so multiple tunneling contracts and simultaneous work must be performed to minimize overall construction duration.

Based on comparison of the advantages and disadvantages of both the TBM methodology and the Sequential Excavation method, the Sequential Excavation technique has been selected as the optimum option for the construction of the Silver Line Phase III Project.

Contract Packaging

Currently, the project is being set up to be procured in six construction packages. Package 1 is being referred to as the Early Construction Package and will include utility relocations, bridge rehabilitation/modifications necessary for the surface connection from Phase III to Phase I, and structure demolition.

Package 2 will include the cut and cover work necessary to make the connection of the Phase III work to the Phase II tunnel in Atlantic Avenue, underpinning and mining of the tunnels under the NESI pipe, underpinning and mining under the Central Artery Tunnel, and mining the tunnels to the eastern limit of the Chinatown Station eastern vertical circulation cut and cover pit.

Package 3 will include construction of the eastern and western vertical circulation cut and cover pits, underpinning and mining under the Orange Line tunnels, structural work necessary to connect the existing Orange Line Station to the new Silver Line Station at Chinatown, and all the architectural, mechanical, electrical, HVAC, plumbing, and vertical circulation elements for the station.

Package 4 will include the mining of the tunnels from the western edge of the west vertical circulation core of Chinatown Station through the cavern tunnels of the Boylston Station, construction of the cut and cover pits necessary for the mezzanine

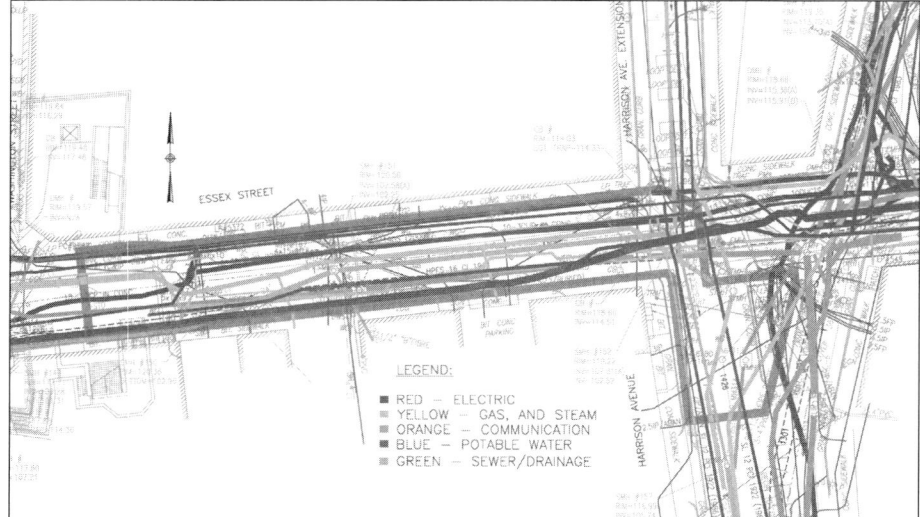

Figure 7. Existing utilities within the Essex/Harrison Streets intersection at the east vertical circulation core of Chinatown Station

connection of Boylston Green Line Station to the new Silver Line Station, construction of the western cut and cover pit to accommodate and construct the new Charles Street entrance, mining of the turn-around loop, and all architectural, mechanical, electrical, HVAC, plumbing and vertical circulation elements for the new Boylston Station.

Package 5 will consist of the mining of the running tunnels from the southern edge of the Charles Street entrance pit to the limit of the cut and cover section in Charles Street, construction of the portal and boat section, and construction of the underground traction power substation and tunnel ventilation structure.

Package 6 will include all of the transit and tunnel facilities such as emergency tunnel lighting and ventilation, communications and signals, traction power substation, and overhead contact system. All communications and station facilities will then be carried back to the Operations Control Center.

CONSTRUCTION IMPACTS AND MITIGATION

Existing Utilities

Due to the extensive evolution and development of downtown Boston over the last few centuries, utilities have been abandoned, rehabilitated, replaced, and relocated many times within the narrow public rights of way. Figure 7 shows the overlay of existing utilities within the limits of the east vertical circulation core of Chinatown Station.

An extensive part of the Silver Line Phase III construction will be the necessary coordination with the local utility companies. The goal would be to have the utility companies permanently relocate their facilities outside of the cut and cover construction. For those that cannot be relocated one-time outside of the construction will have to develop and follow a detailed phasing plan of a series of temporary relocations.

Figure 8. Intersection of Essex and Washington Streets

Figure 9. Boylston Street across from Emerson College's Little Building

Maintenance and Protection of Vehicular/Pedestrian Traffic

The Silver Line Phase III project is being constructed in a densely urbanized area of Boston and the corridor is also extremely narrow. The cut and cover areas to facilitate the mining operations and the physical connections to the existing stations were minimized to reduce the amount of surface disruption. Figures 8 and 9 show the corridor along Essex Street and along Boylston Street respectively.

The deep cut and cover excavations are being accomplished using slurry wall construction down to rock in order to seal off the groundwater. The excavations at Chinatown and at the west circulation pit/new entrance for Boylston Station will both be in excess of one hundred (100) feet deep. Vehicular detour routes are being planned in conjunction with Boston's Department of Transportation while the slurry walls are being installed. Once installed, the excavations will be decked over and the work will proceed from below. Vehicular traffic will then be allowed to travel through the corridor.

Access to the businesses for deliveries will be maintained throughout construction as will pedestrian access through the corridor.

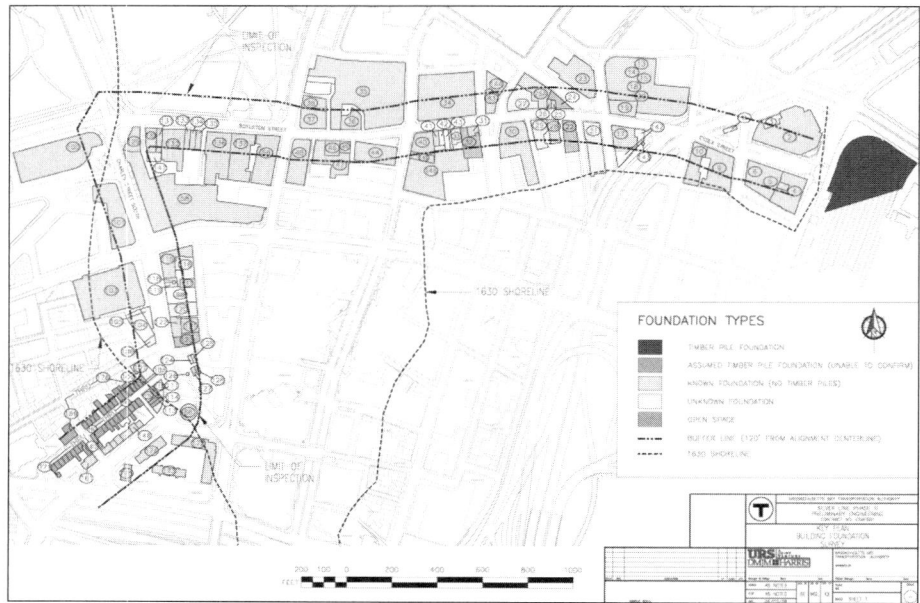

Figure 10. Building foundation types along the Silver Line Phase III Alignment

Settlement to Historic Buildings and the Existing Green Line

The Silver Line Phase III corridor along Essex, Boylston and Charles Streets is lined with older, and in most cases historic, buildings. Most of the corridor falls within the shoreline of the 1630 Shawmut Peninsula which is part of the original land mass of Boston. Most of buildings that were constructed on the peninsula were founded on stone or concrete spread type footings. A small portion of the turn-around loop and the Tremont Street portal and boat section are being constructed "off" of the peninsula. The Bay Village community (shown outside the 1630 shoreline) is assumed to have been founded on timber piles making them much more susceptible to damage from a drop in the groundwater table. See Figure 10.

An extensive use of soil improvement techniques will be employed where necessary to deal with the sequential excavation operations. Although mining provides better control of face loss, and therefore ground settlement at the surface, it will require a significant use of soil improvement. The use of horizontal jet grouting is also being considered due to the inability to use more conventional jet grouting techniques from the surface because of the utilities.

All of the buildings along the corridor will be outfitted with instrumentation and will be closely monitored during the mining of the tunnels and the construction of the cut and cover pits. The contract provisions will provide for certain thresholds of ground movement that may not be exceeded. If the limiting values are met, then work will be required to stop and additional mitigation measures employed.

The existing Green Line Tunnel runs along Boylston Street on the north side of the street adjacent to the Boston Common. Put into service in the early 1900s, it is America's oldest subway tunnel. The tunnel was constructed using cut and cover techniques and it is comprised of individual concrete encased frames. Although the tunnel is ridged in the traverse direction, it will be susceptible to differential settlement longitudinally along its alignment. In addition to extensive soil improvement in the area of the

Boylston Station tunnels, the existing Green Line tunnel will be fitted with instrumentation and monitored throughout the mining operations. Compensation grouting will also be used if necessary during construction. Further geotechnical evaluation will be performed during Final Design to assess the need for direct underpinning of the Green Line tunnel.

SUMMARY

The Silver Line is Boston's newest transit line and its first Bus Rapid Transit Line. Phase I serving the Dudley/Roxbury area, providing service to downtown via the Washington Street corridor, has been in operation since 2002. Phase II providing service from South Station out to points along Boston's developing waterfront and to Logan Airport as been in service since 2004. The Phase III project will provide the final connection between Phase I and Phase II and create a "one-seat" continuous ride from Dudley to Logan Airport.

The tunnel alignment of Phase III goes through the heart of downtown Boston between 60 and 100 feet below the street surface. Buildings, including some that are on the National Register of Historic Places, line both sides of the corridor. The deep cut and cover sections necessary to support the mining operations and to provide the cores for the vertical circulation connections to the very deep new stations will be constructed in a busy urban environment. Mitigation strategies will be implemented to protect the residences and businesses within the construction areas.

Nearly the entire alignment will pass through the tills, marine clays and sands within the 1630 shoreline of the Shawmut Peninsula. The portal and boat section to be constructed in Tremont Street will fall within the beginning of fills and organic soils used to expand the developable areas of Boston. The groundwater levels measured in both shallow and deep observation wells installed along the alignment indicate that the entire project, except for the boat section and short cut and cover section in Tremont Street, will be constructed below the groundwater table.

The urban construction environment, proximity to historic buildings and narrow right-of-way lead to the decision to use sequential excavation (NATM) tunneling techniques in lieu of a tunnel boring machine. Significant ground improvement and underpinning will be necessary to pass beneath the existing underground structures such as the NESI sewer, Central Artery tunnel, and Orange Line Chinatown and Green Line Boylston stations. Tunneling techniques such as horizontal jet grouting and face reinforcement/stabilization are being evaluated for use on the project.

Working under the guidelines for eligibility in the Federal Transit Administration's New Starts Funding Program, the engineering for Silver Line Phase III is proceeding for a construction start in late 2010/early 2011 and revenue service to begin at the end of 2016.

PART 12

New Projects 2

Chairs

John Kennedy
Atkinson Construction

David Young
Hatch Mott MacDonald

DESIGN OF NATM TUNNELS AND STATIONS OF SILVER LINE PHASE III PROJECT IN BOSTON

Verya Nasri ▪ AECOM

Kosmas Vrouvlianis ▪ AECOM

Irwan Halim ▪ URS Corporation

ABSTRACT

The Silver Line Phase III Project is the final one mile underground segment of the new Bus Rapid Transit in downtown Boston. The project is expected to significantly reduce existing and anticipated traffic congestion. The twin one-lane tunnels and two station platforms will be built by NATM in Boston saturated soft clay, sand and till layers in proximity to several historical buildings, underpinning two existing cut and cover subway stations and several major utilities in densely built urban area. This paper presents the analysis and design of the NATM tunnels and stations detailing the excavation sequence and support system and various ground improvement and pre-support techniques used in the preliminary and final design of the project.

INTRODUCTION

Silver Line is the Massachusetts Bay Transportation Authority's (MBTA) fifth and most recently constructed transit line. It is a Bus Rapid Transit (BRT) line that operates articulated buses in the City of Boston, Massachusetts. The plan called for the Silver Line to be constructed in three phases, of which Phases I and II have already been constructed and Phase III is currently under design (Figure 1). Phase I is a 2.2 mile surface BRT line which opened for service on July 2002 and operates between Dudley Square and Downtown Boston. Phase II opened on December 2004, and consists of a 1.5 mile underground segment from Boston's South Station to World Trade Center, continuing on to Logan Airport through the Ted Williams Tunnel, and as a surface route to South Boston. The currently under design Phase III is an underground segment linking the existing Phase I and II and resulting in a continuous line from Dudley Square to Logan Airport and South Boston.

Silver Line Phase III segment is approximately 1 mile, and runs through one of Boston's oldest and most congested areas and is flanked on both sides by many historic buildings and landmarks. Most of the city's land is man-made. Figure 2 shows the original shoreline of 1630 with Shawmut Peninsula dividing the Back Bay from the South Bay. Since 1643, various land-making projects were undertaken and new land was created, transforming the original shoreline to its present shape. For the most part, the Phase III segment of the alignment is located within the original Shawmut Peninsula, with the exception of the south west segment which extends outside the original shore line.

The project consists of a series of twin cut-and-cover and mined tunnels and two underground stations: Boylston Station and Chinatown Station (Figure 3).The two running tunnels are identified as inbound and outbound. The inbound tunnel is the tunnel that carries a vehicle traveling towards South Station, while the outbound tunnel carries a vehicle traveling in the direction away from South Station. The alignment consists of

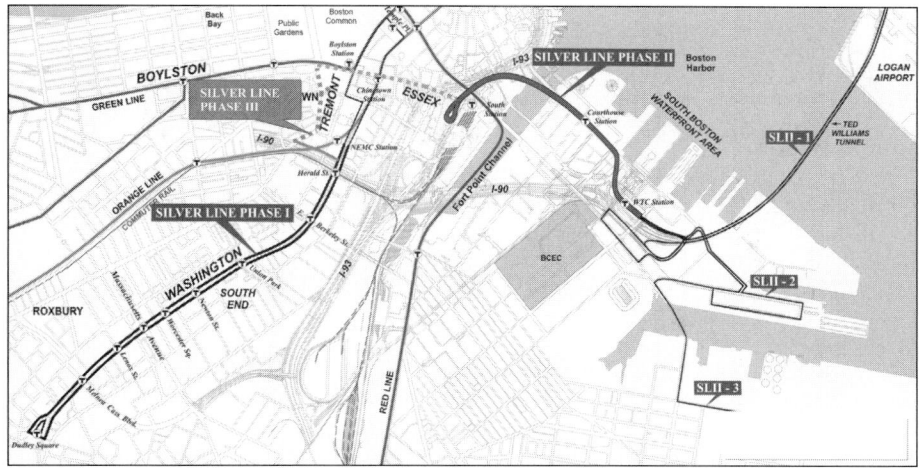

Figure 1. Existing Silver Line Phase I & II and proposed Silver Line Phase III

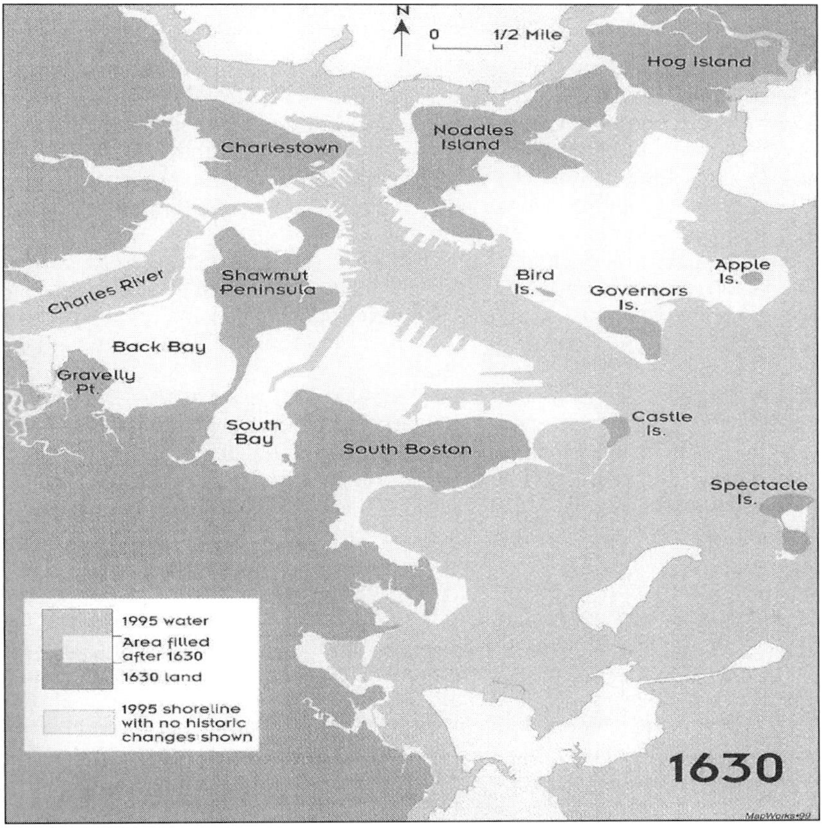

Figure 2. Original 1630 Boston shoreline

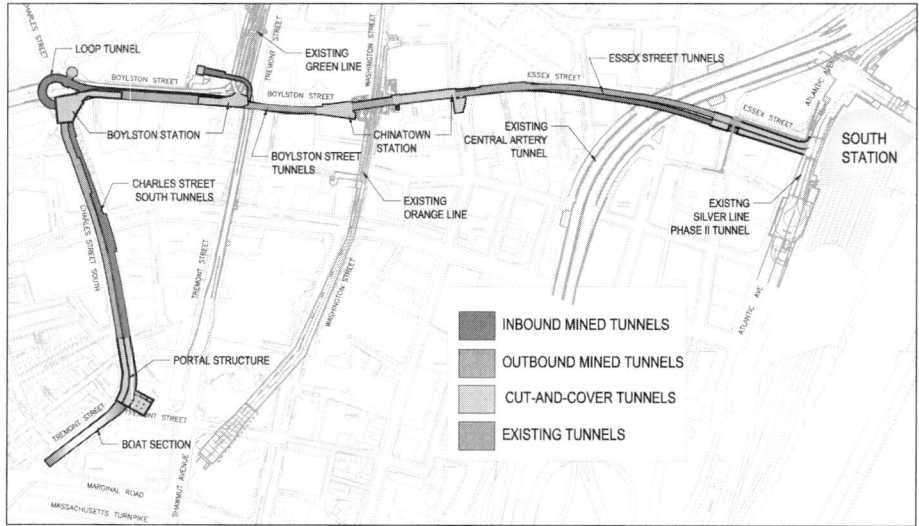

Figure 3. Silver Line Phase III

two parts: the Core segment of the alignment, which connects to the existing Phase II tunnel at Atlantic Avenue, runs under Essex Street and Boylston Street, and terminates in a Loop tunnel at the intersection of Boylston and Charles Streets. The Loop tunnel allows the outbound vehicle to reverse direction onto the inbound tunnel. The Charles Street South segment of the alignment begins at the intersection of Boylston and Charles Streets and continues south under Charles Street South and Tremont Street, to portal on Marginal Road.

Mined tunnels are constructed using the NATM and vary in size, depending on their function (Figure 4). The running tunnel is the smallest in cross section of all mined tunnels and the most dominant in use. The blister tunnel is a wider tunnel section providing an additional lane to allow a disabled vehicle to park. The platform tunnel is the widest tunnel section used at the two stations to provide the extra width required for the station platforms. All mined tunnels include a shotcrete initial liner, a waterproofing membrane and a cast-in-place final liner. In an effort to maintain geometric consistency between the different tunnel sizes, and to minimize variations in the formwork of the final liner, the running tunnel geometry is used as the basis to generate the larger blister and platform tunnel sections. Maintaining the sidewall geometry of the running tunnel and revising the crown and invert geometries, the larger tunnel sections are generated by varying the spring line width.

Cut-and-Cover tunnels are constructed using 3 ft thick slurry walls serving as excavation support and permanent structural walls. To minimize groundwater ingress into the tunnels through the slurry wall joints, an internal cast-in-place wall is poured against the slurry wall. A waterproofing membrane is applied between the slurry wall and the interior cast-in-place wall and the outside faces of the invert and roof slabs.

The two stations are constructed by both the Cut-and-Cover method and the NATM. The Cut-and-Cover portions of the stations house the vertical circulation elements (escalators, elevators, staircases) and various ancillary spaces, while the mined portion provides the space for the inbound and outbound platforms.

Essex Street and Boylston Street are among the oldest streets in Boston. They are located on the original Shawmut Peninsula and carry a massive amount of underground

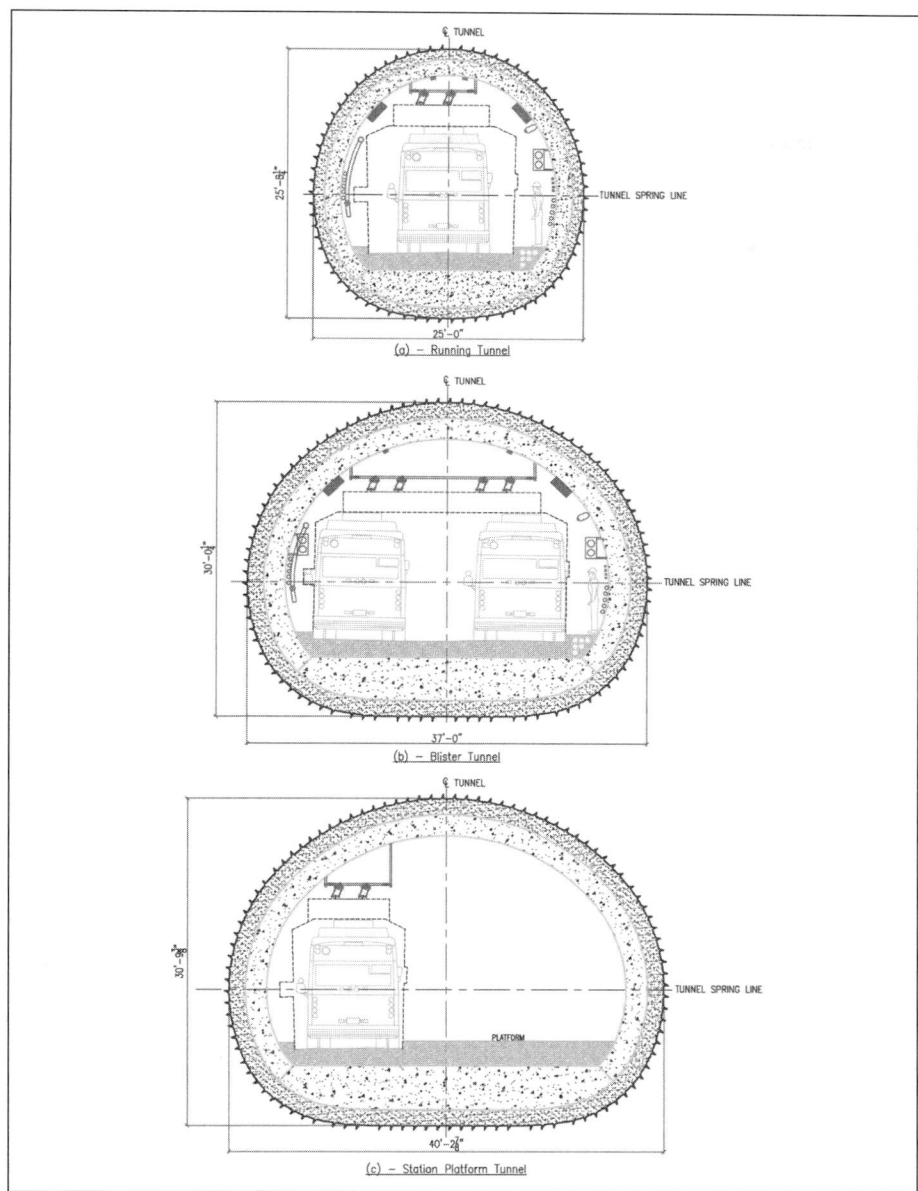

Figure 4. Mined tunnel sections

utilities, new, old and abandoned. In the areas of cut-and-cover construction, these utilities will be removed, relocated, or temporarily supported in place to clear the space allowing construction of the cut-and-cover tunnels and station structures. In areas of mined tunneling, a minimum amount of utility relocation will be expected, as mined tunnels are located below the existing utilities. It should be anticipated, however, that underground obstructions will be encountered during mining operations.

Because of close proximity of the alignment to several historical buildings supported on timber piles and the potential of ground settlements due to lowering of the ground water table, the ground water table shall not be lowered during construction and measures to prevent lowering of the ground water table are incorporated into the design.

GEOLOGIC ENVIRONMENT

The project site is located in the structural and topographic depression known as the Boston Basin at the eastern portion of the Appalachian orogenic zone. The Basin is a roughly triangular-shaped downfaulted area of Precambrian to Cambrian age sedimentary, metamorphic, and volcanic rocks, which are east-northeast trending and widening toward the coast. The upper zone of rock within the Boston Basin is a sequence of sedimentary rocks referred to as the Boston Bay Group (Barosh et al., 1989). The predominant bedrock in the project area is the Cambridge argillite, which is a gray, very thinly and steeply bedded, fine-grained sedimentary rock composed of clay and silt-sized particles that have undergone low-grade metamorphism.

The project area was extensively glaciated during the Wisconsin Period. Repeated retreats and re-advances of the glacier between about 18,000 and 12,000 years ago fractured and scoured the bedrock and left complex deposit of soils including basal tills, moraines, drumlins, and outwash deposits above the bedrock. During glacial retreats, fine-grained soils, including fine sands, silts, and clays, were carried to the ocean by rivers and streams and deposited in quiet marine waters, including the project area. This process created a marine clay deposit known locally as the Boston Blue Clay. This clay often contains interbedded layers of fine sand and silty fine sand. More recent tidal marsh deposits consisting of organic silt, silty fine sand and peat overlie the glacial and marine deposits. Man-made fills were placed over the natural soils to reclaim low-lying and/or submerged areas, particularly as the City of Boston expanded beyond its colonial-era shoreline.

SUBSURFACE CONDITIONS

The subsurface profile from investigation along the Silver Line Phase III tunnel is shown on Figure 5, also depicting the current inbound and outbound tunnel alignments. Starting at the ground surface, the profile consists of the following layers: fill, organic soil (in limited areas outside the original Shawmut Peninsula), marine clay and sand, glacial till, and bedrock. Marine clay and sand layers were typically encountered beneath the fill and/or organic soil (when present). This layer ranged from about 11 feet to 117 feet thick, generally became thinner as the tunnel alignment progressed from its starting point at Marginal Road to its end point at Atlantic Avenue. In general, the individual sub-layers of clay and sand were discrete and relatively homogeneous. However, finely stratified clay and sand, consisting mostly of clay but with sand layers a few millimeters thick up to several inches thick, were encountered in several areas.

Glacial till was encountered beneath the marine sand and clay, and overlying bedrock, at most locations. The maximum till thickness was about 48 feet, and generally the top of the till layer rose in elevation, and the till thickness increased, as the tunnel alignment progressed from its starting point at Marginal Road to its end at Atlantic Avenue. The depth to bedrock ranged from 72 to 134 feet, generally decreased as the tunnel alignment progressed from its starting point at Marginal Road to its end point at Atlantic Avenue.

From Figure 5, the mined tunnels will mainly be excavated in the glacial tills, marine clay and sand, or stratified clay and sand. A small portion of the inbound tunnel will be excavated in the argillite bedrock. The following paragraphs contain descriptions of pertinent soils and rock properties along the tunnel alignments.

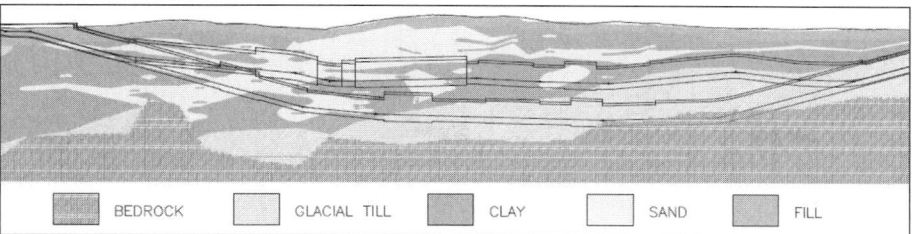

Figure 5. Interpreted geological profile along the alignment

The glacial till in the project area is typically dense to very dense, and consists of a wide ranged mixture of clay, silt, sand, and gravel with occasional cobbles and boulders. The fines content of the till mostly ranged from about 20 to 60 percent, with an average of about 44 percent, indicating a mostly cohesive nature of the soils. Most of the fines consist of non-plastic or low-plasticity silt and clay. The measured hydraulic conductivity in boreholes during drilling ranged from the order of 10^{-6} cm/sec to as high as 10^{-3} cm/sec, indicating occurrence of high permeability zones in the otherwise low permeability deposit. The bedrock in the project area consists of medium hard to soft, slightly to extremely fractured, and weathered argillite. The degree of fracturing and weathering of the bedrock generally decreases with depth.

The marine sand in the project area consists mostly of narrowly graded fine- to very fine-grained silty sand with varying amounts of non-plastic to slightly plastic fines. The amount of fines in the deposit mostly ranged between 10 to 30 percent with an average of 19 percent. A few occasional thin interbedded layers of plastic silt/clay are present within the deposit. The sand is typically medium dense to dense, with density generally increasing with depth. The measured hydraulic conductivity in boreholes during drilling ranged from the order of 10^{-6} cm/sec to as high as 10^{-3} cm/sec, which is similar to the range exhibited by the glacial till.

The marine clay in the project area typically consists of medium plasticity, soft to medium stiff clay, frequently containing layers of fine silty sand ranging from a few millimeters to several inches thick. Figure 6 shows the clay undrained strength profiles with depth along part of the tunnel alignment. As seen from the profile, the upper portion of the clay is generally over-consolidated due to desiccation, the over-consolidation ratio decreases with depth, and becoming close to normally consolidated in some of the deepest sections of the layer. A zone of particularly soft, sensitive clay, with standard penetration test (SPT) N-values of 3 or less, and liquidity index values of greater than 1, was encountered near the bottom half of the layer along the Boylston Street section of the tunnel alignment.

The groundwater levels measured in shallow observation wells installed in the fill and upper marine clay and sand layers along the project alignment typically ranged between 10 to 30 feet below existing ground surface, the variation of which generally attributed to the local drainage conditions. The groundwater levels measured in the deeper observation wells installed in the lower portion of clay, glacial till, and bedrock were typically several feet lower than those measured in the shallow wells, indicating a slight downward flow gradient. Therefore, except for the boat section and short cut-and-cover tunnel section on the Charles Street South end of the alignment, the remaining Silver Line Phase III tunnels are completely below the groundwater table.

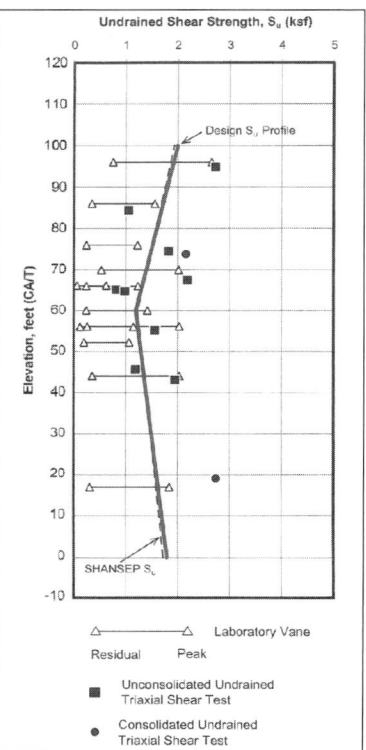

Figure 6. Marine clay undrained shear strength along Charles Street South

BASIC TUNNEL DESIGN CONSIDERATIONS

The anticipated subsurface conditions and the stringent project settlement criteria create difficult conditions for the design and construction of the tunnels. The soft deposits at the project site with their low shear strength can experience large deformations during tunneling. The progressive stress release at the excavation face generates a large plastic zone resulting in instability ahead of the face, and which does not allow the formation of arching effect. Therefore, application of various pre-support techniques ahead of the advancing tunnel face is proposed to limit the ground deformation.

Minimum surface settlement, minimum disturbance to the street traffic and minimum impact on the operation of the existing underground tracks and roadways were the main requirements for the design. The critical design criteria consisted of strict limitation of surface subsidence due to tunneling at the proximity of existing buildings foundations. Based on these criteria the angular distortion at these locations can not exceed $1/1000$, and no surface point shall subside more than one inch below its pre-construction location. Presence of important mid-rise historical buildings supported on shallow spread footings in the zone of mining influence dictated such strict settlement criteria.

The type of ground in which most of the tunnels will be mined is classified as soft clay and sand with low strength characteristics. Different ground improvement techniques and various pre-support methods were considered in initial phases of design. Ground improvement approaches such as massive jet grouting and ground freezing from the ground surface were excluded due to their incompatibility with the densely distributed utilities, heavy street traffic and other site constraints.

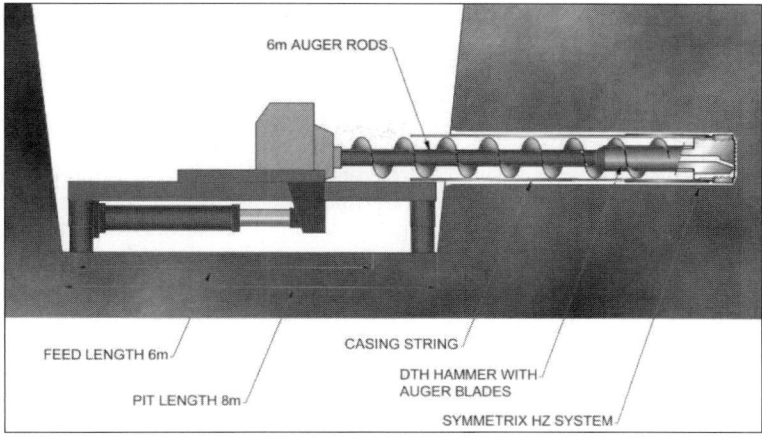

Figure 7. Atlas Copco's drilling equipment for the installation of pipe roof umbrella

Use of pipe roof umbrella, forepoling and horizontal jet grout arch techniques as pre-support methods were studied in detail. The pipe roof umbrella concept was selected for crossing the existing Orange Line Subway and also an existing 6 ft diameter concrete sewer pipe and to achieve the above described settlement criteria. A pipe drilling technology was recommended to minimize installation induced soil disturbance, excess pore water pressure and surface settlement. This technique combines drilling with continuous controlled jacking and auguring (Figure 7). The forepoling method was used for the sections in clay and where the thickness of the till layer cover was considered to be insufficient. For mining under the saturated sand layers located at the crown of excavation, a horizontal jet grout pre-support arch was designed to provide the necessary preconfinement action.

To limit the face extrusion and its resulting ground deformation and to increase the excavation stand up time, fiberglass face reinforcement bars were designed as an element of the pre-support system. The reinforced core in the excavation face reduces ground deformations including extrusion, convergence and pre-convergence components. The design provides fiberglass face reinforcement in combination with one of the three above-mentioned different types of pre-support methods including pipe roof umbrella, forepoling and horizontal jet grout arch techniques along the entire alignment except where the majority of the excavation face consists of the till layer with higher strength parameters.

Continuous and swift excavation and support installation combined with an appropriate staging and support system is necessary to minimize short and long-term deformations. The tunnels will be constructed using available pre-support techniques and fiber reinforced shotcrete and lattice girders as an initial support lining to provide required preconfinement or confinement role. The final lining will be installed using cast in place concrete.

The sequential excavation method has been selected as a means of progressively excavating and supporting the ground. The running tunnels and station platforms are to be excavated with two headings in accordance with the specified sequences designed to stabilize the ground in limited excavation rounds. The excavation sequences of specified round lengths consist of application of steel fiber-reinforced flashcrete to newly exposed surfaces, installation of lattice girders at predetermined intervals, and application of fiber reinforced shotcrete support with variable thicknesses depending upon final opening dimensions.

Monitoring of surface movements will be conducted using surface settlement points installed in a series of arrays perpendicular to tunnel centerline across the expected settlement trough. In addition, inclinometers, extensometers and piezometers will be installed to detect horizontal and vertical subsurface movements and ground water level. Total station technology will be used to determine buildings deformation in the zone of influence of tunneling operation. Monitoring points and tiltmeters will be installed to observe any building movement. In-tunnel monitoring will primarily comprise convergence monitoring of the new and existing tunnels and deformation and stress measurement of liner using strain gauges and concrete and soil pressure cells.

SUPPORT SYSTEMS

Conventional NATM with top heading and bench and shotcrete and lattice girder is used for mining in the till layer when the thickness of the till layer cover is sufficient (more than quarter of the excavation span). Forepoling is added when the thickness of the till layer cover is not sufficient (less than quarter of the excavation span). Fiberglass face reinforcement is included when more than a half of the excavated top heading face is in clay. Horizontal jet grout arch combined with fiberglass face reinforcement is used when crossing the saturated sand layer or clay layer with large sand lenses. Pipe roof method is considered for crossing a 6 ft diameter concrete sewer pipe and also the existing Orange Line Subway at Chinatown station platform tunnel.

PRE-SUPPORT ELEMENTS

The typical pre-support systems used for the running tunnels and the platform stations in order to ensure the stability of the excavation and reduce the ground deformation are comprised of the following prime elements:

- Pipe roof umbrella
- Forepoling
- Horizontal jet grouting arch
- Fiberglass face reinforcement bars

Pipe Roof Umbrella

In recognition of the poor quality of the ground surrounding the tunnels, the limited cover above the tunnels and strict ground deformation criteria, a pre-support system will form an integrated part of the tunnel construction for crossing under the Orange Line Subway at Chinatown station platform tunnel and also for under passing a 6 ft diameter concrete sewer pipe. The steel pipes will be installed horizontally into the ground with arch-shaped arrangements following the geometry of the tunnel for running tunnels when crossing the sewer and for the platform tunnels when crossing the Orange Line.

A rotary percussive means of drilling will be used which allows drilling of both clay and any encountered obstructions. The system will consist of a DTH hammer, percussive/rotary bit, concentric reamer, casing shoe, and flight drill rods. The drill rig will be able to provide jacking force to the casings, rotation to the drill rods, and air through the drill rods to the DTH hammer. It will also allow expulsion and disposal of cuttings. In order to maintain straightness, the pipes are installed by a method that both pulls and pushes the casings. The system is a symmetrical circular design without eccentric movement or other protrusions which may cause deviation. The steel pipes will form reinforcement within the ground, stabilizing the soil above the tunnel roof and in the shoulder area ahead of the advancing tunnel face (Figure 7; Atlas Copco method).

The installation of the pipe roof system is simple and quick and has the following effects:
- Reducing the ground loss by creating a boundary of stiff material between the soil and the tunnel core;
- Stabilizing the ground in the area of newly excavated and yet unsupported excavation round;
- Distributing loads acting upon the tunnel roof above the freshly excavated round into the in-situ ground ahead of the face and the lined tunnel behind the face;
- Confinement of the tunnel excavation area ahead of the tunnel face;
- Enhancing tunnel face stability by reducing loads acting upon the tunnel face; and
- Reducing surface settlement.

In addition to the above aspects, the pipe roof installation provides additional ground exploration ahead of the tunnel construction face. This is particularly important when the roof of the progressing tunnel reaches the clay-sand or clay-fill interface or the tunnel intersects water saturated silt or sand lenses. The augering for the pre-support will be used as part of a probing system investigating the soil conditions ahead of the progressing face.

In the current case with shallow cover, one foot diameter steel pipes are used. To determine their thickness, it is assumed that the full overburden pressure is applied on the pipe roof in order to design a stiff pre-support to limit the ground deformation.

Forpoling

The Forepoling technique is a method of pre-reinforcing the poor ground ahead of the tunnel face providing the stability of the excavation sequence for the time necessary to install the initial support. The implementation of the technique consists of the construction of a canopy of sub-horizontal steel pipes ahead and above the excavation face, which acts as an in-situ support placed ahead of the excavation. Forepoling is generally used for poor ground, shallow cover, or for reducing the surface settlement.

The design requires installing 4" diameter and 40 ft long steel pipes at 1 ft spacing and 5 degree angle with horizontal to form a truncated cone shape with 12 ft overlapping of consecutive forepoling rounds to ensure the stability of the excavation.

Despite the significant progress in numerical analysis methods in the recent years, the design of a forepoling pre-support system is still based on empirical considerations or simplified schemes. To determine the required steel pipe section, the forpole is considered to be a continuous beam supported on the installed initial liner lattice girders and the ground ahead of the excavation face. The computation is carried out for the most critical stage with the longest span, which is just before the installation of the lattice girder. The load applied on the steel pipes is assumed to be half of the total vertical load given by Terzaghi's formula. Forepoling does not form an arch around the excavation and the cement grout filling and surrounding the pipe is not normally considered in the calculation.

Fiberglass Face Reinforcement Bars

Due to de-confinement processes occurring at the tunnel face, the ground tends to extrude from the tunnel face into the tunnel opening. Particularly soft soil has the tendency to protrude through the face into the tunnel. The use of a dome shape face and steel fiber reinforced shotcrete support in more competent soils increases the face stability. However, for soft soils (as anticipated at the project site), additional improvement of the ground shear strength will be required to prevent excessive deformation of clay and sand layers and to further limit the ground extrusion at the face and pre convergence ahead of the face.

The technique proposed for this project improves the strength and deformation characteristics of the ground in the advance core of the tunnel and results in the development of effective preconfinement action. The proposed method consists of using fiberglass bars for the face reinforcement. The fiberglass material combines high tensile and low shear strength properties. It is easy to cut reinforcement made of this material with the same bucket used for tunneling excavation.

The use of fiberglass reinforcement elements such as bars or pipes has been proven to be a reliable means of soil reinforcement in the tunnel face in numerous projects (Lunardi, 2000). This method is normally used in combination with full face excavation which allows rapid closure of the initial lining ring. The completion of the initial lining ring close to the advance face contributes significantly to the reduction of surface settlement.

Conventional drilling techniques will be used with standard drilling equipment. The borings carried out for the face reinforcement will be closely observed and used as part of the pre-probing system that should be implemented during the tunnel construction.

40 ft long fiberglass bars with 12 ft overlap and 3 ft by 3 ft spacing are used to reinforce the excavation face. In terms of analysis, the action of the fiberglass bars is considered using an equivalent applied pressure in an axisymmetric analysis. However, numerical 3D simulation taking into account each bolt individually as well as a realistic behavior of the grout-ground interface is often the only approach that leads to a reasonable design.

Horizontal Jet Grouting Arch

This method consists of an arch of sub-horizontal jet-grouting columns constructed at the crown, shoulders and sidewalls of the tunnel. The jet-grouting arch supports the ground during the excavation in soils with poor cohesion and distributes the stresses applied on the initial and final liners. Horizontal jet-grouting is used in urban tunneling to reduce the surface settlements and prevent damage to buildings and infrastructures.

If the stress in the ground is considerably higher than the strength of the soil even in the zone around the face, an arch effect cannot be formed because the ground does not have sufficient residual strength. The deformation becomes unacceptable developing immediately into the failure range resulting in major instability such as failure of the face and collapse of the cavity without allowing time for installation of the conventional support system. In this case, ground improvement operation must be launched ahead of the face to develop preconfinement action capable of creating an artificial arch effect. A continuous pre-support of horizontal jet grouting arch is designed for crossing the saturated sand layers and clay layers with large sand lenses.

The design requires installing a row of 40 ft long and 24" diameter jet grout columns at 16" spacing and 5 degree angle with horizontal to form a truncated cone shape with 12 ft overlapping of consecutive jet grouting rounds providing an 18" thick pre-support arch around the crown, shoulders and sidewalls of excavation (Figure 8).

NUMERICAL ANALYSES AND MODELLING

The three layers of sand, clay, and till along with the miscellaneous fill were modeled in the finite element analysis. The Hardening Soil (H-S) model (Schanz, 1999) was used for modeling sand, clay, and till strata. H-S model is an advanced model with stress-dependant soil stiffness using the following three different input stiffnesses: the triaxial loading stiffness (E_{50}), the triaxial unloading stiffness (E_{ur}) and the oedometer loading stiffness (E_{oed}). The limiting states of stress are defined by the friction angle

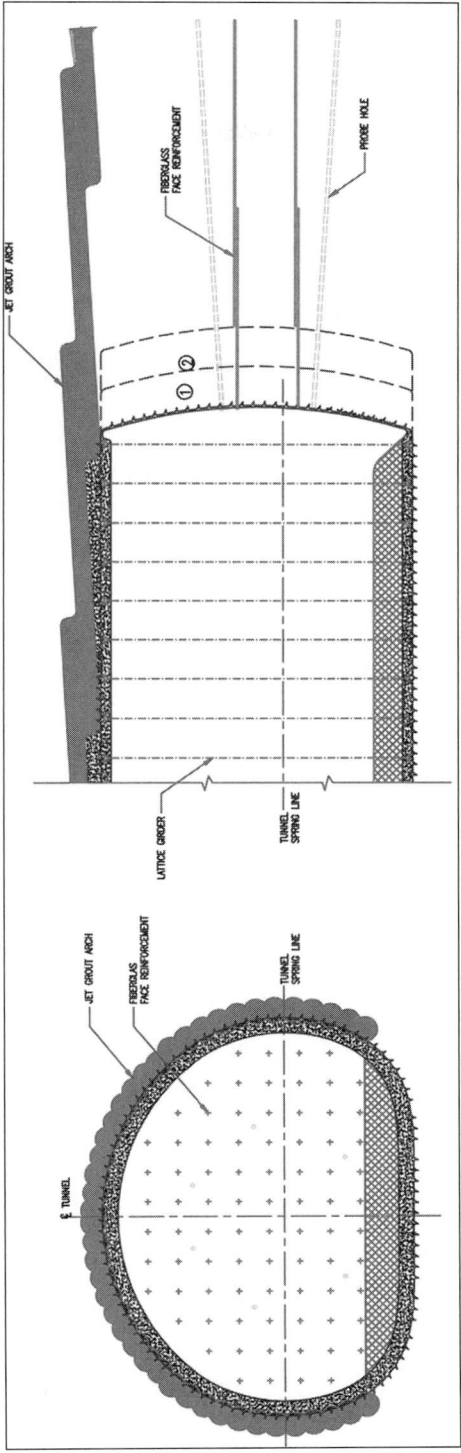

Figure 8. Mined platform station with horizontal jet grouting arch and fiberglass face reinforcement bars

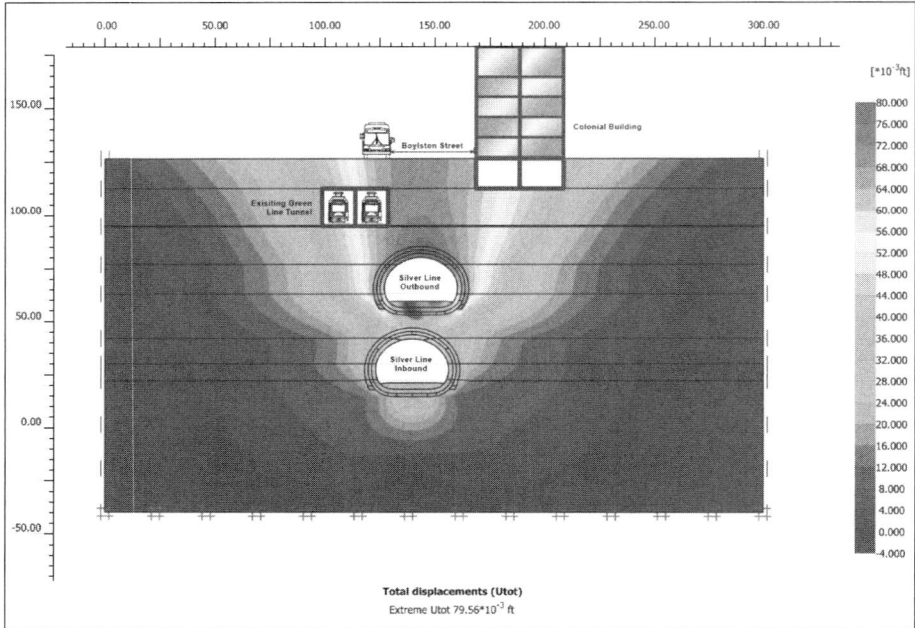

Figure 9. Example of PLAXIS deformation results for horizontal jet grouting arch pre-support system

(ϕ), cohesion (c) and dilatancy angle (Ψ). The relatively simple Mohr-Coulomb model tends to overestimate soil dilatancy. For miscellaneous fill, Mohr-Coulomb soil model was used in the analysis. Material properties used for analysis of Silver Line Phase III tunnels were obtained from GEI Consultants, Inc. Geotechnical Interpretive Report.

The excavation and support of various elements of the project such as the running tunnel, blister tunnel and platform tunnel were modeled according to the construction stages of the existing structures and new tunnels using PLAXIS software (Brinkgreve, 2002). As an example, the stages used to model the construction of platform tunnels are as follows:

1. Initialization of stresses: in this stage the soil layers and ground water table are defined and vertical and horizontal stresses are generated.
2. Installation of building basements and existing subway: the basements of adjacent buildings and the existing subway are excavated and their structures are activated.
3. The displacements from previous stage are set to zero at this stage, which marks the start of Silver Line construction. Excavation of lower platform tunnel: the horizontal jet grout arch is installed and the full face excavation is performed with 50% relaxation of the initial stresses at the excavation boundary. Undrained analysis is used for this stage.
4. Installation of initial liner: shotcrete initial liner is activated and 100% of the initial stresses at the excavation boundary are released.
5. Installation of final liner for the lower tunnel.
6. Repeating stages 3, 4, and 5 for the upper tunnel.

7. Consolidation for 100 years: a 100-year consolidation was performed to calculate long-term ground movements and stresses in surrounding soil and liner.

The tensile stresses in the existing tunnel liner were checked against the rupture modulus of concrete after applying the appropriate load and resistance factors. The analysis indicates that the surface settlement and angular distortion are within the acceptable limits (Figure 9).

CONCLUSIONS

A comprehensive geotechnical investigation was carried out in order to identify critical soil parameters required for design of the NATM tunnels. The interpretation of the results of these tests was discussed in this paper.

The methodology applied for selection of optimum construction techniques and their complexities were discussed. The tunneling method included application of NATM combined with the use of various pre-support systems such as pipe roof umbrella, forepoling, horizontal jet grouting arch, and fiberglass face reinforcement bars to minimize the excavation induced surface settlement.

To verify the efficiency of the applied construction technique, series of finite element analyses were performed to investigate: a) the initial state of stress in soil due to construction of the existing underground structures, b) the soil-structure interaction during various stages of NATM tunnels construction, and c) the structural response of the existing underground and aboveground structures.

REFERENCES

Barosh et al., 1989. Geology of the Boston Basin & Vicinity. Civil Engineering Practice, Journal of the Boston Society of Civil Engineers, Volume 4, No. 1, pp 39–52.

Brinkgreve, R. B. J., 2002. PLAXIS 2D Version 8 Manual. Balkema Publishers, the Netherlands.

GEI Consultants, Inc., 2008. Geotechnical Interpretive Report, Preliminary Engineering Submittal, MBTA Silver Line—Phase III, Boston, Massachusetts.

Lunardi, P., 2000. ADECO-RS Approach. T&T International Special Supplement.

Nasri, V. and Ayoubian, A. 2007. Tunneling Alternatives for Subway Connection in Downtown Chicago. Rapid Excavation Tunneling Conference 2007, Toronto, Canada, June 10–13, 2007, pp. 56–68.

Schanz, T., Vermeer, P. A., and Bonnier, P. G., 1999. The hardening soil model formulation and verification. Proc. Plaxis Symposium Beyond 2000 in Computational Geotechnics, Belkema, Amsterdam, the Netherlands, pp. 281–296.

GEOTECHNICAL AND STRUCTURAL DESIGN CHALLENGES OF THE FREMONT CENTRAL PARK SUBWAY FOR THE BART WARM SPRINGS EXTENSION

Mitchell L. Fong ▪ PB Americas

M. Shing Owyang ▪ PB Americas

Thomas S. Lee ▪ PB Americas

ABSTRACT

The Bay Area Rapid Transit (BART) District will be extending BART service south along an 8.7-km (5.4-mile) -long corridor in the city of Fremont, California as part of the Warm Springs Extension (WSX). The WSX project will link the existing BART system to the future BART to San Jose extension. A segment of WSX is located in Fremont Central Park (referred to as the Fremont Central Park Subway Project). The Fremont Central Park Subway is considered the most complicated and geotechnically challenging portion of the project with an underground subway box being constructed near the active Hayward Fault which is capable of ground motions up to 0.75 g and is located in an aquifer which is used for water storage and supply. Soil-structure interaction analysis was used in the design of the subway box to resist deformations due to earthquake ground motions. A water-tight excavation support system has been designed to allow subway box construction while minimizing impacts to the aquifer.

INTRODUCTION

The Bay Area Rapid Transit (BART) is a heavy rail electric train system that has provided the Bay Area congestion relief for over 30 years. BART plans to extend its service south along an 8.7-km (5.4-mile) -long corridor to the Warm Springs District in Fremont, California as part of the Warm Springs Extension (WSX). The WSX project has been divided into two contracts with different delivery methods. A convention design/bid/build contract was selected for the Fremont Central Park Subway (FCPS) which begins on the south side of Walnut Ave and ends north of Paseo Padre Parkway and is about 2.3-km (1.4 mile) long. FCPS includes the subway box shell, subway transition structures, ventilation structures and relocated park facilities. The follow-on contract, Line, Track, Station, and Systems (LTSS) will be delivered using a design/build (D/B) approach. This D/B contract includes the trackway between Fremont Station and Walnut Ave, Warm Springs Station, and installing systems and tracks from the Fremont BART station to the tail tracks north of Mission Boulevard. Figure 1 shows the entire WSX and the FCPS project.

In order to minimize the impact to the Fremont Central Park, a subway was selected. Within the park, the subway alignment will cross under a portion the east lobe of Lake Elizabeth, one major city street—Stevenson Boulevard, and the Union Pacific railroad (UPRR) tracks. A shoo-fly will be built for the UPRR crossing so that UP operations will not be interrupted.

The 2.3-km (1.4-mile) -long FCPS is dominated by two significant geologic features, the Hayward Fault and the Above Hayward Fault (AHF) aquifer. The alignment

CHALLENGES OF THE FREMONT CENTRAL PARK SUBWAY

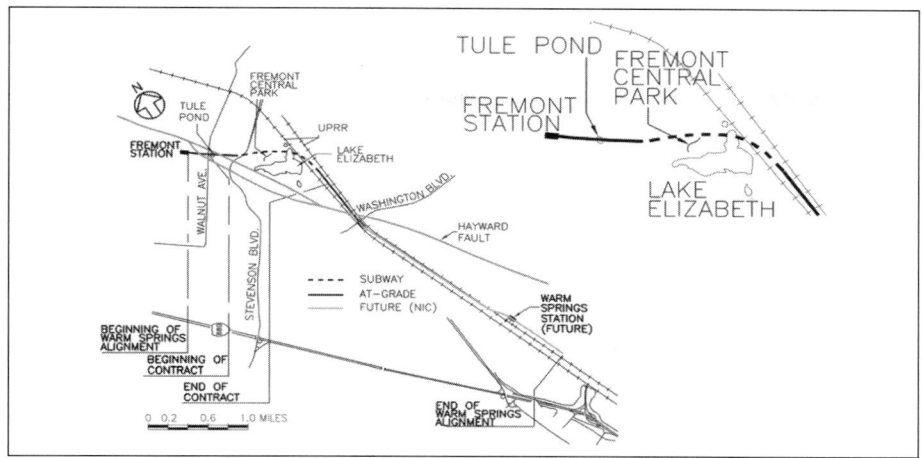

Figure 1. BART WSX & Fremont Central Park subway alignment

crosses the seismically active Hayward Fault twice, once in Tule Pond and the other at Washington Blvd. Extensive previous fault investigations have pinpointed the location of the Hayward fault in Tule Pond and an embankment to accommodate the anticipated seismic displacement is planned. The other fault crossing is at grade and within the design/build portion of the project. The second geologic feature is the AHF aquifer which provides a significant source and storage of water for the local water district. The underground subway will penetrate this aquifer and is up to 11 m (36 ft) below the groundwater table.

GEOLOGICAL, SEISMOLOGICAL, AND HYDROGEOLOGIC SETTING

The San Francisco Bay Area is located in the California Coast Range Physiographic Province which is characterized by northwest-southeast trending valleys and ridges. These features are controlled by folds and faults that resulted from the collision of the Farallon and North American plates and subsequent strike-slip faulting along the San Andreas fault zone. The Bay Area experienced uplift and faulting in several episodes during late Tertiary time (about 25 to 2 million years ago) that produced a series of northwest-trending valleys and mountain ranges, including the Berkeley Hills, the San Francisco Peninsula, and the intervening San Francisco Bay.

The active faults that have the most impact to the project are the Hayward Fault, the San Andreas Fault, and the Calaveras Fault. These faults have caused severe ground shaking at the site and the Hayward Fault has the highest probability of generating a magnitude 6.7 or greater earthquake before 2031. Historically, the most significant earthquake near the WSX alignment occurred in 1868 along the southern section of the Hayward Fault. There is about a 60% chance that the next magnitude 6.7 or greater earthquake in the Bay Area will occur in the next 30 years with a seismic ground acceleration of 0.75 g.

The FCPS project is located within the Niles Cone Subbasin of the Santa Clara Valley Groundwater Basin. The Nile Cone Subbasin is further divided into the Above Hayward Fault (AHF) aquifer and the Below Hayward Fault (BHF) aquifer. The Hayward Fault provides a low permeability boundary within the aquifer. Groundwater elevations have varied historically from 15.3 m (50 ft) above mean sea level (MSL) to 3.1 m (10 ft) above MSL.

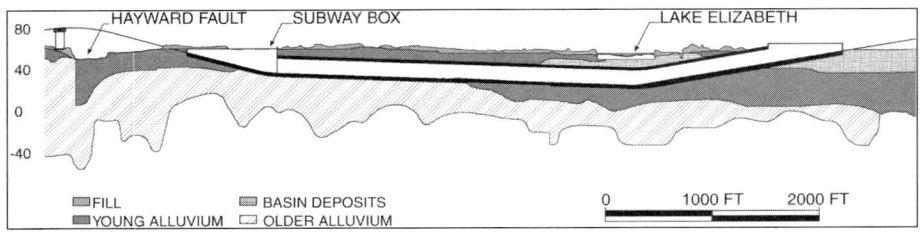

Figure 2. Inferred geologic profile with FCPS box

Geotechnical investigations conducted for this project consisted of mud rotary borings with soil sampling and field testing including Standard Penetration Test (SPT), Cone Penetration Tests (CPT), trenching, monitoring wells, and pump tests. A generalized geologic subsurface profile is presented in Figure 2. Geologic formations with soil types as inferred by the geotechnical investigations at the project site include:

1. Fill—Roadway or railroad embankments, varies from silty clay to track ballast,
2. Basin Deposits—Holocene localized marsh deposits of soft to firm clay, organic clay, and loose to medium dense sands and silts,
3. Young Alluvium—Holocene age alluvial fan deposits and Holocene alluvial deposits, primarily fine-grained, some sand lenses and
4. Older Alluvium—Late Pleistocene to early Holocene alluvial deposits, primary coarse-grained within the unconfined AHF aquifer.

GEOTECHNICAL DESIGN CHALLENGES

Geotechnical design challenges confronting the FCPS Project include the following:

- Selection of a temporary cutoff wall system to minimize dewatering in an existing aquifer,
- Design of the bottom plug to counteract uplift hydrostatic pressures during construction of the subway box, and
- Selection of the most appropriate groundwater table and shear resistance mobilized between CDSM cut off wall and soil/grout plugs.

Temporary Cutoff Wall for Subway Box Excavation Below Grade

Due to the relatively shallow depth below existing ground surface and short length of the subway box, 1.9-km (1.2-mile), cut-and-cover techniques were selected as the most cost effective construction method through Fremont Central Park. In order to limit the surface disruption to Fremont Central Park and the public, cutoff walls were considered over a slope back excavation.

As shown in Figure 2, the subway box will encounter four different soil formations, Fill, Basin Deposits, Young Alluvium, and Older Alluvium and there is significant variation in the composition, consistency, and engineering properties of the subsurface soils encountered. The soils predominately consist of sands, silts, and clays. About 40% of the subway box in the north penetrates into the Older Alluvium which consists of dense to very dense sands and gravels and is very permeable. As the subway box progresses south and under Lake Elizabeth, the very soft to medium stiff clays in

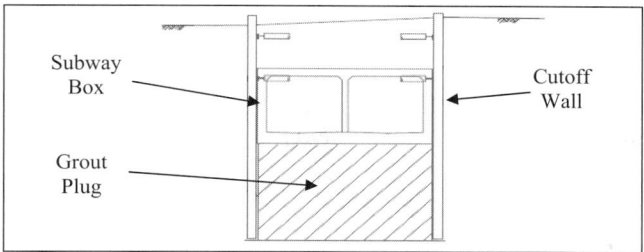

Figure 3. Cutoff walls and grout plug for construction of the subway box

the Basin Deposits are encountered. These soils are less permeable but have some discontinuous silty to clayey sand lenses which are more permeable.

The bottom of the subway box at its deepest section just south of Lake Elizabeth is about 11.6 m (38 ft) below the existing ground surface. To minimize dewatering of the AHF aquifer during excavation for the subway box, watertight cutoff walls and a bottom plug at the base of the subway box excavation will be required to limit groundwater inflow and resist large hydrostatic uplift forces.

The types of watertight cutoff walls considered for construction of the subway box included cement deep soil mixing (CDSM) walls, diaphragm walls (slurry walls), secant pile walls, and ground freezing. The pros and cons of these four options are shown in Table 1 below. Based on evaluation of the different methods and their applicability to the subway construction, it was concluded CDSM walls is the most practical option to collectively cut off the water inflow into the subway box excavation. The impact of the cutoff walls to the water supply aquifer is considered minimal due to the shallow depth of the walls—±18 m (60 feet). A grout plug was selected to provide a bottom seal and can be constructed using either jet grouting or a combination of jet grouting and CDSM methods. Figure 3 shows a typical cross section of the temporary excavation support.

Cement Deep Soil Mixing (CDSM) Method. CDSM is a method of deep in situ soil improvement in which cement is injected and mixed into the soil using an auger. CDSM involves the mixing of cement slurry with in situ soil to construct a continuous and practically waterproof wall made of a line of individual columns overlapping with each other. The overlapping columns reduce the event of leakage through the CDSM wall. Every other column is structurally reinforced with vertical steel H-piles (soldier piles) that are inserted into the soil-cement mixture while the mix is still viscous (i.e., before it sets and hardens). Figure 4 shows the sequence of installation for a 3-auger rig. During excavation, the walls will require lateral bracing members such as struts, composed of steel H-beams or steel pipes that span across the width of the excavation, or tieback anchors placed in drilled holes through the walls and into the earth behind the walls and grouted to provide an anchor from outside the walls.

Construction of CDSM walls does not require full soil replacement, thus reducing both "binder" and spoil volumes. The spoil can be handled as a solid waste and reused for landscaping after drying of the excess water. This type of wall is applicable for groundwater cutoff in the soil conditions encountered at the WSX project site where the walls are expected to be as deep as 18 m (60 feet) below ground surface.

CDSM walls are considered to be the best technical and most practical solution and will provide the required goals of water cut-off and structural support for the WSX subway box construction. The CDSM walls will be reinforced with wide flange soldier piles, and will require one to two levels of structural bracing. The CDSM technology will provide good wall water-tightness and can be used not only for the temporary wall

Table 1. Pros and cons of various cutoff wall alternatives

Wall Construction Option	Relevant Characteristics	
	Pros	Cons
Diaphragm Walls	Common technology	High mobilization cost
	Local experience in Bay Area	Possible different equipment for walls and base plug
	Vertical alignment easier to control, depending upon equipment used	Risk of side wall collapse during construction
	Can be used as a single structural wall	Risk of bentonite slurry loss in high permeability ground
	Fewer joints between structural elements	Requires large work area for equipment
	Well suited to planned depths	Underground utilities conflicts problematic
	Boulder obstructions easier to excavate	Spoils contaminated with bentonite slurry
	Provides a smooth wall for waterproofing installation	
Secant Piles	Common technology	Moderate mobilization cost
	Local experience in Bay Area	Frequent joints
	Well suited to planned depths	Vertical alignment more difficult
		Different equipment for walls and base plug
		Risk of slurry loss in high permeability ground
		Vertical misalignment leads to open "windows" between adjacent columns
		Underground utilities conflicts problematic
		Produces a rough irregular surface for waterproofing installation
CDSM Walls	Same equipment for both walls and base plug, if CDSM used for plug	High mobilization cost
	Local experience in Bay Area	Relatively new technology
	Good overlap between adjacent column elements	Risk of heterogeneous mix
	Generally considered to have low unit construction cost	Requires large work area for equipment
	Well suited to planned depths	Underground utilities conflicts problematic
	Full soil replacement not required	Boulder obstructions potentially problematic
	In-situ quality verifiable	
	Provides a relatively smooth wall for waterproofing installation	
Ground Freezing	Could use the same equipment as for a frozen soil plug.	High mobilization costs
	Applicable to a variety of soil types.	Linear facility requires multiple freeze plants
	Can maintain a continuous freeze as long as desired	Cost to maintain frozen ground for long periods of time
	Does not require the injection of any chemicals or cements into the ground with a potential to contaminate the ground water as a drinking water source.	Exposed frozen wall during construction may require insulation to prevent thawing and degradation during the process of excavation and box construction. Requires substantial cost of energy to maintain frozen ground.
		Limited previous applications in Bay Area
		Requires instrumentation temperature monitoring for ice wall closure

CHALLENGES OF THE FREMONT CENTRAL PARK SUBWAY

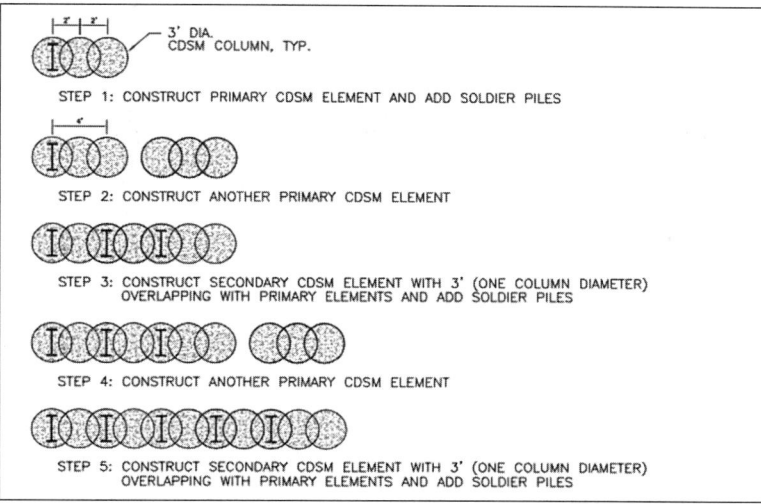

Figure 4. Installation sequence for a 3-auger CDSM wall

support, but also for the bottom grout plug. In addition, this method of soil improvement has been successfully used throughout the Bay Area.

Grout Plug to Counteract Uplift Hydrostatic Pressures

Due to a high differential hydrostatic head between the bottom excavation for the subway box and the design groundwater table, upward seepage through the excavation bottom will occur and substantial uplift pressures will be developed at the base of the excavation upon removal of overburden soils. These uplift pressures must be resisted by the counter weight of a clayey soil (sometimes called soil plug) located below the box. If the weight of the soil plug is not heavy enough to counteract the uplift forces, the soil will lose strength and eventually become "quick" and begin to boil.

The factor of safety (FS) against bottom heave of the soil plug can be computed by the following:

$$FS = \frac{\gamma_t T_{soil} A_b + \tau A_p}{\gamma_w H_w A_b} \tag{1}$$

where
  τ = side friction for soil plug,
  γ_t = unit weight of the soil plug,
  γ_w = unit weight of water,
  T_{soil} = thickness of the soil plug,
  A_b = soil plug base area,
  A_p = soil plug perimeter area, and
  H_w = differential hydrostatic head.

Figure 5 shows a sketch of the soil plug at Station 2286+50 showing how the FS against the bottom instability is determined. At this location, the differential hydrostatic head was assumed to be T + 6.7 m (22 ft) and equal to 12.8 m (42 ft) with T = 6.1 m (20 feet), whereas the side friction for soil plug was assumed to be 480 psf, and A_b and

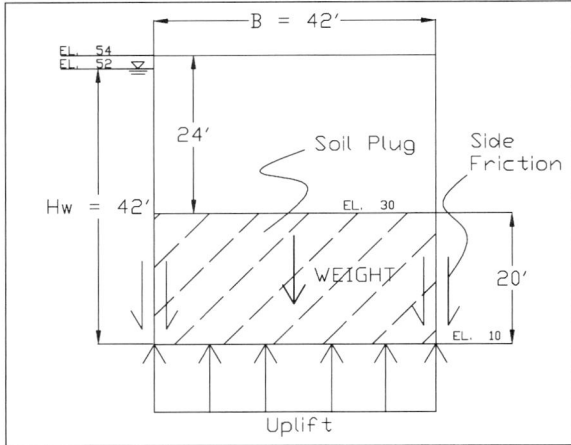

Figure 5. Soil plug at Station 2286+50

A_p were respectively assumed to be 12.8 m (42 ft) and 12.1 m (40 ft). A minimum FS of 1.5 was required bottom stability with a soil plug.

In the case where the soil plug does not provide adequate stability, a grout plug is needed. A typical section for the temporary excavation support with a grout plug is shown in Figure 6. Stability analyses of the excavation bottom were conducted to determine base rupture (blow-ins) caused by water pressure acting on the bottom plug. The side friction of the grout plug was assumed to be mobilized at the interface between the cutoff wall and the plug. A FS of 1.5 was also selected against bottom heave using a grout plug to account for soil/grout variability (in terms of consistency and shear resistance) within the Young Alluvium. A lower factor of safety of 1.2 may also be considered if the perimeter shear is ignored.

Using equation (2) below, the calculated grout plug thickness ranged from 3.1 to 7.3 m (10 to 24 ft) along the proposed WSX subway alignment.

$$FS = \frac{\gamma_{gp} T_{gp} A_b + 2 T_{gp} \tau_{gp}}{\gamma_w (H_w + T_{gp}) A_b} \qquad (2)$$

τ_{gp} = mobilized shear for the grout plug,
γ_{gp} = unit weight of the grout plug,
T_{gp} = thickness of the grout plug,
A_b = grout plug base area,
A_p = grout plug perimeter area, and
H_w = differential hydrostatic head.

Analyses indicated two design parameters, groundwater elevation and shear resistance mobilized between the soil/grout plug and the CDSM walls, governed the design of the soil plug to resist the uplift pressure. The design groundwater elevations mentioned above were based on the groundwater measured from the monitoring readings. For design purposes, design groundwater tables of El. 15.2 m (50 ft) in the northern subway section and El. 15.9 m (52 ft) in the southern subway section were used for conditions during construction. These groundwater tables account for groundwater levels during wet seasons and considered reasonable for design. However, the design shear resistance mobilized between the soil/grout plugs and the CDSM walls

CHALLENGES OF THE FREMONT CENTRAL PARK SUBWAY

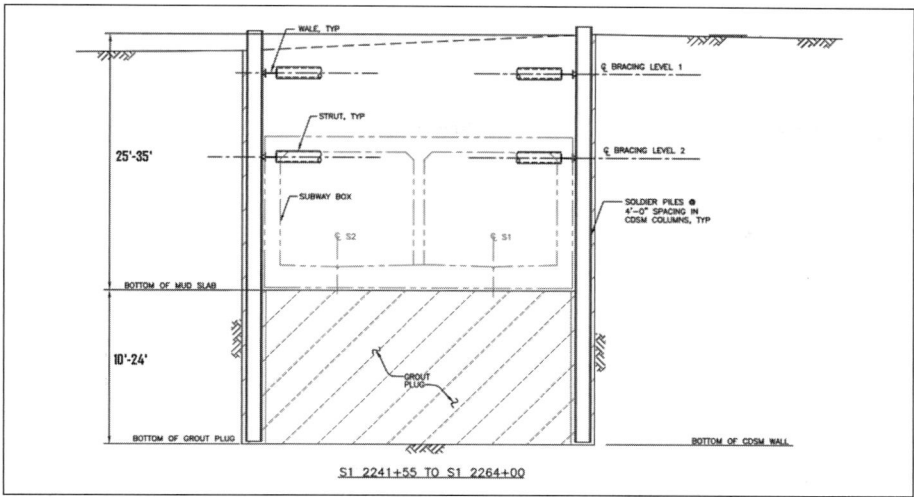

Figure 6. Typical subway box section with grout plug

was typically assumed to be about 20% to 33% of the undrained shear strength of the soil/grout plugs which may vary from locations to locations. To ensure that this shear resistance is mobilized, the contract documents specify that the minimum average undrained shear strength of the grout plug should be at least 690 kPa (100 psi) at 28 days, and that the grout plug must be in direct contact with the in-place CDSM walls. Because of the uncertainty in the design mobilized shear resistance, a factor of safety of 1.5 was assumed in the analysis.

After excavation to the planned invert, a minimum 10-cm (4-in) mud slab will be poured over the foundation subgrade to provide a firm working surface. This slab will also help prevent loosening and boiling of the subgrade. For long term stability, the weight of the backfill soil and subway box is adequate to resist uplift hydrostatic forces from the 500-year flood event without including the weight of the grout plug. Therefore, the grout plug is only needed during construction as the completed structure and backfill have sufficient weight to resist buoyancy.

STRUCTURAL DESIGN CHALLENGES

The WSX FCPS is a reinforced concrete structure designed to resist extremely high seismically induced forces due to earthquake ground motions. In addition, the structure is designed to resist buoyancy or flotation under a 500-year flood event.

The FCPS was structurally designed in accordance with BART Facility Standards (BFS) release 1.2, California Building Codes, ACI codes, AISCI codes, and AREMA requirements. Design loads for the subway structure include dead load, live load (HS20-44 wheel loads, Cooper E80 train loading), lateral loads, buoyancy forces and earthquake loads. The vertical earthquake force was taken as ⅔ of the horizontal peak ground acceleration (PGA = 0.75 g) multiplied by the dead load. The transverse earthquake force was determined from the racking analysis.

Site Specific Design Response Spectra

For the design of above ground structures, the design response spectra as shown in Figure 7 were used to generate the seismic inertial forces:

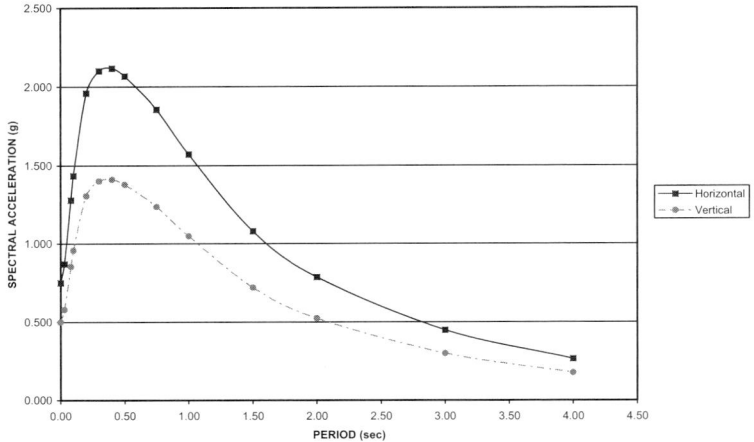

Figure 7. Design response spectra

Load Combinations

Structural components were designed to have adequate strength to resist the most critical effects resulting from various load combinations. For the cut-and-cover subway box, the following load combinations were used:

$U_1 = 1.4D + 1.7L + 1.7H + 1.7B$

$U_2 = 1.0D_{500} + 1.0L + 1.0H_{500} + 1.0B_{500}$

$U_3 = 1.0[D + L + H + B + (EQT + 0.4EQV) / R]$

$U_4 = 1.0[D + L + H + B + (0.4EQT + EQV) / R]$

where D = dead load, L = live load, H = lateral earth pressure and ground water pressure, B = buoyancy force, EQT = elastic transverse earthquake force, EQV = elastic vertical earthquake force, R = response modification coefficient.

Load combination U_1 was taken directly from BFS 1.2, facility design criteria. Load factors in load combination U_1 are higher than current industry standards. Load combinations U_3 and U_4 were also based on BFS 1.2.

Racking Analyses

Free field ground displacements due to earthquake ground motions were obtained from the analyses using SHAKE computer program. Results from SHAKE determined the relative free field ground displacements of the subway box that varies from 0.19 to 1.20 inches.

Transverse displacements (side-sway) of the structure, due to differential ground deformation at each level of its depth during seismic ground motion, were determined using soil-structure interaction analysis. For the FCPS project, the soil profile and its engineering properties vary significantly from station to station with soft soils such as the Basin Deposits and Young Alluvium to stiff to very stiff and dense to very dense Older Alluvium. Four stations were selected for racking analyses as follows: 2242+00, 2252+00, 2274+00, and 2282+00.

For the analysis of subway box at each location, three computer models were created using SAP2000 finite element computer program.

Model 1: This model represents the in-situ condition without the presence of the subway box. This model is used to determine the scaling factor of the ground displacement at the location of the subway box.
Model 2: This model is based on Model 1 but with the presence of the subway box. This model is used to determine the racking displacement of the subway box.
Model 3: This is a stick model represents the reinforced concrete subway box. The racking displacement obtained from model 2 was applied to this model in order to determine the member forces.

The ground displacement at each selected location was then used and applied to Model 1 as the boundary loading to obtain the free-field deflection which was in turn used to determine the scaling factor to account for the loss of the boundary displacement when transmitting through the soil (the boundary displacement should be maintained through out the entire model because the ground displacement should be the same everywhere. However, this condition cannot be maintained due to the drawback of computer modeling with highly compressible materials). It was also applied to Model 2 as a boundary loading to obtain the deflections in the subway box with soil-structure interaction. The differential deflection between the top and bottom the subway box was then calculated and modified by the scaling factor obtained from Model 1 before applying to Model 3 to determine the member forces.

Three-Dimensional Finite Element Model

When generating computer Model 1, the following material properties and assumptions or bases were incorporated into the model.

Material properties. Soil properties include unit weight, Poisson's ratio and modulus of elasticity

Assumptions. In the computer model, the soil was modeled by using shell elements; free field deflections were applied at the vertical boundaries; relative deflections were taken at the location of the structure; soil properties were considered by using soil layers that varied by depth; and each soil layer was considered uniform in thickness with the same properties.

Model 2 was based on Model 1 with the inclusion of the subway box in the soils. The relative deflection between the top and bottom of the box structure generated from this analysis was applied to model 3 after adjustment with the scaling factor.

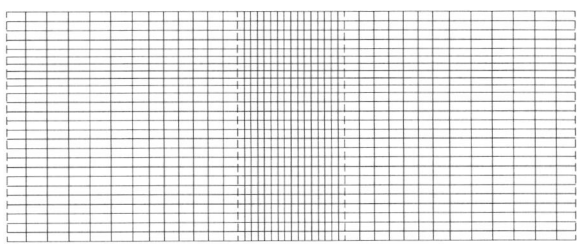

Figure 8. Mathematical Model 1

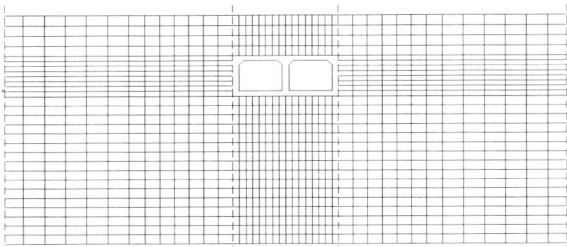

Figure 9. Mathematical Model 2

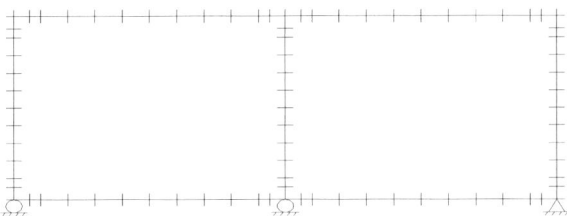

Figure 10. Mathematical Model 3

In addition to the assumptions or bases used in Model 1, the following bases were also used in developing this model:
- The top and bottom slabs and exterior walls were modeled using shell elements.
- The center wall was modeled using frame elements.
- For the walls, 50% of the gross moment of inertia ($0.5I_g$) was used and for slabs, the gross moment of inertia (I_g) was used.

Model 3 was used to determine the member forces in the subway box due to the relative displacements from racking analysis.

The following bases were used in the development of this model:
- Structure was modeled using frame elements representing the centerline of the members with unit length.
- The support boundary conditions were pinned at one exterior wall with rollers at the remaining two walls.
- For walls, 50% of the gross moment of inertia ($0.5I_g$) was used and for slabs, the gross moment of inertia (I_g) was used.

Results of Racking Analyses

The results from racking analysis at the four selected locations are shown in the Table 2. Due to the difference in soil conditions and its properties, the displacements obtained vary from station to station. The free-field displacements at Station 2242+00 and Station 2252+00 are a little bit smaller than those of the structures. This phenomenon can be explained by the fact that the ground surrounding the structure has a cavity in it (i.e., a perforated ground). A perforated ground, compared to the non-perforated ground in the free field, has a lower stiffness in resisting shear distortion and thus will distort more than the non-perforated ground. (Wang, 1993).

Table 2. Differential displacements

Station	$\Delta_{free\ field}$ cm (in)	$\Delta_{racking}$ cm (in)
2242+00	0.48 (0.19)	0.61 (0.24)
2252+00	1.47 (0.58)	1.63 (0.64)
2274+00	4.95 (1.95)	4.04 (1.59)
2282+00	3.05 (1.20)	1.57 (0.62)

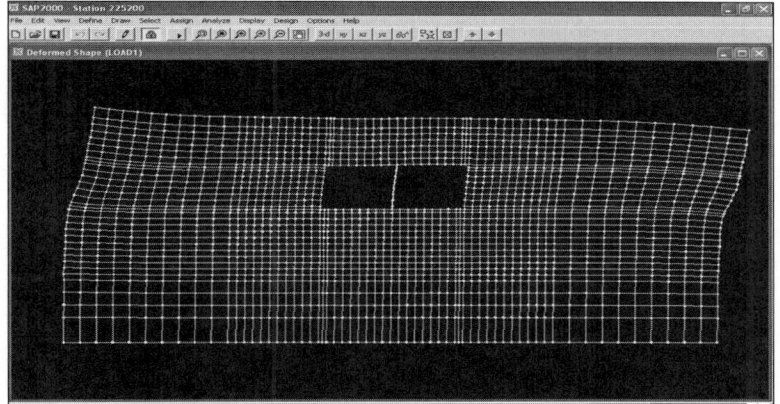

Figure 11. Typical deformation of soil-structure interaction

The free-field displacements (without soil-structure interaction), Δ_{free}, at Stations 2242+00 and 2252+00 are much smaller than that of Stations 2274+00 and 2284+00 because of stiffer/denser soils (Older Alluvium). On the other hand, the free-field displacement at Station 2274+00 is much greater due to its location underneath the Lake Elizabeth, which is underlain by soft Basin Deposits. At Station 2284+00, the free-field displacement is also quite large because it has shallow overburden in soft soils (Basin Deposits, and Young Alluvium). Therefore, structural design using free-field displacements without taking into consideration of soil-structure interaction effect may over- or under-estimate the demands, thereby unduly over- and under-design of the subway structure, respectively.

A typical deformation of Model 2 at Station 2252+00 from the racking analysis is shown in Figure 11. The differential displacements obtained from racking analyses were then used and applied to Model 3 to determine the member forces. A typical deformation of Model 3 at Station 2252+00 of the subway box subjected to racking displacement is shown in Figure 12. The member moments of the subway box at Station 2252+00 corresponding to the locations shown in Figure 13 are presented in Table 3.

Longitudinal Seismic Analysis

During an earthquake, the ground will be deformed and any structure built on or in the ground will be forced into a distorted shape, takes the form of a so-called snake motion. The forces induced in the structure during seismic ground motions depend on several parameters, e.g., soil properties, ground accelerations; shear wave velocity, cross sectional properties of the structure, elevation of the structure below the ground surface, and the direction of wave propagation relative to the structure.

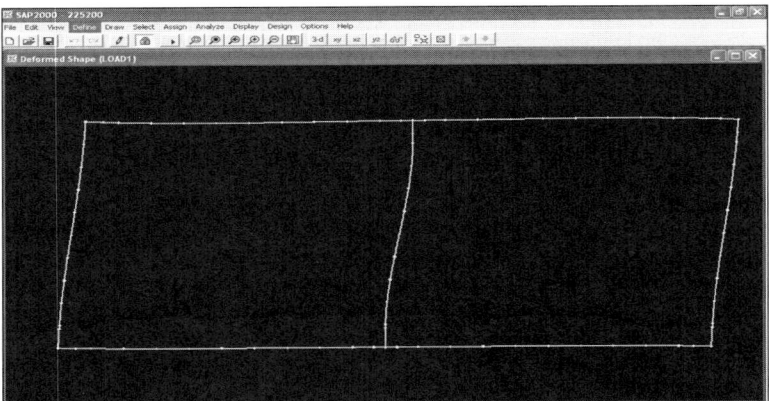

Figure 12. Typical deformation of subway box subjected to racking displacement

Table 3. Racking moments in subway box at Station 2252+00

Location	Racking Moment (ft-kips)
M1	119.6
M2	−47.4
M3	−47.2
M4	−119.4
M5	−119.6
M6	94.6
M7	119.4
M8	134.5
M9	93.2
M10	134.2
M11	−134.5
M12	−46.7
M13	−46.5
M14	−134.2

The analytical procedure used for estimating forces experienced by the subway box that conforms to the ground motions during seismic excitation is based on the theory of wave propagation in an infinite, homogeneous, isotropic, elastic medium, together with the theory of a beam on elastic foundation. Hence, from the imposed ground displacements, the induced bending moment, shear force, lateral load, and axial force in the subway box during earthquake ground motions can be approximately determined by the following equations:

Moment = $E \: I \: d^2u / dx^2$

Shear = $E \: I \: d^3u / dx^3$

Load = $E \: I \: d^4u / dx^4$

Axial Force = $E \: A_c \: d^2\sigma / dx^2$

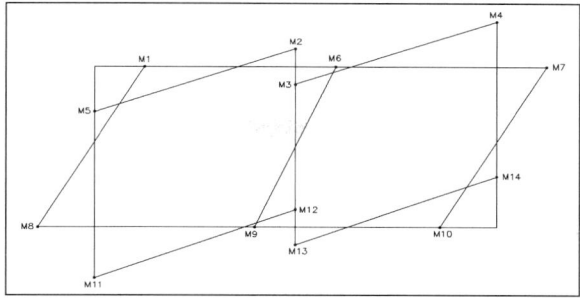

Figure 13. Single line diagram of subway box (Model 3)

where E = modulus of elasticity of subway box material, I = moment of inertia of subway box cross section, u = maximum amplitude displacement at depth z (below ground surface), A_c = subway box cross sectional area, and σ = actual axial deformation.

By solving the differential equations shown above and make provision for the direction of wave propagation, the maximum bending moment, maximum shear force, the equivalent load density necessary to cause the curvature and the maximum axial force in the subway box, can be determined.

The forces obtained in the subway box from the approach described above are the elastic un-reduced forces. In order to account for the inelastic behavior of the subway box, the seismically induced elastic forces should be reduced for design by a response modification coefficient (R).

Seismic induced bending moment and forces calculated from equations shown above are based on no soil-structure interaction and may be used as a "back-envelope" solution. In reality, the seismic induced bending moment and forces are smaller as the subway structure is much stiffer than the surrounding medium. This is because the surrounding medium, will distort less than or slightly more than the free field deformations, and there will be soil-structure interaction. This interaction can be taken into account if it is assumed that the structure behaves as an elastic beam supported on an elastic foundation and the soil provides a support that can be idealized as a series of linear elastic springs.

As shown in Table 4, results of the longitudinal seismic analyses show that the Demand/Capacity (D/C) ratio is much less than 1.0, indicating that the subway box will behave close to the elastic limit with very little demand into the inelastic behavior under a strong ground motion. Hence, the structure will be structurally intact during the design earthquake motion of 0.75g with very little, minor cracks, if any. No major damage is expected for the subway structure during the design earthquake.

CONCLUSIONS

- Two major geologic features at the Fremont Central Park subway project site have a significant impact on the geotechnical design of the project, namely the Hayward Fault and the AHF aquifer. The Hayward Fault crosses the alignment at two locations and the underground subway box structure penetrates and sits in the main water bearing formation of the AHF aquifer.

- The subway box will be constructed using cut-and-cover technique. A water tight temporary excavation support system such as CDSM walls will have to be used to minimize the impact to the AHF aquifer and limit dewatering during

Table 4. D/C ratio for vertical bending due to horizontal shear wave

Subway Section	Station	Elastic Forces		Design Forces		Capacity	Max. D/C Ratio
		Moment ft-kips	Axial Force kips	Moment ft-kips	Axial Force kips	Moment ft-kips	
1	2257+60	53,332	4,060	35,554	2,706	n/c	n/c
2	2257+60 to 2277+50	60,711	5,125	40,474	3,417	91,868	0.44
3	2282+10	47,204	3,076	31,469	2,050	n/c	n/c
4	2282+10 and beyond	110,695	4,239	73,797	2,826	95,626	0.77

Force Reduction Factor, R = 1.5
n/c = not calculated. The ratio will be lot less than the other two box sections by observation.

construction of the subway box. A grout plug will be used to seal off the bottom of the temporary excavation and balance high hydrostatic pressures.

- Instability of excavation bottom due to differential hydrostatic uplift should always be evaluated. This is the fundamental mode of instability for mixed granular and cohesive plugs with discontinuities. If the factor of safety is less than 1.0, a risk of a bottom blow out exists, which should be rectified by constructing a grout plug to resist the uplift.
- In the analysis of the bottom excavation stability subject to uplift, the design ground water table and the shear resistance mobilized between the cutoff walls and the soil/grout plugs are the key parameters governing the design. A higher factor of safety may deem to be necessary if these two parameters cannot be determined with confidence in the design phase.
- Racking analyses of underground structures such as subways, tunnels, and underground stations are essential for proper design of the structures which will have sufficient strength to resist the deformation associated with the ground displacements during seismic ground motions.
- The longitudinal seismic analyses have indicated the D/C ratio for the subway structure is much less than 1.0 at a low response modification coefficient of 1.5. This suggests that the structure will behave close to the elastic limit with very little demand into the inelastic behavior under a strong earthquake ground motion.

ACKNOWLEDGMENTS

The authors are thankful to BART and PB Americas, Inc for their support of publishing this paper. The conclusions of this paper reflect the views and opinions of the authors, and do not necessarily reflect official views and opinions of BART and PB Americas, Inc.

REFERENCE

Wang, Joe (1993). Seismic Design of Tunnels—A Simple State of the Art Design Approach, 1991 William Barclay Parsons Fellowship, Parsons Brinckerhoff, Monograph 7, June.

ATLANTA NORTH-SOUTH TUNNEL

Samer Sadek ▪ HNTB Corporation

Hugh Caspe ▪ HNTB Corporation

Tim Heilmeier ▪ HNTB Corporation

Darryl D. VanMeter ▪ Georgia Department of Transportation

John D. Hancock ▪ Georgia Department of Transportation

ABSTRACT

The Georgia Department of Transportation (GDOT) undertook an exploratory feasibility study to one of the most ambitious tunneling projects in the United States. HNTB Corporation was retained to perform the study, deemed the Atlanta North-South Tunnel, which is approximately 7 miles long and consisting of four lanes (or six lanes) starting at the southern terminus of SR 400 extending to the south connecting with the northern terminus of I-675, providing substantial relief to the Downtown Connector (I-75/I-85), the most congested portion of Atlanta's freeway system. This paper summarizes the analysis performed during the first phase of this challenging assignment which, if constructed, would be one of the longest roadway tunnels ever built in the United States.

INTRODUCTION

The proposed Atlanta North-South Tunnel project would connect the SR 400/I-85 interchange to the north with the I-675/I-285 interchange to the south (see Figure 1). The total length of the project would be approximately 12 miles, of which about 7 miles would be constructed as a tunnel. The primary purpose of the project is to reduce congestion on the Downtown Connector, one of the most congested portions of the freeway system in and around Atlanta. The construction project is currently not programmed into the Georgia Transportation Improvement Plan (TIP) and has no dedicated source of funding.

This paper summarizes the findings for Phase 1 of the feasibility study for the proposed Atlanta North-South Tunnel project and explains the methodologies and processes adopted for the study. The purpose of the study is to evaluate the project constructability, travel demand, revenue potential, costs and overall feasibility to see if the project has viability as a stand-alone, toll-financed project. The study was performed by HNTB Corporation as a Task Order under a contract with the Georgia Department of Transportation (GDOT). This Task Order is unique in that a "Hold Point" has been built into the contract, dividing the contract into two phases. The Hold Point denotes the completion of Phase 1, which intends to identify the transportation system benefits, the general revenue potential of the facility and to devise a cost estimate for the facility in order to assess project feasibility potential. When the Department reviews the findings and makes the determination to proceed to Phase 2, further refinement and analysis will be performed, in line with GDOT Executive Management directives and additional stakeholder input.

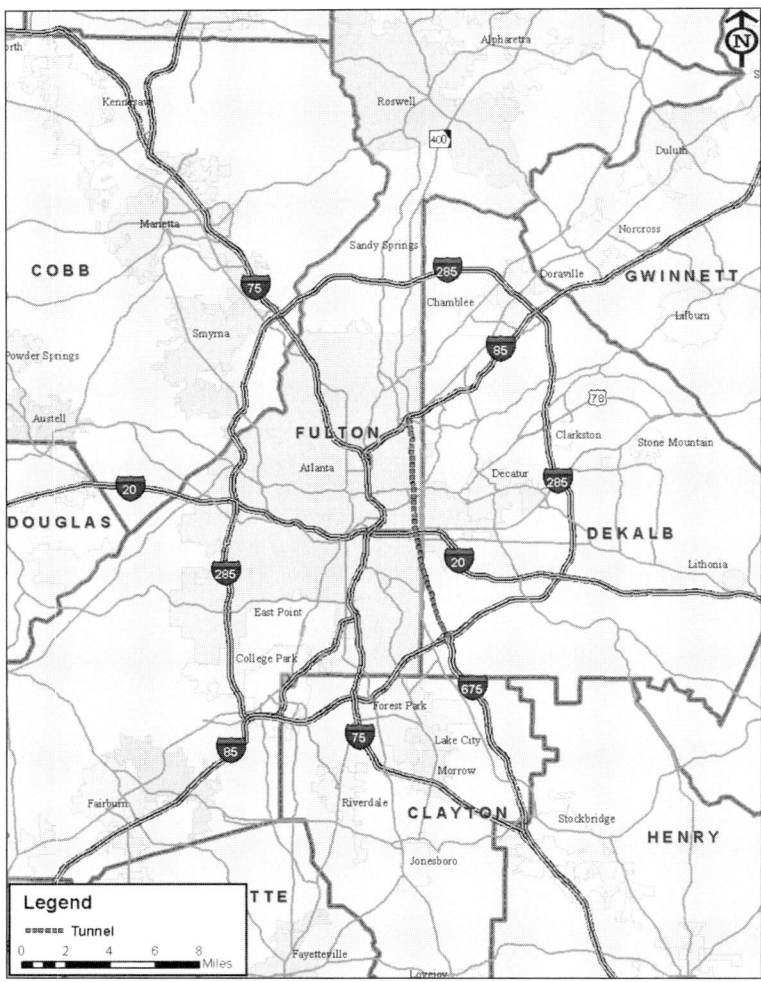

Figure 1. Project location

PROJECT BACKGROUND

The idea of providing a direct connection from SR 400/I-85 to I-675/I-285 to relieve the congestion on the Downtown Connector was first proposed as a surface toll road in 1970 by Wilbur Smith Associates in the original Atlanta Urban Area Tollways study. This surface roadway was never constructed due to environmental constraints. In November 2006, the Reason Foundation published the Galvin Mobility project paper: *Reducing Congestion in Atlanta: A Bold New Approach to Increasing Mobility.* In this report, author Robert Poole proposed four major projects that would offer the needed amount of new freeway capacity to reduce congestion, and one of the extremely ambitious and costly proposals included constructing the link between SR 400/I-85 and I-675/I-285 as a double-decked tunnel. Mr. Poole stated in his paper that constructing this critical transportation link underground as a tunnel is the best way to provide the needed capacity while minimizing environmental impacts and protecting Atlanta's neighborhoods. This tunnel would provide a north-south alternative running the full length of the Downtown Connector, the most congested portion of the freeway system, with the possibility of an intermediate access to Downtown Atlanta at I-20.

Under a pressing need for mobility options in Atlanta, the Georgia Department of Transportation Office of Innovative Program Delivery was charged with performing the exploratory feasibility studies on the Atlanta North-South Tunnel. In November, 2007, GDOT executed a Task Order and charged HNTB Corporation to evaluate the constructability and feasibility of the Atlanta North-South Tunnel.

SCOPE OF WORK—PHASE I STUDY

The Task Order has been structured to include two different phases with the intention that GDOT could objectively analyze the tunnel concept with current state of the practice knowledge and study findings at the conclusion of Phase 1 and be in a position to either authorize additional studies or terminate activities if the tunnel appeared infeasible. The primary emphasis of the Tunnel Feasibility study is to identify project feasibility, including fatal flaws that might exist (ability to be financed, environmental, unsuitable geology, no travel demand, constructability, etc.), geometry constraints and feasible horizontal alignment, profile and cross sections of the facility, locations of the termini and/or the location of intermediate access and demand for the facility if implemented as a toll tunnel. Phase 1 is expected to conclude when the conceptual project finances could be established by evaluating the potential revenue vs. the feasibility level cost estimate.

The framework of Phase 1 is to establish strategic goals and objectives, as well as to perform high level examination of alternatives.

With the aim of providing a meaningful feasibility level cost estimate, several different tunnel alternates were identified, bringing focus to the study. The different alternates are listed below:

- Alternate 1A—Twin tunnels pipeline (no intermediate access) with two 12-foot lanes in each direction (see Figure 2)
- Alternate 1B—Twin tunnels with access/interchange at I-20 with two 12-foot lanes in each direction
- Alternate 2A—Twin tunnels pipeline with three 12-foot lanes in each direction
- Alternate 2B—Twin tunnels with access/interchange at I-20 with three 12-foot lanes in each direction
- Alternate 3A—Single-bore stacked (over/under) tunnel pipeline with two 12-foot lanes in each direction (see Figure 3)
- Alternate 3B—Single-bore stacked tunnel with access/interchange at I-20 with two 12-foot lane in each direction
- Alternate 4A—Single-bore stacked tunnel pipeline with three 12-foot lanes in each direction
- Alternate 4B—Single-bore stacked tunnel with access/interchange at I-20 with three 12-foot lanes in each direction

Amongst the aforementioned alternates, Alternate 1A was established as the base case. The Atlanta North-South Tunnel Feasibility Study adopted a modularized approach. The costs of alignments and access points were compared with their respective traffic demand and revenue potential. The specific project features were modularized and weighed in a cost/benefit manner in order to provide the most cost effective and financially feasible solution.

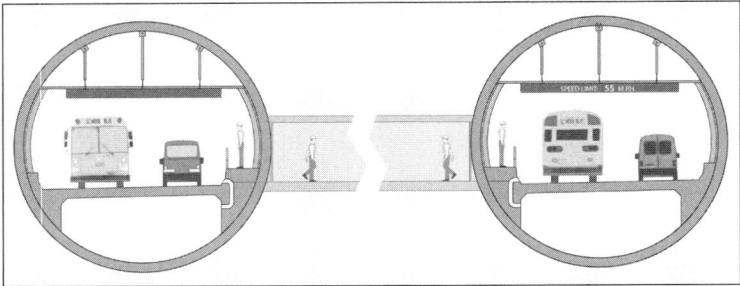

Figure 2. Twin tunnel option

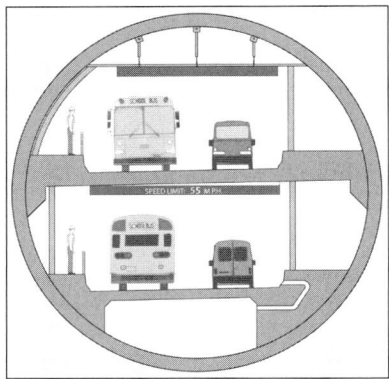

Figure 3. Stacked tunnel option

ALIGNMENT CONSIDERATION

The project team established preliminary design criteria for the tunnel to establish the tunnel geometry. The design criteria assumed basic requirements such speed limit, shoulder width, lane width, clearances, and based on those basic geometrical parameters, the tunnel section was established for each option, assuming a circular tunnel. The circular tunnel section was established as a base assumption for phase 1 study and this is primarily driven by the extensive length of the tunnel (7 miles). Conceptual horizontal and vertical alignment was later established to provide adequate rock cover for the main tunnel, as such, the TBM excavation for the main tunnel was assumed to be fully in rock.

GEOTECHNICAL ASSESSMENT BASIS

The geotechnical study conducted to date has consisted solely of a review of existing information, no new alignment specific field investigations were conducted to verify the conditions presented in the reviewed materials.

These conditions include top of rock elevation, type and condition of rock, groundwater, and rock properties. Most of the borings in the reviewed investigations were terminated at auger refusal, typically defined as the point at which the cutting head can no longer penetrate the materials, or when penetration reaches a defined low rate. As such, these depths may not represent the actual top of rock elevations.

The majority of the geotechnical information available for review from Bridge Foundation Investigations (BFI) and commercial/residential developments were from projects located at and around the SR 400/I-85 Interchange. A small number of other projects were reviewed which were located elsewhere along or adjacent to the proposed alignment.

Adjacent waste management tunnel projects which were reviewed included the Chattahoochee Tunnel, Nancy Creek Tunnel, West Side CSO projects, Intrenchment Creek Tunnel, Custer Avenue CSO Tunnel and Three River Tunnel. These tunnels are of considerable distance from the proposed alignment, ½ mile or more. Both the Intrenchment Creek and Three River tunnels were constructed in the early 1980s and as such limited information was available for review.

The proposed alignment, specifically at the potential intermediate access at I-20 and the southern portal location, will require additional investigations to verify and define the conditions present. Based on the materials reviewed to date, the closest geotechnical boring information to the southern portal is located about ½ mile to the south of the portal.

GEOTECHNICAL CONDITIONS FOR THE RUNNING TUNNEL

The following is a general geologic description of the project area.

Area Geology

The proposed Atlanta North-South Tunnel is located within the Piedmont region in the northeastern portion of Georgia. The Piedmont region in this area generally consists of metamorphic rocks with isolated areas of granitic intrusions. The rocks are Ordovician aged, greater than 200 million years, and have undergone significant periods of structural deformation resulting in faulting, folding and fracturing.

The Brevard Zone is the major structural fault in the Atlanta area, and is located to the north of the proposed Atlanta North-South Tunnel alignment. The Brevard Zone is a series of shear faults, which can be delineated from Alabama to North Carolina. The latest movement of the Brevard occurred at least 325 million years ago. Numerous other folding events and igneous intrusions have occurred subsequent to the formation of the Brevard Fault. Geologic mapping indicates that the Brevard Fault and the rock formations within the project area generally strike in a northeast to easterly direction and dip towards the east, southeast and south. Additionally, an unnamed thrust fault is shown in geologic mapping and could possibly cross the extreme northern portion of the proposed Tunnel alignment.

Presently no active faults are located within the project area. However, based on historical events, the Atlanta region has been affected by seismic events that took place elsewhere within the Southeastern United States. Specifically, the New Madrid earthquake series (1811–1812) and the Charleston earthquake (1886) resulted in significant ground movements and property damage in portions of Georgia.

Another prevalent geologic feature common in the Atlanta region is lineaments. Lineaments are long valleys or draws which are related to underlying areas of possibly highly jointed or fractured rock. Lineaments can be associated with existing or prior waterway courses. The fractured zones are typically cemented at deeper levels and consist of weathered and fractured rock at shallower depths. These fractured areas can also be water bearing. Some masking of the surface features indicative of lineaments can occur due to infilling with alluvial soils in the vicinity of larger waterways.

Bedrock

The geologic information reviewed to date indicates that the rock formations along the proposed alignment, extending from the north to the south, consist of metamorphic rock with the possibility of isolated igneous intrusions. The mapping indicates that the Tunnel alignment will run roughly parallel to the prevailing dip direction of the rock—south–southeast; and at an angle to the prevailing strike—northeast. Along the alignment, five major rock formations are likely to be noted. The rock formations along the alignment are:

- Clairmont Formation (OZcm)—Light gray to bluish-gray granitic feldspar gneiss
- Wahoo Creek Formation (OZw)—The most common lithology is biotite gneiss. The formation also contains feldspar gneiss, amphibolite, hornblende and quartzite.
- Stonewall Gneiss (OZs)—Gray to grayish-brown, schistose feldspar gneiss
- Clarkston Formation (OZcl)—Generally quartz schist with lesser amounts of amphibolite and gneiss
- Informal Mixed Unit (OZm)—A mix of muscovite schist, plagioclase, gneiss and amphibolite

A portion of the geologic map and cross section with the approximate tunnel alignment indicated is shown in the following Figure 4.

In general, the main tunnel alignment is expected to encounter:

1. From the northern portal extending about 3.5 miles to the south, the Tunnel passes through rock from the Clairmont, Wahoo Creek and Stonewall Gneiss formations. These rocks have generally been described as a medium to fine grained gneiss which can vary from strongly to weakly foliated. The most common minerals in the gneiss are quartz and feldspar, with lesser concentrations of biotite, muscovite, chlorite, and amphibolite. Local areas of the rock can be classified as amphibolite or schist and typically exist as small lenses and layers within the rock mass.

2. From approximately 3.5 miles south of the northern portal extending to the proposed southern portal, the proposed Tunnel passes through rock from the Clarkston and Informal Mixed formations. Generally, the rocks in these formations have been described as gneiss, granitic gneiss, granite, biotite gneiss, amphibolite and mica schists. The transition from one rock type to another is generally gradual, but some dramatic transitions are possible in areas of heavy folding.

The rocks along the alignment generally dip to the southeast at angles of 9 to 20 degrees. The dip can vary by as much as 35 degrees in areas of heavy folding leading to areas in which the rock dips back towards the northwest for short stretches.

CONSTRUCTION STAGING/SCHEDULING

For the purpose of cost estimating, some major assumptions were made for construction staging:

- The Northern ramps at SR 400 interchange and I-20 ramps (if opted) will be constructed all as one independent construction package
- Twin Tunnels will be constructed as one construction package utilizing two TBMs
- Stacked tunnel alternates will be construction utilizing one TBM (Utilizing two TBMs will be further analyzed in phase 2 of the study)

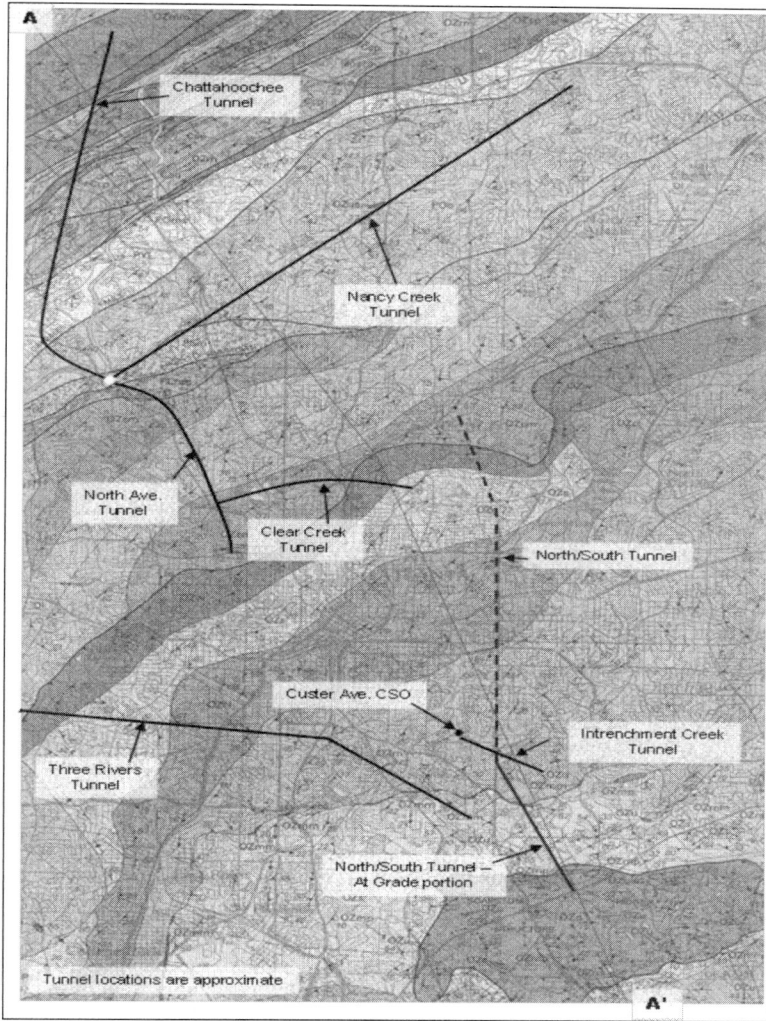

Reproduced from the "Geologic Map of the Atlanta 30' × 60' Quadrangle, Georgia," 2003.

Figure 4. Regional geology and approximate tunnel location / other constructed tunnels

- Construction of the ventilation shaft and connecting ducts to the tunnel will be part of the main tunnel contract
- Construction of surface structures related to roadway approaches for I-20 interchange will be in a separate construction package.
- South approaches and roadway work associated with Southern terminus of the project will be constructed as an independent construction package.
- Over excavation of the main tunnels at ramp intersection will be part of the main tunnel construction

Table 1. Required TBM diameter for all options included in the study

Option	Tunnel Diameter
Alternate 1A and 1B	40'-3"
Alternate 2A and 2B	53'-6"
Alternate 3A and 3B	54'-6"
Alternate 4A and 4B	64'-9"

RECOMMENDED TUNNELING METHODS FOR THE NORTH-SOUTH TUNNEL

The alignment used in this conceptual study indicates the mainline tunnels (excluding exit and entrance ramps) is approximately 7.1 miles. This length is considered a long tunnel, and TBM construction will be the selected method of construction for the main tunnel section. For the ramps, the construction is assumed to be combination of TBM, SEM and cut & cover. TBMs are assumed to be utilized in the long ramps that have curvatures that can be negotiated by TBMs. For portions of the ramps that are considered short or have a small radius of curvature, SEM methods are assumed to be the most cost effective method of construction. Transition sections between mined section and the surface roads were assumed as cut and cover. The limit between cut and cover and a mined section (whether SEM or TBM) was identified as the point at which there was sufficient rock cover to permit utilizing SEM or TBM. For the purpose of the conceptual estimate, one tunnel diameter of rock cover for the main tunnel TBM and approximately ½ of the tunnel diameter for the SEM approaches were identified as the limits.

The study considered various alternatives as combinations of: two lanes, three lanes, stacked tunnel and twin tunnel options. The diameter of the TBM sections varied from 40'-3" for Alternate 1 to 64'-9" for Alternate 4, see Table 1. Those diameters are considered to be extremely large for TBMs, especially when considering the rock conditions in Atlanta. Two manufacturers were consulted Herrenknecht and Robbins during the preparation of the cost estimate. Both manufacturers have produced TBMs used to excavate tunnels in the Atlanta area. Both agreed that the current state of the art does not support the manufacturing of a 64'-9" tunnel, the design team excluded this option as an infeasible option and accordingly the cost estimate was not prepared for Alternate 4. For Alternate 1 both manufacturers submitted a price for the TBM and for Alternate 2 and 3 only one manufacturer submitted TBM price accordingly, construction cost estimates for alternates 1, 2 and 3 were prepared.

ESTIMATE COMPONENTS

The cost estimate considered all the main tunnel components and associated costs including:

- Tunnels (bored, mined, cut-and-cover box, U-section) including excavation, support and structure
- Surface Structures (buildings and facilities)
- Fire-Life Safety Elements (exits, passages and vent buildings)
- Mechanical Components
- Architecture and Tunnel Finishes
- Electrical Components
- System Components
- Utility Relocation

Table 2. Total project cost for different alternates (in $million)

Alternate	1A	1B	2A	2B	3A	3B
Total escalated project cost	$4,860	$6,153	$6,257	$7,680	$5,204	$6,599

- Traffic Control During Construction
- Right-Of-Way (land acquisition costs)

The estimate for each of those elements is developed as an independent estimate using previous experience in tunnel facilities and some basic design assumptions as described throughout this report. Available construction cost data for similar projects was used to identify material costs, production rates, labor costs and all other cost components. Roadway work and surface structures were estimated using GDOT standard cost estimate spread sheets.

The cost of the main line tunnel construction, which comprises the lion's share of the total cost, was developed with a special consideration for its uniqueness in terms of TBM size and tunnel length. A specialty sub-consultant (Don Hilton and Associates) was hired to estimate the construction cost for the mined and the TBM tunnels.

CONCLUSION

The proposed Atlanta North-South Tunnel project would connect the SR 400/I-85 interchange to the north with the I-675/I-285 interchange to the south. The total length of the project would be approximately 12 miles, of which about 7 miles would be constructed as a tunnel. The traffic analysis confirmed that the project will reduce congestion on the Downtown Connector, one of the most congested portions of the freeway system in and around Atlanta. This substantial benefit for the freeway system in Atlanta will come at cost. The current project study estimates vary between $4.9 Billion to $7.7 Billion depending on the selected alternative. The most feasible option from cost/benefit prospective is Option 1A, (twin tunnel alternative with two lanes each direction with no I-20 connection). Based on this early assessment and the current pre-conceptual analysis, each of the presented options will have a financial gap that is created due to the revenue generated from tolling is being less than the project cost. This paper summarizes the findings for Phase 1 of the feasibility study for the proposed Atlanta North-South Tunnel project and explains the methodologies and processes adopted for the study.

PROPOSED CONTRACTING PRACTICES FOR THE CALTRAIN DOWNTOWN EXTENSION

Derek J. Penrice ▪ Hatch Mott MacDonald

Bradford F. Townsend ▪ Hatch Mott MacDonald

ABSTRACT

The Caltrain Downtown Extension (DTX) project will extend commuter and future statewide high-speed rail service into the Transbay Transit Center, which will be located in the business district in downtown San Francisco. The DTX project includes 1.5 miles of underground construction, comprising complex sections of mined and cut-and-cover tunnel.

As part of the DTX preliminary engineering process, development of appropriate contracting practices, including contract packaging, procurement methods and contract terms for the underground construction, has been initiated. This paper provides a summary of the evaluations undertaken, preliminary conclusions reached, and proposed next steps to promote industry interest and maximize bid competition.

INTRODUCTION

The Transbay Transit Center (TTC) Program is a landmark undertaking for the San Francisco Area. The Program will enhance transit service to the region by improving the bus and rail connectivity of eight major transit providers at one multi-modal facility in downtown San Francisco, resulting in the largest such facility in the United States west of New York City. The design, construction and subsequent operation and maintenance of the Program is headed by the Transbay Joint Powers Authority (TJPA), a collaboration of Bay Area government and transportation agencies. The DTX project is a critical component of the Program and comprises the following infrastructure:

- A 1.5-mile below-grade extension of Caltrain commuter rail service along Townsend Street and Second Street to the TTC
- A subsurface station at the intersection of Fourth Street and Townsend Street
- Improvements to the existing Fourth and King Street surface station including reconfiguration and improved functional layout of the existing Caltrain Yard

These major project elements are indicated in Figure 1.

The DTX project is currently in Preliminary Engineering, with a 30% level of design completion expected in early 2010. As part of the preliminary engineering phase, Hatch Mott MacDonald, as part of the Program Management/Program Controls (PMPC) team, has been tasked with the development of the DTX Contract Packaging Strategy.

The Contract Packaging Strategy defines how the scope of the DTX project will be divided into specific construction contracts, how those contracts will be sequenced, which allows the project construction schedule to be developed, and how the contracts will be procured and subsequently managed during construction. The report also provides a strategy for promoting construction industry awareness of the DTX project with the goal of maximizing bid competition.

CONTRACTING PRACTICES FOR THE CALTRAIN DOWNTOWN EXTENSION

Figure 1. TTC program elements and location plan

The Contract Packaging Strategy is a key project document which provides a roadmap for the successful execution of the DTX project. Design, procurement and construction will be developed around the contract packages presented in the report.

The report is a living document, which will be updated on a periodic basis. This paper provides a summary of the evaluations undertaken and conclusions reached to date.

Project Alignment and Scope

The DTX alignment begins on the existing Caltrain mainline on the approach to the Fourth and King Street surface station, the operator's current San Francisco terminus. At the limits of Channel Street and Seventh Street approximately (lower left of Figure 1), a two-track lead begins its descent from grade to fully underground by means of an open-cut U-wall, approximately 1,960 feet in length. While in the open-cut section, the alignment curves east and widens to a three-track rail system leading to a new underground station at the intersection of Fourth and Townsend streets. The alignment thereafter continues due east under Townsend Street using cut-and-cover methods to an interface with a mined tunnel at Clarence Place. The total length of cut-and-cover tunnel in this segment including the Fourth and Townsend Street Station is approximately 2,290 feet.

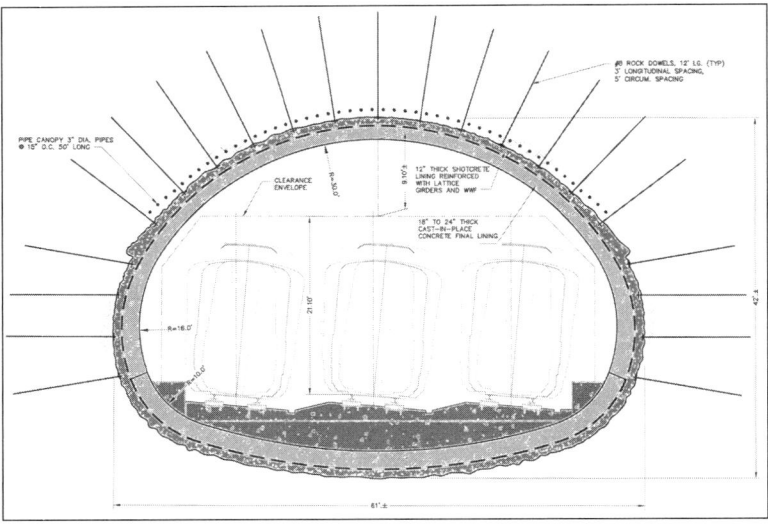

Figure 2. Mined tunnel typical cross section

From the cut–and-cover tunnel interface at Clarence Place, the alignment curves north in a mined tunnel, to be constructed using sequential excavation methods, to Second Street, passing directly beneath twelve historic buildings. The buildings are typically low-rise industrial, commercial and residential structures which contribute to a historic district, having survived the major 1906 earthquake and subsequent fire which devastated much of the city. The mined tunnel continues on Second Street to an interface with a second section of cut-and-cover tunnel at Clementina Street approximately. The mined tunnel has an overall length of 3,210 feet, and its cross section will be approximately 61 feet in width and 42 feet high. Due to the size of the cross section, it is currently proposed that the cross section be excavated in a series of seven drifts, three top heading, three bench drifts, and a single invert. A typical cross section of the mined tunnel is shown in Figure 2.

From the northern mined tunnel interface at Clementina Street, the DTX alignment gradually turns to the northeast over a length of 680 feet to connect to the west end of the TTC. At this point, the DTX flares from three to six rail tracks to align with the TTC platform tracks, producing a structure which varies to a maximum width of 175 feet. As a result of the structure size, required span lengths, complex track geometry and prevailing ground conditions, this section of the alignment will also be constructed using cut-and-cover methods.

In addition to the underground structures, the scope of the DTX project includes site preparation including utility relocation and building demolition, as well as core systems—track, overhead contact system, signals and communications, mechanical, electrical and plumbing work, and finishes.

CONTRACT PACKAGING

The primary goal of the Contract Packaging Strategy is to develop a strategy which allows the DTX project to be constructed in the shortest possible timeframe. The benefits of this approach are to minimize the impacts of escalation and time-dependent costs such as program and construction management, and to allow DTX and the

Transit Center to open for revenue service at the earliest opportunity, in turn promoting faster repayment of federal and other loans.

The Baseline Budget for the construction of DTX was established in March 2008 at $1.22 billion. This figure is inclusive of contingencies, but exclusive of soft costs and right-of-way acquisition. This value was considered by the project team to be too large to be procured as a single construction contract. Therefore, the scope of the DTX project was examined to determine how it could best be divided into several discrete and manageable construction contracts. Like a jigsaw puzzle, the DTX project was broken into as many separate scope elements as could be identified. These elements were then combined to generate an 'optimum' number of contract packages. The team sought to avoid either too many or too few packages, either of which could be detrimental to the success of the project as follows:

- Too few contracts—Individual contract packages will become too large for any single contractor to bond. Contractors will have to combine in joint ventures, thereby limiting the number of qualified bidders, and reducing bid competition.
- Too many contracts—Given the constraints of the project alignment and the limited number of construction staging areas, too many contracts will yield too many project interfaces, thereby setting conditions for disputes and claims arising from site possession and prior workmanship.

A number of permutations for packaging the construction of the DTX project were studied using the following parameters to guide the decision-making process:

- Construction method—Are the proposed methods of construction for project elements similar? Cut-and-cover structures, mined tunnel, and surface rail works are three distinctly different construction operations, with contractors who specialize in each operation.
- Construction location—Are project elements whose work is of a similar nature adjacent to each other or geographically isolated? Similar work in adjacent areas is commonly combined.
- Construction schedule—Does the order in which the work is anticipated to be progressed contribute to the packaging of particular work elements?
- Contract value—Does the combination of various work elements result in overly large contract values?
- Project funding constraints—Are there limitations on the release of capital funding commitments that may dictate when particular work elements can be performed? Such constraints may dictate how the overall project can be packaged and scheduled.
- Logistical Constraints—Can the desired number of contract packages be supported by an adequate number of dedicated construction staging and laydown areas?

It became particularly evident for the larger civil contracts involving the mined tunnel and cut-and-cover tunnel construction that several of these key parameters would be in direct conflict.

One of the goals of the packaging strategy had been to limit the maximum value of any individual contract to approximately $250 million, which has become a North American industry rule-of-thumb figure as the maximum bid value for a single contractor. The underground structures on the DTX alignment would have to be broken into at least three contracts to accommodate this requirement. To accommodate the desired schedule for construction, these contracts would also have to be in construction concurrently. Therefore, each contract would require its own dedicated staging areas.

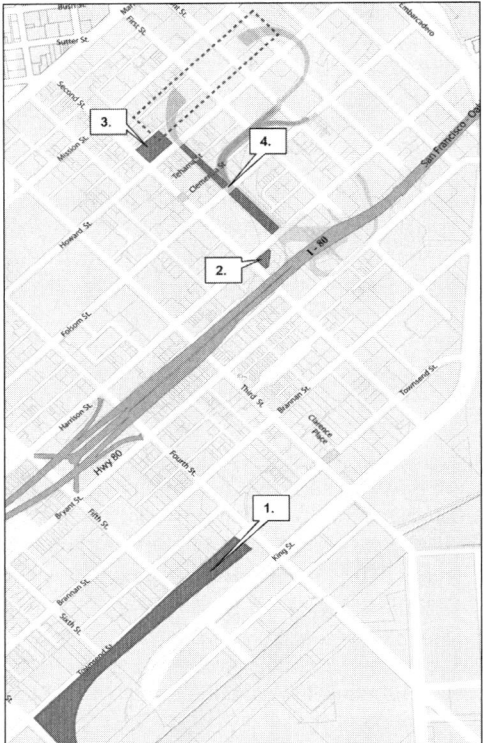

Figure 3. DTX construction staging areas

As with any urban setting, space for staging construction is at a premium. A primary consideration for the development of the DTX contract packages is the scarcity of construction staging areas, and how the available areas could best be utilized to support concurrent construction of the major civil works. The currently identified staging and laydown areas for the DTX project are indicated in Figure 3.

The primary at-grade staging areas available for DTX include the sites listed below. The numbers in Figure 3 correspond with those in the bulleted list.

1. The existing Caltrain Yard over the footprint of the DTX construction and the corner at Seventh and Townsend streets
2. Parcel 'Q,' located at the southeast corner of the intersection of Second and Harrison streets.
3. The area bounded by Natoma, Second and Howard streets
4. The area underneath the proposed bus ramps to the TTC

Additional on-street surface staging areas will be created during construction of the cut-and-cover sections. The City of San Francisco has indicated that Townsend Street can become a two-lane, one-way street during construction. Therefore, up to half the width of each roadway will be available at any given time for the contractors' use. It is also anticipated that a below-grade staging area would be created within the footprint of the cut-and-cover tunnel excavation on Townsend Street between Third Street and the mined tunnel portal.

Table 1. DTX contract packages and construction values

Contract Designation	Construction Value ($ million)
A.101 Interim Caltrain Yard	$3
A.201 Building Demolition	$17
A.301 Utility Relocation Townsend & Second steets	$73
C.101 Fourth and King Station/Final Caltrain Yard	$139
C.201 Cut-and-Cover Tunnel Townsend Street	$300
C.301 Mined Tunnel and Cut-and-Cover Tunnel Second Street	$555
S.101 Track	$68
S.201 Systems—Tunnel	$55
S.301 Systems—Buildings	$11
F.101 Finishes	$2

It is recognized that further property acquisition will be necessary to support the construction of the projects ventilation structures. These sites will also be used in the interim to provide additional construction staging areas. While specific properties have been identified for acquisition, they have not been identified in this paper as the property acquisition will require an addendum to the project Final Environmental Impact Statement/Environmental Impact Report.

Contract Packaging—Outcome

Ten construction contracts have been identified in the DTX Contract Packaging Strategy using the previously defined parameters as a guideline. The proposed packages have been broken into four categories: advance works, civil, core systems, and finishes contracts. These vary in size and value from minor site preparation contracts to the major underground civil works packages. No specific procurement, third party or force account contracts have yet been considered. The contracts and their anticipated construction values are shown in Table 1. The figures presented are in year 2008 dollars inclusive of various contingencies, but excluding soft costs.

During the evaluation of potential contract packages, it was immediately apparent that due to the size of the mined tunnel cross section and the complexity of the construction—overlying historic structures, proximity to the foundations of the I-80 Bay Bridge approach viaduct, and challenging ground conditions—progress on the mined tunnel would be slow. Correspondingly, the mined tunnel will undoubtedly form a major component of the critical path of the DTX project construction schedule. Therefore, minimizing the overall project schedule duration required seeking a strategy for packaging the DTX project which provided early access to the mined tunnel and which allowed the tunnel to be advanced from a minimum of two faces.

Therefore, a concept was developed for the construction of the underground structures which included three distinct contracts comprising two separate cut-and-cover tunnel contracts on Townsend Street and Second Street, respectively, and a third contract for the mined tunnel. However, when a site which was assumed to be available to stage the construction of the mined tunnel adjacent to the north portal was lost to a developer, this concept became unworkable from a logistical perspective. The cut-and-cover tunnel on Second Street and the mined tunnel were subsequently combined into a single contract to provide sufficient staging and laydown area at the north end of the mined tunnel.

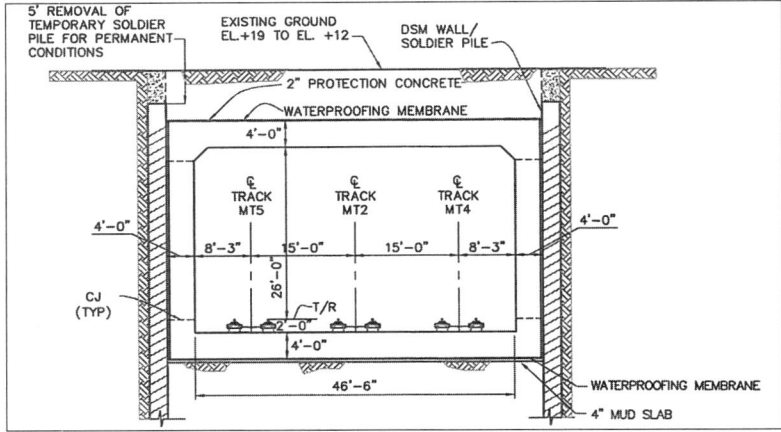

Figure 4. Townsend Street cut-and-cover tunnel typical cross section

This has resulted in two major underground construction contracts. While the majority of the construction of the two contracts will be underway concurrently, the bid and award period for each contract will be staggered to maximize opportunities for bidders. The scope of the underground construction contracts and reasoning for the extent of each contract package is summarized below.

C.201 Townsend Street Cut-and-Cover Tunnel

This scope of this contract package combines three specific elements of the DTX project:

- Retained cut (U-wall), 1,960 feet in length
- Fourth and Townsend Street Station structure, 800 feet in length
- Townsend Street Cut-and-Cover Tunnel, 1,010 feet in length

These elements were combined due to the similar nature of the work proposed: each of the structures comprises cast in place concrete construction—undertaken within the limits of a temporary support of excavation system comprising deep soil mix (DSM) walls, reinforced with structural steel shapes, each is geographical adjacent to the other elements, and importantly the opportunity to support the construction of each exists from the available staging area within the existing Caltrain Yard.

A section of the Townsend Street cut-and-cover tunnel indicating the general nature of the construction on Townsend Street is shown in Figure 4.

While the resulting contract value, at $300 million exceeds the desirable maximum figure, it is not considered to be excessive.

C.301 Mined Tunnel and Second Street Cut-and-Cover Tunnel

This scope of this contract package combines several elements of the DTX project:

- Townsend Street Cut-and-Cover Tunnel stub, 200 feet in length
- Mined Tunnel, 3,200 feet in length
- Second Street Cut-and-Cover Tunnel, 680 feet in length
- Ventilation structures on Townsend and Second streets

As mentioned previously, the mined tunnel and cut-and-cover tunnel on Second Street were combined into a single contract to provide access and staging area at each end of the mined tunnel. The decision to combine these elements was based upon construction schedule and logistical constraints.

A stub of cut-and-cover tunnel on Townsend Street is included as part of this contract as an access pit to provide early access to the southern mined tunnel portal, and also to provide direct connectivity to proposed staging areas.

In addition to the mined tunnel and cut-and-cover tunnel on Second Street, this contract package also includes the construction of ventilation structures on Townsend and Second Streets. As the permanent sites of the ventilation structures will be used to support the mined tunnel construction, the construction of the ventilation structures cannot commence until such time as the mined tunnel contractor has vacated the sites. Any delay in the mined tunnel construction will adversely affect the bid, award and construction of the ventilation structures, and will ultimately delay the opening of DTX to revenue service. Therefore, these elements have been combined into one single construction contract. This allows the timing of the ventilation structure construction to be determined by the mined tunnel contractor.

The construction value for contract C.301 Mined Tunnel and Second Street Cut-and-Cover Tunnel is approximately $555 million. This exceeds the theoretical maximum contract figure by a considerable margin. There is no doubt that many contractors will be forced to joint venture to bid this contract.

While the contract value is larger than initially desired, the contract features distinct and different work elements—the mined tunnel and cut-and-cover tunnel, each of which has significant value. It has been assumed that joint ventures would form on that basis, with one contractor performing the mining operation, and the second contractor performing the cut-and-cover construction. As the project continues to develop, opportunities to reduce the contract value will be sought.

PROCUREMENT

As the selection of a procurement method has a profound effect on the scope of work for the preliminary and final engineering design for that particular contract, another goal of the Contract Packaging Strategy is to determine a preferred procurement method for each of the identified contract packages at the earliest opportunity.

Determining the most appropriate procurement method for each of the contract packages required closely examining the project goals in terms of key parameters including cost, schedule, quality, and risk. While no procurement method is perfect, and compromise is often necessary, in keeping with the objective of constructing the DTX project in the shortest possible timeframe, the following procurement methods have been considered:

- Design-Bid-Build (DBB): This is the traditional contracting method used for public works construction in the United States. The owner engages an engineer to complete a design, comprising drawings and specifications. The completed design is bid by contractors in a competitive process, with the lowest bidder normally retained to construct the project.
- Design-Build (DB): This delivery method has been used overseas for some time, and while its use is increasing in the United States, examples of underground construction procured using DB are still scarce. Under a DB approach, the owner engages an engineer to prepare design criteria, reference drawings and specifications. The reference design is bid by contractors in a competitive process, with the lowest bidder normally retained to complete the final design and thereafter construct the project.

- Construction Manager/General Contractor (CM/GC): This approach is commonly used in the building industry, and has had some limited application in the underground construction industry. While there are a number of permutations to this approach, in general the owner engages an engineer to complete a design, comprising drawings and specifications. While design is in progress, the owner engages the services of the CM/GC through a process involving consideration of qualifications and technical approach as well as bid price on a partially complete design. Thereafter, the CM/GC contributes through constructability review and proposed means and methods to the streamlining and completion of design. Once design is complete the owner and CM/GC negotiate a price for the construction of the project.

A brief comparison of the advantages and limitations of these procurement methods is provided in Table 2.

Based upon the current Program Schedule, it is expected that the final design of the DTX project will commence in early 2010, once full funding for the project has been identified. Construction of the DTX advance contract packages is scheduled to proceed in 2011, with civil and systems packages following in sequence thereafter.

At this stage, the comparison of the delivery methods has focused on DB and DBB, as these require a significantly different design approach. Conversely, the decision to revert to a CM/GC type contract can be made while developing the design under a DBB approach.

As mentioned previously, a principal reason for the consideration of alternative procurement for the DTX, such as DB, is to take advantage of anticipated savings in construction schedule, which arises in part from the integration and overlap of the design and construction processes. Reductions in the projects construction timeframe will yield savings in project escalation costs, and time-dependent costs, such as program and construction management. As the project's construction phase is still several years distant, and is expected to last approximately nine years, these costs contribute significantly to the overall Program budget.

Furthermore, for structures where the contractors means and methods of construction impact the design of temporary and permanent structures, such as the DTX mined tunnel and cut-and-cover tunnel sections, a DB approach where the designer and contractor work together as an integral team has obvious advantages from a design efficiency perspective.

Conversely, there are a number of risks specific to the DTX construction which would have to be resolved prior to the implementation of a DB procurement. These include finalizing agreements with property owners on the alignment, with a myriad of utility agencies, and other project stakeholders including city and state agencies. These aspects of the project, which are beyond the ability of the contractor to control, must be resolved prior to contract award for a DB contract, otherwise construction delays are inevitable. However, these risks to the schedule are identified and mitigations, through development of memorandums of agreement, are in progress.

The current schedule for DTX is based upon a critical assumption that all funding for the project will be in place by the start of 2010, as full funding is necessary for the project to continue into final design and construction. This is considered to be the most significant risk to the project schedule.

Therefore the primary consideration in the selection of procurement method is the timing of the availability of the project funding, and the impacts to the project should funding not become available as anticipated.

Were the design advanced in a DB format, and funding delayed, the completed reference design prepared would effectively be shelved until such time as funding for construction becomes available. Conversely, with a DBB procurement, should funding for construction not be available as scheduled, design could continue in an advanced

Table 2. Comparison of procurement methods

Design-Bid-Build	Design-Build	CM/GC
Advantages	**Advantages**	**Advantages**
■ Traditional practice, well understood by owners, designers and contractors ■ Owner retains control of design process ■ Projects can be thoroughly planned before contact award, thus producing the most accurate estimate of final project costs ■ Owner retains project risk for unforeseen ground conditions	■ Offers potential savings in project schedule, yielding project cost savings through reduced escalation and time dependent costs ■ Provides opportunity for construction innovation ■ Integration of designer and contractor leads to efficiencies through a more focused design effort and on board constructability review ■ Transfers risk for defective design and design changes from owner to contractor	■ Contractor qualifications and approach can be include as part of selection process ■ Selection process is complete before completion of design, offering reduction in overall schedule ■ Contractor provides design phase services including constructability review ■ Allows fast-tracking of project elements ■ Lower potential for cost growth
Disadvantages	**Disadvantages**	**Disadvantages**
■ Overall schedule is longer due to separation of design and construction process ■ Contractor selected late in project; there is often no contractor input during the design phase ■ Owner uncertainty over whether project will be within budget until bid opening ■ Potential for cost growth during construction from change orders and differing site conditions	■ Initial price is typically higher as contractor is bidding on an incomplete design. ■ Owner relinquishes control of design process ■ Owner must know exactly what it wants at an early stage in the project; owner-initiated design changes occurring during final design can have profound impacts on project cost and schedule ■ Owner still bears risk for differing site conditions ■ All third party/stakeholder/utility requirements must be identified and resolved by the owner in advance of construction	■ Initial price is typically higher as contractor is bidding on an incomplete design. ■ Potential for cost growth as design is finalized ■ Schedule advantages may be lost during negotiation of price ■ Potential for reductions in quality to maintain negotiated price ■ May require specific legislation to be adopted

preliminary engineering stage. The additional time would be spent advancing the design, resulting in better defined risks, and refined quantities, thereby promoting better contract pricing.

While there is opportunity to revert from one process to another, this needs to happen sooner in the design process rather than later if any benefits of a DB approach are to be realized. Once the design is advanced beyond a preliminary design concept, there is reduced opportunity for innovation, and the time required to complete the design by a DBB approach becomes comparable with DB, so any schedule advantage of the latter approach is effectively lost.

As the current limitation on funding essentially allows the design to be completed under a DBB within the same timeframe as a DB package, and the former offers less risk, the recommendation for procurement of the majority of DTX contract packages is for DBB.

Should funding for DTX become available earlier than scheduled, this decision will be re-evaluated. A decision on whether to adopt a CM/GC approach will be made at a later stage in the design development.

MARKET CONDITIONS & CONTRACTOR OUTREACH

It is fundamental to the success of the DTX project that sufficient qualified contractors are attracted to bid for each of the identified contract packages—in particular the larger utilities, tunneling, track, and systems contracts.

Potentially, there are a number of large, complex tunnel and rail projects, both local and regional, which may be procured or ongoing within a similar timeframe to the proposed DTX construction. Therefore, as a new and relatively unknown owner, it is imperative that the TJPA maximizes contractor awareness of the project and incentivizes bidding conditions such that the DTX contract packages are enticing in a marketplace where contractors may bid selectively.

The basic outline of a Contractor Outreach program has been developed to promote the project and help maximize the number of bids and is included within the Contract Packaging Strategy.

The initial phase of the Contractor Outreach program involves promoting contractor awareness of the DTX project through the publication of articles in trade magazines and the preparation and presentation of project-specific papers at regional and national conferences. Local contractors associations will be solicited to provide presentations and project updates on a regular basis.

In addition, a substantial mailing list of contractors who may be qualified to bid on parts of the DTX project will be developed, based upon established lists of prequalified contractors maintained by the City of San Francisco, local and regional transit agencies, and utility providers and specific contacts made through contracting organizations. These contractors will be sent regular DTX project fact sheets advising of project progress and potential opportunities for bidding purposes. This is a low cost method of advertising the project to a substantial audience.

Ultimately, the TJPA's goal is to provide opportunities for contractors to participate in the development of the project, and in particular to provide input to the contract provisions. The participation effort will involve the following:

- Conduct informal interviews with potential prime contractors to get their impressions of the project.
- Conduct workshops/roundtable sessions with representatives of multiple contractors related to contract provisions—labor agreements, bonding requirements, escalation clause, payment provisions, limitations on retention, etc., or other items raised as concerns during the interview process.
- Provide/sponsor networking events on an annual basis, with invites sent to contractors on the mailing list. Each event will allow time for a presentation of the project status, schedule and discussing specific technical aspects, and provide sufficient time for questions and answers. A part of the event will be dedicated to allowing small businesses and prime contractors to network.

As part of the Contractor Outreach program, the TJPA will seek opportunities for local and disadvantaged businesses, consistent with the TJPA's Disadvantaged Business Enterprise (DBE) Program. The outreach process for disadvantaged and small businesses will be similar to that described for prime contractors.

No DBE participation requirements will be set for any of the DTX contracts. However, in each case DBE participation goals will be set, and bidders will be encouraged to achieve the specified goal. However, compliance with the participation goal will not be a condition of contract award. Participation goals will be established on a contract-by-contract basis. For specialist items, such as the mined tunnel contract, it is anticipated that participation goals will be nominal.

CONCLUSIONS

Based upon the evaluations undertaken for the DTX project thus far, the following can be concluded:

- A Contract Packaging Strategy is an important tool for owners and engineers to successfully plan and execute complex programs. The report provides a structure for planning, scheduling, budgeting, and controlling the program.
- The importance of identifying adequate areas for construction laydown and support during the environmental phase of a project cannot be understated. A major constraint to achieving the completion of the DTX project in the shortest possible timeframe is the availability of staging areas. While additional properties to support the DTX construction are being sought, this constraint has currently resulted in the development of contract packages whose values are larger than desirable, as shown in Table 1.
- The timing of the availability of funding is a key factor in the determination of the most appropriate procurement method for the DTX project.
- In today's bidding climate, with finite resources in the tunneling industry and increasing demand for those resources, it is critical that Owners to accomplish effective outreach program to maximize project awareness and subsequently stimulate bid competition.

NEXT STEPS

As the DTX design progresses, and the construction schedule of DTX and of adjacent projects, such as the proposed electrification of the Caltrain Peninsula Corridor service, are coordinated and additional land parcels are acquired, it may become advantageous to combine elements currently indicated as discrete contracts or, conversely, to separate elements included within a single contract package. Revisions to the report, which will address any modified scope and budget for contract packages, will be prepared and issued as necessary.

ACKNOWLEDGMENTS

The authors would like to thank the Transbay Joint Powers Authority for the permission to publish this paper. In addition, the contributions of other key members of the project team are gratefully acknowledged: URS—prime consultant, PMPC team; EPC Consultants, PMPC team; Parsons—prime consultant, cut-and-cover tunnel design; Jacobs Associates—mined tunnel design; and Arup—geotechnical investigations.

This paper summarizes the preferred scenario for packaging the DTX project. This scenario was presented and discussed at a DTX Contracting Strategies Workshop held on September 11 and 12, 2008. In addition to the TJPA, the PMPC and design team, staff from Caltrain, the San Francisco County Transportation Authority, and the San Francisco City Attorney's Office attended the workshop. The authors gratefully acknowledge the contributions of these agencies.

PART 13

New York City

Chairs
Berry Roberts
Parsons Brinckerhoff

Don Hickey
Judlau

ALTERNATIVE FINAL CAVERN LININGS FOR THE EAST SIDE ACCESS TRANSIT PROJECT

Colin Barratt ▪ Parsons Brinckerhoff

William Cao ▪ Parsons Brinckerhoff

ABSTRACT

The East Side Access (ESA) project is the largest underground rail infrastructure project under construction in New York City, USA. The Manhattan segment of the project is to be mined through rock and consists of 4.1 miles of Tunnel Boring Machine tunnels for track work and mined excavations to house rail switches, air plenums, and other rail facilities. This paper considers different methods for concrete lining, such as the use of fixed forms, adjustable forms, and pneumatically applied shotcrete. The advantages and disadvantages of these methods are discussed and the basis for considering one over the other are illustrated. Further discussions on design development and preparation of documents to facilitate quality and constructability in the field are presented.

INTRODUCTION

The East Side Access Project will provide the Long Island Rail Road (LIRR) with a second terminal station in New York City at Grand Central Terminal (GCT) in addition to the existing Penn Station Terminal (Figure 1). The project will divert main line trains from Queen's Sunnyside Yard through two existing, currently unused tubes within the existing four tube 63rd Street tunnel, under the East River and in to GCT Manhattan. At GCT, the existing Madison Yard in the lower level of Metro North Railroad (MNR) train-shed will be cleared for a new concourse and will connect to the new GCT Station Caverns via inclined escalator shafts.

General Geology and Groundwater of the Project Site

The rocks underlying Manhattan belong to the New England Upland, locally known as part of the Manhattan Prong consisting of schist, gneiss and marble. The tunnels and caverns generally pass through the schist. Water levels measured in borehole standpipes are approximately 15 feet below street level. The new tunnels and caverns are between 90 and 170 feet below street level.

MANHATTAN NEW CONSTRUCTION

Manhattan Alignment

Manhattan new construction comprises three major segments; the Approach Tunnels, Grand Central Terminal Station Caverns, and the Tail Tracks (Figure 2). The construction of the Manhattan tunnels begins with the Approach Tunnels at the western terminus of the existing 63rd Street Tunnel located beneath the intersection of East 63rd Street and Second Avenue. The new construction will extend the existing two tunnels that bifurcate to four through two side-by-side interlocking wye caverns (GCT 5). The tunnels will then descend and ascend in pairs. The upper tunnels further bifurcate into four upper tunnels and likewise the lower tunnels bifurcate into four lower

ALTERNATIVE FINAL CAVERN LININGS FOR THE EAST SIDE

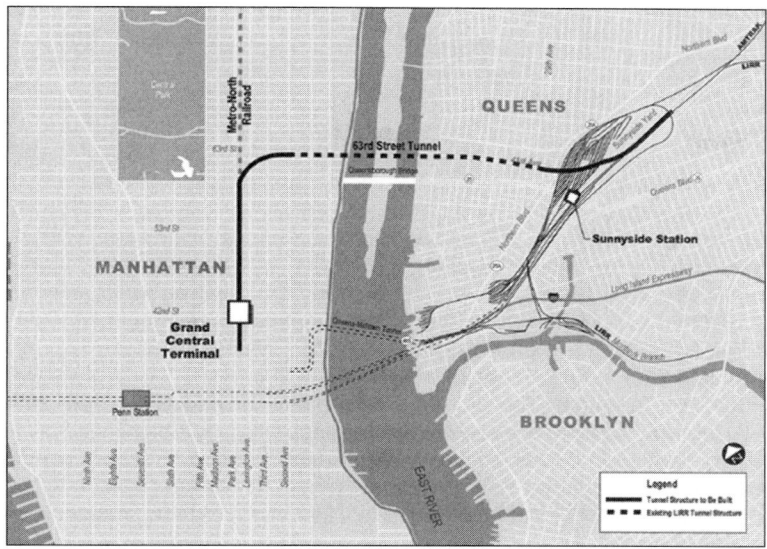

Figure 1. General map of project

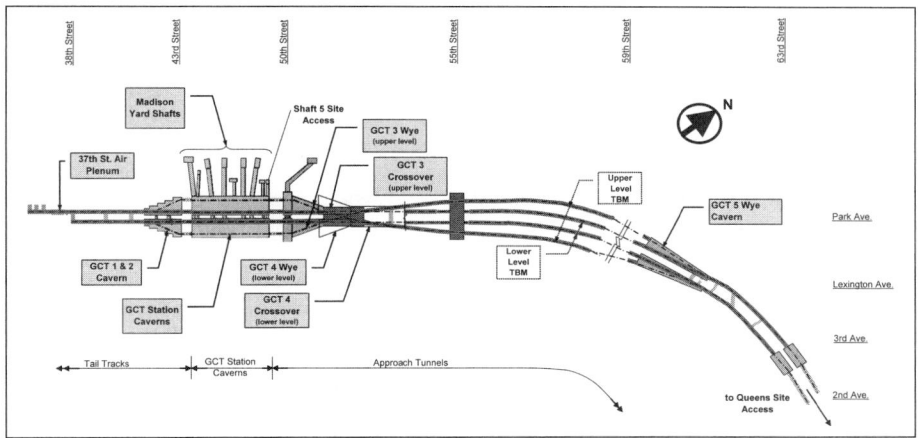

Figure 2. Manhattan alignment

tunnels through two pairs of interlocking wye caverns (GCT 3 and GCT 4, respectively) stacked one over the other. The eight tunnels then enter into two new GCT Station Caverns located below Park Avenue between 49th and 44th Street. On exiting the Station Caverns below 44th Street and Park Ave, the eight tunnels merge into four within two side-by-side three level interlocking caverns (GCT 1 and 2). The tunnels then follow Park Avenue and terminate at 37th/38th Street where a ventilation plant and sump pump is to be constructed.

Manhattan Caverns

Manhattan Approach Tunnel Caverns. The GCT 3, GCT 4, and GCT 5 Wye caverns are pairs of caverns each with a length of 320, 320, and 430 feet respectively. Additionally, GCT 3 and GCT 4 Crossover Caverns are the interlocking structures for the respective Wye caverns, and are each with a length of 320 and 390 feet long, respectively. These caverns are single level, shotcrete lined chambers which vary in cross section, ranging from 22 to 52 feet width by 23 to 33 feet height.

GCT Station Caverns. The main caverns are two 60 feet wide by 70 feet high, three level caverns, each 1,144 feet long with a constant cross section. The cast-in-place concrete arch roof lining is self supported directly on rock ledges, and provides for an overhead crane structure during construction. The internal structure of walls and floor beams are to be built and connected to the arch in a subsequent contract.

Tail Track Interlocking Caverns. The GCT 1&2 Wye Caverns are a pair of three-leveled structures running 326 feet long and varying in cross section from 30 to 51 feet wide by 60 to 65 feet high. The arch is a shotcrete liner structure to be constructed before the internal walls and slabs. It will be supported by bolting directly into the overlying rock until the internal walls are constructed in the subsequent contract. The Ventilation facility at 37th / 38th Street is an enlarged TBM tunnel lined with reinforced shotcrete and connected to the street level by a vertical air shaft excavated by raised bores.

TUNNELS AND CAVERNS EXCAVATION AND LINING CONSTRUCTION LOGISTICS

Site access is via Queens through the existing 63rd Street Tunnel. Initially all equipment, work crews, ventilation and construction utility delivery, and muck removal, are to be through this access point. Once mined, the upper TBM bores, in addition with the raised bores at 37th Street, provide a second temporary ventilation point. Upon completing the service shafts and access tunnels below 49th Street within Madison Yard, this access becomes available for work crews entering from the GCT Station Caverns. However, the only entry point for heavy equipment and muck removal still remains through the Queens Access.

The choice of lining the caverns along the alignment is dictated by two primary factors; access to downstream construction activities and underground equipment storage, and cross sectional geometry and function. Initially, the construction schedule supports the continuous operation of the Tunnel Boring Machines (TBM), opening up only as much of the Wye caverns required to re-launch the TBM along its new alignment. Excavation begins with two TBM bores of the upper level and inner tracks. Upon completion of the upper TBM drives, temporary ventilation system can be set up and work crews can access the site via 37th Street and Shaft # 5 (49th Street), respectively. Excavation of the GCT Stations Caverns and GCT 1&2 Wye Caverns can begin, while the facilities in the upper level tunnel portions north of 49th Street can be enlarged and lined. Additionally, the two TBM's reposition from the upper level and begins to mine the lower level tunnels. The excavated muck from all tunnel excavation and cavern enlargements is conveyed to the Queens Site Access via the newly bored upper level tunnels. Additionally, the previously bored lower level caverns between 49th Street and Queens can also be enlarged and lined as the TBM clears the area (Figure 3).

CONCRETE LINING CONSTRUCTION METHODS

The three primary concrete lining construction methods considered for the Manhattan alignment facilities were (1) fixed formwork cast-in-place concrete, (2) adjustable formwork cast-in-place concrete and (3) pneumatically applied shotcrete

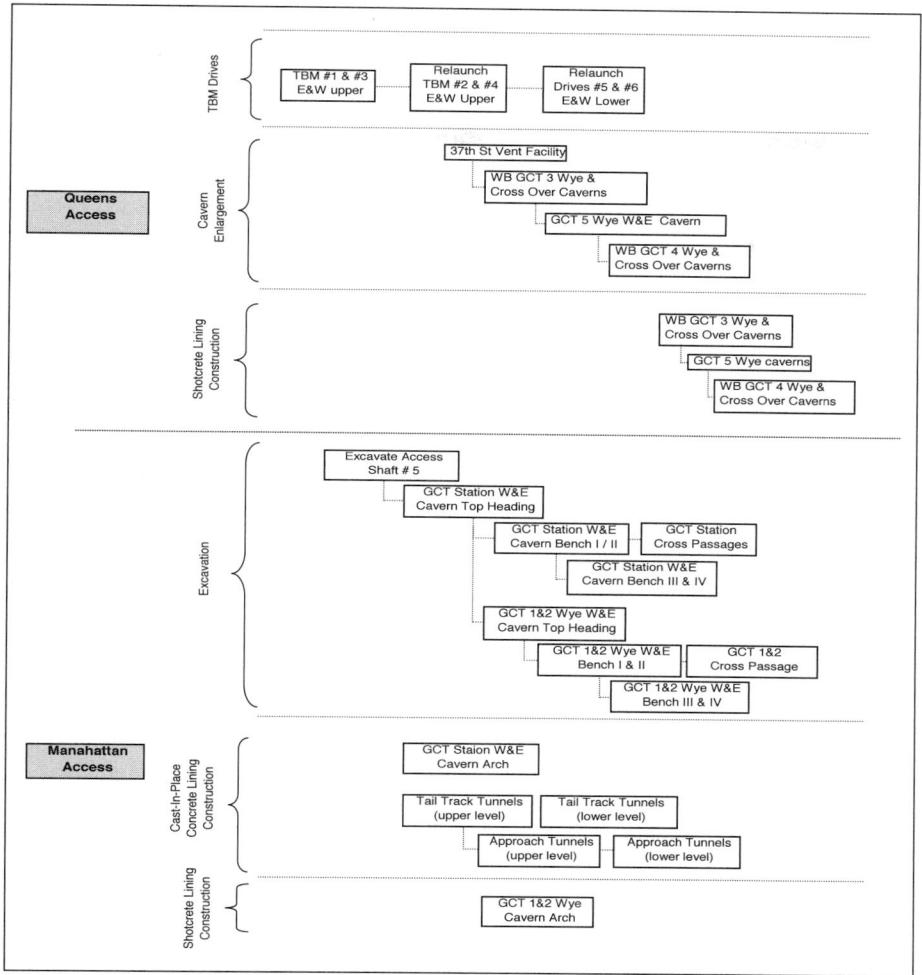

Figure 3. Construction logic diagram

construction. Each method has its own advantages and disadvantages and the choice of system is often dictated to by site specific constraints of the project.

Fixed Formwork Cast-In-Place Construction

Fixed formwork has a constant shape with steel plates shaped to the final profile of the lining. The plates are framed on a carriage that travels the length of the cavern on rails that are self propelled or pulled by winch. The assembly has some moving parts, often hinged at the shoulders and with a degree of vertical movement to facilitate formwork erecting and stripping. It can be quick and cost effective for caverns with a constant cross section and of sufficient length to enable repeat use of the formwork system. Caverns with varying geometry can utilize fixed formwork, however, it will require a number of different forms and carriages. The fixed formwork will be used in series, where the cross section changes over a series of steps to form "stepped

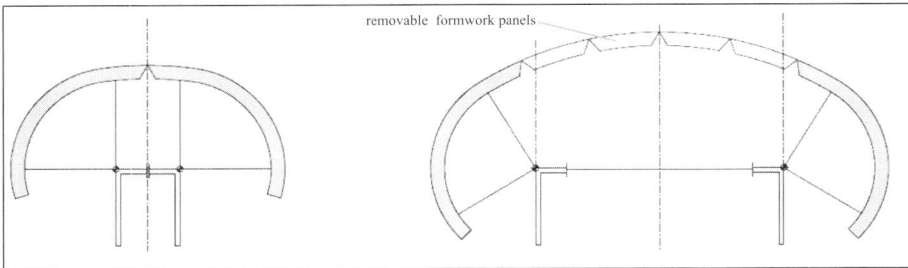

Figure 4. Schematic of an adjustable form system

plate" construction. Greater reuse of wall forms can be achieved by keeping the shoulder radius constant and reusing them with different arch radius forms. However, when keeping the wall radius constant while varying the cavern cross section, there is a compromise to the optimal profile of the cross sectional geometry, and quantities of reinforcement and concrete may be increased as required.

Adjustable Formwork Cast-In-Place Construction

Adjustable formwork is similar to fixed formwork assembly and is often used for caverns of variable cross section (Figure 4). The carriage that carries the forms has a series of hydraulically operated telescopic struts that rotate and lift the formwork panels into place. The carriage consists of two parts which move horizontally relative to each other. When combined with formwork panels of constant radii that can be added and removed, the system can be adjusted to the size of cavern. The carriage can be further extended with outriggers. The constant arch and wall radii, when adjusted incrementally between pours, creates a smooth uninterrupted profile along the cavern length. Fewer carriages are required than with fixed formwork, reducing the amount of time required for mobilization and demobilization of equipment, and occupying less underground storage space. The smooth transition alleviates the need for end walls, and the activity of drilling and shooting out the rock profile can advance in a single direction from the smallest to largest cavern without interruptions, as oppose to the incremental process that would be required for the stepped plate construction when using fixed forms. A similar compromise to the optimal arch profile, however, still occurs when using a constant radius for the arch and walls. The profile is defined by the spatial train clearance requirements at the beginning and end stations of the cavern, while the intermediate cross sections are a smooth interpolation of the two profiles. This method yields a relatively flat crown arch as the cavern widens, thus, requiring more reinforcement and/or concrete during design.

Pneumatically Applied Shotcrete Construction

Shotcrete is the process of pneumatically spraying the concrete directly onto the structural surface. The concrete is conveyed through hoses with compressed air and expelled through a nozzle at a high velocity in which the concrete is consolidated upon impact with the surface. The final profile is sprayed to a set of pins with no limitation on variation. For caverns of variable cross section, the optimum profile can be maintained at any location along the cavern's length. Maintaining the optimum profile requires less concrete and reinforcement to be used during in the works. The equipment required for the operation is relatively small and highly mobile, making it ideal for use where under ground space is limited. The main disadvantage of shotcrete compared to cast-in-place concrete is the need for a skilled workman proficiency in the art of shotcrete and the difficulty in achieving a smooth finished surface similar in quality to formwork.

METHOD SELECTION FOR CONCRETE LINING

Manhattan Approach Tunnel Caverns

The Manhattan Approach Tunnel Caverns are the gateway to the Manhattan part of the project. During the earlier stages of construction, space availability is limited until multiple tunnels are bored and the caverns are mined and enlarged. Blocking either the GCT 5 East or West Wye caverns will stop construction activities further along the alignment to the respective east or west portion of the works. Blocking GCT 3 or 4 Crossover caverns would stop construction further along the alignment to the upper or lower portion of the works respectively. Selecting for the shotcrete construction method will utilize smaller and more mobile equipment, thus enabling other activities further along the alignment to continue and occur simultaneously. The use of any form carriages, particularly at the throat of the Wye caverns, would restrict movement of materials and equipment to areas further down the alignment. Additionally, the various shapes of all the facilities in the approach tunnels further dictates the efficiency and feasibility of selecting shotcrete as the construction method.

GCT Station Caverns

For the GCT Station Caverns, access for work crews becomes available at Shaft #5 (49th Street) and an alternative source for underground ventilation becomes available from 37th Street, allowing the construction of the GCT Station Caverns to operate independently of other construction activities in the Tail Track and Approach Tunnel areas. The excavated width of 60 feet for the GCT Caverns will be sufficient to house a formwork carriage that will allow access through the site to access the Tail Track area beyond. Additionally, the constant cross sectional geometry for over 1,140 feet allows for repeated and continual use of the same formwork. Furthermore, the GCT Station Caverns will be used for public services and thus a quality finish to the structure will be required. Selection of a cast-in-place fixed form construction method for the GCT Station Caverns would be the most economical solution.

Tail Track Tunnels

The Tail Tracks are at the end of the alignment, and lining the GCT 1&2 Wye caverns have less impact on other activities. The relatively short length of the caverns, along with the varying geometry, would have required specific formwork systems to be made, lending cast-in-place construction method to be a more costly solution. The decision to use shotcrete for the Tail Track structures was thus a function of economy of scale and the availability of trained crews already available on site.

CAST-IN-PLACE CONCRETE CAVERN LINING DESIGN

GCT Station Cavern Arch: Fixed Formwork Cast-In-Place Concrete

With a constant cross section of over eleven hundred feet long for each cavern, the design approach was to optimize the arch profile from both a materials and formwork turnaround perspective.

In the transverse direction the intrados was an architecturally derived profile to accommodate overhead platform air plenums and catwalk hung from the crown. The

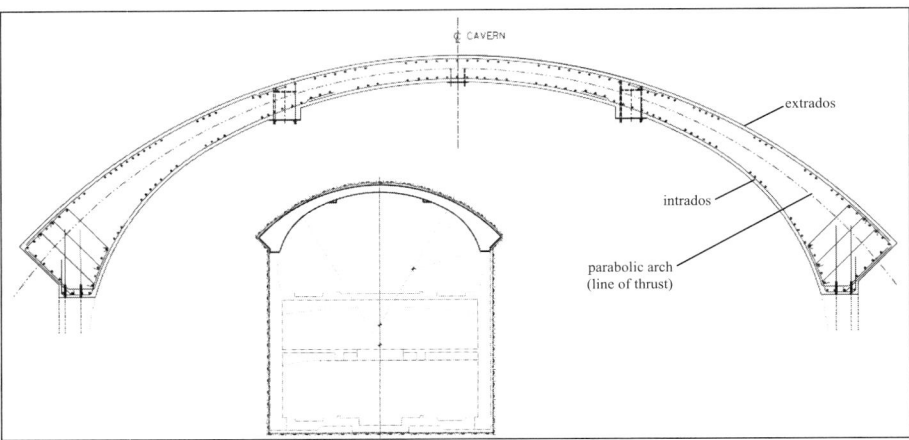

Figure 5. GCT Station cavern arch reinforcement cross section

extrados was profiled to keep the arch axis coincident as practically possible to the line of thrust for the weight of the wet concrete. This was done through trial and adjustment by fixing the crown thickness and determining the associated extrados to give an axis in the shape of a parabolic arch. In turn the relationship between the intrados and extrados fixed the thickness of the arch at the springing. By keeping the axis coincident with the line of thrust, the formwork strip time could be reduced to when the concrete has gained a minimum compressive strength of 1000 psi. The main reinforcement was then derived from influence lines and load combinations of rock wedge size and locations with and without hydrostatic loads. The maximum positive bending moment in the crown occurring when the middle third of the arch is loaded and the maximum negative bending moment at a springing occurring when four-tenths of the span adjacent to the springing is loaded or thereabouts (Figure 5).

To further enhance the production cycle the reinforcement was detailed as a series of 10 by 30 foot mats that could be prefabricated, delivered and mechanically lifted in to place. Details were provided of anchors that could be incorporated in to the waterproofing system that would hold the mats in place, anchored to the rock. This allowed the reinforcement mats to go up independent of the formwork. To complement the mats, the reinforcement along the springing was detailed in cages that could be prefabricated and lifted in to place against the mats along the extrados before the intrados mats were placed. To support using two 25 foot long formwork carriages operating in series, the arch shrinkage/distribution reinforcement was designed to control cracking for a pour length of fifty foot.

SHOTCRETE CAVERN LINING DESIGN AND DETAILING

Train Interlocking Wye and Crossover Caverns GCT 1&2, 3, 4 and 5: Shotcrete as a Final Lining

With no constraints on the geometric profile as the cavern width varies along the alignment of the cavern, the design approach was to derive a simplified relationship that could be used in the field between the cavern cross sectional profile and the cavern width. The cavern width was set out to the tracks as they diverged from the turnout or car body clearance curve through the turnout. The latter was derived based on center and end train car excess. The simplified relationships were linear between the cavern

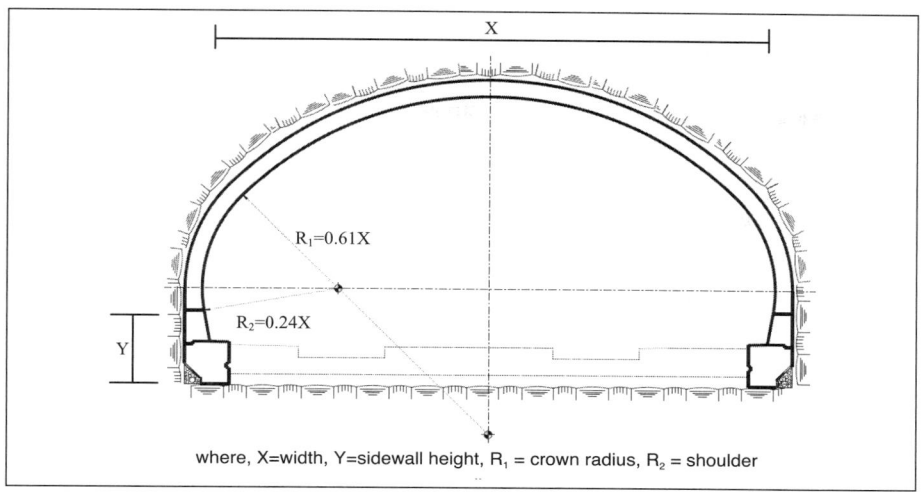

Figure 6. GCT 3, 4 5 wye caverns and GCT 3 and 4 crossover caverns

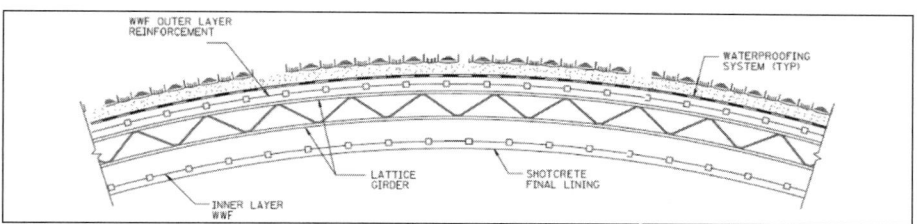

Figure 7. Shotcrete—Typical crown reinforcement detail

width and the arch and wall radii respectively (Figure 6). The width of cross section one third the way in from the largest cross section was taken as a base and the optimal profile starting with a semi-circle and then expanding the arch radii and reducing the wall radii to bring the crown down in order to reduce the quantity of excavation whilst maintaining the clearance envelope around the trains. To maintain the linear relationship, but to save the crown height from dropping within the train clearance envelope as the cavern width got smaller the height between track and cavern spring-line was linearly adjusted along the cavern length. The latter also allowed the crown height to be locally adjusted to accommodate jet fans without raising the height of the cavern along its entire length.

To enhance quality and production in the field the reinforcement detailing for the shotcrete part of the works was kept as uniform as the changing profile would allow. Welded wire reinforcing mats that could be electrically resistance welded and pre-profiled to their final shapes was detailed. Bar sizes and spacing were kept uniform within a cross section only stepping down in size along the cavern length as the caverns got smaller. With a limiting bar size for electrically resistance welding of a #5 bar and to reduce shadow from the pneumatically applied shotcrete during placement, bar sizes were all kept below #5, choosing to go to a 4" spacing or locally using double layers in the crown where required structurally.

For a platform to fix and adjust the lattice girders to their final position, a cast-in-place wall with a straight face was detailed (Figure 8). The straight face was to enable

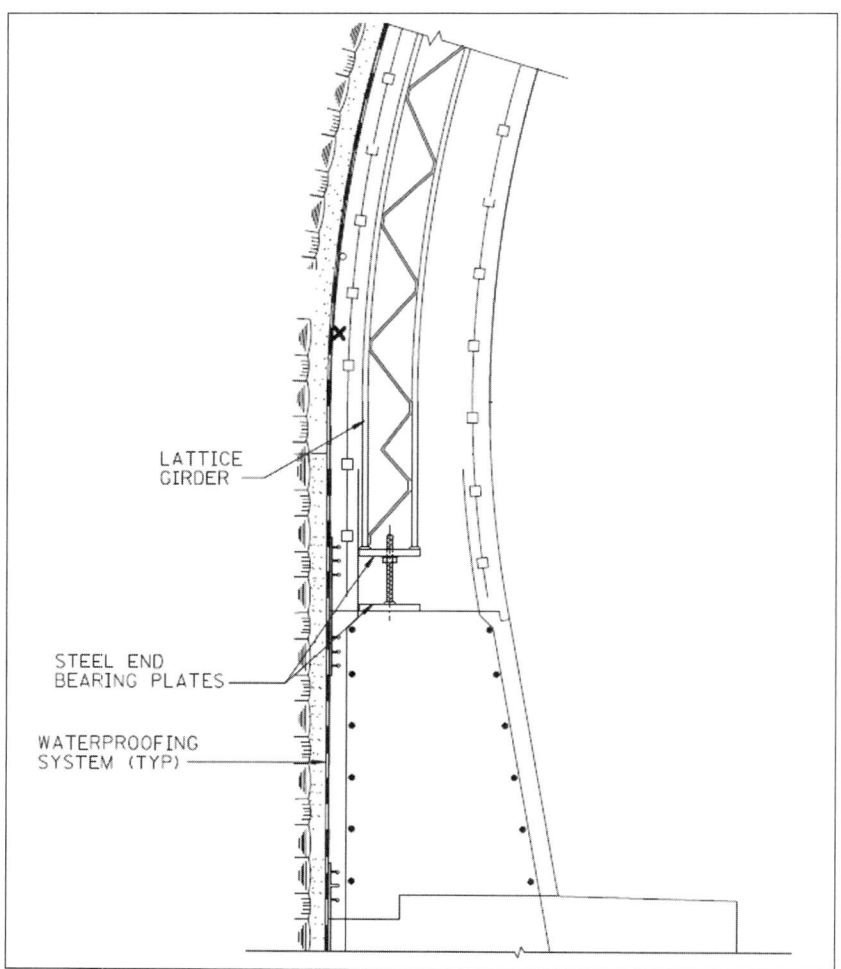

Figure 8. Shotcrete—Typical springing detail

simple building forms to be used. The same anchor details as for the GCT Station Cavern arch were provided to support the mats prior to applying the shotcrete. The intention was to do as much pre-installation to support the shotcrete operation as possible. Leaving the shotcrete process once started as a discrete number of repetitive steps.

To enable cavern interfaces to be shotcrete, the detailing of the reinforcement mats and lattice girders was kept uniform and simple (Figure 9). Both a portion of the connecting cavern and the cavern itself were detailed so as to be shotcrete in one operation. The lattice girders and mats, and the connecting bars were arranged so as to be lifted and install as part of the shotcrete cycle.

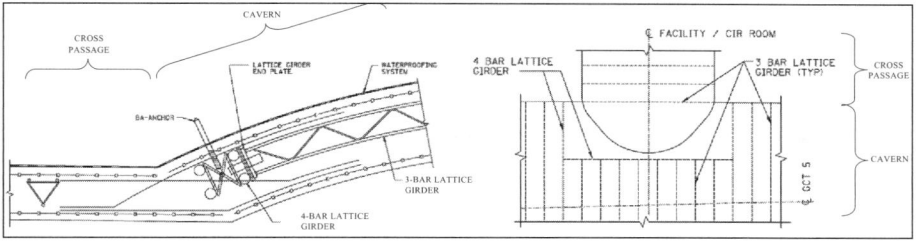

Figure 9. Shotcrete—Cavern interface detail

CONCLUSION

The methods considered for lining the ESA Manhattan alignment structures are dictated by the size and configuration of the structures, site constraints, and construction sequence. The three construction methods explored on the ESA project have advantages and disadvantages to both design and construction. Facilities located at earlier stations along the alignment may bottleneck construction activities. Additionally, some of these facilities continually vary in geometry. Thus a shotcrete method for the Approach Tunnel Caverns was selected to be the most feasible and efficient method. Whereas, facilities located towards the end of the Manhattan alignment, such as the GCT Station Caverns, has a constant geometry and is large enough to allow for access through the work site, thus, lending the design and construction method to be cast-in-place.

ACKNOWLEDGMENTS

The authors would like to thank the MTACC and the GEC Design Team for their contributions throughout the project.

BIBLIOGRAPHY

Smith, D.I., Sadek, S., Liu, J., (2008), "East Side Access—Segmental Lining Design Challenges For Closely Spaced and Shallow Tunnels," North American Tunneling 2008 Proceedings, pp. 288–293

Martin, B., Smith, D., (2005), "Queens Bored Tunnels, East Side Access, New York," 2005 RETC Proceedings, Chapter 104.

Munfah, N., (2001), "Connecting Long Island Rail Road To Grand Central Terminal In Midtown Manhattan," 2001 RETC Proceedings, Chapter 71.

Welded Wire Institute, Inc., (2008), "Manual of Standard Practice—Structural Welded Wire Reinforcement," 8th ed., Reported by WWR-500, WWR 500-R08

American Concrete Institute, (1995), "Specification for Shotcrete," Reported by ACI Committee 506, ACI 506.2-95.

American Concrete Institute, (1999), "Building Code Requirements for Structural Concrete," Reported by ACI Committee 318, ACI 318–99.

CONTINUING THE LEGACY: AN UPDATE ON THE CONSTRUCTION OF THE NEW SECOND AVENUE SUBWAY

Jaidev Sankar ■ AECOM Transportation

Christopher K. Bennett ■ AECOM Transportation

David Caiden ■ Arup

Anil Parikh ■ MTA Capital Construction

Thomas F. Peyton ■ Parsons Brinckerhoff

ABSTRACT

The dream of constructing the new Second Avenue Subway in New York has finally become reality as the first construction contract package is currently underway. This first contract requires the construction a TBM Launch Box in the middle of Second Avenue from which a Tunnel Boring Machine (TBM) will bore two tunnels south to 63rd Street. This paper will discuss the design of this contract and how construction has progressed including preparation, site constraints, utility construction and community impacts while highlighting the challenges of working in an urban environment. It will also discuss moving forward on this major project outlining the remaining contract packages including their scope and status.

HISTORY AND CONTEXT

The construction of a new Second Avenue Subway has had a long history that has been covered in many papers previously and will not be discussed at length herein. Only a brief summary of this history and current state of the project follows.

The history of the Second Avenue Subway Project spans approximately 80 years. First conceived in 1929, planning and preliminary design had progressed to a point where the existing elevated lines on 2nd and 3rd Avenues were demolished in 1942 and 1955, respectively. But priorities changed during World War II and the Korean War, which prevented the real design effort from starting until the late 1960s when DeLeuw Cather started a design for the full length Second Avenue Subway. Work had started and was progressing on three sections of the new tunnel when the New York City Financial crisis once again stopped construction in the late 1970s. The three sections located between 110th and 120th Street (Sect. 13), 99th and 105th (Sect. 11) and at Chatham Square (Sect. 5) near the entrance to the Manhattan Bridge in Chinatown were completed and closed up.

In 1995, The Manhattan East Side Alternatives study revived the concept of a new subway serving the East Side of Manhattan. In the intervening years between the work stoppage and the MESA Study, work progressed on the 63rd Street Tunnel Project which connected Queens and Manhattan via the F Line. The tunnel is actually an example of the advance planning that took place during the design for the proposed Second Avenue Subway. The tunnel section is composed of two levels with two tracks each level. The southern two tracks serve the F Line in an over-under configuration.

The northern pair of tracks was built to specifically carry the Second Avenue trains west to the Q line at 57th Street.

With a population that is projected to grow and only the Lexington Avenue line serving the East side, the need for a new Second Avenue Subway is greater than ever. The Lexington Avenue Line is already overcrowded and carries approximately 1.3 million riders daily—more than the combined ridership of San Francisco, Chicago, and Boston's entire transit systems. The burden on the Lexington Avenue line will only become greater when the East Side Access Project is completed. This will bring riders from Long Island to Grand Central Terminal discharging more commuters to the Lexington Avenue Line.

EXPLANATION OF PROJECT PHASING

The entire Second Avenue Subway Project has a price tag of over $16 billion. Because of limited funding availability and impacts that such a large program would have on the urban environment, the project was broken into four phases each with estimated construction costs of around $4 billion. The intent of the phased construction is to provide viable new service to the Upper East Side beginning with an initial operating segment which can then be expanded to increase the catchment area. The Phase 1 initial operating segment will also bring early revenue service and will serve as an extension of the existing Q train.

After the completion of Preliminary Engineering of the full build Second Avenue Subway line, which is envisioned to have 16 stations running from 125th Street in Upper Manhattan to Hanover Square in Lower Manhattan, Final Engineering was started on Phase 1 in the spring of 2006. The owner, Metropolitan Transportation Authority Capital Construction (MTACC), awarded the design contract to a joint venture of DMJM Harris*Arup (DMJM Harris is now AECOM Transportation).

Phase 1 is the construction of the new 72nd Street, 86th Street and 96th Street stations and includes the refurbishment of the existing station at 63rd Street/Lexington Avenue (See Figure 1). It also includes a tunneling contract for the construction of the running tunnels between stations. The remaining three Phases are:

- Phase 2—northern stations from 106th Street to 125th Street.
- Phase 3—midtown stations from Houston Street to 55th Street.
- Phase 4—downtown stations from Hanover Square to Grand Street.

The entire new Second Avenue Subway is a two-track system with local stops only.

PACKAGING

The packaging of the contracts was a much deliberated issue. The original plan was to package each station—at 96th Street, 72nd Street, 86th Street and 63rd Street—into its own, large contract with one tunneling contract and on systems contract for the entire Phase 1 segment. After reviewing the local market conditions and the vast competition for design and construction resources from other mega-projects currently underway in the New York Metropolitan area, it was decided to break the major station contracts into smaller contracts which is hoped to lead to more bidders and therefore more competitive bids.

Currently there are 11 contract packages with like trades grouped into contracts.

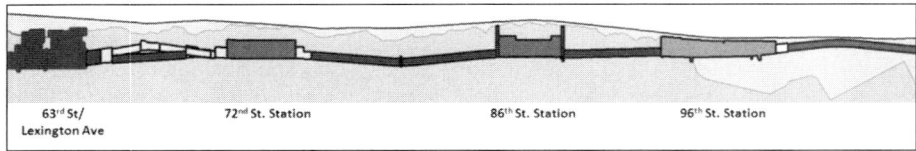

Figure 1. Profile of Phase 1

CURRENT STATE OF DESIGN AND CONSTRUCTION

Contract 1 for the construction of the launch box and the TBM tunnels was awarded in April 2007. At the time of writing, work is progressing with the installation of secant and slurry wall earth support structures which is discussed later in this paper.

Design of Contract 2A is complete and an award of the construction contract is expected in spring 2009. This contract includes the completion of the northern section of the major earth support structure and utility relocation work between 95th Street and 99th Street.

The first of two contracts for the 86th Street Station (Contract 5A) is expected to be advertised in winter 2009. This contract will consist of the utility relocations and cut-and-cover excavation of two starter excavations at 86th Street and 83rd Street from which the follow-on contractor will begin the cavern mining operation. Award of this contract is expected in summer 2009.

The remaining contracts are in various stages of completion with 63rd and 72nd Street design extending into late 2009 when all design work will be completed.

EXISTING CONDITIONS

Second Avenue is a major north-south route on the east side of Manhattan and runs through numerous communities and therefore community board districts.

The avenue is 18 m (60 ft) wide with six traffic lanes with two parking lanes on each side of the street flanked by six meter (20 ft) wide sidewalks. The street is lined with residences and businesses in buildings that range from four-story unreinforced masonry structures to brand new steel and concrete high-rise structures. In addition, many buildings have access doors to vaults that encroach into the sidewalk.

At 96th Street, a park was taken temporarily to be used as a construction yard for Contract 1 and the 96th Street Station Contracts. This enables the contractors to minimize the amount of street and sidewalk takings for laydown areas and provide space for the equipment required to support the TBM operations.

Coordination with the numerous City agencies and utility owners is a significant challenge for the project. These include New York City Dept. of Transportation, NYC Dept. of Environmental Protection, NYPD, NYFD, Sanitation Department, the Parks Department NYC DEP and Consolidated Edison to name only a few.

CONTRACT 1 (C-26002)

Contract 1 was developed primarily as a tunneling contract to be undertaken using a Tunnel Boring Machine (TBM). During the design phase, many options were considered for location of the TBM launch box and the length and directions of the tunnel drives. The evaluation of the options included schedule, contract packaging, consideration of the geologic conditions and the presence of existing obstructions such as existing subway lines.

Once the station locations within Phase 1 were finalized, it was decided that the TBM launch box could double as the southern portion of 96th Street Station where the

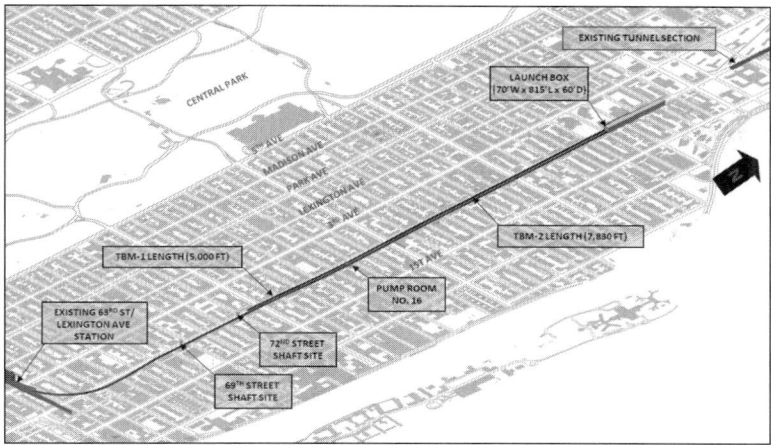

Figure 2. Contract 1 isometric

rock /soil interface exists. This also brings economy to the first Phase because the TBM launch box will eventually become a part of the station.

Scope of Work

Contract 1 is the construction of a TBM Launch Box between 92nd and 95th Streets, the Construction of two circular shafts at 69th and 72nd Streets and the boring of two tunnels using a Tunnel Boring Machine (TBM) from 92nd to 63rd Streets (see Figure 2). The work entails major utility relocations, maintenance and protection of traffic and the construction of a sump at 78th Street.

The Contract was advertised for bid in the fall of 2006 and was awarded to a tri-venture of Skanska USA, Schiavone Construction and J.F. Shea Construction in March 2007 as a low-bid lump sum procurement totaling $337,025,000. This includes a unit price item for fissure grouting in the TBM tunnels which is priced for 15 separate fissure grouting events.

Constraints

To mitigate impacts to the surrounding community, surface work can only be performed Monday through Friday between the hours of 7AM and 10PM and on Saturday between 10AM and 7PM. Underground work can be performed 24 hours a day, 7 days a week.

The Contractor must provide a minimum of four lanes of traffic at all times and sidewalks must be maintained at a minimum width of 2.1 m (7 ft). Pedestrian and emergency vehicle access must also be provided at all times for all buildings.

Noise and vibrations limits are specified in the Contract as are settlement and air quality requirements.

TBM Launch Box

Since the final station structure is shallow, the method of construction for the TBM Launch box is cut-and-cover. The launch box is approximately 248.4 m (815 ft) long and situated in the middle of Second Avenue between 92nd and 95th Streets. The depth of excavation varies between 15.2 m (50 ft) to 18.3 m (65 ft) below grade and involves removal of rock and soil. The top of rock dips down dramatically from 92nd Street heading north towards 93rd Street and is deep between 93rd and 95th Streets.

Utilities. The construction of the launch box requires a large-scale relocation of utilities. This proved to be very challenging because of the number and configuration of the utilities which required careful staging to allow the construction of the support of excavation (SOE) walls and the deck structure. Numerous utility agencies were involved and meetings were held with them on a bi-weekly basis to review the progress of the job, to clarify requirements and to resolve problems that arose as construction proceeded. In many cases, representatives of the utility were either given an office onsite or came to the job site regularly to assist with the relocations and helped expedite resolution to field issues as they arose.

The requirement to keep Second Avenue open to a minimum of four lanes of traffic dictated that the launch box be built one half at a time—east side and west side. The Contractor chose to begin utility relocation work on the west side then move to the east side and repeated this sequence to construct the SOE walls and deck structure. This required the sidewalk widths to be reduced from six meters (20 ft) wide to 2.1 m (7 ft) wide.

Once the SOE walls and deck are installed, the Contractor can work underneath the deck while traffic and pedestrians travel above.

Secant Pile Walls. Where the top of rock is high between 92nd and 93rd Streets, the SOE wall functions solely as support of the excavation and decking and is temporary to the construction of the permanent 96th Street Station structure. Since these walls are temporary, only a suggested design of 1066 mm (42") diameter secant pile walls was shown in the Contract. The ultimate selection of the wall type was left to the Contractor as part of their means and methods.

Contractor's Design. At the time of writing the Contractor installed all of the secant piles on the west side of Second Avenue which varied from approximately six meters (20 ft) to 21.3 m (70 ft) in depth. The Contractor installed 1130 mm (46 ½") diameter secant piles using Bauer BG-40 and BG-28 rigs and reinforced the secondary piles with a steel core beam which varied in size according to the depth of the piles. The secant piles have a concrete compressive strength of 27.6 MPa (4,000 psi) for all piles.

The layout of the secant pile wall in the Contract drawings showed a shift of a portion of the western wall to avoid the relocation of a major communications ductbank. The Contractor was able to work with the utility company to shift the ductbank slightly to the west thereby allowing a straight alignment of the secant pile wall on the west side. In addition, the Contractor chose to use a cast-in-place concrete wall instead of secant pile walls for the south bulkhead wall since the wall is shallow and to facilitate utility penetrations.

The Contractor is casting a concrete wale at the top of the secant pile walls that is also used to support the roadway deck beams.

Slurry Walls. Between 93rd and 95th Streets, the SOE walls are constructed in soil consisting of fill over a 3 m (9.8 ft) thick deposit of organics over sand/silty sand, over varved silt and clay. Slurry walls were mandated as the SOE in the Contract because these reinforced concrete slurry walls will also be the permanent walls for the 96th Street Station. The slurry walls are excavated to a maximum depth of approximately 27.4 m (90 ft) and are also used as ground water cut-off walls. Near 93rd Street, a 0.3 m (one foot) minimum rock socket is required where the slurry walls are founded on rock.

The Contract Drawings show the slurry wall reinforcement cages as one cage the width of the excavated panel, typically six meters (20 ft), requiring splicing the cages as they are lowered into the excavation. Three tremie pipes were also shown and the reinforcement cages were detailed to allow space for the pipes.

Figure 3. Slurry wall reinforcement cage lift

Since the slurry walls will become the permanent walls for the future 96th Street Station, reinforcement bar couplers were specified to allow the connection of the future invert slab, mezzanine beams and slab and roof slab to the walls to form the final station structure by another Contractor.

Contractor's Design. To facilitate the Contractor's preference for more headroom for TBM operations, the Contractor modified the design of the excavation support by changing the number and levels of the bracing in the Contract. This was carried out by the Contractor's design engineer and required reanalysis of the slurry walls including wall reinforcement.

In addition, the Contractor decided to split the slurry wall reinforcement in plan into two separate cages that are each the full height of the slurry wall panels rather than splice full width cages on-site. This required the re-detailing of the reinforcement and shifting certain reinforcement bars to make space for the tremie pipes. The Contractor chose to use two tremie pipes due to the difficulties of bringing three concrete trucks into the site with limited space. This redesign of the slurry walls required a lot of interaction between the design team and the Contractor to agree on a mutually acceptable solution so that the work in Contract 1 can proceed as efficiently as possible without adversely impacting the future station contracts.

The reinforcement cages were fabricated on-site and were lowered into the slurry wall trenches using two cranes which required a highly coordinated effort to ensure the stability of the each cage is maintained as it is lifted from a horizontal position from a flat bed trailer to the vertical position (see Figure 3).

The slurry walls on the west side of Second Avenue were constructed primarily using a Cassagrande rig with clamshell on a Kelly Bar and Liebherr 855 and 885 rigs with clamshell on a cable. In areas where the top of rock is high, Bauer rigs, BG-40 and BG-28, were sometimes used to auger the rock to achieve the required rock sockets instead of flat or star chisels. The sequencing of the panels was in part dictated by the presence of utility manholes and the staging of traffic across the side streets.

TBM Tunnels

Two tunnels are to be bored by TBM with an internal diameter of six meters (19.75 ft) which will serve as the running tunnels between the stations. The TBM drive lengths and sequencing were driven by the Phase 1 schedule.

In order to coordinate the award of the station contracts and compress the construction schedule of Phase 1, only one TBM tunnel is driven through the area of 72nd Street Station to the existing bellmouth under 63rd Street. The second TBM tunnel drive is terminated north of the 72nd Street Station limits allowing the award of the station contract once the first TBM drive is completed.

Moreover, where the TBM tunnels will be enlarged to construct future station and crossover caverns, the TBM tunnels receive initial support only in Contract 1. Where the TBM tunnels remain intact and will be permanent running tunnels, a concrete final lining will be installed that is reinforced with steel fibers rather than rebar to facilitate construction and improve durability.

The first TBM drive will start from the southern end of the launch box under the east side of Second Avenue and proceed south generally on a straight alignment until 69th Street where the TBM proceeds on a curve with a radius of 195 m (640 ft) finally holing-through at the exiting bellmouth structure under 63rd Street. This first drive is approximately 2,377 m (7,800 ft) long. Once the TBM is dismantled and brought back to the launch box and reassembled, the second TBM drive under the west side of Second Avenue will proceed generally straight south to 73rd Street. The length of the second drive is approximately 1,524 m (5,000 ft). The depth of the drives varies from 13.7 m (45 ft) to 25.9 m (85 ft) with a minimum rock cover of 4.6 m (15 ft).

The geology of the drives is through rock that is typically strong (generally between 34.5 MPa–82.7 MPa) and competent gneiss and schist with pegmatite with faults and shear zones.

In the original bid documents, a precast concrete segmental lining was shown as the permanent tunnel lining. During the Bid Phase, an option was introduced for a cast-in-place concrete lining at the request of the bidders to allow for a more competitive bid and to account for local market preferences.

The Contractor has chosen to refurbish a main-beam Robbins TBM and to use a cast-in-place lining with a thermoplastic sheet waterproofing membrane. The initial support of the lined tunnel zones is the Contractor's design.

Construction Shafts

To shorten the duration of the 72nd Street Station Cavern excavation contract, two construction access shafts were included in Contract 1 and can be completed off the schedule's critical path. These shafts are located on the east side of Second Avenue near 69th Street and 72nd Street and are approximately 15.2 m (50 ft) and 12.2 m (40 ft) deep respectively.

In the Contract drawings the two shafts originally had a 9.1 m (30 ft) internal diameter, but this was revised to 8.5 m (28 ft) to avoid certain utility relocations in the middle of Second Avenue. These shafts are to be constructed to the level of the top of the future 72nd Street Station cavern and require 3.7 m (12 ft) to 4.6 m (15 ft) of excavation in fill then drilling and blasting to remove rock.

Community

Construction on Second Avenue in the middle of a vibrant community requires consideration of the surrounding residences and businesses. The construction team has regular interaction with the residents and business owners which can be as simple as a question from a local passerby to a formal scheduled briefing with the residents of

Figure 4. Maintaining access to businesses and pedestrians

an apartment building. The team has made an extra effort to work with the community to minimize impacts by placing limits on noise from construction equipment, maintaining access to businesses on Second Avenue within the zone of construction, and to sign the sidewalks to indicate the pathways to get to businesses and to direct pedestrians (See Figure 4). Working with the MTA, a community liaison based in the field office responds to questions or requests that come in through a dedicated telephone line.

96TH STREET STATION CONTRACTS

96th Street Station includes two contract packages for the construction of the new 96th Street Station. The work entails the construction of the main station box between 92nd and 99th Street, refurbishment of the existing Section 11 tunnel, two ancillary structures and three entrances. The scope of work also includes major utility relocations, underpinning of existing structures for the construction of the entrances, demolition of existing structures and construction of new ancillary structures, and mechanical, electrical, plumbing and finishes for the permanent station and tunnels between 86th Street Station and 105th Street.

Contract 2A (C-26005)

Contract 2A is the first of the station contracts to be issued for bid. This is a heavy civil contract that includes the construction of approximately 670 m (2,200 ft) of permanent reinforced concrete slurry walls from the north end of Contract 1 at 95th Street to the existing Section 11 tunnels at 99th Street and the station invert slab from 93rd Street to 99th Street (See Figure 5). Also included are the SOE walls and excavation for the Ancillary 1 Structure at 93rd Street, Ancillary 2 Structure at 97th Street, Entrances 1 and 2 at 94th Street and Entrance 3 at 96th Street and the roadway decking structure. For Entrance 1, underpinning of a portion of an existing building is required to construct the new entrance within the building at the southwest corner of Second Avenue and 94th Street.

For Entrances 1 and 2, the contractor will be required to demolish knock-out panels that are built as part of the slurry walls constructed in Contract 1 and connect to slurry wall stub panels to complete the entrance SOE walls.

Due to the proximity of the ancillary structures to adjacent buildings, secant pile walls are specified in the contract for the support of excavation walls for Ancillary 1 and Ancillary 2 which will interface with the main station walls. Since the secant piles are

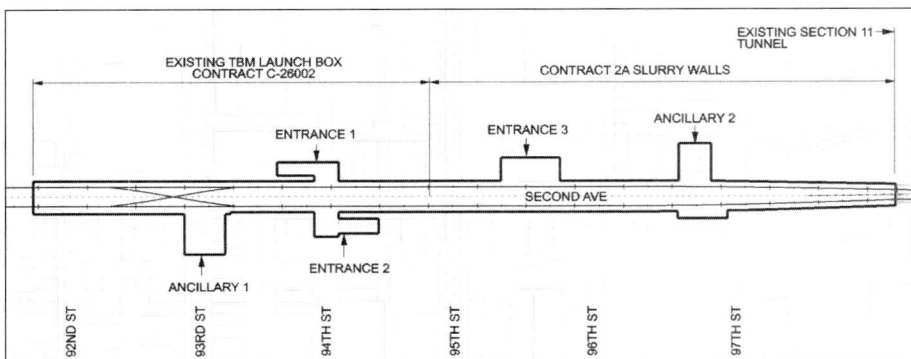

Figure 5. Contract 2A plan

cased during excavation and the excavations are relatively small, the impacts to the buildings are anticipated to be small. During the bid period, Contractors asked if slurry walls could be used in lieu of secant pile walls and this was accepted as an option for the walls that were not adjacent to the existing buildings, provided that the building movement limits set forth in the Contract documents are not to be exceeded.

Because of an abandoned buried manufactured gas plant near the north of the project site, it is expected that a portion of the excavated soil will contain contaminants that will have to be disposed of according to environmental regulations for gas contaminants.

The packaging of the contracts creates multiple interface points. Contract 2A is required to take over the launch box area and traffic deck constructed by the Contract 1 Contractor and take over the building monitoring and groundwater systems then hand over the completed works to the Authority for handover to the next Contractor for Contract 2B.

Contract 2B (C-26010)

The work in Contract 2B consists of the completion of the 96th Street Station which includes the construction of the ancillary and entrance structures, the final station structure including placement of the mezzanine and roof structures, completion of the invert slab for the station between 92nd and 93rd Street, station finishes including mechanical, electrical, plumbing and architectural work and restoration of the utilities and street. Also included are the refurbishments of the existing Section 11 tunnel between 99th and 105th Streets and the new tunnels between 86th Street Station and 96th Street Station. The Contract requires the connection to structures built by the Contract 1 and 2A Contractors to complete the station.

63RD STREET STATION CONTRACT

Contract 3 is the refurbishment of the existing 63rd Street/Lexington Avenue Station originally constructed in the early 1980s. This includes the opening of existing entrances that were constructed to facilitate the connection to the new Second Avenue Subway line but were left closed.

The station was constructed as a two-by-two stacked station with two tracks on the upper level and two on the lower with the southern half of the station currently serving F line trains. The northern half of the station, which currently serves as non-revenue off-peak train storage for the Q line trains, will be refurbished and serve the Second Avenue Subway line allowing revenue Q trains to pick-up and discharge passengers.

The connection to the new Second Avenue Line is accomplished by the use of a reinforced concrete stub tunnel that was constructed with the original station which curves to the north as the tunnel proceeds east towards Second Avenue. This vertically stacked stub tunnel ends at a rock face.

The scope of work for Contract 3 consists of the installation of new elevators, ventilation and electrical work, architectural finishes and structural modifications to the existing station.

72ND STREET STATION CONTRACTS

The contracts for the 72nd Street station include the construction of the new 72nd Street Station and the connection to the existing stub tunnel structure under 63rd Street. The station and connection to the stub tunnel structure are mined caverns given the shallow depth of rock which has the benefit of minimizing impacts of the construction to the surrounding community.

72nd Street Station was originally envisioned as a three-track station for flexibility of operations given that the extension of the Q line via the existing 63rd Street/ Lexington Avenue Station and the dedicated Second Avenue T line converge at 72nd Street station. After a further operational study leading to modifications to the cavern configurations south of 72nd Street Station, the station was modified to be a two-track station with a center platform without compromising the flexibility of train movements. This change also leads to cost savings because of the smaller cavern shapes associated with a two-track station configuration.

72nd Street Station is approximately 302 m (990 ft) long and located between 69th and 72nd Streets and favors the west side of Second Avenue. The station complex includes two crossover caverns on each end of the station, three entrances and two ancillary structures which will require modifications and underpinning of existing structures.

The station cavern is approximately 21.3 m (70 ft) wide and 15.2 m (50 ft) high with a minimum rock cover of 9.1 m (30 ft). The caverns are constructed by drill-and-blast and receive an initial shotcrete and final CIP concrete lining. The two track caverns south of 72nd Street range from 9.1 m (30 ft) to 14.6 m (48 ft) spans and make provision for the future extension south for Phase 3 without requiring major service disruptions for the construction. The configuration of the station and caverns is shown in Figure 6.

Access to the cavern during construction is through the two shafts at 69th and 72nd Streets that were started in Contract 1. These two circular shafts are located on the east side of Second Avenue and enable the station Contractor to start work on the shaft sinking while avoiding any further utility relocations in these areas.

The connection to the existing stub tunnel south of 72nd Street station is a combination of TBM tunnels, single track mined tunnels and two track caverns. The two tracks at the station are side-by-side; therefore the geometry of the tracks to the station must take the tracks from the vertical stacked stub tunnel to a horizontal side-by-side position. This track geometry in addition to special track work required to connect to the future Phase 3 tunnels which take the T line south leads to some challenging cavern configurations.

86TH STREET STATION CONTRACTS

86th Street Station is approximately 290 m (950 ft) long and is located between 83rd Street and mid-block between 86th and 87th Street and is centered on Second

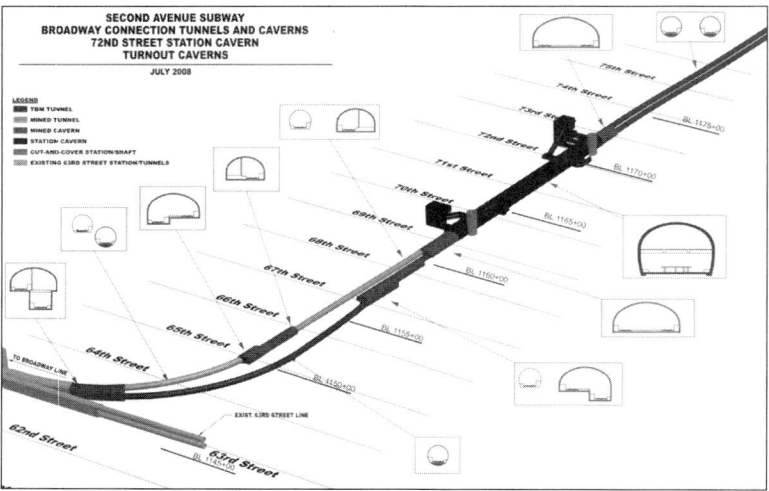

Figure 6. Contract 4 isometric

Avenue. The shallow top of rock enabled 86th Street Station to be designed as a mined cavern with two cut-and-cover shafts at the north and south ends. The shafts serve as the construction access for the mining operations and will ultimately house ancillary spaces.

The start of construction for 86th Street Station is constrained by the TBM tunneling that is part of Contract 1. Both TBM tunnels pass through the area of this station, therefore the TBM tunnels and operations must be complete before the 86th Street Station Contractor can fully sink the two access shafts and begin drilling and blasting operations.

86th Street Station will have two entrances and two ancillary structures and will require modifications and underpinning of existing buildings.

Finding adjacent properties to house the required spaces to serve the station was challenging. Limited real estate takings led to the relocation of space to below ground within the station with some of the ancillary space incorporated into the two cut-and-cover shafts at the north and south ends. Another way space was developed is through the use of two cavern sections—a high-roofed ancillary section and a low-roofed public section. The public cavern shape runs through the length of the platform and is approximately 13 m (43'-6") high. To gain additional space the cavern height is increased to 16.5 m (54'-3") which is used to house additional ancillary space (See Figure 7).

The mining operations for the caverns will commence at the north and south shafts and will involve excavation sequences incorporating two existing TBM tunnel bores. The cavern is supported by initial support consisting of bolts, shotcrete and ribs and a final lining of reinforced cast-in-place concrete and waterproofing.

CONCLUSIONS

Construction of any kind in New York City is always challenging. The communities that line the Second Avenue Subway alignment are vibrant and filled with residents and numerous businesses all wondering how their lives will be impacted by the construction of this new line. Keeping residents informed through continuous communication and working with businesses that are directly impacted by the construction activities helps to alleviate some of their concerns and to minimize disruptions. Also, working closely

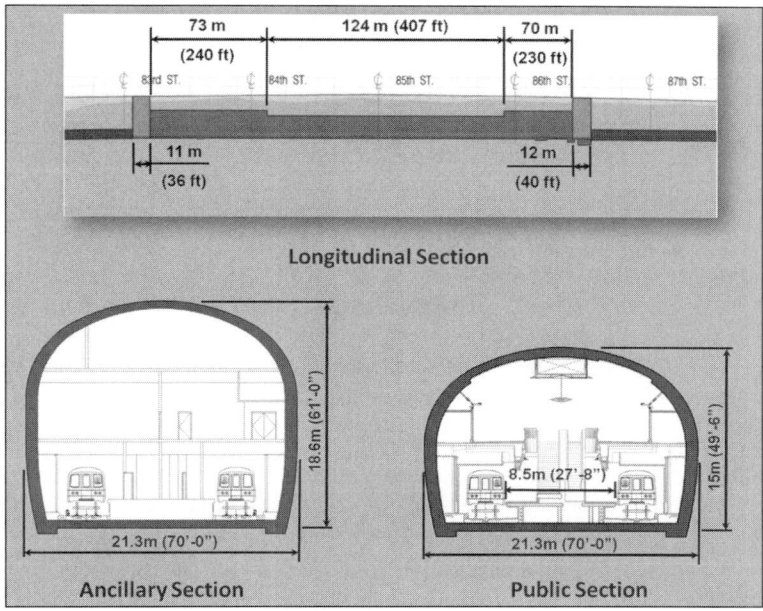

Figure 7. Sections and profile of Contract 5 caverns

with the various City agencies to understand their requirements prior to work commencing has also helped to minimize the surprises during construction.

The experiences during the design and construction of Contract 1 are being considered while developing the subsequent contracts to adapt the future contracts to local conditions.

With construction underway and subsequent contracts to follow in the near future, the Second Avenue Subway is moving closer to a reality.

Views and/or opinions expressed in this paper do not necessarily reflect the views and opinions of the MTA and/or its Agencies.

NO. 7 SUBWAY EXTENSION CROSSING UNDER AN EXISTING SUBWAY STATION: CHALLENGES AND INTEGRATION OF UNDERPINNING INTO THE DESIGN OF NEW TUNNELS

Aram Grigoryan ▪ PB Americas

ABSTRACT

This paper addresses the challenges of designing the underpinning of major subway station with 4 operating tracks. No. 7 subway extension crosses under the existing 42nd street subway station on the 8th Avenue line. Understanding the history of the original construction, of the current structures and their condition were instrumental in developing a feasible underpinning design that is integrated with the design of the new tunnels. The existing 8th Avenue subway station has a lower abandoned level. Several alignments and corresponding underpinning schemes were studied. The design allows new tunnel construction under the station without interruption of station operations.

INTRODUCTION

All public agencies in charge of mass transit maintenance and construction are facing new challenges as they develop public infrastructure that is increasingly complex. The existing transit structures and surrounding aging infrastructure present numerous challenges, constraints and limitations.

New York City Transit (NYCT) operates the largest transit system in the world. NYCT engineers, Consulting Engineers and Contractors face a dual challenge of improving and expanding the existing infrastructure to meet increasing demand, while simultaneously maintaining, repairing and rehabilitating existing infrastructure.

No.7 subway line extension project is undertaken by The Metropolitan Transportation Authority (MTA) and the City of New York department of City Planning to promote the transit-oriented redevelopment of the Hudson Yards area, which extends from West 28th Street on the south, West 43rd Street on the north and the Hudson River Park on the west. The key element of the redevelopment is construction and operation of an extension of the No. 7 Subway Line to serve the Hudson Yards.

The concept of extending the No. 7 line westward was considered in the past. Subway service began in 1915 between Grand Central and Jackson Avenue, to Corona in 1917, and Flushing in 1928. An extension to Times Square opened in 1927. Tail tracks of the Times Square Station have reached the middle of Eighth Avenue. At that time, in the mid-1920s, plans were developed for extending the line to Twelfth Avenue.

The Independent (IND) subway system's 42nd street station was opened shortly after—in 1932. The IND lower level is a single southbound track (Track D3) on the same elevation as the end of the No. 7 line tail tracks, and it connects with the regular tracks at both ends within a few blocks south and north of the station.

In the late 1970s the extension was considered in connection with the development of the Jacob K. Javits Center. At that time the estimated cost of $100,000,000 was considered to be excessive.

Several additional studies were conducted in the 1980s and 1990s.

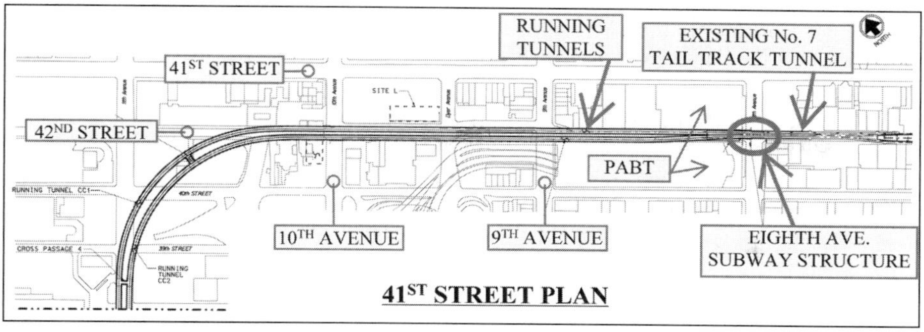

NO. 7 SUBWAY EXTENSION CROSSING UNDER EXISTING SUBWAY STATION 803

Figure 1. No. 7 Line Extension—Part Plan

Finally, in 2002 the MTA undertook a study to prepare an environmental impact report and preliminary project design. An additional incentive was to boost the city's bid for the 2012 Olympics.

The No. 7 line extension project will extend the existing line by approximately 2.4 kilometers (1.5 miles) from Times Square at 41st street and Eighth Avenue southwest to the Jacob K. Javits Convention Center at 11th Avenue and 34th Street.

In the vicinity of the 8th Avenue (see Figure 1) the proposed No. 7 subway line extension tunnel:

- Connects with the existing tail track tunnels on east side. The existing tail track tunnels are lowered to accommodate the modified vertical alignment
- Passes under the existing NYCT 42nd Street/8th Avenue subway station structure
- Passes under the Interface of Port Authority Bus Terminal (PABT) and NYCT subway station structure

EXISTING SUBWAY COMPLEX

42nd Street/Times Square Subway Complex is the largest of the entire NYCT system in passenger volume and size.

The complex includes the following subway stations (see Figure 2):

1. **IRT 42nd Street shuttle "Times Square"**—on West 42nd Street at Broadway; opened 10/27/1904.
2. **BMT Broadway Line "Times Square-42nd Street"**—on Broadway between West 40th and West 42nd Streets; opened 4/14/1918.
3. **IRT West Side IRT "Times Square-42nd Street"**—on 7th Ave between West 42nd and West 40th Streets; opened 7/1/1918.
4. **IRT Flushing Line "Times Square"**—West 41st Street between Broadway and 7th Avenues; opened 3/22/1926.
5. **IND 8th Ave Line "42nd Street-Port Authority Bus Terminal**—on 8th Ave between West 40th and West 44th Streets; opened 9/10/1932.

IRT Flushing Line "Times Square" is the southern Terminal of the #7 line.

The tail tracks extend to the IND 8th Ave line, but do not physically connected to the lower level (see Figure 3). The incompatibility of the subway car and tunnel dimensions would prohibit this connection.

IND 8th Ave Line "42nd Street-Port Authority Bus Terminal services the IND A/C/E lines and has 2 platforms and 4 tracks, plus an abandoned platform on the lower

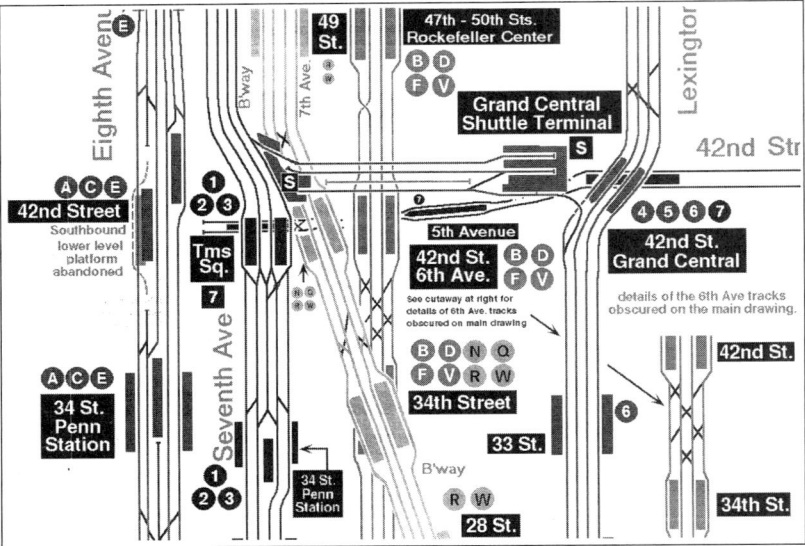

Figure 2. 42nd Street/Times Square Subway Complex

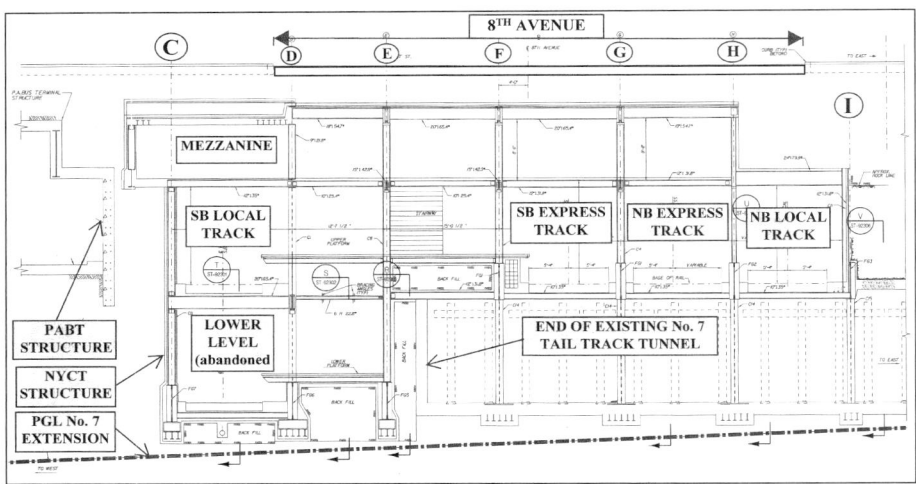

Figure 3. Section through existing structures under the 8th Avenue

level (see Figure 3). The lower level was in use only from 1959 to 1981; it is 366 meters (1,200 feet) in length and has only one track and side platform. This track is aligned with the local southbound IND track above. The station has been repaired and renovated in 2004. PB was the design consultant in Joint Venture with Dattner Architects. It has been the author's privilege to serve as a Structural Engineer and PB's Project Manager during the design, and to provide construction support services to NYCT and the contractor (Judlau) during the construction. During the structural inspection of the lower level we have observed that horizontal bracing members (6" I-beams) have buckled. We investigated various possibilities trying to determine what might have caused these members

NO. 7 SUBWAY EXTENSION CROSSING UNDER EXISTING SUBWAY STATION   805

to buckle. There were no obvious reasons such as signs of fire or impact damage. All the buckled members were concentrated at one location. After we have studied the surrounding structures and superimposed the drawings of the 8th Avenue station and of the No. 7 tail track, we were able to conclude that it was an effect of interaction of two underground structures that caused the buckling. At that time we have developed cost effective design solution and provided additional bracing to stabilize the existing structure. Being familiar with the existing 8th Avenue structure was very helpful for understanding the structural challenges associated with the No. 7 line extension under the 8th Avenue.

One of the challenging requirements of the No 7 line extension project was developing a feasible and constructible design that would allow extending the No. 7 line to cross the 8th avenue line subway structure below 4 operating tracks and passenger platforms.

The structural challenge is to develop a feasible and cost effective method that would satisfy numerous technical challenges, operational and logistical constraints, (delivery of materials and people, removal of muck and debris, working under 8th Avenue, and directly under 4 operating subway lines, near the foundations of the PABT and under the PABT structure). We began with thoroughly studying the history of the structure and exploring the reasoning behind various design concepts and design solutions that the original designers have developed. Seemingly "small" details of the existing structure elements, the history and sequence of the original construction might define and significantly affect the entire underpinning concept.

NO. 7 EXTENSION DESIGN AND UNDERPINNING CHALLENGES

Alignment

Various alignment options were studied. Horizontal and vertical alignments were controlled by numerous factors and constraints along the entire project. Some of the factors were geotechnical conditions and existing structures and foundations along and above the proposed tunnels. These factors and the alignment development are not discussed in this paper.

In the vicinity of the 8th Avenue the horizontal alignment was controlled by the existing tail track structure; it had to meet the plan and profile of the existing No. 7 terminal station.

In the same area the vertical alignment options were:

- **"Deep option" Alignment** that would allow continuous tunneling under the 8th Avenue using Tunnel Boring Machine (TBM). The alignment would allow the new tunnels to pass under the existing lower abandoned level of the 8th Avenue Station.
- **"High option" Alignment** that would allow the new tunnels to pass directly under the existing 8th Avenue structure minimizing excavation and modification of the tail tracks.
- **"Medium option" Alignment** that would pass through the lower level and would require modification and lowering of the tail track structure.

A variation of "Medium option" alignment was selected for the final design.

Underpinning

The No. 7 line extension tunnel interferes with the 8th Avenue subway lower level tunnel, and requires the lower level to be demolished. The remaining structure must be reframed and re-supported within the new construction limits. The construction of the new tunnel structure under the existing subway structure requires extensive underpinning and careful construction phasing.

After considering various design concepts it became apparent that the new structure configuration will greatly depend on the underpinning method and construction sequence. Moreover, the final structure will be an integral part of the underpinning and load transfer framing, thereby making it difficult to separate the temporary works, usually handled by the contractor, from the final structure framing design. The underpinning concepts and construction methods that were developed will integrate the new structure, parts of the temporary structures, and the modified existing structure.

In the vicinity of the 8th Avenue there are three distinct structural zones that require customized underpinning design concepts to address specific challenges and to develop a feasible and constructible design (see Figure 3):

Zone 1: Interface of the PABT Bus Terminal and the west side of the NYCT subway.
Zone 2: Column lines C-D-E within the abandoned lower level of the 8th Avenue subway and under the operating IND southbound local track and platform.
Zone 3: Column lines F-G-H-I within the existing tail track and under the two operating IND express tracks and one local northbound track and corresponding platforms.

These three zones differ from each other by their structural framing and details, load magnitude and load distribution, foundation types, and consequently, did require customized design and underpinning methods. Consistent underpinning design criteria were developed to unify various underpinning design concepts and construction methods. One of the main requirements was to develop feasible and constructible methods and procedures for temporary supporting the existing structures, transferring loads from existing structures to temporary support structures, and transferring loads from temporary support structures to permanent support structures. The methods of load transfer may vary and may include hydraulic jacks, screw jacks, wedges, beveled shims, etc. depending on load magnitude and actual framing. All temporary and permanent supporting structural members must be preloaded by jacking, wedging and shimming as required to prevent future deflections at existing support locations prior to removal of any temporary or permanent supporting structural members. The ultimate goal of the underpinning and load transfer is to protect the existing structure by stabilizing the existing geometry and to prevent the existing structure from additional deformations and deflections. Since the underpinning work is to be performed under four operating tracks, the transfer of loads from structures that support existing tracks shall be performed while there is no train load on the particular structure. Normally this is achieved by performing the actual load transfer under flagging when the train is temporarily stopped by NYCT "flaggers" for a short time, or during General Orders (GO's) when the particular line is closed for a weekend. The GO's are costly, therefore the design was specifically "fine tuned" to eliminate or minimize GO's.

The existence of the 8th Avenue abandoned lower level and unused tail tracks of No.7 Line presented opportunities for the construction staging and underpinning. These spaces, if used properly, might help to eliminate or minimize the impact of the construction on the existing station and train operations above.

The existing transfer girders FG are critical elements of the existing structural framing. We used these girders in designing the underpinning and load transfer procedures. The FG girders are built-up riveted sections comprised of web plates, flange angles, and cover plates. The challenging task of transferring loads from the existing columns to temporary supports is further complicated because the plate girders require additional bearing stiffeners at temporary support locations. Providing bearing stiffeners might have been the obvious choice under the different circumstances; however it is not viable due to cost considerations and significant impact on service. Installing new bearing stiffeners would require taking an operating track out of service, removing part of the concrete track structure and concrete encasement of the girder, installing stiffeners and rebuilding the structure to its original condition.

The alternative design concept was developed to transfer loads from the FG girders to temporary supports by combining several load transfer paths:

- Through distribution beams that engage sufficient number of girder rivets that connect bottom flanges to the web.
- With temporary track support framing to support track live load during the underpinning.
- Engaging the existing FG girders' "built-in" hidden strengths. The original designers have provided structural elements and details for so-called "future" columns, as indicated on archive drawings.

"Future" columns. We have noticed that the existing structure drawings show strategically located columns under the transfer girders FG. These columns are shown in dashed lines and are referred to as "future" columns. Apparently the IND designers have envisioned some kind of No. 7 line extension to the west side. The "future" columns are strategically positioned to accommodate two IRT tracks. This Interesting fact suggests that IND designers made some allowances in their design for underpinning and construction under the lower level. This was an intriguing discovery since it is suggested (but not confirmed) by some transit historians that the IND was built deliberately on two levels so that any extension of the IRT No. 7 Line would be impossible or very difficult. The "future" columns are positioned very close to the existing columns, making it questionable if there is sufficient room to accommodate the temporary column cap plate. The actual dimensions need to be verified in the field. During the design it is prudent to assume that there might be insufficient space, and to develop an alternative or additional support and load transfer method.

It was not possible to take full advantage of these "future" column provisions because the original designers envisioned the future tunnels passing under the lower level. The concept was based on deep vertical alignment that would keep the existing lower level structure. The current No. 7 Line extension design alignment passes through the lower level requiring partial removal of some of the FG girders and significant structural modifications.

The future column locations were used in our design as part of load pick up and load transfer points for underpinning. This original solution was not obvious, and we have shared the information with the contractor during the early construction stages to assist in developing the feasible underpinning method.

Zone 1: PABT and NYCT Interface

The proposed new No. 7 subway line extension tunnels pass under the interface of the PABT and the NYCT structures (see Figure 4). The PABT main structure at the interface has a structural support framing system that, as it appears from the PABT extension drawings dated 1975, is not significantly affected at the interface by the new tunnels. The PABT main supporting columns are beyond the footprint of the proposed tunnels. There is a 1'-6" thick underground concrete wall along the east edge of the PABT bus ramp that was built around 1975. The wall serves as a retaining wall for the bus ramp and does not support the PABT main framing. Part of the underground wall rests on rock, and part of the wall is supported by the NYCT ejector pit structure.

It appears that the wall confines the rock foundation that supports the cantilevered part of the NYCT mezzanine structure.

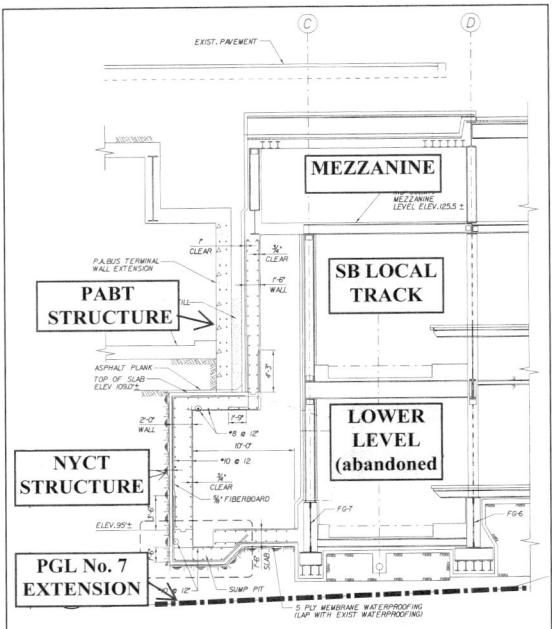

Figure 4. Section at the interface of NYCT and PABT structures

Depending on the condition of the foundation rock under the mezzanine cantilever structure, the wall might be supporting horizontal components of the superimposed column vertical loads of the mezzanine structure that overhangs over the subway wall.

Hence, an adequate structural support of the PABT wall extension—both vertical and horizontal—is critical for ensuring the safety and structural integrity of the NYCT 8th Avenue subway structure.

NYCT 8th Avenue Subway Station structure west exterior wall runs in north-south direction under the west side of the 8th Avenue. The structure has three levels (see Figure 4):

- Mezzanine level that is approximately 2 meters (7 feet) below the street surface;
- Main operating track level;
- Abandoned lower track level.

The subway structure west wall consists of steel columns spaced at 1.5 meters (5 feet) off centers, and concrete jack arches spanning horizontally between the steel columns.

The part of the mezzanine structure that extends from column line 20 to column line 23, overhangs over the main exterior subway wall below. The overhang part of the existing mezzanine is located over the footprint of proposed new tunnels. The mezzanine overhang structure houses pipe chamber, toilets, ejector shaft, and ejector pit. The bottom of the ejector pit is located approximately at the elevation of the abandoned lower level track. The overhanging part of the mezzanine exterior wall at the south end is supported on the ejector pit structure. To the north of the ejector pit structure the mezzanine overhang rests on a narrow strip of rock that is approximately 4.5 meters (15 feet) wide. It is a remaining part of the rock substrate that has remained after the

NO. 7 SUBWAY EXTENSION CROSSING UNDER EXISTING SUBWAY STATION 809

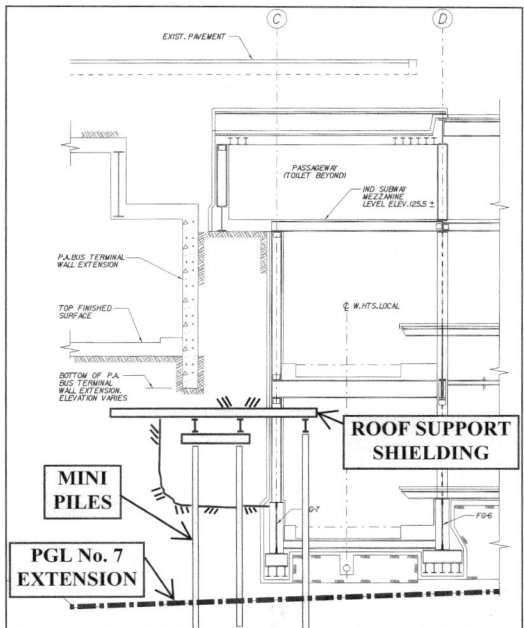

Figure 5. Section at NYCT and PABT interface—underpinning work in progress

excavation and construction of the NYCT structure on the east side, and PABT structure on the west side.

Two construction alternatives were considered in developing the underpinning method and final design under the interface: first, for the construction progressing from east to west, and second, for the construction progressing from west to east. Direction of the construction does impact the construction method since the adjacent construction zones are interrelated, and the underpinning of these zones needs to be coordinated.

To the west of the interface the construction will be done in the open cut and braced excavation with a decking over the excavation in order to support and maintain PABT bus ramps. The open excavation will terminate near the PABT and NYCT interface, approximately 5 meters (16 feet) to the west from the NYCT west exterior wall. The underpinning of line C in zone 2 is critical for the completion of the underpinning and new construction of zone 1 under the PABT.

In order to construct the proposed new tunnels at the interface, the existing PABT underground wall and the NYCT mezzanine overhang structures shall be underpinned. The key element of the underpinning design was the roof support shielding that consists of closely spaced 20 mm diameter (8 inches) structural pipes reinforced with rails or bars, and filled with concrete or grout (see Figure 5).

Zone 2: Lines C–D–E

The existing structure framing within the zone 2 (column lines C, D, and E) varies and is different for each of the column lines. Consequently, the underpinning methods, sequence, and details vary from one column line to other. The underpinning design and the final structure were developed to provide maximum flexibility during the construction. Each of these column lines can be underpinned and the new structure can be built independently of each other. The direction of the construction does not affect the

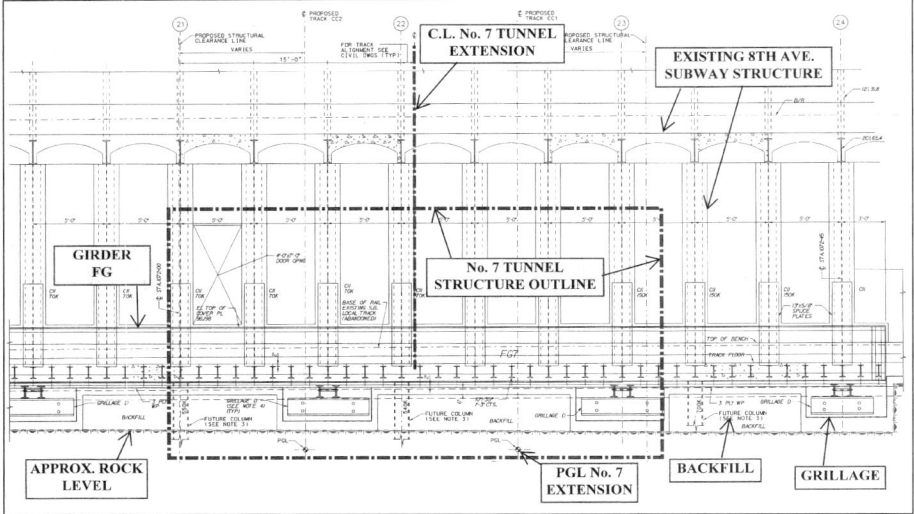

Figure 6. Section along column line C. Note columns spaced at 1.5 m (5 feet).

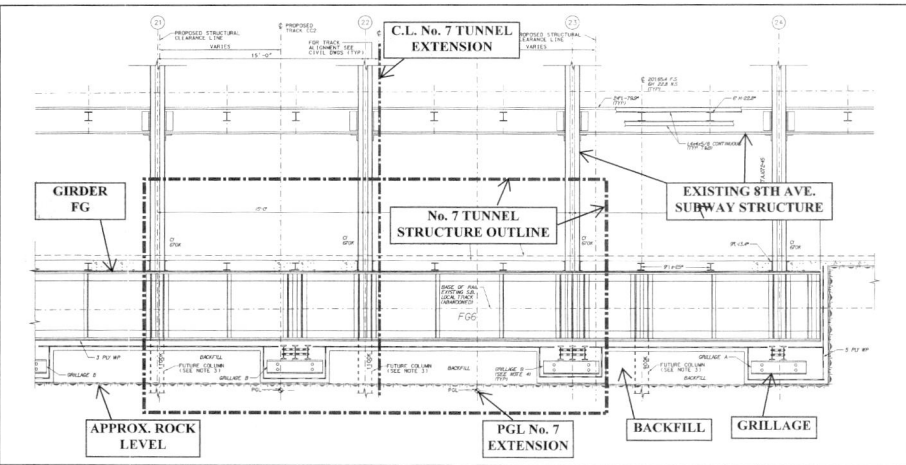

Figure 7. Section along column line D. Note columns spaced at 4.5 m (15 feet).

design. Moreover, the final structure is designed to be flexible to allow the contractor to modify the underpinning and construction methods as needed to best utilize the contractor's preferences and to respond to the field conditions. This flexibility was provided without sacrificing the quality of underpinning work. Existing structure integrity and public safety were paramount of the criteria developed for all underpinning procedures.

Wherever it was advantageous, the existing structure was used as an integral part of the construction method and of the new construction. The structural support system of these three column lines have some similarities that were noticed while studying the existing structure drawings.

The column lines are supported by transfer girders FG (see Figures 6 and 7.)

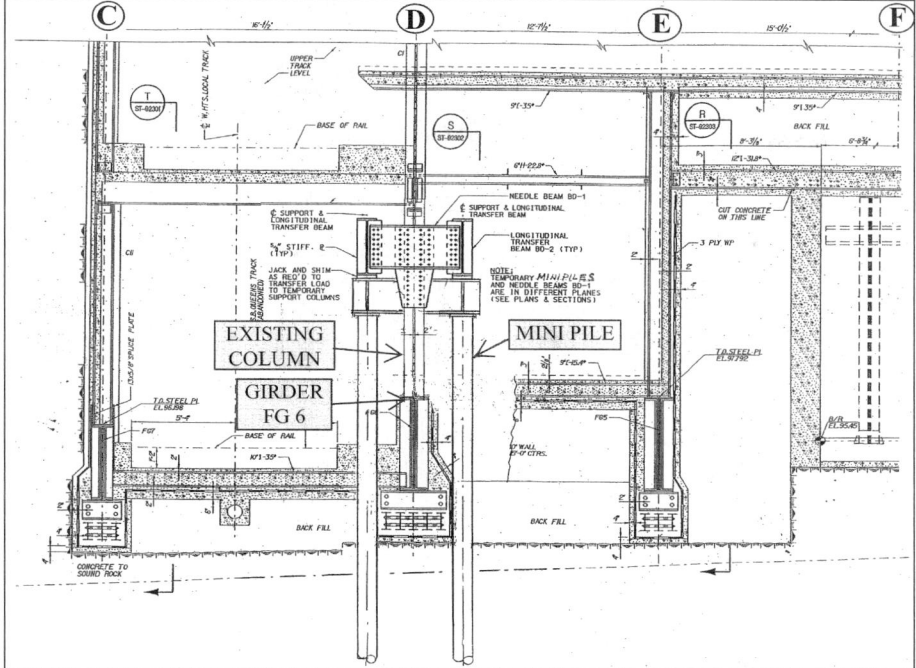

Figure 8. Section showing underpinning of line D in progress

The floor girders are, in turn, supported on grillages founded on rock. It appears that during the construction of the original structure the rock was excavated down to the base of the grillages, and the space between the grillages was backfilled after the framing was built. Based on this observation (subject to field verification) it was deemed possible to demolish and excavate and remove the backfill to the top of the rock without installing major temporary support structures. We have performed computer analysis of the existing framing and have verified that the girders FG were originally designed to support the applied loads and to span between the grillages.

Figure 8 and Figure 9 show samples of underpinning work in progress.
Figure 10 shows the completed new structure at line D.

Zone 3: Lines F–G–H–I

The width of the new tunnels in zone 3 is designed to match the existing No. 7 line tail track structure; however the new construction requires underpinning and replacement of the existing center columns, and underpinning and extension of the existing tunnel exterior walls (see Figure 11). The work in this zone requires careful construction staging and load transfer procedures to ensure the existing structure integrity and safe and uninterrupted train operations.

Figure 12 shows the completed design of the No. 7 Line structure under the 8th Avenue.

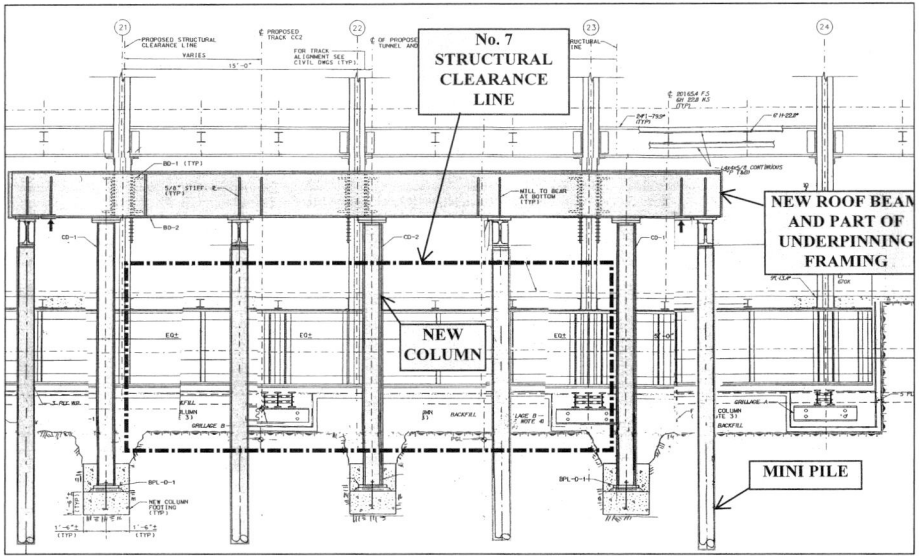

Figure 9. Section showing underpinning of line D in progress

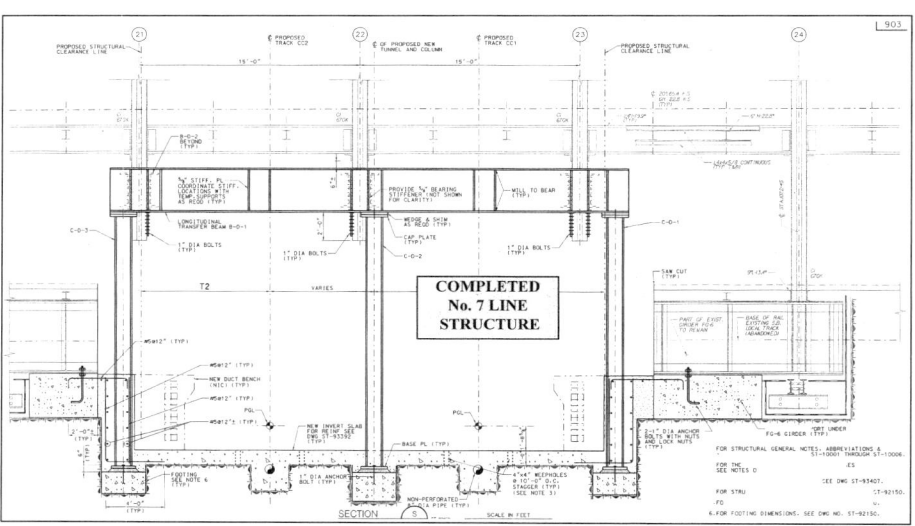

Figure 10. Section showing completed structure at line D

NO. 7 SUBWAY EXTENSION CROSSING UNDER EXISTING SUBWAY STATION   813

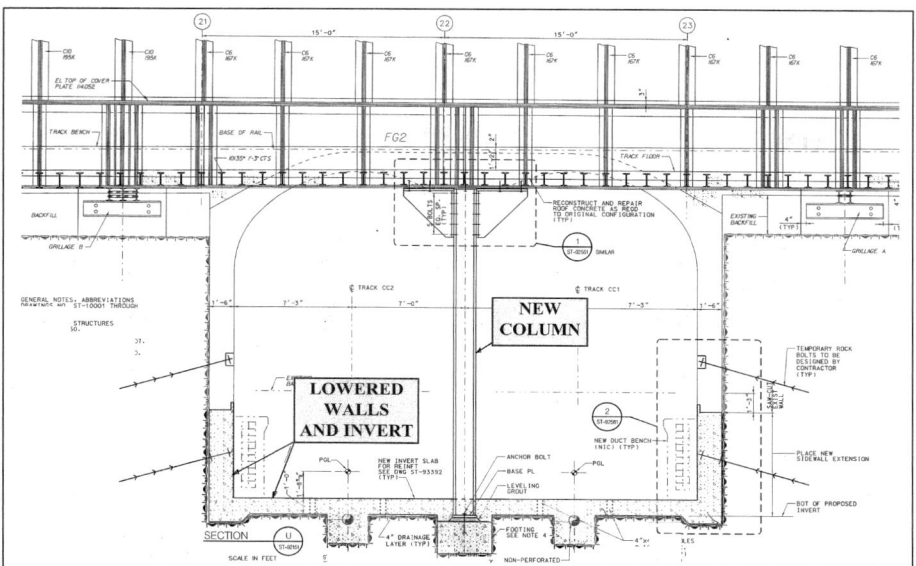

Figure 11. Section showing completed structure with lowered tail track tunnels

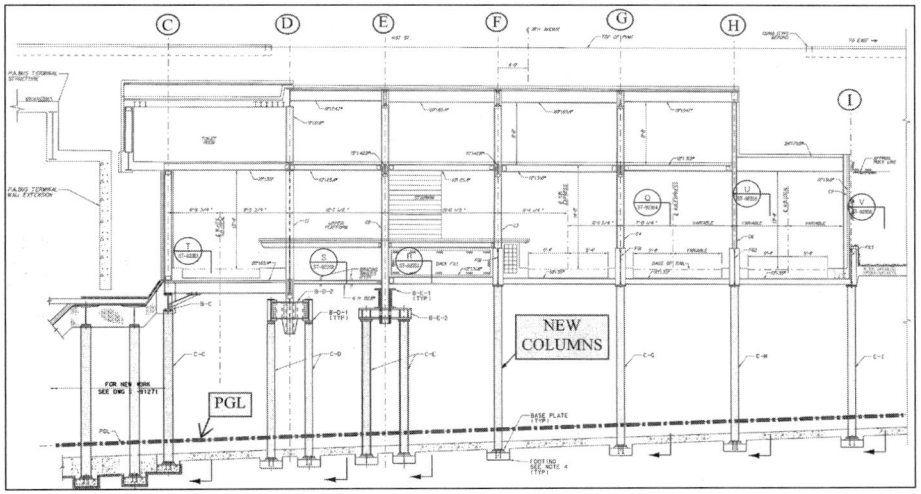

Figure 12. Section showing completed structure under the 8th avenue

RAILROAD INTERFACE MANAGEMENT FOR MTA EAST SIDE ACCESS PROJECT TUNNELS AND STRUCTURES

Michael A. Piepenburg ▪ Hatch Mott MacDonald

Daniel A. Louis ▪ MTACC/URS Corp

Robert Magnifico ▪ Metropolitan Transit Authority

Augustine Juliano ▪ Amtrak

ABSTRACT

The East Side Access (ESA) Project will extend Long Island Rail Road (LIRR) into midtown Manhattan via 11 km of tunnels, a new terminal station and associated support facilities. In addition to the hard-rock tunneling currently underway in Manhattan, soft-ground tunnels in Queens will traverse beneath Amtrak's Sunnyside Yard and Harold Interlocking, the busiest rail interlocking in North America. This paper details ESA's coordinated approach to interfacing with both Amtrak and LIRR with regard to the planned tunnels, shafts and related construction beneath or adjacent to the operating railroads. This interface includes integration of railroad staff, extensive presentations, reviews and agreements with the railroads regarding settlement mitigation, real-time automated geo-instrumentation monitoring, ground improvement, and geotechnical/environmental controls.

Introduction and Summary of Project and Project Purpose

The Long Island Rail Road (LIRR) transports over 270,000 passengers traveling on over 700 trains into and out of Manhattan's Penn Station on a daily basis. To improve service into Manhattan, a new extension, the East Side Access (ESA) Project is being built to carry traffic to and from Grand Central Terminal. The primary components of the ESA Project are shown on Figure 1 and consist of 31,000 feet of stacked twin tunnels mined beneath Manhattan through hard metamorphic rock and four tunnels (totaling 10,500 feet in length) mined in unconsolidated glacial soils beneath Amtrak's Sunnyside Yard and LIRR's Harold Interlocking in Queens. The two sets of new tunnels will be connected by a set of existing tunnels located beneath the East River and constructed in the mid- to late-1970s.

The Queens area tunnels will be constructed from a large open-cut excavation and will be advanced beneath the six-track Harold Interlocking mainline capable of speeds up to 60 mph, Amtrak's Sunnyside rail-yard, plus additional service tracks for train washing, maintenance, and turn-arounds. In addition to the tunnels, three TBM reception pits, three ventilation/emergency exit structures, and smaller structures will be constructed from the surface in and around the active railroad.

The primary stakeholders for the Queens-area tunnels are the LIRR and Amtrak. In addition, the Sunnyside railyard is used for servicing and mid-day storage of trains from New Jersey Transit (NJT). Electric power for the railroads is provided through overhead catenary lines for Amtrak and NJT, and by third rail for LIRR. Both railroads, serving as passenger-carriers are subject to similar Federal Railroad Administration (FRA) regulations with regards to track safety standards and railroad work place safety.

Figure 1. Location of the East Side access project in New York City

Both railroads are heavily traveled, with high-speed rail lines with two daily rush hour schedules. In addition, both railroads have stringent safety requirements to protect passengers, railroad employees, non-railroad contractors, and railroad infrastructure.

Project Description

The ESA Project is projected to be completed in four, overlapping stages with major completion estimated in 2016. Each stage has a defined set of track outages and train operational routings through Harold Interlocking. These routings keep train services away from areas in which construction of structures and railroad facilities is ongoing. The Queens-area tunnels and ancillary structures described in this paper will largely be constructed in the first two stages.

Among the largest and most complex contracts and the ones with the most significant railroad interfaces are:

- Queens-area Geotechnical Investigation
- Contract CH053—Harold Structures—Part 1
- Contract CQ031—Queens Bored Tunnels and Structures

Major Queens-area work elements shown on Figure 2 are described in the following paragraphs and will be used to illustrate the surface and subsurface interface issues with the railroads.

Queens-area Geotechnical Investigation. The geotechnical investigation for the Queens-area projects was performed between 1999 and 2006 and consisted of advancing over 300 geotechnical borings. Of the borings, approximately 250 were advanced in and around the railyard and mainline tracks.

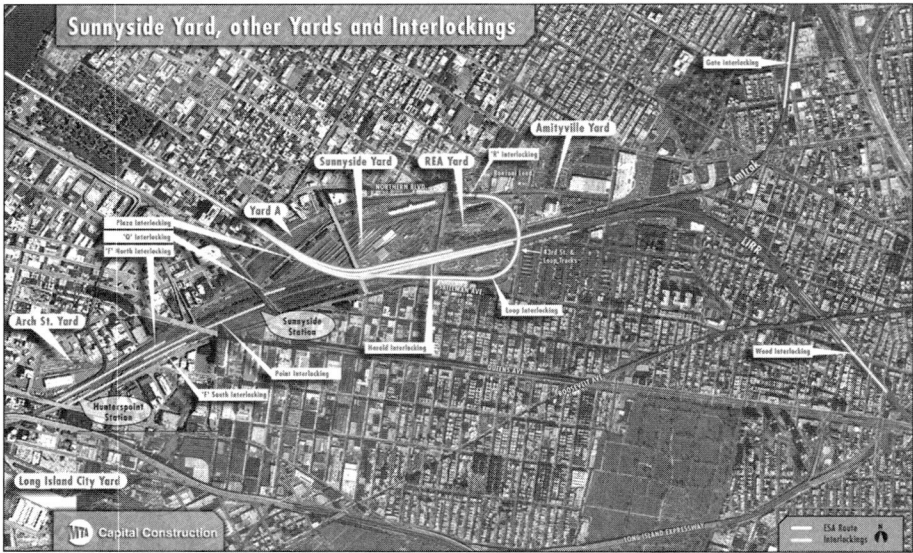

Figure 2. Sunnyside Yard and Harold interlocking in Queens, New York

Figure 3. Subsurface exploration drilling in Sunnyside yard

The soil borings were drilled with hollow-stem augers and rotary wash equipment and samples collected with split-spoon samplers or thin-walled samplers. Selected borings were advanced into rock and continuous lengths of rock core were collected with wireline diamond-bit drilling equipment. Monitoring wells were installed in selected boreholes to monitor groundwater levels.

In Sunnyside Yard, the drilling was performed using drilling rigs mounted on trucks equipped with "Hi-Rail" equipment as shown in Figure 3. More commonly, rubber-tired truck-mounted drill rigs were used to advance borings along roadways and open areas away from the tracks.

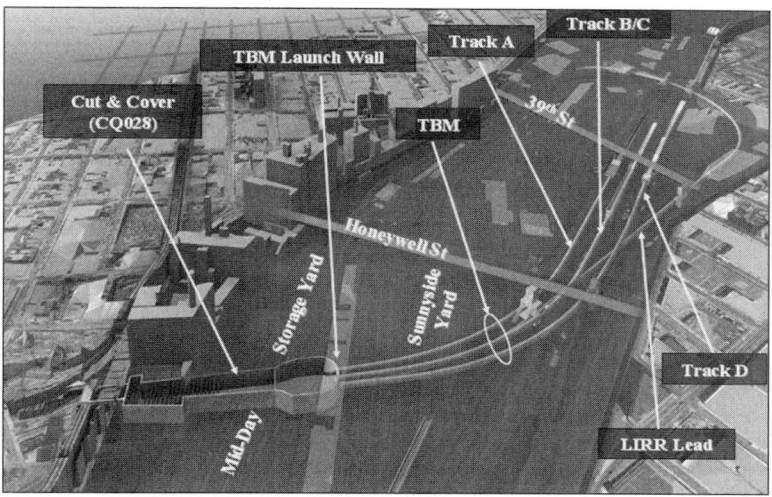

Figure 4. Contract CQ031—Queens bored tunnels and structures (view direction to the northeast—Manhattan is to the lower left)

Contract CH053, Harold Structures, Part 1. Contract CH053 is among the first major projects to construct elements directly within Harold Interlocking and Sunnyside Yard. The major work elements of this contract include:

- Construction of a series of retaining walls
- Construction of over a dozen microtunneling jacking and reception pits
- Construction of nine 60-inch diameter microtunnels ranging in length from 51 to 306 feet and at depths between six and 30 feet.
- Installation of new catenary poles, signal towers and other utilities.
- Installation of geotechnical instrumentation on sensitive structures.

Contract CH053 was awarded to Perini in December 2007 and the work is projected for completion in early 2010.

Contract CQ031, Queens Bored Tunnels and Structures. The key elements of this contract are shown on Figure 4 which includes four 22.5-foot outside-diameter tunnels, designated Tunnels A, B/C, D, and Yard Lead (YL). All four of the tunnels will be excavated beneath the Sunnyside Yard and three of tunnels (Tracks B/C, D and YL) will also cross beneath the Harold Interlocking.

The tunnels will be advanced using two slurry Tunnel Boring Machines (TBMs), which will access the tunnels through the Open Cut Excavation and be removed from three reception pits constructed as part of this contract. The first TBM will be used to advance Tunnel A and D, the second TBM will excavate the Yard Lead Tunnel and then Tunnel B/C. The second TBM will remain in place upon completion of Tunnel B/C and the skin will be removed in a subsequent contract.

Negotiations with proposal teams are underway and a Notice of Award is anticipated in April 2009. Contract CQ031 is scheduled to be completed in May 2012.

Common Railroad Interface Items

Regardless whether the new construction is surface- or subsurface-based, all the ESA Project work is subject to the following railroad-generated restrictions and concerns:

Safety. Safety and protection of passengers, railroad employees, contractors, and equipment is paramount. Contractors and their equipment cannot occupy the zone within 25 feet of the track centerline without railroad approved plans, controlled access and protection provided by flagmen, and decommissioning or de-energizing of electrical lines by railroad electrical personnel.

Per specifications, all on-site workers must annually attend and pass the Railway Worker Protection (RWP) safety programs offered by the railroads. The specifications also require that worker safety be addressed and described in the task specific plans work plans submitted to and reviewed by the railroads prior to task performance.

No Service Impacts. Delays to passenger service, especially rush hour service, are not acceptable. Reduction of track speed due to delays caused by contractor equipment within track fouling limits or damage to track, ballast, or infrastucture is also not acceptable.

Limited Access. Access to track or the area around selected tracks is limited and require full coordination. The railroads agreed to make track available, pending approval of the site access plan coordinator and subject to other work activities and the planning of specific tasks to contractors at the following general times:

- Between rush hours from approx. 10:30 a.m. to 3:00 p.m. (4.5 hours)
- Overnight between approx. 10:00 p.m. to 4:00 a.m. (6 hours)
- Weekends between approx. 10:00 p.m. Friday night and 5:00 a.m. Monday morning (55 hours).

Additionally, the railroads indicated that approximately one out of every five contractor track access requests will be denied due to conflicts with planned railroad activities or protection staffing issues and that some of the accepted planned outages will be cancelled at the last moment due to railroad service disruptions or emergencies.

Track access restrictions and limitations, as described above, are included in the Contract CQ031 and CH053 specifications. The contractor is to include and plan for up to twenty late notice or emergency denials of previously accepted track outages per year.

Settlement Control. Tunneling and excavation-related surface settlement of track, switches, and surface structures must be controlled and the limits of track lateral and differential settlement between rails must controlled and be kept within the FRA limits. "Traffic light" alert levels, which trigger implementation of agreed action plans, are shown on Table 1.

Prior to the start of tunneling, geotechnical instrumentation will be installed along the yard and mainline track alignments and on critical structures and baseline reading levels established. The type and projected number (in parenthesis) of geotechnical instruments to be installed at the surface along the CH053 and CQ031 alignments are listed below:

- Borehole Extensometers (14)
- Inclinometers placed in Soil and Slurry Walls (15)
- Horizontal Inclinometers (4)
- Liquid Level Gages (45)
- Deep Bench Marks (19)
- Open Standpipe Piezometers (16)
- Structure Monitoring Point (109)

Table 1. Railroad infrastructure response ranges

		Values For Green Range (in)		Values For Yellow Range (in)		Values For Red Range (in)	
		Tracks	Turnouts & Switches	Tracks	Turnouts & Switches	Tracks	Turnouts & Switches
Track Surface	The runoff in any 31' of rail at the end of a raise may not be more than	0 to 1-⅛	0 to 1-⅛	1-⅛ to 1-½	1-⅛ to 1-½	1-½ to 2	1-½ to 2
	The deviation from uniform profile on either rail at the mid-ordinate of a 62' chord may not be more than	0 to 1-¼	0 to ⅞	1-¼ to 1-⅝	⅞ to 1-⅛	1-⅝ to 2-¼	1-⅛ to 2-¼
	The deviation from zero crosslevel at any point on a tangent may not be more than	0 to ⅞	0 to ⅝	⅞ to 1-¼	⅝ to ⅞	1-¼ to 1-¾	⅞ to 1-¾
	The reverse elevation on curves may not be more than	0 to 1	0 to ⅝	1 to 1-⅜	⅝ to ⅞	1-⅜ to 1-¾	⅞ to 1-¾
	The difference in crosslevel between any two points less than 62' apart may not be more than	0 to 1-⅛	0 to ¾	1-⅛ to 1-½	¾ to 1	1-½ to 2	1 to 2
Alignment	The deviation of the mid-ordinate from a 62' chord may not be more than	0 to ⅞	0 to ⅝	⅞ to 1-¼	⅝ to ⅞	1-¼ to 1-¾	⅞ to 1-¾
	For curved track, the deviation of the mid-ordinate from a 31' chord may not be more than	0 to ⅝	0 to ½	⅝ to ⅞	½ to ⅝	⅞ to 1-¼	⅝ to 1-¼

- Surface Settlement Monitoring Points—Type 1 (200)
- Surface Settlement Monitoring Points—Type 2 (11)
- Survey Prisms mounted on Rails (572)
- Biaxial Tiltmeters (102)
- Probe Extensometer (2)
- Track Dynamic Profile Monitoring Points (1,679) located beneath each rail

The CH053 instrumentation is being installed by a specialty subcontractor retained by the Contractor. The number of instruments on CQ031 is large enough to justify the establishment of a separate contract to procure and install the instruments. This contract will be awarded in April 2009.

Upon acceptance by the Construction Manager, the instrumentation will be surveyed and monitored by specialty subcontractors retained by the MTA. The instrumentation reading intervals are established in the specifications and will be adjusted during the work in response to observed results and input from the railroads.

The instrument installation and geo-instrumentation monitoring contractors must comply with the safety and access requirements applied to the major contractors and described above. The majority of the instruments can be installed with hand tools under railroad foul-time, which reduces the need for track outages. Other instruments, such as inclinometers or extensometers, will require track outages to allow for a drill rig to access the site and advance a borehole.

To reduce the amount of track access time, many of the on-track settlement monitoring instruments, especially the survey prisms mounted on the rails will be read with an automated total station located outside the railroad fouling limits. The data collected in this manner will be forwarded on a "real-time" basis that is made accessible to the construction manager, contractor, and pre-selected members of the railroad engineering staff.

The specifications also establish "alert" and "review levels" for the instrumentation. When these levels of settlement are reached, the contractor, construction manager and railroad personnel will be notified via the automated monitoring system and will are to respond according to a pre-submitted and approved action plan. Preliminary plans are being developed at this time by the railroads and the MTA and will be expanded prior to start of tunneling by the contractor in conjunction with the railroads.

Railroad Interface. In addition to these constraints, the railroads closely scrutinize all construction that may impact their operations. Furthermore, railroad organizations are complex to outsiders with each railroad having numerous divisions (engineering, operations, signals, structures, etc.) with responsibilities, schedules, and future construction plans that often overlap or conflict even within a single area like Sunnyside Yard or Harold Interlocking. As a result, frequent and early planning, communication and concurrence with railroad field- and office-level decision-makers from all divisions is imperative to the project's success.

Lastly, the costs for all railroad support of on-site contractors and any subsequent railroad infrastructure repair or rehabilitation work will be borne by the MTA. The costs must be estimated, budgeted, and managed. The estimating process is also affected by railroad worker union requirements and rules which are different then ESA's construction contracts.

Railroad Liaison Personnel

Among the first undertakings of the MTA for the ESA Project was the establishment of a lead liaison from both Amtrak and the LIRR. These two individuals (including one of the co-authors), each with nearly 30 years of a variety of experience with their respective railroads, remain as employees of their railroad but their salaries are paid by the MTA. Their work efforts are entirely dedicated to the various contracts within the ESA Project.

As the work has increased, each lead liaison has brought on other railroad staff to support ESA's efforts. The liaison personnel attend design and construction meetings, understand the meeting issues, and then respond with the railroad's viewpoints, requirements, or restrictions or forward this information to the appropriate parties within their organization.

To help keep the liaison personnel accessible to the project, office space has been made available in the Construction Manager's Queens Area field office. The ease of access to these liaison personnel has helped quickly resolve many technical issues. Because the railroad and construction management staff are in one location, all parties receive the same information at the same time and can work together to quickly develop a unified response to problems.

Additional former railroad staff, supported by design and construction consultants, have been retained to form ESA's own Railroad Construction department, which is also located in the Construction Manager's Queens Area field office. This department is responsible for interfacing with the railroads on a daily basis and coordinating railroad flaggers with designer and contractor staff. The ESA Railroad Construction staff is also responsible for developing cost and schedule estimates for railroad direct construction work, including equipment and staffing, and tracking these costs. The ESA Railroad Construction department also played a key role helping the geotechnical exploration staff prepare plan documents for submittal to the railroads and act as a liaison with the railroads.

Interactions with the Railroads During Design

In addition to frequent updates and meetings about the ESA Project design, railroad staff were also kept informed and educated through a series of presentations. The presentations were often topic-specific (microtunneling, instrumentation, tunnel boring machines) and were often made several times to the various divisions and staff levels within the railroad. Key railroad staff and the railroad liaison personnel were also invited to participate in risk workshops.

Supplementing the presentations, railroad staff were also invited to attend field trips to a microtunneling project, an EPBM-driven tunnel project in northern New Jersey, and a slurry TBM driven tunnel project in Pittsburgh. The field trips were well received by the railroad staff who noted that many pre-conceived ideas were quickly clarified or corrected by the site visit.

The lessons learned from earlier contracts were also incorporated into the design of later-stage contacts. As part of the CH053 instrumentation installation, a field demonstration was given to the railroads to show how the instruments were to be installed, the tools and rate of the installation, and to verify if the rail-mounted survey-prisms could be located to face away from the train operators. Photographs of the installation demonstration for the Track dynamic Profile Monitoring Points are shown in Figure 5. The railroad input from this demonstration has been included into the CQ031 geotechnical instrumentation contract design.

Interactions with the Railroads During the Geotechnical Site Investigation

During the geotechnical subsurface investigation, the geotechnical and survey staff met weekly with the railroad liaison personnel and the ESA Railroad Construction department staff. At these meetings, a "two-week look-ahead" activity schedule was presented by each of the design disciplines to the railroad. The activity schedule supplemented a work plan prepared by the design field staff (with assistance from the

Figure 5. Field demonstration of installation of Track Dynamic Profile monitoring points

ESA Railroad Construction department staff) and submitted earlier in the week to the railroads. The work plan included:

- The type and location of the activities, including the number of tracks requested to be taken out of service,
- The type of equipment to be used for the activities,
- The estimated time and duration of the activities, and
- The requested railroad support services such as flagmen or electric personnel

The railroads would provide an initial review of the requests at the meeting and, where necessary, identify areas such as tight overhead clearances and spacing that may require modifications to the drill rig, and suggest an alternate approach to the request.

Following the meeting, the railroads would then evaluate all requests and consider them to determine the needed railroad resources. This railroad review process is increased in complexity because jurisdiction of Harold Interlocking is mixed between Amtrak and LIRR. An example of this is the drilling of a subsurface investigation boreholes located between LIRR and Amtrak infrastructure. At such overlapping sites, a total of nine flagmen and electrical personnel were required to protect a drilling crew.

Once the work plan was approved, a drill rig or survey crew would meet a railroad flagman at a pre-determined location and time to receive the site- and task-specific railroad safety briefing. During the briefing, the railroad's electrical personnel would start the process of de-energizing the third rail or catenary.

Drilling was typically performed by rubber-tired truck- or all terrain vehicle-mounted drill rigs. Where these vehicles could not be used, a truck-mounted drill rig with Hi-Rail attachments that permitted on-rail travel was used. Upon passing a safety inspection by the railroad, the drill rig was driven on the rails to the site by the driller, with a railroad employee serving as a pilot. The inspection and pilot requirement is mandated by the railroads for all contractor-operated Hi-Rail equipped vehicles and is included in ESA's specifications.

Even with an approved plan, due to competing railroad interests and projects, it is not uncommon for the field work to take longer than expected. This often requires flexibility from the contractor and ESA personnel.

One additional construction activity that began with the subsurface investigation and is now incorporated into all subsequent construction contracts is the use of a 6-foot deep hand- (or vacuum-) excavated probe hole advanced prior to drilling to find any utilities. This practice (now a project-wide requirement) came as the result of several smaller utility-related incidents that fortunately did not cause injury or equipment damage.

Interactions with the Railroad During the Construction

The ESA specifications require the contractor to prepare work plans for Construction Manager and Railroad review. The work plans are submitted in an eight-week, four-week and two-week "look-ahead" format and identify needs for pilots, electrical personnel and railroad protection requirements. The two week "look-ahead" work plan provides the greatest level of detail identifying the work activities, equipment and schedule.

In addition to the "look-ahead" meetings, the contractor, Railroads, ESA Railroad Construction department staff and Construction Manager meet on a weekly basis to review the past week's performance and upcoming activities. During the mining of the tunnels, these meetings will be held daily at the start of the day shift and will also include the design professional and geotechnical instrumentation monitoring contractor. During this meeting, the instrumentation readings and the TBM operation data readings (torque, thrust, slurry pressures, etc.) from the previous day will be presented. Based on this information, combined with surface observations of the rail, track, and ballast by the railroads, the TBM performance will be evaluated and adjustments to the mining operations or monitoring plans will be made.

SUMMARY

The Railroad Interface Management program described herein has been an essential element to the overall ESA construction program. It will be further refined and developed as the Queens-area tunnel work commences. This coordinated approach to interfaces, with the key Railroad stakeholders has been well-received by all participants and will serve as a framework for future such projects.

ACKNOWLEDGMENTS

The authors would like to thank Ms. Carroll Stewart for her input regarding the geotechnical investigation and to Messrs. Alan Paskoff, David Smith, Robert Spero, Andrew Thompson and Kevin Tomlinson for their efforts and comments as contributors and reviewers.

CONSTRUCTION OF THE MCUA TUNNEL AND FORCE MAINS UNDER THE RARITAN RIVER, NEW JERSEY: A CASE HISTORY

Bob Rautenberg ▪ Kenny Construction Company

Julian Prada ▪ Hatch Mott MacDonald

Frank Perrone ▪ Hatch Mott MacDonald

Donato Tanzi ▪ Middlesex County Utilities Authority

ABSTRACT

The Construction of Tunnel and Edison Force Mains contract involves the installation of two new force mains to provide a redundant means for sewage conveyance from Middlesex County Utility Authority's (MCUA) Edison Pump Station on the northern shore of the Raritan River to the Central Wastewater Treatment Plant on the southern shore. The force mains will be installed in a 1,192 m (3,910 lineal foot) long tunnel with a 4.09 m (13'-5") inside diameter, fully gasketed precast concrete segmental lining, which is to be tunneled underneath the Raritan River utilizing an Earth Pressure Balance TBM. The tunnel is anticipated to encounter flowing soils near and beneath the river.

Two new 60-inch diameter Centrifugal Cast Fiberglass Reinforced Polymer Mortar force mains will be installed within the tunnel leaving an open utility corridor above the force mains for future consideration. The project also includes the design and construction of two slurry wall shafts with associated piping. The paper will provide an overview of design and provide a case history of the construction of the tunnel and highlights of the construction methods implemented to date.

INTRODUCTION

MCUA Facilities and Service Area

The Middlesex County Utilities Authority (MCUA) is located in Sayreville New Jersey. MCUA owns the Central Wastewater Treatment Plant (CWWTP) which is designed to process 147 MGD and services Middlesex, Somerset and Union counties. The service area includes roughly 750,000 people and 1,500 industrial and commercial facilities.

The MCUA Tunnel and Edison Force Mains Project (EFM) is part of $70.8 million treatment works improvement project that will provide a redundant means of sewage conveyance from MCUA's Edison Pump Station (EPS) to the CWWTP. Prior to the EFM Project, MCUA conveyed 28 MGD of sewage on a daily basis with an 86 MGD peak via an existing 60-inch diameter pipe under the Raritan River known as the Arsenal Force Main. In March, 2003 MCUA experienced a failure of a 102 inch Precast Concrete Cylinder Pipe Sayreville force main which resulted in the New Jersey DEP issuing an Administrative Consent Order (ACO) with various conditions, one of which being the construction of a new supplemental Edison force main.

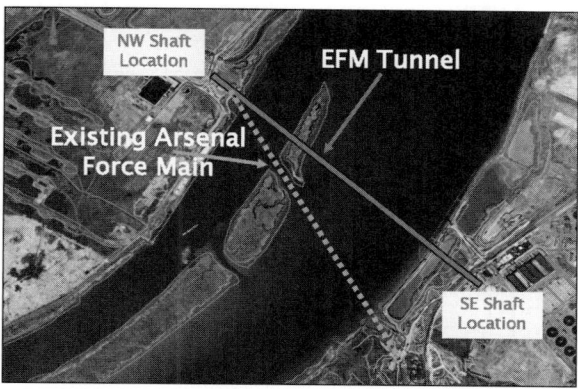

Figure 1. Aerial plan of project

Hatch Mott MacDonald (HMM) performed design and construction phase services for the EFM Project. Following pre-qualification and bid selection, the EFM Project was awarded to Kenny Construction Company (KCC) with a low bid of $45,150,000. Notice to Proceed was given to the contractor, KCC, on July 10, 2007 and Final Completion is anticipated to be November 5, 2009. At the time of submission of this paper tunnel excavation was 55% complete and breakthrough is anticipated in early to mid February 2009.

PROJECT DESCRIPTION

The EFM Project involves the design and construction of two 1,524 mm (60 in) diameter force mains within a 1,192 m (3,910 ft), 4.09 m (13'-5") internal diameter tunnel constructed beneath the Raritan River with a fully gasketed precast concrete segmental lining. The project also includes the design and construction of two slurry wall shafts and associated piping.

The tunnel drive heads on a northwest bearing from the Southeast launch shaft and main work site, located on the southern shore of the Raritan River in the Borough of Sayreville to the Northwest reception shaft, which is located at the EPS on the northern shore of the Raritan River in the Township of Woodbridge. An aerial plan of the project is shown in Figure 1.

Regional Geology

The project is located in the Coastal Plain Physiographic Province, a region characterized by flat to gently undulating topography underlain by soil zones that gently dip towards the east. The surface expressions of the contact between the sediments of the Coastal Plain and the hard bedrock of the Piedmont Physiographic Province are located a few miles west of the alignment. Underground, this contact slowly deepens towards the southeast. The terminal moraine of the most recent Wisconsin Age glaciations terminates within one or two miles north of the site giving rise to diverse and challenging subsurface conditions. This was seen at the North West reception shaft where slurry wall construction encountered glacial outwash and diabase bedrock.

Table 1. Description of anticipated geologic units

Geologic Unit	Symbol	Description
Artificial Fill	Af	Heterogeneous mixture of sand, silt, clay, gravel, dredge spoil, and man-made materials
Estuarine Deposits	Qm	Very soft to soft organic clay and silt
Raritan Terrace Deposit	Qrt	Dense fine to coarse grained clean to silty sand and gravel
Raritan Formation	Krw	Woodbridge Clay member: Medium stiff to hard silty clay and clayey silt
Raritan Formation	Krf	Farington Sand member: very dense fine to coarse clean sand
Weathered Diabase	Kdw	Weathered rock having soil-like consistency with remnant rock pieces
Diabase	Kd	Hard, fine grained igneous rock exhibiting RQD's in excess of 70% and UCS of 50,000 psi.

Project Geotechnical Conditions

A geotechnical data report (GDR) and geotechnical baseline report (GBR) was prepared by HMM as part of the Contract Documents for the EFM Project. A phased geotechnical exploration and laboratory testing program was undertaken along the proposed alignment during the design phase to characterize the subsurface conditions and supplement available geologic information. Based on the findings a GDR was prepared identifying the subsurface conditions and geologic units along the project alignment. The geologic units identified along the alignment are summarized in Table 1.

The GBR divided the tunnel alignment into six reaches based on subsurface conditions encountered within the tunnel horizon and their anticipated effects on construction including full face to mixed face conditions within varying reaches. To date, the excavated muck has been consistent with the baseline conditions established in HMM's GBR. The predominate geologic units within the tunnel horizon through the first three reaches consisted of Woodbridge Clay of the Raritan Formation (Krw) and estuarine deposits (Qm) consisting of soft organic clays and silts. The generalized project profile is summarized in Figure 2.

PROJECT DESIGN

Challenges

In addition to the challenging ground conditions, the EFM Project includes a number of technical challenges that needed to be addressed during construction. These include groundwater control at the soil rock interface for shaft construction, tunneling beneath three (3) MCUA force mains that provide the flow to the plant, tunneling beneath an active railroad siding and a lagoon from a former lead processing plant as well as abandoned timber piles within the Raritan River. To mitigate these risks HMM required borings within the slurry wall footprint to confirm rock head, designed an instrumentation and monitoring plan to be implemented during construction as well as included provisions for pile removal and railroad crossing requirements within the project specification.

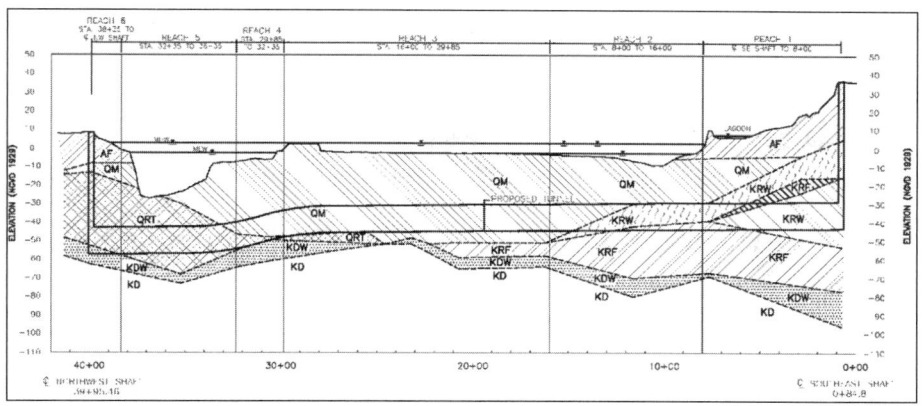

Figure 2. Project profile

Tunnel

As the geotechnical investigation encountered the presence of soft ground deposits throughout the alignment, a potential for flowing soils in the tunnel heading under atmospheric pressure was identified, which led to the specified requirement for a pressurized face Tunnel Boring Machine of either EPB or slurry capabilities to provide active face support and control groundwater during tunneling. The Contract Documents established the Contractor was responsible for determining the diameter of the overall tunnel excavation subject to satisfying the minimum internal diameter of 3,911 mm (154 in) including a maximum overcut dimension of 152 mm (6 in) between tunnel lining extrados and tailshield extrados in the radial dimension specified maximum requirements and making due allowance for tolerance on segmental lining construction and installing the force main pipes.

In terms of lining, the tunnel lining was designed as a 228.6 mm (9 in) thick bolted, double gasketed, precast concrete segmental lining that was to be fully grouted in place. One ring consists of six separate double tapered concrete segments. As the design alignment of the tunnel has a vertical curve in order to negotiate a U.S. Army Corps of Engineer shipping channel in the Raritan River while maintaining sufficient ground cover, allowances were made for installing straight pipe segments on a curved alignment while ensuring the minimum concrete thickness is provided for corrosion protection.

Shafts

To meet both temporary tunnel construction and permanent equipment layout requirements in the shafts, slurry walls were designed to serve as both the support of excavation during shaft sinking operations and the final permanent lining. Dewatering was not permitted for shaft excavation purposes.

The Southeast shaft was designed to be an octagonally shaped approximately 26.8 m (88 ft) deep shaft with a minimum diameter of 8.5 m (28 ft). This shaft was required to be the launch shaft for the TBM and the main work site for the tunnel operations. The toe of the Southeast shaft slurry walls were designed to be embedded beyond the base of excavation to provide adequate base stability and seepage control.

The Northwest Shaft was also designed to be octagonal shaped with a minimum diameter of 8.5 m (28 ft) and approximately 21.3 m (70 ft) deep. This shaft will be used for TBM removal. As the bedrock was found during the geotechnical investigation at the shaft location, the toe of the slurry walls were designed to be seated on the top of rock and the base slab of the shaft is to be secured using rock anchors grouted into the bedrock.

TUNNELING EQUIPMENT

The TBM selected by KCC was an EPB machine manufactured by Lovat Tunnel Equipment, Inc. which was originally manufactured in 2002 as one of four identical TBM's used to construct the East Central Interceptor Sewer (ECIS) in Los Angeles, California. Upon award of the project, the TBM was transported from California to Milwaukee, Wisconsin where it was given a complete refurbishment by KCC forces prior to being shipped to the jobsite.

The 4.69 m (15.4 ft) Lovat RME185SE is capable of mining in either Open or Closed Mode (EPB). The TBM is equipped with eight flood doors and a 724 mm (2.4 ft) diameter by 10.4 m (34 ft) long screw conveyor to control muck flow. Injection ports for ground conditioning are located on the face of the machine as well as in the forward chamber and screw conveyor. To facilitate ground control and annular grouting, three rows of wire brushes were installed in the TBM tail shield.

The TBM is equipped with two 300 kW electric water-cooled motors that power ten hydraulic drive motors capable of producing 3,506 kN.m (2,585,889 ft-lbs) of torque at optimal efficiency. Propulsion is accomplished with eighteen, 115-tonne (130 ton) cylinders with a maximum stroke of 2,250 mm (7.4 ft).

The TBM trailing system is approximately 83 m (272 ft) long and consists of 16 gantry cars that support the trailing conveyor belt as well as all electrical, grouting, dewatering and ventilation system components. In addition to the gantries, the trailing system includes a segment handling system that consists of a railing up car, a segment handling monorail, two segment unloading cars and a primary belt conveyor. Erection of precast concrete segments was accomplished with a ring-type segment erector mounted within the TBM tail shield.

Operation and control of the TBM is accomplished by an on board Programmable Logic Controller (PLC) which has the capability to read information from various sensors on the TBM as well as store and transmit the information to computers located in the field offices. The PLC is also linked to an onboard TACS System which displays TBM positioning in real time and calculates ring sequencing and positioning.

To remove muck from the tunnel, KCC chose to once again use equipment that was previously used on the ECIS Project and designed to work with this TBM. Four 10 m^3 (13 yd^3) capacity lift off type muck cars were moved in and out of the tunnel by an 18.1 tonne (20 ton) Goodman locomotive. In addition to the four muck cars, a Lovat grout car and two Lovat segment cars were also included with each train. A Liebherr 883 crane was used for hoisting and dumping muck cars that weighed approximately 21.8 tonnes (24 tons). The crane also served as the primary shaft crane for lowering segments, supplies and personnel.

To supply grout for annulus grouting, a Mobil Mat Mo 30 plant was purchased from Advanced Concrete Technologies and erected adjacent to the SE Shaft. The MobilMat plant has a 0.50-m^3 (0.65 yd^3) batch capacity counter current mixer and is supplied by two 350 bbls capacity silos, one for cement and the other for fly ash, and a 13.6-ton (15 ton) capacity aggregate hopper. Once a batch was completed, the grout was discharged down a 10 in drop-pipe and into a 4.6 m^3 (6 yd^3) hopper where it was held for discharge into the 4 m^3 (5.2 yd^3) Lovat grout car. Annulus grouting was accomplished using a Putzmeister KOV 550 Duo grout pump. Grout was transported to the heading, hooked up to the grout pump and pumped into grout ports on the segments via two, 2-in grout hoses.

PROJECT CONSTRUCTION

In September of 2007, KCC began its on-site mobilization. Contract Documents dictated that prior to the start of construction at both shafts, a series of boreholes were to be drilled along the centerline of the proposed slurry wall panels to verify the depth of bedrock. Additionally, a series of inclinometers, piezometers and settlement markers were installed by Jersey Boring and Drilling and subsequently monitored during the various stages of construction by KCC. To date the instrumentation installed has been monitored continuously and construction has progressed without incident. The following sections describe the shaft and tunnel construction performed to date.

SE Shaft Construction

Due to the space required to launch the TBM and set up the mining operations, KCC worked with its slurry wall subcontractor Bencor Corporation to redesign the shaft to a fourteen-panel shaft with an inside clear distance of 11.28 m (37 ft). In addition to the change in size, the depth of the wall was shortened by 1.52 m (5 ft) and the jet grouting of the base slab was eliminated. In lieu of the jet grouting, KCC proposed excavating the bottom 5.2 m (17 ft) of shaft under water and constructing the final base slab as a tremie slab.

KCC constructed the guide walls for the slurry walls and Bencor mobilized on site to begin excavation of the first panel on December 6, 2007. As previously mentioned, the shaft was made up of fourteen panels, ten of which were considered primary panels and four that were considered closing panels. The primary panels typically consisted of three "bites" that were excavated individually and adjacent to each other and then concreted as a single monolithic panel. A closing panel was an individual panel that was excavated between two primary panels and included a minimum of 152-mm (6 in) overbite into each panel as a keyway to insure water tightness.

The excavation of each panel was carried out using a Casagrande K3L Hydromill (Figure 3) suspended from a Liebherr 893, 90.7 tonne (100 ton) crawler crane. The K3L hydromill was equipped with two hydraulically driven cutting wheels equipped with heavy-duty carbide tipped teeth that rotated in opposite direction. Each "bite" was 3.14 m long by 0.91 m wide (10.3 ft by 3 ft). Concurrent to the panel excavation, KCC forces were tying reinforcing steel cages in an area adjacent to the shaft. As the excavation of a primary panel was completed, an 18.3 m (60 ft) long cage was carried to the shaft and suspended into the excavation. A second 13.7 m (45 ft) cage was then hoisted and spliced to the first cage and the entire 31 m (102 ft) long cage was lowered and suspended to the design elevation within the excavated panel. With the reinforcement positioned, tremie pipes were installed and the concrete operation commenced. Typically, two tremie pipes were installed on each primary panel and one tremie pipe was installed for a closing panel. Superplasticized concrete was placed and depth measurements were taken to determine when and how many tremie pipes were to be removed. Bencor completed the SE slurry wall on January 30, 2008 and immediately began breaking down equipment and mobilization to NW Shaft.

Mass excavation of the SE Shaft was carried out using a Caterpillar 307 excavator to trim, excavate and pile up the material and then used a 3.06 m³ (4.0 yd³) clamshell suspended from the Liebherr 883 to remove the material from the shaft. The excavation was carried out in 3.0 m (10 ft) lifts. Following the excavation of a lift, the walls were pressure washed with water and a 51 mm (2 in) layer of un-reinforced shotcrete was applied as the finished surface. Excavation and shotcrete proceeded in such a manner until a depth of 21.3 m (70 ft) was reached. During the redesign phase of the shaft, it was determined that the excavation could be safely completed to this depth without the underlying granular materials heaving upwards. Once the excavation reached this depth, the shaft was flooded to within 7.3 m (24 ft) of the surface. At this point the

Figure 3. Casagrande K3L Hydromill

remaining 5.1 m (17 ft) of excavation was completed solely by the Liebherr crane and the clamshell. The depth of the excavation was constantly monitored until the final excavation depth of 26.4 m (86.5 ft) was reached.

While the shaft was flooded with over 60 feet of water, Commercial Diving Services was brought in to inspect and assist in the final construction of the base slab. During the initial dives, elevations were verified and the walls were checked for cleanliness and keyway boxes that had been formed into the slurry panels were cleaned and threaded shear reinforcement dowels installed. At this point, the base slab reinforcement which had been pre-tied to a fabricated steel framework was lowered into the shaft and suspended from the surface by chains. The divers were then used to adjust the elevation of the reinforcement mat to its final position.

With the reinforcing mat positioned, five 10-in. tremie pipes were installed from the surface and the concrete slab was poured. During the tremie operation, soundings were performed in various location throughout the shaft and the divers performed a final elevation check. A total of 128 m^3 (167 yd^3) was placed in the 1.4 m (4.5 ft) thick base slab. After the concrete achieved a strength of 27.5 MPa (4,000 psi) the shaft was dewatered, all sediment was removed and shotcrete was placed on the remaining 5.1 m (17 ft) of exposed slurry wall.

NW Shaft Excavation

Excavation of the NW Shaft slurry wall began on February 11, 2008 and was completed in one month. With the geology at the NW Shaft being comprised entirely of sands and gravels, the excavation of each "bite" was typically completed with the hydromill in a matter of hours but underlying the sands and gravels was diabase bedrock with a compressive strength of 345 MPa (50,000 psi). To ensure a groundwater seal at the soil-rock interface, the redesign called for the toe of the wall to be keyed into rock 152 mm (6 in) which at times, rock excavation took an entire shift to accomplish.

Mass excavation of the NW Shaft was accomplished in a similar manner as the SE Shaft with the exception of a smaller crane being implemented. Upon completion of the excavation, the shaft was abandoned until the expected arrival of the TBM.

SE Shaft TBM Set-up and Launch

Setup for launching the TBM included pouring a, 0.76 m (2.5 ft) thick launch wall with a circular opening that was 0.75 mm (3 in) larger than the diameter of the TBM.

Figure 4. Cutterhead being lowered in the shaft

A secondary floor slab was cast on top of the tremie base slab that included a steel launch frame. For groundwater control, a 3-ply rubber seal was installed on the face of the launch wall which was designed to allow the TBM to pass through and form a seal around the outside perimeter and then around the first precast concrete ring as the TBM mined through the slurry wall.

Due to the size of the SE Shaft, launching the TBM required a complex and detailed launch plan that would minimize the movement of equipment and maximize the use of the limited space. On July 14, 2008, a 453.5 tonne (500ton) DeMag crane was mobilized on site and by the end of the day, the forward shell, which included the cutterhead and a shortened screw conveyor, and the stationary shell were both in position at the bottom of the shaft (Figures 4 and 5). During the refurbishment of the TBM, supports were fabricated that allowed the stacking of key components which eliminated the need for temporary hoses and cables. One support was built for the electrical controls, another for the ground conditioning equipment and orfinal for the grease systems. As the TBM advanced, sections of trailing gantry could be lowered to the bottom of the shaft and the components simply unbolted off of the supports and positioned onto the gantry section.

On August 11, 2008, the TBM began mining through the slurry wall. After cutting approximately 6.5 m (2 ft) the entire cutterhead and screw conveyor was filled with bentonite slurry in order to create EPB pressure on the machine and prepare for breaking through the slurry wall. With the seal in place, minimal groundwater infiltration was encountered during the break through. To propel the TBM, concrete thrust blocks were installed behind the machine every time it advanced an additional 1.22 m (4 ft). The following items describe the launch of the machine:

- TBM advanced until the portal seal was at the end of the intermediate shell
- Thrust blocks were removed from the shaft
- The shortened screw conveyor was removed and lengthened
- Ring erector installed and tail shield added
- Thrust blocks were put back in place and mining resumed.
- At the 26 m (85 ft) mark mining operations were stopped.
- Segment handling system as well as the first two gantry sections were installed
- Mining resumed, with gantry sections added every 10 m (33 ft)

Figure 5. TBM assembly

Figure 6. Shaft layout prior to car passer installation

- Mining continued until the TBM had advanced 177 m (580 ft) at which time a car passer was installed at the base of the shaft along with an equilateral switch within the tunnel

Tunneling Progress

Prior to installing the switch and car passer, the initial mining production was limited due to size of the SE Shaft. With limited space, the mining had to be accomplished with only two muck boxes. In addition, following each mining cycle, both muck boxes and their chassis had to be removed from the shaft so that the two segment cars and the grout car could be lowered and loaded (Figure 6). Under these limitations, a best shift of 6 rings was achieved.

At the time of this writing, the drive is 55% complete and has excavated though Reaches 1 and 2 which varied from a full face of clay (Krw) to a mixed face of clay and granular materials (Krf). The machine is presently in Reach 3 which consists of very soft to soft, "weight of rod," organic clay and silt (Qm). To date we have achieved a best shift of 11 rings and a best day of 19 rings.

As the TBM advanced through Reach 1 and 2, ground conditioning included both foam injection through the three (3) ports on the face and bentonite slurry through the perimeter port on the cutting head. EPB pressures have been maintained at approximately 1.5 bar (22 psi). As the TBM began entering Reach 3 of the drive, the foam injection rates had to be varied and polymer had to be injected with the foam to optimize the soil conditions and maintain the EPB pressures.

CONCLUSIONS

Like most tunnel projects, the EFM project has a number of challenges that needed to be overcome. To date, KCC has successfully met the challenges presented and is moving forward with mining operations towards the successful completion of the project. The TBM has been successfully launched and 55% of the drive has been mined. Further challenges that are upcoming are the completion of the mining in the "weight of rod" ground, negotiating the TBM through a vertical curve under the main shipping channel with one diameter of cover and break though of the machine at the NW shaft. Upon completion, the force mains will be installed to enable MCUA to have redundant means to convey sewage under the Raritan River.

ACKNOWLEDGMENTS

The Authors would like to thank the Middlesex County Utilities Authority for the permission to publish this paper and also acknowledges the contributions of Ted Budd (KCC), Peter Kocsik (HMM), Colin Lawrence (HMM) and Andy Thompson (HMM).

REFERENCES

Geotechnical Baseline Report: Edison Force Main /Edison Pump Station Upgrade Project, Prepared by Hatch Mott MacDonald for the Middlesex County Utilities Authority, November 2006.

Geotechnical Data Report: Edison Force Main /Edison Pump Station Upgrade, Prepared by Hatch Mott MacDonald for the Middlesex County Utilities Authority, August 2006.

PART 14

Risk Management

Chairs

Mike Ryan
Schiavone

Robert Goodfellow
Black & Veatch

HINDSIGHT IS 20/20—REVERSE ENGINEERING TUNNEL RISK ANALYSES

Lee W. Abramson ■ Hatch Mott MacDonald

ABSTRACT

Performing risk analyses on tunnel projects has become common place and on many projects it is required from an owner or regulatory perspective. After the project is planned, designed and constructed, the data concerning identified risks, assumptions, costs, probabilities and mitigations is generally long forgotten and buried in some file somewhere. A study was performed comparing predicted risks versus actual events on several recently constructed tunnel projects. Probabilities and costs for selectively common risks on tunnel projects have been studied and re-evaluated. Conclusions are presented that can guide future projects and the veracity of the tunnel risk analyses performed.

INTRODUCTION

Tunneling is a risky business. More than ever, attention is being paid to the risks associated with tunneled facilities and implications on project or program cost and schedule. Questions about why tunneling projects notoriously exceed budget numbers abound. Evaluations in retrospect and going forward have focused on the wide breadth of inherent risks associated with tunneling. More and more, these risks are being identified, studied and managed to a higher degree. As more and more scrutiny pervades the industry, little value is being obtained and leveraged on future projects. The industry seems to be headed in a direction of making the same mistakes over and over.

The notion of risk evaluation has been around for decades and received wide notoriety with the U.S. Space Program and ambitious projects in the energy industry such as hydropower and nuclear power in the 1960s and 1970s (Stewart and Melcher, 1997). This led to the evolution of tunnel risk analyses with more rigorous identification of the unique risks in tunneling projects impacting cost and schedule. As tunneling risks became more readily identified, more attention has been paid to the frequency and impact of the occurrences unique to tunneling (BTS, 2003). Currently, few tunnel projects are being planned, designed or constructed without a formal risk register that identifies the risks, mitigates the risks and quantifies the probable impacts of risks that cannot be mitigated. In hind sight on many recent projects, events that brought projects to "its knees."were not addressed in risk assessments or were not dealt with to adequately mitigate the risk of these events.

The industry has come a long way assessing the risks on tunnel projects but significant work remains to be done. While many funding agencies such as the U. S. Federal Highway Administration, Federal Transit Administration (USFTA, 2007) and Federal Aviation Administration have embraced the risk assessment processes, one burning omission remains—a rigorous post-project review of the risk evaluation process relative to what actually happened versus the risks anticipated and their relative impact on the costs and schedules of the projects. The information gleaned would provide very insightful information for future risk analyses particularly related to major cost and schedule impacts of recurrent risks on tunnel projects for future planning, design and construction. This information could also be used to identify and prioritize the most

Table 1. Mechanized tunnel risks

Project	Location	Diameter (feet)	Length (miles)	Geology	Risks
West Area CSO Tunnel	Atlanta, GA	27.0	8.5	Gneiss Schist Granite Mylonite	■ Poor rock conditions ■ TBM bull gear failure ■ One year delay
Detroit River Outfall Tunnel	Detroit, MI	21.0	1.2	Limestone	■ Tunnel flooded ■ TBM abandoned
MWD Inland Feeder Arrowhead Tunnels	San Bernadino, CA	18.0	9.5	Granite Gneiss	■ Excessive water inflow ■ Tunnel contract terminated
King County Mercer Street Tunnel	Seattle, WA	16.0	1.2	Glacial soils	■ Excessive wear of TBM face ■ One month delay ■ $600k claim

significant risks so that energies are spent on the most important factors and relatively minor issues can be dealt with systematically and with less effort. Technical aspects of the tunneling projects that should be evaluated include geotechnical conditions, size and length of the tunnels, capability and operation of the TBMs, initial support methods, final support/lining methods and other issues that cause risk (Abramson, 1998).

Isaksson (2002) proposed a very theoretical model for estimating time and cost risk evaluations on tunneling projects relying on probabilistic methods. Her model however did not tie into risk assessments per se, dealt primarily with geotechnical variations over various reaches of tunnels, did not utilize commonly accepted cost estimating and scheduling tools, and did not consider correlation of the multitude of other variables that impact tunnel project costs and schedules. Further, her thesis looked back on two specific tunnel projects—the Grauholz tunnel in Switzerland and the South Marginal Zone of the Halland Ridge Tunnel in Sweden and demonstrated with her model that given the multitude of possibilities and probabilities of every possible event occurring on a tunnel project, the actual cost and time of those projects fell within the probable ranges of cost and time calculated by the probabilistic model, respectively.

CURRENT EXAMPLES OF TUNNELING RISKS

There are several mechanized tunnel projects currently underway or recently completed that beg for attention regarding the veracity of the risk registers prepared during planning and design and the outcomes of these projects. Some examples from published literature are given in Table 1.

Some risks anticipated for the Atlanta project and the probabilistic cost studies for the Seattle project are highlighted below. Additionally, a parametric sensitivity evaluation is shown to demonstrate the importance of risk on schedule and cost.

CITY OF ATLANTA WEST AREA CSO TUNNEL PROJECT

The City of Atlanta West Area CSO Tunnel had several risks (Hutton et al., 2007). Summary details regarding this project are as follows:

- Eight and one half miles of 27-foot-diameter TBM driven rock tunnel
- Two new TBMs used with rock reinforcement support

Figure 1. Vertical muck conveyor

- A few short drill and blast sections
- Partial concrete lining in poor rock conditions and/or excessive water inflow
- Three large intakes in active CSO channels
- 85 MGD underground pump station and shaft

A risk assessment was prepared for this project as on similar projects. Risks identified were mitigated during design as much as possible. Most remaining risks had low to moderate ratings. Risks deemed to have high likelihoods primarily dealt with potential issues related to ground behavior and groundwater. Risks deemed to have the most potential severity also dealt with ground behavior and groundwater as well as safety. Per Hutton et al. (2007), more than expected steel rib support was required, blocky ground was responsible for damage to the vertical conveyor (Figure 1) at least twice and the cutterhead drive pinions and bull gear failed on one of the TBMs. Also, additional steel mesh was added to ground support measures to address "flaking of the rock in the tunnel crown, even in areas of Type A-classified ground" (Figure 2). Further, "...overall excavation rates were found to be slower than anticipated due to unexpected mechanical downtime and worse than expected ground conditions." While risk assessments did address these potential risks, they were deemed to be moderate and similar to a host of others. Issues with the vertical conveyor were not addressed in the risk assessments.

KING COUNTY (SEATTLE) MERCER STREET TUNNEL

The King County Mercer Street Tunnel Project in Seattle was part of the Denny Way/Lake Union CSO Control project. The Mercer Street Tunnel is 16 feet O.D. and 1.2 miles long through glacial soils including clays, silts, sands and cobbles below the

Figure 2. Atlanta West CSO tunnel steel rib supports

Table 2. Mercer Street Tunnel methodology study

Alternative	5th Percentile Cost	95th Percentile Cost	Range
Open face—one pass	$17.8 M	$25.8 M	$8.5 M
Open face—two pass	$21.3 M	$28.5 M	$7.2 M
EPB—one pass	$18.3 M	$24.1 M	$5.8 M
EPB—two pass	$22.4 M	$28.4 M	$6.0 M

groundwater table (Abramson et al, 2002). During preliminary design, a probable cost evaluation was carried out for alternative construction methods including:

- Open face—single pass lining
- Open face—two pass lining
- EPB TBM—single pass lining
- EPB TBM—two pass lining

For each method, low, best and high potential costs were evaluated via a Monte Carlo simulation of variable base cost items (Montgomery Watson, 1998). Additionally, potential "problems."that could occur were also evaluated as follows:

- Boulders encountered
- Running sands encountered
- Difficulties with tunnel water discharge occurred
- Main water supply arterial pipe near tunnel damaged

Figure 3. Mercer Street tunnel EPB TBM

- Ground settlement occurred
- Tunnel steering/alignment difficulties occurred

These risks were given a triangular probability distribution for low, medium and high values for the number events and the consequent costs of those events.

From these analyses, the following was concluded (Table 2).

It was concluded from the probable ranges in cost that "Based on the current state of knowledge which has been incorporated into the model, the EPB methodology, though not always the least expensive alternative, minimizes risk when compared to the open face methodology."

Subsequent to this conclusion it was noted that "One key issue which seems to be important in all of the above alternatives concerns the presence or absence of running sands.."A sensitivity analysis was performed considering the low, medium and high number of times running sands were encountered of 0, 10 and 30 versus 0, 5, and 15 versus 0, 15 and 45, respectively. For the latter, EPB methodology became the preferred option in most trials. That may seem obvious today but keep in mind that EPBs were still somewhat new in the U. S. market in 1997 whereas today many tunnels have utilized this method and it is much better proven and accepted.

In the end, project cost values were as follows:

- Final design engineer's estimate $39.4 M
- Low Bid $29.5 M
- As-constructed $30.1 M

The primary issue during construction was encountering a cluster of bouldery soils and excessive wear and tear on the EPB TBM face and head (Figure 3). The contractor ground through this zone into more cohesive materials and refaced the head of the machine and retooled it. More heavy use of soil conditioners may have lessoned wear and tear on the TBM thereby calling into question whether this was a differing site

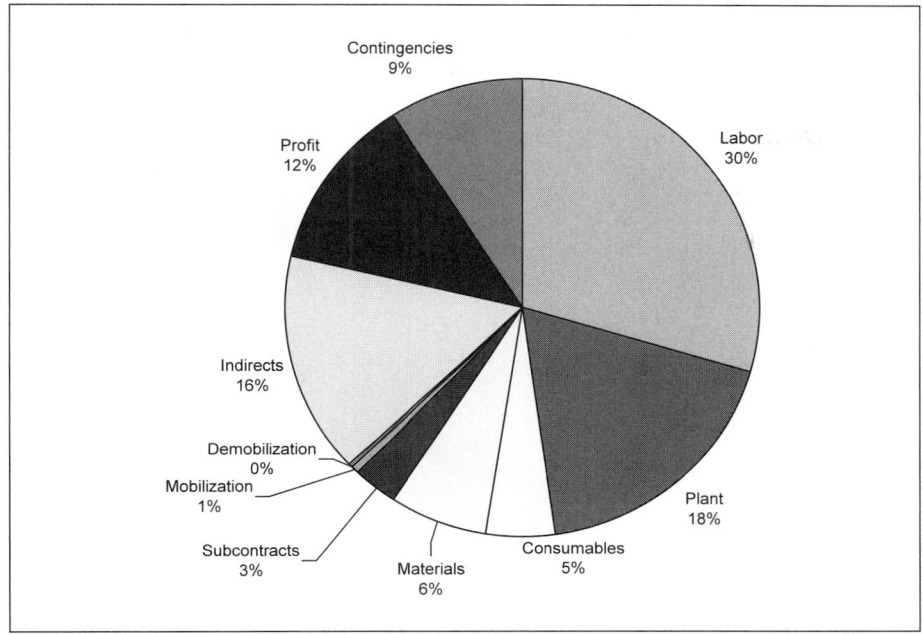

Figure 4. Tunnel cost estimate—base case

condition or poor machine operation. However, the owner agreed to fund rehabilitation of the machine at a cost of approximately $600k.

PARAMETRIC COST ESTIMATING STUDIES AND SENSITIVITY ANALYSES

Computerized computer programs now make it fairly simple for an experienced tunnel engineer to estimate the costs of tunnel projects as well as perform parametric studies such as that done for the California High Speed Rail Program (Abramson and Crawley, 1995). A Microsoft Access Database cost estimating program was used to perform parametric studies for a prototypical soft ground EPB TBM tunnel with precast concrete segmental lining. This tunnel had the following features:

- Subaqueous Tunnel
- Very soft soils
- Finished Diameter—10.33 ft
- Length—9,440 ft
- Average Progress Rate—33 ft/day
- Base Cost Estimate—$72.6 million
- Profit—15%
- Contingency—10%

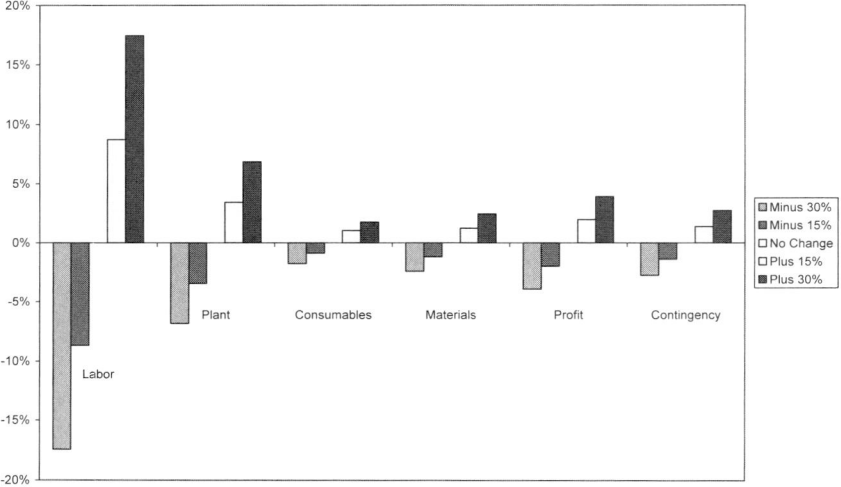

Figure 5. Tunnel cost variations

In summary, the base cost included items for:
- Labor
- Plant
- Consumables
- Materials
- Subcontracts
- Mobilization
- Demobilization
- Indirect Costs
- Profit
- Contingencies

The distribution of these costs for the base case is shown in Figure 4.

Next, a sensitivity evaluation was performed by varying each of these costs by plus 15 and 30 percent and minus 15 and 30 percent. The biggest sensitivity was on the largest cost contributor, labor. When labor costs increase or decrease by 30 percent, total costs can vary by plus/minus 17 percent (Figure 5).

This would be the case for instance if TBM productivity was significantly slower than expected. Seventeen percent represents approximately $10 to 15 million. Many contractors would put a claim in for that whether justified or not. If labor costs were to jump to double the base cost estimate, total project cost would jump by about 29 percent or $20 to 25 million (Figure 6).

CONCLUSIONS

The author believes that of the number of risks evaluated on tunneling projects, only a handful have dramatic effects on cost and schedule. The most serious risks on tunnel projects should be identified based on a review of case histories and independent cost and schedule post mortems. This information should be utilized to develop

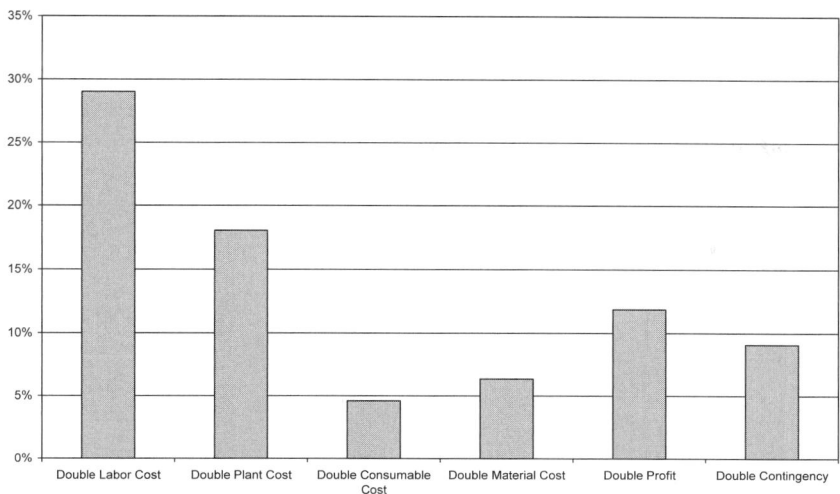

Figure 6. Tunnel costs when something doubles

guidelines for future risk analyses including a catalogue of characteristics and ranges (low, medium and high) and probabilistic distributions of cost and schedule impacts of various geotechnical, equipment, support and other risk types based on recent mechanized tunnel projects. A living document could be created as a data base with updates in the future based on new information.

ACKNOWLEDGMENTS

The author is grateful to his colleagues at City of Atlanta, GA, King County, WA, Jordan Jones and Goulding, Delon Hampton Associates, Obayashi Corp. and Hatch Mott MacDonald for their participation and involvement in the projects and information presented in this paper.

REFERENCES

Abramson, L., Cochran, J., Handewith, H. and MacBriar, T., 2002, Predicted and Actual Risks in Construction of the Mercer Street Tunnel, L. Ozdemir, editor, Proceedings: North American Tunneling Conference, p. 211–218.

Abramson, L.W. and Crawley, J. E., 1995, High Speed Rail Tunnels in California, in G. E. Williamson and I. M. Gowring, editors, Proceedings: Rapid Excavation and Tunneling Conference, p. 575–594.

Abramson, L.W., 1998, Root Causes of Disputes on Tunnel Construction Projects, in L. Ozdemir, editor, Proceedings: North American Tunneling Conference, p. 55–62.

Abramson, L.W., 1992, Factors Related to Water Tunnel Design, William Barclay Parsons Fellowship Program, Parsons Brinckerhoff Quade and Douglas, Inc., July.

British Tunnelling Society, 2003, Joint Code of Practice for Risk Management of Tunnel Works in the UK, Prepared Jointly by the Association of British Insurers and The British Tunnelling Society, First Edition, September, 20 pp.

Hutton, R. Liebno, D. and Nonaka, T., 2007, Construction of the West Area CSO Tunnels and Pumping Station, in M. Traylor and J.W. Townsend, editors, Proceedings: Rapid Excavation and Tunneling Conference, p. 1064-1078.

Isaksson, T., 2002, Model for Estimation of Time and Cost Based on Risk Evaluation Applied on Tunnel Projects, Doctoral Thesis, Division of Soil and Rock Mechanics, Royal Institute of Technology, Stockholm, Sweden.

Montgomery Watson, 1998, Draft Mercer Street Tunnel Risk Analysis, Memo to King County J. Cochran, C. Locke and L. Abramson, March 24.

Stewart, M. and Melchers, R., 1997, Probabilistic Risk Assessment of Engineering Systems, Publisher Chapman and Hall, 274 pp.

U.S. Federal Transit Administration, 2007, Contractor Performance Assessment Report, Office of Planning and Environment, U. S. Department of Transportation, September, 211 pp.

GETTING THE ENGINEER'S ESTIMATE RIGHT

Tom Martin ▪ Parsons Brinckerhoff

Joe O'Carroll ▪ Parsons Brinckerhoff

Keith Caro ▪ Parsons Brinckerhoff

Tom Peyton ▪ Parsons Brinckerhoff

ABSTRACT

Mega projects present unique challenges with regard to estimating costs. With such projects, which may span decades, these challenges often begin as soon as a project is conceived. The importance of having a sound, resource based estimate available early in the process cannot be overstated because it provides the basis for ongoing project decisions. For the final phase of providing an updated cost estimate, often while bidding by contractors is underway, the Engineer's Estimate (EE) must incorporate present market conditions; account for possible volatile material price swings; be based on actual labor and fringe costs; rely on meaningful productivity rates; and accommodate any recent technology advances while identifying and allowing for risk factors. This paper explains the benefits associated with consultants adopting a contractor's estimating approach when preparing the Engineer's Estimate on megaprojects like the various tunneling projects in New York City.

INTRODUCTION

As soon as the idea for a new project is conceived, the next logical question will be "How much is it going to cost?" In a rapidly changing world, there is not an endless supply of funds allowing every worthwhile project to be built. Instead, new projects must compete for available funds or find other funding paths such as grants, bond issues or Bid-Build-Operate while demonstrating they are viable enough to be a "paying proposition" meaning they will generate adequate revenue to pay-back the initial investment in a reasonable period of time.

At numerous stages before the work scope is approved for construction, the project cost will be estimated, modifications will be made and new or revised design costs considered, all with the purpose of getting the best project for the least cost. When sound resource based cost estimating methods are adopted early in the progression of a project from the "Can we build this?" to "Let's go to bid and find out what the cost is going to be," the ability to provide a sound cost estimate is fundamental. This procedure must allow for design refinements, incorporate current construction procedures, take into account work and safety rules and regulations and base costs on similar historical production rates while complying with local manning requirements. Consideration or incorporation of these variables gives a project owner a level of confidence that the overall, final cost will not exceed the budget amount he has earmarked for the completed work.

ESTIMATING COSTS FROM THE BOTTOM UP

In Parsons Brinckerhoff' Geotechnical and Tunneling Technical Excellence Center (PBGT) the method used to develop reliable cost estimates is to use experienced estimators with a contracting background in conjunction with industry recognized heavy construction estimating software to produce the appropriate level of cost for the work as the design progresses. From the initial collection of the facts that affect the project; e.g., type of work; wage and benefit costs; project duration; ground conditions; construction staging and access; groundwater issues; similarity to previous projects completed a sound basis is very important. Using the facts or answers to similar questions, a PBGT estimate is initiated based on the collected data and, at least during the preliminary stages, may include possible alternatives or "what if" scenarios.

Often the initial conceptual cost estimate sets the milestone against which all future re-estimates are compared and/or measured. It is conceivable that some worthwhile projects have been scuttled because an unrealistically high early cost estimate meant the work was judged to be too costly and something the owner could not afford. Whether a low conceptual estimate causing a marginal project to remain valid for further consideration, or a high initial estimate that results in immediate cancellation of the project before additional design or investigation is done, has the most negative impact is an argument that will continue to be debated. PBGT's contractor style approach to cost estimating will provide Owners with a greater confidence in the cost estimate during the early stages of their project.

APPROACH

PBGT's approach is to develop a cost estimate from the "bottom up" at the earliest possible time. That means each estimate begins at the lowest level and is refined toward the total cost using a comprehensive, contractor type estimating format which takes into consideration union agreements, wage and fringe packages, insurance costs, manning requirements, safety issues and crew makeup, all based on experience. When the work will be non-union then differences in work rules and wage packages are carefully considered by the estimator when he sets up the project data base which includes predicted labor costs, equipment, permanent material, supplies and subcontractors as well as taxes and insurance allowances. Existing data base component costs are evaluated and modified where necessary using factors applicable for the location and type of work being estimated.

In addition to relying on estimators with decades of experience bidding similar work for contractors, PBGT maintains cost data bases for tunnels, transit systems, power schemes, water and highway projects and even special projects. Using HCSS[*] software specifically developed for contractor use in estimating work, crews, equipment, permanent material, supplies and subcontractor costs can be sourced from the data base and entered with calculated or known production rates to develop a complete cost estimate. While in the early stages of a project, the required detail will not be as great as that needed for the final Engineer's Estimate, generally due at the time of bidding, the effort will always be more meaningful than using previous line-item bid prices or relying on one of the commercially available estimating guides that provide unit costs and then using these prices for forecast the project cost. Sound reliable cost estimates are necessary to provide the owner and designer confidence planned work can be completed for the funds available.

---

[*] HCSS HeavyBid is a Copyrighted product of Heavy Construction Systems Specialists, Inc. Houston, TX.

QUANTITY TAKEOFF

To understand the different work components needed for completion, takeoff of individual quantities such as excavation, concrete, formwork, backfill, etc., are required which are then used to develop pricing and an overall duration for the work. While some methods may generalize this important step, the PBGT estimating approach prefers to invest the time and effort up front to get meaningful quantity summaries which can then be used by the estimator to determine the true cost of the work. A quantity summary is also a critical component of the estimate when design changes are considered or implemented. It is difficult to accurately predict the net impact of a new design unless a comparison between "before" and "after" quantities and the costs can be provided to the owner and designer.

A summary of work items, developed for bidding purposes, allows grouping of quantities into logical cost centers. While frequently the bid item will end up a "Lump Sum" price, none-the-less, the components that make up that item need to be measured and costs applied to develop a total direct cost for every bid item. The decision to use unit prices vs. lump sum costs is often influenced by the designer, owner and sometimes the estimating team to provide a package contractors are comfortable bidding. Too much reliance on 'all inclusive bid items' can give the appearance an Owner might be requiring unreasonable performance from the successful bidder, even to the extreme case where some projects have failed to receive competitive bids. This is a situation which can be improved by giving contractors at least a total package cost range which represents what competent, qualified firms would be expected to bid for the work scope listed. By providing this range during the advertising of the work, unqualified or smaller firms can check to see if the expected price is within their bonding capacity and then make a decision to joint venture the work or drop out of the competitive bid process. The value of a 90% complete estimate including quantity takeoff, crew makeup, productions, material and supply costs, cannot be over stated. Each stage of developing a project estimate should increase the confidence in the cost of the work and refine that project cost toward what will be submitted by the contractors when the bidding phase is completed.

METHODS, PRODUCTIONS AND SEQUENCES

Once the quantities are established, the PBGT estimator accesses previously established data bases and selects crews, material and supply costs as well as the needed equipment to complete each item of work. When the work scope involves components which can be broken down into steps or parts, the estimator will begin with the most basic operation and work his way to a complete bid item or project cost. This is where the experience of the estimator pays the highest dividend. The HCSS system allows additional activities to be inserted into each cost item if a step was missed, overlooked during the initial setup, or the scope or construction method is modified.

For tunnels which include drill and blast (conventional excavation) rounds of various lengths may be laid out to determine the total cycle time including the number of holes, drilling times, loading or charge time, smoke delays, mucking and support installation allowances. Sufficient detail is available to calculate the total amount of powder and caps necessary as well a the number of man hours in the estimate either by bid item or craft. Such summaries are beneficial for comparisons with the contractor's estimate when the bid review is undertaken to make sure the low bidder has no errors which could result in rejection of his bid.

If a tunnel boring machine (TBM) is the preferred method for excavating a tunnel, predicted advance rates, cutterhead rpm, number of cutters installed on the cutterhead and cost per cubic yard will be requested from the machine manufacturers and used to calculate the basic cost of the tunnel. TBM and backup purchase costs, estimated freight to site, spare main bearing allowances and even the machine delivery schedule for budgetary purposes are discussed with the companies that provide TBMs so the owner has a valid cost basis at bid time.

For other cost components such as temporary support systems, precast segments, steel ribs, etc., previous actual or adjusted costs are used whenever possible so the cost in the Engineer's Estimate represents the most up-to-date information available at the time of bid.

DETAILED ESTIMATE INFORMATION

The true measure of the Engineer's estimate is not just the ability to provide a check of the bid price, but also to present the information contained in that estimate in a reasonable format for reference and review. Without the capability to review and modify items that make up the cost components, the final number becomes just that, simply a number.

A comprehensive estimate should provide:

- Summary of Bid Items with unit costs, margin, indirect and balanced bid
- Direct Cost report
- Crew Summary
- Equipment Report
- Permanent or Construction Material Listing
- Taxes and Fringe Summary
- Workers Comp
- Per Diem Report
- Cash Flow Report
- Notes
- Bid Schedule in Excel format

Just a few of the many reports and summaries that can be generated and made available to the owner are included for reference. The report examples that follow include:

- Estimate Summary—Costs & Prices
- Direct Cost Report
- Bid Item Summary in Excel format

INDIRECT COSTS AND PROFIT

In some low level estimates the allowances for indirect costs (or overhead) and profit will merely be a factor on total cost or a percentage of labor. Typically, as the estimate is refined, a predicted indirect cost may be calculated to verify to the owner that the amount included represents a viable cost. When necessary, the same amount of detail can be provided for indirect costs as is done for direct costs.

The margin allowance generally starts out high in earlier estimates and is reduced as the design details become more defined. This procedure is often used by contractors in determining the amount of profit they seek from a project, so PBGT estimators are following the same procedure as the construction industry, one that adjusts for marketplace competition, risky work, concerns over ground conditions, difficult contract terms and conditions, etc. The advantage of a "bottoms up" estimate over unit pricing

Parsons Brinckerhoff
CM009042
Keith Caro - PB

ESTIMATE SUMMARY - COSTS & BID PRICES

Page 1
01/06/2009 14:51

Bid#	Client# / Bid Description	Quantity	Unit	Manhours	Direct Labor	Perm Matl	Constr Matl	Equip-Ment	Sub-Contr	Direct Total	Indirect Charge	Total Cost	Total Cost Unit Price	Markup	------Balanced Bid------ Total	Balanced Bid Unit Price	Bid Price	Bid Total
1000	All Required Work Except the Bid Item Lis	1.00	LS															
1001	Assemble TBM's	1.00	LS	38,640 38,640.00	3,389,441		338,944	292,236		4,020,621	4,570,177	8,590,798	8,590,797.85	2,449,407 28.5%	11,040,205	11,040,204.74	11,040,204.74	11,040,204.74
1002	Temporary Facilities & Utilties	1.00	LS	28,832 28,832.00	2,504,448		1,572,992	231,210		4,308,650	1,394,994	5,703,644	5,703,643.61	1,809,859 31.7%	7,513,503	7,513,503.03	7,513,503.03	7,513,503.03
1003	Temporary Medium Voltage Substation	1.00	LS						5,905,891	5,905,891		5,905,891	5,905,891.00		5,905,891	5,905,891.00	5,905,891.00	5,905,891.00
1004	Partial Disassembly and Reassembly of TBM	1.00	LS	47,520 47,520.00	4,214,112		421,411	965,997		5,601,521	5,682,129	11,283,650	11,283,650.26	3,045,363 27.0%	14,329,013	14,329,012.94	14,329,012.94	14,329,012.94
1005	Final Disassembly & Removal ofTBM's	1.00	LS	52,800 52,800.00	4,682,347		468,235	1,073,330		6,223,912	6,313,477	12,537,389	12,537,389.17	3,383,736 27.0%	15,921,125	15,921,125.48	15,921,125.48	15,921,125.48
1008	Resident Engineers Main Office	1.00	LS				656,850			656,850		656,850	656,850.00		656,850	656,850.00	656,850.00	656,850.00
1009	Maintain RE Office	46.00	MTHS				992,473			992,473		992,473	21,575.50		992,473	21,575.50	21,575.50	992,473.00
1010	Excav & Support Approach Tunnels	309.00	LF	30,848 99.83	2,796,540	287,876	1,120,988	338,842	164,016	4,708,262	3,770,735	8,478,996	27,440.12	2,020,942 23.8%	10,499,939	33,980.38	33,980.38	10,499,937.42
1016	Excav & Support Assembly Chamber	80.00	LF	26,512 331.40	2,383,957	502,836	782,733	260,574	150,048	4,080,150	3,214,426	7,294,576	91,182.20	1,722,786 23.6%	9,017,362	112,717.03	112,717.03	9,017,362.40
1017	Excav & Support Starter Tunnel Class I	40.00	LF	2,624 65.60	236,122	35,779	92,989	24,386	21,139	410,415	318,377	728,792	18,219.80	170,636 23.4%	899,428	22,485.69	22,485.69	899,427.60
1019	Excav & Support Starter Tunnel Class III	120.00	LF	12,880 107.33	1,158,636	141,689	431,773	122,780	64,598	1,919,477	1,562,256	3,481,732	29,014.44	837,298 24.0%	4,319,030	35,991.92	35,991.92	4,319,030.40
1022	GCT # 5 Excavation & Support	1,000.00	LF															
1023	Enlarge & Support WYE Cavern No. 1 WB	250.00	LF	21,640 86.56	1,940,427	147,600	495,706	176,481	66,730	2,826,943	2,616,389	5,443,332	21,773.33	1,402,265 25.8%	6,845,597	27,382.39	27,382.39	6,845,597.50
1024	Enlarge & Support WYE Cavern No.2 WB	70.00	LF	10,712 153.03	960,781	65,269	254,309	86,054	39,571	1,405,985	1,295,477	2,701,461	38,592.30	694,316 25.7%	3,395,778	48,511.11	48,511.11	3,395,777.70
1025	Enlarge & Support Wye Cavern No. 3 WB	180.00	LF	39,776 220.98	3,565,881	248,400	1,071,529	328,401	157,104	5,371,315	4,808,081	10,179,396	56,552.20	2,576,913 25.3%	12,756,309	70,868.38	70,868.38	12,756,308.40
1028	Enlarge & Support WYE Cavern No. 1 EB	250.00	LF	21,640 86.56	1,940,427	147,070	495,706	176,481	66,730	2,826,413	2,616,389	5,442,802	21,771.21	1,402,265 25.8%	6,845,067	27,380.27	27,380.27	6,845,067.50
1029		70.00	LF	10,712	960,781	65,269	254,309	86,054	39,571	1,405,985	1,295,477	2,701,461	38,592.30	694,316	3,395,778	48,511.11	48,511.11	3,395,777.70

Figure 1. HCSS example report

RISK MANAGEMENT

```
Parsons Brinckerhoff                                                              Page 6
000100          HCSS EXAMPLE SPREADSHEET                                    01/09/2009 8:02
Keith Caro - PB                            DIRECT COST REPORT
```

Activity Resource	Desc	Quantity Pcs Unit	Unit Cost	Perm Labor	Constr Matl/Ex	Equip Ment	Sub-Contrac	Total

BID ITEM = 1014   Land Item   SCHEDULE: 1 92 100
Description = Excav & Support App Tunnels Type 1A   Unit = BCY   Takeoff Quan: 1,266.000   Engr Quan: 0.000

D100	SUBS & MATERIALS		Quan: 1,298.00 BC Hrs/Shft: 8.00			WCWC1		
2CS02	Shotcrete - Wet	452.00 CY	103.809	46,922				46,922
2CT02	Accelerator	4,520.00 LB	1.191	5,384				5,384
2CT10	Steel Fibers	33,900.00 LB	0.596	20,225				20,225
2STS027	Lattice Girders	10,059.00 LB	0.896	9,013				9,013
2TB009	#9 Rock Bolts	3,044.00 LF	1.792	5,455				5,455
2TB009A	#9 - Nut/Washer/Plate	304.00 EA	11.222	3,412				3,412
2TB098	Rock Bolt Resin	3,044.00 LF	1.016	3,093				3,093
3DA002	Drill Bits & Steel	23,414.00 DLF	0.546		12,793			12,793
3E030	Powder	6,647.00 LB	12.020		79,897			79,897
3E050	Blasting Caps	4,074.00 EA	13.112		53,421			53,421
3E060	Blasting Supplies	24.00 RND	418.514		10,044			10,044
3E091	Powder Delivery Char	24.00 DAY	3,551.362		85,233			85,233
4MH015VK	Muck Haul - Truck	1,533.00 BCY	28.800			44,150		44,150
$379,042.07			[ ]	93,503	241,389	44,150		379,042
				72.04	185.97	34.01		292.02

D200	EXCAVATE APPROACH TUNNELS CL		Quan: 1,298.00 BC Hrs/Shft: 8.00 Cal L10 WCWC1					
ERDT10	DRILL/ BLAST HEADING TUNNEL	576.00 CH	Prod: 72.0000 S		Lab Pcs: 16.00		Eqp Pcs: 23.00	
3ADDSTS10	10% small tools & su	835,306.55 LAB$	0.100		83,531			83,531
8DRL0150	TNL JUMB HYD 2 D 1.00	576.00 HR	18.590			10,708		10,708
8DRL0210	JACKLEG/SINKER/ 1.00	576.00 HR	0.575			331		331
8LDW0050	CAT 966 4-1/4 CY 20 1.00	576.00 HR	52.020			29,964		29,964
8RAL025	25 T DSL LOCOMO 2.00	1,152.00 HR	12.737			14,673		14,673
8RAL070	MUCK CAR 10 CY 10.00	5,760.00 HR	2.140			12,326		12,326
8RAL080	FLAT RAIL CAR 2.00	1,152.00 HR	2.140			2,465		2,465
8RAL100	MANTRIP RAIL CA 1.00	576.00 HR	2.140			1,233		1,233
8RAL140	6 CY AGITATOR C 1.00	576.00 HR	3.886			2,238		2,238
8RAL200	CALIFORNIA SWIT 1.00	576.00 HR	0.967			557		557
8SHT030	CAT 950 LDR SHTC 1.00	576.00 HR	8.280			4,769		4,769
8VNT100	VNT FAN 200HP 15 2.00	1,152.00 HR	24.370			28,074		28,074
LA001	Walker 1.00	576.00 MH	42.408	59,911				59,911
LA003	Shifter (Heading Fore 1.00	576.00 MH	39.227	55,691				55,691
LA007	Miner - Driller 2.00	1,152.00 MH	35.340	101,065				101,065
LA013	Miner - Chuck Tender 2.00	1,152.00 MH	35.340	101,065				101,065
LA014	Miner - Powder Carri 2.00	1,152.00 MH	35.340	101,065				101,065
LA032	Miner - Electrician 1.00	576.00 MH	35.340	50,532				50,532
LA034	Miner - Brakeman 2.00	1,152.00 MH	35.340	101,065				101,065
OP032	Tunnel Mucker Opera 1.00	576.00 MH	47.710	55,859				55,859
OP048	Locomotive Op (>10 2.00	1,152.00 MH	41.010	101,365				101,365
OP051	Heading Mechanic - L 1.00	576.00 MH	44.840	51,829				51,829
OP069	Face Jumbo Operator 1.00	576.00 MH	47.710	55,859				55,859
OP090	Shotcrete Machine Op	0.00 MH	47.710					

Figure 2. Direct cost report

GETTING THE ENGINEER'S ESTIMATE RIGHT

Bidltem	Bid Description	Units	Quantity	Labor	Perm Matl	Constr Matl	Equip	Sub	Direct Total	Indirect	Balanced Bid
	100 VARIABLE CONTRACTOR INDIRECTS	MTH	46	$15,661,062	$0	$1,706,606	$603,870	$26,744	$27,641,640	$0	$29,443,787
1000	All Required Work Except the Bid Item Listed Below	LS	1						$0	$3,621,428	$0
1001	Assemble TBM's	LS	1	$1,726,368	$0	$338,944	$292,236	$0	$4,020,621	$3,621,428	$11,040,205
1002	Temporary Facilities & Utilities	LS	1	$1,252,382	$0	$1,572,992	$231,210	$0	$4,308,650	$693,966	$7,513,503
1003	Temporary Medium Voltage Substation	LS	1					$5,905,891	$5,905,891	$0	$5,905,891
1004	Partial Disassembly and Reassembly of TBM's	LS	1	$2,042,512	$0	$421,411	$965,997	$0	$5,601,521	$4,502,544	$14,329,013
1005	Final Disassembly & Removal of TBM's	LS	1	$2,269,458	$0	$468,235	$1,073,330	$0	$6,223,912	$5,002,826	$15,921,125
1008	Resident Engineers Main Office	LS	1			$656,850			$656,850	$0	$656,850
1009	Maintain RE Office	MTHS	46			$992,473			$992,473	$0	$992,473
1010	Excav & Support Approach Tunnels	LF	309	$1,347,360	$287,876	$1,120,988	$338,842	$164,016	$4,708,262	$2,987,946	$10,499,939
1014	Excav & Support App Tunnels Type 1A	BCY	1,266	$427,781	$93,503	$330,156	$108,298	$44,150	$1,463,784	$0	$0
1015	Excav & Support App Tunnels Type 1B	BCY	3,228	$919,579	$194,373	$790,832	$230,545	$119,866	$3,244,477	$0	$5,088,605
1016	Excav & Support Assembly Chamber	LF	80	$1,160,814	$502,836	$782,733	$260,574	$150,048	$4,080,150	$2,547,125	$9,017,362
1017	Excav & Support Starter Tunnel Class I	LF	40	$114,673	$35,779	$92,989	$24,386	$21,139	$410,415	$252,283	$899,428
1019	Excav & Support Starter Tunnel Class III	LF	120	$563,353	$141,689	$431,773	$122,780	$64,598	$1,919,477	$1,237,938	$4,319,030
1022	GCT # 5 Excavation & Support	LF	1,000						$0	$0	$0
1023	Enlarge & Support WYE Cavern No. 1 WB	LF	250	$936,080	$147,600	$495,706	$176,481	$66,730	$2,826,943	$2,073,238	$6,845,597
1024	Enlarge & Support WYE Cavern No.2 WB	LF	70	$463,219	$65,269	$254,309	$86,054	$39,571	$1,405,985	$1,026,641	$3,395,778
1025	Enlarge & Support Wye Cavern No. 3 WB	LF	180	$1,721,044	$248,400	$1,071,529	$328,401	$157,104	$5,371,315	$3,809,944	$12,756,309
1028	Enlarge & Support WYE Cavern No. 1 EB	LF	250	$936,080	$147,070	$495,706	$176,481	$66,730	$2,826,413	$2,073,238	$6,845,067
1029	Enlarge & Support WYE Cavern No.2 EB	LF	70	$463,219	$65,269	$254,309	$86,054	$39,571	$1,405,985	$1,026,541	$3,395,778
1030	Enlarge & Support Wye Cavern No. 3 EB	LF	180	$1,721,044	$248,400	$1,071,529	$328,401	$157,104	$5,371,315	$3,809,944	$12,756,309
1044	GCT # 3 Excavation & Support	LF	659						$0	$0	$0
1047	Enlarge & Support WYE Cavern No. 1 WB	LF	125	$524,303	$64,223	$263,924	$99,099	$30,586	$1,544,565	$1,161,113	$3,795,206
1048	Enlarge & Support WYE Cavern No. 2 WB	LF	67	$403,613	$29,670	$220,373	$77,062	$28,253	$1,191,595	$893,473	$2,923,455
1049	Enlarge & Support WYE Cavern No. 3 WB	LF	137	$919,753	$62,604	$569,180	$175,842	$105,034	$2,818,172	$2,035,934	$6,764,520
1052	Enlarge & Support WYE Cavern No. 1 EB	LF	125	$524,303	$64,223	$263,924	$99,099	$30,586	$1,544,565	$1,161,113	$3,795,206
1053	Enlarge & Support WYE Cavern No. 2 EB	LF	67	$403,613	$29,670	$220,373	$77,062	$28,253	$1,191,595	$893,473	$2,923,455
1054	Enlarge & Support WYE Cavern No. 3 EB	LF	137	$919,753	$62,604	$569,180	$175,842	$105,034	$2,818,172	$2,035,934	$6,764,520
1070	Drilling Ahead of Tunnel Face	LF	2,674	$25,531		$6,729	$12,075		$71,480	$56,282	$180,574
1071	Grouting Ahead of the Tunnel Face	SACK	4,467	$257,205	$28,802	$53,067	$121,644		$732,179	$566,987	$1,831,196
1072	Drill & Grout within Excav Tunnel	SACK	5,793	$189,122	$115,860	$40,602	$89,444		$636,102	$416,902	$1,444,203
1073	Additional Steel Fiber Reinforced Shotcrete	CY	250	$101,067	$80,233	$20,703	$23,853	$105,034	$331,816	$221,197	$760,573
1074	Additional Rock Dowels # 8 X 10 ft	EA	520	$18,950	$20,439	$6,723	$4,472		$70,452	$41,474	$150,844
1076	Additional Rock Bolts # 9 X 12 ft	EA	125	$25,267	$5,615	$5,995	$5,963		$69,331	$55,299	$176,520
1078	Additional GRP Dowels 1" x 10 ft	EA	360	$15,130	$26,396	$5,089	$7,156		$69,855	$33,352	$134,503
1080	Additional Steel Sets Installed	EA	40	$37,824	$54,469	$7,804	$17,889		$158,201	$83,380	$319,821
1081	Additional Mine Straps Installed	EA	400		$42,616				$42,616	$0	$42,616
1082	Spiles # 11 X 12 ft	EA	1,000	$44,217	$45,894	$15,614	$10,436		$162,518	$96,774	$350,100
1101	Deep Bench Marks Surface	EA	8					$160,000	$160,000	$0	$160,000
1102	Open Standpipe Piezometers	EA	7					$140,000	$140,000	$0	$140,000
1103	Surface Settlement Pt Type 1 in pavement	EA	15					$1,800	$1,800	$0	$1,800
1104	Surface Settlement Pt Type 2 in Masonary	EA	78					$14,040	$14,040	$0	$14,040
1105	Prism for auto total station building facades	EA	322					$48,300	$48,300	$0	$48,300
1106	Monitirring point vrertical masonry	EA	292					$14,600	$14,600	$0	$14,600
1107	Electric Tiltmeters columns	EA	56					$132,976	$132,976	$0	$132,976
1108	Inclinometer - Surface	EA	5					$202,988	$202,988	$0	$202,988
1109	MPBX Surface	EA	5					$72,983	$72,983	$0	$72,983
1110	Portable Siesmographs	EA	30					$120,000	$120,000	$0	$120,000
1114	Auto Motorized Total Stations Building Facades	EA	12					$427,167	$427,167	$0	$427,167
1117	Horizontal MPBX New Tunnels	EA	112					$11,200	$11,200	$0	$11,200
1118	High Frequency Geophones	EA	14					$204,362	$204,362	$0	$204,362
		EA	17					$34,000	$34,000	$0	$34,000

Figure 3. Bid item summary in Excel format

cost is readily apparent because the profit amount embedded in a line item estimate is sometimes difficult to determine and open to speculation or challenge.

Similarly, when estimating overhead costs for a project, a complete breakdown of all components may need to be produced. Indirect costs, based on known rates for the geographical area, adjusted as necessary, will be provided as separate cost items and spread over the directs costs. Typically this level of effort may only be justified on larger projects and could include allowances for special transportation, accommodations, supplies, per diem, etc., all items of cost which the owner might need knowledge of prior to soliciting bids for the work.

Contingency or an allowance for unfavorable, unknown conditions or situations, are best left out of the basic estimate. When the estimator embeds contingency amounts in the direct costs, whether by altering production rates or bumping up component costs, it becomes very difficult to quantify just what impact those allowances have on final cost. Similarly when conditions change such that the contingency allowance should be removed or reduced, the task to modify the estimate and remove embedded contingencies can be both time consuming and risky. Keeping project costs based on known productions and separating contingency allowances which can then be entered as a one or more line item cost before arriving at the final cost draws attention to the amount of added money, and makes it easily removed or modified if conditions change. Owners should understand that some contingency allowance may be valid depending on the pre-bid information available. The decision to include reasonable amounts does not invalid the Engineer's estimate or the contractor's bid.

MODIFICATIONS AND UPDATES

The benefit of computer-based software is readily apparent when a revised cost is needed due to design issues, alignment changes, different excavation methods or other major modifications to the work are necessary. In the PBGT cost estimating system, the use of crews and production rates provides the duration for each task and it is simple to provide the updated cost estimate by using new or adjusted factors. In actual use, separate alignments that contain the same end points but present distinct alternative routes, can be readily compared to determine the least expensive option. Because an entire new estimate can be developed by copying any existing estimate and then changing only those parameters which are affected by the new information or design data, the estimator is able to quickly provide alternative cost scenarios.

Another simple and quick modification is revising the labor rates for a project. If the labor and/or fringes that an estimate is based on must be modified, the data base values can be updated by setting up an excel file which contains the new data and importing those changes directly into the estimate. The software will then update all or a single item of the estimate as required using the new information. The result is a revised estimated cost using the modified labor and/or fringe values. Similar modifications for changes in equipment rent or operating costs, subcontract prices or material costs are accommodated just as easily.

ENGINEER'S ESTIMATE

The true test for any estimating program is to measure it against actual bid results under competitive market conditions. The goal is not to have a low Engineer's Estimate which will cause the owner to question why he must pay a higher price for the work. Similarly, if the Engineer's Estimate comes in significantly higher than all the bids, then the owner may challenge that estimate as being too conservative and not realistic. Ideally the Engineer's Estimate should aim to be the second lowest cost once bids are received. Such a ranking will confirm PBGT's interim pricing was sound and the project will likely have a final cost close to the Engineer's estimate. As all contractors and most owners understand, the price on the day bids are received is not necessarily the final cost, that amount is only known when all work is complete and accepted, claims or disputes have been settled and final payment made. The period between bid-award and final acceptance can be a minefield filled with changes, claims and counterclaims, even serious disputes some of which may lead to legal remedies before final settlement is reached and the true cost is known.

On a major project in New York City, bid dated May 16, 2006, Contract CM009, Manhattan Tunnels, the bids were received by MTA LIRR. The low bid on this contract was accepted and the work is underway as this paper was being prepared. The PB provided Engineer's Estimate was approximately 12.8% higher than the low bid, typically a greater margin than desired, however in this case it is believed the Contractor Joint Venture was more aggressive and wanted the work enough to break into a new market territory to tighten costs and/or margins. Of course there could be a host of other reasons that produced this difference as well. PB was approximately 2.8% under the second bid received which is the ideal result and confirms the validity of the PB Engineer's Estimate.

SUMMARY

The risk of not "Getting the Engineer's Estimate right" is something that each consultant and owner should consider early in the life of any proposed project. Owners should demand that whatever method the consultant uses to forecast costs, whether in the early stages or for the Engineer's Estimate, the costs be based on up-to-date information, take into account all appropriate work rules and regulations, and represent methods for completing the work on time for the amount indicated. The cost estimate should also provide overall and individual item cost summaries to a reasonable degree of detail.

Competent estimators using software that is designed for the specific purpose will provide traceability and comparison for items that may later come into question. The ability to modify and re-estimate the entire work scope when new or different information is developed will save time and provide the needed confidence in any cost estimate produced. Estimators must endeavor to keep current with methods, procedures and regulations that affect the cost of new work. Failure to do so will inevitably widen the gap between the Engineer's Estimate and the prices contractors submit for actually completing the work scope. When this gap is excessive, a loss of confidence occurs which may be difficult or even impossible to remedy in the time available.

Use the best products available, employ those who have "been there, done that" whenever possible and give the owner what he needs, a reliable cost estimate that factors in all conditions, variables, keeps contingencies to a minimum, and will withstand close inspection of content. To do less is to shortchange the owner, the designer and the future successful contractor.

While there is currently no requirement for the Engineer's Estimate to be archived and referred to during disputes or claims, a detailed cost estimate produced at the time of bidding would be useful to support productions, manpower requirements, etc. when challenges by the contractor occur. To have this information considered would require modification of most current specifications which permit escrowing of only the contractor's bid documents. Such supplemental information, provided it had been developed along the same lines as the contractor, could also provide another basis for costing extra work independent of the contractor's Escrow Bid Documents.

TRANSFER OF A PROJECT RISK REGISTER FROM DESIGN INTO CONSTRUCTION: LESSONS LEARNED FROM THE WSSC BI-COUNTY WATER TUNNEL PROJECT

R.J.F. Goodfellow ■ Black & Veatch

P.J. Headland ■ Black & Veatch

ABSTRACT

Many underground projects now include preparation of a Risk Register through planning and design as a tool for project and risk management. There is still significant discussion within the industry as to whether and how to best present the Risk Register through the bidding process and the construction phase of a project. While the ITIG Code of Practice for Risk Management of Tunnel Works recommends this approach of continuity in risk management procedures, it is silent on the specifics of how this should be achieved.

This paper will explore the perceived benefits and dis-benefits of including the project Risk Register with bid documents; format and content that should be transferred and lessons learned using the case history of bidding and early construction phase of the Bi-County Water Tunnel in Montgomery County, MD.

INTRODUCTION

The use of a Risk Register during the planning and design of major underground projects is becoming standard practice in North America. The introduction of the Joint Code of Practice for Tunnel Works in the UK (2003) and its international counterpart published by the International Tunnel Insurance Group (ITIG) in 2006 (referred to henceforth as "the Code") has encouraged this development in risk management practice.

As is commonplace in the development of an area of practice, there has been a series of lessons learned in application and practical experience within the design and consulting fraternity. The use and transfer of existing Risk Registers into the construction phase and in particular the debate on what information, if any, is transferred to the Contractor has not reached any consensus to date. The debate continues to present reasoned argument for both extremes of transmitting all or transmitting no data:

Transmitting *all* Data—transmitting all data to the Contractor can be justified as an extension of the full disclosure of all information, including geotechnical information, so that the Contractor can produce the most informed bid without fear of contradiction of other information. The Risk Register forms the road map for where risk has been mitigated and where the residual risk is allocated within the documents.

Transmitting *no* Data—the argument for transmitting no design information to the Contractor is equally reasoned, suggesting that the design decisions are made for good reason and that the Contractor has no need or justification to know reasons why or be given the opportunity to revisit these decisions.

The robust format and use of a Risk Register in managing project risk is useful in many ways during planning and design. Black & Veatch has been working diligently to maintain those advantages to the project by continuing usage of the Risk Register through the construction phase encouraging cooperation and coordination between the Owner and Contractor.

It is proposed that the Risk Register be transferred to the selected Contractor for information as an early partnering activity. This register is intended to form a working document both identifying where risk mitigation and allocation of risk lies within the Contract Documents and to provide notification of changes in risk profile due to construction events. It is important to note that while this early notice is, in part mimicking clauses within the NEC3 Form of Contract, the Risk Register activities are not demanded by the Contract, nor is the Risk Register a Contract Document. Thus, changes to the Risk Register do not necessarily define changed conditions but are indicative of the willingness of the Contractual parties to collaborate for the benefit of the project and give some structure and focus to partnering activities—not just platitudes and abstract meetings.

The Introduction of Risk Registers

It is important to note that Risk Register's existed in design and construction practice before the Code was published. Indeed, the FTA has required a Risk Register through planning and design for several years. The introduction of the Code though was an attempt to provide a clear framework for management of risk. In the first section of the Code its stated objective is "to promote and secure best practice for the minimization and management of risks associated with the design and construction of tunnels, caverns, shafts, and associated underground structures..." (ITIG; Section 1.1).

The Code encourages risk assessments at every stage throughout a project life: development (or planning), procurement, design, and construction. The Code also states that "risk assessments required at each stage of a project shall be summarized in appropriate Risk Registers" (ITIG; Section 4.2.2). The Code goes on to describe the role of a Risk Register in managing project risk. "Risk Registers shall be live documents that are continually reviewed and revised as appropriate" (ITIG; Section 4.3.2) meaning that the frequency of review should be commensurate with the activities undertaken, for example, quarterly review is meaningless if a construction activity is planned to take 6 weeks; more frequent reviews are needed for that specific activity if the effectiveness of mitigation measures is to be assessed properly.

A Risk Register "shall identify hazards, consequent risks, mitigation and contingency measures, proposed actions, responsibilities, critical dates for completion of actions and when required actions have been closed out" (ITIG; Section 4.3.2). In this way, the Code lays down a clear framework for systematic use of the Risk Register as a project management tool that links together engineering practice, technical progress through design and construction with schedule, and assesses the risk to timely and cost effectively complete the project.

"Compliance with the Code from the insurer's perspective has been the subject of much discussion. The outcome of much practical project experience internationally has seen the development of two phases of communication: A first phase (Phase I) that provides assessment of the intentions of the Contractor seeking insurance and the definition of insurance coverage and limits. This is followed by an auditing phase (Phase II) throughout construction that simply verifies that the Contractor is carrying out the actions promised during the Phase I assessment.

Concept of Risk Management and Project Management

The use of a Risk Register as a project management tool is something that cannot be achieved with any format of register. The Risk Register format must be sufficiently robust and contain several key attributes to be used in this way. The system of application is described briefly herein and the reader is referred to Goodfellow and Mellors

(RETC, 2007) for a more detailed review of how the Risk Register can be formatted to meet the Code requirements and provide a strong project management tool.

A properly formatted Risk Register must contain a clear presentation of the hazard, cause of the hazard, and show an initial assessment of the risk. For a better audit and appreciation of progress made in risk mitigation the Risk Register should continue to display this initial risk score.

Management of risk then begins with the repeated cycle of mitigation and reassessment. A current snapshot of project risk is provided with a residual risk assessment after consideration of mitigation measures. A suggested format of Risk Register can be seen in two sections below (Table 1 and Table 2).

The concept of using Risk Register's as a project management tool relies on the consistent assessment of project actions with reference to risk mitigation. If actions are planned, logged, and re-assessed with reference to their impact on particular project risks then the register accurately reflects the current project status.

One important aspect of the management of risk is that after risks are avoided, if possible, and then mitigated to levels as low as reasonably practicable (ALARP); the residual risk must be allocated in the Contract Documents. This is noted on the Risk Register with referencing of particular project risks to either specification or contract sections. This referencing is important when transferring the Risk Register to the Contractor during the construction phase in design-bid-build projects in order to directly link the specific risk and where the risk is considered in the Contract Documents.

Transfer Risk Register to Construction Phase

The Code of Practice asks for the production of a construction stage Risk Register and defines this document as: "A register that records all project related risks identified for the construction stage of the project and includes and identifies the project-related risks brought forward from the client's pre-contract Risk Register..." (ITIG, Appendix A).

The use of a Risk Register during construction has been considered elsewhere in the World. One example is the New Engineering Contract. The NEC 3 Conditions of Contract consider Risk Register's in the Early Warning provisions. An extract from this clause reads as follows: "The Contractor may give an early warning by notifying the Project Manager of any other matter which could increase his total cost. The Project Manager enters Early Warning matters into the Risk Register." In this context, the Project Manager is the Owner's Representative.

The NEC goes on to say that "the Project Manager revises the Risk Register to record the decisions made at each risk reduction meeting and issues the revised Risk Register to the Contractor."

The concept of contractual obligation to partner for the betterment of the overall project is intriguing, as is the use of the Risk Register as a tool to achieve this aim. It is the opinion of the authors that these objectives are in agreement with the overall aims and intents of the Code insofar as the Code seeks to provide a general reduction in the risk profile of the project.

Moreover, the use of a Risk Register by insurers leads to much more precise and clear exclusions, preventing the same long litigation seen, for example, in the recent Sarnia Crossing final settlement from insurance payment disputes from the early 1990s (Tunneltalk.com, 2008). This would provide a similar objective to a GBR that is principally designed to prevent lengthy disputes on differing site conditions and geotechnical issues.

It is the belief of the authors that the Risk Register should not be a Contract Document as it should not detract from the Specifications and other provisions in the Contract. Provided for information, however, the Risk Register transferred to the Contractor should contain all hazards that pertain to design, withholding only those

Table 1. Risk register example: Hazard identification and initial assessment

Hazard	Cause of Hazard	Potential Consequence	Risk likelihood	Risk Consequence						Risk Score
				Financial	Project Schedule	Corporate Reputation	Regulatory/ Legal	Health and Safety	Environment	
Bid Price from Contractor higher than budget allowance of Owner	Insufficient bid Competition	Schedule delay and cost increase	4	4	3					16

Table 2. Risk register example: Risk management and continuing assessment

Control Measures Implemented (actually in place at time of assessment)	Indicators or Metrics (Measuring the effect of Control Measures)	Residual Likelihood—After Mitigation	Residual Consequence – Once Controls in Place						Residual Risk Score—After Mitigation Action	Action Item for Risk Mitigation	Action Item Completion Date (Target Date)	Risk Owner (Name of individual)
			Financial	Project Schedule	Corporate Reputation	Regulatory/ Legal	Health and Safety	Environment				
Personal contact from Design team and advertising of project ahead of bid date	Letters of interest/ response from Contractor Community	2	3						8	Pre-qualification of Contractors	February 15	J. Smith

hazards that are political or commercially sensitive and refer to planning and project feasibility.

Disposition of the hazards contained in the Risk Register should not only designate the responsible party for any residual risk but should also reference where in the Contract Documents a particular risk is considered. The Risk Register not being part of the Contract prevents confusion and also prevents inconsistency with the Contract becoming the basis for any claim.

The Case History below shows an example that may be the first example of the transfer and consideration of a Risk Register from planning through design and into Construction with a meaningful role within the Contract Documents. While at the time of writing, this process is only beginning, it is expected that the use of a Risk Register alongside the other contractual tools to actively and collaboratively manage project risk will benefit the underground industry and provide a more solid basis for future partnering efforts between Owner and Contractor.

CASE HISTORY—WSSC; BI-COUNTY TUNNEL PROJECT

The Bi-County Water Tunnel project includes approximately 5.3 miles of 10-foot diameter deep rock tunnel, located in Montgomery County, Maryland. The water tunnel follows an alignment which runs from Shaft S-1, north of Tuckerman Lane just east of Interstate I-270 and terminating at Shaft S-4 at the intersection of Stoneybrook Drive and Beach Drive (Figure 1). The tunnel is specified to be bored using a Tunnel Boring Machine (TBM), with a minimum diameter of 10 feet. The finished tunnel will contain a grouted-in-place 84-inch ID steel water transmission pipe.

The tunnel alignment is located in rock at depths ranging from 90 feet to 275 feet below ground surface with surface access via three vertical shafts (S-1, S-3, & S-4). Small sections of open-cut pipeline and valve vaults are to be constructed to facilitate tie-ins at each end of the alignment.

The project site lies within the Maryland Piedmont Physiographic Province, northwest of the contact between the crystalline Piedmont rocks to the west and the younger Coastal Plain sediments to the east. The bedrock consists of metamorphosed sedimentary and igneous lithologies including the Sykesville Formation (metamorphic sequence of gneiss and schist) which has been intruded by the Georgetown Intrusive Suite (quartz tonalite with some schist and amphibolites) along central portions of the project alignment.

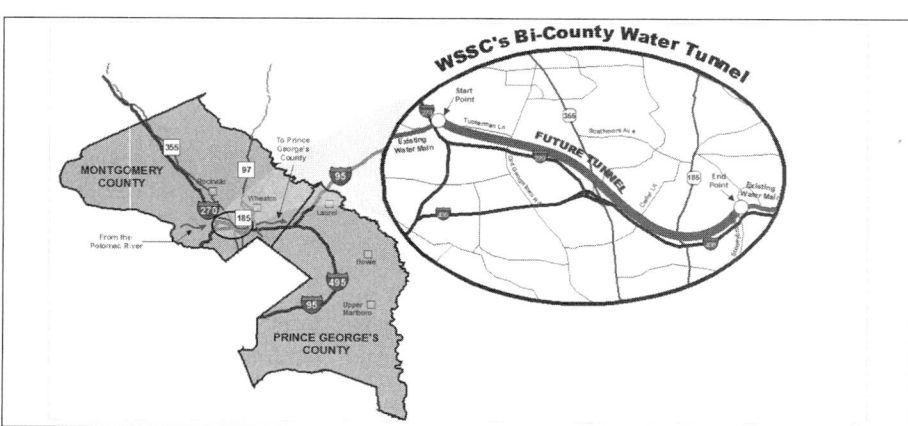

Figure 1

TRANSFER OF A RISK REGISTER FROM DESIGN INTO CONSTRUCTION   859

The use of the Risk Register during construction is designated in the Contract in the Information for Bidders section of the Conditions of Contract. The clause reads as follows:

> "The use of a Risk Register as a tool for risk management during construction is part of an expected open and cooperative relationship between the Owner and Contractor. A risk reduction review will be a part of each progress meeting, but each party may call a risk reduction meeting at any time and request the presence of the other contractual parties to attend. Within this meeting, the Owner and Contractor will discuss and cooperate to achieve:
>
> - Making or considering proposals for how the effect of the registered risks can be avoided or reduced;
> - Seeking solutions that will bring advantage to all those who will be affected;
> - Deciding on the actions which will be taken and who, in accordance with this contract, will take them; and
> - Deciding which risks have now been avoided or have passed and can be removed from the Risk Register."

While it was decided that the Risk Register would not be a Contract Document it was considered beneficial to the Project and to the partnering process to include the Risk Register and an active regular discussion about project risk and specific hazards associated with upcoming activities. Monitoring and management of risk by use of the Risk Register is anticipated to be beneficial through the construction phase. This initial foray into the transfer of risk registers is expected to provide many lessons for the industry and these lessons will be shared in a follow-up paper at the conclusion of construction.

CONCLUSIONS

The following conclusions have been drawn from our experience thus far in the project.

- The Risk Register was helpful in alignment selection during the planning phase and helped the design team in management of risk through the detailed design phase of the Bi-County Tunnel project.
- The use of a Risk Register alongside the other contractual tools to actively and collaboratively manage project risk will benefit the underground industry and provide a more solid basis for future partnering efforts between Owner and Contractor.
- Prudent use of a Risk Register allows auditing of the project at any phase from political, commercial or insurance agents. This allows the sometimes arcane world of underground engineering to be better understood.
- The Risk Register should be transferred in some form into the construction phase of a project.
- During the construction phase the Risk Register must be actively reviewed and managed on a regular basis (as least monthly) to be effective;
- The Risk Register should not be transferred as a Contract Document.
- The Risk Register should allow precise limits and definition of insurance requirements, preventing lengthy disputes between insurer and project participants.
- The Risk Register should form a focal point of closer collaboration between Contract Entities that forms part of a meaningful partnering process.

ACKNOWLEDGMENTS

The help of WSSC and the Project Team has been valuable in generation of this paper. Gathering opinion on implementation of the Code has led to the proposed considered course of action. Personal correspondence and discussion with the tunnel industry on the implementation and transfer of Risk Registers is considered an important education tool for the authors and for this open interaction with our colleagues, we are very grateful.

REFERENCES

The Joint Code of Practice for Risk Management of Tunnel Works in the UK. Mellors, T.W. and Southcott, D eds. British Tunnelling Society and Association of British Insurers, 2003.
A Code of Practice for Risk Management of Tunnel Works, ITIG, 2006.
Cracking the Code—Assessing Implementation in the United States of the Codes of Practice for Risk Management of Tunnel Works. Goodfellow, R. J. F. and Mellors, T.W., RETC Proceedings, 2007.
NEC 3 Engineering & Construction Contract, Thomas Telford, June 2005.
Supreme Court Finds for the Engineers. Tunnel Talk.com, December 2008.

THE DELIVERY OF UNDERGROUND CONSTRUCTION PROJECTS IN THE UK: A REVIEW OF GOOD PRACTICE

Andy Alder ▪ Halcrow Group Ltd

Mike King ▪ Halcrow Group Ltd

ABSTRACT

The aim of this paper is to provide an overview and details of good practice, from a United Kingdom perspective, in the development and implementation of major tunnel and underground infrastructure projects. In doing so, it draws on a broad base of project management theory.

A central theme is that successful project delivery is dependent on detailed planning of the complete project delivery process, forming an effective team that works well together and the proper management of risk. The role of partnering in forming a strong project coalition is described. Particular emphasis is given to the integration of design and construction, and the role of the designer throughout the construction phase.

The implementation of risk management in the design and construction of tunnel works is discussed, and recommendations are made on how risk, value and constructability can be integrated in project delivery.

Recent practice in project procurement, particularly in the use of innovative collaborative forms of contract are described, and details of particular provisions for joint management of risk and problem resolution are provided.

Finally, the relevance of the ITIG Code of Practice for Risk Management of Tunnel Works to good practice in project delivery is highlighted throughout the paper.

INTRODUCTION

The aim of this paper is to provide an overview and details of good practice, from a United Kingdom perspective, in the development and implementation of major tunnel and underground infrastructure projects. In doing so, it draws on a broad base of project management theory.

In preparing this paper reference was also made to recent guidance published in the United States on the management of underground construction projects (Edgerton, 2008). Where UK practice offers further insight to or potential development of US guidance these aspects are explored in some detail in this paper.

A key message from both the UK and US is that the success of an underground project depends heavily on good planning, and having the right team in place to deliver that plan. Furthermore, effective teamwork and management processes to manage and control risk are vital to successful underground project delivery.

PLANNING

Effective planning is vital to underground construction projects in the broadest sense, incorporating feasibility studies, design, scoping, construction planning and scheduling and development procurement strategy. Good planning at the design stage reduces the future contractor's risks, and should give rise to more accurate bidding. It is generally accepted that the client (owner) should obtain the land that is required for the project, including suitable and sufficient temporary worksites to support the envisaged construction methodology, so that the contractor has certainty regarding land availability in his tender.

In a similar manner, obtaining the necessary approval and environmental permits, such as traffic management plans, represents a significant risk to the contractor. Obtaining these approvals, or at least establishing the ground rules for issues such as allowable working hours or acceptable noise levels, reduces uncertainty at bidding stage and allows an earlier construction start. On major UK projects (such as the Docklands Light Railway Extensions and Crossrail Project in the UK) this has been achieved through the use of a Code of Construction Practice agreed with the relevant authorities or agencies at the design stage. This does require construction expertise in the development stage of the project so that environmental impacts are properly assessed and verified, and that commitments made or agreements reached are supportive of efficient construction. In addition, having the design and construction methodology developed concurrently pre-Tender is essential to demonstrate the feasibility of the scheme and to confirm the adequacy of the provisions of the contract.

Consultation with project stakeholders, both internal and external to the project organisations, should be undertaken at an early stage. This provides the opportunity to build important relationships that will assist in project delivery, and if necessary will assist in resolving difficulties or differences of opinion as the project develops. Furthermore, early consultation allows problems to be identified at a stage when there is greatest opportunity to tackle them, and maximum scope to make changes to avoid problems with minimum impact on the project and minimum abortive costs.

Many UK agencies and government bodies (e.g. Office of Government Commerce, 2001 and Network Rail) require Stage Gates within the project development process so that there are defined points where the project is critically reviewed from a number of perspectives, including: fitness of purpose and acceptability of project proposals; cost estimates, budget allowance and value for money; schedule performance of the previous stage and planning and management arrangements for the subsequent stage; and management of risk. For a Design-Bid-Build procurement route the stages during the construction lifecycle would typically be as follows. Assessing performance in use and providing feedback to following projects is a valuable part of the process.

1. Project Development Stage (assessment and evaluation of project options)
2. Preliminary Design
3. Detailed Design
4. Construction Contract Procurement
5. Construction
6. Testing and Commissioning
7. Handover and Project Close-out

Adequate time needs to be allowed for these reviews so that they are undertaken with the necessary degree of rigour. Occasionally sufficient time is not allowed, or the review period is reduced to correct schedule slippage, and this tends to reduce the value of the reviews. Peer reviews or expert reviews are frequently carried out at Stage Gates, and many agencies require independent design checking as part of the conclusion of the detailed design Stage.

Various beneficial project activities can be effectively aligned with the Stage Gates, for example:

- Value can be facilitated by establishing Value Management and Value Engineering workshops at the start of each stage, to provide direction to the design process. Value Management is undertaken at the earliest stages and identifies the required function and subsequently shifts its focus to the selection of the project option that best balances function and price, whereas Value Engineering, which looks at detailed engineering options to select the most efficient design, is introduced at the preliminary and detailed design stages. In this manner Value Engineering forms an integral part of the design process and is hence efficiently used by the design team. (Connaughton and Green, 1996)

- For underground projects, in particular, constructability must form part of the design development with the construction methodology developed alongside the design to ensure that the project can be constructed economically and safely. Constructability reviews can be undertaken at the conclusion of each Stage, to formally verify that constructability has been properly considered.
- Risk Workshops are often used at the start of each Stage to identify the risks and proactively develop mitigations that should be implemented in the forthcoming Stage of the project. Risk reviews are then carried out at the conclusion of each Stage, again to verify that mitigations are incorporated and have not unintentionally adversely altered the risk profile. (Godfrey, 1996)

Procurement Strategy

The procurement strategy needs to be designed to deliver the client's objectives for the project. At an early stage, the relative importance of the primary objectives needs to be established, although the fact that all of these elements are important will make this difficult:

- Quality. Using the word in a broad sense, how much flexibility the client wishes to give the Contractor in developing design or construction proposals.
- Time. The need for early completion of the project, often to meet a predicted demand or to coincide with other elements of a programme.
- Cost. The relative importance of either lowest cost, or greatest cost certainty. Source of finance from public or private sector.
- Risk. Client's attitude to risk, and the relative desire to have certainty of project outcome. This will need to be balanced against potentially lower costs if the client is willing to share risk.

A detailed consideration of these issues, often through a series of workshops, will determine the preferred procurement method, including the finance arrangement, packaging and scope of contract and the form of contract to be used. It is important to define these at a reasonably early stage in the project development since they effect the design and project documents in numerous and complex ways. Late changes in the procurement strategy introduce the risk of inconsistencies in the Tender Documents, which can delay the construction start or give the potential for disputes and claims.

The project schedule often drives the procurement strategy, and for underground construction projects, in particular, consideration is frequently given to purchasing equipment prior to contract award. Purchase of the Tunnel Boring Machine (TBM) without involvement of the Contractor is not recommended, since the Contractor needs to be fully satisfied with the piece of equipment that will drive the critical path of the construction programme (on many underground projects the critical path runs through the tunnel boring). The procurement of the TBM requires some alignment design, the tunnel segment design (including special segments at cross-passages) and the casting of trial segments. Furthermore, to obtain the benefit of early TBM procurement there is likely to be a need to start the launch pit detailed design and construction. For design-bid-build projects these issues are more readily addressed, but with design and build they are more difficult. However, success has been achieved even on Private Finance/Design and Build projects when the preferred Contractor has been allowed to commence design and TBM procurement in advance of contract award (for example the DLR Woolwich Arsenal Extension, UK which was completed 7 weeks early).

Consideration may be given to other advance contracts, either to expedite construction or to assist in management of risk. Typically these include demolition and site clearance, utilities diversions and enabling works such as road or rail diversions.

TEAMWORK AND INTEGRATION OF DESIGN AND CONSTRUCTION

Relationships within the Project Team are critical to success. Partnering has the potential to reduce costs and shorten project durations through cooperative working across the project team. Within the UK, guidance suggests that there are three features critical to the success of project-based partnering (Bennett and Hayes, 1995):

- **Mutual objectives**. Partnering teams **agree mutual objectives**, designed to give everyone in the project team a realistic expectation that their work will be successful in business terms. It is more efficient to concentrate on getting the work done than to argue over who is to blame for problems.
- **Agreeing the decision-making systems** and cooperative problem resolution. Deal with problems quickly, based on a cooperative search for solutions. In solving problems partnering teams accept joint responsibility for their projects and their outcomes.
- **Measurement of performance** and actively seek **continuous improvement**.

Clearly to enable this it is vital that the selection of project partners is undertaken thoroughly. Relationships and trust are very important, and time needs to be invested in workshops and, much more importantly, through working together on the projects. This investment of time to build relationships means that staff continuity is necessary to maximise the benefit. Changing key project team members is likely to be to the detriment of the project.

The UK Government Construction Task Force (the Egan Report, Construction Task Force, 1998) promoted 'rethinking construction' based on partnering and lean production. One of the recommendations was the integration of the supply chain; in practice bringing together clients, designers, contractors and suppliers for efficient delivery of projects. This is particularly important in underground projects since the design is heavily dependent on the construction process, and hence construction expertise is required in the design stage. Having that construction input assists in buildability reviews and in risk management, with a team more able to identify risks and develop effective mitigation strategies and plans. In addition, costs are heavily dependent on the construction method since the costs of labour, preliminaries and the equipment and temporary works are significant. It follows that better definition of these, through having construction expertise available at the design stage, will result in more accurate cost estimates.

Having construction expertise in the design team has been achieved in various ways in the UK. Formal methods such as design and build, and Early Contractor Involvement allow the Contractor who will ultimately build the project to be working alongside the designer. Where the procurement route does not permit this, designers and clients have appointed either a contractor or a specialist construction consultant to provide advice and develop construction planning during the design stage (for example DART Underground Project, Dublin, Ireland and Tottenham Court Road Station Upgrade, London, UK).

Integration of Design and Construction Process: Temporary Works

On of the most important interfaces between the design and construction teams is the issue of temporary works. For cut and cover tunnel works, the design of the permanent works is heavily dependent on the designer's assumptions on propping locations and stiffness, and the same is true in many other aspects of underground construction. There are two primary rules that must be followed:

- The responsibilities for the design of temporary works must be clearly set out.
- The designer's assumptions on the construction sequence and associated temporary works need to be clearly stated.

On the Docklands Light Railway Extension to Woolwich, in London, UK, this was addressed by the designer preparing detailed construction sequence drawings and temporary works design requirements. In addition, the contractor's temporary works designs were independently checked by the designer. This gave the dual benefit of a full design check of these critical items and a closing of the loop between the contractor's temporary works and the designer's permanent works designs. Most forms of UK contract allow for the contractor's temporary works proposals to be issued to the Project Manager/Engineer for acceptance.

On the Tottenham Court Road Station Upgrade Project in London, UK, this issue was handled slightly differently. This project was heavily constrained by the existing station structures which could be adversely affected by the construction works. In this case the designer developed the design of all of the major temporary works, which the contractor was then required to review, check and adopt. If changes were required to suit the contractor's proposed methods they were to be submitted back to the designer for revision.

RISK MANAGEMENT

The essence of good risk management is the integration across all aspects of the design and project development process. Hazards identified in relation to the construction method can be eliminated or mitigated through the design and planning, and can be revealed if the right considerations are made at the early stage. This requires that a team with the appropriate competence across all aspects of the project are in place from early on. Planning of the project and its implementation, design and the various surveys, investigations and environmental studies are all inter-related and interactive.

The management of risk requires cooperation between all parties, and crucially this is supported by the equitable allocation of risk. Where the whole team can see the joint benefits of taking proactive coordinated action to address risks across the project, they will be more motivated to do so. Different participants may be able to assist each other in tackling risks that they are not necessarily owners of or contractually responsible for, but which they are able to positively influence. If other team members are reciprocating, then the synergy and collaboration of the team will allow the risk profile to be reduced. All team members are likely to get benefit from understanding the full set of risks that face the project. Again, this process is supported by equitable allocation of risk so that all participants see the benefit to them and to others. The approach to risk management and the contract conditions need to be consistent to support this process.

Team-based risk management, with good communication, maximises the chances of meeting the following objectives:

- All significant risks are identified.
- Management of risks is efficient: that the best party to manage a risk has clear ownership of that risk.

- Mitigation measures are coordinated and complementary, and where possible mitigation measures are selected that address several risks.

The ITIG Code of Practice (The International Tunnel Insurance Group, 2006) requires the use of a formalised Risk Management procedure as a means of documenting formally the identification, evaluation and allocation of risks. This is commonly achieved through a Risk Register to identify and clarify ownership of risks which shall detail clearly and concisely how the risks are to be allocated, controlled, mitigated and managed. Greatest benefit is achieved if a common Register is used across the team, so that team members know not only the risks that they are assigned, but where ownership of other risks lies. The Risk Register should also include the risk materialisation date (i.e. the date when the risk is expected to occur if unmitigated) and required action date for mitigation plans.

The approach to health and safety risks is similar to that for risks generally, but in the UK and Europe special emphasis is placed on risk management to reduce health and safety risks (Health and Safety Executive, 2007). Each designer, or design team, should use a Risk Register as a live document during the design development to identify and designs out risks to such that they are As Low as Reasonably Practicable (ALARP), and to provide evidence that they are doing so. This is achieved through the use of the hierarchy of risk control: taking action to eliminate and then reduce risks, before resorting to protection against the risk.

Risk workshops are held at the start and end of each design stage, and designers develop their risk assessments and risk mitigation plans during the design process. Where assumptions are needed to allow a design to progress they are identified in a separate Assumptions Log so that they can be validated at a later stage. Communication of the residual risks to the contractor is important, and is achieved through notation on the drawing as well as through the Risk Register. It has been found useful to use workshops to identify the "Designer's Critical Items" (i.e. items of design that the designer is most concerned about, for example workmanship of joint in SEM excavation) to ensure that the contractor's attention is drawn to them and the need for specific controls on materials and workmanship. These can assist the development of Inspection and Test Plans for the works.

The ITIG Code of Practice requires that provision is made for the appointment of an identified individual(s) who is/are suitably competent (i.e. qualified and experienced) in risk management practices and responsible for the identification and coordination of hazards and associated risks and the development and preparation of appropriate risk assessments, Risk Registers and risk management plans for all stages of the project. Benefit has been had in the UK from having a dedicated Risk Manager on major projects to ensure that Risk Management is proactively undertaken. Furthermore, there are legislative requirements for a Coordinator to ensure that health and safety risks are being managed and coordinated across the project, and this has been beneficial in maximising the health and safety performance of projects.

It has been suggested that generic risk acceptance criteria are selected for a particular project. This is not recommended. Rather, each risk should be considered in its own right to economically reduce each risk to the lowest practicable level.

Ground and Groundwater Conditions

Ground and groundwater conditions are, without doubt, the greatest risk to tunnelling. The decision on who should carry the risk of unforeseen ground and groundwater conditions is a difficult decision that needs considerable thought. Recent practice on some projects in the UK has been that clients need cost certainty and are not willing to retain ground risk, and hence transfer it to Contractor. However, the client (with the support of his engineering adviser) has more time to consider and investigate the ground and define the expected conditions. In any case, adequate site investigation prior to construction is essential.

The use of a Geotechnical Baseline Report has been suggested for many years, and enables clear risk allocation. In line with the argument above, equitable risk sharing from the outset will tend to encourage collaboration in the pursuit of project delivery across the team. The use of a Geotechnical Baseline Report shows a professional equitable attitude from the client, and in turn, the Contractor should provide details of the methods that he proposes to use for the Client's engineering advisers to review and identify any potential effects on the Client's interests. This promotes shared understanding of the project and potential problems, which provides a good foundation for dealing with problems should they arise.

This approach is consistent with ITIG Code of Practice. This recommends that contract documentation shall include Ground Reference Conditions prepared by the Client or shall require each tenderer to submit with their tender their own assessment of Ground Reference Conditions. When prepared by the Client, the Ground Reference Conditions are issued to tenderers as information on which tenders shall be based. When prepared by a tenderer, the Ground Reference Conditions shall be used by the Client in the tender assessment process. Ground Reference Conditions (prepared either by the Client or by a tenderer) are intended to form part of the Contract and provide the basis for comparison with ground conditions actually encountered when evaluating payments due.

There are situations where a Geotechnical Baseline Report may not be appropriate. Examples would be where the client transfers all of the ground risk to the contractor due to his approach to risk, or due to the procurement approach (e.g. in a PPP Contract). However, in these situations all factual and interpretative information needs to be provided to the Tenderers, but as non-contractual information with factual and interpretative data clearly differentiated.

The importance of comprehensive Ground Investigation has been noted. The Ground Investigation needs to cover not just design parameters, but parameters of construction importance such as permeability data to design dewatering systems, grading for ground treatment, and abrasivity parameters for TBM specification and performance prediction. This is another reason why construction expertise is needed at the design stage.

Ground Settlement and Building Damage

Ground settlement predictions are typically undertaken during the design stage, followed by assessments of building and utilities damage, and damage to other infrastructure. In line with recommendations above, this is essential so that mitigation requirements can be developed and the necessary land, and possibly environmental approvals, obtained prior to seeking tenders.

Following selection of the contractor, it is important that the settlement assessments are reviewed and updated to take into account the exact methods selected by the contractor.

The appropriate risk owner for building damage due to ground settlement is another complex subject, and is affected by the ground conditions and contractor's method and workmanship. It is even more complex when multiple contractors could be working at one location on major projects. The right approach will necessarily depend on individual circumstances, considering also the insurance arrangements.

CONTRACT DOCUMENTS AND PROCEDURES

Within the UK a major industry review in 1994 recommended development of improved contract practices to improve efficiency and productivity (Latham, 1994). The

Table 1. Engineering and construction contract payment options

Contract Payment Option	Risks/Benefits	
	Contractor	Client
Priced with activity schedule	Carries financial risk as a fixed price.	Facilitates tender comparisons.
Priced with quantities	Carries some financial risk	Carries risk of variation in quantities.
Target cost with activity schedule	Financial pain/gain share. Risk sharing.	Financial pain/gain share. risk sharing.
Target cost with quantities	Financial pain/gain share. Risk sharing.	Financial pain/gain share. Risk sharing. Carries risk of variation in quantities.
Cost reimbursable	Least financial risk.	Outturn cost uncertainty.

Engineering and Construction Contract (ECC, now in its 3rd Edition) was developed in response and is now endorsed by the UK government for public sector procurement (Institution of Civil Engineers, 2005). This is a manual for good management of the project as well as being the contract conditions. The first clause is that the parties shall act in a spirit of mutual trust and co-operation.

The ECC permits a range of payment options, including Fixed Price, Target Cost and Cost Reimbursable. Key features of these payment mechanism in terms of risk are set out in Table 1.

A particular part of the ECC that promotes good management and cooperation in problem resolution is the Early Warning procedure, which is set out as follows:

- The Contractor or Project Manager are to notify each other as soon as either becomes aware of any matter which could increase costs to the client or to the contractor (whether these are met by the client or not), delay completion or meeting milestones, or impair the performance of the works in use.
- Early Warnings are entered into the Risk Register. A risk reduction meeting is held, where the parties cooperate in considering proposals to avoid or reduce the risk, and to seek solutions that will bring advantage to all who may be affected. Decisions are made on actions to be taken and who will take them.
- The Risk Register is updated to record the decisions made, and changes necessary to the Works Information are instructed at the same time.

Similarly there is a strong emphasis on planning. The programme that the Contractor is required to submit is also expected to include:

- The order and timing of work to be undertaken by the Contractor, and by others who are part of the project. This allows the Project Manager to make arrangements for coordination across the project.
- Dates when site access is required, and when equipment, materials and information to be furnished by the Client are required. Again this supports good planning.
- For each operation, a statement is supplied as to how the Contractor plans to do the work and the principal equipment and other resources that he plans to use. This is consistent with ITIG Code of Practice which requires that prior to commencement of construction of an element the Contractor provides Method Statements, Inspection and Test Plans and Risk Assessments.

Risk allocation is explicitly stated in the Contract and variations or claims are termed Compensation Events. The procedure for dealing with compensation events is proactive:

- If the Project Manager gives an instruction for a change then the Compensation Event and request for quotation are given at the same time as the instruction. The Contractor must implement the change and the Project Manager must respond to the Contractor's quotation within 2 weeks of receipt. If there are uncertainties about the change required, the Project Manager is required to state the assumptions on which the Contractor should proceed.
- Similarly if the Contractor becomes aware of an event which he believes is a compensation event he notifies the Project Manager. The Project Manager has a maximum of 3 weeks to respond.

The ways for dealing with the change are developed collaboratively, with the Project Manager and the Contractor both encouraged to propose effective ways to progress. In addition, the need for timely decisions is emphasised by the provision that failure of the Project Manager to accept a compensation event and quotation within the defined timescales is deemed as acceptance. Throughout the ECC strict timescales are imposed to promote swift communication between the parties.

CONTINUITY OF DESIGN/SITE SUPERVISION

As with all projects, defects in materials and workmanship can lead to loss of performance of the permanent works. Events at Heathrow Express have shown that good quality control is also necessary for the safe execution of tunnel works (Health and Safety Executive, 2000). It is therefore important that the quality of materials and workmanship are supervised without commercial or progress pressures. Traditionally this has been undertaken by supervision by the client's Engineer, although recently projects in the UK have used Contractor self-certification. In the case of self-certification it is essential that the Quality Control function is independent of the production team, and it is essential that audits are undertaken to demonstrate this. It would seem appropriate that the designer, with experience in the works to know what is important to monitor, undertakes that audit function.

For any construction process, Method Statements and Inspection and Test Plans are important to ensure that the critical parameters are clearly identified and monitored in such a way as to be able to be confirmed by audit that they are in compliance with the requirements of the design and the contract. Again the ITIG Code of Practice sets out important requirements. It requires that the Method Statements and Inspection and Test Plans detail the monitoring and checking that is to be carried out, by whom and at what intervals, and quality records are to be produced and provided accordingly. In the case of a self-certification contract, the contractor is additionally required to demonstrate how he will control and maintain the independent supervision of the construction checking process.

The design of a tunnel project, with its dependence on the ground conditions, is necessarily a prediction. Monitoring and observation of the ground and groundwater conditions and the performance of both the permanent and temporary works during the actual execution is essential, with feedback provided to the design team to confirm performance, optimise future designs, or to adjust the current design if necessary to reflect changed site conditions. This reinforces the ongoing role of the designer during the construction phase, particularly in changeable ground conditions. This ongoing design role and the need for monitoring are also emphasised by the ITIG Code of Practice. In the UK there is a tendency for division to be introduced between design and construction support, partly due to the length of time that major projects develop over, and partly due to various procurement rules. As discussed above, this division can be to the detriment of project success and needs to be dealt with in the project procurement strategy.

CONCLUSION

This paper has provided an overview of good practice in the UK in tunnel project delivery, with reference to wider aspects of best practice. It is useful to note that the ITIG Joint Code of Practice supports and reinforces this good practice in many areas, and compliance with the UK version of the Code is becoming a requirement of obtaining insurance in the UK market (for example on the DLR Woolwich Arsenal Extension, London, UK).

As stated in the introduction, there are three conditions that are essential to provide the environment for underground project success: good planning; formation and development of a strong and effective project team; and managing the risks. These are supported by appropriate procurement and contract conditions and equitable sharing of project risks. Continuity of the design presence throughout the duration of the project is also essential to the success of an underground construction project and to the continuous improvement of the industry through performance feedback.

REFERENCES

Bennett, J. and Hayes, S. L., 1995, Trusting the Team: The Best Practice Guide to Partnering in Construction, Thomas Telford, UK.
Connaughton, J. N. and Green, S. D., 1996, CIRIA Special Report 129 Value Management in Construction, CIRIA, UK.
Construction Task Force, 1998, Rethinking Construction, HMSO, UK.
Edgerton, W. W. (Ed), 2008, Recommended Contract Practices for Underground Construction, SME, US.
Godfrey, P., 1996, CIRIA Special Report 125 Control of risk: a guide to the systematic management of risk from construction, CIRIA, UK:
Health and Safety Executive, 2000, The Collapse of NATM Tunnels at Heathrow Airport, HSE Books, UK.
Health and Safety Executive, 2007, Managing health and safety in construction: Construction (Design and Management) Regulations 2007. (CDM) Approved Code of Practice, HSE Books, UK.
Institution of Civil Engineers, 2005, Engineering and Construction Contract June 2005, Thomas Telford, UK.
The International Tunnel Insurance Group, 2006, A Code of Practice for Risk Management of Tunnel Works.
Latham, Sir M., 1994, Constructing the Team, HMSO, UK
Network Rail, Guide to Railway Investment Projects (GRIP), Network Rail, UK
Office of Government Commerce (OGC), 2001, Gateway Review Leadership Guide, HMSO, UK.

SELECTED READING

Muir Wood, A., 2000, Tunnelling: Management by Design, E & F N Spon, UK

USING RISK ANALYSIS TO SUPPORT DECISION MAKING ON THE CENTRAL SUBWAY PROJECT

Joe O'Carroll ▪ Parsons Brinckerhoff

Noel Berry ▪ Parsons Brinckerhoff

Arthur Wong ▪ San Francisco Municipal Transportation Agency

ABSTRACT

What decision making tools and methods are available to a project owner when there is no clear differentiator between construction alternatives? A team led by a joint venture of Parsons Brinckerhoff and PGH Wong Engineering (PB/Wong) has been providing preliminary engineering services to the San Francisco Municipal Transportation Agency (SFMTA). The SFMTA and PB/Wong team developed several alternative construction methods for the tunnels and stations of the San Francisco Central Subway Project, none of which could be judged clearly better than another using traditional approaches and metrics. The team employed a risk assessment and analysis frameworks as a means of comparative analysis, drawing upon the knowledge of industry experts in the geotechnical, tunneling and construction fields to assist in the effort. This paper presents the approach and results of a cost and schedule risk analysis for two such analyses designed to answer two different questions regarding the advisability of different construction approaches. These assessments paved the way for key decision making in the preliminary engineering phase of the project and formed the foundation for continuing risk management efforts on the project. The focus of this paper is primarily the first two stages of risk management—identification and assessment—and some of the organizational dynamic issues that arise in such efforts.

INTRODUCTION

The Central Subway Project is a 1.7 mile surface/subsurface extension of the Muni Metro Third Street Light Rail Project. The Locally Preferred Alternative Alignment, continues as in-street surface guideway from the Initial Operating Segment northerly terminus at Fourth and King Streets, to a double track portal between Bryant and Harrison Streets where the surface alignment transitions to a subway configuration. The project continues north in subway under Fourth Street to the Moscone Station (MOS) between Folsom and Howard Streets, continuing on Fourth Street under Market Street to Stockton Street, below the existing Muni and BART tunnels. North of Market Street, the subway will enter the Union Square/Market Street Station (UMS) located between Market Street and Geary Street. The south concourse of UMS will connect directly to the mezzanine of the Muni/BART Powell Street Station. The north concourse will connect to Union Square and Geary Street. The subway will then continue under Stockton Street to the Chinatown Station (CTS) between Clay Street and Jackson Street with a crossover cavern (CXO) located south of Chinatown station and tail tracks to the north of the station to the project's operational limit near Jackson Street. Non-operational tunnels will extend under Stockton Street to Columbus Avenue in the vicinity of Washington Square. The Central Subway project requires a major tunneling operation to be conducted from beneath the I-80 Freeway on Fourth Street between Harrison Street and Bryant Street. Utility relocations, maintenance of traffic and co-ordination with the proposed Transbay Bus Depot, protection of the Caltrans structures during excavation of the tunnel launch box all required careful consideration. Hazards associated with the horizontal and vertical profiles of the guideway tunnels, excavation beneath the Bay Area Rapid Transit (BART) and Muni tunnels on Market Street, sensitive structures and utilities above the tunnels along the entire alignment coupled with complex interfaces with stations construction were identified, evaluated, quantified and mitigated as far a reasonably practical during this preliminary engineering phase of the project.

Construction of the new subway in this densely developed urban environment of downtown San Francisco presented the project team with many political, economic and technical challenges that impart risk to the financial certainty of the project and the ability to deliver the project within the proposed schedule. The SFMTA recognize that there are significant advantages and benefits to the local communities to open the Central Subway as early as possible and, in scheduling the project, aimed to achieve a balance between the design schedule and the construction schedule (i.e., allow enough time for design but begin construction as quickly as possible) that provides assurance in projections of project development and service start up. The timing for construction of the guideway tunnels relative to stations construction and the approach to the design and construction of Union Square/Market Street Station (UMS) were considered as the two most critical components in optimizing the construction schedule and achieving on-time completion of the project and were therefore considered deserving of in-depth risk and opportunity assessments. Constructability of the station platform and the crossover caverns in Chinatown by sequential excavation methods (SEM) post guideway tunnel construction was also included in the risk assessment.

The approach to the design of UMS was identified as a critical component in identifying the program risks, thus deserving of an in-depth risk and opportunity assessment. Three design alternatives were considered in this assessment:

1. A top down deep cut and cover station design with the primary ground support system comprising inclined secant piles.
2. A mined cavern station utilizing a sequential excavation approach with a shotcrete lining as the primary ground support system.
3. A "hybrid" design for the primary ground support system consisting of shorter vertical secant piles in conjunction with a "bulb out" excavation at the guideway and passenger platform elevation.

UNION SQUARE/MARKET STREET STATION (UMS) RISK/OPPORTUNITY ASSESSMENT

UMS is a complex underground structure with many design and construction challenges. Maintenance of traffic, close proximity to buildings, its location in a busy commercial district, and challenging geotechnical conditions make the station sensitive to construction means and methods. Risks of failure of the excavation and ground support systems, impacts to property and businesses, potential increases in costs, schedule delays, and the risk of not meeting design, operational, maintainability, and quality were all risks considered in the assessment. Hazards identified by the workshop participants included:

- Potential safety, security and environmental impacts.
- Public impacts including traffic, business disruption, noise, and dust.
- Geotechnical uncertainty, obstructions, differing site conditions, and ground water.

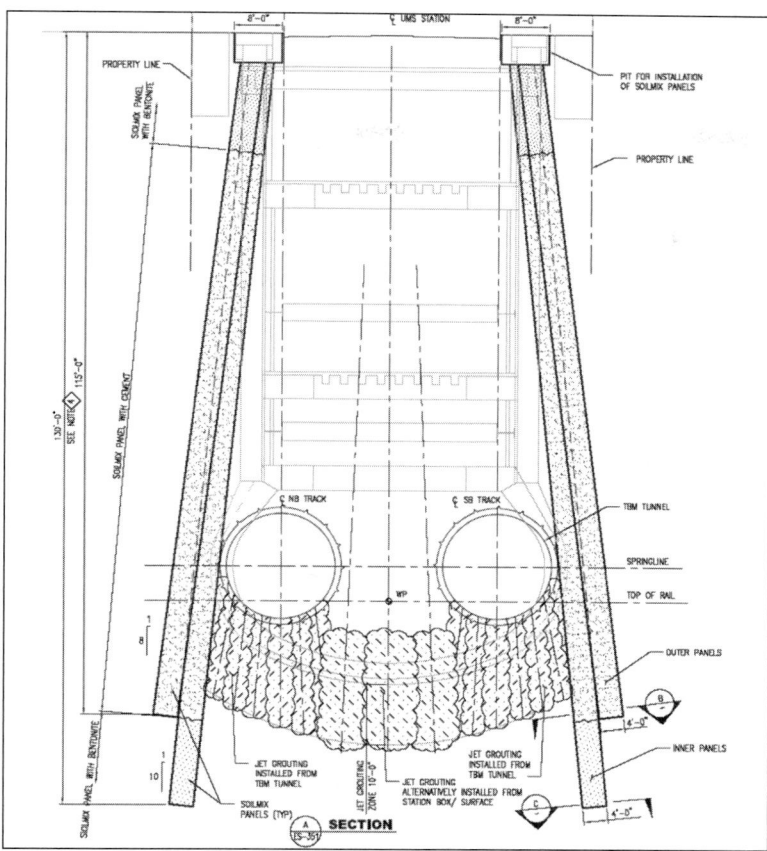

Figure 1. UMS station—hybrid design

- Design complications (ground support, site constraints, access, etc.).
- Plant equipment availability and performance.
- Materials supplies and logistical challenges.

The risk assessment and analysis was comprised of three stages; (1) a workshop that identified and evaluated hazards for each alternative; (2) an analysis of risks associated with cost; and (3) an analysis of risk and uncertainty associated with proposed construction schedules for each alternative. The overall risk management approach is indicated in Figure 4. This paper focuses on the first three stages, Identification, Assessment and Analysis.

The elements of the UMS Station risk assessment were:

- Identify and document potential hazards associated with construction of each of the three alternatives.
- Translate these hazards into risks based upon their probability of occurrence and potential cost and schedule impact should they occur.
- Provide a quantitative assessment of the risks in terms of impact on cost and schedule.
- Identify the most appropriate mitigation measures for each risk identified.

874 RISK MANAGEMENT

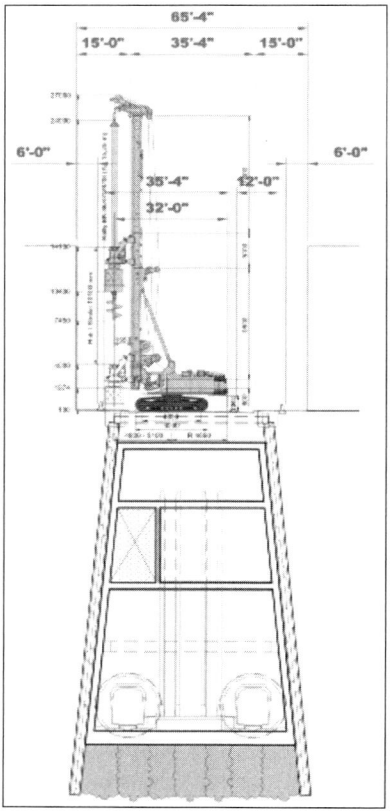

Figure 2. UMS station—deep cut and cover design

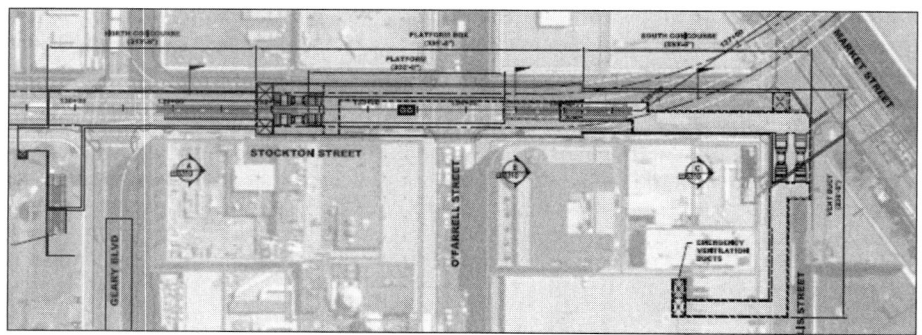

Figure 3. UMS station—layout and location

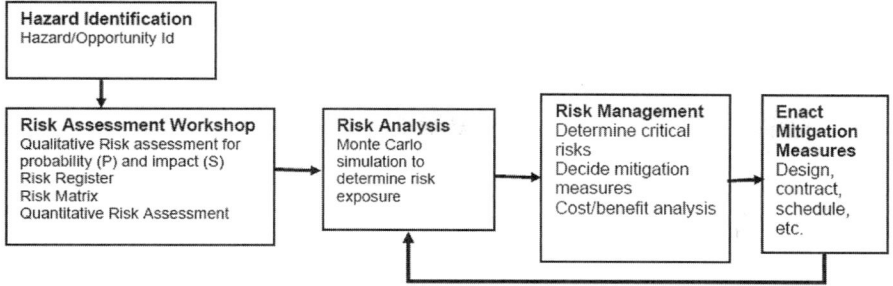

Figure 4. Stages and associated activities in risk management

- Assess the uncertainty in the current construction schedule for UMS Station based upon an assessment of the most likely, optimistic and pessimistic durations for each of the activities in the schedule.
- Aggregate individual assessments to create probable ranges of unmitigated cost and schedule impacts and compare these with contingencies in the current capital cost estimate.

WORKSHOP ACTIVITIES

The general outline of the risk assessment procedure was to identify hazards, define the hazard as specifically as possible, translate the hazard into actionable risk statements and evaluate the risk in qualitative and, where possible, quantitative terms. Workshop participants "walked through" construction of the station alternatives identifying potential hazards based upon their knowledge of the project, personal experience, and expert judgment. Participants were encouraged not to be hindered by discussions of possible mitigations at this stage—as discussed later, thinking or discussing mitigations at this stage may lead to premature exclusion of certain risks if participants feel that a particular issue is easily dealt with. Participants then defined the hazard in specific terms with regards to the event itself, and its location or source. Hazards were then translated into actionable risk statements. For our purposes, a hazard is differentiated from a risk statement in that the latter contains enough detail to assess both a likelihood of occurrence and severity of impact. A unique tracking identification number was assigned to each risk. The tracking number identified the construction method and was numerically sequenced. The risk was then evaluated in qualitative terms by specifying the likelihood of the event and the severity of the impact should the event occur. Once a qualitative ranking for a risk is determined the team will move on to a quantitative definition, ultimately defining the risk in terms of numeric ranges for potential impact and probability of occurrence. The impact assessment is made separately from the probability assessment to prevent conflation of the two. The final risk exposure of the project due to an individual risk is determined using Monte Carlo simulations and is the product of the Probability (P) and Severity of the Impact (S).

Not every risk can be assessed in terms of time or dollars. Certain risks such as safety of construction personnel, quality of life for residents, impact on business owners in close proximity to the construction site, political and media pressures may result in significant exposure that, though not directly measurable, still needs to be addressed and managed. For such risks a qualitative ranking system together with objective criteria for each rank is established prior to assessment. The impact criteria must be clear and objectively determined to ensure what every assessor understands, for example, what exactly assessing a risk as a '3' in safety and health means.

RISK REGISTER

Risk registers were developed containing the identified risk event, the likelihood it will occur, the estimated severity of impact should the event occur, and mitigation measures for those risks judged to require mitigation. For each identified risk, contractual and design and construction measures that should be incorporated into the contract documents to reduce risk exposure are generally proposed. The UMS risk register was used to guide the design team in mitigating risks to the selected alternative. Each risk should be assigned a unique identification number prior to the management stage.

RISK ANALYSIS

During the risk analysis stage, the cost and schedule risks specified in qualitative terms in the assessment are quantified and their relationship to one another, if any, determined. The risk analysis component of this assessment quantified and structured the cost and schedule risks into a comprehensive model for the purpose of determining the overall risk exposure associated with a particular method of construction. Both the probability and impact are quantified, that is, the qualitative descriptions are replaced with numerical ranges representing the probability of occurrence and likely impact to the project should the event occur. Risks are then grouped and analyzed for the alternative as a whole using Monte Carlo simulations to determine the most likely outcomes. As noted earlier, many impact categories such as safety and health, reputation etc, can not be specified in quantitative terms and must be maintained in qualitative terms.

Estimation of cost impact performed in a workshop setting without more detailed analysis is not absolute. Participants in the assessment are guided, in qualitative terms, as to what represents a low, medium and high risk by categorizing the potential impacts. Qualitative is not a synonym for subjective: the criteria for judging the severity of a risk should be clear and unambiguous to all assessors. For the purposes of the UMS Station construction risk assessment a low risk was considered as any individual event that had a potential cost impact of less than $1 million and less than 20% probability of occurrence. A high risk was considered as any event that had a potential impact of more than $5 million and a probability of occurrence of greater than 80%. Most risks were of course some combination, e.g., high probability and low impact, high potential impact but low probability, etc.

The assessment form allowed participants in the assessment to be more specific in their assessment between these ranges if the data existed to support it or if they were confident enough to narrow the range and/or provide the most likely value.

APPROACH

A range of possible impact was determined together with a probability of occurrence for each cost and schedule risk. Risks that appear in more than one alternative may have different impact ranges and/or probabilities associated with them under different alternatives. For example, certain surface effect risks having to do with business disruption appear in all three alternatives, but their impact or probability of occurrence are higher in Alternative 1 and 3 than in Alternative 2 because businesses are subject to interruption for longer periods of time under alternatives 1 and 3. Consequently, the project's risk exposure due to these events is greater under these alternatives. The identified risks and their attendant impact ranges and probabilities were collected for each alternative and a Monte Carlo simulation of 4,000 to 5,000 cycles was run for

Individual Cost and Schedule Risk Events

Many risks have both a discrete cost component and a schedule component. For example, the risk *'Limited number of NATM/SEM-experienced contractors leads to delay in contract award and/or increase in bid costs'* has both a discrete cost component and schedule component. The discrete cost component arises from the assumption that the contractors bidding on the project know that their competition is limited and as such they can submit higher bids than they would in a more competitive environment. Therefore, there will be an up-front (discrete) increase in cost if this is the environment at the time of bidding. In addition, it is assumed that in such an environment, the contractor will have trouble obtaining experienced personnel and as such schedule delays are more likely as less experienced crews have difficulty operating at the productivity levels assumed in the schedule. What assumptions are made regarding this division between up front and delay costs, ramp-up rates for inexperienced crews etc are all important, but the particulars of each are less so than consistent application in a comparative analysis such as this. When risk assessment and analysis is deployed to choose between alternatives the absolute dollar value of risk or schedule delay is not the issue (one of the alternatives will be used), what is absolutely critical is that assumptions be consistent across all alternatives. This is easier with discrete cost side risks than it is for schedule risks.

A schedule risk analysis is complicated by the fact it is not just the 'size' of the impact or probability of a risk that matters, but where in the schedule the risk event may occur. The relationship between the timing of the risk event and amount of float in the schedule at that point may matter more than either the probability of a risk event or its potential impact. When comparing alternative construction methodologies with different schedules this is another area that must be considered—where in the schedule do we think this risk event might occur? Where the risk event is attached in the schedule can make a large difference in the results extracted from Monte Carlo simulations. If the risk is associated with a specific activity then it is fairly straight forward and it can be attached it in the same way to the same activity in all the schedules you are comparing. Other, more nebulous risks, such as 'store owners protest continued work' are more difficult.

Risk managers and project leaders should be aware that all prior efforts to be conservative with regards to assessing the schedule risk exposure can be nullified at this stage. It is quite possible to have a very conservative (i.e., pessimistic) risk register come out of the assessment phase and yet end up with highly optimistic analytic results. If all individual risk events are placed in areas of the schedule where float is higher than the expected value of the risk, then you can expect that the results of the simulation will indicate a better than even chance of completing the work on time. If these same risks are placed on or near the critical path (assuming of course that there is more than one reasonable point in the schedule where such an event would occur), results can be radically different.

If, on one alternative, risk events are consistently placed optimistically ('they'll occur when we'll have the most time to deal with them') while on another they are placed pessimistically ('they'll occur on the critical path'), then the same underlying risk register that came out of assessment can lead to two very different conclusions. This is not to say that one alternative should be unfairly 'punished' in order to keep things even. Certain construction methodologies will be more appropriate under certain situations and one indication may be that there is more room for error in the schedule.

Duration Uncertainty Around Planned Activities

In addition to identifying and assessing individual risk events, which may or may not occur, workshop participants also assessed the uncertainty surrounding the planned duration of scheduled activities. Schedules for each alternative had been prepared during earlier efforts to identify the best approach to the tunnel and station construction. The assessors were asked to revisit these schedules go through each activity and establishing upper (pessimistic) and lower (optimistic) bounds on the amount of time an activity would take. They were also asked to give a 'Most Likely' time to completion. In theory, the most likely time should have corresponded with the duration in the original schedule. This was not often the case however. In some situations the market realities that undergirded some of the assumptions in the schedule had changed, in others state of geotechnical information had increased. In the majority of instances where there was a difference, they believed that the current schedule underestimated the amount of time the activity would take. This is not an uncommon occurrence and in addition to the factors mentioned above owes something to the organizational dynamics of large projects. At the beginning, when different alternatives are being considered, those with the most experience in a particular construction methodology are sought out. Many times however, those with the most experience in any discipline are also the ones most optimistic about it. This in turn may lead to preliminary schedules which are themselves optimistic, populated with activities with high production rates and low durations.

Taking Advantage of Built-in Biases to Increase Objectivity

As with most risk management efforts, the 'soft-skills' of organizational management were as critical to success as the technical details such as distribution types and sampling rates. As mentioned in the previous paragraph, those with the most knowledge of a particular construction methodology tend to be its biggest proponents. So the question confronting the risk management team was how to incorporate all the years of specialist knowledge the expert panel possessed without arriving at a situation where every methodology was 'above average.' Another issue that generally presents itself is a type of anticipatory mitigation. The specialist thinks of a risk, but then thinks about what they would do, and then either discounts the risk or never includes it on the register. 'The risk has been eliminated.' Risks that never make it on the register or are discounted from the beginning may distort and management efforts for the life of the project.

The situation was made more complicated by the practical realities that a certain amount of information needed to be collected in the limited time the expert panel was convened. There was not on this project, and generally is not on most projects, time for endless back and forth between members backing differing construction methodologies.

The approach taken was to break up the members into teams with each team containing both 'neutral' members with general knowledge and experience in *all* the methodologies and members with specialist knowledge of *one* of the methodologies. The risk assessment process was then divided into two stages, identification and assessment. Teams with members who were specialists in, for example, mining were asked to identity all the possible problems with a cut-and-cover approach and hybrid approach. Conversely, teams with cut-and-cover expertise were asked to identify all risks associated with mining. Generalist members of the team served as 'ballast,' identifying some risks that the specialists may have missed and limiting any excesses that might occur with this dynamic.

In the second stage, the risks registers compiled in the first were exchanged, to take advantage of the specialist's knowledge. We were primarily interested in questions of impact for the identified risks, not their potential existence. In other words, we wanted the teams to come at the identified risks assuming the event had happen and then determine how the event might impact the project. When questions of impact and probability (existence) are considered together the two aspects are often conflated with proponents of a particular methodology often understating the actual risk exposure because they believe the event is unlikely to occur or easily mitigated.

The results using this approach both in identifying individual risk events and assessing the earlier schedules with respect to duration uncertainty were fundamental to determining the best approach to the UMS construction method.

Results

The Risk/Opportunity Assessment lead to recommending Alternative No. 3 as the preferred design. It also highlighted a number of issues that needed to be addressed in the management plan including such items as delays in 'bulb out' excavation possibly leading to delays in closure of the invert ring and questions surrounding the amount of jet grouting required that would need to be analyzed in greater detail. It also highlighted the necessity of revisiting assumptions about activity durations made early in the process. The duration uncertainty surrounding key activities was one of the drivers in selecting the best alternative. If basic questions concerning how long an activity would take had not been revisited this could very well have led to a different, and potentially less preferable, alternative being selected.

GUIDEWAY TUNNEL RISK/OPPORTUNITY ASSESSMENT

The purpose of the guideway tunnel risk and opportunity assessment was to:
- Determine, from a risk perspective, the feasibility of constructing the guideway tunnels in advance of the stations.
- Determine which elements of the stations needed to be advanced into final design in conjunction with the guideway tunnel to support 'early' construction of the guideway tunnel.
- Expand the project risk register to include specific hazards related to construction of the guideway tunnel (assuming tunnel construction precedes station construction).
- Quantify the identified cost and schedule risks.

The process for the guideway tunnel risk assessment mirrored that described above for UMS. The outcome of the risk assessment concluded that from a risk perspective construction of the guideway tunnels in advance of the stations was technically feasible. Utility diversions are required at the tunnel boring machine launch box (TLB) and, as required, at MOS and UMS to support 'early' construction of the guideway tunnel. Headwalls at MOS and ground improvement for the platform cavern at UMS are also required before arrival of the TBMs.

The opportunity element of the assessment demonstrated that constructing the tunnels early in the project in advance of station construction benefits the overall project by providing:
- Greater knowledge of geotechnical and groundwater conditions particularly at station locations.
- Flexibility in removing excavated material from stations.
- Flexibility in avoiding future contract interface issues. additional access to performing ground improvement needed at station interfaces, and
- A major cost component to be funded early in the project thus reducing the impact of escalation on the overall project budget.

The results of the assessment process further concluded that constructing CTS platform and crossover caverns with the guideway tunnel linings in place ahead of the cavern excavation was also technically feasible. Risk Mitigation efforts that were taken during completion of preliminary engineering included

- Further geotechnical information at particular locations along the alignment and to clarify uncertainty of the Alluvium/Colma contact with respect to the tunnel crown.
- Obtaining more information regarding groundwater at the invert of the platform caverns and crossover cavern.
- Advancement of the utility relocation designs to allow commencement of utility relocations as early as possible.
- Working closely with Bay Area Rapid Transit (BART) on protection measures including provision of real time instrumentation monitoring in the BART tunnels and specifying continuous excavation of the guideway tunnels beneath the BART and Muni tunnels on Market Street.
- Commencement of pre-construction surveys to minimize risk of property owners claiming adverse effects or diminished seismic resistance of their facilities.

SUMMARY

Risk assessments performed by the project team included impact analyses on both cost and schedule that provided the basis for critical decision making on the chosen alignment, depth of tunneling, extent of geotechnical investigations performed, construction methodology for the stations and the sequence of construction on the project. The results of these risk assessments were documented in the project risk register. Comprehensive measures to eliminate or minimize the potential impacts of these risks to the project were identified. The project team then proceeded to implement these mitigation measures as far as reasonably practical within the limits of preliminary engineering. The targeted risk assessments and registers that were developed were used to support the overall FTA Capital Project Risk Assessment.

Schedule risk analyses performed by the project team as part of the preliminary engineering risk assessments resulted in a selected construction sequence for the overall project and selected construction methodologies for key elements of the project such as Union Square/Market Street Station, Chinatown Station and the guideway tunnels. The outcome of these risk assessments will allow construction to commence as early as possible and will minimize the potential that conflict will arise between separate contractors working concurrently in the same work space. Conflicts result in inefficiencies that delay achievement of key milestones and increase the risk of claims attributed to delays or other impacts that increase cost. This risk mitigation was instrumental in determining the proposed contract packaging strategy.

Other hazards commonly associated with most urban subway construction projects also present themselves as specific risks to the Central Subway. These include public safety, security and environmental impacts, disruption to traffic, transit and business, geotechnical uncertainty, obstructions, differing site conditions, adverse groundwater conditions; design complications (ground support, site constraints, excavation sequencing, access, etc.); plant equipment availability and performance and materials supplies and logistical challenges at each of the construction sites .To mitigate these risks it is proposed to include differing site condition clauses and geotechnical baseline reports in the contract documents to minimize bidder contingencies. Although changed conditions during construction can result in a drawdown of the 'contingency' provided in the estimate, it is generally accepted that it is more cost effective to pay for conditions only if they are encountered, rather than putting the risk on the bidders, who will include contingencies in the base price.

SHORT TUNNELS = HIGH RISK?: PIPELINE CONSTRUCTION INVOLVING OPEN-CUT AND TUNNEL SEGMENTS

Michael Gilbert ■ CDM

Michael Schultz ■ CDM

ABSTRACT

In the world of tunneling the glamour is in the large diameter, long tunnels, but perhaps the number of projects for small diameter and short length tunnels are greater and generally have higher risk factors. Reasons for these higher risks include both the "boiler plate" contract provisions used on these types of contracts, the technical specifications related to the tunnel work (versus the pipeline construction), the bid practices including measurement and payment and the construction. This paper reviews and comments on some of the key issues the authors have observed in recent major pipeline projects that involve a small percentage of tunneling footage of the total contract.

INTRODUCTION

This paper addresses the risk issues associated with the small tunnels that are a part of large sewer and water main conveyance pipeline projects that are primarily installed using open trench methods of construction.

This paper includes data from 13 projects that include a total of 76 short tunnels and/or tunnel sections. These projects include both sewer and water main pipelines ranging in length from 380 m to 53,600 m and excavated diameters of 1,240 mm to 2,540 mm. The criteria used to select projects for the database included: (1) construction completion within the last 5 years; (2) the majority of the pipeline footage or cost of the overall project was greater than half of the project total as compared to the tunnel length; (3) tunnel work was completed by either conventional tunneling or pipe jacking; and (4) the work was done in one of the 48 contiguous states. Horizontal directionally drilled (HDD) pipeline projects were not included.

The projects data are summarized in Table 1. These projects all have in common elements; most are short tunnels that are driven under RR tracks, heavily traveled roadways or bodies of water. Generally the vertical and horizontal alignments have been set within relatively small limits to meet operational requirements. Tunnel crossings were completed in soft ground, rock, and mixed face conditions. The methods of excavation in soil included either a closed-face microtunnel boring machine (MTBM), or the use of an open-face shield for either pipe jacking or conventional tunneling. In rock either a MTBM or drill and blast methods were used. Initial support systems consisted of liner plate or rib and lagging support in soil and rock bolts, or liner plate, ribs and lagging in the rock tunnels depending on the rock quality.

COST AND RISK OF SHORT TUNNELS

There are several varied reasons for the high potential of an element of risk occurring such as the location of these tunnels. Usually they are crossing highways, railroad

RISK MANAGEMENT

Table 1. Summary of pipeline projects with tunnel crossings

Project Name	Project Bid Cost, Million Dollars	Total No. of Trenchless Crossings	Total Crossings Cost as % of Total Bid Cost	Crossing Cost as % of Total Bid Cost	Total Project Length, Meters	Crossing Length, meters	Total Crossing Length as % of Total Project	Individual Crossing as % of Total Project Length	Excavated Tunnel Dia. mm	Method of Const	Ground Conditions	Magnitude of Claim relative to Tunnel Cost Item	Cost Ratio of Tunnel unit cost / Trench unit cost	Comments
NC1	$13.3	4	9.7%	1.6 to 3.3%	8760	26 to 69	2.1%	0.3 to 0.8%	1850	PJ	S	None	4.5 to 5.8	Subsurface conditions consistent with face of silty sand - residual soils.
NC2	$28.9	3	53.5%	2.4 to 48.3%	4300	83 to 798	22.0%	1.5 to 18.5%	2250	T	MF	1-Moderate	2.5 to 4.4	Conditions vary from mixed face of residual soil over partially weathered rock to solid face of hard rock.
NC3	$13.7	2	17.3%	8.0 to 9.3%	3610	22 to 228	7.8%	1.5 and 6.3%	2450	T	MF	None	1.6 and 6.0	Residual soils consisting of sand, clayey sand with boulder size rock in weathered zone above intact rock.
NC4	$39.7	11	20.0%	0.3 to 10.0%	11060	17 to 803	12.5%	0.1 to 7.3%	1850	PJ-7 T-4	S - 8 MF - 2 R - 1	1- Moderate and 1-Major	1.5 to 2.8	Tunnel costs do not include shafts. Conditions at each crossing varied from sandy silt to partially weathered rock to intact rock. One claim settled by arbitrator and second by owner. Owner settled claim was result of two borings showing decomposed rock, actual conditions were solid intact rock.
CO1	$65.0	16	3.4%	0.1 to 0.3%	53650	20 to 112	1.2%	<0.1 to 0.2%	1850	PJ	S - 8 MF - 5 R -3	1-Moderate	1.4 to 6.1	Two projects awarded, first was for tunnel excavation and lining support using jacked casing. Second was for installation of the pipeline for the entire project. Tunnel costs combines cost of tunnel, pipe and grout filling of annular void. Claim was made against the all crossings not one single crossing
PJ - 3	$44.0	15	11.0%	0.4 to 1.4%	22180	24 to 127	4.8%	0.1 to 0.6%	1240 to 2000	PJ -3 T - 12	S - 11 MF - 2 R - 2	None	1.2 to 2.8	Project awarded in several contracts for materials, tunneling and open trench pipeline work. Tunneling conditions ranged from soil consisting of sand, sandy clay and clay typically dense and moist to dry. Rock when encountered was sandstone and claystone and generally soft and dry.
FL1	$18.4	10	31.9%	1.4 to 5.5%	10240	31 to 120	6.5%	0.3 to 1.2%	2450	T	S - 9 MF - 1	None	4.4 to 8.9	Contractors used several methods of ground stabilization including; closed face TBM, dewatering, permeations grouting, and compressed air at various crossings. Soil conditions generally ranged from medium dense to dense sand, sometimes cemented and some clayey sand
FL2	$24.0	9	49.0%	0.5 to 20.7%	10340	18 to 291	10.7%	0.2 to 2.8%	1850 to 2540	PJ	S - 5 MF - 4	None	2.9 to 12.9	Contractors used several methods of ground stabilization including, closed face TBM, dewatering, permeations grouting, and compressed air at various crossings.
OH1	$12.0	1	8.2%	8.2%	1710	119	7.0%	7.0%	2150	PJ	S	1-Moderate	1.2	Initial start encountered nest of boulders and contractor changed to open face shield from MTBM
CA1	$9.4	3	12.7%	1.7 to 8.6%	4240	33 to 149	5.4%	0.8 to 3.5%	2000	PJ	S	Minor	2.2 to 2.7	Claim was made regarding the installation of the pipe within the tunnel. No claims on the direct tunnel work.
MO1	----	1	----	----	380	196	51.5%	51.5%	1250	PJ	S	None	----	Private owner who selected contractor
NJ1	$9.7	1	13.4%	13.4%	1450	84	5.8%	5.8%	2300	T	S and MF	1-Major	2.5	Contractor claimed work could not be done to the tolerances specified, work removed from contract.
SC1	$7.6	1	16.8%	16.8%	----	158	----	----	1550	PJ	R	1-Moderate	----	Project included other work in addition to the tunnel. Claim based on the hardness of the rock causing additional disc cutter changes.

NOTES:
Project Name column identifies the state and number of the project within the state, the subset identifies the number of the crossing within the contract.
Methods of Construction are identified as: PJ = pipe jacking with either shield or microtunnel boring machine and T = tunnel with either steel liner plate or ribs & lagging.
Ground Conditions are identified as: S = soft ground; MF = mixed face of soil and rock or partially weathered rock; and, R = rock
Refer to Table 2 for description of the magnitude of the claim as defined for this paper.

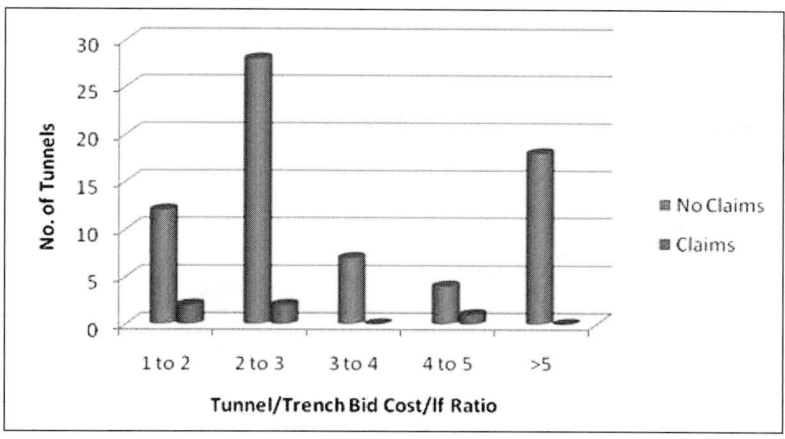

Chart 1. Tunnel tunnel/trenchy bid cost/IF ratio

or rivers where rescue shafts are not an option. Because of the tunnel size the financial impact is often significant, especially in relative terms to the tunnel cost. The reasons for the higher potential risk start with the use of standard "boiler plate" general conditions and specifications designed for use on pipeline distribution, collection and consolidation projects but are in some way adapted for use on the tunnel work. Bidding the tunnel work as lump sum items of a larger project, the extent of the exploration program, the use of the GBR and the ability to pre-qualify tunneling subcontractors are all potential pitfalls. In addition there is the consideration of the means and methods available to the tunneling subcontractor with regards to: experience with particular methods, and funding to perform specific tasks with the most advantageous equipment.

Defining the magnitude of a claim on a tunnel project could be quantified similar to the cost of a risk as it is done in a tunnel-risk workshop. We recognize that the cost and associated ranking varies from job to job and owner to owner, as it should. For this paper, in the cases where a claim has been made the magnitude of the claim is defined in Table 2. This ranking is only to show some relative size of a claim to overall tunnel cost.

Of the 76 crossings evaluated, a total of 8 claims involving 7 of the projects were made based on the encountered tunneling conditions. Based on the definitions presented in Table 2 one claim was minor; five claims were moderate; no claims were significant and two were major claims. It should be noted that the claim value does not represent the final agreed upon compensation in any of the cases, either resolved or still pending.

Overall these claims range from 4% to 277% of the initial bid price for the tunnel work in a particular contract. The claims pertain to items that ranged from delays caused by permitting issue to change conditions due to differences between the subsurface characteristics as seen in borings compared to the conditions during construction. One claim regarding a permit delay was with regards to several of the crossings in the contract, and not specific to any one particular crossing. For this paper that claim was considered as "moderate" as defined in Table 2 when taking into account the total cost of the crossings affected by the permit delay.

The following is a discussion of those claims and present methods of mitigating claims for future short tunnels in pipeline projects.

Table 2. Claim magnitude

Claim Magnitude	Relative Value of Claim
Minor	Less than 20% of the tunnel crossing bid item
Moderate	21% to 50% of tunnel crossing bid item
Significant	51% to 100% of tunnel crossing bid item
Major	Exceeds the tunnel bid cost item

DIVISION 1 SPECIFICATIONS

Owners for many of the listed projects are small cities or town, utility districts or regional townships. Often Owners are not familiar with tunnel construction. Owners often compare the unit price of a tunnel versus an open cut pipeline. The average cost ratio in the database presented in Table 1 shows that tunneling is 3.7 times the cost of open cut work per unit of length and ranges from 1.2 to 12.9 times the open cut cost. Of interest is that for the crossings which involved a claim, the tunnel to open trench cost ratio averaged 2.5 times and ranged from 1.2 to 4.4. If not well versed in the details of tunnel construction the natural instinct is to only be aware of a cost difference for the tunnel and ask questions why the cost is high.

Resistance within the owning authority to change of the Division 1 documents. The Division 1 specification is very much boiler plate for pipeline projects. Unfortunately the tunnel world and the method of doing work result in different costs, different work schedules and different risks than those same items in open trench pipeline work.

This only means that the contractor has to recuperate costs on a project is by the differing site conditions clause.

Rental cost of TBM. The average length of a crossing of the Table 1 data is approximately 85 m. Most of the projects had multiple crossings requiring several site remobilizations. As a result the tunneling work usually extended 2 or 3 months as a minimum and in some projects much longer. The rental cost of this equipment has to be factored into the bid because there is no other means of paying for the equipment. While this may be a small percentage of the overall project it is a significant cost to the subcontractor performing the tunneling. Often the result is a decision to reduce costs by utilizing an older machine that is less reliable, resulting in poorer than average utilization rates for the TBM or a decision to use just a shield and rely on ground modification where potential poor tunneling conditions exist. TBM "availability" on some of the projects has been less than 50%. Availability is defined as the time the machine can be used to actually penetrate into the tunneling media as a percentage of the total time the machine is in the ground. The availability time times the utilization percentage time yields the TBM usage. With such low availability time it is impossible to come close to a typical TBM usage value that would be anticipated on a major tunnel project.

As a result of this low availability, coupled with a machine that is probably not ideally used for the anticipated or encountered conditions, progress is slower and repair costs are higher. This issue is a part of the basis of claims that were made on projects #2 and #4.

High cost of delays and work hours. The same risks involved with major tunnel projects regarding limited choices of the "critical path," i.e. work stoppages, are costlier than that for other types of construction exist for the short tunnels on pipeline work.

A part of the reason for the claim on the CO1 project was the delay in permits that the contractor encountered. As a result he could not schedule the tunnel work to proceed in accordance with the schedule that was submitted. This permitting issue involved several locations and each requiring its own permit. Whereas a major tunnel project would have required a single permit. This cause for delay is far less common for a major tunnel project because that tunnel is considered one element whereas for this project each crossing was considered individually by the governing agencies.

Work hours are usually limited to daylight hours during a pipeline project. While this does not impose a significant burden on trench work it does have significant consequences on tunnel work. Most tunneling operations would prefer to work two shifts and either five or six days and sometimes seven days a week. Most often the owner has to deal with the concerns of the nearby residences and the noise at the shaft during the night. There is down time associated with bulkhead work to secure the heading each day, the potential of excessive settlement at the stopping point and in pipe jacking operations, and the risk of a stuck shield and MTBM.

Typically noise presents a large issue. We have measured and found that the noise level at the shaft is usually below the allowable levels for construction. However, because the construction activity is at night, consideration of a noise barrier may be considered as an alternative that would allow longer hours of operation. Again cost becomes an issue. When there are several working tunnel shafts on a project each one has to be evaluated on its own conditions: proximity to residential houses; engineering time to monitor and evaluate the conditions; and appeal to the governing town or city board. Therefore, the engineering cost is increased to duplicate a task several times that may only occur once or twice on a major tunnel project. And on the major tunnel project the percentage cost to the job of making modifications to minimize disturbance to neighborhood is smaller.

Conflicts. In many cases a geotechnical consultant is part of the design team for a pipeline project. They oversee field explorations, perform the laboratory testing and prepare both data reports and a geotechnical report. Typical issues with this report follow.

A report often contains both design and construction recommendations and often becomes part of the contract documents. The result can be misleading to a contractor's interpretation of a design consideration or recommendation as a reason to alter a selected mean and method of construction.

The report is prepared long before the design is finalized and often dated because of changes made during the design. As a result statements and recommendations in some cases do not agree with specifications. This conflict can cause confusion both during design as well as during construction. Typical of these conflicting or apparent conflicting statements is a geotechnical report stating several potential viable methods of tunneling. As the design progresses, one or more of those methods are no longer considered viable. A bidder reads the report and assumes a tunneling method that is not allowed. One project not listed was delayed because of such concerns raised during the bidding period resulting in extending the bid period by a couple of weeks.

The lesson learned here is in general accordance with the "Geotechnical Baseline Reports for Construction." Only a memorandum for geotechnical design should be prepared not a report and it should be labeled as such and not included in contract documents.

QUALIFICATIONS

The availability of an experienced specialty subcontractor to perform the tunnel work on a project is often limited. This becomes a multi-faceted problem. In many cases the dollar magnitude of tunnel work on a project is too small to be cost effective for a firm to bid. As a result, a tunnel crossing portion of a project that has any special requirements such as grouting, compressed air, ground freezing is either going to be very expensive or an attempt at a different means of maintaining a stable heading will be attempted.

Several of the projects listed encountered problems because the local ground conditions encountered were significantly different from the typical conditions encountered by the tunneling contractor in a different part of the country. Tunneling in stiff silty clay and soft limestone is entirely different than tunneling in residual soils consisting of sandy silt and clay with partial weathered rock. For these short tunnels the tunneling risk is reduced by using an open face shield and tunnel using liner plates rather than jacking a casing.

Every tunnel project has a learning curve at the start. All but 5 of the tunnels that comprise the database are less than 200 m in length and 75% of them are less than 100 m in length. As a result of working on the steep portion of a the curve for most of the tunnel the overall production rate is typically about 0.3m /hr for hand mined tunnels and 1.3 m/hr for pipe jacking.

For longer tunnel projects this is also the time when equipment is brought into the tunnel to allow for a more efficient excavation and installation process. That is not the case for the short tunnels. Depending on the size of a tunnel and the complexity of the machinery being used that learning process can easily take over 100 to 200 meters of tunneling. Those inherited delays plus the experience of the tunnel contractor with the ground conditions will affect schedule and can result in claims.

The ability to qualify contractors to perform tunnel work is limited and in many states not allowed at all. Where qualifications are allowed it is often difficult to disqualify a contractor based on technical issues and the risk of delaying a project in an attempt to start such proceedings is high.

The claims associated with projects NC2 and NC4 are partly due to the limited experience that the contractor had with the given ground conditions.

The local experience factor also shows a difference in advancement rates. Similar conditions in the residual soils in projects NC1, NC2, NC3, NC4 and SC1, the "local" tunneling contractor typically advanced 6 to 7 meters per shift in hand mined tunnels and more when the conditions allowed for use of MTBM whereas the hand mining rate of advance of an "out of town" contractor was 1 to 1.3 meters per shift. Contributing factors to this slower advancement rate were: the learning curve regarding typical ground behavior of residual soil and rock conditions, and selection of excavation equipment.

PRICING

A common bid format for pipeline projects is to require a lump sum price for each crossing. That lump sum price often includes the shafts, tunneling, ground modification such as grouting or dewatering, installation of a carrier pipe within the initial lining system, filling the annular void, and, in some cases, geotechnical instrumentation. As shown in Figure 2, the resulting difference in cost per unit of length of tunnel work is significantly more costly. Also there is a slight trend to indicate that if that ratio is too small claims are likely.

With the exception of rock tunnels, projects listed in Table 1 which were in residual soils or were over 300 m in length were bid as lump sum items. In a couple of cases the shafts were separate items. The range of low bids for these projects was $7.6 million to $65 million and averaged just over $24 million.

Considering that a typical pipeline project can have upwards of 30 different bid line items the desire to keep the number of bid items down can be understood. Of the projects listed, 97% of the crossings are each less than 3% of the total project cost. However, there are several unknown quantities in a tunnel construction regardless of the tunnel size and length. Item such as number of cobbles or boulders, percentage of the tunnel that will be in mixed face, need for permeation grouting, or how much groundwater will be pumped are all indeterminate.

In many cases some of these issues are more difficult to handle because the project is too small to justify certain equipment. Consider a 50 m long 2 m diameter tunnel in sand under 5 m of water pressure. During the design phase does this condition warrant a slurry MTBM be specified as the only means of tunneling? By only requiring that type of equipment several potential bidders may not bid because of lack of experience or crew for such equipment. If used, by the time the crew is knowledgeable of the ground behavior and through the learning curve process and ready to get into full production the tunnel is completed.

A viable alternative is to specify permeation grouting to stabilize the heading. A lump sum bid for tunneling puts the entire risk on the contractor. What can be expected and experienced is that the initial bid is low and optimistic of the conditions and the claim is high. Even negotiations will start with a high lump sum price by the contractor that includes high contingencies to account for the unknowns in grout quantities and affects on tunnel production rates. If the bid documents include line items for mobilization, drilling and grout consumption there is a sharing of the risk.

The argument against the additional bid items is that for 78% of the crossings the bid price only represented a cost of less than 3% of the total project. While none of the claims presented in this database are directly related to a lump sum bid it is our opinion that if unit pricing was used for items such as M&D to cover the cost of the TBM, specialty work such as grouting and different support types that this results in a risk sharing between contractor and owner the overall cost would be less and probably fewer claims. This unit pricing approach reduced the number of claims received on projects NC1, NC2, NC3, and NC4 because the change was addressed in the contract documents. The contractor had a bid for the additional cost of tunnel support and was aware that the risk of encountering poor ground would be shared.

EXPLORATIONS

In a large tunnel project the exploration and laboratory testing programs are developed for the purpose of understanding the subsurface conditions. Total boring footage is usually on the flat portion of the "knee" of the plotted data of linear footage versus "Changes Requested" (USNCTT 1984). This is about a ratio of 1.0 to 1.5 boring footage to tunnel length. These borings are mostly drilled to a depth below proposed tunnel invert. As a result a subsurface profile is usually developed for design purposes. That profile is based on engineering judgment utilizing data in the vicinity of the actual borings not just borings at the stationing of the profile.

This is not the case for each of these pipeline projects. Most of the tunnel crossings are located at areas that are not accessible for borings during design or have a significant price tag associated with the drilling and therefore, a decision was made not to drill. The former is typical of a highway or railroad crossing and the latter for a river crossing where a barge would be required to access the site.

While the boring spacing at these crossings is often well within acceptable limits, 150 m or less, the overall quantity of explorations at a crossing is usually limited to 2 or 3 borings. Other adjacent borings along the alignments were often drilled to a shorter depth and therefore the population of the data base is limited. This can result in misleading information depending upon the geology of the area.

Composition of glacial till soils in the northern parts of the country can vary. This was the case in project OH1 that resulted in a claim. As shown in Figure 1 rapid changes in residual soil and weathered rock interface as were the conditions in projects NC1, NC2, NC3, NC4 and SC1 can result in misleading information regarding subsurface tunneling conditions. Figure 1 is typical of the rapid change in ground conditions in a residual soil and rock geology. In this case a boring, shown at the vertical red line in Figure 1, indicated solid rock and less than 60 cm away the rock was weathered to a depth below the tunnel invert and required a more robust support system.

In this particular case, because the bid had unit prices for the different support types there was no claim. At another crossing in similar geology, two borings taken during design indicated residual soils and partially weathered rock. The crossing encountered

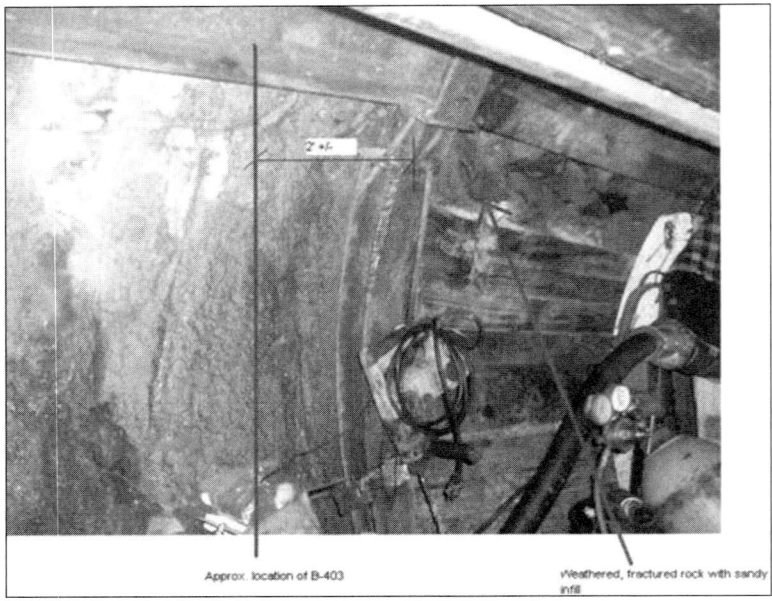

Figure 1. Subsurface exploration actual conditions

a full face of rock for the entire distance. Since it was a highway crossing the state DOT required that the tunnel be supported with liner plates, this resulted in a major claim.

The reverse conditions can also be encountered. Highway cuts in rock are blasted and the rock below grade is often a poorer quality than what is shown of core taken on the side of the traveled way. The poorer rock quality will have an effect on the required tunnel support with regards to tunneling behavior and on one crossing resulted in a significant claim.

TECHNICAL SPECIFICATIONS AND GBR

The risk involved with technical specification and the GBR is raised because of limited data when dealing with only a few tunnel crossings and the population of the data base to establish parameter values and also the apparent or limited understanding of the purpose of the GBR by the bidders.

Project SC1 had 14 borings drilled as part of the overall exploration program. The primary purpose was of these explorations and geophysical survey was to site the tunnel alignment in sound rock and avoid zones of decomposed rock. Achieving that goal was successful. However, of the 14 borings only 5 were within 50 feet of centerline of the final alignment. There were only a limited number of samples that were tested for hardness and abrasion. The number of tests performed 6 Unconfined Compressive Strength (UCS) and 3 Cerchar Abrasion Index (CAI) for 480 lf of tunnel is a good ratio of tests per foot of tunnel. However, it is a limited database. The high end of the parameter values was used as baseline values. A claim was submitted because of excessive cutter wear. Additional testing submitted as part of the claim showed similar strength test values but a significantly higher CAI value.

In many cases the pipeline design engineer may be familiar with the concept of a Geotechnical Baseline Report (GBR) but not deem it appropriate for a pipeline project. Even in the projects where the GBR document is used or baseline type of statements are included in the technical specification the implication of those statements are not understood by all the bidders. It is not uncommon that we see a wider spread in the bids when there is a tunnel crossing involved. It is also not uncommon that we see a low unit or lump sum price for the tunnel portion of a project be assembled on assumptions that do not meet specification criteria or take into account the ability to handle the given ground conditions as stated in the GBR or similar types of statements.

As an example, the following text is from a tunnel specification where no GBR was prepared. The text is providing an interpretation of the ground conditions based on the available subsurface data and experience in the part of the country where the project is currently on-going.

> "The Contractor shall anticipate that 100 feet of tunnel length will encounter decomposed rock during the tunneling operation at or above the tunnel liner plate invert elevation. The additional cost for rock excavation of these 100 feet of tunnel partially weathered rock shall be included in the lump sum cost for tunneling."

The intent of this statement and similar statements regarding groundwater inflow were to provide all bidders with the same interpretation of ground conditions with regards to the probability of encountering rock and not to assume that the entire tunnel will be excavated in sand.

For a project that is not listed which included two tunnel crossings, a total of 12 bids were received. The low lump sum bid for each crossing was the overall low bidder. Bids ranged from $850/m to $2,330/m for one crossing and $890/m to $1,940/m for the other crossing. The average bid for the two crossings was $1,510/m and $1,310/m and the engineer's estimates were $1,670/m and $1,740/m. The high total bid on the project was almost twice the bid of the low bid. The tunnel bid range ratio was 2.25 and 2.75 from high to low for the two tunnel items.

The two lowest total contract bids were within 3% of each other and 10 of the 12 bids were within 15% of the third place bid. However, only 8 of the bids were within 15% of the third place tunnel bid and the low bid was almost 40% less than the third place bid.

Our interpretation is that the two low bids were aggressive overall and more so with the tunnel costs. The majority of the bids were very closely grouped which is an indication that the specifications were understood in general but more risky bidding on the tunnel portion which represented slightly more than 20% of the total bid.

SPECIALITY WORK

One of the major risk issues is the tunnel location. The consequence of a stuck TBM in a location where a rescue shaft is not possible is very expensive. These short tunnel crossings are most commonly located in below highways or railroad tracks. An option of relocating the alignment that may exist on a large tunnel project seldom exists for these tunnel crossings. Performance limitations imposed by the stakeholder—railroad or state or federal highway transportation department often further increase the cost. That is not to say that the need to mitigate ground movement is not necessary. It is just a comparison of the tunnels. It is not uncommon that the vertical alignment of the large conveyance or CSO tunnel project will be deeper than then these small tunnels that are often consolidation pipelines. As a result the geometry often works in favor of reduced ground distortion with the larger tunnels.

Railroad crossings usually require that the carrier pipe be encased. If the carrier pipe is 1,830 mm or larger it usually means that the crossing has to be excavated using conventional tunneling. This method of trenchless construction is required to maintain line and grade of the carrier pipe and account for construction tolerances as well as carrier bedding within the casing and availability of casing diameter sizes. Transportation agencies often have similar requirements.

Evaluation of the subsurface data with regards to tunneling behavior of the soil, probability of encountering cobbles or boulders has to be made. If microtunneling is still viable, that is a method that can be allowed. The other alternative is to tunnel using a shield with appropriate shelving and ground modification to maintain face stability.

Issues can arise here when the general contractor is negotiating with the tunnel subcontractor and also a grouting subcontractor. Who is working for whom? Who has what responsibility to perform the work as specified? What is considered acceptable with regards to the amount of ground modification that is done? Is the contract written such that the all of the risk is on the contractor who then puts it onto the specialty subcontractor?

The design team for a pipeline project may have an engineer assigned to the project that has some experience with tunnel crossings. However, it is unusual that such person has the time to keep up with the changes that are occurring in the tunneling industry. A combination of current experience in tunneling, understanding of geotechnical engineering and tunneling behavior of the subsurface material, and time to write the appropriate tunnel specification are required. The engineer experienced in pipeline design and hydraulics will seldom have this expertise also.

Usually there is little design time available to develop a detailed analysis of a specific issue concerning a tunnel crossing. As an example, a design issue such as a probable need to use a chemical grout or micro-fine cement grout to modify the ground to improve face stability is resolved by reviewing a few nearby soil gradation curves and making modifications to an existing grouting specification. However, for a tunnel project, attention to the need to obtain the necessary laboratory testing, an analysis of the grout hole spacing and grouting pressures is performed, and quantity of drilling and grout take is calculated. Percentage wise grouting for a 100 m long tunnel crossing may be 50% or more of the total crossing bid whereas the same 100 m long segment of a 1.6 km long tunnel is probably less than 5% of the total tunnel cost.

The specification needs to provide for an adequate means and mechanism in payment provisions to make fair adjustments to the contract. This often will mean that additional line items for a crossing will be required. The most common method of ground modification will be grouting and there may be some passive dewatering if a horizontal drain can be installed and the soil is sufficiently permeable to allow for drainage to occur. We recommend that bid items be provided for both drilling and grout take quantity. With careful monitoring and mapping of the location, injection pressure and quantity of grout pumped, a tunnel inspector can determine the effectiveness of the grouting program. The ability to make modifications to the program can be made using the same unit prices. The technical specification should also provide a criterion that can be clearly defined in the specification and measured during the tunneling process.

Because of the small diameter of the tunnels, cobbles (up to 300 mm) can be an issue similar to boulders. We recommend that a unit price for cobbles, boulders or mixed face (soil/rock) be paid for as a separate line item.

Dewatering, if used, should be designed by the contractor to a set of criteria in the specification similar to what would be presented in a GBR if that document is not provided in the contract documents. Typically that information would provide; design groundwater level, a soil permeability, and potentially an estimate of groundwater pump rate to achieve the lowering of the groundwater to a specified elevation.

Prequalification of tunnel contractors is an issue that has to be addressed on a state by state basis. The ability to prequalify a firm is a plus if allowed. Often the owner is limited to prequalification only to ensure that the contractor is solvent. In addition because the tunneling work associated with a pipeline project is small (75% of the projects listed in Table 1 had tunnel costs that were 20% or less of the total bid cost for the project) it is difficult to prequalify only subcontractors. To win the bid, the general contractor typically wants to utilize local subcontractors to the extent possible primarily because of cost associated with crew per diems. The pool of local qualified tunneling experience in some cases is limited. In other cases the general contractor is expanding territories and utilizing tunneling subcontractors that they are familiar with but who are not familiar with the nuances of tunneling in different parts of the country.

The ability to technically prequalify needs to take into account the ground conditions, methods of tunneling and any modifications to the ground. It also has to take into account that in many cases the type of tunneling that is required does not require use of a modern TBM and therefore current experience with modern methods of tunneling are not essential and could result in disqualifying capable tunnel contractors. The technical prequalification should be specific to the means and methods allowed in the contract documents and selected by the bidder.

We recommend that the bid documents for the three low bidders be put in escrow and the documents of the contractor awarded the project be held in escrow and those unopened documents be returned to the other bidders once the contract is signed.

Establishment of Baseline Conditions either by using a GBR or establishing the interpreted conditions in the technical specifications. In many cases there is a pre-bid conference for the bidders. This is a good opportunity for the engineer to explain to the bidders what those conditions are and what they mean to the contractor with regards to his bid. This presentation should be done by an engineer who is familiar with the project and with tunneling.

Proactive construction management is the best method of controlling costs on the project. Specifically the tunnel crossing work should be inspected on a full time continuous basis by engineers fully knowledgeable of tunnel work and how the ground conditions can and will affect the work being performed. It may not reduce the number of claims but it will help in the resolution of what portion of a claim is fair. This management involves a proper level of inspection and record keeping. While this may appear to be an expensive item to the owner, the reality is that this specialty staff is only required on the job during the active tunnel work, not necessarily during shaft construction nor when the carrier pipe is being installed in the tunnel. Considering that most of these tunnels are less than 200 m is length this tunnel inspection time is usually about two to three months.

SUMMARY OF LESSONS LEARNED

The lessons that we have learned from these projects can be summarized as follows:

1. While the chances of getting Division 1 specifications changed for a pipeline project are small there is often some agreement with the owner to allow limited supplements to these various items. If this can be agreed upon making such changes as the project warrants should be performed.
2. Consider adding appropriate bid items as part of the tunnel crossings that share the risk, such as chemical grouting rather than including it as part of a lump sum bid. The subsurface data base may be too limited if borings at shaft locations are considered for a crossing evaluation.
3. When developing an exploration program consider the number and depth of borings adjacent to a tunnel crossing, not just the data that will be obtained at shaft locations.
4. Inclusion of both a Geotechnical Data Report and Geotechnical Baseline Report that addresses each crossing in a pipeline project is an excellent method of leveling the field during the bid process as well as for evaluating claims.
5. Take into account the level of sophistication required of the tunneling and experience of the small tunnel sub-contractor when specifying tunneling qualifications.

PART 15

SEM

Chairs

Jean Habimana
Jacobs Engineering Group

Scott Wimmer
Kiewit

CASE HISTORY OF THE WACHOVIA–KNIGHT THEATER PEDESTRIAN TUNNELS

Eric Eisold ▪ Bradshaw Construction Corp.

Prakash Donde ▪ Jenny Engineering Corp.

Ivona Tarchala ▪ Jenny Engineering Corp.

ABSTRACT

The Wachovia Headquarters and Arts and Cultural Campus project in Charlotte, NC includes a 1.5 million square foot, 48-story office tower, two museums, an Afro-American cultural center and a 1,200 seat theater for the performing arts. Public access to the Cultural Campus is through two pedestrian tunnels constructed under city streets connecting to parking garages. The 5.7 m wide by 4.7 m high mined tunnels were procured through a design-build contract. The tunnels were designed as initial/final shotcrete linings with a sprayed water barrier sandwiched between layers of shotcrete. In addition, the mix design was supplemented with a cementitious admixture to develop internal water resistant properties in the shotcrete.

DESCRIPTION OF PROJECT

The design-build procurement for the two tunnels was accomplished by Rodgers Builders Inc. of Charlotte, NC (general contractor). The overall design of the project was led by TVS of Atlanta, GA (architect). At the time, the only parameters for the tunnels were that they had to be mined, waterproofed and must accommodate a finish box shaped cross section of 4.1 m wide by 2.8 m high. After a process of several technical and price proposals, the team of Bradshaw Construction Corp. (BCC) and Jenny Engineering Corp. (JEC) was selected. The tunnels were constructed under South Tryon Street (29.1 m) and First Street (20.6 m) in downtown Charlotte, NC as shown on Figure 1. Cover over the tunnels ranged from 4.6 m to 6.4 m.

SUBSURFACE CONDITIONS

Charlotte, NC is located in the Piedmont Physiographic Province which is characterized by the in-place weathering of the underlying igneous and metamorphic bedrock into blocky and irregular layers of weathered, partially weathered rock and residual soils. Groundwater is usually found at stream depths near valleys and at higher elevations under hills, and may be perched on discontinuous layers of less permeable materials. Flows into excavations vary according to rainfall, depth, and lateral and vertical hydraulic conductivity. Stand-up time of the residual soils varies from slow raveling to flowing ground depending on the density, groundwater, and grain size distribution. Man-made fills and utility trenches typically complicate these natural conditions even further.

At the job site, the Tryon Street tunnel was located almost entirely in slow raveling residual soils and partially weathered rock. The First Street tunnel was situated mostly in slightly weathered granitic rock overlain by moderately weathered rock near the top

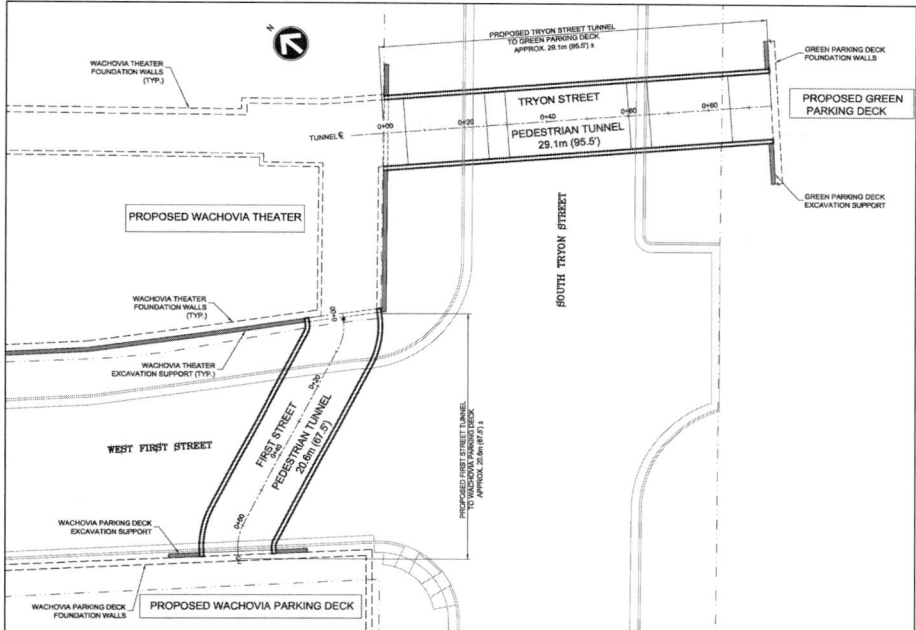

Figure 1. Tunnel plans

of the tunnel. Groundwater was not an issue as the surrounding area in downtown Charlotte is more or less permanently dewatered by sump pumps in the underground parking garages and lower levels of the buildings. Downward percolation from wet weather events and leaking water lines was somewhat of a nuisance, especially with the shotcrete as will be presented later in this report. Profiles of the tunnels are shown on Figures 2 and 3.

DESIGN

The design concept was developed to:
- Meet the project requirements
- Fit the tunnel contractor's preferred means and methods
- Provide the lowest cost

A horseshoe shaped cross section was developed to accommodate the finish walkway. The structural initial/final tunnel lining consisted of a shotcrete shell and a cast-in-place concrete lining. Waterproofing consisted of a concrete admixture which intensifies and prolongs the hydration process while growing silica based needle shaped crystals that fill the voids and pores in concrete. The admixture is designed to endure repeated periods of wetting and drying by growing new crystals as new moisture is introduced during the life of the concrete. In addition to the admixture, a sprayed-on waterproofing membrane was applied between layers of shotcrete.

The tunnel contractor, experienced with shotcrete SEM construction, preferred this approach because of its flexibility for accommodating different ground types and shapes, controlling subsidence, and compatibility with various waterproofing systems.

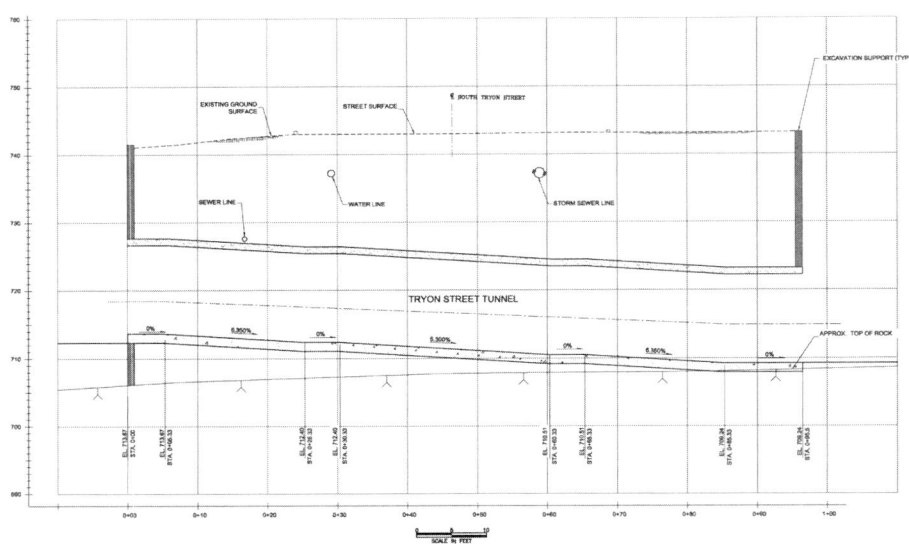

Figure 2. Tryon Street tunnel profile

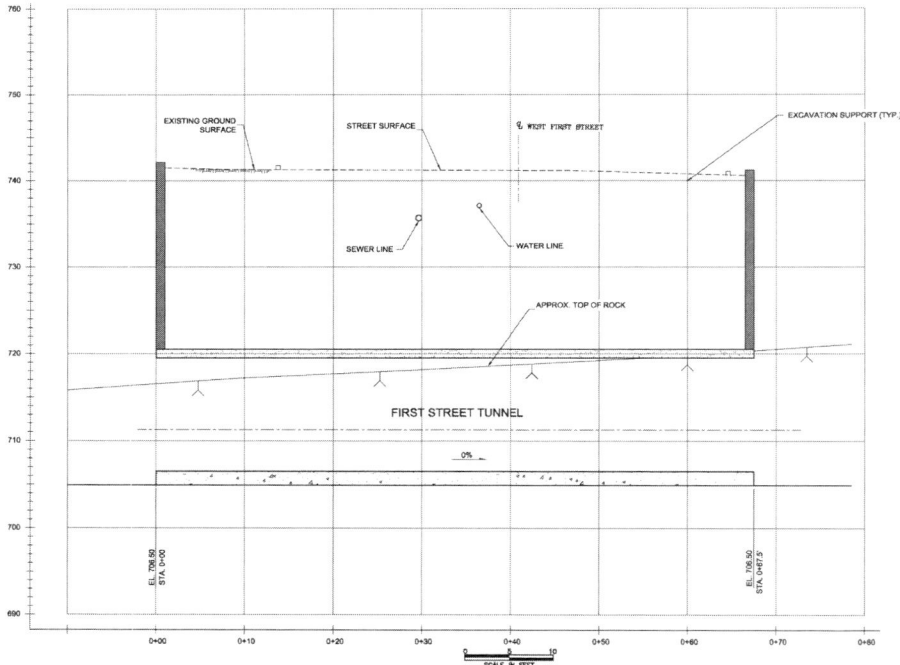

Figure 3. First Street tunnel profile

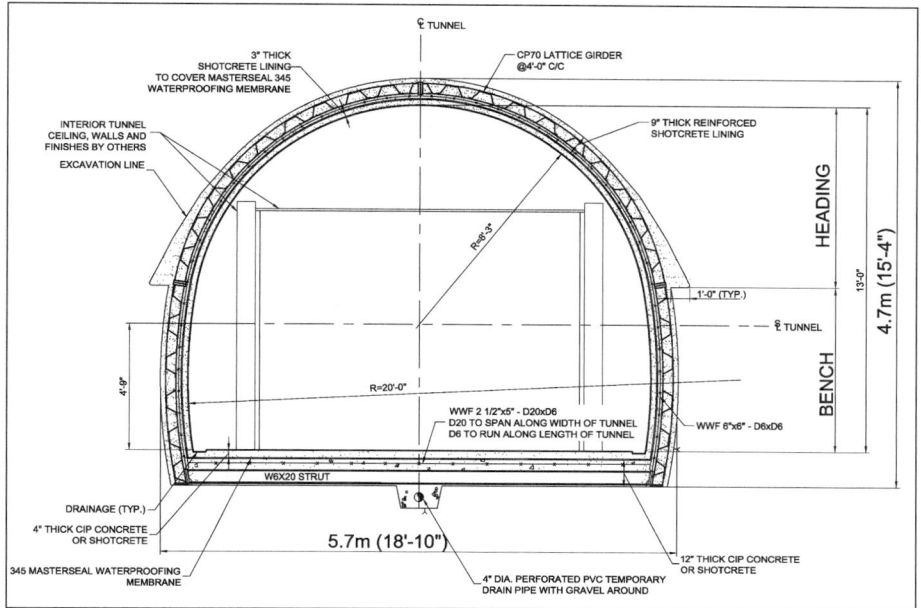

Figure 4. Typical tunnels cross section

The capability of constructing a flat cast-in-place concrete bottom was especially attractive given the floor plan of the walkway finish tunnel.

Inasmuch as the procurement was competitive for design and price, the team endeavored to come up with a cost effective solution that met the project requirements. Given the site constraints, subsurface conditions, and short lengths, a shield driven tunnel was not practical. Hand-mining was clearly the best alternative. A schedule analysis indicated that concurrent mining and supporting two shotcrete SEM tunnels through the variable ground conditions would be optimal, especially if the waterproofing and final load support could be constructed in flexible overlapping construction activities at different locations. In addition, a flat invert eliminated the need for placing a curved shotcrete invert and associated extra concrete, spoils, etc. Other approaches considered were cast-in-place or shotcrete final linings placed against an HDPE or PVC barrier type waterproofing system attached by hand to the initial shotcrete shell.

The tunnels were designed with a horse-shoe shaped configuration with a cross section of 4.7 m (15'-4") high and 5.7 m (18'-10") wide. The dimensions of the finished tunnels were 4.0 m (13'-0") high and 5.1 m (16'-10") wide. The structural arch of the tunnels consisted of a 229 mm (9") thick reinforced shotcrete shell and a 305 mm (12") thick cast-in-place reinforced concrete slab as indicated in Figure 4.

The Tryon St. Tunnel was 29.1 m (95.5') long with a ground cover of approximately 6.4 m (21'). Based on the borings taken at site for design purposes, the entire section of the tunnel was located in the soft ground. The borings also indicated that the top of rock would approximately be at the invert of excavated tunnel as shown in Figure 2, and the ground water elevation would be just above the tunnel spring line. The First St. tunnel was 20.6 m (67.5') long with a ground cover varying from 4.6 m (15') to 6.1 m (20'). Although the two tunnels were in close proximity of each other the borings indicated that the top of rock for the First St Tunnel would be located between the tunnel spring line and the top of the tunnel as illustrated in Figure 3. Accordingly, the design

Table 1. Geotechnical parameters used in beam-spring analytical model

Soil			
Unit weight	Internal Friction Angle	Modulus of Subgrade Reaction	Average SPT N-value
19.6 kN/m^2/m (125 psf/ft)	30°	62.8 kN/m^2/m (400 psf/ft)	15
Rock			
Unit weight	Internal Friction Angle	Modulus of Subgrade Reaction	Average RQD/REC
23.6 kN/m^2/m (150 psf/ft)	36°	125.6 kN/m^2/m (800 psf/ft)	64% / 72%

Table 2. Results of analyses

Location	Max. Bending Moment	Max. Axial Compression	Shear (*at distance 'd' from face of the tunnel lining)
Tunnel lining at crown	8.8 kN-m (6.5 kip-ft)	80 kN (18 kips)	9 kN (2 kip)
Tunnel lining at springline	8.7 kN-m (6.4 kip-ft)	111 kN (25 kips)	13 kN (3 kip)
Tunnel lining at floor	8.1 kN-m (6.0 kip-ft)	85 kN (19 kips)	18 kN (4 kips)
Tunnel floor slab	38.0 kN-m (28 kip-ft)	31 kN (7 kips)	31 kN (7 kips*)

of this tunnel was performed for the mixed face conditions with the groundwater table at the tunnel crown. The design was conducted with geotechnical parameters as summarized in Table 1.

To perform the structural analysis the tunnel lining was divided in to a number of small beam elements. At the nodal points of the beam elements, springs were placed to represent the ground. The springs were assigned elastic properties derived from the modulus of subgrade reactions. The analytical model thus represented the tunnel surrounded by the ground thus permitting a soil-structure interaction to occur. The values of modulus of subgrade reactions used in analyses are presented in Table 1. The tunnels were analyzed utilizing STAAD Pro structural analysis software. The analyses were performed for load cases which included overburden loads from the ground, lateral soil pressures, and hydrostatic pressures. In addition to these loads AASHTO H20-44 highway load and an 80 tonne (180,000 lb) crane load were also included in analyses. The crane load was present from the crane to be used at the site for construction activities. After performing analyses for various loading cases, internal forces in the linings were obtained. The results of the analyses are as summarized below in Table 2.

The design was performed using the strength design and allowable stress design methods. The shotcrete and concrete both with a strength (f'c) of 27.8 MPa (4,000 psi) was used for the tunnel structure. Welded wire fabric was provided as the tunnel lining reinforcement as shown in Figure 5 and Figure 6. The lattice girders were primarily provided to support the tunnel excavation although they contributed a small amount of reinforcement value to the lining also.

The sequence of construction was developed using what is known as the Sequential Excavation Method (SEM) or NATM. Figure 7 shows how the excavation

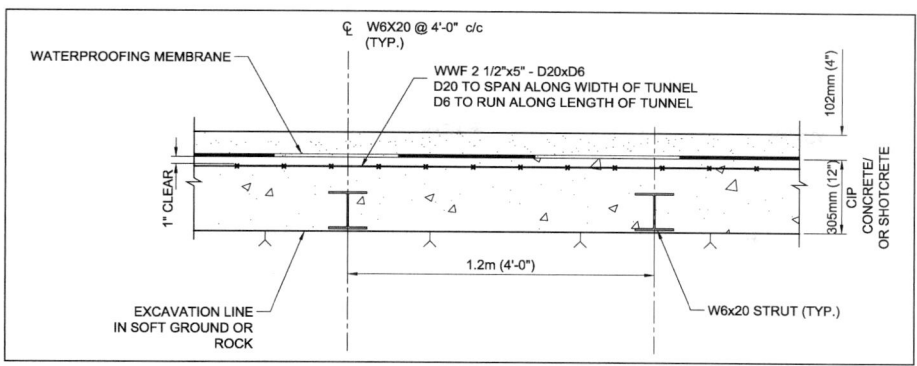

Figure 5. Tunnel base slab detail

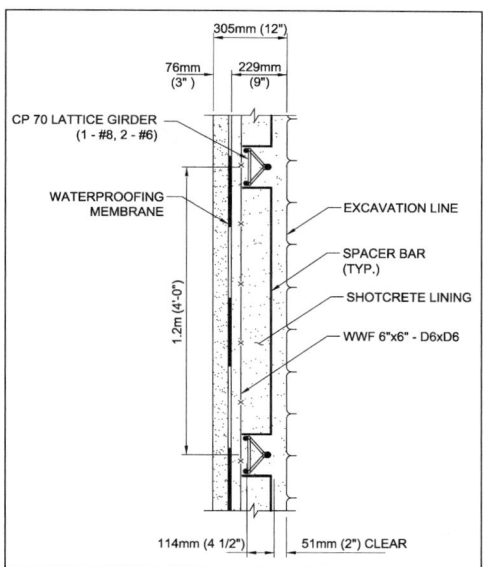

Figure 6. Tunnel lining detail

was to be performed. It was required that the top heading had to be at least two rounds ahead of the bench all the time during excavation of the tunnel. The photo in Figure 8 indicates the excavated heading with the bench two rounds behind it. The tunnel cross section was divided approximately in two sections with 2.4 m (7'-9") high heading and 2.3 m (7'-7") high bench. Excavation rounds were limited to a length of 1.2 m (4'). A lattice girder was placed at the face of each heading to provide immediate ground support at the face of the tunnel. The 305 mm (1'-0") wide "elephant foot" as shown in Figure 9 was provided at the bottom of the top heading during excavation. That extra wide footing ensured that the support for the heading was not undermined during the bench removal later on.

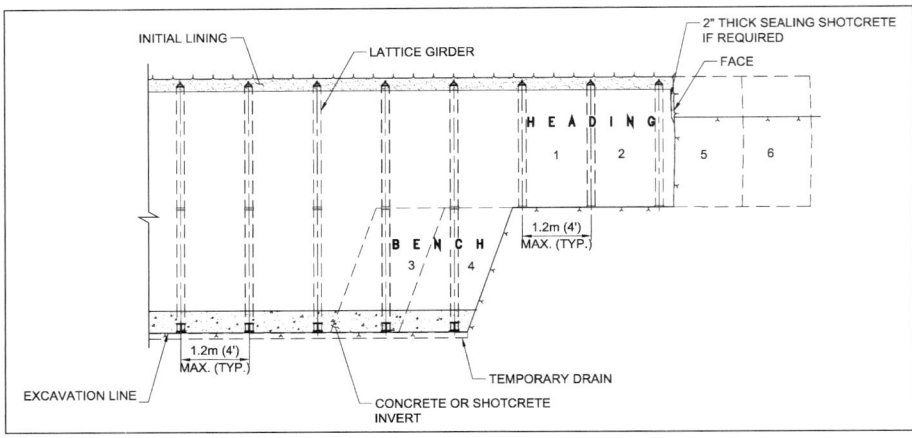

Figure 7. Construction sequence

Figure 8. Photo illustrating Tryon St. tunnel excavation

WATERPROOFING

The tunnels were designed to be watertight structures. Several systems of waterproofing were considered for the project including HDPE membrane, PVC membrane, double-bonded spray applied cementitious membrane and cementitious concrete admixtures. After evaluating the cost and constructability issues of these methods, a combination of two systems was selected to achieve the specified and desired results. It was decided to use a cementitious admixture system known as Krystol Internal Membrane (KIM), manufactured by Kryton, in the structural shotcrete walls and invert concrete slab of the tunnels and then apply a double-bonded "MASTERSEAL® 345" product on the inside surface of the walls as a secondary waterproofing protection.

The KIM material was mixed with the concrete/shotcrete to enhance internal waterproofing characteristics of the structural concrete lining. The KIM admixture reacts with water in the mix first and later with water seeping through the applied shotcrete and concrete. In reacting with water it develops millions of needle-like crystals. These crystals over a period of time continue to grow, filling the pores and cracks in concrete, and thus permanently block the pathways of water. The crystals also densify the concrete

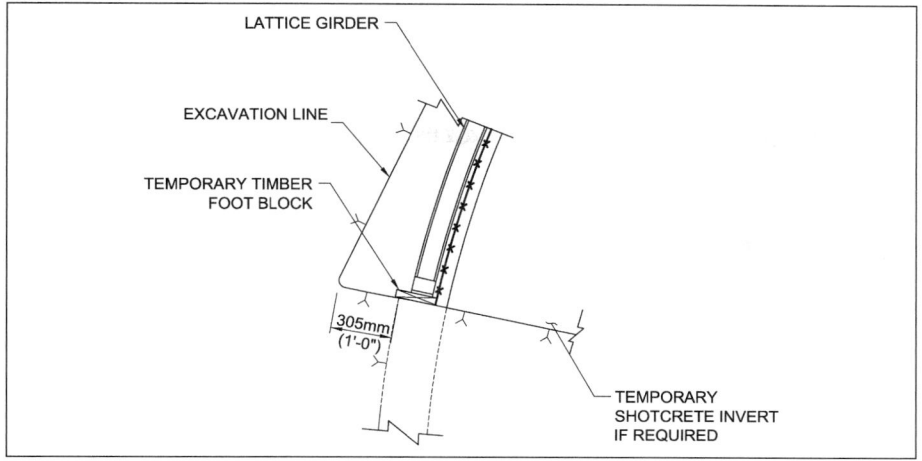

Figure 9. "Elephant foot" detail

over a period of time. Five kg (11 lb) of KIM per 0.8 m³ (1 cy) of the concrete mix was used for the tunnel slabs and six kg (13.5 lb) per 0.8 m³ (1 cy) of shotcrete mixture for the tunnel walls. After completion of the structural tunnel lining, which contained the KIM admixture, a second waterproofing system, MASTERSEAL® 345, was applied to the inside surface of the tunnels. After the application of the MASTERSEAL product a 76 mm to 102 mm (3" to 4") thick shotcrete layer was installed. The MASTERSEAL product structurally bonds to surfaces on both sides and produces a composite structural thickness. By doing so it does not produce weakness in the lining which is inherent in the application of most other types of waterproofing membranes And for that reason, they are always applied on the outside of structural walls. However, the double-bonding property of the MASTESEAL product allows application of waterproofing on the inside of structural walls. This method of waterproofing application on the inside surface is also called "negative side waterproofing."

CONSTRUCTION

Excavation support for the construction of the theater building consisted of drilled piles, timber lagging and tiebacks designed by another subcontractor. The design of the excavation support was accomplished concurrent with the tunnel design to space the piles favorably at the tunnel eyes and avoid interference between the tiebacks and the tunnel. The tunnel eyes were supported with pre-drilled forepoling consisting of #8 rebar on 0.30 m (4 ft) spacing over the crown. The soldier piles were stitched together by horizontal walers welded over the crown and further braced by diagonal beams over the shoulders of the tunnels.

Excavation was accomplished using a Grant 16,000 kg electric powered excavator and a Komatsu 4,000 kg diesel excavator. Line and grade were maintained with a conventional pipeline laser. Mucking was accomplished with track mounted skid-steer loaders. Spoils were removed from the site by using a 1 cubic meter skip-pan lifted by the general contractor using a tower crane. The concrete was produced by a local ready mix supplier and delivered to the job site in mixer trucks. It was pumped from the surface staging area to the headings through 76 mm (3 in) concrete pipe. Shotcrete was applied manually through a 51 mm (2 in) nozzle where an alkali free accelerator

(Meyco SA 160) was added at the nozzleman's control. Tunnel construction was performed on a one work shift per day basis and coordinated closely with the general contractor's schedule. Foundations and walls for the theater building were constructed simultaneously with the pedestrian tunnels which led to numerous challenges associated with work areas and coordination with other trades.

Both tunnels terminated against existing underground parking garages. Several tiebacks were encountered from the existing structures and they were cut off. Waterproofing the connections was accomplished by placing a mastic-type hydrophilic waterstop onto the existing concrete and spraying/pouring the new concrete directly against the existing structures. At the interfaces with the new theater construction, the waterproofing membrane that was placed against the soldier piles and lagging was wrapped into the tunnel over the first layer of shotcrete and then sandwiched in with the sprayed-on waterproofing. Three of the four connections required remedial action to stop leaks by drilling and grouting through the shotcrete/concrete.

Tryon Street Tunnel

After turning the eye, the excavation sequence started by drilling in 1.83 m (6 ft) long pre-supports consisting of #8 rebar at 0.3 m (1 ft) spacing over the arch with jackleg type drills. The top heading was advanced about 1.5 m (5 ft), far enough to place the arch lattice girder segments and welded wire fabric. Another top heading was then advanced. One or two bottom headings were then completed and the invert supported with steel beams. Some rock seams were encountered that required hammering with a hydraulic powered bull-point mounted on the excavator. The stand-up time was good and allowed for the advances to be completed without any shotcrete applications until the excavation sequence was done and all the steel supports were erected. The initial shotcrete was applied towards the end of shift and allowed to set overnight. This sequence was continued until reaching the end of the tunnel which terminated against wall of an existing underground parking garage. A thin coating of gunite was then applied to the surface before applying a 4 mm layer of the sprayed-on waterproofing. A 76 mm (3 in) layer of gunite was then placed over the waterproofing barrier including the KIM additive. The floor slab reinforcement was placed and the concrete was placed by pumping and finished using a vibratory truss screed.

During tunneling, groundwater seepage was encountered that persisted even after shotcreting. However, over a period of time these leaks slowly sealed to moisture patches and then dried up. After the sprayed-on waterproofing was applied, the tunnel arch and walls were completely dry. See Figure 10. After the final gunite layer was placed and the floor slab was poured, there were no leaks or seepage in the tunnel whatsoever. See Figure 11. That was in late April 2008.

After a few wet weather events, some seepage was observed in late May 2008 consisting of two minor dripping leaks and some wet spots and discolored dry spots. These were chipped out and patched with Kryton's patching compounds under the direction of Kryton's technical representative. In early July 2008, a few more wet patches appeared after heavy rains and were chipped and patched. In late July, several more patches appeared and the general contractor drilled and grouted the shotcrete without consulting the tunnel contractor. In mid-August 2008, the tunnel contractor chipped and patched all the areas where the general contractor had tried drilling and grouting in accordance with Kryton's recommended procedures. Over 20 cm (8 in) of rain fell on Charlotte between August 26 and 27, and an inspection of the tunnel on September 2, 2008 revealed no visible leaks. However, just as a precaution, the tunnel contractor installed dimpled plastic panning material in the arch full length in both tunnels so as to divert any possible future seepage to the space behind the concrete masonry walls of

Figure 10. Tryon Street tunnel after spraying waterproofing on sides and arch

Figure 11. Tryon Street tunnel after pouring floor slab

the finish walkway. Almost all of the wet spots causing the problem were found between 18 m (60 ft) and 27 m (90 ft) in from the start of the tunnel. See Figure 12.

First Street Tunnel

After turning the eye, the excavation sequence started by drilling in 1.83 m (6 ft) m long pre-supports consisting of #8 rebar at 0.3 m (1 ft) spacing over the arch with jack-leg type drills. The First Street tunnel was started when the Tryon Street tunnel was in about 4.9 m (16 ft). By sharing resources, the crew work was optimized and the construction schedule was accelerated. After consulting with the tunnel engineer, it was determined that the First Street tunnel could be advanced by driving the top heading to completion and then taking out the bench. Most of the tunnel was in rock and excavation was accomplished by conventional drill-and-blast methods. Drilling was achieved with jack-leg type drills and explosives were supplied by Dyno Nobel. A packaged extra gelatin dynamite (Unimax) was detonated with long period delays attached to primacord. The powder factor was typically between 2 and 4 and no more than 2.2 kg (4.8 lb)

Figure 12. Tryon Street seepage area

Figure 13. Tryon Street tunnel panning installation

was used per delay. Blasting mats were placed over the tunnel portal before each blasting event and no fly-rock escaped the blast zone or caused damage. Sauls Seismic, Inc. provided blasting consulting services and vibration monitoring. Seismographs were placed near the closest structures and reports were provided after each blast. USBM Report 8507 vibration criteria were used to assess blasting vibrations and no measurements exceeded the limits. No complaint or damage was reported.

During tunnel excavation, groundwater flows and seepage were encountered. This phenomenon was much more evident than the experience with the Tryon Street tunnel. A leaking water line was discovered approximately 6 m (20 ft) from centerline at about the mid-point of the alignment. It wasn't repaired until the tunnel was completed. However, the seepage through the first layer of shotcrete subsided quickly and the sprayed-on waterproofing was applied on dry and moist surfaces meeting the manufacturer's criteria. Remarkably, very few problems with the waterproofing on this tunnel were experienced. Nevertheless, the same dimpled plastic panning was installed as in the Tryon Street tunnel as an extra precaution. See Figure 13.

Quality Control

The tunnel design drawings and specifications prepared by the design/build team had to be approved by the City of Charlotte, the architect/engineer (TVS), the general contractor (Rodgers Builders), and the testing and inspection consultant (SME). A separate drawing was included in the tunnel design submittal to conform to the Special Inspection provisions of the North Carolina State Building Code.

The shotcrete and concrete mix designs were developed and tested well in advance of construction. Pre-construction testing included overhead and sidewall panels that were cored and tested. The concrete supplier, Concrete Supply, Inc. was especially helpful in working through the mix design process and construction phase using the KIM additive. Additional panels were constructed and tested by SME during construction as well as conventional concrete cylinders. All shotcrete was placed by an ACI certified nozzleman.

The tunnel designer (Jenny Engineering Corp.) made periodic site visits and inspections during construction. They performed Schmidt hammer and Windsor probe tests, met with the project architects and engineers, and generally observed construction procedures. It helped that both Bradshaw and Jenny had a long history of working on these types of design-build projects and understood each other's concerns. Decisions on design and construction elements were made quickly and within the parameters of the overall project requirements.

LESSONS LEARNED

The practice of waterproofing with the KIM additive in shotcrete proved to be valid to the extent that some gouging and patching may be required under minimal hydrostatic pressures. Neither of the tunnels experienced consistent hydrostatic pressures over the crown of the tunnel, so the performance of this method has yet to be seen at deeper applications below the groundwater table. Based on the performance of the of the first layer of shotcrete in the wet ground at the First Street tunnel, it appears that the KIM ingredient actually performs better and faster under continuous wet conditions as the presence of moisture allows for the hydration the crystals in the KIM. The behavior at the Tryon Street tunnel indicates that the KIM additive takes longer to develop waterproof properties in shotcrete in the presence of intermittent wetting and drying cycles. Our experience was that KIM does extend the curing time of concrete, so that effect should be considered for future applications. The product support we received was excellent. More study on this should be performed for shotcrete applications.

Based on the localized extent of problems associated with seepage, it appears that the secondary means of waterproofing, sprayed-on water barrier, failed at specific areas. Perhaps this was due to localized under-application and/or mixing and spraying variables. In any event, the quantity of materials used exceeded the manufacturer's guidelines. In hind-sight, we aimed at the recommended thickness and probably should have doubled it. The product sprayed nicely and stuck firmly to the smoothing gunite substrate and held the next layer of gunite without any problems. This material appears to be a good waterproofing product and may have performed better with more training, supervision and experience.

BOGGO ROAD BUSWAY PROJECT, BRISBANE, AUSTRALIA

Ted Nye ■ Sinclair Knight Merz

Maxwell Kitson ■ Sinclair Knight Merz

Ravin Chinniah ■ Sinclair Knight Merz

INTRODUCTION

This paper describes a number of aspects of the design and construction of a driven tunnel which forms part of the Boggo Road Busway project. The driven tunnel is 430m long with an excavated width of 15m and a full tunnel excavation height of 8m. The first section of driven tunnel was excavated under the heritage listed Boggo Road Jail for a length of 120m. The main jail buildings and perimeter walls were built in 1908 and are of brick construction.

Ground cover over the tunnel beneath the jail site varied from 5.5m to 8m and as a consequence the predicted settlement values and their potential to cause damage to the buildings were critical to determining and obtaining approval for the selected tunnel alignment.

The geology along the driven tunnel alignment was very variable, providing mixed face tunneling conditions and ranging from surface residual soils to high strength rock at depth. The maximum ground cover above the driven tunnel is 20m.

DESCRIPTION OF PROJECT

The A$326 million Boggo Road Busway a dedicated public transport link to the South Eastern and future Eastern Busway at Buranda and provides the connection to Eastern and Southern Busway corridors to the University of Queensland with a total travel length of 2km.

The Boggo Road Busway commences at the new PA Hospital busway station and continues through a cut and cover underpass beneath the Queensland Rail lines, then travels parallel to the railway lines to reach the new Park Road busway station and existing Park Road railway station. From this point the busway goes underground through a 630m long tunnel that passes beneath the Boggo Road Jail, Gair Park and Annerley Road before emerging at the new 330m span cable stayed Eleanor Schonell Bridge in Dutton Park which then connects across the river to the University of Queensland.

At its eastern end the Boggo Road Busway connects to the South Eastern Busway via the first section of the Eastern Busway. This section commences from the PA Hospital busway station then travels over Ipswich Road and through a cut and cover tunnel under the Pacific Motorway.

The Boggo Road Busway is expected to be completed in mid 2009 and will be used by 600 buses daily (this equates to around 13,000 passengers).

The Boggo Road Busway project and the first stage of the South Eastern Busway has been designed and constructed using the Alliance method of project delivery. The Alliance team members are Department of Main Roads/Queensland Transport (the

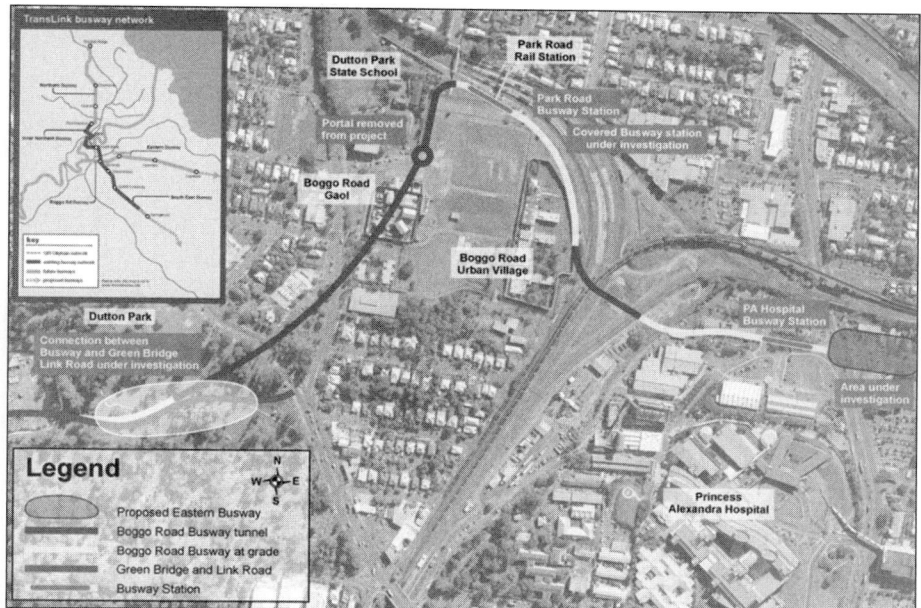

Figure 1. Boggo Road Busway route alignment

client), Thiess Contractors (the builder) and Sinclair Knight Merz (the designer). The route alignment is shown in Figure 1.

DESCRIPTION OF DRIVEN TUNNEL

The busway tunnel's full length is 630m and consists of 430m of driven tunnel with cut and over tunnels at both ends of the driven tunnel. The north end the cut and cover tunnel is 130m long and includes a 34m diameter bus turnaround area at the driven tunnel north portal. At the south end of the driven tunnel the cut and cover section is 70m long and daylights into Dutton Park.

The driven tunnel has a horseshoe shaped profile with a final excavated width and height of 15m and 8m respectively.

The busway tunnel profile includes 2 × 3.5m bus lanes with 1.6m shoulders on both sides of the busway together with a 1.5m wide emergency egress passage along one side of the tunnel. The road pavement to tunnel crown is around 7m in height. Four jet fans niches which required over excavation of the tunnel crown to accommodate three 1.2m diameter jet fans at each fan niche location. The jet fan niches were completed outside the jail buildings within the driven tunnel.

The final lining of the driven tunnel consists of a 300mm in-situ reinforced concrete for the main running tunnel. At the fan niches the final lining consists of a pattern of permanent rock bolts and a 250mm steel fiber reinforced shotcrete. The waterproofing membrane consist of 2mm polyethylene welded sheet over the arch and upper wall of the standard tunnel profile and a 4mm thick spray on membrane over the arch and upper wall at the fan niche locations. The driven tunnel has been designed to be fully drained and has a "no fines" concrete drainage layer under the reinforced concrete road pavement.

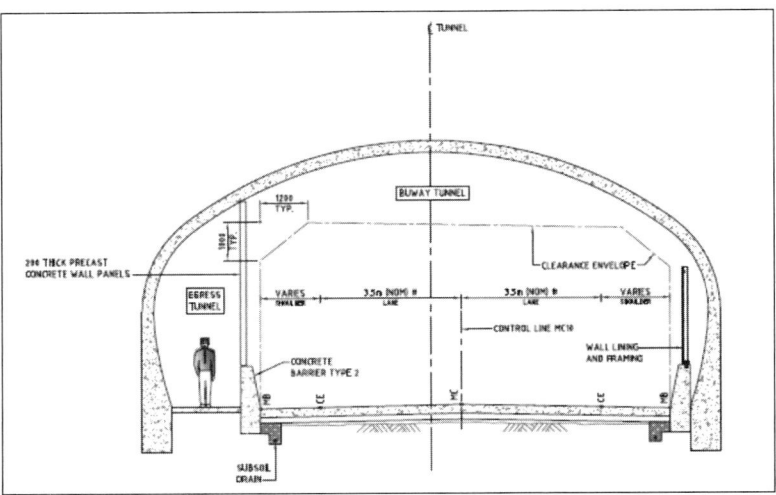

Figure 2. Standard tunnel section through completed tunnel

GEOLOGY

Regional Geology

The regional geology comprises Tertiary Brisbane Tuff Formation typically comprising "rhyolitic tuff," (similar to the Kangaroo Point Cliffs), breccia and minor sandstone and shale that were derived from volcanic activity to the north, overlying the Tingalpa Formation of carbonaceous shale, sandstone, conglomerate and coal. Both formations generally have deep weathering and variable strength profile, with many zones of completely weathered rock with soil like properties as well as zones of high strength.

Geotechnical Investigation for the Tunnel

The investigation for the driven tunnel required the drilling of 19 boreholes with 5 seismic traverses and test pits for inspecting the historical Boggo Road Goal building footings. Laboratory testing generally comprised UCS and aggressivity testing of soil and groundwater. Point Load Testing of core samples was carried out during the site investigation and also on rock samples taken during tunnel excavation.

The geological profile typically comprised soil/fill up to 2m depth over residual soil up to 4.5m depth over weathered rock from the Brisbane Tuff Formation. The Brisbane Tuff Formation generally comprised Breccia (2m to 6m thick) over the both non-welded (Claystone Tuff) very low to low strength (0.5 to 5m thick) and Welded Tuff (Ignimbrite) that was high to very high strength and jointed with clay seams (ranged in thickness up to 10m thick). The Brisbane Tuff (welded) joint system is considered random. Underlying the Brisbane Tuff the Tingalpa Formation of very low to medium strength, comprising Siltstone/Sandstone and Conglomerate.

Ground water monitoring peizometers were installed and packer tests were performed during the site investigation. The groundwater encountered comprised a perched water table (at approximately 8m depth) and a deeper confined aquifer within the Tingalpa formation (at 10m to 16m depth). Actual ground water inflows encountered during tunneling were for practical purposes insignificant.

Geological Model

The geological profile for both the rock type and rock strength varies considerably along the 430m length of the driven tunnel and across the tunnel section. Developing the geological model profile is quite difficult because of the varying geology. To develop the model it required a good understanding of the local geological history as well as the ability to extrapolate the data in three dimensions. It was realized in the design development stage that the geological model would have a significant impact on the selection of tunnel support types along the tunnel and hence the actual cost of the tunnel.

There are industry recognized classification systems which also provide recommended ground support for a given rock mass classification. These include the Q and the Rock Mass Rating (RMR) systems. An initial assessment using both these systems was carried out with the RMR system classifying the ground along the tunnel alignment in the range of poor to good rock conditions.

The geological model purpose is to provide a prediction of the variation of the rock strata along, across and above the tunnel. Rock strength data was superimposed on the geological model. It was this latter information that was used to develop a very simple site specific rock support classification system and consequently this was used to determine the extent of each support type along the tunnel. For example, the fully encapsulated resin rock bolts could not be used as initial ground support if the UCS of the rock was less than 10MPa as this would compromise the desired bond strength along the rock bolt.

Rock strength data principally consisted of Point Load Test (PLT) and laboratory Unconfined Compression Tests (UCS) results. On this project a multiplication factor of between 17 (welded tuff) and 20 was used to estimate the UCS of the rock from the much larger test sample of PLT results.

In the design phase of this project careful assessment of the ground support requirements were assessed over very short lengths of the tunnel due to the wide range of geotechnical parameters. The geological model and the superimposed strength data were continuously reviewed throughout the design and subsequent construction phase. This review resulted in two additional boreholes in Annerley Road (after tunnel construction had commenced), daily geological mapping data from the tunnel, additional field PLTs on rock samples taken from the tunnel and regular pullout tests on rock bolts in the tunnel walls and crown as tunnel excavation progressed.

INITIAL TUNNEL GROUND SUPPORT

The design of the initial ground support for the tunnel is briefly described below. During construction very little modifications were made to the initial design. One significant change during construction was the deletion of steel fibers from the shotcrete for Support Types 1 to 3. Variations in the length of the support types from the original design estimates occurred for along the tunnel outside the jail section.

Support Type 1—3.6m long 21mm dia, M24 thread, 28mm hole, 310kN ultimate capacity fully resin encapsulated rock bolts on a 1.5m grid, 50mm thickness shotcrete.
To be used in "good rock" where >80% of rock over the arch is high strength.

Support Type 2—as above but 1.2m grid and with 100mm shotcrete.
To be used in "fair rock" where >50% of rock over arch is high strength.

Support Type 3—as above but 150mm shotcrete.
To be used in "poor rock" where > 80% medium strength rock over tunnel arch and UCS > 10MPa.

Table 1. Initial tunnel support—predicted and actual

Support Type	% Predicted	% Actual
1	20	5
2	16	47
3	5	0
4	32	21
5	27	27

Support Type 4—Triangular profile lattice girder at 1.2m spacing (170mm deep section) with 300mm shotcrete.
To be used in "very poor rock" where UCS less than 10MPa.
Support Type 5—140mm dia Alwag canopy tubes at 500mm spacing (27 no.) over tunnel arch.
The canopy tubes are 12m long with 3m over lap, plus triangular lattice girder 170mm deep (with 350mm shotcrete). Lower cover tunneling under Boggo Road Jail. (The tunnel width was kept constant, i.e., the canopy tubes were only angled up vertically not horizontally).

The estimate of support for each type and the actual percentage used is given in Table 1.

During construction Type 5 support under the jail was not modified. Type 4 had the shotcrete thickness reduced to 200mm. Shotcrete for Types 1 to 3 support was to have steel fibres at $45kg/m^3$, however, this was taken out very early in construction. Type 2 modified was used for the majority of the length of tunnel beyond the jail. The fan niches have the same arch profile as the standard tunnel profile therefore the same arch support. The fan niches are 19m long including the 10% graded transitions required for efficient operation of the jet fans.

Under the jail the design intent was to provide initial tunnel support that would allow the minimum amount of ground relaxation. This was achieved by the forward installation of canopy tubes and fibre glass dowels and by the early application of the 350mm shotcrete over the arch. Early strength of the shotcrete was critical to the success of this approach. The following table has been taken from the shotcrete specification and gives the strength requirement of the shotcrete from as early as 3 hours after application together with the test used to measure the shotcrete strength (see Table 2).

One of the options considered for the final lining during design development was to use a combined lattice girder and shotcrete lining and not have an in-situ concrete lining. There were potential cost savings and productivity gains with this option. Shotcrete shadow trials on test panels with sections of lattice girders did not prove satisfactory because it could not be demonstrated that the reinforcement would be fully embedded in shotcrete. This is potentially a long term durability issue.

DISPLACEMENT MONITORING

The monitoring instrumentation program for the tunnel was quite comprehensive and included the following:

1. Surface settlement readings of the building and ground surface was completed on predetermined gridlines to 15m either side of the tunnel centreline. General a 5m by 5m grid on open ground. Survey leveling points were anchored into the ground on 1m long rods or attached to the building walls.

Table 2. Type 5 support, minimum shotcrete strengths

At Time	Minimum Strength	Test Applied
3 hours	1 MPa	Initial Strength - Meyco Penetrometer
12 hours	6 MPa	Sprayed Beam Compression (ASTMC116)
24 hours	18 MPa	core compressive strength (AS1012)
3 days	26 MPa	ditto above
7 days	35 MPa	ditto above
28 days	40 MPa	ditto above

2. Settlement monitoring within the tunnel was completed on the side walls at springline level and in the tunnel crown (with optical targets) at specified spacing and installed within a minimum distance from the tunnel face.
3. Convergence tape readings were completed in the tunnel at specified spacing and installed within a minimum specified distance of the tunnel face measuring across the tunnel and to the tunnel crown.
4. Additional building settlement monitoring using electronic tilt beams.

Standard survey leveling would be accurate to ±1mm or better while optical targets in the tunnel crown would be accurate to approximately ±2mm. Convergence tape readings were considered the most repeatable and are probably accurate to ±0.1mm. The electronic beams, each 2m long each, and which were attached as continuous strings to the "rigid" building structures at footing level showed very little relative cumulative movement. The movements were so small it would be difficult to separate out instrument error from actual tilt movement.

The frequency of readings was generally daily under the jail and within two tunnel diameters of the tunnel face. Beyond the jail the frequency of readings was reduced and latter also allowed a reduced level of monitoring installation.

Pre construction and post construction dilapidation surveys of the jail structures were carried out by an independent consultant engage by the Alliance. Existing cracks were monitored visually with tell tales or remotely with electronic displacement gauges. During construction regular inspections of the buildings above the tunnel alignment were carried out by both representatives of design and construction teams.

HERITAGE BUILDING ANALYSIS

The predicted settlement and settlement trough profile and width were used in subsequent 3D FE analyses of the north and south perimeter walls and of one of the three storey high Cell Block buildings.

The FE modeling of the building used solid 8-node hexahedron "brick" elements for the internal and external walls, slabs and strip footings. The structural analysis FE program STRAND7 was used to analyze potential settlement damage to the buildings. The loading for the model was from the building self weight. The foot print size of the building measures 17.5m (across the tunnel alignment) and 12.5m (along the tunnel alignment).

The building is constructed of mainly solid brick work (up to 500mm thickness) with unreinforced concrete lintel beams above and below openings in walls for windows and doorways with very shallow depth unreinforced concrete strip footings and two levels of reinforced concrete floor slabs.

From the 2D tunnel FE analysis and the location of the building with respect to the tunnel centerline settlements at the building corners were predicted to be 0.5mm, 3mm, 2.5mm and 6mm.

Figure 3. Tunnel alignment under the Boggo Road Jail

The building FE model ground stiff stiffness was developed using these displacements, where all four corners of the building are supported on springs and the spring values were adjusted by trial and error to achieve the prescribed displacements at the building corners. The span sections of the footings/external walls were unsupported, which provides an assume representation of the ground structure interaction after the tunnel excavation has traversed past the building.

The building is displaced differentially due to the differentially settlements at the corners and this causes the building to deflect, warp and twist. The result of these effects is to cause the stresses to develop on the surfaces and at the centroid of the "brick" elements.

The flexural tensile stress limits adopted (using AS3700) were 0.2MPa for the brick/mortar bond. For unreinforced concrete the stress limit taken was 2.68MPa.

As expected, maximum stresses developed at mid span locations and were in the range along the strip footings for example of 2MPa to 5MPa. Stress concentrations occurred around opens in walls, for example at the front of the building around the main doorway openings, surface stresses were in the range of 0.13MPa to 1.36MPa.

The conclusion from these analyses was that some minor cracking could occur. The issue becomes will the cracking be evenly spread or be concentrated at limited locations, in which case the cracks would be clearly visible. While grouting of wide cracks is possible it is not a desirable outcome for heritage listed buildings. The final recommended remedial measures agreed to was to recover all settlement after the tunnel had traversed through the site. The effected building element or perimeter wall section was to be underpinned and then jacked back into its original position and in the process close any cracks that had developed. This contingency measure was not required. Figure 3 is an aerial of the Boggo Road Jail with the tunnel alignment indicated by the shading.

SURFACE SETTLEMENT

Previous experience on other busway tunnels in the Brisbane area had demonstrated that surface settlements in the range of 6mm to 20mm could be achieved. However, in two cases, including Buranda referred to in more detail below, the designer and constructor had chosen to first construct small side drift tunnels with the purpose of providing a stiff bearing/footing for the tunnel arch. At the Vulture Street Busway tunnel these side drifts were formed by 3m by 3m excavations along both sides of the tunnel which were subsequently partially backfilled with mass concrete.

In another case at the 22m wide Y-junction at Vulture Street, with 10m of ground cover recorded a surface settlement of 10mm with a preinstalled surface grid pattern of cables, lattice girders and a 500mm shotcrete arch thickness.

Through our alliance partner, the Construction Team were able to provide all of the settlement data for the Buranda Busway tunnel which was completed in 1999. The tunnel was excavated under an operating railway line and its initial excavation span was 19m, reducing to 14m and with just 3m of ground cover. The maximum settlement recorded on this site just over 20mm. The method of tunnel support was canopy tubes, steel lattice girders and shotcrete. The lattice girders were considerably heavier than those used at Boggo while the shotcrete was nominally 100mm thinner.

The criteria adopted for the heritage listed buildings was a maximum settlement of 10mm with a maximum surface slope of 1:1000.

Considerable effort was made in reassessing the Baranda results with the purpose of developing a model with known settlement results and then using this model at Boggo. This comparison included for both sites the geological model, the construction methodology and the rock properties (predominately Point Load Test data).

This approach was partially successful, the main concern was that the finite element model consistently underestimated the actual surface settlement at Baranda. Without knowing the actual details of construction at Buranda (i.e., in particular the timing of support installation) it was difficult to know the reason for this discrepancy.

This is a fundamental problem with theoretical analysis which from our experience has to be over laid with considerable experience and judgment. The experience part is understanding tunnel construction and the relative contribution of support types to and their timing of installation whether they be forward steel canopy tubes, fiberglass face dowels, lattice girders or shotcrete. Without this knowledge theoretical analyses can be very miss leading.

The differences and later confirmed at Boggo was probably the actual stiffer (i.e., thicker shotcrete) and very earlier installation of ground support installation, particularly the shotcrete and its early setting time relative to the excavation cycle advance rate. The full 350mm initial shotcrete thickness over the tunnel crown was achieved very close to the tunnel face at Boggo and combined with pre-installed canopy tubes, fibre glass face dowels and arch footing (local widening where required) were the main influencing factors controlling the magnitude of surface settlement. This would explain the very close comparison between the 2D finite element model used and the actual settlement results. The finite element model (developed using Phases2D) gave predictions of around 7mm with instantaneous support installation. The predicted settlement was 10mm and the actual settlement along the centre line of the tunnel under the jail ranged between 7mm and 12mm. The difference between the upper and lower values can be attributed to the variance in the geological profile.

DESIGN AND CONSTRUCTION EXPECTATIONS

The following items summarize some of the significant aspects of the tunnel design and construction approach adopted for this project and served as a good reminder to this approach to both the construction and design teams.

1. That the Construction Team will construct the tunnel according to the specifications and drawings. In particular following the construction sequences as noted on the drawings for Type 5 support (canopy tubes, lattice girders, shotcrete and face dowels) under the jail and Types 1 to 4 further along the tunnel.
2. The Permit to Excavate (PtE) procedure adopted on this tunnel site will be effective because the participants believe in the process (the PtE process is that the builder, designer and geotechnical engineer review the previous days construction/monitoring/geological data and then agree on the tunnel support and construction methodology to be adopted over the next 24 hours).
3. Settlement predictions made under the jail by the Design Team (around 10mm) rely on a high level of construction supervision to ensure that the design drawings are strictly adhered to.
4. That the buildings of the Boggo Road Jail are sensitive to surface settlement. The design and construction methodology have been developed together to minimise any adverse impact on these buildings.
5. The geology is extremely variable. However, the ground will at worst behave in the tunnel face as a very weak rock and not as a soil.
6. There are no significant ground water issues. Firstly, there will be no impact on settlement above the tunnel due to draw down of a water table and secondly that ground water will not be a significant factor influencing the stability of the tunnel face during construction.
7. There are 5 initial ground support types. There is sufficient flexibility within and between these support types to handle all of the expected ground conditions. The possibility of installing 12m long face nails is an option for all of the support types. Full face and half face excavation sequencing are also feasible when required.
8. That fibre glass dowels at the tunnel face are a more robust method of ensuring face stability than shotcrete alone.
9. The geological model can be improved as tunneling progresses. This will require ongoing face mapping and rock strength classification supplemented by the interpretation of drilling rates (when available) from canopy tube and fiber glass face dowel installations. Supplementary probe drilling will only be carried out if these methods prove inadequate.
10. The final lining will consist of an in-situ steel reinforced concrete arch with a waterproofing membrane. Most ground movement will have already occurred prior to installation of the final concrete lining.

CONSTRUCTION

The tunnel excavation was carried out using an ATM105 road header in two stages (full heading and benching).

Excavation of the tunnel under the Boggo Road Jail was the most challenging design and construction aspect of the project. Despite a slower than expected startup, the 120m long canopy tube section was completed on budget and time. Rates of 10 × 12m long tubes drilled and installed per 10 hour working shift were achieved towards the end of this section of the tunnel. Canopy tubes were installed with a Tamrock Minimatic

Figure 4. Tamrock Minimatic two boom jumbo drill in action

two boom jumbo drill. Each tube consists of 4 × 3m long sections. The initial 3m section has a sacrificial drill bit. Alignment accuracy was very good. This good accuracy results in a neat excavation profile below the canopy tube array and subsequent installation of the lattice girders and shotcrete becomes more efficient in both time and in reduced material wastage. Figure 4 shows the jumbo drill during installation of the canopy tubes.

For the final arch concrete lining reinforcement installation an innovative approach was adopted. Self supporting lattice girders were installed at 1800mm centers prior other reinforcement. This approach negated the need to penetrate the waterproofing membrane to support the reinforcement. The arch formwork length was 9m.

ACKNOWLEDGMENTS

For permission to publish this paper and creating a great working environment we thank and acknowledge our Alliance partners, the Department Main Roads and Queensland Transport, Thiess and finally our work colleagues at Sinclair Knight Merz.

CONCLUSIONS

1. The Alliance approach has proved to be a successful delivery method for the driven tunnel and for the Boggo Road Busway Project as a whole.
2. During the initial design phase considerable effort, over a 6 month period, was devoted to developing the complex geological model and in estimating surface settlement under the Boggo Road Jail. This effort has proven to be particularly worthwhile, firstly, to provide design and construction validity to the Alliance design team and management and secondly, as a demonstration to outside stakeholders of the Alliance's commitment to satisfying their expectations.
3. The daily PtE process helped develop a stronger relationship between the Construction and Design Teams. This good relationship led to open and free discussion and to the adoption of a number of innovations during the construction of the tunnel.
4. Ground behavior during tunneling was as predicted with no damage to the heritage listed buildings above the tunnel.
5. The combination of the Alliance approach and the PtE process has delivered the driven tunnel ahead of program and under budget.

LOOSENING AND FACE STABILITY WITH SHALLOW OVERBURDEN IN THE "SITINA TUNNEL," BRATISLAVA, SLOVAKIA

Chikaosa Tanimoto ▪ Taisei Corp.

Kimikazu Tsusaka ▪ Japan Atomic Energy Agency

Toshihiko Aoki ▪ Taisei Corp.

Masahiro Iwano ▪ Taisei Corp.

INTRODUCTION

Bratislava is located at the intersection of important traffic routes, including motorways, roads, railways, air traffic, trading, and inland waterway with an important harbour. The construction of the D2 Motorway Bratislava, Lamacska cesta Stare grunty was the last missing section of D2 Motorway and its completion linked the Czech Republic through Bratislava to the Hungarian Republic. Traffic was obliged until its completion to use the Lamacska Road and Mlynska Dolina, which were the most heavily trafficked roads in the city. The section of D2 Motorway to be completed in 2006 was 3,300m in length, of which the SITINA Tunnel was approximately half the length. After putting into operation the transit traffic (motorway) and urban traffic was separated. Transit traffic by passing the town was eased. The junctions D2 with Lamacska Road (Harmincova) and Mlynska Dolina also allowed better use of the motorway for the urban traffic.

The tunnel construction was specified by the features with the highly weathered and thermally altered granitic rock formation with the overburden of Quaternary sediment of 2–4 m thick, very thin overburden, severely controlled deformation limit and environmental protection. The tunnel had to be excavated beneath and nearby the academic complex area and the zoo, where sensitive facilities exist and animals live in. The tunnel consisted of two tunnels (Tunnel A and B) with the length of 1,415 and 1,440 m, respectively. The overburden thickness was in the range from 10–30 m. The standard cross-section and the one with invert arch were 79.5–81.5 m^2 and 92.1–93.3 m^2. The excavation took 21 months from September 2003 to May 2005.

The construction was based on the concept of NATM. The SITINA Tunnel was the third application of NATM in Slovakia, following the BRANKISKO Tunnel (commenced in July 2003) as the first and the HORELICA Tunnel (in October 2004) as the second NATM practice. To protect the surrounding structures and citizens from tunneling operations, seismic monitoring was performed in association with blasting. The allowable limit of blasting vibration was fixed to be under the particle velocity of 15 mm/sec. The tunnel excavation cleared this limit as expected before construction.

In driving through the difficult geological situation the excavation work hit 49 faults, and some of them occasionally caused serious face instability. In principle, the rock formation has been affected by the high hydrothermal effect and the resultant clay filling along faults and joints. The auxiliary reinforcement with the 4–8 m long pipe roofing was needed to protect the vicinity of mining face area. This operation extraordinarily required additional works and time which had not been expected before construction.

The characteristic manner of deformation in the vertically jointed rock formation and quantifying the loosening were presented as well as the lessons obtained in NATM.

SUMMARY OF SITINA TUNNEL PROJECT

Geological Conditions

The SITINA Tunnel was excavated through the granitic formation of Sitina Hill found at the end of the Low Carpathians, at the contact with the Lamacsky Pass. From lithological aspect, the structure of the Sitina formation was simple. The surface was covered by anthropogenic fill, with a continuous layer of Quaternary sediments. The thickness of this layer usually varied from 2 to 4 m. The surface of this granitic formation was heavily weathered. The granitic formation consisted of muscovite-biotite granite and granodiorite with frequent occurrence of pegmatite veins, and interbeds of crystalline schist – biotite paragneiss. Regarding the physical condition, the rock was highly heterogeneous, ranging from relatively sound rock to intensively faulted and deeply weathered interbeds of semi-rock to the ground. The overburden of both tunnels was very shallow in the entire tunnel length with maximum value about 30 m and local minimum about 10 m. The most significant feature of tunnel construction was the inhomogeneity of the rock mass resulting from various faulting. The most representative appearance of faulting of the formation was in the existence of irregular zones with varying thickness amounting to several meters. The rock mass in the vicinity of the faults were significantly altered, locally intensively mylonized (crystalline schist above all). Occurrence of rock blocks, falling down from the excavating face and roof, was frequent at crossing of geological boundary. The rock mass was dry, and only groundwater of rainfall origin could be encountered, while its presence was restricted to the locations of the weathered rock zone.

Support Classes

The construction of the tunnel started in the summer of 2003, by the erection of temporary tunnel portals in the south in Mlynska Dolina, and in the north in the premises of the Slovak Academy of Sciences (SAV). The excavation of both tunnels started from the southern portal in Mlynska Dolina at the end of 2003. In 2004, an initial 30 m section of Tunnel A was excavated from the northern portal, and Tunnel B excavation started. The tunnel was broken through in February, 2005. Five support classes were specified in the tender documents for the tunnel excavation and support, with the Class V subdivided to Class Va for faulted rock, and Vb for the portal sections. The tender documents divided the rock condition along the center lines of the two tunnels into homogeneous sections with corresponding support classes assigned. The division was based on the results of the geological and hydrogeological investigation, as well as other relevant data, e.g., the overburden depth.

The modified design shows in Figure 1. It left the face division unchanged, i.e., the top heading, bench and invert (if required), but it changed the height of the individual parts. To allow utilization of tunneling equipment, an equal height of the top heading was adopted for all support classes. In addition, the invert structure for support Classes III and IV is considered as an option. The decision is made on the basis of the measured deformation values, and the assessment of particular conditions in the given section. Generally, the invert was placed in the portal sections and sections passing through tectonically disturbed zones. And, it was estimated that about 20% of the length of tunnels would have the invert. According to the modified design, the pre-support system consisted of 4 m long steel piles for Classes III and IV, and 12 m long, 114 mm-diameter micropiles for Classes Va and Vb in sections passing through tectonic disturbances

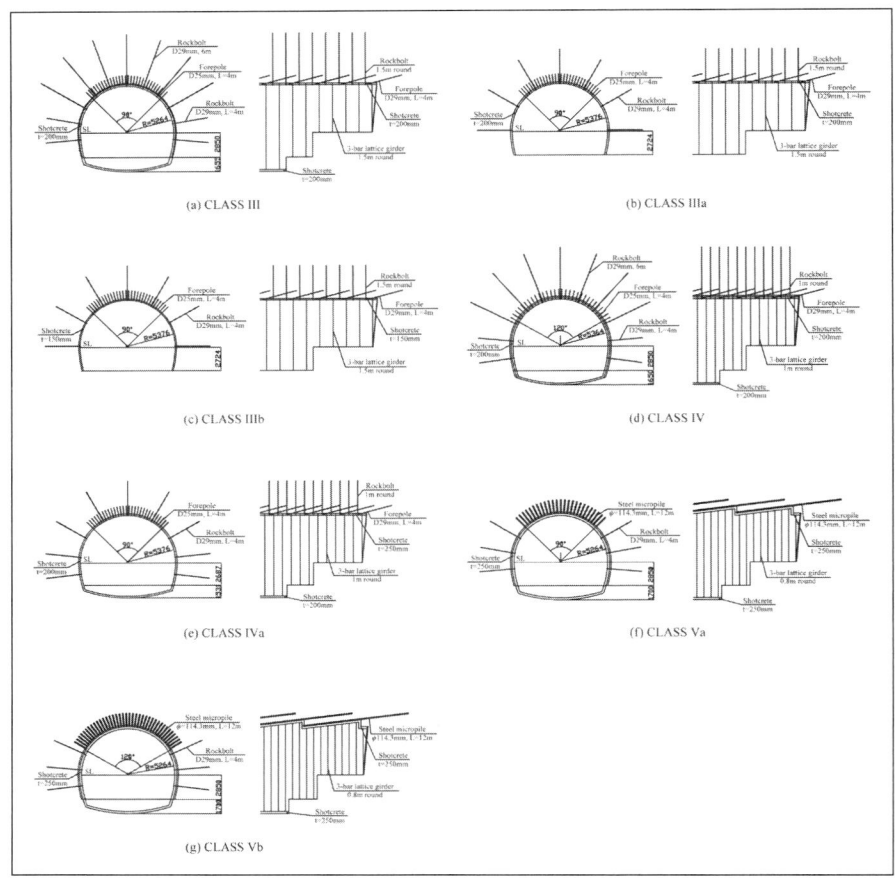

Figure 1. Support classes in SITINA Tunnel

and the portal sections. In addition to shotcrete and support wedges, the face was supported by IBO anchors when needed. The primary shotcrete lining was designed with the thickness ranging from 100 to 250 mm, reinforced with one or two layers of steel mesh and the arcus lattice girders. Another element is the systematic radial anchoring using 4 and 6 m long SN anchors. The round length was supposed to vary from 0.8 to 3.0 m, depending on the classes but in the actual construction the maximum length of a round was 1.5 m. This limit caused additional cost and time. Even the modified design is, however, changed in the progress of excavation, according to the actually encountered geology. The support Classes III and IV with the round lengths of 1.5 and 1.0 m are the most frequent. The two classes were modified during the excavation with respect to the measured deformations of excavated opening; reducing numbers and lengths of the radial rockbolts was supervised by the consultants. The repeated rockfall from the roof and face even in conditions where they had not been expected to be subjected to additional means of investigation, had to be solved by the contractors, with the intensive cares in improving safety and saving time and cost. The solution was found in the systematic reinforcement of the roof and of the excavation face. 8 m long self-drilling rock bolts were employed in addition to 4 m long plies, with grouting ahead of the face.

ESTIMATION OF ROCK MASS BEHAVIOR BASED ON CONVERGENCE MEASUREMENT

Figure 2 shows overburden height, sections of cave-in, driving direction, installed support class, rock type on a face, crown settlement, half value of convergence, rock behavior class and joint frequency on a face in the entire length of both tunnels. In the sections of cave-in No.1, 2, 4, 6–8, heavily jointed granodiorite distributed on the face and it occurred after blasting 1.0–1.5 m round. On the other hand, in the section of cave-in No.3, heavily mylonized granite widely distributed on the face and the excavation work with machine needed to be divided into three stages due to the face instability. And, the section of cave-in No.5, heavily jointed granodiorite distributed on

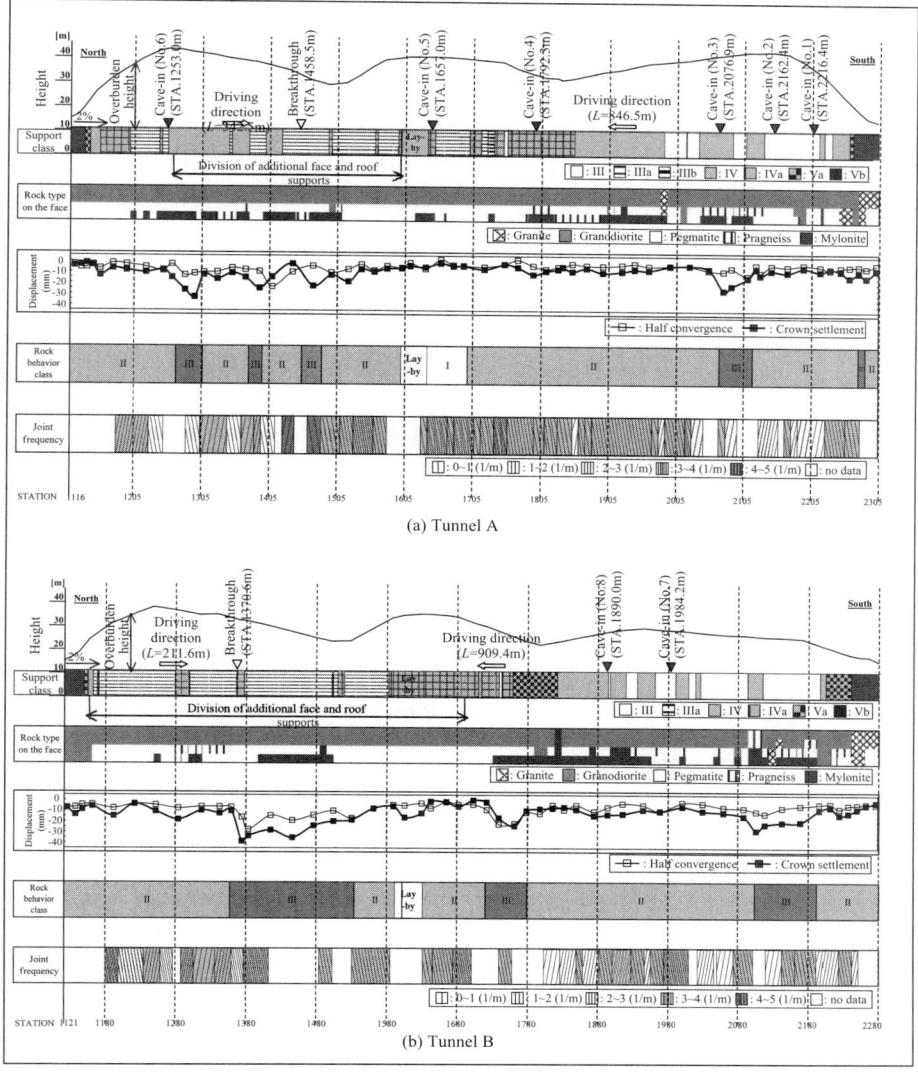

Figure 2. Tunnel profile

(a) Cave-in No. 3

(b) Cave-in No. 5

Photograph 1. Face condition of cave-ins

the face and it occurred after blasting for widening standard 11 m span into 14 m span of a lay-by section. In addition to the support elements in the standard support classes from III to Vb shown in Figure 1, fully grouted (untensioned) rockbolts of 8 m long and 32 mm diameter for face support were installed at 4 m round, giving an overlap of 4 m. Also, fully grouted (untensioned) rockbolts of 8 m long and 32 mm diameter and those of 4 m long and 25 mm diameter were installed alternatively for roof support in the divisions between STA.1,632 m and 1,272.5 m (the total distance is 359.5 m) in Tunnel A and between STA.1,671.6 m and 1,156.4 m (that is 515.2 m) in Tunnel B, so that no significant cave-in occurred in both of the divisions. Face condition in the sections of cave-in No. 3, 5, 6, and 8 are shown in Photograph 1.

One of the authors had categorized hundreds of the convergence curves observed in over 60 projects of 12 m span tunnels into the five classes (Tanimoto et al. 1987; Tanimoto et al. 1988). The classification was expressed by the relationships among the initial deformation rate, the final deformation, the support pressure, and Terzaghi's rock load as shown in Table 1. Also, it was clarified that the outer diameter of inelastic zone from the center of a tunnel was equal to twice the relative distance from the face to the observed section at the final value of convergence by means of a numerical analysis on the assumption of the elastic-perfectly plastic stress-strain relationship. It was expressed by the following equation (Tanimoto et al. 1988).

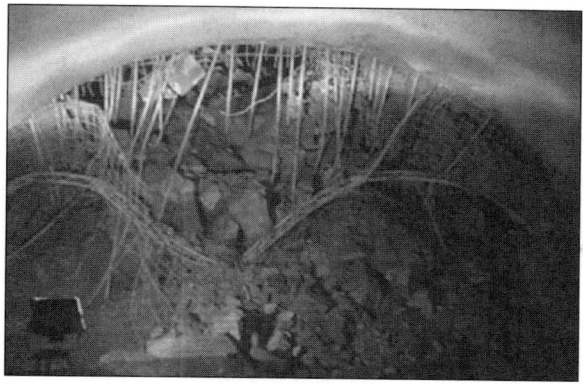

(a) Cave-in No. 6

(b) Cave-in No. 8

Photograph 1.　Face condition of cave-ins (continued)

$$W_p = \frac{L_F/2 - D}{2} \quad (1)$$

where W_p (m) is the width of inelastic zone from tunnel wall, L_F (m) is the relative distance from the face to the observed section at the final value of deformation, and D (m) is the excavated diameter. Also, the initial deformation rate was calculated by the following equation. For example, in the case of 10 m in tunnel diameter, if deformation converges at 20 m in the relative distance from a face, the width of inelastic zone around a tunnel is estimated to be 0 m. Namely, the rock deformation should be subjected to elastic behavior. Therefore, it is possible to distinguish between elastic behavior and inelastic one from the viewpoint of predominant rock deformation caused by tunnel excavation.

$$\frac{dU}{dL} = \frac{0.5 \times \Delta D_{L=0.3D}}{L_{0.3D} - L_0} \times \frac{A_0}{A_1} \quad (2)$$

where dU/dL (mm/m) is the initial deformation rate, $\Delta D_{L=0.3D}$ (m) is the value of convergence after a face advances to 0.3D, $L_{0.3D}$ (m) is the relative distance from the observed section to a face at the observation of $\Delta D_{L=0.3D}$, L_0 (m) is the relative distance

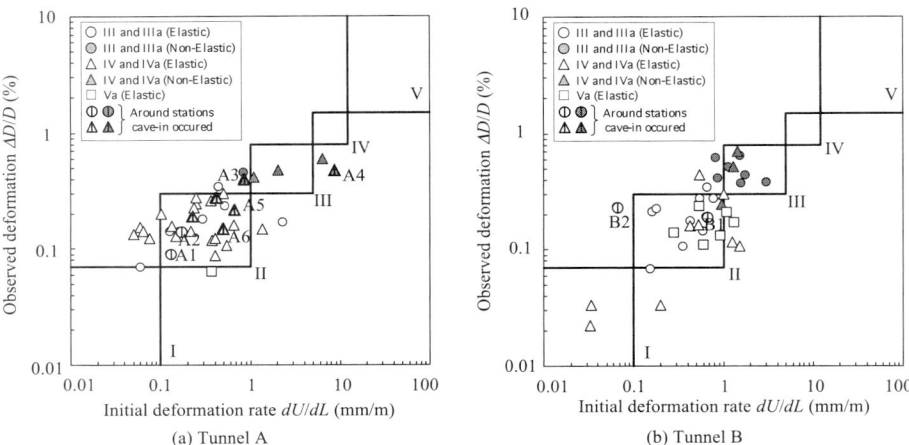

Figure 3. Relationship between initial deformation rate and final deformation in Tunnel A and B

from the observed section to a face at the beginning of the observation (m), A_0 (m²) is the total excavated area of the top heading and bench, and A_1 (m²) is the excavated area of the top headings. In the equation, there are the reasons why the authors focused on the value of deformation after a face advances to 0.3D. First, the round of excavation can be approximately equal to about 3 m in the good rock quality condition. Second, the half-dome action of a cutting face considerably reduces beyond its distance. Third, the additional supports are afforded to install in case that large deformation occurs. Also, excavated diameter means a diameter of a circle whose area is equal to the excavated cross-sectional area of a tunnel. In the SITINA Tunnel, it was the range of 10 to 11 m.

Both convergence and crown settlement were measured in 62 and 54 sections in Tunnel A and B, respectively. In Figure 3, the relationship between the initial deformation rate and the final observed deformation was plotted on the Tanimoto's Classification shown in Table 1. The plotted data in the figure are the results from crown settlement measurements whose initial value was measured within 3 m in the relative distance from a face. Based on Equation (1), white and gray marks mean elastic and inelastic behavior, respectively. The results from the SITINA Tunnel met with the Tanimoto's Classification well, namely the relation between the converged deformation and the initial deformation rate is clear enough to estimate actual support requirement through convergence measure. Also, it can be seen that the inelastic zone develops around tunnel wall when the initial deformation rate beyond 1.00 mm/m was observed. As a result, Class I, II and III on the classification were 5%, 82% and 13% in Tunnel A, and 0%, 70% and 30% in Tunnel B in length to the whole except for lay-by divisions, respectively. Consequently, it was clarified that the extent of loosening zone was less than 1 m in the 70–80% of the total length in both Tunnel A and B.

JOINT ORIENTATION ON A FACE

In every round, rock condition such as rock type, joint orientation, joint frequency and the amount of water inflow were observed in a face observation report and a face photograph was taken by a digital camera at 5–6 m behind a face. By using these materials about a face, joint orientation and frequency were evaluated at 60 and 46 sections in Tunnel A and B, respectively. Generally, the joint frequency can be defined

Table 1. Rock behavior classification on convergence measurement

Class	Support Load	Competence Factor	Initial Deformation Rate (mm/m)	Observed Deformation ΔD/D (%)
I	Slight	over 1.5	less than 0.1	less than 0.07
II	Medium	1.0–1.5	0.1–1	0.07–0.3
III	Heavy	0.75–1.0	1–5	0.3–0.8
IV	Very heavy	0.5–0.75	5–12	0.8–1.5
V	Extremely heavy	less than 0.5	over 12	over 1.5

(Tanimoto et al. 1987)

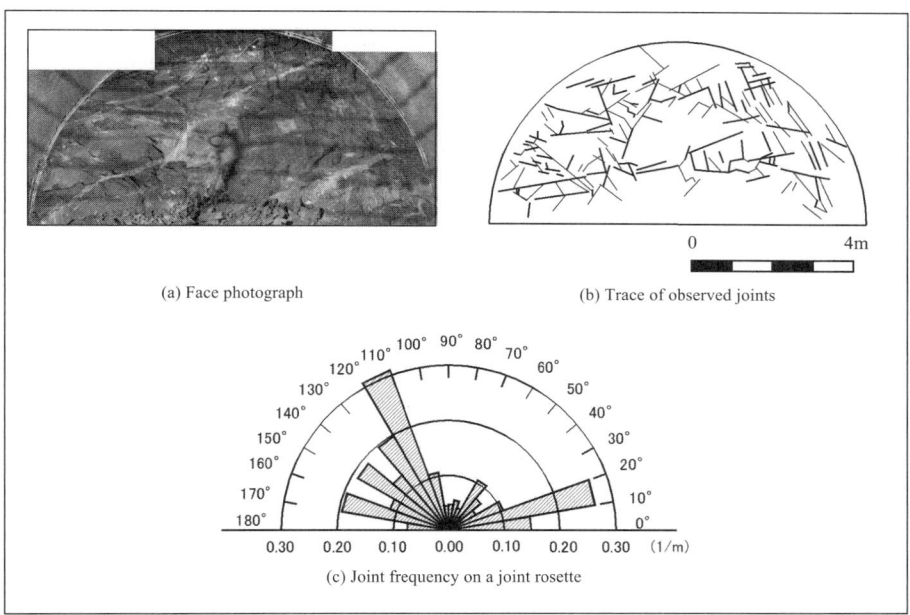

(a) Face photograph

(b) Trace of observed joints

(c) Joint frequency on a joint rosette

Figure 4. Joint distribution at STA. 1801.3 m in Tunnel A

as number joint per liner meter along a scan-line. This definition can be extended to the description in either two or three dimensional manner, namely a total length of joints per unit area [m/m^2] or total area of joint planes per unit volume [m^2/m^3]. Both units are attributed to be joints per meter [1/m]. In this study, it is calculated by dividing the total length of joints into the area of the observed face.

Figure 4 (a) is the face photograph at STA.1,801.3 m in Tunnel A. Figure 4 (b) is obtained by tracing the observed joints on Figure 4 (a) on a PC screen. And, Figure 4 (c) is a joint rosette expressed the respective joint frequency every 10 degrees in counterclockwise from spring line. There are 2 sets with higher frequency in the range of 10–20 degrees and 110–120 degrees, which are named Set I and II, respectively. And, based on a face observation report, the strike and dip of these sets were estimated and joint distribution on a face was distinguished between dip slope and opposite dip. As a result, Set I and II was dip slope and opposite dip, respectively, and the joint distribution on the face at STA.1,801.3 m was evaluated as dip slope since at least one of two joint sets was dip slope. The results from 106 sections in both tunnels were illustrated

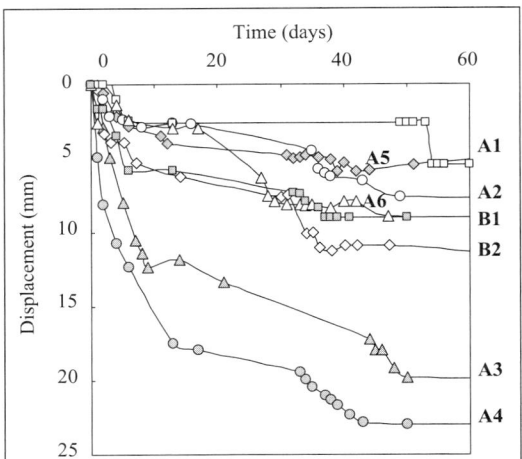

Figure 5. Deformation curves around stations of cave-in

in the bottom band chart as "Joint frequency" in Figure 2. Assuming the result from each face to be the representative condition in the range of 10 m in driving distance, joint frequency on a face is expressed by the extent of density of line and dip slope is colored by gray. In the figure, joint distribution on a face was dip slope in 49 of 60 and 31 of 46 faces in Tunnel A and B, respectively. Therefore, it was concluded that the joint distribution on a face encountered in a large part of the tunnels was one of the predominate causes of face instability.

ALLOWABLE LIMIT OF DEFORMATION IN THE SITINA TUNNEL

Crown settlements observed around sections of cave-in are shown in Figure 5. The results from these curves observed in the sections where cave-in occurred are summarized in Table 2. The initial deformation rate is 0.83 m and 8.17 mm/m in section A3 and A4, respectively. These sections were located at 11 m and 2 m behind cave-in No.3, respectively. Based on Equation 2, the extent of loosening is calculated in the range of 1–2 m and the rock behavior is supposed to be subjected to inelastic behavior. On the other hand, concerning section A1, 2, and 5–8, the rate and the observed deformation are in the range of 0.07–0.65 mm/m and 0.09–0.23%, respectively. According to Equation 2, the extent of loosening is less than 1 m. Based on the results from measurement of crown settlement and situation of cave-in, the rock deformation mainly depends on mylonite material of rock in the section of cave-in No.3. On the other hand, that significantly depends on frequency and orientation of joints in the sections of the other cave-ins.

To allow or not to allow a certain amount of deformation in NATM has raised a long-term discussion among many tunnellers in the world. In the case of driving through earthen (soil) ground, in general, a rather high amount of deformation (up to 1–2% of its original diameter) can be accepted so as to minimize support load, but in rock tunneling, the amount of allowable deformation is quite limited to be as less as possible so as to maintain the potential bearing capacity of rock mass (usually in the range of 0.2 ~ 0.5% of deformation to be defined by $\Delta D/D$; where D is tunnel diameter, and ΔD is convergence). It is a key issue to avoid loosening in rock at anytime. However, it is also difficult to prevent loosening from the excavation process. The quality of a tunnel highly depends on the control of loosening.

Table 2. Results of observed deformation around stations of cave-in

Number of Curve	Observed Station (m)	Support Class	Initial Deformation Rate (mm/m)	Observed Deformation (%)	Cave-in Nearby	Magnitude of Cave-in (m³)	Relative Distance from Observed Station (m)
A1	2193.89	III	0.13	0.09	1	11.4	20
A2	2145.89	III	0.17	0.14	2	31.8	16
A3	2088.64	III	0.83	0.4	3	84.0	11
A4	2074.90	IV	8.17	0.47	3	84.0	2
A5	1796.34	IVa	0.65	0.22	4	7.9	4
A6	1245.49	IV	0.48	0.15	6	47.0	2
B1	1969.24	III	0.65	0.19	7	17.5	15
B2	1916.63	III	0.07	0.23	8	48.5	26

In the past, many laboratory tests have been carried out on rock mass with joints (jointed rock) particularly under the constant confining pressure, namely in the state of ordinary triaxial condition, but there are quite few of the examples which are subjected to constant dilatancy (allowance of joint aperture). Figure 6 shows some experimental results by Tanimoto and his colleagues through the direct shear test under constant dilatancy. The higher the roughness of rock joint plane is, the more the shear strength is generated along a joint. Contrarily, when large deformation has been accepted, the loss of potential bearing capacity of rock itself should be amazingly high. Its magnitude depends on the degree of roughness on joint surface (plane). Experimental results suggest that, when allowable dilatancy is less than 0.5 mm per joint for slightly smooth joint, more than 50% of the ultimate shear strength can be mobilized at shearing, but more than 70–80% of the potential strength has been lost for allowing 0.5–1.0 mm dilatancy per joint. (In the case of Roughness Ms = 0.083 in Figure 6 (a)) For moderately jointed rock condition with medium roughness, the magnitude of lost strength is more obvious, namely 60–70% and more than 85–90% for allowing dilatancies of 0.5 mm and 1.0 mm (per joint), respectively (In the case of Roughness Ms = 0.137 in Figure 6 (b)).

Concerning "loosening" and "confinement," the scientific explanation to them are still rather complicate and has not been clarified yet, but through experience in the past it is extremely essential to minimize deformation and loosening at anytime in tunnel driving. There is no necessary deformation, but only "unavoidable" one. How much is unavoidable? That depends on individual conditions. Based on the experimental results in Figure 6, sudden loss of mobilized shear stress in rock with small extent of deformation induces large-scale rock behavior such as cave-in because rock behavior in tunneling through jointed rock is strongly subjected to shear deformation and dilatancy on joints.

In the case of the SITINA Tunnel, the allowable limit of deformation is considered to be quite small because of the nature of heavily faulted and weathered granitic rock. The range of the observed deformation (crown settlement) around the sections of cave-in except for No.3 was 0.09–0.23% as shown in Table 2. Also, the maximum deformation observed in the sections where the rock behavior is supposed to be elastic deformation (crown settlement) was 0.39% as shown in Figure 3. Therefore, the allowable limit of deformation would be in the range of 0.3–0.4% in tunnel diameter.

CONCLUSIONS

The SITINA Tunnel was constructed as the third application of NATM in Slovakia. It consisted of two tunnels with the lengths of 1,415 m and 1,440 m, respectively (Tunnel A and B). It was excavated for a double-track motorway tunnel and the cross-sections

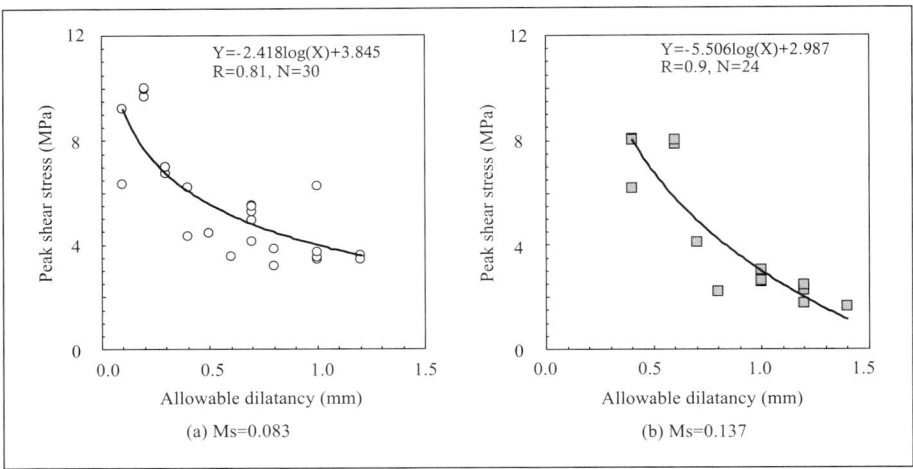

Figure 6. Relationship between peak shear strength and allowable dilatancy

varied in the range of 79–94 m^2. The overburden of both tunnels was very thin along the entire length, with the thickness from 10 to 30 m. The tunnels were driven through the highly weathered and thermally altered granitic rock formation. The excavation work encountered 49 faults, and some of them occasionally caused serious face instability. The rock masses in vicinity of the faults were significantly altered, locally and intensively mylonized. A happening of rock block spalling off from both a face and roof was frequent. And, the remarkable cave-in occurred six times in Tunnel A and twice in Tunnel B. The authors analyzed the joint orientation and frequency on a face by using a face observational report and digital photographs of a face. Also, allowable limit of deformation was estimated through convergence curves based on concept that sudden loss of mobilized shear stress in rock with small extent of deformation induces large-scale rock behavior such as cave-in. As a result, it was concluded that the predominant cause of face instability was considered as the joint distribution on a face. And, based on the measurement of crown settlement observed in the vicinity of the sections where cave-in occurred, allowable limit of deformation could be in the range of 0.3–0.4% in tunnel diameter in the SITINA Tunnel.

REFERENCES

Tanimoto, C., Hata, S., Fujiwara, T., Yoshioka, H., and Michihiro, K. 1987. Relationship between deformation and support pressure in tunnelling through overstressed rock, Proceedings of the 6th International Congress on Rock Mechanics, 1271–1274, Balkema, Rotterdam.

Tanimoto, C., Fujiwara, T., Yoshioka, H., Hata, K. and Michihiro, K. 1988. Determination of rock mass strength through convergence measurements in tunneling, Proceedings of the 2nd International Symposium on Field Measurements in Geomechanics, 1069–1078, Balkema, Rotterdam.

INNOVATIVE NATM—DESIGN FOR A LARGE SHALLOW CAVERN AT STANFORD

Thomas Marcher ▪ ILF Consulting Engineers

Max John ▪ ILF Consulting Engineers

Steffen Matthei ▪ ILF Consulting Engineers

Zuzana Skovajsova ▪ ILF Consulting Engineers

ABSTRACT

Stanford Linear Accelerator Center (SLAC) is located in Menlo Park, CA. In order to allow for enhanced experiments, the current facilities required expansion. Tunnels were excavated through very weak sedimentary rock interspersed with uncemented zones (Ladera Sandstone) employing the principles of the NATM. The overburden varied from 10 to 80 feet.

The paper addresses the characterization of Ladera Sandstone that can exhibit the properties of dense sand and weak rock. The paper focuses on the challenges that were encountered during the NATM initial support design and the construction phase including the geotechnical instrumentation, monitoring and design verification process in particular the excavation of the junction between access tunnel and the FEH cavern (49 ft wide by 32 ft high).

PROJECT DESCRIPTION

The Stanford Linear Accelerator Center (SLAC) is a national research facility operated by Stanford University for the United States Department of Energy (USDOE). SLAC is responsible for construction of the LINAC Coherent Light Source (LCLS) Project which will involve the projection of a new type of X-Ray through a series of tunnels and underground chambers. The project is located at the Stanford Linear Accelerator Center in Menlo Park California, as shown in Figure 1. The new facility includes a series of tunnels and underground caverns about 2,200 feet in length which extends to the east of the existing research facility connected with a new Beam Transport Hall, which is a box structure constructed across the existing Research Yard.

Underground structures consist of the Undulator Tunnel, Access Tunnel, X-Ray Tunnel and FEH Cavern. The three tunnels have a horseshoe shaped cross section approximately 20 feet high and 21 feet wide. The cavern is also horseshoe shaped with a cross section of 32 feet high and 49 feet wide. During construction an additional temporary Access Tunnel to X-Ray Tunnel was added.

DESIGN

Contractual Design

The contractual tunnel design was delivered by Jacobs Engineers (Halim et al., 2007). During the contractual design, NATM was selected as the most appropriate

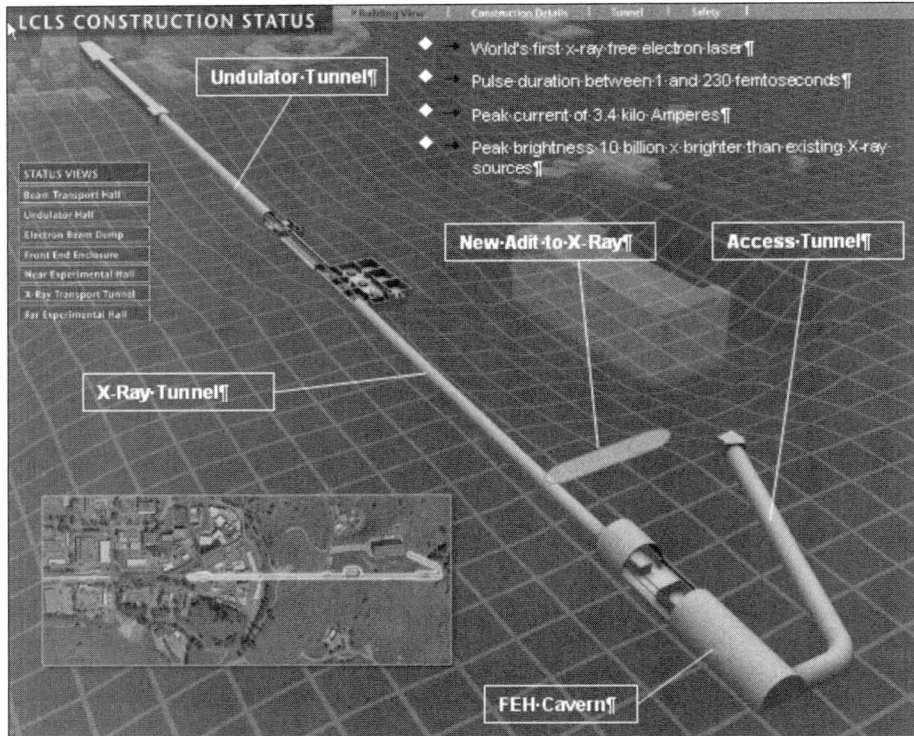

Figure 1. Layout of the LCLS construction site

excavation and support approach. Using NATM a balance is sought between preserving the inherent strength of rock mass while allowing controlled deformations in order to release high stresses at the boundary of the opening. The support resistance of the surrounding ground mass is achieved by using an initial support system. The initial shotcrete lining is flexible and allows for limited rock deformations until equilibrium is reached. In order to prevent loss of strength or high surface settlements, excessive deformation has to be avoided.

The tunnel cross sections and dimensions are based on SLAC's technical requirements. All tunnels and the cavern were designed using a shotcrete lining consisting of an initial support followed by a final shotcrete lining. The initial support system consisted of a combination of synthetic fiber reinforced shotcrete (SFR) and lattice girders. For the cavern, permanent rock dowels were added in the design. The tunnels and cavern were excavated using multiple headings. In areas where there is shallow rock cover, forepoling, with either grouted pipe arches or canopy tubes, was planned. The design criteria and specifications limited the excavation width and height to a maximum of 18 ft. The final support incorporates the initial support as described above and additional plain shotcrete including welded wire mesh reinforcement. Overall thickness of the composite lining is 12 inches for tunnels and 15 inches for the cavern. The reinforced cast-in-place invert slab is connected to the tunnel lining. The construction sequence of the excavation from the Access Tunnel into the FEH Cavern was not addressed in the original design.

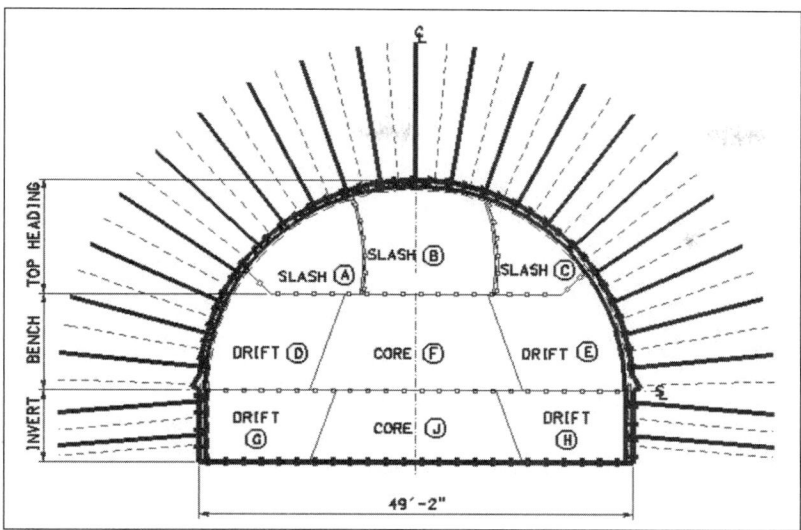

Figure 2. FEH Cavern

Construction Design

Tunnel excavation and ground support work was carried out by Affholder, Inc. under the General Contract by Turner Construction. ILF Consultants was awarded the detail design by the Tunnel Contractor.

It was ILF's task to provide a solution for constructing the intersection that was in accordance with the client's original design. For the construction stages, the excavation sequences were adjusted to accommodate the equipment used by the Tunnel Contractor (Affholder Inc.). These adjustments were minor and had little impact on construction of the three tunnels. However, the elevation of the bench excavation had to be raised to enable the equipment to reach the crown.

By raising the top heading invert to accommodate the maximum reach of available equipment, the cross section had to be subdivided in three stages: top heading, bench and invert resulting in nine headings (Figure 2). Three slashes in top heading were chosen to accommodate the maximum width limitation, three drifts in bench and three in the invert, all with a maximum height limitation of 13 feet.

The excavation from the Access Tunnel into the FEH Cavern was technically complex because the Access Tunnel top heading excavation was lower than FEH Cavern top heading. After discussion with the owner and his designer, it was decided that the excavation sequence for the intersection of the Access Tunnel and FEH Cavern was the contractor's responsibility. To simplify the excavation process a temporary ramp was employed for the connection to the FEH Cavern top heading as shown in Figure 3, to facilitate the Contractor's equipment, an Alpine Roadheader-AM 50.

Another challenge was the need of different support systems for the two geological scenarios as specified in the GBR (2006) which range from medium cemented Sandstone (best scenario) to uncemented Sandstone (worst case scenario). Also, the alignment at the Access Tunnel/ FEH Cavern intersection consists of a horizontal curve which added complexity to the excavation scheme.

The ramp of the Access Tunnel starts 25 feet outside the cavern wall and the cross section was uniformly widened as it was advanced. In order to keep the excavation at a minimum, the ramp was sloped at the maximum grade of 25%. For tunnel excavations

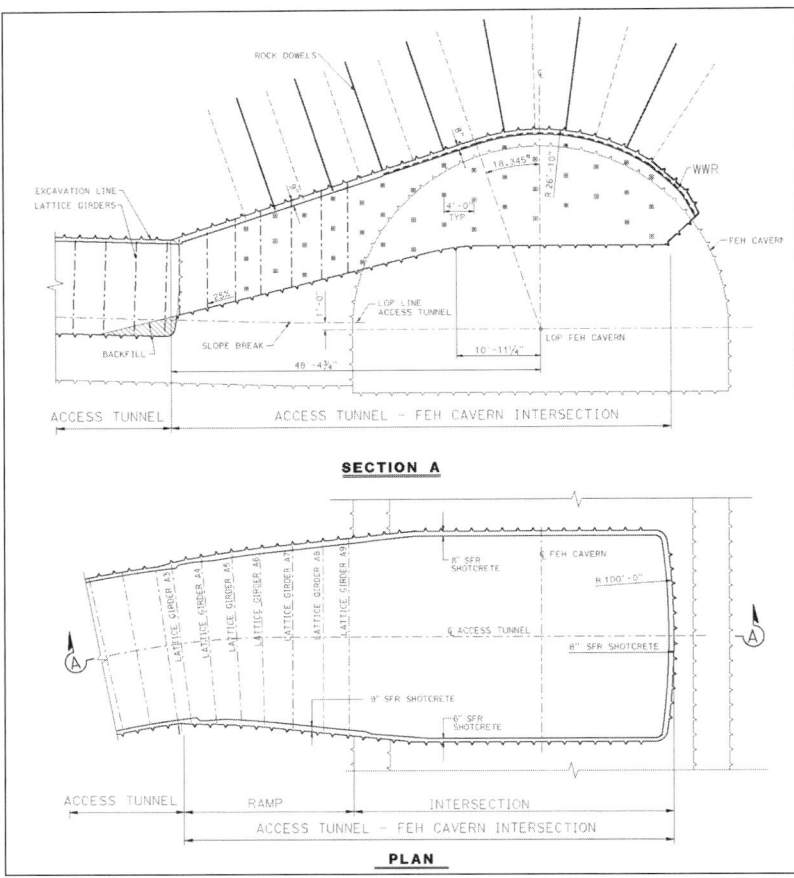

Figure 3. Access tunnel—FEH Cavern intersection initial lining

in standard ground conditions, the initial support consisted of synthetic fiber reinforced shotcrete, lattice girders spaced according to the advance length and fully grouted galvanized rock dowels. Top heading excavation consisted of 4-feet advance increments. The cross section gradually changed from horseshoe shape to almost rectangular with slightly curved walls and ceiling. The reason for the rectangular cross section geometry was to minimize the disturbance to the surrounding ground caused by temporary dowel installations and over-excavation of the cavern standard cross section. The flat roof allowed for permanent, fully grouted galvanized rock dowels to be employed during excavation. Fiberglass dowels were installed in the side walls to stabilize the opening and to provide face stability for the FEH Cavern excavation. Steel lattice girders were omitted at the intersection since they would have to be removed at the sidewalls in any case. The nearly flat roof required additional reinforcement measures to account for the bending moments at the tunnel crown.

The maximum width of the cavern intersection section is 28.5 feet and the maximum height is almost 16 feet. Because the intersection is essentially T-shaped, the roadheader had to make a 90 degree turn. Therefore, the width of the excavation was governed by the roadheader length. Due to the opening size limitations, the intersection was excavated in two slashes. The advance length determined the timing for the

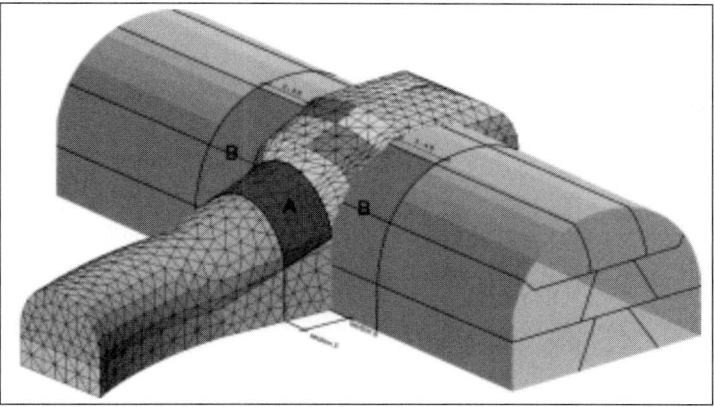

Figure 4. 3-Dimensional view of the intersection

initial lining installation. Loads develop with distance from the face, thus determining the standup time of the free span. The dowels needed to be installed within 24 hours after the excavation was completed. An 8 inch thick shotcrete lining was designed for the intersection, using one layer of wire mesh on the inside surface. Minimal cover of the wire mesh was 2 inches. Because of the use of wire mesh together with fiber reinforced shotcrete, the shotcrete needed to be carefully applied from all sides so as to completely cover the wire mesh and to prevent voids and guarantee full bonding.

GEOLOGY

Two subsurface investigations were carried out in 2003 and 2004. The explorations consisted of drilling, logging, and sampling, conducting in-situ testing, and performing laboratory testing of the cores. In-situ geophysical and pressure-meter testing were conducted in the borings during the 2004 field exploration.

The underground excavation was expected to encounter the Miocene-age Ladera Sandstone over its entire length. The Ladera Sandstone is an intermediate geo-material; a fine to medium grained silty, friable sandstone that includes interbeds of siltstone, with a strength that varies according to the degree of cementation. The sandstone is described using terms like weakly to strongly cemented or uncemented to well cemented. For baseline purposes, the least cemented Ladera Sandstone was described as a very dense granular soil. Bedding thickness was reportedly thin to massive, but bedding planes are not a source of weakness in the rock mass.

Generally, the sandstone near the ground surface is either uncemented or weakly cemented, while the cementation is weak to moderate as depth increases. Cementation in the sandstone is based on a combination of calcium carbonate, silicate and clay. Core recovery in the Ladera Sandstone is typically greater than 80 percent. The sandstone grains are made up of quartz, feldspar and lithic (rock) grains, which are a combination of minerals.

Groundwater at the site lies beneath the tunnel and cavern and the elevation varies from 10 feet to 60 feet below the invert slab.

No major faults have been discovered or mapped on the SLAC site nor have any active faults been located. The nearest of the known active faults, the peninsula section of the San Andreas fault, passes approximately 2.5 miles southwest of the site. The site is located within the Uniform Building Code Zone 4, during a major earthquake on any one of the nearby active faults, the site may experience strong ground shaking.

According to the project design criteria seismic loading has not to be considered for the initial support design, but for the permanent structure.

GROUND CHARACTERIZATION

Intact Strength Characterization

There is no consistent classification system to describe the degree of cementation in sandstone, hence following procedure has been used (GBR, 2006):

- the least cemented Ladera Sandstone will be uncemented and described using soil classification terminology as a very dense granular soil.
- the highly cemented Ladera Sandstone is classified as weak rock using terminology defined by the International Society of Rock Mechanics, (1981).

According to the GBR, (2006) it is expected that weathering and lack of cementation appear randomly throughout the rock mass.

Methodology for Determining Rock Mass Strength

A direct determination of rock mass strength properties on a true scale for tunnel excavation problems is not available. The usual approach is down-scaling the strength of the intact rock taking into account rock mass jointing. The mechanical and hydraulic properties shall be based on data which describe the ground at a scale proportionate to the volume of rock affected by the tunnel structure.

Laboratory tests provide data of the selected rock matrix, without effects of discontinuities and other defects. The characteristics of the rock mass in general can be determined with in-situ field tests; however the properties for the rock mass are to be determined with regard to the scale of discontinuities and heterogeneities related to the tunnel dimensions.

Usually properties for the rock mass are determined by indirect methods. The application of indirect methods considers either an empirical scaling of existing lab- and in-situ data or the use of empirical classification systems or both. Due to the subjective interpretations possible it is necessary to cross check between data obtained from different methods.

The rock mass rating by Bieniawski, Z. T. (1974) provides a method of determination of the properties of the rock mass. The concurrent version used is the RMR_{89} from Bieniawski, Z. T. (1989). The RMR is derived from 5 characteristic rock mass parameters and an adjustment factor dependent on azimuth and dip of the discontinuities. More than 70% of the factors depend on discontinuities and 15% on intact rock properties and 15% on hydro-geological parameter. It has to be noted that these factors do not take into account the state of stress of the rock mass at tunnel depth nor the deformability of intact rock. Hence, this method was considered not to be appropriate for the Ladera Sandstone.

The "Q" Rock Mass Rating System originally proposed by Barton et al., (1974) also considers intact rock and rock mass properties, including RQD (Deere, 1989). Previous experience in Ladera Sandstone indicated that Q-rating is inconsistent with observed ground behavior and is considered non-representative of the conditions to be encountered (GBR, 2006).

The most decisive parameter of the Hoek and Brown Method is the so called Geotechnical Strength Index (GSI). The GSI value is a classification of the rock structure and conditions of discontinuities. The mechanical properties are determined on the basis of an empirical formula by Hoek, E. et al., (2002).

When using the Hoek Brown Criterion (Hoek et al. 2002) difficulties remain in defining corresponding friction angle and cohesive strength for a given rock mass

Table 1. Characteristic strength parameters for Ladera Sandstone

			MN/m²	0.8–1.2
Input		UCS	PSI	xxx
		mi	—	11–13
		GSI	—	40–50
		D	—	0–0.1
Result	Friction Angle	φ	[°]	30–35
	Cohesion	c	[kN/m²] [psf]	25–75 522–1566

strength. Rocscience offers tools to determine corresponding angles of friction and cohesion strengths. The fitting process involves balancing the areas above and below the Mohr-Coulomb strength envelope.

Hoek, E. (1990) and Hoek, E. (1994) discussed the determination of corresponding friction angles and cohesive strengths for various practical situations. These formulas are based upon tangents on the Mohr envelope derived by Bray. Hoek, E. (1990) and Hoek, E. (1994) suggested that the cohesive strength determined by fitting a tangent to the curvilinear Mohr envelope is an upper bound value and may only give optimistic results in stability analyses.

It has to be emphasized that the upper limit of confining stress over which the relationship between the Hoek-Brown and the Mohr-Coulomb criteria is considered, has to be determined for each individual case.

Characteristic Rock Mass Strength of Ladera Sandstone

The properties are defined on the safe side because they are used for limit state analyses. The characteristic rock mass strength parameters are based on input parameters also listed in Table 1.

UCS describes the uniaxial compressive strength of the intact rock. GSI indicates the structural composition of rock mass. The Hoek Brown constant mi depends on the rock lithology. The disturbance factor D depends on the degree of disturbance which is mainly depending on the type of excavation (TBM driven or drill and blast), on the rock mass quality and the level of on-site excavation technology.

Taking into account the depth of the cavern excavation $k0 = 0.5$ can be used to define the confining in-situ pressure.

Deformability Characterization of Ladera Sandstone

There are several methods of determining the stiffness. In order to determine a realistic stiffness modulus of the rock mass the scale of the tunnel opening has to be taken into account. Apart from direct in-situ measurements (i.e., pressure-meter tests, dilatometer, etc.) several methods have been proposed to derive the rock mass deformation modulus. These methods use either intact rock and discontinuity properties or are derived from classification schemes.

The Rocscience Software calculates the deformation modulus of the rock mass using the generalized Hoek-Diederichs formulas (Hoek 2006). At various projects it was found that results using Hoek-Diederichs formulas lead to an overestimation of the rock mass stiffness compared to dilatometer results. Various studies clearly indicate the limits of rock mass estimation methods, i.e., Edelbro, C. et al. (2006) and Romana Ruiz, M. (2002).

Considering the aspects cited in Romana Ruiz, M. (2002), it is proposed to take results of various empirical methods into account and to define characteristic values

Table 2. Worst case strength parameters for Ladera Sandstone

			MN/m²	
		UCS	PSI	<0.8
Input		mi	—	11–13
		GSI	—	15–30
		D	—	0.1–0.2
Result	Friction Angle	φ	[°]	30–35
	Cohesion	c	[kN/m²]	0
			[psf]	0

for the rock mass properties on the basis of a engineering judgment regarding the conclusiveness of results between the various rock types (weaker rock shall have lower parameters).

Hence, characteristic Young's Modulus has been derived based on geotechnical judgment in the range of: $E = 300 - 1000$ MN/m². The Poisson's ratio is chosen to be in the range of 0.15 to 0.3.

Uncemented Zones in Ladera Sandstone

Weathering and/or lack of cementation are anticipated to affect the behavior of the Ladera Sandstone. These "Soil Tunneling Conditions" are encountered in limited areas, which have on previous excavations termed "running sands" (GBR 2006).

The evaluation of characteristic parameters for uncemented Ladera Sand is based on a Soil Classification Terminology for a very dense granular soil. The stiffness modulus is in the range of $E = 100 - 250$ MN/m², the Poisson's ratio is chosen to be in the range of 0.25 to 0.3. The Ladera Sandstone strength parameters for worst case scenario are summarized in Table 2.

GEOTECHNICAL DESIGN

Rock Mass Behavior

Tunnelman's Classification System, developed for classifying tunneling conditions in soil, has been used to describe the ground behavior during previous tunneling at SLAC. The ground behavior during tunnel excavation has been expected to be firm to slow raveling for 85% of the project, and fast-raveling to running for 15% of the project GBR, (2006).

By considering the boundary conditions such as virgin stress, orientation of the opening with regard to geologic features, size and shape of the opening, Rock Mass Behavior Types are defined according to Guideline for the Geomechanical Design of Underground Structures with Conventional Excavation (2006).

- Behavior Type 7: Shear failure under low confining pressure, i.e., potential for excessive overbreak and progressive shear failure with the development of chimney type failure, caused mainly due to low horizontal stresses
- Behavior Type 8: Ravelling ground, i.e., flow of cohesionless dry or moist, intensely fractured rock or soil

For the unsupported tunnel in Ladera Sandstone, the failure mode during tunnel excavation is expected to be stress-induced (plastic failure or yielding of the rock mass). Failure modes controlled by discontinuities are not expected. Hence, a continuum approach using FE-models is used to analyze the support requirements.

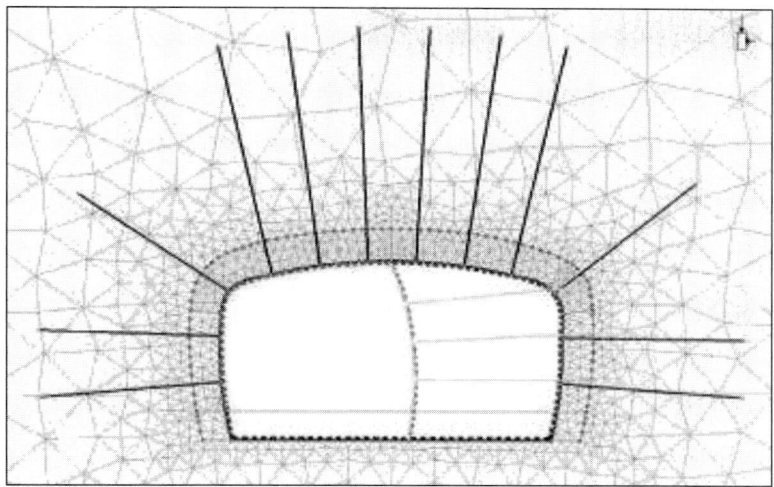

Figure 5. 2D FE-Model of the Section with worst case conditions (max. overburden)

Design Analysis

The geometry of the FE-models and the calculation sequences are in accordance with the detailed construction drawings. The overburden has a maximum height of 75 ft (22.9 m) above the FEH Cavern crown and this section was used for the analyses.

Because the intersection between the Access Tunnel and FEH Cavern is complex, the analyses were conducted using 3-dimensional calculations (Figure 4) employing midasGTS by TNO DIANA. Additional 2-dimensional calculations are carried out at critical cross sections (Figure 5) using the PHASE2 program by Rocscience. The 3D analyses used characteristic design values but, for the most critical cross sections, worst case parameters were used.

The vertical in situ stress was calculated by multiplying the overburden height by the unit weight of the Ladera formation. For calculation of the horizontal in situ stress, a factor $k0 = 0.5$ was applied.

The design analysis started by defining the primary stress field and calculation phases for all stages of excavation together with the corresponding support measures. The initial lining was stressed as a result of tunnel excavation and was deformed as a result of the primary stress state, the strength and the stiffness of the rock mass. The ground surrounding the tunnel was also considered as a load carrying element in the FE analysis and was modeled as a continuum using the isotropic model of Mohr-Coulomb with linear elasticity and perfect plasticity.

As a result of the excavation, deformations occur in the ground ahead the tunnel face. After the excavation and the application of the initial lining, further deformations occur, causing stresses in the shotcrete.

The following construction sequences were used for the 3-dimensional calculations:

- Access Tunnel Top Heading Excavation and Support
- Access Tunnel Ramp Excavation and Support up to FEH Cavern wall
- Intersection Top Heading Excavation and Support
- FEH Cavern Top Heading Excavation and Support
- Intersection Bench Excavation and Support
- FEH Cavern Bench Excavation and Support

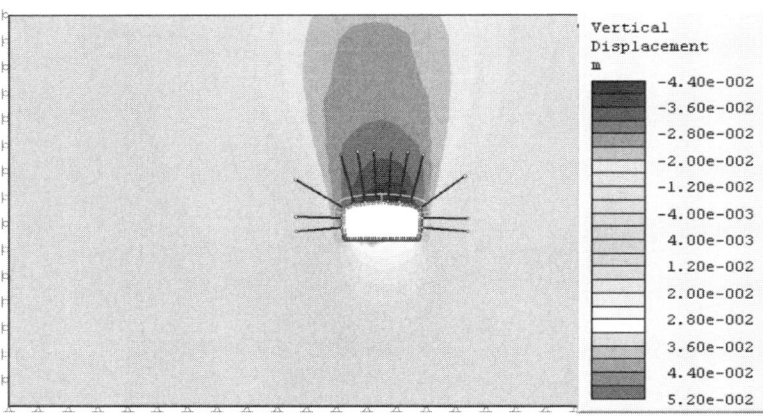

Figure 6. Results of 2D analysis intersection (vertical displacements)

- Access Tunnel Bench Excavation and Support
- Intersection Invert Excavation and Support
- FEH Cavern Invert Excavation and Support

The results of calculations indicated that redistribution of the in-situ stress field caused by the excavation was limited to about 1.0–1.5 tunnel diameter from the opening. Beyond this area, both the vertical and horizontal stresses approached the respective in-situ values. The plastic zones were limited to the zones near the sides of the cavern for average ground conditions. In the worst case scenario, the plastic zones were more pronounced, and dowels were therefore needed to extend beyond the plastic zones.

DESIGN VERIFICATION

Design Verification Process

NATM support measures generally include standard support categories which correspond to anticipated rock mass behavior and contingency measures used supplementary as required locally for specific ground conditions or if tunnel lining convergence exceeds predefined warning levels.

The support measures shall provide sufficient flexibility during construction to allow adaptation to actual conditions. Appropriate selection of support categories during construction requires comparison and matching of encountered rock mass description ground behavior to anticipated behavior for each support category.

In the course of the design verification process the support measures have to be continuously reviewed and updated. The interpretation of data according to the Observational Method is the main key to achieve this goal.

Geotechnical Monitoring

To ensure the actual deformations during the excavation of Intersection between Access Tunnel and FEH Cavern would not exceed the defined limits, an extensive geotechnical monitoring program was established in the Tunnel Contract.

The program consisted of different types of instrumentation, including:

- Surface settlement control (GDMP = Ground Deformation Monitoring Points)

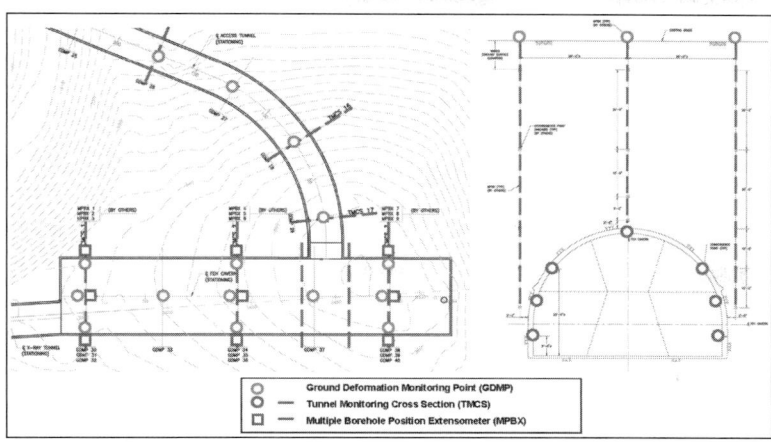

Figure 7. Geotechnical monitoring program at access tunnel and FEH Cavern

- Borehole Extensometer (MPBX = Multiple Position Borehole Extensometer)
- Tunnel Monitoring Cross Sections within the Tunnel/Cavern (TMCS)

The frequency of the geotechnical readings was related to the distance of the instrument to the face and defined as follows:

- Distance < 1.0 tunnel diameter → 2 times a day
- Distance < 2.5 tunnel diameter → daily
- Distance > 2.5 tunnel diameter → weekly

The arrangement of the geotechnical instrumentation is shown in Figure 7.

Geological Mapping

Furthermore, during construction, Geological Mapping was performed by the ILF Geotechnical Engineer. A geological profile at the Intersection of Access Tunnel and FEH Cavern is shown in Figure 8.

The encountered rock mass consisted mainly of interlaying fine Sandstone (grey-yellow-brown) and medium grained Sandstone (red-brown) layers with several small, hard Siltstone (grey) interlayer with almost horizontal massive/compact bedding. Strata with small bedding width and changing grades of weathering between fine to medium Sandstone, resulted in isolated small overbreaks at the crown of top heading excavation.

For the description of the in-situ rock strength, the Weathering Index and Rock Hardness according to the ISRM (1981) were used. The observed grade of rock mass weathering was mainly between V3–V4 which can be described as "highly" (V4) to "moderately" (V3) weathered rock with a distribution of 60% (V4) and 40% (V3) at the face. The fine grey Siltstone layers usually tended to the class V3 and the yellow and medium grained Sandstone layers can be classified as V4. The fine to medium grained Sandstone, as a main component, showed varying weathering indexes from V3 to V4, depending on the grade of cementation. The rock hardness was estimated in the field with the so-called "hammer blow method." Specimens with weathering index V3 could be broken only with heavy hammer blows, whereas specimens of V4 broke with light hammer blow.

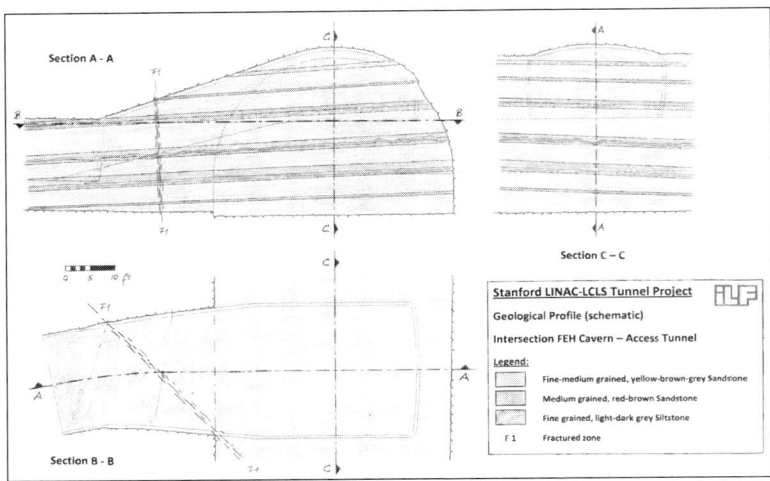

Figure 8. Geological profile at intersection access tunnel and FEH Cavern

Isolated grades of V2 (dark-brown medium Sandstone with calcitic bedding planes) and V5 (fractured zones with calcitic/clayey healed joints) also occurred but did not influence the stability.

Verification and Construction Adjustments

The readings of the MPBX deformations showed the most accurate results. Also, with the early installation of the Extensometer, the deformations could be controlled ahead of the excavation. Most of the readings for surface settlement pints showed variations and inconsistencies. Thus the evaluation was suspect.

An example of a Monitoring Sheet for the FEH MPBX is shown in Figure 9 and Figure 10 demonstrates an example of the TMCS/CMCS inside of the FEH Cavern, next to the Intersection.

Table 3 summarizes the anticipated deformations in comparison to the onsite monitored deformations.

The measured deformations at the MPBX are considerably smaller and the CMCS deformations are in the range of anticipated values for characteristic parameters. It was also found that the settlements ahead of excavation were limited and the deformations leveled off quickly after the excavation had passed and the support was in place.

With this information at hand, it was possible to agree to on-site relaxations of the contract requirements. The following adjustments were agreed between the Client, CM/GC and the Contractor/ILF:

- Larger excavation sections for FEH top heading excavation (open of slash A and B together, followed by slash C)
- Relaxation of the 18 ft restriction of excavation width at the FEH Cavern
- Change to full face excavation for FEH top heading
- Early start of bench excavation of FEH East drift

DESIGN FOR A LARGE SHALLOW CAVERN AT STANFORD 939

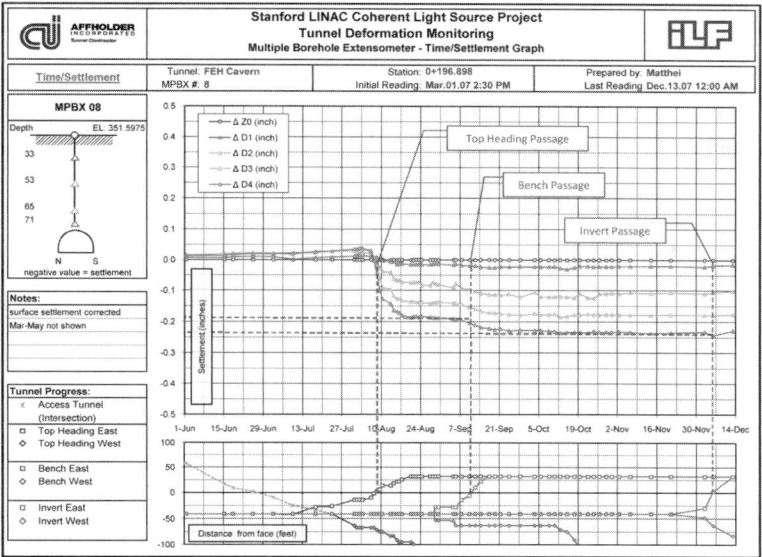

Figure 9. MPBX deformation diagram at the FEH Cavern

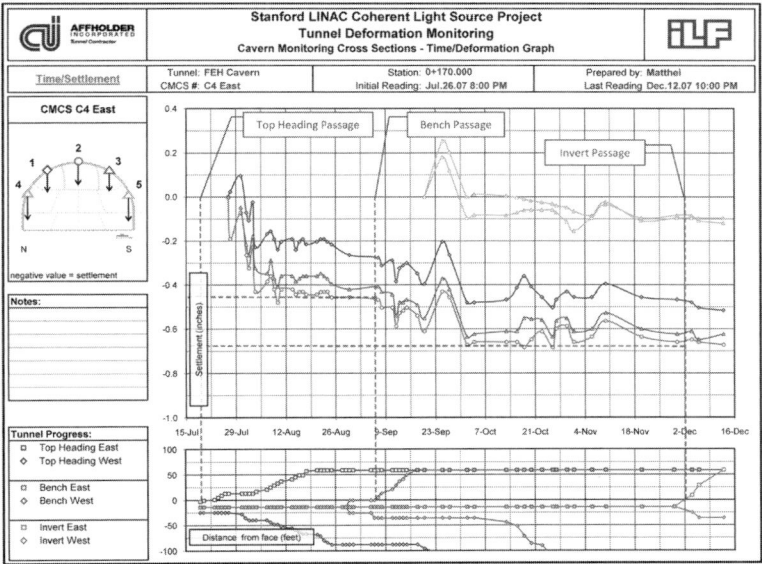

Figure 10. CMCS deformation diagram at the FEH Cavern

Table 3. Anticipated deformations at design stage and real encountered

	Max. Deformations (Tunnel Crown only)			
Excavation Sequence	Design Average	Design Worst Case	On-Site— MPBX Full Face	On Site— CMCS Full Face
Top Heading	0.47"	1.54"	0.18"	0.45"
Bench	0.63"	2.52"	0.24"	0.68"
Invert	0.67"	3.27"	0.25"	0.70"

CONSTRUCTION DOCUMENTATION

The construction of the Intersection started at the end of June 2007 with the partial excavation (left and right side separated) of the ramp to the Top Heading section of the FEH Cavern. The monitoring readings showed very little deformation and therefore it was possible to accommodate adjustments during construction to expedite excavation. In no case additional measures were needed. Details on excavation scheme, design adaptation and adjustments have been described in paper Sander, H. J. et al. (2008) and are not repeated here. The excavation work was finished by February 2008.

SUMMARY

For rock mass characterization, the results of various empirical methods have been taken into account to define characteristic values for the rock mass properties, which were then adjusted according to engineering judgment. Although the parameters have been defined on the safe side for the limit state analyses, it has turned out that the onsite rock mass strength and stiffness was higher than the anticipated properties.

During construction no additional support was required because no large lenses of uncemented sandstone have been encountered.

The final alignment of the intersection of Access Tunnel/ FEH Cavern required unique construction details.

ACKNOWLEDGMENTS

The authors would like to thank Affholder Inc. for the opportunity to work on this project and for their timely support and enthusiastic cooperation during the construction phase.

The contents of the paper reflect the views of the authors who are responsible for the facts and accuracy of the data presented herein. The contents do not necessarily reflect the official views or policies of SLAC of Stanford University as the owner.

REFERENCES

Halim, J., Vincent, F., Taylor, J. 2007. *NATM Design for Stanford LINAC Coherent Light Source Tunnels*, RETC Proceedings.

The Stanford Linear Accelerator Center—LCLS Project Geotechnical Baseline Report, Issued 3/16/06, Rev. 2.

International Society of Rock Mechanics Commission on Standardization of Laboratory and Field Tests. 1981, *Rock Characterization, Testing and* Monitoring, ISRM Suggested Methods. E.T. Brown editor, Pergamon Press.

Bieniawski, Z.T. 1974. *Geomechanics classification of rock masses and its application in tunneling*, Proc. of the 3rd International Congress on Rock Mechanics, Denver, 27–32.

Bieniawski, Z.T. 1989. *Engineering Rock Mass Classification: A Manual*. Wiley, New York, 205–219.
Barton, N. Lien, R., and Lunde, J. 1974. *Engineering Classification of Rock Masses for the Design of Tunnel Support*, Rock Mechanics, Vol. 6, No. 4.
Hoek, E., Carranza-Torres, C. and Corkum, B. 2002. *Hoek-Brown criterion—2002 edition*. Proc. NARMS-TAC Conference, Toronto, 1, 267–273.
RocData v4 and RocLab v1, Rocscience Software.
Hoek, E.1990. *Estimating Mohr-Coulomb friction and cohesion values from the Hoek-Brown failure criterion*. International Journal of Rock Mechanics and Mining Sciences & Geomechanics Abstracts, 12 (3), 227–229.
Hoek, E. 1994. *Strength of Rock and Rock Masses*. ISRM News Journal, 2 (2), 4–16.
Hoek, E and Diederichs, M.S. 2006. *Empirical estimation of rock mass modulus*. International Journal of Rock Mechanics and Mining Sciences, 43, 203–215.
Edelbro, C., Sjöberg, J., Nordlund, E. 2006. *A quantitative comparison of strength criteria for hard rock masses*. Tunelling and Underground Space Technology, 22, 57-68
Romana Ruiz, M. 2002. *Determination of deformation modulus of rock masses by means of geomechanical classifications*. In: Eurock Symposium, Madeira, Spain
Austrian Society for Geomechanics. 2006. *Guideline for the Geomechanical Design of Underground Structures with Conventional Excavation, Ground characterization and coherent procedure for the determination of excavation and support during design and construction*, Salzburg.
Sander H.J., Matthei S., Skovajsova Z., John M., Marcher T. 2008. *Stanford LCLS Project—Detailed Construction Design for the Intersection between Access Tunnel and FEH*. Proceedings of NAT Conference, San Francisco.

ADECO AS AN ALTERNATIVE TO NATM: HOW IT WORKS, WHY IT WORKS

Fulvio Tonon ▪ University of Texas

ABSTRACT

And Rabcewicz said *"tunnels should be driven full face whenever possible."* ADECO, which stands for "Analysis of Controlled Deformations in tunnels," now allows us to fulfill Rabcewicz's dream in any stress-strain condition. In order to achieve that dream and its consequent control over cost and schedule, however, NATM must be abandoned for the ADECO. The paper presents the basic concepts in the ADECO approach to design, construction and monitoring of tunnels together with some case histories, including: full face excavation for Cassia tunnel (width of 22 m, height of 14 m) in sands and silts under 5 m cover below an archeological area in Rome, Italy; Tartaguille tunnel (face area > 100 m^2) advanced full face in highly swelling and squeezing ground under 100 m cover where NATM led to catastrophic failure, France; and 80 km of tunnels (face area > 100 m^2) advanced full face in highly squeezing/swelling ground under 500 m cover for the high-speed railway line between Bologne and Florence, Italy (turnkey contract).

INTRODUCTION

Several generations of NATM (New Austrian Tunneling Method) consultants have us believe that NATM necessarily uses sequential excavation. Was this the original Rabcewicz's intent? On the other hand, in some countries such as the US, sequential excavation is currently used to indicate soft ground tunneling without a tunnel boring machine (Romero, 2002). Many points of view on and definitions of the NATM have been proposed (Kovari, 1994) and reviewed by Karakuş and Fowell (2004). Brown (1990) and Romero (2002) suggest to differentiate NATM philosophy:

- The strength of the ground around a tunnel is deliberately mobilized to the maximum extent possible.
- Mobilization of ground strength is achieved by allowing controlled deformation of the ground.
- Initial primary support is installed having load-deformation characteristics appropriate to the ground conditions, and installation is timed with respect to ground deformations.
- Instrumentation is installed to monitor deformations in the initial support system, as well as to form the basis of varying the initial support design and the sequence of excavation.

from NATM construction method:

- The tunnel is sequentially excavated and supported, and the excavation sequences can be varied.
- The initial ground support is provided by shotcrete in combination with fiber or welded-wire fabric reinforcement, steel arches (usually lattice girders), and sometimes ground reinforcement (e.g., soil nails, spiling).
- The permanent support is usually (but not always) a cast in place lining.

ADECO AS AN ALTERNATIVE TO NATM 943

This paper traces the history of the sequential excavation, NATM (as first conceived) and ADECO (Analysis of Controlled DEformations) with the aim of shedding light on the *unavoidable* use of sequential excavation in "soft ground," and of highlighting advances in tunnel design and construction that have occurred in Europe after and as alternative to the NATM.

SEQUENTIAL EXCAVATION: A 200 YEAR OLD APPROACH

In his 1963 book entitled "The History of Tunneling," G.E. Sandström talks about the tunneling methods devised when the canal era and the railroad era developed in the first half of the 1800s: yes, this is 200 years ago! Since the book was published in 1963 and Rabcewicz's papers on NATM were published in late 1964 and early 1965, there is little doubt that what Sandström describes are methods that preceded the NATM. Let's here from Sandström (pages 113 and ff):

"*An old-time mining tunnel, or drift, seldom exceeded an area of 10 × 10 ft, whereas a single-track railway tunnel used to be given an area of 16 × 22 ft., and a double track 28 × 22 ft. (modern tunnels are larger). The conventional practice used to be to advance a small pilot heading first in the forepoling manner described—if in heavy ground—and subsequently expand it to full size in some other way.*

The method of breaking out from a safe, wholly enclosed pilot tunnel is one of the central problems in tunneling and was endlessly debated throughout the last century. As a matter of fact it is still an issue that has to be argued as a preliminary to any tunneling scheme, because if it is not correctly settled beforehand men will lose their lives and the contractor his capital.

During the last century, a number of different tunneling systems were evolved which derived their names from their national origin. These were the English system, the Belgian System, the Austrian system, German (actually French) system, and the Italian so-called Cristina system. The Americans also laid claim to an independent system."

And on page 130: "*..., the interesting feature of these early American railway tunnels is that most of them were driven full face, i.e., the entire tunnel area was excavated, although in poor ground the top half was taken out to the full width and the roof secured with rafter timbering and lagged.*"

The methods are illustrated in Figure 1 through Figure 4, and the reader is referred to Sandström's book for excellent details.

Take home:

- The "sequential excavation method" is 200 years old and was well known when the NATM was coined in 1964.
- The "sequential excavation method" was developed 200 years ago by miners that had to adapt their mining techniques to the needs of civil engineering works.
- Power is defined as work/time, i.e., (ability to do work)/time.
- When the "sequential excavation method" was devised, tunnels were driven without electricity and compressed air, i.e., the available power was very small, mainly manpower.
- Breaking out from the pilot tunnel is one of the central problems in tunneling; if it is not correctly settled beforehand men will lose their lives and the contractor his capital.
- Early American tunneling was full face.

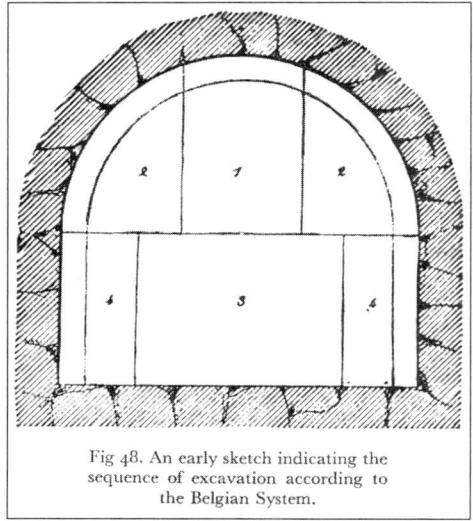

Figure 1. Belgian system used in the 1800s. From Sandström (1963).

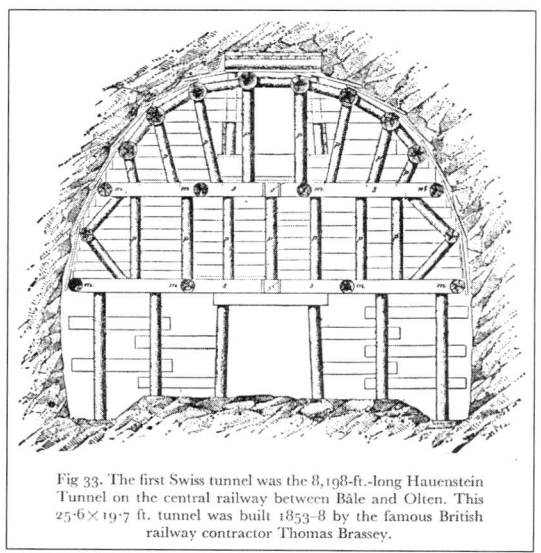

Figure 2. British system used in the 1800s. From Sandström (1963).

AND RABCEWICZ SAID "*TUNNELS SHOULD BE DRIVEN FULL FACE WHENEVER POSSIBLE*"

In his abstract to the first 1964 paper on NATM, Rabcewicz refers to the NATM as: "*a new method consisting of a thin sprayed concrete lining, closed at the earliest possible moment by an invert to a complete ring—called "an auxiliary arch"—the deformation of which is measured as a function of time until equilibrium is obtained.*" In the paper, page 454, Rabcewicz states that "*One of the most important advantages of*

Figure 3. German system used in the 1800s. From Sandström (1963).

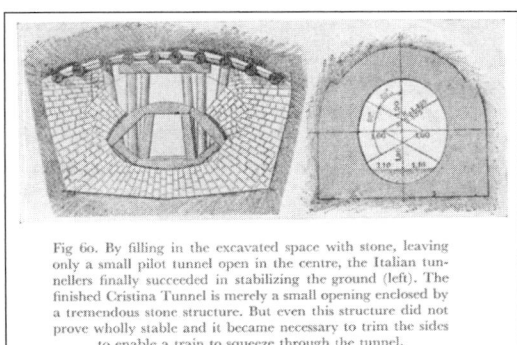

Figure 4. Cristina (Italian) system used in the 1800s. From Sandström (1963).

steel supports is that they allow tunnels to be driven full face to very large cross sections. The resulting unrestricted working area enables powerful drilling and mucking equipment to be used, increasing the rate of advance and reducing costs. Nowadays, dividing the face into headings which are subsequently widened is used only under unfavourable geological conditions." On page 457, Rabcewicz continues on this topic: *"There are still some difficulties to be overcome in normal methods of construction, as inverts are still usually built last of all, leaving the roof and sidewalls of the lining to deform at will. In the meantime, experience has taught us that it is by far more advantageous from all points of view, and frequently even imperative, to close a lining to a complete ring at a short distance behind the face as soon as possible. To comply with this requirement, tunnels should be driven full face whenever possible, although this cannot always be done, particularly in bad ground, where it often becomes necessary to resort to heading and benching. In the most difficult cases it may even be necessary to drive a pilot heading before opening it out to full section. An auxiliary arch executed in the upper heading (Belgian roof arch) though fairly effectively preventing roof loosening, represents an intermediate construction stage, which is still subject to lateral deformation. Such instability has to be removed as soon as possible by excavating the bench and closing the lining by an invert."*

Take home:

- NATM has nothing to do with sequential excavation.
- Rabcewicz realized that tunnels should be driven full face.
- Rabcewicz realized that full face allows use of large equipment i.e., large power at the face, which translates into fast tunnel advance and reduced costs.
- Rabcewicz never cared about nor mentioned the ground ahead of the tunnel face or ground support/reinforcement ahead of the tunnel face.
- Rabcewicz wanted but could not find a way to advance full face in difficult stress-strain conditions. His inability to proceed full face in all stress-strain conditions in 1964 was caused by a technological limitation in the normal methods of construction of those days.

QUANTIFICATION OF PRE-CONVERGENCE

Let us establish the nomenclature illustrated in Figure 5. In 1982, Panet and Guenot (1982) quantified the radial displacement of the ground at the future tunnel perimeter that occurs ahead of the tunnel face (preconvergence) in an unlined tunnel (Figure 6). At the face, about 30% of the final convergence has already occurred. Other researchers have quantified the preconvergence and convergence with and without the effect of the installed lining (e.g., Corbetta et al. (1991); Bernaud e Rousset (1992), (1996); Nguyen-Minh (1994); Nguyen-Minh et al. (1995); Nguyen-Minh and Guo (1993.a; 1993.b; 1996) and Guo (1995)). In particular, these studies show that a stiff lining may significantly reduce the convergence at the face, and thus preconvergence.

ITALIAN ADVANCES IN PRE-SUPPORT

The micropile umbrella-arch (also known as pipe-arch umbrella) consists of subhorizontal micropiles made up of steel pipes grouted in place at high pressure to improve the ground all around the perimeter of the excavation. In 1975, micropiles at different angles were used to tunnel through a collapsed zone (Carrieri et al. 2002), and in 1976 the first umbrella was designed as integral part of the support system for the S. Bernardino tunnel along the Genova-Ventimiglia railway line (Piepoli, 1976). By

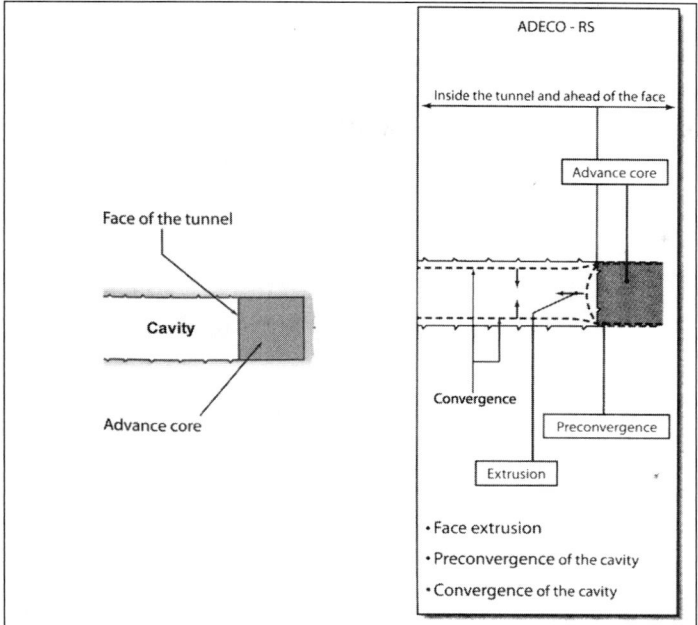

Figure 5. Nomenclature. After Lunardi (2008).

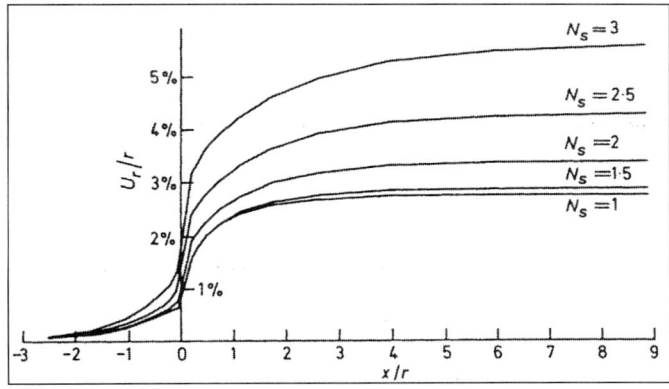

Figure 6. Preconvergence and convergence vs. distance to the tunnel face for tunnels in clays, undrained conditions. $N_s = p_0/s_u$; p_0 = in situ hydrostatic stress, s_u = undrained shear strength. After Panet and Guenot (1982).

1982, 15 tunnels in Italy had been driven by using a micropile umbrella (Barisone et al., 1982). Unfortunately, in many countries a pipe-arch umbrella is erroneously thought of being part of the NATM. In Italy, other major technological advances were made in the 1980s as a consequence of Lunardi's basic observations on and improved understanding of tunneling. Let's see what they were.

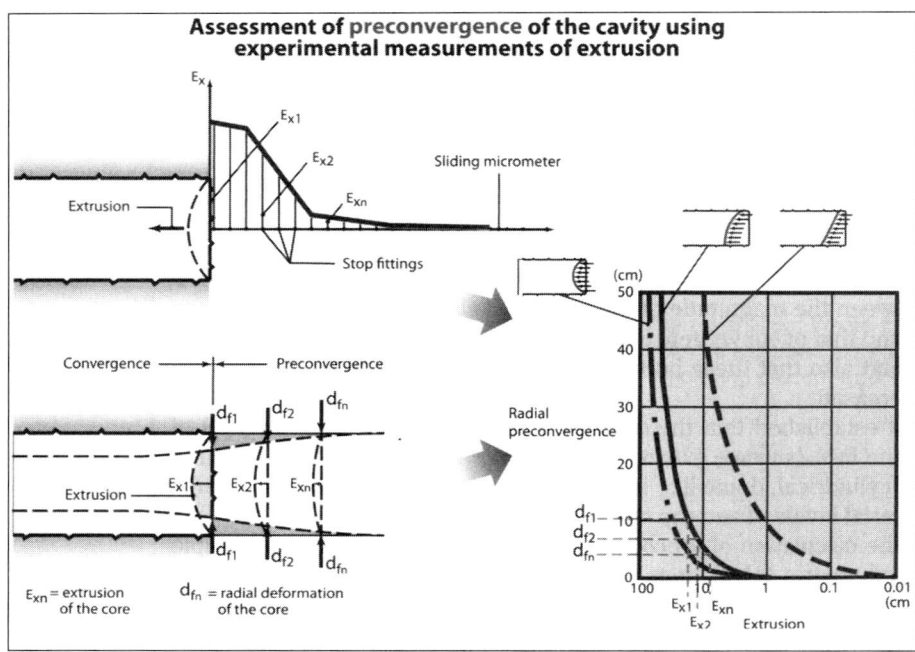

Figure 7. Measurement of extrusion with sliding micrometer and relationship between extrusion and preconvergence. After Lunardi (2008).

LUNARDI'S BASIC OBSERVATIONS ON TUNNEL BEHAVIOR

The same way as Rabcewicz conceived of the NATM in the 1960s by observing tunnel behavior, in the 1970–80s Lunardi made the following basic observations in the tunnels that he designed and/or built:

1. Convergence (radial displacement of cavity wall, Figure 5) is only the last manifestation of ground deformation. The convergence is always preceded by and is the effect of the deformation of the advance core: preconvergence = radial displacement of ground at the future tunnel perimeter, and extrusion = horizontal displacement of the core.

2. Extrusion can be measured *in situ* and is related one-to-one with the preconvergence (Figure 7)

3. Everything else being the same, the deformation (convergence) of the cavity increases as the speed of tunnel advance decreases. This is illustrated in Figure 8, which gives the convergence measured in the calcshists of the Frejus tunnel. When the tunnel advanced 100 m/month (Section 6), the convergence in the cavity was three times as large as the convergence measured when the tunnel advanced 200 m/month. When advancing 100 m/month, it was observed that the ground in the tunnel core deformed much more then when advancing 200 m/month.

4. The collapse of the cavity is always preceded by the collapse of the face-core system (Figure 9).

5. In top-heading and benching, the tunnel face starts at the crown of the top heading and ends at the invert of the bench (Figure 10).

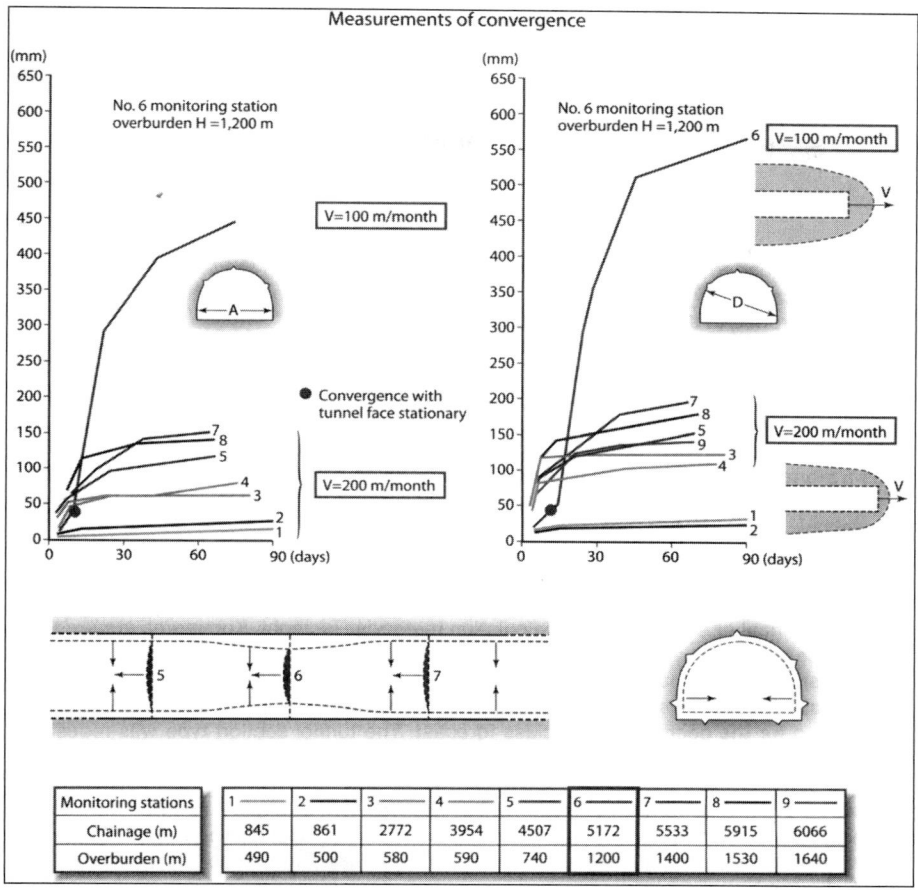

Figure 8. Convergence measurements in the Frejus highway tunnel, 1970s. After Lunardi (2008).

6. The arrival of the tunnel face reduces the confinement in the core and increases the major principal stress, giving rise to three basic face-core behaviors: A = stable; B = stable in the short term; C = unstable (Figure 11).

Take home (Figure 12):
- The ground behavior around the cavity and the convergence in the cavity at a given tunnel chainage X are controlled by the deformation and the behavior of the ground in the tunnel core when excavating the tunnel at chainage X (what Rabcewicz did not understand and could not do in 1960s).
- In difficult stress-strain conditions, counteracting convergence is not feasible. One needs to control preconvergence and extrusion, i.e., the deformations in the core ahead of the tunnel face (what Rabcewicz did not understand and could not do in 1960s).
- Sequential excavation extends the tunnel face even if the top heading is lined (same as Rabcewicz "*An auxiliary arch executed in the upper heading … represents an intermediate construction stage, which is still subject*

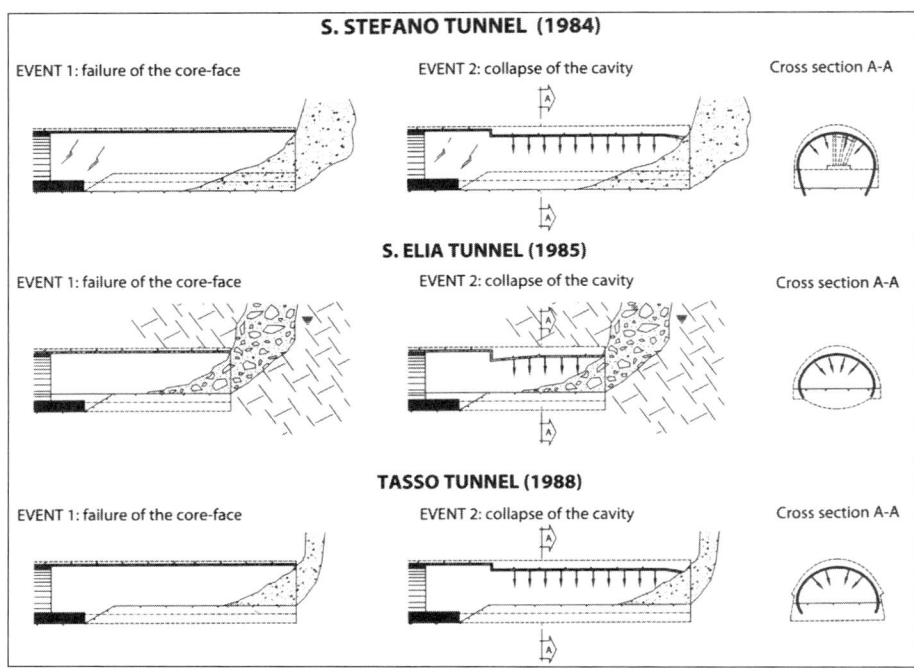

Figure 9. Case histories of tunnel collapses. After Lunardi (2008).

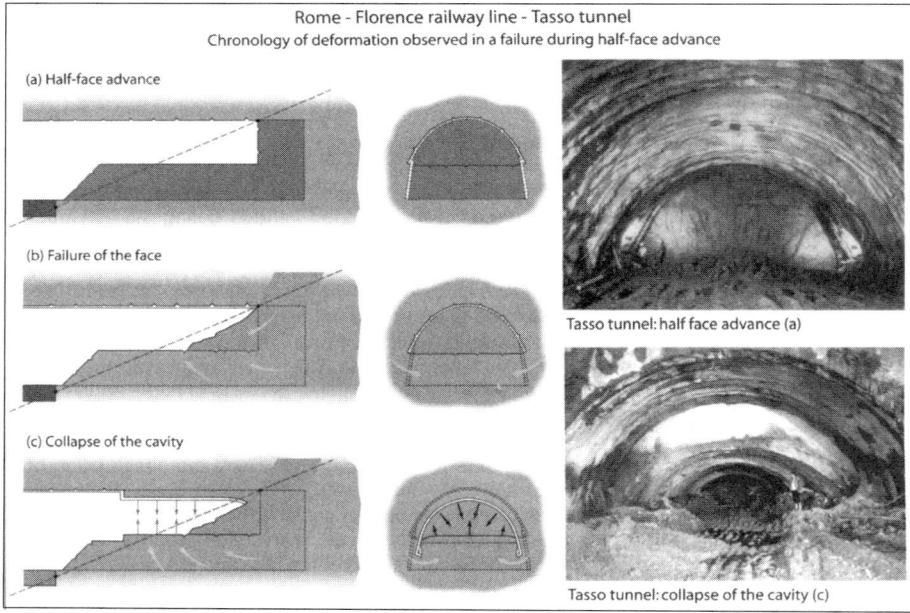

Figure 10. Failure at Tasso tunnel excavated top heading and benching, 1988. Notice 2 m convergence in top heading. After Lunardi (2008).

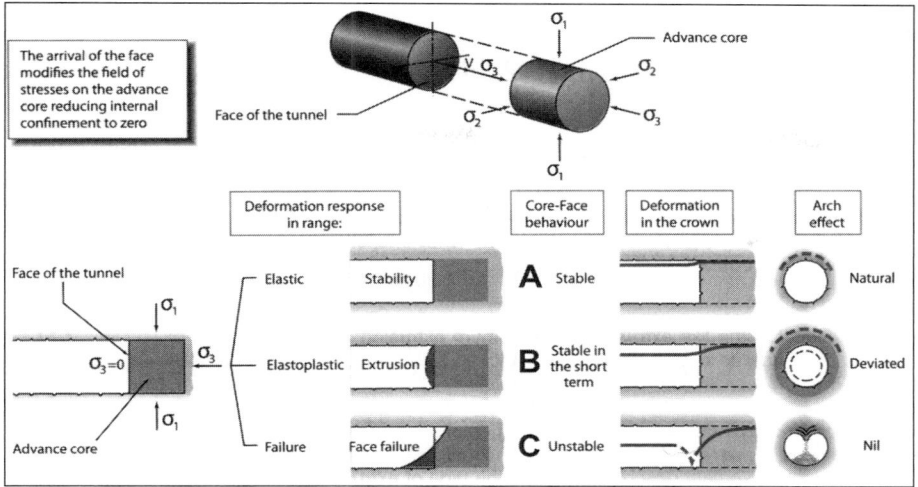

Figure 11. Tunnel behavior categories based on face-core behavior. After Lunardi (2008).

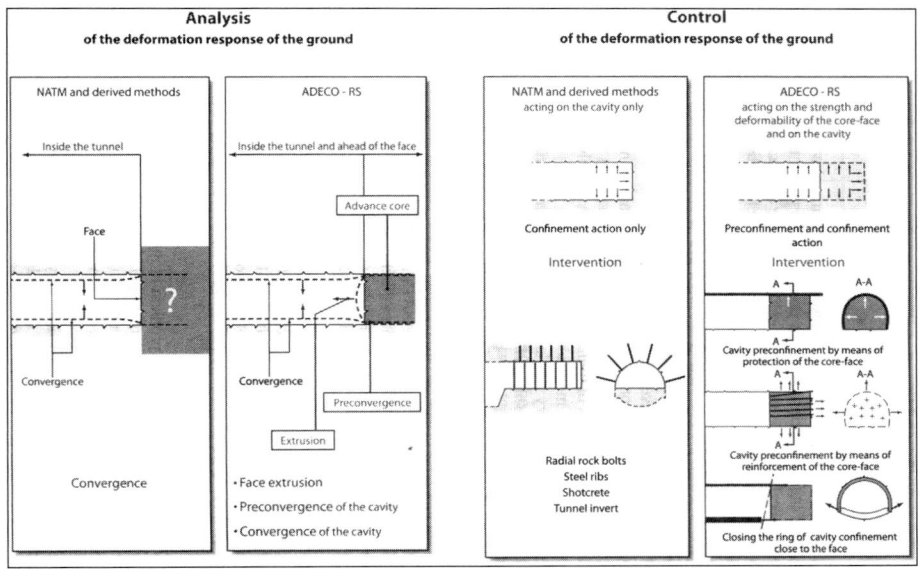

Figure 12. NATM vs. ADECO. After Lunardi (2008).

to lateral deformation") and increases the volume of ground in the core that, by deforming, controls the behavior of the cavity (what Rabcewicz did not understand).

- If the extent of the face and of the core must be minimized, one has to proceed full face (same as Rabcewicz "*tunnels should be driven full face whenever possible*").

These results led Lunardi to the idea of engineering the core in order to use the core as a stabilization method for the cavity, the same way as rockbolts, shotcrete and steel sets are used to stabilize the cavity. The idea was implemented by developing new technologies, such as:

- Sub-horizontal jet-grouting (Campiolo tunnel, 1983).
- Pre-cut with full face excavation (Sibari-Cosenza railway line, 1985, evolution of the pre-decoupage used in the top heading in the Lille Metro, France).
- Fiberglass reinforcement of the core as a construction technology to be used systematically in full-face tunnel advance (1985, high speed railway line between Florence and Rome), and not only as an ad-hoc means to overcome unpredicted tunneling problems.

The ADECO is the culmination of these observations and experiments, and the new technologies introduced with it can thus only be understood and properly used within the context of the ADECO approach.

ADECO APPROACH

The ADECO (Analysis of COntrolled DEformations) workflow is illustrated in Figure 13. In the Diagnosis Phase, the unlined/unreinforced tunnel is modeled in its *in situ* state of stress with the aim of subdividing the entire alignment into the three face/core behavior categories: A, B, and C: these depend on the stress-strain behavior of the core (ground strength, deformability and permeability + *in situ* stress), not only on the ground class. The site investigation must be detailed and informative enough to carry out such quantitative analyses: this clearly defines what the investigation should produce.

In the Therapy phase, the ground is engineered to control the deformations found in the Diagnosis Phase. For tunnel category A, the ground remains in an elastic condition, and one needs to worry about rock block stability (face and cavity) and rock bursts; typically, rock bolts, shotcrete, steel sets and forepoling are used to this effect. In categories B and C yielding occurs in the ground. By looking at the Mohr plane (Figure 14) two courses of action clearly arise:

- Protecting the core by reducing the size of the Mohr circle: this can be achieve either by providing confinement (increasing σ_3) or by reducing the maximum principal stress (reducing σ_1).
- Reinforcing the core, thereby pushing up and tilting upwards the failure envelope.

The rightmost column in Figure 12 depicts the actual implementation of these two ideas as pre-confinement actions. The third line of action consists of controlling the convergence at the face by using the stiffness of the lining (preliminary or even final, if needed), which may also longitudinally confine the core. It is only in this context that the different technologies currently available and listed in Figure 15 take their appropriate role. Notice that the ADECO embraces tunnels excavated with and without a tunnel boring machine.

Once the confinement and preconfinement measures have been chosen, the section is composed both in the transverse and longitudinal directions, and then analyzed. In all cases, full face advance is specified in all stress-strain conditions, thus fulfilling Rabcevicz's dream.

For each section, displacement ranges are predicted in terms of convergence and extrusion (Figure 16). Besides plans and specs, construction guidelines are also produced during the design stage. The construction guidelines are used at the construction site to make prompt decisions based on the displacement readings. If the readings

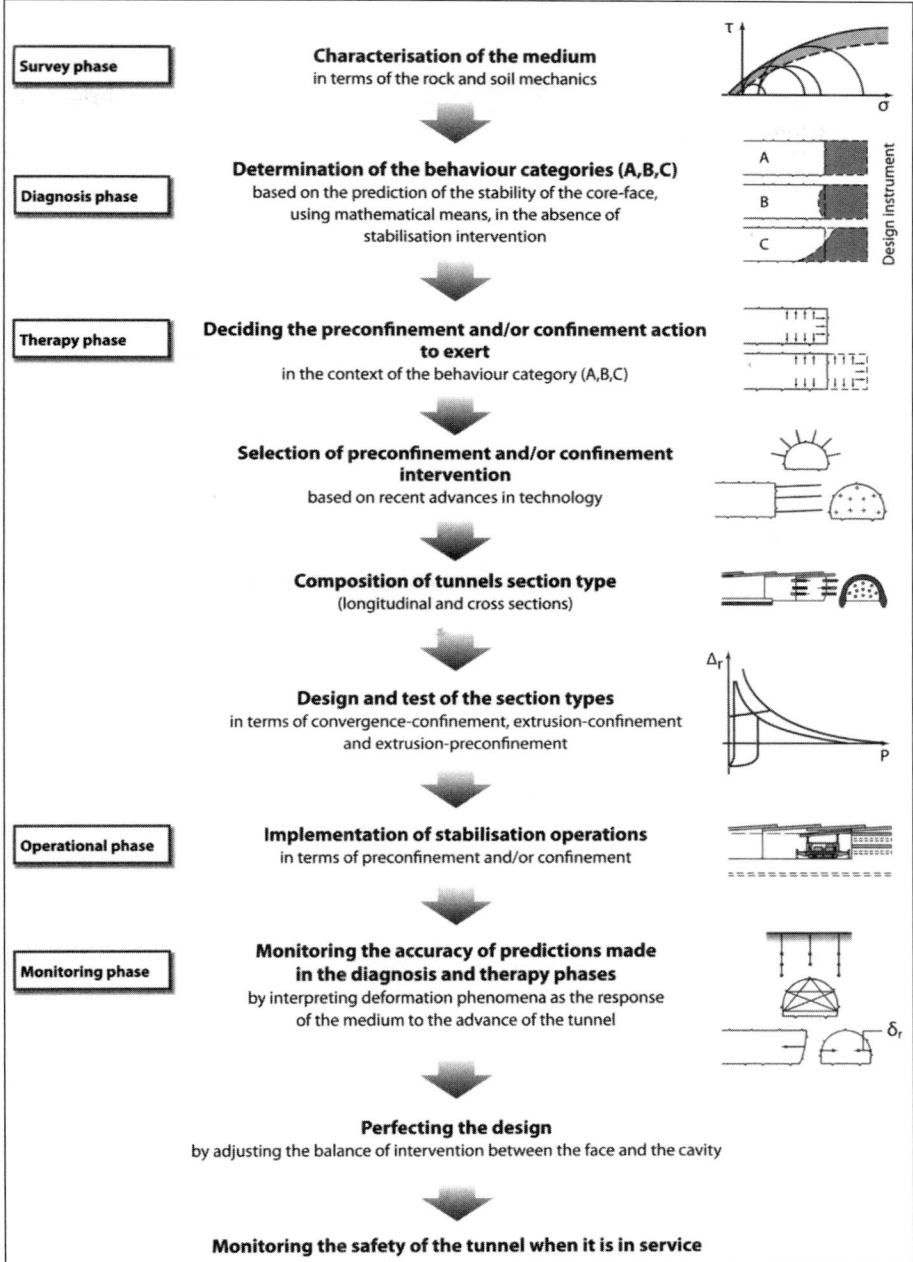

Figure 13. ADECO workflow. After Lunardi (2008).

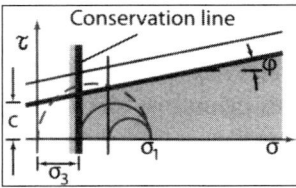

Figure 14. Mohr-plane explanation of approaches to stabilize/stiffen the core. After Lunardi (2008).

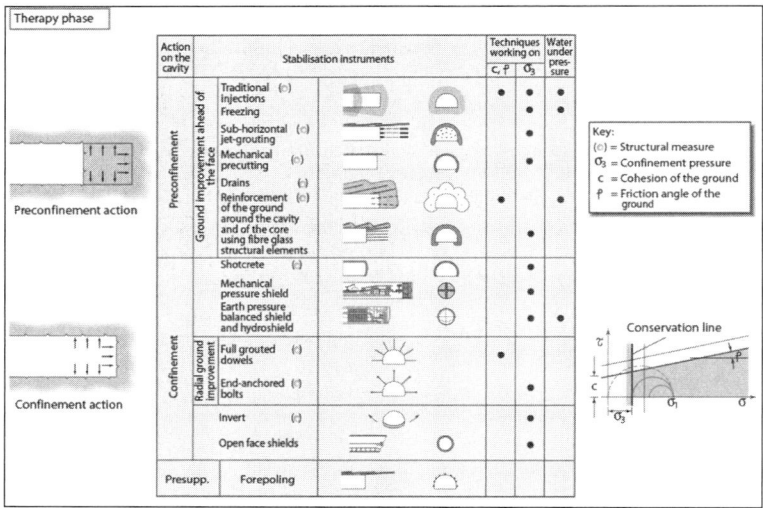

Figure 15. Subdivision of stabilization tools based on their action as pre-confinement or confinement. After Lunardi (2008).

are in the middle of the predicted ranges, then the section in the plans and specs is adopted; if reading values fall to the lower end of the ranges, then the minimum quantities specified in the guidelines are adopted for the stabilization measures (Figure 16). Likewise, if reading values are on the upper end of the ranges, then the maximum quantities specified in the guidelines are adopted. Finally, if the readings are outside the ranges, the guidelines specify the new section to be adopted. In this way, ADECO clearly distinguishes between design and construction stages because no improvisation (design-as-you-go) is adopted during construction.

Monitoring plays a major role in the ADECO, but with two main differences with respect to the NATM:

- In categories B and C, not only convergence but also extrusion is measured because the cause of instability is the deformation of the core.
- Monitoring is used to fine tune the design, not to improvise cavity stabilization measures, so that construction time and cost can be reliably predicted.

Tunnels are thus paid for how much they deform, which, unlike rock mass classifications carried out at the face, is an objective measure void of any interpretation. In addition, in soils, rock mass classifications are inapplicable. Experience in over 500 km

Section Types	Geology	Convergence (cm)	Extrusion (cm)
A	Monte Modino Sandstones	2-3	Negligible
B0		3-5	Negligible
B0V		5-10	< 3
B2	Scaly Clays	8-12	< 6
B2V		6-10	< 5
C2		10-14	< 10
C6		8-12	< 8

Section Types	Intervention	Variabilities		
		Minimum	Nominal	Maximum
C2	Steel rib step	1.2 m	1.0 m	0.8 m
	N° VTR face	50	70	90
	VTR face overl.	10.0 m	12.0 m	14.0 m
	Excavation	14.0 m	12.0 m	10.0 m
	Invert-face (°)	< 2.0∅	< 1.5∅	< 0.5∅
	Crown-face	< 3.0∅	< 5.0∅	< 7.0∅

Figure 16. Displacement predictions and design guidelines. After Lunardi et al. (2008).

of tunnels indicates that, when the ADECO has been adopted and tunnels were paid for how much they deformed, claims have decreased to a minimum.

CASE HISTORIES

Bologne-Florence High Speed Railway, Italy

The largest tunnel construction project ever implemented in the world entailed 84.5 km of running tunnels with a cross-section of 140 m^2 and additional 20 km of service tunnels for a total of about 13 million m^3 of excavated material (Figure 17 through Figure 24). Because of the difficult tunneling conditions, all running tunnels were excavated *without* a tunnel boring machine. Indeed, the route passed through the highly squeezing conditions of the Apennines with covers varying form zero to 550 m. Once the ADECO design was complete, the E 4.209 billion lump sum contract was won by FIAT, who took all risks including the geological risks. Construction started in 1996 and finished in time and on budged in 2006: a maximum of 26 faces were open simultaneously with a production of 1,600 m/mo. Figure 19 through Figure 22 exemplify the case of a C category section designed for highly squeezing scaly clays in the Raticosa tunnel. Scaly clays are extremely sensitive to stress-relief, and lose all of their cohesion if confinement drops to zero. It was thus of utmost importance to pre-confine the core and to adopt a stiff preliminary lining with very stiff final invert to be always kept very close to the tunnel face. Despite the heavy ground improvement and the final invert poured at the tunnel face, production rates were constant and equal to 1.5 m/day: a clear indication of the benefits in using the core as a stabilization measure and of the industrialization achieved with the ADECO.

Figure 25 illustrates the application of the ADECO to a tunnel boring machine (TBM) drive for the 9.26-km long 5.6-m diameter Ginori service tunnel to the Vaglia tunnel. A TBM was chosen because the service tunnel had to be completed rapidly from only one portal, and it was necessary to keep the excavation watertight at all times. The ground, which varied from compact limestone to argillites, was excavated with a Wirth TB 630E/TS double shield TBM equipped with drilling equipment to drill through the cutter head and the shield in order to preconfine and investigate (by georadar) the advance core. Construction finished on time and on budget with an average advance rate of 20 m/day under a maximum water pressure of 5 bar.

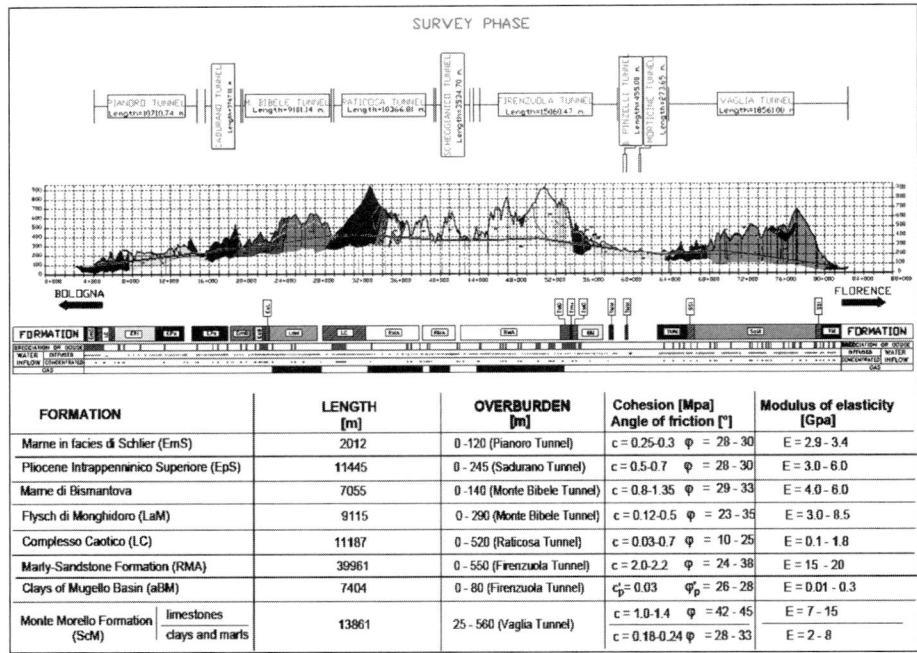

Figure 17. Bologne-Florence high-speed railway tunnels. After Lunardi (2008).

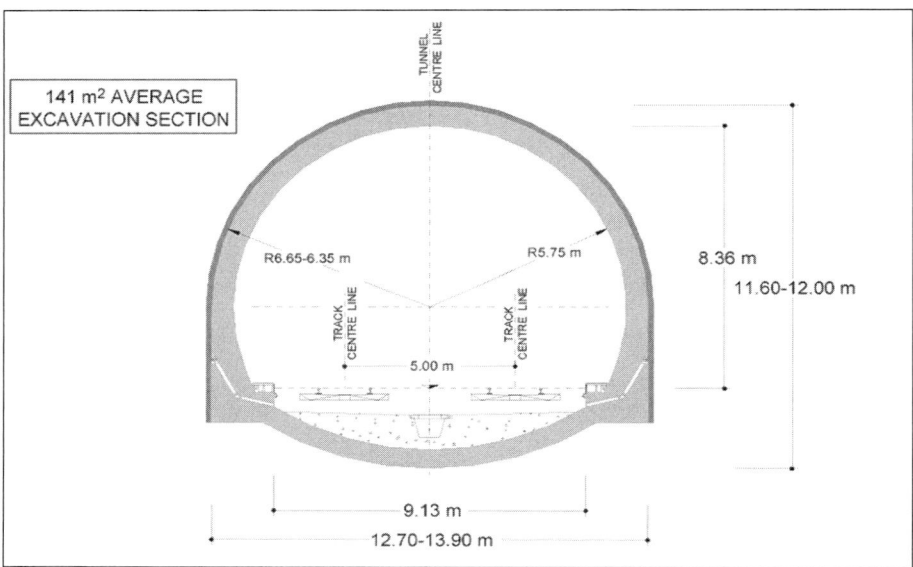

Figure 18. Bologne-Florence high-speed railway: typical cross-section. After Lunardi (2008).

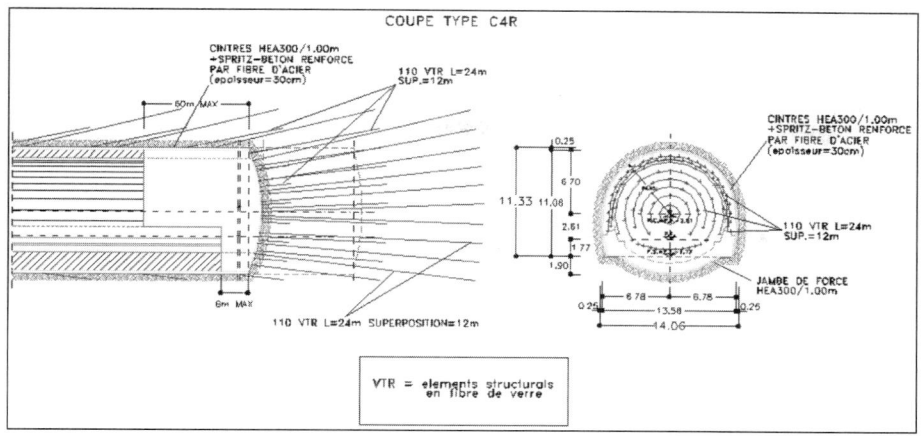

Figure 19. Raticosa tunnel for the Bologne-Florence high-speed railway: Longitudinal and transverse cross-sections for highly squeezing scaly clays. After Lunardi (2008).

Figure 20. Raticosa tunnel for the Bologne-Florence high-speed railway: Full face excavation under 500 m of cover in highly squeezing scaly clays. After Lunardi (2008).

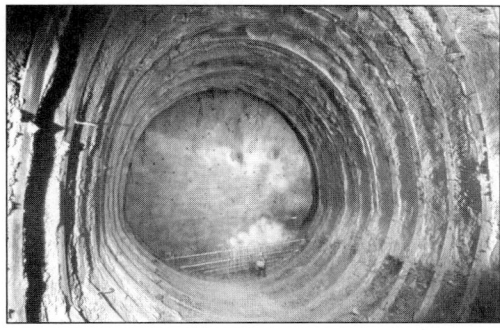

Figure 21. Raticosa tunnel for the Bologne-Florence high-speed railway: Preparing for pouring the final invert under 500 m of cover in highly squeezing scaly clays. After Lunardi (2008).

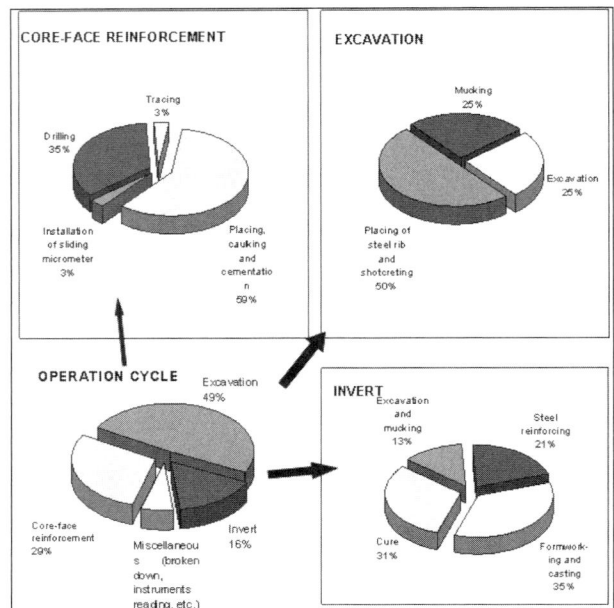

Figure 22. Raticosa tunnel for the Bologne-Florence high-speed railway: Breakdown of construction operations. After Lunardi (2008).

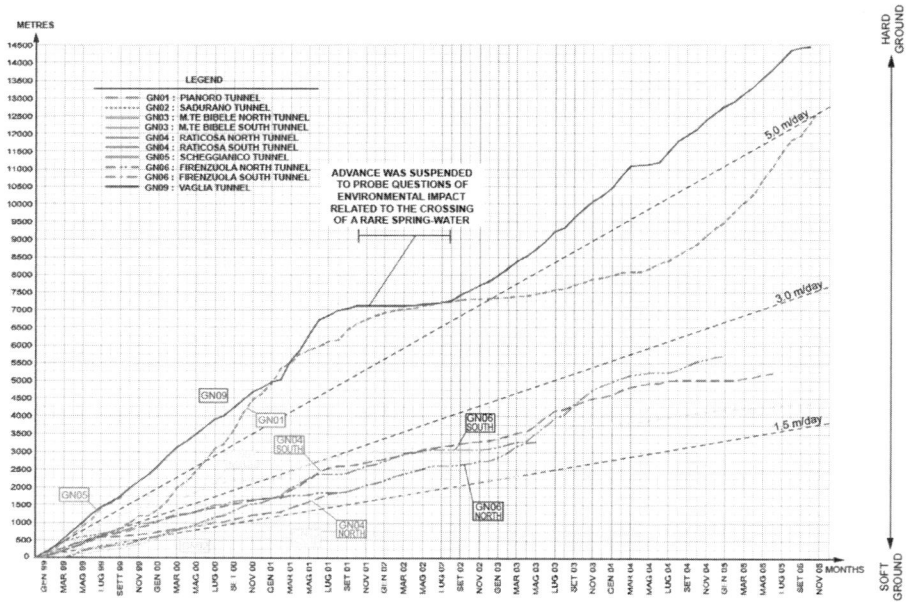

Figure 23. Production data in the Bologne-Florence high-speed railway tunnels. After Lunardi (2008).

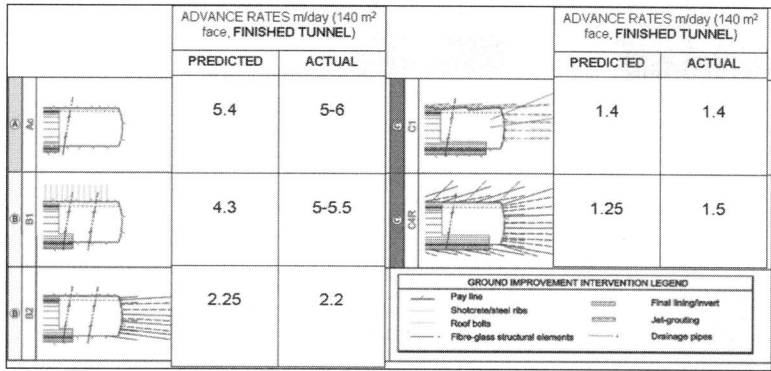

Figure 24. Predicted vs. actual production rates in the Bologne-Florence high-speed railway tunnels. Reconstructed after Lunardi (2008).

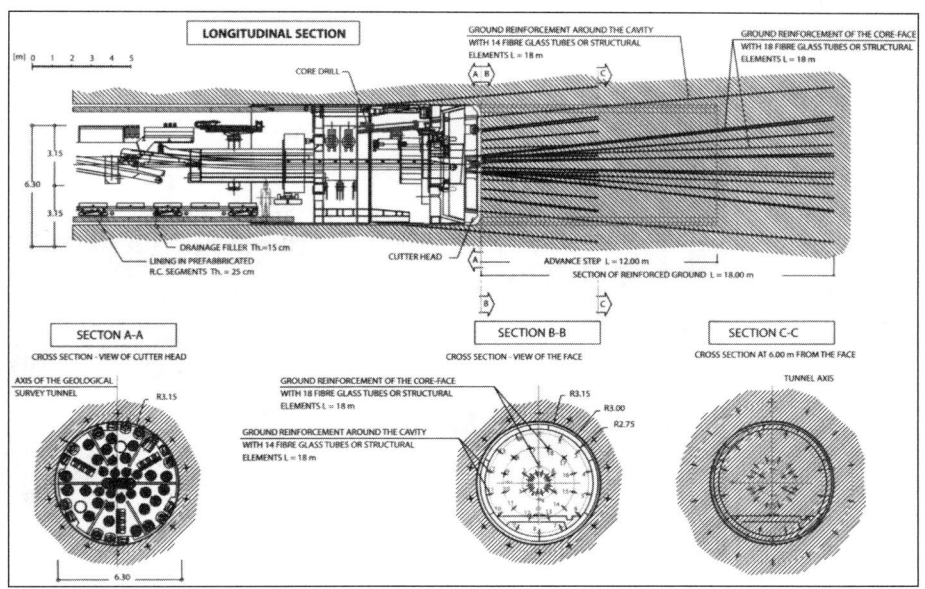

Figure 25. Bologne-Florence high-speed rail: TBM used in the Ginori tunnel. After Lunardi (2008).

Tartaguille Tunnel, France

The Tartaguille tunnel, 2.3 km in length with a cross-section of 180 m², is one of six tunnels on the high speed railway line that connects Lyon to Marseille in France. The tunnel passes through Cretaceous formations, including the lower Stampien marly clays, which are 75% montmorillonite. As depicted in Figure 26, construction started according to the original design in the Stampien clays from the North portal with top heading and benching. The top heading was equipped with temporary invert and composed of 25-cm thick shotcrete and HEB 240 steel sets at 1.5 m spacing that were founded on micropiles and on rock bolts after benching. Construction proceeded with

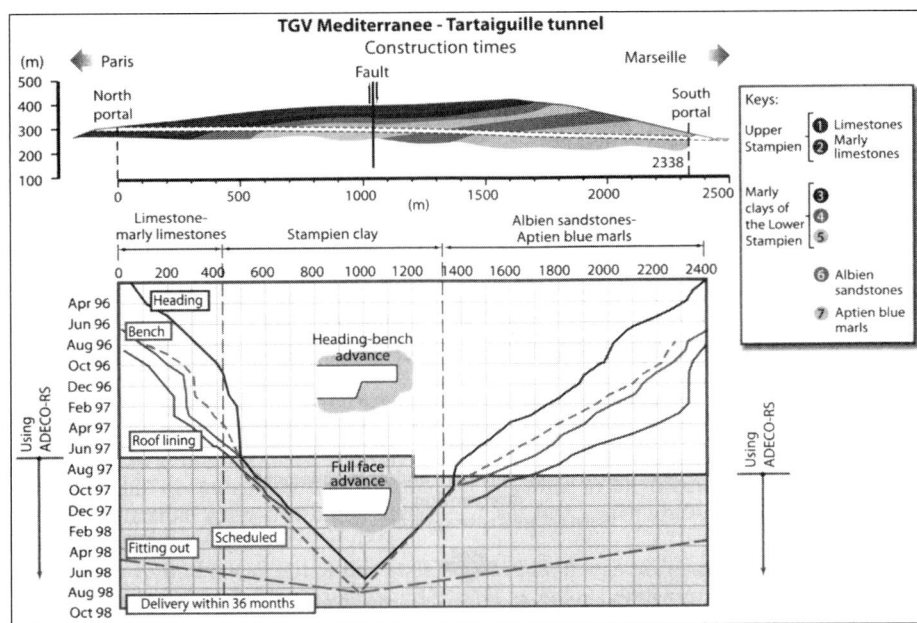

Figure 26. Tartaguille tunnel construction time versus geology. After Lunardi (2008).

great difficulties and it was much slower than anticipated. After clear signs of distress appeared in the primary lining and construction became unsafe, several solutions were proposed by eminent European consultants as illustrated in Figure 27. Only the ADECO proposal adopted full face (180 m²) advance and used the advance core as a stabilization measure; the other solutions tried to counteract or control convergence by advancing with sequential excavation and by installing support and reinforcement in the cavity. The flexible solution is a typical NATM solution. The French Rail (SNCF) decided to adopt the ADECO approach because the ADECO proposal was the only one that promised to finish the tunnel on budget and within schedule. Figure 28 through Figure 34 illustrate the proposed design and some construction phases. Notice:

- The large and powerful equipment deployed at the face
- The large number of workers that can work at the face at the same time
- The steel rib erected at the face with only two connections: this ensures quick installation, and quality control and assurance are much more simplified.
- The sheer stiffness of the preliminary lining and of the final invert meant to avoid (together with fiber glass reinforcement of the core) any decompression of the montmorillonitic clays so as to avoid any swelling.
- Construction schedule included waterproofing and pouring of the invert to the tunnel face without disruption while keeping a constant full face advance of 1.55 m/day

The tunnel was completed one and a half months ahead of schedule and below budget. Figure 27 shows how ADECO yielded constant production rates, whereas the production rates obtained with the sequential excavation were not constant.

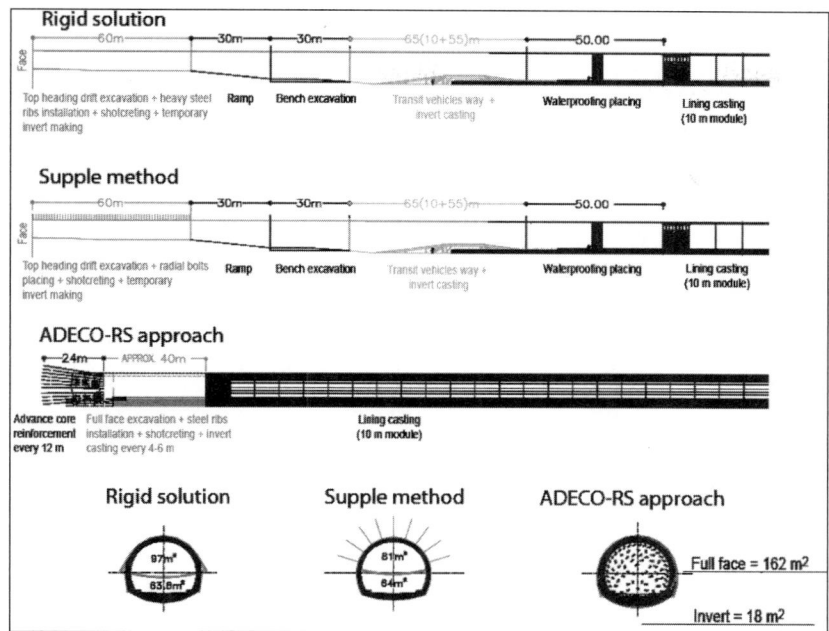

Figure 27. Tartaguille tunnel: The three proposed solutions to advance in the Stampien clays. After Lunardi (2008).

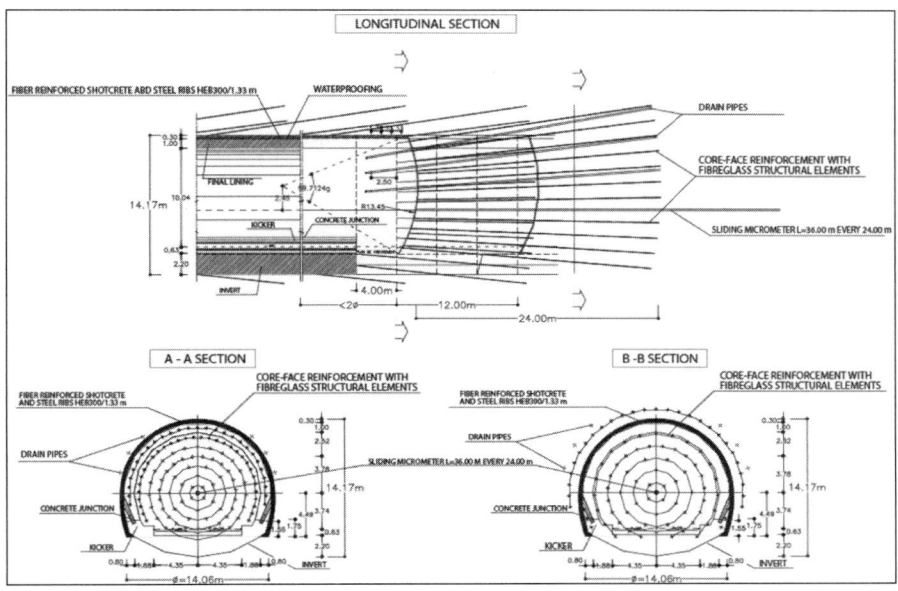

Figure 28. Tartaguille tunnel: Adopted ADECO solution in Stampien clays. After Lunardi (2008).

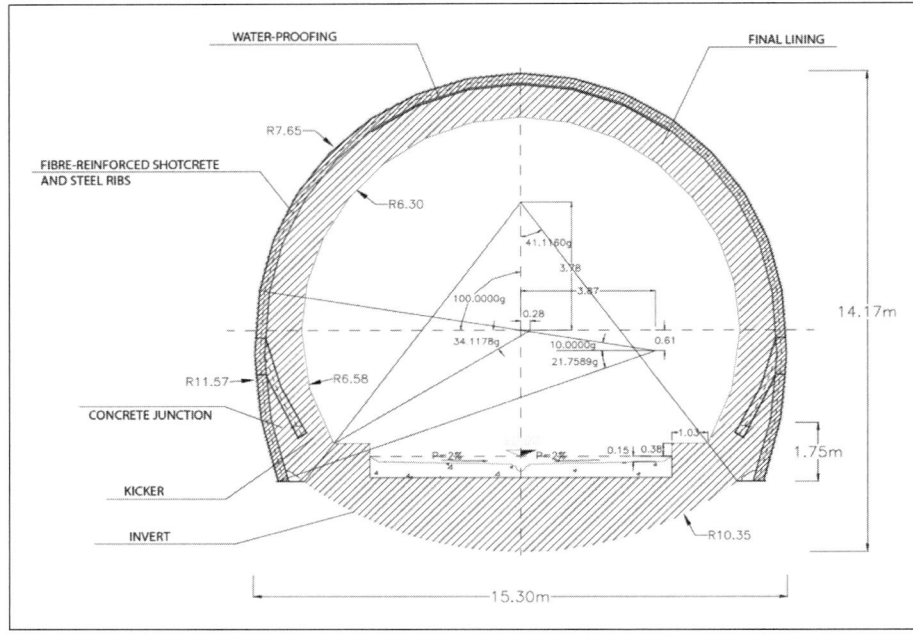

Figure 29. Tartaguille tunnel: Cross section showing primary lining and final lining. After Lunardi (2008).

Figure 30. Tartaguille tunnel: Installation of fiberglass elements in the core. Notice the kickers and the final invert against the face. After Lunardi (2008).

Figure 31. Tartaguille tunnel: Erection of a steel rib. After Lunardi (2008).

Figure 32. Tartaguille tunnel: The steel rib is erected. After Lunardi (2008).

Figure 33. Tartaguille tunnel: Waterproofing and formwork are placed before pouring the final invert against the tunnel face. After Lunardi (2008).

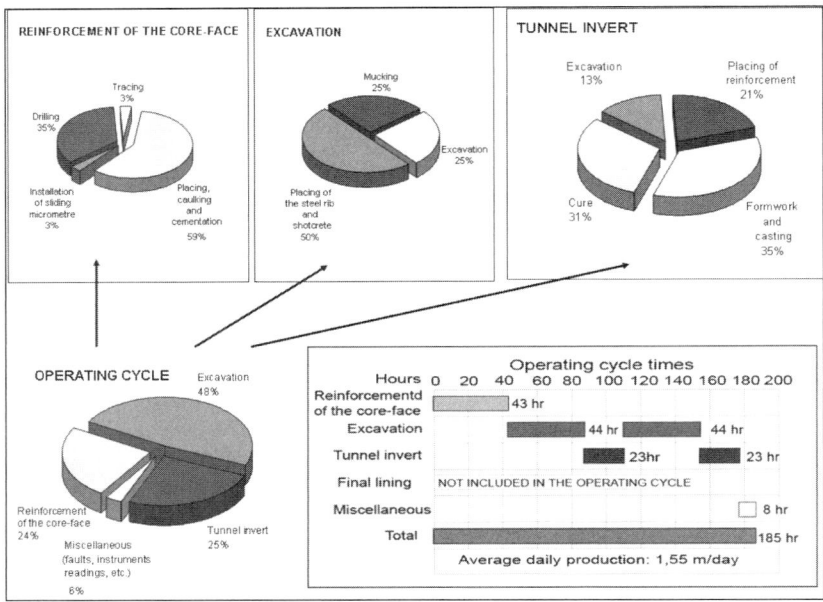

Figure 34. Tartaguille tunnel: Breakdown of construction schedule. After Lunardi (2008).

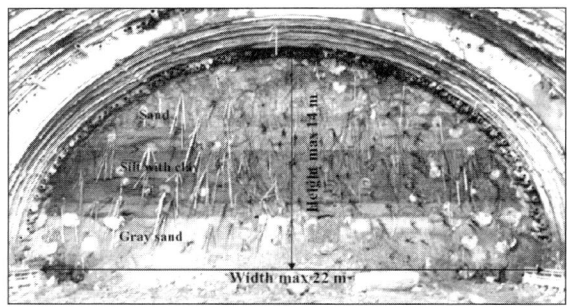

Figure 35. Cassia 1 tunnel: Dimensions and stratigraphy

Cassia 1 Tunnel, Italy

The construction of the "Cassia Tunnel" (outer lane) is part of a larger project for increasing the capacity of the external Ring Road in Rome. The tunnel is 22-m wide and 230-m long, it has 3 traffic lanes plus one emergency lane. The tunnel passes through sands and silts with sand with an overburden of 5 m below a Roman archaeological site. Construction advanced full face under an umbrella of jet grouting, while the core was reinforced and confined by jet-grouting columns (Figure 35 through Figure 37).

The roto-injection technique developed by Trevi and Soilmec for this project uses a double counter-rotating system made up of a rod and a pipe. The internal rod includes the jetting system, equipped with self-drilling monitor. In the umbrella, the pipe remains permanently inserted and works as reinforcement. The spoil is directed into the internal space between the rod and the pipe, which allows one to check and control the flux, thus preventing voids to form in the jet-grouting column.

Figure 36. Cassia 1 tunnel: Jet-grouting columns in the core reinformced with fiberglass elements

Figure 37. Cassia 1 tunnel: Construction stages

ADVANTAGES OF THE ADECO APPROACH OVER SEQUENTIAL EXCAVATION AND NATM

- ADECO fulfills Rabchewicz's dream of advancing full face in all stress-strain conditions, which allows risk, cost and construction time to be minimized.
- Tunnel construction is finally industrialized in all tunneling conditions because tunneling advance is no longer subject to the ground but the ground is made what it needs to be in order to proceed as fast as possible. This is illustrated in Figure 23 and Figure 26, where production rates are constant even in the most difficult stress-strain conditions (highly squeezing, and

Figure 38. Typical transportation means in the early 1800s, when sequential excavation was conceived

squeezing and swelling, respectively). In Figure 26, compare with sequential excavation rates, which are overall much smaller and are not constant.
- Industrialization entails that cost and time can be reliably predicted at the design stage. Figure 24 shows how predicted production rates were maintained during construction of 85 km of tunnels even under the most difficult stress-strain conditions (highly squeezing). Notice that these rates refer to the finished 140 m^2 face tunnel (including final lining), not for top heading, or pilot drift. As stated in the introduction, NATM philosophy entails designing the cavity support/reinforcement based on monitoring results, which means that construction time and cost cannot be predicted.
- Constant production minimizes ground deformation, which minimizes squeezing and thus the loading on the final lining, which becomes cheaper.
- By advancing full face under all conditions, large and powerful equipment can be used, which means that a lot of work can be done in a short time. This reduces cost and construction time.
- By advancing full face and minimizing squeezing, settlements are minimized, which, for example, is of paramount importance in urban area.
- Tunnels construction with and without a tunnel boring machine can be handled within the same approach.

CONCLUSIONS

Sequential excavation was started two hundred years ago; at that time, there was no electricity, horse and buggy were commonly used to move around (Figure 38); ladies wore crinolines and gentlemen wore top hats (Figure 39). As originally conceived by Rabcewicz, the NATM did not necessarily embrace sequential excavation. Rather, Rabcewicz was completely in favor of full face advance but he realized that NATM did not allow him to advance full face in difficult stress-strain conditions. The research and projects carried out by Lunardi indicate the reasons why Rabcewicz could not fulfill his dream in difficult tunneling conditions. He (and all his followers to date):
- Ignored the behavior of the advance core.
- Tried to counteract only convergence, which is the effect, instead of counteracting the cause of instability, i.e., the deformation of the advance core.
- Used deformable linings, which allow the ground to deform and provide negligible confinement to the core.

Figure 39. Opening of the tunnel under the Tames in the early 1800s, when sequential excavation was conceived

Figure 40. 1964 Cadillac Fleetwood 60 Special Sedan, produced when the NATM was conceived

- Let the ground deform and tried to mobilize the strength of the ground. In squeezing conditions, this practice allows the ground to start creeping, which is an irreversible phenomenon and is very difficult (if not impossible) to control by acting only on the cavity.
- Did not have the technology to preconfine the core.

Ironically, continuing using the sequential excavation was a *consequence* of Rabcewicz's choices (not Rabcewicz's choice), which led him (and all of his followers to date) to give up on full face excavation, i.e., Rabcewicz's goal itself.

We now know much more than in 1960s, we have much improved technology (both in design and construction), we can deploy much more computational and construction power, and we have a complete design and construction approach that allows us to advance full face in all stress-strain conditions; it works with and without a tunnel boring machine. This approach has been proven in over 500 km of tunnels, the majority of which in difficult tunneling conditions. As for the United States, proceeding full face is just going back to the roots of early American tunneling. In the end, none of us rides horse and buggy (Figure 38), nor wear crinolines or top hats (Figure 39) anymore. Let's update our tunneling approach as well!

We may still listen to the Beatles, but we do not take the risk and (fuel) cost of driving a 1964 Cadillac Fleetwood (Figure 40) across the US. Why should owners (and, eventually, taxpayers) across the US (and across most of the world) take the risk and pay the cost entailed in a 1964 tunneling approach?

REFERENCES

Barisone, G., Pelizza, S. And Pigorini, B. (1982). Unbrella arch method for tunnelling in difficult conditions—analysis of Italian cases. Proc. IV International Association of Eng. Geol., New Dehli, Vol. IV, Theme 2, 15–27.

Bernaud, D., Rousset, G.; (1992). La "Nouvelle Méthode implicite" pour l'Etude du Dimensionnement des Tunnels. *Rev. Franç. Géotech.* n° 60, pp.5–26 (juillet 1992).

Bernaud, D., Rousset, G.; (1996). The New Implicit Method for Tunnel Analysis (Short Communication). *Int. J. Numerical and Analytical Methods in Geomechanics,* Vol. 20, pp. 673–690.

Brown, E.T. (1981). Putting the NATM into perspective. Tunnels & Tunnelling, November 1981, 13–17.

Carrieri, G., Fiorotto, R. Grasso, P., and Pelizza, S. (2002). Twenty years of experience in the use of the umbrella-arch method of support for tunneling. International Workshop on Micropiles, Venice, Italy, May 30th–June 2nd 2002.

Corbetta, F., Bernaud, D., and Nguyen-Mihn, D.; (1991). Contribution à la Méthode Convergence-Confinement par le Principe de la Similitude. *Rev. Franç. Géotech.* n° 54, pp.5–11 (janvier 1991).

Guo, C.; (1995). *Calcul de Tunnels Profonds Soutenus, Mèthode Stationnaire et Mèthodes Approchèes.* Ph.D. Thesis, ENPC, France.

Karakuş, M. and. Fowell, R.J (2004). An insight into the New Austrian Tunnelling Method (NATM). KAYAMEK'2004-VII. Bölgesel Kaya Mekaniği Sempozyumu / ROCKMEC'2004-VIIth Regional Rock Mechanics Symposium, 2004, Sivas, Türkiye.

Kovári K. (1994). Erroneous concepts behind the New Austrian Tunnelling Method, Tunnels & Tunnelling, November 1994, Vol. 26, 38–42.

Lunardi, P. (2004). Design and Construction of Tunnels. Springer.

Lunardi, P., Cassani, G. and Gatti, M.C. (2008) Design aspects of the construction of the new Apennines crossing on the A1 Milan-Naples motorway: the base tunnel.

Nguyen-Minh, D. and Guo, C.; (1993.a). A Ground-Support Interaction Principle for Constant Rate Advancing Tunnels. *Proc.EUROCK '93,* Lisboa, Portugal, pp. 171–177.

Nguyen-Minh, D. and Guo, C.; (1993.b). Tunnels Driven in Viscoplastic Media. *Geotechnique et Environment Coll. Proc. Franco-Polonias coll.* , Nancy, France.

Nguyen-Minh, D. and Guo, C.; (1996). Recent Progress in Convergence Confinement Method. In *Proc. EUROCK '96* (G. Barla ed.), 2–5 September 1996, Torino, Italy, Vol.2, pp. 855–860. A.A. Balkema, Rotterdam.

Nguyen-Minh, D., Guo, G., Bernaud, D., Rousset, G.; (1995). New Approches to Convergence Confinement Method for Analysis of Deep Supported Tunnels. *Proc. 8th Cong. Int. Soc. Rock Mech.* Tokio 2, pp. 883–887. A.A. Balkema, Rotterdam.

Nguyen-Minh, D.; (1994). New Approaches in the Convergence Confinement Method for Analysis of Deep Supported Tunnels. *Corso su "Trafori Alpini e Gallerie Profonde,"* Politecnico di Milano, Dip. di Ingegneria Strutturale, 19–23 Settembre 1994.

Panet, M., Guenot, A.; (1982). Analysis of Convergence behind the Face of a Tunnel. *Tunnelling '82.* The Institution of Mining and Metallurgy. London.

Piepoli, G. (1976). La nuova galleria S. Bernardino della linea Genova-Ventimiglia. Ingegneria Ferroviaria, 10.

Rabcewicz L. (1964). The New Austrian Tunnelling Method, Part one, Water Power, November 1964, 453–457, Part two, Water Power, December 1964, 511–515

Rabcewicz L. (1965). The New Austrian Tunnelling Method, Part one, Part Three, Water Power, January 1965, 19–24.

Romero V. 2002. NATM in soft-ground: A contradiction of terms? World Tunnelling, 338–343.

Sandström, G.E (1963). The history of tunnelling, underground workings through the ages. Barrie and Rockliff.

PART 16

Shaft

Chairs

Greg Hauser
Jay Dee

Scott Hoffman
Skanska

NEW TECHNOLOGY CHANGES BLIND SHAFT DRILLING

Alan Zeni ▪ Frontier Kemper Constructors

Blind shaft drilling equipment has not changed substantially since the 1970s. Since it resides in a niche market between conventional shaft sinking and raise boring, the relatively small demand has resulted in some stagnation of development. In the 1990s the technology found a larger foothold in foundation and deep piling work for civil construction. WIRTH GmbH from Erkelenz, Germany has been servicing this market for many years with pile top drilling rigs and tools. Frontier-Kemper Constructors, Inc. recently entered into an agreement with WIRTH to develop a new generation of large diameter blind drilling equipment more specifically aimed at mining needs. The, recently completed, DHI-240 is the first rig to apply the most recent technology. This machine made its commercial debut in the fall of 2008. A description of the technology and a short case study of the first project will be described.

THE DHI-240 RIG

The DHI-240 was specifically built for drilling large diameter shafts using the blind drilling method. The blind drilling process is a hybrid of rotary drilling such as is used when drilling for oil, gas or water wells. A typical drilling rig consists of a hoisting mechanism designed to handle the entire weight of the below ground drilling tools and a rotary table to provide the rotating action needed to cut the rock. The third element is weight from heavy tools and the fourth is some method of circulation for removal of the cuttings. The blind drilling method for drilling large diameter shafts uses a very similar mechanism and technique except the tools are much larger and move slower. An oilfield rig is built for speed like a powerful sports car, the blind drilling rig, on the other hand, is more akin to an all wheel drive truck or jeep made for slow going on steep or rough ground, a similar amount of power is applied in a different way to obtain a different result.

The DHI-240 is the product of a joint venture between FKCI and WIRTH GmbH. The specialty drilling equipment such as the rotary table, elevators, traveling block, swivel, kelly and all the bottom hole tools are WIRTH components designed for use on an L-35 MP rig. All of the other components: hoist, mast, substructure, hydraulic drive and control station were designed and built by FKCI. The DHI-240 improved upon the L-35 MP in several important ways:

- The draw-works are driven with a 300 HP variable frequency AC electric motor. This allows the draw-works to be fail safe and easily controlled with dynamic braking, automatic parking and emergency brakes.
- The 382-ton capacity mast is capable of tilting hydraulically to make rigging, tool handling and casing installation easier and safer.
- The rig has the necessary clearance for handling 20 ft. (6.0 meter) diameter drilling assemblies.
- The rig is arranged on a low substructure supported by a concrete foundation. All components pin-in to the substructure for simple and fast assembly.

Figure 1

Figure 2

The sub also carries a sliding table which operates similar to a guillotine allowing the bottom holes tools to land below the rotary table.
- The sliding table can be operated with the main rotary table in place.
- The sliding table, pipe handler, elevators and back leg mast tilt cylinders are operated by a handheld remote control from any location on the rig.
- The freestanding operator's cabin is equipped with control panels for hoisting and rotary functions. It is equipped with electronic touch screen controls and component monitoring.
- The electrical distribution and draw-works power containers reside directly behind the draw-works and cables are made to the proper length and are easily connected.
- Twin 400-HP hydraulic units are used for maximum power to the rotary and provides backup so that production can continue in the event one unit fails. The rig can be operated satisfactorily with one unit. The hydraulic units

Figure 3

Figure 4

are housed in a 40 ft container with quick connections for hydraulic lines conveniently located on the outside of the container.
- Two high pressure 800 CFM x 300 PSI compressors are housed in a 40 ft container. Simple electrical and air connections allow these compressors to be set up quickly. A 1,000 gallon common air receiver simplifies air distribution to the circulation and utility systems and results in more consistent pressure control.

The DHI-240 is capable of blind boring shafts to a maximum 20 ft. (6,096 mm) diameter. It is the only blind drilling rig in the US designed specifically to excavate this size shaft. With a conventional concrete lining of 12" (305 mm) thickness a finished shaft of 18 ft. (5,486 mm) can be constructed without the need for a conventional sinking operation. The ability to excavate using the blind drill can save from 2 to 6 months of construction time on ventilation or production shaft projects.

Figure 5

Figure 6

The large diameter blind shaft drilling tools currently in use in the US are built to work with 13 ⅜" drill pipe which has been a standard since the 1960s. The pipe is 12" ID N-80 API casing with threaded connections designed by Hughes and Drilco Industrial for high torque service. This drill pipe is capable of driving cutter heads up to a maximum of about 16 ft. in diameter. Beyond this limit, the pipe will tighten in the drilling process and cause the connections to fail. The WIRTH 330 mm pipe is built with flanged connections that are much stronger and are capable of withstanding the torque required to drill diameters up to 24 ft. The flanged connection also offers another significant advantage for blind drilling. The pipes that conduct the compressed air to the bottom of the tool assembly are mounted to the outside of the pipe and are connected through the flanges. In addition to being more convenient to use, they eliminate the need for an internal air injection line that interferes with the transport of cuttings to the surface. The superior strength and carrying capacity of the drill pipe are the most important elements that make it possible to drill to 20 ft. diameter.

The WIRTH rotary table has a torque capacity of 300,000 foot-pounds (420 Knm) which can drive the 20 ft. cutterhead up to 17 RPM. The maximum reasonable speed to rotate the 20 ft. head, however, would be no more than 5 or 6 RPM. At that speed and

Figure 7

using approximately 150 tons of thrust weight, the 20 ft. head will advance at approximately .75 to 1 foot per hour.

THE BLACK PANTHER PROJECT

The first application of this machine and technology on a commercial endeavor is the Return Shaft for the Black Panther Mine No. 1 near Oaktown, IN. This shaft is approximately 350 ft. deep and drilled to 20 ft. raw diameter. The shaft will be lined with concrete having a finished size of 18 ft. Until now, this shaft would have to be sunk conventionally. Raise boring is precluded because of the requirement for the shaft to be complete before the underground development reaches the shaft location. The conventional option would require at least 6 months to complete with a crew of about 30 persons. Conventional operations involve drilling, blasting and mucking. It also requires ventilation and water control as well as temporary ground support. In contrast, the blind drilling operation requires 10 to 12 persons, none of whom are required to work underground. No ventilation or temporary ground support is needed because all of the excavation and mucking work is done underwater. The water also provides any ground support needed.

The lining will be cast in place concrete, jump formed from the top down. The drilling water is removed just below the lining so that there is very little if any deterioration of the shaft walls before the concrete is placed.

In addition to the cost savings gained by using only one third of the labor, the drilling operation also shortens the excavation time by at least one half.

Set up of the DHI-240 on the Black Panther site was completed in about 3 weeks including the Christmas holidays. On completion of the collar and concrete foundation, the rig was set-up in about the same time as the conventional sinking equipment.

At this writing, the drilling operation is ongoing. The excavation is expected to take less than one month to complete.

The Black Panther project is the first attempt to blind drill a 20 ft. diameter shaft in the US. This project will be the first benchmark for this technology and will represent a real alternative to conventional sinking for large diameter ventilation shafts. This technology also adds another dimension to the use of blind drilling in the civil construction market, complementing pile top machines and opening up new applications for mechanical shaft drilling.

TAMERLANE HOIST AND VERTICAL BELT PROJECT

Shawn Collins ▪ Frontier Kemper Constructors

BACKGROUND

The history of the Pine Point Mine is a rich one. This property once boasted one of the largest and most profitable lead-zinc mines in Canada's long and storied mining history. The Pine Point mine used open pit technology to extract the ore, but they faced several challenges. In addition to logistics problems with haul distances and metal prices, the mine faced the challenge of a large in flow of water from the Great Slave Lake. Eventually, the Pine Point Mine was closed leaving behind a large deposit of lead-zinc ore. Following the closing of the Pint Point Mine, Tamerlane Ventures Inc. acquired this property which includes approximately 70 million tonnes of lead-zinc reserves. The initial phase of this project will concentrate on the R-190 deposit as a test site to prove their new approach for extracting the ore. The R-190 reserve is believed to have approximately 1,014,000 million tonnes of high grade ore varying from 5% to 52% combined lead-zinc. Once the pilot project proves the approach is valid, the process will be repeated in approximately 33 other ore bodies on the property.

This paper is written to discuss the design process to determine the most effective means of placing production and service equipment in the same shaft. Currently the design work and the mining scheme is complete, but lead-zinc pricing has prevented Tamerlane Ventures Inc. from moving forward with the project.

LOCATION

The property that Tamerlane Ventures Inc. has acquired is the Pine Point Mine in Canada's Northwest Territories, just south of the Great Slave Lake (Figure 1).

FEASIBILITY STUDY

The preliminary studies conducted by Tamerlane Ventures Inc. concluded that an underground mining facility would be the most effective means of extracting the ore, but this approach does present some challenges. Because a total of thirty four separate underground operations would eventually be required to mine the different ore bodies, it was essential to minimize the shaft, equipment, installation and operating costs.

INITIAL DESIGN CONSIDERATIONS

To prevent inflow of ground water, Tamerlane Venture Inc. determined that freezing a perimeter around the ore body would be the best approach. The idea was to sink a single shaft, approximately 185 m (600 ft) deep, to keep the initial cost of the project low. Shaft sinking options included blind boring or conventional shaft sinking depending on the diameter of the shaft the equipment dictated. After sinking the main shaft, a drift would then be driven across the ore body in preparation for a 2.47 m (8 ft) diameter raise bored shaft for ventilation. FKC-Lake Shore's focus on the project was to encorporate production and service equipment into the single shaft.

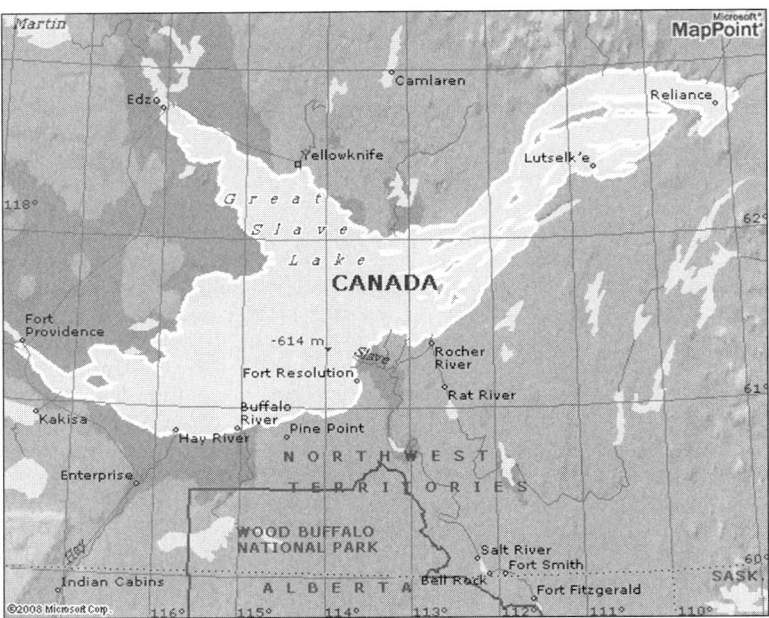

Figure 1. Map of Great Slave Lake and Pine Point

The main challenge to incorporate production and service equipment into a single shaft for this project is exposed by comparing the approach to typical mining or tunneling applications. In typical underground mining applications, the life of mine is substantial. The initial cost of the shaft, production system and service system is relatively small in comparison to the long term operating costs of the mine. However, in this case, the underground operation at each ore body location is approximately one year in duration. Therefore, the shaft, production system and service system costs are relatively large in comparison to the operating costs. Considerations for keeping the initial costs as low as possible were essential. The approach of putting service equipment and production equipment into a single shaft is not new in the tunneling world. Tunneling jobs are typically done from a single launch shaft where equipment and men access the underground workings while muck is removed from the tunnel. For this reason, the proposed approach would seem to more closely resemble a tunneling operation. However, it is different in a couple of important ways. First, in tunneling the shafts are relatively shallow in comparison to this shaft which is to be 600 ft deep. While minimizing the cost of sinking the launch shaft is certainly in the best interest of a tunneling project, it is less critical than this Tamerlane's application. In the majority of examples from tunneling, the shaft and hoisting equipment represent a relatively low cost in comparison to the tunnel excavation. Second, the number of operations make this project different from tunneling. Once the pilot program is successfully completed in the R-190 ore body, the process will be repeated thirty three more times in different ore bodies. With a tunnel, generally one launch shaft is used despite the tunnel length and the process is not repeated multiple times.

To create a successful design, it was necessary to rely upon experience in both mining and tunneling.

PHASE 1—VERTICAL CONVEYOR

Production Hoist vs. Vertical Conveyor

Production hoisting has been around for many years and is a reliable means to remove material from underground; while the vertical conveyor is a more recent technology, but has proven successful for certain applications. Both systems must be considered to understand which one makes the most sense. Below is a list of pros and cons for both system.

Production Hoist

Pros
- High tonnages are possible
- Deep applications are possible
- Technology is proven

Cons
- Shaft steel & skip loading station is required
- Power consumption is high
- Electrical switches are required in the shaft
- Large footprint is required
- Installation is relatively expensive

Vertical Conveyor

Pros
- Shaft steel is not required
- Power consumption is low
- Small footprint is required
- Installation is relatively inexpensive

Cons
- Shaft depth is limited
- Production rates are limited

After examination of the project parameters and discussions with the owner, it was concluded that a vertical conveyor would be ideal. The vertical conveyor would use only a small portion of the shaft for production allowing area for a larger service cage. This would also maximize the area for available for ventilation and minimize the shaft diameter required (Figure 2).

The production design parameters for the R-190 pilot project were as follows:
- Required Production Rate: 220 tph
- Lift Height: 185 m (600 ft)
- Material : Lead-zinc ore
- Material Density: 2 tons/m^3 (130 lbs/ft^3)
- Raw Feed: Constant by hopper

Based on the design criteria above, the belt that was specified was as follows:
- Base Belt : XST 5400 N/mm
- Cleat Type: TC-G 220

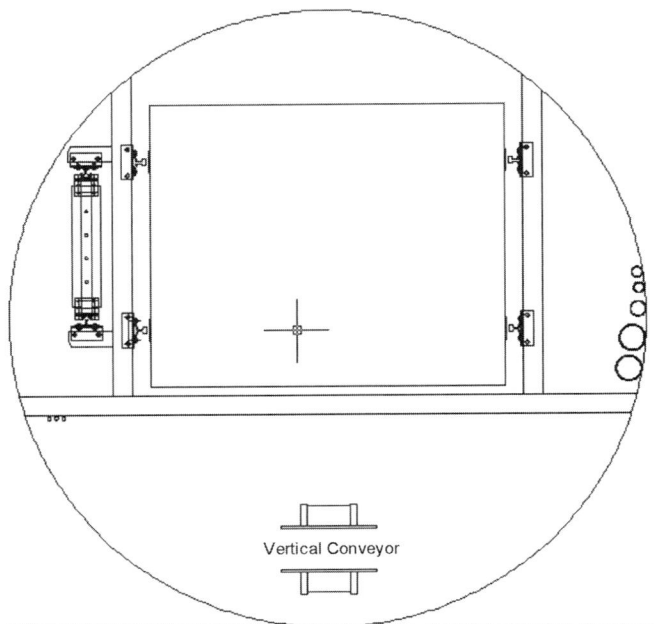

Figure 2. Plan view of 22' diameter shaft

- Cleat Pitch: 333 mm (13.11 in)
- Sidewalls: F240 S
- Overall Belt Width: 1,000 mm (39.37 in)
- Design Fill Factor: 78.89%

General Description

The vertical belt system is laid out in an "S" configuration (Figure 3) with one drive pulley, roller curve pulleys, upper deflection pulley, lower bend pulley, take-up pulley and a lower deflection pulley.

The surface structure incorporates the discharge chute and rock boxes to feed the surface stacking conveyor. The head pulley elevation is approximately 10.8 m (35 ft) above grade. The surface structure is fully enclosed and insulated to survive the harsh conditions in winter.

The underground structure is turned 90 degrees from the top structure, putting a twist in the belt. The mine will provide consistent flow of 220 tph by means of a storage bin and feeding hopper underground. Loading of the material into the vertical belt is handled by a short inclined conveyor.

Belt Specifications

The vertical belt is composed of a single strand of steel cord belting 1,000 mm (39 in) wide constructed with sidewalls along both sides. There is approximately 200 mm (7.87 in) of free space between the edge of the belt and the sidewall. In between the sidewalls, cleats are installed to carry the material. The sidewall and cleat bases are cold vulcanized to base belt, and the cleats are then bolted to the sidewalls as well as the cleat bases.

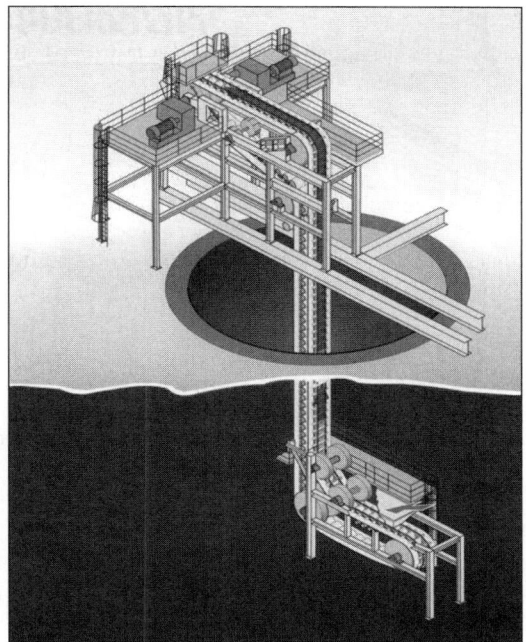

Figure 3

Drive Electrical Control

The vertical conveyor is controlled via the use of a variable frequency drive. The VFD allows for flexibility to adjust speeds for production and inspection. In addition, emergency stop pushbuttons, belt alignment switches and sequence/slip switches are used to ensure safe operation.

Drive Pulley Assembly

The drive pulley assembly is composed of a 1,600 mm (63 in) diameter × 1,100 mm (43.3 in) wide straight face pulley with external pillow block bearings mounted on a common shaft. The pulley has 12.7 mm (½") thick neoprene lagging. The pulley is driven by a right angle gearbox and a DC motor.

Roller Curve Pulleys

There are ten roller curve pulleys in the system. Each roller is 300 mm (11.81 in) diameter × 1,100 mm (43.3 in) wide straight faced with pillow block bearings.

Upper & Lower Deflection Pulleys

The upper deflection pulley turns the empty belt from horizontal to vertical allowing the belt to enter the shaft after traveling past the drive pulley. The upper deflection pulley is 1,600 mm (63 in) diameter and each of its two discs has a 200 mm (7.87 in) face width. Each disc supports the belt free space between the edge of the belt and the sidewall.

The lower deflection pulley, which turns the belt from horizontal to vertical allowing the loaded belt to enter the shaft on its way up, is 1,600 mm (63 in) diameter and each of its two discs has an 200 mm (7.87 in) face width.

Lower Bend Pulley

The lower bend pulley is composed of a 1,600 mm (63 in) diameter × 1,100 mm (43.3 in) wide straight face pulley with external roller bearings mounted on a common shaft. The bend pulley turns the empty belt from vertical to horizontal underground.

Take Up Pulley

The take-up pulley is composed of a 1,600 mm (63 in) diameter × 1,100 mm (43.3″) wide straight face pulley with external pillow block bearings mounted on a common shaft. The pulley is mounted on a screw type take-up frame with 1,220 mm (48 in) of travel for belt tensioning.

Guide Rollers

Four sets of guide rollers, including four rollers per set, are used at the top and bottom of the vertical conveyor to train the belt and prevent it from running off the edge of the pulleys. They do not extend down into the shaft on the surface nor up into the shaft underground, making examination and maintenance convenient.

Surface Structure

Due to weather conditions, the headframe is designed to be totally enclosed and insulated. However, access doors are incorporated to facilitate the repair, removal, and/ or replacement of all major components. One side of the surface structure is removable allowing placement of a belt clamp for initial belt installation and maintenance if it becomes necessary.

Lifting System

A monorail trolley system is designed to facilitate the removal and replacement of major components excluding the vertical belt.

Underground Clean Up Conveyors

A small horizontal cleanup conveyor will catch any spillage from the vertical belt underground and put it onto a small vertical conveyor. The small vertical conveyor will dump the spillage back onto the inclined feed conveyor which will take the spillage back to the feed belt (Figure 4).

PHASE 2—SERVICE HOIST

General Description

After the vertical belt components were established and the area needed for ventilation was finalized by the owner, we turned our attention to the service hoist. Because the service hoist would be the only way to get mining equipment underground, it was necessary to design for both the largest piece dimensionally and also the heaviest piece. In addition, we had to keep in mind that the shaft diameter needed to be minimized as much as possible. The heaviest piece of equipment to be lowered down the shaft was 11,340 kg (25,000 lbs). To optimize the floor space and to make sure the largest piece of equipment would fit inside the cage, the design team decided on a tall cage with an overhead crane, allowing large pieces to be hung vertically in the cage. While the process of hanging a large piece of equipment in the cage is somewhat time consuming,

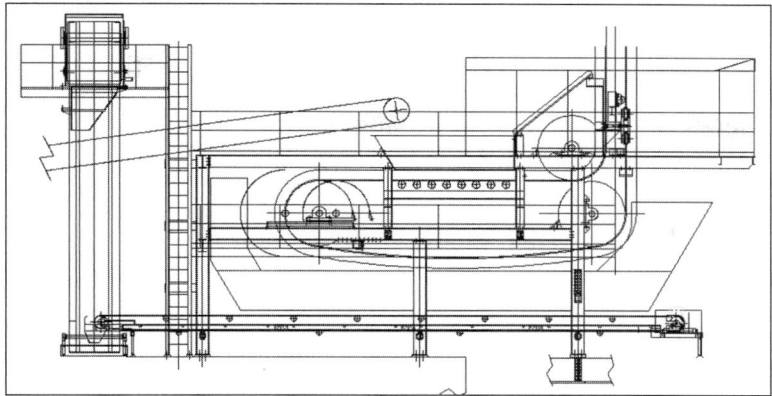

Figure 4. Underground structure including horizontal and vertical cleanup conveyors

it was a good trade off for the owner because they knew it would only be required on startup of the mine and those rare occasions when a major repair was necessary.

A choice then had to be made as to the most appropriate type of hoist for this application: drum winder (no counterweight), over/under, ground mounted friction or tower mounted friction. Each of these hoists can be a good selection under the right circumstances. Here are a few pros and cons for each type, which were taken into consideration before making the final choice.

Drum Winder

Pros

- Design is simple
- Installation is simple
- Headframe and shaft steel are simple because crash timbers and buffers are unnecessary

Cons

- Power consumption is relatively high because there is no counterweight
- Rope safety factors limit load capacity

Over/Under

Pros

- Design is simple
- Power consumption is relatively low due to counterweight
- Headframe and shaft steel are simple because crash timbers and buffers are unnecessary

Cons

- Rope safety factors limit load capacity
- Hoist house cannot be located in the headframe

Ground Mounted Friction

Pros

- Power consumption is relatively low due to counterweight
- Large loads are possible due to multiple ropes
- Deep shafts are possible

Cons

- Design is more complex due to multiple ropes, friction liners and tension ratios to avoid slipping
- Headframe and shaft steel are complicated because crash timbers and/or buffers are needed
- Hoist house cannot be located in the headframe
- Multi rope sheaves are expensive
- Deeper shaft sump is required

Tower Mounted Friction

Pros

- Power consumption is relatively low due to counterweight
- Large loads are possible due to multiple ropes
- Deep shafts are possible
- Sheaves are not required
- Separate hoist house is not required

Cons

- Headframe is more complicated due to mounting the hoist
- Shaft steel is more complicated because of crash timbers and/or buffers
- Deeper shaft sump is required

After evaluating the options, the decision was made to use a tower mounted friction hoist. In part, the cold winters and the desire to enclose the entire headframe drove the decision to move the hoist inside the headframe. Preliminary layouts were completed to make sure that a tower mounted friction hoist could give us the appropriate rope center lines while maintaining the appropriate D/d (Drum diameter to rope diameter) ratio of 80:1. We concluded that it was not only a feasible option, but the best option, so we and moved forward with the details of the design as described below.

Hoist

The hoist is designed for the maximum load condition of 11,340 kg (25,000 lbs) ton running 3.1 m/s (600 fpm) in the shaft. To satisfy the load and speed requirements, the installed power required is 224 kw (300 HP). The hoist is 2.54 m (100 in) in diameter to give the appropriate rope centerlines and D/d ratio for the 32 mm (1.25 in) diameter hoist ropes. For this friction application, the angle of wrap around the hoist drum is 180°. The hoist drum has redundant disc brakes, a right angle gearbox and a DC motor. The disc brakes are spring applied hydraulic release for failsafe operation and each system is cable of holding 125% of the full load capacity. The hydraulic package is designed to release the brakes and includes redundant dump circuits to ensure the brakes are able to set in the event of a clogged dump valve.

Headframe

The headframe will accommodate a friction hoist mounted on a concrete floor in top of the headframe. The main columns of the headframe are designed to rest on the shaft collar with the back legs to rest on individual foundations to be built off of the collar. The headframe is approximately 21.6 m (70 ft) high to the hoist level. Electrical controls are housed on a second level. The headframe is completely enclosed with covers and insulation and includes swinging gates on two sides for access to the cage. One side allows access to the counterweight. The headframe design includes tapered wooden retarders to arrest the cage if it over travels into the headframe.

Cage

The service cage is a key component to the overall success of the approach. The cage has a single deck with an overall height of approximately 11.1 m (36 ft), and has a capacity of 11,340 kg (25,000 lbs) to handle the maximum material load. The cage consists of a man deck and a hinged roof approximately ten feet above the man deck. The most common use for the cage will be man trips, therefore the hinged roof will typically be left in place. However, for large loads the hinged roof can be lifted out of the way by means of the hand crank winch. Once the hinged roof is out of the way, the upper crosshead with overhead crane will be exposed.

The overall weight of the cage is approximately 14,569 kg (33,000 lbs). The man deck is made of plate with 9.5 mm (⅜ in) thick skid rails for ease of loading and unloading supplies. The deck is approximately 3.7 m (12 ft) long × 3.1 m (10 ft) wide, and 3.86 m (12.5 ft) face to face of the guides, with gates on each end for man containment. The cage sides are expanded metal except for the bottom 457 mm (18 in) which are 6.35 mm (¼ in) steel plate (Figure 5).

Guide roller assemblies are used to ensure safe and smooth operation. The cage is equipped with eight units, each unit consisting of three rollers mounted to a common bracket. Guide shoes with removable wear bars are provided as a backup system for the guide rollers.

Counterweight

The counterweight was designed by taking the cage weight and adding 50% of the full load capacity.

Cage Weight + 50% Full Load = Counterweight

14,569 kg (33,000 lbs) + 11,340 kg (25,000 lbs) / 2 = 20,639 kg (45,500 lbs)

The counterweight uses six replaceable wear shoe assemblies for operation on the guide rails.

Shaft Steel & Guide Rails

Bunton beams are attached to the shaft wall with top and bottom brackets which bolt to the shaft wall with screw type concrete anchors. Stub beams for the counterweight guides are welded to the main bunton beams. To decrease the installation time and expense, the bunton sets are on 4.63 m (15 ft) vertical spacing rather than the more typical 3.1 m–3.7 m (10 ft–12 ft) spacing. A typical bunton set, as described, weighs approximately 386 kg (850 lbs). The cage is guided by four 85 lb rail guides, and the counterweight by two 60 lb rail guides.

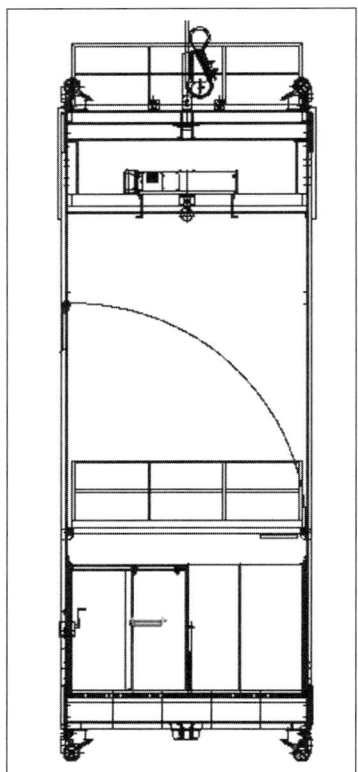

Figure 5. Service cage

Hoist & Balance Ropes

In the past, most friction hoisting applications would have used six strand ropes with two right hand lay and two left hand lay to minimize problems with twisting. For this application, we decided on four 32 mm (1.25 in) NRHD (non-rotating high density) ropes for their resistance to rotation and high breaking strengths. Two 41 mm (1-⅝ in) diameter balance ropes are also required to offset the weight of the four hoist ropes.

CONCLUSION

The unique approach that Tamerlane Ventures Inc. proposed to take in order to extract the ore at Pine Point along with harsh conditions made for a challenging design process. By working closely with the owner we were able to create a successful design scheme that is ready to be implemented. Using freeze technology, Tamerlane Ventures Inc. will be able to avoid the costly water problems that plagued the mine in years past. Using the vertical conveyor, underground mining becomes a cost effective means of extracting ore by minimizing the shaft diameter, installation of shaft equipment and power requirements. The tower mounted friction hoist provides the hoisting capacity necessary for the heaviest piece of mining equipment while maintaining the smallest possible footprint. The multi-functional cage design makes normal man riding convenient and safe, but also allows for the occasional large or heavy loads to be taken

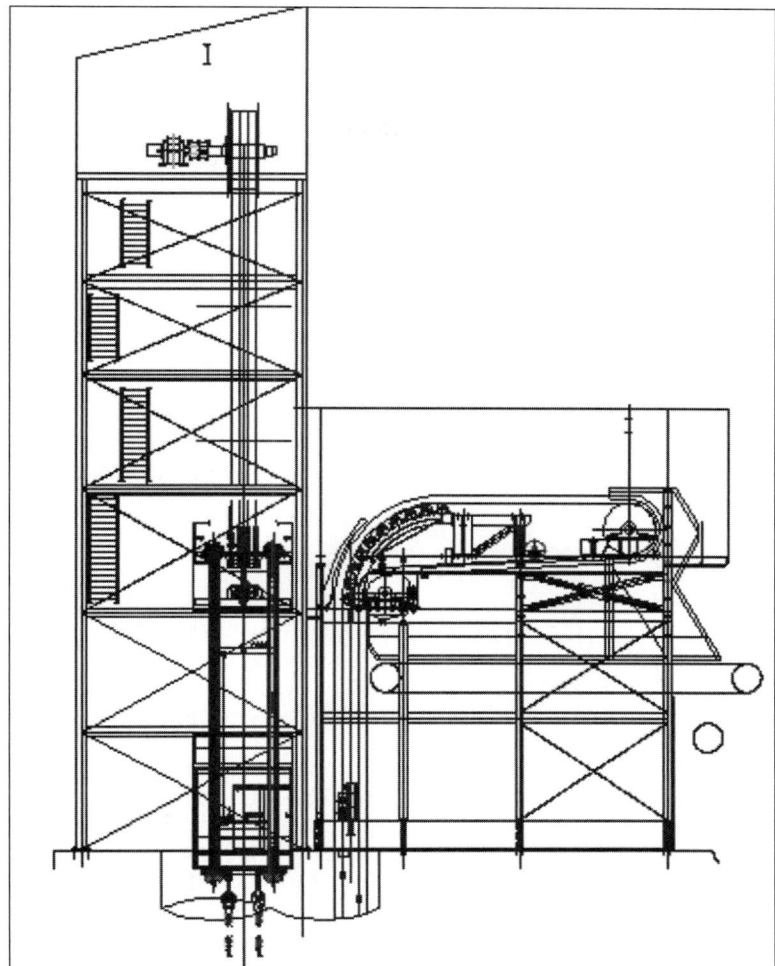

Figure 6. Service hoist and vertical belt headframe

underground. By installing the 12.5 ton overhead crane in the cage, the footprint of the cage floor was optimized affording Tamerlane the opportunity to decrease the shaft diameter to 22'. In the end, effective communication from the owner of their needs along with careful exploration of design options for production and service equipment enabled us to overcome the hurdles and present a whole new approach to mining (Figure 6).

ACKNOWLEDGMENTS

A special thanks to David Swisher and Justin Smoak of Tamerlane Ventures Inc. for their input and dedication to this project. Thank you to the design team at FKC-Lake Shore for commiting to put together a solid, workable design and helping craft this paper.

A SMALL DIAMETER SHAFT DESIGN ALTERNATIVE

Brian E. Gombos ▪ Wade Trim Associates

David D. DiPonio ▪ Wade Trim Associates

ABSTRACT

During early stages of the detailed design for Detroit's Upper Rouge CSO Tunnel Project, a Value Engineering Recommendation suggested the design team investigate alternative concepts to the usual cast-in-place or precast concrete lining system for the Project's many smaller diameter Deaeration Vent Shafts. The resulting design investigation produced not only an alternative lining system, but lead to a proposed non-conventional approach to shaft installation means and methods through soil and rock necessary to construct the permanent shaft lining system. This paper will outline and summarize the technical issues, contractual restrictions, and cost considerations that eventually lead to the final shaft design.

INTRODUCTION

The Rouge River passes by approximately 1.5 million people, through 48 cities, and flows over 203 km (126 mi) before entering the Detroit River. During wet weather events, the river receives pollutants from both storm water runoff and overflow from the sewer systems. The Detroit Water and Sewerage Department (DWSD) has taken the initiative to improve the quality of the Rouge River by implementing the Upper Rouge CSO Tunnel Project (URT).

The URT Project evolved following the enactment of the U.S. Government's Clean Water Act, after the Federal EPA issued their National Combined Sewer Overflow Control Strategy in 1989. In anticipation of regulatory reforms, DWSD created their CSO Team in 1993 to begin planning and evaluating various measures that would lead to regulatory conformance. In 1996, DWSD submitted its Long Term CSO Control Plan (LTCSOCP) which recognized the Detroit and Rouge Rivers as the two receiving waterways that were most affected by the Department's CSO discharges. The LTCSOCP addresses 78 outfalls that have historically discharged around 20 billion gallons of combined sewage into the Detroit and Rouge Rivers annually. The LTCSOCP has been recently updated to include recommendations for future control of wet weather CSO discharges in the Detroit River; however, the URT project represents the final step towards meeting the water quality objectives established by the Clean Water Act along the Rouge River.

Preliminary design of the URT was completed in 2005. Final design began in 2006, following the award to Jacobs Engineering, assisted by Arcadis; Tucker, Young, Jackson, Tull; Multi Tech Resources; SOMAT; SIGMA; and Wade Trim.

The storage tunnel and conveyance system that the URT Project represents was conceived in 1996 and the first contract got underway in 2008. Construction of the URT Project is expected to take seven or eight years to complete.

PROJECT DESCRIPTION

When completed, the Upper Rouge Tunnel (URT) will capture and store up to 210 MG of combined sewage, controlling overflows from 28 outfalls along the Rouge River. The 11.3 km (7 mi) long, 9.1 m (30 ft) diameter, 49 m (160 ft) deep, storage tunnel will receive flow about 55 times per year. Only one of these 55 inflow events will nearly or completely fill the tunnel causing an overflow to the local receiving waters, with a frequency projection of less than one per year on average.

In addition to the mainline storage tunnel, the project includes three in-line shafts; one 125 MGD dewatering pump station; 14 near surface facilities, that include collector sewers, diversion structures, bar racks and flap gates, drop shaft inlet structures, surge conveyance conduits, surge control structures, surge control volume storage shafts, and deaeration vent shafts; 11 off-line drop shafts, deaeration chambers, and adits; and outfall modifications.

The near surface system (Figure 1) functions by diverting flows from the existing system via a diversion structure located downstream of the existing in-system control structures. The flow is then directed through a series of flap gates and through the surface collector conduit to either a tangential vortex or plunge type inlet, after which the flow drops vertically through the drop shaft into the deaereation chamber approximately 46 m (150 ft) below grade where entrained air is allowed to evacuate to the surface through the small diameter deaeration vent pipe. The flow is then transported to the main line storage tunnel via adits that vary in length from several hundred feet to nearly 610 m (2,000 ft).

In addition to the CSO conveyance components, the near surface facility utilizes a series of structures to passively control and partially contain hydraulic surge waves associated with uncontrolled filling of the tunnel. A surge control structure located above the surface collector conduit allows rapidly rising water that reverses direction and travels from the URT back up the drop shaft, generated by a surge wave moving through the system, to travel vertically upwards towards, and often above, grade level. When the water approaches the maximum expected hydraulic grade line due to surge, it flows through a series of weir openings. The water that is able to flow through the openings then plunges nearly 15 m (50 ft) into a surge conduit, where it is directed to the surge control volume shaft. The surge control volume shaft stores the volume of water that is generated by a surge wave that rises above the control elevation of the weirs in the surge control structure. The stored volume is retained in the shaft until the levels in the URT begin to drop, at which time the water is passively reintroduced to the drop shaft through several flap gates.

GROUND CONDITIONS

The ground conditions in southeastern Michigan consist of glacial drift, characterized as clayey tills, outwash sands and gravels, and glaciolacustrine silts and clays that are underlain by glacial till. The underlying bedrock consists of layers of sedimentary rocks comprised of shales, limestones, and dolomites that slope or dip inward from the rim of the Michigan Basin toward the center of the basin. Overlying the bedrock is often a layer of highly over consolidated glacial till locally referred to as "hardpan."

The URT Project alignment and tunnel profile encounters three distinct bedrock formations (Figure 2). In the northern reach of the project, the glacial drift is underlain by the Antrim Shale formation, comprised of hard, brittle, bituminous shale. This formation has historically produced methane gas. Near the center of the project alignment, the bedrock formation underlying the glacial drift transitions to the Traverse Group. This formation has potential for producing both methane and hydrogen sulfide gases. The Traverse Group is directly underlain by the Dundee Limestone formation, comprised of hard limestone and dolomite that has historically produced oil and gas, and often contains groundwater under artesian pressure that can contain dissolved sulfides and methane.

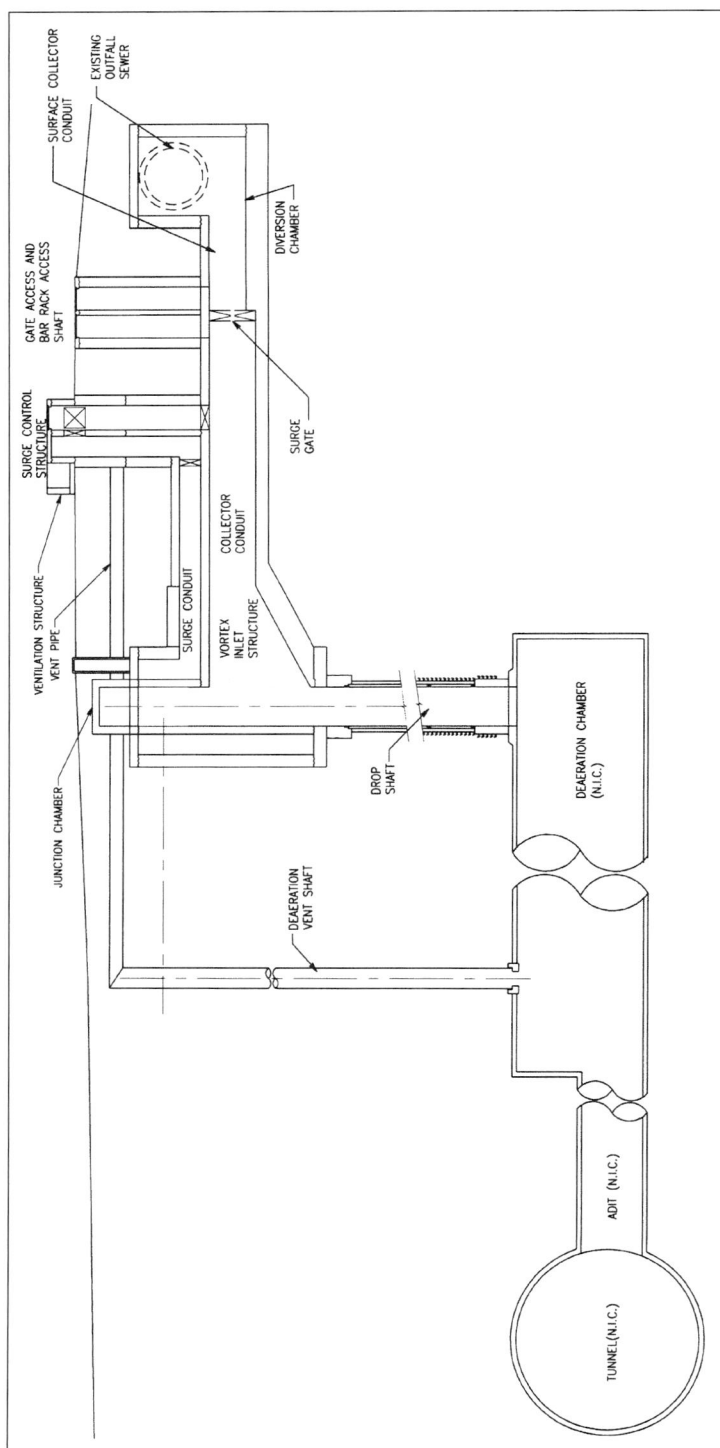

Figure 1. Near surface facility structures key plan

A SMALL DIAMETER SHAFT DESIGN ALTERNATIVE

The long term static groundwater is typically located 3 to 4.5 m (10 to 15 ft) below grade level, and is dependent upon seasonal variations of perched groundwater from the adjacent floodplain and granular surface deposits that are hydraulically influenced by the Rouge River. Additionally, groundwater is also present in silt and sand seams found within the deeper glaciolacustrine clay deposits. Deep granular layers at or near the soil-rock interface may be hydraulically connected to the underlying bedrock aquifers and have greater recharge capacity. The confinement of these layers from the overlying clay often results in artesian groundwater conditions. The underlying bedrock formations are also capable of producing artesian pressure conditions. These formations are hydraulically connected and recharged by higher elevation ground located north of the project area.

The overburden soils along the project alignment are fairly uniform. The surface topography varies by as much as 12 m (40 ft); however, the subsurface strata are fairly uniform, varying in thickness from 11 to 29 m (35 to 95 ft). The surficial fill, consisting typically of stiff silty or sandy clay, varies from several feet to as much as 4.5 m (15 ft). Below the fill is approximately 9 to 21 m (30 to 70 ft) of natural soil deposits consisting generally of silty clay with varying layers of sand and silt. The unconfined compressive strength of the clay varies from approximately 168 kPa (3,500 psf) near the top of the deposit where the clay is characterized as stiff to very stiff, to less than 48 kPa (1,000 psf) for the lower soft clay strata. Below the natural clay soils is the over consolidated clay "hardpan" with thickness ranging between 0 and 6 m (0 and 20 ft). Gravels, boulders, and rock fragments are present at the soil-rock interface that occurs approximately 11 to 29 m (35 to 95 ft) below grade.

OPERATIONAL REQUIREMENTS

Hydraulic and ventilation performance, as well as the configuration of drop shafts at 14 locations, were studied during design development of the near surface facilities and connecting structures which integrate the shallower existing DWSD collection system into the proposed 9.1 m (30 ft) diameter deep storage and conveyance tunnel.

The design team initially presented the concept of combining the drop conduit and ventilation conduit required at each of the drop shaft locations into a single deep shaft excavation large enough in diameter to consolidate both of these vertical structures. The design of a proposed joint conduit arrangement in a single shaft excavation has been accomplished before in other deep storage and conveyance tunnel systems and remains in successful operation at a number of existing CSO facilities without any known problems associated with that design characteristic.

Uniqueness of the Upper Rouge Tunnel System Prevails

One of the underlying operational and design requirements imposed by the DWSD was to make this CSO system *passive*. In other words, it was to be a system without active inflow control gates. The only gates to be incorporated in the system would be for surge or backflow.

The passive control concept and its related hydraulic analysis required the design of a number of large diameter surge control towers which extend 6 to 9 m (20 to 30 ft) above existing ground surface elevation. A hydraulically designed horizontal separation distance between the deaeration vent shaft and the drop shaft was required. The hydraulic designers also insisted that the drop and vent conduits maintain vertical plumbness for their full depth. The final analysis included all of these conditions and resulted in the need for two separate shafts at each of 11 "offset" near surface facility collection locations. The three other collection locations which were "inline" with the main tunnel had their flow and ventilation conduits incorporated into permanent large diameter access shafts.

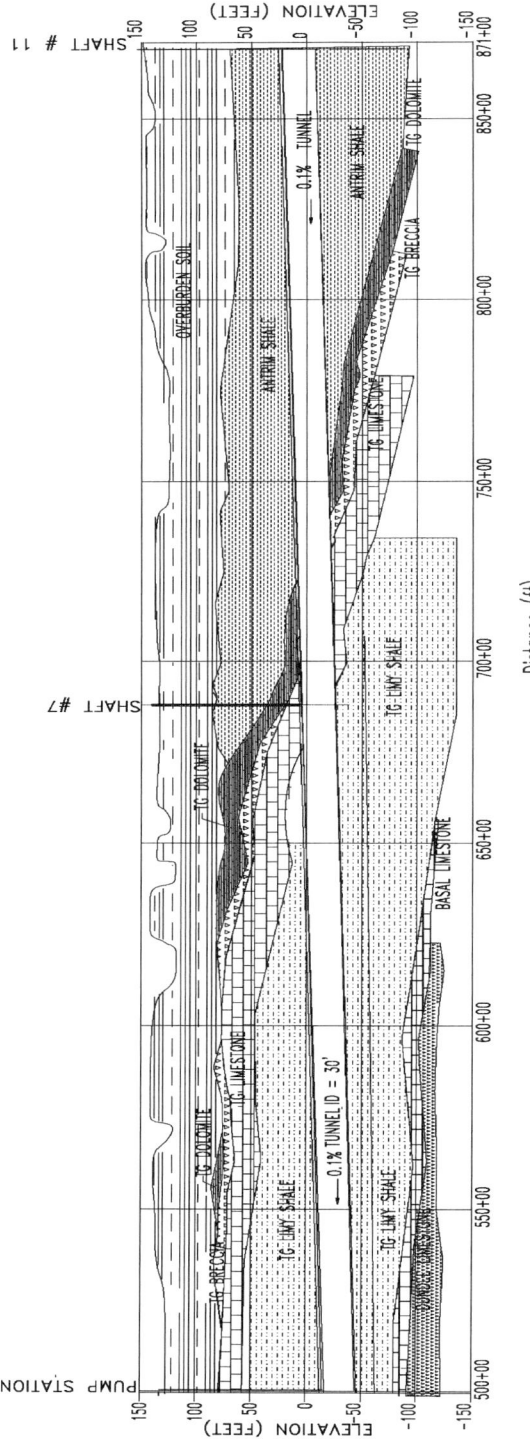

Figure 2. Generalized soil and rock profile

A SMALL DIAMETER SHAFT DESIGN ALTERNATIVE

BASIS OF DESIGN PARAMETERS AND CRITERIA

At a 30% design level, an independent Value Engineering (VE) Study was conducted. Since implementation and operation of "consolidated or combined drop shaft and vent shaft" systems would not occur on the URT Project, the design staff urged the VE team to explore any areas that could result in technically sound innovative approaches to all of the "off-line" small diameter deep shafts, ranging from 1 to 6 m (3 to 20 ft) finished diameter, proposed for the URT Project.

The VE team proposed the use of alternative "thin-walled" pipe material products for both the drop shafts and vent shafts, instead of the more conventional precast or cast-in-place reinforced concrete pipe and lining systems that are ordinarily used. The VE team cited that if a thin walled pipe lining alternative were found to be technically suitable, they would be much easier to install, especially if they were characteristically lighter and manufactured in longer lengths.

Who Builds the "Off-Line" Drop Shafts and Vent Shafts?

Obviously, the Contractor that builds the drop shafts and vent shafts has access to the deaeration chamber and adit below. That is, of course, if those facilities below are contained within the same Contract as the "off-line" shafts.

The design team was urged by DWSD to have this project engage in as much local economic development as possible. Heavy civil construction projects requiring deep underground excavations are considered a highly specialized segment of the industry that is not often found locally. Due to this consideration, a decision was made to keep the tunnel adits and deaeration chambers part of the main tunnel construction contracts and local participation would be found elsewhere within the project. Being conscious of the need to support the local economy, the design team decided to group the "off-line" shafts into each of the respective near surface facility contracts which were anticipated to be performed by smaller Detroit area construction firms.

Therefore, all of the proposed structures, including the horizontal and vertical conduits that connect to the existing outfall sewers on the upstream end and the deaeration chamber on the downstream end, are contained in Contracts separate from the Tunnel Contract (see Figure 1). The interface between these Contracts is at the drop shaft and vent shaft connections to the deaeration chambers. The sequencing and coordination outlined in both Contracts is based upon having the drop shafts and vent shafts constructed prior to the tunnel and the respective adits and deaeration chambers, and thus requires top down construction and precludes a raise bore approach.

CONSTRUCTION METHOD ANALYSIS

The designers for the near surface facilities were encouraged by the VE team's initiative to explore technically sound alternative approaches to the more conventional use of reinforced concrete shaft final linings. The shaft designers investigated a number of "thin walled" pipe materials that could serve as an acceptable final lining for vent shaft and drop shaft risers providing that all concerns and technical considerations were complied with.

Before departing from the conventional top down approaches, the shaft designers were responsible for determining that vertical loadings of precast or cast-in-place concrete on the structures below would not appear to be an issue for any of the diameters of drop shafts and vent shafts on the URT Project. This was necessary since stacking precast concrete riser sections would be the obvious fallback should no alternatives be deemed acceptable. In all cases, due to the interface between the Tunnel Contracts and the Near Surface Facility Contracts, the completed shafts with their final linings are contractually required to be in place and self-supporting prior to the start of all adjoining tunnel, adit, and deaeration chamber excavations.

For any alternatives considered, conclusive evidence would need to be provided to ensure that all construction joints would be as dependable as the gasketed tongue and groove or bell and spigot joint that standard precast RCP provides.

An item of considerable concern to the design team was the need for continuous personnel entry into the shaft throughout every phase of construction, i.e., excavation, ground support, and final lining installation. During interviews, experienced members of the construction industry indicated that a 0.5 to 1 m radial annulus would be required to be added to the shaft excavation diameter to accommodate personnel access and to maintain plumbness or required vertical tolerances.

As the designers conducted their in-depth analysis of small diameter shaft lining designs and related construction alternatives, they found there was significant potential to safely install the appropriately sized final lining with a number of pre-fabricated products or materials with limited personnel entry required (see Table 1).

Design Alternative Limited to Vent Shafts…But Could Include Drop Shafts!

Eventually, the shaft design analysis was limited to deareation vent shafts that were hydraulically sized at a 1.7 m (5.5 ft) finished diameter since it was assumed they would provide the greatest economy of savings. There was testimony, however, from representatives of contracting companies in the mining industry who specialize in deep shaft drilling for public works projects that insisted they could be competitive and technically compliant on larger diameters up to 6 m (20 ft), which would have also included most of the drop shafts. Since the design team was more familiar with conventional means and methods for larger shaft sizes, it was decided to limit their "formal shaft design" efforts to vent shafts that would be "blind drilled" and proceed with designing the drop shafts, considering more conventional "top down" construction methods.

There would be Contract provisions, however, to permit the successful Contractor, on a post-bid basis, the opportunity to propose technically and economically acceptable alternatives to the design of the drop shafts, if they desired to do so. The design team recognized that exercising that option might invite the Contractors to propose similar "blind drilled" means and methods used to construct the larger drop shafts. The designers assumed that any Contractor proposed alternatives would likely be formulated to have their designs conform more closely to those of the smaller vent shafts and potentially generate an even greater economy of savings to the Project, due to repetition of similar work and an inherent economy of scale.

The initial design phase of the vent shafts involved the assessment of ten different shaft lining systems in order to arrive at the final design. All were considered constructible. Half were determined to be not technically viable or had installation issues. The two lining systems that the design team felt were the most technically preferable were:

- Cast-in-place reinforced concrete
- Steel casing liner with a corrosion protection system

In the final analysis, the cast-in-place reinforced concrete lining, which was assumed to be placed by way of a slip lining operation, was estimated by the designers to cost more per vertical foot than the "blind drilled" steel casing liner system permanently installed with a corrosion protection coating. Therefore, a consistent rapid excavation and lining process for all 11 deareation vent shafts on the Project, averaging a depth of 46 m (150 ft) each, was expected to yield a significant cost savings to the Project.

Table 1. Final lining selection matrix

Upper Rouge CSO Tunnel Project 5' 6" (66") ID Vent Shaft Pipe Analysis							
Pipe Material	Weight (lb/ft)	Laying Length (ft)	Economy	Efficiency	Safety	Comments
RCP	1830	6, 8	2	1	1	Standard C-76-III RCP manhole pipe
CMP (plain)	157	20	1	1	1	10 GA Standard galvanized **(determined not technically viable)**
CMP (bituminous)	177	20	1	1	1	10 GA with bituminous coating **(determined not technically viable)**
CMP (polymer)	160 +/-	20	1	1	1	10 GA with polymer coating **(determined not technically viable)**
CCFRPM (Hobas)	317	4, 8, 10, 20	1	1	1	Add'l cost per foot for lengths smaller than 20 foot lengths **(installation issues)**
AWWA C301(E)	1470	16	3	1	2	PCCP--precast concrete cylinder pipe (pressure pipe) **(installation issues)**
CIP Concrete (R)	by design	full shaft depth	3	2	2	continuous "slipline" operation
CIP Concrete (P)	by design	full shaft depth	3	2	2	continuous "slipline" operation
Steel caisson	by design	full shaft depth	2	3	3	"Blind Drill" liner installed with Corrosion Protection System
Steel pipe/ drilled	by design	full shaft depth	2	3	3	"Blind Drill" excavated and lined w/ permanent carbon steel casing-- not finished

KEY: 1=Least Measure 3=Greatest Measure
Economy=Installation Costs
Efficiency=Quality, Schedule, handling
Safety=Personnel entry

DETAILED DESIGN CONSIDERATIONS

The design of the deaeration vent shafts' initial support and final lining was required to accommodate the contractual restrictions, limited personnel entry, and necessary coordination associated with the staged bidding schedule of eight independent contracts. In addition, the design of the final lining was to provide for a design life of 100 years, while being exposed to an aggressive environment, including alternating wet-dry cycles, dissolved gas in soil and rock, chemical and biological reactions prevalent in sewers, as well as structural loading, including the static pressure due to soil and groundwater, and extreme transient pressures generated by surge waves.

To allow flexibility in means and methods and to promote economy during the bidding phase, the design of the initial support of the overburden soil is a Contract requirement for the Contractor. The Contract, however, provides a recommended method of shaft support and final lining installation that includes certain specific design features to ensure excavation stability during excavation and installation of the final lining. These include Contractor design of all temporary excavation support systems conforming to detailed design criteria presented as part of the Contract Documents; pre-excavation grouting; minimum embedment of the excavation support system into competent rock; and construction of the vent shafts without dewatering.

Several types of technically and economically viable temporary excavation support systems were evaluated for support of the overburden soil. These included a drilled steel casing, steel ribs with timber lagging, and liner plates. The evaluation included preliminary analysis of these systems considering both a dewatered excavation, where possible, as well as a fully submerged condition. The temporary drilled steel casing was ultimately recommended as an economical and technically sound ground support method that does not require dewatering of the shaft excavation, limits potential problems at the often highly permeable soil-rock interface, and limits construction personnel entry to the shaft.

The general concept for temporary support of shaft excavation and installation of the deaeration vent shaft final lining (Figure 3) anticipates that a temporary drilled steel casing will be installed to support the overburden soil while the shaft is being excavated. It was envisaged during design that the drilled casing will be installed through the overburden and terminate in competent bedrock, below the soil-rock interface that exhibits high hydraulic conductivity. The excavation design will likely include pre-excavation grouting of the soil-rock interface zone to ensure adequate and uniform seating of the casing and reduction in hydraulic conductivity of the fractured rock zone. Following installation of the drilled casing, the overburden soils will be excavated by soft ground flight auger from inside the casing to the top of rock. Upon reaching the rock horizon, the excavation will continue to the terminal elevation by changing to a blind drilling operation utilizing reverse circulation technology. Upon completion of the excavation, the vent shaft final lining will be sequentially submerged into final position by introducing ballast fluid into the field assembled liner. Following installation of the ventilation liner to final position and alignment, the annular space between the final lining and excavation limits will be filled with grout through a series of grout placement pipes that are fabricated to be integral with the final lining assembly.

The structural design of the thin walled cylindrical shaft lining included evaluation of external loads, including earth pressure in the overburden soils, hydrostatic pressure due to groundwater present in the soil and rock, construction surcharge loads, distortion during handling, and unequal grouting pressures. The lining thickness was governed by the structural requirements of the lining to resist buckling. Internal loading resulting from a transient surge wave with pressure head exceeding 43 m (140 ft) above the crown of the deaeration chamber was also accounted for in the design of the lining, connections, and appurtenances. Additionally, the analysis and design included the affects of initial deformation by assuming deviation from the Contract tolerances for verticality and radial out-of-roundness.

The design and detailing of the liner construction incorporates measures to minimize field fabrication. The liner is designed to be installed by field-assembling several prefabricated sections. The length of the recommended component pipe sections was selected to minimize the number of joints required to be constructed in the field, yet allow the maximum length of pipe that is practical for transportation to the construction site. In addition, to remain consistent with the basic design assumption that man entry into the shaft excavation was to be minimized, the detailing includes features such as centralizers (to maintain concentricity within the shaft excavation) and integral grout tubes, installed prior to sinking the shaft liner that allow primary, secondary, and remedial grouting of the annulus. All stiffener rings or bands, gussets, centralizers, and flanges for field connections are intended to be shop-welded. The minimal number of liner connections necessary to be made in the field is intended to be bolted, gasketed, stiffened flange connections.

To accommodate known Contract interfaces at the deaeration chamber, as well as coordination that likely will need to occur between near surface facility subcontractors that are installing the shafts and the shallow horizontal runs of ventilation piping, temporary bulkheads were incorporated at the top and bottom terminal locations of the shaft liner as temporary isolation measures. The bulkhead at the base of the liner, at the interface with the top of the deaeration chamber which is the interface between two

A SMALL DIAMETER SHAFT DESIGN ALTERNATIVE

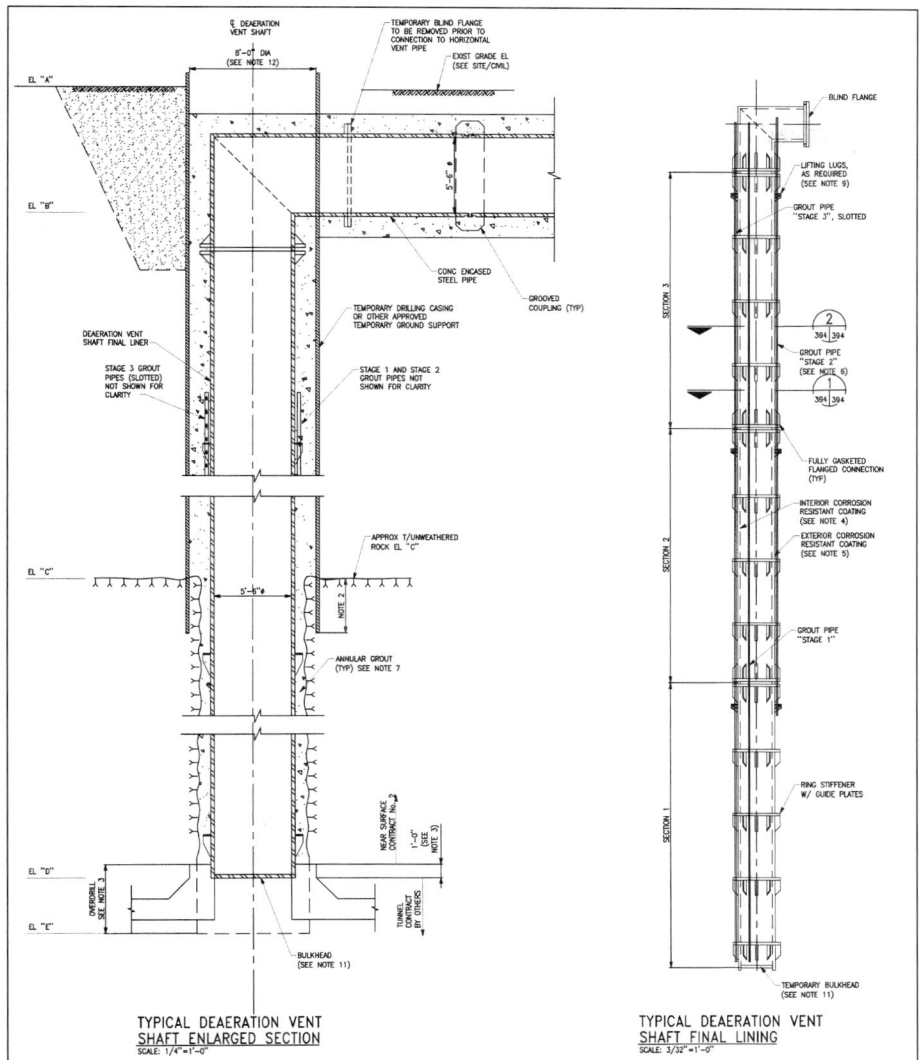

Figure 3. Typical deaeration vent shaft section

construction Contracts, is designed to be a water tight bulkhead resisting long-term hydrostatic and artesian pressure acting from the exterior, as well as hydrostatic pressure on the interior resulting from the liner being potentially flooded at the time of excavation of the deaeration chamber. In addition to the bulkhead, the Contract requires that the shaft excavation extend 1 m (3 ft) below the final steel lining and be filled with controlled low strength material (CLSM) to aid the tunnel contractor in locating and subsequently removing the bulkhead. The bulkhead will be removed by the tunnel contractor that constructs the deaeration chamber and adits for the respective shaft location. It is anticipated that the upper limit of work for the subcontractor installing the shaft will be at the top of the shaft final lining, and that a different, more conventional underground or process piping contractor will complete the remaining horizontal run of ventilation pipe that extends from the top of the deaeration ventilation shaft to the top of the drop shaft, which then continues to the terminal venting chamber located above the surge control structure. The Contract requires a 90 degree fabricated fitting or elbow with bolted connection at the top of the deaeration vent shaft, with a bolted blind flange or similar bulkhead at the transition to the horizontal section of ventilation piping. This is intended to provide temporary isolation of the shaft work and closure of the shaft to prevent accidental entry of personnel, material, debris, or surface water. Additionally, although the top of the ventilation shaft will be below grade by design, the bolted connection will facilitate easier removal should maintenance access be required in the future.

Durability of the final lining system was identified as critical design criteria to achieve the intended service life of 100 years with minimal maintenance or repair. Chemical and biological constituents of the CSO flow can create generally aggressive environments to unprotected ferrous metals. In addition, the alternating wet and dry cycles created by tunnel filling and ventilation, and the occasional surge wave, increase the potential for corrosion. Surging water traveling through the deaeration ventilation shaft under high pressure at a relatively high velocity, likely carrying suspended solids, also generates an abrasive effect on the steel substrate. To protect the steel lining system from the effects of corrosion and abrasion, a high build elastomeric polyurethane coating system was specified. The exterior surface of the steel lining system, although subject to less potential for corrosion because it will be permanently installed in soil and rock and within the grouted annulus between the excavation support system and final liner, was also specified to receive the high build coating system for greater long term durability. The coatings are envisaged to be shop-applied under strict quality control with field application only at the joints and as required for touch up.

CONSTRUCTION TOLERANCES

Maintaining the ventilation shaft tolerances for shaft verticality, as well as vertical and horizontal control, is critical for the Contract interfaces at the deaeration chamber. The near surface facility Contract under which the shafts are to be constructed requires maximum allowable deviation from plumb for the final lining of 51 mm (2 in) per 30.5 m (100 ft) with a cumulative total deviation in the overall depth of the shaft not to exceed 76 mm (3 in). The deviation from horizontal location at the deaeration chamber is specified not to exceed 100 mm (4 in). The terminal location of the steel final lining at the bottom of the shaft was intentionally designed to include a 300 mm (12 in) overlap with the contract limits of the deaeration chamber final lining. The tolerances for vertical location of the bottom of the ventilation shaft final lining are plus or minus 25 mm (1 in). Tolerances for radial out-of-roundness are specified to be plus or minus 13 mm (0.5 in) at any location. The Contract will require a rigorous geotechnical instrumentation and monitoring program to monitor soil, rock, utility, and structure deformations; groundwater observations; and vibration.

PLANNED BIDDING SCHEDULE

The 11.3 km (7 mi) long URT Project consists of eight construction contracts with two tunnel contracts of equal length, five near surface facility contracts, and a pump station contract. Procurement began in 2008 getting work underway on the downstream tunnel contract and first near surface facility contract. Procurement is expected to continue during 2009 through 2011 for the second tunnel contract and remaining near surface facility contracts. The pump station contract is scheduled to be bid in 2013.

CONCLUSION

The URT design effort, initiated during the early phases of engineering with the help of a Value Engineering Recommendation, produced not only an alternative lining system, but one which is somewhat non-conventional from a civil public works perspective. This concept incorporated the mining industry's proven approach to shaft installation means and methods through soil and rock necessary to construct a complete permanent shaft lining system from the surface.

At the conclusion of the analyses, the Wade Trim design team discovered that blind shaft drilling in conjunction with the advancement of a steel casing liner permanently installed with a corrosion protection system was judged to be the most efficient and economical, while providing the safest construction method of all lining systems considered. Wade Trim's findings were then endorsed by the Project Design Manager and communicated to all Near Surface Facility design teams.

The result of this effort was accepted by the lead consultant, Jacobs Engineering, and DWSD and is reflected in the final design of the deaeration vent shafts for Contract PC-766 of the Upper Rouge CSO Tunnel Project.

The deaeration vent shaft design suggesting blind shaft drilling with steel final linings for DWSD Contract PC-766 is expected to provide the Project with considerable schedule, cost, and risk reduction advantages.

ACKNOWLEDGMENTS

The authors wish to acknowledge and extend their appreciation to their client, Jacobs Engineering, and the project's owner, the Detroit Water and Sewerage Department.

REFERENCES

Parsons Brinckerhoff Michigan, Inc. 2005. Upper Rouge River CSO Tunnel Basis of Design Report. Prepared for the City of Detroit, Water and Sewerage Department, April 2005.

Parsons Brinckerhoff Michigan, Inc. 2005. Upper Rouge River CSO Tunnel Basis of Design Geotechnical Interpretive Report. Prepared for the City of Detroit, Water and Sewerage Department, August 2005.

Jacobs Engineering 2006. Upper Rouge CSO Tunnel Supplemental Basis of Design Report. Prepared for the City of Detroit, Water and Sewerage Department, September 2006.

Jacobs Engineering 2007. Upper Rouge CSO Tunnel Geotechnical Interpretive Report. Prepared for the City of Detroit, Water and Sewerage Department, September 2006.

KANSAS RIVER TUNNEL SHAFT DRILLING

Clay Haynes ▪ Black & Veatch

Cary Hirner ▪ Black & Veatch

Clay Griffith ▪ Southland Contracting, Inc.

ABSTRACT

While the Kansas River Tunnel is a relatively short, small diameter tunnel, the shaft excavation methods implemented were long and large on innovation and use of reverse circulation drilling equipment. ATS Drilling, L.P. used mechanical excavation techniques for both soil and rock segments of the 42.7 m (140 feet) and 38.1 m (125 feet) deep shafts that were 2.9 m (9.5 feet) and 4.1m (13.5 feet) diameter, respectively.

INTRODUCTION

Water District No. 1 of Johnson County, Kansas (Water One) is a quasi-municipal government, similar to a city government. Water One's sole purpose is to supply potable water to eastern Johnson County, Kansas. Formed in 1957, Water One covers 69,927 hectare (270 square miles) of service area; serves over 380,000 customers; supplies water to 15 cities; and has a current treatment capacity of 473,110 l/min (180 million gallons per day (mgd)). The Water One customer service area is depicted on Figure 1.

To meet future water demands in Johnson County, Water One implemented the Phase V program which is projected to be completed in 2009. The Phase V program consists of the following major elements:

- 78,850 l/min (30 mgd) collector well
- Raw water pipeline from collector wells to the water treatment plant
- 78,850 l/min (30 mgd) softening with membrane filtration water treatment plant
- 29 km (18 miles) of open cut 60-inch steel transmission main
- 426.8 m (1,400 feet) long, 60-inch steel lined tunnel

The Kansas River Tunnel is a critical trenchless crossing to convey treated water to Water One's customer base as depicted on Figure 2.

REGIONAL AND LOCAL GEOLOGIC CONDITIONS

Johnson County is located in eastern Kansas bordering the state of Missouri and the cities of Kansas City, Kansas and Kansas City, Missouri. The bedrock beneath Johnson is Pennsylvanian age sedimentary rock consisting primarily of shale, limestone and sandstone.

The bedrock at the Kansas River Tunnel consists of a few feet of decomposed Galesburg Shale, Bethany Falls Limestone, Hushpuckney Shale, Middle Creek Limestone, Ladore Shale, Sniabar Limestone, Mound City Shale, Critzer Limestone and Tacket Shale depicted on Figure 3.

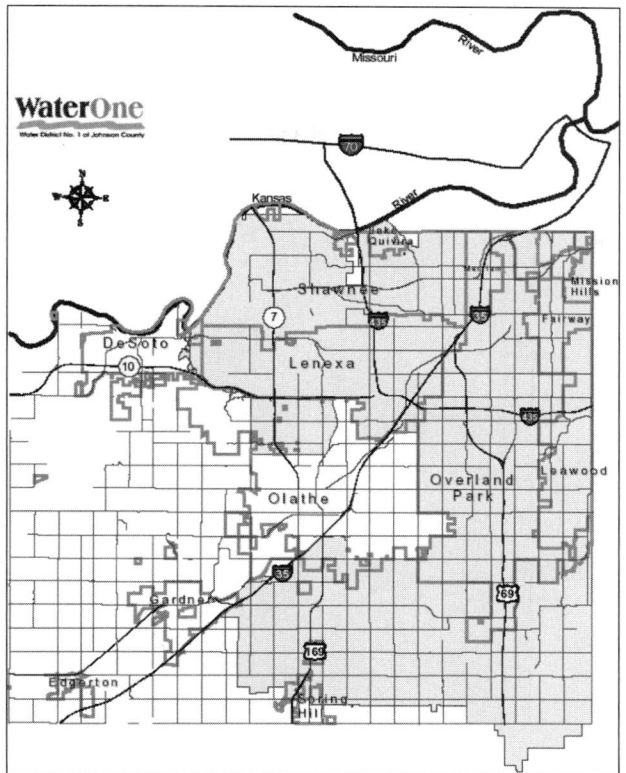

Figure 1. Water One customer service area

DECISION MAKING—CROSSING METHOD

Aesthetic, cost and reliability considerations immediately eliminated an aerial crossing from consideration. A comprehensive decision making process was employed to determine the acceptable construction method for the Kansas River crossing. Dreding, horizontal directionally drilling (HDD), microtunneling and tunneling in bedrock were evaluated.

There were several reliability concerns with dredging. The first major concern was that ongoing sand dredging operations in the river could damage the proposed transmission main. The second major concern was that the steep banks may fail if the river migrated or floods damaged the river banks. For these reasons, this method was eliminated.

The proposed main diameter of 1.5 m (60-inch) would be nearly a world record diameter for a HDD installation. Specifying a 1.5 m (60-inch) diameter HDD main would severely limit the contracting pool. Hydraulic considerations would necessitate two 1.2 m (48-inch) transmission mains as an alternative to a single 1.5 m (60-inch) main. The geology and geometry of the river crossing would require the HDD installation to be excavated in the bedrock. Discussions with prominent HDD contractors convinced us that two 1.2 m (48-inch) HDD mains in bedrock would be cost-prohibitive.

Microtunneling in either the bedrock or the soil was judged to be cost-prohibitive. Thus, tunneling was selected as the crossing method.

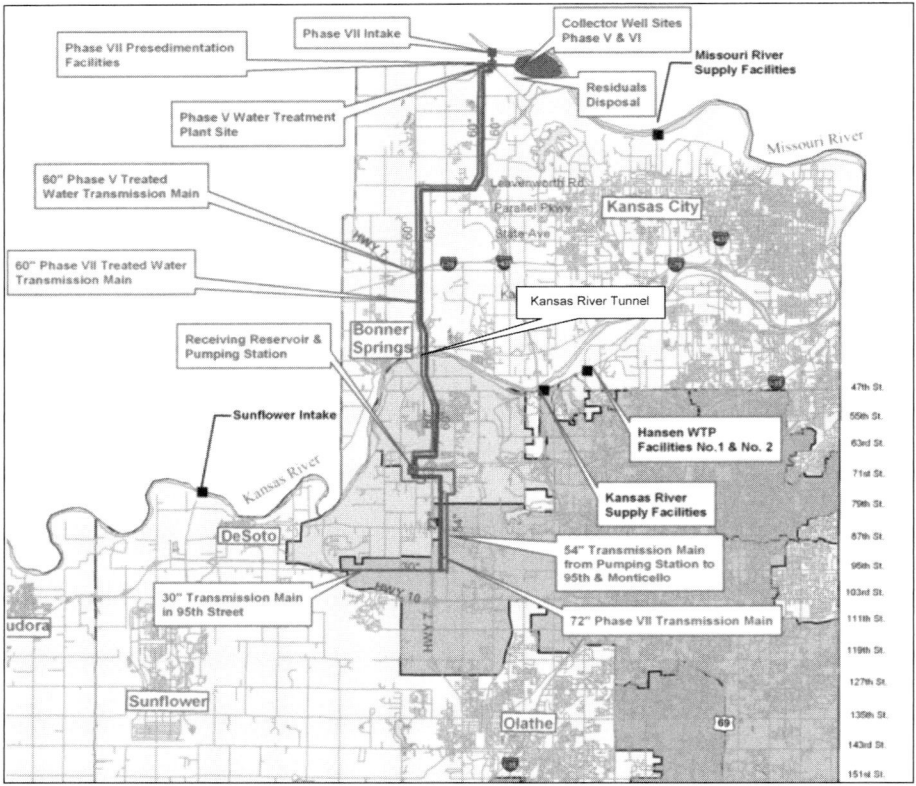

Figure 2. Phase V program components

GEOTECHNICAL INVESTIGATION

After the construction method was selected, Black & Veatch (B&V) conducted two land-based borings in close proximity to the proposed shaft locations. B&V drilled the borings approximately 24.4 m (80 feet) into the bedrock on each side of the river. NQ size core 47.6 mm (1.875 inches) core was taken in the bedrock. Hydraulic packer tests were conducted on the bedrock to determine permeability.

While drilling the two land-based borings, B&V discovered that the Bethany Fallls Limestone, the first competent bedrock formation encountered, occurred at approximately the same elevation on both sides of the river. B&V mobilized a barge to drill boreholes within the river to verify that the Bethany Falls Limestone was continuous across the river length and no deep erosion channel was present. Five borings were drilled at roughly equal spacing across the river coring into the Bethany Falls Limestone approximately 3 m (10 feet) at each location.

Due to budget constraints, no testing was performed on the core before selecting the proposed tunnel envelope.

VERTICAL ALIGNMENT

Considering risk exposure, Owner risk tolerance, and our knowledge of formations in the area, the Critzer Formation was selected for the tunnel envelope.

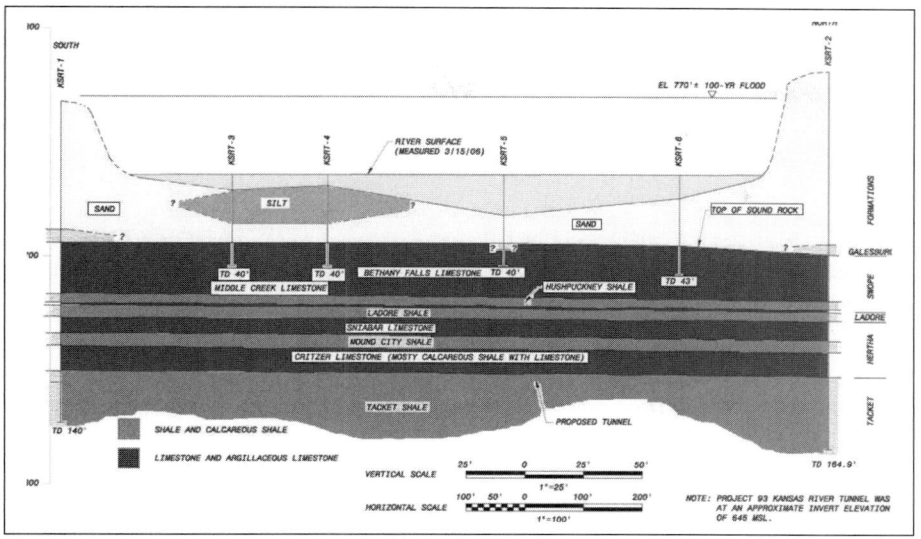

Figure 3. Geologic cross section

ASSUMED SHAFT CONSTRUCTION METHODS

Geologic conditions were a primary factor in evaluating permissible shaft construction methods. However, adjacent infrastructure on the north side of the river had an equally profound influence on the proposed shaft construction methods. The City of Bonner Springs, Kansas groundwater well field for drinking water was within 61 m (200 feet) of our proposed shaft location. Dewatering of the alluvial aquifer in the area could not be tolerated.

Knowing that the north shaft's initial support system in the soil would need to be watertight and realizing that this would make the initial support costly, it was decided that the north shaft would be the smaller retrieval or exit shaft. Black & Veatch assumed that the contractor could construct an enlarged chamber in the base of the shaft in the bedrock to facilitate rotation of their tunnel boring machine (TBM) segments to a vertical orientation for removal. During design a minimum diameter of 2.9 m (9 ft-6 inches) was established for the north shaft but allowed the contractor to construct it to larger dimensions to accommodate his specific means and methods.

Shaft drilling with a steel casing initial support system was specified in the soil section of the north shaft and an initial support system of rockbolts, mesh and shotcrete in the rock section. Black & Veatch anticipated the contractor would employ drill and blast excavation methods in the rock section of the north shaft, although other mechanical methods of shaft excavation were not specifically prohibited. The soil and rock were approximately 24.4 m (80 feet) and 18.3 m (60 feet) thick, respectively at the north shaft.

No well fields or critical infrastructure were located near the shaft on the south side of the river. From construction experience on a shaft in the Kansas River floodplain a few miles downstream and nearby water well data, B&V knew that dewatering the soil would not be economically feasible at this location. The soil and rock were both approximately 18.3 m (60 feet) thick, respectively, on the south side of the river. B&V believed that a two tier sheet piling system with watertight interlocks for sheeting below the water table would be a typical initial support system. While B&V suspected that the

Table 1. Construction bids

Contractor	Amount
Southland Contracting, Inc.	$6,719,727
Kissick Construction Co.	$7,944,786
Affholder, Inc.	$8,890,000

highly decomposed Galesburg Shale would provide an effective seal at the soil/rock interface, B&V did provide a third sheet pile inset that could be tremie grouted at the soil/rock interface to seal off groundwater inflows. B&V also provided a value engineering clause for the contractor to propose an alternative initial support method.

CONSTRUCTION BIDS

The project was bid on February 27, 2007. The bid results were as shown in Table 1.

The Engineer's Opinion of Probable Construction Cost was $8,425,000. The project was awarded to the lowest, responsive bidder, Southland Contracting, Inc.

CONTRACTOR'S PROPOSED CHANGES TO SHAFT CONSTRUCTION— NORTH SHAFT

Southland proposed, through a subcontract to ATS Drilling, L.P., to drill the soil section and install a steel casing as initial support as envisaged in the original design. As previously noted, B&V assumed that the contractor would drill and blast the rock section of the north shaft. However, ATS Drilling, L.P. proposed to use reverse circulation drilling techniques to excavate the rock section. B&V was somewhat skeptical and supplied the contractor cores of the Bethany Falls Limestone, some of the hardest rock in the rock section, for testing. The Contractor tested the rock and confirmed his intention to excavate the shaft using reverse circulation drilling equipment.

ATS, L.P. began drilling the soil section of the north shaft on July 24, 2007. They finished drilling the soil section on July 29, 2007 taking approximately 72 hours to drill and support the 24.4 m (80 foot) soil section. ATS, L.P. began drilling the rock section of the shaft on July 30, 2007 and finished drilling the rock section at the end of the day on August 7, 2007. Discounting a two day break between start and finish, ATS, L.P. drilled the 2.9 m (9 ft-6 inches) diameter, 18.3 m (60 ft) rock portion of the shaft in approximately 96 hours. This equates to an approximately (0.19 m/hr) 0.625 ft / hour rolling average advance rate.

CONTRACTOR'S PROPOSED CHANGES TO SHAFT CONSTRUCTION— SOUTH SHAFT

ATS, L.P. did propose significant changes to the South Shaft design. ATS, L.P. proposed to use a single stage circular steel casing as the initial support for the 18.3 m (60 ft) deep soil portion of the shaft. ATS, L.P. had problems with getting the 4.2 m (13 ft. – 10 inch) diameter steel casing to advance into the ground. ATS, L.P. installed secant piles outside the steel casing and excavated the inner material in the wet. The 4.2 m (13 ft -10 inch) diameter casing was lowered into the steel casing and the annular space between the secant piles and steel casing was tremie grouted.

Reverse circulation drilling of the 4.1 m (13 ft-6 inch) diameter shaft began on December 8, 2007. ATS, L.P. worked 24 hour days to December 15, 2007 when a gear box at the drill rig drive unit seized. The parts for the Wirth drill rig had to be shipped from Germany. ATS, L.P. elected to resume drilling after Christmas. Drilling resumed on December 26, 2007 and was completed during daylight hours on January 5, 2008.

Figure 4. North Shaft drill string

Counting all breakdowns, ATS, L.P. worked approximately 192 hours from December 8 to December 15, 2007, and approximately 252 hours from December 26 to January 5, 2008 on drilling the rock section of the shaft. The total time expended drilling the 18.3 m (60 ft) long rock section of shaft excavation was approximately 444 hours. This equates to a rolling average advance rate of approximately 0.04 m/hr (0.135 ft / hour) for the 4.1 m (13 ft-6 inch) excavated diameter shaft.

THE REST OF THE STORY

Southland Contracting, Inc. (SCI) self-performed most of the remaining work on the project. SCI excavated approximately 426.8 m (1,400 feet) of tunnel using a tunnel boring machine of their design that employed a gripper ring for thrust reaction. The tunnel was supported with a combination of rock bolts and mesh and half-circle corrugated metal segments.

SCI lowered the 6.1 m (20 ft.) long, 1.5 m (60-inch) internal diameter (ID) pipe in a vertical orientation down the south shaft and rotated the pipe horizontal for transport into the tunnel. The steel pipe was blocked in place using steel screw jacks in the upper pipe section. The steel pipe joints were lap welded and tested using the magnetic particle method. A sand-cement grout was placed in several lifts in the annular space between the ground and the steel pipe.

The 1.5 m (60-inch) ID steel pipe were installed in a vertical orientation in the shafts, lap welded internally and externally, tested and backfilled with concrete. Short sections of open cut pipeline were constructed from the shafts to a short distance outside of the shaft perimeter (less than 61 m (200 ft) to allow sufficient buffer space between the tunnel contractor and adjoining open cut contractor.

SCI passed the hydrostatic test on the main on November 23, 2008, a full month ahead of schedule and on budget. Black & Veatch anticipates that the tunnel will be flushed and disinfected after adjoining open cut contracts are completed in the first quarter of 2009.

CONCLUSION

A project does not have to be extremely large for innovative techniques to be employed. In fact, smaller projects are probably the preferred venue for innovation because the consequences of adversity or outright failure are less severe.

Figure 5. South Shaft drill string

ATS, L.P. should be commended for their grit and determination in completing the south shaft using the chosen technique. Using a reverse circulation drill rig in December and January in Kansas is quite a challenge given the harsh winter weather. ATS, L.P. has gained considerable knowledge with this method.

SCI redoubled their efforts to make up schedule consumed by the shaft excavation phase. They finished ahead of schedule and months ahead of the adjoining open cut contracts.

The reverse circulation drilling rates in the rock section of the 2.9 m (9 ft-6 inch) diameter north shaft were comparable with drill and blast shaft excavation advance rates. The reverse circulation drilling rates in the rock section of the 4.1 m (13 ft-6 inch) south shaft were much lower than the advance rates for a typical drill and blast shaft excavation.

ACKNOWLEDGMENT

The authors would like to thank Dan Smith and Brent Lawler for their contributions to this paper.

DESIGN CONSIDERATIONS FOR THE USE OF SLURRY WALLS AS PERMANENT WALLS FOR DEEP RECTANGULAR SHAFT STRUCTURES IN SEISMIC AREAS—SILICON VALLEY RAPID TRANSIT PROJECT

Michael J. Lehnen ▪ Hatch Mott McDonald

Ching Wu ▪ Sunrise Pacific, Inc.

Mike Wongkaew ▪ Hatch Mott MacDonald

James Chai ▪ Santa Clara Valley Transportation Authority

ABSTRACT

Slurry diaphragm walls are commonly used as both temporary excavation support walls, and permanent structural walls, for circular shafts and long rectangular structures (such as underground subway stations); however, they are not commonly used for more compact rectangular or square shafts. In these instances, a second wall is typically cast inside the slurry walls to form the permanent structure. This paper discusses the approach used on the Silicon Valley Rapid Transit Project in California to analyze these compact rectangular types of structures for static and seismic loading, and to determine whether or not slurry walls can be confidently used as both the temporary and permanent walls for two tunnel ventilation shafts.

INTRODUCTION AND PROJECT DESCRIPTION

The Santa Clara Valley Transportation Authority (VTA) intends to construct the Silicon Valley Rapid Transit (SVRT) Project in Santa Clara County, California. This project consists of a 26.2-km (16.3-mile) long extension of the Bay Area Rapid Transit (BART) heavy rail rapid transit system from the planned Warms Springs Extension in Fremont on the east side of the San Francisco Bay, through San Jose—the heart of Silicon Valley—to the city of Santa Clara. The existing BART system is shown in Figure 1, and the limits of the SVRT Project in Figure 2.

The alignment includes six proposed stations (three above-grade and three below-grade), an optional future station, and vehicle storage and maintenance facilities.

Shown in Figure 3 is the approximately 5-mile Tunnel Segment alignment through downtown San Jose, which is comprised of twin circular 17'-10" ID bored tunnels, and various cut-and-cover facilities, which are listed below:

- Stations:
  - Downtown San Jose Station and Crossover
  - Alum Rock Station
  - Diridon / Arena Station
- Portals:
  - East Portal
  - West Portal

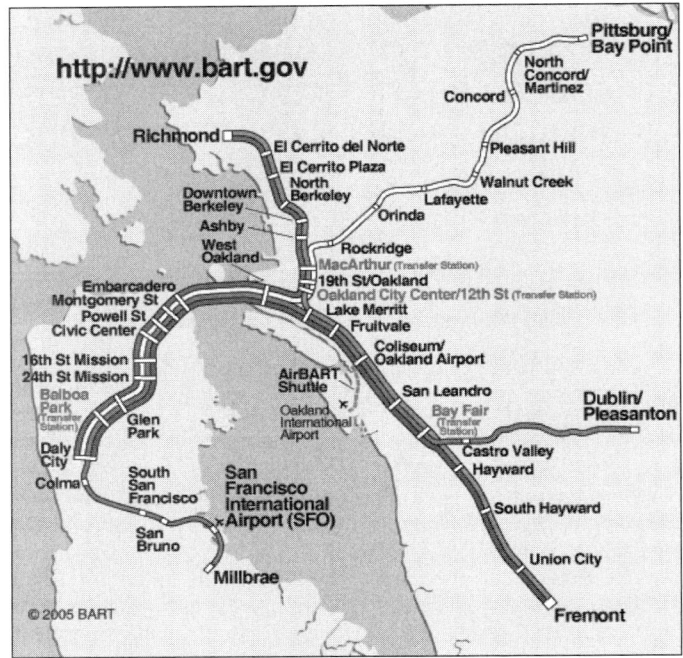

Figure 1. Existing BART system

- Mid-Tunnel Ventilation Structures:
  - East Vent Structure FSS
  - West Vent Structure STS

The SVRT project has recently completed the '65% Engineering' phase, and is currently in the 'Engineering Readiness Work' phase in anticipation of the commencement of final design.

The focus of this paper is the analysis and design work performed to date on the two mid-tunnel ventilation structures, whose locations are highlighted in Figure 3.

MID-TUNNEL VENTILATION STRUCTURES

The mid-tunnel vent structures are comprised of two main elements: a compact rectangular vent shaft situated around and above the bored tunnels and extending upward below city streets, and a headhouse connected to the shaft and located off the street. For the current 65% design, the excavation depth is approximately 26.7 m (87.5 ft) for FSS and 27.3 m (89.5 ft) for STS. Excavation depth for the basement/ equipment room under the headhouse is approximately 6.4 m (21 ft) at both locations. To provide adequate groundwater cutoff, the excavation support walls extend to a depth of approximately 61 m (200 ft).

Figure 4 shows the current general layout, longitudinal and transverse sections of Vent Structure STS. The general configuration of vent structure FSS is similar.

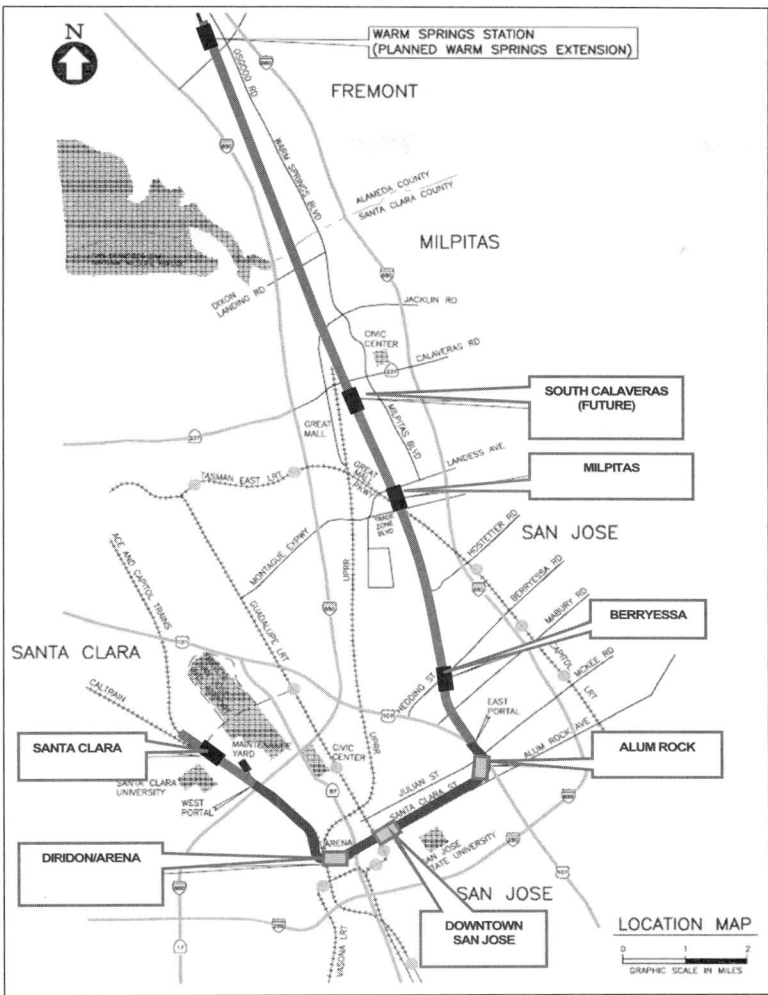

Figure 2. Silicon Valley Rapid Transit (SVRT) project alignment

GEOLOGICAL DESCRIPTION

Geotechnical investigations have established that the soils occurring along the entire tunnel alignment can, for the purposes of engineering behavior, be grouped in two principal types. These are:

- Stiff fine-grained soils (Classifications CL, ML, CH)
- Dense Silty Sands and Gravels (Classifications SP, SW, SC, GP, GW, GC, GM)

The subsurface conditions along the alignment are, in general, similar at the various locations. Generally the stiff fine-grained soils (silty clays), with some sand layers or lenses, predominate over most of the planned excavation depth. Silty sands and

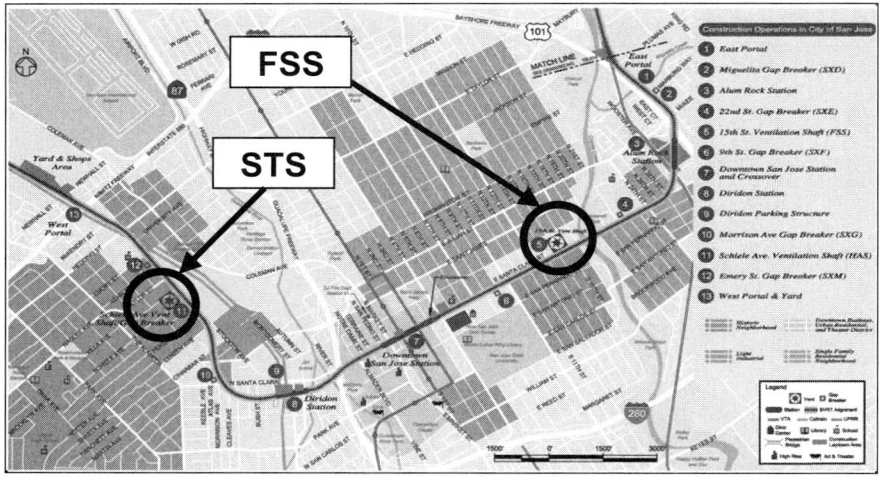

Figure 3. Tunnel segment of the SVRT project

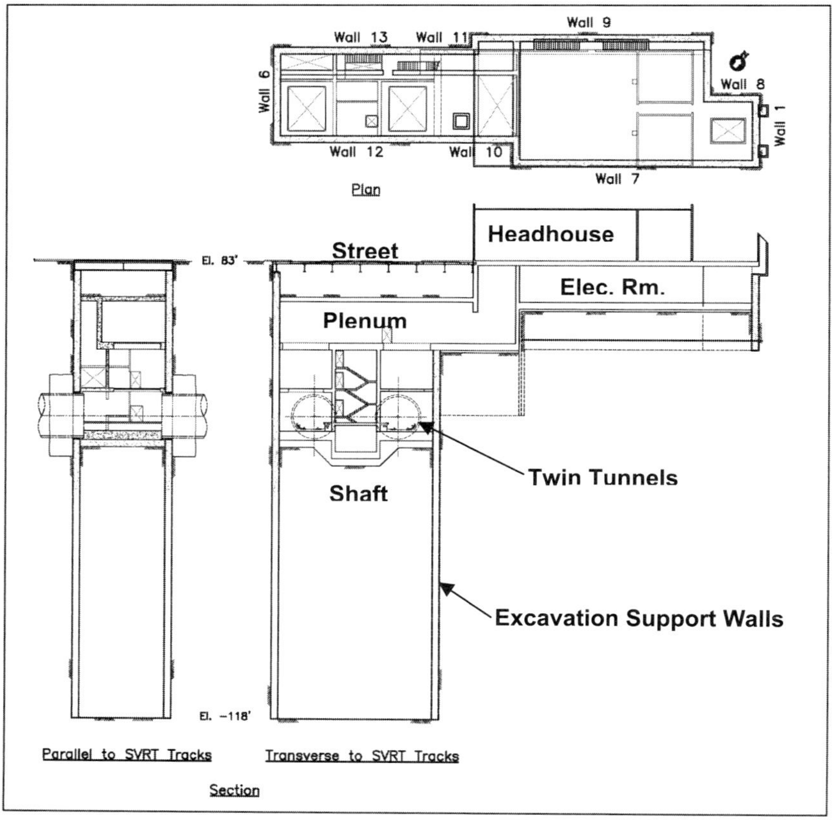

Figure 4. General arrangement of STS ventilation structure

gravels occur as the level of the bottom of vent structure excavations are approached. Fine soils occur again at greater depth, but are more variable in elevation and thickness than in the upper part of the soil profile, and include more frequent and at times thicker layers of sand.

Although local variations occur, it has been assessed that the clayey soils can be conservatively assumed to have undrained shear strength of about 48 kPa (1,000 psf) above a depth of 6.1 m (20 ft), increasing to 115 kPa (2,400 psf) and higher below a depth of 30.5 m (100 ft). The clays are lightly overconsolidated, with approximate overconsolidation ratios of 4 to 7 in the desiccated crust above the level of the water table, reducing to an average of about 2 to 3 at depths exceeding 9.1 m (30 ft). The total unit weight averages about 18.8 kN/m^3 (122 pcf) above a depth of 10.7 m (35 ft), and about 19.8 kN/m^3 (126 pcf) below that depth. Maximum ground water table is approximately 0.9 m (3 ft) bgs.

The sands and gravels are generally dense to very dense, with estimated friction angle of 33° or higher.

STRUCTURAL CONFIGURATION AND DESIGN CONSIDERATIONS

During the Preliminary Engineering (PE) phase of the project, and based on known ground conditions at that time, the two mid-tunnel vent structures were designed as more traditional 'two-pass' systems comprised of a temporary excavation support system, followed by construction of the permanent cast-in-place reinforced concrete structures after completion of bored tunneling operations. The temporary excavation support was comprised of soil-cement walls with overlapping 0.9 m (3-ft.) diameter piles spaced at 0.6 m (2-ft.) centers and with steel soldier piles embedded in every other pile. Due to the inability of current soil-cement technology in North America to reach the depths required to provide groundwater cutoff, the intent was to 'toe' the soil-cement walls into a block of treated ground located beneath the vent structure. Jet grouting was the assumed method of ground treatment for the PE design.

As the design progressed in the early stages of the 65% Engineering phase of the project, the feasibility of using slurry walls as both the temporary excavation support and permanent walls at other underground structures along the tunnel alignment was investigated. This led to the undertaking of a similar study for the two mid-tunnel vent structures.

Initially, three options were evaluated and compared for the vent structures, as listed below:

- PE 'Two-Pass' Design as described above, with soil-cement walls toed into a block of treated ground as temporary excavation support, followed by construction of the permanent cast-in-place reinforced concrete structure
- Alternate 'Two-Pass' Design, with deeper slurry walls as temporary excavation support only, and no ground treatment, followed by construction of the permanent structure
- 'One-Pass' Design, with slurry walls acting as both the temporary excavation support walls, and the permanent walls of the structure

For all options considered, the excavation was assumed to be supported by internal bracing (walers and struts).

For the slurry wall options, the walls are intended to be reinforced concrete, and not soldier pile tremie concrete (SPTC) walls.

Due to the depth requirements for these walls, they would most likely be constructed utilizing a hydromill system, and not a clamshell system. A typical hydromill is shown in Figure 5.

Figure 5. Typical slurry wall hydromill rig

The final result of this study was the recommendation to utilize the slurry walls as both the temporary excavation support and permanent walls for the two mid-tunnel vent structures.

The major design considerations leading to this recommendation are discussed in the following sections.

DESIGN IN HIGH SEISMIC ZONE

Slurry walls have been used as permanent walls in other high seismicity areas outside the US. Several types of slurry wall panel joints have been developed to provide continuity between the panels; however, there have been inherent construction difficulties and limited available data to confirm actual seismic performance of these types of joints. Therefore, for this project, it is assumed that continuity of reinforcement between the individual slurry wall panels will not be provided.

The assumption of unreinforced panel joints led to concerns about the magnitude of the stresses across these joints, and to the approach that should be taken to properly analyze the seismic effects on the vent structures.

The design earthquake effect on a long underground structure was preliminarily analyzed by imposing an earthquake-induced racking deformation on the structure. Using racking ratios similar to those applied to the other cut-and-cover structures on the project (portals and stations), which are rectangular in geometry but much longer than the vent shafts, resulted in extremely high and unrealistic in-plane shear stresses between the individual slurry wall panels.

Therefore, a literature review was undertaken to better understand how structures of a more compact rectangular geometry similar to the mid-tunnel vent shafts are typically analyzed in areas of high seismicity in North America. However, no such similar structures were found.

SEISMIC DESIGN CRITERIA

In addition to meeting the federal, state and local requirements, the design of the ventilation structures is in accordance with the SVRT Project Central Area Guideway Design Criteria Manual (HMM/B 2007) which combines BART Facilities Standards (BFS) Release 1.2 with BART-approved Requests for Variance.

Design Earthquake

The design ground motion is the higher of a site specific 10% in 50-year probabilistic analysis ground motion or the median deterministic ground motions from San Andreas magnitude 8.0 event, Hayward magnitude 7.25 event or Calaveras magnitude 7.0 event. The resulting peak ground acceleration is 0.55 g for permanent structures. This level of earthquake would be perceived as a "severe shaking" according to USGS scale with an instrumental intensity level of "VIII." For temporary structures, the horizontal peak ground acceleration is 0.34 g, which corresponds to a probability of exceedance of 10% in 10 years. The vertical peak ground accelerations are 0.49 g and 0.28 g, for permanent and temporary structures, respectively.

Seismic Effects on Underground Structures

The seismic effects on underground structures take the form of deformations that cannot be changed significantly by stiffening the structures (BART 2004). The structure should instead be designed and detailed to withstand the imposed deformations without losing the capacity to carry applied loads and to meet the performance goals for the structure. These goals are continued safe operation following the design ground motions and prevention of excessive cracking that will lead to unacceptable levels of water infiltration.

The deformations will depend on soil properties, particle velocity, shear wave propagation velocity and relative stiffness of the soil and the structure. Structural displacements can be determined from soil-structure interaction analysis. BFS R1.2 states that free-field displacements and soil properties for soil-structure interaction analysis can be determined using a computer program such as SHAKE; however, the method for performing the soil-structure interaction analysis of the structure and surrounding soil is not prescribed in BFS R1.2.

Enhanced Performance Goal

The project requirement for revenue structures is not only to ensure life safety but also "to be capable of being operational within 72 hours" following the design earthquake. To achieve this enhanced performance goal, a series of structural acceptance criteria are stipulated.

Structural Acceptance Criteria for Flexure

BFS R1.2 specifies that strains in exterior walls shall be below ⅔ of the ultimate strains. For Grade 60 reinforcement, the ultimate strains (ε_{su}) are specified as:

ε_{su} = 0.09 for #10 bars (d_b = 29 mm) and smaller

ε_{su} = 0.06 for #11 bars (d_b = 36 mm) and larger

For concrete, the allowable strain (ε_u) is defined as ⅔ the value calculated as per Section 3.2 of Caltrans Seismic Design Criteria, which is based on Mander's concrete stress-strain model for confined concrete.

The seismic loads are included in the following combination of loads:

U ≥ 1.0 (D + L + H + B + EQ)

For ductile flexural members, which are allowed to undergo inelastic deformation under the load combination above,

$\mu_{\Delta c} \geq 1.5 \geq \mu_{\Delta d}$

where $\mu_{\Delta c}$ and $\mu_{\Delta d}$ are the lateral displacement ductility capacity and demand, respectively.

Structural Acceptance Criteria for Shear

The project design criteria do not permit any ductility ($\mu_{\Delta d} < 1$) for shear and therefore require that the ultimate shear strength of the members with the use of appropriate strength reduction factor exceeds the shear demand computed in accordance with the loading combination above.

SEISMIC SOIL-STRUCTURE INTERACTION ANALYSIS

With regard to different soil-structure interaction (SSI) analysis methods for determining the seismic racking demand of vent structures, the project employs an iterative equivalent linear SSI analysis, i.e., SHAKE and SASSI computer programs, to determine global racking demand on the structure and local nonlinear pseudo-static push over analysis to assess the structural performance. Even though equivalent linear analysis does not simulate local nonlinear structural behavior in detail, the iterative procedure adjusts structural stiffness according to local nonlinear structural behavior, and approximates the global nonlinear behavior reasonably. Neglecting local non-linear soil behavior should have minor effect on the global results because the extent of local non-linear behavior is small compared with the size of the structure. Furthermore, neglecting local non-linear soil behavior is on the conservative side, so if satisfactory behavior can be demonstrated with this method, we can be confident in the actual performance of the structure during a seismic event.

Site response analyses were performed using SHAKE software to determine free-field ground motions and strain compatible soil properties at various depths. Acceleration time histories from three seismic events (Kobe Nishi-Akashi, Kocaeli Izmit and Loma Prieta—Coyote Lake Dam SW Abutment) in fault-normal, fault parallel and vertical directions were modified to produce spectral accelerations that match the project design spectral acceleration derived for NEHRP soil type "C." These spectrum-compatible time histories were then used in conjunction with the strain-dependent dynamic soil properties as the input into SHAKE software to conduct a site response analysis and compute the iterated strain compatible soil properties and ground motions.

Subsequently, a preliminary soil-structure interaction analysis of the STS Ventilation Structure was carried out using SASSI software to evaluate the seismic effects on the structure. Multiple parametric SASSI runs were made for the three spectrum-compatible time histories using lower bound, average and upper bound soil properties. The most conservative results were obtained using lower bound (geometric mean minus one standard deviation) soil properties and Kocaeli time histories. Three SASSI iterations have been performed thus far. The results were used in developing the seismic design approaches discussed below.

SEISMIC DESIGN APPROACH FOR SLURRY WALLS IN THE OUT-OF-PLANE DIRECTION

Similar to cut-and-cover rectangular box structures subject to transverse seismic deformation, the ground motion perpendicular to a slurry wall of the ventilation structures will cause racking of the wall-slab structural framing system. The moment at slab-to-wall joints will be high and flexural cracks will form. Reinforced concrete members can generally be made flexurally ductile to stably sustain cyclic moment demand corresponding to its plastic hinging moment capacity through the use of confinement reinforcement such as the prescriptive requirements of BFS R1.2. The structural acceptance criteria including the strain limitation to ⅔ of the ultimate value and the flexural ductility demand limitation of 1.5 would limit the crack widths and level of degradation. As a comparison, Ordinary Moment Frames designed per Chapter 1 through 18 of ACI 318, i.e., without seismic provisions, generally have a ductility capacity of 3. Thus, the ductility demand limit of 1.5 required for this project should provide a very high confidence that the structure integrity will be maintained and the structure will be capable of being operational within 72 hours.

The strain criteria can be met through increasing the amount of flexural reinforcement. The ductility capacity requirement can be met by increasing the amount of confinement reinforcement. Nonlinear pushover analysis will be performed during the next design phase to confirm the displacement ductility demand and capacity. In the unlikely scenario that flexural reinforcement in the slurry wall exceeds the maximum limit permitted by the design code, a thicker reinforced concrete section may be necessary. For slurry wall to be used as a single-pass and permanent wall of the ventilation structure, localized thickening may be accomplished using second-pass reinforced concrete interior lining made composite with the slurry wall. This approach should be more cost effective than the use of thicker slurry wall.

There will also be shear demand associated with the racking deformation of the slurry wall-slab system in the out-of-plane direction. The project design criteria do not permit shear ductility (i.e., $\mu_{\Delta d} \leq 1$) and further require that the shear capacity of reinforced concrete slurry wall exceeds the shear demand corresponding to the plastic-hinging moment capacity of the wall with an overstrength factor of 1.25. Shear reinforcement will be provided to meet this demand and therefore shear cracking in the slurry wall due to out-of-plane deformation is not anticipated.

Potentially, flexural cracks will occur near wall-to-base slab and wall-to-roof slab connections as a result of the design earthquake. However, the structure will be stable. After the earthquake, if localized flexural cracks near wall-to-base slab and wall-to-roof joints do occur, they will have to be repaired by grouting and/or concrete patching to stop water leakage.

SEISMIC DESIGN APPROACH FOR SLURRY WALLS IN THE IN-PLANE DIRECTION

Ground motion parallel to the slurry walls and the torsional response of the entire ventilation structure generally cause in-plane deformation of the walls. The total in-plane deformation comprises both flexural and shear components. The contribution of flexural and shear to the overall deformation varies with the wall length to height ratio.

As emphasized earlier, reinforced concrete walls have very low to negligible shear ductility. Further, panelized slurry wall construction makes it difficult to provide reinforcement continuity across the slurry wall joints. Because of these factors, emphasis of the post-processing of the SASSI analysis results was placed on obtaining the distribution and magnitude of in-plane shear strain and stress in the slurry walls.

Figure 6 illustrates approximately the distribution of shear demand-to-capacity ratio in exterior walls where slurry walls will be used as a permanent structure. For evaluation of shear crack in concrete, the shear "capacity" is taken as $3.5\sqrt{f_c'}$(psi) without the use of strength reduction factor (Nilson and Winter 1991). This value, 1.52 MPa (221 psi) for concrete with specified compressive strength of 27.6 MPa (4,000 psi), corresponds to ultimate shear strength of plain concrete. It is used in this context to demarcate the level of shear stress above which shear cracking of plain concrete is very likely. The use of plain concrete's strength is to capture the weakest link in the slurry wall at its vertical joints. It should be noted that the nominal shear strength of reinforced concrete per the ACI 318-08 design code is ($\phi = 0.75$)$\cdot 10\sqrt{f_c'}$ (psi), which is 3.27 MPa (474 psi).

In Figure 6, the unshaded area indicates a zone with no cracks, i.e., shear demand is less than the shear capacity of unreinforced concrete. Shaded area indicates region where, according to SASSI soil-structure interaction analysis, the shear stress exceeds the ultimate shear strength of plain concrete. Unreinforced vertical slurry wall joints located in the shaded regions would likely experience cracking.

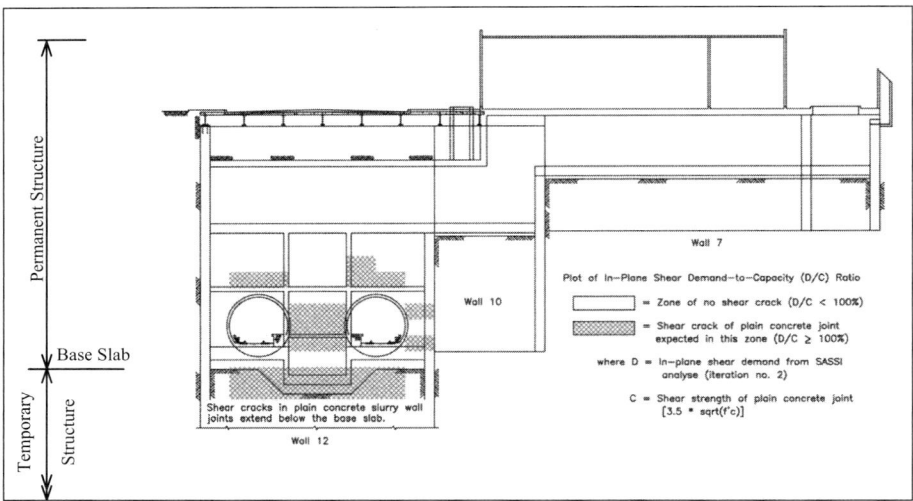

Figure 6. Shear demand-to-capacity (D/C) ratio in the south exterior walls from SASSI seismic soil-structure interaction analysis

From Figure 6, it is predicted that some portions of the ventilation structures below the plenum will experience high seismic shear demand as compared to the available shear strength of the plain concrete joints of the slurry walls. Consequently, localized cracks and some slippage of slurry wall joints during cyclic deformation induced by a design seismic event are anticipated. Although the ventilation structures will be stable and capable of being operational within 72 hours, the possibility of localized cracks and slippage in the slurry wall panel joints in the lower portion of the shaft raise two concerns. First, since the region of high shear stress is below the ground water table, water leakage may occur. Second, since the slabs are framed into the slurry walls, the slippage along the slurry wall panel joints may impose high shear demand at the slab-to-slurry wall connections. However, because the potential shear cracks do not extend to the full height of the slurry wall joints, the remaining uncracked segment of the joints should provide some restraints and reduce the potential impacts of the joint slippage on slab-to-slurry wall connections. Slurry wall below the base slab may also have cracks at joints but they are not part of the permanent structure and should not be of concern.

Figure 6 also illustrates a key benefit of performing three-dimensional seismic soil-structure interaction analysis in elucidating response of underground structures with complex geometry. The occurrence of high in-plane shear stress at the transition from the deep shaft portion to the shallower headhouse portion as observed in Figure 6 is indicative of stress concentration caused by an abrupt change in structural stiffness. In the case of the ventilation structures in this project, the offset shaft and headhouse configuration was necessitated by site constraints. This was exacerbated by the locations of tunnel openings in the slurry walls.

Three-dimensional seismic SSI analysis is also a useful tool in refining the design. It was found that by lowering the bottom elevation of "Wall 10" to approximately the bottom of invert slab elevation, an alternative stress path would be created for the shear transfer between the deep shaft and the headhouse. Consequently, the high shear stress zone is shifted away from the region of the slurry wall between the tunnel eyes and slightly above the tunnel eyes (see Figure 6), which will be a part of the permanent structure, to the buried and deepened portion of "Wall 10" which will not be critical to the serviceability of the ventilation structure.

APPROACHES FOR CRACKS AND WATER SEEPAGE MITIGATION

Cracks (due to flexure and shear) and water seepage are inherent characteristics of underground concrete structures. Unlike above ground structures where cyclic shear deformation of walls may lead to gradual degradation, increasing drift and eventual instability of the entire structure, underground structures cannot overturn because they are surrounded and confined by the ground. Even with shear cracks at plain concrete slurry wall joints, the underground ventilation structures are stable, provided that the slurry wall panels are designed so that they will perform their primary function of retaining soil and span vertically between the slab levels during and after the seismic event. The effect of shear cracks in plain concrete slurry wall panel joints and possible degradation of the slurry wall needs to be considered.

Post-Earthquake Repair

Since liquefaction or fault rupture is not anticipated at the two ventilation structure sites, once the seismic ground motion stops the surrounding ground and the embedded ventilation structure will return to its original configuration. Most of the cracks due to flexure or shear will likely be closed, although some cracks may remain open and seepage may continue. Post earthquake grouting of cracks may be required. The water that may have seeped into the ventilation structure will need to be collected and pumped out. Water collection system and sump pumps are standard features of any underground structure, and are included in the STS and FSS vent structures.

Drained Cavity Walls

In areas where leakage is prohibited or objectionable, such as the electrical rooms, drained cavity walls such as schematically shown in Figure 7 can be constructed. This inner false wall, which may be of masonry, reinforced concrete, or metal panel construction, is generally non-load bearing. Any seepage that penetrates the exterior slurry wall is collected within the cavity between the exterior wall and the inner false wall and is discharged to a sump. Alternatively, electrical equipment can be mounted on brackets that are structurally connected to the walls and provide a sufficient gap between the walls and the equipment.

APPROACHES FOR ACCOMMODATING SLURRY WALL JOINT SLIPPAGE

Recall that slippage of slurry wall vertical joints may potentially cause some distress in the slab-to-wall connections. Accordingly, it may be advantageous not to construct a moment connection at intermediate slab-to-wall joints. This may be accomplished through seat-type connection where a series of corbels are constructed on the slurry walls to support the intermediate slabs. The corbels should not extend horizontally across slurry wall vertical joints. If this approach is used, the SSI analysis needs to account for the seat-type connections.

The roof slab to slurry wall connection should be a moment connection to minimize potential for ground water seepage. For the ventilation structures in this project, the slurry wall in this zone is above the zone of expected high shear stress. Thus, cracks and slippage at slurry wall joints are not anticipated and the moment connection should be feasible.

These issues will be analyzed further during the next phase of design.

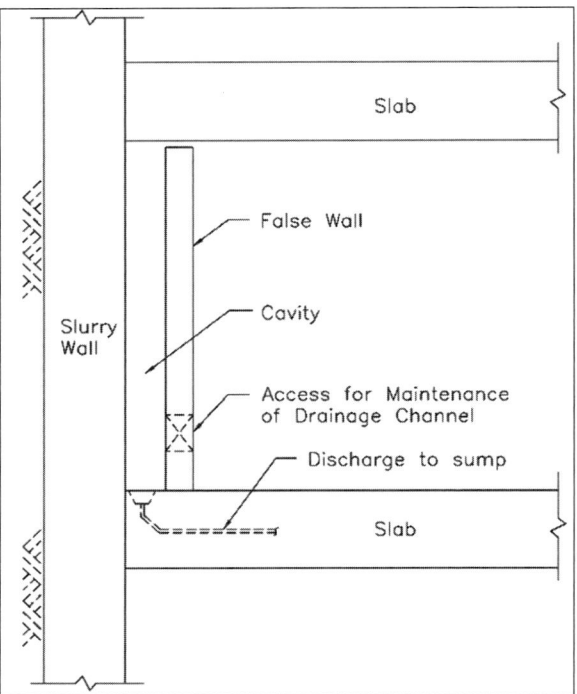

Figure 7. Drained cavity wall system

LOCALIZED STRENGTHENING AT TUNNEL EYES

The penetration of the bored tunnels through the slurry wall panels will cause discontinuity and stress concentrations. For this project, to be determined by nonlinear pushover analysis during the next design phase, localized strengthening at this location may be made. A feasible scheme is illustrated in Figure 8, in which a cast-in-place reinforced concrete wall is constructed against the inside face of the slurry wall. Connections would be made between the cast-in-place wall and the longitudinal slurry walls (parallel to the bored tunnels and tracks), the base slab below and the intermediate floor slab above. The cast-in-place concrete wall would be designed for full static and seismic loads. The slurry wall panels transverse to the bored tunnels in this region would serve only as a temporary support.

NO SHEAR CRACKS IN MODERATE EARTHQUAKE

The preceding analysis is for the project design earthquake with an approximate recurrence of one in 500 years, which would result in a peak ground acceleration of 0.55 g and be perceived as "severe shaking" according to the USGS scale. An event such as this may occur once in a lifetime or less frequent.

However, during a moderate but more probable earthquake (such as the one that would be perceived as "strong shaking" according to the USGS scale with a corresponding peak ground acceleration of 0.18 g and an instrumental intensity of VI at the SVRT site), the plain concrete slurry wall panel joints are not expected to crack.

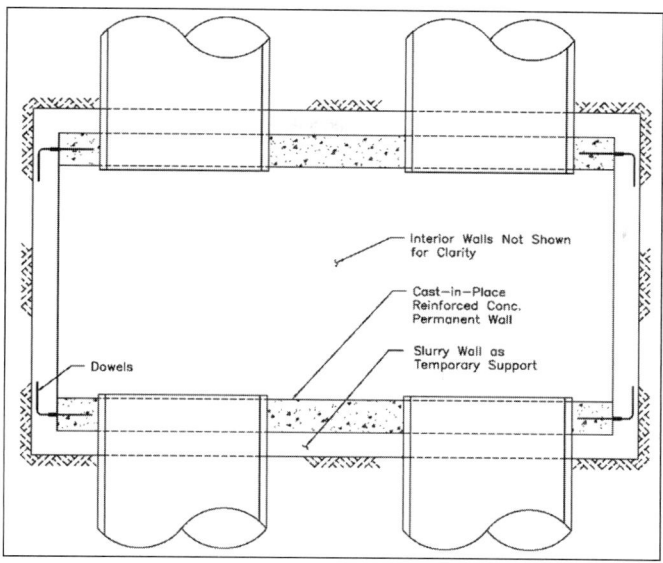

Figure 8. Localized strengthening at tunnel eyes

OTHER CONSIDERATIONS

Not addressed in this paper but also considered during development of the design were the following:

- Potential Ground Displacements Near Excavations
- Environmental Considerations:
  - Spoil Volume and Transportation of Excavated Material
  - Cross Contamination of Deep Aquifer
- Constructability:
  - Stability of Wall During Construction
  - Potential Obstructions
  - Construction Tolerances
  - Compatibility with Bored Tunnel Operations
  - Construction Management and QC
  - Interference with City and Private Properties and Utilities
  - Case Histories and Past Performance
- Risk
- Cost and Schedule

CONCLUSIONS

As of this writing, the SVRT project has recently completed the '65% Engineering' phase. Further refinement and finalization of the design remains; however, results of preliminary analyses performed to date indicate that the slurry walls can reasonably be expected to properly function as both the temporary and permanent exterior walls of the vent structures. The following approaches have been used on the SVRT project, and are recommended for the seismic design of compact rectangular shafts using slurry walls as permanent walls:

1. The configuration and sizing of interior structural components and the design of slurry walls should satisfy all the Project's design criteria requirements for static loads.
2. The resulting design should be fully evaluated for seismic effects through the use of seismic soil-structure interaction analyses such as the combination of SHAKE and SASSI software. Configurations that would create stress concentrations, such as offset deeper and shallower portion or abrupt change in structural stiffness, should be avoided if practical. If unavoidable, the seismic soil-structure analysis should be three-dimensional.
3. The slurry wall in the out-of-plane direction should be designed as ductile flexural members.
    a. With ductile design, ensure that the structure is stable and capable of returning to service in 72 hours (or as mandated by project-specific criteria).
    b. Flexural cracks and plastic hinging are acceptable provided the tensile strain of steel reinforcement and compressive strain of concrete are limited to ⅔ of the ultimate values and the ductility demand is limited to 1.5. Prescriptive minimum longitudinal and confinement reinforcement may be imposed.
    c. Shear ductility is not permitted. The shear strength of slurry wall and slabs, with appropriate strength reduction factor, should not be less than the plastic hinging moment capacity of the slurry wall with an overstrength factor of 1.25.
    d. The structural acceptance criteria (strain limits and ductility demand limit) are intended to minimize flexural crack widths. Nevertheless, post-earthquake repair of flexural cracks by grouting and concrete patching to stop water leakage may be required and plan for such repair should be developed.
4. The primary load effect due to seismically induced in-plane deformation of the slurry wall will be shear. Shear cracks in plain concrete joints of the slurry walls should be evaluated. Slippage of cracked joints may occur during the earthquake.
    a. Underground structures, being embedded in the ground, are inherently stable. By limiting shear cracks only to the slurry wall joints, the structure should be capable of returning to service in 72 hours.
    b. Each slurry wall panel should be designed such that the shear strength with appropriate strength reduction factor, exceed the seismic shear demand. By maintaining the integrity of the slurry wall between the panel joints, the slurry wall should continue to perform its primary function of retaining the soil and spanning vertically between the slabs during the seismic event.
    c. After the earthquake, no permanent deformation is expected and most joints should be closed. Nevertheless, post-earthquake repair of shear cracks in the vicinity of plain concrete joints by grouting and concrete patching to stop water leakage may be required and plan for such repair should be developed.

d. Drained cavity wall should be considered where leakage is prohibited or objectionable.
e. To accommodate slippage of slurry wall vertical joints and minimize consequential localized high shear stress at the slab-to-slurry wall connections, the use of seat type connection where a series of corbels are constructed to support the intermediate slabs may be considered.
f. The roof slab-to-slurry wall connection should be a moment connection to minimize ground water seepage. For the ventilation structures of this project, the slurry wall in this zone is not subjected to high in-plane shear stress and cracks and slippage at the slurry wall panel joints are not expected.

5. To determine the displacement ductility demand and capacity, strains in concrete and reinforcing steel, as well as to determine the need for localized strengthening at tunnel eyes and confirm compliance with project structural acceptance criteria, nonlinear pushover analysis of the structure should be performed.

ACKNOWLEDGMENT

The Authors wish to acknowledge the Santa Clara Valley Transportation Authority for allowing us to present the information contained in this paper.

REFERENCES

HMM/Bechtel JV (2007), Central Area Guideway Design Criteria Manual, Design Criteria No. P0503-D300-DC-DE-001, Rev. 1, Silicon Valley Rapid Transit Project.
Bay Area Rapid Transit Authority (2004), BART Facilities Standards, Release 1.2.
Nilson and Winter (1991), Design of Concrete Structures, 11th Edition, McGraw Hill, p.123.

PART 17

Slurry and EPB 1

Chairs

Darrel Liebno
Obayashi

Luminita Calin
Kenny Construction

SELECTION, DESIGN, AND PROCUREMENT OF NORTH AMERICA'S LARGEST MIXSHIELD TBM FOR PORTLAND, OREGON'S EAST SIDE CSO TUNNEL

Christof Metzger ▪ Kiewit–Bilfinger Berger Joint Venture

Greg Colzani ▪ Jacobs Associates

Gary Irwin ▪ City of Portland Oregon, Bureau of Environmental Services

ABSTRACT

This paper discusses the process utilized by the Kiewit–Bilfinger Berger Joint Venture (Contractor) to select, design, and procure the largest mixshield tunnel boring machine (TBM) ever implemented in North America. The mixshield will excavate the 8,534 linear m × 6.7 m ID (28,000 linear ft × 22 ft ID) tunnel for the City of Portland Oregon's, Bureau of Environmental Services (Owner) East Side CSO Tunnel Project. TBM selection was conducted under a preconstruction contract in which the Contractor, Owner, Construction Management Consultant and Designer collaborated on TBM selection prior to commencement of construction. This paper focuses on TBM design considerations, including the use of one versus two machines, machine configuration, extended-wear protection, hyperbaric interventions, boulder digestion, and erection of a steel-fiber-reinforced segmental lining.

INTRODUCTION

The City of Portland entered into an Amended Stipulation and Final Order (ASFO) agreement with the Oregon Department of Environmental Quality in August 1994. The agreement requires the City to control its 55 combined sewer overflows (CSOs) by 2011 with interim deadlines imposed to complete specific portions of the work prior to that date. The purpose of this project is to reduce the frequency and volume of combined sewer overflows from the drainage areas on the east side of the Willamette River. In accordance with the ASFO, the facilities associated with the East Side Combined Sewer Overflow (ESCSO) Tunnel Project must be constructed and operational by December 1, 2011. Otherwise, substantial fines may be imposed.

The East Side CSO Tunnel and associated shafts and pipelines are the largest and final major element of this CSO Program (Figure 1). The tunnel will collect and intercept overflows from 14 existing combined sewer outfalls that discharge to the Willamette River. It will extend for approximately 9,144 linear m (30,000 linear ft) from the vicinity of SE 17th Avenue and SE Mcloughlin Boulevard, proceeding north and paralleling the east bank of the Willamette River to a connection with the recently completed and operational West Side CSO Tunnel at the confluent structure on Swan Island. The tunnel being excavated by a single-slurry mixshield TBM (SPBM) has a 6.7-m (22-ft) inside diameter (ID) and depths ranging from 26 to 50 m (85 to 165 ft); it will intercept six drop/access multiuse shafts along the alignment (see Figure 2 for an overview of the alignment). The shafts will range in depth from approximately 37 to 53 m (120 to 175 ft), with IDs of up to approximately 17.7 m (58 ft). The main mining shaft is the Opera Shaft. From the Opera Shaft, the TBM will excavate the 6,096-m-long (20,000-ft-long)

Figure 1. Typical cross section of the ESCSO tunnel

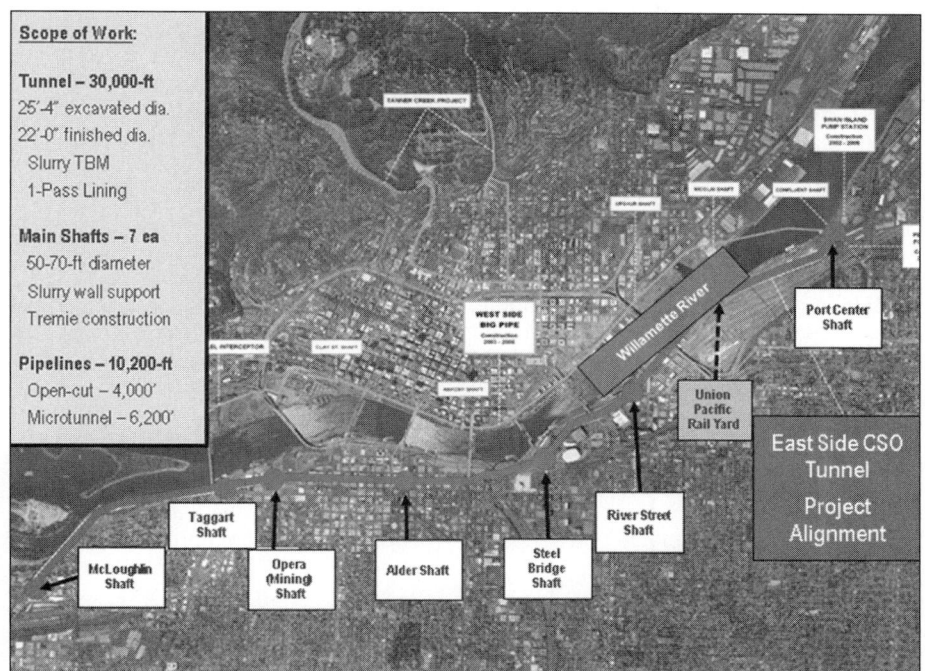

Figure 2. ESCSO project map

north drive. The TBM will be removed at the Port Center Way Shaft, be refurbished and then relaunched from the Opera Shaft to complete the remaining 3,048-m-long (10,000-ft-long) south drive. Tunnel and shaft construction will encounter soft-ground conditions consisting of artificial fill, sand/silt alluvium, gravel alluvium with random large cobbles and boulders, and Troutdale Formation consisting of tightly packed gravels with interbedded sand, silt, and clay lenses. Groundwater levels vary seasonally from approximately 15.2 to 38 linear m (50 to 125 linear ft) above the tunnel crown. The tunnel system will function in both conveyance and storage modes.

The Contractor, Kiewit–Bilfinger Berger JV (KBB), was selected via a qualifications-based selection process by the City of Portland, Oregon's Bureau of Environmental Services, to complete the project. The work is being performed under a two-phase contract—a preconstruction consulting contract followed by a cost-reimbursable plus a fixed-fee construction contract. During the preconstruction phase of this contract, TBM evaluation, specification, and procurement was one of the major contract deliverables. The remainder of this paper will discuss the TBM evaluation process, TBM procurement, TBM and physical plant design, and TBM performance to date.

TBM EVALUATION

As part of preconstruction phase, KBB was required to perform an evaluation and cost comparison of using one versus two TBMs for the East Side CSO Project. This methodology evaluation and cost comparison was prepared based on the 60% design. Key parameters of the analysis are included in Table 1.

The cost comparison methodology included the following study assumptions:

Mining Shaft Construction

1 TBM: Single Mining Shaft
2 TBMs: Enlarged Mining Shaft, additional work includes:

- Slurry wall (2,648 m^2, 28,500 ft^2)
- Excavation (7,561 m^3, 9,890 yd^3)
- Concrete (1,013 m^3, 1,325 yd^3)

Tunnel Setup/Removal

1 TBM: TBM removal at Port Center Shaft.
        Major overhaul of TBM at OMSI prior to south tunnel drive.
2 TBMs: Installation of second TBM drive support equipment, including slurry separation plant, annular grout plant, and shaft utilities.

TBM Operations

1 TBM: Linear tunnel operation, including use of shaft, service, and support crews.
2 TBMs: Share service and support crews during concurrent north and south tunnel drives.

Table 1. Key parameters

Tunneling Basis	1 TBM	2 TBMs
TBM type/condition	Slurry, new	Slurry, new
Mining site location	OMSI	OMSI
Mining shafts	1 each	2 each
Muck disposal	Barge conveyor	Barge conveyor
Segmental Liner	Off-site plant	Off-site plant

Table 2. Order-of-magnitude cost comparison between 1 and 2 TBMs

Work Activity	1 TBM	2 TBMs
Mining shaft construction	—	1.6×
Tunnel set-up removal	—	1.06×
TBM operations	1.05×	—
Tunneling equipment	—	1.59×
Schedule-related costs	1.2×	—
Total Cost	—	1.18×

Table 3. Project schedule impact analysis

ESCSO Project Schedule	1 TBM	2 TBMs
Project start	March 2006	March 2006
Project finish	September 2011	October 2010
Total Duration	65 months	54 months
Completion Date float, ASFO	3 months	14 months

Tunneling Equipment
1 TBM: Single boring machine and support equipment.
2 TBMs: Two complete (new) boring machines, including additional slurry separation plant, annular grout plant, rolling stock, linear plant, and electrical equipment.

Schedule-Related Costs
1 TBM: Linear tunnel operation, including use of single tunnel supervision team. Supervisory and non-reimbursable costs for increased overall project schedule.
2 TBMs: Additional relocation costs and planning time associated with second TBM operation (tunnel supervision).
Reduced supervisory and other non-reimbursable costs associated with shortened overall project schedule.

The analysis projected that the use of one TBM offered the project approximately $20.0 million in cost savings. An order of magnitude cost comparison of one TBM versus two TBMs is shown in Table 2. In addition, it is recognized that the use of one TBM would significantly lessen the impacts from construction on the community during the period of tunnel excavation as a result of less truck and construction traffic, less noise, and fewer workers being required at the mining site.

However, it is acknowledged that while more cost effective, the use of a single TBM would increase the overall construction schedule. While either approach (one or two machines) would allow project completion in advance of the December 2011 ASFO date, the use of one TBM would reduce available schedule float. The project schedule impact analysis is shown in Table 3.

While the use of one TBM would increase project schedule risk, the project team identified a number of identifiable measures to mitigate this risk and to consider as contingency plans in the event that unforeseen conditions or other production delays arose affecting the project's critical path. These include:

- TBM procurement premiums: including design, spare parts, and manufacturer's guarantees to ensure maximum TBM performance.

Table 4. Mitigation cost comparison

Item	1 TBM	Mitigation Cost
Cost savings on use of 1 TBM	($20.6 million)	
Mitigation measures		
TBM upgrades, guarantees		$1.5 million
Working 7 days/week		$10.0 million
2nd boring machine		$9.0 million
Totals	($20.6 million)	$20.5 million

- Reasonable production rates: use of productivity rates that offer the possibility of an improved schedule.
- TBM operation/refurbishment: regular scheduled events of downtime to replace parts and perform major overhaul of the TBM.
- Working days: increasing from five work days per week to seven days per week to account for productivity issues or to make up lost time.
- 1.5 TBM: if required, procurement and installation of a second boring machine during operation of the first machine to mitigate schedule time.
- Phased completion: putting the north tunnel drive on line for receiving overflows and providing storage separate from the south tunnel, thereby increasing the available schedule float for the majority of the alignment.

A comparison between (1) the cost savings of using one TBM and (2) being required to perform the above schedule mitigation costs indicates that even under a worst-case scenario of performing multiple mitigation measures, the use of a single machine does not adversely affect the overall project cost (see Table 4).

TBM PROCUREMENT

In September 2005, separate technical and commercial request for proposals were sent to four potential TBM suppliers (Herrenknecht, Lovat, NFM/Wirth, Robbins/Mitsubishi). The technical request for proposal established a procurement process as follows: (1) TBM supplier pre-proposal conference; (2) submittal of each TBM supplier's technical proposal; (3) TBM supplier presentation; (4) short list of suppliers; (5) submittal of final technical and commercial price proposals; (6) selection of TBM supplier and negotiation; (7) execution of a letter of intent.

The abbreviated general scope of design included:

- Design and manufacture of one/up to two slurry shield(s), and associated backup equipment.
- Complete assembly and workshop testing in accordance with an agreed-upon schedule to fully demonstrate all functions of the equipment.
- Supply of hazard and risk assessments schedules.
- Safe and robust design of the slurry shield, with special consideration given to the torque of the cutterhead and an appropriate design of the main bearings, propelling cylinders, and shield. Heavy-duty construction for the stone crusher and hard welding to minimize the wear of the structure and tools shall be taken into account.
- The diameter of the main shield and cutterhead is to be designed to ensure that boring along the tunnel line with the tolerance requirements specified for

the correction curve of 244 linear m (800 linear ft) can be implemented with the minimum possible over excavation.
- External loads (earth pressure, water pressure, foundations, overloads, etc.), piston pressure, and forces from the shield controlling are taken into account in the dimensioning of the elements of the TBM. Design calculations of the tail shield showing that deformation of the shield does not occur during the tunnel driving.
- The design of the shield was to be submitted and approved not later than two months after the contract had been awarded or no later than two months prior to the start of manufacturing.

A comprehensive project team composed of KBB project specialists and Owner's specialists was formed to evaluate manufacturer proposals. (Jacobs Associates provided the Owner with Construction Management Services.) A total of three technical proposals were submitted by Herrenknecht Tunneling Systems, USA (HTS), NFM/Wirth and Robbins/Mitsubishi. Individual meetings were held with each supplier, at which they presented their technical solution followed by an opportunity for the project team to ask technical questions. In addition, an initial discussion of KBB's Commercial Request for Proposal was conducted. After evaluating each technical proposal and each TBM supplier presentation, the project team determined that HTS was fully responsive to the requirements established by the contract documents and KBB's technical request for proposal and provided the best technical approach to the project requirements. Upon completion of successful negotiations with HTS, a letter of intent to proceed with TBM design and procurement of long-lead items and materials was issued several months prior to the construction notice to proceed.

PLANT DESIGN

The performance of the tunneling operation on the East Side CSO Project will largely determine the final outcome of project cost, schedule, and overall job success. Therefore, successful completion of the tunnel work is essential and must be based on detailed plans, technical competence, contractor commitment, and mitigation of risks. KBB benefitted from the synergic effects of the local knowledge and the technical experience for slurry operations of the joint venture partners. During the preconstruction stage, Phase 1 of the project, partnering with the Owner was established, which is another leading indicator of a project's success.

The preconstruction stage gave the Contractor and Owner the advantage of developing the project planning from a 30% status to a 100% status. During this stage, detailed planning was done by several task groups and included the design for major plant equipment such as the separation plant and the tunnel boring machine (TBM). Task groups worked out concepts for a number of key components, some of which are:
- Ground modification and stabilization
- Microtunneling
- Segmental lining
- Shaft work
- Main tunnel

TBM Design

There are characteristics of each project that will require the TBM to have certain features to complete the work. In the case of the ESCSO, the soil conditions are

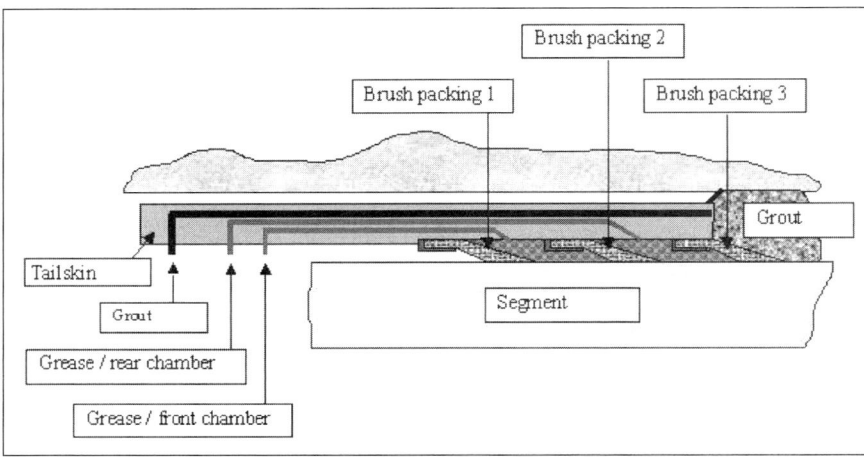

Figure 3. Triple brush system (seal brushes) and annulus grout system to prevent soil loss around lining

anticipated to open-graded, abrasive sands and to contain cobbles and boulders. As such, the team designed the TBM around the following key parameters:

- Ability to control permeable soils working under significant water pressure
- Resistance to extreme wear conditions
- Capacity to handle and crush a large number of cobbles and boulders
- Facilities to access and work under compressed air with slurry chamber

Each of these key technical design issues are addressed below.

Ability to Control Soils Under Large Water Pressures

The slurry TBM must be able to fully withstand up to 3.85 bar (56 psi) of both water and soil pressure. Risks associated with this include ground loss, settlements, and potential structure damage. Features that were incorporated into the TBM include:

TBM Component	Feature
Shield Structures	Welded steel shield structure, designed for operating load of 5.0 bar (72.5 psi).
Tail Seal Brushes	Three rows to prevent water inflow and allow efficient annulus grouting (Figure 3).
Annulus Grout Equipment	Four independent pumping systems to ensure full-time grout capabilities. Grouting to be performed through the tail shield.
Air Bubble Technology	Redundant, independent system of face and soil stability control exerting regulated pressure on bentonite suspension bearing on tunnel face.
Pressure Bulkhead	Installed in front shield, immediate activation, capable of 4.0 bar (58 psi).
Data Acquisition	Instantaneous data regarding face pressure, bentonite flow, and machine torque to immediately detect unexpected ground loss or instability.

Resistance to Extreme Wear Conditions

It was expected that the soil formation in Portland (Troutdale Formation and gravel alluvium, as described above) will cause significant wear on all elements of the TBM, including the cutterhead, shield assembly, and overall structure. Risks from this include cutterhead damage, increased tool usage, and shield wear. Features that were incorporated into the TBM to account for these issues include:

TBM Component	Feature
Shield Structures	Extensive hard facing of front shield at the cutterhead.
Cutterhead	Extensive hard facing on periphery and additional steel plating.
Wear Protection Tools	Electronic detection of tool wear through induction coils and cables; abrasion detection of tools with switches and LEDs.
Scrapers	Carbide cutting edges, high-strength steel bodies.
Bucket Lips	Hardfacing of front with carbide inserts at rear.

Capacity to Handle and Crush Large Numbers of Boulders

Risks from cobbles and boulders include cutterhead damage, plugged slurry lines, and loss of face pressure. Features that were incorporated into the TBM to account for these issues include:

TBM Component	Feature
Closed-faced Cutterhead	Approximately 70% of face closed to minimize boulder rotation (Figure 4).
Disc Cutters	Full-face of disc cutters. Use 432-mm (17-in.) twin disc cutters to achieve longer life, combined with single cutters in the center.
Stone Crusher	Capable of crushing 813 mm (32-in.) boulders with "upsized" design and hardfacing to account for frequency of cobbles and boulders (Figure 5).
Suction Grill	Upsized design to account for frequency of cobbles and boulders.
Air Bubble Control	Use of compressed air immediately replaces pressure lost from boulders, ensuring face stability.

Facilities to Access and Work Under Compressed Air

There is no question that interventions into the cutterhead chamber are required within a long drive such as the ESCSO Project. These interventions must be performed under compressed air. Risks include employee safety, face stability, and increased downtime. Features that were incorporated into the TBM to account for these issues include:

TBM Component	Feature
Man Locks	Twin, independent chamber locks designed for 4.5 bar (65.2 psi), with an arrangement to allow use during cutterhead rotation (Figure 6).
Rear-loaded Cutting Tools	Safe access from within the working chamber.

Attention was paid to the design of cutters and bucket lips. The design of the bucket lips was changed, after the experience with the first drive section from Opera Shaft to Alder Shaft, in order to provide an overcut of 12.7 mm (0.5 in.). The change

Figure 4. Cutterhead design for the challenging ground conditions

Figure 5. Rock crusher with 152 mm (6 in.) opening of suction grill

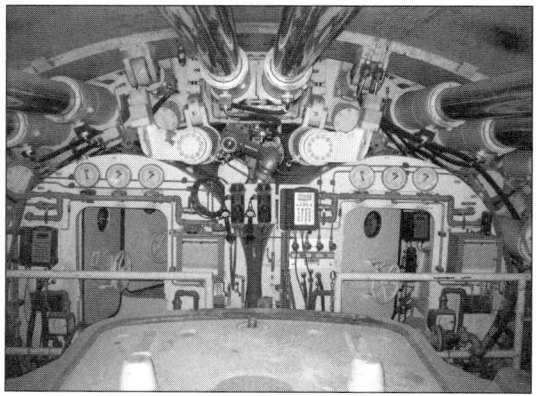

Figure 6. Two independent man locks

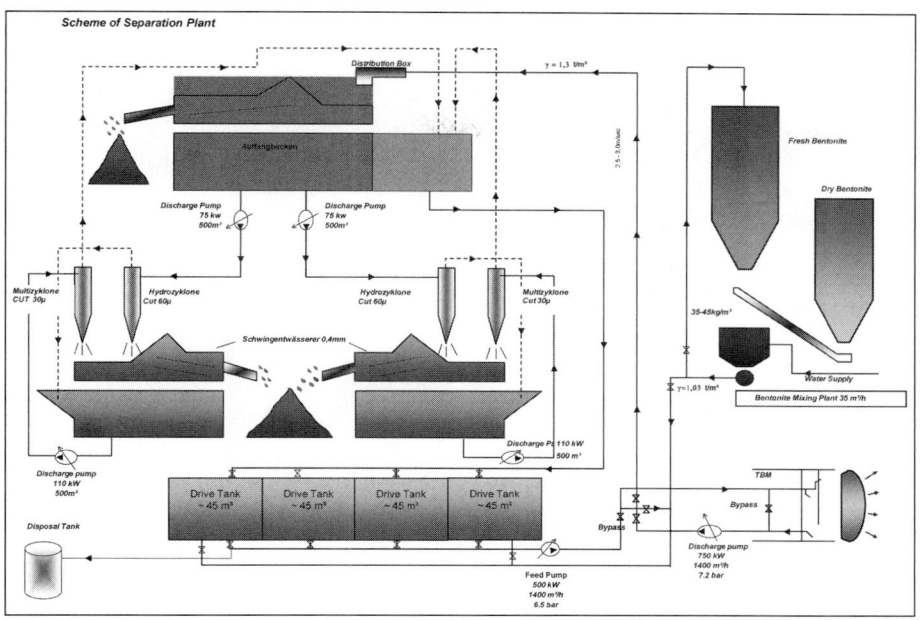

Figure 7. Schematic of separation plant, ESCSO Portland

allowed us to drive the distances from Alder Shaft to Steel Bridge Shaft (1,350 linear m [4,500 linear ft]) and from Steel Bridge Shaft to River Street Shaft (1,050 linear m [3,500 linear ft]) without changing any tools in interventions. It was a more economical and risk-free approach to change tools in the shafts under atmospheric conditions.

Slurry Treatment Plant

For the East Side CSO Project, we utilized a modular slurry treatment system that is easily installed and capable of being upsized quickly and efficiently. The plant is identical to the system used on a number of past projects.

Separation plants in tunneling operations must be designed with the utmost flexibility regarding capacity and muck characteristic fluctuations (Figure 7). Otherwise, the bottlenecks in the separation plant process can greatly restrict the rate of TBM advance. For the ESCSO Project, we designed a separation plant with a total of four process stages to separate the muck from the return bentonite slurry. For the sake of operational reliability and flexibility, we designed two subsystems operating in parallel with a maximum capacity of 700 m³/h (24,720 ft³/hr) each.

The surface-mounted separation plant is fed by a 356-mm-diameter (14-in.-diameter) pipe coming from the TBM. The distribution system at the head of the slurry separation plant required special attention. The majority of the tunnel is in the Troutdale Formation, which has a high percentage of cobbles and gravel. A flow velocity of 3.5 linear m/sec (12 linear ft/sec) is required to prevent cobbles from settling and plugging up the slurry pipe. Past experience has shown that serious clogging can occur at the area where the slurry flow is split and distributed to multiple slurry separation subsystems because the slurry velocity needs to be maintained and pipe diameters need to be decreased. The solution was to install an impact box and a heavy-duty vibrating screen at the head end of the plant. The high velocity slurry discharge from the pipe hits an impact area inside the box, and then rains down onto the screen. The double-deck screen rated at

Figure 8. Left: Preclassifying stage; Right: Secondary hydrocyclone stage

1,400 m³/hr (49,441 ft³/hr) scalps off the cobbles and gravels, and the underflow is then evenly distributed to the two 700 m³/hr (24,270 ft³/hr) slurry separation subsystems.

The following process stages were incorporated (Figure 8):

1. Preclassifying of the coarse particles and agglomerates by bar size and screen with an opening of 45 × 45 mm (1.8 × 1.8 in.) at the upper deck and 4 × 4 mm (0.16 × 0.16 in.) at the lower deck.
2. First cyclone stage with a cut of 60 mm (2.36 in.).
3. Secondary hydrocyclone stage with clustered hydrocyclones and a cut of 40 mm (1.57 in.).
4. Dewatering screens for hydrocyclone underflows with an opening of 0.2 mm (0.008 in.).
5. Fine particle separation with centrifuge.

The capacities of all subsystems are sized to assure a smooth and trouble-free overall operation. The preclassifying screen is a proven linear motion screen type with a polyurethane deck. The screen overflow is directly discharged to the muck conveyor belt system; the underflow is pumped via the screen launder to the first hydrocyclone stage consisting of two hydrocyclones per line. The hydrocyclones are installed above the dewatering screen. The hydrocyclones are equipped with a vacuumatic-controlled underflow discharge pocket. The underflow discharge pocket, in combination with the valve installed in the overflow system, allows a precise adjustment of the required solids concentration in the underflow and, thereby, in the feed to the subsequent dewatering screen. The dewatered muck is then discharged via a chute to the muck conveyor belt system for heap dumping.

The underflow of the secondary hydrocyclone stage also is equipped with vacuumatic-controlled underflow discharge pockets.

The underflow is discharged to a dewatering screen. The overflow of the secondary hydrocyclone stage is piped directly to one of the slurry storage tanks for recirculation to the TBM.

The dewatered solids are removed via conveyor belt to a temporary muck pile before final disposal. The drives of the dewatering screens can be manually adjusted

Table 5. Installed booster pump stations for the ESCSO Tunnel

Slurry Supply System			
Location	No.	Capacity	Size
OMSI–Slurry Plant	1	23,091 L/min (6,100 gal/min)	500 kW
Steel Bridge Shaft	1	23,091 L/min (6,100 gal/min)	500 kW
Slurry Return System			
Location	No.	Capacity	Size
TBM	1	23,091 L/min (6,100 gal/min)	750 kW
Opera Shaft Bottom	1	23,091 L/min (6,100 gal/min)	750 kW
Steel Bridge Shaft	1	23,091 L/min (6,100 gal/min)	750 kW
River St. Shaft	1	23,091 L/min (6,100 gal/min)	750 kW
Albina	1	23,091 L/min (6,100 gal/min)	750 kW

Figure 9. Booster pump station in the tunnel

by a frequency controller to optimize the dewatering efficiency and regulate for grain size and amount of sludge.

The main drive of the ESCSO Project is in the Troutdale Formation. The high-performance Hiller centrifuge DP-84 can handle the Troutdale without any problem by separating 40 m^3/hr (1,412 ft^3/hr) of disposed slurry. In distances where the alignment of the tunnel is located in the alluvium (silt), an additional mobile centrifuge was rented.

The long tunnel requires a series of booster pumps to move the slurry (Table 5; Figure 9). A single supply booster pump is required at about 3,650 m (12,000 ft) inside the tunnel. Return slurry booster pumps are required about every 1,200 m (4,000 ft).

Bentonite

Another focus at the separation plant was to evaluate which supply of bentonite would work best for the existing ground. Analysis of the Troutdale Formation and gravel alluvium showed that the system will be effective in removing at least 99.00% of the bentonite from the tunnel spoils. In the sand/silt alluvium, it is difficult to achieve less than

Figure 10a. Separation of Troutdale material   Figure 10b. Fines and centrifuge material

2% retained bentonite without increased centrifuge efforts. When the amount of retained bentonite reaches 2%, there are disposal premiums and barge stability problems.

A long tunnel such as the ESCSO Project, which is driven in differing ground conditions, clearly shows the advantages and disadvantages of a slurry TBM. In the sand and cobble formations, the material is separated as shown in Figure 10a); whereas in alluvium material several adjustments, including flocculants and additional centrifuge operation, were necessary to prevent extra days (see Figure 10b).

Piping System

The length of the tunnel, coupled with the fluid volume capacity required to support the anticipated mining rates, required 355.6 mm (14 in.) slurry pipes with a thickness of 12.7 mm (0.5 in.) for the discharge pipe, and 406.4 mm (16 in.) with a thickness of 9.5 mm (0.375 in.) for the fresh bentonite supply. Pipe maintenance and rotation are regularly scheduled to evenly distribute wear on the discharge pipe. This process will maximize the life of the pipe and overall cost efficiency.

Segmental Lining

The decision was made to build about 85% of the tunnel with steel-fiber-reinforced segments. The remaining 15% are to be installed in soil conditions that produce loads during seismic events greater than the steel-fiber-reinforced segments could adequately support. This 15% will be built with standard reinforced segments. This is the first use of a "wholly steel fiber reinforced on pass lining system" in the United States. Therefore, extensive research and calculations were performed in the preconstruction stage of the project. Full-scale tests were performed to guarantee the success of this approach.

The preliminary design had foreseen the use of a standard reinforced ring with six segments plus one keystone. The usage of steel-fiber-reinforced segments required an adjustment to seven segments plus one keystone to reduce bending moments during handling. The original production was based on 64 segments per day, using a single shift operation and eight sets of molds. The production schedule showed 912 production days. Better-than-anticipated performance of the TBM from the start of the regular drive required an increase in production to 144 segments a day by using two additional sets of molds and casting twice a day. This shortened the production period to 450 days. The segment design is summarized in Figure 11; Figure 12 gives the tunnel liner ring life cycle; Figure 13 illustrates the ESCSO segment storage.

MIXSHIELD TBM FOR PORTLAND'S EAST SIDE CSO TUNNEL

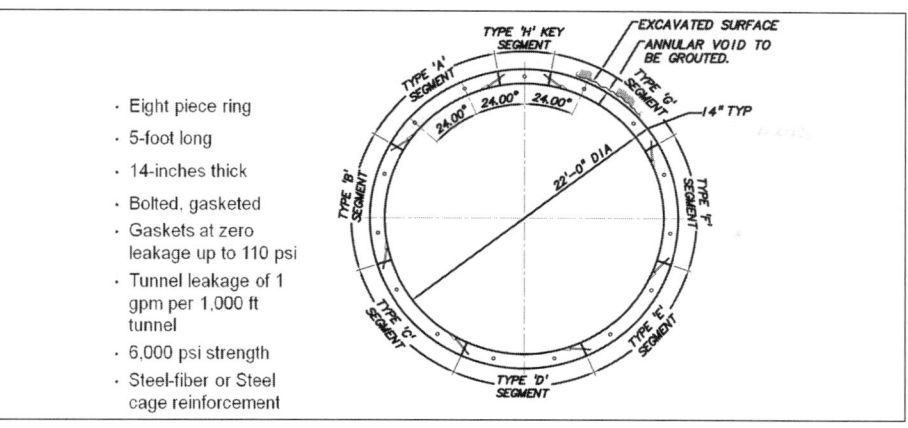

- Eight piece ring
- 5-foot long
- 14-inches thick
- Bolted, gasketed
- Gaskets at zero leakage up to 110 psi
- Tunnel leakage of 1 gpm per 1,000 ft tunnel
- 6,000 psi strength
- Steel-fiber or Steel cage reinforcement

Figure 11. Segmental lining design

Figure 12. Life of a segment ring

Figure 13. Segment storage of the ESCSO project

The use of steel-fiber reinforcement saved the project about $4 million. Nevertheless, some disadvantages exist:

- Increased wear at the batch plant
- Increased factory maintenance requirements
- Extensive quality tests and quality assurance

The positives of lining installation, performance, and cost saving greatly outweigh the negatives and have proven to be one of the important contributors to the success of the project.

SUMMARY

The successful completion of the design of the operational plant, as well as putting the right team together, allowed Kiewit–Bilfinger Berger to achieve the following maximal performance rates:

- Best shift: 13.7 m; 45 ft (9 rings)
- Best 24-hour period: 38 m; 125 ft (25 rings)
- Best 5-day period: 160 m; 525 ft (105 rings)
- Best 28-day period: 535 m; 1,755 ft (351 rings)

These are the best performances as of January 2009. We expect to continually improve as we work towards completion of the North Tunnel Drive in 2009 and the South Tunnel Drive in 2010.

CONSTRUCTION OF DRILLED SHAFTS FOR THE UPPER NORTHWEST INTERCEPTOR, SECTIONS 1&2 PROJECT— SACRAMENTO, CA

Jeremy Theys ▪ Traylor Brothers, Inc.

Chris Schäfer ▪ Haywood Baker, Inc.

The Sacramento Regional County Sanitation District (SRCSD) is improving the sewer system in the Natomas area of Sacramento, CA. The Upper Northwest Interceptor (UNWI) system is an ongoing series of tunnel pipeline projects that will provide sewer conveyance for all residents of northeast Sacramento County to the New Natomas Pumping Station, located North of downtown Sacramento, also in the city of Sacramento limits. The New Natomas Pumping Station then carries wastewater flows to the Sacramento Regional Wastewater Treatment Plant in Elk Grove via the existing Lower Northwest Interceptor. The Joint Venture of Traylor Brothers, inc. and J.F. Shea (TSJV) were awarded the UNWI 1&2 project in late September, 2007.

The planned 5.8 km (3.65 mile) construction route for UNWI 1&2 begins at the New Natomas Pump Station and runs north along the East Drainage Canal to the beginning of UNWI 3&4 at Bridgecross Drive. UNWI 1&2 includes approximately 5.8 km (19,240 feet) of 3.65 m (144 inch) inside diameter pipe using Earth Pressure Balance Tunneling Methods. The project also includes 16.76 m (55 feet) of 3.04 m (120-inch) diameter interceptor pipeline, twenty three access manholes spaced approximately 305 m (1,000 feet) apart along the alignment, a 2.13 m (84-inch) by 3.65 m (144-inch) transition structure, a 3.65 m (144-inch) by 3.04 m (120-inch) transition structure, and the connection of three existing sanitary sewer lines to the interceptor.

ENVIRONMENTAL CONSIDERATIONS

The project is located entirely in the Natomas Basin located north of downtown Sacramento. The entire tunnel alignment lies inside the habitat for several threatened or protected animal species, including primarily the Giant Garter Snake (protected under the Endangered Species Act), Swainson's Hawk, and Burrowing Owls. Due to the endangered species present along the alignment, the project is to be constructed according to the Natomas Basin Habitat Control Plan. This plan was created to detail the take avoidance, impact minimization, and mitigation measures necessary to limit the construction impacts upon the species. Due to the nesting and hibernating periods of the various protected species, the construction season for any work in the habitat area of the protected species has to occur between May 1st and October 1st, 2008. This resulted in an allowable 20-week working period, which created a very tight schedule to perform all of the ancillary activities to the tunnel, namely jet grouting and shaft drilling activities. The only preliminary activities that could take place were general surveying and layout of the manholes and site extents. Any ground disturbing activities could not begin until May 1st 2008. Also due to the tunnel alignment running through a residential area, the shaft construction activities had to take place between 7AM and 7PM each day. 24-hour working periods were not allowed.

PROJECT GEOTECHNICAL CONDITIONS

The UNWI 1&2 alignment lays in an area of generally flat ground within the southern Sacramento Valley on the north side of the city of Sacramento in Natomas. The topography along the alignment varies from approximately 6.01 m (20 feet) in elevation at its highest point to zero feet in elevation at its lowest point at the north end of the project. Before the project bid, a total of 49 exploratory borings were taken along the tunnel alignment to identify the soils to be expected throughout the project. A study of the borings results in a generalization of the soils present. They indicated four main geologic units which include the Upper Clay, the Upper Sand, the Lower Clay, and the Lower Sand. A thin layer of man-made fill lies at the surface along the alignment. For the majority of the tunnel alignment, the lower section of the tunnel profile lies in the lower clay, while the upper portion of the tunnel face is the upper sand (silty sand). There are no boulders expected throughout the tunnel alignment. The geologic conditions seem ideal for Earth Pressure Balance Tunneling methods, as well as great conditions for drilling the manhole shafts. The groundwater level generally exists around elevation −1 to −4 (feet). Due to the presence of the groundwater, all of the shafts were be drilled in the wet, using a slurry to aid in the support of the existing soils during excavation. Jet grouting was also specified to be performed in advance of the shaft drilling.

JET GROUTING FOR MANHOLE CONSTRUCTION

Jet grouting was required at each of the twenty manholes to be constructed along the alignment of the Upper Northwest Interceptor (UNWI) Sewer Tunnel in Sacramento, CA. A 6.09 m by 6.09 m by 6.55 m (20 feet by 20 feet by 21.5 feet) deep block of jet grout was installed, into which the steel riser casing was set and through which the tunnel boring machine will bore. The jet grouting will allow for a relatively water-free environment in which the connection can be constructed. Similar to the jet grout manhole connection blocks, jet grout was also installed to connect three existing lateral sewers to the new tunnel.

Jet Grouting Procedure

Jet grouting is a versatile grouting procedure that creates in situ geometries of soil-cement known as soilcrete. A jet grout monitor is advanced through a borehole to the limits of the treatment zone. As the monitor is raised and rotated, cutting fluid introduced through the monitor at high velocity erodes the surrounding soils, mixing it with grout slurry. Excess slurry returns to the surface through the borehole annulus, and is directed in to a containment pit where it is allowed to cure before removal. Jet Grouting is effective across the widest range of soil types, of any grouting system, including silts and certain clays. Because it is an erosion-based system, soil erodibility plays a major role in predicting geometry, quality and production. Cohesion less soil is typically more erodible than cohesive soils. Since the geometry and physical properties of the soilcrete are engineered, the degree of improvement can be readily predicted.

There are three traditional jet grouting systems. The in situ soil, the application, and the required strength of the soilcrete determine which system is to be used. However, any system can be used for almost any application, provided that the right design and operating procedures are used. The visual representation of the jet grouting types is shown in Figure 1.

Single Fluid Jet Grouting (Soilcrete S)

Grout is pumped through the rod and exits the horizontal nozzle(s) in the monitor at high velocity. This energy breaks down the soil matrix and replaces it with a mixture

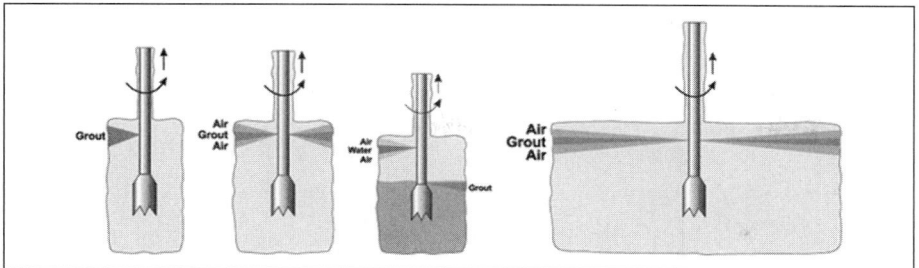

Figure 1. Jet grouting systems. Left to right: single fluid, double fluid, triple fluid, and SuperJet.

of grout slurry and in situ soil. Single fluid jet grouting is most effective in cohesionless soils.

Double Fluid Jet Grouting (Soilcrete D)

A two phase internal fluid system is employed for the separate supply of grout and air down to different, concentric nozzles. The grout erodes in the same effect and for the same purpose as with Single Fluid. Erosion efficiency is increased by shrouding the grout jet with air. Soilcrete columns with diameters over 3 ft can be achieved in medium to dense soils, and more than 1.82 meters (6 feet) in loose soils. The double fluid system is more effective in cohesive soils than the single fluid system.

Triple Fluid Jet Grouting (Soilcrete T)

Grout, air and water are pumped through different lines to the monitor. Coaxial air and high-velocity water form the erosion medium. Grout emerges at a lower velocity from separate nozzle(s) below the erosion jet(s). This separates the erosion process from the grouting process and tends to yield a higher quality soilcrete. Triple fluid jet grouting is the most effective system for cohesive soils.

SuperJet Grouting

SuperJet grouting is a modified double fluid jet grouting system that takes advantage of tooling design efficiencies and increased energy to create high-quality, large diameter soilcrete elements. It is effective in most soil types and is best when applied for bottom seals, mass stabilization and surgical treatment applications.

Grout, air and drilling fluid are pumped through separate chambers in the drill string. Upon reaching the design drill depth, jet grouting is initiated with high velocity, coaxial air and grout slurry to erode and mix with the soil, while the pumping of drilling fluid is ceased. This system uses opposing nozzles and a highly sophisticated jetting monitor specifically designed for focus of the injection media. Using very slow rotation and lift, soilcrete column diameters of 3–5 m (10–16 feet) can be achieved. This is the most effective system for mass stabilization application or where surgical treatment is necessary. There are more variations of these systems than there are systems themselves, but in most cases they are a "bottom-up" process. That is to say, they use hydraulic rotary drilling to reach the design depth, and at that point initiate jet grouting parameters and procedures to create a cementitious soil matrix commonly called soilcrete. Figure 2 shows the Super Jet grouting process.

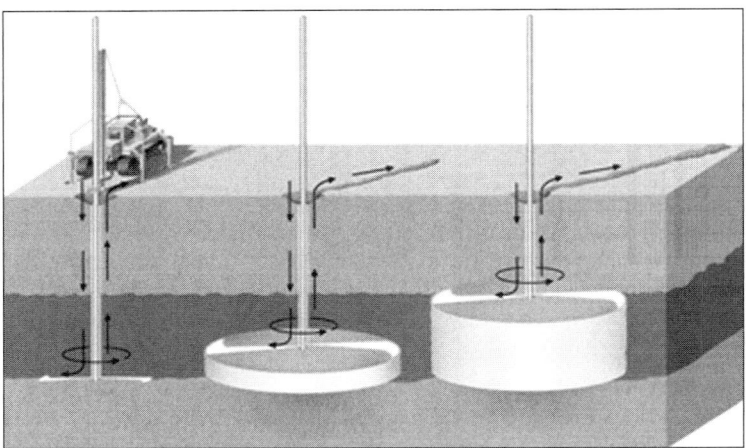

Figure 2. SuperJet grouting process

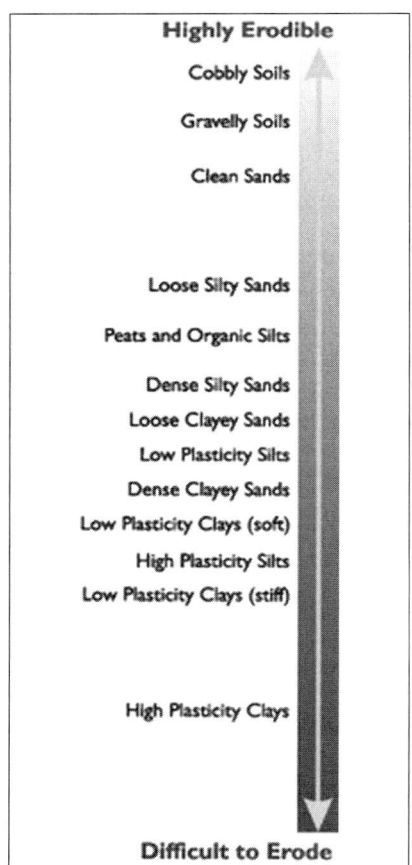

Figure 3. Soil erodibility scale for jet grouting

Jet Grouting Equipment

The jet grouting operations consist of the on-site drilling and grouting activities as well as the delivery of cement to the site. The jet grouting plant consists of four major components:

- Cement Grout Batch Plant
- Jet Grout Pump
- Drill Rig
- Quality Control System

The cement grout batch plant consists of a storage silo equipped with variable speed auger and a jet valve grout mixer. The nozzle size, water flow rate, and cement feed rate are adjusted to obtain the necessary specific gravity mix to be used. The mixed grout is temporarily stored in a holding tank and then transferred to the high pressure pump upon demand. The specific gravity (water to cement ratio) is monitored by an electronic mass flow meter. The mass flow meter is verified with a mud balance several times throughout the shift. The high pressure grout pump is mounted on a skid with other auxiliary equipment. In addition an air compressor is located near the drill to supply compressed air. Through a series of hoses and slick line the grout is delivered to the drill rig. The rig contains a rotary head coupled with automatic chucks mounted on a crowd winch system which runs along the mast. With two lattice mast extensions the drill rig is capable of reaching the target treatment depth with a single stroke. The swivel is located at the top of the drill steel. This connects the high pressure grout, drilling fluid and air hoses to the drill steel and directs the fluids into the appropriate annulus of the drill steel. Additionally, it allows the rotation of the drill steel separate from the hose connections. The drill steel is composed of multiple chambers, each of which conveys an individual fluid (high pressure grout and air). At the base of the drill steel the fluids exit through the jet grout monitor. The monitor is a proprietary piece of equipment that transfers the flow of the grout into a highly concentrated stream of grout perpendicular to the axis of the drill steel. The monitor also provides pathways that allow the air to "shroud" the stream of high velocity grout, and allows the drilling fluid to pass through the drill bit end of the monitor.

Jet Grout Design

The jet grout parameters were chosen to generate soil-cement columns of approximately 1.83 m (6 feet) diameters and 3.35 m (11 feet) diameters, respectively. These parameters and installation methods were verified in the test section prior to production work. At each manhole one jet grout core was taken to verify column diameters. Wet samples and core samples were tested for unconfined compressive strength at a local laboratory.

The strength of the soil-cement is highly dependent upon the in-situ soils. With the large variance in soil conditions throughout the work area, substantial variations in strength were predicted. The jet grouting design criteria called for a minimum strength of 1,034 kPa (150 psi) after 28 days. The geotechnical contractor's initial target strength was 1,034 to 1,723 kPa (150 to 250 psi) to ensure the minimum strengths were met. The cement content was adjusted throughout the project in order to keep the unconfined compressive strength of the soilcrete in the desired range. A jet grout test section was contractually required to be built, for proof testing of the operation. During the test section operation wet samples and core samples were taken for laboratory testing. Initial jet grout strengths resulted in the range of 3,447 to 5,515 kPa (500 to 800 psi), much harder than the desired target strength levels. This harder jet grout block created early problems for the shaft drilling operation until the compressive strength was brought down closer to the target.

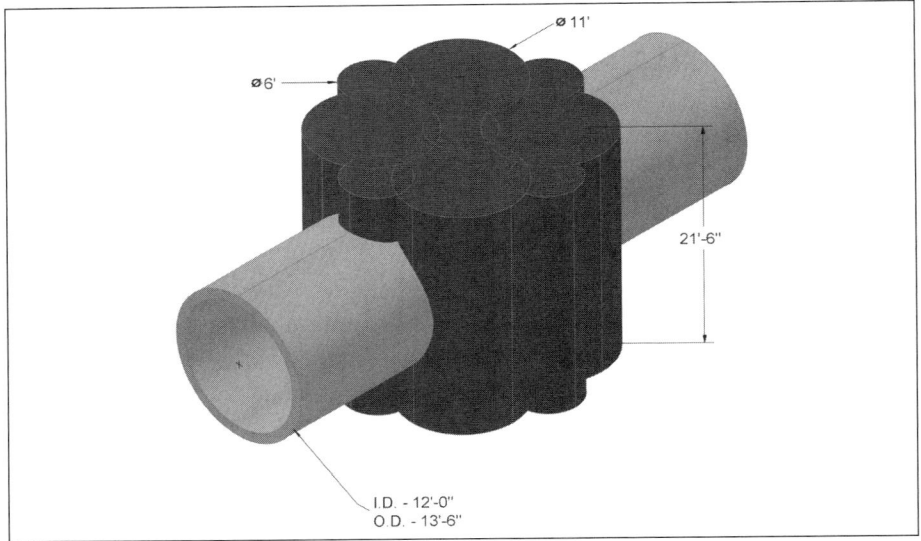

Figure 4. Typical jet grout block for manhole connection to 3.65 m (144 inch) internal diameter UNWI pipeline

A jet grout block was required to be installed at each access manhole location to reduce permeability and increase the compressive strength of the soils. The design of the access shafts for the project consisted of a 1.83 m (72-inch) internal diameter manholes at all locations. In order to provide a dry and stable environment to make the manhole connection, the jet grout blocks are designed to be 6.09 m (20 feet) wide by 6.09 m (20 feet) long. They are also 6.09 m (20 feet) tall, with the tunnel situated in the lower portion with 0.61 m (2 feet) of stabilized ground below the tunnel profile. TSJV increased the jet grout block height to 6.70 m (22 feet) tall due to the desire to have 1.22 m (4 feet) of jet grouted soils around any portion of the work as a buffer. The grout block was to be installed at a specific depth to encapsulate the tunnel and the bottom of the shaft at their intersection point with the intention of providing a dry environment during the performance of the shaft/tunnel connection work. The jet grout columns and tunnel are represented in Figure 4.

Hayward Baker, Inc. (HBI) was awarded the subcontract in early 2008 for the installation of all the jet grouting on the project. For the manholes, the jet grout block was composed of an array of 3.35 m (11 feet) diameter and 1.83 m (6 feet) diameter jet grout columns formed in rows adjacent to each other, with overlapping extents. There are four 3.35 m (11 feet) diameter columns at each shaft location in a 2×2 grid on 2.44 to 2.59 m (8 to 8.5 feet) centers and 1.52 m (5 feet) diameter columns on 1.22 m (4 feet) spacing to act to fill suspected holes of the larger column grid pattern. The column layout is shown in Figure 5.

Jet Grouting Parameters

The jet grout injection parameters were determined from a test program that was performed at the first manhole location where the jet grouting was to be executed. In order to obtain the desired range of compressive strengths of the in situ soilcrete, a neat cement grout with a specific gravity of 1.45 is used for the 1.52 m (5 feet)

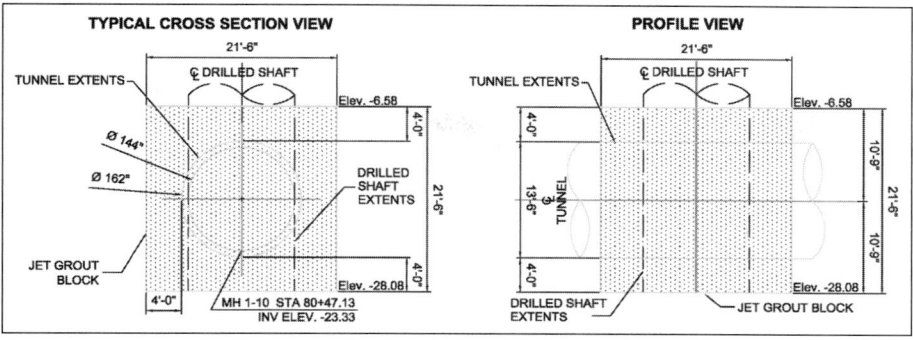

Figure 5. Typical manhole jet grout block

diameter columns and a 1.52 specific gravity is used for the 3.35 m (11 feet) diameter columns. This corresponds to a 1.39 water to cement ratio by weight (58.1% water, 41.9% cement) and a 1:1 water to cement ratio by weight (50% water 50% cement). The injection pressure is approximately 31,000 kPa (4,500 psi) and the grout flow rate of 423 liters/minute (112 gpm), air pressure is maintained at 689 kPa (100 psi) pressure and volume output at 5,663 lpm (200 cfm), and the injection rod rotation was set at 15 rpm. The withdrawal rate was 10 cm (3.93 inches) per minute or the 3.35 m (11 feet) diameter columns, while it increased to 40 cm (15.75 inches) per minute for the 1.52 m (5 feet) diameter columns.

DRILLED MANHOLE SHAFTS

The purpose of interceptor manholes is to allow access to the interceptor (tunnel) for future inspection and maintenance, once the tunnel is complete and operational. Based on maintenance and operations requirements all of the manholes were located on 305 m (1,000 feet) spacing wherever feasible. The project specified the use of blind augured shaft drilling techniques to drill the twenty manhole shafts along the tunnel alignment. For the connection of the three existing sanitary sewers to the interceptor tunnel, we chose to also drill the three access shafts for either microtunneling or conventional hand or auger mining techniques. There were a total number of twenty three manhole shafts to be excavated in a twenty week timeframe. The shafts had to be completed in the 2008 construction season as the tunneling portion of the project was scheduled to begin in the winter. The shafts will also be used for maintenance of the TBM cutterhead during the drive. After the TBM maintenance is complete and the TBM passes through the shaft, the construction of a cast-in-place concrete collar will be performed to allow transfer of the tunnel loads when the tunnel/manhole connection is performed. This exterior collar takes the place of any necessary interior support bracing of the tunnel lining precast concrete segments. Once the concrete collar has achieved strength, the tunnel lining circular connection will be water-jetted and the circular hole will be opened to the tunnel. Once the opening is completed, the shaft can be used during the tunnel drive to supply tunnel ventilation air from the surface, and provide power for the tunnel conveyor booster drive units. Ultimately, the precast concrete manhole sections will be stacked and backfilled to create the final manhole structure.

Manhole Construction

The shafts for the manholes were excavated using the cased blind auger shaft drilling method. The general stages of the construction are as follows:

- Crane set up
- Install upper casing
- Drill pilot hole
- Ream hole to 3.43 m (11' 3") diameter
- De-sand the reamed hole
- Place CMP casing in reamed hole
- Place 3,447 (500 psi) flowable fill concrete into bottom of cased hole
- Place 13,789 kPa (2,000 psi) annulus concrete (grout) around casing
- Remove upper casing
- Move to the next drilled shaft site

A heavyweight steel upper casing was fabricated from 1.27 cm (½") rolled plates to create an upper drill casing, which holds back the man-made fill zone at the surface during the shaft excavation process. The upper casing was faced with carbide tipped teeth welded to the lower edge. This 3.96 m (13 feet) long upper casing was spun into the ground (using the casing spinner) until only a 1.2–1.5 m (4–5 feet) long section was left above ground. This above ground section of casing acted as a safety mechanism and also allowed a high level of drilling fluid to be used in case of shaft instability that was found. A bentonite and water mixture was used to aid the carbide teeth to cut through the ground while the upper casing was spun. Without the slurry, high torque was witnessed while installing the casing.

After spinning in the upper casing, the shaft was then drilled to the design depth at a diameter of 3.43 m (11'3"). A gap of approximately 15.2 cm (6 inches) wide was left for grout between the lower casing and the excavated soil. Twelve gauge (0.109" wall thickness) corrugated metal pipe (CMP) is used as the lower casing and temporary shaft support for the shafts. All the shafts used a standard CMP casing diameter of 3.05 m (120 inches) inside diameter. The CMP casing is installed (hung) inside the drilled shaft and stopped .45 m (1.5 feet) above the tunnel at each manhole location (measured from the top outside edge of the precast concrete segmental liner elevation). The lengths of the shafts range due to the differing surface elevation and gradual rise of the tunnel as the alignment proceeds from the south to the north. The deepest shafts are at the beginning of the alignments and were roughly 13.1 m (43 feet) deep to final drilled depth. The shortest shafts at the north end are only in the range of 6.09 m (20 feet) deep. The construction sequence of the shafts (cross-sectional view) is shown in Figure 6 (MH 1-2 used for this example).

Drilling Equipment

The drill table used was a "471 Drill Attachment w/o Insert," manufactured by the Steven M. Hain Company. The drill table is rated to supply 223.7 Kn-m (165,000 ft. lbs) of torque, and was repowered with a John Deere Tier 3 engine by the Traylor Brothers inc. Equipment Division to meet the new California air quality requirements. The drill table was attached to a Liebherr HS 853 HD Hydraulic Crawler Crane with 32.0 m (105 feet) of boom. This crane would eventually act as the main shaft servicing crane for the tunneling portion of the work once the shaft drilling portion of the work was completed. A single 19.8 m (65 feet) long Kelly Bar performed the drilling work and was long enough to reach the deepest depths needed for the shaft locations on the project (longest depth roughly 13.1 m (43 feet) at MH 1-3). A specialty upper casing spinner with fabricated arms attached to the Kelly Bar and allowed for installation of the large diameter upper steel casing. This assembly is simply a four-arm device that attached to the Kelly Bar and sit in the fabricated pockets in the reinforced area of the upper casing.

DRILLED SHAFTS FOR THE UPPER NORTHWEST INTERCEPTOR

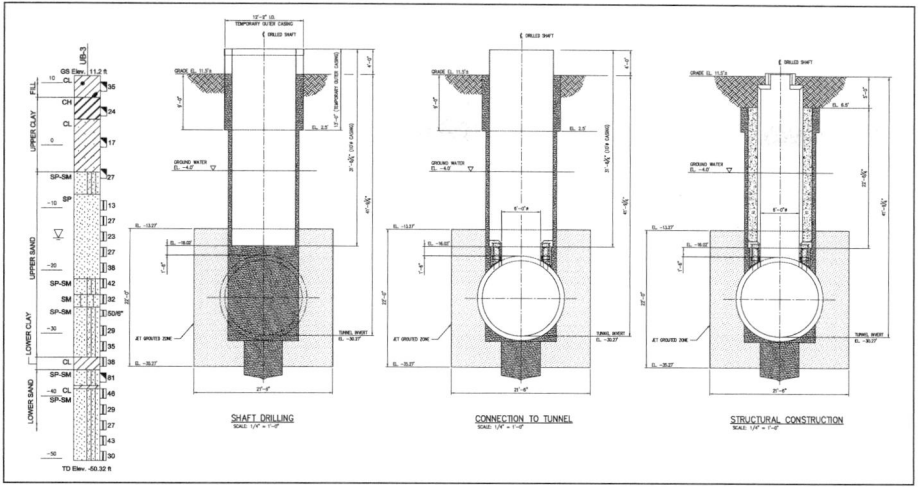

Figure 6. Manhole drilling sections (example from MH 1-2)

Drill Tools

Multiple drill tools were used to excavate the shafts. The initial reaming hole was drilled using a single flight 72" diameter auger with cutting edge. This auger creates a pilot hole to act as a centralized hole for the main excavation bucket and auger. This auger was found to adequately excavate the jet grout block to the final manhole depth. The main shaft drilling is performed with a 72" diameter XXHD Earth bucket with reaming arms, manufactured by the Steven Hain Company (shown in Figure 7). The bucket uses 2 swing-arm reamers, which allowed for excavation of the shaft to three different diameters. The bucket was fitted with cutting teeth and inner seal flaps, to prevent the loss of wet material as the bucket was extracted from the hole. The 2 settings of the adjustable arms could excavate to the following different diameters: 11'-3" diameter ream (lower casing drilling—1'-1" larger diameter), 11'-10" diameter ream (excavation inside the upper casing), and 12' 9" diameter ream (upper casing drilling—6" larger diameter than O.D.).

The swing arm reamers used during the first two shafts originally had blade type pick teeth that were intended to scrape the ground and excavate as the bucket lowers. Due to the high compressive strength of the jet grout, it was found that the teeth did not penetrate the jet grouted soils adequately, and simply wore away as the shaft was excavated at a very slow vertical penetration rate. The excavation of the jet grout block took approximately twenty hours for the bottom 4.87 m (16 feet) depth of full diameter excavation. TSJV ordered a different set of new reaming arms made for rock excavation that utilized carbide tipped bits (Kennametal C4 bits) from the Steven Hain Company. The revised arms took about two weeks to fabricate and deliver to the job site. The new arms were used from the third manhole (MH 1-4) onward, and worked with great success. The blade style teeth could have worked better if more vertical weight could have been applied to the drill string, in order to "load" the blade style teeth more and thus cause better penetration. The drill setup used was limited due to the size of the crane and thus the associated line pull limitations.

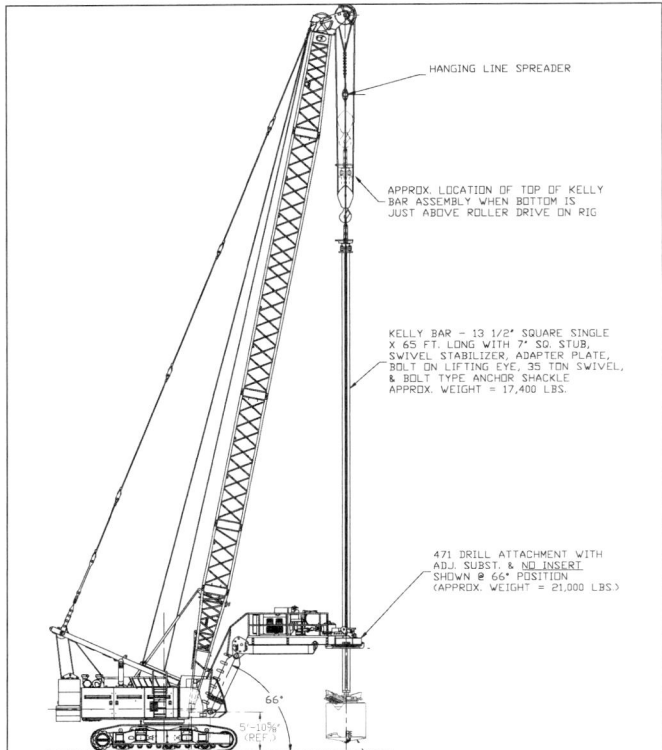

Figure 7. Shaft drilling equipment

Drilling Fluid and Handling Equipment

The blind-auger drilled shaft methodology requires the use of drill slurry, or "mud." The purpose of the drilling fluid has four primary benefits:

- Remove cuttings from the hole
- Stabilize the formation
- Lubricate and cool the drill bits and tools
- Suspend the cuttings during static periods

The primary use of the drill fluid is to provide stabilization of the formation (provide high internal pressure at or above the pressure of the soils that form the shaft walls) until a permanent casing can be installed. Without the drilling mud, the hole can become enlarged causing much difficulty to stabilize.

A drilling fluid engineer from Mi SWACO worked with TSJV for the first 2 weeks of the shaft drilling to develop the correct drilling fluid for the formation. The initial drilling fluid used in the shaft drilling process was the time-tested basic drilling mud comprised of bentonite and water. The typical bentonite slurry mixes were comprised of a range of bentonite power added to water at a concentration of 3.1 to 5.4 kg (7 to 12 pounds) per bbl (42 gallons). The high yield Wyoming bentonite powder, which swells from five to ten times its original volume when hydrated with water has a unique rheological property, turning to a low viscosity liquid when agitated while drilling, but provides stiffness, or high viscosity, when at rest. The stiffening action help hold any of the soils in

Figure 8. Shaft drilling setup

suspension, and also acts to coat the walls of the excavation to provide somewhat of a barrier to ground water transfer into the shaft excavation. The barrier formed is referred to as a filter cake. In many ground conditions, the filter cake is necessary in order to prevent drill fluid loss into the surrounding ground formation. For the UNWI project, the majority of the shafts are drilled through silty sands and low plasticity clays, which are not prone to abnormally high transmission of fluids.

TSJV acquired a large supply of synthetic polymers as well as the high yield bentonite powder, to supplement the slurry if the ground conditions deteriorated to the point where additional additives are necessary to keep the hole open. The synthetic polymers (polyacrylamides) can supplement the bentonite mixture or can be used on their own with water, to modify the viscosity of the drill fluid. A 75,000 liter (20,000 gallon) portable tank was used for all batching of the slurry and recirculation.

Once the shaft is drilled to depth in-the-wet, a desander and slurry pump was utilized to re-circulate the slurry and remove all of the fines that were suspended in the slurry. This created a clean bottom to the shaft, which ensured the placement of the backfill materials would not be compromised. The desander or hydrocyclone removed the fines from the slurry and preserved the integrity of the drilling fluid, so it could be reused over and over. Any sand or other particles were removed and trucked off site with the drilling spoils from the shaft drilling excavation. The re-use of the slurry was a huge cost savings to the project covered any costs necessary to run the equipment including the labor that was involved.

Polymer Slurry

After the first 5 manhole shafts were drilled, the site extents for each manhole shaft shrunk progressively smaller in dimensions, down to about a 12.1 m by 30.5 m (40 feet by 100 feet) area. The small site plan proved to be too limiting to house all the drilling and slurry recirculation equipment and operate in a safe manner. TSJV decided to try the use of strictly polymer drill fluid in lieu of the standard bentonite-based polymer. Based on guidance by the drilling fluid engineer, it was advised that we utilize the use of polymer slurries to make the operation simpler and more streamlined. There are two main types of polymers used in drilling applications, poly anionic cellulose (PAC's) and partially hydrated polyacrylamides (PHPA's). Both types of polymer provide similar advantages of a drilling fluid, with the PAC's primarily being powdered polymer, while the PHPA's are primarily pre-hydrated polymers which are incorporated into the fluid more rapidly. The PHPA's do not hold solids in suspension; they coat the particles and allow them to rapidly settle to the bottom of the excavation. The PAC's provide thicker slurry and provide better filtration control at the interface between the drill fluid and the ground formation.

Due to relatively low pressures due to the shallow depths of the shafts, a dense drill fluid was not necessary to provide formation stabilization. The polymer chosen to be used was Poly Plus 2000, a PHPA product. This polymer is a high molecular weight, anionic liquid designed to act as a viscofier, friction reducer, and flocculant. The polymer also provides encapsulation of the drill cuttings and formation stability control. The Poly Plus 2000 provided all the properties necessary to successfully drill the shafts in a very short time frame. One of the main advantages of the polymer is that the drill cuttings left in suspension would fall out of the solution if left un-agitated overnight. Once the shaft was drilled to depth, the crew could come in the next morning and remove the drill fines left in the fluid with a small diameter clean-out bucket in a very short time frame. This left a full shaft of fairly clean drill fluid, capable of re-use on the next manhole shaft. The remaining polymer slurry was transferred into a 15,100 liter (4,000 gallon) tanker truck for use on the next shaft. When the viscosity of the slurry was found to be too low (higher torque also noted by the drill rig operator), more Poly Plus 2000 would be added to the slurry. To add more polymer material, TSJV simply hung a 18.9 liter (5 gallon) bucket of the polymer from the shaft casing, and allowed the polymer to drip into the hole once a small hole was made in the bottom of the bucket. This allowed for slow addition of the product into the active slurry system.

With the use of the polymers, it was very important to watch the chemical makeup of the slurry. Any pH values higher than 10.3 would cause the polymers to start to disintegrate. This action was found to occur whenever the shaft depth would reach the jet grouted layer. The very basic nature of the cement caused the pH to rise drastically. It was found that the addition of Sodium Bicarbonate (baking soda) in high quantities would keep the pH in the range of 8–10.

Concrete Placing Equipment

Once the shaft was drilled to depth and the lower steel casing was installed, and the lower portion of the shaft was backfilled with flowable fill (500 psi compressive strength at 28 days age). All flowable fill installation and casing backfilling took place while the shaft remained full of drilling fluid slurry. This provided internal pressure to counter any pressures developed during the backfill grouting process, and also prevented any ground loss into the excavation. The services of a 28 meter (91 feet) long boom concrete pump truck were utilized with a long 12.7 cm (5 inch) diameter pipe to act as a tremie pipe to place flowable fill to create the bottom seal as detailed in Figure 6. This same equipment would be used to backfill the CMP casing, with a slightly smaller drop pipe to fit in the 15.2 cm (6 inch) wide annulus. The backfill concrete mix used was a 13,790 kPa (2,000 psi) (28 day strength) concrete using anti-washout concrete admixture to promote the gain of proper desired strength in under water submerged conditions.

Shaft Re-Drilling Operation

Once all the shafts were drilled and cased, the drill rig was used to perform a re-drilling operation in preparation for the first use of the shafts, the maintenance of the TBM cutterhead tools. In order to minimize downtime while changing the tools on the cutterhead, TSJV plans to have a crew working from inside the drilled shafts in front of the TBM cutterhead. In order to allow access, the shafts were re-drilled, removing the center section of the flowable fill material down to a depth that is 0.91 m (3 feet) below the nose of the TBM face. The hole was drilled out to a 1.8 m (6-feet) diameter to allowable a reasonable working cavern for two workers changing ripper teeth or welding the

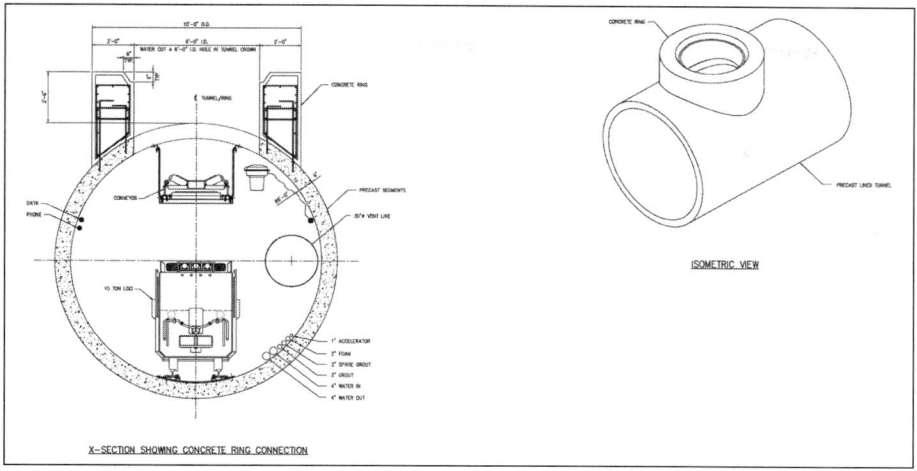

Figure 9. Standard manhole connection collar drawing

cutterhead structure itself. After the re-drilling was finished, clean sand was dumped into the shaft to fill up the hole to the top elevation of the original flowable fill. When the TBM enters the shaft location, forward advance will cease while allowing the rotating cutterhead to ingest the flowing sand. A clean hole should be left to allow man access to the face.

The re-drilling shaft phase proceeded very quickly, with the completion of two shafts per twelve hour working shift. The majority of the time was spent moving the drill rig and crane mats between shafts, along with the spoil disposal.

Access Manhole Connections to the Tunnel

Note—At the time of this writing, the tunneling process is underway, but has not proceeded to a manhole shaft location.

The shafts are to be used during maintenance periods on the TBM cutterhead throughout the tunneling operation. After the tunnel boring machine advances a distance in which the trailing gear is clear of the shaft location, the connection to the tunnel will be executed to facilitate usage of the shaft for tunnel ventilation. In order to make the cut in the precast concrete tunnel liner, it is mandatory to develop a permanent structure element that would bear the tunnel liner loading after the opening is made. A cast in place reinforced concrete collar will be poured at the shaft/tunnel connection before cutting the precast concrete segmental liner. The contract specifications required the inside diameter of the access manholes to be 1.82 m (6 feet). The inside diameter of the cast in place collar will be held at the same diameter. The outside diameter of the collar is limited to a maximum of the 3.05 m (10 foot) inside diameter of the CMP shaft casing. To install the collar, the top of the tunnel will be uncovered and cleaned from inside the shaft. The tunnel liner surface will be roughened and the collar reinforcement will be installed in place. The collar will be anchored to the liner with rebar dowels that will be installed around the perimeter of the collar. A custom made formwork will be lined inside with T-lock PVC liner before the concrete is placed. The collar will then be cast in place and given adequate cure time before making the connection to the tunnel liner. The disk-shaped portion of the precast concrete segments will be cut and removed using hydro-blasting methods. Figure 9 shows the standard connection details.

Final Manhole Risers

The final manhole structure will be comprised of T-lock PVC lined precast concrete manhole sections, manufactured out of calcareous concrete. TSJV will stack the sections of the manhole to the surface and make all the final T-lock welding joints to create a pinhole free liner, to prevent corrosion once the sewer is in service.

Manhole Construction Summary

The ever-changing environmental constraints on the project have caused a great deal of headaches and logistical nightmares. TSJV had the site access cut off to two different manholes due to the presence of Burrowing Owl nests once the construction season was underway, and one shaft was moved over 200 feet to allow construction without disturbance to the nesting owls. The extensive precautionary measures and procedures to follow to prevent the harm of any Giant Garter Snakes caused a large expenditure of labor and materials to satisfy the local regulatory agencies for this matter. The owner and TSJV both worked hard to ensure there were not delays to the project. Numerous change orders were granted to cover the environmental impacts during construction of the drilled shafts.

The drilling of all 23 manhole shafts was completed by early September 2008. After initial shafts took four to five days each to drill to depth, the later shafts were completed in two to three days due to fine tuning the procedures and the use of polymer drill fluid. The act of using the polymer in lieu of the bentonite slurry saved a full day per manhole location due to not having to recirculate the slurry through a de-sanding facility for eight to ten hours to clean it for re-use.

PORT AUTHORITY OF ALLEGHANY COUNTY NORTH SHORE CONNECTOR PROJECT TUNNELS AND STATION SHELL CASE HISTORY CONTRACTS 003 AND 006

Paul Zick ▪ North Shore Constructors

ABSTRACT

The Port Authority of Allegheny County North Shore Connector (NSC) Project involves TBM excavation of twin 690 m (2,240 lf) precast segment lined light rail tunnels through soil and rock under the Allegheny River and a narrow, downtown city street. The $435 M project is funded 80% ($348 M) by the Federal Government, 17% ($72.5 M) by the State of Pennsylvania and 3% ($14.5 M) by Allegheny County. Along the tunnel alignment are a historic building on shallow foundations, modern buildings on deep foundations, and one building constructed directly over the alignment. The project includes construction of twin 369 m (1200 lf) cut and cover concrete box tunnels and an underground station shell. This paper provides an overview of general site conditions, current project status and construction challenges for the Twin Tunnels and North Shore Light Rail Station.

INTRODUCTION

Downtown Pittsburgh is southwestern Pennsylvania's central business district, as well as its cultural and recreational capital. This urban area of Pittsburgh is known as the "Golden Triangle." The North Shore Connector project (Figure 1) will extend the current 62 km (38 miles) Light Rail Transit System, known as the "T," 2 km (1.2 miles) from the Gateway Subway Station underneath Stanwix Street and the Allegheny River in twin bored tunnels to the North Shore. Remaining underground on the North Shore, the alignment travels adjacent to Mazeroski Way to an underground station near PNC Park (Home of the Pittsburgh Pirates). The alignment continues westerly below grade adjacent to Reedsdale Street, then transitions to an elevated structure near Art Rooney Avenue, terminating in a station at Allegheny Avenue on the western edge of Heinz Field (Home of the Pittsburgh Steelers).

The extended light rail will benefit the North Shore's residents, sports stadiums, businesses, museums, as well as Pittsburgh's first casino now under construction. The North Shore Connector will also enable future extensions of the T to other destinations within Allegheny County.

PROJECT GEOTECHNICAL SETTING AND GROUND CONDITIONS

The project is located in the Allegheny River valley at the confluence of the Allegheny and Monongahela Rivers where they merge to form the Ohio River. The alignment is located in the Pittsburgh Low Plateau section. Natural soil deposits are referred to as valley fill, and include moderately to poorly sorted clay, silt, sand, and gravel with some cobbles and boulders.

Two distinct units of valley fill deposits have been encountered during tunneling, the lower unit, Fluvioglacial deposits, and the upper unit, alluvial deposits. Fluvioglacial

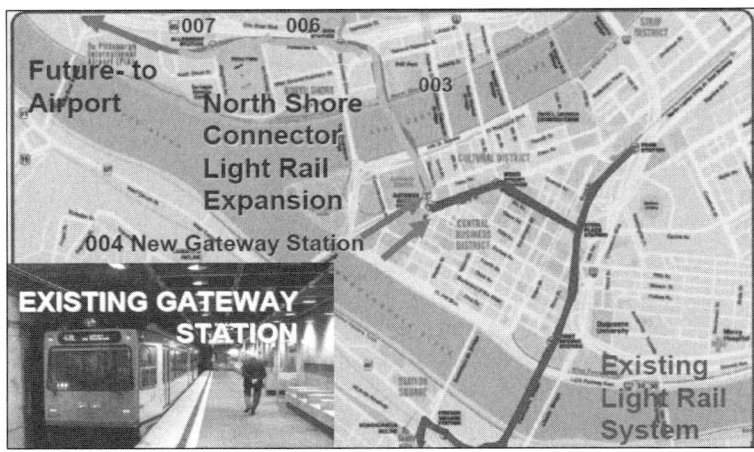

Figure 1. North Shore connector

deposits are generally coarse-grained and consist of mostly sand and gravel from melting glaciers, and the alluvial deposits are more fine grained and typically consist of clay, fine silt, and sand.

Beneath the fluvioglacial and alluvial deposits is Bedrock. Bedrock is generally horizontal with some dipping of less than 2 degrees toward the southeast. Some faults occur in this section, but are generally not readily apparent topographically. Joints are obvious in some areas. Neither faults nor joints caused any problems during mining the two bores.

Bedrock bedding sequence encountered in the Allegheny River crossing (Figure 2) consists of an upper layer of silty shale with uniaxial compressive strengths (UCS) ranging from 37 to 81 MPa (5,400 to 11,800 psi) overlying a 3 to 6 m (9 to 20-ft) thick layer of soft calcareous claystone with UCS of 1.4 to 30 MPa (200 to 4,400 psi.) A thin, .3 to .6 m (1 to 2 ft) -thick, discontinuous layer of hard limestone with UCS of 16 to 78 MPa (2,400 to 12,700 psi), overlies the claystone, except beneath the river, where an up to 150 mm (6-inch) thick, discontinuous, bony coal seam is present over the claystone. Area experience indicates that the strength of this coal seam is around 7 MPa (1000 psi). The next lower rock unit is a 3–4 m (10 to 14 ft) thick layer of thinly bedded, blocky to massive, siltstone with hard calcareous inclusions having UCS between 37 to 76 MPa (3,400 and 11,000 psi). The lower most rock unit encountered is a fine-grained thinly to medium bedded, blocky to massive sandstone with UCS ranging from 29 to 83 MPa (3,400 to 12,000 psi).

Ground water was a factor in all excavations. The valley fill deposits provided a relatively permeable aquifer that was continuously recharged by the adjacent rivers, and the aquifer generally responds quickly to changes in river pool elevation.

GROUND IMPROVEMENTS BEFORE AND DURING TUNNELING

On the North side of the river, the tunnels passed beneath the newly constructed Equitable Resources Building. On the South side of the river, the tunnels passed beneath two major roadways (10th Street Bypass and Fort Duquesne Boulevard), and adjacent to the historic Joseph Horne Building, a 150 year old structure constructed with shallow spread footings at an elevation above the tunnel spring line.

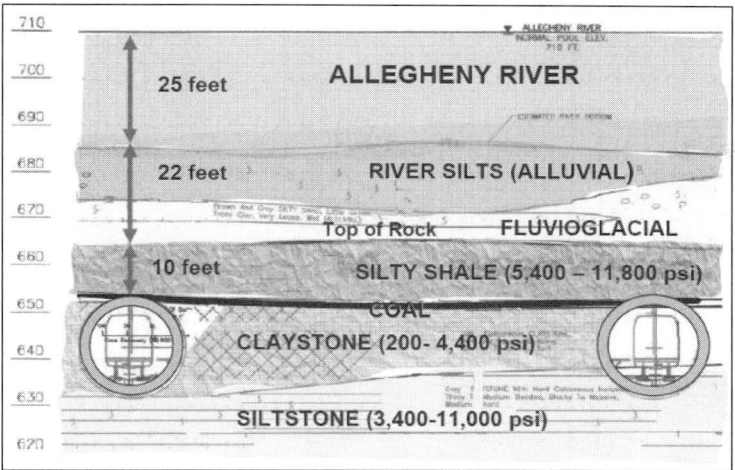

Figure 2. River crossing

Ground improvement, consisting of jet grouting, was designed for four locations. The 10th Street Bypass, Launch Pit Exit block, Receiving Pit Entrance Block and 144 m (468 lf) of Stanwix Street in front of the Joseph Horne Building. The Launch Pit exit block and Receiving Pit entrance blocks were designed as gravity structures to be incorporated into the support of excavation. These areas were constructed with 100% jet grout coverage at a minimum 1 MPa (150 psi). .

On Stanwix Street, a 144 m (468 foot) jet grout zone was constructed from the Receiving Pit northward along the right tunnel which was adjacent to Joseph Horne Building. Jet grout ground improvement was specified in lieu of structural underpinning to be less intrusive on the Horne Building occupants and to minimize angular distortion due to ground loss. Building settlements during underpinning were expected to be in the 10 to 15 mm range. The goal was to keep angular distortions in the range of 1/600 or less so that damage, if it developed, would be substantially limited. Maximum centerline settlement for both the first drive through an unimproved zone and the second drive was (5 mm) .2 in. Angular distortion was essentially zero.

The originally designed trapezoidal jet grouted zone surrounding the tunnel, was modified using stepped columns (Figure 3) to create a shroud around the tunnel that extends 2 m (6 feet) beyond the limits of the tunnel bore above spring line with the two outside columns extending from spring line to 2 m (6 feet) below the invert. In addition to protecting the Horne building foundations, the close proximity of the bored tunnels to each other required ground stabilization for the pillar between them. Also the lining in the first tunnel had to be protected from damage during the boring of the second tunnel. Temporary steel ribs were installed in the completed tunnel until the adjacent tunnel precast segments were grouted. Treated material in this zone required a minimum 1 MPa (150 psi) compressive strength. This serves to strengthen the pillar between the west tunnels (driven first) and the east tunnel (driven second).

On the South side of the Allegheny River, jet grouting was planned at three retaining walls along Fort Duquesne Boulevard and the 10th Street Bypass (Figure 4). Foundation piles supporting the retaining walls were to be removed to facilitate tunneling. Portions of the retaining walls were removed to facilitate pile extraction. Jet grouting beneath the retaining walls was planned to form a foundation for reconstructed portions of the retaining walls. However, the piles were found to be shorter than anticipated.

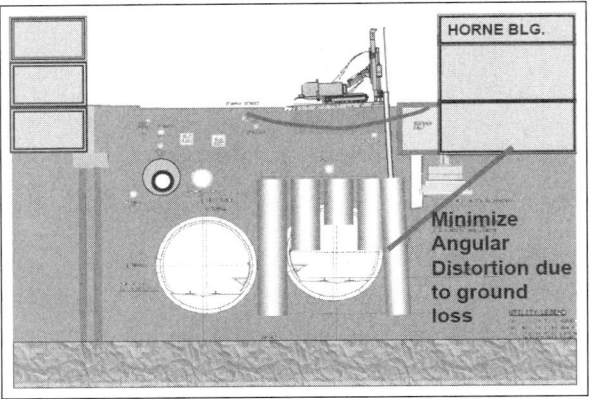

Figure 3. Stanwix Street jet grouting

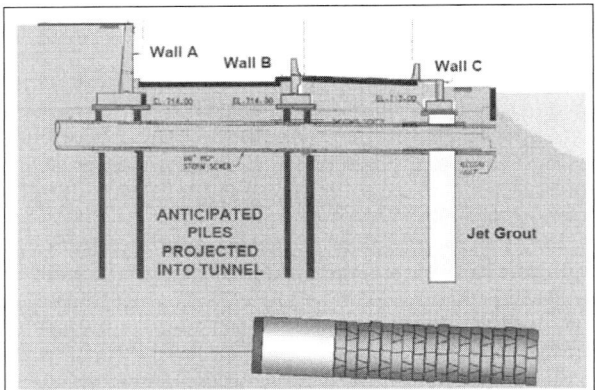

Figure 4. 10th Street bypass

Only the most northern wall (C) was completely removed and piles extracted. A zone equal to the pile foot print was then jet grouted. Selected piles under Walls A and B were checked for depth. The tips of these piles were found to be out side the tunnel excavation window significantly reducing the treated quantity. The area under wall (C) that was treated had 100% coverage at a minimum 4.1 MPa (600 psi) 28 day compressive strength.

A high capacity cooling well is located at the North West Corner of the Joseph Horne Building. Ground modification designed for the northern most 21.5 m (70 lf) of the east tunnel bore along the Joseph Horne Building consisted of injecting low mobility compensation grouting as the tunnel advanced. Compensation grouting was selected over jet grouting to minimize the potential for interfering with the water well. Subcontractor Moretrench America Corp, injected low mobility sanded grout in pre-drilled holes consolidating the ground above the tunnel quarter arch. The grout compensated for surface movements due to ground loss as the TBM advanced through this tunnel reach.

On the North side the first tunnel passed between first and second column lines of the Equitable Resources Building. Compensation grouting was specified to reduce

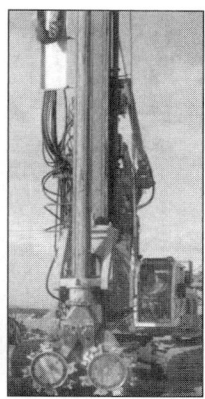

Figure 5. CDSM wall launch pit

Figure 6. CDSM mixing rig

potential damage to the parking garage on grade basement slab from tunneling-related ground movements. Actual settlement was less than 10 mm (⅜"). Only minimal grouting was required to fill small voids that develop beneath the slab as soil below the slab settled.

SUPPORT OF EXCAVATION (SOE)

Three methods were used to provide support for the cut and cover excavations. The first system, Cement Deep Soil Mixing Walls (CDSM) was the primary perimeter element used to support the Launch Pit (Figure 5), Receiving Pit, and cut and cover tunnels excavation. The second system, slurry walls with a cast in place roof were used to construct the North Side Station structural shell by the top down method. The station platform is about ± 15 m (49 feet) below the ground level. At one point, the cut-and-cover tunnels pass underneath an aerial roadway section (SR 65). Two bents of this aerial section required extensive reinforced concrete post tension underpinning structures, (two mini bridges to hold up the bridge). The third system, soldier piles, whalers, struts, and lagging were used to support the Boat Section excavation where the tunnels emerge from below grade to transition onto an aerial structure that will be located on the north side of Heinz Field.

In the CDSM method, subcontractor, Nicholson Construction Company used a base machine with 15 m (70 ft) Kelly bar and twin hydraulically driven mixing heads (Figure 6). The walls are constructed using rectangular elements consisting of cement grout mixed with in-situ soils. The rectangular elements are formed vertically by first mixing bentonite slurry with the in-situ soils from grade to the top of rock. Mixing is from the top down. Upon reaching the top of rock a 1 m (3 ft) deep socket was cut into the foundation rock to provide ground water cut off. Slag/bentonite grout was then injected into mixing heads blending the grout with the in-situ soils as the mixing head is with drawn. The slag/bentonite mixture develops a product with a minimum 1 MPa (150 psi) compressive strength. Steel soldier piles were then inserted with the aid of a vibratory hammer while the mixture was still fluid. Fiberglass wide flange beams were used in place of steel to form a soft eye in the sections of wall where the TBM will break out and in of the Launch and Receiving Pits. Wide flange beams and pipe struts completed the internal support.

At two (2) locations under existing highways headroom was to low to accommodate the CDSM equipment. In the low head room areas, the CDSM method was modified

Figure 7. CB wall

Figure 8. SR 65 Instrumentation

to a Cement Bentonite (CB) system using a low headroom excavating grab (Figure 7). The CB walls were excavated similarly to Slurry wall construction, completely removing all in place materials. At completion of excavation, an air lift pump was inserted into the cut and the bentonite slurry was mixed with cement grout.

After mixing, steel soldier piles were inserted. The resulting mixture produced a product with minimum 1 MPa (150 psi) 28 day strength. At the Receiving Pit, the CB method was further modified by excavating approximately 60% of the panel with the grab, then mixing the remaining in place situ soil materials with the bentonite slurry and cement grouts.

INSTRUMENTATION MONITORING

An extensive Instrumentation and monitoring program was put in place before excavation for SR 65 underpinning (Figure 8). The program includes monitoring vertical and horizontal movements of structures; strains in the SR 65 highway bridge bents; tilt of highway bridge bents; groundwater levels; and construction-related vibrations.

The instrumentation monitoring and reporting program was designed to alert the Contractor and Engineer when construction activities could potentially impact critical

Figure 9. Herrenknecht 6.92 meters slurry pressure balance tunnel boring machine

structures. This system allows adjustment of construction procedures or methods before significantly impacting structures and utilities. Data is collected by a remotely read continuously operating data logger. Data is transmitted automatically to a collection computer in Boston where it is constantly evaluated. If initial response values are exceeded, the project site is immediately notified. Reports of all instruments are produced daily.

TUNNEL EXCAVATION

The Joint Venture selected a Herrenknecht 6.92 meter Slurry Pressure Balance Tunnel Boring Machine (SPBTBM) to bore the twin tunnels (Figure 9). It was determined that the pressurized, closed face Slurry TBM system would best handle the varied mixed face conditions found on this project. Using the Slurry TBM system, a 1.5% face loss was anticipated for design in the fluvioglacial deposits and a 1% face loss was anticipated for design in bedrock. The actual settlement trough for both tunnels was well within these parameters.

The TBM is a Mixshield TBM capable of both soft ground and rock excavation. Operating maximum face pressure is 3.0 bars. The cutter head is equipped with 24 twin disk 430 mm (17-in) cutters, 20 scrapers and bucket lips. The cutterhead is also equipped with 3 wear detection devices. The TBM is designed to navigate a 176 m (574 ft) min radius curve, with the ability to correct alignment at a 141 m (458 ft) radius. The articulated shield is 8.3 m (27.1 ft) in length and is equipped with provisions to drill probe and grout holes. The Tail Shield is equipped with three (3) rows of stainless steel brushes. Overall length of the TBM and trailing gear is 47 m (151 ft) with the weight of the TBM at 490 tons.

Twenty four (24) ea × 260 mm (10.2 in) thrust cylinders that develop 4,200 tons of advance force propel the TBM. Maximum stroke is 1,750 mm (5.74 ft). The (6.92 m) 22.8 foot diameter cutter head is powered by 8ea -101hp, water cooled, VFD controlled electric drive motors. The drives develop 4,203 lb-ft of torque. Direction of rotation can be left or right with a rotational speed of 0–2.7 rpm. The main bearing is 8.5 feet in diameter.

The tunnels are lined immediately behind the TBM with a 1.25 m (4 ft) long tapered precast segment ring (Figure 10). Inside diameter of each erected 4 foot ring is 6.15 m (20 ft) with an outside diameter of 6.72 m (21'-10"). The tapered rings are made up of

Figure 10. Precast segments

6 pieces and a key and were manufactured in two size tapers 41 and 57 mm (1-⅝" and 2-¼"). The ring thickness is 280 mm (11 in).

The TBM is equipped with a double chamber man lock to accommodate up to 4 persons and single material lock (Figure 11). Maximum operating pressure is 3 bars. During the mining, 125 interventions were made to service the cutter head and clear the excavation chamber. Face pressure range between 1.41 and 1.54 bar.

The machine is also equipped with a hydraulic jaw crusher (Figure 12) capable of reducing 600 mm (2 ft) rocks to the 140 mm (5.51 in) maximum size that can pass through the slurry material handling system. The slurry system (Fig 13) has a capacity of 1,000 m³/hr (1,270 cy/hr) using a 350 kw (460 hp) main slurry pump located on the first back up gantry, two (2) 350 kw (460 hp) in line pumps (one at the base of the launch pit and one at the receiving pit—675 m (2,200 lf)) and a 402 kw (536 hp) feed pump. Slurry lines are 305 mm (12 in).

Slurry enters the plant at the top (Figure 14), passing over a 50 mm scalping screen, flow then splits into two parallel circuits consisting of secondary screens, 250 mm cyclones, 75 mm cyclones, and twin 12 tn/hr centrifuges.

The maximum tunnel gradient is plus and minus 8.15%. To accommodate the steep grade, rubber tired diesel multipurpose vehicles (MPV) (Figure 15) were used to transport segments and supplies. to the heading. Segment annular backfilling grout and accelerator were pumped directly from the portal batch plant to a holding tank on the back up equipment. The grout was then pumped through multiple points in the tail shield behind the segments as the TBM advanced. The annular grout components were cement, bentonite and flyash. Personnel were transported in separate diesel vehicles.

TBM MINING

The TBM arrived on site August of 2007 and, to save time, was assembled on the surface concurrent with final launch pit excavation. The assembled 490 ton TBM was lifted, transported 43 m (140 ft) to the launch pit, then lowered 15 m (70 ft) by gantry crane to the invert (Figure 15).

Figure 11. Double chamber man lock

Figure 12. Hydraulic jaw crusher

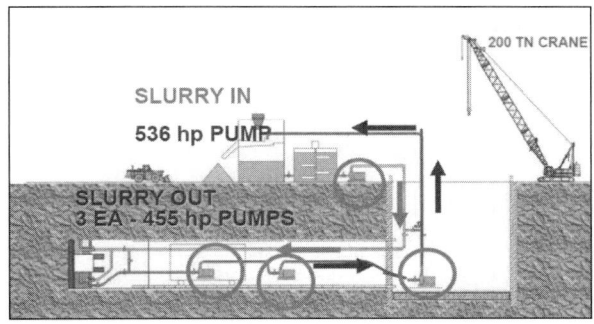

Figure 13. Slurry system

The initial 43 m (140 ft) of mining started January 14th, 2008 with two of the 3 tailing decks in place. At completion of the initial mining the third deck was added and production mining began March 14th 2008. Mining the west (left) tunnel first, the TBM proceeded southward under the Equitable Resources Building and Allegheny River to the receiving pit. The first tunnel hole through was July 11, 2008 (Figure 17). Upon reaching the Receiving pit, the TBM was rotated 180°. Tunneling resumed September 3, 2008 and returned to the Launch Pit located January 15th 2009 (Figure 17).

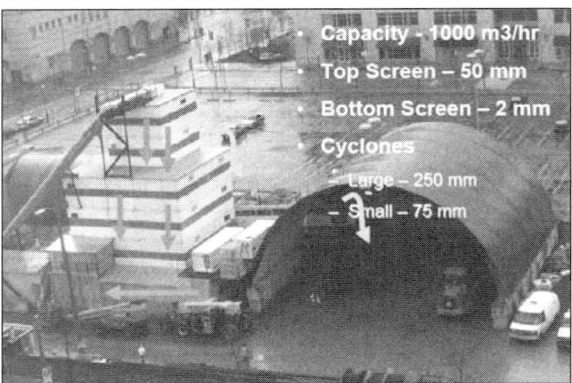

Figure 14. Separation plant

Figure 15. Multipurpose vehicles (MPV)

Figure 16. Gantry crane

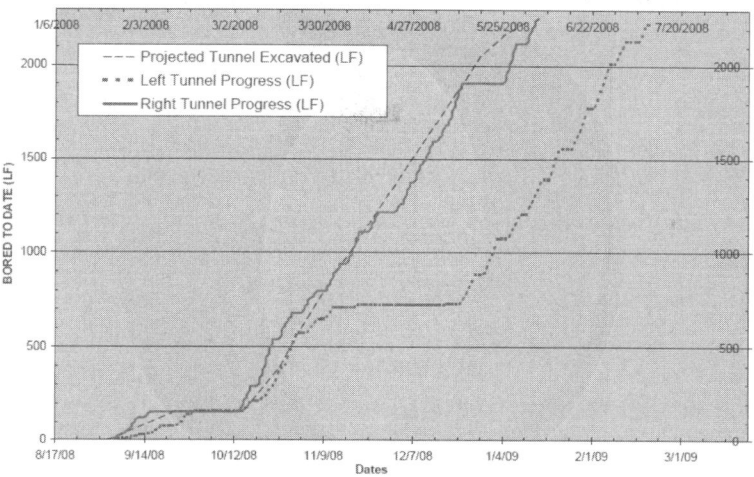

Figure 17. Tunnel production

CHALLENGES

During the first drive, a collection of wood mixed with a sticky material plugged the TBM excavation chamber. The plug occurred while the TBM was making the transition from the soft ground to a full face of rock. An intervention at 1.8 bar was required to clear the plug. The TBM, however, was found to be located directly under the North River sheet piling wall. Additional investigation found the wall sheet piling was within a foot of the TBM crown and that instead of crossing the tunnel perpendicular, one leg ran parallel to the TBM crown for the length of the machine. The close proximity of the sheet piling to the TBM made it impossible to seal the Fluvioglacial formation and pressurize the face for the intervention. The passage created by the sheet pile allowed the air to escape directly to the river. Grouting of the formation was required before an intervention could be made.

SUMMARY AND CONCLUSIONS

In mining mixed face tunnel alignments, transitions from soft ground to rock and rock to soft ground are the most challenging. The Slurry TBM proved to be the correct choice of TBM for mining the fluvioglacial and rock formations under the Allegheny River. With the exception of the River Wall sheet pile encounter, transitions went smoothly. The construction challenges to underground construction in a mixed ground condition require adaptation of methods and equipment to accommodate the unusual technical problems as well as co existing with the special needs of the public.

REFERENCES

Port Authority of Allegheny County. 2007. North Shore Connector; Project Information.
Boscardin, M D, Roy, P.A., Miller, A.J., DiRocco, K.J., 2007. "Designing to Protect Adjacent Structures during Tunneling in an Urban Environment" RETC 2007 Proceedings pp 70–79.
North Shore Connector Contract No. NSC-003/006 November 2005 "Geotechnical Baseline Report" DMJM HARRIS/AECOM.

CONSTRUCTION OF THE NORTH DORCHESTER BAY CSO STORAGE TUNNEL IN BOSTON

J. Davies ▪ Hatch Mott MacDonald

K. Chin ▪ Massachusetts Water Resources Authority

J. Ohnigian ▪ Shaw Environmental & Infrastructure

J. Stokes ▪ M.L. Shank

ABSTRACT

This paper describes the successful management and construction of the 5.18 m (17 ft) diameter, segmentally lined, North Dorchester Bay CSO storage tunnel in Boston, MA. This 3,300 m (10,830 ft) long tunnel was mined through variable soft ground with only 6 m (20 ft) of cover, using an earth pressure balance tunnel boring machine. These extensive data collected from the projects unique automated and internet accessible geotechnical and TBM monitoring system, will be examined and used to explore the relationships between geological conditions, TBM operating parameters and measured settlements. The paper will also discuss the use of screw revolutions as an indicator of muck volume.

INTRODUCTION

As the largest single project in the Massachusetts Water Resources Authority (MWRA) combined sewer outfall (CSO) control program, the North Dorchester CSO Project will virtually eliminate combined sewer overflows and storm water discharges to the beaches in South Boston. For decades, seven outfalls along the South Boston waterfront have discharged combined sewer overflows approximately 20 times a year, as well as storm water in every storm. These discharges have contributed to water quality standards violations and beach closings. Once the project is completed combined sewer overflow and storm water will be diverted into the tunnel for storage. After the storms, this stored water will be pumped from the tunnel to a local sewer interceptor by a 37.8 million L a day (10 mgd) dewatering pump station and 0.6 m (24-inch) diameter force main. The storage capacity of the tunnel is approximately 70 million L (18.5 million gallons), which will effectively eliminate CSO discharges into the North Dorchester Bay for up to a 25 year storm and eliminate storm water discharges into Pleasure Bay.

The 5.91 m (19 ft 4 ¾ inches) excavated diameter segmentally lined tunnel was excavated with an earth pressure balance (EPB) tunnel boring machine (TBM) through glacial and post-glacial deposits consisting of clay, sand, gravel, till, including boulders and organics. This tunnel is 3,300 m (10,830 ft) long and follows the coastline to intersect the seven existing CSOs, as shown on Figure 1. The alignment is very shallow with only 6 m (20 ft) of cover with the tightest curve having a radius of 183 m (600 ft). In addition to the tunnel, a diversion structure and a drop shaft have been constructed at each of the existing CSOs, together with a number of small diameter near surface connecting pipes.

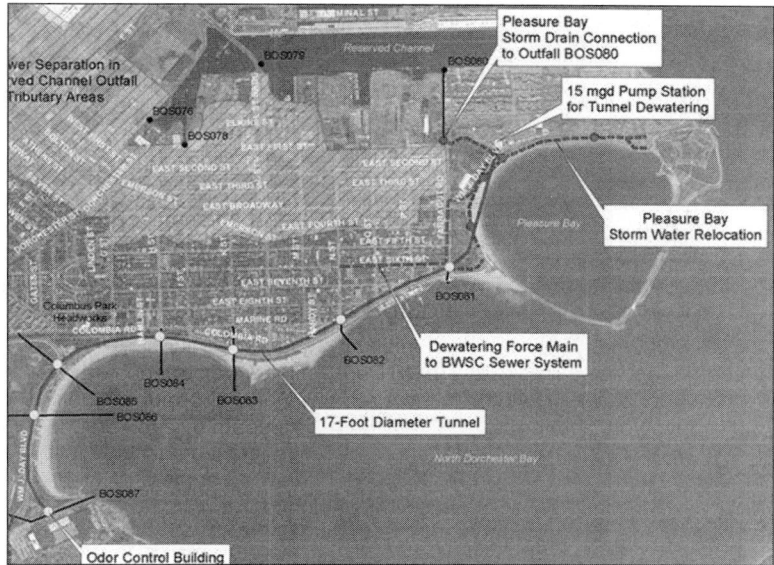

Figure 1. Tunnel alignment and location plan

The MWRA hired the Hatch Mott MacDonald / Shaw Environmental & Infrastructure Joint Venture to be Construction Manager (CM) for this project. The CM has the prime responsibility for the construction management, resident inspection and all other construction support activities during the construction and start phrase of the entire North Dorchester Bay (NDB) CSO Project.

The tunnel mining commenced on October 17, 2007 and the TBM holed through on August 13 2008 approximately six months ahead of schedule and without any claims.

CONTRACTOR/OWNER/CM PARTNERSHIP

In recent years, it has been a trend for owners to share the project risks in an effort to reduce the cost of contingencies added to the bid price. Like most current tunnel construction contracts, the contract document was developed based on the Underground Technology Research Council; (UTRC) guidelines which recommend Escrow Bid Documents, a Geotechnical Baseline Report (GBR), a Disputes Review Board (DBR) etc. The intent was to provide a clear and fair document to allocate risk among all project participants and to facilitate dispute resolution when both parties could not agree. In addition, the MWRA implemented a risk assessment program during the design to reduce potential problems during the construction. The project also utilized lessons learned on recent projects and continuous peer review during the design by a team of outside tunnel experts and the CM.

Since this project is under court order, the completion day of the contract cannot be moved. The project was therefore advertised in April, 2006 and the bids were received in June 2006. The three responsive bidders were:

- Shank/Balfour Beatty / Barletta Heavy Construction JV—$146 million
- J.F. Shea Construction, Inc.—$166 million
- Cashman/Traylor Brothers JV—$238 million

The lowest bidder was $32 million below the engineer's estimate of $178 million and $20 million lower then the second lowest bidder. The difference in bid prices did create some initial concerns, which prompted a meeting between the MWRA and the owner and project director of Shank Construction, Mr. Mike Shank. It was confirmed during this meeting that the bid covered all the work that was required in the contract and that Shank Construction attributed at least part of the savings to the owners approach to the contract documents. The contractor's approach on risk management was also discussed along with the importance of fully engaging the MWRA, so that they would understand his approach to completing the project on time.

Following the award of the contract and issuing the notice to proceed, the entire project team gathered together. The team consisted of the MWRA, the Construction Management, the Design Engineer (DE) and the contractor. During this initial meeting, Mr. Shank offered the idea of a "partnering" concept between the owner, the CM and the contractor. The concept of "partnering" had been around for sometime but the implementation in construction has been patchy with varying degrees of success. Although its adoption is more an exception than the norm, the benefits to the project were carefully considered. After several exchanges of ideas, the MWRA embraced the concept. It was also agreed that no formal partnering document was needed and that an old fashion "hand shake" was sufficient. However, a number of key critical factors to successful partnering were identified and these included:

- Top management support
- Open communication
- Effective coordination
- Mutual trust.

The end result of "partnering" was very successful due to the co-operation of the designer, CM, owner and contractor. The tunnel construction was completed six months ahead of schedule and without any claims. There were also no construction related change orders. The only change order being for additional work requested by the owner.

It was felt that the "partnering" approach helped the team get to know each other and develop a shared approach and ownership of the project. Through this, a good working relationship was developed which led to numerous informal and formal meetings where ideas, data and information were shared and discussed. These discussions often led to a better or more practical approach. An example of this was the brainstorming and discussion that took place on the real-time data produced by the TBM data logging system following excessive settlement. In each case, the aim was to identify the cause or potential causes, so that we could implement changes to prevent it occurring again, rather than assign blame.

The handling of construction safety was also an area of successful corporation where the owner / CM gave conditional approval for a trial of a contractor / union proposal. This proposal identified safety representatives from a pool of experienced local 88 personnel. These safety miners were given the authority to not only police but to do physical repairs and installation of safety equipment. This proposal got excellent support from all parties including the union management and was permanently adopted resulting in a successful and proactive contract safety program.

GROUND CONDITIONS & RISK MANAGEMENT

MWRA implemented an extensive risk assessment program during design to manage and mitigate potential issues during tunneling. Issues such as potential boulder and historic seawall obstructions, excessive ground deformation, protection of existing utilities and muck disposal were handled during the design though extensive subsurface

Figure 2. Installation of the cutter wheel

investigations. These investigations also included rotosonic borings to enable collection of larger diameter intact samples of granular soils. The DE, Parsons Brinkerhoff/ Metcalf & Eddy, prepared a geological section of the tunnel alignment, which showed over 65 borings at an average spacing of a little over 50 m (160 ft).

Based on this very detailed geotechnical information, the ground conditions along the alignment were estimated to be 71% clays, 13% sands 6% gravels, 3% sands and gravels and 7% organics. Throughout the mining process, the tunnel shift inspectors retrieved 'grab' samples of the different soils from the TBM conveyor belt using sealable plastic bags. These samples were laboratory tested on site, and the results were compared to the ground conditions described in the contract documents. The conclusion was that investment in upfront geotechnical investigation paid off and that the profile was a very accurate representation of the ground encountered.

TBM LAUNCH AND RETRIEVAL SHAFTS

The design required slurry walled shafts for both the mining and retrieval of the TBM. The mining shaft was located within the Conley Terminal work site and had a nominal diameter of 15.24 m (50 ft). Due to the possibility that dewatering the area could lead to the migration of hydrocarbon contamination, the slurry wall penetrated down 38.4 m (126 ft) to bedrock, thereby providing additional groundwater cut off. The tunnel invert at this location was approximately 15.24 m (50 ft) below ground surface. The contractor redesigned this shaft by extensively revising and reducing the wall reinforcement and hence owner cost. He also proposed introducing a floating slab at both the mining and retrieval shafts. The proposed changes at the mining shaft were not accepted due to the contamination concerns; however the revisions to the retrieval shaft were adopted, generating savings for the owner. The receiving shaft was constructed at the BOS 087 outfall and was 12.2 m (40 ft) in diameter and approximately 13.7 m (45 ft) deep, which was about 1.5 m (5 ft) below the tunnel invert. A block of ground on the outside of the shafts was jet grouted from the surface to simplify the TBM launch and retrieval.

Figure 3. Segment being installed

TUNNEL CONSTRUCTION

In September 2007, the Hitachi Zosen TBM arrived at the Mass Port Container facility, which is adjacent to the Conley Terminal Worksite. The TBM and its back-up equipment were offloaded from the container ship and transported to the assembly area within 24 hours. Following assembly and testing, initial mining commenced on October 17, 2007. The TBM had an excavation diameter of 5.91 m (19 ft 4 ¾ inches), the outer shield diameter was 5.86 m (19 ft 2 ¾ inches), which gave an overcut with the diameter of 50 mm (2 inches). The TBM was equipped with a 700 mm (27 ½ inches) diameter ribbon screw, which was designed to accommodate boulders of up to 450 mm (17 ¾ inches). The cutter head was dressed with 64 cutter bits and 12 trim bits. The face was capable of being equipped with disk cutters, however these were not installed. The backfill segment grouting was done through three 1 ½ inch diameter grout pipes, which extended through the wire brush tail seals.

The TBM's best overall advance rates on regular nine hour shifts were 17 m (56 ft) /shift, 31.7 m (104 ft) /day and 146.3 m (480 ft) /week. During continuous uninterrupted mining adjacent to sensitive buried infrastructure the advance rates peaked at 19.5 m (64 ft) /shift, 36.6 m (120 ft) /day with a weekly maximum of 159.7 (524 ft).

The DE's original design for the tunnel lining utilized differently tapered rings for the vertical and horizontal rings and had a requirement to always have the key segment above the tunnel springline. This resulted in a very cumbersome design, which utilized numerous different segment shapes and sizes. The contractor therefore hired CSI-Hanson, to redesign the lining to use a streamlined easy to handle six segment universal ring. This simplified topside logistics and handling greatly as well as enhancing QA/QC via simplification. The 1.2 m (4 ft) wide rings, 0.25 m (10 inch) thick rings were built with a taper that allowed a minimum radius of 180 m (590 ft). Production and delivery of segments commenced in October 2007 and were completed in August 2008. Unfortunately due to the exceptional mining progress, a supply shortfall occurred when CSI-Hanson's supplier of the rebar components could not increase their production sufficiently, to keep up with the mining requirements. The tunnel contractor opted to stop mining from April 16 2008 until May 19, 2008 to allow CSI-Hanson JV sufficient time to rebuild inventory. The rings were double gasketed with both an ethylene propylene diene M-class (EPDM) and hydrophilic gasket. The green hydrophilic gasket was installed on the outside of the larger EPDM gasket as can be seen in the Figure 3.

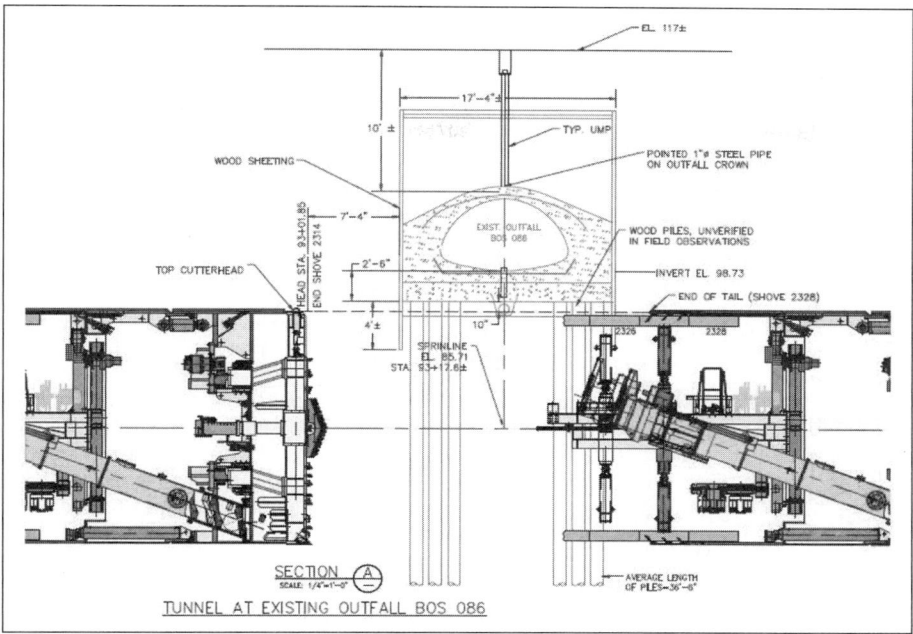

Figure 4. A cross section of the 086 outfall showing the TBM approaching and leaving

The combination of the two gaskets proved to be very successful and the tunnel lining easily achieved the specified criteria of 18.9 L (5 gals) /min/304.8 m (1,000 ft) of tunnel.

Soil Conditioning

The success of the EPB process relies on the successful transformation of the mined soil into a paste. This was achieved by adding a variety of conditioning agents. The conditioning agent dosage rate was monitored continuously and adjusted according to the observed soil type. On a few occasions, water alone was used as an conditioning agent. The cohesive soils, which formed the majority of the soils, were treated with foam and an anti-clay additive so that the dense cohesive lumps were broken down into a smooth paste. The granular soils were treated with a polymer to increase the viscosity of the muck and prevent uncontrolled flow of fluid spoil through the screw conveyor and guillotine gate. In addition to the normal polymer, the contractor also used Baroid Fuse-It, as an aid in start-up at the beginning of a tunneling day.

Mining Below the Existing Outfalls

Due to alignment constraints, the TBM had to pass below the existing BOS085 and BOS086 outfalls with as little as 0.25 m (10 inches) of clearance, as shown in Figure 4. This was further complicated by the outfalls being supported on timber piles. The outfalls were carefully instrumented and continuous mining was both specified and carried out. The TBM carefully mined through the timber piles without difficulty and caused no damage to the outfall structure. The monitoring showed that the outfalls started moving some distance ahead of the TBM, initially heaved a little, and then settled somewhat resulting in a maximum final differential of –5.93 mm (¼ inch).

RESCUE SHAFT CONSTRUCTION

After the contract award, the Boston Fire Department required a second means of tunnel egress in the event of an emergency evacuation during mining. It was therefore decided to construct a rescue shaft approximately halfway along the tunnel horizontal alignment. This shaft consisted of a 1.82 m (72 inch) diameter hand mined shaft through jet grouted soils to the top of the tunnel. The initial ground support was steel liner plate. A 1.52 m (60 inch) diameter ductile iron pipe final liner was then installed and grouted in place. An A316 Stainless Steel Access Hatch and Frame was built into the top of the tunnel liner. The jet grouted zone also provided a stable tunnel face, which permitted manned inspection and intervention of the TBM cutter head without the need for compressed air.

ADIT CONSTRUCTION

The contractor opted to jack steel pipes through the ground to provide initial ground support to construct the five drop shaft connections. Following the probing of each adit, to check ground conditions and water inflow, adit openings were line cored through the sidewall of the tunnel. Temporary support of the tunnel liner openings was provided by four steel rings at each adit location. The steel pipes where then jacked horizontally across to the drop shaft casings which had been installed from the surface. Reinforced concrete pipes within stainless steel collars were installed then, all voids were grouted to complete the connections. Four of the five adits were constructed in the Boston Blue clay, without any additional grouting, the 085 adit area was jet grouted from the surface to improve ground conditions prior to the start of TBM mining.

GEOTECHNICAL AND TBM MONITORING SYSTEM

During the peer review process, the CM recommended that the geotechnical and TBM monitoring system be automated and transferred onto a password protected website. This improved the project team's access to these data significantly, and would enable them to examine both live data and historic records from any internet connection. A very large quantity of information was gathered and analyzed during the project, however only four areas are discussed in this paper.

Water Level Readings

Deep observation wells were installed at specific locations, such as around the mining shaft. These confirmed both season and tidal variations of water level as expected. Piezometers were also installed throughout the length of the tunnel alignment, to monitor the influence of the TBM on the water pressure within the surrounding soil. These picked up significant changes in water level as the TBM approached and past. During mining, when the TBM had greater face pressures the ground water levels would increase, and then reduce during ring build or for shift change. These data logged piezometers began recording an increase in water pressure between 0.6 m (2 ft) and 30.50 m (100 ft), in advance of the TBM. Some of the water level increases were gradual where as others were very rapid. For example, some increased 0.91 m (3 ft) over 48 hours, where others increased over 4.9 m (16 ft) in 40 minutes. Generally, the increase would peak as the TBM passed beneath the piezometer; however some did not peak until the TBM was well past the piezometer. After the peak, the readings would gradually decrease back to a level which was slightly higher than the water level before the TBM had passed.

Figure 5 shows the results from a typical piezometer. From this and other results it was clear that the degree of hydraulic connectivity varied considerably and was difficult

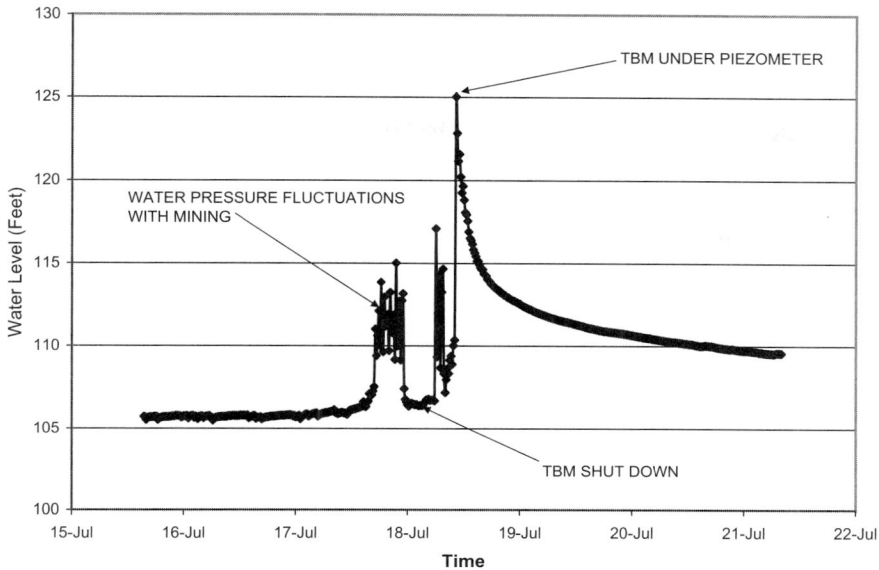

Figure 5. Water level fluctuation as the TBM passed a piezometer

to predict. As shown in the figure, the water level responded closely to the TBM face pressure as the TBM approached. Initially, there was also concern that the increase in water pressure could result in increased settlement. However, this did not appear to be the case perhaps due to the fairly rapid dissipation of pressure.

Settlement Monitoring

A comprehensive settlement monitoring program was instigated, which included the installation of monitoring points along the alignment, on buildings and on buried utilities. Extensometer arrays were also installed to measure subsurface movements. The settlement points typically consisted of masonry nails installed in rows, oriented perpendicular to the tunnel alignment spaced at 15.24 m (50 ft) centers. Each row consisted of seven points, spaced at 7.62 m (25 ft) centers, extending 22.86 m (75 ft) left and right of tunnel centerline. The surface settlement points were read manually and uploaded to the website along with the information collected automatically from the data logged extensometers.

The specification included maximum specified settlements, such as 19 mm (0.75 inches) for high pressure gas lines. There were also corresponding action levels at which point the TBM operator would change the TBM parameters to try and reduce the settlement or heave as appropriate. This approach was very successful and the final settlements were typically low, with average settlements of 16.5 mm (⅔ inch) in the clays and 34 mm (1 ⅓ inches) in the sands. The overall average for the entire project was 22.6 mm (⅞ inches) of settlement. Figures 6 and 7 illustrate typical settlement troughs for the clays and sands, respectively. In each case five different data sets have been plotted, which illustrate the settlement trough as the TBM approached and past. For example, "settlement 92+50 + 13" gives the settlement trough when the TBM was 13 ft in advance of the monitoring point, where as "settlement 92 + 50 –87" represents the trough when the TBM has past the monitoring points by 87 ft.

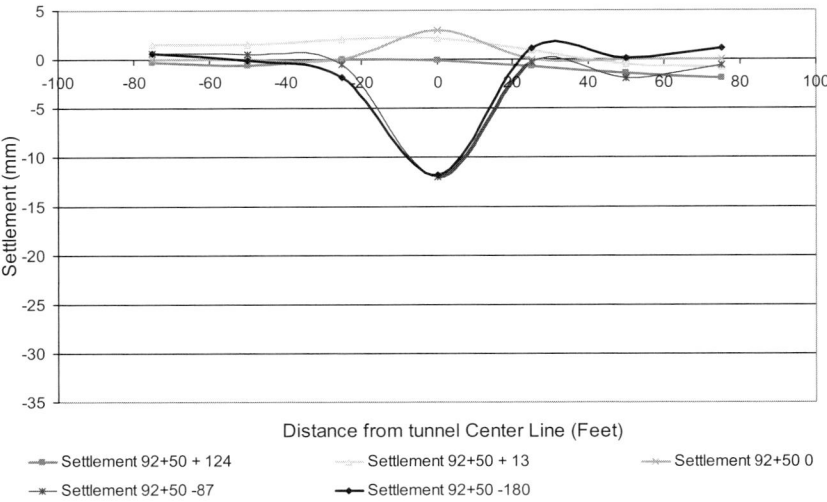

Figure 6. Typical settlement curve in clay

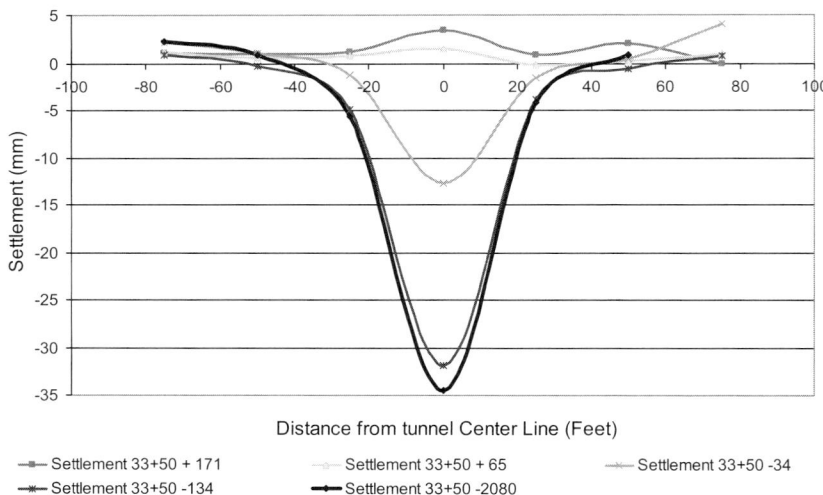

Figure 7. Typical settlement curve in sand

The settlement curve data above were also examined in detail for all the alignment and theoretical curves fitted to these data points. From these we were able to obtain an average face loss per material, as shown in Table 1.

Excessive settlement occurred on only three occasions. The first event occurred early in the drive when a mechanical issue led to a loss of ground and the creation of a sink hole in a beach area. The second area was relatively minor in extent when three inches of settlement produced cracking in the hard landscaping. The third event was a little more significant and occurred when the EPB was not maintained in a sand and gravel over a three day weekend and then during start-up the TBM screw blocked.

Table 1. Average face loss per soil type

Soil Type	Average Face Loss
Clay and Clay with Organics, Highly Plastic Clay, Moderately to highly plastic clay	0.50%
Clayey Sand and Silt Clay, Sandy Clay and Gravel	0.75%
Silty Sand, Silty Clay with Silty Sand Pockets	1.14%

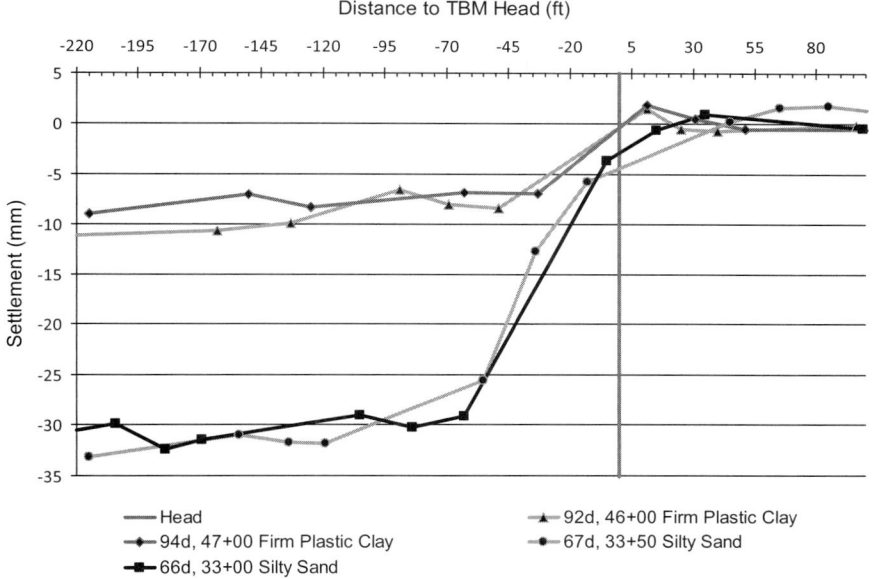

Figure 8. Typical settlement plotted against the distance to the TBM head for different materials

The TBM position was also examined in relation to the observed settlement. Typically, the TBM caused heave ahead of it, which was often over 9.14 m (30 ft) down station. Settlement would then follow as the TBM mined under the reference point. However, the behavior of the sands and clays were different. The clays typically had sufficient stand up time for the annular overcut to stay open over the length of the TBM. This void was then filled with tail seal grouting resulting in small settlements. On the other hand, the sands and gravels almost immediately collapsed onto the TBM shield after the TBM face had passed. This loss of the overcut resulted in the higher observed face losses and surface settlement of these materials. Figure 8 shows monitoring data for four discrete points plotted against the distance to the TBM head. Two of the points were in firm plastic clays and recorded overall settlement of around 10 mm, where as the other two were in silty sands and recorded settlements of between 30 and 35 mm.

Screw Revolutions as an Indicator of Muck Volume

At the start of the project, an automated muck car weighing system was installed. This did not prove reliable and also had the disadvantage that these data were not available during the mining cycle. This approach was therefore abandoned in favor

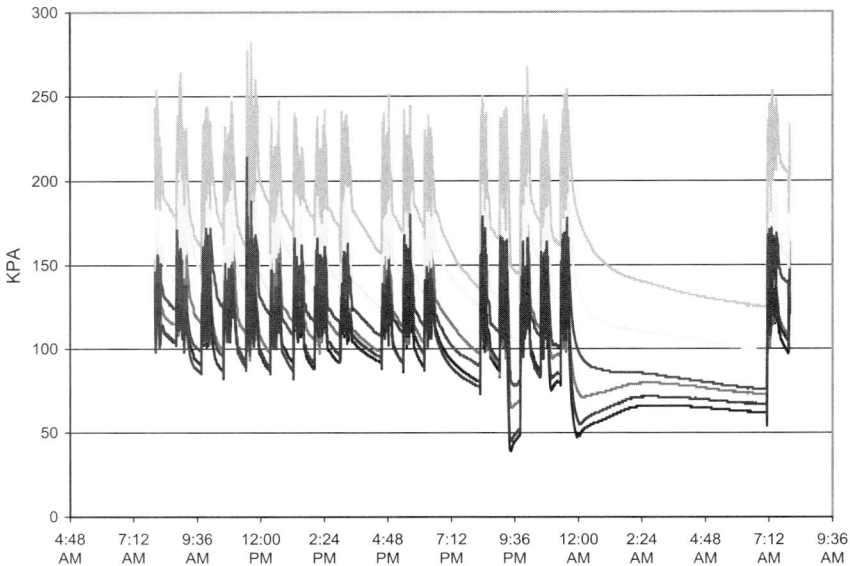

Figure 9. Typical bulkhead pressures for a 24 hr period

of monitoring the number of screw revolutions. The TBM data system was therefore configured to show a live graph of screw revolutions versus advance rate, together with a predetermined upper and lower boundary line. A number of variables existed in the calculations that had the potential to cause errors. However, the graphic presentation was found to be a useful tool to monitor the rate of muck generation and hence give an early warning of over mining or under mining. Under normal operating conditions the number of screw revolutions per advance was found to be relatively consistent within each geological unit. As the materials varied so did the revolutions, with between 15 and 25% extra screw rotations required for the clays or clay mixtures over the sands. As the clay became drier and stickier the number of revolutions went up.

Bulkhead Pressures

Generally, bulkhead pressures were consistent with the mining cycle. They increased during mining and decreased during ring builds or periods of shutdown as shown in Figure 9. When extended shut downs occurred say for a weekend or longer, the bulkhead pressures generally would equalize, and return to a pressure similar to the pressure of groundwater in that area. This equalization did not appear to have a negative impact on the amounts of settlement. Start up was often loose but not uncontrollable, especially if the start up specific polymers were used.

There were six pressure sensors on the face of the machine, separated in pairs at three different heights. During mining it was noted that the pairs of sensors at the same elevation would generally show similar but not identical readings. It was also noted that gauges at different elevations often recorded similar pressures. In was concluded that the material within the plenum was acting more as a solid than a liquid with localized pressure variations. On a few occasions the pressures recorded at the same elevation become very closely paired, and gauges at different elevations showed an increase in pressure with depth. When this happened it suggested that the material was acting more like a liquid. If this phenomenon was observed, particular care was taken to ensure no loss of face support.

CONCLUSION

The Hitachi Zosen TBM and the contractor's methods of operation proved to be very successful at mining the tunnel. The cutter head tool types performed well in dealing with timber piles and boulders, and whilst mining through the jet grout and concrete at the shafts. The average settlement on the project was also generally below the contract limits, particularly in the clays. However, future designs of this type of TBM should at least consider the inclusion of injection ports around the shield sections to allow void filling with a bentonite slurry. If the TBM had this provision, it may have been possible to reduce the settlement in the sands, significantly. The average advance rate for the tunnel was around 24.4 m (80 ft) a day, which was considerably faster than originally anticipated and resulted in the early breakthrough of the tunnel.

The automated monitoring system worked very well and provided the team with ready access to live data. This enabled decisions to be made earlier and with more confidence. The use of screw revolutions to monitor over and under mining also proved to be a valuable indicator and is something that may be worth further development.

The universal ring worked very well and was a significant improvement over the original design. However, when negotiating the 183 m (600 ft) radius curve having a lining set up which would allow the TBM to steer a curve tighter than 180 m (590 ft) would have allowed for better alignment control. The decision to pipe jack the adits also proved extremely successful with good ground support and good protection for the workers.

The owner's investment in a detailed geotechnical investigation enabled the DE to prepare a very accurate longitudinal profile of the tunnel alignment. By defining the geotechnical conditions in this way, day to day decisions regarding the mining through these materials could be made with more confidence. This improved productivity and reduced the settlements observed. The early involvement of the CM and the owners desire to provide a clear and fair set of bid documents enabled the project to be completed significantly below the engineers estimate. Finally the partnering between the MWRA, the CM, the DE and the contractor proved very worthwhile and was perhaps the key to the success of the project.

ACKNOWLEDGMENTS

The authors would like to acknowledge the efforts of many colleagues who played a part in the work described here and contributed to the success of the project. In particular the support of the client's senior management was vital to the success of the partnering. We would also like to thank Jeb Pittsinger for his assistance in preparing the illustrations and figures.

HIGH RISK TUNNELING ADJACENT TO LARGE WATER TANK ON THE UNWI SECTIONS 3&4 PROJECT

Andrew Finney ▪ CH2M HILL

John Wong ▪ Sacramento County Sanitation District

Craig Vandaelle ▪ Michels Tunneling

Cody Painter ▪ PB Americas

Steve Cano ▪ KBR

INTRODUCTION

The Sacramento Regional County Sanitation District (SRCSD) serves unincorporated areas of Sacramento County, parts of the cities of Sacramento and Folsom, and the cities of Citrus Heights, West Sacramento, Rancho Cordova, and Elk Grove. SRCSD maintains 145 kilometers (km) (90 miles) of interceptors that convey a combined total of 0.6 million meters^3 per day (160 million gallons per day) or more of wastewater, with additional miles of sewer interceptors that are scheduled to be built by 2020. Wastewater is conveyed to the Sacramento Regional Wastewater Treatment Plant in the southwestern portion of the County, also managed and operated by SRCSD, in interceptors with diameters as large as 3.0 m (120 inches.)

As part of its overall expansion program, SRCSD has constructed a number of major interceptors over the last few years with tunneling methods due to concerns associated with either the impacts of open cut construction in an urban environment (traffic, safety, noise, environmental concerns) or the need to achieve a crossing of a stream, river, or transportation corridor.

This paper focuses on a specific issue that affected construction of the Upper Northwest Interceptor (UNWI) Sections 3&4 project, located at the northern limit of the SRCSD service boundary and within the limits of the City of Sacramento. This project involved the construction of 3.2 km (2 miles) of 2.1-meter (84-inch) and 2.4 km (1.5 miles) of 1.7-meter (66-inch) diameter reinforced concrete pipe (RCP) using MTBM, EPBM, and 3.3-meter (130-inch) two-pass TBM methods. The owner elected to install the RCP using tunneling for a number of reasons including a desire to avoid traffic impacts in a dense urban corridor, to minimize disturbance to the neighborhood, and to avoid potential dewatering-induced settlement issues along the western one-third of the project. For the western portion of the alignment the 2.1-meter (84-inch) reinforced concrete interceptor pipe was to be installed using a 2.7-meter (105-inch) earth pressure balance machine (EPBM). No external dewatering of shafts or tunnels was permitted for reasons of concern about settlement, discussed previously. At the intersection of Elkhorn Boulevard and Natomas Boulevard (at the extreme northwest corner of the project) the 2.7-meter (105-inch) EPBM tunneled past a recently constructed 11-million liter (3-million gallon) water storage reservoir, passing within 3 meters (10 feet) of the tank's shallow foundation while tunneling through clean sands with a shallow groundwater table. The paper will discuss the implemented solution to mitigate additional settlement risk that was developed as a joint collaboration between the owner, CM, designer, and contractor.

CHANGES DURING DESIGN

The designer, CH2M HILL, was aware of the planned construction of the water reservoir by the City of Sacramento during the UNWI Sections 3&4 design. The prestressed concrete reservoir, which has a diameter of approximately 41 meters (135 feet) and a height of 11 meters (35 feet), was originally designed with a 305-milimeter (12-inch) precast concrete pile foundation bearing approximately 14 meters (45 feet) below ground surface (bgs.) During the concurrent construction of the reservoir in 2006 and the advancement from 90-percent to 100-percent design of the UNWI Sections project, the tank contractor was unable to advance the concrete piles to their specified minimum tip depth. The City attempted to preauger to permit further penetration and considered switching to an H-pile or auger cast pile foundation system. Eventually the City elected to modify the reservoir foundation to include a 1.5-meter (5-foot) overexcavation and replacement of native soils beneath a revised 600-milimeter (2-foot) thick concrete slab foundation. The overexcavated zone (to be backfilled with aggregate base course) extended approximately 3 meters (10 feet) beyond the footprint of the reservoir to within 0.9 meters (3 feet) of the centerline of the UNWI 3&4 interceptor (see Figures 1 and 2).

The designer and the owner (SRCSD) recognized that the redesign of the reservoir to a shallow foundation system introduced an additional level of risk to the tunneled interceptor construction, as prestressed concrete cylinder tanks are recognized as having a low tolerance for differential settlement. The final design interceptor alignment was constrained to the north by a planned installation of 69 kilovolt power lines (limiting future overhead working room) which were themselves constrained by the northern public utility easement on the south side of Elkhorn Boulevard. Vertical alignment of the gravity interceptor was also constrained by the connection with the completed upstream segment of interceptor pipe and downstream by the connection to the Natomas Pump Station. Settlement calculations performed by the designer indicated that the risk of settlement damage from the tunneling was low. Therefore the designer and owner elected to maintain the current vertical and horizontal alignment and implemented additional risk management measures to satisfy both the City and SRCSD, which included the following revised contractual requirements:

- The EPBM drive past the reservoir could not be completed as the first EPBM drive of the project to allow time to develop familiarity with the EPBM and the ground conditions.
- The EPBM drive past the reservoir must be completed continuously to avoid issues associated with start up and shut down of the EPBM.
- Install additional surface and subsurface settlement monitoring points and inclinometers along the alignment on either side of the point of least clearance between the reservoir foundation and the interceptor alignment.
- Establish survey pins around the perimeter of the concrete reservoir foundation to allow monitoring during both the initial filling of the reservoir (which was to occur during the UNWI Sections 3&4 construction but before the adjacent EPBM drive) and the subsequent tunneling activities.

In addition, several threshold settlement values were contractually established to alert the team to potential issues with the mining operation prior to passing the reservoir. A significant portion of the pre-award conference was devoted to discussion of these potential tunneling issues associated with the proximity of the Elkhorn Reservoir.

ADDITIONAL REQUIREMENTS AT THE START OF CONSTRUCTION

The project was awarded to Michels Pipeline Construction (Michels) of Brownsville, WI in January of 2007. Once the project was underway, the City of Sacramento revised

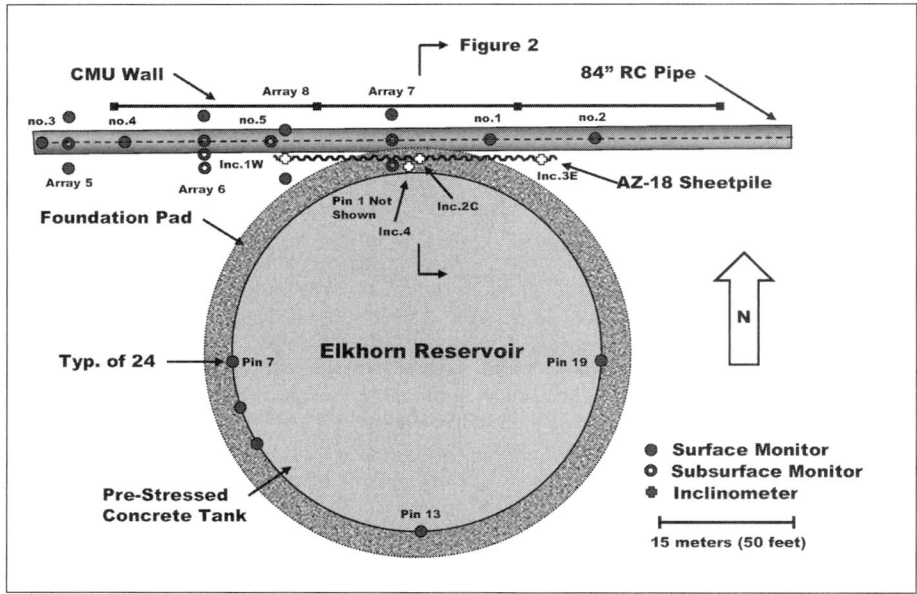

Figure 1. Plan view of Elkhorn Reservoir and the UNWI Sections 3&4 tunneled pipe

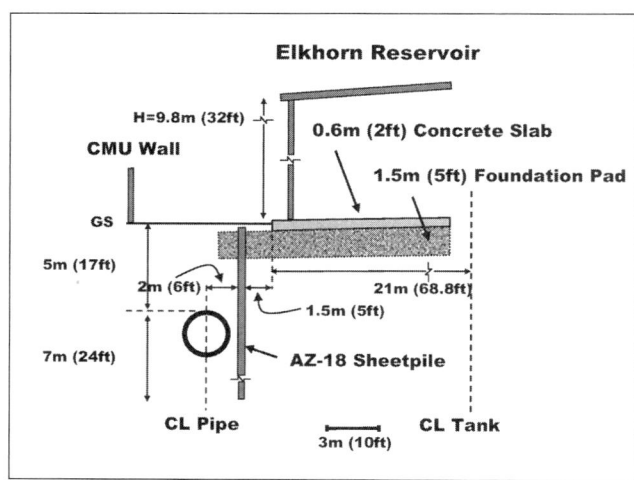

Figure 2. Section view of Elkhorn Reservoir and the UNWI Sections 3&4 tunneled pipe

the right of way easement conditions to include the development of an additional mitigation action plan in the event settlement of 2.5 millimeters (0.1 inches) of settlement was observed at the base of the tank foundation. In light of the reduced threshold settlement and the existing subsurface conditions, additional discussion occurred among the parties at the initial project partnering meeting. Specific concerns included:

- Subsurface conditions within the tunnel zone were baselined as "predominantly silty sand, with minor clean, well-graded sand, clayey sand and well-graded gravel constituents." This material would "behave as a flowing soil if unsupported." Above the tunneled zone was a layer of firm lean and fat clay and beneath the tunnel zone was older partially-cemented sandy alluvium of the Riverbank Formation. Blow counts in the Riverbank formation routinely exceeded 50 to 100 blows per foot.

- In addition to the proximity of the reservoir to the center of the tunnel, the reservoir was also very near the launch shaft for the drive, increasing the risk of initial start up difficulties with the tunneling operations.
- The recent construction of a cased water line crossing of a drainage canal to the west of the reservoir site experienced major ground loss which significantly disturbed the ground in the area and highlighted the difficulty posed by the subsurface soil and groundwater conditions.

The group agreed that additional mitigation of potential settlement risk would be of benefit to both the owner and contractor. The following four options were proposed by the contractor to further satisfy the City of Sacramento's requirements:

1. A secant pile wall installed between the reservoir and the outside edge of the tunnel. This option was eliminated based on cost and practicality of completing the work on time.
2. Consolidation grouting from the ground surface to improve the tunneled zone and the ground between the tunnel and the reservoir. This method was eliminated because of variation in the existing soil conditions and a lack of assurance that ground improvement would be continuous throughout the area required for the protection of the reservoir. There was also concern that the EPBM would have difficulty excavating the improved ground.
3. A series of interlocking jet grouted columns placed between the reservoir and the outside edge of the tunnel. The columns would also be installed to a depth sufficient to toe into the undisturbed ground beneath the tunnel reservoir. This method was eliminated because any inconsistency in the column wall could allow the flow of ground from beneath the reservoir to the tunnel, and because of the potential for a portion of the column to intrude into the tunnel zone impeding the progress of the EPBM and potentially resulting in over-mining of adjacent untreated ground.
4. Lastly, a steel sheet pile wall installed halfway between the closest part of the reservoir and the outside edge of the tunnel. The wall would be installed to a depth sufficient to toe into the undisturbed ground beneath the tunnel reservoir. After consideration of all options, it was agreed that the steel sheet pile wall was the most viable option for the protection of the water reservoir.

The designer, working with the construction team developed a sheet pile wall that was capable of retaining the soils beneath the tank in the event of a loss of support from the soils within the tunneled zone. The computer program PYWALL v. 2.0 was used to estimate pile deflection under the imposed lateral pressure of the tank (using a Boussinesq distribution) and determine a required wall depth of 13 meters (42 feet). The wall was specified to be installed within a 1.5-meter (5-foot) excavation to avoid installing the sheets through the aggregate foundation pad and to provide a vibration air-gap between the installation operations and the foundation pad materials. Geotechnical instrumentation was also included with the sheetpiles in the form of inclinometer casings grouted into 5-inch square steel tubes welded to, and installed with, the sheet piles.

Figure 3. Progressive vibratory installation of sheet piles at the reservoir

Figure 4. Outer inclinometer casing prior to sheet pile installation

IMPLEMENTATION OF RESERVOIR PROTECTION/MONITORING PROGRAM

Installation of the AZ-18 sheet piles was completed without incident using a vibratory driver mounted on an ABI TM 18/22. In order to reduce the risk of vibration damage the contractor pre-drilled every other interlock to a depth of within 1.5 meters (5 feet) of the pile tip with a 405-millimeter (16-inch) auger prior to advancing the sheets. Pre-drilling was limited to a maximum length of approximately 3 meters (10 feet) in advance of the sheet piles to avoid initiating soil movement toward the loosened soil zone and to avoid reconsolidation of the augered zone. Figure 3 shows the incremental vibratory installation, while Figure 4 shows the outer inclinometer casing attached to the sheet piles.

The monitoring program included the use of inclinometers, surface settlement points (surveyor's nail or scribed mark), and subsurface monitoring points (#4 rebar installed in a cased borehole and driven to a point 1.5 meters (5 feet) above the tunnel crown). Instrument locations are shown in Figure 1. All instruments were installed and a baseline established prior to the first filling of the reservoir (in the case of the tank monitoring points), and prior to the start of the affected drive (for the remaining points).

TUNNELING AND GROUTING OPERATIONS

The 8.5-meter (28-foot) deep jacking shaft was constructed of driven AZ-18 sheet piles with a 1.5-meter (5-foot) reinforced concrete shaft plug and associated vertical hold-downs. The contractor elected to jet grout the soils beyond the tunnel eye and poured an un-reinforced concrete soft eye while simultaneously lifting the sheets within the tunnel eye in order to mitigate the risk of ground loss during EPBM launch.

Table 1. Time–distance information for the EPBM drive

Date and Time	Monitoring Point	Location of EPBM Face (station in feet)	Advance Duration (minutes)
4/22/08 10:00AM	SMP-3	212+20	0
4/22/08 10:00AM	Array 5	212+30	0
4/22/08 5:30PM	SMP-4	212+50	204
4/22/08 10:30PM	Array 6	212+80	358
4/22/08 3:20AM	SMP-5	213+05	185
4/22/08 4:30AM	Suppl. Array	213+10	48
4/23/08 12:45PM	Array 7	213+50	375
4/23/08 9:10PM	SMP-1	213+86	319
4/24/08 4:10AM	SMP-2	214+25	173

Jacking commenced on April 22, 2008. Reported surface settlement of 10 millimeters (0.4 inches) at SMP-3 immediately following successful launch indicated that the launch ground control measures were successful at preventing significant settlement. After successful launch, advance rates ranged from 22 to 45 millimeters per minute (0.9 to 1.8 inches per minute.) Table 1 presents a summary of time-distance milestones for the face of the EPBM as they relate to monitoring points identified on Figure 1.

At the completion of the drive, contact grouting of the annular space was completed on May 2, 2008 in order to reduce post construction settlement associated with the partial collapse of the overcut annulus. Approximately 38 meters^3 (50 cubic yards) of grout was pumped along approximately 90 meters (300 feet) of alignment. Surface returns of grout occurred along preferential pathways at monitoring array 5 and inclinometer 3E and surface heave of the paved access to the reservoir was observed during grouting.

OBSERVATIONS

The observed response of the ground, tank, and sheet piles are discussed in the following sections, organized by monitoring device.

Inclinometers

Four inclinometers were installed to monitor lateral movement adjacent to the reservoir. Three of the inclinometers (nos. 1W, 2C, & 3E) were installed on the sheet piles as shown in Figure 4; the remaining inclinometer (no. 4) was installed in a borehole between the sheet piles and the reservoir as shown in Figure 1. Ultimately, no movement was detected in inclinometer no. 4, as anticipated. Figures 5 and 6 present the inclinometer response for inclinometers 1W and 2C, respectively. The response of inclinometer 3E is not depicted, although the deflection was similar in magnitude to that of 2C. It is observed that the lateral deflection toward the tunneled pipe during and after tunneling was extremely small (less than 0.8 millimeters [0.03 inches]) and that the embedment depth selected was appropriate.

Of greater note was the deflection toward the tank (approaching 5 millimeters [0.2 inch]), observed on May 8, 2008 and likely the result of pressurized contact grouting operations conducted on May 2, 2008. All observed movement is considered insignificant to the reservoir and unlikely to result in surface manifestation of displacement.

Surface and Subsurface Monitoring Points

Eleven surface monitoring points (SMP) and 6 subsurface monitoring points (SSM) were installed in the vicinity of the Elkhorn Reservoir along the UNWI Sections

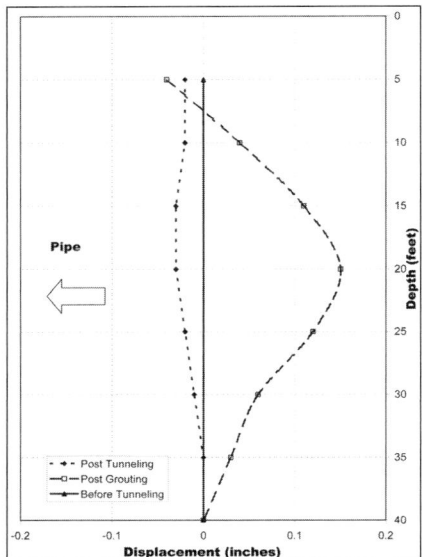 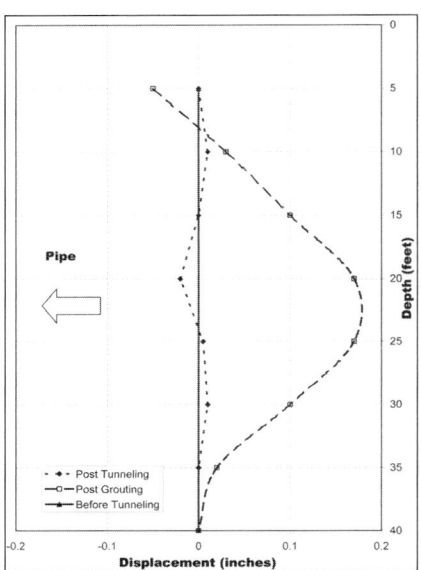

Figures 5 & 6. Measurements along North-South axes of inclinometer 1W (left) & 2C (right)

3&4 alignment as shown in Figure 1. The response of select surface monitoring points located outside the limits of the sheet pile wall are presented in Figure 7, while Figure 8 presents the response of select points located adjacent to the portion of the alignment where the sheet pile was installed. Also presented on Figures 7 and 8 are predicted surface settlement values computed using the method of New & O'Reilly, 1992 with an estimated half width of 2.5i = 6 meters (20 feet) and a trough area equal to approximately 4% of the tunnel face area. For the sheet pile case presented in Figure 8 the increased predicted settlement is the result of the mechanical limit placed on the settlement trough by the sheet pile. It was assumed that the area of the settlement trough that was intercepted by the sheet pile would be redistributed to the remaining trough, thereby increasing overall settlement by approximately 25%. Both cases considered indicate reasonable agreement between the measured values and the predicted values. It should be noted that these are simple approximations, based on the assumption that advance and spoil feed rates were identical at all times during the drive. It was also observed from the settlement data that all settlement occurred after the passage of the EPBM face, likely the result of careful maintenance of positive face pressure at all times during the drive which prevented significant loss of ground in advance of the face.

Subsurface settlement results, while not plotted, indicate a similar relationship. SSM 5 and SSM 6, both located before the sheetpile at points approximately 1.5 meters (5 feet) above the tunnel crown both reported 63 millimeters (2.5 inches) of settlement. SSM 7, located within the sheetpile zone at a point approximately 1.5 meters (5 feet) above the tunnel crown reported 86 millimeters (3.4 inches) of settlement, approximately 35% greater settlement than the corresponding points before the sheets.

Tank Monitoring Points

As shown in Figure 1, 24 monitoring points were established around the concrete foundation of the Elkhorn Reservoir prior to the start of adjacent construction. Given the presence of lean and fat clays in the near-surface soils the team recognized the

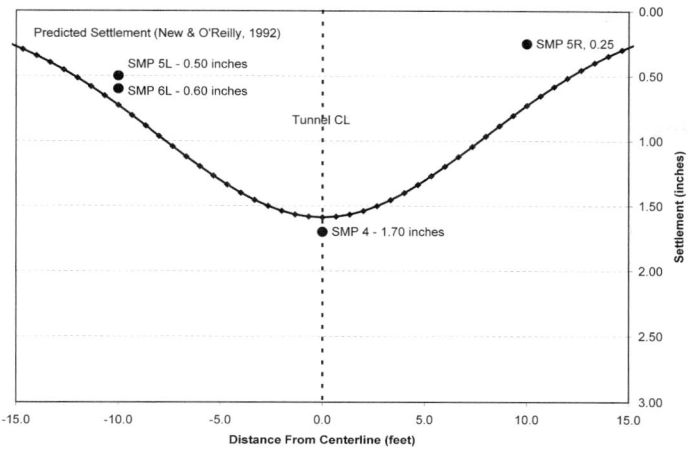

Figure 7. Comparison of measured and predicted settlement prior to the sheet pile wall

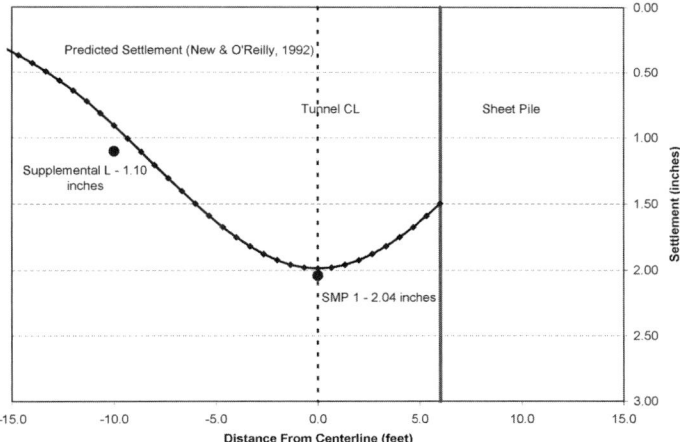

Figure 8. Comparison of measured and predicted settlement adjacent to the sheet pile wall

potential for time dependent settlement of the reservoir upon first filling. Figure 9 presents the vertical deflections of the 4 cardinal monitoring pins over the duration of the project with specific identified milestones (refer to Figure 1 for pin locations).

It is observed from Figure 9 that the reservoir immediately developed a preferential settlement to the north after first filling, as the observed settlement at Pin 1 (north) on March 1, 2008 (prior to tunneling) was 125% greater than the observed settlement at Pin 13 (south). However, the overall differential settlement of the reservoir in the north-south axis of 18 millimeters (0.7 inches) is not considered to be a concern. The construction activities of sheet pile installation, tunneling, and contact grouting do not appear to have affected the settlement of the tank. Instead any response to these activities is masked by the relatively elastic rebound of the tank upon emptying, which was performed prior to installation of the sheet piles and again prior to the start of tunneling

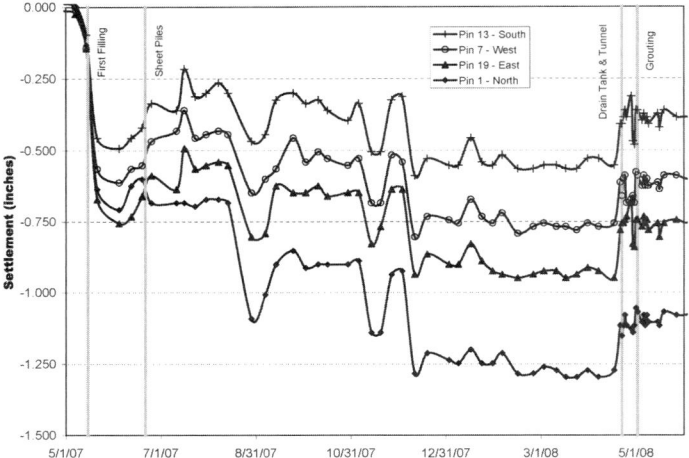

Figure 9. Reported settlement at the four cardinal points of the Elkhorn Reservoir foundation

activities. Without the establishment of a baseline set of measurements well in advance of the tunneling activities it could be perceived that the preferential settlement of the northern limit of the tank was potentially related to the UNWI Sections 3&4 project.

CONCLUSIONS

The protection of the City of Sacramento's Elkhorn Reservoir was a collaborative effort between the owner, the construction manager, the designer, and the contractor. Each party recognized the risks posed by the last minute tank design revisions and worked to mitigate risk and maximize project success.

Specific observations include the following

- Highlights the value of baseline measurements with regard to settlement of the tank. The tank developed a preferred direction of settlement that if undocumented may have led others to the conclusion that the UNWI Sections 3&4 project may be responsible.

- Draws attention to the need for careful pressure monitoring during annular space grouting. Grouting operations always have a risk of frac-out, and as was observed here, imposing significant pressures on adjacent facilities.

- By discussing concerns early and openly during project partnering the ultimate solution met both the needs of the owner and satisfied the liability concerns of the contractor.

REFERENCE

New B.M & O'Reilly, M.P., 1992: "Tunneling Induced Ground Movements; Predicting Their Magnitude and Effects," Ground Movements and Structures. Geddes ed, Pentech Press, London, pp. 671–697.

CONSTRUCTION WORKS OF LARGE-SECTION VERTICALLY PARALLEL TWIN TUNNELS IN CLOSE PROXIMITY

Kentaro Kuraji ▪ Metropolitan Expressway Co.

Masami Morita ▪ Metropolitan Expressway Co.

Naoyuki Araki ▪ SHIMIZU Corp.

Yoshihir Taniguchi ▪ SHIMIZU Corp.

INTRODUCTION

Background

Tokyo, the capital of Japan, is one of the largest cities of the world. As many as 33 million people live in Tokyo and three neighboring prefectures although the city occupies only 0.6% of the area of the country. The network of Metropolitan Expressways serves as a major artery for daily physical distribution in Tokyo. The Central Circular Route of the Metropolitan Expressway (MECCR) is located innermost among the three circular routes in Tokyo (Figure 1). Once completed, MECCR is expected to drastically mitigate traffic congestion in the central Tokyo area and provide drivers wider varieties of options in route selection. Alleviating congestion is estimated to reduce 400,000 tons of carbon dioxide emissions annually.

The eastern and northern sections and part of the western section of MECCR have been completed with the total length reaching 33 km. The western section will be extended southward by approximately 4 km in the spring of 2010. An 11-km-long western section of MECCR that is currently being constructed is composed mostly of tunnels.

A large number of power supply, gas supply, telephone, drinking water and sewerage pipes have been placed under the Yamatedori avenue along which MECCR is being constructed. The shield tunneling method has therefore been adopted for constructing tunnels along the route. The method involves the advancement of a shield tunnel boring machine (TBM) through the ground simultaneously assembling segmental rings within the shield. The method not only enables the building of highway tunnels more quickly and efficiently than other methods but also minimizes the effects of construction work on the surrounding area and the environment. Constructing the tunnel using the conventional cut-and-cover method might have required large amounts of time and cost because relocating the numerous objects buried under the Yamatedori avenue might have been necessary during tunnel construction. It was also considered desirable to place the tunnel at a depth of 30 m or more from surface to prevent tunnel construction from adversely affecting railways because the tunnel was expected to cross eleven railways and subways. This is one of the reasons for selecting shield tunneling as an effective method. The shield tunneling method was finally adopted for a length of 7 km in the 11-km western section of MECCR.

The shield tunneling method was also employed in the Ohashi Tunnel located southernmost in the 11-km western section of MECCR. This paper describes the construction of the Ohashi Shield Tunnel.

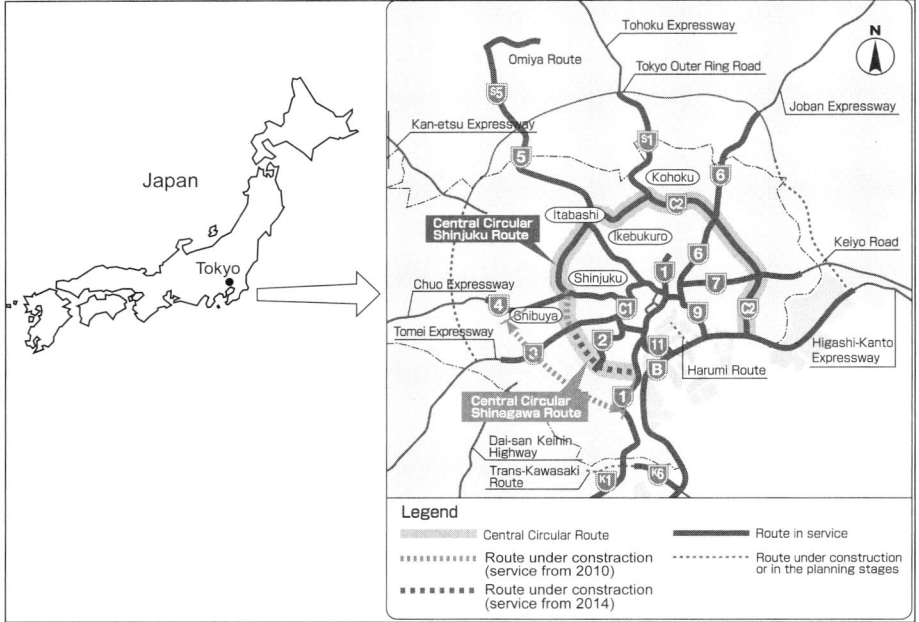

Figure 1. Metropolitan Expressway Central Circular Route (MECCR)

OHASHI SHIELD TUNNEL

Outline of the Ohashi Shield Tunnel

The Ohashi Shield Tunnel is a highway tunnel with a length of 431.7 m. The combined total length of upper and lower tunnels is 863.4 m. The upper tunnel is southbound and the lower tunnel northbound. The shield tunnelling method was adopted for construction to reduce the construction period and minimize the effects on the surrounding environment. The outer diameter of the shield tunnel is 12.65 m. The shield TBM used the Matsumizaka shaft under the Yamatedori avenue and the Ohashi shaft located beside National Highway 246 (Figure 2).

The shield TBM was first used to excavate the upper tunnel from the Matsumizaka shaft to the Ohashi shaft. Then, the TBM that reached the Ohashi shaft was jacked down to the level of the lower tunnel and completely turned around. The lower tunnel was excavated from the Ohashi shaft to the Matsumizaka shaft. The TBM, upon return to the Matsumizaka shaft, was dismantled and removed.

A 5.1-m-diameter vertical tunnel (Escape Shaft) for accommodating an emergency exit, draining water during the fighting of tunnel fire and pumping seepage was constructed near the midpoint in the Ohashi Shield Tunnel (Figure 3). The vertical tunnel was connected to the Ohashi Shield Tunnel without using any trenches.

A 4-km section in the western section of MECCR including the Ohashi Shield Tunnel is scheduled to be completed by the end of March 2010. When MECCR is extended southward, some of the segmental rings will be removed from the Ohashi Shield Tunnel to increase the road width (Figure 4). The southern section of MECCR is scheduled to be completed in 2014.

LARGE-SECTION VERTICALLY PARALLEL TWIN TUNNELS

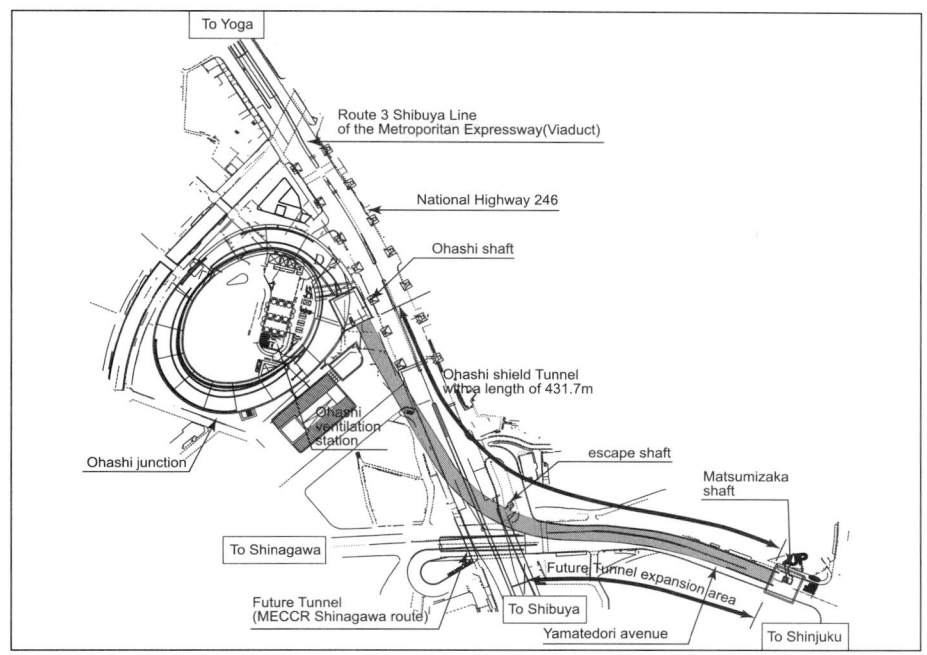

Figure 2. Ohashi Shield Tunnel and shafts

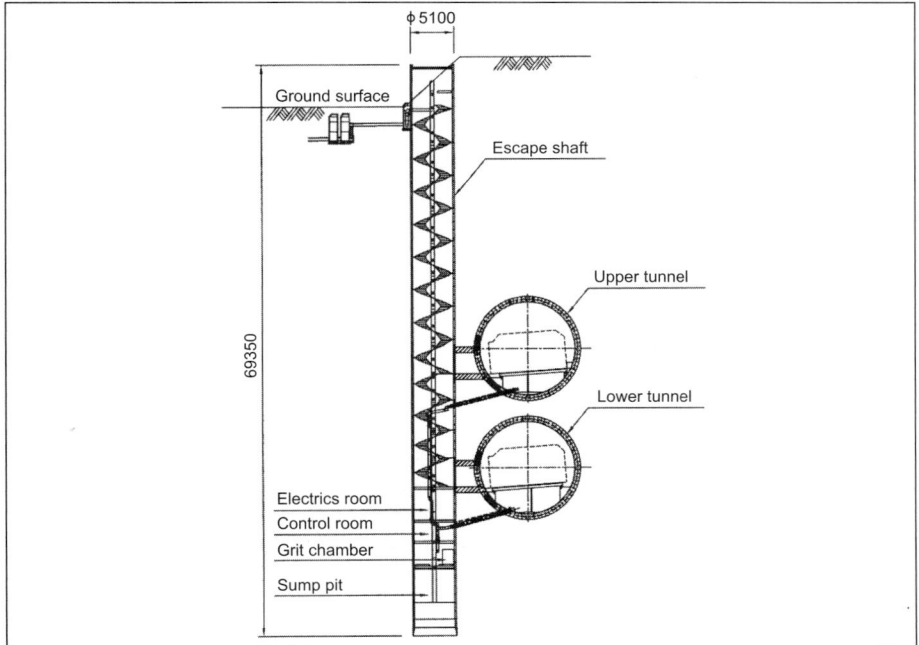

Figure 3. Cross Section of the connection of Escape Shaft

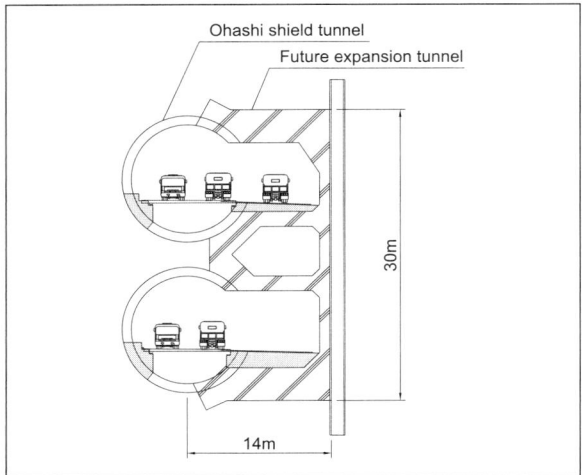

Figure 4. Image of future expansion tunnel

Alignment

The horizontal alignments of the upper and lower tunnels constituting the Ohashi Shield Tunnel are more or less the same with each other. The minimum horizontal and vertical curve radii are 123.5 and 476.5 m, respectively, both near the center of the tunnel.

The highway will carry the traffic only on one lane when it will be opened to traffic. The highway, however, has a width sufficient to carry traffic on two lanes (two 3.25-m-wide lanes and a 1.25-m-wide shoulder lane) throughout to carry out repair or maintenance work without suspending traffic.

Adjacent Structures

Numerous important structures are located right above the Ohashi Shield Tunnel. The major adjacent structures are listed below with the minimum clearance.

- Eighteen piers and groups of foundation piles under a viaduct (Route 3 Shibuya Line of the Metropolitan Expressway) with a minimum clearance of 3.1 m
- Subway (Tokyu Den-entoshi line) with a minimum clearance of 6.5 m
- Utility duct (Shibuya utility duct) with a minimum clearance of 15.5 m

Soil Properties

The geological conditions at the tunnel site are shown in Figure 6. Buried soil and surface soil, the loam layer of the Kanto region and the Tokyo group composed of sand, clay and gravel are distributed from surface on the Kazusa group mainly composed of consolidated silt (Kc layer). Soft alluvial clay layers are distributed in a dissected valley. The shield TBM excavates consolidated silt layers (Kc layers) throughout. The cohesion of the Kc layer is 560 kN/m^2 and the angle of internal friction is 24 degrees. Thus, the ground is stable. Sandy layers with confined water, however, also exist.

Adjacent structures are supported on the sandy gravel layer (Tog layer) above the Kc layer. The Tog layer is an unconsolidated layer that deposited in shallow inland sea

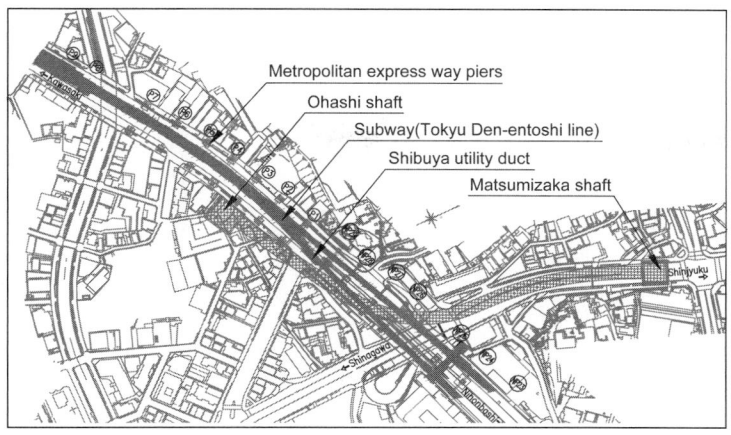
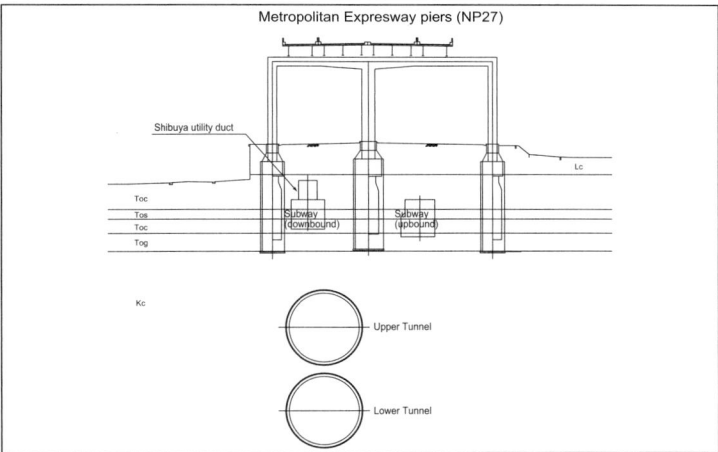

Figure 5. Adjacent structures (plan and cross section)

and mainly composed of gravel. The adjacent structures were therefore likely to settle due to soil settlement during the advancement of the shield TBM. Figure 6 and Table 1 show the geological profile and a list of soil constants, respectively.

Segmental Rings

In the Ohashi Shield Tunnel, 681 steel segmental rings and 345 ductile segmental rings both with an outer diameter of 12.65 m were adopted. Steel segmental rings were used in the section where a new tunnel will be connected in the future with the extension of the highway southward (in the section where some of the segmental rings will be removed).

A vertical tunnel (Escape Shaft) to accommodate emergency stairways, drain water during fire fighting and pump seepage was connected to the Ohashi Shield Tunnel without using any trenches. Some segments were removed at the connection to create an opening. In order to reinforce the area around the removed segments, steel segments with a width of 900 mm and a girder depth of 700 mm were adopted. Ductile segments were employed in other ordinary sections. The width of a ductile segment

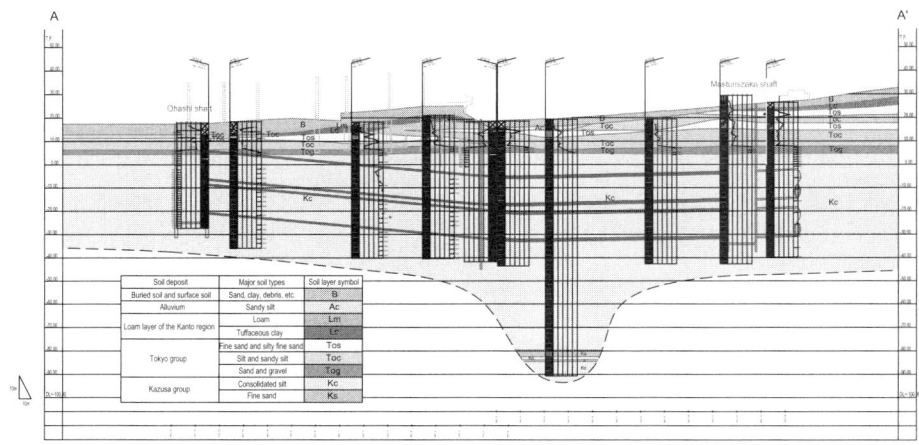

Figure 6. Geological profile

Table 1. List of soil constants

Soil Layer Symbol	N-Value	Unit Weight γ (kN/m³)	Angle of Internal Friction ϕ (degrees)	Cohesion C (kN/m²)	Modulus of Deformation E(MPa)
B	8	13.5	10	52	4.8
Ac	3	15.0	0	15	2.1
Lm	2	13.5	10	52	4.8
Lc	4	15.5	11	58	9.1
Upper Toc	9	19.0	11	42	7.2
Lower Toc	9	18.0	16	51	11.6
Tos	17	19.0	34	18	16.0
Tog	50	22.5	42	0	187.0
Kc	50	19.0	24	560	527.0

was set at 900 mm in sharp curves to enable construction in sharp curves and at 1,200 mm in straight sections.

Since the steel segments were expected to be combined with the reinforced concrete tunnel to form a tunnel of large cross section, the girder depth was set at an extremely high level of 900 mm (Figure 7). The extended highway will be connected (segments will be removed) without regulating traffic. The steel segments should be preferably connected with the reinforced concrete framework of the extended highway outside the Ohashi Shield Tunnel. To that end, metal fittings were embedded on the outer side of steel segments for connection at the connection with the extended highway (Figure 8).

Steel segments have a maximum arch length of 7.2 m, width of 750 mm and a maximum weight of 8.4 tons (Figure 9).

The meandering of the shield tunnel was corrected by adjusting the positions of tapered segments with the maximum width. In the area where expansion would be required, however, correcting the meandering of the tunnel using tapered segments was impossible because the segmental rings were fixed in the reinforced sections. The

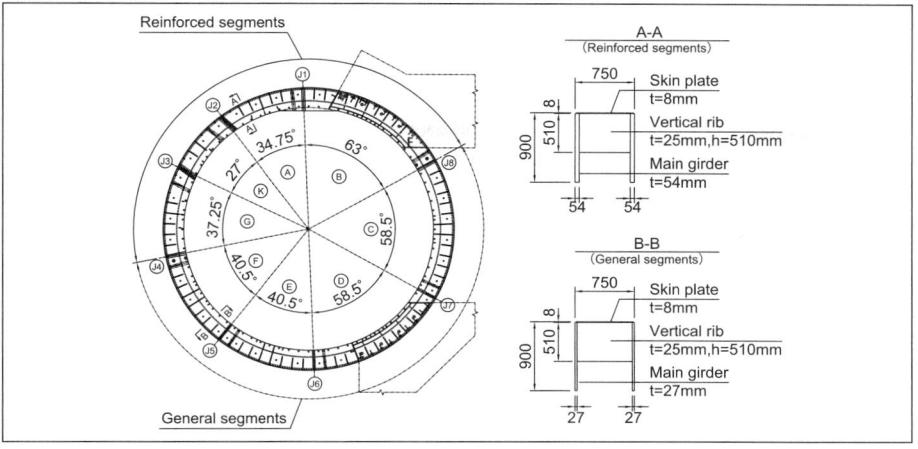

Figure 7. Steel segmental ring

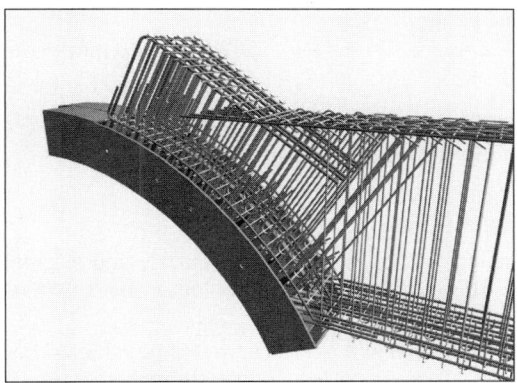

Figure 8. Detailed view of the connection with the future expansion tunnel

Figure 9. General view of steel segmental ring

problem was solved by setting a taper plate ring composed of 10 equal-size segments with a thickness of 30 to 60 mm between rings.

Shield Tunnel Boring Machine (TBM)

In the Ohashi Shield Tunnel, construction took place in a horizontal curve with a radius of 123.5 m and a vertical curve with a radius of 476.5 m in the sharp curve. A shield TBM with articulated equipment was therefore adopted. The specifications for the shield TBM used in the Ohashi Shield Tunnel are described below.

- Dimensions: Outer diameter: 12.940 m. Length: 11.105 m
- Type: Slurry shield
- Thrust: 4,000 kN × 1,850 mm × 40 = 160,000 kN-m
- Rate of shield jack advance: 30 mm/min
- Cutter torque: Maximum: 29,009 kN-m
- Cutter speed: 0.37 r.p.m
- Articulated equipment: Spherical X-shaped articulated equipment
- Maximum angles of articulation: 3.2 degrees horizontally, and 0.5 degree vertically
- Copy jack: 362 kN × 150 mm × four
- Tail seal: Wire brush type in three layers
- Erector speed: 0 to 1.5 r.p.m
- Maximum load carried by erector: 9.0 tons

Technical Problems in Construction

The shield tunneling method is effective for quickly and efficiently constructing tunnels in urban areas. Solving four technical problems was required at the site to adopt the shield tunneling method.

- Excavating the tunnel in a sharp curve using a shield TBM with a diameter of approximately 13 m (Construction in a sharp curve)
- Advancing the shield TBM right below the piers and foundation piles of a viaduct, subway, utility duct and other key structures (Construction in the vicinity of key structures)
- Lowering a large shield TBM from the level of the upper tunnel to that of the lower tunnel and completely turning it around (Jacking down and U-turn of a large shield TBM)
- A minimum clearance of 1.4 m between upper and lower tunnels (Soil stabilization between upper and lower tunnels)

These technical problems and their solutions are described below.

SOLUTION OF TECHNICAL PROBLEMS

Construction in a Sharp Curve

A minimum curve radius of 123.5 m in the Ohashi Shield Tunnel meant a serious problem for the shield TBM with an outer diameter of approximately 13 m. The shield

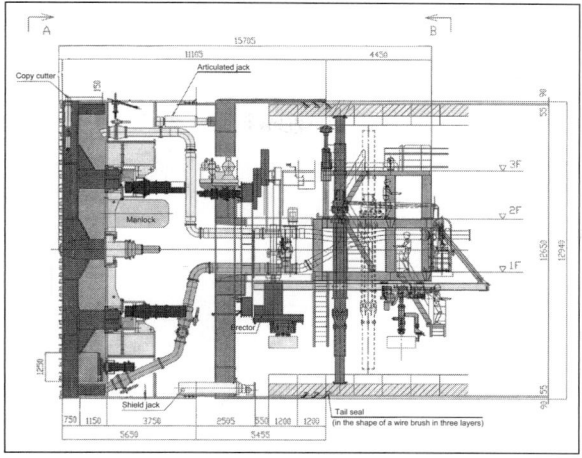

Figure 10. TBM (top: Structural drawing; bottom: general view)

TBM was therefore equipped with an articulated device, and ductile segments with a narrow width of 900 mm were adopted in the sharp curve as described earlier.

While excavating the tunnel in the sharp curve, higher jack thrust was applied on the convex side of the curve than on the concave side. Then, eccentric loading was expected to act along the tunnel axis. The safety of segmental rings against the eccentric loading due to shield jack thrust was verified by making analysis along the tunnel axis.

Segmental rings were expected to be subjected to soil reaction perpendicular to the tunnel axis owing to the jack thrust while the shield TBM was advancing in the sharp curve. It was therefore also necessary to verify safety in the direction perpendicular to the tunnel axis by applying the soil reaction obtained in the analysis along the tunnel axis to a model perpendicular to the tunnel axis as lateral pressure. The models for analysis along and perpendicular to the tunnel axis are described below.

1. Model for analysis along the tunnel axis (Figure 11)
    - A model for structural analysis along the tunnel axis was defined as an axial beam-spring model.
    - The planar shape of the curve was reflected in the axial beam-spring model.

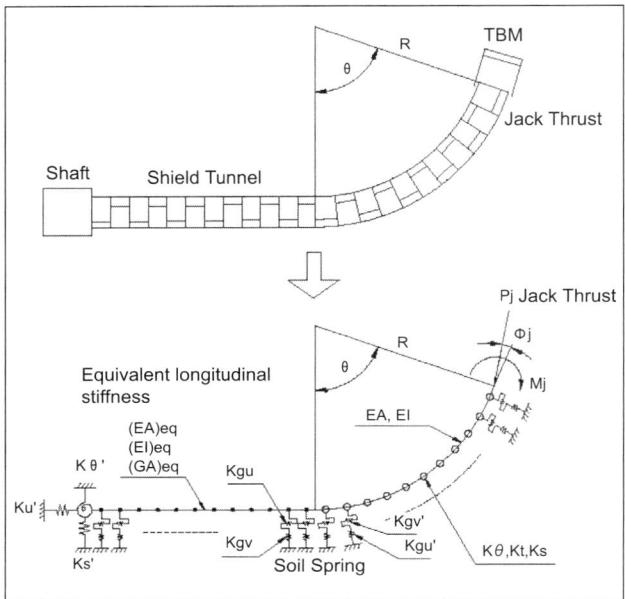

Figure 11. Model for analysis along the tunnel axis

- Segments were represented by beams and ring joints by springs (rotational, shear and axial springs).
- Moments induced by the jack thrust and eccentric loading were applied at the end of the axial beam-spring model.
- The ground was modeled using springs working along and perpendicular to the tunnel axis.
- The stiffness of backfill material was evaluated with time.
- The model was used to calculate the sectional forces along the tunnel axis and to verify the stress of the ring joint.

2. Model for analysis perpendicular to the tunnel axis (Figure 12)
   - A segmental ring model was created by representing the segmental joints as rotational springs and ring joints as shear springs.
   - The model was subjected to permanent design load (soil pressure, water pressure, weight of the segment) and the lateral pressure obtained in the analysis along the tunnel axis (soil reaction) to calculate sectional forces.
   - The calculated sectional forces were used to verify stresses in the main cross section, segmental joint and ring joint.

As a result of analysis, it was found that stress exceeded the design level only in the bolts of ring joints in tension. All of the 80 M30 bolts with an yield strength of 940 N/mm^2 or higher were therefore replaced with M36 bolts to reinforce segmental rings.

The stroke of the copy cutter was set to be 40 mm as the overbreak for construction in the sharp curve was calculated to be 29 mm. The volume of backfilling was maintained at 120% of the volume that took the overbreak into consideration. The backfilling pressure was kept at the same level as in straight sections. As a result, no backfill materials entered the cutting face or leaked through the tail seal.

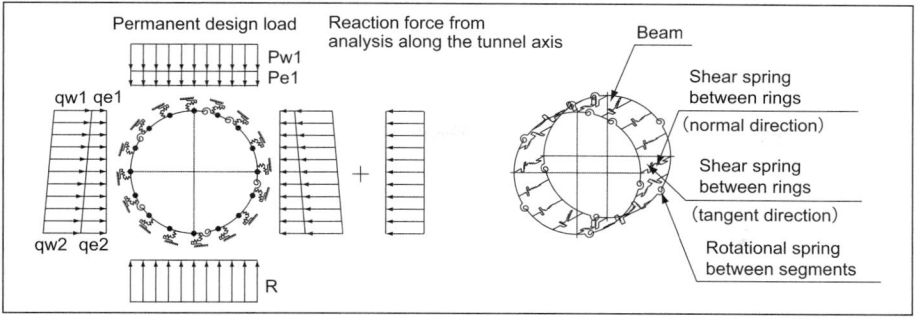

Figure 12. Model for analysis to the tunnel axis

In the sharp curve, segmental rings were assembled while confirming the tail clearance to prevent segmental rings from contacting the tail plate of the shield TBM. The shield TBM was articulated gradually in several rings while measuring the stroke of the articulated jack. The rolling of the shield TBM was controlled by adjusting the direction of cutter rotation according to the degree of rolling.

The above measures made possible the completion of construction of both upper and lower shield tunnels in the sharp curve without any trouble.

Construction in the Vicinity of Key Structures

In the Ohashi Shield Tunnel, the shield TBM advanced right below the key structures such as 18 piers and a group of foundation piles of a viaduct, subway and utility duct. Confirming the effects of construction on the structures was therefore required before the commencement of work. The subway and utility duct in particular were located right above a sharply curved section with great overbreak. The effects of the advancement of the shield TBM were of concern.

The minimum clearance between the upper tunnel and the foundation piles was only 3.1 m and that between the tunnel and the subway was only 6.5 m. The structures were supported on the Tog layer above the Kc layer. The ground movements expected during the advancement of the shield TBM were likely to affect the ground that supports the adjacent structures.

A shield TBM of large cross section had never advanced right below adjacent structures in a sharp curve both ways to construct upper and lower tunnels at any other construction sites. Then, the effects on and the safety of adjacent structures were verified in advance by finite element analysis. Figure 13 shows an FEM analysis model.

Measurements were taken as described below to verify the agreement of FEM analysis results with actual ground movements (calibration), identify the displacements of adjacent structures and of the ground near the adjacent structures and verify whether the slurry at the cutting face leaked to the Tog layer or not.

1. Measurements in boreholes
   - Preliminary measurement (in three cross sections): Differential settlement gauges, pressure gauges for measuring settlement and multistage inclinometers
   Objective: To calibrate analysis parameters such as stress relief ratio
   - Measurement at the locations of adjacent structures (in two cross sections): Differential settlement gauges, pressure gauges for measuring settlement and multistage inclinometers

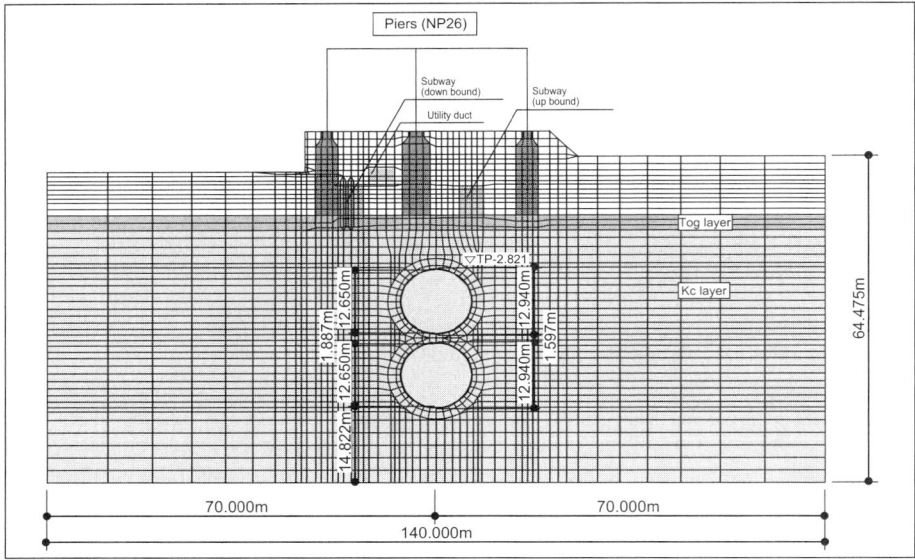

Figure 13. FEM analysis model

      Objective: To identify the displacement of the ground in the vicinity of adjacent structures
- Measurement of pore water pressure (at two locations): Pore water pressure gauges
Objective: To verify whether the slurry at the cutting face leaked to the Tog layer or not
2. Measurement of displacements of adjacent structures
   - Eighteen piers of a viaduct: Horizontal displacement, vertical displacement and inclination
   - Subway (470 m each for up- and down-bound lines): Horizontal displacement, vertical displacement, inclination, crack displacement and surface strain
   - Utility duct (with a length of 140 m): Vertical displacement and inclination
   - Two abutments: Horizontal displacement and vertical displacement
   - Earth retaining wall for national highway: Horizontal displacement and vertical displacement

The soil constants and the stress relief during the advancement of the shield TBM that were used for FEM analysis were calibrated based on the measurements. As a result, it was found that the construction work would not cause any damage to adjacent structures. The upper tunnel was completed while measuring the displacements of adjacent structures and ground, and the pore water pressure in the Tog layer.

    When the lower tunnel was constructed, the values input for FEM analysis were adjusted again using the actual displacements of adjacent structures obtained during the construction of the upper tunnel. As a result, it was found that the excavation of the lower tunnel was nearly unlikely to affect the adjacent structures as well as the construction of the upper tunnel. The lower tunnel was constructed while taking

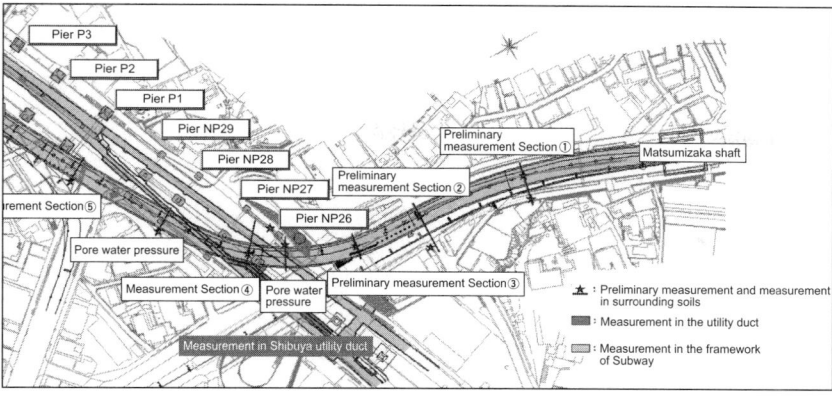

Figure 14. Plan of measurement points

measurements for the adjacent structures as during the construction of the upper tunnel. As of January 2009, or the point six months after the completion of tunneling when the displacements due to the advancement of the shield TBM in the upper and lower tunnels may have ceased, no deformation has been found in any adjacent structures.

Shown below are examples of measurements (Figure 15). The measurements include seasonal changes in the piers of the viaduct, subway lines and utility duct. The adjacent structures that settled due to the excavation of the upper tunnel later thrusted upward. Thus, the volume of settlement due to the excavation of the upper tunnel was not explicitly identified.

The measurements of the maximum settlement taken at the end of September, 2009, or the time when displacements might have discontinued, were compared with the values calculated in FEM analysis (Table 2).

Jacking Down and U-turn of a Large Shield TBM

For excavating the upper and lower tunnels of the Ohashi Shield Tunnel, it was necessary to lower a 2,100-ton shield TBM from the upper to lower stage and turn it around upon its arrival at the shaft. A flow of steps from the placement of temporary work materials to the jacking down and U-turn of the shield TBM is shown below (Figure16).

- Step 1 Install a pedestal and advance the shield TBM to the shaft
- Step 2 Push and move laterally the shield TBM (one month)
- Step 3 Install jack supporting beams and jack up the shield TBM (two months)
- Step 4 Remove the pedestal and jack down the shield TBM (one month)
- Step 5 Completely turn around and move laterally the shield TBM (0.5 month)
- Step 6 Prepare for resuming the advancement of the shield TBM (set up the cutting face and assemble a reaction pedestal) (3.5 months)

Approximately eight months were required from the arrival of the shield TBM at the shaft to the resumption of advance.

Eight steel I-girders (3000 × 700) were installed above the framework of the Ohashi shaft.

The shield TBM was lifted using 14 400-ton hydraulic jacks installed on the I-girders. Fourteen prestressing strands composed of 27 18-mm-diameter wires were used to lift the shield TBM.

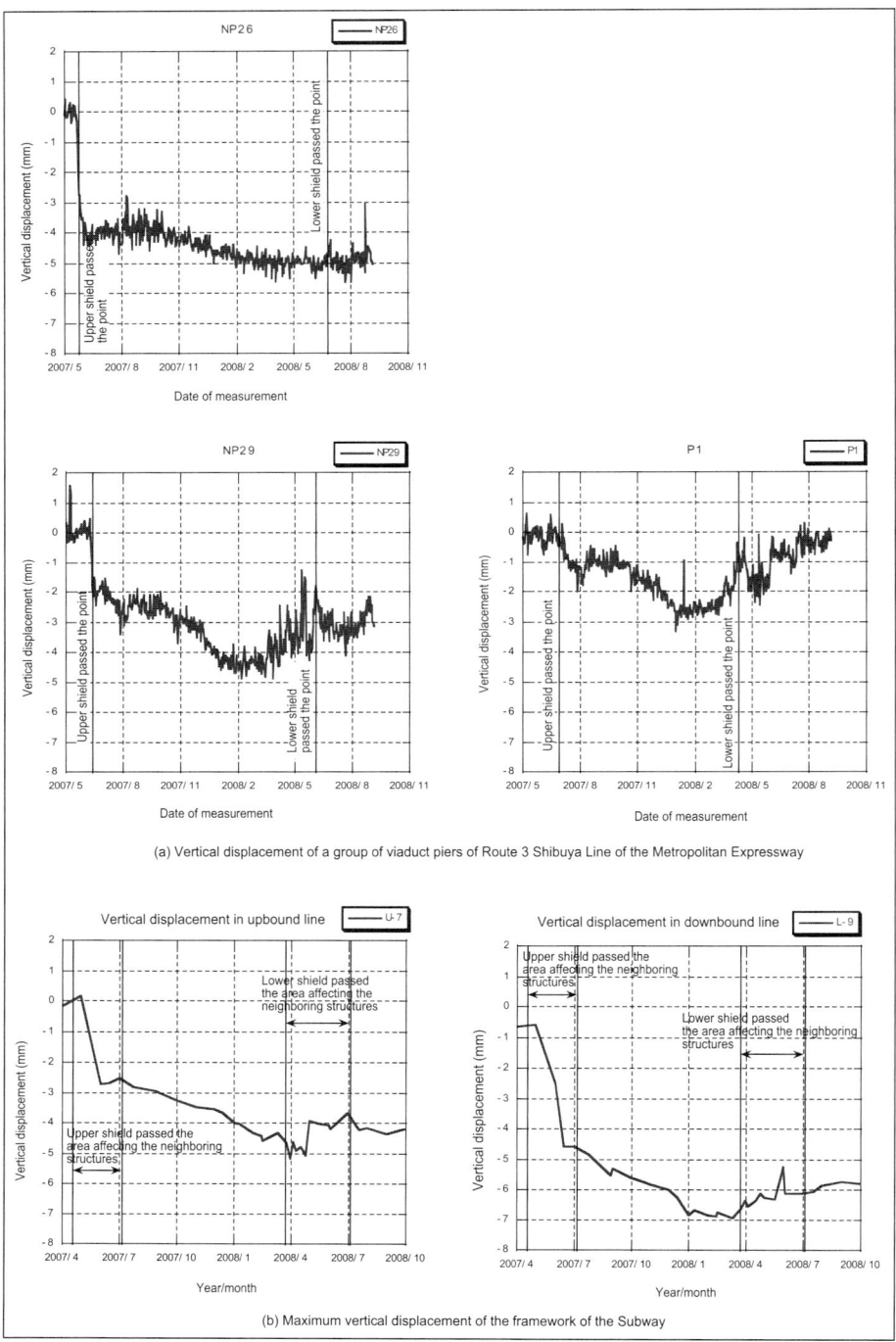

Figure 15. Examples of measurements

Table 2. FEM analysis results and measurements

	Maximum Settlement at the Completion of Lower Tunnel Excavation (mm)		
	FEM	Measurement (September, 2009)	Variance
Pier of the viaduct (NP26)	5.4	5.0	0.4
Pier of the viaduct (NP27)	6.2	6.0	0.2
Pier of the viaduct (NP28)	6.8	6.6	0.2
Subway (upbound line)	4.6	4.2	0.4
Subway (downbound line)	5.1	5.5	−0.4
Utility duct	5.9	5.7	0.2

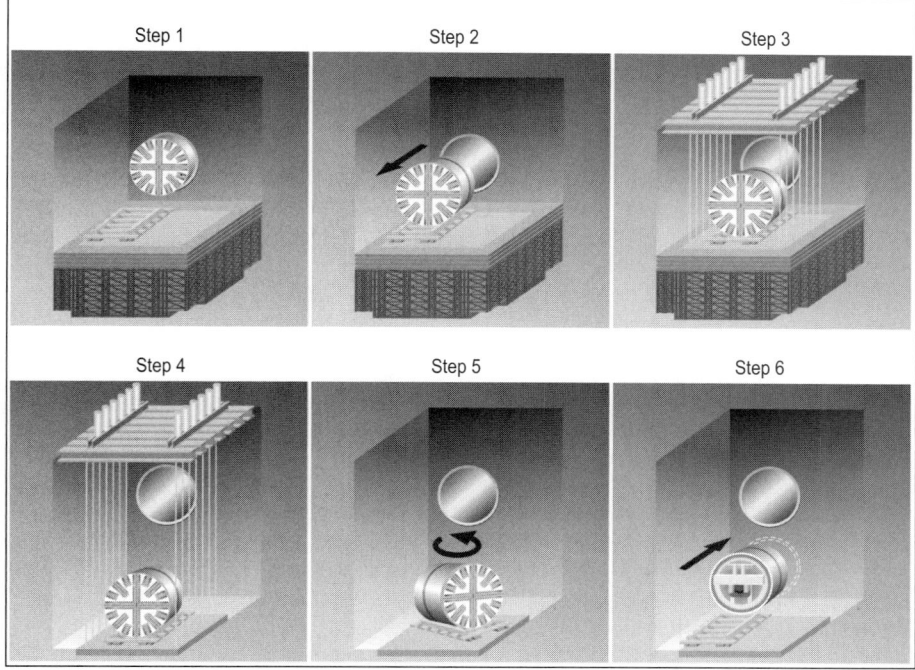

Figure. 16 Jacking down and U-turn

In order to dismantle the pedestal installed earlier to push the shield TBM on the upper stage into the Ohashi shaft, the shield TBM was first jacked up nine meters. The jacks were operated 28 times for approximately three hours. After the pedestal was removed, the shield TBM was jacked down 23 m. The jacks were operated 88 times for approximately five hours. When the shield TBM was lifted or lowered using hydraulic jacks, the stroke of the jacks was centrally controlled via a control panel to balance the shield TBM.

The ball slider method was adopted to turning around the shield TBM. The ball slider is a box-shaped equipment containing steel balls used for shifting heavy weight load placed on it. Sixty 90-mm-diameter steel balls rolled in each two-tier steel box to move the shield TBM. Twenty eight ball sliders were placed under the platform

Figure 17. Detailed view of a ball slider

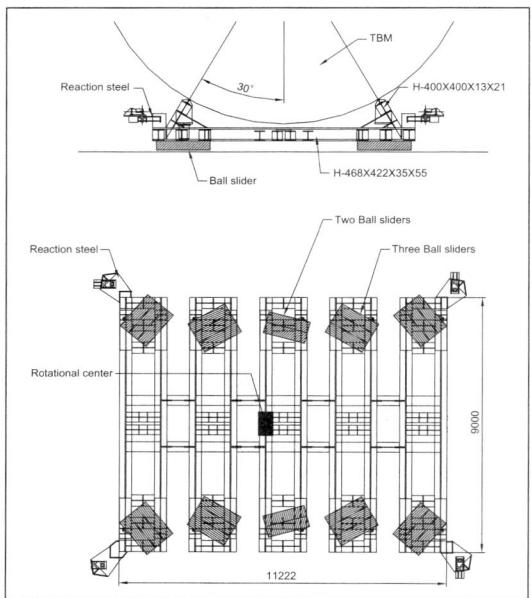

Figure 18. Positions of ball sliders

supporting the shield TBM in the directions of movement and rotation. Figure 17 shows a detailed view of a ball slider and Figure 18 shows the positions of ball sliders when rotating the shield TBM.

The top surface of the deck in the Ohashi shaft had to be flat to enable the ball sliders to move smoothly. To that end, mortar was placed above the deck for leveling and 25-mm-thick steel plates were laid over the deck.

Reaction steel plates were installed at four locations on the side wall of the shaft to carry the reaction when rotating the TBM platform. The reaction steel plates were connected to the four corners of the platform with high strength steel rods, and jacks with a combined traction of 200 tons were employed to rotate the shield TBM (Figure.19). A stroke of the jacks was only 200 mm. The jacks therefore had to be replaced a total of 135 times. The jacks were operated for approximately eight hours. The U-turn of the shield TBM took two days.

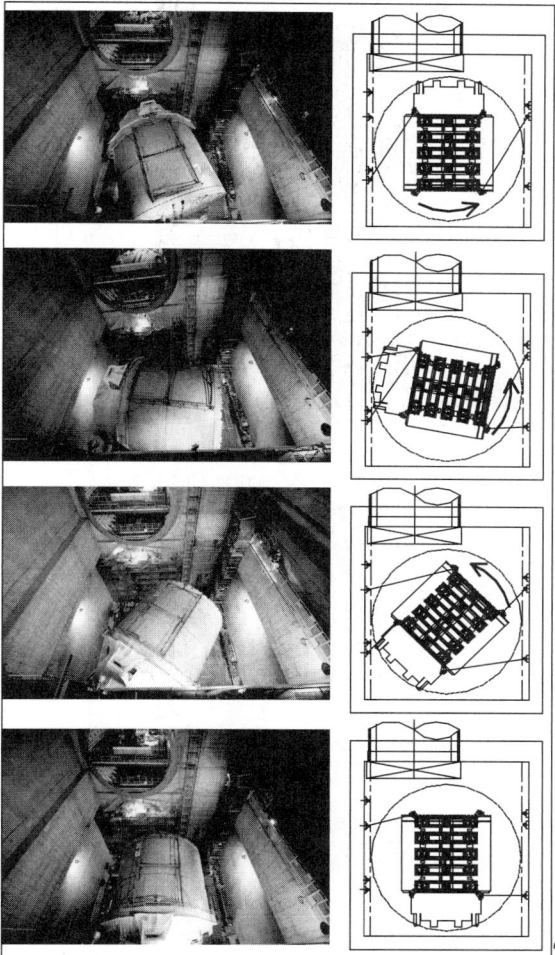

Figure 19. U-turn of the shield TBM

Soil Stabilization Between Upper and Lower Tunnels

The clearance between the upper and lower tunnels was extremely small ranging from 1.4 m to 4.3 m. The excavation of the upper tunnel was followed by the excavation of the lower tunnel.

The soil between the two tunnels had to be stable while the shield TBM was advancing through the lower tunnel in particular. The collapse of the soil between the two tunnels would cause the upper tunnel to lose the bearing ground, settle greatly and suffer damage, and have serious effects on adjacent structures. The stability of the soil between the upper and lower tunnels was therefore verified using the FEM analysis.

As for the mechanical properties of mudstone, mudstone is assumed to be non-linear elastic according to the stress level. The non-linear behavior of mudstone is modeled using R, a non-dimensional parameter that is called fracture severity. (The fracture severity method was proposed by the Central Research Institute of Electric Power Industry.) R shows the stress margin ratio against failure under the generated

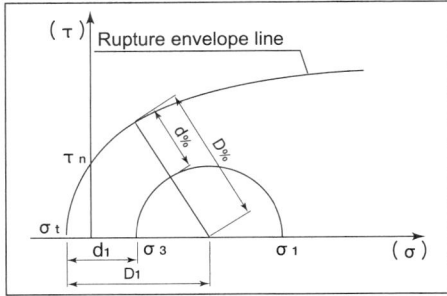

Figure 20. Failure criteria for mudstone

stress condition. R is defined by equation (1) and indicates the limit of linear elasticity. The range 0 < R < 1 indicates non-linear elasticity and the modulus of deformation E decreases according to R. The parameter Fs is related to R as the inverse of stress level and represents the local stress safety factor. Both R and Fs are regarded as state parameters that indicate stability during excavation.

$$R = k \cdot min(d_1/D_1, d_2/D_2) \quad (1)$$

$$F_s = min\{D_1/(D_1 - d_1), D_2/(D_2 - d_2)\} \quad (2)$$

$$R \geq 1 \quad :E = E_0 \quad (3)$$

$$0 < R < 1 \quad :(E - E_f)/(E_0 - E_f) = R^{1/a} \quad (4)$$

$$R = 0 \quad :E = E_f \quad (5)$$

E_0 : Initial modulus of deformation

E_f : final modulus of deformation

a : coefficient of envelope line

where k is a constant that indicates the limit of linear elasticity (yield strength = peak strength x (k − 1) / k) and k is 2 to 3 on the mudstone.

Based on the analysis results, two methods are used for judging stability. The zone in which both judgements are not satisfied is defined as the "loosened zone." The stability judgments are shown below.

Judging from how far the stress circle is from the rupture envelope line;

$F_s \geq 1.5$

$R \leq k \times \frac{1}{3}$

As an analysis result, the distribution of coefficient of loosening R (at the completion of the bottom tunnel) is shown in Figure 21. Coefficient of loosening R exceeded 1.0 throughout the Kc layer in the vicinity of the tunnel at the end of the excavation of the lower shield tunnel. Then, it was determined that the soil between the two tunnels was in the elastic zone at the end of the excavation of the lower tunnel.

Safety was confirmed by FEM analysis. Full attention was paid to slurry pressure at the cutting face, amount of excavated material and volume of backfill grouting when advancing the shield TBM in the lower tunnel. The vertical displacement during the advancement of the shield TBM in the lower tunnel was approximately one millimeter upward, nearly the same as the analysis value.

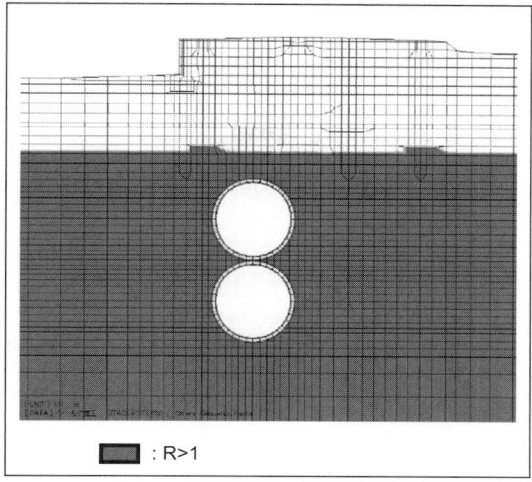

Figure 21. Distribution of coefficient of loosening R (at the completion of the bottom tunnel)

CONCLUSIONS

The shield tunneling method was adopted to excavate the Ohashi Shield Tunnel to reduce the construction period and save the construction cost. It was necessary to advance the shield TBM under the piers of an existing expressway and subway lines in a sharp curve with a radius of 123.5 m. In the shaft, lowering the 2,100-ton shield TBM using hydraulic jacks and turning it around were also required to excavate the lower tunnel. The clearance between the upper and lower tunnels was only 1.4 m. These technical problems were solved by using the method described in detail in this paper. Then, it became possible to adopt the shield tunneling method at this construction site. The assembly of the shield TBM was started in the Matsumizaka shaft in January 2006 and the dismantling of the TBM was completed in the same shaft in December 2008. Interior finish work will be carried out to open the tunnel to traffic in March 2010.

REFERENCES

1) Atsushi KOIZUMI & Hirotomo MURAKAMI: Design method of segments at a sharply curved section, Proceedings of Japan Society of Civil Engineers, No.448/III-19, pp.111-120, Jun 1992.
2) Hayashl M. & Hibino S: Progressive analysis of relaxation around under ground excavation space, Central Research Institute of Electric Power Industry Report, 1968.

A PRACTICAL APPROACH FOR PRECAST CONCRETE SEGMENTAL RING SELECTION

Steve Skelhorn ▪ McNally Construction Inc.

Laura McNally ▪ McNally Construction Inc.

ABSTRACT

Over the recent years, major advances have been made in design of precast concrete segmental tunnel liner. Starting from basic, parallel sided rings, we now have state of the art, universal rings incorporating custom tapers to suit individual requirements. At planning stage, the ring diameter is normally fixed and the ring geometry is one of the first items to be assessed. Obvious factors such as TBM layout and curve radius can constrain geometry; however, there others factors that should be considered in order to optimize production.

This paper takes a look at the history of the segmental rings and criteria for ring selection, including: general ring geometry;; individual segment geometry; ring taper, and ring width. This paper also explores optimization of ring with a focus on constructability.

HISTORY

Tunnels have been under construction throughout the history of man, with the first tunnels built in prehistoric times, probably as a means of enlarging cave dwellings. Ancient Babylonians, circa 2180 BC, constructed the first recorded tunnel under the river Euphrates. Used to connect the Royal Palace with the temple, the tunnel involved the use of copper tools and reed drills and removed rock using fire and rock quenching, often using wine rather than water to quench and spall the rock. Technically this was not the first sub aqueous tunnel as the entire river was diverted out of the way prior to tunnelling; presumably budget and schedules were not of major concern in those days.

These early tunnels were driven through sound rock and required no lining. As tunnelling advanced, and was completed in more challenging ground conditions, various types of support systems were developed, including segmental liners and simple rib and lagging liners. The focus of this paper is on the evolution and design of segmental lining systems.

EARLY SEGMENTAL TUNNEL LININGS

Early tunnel lining materials consisted of brick or masonry blocks. The first tunnel across the River Thames (London, England) was the Rotherhithe tunnel, constructed by Marc Brunel between 1825 and 1841, the tunnel demonstrated the first use of a shield and was lined with brick masonry.

The Simplon tunnel, built at the turn of the 20th century, required a masonry liner of up to 10 feet thick to safeguard against rock bursts. The 12 mile long tunnel reached maximum cover of 7,000 feet as it passed under the Swiss Alps.

As tunnelling methods developed, so did the type of lining, allowing tunnels to be built through increasingly more challenging ground. In 1869, Peter Barlow and James

PRACTICAL PRECAST CONCRETE SEGMENTAL RING SELECTION

Figure 1. Brunel's Thames tunnel

Advancing the Shield   Erecting the Segments

Figure 2. Greathead's shield

Greathead, pioneered the shield and the segmental liner. The "great head shield" used for the second Thames tunnel was so successful, its design remained unaltered for almost 75 years.

The Shield allowed the erection of precast segments as a support system. These were cast iron rings, consisting of up to 10 segments per ring. The liners were so successful that they were used for the majority of the London Underground System constructed between 1854 and 1906.

Cast iron segmental liners, or spherical graphite iron (SGI) are still used today. Casting tolerance combined with machined faces allows some intricate shapes to be formed and many of the London Underground escalator shafts were constructed this way.

Figure 3. SGI escalator shaft

Figure 4. Standard bolted segmental liner

PRECAST CONCRETE SEGMENTAL LINERS

Since the 1960s, concrete liners have been used in many small diameter tunnels in Europe. The liners were designed to be constructed by hand; consequently the segment width was restricted to around 600 mm due to weight. These generally formed the primary liner with a secondary liner consisting of in-situ concrete or infill panels.

With the development of tunnel boring machines came the introduction of mechanical ring erectors, which eliminated size constraints related to segment weight This allowed design of a solid (non-paneled) ring that was sufficient for use as a final lining, leading to a one pass tunnelling approach. Ring widths could also be increased and consequently standard rings of up to 1m in width became available.

Individual segments are normally constructed using bolted connections. There are some exceptions to this, notably, in the case of wedge block linings, segments are expanded against the ground with the installation of a tapered key which compresses the ring negating the need for bolts. This type of liner is only suitable in stable ground.

Waterproofing of the constructed tunnel was always an issue, as achieving an effective seal between the segments was difficult. Initial ring designs incorporated tarred yarn (tar impregnated hemp rope) between each of the segments to temporarily seal the gap between adjacent segments and retain backfill grout. The joints between

Figure 5. Wedge block liner

segments were sealed with a caulking compound after completion of the tunnel, which increased the overall project schedule. Over time, the yarn was substituted with a permanent hydrophilic gasket. This was set into a groove cast into the ring. These proved partly successful, but many clients still demanded secondary caulking of the joints.

Ring geometry at this stage consisted of a number of parallel segments, two top segments (segments either side of the key) and a key segment. Keys were often parallel sided consisting of a rectangular keystone approximately 150 mm wide. Construction required the two top plates to be physically lifted to allow the key to be installed. This was fine in good ground with hand built rings, but was not practical for mechanised tunnelling. Subsequently the parallel key was replaced with a trapezoidal shaped key, which allowed a machine ram to push the key into place. An added advantage of this approach was that the gaskets were not in contact until the key segment was slid into place, and tear-out of the gasket was greatly reduced.

THE TRAPEZOIDAL RING

In the early 1980s, designers and contractors began to experiment with trapezoidal segments. The change was driven by the need to improve water tightness of the lining system, and the substitution of rubber gaskets for previously used hydrophilic gaskets. With a rubber gasket, there was a need to slide each segment into place before contact was made with between gaskets. This was not possible with parallel segments, where it was necessary for the TBM erector to compress the radial gaskets to align the bolt holes. With trapezoidal segments, the gasket does not make contact until the segment bolts holes are aligned.

The initial trapezoidal ring consisted of 6 identical trapezoidal segments. These could essentially be thought of as three keys and three counter keys. One of the disadvantages of this type was in the building of the ring. Starting with a counter key segment, the build sequence required a "key" segment to be built on either side of the key before installing the other two counter keys. This resulted in difficultly compressing all the gaskets and ensuring water tightness of the liner.

Bolting of the rings was also an issue. One pass rings had bolted connections for the radial and circumferential joints. To make this practical, bolt pockets with curved holes were provided in each segment and curved "banana bolts" were used for the connection. These were effective; however, to provide sufficient clearance for the threaded section it was necessary to oversize the hole. This inevitably led to movement of the segments, an increase in lipping of the joints, and overall liner quality concerns.

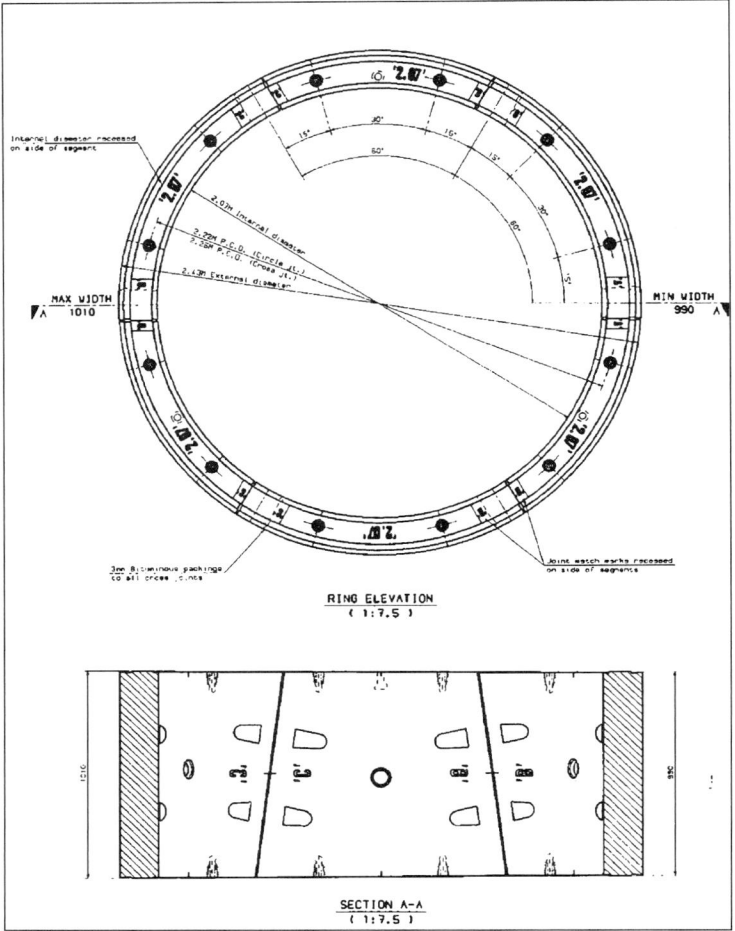

Figure 6. Trapezoidal segment layout

Over the course of several projects, the system quickly advanced culminating in the modern universal ring used today.

TAPERED RING

One of the issues with any segmental ring is negotiating curves. With a non tapered ring, it was necessary to pack the longitudinal joints. This would accommodate curves in the tunnel; however it created ring build quality problems. When a parallel ring is packed on one edge, the ring loses its plane; for example, if the left side of a ring is packed the sides of the ring will project further forward than the mean of the top and bottom. Within a segmental ring, this causes a geometrical shift and the ring looses it circularity. This in turn makes it harder to build subsequent rings.

To counteract this, extensive surveys and checks were required to the liners during and after construction to allow appropriate adjustments through packing of the rings. A tunnel engineer would typically carry out a series of checks as detailed below:

Figure 7. Tapered ring—Sheppard Subway

- Check the laser orientation and record the position of the TBM
- Extend "square marks" along the tunnel wall by measuring forward from previous marks with periodic checks using an optical square from the laser or by means of a 90 degree eyepiece from the theodolite.
- Measure the square of the ring relative to the theoretical alignment. This would be done by using a square mark for the horizontal lead and by means of a plumb bob for the vertical.
- Measure the ring build diameter both horizontally and vertically to determine the circularity
- Measure the plane on the ring by means of a horizontal straight edge combined with a plumb bob.
- Calculate the theoretical leads and advise the lead miner of the packing required to both steer the ring and to mitigate plane issues.

This was obviously a lengthy procedure and required a full time engineer at the TBM. In addition, any packing to the joints would introduce a weak point in the gasket system and could lead to leaks and the need for additional waterproofing of the finished tunnel.

The tapered ring was developed using a standard ring with a single key, and introducing a taper to the ring width. Tapers varied but were generally between 10 and 20 mm. A nominal 1m wide ring with a 20 mm taper would therefore consist of segments ranging in width from 990 mm to 1010 mm; fundamentally, the tunnel could now be steered by pointing the ring (directing the taper) in the desired direction, while maintaining the plane of the rings.

With a tapered ring, the geometry of each individual segment is different. This introduces complexity in segment casting, with individual moulds required for each segment. The tapered trapezoidal ring mitigates this issue to some degree. With all the segments trapezoidal, this ring consists of three pairs of identical segments.

Ring building typically starts with a segment in the invert and ends with the final plate or key placed above spring line. With a tapered ring, the required alignment of the taper is determined at the end of each mining cycle by measuring the gap between the last ring built and TBM. It is not possible to determine the orientation until the TBM has completed mining, as it is dependent on steering of the TBM.

A build sequence starting in the invert was driven by the method of operating the TBM segment erectors, combined with most owners specifying that the key needed to

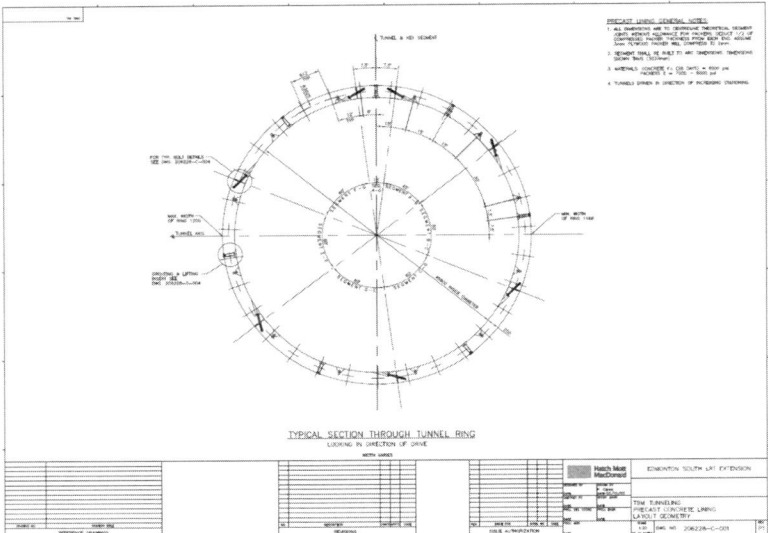

Figure 8. Tapered precast liner—Edmonton SLRT

be built above spring line. The origin of this requirement is not really clear; however, discussions with people from the industry have found that it was thought to provide better build quality. Subsequent projects with the key built anywhere around the circle have disproven this assumption. It should be noted that in the case of a full trapezoidal ring design there are actually three keys, so this specified requirement does not apply.

A key issue with restricting placement of the key segment to above spring line is the impact on delivery of segments to the TBM. This is a greater issue with small diameter tunnels, where there is generally insufficient clearance to rearrange the segments within the machine in order to provide the segment with the required taper to start the ring build. Consequently the delivery order of segments is critical but in many cases is not practical to make adjustments during operation without causing delays due to the length of the tunnel drive

UNIVERSAL RING

Modern tapered segmental rings have mitigated many of the early issues. They generally consist of two trapezoidal key segments and a number of rhomboid side segments (normally between 4 and 6) to make the full ring. The usual configuration is for the keys, referred to as a key and counter key, to be opposite one another within the ring (i.e., 180 degrees). Ring erection commences with the key, follows with placement of two (or three, depending on design) rhomboid segments on either side, and finishes with the counter key.

Some projects have been carried out with the two keys adjacent to one another in the ring; however, this configuration restricts the building process to one direction only. In this case, ring building must start with the counter key, and work sequentially around the circle, ending with the key. For a ring design with the key to the right of counter key, the build will always need to be carried out in a counter clockwise direction. A consequence of this approach is that, with the erector always acting on the same direction, it can introduce a roll to the tunnel.

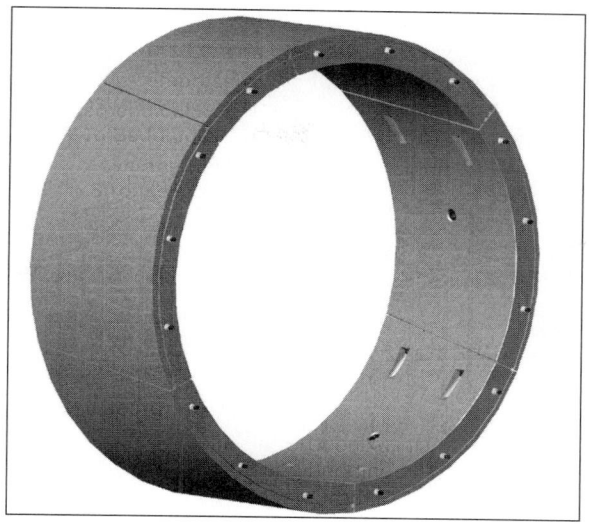

Figure 9. The universal ring

A universal ring has the great advantage that the build process always starts with the same segment (the counter key) for all rings. Depending upon the required orientation of the ring taper, this segment may be built anywhere around the circle. Taking advantage of this factor, it has been possible to reduce the key size, which increases ease of construction. It also reduces the recoil distance (clearance) required and consequently reduces TBM ram stroke and tail can length.

Modern rings have replaced the old style "banana bolts" with straight spear bolts that screw in to cast in plastic inserts and have introduced the option of using dowels for longitudinal joints. These changes have greatly improved the build quality and ease of construction of the liner. Other advancements, such as double gasketed systems and hydrophilic gaskets for use in extreme water head areas have also improved the quality of the final product.

RING SELECTION—SOME FACTORS FOR CONSIDERATION

There are several criteria to consider when selecting the optimum ring configuration. A review of recent segmental tunnel projects will reveal that many different variations of the universal ring are being used. This section provides a discussion on the key factors to be considered. It is important to note that there are no right and wrong choices, and many decisions will be a matter of preference.

Ring Width. (Width of the segments along the tunnel). The optimum ring width is the subject of much debate. Logically, the wider each ring, the more production will be achieved for a set number of rings in a shift. As tunneling is generally cyclic, this is an important factor, especially as tunnel length increases. Additionally by increasing the width, the total number of rings required for the tunnel is reduced. This can be an important factor of the ring joints and bolt pockets are to be patched at the end of the day.

The ring width affects the design of the TBM and tunneling system in a couple of ways. First, the design the TBM, such as tail can length and clearance through the trailing gear must be considered. Back up equipment to accommodate the ring width and resultant excavated material per cycle is impacted—such as muck handling equipment and grout quantities. The width also directly affects the taper required to navigate

Figure 10. Segment storage—Toronto, Ontario

curves; as ring width increases, amount of taper required to achieve the same curve radius increases.

There is no definitive consensus on an ideal ring width, however, analysis of projects to date suggests an optimum width between 30% and 40% of the finished diameter.

Ring taper. The taper is generally determined by a combination of the tunnel alignment and width of ring. The ring taper and ring width should always be considered together. The minimum curve radius on the project will dictate the taper of the rings required. As a rule of thumb, it is preferable to incorporate a taper equaling twice the calculated taper required to negotiate the tightest curve, to allow for steering adjustments. There are issues with providing too much taper including build problems within the tail can and a tendency for adjacent rings to move sideways as thrust imposed producing lipping between rings. As the positioning of the segments within a completed ring is incremental, a compromise must always be met to position the segments in the best fit orientation. One solution is to provide standard tapers with shorter rings for the curve.

One of the factors to consider is where on the circle the taper should be. This is not of major importance but for simplicity in ring selection, it is suggested that the widest part of the ring should be the counter key (first segment built). This has advantages in ring selection and also means the key is the narrowest segment, assisting in the placement.

End purpose—Type of concrete—surface finish / durability. Many clients are now seeking more durable products, with longer design life. For sanitary sewers, some owners are now specifying secondary linings for corrosion protection. In lieu of installing a separate pipe within the completed tunnel, several methods, such as spray on linings or secondary PVC membranes have been tried with varying degrees of success. Any secondary lining will extend the schedule and negates one of the advantages of a one pass lining system. Recent developments in sulphate resistant concrete and polymer concrete provide an alternative, as well as fundamental system designs providing improved ventilation. For tunnels that do require a secondary lining, minimizing the number of bolt pockets and mitigating steps and lips is a must.

Another consideration is the use of steel fibre reinforcement. The majority of concrete segments cast to date have used a traditional steel reinforcement cage. Modern developments with steel fibres have allowed the deletion of this cage, which reduces segment manufacturing costs, improves durability of the finished product, and mitigates the potential for damage during handling.

CONSTRUCTABILITY

There have always been concerns with commencing the ring build above spring line. For this reason, some people have elected to provide left and right rings with opposite taper orientation, or rings that can be rotated within the TBM. The latter option is only possible with a standard universal ring with a key and counter key that are the same size. Although these systems do increase the odds that the first segment will be in the invert, it does not guarantee it, and can lead to other issues. Increasing the number of variables in the ring build process does increase the possibility of human error, and potential for ring build quality problems as a result of improper ring orientation.

RING BUILDING

Modern universal rings make ring building simple compared to the early systems. Tunnel liners are constructed to follow the TBM, reducing the number of measurements and calculations required. In the early days of the tapered ring, tunnel engineers continued with their extensive measurement and checking regimes, however it very quickly became apparent that most of these were unnecessary. The method of installation now requires a check of the gap between the segment and the tail can. The maximum taper width is then placed at that location. Improvements in guidance systems have also simplified the process, especially with the addition of ring placement software, which allows the placement of the next segment to be determined at the push of a button.

The ring plane, square and circularity and now generally less important and in some cases are not measured at all. Improvements in tolerances of segment casting, bolting systems, TBM erector systems and guidance system have all contributed to creating a better finished product.

CONCLUSIONS

At this stage there are no hard and fast conclusions when it comes to the optimum ring; however, the Authors consider the Universal Ring to be the ideal lining system for modern tunneling. Extensive research into the modern systems would provide more insight, but current systems are advancing at such a rate that such a study would be outdated before it was completed .Future developments on TBM design, concrete materials, casting techniques and automated build systems will inevitably develop things further, and exciting developments such as extruded liners, laser guided erectors, expandable liners for EPB tunnels and polymer concrete segments are on the horizon.

ACKNOWLEDGMENTS

The Authors would like to thank the following for their assistance in the preparation of the paper:
- Chris Smith—CRS Consultants
- Hatch Mott Macdonald.

BIBLIOGRAPHY

British Transport Museum—London England.
London Under London—A subterranean guide—Trench / Hillman.

PART 18

Slurry and EPB 2

Chairs

Laura McNally
McNally Construction

Steve Skelhorn
McNally Construction

BIG WALNUT OUTFALL AUGMENTATION SEWER—PART II: TBM CASE HISTORY

Tom Szaraz ▪ McNally Kiewit JV

Gary Bulla ▪ McNally Kiewit JV

Chris Smith ▪ CRS Consultants

ABSTRACT

Part of a major sewer expansion for the City of Columbus, Ohio. The Big Walnut Outfall Augmentation Sewer—Part II (BWOAS) consists of 13,200 feet of tunnel of 12-foot tunnel passing mainly through glacial deposits with a water head of up to 2 bar. A Lovat RME167SE was used to drive this tunnel with steel fiber reinforced precast segmental liner used for support. This paper will discuss the concepts for tunnel support and TBM selection, the problems encountered with the TBM sealing systems and the solutions used to repair the machine.

INTRODUCTION

The project is located on the south side of Columbus, Ohio on the Northwest corner of Alum Creek Drive and State Route 317 to a location just south of the Alum Creek Drive and Groveport Road intersection. The project location and layout is shown in Figure 1. Included with the tunnel other major work involved approximately 5,000 lineal-feet of 42″ sanitary sewer in open cut; two gated by-pass structures; one flow control structure; tunnel connect structure and associated shaft work.

The short term goal of this tunnel is to relieve an existing 108-inch sewer during times of surcharge (480 MGD) by providing storage in both the BWOAS Part 2 tunnel, into the BWARI Part 1 tunnel, recently installed by Jay Dee/Michel's, then on to the newly expanded Southerly Wastewater Treatment Plant. In the future it will function as conveyance and storage from future tunnels slated to be installed over the next 30 years. The project is located in an industrial warehouse area situated in between City of Columbus and Rickenbacker Airport; which was primarily used for cargo and military flights.

STEEL FIBER SEGMENTS

The original lining procurement strategy had been to purchase the PCC linings from a specialist manufacture already set up in the area, however due to schedule limitations and other considerations it was not possible to utilize this supplier. McNally already had experience in manufacturing segmental linings on site and had set up NASCO (North American Segment Company) to manage the work. NASCO had successfully cast the rings for the Alki Project in Seattle with the assistance of a UK company using well proven techniques.

The decision was therefore made to set up a plant to make the linings for the BWOAS II project using NASCO as the management vehicle. For this project, NASCO enlisted the services of Chris Smith of CRS Engineering Consultants. Mr. Smith has

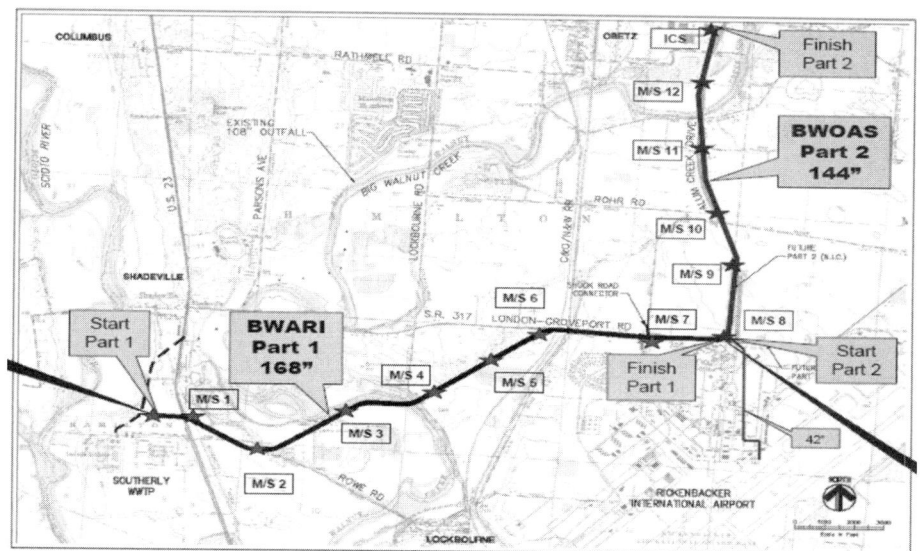

Figure 1. Project location and layout

had over 25 years of experience in the supplying tunnel lining systems, with experience in the lining design, mould design, manufacturing techniques and concrete mix designs. He was also the expert tasked by the UK company to work with NASCO on the Alki project. Mr. Smith was the primary plant manager of the segment production facility located Mount Vernon, OH. And was responsible for all aspects of the NASCO operation and directed daily operations at the plant as well as ensuring product quality.

Mr. Smith had recently worked on a number of projects including the Channel Tunnel Rail Link in the UK where the linings had been manufactured without rebar, using structural steel fibers as the only reinforcement. These rings had proved very successful and cost effective, reducing damage in the tunnel by a significant amount and resulting in less joint leakage. Using his design experience gained from these projects he suggested that it should be possible to design and manufacture a fiber reinforced segmental lining system for BWOAS II.

The lining system had already been designed with traditional rebar reinforcement by Hatch Mott MacDonald. McNally asked Hatch Mott MacDonald to carry out a redesign using structural steel fibers only. The geometry had been fixed to suit the TBM which was already well advanced in the manufacturing process. Due to this and project time constraints it was not possible to change the ring geometry at this stage. A steel fiber design was found to be satisfactory for the permanent ground loads but there were concerns over the localized torsional stresses that might be induced in the segments due to the angle of the universal rings longitudinal joints. It was therefore decided to incorporate a minimal rebar cage around the periphery of each segment to eliminate any risk from these stresses.

The final lining was of the "Universal" configuration with 4 No. 67.5° and 2 No 45° segments, the internal diameter was 144" with a thickness of 9" and a nominal width of 60". The ring was tapered 1 ½" across its diameter to allow for steering corrections. The longitudinal joint fixings were angled spearbolts incorporating a compressible washer to allow for flex in the ring and the circle joint incorporated Dowelock dowel fixings. Handling at the TBM and grouting was carried out through a cast in plastic socket

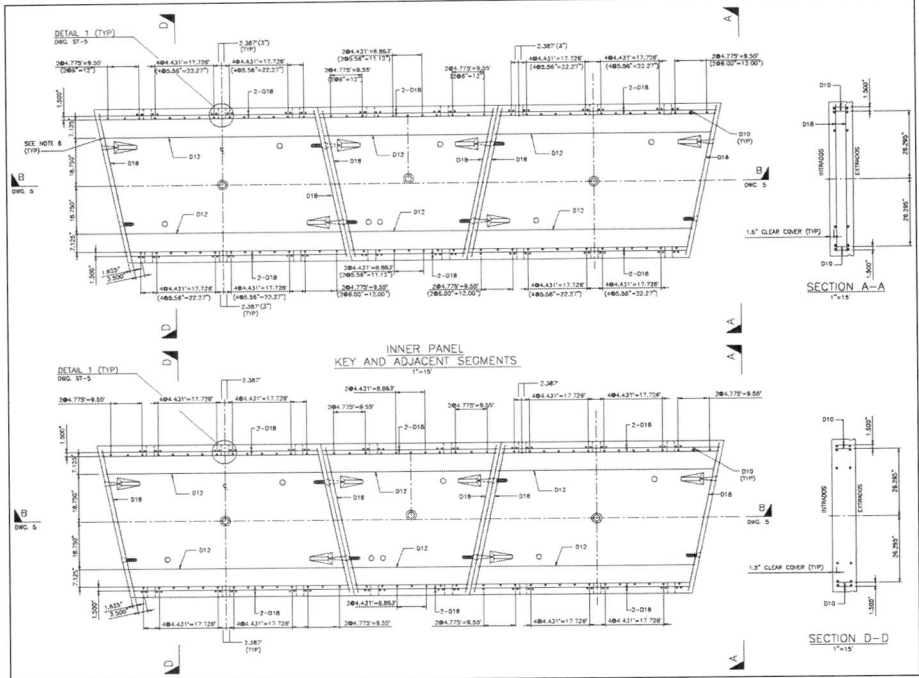

Figure 2. Precast segment drawing

incorporating a Non Return Valve. Every segment was also supplied with a compressible sealing gasket from Trelleborg.

Although Mr. Smith had experience in the manufacture of steel fiber reinforced linings this was the first project to use them in a permanent segmental lining in the United States. Extensive work was carried out to add relevant requirements to the specification and to expand the testing requirements originally laid down, in order to ensure that the steel fiber concrete not only met the contract requirements but that the City of Columbus could be satisfied as to the structural performance of the lining. All parties to the contract greatly assisted in this process with grateful acknowledgement to their help in moving the design forward going to H.R.Gray; URS Corporation; and Lachel & Associates Inc.

The plant at Mount Vernon was adjacent to an existing precast facility own by United Precast, Inc. (UPI). They had extensive local contacts and a well trained work force, and they were sourced by NASCO for their expertise, supplying NASCO with concrete, skilled production labor, concrete quality control technicians and general onsite supervision.

Control of fiber reinforced concrete is an undertaking that requires careful supervision and testing. All normal concrete tests were carried out on a regular basis at UPI's laboratory. Additional tests were also carried out on the wet mix and on the finished segments to ensure adequate dosage and distribution of the fiber. Flexural beams were tested regularly to ASTM 1609 in accordance with an agreed schedule at the University of Michigan's laboratory at Ann Arbor.

The moulds used in this project were designed by CBE Tunnels Engineering/ Manufacturing of France. CBE has over 25 years of experience in the design and

Figure 3. Mould layout

manufacturing of steel moulds and the supply of lifting and handling equipment for the production of concrete segments. Every mould had a pair of air vibrators together with a steam curing distribution outlet; these could be individually controlled as required. The moulds were designed to incorporate the different plastic inserts required for grouting, handling and erection in the tunnel.

NASCO together with UPI's technicians developed a mix that utilized the locally available materials, suited the existing batch plant's process controls, and could be efficiently handled from the batch plant and distributed into the moulds, as well as meeting all the contract requirements. The final mix was designed to have a 0.38 water cement ratio and compressive strength of 6,000 psi minimum at 28-days. The actual average of the 28-day cylinder breaks was 8,826 psi. However, a reduction in compressive strength which would have been an advantage when using fiber reinforcement was limited by durability considerations and the minimum cementitious content requirement of the specification. The mix was designed to include 55 lbs. of fibers reinforcement per cubic yard of concrete.

The segments were produced using two shifts together with steam curing to speed the concrete curing process. The required schedule and logistics of this contract were such that a carousel installation was not cost effective and the plant layout was tailored to suit the existing warehouse used for the production of the segments.

The segments were then removed from the moulds and put into secondary curing bays for five days. Each bay was individually control for moisture level with a warm water mist spray system. After the curing process was completed, gaskets and packing were applied to the segments and they were then stored until the jobsite needed them for the tunnel.

TBM

A Lovat Model RME167SE was chosen for the project. This is a self-contained tunneling system complete with screw conveyor and a lining erection system. As typical with a Lovat there are three main sections; a forward shell which is made up of the cuttinghead and main drive system; the stationary shell with the operating controls, electric and hydraulic power units, propulsion system and conveyor system; and the trailing shield which houses the lining erection equipment. A more detail description of the TBM systems and the project were previously presented at the 2007 RETC.

Figure 4. Steam curing

Figure 5. Segment storage

MERKEL SEALS

The Merkel seals are part of the forward shell assemble and can only be accessed from within the plenum. There is an inner and outer set of seals. Each seal cavity is made up of four Merkel seals and three spacer rings with an outer retaining cover holding the seals in place. The Merkel seals protect and lubricate the main bearing on the TBM. The seal assembly is shown in figure 6. Three cavities exist within the seal assembly. The forward most cavity or "G" cavity is for HWB grease. This cavity receives a constant flow of grease to protect against mined material from entering the seal assembly. The CV1 and CV2 cavities receive a constant flow of 80/90 oil. The two CV cavities are not pressurized during open mode mining. When pressure in the plenum reaches 2.2 bar EPB pressure stage 1 initializes and the CV1 cavity is pressurized to 1.7 bar while the CV2 cavity remains at 0 bar. During stage 2 the CV1 cavity pressurizes to 3.4 bar and the CV2 cavity pressurizes to 1.7 bar.

Fifty-feet into the initial launch of the TBM from shaft 8 problems were encountered with the seal system. Water was discovered in the 80/90 oil tank; upon further investigation HBW grease was found in the CV1 and CV2 return lines on the outside diameter (OD) set of seals. A Lovat technician was dispatch from Toronto to site. The plenum was clean and the OD retaining ring removed. The forward seal was torn into

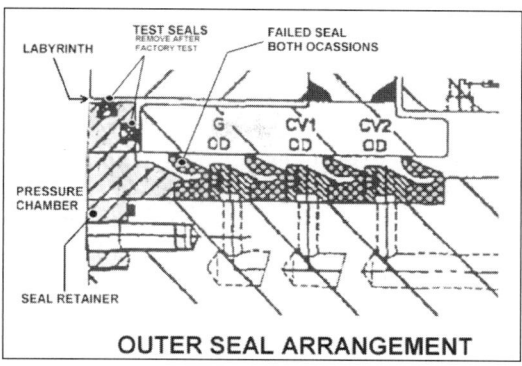

Figure 6. Mekel seal assembly

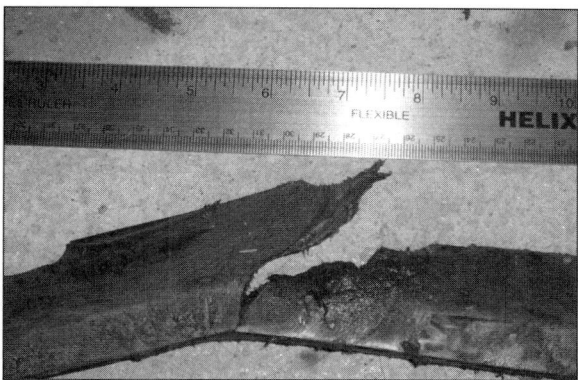

Figure 7. Damaged Merkel seal

two pieces and several uniform scallop shaped cuts were in the seal. The remaining spacers and seals were removed from the cavity and inspected. The cavity and seals were cleaned along with all of the feed and return lines. Figure 7 shows the damaged seal. A new seal was spliced and all of the remaining original seals and spacer rings were installed. The system was charged and mining resumed.

After 100-feet into the initial launch water was again discovered in the 80/90 tank along with HBW in the CV OD return lines. While removing the seals and spacers it was discovered that the seals had twisted and were not properly seated with the spacer rings during installation. Although no seals or spacers were damaged the entire system had to be cleaned and flushed. It was determined that the seals being replaced with the forward shell in a vertical position caused the seals to unseat from the spacers during installation. This process is done in the factory with the forward shell laying horizontal on the shop floor.

As we found out with this second system failure there were no procedures for replacing the Merkel seals underground. Through trial and error and several meetings Paul Cott with Lovat came up with a procedure for replacing the seals and spacers while maintaining the required tolerances in the cavity. Mr. Cott fabricated several special rods designed to ensure that the seal did not twist while being installed. Further, he fabricated several different spacer clips that bolted to the machine using the outer

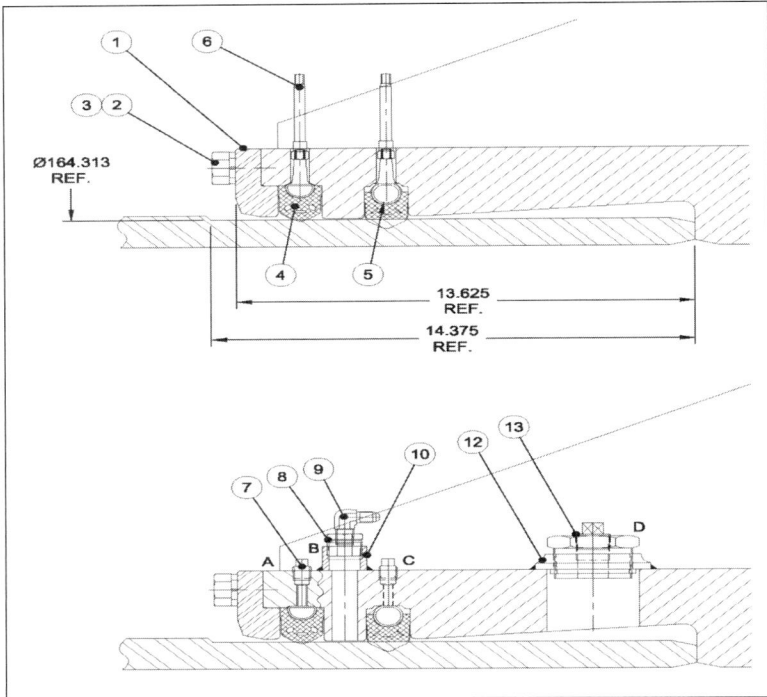

Figure 8. Articulation assembly

retaining ring holes. The spacer clips were designed at different lengths to hold the spacer rings to their designed location within the seal cavity. Although the process was extremely time consuming it proved to be effective. After the system was reassembled, static and dynamic testing was carried out to ensure no problems with the seals. A total of two months were lost due to the problems with the Merkel seals.

ARTICULATION SEALS

To allow for additional steering of the TBM, the forward shell can be articulated up to 2° relative to the stationary shell and the trailing shield in any direction. The twelve articulation cylinders are arranged in four groups of four. Total articulation cylinder capacity at maximum pressure of 5,000 psi equals 1,500 tons of thrust.

The designed articulation seals are an air energized system made up of two rubber Buna-N 90-durometer seals. Each seal had a separate ⅞" inflatable tube to press the Buna-N seals against the inside diameter of the outer shell. A gauged air system was connected to the tubes through valve stems with an operating pressure of 30 psi. EP2 grease was fed to ports between the seals by an automatic grease system. Figure 8 shows the articulation assembly. The forward seal in the system is the only one that can be replaced without separating the TBM. Figure 9 shows the supplied seal and tube.

1,300-feet into the drive a 7" section of the forward articulation seal was found in the bottom of the stationary shell. A Lovat technician was dispatched to site with a replacement seal and tube. The retainer ring was removed and the replacement seal installed. The seal and tube had to be threaded around the articulation cylinders and spliced together. Several attempts with different Loctite primers and glues failed to hold

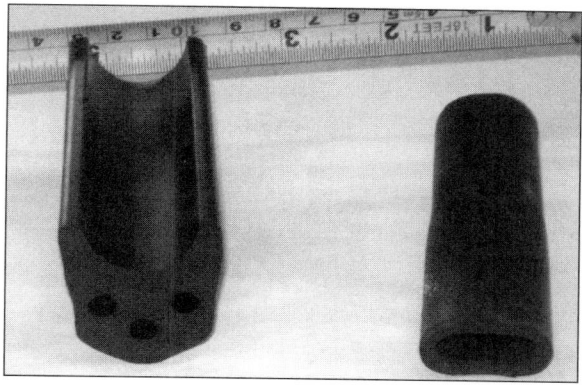

Figure 9. Articulation seals

the seal and tube together during inflation. Loctite 415 glue was eventually successful for both the seal and tube. The tube was inflated to 30 psi and held.

TBM mining resumed for 25-feet before the forward seal failed again. The forward seal was replaced and failed again after 75-feet of mining. The seal was not damaged; however the tube was torn into several pieces. An inspection of the rear seal revealed that it was torn in several places as well. The forward seal and a new tube were installed. The "D" ports as shown in figure 8 were plumed into the tail grease system. The EP2 grease system was augmented with additional ports to feed additional grease between the seals and into the "C" ports above the rear seal. This system worked well for 550-feet when the tube blew out of the forward seal. The inflow could be controlled with increased grease flow and shaft 9 was 400-feet from our current position it was decided not to replace the seal until the TBM was at the shaft.

After the second articulation seal failure several planning meetings were held with Lovat to discuss different options for this system. Larry Snyder from Snyder Engineering LLC was hired to investigate options for possible modifications to the seal system. Several options were considered including expanding the 10-foot diameter augured cased shaft to remove and repair the TBM at Lovat's shop in Toronto.

It was decided that the best option was to repair the articulation seals underground. The shaft was excavated to a 16-foot diameter by 3-foot long slot to provided complete access to the exterior of the TBM. The excavation was initially going to be supported with liner plate. The geology of the tunnel envelope through the shaft was a stiff glacial till changing to till with sand layers at the bottom of the shaft. It was found that the jet grouting performed by Nicholson Construction Company along with the stiff till was stable enough to stand on its own. The TBM was drove through the shaft until the articulation joint was centered in the 3-foot channel. The pins were removed from the articulation rams and a series of three different length Dutchmen were used to push the stationary shell back into the tunnel. Figure 10 shows the TBM separated in shaft 9. A channel that was cut into the bottom of the shaft was cleaned out and a sump and pump were installed to remove and water from the joint.

It was decided to move away from the original air energized seal system to a solid set of rubber Buna-N 70-durometer seals. The seals are shown in figure 11. The repair included cutting ¾" deep by 3" long serrations every 18" along the edge of the forward shell. This was done to help relieve any trapped soil between the two shells. After the seals were installed the joint was closed with the propulsion jacks. The articulation rams were reconnected along with all of the removed systems.

Figure 10. TBM separated at Shaft 9

Figure 11. Replacement articulation seal

The TBM mined forward and was positioned so that the tail seal brushes could be exposed for inspection. After seeing the amount of repair need to the first two rows of brushes the trailing shield was separated from the concrete segments. All three rows of tail brushes were replaced in shaft 9. The repairs to the articulation and tail seals took seven weeks.

The TBM was re-launched out of shaft 9 and after 590-feet of mining a 20" piece of the forward seal was found in the bottom of the stationary shell. Another seal was ordered from Lovat. HBW grease was pumped between the seals along with tail seal grease pumped behind the rear seal. The TBM was advanced 135-feet until the new seal arrived. After inspecting the damaged seal it was determined that the 70-durometer rubber was too soft. The replacement 70-durometer seal was installed and a 90-durometer seal was ordered for future replacement. During an inspection of the rear seal it was discovered that a 6" piece was missing. Mining resumed for 65-feet and was stopped when the new 90-durometer seal arrived on site. The seal was replaced and a separate HBW pump was installed for adding grease to the rear seal. This worked well for over 1,000-feet until the forward seal failed again. After the seventh repair was completed the seals lasted for the remainder of the project.

CLOSING REMARKS

The first use of fiber reinforced concrete segments in the United States. The fiber reinforced segments provided the City of Columbus with a superior product which was better than specified. In addition, the rings provided greater durability with significantly less damages during handling and installation.

In spite of all of the problems with the TBM sealing system and the fact that all of the seals were replaced underground the project was completed under budget and early.

ACKNOWLEDGMENTS

The author would like to acknowledge the following companies that contributed to the success of the project: The McNally and Kiewit Group of companies, Hatch Mott MacDonald, CBE Tunnels Snyder Engineering LLC, United Precast, Inc., Elio Calabrese Construction, Inc., Lachel & Associates Inc, H.R. Gray, Nicholson Construction Company, Case Foundation and Soletanche Inc. The assistance of Chris Smith, CRS Engineering Consultants and Gary Bulla in preparing this document is greatly appreciated.

EPB TUNNELLING THROUGH COHESIONLESS SATURATED GROUND UNDER VERY SHALLOW COVER— PERTH NEW METRORAIL CITY PROJECT

Hiroshi Yamazaki ▪ Leighton Contractors

Oskar Sigl ▪ Geoconsult Asia Singapore Pte Ltd.

Fumihiro Aikawa ▪ Leighton–Kumagai Joint Venture

Raghvendra Bhargava ▪ Parsons Brinckerhoff

ABSTRACT

The paper describes prediction, control and validation of face pressures, and special TBM features for the bored tunnels of the New MetroRail City Project, Perth, Australia. The selected EPB-TBM passed below a heritage bridge and the operational Perth Railway station in ground conditions consisting of fully saturated uniform sand combined with shallow overburden and a horizontal alignment radius of only 135m. Therefore, strict control of volume loss, TBM operations and ground movements was vital requiring continuous evaluation and adjustments to TBM operation parameters. The paper discusses the effectiveness of applied TBM face pressures and the used foam/polymer soil conditioning which was measured against achieved volume loss and the predicted range of ground settlements.

INTRODUCTION

The Perth CBD section of the New MetroRail City Project comprises twin bored tunnels 1.45km. combined length, two underground stations, cut & cover tunnels and dive structures on either ends of the bored tunnels. As shown in Figure 1, the bored tunnel works were divided into the northern and southern sections. The northern section was located in very challenging conditions of shallow sandy overburden, full ground water head, rising alignment, crossing under a heritage bridge and the operational Perth main railway station. The horizontal alignment radius below the station's railway tracks and platforms was as tight as 135m. In order to address these specific and demanding conditions, the tunnels were constructed using earth pressure balanced tunnel boring machine (EPB-TBM). The tunnels have finished internal diameter of 6.16m and 275mm thick segmental lining rings of 1,000 and 1,200mm widths.

The specific alignment and geotechnical conditions, and presence of critical civil infrastructures in very close proximity along the northern section, required an integrated design and construction process. The process involved a series of calculations and numerical analyses to derive theoretical TBM face support pressures, consideration of respective influence of TBM operation control parameters, use of foam/polymer to condition the saturated sandy soil, and predictions of ground settlements at every 20m sections along the tunnel alignment. During construction, the predicted face support pressures and ground settlements were continuously compared with actually applied TBM pressures and measured ground settlements from instrumentation monitoring sections installed at every 20m along the tunnel route.

EPB TUNNELLING THROUGH COHESIONLESS SATURATED GROUND 1125

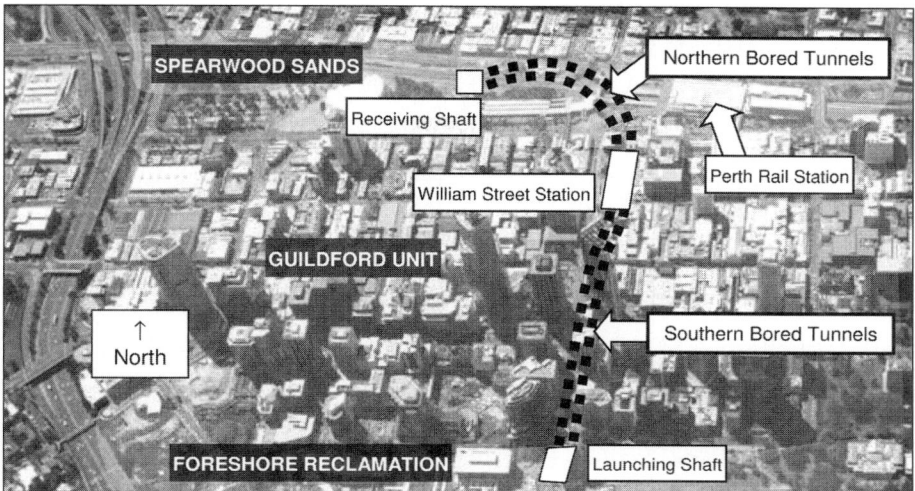

Figure 1. Bored tunnel alignment overview (New Metrorail City Project-Perth)

ALIGNMENT AND GEOTECHNICAL CONDITIONS

As shown in Figure 1, the northern bored tunnel section commences at the north head wall of William Street Station, and passes below the heritage listed Horseshoe Bridge. Soon after crossing the bridge, the tunnel alignment goes within the rail reserve and passes underneath the railway tracks and platforms at a very sharp radius of 135 m, turning further into Roe Street and arriving at the TBM receiving shaft. The vertical alignment along the northern section rises steadily with overburden cover reducing from 10.2 to 3.25m at the receiving shaft, which is less than 0.5 times the excavation diameter of tunnel.

The ground conditions around TBM tunnels of northern section mainly comprises of the Spearwood Sand (SS), which is a poorly graded medium dense to dense quartz sand. In the tunnel invert firm to stiff sandy clay of Guildford formation occur (UGU/GFU). From particle size gradation curves shown in Figure 2, a uniformity coefficient $C_u = 4$ and a coefficient of gradation $C_g = 1.6$ was determined. From tunnelling perspective, the spearwood sand is practically pure sand with full groundwater pressure. There is potential for lenses of peaty soils within it, and occasional cemented zones of considerably higher strength. Towards the end section of tunnel, the vertical alignment rises at an acute angle to the predominant bedding of sand and clay layers, thereby increasing the potential for non-homogeneity within the tunnel face.

The tunnel section passes through two ground water regimes i.e., from an underdrained hydraulic profile below the predominantly clayey layers on the south side of the Perth Rail Yard close to the existing railway station to hydrostatic ground water regime on the north side close to the receiving shaft. It is likely that these changes in hydraulic heads resulted from the relative differences in permeability of the sand and clay materials across this tunnel section. The operation of boring the tunnel may connect some of these systems and could result in flow along the interface between the lining and the soil. In the event the TBM face is opened, risks related to the potential for a zone or lenses with significantly higher permeability to be exposed in the face was identified in the geotechnical assessment. As a conservative assumption, dictated by the prevailing sandy ground conditions, full groundwater pressures had to be assumed all along the bored tunnel alignment.

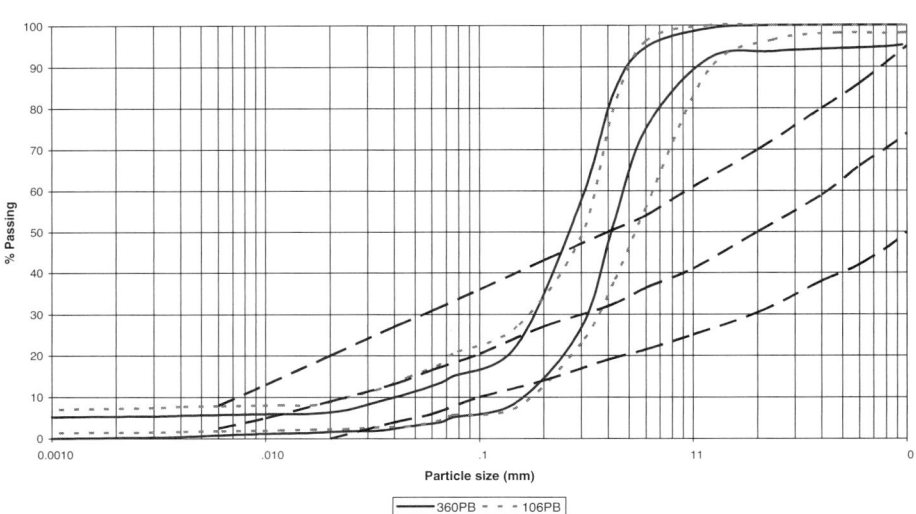

Figure 2. Typical grain size distribution of soils within the northern TBM section

TBM SELECTION AND SPECIAL FEATURES

In the Northern bored tunnel section, the alignment under-passed the operational Perth main rail station, heritage listed Horse Shoe bridge and had to tunnel through the very shallow overburden of saturated sand at the same time. The overburden reduced from about 1.5 times the excavation diameter to a minimum of only half a tunnel diameter.

In the southern section, the bored tunnels are located directly below William Street, Perth's most busy road with TBM operations taking place in the immediate vicinity of high rise buildings. Since the William Station is off-set from the William Street, the bored tunnel alignment has to eventually leave the road reserve and slew underneath a group of buildings on shallow foundations.

Based on assessment of geotechnical conditions along both the northern and southern sections, a closed TBM shield machine with active face pressure control was required. Since the available site areas for a slurry plant were extremely limited, the use of a slurry TBM was ruled out and an earth pressure balanced TBM (EPB-TBM) selected instead. In addition, previous successful records of the Contractor using EPB-TBMs in sandy ground conditions were considered in this decision too.

An EPB-TBM provides a continuous active face support during advance by maintaining specified pressures in the earth pressure chamber using a bulk head and a screw conveyor system for muck removal under pressurised condition. In addition, a pressurised tail void grouting system for backfilling the void between the extrados of the segmental lining and the excavation circumference significantly reduces surface settlements. The excavation diameter of the TBM was 6,900mm. The TBM was built to accommodate a segmental lining with 6,160mm internal diameter and 275mm thickness. The segment rings were produced in two different ring lengths of 1,000mm and 1,200mm. The shorter rings were used to negotiate the 135m alignment curve.

The main tunnelling risks were face instability, excessive accidental ground loss & settlements, and consequential impacts on existing buildings and rail infrastructures. Face instabilities were mainly expected due to variable geotechnical conditions of

Table 1. Main features of selected EPB TBM

Feature	Details
Manufacturer	Mitsubishi Heavy Industries, Kobe, Japan
Installed cutter head power	540 kW
Excavation diameter/shield length	Φ = 6,900mm / L_{shield} = 8,480mm
Number of cutters	typically 80 nos. drag bits, no discs
Thrust Force maximum	20 × 2,000 kN = 40,000 kN
Articulation jacks	8 × 3,500 kN
Maximum torque	5,485 kNm (100%); 6,582 kNm (120%)
Segmental lining	6,160mm I.D., 275mm thick, 1,000mm/1,200mm width, 5+1key
Special features	3×anchor detectors, knife bits, copy cutters, ENZAN "Arigataya" TBM monitoring & survey system

mixed sandy and clayey soils in the tunnel face. The variable permeability and differential water pressure head in these layers had potential to cause flowing ground in front of the cutter head in the event of incorrect prediction and application of face pressures and lack of control of TBM operations

The risk of excessive accidental ground loss & settlements was allocated due to sudden loss of face pressure in the event of encountering known tunnelling hazards specific for this area such as poorly backfilled old wells, sudden and forceful pull-out of abandoned ground anchors, wooden/sheet piles used for support of old utilities, as well as any incident of blow out in extremely shallow sandy overburden. The consequences were very serious as tunnelling operations were located in the busiest part of the Perth CBD, adjacent and beneath some of the major and sensitive buildings, and critical railway infrastructure.

As a consequence a rigorous risk mitigation exercise was undertaken and following special TBM features were determined to mitigate the risks and minimize the consequences of excessive ground loss and settlements:

- *a newly developed detector* comprising a custom shaped hydraulic cylindrical shaft protruding out of the TBM cutter head and fitted with a hydraulic pressure sensor. If cutter head comes in contact with an obstacle the hydraulic cylinder would be pushed backwards causing increasing pressure in the hydraulic system, and an ALERT would appear on the control display. This simple mechanical arrangement is very simple, reliable and robust against false alarms. The TBM cutter head had 3 anchor detectors installed at 120° close to the circumference.
- *special cutters* in addition to the ordinary drag bits for excavation, heavy duty knife cutters were installed along the cutter head circumference to cut through obstacles such as abandoned ground anchors, old well casings etc., thus minimizing the risk of damages to the cutter head. This cutter configuration has already been used successfully by Kumagai on previous projects in Japan.
- *copy cutters and shield articulation* for steering TBM in curved alignment. The TBM had to negotiate a very tight 135m radius curve with less than one tunnel diameter soil overburden below operational Perth rail station, the overcut was localized using hydraulically operated copy cutters. Shield articulation system consisted of a number of separately operated hydraulic jacks installed between the front and the tail shield. By extending and

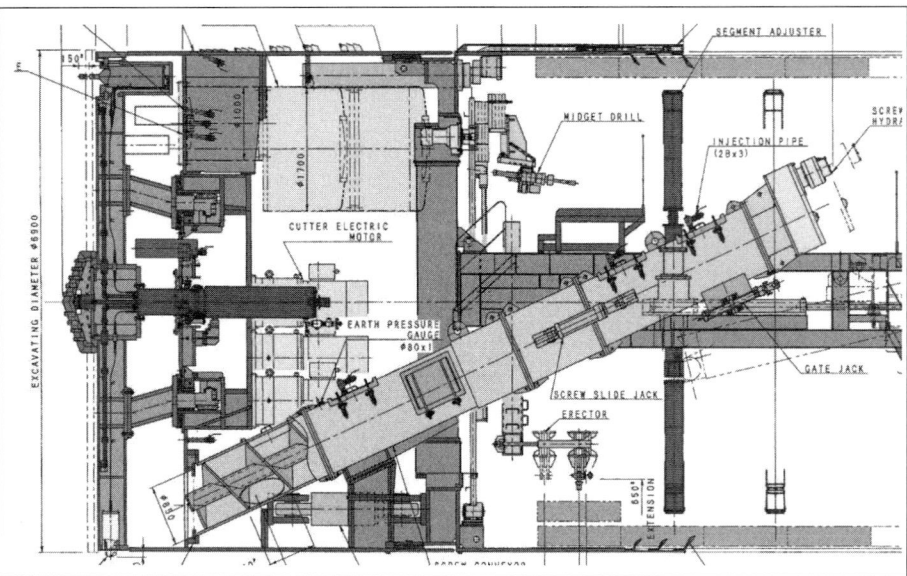

Figure 3. EPB tunnel boring machine—longitudinal section

retracting the articulation jacks the tail shield axis was angled to better fit the alignment curve by reducing length of rigid shield body.

- *simultaneous tail void back fill grout system* to ensure that the space between the extrados of the segmental lining and the line of excavation is pressurised and fully backfilled as the TBM is moving along. The system uses two injection lines per grout port. One line for the grout and the second for the accelerator. The TBM was equipped with grout ports through the TBM tail shield with advanced flushing system "TAC 3rd." This system flushes the grout port automatically in order to avoid blockages by accelerated grout when grout operation is suspended.

- *advanced TBM Operation monitoring system,* the TBM operation monitoring and automatic survey system, ENZAN "Arigataya" was installed. The system monitored and recorded more than 150 items every four times per second. Data was displayed on the operation panel in the TBM operator's cabin and the TBM monitoring room. The system produced a detailed operation report for every ring and every shift for record and submission to the Client.

- *other TBM operation and control systems included,* muck volume measuring system, internal grouting ports in the crown of TBM shield, compressed air work facility and a fully operational airlock for emergency intervention into the cutter head.

Monitoring of the data obtained by these systems was integrated into TBM monitoring system allowing continuous check and adjustments, particularly extracted soil volume against the theoretical volume, face pressures, TBM position, alignment and steering data.

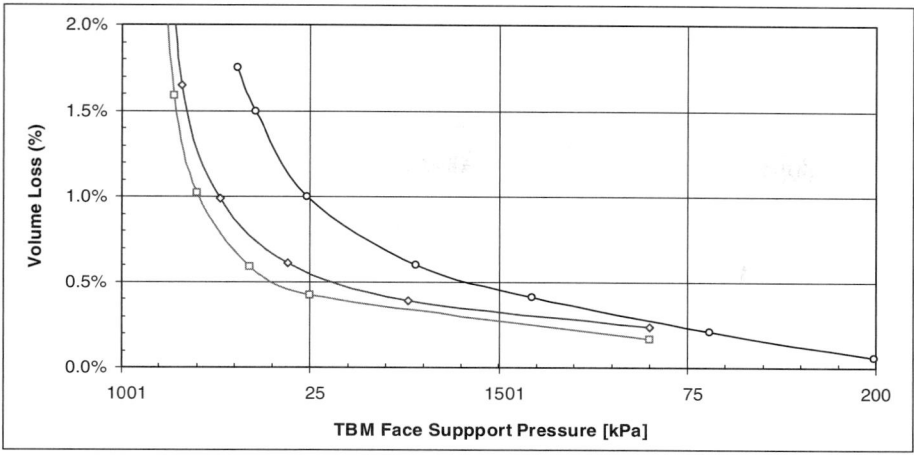

Figure 4. Volume loss versus face pressures—bored tunnel north

VOLUME LOSS ANALYSES AND TBM FACE PRESSURE CALCULATIONS

Detailed analyses of ground volume loss, ground movements and calculations of TBM face pressures were vital design elements to ensure safe tunnelling, gain confidence of the public and the asset holders; and to obtain approvals from utility authorities.

The design works comprised following sequence of analyses and assessment.

Volume Loss Analyses

A series of FE numerical analyses (using Plaxis v.8.2) were carried out to assess the magnitude of ground volume losses expected to occur in response to a range of applied TBM face pressures. The analyses sections modelled representative tunnel geometry, alignment, overburden and geotechnical conditions, the TBM/Shield cross section dimensions and features. The main objectives of the analyses were to determine:

- deformation response of soil around tunnels under a range of applied face pressures
- characteristic relationships between volume loss, face pressures and surface settlement
- typical values of design volume loss and settlement trough parameters

The results of the numerical analyses were plotted to obtain relationship between volume loss and TBM face support pressure for 3 expected sequences of soil profiles along the northern bored tunnel section are shown in Figure 4.

In addition to the expected trend of increasing volume loss with decreasing face support pressure, the Figure 4 also shows that below a face pressure of 115 to 125kPa range, the volume loss tends to increase exponentially indicating excessive deformations around the tunnel opening and thereby providing a lower-bound limit face pressure value for the respective geotechnical and overburden conditions. Above this pressure range the volume loss response is quasi linear.

During the exercise of determining a recommendation for the required TBM face pressures such limit considerations are of particular relevance in order to assess risk margins. The above relationship was cross checked using analytical methods of deformations around a circular hole in an elasto-plastic continuum. The results from the two methods showed good correlation and reassured confidence in the numerical analysis models. The results were used to determine design volume loss percentages for practical range of applied face pressures and settlement trough parameters, which were used as input parameters to derive settlement profiles as described later in the "Ground Movement Analyses" section.

Adopted Volume Loss Values

The volume loss face pressure relationship presented in Figure 4 was used in conjunction with ground water pressures along the tunnel axis to derive design volume loss values. Groundwater monitoring results along the northern bored tunnel section suggested mean water pressure values of around 100kPa.

Considering the target value for the TBM face pressure of about 50% higher than the water pressure, the TBM should be driven at a target face support pressure of about 150kPa. Practical limitations in the controls of TBM operation naturally cause variations in the actually applied face pressures.

Considering this influence, several scenarios were studied and following design volume losses were obtained:

- Driving the TBM at about 85% of the target face pressure value (15% face pressure drop) yields excavation deformations such that a volume loss of about 0.6% would occur. Therefore, a volume loss value of 0.6% along with a trough width parameter of k= 0.40 are adopted as regular design values. This case was called the 'Design Case.'

- In order to limit construction induced damages to buildings within "slight" damage category, and for structures of heritage value to "very slight" damage category, some building required volume loss to not exceed 0.40%. This was achieved by implementing ground treatment. This case was called the 'Ground Treatment Case.'

- At TBM launching and TBM arrival shafts, an increased volume loss of 1.0% was considered for an alignment stretch of 20m (about 2 times the length of the TBM shield). This case was called the 'Launching/Arrival Case.'

- An exceptional case scenario representing accidental situation causing a volume loss of 1.5% was investigated too. This exceptional scenario was used to ascertain appropriate contingency and emergency procedures for implementation in the relevant safety plans. No design was carried out for this scenario.

Additional Volume Loss in Curved alignment

The radii of horizontal alignment of tunnels varied from 135m to 510m. The influence of steering of TBM in curved sections was included using geometry of overcut excavation profiles considering radius of curve, diameter of shield, length of articulated shield body. This represents the worst case scenario as elastic/plastic soil behaviour and bentonite injection along the TBM shield skin are disregarded. The following relationship was used:

$$\Delta V_{S-CURVE} = \frac{\sqrt{\left(R+\frac{D}{2}\right)^2 + \left(\frac{L_s}{2}\right)^2} - \left(R-\frac{D}{2}\right)}{D}$$

R = Alignment radius
D = Diameter of shield

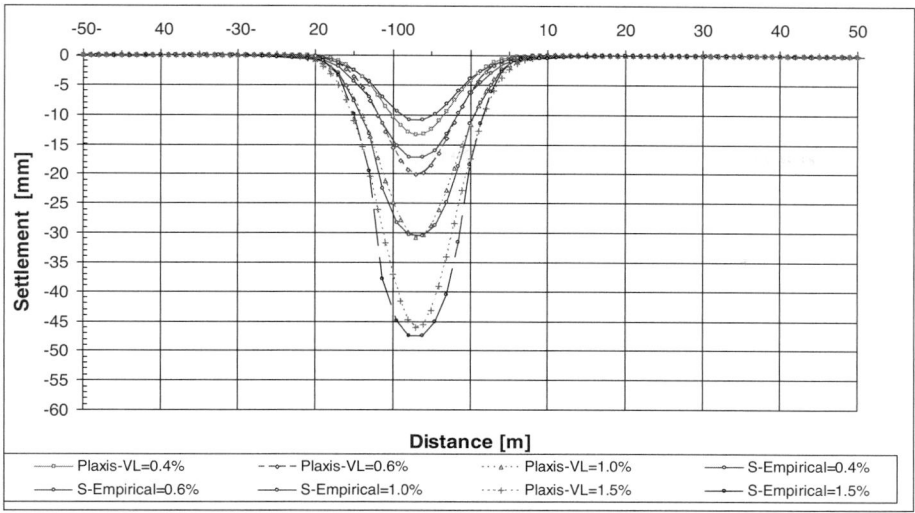

Figure 5. Transverse settlement profiles—validation of numerical analyses with semi-empirical method

L_s = Length of rigid shield body considering articulation
The geometrical additional volume loss during the TBM negotiating the tightest curve (R=135m) and the regular curve (R=510m) are additional 0.26% and 0.07% respectively. Along the vertical alignment, the proposed bored tunnels have radii ranging from 2,800m to 3,000m. These curvatures are considered large enough to result only in insignificant values of additional 'curvature' volume loss.

Ground Movement Analyses

Ground movement analyses comprised derivation of longitudinal and transversal settlement profiles, and calculation of horizontal ground displacements and strains. The design volume loss values obtained from volume loss analyses were used as a guide to determine the range of volume loss for which ground settlements analyses was undertaken.

A number of numerical calculations modelling volume loss values ranging from 0.25% to 2.0% were carried out using Plaxis V8.2. To compare and validate the ground settlements obtained from numerical analyses, equivalent number of semi-empirical calculations using approaches proposed by R.J. Mair, et al 1996, O'Reilly and New, 1982 and 1991 were performed.

Analyses of transversal settlement profiles obtained from numerical and semi-empirical methods were carried out for a number of sections with identical geotechnical, alignment and overburden conditions.

Volume loss values of 0.4%, 0.6%, 1.0% and 1.5% were used to derive corresponding settlement profiles. These values represent volume loss percentage for the 4 scenarios of ground conditions and TBM operations discussed in the previous section.

Figure 5 presents the plots comparing the settlement profiles. The graphs show reasonable good correlations between the results of numerical analyses and semi-empirical methods. The results of these analyses were used to determine zones of influence of tunnel construction and assessment of damage to the buildings and utilities based on the limiting tensile strain approach.

Additional Assumptions in Settlement Analyses

The above discussed tunnel excavation induced ground displacements are immediate settlements. Other influences such as effects of groundwater lowering and long-term consolidation have been excluded based on following assumptions:

- water inflow into the tunnel is largely prevented or at least significantly reduced by maintaining adequate face pressures in the range of hydrostatic pressure.
- an effective soil plug is maintained within the screw conveyor of the EPB TBM by using appropriate soil conditioning agents such as foam and/or polymer.
- after installation of pre-cast segments with permanent waterproofing is ensured by sealing gaskets, preventing water inflow through the segment joints.
- The tunnel overburden and cross section is within the sand with medium stiff to stiff clay in the invert, any presence of soft clays that may cause long-term consolidation settlement is minimal and localized.

Face Stability and TBM Support Pressure

In order to achieve the design volume loss values, the applied face support pressure must ensure stability of the tunnel face. This required calculations of TBM face pressure ranges to ensure that the volume loss scenarios used in deriving ground movements are not exceeded and stability of tunnel face is maintained.

Since alignment of northern bored tunnel is within the Spearwood Sands, face stability calculations under effective stress (drained) conditions using approach proposed by Anagnostou and Kovari (1996) were carried out.

The limit equilibrium of a sliding wedge at the face loaded by a prismatic body was investigated for a number of sections along the tunnel route. Geotechnical layering, overburden conditions, surface surcharge loads, mixed face soil stratification and seepage pressures were included in the calculations according to the Anagnostou and Kovari (1996) approach.

Figure 6 presents plots of face pressure calculated from numerical analyses, analytical calculations, face stability and ground movement considerations.

The comparison of calculated face pressures from above discussed design considerations and approaches is of very practical use in quantification of ranges of support pressures to control volume loss, ground surface settlements and stability of tunnel face.

Soil Conditioning—Use of foam and Polymer

The northern bored tunnel was excavated in fully saturated sand with very little fine content and high permeability. This resulted in a prognosed lack of plastic consistency and poor workability of earth material in the TBM chamber. Plastic consistency and good workability are critical for efficient application and control of TBM face pressures, extraction of material through screw conveyer and overall efficiency of TBM operations. Due to higher permeability of sand and full ground water pressure, there was a serious concern on binding the sandy soil and water together. The shallow overburden and tunnelling beneath critical railway infrastructure further added to the need to investigate soil conditioning requirements in greater detail.

The particle size gradations (sieve analyses) of the ground materials within the tunnel cross section depth were undertaken. As shown in particle size gradation curves in Figure 2, a significant portion of the soil mass lies below the upper and middle dashed lines, which indicates that conditioning of soil was necessary to achieve

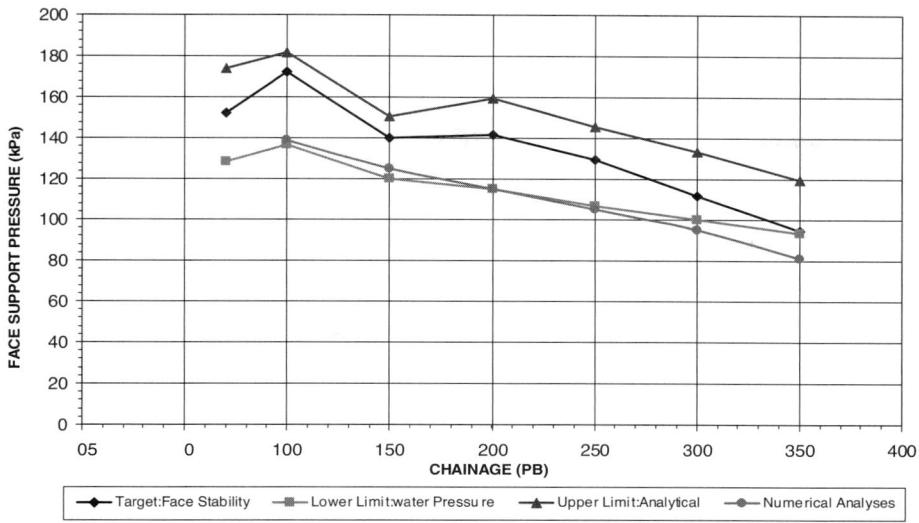

Figure 6. Comparison of calculated face pressures from different approaches

required consistency and ensure uniformity of face pressure application. The uniformity coefficient C_u of about 4 and a coefficient of gradation C_g of about 1.6 were determined from characteristic grading sizes of particle size distribution curves in Figure 2. The injection of water alone was not considered an adequate means and the need for conditioning was judged critical. The positive impacts of soil conditioning in reducing TBM torque, wear and tear of cutterhead and screw conveyer were considered as additional benefits for adopting conditioning of soil.

A series of conditioning mix trials and tests were conducted to determine optimum combination of following key parameters of the foam mix:

- CF: Ratio of the Concentrate to be added with water to make the foam solution.
- CA: Ratio of the foam Additive to be added with water to make the reinforced foam solution.
- FER: Foam Expansion Rate: Ratio of compressed air added to the foam solution to make the foam
- FIR: Foam Injection Rate: Ratio of volume of foam injected to the volume of ground excavated.

The initial on-site trials and tests consisted of taking soil from William street station in the tunnel eye horizon. The effectiveness of foam greatly depends on its chemical composition, injection rate and foam expansion ratio. A special conditioning agent was produced by mixing foam with polymer, and a purpose built computerized foam injection system was installed to allow selection of foam mix or polymer. The mix and operation parameters were defined prior to commencement of tunnelling.

The concentrated foam solution was mixed with water to predetermined dosages to make a foaming solution. The pumps fed the solution through to the foam generator where it was mixed with compressed air to create the foam at the required consistency, which was then injected into the cutter face.

The pressure of the bubbles in the foam was controlled through the pressure of the compressed air mixed with the solution and set to match the earth pressures to prevent

Table 2. Typical foam and polymer components for soil conditioning

Ground	CF (per 1m³ solution)	CA (per 1m³ solution)	FER	FIR (per 1m³ soil)
Silt	30 litre	None	8	20%
Silty sand	20 litre	6 litre	6	20%–35%
Sand	20 litre	6 litre	6	35%< Polymer mix

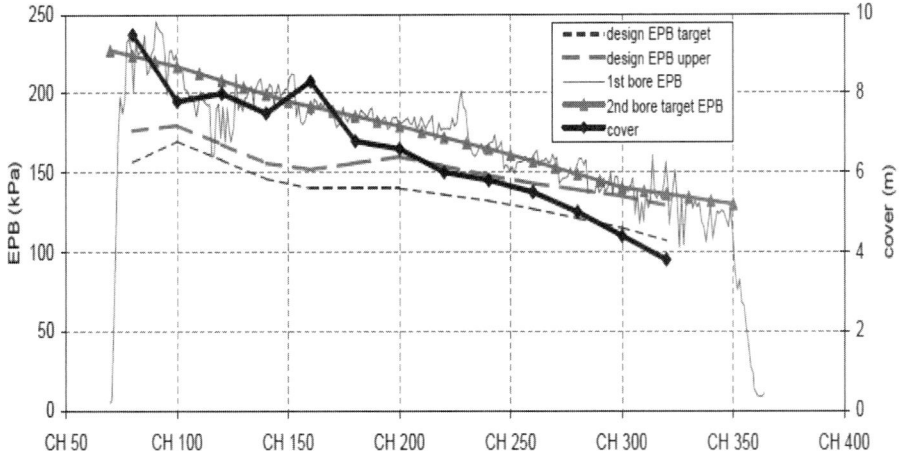

Figure 7. Actually applied vs. calculated TBM face pressures and overburden pressure

the shrinkage of the bubbles due to earth pressure. The bubbles in the foam penetrate the soil structure and aid in the movement of the soil particles over one another thereby increasing the plasticity. They also block water from moving within the pores of the soil structure and also reduce the friction created between soil particles and the cutter tools.

Table 2 is listing the key parameters of the foam mix actually used for the different types of ground material encountered.

VALIDATION OF MONITORED AND PREDICTED DATA

During tunnel excavation predicted face pressure and expected ground settlements were continuously compared against actually applied TBM pressures and actually monitored ground movements.

Figure 7 presents comparison of applied and predicted TBM face pressures along the alignment of the tunnel. The earth pressure corresponding to the overburden cover above the tunnel crown is also shown for reference.

Figure 7 shows that the actually applied face pressure accurately followed the recommended pressure values and its fluctuations were relatively low and acceptable.

Figure 8 presents comparison of predicted and actual ground settlements measured during construction. Ground settlements were monitored at 20m intervals along the tunnel alignment. The graphs in Figure 8 below show the predicted ground settlement troughs for volume loss values of 0.40% and 0.60% and overlain by actually monitored settlement troughs for the two EPB TBM drives (1st and 2nd bore).

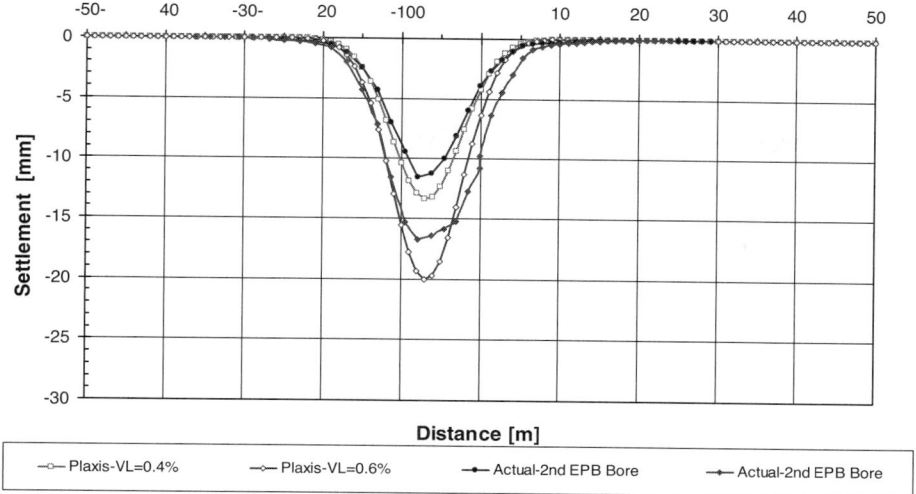

Figure 8. Predicted settlement troughs together with actually recorded surface settlements

From actually recorded settlement troughs of the two TBM bores, volume losses values of approximately 0.36% and 0.50% were derived. These values are in excellent agreement with the results of numerical volume loss analyses carried out before the commencement of TBM tunnelling.

CONCLUSIONS

The twin bored tunnels of the Perth Metro Rail project were successfully constructed under very challenging conditions of saturated sandy ground, shallow overburden, passing below very critical live railway infrastructure causing only relatively small settlement.

The requirement of controlling ground movements and construction impacts on buildings and infrastructures below the "slight" damage category was met. These goals, critical on a project wide scale, were achieved by a combination of factors. These factors included accurate prognosis of required TBM face support pressures through detailed design analyses and calculations, continuous evaluation of applied face pressures, tightly managed control of TBM operation parameters and monitoring of ground movements during construction that enabled timely adjustments to TBM operation parameters.

The chosen approach of deriving the TBM operation parameters were proven very useful and fit for practical application and swift adjustments to re-run predictions based on conditions encountered during TBM driving.

This process included:
- assessing potential volume loss values as a function of applied face pressures using 2D numerical analysis models. Although considered conservative, the approach was proven feasible in practical terms and an adequate design tool to assess critically lower-bound face pressure levels.
- The approach of deriving the actually recommended TBM face pressures from analytical face stability analyses proofed to be very reliable, quick and accurate with adequate safety margins even in sandy conditions with low overburden.

In addition to the design approach and prediction, the use of optimum designed mixture of foam and polymers achieved effective conditioning of high permeable sandy soil under full ground water head. Soil conditioned in this manner helped to apply and maintain TBM support pressures consistently, and allowed controlled removal of excavated material through the screw conveyer.

The predicted volume losses, settlements and face pressure compared well with actual values experienced and monitored during tunnel construction.

Summarizing, it has to be mentioned that the successful driving of the EBP-TBMs on this project through fully saturated uniform sand in very low overburden condition under critical and extremely sensitive existing infrastructure, manifests a tremendous achievement for which it is difficult to find comparative examples in the literature.

REFERENCES

Public Transport Authority (2004) New Metrorail City Project—Contract 27/03: Scope of Works and Technical Criteria, Government of Western Australia, February 2004.

N. Loganathan, J. O'Carrol, R. Flanagan, T. Boon Tee (2005); EPB TBM Tunnelling in Singapore Old Alluvium, 2005 RETC Proceedings, Chapter 28 pp 348–356.

Mair, R.J., Taylor, R.N. and Burland, J.B. (1996); Prediction of ground movements and assessment of risk of building damage due to bored tunnelling, Geotechnical Aspects of Underground Construction in Soft Ground, 1996 Balkema, Rotterdam. ISBN 9054108568.

O'Reilly, M.P. and New, B.M. (1982); Settlement above tunnels in the UK—Their magnitude and prediction, Proceedings Tunnelling '82 Symposium, Institution of Mining and Metallurgy, London (1982), pp 173–181.

New, B.M. and O'Reilly, M.P. (1991); Tunnelling induced ground movements predicting their magnitude and effects.

Peck, R. B. (1969); Deep excavations and tunnelling in soft ground, Proceedings 7th international conference on soil mechanics and foundation engineering, Mexico.

New, B.M. and Bowers, K.H. (1994); Ground Movement Model Validation at the Heathrow Express Trial Tunnel; Tunnelling 94 pp 301–329, Chapman & Hall, ISBN0412598604.

Anagostou, G. and Kovari, K. (1996); Face Stability in Slurry and EPB Shield Tunnelling; Tunnels & Tunnelling; December 1996.

Davis E.H., Gunn M.J., Mair R.J. and Senevirantes H.N. (1980); The stability of shallow tunnels and underground openings in cohesive material; Geotechnique 30 (1980), No. 4, 397–416.

Timoshenko, S.P. (1953); History of strength of materials; McGraw-Hill, London 1953.

Florence A.L. and Schwer L.E. (1978); Axisymmetric compression of Mohr-Coulomb medium around a circular hole, Int.J.Num.Anal.Meth.Geomech, Vol 2, pp 367–379, 1978.

McGough, P. and Williams, M. (2007);Geotechnical instrumentation and monitoring, Proceedings of seminar on New MetroRail City Project—Tunnelling and underground structures.

Johnson, I., Gibson S. and Aikawa, F. (2007); Segmental tunnel lining, Proceedings of seminar on New MetroRail City Project—Tunnelling and underground structures.

Williams, M. and Nobes, C. (2007); Compensation grouting for building protection, Proceedings of seminar on New MetroRail City Project—Tunnelling and underground structures.

SAO PAULO METRO PROJECT—CONTROL OF SETTLEMENTS IN VARIABLE SOIL CONDITIONS THROUGH EPB PRESSURE AND BICOMPONENT BACKFILL GROUT

Lorenzo Pellegrini ▪ SELI S.p.A.

Pietro Perruzza ▪ SELI S.p.A.

ABSTRACT

The respect of EPB reference pressure during TBM advance and hyperbaric activities, the complete and effective filling of the annular gap between tunnel lining extrados and the excavation section and a well organized monitoring system, to control surface settlements and ground distortion, are items of utmost importance in EPB tunnelling process.

The paper describes how these concepts have been applied while excavating in the difficult underground conditions found at Sao Paulo (Brazil) during the execution of the Sao Paulo Metro Line 4-Lot 1.

In this project, the annular gap has been filled by two components type, cement grouting. The back feed of the extensive monitoring campaign carried out to check the surface settlements during the excavation and hyperbaric maintenance operations will also be described in the paper.

INTRODUCTION

The execution of a Metro tunnel in a city such as Sao Paulo do Brazil is a big challenge either due to technical or economical reasons. Sao Paulo, at the moment, is the financial and economical capital of Latin America and contends to Mexico City and NY the lead as the most populated city of the continent.

The city has grown very fast in the last 40 years reaching the number of 18 ML people in between Sao Paulo and the suburb areas (Grande Sao Paulo). Although the first metro project dates late 60s, same as Mexico City, from there on Sao Paulo has seen the execution of only 61km of Metro lines compared to the 202km of Mexico DF.

The Line 4 project connects the west side of the city (Villa Sonia) with the centre (Luz) interconnecting tree of the four existing lines (see Figure 1); Blue (Line1), Green (Line 2) and Red (Line 3). The new line crosses line 2 (6m below it) close to Paulista station and crosses line 3 inside Republica station. Line 1 is intercepted at the far end, at Luz station.

The city is characterized by continuous follow of modern high buildings (over 40 stores) and old low private houses (two or three stores).

The Sao Paulo Metro Company awarded to the local Joint Venture CVA (Consorcio Via Amarela) the execution of 12,8km of tunnel, 12 stations, and 12 ventilation shafts. Half of the total length of tunnel is excavated by TBM while the rest by NATM. CVA is a Joint Venture leaded by CBPO Odebrecht and includes the fifth bigger company of Brazil. The JV has contracted SELI as supplier of the equipment and to assist the Company for the excavation of the tunnel by TBM.

Figure 1. Dense populated area—view from site installation

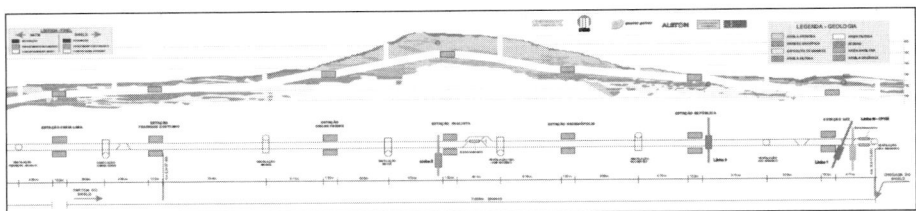

Figure 1A. Geological profile Sao Paulo Metro Line 4—Lot 1

GEOLOGY DESCRIPTION

The area interested by the Metro Line 4 Project (see Figure 1.A), is characterized by four different geological formations:

- Gneiss rock basement
- Tertiary sediments characteristics of Sao Paulo region
- Quaternary alluvial deposits
- Filling basement material

More specifically the TBM has operated in three different geological formations:

- Soil derived from the alteration of gneiss of the basement encountered in a limited area of the Faria Lima region
- Soil of the Resende formation characterized by high to medium plasticity, clay and sandy layers inter-bedded, incorporating gravel
- Soil of the Sao Paulo formation characterized by intercalation of red clay layers having medium to hard consistency and red fine to coarse sand layers

Soil Conditioning

As it is well know to TBM EPB operators, a good soil conditioning is of outmost importance in the operation of the TBM, either to easily muck out the excavated material (see Figure 2) and to maintain homogeneous EPB pressure in the tunnel face.

For the Sao Paulo project two special foams have been developed in the Innotek factory, in Italy, according to the specific geology to be encountered. The first one,

Figure 2. Conditioned muck at disposal area

Inntens TK70, have been used in sandy grounds, while the second, Inntens TK57, has been used where saturated clay were predominant.

During the excavation, the conditioning parameters, FER, FIR and conditioning agent concentration, have been modified according to the characteristics of the materials being excavated, in particular to their plasticity and fineness modulus.

Equipment Description

The execution of the Metro tunnel must avoid interfering with existing transportation system in order to minimize the risk of paralyzing the city. More specifically surface settlements and distortion have to be minimized. Therefore the choice of the excavation system has been oriented to an EPB TBM and the related simultaneous backfilling system to the bi-component cement grout.

The TBM selected is the Herrenknecht S-336 EPB, sizing 9,5m diameter. The backup and all the relevant auxiliary equipment are SELI.

The bi-component grouting system is composed by six eccentric screw pumps having a capacity of 200l/min each to supply grout liquid A to the tailskin injection ports and six eccentric screw pumps each with capacity of 20l/min to supply the accelerator (liquid B) to the tailskin injection port. All pumps are feed through FVD. The grout system can operate in manual, semi automatic or fully automatic mode, setting all relevant parameters on a touch screen panel.

A dedicated PLC, receiving the input commands from the grouting operator panel and the back feed information from the flow meters and the pressure transducers, installed on each grouting lines, ensure that the system works according to the values set. The grout is blended at the portal area by an automatic batching plant.

The component A and B of the grout are pumped from the portal area to the respective tanks on board of the TBM backup through 1.¼" and 1" pipe respectively. A maximum distance of 3,5km have been reached with a pressure of 80 bar measured at the portal.

Work Execution

To avoid surface settlements or heave, during the TBM advancing, reference EPB and grout injection pressure have to be strictly followed.

For the EPB pressure, two reference values have to be provided by the designer for the whole tunnel length:

- Ground stability pressure
- Minimum ground deformation pressure

The first value ensures the stability of the tunnel face and the second has the goal of minimize the deformations of the ground before and after the TBM approaches the specific location.

While the first point is the most important in terms of safety of the people leaving in the buildings on the surface and the users of all surface and underground structure, is the second point the most interesting in terms of fine tuning of the TBM.

The TBM has seven EPB sensors distributed at various heights inside the working chamber.

Playing with the screw conveyor, the advance speed, and the flux of soil conditioning agents, the TBM operator has to keep the pressure on sensor no.1 (the top one) as close as possible to the reference pressure. Moreover, he has to ensure that the distribution of pressure inside the working chamber is rising with the depth with a gamma factor ranging in between 1,2 and 1,4 ton/m^3.

The grouting pressure is a consequence of the EPB pressure. To ensure that the grout will fill completely the cavity behind the rings, a pressure slightly higher than the EPB pressure has to be applied.

Since this pressure is read at some distance from the delivery point, this concept has to be interpreted by the operator of the TBM according to the specific grouting system adopted. The reason is that grouting pressure reading may suffer other influences apart from the EPB pressure at the TBM working chamber, such as: grout flow (depending to the TBM advance speed); grout line length, section and obstruction; grout viscosity etc.

The TBM Manager, the Monitoring Manager and the Design Engineer have to continuously share all the information in order to optimize the EPB and grout pressure to be applied so to minimize the final ground distortion.

For the Sao Paulo Metro Line 4 Lot 1, the TBM parameters have been supplied by the Company Figueredo Ferraz. Latina Company is in charge of the monitoring, while SELI manages the TBM excavation.

Ground stability pressure at the tunnel face has been calculated following the method suggested by Anagnoustou and Kovari (1996) at each 50m, starting from the information available at the beginning of the project such as geological description, water table level and surface loads.

The foreseen ground deformation has been calculated starting from the methods suggested by Peck (1969). Peck method has been chosen due to its simplicity that allows a rapid and simple evaluation of the settlements in a continuous form all along the tunnel alignment. With the retro-analysis of the residual settlements behind the TBM shield, the method is continuously re-calibrated for the evaluation of the future distortion while the TBM advances in different geology.

Back Feed from Monitoring

While TBM advances in ground a huge amount of information is stored by the computer connected to the TBM PLC. Some of this information has to be re-interpreted with the support of the back feed of the surface monitoring to optimize the EPB operations.

To be able to do this work in real time we create a figure in the job site, the ATO or Site Technical Assistant. The ATO receives all the information from the TBM and the monitoring and makes the first screaming, 24hr per day.

As soon as the ATO detects a deviation from the normal behavior of the TBM or monitoring parameters or a deviation from the designed parameters, he informs immediately the TBM Manager.

The TBM Manager will immediately take the pr-established actions to bring the situation inside the normal range of work (e.g. lifting up EPB and/or grouting pressure). If these actions will not succeed then the TBM Manager contacts the Designer that will recalculate the proper pressures and foreseen settlements on the base of the new information available.

Hyperbaric Activities

As well as reference pressure, need for the tunnel excavation, the Designer supplies the relative reference pressure for hyperbaric activity. Hyperbaric operations are needed in order to check the cutter head conditions and to carry out the maintenance of the cutting tools.

SELI's policy is to keep continuously under control the cutter head condition. For this reason a very rigid plan of programmed maintenance has been installed. This plan foresees two hyperbaric interventions per week to control the CH conditions.

Programmed hyperbaric interventions where carried out usually every Tuesday and Friday. Dates may vary according to the geological ground conditions of the particular location or closeness to dangerous structures, such as tunnels, high buildings or deep foundations. Up to date 63 hyperbaric interventions have been carried out with pressure ranging in between 0,8 and 2,5 bar.

Thanks to this plan we were able to keep the cutter head in perfect conditions reducing to zero any time lost associated to CH wearing. We believe that the exceptional performance achieved by our TBM (such as 80mm/min advance speed) have been reached also thanks to the strict plan of CH maintenance.

Monitoring Instrumentation

The instrumentation chosen for the monitoring of the ground response to the TBM advance, have been the most simple and with faster reply as possible for an urban environment such as Sao Paulo city. Therefore such instrumentation consists in:

- Ground marker
- Tassometers
- Convergence anchoring pin and levelling
- Clinometers
- Inclinometer
- Piezometers

Monitoring sections have been located at each 25m of the tunnel alignment, for a total number of 200 sections. Four different sections have been prepared, called type A, B, C and D. Sections type A and B are used to monitor the ground as well the aerial and underground structure. Sections type C and D are used to monitor the ground behaviour only. Sections A, B and C give information of the average behaviour of the ground all along the tunnel length. Section D is applied in the critic zones where a more detailed monitoring campaign is recommended.

Section A consists of five Ground Markers M (see Figure 3), three Tassometers T and four Convergence Pins P; B consists of three M, one T, and four P; C consists of three M and finally sections D consists of two T only.

Frequency of Readings

Sections type A, B and C frequency of reading is shown in the following diagram no.1. The point "zero" represents the position of the TBM cutter head and the other numbers represents the distance in meters ahead and behind the TBM.

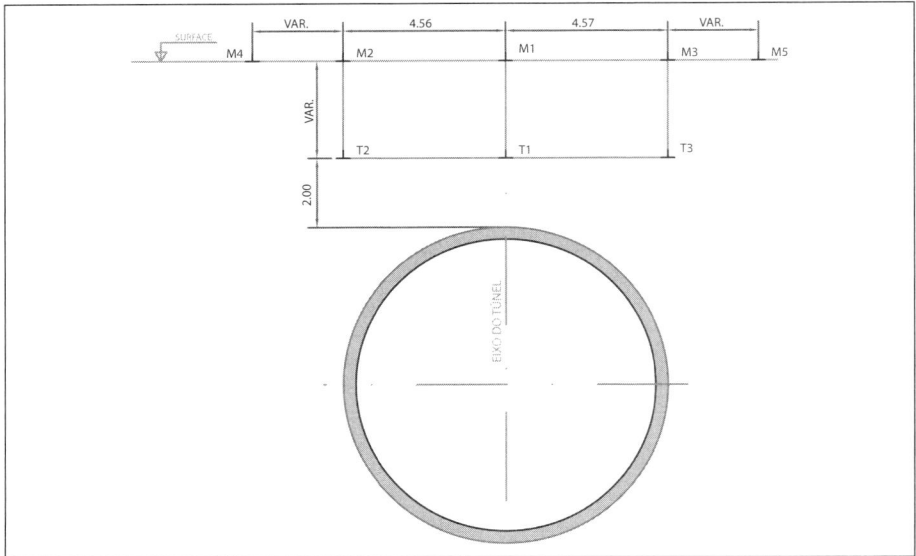

Figure 3. Section type A

Different frequency may be applied according to the needs of special situations such as detected ground movement, proximity of infrastructure or buildings etc.
Section type D frequency of reading is shown in diagram no.2.

Alert Levels

Four reference levels of alert have been set according to settlement detected by the surface monitoring. At each level different pre-established interventions have to be followed until the situation is considered under control again. The interventions going from lifting up, step by step, the EPB and grouting pressure (without stopping the mining activities) to a more drastic solutions such as stoppage of the excavation and post grouting or soil consolidation interventions.

At the beginning of the project an average settlement corresponding to a lost of ground volume ranging in between 0,3 and 0,7% have been considered absolutely acceptable. It is clear that depending on the local geology, water table level, and tunnel coverage, different settlements may occur for a same value of volume lost. First level of alert corresponds to settlements induced by a volume lost higher than 0,7%. Second, third and fourth level correspond respectively to volume lost of 1%, 1,5% and 2,5%.

At the present stage of the work average settlements corresponding to 0,17% of volume lost have been recorded. Average settlement is 6 mm while average distortion is 1:1700.

The normal behaviour of the ground observed while the TBM passes through a monitoring section has been as follows. Starting from 20m ahead of the TBM cutter head, the soil starts to lift few decimals of mm. Once the forward shield reaches the monitoring section is recorded the faster settlement which occurs until the tail skin after reaches the same section. At that point settlement tends to stabilize (see Figure 4).

A completed stabilization of the settlements is reached about 60m behind the cutter head position, two or three days after the TBM passed below the specific section.

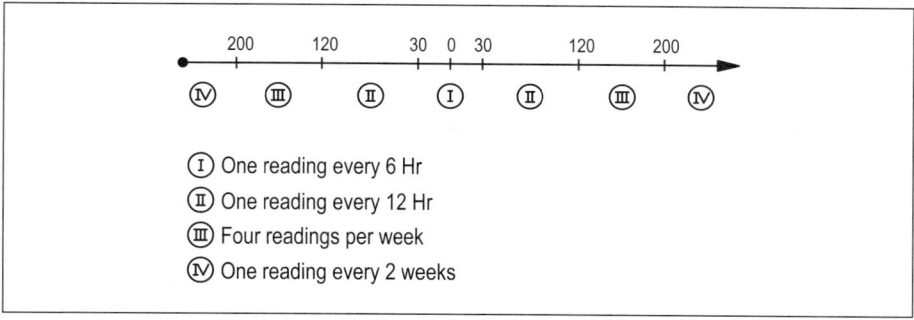

Diagram 1. Frequency of reading for section type A, B, C

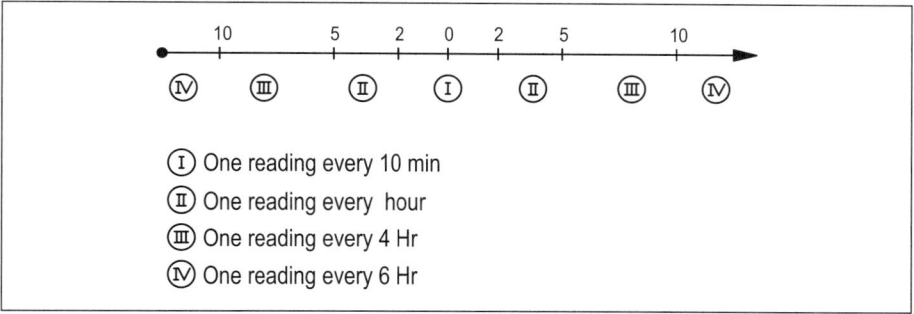

Diagram 2. Frequency of reading for section type D

Bi-component Grouting System

The complete and effective filling of the annular gap is of utmost importance in EBP tunnel excavation. The complete annular gap filling mitigates the tunnel excavation induced settlement and is also the mechanism whereby ground and other loads are applied in controlled mode to the tunnel lining. The continuous backfilling provides an impermeable membrane all around the tunnel lining that, jointly with segment gaskets, enhances waterproofing of the tunnel.

Simultaneous two component backfilling grouting, abbreviation 2CBG, from injection pipes located on the TBM tailskin, has great effectiveness. This system uses two different liquids: liquid A composed by cement, bentonite, water, and retarder; and liquid B or liquid sodium silicate.

Few seconds after mixing the two liquid components, a semi-solid jelly consistency material, in plastic state, is formed. The cement grout keeps this jelly state for about half an hour, then start hardening and reaches 0,05–0,1Mpa in one hour. The strength at this stage depends on cement and sodium silicate dosage.

Grouting pressure starts acting on the extrados of lining immediately after the passage of the tail shield. Thanks to the liquid/plastic state of the grout mixture the grouting pressure distribution becomes quickly uniform moving away to the grouting ports.

After hardening, the grout holds the earth pressure and the ground water pressure, and conveys the load to the tunnel lining.

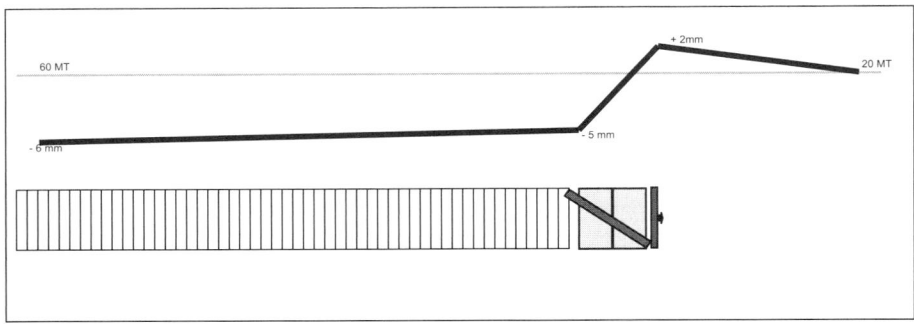

Figure 4. Surface ground movement while TBM approaches a monitoring section

Comparison Between The Mortar Backfilling and Two Component Backfilling Grouting

For many years, in Europe, where EPB TBMs were used, traditional mortar cement has been adopted. Even if there is no doubt that traditional mortar can fulfil its task, we consider that 2CBG bring more vantages. Below are listed some of them compared to traditional mortar.

The mortar backfilling grout requires 4 to 5 hours to set and takes long time, 12 to 24 hours, to achieve a compressive strength between 0,05 and 0,1MPa. During this time, the grout can be washed away by groundwater. Meantime the precast lining floats inside the mortar grout and it can easily lower especially in permeable soil above the water table.

The 2CBG take 6 to 12sec to gel and approximately 1hour to harden reaching up to 0,05 to 0,1MPa; the gel time can be chemically varied directly onboard the TBM, changing the percentage of silicate injected. The gelled material gives an immediate support to precast lining avoiding sinking or floating of the ring, also in impermeable ground.

Traditional mortar requires a sophisticated batching plant for its preparation, heavy wagons for the transport, and a double system of pumps to transfer the mortar from the wagon to the storage tank and from the tank to the tail skin. The mortar easily clogs the pipes of the tail skin, so the lines have to be cleaned frequently. It loses its workability easily even using chemically retarded, especially during prolonged excavation stops, therefore it cannot be kept inside the storage tank for long time.

The 2CBG system requires a very easy colloidal mixing plant for the component A only. The two components are transferred separately directly from the external plant to the storage tank onboard the TBM through two dedicated pipelines.

The cement/bentonite slurry is injected inside the annular gap by six screw pumps assembled under the storage thank, the accelerator is added to component A by six screw pumps as well. The component A doesn't clog easily the lines and the pipes are easy to be cleaned by water. The cement slurry is chemically retarded and can be kept inside the storage thank for more than 8 hours without sedimentation.

Pressure control of traditional mortar grout is very difficult to be carried out, since pistons pumps always make pecks of pressure while pumping. With 2CBG the flow is absolutely constant and so the pressure read at the pressure transducers.

Using the traditional mortar grout is difficult to control the ground volume lost due to TBM advance, while with the 2CBG, in the Sao Paulo Metro Line 4 Project, the average ground volume lost has been kept lower than 0,4%.

Choosing the Back Filling Grouting System

After the experience had in Vancouver Metro Project and the good results obtained in terms of filling ratio, residual settlements and ground volume lost, SELI proposed to replace the traditional mortar backfilling system with two component backfilling grout system for the Sao Paulo Metro Line 4 Project.

In Sao Paulo the 2CBG system looked even more appropriate to be applied due the local geological formations, the presence of a watertable constantly above the tunnel crown, the overburden and the importance of the structures along the project and, consequently, the low residual settlements and distortions to be maintained during the whole TBM progress. Some times technical choices, such as this one, are in conflict with project budgets.

Actually, after laboratory trials and the search of the proper materials on the local market, we came to a grout materials prize higher than the traditional mortar. But grouting costs is not only to be considered as material costs. Comparing equipment costs and more specifically TBM excavation time lost, due to grout lines clogging, of two precedent SELI's Projects (Athens Attiko Metro and Vancouver Projects), we find out that the 2CBG is far better than traditional mortar grouting even in terms of final project cost.

The use of this type of grout provides greater flexibility to the TBM operator. Varying the ratio of two components A:B, the setting time of the grout can be varied directly onboard the TBM giving to the operator the possibility to control how the grout material will penetrate the annular gap and the ground.

Grouting Process

As we said the effectiveness of 2CBG system, filling the ring annulus void, is higher than using the traditional mortar grouting method as the shorter setting time of the first helps to minimize the lost of grout diluted or segregated by the ground water.

The ring annulus void is grouted through injection ports located at the end of the TBM tailskin perimeter using constant pressure grout injection. On our TBM there are six available injection ports in total, and during the regular mining process all the ports are used.

The injection of component A starts with the beginning of the TBM advance cycle. The retarded grout (liquid A) is pumped via the 2" line from the storage tank, onboard the backup, to the TBM tailskin.

Following a minimal delay (e.g. 10 seconds time) the accelerator (liquid B) is pumped via the 1", reduced to ⅜", line to the injection nozzle. The accelerator flows through separate hose contained in the elliptical shape pipe of the component A in the tailskin. At the end of liquid B hose is located the injection nozzle. It is only at that point, located at the tailskin far end that the accelerator and retarded grout get in contact mixing together forming the gel.

2CBG Specification

The 2 component backfilling grouting performances haven't been standardized yet. According to the experience held in other projects we have been able to fix some characteristics and parameters that we want to be respected for the fresh component A and the gelled grout. The sodium silicate characteristics vary according to the gel time required the final compressive strength, the pumping distance and the material available locally.

The component A has been designed to match the following characteristics:

- Efficient flow ability
- Early generation of strength

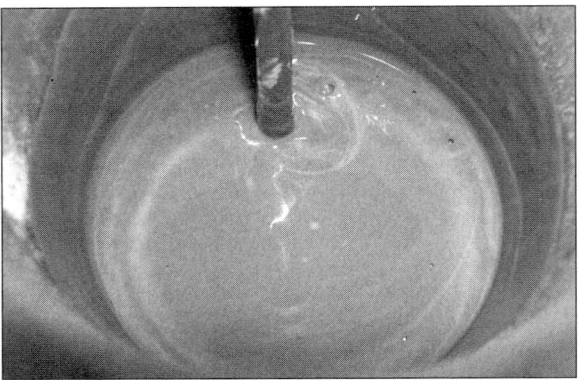

Figure 5. Component A after mixing

- Waterproofing
- Resistance to segregation
- Ability to be transported (pumped) at long distance
- Resistance to dilution by groundwater
- Very low volume reduction

The following values have been considered crucial to be achieved for the correct backfilling grouting:

- Marsh viscosity > 35sec <45sec
- Lost of viscosity after 4h < 0,5%
- Bleeding:
  - After 1h < 0,5%
  - After 2h < 1%
  - After 24h < 4%
- Initial setting time > 24h
- Gel time > 6sec <12sec
- Sodium silicate dosage > 6% calculated on the mix weight
- Compressive strength:
  - After 1h > 0,1Mpa
  - After 24h > 0,5Mpa
  - After 28days > 2,5Mpa

Mix Design for Component A and Accelerator Characteristics

The mixture of the tixotropic gel grout has been designed for the characteristics above mentioned using cement, bentonite, liquid retarder and sodium silicate all available on the local market (see Figure 5).

Cement. Different classes and types of cement have been tested with particular interest in finding the best solution in terms of gel time, early and late compressive strength, sodium silicate consumption and, of course, prize. Cement type V ARI RS from Votoran has been chosen; this cement is early strength sulphate resistant Portland cement.

Bentonite. Scope of the bentonite is to stabilize the slurry, reducing the bleeding, and increasing the viscosity. Therefore drilling grade bentonite (following the API specification 13/A—ISO 13500) has been added in the component A blend.

Liquid retarder. In order to extend the component A setting time a natural sugar solution has been added in the mix as retarder.

Water. Water from local artesian wells has been utilized.

Accelerator. To gel the component A, achieving the specifications previously indicated, we have tested many liquid sodium silicate accelerator.

When it is added to Liquid A, the sodium silicate quickly causes the jellification of the water content in the mixture A itself. The gelling time, that is the time needed to form the gel within the mixture A, depends on various factors, where the most important are: the cement type, the water/cement ratio, the amount and type of silicate used and the humidity and temperature of the environment.

For our mixture we found out that the proper sodium silicate has the following characteristics:

- Density @ 20°C,   38° Baumè
- Viscosity @ 20°C   40 Cp
- Content of solid   34,4%
- SiO_2/Na_2O ratio   3,22
- @ 20°C   1,355 t/m^3

Trial Mix Test Results

The tests performed in laboratory are not standardized yet and they are empirical but at the same time are very practical and give the idea of what happen inside the annular gap when the 2CBG is injected.

Here below are reported the results obtained using the materials adopted for the component A of the Sao Paulo Metro Line 4 Project:

- Marsh viscosity—viscosity of component A, measured @ 20°C with Marsh cone having an opening of 4,76mm, range between 38 and 40sec
- Bleeding—bleeding measured in the laboratory, using a 250cc glass cylinder (as Figure 6 shows), gave the following results:
  – After 1h is none
  – After 2h < 1%
  – After 24h < 4%
- Setting time—no component A setting has been measured in 24h
- Gelling time—the gelling time has been determined just mixing the component A with component B sodium silicate, by two plastic glasses (see Figure 7). The procedure has been repeated with different accelerator percentage until the specified gelling time has been achieved (see diagram 3). Eight to nine seconds gelling time have been reached adding 7% of sodium silicate to the component A.
- Compressive strength—due to the difficulty to measure the compressive strength of the jelly material and to the lack of specifications, we have utilized a no-standard method that is empirical but effective. In order to be able to measure a compressive strength of 0,1Mpa, we have modified the Vicat apparatus applying on the top of it a weight able to exercise directly on the consistency plunger and consequently on the component A being hardened, a pressure equivalent to 1kg/cm^2 (see Figure 9). 0,1Mpa compressive

Figure 6. Component A bleeding after 24h

Figure 7. Component A gelled in the glass during the mixing

strength is achieved when the Vicat consistency plunger doesn't penetrate the sample as shown in the Figure 10.

To verify the compressive strength after 2Hr and 28 days, cubic samples, 100×100×100mm, have been prepared and cured at 20°C and >90% humidity as shown in Figures 11 and 12.

Summarizing, the required compressive strength values have been achieved adding to the component A 7% of 38° Baumè sodium silicate in weight. Following are listed the compressive strength obtained:

- after 1h   > 0,1Mpa
- after 2h   0,7Mpa
- after 28days 3,8Mpa

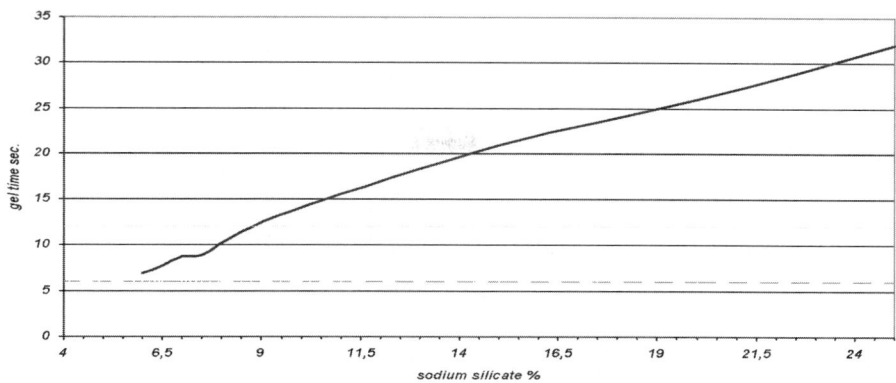

Diagram 3. Grout gel time vs. percentage of sodium silicate

Figure 8. Component A gelled

Figure 9. Modified Vicat apparatus

Figure 10. 0,1Mpa compressive strength is achieved

Figure 11. 2CBG hardened removed from the glasses

Figure 12. 28 days compressive strength

Figure 13. Hardened backfilling cores

Figure 14. Component A turbomixer

Two Component Batching and Delivering Plant

Component liquid A of the grout is batched on the surface in the dedicated mixing plant designed by SELI and fabricated by Innotek Italia (see Figures 14 and 15).

The component A is pumped by duplex pump into a tank onboard the TBM backup gantry, through a dedicated 1¼" pipeline installed along the entire tunnel length (see Figure 16).

The sodium silicate (accelerator liquid B) is stored on the surface (as well onboard the TBM) and pumped by single stroke vertical pump inside the tunnel to the TBM backup gantry through a dedicate 1" pipeline installed along the tunnel (see Figure 17). Two more pipelines 1¼" and 1" are installed as spare.

The two components are mixed together only at the six injection ports in the tail-skin of the TBM (see Figure 18).

Figure 15. Component A agitator

Figure 16. Pumping station for component A

Figure 17. Component B—sodium silicate—pumping station

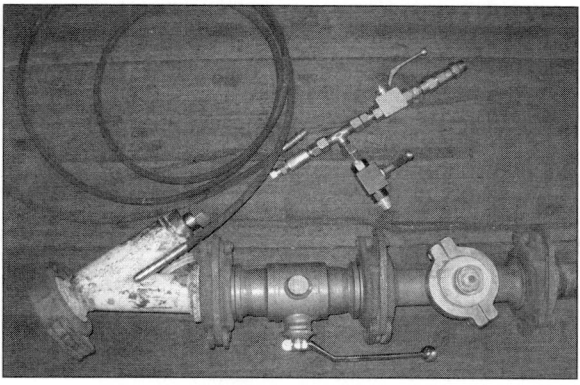

Figure 18. Component A and B grouting line plus flushing port

CONCLUSION

SELI Company have been contracted by CVA (Consorcio Via Amarela) Joint Venture to supply and manage a 9,5m diameter EPB for the excavation of Lot 1 of the Sao Paulo Metro Line 4 Project.

An extensive monitoring system both for the TBM parameters and for the ground distortion has been implemented. Due to the restriction of settlements imposed by the particular urban environment such as the city of Sao Paulo, SELI decided to install onboard the TBM the two component back filling grouting system.

The system, already tested in previous projects run by SELI, results much better in terms of final soil volume lost and so in terms of residual settlements.

Thanks to the rapid share of information coming from the TBM and the back feed of the monitoring, the site personnel could quickly take all the proper actions to maintain the final ground settlements and distortion inside the foreseen limits.

PLANNING AND PREPARATION FOR TUNNELING AT BRIGHTWATER WEST

Mina M. Shinouda ▪ JAY DEE Contractors Inc.

Glen Frank ▪ JAY DEE Contractors Inc.

Greg Hauser ▪ JAY DEE Contractors Inc.

ABSTRACT

Brightwater West (BT4) represents the state of the art in utility tunneling, namely a long relatively small diameter, soft ground tunnel, with no intermediate shafts, under significant groundwater pressures, requiring very precise survey control in order to hit a small exit window. The tunnel is over 6.4 km (4 miles) in length, is expected to encounter active earth pressures of over 5 bars in glacial geology, and a planned hole through into a shaft eye constructed at 45.7 meter (150 feet) below the water table. Despite the fact that all of these challenges have been previously overcome in larger diameter tunnels, the technical solutions are much more challenging in a smaller diameter due to the lack of space available for implementing the equipment and techniques required. In addition to the technical challenges, the project is faced with many of the constraints designed to minimize the impact of these types of projects on the neighboring community and environment. The work shaft for BT4 is within 15.2 m (50 feet) of Puget Sound and is on a site that was previously occupied by petroleum storage tanks, and all of the tunnel muck has to be removed from the site by barge as trucking is not allowed by contract. The paper addresses the work performed during the preparation for tunneling stage of this project both in assembling the TBM and in preparing the site for its arrival.

INTRODUCTION

This paper describes the work required to prepare for the tunnel excavation on Brightwater West, including the preparation of the launch shaft and main work site as well as the composition of compromises that made up the Tunnel Boring Machine (TBM) that is excavating the tunnel.

PROJECT DESCRIPTION

Brightwater West Project (BT4) is part of the overall Brightwater Conveyance System in King County, Washington. The project was awarded to Jay Dee–Coluccio–Taisei JV (JCT) at the beginning of 2007 and the Notice To Proceed (NTP) was issued on February 20th, 2007. BT4 comprises of 6,400 m (21,000 ft.) of 3.96 m (13 ft.) finish tunnel, 152.4 m (500 ft.) of 2.1 m (84 in.) microtunnel, sampling faculty structure, and a metering vault. The portal jobsite is located in the city of shoreline adjacent to the Puget Sound, Paramount Petroleum, and BNSF railroad. The project is expected to be completed in early 2010. Figure 1 shows the location of BT4 in respect to the Brightwater Conveyance System.

PLANNING AND PREPARATION FOR TUNNELING AT BRIGHTWATER WEST 1155

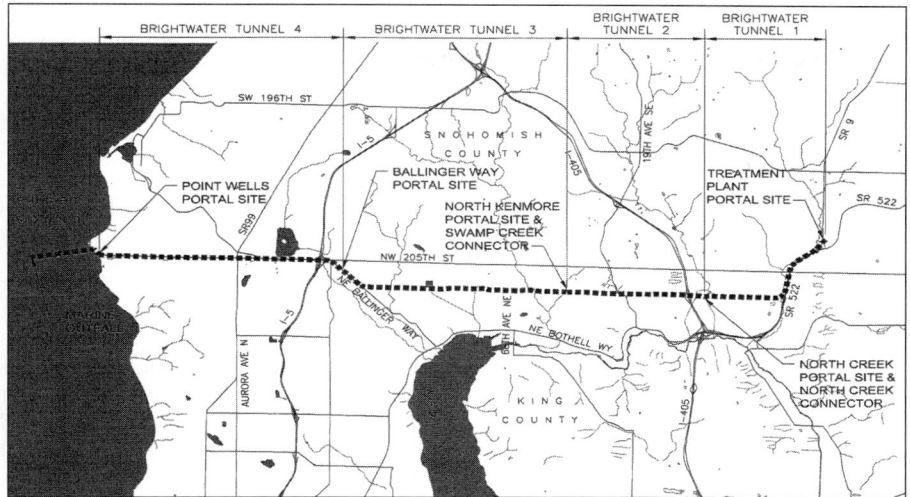

Figure 1. Brightwater conveyance system

Geologic Conditions

The project is located within the Puget Trough, which is a structural basin located between the Olympic and Cascade Mountains, formed by the Juan de Fuca oceanic plate being thrust beneath the North American Continental plate. The bedrock contact is over 305 meter (1,000 feet) below the surface, and is overlain by glacial and non-glacial sediment through which the tunnel will be constructed.

The geologic history of the project site is dominated by at least six different episodes of advance-retreat cycles of continental glaciers during the Pleistocene era. Each of these glacial advances partially eroded the pre-existing stratigraphy, and deposited a fresh sequence of sediment.

The stratigraphy along BT4 is complex due to the multiple erosion/deposition cycles that have occurred during the time that these materials were at the surface, and the orientation between the depositional glaciers, rivers, streams, and lakes, and the tunnel alignment, which is generally perpendicular.

During the last glacial period, which ended about 10,000 years ago in the project area, large quantities of sediments ranging from clay-sized to large boulder-sized were deposited in the Puget Trough. Each depositional event was followed by one of erosion during which large and small channels, ravines and valleys were incised into the previously deposited sediments. Subsequent deposition either filled or partly filled those channels, ravines and valleys, then the process was repeated again and again Consequently, many if not all of the formations are only remnants, and refilled channels are common.

The depositional environment for the soils expected to be encountered during the excavation of BT4 is of two basic types. The soils expected in the tunnel envelope west of Station 636+00 will consist of alluvial sands, gravels, silts and clays, and lacustrine clays deposited during the interglacial period between glacial advances. The soils expected to be encountered east of Station 636+00 are primarily glacial and glaciofluvial silts, clays, and sands. See Figure 2.

The soil deposits at tunnel depth are, for the most part, over 45.7 meter (150 feet) below the current ground surface, and all of the soils have been subjected to additional consolidating pressures from the weight of thousands of feet of overlying ice in the recent geologic past. The alluvial clays and silts are generally very stiff to hard, and the glacially deposited clays and silts are harder yet. The sands, which are typically fine-grained and the gravels, which are rare, are generally dense to very dense.

The Geotechnical Baseline Report for the project indicates that there will be 70 boulders encountered by the TBM. The majority of these boulders are expected during the portion of the tunnel where the depositional environment transitions between Alluvial/Lacustrine to Glacial/Glacialfluvio.

Groundwater. The glacial advance and retreat coupled with the interglacial periods of alluvial erosion and deposition as well as lacustrine deposition, created a sequence of aquifers and aquitards with varying thicknesses and lateral continuity.

The western half of the project appears to contain three basic aquifers. The upper aquifer is primarily coarse sand and gravel and is an important source of water for both private and municipal use. The lower aquifer is also primarily coarse sand and gravel is hydraulically connected to Puget Sound. The tunnel envelope is contained in the middle aquifer, which is primarily made up of fine sand with some lenses of coarser sand and gravel. The groundwater heads at springline for the first (western-most) half of the tunnel are expected to be range from 6 to 15 m (20 to 50 ft), with the majority of the 3,000 lineal meters (10,000 feet) having hydrostatic pressures below 15 psi.

An abrupt change in the groundwater head is expected approximately half way through the tunnel drive. The hydrostatic pressures are expected to rapidly increase from approximately 15 psi to 75 psi as the tunnel envelope transitions into an aquifer dominated by the much more complex glacial deposits. The groundwater head is expected to be above 45.7 meter (150 feet) for the majority of the second (eastern-most) half of the tunnel drive.

Ground Classification. Approximately one-half of the drive is expected to be in fine sand with lenses of silt and/or clay and is expected to behave as Flowing Ground. Approximately one-third of the drive will be very stiff or hard clay with thin lenses of fine sand and is expected to behave as Slow to Fast Raveling, although some areas should behave as Firm especially where the hydrostatic pressures are low. The remaining portion of the drive is expected to be fast raveling in the low hydrostatic head portion of the tunnel and flowing in the high hydrostatic head portion of the tunnel.

Railroad Crossing Within 100 Feet of Launch Shaft

The project had a very challenging start where the launch face station was about 30.5 m (100 ft.) away from the Burlington Northern Santa Fe (BNSF) railroad right-of-way (ROW), also, the portion of tunnel under the railroad was on a 304.8 m (1,000 ft.) radius curve. Moreover, it was not possible to assemble the entire TBM before launch due to shaft length restrictions, thus the assembly of the TBM had to be staged as the mining progressed under the BNSF ROW.

BNSF has a very tight contract with King County (KC) which imposes a significant liquidated damage amount in case of train delays. AS a result, the contract documents have very strict criteria for any contractual activities under or within 7.6 m (25 ft.) of the railroad. Three major components of these criteria are working 24 hours while advancing under the railroad ROW, maximum permissible track movement of 6.35 mm (one quarter of an inch), and the presence of a railroad flagger at all times.

The crown of the tunnel was about 6.1 meter (20 feet) below the tracks which was a borderline of being a critical depth. An extensive array of settlement monitoring points was installed on the tracks and ROW, this array was monitored every two hours during mining. Any personal working within the railroad ROW had to take BNSF specific safety training.

Although JCT has an experienced crew, any new TBM has its unique maneuvering behavior. Due to the fact that the TBM was not fully assembled in the shaft, some of

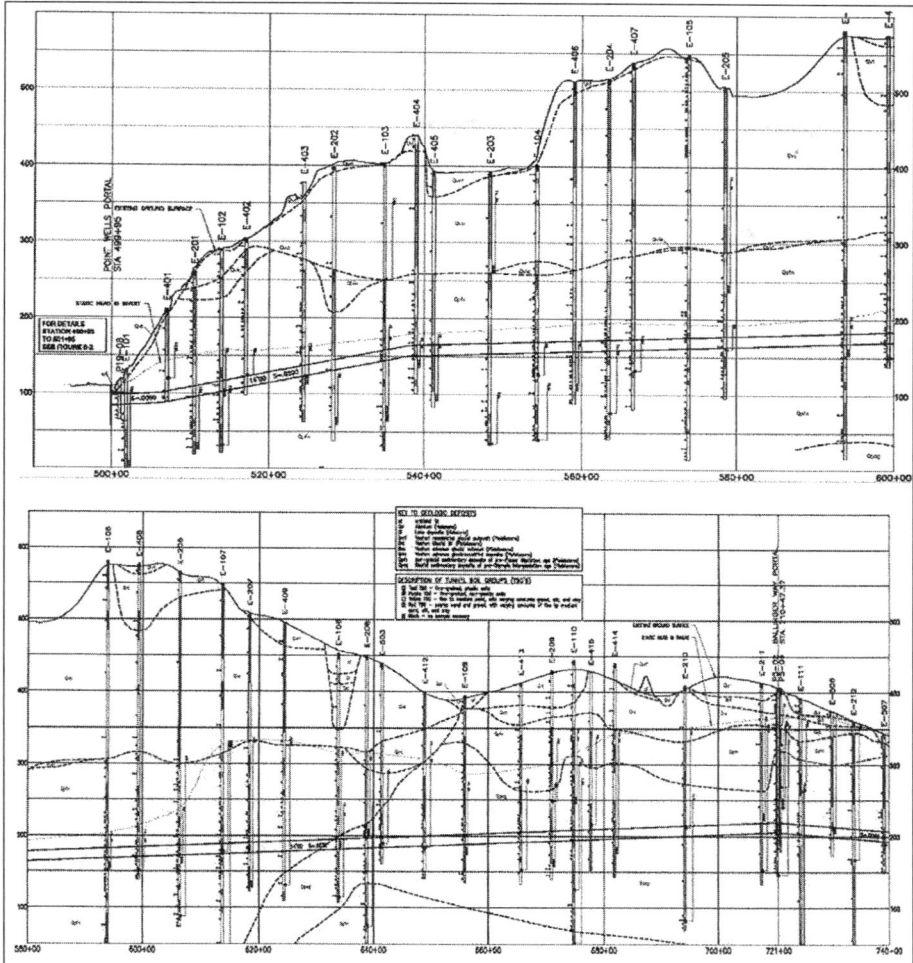

Figure 2. Brightwater West geological profile

the systems were not fully tested before launch one of which was the tail void grouting system. Never the less, passing under the rail road was successful with very minimal ground movements that were well under the permissible values.

No Real Possibility for Intermediate Shafts

The tunnel alignment starts at Point Wells portal shaft and goes eastward passing under BNSF railroad, residential area, local highway (US 99), commercial area, Interstate highway (I-5), and ending at Ballinger Way shaft. There are no intermediate shafts allowed in the contract, in actual fact, there is no real possibility in finding a suitable location for any shafts along the whole alignment. This fact introduced several challenges to project, including survey accuracy, tunnel ventilation, and cutterhead maintenance.

The exit window at Ballinger Way shaft has a very small tolerance of 7.6 cm (3 in.) which requires very accurate survey. Since there is no possibility in tying the tunnel survey to the surface survey along the alignment to create a closed loop traverse, a repeated open loop traverse cycles are being performed regularly from the launch shaft to the heading to weed out errors. In addition, a Gyroscope is being employed to check the tunnel bearing.

A positive displacement ventilation system is used to supply the heading with fresh air, comprising of 122 cm (48 in.) bagline and two 100 hp fans at the shaft with a booster fan to be installed at the midpoint of the drive. An additional smaller ventilation line will be installed from the shaft to the California switch in the middle of the tunnel to compensate for the extra locomotive.

Although it is a common practice to change the worn-out cutting tools (rippers and disk cutters) and scrappers from the inside of the TBM, the nosecone usually has to be replaced from the outside. Since there is no intermediate shaft, this was not an option, thus extreme measures were taken to allow changing the nosecone from the inside of BT4 TBM. While being quite a cumbersome task, it gives the assurance of the possibility of maintaining the entire face of the cutterhead if needed.

High Probability of an Intervention at Above 4 Bar

JCT will need to access the cutterhead chamber to change teeth and do required maintenance at least every 1,524 lineal meter (5,000 feet) of tunnel advance. Since over 2,740 lineal meters (9,000 feet) of the tunnel is more than 30.5 meters (100 feet) below the water table at least one of these "interventions" will be done with very high static water pressures. It is anticipated that this (these) interventions will require hyperbaric pressures of above 50 psig, which is the highest pressure allowed for hyperbaric work in the state of Washington.

In order to ensure that any required work can be completed, JCT was required to obtain 5 variances from the Washington Administrative Code (WAC), as well as mobilize and install a significant amount of infrastructure which would be required to support hyperbaric work at such high pressures. This infrastructure included the following:

- A medical lock manufactured to allow an injured workman to be treated by a doctor under pressure.
- A transfer lock designed to transport an injured man and an attendant from the heading to the medical lock under pressure.
- A 4 chamber lock integral to the TBM in the heading, allowing hyperbaric work in the heading to progress concurrently with decompression of the previous crew.

Challenging Exit Shaft Scenario

The exit shaft for the TBM was constructed by the Central Contractor at the Ballinger Way Portal. The shaft is 60.9 meters (200 feet) deep with static groundwater levels of more than 45.7 meters (150 feet) above the tunnel invert and was constructed using ground freezing as the primary means of earth support. At the time of this writing it is expected that the freeze plant will be de-energized and the ground allowed to thaw prior to the arrival of the TBM. The unknown expected behavior of the ground that has been frozen then thawed is a significant concern for JCT in planning the final few feet of the BT4 tunnel.

JCT will design and manufacture a seal to facilitate the TBM entering this shaft, which will require that the TBM enters the shaft within 7.6 centimeter (3 inches) of the planned location. Due to the fact that there are no intermediate shafts nor provisions for checking the survey from the surface anywhere along the alignment, JCT will utilize several gyro based survey to get the most accurate location of the TBM possible prior to reaching the exit shaft.

No Trucking of Tunnel Muck

With few exceptions, excavated spoils from the jobsite are not permitted to be trucked out. The contract mandates the haul out of spoils by barges via the adjacent Puget Sound. Since there was an extensive amount of work needed to be done to get the barge hauling system in place, a limited number of trucks was allowed at the early stage of the project, but after a specific cutoff date, only contaminated spoils were to be trucked out.

An existing pier and wharf, owned by Paramount Petroleum, was the only way available to load the barges. An elaborate conveyor belt system was developed to transfer the muck from the jobsite to the barge docking location at the wharf. Some restrictions were imposed by Paramount Petroleum for the use of their pier and wharf. These restriction and the details of the conveyor belt system are presented in a later portion of this article.

The hauling of the spoils by barges significantly limited the number of disposal sites that can be used for dumping the materials. Also the number of days available for barge loading was limited by the Puget Sound tides, weather, and Paramount Petroleum marine traffic, this mandated that the muck bin would have enough capacity to hold a volume of muck produced by a 7 to 10 days of mining.

LAUNCH SHAFT SITE DESCRIPTION

The Point Wells Portal where the main work shaft is located on an approximately seven acre site bounded by Puget Sound on the West and South, by the Burlington Northern Santa Fe (BNSF) north/south mainline railroad on the east, and by the Point Wells bulk fuel terminal to the North.

Site Constraints

The BT4 portal jobsite is separated from the entry road by BNSF railroad with the elevation of the road about 9.1 meters (30 feet) higher than the general site elevation. The only access to the Jobsite is an elevated trestle located at the north end of jobsite. This trestle is bridging over the railroad tracks and sloping down to the site elevation. The trestle is owned by Paramount Petroleum (PP) and also serves as an access to their facilities, an easement has been obtained by KC to use it for the duration of the project. PP falls under the jurisdiction of the Department of Homeland Security, thus a security gate is located at the entrance of their property before the trestle. All JCT employees have to obtain access cards for that gate and truck traffic is strictly regulated. The contract limits the loads on the trestle to one truck at any given time.

The main mining site is adjacent to a residential area which was a fact that obligated KC to mandate some restrictions to the jobsite activities. To eliminate any disturbance to the residents, the contract dictates the working hours, truck traffic hours, noise levels, and jobsite lighting.

The job site working hours are 7:00 AM to 10:00 PM during weekdays and 9:00 AM to 10:00 PM on Saturday. The only exception to that is tunneling and tunneling support services which is allowed 24 hours Monday to Saturday. No work is allowed on Sundays except for mining under the railroad.

Truck traffic was allowed during the hours of 9:00 AM to 4:00 PM and 6:00 PM to 10:00 PM on weekdays and 9:00 AM to 10:00 PM on Saturday. Due to the traffic congestion in the Seattle area, these truck traffic hours proved to be challenging especially

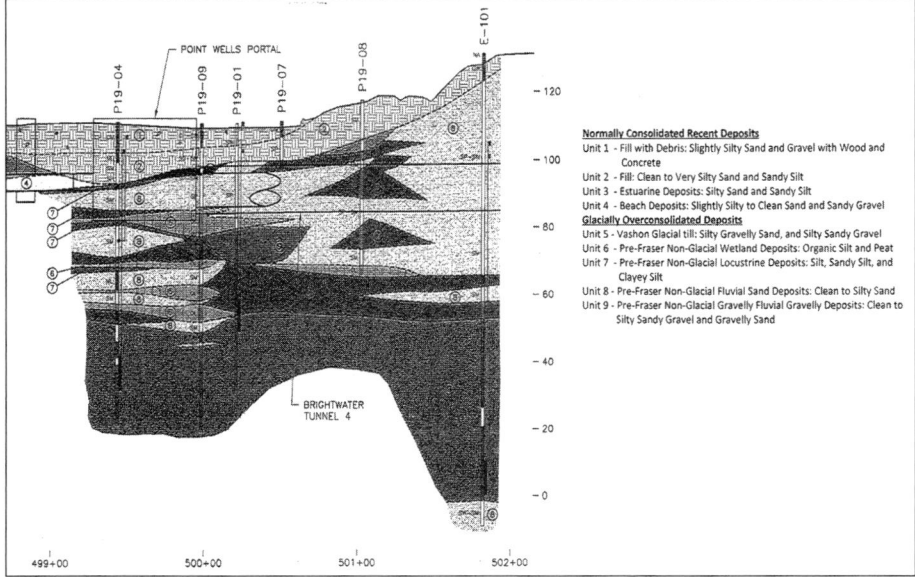

Figure 3. Point Wells Portal geological profile

since the contract does not allow for truck queuing on public roads anywhere within four mile radius of the jobsite.

The contract mandates strict standards for the noise levels reaching the neighboring residents and required noise readings for each major construction activity or after the addition of major equipment. The specified standards were hard to meet due to the preexistent railroad noise levels. A three-dimensional computer model has been developed by JCT, using SoundPLAN software, to identify a solution for this problem and few noise mitigation methods are implemented. A six meters (20 feet) high sound wall is installed at the east and south sides of the jobsite, work is scheduled to avoid simultaneous activities generation high combine noise, strobe light and flag person are used to backup equipment instead of backup alarms, and equipments are not allowed to be left idle for more than 5 minutes.

In order for the residents not to be impacted by light pollution, all lights are shielded, exterior construction lighting is directed downward, light poles are limited to 9.1 meters (30 feet), and pole mounted luminaries are independently controlled to turn off as needed.

A truck wheel wash is installed at the exit of the site to avoid any debris from ending up on the neighboring roads.

Geologic Conditions

The geology of the Point Wells Portal consists of approximately three meter (10 feet) of fill overlying approximately 13.5 meters (44 feet) of dense fine silty sand with layers of stiff to hard silt and peat, with a continuous 1.2 meters (4 feet) thick layer of silt and peat at the bottom. This sand and silt unit is underlain by a coarse sand and gravel. This sequence is shown in Figure 3.

The fill and upper 12.2 meters (40 feet) of the sand and silt unit act as a single aquifer in which the water table is typically at the surface. The bottom 1.2 meters (4 feet) of

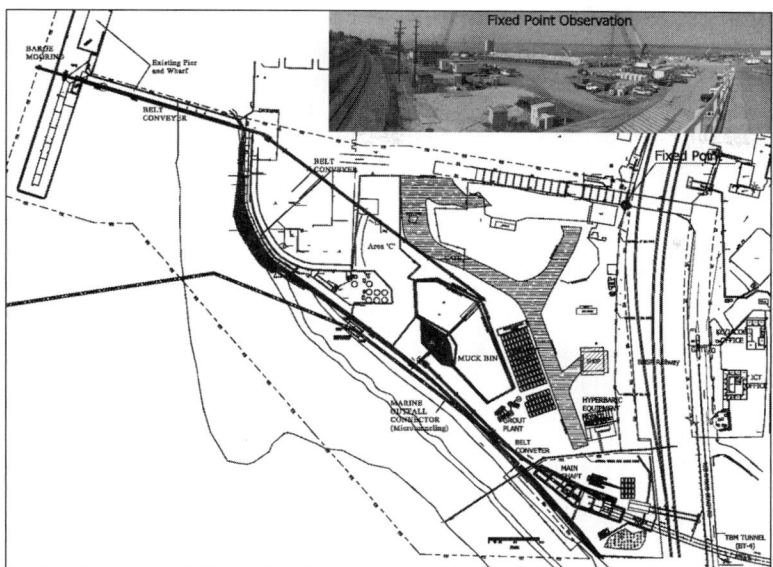

Figure 4. Point Wells jobsite layout

the sand and silt unit acts as an aquitard hydraulically separating the underlying coarse sand and gravel unit, which is slightly artesian at the Point Wells Portal location.

Petroleum Contamination

The site was historically used as part of the fuel terminal and for other industrial activities, and 229.4 cubic meters (300 cubic yards) of problem waste was expected to be encountered during the excavations on the site.

As the excavations were made it became apparent that significantly more than 229.4 cubic meters (300 cubic yards) of material were impacted as almost all of the material within 4.6 meters (15 feet) of the ground surface and some soils even deeper, showed signs of petroleum contamination. At the time of this writing, the total amount of petroleum impacted soils hauled off-site was approximately 9,000 tons.

Area C

The Marine Outfall is a pipeline extended about 1.62 kilometer (one mile) into the Puget Sound and serves as the discharge end of the Brightwater Conveyance System. This project is under a different contract and is being done by others. This pipeline is linked to BT4 tunnel with the Marine Outfall Connector (MOC); a short micro-tunnel installed by JCT at the beginning of the project.

A one acre piece of land, designated in the contract as Area 'C', was cutout of the Point Wells Portal (PWP) and relinquished to the Marine Outfall Contractor (MC) to serve as his jobsite. Area 'C' is located at the north-west corner of the PWP and was handed over after 400 days from the Notice To Proceed (NTP). JCT is required by contract to provide all utilities to MC. At the time of writing this article, MC is still in possession of Area 'C'. Area 'C' is shown in Figure 4.

Existing Pier and Wharf

The existing pier and wharf structures used for barge loading are located within a Maritime Security System area (MARSEC) and regulated by the U.S. Coast Guard. Due to the country current heightened security alert, accessing any MARSEC areas require cretin credentials which presented some restrictions on the barge loading operation. Limited number of personal are allowed access and they had to pass a background check by the Homeland Security. Also, MARSEC gates are opened to entry and leave only and remain closed and locked at all other times. Another limitation was the allowed loads on the pier and wharf to preserve the integrity of the structures. The Existing Pier and Wharf are shown in Figure 4.

In order to facilitate the installation of the Muck Handling System (MHS), some of the existing structures on the pier had to be either modified or removed. This required permission and coordination with Paramount Petroleum (PP) and imposed extra expenses on JCT. Furthermore, PP required a 3 meter (10 ft.) wide by 3.65 meter (12 ft.) high access to all areas on the pier and wharf at all times. The pier and wharf have to be restored to their original configuration at the end of the job.

An extensive survey of the pier, wharf, supporting piles, and wharf fenders was done by JCT and witnessed by PP. Some fenders were deteriorated and had to be fixed by JCT before use.

ON-SITE PREPATORY WORK

Environmental Mitigation and Environmental Controls

JCT was required to employ the services of an environmental manager to ensure that the negative environmental impact of the project was kept to a minimum. Kroner Environmental Services Inc. served this capacity, which included all aspects of environmental work including:

- Prevention of degrading the environment due to the spilling of construction materials.
- Ensuring that the construction noise does not constitute a nuisance to the local community.
- Ensuring that the construction lighting does not constitute a nuisance to the local community.
- Providing all monitoring required to ensure that the items above are accomplished.

In addition to the work touched on above, JCT is also required to treat and monitor all of the water that leaves the site in order to maintain water quality. This includes the water that is discharge to Puget Sound as well as the water that is discharged into the sewer.

Surface Water Discharge. Collected storm water "water that originates as rain" is discharged to Puget Sound using a level spreader constructed just outside the seawall that bounds the west side of the project site. Only storm water that meets the National Pollution Discharge Elimination System permit requirements can be discharged from the site.

The primary water qualities of concern in the surface water run-off from the site are turbidity (suspended sediments) as well as pH and petroleum hydrocarbons both during initial setup of the site and the more long term tunnel construction period.

Process-Water Discharge. Process water is any water which originates as potable water, and any rainwater that comes into direct contact with treated tunnel muck or uncured concrete, and groundwater seepage from the Point Wells Portal site.

The primary contaminates of concern during the initial design of the process water treatment system were sediment, oil, and acidity/alkalinity. Subsequent to the start of construction concern developed about the salinity of the process water discharged from the site and JCT modified the treatment plant and procedures to comply with the additional requirements concerning saline content.

Utility Crossing

Since there were no utilities extended to PWP site at the time of awarding the contract, JCT was required to do so by the contract. As mentioned before, the jobsite is separated from the neighboring area by BNSF railroad which mandated an underground crossing for the utilities. Two 50.8 cm (20 in.) casings were to be installed under the railroad in which utility pipes are fitted.

The alignment of the casings had a gradient of 8% and each pass was approximately 36.6 m. (120 ft.) long. The depth of the casings at its shallowest point under the railroad was 1.8 m. (six ft.)

Pipe Ramming method was chosen to install the casings. The drive pit for the ram operation of both casings was 12.2 by 6.1 m. (40 by 20 ft.) and was located in the PWP site west of the railroad tracks. Two 1.5 m. (5 ft.) diameter drilled steel caissons were installed on the east side railroad right-of-way to serve as exit shafts. After ramming the casings, the soil inside them was bored and disposed off. Utility pipes were installed inside the casings and separated apart using spacers and the annulus area was filled with sand. Figure 5 shows the configuration of utility pipes inside the casings.

An array of settlement monitoring points was installed on the railroad and closely monitored during the ramming operation. A BNSF flagger was present at all times and ramming was halted during the passing of a train. No work under BNSF railroad was allowed during the fourth quarter of the calendar year.

Marine Outfall Connector (MOC) Construction

The MOC is approximately 152.4 m (500 ft.) long pipeline, having an inside diameter of 2.1 m. (84 in.), that connects the BT4 main tunnel to the marine outfall pipeline constructed by others. This MOC crosses under an existing seawall, an abandoned outfall, and an active storm drain line. A slurry pressure balance Micro-Tunneling Machine (MTBM) was employed to perform the job. The MOC was to be completed before the start of the main tunneling of BT4. MOC alignment is shown in Figure 4.

The north end of the main shaft was used to launch the MTBM. A temporary sheet pile wall was installed approximately 11 m (36 ft.) from the north end of the main shaft forming an 11 by 8.5 m (36 by 28 ft.) microtunneling launch pit. Jet grouting was performed to establish a tunneling eye for the MTBM.

The launch pit was excavated to a depth of about 9.1 m (30 ft.) from the ground surface. During the pit excavation, a number of boulders were encountered which superseded the amount and size of boulders specified in the contract document. A Differing Site Conditions (DSC) was issued by JCT involving the change of the MTBM face configuration to accommodate the possibility of mining through this different ground conditions. This DSC was acknowledged by KC and JCT was awarded the appropriate compensation.

A receiving shaft was installed on the Puget Sound Beach, adjacent to the jobsite, about 152.4 m (500 ft.) north of the launch pit. This shaft acted as the MOC terminus point and the Marine Outfall start point and was handed-over to the MC with Area 'C'.

The MTBM was equipped with a jacking system that has a thrust force of 1,200 ton. Direct Jack Micro-tunneling method was applied using Fiberglass Reinforced Polymer Pipe (FRPP), the use of this type of pipe was required by the contract. Since the jacking

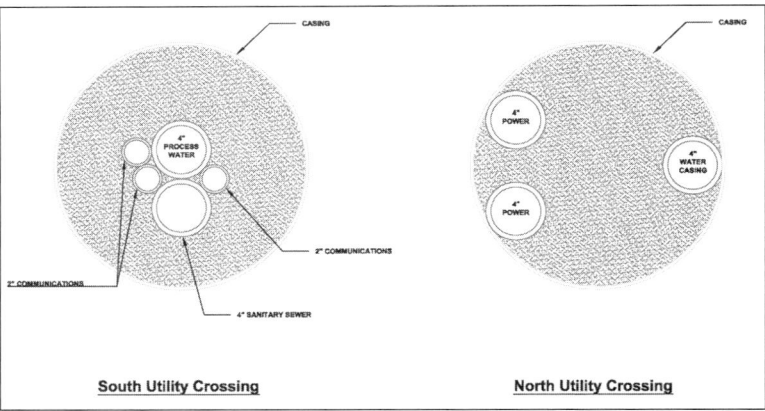

Figure 5. Railroad utility crossings

forces were more than the recommended capacity of the FRPP, an Intermediate Jacking Station (IJS) was used for the second half of the drive.

During the mining of the MOC contaminated materials were encountered resulting from hitting the Seawall. According to the contract, the Seawall was not existent at the MOC depth; this triggered another DSC for delays and cost of disposal of the contaminated materials which was granted by KC.

Muck Handling System

The Muck Handling System is defined here as the means used to handle the excavated muck from the instant it leaves the TBM till its offsite disposal. The TBM muck handling equipments comprise of a screw conveyor that discharges the muck out of the working chamber onto a belt conveyor which moves the muck to its disposal location on the TBM gantry, more detailed description of the TBM components will be presented in a later stage. A layout of the entire Muck Handling System. is shown in Figure 4.

Tunnel Muck Handling. The muck is transported from the TBM to the shaft in muck boxes that are staged within the gantry sections during the mining cycle. A total of five muck boxes are used each mining cycle, each having a capacity of 8.8 cubic-meter (11.5 cubic yards). The muck boxes are hauled through the tunnel by a 20 ton locomotive. Once the boxes reach the shaft, they are hoisted up one at a time to be dumped in a hopper located at the top of the shaft.

Tipping Frame and Muck Hopper at Main Shaft. A tipping frame is used to empty the muck boxes in the shaft hopper. This tipping frame is mounted on top of the hopper and supported by a steel structure that is anchored to the shaft's bottom slab and side. The hopper has a capacity of 23 cubic meter (30 cubic yards) and equipped with a belt feeder at the bottom.

Belt Conveyor from Launch Shaft to Muck Bin. Due to noise and space restrictions, an array of conveyor belts was chosen to move the muck from the shaft hopper to the muck bin which is located about 61 m (200 ft.) away from the shaft. In order to efficiently utilize the entire muck bin area, a stacker is used to pile the muck in different areas of the muck bin. This system is in continuous operation during mining.

Since the belt feeder at the bottom of the shaft hopper is lower than the top shaft elevation, a slightly inclined belt conveyor is installed within the shaft area, along the east shaft wall, to transfer the muck to an overland belt conveyor which in turn shuttles the muck to the muck bin stacker. The overland belt conveyor has a three meter (10 ft.) bottom clearance to allow for equipment traffic around the shaft.

Muck Bin and Tunnel Belt Feeder. The muck bin has a kidney like shape to allow for maximum utilization of the stacker radial motion. The sides of the muck bin are constructed of ecology blocks stacked on top of each other to form retaining walls that are.2.4 m. and 4.3 m (8 ft. and 14 ft.) at the lowest and highest points respectively. A significant length of the sides are double walled mainly for stability, but also a perforated pipe is placed in the gap between the two walls and used for water drainage. The muck bin has an approximate capacity of 4,590 cubic meters (6,000 cubic yards).

A short belt feeder is installed to move the muck out of the muck bin. This belt feeder is completely enclosed in a connex box and fed via a hopper, thus entitled "*Tunnel* Belt Feeder." During the loading of a barge, an excavator is used to feed the muck to the tunnel belt feeder.

Belt Conveyor from Muck Bin to Barge. A second conveyor system is used to move the muck from the muck bin to the barge loading area. This system comprises of an overland conveyor crossing Area 'C', a low-head conveyor along the existing pier, and a stacker on the wharf used to load the barges. This system is operated only during loading a barge.

The overland conveyor was elevated to 4.3 m. (14 ft.) above ground to allow for equipment traffic for the Marine Outfall Contractor (MC). In addition to that, one of the conveyor supports in Area 'C' had to be relocated to permit maximum utilization of Area 'C' by MC. Modifications had to be made to the wooden shelter on the pier to accommodate the low-head conveyor, the wooden shelter is being used to keep the muck from blowing off the conveyor belt to the water.

A 3 meter (10 ft.) wide by 3.65 meter high (12 ft.) high path along the pier and wharf has to be maintained at all times for paramount petroleum (PP) access. This mandated the location of the stacker on the pier as well as the minimum height that it can be lowered to.

Barge Loading/Offloading and Disposal. Currently, a barge is being loaded twice to three times a week to keep up with the mining production rate. Each barge has a capacity of 1,800 tons and takes about 8 to 10 hours to load. Loaded barges travel across the Puget Sound, using a tug boat, to the final disposal site where it is unloaded, cleaned and prepared for the next loading cycle. Precautions are being made to avoid spillage of muck in Puget Sound during transportation.

Shaft Construction

A total of four shafts were required at the Point Wells Portal site namely:

- A launch shaft was constructed on site for two pipe rammed crossings of the railroad. These 50.8 meter (20 inch) steel pipes were launched from a 12 × 6 m (40 × 20 ft) shaft that was three meter (10 feet) deep, which was a braced excavation with sheets and a drained slab bottom.

- A recovery shaft was constructed just above the high tide line on the beach for the 2.1 meter (84 inch) MOC. This shaft was 9.7 meter (32 feet) long and 4.9 meter (16 feet) wide and 7.6 meter (25 feet) deep and was originally designed as a flooded shaft but was changed to an uplift resisting design prior to construction. The final shaft was a braced excavation with steel sheets,with a concrete plug to resist uplift.

- A launch shaft was constructed for the MOC. This shaft was 9.1 meter (30 feet) long, 7.6 meter (25 feet) wide and 7.3 meter (24 feet) deep, and was constructed as a braced excavation with sheet pile walls and a jet grout plug to resist uplift. This shaft was incorporated into the main shaft prior to the launch of the main TBM.

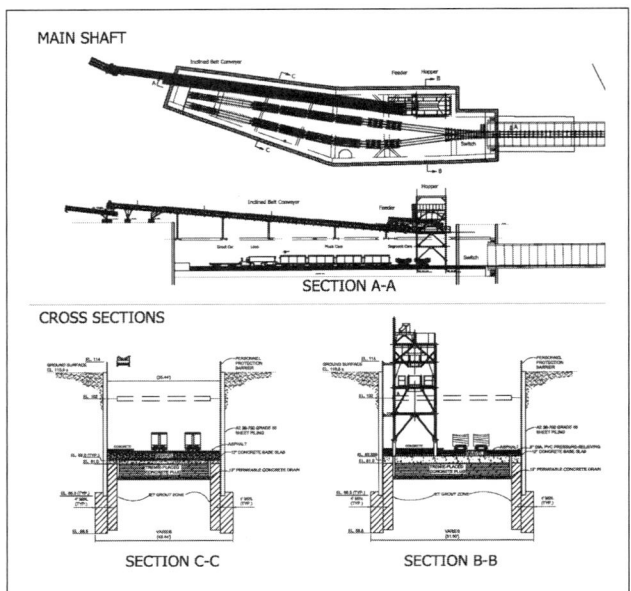

Figure 6. Main shaft configuration

- The main work shaft for the project was constructed at the Point Wells Portal location and is within 30.48 meter (100 feet) of the waters of Puget Sound. The finished shaft was approximately 59 meter (195 feet) long, 15.2 meter (50 feet) wide, and 9.1 meter (30 feet) deep, putting the invert 7.6 meters (25 feet) away from and 7.6 meters (25 feet) below Puget Sound at high tide.

Main Shaft Construction. JCT decided to build a large shaft at the site to facilitate an eventless crossing of the heavily used BNSF railroad, which began less than 18.3 meter (60 feet) from the launch seal. This shaft was large enough to encompass the sampling facility with the 1.8 m (72 in.) gate valve, the meter vault, the 1.8 m (72 in.) steel pipe between them, and the transition from 1.8 m (72 in.) steel pipe to the 2.1 m (84 in.) MOC.

Groundwater was expected to be at or near the ground surface and no dewatering was allowed due to the potential for the spreading of subsurface contamination plumes known to exist near the site. The general plan for constructing the shaft was to drive sheets around the periphery, then install a jet grout plug under the bottom of the shaft to resist the uplift pressure. The planned jet grout plug was 7.6 meter (25 feet) thick and extended under the bottoms of the sheets in order to mobilize the steel weight in resisting the uplift. The planned construction included a jet grout "eyes" for the launch of the MOC slurry MTBM and the main TBM, as well as the uplift resisting block. The total installed volume of jet grout planned was over 6,880 cubic meters (9,000 cubic yards).

Due to significant overruns in the cost of jet grouting primarily associated with the quantity and disposal cost of the jet grout spoil, the plan was changed. The presence of a continuous layer of silt with very low vertical permeability allowed the shaft design to be changed to allow the uplift pressure to be drained through the bottom slab. See Figure 6. This change reduced the overall amount of jet grout required to less than 3,820 cubic meters (5,000 cubic yards).

The launch shaft for the MOC was integral to the main shaft and the overall shaft excavation was staged to facilitate the installation of the MOC pipeline as the main shaft was being excavated. The shaft was excavated down to the strut elevation first and the struts and walers were installed. Then the southern half of the shaft was excavated down to the level where uplift pressures on underlying low permeability soil layers would have required that the shaft be flooded (approximately 20 feet). The muck from the southern portion of the shaft was stockpiled in the northern portion of the shaft in order to provide the required overburden to resist the jacking loads from the MOC installation.

After the MOC was completed, the northern section of the shaft was excavated down to 6.1 m (20 ft.), and the shaft was flooded for the remainder of the excavation. The shaft was excavated down to a level approximately 2.1 m (7 feet) below the final grade in the wet. A 1.2 meter (4 foot) tremied slab was installed across the entire shaft bottom except for the outer one meter (3 feet) which had been jet grouted all the way up to the bottom of the final working slab elevation. Once the tremie slab was installed the shaft was pumped out and the drain pipes were installed.

The final one meter (3 feet) of shaft bottom consisted of steelwork encased in concrete for anchoring the jacking frame for the main TBM launch in the southern portion of the shaft, and one meter (3 feet) of crushed rock in the northern portion of the main shaft. This was the shaft bottom used for the assembly and launch of the TBM, and for the first 122 m (400 ft.) of mining. Once the TBM was completely turned under, the shaft was setup for production mining with the installation of a working slab with a switch and rail in the bottom and a car dumping feeder hopper and truss conveyor above the struts and walers.

High pH Material Disposal

On Sunday September 9th, 2007, the lead story on the Local News page of the Seattle Times was entitled "Some tunnel debris polluted," the story went on to describe how high pH jet grout spoil had led to the contamination of some surface water at a disposal site and a $2.4 Million cleanup was required. On Thursday September 13th, 2007, JCT began jet grouting the eye for the MTBM launch for the MOC pipeline installation. After several weeks of attempting to find a disposal site for the jet grout spoil (the planned site had backed out soon after the paper had published the story), the Point Wells Portal site was inundated with the high pH jet grout spoil, and JCT was forced to dispose of the spoil to the local waste authority, which used it as daily cover at the municipal solid waste landfill. The disposal cost was increased by a factor of 4 and the design and excavation of the main work shaft was modified as a result of this increased cost.

OFF-SITE PREPATORY WORK—TBM DESIGN AND MANUFACTURE

General TBM Description

The Tunnel Boring Machine (TBM) in use to mine BT4 is an Earth Pressure balance (EPB), mixed face TBM manufactured by LOVAT Inc. The TBM is approximately nine meters (30 ft.) long and has a shield diameter of 4.65 m (183.3 in.). Numerous supporting systems are mounted on trailing gantries giving the machine an approximate total length of 85.3 m (280 ft.) The TBM is designed to mine on a minimum radius of 250 m (820 ft.) and efficiently perform under 5 bar gauge pressure. Figure 7 shows the general layout of the TBM.

Four electric water-cooled motors, having a combined 1,200 HP, are used to drive the ctterhead and allowing for a maximum torque and speed of 5,288 kNm (3.9 million lb.ft.) and 3.4 rpm respectively. Sixteen propulsion jacks are used to advance the machine with a total thrust force of 2,000 tons. The TBM has a maximum penetration rate of 20.3 cm/min (8 in/min).

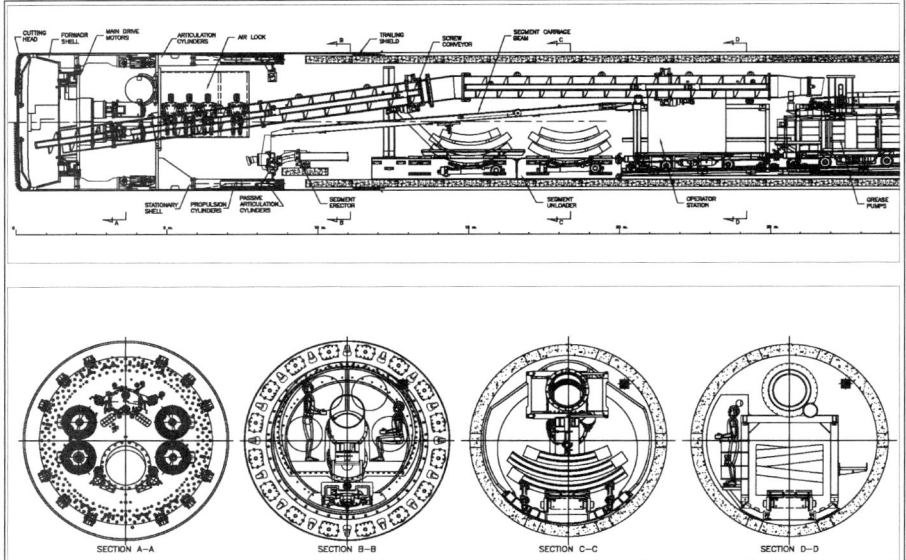

Figure 7. Brightwater West tunnel boring machine

The variable speed bi-directional cutterhead is equipped with eight isolation doors and four spokes. The face is dressed with 33 ripper type cutting teeth, three of which are wear indicator cutters and two are extendable ones. The ground conditions along the alignment of the tunnel does not necessarily require the use of disc cutters, never the less the cutterhead is designed to allow mounting disk cutters if needed. The nosecone is designed to be replaced from the inside of the machine. Figure 8 shows the configuration of the cutterhead.

The TBM is equipped with a soil conditioning system that can disperse ground conditioning materials in six different locations. Five nozzles are fitted on the front of the cutterhead to allow for a full coverage of the face and one nozzle is mounted on the side of the cutterhead.

The annulus area behind the segments is filled with two component grout mix to aid in supporting the precast concrete tunnel liner. The contract mandated grouting through the tail can at a rate equivalent to the TBM advance speed. No grouting is allowed through the segments. The grout system is designed to convey the two parts of the grout mix separately from the holding tanks to the heading where it is injected in the annulus gap via pipes embedded in the tail can skin.

The excavated muck is moved from the cutterhead chamber via a screw conveyor to the trailing belt conveyor which in turn dumps it in muck boxes to be taken to the main shaft for disposal. The screw conveyor comprised of two sections, both having a diameter of 91.4 cm (36 in.) and designed to accommodate a 30.5 cm (12 in.) size boulders. The first section is 12.5 meter (41 ft.) long and fitted on an angle, the second section is 11.9 meter (39 ft.) long and fitted horizontally. The extensive length of the screw is to allow dissipating the EPB pressure when mining under 5 bars. Both sections of the screw have variable and reversible speeds and have separate controls. The belt conveyor is mounted on the trailing gear and has an approximate length of 37.2 m (122 ft.).

The TBM is also equipped with a twin man airlock located in the stationary shield to be used for hyperbaric cutterhead interventions. TACs tunnel guidance system is used to maintain the project alignment.

Figure 8. Cutterhead configuration

Hyperbaric

The location and size of the airlock specified for the TBM was the overriding requirement that drove the design and manufacture of the tunneling machine. The requirement that the TBM be equipped with a manlock inside of the stationary shield that was capable of supporting hyperbaric work at 5 bar gauge resulted in a significant challenge for JCT.

According to the Washington Administrative Code no hyperbaric work above 50 psi is allowed during underground or tunnel construction. The expected hydrostatic pressure at springline for nearly half the tunnel length is expected to be above 50 psi.

JCT is pursuing 5 separate variances from the code. The resulting airlock design, coupled with the relatively high earth pressures expected, resulted in several major components being "non-standard" for a TBM of this size, which required extra time and effort during TBM design and manufacture.

Electric over Hydraulic

The required size and location of the airlock prevented the operators station from being located where is would have normally been located in TBMs of this size, and thus required an electric over hydraulic control system. This change not only increased the complexity of the design dramatically, but also made the TBM much harder to operate and maintain.

Pressurized Motor-room

The required size and location of the airlock required that the drive motors and articulation jacks be located inside of a separate section of the TBM which would need to be pressurized during hyperbaric work in the cutterhead. This modification required that these components and much of the equipments associated with these system had to be capable of withstanding a pressures in excess of 5 bar as well as being capable and classified for operation in a potentially gassy environment.

SUMMARY

At the time of writing this article, JCT has completed installing the MOC pipeline, the railroad crossing, surface & process water discharge systems, and the muck handling system. The main shaft has been completed and the mining of the main BT4 tunnel is in process, 1,097 m (3,660 ft.) of the tunnel has been mined and lined. The best week of mining production was 161.f m (530 ft.) and the best day was 36.6 m (120 ft.).

ACKNOWLEDGMENTS

The authors would like to acknowledge all of the employees of JCT, who's hard work and diligence has resulted in a project that we all can be proud of. We would also like to thank Jay Dee Contractors, Inc., Frank Coluccio Construction Co., and Taisei Construction Corp. for their valuable insight and support.

REFERENCES

Brightwater Conveyance System West Contract—Brightwater Tunnel, Section 4—King County, WA, Contract C00007C06, July, 2006.
Pre-Bid Engineering Geologic Evaluation of Subsurface Conditions for the Tunnels and Shafts of the West Contract Brighwater Conveyance System King County, Washington—Jim Irish, August, 2006.
Washington Administrative Code chapter 296—www.Lni.wa.gov/LawRule/

BRIGHTWATER EAST—A CASE HISTORY

Luminita Calin ▪ Kenny Construction Company

Tony Hupfauf ▪ Kenny Construction Company

ABSTRACT

King County (Seattle, WA) is on a monumental task of building the new Brightwater System complete with a new treatment plant, three main conveyance tunnels, a new outfall into Puget Sound and other ancillary works. This paper will describe the first conveyance tunnel, Brightwater East currently under construction by the Joint Venture of Kenny (Sponsor) / Shea / Traylor that is scheduled for completion in early 2010. The project located in Bothell, WA (north of Seattle) is in both King and Snohomish Counties.

It was the first of the major tunnel projects that was scheduled by King County for the Brightwater System. The East Contract consists of the following major elements: 14,000 ft of 19′-2″ EPBTBM mined tunnel using 16′-8″ ID bolted, gasketed precast concrete segments for a primary liner; installing and encasing 14,200 ft each of 48″, 66″, 27″, 84″ in diameter steel pipes inside the tunnel along with three runs of fiber optic cable; 2,430 ft of 72″ in diameter microtunnel including three caisson shafts and associated structures; one Intercepting Structure (IS) shell to mine from that is 74 ft deep and 74 ft finished in diameter with 130 ft deep slurry diaphragm walls, tremie slab and final concrete wall lining; one Influent Pump Station shell (IPS) 83 ft deep, twin 77 ft inside diameter cells, with 160 ft deep slurry diaphragm walls, tremie slab, and final lining; two short 12 ft in diameter inter-shaft connector tunnels; one extraction shaft 40 ft deep by 40 ft wide and 140 ft long for connection to the new treatment plant piping.

BRIGHTWATER SYSTEM

The King County regional wastewater conveyance and treatment system serves 17 cities, 16 sewer districts and the Muckleshoot Indian Tribe. Brightwater is the third regional treatment plant which will serve the area north of Seattle. Construction started in 2006 and is scheduled to be completed in 2010, with operations starting in 2011. All three treatment plants combined will serve more than 1.4 million people.

The *Treatment Plant* at startup in 2011 will have capacities to treat approximately 36 million gallons of wastewater per day. Due to the rapid growth in the region the capacity can be increased through expansion of the plant to up to 130 million gallons per day. King County expects wastewater to reach that magnitude by 2040.

The *Conveyance System* consists of four major tunnel drives, two microtunnel drives and a pump station. The East Contract TBM was launched from the North Creek Portal and headed east for breakthrough at the Treatment plant. The TBMs on the Central Contract were launched from the North Kenmore Portal site going East and West. And the fourth and last TBM to be launched started from the Marine Outfall and headed east.

The *Marine Outfall* is located on the Puget Sound Shore line approximately 20 miles north of Seattle. The total outfall length is about 5,000 ft long, with a diffuser length of 250 ft consisting of two 63″ pipes at a depth of 600 ft below sea level.

PROJECT DESCRIPTION

The Brightwater East tunnel project was bid on October 20, 2005, by six bidders that submitted the following tenders:

1. Kenny (Sponsor) / Shea / Traylor, JV $130,648,750
2. Jay Dee / Coluccio, JV $144,765,000
3. Impregilo / Healy Brightwater JV $151,683,000
4. Obayashi Corporation $152,500,000
5. Kiewit / Bilfiger Berger, JV $174,224,000
6. VINCI / Parsons RCI / Frontier-Kemper, JV $174,646,000

The Engineer's estimate was $188,740,000. KST's low tender was 10% below the second bidder and 31% below the Engineer's estimate. The mandated duration of the contract is 1,400 calendar days and includes liquidated damages of $24,000 per calendar day for failure to achieve Milestone 1 and $11,000 per calendar day for failure to achieve Milestone 2.

The Brightwater East project is located about 15 miles north of Seattle at the north end of Lake Washington. The tunnel extends from the east end of Bothell, where it connects to the future IPS pump station and the Central Contract, to the north end of Woodinville where the connection to the treatment plant will be made (see Figure 1). The tunnel alignment is 14,000 ft long and consists of eight horizontal curves with the majority of them having a radius of 1,000 ft. The tunnel alignment has a steep incline of 2.8% in the last 3,000 ft. The tunnel is mined from launch until breakthrough under water pressures ranging from 25 psi to 48 psi.

GEOLOGY

The Brightwater System lies within the Puget Sound Trough, a structural basin between the Olympia Mountains and the Cascade Mountains. The bedrock is very deep and it is not encountered during the mining of the tunnels. The bedrock is overlain

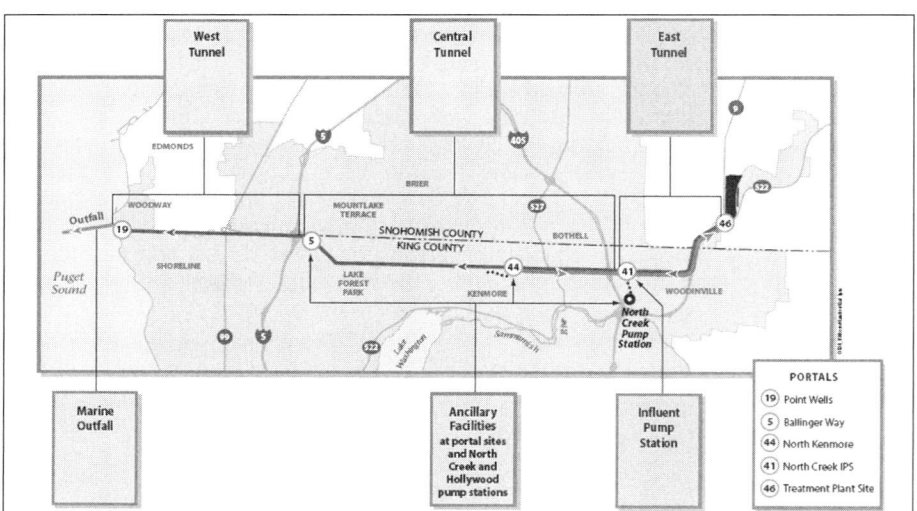

Figure 1.   Project location (King County)

Figure 2. Brightwater East Lovat TBM

by glacial and non-glacial sediments deposited during a succession of several continental glaciations. Each glaciation partially eroded the pre-existing ground surface and then deposited a new sequence of sediments over the land.

Recent (Holocene) deposits, generally composed of silt, sand and silty sand with gravel deposits present in the recent alluvium were only encountered at the North Creek Portal site during the slurry wall and the excavation of the two shafts.

Tunneling through the North Creek Valley, the first 1,200 ft of tunnel, was a very critical and very closely scrutinized operation, due to very low cover and mining through normally consolidated soils, at times through partial or even full face low shear strength, high compressibility peat and a Contract requirement to perform an intervention with a high water table.

The remainder of the tunnel was excavated through silts and sands, rarely clay. Ground conditioning played a very important role by conditioning the soil for torque and wear control while mining this difficult geology. The conditioned soil caused extreme cost overruns due to the water content and borderline pH. This was further aspirated by rising trucking costs and environmental agencies tightening the regulations and putting pressure on available dump sites. The end result caused the trucking costs to triple due to the rising dump fees, longer distances to haul and higher trucking rates due to the fuel crisis. Soil disposal has become a major cost and risk factor in EPB and slurry tunneling, especially in environmentally sensitive regions such as the Seattle area.

TUNNEL BORING MACHINE

The TBM used for mining the Brightwater East tunnel is a new Lovat EPBTBM (see Figure 2). The manufacture was carried out by Lovat forces in Toronto, Canada.

General Description. The Lovat Tunnel Boring Machine Model RME229SE 22900 is a self-contained tunneling system, complete with a double screw, dual two-compartment airlocks for compressed air interventions, muck conveyor, lining erection system for the precast concrete tunnel liner. The typical Lovat integral power packs had to be moved back on the trailing gear due to the screw and dual airlocks. The Lovat TBM is made up of three main sections: a forward shell which contains the cutterhead, the main drive and a manlock vessel; a stationary shell which houses the screw from the cutterhead chamber, the operating controls, a portion of the electro/

Figure 3. CSI-Hanson manufacturing plant in Tacoma

hydraulic power units, propulsion system and segment erection system; and a trailing shield which contains the lining erection equipment and three sets of tail seal brushes.

Hydraulic power is used throughout the machine for excavation, conveying and lining erection. Forward propulsion is powered by hydraulic cylinders, arranged in a ring around the inside diameter of the stationary shell, which act between the stationary shell and the tunnel lining.

Steering is assisted by an active articulation system connecting the forward shell to the stationary shell. The cutterhead could articulate to 1.3° in any direction enabling the machine to negotiate curves both vertically and horizontally. A passive articulation system was mandated by the specifications and installed. KST didn't desire to have the passive articulation installed based on past experiences in soils not as challenging as Brightwater. Needless to say, the request was rejected and the passive articulation proved to be the largest contributor to downtime on the project.

All machine controls, instrumentation and monitoring devices are centralized on the operator's control panel that is built into the stationary shell. The operator's control panel is placed in a position that maximized the view of the cutterhead discharge and lining erection area. The large dual airlock caused an obstructed view of the opposite side erection area requiring a second erection operation station.

For safe operation, the TBM is manufactured with an electrical system suitable for Class I, Division II as required on the Brightwater East project due the tunnel being classified as "Potentially Gassy." An extensive gas detection system is included to warn of the presence of methane gas, with an automatic shutdown of the TBM at a preset level of explosive gas concentration.

The TBM is equipped with a customized lining erection system specifically designed for the six-piece (2 trapezoids and 4 parallelograms) gasketed and bolted precast primary liner. The segments for Brightwater East, as well as all other Brightwater tunnel projects, were manufactured in Tacoma by the Brightwater JV, CSI-Hanson (see Figure 3). The segments performed admirably. There wasn't one segment ring in the entire drive that was out of tolerance. In addition, the total infiltration for the entire tunnel was less than 7.5 gpm, well within the specifications (see Figure 4).

To allow cutterhead inspections under water pressures ranging from 25 psi to 48 psi the machine is equipped with a twin manlock vessel (2 stage type, 2 chambers side-by-side). The vessel is designed for pressures up to 73 psi and was tested to 87 psi.

A summary of the main technical specifications of the TBMs is as shown in Table 1.

Figure 4. Segments installed in the tunnel

Lovat Data Logging System. The TBM is equipped with a Programmable Logic Controller (PLC). The PLC is used to control the machine and / or to read information from sensors.

Any information in the PLC is sent to and displayed with Human Machine Interface software (HMI). The HMI software, which must be run on a PC (Windows) format, can also be used to record information.

The system features one terminal in the TBM (LCD display) and one desktop PC station at the surface. The two units are connected by a fiberoptic cable link and two external Short Haul Modems. The surface computer records the data in a SQL database. Through a web-based application called Bizware the SQL server can be accessed via the Internet. This allows everyone in the supervision team to troubleshoot issues or to track the TBM's mining progress from any remote location (see Table 2).

There are basically two applications for the collected telemetry data from the TBM. One is to track the mining progress and troubleshoot problems in real time via Bizware from any remote location. The other application is to generate ring reports through a program in VBA on the local network. It is required to have values for EPB pressures, grout pressures, GCS flow rates, guidance information, muck volume, muck weight, main cylinders extension and active articulation cylinders extension on the ring report. The VBA program computes the ring reports in semi-automated mode. Grout volume, muck weight and muck volume is retrieved from shift reports. The remaining information is retrieved from the SQL server, averaged for each ring and then merged with information retrieved from the shift reports.

GROUND IMPROVEMENT—JET GROUTING

A very limited program of jet grouting was mandated by the Contract in order to improve the stability of the ground during the breaking-out of the EPBTBM from the IS mining shaft due to high water table and difficult ground conditions. Other jet grout blocks were used for breaking-out of the same shaft by Northwest Boring, our micro-tunnel Subcontractor, for the first of three drives for the 72" in diameter North Creek Connector, for two inter-shaft connector tunnels between the IS and IPS shafts and for a future penetration into the shaft by the Brightwater Central tunnel. The jet grouting was performed by Hayward-Baker and proved to be very successful.

Table 1. Summary of the Lovat EPB-TBM technical specifications

LOVAT—RME229SE Series 22900	
Basic Dimensions of TBM:	
Cut Diameter	5852 mm
Bore Diameter	5827 mm
Shield Diameter	5814 mm
TBM Length	9.4 m
TBM + Back-up Length	104.0 m
TBM Weight	496 ton
TBM + Back-up Weight	830 ton
Cutters:	
Nose Cone	carbide inserts, steel construction
Disc cutters	interchangeable with backloading ripper teeth
Lovat Ripper teeth	carbide inserts and hardfacing
Lovat Scraper teeth	carbide inserts and hardfacing
Advance Rate:	
Maximum Rate of advance	10 cm/min
Cutterhead Power System:	4 × 400 HP electric motors
Cutterhead Drive System:	6 hydraulic motors, variable displacement type
Maximum Torque	660 tonne @ 0.0 to 1.77 rpm
Minimum Torque	290 tonne @ 4.0 rpm
Main Bearing:	Triple axial roller type
Shield Propulsion system:	
Propulsion cylinders	16 cylinders @ 280 tonne / ea
Total maximum thrust	4480 tonne @ 340 bar
Propulsion stroke	2300 mm
Electrical—Summary of installed power:	**2160 kW**
Cutterhead	1200 kW
Propulsion / Erection	225 kW
Screw conveyor	225 kW
Belt conveyors	112 kW

CAISSONS AND MICROTUNNELING

Part of the Brightwater East project included three microtunnel drives with a combined length of 2,430 ft of 72″ pipe. Drive A was launched from the IS structure at the North Creek Portal site. Drive B and C were launched from a 28′ diameter and 65′ deep caisson. All drives were received at 20′ diameter caisson structures with a depth of approximately 65′. The microtunnel work was conducted by Northwest Boring and the caissons were constructed by KST.

Table 2. Lovat data logging system

Automatic Recording of Information	
Cutterhead RPM	Total TBM Electric Current Draw
Cutterhead Torque	Main Drive Motor Electric Current Draw
Cutterhead Direction	
	Main Bearing Lubrication Flow and Pressure
Articulation Cylinder Extension	Sealing Systems Lubrication Flow and Pressure
Flood Control Door Opening	Hydraulic Oil Temperature and Level
TBM Operating Hours	
Propulsion Cylinder Extension	Earth Pressure Sensor Output
Propulsion Cylinder Extension Speed	Ground Conditioning System Output
Propulsion Cylinder Pressure	Environmental Monitoring System
Centre of Thrust	Guidance System Output
Screw Conveyor RPM	Grout Injection Pressure
Screw Conveyor Torque	Grout Injection Flow Rate
Screw Conveyor Internal Pressure	
Screw Conveyor Guillotine Door Opening	
Screw Conveyor Caliper Door Opening	
Belt Conveyor Drive Motor Hydraulic Pressure	

All caissons were excavated under water. Due to the high ground water table, tremie slabs of up to 16 ft of thickness had to be poured to prevent the structures from floating. At break-ins of the microtunnel Northwest Boring elected to flood the retrieval shaft in order to prevent damage to the adjacent structures through possible ground inflow.

During the sinking of the caissons (see Figure 5) bentonite slurry had to be pumped along the perimeter to reduce friction in sand layers. In some other layers, which consisted mainly of peat, the structure would sink during excavation several inches per hour.

SHAFTING

Slurry Wall Shafts at North Creek Portal Site

The Contract Documents required the installation of two slurry walls, one for the IS mining shaft and one for the IPS shaft for the future pump station to be built under a separate contract located at the North Creek Portal, the west end of the project. The slurry wall work was performed by Bencor from Dallas, Texas (see Figure 5). There were problems from the start when the Operating Engineers union called a County-wide strike against the ready-mix concrete suppliers that shut down concrete deliveries to the project in spite of a Project Labor Agreement. In spite of the strike, Bencor worked extended hours and days to make up most of the time but fell further behind due to a failure of a critical "Wye" panel at the common wall intersection between the two lobes of the figure eight shaft. This failure was due to a Differing Site Condition but caused additional schedule slippage.

Figure 5. Manhole 1 caisson operation

Figure 6. Start of the slurry wall operation at the IS shaft

Circular Shafts

The IS slurry wall is 130 ft deep with 4 ft thick heavily reinforced panels. The circular excavation inside the slurry wall is 74 ft in inside diameter and 74 ft deep. The IS shaft (see Figure 7) served as the TBM launch chamber for the Brightwater East tunnel and the North Creek Connector microtunnel and will receive the TBM from the adjoining Brightwater Central tunnel project. Several interesting features during the construction were the utilization of fiberglass reinforcement in order to accommodate the future penetrations through the slurry wall, and the excavation "in the wet" and the underwater pour of the tremie slab.

The second slurry wall was for the IPS shaft (see Figure 7) that was a double lobe (figure eight) shaft with a common wall. The slurry wall panels were 160 ft deep, 4 ft

Figure 7. Influent structure (IS) and influent pump station (IPS) shafts

thick and each of the lobes had an inside diameter of 77 ft. The excavation inside the lobes was 83 ft deep. After the excavation in the wet to the final depth, a 15 ft thick tremie slab, keyed into the walls was poured in each of the lobes using the same system as used in the IS.

Rectangular Shaft

The Treatment Plant Portal (TPP) shaft is a combination rectangular shaft and ellipse, with internal steel bracing supporting the 60 ft deep soldier piles and wood lagging. This shaft was used for the TBM extraction at the north terminus of the project. A unique feature of the 140 ft long by 40 ft wide shaft was the use of double elliptical walers and an internal tie rod system to form a 62 ft long free span for the placement of 4 runs of 59 ft long steel pipes that will be placed in the tunnel (see Figure 8).

The opposite end of this shaft serves as a connection to the treatment plant. Once the tunnel is entirely filled with grout the connection to the treatment plant will be established and the shaft will be backfilled.

TUNNELING

The Lovat TBM was originally scheduled to be launched from the IS shaft for the 14,000 ft long drive in June 2007. Due to a critical main bearing delay at Rotek the TBM launch had to be rescheduled for September 2007. This was handled as a Force Majeure event without cost.

After being inserted into the portal seals, the TBM successfully penetrated the 4-foot thick slurry wall of the IS structure and through the jet grouted block without a hitch. The first 1,200 ft of tunnel was through the critical North Creek Valley with only 45 ft of overburden under a busy road with sewers, water mains, and other utilities running within 15 ft parallel to the tunnel. In addition, the geology for this portion of the drive consisted either of a full face of peat, or a mixed face of peat and fine sand with

Figure 8. Treatment plant portal (TPP) shaft

the water table only a few feet from the top of the ground. This constituted a challenge to the TBM operators. Either low or high deviations from the targeted EPB pressures would lead to settlement. The mining through this section was very successful without major incidents and kept the settlement in the range of less than one inch.

For this section of the tunnel two cutterhead inspections under compressed air were specified by Contract and scheduled. At both interventions the face was sealed with bentonite, yet during the attempt to fill the plenum with compressed air the air communicated through the peat to the surface and was impossible to maintain a constant air pressure as evidenced by leaking air to the surface and the rising of the asphalt pavement. Both interventions had to be aborted.

After mining through this critical section, the operation was changed to 24 hours per day and 6 days a week to make up schedule due to the strike and late delivery of the TBM. Two production shifts started at 6:00 am and lasted until midnight. An overlapping maintenance shift started at 11:00 pm and finished at 7:00 am the next day. This arrangement resulted in a mining progress of up to 1,900 feet per month, with the last six months averaging 1,600 feet per month including maintenance and interventions.

Throughout the remaining time of the tunneling additional nine interventions with working chamber pressures up to 48 psi took place. Including the last intervention which took place less than 3,000 ft before to hole-out, only minor wear on the cutterhead was reported. Unfortunately, very abrasive soils must have been encountered during the run to hole-out that caused a great deal of wear to the cutterhead, but fortunately the wear didn't cause any structural damage and the drive was completed successfully.

On November 14, 2008, the TBM holed-out within one inch of line and grade at the Treatment Plant shaft and the accelerated production allowed the TBM to complete the drive on schedule (see Figure 9).

The Brightwater East project has proven to be a very challenging project due to the soil conditions, a very prescriptive specification that limited the Contractor's ability to control means and methods, a well crafted Geotechnical Baseline, a restrictive PLA and the Owner's first OCIP.

Figure 9. TBM holing-out at the TPP shaft

REFERENCES

King County. Figure 1, Project location.

King County, Department of Natural Resources and Parks, Wastewater Treatment Division, 2005. Brightwater Conveyance System—East Contract, Contract Documents.

Lovat Inc., 2006. Technical specifications Lovat RMP185SE Tunnel Boring Machine, Kenny / Shea / Traylor, Joint Venture, Brightwater East.

GOTTHARD-BASE TUNNEL, SECTION FAIDO PREVIOUS EXPERIENCE WITH THE USE OF THE TBM

Martin Herrenknecht ▪ Herrenknecht AG

Olivier Böckli ▪ Implenia Bau AG

Karin Bäppler ▪ Herrenknecht AG

ABSTRACT

After the successful breakthrough at the end of 2006 of both TBMs into the multifunctional site Faido (MFS), in the middle of 2007 the two tunnel boring machines resumed the 12 km long section between Faido and Sedrun. In the roughly a 10 month interval both approximately 465 m long tunnel boring machines were partly dismantled, pushed through the 2.5 km long MFS, rebuilt for the new larger diameter of 9.5 m and reassembled to be ready for operation.

In the rebuilding phase, improvements were made to the equipment, especially as a result of experience gained from the previous 14 km long drive in the section Bodio and taking into account the latest information regarding the upcoming 14 km section Faido.

Both TBM drives are currently in the rock formations of the Lucomagno gneiss, which stretch over a section of 3.5 km from the MFS Faido to the predicted Piora Syncline. This section is divided into two fold axes of the Chièra Synform, where the Lucomagno gneiss pass from a sub-horizontal into vertical bedding. The drives, with current rock overburdens exceeding 1,500 m and, as known from the excavation phase of the MFS, in part highly disturbed rock formations, are characterized by the locally intense rock squeezing conditions with large convergences in the vicinity of the driving TBM as well as backwards due to the interaction of the two drives. High expenditure for the support measures, reduced driving performances together with extensive extra work associated with deadlines and increased costs are the result.

In the paper the previous experience with the TBM drives in the section Faido and the challenges still awaiting us are presented and discussed.

INTRODUCTION

The sections Bodio and Faido (Figure 1) are the southernmost sections of the altogether 5 main contract sections of the nearly 57 km long Gotthard Base Tunnel. After the successful breakthrough at the end of 2006 of both TBMs of the section Bodio into the multifunctional site Faido (MFS), in the middle of 2007 the two tunnel boring machines resumed the 12 km long section between Faido and Sedrun. In the roughly 10 month interval the approximately 465 m long tunnel boring machines were partly dismantled, pushed through the 2.5 km long MFS, rebuilt for the new larger diameter of 9.5 m and reassembled to be ready for operation. In the rebuilding phase, improvements were made to the equipment, especially as a result of experience gained from the previous 14 km long drive in the section Bodio and taking into account the latest information regarding the upcoming 12 km section Faido. The TBM West has passed the rock formations of the Lucomagno gneiss, which stretch over a section of 3.5 km

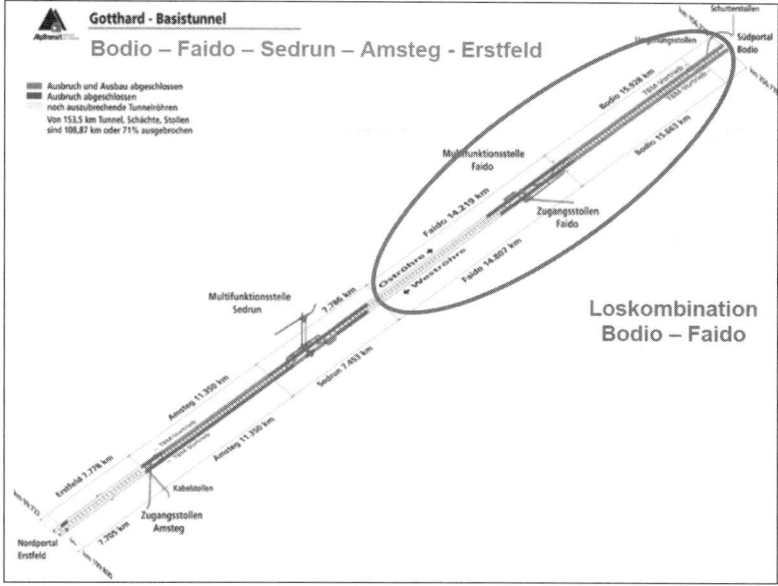

Figure 1. Overall survey Gotthard Base Tunnel of the ATG

from the MFS Faido to the predicted Piora Syncline, in which the TBM drove in at the end of January 2009. The TBM East has successfully passed the Piora Syncline and is currently in the following Medelser Granite..The section of the Lucomagno gneiss is divided into two by the fold axes of the Chièra Synform, where the Lucomagno gneiss pass from a sub-horizontal into sub-vertical bedding. This drive was up to date characterized by the locally intense rock squeezing conditions with large convergences in the vicinity of the driving TBM as well as backwards due to the interaction of the two drives. High expenditure for the support measures, reduced driving performances together with extensive extra work associated with deadlines and increased costs are the result.

In the following, the previous experience with the TBM drives in the section Faido and the challenges still awaiting are discussed.

The TBM East excavated nearly 4,400 m until the beginning of 2009 within about 16 months of driving time, the TBM West nearly 3,600 m within about 14 months. After the summer break 2008, the TBM East reached the so-called parking position in front of the Piora Syncline, subsequently, various additional exploration measures were taken, before the Piora Syncline was successfully penetrated with the TBM revised in the parking position.

PROJECT DESCRIPTION

With AlpTransit Gotthard, a future-oriented flat railway through the Alps is being built. The Base Tunnel at the Gotthard is the core of the new railway connection. The tunnel, being the longest one in the world with its 57 km, will be put into service at the end of 2017. The pioneer work of the 21st century will lead to a prominent improvement of travel and transport possibilities of in the heart of Europe.

With 550 m above sea level, the highest apex of the new flat railway is at the same level as the city of Bern. For comparison: The apex of the existing mountain stretch is located at 1,150 m above sea level. The route through Switzerland will become flatter and shorter by 40 km. The flat railway permits to drive goods trains that are longer compared to those used today and can be twice as heavy (4,000 t instead of 2,000 t today). The fastest goods trains will drive at a speed of up to 160 km/h—twice as fast as today. The slopes and tight curve radii make the employment of such trains impossible on the existing lines in the Alp region. After the construction of the flat railway, less locomotives, personnel and energy will be required for the same transported quantity.

During the planning of the construction, an answer has been given to the question of when, where and in what sequence the construction has to be accomplished in order to optimise time and costs of construction. The concept for the Gotthard Base Tunnel provides a simultaneous advance in five parts of different lengths. These are referred to as sections. In the construction project, two variants have been elaborated which differ from each other as regards the advance method, i.e. tunnel boring machine or drill and blast.

The nearly 8 km long Erstfeld section is the northernmost part of the Gotthard Base Tunnel. It also includes an underground junction to permit a future extension of the tunnel towards the north without interrupting the operation. The tunnel is built in an open cut in the first part that will be filled up again after works have been completed. Currently, two Herrenknecht tunnel boring machines also drive in the Erstfeld section, which have been employed before in the Amsteg section.

The slightly more than 11 km long Amsteg section is the second section from the north. In drill and blast operation, a 1.8 km long access gallery and a construction gallery have been excavated to offer access to the two tunnel tubes and the assembling caverns. In 2003, two Herrenknecht tunnel boring machines (currently revised and employed in the Erstfeld section) started from the cavern towards the Sedrun section border, at the end of 2007, the drive works could be successfully completed with the breakthrough to the Sedrun section.

The Sedrun section is developed via a 1 km long access gallery and two 800 m deep shafts. In this section, one of the two multifunctional sites is constructed which will accommodate technical railway facilities, but also emergency stop and change-of-gauge stations. The excavation of the tunnel tubes towards the north and south was started in 2004 and is performed by conventional drill and blast operation. Tunnel boring machines could not be employed due to the geological conditions. Currently, the drives towards the south as well as the interior works of the excavated areas are still under progress.

The Bodio and Faido sections are prepared by the Consorzio TAT (Joint Venture) in a nearly 30 km long combination of contract sections. The Faido section is developed via a 2.7 km long access gallery with a slope of up to 13%, and its constructional-logistics is coupled to the Bodio section. In the Faido section, the second multifunctional site is located. Due to the extraordinarily difficult geological conditions, this section had to be partially shifted towards the south. The two tunnel boring machines arriving from Bodio have been, upon their arrival, revised and restructured before they tackled the 12 km long drive towards the Sedrun section border. With a length of slightly more than 15 km, the Bodio section is the longest section of the Gotthard Base Tunnel. The first tunnel meters were constructed in an open cut, followed by a stretch of soft rock and finally strong rock permitting the drive with tunnel boring machines. A bypass gallery was built in the portal zone which permitted a quicker development of the underground assembly caverns. At the beginning of 2003, two tunnel boring machines started the drive towards Faido from the caverns and could complete it at the end of 2006 with their arrival at the multifunctional site Faido.

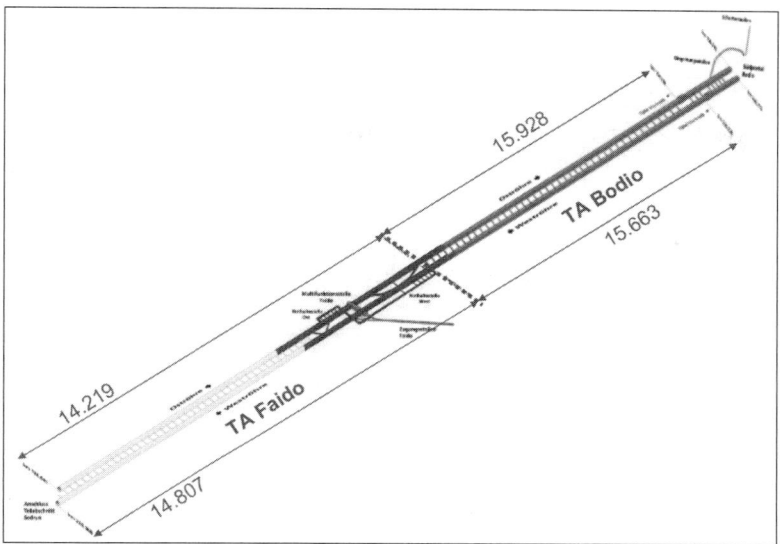

Figure 2. Overview driving status contract sections Bodio and Faido

CONTRACT SECTION 452 (FAIDO)

The technical and programmatic basic conditions of the 14.5 km long contract section Faido represent extraordinary challenges to the design and coordination of the individual construction activities. In the single-track tube east, in parallel to the actual driving works, cross passages are broken out by drill&+blast at a distance of 312 m each, as was already done in the Bodio section, and the definite inner vault is covered with concrete slightly set back. The so-called "worms" (600 m long installations for preparing the inner vault) will also tackle the inner vaults in the single-track tunnels in the Faido section (Figure 2), after the MFS extending over approx. 2.5 km has been covered with concrete in autumn 2009.

Apart from the excavation diameter, the single-track tunnels to be prepared in the Faido section correspond to those that have been excavated in the Bodio section (see Lecture STC 2005, "Bodio Lotto 554—Experience after half of the excavation"). Compared to the technical basic conditions and challenges to the TBMs that are already demanding in the Bodio section, in the Faido section, there are in addition overburdens of up to 2,500 m with correspondingly large primary and secondary states of stress (due to overburden and tectonics), higher rock temperatures (according to predictions up to max. 55°C), the demanding Piora zone as well as extraordinarily long driveways across the Bodio section (15–30 km).

The TBM driving stretches in the Faido section represent those of the complete project of the Gotthard Base Tunnel involving the highest geological and project-related risks. In the project planning phase, in particular for the Piora Basin to be penetrated, extensive previous explorations had to be carried out to be able to prove the technical feasibility of the demanding overall project.

It should be basically mentioned that in the south sections of the Gotthard Base Tunnel, up to now relatively great differences between the geological prediction and the geological finding occurred, in particular as concerns the rock behaviour and the corresponding constructional relevance (finding worse than predicted).

Figure 3. Consorzio TAT

THE CONSORZIO TAT, BODIO + FAIDO

The Consorzio TAT (Figure 3), consisting of the partner companies Implenia Bau AG (Switzerland)—sponsor of the Joint Venture, Alpine Bau GmbH (Austria), CSC Impresa Costruzioni SA (Switzerland), Hochtief AG (Germany) and Impregilo SpA (Italy) has received in the year 2001from the client Alptransit Gotthard AG (ATG) the contract for executing the two southern contract sections Bodio and Faido. For accomplishing the contract works, the internationally composed working group under Swiss management currently employs on site altogether approx. 700 employees, with nearly 90 directors, 490 industrial employees and nearly 120 subcontractor representatives. About ⅔ of the complete workforce are allotted to the contract section Faido.

The Company Herrenknecht AG (Germany) is the manufacturer and supplier of the 2 identical tunnel boring machines that are employed for driving in the two parallel, altogether 30 km long tunnel by the Consorzio TAT.

GEOLOGY
TECTONIC SURVEY

The Alps to be traversed by the Gotthard Base Tunnel practically at right angles (Figure 4) are subjected to high stresses for plate-tectonic reasons. Still today, the Alps "grow" by about 1.2 mm per year. The resulting ground pressures due to tectonic conditions are overburdened by partly massive stresses due to the rock overlap of up to 2,500 m (Figure 5). Lateral pressure values of λ 1.2 cannot be excluded. The consequences of these complex states of stress can lead to constructionally, extremely demanding basic conditions requiring extensive know-how as well as an enormous commitment of the executors if they occur.

GEOLOGICAL LONGITUDINAL SECTION

In the TBM drive of the Faido section, corresponding to the geological prediction, the following geological formations (Figure 6) have to be penetrated chronologically from the MFS:

- Lucomagno gneiss, sub-horizontal bedding
- Lucomagno gneiss, sub-vertical bedding
- Piora Syncline
- Medelser Granite
- Striated gneiss

In connection with the experience gathered up to know, the sections in the Lucomagno gneiss and the transition zone Chièra Synform, the Piora Syncline and the Medelser Granite will be discussed below.

Figure 4. Folding of the Alps

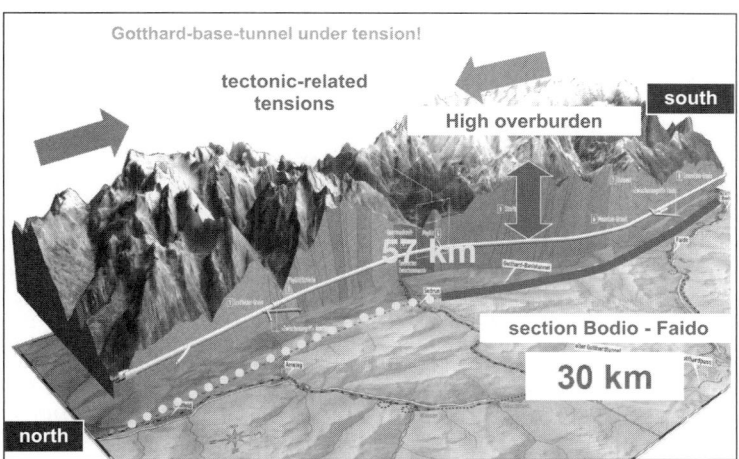

Figure 5. Survey Gotthard base tunnel

PREVIOUS EXPERIENCE WITH THE TBM DRIVES IN THE FAIDO SECTION
PROJECT- AND GEOLOGY-RELATED BASIC CONDITIONS

Two parallel tunnel tubes are excavated with TBMs with a diameter of currently 9.5 m (Figure 7). The distance between the two tunnels is about 32 m (distance between axes 40 m), i.e. just over three times the boring diameter.

Currently, the leading drive of the east tube has an advance of nearly 1 km over the trailing drive of the west tube.

Fault-zones have been driven, in contrast to the prediction, to the tunnel axis in a very slow manner, whereby these remain within the extraction lines over relatively long milling paths. This leads, among others, to the fact that even faults in a dm range

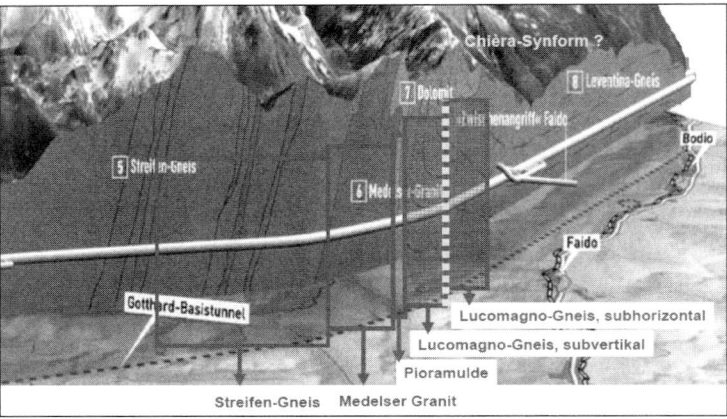

Figure 6. Detail of geological longitudinal section

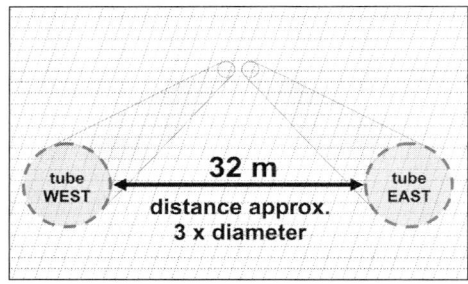

Figure 7. Situation single-track tunnels

become constructionally relevant as these remain for more than 30–40 m in the full section and thus have a negative influence on the excavated cavity.

EXPERIENCE IN THE SUB-HORIZONTALLY BEDDED LUCOMAGNO GNEISS

States of Stress

In the schematic rock cube (Figure 8), the horizontal bedding of the Lucomagno gneiss as well as the tunnel axis extending in parallel thereto are indicated. Due to the high ground pressures from overburden and tectonics, the horizontal bedding tends to buckle which would lead, in particular in the working zone L1* (Figure 9), to cave-ins directly behind the flight conveyor (Figures 9 and 10), making it impossible to prepare the excavation support during driving and leading to bottom upheaval and convergences. This resulted in construction sequences disturbed by these extraordinary incidences and, as a consequence, reduced driving performances.

Interaction East Tube–West Tube

By the drive of the leading TBM, a loosening body or de-strengthening zone formed around the full section (Figure 11). Favoured by large overburdens, possibly

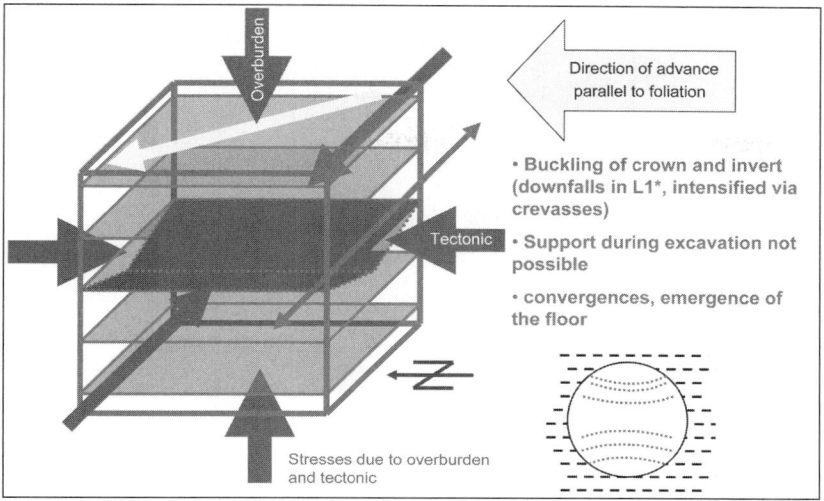

Figure 8. Schematic rock cube

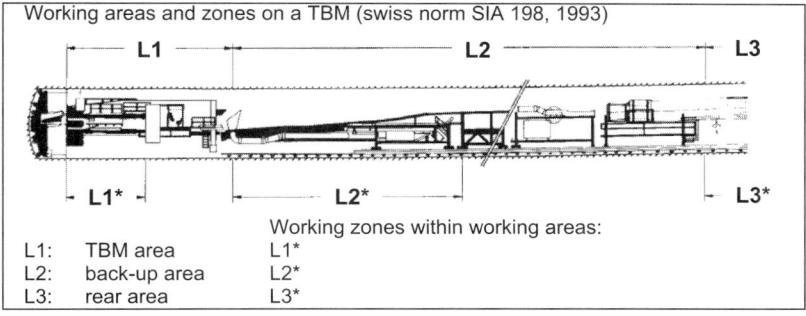

Figure 9. Definition of the TBM (tunnel boring machine) working areas and the back up

Figure 10. Caving in the L1*

Figure 11. Overbreak above the TBM shield

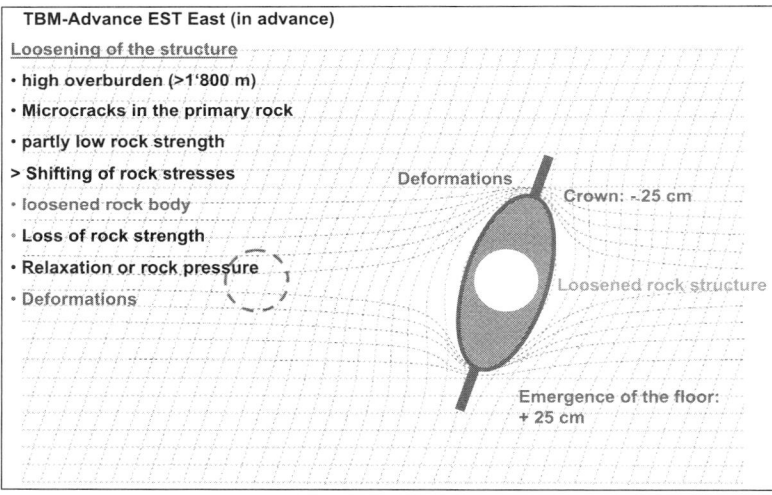

Figure 12. Loosening body leading drive

microcracks in the rock, generally faulty rock conditions and sometimes low rock strengths, deformations formed thereby which amounted to up to 25 cm in the roof and in the bottom.

At the time of the passage of the trailing drive, the single-track tube of the leading drive is secured with roof bolts, ring beams and shotcrete. The deformations that occurred in the west trailing drive were again up to 25 cm in the roof, in the bottom upheavals of up to 75 cm occurred, which was additionally favoured by corresponding preloads and shifting processes as a result of the leading drive (Figure 13). The convergences lead on the one hand to material breaking down onto the trailer, and on the other hand to geometrical conflicts, respectively to collisions of the back-up construction with the deformed excavation support. The individual geometrical "conflict areas" (Figure 14) were removed mainly by hand with the aid of pneumatic rock drills.

Moreover, the trailing drive caused again additional deformations in the already supported single-track tube of the east leading drive of further approx. 5–10 cm in the

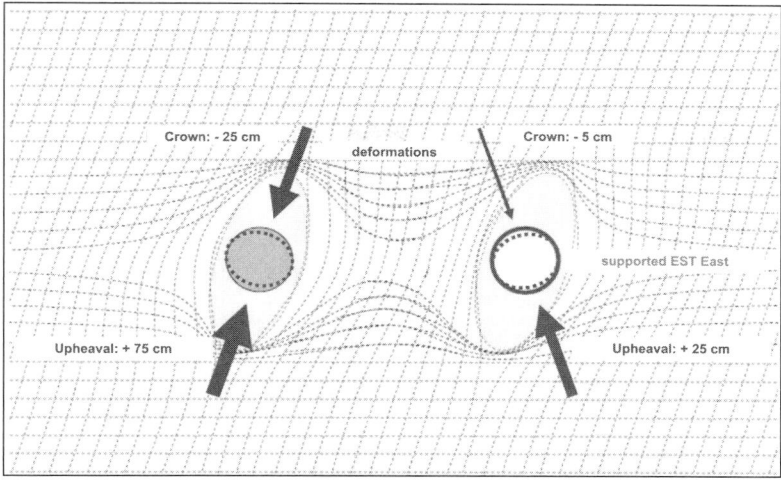

Figure 13. Effects trailing drive

Figure 14. Conflict areas convergence—NLK

roof as well as bottom upheavals of up to 30 cm. Figure 15 represents the chronological influence of the west trailing drive on the already excavated and supported east tube of the leading drive over a period of 120 days. The observation relates to a section of the east tube between tunnel meters 18,250–18,650. The bottom upheavals extended over an area of more than 250 m. With an admissible dimension of 70 mm for occurring bottom upheavals, this phenomenon lead to an area to be redeveloped of more than 200 m and a section where the bottoms have to be completely excavated and rebuilt of approx. 150 m.

The redevelopment of the bottom blocks consisting, among others, of an extensive bottom anchoring, will lead to a partial blocking of the section concerned, respectively to a restricted single-track operation. A bottom excavation necessarily leads to a complete blocking of the tunnel section concerned, as was analogously already done in the Bodio section.

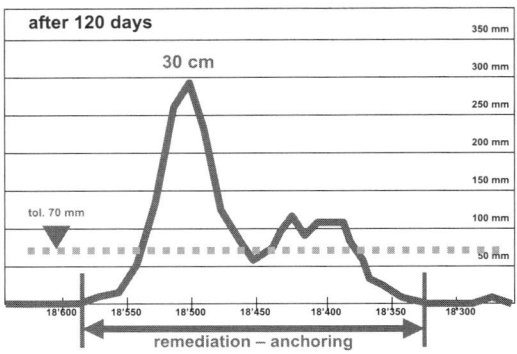

Figure 15. Chronological influence trailing and leading tube

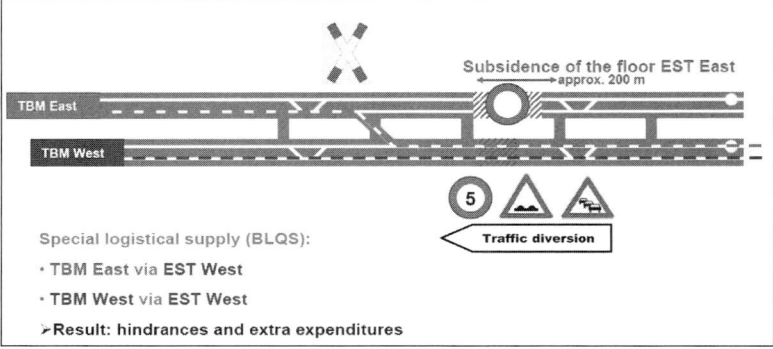

Figure 16. Assisting construction measures constructional-logistic cross passage

It goes without saying that a bottom excavation has serious consequences for the rail operation. The execution of the upcoming bottom excavation of about 150 m in the east tube would necessarily lead to a stoppage of the TBM East drive concerned for 5–6 months without additional assisting building activities. By designing a cross passage that is oblique instead of orthogonal with respect to a constructional-logistic cross passage practicable on one track, the production in both TBM drives can still be maintained, though in a restricted form, taking into consideration impediments (Figure 16).

Convergence Slots in the Excavation Support

In the L1 area, the deformations that occurred lead to the TH (Figure 21) arched support being pushed in and thereby to local failure of the shotcrete lining and a distinctive reduction in the cross-section. To avoid the local destruction of the shotcrete lining when the TH arched supports are pushed in, up to 6 convergence slots were arranged in the periphery (Figure 17). The radial pushing-in forces occurring due to cross-section deformations lead to a compression of the convergence slots. Thereby, partly massive convergences can be absorbed in the periphery of the excavation support system without the shotcrete lining being locally destructed.

Figure 17. Injected convergence slots in the L1 area

EXPERIENCE IN THE TRANSITION ZONE CHIÈRA SYNFORM

The Chièra Synform is the transition zone between the sub-horizontally and sub-vertically bedded Lucomagno gneiss (Figure 18). Geologically, the Chièra Synform becomes manifest in a highly folded transition zone. The geometric extension of this zone characterized by foldings and parasitic folds, involving corresponding constructional consequences, was about 500 m.

Before the last Easter holidays, a kakiritic fault-zone, additionally aggravated by the basic conditions prevailing in the transition zone, lead to a temporary jamming of the TBM East. Local water penetration additionally led to a partial washing out of material which further increased the pressure on the TBM shield in the lower half of the cross-section and finally resulting in a jamming of the TBM shield. Before the decision was made to expose the TBM shield by creating a roof, laterally of the TBM shield, large-caliber pressure releasing drillings were performed with the horizontal drill feeds provided for advanced injection drilling, which remarkably showed the desired success and permitted the TBM to continue without any further assisting construction activities

Exceeding the Maximum Shield Load

The deformations of the rock initiated by the shifting of stress also have a direct influence on the tunnel boring machine. The roof shield can be traversed via 2 roof cylinders. On the basis of the experience gained in the first meters of excavation, 4 "Enerpac" cylinders were additionally retrofitted in the top section of the shield, which on the one hand permitted to measure the existing shield load, and on the other hand decisively assisted in maintaining the controllability of the TBM by correspondingly increasing the pushing-in resistance. By measuring the prevailing hydraulic pressure of the 6 cylinders (2 roof cylinders and 4 "Enerpac" cylinders), direct conclusions could be drawn to the effectively prevailing loosening and/or rock pressure. The shield skin is structurally designed for a pressure of 30 to/m2. By observing the pressure on the roof shield and thereby indirectly the prevailing loosening pressure or rock pressure, measures for preventing serious machine damage could be initiated in time and thus the probability of occurring damage, such as breakaways of the drive pinions of the main drive or cracks in the shield structure, as they occurred in the Bodio section, can be reduced (Figure 19).

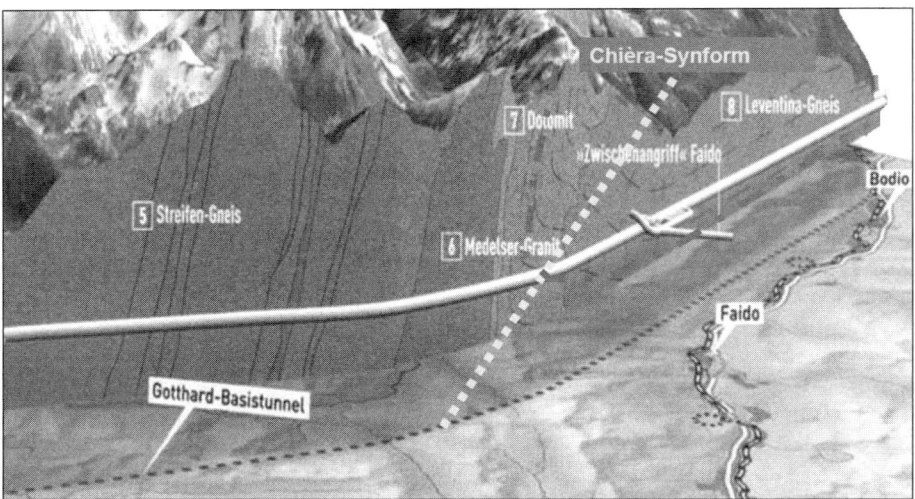

Figure 18. Chièra Synform

After the evaluation of the measuring results of the "Enerpac" cylinders from this extensive measuring campaign, these cylinders were removed again. Corresponding to the rock-mechanical principles, there is, as it its generally known, an interaction between admissible shiftings and building-up pressure of the rock in place. If no or only slight deformations are admitted (rigid shield, rigid excavation), very high pressure can possibly form, however, if deformations are admitted (movable shield (Figure 20), flexible excavation (Figure 21), a lower pressure is built up. As due to the employment of the measuring cylinders, the shield necessarily had to be fixed at a pre-determined position, the movability of the shield was reduced thereby, which was not desirable in squeezing zones. With a drill diameter of 9.5 m, the TBM shield is currently able to diminish by 15 cm in the roof with respect to the milled diameter to thereby minimize a temporary pressure build-up as a result of convergence of the rock in place in the TBM shield area.

EXPERIENCE IN THE SUB-VERTICALLY BEDDED LUCOMAGNO GNEISS

States of Stress

In the schematic rock cube (Figure 22), the vertical bedding of the Lucomagno gneiss as well as the tunnel axis extending at approx. 30–40° with respect thereto are indicated. Due to the high ground pressures from overburden and tectonics, the vertical bedding tends to buckle, leading to lateral buckling of the paraments and to a destabilization of the tunnel face.

Block-like, Instable Tunnel Face

Resulting issues are primary block-like and instable tunnel faces, impact damages at the cutterhead and the belt conveyor systems caused thereby as well as defects in the loading and conveying systems. The corresponding phenomena have been taken into consideration in the new cutterhead design for the Faido section. In part, these phenomena are, however, superimposed by the consequences of the high rock pressure. Relevant changes at the cutterhead design were as follows:

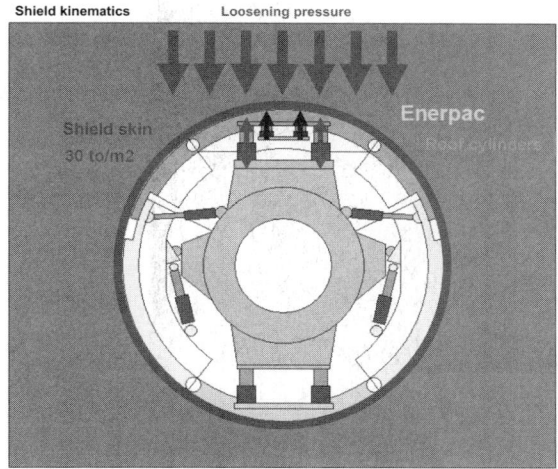

Figure 19. Observation of the shield load

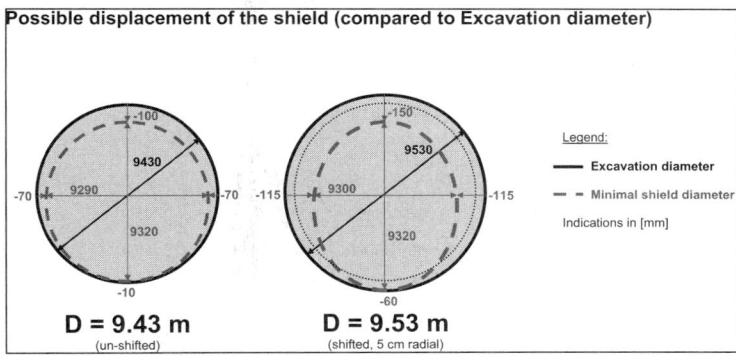

Figure 20. Displaceable shield

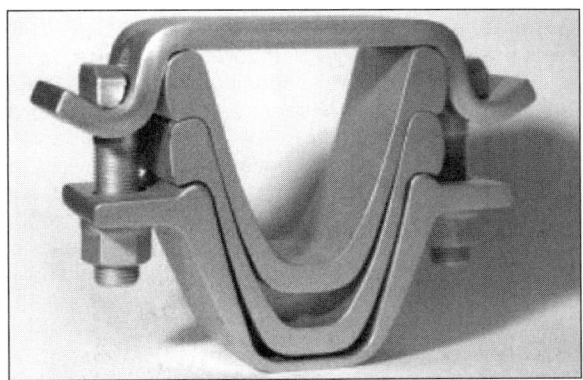

Figure 21. Flexible lining (TH ring beams)

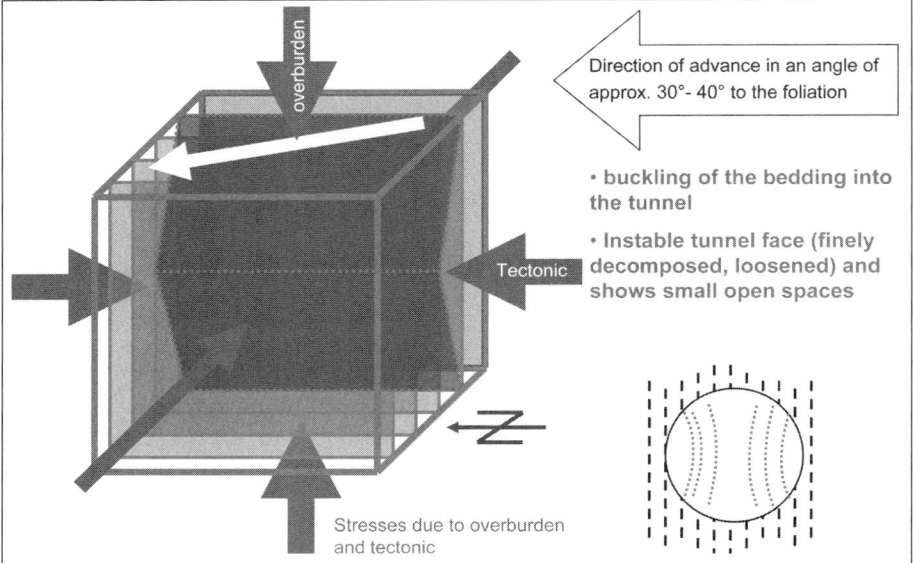

Figure 22. Schematic rock cube

- Spirally arranged roller bits (in contrast to a spoke-like arrangement for the Bodio section)
- Increase of the number of bucket channels from 8 to 12 pcs.
- Reduction of the individual bucket segment size
- Increase of the radius of the vertical curve of the cutterhead

In spite of optimisations of the cutterhead, phenomena such as a block-like and instable tunnel face, necessarily lead to damages at the drive systems, though these can be minimized by corresponding measures.

EXPERIENCE FROM THE PIORA SYNCLINE

The Piora Syncline (Figure 23) is the best explored section of the complete Gotthard Base Tunnel. Already more than 10 years ago (1993–1998), a 5 km long sounding gallery was prepared for the exploration of this "Pièce de Résistence," resp. for the positioning of the corresponding sounding drills. These extensive geological clarifications in the run-up to the project served among others to prove the technical feasibility of the structure of the century. In the process, the formation of the nearly 200 m large Piora Syncline was of essential relevance. In the "worst-case scenario," it had to be expected that an approximately 2,500 m high column of a sugar-like dolomite projected into the tunnel path, with a potential water column of 1,500 m (corresponds to a water pressure of 150 bar). If this had occurred, about 3 years of additional time of construction and additional costs of hundreds of millions would have to be expected for the approx. 150 m long section, with extensive rock improvement measures (rock injections).

In one sounding drill from the sounding gallery Piora that drilled the sugar-like dolomite, at that time an incident with a massive falling-in of material occurred. Within 24 hours, some thousand m3 of mud were washed into the sounding gallery through

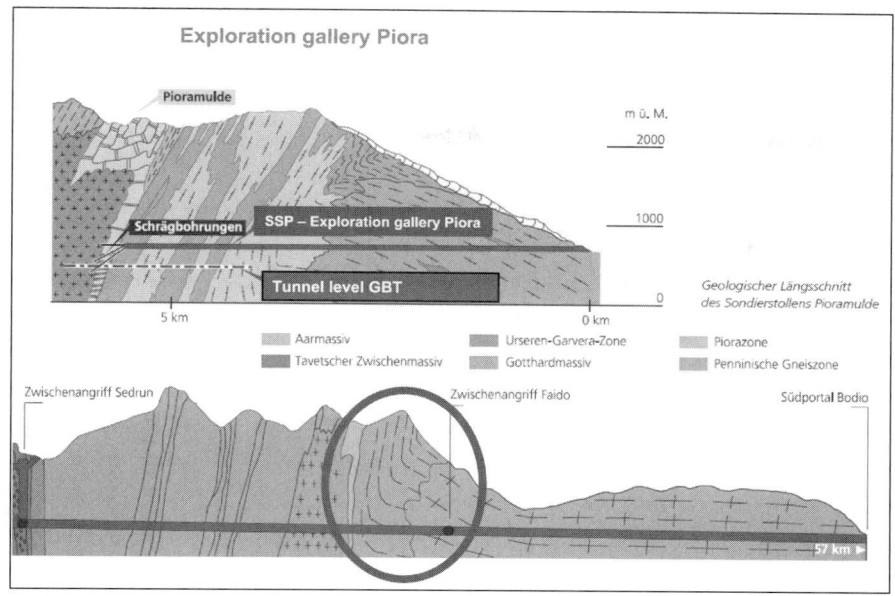

Figure 23. Exploration gallery Piora

Figure 24. Water inrush previous exploration sounding gallery Piora

the shank of the sounding drill, filling half of the TBM (Figure 21). The necessity of a functioning preventer system became obvious in a dramatic way. The 5 sounding drills excavated later from the sounding gallery (Figure 22) through the Quartenschiefer into the Piora Syncline finally showed that the sugar-like dolomite does not project into the path of the Gotthard Base Tunnel, but is "shielded" by a so-called cap rock between the exploration gallery and the Gotthard Base Tunnel alignment level.

Correspondingly, the probability of a geological incidence during the excavation could be classified to be extremely low. However, as the consequences can be catastrophic if something like this happens indeed, extensive planning of measures was

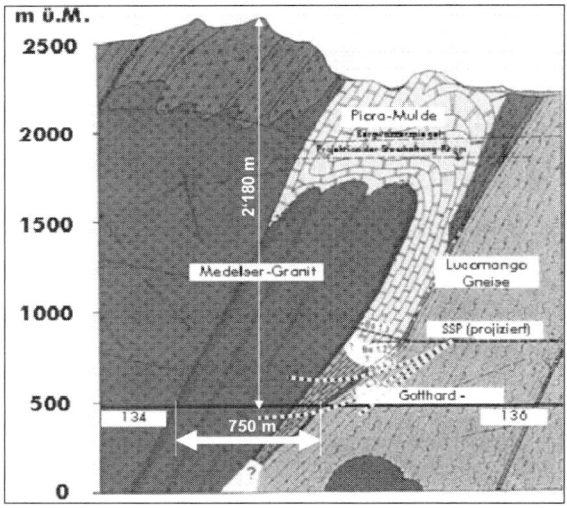

Figure 25. Piora Basin

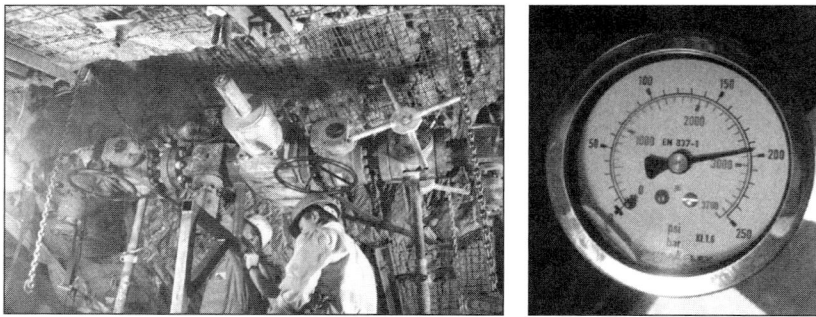

Figure 26. Preventer on TBM, test pressure standpipe 200 bar

elaborated in the run-up, which among others resulted in reserved decisions for the purpose of reducing the reaction time. The penetration of the Piora Basin was accompanied in the run-up by a core drilling with blowout prevention along the entire section to be able to eliminate even the small still remaining residual risk. The exploration drillings were executed with a blowout preventer, the corresponding technique originates from oil industry. When a reservoir (crude oil, natural gas) is opened for the first time, an escape of oil or gas partly under high pressure has to be prevented with a special protecting device (so-called "preventer"). The preventers employed by TAT resist a water pressure of up to 200 bar (Figure 26).

Before the TBM East entered the Piora Syncline, the drive system and in particular the cutterhead were subjected to thorough revision in the so-called parking position. From this parking position, a core drilling with blowout prevention was performed from the TBM along the entire section in the Piora Syncline. It could confirm the results of the previous exploration of the sounding gallery.

Figure 27. TBM excavation material from the Piora

Within 10 days of excavation, then the approx. 180 m long stretch in the marbled dolomite of the Piora Syncline could be successfully penetrated with the first TBM (Figure 27).

EXPERIENCE IN THE MEDELSER GRANITE

The Piora Syncline is followed by the geological formation of the Medelser Granite. The hard metamorphic rock (strength > 130 N/mm^2), combined with the high overburden of approx. 2,500 m, lead to rock-burst like incidents and massive tunnel face instabilities. Over the majority of the hitherto driven approx. 600 m long stretch in the Medelser Granite, on less than 10% of the tunnel face, chisel traces were visible (Figure 28), the rest of the tunnel face broke out in coarse blocks. These blocks are crashed and pulverized as in an impact crusher in front of the cutterhead, leading to massive wear and sometimes impact damages at the cutterhead. The revision intervals of the cutterhead (Figure 29) doubled from a drive stretch of 3–4 km up to then to only 1–2 km according to experience.

OPTIMIZATIONS TBM

The milling operation lasting now for nearly 6 years has partly lead to extensive adjustments and optimizations at both drive systems. The geological predictions resulted in a profile of requirements and performance standards in the service contract in the tender phase (Figure 30). The realization in this respect, including the experience and prior art, finally leads to the acquisition and construction of the corresponding drive systems. On the basis of the geological finding significantly deviating from the prediction and of the experience gathered during the drive activities, performance optimizations (shown in green) are successively realized. Corresponding investment costs (shown in red) have to be set for these. It is in the nature of things that aspects that contain a lot of room for optimisation are initially realized at relatively low costs, and vice-versa those aspects that only contain little room for optimisation are realized at the end at relatively high investment costs (Figure 31).

In the progress of the TBM drives up to now, diverse, partly extensive adjustments have therefore been made: With the adoption of the performance drive in spring 2003, the drive systems have been sent to tackle the 30 km. Due to unexpected rock

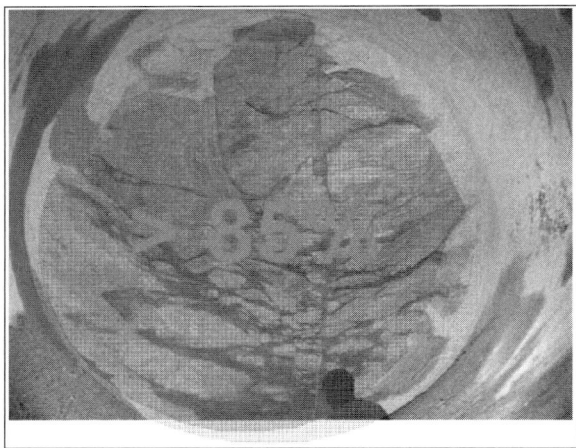

Figure 28. Tunnel face zone in front of and behind Piora

Figure 29. TBM cutterhead revision

conditions, already at an early stage a substantial alteration of the area L1* came into mind, followed by diverse minor and major alteration measures. Other essential adjustments were realized in the alteration of the drive systems in the MFS Faido, where, apart from the enlargement of the drill diameter from 8.9 m to 9.4 m (shifted 9.5 m), mainly the experience of the drives in the milling sections, each having a length of 12 km, of the Bodio section have been realized.

After the recommissioning of the drive systems last year, the upcoming section will also lead to further adjustments, depending on the findings and experience. According to experience, however, the corresponding performance optimizations will presumably only be marginal, however, the corresponding investment costs will be relatively high.

Below, some of the relevant changes and adjustments to the drive systems are listed:

2004
- Rebuilding of the sole vault lining
- Change-over from conveyor belt concrete to pumped concrete
- Employment of mobile sole concrete pumps
- Removal of sole ridding belt conveyor—rebuilding lowerable back-up belt conveyor
- Shifting roller bit in the gauge range
- Change-over shift system from 4/3 to 3/2-operation

2005
- Definite rebuilding L1 area TBM East and West
- Replacement telescopic feeds > rigid feeds L1
- Fixed installation sole concrete pumps in the trailer
- Change-over surveying system (laser); above > below back-up
- Doubling of cutterhead dedusting performance
- Extension transfer station shotcrete (enlargement of intermediate storage sites on the back-up)

2006 (mainly in connection with the rebuilding for the Faido section)
- Reinforcement vertical support, dust wall
- Enlargement drilled diameter Ø 8.9 m > Ø 9.4 m (with shifting Ø 9.5 m)
- Increase of number of gauge cutters
- Spirally arranged cutters
- Intensification cutterhead sprinkler
- Increase of number of bucket channels
- Reinforcement vertical support
- Reinforcement dust wall
- Replacement of steel with aluminum feeds
- Enlargement of intermediate storage sites
- Commissioning of mud treatment
- Increase of refrigerating capacity
- Commissioning Secatol agitating units concrete

2007
- Commissioning of the preventer-protected previous exploration drilling appliance
- Commissioning of measuring and pressure system roof shield
- Adjustments back-up area as a result of convergence phenomena

2008
- Rebuilding in the back-up area as a result of convergence phenomena
- Optimisation of the shield positioning capacity

TBM Capacities and Personnel

The realized measures of optimisation basically permitted to technically successfully master the current sections that presented, in contrast to the prediction, multiply disturbed conditions (in the single-track tunnel West over more than 500 m at a stretch).

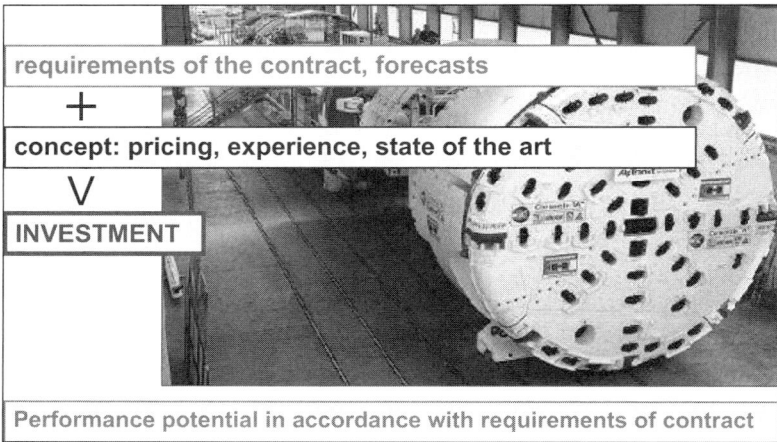

Figure 30. Concept drive systems

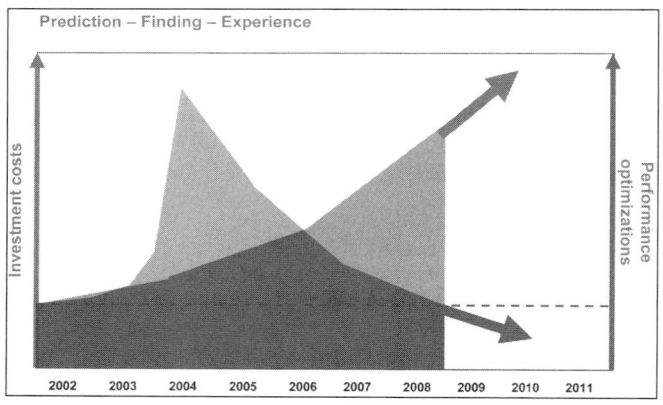

Figure 31. Prediction–Finding–Experience

The personnel and the drive systems thereby impressively demonstrated that their performance profile is essentially higher than the requirement of the service contract, and this proved to be absolutely necessary from the development corresponding to the geological finding greatly deviating from the geological prediction.

CONCLUSIONS AND FINAL REMARKS
GENERAL

In the structure of the century, the Gotthard Base Tunnel, in various fields constructionally uncharted trails were entered, such as e.g., a TBM drive with a diameter > 9 m in crystalline hard rock at overburdens of more than 2,500 m, handling of convergences of a diameter of nearly 1 m in the L1 and L2 areas, massive excavation support in L1* with TH44 arched supports, surrounding system anchorage and up to 5 m^3 of shotcrete per running meter, combined with stress controllers.

Moreover, preventer-protected core and impact drillings were performed for the first time in the TBM drive.

Added are logistics and underground services representing more than 50% of the complete personnel in the execution. However, it turned out so far that the selected concepts basically work, which was impressively confirmed by the TBM West with a driving performance of more than 700 m, notabene 22 km inside the rock in November 2008.

CONCLUSIONS

In connection with the TBM drive of a long tunnel (30 km) with a large overburden (>2,500 m) and in a tectonically active rock mass (folding of the Alps), one can draw the following conclusions:

Despite extensive clarifications in the run-up to the project, there can be a great difference between geological prediction and geological finding (e.g., rock class distribution). The rock behaviour and the hazard scenarios can prove to be less favourable than expected, which on the one hand shows in the constructional relevance (e.g., separate operations of drive and support), and on the other hand makes impossible an optimal use of the drive systems designed according to the hazard scenarios dominating the service contract.

The constructional relevance can change very quickly on site, correspondingly the mountain only forgives faults in exceptional cases and sometimes requires quick decisions of all persons involved in the project and the prompt realization of immediate measures.

In case of fault-zones and low rock strengths, extensive plastic zones can form, the shifting of stress can moreover be effected over more than 3 times the drill diameter.

The employed tunnel boring machines manufactured by Herrenknecht proved, however, that they are in a position to master technically essentially more critical situations than were provided in the service contract.

The construction of the employed TBMs and trailers are subject to extensive adjustments (among others due to the extraordinary conditions) and optimizations during the drive for more than altogether 8 years and nearly 30 km each.

Extraordinary conditions can additionally aggravate the already very demanding technical and logistical challenges.

A close and constructive cooperation between client, author of the project, supervisor of works and enterprise is of essential importance for the success of the project of the structure of the century.

Project, contract and remuneration adjustments are almost inevitable in the sense of a project-oriented overall consideration of optimizations. The realization of these non-contractual adjustments requires a constructive and fair approach by all persons involved in the project.

FINAL REMARKS

The design of the current TBM drives in the Faido section would, from today's point of view, be connected with hardly conquerable contractual and operative problems without the contract section combination with the Bodio section, correspondingly, the offered variant of a contract section combination of Bodio–Faido proved to be the only possible way.

A contract management that is consequent and fair from the beginning of the construction works is of essential and vital importance for the clients as well as for the enterprise and the success of the project of the construction site of the century, the Gotthard Base Tunnel—combination of contract sections Bodio–Faido.

Solutions that can be timely and technically realized are, at the present stage of the project of the contract section combination, no longer possible without above-average professionally and socially competent commitment at all stages, with an optimum utilization of existing scopes for development, or scopes for development still to be created.

At present, in both drives still nearly 6 or 7 km towards the northern section border to the Sedrun section have to be worked. We expect the breakthrough for the second half of the year 2010. In parallel to the TBM drive, the excavation and lining of the cross passages, the rehabilitation of the squeezing zones of the TBM stretches, the preparation of the inner shell of the single-track tubes, as well as the interior works of the underground station, the multifunctional site Faido, approx. in the middle of the altogether nearly 30 km drive section from Bodio, will be done.

The client assumes today that the complete section will be commissioned in 2017. In this respect, in particular with respect to term optimizations across the contract sections, great efforts are still required.

KEY DATA

Gotthard Base Tunnel South (GBTS)
Contract Section 554 Bodio, Contract Section 452 Faido

Region

Switzerland, Ticino

Client

AlpTransit Gotthard AG, Lucerne
(Affiliated company of the Swiss Federal Railways SBB)

Planning

Ingenieurgemeinschaft Gotthard Basistunnel Süd (IG GBTS)
consisting of:

- Lombardi SA, Ingegneri Consulenti, Minusio
- Amberg Engineering AG, Regensdorf
- PÖYRY AG, Zurich

Execution

Consorzio TAT, consisting of:

- Implenia Bau AG, Aarau (CH), sponsor of the Joint Venture
- Hochtief AG, Essen (D)
- Alpine Bau GmbH, Salzburg (A)
- CSC Impresa Costruzioni SA, Lugano (CH)
- Impregilo SpA, Milano (I)

Key Data

Time of construction rough work:	2001–2015
Original final construction costs*:	1,576 Mio. CHF (1,414 Mio. US$)
Currently pred. final construction costs*:	2,100 Mio. CHF (1,884 Mio. US$)
Length:	2 × 30 km
Max. overburden:	2.500 m
Excavation diameter Bodio section:	8.83m / 8.93m (shifted)
Faido section:	9.43 m / 9.53 m (shifted)
Incline:	6.67 ‰

Employed TBMs

Type:	Herrenknecht S-210, S-211
Performance Cutterhead Drive:	3,500 kW
Equipment:	66 pcs. roller bit (17 inch)
Length of the trailer construction:	465 m

Particulars

- Interior works in parallel to the drive works
- All transfer of material and persons in rail operation
- Processing of the excavation material to broken aggregates for concrete production

* without inflation, base 2001

BIBLIOGRAPHY

Lecture "Bodio Lotto 554—Experience after half of the drive" at the occasion of the Swiss Tunnel Congress STC 2005, Dipl.-Ing. Böckli Olivier.
"Die neue Gotthardbahn." AlpTransit Gotthard AG, 2005.
Lecture "Gotthard Base Tunnel South, Faido Section, Previous Experience with the TBM drive" at the occasion of the Swiss Tunnel Congress STC 2008, Dipl.-Ing. Böckli Olivier.

PART 19

TBM Case Studies 1

Chairs
Shemek Oginski
J.F. Shea

Carl Christensen
J.F. Shea

TBM TUNNELING AT THE ASHLU HYDROPOWER PROJECT, SQUAMISH, BC

Serge Moalli ▪ Frontier Kemper Constructors ULC

Steve Redmond ▪ Frontier Kemper Constructors ULC

Dean Brox ▪ Hatch Mott MacDonald

Peter Procter ▪ Hatch Mott MacDonald

Daniel Jezek ▪ Hatch Mott MacDonald

Michelle van der Pouw Kraan ▪ Hatch Mott MacDonald

Richard Blanchet ▪ Innergex Renewable Energy Inc.

Robert Kulka ▪ Innergex Renewable Energy Inc.

ABSTRACT

The Ashlu Creek Hydro-electric project, located north of Squamish, BC, at the confluence of Ashlu Creek and the Squamish River, comprises a 49 MW run-of-river project and features a 4.08 m diameter, 4.4 km power tunnel and a 128 m deep drop shaft. The Ashlu Creek Hydro-electric project is being developed by Innergex under Ashlu Creek Investments Limited. Site civil works are being completed by Ledcor CMI of Vancouver with tunnel construction by Frontier Kemper Constructors ULC under a lump sum design-build contract. Site works got underway in September 2006 with the construction of the tunnel portal and laydown area. These site works required rock slope stabilization over a 50 m rock slope to secure the laydown area as well as the construction of a tail track muck tipping station and structural wall to facilitate TBM excavation and spoil removal. Starter tunnel construction commenced in early 2007 and TBM tunneling started in the late spring of 2007 and is planned to continue until early 2009. The power tunnel is being excavated using a 1,200 kW Wirth open-type hard rock TBM dressed with 31–17″ cutters. The TBM featured two fixed mounted rock bolt drills behind the finger shield: these were removed and replaced with jacklegs to improve access. A probe drill at the end of the main beam is also present. The geology along the 4.4 km power tunnel comprises very strong and generally massive granitic bedrock with rock strength greater than 225 MPa. A unique feature of the project is that no subsurface investigations have been completed for the tunnel however extensive bedrock outcrops are present along the Ashlu Creek valley. A historical quarry exists approximately mid-way along the tunnel alignment that has exposed good quality granitic bedrock. TBM tunneling to date has achieved advance rates of 10–20 m/day with a maximum of 23 m and has encountered all of the anticipated six major fault zones where the installation of various standard high capacity tunnel support was successful to facilitate TBM advance through these sections.

Figure 1. Ashlu Creek and intake

INTRODUCTION

The Ashlu Creek Hydropower Project is currently under construction near Squamish, BC some 120 km north of Vancouver in the Sea to Sky corridor.

The hydropower project is being developed by Ashlu Creek Investments Limited, of Innergex Renewable Energy Inc. of Montreal. The project comprises a 49 MW run-of-river hydropower project that includes a 4.4 km power tunnel and a 128 m drop shaft. The project also includes an inflatable weir intake structure and a surface powerhouse. Figure 1 shows the intake area of Ashlu Creek.

The surface civil works are being completed by Ledcor CMI of Vancouver. The underground works are being completed under a fixed price lump sum contract by Frontier Kemper Constructors ULC. The underground works designer is Hatch Mott MacDonald of Vancouver and RSW/Golder is responsible for the design of the intake and powerhouse components.

The power tunnel is being excavated by tunnel boring machine (TBM) and the drop shaft will be excavated by raisebore drilling. Construction commenced at the TBM portal in late 2006 with rock excavation and slope stabilization works. TBM excavation commenced in late May 2007 and is planned to be completed in February of 2009. TBM tunneling to date has achieved advance rates of 10–20 m/day with a maximum of 24 m and has encountered seven major fault zones. Three of these required the installation of standard high capacity tunnel support to facilitate TBM advance through these sections.

PROJECT LOCATION AND LAYOUT

The Ashlu Creek Hydropower Project is located approximately 40 km northwest of the town of Squamish, BC which is about 120 km north of Vancouver, BC. The Ashlu Creek Hydropower Project is sited on the Ashlu Creek, a tributary of the Squamish River.

The hydropower project is a run-of-river project whereby water is pooled behind a small inflatable intake structure and diverted via a 128 m drop shaft and along a 4.4 km power tunnel. The design flow for the project is 29 m^3/s and the total head is 190 m. The project layout is shown in Figure 2.

The power tunnel is a dead-end tunnel alignment with a grade of +1% for the first 2 km, +1.5% for 500 m and 2% for the last 1.94 km. The tunnel alignment contains one

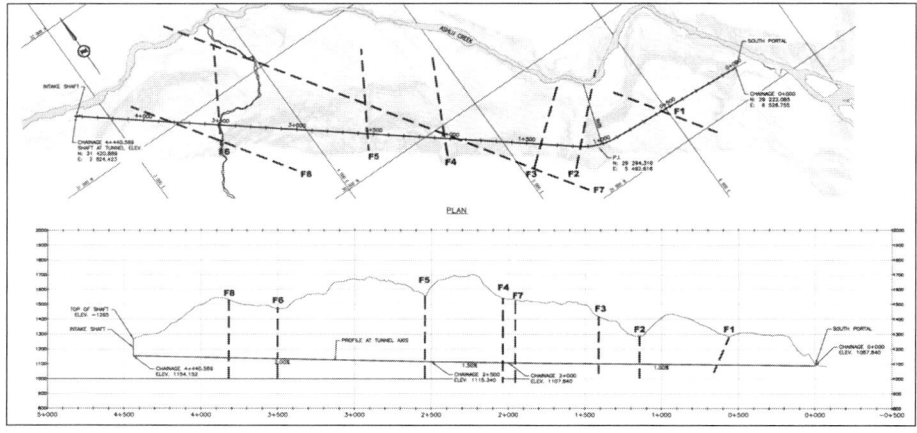

Figure 2. Power tunnel plan and profile

Figure 3. Exposed granitic bedrock at quarry

horizontal curve with a radius of 450 m. The maximum cover along the tunnel is about 600 m over a 1 km section in the middle of the alignment.

TUNNELING CONDITIONS

Geology at the project site comprises very strong granitic bedrock as exposed along the sidewalls of the Ashlu Creek valley. The anticipated quality of rock along the alignment can be appreciated from a rock quarry present along tunnel alignment indicating massive, widely fractured granitic bedrock. Figure 3 shows a picture of the exposed rock at the quarry.

Six major fault zones were originally inferred to range from 15 m up to 25 m in width. These features were inferred from topographic depressions/gulleys along the alignment as well as distinct changes in the creek alignment and are included in Figure 2. Rock strength was estimated to be approximately 225 Mpa based on a single laboratory test prior to construction.

Figure 4. Portal rock stabilization

PORTAL CONSTRUCTION FOR TBM SET-UP

The tunnel portal was sited at the base of a very high natural rock slope with an area of about 1.1 hectares. Given the restrictive area available it was necessary to design a TBM portal laydown with a curved tail-track arrangement that would facilitate TBM assembly as well as muck tipping during tunnel excavation. Such restrictive areas for tunnel portals are common among the steep side slopes in narrow glacial valleys in BC.

The construction of the TBM portal required the excavation of 15 m of loose large rock blocks at the base of the portal slope as well as excavation of massive bedrock. Some of the large blocks required to be blasted. Several loose blocks were present along the natural portal slope that required scaling followed by rock stabilization. Several high capacity rock bolts were installed along with slope mesh over extensive areas with limited areas of shotcreting to secure the portal for tunnel construction. Rock excavation was completed by Ledcor CMI and BAT construction completed the slope stabilization works. A limited amount of drainage works was also completed to divert surface flows along gulleys during heavy precipitation events. Figure 4 presents some of the rock stabilization works at the TBM portal.

TBM TUNNELLING

Drill and Blast Starter Tunnel

A TBM starter tunnel was excavated to a length of 30 m to facilitate the starting of TBM excavation. The 5 m wide starter tunnel was excavated by RokTek Services of Prince George, BC using a single boom jumbo as shown in Figure 5. Relatively good quality rock conditions were encountered throughout the starter tunnel and it was supported with spot rock bolts and fully shotcreted for safety purposes for the installation of the concrete starting cradle and start of TBM tunnel excavation.

TBM and Tunneling Equipment

The TBM selected for the project was a Wirth 400/450 III 1200 kW open type TBM dressed with 31–17" cutters. The TBM has four double-shoe sidewall grippers behind the cutterhead and four single-shoe sidewall grippers further behind as shown in Figure 6.

Figure 5. Drill and blast starter tunnel

Figure 6. Wirth open beam TBM at portal

TBM mucking was done using diesel locomotives and 4–10 m^3 muck wagons. One of the key scheduling challenges with the use of the TBM for the excavation of the power tunnel was the desire to install the main high-voltage power connection to the project site in order to provide grid power for the TBM without the need for generators.

The main 400 kVA transformer for delivering power from the powerhouse to the BC Hydro grid was transported to the project site from Texas, USA via road and barge to Squamish and finally by special multi-wheel flatbed trailer in time to allow for the start of TBM excavation without the need for generators.

Encountered Conditions

Fault Zone F1 was encountered at Ch ~ 640 m which was about 150 m prior to the expected location of this fault and thus this feature was inferred to be dipping. The rock conditions encountered at Fault F1 comprise closely fractured, medium strong rock slabs with no presence of weak fault material. This fault zone was oriented sub-vertically and intersected at approximately 45 degrees towards the tunnel axis as anticipated. Figure 7 illustrates rock conditions at Fault F1.

Fault F2 was encountered at CH ~ 1,035 m for a width of about 85 m over which variably altered rock containing weakly healed mineral coatings/infilling was observed.

Figure 7. Rock conditions at Fault F1

Figure 8. Fault F2 conditions

X-ray diffraction (XRD) testing of this coating/infilling confirmed the presence of laumontite which is a zeolite mineral with swelling characteristics. Figure 8 illustrates the closely spaced fractured associated with Fault F2. An unexpected geological fault designated Fault F7, was encountered over a length of 46 m at about CH 1950 m. This feature proved to be of the same west-northwest orientation as Fault F1, which is also the overall orientation of Ashlu Creek. Upon intersection of Fault F7 an evaluation was undertaken to consider whether an additional similarly oriented fault could be present along the remaining tunnel alignment. From these known faults it was postulated that an additional fault, Fault F8, could be anticipated with the same west-northwest orientation as a repeating structural feature with a perpendicular offset spacing similar to that between F1 and F7 of about 600 m. Fault F8 was thus anticipated to be intersected at about CH 3800 m as shown in Figure 2. However, at the time of writing TBM excavation has surpassed this location without the intersection of any major fault zone.

Geological and geotechnical mapping has been carried out in stages (non-concurrently) during tunnel excavation. At the time of writing the main rock fractures that have been encountered (to CH ~ 2,225 m) are presented in the stereonet of Figure 9 and include a total of three (3) main fractures and seven (7) minor sets. The encountered number of main fractures sets and their orientations are generally consistent

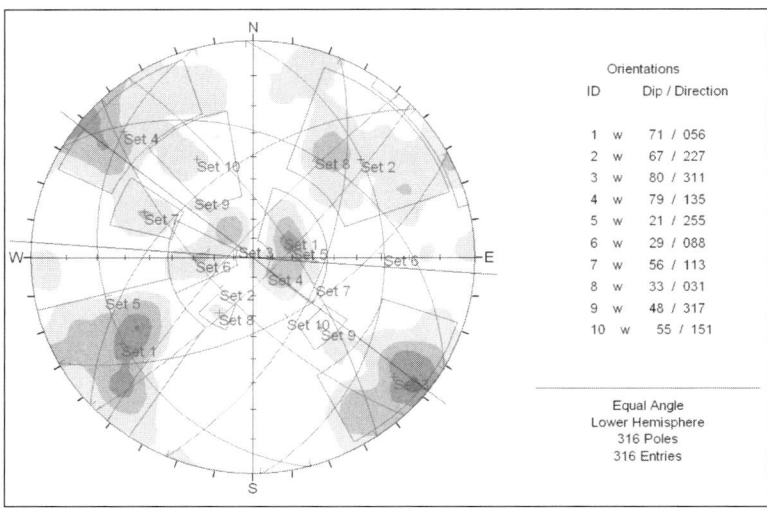

Figure 9. Stereonet of main fracture sets

with pre-construction mapping data that was collected from outcrops along the valley. Fracture spacing has generally ranged from 0.5 m to 2.0 m.

Updated estimates of rock strength were obtained from cores drilled from block samples extracted from the tunnel at CH ~ 1,300 m. The results of uniaxial compressive strength (UCS) tests indicated rock strengths ranging from 160 MPa to 225 MPa with an average of about 175 MPa. Groundwater inflows have been relatively consistent at approximately < 6 l/s (90 US gpm) so far and below predicted amounts.

Inferred Overstressing

Overstressing is inferred to have occurred at approximate CH ~ 1,300 m under a cover of about 285 m. These conditions are believed to have occurred as a result of the presence of variably oriented, closely spaced fractures that are weakly healed with laumontite as confirmed from XRD testing. The overstressing is concentrated along the right-hand haunch area between the 12 o'clock and 2 o'clock positions which is consistent with stress concentrations that are expected to be parallel to the steep sidewalls of the Ashlu Valley. The occurrence of overstressing was investigated for the deeper sections of the tunnel in order to evaluate the potential extent of overstressing and failed rock requiring support. Back-analyses were first completed to simulate the observed extent of overstressing at shallower depths. Figure 10 presents the results of predicted overstressing under the maximum cover of 575 m and indicates possible failure of rock to extend to about 1.2 m along the haunch area.

This occurrence of overstressing has been observed under the maximum cover in moderately altered rock near CH 2,650 m where the drilling of rock bolts for pinning of the steel ribs encountered "soft" penetration over a depth of less than 1.0 m before "hard" penetration. This information appears to be consistent with the extent of predicted overstress for these rock conditions. The occurrence of this overstressing is shown in Figure 11. These support measures has been successfully applied by the drilling and installation of rock bolts through the TBM finger-shield to provide early support and have been effective in preventing the dislodgement of rock slabs onto the TBM.

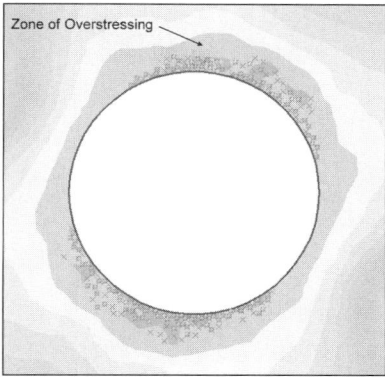

Figure 10. Predicted overstressing at maximum cover

Figure 11. Support for overstressed conditions

TBM Excavation Progress

TBM excavation was done with 2 shifts—7 days per week. Excavation progress has varied from 10–20 m/day with a maximum daily progress of 23 m. TBM excavation progress through the difficult fault zones where heavy capacity support comprising steel sets were installed achieved typical progress rates of 3 to 5 m/day.

TBM Tunnel Support

The TBM tunnel has been excavated through mainly good quality rock conditions to date where the tunnel is stable and tunnel support has not been needed. Where rock support was needed for blocky rock conditions, pattern rock bolts in conjunction with welded wire mesh and/or C100 channels were installed. Figure 12 illustrates a typical example of pattern rock bolts and mesh support in the TBM Tunnel.

Tunnel support for fault zones has comprised W100 x 19 steel sets in conjunction with mesh as initial support prior the application of shotcrete as part of the final support and lining. Figure 13 shows the steel sets installed at Fault F7.

Figure 14 presents a summary pie chart of the tunnel support installed to CH ~ 2,225 m.

Figure 12. Typical tunnel support

Figure 13. Steel set support at Fault F7

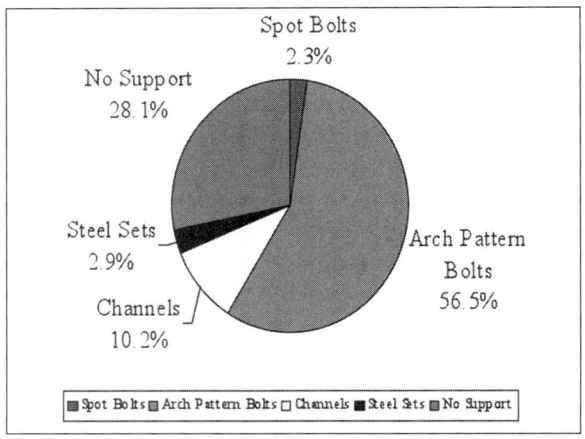

Figure 14. Summary of tunnel support

Figure 15. Muck tipping station

Final Support and Lining Requirements

The final tunnel support and lining requirements were evaluated over each section of the tunnel during excavation. These requirements typically comprised shotcrete reinforced with mesh where fault and highly fractured zones were intersected. Grouting will be carried out at fault zones where voids have formed behind the steel sets in conjunction with the final shotcrete lining.

Spoil Disposal

A spoil tipping station was constructed immediately outside the TBM portal to allow easy discharge of TBM tunnel muck to be picked up by the prime contractor and hauled away to the spoil disposal site located half way up the valley towards the intake.

The TBM muck tipping station was designed as a split level (two-bench) station that required scaling and stabilization of a high rock slope area as well as re-alignment of a Forest Services Road that was required to be maintained and open for public use. The TBM muck tipping station comprised a 6 m high loc-block wall assembly with a length of 50 m. The TBM muck tipping station contains dual hydraulic jacks/tipping rams to allow for the discharge of two muck wagons at a time. The wall was anchored into bedrock with grouted rock bolts. Figure 15 shows the TBM tipping muck station.

ACKNOWLEDGMENTS

The authors gratefully acknowledge the permission of Innergex Renewable Energy Inc. to publish this paper. Geological and geotechnical/rock support mapping of the tunnel was carried out by Daniel Jezek, P.Eng., and Michelle van der Proun Kraan, a UBC Geological Engineering summer student employed by Hatch Mott MacDonald during the summer of 2008 and 2009 university school year.

TBM AND NATM COMBINED SOLUTION FOR A VERY DEEP TUNNEL—THE "PAJARES" CASE

Enrique Fernández ▪ Dragados S.A.

Paz Navarro ▪ Dragados S.A.

Alejandro Sanz ▪ Dragados S.A.

INTRODUCTION

The high speed rail link between Madrid and Asturias, in the North of Spain, has encountered in the Cantabrian range the biggest challenge of the whole line, a very deep tunnel, 25 km long through a foreseen difficult geology. This tunnel was divided in the following four lots for construction:

- Lot 1: Pola de Gordón- Folledo. L=10,7 km
- Lot 2: Folledo-Viadangos. L=4,166 m
- Lot 3: Viadangos- Los Pontones. East tube. L=10,4 km
- Lot 4: Viadangos- Los Pontones. West tube. L=10,4 km.

Dragados S.A. leading a JV was contracted to design and built the Lot 2 on the captioned Pajares High Speed Railway Tunnel. The Lot 2, comprising the middle third section of the 25 km long tunnel has an additional challenge: the construction of one access from the surface to allow it own construction. Feasibility reports and preliminary design do not foresee the use of TBM due to the risk of squeezing ground at the deepest area of the tunnel. On the other hand, conventional NATM excavation of the twin tunnels and ancillary works, including access shafts were against the expected construction schedule by the client, due to the poor ground foresee at the early stage.

PRELIMINARY DESIGN

The preliminary design, for bid purpose, considered two access shafts, 580 m deep and 9 m diameter, which would have given access to a cavern located across the tunnels axis. This cavern would have been used to excavate 2 × 2 km tunnels, one for each direction, by NATM. Above the cavern, a net of galleries allows the ground improvement at cavern area by top down long fiberglass grouted bolts. The design was based on geotechnical studies that expect the existence of very low geotechnical quality ground, San Emiliano shales, in the central cavern zone. Also Lot 1 expect similar ground conditions in the last 3 km and consequently a decline tunnel to accede at sta. 7,000 was considered to execute these 3 km by NATM as well.

This technically complex design would have put the deadline which in the contract was 5 years at risk, so the construction design main challenge should found a technical solution to fulfill the contractual schedule.

CONSTRUCTION DESIGN

The first activity on site following the contract award was the execution of very deep core logs in order to verify the ground condition at the tunnel elevation. These logs give

Figure 1. Access ramp tunnel portal

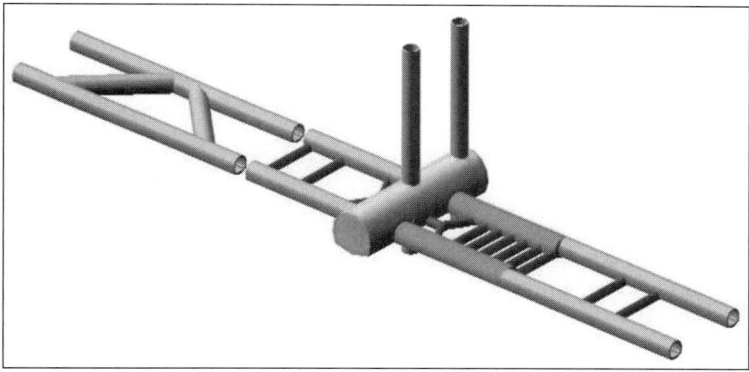

Figure 2. Preliminary design sketch

tremendous valuable information related to squeeze ground conditions. Based in this information a completely different design than the preliminary one was made in order to accomplish with the contractual schedule.

The proposed and approved solution consists of, instead of the Viadangos access shafts, a decline access tunnel to excavate the East tube, 5.466,55 m length and 6,2% downward, that allows the access to the main tunnels, at the beginning of the contracted stretch. This access ramp tunnel portal is located in a village called Buiza very close to the Lot 1 ramp portal, captioned above. A Double shield TBM was selected to built this ramp.

From the access ramp tunnel, the captioned TBM Double shield, proceed driving the East tube using a transition curve onto the main alignment and leaving a short section of tunnel to the Lots 1/ 2 contract boundary. The connection between the Access Tunnel and the Main Tunnel requires the construction of a large-span Bifurcation Chamber ("Bifurcation East") and a short mined tunnel, which has been constructed using the New Austrian Tunneling Method.

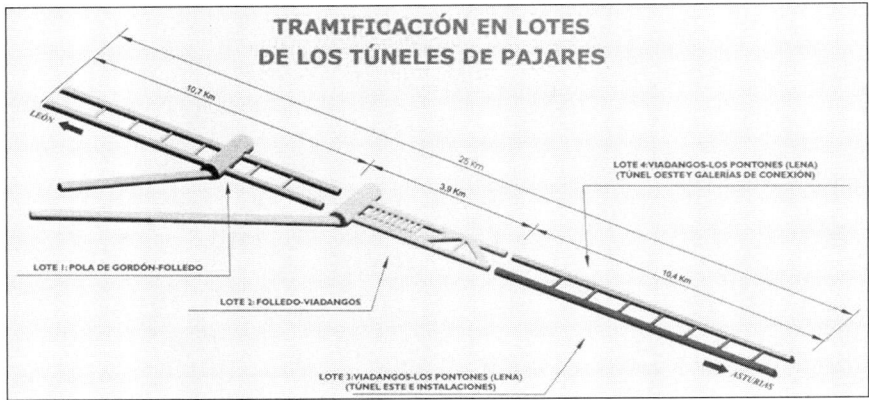

Figure 3. Construction design sketch, lot division

In the Construction Design, the excavation of both main tubes was expected to be carried out by the same TBM, which required a complex execution process sequence in the rock mass, 800 m deep:

- TBM dismantling in a cavern after finishing the first tube (east tube)
- Execution of a big cavern for the TBM assembly to excavate the second tube (west tube)
- TBM assembly inside the cavern and excavation of west tube

Furthermore, cross-passages were designed to be executed between the main tubes every 400 m, as galleries that accommodate the signaling and electrical equipments. Also a preferential stop point next to the junction between the access gallery and the main east tunnel has been designed. This preferential stop point is 400 m long and is composed by a third tube located between the two main tubes, with a smaller cross section, and connected to them with 25 m long cross-passages.

GEOLOGY AND HYDROLOGY

Access Ramp Tunnel and Main Tunnel Geology

The Cantabrian Range is mainly the result of an upsurge by compression of faults and areas of hercynian nature. This compression was caused by the rising of the Pyrenean and Betica mountain ranges during the Alpine orogeny, giving rise to the current formation and very complex geological structure.

In the area of the Pajares Tunnels corridor there are materials which form a very complete sequence of the Palaeozoic era.

The hercynian processes, which stretched greater intensity during the Superior Carboniferous stage, moulded these materials until their subsequent reactivation, in a highly complex orogenic dynamic which gave rise to the strong deformation of these materials and their low ensuing geomechanical quality. Different structural units appear in the area in accordance with type of deformation:

- Region of Folds and Beds (Juliver,1967), characterized by the absence of metamorphisms and the development of overthrust faults and their associated folds.

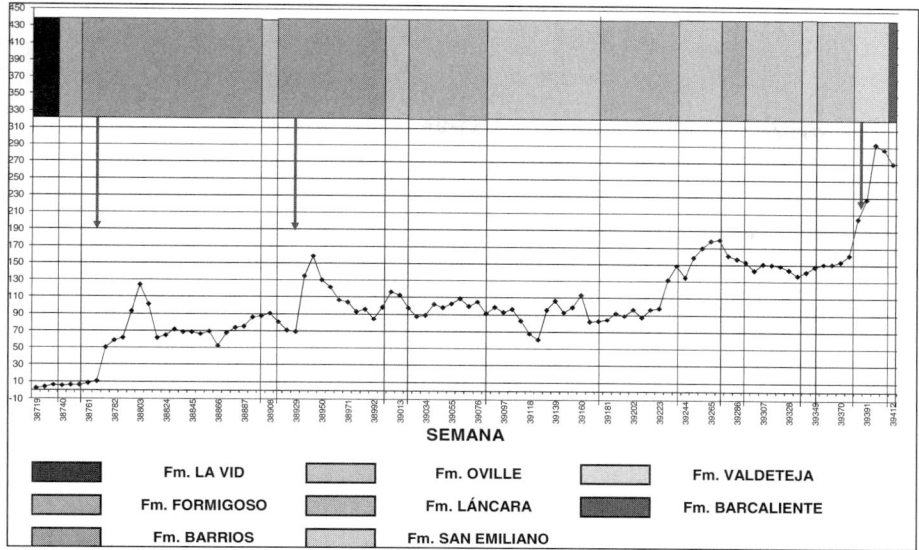

Figure 4. Geological and hydrogeological profile after the construction of the access ramp tunnel (Average week flows and geological formations)

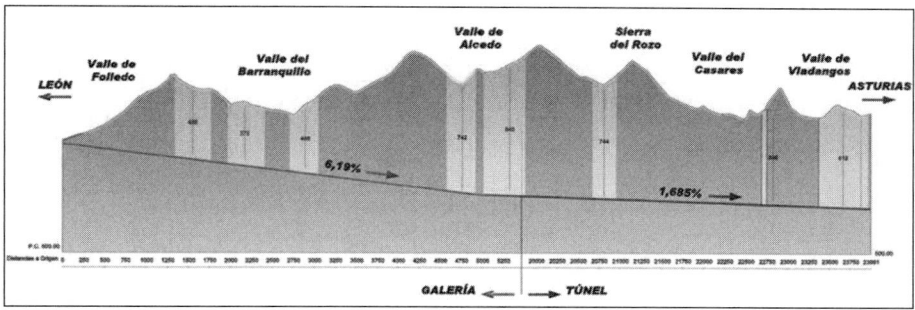

Figure 5. Access ramp and East main tunnel stretch Lot 2 longitudinal profile

- Central Carboniferous basin lying below the two units mentioned above.

The access ramp tunnel has been bored through numerous formations:

- Slate-sandstone formations: Oville, Formigoso, La Vid, San Pedro
- Sandstone-quartzite formations: Barrios
- Carbonated formations: Láncara, La Vid

Two different aquifers have been encountered during the excavation depending on the lithological characteristics of the bored materials:

- Detritic: mainly made of quartzitic materials from the Barrios formation. These materials are related directly with the fracturation and weathering rates
- Carbonated: essentially composed by calcareous materials from Grupo La Vid and Láncara, Valdeteja and Barcaliente formations.

Bifurcation Cavern Geology

The Láncara Formation is the predominant geological unit at the location of the Bifurcation Chamber. The formation mainly consists of dolomite, red limestone and grey limestone.

After the construction of the access ramp tunnel, three exploration borings were drilled in the bifurcation area to evaluate the geological situation. Massive, homogeneus and competent grey limestone and dark dolomite displaying high strength were encountered. One boring encountered dark dolomites and dark shale alternately, 50% proportions for each of the two types. Similar material was encountered during the investigations at the cross-passages with corrected RMR values in the range of 50 to 62.

Geological and Hydrogeological Problems

From the beginning of the project the technicians were conscious of the difficulties of the tunnel excavation through the Cantabrian Range. This matter forced the excavation of the tunnels with the help of intermediate tunnels, to guarantee a higher performance and to assure the project feasibility in case of serious incidents which could occur and limit the advance in any of the portals.

Different problems have turned up during the works, in the four Lots, most of them caused by geological or hydrogeological reasons. In Lot 2 main issues were:

- Floods caused by water leaks in fractured and/or sandy quartzite formations described in section 4.4 of this document
- Operation and maintenance logistic problems during the access ramp tunnel excavation with a 6,2% downward slope. This problem is developed in the section 6.2
- The presence or deflagrate gases presence in the slate formations of the Carboniferous stage. Gas concentration has appeared in the contact between slate and carbonate formations. Gas has been encountered in the "Grupo La Vid" (limestone) and "San Emiliano" (slate) formations in the access ramp tunnel and in the main tunnel. Both the TBM and the ventilation had been configured to detect and decrease gas presence.

Floods Caused by Water Entries in Fractured and/or Sandy Quartzite Formations

During the excavation of the intermediate access ramp tunnel some floods occurred inside the shield. The most outstanding flood occurred on 11 August 2006.

The flood occurred when the cutterhead started the excavation of a sandy fracture in the Quarzite formation. This produced a huge water flow and sand slide towards the shield. Neither the dewatering system nor the pipe net could bear the huge quantity of water and sand which flooded so fast into the TBM.

The water stretched a height of 7 m and the sand tongue filled up the whole shield. Once the water and the sediments were stabilized the cleaning and dewatering works were started. The sediments and muck handling were made by screws, auxiliary conveyor belts and mud pumps.

After the cleaning and the repairs and reconstruction of almost the whole TBM electrical/electronic systems were finished, the production was resumed 25 days after the incidence.

CONSTRUCTION PROCEDURE

The access ramp tunnel has been excavated by a double shield TBM, with a diameter of 10.120 m, in order to avoid the shaft construction risks and make the evacuation

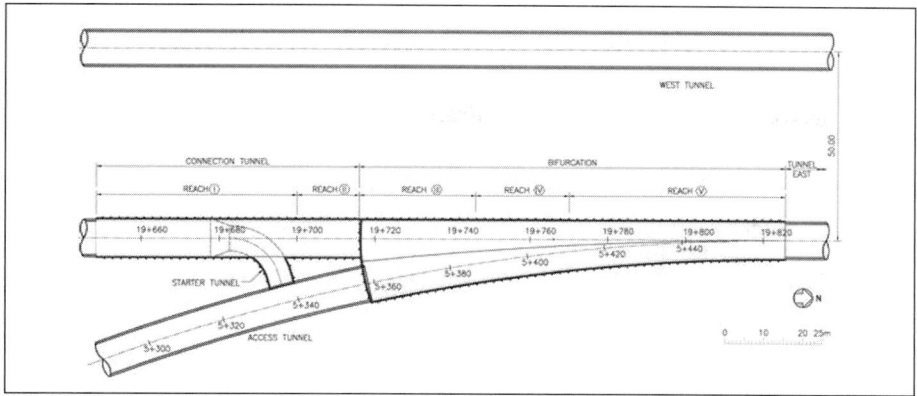

Figure 6. East bifurcation stretches

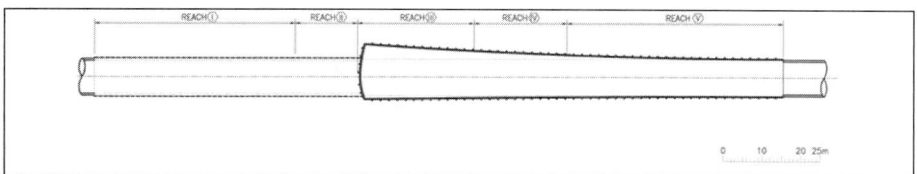

Figure 7. Longitudinal bifurcation profile

in case of emergency. The TBM proceeds driving the East main tunnel stretch using a transition curve onto the main alignment. Once the first drive was completed and the squeezing effect along the main tunnel alignment was quite nil, was taken the decision by the site staff to bore the west tube by using the TBM which excavated the same tube from the south portal along the Lot 1 alignment. With this decision, the construction of the assembly chamber as well as the two bifurcation chambers required to move the TBM from the east tube to the west tube was not required, saving a substantial time in the overall schedule.

The starter tunnel that connects the access tunnel with the main tunnel, the connection tunnel between lots 1 and 2 and the bifurcation chamber are being excavated by NATM method and conventional excavation means: drill and blast method.

To the date, 15 January 2009, the starter and the connection tunnels have been already excavated and supported, by drill and blast, and the initial lining is placed (4 m steel bolts, 6–12 m selfdrilling bolts, shotcrete etc.).

The final lining is going to be executed in the following days. The plant used for the excavation and support placing of this stretch executed by conventional means is the following: jumbo Atlas Copco L2C, Normet 9810 platform, LHD Wagner ST-5B and shotcrete robot PM-407. The project designs define the bifurcation in Figures 6, 7, and 8.

The bifurcation excavation sequence is as follows:

- Starter Tunnel and part of Stretch I heading south
- Remaining parts of Stretch I and Stretch II (Connection Tunnel)
- Top Heading Sidewall in Stretch III

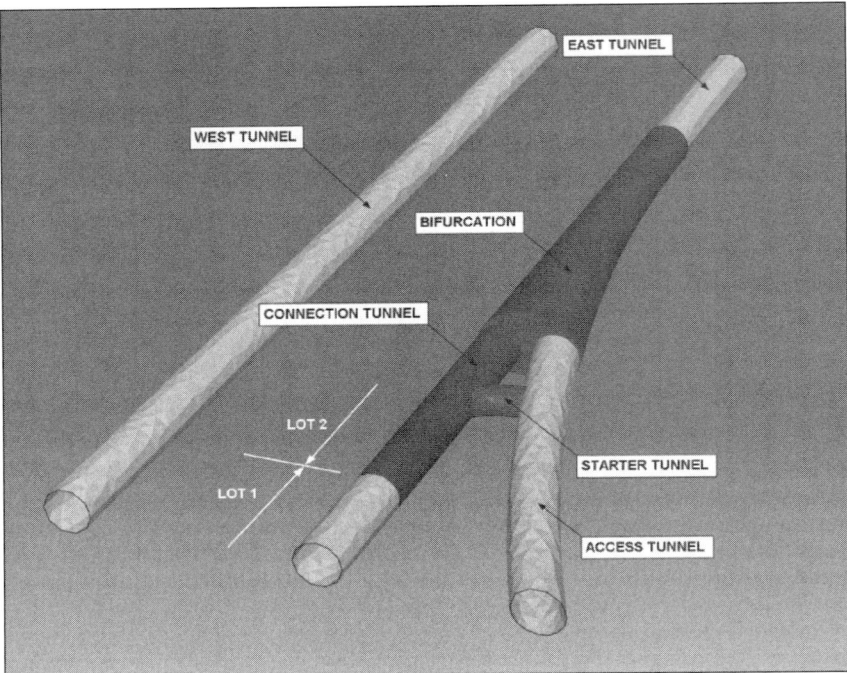

Figure 8. 3D View of tunnel excavations

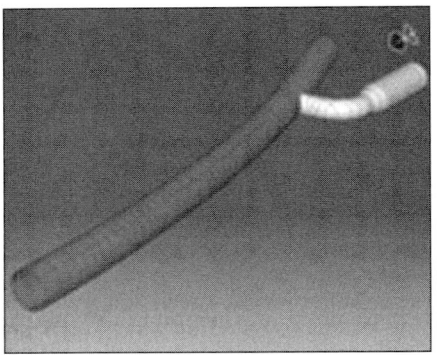 

Figure 9. Excavation of starter tunnel and Stretch I heading south; excavation of Stretch I and Stretch II (connection tunnel)

- Top Heading Niche (Breakout into the TBM Access Tunnel)
- Top Heading in Stretch IV
- Center Top Heading in Stretch III
- Bench/Invert of Stretch III (Sidewall and center) and Stretch IV

TBM AND NATM COMBINED SOLUTION FOR A VERY DEEP TUNNEL   1225

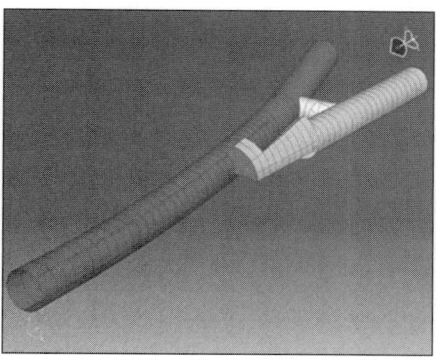 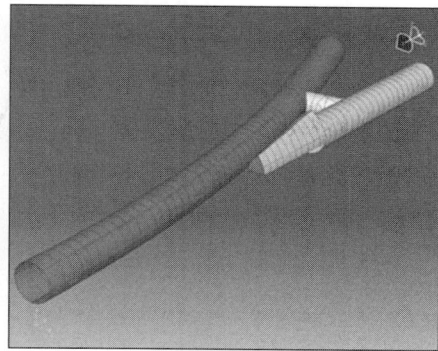

Figure 10.   Excavation of Stretch III top heading (sidewall); excavation of niche top heading

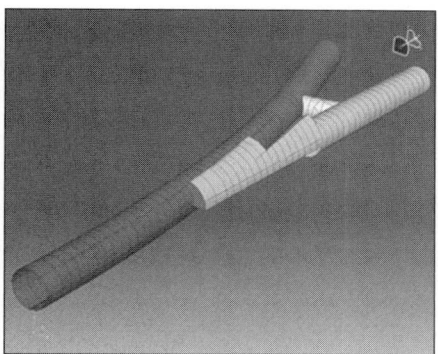 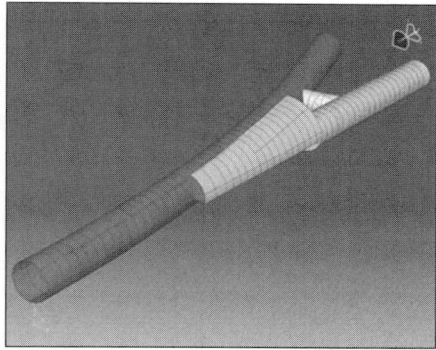

Figure 11.   Excavation of Stretch IV top heading; excavation of Stretch III center top heading

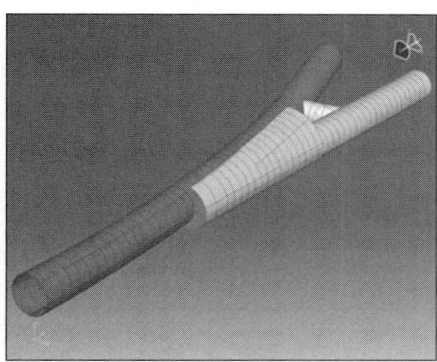 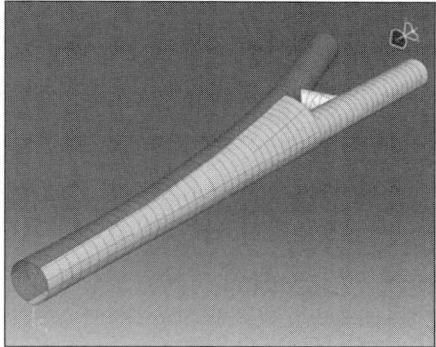

Figure 12.   Excavation of Stretch III and IV bench/invert; excavation of Stretch V top heading and bench/invert

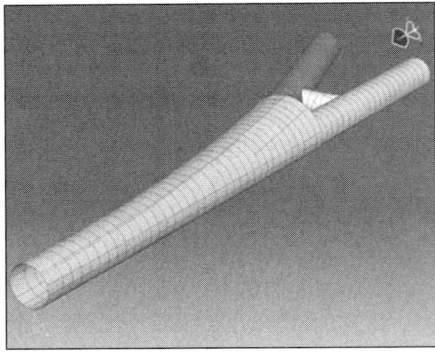

Figure 13. Excavation of remaining parts in Stretches III, IV and V

- West side of Stretch V
- Remaining segmental linings and parts on the east side of the bifurcation for Stretches III, IV and V.

TBM

TBM Main Characteristics

The TBM type chosen to excavate the "Folledo- Viadangos" stretch is a new generation double shield TBM.

The geologic formations of the alignment are very favorable for the use of a double shield TBM due to two basic factors:

- Compressive rock strength medium or low
- Low abrasivity

There had been foreseen circumstances that could cause delays during the excavation; some of them are pointed in the following lines:

- Excessive convergences that could cause squeezing problems
- Heading instability in fracture zones
- Probability of presence of karstic zones

In accordance with the geologic conditions detected, the TBM design has fulfilled its two most important aims:

- Keep excavation rhythms elevated
- Overcome the foreseen problematic zones with less delay

This new TBM, is designed to bore through hard rock with a high thrust per disc cutter and an elevated power due to the "gripper system" that works in the rear shield, with an elevated torque, that the cutterhead develops, allowing its advance with security in unstable grounds and in soft or low cohesive ground.

The TBM was designed with the minimum possible length, compatible with the foreseen lining and additional disc cutters were placed, up to 300 mm. to increase, if required, the free space between the shield and the rock, adapting the possible convergences and avoiding the TBM squeeze. The conicity, the difference between the cutterhead and the tail diameters, collaborates on the decrease of the squeeze risk.

Figure 14. Pajares Lot 2 double shield TBM

Figure 15. Pajares precast segments (500/ 600 mm thicknesses)

It is important that in every type of ground the lining may be placed always under the protection of the rear shield, this allows for maximum security, almost total security for the works.

The segments have two different thicknesses 500 and 600 mm, with variable steel quantities and concrete strength, according to the bored rock quality and the existing overburden.

The double shield TBM allows simultaneous excavation and segment placement achieving elevated performances.

The main TBM characteristics are shown in Table 1.

Operation and Maintenance Logistic Problems During the Access Ramp Tunnel Excavation with a 6,2% Downward Slope

The excavation of the intermediate access ramp tunnel of Buiza has been one of the challenges of the Pajares Tunnels Project, with a 6,2% downward slope and 5,45 km length, its construction has represented a singular logistic and operational problem for the TBM tunnel. Along the ramp length there is a 310 m drop between the portal and the end of the gallery.

Table 1. TBM characteristics

TBM	
TBM Type	Double shield
Back-up type	Open, over rail
Length	12,5 m
TBM+ back-up length	125 m
TBM weight	1.550 t
TBM+ back-up weight	2.400 t
Lining	
Type	Precast segments
External diameter	9,700 mm
Internal diameter	8,500/ 8,700 mm
Segment length	1,500 mm
Segment thickness	600 / 500 mm.
Ring	6+1
Cuttinghead	
Excavation diameter	10,120 mm.
Disc cutters	17" (432 mm)
Thrust	17,088 kN
Diameter overcut	300 mm.
Weight	230 t
Cuttinghead drive	
Type	Electric, variable frequency drive
Power	5,600 kW (16 × 350 kW)
Speed	0–6 rpm
Maximum torque	23,051 kN.m/0–225 rpm.
Torque with maximum speed	8,538 kN.m/ 6 rpm
Peak starting torque	30,428 kN.m
Main thrust	
Cylinders number	18
Stroke	1,620 mm
Thrust 100 bar	22,195 mm
Thrust 300 bar	67,858 mm
Exceptional thrust 460 bar	104,050 kN
Power	200

The TBM Herrenknecht S-281 had to adequate its equipments to the tunnel slope. The main mechanic design singularities for the TBM are detailed in the following points:

- Flexion overstresses caused by the loads normal component on cranes and suspension elements in shield and back-up (containers unload, segment cranes, erector, etc.)
  - In the case of the segment erector
    - The weight horizontal component causes static loads excess over the main bearing
    - Working in flexion causes often gear reducers and motors brake down
    - Each segments weight normal component causes vacuum failures
    - Often damages caused by fatigue in the vacuum plates
- Special brakes for load longitudinal displacements
  - Due to the high slope some problems emerged in the brake systems in the longitudinal movement devices, the most important problem was the segment crane

- The segment crane had 8 movement engines with brake (instead of the 4 foreseen engines without brakes) and double emergency brake device and even though some incidents occurred
- Conditions in hydraulic devices due to the slope levels variation
  - Frequent problems caused by a capacity reduction in hydraulic tanks, water or gasoil tanks and mortar, gravel and cement (capacity reduction in foreseen storage caused by the slope effect up to 15%)
  - Frequent stops due to hydraulic failures
  - Hydraulic cylinders and control electrovalves
- Incidence in the main bearing caused by the slope change
  - The TBM excavated a transition between the access ramp tunnel and the main tunnel, this produced a slope change from 6,2% to 1,7%, always downward. This slope change caused problems in the main bearing
  - During the access gallery excavation all the impurities and metallic particles were accumulated in the front part of the 16 main bearing reducers, and due to the slope change all these impurities came into the hydraulic system circuits causing many problems and breakdowns.

FACILITIES

The use of a TBM supposes a continuous working mode of excavation with high rates of advance, and even more using a Double Shield TBM. This continuous work needs a continuous supply from the outside, with the adequate lines to join the TBM and the facilities outside the tunnel.

In this case, for the TBM supply and feeding, different equipments were installed near the portal with the corresponding lines extended along the gallery until the TBM. These facilities and networks, described next, were calculated based on the needs of the excavation equipment and have been influenced by the significant downward tunnel slope, 6.2% in the first 5 km, as well as the hydrogeological conditions of the rock massif.

Facilities Outside the Tunnel

Facilities were located near the portal, arranging the main following equipments:
- Segments Stock with capacity to store 700 rings
- 40-ton gantry crane for segments loading and storage
- 40-ton gantry crane for loading materials into the trains
- Electrical substation with a total capacity of 17,500 kVA, where 2,500 kVA targeted to the outside facilities and the rest to the force tunnel line.
- Ventilation equipment consisting of two 740 HP fans connected in series provided with snow protection
- Conveyor belt drive and belt extensor
- Secondary transfer conveyor belt to the final disposal
- Mortar Plant for mortar backfill grouting manufacture
- Dewatering equipment, lines and water treatment
- Warehouse and Workshop

Figure 16. Tunnel portal with the main service lines and services: ventilation, conveyor belt, electrical power, light, water in and dewatering lines

Inner Facilities and Networks

As the tunnel advanced the networks and lines needed to operate the TBM were prolonged to provide a continuous supply. The following lines were needed during the works:

- Force tunnel line: electrical 20 kV line for the TBM supply and the inner facilities as dewatering intermediate stations, boosters, lighting… As the TBM drives and other equipments work at low voltage, different transformers in the back-up and at every booster station were used.
- Low voltage electric network for lighting, motor control, ground monitoring and grounding wires.
- Ventilation 2.5 m diameter flexible duct for air inflow.
- Dual railroad track system separated 1,000 mm both tracks and railroad tracks for movement of trains in both senses and to allow circulation of big platforms over the central tracks.
- Telephone network to ensure communication within the tunnel and the portal.
- Water line supply
- Dewatering system comprising 5 pumping intermediate stations and lines
- Conveyor belt for muck hauling

Dewatering System

As a deep tunnel with an overburden around 1,000 m and due to the hydrogeological conditions of the rock massif, it was expected to find significant water inflows at high pressure with the subsequent danger of TBM flooding.

Therefore, to ensure dewatering of the TBM it was a prerequisite for the execution of this tunnel, as taking into account the important downward tunnel slope and anticipated water inflows, located at about 350 liters per second, the TBM could have flooded in a short time if a failure in the dewatering system occurred.

The solution adopted was a phased pumping, with 5 intermediate pumping stations. Each pumping station drove water to the next from the station above, and also the water going down on the floor to the tunnel face by using a special metal empty precast invert segment that picked the water up and sent it to the nearest pump station.

Figure 17. Intermediate dewatering pump station comprising: 3 water tanks, 1,000 kVA transformer, 1,000 kVA in generator sets and water pumps

Proceeding this way a gradual dewatering made step by step was obtained, adapted to changing conditions of water inflow and not entirely concentrated on the TBM, thus minimizing the risk of TBM flooding because it was not allowed to stretch the front of all the water that leaked into the tunnel.

Each pumping station comprised of the following, placed in metal structures on the side of the tunnel so as not to disturb the movement of trains:

- Three 18 c. m. capacity metal tanks.
- Group of 3 pumps with a capacity of 90 liters per second and 100 meters high, placed in each tank.
- Three parallel dewatering metal 300 mm diameter lines of connection between pumping stations, with another 300 mm line and two 200 mm lines of connection between pumping stations and metal empty precast invert segment, besides other auxiliary lines, making a total of over 60 km of drainage pipes.
- A 1,000 kVA transformation center 20 kV / 0.38 kV for each pumping station.
- Two units of emergency 500 KVA generators in each pumping station to ensure dewatering in case of power outages.

In addition, 22 metal empty precast invert segments were used to collect water that seeped through the walls of the tunnel and moved toward the face, each with its corresponding pump 110 liters per second and 80 m high capacity.

These empty precast invert segments were open at the top to facilitate the collection of water inside. It represented a weakness during the excavation, because after their placement, the TBM had to lean on them with auxiliary thrust cylinders. To fix it there were reinforced with steel beams that were removed when the TBM was far enough not to push forward the efforts to those empty segments.

As more significant information of the dewatering system, 1,000 kW on electric water pumps installed on the TBM and 4,000 kW more in water pumps along the tunnel, with 18,500 kW in transformers and 5,300 kW of them secured by 10 generators throughout the tunnel. Also 60 km of steel pipe (20 km 200 mm pipe and 40 km 300 mm pipe), 22 "empty precast invert segments" with a capacity of 4.5 m^3 installed. The dewatering system was able to evacuate peaks flow up to 600 l/s.

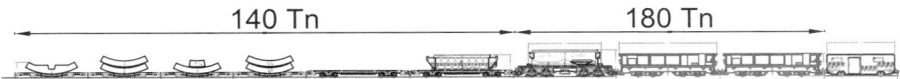

Figure 18. Train composition for material transport

Figure 19. Train at the portal (left) and precast concrete segments in cars (right)

Materials Transport

Supply trains were used for material transport throughout the tunnel. The muck transport in TBM excavation may be done by muck cars or by a conveyor belt along the tunnel but the muck cars have the inconvenience of having the need of powerful locomotives to move the train along the tunnel, with high fumes emission and it has the inconvenient of being a discontinuous transport system. In addition, due to the important slope of 6.2% and water infiltrations in the gallery, adherence between rails and train wheels was difficult so, trains were used only for material transport and the muck hauling was carried out by means of a conveyor belt.

The trains included fixed-body segment cars for segment rings transport, mortar tanks cars (w/mixer), personnel car and one flat car, corresponding to one TBM advance stroke.

It was necessary to go down a slope of 6.2% with a maximum load of up to 320 T, so the main problem was more to curb its descent that the traction needed to return to the surface without materials charge. Given the peculiarity of this situation it was decided to have 2 diesel hydrostatic locomotives SCHÖMA CHL-350BB (510 HP each) and 1 hydraulic tandem CHL-200 for each composition, which makes a total of 12 bogies powered by 12 hydraulic motors by train. Both locomotives and tandem were synchronized to avoid disorders during the march.

Given the complexity of slowing down loaded trains with slopes of 6.2%, locomotives had a triple braking system to work with the diesel engine machines: a brake pneumatic system, a hydraulic brake system and a Voïth brake hydraulic retarder.

In order to improve security conditions, a complementary maintenance of railroad tracks plan was established, as well as protocol sandblasting "dry" silica sand in tracks, essential to ensure both traction and braking of trains.

Muck Transport: Conveyor Belt

The muck was transported by means of a conveyor belt along the tunnel with a band reserve to allow its automatic enlargement while the tunnel was being bored.

TBM AND NATM COMBINED SOLUTION FOR A VERY DEEP TUNNEL 1233

Figure 20. Main conveyor belt with its band extensor at the portal

The conveyor belt was chosen in accordance with the material grading, tunnel geometry and estimated excavation rates. The reel was together with the TBM back-up and advanced at the same time.

A band extensor was placed at the portal to allow the advance of the tunnel excavation without carrying out any additional intervention. The band reserve was automatic and while the tunnel advanced, the tensor car allowed the movement of an equivalent belt length. Tensions caused by both slope and alignment of the tunnel, especially in the intersection between the main tunnel and gallery made operation difficult and needed the help of boosters, located at key points to ensure the adequate tension distribution.

The important length and slope of the tunnel required that the conveyor belt drive had to operate a total length of 19 km with a peak elevation between its tail in the TBM and the main conveyor belt drive at the portal of 400 m. Only the weight of that belt represented 400 T, which had to be added to the weight of the muck transported (300–350 T/pull length).

As the tunnel grows, the transmission capacity was increased connecting transmission intermediate unities (boosters) in intermediate points of the tunnel. Considering the tunnel length and its curves, to evacuate the 1,000 T/h were installed 6 simple boosters 600 HP in the upper band to increase traction and allow the muck handling from the heading without losing the alignment or performance because of the curves. To guarantee the boosters electric supply the TBM medium voltage force line was employed and a transformer station was placed nearby each booster. During the execution, it has been a big deal to make work all the boosters and the main conveyor belt drive in synchronized. Experience has shown in this case that the speed of the TBM and therefore the tunnel excavation rate has been significantly dependent on the ability of the conveyor belt, rather than other factors foreseen a priori.

LESSONS LEARNED

From the construction of the Pajares lot 2 contract several interesting lessons has been learned. From one side and once again, a better known of the ground conditions to be encountered along the alignment is provided on the bid documents or during the design stage, will revert in less stoppage or delays during construction. If it's true that deep bore logs are not always easy to be done, nor to read the results, in this specific case the squeezing expected ground represent a key decision in terms of use TBM or NATM excavation method, and the TBM selection which was finally decided was confirmed during construction.

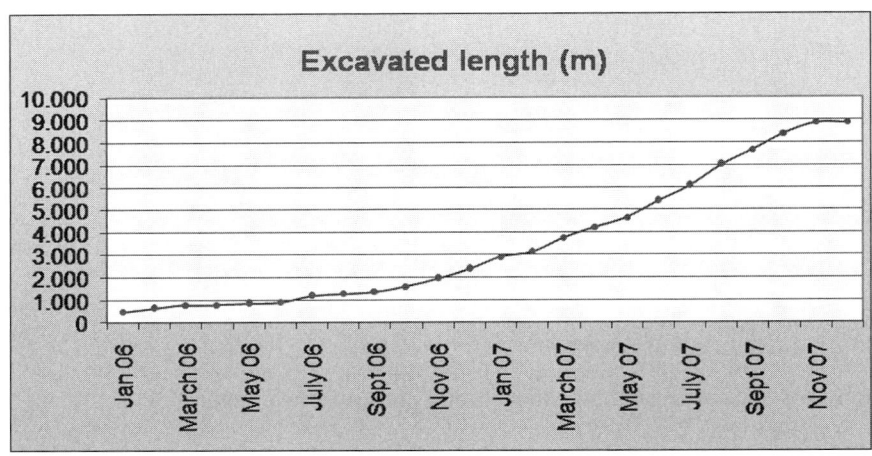

Figure 21

The challenging ramp built to accede to the main tunnel was well performed but require a very heavy conceptual design in terms of TBM itself and facilities as has been described above.

Productions achieved along the ramp and main tube shows the adequacy of the choosen method.

Bifurcation chamber shows that TBM and NATM can and must cooperate together in tunnel designs as the most cost effective way to develop Mega projects as Pajares it is.

Finally, last but not least, Client decision to divide the project in five contracts, by using 5 TBM's has been a conservative but realistic decision to accomplish the very thigh construction schedule. Problems faced during construction, as described above, as they face in the other contracts, but the existence of 5 machines give to the client an extra tranquility to see it project completed for the benefit of the Spanish citizens.

BIBLIOGRAPHY

On site information and documentation.
Ministerio Fomento (2003). *"Proyecto Básico de los Túneles de Pajares."*
Ministerio Fomento (2004). *"Proyecto y obra de plataforma de la Línea de Alta Velocidad León-Asturias. Tramo: Túneles de Pajares (Lote 2)."*
Bifurcation East Report. Dr.G.Sauer Corporation.
"Los Túneles de Pajares" Raúl Mínguez Bailo. Revista de Obras Públicas Nov 2005.
"Los Túneles de Pajares" Carlos Díez Arroyo, Raúl Mínguez Bailo. Documentación Máster AETOS 2007–2008.
"Logística y explotación de una galería con pendiente descendente del 6,2%. L.A.V. León-Asturias. Túneles de Pajares. Lote 2" Juan Luis Magro. 24 Abril 2008.
"Evacuación de los caudales de infiltración durante la ejecución de la obra" Juan Luis Magro, Ernesto Lago (DRAGADOS).

8 M DIAMETER 7 KM LONG BELES TAILRACE TUNNEL (ETHIOPIA) BORED AND LINED IN BASALTIC FORMATIONS IN LESS THAN 12 MONTHS

Antonio Raschillà ▪ S.E.L.I. SpA

Francesco Bartimoccia ▪ S.E.L.I. SpA

ABSTRACT

On the 2nd of June 2007 SELI S.p.A. started the excavation of Beles Tailrace Tunnel (Ethiopia). The new 8,07 m diameter DS Universal TBM utilized for the tunnel excavation was manufactured by SELI S.p.A. and allowed the installation of 30 cm thick concrete lining, contemporaneously with the excavation of 7,200 m of tunnel.

The greatest part of the tunnel alignment was excavated in hard and sound basaltic rocks.

The paper describes the main steps of the tunnelling operations which led to a remarkably rapid excavation, with a peak production of 36 m / day and a final breakthrough achieved 46 days ahead of the contractual deadline.

Furthermore the paper describes the work and logistic organization that was set up to efficiently perform the mechanized excavation in the very remote area of the project.

THE PROJECT

Beles project is one the multi-purpose projects, currently underway, to develop the resources in Ethiopia.

The project area is located in the Amhara Region (north western of Ethiopia), about 150 Km to the Bahir Dar town (see Figure 1). The Beles Multipurpose Project, committed to Salini Costruttori S.p.A., will be a single stage power plant with a total installed capacity of 460 MW.

The project, located in the South-Western bank of the Lake Tana, develops completely underground for a total length of about 20 Km, and consists of mainly:

- 12 Km Headrace Tunnel (which will conveys water from Lake Tana into the penstock of the underground powerhouse)
- 7,2 Km Tailrace Tunnel (which will discharges water into the Jehana River, a small tributary of the Beles River)
- Underground Powerhouse with n. 4 Francis type turbines of 115 MW each and Transformer system
- 270 m Penstock shaft
- 90 m Surge Shaft

THE TRACING AND GEOLOGY

The SE-NW running tunnel is divided in a Tailrace Tunnel and Headrace Tunnel, with a total excavation length of nearly 19 Km.

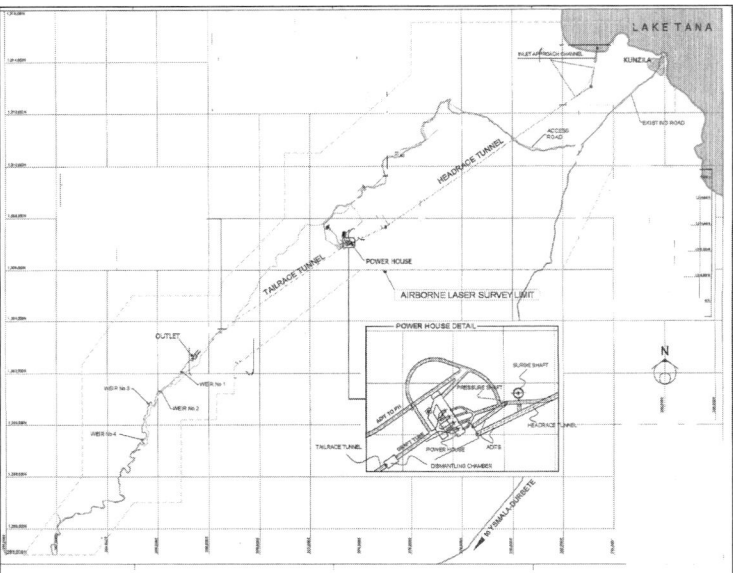

Figure 1. General map of the project

The Tailrace Tunnel runs from the Powerhouse to the outlet portal for 7200 m with a slope of 0,15%, conveying the discharge water to the Jehana River (see Figure 2).

Excavation started from the outlet towards the Power House. The geological conditions are described in TBM advance direction, as follows:

From chainage 7200 to 5000 the tunnel penetrated a brecciated Basalt formation, which was composed of plateau flow basalt and pyroclastica (agglomerates and tuffites). The overburden in that range extended up to 120 m. Excavation conditions was good to fair for most of this TBM section, depending on the fracture intensity.

From chainage 5000 to 0 the tunnel section contained a series of aphanitic Basalt layers with some intercalations of brecciated basalt layers at an overburden from 250–370 m. The rock mass crossed was good and favourable tunnelling conditions (see Figure 3).

The complete Tailrace tunnel was expected to intersect about 17 major lineaments; these zones were described with locally poor tunnelling conditions. The lineaments intersect the alignment most of the times at obtuse angle and almost vertically (see Graph 1).

THE WORK AND METHOD OF EXCAVATION

For the excavation and lining of the about 7,2 Km of Tailrace Tunnel was utilized a Double Shield TBM system designed and built by SELI Tecnologie. The tunnelling system executed all the following operations:
- boring
- erection of the precast lining
- filling of the annular gap between the excavation and the precast lining by pea-gravel.
- grouting of the rock around the lining

BELES TAILRACE TUNNEL (ETHIOPIA) IN BASALTIC FORMATIONS

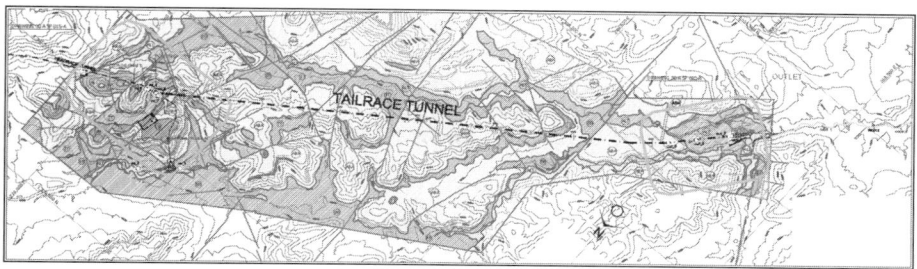

Figure 2. Tracing of Tailrace Tunnel

Figure 3. Front face

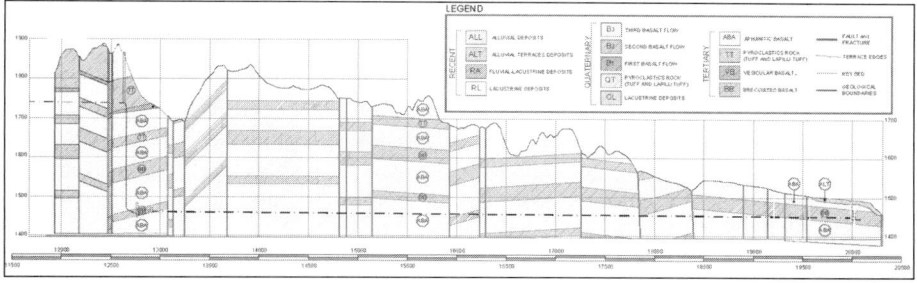

Graph 1. Geologic profile

TBM and Back-up System Description

TBM, back-up system and rolling stock were designed and manufactured by **SELI Tecnologie** (see Figure 4) in the Italian factory of Aprilia (Rome) to ensure the maximum advance rates under variable ground condition through the complete mechanisation of all operations related to handling and installing precast segment, loading and exchanging muck trains and servicing the various TBM power units mounted on the rolling platforms. The following table resumes TBM specifications.

TBM Type	Double Shield Universal
Cutterhead excavation diameter	8,070 mm (new cutters)
TBM Shield Length (Including tail shield)	11,87 mt (TBM closed)
TBM & Back-Up Length	approx. 125 mt
Cutterhead and Cutters	
Cutterhead structure	No. 6 parts—heavy structure bolted
Number of cutting tools	52
Cutter type	17 inch disc—backloading and recessed
Cutterhead Thrust	
Thrust Cylinders	n. 12—345 bar max pressure
Auxiliary Thrust Cylinders	n. 14—345 bar max pressure
Maximum advance speed	6 m/h
Cutterhead Drive	
Type	Electric
Cutterhead Power System	8 × 315 kw Electric water cooled motors
Maximum Torque	5,250 kNm
Cutterhead Speed	0 to 6 rpm
Segment Erector	6 movements radio controlled
Conveyor System	
Width	1,000 mm
Capacity	550 t/h
Machine Weight	Approx. 600 t
Back-Up Components—Auxiliary Equipment	■ ZED system ■ Gas monitoring and alarm system ■ Pea gravel storing and pumping system ■ cement grouting storing, mixing and grouting system ■ emergency diesel generator ■ close circuit TV camera system ■ rock drilling equipment ■ PLC system (see Figure 5)

TBM was also equipped with a special Rock Drilling Equipment to probe in front of the Cutterhead and from the Gripper shield, which has 18 different holes (internal diameter 102 mm) where, by using a special support mounted on the erector, probe drilling can be carried out.

An automatic recording device of the latest generation was installed on the TBM. Excavation parameters were visualized on a monitor in the operator cabin. The computer mounted in the operator cabin recorded and stored all the data by means of datalogging system.

Figure 4. TBM (called Hiwot) in Aprilia factory (Rome)

Figure 5. PLC system

The recorded parameters were:
- TBM total thrust and pressure thrust on each thrust cylinder
- Penetration rate
- Cutterhead torque
- Cutterhead revolution per minutes
- Electric power consumption for each electric motor
- Stroke advance
- Steering parameters

Figure 6. Tunnel portal

Tunnel Transport System

Tunnel excavation and segments installation, in DS Mode, are performed concurrently, except under unfavourable geological conditions.

The rolling stock followed SELI philosophy and design tradition, that has proved to be successful in more than 500 Km of tunnels built.

The choice of muck cars (fix body type) allowed the highest ratio among the transported weight and the own weight of the muck car. Furthermore, all lateral internal surface was smooth and didn't create any obstacle to the operations of unload of the muck.

The wheels had large diameter and high capacity bearing for long life and low rolling friction.

Each Tailrace tunnel trains included:

- 13 fixed-body muck cars for a total capacity of 156 m^3 of muck (see Figure 6)
- 1 pea-gravel car transporting a silos of 5,3 m^3 of capacity
- 3 flatbed cars for the transportation of precast segments
- 1 personnel car having 16 seats capacity

The number of trains utilized was ⅔ up to chainage 3500 and n° 4 from chainage 3500 to the end of the tunnel.

A fully automatic hydraulic muck car tippler (able to unload one muck car each 2 minutes and full train in 26 minutes) was located outside the tunnel portal. The tippler was complete with pusher for the train shunting and therefore there was no need of the locomotive to unload the train.

Lining

The lining was an open type parallel ring rhomboidal segmental system, with the following characteristics:

- External diam. 7800 mm
- Internal diam. 7200 mm
- Lining thickness 300 mm
- Segment ring width 1500 m

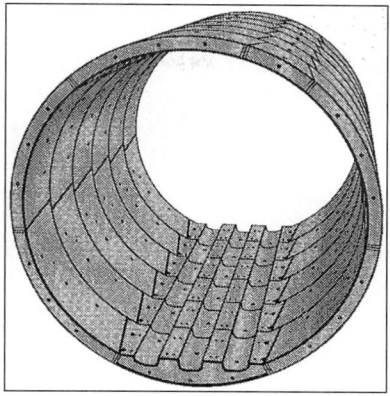

Figure 7. Lining system

- N.6 segments + n.1 key stone
- Invert with flat tracks and 3 central trenches
- Dowel connectors in the circumferential joints
- Grouting holes

For assembling of the segment ring, the invert was installed first with the opportune dowels of the circumferential joint. Next step, the both left and right bottom sidewall segments was installed and kept in position by means of the auxiliary thrust cylinders. Afterward the both left and right top sidewall and as the last elements of the ring the roof segment and key stone were installed (see Figure 7).

Dowel connectors (three for each segment made by extruded polyetilene) had no structural function, but were installed just like simple connecting and centring devices.

Pea Gravel Injection

During excavation "pea gravel" was injected into the annular gap between rock mass and the lining. The size range of the pea-gravel was 8–12 mm and it has been dimensioned to allow its injection by using n° 2 air pumps mounted on the deck n° 3. A dedicated line arrived to the deck n° 1 where, using flexible pipes, pea-gravel was injected behind the second-last completely assembled ring.

Immediately after the injection the grouting holes was plugged with suitable wooden plugs to prevent escaping of pea-gravel.

Contact Grouting

Contact grouting was performed all along the tunnel (see figure 8) to grout the pea-gravel in the annular gap, following the excavation behind the back-up (see figure 9).

Grouting was filled up also the small space between segment and segment along the joint where was present any opening.

Grouting plant was installed at the Tunnel Portal and the grout mixture was pumped inside the tunnel by using n° 2 lines of galvanized steel pipes.

Grout equipment was composed by:

- turbomixer having capacity of 1500 lt and production rate of 15 mc/h
- agitator having capacity of 4000 lt
- automatic dosing system for additives

Figure 8. Tunnel inside sight

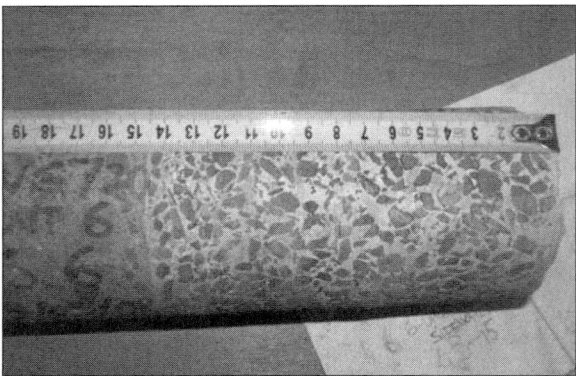

Figure 9. Core drilling example

- automatic dosing system for bentonite
- n° 2 pumps, maximum flow rate of 100 lt/min and maximum pressure of 200 bar

During the normal activities (excavation and regripping), in the range of the back-up decks, the chalking grooves was closed with mortar and all necessary repairs were done.

LOGISTIC AND WORK ORGANIZATION

Due to remote location of the project, it was necessary to overcome several logistic difficulties.

The shipment of the TBM, back-up, rolling stock and all the other spare parts and materials was carried on from Italy to Gibuti and from Gibuti to the job site (in Bahir Dar area) through about 1000 Kms of difficult roads (see Figures 10, 11, 12), especially during the rainy periods.

Particular attention was done in the Nile Bridge (see Figure 13) crossing where all necessary checks of the structure took place before the passage of the main TBM components.

Figure 10. Portal area

Figure 11. Road from Gibuti to site

Figure 12. Road from Gibuti to site

Figure 13. Nile bridge crossing

The waiting to receive any shipment on site was about 4–5 months from purchase order.

For this reason it was indispensable to plan possible spare parts as providently as possible, with considerable advance.

It was also necessary to face low level of specialization of local manpower; a specialized SELI staff with adequate experience supervised all tunnel boring operations training local workers to assure the level of reached quality and productivity.

Tunnelling operations was carried out 7 days/week in 3 shifts of 8 hours each, with shift changing at the front and a scheduled maintenance shift inside the tunnel of about 4 hours in the morning. The maintenance operations consisted mainly in cutterhead inspection and cutter changes as required, supply and extension of tunnel services, eventual probe drilling operations, maintenance pea-gravel lines. When special maintenance, repairs or probe drilling was required, the maintenance duration was extended beyond the normal 4 hours time.

TBM PERFORMANCE

The excavation of the tunnel started on the 2nd June 2007 and was completed the 31st May 2008, after less then one year at an average advance rate of about 20 metres per days.

The maximum daily advance rate of 36 metres (24 rings) was achieved in January '08, while the best weekly advance rate was 189 metres achieved on the 27th week of excavation (see Graph 2).

The best month, as shown in the graph 3, was December '07 with 724 metres of excavation during 29 days of work (excluding 25th and 31st of December) at an average rate of 25 metres per day.

Graph 2 shows the very short period of the learning phase.

Learning curve, usually long 2–3 months, was limited at less that one month. This important target was obtained through an easy and rapid interaction expatriate—local personnel and an optimization of the necessary equipments.

The Graph 4 shows the constant advance of the excavation production (upper line) related with the contractual production (lower line) calculated on the basis of the RMR (Bieniawski) of the rock mass encountered. TBM completed the tunnel with 46 days of advance on contractual time.

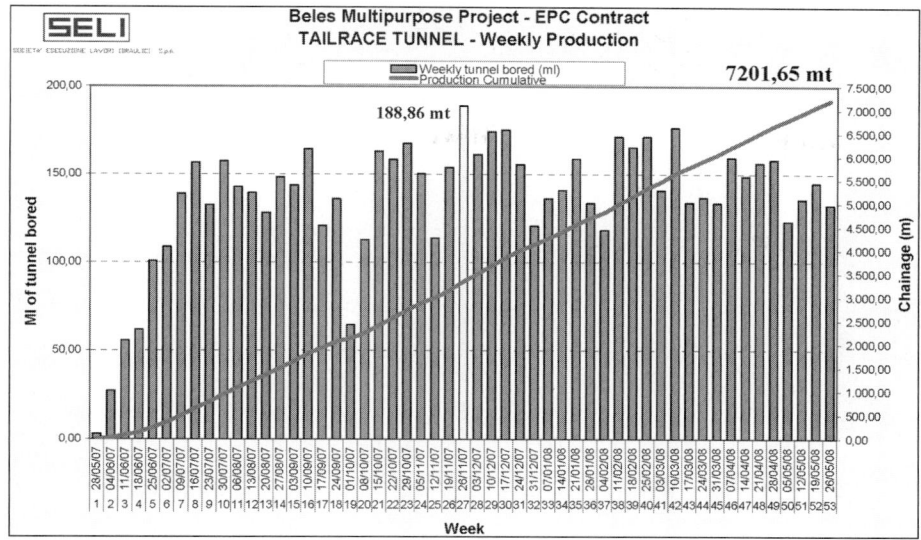

Graph 2. Weekly productions of the TBM DS 0807 114 SELI

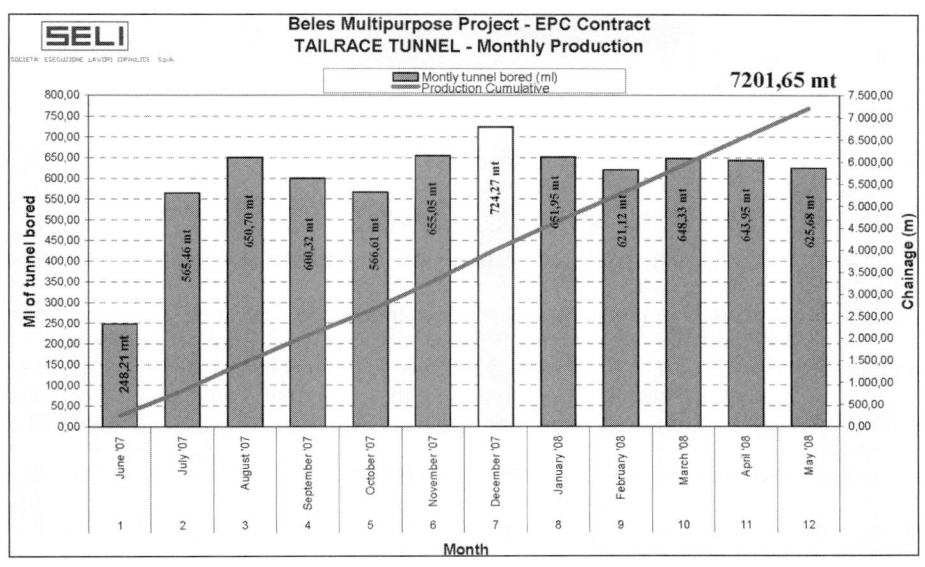

Graph 3. Monthly productions of the TBM DS 0807 114 SELI

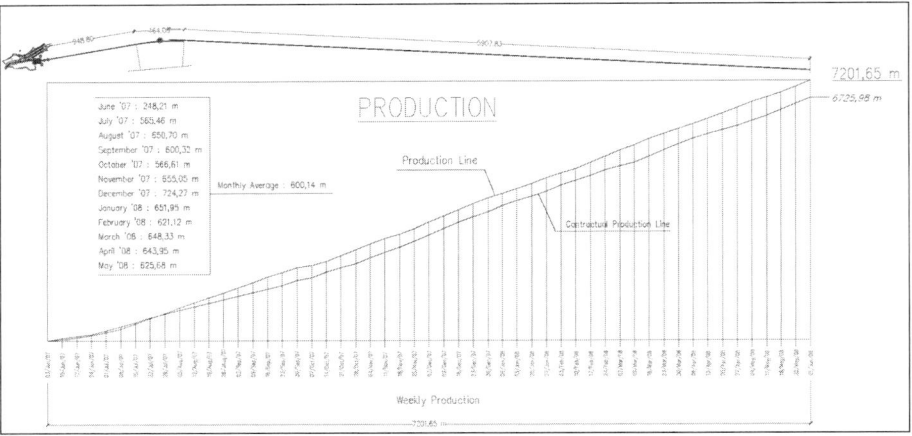

Graph 4. Production lines of the TBM DS 0807 114 SELI

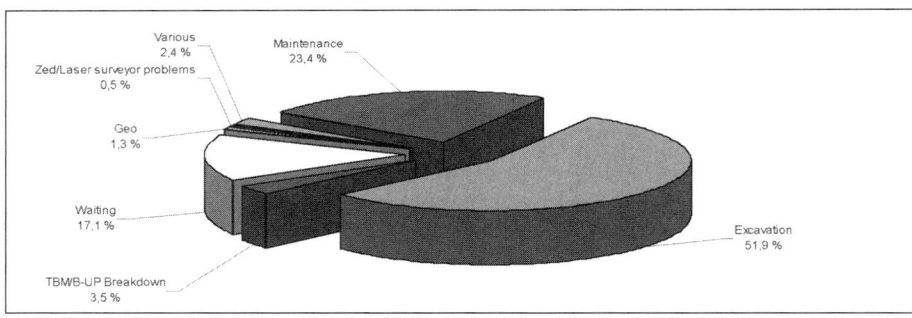

Graph 5. TBM DS 0807 114 SELI—timing analysis

TIMING ANALYSIS

Graph 5 shows the good efficiency reached from the TBM (51,9 % included regripping time and ordinary waiting train).

It was the result of a punctual daily maintenance (23%).

In the "Maintenance" and "Waiting" items, particularly main point was time lost for Pea-gravel injection lines. Substantial delays were imputable to the necessary daily maintenance of the injection lines (to guarantee the good operation of the system) due to the characteristics of the material. Non available fluvial gravel in situ has imposed a sharpy material from quarry with consequent easy deterioration of the lines.

Other important parameter was "TBM breakdown" (3,5 %), a low value considering the rock hardness (with the consequent problems relative to the vibrations) and TBM adjustment during learning phase.

CONCLUSIONS

Main problems encountered during the project were difficult logistic situation, especially for supplying materials and a local manpower with lack of specialization.

Figure 14. Breakthrough of the TBM

It was possible to obtain good results through a sharp design and planning of all the activities, with the employment of specialized SELI human resources, who has organized in the better way all the work and has instructed the local manpower.

Key factor was a careful planning and equipments employment, with consequent optimization of the work phases during the whole work.

CONSTRUCTION OF LOUISVILLE WATER COMPANY'S RIVERBANK FILTRATION TUNNEL AND PUMP STATION PROJECT

S. Holtermann ▪ Jordan, Jones & Goulding, Inc.

W. Klecan ▪ Jordan, Jones & Goulding, Inc.

K. Ball ▪ Louisville Water Company

ABSTRACT

Through the process of Riverbank Filtration (RBF), the Louisville Water Company plans to obtain a higher quality source of water for their B.E. Payne Water Treatment Plant (WTP) by means of four collector wells located in the alluvial aquifer along the bank of the Ohio River. The collector wells are connected to a 2,850 L/s (65 mgd) pumping station by a 2,362 m (7,750 ft) long, 3 m (10 ft) finished diameter tunnel located in limestone and shale bedrock, which is approximately 45 m (150 ft) below the surface. High tunnel boring machine (TBM) production rates were achieved in this moderately weak bedrock; most of the tunnel only required support in the crown. An unanticipated water-bearing fault and unanticipated ground behavior proved challenging and required quick thinking to minimize impacts to construction. Tunneling conditions are discussed in detail, together with the methods used for overcoming the groundwater and ground behavior challenges. The tunnel is expected to be completed in April 2009.

PROJECT OVERVIEW

Purpose of the Project

The purpose of the Riverbank Filtration (RBF) Tunnel and Pump Station project is to enable the Louisville Water Company to meet US Environmental Protection Agency (USEPA) drinking water regulations aimed at balancing the risks of microbial disease occurrences and the health issues associated with exposure to potentially harmful compounds formed during drinking water disinfection. These regulations require water utilities to balance the risks associated with pathogens, synthetic chemicals, radionuclides, natural occurring compounds and various by-products associated with disinfection. In addition, the Louisville Water Company is committed to providing its customers with a high quality of water that is free of taste, odor and color, which requires an advanced technology to remove non-regulated contaminants such as algae that can cause these problems.

RBF is a purification process that uses the natural filtering processes of the riverbank to remove many of the particles and contaminants from the raw river water. The process of RBF utilizes the sorptive, filtering and biological processes of the natural occurring riverbank to remove sediment, pathogens and organic chemicals from the river water. The natural filtering system reduces the threat of Zebra mussels and Asiatic clams, as well as the risks associated with hazardous chemical spills and herbicides and pesticides associated with the current surface water intake approach utilized at the B.E. Payne WTP. The treatment process is highly reliable in that there are no mechanical or chemical elements to fail.

LOUISVILLE'S RIVERBANK FILTRATION TUNNEL AND PUMP STATION 1249

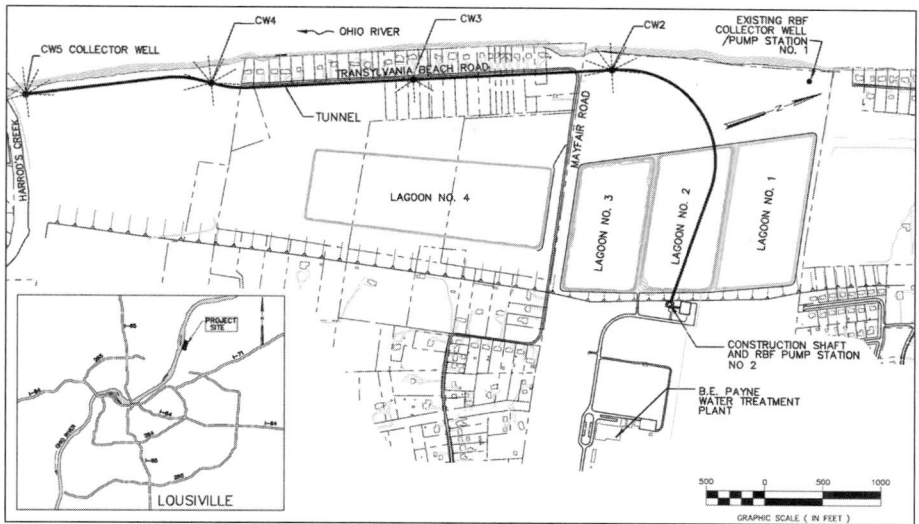

Figure 1. General site plan

By utilizing the sand and gravel layers under the Ohio River as a natural filter to remove turbidity and synthetic organics, RBF generates water quality close to that of finished water prior to entering the plant. The water resulting from RBF offers many benefits, such as low disinfection by-products in the finished water, therefore reducing potential water quality problems in the distribution system, as well as more stable temperature characteristics producing cooler, more aesthetically pleasing water in the summer and warmer water in the winter. This may result in fewer water main breaks in the distribution system due to less significant swings in the water temperature.

The Louisville Water Company constructed and placed in service a 656 L/s (15 mgd) demonstration well (Collector Well No. 1) in 1999, as shown in Figure 1. This well has been very successful and serves as a collection model for the construction of the four additional collector wells associated with the RBF Tunnel and Pump Station project.

Scope of the Proposed Construction

The Louisville Water Company selected a design which utilizes a series of four horizontal collector wells linked to a single pump station by means of collector well drop shafts and a hard rock tunnel. Given the selected design, the scope of the construction project can be broken out into five main components: construction shaft/pump station shaft, hard rock tunnel, collector well, drop shaft and pump station.

Bid and Award

Advertisement-for-Bid took place in August 2006 for a November 8, 2006 bid opening. Mole Constructors, Inc. (MCI) was awarded the project November 14, 2006 with a bid of $34 million. Notice-to-proceed was issued for March 19, 2007 with work to be completed within 900 days, or by September 3, 2009.

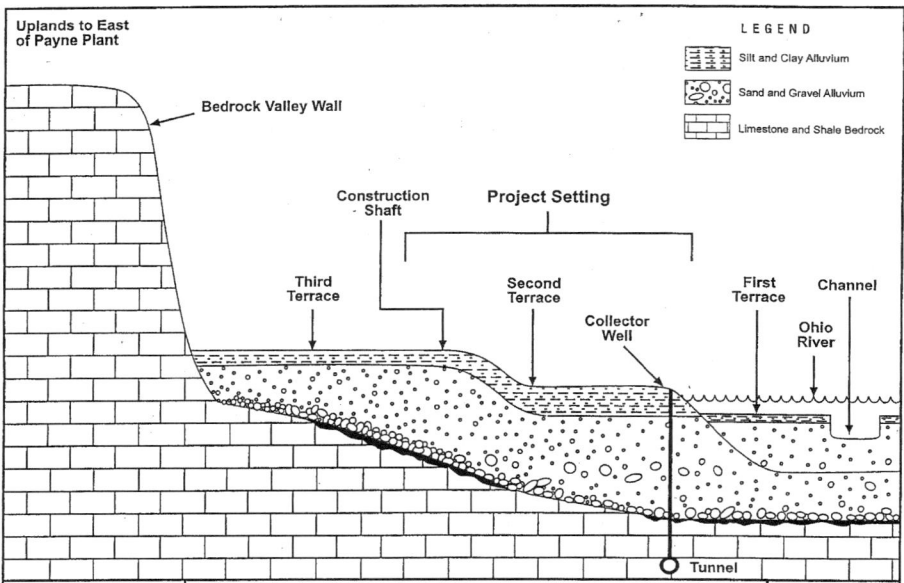

Figure 2. Conceptual profile of the Ohio River terraces

SITE PHYSIOGRAPHY AND GEOLOGY

Alluvial Aquifer

The site is located on the alluvial terraces of the eastern bank of the Ohio River. The alluvial terraces are bordered to the east by limestone and shale cliffs, as shown in Figure 2. The Ohio River has three terraces and an alluvial channel. The first and youngest terrace has the lowest elevation and was flooded by the lock and dam system. The second terrace forms the lowlands along the river and the site for all the collector wells. The third terrace is the highest and oldest; formed before the last ice age when the Ohio River downcut a wide floodplain through the surrounding bedrock uplands and deposited large amounts of alluvium in the floodplain. The top layers of the alluvial terraces contain recent floodplain deposits, consisting of clays, silts and thin fine sand horizons, and organics, with the occasional log or stump. Sands and gravels with the occasional cobble or silt layer underlie the recent floodplain deposits and extend down to the top of the bedrock. The water table rises and falls with the Ohio River. Pumping tests have shown the sand and gravel aquifer to have a transmissivity of about 36.1 m^2/day per metre (300,000 gpd/ft) and a storage coefficient of around 5×10^{-4}, and therefore capable of yielding a large volume of water.

Bedrock

The bedrock consists of Upper Ordovician limestones and interbedded shales, gently dipping to the southeast. The top of bedrock below the collector wells is relatively uniform as can be seen in Figure 3. At the location of the construction shaft, the top of bedrock occurs at an average elevation 106 m (355 ft).

The bedrock has four distinctive lithological units:

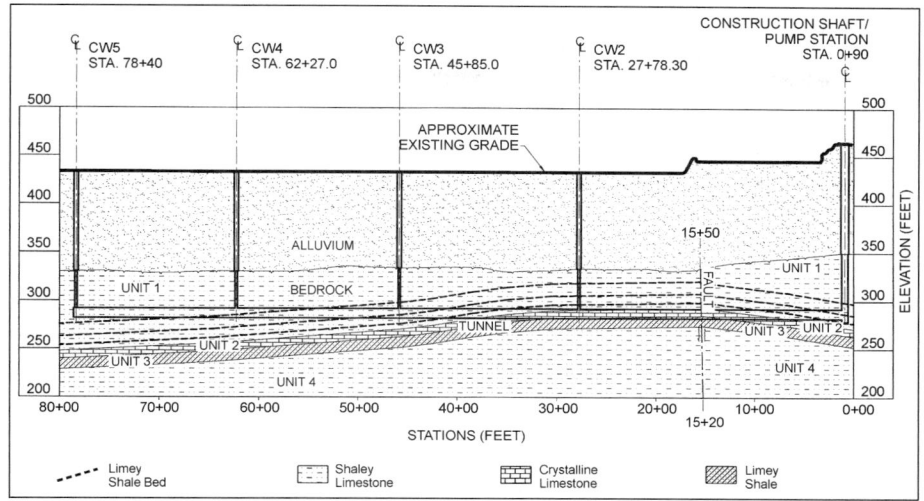

Figure 3. Geological profile

- Unit 1—Shaley limestone with three major shale layers ranging in thickness from 15 to 30 cm (0.5 to 1 ft) thick that are traceable across the entire length of the tunnel.
- Unit 2—Crystalline and bioturbated limestone.
- Unit 3—Predominately shale with occasional layer of bioturbated limestone.
- Unit 4—Shaley limestone.

Most of the tunnel is located in Units 1 and 2, as shown in Figure 3. Unit 3 is present in the invert of the tunnel approximately between Sta.15+00 and Sta.30+00. Unit 4 remains below the tunnel.

Structural Geology

Several major faults are present southwest of the Louisville area, trending in a northeast to southwest direction. The fault encountered in the tunnel at Sta.15+20 was the reverse type with a 0.3 m (1 foot) displacement, striking about 016 degrees and dipping about 20 degrees towards the southeast. Occasional high angle joints are visible in some of the road cutting around Louisville; however, none were encountered in the tunnel.

Groundwater in the Bedrock

The rock material is practically impermeable. Structural discontinuities, such as faults, bedding plane fractures and joints, could contain significant amounts of water and inflow rates may remain constant, especially if continuously recharged by the overlying alluvial aquifer.

During the excavation of the tunnel, a single 3.8 cm (1.5 inch) diameter bolt hole at Sta.14+33 produced approximately 2.2 L/s (35 gpm). At approximate Sta.15+20, the water inflow increased significantly as the tunnel intercepted the fault. The inflow from the fault produced approximately 19 L/s (300 gpm) and remained close to this rate throughout the rest of mining. There were four other areas in the tunnel where groundwater was encountered in the rock bolt holes. The stations where these inflows were located are: 46+13 to 46+38, 49+66 to 49+80, 54+31 to 54+68, and 56+03 to 56+31. The inflow volumes ranged from a steady drip to about 3 L/s (50 gpm). The bolt holes were either grouted (large inflows) or plugged with resin (small inflows). The bolt holes that produced the inflows were likely to have tapped into one of the fault splays, though the fault never crossed the tunnel alignment beyond Sta.15+50. Water pressure in one of the flow producing bolt holes was about 358.5 kPa (52 psi), equivalent to a 36.5 m (120 ft) high column of water and close to the water level reported in some of the nearby boreholes.

Natural Gas

Small amounts of natural gas have been commercially produced in the vicinity of Louisville for many decades and pockets of "nuisance" gas are sometimes encountered in geotechnical borings and water wells in the region. Natural gas with a pressure of 861.8 kPa (125 psia) was discovered during the first stage of site investigations in a borehole located about 457.2 m (1,500 ft) from the southern end of the tunnel. The boring was converted to a well, which was estimated to have produced an average of 637.1 m^3/day (22,500 ft^3/day) over a 4.5 day period before being shut in. The pressure in the well then returned to 861.8 kPa (125 psia) within 11 hours. The well was grouted and no further testing was performed. Gas was also discovered in a borehole at Cox Park, approximately 6.4 km (4 miles) from the southern end of the tunnel. No testing was performed and the boring was grouted.

During the second site investigation phase, core borings were drilled along the length of the tunnel at approximately 60.9 m (200-foot) intervals. No gas was encountered in any of these borings. Since the limestone and shale bedrock in the area is known to produce gas, the tunnel was classified as "potentially gassy."

BEDROCK GEOTECHNICAL PROPERTIES

Typical geotechnical properties of the sub-horizontal stratified bedrock along the tunnel alignment are as follows:

- Core recovery is generally good to excellent and Rock Quality Designation (RQD) results mostly in the 75 to 100 range.
- The bedrock exhibits anisotropic compressive strength properties. The Unconfined Compressive Strength (UCS) tends to be the highest when the sample is loaded in the direction perpendicular to the bedding. The UCS in the direction parallel to the bedding is about 60 percent of the UCS in the perpendicular direction. In most cases, the lowest UCS probably occurs when the applied load is orientated at an angle to the bedding, although no tests were performed to verify this case. Table 1 shows typical UCS values for the lithological units 1, 2 and 3.
- The shale layers contain clay minerals that swell and disintegrate when immersed in water. The rate of disintegration depends on the clay content. When performing Free Swell Tests on representative samples of shale, the majority of the shale samples began to break apart within 30 minutes of starting the test, such that the longitudinal swelling could no longer be measured. The change in diameter during the tests ranged from 0.2 to 0.5 percent. Full reaction to diametrical swelling occurred between 4 and 12 hours for most of the samples tested.

Table 1. Unconfined compressive strength data

Units	Unconfined Compressive Strength psi (MPa)						
	Perpendicular To Bedding			Parallel To Bedding			
	Average	Average Strongest 10%	Average Weakest 10%	Average	Average Strongest 10%	Average Weakest 10%	
1 and 2	12,500 (86)	17,500 (120)	7,600 (52)	8,300 (57)	14,400 (99)	2,200 (15)
3	6,900 (47)	12,000 (83)	1,900 (13)	4,100 (28)	9,100 (63)	100 (0.7)

TUNNEL GROUND SUPPORT DESIGN

Initial Ground Support Design

The ground support for this circular tunnel opening in moderately weak, sub-horizontal sedimentary strata with fairly closely spaced bedding planes was designed with the aid of Phase^2 and UNWEDGE analytical software from Rocscience. The finite element program Phase^2 was initially used to determine if stress induced failures around the tunnel are likely and the probable extent of the failure zone. Results of the Phase^2 analysis indicated that stress induced failure was unlikely for the rock compressive strengths (average weakest 10 percent values) shown in Table 1, under an overburden stress equal to the weight of the overlying soil and rock material, and an assumed in-situ horizontal stress of three times the overburden stress. However, small rock wedges could still fall from the upper sidewalls if the tunnel encountered high angle joints. Larger rock wedges could develop in sections of the tunnel where the thicker shale layers intercept the tunnel crown. The size of potentially unstable blocks was determined using the UNWEDGE software and also the length, capacity and configuration of rock bolts needed to support these blocks.

It was determined that a 1.2 m (5 ft-9 in) long rock bolt, located each side of the tunnel centerline in the upper sidewall and spaced 1.2m (4 ft) apart longitudinally would be sufficient in most cases to maintain a stable excavation. In sections where the shale intercepts the crown, a four bolt pattern in the crown and upper sidewalls, spaced 0.91m (3 ft) longitudinally with welded wire fabric and steel straps would be needed. The analysis concluded that provided the tunnel stayed above the predominately shale Unit 3, the entire tunnel could be supported with rock bolts and heavier support (such as steel circular ribs) was not necessary.

Final Concrete Lining

The interbedded weak shales would deteriorate and could compromise tunnel stability. Therefore, a final concrete lining was designed to ensure long term stability. The concrete lining is minimum 24.5 cm (10 inches) thick based on a 27.5 MPa (4,000 psi) minimum compressive strength at 28 days and unreinforced, except for the dropshaft-tunnel junctions and future connection-tunnel junctions.

CONSTRUCTION OVERVIEW

Construction/Pump Station Shaft

The design of the Construction/Pump Station Shaft, as indicative of its name, provided for the dual role of the structure. It would first serve as the tunnel construction shaft and then as the permanent pump station shaft. A concrete diaphragm wall was selected by the designer to account for the challenges involved in shaft construction in the presence of an aquifer as well as to enable the shaft to be socketed into rock. Another reason for the selection of a diaphragm wall was that the geotechnical investigation indicated the potential for karstic features and a top of rock which sloped across the shaft footprint.

The construction of the 12.7 m (41.5 ft) diameter diaphragm wall consisted of four primary and four secondary reinforced concrete panels 0.9m (3 ft) thick by 37.8m (124 ft) in height. MCI selected Bencor Corporation of America to perform the work. The four primary panels were constructed with four smaller secondary panels interlocked between them to complete the shaft structure. The majority of the excavation was performed by utilizing a hydromill, suspended from a crawler crane, and bentonite slurry mixing and desanding plants.

Upon completion of the shaft collar, excavation activities to muck out the shaft down to top of rock commenced. Once the overburden was excavated and a shotcrete lining applied to the diaphragm wall, conventional drill/blast work was performed to excavate the rock to the design elevation near the base of the wall structure. At this point, the diameter of the shaft was decreased to 7.6 m (25 ft) and drill/blast work continued in order to reach invert elevation at a total shaft depth of approximately 56.9 m (187 ft). Shotcrete was utilized as the permanent lining through the smaller diameter rock portion of the shaft.

Tunnel

Tunnel construction consists of a 2,362 m (7,750 ft) long tunnel with a finished cast-in-place concrete lining of 3 m (10-ft). The tunnel will serve to convey bank-filtered raw water collected from each of the four collector wells to the pump station shaft for eventual lifting and conveyance to the WTP. The tunnel was designed to allow for future expansion with two future tie-in connection bulkheads located near the tangent after the initial curve and the end of the tunnel.

In preparation for assembly of the 3.6 m (12 ft) diameter tunnel boring machine, a 40.5 m (133 ft) long, 4.2 m (14 ft) horseshoe drill/blast starter tunnel was excavated from the construction shaft. A similar 9 m (30 ft) long tail tunnel was also excavated. The main beam was set into the shaft in mid-February 2008 and assembly completed and excavation initiated on March 5, 2008. Muck conveyance was accomplished using 12.1 t (12 ton) electric locomotives and 7.3 cubic meters (10 cubic yard) muck cars. Excavation was halted temporally at the end of March to complete the assembly of the machine, install a car pass system in the starter tunnel to facilitate a two train system and install a guide rail system in the shaft. Mining resumed the second week in April 2008, and was completed on October 9, 2008.

During the course of tunnel excavation, several issues were encountered. The issues included an unanticipated water bearing fault feature, unanticipated ground behavior, and several water bearing seams encountered during rock bolt installation. Tunneling conditions are discussed in greater detail, along with the methods used to overcome the more difficult ground conditions, in the Tunnel Construction Challenges section.

With the completion of mining, the machine was disassembled at the terminus of the tunnel and transported back to the pump station shaft for extraction. Upon removal of the machine in late November 2008, the concrete lining operation was initiated. Initial placements occurred in late December, with production placements starting on January 22, 2009. The lining operation will utilize a 3 m (10-ft) diameter steel tunnel forming system with 73.1 linear metres (240 linear ft) of 1.5 m (5-ft) section forms. Placements of up to 64.0 linear metres (210 linear ft) are anticipated to be made daily on a five day per week schedule. The lining will be reinforced only at future tie-in locations and

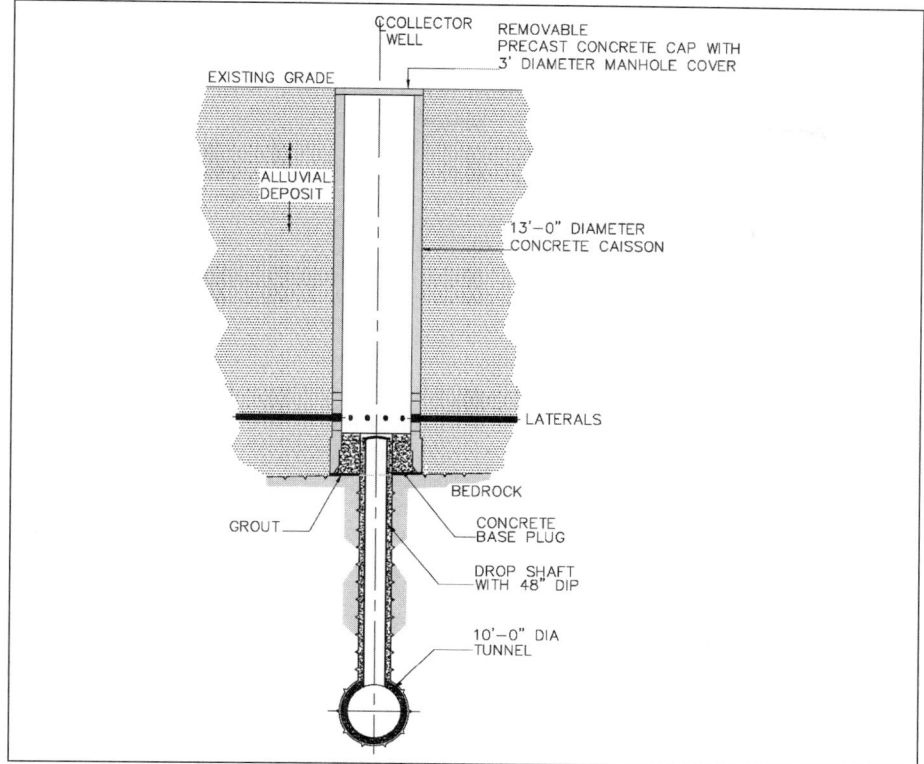

Figure 4. Collector well and drop shaft profile

collector well drop pipe tie-in locations. Lining and contact grouting are scheduled to be completed in April 2009.

Collector Wells

Collector well construction consists of four 4 m (13 ft) inside diameter caisson structures reaching from the surface to top of rock; a depth ranging from approximately 27.4 to 33.5 metres (90 to 110 ft). MCI selected Ranney Collector Wells, a Division of Reynolds, Inc., to perform the work. The structures are built of cast-in-place reinforced concrete in lifts typically 3.6 metres (12-ft) in height. The lifts are sunk through the combination of clam shell excavation and a pull down system. Incorporated in one of the early lifts is a set of collector lateral ports with blind flanges. The height of the final lift is adjusted in the field in order for the structure height to match existing ground elevation.

The caisson structures also serve to provide access for the installation and development of the 30 cm (12 inch) diameter stainless steel lateral well screens which radiate from the caisson into the surrounding aquifer and serve as the collection devices for the bank-filtered raw water. The length of installed screen on the project is between 57.9 and 79.2 linear metres (190 and 260 linear ft) per lateral. A plan view of the layout of the laterals and the locations of the collector wells is shown on Figure 1.

As the bank filtered water is collected into the caisson through the laterals, a concrete encased 121.9 cm (48 inch) diameter drop pipe conveys the raw water from the invert of the collector caisson down to tunnel elevation. Figure 4 provides a profile of a typical collector well and drop shaft on the project. The average total raw water production of the four collector wells is expected to be between 1,970 and 2,630 L/s (45 and 60 MGD).

Drop Shafts

Drop shaft construction consists of the excavation of the shaft and installation of a ductile iron pipe string between the invert of the collector well and the crown of the tunnel. The blind bore method of excavation was selected with MCI selecting Long Foundation Drilling Company to perform the work.

Upon completion of the collector caisson, the blind bore was drilled through the bottom of the caisson. The depth of the blind bore varied slightly from caisson to caisson and was typically around 15.2 m (50 ft).

Pump Station

Construction of the pump station is scheduled to begin following completion of the underground work. MCI selected Reynolds, Inc. to perform the work. The pump station will consist of a structure providing the housing of four turbine pumps generating a collective capacity of 2,850 L/s (65 MGD). The flow will tie into the existing raw water main to the WTP. The collector well-supplied raw water is not planned to be used to expand the capacity of the WTP, but rather replace the current water being drawn from a conventional intake in the Ohio River. This conventional intake and pump station will remain on stand-by and utilized if the tunnel were ever needed to be taken off line. Construction of the pump station is scheduled to take approximately 10 months to complete.

TBM TYPE AND PERFORMANCE

Equipment

The tunnel was excavated with a 3.6 m (12-ft diameter) Robbins Series MB 105-144-5 rock TBM. The machine had last been operational in Birmingham, Alabama on a project completed in 2001. Per information submitted on the machine by MCI, the machine generates 3.5 MN (792,000 pounds) of thrust at 24.0 MPa (3,500 psi) under recommended operational parameters. Each of the five drive motors generates 150 horsepower, which turns the head at 6.7 revolutions per minute at approximately 81,252.5 kgf-m (587,700 foot-pounds) of torque. Six radial buckets collect material generated by 26 cutters; two twin 30.48 cm (12 inch) center cutters, eighteen 35.5 cm (14 inch) face cutters, and six 35.5 cm (14 inch) gauge cutters on nominal 7.62 cm (3 inch) spacing. The boring stroke was 1.2 m (4 ft) with a gripper stroke of 45.7 cm (18 inches). The machine utilized two grippers capable of exerting 396-tons of pressure on the tunnel wall. The machine operated on a 4160 VAC/3phase/60 hz power feed from a drop location near the construction shaft.

The machine utilized a center mounted 56 cm (22 inch) wide conveyor capable of handling 110.8 cubic metres (145 cubic yards) of material an hour. The machine was capable of 230 metres (750 ft) radius curves which allowed the 243.8 m (800 ft) radius curves required on the project to be negotiated. Two stoper drills positioned on drilling decks were utilized to install rock bolts. A 120-degree roof shield with finger mounts was employed; fingers were not utilized during the mining. Trailing gear included an electrical sled and seven platform skid type gantry sections with 56 cm (22 inch) wide belt conveyor capability in order to facilitate muck handling and train length. The air scrubber equipment utilized a machine mounted suction system powered by fans on the surface.

LOUISVILLE'S RIVERBANK FILTRATION TUNNEL AND PUMP STATION   1257

The muck handling system included 12-ton battery operated locomotives. Two locis and 14 muck cars were typically operated in the tunnel. The muck cars were outfitted to utilize a shaft mounted guide rail system. Back-up batteries were available via two battery chargers located on the surface.

The guidance system for the machine utilized a laser and grid system. The TBM operator maintained line and grade through the coordination of cut sheets, as necessary, and targeting of the front and rear grid display panels.

Performance

The machine performed well and mined very effectively through the rock. Under normal conditions, a typical 1.2 m (4 ft) push took approximately 20 minutes resulting in a typical mining rate of 6.1 cm (2.4 inches) per minute. No major mechanical problems were experienced and cutter changes were minimal.

Production

From the start of mining MCI averaged approximately 21.0 m (69 ft) per production day when the machine engaged in mining activity. After the machine was fully assembled and configured with the final gantry section at approximate Sta.19+32, mining averaged approximately 23.7 m (78 ft) per production day. These rates do not include work days when mining was not performed and do include the production days impacted due to water feature inflow. Maximum single shift production accounted for 25.6 m (84 ft) with the maximum single day production figure reaching 43.8 m (144 ft). Typical operations consisted of two pushes concluding with the release of the muck train back to the shaft for handling. Train return times varied due to distance from the shaft, condition of the rail line and the availability of the replacement train. Taking into consideration the train return time and machine performance, a typical production effort generated 1.8 to 2.4 m (6 to 8 ft) per hour during mining operations.

TUNNEL CONSTRUCTION CHALLANGES

Water Features

The fault feature encountered at Sta.15+20 carried an inflow of approximately 19 L/s (300 gpm) and was first encountered at the face of the machine. The machine was able to excavate through the feature, but the water flushed fines off of the belts and generated invert problems with the accumulation of muck. Production impacts were initially significant and then lessened as the machine advanced through the fault zone. After fully clearing the zone at Sta.15+50, the production impacts were minimized with the installation of an upstream and downstream sump and pump system bracketing the feature and discharging water to the shaft sump directly. Until the up-station sump was constructed, due to the absence of tunnel grade, the water inflow could follow the machines advancement and continue to impact production.

A secondary impact of the feature inflow was the presence of the Unit 3 Limey Shale material in the invert through this section of the tunnel as identified in Figure 3. This created the conditions for invert degradation and eventual rail bedding issues and derail events. This situation was mitigated through the resetting and pining of the impacted rail sections with the incorporation of stone bedding material.

In addition to the Sta.15+20 feature inflow event, three of the five areas previously referenced required the temporary halting of mining operations in order to mobilize grouting equipment. In one of the three cases, several days were spent on grouting operations as the zone was slowly mined in a process of grouting, waiting for set, and upon resumption of mining operations encountering additional inflows which required grouting.

Figure 5. Example of slabbing after scaling was performed

UNEXPECTED GROUND BEHAVIOR AND INITIAL GROUND SUPPORT MODIFICATIONS

For about the first 300 meters (1,000 ft) after excavation was initiated, the ground was stable and only showed minor flaking at some bedding plane transitions. As excavation approached Sta.18+00, it was noticed that movement was occurring along some of the bedding planes near the crown, at about 60 m (200 ft) behind the tunnel face. During the next 300 m (1,000 ft) of excavation, minor spalling began occurring in the crown along some of the excavated sections extending back to about Sta.20+00, with a number of slabs falling out due to shear/buckling type failure of the closely bedded rock layers in the crown, as shown in Figure 5. A few of these slabs were between 1 metre long, about 45 cm (1.5 ft) wide and up to 15 cm (6 inches) thick.

As tunneling continued further, the slabbing became more extensive, raising safety questions and concern over the stability of the crown. Tunneling was temporally suspended while additional support was installed in the excavated portion, extending back to the TBM Starter Tunnel. The initial ground support was modified to a four bolt pattern in the crown and sidewalls with welded wire fabric, as shown in Figures 6 and 7, and installed throughout the remaining section of tunnel.

Most of the crown instability occurred in the section of tunnel that runs parallel to the river and happens to coincide with the general strike direction of the major faults in the region and approximately perpendicular to the direction of the maximum regional principal horizontal stress.

Other Ground Support Modifications

Additional ground support utilizing rolled channel was required in the fault zone, where the fault rose in the sidewalls and traversed the crown of the tunnel. The channel was installed in two piece sets of 2.4 m (8 ft) lengths as shown in Figure 8. In addition, straps and spot bolts were installed as necessary to provide support.

Also some areas where slabbing had occurred, several pieces of channel with mesh were installed as shown in Figure 9, prior to installing two additional crown bolts to supplement the modified pattern.

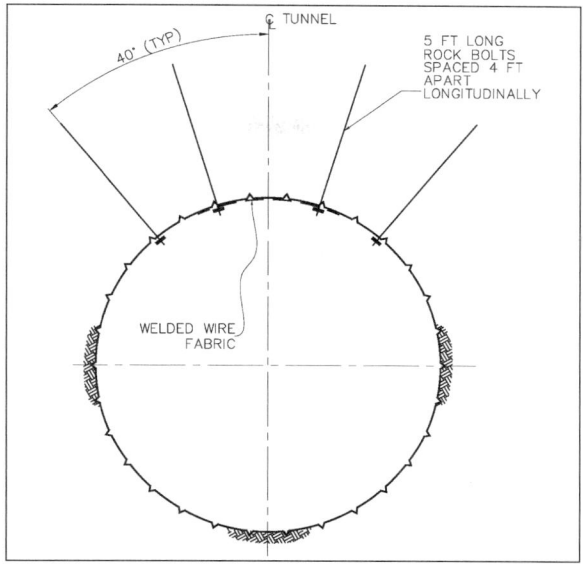

Figure 6. Modified ground support pattern

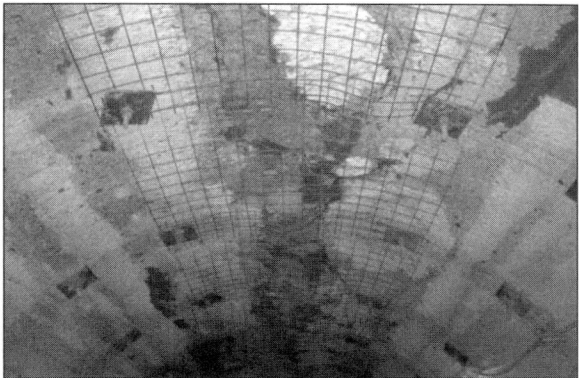

Figure 7. Installed modified ground support pattern

COMPLETING CONSTRUCTION

Upon completion of the underground construction, the tunnel and shaft areas will have a cleaning solution applied prior to acceptance for raw water conveyance. Similarly, collector well caisson structures and collector laterals will also be cleaned prior to acceptance and the tunnel becoming operational. This effort will occur prior to start-up operations at the pump station.

Figure 8. Fault feature ground support

Figure 9. Modified ground support with channel

SUMMARY

With a review of the tunnel construction challenges discussed above, it becomes clear that the unanticipated groundwater inflow events and unanticipated ground behavior required quick thinking to minimize impacts to construction. Through the cooperative efforts of the construction team and the Louisville Water Company, these challenges were overcome while minimizing impacts to the construction, and enabled the successful completion of the mining phase of the project.

TECHNICAL CONSIDERATIONS FOR TBM TUNNELING IN THE ANDES

Dean Brox ▪ Hatch Mott MacDonald

Guido Venturini ▪ Sea Consulting

ABSTRACT

The use of Tunnel Boring Machines (TBMs) for tunnel construction of long tunnels in the Andes of South America has been of mixed success in the past due to a variety of reasons and key lessons can be recognized. A review of past TBM tunnel projects in the Andes has unveiled some interesting findings for different TBM tunnel approaches with both world records of progress as well as nearly in-completed projects with machines being abandoned. Key considerations for the use of TBMs for tunnel construction in the Andes include geological issues; (faults, rock type, rock strength, rock abrasivity, durability, groundwater inflows), depth of cover and the potential for overstressing/ rockbursts, site access and terrain, portal locations, intermediate access possibilities, minimum tunnel size, support requirements, contractor and labour experience, and project schedule demands. TBMs offer highly recognized schedule benefits (high progress, no concrete lining) for a project if intermediate access adits are limited as well as hydraulic benefits for water conveyance projects. Minimum specifications for TBMs can be easily developed to suit the unique requirements of any project so as long as all factors are carefully considered. The overall conditions and factors associated with some projects may suggest that it is not appropriate to use TBMs due to a perceived high risk. Finally, access for geotechnical investigations by drilling in the commonly steep terrain of the Andes is typically difficult and often not impossible. Alternative approaches for geotechnical investigations for TBM tunnels including comprehensive mapping, high-resolution airborne electric-magnetics for delineation of possible fault zones, and rock block sampling is presented and discussed.

INTRODUCTION

The technical considerations for the use of tunnel boring machines (TBMs) in the Andes of South America are different unique from than those for other mountainous regions due to unique conditions associated within the geological conditions of the Andes and the tunneling practices in these countries comprising these regions.

Several TBM tunneling projects have been completed in the Andes with mixed success dating from the 1980s and several new hydro and mine access projects are currently contemplating the use of TBMs. The use of TBMs in the Andes appears to have a negative connotation due to the mixed past success even though most projects were completed within or shortly thereafter the respective construction schedules. Two TBMs were abandoned in the Andes from a total estimated 12 projects.

This paper outlines the key technical issues that need to be considered for the use of TBMs in the Andes, the advantages and disadvantages of the use of TBMs, some minimum requirements for their use, and some unique approaches for geotechnical investigations required for the their use. Figure 1 shows a typical small diameter TBM tunnel.

Figure 1. Small size TBM tunnel

HISTORICAL USE OF TBMs IN THE ANDES

Several long and small size tunnels have been successfully constructed in South America using TBMs since the late 1970s for water supply and hydroelectric projects. There however exists a misunderstanding that there are several tunneling projects in South America with unsuccessful applications of TBMs. Table 1 presents a summary of relevant information from several completed and ongoing TBM tunnelling projects in South America. Additional TBM tunnel projects may have been completed in the Andes but have not been identified by the authors. Many of these projects faced problematic geological conditions and associated delays however several of these projects were completed successfully in terms of schedule and costs using TBMs.

The first use of a TBM in the Andes is believed to have been in 1976 for the 24 km Yacambu Quibor Water Tunnel Project in Venezuela. This project was associated with low strength phyllite rock of about 15 MPa with rock cover over 1,200 m that resulted in significant deformation and squeezing that caused one of two TBMs to be removed in 1987 after about 1.5 km of progress and suspension of works. In comparison TBM tunneling was competed successfully for the 6.5 km headrace tunnel of the Carhuaquero hydropower project in Peru. The Misiscuni water diversion project in Bolivia however faced extreme geological conditions with weak and loose materials combined with high groundwater inflows associated with a 700 m wide fault zone under high cover where TBM progress was very poor. These conditions caused the contractor to leave the project that was completed by the TBM supplier. TBM progress rates of 840 m/month were however achieved during some of the project. It should be noted that the experience of the contractor with the use of a TBM was believed to be very limited for this project.

In 1992, the 11 km Rio Blanco Water Diversion Tunnel completed at the El Teniente Mine in Chile using an under powered 5.7 m diameter TBM was used that resulted in initially low rates of progress for the competent and very strong dioritic rock. Upgrades were completed on the TBM during the early stages of the project that resulted in sustained production rates of 30 m/day.

At the Pappallacta hydropower project in Ecuador the TBM completed 5.7 km of 6.2 km tunnel before encountering a major inrush of 1,200 m^3 of loose material requiring the remainder of the tunnel to be excavated by drill and blast. Double shield TBMs in conjunction with pre-cast concrete segments were used for two of the listed projects where highly mixed rock conditions were present and world record rates of progress of over 1,800 m/month were achieved at Manubi. The Yuncan hydropower project in Peru is the second project in the Andes where a TBM was buried and abandoned. One of the two TBMs at Yuncan was abandoned upon intersection of a 10 m fault with significant

Figure 2. High deformation at Yacambu Quibor

Table 1. Past TBM projects in the Andes

Name	Location	Year[2]	Length (km)	Size (m)
Yacambu Quibor	Venezuela	1975–2008	24	4.5
Carhuaquero	Peru	1990–1992	6.52	3.8
Rosales[1]	Columbia	1991–1992	9.1	3.5
Rio Blanco	Chile	1992–1993	11.0	5.7
Pappallacta	Ecuador	1988–1990	6.2	3.2
Misicuni	Bolivia	1998–2003	19.5	3.5
Chimay	Peru	1999–2001	9.6	5.7
Manubi[1]	Ecuador	2000–2002	11.4	4.0
Yuncan	Peru	2000–2003	6.7	4.1
San Francisco	Ecuador	2006–2007	9.7	7.1
Olmos	Peru	2008–2009	13.8	5.3
Los Bronces[1]	Chile	2009–2011	8.0	4.2

[1] Double Shield TBMs used
[2] Year of Start of Project

water inflows and loose material and required a mine-bypass around the TBM. The San Francisco Hydropower project in Ecuador is the last completed TBM tunnel project in the Andes. TBM progress was reported as good at about 25 m/day over the entire project with out any major delays. Figure 3 shows the portal set up area for the San Francisco TBM tunnel project.

The Olmos Water Project in Peru is currently underway and has experienced challenges associated with high rock cover of almost 2,000 m that has required routine high capacity support that has impacted TBM progress to less than 8 m/day until modifications were made to the TBM including the adoption of the McNally TBM Support System™ that has subsequently allowed for progress to be sustained at almost 20 m/day. Figure 4 shows an example of this tunnel support that effectively contains overstressed conditions comprising slabs and blocks.

Future projects that are planned with TBMs in the Andes include the 4.5 m diameter, 8 km, Los Bronces mine access tunnel in Chile as well as other possible hydropower tunnels in Chile and Peru and mine access tunnels in Chile.

Figure 3. San Francisco TBM portal set up

KEY CONSIDERATIONS

Geological Conditions

The Andes of South America are a relatively young geological environment with highly variable rock conditions from poorly indurated and low durability volcanic tuffs and/or highly altered andesites to extremely strong (> 300 MPa) and highly abrasive andesites that present some unique challenges for consideration for the planning of long tunnels. These series of volcanic rocks have been extremely folded and faulted in some areas. Figure 5 shows an example of differentially eroded interlayered volcanic tuffs and andesites that have been folded upright at moderate elevation of 2,000 m in the Andes. Deleterious minerals including zeolites containing smectites, gypsum/anhydrite, or vein filled laumontite are also commonly present within the volcanic bedrock. The main concern of deleterious minerals is their susceptibility to scour and/or erosion in unlined water conveyance tunnels. However, deterioration can occur during construction and cause problems with instability and tunnel floor softening (Castro et al. 2003). The presence of deleterious minerals will also typically result in moderate to low rock strengths that can lead to overstressing even under moderate rock cover.

In the northern tropical areas of South America the Andes mountains are thoroughly covered with dense vegetation that do not provide much exposure of bedrock whereas in the southern temperate areas there exists significant bedrock exposures along steeply sloped terrain.

TBMs are most appropriately applied in homogeneous rock conditions that are conducive for excavation including very strong rock varying from 150 MPa to 250 MPa. Extremely strong, massive (widely jointed), and abrasive rock will impact TBM progress however larger (19" = 483 mm) cutters in conjunction with high-power capacity can result in attractive TBM progress rates.

TBMs are also most appropriately applied along tunnel alignments where there exists a relatively low percentage of poor quality rock associated with faults/shears and/or highly altered rock.

TBMs are not ideally suited where significant and sustained groundwater inflows may occur associated with permeable fractured or porous rock where large groundwater storage exists since unmanageable groundwater inflows will impact TBM operations.

Figure 4. McNally TBM Support System™

Figure 5. Differentially eroded tuffs

Depth of Cover/Potential Overstressing

One of key considerations for the use of TBMs is the depth of rock cover along the tunnel alignment and the potential for overstressing. Overstressing will occur under the following conditions:

- High rock cover;
- Low/moderate rock strength, and;
- High in situ stresses.

Evaluation of the potential for overstressing requires knowledge of the uniaxial compressive strength and in situ stresses. Estimates of rock strength can be made by field observations at bedrock outcrops and knowledge of the type of rock and the presence of alteration. Laboratory testing of rock block or drillcore samples should ideally be undertaken.

Knowledge of in situ stresses is difficult especially in steep mountainous terrain and in particular along plate tectonic subduction zones that are associated along the Andes. Stress testing results within or near to tectonic subduction zones have commonly indicated stress ratios well in excess of unity (1.0) and as high as k=2 for moderate depths (500 m to 1,000 m). Testing results with k> 3 to 4 have also been noted. The

Figure 6. Overstressing and fall out

presence of regional faults can also have a significant impact on the orientation and magnitude of local in situ stresses.

In the absence of site-specific stress data the evaluation of the potential for overstressing or initiation of spalling and damage of brittle rock can be based on the empirical findings of observed overstressing in deep tunnels. (Diedrichs,2007). Spalling has been observed to occur where the uniaxial stress level (σ_v) is more than 0.33 to 0.50 of the laboratory compressive strength. (σ_v/UCS > 0.33–0.50). A simplified evaluation of the extent of spalling for assumed or known in situ stresses or the stress ratio (for an observed amount of spalling) can also be estimated using the empirical relationship:

$$r/R = 0.5*(UCS/\sigma_{max} + 1)$$

where r = radius depth of spalling/fall-out, R = tunnel diameter. A comprehensive assessment of the extent of overstressing should however be carried out using readily available software programs such as Phases or FLAC.

TBMs are not typically considered to be appropriate for long deep tunnels where extensive lengths may be subject to overstressing due to low rock strengths and/or high in situ stresses. De-stressing ahead of the advancing tunnel face does not occur for TBM excavation but does for drill and blast tunnels and significant amounts of high capacity tunnel support will be required to be installed close behind the cutterhead in a protected area for workers.

Under such high stress conditions there also exists the potential for the occurrence of rockbursts impacting worker safety and where there is limited protection around the cutterhead area of the TBM unless specific medications have been made for this purpose. Figure 6 shows an example of overstressing and resulting rock fall out.

Site Access and Terrain

Appropriate site access and terrain with low-grade roads must be considered to allow for the practical mobilization of TBM equipment. The weight of large size TBMs (>8–10 m) can exceed 130 tonnes and special low-boy access vehicles are typically required to bring TBMs to portal areas for assembly. Alternatively, the maximum payload for high capacity helicopters (Mi26) is limited to 20 tonnes and therefore restricts the use of only small TBMs in very remote locations.

Remote mountainous sites in the Andes also present challenges for the transport of replacement parts such as hydraulic cylinders and cutters. The replacement and transport of main bearings is also a challenge to South America and result in several months of delays and downtime if such an event were to occur.

Portal Locations

Practical locations with sufficient area must exist that facilitate the assembly of TBMs unless large span caverns/chambers can be excavated to allow for the assembly of TBMs and for the starter tunnel. Tunnel portal are always sited at the base of slopes where rockfall and/or avalanche hazards may exist. The site laydown for a TBM is much larger than that for a drill and blat operation and therefore there is greater risk for rockfall/avalanches to impact TBM operations during construction.

Contractor Experience

There exist very few tunnel contractors in South America with good TBM experience. There exists a long history of mining in South America where contractors have achieved high production rates for drill and blast excavation. Accordingly, TBMs have mainly been used in South America in joint venture with specialist TBM contractors from Europe.

With the introduction and experience of user- friendly TBM tunnel support systems through challenging rock conditions there may be an increased use of TBMs for future tunnels in the Andes.

Available Intermediate Access Adits

Where a site provides the topographic possibly of intermediate access adits then the use of TBMs is of little advantage since excavation from multiple adits can result in a similar if not shorter, schedule, with less overall risk since having multiple headings can always contribute to the overall project progress. Figure 7 shows a typical intermediate access adit in gently sloping mountainous terrain where access was possible.

Project Schedule and Completion

Some tunnel projects are revenue-based projects such as hydro and or mine access tunnels that demand early completion or face major penalties. These projects may mandate the use of TBMs in order to provide an overall shorter construction schedule. In some cases it may be attractive for owners to consider to pre-purchase TBMs for an earlier start of tunnel excavation rather than wait the full procurement period for a TBM (typically 12–16 months) after the award of contract which is one of the main disadvantages for the use of TBMs. The recent surge in projects around the world has however also meant a long procurement period for drill and blast equipment. The availability of used TBMs for early procurement is a key advantage for project schedule.

Minimum TBM Size

One of the key issues with regards to the application of TBMs is the minimum acceptable tunnel diameter to meet both the minimum hydraulic requirements as well as for practical construction for the effective installation of initial and final tunnel. An evaluation of the minimum acceptable TBM diameter for each of the three long power tunnels should be carried out recognizing the minimum hydraulic requirements, the maximum anticipated initial tunnel support requirements (Class 5 support with shotcrete), maximum anticipated deformation/closure under weak rock conditions, minimum practical clearance requirements for effective/optimal progress based on precedent practice, and maximum final support/lining requirements. This assessment will indicate that it is prudent to oversize the TBM diameter well above the minimum hydraulic requirements

Figure 7. Intermediate access adit for long tunnel

and will serve as a very useful comparison to any proposals from the EPC contractors who may wish to propose TBM excavation and have not considered all the technical requirements for the project. A TBM diameter of about 4 m is considered to be practical based on the above mentioned criteria. Figure 8 shows the minimize space for the installation of heavy steel rib support for poor rock conditions for a 4.0 m TBM.

Final Tunnel Support and Lining

The final support and lining requirements for the proposed power tunnels are subject to the durability of the encountered rock conditions as well as the performance of the initial support installed and overall stability of the tunnels after excavation. The durability of volcanic rocks common throughout the Andes is uncertain and is of particular importance in terms of their acceptability to remain unlined/protected for long-term serviceability of the tunnels. The durability of volcanic rocks can be initially evaluated from the results of rock strength and petrographic testing however further durability/ slaking potential testing may be appropriate if there exists rock units of low strength and suspected limited durability where zeolites and/or other vein infilling materials may be present.

The decision-making process for final support and lining for water conveyance tunnels should only be made after an appropriate time after excavation and initial support in order that the performance of the tunnels can be evaluated and the encountered rock conditions have been exposed to any possible effects of humidity. The decision-making process for final support and lining should be made by the Owner's representative during regular and routine site inspections during tunnel excavation such that instructions can be provided from the construction management team to the tunnel contractor to complete the works in a timely manner and concurrently with ongoing tunnel excavation and rather not at the completion of all tunnel excavation. With this approach it is necessary to have unit rates from the tunnel contractor for various forms of final support and lining.

It will be necessary that shotcrete and/or concrete linings be placed over areas where low strength and/or altered/non-durable rock is encountered that can be subjected to scour/erosion. An ongoing evaluation of rock durability will be required during tunnel excavation to further assess the durability of all encountered rock units. Figure 9 shows the application of shotcrete as part of final support/lining requirements for a small size TBM tunnel for water conveyance.

Figure 8. Support Installation in small TBM tunnel

Figure 9. Final support as shotcreting

The 5.7 m diameter, 11 km TBM excavated Rio Blanco diversion tunnel constructed for Codelco's El Teniente Mine in 1992 represents an interesting case history with regards to final support and lining requirements. The geology along the tunnel alignment is believed to have been andesites that were of good quality as indicated by the minimum support requirements during excavation. However, shortly after conveyance of water through the tunnel severe problems occurred due to deterioration of the andesite rock and it was subsequently recognized that zeolites containing swelling clays were present within the andesite. Similar adverse mineralogical conditions were identified during the construction of the 45 km water transfer tunnel of the Lesotho Highlands Water Project in Southern Africa during the mid-1990s that led to the decision to place concrete lining for the entire length of the transfer tunnel. These two case histories provide an important lesson to be learned that adequate petrographic and other associated rock testing should be completed and carefully evaluated prior to construction to identify all final support and lining requirements. Significant increases to lining requirements and/or changes during construction can typically lead to major delays and cost overruns.

ADVANTAGES AND DISADVANTAGES OF TBMs

The main advantages for the use of TBMs in the Andes are as follows:

- significantly higher and sustainable progress rates for generally good quality hard rock conditions
- less rock support due to less damage caused to tunnel profile
- long single drives where no intermediate access adits are possible in steep terrain
- lower ventilation requirements allowing smaller tunnels to be constructed
- improved health conditions for workers without exposure to blast smoke/fumes

Where intermediate access adits are not available due to steep topographic terrain and/or environmental reasons, TBMs are the common choice.

For hydropower projects the use of TBMs for water conveyance has significant advantages due to the following:

- improved hydraulic performance in terms of lower headlosses for circular TBM tunnels
- lower cost small size TBM tunnels for minimum hydraulic flows versus over-sized drill and blast tunnels
- schedule and cost savings since no need for concrete invert

Figure 10 shows the extensive work required associated with concrete invert for a drill and blast hydropower tunnel.

Hydraulic headloss reductions of 23% and 67% are typically associated with unlined TBM circular tunnels over shotcrete lined and unlined drill and blast tunnels respectively (Benson, 1986).

The main disadvantages for the use of TBMs in the Andes are as follows:

- immediate stress relaxation and overstressing behind the cutterhead requiring early support and protection of workers
- limited space available for the installation of high capacity tunnel support if very poor geological conditions encountered
- potential squeezing of TBMs at major fault zones
- limited space available for pre-excavation grouting to reduce groundwater inflows
- long procurement time of 12–16 months for new TBMs or 8–9 months for used TBMs

The advent of the McNally TBM Tunnel System™ is however now recognized as an effective solution to allow tunnel contractors to safely support overstressed rock in deep tunnels.

USE OF SHIELDED TBMs

Given the typical complexity of the geology throughout many areas of the Andes it may be appropriate to consider the use of shielded TBMs in conjunction with pre-fabricated support components such as pre-cast concrete segments. This overall excavation and support approach has many benefits over a non-shielded TBM approach followed by final support where extensive sections of tunnels either need early high capacity support for stability or additional final support due to durability concerns. One-pass pre-fabricated supports are commonly used for "soft ground." or non-rock tunnels such as for metro tunnels in urban areas. Significant schedule and associated cost

Figure 10. Concrete invert works for D&B tunnel

benefits can actually be realized by adopting a similar one-pass support approach if extensive sections are anticipated to require support. A key concern with the application of shielded TBMs is their greater susceptibility to squeezing and greater limitations to install flexible types of support. TBM manufacturers are however now incorporating greater flexibility in the design of shielded TBMs to allow for forward probing and grouting through purpose built ports within the shield and also to allow for overcutting capabilities to handle a limited amount of squeezing. The use of shielded TBMs in the Andes requires a detailed evaluation of the risks of probable weak rock and faults under high cover.

MINIMUM SPECIFICATIONS FOR TBMs

Precedent practice in the tunnelling industry has demonstrated that it is not prudent to over-specify the requirements for TBMs such that most of the risks associated with the selection of any type or make of TBM remains with the tunnel contractor.

The minimum requirements for the application of TBMs on any given tunnel project should be based on consideration of the following key issues:

- Distribution of main rock types, strength, quality and durability;
- Quantification of number and extent of major fault and shear zones;
- Presence of weak rock units and potential for overstressing and squeezing conditions;
- Installation requirements for initial tunnel support, and;
- Final support and lining requirements.

A prudent list of minimum requirements are as follows:

- Minimum power requirement to effectively excavate strongest rock units (UCS > 250 MPa);
- Over-cut capability with Reamer Cutter to prevent squeezing of cutterhead in weak rock conditions and facilitate timely changing of gauge cutters;
- Fixed mounted rock bolting drills on either side of TBM close behind cutterhead;
- Mechanical ring-arm erector to facilitate installation of steel rib supports;
- Fix mounted probe drill capable of drilling ahead to a minimum of 30 m in the strongest rock unit through port facility in cutterhead, and;

- Reinforced canopy shield extending 180 degrees over top of TBM.

The above listed minimum requirements pertain to the use of an open-type main beam TBM. Since the mid-1990s a limited number of tunnel contractors have adopted the use of double-shielded and single-shielded TBMs with the concurrent installation of pre-cast concrete segments and/or steel liner plates when a large proportion of poor rock conditions was inferred to be expected along the given tunnel alignment. The main advantage of these types of TBM is that they can operate as a open-type main beam TBM (with sidewall grippers) for good rock conditions but also install full profile segmental support (with rear thrusters) for poor rock conditions.

GEOTECHNICAL INVESTIGATIONS

An appropriate level of geotechnical investigations should be completed prior to the consideration of the use of TBMs. The typical tasks to be undertaken should include the following:

- Geological mapping/evaluation
- Identification of main faults/shear zones
- Seismic surveys at portals
- Long horizontal drillholes at tunnel portals or into side valleys
- Short (Hilti-type) drillholes to obtain core samples for testing
- Rock testing for strength, petrology and abrasivity from cores and blocks
- Evaluation of the distribution of rock quality along the tunnel alignment

Geological mapping and evaluation is the first task that should be undertaken. This work can comprise a desk review of existing literature and historical reports and investigations. The key requirement as part of this task is the identification of faults along the tunnel alignment and then field proofing of suspected features. The confirmation of such features may be difficult due to thick overburden in tropical areas or access in steep mountainous terrain. A fairly new method for the evaluation of fault zones is through the use of high-resolution airborne electric-magnetics. This relatively new method is attracting significant recognition as it is able to delineate lineaments in difficult terrain and under deep/thick overburden that may represent faults and sources of groundwater that are important to identify for tunnel construction.

In the event that colluvium or rock blocks are present around a proposed portal site it is prudent to undertake seismic surveys to determine the depth to bedrock for the design of a portal and excavation and stabilization requirements.

One of the most important requirements for the assessment of TBMs is the undertaking of a comprehensive rock-testing program. Rock parameters in terms of uniaxial compressive strength (UCS), tensile strength (Brazilian), as well as petrology (percentage of hard minerals) and abrasivity (Cerchar index) represent the key parameters that need to be characterized for the application of TBMs. Additional rock testing includes punch penetration as well Drilling Rate Index (DRI) and Cutter Life Index (CLI) that are only performed at the University of Trondheim in Norway. Extremely high rock strengths (> 250 MPa) will result in slow TBM penetration rates. Conversely, low rock strengths with high rock cover can result in extensive overstressing and the potential for rockbursts. Petrographic thin section analyses serve to define the mineral constituents and percentage of overall hard minerals (> Moh 6.5) that can also have a dramatic impact on TBM penetration. Petrographic testing will also identify the presence of rock alteration that is usually associated with a significant loss of strength. Rock abrasivity in terms of the Cerchar abrasivity Index (CAI) has become a recognized parameter that can be correlated to TBM cutter consumption and is usually related to the amount of

Figure 11. Rock block sampling

hard mineral content. Figure 11 shows a typical large rock block sample that was collected at high elevation as part of a hydropower project in Chile that was drilled and tested for TBM rock parameters.

In the steep terrain mountainous areas of the Andes it may not be possible to complete geotechnical drilling investigations due to access. As an alternative it is appropriate to identify representative rock block samples that can be collected and transported to a laboratory where core samples can be drilled and tested under standard procedures. The southern areas of the Andes have been subjected to glaciation where surface rock blocks are not weathered and therefore can be considered to be representative of deep in situ rock conditions. Another alternative to traditional deep drilling is shallow drilling of holes using a Hilt-type drilling machine into rock outcrops to obtain core samples for laboratory testing as shown in Figure 12.

Rock block samples can be collected from along steep mountain slopes where it is difficult to access and sample at higher elevations along tunnel alignments. A rock block sampling and testing program is significant less expensive than geotechnical drilling with helicopter support.

The importance of providing EPC bidders with representative and technically correct rock testing data cannot be over-emphasized. All sample preparation and testing should be completed at an accredited rock testing laboratory and only by certified testing technicians. The specifications for the rock-testing machine should be carefully reviewed as part of the laboratory selection process. A site visit should be made to the proposed laboratory prior to selection for all testing and a quality assurance visit should be made to witness the first uniaxial compressive strength (UCS) test. Photographs prior to and after failure along with failure descriptions should be provided with each UCS test result. A limited number of check tests should be completed during the early stages of the testing program at a different laboratory for an early comparison to confirm that all testing is valid and representative. Photomicrographs should be provided for each of the petrographic thin section analyses. The Rockwell Hardness Number of the testing pins used as part of the rock abrasivity testing should be noted on each of the CAI test results as different testing pins are used at various laboratories.

CONCLUSIONS

The applicability of TBMs is the Andes for major tunnel projects requires a careful evaluation of several key considerations, some of which are technical, and some that are non-technical. The key lessons that can be recognized from past tunnel projects in

Figure 12. Hilti-drilling into outcrop

the Andes include that every tunnel project and site location is unique in terms of geology, access, terrain/cover, experience of candidate contractors, and project schedule demands. A comprehensive evaluation of the anticipated geological conditions including probable fault zones, rock strength and mineralogy/alteration and the potential for overtressing should be carried as part of any assessment for the application of TBMs.

TBMs do not represent the best solution for all major and long tunnel projects where high-risk geological conditions are anticipated or where intermediate access adits are available for well-experienced drill and blast contractors.

TBMs have however contributed significantly to the completion of several major tunnel projects in the Andes despite encountering challenging ground conditions. TBMs remain to represent the most schedule-effective approach for the construction of long tunnels where intermediate access adits are not available and so as long as there does not exist a large portion of poor quality ground.

REFERENCES

Benson, R. Design of Steel Lined Penstocks and Pressure Tunnels. Tunnelling Association of Canada Conference, 1986.

Castro, S.O. Van Sint, J.M., Gonzalez, R.R. Lois, P.V. Velasco, L.E. Dealing with Expansive Rocks in the Los Quilos and Chacabuquito Water Tunnels—Andes Mountains of Central Chile. ISRM Congress 2003.

Diederichs, M.S. 2007. Mechanistic interpretation and practical application of damage and spalling prediction criteria for deep tunneling. *Canadian Geotechnical Journal*, 44(9), 1082–1116.

ACKNOWLEDGMENTS

The authors wish to acknowledge contributions of information regarding the history of past TBM projects in the Andes from Evert Hoek, Canada, Julio Castilla of ARPL, Peru, Pablo Lopez of URS, La Paz, and Carlos Filho of Odebrecht, Sao Paulo as well as Mike McNally of C&M McNally Engineering Corporation, Canada, for discussions on the McNally TBM Support System™. The authors acknowledge Renzo Valentino Cardoza of Pacific Hydro for his contribution to this paper.

ROBBINS 10M DOUBLE SHIELD TUNNEL BORING MACHINES ON SRISAILAM LEFT BANK CANAL TUNNEL SCHEME, ALIMINETI MADHAVA REDDY PROJECT, ANDHRA PRADESH, INDIA

William Brundan ▪ The Robbins Co.

INTRODUCTION

The Srisailam Left Bank Canal (SLBC) Tunnel Scheme is part of the Alimineti Madhava Reddy Project in Andhra Pradesh, South India. This scheme has been planned since 1983 and will provide irrigation facilities to drought-ridden areas in Andhra Pradesh. Water will be transferred from the Srisailam Reservoir during the monsoon months to 0.3 million acres of farmland and will also provide drinking water to villages en-route.

The paper will address the design and onsite assembly of the Robbins Double Shield TBMs for Tunnel-1 of the SLBC tunnel scheme from planning to start-up of the machines. Also, it will provide a current status of the project and describe the advantages of a site assembly. Problems encountered to date and the solutions adopted will also be described. The project comprises over 50 km of tunnels and involves construction of the world's longest tunnel without intermediate access. The TBM tunnels at 43.9 km in length and 10.0 m diameter will cross the Amrabad plateau, containing a wild life sanctuary and India's largest tiger reserve, the Rajiv Gandhi.

Jaiprakash elected to excavate the tunnel using two tunnel boring machines, rather than by drill and blast, in order to alleviate surface disturbances to environment and the tiger reserve, wild life sanctuary, and protected forestry areas.

The Project was awarded to the Main Contractor Jaiprakash Associates Ltd with The Robbins Company as the Tunnel Boring Machine (TBM) supplier in August 2005. The Robbins Company supplied the TBMs along with two continuous conveyor systems and all spares and consumables to complete the tunnel excavation.

PROJECT DESCRIPTION

The tunnel is altogether 43.9 km in length and is exceptional because there is no possibility of intermediate access. Two new tunnel boring machines were chosen by Jaiprakash, to have any prospect of completing the project in the required 60 months. The TBMs are able to install a segment final lining simultaneously with excavation.

The tunnel is lined using precast concrete reinforced segments. The choice of the TBM was primarily made to be double shield, so that lining could be installed in parallel to excavation and therefore obtain best possible production outputs. The segments for the 10.0 m excavated diameter tunnel (face area 78.5 m^2) are 300 mm thick with internal lined diameter of 9.2 m.

Each tunnel drive of up to 22 km is carried out from the respective excavated portal areas at the outlet end and inlet end. Portal surface areas were maximized so that the full length of the TBM could be assembled for ease of launching the machine. This was straight forward for Jaiprakash at the outlet as they had no space limitations and the portal was designed to be 45 m wide and 160 m long. But at the inlet because excavation of the portal area was known to be far more challenging, the portal area had to be reduced to 120 m by 45 m and protected by a bund wall as the site is below the FWL of the reservoir. The site is still under construction by Jaiprakash. Assembly of the inlet TBM is only expected to commence in May 2009.

Excavated spoil from the tunnels is transported by a Robbins Continuous Conveyor system from the TBM to the surface spoil handling area where trucks are loaded to move material to spoil dumps. The excavated granite, from the TBM drive and from the portal excavation at the outlet, is being recycled for use as concrete aggregate and peagravel by Jaiprakash. The same is planned at the inlet where the quartzites also make good aggregate.

PROJECT OPERATION

Robbins is engaged to operate the TBMs for Jaiprakash and provide experienced supervision to work with the Jaiprakash workforce. Robbins maintains at site a team of about 18 persons, composed of both Indian Nationals and Expatriates, to operate the TBM, carry out maintenance, and to provide supervision of cutter changing and the tunnelling operation.

PROJECT ALIGNMENT

Initially the tunnel had been designed to be straight over its entire length but after survey of the surface and detailed investigation Jaiprakash had to revise the alignment and add some large radius curves to avoid a deep valley where the TBM would have day-lighted.

The high adopted radius of 8,000 m in effect allows the non tapered segment ring to be used throughout with minimal corrective measures and packing.

SITE INVESTIGATION AND GEOLOGY

The geology of the tunnel has been interpreted mainly from surface mapping and aerial photography. Jaiprakash had completed walkover of the surface and survey of all river valleys above the tunnel. One such steep valley affected the tunnel and resulted in the realignment of the project.

Nearby underground power stations on the Srisailam reservoir and some 5 km from the tunnel inlet also have given extensive information about the quartzites.

The geology is generally very stable, unlike the conditions in North of the country, part of the ancient South Indian Peninsular Shield, and consists of two main rock types: quartzites and granite.

Normal rock properties of Srisailam quartzites:
- Density 2.53 to 2.80 g/cm^3
- Compressive strength 850 to 2,650 kg/cm^2 (83 to 260 MPa)

Normal rock properties of granites:
- Density 2.62 to 2.86 gm/cm^3
- Compressive strength 1,000 to 2,300 kg/cm^2 (98 to 225 MPa)

From tunnel chainage 0 m at the inlet portal to tunnel chainage 26,450 m this consists of quartzites and thin interbedded layers of shale. From tunnel chainage 26,450 m to the outlet portal at tunnel chainage 43,950 m a length of 15,500 m of granite will be encountered (refer to Figure 1). The expected proportion of quartzite to granite

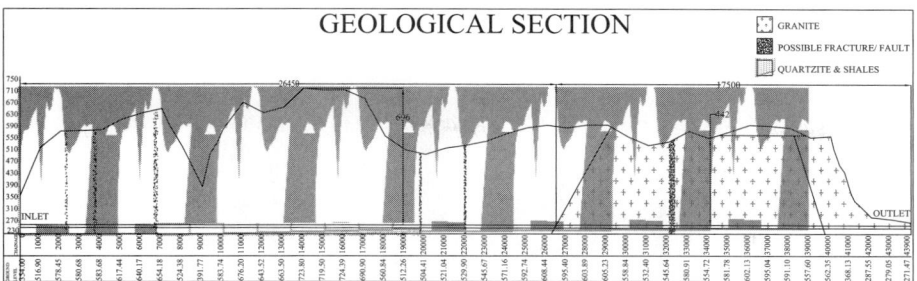

Figure 1. Geological section

Figure 2. Granite geology

depends on the orientation of the granite to quartzite transition currently projected from geological mapping carried out on the surface.

The quartzites, over 60% of the tunnel, are metamorphic and have very high quartz content. Their significant abrasiveness and high strength give the most concern for cutter wear and costs. The expected physical properties of the rocks are given above. Results from tests carried out on the quartzites and from information from the power station at Srisailam have shown that compressive strengths of the quartzite can go as high as 4,588 kg/cm^2 (450 MPa).

The granites for 40% of the tunnel are generally expected to be consistent in nature and are less cause for concern. Figure 2 shows typical granite geology at the assembly portal.

The quartzite section also can be very blocky in nature and includes thin and weak interbedded shales. The layering of quartzite and shales will tend to cause overbreak in the crown and shoulder of the tunnel. A good impression of the quartzite geology can be seen in the portal area photograph (see Figure 3).

The granite section encountered by the TBM up to December 2008 has been extremely variable in nature and rarely presenting a massive clean face. Many joints effect the way the face stands up to the action of the disc cutters, certainly giving increased wear and damage through impact loadings.

Figure 3. Quartzite geology at the inlet portal

TBM CHOICE

Double shields were eventually finalized as the best choice for the tunnel excavation by Jaiprakash. This was after detailed discussion and analysis. Consideration of using Open TBMs, or one Open and one shielded machine was made, but the simplest solution of using two double shields was finally confirmed.

The main reasons for the choice were:

- Erection of segment lining simultaneously with excavation
- 60% of the tunnel in very blocky and layered shale and quartzite
- No temporary support required
- Temporary support difficult to quantify and estimate
- Concrete lining after excavation or even in parallel with excavation eliminated
- Optimum programme as no delay in follow on concrete lining
- Various fault and shear zones expected
- Hydraulic design requiring a lining
- Two TBMs of the same type

TBM DESCRIPTION AND MAIN DETAILS

The key elements of the TBMs are as follows:

- Cutterhead
- Main bearing and seal
- Forward shield
- Main drive
- Telescopic shield
- Gripper shield
- Tail shield
- TBM conveyor

Cutterhead. The cutterhead design is similar to that used on many Robbins hard rock TBMs (see Figure 4). The cutterhead consists of a six piece bolted and doweled heavy duty structure for hard rock. The cutterhead is made from heavy steel plates,

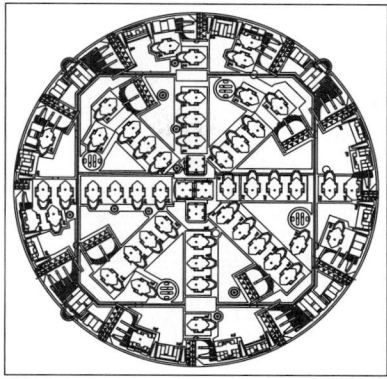

Figure 4. Cutterhead front view

internally reinforced with integral buckets to remove the rock cuttings. The buckets pick up spoil and deliver this by gravity to the central TBM conveyor situated within the cutterhead support. The total weight of the assembled cutterhead is in excess of 300 tonnes. The cutterhead includes:

- Twelve muck buckets with radial face and gauge openings
- Abrasion resistant boltable bucket teeth or lips
- Grill bars
- Abrasion resistant carbide buttons
- Abrasion resistant faceplate and gauge plates

The cutterhead is the mounting structure for the 67 main replaceable cutter rings. An additional three cutter housings of 17-inch (432 mm) diameter can be used if required for overboring and increasing the diameter from 10.0 m to 10.2 m. The 67 disc cutters consist of:

- 8 twin disc centre cutters of 17-inch (432 mm) diameter
- 48 number face cutters of 19-inch (483 mm) diameter
- 11 number gauge cutters also of 19-inch (483 mm) diameter

The TBM and cutterhead is designed for the Series 19-inch (483 mm) disc cutters. To further cope with the hard and abrasive rock expected on the project the 19-inch hubs are actually mounted with 20-inch (508 mm) rims. This allows an additional 80% of rim material for wear and gives additional cutter life of approximately 20%. Initial results from the outlet TBM after some 2 km of excavation is showing that the special 19-inch/20-inch cutters are performing better than expected and certainly warrant the development of this cutter size.

Main bearing, seal and ring gear. The main bearing assembly is a tri-axis bearing that accepts both the radial and thrust loadings associated with the cutterhead operation. Tri-cylindrical roller bearings are mounted on the main bearing and held in place by the cutterhead adaptor. The ring gear is an externally toothed gear mounted to the cutterhead adaptor and secured in place by studs, hex nuts and washers. An inner and outer seal is mounted to the non-rotating structural components with the lips of the seals providing sealing action against the rotating components. The seals are orientated so that the lips prevent entrance of contaminants into the cavities of the bearing.

Front shield and cutterhead support. The front shield is fabricated in four segments weighing some 215t and is designed as the primary support structure of the

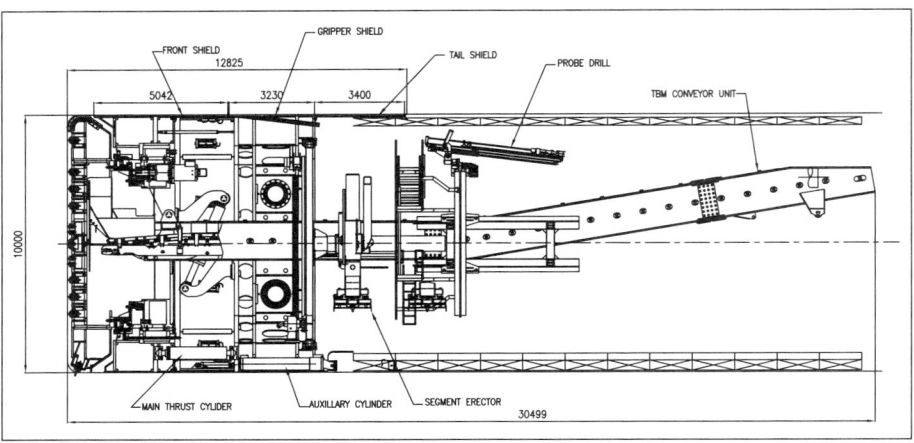

Figure 5. TBM shields and layout

machine. The forward shield supports the weight of the cutterhead assembly and also provides mechanisms for stabilizing it against movement within the bored profile of the tunnel. The forward shield provides the primary structure for mounting the cutterhead assembly and the mounting base for the main drive motors, and a machined and drilled structure for mounting the main bearing assembly. In addition the shield includes provisions for mounting of the conveyor, muck chute and stabilizer cylinders and shoes.

Main drive. The cutterhead is driven by twelve 315 kW, 690-volt, 3 phase, 50 Hz, water cooled, VFD electric motors. Planetary gear reducers from the motors drive the common main ring gear. The water cooled motors should be supplied with a minimum of 20 lpm water each at a temperature not exceeding 21 degrees centigrade.

Telescopic section. This is the intermediate section between the front shield and the gripper shield and comprises an outer shield section connected to the front shield and an inner section connected to the gripper shield. A total of 16 articulation cylinders are mounted between the telescoping shield and the gripper shield.

Gripper shield. This shield is the supporting structure for the two double acting gripper cylinders which develop the hydraulic force required to hold the gripper shoes tightly against the sides of the tunnel to react to the torque and propel forces needed for boring and cutting rock. Twelve main thrust cylinders are installed between the gripper shield and the front shield to generate the hydraulic thrust necessary for propelling and steering the machine. Two torque cylinders also installed between the front and gripper shields provide roll control and also vertical steering of the cutterhead.

Tail Shield. The rear shield contains 19 auxiliary propulsion cylinders which react against the tunnel lining segments. These cylinders have three functions:

- Moving the gripper shield during the reset cycle
- Securely holding the segment lining in place during the build cycle
- Acting as a strut between the gripper assembly and the segment lining and providing further reaction force for the thrust system

The tail shield is open bottomed to allow the segments to be placed directly on the excavated rock. The opening covers an angle of 100 degrees.

Dimensions of the shields and layout of the components described above are given in Figure 5.

Further details are given in Table 1.

Table 1. TBM specification

Technical Specification TBM 322-317 and TBM 322-318	
Cutterhead	
Nominal diameter	10.0m new cutters
Overbore	10.20m
Type	Flat face design
Muck buckets	12
Cutters	
Type	Front, Backloading
Size, face and gauge	Series 19" (483 mm) diameter
Centre cutter	Series 17" (432 mm) diameter
Number	67
Max recommended load per cutter	312 kN
TBM Shields	
Number	3 number
TBM overall length	12.8m
Shield length	11.6m
Tail shield type	Open bottom
Cutterhead Drive	
Type	VFD electric motors
Number of motors	12 × 315 kW
Maximum Power	3780 kW
Speed, constant torque range	0 to 3.66 rpm
Cutterhead Torque	9864 kNm
Speed, constant power range	3.40 to 8.05 rpm
Cutterhead Torque	4484 kNm
Main Thrust	
Main thrust	34392 kN at 228 bar
Max recommended cutterhead thrust	20904 kN
Number of cylinders	12, (400 × 280 × 1700 mm)
Operating pressure	345 bar
Auxiliary Thrust	
Auxiliary thrust	58,190 kN @ 244 bar
Number of cylinders	19, (400 × 330 × 2250)
Operating pressure	345 bar
Grippers, 2 Nos.	
Gripper pad pressure	6.0 Mpa
Anchoring force per pad	55,245 @ 313 bar
TBM Conveyor	
Width	1372 mm
Capacity	1500 tph
Erector	
Type	Vacuum pad
Hydraulic System	
Operating pressure	350 bar
Oil tank capacity	9 m^3
Electrical System	
Cutterhead drive	VFD System
Primary supply voltage	22,000 V
Transformer	2 × 3000 KVA (690 V)
Secondary voltage	690 V, 3 phase
Auxiliary Transformer	1 × 2000 KVA (400 V)
Secondary voltage	400 V, 3 phase, 50 Hz
Machine Weight	
Total	1400 t

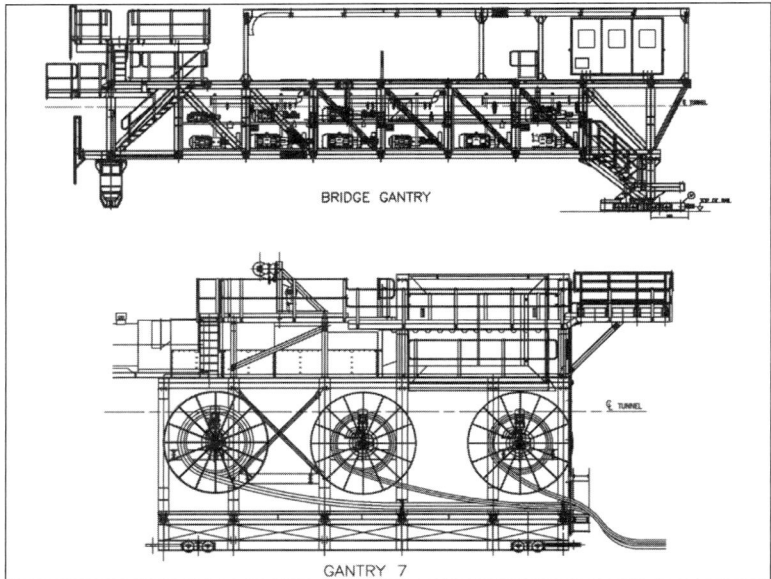

Figure 6. Details of back-up decks 0 and 7

BACK-UP DESCRIPTION AND DETAILS

The back-up deck system consists of strong rolling decks numbered from 0 to 7 and a towed California crossing which allows two trains to pass. The back-up has been designed to run on an outer rail which is recycled once this is exposed at the rear of the equipment. This back-up structure was all fabricated in India.

The back-up is designed as a two push system—one supply train can deliver enough materials and segments for two complete excavation cycles and this can all be unloaded in one fast operation. This system was adopted because of the great length of the tunnel drives and the time for a supply train to reach the TBM—greater than 1.25 hours of travelling time in one direction at 20 km length.

The back-up decks have two main levels and carry all the necessary equipment for running the TBM:

- Bridge Gantry or Deck 0—TBM conveyor, transfer conveyor and bridge gantry and the towing connection between TBM and back-up. The TBM drive hydraulic power packs are mounted on the middle deck. Under the bridge section rail track is installed in up to 9m lengths on which the following backup decks travel. Figures 6 and 7 illustrate this gantry.
- Deck 1 and Deck 2—Electrical cabinets, transformers and VFD equipment are mounted on the top deck. Segment handling equipment on the lower deck.
- Deck 3—Invert and secondary grouting mixers, pumps and material handling equipment
- Deck 4a and b—Pea gravel injection equipment, storage hoppers and handling system
- Deck 5—Water supply and dewatering system. Emergency refuge chambers
- Deck 6—Air compressors, high voltage cable reel and ventilation equipment

Figure 7. Deck 0

Figure 8. Deck 7

- Deck 7—Ventilation equipment and dust scrubber system, duct magazine and hose reels
- California crossing—Towed by the backup and running on the backup outer rails. Rail recycling carried out at the rear of the California. Figures 6 to 8 illustrate decks 0 and 7.

Back-up decks in all are some 120 m long and weigh unloaded 450 t.

SEGMENT HANDLING SYSTEM

Segments are transported into the back-up by train and carried on four Muhlhauser flat cars (refer to Figure 9). Four segment cars are unloaded in one operation by segment lifters at back-up decks 1 and 2 thus allowing the train to move back in a matter of minutes to unload grout and peagravel on decks 3 and 4.

A segment crane rated to lift segments in pairs then moves the segments to the segment feeder and the vacuum erector. The vacuum erector is rated to lift 10 tonnes (refer to Figure 10).

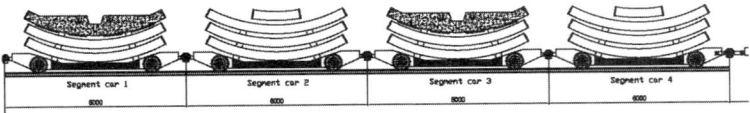

Figure 9. Segment layout on flat car

Figure 10. Segment crane

ROLLING STOCK DETAILS

The TBM is supplied by tunnel trains, Figure 11, which carry all the materials required to support the excavation process. The trains are 65 m in length and weigh when loaded close to 200 tonnes. They include mainly segments, grout materials and pea gravel cars.

CONVEYOR SYSTEM

The conveyor system in itself is a major component of the tunnel. Each system has been designed for an operating length of 22.5 km and is VFD controlled. As a result of this length the conveyor has been divided into two separate runs in each tunnel section.

Run 1 for 11.25 km consists of a portal drive and loop cassette system and one booster drive at 5.626 km, Run 2 is a repeat of Run 1, the difference being that the loop cassette system must here be installed inside the tunnel.

At the outlet portal also included as part of the system are an incline conveyor at gradient 1:5.5 and a stacker conveyor. The stacker allows a stockpile of some 9 m height over an arc of 45 degrees. Some 5,000 m^3 can be stockpiled here waiting for muck removal by truck.

Figures 12 and 13 show the portal loop cassette, the incline conveyor and stacker. Summary specification of the conveyor:

- Per tunnel length of 22,500 m—2 conveyors of 11,250 m
- Drive unit for 11,250 m—2 × 300 KW
- Booster unit for 11,250 m—2 × 300 KW
- Belt capacity—800 TPH
- Belt speed—3.0 m/s
- Belt width—914 mm
- Cassette capacity—600 m

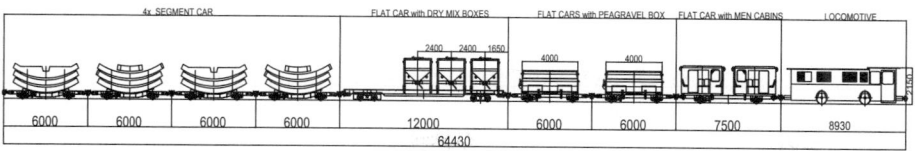

Figure 11. Train layout

Figure 12. Photograph of loop cassette at portal

Figure 13. Photograph of incline and stacker conveyor

Table 2. Summary of segment lining

Description	Bolted Concrete Lining
Concrete grade	50 N/mm^2
Tunnel Alignment	Straight on plan with gradient of 1:3200
Segment type	6 straight segments and a wedged key with 1:7 taper
Ring width	1600 mm
Excavated diameter of tunnel	10000 mm
Internal diameter	9200 mm
Segment thickness	300 mm
Gasket	None
Circle joints	3 Dowels in invert segment and 16 spear bolts in all other segments
Radial joints	14 Spear bolts
Bolt specification	M20 by 375mm long grade 4.6

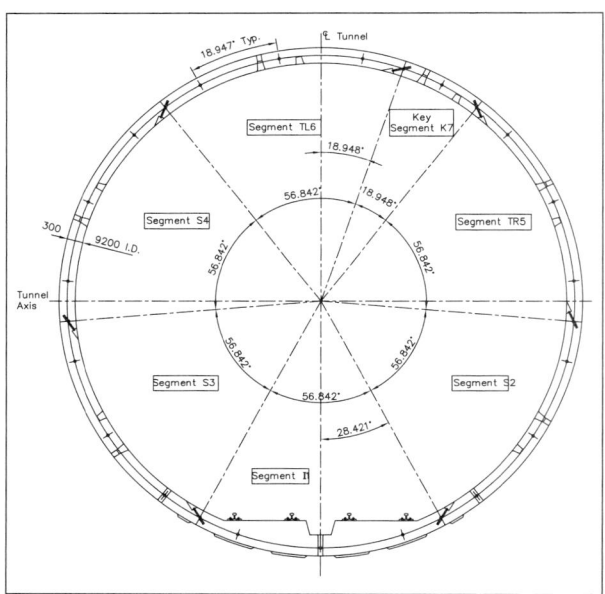

Figure 14. Segment geometry and cross section

SEGMENT DESIGN

Segment design was carried out by MottMacdonald of the UK for Jaiprakash with coordination from The Robbins Company (refer to Figure 14). Design concluded with a 1.6 m length segment and a 6+1 element ring. The segment was designed with reinforcing of 95 kg/m^3 and ring thickness of 300 mm. The ring is bolted parallel sided, with no taper. There is also no gasket. A specification summary of the ring is given in Table 2.

The invert segment has external supporting pads cast onto the extrados which allows the segment to be placed directly onto the invert. In each segment there are two erector shear pin holes with one of these being used for pea gravel injection and grouting purposes. Additional grout holes are also available in the invert and lower two segments.

The full ring weighs some 40 tonnes, with the flat invert segment just below 10 t in weight.

The segments are produced by in a static segment factory situated near the tunnel portals. Layout of the factory was designed by Jaiprakash with specialist equipment and moulds coming from CBE of France.

PROBE DRILLING

Probing equipment is mounted at the rear of the tailshield so that probe drilling if required for grouting can be carried out over some 300 degrees of the tunnel circumference. Twenty four drill holes are installed through the gripper shield and gripper shoes and inclined outwards at approximately seven degrees. In addition it is also possible to drill through the face in six positions if required.

Probe hole layout has been designed so that fans or arrays of grout holes can be drilled when required to improve conditions ahead of the face in poor ground or wet conditions. The drill hole collars in the ground at approximately 6.2 m from the tunnel face. It must be said that it is not expected throughout the entire tunnel to meet these poor conditions.

Probe drilling equipment has been specified to again cope with the very hard and abrasive conditions.

ANNULUS PEAGRAVEL INJECTION AND GROUTING

Peagravel is injected from deck 4 on the back-up to the front segment rings. Injection through the segments, on rings 2 and 3, has proven to be most successful in filling the annulus. Two Aliva 263 pumps are used to give more than 200% necessary injection capability.

Peagravel is produced on site from crushed and screened granite produced initially from excavations for the portal site, and subsequently from tunnel spoil. Significant wear to the equipment and delivery pipelines does occur as a result of the coarse peagravel.

Grouting is carried out initially to securely bed the invert segment as the first stage and then some 40 rings back to fill any voids left in the peagravel and in the crown of the tunnel.

Two sets of grouting equipment, one for each stage, are mounted on the back-up.

VENTILATION

The tunnel ventilation system from the portal to the rear of the TBM back-up is forced from a bank of three 355 kW fans situated at the portal. These force air into a 3,100 mm diameter ducting.

The magazine cassette for this large diameter duct is transported by special rail car to the rear back-up decks where special handling equipment is installed to lift the cassette into position. The duct cassette handles 100 m lengths of ducting.

On the TBM back-up two ventilation systems, one forcing and one extracting via a scrubber system, distribute fresh air around the back-up decks.

Fan specifications:

- Main portal fans—3 × 355 kW VFD fans
- TBM fan—2 × 55 kW
- TBM scrubber—1,000 m^3/min dry dedusting unit and 2 × 75 kW fans

All the ventilation equipment was provided by ECE Cougemacoustic of France.

COOLING

The 12 drive motors on the cutterhead are water cooled. Cold water is provided from a surface chilling plant, at between 21 and 25 degrees C and is pumped into the tunnel and TBM through an insulated 150 mm pipeline. The TBM can use up to 1,200 LPM of water in total, though the drive motor cooling circuit requires some 240 LPM minimum. As water temperature increases, required quantity of water for cooling the TBM drives increases.

Despite the high temperatures of an Indian summer, in excess of 40 degrees C for more than four months, no other cooling measures have yet been required to control the environment in the tunnel. So far the ventilation system provided is working effectively. As the tunnel gets longer this situation may have to be reviewed.

EMERGENCY CHAMBERS

Safety concerns in a tunnel of this length certainly necessitate refuge chambers able to accommodate the full working shift. Two separate pressurized containers provide capacity for 32 people and are included within the back-up. The containers are connected with a compressed air line from the surface. Should emergency occur, compressed air is stored in cylinders for 32 persons sufficient for a two hour period should pressurization not be possible.

SITE ASSEMBLY

Both machines were planned to be delivered in a very short time frame. The first components of the TBMs were to be shipped within eight months of finalizing the order and all components for the first TBM were delivered by month 13. Due to this very short start up programme Robbins had no hesitation in adopting the "Onsite first time assembly" (OFTA) policy. Under this policy components from all around the world were brought to site, and only assembled together for the first time together at the project site. Individual assemblies such as the probe drill and erector were pre-assembled and tested at their respective manufacture locations.

Shipping dates as required under the contract were therefore successfully achieved but this was only possible with OFTA. Time savings estimated to be some 4 to 5 months on the delivery schedule were achieved, making significant cost savings in shipping and handling compared to a full factory assembly.

PROJECT STATUS

The first TBM started the commissioning process during April 2008. Although assembly officially started on 19th November 2007, work only started to really gain momentum in January 2008. This meant that assembly, testing and commissioning took a period of four months. Excavation commenced on 19th May 2008.

The second TBM, although mostly delivered to site during 2007, will only start to be assembled in May 2009 because of the considerable problems that have been encountered by Jaiprakash in obtaining access to land necessary for commencing excavation of the inlet portal.

Production outputs to date for the TBM have been improving, as the significant learning curve with a labour force inexperienced in TBM work has been overcome. Excavation outputs on a weekly and monthly basis are given up to the 20th December 2008 in Figures 15 and 16.

Problems with water and blocky granite have now also been overcome and target production rates in excess of 400 m per month are expected to be reached early in the New Year.

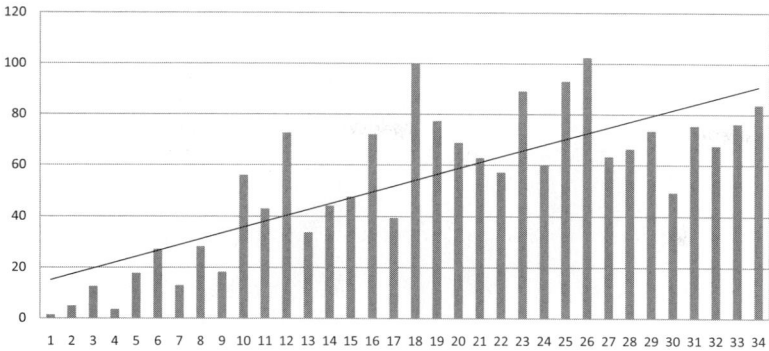

Figure 15. Weekly production rates

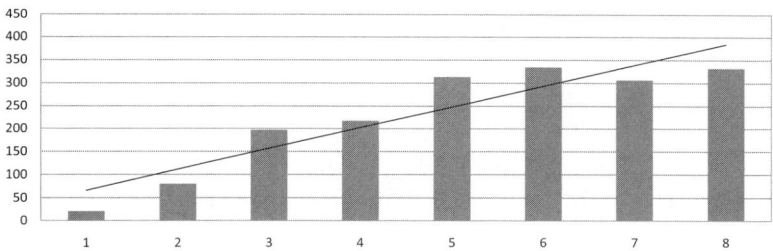

Figure 16. Monthly production rates

PROBLEMS ENCOUNTERED AND ADOPTED SOLUTIONS

Water in the face. The granite section of the tunnel was expected to be very dry with little or no water. This proved not to be the case! In blocky and fractured sections and also associated with dolerite dykes significant water flowed into the face. This not a concern to the TBM, but when picked up by the cutterhead buckets and deposited on the conveyor caused real problems with the conveyor scrapers, as a paste of highly abrasive sand caused rapid wear to the scrapers. Water draining from the TBM belt also tended to pour onto the segment erector causing mechanical problems. Many types of scraper were tried until the best was found (see Figure 17).

Boulders damaging the conveyor. Face conditions often are extremely blocky, giving rise to large blocks and boulders in front of the cutterhead. Boulders often managed to pass through the muck buckets and end up on the conveyor belt (see Figure 18).

Occasional boulders would block in transfer hoppers and point load the conveyor belt causing massive tears in the belt (see Figure 19). Boulders also would roll back and drop off the incline and stacker conveyor. The solution proved to be in reducing the spacing of grizzly bars on the muck buckets and adding additional bars so the boulders could not pass onto the conveyor system (see Figure 20).

Water on the conveyor. Part of the conveyor system from the end of the portal area rises up a gradient of 1:5.5 (10 degrees) to the surface. The incline designed as part of the main conveyor system, when water was present, caused excavated material to flow back down the belt resulting in extensive spillage (see Figure 21). The incline conveyor could not be changed, but the stacker angle was reduced. Water cannot be

Figure 17. Scrapers replaced many times

Figure 18. Boulders!

stopped at the face so this problem is difficult to solve fully, but reducing boulder size by adding more grizzly bars has certainly helped.

Power outages. The project site is meant to run from a main electricity supply but does have full generator backup. Regular power outages occur which cause sudden stops to boring. In some weeks up to 30 power cuts have been recorded. A solution for this problem hasn't yet been found!

Peagravel injection. Initially it was planned to inject peagravel from rings 6 and 7, some 10 m from the tailshield. This assumed that peagravel would flow in the annulus outside the segment lining. This proved not to work and pea gravel injection had to move right up to ring 2, almost at the tail shield itself. The peagravel manufactured from crushed granite and even tunnel spoil is coarse and not ideal. In addition to excessive wear caused by the gravel it certainly doesn't flow properly outside the segments.

Setting invert segment. The tail shield is designed to be open bottom to allow the invert segment to be placed directly onto the excavated rock. This in practice initially did not work and packing had to be placed beneath the segment. The solution proved to be to increase the opening allowing the interaction between the lower three segments and the tailshield to be minimized.

Figure 19. Belt tears

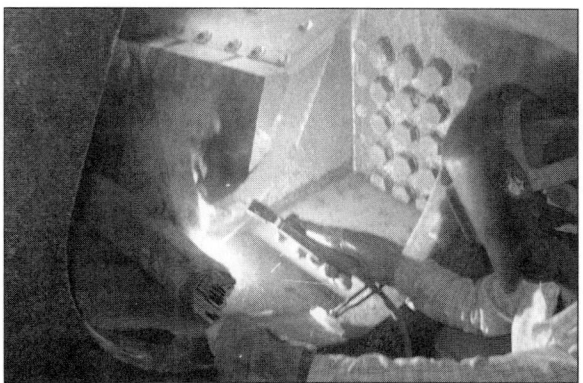

Figure 20. Installing extra grizzly bars on muck buckets

Local labour force. The local Indian labour force is not experienced in TBM work—of course this was to be expected but training and gaining experience certainly lengthened the learning curve process. It must be said that although generally inexperienced the workforce has been keen and very willing to learn. On future projects Robbins has put together a detailed training course which takes the Contractor's personnel through four stages of learning from general tunnelling to looking in detail at electrical and hydraulic circuitry and studying the operation and maintenance manuals of the equipment.

Figure 21. Water and boulders on the conveyor

OTHER PROJECTS IN ANDHRA PRADESH

A third Robbins Double Shield will also commence operation during 2009 on the other bank of the Srisailam reservoir. This machine is almost identical to the two at the Project described above and is currently under assembly. Three TBMs of 10.0 m diameter operating in the State of Andhra Pradesh illustrate well how fast the tunneling market is developing in India.

ACKNOWLEDGMENT

Thanks must be expressed to the site teams currently based at Devarakonda and led by personnel from Jaiprakash Associates Ltd and The Robbins Company for their dedication and hard work in getting the first TBM commissioned and under way.

PART 20

TBM Case Studies 2

Chairs

Eric Eisold
Bradshaw Construction Corp.

Brett Zernich
Traylor Brothers Inc.

MADIQ TUNNEL, LEBANON: TBM TUNNELING VS. KARST GEOLOGY

William D. Leech ▪ MWH Americas, Inc.

Issam Bou Jaoude ▪ Baresel AG/Al Tajj, JV

Nicholas Ghanem ▪ Tajj Est.

ABSTRACT

Lebanon used a TBM for the first time to excavate a tunnel in the country. A main beam TBM, built in 1979, bored the 3.8 m (12.5-ft) diameter by 4,020 m (2.5-miles) Madiq Tunnel. The TBM bored through karstic terrain of dolomitic limestone, limestone and marly-limestone rock having near horizontal beds with more than twelve faults crossing the axis of the tunnel. The TBM also encountered more than six major karst features, filled with various combinations of boulders, clay, void space and water. The excavation through these collapsed-breccia zones drastically slowed the TBM's progress. When complete, which the Madiq Tunnel is a part, the Kesrouane Water Project along the coast will bring fresh and irrigation waters to the parched and heavily populated region north of Beirut. The paper will describe the tunnel geologic features, karst genius, TBM upgrades and performance, and TBM/karst difficult situations.

INTRODUCTION

The Lebanese Ministry of Hydraulic and Electric Resources, the Council for Development and Reconstruction (CDR), and the Kesrouane Water Authority are constructing the Kesrouane Coastal Area Water Supply Project. The Madiq Tunnel is a major part of the Project. The Japanese Bank of International Cooperation is funding the Project. When finished the Project along Lebanon's coast will ultimately bring fresh and irrigation waters to a parched and heavily populated region. The entire Project aims to enhance the country's water network with a system of catchment works, pumping stations and distribution tunnels that will serve 252,000 customers by 2020. Once complete, the Project will supply up to 89,000 cubic meters (72 acre-ft) of clean water per day, sourced from the valley of Nahr Ibrahim.

The Project involves the construction of nine pumping stations, 29 reservoirs, and 51-km of transmission pipelines. Included in the Project is the 4,020-m (13,200-ft) Madiq Tunnel, which will have a 3.8-m (12.5-ft) excavated and finished diameter for the water conveyance of both domestic and irrigation waters.

The Madiq Tunnel is located on west flank of Mount Lebanon in the Casa of Kesrouane with the portals in the valley of Nahr Ibrahim, approximately 40-km (25-miles) north of Beirut.

In August 2006, work on the Project was halted because of hostilities in the region. Lebanon's infrastructure sustained more than $2.5 billion USD in damages during the July 2006 conflict with Israel. Most of Lebanon's construction projects were stopped or abandoned altogether. When the war started, the Tunnel Contractor was preparing to launch the tunnel boring machine (TBM) at the downstream west portal (Figure 1). The tunneling labor crews and expatriate supervision left the worksite immediately.

Figure 1. TBM assembly for launch

From the conflict no damage was sustained to either the TBM or other tunneling equipment, but the Project was delayed about four and a half months. In December 2006, labor crews and expatriate supervision returned to the tunnel project site to resume the launch of the TBM and to begin boring operations.

In mid September 2008 the Tunnel Contractor "holed-thru" Madiq TBM drive. At the time of publishing this technical paper, Tunnel Contractor has nearly completed the concreting/shotcreting, equipping and finishing of the Madiq Tunnel.

TUNNEL CONTRIBUTORS

From 1995 to 1997 Lebanon's Bureau Technique pour le Development (BTD) in association with SIMECSOL a French designer performed the design of the Madiq Tunnel. The construction of the Madiq Tunnel is part of the CDR Contract 2054 of the water scheme. On 14 June 2001, CDR awarded the tunnel contract to Baresel AG–Al Tajj, a joint venture. CDR selected DAHNT-MWH joint venture of Dar Al Handasah Nazih Taleb & Partners of Lebanon and Montgomery Watson Harza of the United Kingdom to perform construction management of the Project. Baresel provided the expatiate supervision (Site Project Manager, Tunnel Superintendent, Tunnel Walkers, Tunnel Shifters, TBM Operators, TBM Mechanics/Electricians and Master Mechanic) for the tunnel excavation. The Tunnel Resident Engineer for the excavation phase came from MWH.

REGIONAL GEOLOGY

Lebanon is located on the eastern coast of the Mediterranean Sea, along the Dead Sea Transform fault system. The Dead Sea Transform fault system in Lebanon has several surface expressions, represented in major faults (Yammouneh, Roum, Hasbaya, Rashaya and Serghaya faults), in uplifts as high mountainous terrain (Mount Lebanon and Anti Lebanon (Figure 2)), and from the seismic activity record. Both Mount Lebanon and Anti- Lebanon are trending SSW-NNE, and the Bekaa valley separates them. Recent work has categorized the Lebanese section of the Dead Sea Transform fault as being a strong seismic activity zone. The structural framework of the Lebanese restraining bend is quite similar to the structural framework of the San Andreas Fault system in California, USA.

Lebanon has more than 65% of its surface area covered with karst terrains (Figure 2). This area is approximately 5550 Km^2 (2140 sq. mi.) of surface coverage.

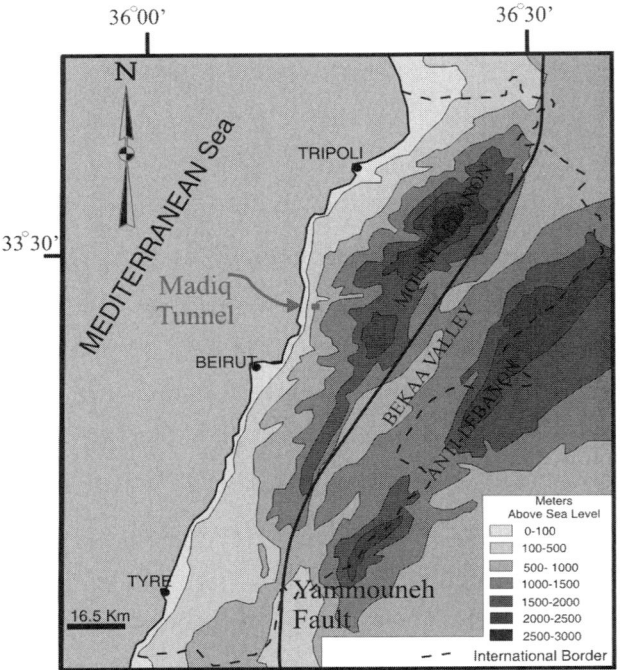

Figure 2. Topographic map of Lebanon showing the location of the Yammouneh Fault and the location of the study are on the northern coast of Lebanon

However, the presence of the Lebanese Restraining Bend has created a structural framework in Lebanon that has its imprints on a lot of features, specifically karst developmenit and morphology. The inclination of bedding, faulting and jointing are all structural elements with their imprinted on the development of karstic features in Lebanon. However, one can not rule out the effect of lithology and the lithological units, in which the karst has developed. This is observed in the shape of the caves and shapes of the cave passages.

PROJECT GEOLOGY

The Madiq tunnel was bored solely in the Sannine Formation. This formation is composed mainly of massive to thinly bedded dolomitic limestone, limestone, marly limestone and marls with chert bands, lenses, and beds. This formation encompasses the second major aquifer in Lebanon, and provides one of the major water sources for the country. This aquifer is a well developed karstified aquifer permitting high infiltration of water. The inflation rate in the Sannine Formation reaches 60%. This Sannine Formation overlies the Hammana Formation (interbeds of marl and limestone with some underlying volcanics). The Hammana is mainly an aquiclude (Figure 4). The Sannine Formation and its underlying Hammana Formation are gently dipping towards the west with bedding dips around 15 degrees.

Water infiltrating into this Sannine karst aquifer will follow both a vertical decent along fractures and a horizontal sense along bedding discontinuities until it reaches the boundary between the Sannine aquifer and the Hammana aquiclude. Then it will start

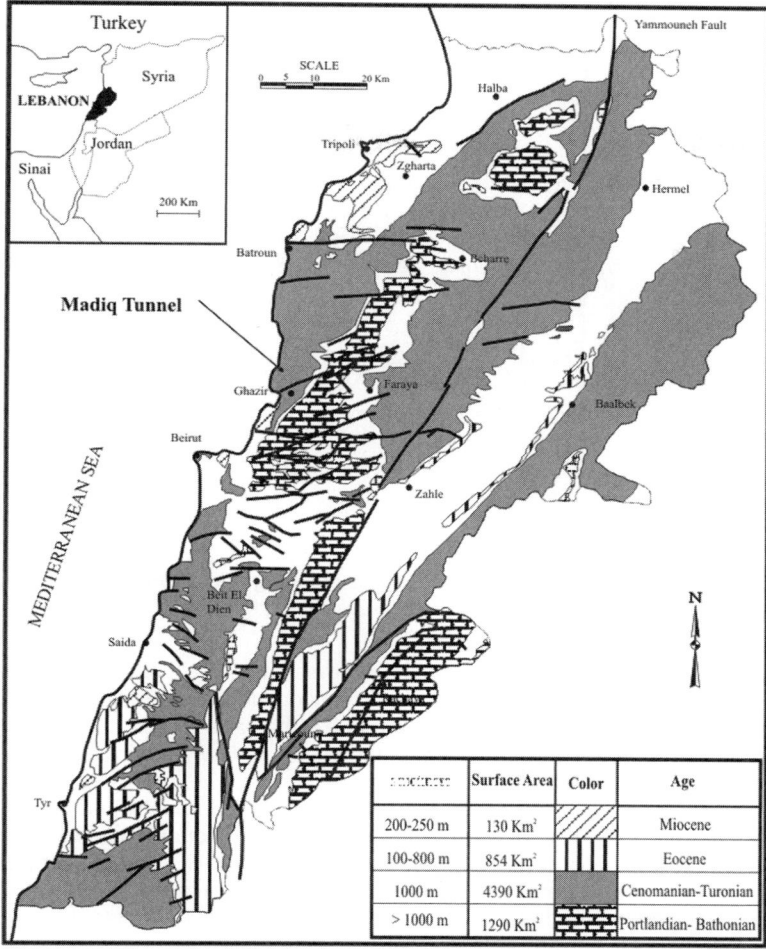

Figure 3. Simplified karstic and structural map of Lebanon showing the location of the tunnel on the northern coast

following an almost horizontal direction in the same decent as the bedding dip towards the sea.

The water table is expected to be at a depth of approximately 75 m (250-ft) below the areas at the north eastern end of the tunnel and increases as we go toward the south western end.

The Sannine Formation encountered during TBM excavation can be broadly grouped into three zones (Figure 5).

Zone 1 is composed of hard to very hard cherty limestone and hard cherty dolomitic limestone. Color ranges between light to dark gray and sometimes light brown. Chert in the form of layers, lenses and nodules are present in large quantities. These cause wear and tear to the cutter heads and hazardous dust to the workers. This zone is well karstified with various types of karstic cavities present as pointed out in the next section.

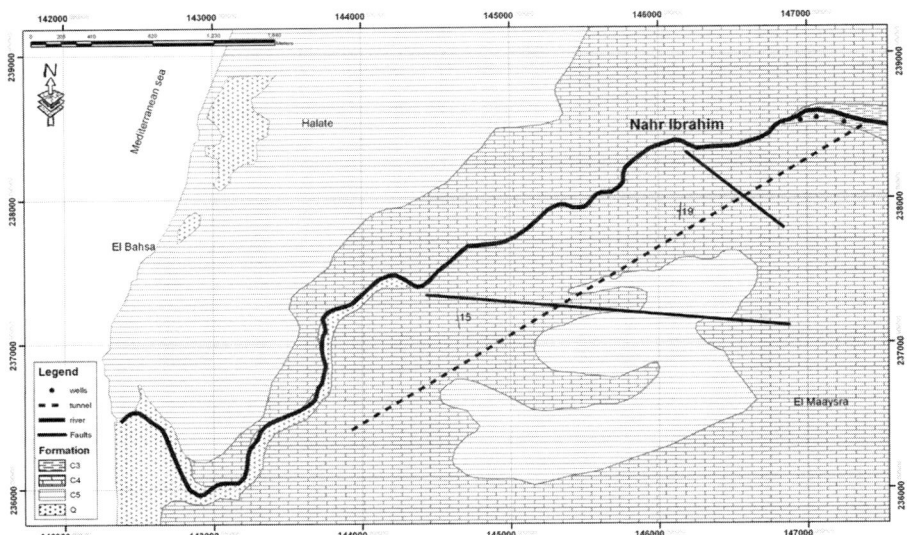

Figure 4. Simplified geological map of the area showing the inclination of bedding of the Sannine Formation

Zone 2 is mainly composed of hard marly limestone that is light to dark bluish gray in color. It is less karstified but few seepage zones occur along the fault/fracture zones and along fault planes. Examples are present at Chainage 2+360 m (7741-ft). Chert bands, lenses and nodules are also present in the zone. Few karstic cavities are present in this zone but not as intense as Zone 1.

Zone 3 is composed mainly of interbeds of hard fossileferous limestone and green and blue soft marl. The thickness of the fossileferous limetones beds decreases and the thickness of the marly beds increases as we go towards the end of the tunnel. The color of the limestone is light brownish to light gray. Small karstic cavities filled with clay and sand are encountered in this zone. Small seepage points are present at the intersection of faults/fractures discontinuities and bedding plane discontinuities. Example of those seepages is present at 2360 m.

The fracture/fault discontinuities present in the tunnel are shown on the section in Figure 5. The dominant orientation encountered is approximately E-W with dominantly strike slip faulting observed as obvious from slickenlines present on the fault planes. The zone of fracturing mainly due to faulting has a range between 10 cm (4-in.) and 20 m (65-ft). Fault zones are either lithified sometimes harder than the rock itself and sometimes they are fractured with no filling other are fractured with soft red clay as matrix.

KARST GENIUS

Karst is a geologic terrain that includes the subsurface caves and/or an aquifer as a single system, generally in limestone and/or dolomite, in which the cavity is formed chiefly by the dissolving of the surrounding rock, and which may be characterized by sinkholes, sinking streams, closed depressions, subterranean drainage, and caves.

Fault is crack or fracture in the earth's crush in which relative rock and/or soil movement has taken place (inactive), or continues to take place (active). Faults have up and down movements (normal and reverse faults) and/or sideway movements (left and

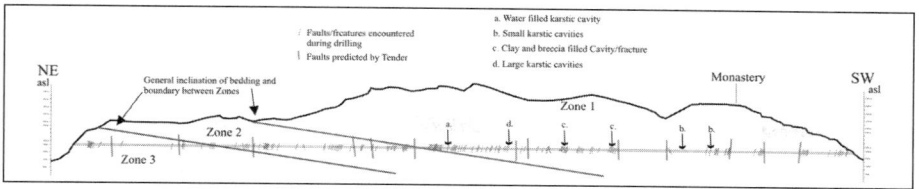

Figure 5. Simplified longitudinal section along the tunnel showing the predicted and the encountered faults/fractures along with the three lithological zones encountered

right lateral faults). The movement displacement may be of a few centimeters (inches) or many kilometers (miles). Faults can have thicknesses of a hairline or of many meters (feet) of width, or are composed of a massive series of faults over kilometers in width. Karst caves form from fault planes that commonly guide vertically or have sub-vertical shafts. The planes can have guiding sub-horizontal or oblique karst passages within their confines. Faults can be filled with clay "gouge," sand, gravel and boulders of rock, and/or water; or can be open voids, but rarely.

Breccias are produced in several geologic processes: a clastic sedimentary rock composed of clasts that is fragments in a consolidated matrix breccia that is in rock fall breccia, collapse breccia that is in karst areas and other geotechnical processes. Breccias are masses of rock composed of angular to rounded fragments. The rocks composed of limestone/dolomite that have accumulated by solution of surrounding or underlying carbonate limestone or dolomite are formed as breccias resulting from the collapse of the roof of a karst cave, of an underlying karst cave, or of an overhanging ledge.

Fault Breccia is an assemblage of broken rock fragments frequently found along faults. The fragments may vary in size from millimeters (inches) to meters (feet).

Disconformity is a break in the stratigraphic record between parallel layers of sedimentary rocks, which represent a period of erosion or non-deposition. Joints are naturally formed hairline or open cracks in rock without lateral or vertical components of rock movement. Joint set is a group of systematic joints.

Collapse and/or solution breccia is a mass of fragments resulting from the roof falling into a karst cavity created by dissolution of the surrounding buried beds and faults and joints in the limestone/dolomite rock beds.

Clay Filling is the time interval between end of the phreatic (ground water and calcium/magnesium) solution of a karst cave and the beginning of clay deposition of mudstone or mud within in the dissolved limestone/dolomite rock bed.

Collapse Chamber is an underground chamber containing notable quantities of collapsed material. The term is commonly abused in describing the origin of karst cave chambers floored by collapse debris. Wall and roof collapses are common modifying processes in larger chambers. Such a collapse cannot form a chamber, as it can only take place into a preexisting cavity.

KARST CAVITIES ENCOUNTERED

Four types of karstic cavities were encountered during boring the tunnel (Figure 5 & 6). The types were characterized based on their size, location with respect to tunnel. The cavities contained combinations of a breccia-clasts/stiff-clay matrix, sands and gravels, air space and ground water. All the karstic features were associated with fault and/or fracture zones.

Water filled karstic cavities were fractures controlled cavities that are either filled water or water flowed through them (Figure 6a). The best example from this type is

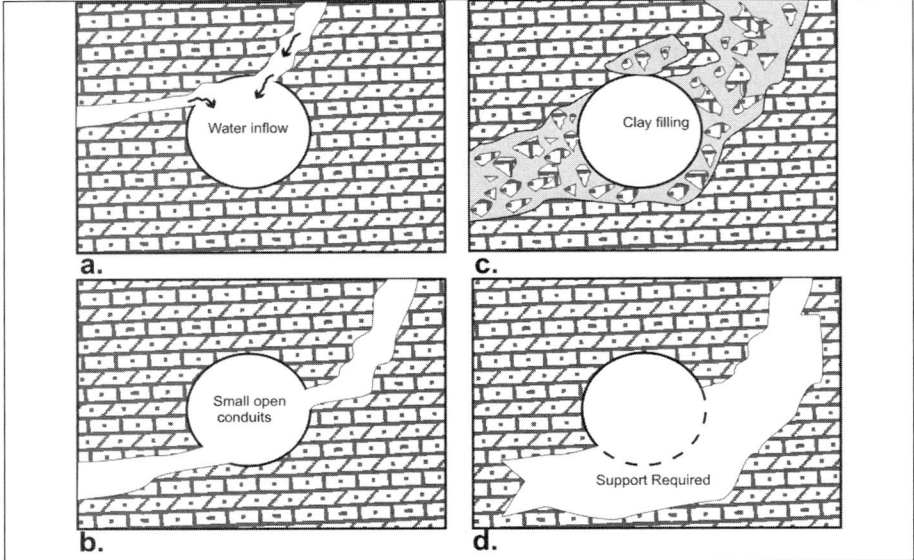

Figure 6. The different types of karstic cavities encountered during tunneling. a. Water filled karstic cavities that resulted in major ground water inflow into the tunnel. b. Small karstic openings with no major problems to TBM boring, c. Karstic cavities/fracture zones filled with clay and breccia of various sizes, which caused major problems to TBM excavation, and d. Large karstic cavities that require filling and support to allow the TBM to pass.

the major water filled karstic cavity encountered at Chainage 2+098 m (6881-ft). The cavity when hit by the TBM caused substantial flooding within the tunnel. Upon impact the ground water inflow measured as high as 60 l/s (950-GPM), which was during the wetter winter season. Finally, the inflow reduced down to approximately 3 l/s (50-GPM) during the drier summer season. This situation is very typical for underground karstic drainage.

Small karstic cavities indicated the formation by hairline fractures and/or beds, which were simple karstic conduits (Figure 6b). Excavating through them required little or no additional ground support. They were either small enough and/or positioned in a way that they do not affect the progress of the TBM. These cavities were scattered all along the length of tunnel. They were sometimes filled with clay and sand. Sometimes they contained small quantities of ground water. Examples of this type are the cavities that were located at a Chainage 0+725 m (2378-ft) from the south portal (Figure 7).

Karstic cavities/fracture zones filled with clay and breccia (Figure 6c) were fracture controlled cavities filled with reddish-brown, stiff clay as matrix along with small to large breccia fragments or clasts from the surrounding rocks. Examples of such karst / fracture zones were at Chainage 1+250 m (4100-ft) and 1+578 m (5176-ft) (Figure 8). These chainage locations were major excavation stoppage and delay points in advance. The karstic/fracture zones caused the TBM to "stick" and "sink" within the cavities. These situations occurred early in the tunnel excavation progress.

Large karstic cavities (Figure 6d) were large enough to require the tunnel invert and sidewalls to be filled and supported in order for the TBM to pass across them. The large cavities were below or on the sides of the boring machine. Example of this type was the cavity encountered at Chainage 1+780 m (5838-ft) (Figure 7).

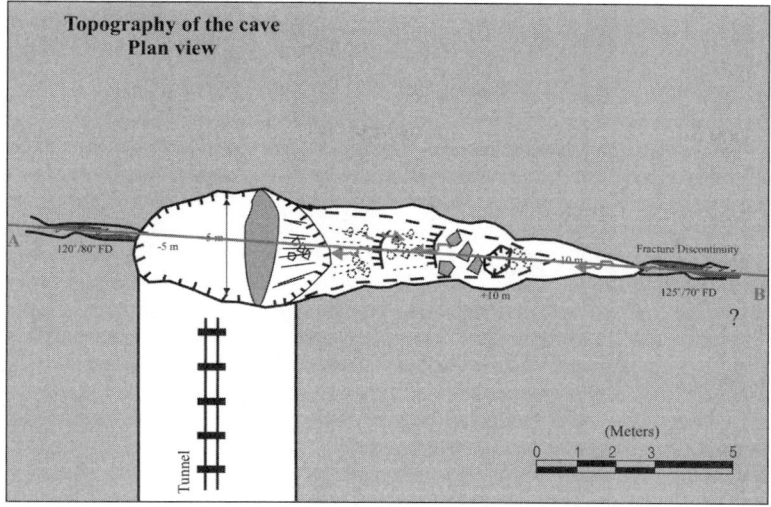
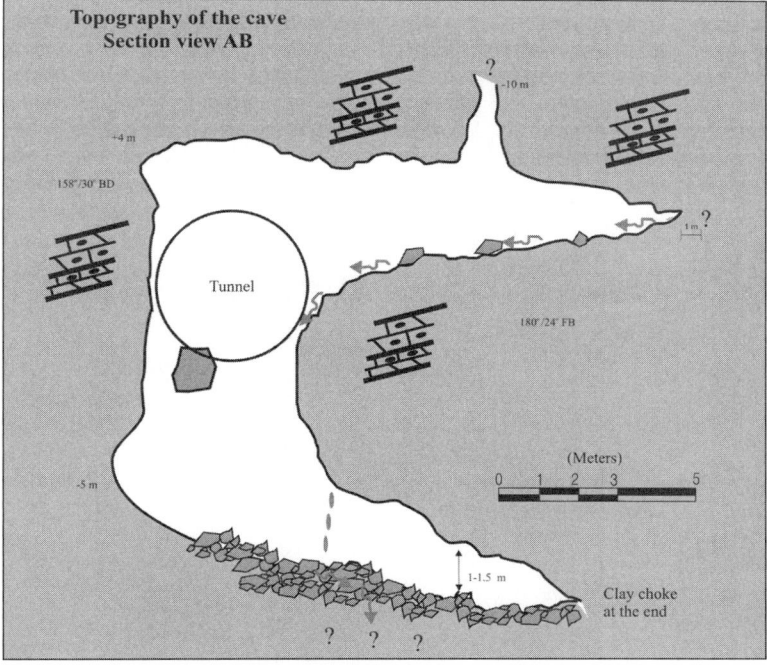

Figure 7. Large karstic cavity at Chainage 1+780 m that required invert filling and sidewall support before continuing TBM excavation

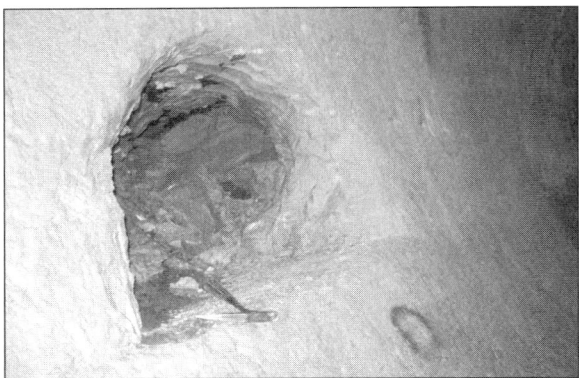

Figure 8. Small karstic cavity at Chainage 0+725 m that caused no problem to TBM excavation

Figure 9. Large karstic cavity in a fracture zone at Chainage 1+578 m. The cavity was filled with clay and breccia boulders of various sizes, which caused a major recovery problem and delay to TBM excavation.

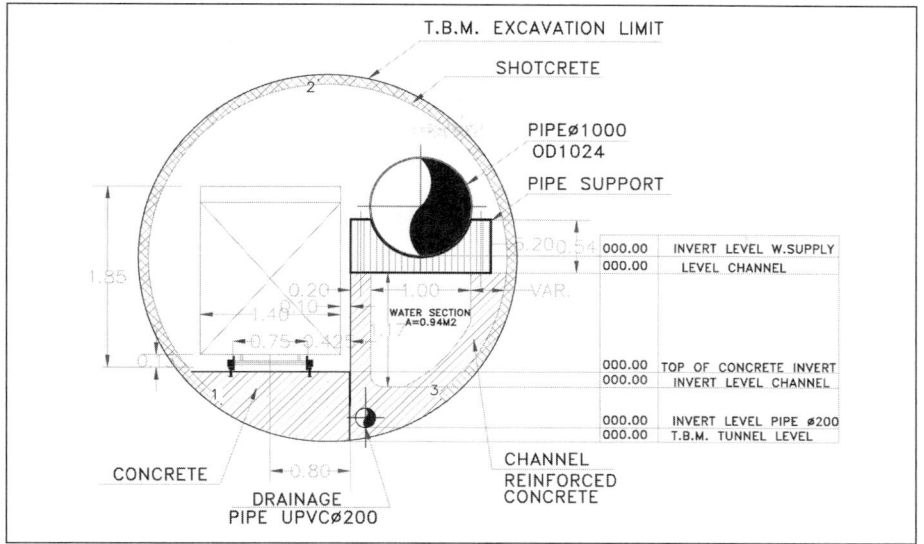

Figure 10. Finished cross-section of Madiq Tunnel

TUNNEL DESIGN

The Madiq Tunnel design comprised a tunnel containing a portable water pipeline, which started approximately 500 m (1650-ft) downstream of the Madiq Spring on the north facing slope of the Nahr Ibrahim valley to south facing slope below the first slope below the first bend of the road leading up the Nahr Ibrahim valley. An irrigation channel was design set inside the tunnel to convey valley stream water to Jabil and Kesrouane costal area. In addition, rail conveyance for carrying men and equipment was part of the design.

The design was prepared specifically for a TBM to perform the tunnel excavation. In addition to TBM excavation, a construction contact was assembled to include ground support, ground treatment, geotechnical ground monitoring, probing and pre- and post-grouting provisions, and ground water drainage design.

The finished cross-section of the tunnel (Figure 10) was designed to consider the following constraints:

- Maximum water flow in the irrigation channel of 2.15 m^3/s (34,000-GPM).
- Constant slope of the irrigation channel and tunnel of 0.35%, with a concrete formwork tolerance of 1 cm (¼ in.).
- A 185 cm (6'-1") tall person standing on a concrete footpath.
- The ability to carry and change out pipe sections inside the tunnel.
- An allowable deviation of 5 cm (2-in.) from the TBM tunnel theoretical center and spring lines from right and left, and up and down respectively.
- Minimum thickness of 20 cm (8-in.) of concrete channel lining.
- The irrigation channel can be easily inspected by removing the top precast concrete slabs.

Based on the above constraints, the tunnel internal diameter had to be greater than 3 m (10-ft). Considering that the heaviest ground support could be 15 cm (6-in.), the excavation diameter had to be 3.8 m (12.5-ft) or greater.

TUNNEL PLANT SELECTION

A Robbins main beam gripper, partially shielded, TBM model 116-189, originally manufactured in 1979, featuring an enlarged cutterhead from 3.5-m (11.5-ft) to 3.8 m (12.5-ft) in diameter, bored the tunnel's karstic sedimentary rock. In some locations of the tunnel the limestone—dolomite rock ranged from 20 to 60 MPa (2900 to 8700-psi) in unconfined compressive strength. The TBM, using a combination of 35.5-cm (14-in.) face and gage cutters with 30.5 cm (12-in.) center cutters bored the tunnel upstream at a 0.35% grade. The tunnel was bored from a single heading by two 12-hour per shift production crews, six days per week.

When required, the initial and final ground support was one or combination of the six following elements: friction anchors inserted into 36-mm (1.5-in.) diameter by 150 to 200 cm (5 to 6.5 -ft) long drill hole, cement grouted thread bar rockbolts/dowels having 20-mm (No. 6) bar diameter by 150 to 300 cm (5 to 10-ft) long, 100 × 100 × 6 (4×4-in.-6g) welded wire fabric, UPN 100 (C6x10.5) channel steel arch segments; HEA 100 (WF6×20) arch and invert steel beams, and dry-mix shotcrete without fibers.

For muck haulage, two 14-tonne 750-mm (30-in.) gauge Ageve DHL 60 M14 mining locomotives were used initially. They were replaced by two 135-kW, 25-tonne GIA Industries DHD 25 (193-bHP) locomotives that pull 16 Lewa muck cars with capacity for five cubic meters. Four 5.5 m-long 2-tonne Mühlhäuser flat cars, two 3.5-tonne 12-man capacity Mühlhäuser man cars, one 2.5-tonne shotcrete car with conveyor belt to an Aliva 252 shotcrete machine, and two 5-tonne Mühlhäuser batch cars made up the balance of railroad equipment fleet.

Because electrical power from Lebanon's grid is not reliable on 24/7 basis, four Stanford generators with Cummings diesel engines delivered 500-kVA each to a Waigel 400/6000 volt up step transformer for the TBM's electrical power.

TBM REMEDIES THROUGH KARST SECTIONS

After the TBM stoppage at Chainage 1+250 (4100-ft), the Construction Manager directed the Contractor to prepare Methods and Means Statements for karst ground recoveries. The instructions were to anticipate and recover in a timely and efficient manner from the work stoppages when the TBM encountered any massive karst/fractured zone cavities. The Contractor did develop and submit well though-out directives and sketches in dealing with each situation. The Construction Manager reviewed and accepted each one of the statements with little or no "exception" comments. Below are five briefs of the Methods and Means Statements dealing with the actual situations and possible scenarios.

Three of the situations and recoveries actually occurred and were successful in their recoveries. The other two were scenarios that never materialized.

1. Massive karst / fracture zone in stiff clay and boulder matrix.
   - Pull back TBM cutterhead to leave work space between the cutter and karst cavity-partial rock face.
   - Apply dry-mix shotcrete to the bored and cavity surfaces in the front of TBM and on top the exposed ground to reduce the risk to exposed personnel and equipment working in front of the TBM from unexpected rockfalls.
   - Slit large boulders with rock chippers. Remove the boulders and stiff clay inside the bore's neat line by hand through the TBM small openings the cutterhead. Note: For this TBM the cutters were front loading, which provided some small openings to the face.
   - Place 22 mm (No. 6) reinforcing steel and welded-wire fabric mesh.

- Fill the invert with shotcrete to the bore's neat line. Allow the shotcrete to gain strength.
- Push/drive the TBM across the reinforced concrete pad in the invert.
- Apply only slight force to the TBM grippers, and also position the grippers to avoid the karst areas in the sidewalls.
- Bring the TBM back up to grade and to line gradually: 50 cm (1.5-ft) in 100 m (330-ft).

2. Partial karst / fracture zone with boulders, little clay and some ground water.
   - Pull back TBM cutterhead to leave work space between the cutter and karst cavity-partial rock face.
   - Apply shotcrete to the bored and cavity surfaces in the front of TBM and on top the exposed ground to reduce the risk to exposed personnel and equipment working in front of the TBM from unexpected rockfalls
   - Drain the water flow by placing and shotcreting in place three rubber 5 cm (2-in.) hoses at the origin of drainage in the karst cavity. Bring the hoses through the TBM below the invert. Allow water to drain into the invert.
   - Reposition boulders and fill with boulders by hand the void space below the bottom of the cavity to 45 cm (18-in.) below the invert.
   - Remove any excessive boulders by hand through the TBM small openings the cutterhead.
   - Place first expanded metal sheet, followed by 22 mm (No. 6) reinforcing steel and welded-wire fabric mesh.
   - Fill the invert with shotcrete to the bore's neat line. Note: The expanded metal placed first will bridge and prevent the shotcrete going deeper into the cavity.
   - Allow the shotcrete to gain strength.
   - Push/drive the TBM across the reinforced concrete pad the invert.
   - Position the TBM grippers to avoid the karst voids in the sidewalls.

3. Large karst/fracture zone with large ground water inflows, numerous boulders and no clay and sand.
   - Pull back TBM cutterhead to leave work space between the cutter and karst cavity-partial rock face.
   - Scale loose rock in crown and sidewalls to springline.
   - Remove all boulders that are in the way of the following activities.
   - Chip rock back to form a rock channel beyond bore's neat line from water flow's source to the invert. Size the channel to fit a 15cm (6-in.) pipe. Attach a curved pipe inside the channel with steel straps and anchor bolts. Note: The first law of dewatering is to get the ground water into a pipe. When in a pipe the water can be controlled.
   - Hang heavy reinforced plastic sheeting over cavity that is making water. Attach sheeting with Ramset nails and washers.
   - Hang and attach welded-wire fabric using anchor bolts and washer plates
   - Shotcrete sheeting and welded-wire fabric seal off water leakage
   - Jam in wood railroad ties into another cavity for the TBM grippers push out against.
   - Shotcrete all potentially dangerous rock surfaces.

- During the above time period, install 15 cm (6-in.) Victaulic grooved and coupling pipe for length.
- Dig sumps and set submersible pumps at key locations along the tunnel. Attach piping to pumps.
- Push forward slow and gently with low speed cutterhead rotation
- Drive the TBM beyond the end of the invert pipe section.
- Dig a sump at its pipe end. Install a submersible pump.
- Continue driving the tunnel with the TBM making gradual adjustment to line and grade.

4. Karst / fracture zone that produces flowing debris against the cutterhead. (This scenario never occurred along the TBM drive. The possibility might have occurred.)
   - Install submersible pumps in the invert behind the cutterhead attempting to lower flow rate within the karst zone. Dewatering may take a considerable amount of time, but in most cases flow rates will diminish relatively fast over a short period of time.
   - Pull back TBM cutterhead to leave work space between the cutter and karst cavity-partial rock face.
   - Install vacuum equipment to suck slimy clay and water from the invert in front and behind the cutterhead. Transfer the material to a muck car or a job-made tanker car. Keep the clay and water in suspension with additional water and a grout mixer arrangement attached to the muck or tanker car.
   - Once the clay is removed from the cavity and most of the ground water is drawn, slit and remove the boulders that are in the way.
   - Follow the methods and means cited in Items 1and 3 above.

5. Large grotto type-cavern scenario void of boulders, clay and water. (This scenario never occurred along the TBM drive. The possibility might have occurred.)
   - Pull back TBM cutterhead to leave work space between the cutter and karst grotto- cavern edge.
   - Scale loose rock in crown and sidewalls to springline.
   - Install safety catch net below the tunnel invert elevation and/or safety personnel harness life line above the crown elevation across the karst cavern.
   - Survey the dimensions of the cavern: distance, width, depth and height using a Distometer.
   - Prepare geotechnical and engineering report defining the key parameters: Evaluate karst genies from geologic data. Estimate the implications and risk in crossing the cavern. Evaluate the possibilities of failing weak rock, big boulders, and rock slides from higher cavern elevations. And evaluate water flow.
   - Decide on detailed the methods and meads, and structural engineering analyses to cross the cavern.
   - Detail the engineering structural design for a) the TBM bearing loads and structures on the tunnel invert "bridge" b) the lateral components and side walls of the tunnel to support the force of the TBM grippers, and c) the

crown section to maintain the round cross section of the tunnel and to deflect and resist impact of any future rock falls and slides.
- Execute of the cavern "bridge" construction. (The details of construction are not included in this paper because their narrative length and many detailed concept sketches.)

Provisions for probing ahead and backfilling karst cavities with grout were contained in the contract. After the recovery from the karst / fracture zone at Chainage 1+250 m (4100-ft), the Construction Manager directed the Contractor to probe ahead. After analyzing the results and time of the probe, the Construction Manager discontinued further probing.

The Contractor did not have the correct set-up or drill to probe. Probing could only be done outside the cutterhead and a very high angle from bore's centerline. Probing could only reach-out to 30 m (100-ft) beyond the tunnel face. Many probes failed to reach the 30 m (100-ft) mark; therefore, probe holes had to be redrilled. Probing became the critical-path activity: the tunnel could not be advanced with the TBM. Probing took to long: 30 m (100-ft) in 3 days to probe.

Both the Contractor and the Construction Manager agreed that the experienced TBM Operators and Walkers would realize that karst/fracture zone was being encountered the by the reduction of trust force (hydraulic gage pressure), the binding of the cutterhead, and the change of muck material on the conveyor belts. They would immediately slow down cutterwheel's rotation and back off the TBM trust, and follow the methods and means stated above.

CONCLUSIONS

The TBM holed-thru on 17 September 2008. The excavation drive took 92 weeks to finish from turning-the-eye to hole-thru.

Drill-and-blast excavation from two headings (north and south portal) would have taken less time to complete the 4,020 m (2.5-mile) drive. The drill-and-blast method is much more versatile in dealing karst / fracture zone ground. Also, the method is much easier to setup. A TBM refurbishment and site setup is much expensive and time consuming than drill-and-blast equipment and preparation. Based on this tunnel's distance and ground conditions, the costs and time to complete the tunnel favors drill-and-blast excavation over TBM excavation.

Assuming a longer drive and more similar karst / fracture zones were envisioned, a double-shielded TBM would be the likely TBM to be selected. The TBM is shielded for better personnel protection. It allows trusting be done from steels ribs and wood lagging when the grippers will not grip to advance forward.

Fortunately, the karst / fracture zone conditions allowed the TBM to be retracted enough. Personnel, small tools and supplies could work between the cutterhead and the face to facilitate the TBM's recoveries.

ONSITE ASSEMBLY AND HARD ROCK TUNNELING AT THE JINPING-II HYDROPOWER STATION POWER TUNNEL PROJECT

Stephen M. Smading ▪ The Robbins Company

Joe Roby ▪ The Robbins Company

Desiree Willis ▪ The Robbins Company

ABSTRACT

Unique onsite assembly of a 12.43 m Main Beam TBM and back-up system was completed on September 18, 2008 in the remote mountains of the Sichuan province of China. The equipment was assembled onsite, without previously having been assembled and tested in a factory, utilizing a method called Onsite First Time Assembly (OFTA). The Jinping-II hydroelectric project features four parallel headrace tunnels approximately 18 km long, two of which will be excavated by TBMs and two by drill and blast. A nearby fifth tunnel is being excavated by a 7.2 m TBM to draw down ground water in advance of excavating the headrace tunnels.

This paper addresses the following topics:

- Project description including geology and terrain;
- The decision process leading to onsite first time assembly, the assembly process and logistics, and lessons learned;
- Design features of the TBM and back-up system; and
- Operational history from startup to the writing of the paper.

PROJECT DESCRIPTION

Jinping-II will be the largest power station (see Figure 1) in an ambitious 21-station mega project for owner Ertan Hydropower Development Co. Ltd. (EHDC). The total 21-station project will harness power from the Yalong River for China's West to East Electricity Transmission Project. EHDC began the scheme in 1991, constructing the Ertan Hydroelectric Project in the west of Sichuan Province. The project was officially completed in 2000 with an installed capacity of 3,300 MW. Additional projects in the lower reaches of the Yalong are planned for completion before 2015, including Jinping-I (3,600 MW), Jinping-II (4,800 MW), Guandi (2,400 MW) and Tongzilin (600 MW). All remaining projects are to be finished by 2025. Currently, three power stations are being built or are already online (Ertan, Jinping-I, and Jinping-II), while the others are in various preparatory stages.

Power from these stations and other resources in the west will be transmitted to Guangdong, Jiangsu, and Zhejiang Provinces, as well as the cities of Shanghai, Beijing, Tianjin, and other eastern locations, where electricity is now in short supply. The entire scheme is envisaged to go online in 2030 and will have a generation capacity of close to 30 GW.

HARD ROCK TUNNELING AT THE JINPING-II HYDROPOWER STATION

Figure 1. Jinping-II underground power station

Preliminary geological surveys, feasibility studies, and necessary approvals for the large scale development of Jinping-II and other stations have been ongoing for the past 40 years. In 2003, work began on the 62 km long main road leading up to the Jinping-II jobsite.

The Jinping-II site is unique in that it will utilize a 180 degree natural hairpin bend in the Yalong River, a tributary of the Yangtze, to generate a multi-year average annual generation of 24.23 TWh. From the intake structure, the river flows northward before turning abruptly southward as it flows around Jin Ping Mountain. The distance along the river from intake to outlet is approximately 150 km during which the river drops 310 m. From intake structures near Jingfeng Bridge water will flow through the four Jinping headrace tunnels downgrade at 3.65% to the underground Dashuigou powerhouse. The powerhouse will utilize eight 600 MW turbine generators for a total generating capacity of 4,800 MW. The four (4) parallel headrace tunnels, with an average length of 16.6km, are separated by 60m from centerline to centerline. Two access tunnels and a drainage tunnel run parallel to the headrace tunnels on the southern side.

Ertan Hydropower, the owner, split the tunneling contracts in two. One contract was let for headrace tunnel Nos. 1 & 2, and a separate contract was let for tunnel nos. 3 & 4. The tender documents specified that two 12.4 m diameter TBMs would excavate 16.7 km long sections of headrace tunnel Nos. 1 and 3. As a result, each of the construction contracts includes one TBM bored tunnel and one drill and blast excavated tunnel. China Railway 18th Bureau (Group) Co Ltd. won the construction contract for headrace tunnel Nos. 1 and 2, while China Railway 13th Bureau (Group) Co Ltd. won the contract to construct headrace tunnel Nos. 3 and 4.

Parallel to the headrace tunnels is the 15.3 km long dewatering tunnel which is being excavated under separate contract by Beijing Vibroflotation Engineering Company (BVEC) with a 7.2 m diameter TBM. This tunnel is being excavated ahead of the four headrace tunnels in order to reduce the water inflow in the headrace tunnels well below the 5 m^3/s otherwise expected (see Figure 2).

GEOLOGY AND TUNNEL ALIGNMENT

All four tunnels are located on the slopes of Jinping Mountain in reportedly stable geology consisting of massive to blocky marble with limestone, sandstone, slate and chlorite schist with unconfined compressive strength (UCS) of between 50 and 85 MPa. A high overburden, with over 70% of the cover greater than 1,500 m and a maximum

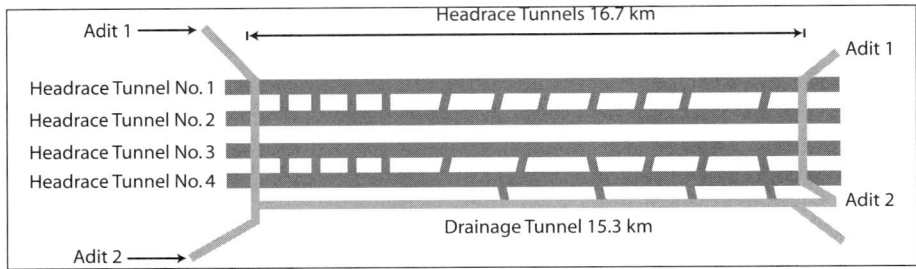

Figure 2. Jinping-II project layout

Figure 3. High cover at Jinping-II jobsite

of 2,525 m, creates a risk of squeezing ground and rock bursts (see Figure 3). Pre-excavation core tests typified rock in the tunnel as:

- Class II (RMR 61 to 80): 29.1% of tunnel
- Class III (RMR 41 to 60): 53.6% of tunnel
- Classes IV and V (RMR < 41): 17.3% of tunnel

Though the rock should provide relatively good conditions for excavation, there are several challenges to overcome. One is the potential for sudden inundation of the tunnel during the excavation work. Underground water in the vicinity is reportedly conveyed by fissures and a network of channels with a continuous water source, resulting in the possibility of high pressure and large flow rates. Core tests revealed a potential for steady flows in the range of 2 to $3m^3/s$ with maximum water flows of up to $5m^3/s$ (Wu & Huang, 2008).

Another challenge is rock bursts, which may occur as a result of the high in-situ stress caused by high cover. Again, according to Wu & Huang, measured maximum major principal stress is approximately 42 MPa vertical, indicating that gravity stress dominates. They reported that the major and minor principal stresses could reach 63 MPa and 26 MPa respectively in the headrace tunnel at the point of maximum overburden. Severe rock bursts occurred during excavation of the access tunnels and an adit; therefore, some rock bursting is expected during construction of the headrace tunnels.

In response to the core test results and high cover, an aggressive ground support program has been developed with various support designs specified based on the rock mass classification. In relatively stable rock, support is minimal including sparse rock bolts. In rock mass Class III, systematic rock bolts up to 6 m long are installed, as well as steel-fiber reinforced shotcrete. Class IV and V sections are also stabilized with rock bolts and reinforced shotcrete, and final lining will include concrete up to 70 cm thick.

Measures to Handle Ground Water

During excavation of the headrace tunnels, the contractor will attempt to reduce and control volume water inflows using several approaches:

Pre-excavation Draining. This plan specifies dewatering the mountain by draining water into the 7.2 m dewatering tunnel, which is being excavated by TBM in advance of excavating the headrace tunnels.

Pre-excavation Probing. Rock drills are employed to drill ahead of the TBM, probing for changing geological and hydrological conditions. Information so gleaned will be used to specify pre-excavation rock consolidation and water cutoff grouting programs, as well as to anticipate near-future rock support measures for safe tunneling. It is imperative that any incoming water flow be limited to allow continued excavation by the TBM.

Post-excavation Draining. The construction design, including the TBM design, allows for large volumes of water to be drained through the bored headrace tunnel as they are excavated, minimizing impact on excavation logistics and TBM operations.

Controlling. The concept for this step is to give the constructor the ability to control the rate at which the groundwater is drained into the tunnel, from every point in the excavated tunnel. In this way, it is hoped that water can be allowed to flow into the bored tunnel to the maximum allowable volume rate which will allow continued TBM operations. Ideally, if successful, the system would permit the constructor to drain where and when necessary to maintain operations. This will require, of course, high quality water cutoff grouting, drain pipes and valves.

Rock Bursts

The high in-situ stress along the headrace tunnel can cause rock bursts during excavation. Measured stresses may reach 63 MPa at the site of maximum overburden. Several measures have been specified by the project owner to reduce the potential for rock bursts during headrace tunnel excavation:

TBM Usage. Headrace tunnel Nos. 1 and 3 will be excavated by TBM to a total length of about 16.7 km. The rock mass surrounding a TBM-bored tunnel is disturbed less than it is with drill and blast excavation. It is hoped that the use of TBMs on two of the four headrace tunnels may reduce the rock stresses enough to reduce the probability of rock bursts somewhat in all four tunnels.

Reinforcement of the Surrounding Rock Masses. Maintaining as much of the rock in place as possible after excavation (i.e. minimizing over break or rock fall) results in better total rock support through the formation of a natural arch and reduces post excavation stress. Rock support has been design to keep as much rock in place as possible:

    a. shotcrete or steel fiber reinforced shotcrete applied immediately after the excavation

    b. patterned rockbolts to prevent the loss of rock blocks and slabs

    c. wire mesh or steel ribs

Figure 4. Partial shop assembly

ONSITE FIRST TIME ASSEMBLY (OFTA)

Onsite First Time Assembly (OFTA) was selected for the 12.4 m TBM due to fast track project scheduling and a limited seasonal window for delivery to site by river. The OFTA process, developed by The Robbins Company, allows machines to be assembled at the jobsite without need of pre-assembly in a manufacturing facility. The process was first utilized in 2006 on the 14.4 m diameter TBM at the Niagara Tunnel Project—the world's largest hard rock TBM. OFTA has since been used on several projects around the world, resulting in reduced TBM startup schedules and cost savings due to decreased shipping costs and man hours.

OFTA was identified as essential for the Jinping project because it would enable early shipment of large components of the TBM. Rapid shipment of the large components was needed in order for them to be moved via barge on the Yangzi River before the onset of the low water season between November and April. The area sees vastly different seasonal rainfall, with the May to October rainy season accounting for as much as 95% of annual rainfall.

All of the heavy structural parts of the TBM were manufactured in a facility located in the city of Dalian in Northeast China. Under the original site assembly plan, pre-assembly of some TBM components was to have begun on site in late November 2007. That assembly schedule required that all of the parts arrive at the Le Shan dock near the city of Chengdu on the Yangtze River in early November 2007 before the low water season started. However, by the end of the summer of 2007 the original assembly schedule was delayed because the site was not ready to receive the equipment. Additionally, the Yangtze River experienced unusually heavy flows that year. For these reasons the decision was taken to partially assemble some of the critical parts in the Dalian factory before shipping. The main bearing, gear, and pinions were installed in the cutterhead support so the ring gear—pinion mesh could be verified. Later, the muck chute, side supports, roof support, and front support were attached. The remaining components were assembled for the first time on site (See Figure 4).

At the end of 2007, all of the heavy structures were loaded on a barge, shipped up the river, and placed in a storage yard near Chengdu until the job site was ready to receive them.

Though all of the structural components of the TBM and backup were manufactured in China, sub-systems such as hydraulic, lubrication, water, electrical, and ventilation were manufactured and tested in facilities in the USA or Europe before being shipped to the site.

Figure 5. TBM assembly in the chamber

Key components of a successful OFTA program include:
- Quality control of component manufacture to ensure proper fit up at site
- Absolute control of the total system bill-of-materials, to ensure that everything required for the system is sent to the job site
- Logistical planning and control, to ensure that everything arrives at the job site in the order that it is required for efficient assembly and use of storage space
- Resources planning, to ensure that all tools and personnel of every type and quantity required for assembly are on site when required
- Advance alternative recovery planning, in order to be ready to react quickly to possible failures in any of the above steps

Much of the challenge of the assembly was a result of the remote location. Once at the site, the 12.4 m machine was erected in an underground assembly chamber measuring 20 m wide × 26 m high. Limited space required that many of the smaller TBM components, the parts imported from outside China, and all of the back-up structures be staged about 80 km away in the town of Manshuiwan, where warehouse space and a large outdoor yard were provided by Ertan.

Every morning a coordination meeting was convened to plan which parts should be sent to the site for the next day's scheduled work. The designated parts were loaded on trucks that day and sent to the assembly chamber, arriving later that evening to be available for the next morning's assembly (See Figure 5).

Because of the remote location and because the TBM had not been previously assembled, it was necessary to equip several shipping containers as workshops. A hydraulic workshop was set up with the hose ends and adapter fittings needed, as well as a high production hose crimping machine (See Figure 6). Similarly, an electrical container, a tool container, a workshop container, and an office container were mobilized in the assembly chamber.

Assembly of the TBM and back-up system began in July 2008 and finished on September 17, a schedule comparable to that for site assembly of a large diameter TBM which has been pre-assembled in the factory. Crews then walked the TBM and the first three back-up gantries 200 m forward from the assembly chamber to a launch chamber. The vacated assembly chamber was then used to erect the conveyor system and six more back-up gantries.

In general, the assembly sequence proceeded according to the plan, with one major exception. Early in the assembly program, it was discovered that the gripper

Figure 6. Hydraulic workshop

Figure 7. Portable boring machine

carrier bushings had not been finish machined in the manufacturing facility in Dalian. Shipping the carrier to the nearest machine shop in Chengdu for repair would have been the preferred way to solve the problem. However, this was impossible because of damage to machine tools and factories resulting from the severe earthquake that hit Sichuan province in May 2008. Instead, a contractor in Shanghai was brought to site with a portable boring machine and the gripper carrier bushings were line bored in 3 days. (See Figure 7).

Another major difficulty was the lack of skilled local workers. For this reason, intense supervision and training of these workers was necessary to ensure the quality of the final product. Robbins had as many as 16 supervisory personnel from the USA and Europe and 26 engineers, mechanics, and electricians from Robbins (China) Underground Equipment Co., Ltd. at the peak of the assembly effort.

The equipment was successfully assembled and launched in only three months, despite record snowstorms that caused major delays, as well as 2008's magnitude 8 earthquake centered near Chengdu, which caused heavy road damage and further delays to the schedule (see Figure 8).

Figure 8. Fully assembled 12.43 m diameter TBM

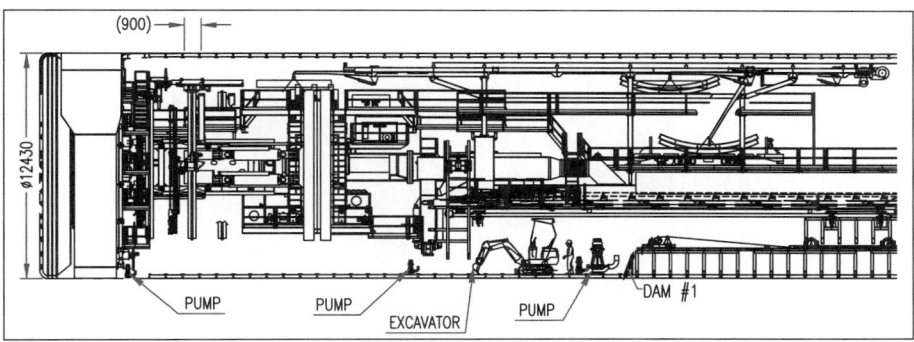

Figure 9. TBM general assembly

TBM FEATURES AND DESIGN CRITERIA

Robbins specially designed the 12.43 m TBM for high water inflows and difficult ground conditions (see Figure 9). Several measures are being taken to address the possibility of deep, flowing water in the invert under the TBM. With the exception of the cutterhead and cutterhead support, the lowest parts of the TBM, back-up, and continuous conveyor systems are 1.5 m above the tunnel invert. In addition, the tunnel train track is assembled on a continuously installed steel framework, also 1.5 m above the tunnel invert. Keeping all of the equipment 1.5 m above the tunnel invert allows a water inflow of approximately 4,000 liters per second to pass under the boring equipment and trains with minimum impact on tunnel excavation.

Primary rock support activities are performed from platforms on top of the TBM. Ring beams are delivered in the top of the tunnel, through the back-up and over the top of the TBM main beam to the ring beam erector. A panel erector can install specially designed steel panels over fissures in the rock where water is entering at high pressure, in order to deflect and redirect the water spray.

Moveable steel dams can be placed in the invert just behind the TBM and dewatering pumps are available to relay water from the cutterhead support area to the end of the back-up to keep the water level as low as possible under the TBM, in the primary tunnel working area.

Figure 10. Back-up system with steel canopies

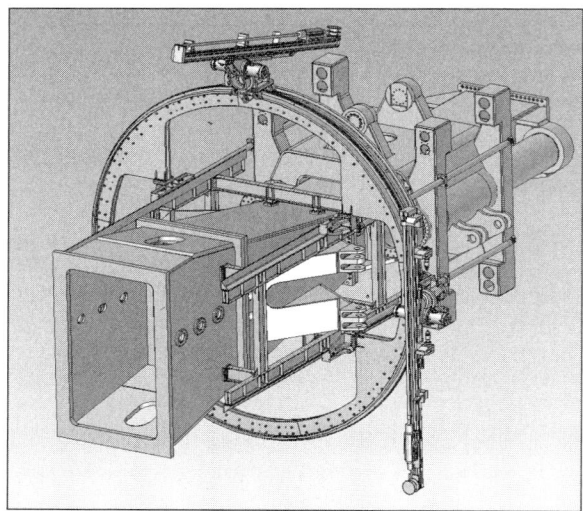

Figure 11. Rock drills on TBM (L1 zone) are used for both rock bolting and probing

The working decks on top of the backup are covered by steel canopies to protect personnel from high pressure or high flow water (see Figure 10).

Rock bolting is done in two locations on the TBM. The L1 zone, located just behind the cutterhead support, has two drills and the L2 zone, on the backup just behind the bridge conveyor, has two more drills. Shotcrete can be applied both in the L1 and L2 areas. In L1 a single robot is used for emergency application of shotcrete. Production shotcreting is done in the L2 area with two robots, one on each side of the backup. The L2 robots have an axial stroke of 12 m and a pumping capacity of 25 m^3/h each. The backup gantries where the L2 drills and shotcrete robots are located are configured as 6m diameter steel tubes. All shotcreting and drilling takes place on the outside of the tubes to protect the facilities on the inside of the tubes and to allow free passage of workers and materials during ground support activities (see Figures 11, 12 and 13).

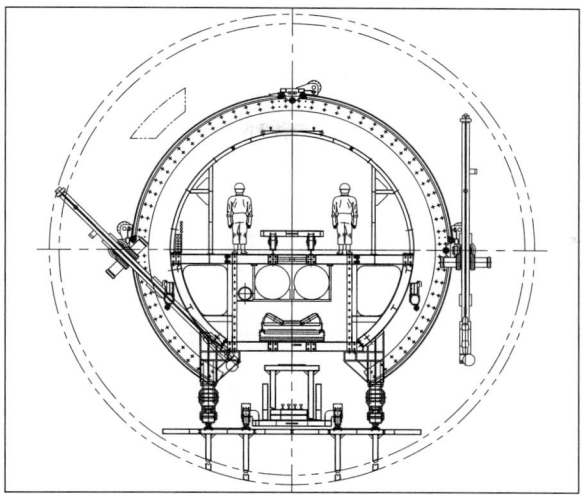

Figure 12. Additional rock drills are on the backup for secondary bolting, outside of 6 m tube

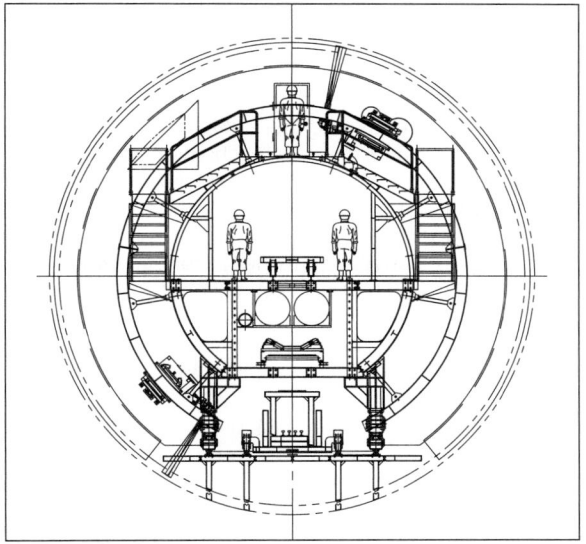

Figure 13. Shotcrete robots work outside of the totally enclosed central working area

Following the L2 zone rock support equipment decks are several three-level backup decks on which various equipment is mounted and various workstations are located (see Figure 14).

The TBM is a Robbins HP-TBM (High Performance TBM) which combines very heavy structural steel components, a very high capacity 3-axis/3-roller main bearing, high thrust and high power. Cutterhead has 4,410 kW of power and is fitted with 19-inch

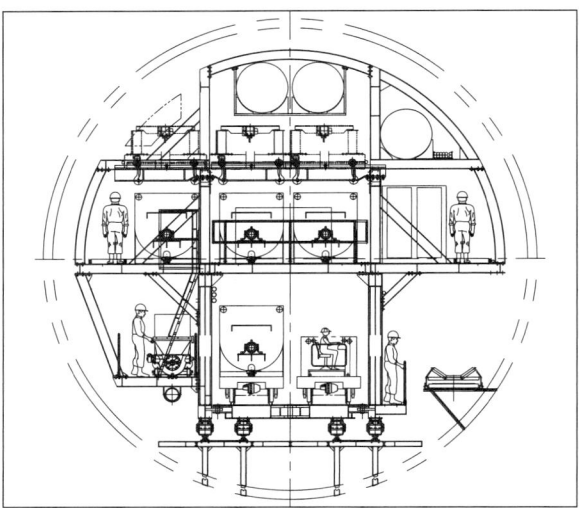

Figure 14. Typical back-up cross section

back loading cutters with two extra housings installed in the gage area for overboring in the event squeezing ground is encountered (see Table 1).

MUCKING SYSTEMS

In anticipation of high water inflows, the conveyors on the TBM and backup are designed to be completely horizontal to minimize the high spillage rates associated with inclined conveyors carrying muck and large amounts of water. The bridge conveyor located just behind the TBM conveyor is straight for two thirds of its length and then curves to the side to discharge directly into the advancing tailpiece on the right hand side of the tunnel. Curving the conveyor was necessary to eliminate the usual transfer conveyor between the bridge conveyor and tunnel conveyor. With the elimination of the transfer conveyor it was possible to keep the bridge conveyor completely flat.

Muck is transported from the TBM by a continuous conveyor system which will eventually be 16.7 km (See Figure 15) in length. The steel cable core conveyor belt system utilizes a 1,200 kW main drive with a 1,200 kW booster drive, which will be installed at the midpoint of the tunnel. The TBM tunnel No. 1 conveyor capacity is large, at 1,800 tons/hour, in order to be able to handle crushed rock from the adjacent drill and blast tunnel No.2 in addition to the TBM generated muck from Tunnel No.1. The tunnel conveyor discharges to a series of conveyors intended to handle muck from all four headrace tunnels and the dewatering tunnel. Final disposal is in a deep valley 7 km from the portal. The dewatering tunnel utilizes a similar steel cable belt system 15.4 km in length.

OPERATIONAL HISTORY

The assembly of both 12.4 m machines was completed in the autumn of 2008, while the 7.2 m machine was launched earlier, in May 2008. As of late October, the Robbins machine at headrace tunnel No. 1 was undergoing testing and had advanced more than 300 m of its 2,000 m long commissioning bore. Increased ground support was required at the interface between the starting chamber and the bored tunnel and

Table 1. TBM general specifications

Year of Manufacture	2008
Machine Diameter (new cutters)	12.42 meters (40.7 ft)
Cutters	
Face/Gage	Series 19 (482.6 mm)
Center	Series 17 (431.8mm)
Number of Disc Cutters (overcut not included)	78
Number of Disc Cutters Overcut	2
Maximum Recommended Individual Cutter Load	267 kN (60,000 lbs.)
Cutterhead	
Recommended Normal Operating Thrust	20,826 kN (4,681,871 lbs.)
Cutterhead Drive	Electric motors/safe sets, gear reducers
Cutterhead Power	4,410 kW (14 x 422.4 HP) (5,914 HP)
Cutterhead Speed	0-5.61 rpm
Approximate Torque (low speed) 0-2.55rpm	16,519 kNm
Approximate Torque (high speed) 5.61 rpm	7,509 kNm
Thrust Cylinder Boring Stroke	1,884 mm (74.2 inches)
Hydraulic System	225 kW (300 HP)
System Operating Pressure at Maximum Recommended Cutterhead Thrust	300 bar (4,351 psi)
Maximum System Pressure	345 bar (5,000 psi)
Electrical System	
Motor Circuit	690 VAC 3-phase, 50 Hz
Lighting System/Control System	230VAC/24 VDC
Transformer Size	2 x 3000 kVA, 1 x 2000 kVA
Primary Voltage	20,000 V 50 Hz
Secondary Voltage	690 VAC drive motors, 400 VAC hydraulic pump motors
Machine Conveyor	
Width	1,370 mm (54 inches)
TBM Weight (approximately)	1,256 metric tons, excluding drilling equipment

Figure 15. Tunnel conveyor

took some time to be agreed. The resulting design included ring beam installation every 900 mm and a 17-bolt pattern of rock bolts every 1.5 m. Progress has been slow to date due to very poor rock conditions. The face is fractured and collapses. Similar rock conditions were present during the first 1.5 km of the dewatering tunnel then improved. It is hoped that conditions will also improve in the head race tunnels in the near future.

Excavation at the dewatering tunnel has advanced a total of 2,890 m as of January 2009, at rates of up to 50 m per day. Boring is done in two 10-hour shifts with a 4-hour maintenance shift. Operations at the drill and blast tunnels were also underway and had advanced approximately 2 km in headrace tunnel Nos. 2 and 4.

The TBM for the dewatering tunnel is expected to finish in late 2009/early 2010, while the machine at headrace tunnel No. 1 is slated for a mid-2012 breakthrough.

CONCLUSIONS

Excavation of the Jinping II headrace tunnels presents many formidable challenges. A condensed construction schedule required a new approach to TBM design and manufacture which resulted in the use of OFTA for rapid launch of the machine, shaving months off the schedule. The extremely high water inflow potential required new methods. Some of the bold new methods planned to battle the water are untested. Tunneling under more than 2 km of cover and the attendant rock stresses and potential for spalling and rock burst would be very challenging, even in the absence of water. At Jinping the extremely remote site location, high water pressure and inflow potential, and 2 km of cover combine to make it one of the most challenging tunneling projects of the day. Regardless of the tunneling production rates finally achieved on the project, one outcome is inevitable; lessons will be learned on this project which will make possible future projects in mountainous regions of China, and even larger ranges such as the Himalayas and the Andes.

REFERENCES

Wu S. and Huang Z. 2008. *Tunnel Vision at Jinping II. Int. Water Power & Dam Construction.* www.waterpowermagazine.com/story.asp?storyCode=2050980. Retrieved January 12, 2009.

Ertan Hydropower Plant, Yalong River, China. Power Technology.com http://www.power-technology.com/projects/ertan/ Retrived January 17, 2009.

DOUBLE SHIELD TBM IN CHALLENGING, DIFFICULT GROUND CONDITIONS—A CASE STUDY FROM ZAGROS LONG WATER TRANSFER TUNNEL, IRAN

Jafar Khademi Hamidi ▪ Amirkabir University of Technology

Kourosh Shahriar ▪ Amirkabir University of Technology

Jamal Rostami ▪ Pennsylvania State University

ABSTRACT

The Zagros long tunnel is one of the main components of Sirvan water transfer project in western Iran. It is approximately 48 km in length and consisted of two lots. The second 26 km lot of this tunnel is under construction by a 6.73 m diameter double shield (DS) TBM. The tunnel passes through several formations with wide range rock mass qualities. During the tunnel alignment, changes in rock quality were highly frequent from poor to very good. The encountered geological conditions required TBM operation to change frequently from hard rock to soft, dry to flowing, sticky to nonsticky ground (and vice versa), more often than expected. In the course of tunneling, the machine has encountered nearly many adverse geologic conditions. The most important problems have been sudden high volume water inflow into the tunnel, sticky ground, and gas seepage, all of which resulted in reduced TBM advance rates. This study will highlight two difficult ground conditions including large water ingress and sticky ground and also the related problems encountered by the TBM along 8 km of excavated tunnel. In sticky grounds, cleaning of the clogged cutterhead caused many delays and contributed to additional downtime. Application of some operational modifications, including decreased TBM thrust showed satisfactory results in such cases. While reduced thrust resulted in reduced penetration rate, but the overall TBM advance was greatly improved by the increase utilization rate and elimination of down time related to cleaning of the head. To cope with high water inflow, application of preventive measures such as pre-grouting has proved to be a more efficient solution. The sealing of ground water inflow also decreased seepage of dangerous H_2S gas and other chemicals such as hydrocarbons accompanying the groundwater.

INTRODUCTION

There are many circumstances when the selected tunnel alignment goes through frequently changing and difficult ground conditions due to availability of land at portals, right of way issues, topographical constraints, accessibility of the site, lack infrastructure, or simply vertical elevation and grade of the tunnel. INI such cases the geological conditions are not the controlling factor in selection of the tunnel route and become facts in the ground that need to be dealt with. In difficult grounds problems such as fault zones with water bearing gouge, high volume of groundwater inflow, gassy conditions, sticky grounds, tunnel wall and face instabilities in running or blocky grounds, hard and abrasive rock, and potential for high convergence in tunnel can be anticipated. This issue was addressed by Barla and Pelizza (2000), where they warn that occurrence of each of these conditions individually or in a combination without warning along

Figure 1. Herrenknecht DS TBM model S-157

the tunnel alignment, may dramatically reduce the average rate of TBM advance. In projects involving challenging conditions, machine manufacturer should provide the contractor with a machine capable of applying many remedial corrections to best handle the adverse and difficult geological conditions, as they are encountered on tunnel alignment.

The 48 km long Zagros tunnel is one of the longest water transfer project located in Zagros mountain range in western Iran. The second lot of this tunnel with the length of 26 km is currently under construction. To date, eight kilometers of the bored tunnel has been completed by a 6.73 m diameter Herrenknecht double shield (DS) TBM which has provided some invaluable experiences of TBM tunneling in difficult ground conditions. The machine has encountered frequently changing geological and geotechnical conditions along the tunnel alignment and nearly all the above mentioned difficult ground conditions. Challenges in construction of the Zagros water transfer tunnel were addressed elsewhere by Khademi Hamidi et al. (2008), Shahriar et al. (2008) and Mirmehrabi et al. (2008). This paper will focus on two problems including high water ingress and sticky ground which drastically affected the TBM advance in this tunneling project.

PROJECT DESCRIPTION AND GEOLOGY

The Zagros Tunnel project involves design and construction of two conveyance tunnels to transfer about 70 m^3/s of water in western Iran. The second tunnel is approximately 26 km long and 6.73 m in diameter currently under construction using a double shield (DS) TBM. The DS TBM was supplied by the Herrenknecht AG and the backup is designed and fabricated by SELI Tecnologie Company with an overall length of about 180 m. Figure 1 shows the DS TBM used in Zagros water transfer tunnel. The TBM specification is also given in Table 1.

The tunnel alignment intersects a variety of geological formations, including Pabdeh, Gurpi and Illam in which most of the tunnel route is located. The oldest geologic unit along the tunnel alignment is brownish gray limestone of Illam formation. Gurpi formation consists of combination of limy shale and argillaceous limestone. Some layers of Illam in transition zone with Gurpi formation are rich with pyrites. The youngest unit is Pabdeh formation which consisted of combination of dark gray limy shale and greenish gray argillaceous limestone. The rock mass quality along the tunnel alignment varies from very poor to good. The maximum depth of tunnel is 1,000 with the average depth equal 400 m. The groundwater level varies from 30 to 340 m above

Table 1. The specifications of the Herrenknecht S-157 DS TBM

Machine diameter (with new cutters)	6,730 mm
Number of cutters	43
Cutter diameter	432 mm (17")
Average cutter spacing	90 mm
Cutterhead torque (maximum)	4,747 kNm
Cutterhead power	2,100 kW
Thrust force	29,038 kN
Rotational speed	0–9 rpm

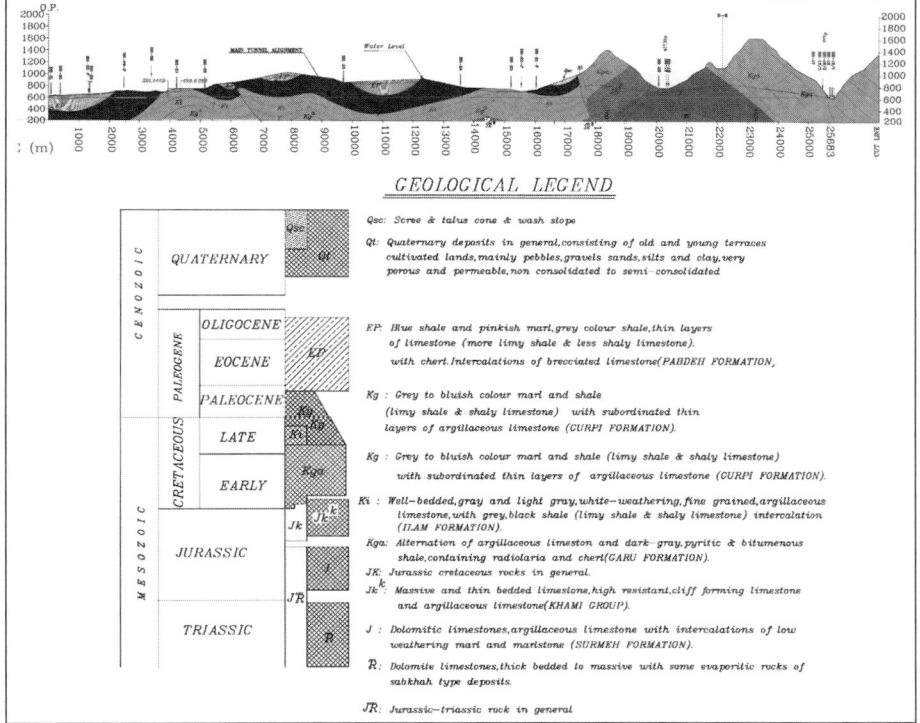

Figure 2. Longitudinal geological profile of Zagros Tunnel (lot 2) (Imensazan Cons. Engrs. 2005)

the tunnel crown. The tunnel will be excavated at a grade of 0.08% and finished with a concrete segmental lining to a diameter of 6 m. The typical geological and geotechnical conditions of the project are illustrated in Figure 2 with the longitudinal profile of lot 2.

Challenges associated with the design and construction of the tunnel mostly include large amounts of water inflow into the tunnel, release of gases dissolved in groundwater in Illam formation, and sticky ground in Gurpi formation.

For the 8 km of tunnel completed to date, wide ranges of TBM performance parameters have been recorded. Table 2 and 3 contain the summary of TBM performance during the project life and the year 2006, respectively. The work was based on 3 shifts a day and 5.5 days a week schedule.

Table 2. The best records of TBM performance parameters during the project life

Maximum daily advance rate (m)	34.78
Maximum weekly advance rate (m)	190.53
Maximum monthly advance rate (m)	757.95
Maximum rate of penetration (m/h)	3.63
Maximum daily utilization factor (%)	57

Table 3. The TBM performance parameters during the year 2006

Maximum daily advance rate (m)	30.55
Average daily advance rate (m)	10.30
Maximum weekly advance rate (m)	177.08
Average weekly advance rate (m)	45.86
Maximum monthly advance rate (m)	591.41
Average monthly advance rate (m)	354.83
Maximum rate of penetration (m/h)	3.63
Average rate of penetration (m/h)	2.75
Maximum utilization factor (%)	44
Average utilization for the period (%)	15.6

TUNNELING IN DIFFICULT GROUND CONDITIONS

Tunnel boring machine can be a magnificent tunneling device under optimal ground conditions. As was stated by Deere (1981), such ideal conditions for TBM tunneling would include good rock material characteristics (e.g., compact, uniform; low to medium strength, hardness, and abrasiveness), good rock mass characteristics (i.e., few joints and shears; unweathered and absence of faults) and good geological-environmental characteristics (low groundwater level, moderate in-situ stress and absence of gas) in which the TBM can operate most effectively. But experience has shown that such idealized rock conditions are rarely encountered at the same time in nature. Difficult or adverse geological conditions are defined where the geological conditions are such that the same TBM cannot work in the operational modes for which it was designed and manufactured. In such cases the advance of the TBM is significantly reduced, while in extreme conditions, tunneling comes to a complete halt. The remedial often innovative measures, on the TBMs to handle difficult ground conditions are becoming fashionable in common practices. Some of the most frequent difficult ground conditions and their common mitigation measures for TBM tunneling are listed in Table 4. However, there have been many TBM tunneling case histories in such conditions in the world in which stop and change of the tunneling technique was the only possible solution (Barton 2000; Khademi Hamidi 2006; Shahriar et al. 2008).

PROBLEMS ENCOUNTERED BY TBM DURING ZAGROS TUNNEL EXCAVATION

Sticky Ground

In highly adhesive clayey formations stickiness causes difficulties using tunnel boring machines. Sticky clay can adversely affect the rate of production by clogging moving parts of the TBM and by adhering to the exposed steel such as the surfaces of cutterhead, disc cutters, bucket scrapers etc. Based on the engineering geological report and other available information, encountering sticky ground condition was

Table 4. Main difficult ground conditions and their common mitigation measures (after Shahriar et al. 2008)

Difficult ground conditions	Mitigation measures
Water inflow	■ Probe drilling ■ Drainage ■ Pre-grouting or grout injection ahead of the face ■ Ground Freezing ■ Use of shielded TBMs with bulkhead or pressurized face
Wall instability (often encountered by open TBM)	■ Use of support systems such as steel arches, shotcrete installed behind cutterhead ■ Pretreatment by injection holes ■ Tunnel lining with precast concrete segments ■ Use of shielded TBMs
Face instability	■ Using fiberglass rock bolt (blocky ground) ■ Use of shotcrete (for soft and raveling ground) ■ Creating artificial face ■ Using grill bars in cutterhead ■ Consolidation grouting
Fault zones	■ Probe drilling ■ Ground improvement ■ Segmental lining ■ Ground Freezing ■ Drilling drainage holes (high water pressure present) ■ Use of shielded TBMs
Karstic voids	■ Drilling drainage holes ■ Filling the karstic voids
Squeezing	■ Over excavation or additional circumferential cutting ■ Use of lubricator such as bentonite, grease ■ Prevention of machine break downs ■ Re-scheduling machine maintenance to avoid long stoppage ■ Use of auxiliary thrust system
Gassy ground	■ Use of gas monitoring and detector systems with regard to gas type (i.e., toxic, flammable etc.) ■ Drilling de-gassing holes in the area of the potential gas source ■ Use of sufficient ventilation systems to provide fresh air to the TBM and face area ■ Use of gas mask ■ Neutralizing chemical compounds or absorbent material

anticipated to occur within the Gurpi formation. The Gurpi formation was alternation of clayey limestone and thick shale layers. XRD analysis showed that the clay in shale layers consists of illite and montmorillonite. Such high clay content rock layers caused serious problems with TBM advance in such a manner that some hours of working shift were being allocated to cleaning of clogged cutterhead, buckets and discs. Figures 3 shows TBM cutters and cutterhead scrapers were clogged in sticky ground in Gurpi formation (chainage 4+927 km).

As it can be seen in the Figure 3, clogging of cutterhead scrapers leads to decreased capacity of mucking by the cutterhead bucket. Accordingly, the balance between the volume of muck excavated from the tunnel face and muck enter the bucket is disturbed. So, the excavated muck will remain between cutterhead and face. It will cause the thrust and torque of the TBM increase. In such conditions, operator had to stop the machine and restart after cleaning of bucket chutes and scraper openings. With short

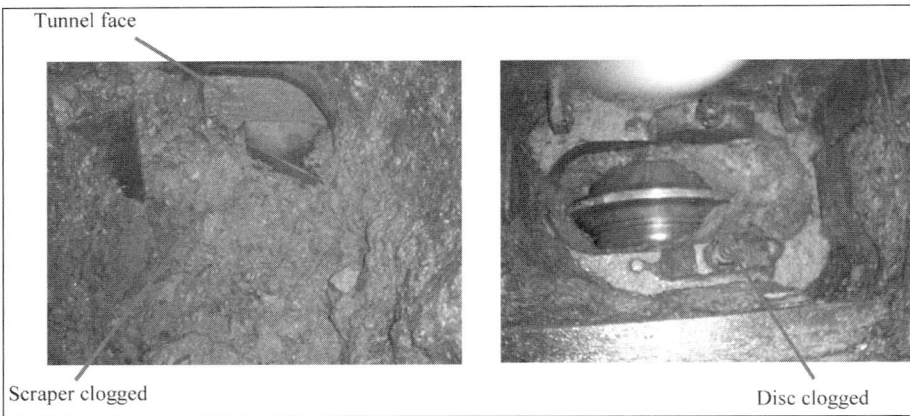

Figure 3. TBM components clogged in sticky ground in Gurpi formation

intervals between stops and frequent cleaning of the buckets, the TBM advance rate was increased.

In weak and sticky grounds, typically the torque capacity of the machine is often the limiting factor regarding penetration rate. This limit can be overcome by using TBMs with variable cutterhead rotational speed and high powered motors. Also, sticky clay can block disc cutters from rotating, so they show one-sided wear or can develop a flat blade. In general, the most common consequences related to muck stuck to TBM cutterhead and its components include:

- Cutterhead clogging makes it heavier and more susceptible to TBM diversion, in the form of diving, during excavation
- Cutterhead clogging leads to increased TBM torque. With regard to machine torque limiting factor, the operator has to inevitably stop the machine
- The scraper clogging prevents the muck from entering the cutterhead. So, the excavated muck will remain in face. This on one hand leads to increased TBM thrust for the given penetration rate and on the other hand leads to one-sided disc wear in abrasive ground
- Cohesive muck makes it sticky and consequently difficult to be removed from belt conveyor. Such a condition also causes difficulties for muck discharge in the transfer points and in muck cars at the end of back-up (Figure 4).

Figure 5 illustrates the monthly distribution of performance and downtimes of the TBM working in sticky clays of Gurpi formation. As it can be observed from the pie chart, cleaning of cutterhead and its components accounts for approximately 25% of total shift time.

There are many case histories of TBM tunneling in all over the world in which presence of soft shaley and sticky ground along the tunnel alignment has resulted in considerable decrease of TBM advance rate. This problem has been more sever in projects where most of tunnel alignment is through hard rock and the TBM was selected for excavation in hard rock (e.g., Zagros tunnel). Obviously these machines are not suitable for incidents of soft sticky ground.

In general, there are two different strategies for coping with sticky (or generally each difficult) ground conditions. The first one = is actually the well-known statement that: "*An ounce of prevention is worth a pound of cure.*" This is related to sufficient

Figure 4. Sticky muck carried on the belt conveyor

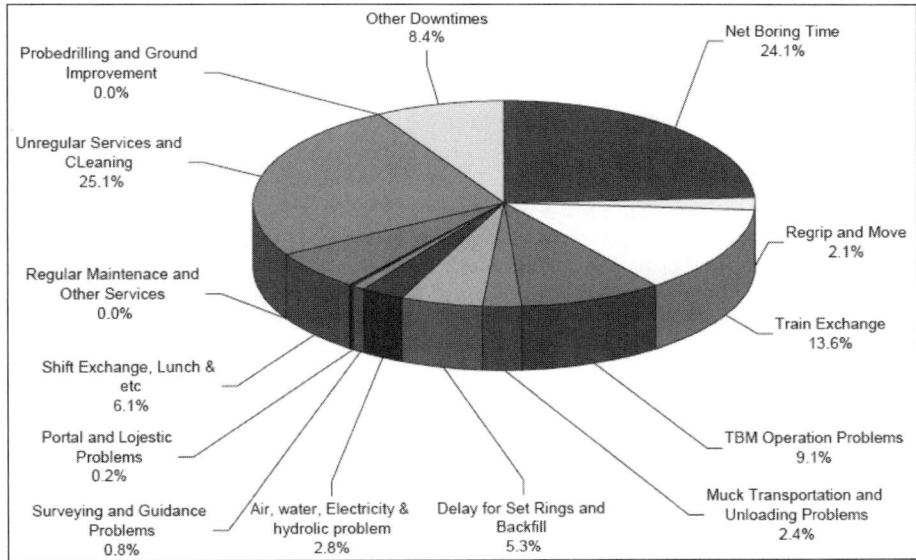

Figure 5. Detailed TBM time analysis for the tunnel in Gurpi clay

site investigations to know every aspect of the geological profile as much as possible. As was addressed by Barla and pelizza (2000) prior understanding of the geological and geotechnical conditions of the site is fundamental for development of underground works. With sufficient information obtained in a correct manner from the tunnel path ahead, the most appropriate TBM will be selected in which all precautionary measures required for mitigation of adverse conditions are supplied by the machine manufacturer at the design stage.

One of these measures is to provide enough water and foam injection nuzzles for cooling discs and controlling ground stickiness, respectively. Alternatively, probe drill installation on TBM for early detection of fault zones is examples of this kind of strategy. In Zagros tunnel, lack of pre-construction exploration, due to the nature of the design-build contract has prevented designers from coping with the presence of sticky ground conditions. Hence, precautionary measures, for example providing enough injection nuzzles at the face to cope with possible sticky ground had not been envisioned at the machine procurement stage. In the case of Zagros tunnel, since presence of severe sticky ground was not anticipated, the tunneling engineers searched for some practical solutions after the fact. The recommended measures to cope with sticky ground in this project included:

- Use of high pressure water jet in cutterhead where the exposed surfaces are prone to be blocked with sticky clay
- Use of ground conditioning techniques such as foam injection
- Use of some operational modification, for instance decreasing the applied thrust

Use of water jet and foam were tested with no significant level success. This was due to insufficient number of nuzzles for water jet and foam injection (i.e., only four nuzzles for water jet and two for foam were available) on the cutterhead. Therefore, it was decided to change TBM operational parameters to cope with sticky ground. This was intended to decrease the volume of the muck that enters the cutterhead bucket. For this purpose, the penetration rate was reduced by using lower machine thrust. It should be noted that decreasing the TBM thrust together with fixed or higher cutterhead rotational speed lead to cutterhead buckets clogging in higher time intervals. Accordingly, the TBM utilization was much increased and downtime related to cutterhead inspection and cleaning was considerably decreased.

Large Water Inflow

Groundwater is usually a major consideration in most tunneling and underground construction operation. The water inflow in tunnel affects not only the construction schedule, but also stability and safety. Since the beginning of Zagros tunnel the TBM has twice experienced huge amount of groundwater flowing into the tunnel. The first incident was at the beginning of Ilam formation where the maximum flow of 320 lit/s was recorded at the tunnel portal at around chainage 4+428 km.

Having passed the Ilam limy/carbonitic formation, the TBM reached the limy and shaly formation of Pabdeh and Gurpi, it had been predicted that water inflow would be decreased, due to lower permeability estimates within these formations based on the given hydrogeological studies. However, everything was unexpectedly changed when the tunnel reached the unit K_{Gu}^4 in Gurpi formation at the chainage of 7+600 km. While the TBM was going through this unit, the water flow at the tunnel portal recorded at the maximum value of 700 lit/s, which was an exceptional and inconceivable value during project life to date. According to the engineering geological information, the tunnel face in this unit passes through the syncline S.5 which due to presence of water pathways including tension fractures and joints has created a local pressurized aquifer with high permeability. Figure 6 shows the cores taken from borehole BH-C in this unit during site investigation. It is seen in the Figure that the cores were of small pieces indicating the unit K_{Gu}^4, which is a severely fractured media.

Figure 7 shows huge amount of water flowing within tunnel and the tunnel portal. The water discharge rate from tunnel portal in a selected period is shown in Figure 8. As it can be seen in this, figure, the water flow had a rising trend.

The substantial water inflow into tunnel has caused a number of problems such as face instability, disturbing segment erection and backfill grouting, material inrush into the front and telescopic shield, pea gravel inrush into TBM from annular space behind the lining, disturbing track-laying practice ahead of back-up system, etc. which all in all has led to low TBM utilization. Groundwater may find paths through tunnel face or walls to enter into the tunnel. With tunnel excavated by an open TBM, water leakage through tunnel roof can endanger personnel and electrical installations exposed inside

DOUBLE SHIELD TBM IN CHALLENGING GROUND CONDITIONS

Figure 6. Core box obtained from borehole BH-C in unit K_{Gu}^4

Figure 7. Large water flow in tunnel in Gurpi formation (chainage 7+600 km)

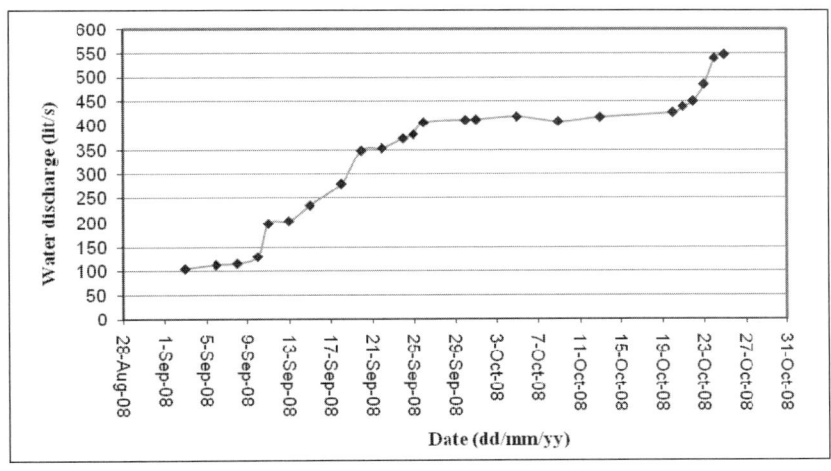

Figure 8. Water discharge rate from tunnel portal for a selected period in Gurpi formation (Ch: 7+450 to 8+130 km)

the machine. With a DS TBM, the shield prevents water flow to directly leak through tunnel roof, although some water may leak through telescopic joint or gripper openings. One of the most important water-related problems which caused much downtime with excavation was related to shield cleaning. This was due to excavated muck and pea gravel moving into front shield and tail shield, respectively. At the tunnel face, big flow normally was accompanied by excavated material and muddy water entered the shield through cutterhead muck ring where the muck was in turn transmitted onto the belt conveyor. Some of this water came directly through cutterhead. In addition to the material overflow from the belt conveyor along with water, some of material came into the shield through the openings beneath the belt conveyor. This material therefore, was deposited on the shield invert and could jam the two shields by filling the gap in the telescopic part of the machine. The sketch shown in Figure 9 provides a simplified illustration of how the water flow transfers fine muck and particles into the shield. Within telescopic shield, this material causes problems with the action of the thrust cylinders. Some of muck particles and fines in face found their path through the gap beneath the TBM shield to tail shield. Figure 10 is a schematic drawing of water containing fine muck flowing toward the tail shield. To prevent fines from jamming the shields, the tail shield was fitted with steel brush seals.

The jammed muck in telescopic joint and tail shield required cleaning of shield which usually took several hours in every shift. The distribution of TBM downtimes in these conditions, including shield cleaning in telescopic joint and tail shield was illustrated in Figure 11. As it can be seen, about 29% of the shift time was allocated to shield cleaning.

Another problem arising from presence of high water pressure surrounding the tunnel is related to pea-gravel and backfill grouting. High pressure water caused pea-gravel to rush into TBM shield where the invert segment was installed in tail shield. Pea-gravel had to be cleaned before segment installation which usually caused loss of several hours in every shift and consequently machine downtime. To solve this problem in the tunnel, coarse-gravel was used with acceptable results.

High water inflow toward the tunnel also had environmental effects regarding to groundwater table falling and discharge rate of springs around the tunnel alignment being effected. Table 5 gives the discharge rate of some of these springs compared before and after tunnel construction. As shown in the Table most of these springs were completely dried even at the distance more than 4 km from tunnel axis.

To mitigate groundwater related problems in the project, two strategies were taken into consideration. The first one was to collect inflow water in certain points in tunnel and discharge the collected water using pumps and pipe lines. It was decided to use special pumps such as slurry pumps which are designed to transfer water containing solid materials such as sand, mud and gravel with different sizes. In Zagros tunnel, because of large water flow in tunnel (i.e., sometimes up to 800 lit/sec), this strategy was somewhat unsuccessful. Hence, the owner directed the contractor to implement preventive measures such as pre-grouting, sealing of segment by using rubber gaskets, and possibly ground freezing. Using these precautionary measures could facilitate and improve tunnel construction, and maintain surrounding groundwater table at desired elevations. In some special conditions where the groundwater contained undesirable or dangerous chemicals or gases (e.g., conditions present in Zagros tunnel), precautionary measures such as grouting is recommended to minimize safety hazards to personnel. Thus by implementing measures to reduce the volume of water inflow, which had to be collected, pumped out, and put through chemical treatment could be significantly reduced. Therefore, pre-injection of cementatious grouting ahead of the heading was selected as the best practice in Zagros tunnel was to use.

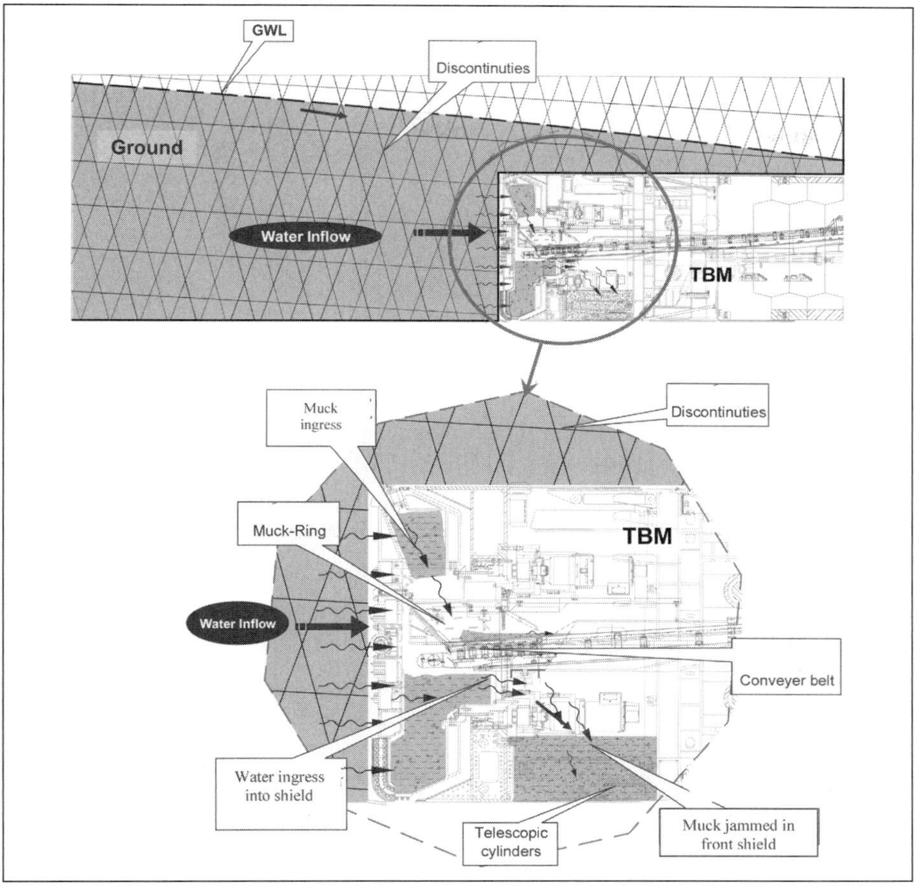

DOUBLE SHIELD TBM IN CHALLENGING GROUND CONDITIONS 1331

Figure 9. Water flow transferring muck and fine material into TBM shield (after Sahel Cons. Engrs. 2008)

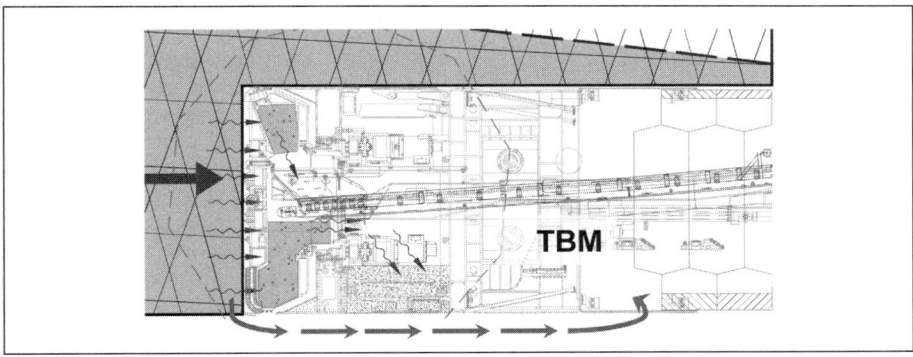

Figure 10. Flow of water containing fine muck towards the tail shield (after Sahel Cons. Engrs. 2008)

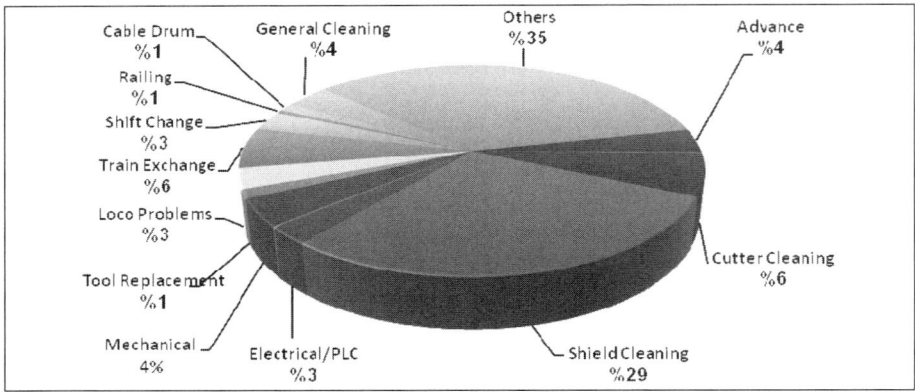

Figure 11. Distribution of TBM downtimes under condition of large water inflow

Table 5. Discharge rate of some springs around tunnel before and after construction

Spring	Discharge Rate (lit/s)		Distance From Tunnel Axis (m)
	Before Construction	After Construction	
Jalekouse 1	3-10	dry	250
Jalekouse 2	0.5	dry	200
Aspar 1	3-20	dry	1450
Aspar 2	2-10	dry	1450
Abdalan 1	0.5	dry	4000
Abdalan 2	15	4.5	4000
Abdalan 3	10	4.5	4000

CONCLUSIONS AND REMARKS

Difficult ground conditions including sticky ground and large water inflow occurred in Zagros long tunnel. Although occurrence of such ground conditions were expected to some extent during site investigation and TBM design, the magnitude of problem and its related hazards was certainly not anticipated. During the tunneling, changes in rock quality were frequent and ranged from very poor to very good. The encountered geological conditions required TBM operation to change frequently from hard rock to soft, dry to flowing, sticky to nonsticky ground, more often than expected.

Unexpectedly high clay content and high permeability of the fractured rock in Gurpi formation, presented many challenges and difficult ground conditions including high water ingress, tunnel face instability, sticky ground. These factors have contributed to lower TBM utilization and advance rate.

In sticky ground condition where cutterhead clogging caused much downtime for machine, application of some operational modifications such as decreased TBM applied thrust showed satisfactory results. Although the average penetration rate became lower for the modified operational measures, but the overall TBM advance was greatly improved. With high water inflow condition, application of preventive measures such as pre-grouting rather than mitigation measures such as pumping seems to provide more satisfactory results. Pre-grouting of the ground to prevent water ingress as opposed to simple drainage and pumping of water is highly recommended. This is primarily because seepage of high volume of water involved many other problems such as environmental impacts related to decreased flow of the local springs, change in the groundwater table, introduction dangerous gases including H_2S and other chemicals such as hydrocarbons into the tunnel environment, and finally the issue of chemical treatment and discharge of contaminated water.

It is obvious that conducting a sufficient site investigation can reduce the potential of any unexpected adverse geological conditions in tunneling projects. Moreover, the exploration for TBM tunneling should be somewhat more extensive than conventional tunneling. Compatibility of the selected TBM with ground conditions along the tunnel alignment is a key factor for a successful operation. This refers to installation of appropriate systems on the machine to handle adverse ground conditions when there is possibility of encountering less than favorable conditions along the tunnel.

REFERENCES

Barla, G., Pelizza, S. 2000. TBM tunneling in difficult ground conditions. In: Geoeng, Melbourne, Australia.

Barton, N., 2000. TBM Tunneling in Jointed and Faulted Rock. Balkema, Brookfield, p. 173.

Deere, D.U. 1981. Adverse geology and TBM tunneling problems. Proc. Rapid Excavation and Tunneling Conference (RETC), San Francisco, pp. 574–85.

Imensazan Consulting Engrs., 2005. Nosoud Tunnel construction method. Report: IMN-10031-MCNP-OR-GE-TR-111-00.

Khademi Hamidi, J. 2006. A methodology for rock TBM selection according to geotechnical conditions. MSc thesis, Amirkabir University of Technology (Tehran Polytechnic), Iran.

Khademi Hamidi, J., Bejari, H., Shahriar, K. 2008. Assessment of ground squeezing and ground pressure imposed on TBM shield. Proc. 12th International Conference of International Association for Computer Methods and Advances in Geomechanics, Goa, India.

Mirmehrabi, H., Hassanpour, J., Morsali, M., Tarigh Azali, S. 2008. Experiences gained from gas and water inflow toward the tunnel, case study: Aspar anticline, Kermanshah, Iran. Proc. 5th Asian Rock Mechanics Symposium, Tehran, Iran

Sahel Consulting Engrs., 2008. Practical solutions for mitigating the muck inrush into TBM shield. Report: SCE 2026 UNGR OFC EG RP 05 D0.

Shahriar, K., Sharifzadeh M., Khademi Hamidi J. 2008. Geotechnical risk assessment based approach for rock TBM selection in difficult ground conditions. Tunnelling and Underground Space Technology, Vol. 23, pp. 318–25.

IMPACTS OF GROUND CONVERGENCE ON TBM PERFORMANCE IN GHOMROUD TUNNEL

Ebrahim Farrokh ▪ Pennsylvania State University

Jamal Rostami ▪ Pennsylvania State University

ABSTRACT

Ghomroud water conveyance tunnel project is under construction using shield Tunnel Boring Machines (TBM) for a total length in excess of 50 km. Phase 1 of the project included 36 km of tunnel divided into four lots (1, 2, 3, and 4) of nine km each. Lots 3 and 4 of this tunnel for a total length of 18 km were combined into one award and have been constructed using a double shield (DS) TBM or one path system. Extreme ground convergence in some sections of the tunnel has stalled TBM performance in early reaches, due to face collapse and shield jammings. This paper presents the latest update on the tunnel progress and impact of ground convergence on machine performance based on the information obtained from field observations. Also, the method and results of tunnel convergence measurements were studied in conjunction with geological parameters and an attempt was made to correlate TBM operational parameters and ground convergence. The result of the analysis indicates a good correlation between machine's operational parameter and tunnel convergence and can be used as an indicator of the potential for high rates of convergence. An early warning on ground convergence is essential for taking precautionary measures to avoid TBM jamming and related long delays and costs.

INTRODUCTION

TBMs have been used for excavation of water tunnels because of their high excavation rate and circular tunnel profile which is the preferred choice for such applications. Range of application of TBMs has expanded considerably in past two decades. Although a more comprehensive tunneling machine classification and selection chart has been developed by ITA/IATES, for simplicity, it is possible to classify the rock TBMs into three classes according to Barla and Pelizza 2001. This includes:

1. Open type or unshielded TBM (for hard and sound rock formations consisting rock classes I and II according to RMR classification).

2. Single shield TBM (for weak to very weak rock formations in relatively short tunnels).

3. Double shield TBM (for weak to hard rock formations in relatively long tunnels).

The use of shielded TBM allows the machine to mine through weak grounds and adverse geological conditions. However, TBM may get stuck (including shield jamming and cutterhead blocking) in the complicated geological structures, especially under high cover in weak rocks. This could cause major delays and impose a heavy and expensive burden on the tunneling operation.

Machine stoppage in adverse ground is bad news in many respects as it was reported in some tunnel projects such as Evinos-Mornos tunnel in Greece (Grandori et al. 1995) and Plave II in Slovenia (Guetter et al. 2001). Obviously, the passage of time has a negative impact on the situation. Also, the process of releasing the machine is very labor intensive and thus is very slow and dangerous. Therefore, examining the possibility of machine jamming due to adverse ground condition is an important step in tunneling operation involving the use of shielded machines. When such conditions are encountered, work schedule should be modified to expedite efficient and rapid tunneling through such ground and performing more extensive maintenance in stable grounds. This paper presents a case study of Ghomroud tunnel project in central Iran where study of ground convergence and machine parameters has led to development of a method to evaluate ground conditions from within the machine.

In previous articles by the same authors (Farrokh and Rostami 2007, 2008), the project was described and a discussion of the geology and machine performance was offered while the main focus of the paper was to explore relationships between tunnel convergence and chip size as well as TBM operational parameters in different sections of tunnel. This included marginal to high squeezing conditions. The results indicated that analysis of machine's operational parameters as well as grain size distributions can be used as an indicator of high deformations in the wall and potentials for ground squeezing. This article is a follow up with focus on specific areas of tunnel near to TBM jamming through observation of TBM parameters and their variations leading to the jamming. The conditions of these areas are typical of adverse geology in tunneling operations. Although the study is in its preliminary stage, the results to date seem to be promising. The outcome could allow the operators to identify potential for high convergence before the TBM is jammed by squeezing ground or a face collapse is encountered.

PROJECT DESCRIPTIONS AND GEOLOGY

The 36 km long phase one of Ghomroud water conveyance tunnel from Dez river to Golpayegan dam reservoir is near completion with excavated diameter of 4.5 m. A brief overview of the project, geology, and machine specifications has been provided in previous publications by the authors and is not repeated here (Farrokh and Rostami 2007, 2008). The tunnel is lined with hexagonal pre-cast concrete segments to an ID of 3.8 m. To date, over 18 km of tunnel in Lot 3&4 has been successfully excavated and the TBM continues to work as the share of the current contractor has been increased and they are allowed to continue tunneling until they meet up with the other tunneling operation underway in the opposite direction.

The tunnel is driven in Sannandaj-Sirjan geological zone of central Iran. This zone consists of series of asymmetric foldings and faults and has experienced mild to high metamorphisms which have caused schistosity and recristalization of minerals. The formations comprise massive limestone and dolomite, as well as slate, schist, metamorphic shale, and sandstone units. These rock types have similar geomechanical properties. The rock types along tunnel alignment are classified into four major categories as shown in Figure 1 (SCE 2003) and summarized as follows:

   I. Highly foliated and schistose rock types (schist, slate). This category covers more than 70% of the tunnel length. In general, they easily separate along the foliation planes, which are highly persistent. The uniaxial compressive strength of this formation was measured at a range of 30 to 60 MPa.

   II. Massive rock types (limestone and sandstone units). Intact strength of these rocks range between 40 and 80 MPa.

   III. Quartzite and quartz veins. These veins can be seen randomly in different parts of Jurassic formations. The maximum thickness of these veins reaches

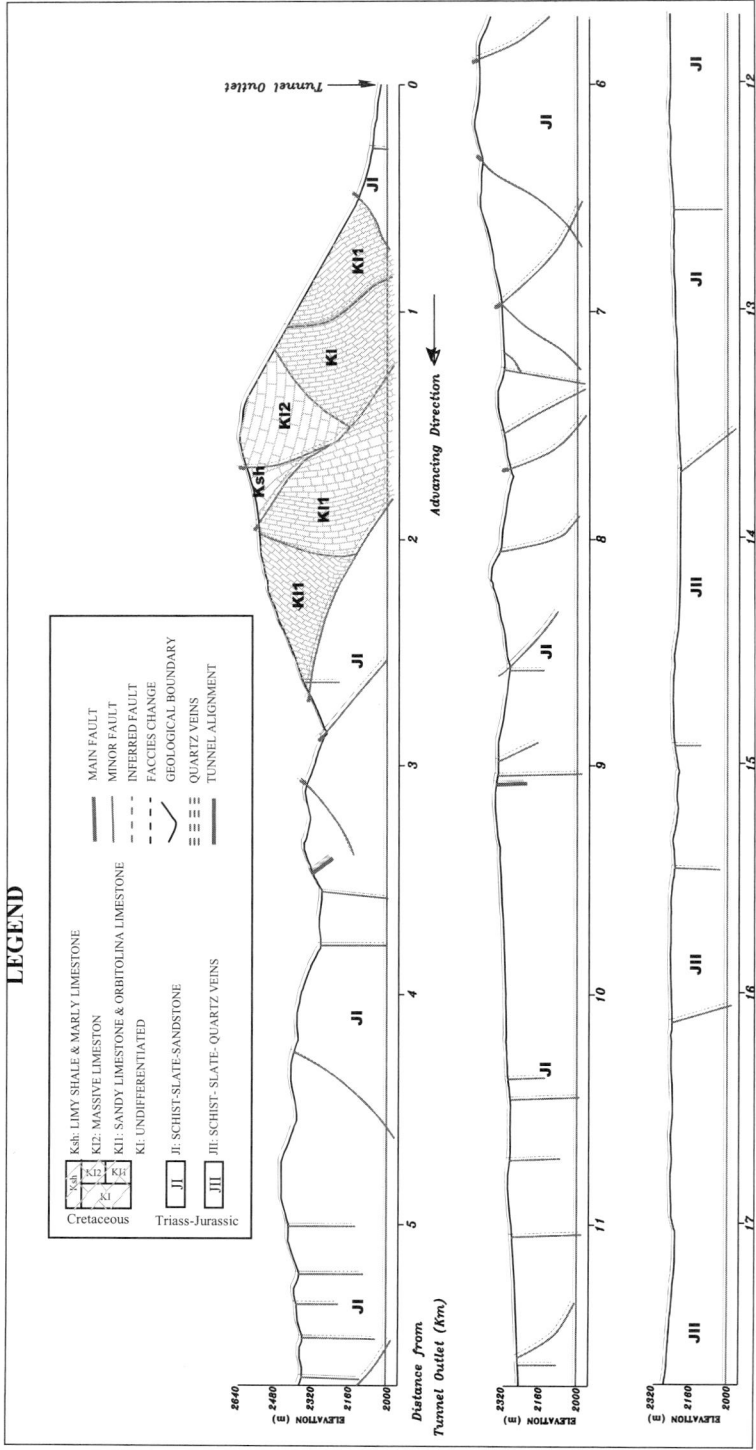

Figure 1. Longitudinal geological section along the tunnel (SCE 2003)

up to 80 m. Their strength reaches over 100 MPa. The origin of these veins is mainly from pegmatite formations in the area.

IV. Crushed rocks (shear zones). Tectonic activities have caused formation of various faults with thick shear and crushed zones along the tunnel alignment. Most of these faults are of reversed type. In addition some normal and strike slip faults have been observed in the area. Based on the geological studies, about 20 faults were recognized along the tunnel alignment with width ranging from 10 to 50 m. The rock mass is classified as "very poor" by standard systems.

As mentioned, the intact rock strength of the formations are generally medium to low and have low abrasivity, except for quartzite and quartz veins, which are not long sections of the tunnel. Joints, fractures and shear zones have reduced the strength of rock mass where they are present. The majority of the rock masses are classified as weak to fair quality.

Permeability of the rock masses is generally low (with hydraulic conductivity below 10–7 m/s). Ground water condition along the tunnel can generally be considered as dry to wet, except for crushed zones where higher rock mass permeability is encountered and water dripping can be spotted. The tunnel overburden is mostly over 200 m and ranges from 10 to 650 m for Jurassic and 200 to 650 for Cretaceous units. Therefore, gravitational stresses are expected to be up to 17.5 MPa. Table 1 is a summary of rock mass properties.

TUNNELLING OPERATION

The prime contractor for the design-built project was Ghorb Khatam and two of the subsidiaries Ghaem Co. and Sahel Co. were selected to perform as prime contractor and engineer/consultant, respectively. SELI of Italy was later subcontracted to assist in set up and operation of the TBM. Tunneling of Ghomroud tunnel was commenced in fall of 2002. About 380 m was excavated by drill and blast method, as the starter tunnel while the TBM was being manufactured and mobilized. The Double Shield TBM manufactured by Wirth AG of Germany and arrived at the site in fall of 2003 and was put to work in early 2004. The remained length of the contract was mined by the TBM and the excavation of lots 3 & 4 has been successfully completed in spring 2008.

In the TBM driven section, machine has had several stops including a major delay due to jamming in squeezing grounds and also for face collapses (see Figure 2). This is in contrast with rather good performances in other sections of the tunnel where advances of up to 53 m/day and 1000 m/month have been recorded (Table 2).

TUNNELING DOWNTIME CAUSED BY GEOLOGICAL CONDITIONS

Tunneling operation has encountered adverse geological conditions at several locations, which has caused TBM jamming (cutter head blocking and shield jamming). The geological causes of TBM stoppages and long delays may be divided into two main categories: 1. Ground squeezing/jamming, 2. Face collapse.

Ground Squeezing

As stated by the International Society of Rock Mechanics (ISRM) (Barla 2001), Squeezing rock is the time dependent large deformation, which occurs around the tunnel and is essentially associated with creep, caused by exceeding shear stress. Deformation may terminate during construction or continue over a long period of time. The magnitude of tunnel convergence, the rate of deformation and the extent of the yielding zone around the tunnel, depends on the geological and geotechnical

Table 1. Lithology description and properties

	Excavation method	TBM			D&B
	Excavated Length (Km)	17.6			0.4
	Length (Km) Tunnel Route	1.3	14.85	1.25	0.6
	Lithology Lable	JII	JI Advancing direction ←	KI JII	JI JIII

Lithology	JI		KI	KIII	JII			JIII
Description	Sandstone (JIa) Shale-Slate-Schist (JIb)		Limestone and Dolomite	Crushed zones in KI unit	Ghraphite schist (JIIa) Qaurtz schist (JIIb) Qaurtzite (JIIc)			Crushed zones in JI and JII units
Overburden height max. (m)	650		650	650	200			400
Lithology length along the tunnel (m)	14706		1125	135	1300			690
Percentage of Lithology length along the tunnel	82		6	1	7			4
Rock units	JIb	JIa	—	—	JIIc	JIIb	JIIa	—
σ_{ci} (MPa)	40	60	75	50	100	50	45	30
m_i	9	19	13	13	20	7	7	7
GSI	35	50	52	20	52	33	33	25

* σ_{ci} =Average intact rock compressive strength, m_i = Hoek-Brown constant, and GSI=Average geological strength index obtained from site investigation

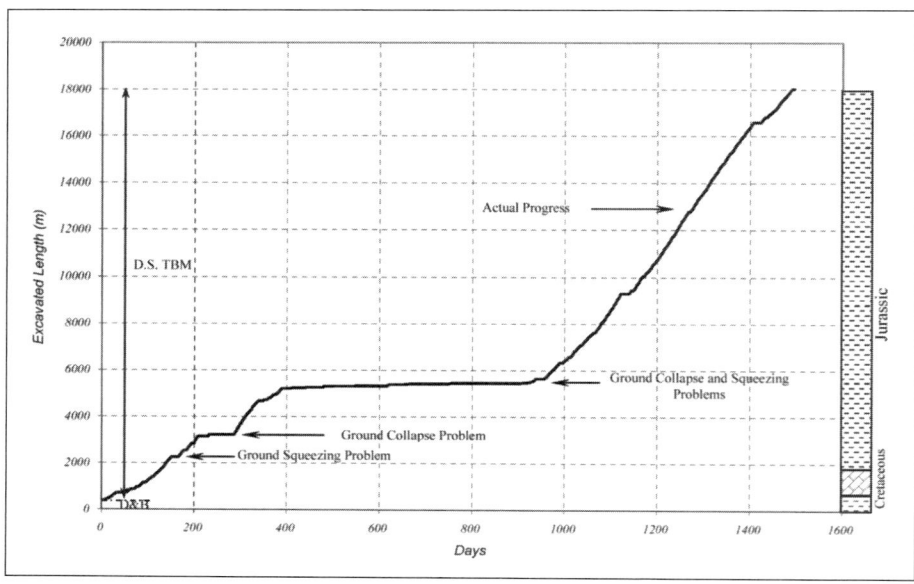

Figure 2. TBM progress and encountered geological difficulties

Table 2. TBM performance parameters

Parameters	
Date	2002–2008
Length (m)	17956
Excavated Diameter (m)	4.5
Best Day Advance (m)	52.7
Best Month Advance (m)	1000.3
Best Week Advance (m)	251.8
Daily Average (including stoppages) (m)	13.1
Daily Average (without long delays/stoppages) (m)	24
Penetration Rate Range (m/h)	0.6–6.95
Utilization (including stoppages) (%)	12
Utilization (without stoppages) (%)	22

conditions, the in-situ state of stress relative to rock mass strength, groundwater condition and pore pressure, and the rock mass properties. Gneiss, micaschists, calcschists (typical of contact/shear zones and faults), and claystones are numbers of rock complexes where squeezing will occur, provided that loading conditions necessary for the squeezing are present.

On the basis of the available data, it is possible to study squeezing behavior according to the semi empirical criteria proposed by Hoek (2001) for each tunnel section and the corresponding depth (H). Hoek (2001) used the ratio of the rock mass uniaxial compressive strength (σ_{cm}) to the in situ stress (σ_0) as an indicator of potential tunnel squeezing problems. In particular, Hoek and Marinos (2000) showed that a plot of tunnel strain (defined as the percentage ratio of radial tunnel wall displacement to tunnel radius) against the ratio σ_{cm}/σ_0 can be used effectively to assess potential tunneling problems under squeezing conditions.

Hoek (2001) by means of axi-symmetric finite element analyses and a range of different rock masses, in situ stresses and support pressures (p_i) suggested the following approximate relationship for the tunnel strain ε_t:

$$\varepsilon_t(\%) = 0.15\left(1 - \frac{p_i}{\sigma_0}\right)\left(\frac{\sigma_{cm}}{\sigma_0}\right)^{-\left(\frac{3p_i/\sigma_0 + 1}{3.8 p_i/\sigma_0 + 0.54}\right)} \tag{1}$$

Where:
p_i = support pressure,
σ_0 = in situ stress,
σ_{cm} = uniaxial compressive strength of rock mass

A possible way to estimate σ_{cm}, which has been proposed by Hoek and Marinos (2000), is to use the following equation:

$$\sigma_{cm} = (0.0034 \cdot m_i^{0.8})\sigma_{ci}\left[1.029 + 0.025 e^{(-0.1 m_i)}\right]^{GSI} \tag{2}$$

Where:
σ_{ci} = uniaxial compressive strength of the intact rock (or laboratory testing),
m_i = Hoek-Brown constant,
GSI = Geological Strength Index

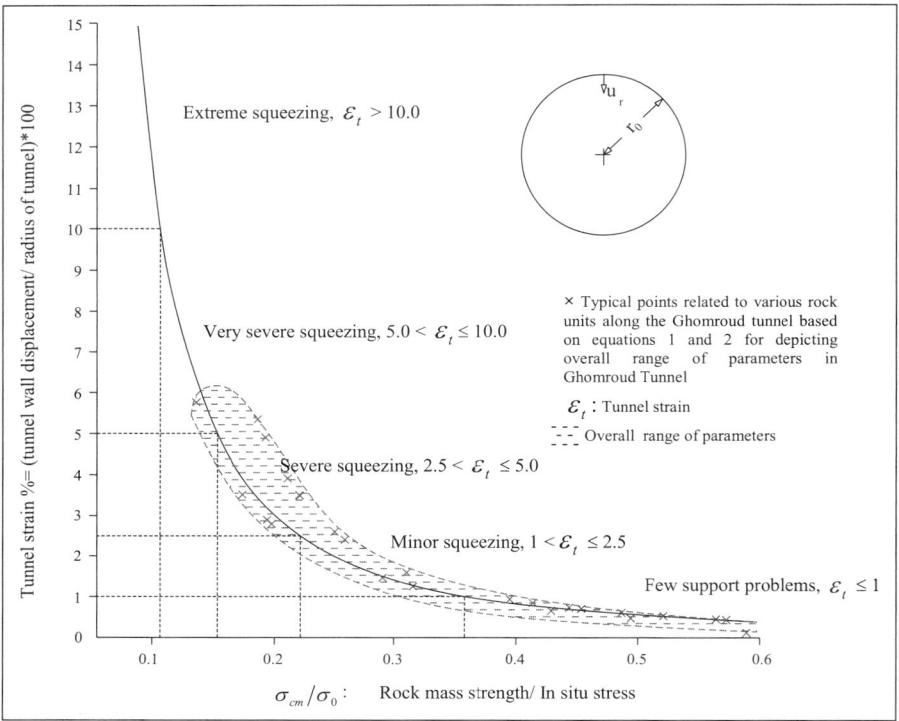

Figure 3. Classification of squeezing behavior according to Hoek (2001) and predicted range of tunnel strain for Ghomroud tunnel

On the basis of the above consideration and study of several case histories of tunnel practices in Venezuela, Taiwan and India (the cases considered, include 16 tunnels in graphitic phyllites, sandstone, shale, slates, fractured quartzite, sheared metabasic rocks and fault zones.), Hoek (2001) suggested Figure 3 to be used as a first estimate of tunnel squeezing problems.

The results of analysis for Ghomroud tunnel using the rock mass properties (Table 1) and tunnel overburden at various stations for convergence prediction based on Equations 1 and 2 (using pi=0 for estimation of max convergence) indicated that most parts of the tunnel alignment have the potential for squeezing and in some parts of the tunnel alignment, possibility of sever squeezing exists. Figure 3 illustrates overall range of tunnel strains predicted by Equations 1 and 2. As can be seen, different sections in the tunnel are anticipated to show various levels of squeezing behavior, which can be categorized from non-squeezing ($\varepsilon_t <1$) to severe squeezing ($\varepsilon_t >2.5$).

In order to quantify and classify the real squeezing behavior, it was necessary to monitor tunnel wall convergence. Tunnel wall convergence was measured by comparing the available annular space around the segmental lining at the end of the tail shield (in vertical direction) and the annular space around the segmental lining in the non-squeezing ground. This was done by measuring the amount of grout backfill used behind the segments to estimate the amount of ground closure and thus reduced volume of annular space between the lining and bored tunnel. For two kilometers of tunnel (Km 30 to 32 of the contract or station 20+00 to 40+00 in the direction of tunneling), wall convergence was directly measured using a scaled instrument in the grout holes inside

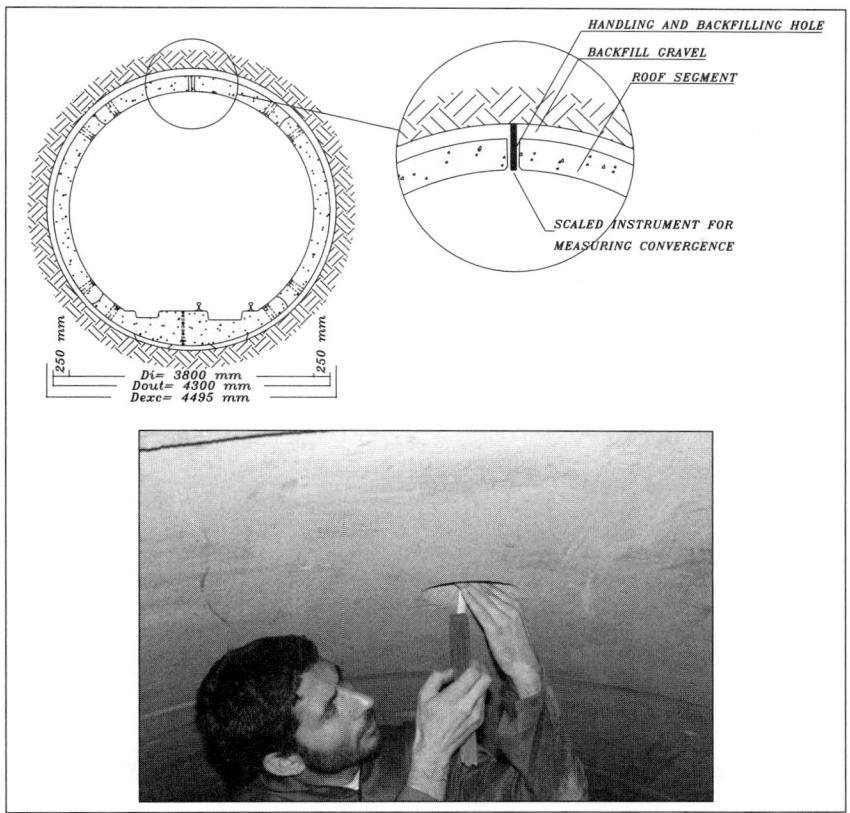

Figure 4. Measurement of convergence by a scaled instrument through the grout holes

the segmental lining (Figure 4). For this purpose, the instrument was used through the grout hole on the crown segment to measure the amount of ground deformation relative to the bored diameter. This was deemed appropriate since the segments were placed directly on tunnel invert and thus the measured distance could represent the amount of vertical convergence relative to the tunnel invert. Figure 5 illustrates the magnitude of measured radial convergence along the different lithological units as well as the tunnel strain classifications based on predictions (by Equation 1). As can be seen, most of the tunnel sections fall within the squeezing ground category (with ε_t >1) as predicted by Hoek formula for squeezing behavior.

Table 3 is the summary of squeezing behavior for the mined tunnel in term of incurred tunnel strain, calculated from the measured radial convergence. These results indicate that about 40 percent of the excavated tunnel is within the squeezing category (ε_t >1) (about 8 percent more than predicted amount—Figure 6).

As can be seen from Figure 5 and Table 3, about 10% of the mined length of tunnel is in severe squeezing category (which is 5 percent greater than the predicted value-Figure 6). Figure 6 shows the amounts of predicted and actual percentage of tunnel length with different ε_t. In these areas, some ground support problems due to high ground squeezing pressures as well as TBM advancing difficulties and shield jamming have been experienced. The amount of actual and predicted length of tunnel in each category is not exactly the same. Existence of the more complicated and relatively

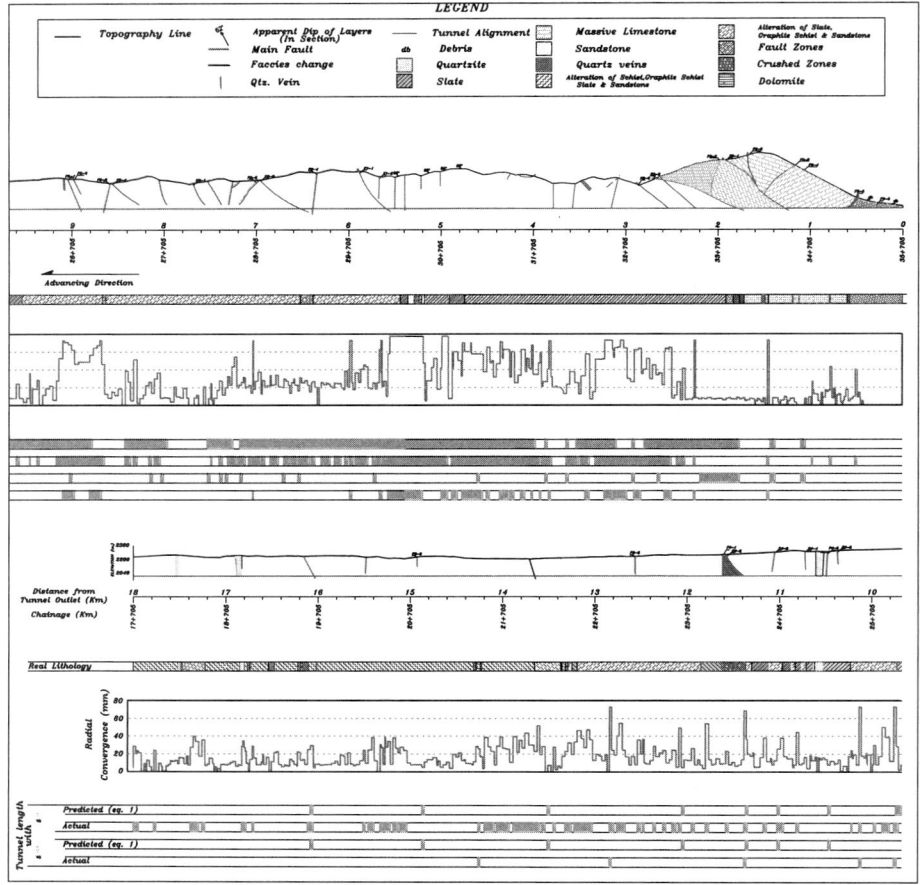

Figure 5. Measured and actual radial convergence along Ghomroud tunnel alignment expressed in term of tunnel strain

weak structures was the main cause of the difference. Therefore, the exact boundaries of areas with sever squeezing behavior could not be delineated.

The rock mass in the shield jamming locations consisted of schist, quartz veins and some soft interbedded Graphite-Schist. In these locations, the strike of the schistosity was parallel to the tunnel axis and this was very instrumental for occurrence of ground squeezing. Moreover, estimated RMR for the rock mass at these section was about 25 to 30, meaning that the rock mass could be considered weak and low in strength. As shown in Figure 7 to 9, with very high ground convergence, the tunnel walls completely confined the TBM shield.

In double shield TBM, the shield is tapered cylinder, meaning that the diameter of the front shield is larger than the tail shield. Table 4 is the summary of shield diameters for Ghomroud double shield TBM. Based on the measurements, the amount of over-excavation was about 145 mm at the tail shield, which means 145 mm of convergence before tail shield could be engulfed and subsequently squeezed by the ground. Therefore, in severely squeezing sections of tunnel where the ground completely confined the shield (Figure 9); the amount of ground displacement had to be over 145 mm

Table 3. Classification of squeezing behavior in term of tunnel strain along the TBM driven length based on the monitored radial convergence

Parameters	Lithology units along the mined tunnel				Sum
	Carbonate units (Limestone and Dolomite)	Crushed zones in carbonate units	Noncarbonate units (schist-slate-shale-sandstone-graphite schist-quartz schist-quartzite)	Crushed zones in noncarbonate units	
Length of the Section m (%)	1125 (6.4)	135 (0.8)	15875 (90.3)	445 (2.5)	17580 (100)
Length of Tunnel in $\varepsilon_t < 1$ m (%)	1079 (6.15)	130 (0.75)	8970 (51)	237 (1.35)	10416 (59.25)
Length in $1 < \varepsilon_t < 2.5$ m (%)	28.5 (0.16)	3 (0.02)	5285.5 (30)	173 (1)	5490 (31.23)
Length in $\varepsilon_t > 2.5$ m (%)	17.5 (0.1)	2 (0.01)	1617.5 (9.2)	35 (0.2)	1672 (9.51)

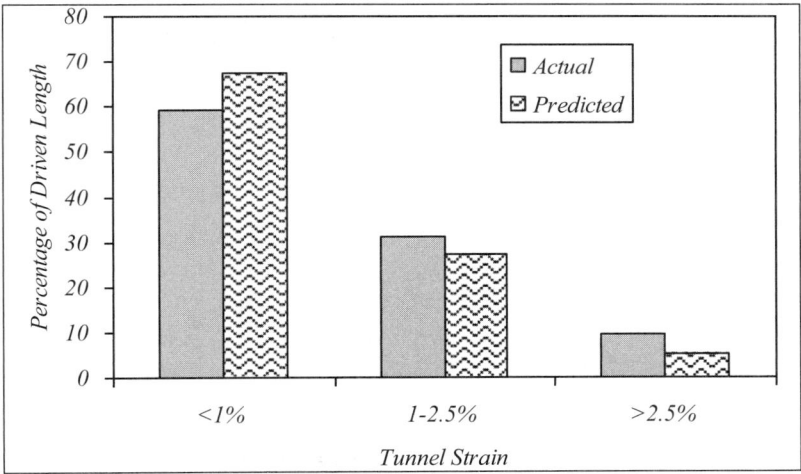

Figure 6. Predicted and actual percentage of driven tunnel length with different ε_t (predicted values obtained from Equation 1)

to cause shield confinement. In such conditions, no backfill gravel could be applied behind the segments because of the limited space between tunnel wall and lining.

There was excess capacity in the auxiliary thrust cylinders to overcome shield friction in squeezing ground, nonetheless, this large capacity could not pull machine through in some tunnel sections. In such conditions, because of the quick and large radial displacement (more than 3 cm) in the vicinity of the tunnel face (immediately behind the cutterhead and within the front shield), the thrust force of the telescopic cylinders could go up to more than 1,500 tons, to release the jammed shield. This is 80% more than the maximum operational thrust force of the machine. In some instances, it was necessary to excavate the ground around the front shield and near the cutter

Figure 7. Ground confinement around the front shield

Figure 8. Hand mining around the front shield and cutter head

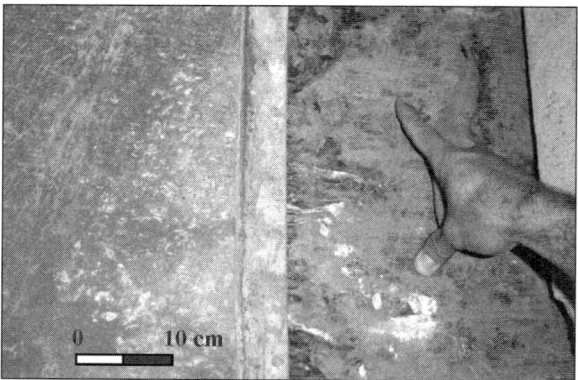

Figure 9. Shield movement marks on the tunnel wall

Table 4. Technical data of Ghomroud D.S. TBM

Boring diameter	New cutters	4,525 mm
	Worn-out cutters	4,495 mm
Over cut over cutter head shield	With new cutters	65 mm
	With worn-out cutters	35 mm
Front shield diameter	Outer diameter	4,460 mm
Tail shield diameter	Outer diameter	4,380 mm
Max over cut measured at the tail shield		145 mm

head to inject grease through the shield to lubricate the contact area of outer shield surface and ground to release the machine. In order to excavate the surrounding ground, two circular positions near to the stabilizers in the front shield was defined by Wirth Company to be cut as window openings for accessing the surrounding ground behind the front shield and near the cutterhead. In some locations, it was necessary to excavate the ground around the tail shield through small opening behind the tail shield, mined to provide access to the ground. For this purpose, in some incidents, some erected segments were removed to provide access to the tail shield. In these situations, the shields were seized by the squeezing ground, with some miner ground intrusions into the shield as shown in Figure 9.

Face/Ground Collapse

Joint surface condition has a very important role in tunnel stability. During the excavation, when some intersecting joints with weak surface conditions were encountered, they produced sizable rock wedges falling out at the face or the walls. This is referred to as fallouts or ground collapses. This process typically occurs in highly jointed rock mass or in fault/crushed zone. Based on the Geological investigations for Ghomroud tunnel it was predicted that the tunnel alignment would intersect several faults and crushed zones. From the time of the last revision of this paper, there has been several face collapses along the tunnel. In some locations, the TBM passed through such zones successfully, while in other locations, the cutter head was blocked. The encountered geological conditions comprised graphite schist with slickensided schistose surfaces and quartz veins, interbedded soft rocks with high percentage of Graphite, and localized wet conditions even dripping water. Existence of the quartz veins in these locations (Figure 10) confirms that the rock mass was highly fractured and these fractures were subsequently filled by siliceous material.

At some locations, the dip direction of these fractures and joints was against the tunnel advancing direction (the situation of drive against dip) and the excavation process was followed by relatively large over breaks (about 1 to 2 m deep) and release of relatively large blocks causing cutterhead jamming. On the other hand, the existence of soft interbedded and slickensided surfaces especially in the case of steeply dipping joints (observed average dip of 60–80 degrees) and relatively low spacing (about 20 cm) caused easy dislocation of the rock blocks.

RMR values for these sections were typically around 20 to 30 range (poor rock mass). Figure 11 shows the rock conditions in one of these locations where additional efforts were required to release the cutterhead. In such cases a bypass was excavated around the cutterhead and the top of the front shield (Figure 12). Then, the collapsed rock blocks on the cutterhead were removed and ground was stabilized by wood timbers and in some instances by shotcrete, the empty cavity above the cutterhead was filled by some light filled bags of rock fragments. The remaining cavities were filled with expandable foams (Figure 13). In a couple of locations, ground squeezing conditions were also experienced while the TBM cutterhead was being released. This

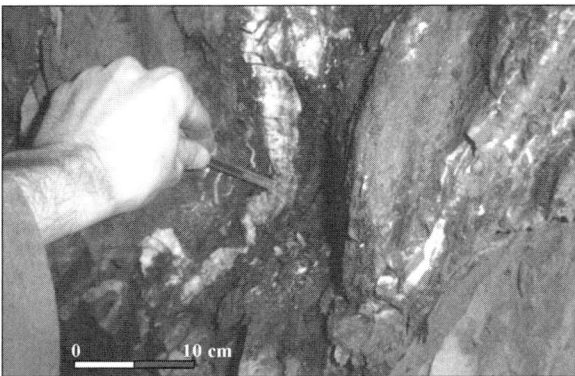

Figure 10. Quartz veins in the rock mass around the collapsing locations

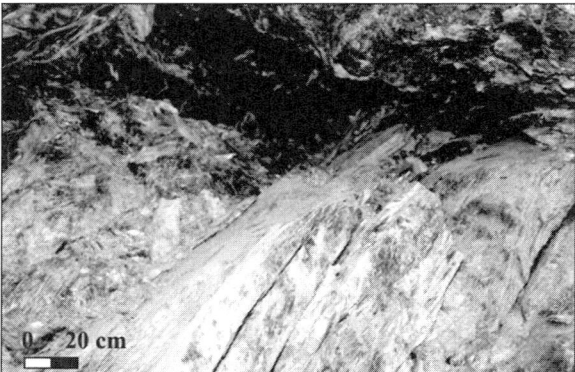

Figure 11. Rock collapse on top of the front shield

Figure 12. Hand mining around the front shield and stabilizing by wood timbers

Figure 13. Filling cavities above the front shield by filler packs and foam

was naturally due to time dependent behavior of the weak rock and high potential of squeezing in such locations.

EVIDENCES OF THE ADVERSE GEOLOGY CONDITIONS DURING THE EXCAVATION

In the Ghomroud tunnel project, some geotechnical investigations and observations were carried out during the construction phase and while TBM was being operated. Some basic activities are summarized as follows:

- Field observation and investigation to prepare various geological maps (structural and engineering geology).
- Drilling probe holes to evaluate ground conditions and rock mass permeability.
- Additional Geomechanical laboratory tests on retrieved samples.

The results of these geotechnical studies show that the geological conditions encountered along the tunnel, especially the structural features observed during construction, were hardly predictable. Therefore it was not possible to find exact locations of crushed and highly squeezing zones from the available maps of pre-bid site investigations. One of the best methods for detail geological investigations during the tunnel excavation was to perform probe drilling ahead of the tunnel face. The main objective of the probe drilling was to detect water and gas, as well as major faults. For this reason, it was necessary to observe drilling rate, examine drill cuttings, and monitor groundwater inflow. However, in a shielded TBM, the probe drilling can only be performed through designated holes in the tail shield. This limited the observation of ground features especially when the rock mass behind the shield had collapsed (Figure 14). In this situation, the only available data from probe drilling was general operational parameters such as feed pressure and drilling rate. Moreover, since the Graphite schist has plastic behavior, there was no notable water flow detected in related zones because of low rock mass permeability. On the other hand, selection of the suitable starting point for probe holes could be crucial. Many approaches were used to predict the location of unfavorable geological conditions by probe drilling during the excavation, before entering such zones. But this proved to be very difficult in the field due to practical restrictions indicating the need to develop alternative methods for predicting the location of unfavorable

Figure 14. Picture of the drilling rods from the cavity on top of the shield

ground. The following section describes one of the methods developed and studied in this project to provide some level of observation on ground conditions from peripheral measurements. The method involves observing TBM operational parameters and subsequent analysis to predict ground behavior at the face and around the front end of the machine.

Evaluation of TBM Operational Parameters

Based on the analysis of the TBM operational parameters such as thrust, torque and RPM, near the TBM jamming locations, some notably different values for these parameters can be identified for the last few boring cycles. It should be mentioned that the behavior of the ground near areas of potential face collapse could not be easily distinguished from squeezing ground (because of plastic behavior of schist). TBM penetration rate and cutterhead coefficient are two important parameters used by many researchers in TBM performance prediction. Composite indices such as specific penetration, "SP" (Alber 2000) and field penetration index "FPI" (Klein et al. 1995) have also been introduced and used by some researchers in order to consider effects of TBM operational parameters on the net penetration rate. In this research both of these indices were examined for assessment of TBM operational parameters. Overall, the SP index (Equation 3) shows better correlation and more consistent trend than does FPI. Therefore this study is focused on results of analysis for SP. The ratio of rolling force to normal force or so-called cutter (or rolling) coefficient (CC) typically shows cutterhead torque for a given thrust. To simplify, it is possible to use torque to thrust ratio (named as Cutterhead Coefficient which is presented in Equation 4) instead of rolling force to normal force ratio (Cutter or Rolling Coefficient).

$$SP = \frac{p}{RPM.F_n} \text{ (Penetration per unit force)} \quad (3)$$

$$CC = \frac{Tq}{Th} \text{ (Cutterhead Coefficient)} \quad (4)$$

Where: Tq is the cutterhead torque in kN-m, Th is the cutterhead thrust in kN, RPM is cutterhead rotational speed in revolution per minute, p is penetration in cm per minute, and F_n is normal force in kN per disc cutter.

IMPACTS OF GROUND CONVERGENCE ON TBM PERFORMANCE

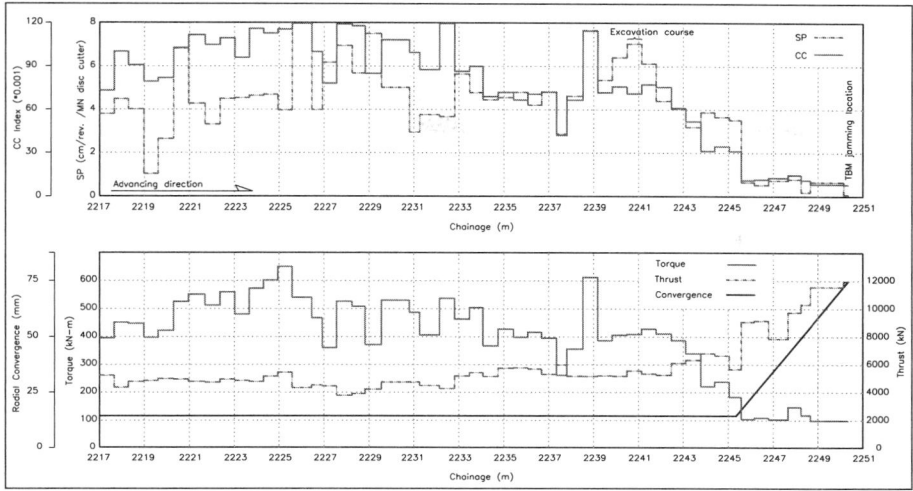

Figure 15. TBM parameters analysis for squeezing ground

The study shows that the ratio between Tq and Th changes and deviates from the predicted trend based on the Cutting Coefficient (ratio of rolling to normal force). This deviation can be attributed to the nature of a shielded TBM, where the total thrust force is not transferred to the cutters because of the imposed frictional forces on the shield, resulting in unusually low Tq/Th ratios. Figure 15 shows the results of the analysis as well as the amount of radial convergence for one situation of squeezing ground. The radial convergence diagram in Figure 15 is measured by comparing the available annular space around the segmental lining (in vertical direction) at the end of the tail shield with non-squeezing ground. This diagram show that excessive convergence, indicated by higher friction and thus lower cutterload, was associated with lower torque on the cutterhead. The excessive convergence was due to the quality of the ground which deteriorated gradually up to jamming location. As shown in Figure 15, when the ground squeezing pressure gradually increases, the frictional force imposed on the outer surface of the shield increases while specific penetration of the cutterhead per revolution and subsequently the cutterhead torque decreases. To compensate for the conditions, operators typically increase thrust, resulting in considerable decrease in CC index value. Therefore, notable drop in CC index corresponds to highly squeezing grounds. These results are consistent with the measurements and observations in other TBM jamming locations along the tunnel.

To compare the validity of the indices, the daily average parameters of SP, CC for the first 5 km of tunnel, in which there were weak to very weak rock masses as well as relatively good rock masses (Limestone) (Figure 17 shows typical tunnel face situations of this area) are shown in diagrams of Figure 16. This area contained important locations where TBM jamming had occurred. TBM jamming locations are easily spotted with high regripping pressure, high difference of max and min values of thrusts and regripping pressures for daily excavation (Figure 16). TBM jamming locations have been shown in diagrams of Figure 16 with dotted straight lines. As it can be seen from Figure 16, the amount of CC values decreased sharply at the jamming locations but SP did not show high sensitivity at these locations.

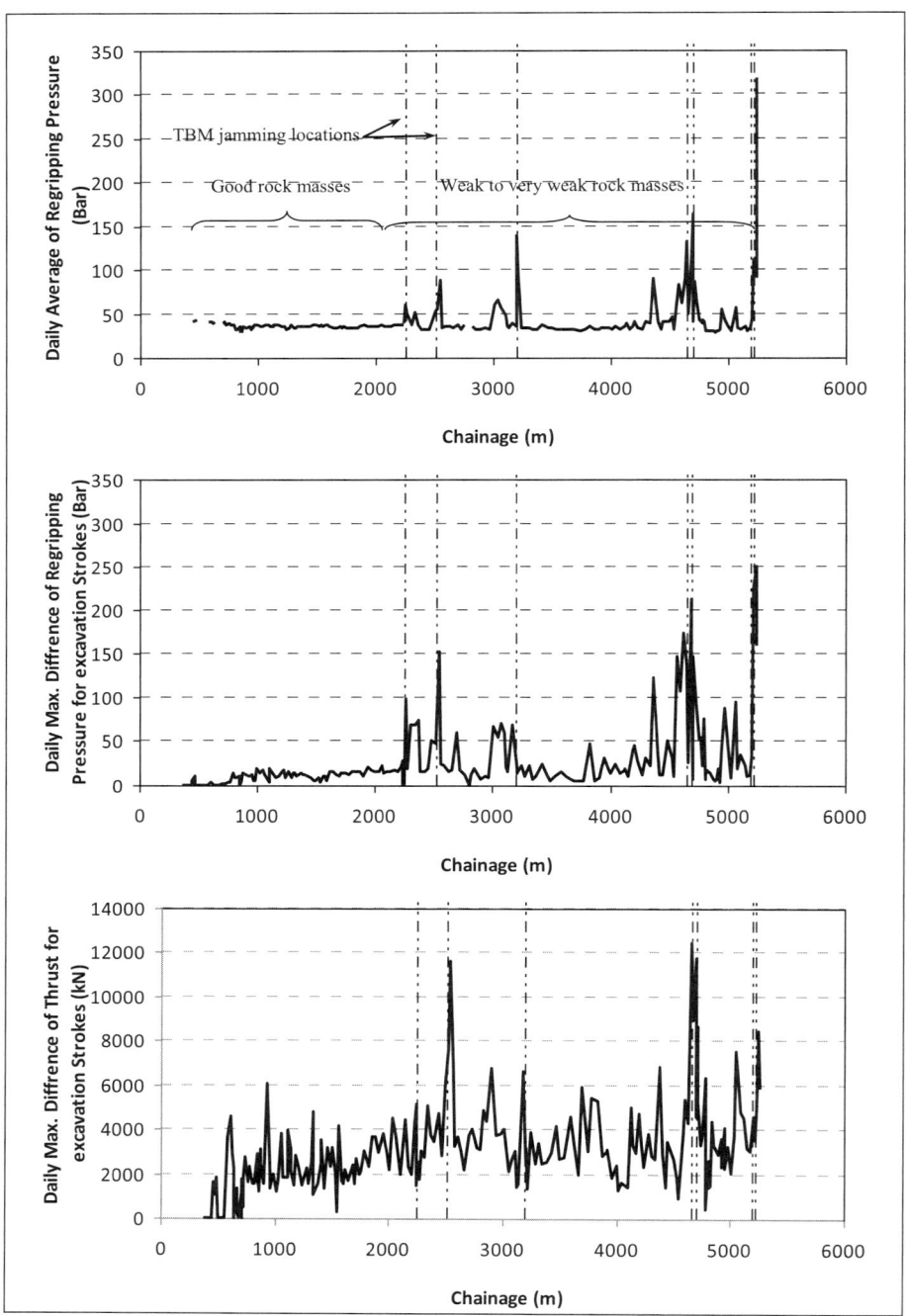

Figure 16. Variations of different parameters for 5 km of tunnel including squeezing ground as well as good rock mass *(continues)*

IMPACTS OF GROUND CONVERGENCE ON TBM PERFORMANCE 1351

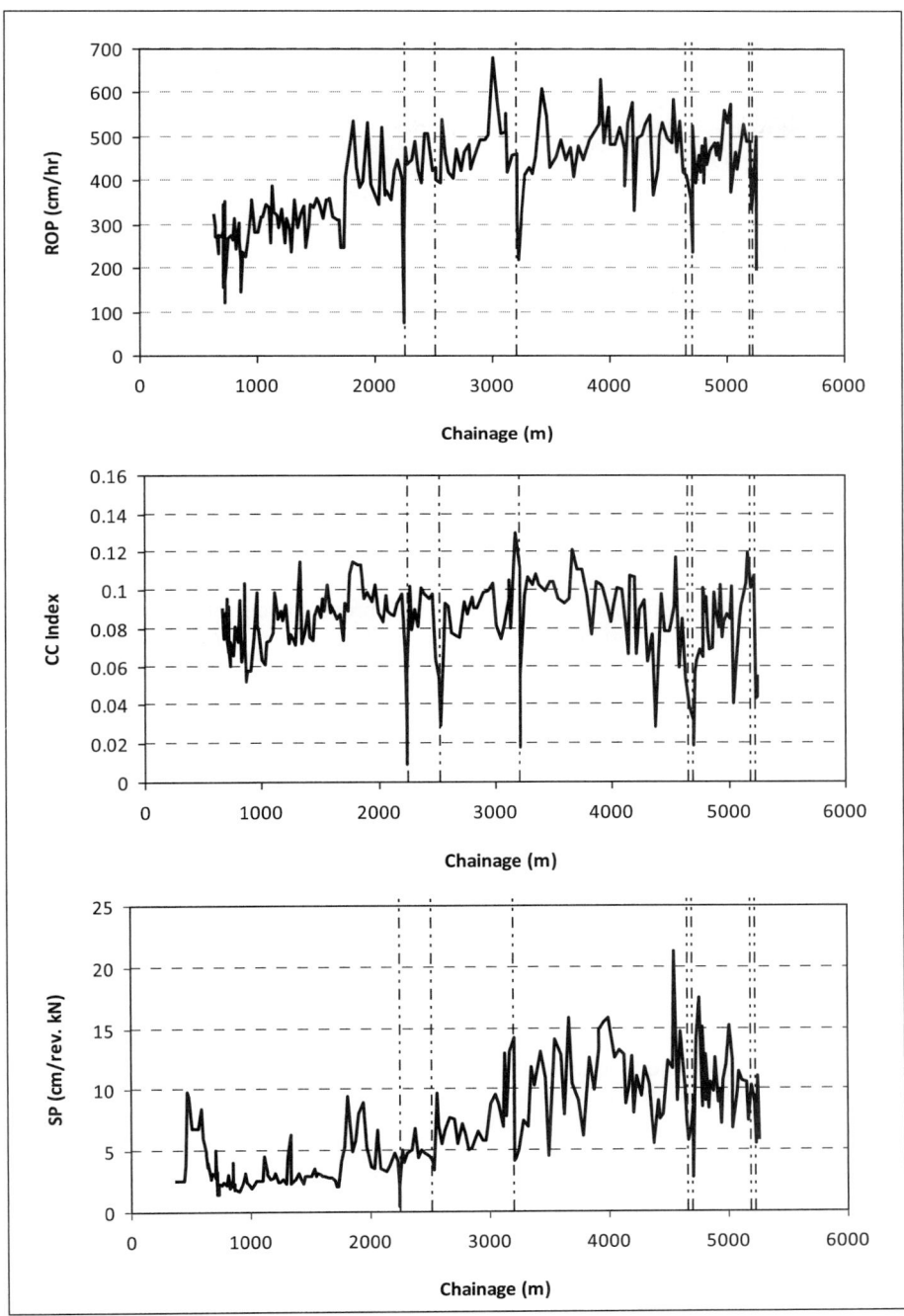

Figure 16. Variations of different parameters for 5 km of tunnel including squeezing ground as well as good rock mass *(continued)*

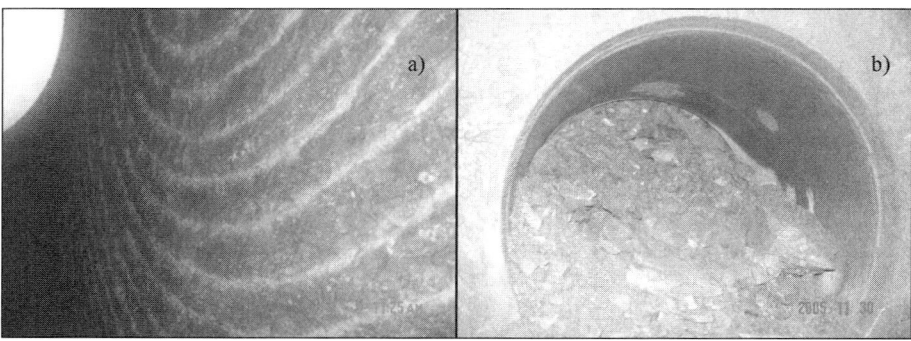

Figure 17. Typical tunnel face conditions of (a) good rock mass (in Limestone), and (b) very weak rock mass (in Graphite Schist)

DISCUSSIONS

Finding evidence of squeezing conditions during the excavation is very important in allowing the operator to prepare for upcoming unfavorable conditions. If known in advance, operators can implement measures to avoid TBM jamming and time-consuming activities of releasing the machine with manual labor. Based on our preliminary findings, TBM operational parameters (TBM thrust, torque) and some composite indexes including SP, CC, and penetration rate (ROP) can play a very important role in predicting potential problems in ground conditions near the sections with high potential of squeezing. Preliminary findings show that unusual increase in regripping pressure, higher than normal thrust level, and sudden decrease in CC and penetration rate are indicative of the upcoming squeezing situations. This could be explained by the fact that in high ground convergence and squeezing ground, ground pressure against the shield increases, leading to shield confinement. In this situation, shield friction increases, resulting in decrease in net cutter load for given applied thrust and therefore the amount of penetration rate decreases. This results in high compensatory thrust pressure from the machine to achieve the least acceptable penetration. In this situation, the contact pressure between cutter head and the tunnel face decreases; which in turn results in lower torque. Increasing thrust and decreasing torque lead to low CC index, which can easily be observed by monitoring machine parameters. On the other hand, increased regripping pressure and difficulty in moving of the tail shield can be attributed to high imposed ground pressure and high amount of frictional forces between the shield and the ground. Using these indexes, early indications and warnings could be generated; machine operational parameter adjusted, and scheduled activities revised to cope with the ground conditions at and ahead of the tunnel face.

CONCLUSIONS

The encountered adverse geological conditions including squeezing ground and face collapses have caused several TBM stoppages and delays during excavation of Ghomroud tunnel lots 3 and 4. Weak intact rock and rock mass and the existence of extensive foliated structures and faults are main effective causes of the TBM jamming. Based on the tunnel convergence measurements, about 10% of the TBM excavated length of the tunnel was in sever squeezing category, while unfavorable orientation of weak structures such as joints, crushed zones, and faults caused several face collapses. Review and analysis of TBM operational parameters shows that, in highly squeezing grounds, machine thrust and regripping pressure has increased gradually while the cutter head torque has decreased. In order to analyze the TBM parameters during the excavation in adverse situations, two indices (Including CC and SP) have been used. These indices demonstrate a considerable decrease in the vicinity of unfavorable geological locations. Although preliminary, these results can provide a useful tool to examine rock quality during tunnel excavation by TBM or shielded machines and develop a measure of potential for occurrence of machine jamming while tunneling is in progress. Once these areas are identified, suitable precautionary measures can be taken to avoid long delay and prevent machine from being held by adverse geological conditions.

REFERENCES

Alber, M., 2000. Advance rates of hard rock TBMs and their effects on projects economics. Tunnelling and Underground Space Tech., Vol. 15, No. 1, pp. 55–84.

Barla, G., 2001. Tunnelling under squeezing rock conditions. In: Kolymbas, D. (ed.), Euro Summerschool on Tunnelling Mechanics, Innsbruck October 08–11, pp. 169–268.

Barla, G., Pelizza, S., 2000. TBM tunnelling in difficult ground conditions. Proc. of GeoEng 2000—Proceedings of the International Conference on Geotechnical & Geological Engineering, Melbourne, 19–24 November 2000, Technomic Publishing Company: Lancaster, pp. 329–354.

Farrokh, E., Rostami, J., 2008. Correlation of tunnel convergence with TBM operational parameters and chip size in the Ghomroud tunnel, Iran. Tunneling and Underground Space Technology, Vol. 23, No. 6, pp. 700–710.

Farrokh, E., Rostami, J., 2007. The relationship between tunnel convergence and TBM operational parameters and chip size for double shield TBMs. RETC, Canada, Chapter 88, pp. 1094–1108.

Grandori, R., Gager, M., Antonini, F., Vigle, A., 1995. Evinos-Mornos tunnel- Greece. Proceedings of the Rapid Excavation & Tunnelling Conference, Chapter 47, pp. 747–767.

Guetter, W., Weber, W., 2001. Two tunnels in totally different geological formations. Proceedings of the Rapid Excavation & Tunnelling Conference, Chapter 21, pp. 241–260.

Hoek, E., 2001. Big tunnels in bad rock. 2000 Terzaghi lecture, The ASCE Journal of Geotechnical and Geoenvironmental Engineering, Vol. 127, No. 9, pp. 726–740.

Hoek E, Marinos, P., 2000. Predicting tunnel squeezing. Tunnels and Tunnelling International. Part 1: November 2000, pp. 45–51; Part 2: December 2000, pp. 34–36.

Klein, S., Schmoll, M., Avery, T., 1995. TBM performance at four hard rock tunnels in California. Proceedings of the Rapid Excavation & Tunnelling Conference, chapter 4, pp. 61–75.

Sahel Consultant Engineers (SCE), 2003. Geological and engineering geological review of Ghomroud tunnel.

TBM DATA MANAGEMENT AND QUALITY ASSURANCE FOR THE BRIGHTWATER CONVEYANCE PROJECT

Jeffrey Mitsopoulos ▪ Jacobs Engineering

Frank Stahl ▪ Babendererde Engineers GmbH

James Wonneberg ▪ Jacobs Engineering

Kenneth Rossi ▪ EPC Consultants, Inc.

ABSTRACT

King County's (KC) Brightwater Conveyance Project in Seattle, Washington involves the construction of approximately 21 kilometers (13 miles) of bored tunnel, in three contracts, with four TBM's which produce large amounts of real-time digital mining data. To facilitate the management of TBM data with a single interface and assist in project Quality Assurance, KC's Construction Manager (Jacobs Engineering) recommended the use of the software TPC (Tunneling Process Control), developed by Babendererde Engineers GmbH.

The paper demonstrates how TPC was used for oversight, analysis and QA on a complex tunneling project and how contemporaneous review and management of data results in a better understanding of TBM performance than can be obtained by retrospective analysis of data.

INTRODUCTION

King County's Brightwater Conveyance System will be a sewage collection and outfall system that ultimately directs treated wastewater by gravity approximately 21 Km (12.5 miles) from a treatment plant in Woodinville, Washington to the system's outfall nearly due west in Puget Sound. Tunnel construction is divided amongst three essentially simultaneously constructed contracts totaling approximately $444 B for four separate tunnels (Figure 1).

The tunnels being constructed have outside diameters ranging from 4.5 m (14'-8") to 5.6 m (18'-4") with a bolted, gasketed, segmental pre-cast concrete design (Figure 2). The BT-1 and BT4 tunnels are being mined by EPB TBM's where as the BT-2 and BT-3 tunnels are being mined by slurry TBM's. The drives are through a highly varied geological region that encounter maximum face pressures from 3.2 Bar (47 psi) to 6.8 Bar (98 psi) through a region of soil that is primarily highly varied glacial till (Figure 3). King County faced with this scenario inherent to the complexities of multiple-Contract Construction Management and technical challenge wrote into its Specification 02310, Tunnel Excavation, provisions for a Contractor provided TBM Data Monitoring System (DMS) that would supply the Construction Management Teams for each project the three following provisions:

1. Record data at maximum time intervals of ten seconds and display in real-time.

2. Store and record data via an automated acquisition system in digital form for later use and retrieval.

TBM DATA MANAGEMENT AND QUALITY ASSURANCE 1355

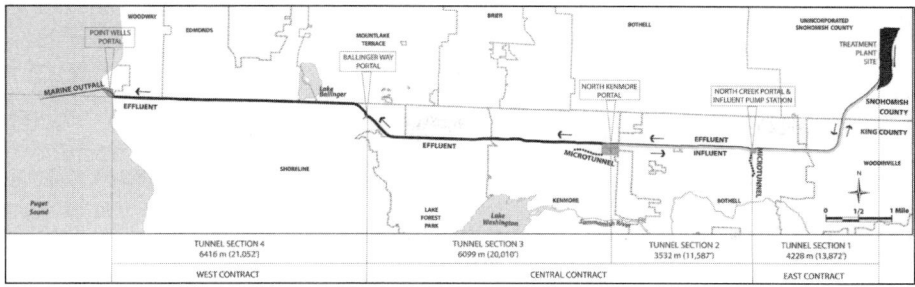

Figure 1. Brightwater conveyance system

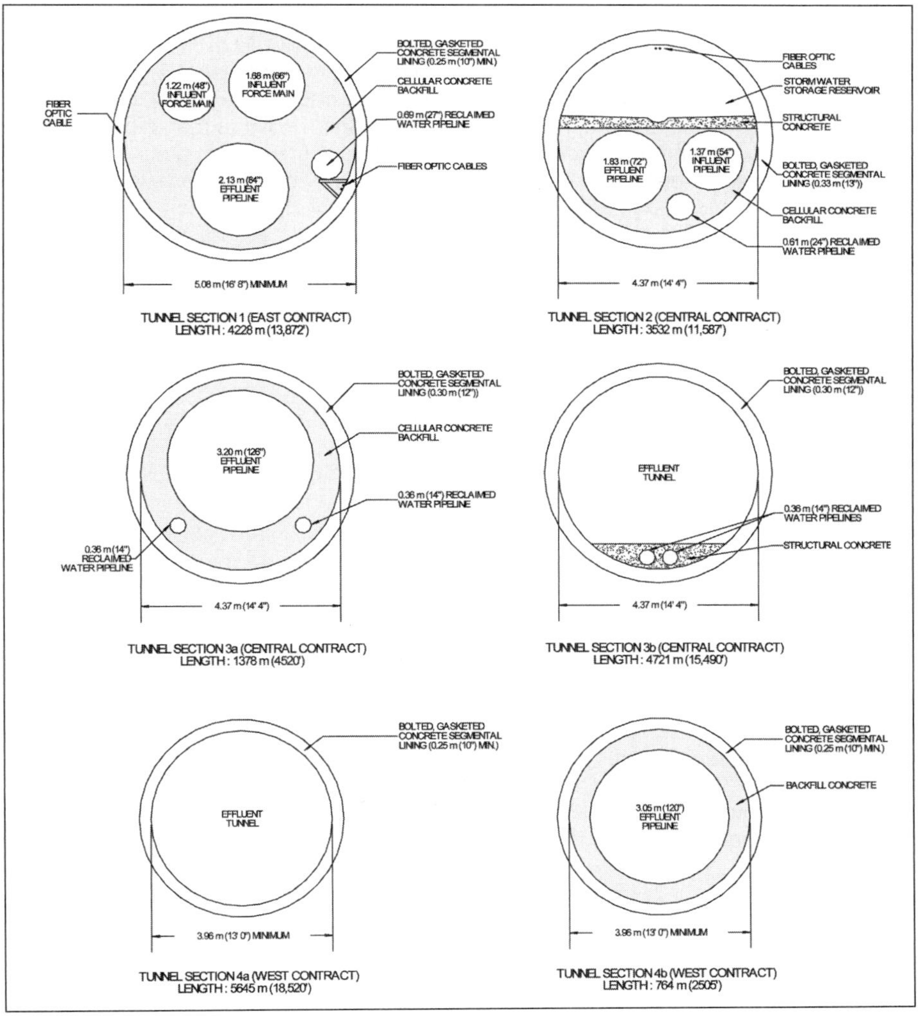

Figure 2. Brightwater conveyance tunnel sections

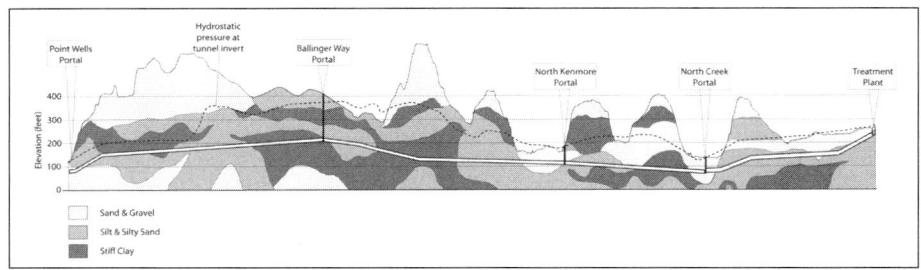

Figure 3. Brightwater conveyance system geologic profile

3. Provide secure Internet-based access to real time data for use by the Project Representative in the Project Representative office.

Given the inevitable Contractor compliance of these provisions the Jacobs Engineering CM (JECM) was to decide on the tool, a TBM Data Manager and Reporter (DMR), to archive, recall, and graphically present the data of the TBM DMS. Number one above required that the project acquire systems and hardware that could at a minimum record 10 second incremental data points, tracking approximately 130 parameters, on three shifts per day, for four separate TBM's; a massive amount of data. In total, the data could populate a database with 1,639,872,000 entries annually. Item two required that JECM obtain a DMR with the capability to accept and subsequently store continuous real-time digital data from the three separate Brightwater Contractors. Finally, the third provision suggested to the JECM to ensure the DMR provide a means to distribute data among the three planned project offices to a central hub server. Chosen by the KCCM to address the three provisions for data acquisition and processing was a software proprietary to Babendererde Engineers GmbH, Germany, called Tunneling Process Control (TPC).

TBM DATA ACQUISITION AND DATABASE MANAGEMENT

DMR Hardware

The recommended hardware implementation strategy for TPC was to set up a separate, off-the-shelf, secure network to host the system. Figure 4 shows a schematic with the key features of the system hardware used for Brightwater. The TPC network captures data from the four separate Contractor maintained TBM DMS systems.

Upon receipt of mining data from the individual TBM's, data is streamed from the two SQL servers and one ftp server to the Main Server via a high speed (T1) internet connection. The Main Server is located at the Construction Management Project Office. Using a Virtual Private Network (VPN) connection, the Main Server can be used to view real time TBM data for any of the four TBMs from a computer anywhere in the world. This is how JECM staff utilize TPC.

Key hardware parameters for the Brightwater TPC system include:

- Servers with 2.0GHzCPUs
- 4GB RAM,
- 100/1000 Base T Ethernet cards,
- wireless networking capability,
- T1 internet connection,
- 5×250GB RAID hard drive configuration.

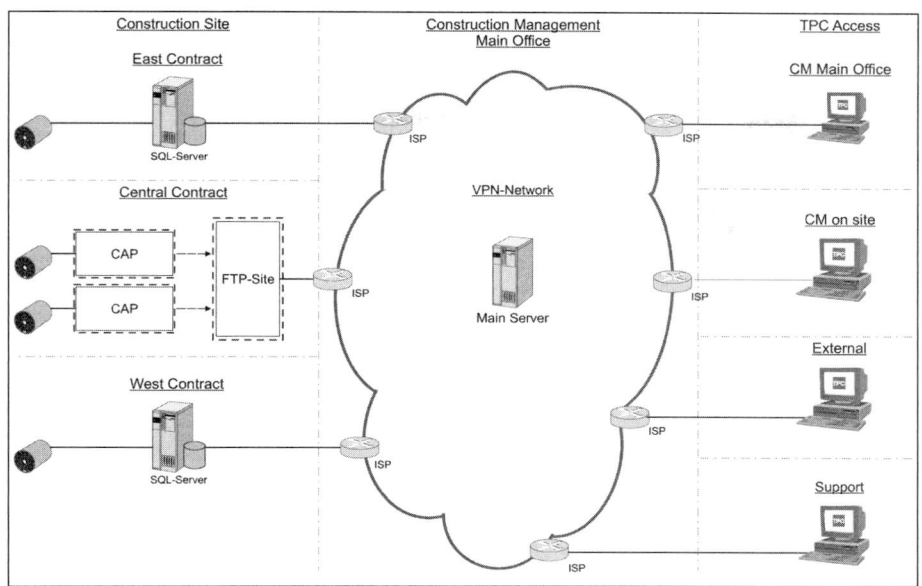

Figure 4. Brightwater DMR hardware

The Central Contract tunnels, BT-2 and BT-3, had a unique situation in that their data access was provided by periodic (20 min) posting of data to an ftp site. TPC was customized for this application to extract data from the ftp site.

DMR Software

TPC has provided the JECM with the ability to log, observe in real time, and review large quantities of multidimensional data, with reasonable processing times and powerful built-in reporting tools. Another key advantage of the system is that it allows data from different TBM types and manufacturers to be collected, managed and analyzed using a single application and interface.

The software eliminates the need for time-consuming data manipulation to review TBM performance. Instead, TPCs customizable querying and reporting tools are used to quickly generate useful and presentable information on an ongoing basis throughout each tunnel drive. These reports can be exported to pdf format, automatically sent by email, or as printed hardcopy through a LAN printer.

The software also allows manually collected data to be added to the digital data retrieved from the TBMs. This data includes, ring damage and repair information, photographs, geotechnical instrumentation data, and classification of TBM down time from field inspectors' observations. All of this data is useful in assessing TBM performance.

OVERSIGHT, CONTEMPORANEOUS REVIEW, AND REPORTING

JECM's goal for the DMR of Oversight, Contemporaneous Review, and Reporting was realized by use of the following features of TPC that allows the simultaneous and continuous monitoring of daily TBM activity on the four Brightwater tunnel drives.

1. Real-time TBM Overview and Mining Parameter Monitoring.
2. Shift Reports

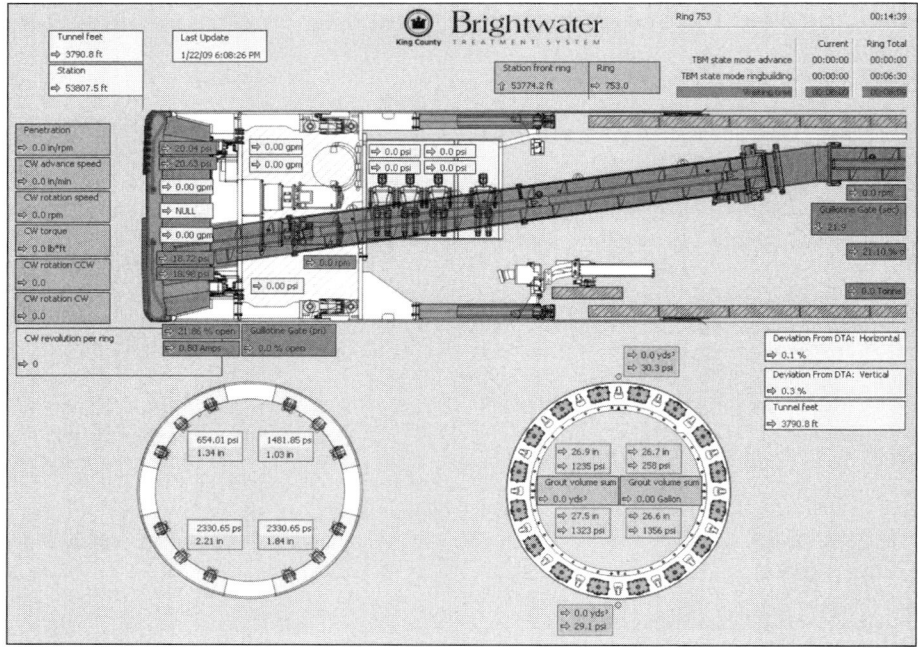

Figure 5. DMR real time viewer—BT-4 TBM

3. Productivity Analysis
4. Ring Reports
5. Multiple Analysis Plotting and Reporting.
6. Project Advance

This DMR's platform provides multiple user accounts offering the above features to all JECM personnel levels, and their specific uses for the project from Tunnel Inspectors to Assistant Resident Engineers to Project Managers. This attribute of the DMR allows for systematic monitoring and reporting activities inherent to a quality assurance program.

Real Time TBM Overview and Mining Parameter Monitoring

TPC gives the JECM User the ability to view, in real-time, at ten second intervals any of the mining parameters that the DMT is cataloging. This User can be the Inspector, the Resident Engineer, the Owner or whoever is enabled as a User, anywhere on the planet with internet access. Figure 5 shows real-time views as the User sees them for the TBM Overview. Views in this feature are refreshed in ten second intervals with the most currently reported data and are customizable to the TBM parameters of specific interest. The real time viewer for Brightwater is configured to show a TBM overview, guidance system, conveying circuit, geology, atmospheric gas monitoring, and User selected parameters.

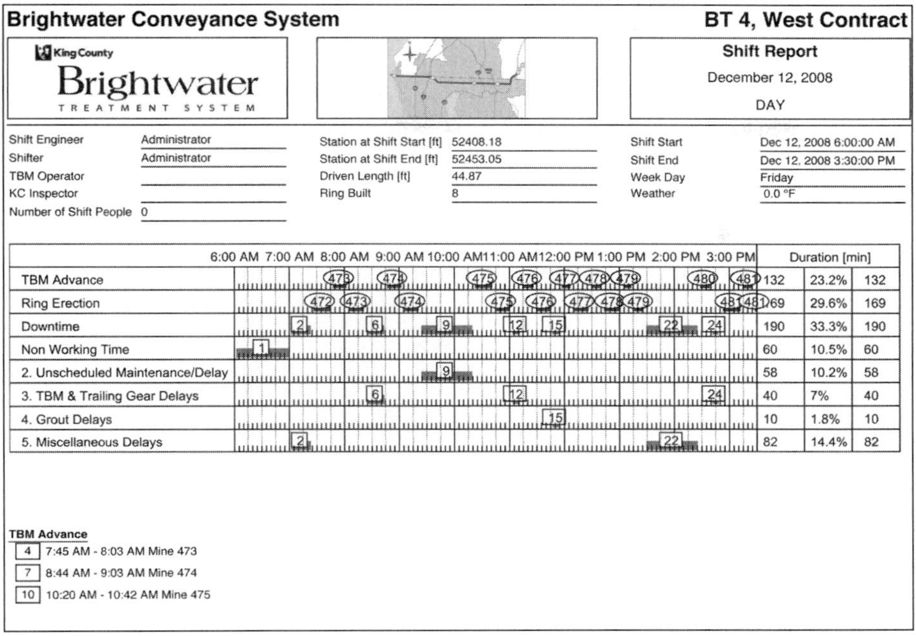

TBM DATA MANAGEMENT AND QUALITY ASSURANCE 1359

Brightwater Conveyance System		BT 4, West Contract	
King County **Brightwater** TREATMENT SYSTEM		Shift Report	
		December 12, 2008	
		DAY	
Shift Engineer	Administrator	Station at Shift Start [ft] 52408.18	Shift Start Dec 12, 2008 6:00:00 AM
Shifter	Administrator	Station at Shift End [ft] 52453.05	Shift End Dec 12, 2008 3:30:00 PM
TBM Operator		Driven Length [ft] 44.87	Week Day Friday
KC Inspector		Ring Built 8	Weather 0.0 °F
Number of Shift People 0			

	6:00 AM – 3:00 PM	Duration [min]		
TBM Advance		132	23.2%	132
Ring Erection		169	29.6%	169
Downtime		190	33.3%	190
Non Working Time		60	10.5%	60
2. Unscheduled Maintenance/Delay		58	10.2%	58
3. TBM & Trailing Gear Delays		40	7%	40
4. Grout Delays		10	1.8%	10
5. Miscellaneous Delays		82	14.4%	82

TBM Advance
- 4 | 7:45 AM - 8:03 AM Mine 473
- 7 | 8:44 AM - 9:03 AM Mine 474
- 10 | 10:20 AM - 10:42 AM Mine 475

Figure 6. DMR shift report

Shift Reporting

The Brightwater Conveyance Project DMR tracks and categorizes TBM activity as Advancing, Ring Building, Nonworking Time, or Downtime over a mining operation shift which was defined by the JECM TPC User. The TPC User aided by an Inspector's daily hand written shift reports would further differentiate Downtime into subcategories pertaining to tunnel mining activity. This step became necessary to discern between Contractor error and tunneling logistics. These subcategories and the associated periods of time become part of the DMRs query-able database. Significant to the separate Brightwater contracts, these descriptions of subcategories were further tailored by the separate contracts as their machines, EPB and Slurry, have different associated causes for delay. Figure 6 shows resultant shift report showing time utilization as percentage of the shift's period. This activity was instituted as JECM daily procedure and has provided a database that is useful when evaluating Contractor mining operation efficiency.

Productivity Analysis

Directly resulting from the Shift Reporting procedure above was the DMRs capability to produce an analysis of Contractor productivity, the Productivity Analysis. Previously mentioned the shift reporting procedure provided a database of mining activity representing shift efficiency based on the categorization of Downtime. At intervals periodic and specific to the separate tunneling contracts, the JECM would produce analysis of the TBM productivity extracted from the query-able database of the DMR. TPC allows the filtering of the data fields for this feature over durations of time specified by the User. Figure 7 and demonstrate an automatically generated report of such

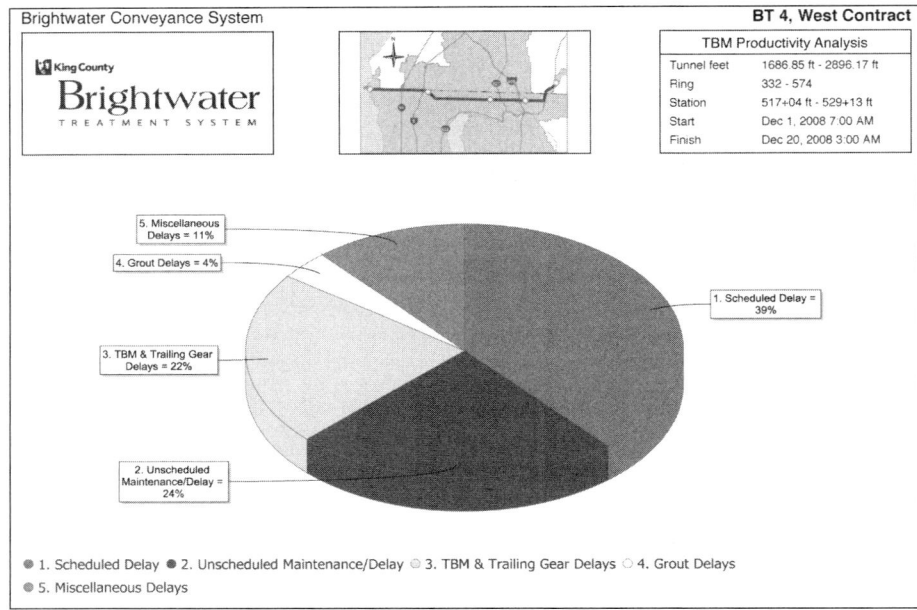

Figure 7. DMR TBM productivity analysis

an analysis over a period defined by the User. Of note in Figure 11 is the breakdown of mining activity downtime for the BT-4 TBM.

Ring Reports

Customized by the JECM to show the selected tunneling parameter summaries for each of the separate mining contracts were the Ring Mining Summaries (Ring Reports). Figure 8 represents a typical Ring Report that is generated at any time for any ring of the tunnel alignment based on the DMR User's search criteria. These graphic representations of a traditional reporting of TBM mining parameters were immediately retrievable for all rings at any point in the project.

Multiple Analysis and Reporting (Special Reports)

The complexities for the TPC system regarding its computational representation of data is too broad for the scope of this paper, however, of note is its ability to conduct compound queries of all compiled mining data over a range defined by time, ring, or station. The results of such queries are the Special Reports which allow the evaluation of essentially any TBM performance or compliance specification. This element of TPC's reporting gives the JECM analysis and graphic representation of multiple mining parameters over a common range. Figure 9 shows such an analysis for tail shield grouting.

Project Advance

At perhaps the highest level of reporting capabilities of this particular TBM DMR was the Project Advance report. Figure 10 exhibits a screen capture of such a report and its demonstration of the plotted comparisons of actual project TBM advance versus the Contractor's planned advance. The feature was primarily used to demonstrate TBM

Brightwater Conveyance System

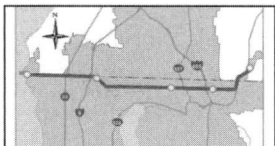

BT 4, West Contract

Ring Mining Summary
No. 725

Station	53672.42
Tunnel feet	3655.72
Advance time	01/16/2009 13:38
	01/16/2009 13:53

1. Cutting wheel (Maximum Values)

RPM	2.67 rpm		CW Rotation	94.2% 5.8%
Thrust	375.58 Ton		Advance Rate	7.0 in/min
Total Advance Pressure	9485.94 psi		Penetration	2.7 in/rpm

2. Face Pressure

	BL	BR	TL	TR			
AVERAGE	32.0	31.0	20.4	27.4 psi		Advance time	7 min/ring
MAXIMUM	49.3	44.1	36.6	41.2 psi		Ring erection time	1 min/ring
MINIMUM	20.2	20.7	14.8	19.3 psi		Waiting time	5 min/ring

3. Jacking Units

Group	max. Advance Pressure	Extension [in]	Extension Start [in]	Extension Difference [in]
A	2,257	87	27	59
B	2,480	86	27	59
C	2,463	87	27	59
D	2,286	87	0	87

4. Grout Injection (Ring n-1)

	1	2	V_{Act}	V_{Target}
V [yd³]	0.88	1.09	1.98	
P_{avg} [psi]	32.62	26.01		
P_{max} [psi]	60.20	50.28		

5. TBM Guidance System

Roll in/ft	0	Deviation From DTA, Horizontal	0.4	%
Pitch in/ft	3	Deviation From DTA, Vertical	0.7	%

6. Installed Ring Data

Station at Ring	53636.4 ft
Station at Ring n-1	53631.4 ft

7. Belt Conveyor

Scale 1	
Scale 2	41.2
Excavated	
Calc. WT.	
Bulk F	

8. Air Monitoring

Gas	Max	Min	Avg
CH4 (%LEL)	0.0	0.0	0.0

9. Soil Conditioning

	Total Gallons
Soap	-24.87
Polymer	11.05

Figure 8. DMR ring report

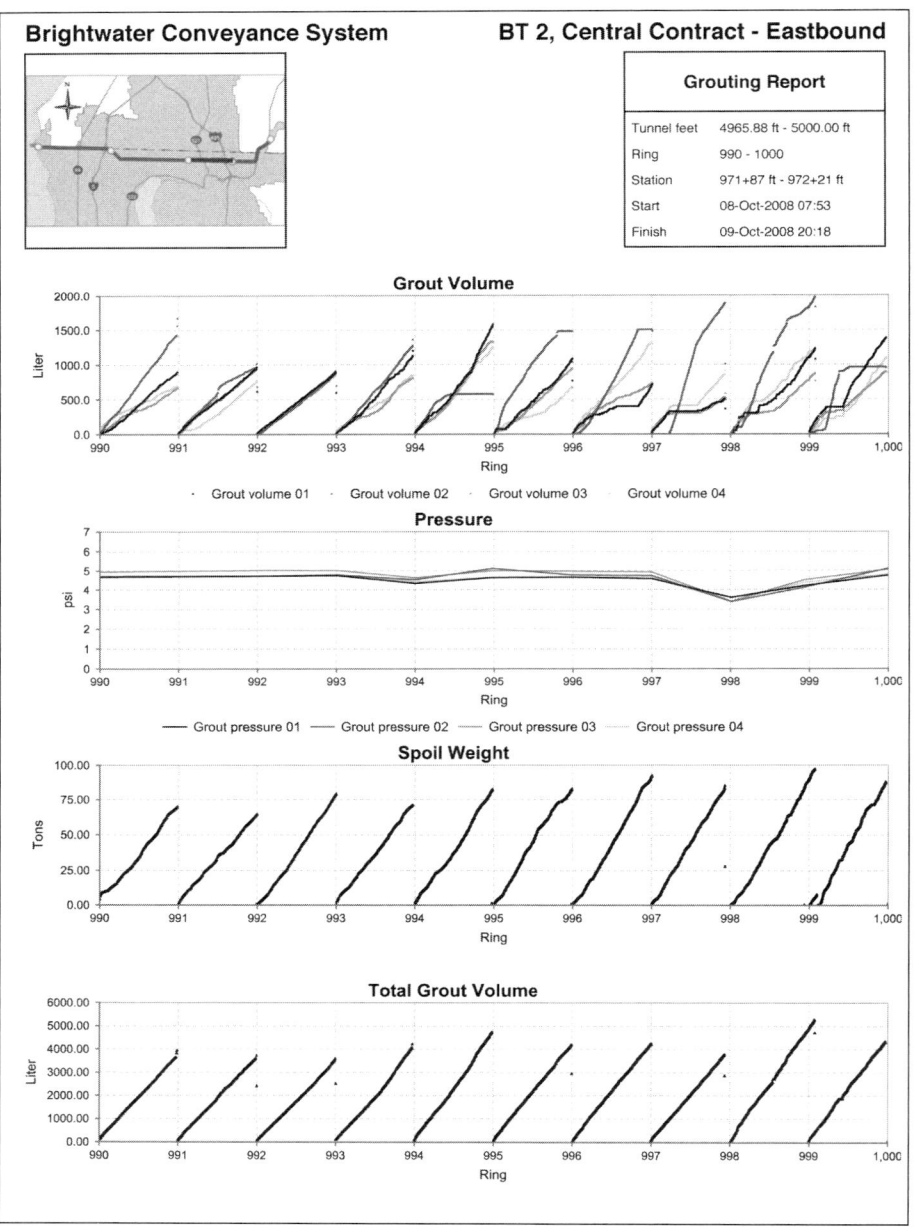

Figure 9. Special reports—grouting

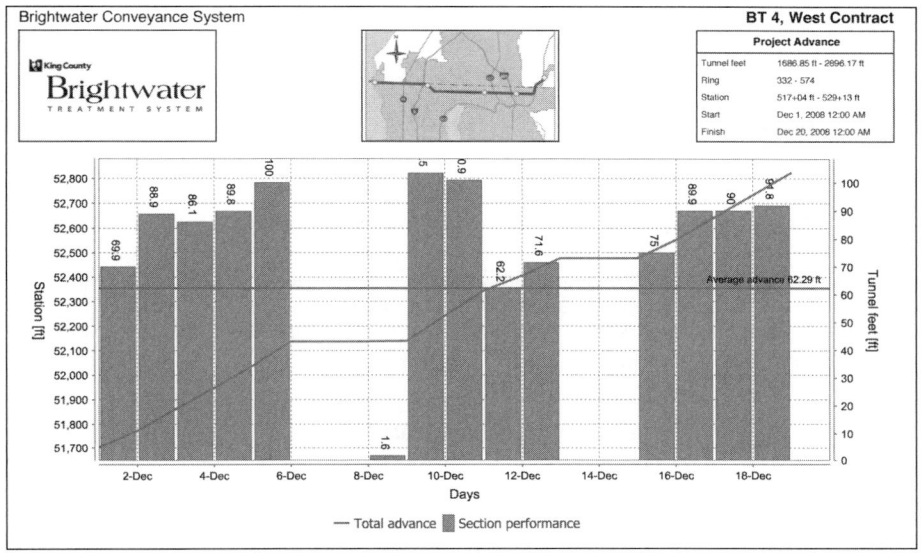

Figure 10. DMR Project Advance Report

progress in a summary level to Owner oversight audiences. Automatically generated from database query of TBM progress and the Contractor's CPM over a User selected range of time, the resulting graphic would display a succinct summary of project timeliness that was immune to debate.

AD HOC IMPLEMENTATION

During the course of the TBM drives opportunities have presented themselves for TBM activity evaluation that were not a routine part of the CM effort but rather an as-needed occurrence. Although some cases of such Ad-Hoc implementation of the Brightwater DMR are still contractually sensitive and can not yet be discussed at great length, several examples can be relayed.

Differing Site Condition

An obvious example of such a use is to evaluate a Contractor claim of Differing Site Condition (DSC) affecting TBM mining activities. Few tools for such a task will be as valuable to an Owner or Construction Manager as a query-able database of all pertinent TBM mining parameters recorded since the inception of a Contract. Ability to graphically present data that has been filtered to distill specific TBM mining parameters over a finite period of a TBM drive is valuable to Owners and the CM's.

Public Outreach/Wow Factor

During the Brightwater Conveyance Project the JECM frequently had the task of demonstrating the TPC DMR to Owner Administrators or as a public outreach. When demonstrating the real-time features of TPC, invariably the response was one tinged with a 'wow' factor. Graphically viewing the streaming data of a TBM mining activity that is ten miles away under two hundred feet of glacial till was always impressive to visitors. A project favorite anecdotal example of public outreach occurred as the JECM was trying to validate the cause of a noise complaint of a Homeowner living above the tunnel alignment. With one JECM team member viewing TPC's Real-time viewer for an instantaneous indication of TBM activity and another JECM team member standing in the subject Homeowner's kitchen on the telephone, the team was indeed able to verify that the source of the noise complaint was the TBM.

TBM Operation Assessment

Used to a lesser extent, as cases for application on the project have thankfully been limited, are evaluations of suspect TBM Operator procedure. In the event that problems develop with the TBM, at the alignment surface, within the tunnel, etc, the DMR provides a means to thoroughly evaluate mining activity conditions and TBM Operator response. Given the dynamics of the DMR such evaluation is capable moments or months after the occurrence.

CONCLUSION

Given the current state of TBM sophistication and the inevitable increase in the application of technology to this industry's method of construction, the experiences of the Brightwater Conveyance Project's DMR can provide assistance to those planning a real-time data base management system for TBM mining parameters.

First, there is value to the Owner or Construction Manager that implements a DMR system that can evaluate data on a contemporaneous basis as opposed to a stored (retrospective) system such as one that would be periodically received in batch data from a Contractor. Primarily, value is achieved by the review of data that is current and part of a project's current state. Data does not have to be retrospectively dissected and reconstructed to show trends and events. The value to the Owner is in quality TBM data management that is efficient, that is timely, and conserves project resources.

Owners should also insist on clear succinct language in the Contract Documents that define the Contractor's requirements for interface with a project's planned DMR. This statement is painfully self explanatory and can basically be applied as a goal of all specification writing. A loophole in this project's specification did allow a Contractor to provide full TBM data access and viewing only via their own proprietary DMS.

Finally, instances in multi-contract TBM construction, such as that with Brightwater, involve individuals and CM practices of wide experience and management technique. With these conditions comes skepticism for the use of a new paradigm such as a project-wide TBM DMR like TPC. Paramount to the DMRs successful implementation is to instill the support of the project stakeholders by outlining the goals, procedures for use, and reporting well ahead of a system's implementation.

CONSTRUCTION OF THE EAST SIDE ACCESS MANHATTAN TUNNELS

Don Hickey ▪ Judlau

The MTA in conjunction with LIRR, are in the process of expanding one of the largest commuter railroads in the country, with over 260,000 passengers a day. The LIRR, which provides 700 passenger trains every 24 hours, can't continue on the present growth course as a joint tenant in Penn Station with Amtrak and NJT. For this reason, in conjunction with the fact that over 53% of LIRR's current riders into Penn Station are ultimately bound for the East Side of Manhattan opening, have created the need for the expansion of the LIRR. The complete expansion will cost approximately 6.3 billion dollars and all of the contracts are to be completed by 2013.

HISTORY

Work on the expansion has been constructed as far back as the 1970s with the East River sunken tubes. Construction in the 1980s consisted of short TBM runs and shaft construction, the bellmouth open cut and rock excavation in the 1990s and early 2000s. The existing work completed has 4 tubes, 2 top and 2 bottom. The top 2 are in use, used for the NYCTA and the bottom 2 tubes would be used for the LIRR. The lower LIRR tunnels were left dead ended at 63rd and 2nd Avenue in Manhattan's East side. The upper tubes were connected into the existing NYCT system.

PROJECT DESCRIPTION

The complete excavation of 25,100 ft of 22' diameter hard rock tunnels with two TBM's. The excavation is completed in 4 tunnels, 2 each top tunnels which are 7,200 ft each and 2 each lower Tunnels which are 5,200 ft each. The TBM excavation begins 1.5 miles from the access shaft located in Long Island City, Queens, traveling through the lower tunnels from Queens to Manhattan. We travel through open cut construction, drill and blast construction, TBM tunnels and sunken tubes through the East River. All equipment and personell have to enter in Queens and travel 1.5 miles to the start of excavation. At the farest distance, it will be 3 miles from the Queens shaft. In 2008, the JV was awarded contract CM019, ESA Cavern Excavation which includes 4 each 1600 ft tunnel drives, 2 on top and 2 on bottom. This changed the sequence of our construction so we now have 4 TBM drives on top and 4 TBM drives on bottom. After the TBM's mined the 1st upper tunnels the TBM's are backed up to 1500 ft to enlarged Wye Caverns and relaunched for the 2nd upper drive. The Wye Caverns are 60' wide by 30 ft high and 400 ft long. The project also includes 2 sets of caverns GCT5 and GCT3, several cross passages, sump chamber, a central instrumentation room and a cross flue. When the 4 upper drives are completed, the TBM will be backed up to GCT5 Wye Cavern and relaunched for lower tunnel drives. After excavation, roughly 50% of the TBM tunnels and all Wye caverns, drill and blast excavations are water proofed and concrete lined. Project Quantities are 352,000 Bcy of TBM excavation, 60,000 Bcy of drill and blast excavation, 50,000 cy of CIP lining. On July 6, 2006, the

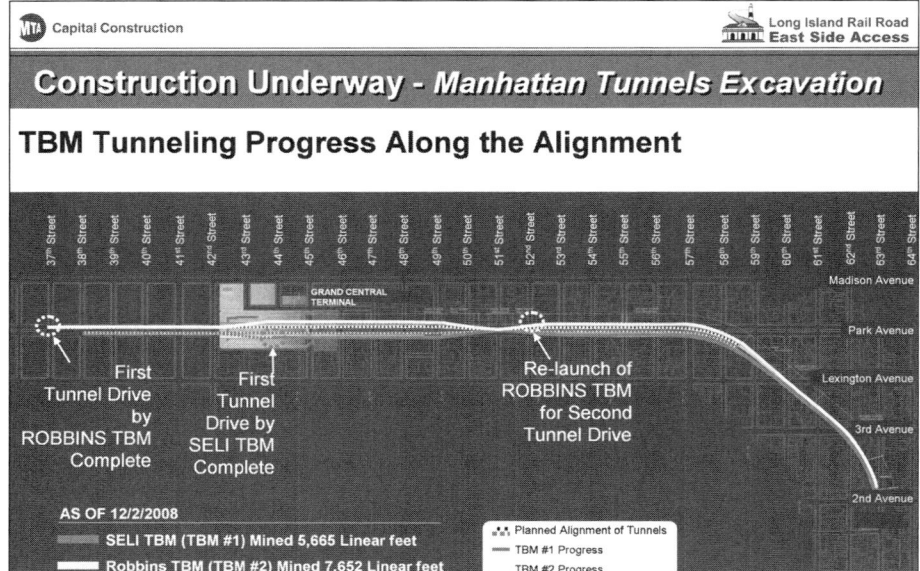

Figure 1. Tunnel alignment over Manhattan map with excavation progress as of 12/02/08 (Courtesy of MTACC)

JV of Dragados USA and Judlau Contracting was awarded CM009 "Construction of the Manhattan Tunnels for $427,000,000.00. Dragados USA is a subsidery of Dragados, Madrid and parent company ACS. Judlau Contracting is NYC local heavy contractor.

The project owner is the LIRR, the project is managed by MTA Capital Construction, the project was designed by General Engineering Consultants comprised of PB, STV and Parson Transportation.

PROJECT GEOLOGY & ALIGNMENTS

Tunnel alignment are being built under existing operating NYCTA subway lines, Metro North Railroad lines and Grand Central Station. Our drill and blast assembly chamber was only 8 ft below an active NYCTA subway line. Grades of new construction range from 1% to 3% up hill and down hill. The prominent geology for this project is metamorphic rock of schist & gneiss in the Hartland Formation.

The average depth below street grade is 110 ft. The upper tunnels drives are about 20 ft higher than the lower tunnel drives.

TUNNEL BORING MACHINES

After the project award, the JV procured 2 each TBM that were required by contract. The JV chose a SELI double shielded TBM with complete back up for EB tunnel and Robbins HP Main Beam TBM for the WB tunnels.

The SELI TBM is a complete double shielded machine without a segment erector. The TBM was designed to be assembled in a small starter tunnel and disassembled in a mined tunnel. The cutterhead is comprised of 1 center section and 4 outer segments that use a bolted connection. The shields are made in section with the top of the forward shield and gripper shield made up of 3 pieces one of which a stripping key. The

Figure 2. SELI TBM being assembled in starter tunnel

Figure 3. Robbins TBM being assembled at the Queens shaft

TBM is 22' diameter powered by 7 each 477 HP VFD which rotate the cutterhead at a maximum 8.0 RPM. The cutterhead has 45 each 19 inch cutters on the back loaded cutterhead. The back up has 10 decks that account for all hydraulic and electric cabinets. Deck 7 has the mobile tail piece for the continuous conveyor.

The Robbins HP Main beam TBM was the second TBM obtained by the JV, TBM 244-313 was designed for assembly and disassembly in a 22' diameter tunnel. The cutterhead assembly is made up of 1 inner section and 4 outer sections. The cutterhead has 46 each 19 inch back loaded cutters. It is powered by 7 each 440 HP electric VFD motors. The 675 ton TBM is capable of providing 3,000,000 lbs of thrust. The cutterhead speed is variable up to 8.4 RPM. The trailing gear has 10 decks which hold the mechanical and electrical power systems as well as the mobile tail piece of the continuous conveyor system.

Both TBM's were outfitted with 2 roof bolters that had the ability to probe as required by contract. The GBR and contract stated anticipated shear zones that would require steel ring so ring beam erectors were provided on both TBM's. Both TBM's also had survey systems provided by PPS.

Figure 4. View shows York Avenue existing tunnel. We have 2 belt storage units for EB and WB TBM tunnels and transfer conveyor to feed fixed tunnel conveyor.

Both TBM's have back loaded cutterheads which is new on TBM's in NYC. It's a safer option but cutter change time is longer because of tighter quarters for cutter crews.

Since the TBM's need to be assembled and disassembled 4 times each, it was important to get a TBM that could be disassembled and moved back through the excavated tunnel for the second launch. Both TBM's have bolted cutterhead for quick disassembled. The SELI TBM shields have a top stripping key that allows us to pull the wedge out that will free up removing the other components of the front telescopic and gripper shields.

The Robbins TBM front shield was designed with 6 inch filler plates that made outer diameter of 6.7 meters while mining. When the TBM run is complete and needs to be retracted, the filler plates are removed hydraulically which reduces the diameter to 6.1 M to walk back through the ring steel. The collapsed TBM will travel on rail gripping dollies. The plates are reinstalled with the quad sections of the cutterhead and the TBM is relaunched.

Both TBM's are supplied with data logging PLC with remote access and displays. You can view mining activities as long as you have internet access and the password.

CONVEYOR SYSTEM

The JV chose a continuous conveyor for muck removal. Both TBM's will have these own individual 36 inch tunnel conveyor system which will travel from the mobile tail piece back to York Avenue where they will join and transfer muck to a fixed length belt in the existing tunnel. The WB tunnel belt will cross over via conveyor to feed the fixed tunnel conveyor that run 7,000 ft to the access shaft in queens. The vertical bucket belt takes the muck up 110 VF to transfer belts over a busy street called Northern Blvd. to a stacker belt 600 ft away. The system driven by VFD electric motors was provided by Robbins Conveyor. The design capacity of the TBM belt is 800 tons per hour. The fixed belt and surface belts are rated for 1,200 tons per hours. The system is completely PLC controlled and interlocked together so they are tied into each TBM. The total belt system is 12,000 ft long at the end of the longest drive.

After the muck gets to the surface it travels across a very busy street called Northern Blvd. in an enclosed box to a surface transfer conveyor to a radical stacker

Figure 5. View showing started tunnel for SELI with gripper wall, poured invert and monorail hoist for assembly

conveyor located in the Sunnyside rail yards. The muck is stockpiled than off loaded via trucks. The material is recycled backfilling numerous sites in NYC.

TBM INSTALLATION

The assembly chambers were built along side each other with a 6 ft rock pillar between both tunnels. The assembly chambers were excavated as small as possible. Blasting was very restricted due to the NYCTA tunnels above us. The WB side had 8 ft of clearance between neat line of excavation and the outside limit of the live subway above us. Each time we blasted, NYCTA trains were stopped while the shot was detonated, and then released after a quick track inspection.

The assembly chambers were roughly 24 ft headings while the connected existing tunnel were 18'-10". We excavated only enough to launch the TBM while the trailing gear was extended into the existing tunnel. We had built the trailing gear or the surface but couldn't install any equipment on the top decks such as the scrubber and fans. We would have to mine in a few hundred feet then stop to reconfigure the trailing gear because of head clearance.

After the drill and blast was completed for assembly chambers, the chambers were prepared with concrete invert and gripper walls. The SELI chamber required the installation of 3 heavy duty monorail beams capable of lifting 25 tons. The segments were lowered in pieces and transported 1 ½ miles to the assembly chamber where the assembly was being performed. The trailing gear was pre built on surface with equipment and lowered into the tunnel and transported into the assembly chambers.

The existing tunnel diameter was 18'-10" so all trailing gear had to be able to fit in the existing tunnel. After we got launched and the trailing gear was inside the assembly chamber we used the same monorail system to raise the scrubber and fan set up on to the top decks of the trailing gear.

The Robbins TBM was built quite different. It was built at the bottom of the shaft with the use of a crane. The only pieces that were transported to the heading and preassembled were the quad sections of the outer cutterhead. The front shield was assembled in the heading so the TBM could drive through it and grab the shield. The gripper shoes were prehung on the gripper walls.

Figure 6. View shows Robbins Assembly Chamber with front shield built and gripper shoe hung waiting on TBM to arrive. TBM will drive through front shield.

Figure 7. View showing the building of the Robbins TBM at the Queens shaft included

The TBM was assembled at the shaft, the inner cutterhead, the cutterhead support with all drive units and the two section main beam. All the auxiliary equipment was also built, rock drills, ring beam erector and all hydraulic hoses and belts were installed at the shaft. When the TBM was completed the bridge conveyor and the trailing gear was attached. The TBM was moved up into the tunnel until the trailing gear was ready to travel into the assembly/launch chamber.

The Robbins rail gripped itself 7,700 ft in 5 days up and down 3% grades with no problems. When it got to the assembly chamber, the TBM walked itself into the front shield comprised of roof support, side supports and vertical shoe and attached. At the same time the gripper shoe were attached and the final connections were made and the TBM was launched.

Each TBM start up had a 3% grade and in a curve, both very challenging starts to get the TBM's collared in while keeping line and grade.

Figure 8. View showing TBM tunnel with ring steel ground support

Figure 9. View Showing TBM mined tunnel

GROUND SUPPORT FOR THE PROJECT

The contract has 3 rock support types, type I which is 4 each 10 ft rock bolts, type II which is channels and 4 each 10 ft rock bolts and type III which is rolled steel sets. Ground supports for the Wye Caverns is 16 ft rock bolts and 4" of initial shotcrete lining.

The project documents indicated shear zones that would require type III ground support as well as when the tunnels crossed under live subway facilities. There are 2 subway systems that cross out tunnel excavation at 53rd Street in Manhattan the crossing is 17–20' above our excavation. At 42nd Street, there is the number 7 Line crossing which is 20' of clearance above our excavation as examples of the many crossings.

CURRENT STATUS

The SELI TBM has mined 5665 feet of tunnel than was stopped and walked back to GCT3 Wye cavern for the next tunnel run of the top tunnels. The SELI TBM is on hold until GCT3 Wye cavern is excavated and a starter tunnel is completed for relaunching.

Figure 10. View showing GCT 3 Wye Cavern excavation

The Robbins TBM mined 7,651 feet of the top tunnel and was walked back to GCT3 WB Wye cavern. Rather than excavating the cavern for relaunch the JV placed a concrete plug and launched the TBM for a 1,500 ft run.

The JV chose to excavate using a Sandvik MT 720 Road header, this is a 135 ton road header. The GCT3 Wye cavern is 31 ft high × 61 ft wide × 320 ft long. We are widening the 22ft TBM tunnel to a cavern. After the cavern is excavated and a starter tunnel is build, the SELI TBM will relaunch for a 3150 ft run to a dead end at 38th Street in Manhattan. The project is currently 45% complete as of December 2008. The Robbins TBM was relaunched for it's second top drive.

INDEX

A
ADECO (Analysis of Controlled Deformations) method, 952–955, 966–967
   advantages over SEM and NATM, 965–968
   case histories, 955–964
   and history of SEM and NATM, 942–952
AECOM, 589
Alafia River Tunnel (Tampa, FL) case study, 228
Alaska Electric Light & Power Company, 12–14
   Lake Dorothy lake tap and tunnel, 12–21
AlpTransit Gotthard Ltd., 471
Anacostia River Projects (Washington, DC), 242–244
   project-specific soil grouping system, 244–249
Andes Mountains
   historical use of TBMs, 1262–1263
   technical considerations for TBM use in, 1261–1274
Apex Tunnel (Nevada), 534
   geologic investigations, 534–547
Arup, 488
ASFINAG
   and Austrian tunnel construction and management, 308–309
   visual and geometric documentation of Tauern Tunnel construction by tunnel scanner, 308–319
Ashlu Creek Hydro-electric project (Squamish, BC), 1208–1210
   TBM tunneling for power tunnel, 1208–1217
Atlanta, Georgia
   North-South Tunnel feasibility study, 757–765
ATS Drilling, L.P., 998–1004
Auckland, New Zealand
   Hobson sewer storage tunnel, 450–456, 458–462
   Rosedale Outfall project, 456–458, 460, 461, 462
Austin, Texas: Waller Creek Tunnel, 87–95
Australia
   EPBM tunneling for New MetroRail City Project (Perth), 1124–1136
   SEM in construction of bus tunnel under road, park, and historic buildings (Brisbane), 906–915
Austria
   NATM in construction of Strenger Tunnel, 98
   visual and geometric documentation of Tauern Tunnel construction by tunnel scanner, 308–319

B
Ballard Siphon Replacement Project (King County, WA), 581
   consideration of microtunneling vs. EPB tunneling, 581–588
BASF Construction Chemicals
   research on soil conditioning for clogging soils in EPBM tunneling, 320–327, 332
   research on soil conditioning for coarse-grained soils with low fines in EPBM tunneling, 327–333
Bauer Machinen GmbH, 370, 374
Bay Area Rapid Transit (California), 742
   Fremont Central Park Subway Project, 742–756
Bay Tunnel Project (California), 60–61, 70–71
   EPBM in construction of, 65–66
   geologic conditions, 61–64
   seismic analysis and design, 66–70
   tunneling conditions, 64
Beacon Hill Station (Seattle, WA), 98
Beles Tailrace Tunnel (Ethiopia), 1235
   rapid excavation by shield tunneling, 1235–1247
Bergen County (New Jersey) Utilities Authority, 558
   1.2-mile CSO microtunneling component underlying highways and railyard, 558–569
Bi-County Water Tunnel (Montgomery County, MD), risk management, 858–859
Big Walnut Outfall Augmentation Sewer (Columbus, OH), 1114
   solutions to TBM sealing system problems, 1114–1123
   steel fiber reinforced precast segmental liner, 1114–1117
Black & Veatch, 854–855, 998–1004
Black Panther Mine No. 1 (Oaktown, IN), 974
   DHI-240 rig in drilling of return shaft, 974
Boggo Road Busway (Brisbane, Australia), 906–907
   SEM in construction of tunnel under road, park, and historic buildings, 906–915

Bologne-Florence High Speed Railway (Italy), 955
application of ADECO, 955–959
Boston, Massachusetts
design, pre-support, and modeling for Silver Line Phase III, 728–741
EPBM in construction of North Dorchester Bay CSO Storage Tunnel, 1062–1073
planning and tunneling method selection for Silver Line Phase III, 713–725
research on soil conditioning for clogging soils in EPBM tunneling (North Dorchester Bay Tunnel), 320–323, 325
Bradshaw Construction Corp., 894
Bratislava, Slovakia
NATM in excavation of motorway tunnel through granitic rock, with numerous faults and thin overburden, and underlying academic complex and zoo, 916–926
Sitina Tunnel, 916
Brazil: Sao Paulo Metro Line, 1137–1153
Brierley Associates, 488
Brightwater Conveyance System (Seattle, WA), 250–251, 297, 1154, 1354
East Tunnel case history, 1171–1181
ground freezing for access shaft excavation in soft ground, 297–307
preparations for EPBM tunneling (BT4), 1154–1170
probabilistic geotechnical baseline assessment compared with actual tunnel face conditions, 251–262
TPC software for TBM data management and quality assurance, 1354–1364
Brisbane, Australia
Boggo Road Busway, 906–907
SEM in construction of tunnel under road, park, and historic buildings (Boggo Road Busway), 906–915
British Columbia: Ashlu Creek Hydro-electric project (Squamish), 1208–1217

C

Caldecott Tunnels fourth bore (Oakland, California), 96–97
NATM design features, 100–107
California
assessment of slurry diaphragm walls for deep rectangular ventilation shafts (Silicon Valley Rapid Transit), 1005–1019
Bay Tunnel Project, 60–71
blind auger drilling in construction of access shafts for Upper Northwest Interceptor (Sacramento), 1037–1050
Caldecott Tunnels fourth bore (Oakland), 96–107
Caltrain Downtown Extension project (San Francisco), 766–777
design and construction challenges for dam outlet tunnel and shaft in mixed ground of hard rock blocks and soft clayey matrix (Los Gatos), 171–184
design/build/maintain rehabilitation project (Iron Mountain Mine), 616–628
Fremont Central Park Subway Project, 742–756
NATM in construction of large shallow cavern in weak sandstone (SLAC), 927–941
planning for Joint Water Pollution Control Plant (Los Angeles County), 700–712
research on soil conditioning for clogging and low-fines soils in EPBM tunneling (Silicon Valley Rapid Transit), 320–333
risk management for EPBM tunneling adjacent to reservoir (Sacramento), 1074–1082
San Vicente Tunnel Project (San Diego County), 195–214
Caltrain Downtown Extension (DTX) project (San Francisco, CA), 766
contract packaging strategy, 766–777
Camp Dresser & McKee, 272
Carbonate sedimentary rocks, 215, 228–230
Alafia River Tunnel case study (Florida), 228
identifying karstic features, 218–220
karst, 216
karst development in south Florida, 216–217
karstic terrane, 216
laboratory testing, 222–225
remote detection of karstic features, 220–222
tunneling considerations, 225–227
Cassia 1 Tunnel (Italy), and application of ADECO, 964–965
Caverns
design and planning of rock caverns for underground water reservoirs (Hong Kong), 438–447, 449
end-user requirements and rock and site characterization for physics research facility in large rock cavern(s) (South Dakota), 638–645
large rock caverns for transit hubs (New York City), 35–47
Cement deep soil mixing (CDSM) method, 745–747
Center for Integrated Facility Engineering (CIFE), 385–386
Central Subway Project (San Francisco, CA), 871–872
risk assessment and analysis in decision making, 872–880
Charleston (South Carolina) Water System, 185–186
response to sand lens in Cooper Marl while constructing shaft for wastewater tunnel, 185, 186–194

INDEX

Charlotte, North Carolina
   SEM with water resistant additions in construction of Wachovia–Knight tunnels, 894–905
   Wachovia–Knight Theater pedestrian tunnels, 894
Chicago, Illinois
   design and planning of Thorn Creek Tunnel, 123–131
   Thornton Composite Reservoir, 123–124
China
   feasibility study of EPBM for construction of transit tunnel passing under airport runway (Shanghai), 463–470
   hard rock tunneling in deep mountain setting with high water inflow potential (Jinping-II), 1308–1320
City Water Tunnel No. 3 (New York City)
   long distance pumping of concrete for lining (Stage 2), 396–413
   Stage 2, 396
City-Tunnel (Leipzig, Germany), 341–343
Clark, Lincoln, and White Pine Counties Groundwater Development Project (Nevada), 534. See also Apex Tunnel
Coastal Projects Pvt. Ltd., 364
Columbus, Ohio
   solutions to TBM sealing system problems (Big Walnut Outfall Augmentation Sewer), 1114–1123
   steel fiber reinforced precast segmental liner (Big Walnut Outfall Augmentation Sewer), 1114–1117
Connecticut: Hampstead Avenue Interceptor Extension (Hartford), 589–601
Construction manager at risk contracts, 508–509
Contracts
   Caltrain DTX contract packaging strategy, 766–777
   construction manager at risk, 508–509
   design-bid-build, 507–508
   design-build (Lake Mead Intake No. 3), 503–515
   Mitholz Project (Switzerland), 4–5
   packaging for Second Avenue Subway (New York City), 790–801
   See also Design and planning
Conveyor systems
   Hallandså Railway Tunnel Project (Sweden), 428, 433–434
   long (19.35 km) extensible system (Andhra Pradesh, India), 363–369
Cut and cover tunnels
   planned for Caltrain Downtown Extension project (San Francisco), 767–768, 772–773
   for planned Silver Line Phase III (Boston, MA), 713–725

Cutter-soil mixing (CSM), 373–375
   in construction of microtunneling shafts for Mokelumne River Crossing (California), 370–382

D

D.C. Water and Sewer Authority, 242
Daniel Island Extension Tunnel (Charleston, SC), 185
   response to sand lens in Cooper Marl while constructing shaft for wastewater tunnel, 185–194
Deep Underground Science and Engineering Laboratory (South Dakota), 638
   end-user requirements and rock and site characterization for large rock cavern facility, 638–645
Design and planning
   Apex Tunnel geologic investigations (Nevada), 534–547
   assessment of slurry diaphragm walls for deep rectangular ventilation shafts (Silicon Valley Rapid Transit), 1005–1019
   Bay Tunnel Project (California), 60–71
   Caldecott Tunnels fourth bore (Oakland, California), 96–107
   caverns for underground water reservoirs (Hong Kong), 438–447, 449
   combined service/production shaft with vertical conveyor and tower-mounted hoist (Pine Point Mine), 975–985
   cost and benefit of second subsurface investigation for Reach 4 Tunnel Project (Nevada), 548–555
   determining feasible construction options for Reach 3 Tunnel Project (Nevada), 525–533
   interactions of Eastside Access Project with Amtrak and Long Island Rail Road regarding work under Queens rail facilities (New York City), 814–823
   Joint Water Pollution Control Plant (Los Angeles County, CA), 700–712
   Kensico-City Tunnel (New York), 108–122
   Lake Mead Intake No. 3 shafts and tunnel (Nevada), 488–502
   Port of Miami Tunnel (Miami, FL), 687–699
   preparations for EPBM tunneling (Brightwater Conveyance System, BT4), 1154–1170
   rock mass characterization of weak sandstone setting for large NATM-constructed cavern (Stanford Linear Accelerator Center), 927, 931–941
   Silver Line Phase III (Boston, MA), 713–725, 728–741
   Thorn Creek Tunnel (Chicago, IL), 123–131

Upper Rouge Tunnel CSO Control Project
(Detroit, MI), 72–85
Waller Creek Tunnel (Austin, Texas),
86–95
Westside Extension Transit Corridor (Los
Angeles, CA), 660–674
Yucca Mountain nuclear waste repository
(Nevada), 516–524
See also Contracts
Design-bid-build contracts, 507–508
for Lake Mead Intake No. 3 (Nevada),
503–515
Detroit, Michigan
Modified Detroit River Outfall No. 2,
675–686
Upper Rouge Tunnel CSO Control Project,
72–85
Devil's Slide Tunnel (California), 97–98
DHI-240 large-diameter blind drilling rig,
970–974
Difficult ground
design and construction challenges for
dam outlet tunnel and shaft in mixed
ground (hard rock blocks and soft
clayey matrix), 171–184
double-shield TBM in highly variable
ground conditions (Zagros water
tunnel), 1321–1333
EPBM tunneling through saturated
ground under very shallow cover and
critical, sensitive infrastructure (Perth,
Australia), 1124–1136
impacts of extreme ground convergence
on TBM performance (Ghomroud
water conveyance tunnel, Iran),
1334–1353
problems and remedies in TBM tunneling
through karstic terrain (Lebanon),
1294–1307
recovery, refurbishing, and revised tunnel
alignment for TBM halted by fluid mud,
151–170
response to sand lens in Cooper Marl
while constructing shaft for wastewater
tunnel (Charleston, SC), 185–194
tunnel shield machines in water tunnel
construction in varied ground
(California), 195–214
tunneling difficulties in carbonate
sedimentary rocks, 215–230
underpinning and excavation works for
Number 7 Line Extension (New York
City), 134–150
See also Ground characterization; Ground
conditions
Dispute resolution (North 27th St. tunnel,
Milwaukee), 286–296
Double-shield TBMs
in construction of Pula Subbaiah Veligonda
Tunnel #2, 363, 364

in construction of Srisailam Left Bank
Canal Tunnel Scheme (Andhra
Pradesh, India), 1275–1291
in highly variable and difficult ground
conditions (Zagros water tunnel),
1321–1333
impacts of extreme ground convergence
on TBM performance (Iran),
1334–1353
and NATM in construction of Pajares High
Speed Railway Tunnel Lot 2 (Spain),
1218–1234
in rapid excavation of Beles Tailrace
Tunnel (Ethiopia), 1235–1247
Drill and blast
compared with tunnel bore extender,
357–362
in construction of Lake Mead Intake shaft,
496–497
in construction of large rock caverns for
transit hubs (New York City), 35–47
evaluation for Thorn Creek Tunnel
(Chicago, IL), 123–131
in excavation of adits, deaeration
chambers, and shafts for Upper Rouge
Tunnel CSO Control Project, 82–85
for Harlem River Crossing utility tunnel
(New York City), 55
in Mitholz Project (Switzerland), 2–11
Drill Tech Drilling and Shoring, Inc., 176, 370,
374–376
DTX. *See* Caltrain Downtown Extension
DUSEL. *See* Deep Underground Science and
Engineering Laboratory

E
Earth pressure balance TBMs
with bicomponent backfill grout for
settlement control in variable soil
conditions (Sao Paulo Metro Line),
1137–1153
Brightwater East case history, 1171–1181
considered for Port of Miami Tunnel
(Florida), 687–699
in construction of Alafia River Tunnel
(Florida), 228
in construction of Bay Tunnel Project, 61,
65–66, 70
in construction of Hobson sewer
storage tunnel and Rosedale Outfall
(Auckland, New Zealand), 450–462
in construction of North Dorchester Bay
CSO Storage Tunnel (Boston, MA),
1062–1073
in construction of Tunnel and Edison Force
Mains Project (New Jersey), 824–833
feasibility study for construction of transit
tunnel passing under airport runway
(Shanghai, China), 463–470
in high risk tunneling adjacent to reservoir
(Sacramento, CA), 1074–1082

preparations for EPBM tunneling
(Brightwater Conveyance System,
BT4), 1154–1170
and probabilistic geotechnical baseline
assessment compared with actual
tunnel face conditions, 250–262
research on soil conditioning for clogging
and low-fines soils in EPBM tunneling,
320–333
tunneling through saturated ground
under very shallow cover and critical,
sensitive infrastructure (Perth,
Australia), 1124–1136
weighed against microtunneling for Ballard
Siphon Replacement Project (King
County, WA), 581–588
East Bay (California) Municipal Utility District,
370
cutter-soil mixing in construction of
microtunneling shafts for Mokelumne
River Crossing (California), 370–382
East Side CSO Tunnel Project (Portland,
OR), 570, 1022
multiple microtunnel drives, 570–580
selection, design, and procurement of
large mixshield TBM, 1022–1036
Eastside Access Project (New York City),
602, 780
alternative concrete linings considered and
selected for caverns, 780–789
interactions with Amtrak and Long Island
Rail Road regarding work under
Queens rail facilities, 814–823
microtunneling for multiple utility crossings
under Harold Interlocking railroad
system, 602–614
rock cavern for transit hub, 36, 38
TBMs in hard rock tunneling, 1365–1372
Edison Force Mains (EFM). *See* Tunnel and
Edison Force Mains Project
EPBMs. *See* Earth pressure balance TBMs
Ertan Hydropower Development Co. Ltd.,
1308–1309
Ethiopia
Beles Tailrace Tunnel, 1235–1247
recovery, refurbishing, and revised tunnel
alignment for TBM halted by fluid mud,
151–170

F
Finite-element modeling. *See under* Modeling
Flatiron West, 176
Florida
Alafia River Tunnel (Tampa), 228
karst development in south Florida,
216–217
proposed Port of Miami Tunnel, 687–699
Florida Department of Transportation District
6, 687–699
Folsom South Canal Connection Project. *See*
Mokelumne River Crossing

Forepoling, 737
Fremont (California) Central Park Subway
Project, 742–744, 755–756
alignment crossing fault and aquifer and
under park and lake, 742–743
geotechnical design challenges, 744–749
structural design challenges, 749–755
Frontier Kemper Constructors, Inc., 970–974,
975, 1208
Fugro West, 370

G
Georgia Department of Transportation, 757
feasibility study for Atlanta North-South
Tunnel, 757–765
Geotechnical baseline reports (GBRs), 232
and construction considerations, 238–239
for East Side CSO Tunnel Project
(Portland, OR), 570–571
establishing baselines, 233–237
ground classification, 237
and groundwater control, 236
Hobson sewer storage tunnel (Auckland,
New Zealand), 454–455
probabilistic geotechnical baseline
assessment compared with actual
tunnel face conditions, 250–262
recommendations, 240–241
in risk management, 867, 888–889, 891
use during construction, 239–240
*Geotechnical Baseline Reports for
Underground Construction*, 232
Germany
finite-element modeling of compensation
grouting (Leipzig City-Tunnel),
341–343
finite-element modeling of compensation
grouting (Limburg railway tunnel),
338–341
Ghaem Co., 1337
Ghomroud water conveyance tunnel (Iran),
1334, 1335–1336
impacts of extreme ground convergence
on TBM performance, 1334–1353
Ghorb Khatam, 1337
Gilgel Gibe II Hydroelectric Power Plant
(Ethiopia), 151
recovery, refurbishing, and revised tunnel
alignment for TBM halted by fluid mud,
151–170
Gotthard Base Tunnel (Switzerland), 2
design and construction of 57-km alpine
rail tunnel, 471–485
past and future geotechnical challenges
and TBM improvements in section
Faido tunneling, 1182–1205
Ground characterization
of challenging mixed ground for Mather
Interceptor (Sacramento, CA),
272–284
and GBRs, 237

probabilistic geotechnical baseline assessment compared with actual tunnel face conditions, 250–262
project-specific soil grouping system (Anacostia River Projects), 244–249
visual and geometric documentation by tunnel scanner (Austria), 308–319
by wireless TRT system ahead of TBM excavation, 263–271
See also Difficult ground
Ground conditions
ground freezing for access shaft excavation in soft ground (Brightwater Conveyance System), 297–307
microtunneling in soft ground with low ground cover under transportation routes and historic buildings (Hartford, CT), 589–601
San Vicente Tunnel Project (San Diego County, CA), 196–199, 208–213
soft ground, gassy ground, tar seeps, and fault crossings for planned Westside subway (Los Angeles), 660–674
tunnel bore extender employing undercutting in soft ground (Uetliberg Tunnel, Switzerland), 348–357
See also Difficult ground
Ground modification
compensation grouting, 334–336
contractor's acceptance of responsibility for grouting (N. 27th St. ISS Ext. tunnel), 286–296
finite-element modeling of compensation grouting, 334–344
ground freezing for access shaft excavation in soft ground (Brightwater Conveyance System), 297–307
research on soil conditioning for clogging soils in EPBM tunneling, 320–327, 332
research on soil conditioning for coarse-grained soils with low fines in EPBM tunneling, 327–333
Groundwater control
contractor's acceptance of responsibility (N. 27th St. ISS Ext. tunnel), 286–296
and geotechnical baseline reports, 236
Grouting
bicomponent backfill grout for settlement control in variable soil conditions (Sao Paulo Metro Line), 1137–1153
contractor's acceptance of responsibility for (N. 27th St. ISS Ext. tunnel), 286–296
finite-element modeling of compensation grouting, 334–344
pre-excavation (Providence, RI), 28–30

H
Hallandså Railway Tunnel Project (Sweden), 416–417
and dual mode hard rock mix-shield TBM, 417–437

Hampstead Avenue Interceptor Extension (Hartford, CT), 589
microtunneling in soft ground with low ground cover under transportation routes and historic buildings, 589–601
Harlem River Crossing utility tunnel (New York City), 48–50
selected tunneling approach, 53–57
site challenges, 51–53
Hartford (Connecticut) Metropolitan District Commission, 589
microtunneling in soft ground with low ground cover under transportation routes and historic buildings, 589–601
Hatch Mott MacDonald, 176–177, 558, 825, 1115
HDR Engineering Inc., 534
Herrenknecht AVN1200TB Microtunneling System, 612
Herrenknecht AVND1800AB MTBM, 558–569
Herrenknecht AVND2000AB, 573
Herrenknecht S-336 EPB, 1139
Hindustan Construction Company, 364
Hitachi Zosen TBM, 1066–1067, 1073
HNTB Corporation, 757
Hobson sewer storage tunnel (Auckland, New Zealand), 450–456, 458–462
Homestake Mine (South Dakota), 638. *See also* Deep Underground Science and Engineering Laboratory
Hong Kong, University of, underground water reservoirs
construction, 447–448
planning, 438–447, 449
Horizontal Directional Drilling (HDD), 48–49, 51

I
Impregilo SpA, 488
India
long (19.35 km) extensible conveyor system (Andhra Pradesh), 363–369
shield tunneling in construction of Srisailam Left Bank Canal Tunnel Scheme (Andhra Pradesh), 1275–1291
International Tunnel Insurance Group. *See* ITIG Code of Practice for Risk Management of Tunnel Works
Iran
double-shield TBM in highly variable and difficult ground conditions (Zagros water tunnel), 1321–1333
impacts of extreme ground convergence on TBM performance (Ghomroud water conveyance tunnel, Iran), 1334–1353
Iron Mountain Mine (California), 616
design/build/maintain mine rehabilitation project, 616–628
Italy
application of ADECO (Bologne-Florence High Speed Railway), 955–959

INDEX 1379

application of ADECO (Cassia 1 Tunnel), 964–965
Bologne-Florence High Speed Railway, 955
ITIG Code of Practice for Risk Management of Tunnel Works, 854, 866–870

J
J. F. Shea, 134
  dispute resolution and acceptance of responsibility for grouting and groundwater control, 286–296
J. S. Redpath Corporation, 12, 13
Jacobs Associates, 176–177, 579
Jacobs Engineering, 986, 997, 1354
Jaiprakash Associates Ltd., 1275
Japan: Ohashi Shield Tunnel (Tokyo), 1084–1101
Jenny Engineering Corp., 894
Jinping-II Hydropower Station headrace tunnels (China), 1308–1309
  hard rock tunneling in deep mountain setting with high water inflow potential, 1308–1320
  onsite first time assembly of TBM, 1308, 1312–1314
Johnson County, Kansas, 998
  shaft excavation by reverse circulation drilling (Kansas River Tunnel), 998–1004
Joint Code of Practice for Tunnel Works in the UK, 854
Joint Water Pollution Control Plant (Los Angeles County, CA), 700
  planning for wastewater tunnels and outfall, 700–712

K
Kansas River Tunnel (Johnson County, KS), 998
  shaft excavation by reverse circulation drilling, 998–1004
Kenny Construction Company, 825–833
Kensico-City Tunnel (New York), 108
  design and planning, 108–122
Kesrouane Water Project (Lebanon). *See* Madiq Tunnel
Kiewit–Bilfinger Berger Joint Venture, 579, 1022
King County, Washington, 581
  consideration of microtunneling vs. EPB tunneling, 581–588
  See also Seattle, Washington
Kleinfelder Inc., 534
Krystol Internal Membrane (KIM), 900–901

L
Lachel Felice & Associates, Inc., 13
Lake Dorothy (Alaska) lake tap and tunnel, 12–21
Lake Mead Intake No. 3 (Nevada)
  design-build contract, 503–515

intake structure design, 500–502
shaft design and construction, 489–499
tunnel design, 499–500
Lebanon: Madiq Tunnel, 1294–1307
Leipzig (Germany) City-Tunnel, 334–337, 341–344
Lenihan Dam (Los Gatos, CA), 171–172
  ground conditions (hard rock blocks and soft clayey matrix), 171, 172–174
  multiple support approaches, 174–175
  outlet tunnel and shaft design and construction, 175–184
Limburg (Germany) railway tunnel, 334–341
Linings
  adjustable formwork cast-in-place concrete considered for Eastside Access Project, 784
  documentation by tunnel scanner of shotcrete and concrete lining construction stages (Tauern Tunnel, Austria), 308–319
  fixed formwork cast-in-place concrete selected for Eastside Access Project, 780–789
  history and ring characteristics of precast concrete segmental liners, 1102–1111
  Phase I Combined Sewer Overflow (Rhode Island), 33
  precast concrete segmental liner for Tunnel and Edison Force Mains Project (New Jersey), 824, 825, 827
  ring selection, 1109–1110
  San Vicente Tunnel Project (California), 202
  segmental lining for Lake Mead Intake tunnel, 499–500
  shotcrete selected for Eastside Access Project, 780–789
  steel fiber reinforced precast segmental liner (Big Walnut Outfall Augmentation Sewer), 1114–1117
  tapered rings, 1106–1108
  trapezoidal rings, 1105–1106
  universal rings, 1108–1109
Los Angeles County, California
  planned Regional Connector (light rail), 669–674
  planned Westside Extension Transit Corridor, 660–669, 673–674
  planning for Joint Water Pollution Control Plant, 700–712
Los Angeles County (California) Metropolitan Transportation Authority, 660–674
Los Gatos, California, 171
  design and construction challenges for dam outlet tunnel and shaft in mixed ground of hard rock blocks and soft clayey matrix, 171–184
Lötschberg Base Tunnel (Switzerland), 2, 471
Louisville (Kentucky) Water Company, 1248
  TBM in construction of Riverbank Filtration Tunnel and Pump Station, 1248–1260
Lovat Inc., 1167

Lovat RME167SE, 1117
Lovat RME185SE, 828
Lovat RME229SE 22900, 1173–1175
Lunardi, Pietro, 948–952

M

M. L. Shank, 26
Madiq Tunnel (Lebanon), 1294
　problems and remedies in TBM tunneling through karstic terrain, 1294–1307
Massachusetts
　design, pre-support, and modeling for Silver Line Phase III (Boston), 728–741
　EPBM in construction of North Dorchester Bay CSO Storage Tunnel (Boston), 1062–1073
　planning and tunneling method selection for Silver Line Phase III (Boston), 713–725
　research on soil conditioning for clogging soils in EPBM tunneling (Boston), 320–323, 325
Massachusetts Water Resources Authority, 1062
MASTERSEAL 345, 900–901
Mather Interceptor (Sacramento County, CA), 272–273
　feasibility evaluation of tunneling methods, 278–284
　ground characterization, 274–278, 284
McNally Kiewit JV, 1114–1117
Menlo Park, California. *See* Stanford Linear Accelerator Center
Metropolitan Transportation Authority Capital Construction (New York City), 134, 602, 612
Miami, Florida
　desired traffic improvement to Port of Miami, 687
　proposed Port of Miami Tunnel, 687–699
Michigan
　Modified Detroit River Outfall No. 2 (Detroit), 675–686
　Upper Rouge Tunnel CSO Control Project (Detroit), 72–85
Microtunneling
　in East Side CSO Tunnel Project (Portland, OR), 570–580
　for multiple utility crossings under Harold Interlocking railroad system (New York City), 602–614
　1.2-mile CSO component underlying highways and railyard (Bergen County, NJ), 558–569
　in soft ground with low ground cover under transportation routes and historic buildings (Hartford, CT), 589–601
　weighed against EPB tunneling for Ballard Siphon Replacement Project (King County, WA), 581–588

Milwaukee (Wisconsin) Metropolitan Sewerage District
　dispute resolution and contractor's acceptance of responsibility for grouting and groundwater control, 290–296
　North 27th St. ISS Extension tunnel, 286–290
Mining
　design/build/maintain mine rehabilitation project (California), 616–628
　TBM and contractor experience problems in construction of drainage tunnel (Papua New Guinea), 629–637
Mitholz Project (Switzerland), 2–4
　contract, 4–5
　drill and blast approach, 7–11
　mobile crusher, 7–9
　rock classes and support, 5–7
Modeling
　finite-element modeling of compensation grouting for City-Tunnel (Leipzig, Germany), 334–337, 341–344
　finite-element modeling of compensation grouting for railway tunnel (Limburg, Germany), 334–341
　finite-element modeling of ground displacement for Fremont Central Park Subway Project (California), 750–753
　finite-element modeling of heritage building under which busway tunnel was constructed (Australia), 911–913
　virtual design and construction, 383–395
Modified Detroit River Outfall No. 2, 675
　slurry TBM tunneling in rock, 675–686
Mokelumne River Crossing (California), 370–373
　cutter-soil mixing in construction of microtunneling shafts, 370, 373–382
Mole Constructors, Inc., 1249
Montgomery County, Maryland, Bi-County Water Tunnel risk management, 858–859

N

Narragansett Bay Commission
Nevada
　design and construction of Lake Mead Intake No. 3 shafts and tunnel, 488–502
　design and future subsurface construction of Yucca Mountain nuclear waste repository, 516–524
　design-build contract for Lake Mead Intake No. 3, 503–515
　feasible construction options for Reach 3 Tunnel Project, 525–533
　geologic investigations for Apex Tunnel, 534–547
　subsurface investigation for Reach 4 Tunnel Project, 548–555

INDEX

subsurface ventilation system for Yucca Mountain nuclear waste repository, 646–658
Systems Conveyance and Operations Program (SCOP), 525
New Austrian Tunneling Method (NATM)
  ADECO as alternative to, 942–968
  and Beacon Hill Station (Seattle, WA), 98
  in construction of large shallow cavern in weak sandstone (SLAC), 927–941
  and design of Caldecott Tunnels fourth bore (Oakland, CA), 96–107
  and Devil's Slide Tunnel (California), 97–98
  in excavation of motorway tunnel through granitic rock, with numerous faults and thin overburden, and underlying academic complex and zoo (Slovakia), 916–926
  for planned Silver Line Phase III (Boston, MA), 713–725, 728–741
  and Rabcewicz, 942–946, 966–967
  and shield tunneling in construction of Pajares High Speed Railway Tunnel Lot 2 (Spain), 1218–1234
  and Strenger Tunnel (Austria), 98
  U.S. and European experience with, 98–99
  See also Sequential Excavation Method
New Jersey
  design and construction to date of Tunnel and Edison Force Mains Project, 824, 825–833
  1.2-mile CSO microtunneling component underlying highways and railyard (Bergen County), 558–569
New MetroRail City Project (Perth, Australia), 1124
  EPBM tunneling through saturated ground under very shallow cover and critical, sensitive infrastructure, 1124–1136
New York City
  alternative concrete linings considered for caverns (Eastside Access Project), 780–789
  contract packaging for Second Avenue Subway, 790–801
  design and planning of Kensico-City water tunnel, 108–122
  Harlem River Crossing utility tunnel, 48–57
  interactions of Eastside Access Project with Amtrak and Long Island Rail Road regarding work under Queens rail facilities, 814–823
  large rock caverns for transit hubs, 35–47
  long distance pumping of concrete for lining City Water Tunnel No. 3 (Stage 2), 396–413
  microtunneling for multiple utility crossings under Harold Interlocking railroad system (Eastside Access Project), 602–614
  TBMs in hard rock tunneling for Eastside Access Project, 1365–1372
  underpinning and excavation works for Number 7 Line Extension, 134–150
  underpinning design challenges for Number 7 Line Extension, 803–813
New York City Metropolitan Transit Authority (NYCMTA), 35–38
New York City Transit, 134
New Zealand
  Hobson sewer storage tunnel (Auckland), 450–456, 458–462
  Rosedale Outfall project (Auckland), 456–458, 460, 461, 462
Norconsult, 13, 19
North 27th St. ISS Extension tunnel (Milwaukee, WI), 286–290
  dispute resolution and contractor's acceptance of responsibility for grouting and groundwater control, 290–296
North American Segment Company (NASCO), 1114–1117
North Carolina
  SEM with water resistant additions in construction of Wachovia–Knight tunnels, 894–905
  Wachovia–Knight Theater pedestrian tunnels (Charlotte), 894
North Dorchester Bay CSO Storage Tunnel (Boston, MA), 1062
  EPBM in construction through variable soft ground with 6 m of cover, 1062–1073
  research on soil conditioning for clogging soils in EPBM tunneling, 320–323, 325
North Pacific Research, 616
North Shore Connector Project (Pittsburgh, PA), 1051
  slurry TBM in construction of light rail tunnel under Allegheny River and city street, 1051–1061
North-South Tunnel (Atlanta, GA), 757
  feasibility study, 757–765
Northeast Remsco Construction, 558–559, 589
Norwegian Lake Tap Method, 12, 18–21
Number 7 Line Extension (New York City), 134–136, 802–803
  rock cavern for transit hub, 35, 38
  underpinning and excavation works, 134–150
  underpinning design challenges, 803–813
Number 10 metro line (Shanghai, China), 463
  feasibility study of EPBM for construction of transit tunnel passing under airport runway, 463–470

O

Oakland, California: Caldecott Tunnels fourth bore, 96–107
Ohashi Shield Tunnel (Tokyo, Japan), 1083

shield tunneling in construction of upper and lower twin highway tunnels with 1.4 m clearance, 1084–1101
Ohio
  solutions to TBM sealing system problems (Big Walnut Outfall Augmentation Sewer), 1114–1123
  steel fiber reinforced precast segmental liner (Big Walnut Outfall Augmentation Sewer), 1114–1117
Ok Tedi Mine (Papua New Guinea), 629
  TBM and contractor-experience problems in construction of drainage tunnel, 629–637
Onsite first time assembly, 1308, 1312–1314
Oregon
  multiple microtunnel drives in East Side CSO Tunnel Project (Portland), 570–580
  selection and design of large mixshield TBM for East Side CSO Tunnel Project (Portland), 1022–1036
Overpeck Valley Parallel Sewer Project (Bergen County, NJ), 558
  1.2-mile CSO microtunneling component underlying highways and railyard, 558–569

P
Pajares High Speed Railway Tunnel Lot 2 (Spain), 1218
  shield tunneling and NATM in construction of Pajares High Speed Railway Tunnel Lot 2 (Spain), 1218–1234
Papua New Guinea, 629
  TBM and contractor-experience problems in construction of drainage tunnel (Ok Tedi Mine), 629–637
Parsons Brinckerhoff, 386, 579
Parsons Infrastructure, 534
Pennsylvania: North Shore Connector Project (Pittsburgh), 1051–1061
Perth, Australia: New MetroRail City Project, 1124–1136
Phase I Combined Sewer Overflow (Providence, RI), 22–24, 34
  adits, 30–33
  main tunnel lining, 33
  pre-excavation grouting, 28–30
  TBM in construction of, 26–27
  work shafts, 24–26
Pine Point Mine (Canada), 975
  design of combined service/production shaft with vertical conveyor and tower-mounted hoist, 975–985
Pittsburgh, Pennsylvania: North Shore Connector Project, 1051–1061
Port of Miami Tunnel (Miami, FL), 687–688
  proposed undersea highway tunnel, 687–699
Portland (Oregon) Bureau of Environmental Services, 570

multiple microtunnel drives in East Side CSO Tunnel Project, 570–580
selection, design, and procurement of large mixshield TBM (East Side CSO Tunnel Project), 1022–1036
Providence, Rhode Island: Phase I Combined Sewer Overflow, 22–24, 34
Public-private partnerships, 687–699
Pula Subbaiah Veligonda Tunnel #2 (Andhra Pradesh, India), 363–364
  long (19.35 km) extensible conveyor system, 363–369

R
Rabcewicz, Ladislaus von, 942–946
Reach 3 Tunnel Project (Nevada), 525–526
  anticipated ground behavior, 528–529
  feasible tunnel construction options, 530–533
  geologic conditions, 527–528
Reach 4 Tunnel Project (Nevada), 548–549
  cost and benefit of second subsurface investigation, 548–555
Regional Connector (Los Angeles County, CA), 669–674
Rhode Island: Providence Combined Sewer Overflow, 22–24, 34
Risk assessment and analysis
  in decision making for Central Subway Project (San Francisco, CA), 871–880
  microtunneling vs. EPB tunneling (King County, WA), 586–587
Risk management
  Bi-County Water Tunnel (Montgomery County, MD), 858–859
  and contracts, 867–869
  engineer's estimates using contractors' estimating approach, 845–853
  for EPBM tunneling adjacent to reservoir (Sacramento, CA), 1074–1082
  and geotechnical baseline reports, 867, 888–889, 891
  and ground and groundwater conditions, 866–867
  and ground settlement, 867
  and high-risk factors in short tunnel construction, 881–891
  and procurement strategy, 863–864
  and project planning, 861–863
  and quality control, 869–870
  risk registers, 854–860, 868
  and stage gates, 862–863
  team-based, 864–866
  UK good practice, 861–870
  See also ITIG Code of Practice for Risk Management of Tunnel Works
Riverbank Filtration Tunnel and Pump Station (Louisville, KY), 1248
  high TBM production rates in moderately weak bedrock, 1248–1260
Robbins Company, 364, 633

INDEX

Robbins Double Shield TBMs, 1275, 1278–1281
Robbins HP Main Beam TBM, 1366–1368
Robbins HP-TBM, 1315–1319
Robbins partially shielded TBM model 116–189, 1304
Robbins Series MB 105-44-5 rock TBM, 1256
Rock caverns. See Caverns
Rock tunneling
 in deep mountain setting with high water inflow potential (China), 1308–1320
 dual mode hard rock mix-shield TBM in construction of Hallandså Railway Tunnel (Sweden), 416–437
 Gotthard Base Tunnel (Switzerland), 471–485
 high TBM production rates in moderately weak bedrock (Louisville, KY), 1248–1260
 past and future geotechnical challenges and TBM improvements in section Faido tunneling (Gotthard Base Tunnel), 1182–1205
 TBMs in hard rock tunneling for Eastside Access Project (New York City), 1365–1372
Rodgers Builders Inc., 894
Rosedale Outfall project (Auckland, New Zealand), 456–458, 460, 461, 462

S

SA Healy, 488
Sacramento (California) Regional County Sanitation District, 272
 blind auger drilling in construction of access shafts (Upper Northwest Interceptor), 1037–1050
 feasibility evaluation of tunneling methods (Mather Interceptor), 272–273, 278–284
 ground characterization (Mather Interceptor), 272–278, 284
 Mather Interceptor, 272–273
 risk management for EPBM tunneling adjacent to reservoir (Upper Northwest Interceptor), 1074–1082
Sahel Co., 1337
San Diego County (California) Water Authority, 195
 tunnel shield machines in water tunnel construction in varied ground (San Vicente Tunnel Project), 195–214
San Francisco, California
 Bay Tunnel Project, 60–71
 risk assessment and analysis in decision making for Central Subway Project, 871–880
San Francisco (California) Public Utilities Commission, 60
San Jose, California
 research on soil conditioning for clogging soils in EPBM tunneling, 320–327, 332
 research on soil conditioning for coarse-grained soils with low fines in EPBM tunneling, 327–333
San Vicente Tunnel Project (San Diego County, CA), 195–196, 213–214
 geology and ground conditions, 196–199, 208–213
 shafts, 199–202
 tunnel lining, 202
 tunnel production, 207–208
 tunnel shield machines, 202–207
Sandstrom, G.E., 943
Santa Clara Valley (California) Water District, 171
Sao Paulo (Brazil) Metro Line, 1137
 bicomponent backfill grout for settlement control in variable soil conditions, 1137–1153
SAP2000 software, 751
Satco, 2
Schiavone, 134
Schiavone/Shea/Frontier-Kemper, 396, 399
Seattle, Washington
 Brightwater East case history, 1171–1181
 ground freezing for access shaft excavation in soft ground (Brightwater Conveyance System), 297–307
 NATM in construction of Beacon Hill Station, 98
 preparations for EPBM tunneling (Brightwater Conveyance System, BT4), 1154–1170
 probabilistic geotechnical baseline assessment compared with actual tunnel face conditions (Brightwater Conveyance System), 251–262
 TPC software for TBM data management and quality assurance (Brightwater Conveyance System), 1354–1364
 See also King County, Washington
Second Avenue Subway (New York City)
 contract packaging, 790–801
 rock cavern for transit hub, 35, 36–37
SELI, 1337
SELI double shielded TBM, 1366–1368
Sequential Excavation Method (SEM)
 in construction of tunnel under road, park, and historic buildings (Brisbane, Australia), 906–915
 in construction of Wachovia–Knight Theater pedestrian tunnels (Charlotte, NC), 894–905
 history of, 943, 966
 planned for Caltrain Downtown Extension (DTX) project (San Francisco), 768, 772–773
 proposed for New York City rock caverns, 47
 See also New Austrian Tunneling Method (NATM)
Seven Train (New York City). See Number 7 Line Extension

Shafts
  assessment of slurry diaphragm walls for deep rectangular ventilation shafts (Silicon Valley Rapid Transit), 1005–1019
  construction (Tunnel and Edison Force Mains Project), 824, 827–828, 829–830
  cutter-soil mixing in construction of microtunneling shafts for Mokelumne River Crossing (California), 370–382
  design and construction (Lake Mead Intake No. 3), 489–499
  design and construction challenges for shaft in mixed ground (hard rock blocks and soft clayey matrix), 171–184
  design of combined service/production shaft with vertical conveyor and tower-mounted hoist (Pine Point Mine), 975–985
  DHI-240 rig in drilling of return shaft for Black Panther Mine (Indiana), 974
  drill and blast in excavation of shafts (Upper Rouge Tunnel CSO Control Project), 82–85
  excavation by reverse circulation drilling (Kansas River Tunnel), 998–1004
  ground freezing for access shaft excavation in soft ground (Brightwater Conveyance System), 297–307
  new-generation large-diameter blind drilling rig (DHI-240), 970–974
  Phase I Combined Sewer Overflow work shafts (Providence, RI), 24–26
  response to sand lens in Cooper Marl while constructing shaft for wastewater tunnel (Charleston, SC), 185–194
  San Vicente Tunnel Project (San Diego County, CA), 199–202
  steel casing liner as alternative to concrete lining for deaeration vent shafts (Upper Rouge Tunnel CSO Control Project), 986–997
SHAKE software, 750
Shanghai (China) No. 10 metro line, 463
  feasibility study of EPBM for construction of transit tunnel passing under airport runway, 463–470
Shank/Balfour Beatty, 23
Shank/Balfour Beatty/Barletta Heavy Construction JV, 1063–1064, 1073
Shea. *See* Schiavone/Shea/Frontier-Kemper; J. F. Shea
Shield tunneling
  in Andes, 1270–1271
  in construction of upper and lower twin highway tunnels with 1.4 m clearance (Tokyo, Japan), 1084–1101
  dual mode hard rock mix-shield TBM in construction of Hallandså Railway Tunnel (Sweden), 416–437
  feasibility study for construction of transit tunnel passing under airport runway (Shanghai, China), 463–470
  problems and remedies in tunneling through karstic terrain (Lebanon), 1294–1307
  in water tunnel construction in varied ground (California), 195–214
  See also Double-shield TBMs
Shotcrete
  dual layers with KIM admixture and MASTERSEAL intermediate layer for waterproofing (Wachovia–Knight Theater), 894–905
  selected for Eastside Access Project, 780–789
  visual and geometric documentation by tunnel scanner of application (Austria), 308–319
Silicon Valley Rapid Transit (California)
  assessment of slurry diaphragm walls for deep rectangular ventilation shafts, 1005–1019
  research on soil conditioning for clogging soils in EPBM tunneling, 320–327, 332
  research on soil conditioning for coarse-grained soils with low fines in EPBM tunneling, 327–333
Silver Line Phase III (Boston, MA), 713, 728–733
  geotechnical investigation, 732–733
  planning and tunneling method selection, 713–725
  pre-support elements, 736–738
  tunnel design considerations, 734–736
Sitina Tunnel (Bratislava, Slovakia), 916
  estimation of rock mass behavior based on convergence measurement, 919–922
  excavation through granitic rock, with numerous faults and thin overburden, and underlying academic complex and zoo, 916–917, 925–926
  NATM in construction of, 916–926
Skanska, 134
SLAC. *See* Stanford Linear Accelerator Center
Slovakia: Sitina Tunnel (Bratislava), 916–926
Slurry TBMs
  considered for Port of Miami Tunnel (Florida), 687–699
  in construction of light rail tunnel under Allegheny River and city street (Pittsburgh, PA), 1051–1061
  in construction of Modified Detroit River Outfall No. 2, 675–686
  expected for construction of Westside subway (Los Angeles), 660–674
  and probabilistic geotechnical baseline assessment compared with actual tunnel face conditions, 250–262

INDEX 1385

selection, design, and procurement of North America's largest mixshield TBM (East Side CSO Tunnel), 1022–1036
Smith, Chris, 1114–1117
South America. *See* Andes Mountains
South Carolina: Daniel Island Extension Tunnel (Charleston), 185–194
South Dakota
 defunct Homestake Mine (Lead), 638
 planning and site characterization for DUSEL research facility (Lead), 638–645
Southern Nevada Water Authority, 488
 and Apex Tunnel geologic investigations, 534–547
 Clark, Lincoln, and White Pine Counties Groundwater Development Project, 534
 design-build contract for Lake Mead Intake No. 3, 503–515
Southland Contracting, Inc., 1003–1004
Spain: Pajares High Speed Railway Tunnel Lot 2, 1218–1234
Squamish, British Columbia: Ashlu Creek Hydro-electric project, 1208–1217
Srisailam Left Bank Canal Tunnel Scheme (Andhra Pradesh, India), 1275
 double-shield TBMs in tunnel excavation, 1275–1291
STAAD Pro structural analysis software, 898
Staheli Trenchless Consultants, 581
Stanford Linear Accelerator Center (California), 927
 NATM in construction of large shallow cavern, 927–931, 940
 rock mass characterization of weak sandstone setting for new cavern, 927, 931–941
S3-II, 134
Strenger Tunnel (Austria), 98
Sweden: Hallandså Railway Tunnel, 416–437
Swiss Federal Railways, 471
Switzerland
 drill and blast in Mitholz Project, 2–11
 Gotthard Base Tunnel, 2, 471–485, 1182–1205
 Lötschberg Base Tunnel, 2, 471
 TBM followed by tunnel bore extender employing undercutting in soft ground (Zurich), 346–362
Systems Conveyance and Operations Program (SCOP; Nevada), 525, 549. *See also* Reach 3 Tunnel Project; Reach 4 Tunnel Project

T

Tamerlane Ventures Inc., 975
Tampa, Florida: Alafia River Tunnel, 228
Tartaguille Tunnel (France), 959
 application of ADECO, 959–964
Tauern Tunnel (Austria), 308–309

visual and geometric documentation by tunnel scanner of construction stages, 308–319
TBE. *See* Tunnel bore extender
TBMs. *See* Tunnel boring machines
Texas: Waller Creek Tunnel (Austin), 87–95
Thorn Creek Tunnel (Chicago, IL), 123–125
 design and planning, 123, 126–131
Tokyo, Japan: Ohashi Shield Tunnel, 1084–1101
TPC (Tunneling Process Control) software, 1354–1364
Trans Hudson Express (THE) Tunnel (New York City)
 rock cavern for transit hub, 35, 36
TRT. *See* Tunnel Reflector Tracing (TRT) system
Tunnel and Edison Force Mains Project (New Jersey), 824–825
 design and construction, to date, of sewer tunnel under Raritan River, 824, 825–833
 shaft construction, 824, 827–828, 829–830
Tunnel bore extender (TBE)
 compared with drill and blast, 357–362
 following TBM and employing undercutting in soft ground (Uetliberg Tunnel), 348–357
Tunnel-boring machines
 advance rates for rock conditions in Kensico-City Tunnel (New York), 113–115
 in construction of mine drainage tunnel (Papua New Guinea), 629–637
 in construction of power tunnel (Ashlu Creek Hydro-electric project), 1208–1217
 in construction of Upper Rouge Tunnel CSO Control Project, 79–82
 dual mode hard rock mix-shield TBM in construction of Hallandså Railway Tunnel (Sweden), 416–437
 followed by tunnel bore extender employing undercutting in soft ground (Zurich), 346–357
 in future subsurface construction of Yucca Mountain nuclear waste repository, 516–524
 in hard rock tunneling in deep mountain setting with high water inflow potential (China), 1308–1320
 in hard rock tunneling for Eastside Access Project (New York City), 1365–1372
 high production rates in moderately weak bedrock (Louisville, KY), 1248–1260
 historical use in Andes, 1262–1263
 problems and remedies in tunneling through karstic terrain (Lebanon), 1294–1307
 recovery, refurbishing, and revised tunnel alignment for TBM halted by fluid mud, 151–170

solutions to problems with sealing systems
(Big Walnut Outfall Augmentation
Sewer), 1114–1123
technical considerations for use in Andes,
1261–1274
TPC software for TBM data management
and quality assurance (Brightwater
Conveyance System), 1354–1364
wireless TRT system for rapid ground
assessment ahead of excavation,
263–271
See also Double-shield TBMs; Earth
pressure balance TBMs; Rock
tunneling; Shield tunneling; Slurry
TBMs; Tunnel bore extender; *and
machines by manufacturers' names*
Tunnel Reflector Tracing (TRT) system,
263–271
Tunneling. *See* Cut and cover tunnels;
Microtunneling; Rock tunneling; Shield
tunneling; Tunnel boring machines
TVS, 894

U

Uetliberg Tunnel (Zurich, Switzerland),
346–348
comparison of tunnel bore extender (TBE)
with drill and blast, 357–362
TBM followed by TBE employing
undercutting in soft ground, 348–357
Underground Technology Research Council,
232
Upper Northwest Interceptor (Sacramento,
CA)
blind auger drilling in construction of
access shafts, 1037–1050
EPBM in construction of, 1037, 1049
risk management for EPBM tunneling
adjacent to reservoir, 1074–1082
Upper Rouge Tunnel CSO Control Project
(Detroit, MI), 72–75, 85, 986–987
drill and blast in excavation of adits,
deaeration chambers, and shafts,
82–85
geotechnical investigations and ground
characterization, 75–79
steel casing liner as alternative to concrete
lining for deaeration vent shafts,
986–997
TBMs in construction of, 79–82
Utilidors, 48, 53

V

Vegas Tunnel Constructors, 488
Virtual design and construction (VDC),
383–395

W

Wachovia–Knight Theater pedestrian tunnels
(Charlotte, NC), 894

SEM with water resistant additions in
construction of, 894–905
Wade Trim Associates, 986, 997
Waller Creek Tunnel (Austin, Texas), 86–87
planning and design, 87–95
Washington, D.C.: Anacostia River Projects,
242–249
Washington (State)
baseline assessment compared with actual
tunnel face conditions (Brightwater
Conveyance System, Seattle),
250–262
Brightwater East case history (Seattle),
1171–1181
consideration of microtunneling vs. EPB
tunneling (King County), 581–588
ground freezing for access shaft
excavation in soft ground (Brightwater
Conveyance System, Seattle),
297–307
NATM in construction of Beacon Hill
Station (Seattle), 98
preparations for EPBM tunneling
(Brightwater Conveyance System,
BT4, Seattle), 1154–1170
TPC software for TBM data management
and quality assurance (Brightwater
Conveyance System, Seattle),
1354–1364
Water One (Johnson County, KS), 998
shaft excavation by reverse circulation
drilling (Kansas River Tunnel),
998–1004
Westside Extension Transit Corridor (Los
Angeles, CA), 660–664
ground conditions and slurry TBMs for
planned subway tunnels, 664–674
Wirth 400/450 III TBM, 1208, 1211
WIRTH GmbH, 970–974
Wisconsin: N. 27th St. ISS Ext. tunnel
(Milwaukee), 286–296

Y

Yucca Mountain nuclear waste repository
(Nevada), 516, 646
design and future subsurface construction,
516–524
development ventilation system, 646–653
emplacement ventilation system, 646–648,
653–658

Z

Zagros Tunnel (Iran)
double-shield TBM in highly variable and
difficult ground conditions, 1322–1333
48 km water transfer tunnel, 1321–1322
Zurich, Switzerland: Uetliberg Tunnel,
346–362